TABLE OF SIGN CONVENTIONS

Reference	g sign	Riemann	Einstein	Spacetime four-dimensional indices
Landau, Lifshitz (1962) "spacelike convention"	+	+	+	latin
Landau, Lifshitz (1971) "timelike convention"	−	+	+	latin
Misner, Thorne, Wheeler (1973: this text)	+	+	+	greek
Adler, Bazin, Schiffer (1965)	−	−	−	greek
Anderson (1967)	−	−	−[b]	greek
Bergmann (1942)	−	−[a]	−	greek
Cartan (1946)		−	−	
Davis (1970)		+	−	latin
Eddington (1922)	−	+	−	greek
Ehlers (1971)	+	+	+	latin
Einstein (1950)	−	+	−	greek
Eisenhart (1926)		+	−	
Fock (1959)	−	−[a]	−	greek
Fokker (1965)	−	−	+	latin
Hawking and Ellis (1973)	+	+	+	latin
Hicks (1965)		+	+	
Infeld, Plebanski (1960)	−	+	−	greek
Lichnerowicz (1955)	−	+	+	greek
McVittie (1956)	−	+	−	greek
Misner (1969a)	+	+	+	greek
Møller (1952)	+	−	−	latin
Pauli (1958)	+	−	−	latin
Penrose (1968)	−	−	−	latin
Pirani (1965)	−	−	−	latin
Robertson, Noonan (1968)	+	+	−	latin
Sachs (1964)	±	+	−	latin
Schild (1967)	−	+	−	latin
Schouten (1954)	−	+		
Schroedinger (1950)	−	+	−	latin
Synge (1960b)	+	+	−	latin
Thorne (1967)	−	+	+	greek
Tolman (1934a)	−	+	−	greek
Trautman (1965)	−	−	−	latin
Weber (1961)	+	+	+	greek
Weinberg (1972)	+	−	−	greek
Weyl (1922)	−	+	+	latin
Wheeler (1964a)	+	+	+	greek

[a] Unusual index positioning on **Riemann** components gives a different sign for $R_{\mu\rho\alpha\beta}$.

[b] Note: his $\kappa < 0$ is the negative of the gravitational constant.

This book follows the "Landau-Lifshitz Spacelike Convention" (LLSC). Arrows below mark signs that are "+" in it. The facing table shows signs that other authors use.

$$+\boldsymbol{g} = -(\boldsymbol{\omega}^0)^2 + (\boldsymbol{\omega}^1)^2 + (\boldsymbol{\omega}^2)^2 + (\boldsymbol{\omega}^3)^2$$

\boldsymbol{g} sign
(col. 2)

$$+\mathscr{R}(\boldsymbol{u}, \boldsymbol{v}) = \nabla_{\boldsymbol{u}} \nabla_{\boldsymbol{v}} - \nabla_{\boldsymbol{v}} \nabla_{\boldsymbol{u}} - \nabla_{[\boldsymbol{u},\boldsymbol{v}]}$$

$$+R^{\mu}{}_{\nu\alpha\beta} = \partial_{\alpha}\Gamma^{\mu}{}_{\nu\beta} - \partial_{\beta}\Gamma^{\mu}{}_{\nu\alpha} + \Gamma^{\mu}{}_{\sigma\alpha}\Gamma^{\sigma}{}_{\nu\beta} - \Gamma^{\mu}{}_{\sigma\beta}\Gamma^{\sigma}{}_{\nu\alpha}$$

Riemann sign
(col. 3)

$$+R_{\mu\nu} = R^{\alpha}{}_{\mu\alpha\nu}$$

quotient of **Einstein**
and **Riemann** signs

$$\textbf{Einstein} = +8\pi\boldsymbol{T}$$

$$G_{\mu\nu} = R_{\mu\nu} - \frac{1}{2}g_{\mu\nu}R = +8\pi T_{\mu\nu}$$

Einstein sign
(col. 4)

$$T_{\hat{0}\hat{0}} = \boldsymbol{T}(\boldsymbol{e}_{\hat{0}}, \boldsymbol{e}_{\hat{0}}) > 0$$

all authors agree
on this "positive
energy density" sign

 The above sign choice for **Riemann** is convenient for coordinate-free methods, as in the curvature operator $\mathscr{R}(\boldsymbol{u}, \boldsymbol{v})$ above, in the curvature 2-forms (equation 14.19), and for matrix computations (exercise 14.9). The definitions of **Ricci** and **Einstein** with the signs adopted above are those that make their eigenvalues (and $R \equiv R^{\mu}{}_{\mu}$) positive for standard spheres with positive definite metrics.

GRAVITATION

The Crab Nebula (NGC 1952), the remains of the supernova of July 1054, an event observed and recorded at the Sung national observatory at K'ai-feng. In the intervening 900 years, the debris from the explosion has moved out about three lightyears; i.e., with a speed about 1/300 of that of light. In 1934 Walter Baade and Fritz Zwicky predicted that neutron stars should be produced in supernova explosions. Among the first half-dozen pulsars found in 1968 was one at the center of the Crab Nebula, pulsing 30 times per second, for which there is today no acceptable explanation other than a spinning neutron star. The Chinese historical record shown here lists unusual astronomical phenomena observed during the Northern Sung dynasty. It comes from the "Journal of Astronomy," part 9, chapter 56, of the *Sung History* (*Sung Shih*), first printed in the 1340's. The photograph of that standard record used in this montage is copyright by, and may not be reproduced without permission of, the Trustees of the British Museum.

GRAVITATION

Charles W. MISNER
UNIVERSITY OF MARYLAND

Kip S. THORNE
CALIFORNIA INSTITUTE OF TECHNOLOGY

John Archibald WHEELER
PRINCETON UNIVERSITY

W. H. FREEMAN AND COMPANY San Francisco

Library of Congress Cataloging in Publication Data

Misner, Charles W. 1932–
 Gravitation.
 Bibliography: p.
 1. Gravitation. 2. Astrophysics. 3. General
relativity (Physics) I. Thorne, Kip S., 1940– joint author.
II. Wheeler, John Archibald, 1911– joint author.
III. Title.
QC178.M57 531'.14 78-156043
ISBN 0-7167-0334-3
ISBN 0-7167-0344-0 (pbk)

10 11 12 13 14 15 16 17 18 19 20 KP 8 9 8 7 6 5 4 3 2 1

We dedicate this book
To our fellow citizens
Who, for love of truth,
Take from their own wants
By taxes and gifts,
And now and then send forth
One of themselves
As dedicated servant,
To forward the search
Into the mysteries and marvelous simplicities
Of this strange and beautiful Universe,
Our home.

PREFACE

This is a textbook on gravitation physics (Einstein's "general relativity" or "geometrodynamics"). It supplies two tracks through the subject. The first track is focused on the key physical ideas. It assumes, as mathematical prerequisite, only vector analysis and simple partial-differential equations. It is suitable for a one-semester course at the junior or senior level or in graduate school; and it constitutes—in the opinion of the authors—the indispensable core of gravitation theory that every advanced student of physics should learn. The Track-1 material is contained in those pages of the book that have a 1 outlined in gray in the upper outside corner, by which the eye of the reader can quickly pick out the Track-1 sections. In the contents, the same purpose is served by a gray bar beside the section, box, or figure number.

The rest of the text builds up Track 1 into Track 2. Readers and teachers are invited to select, as enrichment material, those portions of Track 2 that interest them most. With a few exceptions, any Track-2 chapter can be understood by readers who have studied only the earlier Track-1 material. The exceptions are spelled out explicitly in "dependency statements" located at the beginning of each Track-2 chapter, or at each transition within a chapter from Track 1 to Track 2.

The entire book (all of Track 1 plus all of Track 2) is designed for a rigorous, full-year course at the graduate level, though many teachers of a full-year course may prefer a more leisurely pace that omits some of the Track-2 material. The full book is intended to give a competence in gravitation physics comparable to that which the average Ph.D. has in electromagnetism. When the student achieves this competence, he knows the laws of physics in flat spacetime (Chapters 1–7). He can predict orders of magnitude. He can also calculate using the principal tools of modern differential geometry (Chapters 8–15), and he can predict at all relevant levels of precision. He understands Einstein's geometric framework for physics (Chapters

16–22). He knows the applications of greatest present-day interest: pulsars and neutron stars (Chapters 23–26); cosmology (Chapters 27–30); the Schwarzschild geometry and gravitational collapse (Chapters 31–34); and gravitational waves (Chapters 35–37). He has probed the experimental tests of Einstein's theory (Chapters 38–40). He will be able to read the modern mathematical literature on differential geometry, and also the latest papers in the physics and astrophysics journals about geometrodynamics and its applications. If he wishes to go beyond the field equations, the four major applications, and the tests, he will find at the end of the book (Chapters 41–44) a brief survey of several advanced topics in general relativity. Among the topics touched on here, superspace and quantum geometrodynamics receive special attention. These chapters identify some of the outstanding physical issues and lines of investigation being pursued today.

Whether the department is physics or astrophysics or mathematics, more students than ever ask for more about general relativity than mere conversation. They want to hear its principal theses clearly stated. They want to know how to "work the handles of its information pump" themselves. More universities than ever respond with a serious course in Einstein's standard 1915 geometrodynamics. What a contrast to Maxwell's standard 1864 electrodynamics! In 1897, when Einstein was a student at Zurich, this subject was not on the instructional calendar of even half the universities of Europe.[1] "We waited in vain for an exposition of Maxwell's theory," says one of Einstein's classmates. "Above all it was Einstein who was disappointed,"[2] for he rated electrodynamics as "the most fascinating subject at the time"[3]—as many students rate Einstein's theory today!

Maxwell's theory recalls Einstein's theory in the time it took to win acceptance. Even as late as 1904 a book could appear by so great an investigator as William Thomson, Lord Kelvin, with the words, "The so-called 'electromagnetic theory of light' has not helped us hitherto . . . it seems to me that it is rather a backward step . . . the one thing about it that seems intelligible to me, I do not think is admissible . . . that there should be an electric displacement perpendicular to the line of propagation."[4] Did the pioneer of the Atlantic cable in the end contribute so richly to Maxwell electrodynamics—from units, and principles of measurement, to the theory of waves guided by wires—because of his own early difficulties with the subject? Then there is hope for many who study Einstein's geometrodynamics today! By the 1920's the weight of developments, from Kelvin's cable to Marconi's wireless, from the atom of Rutherford and Bohr to the new technology of high-frequency circuits, had produced general conviction that Maxwell was right. Doubt dwindled. Confidence led to applications, and applications led to confidence.

Many were slow to take up general relativity in the beginning because it seemed to be poor in applications. Einstein's theory attracts the interest of many today because it is rich in applications. No longer is attention confined to three famous but meager tests: the gravitational red shift, the bending of light by the sun, and

[1] G. Holton (1965).
[2] L. Kolbros (1956).
[3] A. Einstein (1949a).
[4] W. Thomson (1904).
Citations for references will be found in the bibliography.

the precession of the perihelion of Mercury around the sun. The combination of radar ranging and general relativity is, step by step, transforming the solar-system celestial mechanics of an older generation to a new subject, with a new level of precision, new kinds of effects, and a new outlook. Pulsars, discovered in 1968, find no acceptable explanation except as the neutron stars predicted in 1934, objects with a central density so high ($\sim 10^{14}\mathrm{g/cm^3}$) that the Einstein predictions of mass differ from the Newtonian predictions by 10 to 100 per cent. About further density increase and a final continued gravitational collapse, Newtonian theory is silent. In contrast, Einstein's standard 1915 geometrodynamics predicted in 1939 the properties of a completely collapsed object, a "frozen star" or "black hole." By 1966 detailed digital calculations were available describing the formation of such an object in the collapse of a star with a white-dwarf core. Today hope to discover the first black hole is not least among the forces propelling more than one research: How does rotation influence the properties of a black hole? What kind of pulse of gravitational radiation comes off when such an object is formed? What spectrum of x-rays emerges when gas from a companion star piles up on its way into a black hole?[5] All such investigations and more base themselves on Schwarzschild's standard 1916 static and spherically symmetric solution of Einstein's field equations, first really understood in the modern sense in 1960, and in 1963 generalized to a black hole endowed with angular momentum.

Beyond solar-system tests and applications of relativity, beyond pulsars, neutron stars, and black holes, beyond geometrostatics (compare electrostatics!) and stationary geometries (compare the magnetic field set up by a steady current!) lies geometrodynamics in the full sense of the word (compare electrodynamics!). Nowhere does Einstein's great conception stand out more clearly than here, that the geometry of space is a new physical entity, with degrees of freedom and a dynamics of its own. Deformations in the geometry of space, he predicted in 1918, can transport energy from place to place. Today, thanks to the initiative of Joseph Weber, detectors of such gravitational radiation have been constructed and exploited to give upper limits to the flux of energy streaming past the earth at selected frequencies. Never before has one realized from how many kinds of processes significant gravitational radiation can be anticipated. Never before has there been more interest in picking up this new kind of signal and using it to diagnose faraway events. Never before has there been such a drive in more than one laboratory to raise instrumental sensitivity until gravitational radiation becomes a workaday new window on the universe.

The expansion of the universe is the greatest of all tests of Einstein's geometrodynamics, and cosmology the greatest of all applications. Making a prediction too fantastic for its author to credit, the theory forecast the expansion years before it was observed (1929). Violating the short time-scale that Hubble gave for the expansion, and in the face of "theories" ("steady state"; "continuous creation") manufactured to welcome and utilize this short time-scale, standard general relativity resolutely persisted in the prediction of a long time-scale, decades before the astro-

[5] As of April 1973, there are significant indications that Cygnus X-1 and other compact x-ray sources may be black holes.

physical discovery (1952) that the Hubble scale of distances and times was wrong, and had to be stretched by a factor of more than five. Disagreeing by a factor of the order of thirty with the average density of mass-energy in the universe deduced from astrophysical evidence as recently as 1958, Einstein's theory now as in the past argues for the higher density, proclaims "the mystery of the missing matter," and encourages astrophysics in a continuing search that year by year turns up new indications of matter in the space between the galaxies. General relativity forecast the primordial cosmic fireball radiation, and even an approximate value for its present temperature, seventeen years before the radiation was discovered. This radiation brings information about the universe when it had a thousand times smaller linear dimensions, and a billion times smaller volume, than it does today. Quasistellar objects, discovered in 1963, supply more detailed information from a more recent era, when the universe had a quarter to half its present linear dimensions. Telling about a stage in the evolution of galaxies and the universe reachable in no other way, these objects are more than beacons to light up the far away and long ago. They put out energy at a rate unparalleled anywhere else in the universe. They eject matter with a surprising directivity. They show a puzzling variation with time, different between the microwave and the visible part of the spectrum. Quasistellar objects on a great scale, and galactic nuclei nearer at hand on a smaller scale, voice a challenge to general relativity: help clear up these mysteries!

If its wealth of applications attracts many young astrophysicists to the study of Einstein's geometrodynamics, the same attraction draws those in the world of physics who are concerned with physical cosmology, experimental general relativity, gravitational radiation, and the properties of objects made out of superdense matter. Of quite another motive for study of the subject, to contemplate Einstein's inspiring vision of geometry as the machinery of physics, we shall say nothing here because it speaks out, we hope, in every chapter of this book.

Why a new book? The new applications of general relativity, with their extraordinary physical interest, outdate excellent textbooks of an earlier era, among them even that great treatise on the subject written by Wolfgang Pauli at the age of twenty-one. In addition, differential geometry has undergone a transformation of outlook that isolates the student who is confined in his training to the traditional tensor calculus of the earlier texts. For him it is difficult or impossible either to read the writings of his up-to-date mathematical colleague or to explain the mathematical content of his physical problem to that friendly source of help. We have not seen any way to meet our responsibilities to our students at our three institutions except by a new exposition, aimed at establishing a solid competence in the subject, contemporary in its mathematics, oriented to the physical and astrophysical applications of greatest present-day interest, and animated by belief in the beauty and simplicity of nature.

High Island *Charles W. Misner*
South Bristol, Maine *Kip S. Thorne*
September 4, 1972 *John Archibald Wheeler*

CONTENTS

BOXES

FIGURES

ACKNOWLEDGMENTS

Deep appreciation goes to all who made this book possible. A colleague gives us a special lecture so that we may adapt it into one of the chapters of this book. Another investigator clears up for us the tangled history of the production of matter out of the vacuum by strong tidal gravitational forces. A distant colleague telephones in references on the absence of any change in physical constants with time. One student provides a problem on the energy density of a null electromagnetic field. Another supplies curves for effective potential as a function of distance. A librarian writes abroad to get us an article in an obscure publication. A secretary who cares types the third revision of a chapter. Editor and illustrator imaginatively solve a puzzling problem of presentation. Repeat in imagination such instances of warm helpfulness and happy good colleagueship times beyond count. Then one has some impression of the immense debt we owe to over a hundred-fifty colleagues. Each face is etched in our mind, and to each our gratitude is heartfelt. Warm thanks we give also to the California Institute of Technology, the Dublin Institute for Advanced Studies, the Institute for Advanced Study at Princeton, Kyoto University, the University of Maryland, Princeton University, and the University of Texas at Austin for hospitality during the writing of this book. We are grateful to the Academy of Sciences of the U.S.S.R., to Moscow University, and to our Soviet colleagues for their hospitality and the opportunity to become better acquainted in June–July 1971 with Soviet work in gravitation physics. For assistance in the research that went into this book we thank the National Science Foundation for grants (GP27304 and 28027 to Caltech; GP17673 and GP8560 to Maryland; and GP3974 and GP7669 to Princeton); the U.S. Air Force Office of Scientific Research (grant AF49-638-1545 to Princeton); the U.S. National Aeronautics and Space Agency (grant NGR 05-002-256 to Caltech, NSG 210-002-010 to Maryland); the Alfred P. Sloan Foundation for a fellowship awarded to one of us (K.S.T.); and the John Simon Guggenheim Memorial Foundation and All Souls College, Oxford, England, for fellowships awarded to another of us (C.W.M.).

GRAVITATION

SPACETIME PHYSICS

*Wherein the reader is led, once quickly (§ 1.1),
then again more slowly, down the highways and
a few byways of Einstein's geometrodynamics—
without benefit of a good mathematical compass.*

GEOMETRODYNAMICS IN BRIEF

§1.1. THE PARABLE OF THE APPLE

> *One day in the year 1666 Newton had gone to the country,*
> *and seeing the fall of an apple, as his niece told me, let himself*
> *be led into a deep meditation on the cause which thus*
> *draws every object along a line whose extension would pass*
> *almost through the center of the Earth.*
>
> VOLTAIRE (1738)

Once upon a time a student lay in a garden under an apple tree reflecting on the difference between Einstein's and Newton's views about gravity. He was startled by the fall of an apple nearby. As he looked at the apple, he noticed ants beginning to run along its surface (Figure 1.1). His curiosity aroused, he thought to investigate the principles of navigation followed by an ant. With his magnifying glass, he noted one track carefully, and, taking his knife, made a cut in the apple skin one mm above the track and another cut one mm below it. He peeled off the resulting little highway of skin and laid it out on the face of his book. The track ran as straight as a laser beam along this highway. No more economical path could the ant have found to cover the ten cm from start to end of that strip of skin. Any zigs and zags or even any smooth bend in the path on its way along the apple peel from starting point to end point would have increased its length.

"What a beautiful geodesic," the student commented.

His eye fell on two ants starting off from a common point P in slightly different directions. Their routes happened to carry them through the region of the dimple at the top of the apple, one on each side of it. Each ant conscientiously pursued

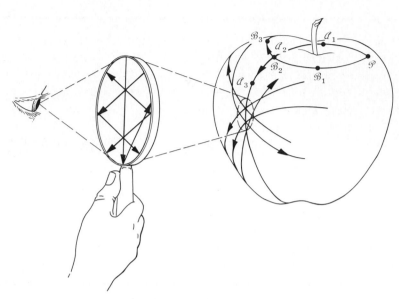

Figure 1.1.

The Riemannian geometry of the spacetime of general relativity is here symbolized by the two-dimensional geometry of the surface of an apple. The geodesic tracks followed by the ants on the apple's surface symbolize the world line followed through spacetime by a free particle. In any sufficiently localized region of spacetime, the geometry can be idealized as flat, as symbolized on the apple's two-dimensional surface by the straight-line course of the tracks viewed in the magnifying glass ("local Lorentz character" of geometry of spacetime). In a region of greater extension, the curvature of the manifold (four-dimensional spacetime in the case of the real physical world; curved two-dimensional geometry in the case of the apple) makes itself felt. Two tracks \mathcal{A} and \mathcal{B}, originally diverging from a common point \mathcal{P}, later approach, cross, and go off in very different directions. In Newtonian theory this effect is ascribed to gravitation acting at a distance from a center of attraction, symbolized here by the stem of the apple. According to Einstein a particle gets its moving orders locally, from the geometry of spacetime right where it is. Its instructions are simple: to follow the straightest possible track (geodesic). Physics is as simple as it could be locally. Only because spacetime is curved in the large do the tracks cross. Geometrodynamics, in brief, is a double story of the effect of geometry on matter (causing originally divergent geodesics to cross) and the effect of matter on geometry (bending of spacetime initiated by concentration of mass, symbolized by effect of stem on nearby surface of apple).

his geodesic. Each went as straight on his strip of appleskin as he possibly could. Yet because of the curvature of the dimple itself, the two tracks not only crossed but emerged in very different directions.

Einstein's local view of physics contrasted with Newton's "action at a distance"

"What happier illustration of Einstein's geometric theory of gravity could one possibly ask?" murmured the student. "The ants move as if they were attracted by the apple stem. One might have believed in a Newtonian force at a distance. Yet from nowhere does an ant get his moving orders except from the local geometry along his track. This is surely Einstein's concept that all physics takes place by 'local action.' What a difference from Newton's 'action at a distance' view of physics! Now I understand better what this book means."

Physics is simple only when analyzed locally

And so saying, he opened his book and read, "Don't try to describe motion relative to faraway objects. *Physics is simple only when analyzed locally*. And locally

the world line that a satellite follows [in spacetime, around the Earth] is already as straight as any world line can be. Forget all this talk about 'deflection' and 'force of gravitation.' I'm inside a spaceship. Or I'm floating outside and near it. Do I feel any 'force of gravitation'? Not at all. Does the spaceship 'feel' such a force? No. Then why talk about it? Recognize that the spaceship and I traverse a region of spacetime free of all force. Acknowledge that the motion through that region is already ideally straight."

The dinner bell was ringing, but still the student sat, musing to himself. "Let me see if I can summarize Einstein's geometric theory of gravity in three ideas: (1) locally, geodesics appear straight; (2) over more extended regions of space and time, geodesics originally receding from each other begin to approach at a rate governed by the curvature of spacetime, and this effect of geometry on matter is what we mean today by that old word 'gravitation'; (3) matter in turn warps geometry. The dimple arises in the apple because the stem is there. I think I see how to put the whole story even more briefly: *Space acts on matter, telling it how to move. In turn, matter reacts back on space, telling it how to curve.* In other words, matter here," he said, rising and picking up the apple by its stem, "curves space here. To produce a curvature in space here is to force a curvature in space there," he went on, as he watched a lingering ant busily following its geodesic a finger's breadth away from the apple's stem. "Thus matter here influences matter there. That is Einstein's explanation for 'gravitation.'"

Then the dinner bell was quiet, and he was gone, with book, magnifying glass—and apple.

Space tells matter how to move

Matter tells space how to curve

§1.2. SPACETIME WITH AND WITHOUT COORDINATES

> *Now it came to me: . . . the independence of the gravitational acceleration from the nature of the falling substance, may be expressed as follows: In a gravitational field (of small spatial extension) things behave as they do in a space free of gravitation. . . . This happened in 1908. Why were another seven years required for the construction of the general theory of relativity? The main reason lies in the fact that it is not so easy to free oneself from the idea that coordinates must have an immediate metrical meaning.*
>
> ALBERT EINSTEIN [in Schilpp (1949), pp. 65–67.]

Nothing is more distressing on first contact with the idea of "curved spacetime" than the fear that every simple means of measurement has lost its power in this unfamiliar context. One thinks of oneself as confronted with the task of measuring the shape of a gigantic and fantastically sculptured iceberg as one stands with a meter stick in a tossing rowboat on the surface of a heaving ocean. Were it the rowboat itself whose shape were to be measured, the procedure would be simple enough. One would draw it up on shore, turn it upside down, and drive tacks in lightly at strategic points here and there on the surface. The measurement of distances from tack to

Problem: how to measure in curved spacetime

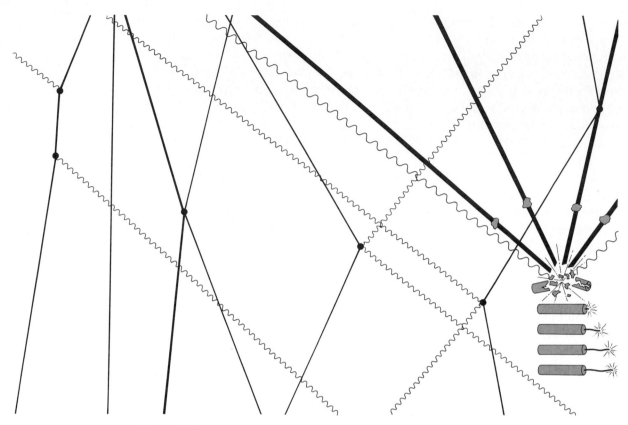

Figure 1.2.

The crossing of straws in a barn full of hay is a symbol for the world lines that fill up spacetime. By their crossings and bends, these world lines mark events with a uniqueness beyond all need of coordinate systems or coordinates. Typical events symbolized in the diagram, from left to right (black dots), are: absorption of a photon; reemission of a photon; collision between a particle and a particle; collision between a photon and a particle; another collision between a photon and a particle; explosion of a firecracker; and collision of a particle from outside with one of the fragments of that firecracker.

Resolution: characterize
events by what happens
there

tack would record and reveal the shape of the surface. The precision could be made arbitrarily great by making the number of tacks arbitrarily large. It takes more daring to think of driving several score pitons into the towering iceberg. But with all the daring in the world, how is one to drive a nail into spacetime to mark a point? Happily, nature provides its own way to localize a point in spacetime, as Einstein was the first to emphasize. Characterize the point by what happens there! Give a point in spacetime the name "event." Where the event lies is defined as clearly and sharply as where two straws cross each other in a barn full of hay (Figure 1.2). To say that the event marks a collision of such and such a photon with such and such a particle is identification enough. The world lines of that photon and that particle are rooted in the past and stretch out into the future. They have a rich texture of connections with nearby world lines. These nearby world lines in turn are linked in a hundred ways with world lines more remote. How then does one tell the location of an event? Tell first what world lines participate in the event. Next follow each

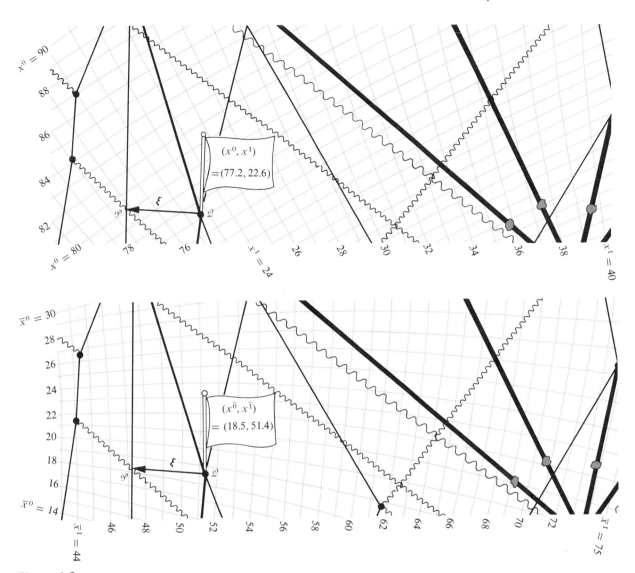

Figure 1.3.

Above: Assigning "telephone numbers" to events by way of a system of coordinates. To say that the coordinate system is "smooth" is to say that events which are almost in the same place have almost the same coordinates. Below: Putting the same set of events into equally good order by way of a different system of coordinates. Picked out specially here are two neighboring events: an event named "$\mathscr{2}$" with coordinates $(x^0, x^1) = (77.2, 22.6)$ and $(x^{\bar{0}}, x^{\bar{1}}) = (18.5, 51.4)$; and an event named "$\mathscr{P}$" with coordinates $(x^0, x^1) = (79.9, 20.1)$ and $(x^{\bar{0}}, x^{\bar{1}}) = (18.4, 47.1)$. Events $\mathscr{2}$ and \mathscr{P} are connected by the separation "vector" $\boldsymbol{\xi}$. (Precise definition of a vector in a curved spacetime demands going to the mathematical limit in which the two points have an indefinitely small separation [N-fold reduction of the separation $\mathscr{P} - \mathscr{2}$], and, in the resultant locally flat space, multiplying the separation up again by the factor N [$\lim N \to \infty$; "tangent space"; "tangent vector"]. Forego here that proper way of stating matters, and forego complete accuracy; hence the quote around the word "vector".) In each coordinate system the separation vector $\boldsymbol{\xi}$ is characterized by "components" (differences in coordinate values between \mathscr{P} and $\mathscr{2}$):

$$(\xi^0, \xi^1) = (79.9 - 77.2, 20.1 - 22.6) = (2.7, -2.5),$$
$$(\xi^{\bar{0}}, \xi^{\bar{1}}) = (18.4 - 18.5, 47.1 - 51.4) = (-0.1, -4.3).$$

See Box 1.1 for further discussion of events, coordinates, and vectors.

of these world lines. Name the additional events that they encounter. These events pick out further world lines. Eventually the whole barn of hay is catalogued. Each event is named. One can find one's way as surely to a given intersection as the city dweller can pick his path to the meeting of St. James Street and Piccadilly. No numbers. No coordinate system. No coordinates.

That most streets in Japan have no names, and most houses no numbers, illustrates one's ability to do without coordinates. One can abandon the names of two world lines as a means to identify the event where they intersect. Just as one could name a Japanese house after its senior occupant, so one can and often does attach arbitrary names to specific events in spacetime, as in Box 1.1.

Coordinates, however, are convenient. How else from the great thick catalog of events, randomly listed, can one easily discover that along a certain world line one will first encounter event Trinity, then Baker, then Mike, then Argus—but not the same events in some permuted order?

To order events, introduce coordinates! (See Figure 1.3.) Coordinates are four indexed numbers per event in spacetime; on a sheet of paper, only two. Trinity acquires coordinates

$$(x^0, x^1, x^2, x^3) = (77, 23, 64, 11).$$

In christening events with coordinates, one demands smoothness but foregoes every thought of mensuration. The four numbers for an event are nothing but an elaborate kind of telephone number. Compare their "telephone" numbers to discover whether two events are neighbors. But do not expect to learn how many meters separate them from the difference in their telephone numbers!

Nothing prevents a subscriber from being served by competing telephone systems, nor an event from being catalogued by alternative coordinate systems (Figure 1.3). Box 1.1 illustrates the relationships between one coordinate system and another, as well as the notation used to denote coordinates and their transformations.

Choose two events, known to be neighbors by the nearness of their coordinate values in a smooth coordinate system. Draw a little arrow from one event to the other. Such an arrow is called a *vector*. (It is a well-defined concept in flat spacetime, or in curved spacetime in the limit of vanishingly small length; for finite lengths in curved spacetime, it must be refined and made precise, under the new name "tangent vector," on which see Chapter 9.) This vector, like events, can be given a name. But whether named "John" or "Charles" or "Kip," it is a unique, well-defined geometrical object. The name is a convenience, but the vector exists even without it.

Just as a quadruple of coordinates

$$(x^0, x^1, x^2, x^3) = (77, 23, 64, 11)$$

is a particularly useful name for the event "Trinity" (it can be used to identify what other events are nearby), so a quadruple of "components"

$$(\xi^0, \xi^1, \xi^2, \xi^3) = (1.2, -0.9, 0, 2.1)$$

The name of an event can even be arbitrary

Coordinates provide a convenient naming system

Coordinates generally do not measure length

Several coordinate systems can be used at once

Vectors

Box 1.1 MATHEMATICAL NOTATION FOR EVENTS, COORDINATES, AND VECTORS

Events are denoted by capital script, one-letter Latin names such as
Sometimes subscripts are used:

$\mathcal{P}, \mathcal{Q}, \mathcal{A}, \mathcal{B}.$
$\mathcal{P}_0, \mathcal{P}_1, \mathcal{B}_6.$

Coordinates of an event \mathcal{P} are denoted by

or by

or more abstractly by
where it is understood that Greek indices can take on any value 0, 1, 2, or 3.

$t(\mathcal{P}), x(\mathcal{P}), y(\mathcal{P}), z(\mathcal{P}),$
$x^0(\mathcal{P}), x^1(\mathcal{P}), x^2(\mathcal{P}),$
$x^3(\mathcal{P}),$
$x^\mu(\mathcal{P})$ or $x^\alpha(\mathcal{P}),$

Time coordinate (when one of the four is picked to play this role)

$x^0(\mathcal{P}).$

Space coordinates are
and are sometimes denoted by
It is to be understood that Latin indices take on values 1, 2, or 3.

$x^1(\mathcal{P}), x^2(\mathcal{P}), x^3(\mathcal{P})$
$x^j(\mathcal{P})$ or $x^k(\mathcal{P})$ or \ldots.

Shorthand notation: One soon tires of writing explicitly the functional dependence of the coordinates, $x^\beta(\mathcal{P})$; so one adopts the shorthand notation
for the coordinates of the event \mathcal{P}, and
for the space coordinates. One even begins to think of x^β as representing the event \mathcal{P} itself, but must remind oneself that the values of x^0, x^1, x^2, x^3 depend not only on the choice of \mathcal{P} but also on the *arbitrary* choice of coordinates!

x^β
x^j

Other coordinates for the same event \mathcal{P} may be denoted

$x^{\bar{\alpha}}(\mathcal{P})$ or just $x^{\bar{\alpha}}$,
$x^{\alpha'}(\mathcal{P})$ or just $x^{\alpha'}$,
$x^{\hat{\alpha}}(\mathcal{P})$ or just $x^{\hat{\alpha}}$.

EXAMPLE: In Figure 1.3 $(x^0, x^1) = (77.2, 22.6)$ and $(x^{\bar{0}}, x^{\bar{1}}) = (18.5, 51.4)$ refer to the *same* event. The bars, primes, and hats distinguish one coordinate system from another; by putting them on the indices rather than on the x's, we simplify later notation.

Transformation from one coordinate system to another is achieved by the four functions

which are denoted more succinctly

$x^{\bar{0}}(x^0, x^1, x^2, x^3),$
$x^{\bar{1}}(x^0, x^1, x^2, x^3),$
$x^{\bar{2}}(x^0, x^1, x^2, x^3),$
$x^{\bar{3}}(x^0, x^1, x^2, x^3),$
$x^{\bar{\alpha}}(x^\beta).$

Separation vector* (little arrow) reaching from one event \mathcal{Q} to neighboring event \mathcal{P} can be denoted abstractly by
It can also be characterized by the coordinate-value differences† between \mathcal{P} and \mathcal{Q} (called "components" of the vector)

\boldsymbol{u} or \boldsymbol{v} or $\boldsymbol{\xi}$, or $\mathcal{P} - \mathcal{Q}$.

$\xi^\alpha \equiv x^\alpha(\mathcal{P}) - x^\alpha(\mathcal{Q}),$
$\xi^{\bar{\alpha}} \equiv x^{\bar{\alpha}}(\mathcal{P}) - x^{\bar{\alpha}}(\mathcal{Q}).$

Transformation of components of a vector from one coordinate system to another is achieved by partial derivatives of transformation equations

since $\xi^{\bar{\alpha}} = x^{\bar{\alpha}}(\mathcal{P}) - x^{\bar{\alpha}}(\mathcal{Q}) = (\partial x^{\bar{\alpha}}/\partial x^\beta)[x^\beta(\mathcal{P}) - x^\beta(\mathcal{Q})].$†

$\xi^{\bar{\alpha}} = \dfrac{\partial x^{\bar{\alpha}}}{\partial x^\beta} \xi^\beta,$

Einstein summation convention is used here:
any index that is repeated in a product is automatically summed on

$\dfrac{\partial x^{\bar{\alpha}}}{\partial x^\beta} \xi^\beta \equiv \displaystyle\sum_{\beta=0}^{3} \dfrac{\partial x^{\bar{\alpha}}}{\partial x^\beta} \xi^\beta.$

*This definition of a vector is valid only in flat spacetime. The refined definition ("tangent vector") in curved spacetime is not spelled out here (see Chapter 9), but flat-geometry ideas apply with good approximation even in a curved geometry, when the two points are sufficiently close.

†These formulas are precisely accurate only when the region of spacetime under consideration is flat and when in addition the coordinates are Lorentzian. Otherwise they are approximate—though they become arbitrarily good when the separation between points and the length of the vector become arbitrarily small.

is a convenient name for the vector "John" that reaches from

$$(x^0, x^1, x^2, x^3) = (77, 23, 64, 11)$$

to

$$(x^0, x^1, x^2, x^3) = (78.2, 22.1, 64.0, 13.1).$$

How to work with the components of a vector is explored in Box 1.1.

Coordinate singularities normally unavoidable

There are many ways in which a coordinate system can be imperfect. Figure 1.4 illustrates a coordinate singularity. For another example of a coordinate singularity, run the eye over the surface of a globe to the North Pole. Note the many meridians that meet there ("collapse of cells of egg crates to zero content"). Can't one do better? Find a single coordinate system that will cover the globe without singularity? A theorem says no. Two is the minimum number of "coordinate patches" required to cover the two-sphere without singularity (Figure 1.5). This circumstance emphasizes anew that points and events are primary, whereas coordinates are a mere bookkeeping device.

Continuity of spacetime

Figures 1.2 and 1.3 show only a few world lines and events. A more detailed diagram would show a maze of world lines and of light rays and the intersections between them. From such a picture, one can in imagination step to the idealized limit: an infinitely dense collection of light rays and of world lines of infinitesimal test particles. With this idealized physical limit, the mathematical concept of a continuous four-dimensional "manifold" (four-dimensional space with certain smoothness properties) has a one-to-one correspondence; and in this limit continuous, differentiable (i.e., smooth) coordinate systems operate. The mathematics then supplies a tool to reason about the physics.

The mathematics of manifolds applied to the physics of spacetime

Dimensionality of spacetime

A simple countdown reveals the dimensionality of the manifold. Take a point \mathscr{P} in an n-dimensional manifold. Its neighborhood is an n-dimensional ball (i.e., the interior of a sphere whose surface has $n - 1$ dimensions). Choose this ball so that its boundary is a smooth manifold. The dimensionality of this manifold is $(n - 1)$. In this $(n - 1)$-dimensional manifold, pick a point \mathscr{Q}. Its neighborhood is an $(n - 1)$-dimensional ball. Choose this ball so that ..., and so on. Eventually one comes by this construction to a manifold that is two-dimensional but is not yet known to be two-dimensional (two-sphere). In this two-dimensional manifold, pick a point \mathscr{M}. Its neighborhood is a two-dimensional ball ("disc"). Choose this disc so that its boundary is a smooth manifold (circle). In this manifold, pick a point \mathscr{N}. Its neighborhood is a one-dimensional ball, but is not yet known to be one-dimensional ("line segment"). The boundaries of this object are two points. This circumstance tells that the intervening manifold is one-dimensional; therefore the previous manifold was two-dimensional; and so on. The dimensionality of the original manifold is equal to the number of points employed in the construction. For spacetime, the dimensionality is 4.

This kind of mathematical reasoning about dimensionality makes good sense at the everyday scale of distances, at atomic distances (10^{-8} cm), at nuclear dimensions (10^{-13} cm), and even at lengths smaller by several powers of ten, if one judges by the concord between prediction and observation in quantum electrodynamics at high

Figure 1.4.

How a mere coordinate singularity arises. Above: A coordinate system becomes *singular* when the "cells in the egg crate" are squashed to zero volume. Below: An example showing such a singularity in the Schwarzschild coordinates r, t often used to describe the geometry around a black hole (Chapter 31). For simplicity the angular coordinates θ, ϕ have been suppressed. The singularity shows itself in two ways. First, all the points along the dotted line, while quite distinct one from another, are designated by the same pair of (r, t) values; namely, $r = 2m$, $t = \infty$. The coordinates provide no way to distinguish these points. Second, the "cells in the egg crate," of which one is shown grey in the diagram, collapse to zero content at the dotted line. In summary, there is nothing strange about the geometry at the dotted line; all the singularity lies in the coordinate system ("poor system of telephone numbers"). No confusion should be permitted to arise from the accidental circumstance that the t coordinate attains an infinite value on the dotted line. No such infinity would occur if t were replaced by the new coordinate \bar{t}, defined by

$$(t/2m) = \tan(\bar{t}/2m).$$

When $t = \infty$, the new coordinate \bar{t} is $\bar{t} = \pi m$. The r, \bar{t} coordinates still provide no way to distinguish the points along the dotted line. They still give "cells in the egg crate" collapsed to zero content along the dotted line.

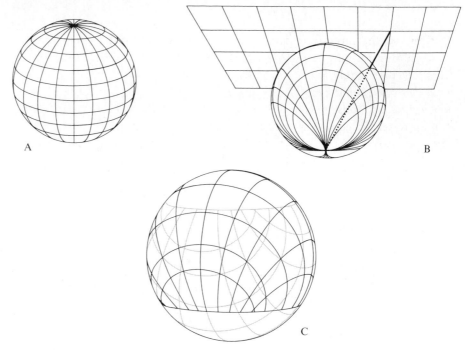

Figure 1.5.
Singularities in familiar coordinates on the two-sphere can be eliminated by covering the sphere with two overlapping coordinate patches. A. Spherical polar coordinates, singular at the North and South Poles, and discontinuous at the international date line. B. Projection of the Euclidean coordinates of the Euclidean two-plane, tangent at the North Pole, onto the sphere via a line running to the South Pole; coordinate singularity at the South Pole. C. Coverage of two-sphere by two overlapping coordinate patches. One, constructed as in B, covers without singularity the northern hemisphere and also the southern tropics down to the Tropic of Capricorn. The other (grey) also covers without singularity all of the tropics and the southern hemisphere besides.

Breakdown in smoothness of spacetime at Planck length

energies (corresponding de Broglie wavelength 10^{-16} cm). Moreover, classical general relativity thinks of the spacetime manifold as a deterministic structure, completely well-defined down to arbitrarily small distances. Not so quantum general relativity or "quantum geometrodynamics." It predicts violent fluctuations in the geometry at distances on the order of the Planck length,

$$L^* = (\hbar G/c^3)^{1/2}$$
$$= [(1.054 \times 10^{-27} \text{ g cm}^2/\text{sec})(6.670 \times 10^{-8} \text{ cm}^3/\text{g sec}^2)]^{1/2} \times$$
$$\times (2.998 \times 10^{10} \text{ cm/sec})^{-3/2} \quad (1.1)$$
$$= 1.616 \times 10^{-33} \text{ cm.}$$

No one has found any way to escape this prediction. As nearly as one can estimate, these fluctuations give space at small distances a "multiply connected" or "foamlike" character. This lack of smoothness may well deprive even the concept of dimensionality itself of any meaning at the Planck scale of distances. The further exploration of this issue takes one to the frontiers of Einstein's theory (Chapter 44).

If spacetime at small distances is far from the mathematical model of a continuous manifold, is there not also at larger distances a wide gap between the mathematical

idealization and the physical reality? The infinitely dense collection of light rays and of world lines of infinitesimal test particles that are to define all the points of the manifold: they surely are beyond practical realization. Nobody has ever found a particle that moves on timelike world lines (finite rest mass) lighter than an electron. A collection of electrons, even if endowed with zero density of charge (e$^+$ and e$^-$ world lines present in equal numbers) will have a density of mass. This density will curve the very manifold under study. Investigation in infinite detail means unlimited density, and unlimited disturbance of the geometry.

Difficulty in defining geometry even at classical distances?

However, to demand investigatability in infinite detail in the sense just described is as out of place in general relativity as it would be in electrodynamics or gas dynamics. Electrodynamics speaks of the strength of the electric and magnetic field at each point in space and at each moment of time. To measure those fields, it is willing to contemplate infinitesimal test particles scattered everywhere as densely as one pleases. However, the test particles do not have to be there at all to give the field reality. The field has everywhere a clear-cut value and goes about its deterministic dynamic evolution willy-nilly and continuously, infinitesimal test particles or no infinitesimal test particles. Similarly with the geometry of space.

No; one must accept geometry at classical distances as meaningful

In conclusion, when one deals with spacetime in the context of classical physics, one accepts (1) the notion of "infinitesimal test particle" and (2) the idealization that the totality of identifiable events forms a four-dimensional continuous manifold. Only at the end of this book will a look be taken at some of the limitations placed by the quantum principle on one's way of speaking about and analyzing spacetime.

§1.3. WEIGHTLESSNESS

"Gravity is a great mystery. Drop a stone. See it fall. Hear it hit. No one understands why." What a misleading statement! Mystery about fall? What else should the stone do except fall? To fall is normal. The abnormality is an object standing in the way of the stone. If one wishes to pursue a "mystery," do not follow the track of the falling stone. Look instead at the impact, and ask what was the force that pushed the stone away from its natural "world line," (i.e., its natural track through space-time). That could lead to an interesting issue of solid-state physics, but that is not the topic of concern here. Fall is. Free fall is synonymous with weightlessness: absence of any force to drive the object away from its normal track through space-time. Travel aboard a freely falling elevator to experience weightlessness. Or travel aboard a spaceship also falling straight toward the Earth. Or, more happily, travel aboard a spaceship in that state of steady fall toward the Earth that marks a circular orbit. In each case one is following a natural track through spacetime.

Free fall is the natural state of motion

The traveler has one chemical composition, the spaceship another; yet they travel together, the traveler weightless in his moving home. Objects of such different nuclear constitution as aluminum and gold fall with accelerations that agree to better than one part in 10^{11}, according to Roll, Krotkov, and Dicke (1964), one of the most important null experiments in all physics (see Figure 1.6). Individual molecules fall in step, too, with macroscopic objects [Estermann, Simpson, and Stern (1938)]; and so do individual neutrons [Dabbs, Harvey, Paya, and Horstmann (1965)], individual

All objects fall with the same acceleration

(continued on page 16)

Figure 1.6.

Principle of the Roll-Krotkov-Dicke experiment, which showed that the gravitational accelerations of gold and aluminum are equal to 1 part in 10^{11} or better (Princeton, 1964). In the upper lefthand corner, equal masses of gold and aluminum hang from a supporting bar. This bar in turn is supported at its midpoint. If both objects fall toward the sun with the same acceleration of $g = 0.59$ cm/sec^2, the bar does not turn. If the Au mass receives a higher acceleration, $g + \delta g$, then the gold end of the bar starts to turn toward the sun in the Earth-fixed frame. Twelve hours later the sun is on the other side, pulling the other way. The alternating torque lends itself to recognition against a background of noise because of its precise 24-hour period. Unhappily, any substantial mass nearby, such as an experimenter, located at M, will produce a torque that swamps the effect sought. Therefore the actual arrangement was as shown in the body of the figure. One gold weight and two aluminum weights were supported at the three corners of a horizontal equilateral triangle, 6 cm on a side (three-fold axis of symmetry, giving zero response to all the simplest nonuniformities in the gravitational field). Also, the observers performed all operations remotely to eliminate their own gravitational effects*. To detect a rotation of the torsion balance as small as $\sim 10^{-9}$ rad without disturbing the balance, Roll, Krotkov, and Dicke reflected a very weak light beam from the optically flat back face of the quartz triangle. The image of the source slit fell on a wire of about the same size as the slit image. The light transmitted past the wire fell on a photomultiplier. A separate oscillator circuit drove the wire back and forth across the image at 3,000 hertz. When the image was centered perfectly, only even harmonics of the oscillation frequency appeared in the light intensity. However, when the image was displaced slightly to one side, the fundamental frequency appeared in the light intensity. The electrical output of the photomultiplier then contained a 3,000-hertz component. The magnitude and sign of this component were determined automatically. Equally automatically a proportional D.C. voltage was applied to the electrodes shown in the diagram. It restored the torsion balance to its zero position. The D.C. voltage required to restore the balance to its zero position was recorded as a measure of the torque acting on the pendulum. This torque was Fourier-analyzed over a period of many days. The magnitude of the Fourier component of 24-hour period indicated a ratio $\delta g/g = (0.96 \pm 1.04) \times 10^{-11}$. Aluminum and gold thus fall with the same acceleration, despite their important differences summarized in the table.

Ratios	Al	Au
$\dfrac{\text{Number of neutrons}}{\text{Number of protons}}$	1.08	1.5
$\dfrac{\text{Mass of kinetic energy of K-electron}}{\text{Rest mass of electron}}$	0.005	0.16
$\dfrac{\text{Electrostatic mass-energy of nucleus}}{\text{Mass of atom}}$	0.001	0.004

The theoretical implications of this experiment will be discussed in greater detail in Chapters 16 and 38.

Braginsky and Panov (1971) at Moscow University performed an experiment identical in principle to that of Dicke-Roll-Krotkov, but with a modified experimental set-up. Comparing the accelerations of platinum and aluminum rather than of gold and aluminum, they say that

$$\delta g/g \lesssim 1 \times 10^{-12}.$$

*Other perturbations had to be, and were, guarded against. (1) A bit of iron on the torsion balance as big as 10^{-3} cm on a side would have contributed, in the Earth's magnetic field, a torque a hundred times greater than the measured torque. (2) The unequal pressure of radiation on the two sides of a mass would have produced an unacceptably large perturbation if the temperature difference between these two sides had exceeded 10^{-4} °K. (3) Gas evolution from one side of a mass would have propelled it like a rocket. If the rate of evolution were as great as 10^{-8} g/day, the calculated force would have been $\sim 10^{-7}$ g cm/sec^2, enough to affect the measurements. (4) The rotation was measured with respect to the pier that supported the equipment. As a guarantee that this pier did not itself rotate, it was anchored to bed rock. (5) Electrostatic forces were eliminated; otherwise they would have perturbed the balance.

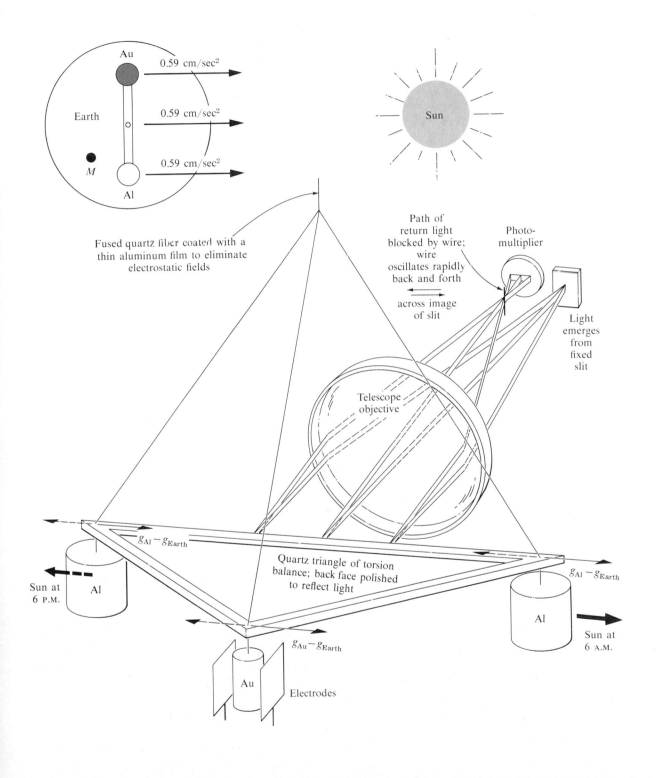

0.59 cm/sec²

0.59 cm/sec²

0.59 cm/sec²

Au

Earth

M

Al

Sun

Fused quartz fiber coated with a
thin aluminum film to eliminate
electrostatic fields

Path of
return light
blocked by wire;
wire
oscillates rapidly
back and forth

across image
of slit

Photo-
multiplier

Light
emerges
from
fixed
slit

Telescope
objective

$g_{Al} - g_{Earth}$

$g_{Al} - g_{Earth}$

Quartz triangle of torsion
balance; back face polished
to reflect light

Sun at
6 P.M.

Al

Al

Sun at
6 A.M.

$g_{Au} - g_{Earth}$

Au

Electrodes

electrons [Witteborn and Fairbank (1967)] and individual mu mesons [Beall (1970)]. What is more, not one of these objects has to see out into space to know how to move.

Contemplate the interior of a spaceship, and a key, penny, nut, and pea by accident or design set free inside. Shielded from all view of the world outside by the walls of the vessel, each object stays at rest relative to the vessel. Or it moves through the room in a straight line with uniform velocity. That is the lesson which experience shouts out.

Forego talk of acceleration! That, paradoxically, is the lesson of the circumstance that "all objects fall with the same acceleration." Whose fault were those accelerations, after all? They came from allowing a groundbased observer into the act. The

Box 1.2 MATERIALS OF THE MOST DIVERSE COMPOSITION FALL WITH THE SAME ACCELERATION ("STANDARD WORLD LINE")

Aristotle: "the downward movement of a mass of gold or lead, or of any other body endowed with weight, is quicker in proportion to its size."

Pre-Galilean literature: metal and wood weights fall at the same rate.

Galileo: (1) "the variation of speed in air between balls of gold, lead, copper, porphyry, and other heavy materials is so slight that in a fall of 100 cubits [about 46 meters] a ball of gold would surely not outstrip one of copper by as much as four fingers. Having observed this, I came to the conclusion that in a medium totally void of resistance all bodies would fall with the same speed." (2) later experiments of greater precision "diluting gravity" and finding same time of descent for different objects along an inclined plane.

Newton: inclined plane replaced by arc of pendulum bob; "time of fall" for bodies of different composition determined by comparing time of oscillation of pendulum bobs of the two materials. Ultimate limit of precision in such experiments limited by problem of determining effective length of each pendulum: (acceleration) = $(2\pi/\text{period})^2$(length).

Lorand von Eötvös, Budapest, 1889 and 1922: compared on the rotating earth the vertical defined by a plumb bob of one material with the vertical defined by a plumb bob of other material. The two hanging masses, by the two unbroken threads that support them, were drawn along identical world lines through spacetime (middle of the laboratory of Eötvös!). If cut free, would they also follow identical tracks through spacetime ("normal world line of test mass")? If so, the acceleration that draws the actual world line from the normal free-fall world line will have a standard value, *a*. The experiment of Eötvös did not try to test agreement on the magnitude of *a* between the two masses. Doing so would have required (1) cutting the threads and (2) following the fall of the two masses. Eötvös renounced this approach in favor of a static observation that he could make with greater precision, comparing the *direction* of *a* for the two masses. The direction of the supporting thread, so his argument ran, reveals the direction in which the mass is being dragged away from its normal world line of "free fall" or "weightlessness." This acceleration is the vectorial resultant of (1) an acceleration of magnitude *g*, directed outward against so-called gravity, and (2) an acceleration directed toward the axis of rotation of the earth, of magnitude $\omega^2 R \sin\theta$ (ω, angular ve-

push of the ground under his feet was driving him away from a natural world line. Through that flaw in his arrangements, he became responsible for all those accelerations. Put him in space and strap rockets to his legs. No difference!* Again the responsibility for what he sees is his. Once more he notes that "all objects fall with

*"No difference" spelled out amounts to Einstein's (1911) principle of the local equivalence between a "gravitational field" and an acceleration: *"We arrive at a very satisfactory interpretation of this law of experience, if we assume that the systems K and K' are physically exactly equivalent, that is, if we assume that we may just as well regard the system K as being in a space free from gravitational fields, if we then regard K as uniformly accelerated. This assumption of exact physical equivalence makes it impossible for us to speak of the absolute acceleration of the system of reference, just as the usual theory of relativity forbids us to talk of the absolute velocity of a system; and it makes the equal falling of all bodies in a gravitational field seem a matter of course."*

locity; R, radius of earth; θ, polar angle measured from North Pole to location of experiment). This centripetal acceleration has a vertical component $-\omega^2 R \sin^2 \theta$ too small to come into discussion. The important component is $\omega^2 R \sin \theta \cos \theta$, directed northward and parallel to the surface of the earth. It deflects the thread by the angle

$$
\frac{\text{horizontal acceleration}}{\text{vertical acceleration}}
$$

$$
= \frac{\omega^2 R \sin \theta \cos \theta}{g}
$$

$$
= \frac{3.4 \text{ cm/sec}^2}{980 \text{ cm/sec}^2} \sin \theta \cos \theta
$$

$$
= 1.7 \times 10^{-3} \text{ radian at } \theta = 45°
$$

from the straight line connecting the center of the earth to the point of support. A difference, δg, of one part in 10^8 between g for the two hanging substances would produce a difference in angle of hang of plumb bobs equal to 1.7×10^{-11} radian at Budapest ($\theta = 42.5°$). Eötvös reported $\delta g/g$ less than a few parts in 10^9.

Roll, Krotkov, and Dicke, Princeton, 1964: employed as fiducial acceleration, not the 1.7 cm/sec² steady horizontal acceleration, produced by the earth's rotation at $\theta = 45°$, but the daily alternating 0.59 cm/sec² produced by the sun's attraction. Reported $|g(\text{Au}) - g(\text{Al})|/g$ less than 1×10^{-11}. See Figure 1.6.

Braginsky and Panov, Moscow, 1971: like Roll, Krotkov, and Dicke, employed Sun's attraction as fiducial acceleration. Reported $|g(\text{Pt}) - g(\text{Al})|/g$ less than 1×10^{-12}.

Beall, 1970: particles that are deflected less by the Earth's or the sun's gravitational field than a photon would be, effectively travel faster than light. If they are charged or have other electromagnetic structure, they would then emit Čerenkov radiation, and reduce their velocity below threshold in less than a micron of travel. The threshold is at energies around 10^3 mc^2. Ultrarelativistic particles in cosmic-ray showers are not easily identified, but observations of 10^{13} eV muons show that muons are not "too light" by as much as 5×10^{-5}. Conversely, a particle P bound more strongly than photons by gravity will transfer the momentum needed to make pair production $\gamma \to P + \bar{P}$ occur within a submicron decay length. The existence of photons with energies above 10^{13} eV shows that e^{\pm} are not "too heavy" by 5 parts in 10^9, μ^{\pm} not by 2 in 10^4, Λ, Ξ^-, Ω^- not by a few per cent.

Figure 1.7.

"Weightlessness" as test for a local inertial frame of reference ("Lorentz frame"). Each spring-driven cannon succeeds in driving its projectile, a steel ball bearing, through the aligned holes in the sheets of lucite, and into the woven-mesh pocket, when the frame of reference is free of rotation and in free fall ("normal world line through spacetime"). A cannon would fail (curved and ricocheting trajectory at bottom of drawing) if the frame were hanging as indicated when the cannon went off ("frame drawn away by pull of rope from its normal world line through spacetime"). Harold Waage at Princeton has constructed such a model for an inertial reference frame with lucite sheets about 1 m square. The "fuses" symbolizing time delay were replaced by electric relays. Penetration fails if the frame (1) rotates, (2) accelerates, or (3) does any combination of the two. It is difficult to cite any easily realizable device that more fully illustrates the meaning of the term "local Lorentz frame."

the same acceleration." Physics looks as complicated to the jet-driven observer as it does to the man on the ground. Rule out both observers to make physics look simple. Instead, travel aboard the freely moving spaceship. Nothing could be more natural than what one sees: every free object moves in a straight line with uniform velocity. This is the way to do physics! Work in a very special coordinate system: a coordinate frame in which one is weightless; *a local inertial frame of reference.* Or calculate how things look in such a frame. Or—if one is constrained to a ground-based frame of reference—use a particle moving so fast, and a path length so limited, that the ideal, freely falling frame of reference and the actual ground-based frame get out of alignment by an amount negligible on the scale of the experiment. [Given a 1,500-m linear accelerator, and a 1 GeV electron, time of flight $\simeq (1.5 \times 10^5 \text{ cm})/$

Eliminate the acceleration by use of a local inertial frame

$(3 \times 10^{10} \text{ cm/sec}) = 0.5 \times 10^{-5} \text{ sec}$; fall in this time $\sim \frac{1}{2}gt^2 = (490 \text{ cm/sec}^2)(0.5 \times 10^{-5} \text{ sec})^2 \simeq 10^{-8} \text{ cm.}]$

In analyzing physics in a local inertial frame of reference, or following an ant on his little section of apple skin, one wins simplicity by foregoing every reference to what is far away. Physics is simple only when viewed locally: that is Einstein's great lesson.

Newton spoke differently: "Absolute space, in its own nature, without relation to anything external, remains always similar and immovable." But how does one give meaning to Newton's absolute space, find its cornerstones, mark out its straight lines? In the real world of gravitation, no particle ever follows one of Newton's straight lines. His ideal geometry is beyond observation. "A comet going past the sun is deviated from an ideal straight line." No. There is no pavement on which to mark out that line. The "ideal straight line" is a myth. It never happened, and it never will.

Newton's absolute space is unobservable, nonexistent

> "It required a severe struggle [for Newton] to arrive at the concept of independent and absolute space, indispensible for the development of theory. . . . Newton's decision was, in the contemporary state of science, the only possible one, and particularly the only fruitful one. But the subsequent development of the problems, proceeding in a roundabout way which no one could then possibly foresee, has shown that the resistance of Leibniz and Huygens, intuitively well-founded but supported by inadequate arguments, was actually justified. . . . It has required no less strenuous exertions subsequently to overcome this concept [of absolute space]"
>
> [A. EINSTEIN (1954)].

What is direct and simple and meaningful, according to Einstein, is the geometry in every local inertial reference frame. There every particle moves in a straight line with uniform velocity. *Define* the local inertial frame so that this simplicity occurs for the first few particles (Figure 1.7). In the frame thus defined, every other free particle is observed also to move in a straight line with uniform velocity. Collision and disintegration processes follow the laws of conservation of momentum and energy of special relativity. That all these miracles come about, as attested by tens of thousands of observations in elementary particle physics, is witness to the inner workings of the machinery of the world. The message is easy to summarize: (1) physics is always and everywhere locally Lorentzian; i.e., locally the laws of special relativity are valid; (2) this simplicity shows most clearly in a local Lorentz frame of reference ("inertial frame of reference"; Figure 1.7); and (3) to test for a local Lorentz frame, test for weightlessness!

But Einstein's local inertial frames exist, are simple

In local inertial frames, physics is Lorentzian

§1.4. LOCAL LORENTZ GEOMETRY, WITH AND WITHOUT COORDINATES

On the surface of an apple within the space of a thumbprint, the geometry is Euclidean (Figure 1.1; the view in the magnifying glass). In spacetime, within a limited region, the geometry is Lorentzian. On the apple the distances between point and point accord with the theorems of Euclid. In spacetime the intervals ("proper distance," "proper time") between event and event satisfy the corresponding theorems of Lorentz-Minkowski geometry (Box 1.3). These theorems lend themselves

Local Lorentz geometry is the spacetime analog of local Euclidean geometry.

(continued on page 23)

**Box 1.3 LOCAL LORENTZ GEOMETRY AND LOCAL EUCLIDEAN GEOMETRY:
WITH AND WITHOUT COORDINATES**

I. Local Euclidean Geometry

What does it mean to say that the geometry of
a tiny thumbprint on the apple is Euclidean?

A. *Coordinate-free language* (Euclid):
Given a line $\mathcal{A}\mathcal{C}$. Extend it by an equal
distance $\mathcal{C}\mathcal{Z}$. Let \mathcal{B} be a point not on $\mathcal{A}\mathcal{Z}$
but equidistant from \mathcal{A} and \mathcal{Z}. Then

$$s_{\mathcal{A}\mathcal{B}}{}^2 = s_{\mathcal{A}\mathcal{C}}{}^2 + s_{\mathcal{B}\mathcal{C}}{}^2.$$

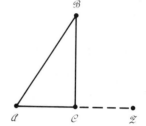

(Theorem of Pythagoras; also other theo-
rems of Euclidean geometry.)

B. *Language of coordinates* (Descartes):
From any point \mathcal{A} to any other point \mathcal{B}
there is a distance s given in suitable (Eucli-
dean) coordinates by

$$s_{\mathcal{A}\mathcal{B}}{}^2 = [x^1(\mathcal{B}) - x^1(\mathcal{A})]^2 + [x^2(\mathcal{B}) - x^2(\mathcal{A})]^2.$$

If one succeeds in finding any coordinate
system where this is true for all points \mathcal{A}
and \mathcal{B} in the thumbprint, then one is guar-
anteed that (i) this coordinate system is
locally Euclidean, and (ii) the geometry of
the apple's surface is locally Euclidean.

II. Local Lorentz Geometry

What does it mean to say that the geometry of
a sufficiently limited region of spacetime in the
real physical world is Lorentzian?

A. *Coordinate-free language* (Robb 1936):
Let $\mathcal{A}\mathcal{Z}$ be the world line of a free particle.
Let \mathcal{B} be an event not on this world line.
Let a light ray from \mathcal{B} strike $\mathcal{A}\mathcal{Z}$ at the
event \mathcal{Q}. Let a light ray take off from such
an earlier event \mathcal{P} along $\mathcal{A}\mathcal{Z}$ that it reaches
\mathcal{B}. Then the proper distance $s_{\mathcal{A}\mathcal{B}}$ (spacelike
separation) or proper time $\tau_{\mathcal{A}\mathcal{B}}$ (timelike
separation) is given by

$$s_{\mathcal{A}\mathcal{B}}{}^2 \equiv -\tau_{\mathcal{A}\mathcal{B}}{}^2 = -\tau_{\mathcal{A}\mathcal{Q}}\tau_{\mathcal{A}\mathcal{P}}.$$

Proof of above criterion for local Lorentz geometry, using coordinate methods in the local Lorentz frame where particle remains at rest:

$$\tau_{\mathcal{A}\mathcal{B}}{}^2 = t^2 - x^2 = (t - x)(t + x)$$
$$= \tau_{\mathcal{A}\mathcal{P}}\tau_{\mathcal{A}\mathcal{Q}}.$$

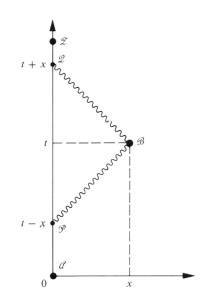

B. *Language of coordinates* (Lorentz, Poincaré, Minkowski, Einstein):

From any event \mathcal{A} to any other nearby event \mathcal{B}, there is a proper distance $s_{\mathcal{A}\mathcal{B}}$ or proper time $\tau_{\mathcal{A}\mathcal{B}}$ given in suitable (local Lorentz) coordinates by

$$\begin{aligned} s_{\mathcal{A}\mathcal{B}}{}^2 = -\tau_{\mathcal{A}\mathcal{B}}{}^2 = &-[x^0(\mathcal{B}) - x^0(\mathcal{A})]^2 \\ &+ [x^1(\mathcal{B}) - x^1(\mathcal{A})]^2 \\ &+ [x^2(\mathcal{B}) - x^2(\mathcal{A})]^2 \\ &+ [x^3(\mathcal{B}) - x^3(\mathcal{A})]^2. \end{aligned}$$

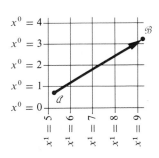

If one succeeds in finding any coordinate system where this is locally true for all neighboring events \mathcal{A} and \mathcal{B}, then one is guaranteed that (i) this coordinate system is locally Lorentzian, and (ii) the geometry of spacetime is locally Lorentzian.

III. Statements of Fact

The geometry of an apple's surface is locally Euclidean everywhere. The geometry of spacetime is locally Lorentzian everywhere.

Box 1.3 (*continued*)

IV. Local Geometry in the Language of Modern Mathematics

A. *The metric for any manifold:*
At each point on the apple, at each event of spacetime, indeed, at each point of any "Riemannian manifold," there exists a geometrical object called the *metric tensor* **g**. It is a machine with two input slots for the insertion of two vectors:

$$\underset{\downarrow}{\text{slot 1}} \quad \underset{\downarrow}{\text{slot 2}}$$

$$\boldsymbol{g}(\quad,\quad).$$

If one inserts the same vector **u** into both slots, one gets out the square of the length of **u**:

$$\boldsymbol{g}(\boldsymbol{u}, \boldsymbol{u}) = \boldsymbol{u}^2.$$

If one inserts two different vectors, **u** and **v** (it matters not in which order!), one gets out a number called the "scalar product of **u** on **v**" and denoted **u · v**:

$$\boldsymbol{g}(\boldsymbol{u}, \boldsymbol{v}) = \boldsymbol{g}(\boldsymbol{v}, \boldsymbol{u}) = \boldsymbol{u} \cdot \boldsymbol{v} = \boldsymbol{v} \cdot \boldsymbol{u}.$$

The metric is a linear machine:

$$\boldsymbol{g}(2\boldsymbol{u} + 3\boldsymbol{w}, \boldsymbol{v}) = 2\boldsymbol{g}(\boldsymbol{u}, \boldsymbol{v}) + 3\boldsymbol{g}(\boldsymbol{w}, \boldsymbol{v}),$$
$$\boldsymbol{g}(\boldsymbol{u}, a\boldsymbol{v} + b\boldsymbol{w}) = a\boldsymbol{g}(\boldsymbol{u}, \boldsymbol{v}) + b\boldsymbol{g}(\boldsymbol{u}, \boldsymbol{w}).$$

Consequently, in a given (arbitrary) coordinate system, its operation on two vectors can be written in terms of their components as a bilinear expression:

$$\boldsymbol{g}(\boldsymbol{u}, \boldsymbol{v}) = g_{\alpha\beta} u^{\alpha} v^{\beta}$$
$$\text{(implied summation on } \alpha, \beta)$$
$$= g_{11} u^1 v^1 + g_{12} u^1 v^2 + g_{21} u^2 v^1 + \cdots.$$

The quantities $g_{\alpha\beta} = g_{\beta\alpha}$ (α and β running from 0 to 3 in spacetime, from 1 to 2 on the apple) are called the "components of **g** in the given coordinate system."

B. *Components of the metric in local Lorentz and local Euclidean frames:*
To connect the metric with our previous descriptions of the local geometry, introduce local Euclidean coordinates (on apple) or local Lorentz coordinates (in spacetime).

Let **ξ** be the separation vector reaching from \mathcal{A} to \mathcal{B}. Its components in the local Euclidean (Lorentz) coordinates are

$$\xi^{\alpha} = x^{\alpha}(\mathcal{B}) - x^{\alpha}(\mathcal{A})$$

(cf. Box 1.1). Then the squared length of $\boldsymbol{u}_{\mathcal{A}\mathcal{B}}$, which is the same as the squared distance from \mathcal{A} to \mathcal{B}, must be (cf. I.B. and II.B. above)

$$\begin{aligned}
\boldsymbol{\xi} \cdot \boldsymbol{\xi} = \boldsymbol{g}(\boldsymbol{\xi}, \boldsymbol{\xi}) &= g_{\alpha\beta} \xi^{\alpha} \xi^{\beta} \\
&= s_{\mathcal{A}\mathcal{B}}{}^2 = (\xi^1)^2 + (\xi^2)^2 \text{ on apple} \\
&= -(\xi^0)^2 + (\xi^1)^2 + (\xi^2)^2 + (\xi^3)^2 \\
&\qquad\qquad\qquad\qquad \text{in spacetime.}
\end{aligned}$$

Consequently, the components of the metric are

$$g_{11} = g_{22} = 1, g_{12} = g_{21} = 0;$$
i.e., $g_{\alpha\beta} = \delta_{\alpha\beta}$ \qquad on apple, in local Euclidean coordinates;

$$g_{00} = -1, g_{0k} = 0, g_{jk} = \delta_{jk}$$
in spacetime, in local Lorentz coordinates.

These special components of the metric in local Lorentz coordinates are written here and hereafter as $g_{\hat{\alpha}\hat{\beta}}$ or $\eta_{\alpha\beta}$, by analogy with the Kronecker delta $\delta_{\alpha\beta}$. In matrix notation:

$$\|g_{\hat{\alpha}\hat{\beta}}\| = \|\eta_{\alpha\beta}\| = \begin{array}{c} \\ \alpha \\ \downarrow \end{array} \begin{array}{c} \overset{\displaystyle \longleftarrow \beta \longrightarrow}{\begin{array}{cccc} 0 & 1 & 2 & 3 \end{array}} \\ \begin{array}{c} 0 \\ 1 \\ 2 \\ 3 \end{array} \left\| \begin{array}{cccc} -1 & 0 & 0 & 0 \\ 0 & 1 & 0 & 0 \\ 0 & 0 & 1 & 0 \\ 0 & 0 & 0 & 1 \end{array} \right\| \end{array}$$

to empirical test in the appropriate, very special coordinate systems: Euclidean coordinates in Euclidean geometry; the natural generalization of Euclidean coordinates (local Lorentz coordinates; local inertial frame) in the local Lorentz geometry of physics. However, the theorems rise above all coordinate systems in their content. They refer to intervals or distances. Those distances no more call on coordinates for their definition in our day than they did in the time of Euclid. Points in the great pile of hay that is spacetime; and distances between these points: that is geometry! State them in the coordinate-free language or in the language of coordinates: they are the same (Box 1.3).

§1.5. TIME

Time is defined so that motion looks simple.

> *Time is awake when all things sleep.*
> *Time stands straight when all things fall.*
> *Time shuts in all and will not be shut.*
> *Is, was, and shall be are Time's children.*
> *O Reasoning, be witness, be stable.*
>
> VYASA, the *Mahabarata* (*ca.* A.D. 400)

Relative to a local Lorentz frame, a free particle "moves in a straight line with uniform velocity." What "straight" means is clear enough in the model inertial reference frame illustrated in Figure 1.7. But where does the "uniform velocity" come in? Or where does "velocity" show itself? There is not even one clock in the drawing!

A more fully developed model of a Lorentz reference frame will have not only holes, as in Fig. 1.7, but also clock-activated shutters over each hole. The projectile can reach its target only if it (1) travels through the correct region in space and (2) gets through that hole in the correct interval of time ("window in time"). How then is time defined? Time is defined so that motion looks simple!

No standard of time is more widely used than the day, the time from one high noon to the next. Take that as standard, however, and one will find every good clock or watch clashing with it, for a simple reason. The Earth spins on its axis and also revolves in orbit about the sun. The motion of the sun across the sky arises from neither effect alone, but from the two in combination, different in magnitude though they are. The fast angular velocity of the Earth on its axis (roughly 366.25 complete turns per year) is wonderfully uniform. Not so the apparent angular velocity of the sun about the center of the Earth (one turn per year). It is greater than average by 2 per cent when the Earth in its orbit (eccentricity 0.017) has come 1 per cent closer than average to the sun (Kepler's law) and lower by 2 per cent when the Earth is 1 per cent further than average from the sun. In the first case, the momentary rate of rotation of the sun across the sky, expressed in turns per year, is approximately

$$366.25 - (1 + 0.02);$$

The time coordinate of a local Lorentz frame is so defined that motion looks simple

in the other,

$$366.25 - (1 - 0.02).$$

Taking the "mean solar day" to contain $24 \times 3,600 = 86,400$ standard seconds, one sees that, when the Earth is 1 per cent closer to (or further from) the sun than average, then the number of standard seconds from one high noon to the next is greater (or less) than normal by

$$\frac{0.02 \text{ (drop in turns per year)}}{365.25 \text{ (turns per year on average)}} \, 86,400 \text{ sec} \sim 4.7 \text{ sec.}$$

This is the bookkeeping on time from noon to noon. No standard of time that varies so much from one month to another is acceptable. If adopted, it would make the speed of light vary from month to month!

This lack of uniformity, once recognized (and it was already recognized by the ancients), forces one to abandon the solar day as the standard of time; that day does not make motion look simple. Turn to a new standard that eliminates the motion of the Earth around the sun and concentrates on the spin of the Earth about its axis: the sidereal day, the time between one arrival of a star at the zenith and the next arrival of that star at the zenith. Good! Or good, so long as one's precision of measurement does not allow one to see changes in the intrinsic angular velocity of the Earth. What clock was so bold as first to challenge the spin of the Earth for accuracy? The machinery of the heavens.

Halley (1693) and later others, including Kant (1754), suspected something was amiss from apparent discrepancies between the paths of totality in eclipses of the sun, as predicted by Newtonian gravitation theory using the standard of time then current, and the location of the sites where ancient Greeks and Romans actually recorded an eclipse on the day in question. The moon casts a moving shadow in space. On the day of a solar eclipse, that shadow paints onto the disk of the spinning Earth a black brush stroke, often thousands of kilometers in length, but of width generally much less than a hundred kilometers. He who spins the globe upon the table and wants to make the shadow fall rightly on it must calculate back meticulously to determine two key items: (1) where the moon is relative to Earth and sun at each moment on the ancient day in question; and (2) how much angle the Earth has turned through from then until now. Take the eclipse of Jan. 14, A.D. 484, as an example (Figure 1.8), and assume the same angular velocity for the Earth in the intervening fifteen centuries as the Earth had in 1900 (astronomical reference point). One comes out wrong. The Earth has to be set back by 30° (or the moon moved from its computed position, or some combination of the two effects) to make the Athens observer fall under the black brush. To catch up those 30° (or less, if part of the effect is due to a slow change in the angular momentum of the moon), the Earth had to turn faster in the past than it does today. Assigning most of the discrepancy to terrestrial spin-down (rate of spin-down compatible with modern atomic-clock evidence), and assuming a uniform rate of slowing from then to now

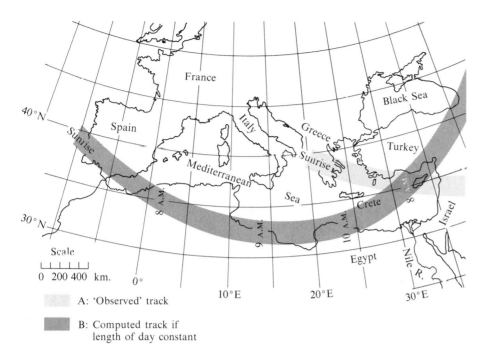

Figure 1.8.

Calculated path of totality for the eclipse of January 14, A.D. 484 (left; calculation based on no spin-down of Earth relative to its 1900 angular velocity) contrasted with the same path as set ahead enough to put the center of totality (at sunrise) at Athens [displacement very close to 30°; actual figure of deceleration adopted in calculations, 32.75 arc sec/(century)2]. This is "undoubtedly the most reliable of all ancient European eclipses," according to Dr. F. R. Stephenson, of the Department of Geophysics and Planetary Physics of the University of Newcastle upon Tyne, who most kindly prepared this diagram especially for this book. He has also sent a passage from the original Greek biography of Proclus of Athens (died at Athens A.D. 485) by Marinus of Naples, reading, "Nor were there portents wanting in the year which preceded his death; for example, such a great eclipse of the Sun that night seemed to fall by day. For a profound darkness arose so that stars even appeared in the sky. This happened in the eastern sky when the Sun dwelt in Capricorn" [from Westermann and Boissonade (1878)].

Does this 30° for this eclipse, together with corresponding amounts for other eclipses, represent the "right" correction? "Right" is no easy word. From one total eclipse of the sun in the Mediterranean area to another is normally many years. The various provinces of the Greek and Roman worlds were far from having a uniform level of peace and settled life, and even farther from having a uniform standard of what it is to observe an eclipse and put it down for posterity. If the scores of records of the past are unhappily fragmentary, even more unhappy has been the willingness of a few uncritical "investigators" in recent times to rush in and identify this and that historical event with this and that calculated eclipse. Fortunately, by now a great literature is available on the secular deceleration of the Earth's rotation, in the highest tradition of critical scholarship, both astronomical and historical. In addition to the books of O. Neugebauer (1959) and Munk and MacDonald (1960), the paper of Curott (1966), and items cited by these workers, the following are key items. (For direction to them, we thank Professor Otto Neugebauer no relation to the other Neugebauer cited below!) For the ancient records, and for calculations of the tracks of ancient eclipses, F. K. Ginzel (1882, 1883, 1884); for an atlas of calculated eclipse tracks, Oppolzer (1887) and Ginzel (1899); and for a critical analysis of the evidence. P. V. Neugebauer (1927, 1929, and 1930). This particular eclipse was chosen rather than any other because of the great reliability of the historical record of it.

(angular velocity correction proportional to first power of elapsed time: angle correction itself proportional to square of elapsed time), one estimates from a correction of

30° or 2 hours 1,500 years ago

the following corrections for intermediate times:

$30°/10^2$, or 1.2 min 150 years ago,

$30°/10^4$, or 0.8 sec 15 years ago.

Thus one sees the downfall of the Earth as a standard of time and its replacement by the orbital motions of the heavenly bodies as a better standard: a standard that does more to "make motion look simple." Astronomical time is itself in turn today being supplanted by atomic time as a standard of reference (see Box 1.4, "Time Today").

Look at a bad clock for a good view of how time is defined. Let t be time on a "good" clock (time coordinate of a local inertial frame); it makes the tracks of free particles through the local region of spacetime look straight. Let $T(t)$ be the reading of the "bad" clock; it makes the world lines of free particles through the local region of spacetime look curved (Figure 1.9). The old value of the acceleration, translated into the new ("bad") time, becomes

$$0 = \frac{d^2x}{dt^2} = \frac{d}{dt}\left(\frac{dT}{dt}\frac{dx}{dT}\right) = \frac{d^2T}{dt^2}\frac{dx}{dT} + \left(\frac{dT}{dt}\right)^2\frac{d^2x}{dT^2}.$$

To explain the apparent accelerations of the particles, the user of the new time introduces a force that one knows to be fictitious:

$$F_x = m\frac{d^2x}{dT^2} = -m\frac{\left(\frac{dx}{dT}\right)\left(\frac{d^2T}{dt^2}\right)}{\left(\frac{dT}{dt}\right)^2}. \tag{1.2}$$

It is clear from this example of a "bad" time that Newton thought of a "good" time when he set up the principle that "Time flows uniformly" ($d^2T/dt^2 = 0$). Time is defined to make motion look simple!

The principle of uniformity, taken by itself, leaves free the scale of the time variable. The quantity $T = at + b$ satisfies the requirement as well as t itself. The history of timekeeping discloses many choices of the unit and origin of time. Each one required some human action to give it sanction, from the fiat of a Pharaoh to the communique of a committee. In this book the amount of time it takes light to travel one centimeter is decreed to be the unit of time. Spacelike intervals and timelike intervals are measured in terms of one and the same geometric unit: the centimeter. Any other decision would complicate in analysis what is simple in nature. No other choice would live up to Minkowski's words, "Henceforth space by itself, and time by itself, are doomed to fade away into mere shadows, and only a kind of union of the two will preserve an independent reality."

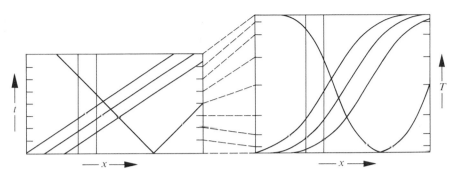

Figure 1.9.
Good clock (left) vs. bad clock (right) as seen in the maps they give of the same free particles moving through the same region of spacetime. The world lines as depicted at the right give the impression that a force is at work. The good definition of time eliminates such fictitious forces. The dashed lines connect corresponding instants on the two time scales.

One can measure time more accurately today than distance. Is that an argument against taking the elementary unit to be the centimeter? No, provided that this definition of the centimeter is accepted: *the geometrodynamic standard centimeter is the fraction*

$$1/(9.460546 \times 10^{17}) \tag{1.3}$$

of the interval between the two "effective equinoxes" that bound the tropical year 1900.0. The tropical year 1900.0 has already been recognized internationally as the fiducial interval by reason of its definiteness and the precision with which it is known. Standards committees have *defined the ephemeris second* so that 31,556,925.974 sec make up that standard interval. Were the speed of light known with perfect precision, the standards committees could have given in the same breath the number of centimeters in the standard interval. But it isn't; it is known to only six decimals. Moreover, the *international centimeter* is defined in terms of the orange-red wavelength of Kr^{86} to only nine decimals (16,507.6373 wavelengths). Yet the standard second is given to 11 decimals. We match the standard second by arbitrarily defining the geometrodynamic standard centimeter so that

$$9.4605460000 \times 10^{17}$$

such centimeters are contained in the standard tropical year 1900.0. The speed of light then becomes exactly

$$\frac{9.4605460000 \times 10^{17}}{31,556,925.974} \text{ geometrodynamic cm/sec.} \tag{1.4}$$

This is compatible with the speed of light, as known in 1967, in units of "international cm/sec":

$$29,979,300,000 \pm 30,000 \text{ international cm/sec.}$$

Box 1.4 TIME TODAY

Prior to 1956 the second was defined as the fraction 1/86,400 of the mean solar day.

From 1956 to 1967 the "second" meant the ephemeris second, defined as the fraction 1/(31,556,925.9747) of the tropical year 00h00m00s December 31, 1899.

Since 1967 the standard second has been the SI (Système International) second, defined as 9,192,631,770 periods of the unperturbed microwave transition between the two hyperfine levels of the ground state of Cs^{133}.

Like the foregoing evolution of the unit for the time *interval,* the evolution of a time *coordinate* has been marked by several stages.

Universal time, UTO, is based on the count of days as they actually occurred historically; in other words, on the actual spin of the earth on its axis; historically, on mean solar time (solar position as corrected by the "equation of time"; i.e., the faster travel of the earth when near the sun than when far from the sun) as determined at Greenwich Observatory.

UT1, the "navigator's time scale," is the same time as corrected for the wobble of the earth on its axis ($\Delta t \sim 0.05$ sec).

UT2 is UT1 as corrected for the periodic fluctuations of unknown origin with periods of one-half year and one year ($\Delta t \sim 0.05$ sec; measured to 3 ms in one day).

Ephemeris Time, ET (as defined by the theory of gravitation and by astronomical observations and calculations), is essentially determined by the orbital motion of the earth around the sun. "Measurement uncertainties limit the realization of accurate ephemeris time to about 0.05 sec for a nine-year average."

Coordinated Universal Time (UTC) is broadcast on stations such as WWV. It was adopted internationally in February 1971 to become effective January 1, 1972. The clock rate is controlled by atomic clocks to be as uniform as possible for one year (atomic time is measured to ~0.1 microsec in 1 min, with diffusion rates of 0.1 microsec per day for ensembles of clocks), but is changed by the infrequent addition or deletion of a second—called a "leap second"—so that UTC never differs more than 0.7 sec from the navigator's time scale, UT1.

Time suspended for a second

Time will stand still throughout the world for one second at midnight, June 30. All radio time signals will insert a "leap second" to bring Greenwich Mean Time into line with the earth's loss of three thousandths of a second a day.

The signal from the Royal Greenwich Observatory to Broadcasting House at midnight GMT (1 am BST July 1) will be six short pips marking the seconds 55 to 60 inclusive, followed by a lengthened signal at the following second to mark the new minute.

THE TIMES
Wednesday
June 21 1972

The foregoing account is abstracted from J. A. Barnes (1971). The following is extracted from a table (not official at time of receipt), kindly supplied by the Time and Frequency Division of the U.S. National Bureau of Standards in Boulder, Colorado.

Timekeeping capabilities of some familiar clocks are as follows:

Tuning fork wrist watch (1960),
 1 min/mo.

Quartz crystal clock (1921–1930),
 1 μsec/day,
 1 sec/yr.

Quartz crystal wrist watch (1971),
 0.2 sec/2 mos.,
 1 sec/yr.

Cesium beam (atomic resonance, Cs^{133}), (1952–1955),
 0.1 μsec/day,
 0.5 μsec/mo.

Rubidium gas cell (Rb^{87} resonance), (1957),
 0.1 μsec/day,
 1–5 μsec/mo.

Hydrogen maser (1960),
 0.01 μsec/2 hr,
 0.1 μsec/day.

Methane stabilized laser (1969),
 0.01 μsec/100 sec.

Recent measurements [Evenson *et al.* (1972)] change the details of the foregoing 1967 argument, but not the principles.

§1.6. CURVATURE

Gravitation seems to have disappeared. Everywhere the geometry of spacetime is locally Lorentzian. And in Lorentz geometry, particles move in a straight line with constant velocity. Where is any gravitational deflection to be seen in that? For answer, turn back to the apple (Figure 1.1). Inspect again the geodesic tracks of the ants on the surface of the apple. Note the reconvergence of two nearby geodesics that originally diverged from a common point. What is the analog in the real world of physics? What analogous concept fits Einstein's injunction that physics is only simple when analyzed locally? Don't look at the distance from the spaceship to the Earth. Look at the distance from the spaceship to a nearby spaceship! Or, to avoid any possible concern about attraction between the two ships, look at two nearby test particles in orbit about the Earth. To avoid distraction by the nonlocal element (the Earth) in the situation, conduct the study in the interior of a spaceship, also in orbit about the Earth. But this region has already been counted as a local inertial frame! What gravitational physics is to be seen there? None. Relative to the spaceship and therefore relative to each other, the two test particles move in a straight line with uniform velocity, to the precision of measurement that is contemplated (see Box 1.5, "Test for Flatness"). Now the key point begins to appear: precision of measurement. Increase it until one begins to discern the gradual acceleration of the test particles away from each other, if they lie along a common radius through the center of the Earth; or toward each other, if their separation lies perpendicular to that line. In Newtonian language, the source of these accelerations is the tide-producing action of the Earth. To the observer in the spaceship, however, no Earth is to be seen. And following Einstein, he knows it is important to analyze motion locally. He represents the separation of the new test particle from the fiducial test particle by the vector $\xi^k (k = 1, 2, 3$; components measured in a local Lorentz frame). For the acceleration of this separation, one knows from Newtonian physics what he will find: if the Cartesian z-axis is in the radial direction, then

Gravitation is manifest in relative acceleration of neighboring test particles

$$\frac{d^2\xi^x}{dt^2} = -\frac{Gm_{\text{conv}}}{c^2 r^3} \xi^x;$$

$$\frac{d^2\xi^y}{dt^2} = -\frac{Gm_{\text{conv}}}{c^2 r^3} \xi^y; \qquad (1.5)$$

$$\frac{d^2\xi^z}{dt^2} = \frac{2Gm_{\text{conv}}}{c^2 r^3} \xi^z.$$

Proof: In Newtonian physics the acceleration of a single particle toward the center of the Earth in conventional units of time is Gm_{conv}/r^2, where G is the Newtonian constant of gravitation, 6.670×10^{-8} cm^3/g sec^2 and m_{conv} is the mass of the Earth in conventional units of grams. In geometric units of time (cm of light-travel time),

the acceleration is Gm_{conv}/c^2r^2. When the two particles are separated by a distance ξ perpendicular to r, the one downward acceleration vector is out of line with the other by the angle ξ/r. Consequently one particle accelerates toward the other by the stated amount. When the separation is parallel to r, the relative acceleration is given by evaluating the Newtonian acceleration at r and at $r + \xi$, and taking the difference (ξ times d/dr) Q.E.D. In conclusion, the "local tide-producing acceleration" of Newtonian gravitation theory provides the local description of gravitation that Einstein bids one to seek.

What has this tide-producing acceleration to do with curvature? (See Box 1.6.) Look again at the apple or, better, at a sphere of radius a (Figure 1.10). The separation of nearby geodesics satisfies the "equation of geodesic deviation,"

Relative acceleration is caused by curvature

$$d^2\xi/ds^2 + R\xi = 0. \tag{1.6}$$

Here $R = 1/a^2$ is the so-called Gaussian curvature of the surface. For the surface of the apple, the same equation applies, with the one difference that the curvature R varies from place to place.

Box 1.5 TEST FOR FLATNESS

1. Specify the extension in space L (cm or m) and extension in time T (cm or m of light travel time) of the region under study.

2. Specify the precision $\delta\xi$ with which one can measure the separation of test particles in this region.

3. Follow the motion of test particles moving along initially parallel world lines through this region of spacetime.

4. When the world lines remain parallel to the precision $\delta\xi$ for all directions of travel, then one says that "in a region so limited and to a precision so specified, spacetime is flat."

EXAMPLE: Region just above the surface of the earth, 100 m \times 100 m \times 100 m (space extension), followed for 10^9 m of light-travel time ($T_{conv} \sim$ 3 sec). Mass of Earth, $m_{conv} = 5.98 \times 10^{27}$ g, $m = (0.742 \times 10^{-28}$ cm/g$) \times (5.98 \times 10^{27}$ g$) =$ 0.444 cm [see eq. (1.12)]. Tide-producing acceleration $R^z{}_{0z0}$ (relative acceleration in z-direction of two test particles initially at rest and separated from each other by 1 cm of vertical elevation) is

$$(d/dr)(m/r^2) = -2m/r^3$$
$$= -0.888 \text{ cm}/(6.37 \times 10^8 \text{ cm})^3$$
$$= -3.44 \times 10^{-27} \text{ cm}^{-2}$$

("cm of relative displacement per cm of light-travel time per cm of light-travel time per cm of vertical separation"). Two test particles with a vertical separation $\xi^z = 10^4$ cm acquire in the time $t = 10^{11}$ cm (difference between time and proper time negligible for such slowly moving test particles) a relative displacement

$$\delta\xi^z = -\tfrac{1}{2}R^z{}_{0z0}t^2\xi^z$$
$$= 1.72 \times 10^{-27} \text{ cm}^{-2}(10^{11} \text{ cm})^2 \, 10^4 \text{ cm}$$
$$= 1.72 \text{ mm}.$$

(Change in relative separation less for other directions of motion). When the minimum uncertainty $\delta\xi$ attainable in a measurement over a 100 m spacing is "worse" than this figure (exceeds 1.72 mm), then to this level of precision the region of spacetime under consideration can be treated as flat. When the uncertainty in measurement is "better" (less) than 1.72 mm, then one must limit attention to a smaller region of space or a shorter interval of time or both, to find a region of spacetime that can be regarded as flat to that precision.

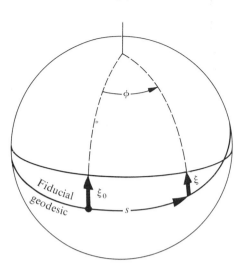

Figure 1.10.

Curvature as manifested in the "acceleration of the separation" of two nearby geodesics. Two geodesics, originally parallel, and separated by the distance ("geodesic deviation") ξ_0, are no longer parallel when followed a distance s. The separation is $\xi = \xi_0 \cos \phi = \xi_0 \cos (s/a)$, where a is the radius of the sphere. The separation follows the equation of simple harmonic motion, $d^2\xi/ds^2 + (1/a^2)\,\xi = 0$ ("equation of geodesic deviation").

The direction of the separation vector, $\boldsymbol{\xi}$, is fixed fully by its orthogonality to the fiducial geodesic. Hence, no reference to the direction of $\boldsymbol{\xi}$ is needed or used in the equation of geodesic deviation; only the magnitude ξ of $\boldsymbol{\xi}$ appears there, and only the magnitude, not direction, of the relative acceleration appears.

In a space of more than two dimensions, an equation of the same general form applies, with several differences. In two dimensions the *direction* of acceleration of one geodesic relative to a nearby, fiducial geodesic is fixed uniquely by the demand that their separation vector, $\boldsymbol{\xi}$, be perpendicular to the fiducial geodesic (see Figure 1.10). Not so in three dimensions or higher. There $\boldsymbol{\xi}$ can remain perpendicular to the fiducial geodesic but rotate about it (Figure 1.11). Thus, to specify the relative acceleration uniquely, one must give not only its magnitude, but also its direction.

The relative acceleration in three dimensions and higher, then, is a vector. Call it "$D^2\boldsymbol{\xi}/ds^2$," and call its four components "$D^2\xi^\alpha/ds^2$." Why the capital D? Why not "$d^2\xi^\alpha/ds^2$"? Because our coordinate system is completely arbitrary (cf. §1.2). The twisting and turning of the coordinate lines can induce changes from point to point in the components ξ^α of $\boldsymbol{\xi}$, even if the vector $\boldsymbol{\xi}$ is not changing at all. Consequently, the accelerations of the components $d^2\xi^\alpha/ds^2$ are generally not equal to the components $D^2\xi^\alpha/ds^2$ of the acceleration!

How, then, in curved spacetime can one determine the components $D^2\xi^\alpha/ds^2$ of the relative acceleration? By a more complicated version of the equation of geodesic deviation (1.6). Differential geometry (Part III of this book) provides us with a geometrical object called the *Riemann curvature tensor,* "**_Riemann._**" **_Riemann_** is

Curvature is characterized by Riemann tensor

(*continued on page 34*)

Figure 1.11.

The separation vector $\boldsymbol{\xi}$ between two geodesics in a curved three-dimensional manifold. Here $\boldsymbol{\xi}$ can not only change its length from point to point, but also rotate at a varying rate about the fiducial geodesic. Consequently, the relative acceleration of the geodesics must be characterized by a direction as well as a magnitude; it must be a vector, $D^2\boldsymbol{\xi}/ds^2$,

Box 1.6 CURVATURE OF WHAT?

Nothing seems more attractive at first glance than the idea that gravitation is a manifestation of the curvature of space (A), and nothing more ridiculous at a second glance (B). How can the tracks of a ball and of a bullet be curved so differently if that curvature arises from the geometry of space? No wonder that great Riemann did not give the world a geometric theory of gravity. Yes, at the age of 28 (June 10, 1854) he gave the world the mathematical machinery to define and calculate curvature (metric and Riemannian geometry). Yes, he spent his dying days at 40 working to find a unified account of electricity and gravitation. But if there was one reason more than any other why he failed to make the decisive connection between gravitation and curvature, it was this, that he thought of space and the curvature of space, not of spacetime and the curvature of spacetime. To make that forward step took the forty years to special relativity (1905: time on the same footing as space) and then another ten years (1915: general relativity). Depicted in spacetime (C), the tracks of ball and bullet appear to have comparable curvature. In fact, however, neither track has any curvature at all. They both look curved in (C) only because one has forgotten that the spacetime they reside in is itself curved—curved precisely enough to make these tracks the straightest lines in existence ("geodesics").

If it is at first satisfying to see curvature, and curvature of spacetime at that, coming to the fore in so direct a way, then a little more reflection produces a renewed sense of concern. Curvature with respect to what? Not with respect to the labo-

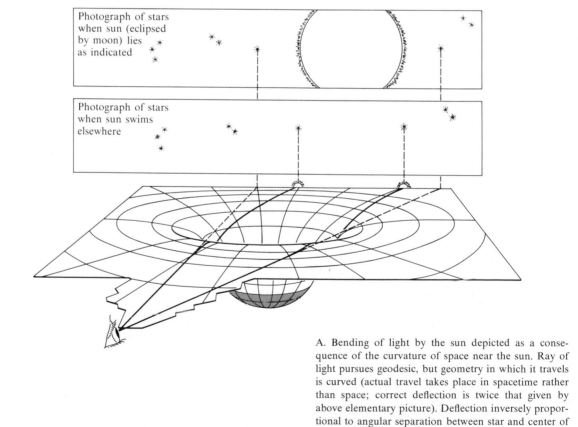

A. Bending of light by the sun depicted as a consequence of the curvature of space near the sun. Ray of light pursues geodesic, but geometry in which it travels is curved (actual travel takes place in spacetime rather than space; correct deflection is twice that given by above elementary picture). Deflection inversely proportional to angular separation between star and center of sun. See Box 40.1 for actual deflections observed at time of an eclipse.

Photograph of stars when sun (eclipsed by moon) lies as indicated

Photograph of stars when sun swims elsewhere

ratory. The earth-bound laboratory has no simple status whatsoever in a proper discussion. First, it is no Lorentz frame. Second, even to mention the earth makes one think of an action-at-a-distance version of gravity (distance from center of earth to ball or bullet). In contrast, it was the whole point of Einstein that physics looks simple only when analyzed locally. To look at local physics, however, means to compare one geodesic of one test particle with geodesics of other test particles traveling (1) nearby with (2) nearly the same directions and (3) nearly the same speeds. Then one can "look at the separations between these nearby test particles and from the second time-rate of change of these separations and the 'equation of geodesic deviation' (equation 1.8) read out the curvature of spacetime."

B. Tracks of ball and bullet through space as seen in laboratory have very different curvatures.

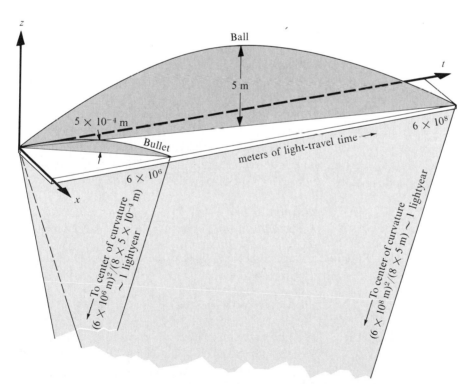

C. Tracks of ball and bullet through spacetime, as recorded in laboratory, have comparable curvatures. Track compared to arc of circle: (radius) = (horizontal distance)2/8 (rise).

the higher-dimensional analog of the Gaussian curvature R of our apple's surface. **Riemann** is the mathematical embodiment of the bends and warps in spacetime. And **Riemann** is the agent by which those bends and warps (curvature of spacetime) produce the relative acceleration of geodesics.

 Riemann, like the metric tensor **g** of Box 1.3, can be thought of as a family of machines, one machine residing at each event in spacetime. Each machine has three slots for the insertion of three vectors:

$$\text{slot 1}\quad\text{slot 2}\quad\text{slot 3}$$
$$\downarrow\qquad\downarrow\qquad\downarrow$$
$$\textbf{Riemann }(\qquad,\qquad,\qquad).$$

Choose a fiducial geodesic (free-particle world line) passing through an event \mathscr{P}, and denote its unit tangent vector (particle 4-velocity) there by

$$\textbf{u} = \textbf{dx}/d\tau; \text{ components, } u^\alpha = dx^\alpha/d\tau. \tag{1.7}$$

Choose another, neighboring geodesic, and denote by $\boldsymbol{\xi}$ its perpendicular separation from the fiducial geodesic. Then insert **u** into the first slot of **Riemann** at \mathscr{P}, $\boldsymbol{\xi}$ into the second slot, and **u** into the third. **Riemann** will grind for awhile; then out will pop a new vector,

$$\textbf{Riemann } (\textbf{u}, \boldsymbol{\xi}, \textbf{u}).$$

Riemann tensor, through equation of geodesic deviation, produces relative accelerations

The equation of geodesic deviation states that this new vector is the negative of the relative acceleration of the two geodesics:

$$D^2\boldsymbol{\xi}/d\tau^2 + \textbf{Riemann } (\textbf{u}, \boldsymbol{\xi}, \textbf{u}) = 0. \tag{1.8}$$

 The Riemann tensor, like the metric tensor (Box 1.3), and like all other tensors, is a linear machine. The vector it puts out is a linear function of each vector inserted into a slot:

$$\textbf{Riemann } (2\textbf{u}, a\textbf{w} + b\textbf{v}, 3\textbf{r})$$
$$= 2 \times a \times 3\, \textbf{Riemann } (\textbf{u}, \textbf{w}, \textbf{r}) + 2 \times b \times 3\, \textbf{Riemann } (\textbf{u}, \textbf{v}, \textbf{r}). \tag{1.9}$$

Consequently, in any coordinate system the components of the vector put out can be written as a "trilinear function" of the components of the vectors put in:

$$\textbf{r} = \textbf{Riemann } (\textbf{u}, \textbf{v}, \textbf{w}) \Longleftrightarrow r^\alpha = R^\alpha{}_{\beta\gamma\delta}\, u^\beta\, v^\gamma\, w^\delta. \tag{1.10}$$

(Here there is an implied summation on the indices β, γ, δ; cf. Box 1.1.) The $4 \times 4 \times 4 \times 4 = 256$ numbers $R^\alpha{}_{\beta\gamma\delta}$ are called the "components of the Riemann tensor in the given coordinate system." In terms of components, the equation of geodesic deviation states

$$\frac{D^2\xi^\alpha}{d\tau^2} + R^\alpha{}_{\beta\gamma\delta}\, \frac{dx^\beta}{d\tau}\, \xi^\gamma\, \frac{dx^\delta}{d\tau} = 0. \tag{1.8'}$$

In Einstein's geometric theory of gravity, this equation of geodesic deviation summarizes the entire effect of geometry on matter. It does for gravitation physics what the Lorentz force equation,

Equation of geodesic deviation is analog of Lorentz force law

$$\frac{D^2 x^\alpha}{d\tau^2} - \frac{e}{m} F^\alpha{}_\beta \frac{dx^\alpha}{d\tau} = 0, \qquad (1.11)$$

does for electromagnetism. See Box 1.7.

The units of measurement of the curvature are cm^{-2} just as well in spacetime as on the surface of the apple. Nothing does so much to make these units stand out clearly as to express mass in "geometrized units":

Geometrized units

$$m(\text{cm}) = (G/c^2) m_{\text{conv}}(\text{g})$$
$$= (0.742 \times 10^{-28} \, \text{cm/g}) m_{\text{conv}}(\text{g}). \qquad (1.12)$$

Box 1.7 EQUATION OF MOTION UNDER THE INFLUENCE OF A GRAVITATIONAL FIELD AND AN ELECTROMAGNETIC FIELD, COMPARED AND CONTRASTED

	Electromagnetism [*Lorentz force, equation (1.11)*]	*Gravitation* [*Equation of geodesic deviation (1.8′)*]
Acceleration is defined for one particle?	Yes	No
Acceleration defined how?	Actual world line compared to world line of uncharged "fiducial" test particle passing through same point with same 4-velocity.	Already an uncharged test particle, which can't accelerate relative to itself! Acceleration measured relative to a nearby test particle as fiduciary standard.
Acceleration depends on all four components of the 4-velocity of the particle?	Yes	Yes
Universal acceleration for all test particles in same locations with same 4-velocity?	No; is proportional to e/m	Yes
Driving field	Electromagnetic field	Riemann curvature tensor
Ostensible number of distinct components of driving field	$4 \times 4 = 16$	$4^4 = 256$
Actual number when allowance is made for symmetries of tensor	6	20
Names for more familiar of these components	3 electric 3 magnetic	6 components of local Newtonian tide-producing acceleration

This conversion from grams to centimeters by means of the ratio

$$G/c^2 = 0.742 \times 10^{-28} \text{ cm/g}$$

is completely analogous to converting from seconds to centimeters by means of the ratio

$$c = \frac{9.4605460000 \times 10^{17} \text{ cm}}{31,556,925.974 \text{ sec}}$$

(see end of §1.5). The sun, which in conventional units has $m_{\text{conv}} = 1.989 \times 10^{33}$ g, has in geometrized units a mass $m = 1.477$ km. Box 1.8 gives further discussion.

Using geometrized units, and using the Newtonian theory of gravity, one can readily evaluate nine of the most interesting components of the Riemann curvature tensor near the Earth or the sun. The method is the gravitational analog of determining the electric field strength by measuring the acceleration of a slowly moving test particle. Consider the separation between the geodesics of two nearby and slowly moving ($v \ll c$) particles at a distance r from the Earth or sun. In the standard, nearly inertial coordinates of celestial mechanics, all components of the 4-velocity of the

Components of Riemann tensor evaluated from relative accelerations of slowly moving particles

Box 1.8 GEOMETRIZED UNITS

Throughout this book, we use "geometrized units," in which the speed of light c, Newton's gravitational constant G, and Boltzman's constant k are all equal to unity. The following alternative ways to express the number 1.0 are of great value:

$$1.0 = c = 2.997930 \cdots \times 10^{10} \text{ cm/sec}$$

$$1.0 = G/c^2 = 0.7425 \times 10^{-28} \text{ cm/g};$$

$$1.0 = G/c^4 = 0.826 \times 10^{-49} \text{ cm/erg};$$

$$1.0 = Gk/c^4 = 1.140 \times 10^{-65} \text{ cm/K};$$

$$1.0 = c^2/G^{1/2} = 3.48 \times 10^{24} \text{ cm/gauss}^{-1}.$$

One can multiply a factor of unity, expressed in any one of these ways, into any term in any equation without affecting the validity of the equation. Thereby one can convert one's units of measure from grams to centimeters to seconds to ergs to For example:

$$\text{Mass of sun} = M_\odot = 1.989 \times 10^{33} \text{ g}$$

$$= (1.989 \times 10^{33} \text{ g}) \times (G/c^2)$$

$$= 1.477 \times 10^5 \text{ cm}$$

$$= (1.989 \times 10^{33} \text{ g}) \times (c^2)$$

$$= 1.788 \times 10^{54} \text{ ergs}.$$

The standard unit, in terms of which everything is measured in this book, is centimeters. However, occasionally conventional units are used; in such cases a subscript "conv" is sometimes, but not always, appended to the quantity measured:

$$M_{\odot\text{conv}} = 1.989 \times 10^{33} \text{ g}.$$

fiducial test particle can be neglected except $dx^0/d\tau = 1$. The space components of the equation of geodesic deviation read

$$d^2\xi^k/d\tau^2 + R^k{}_{\hat{0}j\hat{0}}\xi^j = 0. \tag{1.13}$$

Comparing with the conclusions of Newtonian theory, equations (1.5), we arrive at the following information about the curvature of spacetime near a center of mass:

$$\begin{Vmatrix} R^{\hat{x}}{}_{\hat{0}\hat{x}\hat{0}} & R^{\hat{y}}{}_{\hat{0}\hat{x}\hat{0}} & R^{\hat{z}}{}_{\hat{0}\hat{x}\hat{0}} \\ R^{\hat{x}}{}_{\hat{0}\hat{y}\hat{0}} & R^{\hat{y}}{}_{\hat{0}\hat{y}\hat{0}} & R^{\hat{z}}{}_{\hat{0}\hat{y}\hat{0}} \\ R^{\hat{x}}{}_{\hat{0}\hat{z}\hat{0}} & R^{\hat{y}}{}_{\hat{0}\hat{z}\hat{0}} & R^{\hat{z}}{}_{\hat{0}\hat{z}\hat{0}} \end{Vmatrix} = \begin{Vmatrix} m/r^3 & 0 & 0 \\ 0 & m/r^3 & 0 \\ 0 & 0 & -2m/r^3 \end{Vmatrix} \tag{1.14}$$

(units cm^{-2}). Here and henceforth the caret or "hat" is used to indicate the components of a vector or tensor in a local Lorentz frame of reference ("physical components," as distinguished from components in a general coordinate system). Einstein's theory will determine the values of the other components of curvature (e.g., $R^{\hat{x}}{}_{\hat{z}\hat{x}\hat{z}} = -m/r^3$); but these nine terms are the ones of principal relevance for many applications of gravitation theory. They are analogous to the components of the electric field in the Lorentz equation of motion. Many of the terms not evaluated are analogous to magnetic field components—ordinarily weak unless the source is in rapid motion.

This ends the survey of the effect of geometry on matter ("effect of curvature of apple in causing geodesics to cross"—especially great near the dimple at the top, just as the curvature of spacetime is especially large near a center of gravitational attraction). Now for the effect of matter on geometry ("effect of stem of apple in causing dimple")!

§1.7. EFFECT OF MATTER ON GEOMETRY

> *The weight of any heavy body of known weight at a particular distance from the center of the world varies according to the variation of its distance therefrom; so that as often as it is removed from the center, it becomes heavier, and when brought near to it, is lighter. On this account, the relation of gravity to gravity is as the relation of distance to distance from the center.*
>
> AL KHĀZINĪ (Merv, A.D. 1115), Book of the Balance of Wisdom

Figure 1.12 shows a sphere of the same density, $\rho = 5.52$ g/cm^3, as the average density of the Earth. A hole is bored through this sphere. Two test particles, A and B, execute simple harmonic motion in this hole, with an 84-minute period. Therefore their geodesic separation ξ, however it may be oriented, undergoes a simple periodic motion with the same 84-minute period:

$$d^2\xi^j/d\tau^2 = -\left(\frac{4\pi}{3}\rho\right)\xi^j, \qquad j = x \text{ or } y \text{ or } z. \tag{1.15}$$

Box 1.9 GALILEO GALILEI
Pisa, February 15, 1564—Arcetri, Florence, January 8, 1642

Uffizi Gallery, Florence

*"In questions of science the authority
of a thousand is not worth the humble
reasoning of a single individual."*

GALILEO GALILEI (1632)

*"The spaces described by a body falling from rest
with a uniformly accelerated motion are to each other
as the squares of the time intervals employed in
traversing these distances."*

GALILEO GALILEI (1638)

*"Everything that has been said before and imagined by other people [about the
tides] is in my opinion completely invalid. But among the great men who have
philosophised about this marvellous effect of nature the one who surprised me the
most is Kepler. More than other people he was a person of independent genius,
sharp, and had in his hands the motion of the earth. He later pricked up his ears
and became interested in the action of the moon on the water, and in other occult
phenomena, and similar childishness."*

GALILEO GALILEI (1632)

*"It is a most beautiful and delightful sight to behold [with the new telescope] the
body of the Moon . . . the Moon certainly does not possess a smooth and polished
surface, but one rough and uneven . . . full of vast protuberances, deep chasms
and sinuosities . . . stars in myriads, which have never been seen before and
which surpass the old, previously known, stars in number more than ten times. I
have discovered four planets, neither known nor observed by any one of the
astronomers before my time . . . got rid of disputes about the Galaxy or Milky
Way, and made its nature clear to the very senses, not to say to the
understanding . . . the galaxy is nothing else than a mass of luminous stars
planted together in clusters . . . the number of small ones is quite beyond
determination—the stars which have been called by every one of the astronomers
up to this day nebulous are groups of small stars set thick together in a wonderful
way."*

GALILEO GALILEI IN *SIDEREUS NUNCIUS* (1610)

*"So the principles which are set forth in this treatise will, when taken up by
thoughtful minds, lead to many another more remarkable result; and it is to be
believed that it will be so on account of the nobility of the subject, which is
superior to any other in nature."*

GALILEO GALILEI (1638)

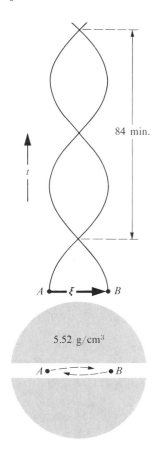

Figure 1.12.

Test particles A and B move up and down a hole bored through the Earth, idealized as of uniform density. At radius r, a particle feels Newtonian acceleration

$$\frac{d^2 r}{d\tau^2} = \frac{1}{c^2} \frac{d^2 r}{dt_{\text{conv}}^2}$$

$$= -\frac{G}{c^2} \frac{(\text{mass inside radius } r)}{r^2}$$

$$= -\left(\frac{G}{r^2 c^2}\right)\left(\frac{4\pi}{3} \rho_{\text{conv}} r^3\right)$$

$$= -\omega^2 r.$$

Consequently, each particle oscillates in simple harmonic motion with precisely the same angular frequency as a satellite, grazing the model Earth, traverses its circular orbit:

$$\omega^2 (\text{cm}^{-2}) = \frac{4\pi}{3} \rho (\text{cm}^{-2}),$$

$$\omega^2_{\text{conv}} (\text{sec}^{-2}) = \frac{4\pi G}{3} \rho_{\text{conv}} (\text{g/cm}^3).$$

Comparing this actual motion with the equation of geodesic deviation (1.13) for slowly moving particles in a nearly inertial frame, we can read off some of the curvature components for the interior of this model Earth.

<div style="text-align: right">The Riemann tensor inside the Earth</div>

$$\begin{Vmatrix} R^{\hat{x}}_{\hat{0}\hat{x}\hat{0}} & R^{\hat{y}}_{\hat{0}\hat{x}\hat{0}} & R^{\hat{z}}_{\hat{0}\hat{x}\hat{0}} \\ R^{\hat{x}}_{\hat{0}\hat{y}\hat{0}} & R^{\hat{y}}_{\hat{0}\hat{y}\hat{0}} & R^{\hat{z}}_{\hat{0}\hat{y}\hat{0}} \\ R^{\hat{x}}_{\hat{0}\hat{z}\hat{0}} & R^{\hat{y}}_{\hat{0}\hat{z}\hat{0}} & R^{\hat{z}}_{\hat{0}\hat{z}\hat{0}} \end{Vmatrix} = (4\pi\rho/3) \begin{Vmatrix} 1 & 0 & 0 \\ 0 & 1 & 0 \\ 0 & 0 & 1 \end{Vmatrix} \qquad (1.16)$$

This example illustrates how the curvature of spacetime is connected to the distribution of matter.

Let a gravitational wave from a supernova pass through the Earth. Idealize the Earth's matter as so nearly incompressible that its density remains practically unchanged. The wave is characterized by ripples in the curvature of spacetime, propagating with the speed of light. The ripples will show up in the components $R^j_{\hat{0}k\hat{0}}$ of the Riemann tensor, and in the relative acceleration of our two test particles. The left side of equation (1.16) will ripple; but the right side will not. Equation (1.16) will break down. No longer will the Riemann curvature be generated directly and solely by the Earth's matter.

<div style="text-align: right">Effect of gravitational wave on Riemann tensor</div>

Nevertheless, Einstein tells us, a part of equation (1.16) is undisturbed by the

waves: its trace

$$R_{\hat{0}\hat{0}} \equiv R^{\hat{x}}{}_{\hat{0}\hat{x}\hat{0}} + R^{\hat{y}}{}_{\hat{0}\hat{y}\hat{0}} + R^{\hat{z}}{}_{\hat{0}\hat{z}\hat{0}} = 4\pi\rho. \qquad (1.17)$$

Even in the vacuum outside the Earth this is valid; there both sides vanish [cf. (1.14)].

More generally, a certain piece of the Riemann tensor, called the *Einstein tensor* and denoted **Einstein** or **G**, is always generated directly by the local distribution of matter. **Einstein** is the geometric object that generalizes $R_{\hat{0}\hat{0}}$, the lefthand side

Einstein tensor introduced

Box 1.10 ISAAC NEWTON
Woolsthorpe, Lincolnshire, England, December 25, 1642—
Kensington, London, March 20, 1726

"The description of right lines and circles, upon which geometry is founded, belongs to mechanics. Geometry does not teach us to draw these lines, but requires them to be drawn."

[FROM P. 1 OF NEWTON'S PREFACE TO THE FIRST (1687) EDITION OF THE *PRINCIPIA*]

"Absolute space, in its own nature, without relation to anything external, remains always similar and immovable
"Absolute, true, and mathematical time, of itself, and from its own nature, flows equably without relation to anything external."

[FROM THE *SCHOLIUM* IN THE *PRINCIPIA*]

"I have not been able to discover the cause of those properties of gravity from phenomena, and I frame no hypotheses; for whatever is not reduced from the phenomena is to be called an hypothesis; and hypotheses . . . have no place in experimental philosophy. . . . And to us it is enough that gravity does really exist, and act according to the laws which we have explained, and abundantly serves to account for all the motions of the celestial bodies, and of our sea."

[FROM THE *GENERAL SCHOLIUM* ADDED AT THE END OF THE THIRD BOOK OF THE *PRINCIPIA* IN THE SECOND EDITION OF 1713; ESPECIALLY FAMOUS FOR THE PHRASE OFTEN QUOTED FROM NEWTON'S ORIGINAL LATIN, *"HYPOTHESES NON FINGO."*]

"And the same year [1665 or 1666] I began to think of gravity extending to the orb of the Moon, and having found out. . . . All this was in the two plague years of 1665 and 1666, for in those days I was in the prime of my age for invention, and minded Mathematicks and Philosophy more than at any time since."

[FROM MEMORANDUM IN NEWTON'S HANDWRITING ABOUT HIS DISCOVERIES ON FLUXIONS, THE BINOMIAL THEOREM, OPTICS, DYNAMICS, AND GRAVITY, BELIEVED TO HAVE BEEN WRITTEN ABOUT 1714, AND FOUND BY ADAMS ABOUT 1887 IN THE "PORTSMOUTH COLLECTION" OF NEWTON PAPERS]

of equation (1.17). Like $R_{\hat{0}\hat{0}}$, **Einstein** is a sort of average of **Riemann** over all directions. Generating **Einstein** and generalizing the righthand side of (1.16) is a geometric object called the *stress-energy tensor* of the matter. It is denoted **T**. No coordinates are need to define **Einstein**, and none to define **T**; like the Riemann tensor, **Riemann**, and the metric tensor, **g**, they exist in the complete absence of coordinates. Moreover, in nature they are always equal, aside from a factor of 8π:

Stress-energy tensor introduced

$$\textbf{\textit{Einstein}} \equiv \textbf{\textit{G}} = 8\pi\textbf{\textit{T}}. \qquad (1.18)$$

"For hypotheses ought . . . to explain the properties of things and not attempt to predetermine them except in so far as they can be an aid to experiments."

[FROM LETTER OF NEWTON TO I. M. PARDIES, 1672, AS QUOTED IN THE CAJORI NOTES AT THE END OF NEWTON (1687), P. 673]

"That one body may act upon another at a distance through a vacuum, without the mediation of any thing else, by and through which their action and force may be conveyed from one to another, is to me so great an absurdity, that I believe no man, who has in philosophical matters a competent faculty of thinking, can ever fall into it."

[PASSAGE OFTEN QUOTED BY MICHAEL FARADAY FROM LETTERS OF NEWTON TO RICHARD BENTLY, 1692–1693, AS QUOTED IN THE NOTES OF THE CAJORI EDITION OF NEWTON (1687), P. 643]

"The attractions of gravity, magnetism, and electricity, reach to very sensible distances, and so have been observed . . . ; and there may be others which reach to so small distances as hitherto escape observation; . . . some force, which in immediate contract is exceeding strong, at small distances performs the chemical operations above-mentioned, and reaches not far from the particles with any sensible effect."

[FROM QUERY 31 AT THE END OF NEWTON'S *OPTICKS* (1730)]

"What is there in places almost empty of matter, and whence is it that the sun and planets gravitate towards one another, without dense matter between them? Whence is it that nature doth nothing in vain; and whence arises all that order and beauty which we see in the world? To what end are comets, and whence is it that planets move all one and the same way in orbs concentrick, while comets move all manner of ways in orbs very excentrick; and what hinders the fixed stars from falling upon one another?"

[FROM QUERY 28]

"He is not eternity or infinity, but eternal and infinite; He is not duration or space, but He endures and is present. He endures forever, and is everywhere present; and by existing always and everywhere, He constitutes duration and space. . . . And thus much concerning God; to discourse of whom from the appearances of things, does certainly belong to natural philosophy."

[FROM THE *GENERAL SCHOLIUM* AT THE END OF THE *PRINCIPIA* (1687)]

Einstein field equation: how matter generates curvature

This *Einstein field equation*, rewritten in terms of components in an arbitrary coordinate system, reads

$$G_{\alpha\beta} = 8\pi T_{\alpha\beta}. \tag{1.19}$$

The Einstein field equation is elegant and rich. No equation of physics can be written more simply. And none contains such a treasure of applications and consequences.

Consequences of Einstein field equation

The field equation shows how the stress-energy of matter generates an average curvature (**Einstein ≡ G**) in its neighborhood. Simultaneously, the field equation is a propagation equation for the remaining, anisotropic part of the curvature: it governs the external spacetime curvature of a static source (Earth); it governs the generation of gravitational waves (ripples in curvature of spacetime) by stress-energy in motion; and it governs the propagation of those waves through the universe. The field equation even contains within itself the equations of motion ("Force =

Box 1.11
ALBERT EINSTEIN
Ulm, Germany,
March 14, 1879—
Princeton, New Jersey,
April 18, 1955

Library of E. T. Hochschule, Zürich

Académie des Sciences, Paris

Archives of California Institute of Technology

mass × acceleration") for the matter whose stress-energy generates the curvature.

Those were some consequences of $\boldsymbol{G} = 8\pi\boldsymbol{T}$. Now for some applications.

The field equation governs the motion of the planets in the solar system; it governs the deflection of light by the sun; it governs the collapse of a star to form a black hole; it determines uniquely the external spacetime geometry of a black hole ("a black hole has no hair"); it governs the evolution of spacetime singularities at the end point of collapse; it governs the expansion and recontraction of the universe. And more; much more.

Applications of Einstein field equation

In order to understand how the simple equation $\boldsymbol{G} = 8\pi\boldsymbol{T}$ can be so all powerful, it is desirable to backtrack, and spend a few chapters rebuilding the entire picture of spacetime, of its curvature, and of its laws, this time with greater care, detail, and mathematics.

Thus ends this survey of the effect of geometry on matter, and the reaction of matter back on geometry, rounding out the parable of the apple.

"What really interests me is whether God had any choice in the creation of the world"

EINSTEIN TO AN ASSISTANT, AS QUOTED BY G. HOLTON (1971), P. 20

"But the years of anxious searching in the dark, with their intense longing, their alternations of confidence and exhaustion, and the final emergence into the light—only those who have experienced it can understand that"

EINSTEIN, AS QUOTED BY M. KLEIN (1971), P. 1315

"Of all the communities available to us there is not one I would want to devote myself to, except for the society of the true searchers, which has very few living members at any time. . ."

EINSTEIN LETTER TO BORN, QUOTED BY BORN (1971), P. 82

"I am studying your great works and—when I get stuck anywhere—now have the pleasure of seeing your friendly young face before me smiling and explaining"

EINSTEIN, LETTER OF MAY 2, 1920, AFTER MEETING NIELS BOHR

"As far as the laws of mathematics refer to reality, they are not certain; and as far as they are certain, they do not refer to reality."

EINSTEIN (1921), P. 28

"The most incomprehensible thing about the world is that it is comprehensible."

EINSTEIN, IN SCHILPP (1949), P. 112

EXERCISES

Exercise 1.1. CURVATURE OF A CYLINDER

Show that the Gaussian curvature R of the surface of a cylinder is zero by showing that geodesics on that surface (unroll!) suffer no geodesic deviation. Give an independent argument for the same conclusion by employing the formula $R = 1/\rho_1\rho_2$, where ρ_1 and ρ_2 are the principal radii of curvature at the point in question with respect to the enveloping Euclidean three-dimensional space.

Exercise 1.2. SPRING TIDE VS. NEAP TIDE

Evaluate (1) in conventional units and (2) in geometrized units the magnitude of the Newtonian tide-producing acceleration $R^m{}_{0n0}(m, n = 1, 2, 3)$ generated at the Earth by (1) the moon ($m_{\mathrm{conv}} = 7.35 \times 10^{25}$ g, $r = 3.84 \times 10^{10}$ cm) and (2) the sun ($m_{\mathrm{conv}} = 1.989 \times 10^{33}$ g, $r = 1.496 \times 10^{13}$ cm). By what factor do you expect spring tides to exceed neap tides?

Exercise 1.3. KEPLER ENCAPSULATED

A small satellite has a circular frequency $\omega(\mathrm{cm}^{-1})$ in an orbit of radius r about a central object of mass $m(\mathrm{cm})$. From the known value of ω, show that it is possible to determine neither r nor m individually, but only the effective "Kepler density" of the object as averaged over a sphere of the same radius as the orbit. Give the formula for ω^2 in terms of this Kepler density.

It is a reminder of the continuity of history that Kepler and Galileo (Box 1.9) wrote back and forth, and that the year that witnessed the death of Galileo saw the birth of Newton (Box 1.10). After Newton the first dramatically new synthesis of the laws of gravitation came from Einstein (Box 1.11).

And what the dead had no speech for, when living,
They can tell you, being dead; the communication
Of the dead is tongued with fire beyond
the language of the living.

T. S. ELIOT, in *LITTLE GIDDING* (1942)

I measured the skies
Now the shadows I measure
Skybound was the mind
Earthbound the body rests

JOHANNES KEPLER, d. November 15, 1630.
He wrote his epitaph in Latin;
it is translated by Coleman (1967), p. 109.

Ubi materia, ibi geometria.
JOHANNES KEPLER

PHYSICS IN FLAT SPACETIME

*Wherein the reader meets an old friend, Special Relativity,
outfitted in new, mod attire, and becomes more
intimately acquainted with her charms.*

CHAPTER **2**

FOUNDATIONS OF SPECIAL RELATIVITY

In geometric and physical applications, it always turns out that a quantity is characterized not only by its tensor order, but also by symmetry.

HERMAN WEYL (1925)

Undoubtedly the most striking development of geometry during the last 2,000 years is the continual expansion of the concept "geometric object." This concept began by comprising only the few curves and surfaces of Greek synthetic geometry; it was stretched, during the Renaissance, to cover the whole domain of those objects defined by analytic geometry; more recently, it has been extended to cover the boundless universe treated by point-set theory.

KARL MENGER, IN SCHILPP (1949), P. 466.

§2.1. OVERVIEW

Curvature in geometry manifests itself as gravitation. Gravitation works on the separation of nearby particle world lines. In turn, particles and other sources of mass-energy cause curvature in the geometry. How does one break into this closed loop of the action of geometry on matter and the reaction of matter on geometry? One can begin no better than by analyzing the motion of particles and the dynamics of fields in a region of spacetime so limited that it can be regarded as flat. (See "Test for Flatness," Box 1.5).

Chapters 2–6 develop this flat-spacetime viewpoint (special relativity). The reader, it is assumed, is already somewhat familiar with special relativity:* 4-vectors in general; the energy-momentum 4-vector; elementary Lorentz transformations; the Lorentz law for the force on a charged particle; at least one look at one equation

Background assumed of reader

*For example, see Goldstein (1959), Leighton (1959), Jackson (1962), or, for the physical perspective presented geometrically, Taylor and Wheeler (1966).

in one book that refers to the electromagnetic field tensor $F_{\mu\nu}$; and the qualitative features of spacetime diagrams, including such points as (1) future and past light cones, (2) causal relationships ("past of," "future of," "neutral," or "in a spacelike relationship to"), (3) Lorentz contraction, (4) time dilation, (5) absence of a universal concept of simultaneity, and (6) the fact that the \bar{t} and \bar{z} axes in Box 2.4 are orthogonal even though they do not look so. If the reader finds anything new in these chapters, it will be: (i) a new viewpoint on special relativity, one emphasizing coordinate-free concepts and notation that generalize readily to curved spacetime ("geometric objects," tensors viewed as machines—treated in Chapters 2–4); or (ii) unfamiliar topics in special relativity, topics crucial to the later exposition of gravitation theory ("stress-energy tensor and conservation laws," Chapter 5; "accelerated observers," Chapter 6).

§2.2.　GEOMETRIC OBJECTS

Every physical quantity can be described by a geometric object

Everything that goes on in spacetime has its geometric description, and almost every one of these descriptions lends itself to ready generalization from flat spacetime to curved spacetime. The greatest of the differences between one geometric object and another is its scope: the individual object (vector) for the momentum of a certain particle at a certain phase in its history, as contrasted to the extended geometric object that describes an electromagnetic field defined throughout space and time ("antisymmetric second-rank tensor field" or, more briefly, "field of 2-forms"). The idea that every physical quantity must be describable by a geometric object, and

All laws of physics can be expressed geometrically

that the laws of physics must all be expressible as geometric relationships between these geometric objects, had its intellectual beginnings in the Erlanger program of Felix Klein (1872), came closer to physics in Einstein's "principle of general covariance" and in the writings of Hermann Weyl (1925), seems to have first been formulated clearly by Veblen and Whitehead (1932), and today pervades relativity theory, both special and general.

A. Nijenhuis (1952) and S.-S. Chern (1960, 1966, 1971) have expounded the mathematical theory of geometric objects. But to understand or do research in geometrodynamics, one need not master this elegant and beautiful subject. One need only know that geometric objects in spacetime are entities that exist independently of coordinate systems or reference frames. A point in spacetime ("*event*") is a geometric object. The arrow linking two neighboring events ("*vector*") is a geometric object in flat spacetime, and its generalization, the "*tangent vector*," is a geometric object even when spacetime is curved. The "*metric*" (machine for producing the squared length of any vector; see Box 1.3) is a geometric object. No coordinates are needed to define any of these concepts.

The next few sections will introduce several geometric objects, and show the roles they play as representatives of physical quantities in flat spacetime.

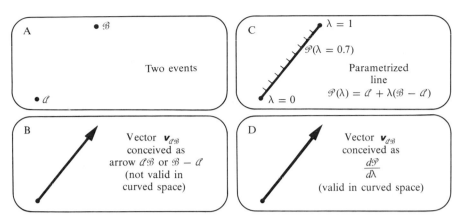

Figure 2.1.
From vector as connector of two points to vector as derivative
("tangent vector"; a local rather than a bilocal concept).

§2.3. VECTORS

Begin with the simplest idea of a vector (Figure 2.1B): an arrow extending from one spacetime event \mathcal{A} ("tail") to another event \mathcal{B} ("tip"). Write this vector as

$$\boldsymbol{v}_{\mathcal{A}\mathcal{B}} = \mathcal{B} - \mathcal{A} \ \text{(or $\mathcal{A}\mathcal{B}$)}.$$

Ways of defining vector:

As arrow

For many purposes (including later generalization to curved spacetime) other completely equivalent ways to think of this vector are more convenient. Represent the arrow by the parametrized straight line $\mathcal{P}(\lambda) = \mathcal{A} + \lambda(\mathcal{B} - \mathcal{A})$, with $\lambda = 0$ the tail of the arrow, and $\lambda = 1$ its tip. Form the derivative of this simple linear expression for $\mathcal{P}(\lambda)$:

As parametrized straight line

$$(d/d\lambda)[\mathcal{A} + \lambda(\mathcal{B} - \mathcal{A})] = \mathcal{B} - \mathcal{A} = \mathcal{P}(1) - \mathcal{P}(0) \equiv (\text{tip}) - (\text{tail}) \equiv \boldsymbol{v}_{\mathcal{A}\mathcal{B}}.$$

This result allows one to replace the idea of a vector as a 2-point object ("bilocal") by the concept of a vector as a 1-point object ("tangent vector"; local):

$$\boldsymbol{v}_{\mathcal{A}\mathcal{B}} = (d\mathcal{P}/d\lambda)_{\lambda = 0}. \tag{2.1}$$

As derivative of point along curve

Example: if $\mathcal{P}(\tau)$ is the straight world line of a free particle, parametrized by its proper time, then the displacement that occurs in a proper time interval of one second gives an arrow $\boldsymbol{u} = \mathcal{P}(1) - \mathcal{P}(0)$. This arrow is easily drawn on a spacetime diagram. It accurately shows the 4-velocity of the particle. However, the derivative formula $\boldsymbol{u} = d\mathcal{P}/d\tau$ for computing the same displacement (1) is more suggestive of the velocity concept and (2) lends itself to the case of accelerated motion. Thus, given a world line $\mathcal{P}(\tau)$ that is not straight, as in Figure 2.2, one must first form $d\mathcal{P}/d\tau$, and only thereafter draw the straight line $\mathcal{P}(0) + \lambda(d\mathcal{P}/d\tau)_0$ of the arrow $\boldsymbol{u} = d\mathcal{P}/d\tau$ to display the 4-velocity \boldsymbol{u}.

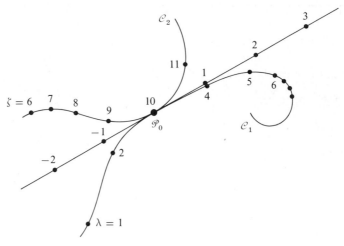

Figure 2.2.

Same tangent vector derived from two very different curves. That parametrized straight line is also drawn which best fits the two curves at \mathscr{P}_0. The tangent vector reaches from 0 to 1 on this straight line.

Components of a vector

The reader may be unfamiliar with this viewpoint. More familiar may be the components of the 4-velocity in a specific Lorentz reference frame:

$$u^0 = \frac{dt}{d\tau} = \frac{1}{\sqrt{1 - v^2}}, \qquad u^j = \frac{dx^j}{d\tau} = \frac{v^j}{\sqrt{1 - v^2}}, \qquad (2.2)$$

where

$$v^j = dx^j/dt = \text{components of "ordinary velocity,"}$$
$$v^2 = (v^x)^2 + (v^y)^2 + (v^z)^2.$$

Even the components (2.2) of 4-velocity may seem slightly unfamiliar if the reader is accustomed to having the fourth component of a vector be multiplied by a factor $i = \sqrt{-1}$. If so, he must adjust himself to new notation. (See "Farewell to 'ict,'" Box 2.1.)

More fundamental than the components of a vector is the vector itself. It is a geometric object with a meaning independent of all coordinates. Thus a particle has a world line $\mathscr{P}(\tau)$, and a 4-velocity $\boldsymbol{u} = d\mathscr{P}/d\tau$, that have nothing to do with any coordinates. Coordinates enter the picture when analysis on a computer is required (rejects vectors; accepts numbers). For this purpose one adopts a Lorentz

Basis vectors

frame with orthonormal basis vectors (Figure 2.3) \boldsymbol{e}_0, \boldsymbol{e}_1, \boldsymbol{e}_2, and \boldsymbol{e}_3. Relative to the origin \mathcal{O} of this frame, the world line has a coordinate description

$$\mathscr{P}(\tau) - \mathcal{O} = x^0(\tau)\boldsymbol{e}_0 + x^1(\tau)\boldsymbol{e}_1 + x^2(\tau)\boldsymbol{e}_2 + x^3(\tau)\boldsymbol{e}_3 = x^\mu(\tau)\boldsymbol{e}_\mu.$$

Expressed relative to the same Lorentz frame, the 4-velocity of the particle is

$$\boldsymbol{u} = d\mathscr{P}/d\tau = (dx^\mu/d\tau)\boldsymbol{e}_\mu = u^0\boldsymbol{e}_0 + u^1\boldsymbol{e}_1 + u^2\boldsymbol{e}_2 + u^3\boldsymbol{e}_3. \qquad (2.3)$$

Box 2.1 FAREWELL TO "ict"

One sometime participant in special relativity will have to be put to the sword: "$x^4 = ict$." This imaginary coordinate was invented to make the geometry of spacetime look formally as little different as possible from the geometry of Euclidean space; to make a Lorentz transformation look on paper like a rotation; and to spare one the distinction that one otherwise is forced to make between quantities with upper indices (such as the components p^μ of the energy-momentum vector) and quantities with lower indices (such as the components p_μ of the energy-momentum 1-form). However, it is no kindness to be spared this latter distinction. Without it, one cannot know whether a vector (§2.3) is meant or the very different geometric object that is a 1-form (§2.5). Moreover, there is a significant difference between an angle on which everything depends periodically (a rotation) and a parameter the increase of which gives rise to ever-growing momentum differences (the "velocity parameter" of a Lorentz transformation; Box 2.4). If the imaginary time-coordinate hides from view the character of the geometric object being dealt with and the nature of the parameter in a transformation, it also does something even more serious: it hides the completely different metric structure (§2.4) of $+++$ geometry and $-+++$ geometry. In Euclidean geometry, when the distance between two points is zero, the two

points must be the same point. In Lorentz-Minkowski geometry, when the interval between two events is zero, one event may be on Earth and the other on a supernova in the galaxy M31, but their separation must be a null ray (piece of a light cone). The backward-pointing light cone at a given event contains all the events by which that event can be influenced. The forward-pointing light cone contains all events that it can influence. The multitude of double light cones taking off from all the events of spacetime forms an interlocking causal structure. This structure makes the machinery of the physical world function as it does (further comments on this structure in Wheeler and Feynman 1945 and 1949 and in Zeeman 1964). If in a region where spacetime is flat, one can hide this structure from view by writing

$$(\Delta s)^2 = (\Delta x^1)^2 + (\Delta x^2)^2 + (\Delta x^3)^2 + (\Delta x^4)^2,$$

with $x^4 = ict$, no one has discovered a way to make an imaginary coordinate work in the general curved spacetime manifold. If "$x^4 = ict$" cannot be used there, it will not be used here. In this chapter and hereafter, as throughout the literature of general relativity, a real time coordinate is used, $x^0 = t = ct_{conv}$ (superscript 0 rather than 4 to avoid any possibility of confusion with the imaginary time coordinate).

The components w^α of any other vector \boldsymbol{w} in this frame are similarly defined as the coefficients in such an expansion,

Expansion of vector in terms of basis

$$\boldsymbol{w} = w^\alpha \boldsymbol{e}_\alpha. \tag{2.4}$$

Notice: the subscript α on \boldsymbol{e}_α tells which vector, not which component!

§2.4. THE METRIC TENSOR

The metric tensor, one recalls from part IV of Box 1.3, is a machine for calculating the squared length of a single vector, or the scalar product of two different vectors.

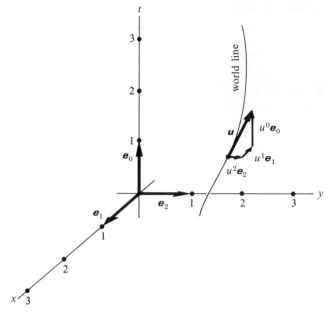

Figure 2.3.

The 4-velocity of a particle in flat spacetime. The 4-velocity \boldsymbol{u} is the unit vector (arrow) tangent to the particle's world line—one tangent vector for each event on the world line. In a specific Lorentz coordinate system, there are basis vectors of unit length, which point along the four coordinate axes: \boldsymbol{e}_0, \boldsymbol{e}_1, \boldsymbol{e}_2, \boldsymbol{e}_3. The 4-velocity, like any vector, can be expressed as a sum of components along the basis vectors:

$$\boldsymbol{u} = u^0\boldsymbol{e}_0 + u^1\boldsymbol{e}_1 + u^2\boldsymbol{e}_2 + u^3\boldsymbol{e}_3 = u^\alpha\boldsymbol{e}_\alpha.$$

Metric defined as machine for computing scalar products of vectors

More precisely, the metric tensor \boldsymbol{g} is a machine with two slots for inserting vectors

$$\underset{\underset{\displaystyle \boldsymbol{g}(\quad,\quad).}{\downarrow\qquad\downarrow}}{\text{slot 1}\ \ \text{slot 2}} \tag{2.5}$$

Upon insertion, the machine spews out a real number:

$$\boldsymbol{g}(\boldsymbol{u},\boldsymbol{v}) = \text{``scalar product of } \boldsymbol{u} \text{ and } \boldsymbol{v}\text{,''} \text{ also denoted } \boldsymbol{u}\cdot\boldsymbol{v}. \tag{2.6}$$
$$\boldsymbol{g}(\boldsymbol{u},\boldsymbol{u}) = \text{``squared length of } \boldsymbol{u}\text{,''} \text{ also denoted } \boldsymbol{u}^2.$$

Moreover, this number is independent of the order in which the vectors are inserted ("symmetry of metric tensor"),

$$\boldsymbol{g}(\boldsymbol{u},\boldsymbol{v}) = \boldsymbol{g}(\boldsymbol{v},\boldsymbol{u}); \tag{2.7}$$

and it is linear in the vectors inserted

$$\boldsymbol{g}(a\boldsymbol{u} + b\boldsymbol{v},\boldsymbol{w}) = \boldsymbol{g}(\boldsymbol{w}, a\boldsymbol{u} + b\boldsymbol{v}) = a\boldsymbol{g}(\boldsymbol{u},\boldsymbol{w}) + b\boldsymbol{g}(\boldsymbol{v},\boldsymbol{w}). \tag{2.8}$$

Because the metric "machine" is linear, one can calculate its output, for any input,

as follows, if one knows only what it does to the basis vectors \boldsymbol{e}_α of a Lorentz frame. (1) Define the symbols ("metric coefficients") $\eta_{\alpha\beta}$ by

$$\eta_{\alpha\beta} \equiv \boldsymbol{g}(\boldsymbol{e}_\alpha, \boldsymbol{e}_\beta) = \boldsymbol{e}_\alpha \cdot \boldsymbol{e}_\beta. \tag{2.9}$$

(2) Calculate their numerical values from the known squared length of the separation vector $\boldsymbol{\xi} = \Delta x^\alpha \boldsymbol{e}_\alpha$ between two events:

$$\begin{aligned} (\Delta s)^2 &= -(\Delta x^0)^2 + (\Delta x^1)^2 + (\Delta x^2)^2 + (\Delta x^3)^2 \\ &= \boldsymbol{g}(\Delta x^\alpha \boldsymbol{e}_\alpha, \Delta x^\beta \boldsymbol{e}_\beta) = \Delta x^\alpha \Delta x^\beta \boldsymbol{g}(\boldsymbol{e}_\alpha, \boldsymbol{e}_\beta) \\ &= \Delta x^\alpha \Delta x^\beta \eta_{\alpha\beta} \quad \text{for every choice of } \Delta x^\alpha \end{aligned}$$

$$\Longrightarrow \|\eta_{\alpha\beta}\| \equiv \begin{Vmatrix} -1 & 0 & 0 & 0 \\ 0 & 1 & 0 & 0 \\ 0 & 0 & 1 & 0 \\ 0 & 0 & 0 & 1 \end{Vmatrix} \text{ in any Lorentz frame.} \tag{2.10}$$

(3) Calculate the scalar product of any two vectors \boldsymbol{u} and \boldsymbol{v} from

$$\begin{aligned} \boldsymbol{u} \cdot \boldsymbol{v} &= \boldsymbol{g}(\boldsymbol{u}, \boldsymbol{v}) = \boldsymbol{g}(u^\alpha \boldsymbol{e}_\alpha, v^\beta \boldsymbol{e}_\beta) = u^\alpha v^\beta \boldsymbol{g}(\boldsymbol{e}_\alpha, \boldsymbol{e}_\beta); \\ \boldsymbol{u} \cdot \boldsymbol{v} &= u^\alpha v^\beta \eta_{\alpha\beta} = -u^0 v^0 + u^1 v^1 + u^2 v^2 + u^3 v^3. \end{aligned} \tag{2.11}$$

That one can classify directions and vectors in spacetime into "*timelike*" (negative squared length), "*spacelike*" (positive squared length), and "*null*" or "*lightlike*" (zero squared length) is made possible by the negative sign on the metric coefficient η_{00}.

Box 2.2 shows applications of the above ideas and notation to two elementary problems in special relativity theory.

§2.5. DIFFERENTIAL FORMS

Vectors and the metric tensor are geometric objects that are already familiar from Chapter 1 and from elementary courses in special relativity. Not so familiar, yet equally important, is a third geometric object: the "*differential form*" or "*1-form.*"

Consider the 4-momentum \boldsymbol{p} of a particle, an electron, for example. To spell out one concept of momentum, start with the 4-velocity, $\boldsymbol{u} = d\mathcal{P}/d\tau$, of this electron ("spacetime displacement per unit of proper time along a straightline approximation of the world line"). This is a vector of unit length. Multiply by the mass m of the particle to obtain the *momentum vector*

$$\boldsymbol{p} = m\boldsymbol{u}.$$

But physics gives also quite another idea of momentum. It associates a de Broglie wave with each particle. Moreover, this wave has the most direct possible physical significance. Diffract this wave from a crystal lattice. From the pattern of diffraction, one can determine not merely the length of the de Broglie waves, but also the pattern in space made by surfaces of equal, integral phase $\phi = 7$, $\phi = 8$, $\phi = 9, \ldots$. This

Box 2.2 WORKED EXERCISES USING THE METRIC

Exercise: Show that the squared length of a test particle's 4-velocity \boldsymbol{u} is -1.
Solution: In any Lorentz frame, using the components (2.2), one calculates as follows

$$\boldsymbol{u}^2 = \boldsymbol{g}(\boldsymbol{u}, \boldsymbol{u}) = u^\alpha u^\beta \eta_{\alpha\beta} = -(u^0)^2 + (u^1)^2 + (u^2)^2 + (u^3)^2$$

$$= -\frac{1}{1 - v^2} + \frac{v^2}{1 - v^2} = -1.$$

Exercise: Show that the rest mass of a particle is related to its energy and momentum by the famous equation

$$(mc^2)^2 = E^2 - (pc)^2$$

or, equivalentiy (geometrized units!),

$$m^2 = E^2 - \boldsymbol{p}^2.$$

First Solution: The 4-momentum is defined by $\boldsymbol{p} = m\boldsymbol{u}$, where \boldsymbol{u} is the 4-velocity and m is the rest mass. Consequently, its squared length is

$$\boldsymbol{p}^2 = m^2 \boldsymbol{u}^2 = -m^2$$

$$= -(mu^0)^2 + m^2\boldsymbol{u}^2 = -\underset{\underset{E^2}{\uparrow}}{\frac{m^2}{1 - v^2}} + \underset{\underset{\boldsymbol{p}^2}{\uparrow}}{\frac{m^2 v^2}{1 - v^2}}.$$

Second Solution: In the frame of the observer, where E and \boldsymbol{p} are measured, the 4-momentum splits into time and space parts as

$$p^0 = E, \qquad p^1 \boldsymbol{e}_1 + p^2 \boldsymbol{e}_2 + p^3 \boldsymbol{e}_3 = \boldsymbol{p};$$

hence, its squared length is

$$\boldsymbol{p}^2 = -E^2 + \boldsymbol{p}^2.$$

But in the particle's rest frame, \boldsymbol{p} splits as

$$p^0 = m, \qquad p^1 = p^2 = p^3 = 0;$$

hence, its squared length is $\boldsymbol{p}^2 = -m^2$. But the squared length is a geometric object defined independently of any coordinate system; so it must be the same by whatever means one calculates it:

$$-\boldsymbol{p}^2 = m^2 = E^2 - \boldsymbol{p}^2.$$

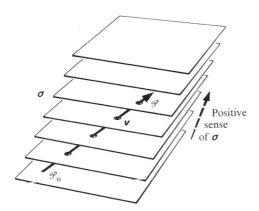

Figure 2.4.

The vector separation $\boldsymbol{v} = \mathscr{P} - \mathscr{P}_0$ between two neighboring events \mathscr{P}_0 and \mathscr{P}; a 1-form $\boldsymbol{\sigma}$; and the piercing of $\boldsymbol{\sigma}$ by \boldsymbol{v} to give the number

$$\langle \boldsymbol{\sigma}, \boldsymbol{v} \rangle = (\text{number of surfaces pierced}) = 4.4$$

(4.4 "bongs of bell"). When $\boldsymbol{\sigma}$ is made of surfaces of constant phase, $\phi = 17$, $\phi = 18$, $\phi = 19, \ldots$ of the de Broglie wave for an electron, then $\langle \boldsymbol{\sigma}, \boldsymbol{v} \rangle$ is the phase difference between the events \mathscr{P}_0 and \mathscr{P}. Note that $\boldsymbol{\sigma}$ is not fully specified by its surfaces; an orientation is also necessary. Which direction from surface to surface is "positive"; i.e., in which direction does ϕ increase?

pattern of surfaces, given a name "$\widetilde{\boldsymbol{k}}$," provides the simplest illustration one can easily find for a 1-form.

The pattern of surfaces in spacetime made by such a 1-form: what is it good for? Take two nearby points in spacetime, \mathscr{P} and \mathscr{P}_0. Run an arrow $\boldsymbol{v} = \mathscr{P} - \mathscr{P}_0$ from \mathscr{P}_0 to \mathscr{P}. It will pierce a certain number of the de Broglie wave's surfaces of integral phase, with a bong of an imaginary bell at each piercing. The number of surfaces pierced (number of "bongs of bell") is denoted

> Vector pierces 1-form

$$\langle \widetilde{\boldsymbol{k}}, \boldsymbol{v} \rangle;$$
$$\text{1-form pierced} \underset{\uparrow}{\rule{0pt}{0pt}} \quad \underset{\uparrow}{\rule{0pt}{0pt}} \text{vector that pierces}$$

in this example it equals the phase difference between tail (\mathscr{P}_0) and tip (\mathscr{P}) of \boldsymbol{v},

$$\langle \widetilde{\boldsymbol{k}}, \boldsymbol{v} \rangle = \phi(\mathscr{P}) - \phi(\mathscr{P}_0).$$

See Figure 2.4.

Normally neither \mathscr{P}_0 nor \mathscr{P} will lie at a point of integral phase. Therefore one can and will imagine, as uniformly interpolated between the surfaces of integral phase, an infinitude of surfaces with all the intermediate phase values. With their aid, the precise value of $\langle \widetilde{\boldsymbol{k}}, \boldsymbol{v} \rangle = \phi(\mathscr{P}) - \phi(\mathscr{P}_0)$ can be determined.

To make the mathematics simple, regard $\widetilde{\boldsymbol{k}}$ not as the global pattern of de Broglie-wave surfaces, but as a local pattern near a specific point in spacetime. Just as the vector $\boldsymbol{u} = d\mathscr{P}/d\tau$ represents the local behavior of a particle's world line (linear approximation to curved line in general), so the 1-form $\widetilde{\boldsymbol{k}}$ represents the local form

> The 1-form viewed as family of flat, equally spaced surfaces

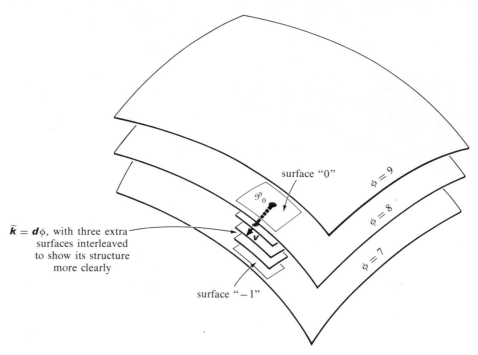

Figure 2.5.

This is a dual-purpose figure. (a) It illustrates the de Broglie wave 1-form $\widetilde{\boldsymbol{k}}$ at an event \mathscr{P}_0 (family of equally spaced, flat surfaces, or "hyperplanes" approximating the surfaces of constant phase). (b) It illustrates the gradient $\boldsymbol{d}\phi$ of the function ϕ (concept defined in §2.6), which is the same oriented family of flat surfaces

$$\widetilde{\boldsymbol{k}} = \boldsymbol{d}\phi.$$

At different events, $\widetilde{\boldsymbol{k}} = \boldsymbol{d}\phi$ is different—different orientation of surfaces and different spacing. The change in ϕ between the tail and tip of the very short vector \boldsymbol{v} is equal to the number of surfaces of $\boldsymbol{d}\phi$ pierced by \boldsymbol{v}, $\langle \boldsymbol{d}\phi, \boldsymbol{v} \rangle$; it equals -0.5 in this figure.

of the de Broglie wave's surfaces (linear approximation; surfaces flat and equally spaced; see Figure 2.5).

Regard the 1-form $\widetilde{\boldsymbol{k}}$ as a machine into which vectors are inserted, and from which numbers emerge. Insertion of \boldsymbol{v} produces as output $\langle \widetilde{\boldsymbol{k}}, \boldsymbol{v} \rangle$. Since the surfaces of $\widetilde{\boldsymbol{k}}$ are flat and equally spaced, the output is a linear function of the input:

$$\langle \widetilde{\boldsymbol{k}}, a\boldsymbol{u} + b\boldsymbol{v} \rangle = a\langle \widetilde{\boldsymbol{k}}, \boldsymbol{u} \rangle + b\langle \widetilde{\boldsymbol{k}}, \boldsymbol{v} \rangle. \tag{2.12a}$$

The 1-form viewed as linear function of vectors

This, in fact, is the mathematical definition of a 1-form: *a 1-form is a linear, real-valued function of vectors;* i.e., a linear machine that takes in a vector and puts out a number. Given the machine $\widetilde{\boldsymbol{k}}$, it is straightforward to draw the corresponding surfaces in spacetime. Pick a point \mathscr{P}_0 at which the machine is to reside. The surface of $\widetilde{\boldsymbol{k}}$ that passes through \mathscr{P}_0 contains points \mathscr{P} for which $\langle \widetilde{\boldsymbol{k}}, \mathscr{P} - \mathscr{P}_0 \rangle = 0$ (no bongs of bell). The other surfaces contain points with $\langle \widetilde{\boldsymbol{k}}, \mathscr{P} - \mathscr{P}_0 \rangle = \pm 1, \pm 2, \pm 3, \ldots$.

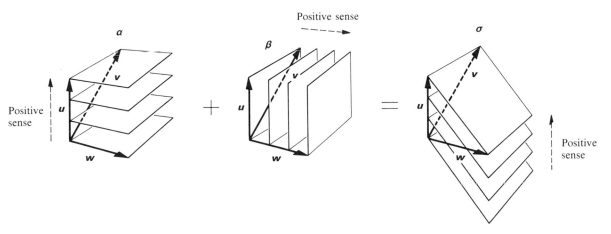

Figure 2.6.

The addition of two 1-forms, $\boldsymbol{\alpha}$ and $\boldsymbol{\beta}$, to produce the 1-form $\boldsymbol{\sigma}$. Required is a pictorial construction that starts from the surfaces of $\boldsymbol{\alpha}$ and $\boldsymbol{\beta}$, e.g., $\langle \boldsymbol{\alpha}, \mathscr{P} - \mathscr{P}_0 \rangle = \cdots -1, 0, 1, 2, \ldots$, and constructs those of $\boldsymbol{\sigma} = \boldsymbol{\alpha} + \boldsymbol{\beta}$. Such a construction, based on linearity (2.12b) of the addition process, is as follows. (1) Pick several vectors $\boldsymbol{u}, \boldsymbol{v}, \ldots$ that lie parallel to the surfaces of $\boldsymbol{\beta}$ (no piercing!), but pierce precisely 3 surfaces of $\boldsymbol{\alpha}$; each of these must then pierce precisely 3 surfaces of $\boldsymbol{\sigma}$:

$$\langle \boldsymbol{\sigma}, \boldsymbol{u} \rangle = \langle \boldsymbol{\alpha} + \boldsymbol{\beta}, \boldsymbol{u} \rangle = \langle \boldsymbol{\alpha}, \boldsymbol{u} \rangle = 3.$$

(2) Pick several other vectors \boldsymbol{w}, \ldots that lie parallel to the surfaces of $\boldsymbol{\alpha}$ but pierce precisely 3 surfaces of $\boldsymbol{\beta}$; these will also pierce precisely 3 surfaces of $\boldsymbol{\sigma}$. (3) Construct that unique family of equally spaced surfaces in which $\boldsymbol{u}, \boldsymbol{v}, \ldots, \boldsymbol{w}, \ldots$ all have their tails on one surface and their tips on the third succeeding surface.

Sometimes 1-forms are denoted by boldface, sans-serif Latin letters with tildes over them, e.g., $\widetilde{\boldsymbol{k}}$; but more often by boldface Greek letters, e.g., $\boldsymbol{\alpha}, \boldsymbol{\beta}, \boldsymbol{\sigma}$. The output of a 1-form $\boldsymbol{\sigma}$, when a vector \boldsymbol{u} is inserted, is called "*the value of $\boldsymbol{\sigma}$ on \boldsymbol{u}*" or "*the contraction of $\boldsymbol{\sigma}$ with \boldsymbol{u}*."

Also, 1-forms, like any other kind of function, can be added. The 1-form $a\boldsymbol{\alpha} + b\boldsymbol{\beta}$ is that machine (family of surfaces) which puts out the following number when a vector \boldsymbol{u} is put in:

Addition of 1-forms

$$\langle a\boldsymbol{\alpha} + b\boldsymbol{\beta}, \boldsymbol{u} \rangle = a\langle \boldsymbol{\alpha}, \boldsymbol{u} \rangle + b\langle \boldsymbol{\beta}, \boldsymbol{u} \rangle. \tag{2.12b}$$

Figure 2.6 depicts this addition in terms of surfaces.

One can verify that the set of all 1-forms at a given event is a "vector space" in the abstract, algebraic sense of the term.

Return to a particle and its de Broglie wave. Just as the arrow $\boldsymbol{p} = md\mathscr{P}/d\tau$ represents the best *linear* approximation to the particle's actual world line near \mathscr{P}_0, so the flat surfaces of the 1-form $\widetilde{\boldsymbol{k}}$ provide the best linear approximation to the curved surfaces of the particle's de Broglie wave, and $\widetilde{\boldsymbol{k}}$ itself is the *linear function* that best approximates the de Broglie phase ϕ near \mathscr{P}_0:

$$\phi(\mathscr{P}) = \phi(\mathscr{P}_0) + \langle \widetilde{\boldsymbol{k}}, \mathscr{P} - \mathscr{P}_0 \rangle$$
$$+ \text{ terms of higher order in } (\mathscr{P} - \mathscr{P}_0). \tag{2.13}$$

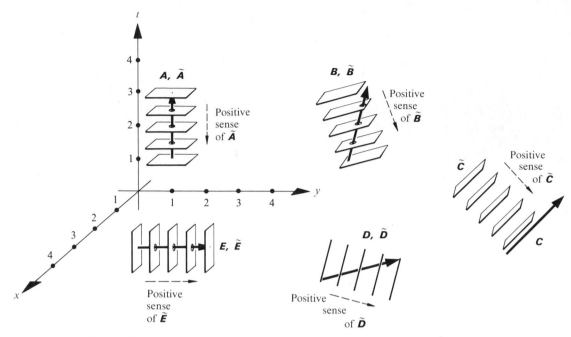

Figure 2.7.

Several vectors, **A**, **B**, **C**, **D**, **E**, and corresponding 1-forms \tilde{A}, \tilde{B}, \tilde{C}, \tilde{D}, \tilde{E}. The process of drawing \tilde{U} corresponding to a given vector **U** is quite simple. (1) Orient the surfaces of \tilde{U} orthogonal to the vector **U**. (Why? Because any vector **V** that is perpendicular to **U** must pierce no surfaces of \tilde{U} ($0 = \mathbf{U} \cdot \mathbf{V} = \langle \tilde{U}, \mathbf{V} \rangle$) and must therefore lie in a surface of \tilde{U}.) (2) Space the surfaces of \tilde{U} so that the number of surfaces pierced by some arbitrary vector **Y** (e.g., **Y** = **U**) is equal to **Y** · **U**.

Note that in the figure the surfaces of \tilde{B} are, indeed, orthogonal to **B**; those of \tilde{C} are, indeed, orthogonal to **C**, etc. If they do not look so, that is because the reader is attributing Euclidean geometry, not Lorentz geometry, to the spacetime diagram. He should recall, for example, that because **C** is a null vector, it is orthogonal to itself ($\mathbf{C} \cdot \mathbf{C} = 0$), so it must itself lie in a surface of the 1-form \tilde{C}. Confused readers may review spacetime diagrams in a more elementary text, e.g., Taylor and Wheeler (1966).

Physical correspondence between 1-forms and vectors

Actually, the de Broglie 1-form \tilde{k} and the momentum vector **p** contain precisely the same information, both physically (via quantum theory) and mathematically. To see their relationship. relabel the surfaces of \tilde{k} by $\hbar \times$ phase, thereby obtaining the *"momentum 1-form"* \tilde{p}. Pierce this 1-form with any vector **v**, and find the result (exercise 2.1) that

$$\mathbf{p} \cdot \mathbf{v} = \langle \tilde{p}, \mathbf{v} \rangle. \tag{2.14}$$

In words: the projection of **v** on the 4-momentum vector **p** equals the number of surfaces it pierces in the 4-momentum 1-form \tilde{p}. *Examples:* Vectors **v** lying in a surface of \tilde{p} (no piercing) are perpendicular to **p** (no projection); **p** itself pierces $\mathbf{p}^2 = -m^2$ surfaces of \tilde{p}.

Mathematical correspondence between 1-forms and vectors

Corresponding to any vector **p** there exists a unique 1-form (linear function of vectors) \tilde{p} defined by equation (2.14). And corresponding to any 1-form \tilde{p}, there exists a unique vector **p** defined by its projections on all other vectors, by equation (2.14). Figure 2.7 shows several vectors and their corresponding 1-forms.

A single physical quantity can be described equally well by a vector \boldsymbol{p} or by the corresponding 1-form $\tilde{\boldsymbol{p}}$. Sometimes the vector description is the simplest and most natural; sometimes the 1-form description is nicer. *Example:* Consider a 1-form representing the march of Lorentz coordinate time toward the future—surfaces $x^0 = \ldots, 7, 8, 9, \ldots$. The corresponding vector points toward the past [see Figure 2.7 or equation (2.14)]; its description of the forward march of time is not so nice!

One often omits the tilde from the 1-form $\tilde{\boldsymbol{p}}$ corresponding to a vector \boldsymbol{p}, and uses the same symbol \boldsymbol{p} for both. Such practice is justified by the unique correspondence (both mathematical and physical) between $\tilde{\boldsymbol{p}}$ and \boldsymbol{p}.

EXERCISE

Exercise 2.1.

Show that equation (2.14) is in accord with the quantum-mechanical properties of a de Broglie wave,

$$\psi = e^{i\phi} = \exp[i(\boldsymbol{k} \cdot \boldsymbol{x} - \omega t)].$$

§2.6. GRADIENTS AND DIRECTIONAL DERIVATIVES

There is no simpler 1-form than the *gradient,* "\boldsymbol{df}," of a function f. Gradient a 1-form? How so? Hasn't one always known the gradient as a vector? Yes, indeed, but only because one was not familiar with the more appropriate 1-form concept. The more familiar gradient is the vector corresponding, via equation (2.14), to the 1-form gradient. The hyperplanes representing \boldsymbol{df} at a point \mathscr{P}_0 are just the level surfaces of f itself, except for flattening and adjustment to equal spacing (Figure 2.5; identify f here with ϕ there). More precisely, they are the level surfaces of the linear function that approximates f in an infinitesimal neighborhood of \mathscr{P}_0.

Gradient of a function as a 1-form

Why the name "gradient"? Because \boldsymbol{df} describes the first order changes in f in the neighborhood of \mathscr{P}_0:

$$f(\mathscr{P}) = f(\mathscr{P}_0) + \langle \boldsymbol{df}, \mathscr{P} - \mathscr{P}_0 \rangle + \text{(nonlinear terms)}. \qquad (2.15)$$

[Compare the fundamental idea of "derivative" of something as "best linear approximation to that something at a point"—an idea that works even for functions whose values and arguments are infinite dimensional vectors! See, e.g., Dieudonné (1960).]

Take any vector \boldsymbol{v}; construct the curve $\mathscr{P}(\lambda)$ defined by $\mathscr{P}(\lambda) - \mathscr{P}_0 = \lambda\boldsymbol{v}$; and differentiate the function f along this curve:

$$\partial_{\boldsymbol{v}} f = (d/d\lambda)_{\lambda=0} f[\mathscr{P}(\lambda)] = (df/d\lambda)_{\mathscr{P}_0}. \qquad (2.16a)$$

The "differential operator,"

$$\partial_{\boldsymbol{v}} = (d/d\lambda)_{\text{at } \lambda = 0, \text{ along curve } \mathscr{P}(\lambda) - \mathscr{P}_0 = \lambda\boldsymbol{v}}, \qquad (2.16b)$$

which does this differentiating, is called the *"directional derivative operator along the vector **v**."* The directional derivative $\partial_v f$ and the gradient df are intimately related, as one sees by applying ∂_v to equation (2.15) and evaluating the result at the point \mathscr{P}_0:

$$\partial_v f = \langle df, d\mathscr{P}/d\lambda \rangle = \langle df, v \rangle. \tag{2.17}$$

This result, expressed in words, is: df is a linear machine for computing the rate of change of f along any desired vector v. Insert v into df; the output ("number of surfaces pierced; number of bongs of bell") is $\partial_v f$—which, for sufficiently small v, is simply the difference in f between tip and tail of v.

§2.7. COORDINATE REPRESENTATION OF GEOMETRIC OBJECTS

In flat spacetime, special attention focuses on Lorentz frames. The coordinates $x^0(\mathscr{P})$, $x^1(\mathscr{P})$, $x^2(\mathscr{P})$, $x^3(\mathscr{P})$ of a Lorentz frame are functions; so their gradients can be calculated. Each of the resulting "basis 1-forms,"

$$\boldsymbol{\omega}^\alpha = \boldsymbol{d}x^\alpha, \tag{2.18}$$

has as its hyperplanes the coordinate surfaces $x^\alpha = \text{const}$; see Figure 2.8. Consequently the basis vector \boldsymbol{e}_α pierces precisely one surface of the basis 1-form $\boldsymbol{\omega}^\alpha$,

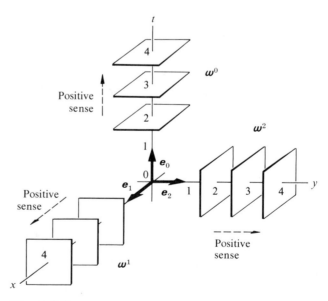

Figure 2.8.
The basis vectors and 1-forms of a particular Lorentz coordinate frame. The basis 1-forms are so laid out that

$$\langle \boldsymbol{\omega}^\alpha, \boldsymbol{e}_\beta \rangle = \delta^\alpha{}_\beta.$$

while the other three basis vectors lie parallel to the surfaces of \boldsymbol{w}^α and thus pierce none:

$$\langle \boldsymbol{w}^\alpha, \boldsymbol{e}_\beta \rangle = \delta^\alpha{}_\beta. \tag{2.19}$$

(One says that the set of basis 1-forms $\{\boldsymbol{w}^\alpha\}$ and the set of basis vectors $\{\boldsymbol{e}_\beta\}$ are the *"duals"* of each other if they have this property.)

Just as arbitrary vectors can be expanded in terms of the basis \boldsymbol{e}_α, $\boldsymbol{v} = v^\alpha \boldsymbol{e}_\alpha$, so arbitrary 1-forms can be expanded in terms of \boldsymbol{w}^β:

Expansion of 1-form in terms of basis

$$\boldsymbol{\sigma} = \sigma_\beta \boldsymbol{w}^\beta. \tag{2.20}$$

The expansion coefficients σ_β are called "the components of $\boldsymbol{\sigma}$ on the basis \boldsymbol{w}^β."

These definitions produce an elegant computational formalism, thus: Calculate how many surfaces of $\boldsymbol{\sigma}$ are pierced by the basis vector \boldsymbol{e}_α; equations (2.19) and (2.20) give the answer:

Calculation and manipulation of vector and 1-form components

$$\langle \boldsymbol{\sigma}, \boldsymbol{e}_\alpha \rangle = \langle \sigma_\beta \boldsymbol{w}^\beta, \boldsymbol{e}_\alpha \rangle = \sigma_\beta \langle \boldsymbol{w}^\beta, \boldsymbol{e}_\alpha \rangle = \sigma_\beta \delta^\beta{}_\alpha;$$

i.e.,

$$\langle \boldsymbol{\sigma}, \boldsymbol{e}_\alpha \rangle = \sigma_\alpha. \tag{2.21a}$$

Similarly, calculate $\langle \boldsymbol{w}^\alpha, \boldsymbol{v} \rangle$ for any vector $\boldsymbol{v} = \boldsymbol{e}_\beta v^\beta$; the result is

$$\langle \boldsymbol{w}^\alpha, \boldsymbol{v} \rangle = v^\alpha. \tag{2.21b}$$

Multiply equation (2.21a) by v^α and sum, or multiply (2.21b) by σ_α and sum; the result in either case is

$$\langle \boldsymbol{\sigma}, \boldsymbol{v} \rangle = \sigma_\alpha v^\alpha. \tag{2.22}$$

This provides a way, using components, to calculate the coordinate-independent value of $\langle \boldsymbol{\sigma}, \boldsymbol{v} \rangle$.

Each Lorentz frame gives a coordinate-dependent representation of any geometric object or relation: \boldsymbol{v} is represented by its components v^α; $\boldsymbol{\sigma}$, by its components σ_α; a point \mathscr{P}, by its coordinates x^α; the relation $\langle \boldsymbol{\sigma}, \boldsymbol{v} \rangle = 17.3$ by $\sigma_\alpha v^\alpha = 17.3$.

To find the coordinate representation of the directional derivative operator $\partial_{\boldsymbol{v}}$, rewrite equation (2.16b) using elementary calculus

$$\partial_{\boldsymbol{v}} = \left(\frac{d}{d\lambda} \right)_{\mathscr{P}_0} = \underbrace{\left(\frac{dx^\alpha}{d\lambda} \right)_{\text{at } \mathscr{P}_0 \text{ along } \mathscr{P}(\lambda) - \mathscr{P}_0 = \lambda \boldsymbol{v}}}_{v^\alpha;\ \text{see equation (2.3)}} \left(\frac{\partial}{\partial x^\alpha} \right);$$

the result is

$$\partial_{\boldsymbol{v}} = v^\alpha \, \partial / \partial x^\alpha. \tag{2.23}$$

Directional derivative in terms of coordinates

In particular, the directional derivative along a basis vector \boldsymbol{e}_α (components $[\boldsymbol{e}_\alpha]^\beta = \langle \boldsymbol{w}^\beta, \boldsymbol{e}_\alpha \rangle = \delta^\beta{}_\alpha$) is

$$\partial_\alpha \equiv \partial_{\boldsymbol{e}_\alpha} = \partial / \partial x^\alpha. \tag{2.24}$$

This should also be obvious from Figure 2.8.

Components of gradient

The components of the gradient 1-form df, which are denoted $f_{,\alpha}$

$$df = f_{,\alpha} \boldsymbol{\omega}^{\alpha}, \tag{2.25a}$$

are calculated easily using the above formulas:

$$
\begin{aligned}
f_{,\alpha} &= \langle df, e_{\alpha} \rangle && \text{[standard way to calculate components; equation (2.21a)]} \\
&= \partial_{\alpha} f && \text{[by relation (2.17) between directional derivative and gradient]} \\
&= \partial f / \partial x^{\alpha} && \text{[by equation (2.24)].}
\end{aligned}
$$

Thus, in agreement with the elementary calculus idea of gradient, the components of df are just the partial derivatives along the coordinate axes:

$$f_{,\alpha} = \partial f / \partial x^{\alpha}; \quad \text{i.e., } df = (\partial f / \partial x^{\alpha}) \, dx^{\alpha}. \tag{2.25b}$$

(Recall: $\boldsymbol{\omega}^{\alpha} = dx^{\alpha}$.) The formula $df = (\partial f / \partial x^{\alpha}) \, dx^{\alpha}$ suggests, correctly, that df is a rigorous version of the "differential" of elementary calculus; see Box 2.3.

Other important coordinate representations for geometric relations are explored in the following exercises.

EXERCISES

Derive the following computationally useful formulas:

**Exercise 2.2. LOWERING INDEX TO GET THE 1-FORM
 CORRESPONDING TO A VECTOR**

The components u_{α} of the 1-form \tilde{u} that corresponds to a vector \boldsymbol{u} can be obtained by "lowering an index" with the metric coefficients $\eta_{\alpha\beta}$:

$$u_{\alpha} = \eta_{\alpha\beta} u^{\beta}; \text{ i.e., } \quad u_0 = -u^0, \, u_k = u^k. \tag{2.26a}$$

Exercise 2.3. RAISING INDEX TO RECOVER THE VECTOR

One can return to the components of \boldsymbol{u} by raising indices,

$$u^{\alpha} = \eta^{\alpha\beta} u_{\beta}; \tag{2.26b}$$

the matrix $\|\eta^{\alpha\beta}\|$ is defined as the inverse of $\|\eta_{\alpha\beta}\|$, and happens to equal $\|\eta_{\alpha\beta}\|$:

$$\eta^{\alpha\beta} \eta_{\beta\gamma} \equiv \delta^{\alpha}{}_{\gamma}; \quad \eta^{\alpha\beta} = \eta_{\alpha\beta} \text{ for all } \alpha, \beta. \tag{2.27}$$

Exercise 2.4. VARIED ROUTES TO THE SCALAR PRODUCT

The scalar product of \boldsymbol{u} with \boldsymbol{v} can be calculated in any of the following ways:

$$\boldsymbol{u} \cdot \boldsymbol{v} = g(\boldsymbol{u}, \boldsymbol{v}) = u^{\alpha} v^{\beta} \eta_{\alpha\beta} = u^{\alpha} v_{\alpha} = u_{\alpha} v_{\beta} \eta^{\alpha\beta}. \tag{2.28}$$

Box 2.3 DIFFERENTIALS

The "exterior derivative" or "gradient" df of a function f is a more rigorous version of the elementary concept of "differential."

In elementary textbooks, one is presented with the differential df as representing "an infinitesimal change in the function $f(\mathscr{P})$" associated with some infinitesimal displacement of the point \mathscr{P}; but one will recall that the displacement of \mathscr{P} is left arbitrary, albeit infinitesimal. Thus df represents a change in f in some unspecified direction.

But this is precisely what the exterior derivative df represents. Choose a particular, infinitesimally long displacement \boldsymbol{v} of the point \mathscr{P}. Let the displacement vector \boldsymbol{v} pierce df to give the number $\langle df, \boldsymbol{v} \rangle = \partial_{\boldsymbol{v}} f$. That number is the change of f in going from the tail of \boldsymbol{v} to its tip. Thus df, before it has been pierced to give a number, represents the change of f in an unspecified direction. The act of piercing df with \boldsymbol{v} is the act of making explicit the direction in which the change is to be measured. The only failing of the textbook presentation, then, was its suggestion that df was a scalar or a number; the explicit recognition of the need for specifying a direction \boldsymbol{v} to reduce df to a number $\langle df, \boldsymbol{v} \rangle$ shows that in fact df is a 1-form, the gradient of f.

§2.8. THE CENTRIFUGE AND THE PHOTON

Vectors, metric, 1-forms, functions, gradients, directional derivatives: all these geometric objects and more are used in flat spacetime to represent physical quantities; and all the laws of physics must be expressible in terms of such geometric objects.

As an example, consider a high-precision redshift experiment that uses the Mössbauer effect (Figure 2.9). The emitter and the absorber of photons are attached to

Geometric objects in action: example of centrifuge and photon

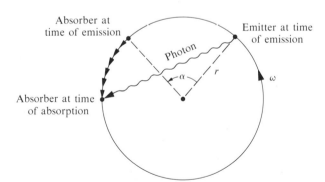

Figure 2.9.
The centrifuge and the photon.

the rim of a centrifuge at points separated by an angle α, as measured in the inertial laboratory. The emitter and absorber are at radius r as measured in the laboratory, and the centrifuge rotates with angular velocity ω. PROBLEM: What is the redshift measured,

$$z = (\lambda_{\text{absorbed}} - \lambda_{\text{emitted}})/\lambda_{\text{emitted}},$$

in terms of ω, r, and α?

SOLUTION: Let \boldsymbol{u}_e be the 4-velocity of the emitter at the event of emission of a given photon; let \boldsymbol{u}_a be the 4-velocity of the absorber at the event of absorption; and let \boldsymbol{p} be the 4-momentum of the photon. All three quantities are vectors defined without reference to coordinates. Equally coordinate-free are the photon energies E_e and E_a measured by emitter and absorber. No coordinates are needed to describe the fact that a specific emitter emitting a specific photon attributes to it the energy E_e; and no coordinates are required in the geometric formula

$$E_e = -\boldsymbol{p} \cdot \boldsymbol{u}_e \tag{2.29}$$

for E_e. [That this formula works can be readily verified by recalling that, in the emitter's frame, $u_e{}^0 = 1$ and $u_e{}^j = 0$; so

$$E_e = -p_\alpha u_e{}^\alpha = -p_0 = +p^0$$

in accordance with the identification "(time component of 4-momentum) = (energy."] Analogous to equation (2.29) is the purely geometric formula

$$E_a = -\boldsymbol{p} \cdot \boldsymbol{u}_a$$

for the absorbed energy.

The ratio of absorbed wavelength to emitted wavelength is the inverse of the energy ratio (since $E = h\nu = hc/\lambda$):

$$\frac{\lambda_a}{\lambda_e} = \frac{E_e}{E_a} = \frac{-\boldsymbol{p} \cdot \boldsymbol{u}_e}{-\boldsymbol{p} \cdot \boldsymbol{u}_a}.$$

This ratio is most readily calculated in the inertial laboratory frame

$$\frac{\lambda_a}{\lambda_e} = \frac{p^0 u_e{}^0 - p^j u_e{}^j}{p^0 u_a{}^0 - p^j u_a{}^j} \equiv \frac{p^0 u_e{}^0 - \boldsymbol{p} \cdot \boldsymbol{u}_e}{p^0 u_a{}^0 - \boldsymbol{p} \cdot \boldsymbol{u}_a}. \tag{2.30}$$

(Here and throughout we use boldface Latin letters for three-dimensional vectors in a given Lorentz frame; and we use the usual notation and formalism of three-dimensional, Euclidean vector analysis to manipulate them.) Because the magnitude of the ordinary velocity of the rim of the centrifuge, $v = \omega r$, is unchanging in time, $u_e{}^0$ and $u_a{}^0$ are equal, and the magnitudes—but not the directions—of \boldsymbol{u}_e and \boldsymbol{u}_a are equal:

$$u_e{}^0 = u_a{}^0 = (1 - v^2)^{-1/2}, \quad |\boldsymbol{u}_e| = |\boldsymbol{u}_a| = v/(1 - v^2)^{1/2}.$$

From the geometry of Figure 2.9, one sees that \boldsymbol{u}_e makes the same angle with \boldsymbol{p} as does \boldsymbol{u}_a. Consequently, $\boldsymbol{p} \cdot \boldsymbol{u}_e = \boldsymbol{p} \cdot \boldsymbol{u}_a$, and $\lambda_{\text{absorbed}}/\lambda_{\text{emitted}} = 1$. *There is no redshift!*

Notice that this solution made no reference whatsoever to Lorentz transformations—they have not even been discussed yet in this book! The power of the geometric, coordinate-free viewpoint is evident!

One must have a variety of coordinate-free contacts between theory and experiment in order to use the geometric viewpoint. One such contact is the equation $E = -\boldsymbol{p} \cdot \boldsymbol{u}$ for the energy of a photon with 4-momentum \boldsymbol{p}, as measured by an observer with 4-velocity \boldsymbol{u}. Verify the following other points of contact.

EXERCISES

Exercise 2.5. ENERGY AND VELOCITY FROM 4-MOMENTUM

A particle of rest mass m and 4-momentum \boldsymbol{p} is examined by an observer with 4-velocity \boldsymbol{u}. Show that just as (a) the energy he measures is

$$E = -\boldsymbol{p} \cdot \boldsymbol{u}; \tag{2.31}$$

so (b) the rest mass he attributes to the particle is

$$m^2 = -\boldsymbol{p}^2; \tag{2.32}$$

(c) the momentum he measures has magnitude

$$|\boldsymbol{p}| = [(\boldsymbol{p} \cdot \boldsymbol{u})^2 + (\boldsymbol{p} \cdot \boldsymbol{p})]^{1/2}; \tag{2.33}$$

(d) the ordinary velocity \boldsymbol{v} he measures has magnitude

$$|\boldsymbol{v}| = \frac{|\boldsymbol{p}|}{E}, \tag{2.34}$$

where $|\boldsymbol{p}|$ and E are as given above; and (e) the 4-vector \boldsymbol{v}, whose components in the observer's Lorentz frame are

$$v^0 = 0, \qquad v^j = (dx^j/dt)_{\text{for particle}} = \text{ordinary velocity},$$

is given by

$$\boldsymbol{v} = \frac{\boldsymbol{p} + (\boldsymbol{p} \cdot \boldsymbol{u})\boldsymbol{u}}{-\boldsymbol{p} \cdot \boldsymbol{u}}. \tag{2.35}$$

Exercise 2.6. TEMPERATURE GRADIENT

To each event \mathcal{Q} inside the sun one attributes a temperature $T(\mathcal{Q})$, the temperature measured by a thermometer at rest in the hot gas there. Then $T(\mathcal{Q})$ is a function; no coordinates are required for its definition and discussion. A cosmic ray from outer space flies through the sun with 4-velocity \boldsymbol{u}. Show that, as measured by the cosmic ray's clock, the time derivative of temperature in its vicinity is

$$dT/d\tau = \partial_{\boldsymbol{u}} T = \langle \boldsymbol{d}T, \boldsymbol{u} \rangle. \tag{2.36}$$

In a local Lorentz frame inside the sun, this equation can be written

$$\frac{dT}{d\tau} = u^\alpha \frac{\partial T}{\partial x^\alpha} = \frac{1}{\sqrt{1 - v^2}} \frac{\partial T}{\partial t} + \frac{v^j}{\sqrt{1 - v^2}} \frac{\partial T}{\partial x^j}. \tag{2.37}$$

Why is this result reasonable?

§2.9. LORENTZ TRANSFORMATIONS

To simplify computations, one often works with the components of vectors and 1-forms, rather than with coordinate-free language. Such component manipulations sometimes involve transformations from one Lorentz frame to another. The reader is already familiar with such Lorentz transformations; but the short review in Box 2.4 will refresh his memory and acquaint him with the notation used in this book.

Lorentz transformations: of coordinates

The key entities in the Lorentz transformation are the matrices $\|\Lambda^{\alpha'}{}_\beta\|$ and $\|\Lambda^{\beta}{}_{\alpha'}\|$; the first transforms coordinates from an unprimed frame to a primed frame, while the second goes from primed to unprimed

$$x^{\alpha'} = \Lambda^{\alpha'}{}_\beta x^\beta, \qquad x^\beta = \Lambda^{\beta}{}_{\alpha'} x^{\alpha'}. \tag{2.38}$$

Since they go in opposite directions, each of the two matrices must be the inverse of the other:

$$\Lambda^{\alpha'}{}_\beta \Lambda^{\beta}{}_{\gamma'} = \delta^{\alpha'}{}_{\gamma'}; \qquad \Lambda^{\beta}{}_{\alpha'} \Lambda^{\alpha'}{}_\gamma = \delta^{\beta}{}_\gamma. \tag{2.39}$$

From the coordinate-independent nature of 4-velocity, $\boldsymbol{u} = (dx^\alpha/d\tau)\boldsymbol{e}_\alpha$, one readily derives the expressions

Of basis vectors

$$\boldsymbol{e}_{\alpha'} = \boldsymbol{e}_\beta \Lambda^{\beta}{}_{\alpha'}, \qquad \boldsymbol{e}_\beta = \boldsymbol{e}_{\alpha'} \Lambda^{\alpha'}{}_\beta \tag{2.40}$$

for the basis vectors of one frame in terms of those of the other; and from other geometric equations, such as

$$\boldsymbol{v} = \boldsymbol{e}_\alpha v^\alpha = \boldsymbol{e}_{\beta'} v^{\beta'},$$
$$\langle \boldsymbol{\sigma}, \boldsymbol{v} \rangle = \sigma_\alpha v^\alpha = \sigma_{\beta'} v^{\beta'},$$
$$\boldsymbol{\sigma} = \sigma_\alpha \boldsymbol{w}^\alpha = \sigma_{\beta'} \boldsymbol{w}^{\beta'},$$

one derives transformation laws

Of basis 1-forms

$$\boldsymbol{w}^{\alpha'} = \Lambda^{\alpha'}{}_\beta \boldsymbol{w}^\beta, \qquad \boldsymbol{w}^\beta = \Lambda^{\beta}{}_{\alpha'} \boldsymbol{w}^{\alpha'}; \tag{2.41}$$

Of components

$$v^{\alpha'} = \Lambda^{\alpha'}{}_\beta v^\beta, \qquad v^\beta = \Lambda^{\beta}{}_{\alpha'} v^{\alpha'}; \tag{2.42}$$

$$\sigma_{\alpha'} = \sigma_\beta \Lambda^{\beta}{}_{\alpha'}, \qquad \sigma_\beta = \sigma_{\alpha'} \Lambda^{\alpha'}{}_\beta. \tag{2.43}$$

One need never memorize the index positions in these transformation laws. One need only line the indices up so that (1) free indices on each side of the equation are in the same position; and (2) summed indices appear once up and once down. Then all will be correct! (Note: the indices on Λ always run "northwest to southeast.")

Box 2.4 LORENTZ TRANSFORMATIONS

Rotation of Frame of Reference by Angle θ in *x-y* Plane

$$\text{Slope } s = \tan\theta; \quad \sin\theta = \frac{s}{(1+s^2)^{1/2}}: \quad \cos\theta = \frac{1}{(1+s^2)^{1/2}}$$

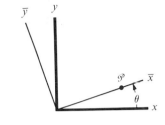

$$t = \bar{t}$$
$$x = \bar{x}\cos\theta - \bar{y}\sin\theta$$
$$y = \bar{x}\sin\theta + \bar{y}\cos\theta$$
$$z = \bar{z}$$

$$\bar{t} = t$$
$$\bar{x} = x\cos\theta + y\sin\theta$$
$$\bar{y} = -x\sin\theta + y\cos\theta$$
$$\bar{z} = z$$

All signs follow from sign of this term. Positive by inspection of point \mathscr{P}.

Combination of Two Such Rotations

$$s = \frac{s_1 + s_2}{1 - s_1 s_2} \quad \text{or} \quad \theta = \theta_1 + \theta_2$$

Boost of Frame of Reference by Velocity Parameter α in *z-t* Plane

$$\text{Velocity } \beta = \tanh\alpha; \quad \sinh\alpha = \frac{\beta}{(1-\beta^2)^{1/2}}; \quad \cosh\alpha = \frac{1}{(1-\beta^2)^{1/2}} = \text{``}\gamma\text{''}$$

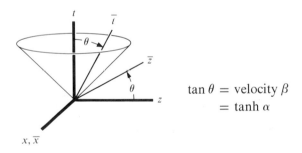

$$\tan\theta = \text{velocity } \beta$$
$$= \tanh\alpha$$

$$t = \bar{t}\cosh\alpha + \bar{z}\sinh\alpha$$
$$x = \bar{x}$$
$$y = \bar{y}$$
$$z = \bar{t}\sinh\alpha + \bar{z}\cosh\alpha$$

$$\bar{t} = t\cosh\alpha - z\sinh\alpha$$
$$\bar{x} = x$$
$$\bar{y} = y$$
$$\bar{z} = -t\sinh\alpha + z\cosh\alpha$$

All signs follow from sign of this term. Positive because object at rest at $\bar{z} = 0$ in rocket frame moves in direction of increasing z in lab frame.

Matrix notation: $\quad x^\mu = \Lambda^\mu{}_{\bar\nu} x^{\bar\nu}, \quad x^{\bar\nu} = \Lambda^{\bar\nu}{}_\mu x^\mu$

$$\|\Lambda^\mu{}_{\bar\nu}\| = \begin{Vmatrix} \cosh\alpha & 0 & 0 & \sinh\alpha \\ 0 & 1 & 0 & 0 \\ 0 & 0 & 1 & 0 \\ \sinh\alpha & 0 & 0 & \cosh\alpha \end{Vmatrix}, \quad \|\Lambda^{\bar\nu}{}_\mu\| = \begin{Vmatrix} \cosh\alpha & 0 & 0 & -\sinh\alpha \\ 0 & 1 & 0 & 0 \\ 0 & 0 & 1 & 0 \\ -\sinh\alpha & 0 & 0 & \cosh\alpha \end{Vmatrix}$$

Box 2.4 (*continued*)

Energy-momentum 4-vector
$$E = \bar{E} \cosh \alpha + p^{\bar{z}} \sinh \alpha$$
$$p^x = p^{\bar{x}}$$
$$p^y = p^{\bar{y}}$$
$$p^z = \bar{E} \sinh \alpha + p^{\bar{z}} \cosh \alpha$$

Charge density-current 4-vector
$$\rho = \bar{\rho} \cosh \alpha + j^{\bar{z}} \sinh \alpha$$
$$j^x = j^{\bar{x}}$$
$$j^y = j^{\bar{y}}$$
$$j^z = \bar{\rho} \sinh \alpha + j^{\bar{z}} \cosh \alpha$$

Aberration, incoming photon:

$$\sin \theta = \frac{-p_\perp}{E} = \frac{(1 - \beta^2)^{1/2} \sin \bar{\theta}}{1 - \beta \cos \bar{\theta}} \qquad \sin \bar{\theta} = \frac{-\bar{p}_\perp}{\bar{E}} = \frac{(1 - \beta^2)^{1/2} \sin \theta}{1 + \beta \cos \theta}$$

$$\cos \theta = \frac{-p^z}{E} = \frac{\cos \bar{\theta} - \beta}{1 - \beta \cos \bar{\theta}} \qquad \cos \bar{\theta} = \frac{-\bar{p}^{\bar{z}}}{\bar{E}} = \frac{\cos \theta + \beta}{1 + \beta \cos \theta}$$

$$\tan (\theta/2) = e^\alpha \tan (\bar{\theta}/2) \qquad \tan (\bar{\theta}/2) = e^{-\alpha} \tan (\theta/2)$$

Combination of Two Boosts in Same Direction

$$\beta = \frac{\beta_1 + \beta_2}{1 + \beta_1 \beta_2} \qquad \text{or} \qquad \alpha = \alpha_1 + \alpha_2.$$

General Combinations of Boosts and Rotations

Spinor formalism of Chapter 41

Poincaré Transformation

$$x^\mu = \Lambda^\mu{}_{\alpha'} x^{\alpha'} + a^\mu.$$

Condition on the Lorentz part of this transformation:

$$ds'^2 = \eta_{\alpha'\beta'} \, dx^{\alpha'} \, dx^{\beta'} = ds^2 = \eta_{\mu\nu} \Lambda^\mu{}_{\alpha'} \Lambda^\nu{}_{\beta'} \, dx^{\alpha'} \, dx^{\beta'}$$

or $\Lambda^T \eta \Lambda = \eta$ (matrix equation, with T indicating "transposed," or rows and columns interchanged).

Effect of transformation on other quantities:

$$u^\mu = \Lambda^\mu{}_{\alpha'} u^{\alpha'} \qquad \text{(4-velocity)} \qquad u_{\alpha'} = u_\mu \Lambda^\mu{}_{\alpha'};$$
$$p^\mu = \Lambda^\mu{}_{\alpha'} p^{\alpha'} \qquad \text{(4-momentum)} \qquad p_{\alpha'} = p_\mu \Lambda^\mu{}_{\alpha'};$$
$$F^{\mu\nu} = \Lambda^\mu{}_{\alpha'} \Lambda^\nu{}_{\beta'} F^{\alpha'\beta'} \qquad \text{(electromagnetic field)} \qquad F_{\alpha'\beta'} = F_{\mu\nu} \Lambda^\mu{}_{\alpha'} \Lambda^\nu{}_{\beta'};$$
$$\boldsymbol{e}_{\alpha'} = \boldsymbol{e}_\mu \Lambda^\mu{}_{\alpha'} \qquad \text{(basis vectors)};$$
$$\boldsymbol{\omega}^{\alpha'} = \Lambda^{\alpha'}{}_\mu \boldsymbol{\omega}^\mu \qquad \text{(basis 1-forms)};$$
$$\boldsymbol{u} = \boldsymbol{e}_{\alpha'} u^{\alpha'} = \boldsymbol{e}_\mu u^\mu = \boldsymbol{u} \qquad \text{(the 4-velocity *vector*)}.$$

Exercise 2.7. BOOST IN AN ARBITRARY DIRECTION

An especially useful Lorentz transformation has the matrix components

$$\Lambda^{0'}{}_{0} = \gamma \equiv \frac{1}{\sqrt{1 - \beta^2}},$$

$$\Lambda^{0'}{}_{j} = \Lambda^{j'}{}_{0} = -\beta\gamma n^{j},$$

$$\Lambda^{j'}{}_{k} = \Lambda^{k'}{}_{j} = (\gamma - 1)n^{j}n^{k} + \delta^{jk}, \qquad (2.44)$$

$$\Lambda^{\mu}{}_{\nu'} = \text{(same as } \Lambda^{\nu'}{}_{\mu} \text{ but with } \beta \text{ replaced by } -\beta),$$

where β, n^1, n^2, and n^3 are parameters, and $\boldsymbol{n}^2 \equiv (n^1)^2 + (n^2)^2 + (n^3)^2 = 1$. Show (a) that this does satisfy the condition $\Lambda^T \eta \Lambda = \eta$ required of a Lorentz transformation (see Box 2.4); (b) that the primed frame moves with ordinary velocity $\beta\boldsymbol{n}$ as seen in the unprimed frame; (c) that the unprimed frame moves with ordinary velocity $-\beta\boldsymbol{n}$ (i.e., $v^{1'} = -\beta n^1$, $v^{2'} = -\beta n^2$, $v^{3'} = -\beta n^3$) as seen in the primed frame; and (d) that for motion in the z direction, the transformation matrices reduce to the familiar form

$$\|\Lambda^{\nu'}{}_{\mu}\| = \begin{Vmatrix} \gamma & 0 & 0 & -\beta\gamma \\ 0 & 1 & 0 & 0 \\ 0 & 0 & 1 & 0 \\ -\beta\gamma & 0 & 0 & \gamma \end{Vmatrix}, \qquad \|\Lambda^{\mu}{}_{\nu'}\| = \begin{Vmatrix} \gamma & 0 & 0 & \beta\gamma \\ 0 & 1 & 0 & 0 \\ 0 & 0 & 1 & 0 \\ \beta\gamma & 0 & 0 & \gamma \end{Vmatrix}. \qquad (2.45)$$

§2.10. COLLISIONS

Whatever the physical entity, whether it is an individual mass in motion, or a torrent of fluid, or a field of force, or the geometry of space itself, it is described in classical general relativity as a geometric object of its own characteristic kind. Each such object is built directly or by abstraction from identifiable points, and needs no coordinates for its representation. It has been seen how this coordinate-free description translates into, and how it can be translated out of, the language of coordinates and components, and how components in a local Lorentz frame transform under a Lorentz transformation. Turn now to two elementary applications of this mathematical machinery to a mass in motion. One has to do with short-range forces (collisions, this section); the other, with the long-range electromagnetic force (Lorentz force law, next chapter).

In a collision, all the change in momentum is concentrated in a time that is short compared to the time of observation. Moreover, the target is typically so small, and quantum mechanics so dominating, that a probabilistic description is the right one. A quantity

Scattering of particles

$$d\sigma = \left(\frac{d\sigma}{d\Omega}\right)_{\theta} d\Omega \qquad (2.46)$$

gives the cross section (cm²) for scattering into the element of solid angle $d\Omega$ at the deflection angle θ; a more complicated expression gives the probability that the

original particle will enter the aperture $d\Omega$ at a given polar angle θ and azimuth ϕ and with energy E to $E + dE$, while simultaneously products of reaction also emerge into specified energy intervals and into specified angular apertures. It would be out of place here to enter into the calculation of such cross sections, though it is a fascinating branch of atomic physics. It is enough to note that the cross section is an area oriented perpendicular to the line of travel of the incident particle. Therefore it is unaffected by any boost of the observer in that direction, provided of course that energies and angles of emergence of the particles are transformed in accordance with the magnitude of that boost ("same events seen in an altered reference system").

Over and above any such detailed account of the encounter as follows from the local dynamic analysis, there stands the law of conservation of energy-momentum:

Conservation of energy-momentum in a collision

$$\sum_{\substack{\text{original} \\ \text{particles, } J}} \boldsymbol{p}_J = \sum_{\substack{\text{final} \\ \text{particles, } K}} \boldsymbol{p}_K \,.$$

(2.47)

Out of this relation, one wins without further analysis such simple results as the following. (1) A photon traveling as a plane wave through empty space cannot split (not true for a focused photon!). (2) When a high-energy electron strikes an electron at rest in an elastic encounter, and the two happen to come off sharing the energy equally, then the angle between their directions of travel is less than the Newtonian value of $90°$, and the deficit gives a simple measure of the energy of the primary. (3) When an electron makes a head-on elastic encounter with a proton, the formula for the fraction of kinetic energy transferred has three rather different limiting forms, according to whether the energy of the electron is nonrelativistic, relativistic, or extreme-relativistic. (4) The threshold for the production of an (e^+, e^-) pair by a photon in the field of force of a massive nucleus is $2m_e$. (5) The threshold for the production of an (e^+, e^-) pair by a photon in an encounter with an electron at rest is $4m_e$ (or $4m_e - \epsilon$ when account is taken of the binding of the $e^+e^-e^-$ system in a very light "molecule"). All these results (topics for independent projects!) and more can be read out of the law of conservation of energy-momentum. For more on this topic, see Blaton (1950), Hagedorn (1964), and Chapter 4 and the last part of Chapter 5 of Sard (1970).

CHAPTER **3**

THE ELECTROMAGNETIC FIELD

*The rotating armatures of every generator and every motor in this
age of electricity are steadily proclaiming the truth of the
relativity theory to all who have ears to hear.*

LEIGH PAGE (1941)

§3.1. THE LORENTZ FORCE AND
THE ELECTROMAGNETIC FIELD TENSOR

At the opposite extreme from an impulsive change of momentum in a collision (the
last topic of Chapter 2) is the gradual change in the momentum of a charged particle
under the action of electric and magnetic forces (the topic treated here).

Let electric and magnetic fields act on a system of charged particles. The accelera-
tions of the particles reveal the electric and magnetic field strengths. In other words,
the Lorentz force law, plus measurements on the components of acceleration of test
particles, can be viewed as defining the components of the electric and magnetic
fields. Once the field components are known from the accelerations of a few test
particles, they can be used to predict the accelerations of other test particles (Box
3.1). Thus the Lorentz force law does double service (1) as definer of fields and (2)
as predicter of motions.

*Here and elsewhere in science, as stressed not least by Henri Poincaré, that view
is out of date which used to say, "Define your terms before you proceed." All the laws
and theories of physics, including the Lorentz force law, have this deep and subtle
character, that they both define the concepts they use (here **B** and **E**) and make
statements about these concepts. Contrariwise, the absence of some body of theory,
law, and principle deprives one of the means properly to define or even to use concepts.
Any forward step in human knowledge is truly creative in this sense: that theory,
concept, law, and method of measurement—forever inseparable—are born into the
world in union.*

Lorentz force as definer of
fields and predicter of
motions

Box 3.1 LORENTZ FORCE LAW AS BOTH DEFINER OF FIELDS AND PREDICTER OF MOTIONS

How one goes about determining the components of the field from measurements of accelerations is not different in principle for electromagnetism and for gravitation. Compare the equations in the two cases:

$$\frac{d^2 x^\alpha}{d\tau^2} = \frac{e}{m} F^\alpha{}_\beta u^\beta \quad \text{in a Lorentz frame,} \quad (1)$$

and

$$\frac{D^2 \xi^\alpha}{d\tau^2} = -R^\alpha{}_{\beta\gamma\delta} u^\beta \xi^\gamma u^\delta \quad \text{in any coordinate system.}$$

$$(2)$$

To make explicit the simpler procedure for electromagnetism will indicate in broad outline how one similarly determines all the components of $R^\alpha{}_{\beta\gamma\delta}$ for gravity. Begin by asking how many test particles one needs to determine the three components of B and the three components of E in the neighborhood under study. For one particle, three components of acceleration are measurable; for a second particle, three more. Enough? No! The information from the one duplicates in part the information from the other. The proof? Whatever the state of motion of the first test particle, pick one's Lorentz frame to be moving the same way. Having zero velocity in this frame, the particle has a zero response to any magnetic field. The electric field alone acts on the particle. The three components of its acceleration give directly the three components E_x, E_y, E_z of the electric field. The second test particle cannot be at rest if it is to do more than duplicate the information provided by the first test particle. Orient the x-axis of the frame

of reference parallel to the direction of motion of this second particle, which will then respond to and measure the components B_y and B_z of the magnetic field. Not so B_x! The acceleration in the x-direction merely remeasures the already once measured E_x. To evaluate B_x, a third test particle is required, but it then gives duplicate information about the other field components. The alternative? Use all N particles simultaneously and on the same democratic footing, both in the evaluation of the six $F_{\alpha\beta}$ and in the testing of the Lorentz force, by applying the method of least squares. Thus, write the discrepancy between predicted and observed acceleration of the Kth particle in the form

$$\dot{u}_\alpha{}^K - \frac{e}{m} F_{\alpha\beta} u^{\beta,K} = \delta a_\alpha{}^K. \quad (3)$$

Take the squared magnitude of this discrepancy and sum over all the particles

$$S = \sum_k \eta^{\alpha\beta} \delta a_\alpha{}^K \delta a^K{}_\beta. \quad (4)$$

In this expression, everything is regarded as known except the six $F_{\alpha\beta}$. Minimize with respect to these six unknowns. In this way, arrive at six equations for the components of B and E. These equations once solved, one goes back to (3) to test the Lorentz force law.

The 6×6 determinant of the coefficients in the equation for the $F_{\alpha\beta}$ automatically vanishes when there are only two test particles. The same line of reasoning permits one to determine the minimum number of test particles required to determine all the components of the Riemann curvature tensor.

The three-dimensional version
of the Lorentz force law

The Lorentz force law, written in familiar three-dimensional notation, with E = electric field, B = magnetic field, v = ordinary velocity of particle, p = momentum of particle, e = charge of particle, reads

$$(dp/dt) = e(E + v \times B). \qquad (3.1)$$

Useful though this version of the equation may be, it is far from the geometric spirit of Einstein. A fully geometric equation will involve the test particle's energy-momentum 4-vector, \mathbf{p}, not just the spatial part p as measured in a specific Lorentz frame; and it will ask for the rate of change of momentum not as measured by a specific Lorentz observer (d/dt), but as measured by the only clock present à priori in the problem: the test particle's own clock ($d/d\tau$). Thus, the lefthand side of a fully geometric equation will read

$$d\mathbf{p}/d\tau = .$$

The righthand side, the Lorentz 4-force, must also be a frame-independent object. It will be linear in the particle's 4-velocity \mathbf{u}, since the frame-dependent expression

$$\frac{dp}{d\tau} = \frac{1}{\sqrt{1 - v^2}} \frac{dp}{dt} = \frac{e}{\sqrt{1 - v^2}} (E + v \times B) = e(u^0 E + \mathbf{u} \times B) \qquad (3.2a)$$

is linear in the components of \mathbf{u}. Consequently, there must be a linear machine named **Faraday**, or **F**, or "electromagnetic field tensor," with a slot into which one inserts the 4-velocity of a test particle. The output of this machine, multiplied by the particle's charge, must be the electromagnetic 4-force that it feels:

Electromagnetic field tensor
defined

Geometrical version of
Lorentz force law

$$d\mathbf{p}/d\tau = e\mathbf{F}(\mathbf{u}). \qquad (3.3)$$

By comparing this geometric equation with the original Lorentz force law (equation 3.2a), and with the companion energy-change law

$$\frac{dp^0}{d\tau} = \frac{1}{\sqrt{1 - v^2}} \frac{dE}{dt} = \frac{1}{\sqrt{1 - v^2}} eE \cdot v = eE \cdot \mathbf{u}, \qquad (3.2b)$$

one can read off the components of **F** in a specific Lorentz frame. The components of $d\mathbf{p}/d\tau$ are $dp^\alpha/d\tau$, and the components of $e\mathbf{F}(\mathbf{u})$ can be written (definition of $F^\alpha{}_\beta$!) $eF^\alpha{}_\beta u^\beta$. Consequently

$$dp^\alpha/d\tau = eF^\alpha{}_\beta u^\beta \qquad (3.4)$$

must reduce to equations (3.2a,b). Indeed it does if one makes the identification

Components of
electromagnetic field tensor

$$\|F^\alpha{}_\beta\| = \begin{array}{c|cccc} & \beta = 0 & \beta = 1 & \beta = 2 & \beta = 3 \\ \hline \alpha = 0 & 0 & E_x & E_y & E_z \\ \alpha = 1 & E_x & 0 & B_z & -B_y \\ \alpha = 2 & E_y & -B_z & 0 & B_x \\ \alpha = 3 & E_z & B_y & -B_x & 0 \end{array} \qquad (3.5)$$

More often seen in the literature are the "covariant components," obtained by lowering an index with the metric components:

$$F_{\alpha\beta} = \eta_{\alpha\gamma}F^{\gamma}{}_{\beta}; \tag{3.6}$$

$$\|F_{\alpha\beta}\| = \begin{Vmatrix} 0 & -E_x & -E_y & -E_z \\ E_x & 0 & B_z & -B \\ E_y & -B_z & 0 & B_x \\ E_z & B_y & -B_x & 0 \end{Vmatrix}. \tag{3.7}$$

This matrix equation demonstrates the unity of the electric and magnetic fields. Neither one by itself, E or B, is a frame-independent, geometric entity. But merged together into a single entity, $F = $ *Faraday*, they acquire a meaning and significance that transcends coordinates and reference frames.

EXERCISE

Exercise 3.1.

Derive equations (3.5) and (3.7) for the components of *Faraday* by comparing (3.4) with (3.2a,b), and by using definition (3.6).

§3.2. TENSORS IN ALL GENERALITY

Examples of tensors

A digression is in order. Now on the scene are several different tensors: the metric tensor g (§2.4), the Riemann curvature tensor *Riemann* (§1.6), the electromagnetic field tensor *Faraday* (§3.1). Each has been defined as a linear machine with input slots for vectors, and with an output that is either a real number, e.g., $g(u, v)$, or a vector, e.g., *Riemann* (u, v, w) and *Faraday* (u).

Should one make a distinction between tensors whose outputs are scalars, and tensors whose outputs are vectors? No! A tensor whose output is a vector can be reinterpreted trivially as one whose output is a scalar. Take, for example, *Faraday* $= F$. Add a new slot for the insertion of an arbitrary 1-form σ, and gears and wheels that guarantee the output

$$F(\sigma, u) = \langle \sigma, F(u) \rangle = \text{real number}. \tag{3.8}$$

Then permit the user to choose whether he inserts only a vector, and gets out the vector $F(\dots, u) = F(u)$, or whether he inserts a form and a vector, and gets out the number $F(\sigma, u)$. The same machine will do both jobs. Moreover, in terms of components in a given Lorentz frame, both jobs are achieved very simply:

$$\begin{aligned} &F(\dots, u) \text{ is a vector with components } F^{\alpha}{}_{\beta}u^{\beta}; \\ &F(\sigma, u) \text{ is the number } \langle \sigma, F(\dots, u) \rangle = \sigma_{\alpha}F^{\alpha}{}_{\beta}u^{\beta}. \end{aligned} \tag{3.9}$$

By analogy, one defines the most general tensor \boldsymbol{H} and its rank $\binom{n}{m}$ as follows: \boldsymbol{H} is a linear machine with n input slots for n 1-forms, and m input slots for m vectors; given the requested input, it puts out a real number denoted

$$\boldsymbol{H}(\underbrace{\boldsymbol{\sigma}, \boldsymbol{\lambda}, \ldots, \boldsymbol{\beta},}_{n \text{ 1-forms}} \underbrace{\boldsymbol{u}, \boldsymbol{v}, \ldots, \boldsymbol{w}}_{m \text{ vectors}}). \tag{3.10}$$

Definition of tensor as linear machine that converts vectors and 1-forms into numbers

For most tensors the output changes when two input vectors are interchanged,

$$\boldsymbol{Riemann}(\boldsymbol{\sigma}, \boldsymbol{u}, \boldsymbol{v}, \boldsymbol{w}) \neq \boldsymbol{Riemann}(\boldsymbol{\sigma}, \boldsymbol{v}, \boldsymbol{u}, \boldsymbol{w}), \tag{3.11}$$

or when two input 1-forms are interchanged.

Choose a specific tensor \boldsymbol{S}, of rank $\binom{2}{1}$ for explicitness. Into the slots of \boldsymbol{S}, insert the basis vectors and 1-forms of a specific Lorentz coordinate frame. The output is a "component of \boldsymbol{S} in that frame":

$$S^{\alpha\beta}{}_{\gamma} \equiv \boldsymbol{S}(\boldsymbol{\omega}^{\alpha}, \boldsymbol{\omega}^{\beta}, \boldsymbol{e}_{\gamma}). \tag{3.12}$$

Components of a tensor

This defines components. Knowing the components in a specific frame, one can easily calculate the output produced from any input forms and vectors:

$$\begin{aligned} \boldsymbol{S}(\boldsymbol{\sigma}, \boldsymbol{\rho}, \boldsymbol{v}) &= \boldsymbol{S}(\sigma_{\alpha}\boldsymbol{\omega}^{\alpha}, \rho_{\beta}\boldsymbol{\omega}^{\beta}, v^{\gamma}\boldsymbol{e}_{\gamma}) = \sigma_{\alpha}\rho_{\beta}v^{\gamma}\boldsymbol{S}(\boldsymbol{\omega}^{\alpha}, \boldsymbol{\omega}^{\beta}, \boldsymbol{e}_{\gamma}) \\ &= S^{\alpha\beta}{}_{\gamma}\sigma_{\alpha}\rho_{\beta}v^{\gamma}. \end{aligned} \tag{3.13}$$

Tensor's machine action expressed in terms of components

And knowing the components of \boldsymbol{S} in one Lorentz frame (unprimed), plus the Lorentz transformation matrices $\|\Lambda^{\alpha'}{}_{\beta}\|$ and $\|\Lambda^{\beta}{}_{\alpha'}\|$ which link that frame with another (primed), one can calculate the components in the new (primed) frame. As shown in exercise 3.2, one need only apply a matrix to each index of \boldsymbol{S}, lining up the matrix indices in the logical manner

$$S^{\mu'\nu'}{}_{\lambda'} = S^{\alpha\beta}{}_{\gamma}\Lambda^{\mu'}{}_{\alpha}\Lambda^{\nu'}{}_{\beta}\Lambda^{\gamma}{}_{\lambda'}. \tag{3.14}$$

Lorentz transformation of components

A slight change of the internal gears and wheels inside the tensor enables one of its 1-form slots to accept a vector. All that is necessary is a mechanism to convert an input vector \boldsymbol{n} into its corresponding 1-form $\tilde{\boldsymbol{n}}$ and then to put that 1-form into the old machinery. Thus, denoting the modified tensor by the same symbol \boldsymbol{S} as was used for the original tensor, one demands

$$\boldsymbol{S}(\boldsymbol{\sigma}, \boldsymbol{n}, \boldsymbol{v}) = \boldsymbol{S}(\boldsymbol{\sigma}, \tilde{\boldsymbol{n}}, \boldsymbol{v}); \tag{3.15}$$

Modifying a tensor to accept either a vector or a 1-form into each slot

or, in component notation

$$S^{\alpha}{}_{\beta\gamma}\sigma_{\alpha}n^{\beta}v^{\gamma} = S^{\alpha\beta}{}_{\gamma}\sigma_{\alpha}n_{\beta}v^{\gamma}. \tag{3.15'}$$

This is achieved if one raises and lowers the indices of \boldsymbol{S} using the components of the metric:

$$S^{\alpha}{}_{\beta\gamma} = \eta_{\beta\mu}S^{\alpha\mu}{}_{\gamma}, \qquad S^{\alpha\mu}{}_{\gamma} = \eta^{\mu\beta}S^{\alpha}{}_{\beta\gamma}. \tag{3.16}$$

Raising and lowering indices

(See exercise 3.3 below.) By using the same symbol \boldsymbol{S} for the original tensor and

the modified tensor, one allows each slot to accept either a 1-form or a vector, so one loses sight of whether \boldsymbol{S} is a $\binom{2}{1}$ tensor, or a $\binom{1}{2}$ tensor, or a $\binom{3}{0}$ tensor, or a $\binom{0}{3}$ tensor; one only distinguishes its total rank, 3. *Terminology:* an "upstairs index" is called "contravariant"; a "downstairs" index is called "covariant." Thus in $S^{\alpha}{}_{\beta\gamma}$, "$\alpha$" is a contravariant index, while "β" and "γ" are covariant indices.

Because tensors are nothing but functions, they can be added (if they have the same rank!) and multiplied by numbers in the usual way: the output of the rank-3 tensor $a\boldsymbol{S} + b\boldsymbol{Q}$, when vectors $\boldsymbol{u}, \boldsymbol{v}, \boldsymbol{w}$ are put in, is

$$(a\boldsymbol{S} + b\boldsymbol{Q})(\boldsymbol{u}, \boldsymbol{v}, \boldsymbol{w}) \equiv a\boldsymbol{S}(\boldsymbol{u}, \boldsymbol{v}, \boldsymbol{w}) + b\boldsymbol{Q}(\boldsymbol{u}, \boldsymbol{v}, \boldsymbol{w}). \tag{3.17}$$

Several other important operations on tensors are explored in the following exercises. They and the results of the exercises will be used freely in the material that follows.

EXERCISES

Exercise 3.2. TRANSFORMATION LAW FOR COMPONENTS OF A TENSOR

From the transformation laws for components of vectors and 1-forms, derive the transformation law (3.14).

Exercise 3.3. RAISING AND LOWERING INDICES

Derive equations (3.16) from equation (3.15′) plus the law $n_\alpha = \eta_{\alpha\beta}n^\beta$ for getting the components of the 1-form $\tilde{\boldsymbol{n}}$ from the components of its corresponding vector \boldsymbol{n}.

Exercise 3.4. TENSOR PRODUCT

Given any two vectors \boldsymbol{u} and \boldsymbol{v}, one defines the second-rank tensor $\boldsymbol{u} \otimes \boldsymbol{v}$ ("tensor product of \boldsymbol{u} with \boldsymbol{v}") to be a machine, with two input slots, whose output is the number

$$(\boldsymbol{u} \otimes \boldsymbol{v})(\boldsymbol{\sigma}, \boldsymbol{\lambda}) = \langle \boldsymbol{\sigma}, \boldsymbol{u}\rangle\langle \boldsymbol{\lambda}, \boldsymbol{v}\rangle \tag{3.18}$$

when 1-forms $\boldsymbol{\sigma}$ and $\boldsymbol{\lambda}$ are inserted. Show that the components of $\boldsymbol{T} = \boldsymbol{u} \otimes \boldsymbol{v}$ are the products of the components of \boldsymbol{u} and \boldsymbol{v}:

$$T^{\alpha\beta} = u^\alpha v^\beta, \qquad T_\alpha{}^\beta = u_\alpha v^\beta, \qquad T_{\alpha\beta} = u_\alpha v_\beta. \tag{3.19}$$

Extend the definition to several vectors and forms,

$$(\boldsymbol{u} \otimes \boldsymbol{v} \otimes \boldsymbol{\beta} \otimes \boldsymbol{w})(\boldsymbol{\sigma}, \boldsymbol{\lambda}, \boldsymbol{n}, \boldsymbol{\zeta}) = \langle \boldsymbol{\sigma}, \boldsymbol{u}\rangle\langle \boldsymbol{\lambda}, \boldsymbol{v}\rangle\langle \boldsymbol{\beta}, \boldsymbol{n}\rangle\langle \boldsymbol{\zeta}, \boldsymbol{w}\rangle, \tag{3.20}$$

and show that the product rule for components still holds:

$$\boldsymbol{S} = \boldsymbol{u} \otimes \boldsymbol{v} \otimes \boldsymbol{\beta} \otimes \boldsymbol{w} \text{ has components}$$
$$S^{\mu\nu}{}_\lambda{}^\zeta = u^\mu v^\nu \beta_\lambda w^\zeta. \tag{3.21}$$

Exercise 3.5. BASIS TENSORS

Show that a tensor \boldsymbol{M} with components $M^{\alpha\beta}{}_\gamma{}^\delta$ in a given Lorentz frame can be reconstructed from its components and from the basis 1-forms and vectors of that frame as follows:

$$\boldsymbol{M} = M^{\alpha\beta}{}_\gamma{}^\delta \boldsymbol{e}_\alpha \otimes \boldsymbol{e}_\beta \otimes \boldsymbol{\omega}^\gamma \otimes \boldsymbol{e}_\delta. \tag{3.22}$$

(For a special case of this, see Box 3.2.)

Box 3.2 THE METRIC IN DIFFERENT LANGUAGES

A. Geometric Language

g is a linear, symmetric machine with two slots for insertion of vectors. When vectors u and v are inserted, the output of g is their scalar product:

$$g(u, v) = u \cdot v.$$

B. Component Language

$\eta_{\mu\nu}$ are the metric components. They are used to calculate the scalar product of two vectors from components in a specific Lorentz frame:

$$u \cdot v = \eta_{\mu\nu} u^\mu v^\nu.$$

C. Coordinate-Based Geometric Language

The metric g can be written, in terms of basis 1-forms of a specific Lorentz frame, as

$$g = \eta_{\mu\nu} \boldsymbol{\omega}^\mu \otimes \boldsymbol{\omega}^\nu = \eta_{\mu\nu} \, \boldsymbol{dx}^\mu \otimes \boldsymbol{dx}^\nu$$

[see equations (2.18) and (3.22)].

D. Connection to the Elementary Concept of Line Element

Box 2.3 demonstrated the correspondence between the gradient \boldsymbol{df} of a function, and the elementary concept df of a differential change of f in some unspecified direction. There is a similar correspondence between the metric, written as $\eta_{\mu\nu} \, \boldsymbol{dx}^\mu \otimes \boldsymbol{dx}^\nu$, and the elementary concept of "line element," written as $ds^2 = \eta_{\mu\nu} \, dx^\mu \, dx^\nu$. This elementary line element, as expounded in many special relativity texts, represents the squared length of the displacement "dx^μ" in an unspecified direction. The metric $\eta_{\mu\nu} \, \boldsymbol{dx}^\mu \otimes \boldsymbol{dx}^\nu$ does the same. Pick a specific infinitesimal displacement vector $\boldsymbol{\xi}$, and insert it into the slots of $\eta_{\mu\nu} \, \boldsymbol{dx}^\mu \otimes \boldsymbol{dx}^\nu$. The output will be $\boldsymbol{\xi}^2 = \eta_{\mu\nu} \xi^\mu \xi^\nu$, the squared length of the displacement. Before $\boldsymbol{\xi}$ is inserted, $\eta_{\mu\nu} \, \boldsymbol{dx}^\mu \otimes \boldsymbol{dx}^\nu$ has the potential to tell the squared length of any vector; the insertion of $\boldsymbol{\xi}$ converts potentiality into actuality: the numerical value of $\boldsymbol{\xi}^2$.

Because the metric $\eta_{\mu\nu} \, \boldsymbol{dx}^\mu \otimes \boldsymbol{dx}^\nu$ and the line element $ds^2 = \eta_{\mu\nu} \, dx^\mu \, dx^\nu$ perform this same function of representing the squared length of an unspecified infinitesimal displacement, there is no conceptual distinction between them. One sometimes uses the symbols \boldsymbol{ds}^2 to denote the metric; one sometimes gets pressed and writes it as $\boldsymbol{ds}^2 = \eta_{\mu\nu} \, \boldsymbol{dx}^\mu \, \boldsymbol{dx}^\nu$, omitting the "$\otimes$"; and one sometimes even gets so pressed as to use nonbold characters, so that no notational distinction remains at all between metric and elementary line element:

$$g = \boldsymbol{ds}^2 = ds^2 = \eta_{\mu\nu} \, dx^\mu \, dx^\nu.$$

Exercise 3.6. *Faraday* MACHINERY AT WORK

An observer with 4-velocity \boldsymbol{u} picks out three directions in spacetime that are orthogonal and purely spatial (no time part) as seen in his frame. Let \boldsymbol{e}_1, \boldsymbol{e}_2, \boldsymbol{e}_3 be unit vectors in those directions and let them be oriented in a righthanded way ($\boldsymbol{e}_1 \cdot \boldsymbol{e}_2 \times \boldsymbol{e}_3 = +1$ in three-dimensional language). Why do the following relations hold?

$$\boldsymbol{e}_j \cdot \boldsymbol{u} = 0, \qquad \boldsymbol{e}_j \cdot \boldsymbol{e}_k = \delta_{jk}.$$

What vectors are to be inserted in the two slots of the electromagnetic field tensor *Faraday* if one wants to get out the electric field along \boldsymbol{e}_j as measured by this observer? What vectors must be inserted to get the magnetic field he measures along \boldsymbol{e}_j?

§3.3. THREE-PLUS-ONE VIEW VERSUS GEOMETRIC VIEW

The power of the geometric view of physics

Great computational and conceptual power resides in Einstein's geometric view of physics. Ideas that seem complex when viewed in the everyday "space-plus-time" or "3 + 1" manner become elegant and simple when viewed as relations between geometric objects in four-dimensional spacetime. Derivations that are difficult in 3 + 1 language simplify in geometric language.

Example of electromagnetism

The electromagnetic field is a good example. In geometric language, it is described by a second-rank, antisymmetric tensor ("2-form") \boldsymbol{F}, which requires no coordinates for its definition. This tensor produces a 4-force on any charged particle given by

$$d\boldsymbol{p}/d\tau = e\boldsymbol{F}(\boldsymbol{u}).$$

It is all so simple!

By contrast, consider the "3 + 1" viewpoint. In a given Lorentz frame, there is an electric field \boldsymbol{E} and a magnetic field \boldsymbol{B}. They push on a particle in accordance with

$$d\boldsymbol{p}/dt = e(\boldsymbol{E} + \boldsymbol{v} \times \boldsymbol{B}).$$

Transformation law for electric and magnetic fields

But the values of \boldsymbol{p}, \boldsymbol{E}, \boldsymbol{v}, and \boldsymbol{B} all change when one passes from the given Lorentz frame to a new one. For example, the electric and magnetic fields viewed from a rocket ship ("barred" frame) are related to those viewed in the laboratory ("unbarred" frame) by

$$\bar{E}_{\parallel} = E_{\parallel}, \qquad \bar{E}_{\perp} = \frac{1}{\sqrt{1 - \beta^2}} (E_{\perp} + \boldsymbol{\beta} \times B_{\perp}),$$

$$\bar{B}_{\parallel} = B_{\parallel}, \qquad \bar{B}_{\perp} = \frac{1}{\sqrt{1 - \beta^2}} (B_{\perp} - \boldsymbol{\beta} \times E_{\perp}). \tag{3.23}$$

(Here "∥" means component along direction of rocket's motion; "⊥" means perpendicular component; and $\beta^j = dx^j{}_{\text{rocket}}/dt$ is the rocket's ordinary velocity.) The analogous transformation laws for the particle's momentum \boldsymbol{p} and ordinary velocity

v, and for the coordinate time *t*, all conspire—as if by magic, it seems, from the 3 + 1 viewpoint—to maintain the validity of the Lorentz force law in all frames.

Not only is the geometric view far simpler than the 3 + 1 view, it can even derive the 3 + 1 equations with greater ease than can the 3 + 1 view itself. Consider, for example, the transformation law (3.23) for the electric and magnetic fields. The geometric view derives it as follows: (1) Orient the axes of the two frames so their relative motion is in the *z*-direction. (2) Perform a simple Lorentz transformation (equation 2.45) on the components of the electromagnetic field tensor:

$$\bar{E}_\parallel = \bar{E}_z = F_{\overline{30}} = \Lambda^\alpha{}_{\overline{3}}\Lambda^\beta{}_{\overline{0}}F_{\alpha\beta} = \gamma^2 F_{30} + \beta^2\gamma^2 F_{03}$$
$$- (1 - \beta^2)\gamma^2 F_{30} = F_{30} = E_z = E_\parallel,$$

$$\bar{E}_x = F_{\overline{10}} = \Lambda^\alpha{}_{\overline{1}}\Lambda^\beta{}_{\overline{0}}F_{\alpha\beta} = \gamma F_{10} + \beta\gamma F_{13} = \gamma(E_x - \beta B_y),$$

(3.24)

etc.

By contrast, the 3 + 1 view shows much more work. A standard approach is based on the Lorentz force law and energy-change law (3.2a,b), written in the slightly modified form

$$m\frac{d^2\bar{x}}{d\tau^2} = e\left(\bar{E}_x\frac{d\bar{t}}{d\tau} + 0\frac{d\bar{x}}{d\tau} + \bar{B}_z\frac{d\bar{y}}{d\tau} - \bar{B}_y\frac{d\bar{z}}{d\tau}\right),$$

(3.25)

... (three additional equations)

It proceeds as follows (details omitted because of their great length!):

(1) Substitute for the $d^2\bar{x}/d\tau^2$, etc., the expression for these quantities in terms of the $d^2x/d\tau^2,\ldots$ (Lorentz transformation).

(2) Substitute for the $d^2x/d\tau^2,\ldots$ the expression for these accelerations in terms of the laboratory **E** and **B** (Lorentz force law).

(3) In these expressions, wherever the components $dx/d\tau$ of the 4-velocity in the laboratory frame appear, substitute expressions in terms of the 4-velocities in the rocket frame (inverse Lorentz transformation).

(4) In (3.25) as thus transformed, demand equality of left and right sides for all values of the $d\bar{x}/d\tau$, etc. (validity for all test particles).

(5) In this way arrive at the expressions (3.23) for the \bar{E} and \bar{B} in terms of the **E** and **B**.

The contrast in difficulty is obvious!

§3.4. MAXWELL'S EQUATIONS

Turn now from the action of the field on a charge, and ask about the action of a charge on the field, or, more generally, ask about the dynamics of the electromagnetic

field, charge or no charge. Begin with the simplest of Maxwell's equations in a specific Lorentz frame, the one that says there are no free magnetic poles:

$$\boldsymbol{\nabla} \cdot \boldsymbol{B} \equiv \operatorname{div} \boldsymbol{B} = \frac{\partial B_x}{\partial x} + \frac{\partial B_y}{\partial y} + \frac{\partial B_z}{\partial z} = 0. \tag{3.26}$$

This statement has to be true in all Lorentz frames. It is therefore true in the rocket frame:

$$\frac{\partial \bar{B}_x}{\partial \bar{x}} + \frac{\partial \bar{B}_y}{\partial \bar{y}} + \frac{\partial \bar{B}_z}{\partial \bar{z}} = 0. \tag{3.27}$$

For an infinitesimal Lorentz transformation in the x-direction (nonrelativistic velocity β), one has (see Box 2.4 and equations 3.23)

$$\bar{B}_x = B_x, \qquad \bar{B}_y = B_y + \beta E_z, \qquad \bar{B}_z = B_z - \beta E_y; \tag{3.28}$$

$$\frac{\partial}{\partial \bar{x}} = \frac{\partial}{\partial x} + \beta \frac{\partial}{\partial t}, \qquad \frac{\partial}{\partial \bar{y}} = \frac{\partial}{\partial y}, \qquad \frac{\partial}{\partial \bar{z}} = \frac{\partial}{\partial z}. \tag{3.29}$$

Substitute into the condition of zero divergence in the rocket frame. Recover the original condition of zero divergence in the laboratory frame, plus the following additional information (requirement for the vanishing of the coefficient of the arbitrary small velocity β):

$$\frac{\partial B_x}{\partial t} + \frac{\partial E_z}{\partial y} - \frac{\partial E_y}{\partial z} = 0. \tag{3.30}$$

Had the velocity of transformation been directed in the y- or z-directions, a similar equation would have been obtained for $\partial B_y/\partial t$ or $\partial B_z/\partial t$. In the language of three-dimensional vectors, these three equations reduce to the one equation

$$\frac{\partial \boldsymbol{B}}{\partial t} + \boldsymbol{\nabla} \times \boldsymbol{E} \equiv \frac{\partial \boldsymbol{B}}{\partial t} + \operatorname{curl} \boldsymbol{E} = 0. \tag{3.31}$$

How beautiful that (1) the principle of covariance (laws of physics are the same in every Lorentz reference system, which is equivalent to the geometric view of physics) plus (2) the principle that magnetic tubes of force never end, gives (3) Maxwell's dynamic law for the time-rate of change of the magnetic field! This suggests that the magnetostatic law $\boldsymbol{\nabla} \cdot \boldsymbol{B} = 0$ and the magnetodynamic law $\partial \boldsymbol{B}/\partial t + \boldsymbol{\nabla} \times \boldsymbol{E} = 0$ must be wrapped up together in a single frame-independent, geometric law. In terms of components of the field tensor \boldsymbol{F}, that geometric law must read

$$F_{\alpha\beta,\gamma} + F_{\beta\gamma,\alpha} + F_{\gamma\alpha,\beta} = 0, \tag{3.32}$$

since this reduces to $\boldsymbol{\nabla} \cdot \boldsymbol{B} = 0$ when one takes $\alpha = 1$, $\beta = 2$, $\gamma = 3$; and it reduces to $\partial \boldsymbol{B}/\partial t + \boldsymbol{\nabla} \times \boldsymbol{E} = 0$ when one sets any index, e.g., α, equal to zero (see exercise 3.7 below). In frame-independent geometric language, this law is written (see §3.5, exercise 3.14, and Chapter 4 for notation)

$$d\boldsymbol{F} = 0, \text{ or, equivalently, } \boldsymbol{\nabla} \cdot {}^{*}\boldsymbol{F} = 0; \qquad (3.33)$$

and it says, "Take the electromagnetic 2-form \boldsymbol{F} (a geometric object defined even in absence of coordinates); from it construct a new geometric object $d\boldsymbol{F}$ (called the "exterior derivative of \boldsymbol{F}"); $d\boldsymbol{F}$ must vanish. The details of this coordinate-free process will be spelled out in exercise 3.15 and in §4.5 (track 2).

Two of Maxwell's equations remain: the electrostatic equation

$$\boldsymbol{\nabla} \cdot \boldsymbol{E} = 4\pi\rho, \qquad (3.34)$$

and the electrodynamic equation

$$\partial \boldsymbol{E}/\partial t - \boldsymbol{\nabla} \times \boldsymbol{B} = -4\pi\boldsymbol{J}. \qquad (3.35)$$

They, like the magnetostatic and magnetodynamic equations, are actually two different parts of a single geometric law. Written in terms of field components, that law says

Electrodynamics and electrostatics unified in one geometric law

$$F^{\alpha\beta}{}_{,\beta} = 4\pi J^{\alpha}, \qquad (3.36)$$

where the components of the "4-current" \boldsymbol{J} are

$$\begin{aligned} J^0 &= \rho = \text{charge density,} \\ (J^1, J^2, J^3) &= \text{components of current density.} \end{aligned} \qquad (3.37)$$

Written in coordinate-free, geometric language, this electrodynamic law says

$$d{}^{*}\boldsymbol{F} = 4\pi {}^{*}\boldsymbol{J} \text{ or, equivalently, } \boldsymbol{\nabla} \cdot \boldsymbol{F} = 4\pi\boldsymbol{J}. \qquad (3.38)$$

(For full discussion, see exercise 3.15 and §4.5, which is on Track 2.)

Exercise 3.7. MAXWELL'S EQUATIONS

Show, by explicit examination of components, that the geometric laws

$$F_{\alpha\beta,\gamma} + F_{\beta\gamma,\alpha} + F_{\gamma\alpha,\beta} = 0, \qquad F^{\alpha\beta}{}_{,\beta} = 4\pi J^{\alpha},$$

do reduce to Maxwell's equations (3.26), (3.31), (3.34), (3.35), as claimed above.

EXERCISE

§3.5 WORKING WITH TENSORS

Another mathematical digression is needed. Given an arbitrary tensor field, \boldsymbol{S}, of arbitrary rank (choose rank = 3 for concreteness), one can construct new tensor fields by a variety of operations.

One operation is the *gradient* $\boldsymbol{\nabla}$. (The symbol d is reserved for gradients of scalars, in which case $\boldsymbol{\nabla} f \equiv df$, and for "exterior derivatives of differential forms;" a Track-2

Ways to produce new tensors from old:

Gradient

concept, on which see §4.5.) Like \mathbf{S}, $\boldsymbol{\nabla}\mathbf{S}$ is a machine. It has four slots, whereas \mathbf{S} has three. It describes how \mathbf{S} changes from point to point. Specifically, if one desires to know how the number $\mathbf{S}(\mathbf{u}, \mathbf{v}, \mathbf{w})$ for *fixed* \mathbf{u}, \mathbf{v}, \mathbf{w} changes under a displacement $\boldsymbol{\xi}$, one inserts \mathbf{u}, \mathbf{v}, \mathbf{w}, $\boldsymbol{\xi}$ into the four slots of $\boldsymbol{\nabla}\mathbf{S}$:

$$\boldsymbol{\nabla}\mathbf{S}(\mathbf{u}, \mathbf{v}, \mathbf{w}, \boldsymbol{\xi}) \equiv \partial_{\boldsymbol{\xi}}\mathbf{S}(\mathbf{u}, \mathbf{v}, \mathbf{w}) \text{ with } \mathbf{u}, \mathbf{v}, \mathbf{w} \text{ fixed}$$
$$\simeq + [\text{value of } \mathbf{S}(\mathbf{u}, \mathbf{v}, \mathbf{w}) \text{ at tip of } \boldsymbol{\xi}] \qquad (3.39)$$
$$- [\text{value of } \mathbf{S}(\mathbf{u}, \mathbf{v}, \mathbf{w}) \text{ at tail of } \boldsymbol{\xi}].$$

In component notation in a Lorentz frame, this says

$$\boldsymbol{\nabla}\mathbf{S}(\mathbf{u}, \mathbf{v}, \mathbf{w}, \boldsymbol{\xi}) \equiv \partial_{\boldsymbol{\xi}}(S_{\alpha\beta\gamma}u^{\alpha}v^{\beta}w^{\gamma}) = \left(\frac{\partial S_{\alpha\beta\gamma}}{\partial x^{\delta}}\xi^{\delta}\right)u^{\alpha}v^{\beta}w^{\gamma}$$

$$= S_{\alpha\beta\gamma,\delta}u^{\alpha}v^{\beta}w^{\gamma}\xi^{\delta}.$$

Thus, the Lorentz-frame components of $\boldsymbol{\nabla}\mathbf{S}$ are nothing but the partial derivatives of the components of \mathbf{S}. Notice that the gradient raises the rank of a tensor by 1 (from 3 to 4 for \mathbf{S}).

Contraction

Contraction is another process that produces a new tensor from an old one. It seals off ("contracts") two of the old tensor's slots, thereby reducing the rank by two. Specifically, if \mathbf{R} is a fourth-rank tensor and \mathbf{M} is obtained by contracting the first and third slots of \mathbf{R}, then the output of \mathbf{M} is given by (definition!)

$$\mathbf{M}(\mathbf{u}, \mathbf{v}) = \sum_{\alpha=0}^{3} \mathbf{R}(\mathbf{e}_{\alpha}, \mathbf{u}, \boldsymbol{\omega}^{\alpha}, \mathbf{v}). \qquad (3.40)$$

Here \mathbf{e}_{α} and $\boldsymbol{\omega}^{\alpha}$ are the basis vectors and 1-forms of a specific but arbitrary Lorentz coordinate frame. It makes no difference which frame is chosen; the result will always be the same (exercise 3.8 below). In terms of components in any Lorentz frame, equation (3.40) says (exercise 3.8)

$$\mathbf{M}(\mathbf{u}, \mathbf{v}) = M_{\mu\nu}u^{\mu}v^{\nu} = R_{\alpha\mu}{}^{\alpha}{}_{\nu}u^{\mu}v^{\nu},$$

so that

$$M_{\mu\nu} = R_{\alpha\mu}{}^{\alpha}{}_{\nu}. \qquad (3.41)$$

Thus, in terms of components, contraction amounts to putting one index up and the other down, and then summing on them.

Divergence

Divergence is a third process for creating new tensors from old. It is accomplished by taking the gradient, then contracting the gradient's slot with one of the original slots:

$$(\text{divergence of } \mathbf{S} \text{ on first slot}) \equiv \boldsymbol{\nabla} \cdot \mathbf{S} \qquad \text{is a machine such that}$$

$$\boldsymbol{\nabla} \cdot \mathbf{S}(\mathbf{u}, \mathbf{v}) = \boldsymbol{\nabla}\mathbf{S}(\boldsymbol{\omega}^{\alpha}, \mathbf{u}, \mathbf{v}, \mathbf{e}_{\alpha}) = S^{\alpha}{}_{\beta\gamma,\alpha}u^{\beta}v^{\gamma}; \qquad (3.42)$$

i.e. $\boldsymbol{\nabla} \cdot \mathbf{S}$ has components $S^{\alpha}{}_{\beta\gamma,\alpha}$.

Transpose is a fourth, rather trivial process for creating new tensors. It merely interchanges two slots:

N obtained by transposing second and third slots of $S \Rightarrow$

$$N(u, v, w) = S(u, w, v). \tag{3.43}$$

Symmetrization and *antisymmetrization* are fifth and sixth processes for producing new tensors from old. A tensor is completely symmetric if its output is unaffected by an interchange of two input vectors or 1-forms:

$$S(u, v, w) = S(v, u, w) = S(v, w, u) = \cdots . \tag{3.44a}$$

It is completely antisymmetric if it reverses sign on each interchange of input

$$S(u, v, w) = -S(v, u, w) = +S(v, w, u) = \cdots . \tag{3.44b}$$

Any tensor can be symmetrized or antisymmetrized by constructing an appropriate linear combination of it and its transposes; see exercise 3.12.

Wedge product is a seventh process for producing new tensors from old. It is merely an antisymmetrized tensor product: given two vectors u and v, their wedge product, the *"bivector"* $u \wedge v$, is defined by

$$u \wedge v \equiv u \otimes v - v \otimes u; \tag{3.45a}$$

similarly, the *"2-form"* $\alpha \wedge \beta$ constructed from two 1-forms is

$$\alpha \wedge \beta \equiv \alpha \otimes \beta - \beta \otimes \alpha. \tag{3.45b}$$

From three vectors u, v, w one constructs the *"trivector"*

$$
\begin{aligned}
u \wedge v \wedge w &\equiv (u \wedge v) \wedge w \equiv u \wedge (v \wedge w) \\
&= u \otimes v \otimes w + \text{terms that guarantee complete antisymmetry} \\
&= u \otimes v \otimes w + v \otimes w \otimes u + w \otimes u \otimes v \\
&\quad - v \otimes u \otimes w - u \otimes w \otimes v - w \otimes v \otimes u.
\end{aligned} \tag{3.45c}
$$

From 1-forms α, β, γ one similarly constructs the "3-forms" $\alpha \wedge \beta \wedge \gamma$. The wedge product gives a simple way to test for coplanarity (linear dependence) of vectors: if u and v are collinear, so $u = av$, then

$$u \wedge v = av \wedge v = 0 \qquad \text{(by antisymmetry of ``}\wedge\text{'').}$$

If w is coplanar with u and v so $w = au + bv$ ("collapsed box"), then

$$w \wedge u \wedge v = au \wedge u \wedge v + bv \wedge u \wedge v = 0.$$

The symbol "\wedge" is called a "hat" or "wedge" or "exterior product sign." Its properties are investigated in Chapter 4.

Taking the dual is an eighth process for constructing new tensors. It plays a fundamental role in Track 2 of this book, but since it is not needed for Track 1, its definition and properties are treated only in the exercises (3.14 and 3.15).

Because the frame-independent geometric notation is somewhat ambiguous (which slots are being contracted? on which slot is the divergence taken? which slots are being transposed?), one often uses component notation to express coordinate-independent, geometric relations between geometric objects. For example,

$$J_{\beta\gamma} = S^{\alpha}{}_{\beta\gamma,\alpha}$$

means "\boldsymbol{J} is a tensor obtained by taking the divergence on the first slot of the tensor \boldsymbol{S}". Also,

$$v^{\gamma} = (F_{\mu\nu}F^{\mu\nu})^{,\gamma} \equiv (F_{\mu\nu}F^{\mu\nu})_{,\beta}\eta^{\beta\gamma}$$

means "\boldsymbol{v} is a vector obtained by (1) constructing the tensor product $\boldsymbol{F} \otimes \boldsymbol{F}$ of \boldsymbol{F} with itself, (2) contracting $\boldsymbol{F} \otimes \boldsymbol{F}$ on its first and third slots, and also on its second and fourth, (3) taking the gradient of the resultant scalar function, (4) converting that gradient, which is a 1-form, into the corresponding vector."

Index gymnastics

"Index gymnastics," the technique of extracting the content from geometric equations by working in component notation and rearranging indices as required, must be mastered if one wishes to do difficult calculations in relativity, special or general. Box 3.3 expounds some of the short cuts in index gymnastics, and exercises 3.8–3.18 offer practice.

EXERCISES

Exercise 3.8. CONTRACTION IS FRAME-INDEPENDENT

Show that contraction, as defined in equation (3.40), does not depend on which Lorentz frame \boldsymbol{e}_{α} and \boldsymbol{w}^{α} are taken from. Also show that equation (3.40) implies

$$\boldsymbol{M}(\boldsymbol{u}, \boldsymbol{v}) = R_{\alpha\mu}{}^{\alpha}{}_{\nu}u^{\mu}v^{\nu}.$$

Exercise 3.9. DIFFERENTIATION

(a) Justify the formula

$$d(u_{\mu}v^{\nu})/d\tau = (du_{\mu}/d\tau)v^{\nu} + u_{\mu}(dv^{\nu}/d\tau).$$

by considering the special case $\mu = 0$, $\nu = 1$.

(b) Explain why

$$(T^{\alpha\beta}v_{\beta})_{,\mu} = T^{\alpha\beta}{}_{,\mu}v_{\beta} + T^{\alpha\beta}v_{\beta,\mu}.$$

Exercise 3.10. MORE DIFFERENTIATION

(a) Justify the formula,

$$d(u^{\mu}u_{\mu})/d\tau = 2u_{\mu}(du^{\mu}/d\tau),$$

by writing out the summation $u^{\mu}u_{\mu} \equiv \eta_{\mu\nu}u^{\mu}u^{\nu}$ explicitly.

(b) Let δ indicate a variation or small change, and justify the formula

$$\delta(F_{\alpha\beta}F^{\alpha\beta}) = 2F_{\alpha\beta}\delta F^{\alpha\beta}.$$

(c) Compute $(F_{\alpha\beta}F^{\alpha\beta})_{,\mu} = ?$

Box 3.3 TECHNIQUES OF INDEX GYMNASTICS

Equation	Name and Discussion
$S^{\alpha}{}_{\beta\gamma} = S(\boldsymbol{\omega}^{\alpha}, \boldsymbol{e}_{\beta}, \boldsymbol{e}_{\gamma})$	Computing components.
$S^{\alpha\beta}{}_{\gamma} - S(\boldsymbol{\omega}^{\alpha}, \boldsymbol{\omega}^{\beta}, \boldsymbol{e}_{\gamma})$	Computing other components.
$\boldsymbol{S} = S^{\alpha}{}_{\beta\gamma}\boldsymbol{e}_{\alpha} \otimes \boldsymbol{\omega}^{\beta} \otimes \boldsymbol{\omega}^{\gamma}$	Reconstructing the rank-$\binom{1}{2}$ version of \boldsymbol{S}.
$\boldsymbol{S} = S^{\alpha\beta\gamma}\boldsymbol{e}_{\alpha} \otimes \boldsymbol{e}_{\beta} \otimes \boldsymbol{e}_{\gamma}$	Reconstructing the rank-$\binom{3}{0}$ version of \boldsymbol{S}. [Recall: one does not usually distinguish between the various versions; see equation (3.15) and associated discussion.]
$S^{\alpha\beta}{}_{\gamma} = \eta^{\beta\mu}S^{\alpha}{}_{\mu\gamma}$	Raising an index.
$S^{\alpha}{}_{\mu\gamma} = \eta_{\mu\beta}S^{\alpha\beta}{}_{\gamma}$	Lowering an index.
$M_{\mu} = S^{\alpha}{}_{\mu\alpha}$	Contraction of \boldsymbol{S} to form a new tensor \boldsymbol{M}.
$T^{\alpha\beta}{}_{\mu\nu} = S^{\alpha\beta}{}_{\mu}M_{\nu}$	Tensor product of \boldsymbol{S} with \boldsymbol{M} to form a new tensor \boldsymbol{T}.
$\boldsymbol{A}^2 = A^{\alpha}A_{\alpha}$	Squared length of vector \boldsymbol{A} produced by forming tensor product $\boldsymbol{A} \otimes \boldsymbol{A}$ and then contracting, which is the same as forming the corresponding 1-form $\widetilde{\boldsymbol{A}}$ and then piercing: $\boldsymbol{A}^2 = \langle \widetilde{\boldsymbol{A}}, \boldsymbol{A} \rangle = A^{\alpha}A_{\alpha}$.
$\eta_{\alpha\beta}\eta^{\beta\gamma} = \delta_{\alpha}{}^{\gamma}$	The matrix formed from the metric's "covariant components," $\|\eta_{\alpha\beta}\|$, is the inverse of that formed from its "contravariant components," $\|\eta^{\alpha\beta}\|$. Equivalently, raising one index of the metric $\eta_{\alpha\beta}$ produces the Kronecker delta.
$S^{\alpha}{}_{\beta\gamma} = N^{\alpha}{}_{\beta,\gamma}$	Gradient of \boldsymbol{N} to form a new tensor \boldsymbol{S}.
$R_{\beta} = N^{\alpha}{}_{\beta,\alpha}$	Divergence of \boldsymbol{N} to form a new tensor \boldsymbol{R}.
$N^{\alpha}{}_{\beta,\gamma} = (\eta_{\beta\mu}N^{\alpha\mu})_{,\gamma} = \eta_{\beta\mu}N^{\alpha\mu}{}_{,\gamma}$	Taking gradients and raising or lowering indices are operations that commute.
$N^{\alpha}{}_{\beta}{}^{,\gamma} \equiv N^{\alpha}{}_{\beta,\mu}\eta^{\mu\gamma}$	Contravariant index on a gradient is obtained by raising covariant index.
$(R_{\alpha}M_{\beta})_{,\gamma} = R_{\alpha,\gamma}M_{\beta} + R_{\alpha}M_{\beta,\gamma}$	Gradient of a tensor product; says $\boldsymbol{\nabla}(\boldsymbol{R} \otimes \boldsymbol{M}) =$ Transpose $(\boldsymbol{\nabla R} \otimes \boldsymbol{M}) + \boldsymbol{R} \otimes \boldsymbol{\nabla M}$.
$G_{\alpha\beta} = F_{[\alpha\beta]} \equiv \frac{1}{2}(F_{\alpha\beta} - F_{\beta\alpha})$	Antisymmetrizing a tensor \boldsymbol{F} to produce a new tensor \boldsymbol{G}.
$H_{\alpha\beta} = F_{(\alpha\beta)} \equiv \frac{1}{2}(F_{\alpha\beta} + F_{\beta\alpha})$	Symmetrizing a tensor \boldsymbol{F} to produce a new tensor \boldsymbol{H}.
$*J_{\alpha\beta\gamma} = J^{\mu}\varepsilon_{\mu\alpha\beta\gamma}$	Forming the rank-3 tensor that is dual to a vector (exercise 3.14).
$*F_{\alpha\beta} = \frac{1}{2}F^{\mu\nu}\varepsilon_{\mu\nu\alpha\beta}$	Forming the antisymmetric rank-2 tensor that is dual to a given antisymmetric rank-2 tensor (exercise 3.14).
$*B_{\alpha} = \frac{1}{6}B^{\lambda\mu\nu}\varepsilon_{\lambda\mu\nu\alpha}$	Forming the 1-form that is dual to an antisymmetric rank-3 tensor (exercise 3.14).

Exercise 3.11. SYMMETRIES

Let $A_{\mu\nu}$ be an antisymmetric tensor so that $A_{\mu\nu} = -A_{\nu\mu}$; and let $S^{\mu\nu}$ be a symmetric tensor so that $S^{\mu\nu} = S^{\nu\mu}$.

(a) Justify the equations $A_{\mu\nu}S^{\mu\nu} = 0$ in two ways: first, by writing out the sum explicitly (all sixteen terms) and showing how the terms in the sum cancel in pairs; second, by giving an argument to justify each equals sign in the following string:

$$A_{\mu\nu}S^{\mu\nu} = -A_{\nu\mu}S^{\mu\nu} = -A_{\nu\mu}S^{\nu\mu} = -A_{\alpha\beta}S^{\alpha\beta} = -A_{\mu\nu}S^{\mu\nu} = 0.$$

(b) Establish the following two identities for any arbitrary tensor $V_{\mu\nu}$:

$$V^{\mu\nu}A_{\mu\nu} = \frac{1}{2}(V^{\mu\nu} - V^{\nu\mu})A_{\mu\nu}, \qquad V^{\mu\nu}S_{\mu\nu} = \frac{1}{2}(V^{\mu\nu} + V^{\nu\mu})S_{\mu\nu}.$$

Exercise 3.12. SYMMETRIZATION AND ANTISYMMETRIZATION

To "symmetrize" a tensor, one averages it with all of its transposes. The components of the new, symmetrized tensor are distinguished by round brackets:

$$V_{(\mu\nu)} \equiv \frac{1}{2}(V_{\mu\nu} + V_{\nu\mu});$$

$$V_{(\mu\nu\lambda)} \equiv \frac{1}{3!}(V_{\mu\nu\lambda} + V_{\nu\lambda\mu} + V_{\lambda\mu\nu} + V_{\nu\mu\lambda} + V_{\mu\lambda\nu} + V_{\lambda\nu\mu}).$$

(3.46)

One "antisymmetrizes" a tensor (square brackets) similarly:

$$V_{[\mu\nu]} \equiv \frac{1}{2}(V_{\mu\nu} - V_{\nu\mu});$$

$$V_{[\mu\nu\lambda]} \equiv \frac{1}{3!}(V_{\mu\nu\lambda} + V_{\nu\lambda\mu} + V_{\lambda\mu\nu} - V_{\nu\mu\lambda} - V_{\mu\lambda\nu} - V_{\lambda\nu\mu}).$$

(3.47)

(a) Show that such symmetrized and antisymmetrized tensors are, indeed, symmetric and antisymmetric under interchange of the vectors inserted into their slots:

$$V_{(\alpha\beta\gamma)}u^{\alpha}v^{\beta}w^{\gamma} = +V_{(\alpha\beta\gamma)}v^{\alpha}u^{\beta}w^{\gamma} = \cdots,$$

$$V_{[\alpha\beta\gamma]}u^{\alpha}v^{\beta}w^{\gamma} = -V_{[\alpha\beta\gamma]}v^{\alpha}u^{\beta}w^{\gamma} = \cdots.$$

(b) Show that a second-rank tensor can be reconstructed from its symmetric and antisymmetric parts,

$$V_{\mu\nu} = V_{(\mu\nu)} + V_{[\mu\nu]},$$

(3.48)

but that a third-rank tensor cannot; $V_{(\alpha\beta\gamma)}$ and $V_{[\alpha\beta\gamma]}$ contain together "less information" than $V_{\alpha\beta\gamma}$. "Young diagrams" (see, e.g., Messiah [1961], appendix D) describe other symmetries, more subtle than these two, which contain the missing information.

(c) Show that the electromagnetic field tensor satisfies

$$F_{(\alpha\beta)} = 0, \qquad F_{\alpha\beta} = F_{[\alpha\beta]}.$$

(3.49a)

(d) Show that Maxwell's "magnetic" equations

$$F_{\alpha\beta,\gamma} + F_{\beta\gamma,\alpha} + F_{\gamma\alpha,\beta} = 0$$

can be rewritten in the form

$$F_{[\alpha\beta,\gamma]} = 0.$$

(3.49b)

Exercise 3.13. LEVI-CIVITA TENSOR

The "Levi-Civita tensor" $\boldsymbol{\varepsilon}$ in spacetime is a fourth-rank, completely antisymmetric tensor:

$$\boldsymbol{\varepsilon}(\boldsymbol{n}, \boldsymbol{u}, \boldsymbol{v}, \boldsymbol{w}) \text{ changes sign when any two of the vectors are interchanged.} \tag{3.50a}$$

Choose an arbitrary but specific Lorentz frame, with \boldsymbol{e}_0 pointing toward the future and with $\boldsymbol{e}_1, \boldsymbol{e}_2, \boldsymbol{e}_3$ a righthanded set of spatial basis vectors. The covariant components of $\boldsymbol{\varepsilon}$ in this frame are

$$\varepsilon_{0123} = \boldsymbol{\varepsilon}(\boldsymbol{e}_0, \boldsymbol{e}_1, \boldsymbol{e}_2, \boldsymbol{e}_3) = +1. \tag{3.50b}$$

[Note: In an n-dimensional space, $\boldsymbol{\varepsilon}$ is the analogous completely antisymmetric rank-n tensor. Its components are

$$\varepsilon_{12\ldots n} = \boldsymbol{\varepsilon}(\boldsymbol{e}_1, \boldsymbol{e}_2, \ldots, \boldsymbol{e}_n) = +1, \tag{3.50c}$$

when computed on a "positively oriented," orthonormal basis $\boldsymbol{e}_1, \ldots, \boldsymbol{e}_n$.]

(a) Use the antisymmetry to show that

$$\varepsilon_{\alpha\beta\gamma\delta} = 0 \text{ unless } \alpha, \beta, \gamma, \delta \text{ are all different}, \tag{3.50d}$$

$$\varepsilon_{\pi_0\pi_1\pi_2\pi_3} = \begin{cases} +1 \text{ for even permutations of } 0, 1, 2, 3, \text{ and} \\ -1 \text{ for odd permutations.} \end{cases} \tag{3.50e}$$

(b) Show that

$$\varepsilon^{\pi_0\pi_1\pi_2\pi_3} = -\varepsilon_{\pi_0\pi_1\pi_2\pi_3}. \tag{3.50f}$$

(c) By means of a Lorentz transformation show that $\varepsilon^{\bar{\alpha}\bar{\beta}\bar{\gamma}\bar{\delta}}$ and $\varepsilon_{\bar{\alpha}\bar{\beta}\bar{\gamma}\bar{\delta}}$ have these same values in any other Lorentz frame with $\boldsymbol{e}_{\bar{0}}$ pointing toward the future and with $\boldsymbol{e}_{\bar{1}}, \boldsymbol{e}_{\bar{2}}, \boldsymbol{e}_{\bar{3}}$ a righthanded set. Hint: show that

$$\varepsilon^{\alpha\beta\gamma\delta} \Lambda^{\bar{0}}{}_\alpha \Lambda^{\bar{1}}{}_\beta \Lambda^{\bar{2}}{}_\gamma \Lambda^{\bar{3}}{}_\delta = -\det|\Lambda^{\bar{\mu}}{}_\nu|; \tag{3.50g}$$

from $\Lambda^T \eta \Lambda = \eta$, show that $\det|\Lambda^{\bar{\mu}}{}_\nu| = \pm 1$; and verify that the determinant is $+1$ for transformations between frames with \boldsymbol{e}_0 and $\boldsymbol{e}_{\bar{0}}$ future-pointing, and with $\boldsymbol{e}_1, \boldsymbol{e}_2, \boldsymbol{e}_3$ and $\boldsymbol{e}_{\bar{1}}, \boldsymbol{e}_{\bar{2}}, \boldsymbol{e}_{\bar{3}}$ righthanded.

(d) What are the components of $\boldsymbol{\varepsilon}$ in a Lorentz frame with past-pointing $\boldsymbol{e}_{\bar{0}}$? with lefthanded $\boldsymbol{e}_{\bar{1}}, \boldsymbol{e}_{\bar{2}}, \boldsymbol{e}_{\bar{3}}$?

(e) From the Levi-Civita tensor, one can construct several "permutation tensors." In index notation:

$$\delta^{\alpha\beta\gamma}{}_{\mu\nu\lambda} \equiv -\varepsilon^{\alpha\beta\gamma\rho}\varepsilon_{\mu\nu\lambda\rho}; \tag{3.50h}$$

$$\delta^{\alpha\beta}{}_{\mu\nu} \equiv \frac{1}{2}\delta^{\alpha\beta\lambda}{}_{\mu\nu\lambda} = -\frac{1}{2}\varepsilon^{\alpha\beta\lambda\rho}\varepsilon_{\mu\nu\lambda\rho}; \tag{3.50i}$$

$$\delta^{\alpha}{}_{\mu} \equiv \frac{1}{3}\delta^{\alpha\beta}{}_{\mu\beta} = \frac{1}{6}\delta^{\alpha\beta\lambda}{}_{\mu\beta\lambda} = -\frac{1}{6}\varepsilon^{\alpha\beta\lambda\rho}\varepsilon_{\mu\beta\lambda\rho}. \tag{3.50j}$$

Show that:

$$\delta^{\alpha\beta\gamma}{}_{\mu\nu\lambda} = \begin{cases} +1 \text{ if } \alpha\beta\gamma \text{ is an even permutation of } \mu\nu\lambda, \\ -1 \text{ if } \alpha\beta\gamma \text{ is an odd permutation of } \mu\nu\lambda, \\ 0 \text{ otherwise}; \end{cases} \tag{3.50k}$$

$$\delta^{\alpha\beta}{}_{\mu\nu} = \delta^{\alpha}{}_{\mu}\delta^{\beta}{}_{\nu} - \delta^{\alpha}{}_{\nu}\delta^{\beta}{}_{\mu}$$

$$= \begin{cases} +1 \text{ if } \alpha\beta \text{ is an even permutation of } \mu\nu, \\ -1 \text{ if } \alpha\beta \text{ is an odd permutation of } \mu\nu, \\ 0 \text{ otherwise;} \end{cases} \tag{3.50l}$$

$$\delta^{\alpha}{}_{\mu} = \begin{cases} +1 \text{ if } \alpha = \mu, \\ 0 \text{ otherwise.} \end{cases} \tag{3.50m}$$

Exercise 3.14. DUALS

From any vector \boldsymbol{J}, any second-rank antisymmetric tensor $\boldsymbol{F}(F_{\alpha\beta} = F_{[\alpha\beta]})$, and any third-rank antisymmetric tensor $\boldsymbol{B}(B_{\alpha\beta\gamma} = B_{[\alpha\beta\gamma]})$, one can construct new tensors defined by

$$*J_{\alpha\beta\gamma} = J^{\mu}\varepsilon_{\mu\alpha\beta\gamma}, \qquad *F_{\alpha\beta} = \frac{1}{2} F^{\mu\nu}\varepsilon_{\mu\nu\alpha\beta}, \qquad *B_{\alpha} = \frac{1}{3!} B^{\lambda\mu\nu}\varepsilon_{\lambda\mu\nu\alpha}. \tag{3.51}$$

One calls $*\boldsymbol{J}$ the *"dual"* of \boldsymbol{J}, $*\boldsymbol{F}$ the dual of \boldsymbol{F}, and $*\boldsymbol{B}$ the dual of \boldsymbol{B}. [A previous and entirely distinct use of the word "dual" (§2.7) called a *set* of basis one-forms $\{\boldsymbol{\omega}^{\alpha}\}$ dual to a *set* of basis vectors $\{\boldsymbol{e}_{\alpha}\}$ if $\langle\boldsymbol{\omega}^{\alpha}, \boldsymbol{e}_{\beta}\rangle = \delta^{\alpha}{}_{\beta}$. Fortunately there are no grounds for confusion between the two types of duality. One relates sets of vectors to sets of 1-forms. The other relates antisymmetric tensors of rank p to antisymmetric tensors of rank $4 - p$.]
 (a) Show that

$$**\boldsymbol{J} = \boldsymbol{J}, \qquad **\boldsymbol{F} = -\boldsymbol{F}, \qquad **\boldsymbol{B} = \boldsymbol{B}. \tag{3.52}$$

so (aside from sign) one can recover any completely antisymmetric tensor \boldsymbol{H} from its dual $*\boldsymbol{H}$ by taking the dual once again, $**\boldsymbol{H}$. This shows that \boldsymbol{H} and $*\boldsymbol{H}$ contain precisely the same information.
 (b) Make explicit this fact of same-information-content by writing out the components $*A^{\alpha\beta\gamma}$ in terms of A^{α}, also $*F^{\alpha\beta}$ in terms of $F^{\alpha\beta}$, also $*B^{\alpha}$ in terms of $B^{\alpha\beta\gamma}$.

Exercise 3.15. GEOMETRIC VERSIONS OF MAXWELL EQUATIONS

Show that, if \boldsymbol{F} is the electromagnetic field tensor, then $\boldsymbol{\nabla} \cdot *\boldsymbol{F} = 0$ is a geometric frame-independent version of the Maxwell equations

$$F_{\alpha\beta,\gamma} + F_{\beta\gamma,\alpha} + F_{\gamma\alpha,\beta} = 0.$$

Similarly show that $\boldsymbol{\nabla} \cdot \boldsymbol{F} = 4\pi\boldsymbol{J}$ (divergence on second slot of \boldsymbol{F}) is a geometric version of $F^{\alpha\beta}{}_{,\beta} = 4\pi J^{\alpha}$.

Exercise 3.16. CHARGE CONSERVATION

From Maxwell's equations $F^{\alpha\beta}{}_{,\beta} = 4\pi J^{\alpha}$, derive the "equation of charge conservation"

$$J^{\alpha}{}_{,\alpha} = 0. \tag{3.53}$$

Show that this equation does, indeed, correspond to conservation of charge. It will be studied further in Chapter 5.

Exercise 3.17. VECTOR POTENTIAL

The vector potential \boldsymbol{A} of electromagnetic theory generates the electromagnetic field tensor via the geometric equation

$$\boldsymbol{F} = -(\text{antisymmetric part of } \boldsymbol{\nabla}\boldsymbol{A}), \tag{3.54}$$

i.e.,

$$F_{\mu\nu} = A_{\nu,\mu} - A_{\mu,\nu}. \tag{3.54'}$$

(a) Show that the electric and magnetic fields in a specific Lorentz frame are given by

$$\boldsymbol{B} = \boldsymbol{\nabla} \times \boldsymbol{A}, \qquad \boldsymbol{E} = -\partial \boldsymbol{A}/\partial t - \boldsymbol{\nabla} A^0. \tag{3.55}$$

(b) Show that \boldsymbol{F} will satisfy Maxwell's equations if and only if \boldsymbol{A} satisfies

$$A^{\alpha,\mu}{}_{,\mu} - A^{\mu}{}_{,\mu}{}^{,\alpha} = -4\pi J^{\alpha}. \tag{3.56}$$

(c) Show that "gauge transformations"

$$\boldsymbol{A}_{\text{NEW}} = \boldsymbol{A}_{\text{OLD}} + \boldsymbol{d}\phi, \qquad \phi = \text{arbitrary function}, \tag{3.57}$$

leave \boldsymbol{F} unaffected.

(d) Show that one can adjust the gauge so that

$$\boldsymbol{\nabla} \cdot \boldsymbol{A} = 0 \qquad \text{("Lorentz gauge")}, \tag{3.58a}$$

$$\Box \boldsymbol{A} = -4\pi \boldsymbol{J}. \tag{3.58b}$$

Here \Box is the wave operator ("d'Alembertian"):

$$\Box \boldsymbol{A} = A^{\alpha,\mu}{}_{,\mu} \boldsymbol{e}_{\alpha}. \tag{3.59}$$

Exercise 3.18. DIVERGENCE OF ELECTROMAGNETIC STRESS-ENERGY TENSOR

From an electromagnetic field tensor \boldsymbol{F}, one constructs a second-rank, symmetric tensor \boldsymbol{T} ("stress-energy tensor," to be studied in Chapter 5) as follows:

$$T^{\mu\nu} = \frac{1}{4\pi} \left(F^{\mu\alpha} F^{\nu}{}_{\alpha} - \frac{1}{4} \eta^{\mu\nu} F_{\alpha\beta} F^{\alpha\beta} \right). \tag{3.60}$$

As an exercise in index gymnastics:

(a) Show that $\boldsymbol{\nabla} \cdot \boldsymbol{T}$ has components

$$T^{\mu\nu}{}_{,\nu} = \frac{1}{4\pi} \left[F^{\mu\alpha}{}_{,\nu} F^{\nu}{}_{\alpha} + F^{\mu\alpha} F^{\nu}{}_{\alpha,\nu} - \frac{1}{2} F_{\alpha\beta}{}^{,\mu} F^{\alpha\beta} \right]. \tag{3.61}$$

(b) Manipulate this expression into the form

$$T_{\mu}{}^{\nu}{}_{,\nu} = \frac{1}{4\pi} \left[-F_{\mu\alpha} F^{\alpha\nu}{}_{,\nu} - \frac{1}{2} F^{\alpha\beta} (F_{\alpha\beta,\mu} + F_{\mu\alpha,\beta} + F_{\beta\mu,\alpha}) \right]; \tag{3.62}$$

note that the first term of (3.62) arises directly from the second term of (3.61).

(c) Use Maxwell's equations to conclude that

$$T^{\mu\nu}{}_{,\nu} = -F^{\mu\alpha} J_{\alpha}. \tag{3.63}$$

CHAPTER 4

ELECTROMAGNETISM AND DIFFERENTIAL FORMS

The ether trembled at his agitations
In a manner so familiar that I only need to say,
In accordance with Clerk Maxwell's six equations
It tickled peoples' optics far away.
You can feel the way it's done,
You may trace them as they run—
dγ by dy *less* dβ *by* dz *is equal KdX/dt . . .*

While the curl of (X, Y, Z) *is the*
minus d/dt *of the vector* (a, b, c).

From *The Revolution of the Corpuscle,*
written by A. A. Robb
(to the tune of *The Interfering Parrot*)
for a dinner of the research students
of the Cavendish Laboratory
in the days of the old mathematics.

This chapter is all Track 2. It is needed as preparation for §§14.5 and 14.6 (computation of curvature using differential forms) and for Chapter 15 (Bianchi identities and boundary of a boundary), but is not needed for the rest of the book.

§4.1. EXTERIOR CALCULUS

Stacks of surfaces, individually or intersecting to make "honeycombs," "egg crates," and other such structures ("differential forms"), give unique insight into the geometry of electromagnetism and gravitation. However, such insight comes at some cost in time. Therefore, most readers should skip this chapter and later material that depends on it during a first reading of this book.

Analytically speaking, differential forms are completely antisymmetric tensors; pictorially speaking, they are intersecting stacks of surfaces. The mathematical formalism for manipulating differential forms with ease, called "exterior calculus," is summarized concisely in Box 4.1; its basic features are illustrated in the rest of this chapter by rewriting electromagnetic theory in its language. An effective way to tackle this chapter might be to (1) scan Box 4.1 to get the flavor of the formalism; (2) read the rest of the chapter in detail; (3) restudy Box 4.1 carefully; (4) get practice in manipulating the formalism by working the exercises.*

(continued on page 99)

*Exterior calculus is treated in greater detail than here by: É. Cartan (1945); de Rham (1955); Nickerson, Spencer, and Steenrod (1959); Hauser (1970); Israel (1970); especially Flanders (1963, relatively easy, with many applications); Spivak (1965, sophomore or junior level, but fully in tune with modern mathematics); H. Cartan (1970); and Choquet-Bruhat (1968a).

**Box 4.1 DIFFERENTIAL FORMS AND
EXTERIOR CALCULUS IN BRIEF**

The fundamental definitions and formulas of exterior calculus are summarized here for ready reference. Each item consists of a general statement (at left of page) plus a leading application (at right of page). This formalism is applicable not only to spacetime, but also to more general geometrical systems (see heading of each section). No attempt is made here to demonstrate the internal consistency of the formalism, nor to derive it from any set of definitions and axioms. For a systematic treatment that does so, see, e.g., Spivak (1965), or Misner and Wheeler (1957).

A. Algebra I (applicable to any vector space)

1. *Basis 1-forms.*
 a. Coordinate basis $\boldsymbol{w}^j = \boldsymbol{d}x^j$
 (j tells which 1-form, not which component).
 b. General basis $\boldsymbol{w}^j = L^j{}_{k'}\,\boldsymbol{d}x^{k'}$.

 An application
 Simple basis 1-forms for analyzing Schwarzschild geometry around static spherically symmetric center of attraction:
 $$\boldsymbol{w}^0 = (1 - 2m/r)^{1/2}\,\boldsymbol{d}t;$$
 $$\boldsymbol{w}^1 = (1 - 2m/r)^{-1/2}\,\boldsymbol{d}r;$$
 $$\boldsymbol{w}^2 = r\,\boldsymbol{d}\theta;$$
 $$\boldsymbol{w}^3 = r\sin\theta\,\boldsymbol{d}\phi.$$

2. *General* p-*form* (*or* p-*vector*) is a completely anti-symmetric tensor of rank $\binom{0}{p}$ [or $\binom{p}{0}$]. It can be expanded in terms of wedge products (see §3.5 and exercise 4.12):

 $$\boldsymbol{\alpha} = \frac{1}{p!}\alpha_{i_1 i_2 \dots i_p}\boldsymbol{w}^{i_1} \wedge \boldsymbol{w}^{i_2} \wedge \cdots \wedge \boldsymbol{w}^{i_p}$$
 $$\equiv \alpha_{|i_1 i_2 \dots i_p|}\boldsymbol{w}^{i_1} \wedge \boldsymbol{w}^{i_2} \wedge \cdots \wedge \boldsymbol{w}^{i_p}.$$

 (Note: Vertical bars around the indices mean summation extends only over $i_1 < i_2 < \cdots < i_p$.)

 Two applications
 Energy-momentum 1-form is of type $\boldsymbol{\alpha} = \alpha_i\boldsymbol{w}^i$ or
 $$\boldsymbol{p} = -E\,\boldsymbol{d}t + p_x\,\boldsymbol{d}x + p_y\,\boldsymbol{d}y + p_z\,\boldsymbol{d}z.$$

 Faraday is a 2-form of type $\boldsymbol{\beta} = \beta_{|\mu\nu|}\boldsymbol{w}^\mu \wedge \boldsymbol{w}^\nu$ or in flat spacetime
 $$\boldsymbol{F} = -E_x\,\boldsymbol{d}t \wedge \boldsymbol{d}x - E_y\,\boldsymbol{d}t \wedge \boldsymbol{d}y - E_z\,\boldsymbol{d}t \wedge \boldsymbol{d}z$$
 $$+ B_x\,\boldsymbol{d}y \wedge \boldsymbol{d}z + B_y\,\boldsymbol{d}z \wedge \boldsymbol{d}x + B_z\,\boldsymbol{d}x \wedge \boldsymbol{d}y$$

Box 4.1 (continued)

3. *Wedge product.*
All familiar rules of addition and multiplication hold, such as

$$(a\boldsymbol{\alpha} + b\boldsymbol{\beta}) \wedge \boldsymbol{\gamma} = a\boldsymbol{\alpha} \wedge \boldsymbol{\gamma} + b\boldsymbol{\beta} \wedge \boldsymbol{\gamma},$$
$$(\boldsymbol{\alpha} \wedge \boldsymbol{\beta}) \wedge \boldsymbol{\gamma} = \boldsymbol{\alpha} \wedge (\boldsymbol{\beta} \wedge \boldsymbol{\gamma}) \equiv \boldsymbol{\alpha} \wedge \boldsymbol{\beta} \wedge \boldsymbol{\gamma},$$

except for a modified commutation law between a *p*-form $\boldsymbol{\alpha}$ and a *q*-form $\boldsymbol{\beta}$:

$$\underset{p}{\boldsymbol{\alpha}} \wedge \underset{q}{\boldsymbol{\beta}} = (-1)^{pq} \underset{q}{\boldsymbol{\beta}} \wedge \underset{p}{\boldsymbol{\alpha}}.$$

Applications to 1-forms $\boldsymbol{\alpha}, \boldsymbol{\beta}$:

$$\boldsymbol{\alpha} \wedge \boldsymbol{\beta} = -\boldsymbol{\beta} \wedge \boldsymbol{\alpha}, \qquad \boldsymbol{\alpha} \wedge \boldsymbol{\alpha} = 0;$$
$$\boldsymbol{\alpha} \wedge \boldsymbol{\beta} = (\alpha_j \boldsymbol{\omega}^j) \wedge (\beta_k \boldsymbol{\omega}^k) = \alpha_j \beta_k \boldsymbol{\omega}^j \wedge \boldsymbol{\omega}^k$$
$$= \frac{1}{2}(\alpha_j \beta_k - \beta_j \alpha_k)\boldsymbol{\omega}^j \wedge \boldsymbol{\omega}^k.$$

4. *Contraction of* p-*form on* p-*vector.*

$$\langle \underset{p}{\boldsymbol{\alpha}}, \underset{p}{\boldsymbol{A}} \rangle$$

$$= \alpha_{|i_1 \dots i_p|} A^{|j_1 \dots j_p|} \underbrace{\langle \boldsymbol{\omega}^{i_1} \wedge \cdots \wedge \boldsymbol{\omega}^{i_p}, \boldsymbol{e}_{j_1} \wedge \cdots \wedge \boldsymbol{e}_{j_p} \rangle}_{[\equiv \delta^{i_1 \dots i_p}_{j_1 \dots j_p} \text{ (see exercises 3.13 and 4.12)}]}$$

$$= \alpha_{|i_1 \dots i_p|} A^{i_1 \dots i_p}.$$

Four applications

a. Contraction of a particle's energy-momentum 1-form $\boldsymbol{p} = p_\alpha \boldsymbol{\omega}^\alpha$ with 4-velocity $\boldsymbol{u} = u^\alpha \boldsymbol{e}_\alpha$ of observer (a 1-vector):

$$-\langle \boldsymbol{p}, \boldsymbol{u} \rangle = -p_\alpha u^\alpha = \text{energy of particle.}$$

b. Contraction of ***Faraday*** 2-form ***F*** with bivector $\delta\mathscr{P} \wedge \Delta\mathscr{P}$ [where $\delta\mathscr{P} = (d\mathscr{P}/d\lambda_1)\Delta\lambda_1$ and $\Delta\mathscr{P} = (d\mathscr{P}/d\lambda_2)\Delta\lambda_2$ are two infinitesimal vectors in a 2-surface $\mathscr{P}(\lambda_1, \lambda_2)$, and the bivector represents the surface element they span] is the magnetic flux $\Phi = \langle \boldsymbol{F}, \delta\mathscr{P} \wedge \Delta\mathscr{P} \rangle$ through that surface element.

c. More generally, a *p*-dimensional parallelepiped with vectors $\boldsymbol{a}_1, \boldsymbol{a}_2, \dots, \boldsymbol{a}_p$ for legs has an oriented volume described by the *"simple"* *p*-vector $\boldsymbol{a}_1 \wedge \boldsymbol{a}_2 \wedge \cdots \boldsymbol{a}_p$ (oriented because interchange of two legs changes its sign). An egg-crate type of structure with walls made from the hyperplanes of *p* different 1-forms $\boldsymbol{\sigma}^1$,

$\sigma^2, \ldots, \sigma^p$ is described by the *"simple"* p-form $\sigma^1 \wedge \sigma^2 \wedge \cdots \wedge \sigma^p$. The number of cells of $\sigma^1 \wedge \sigma^2 \wedge \cdots \wedge \sigma^p$ sliced through by the infinitesimal p-volume $\boldsymbol{a}_1 \wedge \boldsymbol{a}_2 \wedge \cdots \wedge \boldsymbol{a}_p$ is

$$\langle \sigma^1 \wedge \sigma^2 \wedge \cdots \wedge \sigma^p, \boldsymbol{a}_1 \wedge \boldsymbol{a}_2 \wedge \cdots \wedge \boldsymbol{a}_p \rangle.$$

d. The Jacobian determinant of a set of p functions $f^k(x^1, \ldots, x^n)$ with respect to p of their arguments is

$$\left\langle \boldsymbol{df}^1 \wedge \boldsymbol{df}^2 \wedge \cdots \wedge \boldsymbol{df}^p, \frac{\partial \mathscr{P}}{\partial x^1} \wedge \frac{\partial \mathscr{P}}{\partial x^2} \wedge \cdots \wedge \frac{\partial \mathscr{P}}{\partial x^p} \right\rangle$$

$$= \det \left\| \left(\frac{\partial f^k}{\partial x^j} \right) \right\| \equiv \frac{\partial(f^1, f^2, \ldots, f^p)}{\partial(x^1, x^2, \ldots, x^p)}.$$

5. *Simple forms.*
 a. A simple p-form is one that can be written as a wedge product of p 1-forms:

 $$\underset{p}{\sigma} = \underbrace{\boldsymbol{\alpha} \wedge \boldsymbol{\beta} \wedge \cdots \wedge \boldsymbol{\gamma}}_{p \text{ factors}}.$$

 b. A simple p-form $\boldsymbol{\alpha} \wedge \boldsymbol{\beta} \wedge \cdots \wedge \boldsymbol{\gamma}$ is represented by the intersecting families of surfaces of $\boldsymbol{\alpha}, \boldsymbol{\beta}, \ldots, \boldsymbol{\gamma}$ (egg-crate structure) plus a sense of circulation (orientation).

 Applications:
 a. In four dimensions (e.g., spacetime) all 0-forms, 1-forms, 3-forms, and 4-forms are simple. A 2-form \boldsymbol{F} is generally a sum of two simple forms, e.g., $\boldsymbol{F} = -e\,\boldsymbol{dt} \wedge \boldsymbol{dx} + h\,\boldsymbol{dy} \wedge \boldsymbol{dz}$; it is simple if and only if $\boldsymbol{F} \wedge \boldsymbol{F} = 0$.
 b. A set of 1-forms $\boldsymbol{\alpha}, \boldsymbol{\beta}, \ldots, \boldsymbol{\gamma}$ is linearly dependent (one a linear combination of the others) if and only if

 $$\boldsymbol{\alpha} \wedge \boldsymbol{\beta} \wedge \cdots \wedge \boldsymbol{\gamma} = 0 \qquad \text{(egg crate collapsed)}.$$

B. Exterior Derivative (applicable to any "differentiable manifold," with or without metric)

1. \boldsymbol{d} produces a $(p+1)$-form $\boldsymbol{d\sigma}$ from a p-form $\boldsymbol{\sigma}$.
2. Effect of \boldsymbol{d} is defined by induction using the

Box 4.1 (*continued*)

(Chapter 2) definition of df, and f a function (0-form), plus

$$d(\underset{p}{\alpha} \wedge \underset{q}{\beta}) = d\alpha \wedge \beta + (-1)^p \alpha \wedge d\beta,$$

$$d^2 = dd = 0.$$

Two applications

$$d(\alpha \wedge d\beta) = d\alpha \wedge d\beta.$$

For the p-form ϕ, with

$$\phi = \phi_{|i_1\ldots i_p|} dx^{i_1} \wedge \cdots \wedge dx^{i_p},$$

one has (alternative and equivalent definition of $d\phi$)

$$d\phi = d\phi_{|i_1\ldots i_p|} \wedge dx^{i_1} \wedge \cdots \wedge dx^{i_p}.$$

C. Integration (applicable to any "differentiable manifold," with or without metric)

1. *Pictorial interpretation.*

 Text and pictures of Chapter 4 interpret $\int \alpha$ (integral of specified 1-form α along specified curve from specified starting point to specified end point) as "number of α-surfaces pierced on that route"; similarly, they interpret $\int \phi$ (integral of specified 2-form ϕ over specified bit of surface on which there is an assigned sense of circulation or "orientation") as "number of cells of the honeycomb-like structure ϕ cut through by that surface"; similarly for the egg-crate-like structures that represent 3-forms; etc.

2. *Computational rules for integration.*

 To evaluate $\int \alpha$, the integral of a p-form

 $$\alpha = \alpha_{|i_1\ldots i_p|}(x^1, \ldots, x^n) dx^{i_1} \wedge \cdots \wedge dx^{i_p},$$

 over a p-dimensional surface, proceed in two steps.
 a. Substitute a parameterization of the surface,

 $$x^k(\lambda^1, \ldots, \lambda^p)$$

 into α, and collect terms in the form

 $$\alpha = a(\lambda^j) d\lambda^1 \wedge \cdots \wedge d\lambda^p$$

 (this is α viewed as a p-form in the p-dimensional surface);

b. Integrate

$$\int \boldsymbol{\alpha} = \int a(\lambda^j)\, d\lambda^1\, d\lambda^2 \dots d\lambda^p$$

using elementary definition of integration.

Example: See equations (4.12) to (4.14).

3. *The differential geometry of integration.*

Calculate $\int \boldsymbol{\alpha}$ for a p-form $\boldsymbol{\alpha}$ as follows.

a. Choose the p-dimensional surface \mathcal{S} over which to integrate.

b. Represent \mathcal{S} by a parametrization giving the generic point of the surface as a function of the parameters, $\mathcal{P}(\lambda^1, \lambda^2, \dots \lambda^p)$. This fixes the orientation. The same function with $\lambda^1 \leftrightarrow \lambda^2$, $\mathcal{P}(\lambda^2, \lambda^1, \dots, \lambda^p)$, describes a different (i.e., oppositely oriented) surface, $-\mathcal{S}$.

c. The infinitesimal parallelepiped

$$\left(\frac{\partial \mathcal{P}}{\partial \lambda^1}\, \Delta\lambda^1\right) \wedge \left(\frac{\partial \mathcal{P}}{\partial \lambda^2}\, \Delta\lambda^2\right) \wedge \cdots \wedge \left(\frac{\partial \mathcal{P}}{\partial \lambda^p}\, \Delta\lambda^p\right)$$

is tangent to the surface. The number of cells of $\boldsymbol{\alpha}$ it slices is

$$\left\langle \boldsymbol{\alpha}, \frac{\partial \mathcal{P}}{\partial \lambda^1} \wedge \cdots \wedge \frac{\partial \mathcal{P}}{\partial \lambda^p} \right\rangle \Delta\lambda^1 \dots \Delta\lambda^p.$$

This number changes sign if two of the vectors $\partial \mathcal{P}/\partial \lambda^k$ are interchanged, as for an oppositely oriented surface.

d. The above provides an interpretation motivating the definition

$$\int \boldsymbol{\alpha} \equiv \int\!\!\int \cdots \int \left\langle \boldsymbol{\alpha}, \frac{\partial \mathcal{P}}{\partial \lambda^1} \wedge \frac{\partial \mathcal{P}}{\partial \lambda^2} \wedge \cdots \wedge \frac{\partial \mathcal{P}}{\partial \lambda^p} \right\rangle$$
$$d\lambda^1\, d\lambda^2 \dots d\lambda^p.$$

This definition is identified with the computational rule of the preceding section (C.2) in exercise 4.9.

An application

Integrate a gradient \boldsymbol{df} along a curve, $\mathcal{P}(\lambda)$ from $\mathcal{P}(0)$ to $\mathcal{P}(1)$:

$$\int \boldsymbol{df} = \int_0^1 \langle \boldsymbol{df}, d\mathcal{P}/d\lambda \rangle\, d\lambda = \int_0^1 (df/d\lambda)\, d\lambda$$
$$= f[\mathcal{P}(1)] - f[\mathcal{P}(0)].$$

e. Three different uses for symbol "d": *First*, lightface d in explicit derivative expressions such as

Box 4.1 (*continued*)

d/da, or df/da, or $d\mathcal{P}/da$; neither numerator nor denominator alone has any meaning, but only the full string of symbols. *Second*, lightface d inside an integral sign; e.g., $\int\!\int f\,da$. This is an instruction to perform integration, and has no meaning whatsoever without an integral sign; "$\int\dots d\dots$" lives as an indivisible unit. *Third*, sans-serif **d**; e.g., **d** alone, or **d**f, or **d**a. This is an exterior derivative, which converts a p-form into a $(p+1)$-form. Sometimes lightface d is used for the same purpose. Hence, d alone, or df, or dx, is always an exterior derivative unless coupled to an \int sign (*second* use), or coupled to a $/$ sign (*first* use).

4. *The generalized Stokes theorem* (see Box 4.6).
 a. Let $\partial\mathcal{V}$ be the closed p-dimensional boundary of a $(p+1)$-dimensional surface \mathcal{V}. Let $\boldsymbol{\sigma}$ be a p-form defined throughout \mathcal{V}.
 Then

 $$\int_{\mathcal{V}} \boldsymbol{d\sigma} = \int_{\partial\mathcal{V}} \boldsymbol{\sigma}$$

 [integral of p-form $\boldsymbol{\sigma}$ over boundary $\partial\mathcal{V}$ equals integral of $(p+1)$-form $\boldsymbol{d\sigma}$ over interior \mathcal{V}].
 b. For the sign to come out right, orientations of \mathcal{V} and $\partial\mathcal{V}$ must agree in this sense: choose coordinates y^0, y^1, \dots, y^p on a portion of \mathcal{V}, with y^0 specialized so $y^0 \leq 0$ in \mathcal{V}, and $y^0 = 0$ at the boundary $\partial\mathcal{V}$; then the orientation

 $$\frac{\partial\mathcal{P}}{\partial y^0} \wedge \frac{\partial\mathcal{P}}{\partial y^1} \wedge \cdots \wedge \frac{\partial\mathcal{P}}{\partial y^p}$$

 for \mathcal{V} demands the orientation

 $$\frac{\partial\mathcal{P}}{\partial y^1} \wedge \cdots \wedge \frac{\partial\mathcal{P}}{\partial y^p}$$

 for $\partial\mathcal{V}$.
 c. Note: For a nonorientable surface, such as a Möbius strip, where a consistent and continuous choice of orientation is impossible, more intricate mathematics is required to give a definition of "∂" for which the Stokes theorem holds.

 Applications: Includes as special cases all integral theorems for surfaces of arbitrary dimension in spaces of arbitrary dimension, with or without metric, generaliz-

ing all versions of theorems of Stokes and Gauss. Examples:

a. \mathcal{V} a curve, $\partial\mathcal{V}$ its endpoints, $\boldsymbol{\sigma} = f$ a 0-form (function):

$$\int_{\mathcal{V}} \boldsymbol{d}f = \int_0^1 (df/d\lambda)\,d\lambda = \int_{\partial\mathcal{V}} f = f(1) - f(0).$$

b. \mathcal{V} a 2-surface in 3-space, $\partial\mathcal{V}$ its closed-curve boundary, \boldsymbol{v} a 1-form; translated into Euclidean vector notation, the two integrals are

$$\int_{\mathcal{V}} \boldsymbol{d}\boldsymbol{v} = \int_{\mathcal{V}} (\boldsymbol{\nabla} \times v) \cdot d\boldsymbol{S}; \int_{\partial\mathcal{V}} \boldsymbol{v} = \int_{\partial\mathcal{V}} v \cdot d\boldsymbol{l}.$$

c. Other applications in §§5.8, 20.2, 20.3, 20.5, and exercises 4.10, 4.11, 5.2, and below.

D. Algebra II (applicable to any vector space with metric)

1. *Norm of a* p-*form.*

$\|\boldsymbol{\alpha}\|^2 \equiv \alpha_{|i_1...i_p|}\alpha^{i_1...i_p}.$

> *Two applications:* Norm of a 1-form equals its squared length, $\|\boldsymbol{\alpha}\|^2 = \boldsymbol{\alpha} \cdot \boldsymbol{\alpha}$. Norm of electromagnetic 2-form or *Faraday*: $\|\boldsymbol{F}\|^2 = \boldsymbol{B}^2 - \boldsymbol{E}^2$.

2. *Dual of a* p-*form.*

a. In an n-dimensional space, the dual of a p-form $\boldsymbol{\alpha}$ is the $(n - p)$-form $\boldsymbol{*\alpha}$, with components

$$(\boldsymbol{*}\alpha)_{k_1...k_{n-p}} = \alpha^{|i_1...i_p|}\epsilon_{i_1...i_p\,k_1...k_{n-p}}.$$

b. Properties of duals:

$$\boldsymbol{**\alpha} = (-1)^{p-1}\boldsymbol{\alpha} \text{ in spacetime;}$$
$$\boldsymbol{\alpha} \wedge \boldsymbol{*\alpha} = \|\boldsymbol{\alpha}\|^2\boldsymbol{\varepsilon} \text{ in general.}$$

c. Note: the definition of $\boldsymbol{\varepsilon}$ (exercise 3.13) entails choosing an orientation of the space, i.e., deciding which orthonormal bases (1) are "right-handed" and thus (2) have $\boldsymbol{\varepsilon}(\boldsymbol{e}_1,\ldots,\boldsymbol{e}_n) = +1$.

> *Applications*
> a. For f a 0-form, $\boldsymbol{*}f = f\boldsymbol{\varepsilon}$, and $\int\!\!\int fd(\text{volume}) = \int \boldsymbol{*}f.$
> b. Dual of charge-current 1-form \boldsymbol{J} is charge-current 3-form $\boldsymbol{*J}$. The total charge Q in a 3-dimensional hypersurface region \mathcal{S} is
>
> $$Q(\mathcal{S}) = \int_{\mathcal{S}} \boldsymbol{*J}.$$

Box 4.1 (*continued*)

Conservation of charge is stated locally by $d*J = 0$. Stokes' Theorem goes from this differential conservation law to the integral conservation law,

$$0 = \int_{\mathcal{V}} d*J = \int_{\partial\mathcal{V}} *J.$$

This law is of most interest when $\partial\mathcal{V} = \mathcal{S}_2 - \mathcal{S}_1$ consists of the future \mathcal{S}_2 and past \mathcal{S}_1 boundaries of a spacetime region, in which case it states $Q(\mathcal{S}_2) = Q(\mathcal{S}_1)$; see exercise 5.2.

c. Dual of electromagnetic field tensor $F = \textbf{\textit{Faraday}}$ is $*F = \textbf{\textit{Maxwell}}$. From the $d*F = 4\pi *J$ Maxwell equation, find $4\pi Q = 4\pi\int_S *J = \int_S d*F = \int_{\partial S} *F$.

3. *Simple forms revisited.*
 a. The dual of a simple form is simple.
 b. Egg crate of $*\sigma$ is perpendicular to egg crate of $\sigma = \alpha \wedge \beta \wedge \cdots \wedge \mu$ in this sense:
 (1) pick any vector V lying in intersection of surfaces of σ

 $$(\langle \alpha, V \rangle = \langle \beta, V \rangle = \cdots = \langle \mu, V \rangle = 0);$$

 (2) pick any vector W lying in intersection of surfaces of $*\sigma$;
 (3) then V and W are necessarily perpendicular: $V \cdot W = 0$.

 Example: $\sigma = 3\,dt$ is a simple 1-form in spacetime.
 a. $*\sigma = -3\,dx \wedge dy \wedge dz$ is a simple 3-form.
 b. General vector in surfaces of σ is

 $$V = V^x e_x + V^y e_y + V^z e_z.$$

 c. General vector in intersection of surfaces of $*\sigma$ is

 $$W = W^t e_t.$$

 d. $W \cdot V = 0$.

§4.2. ELECTROMAGNETIC 2-FORM AND LORENTZ FORCE

The electromagnetic field tensor, **Faraday** = **F**, is an antisymmetric second-rank tensor (i.e., 2-form). Instead of expanding it in terms of the tensor products of basis 1-forms,

$$\boldsymbol{F} = F_{\alpha\beta}\, \boldsymbol{dx}^\alpha \otimes \boldsymbol{dx}^\beta,$$

the exterior calculus prefers to expand in terms of antisymmetrized tensor products ("*exterior products*," exercise 4.1):

Electromagnetic 2-form expressed in terms of exterior products

$$\boldsymbol{F} = \frac{1}{2} F_{\alpha\beta}\, \boldsymbol{dx}^\alpha \wedge \boldsymbol{dx}^\beta, \tag{4.1}$$

$$\boldsymbol{dx}^\alpha \wedge \boldsymbol{dx}^\beta \equiv \boldsymbol{dx}^\alpha \otimes \boldsymbol{dx}^\beta - \boldsymbol{dx}^\beta \otimes \boldsymbol{dx}^\alpha. \tag{4.2}$$

Any 2-form (antisymmetric, second-rank tensor) can be so expanded. The symbol "∧" is variously called a "wedge," a "hat," or an "exterior product sign"; and $\boldsymbol{dx}^\alpha \wedge \boldsymbol{dx}^\beta$ are the "basis 2-forms" of a given Lorentz frame (see §3.5, exercise 3.12, and Box 4.1).

There is no simpler way to illustrate this 2-form representation of the electromagnetic field than to consider a magnetic field in the x-direction:

$$\begin{aligned} F_{yz} &= -F_{zy} = B_x, \\ \boldsymbol{F} &= B_x\, \boldsymbol{dy} \wedge \boldsymbol{dz}. \end{aligned} \tag{4.3}$$

The 1-form $\boldsymbol{dy} = \operatorname{grad} y$ is the set of surfaces (actually hypersurfaces) $y = 18$ (all t, x, z), $y = 19$ (all t, x, z), $y = 20$ (all t, x, z), etc.; and surfaces uniformly interpolated between them. Similarly for the 1-form \boldsymbol{dz}. The intersection between these two sets of surfaces produces a honeycomb-like structure. That structure becomes a "2-form" when it is supplemented by instructions (see arrows in Figure 4.1) that give a "sense of circulation" to each tube of the honeycomb (order of factors in the "wedge product" of equation 4.2; $\boldsymbol{dy} \wedge \boldsymbol{dz} = -\boldsymbol{dz} \wedge \boldsymbol{dy}$). The 2-form \boldsymbol{F} in the example differs from this "basis 2-form" $\boldsymbol{dy} \wedge \boldsymbol{dz}$ only in this respect, that where $\boldsymbol{dy} \wedge \boldsymbol{dz}$ had one tube, the field 2-form has B_x tubes.

A 2-form as a honeycomb of tubes with a sense of circulation

When one considers a tubular structure that twists and turns on its way through spacetime, one must have more components to describe it. The 2-form for the general electromagnetic field can be written as

$$\begin{aligned} \boldsymbol{F} = {}&E_x\, \boldsymbol{dx} \wedge \boldsymbol{dt} + E_y\, \boldsymbol{dy} \wedge \boldsymbol{dt} + E_z\, \boldsymbol{dz} \wedge \boldsymbol{dt} + B_x\, \boldsymbol{dy} \wedge \boldsymbol{dz} \\ &+ B_y\, \boldsymbol{dz} \wedge \boldsymbol{dx} + B_z\, \boldsymbol{dx} \wedge \boldsymbol{dy} \end{aligned} \tag{4.4}$$

(6 components, 6 basis 2-forms).

A 1-form is a machine to produce a number out of a vector (bongs of a bell as the vector pierces successive surfaces). A 2-form is a machine to produce a number out of an oriented surface (surface with a sense of circulation indicated on it: Figure 4.1, lower right). The meaning is as clear here as it is in elementary magnetism:

A 2-form as a machine to produce a number out of an oriented surface

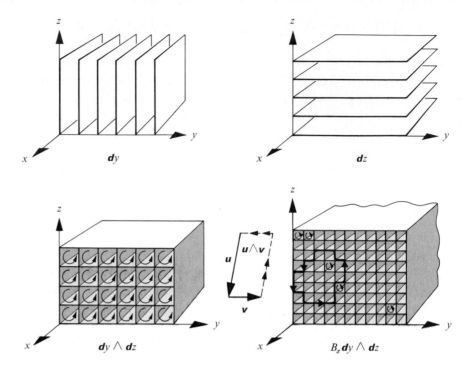

Figure 4.1.
Construction of the 2-form for the electromagnetic field $\boldsymbol{F} = B_x\,\boldsymbol{dy} \wedge \boldsymbol{dz}$ out of the 1-forms \boldsymbol{dy} and \boldsymbol{dz} by "wedge multiplication" (formation of honeycomb-like structure with sense of circulation indicated by arrows). A 2-form is a "machine to construct a number out of an oriented surface" (illustrated by sample surface enclosed by arrows at lower right; number of tubes intersected by this surface is

$$\int_{\text{(this surface)}} \boldsymbol{F} = 18;$$

Faraday's concept of "magnetic flux"). This idea of 2-form machinery can be connected to the "tensor-as-machine" idea of Chapter 3 as follows. The shape of the oriented surface over which one integrates \boldsymbol{F} does not matter, for small surfaces. All that affects $\int\boldsymbol{F}$ is the area of the surface, and its orientation. Choose two vectors, \boldsymbol{u} and \boldsymbol{v}, that lie in the surface. They form two legs of a parallelogram, whose orientation (\boldsymbol{u} followed by \boldsymbol{v}) and area are embodied in the exterior product $\boldsymbol{u} \wedge \boldsymbol{v}$. Adjust the lengths of \boldsymbol{u} and \boldsymbol{v} so their parallelogram, $\boldsymbol{u} \wedge \boldsymbol{v}$, has the same area as the surface of integration. Then

Exercise: derive this result, for an infinitesimal surface $\boldsymbol{u} \wedge \boldsymbol{v}$ and for general \boldsymbol{F}, using the formalism of Box 4.1.

the number of Faraday tubes cut by that surface. The electromagnetic 2-form **F** or **Faraday** described by such a "tubular structure" (suitably abstracted; Box 4.2) has a reality and a location in space that is independent of all coordinate systems and all artificial distinctions between "electric" and "magnetic" fields. Moreover, those tubes provide the most direct geometric representation that anyone has ever been able to give for the machinery by which the electromagnetic field acts on a charged particle. Take a particle of charge e and 4-velocity

$$\boldsymbol{u} = \frac{dx^\alpha}{d\tau}\, \boldsymbol{e}_\alpha. \tag{4.5}$$

Let this particle go through a region where the electromagnetic field is described by the 2-form

$$\boldsymbol{F} = B_x\, \boldsymbol{dy} \wedge \boldsymbol{dz} \tag{4.6}$$

of Figure 4.1. Then the force exerted on the particle (regarded as a 1-form) is the contraction of this 2-form with the 4-velocity (and the charge);

Lorentz force as contraction of electromagnetic 2-form with particle's 4-velocity

$$\dot{\boldsymbol{p}} = d\boldsymbol{p}/d\tau = e\boldsymbol{F}(\boldsymbol{u}) \equiv e\langle \boldsymbol{F}, \boldsymbol{u}\rangle, \tag{4.7}$$

as one sees by direct evaluation, letting the two factors in the 2-form act in turn on the tangent vector \boldsymbol{u}:

$$
\begin{aligned}
\dot{\boldsymbol{p}} &= eB_x\langle \boldsymbol{dy} \wedge \boldsymbol{dz}, \boldsymbol{u}\rangle \\
&= eB_x\{\boldsymbol{dy}\langle \boldsymbol{dz}, \boldsymbol{u}\rangle - \boldsymbol{dz}\langle \boldsymbol{dy}, \boldsymbol{u}\rangle\} \\
&= eB_x\{\boldsymbol{dy}\langle \boldsymbol{dz}, u^z\boldsymbol{e}_z\rangle - \boldsymbol{dz}\langle \boldsymbol{dy}, u^y\boldsymbol{e}_y\rangle\}
\end{aligned}
$$

or

$$\dot{p}_\alpha\, \boldsymbol{dx}^\alpha = eB_x u^z\, \boldsymbol{dy} - eB_x u^y\, \boldsymbol{dz}. \tag{4.8}$$

Comparing coefficients of the separate basis 1-forms on the two sides of this equation, one sees reproduced all the detail of the Lorentz force exerted by the magnetic field B_x:

$$
\begin{aligned}
\dot{p}_y &= \frac{dp_y}{d\tau} = eB_x \frac{dz}{d\tau}, \\[2mm]
\dot{p}_z &= \frac{dp_z}{d\tau} = -eB_x \frac{dy}{d\tau}.
\end{aligned}
\tag{4.9}
$$

By simple extension of this line of reasoning to the general electromagnetic field, one concludes that *the time-rate of change of momentum (1-form) is equal to the charge multiplied by the contraction of the **Faraday** with the 4-velocity.* Figure 4.2 illustrates pictorially how the 2-form, **F**, serves as a machine to produce the 1-form, $\dot{\boldsymbol{p}}$, out of the tangent vector, $e\boldsymbol{u}$.

(continued on page 105)

Box 4.2 ABSTRACTING A 2-FORM FROM THE CONCEPT OF "HONEYCOMB-LIKE STRUCTURE," IN 3-SPACE AND IN SPACETIME

Open up a cardboard carton containing a dozen bottles, and observe the honeycomb structure of intersecting north-south and east-west cardboard separators between the bottles. That honeycomb structure of "tubes" ("channels for bottles") is a fairly apt illustration of a 2-form in the context of everyday 3-space. It yields a number (number of tubes cut) for each choice of smooth element of 2-surface slicing through the three-dimensional structure. However, the intersecting cardboard separators are rather too specific. All that a true 2-form can ever give is the number of tubes sliced through, not the "shape" of the tubes. Slew the carton around on the floor by 45°. Then half the separators run NW-SE and the other half run NE-SW, but through a given bit of 2-surface fixed in 3-space the count of tubes is unchanged. Therefore, one should be careful to make the concept of tubes in the mind's eye abstract enough that one envisages direction of tubes (vertical in the example) and density of tubes, but not any specific location or orientation for the tube walls. Thus all the following representations give one and the same 2-form, $\boldsymbol{\sigma}$:

$$\boldsymbol{\sigma} = B\,\boldsymbol{dx} \wedge \boldsymbol{dy};$$

$$\boldsymbol{\sigma} = B(2\,\boldsymbol{dx}) \wedge \left(\frac{1}{2}\,\boldsymbol{dy}\right)$$

(NS cardboards spaced twice as close as before; EW cardboards spaced twice as wide as before);

$$\boldsymbol{\sigma} = B\,\boldsymbol{d}\left(\frac{x-y}{\sqrt{2}}\right) \wedge \boldsymbol{d}\left(\frac{x+y}{\sqrt{2}}\right)$$

(cardboards rotated through 45°);

$$\boldsymbol{\sigma} = B\,\frac{\alpha\,\boldsymbol{dx} + \beta\,\boldsymbol{dy}}{(\alpha\delta - \beta\gamma)^{1/2}} \wedge \frac{\gamma\,\boldsymbol{dx} + \delta\,\boldsymbol{dy}}{(\alpha\delta - \beta\gamma)^{1/2}}$$

(both orientation and spacing of "cardboards" changing from point to point, with all four functions, α, β, γ, and δ, depending on position).

What has physical reality, and constitutes the real geometric object, is not any one of the 1-forms just encountered individually, but only the 2-form $\boldsymbol{\sigma}$ itself. This circumstance helps to explain why in the physical literature one sometimes refers to "tubes of force" and sometimes to "lines of force." The two terms for the same structure have this in common, that each yields a number when sliced by a bit of surface. The line-of-force picture has the advantage of not imposing on the mind any specific structure of "sheets of cardboard"; that is, any specific decomposition of the 2-form into the product of 1-forms. However, that very feature is also a disadvantage, for in a calculation one often finds it useful to have a well-defined representation of the 2-form as the wedge product of 1-forms. Moreover, the tube picture, abstract though it must be if it is to be truthful, also has this advantage, that the orientation of the elementary tubes (sense of circulation as indicated by arrows in Figures 4.1 and 4.5, for example) lends itself to ready visualization. Let the "walls" of the tubes therefore remain in all pictures drawn in this book as a reminder that 2-forms can be built out of 1-forms; but let it be understood here and hereafter how manyfold are the options for the individual 1-forms!

Turn now from three dimensions to four, and find that the concept of "honeycomb-like structure" must be made still more abstract. In three dimensions the arbitrariness of the decomposition of the 2-form into 1-forms showed in the slant and packing of the "cardboards," but had no effect on the verticality of the "channels for the bottles" ("direction of Faraday lines of force or tubes of

force"); not so in four dimensions, or at least not in the generic case in four dimensions.

In special cases, the story is almost as simple in four dimensions as in three. An example of a special case is once again the 2-form $\boldsymbol{\sigma} = B\,\boldsymbol{dx} \wedge \boldsymbol{dy}$, with all the options for decomposition into 1-forms that have already been mentioned, but with every option giving the same "direction" for the tubes. If the word "direction" now rises in status from "tube walls unpierced by motion in the direction of increasing z" to "tube walls unpierced either by motion in the direction of increasing z, or by motion in the direction of increasing t, or by any linear combination of such motions," that is a natural enough consequence of adding the new dimension. Moreover, the same simplicity prevails for an electromagnetic plane wave. For example, let the wave be advancing in the z-direction, and let the electric polarization point in the x-direction; then for a monochromatic wave, one has

$$E_x = B_y = E_0 \cos \omega(z - t) = -F_{01} = F_{31},$$

and all components distinct from these equal zero. **Faraday** is

$$\begin{aligned} \boldsymbol{F} &= F_{01}\,\boldsymbol{dt} \wedge \boldsymbol{dx} + F_{31}\,\boldsymbol{dz} \wedge \boldsymbol{dx} \\ &= E_0 \cos \omega(z - t)\,\boldsymbol{d}(z - t) \wedge \boldsymbol{dx}, \end{aligned}$$

which is again representable as a single wedge product of two 1-forms.

Not so in general! The general 2-form in four dimensions consists of six distinct wedge products,

$$\boldsymbol{F} = F_{01}\,\boldsymbol{dt} \wedge \boldsymbol{dx} + F_{02}\,\boldsymbol{dt} \wedge \boldsymbol{dy} + \cdots \\ + F_{23}\,\boldsymbol{dy} \wedge \boldsymbol{dz}.$$

It is too much to hope that this expression will reduce in the generic case to a single wedge product of two 1-forms ("*simple*" 2-form). It is not even

true that it will. It is only remarkable that it can be reduced from six exterior products to two (details in exercise 4.1); thus,

$$\boldsymbol{F} = \boldsymbol{n}^1 \wedge \boldsymbol{\xi}^1 + \boldsymbol{n}^2 \wedge \boldsymbol{\xi}^2.$$

Each product $\boldsymbol{n}^i \wedge \boldsymbol{\xi}^i$ individually can be visualized as a honeycomb-like structure like those depicted in Figures 4.1, 4.2, 4.4, and 4.5. Each such structure individually can be pictured as built out of intersecting sheets (1-forms), but with such details as the tilt and packing of these 1-forms abstracted away. Each such structure individually gives a number when sliced by an element of surface. What counts for the 2-form \boldsymbol{F}, however, is neither the number of tubes of $\boldsymbol{n}^1 \wedge \boldsymbol{\xi}^1$ cut by the surface, nor the number of tubes of $\boldsymbol{n}^2 \wedge \boldsymbol{\xi}^2$ cut by the surface, but only the sum of the two. This sum is what is referred to in the text as the "number of tubes of \boldsymbol{F}" cut by the surface. The contribution of either wedge product individually is not well-defined, for a simple reason: the decomposition of a six-wedge-product object into two wedge products, miraculous though it seems, is actually far from unique (details in exercise 4.2).

In keeping with the need to have two products of 1-forms to represent the general 2-form note that the vanishing of \boldsymbol{dF} ("no magnetic charges") does not automatically imply that $\boldsymbol{d}(\boldsymbol{n}^1 \wedge \boldsymbol{\xi}^1)$ or $\boldsymbol{d}(\boldsymbol{n}^2 \wedge \boldsymbol{\xi}^2)$ separately vanish. Note also that any spacelike slice through the general 2-form \boldsymbol{F} (reduction from four dimensions to three) can always be represented in terms of a honeycomb-like structure ("simple" 2-form in three dimensions; Faraday's picture of magnetic tubes of force).

Despite the abstraction that has gone on in seeing in all generality what a 2-form is, there is no bar to continuing to use the term "honeycomb-like structure" in a broadened sense to describe this object; and that is the practice here and hereafter.

Figure 4.2.
The ***Faraday*** or 2-form **F** of the electromagnetic field is a machine to produce a 1-form (the time-rate of change of momentum **ṗ** of a charged particle) out of a tangent vector (product of charge e of the particle and its 4-velocity **u**). In spacetime the general 2-form is the "superposition" (see Box 4.2) of two structures like that illustrated at the top of this diagram, the tubes of the first being tilted and packed as indicated, the tubes of the second being tilted in another direction and having a different packing density.

§4.3. FORMS ILLUMINATE ELECTROMAGNETISM, AND ELECTROMAGNETISM ILLUMINATES FORMS

All electromagnetism allows itself to be summarized in the language of 2-forms, honeycomb-like "structures" (again in the abstract sense of "structure" of Box 4.2) of tubes filling all spacetime, as well when spacetime is curved as when it is flat. In brief, there are two such structures, one **Faraday = F**, the other **Maxwell = *F**, each dual ("perpendicular," the only place where metric need enter the discussion) to the other, each satisfying an elementary equation:

$$\boldsymbol{dF} = 0 \tag{4.10}$$

("no tubes of **Faraday** ever end") and

$$\boldsymbol{d{*}F} = 4\pi \, {*}\boldsymbol{J} \tag{4.11}$$

("the number of tubes of **Maxwell** that end in an elementary volume is equal to the amount of electric charge in that volume"). To see in more detail how this machinery shows up in action, look in turn at: (1) the definition of a 2-form; (2) the appearance of a given electromagnetic field as **Faraday** and as **Maxwell**; (3) the **Maxwell** structure for a point-charge at rest; (4) the same for a point-charge in motion; (5) the nature of the field of a charge that moves uniformly except during a brief instant of acceleration; (6) the **Faraday** structure for the field of an oscillating dipole; (7) the concept of exterior derivative; (8) Maxwell's equations in the language of forms; and (9) the solution of Maxwell's equations in flat spacetime, using a 1-form **A** from which the Liénard-Wiechert 2-form **F** can be calculated via **F = dA**.

Preview of key points in electromagnetism

A 2-form, as illustrated in Figure 4.1, is a machine to construct a number ("net number of tubes cut") out of any "oriented 2-surface" (2-surface with "sense of circulation" marked on it):

A 2-form as machine for number of tubes cut

$$\begin{pmatrix} \text{number} \\ \text{of tubes} \\ \text{cut} \end{pmatrix} = \int_{\text{surface}} \boldsymbol{F}. \tag{4.12}$$

For example, let the 2-form be the one illustrated in Figure 4.1

Number of tubes cut calculated in one example

$$\boldsymbol{F} = B_x \, \boldsymbol{dy} \wedge \boldsymbol{dz},$$

and let the surface of integration be the portion of the surface of the 2-sphere $x^2 + y^2 + z^2 = a^2$, $t = \text{constant}$, bounded between $\theta = 70°$ and $\theta = 110°$ and between $\varphi = 0°$ and $\varphi = 90°$ ("Atlantic region of the tropics"). Write

$$\begin{aligned} y &= a \sin\theta \sin\varphi, \\ z &= a \cos\theta, \\ \boldsymbol{dy} &= a \, (\cos\theta \sin\varphi \, \boldsymbol{d\theta} + \sin\theta \cos\varphi \, \boldsymbol{d\varphi}), \\ \boldsymbol{dz} &= -a \sin\theta \, \boldsymbol{d\theta}, \\ \boldsymbol{dy} \wedge \boldsymbol{dz} &= a^2 \sin^2\theta \cos\varphi \, \boldsymbol{d\theta} \wedge \boldsymbol{d\varphi}. \end{aligned} \tag{4.13}$$

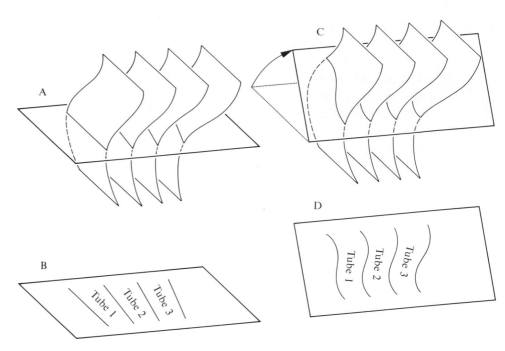

Figure 4.3.

Spacelike slices through **Faraday**, the electromagnetic 2-form, a geometric object, a honeycomb of tubes that pervades all spacetime ("honeycomb" in the abstract sense spelled out more precisely in Box 4.2). The surfaces in the drawing do not look like a 2-form (honeycomb), because the second family of surfaces making up the honeycomb extends in the spatial direction that is suppressed from the drawing. Diagram A shows one spacelike slice through the 2-form (time increases upwards in the diagram). In diagram B, a projection of the 2-form on this spacelike hypersurface gives the Faraday tubes of magnetic force in this three-dimensional geometry (if the suppressed dimension were restored, the tubes would be tubes, not channels between lines). Diagram C shows another spacelike slice (hypersurface of simultaneity for an observer in a different Lorentz frame). Diagram D shows the very different pattern of magnetic tubes in this reference system. The demand that magnetic tubes of force shall not end ($\nabla \cdot B = 0$), repeated over and over for every spacelike slice through **Faraday**, gives everywhere the result $\partial B/\partial t = -\nabla \times E$. Thus (magnetostatics) + (covariance) \longrightarrow (magnetodynamics). Similarly—see Chapters 17 and 21— (geometrostatics) + (covariance) \longrightarrow (geometrodynamics).

The structure $d\theta \wedge d\theta$ looks like a "collapsed egg-crate" (Figure 1.4, upper right) and has zero content, a fact formally evident from the vanishing of $\alpha \wedge \beta = -\beta \wedge \alpha$ when α and β are identical. The result of the integration, assuming constant B_x, is

$$\int_{\text{surface}} F = a^2 B_x \int_{70°}^{110°} \sin^2\theta \, d\theta \int_{0°}^{90°} \cos\varphi \, d\varphi \tag{4.14}$$

It is not so easy to visualize a pure electric field by means of its 2-form F (Figure 4.4, left) as it is to visualize a pure magnetic field by means of its 2-form F (Figures 4.1, 4.2, 4.3). Is there not some way to treat the two fields on more nearly the same footing? Yes, construct the 2-form $*F$ (Figure 4.4, right) that is *dual* ("perpendicular"; Box 4.3; exercise 3.14) to F.

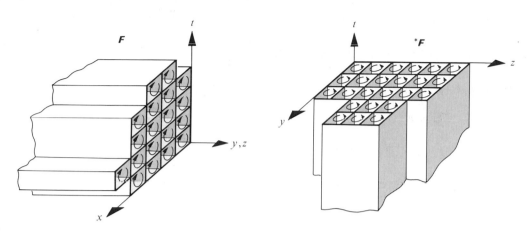

Figure 4.4.
The **Faraday** structure

$$\boldsymbol{F} = \frac{1}{2} F_{\mu\nu} \, \boldsymbol{dx}^{\mu} \wedge \boldsymbol{dx}^{\nu} = \frac{1}{2} F_{01} \, \boldsymbol{dt} \wedge \boldsymbol{dx} + \frac{1}{2} F_{10} \, \boldsymbol{dx} \wedge \boldsymbol{dt} = E_x \, \boldsymbol{dx} \wedge \boldsymbol{dt}$$

associated with an electric field in the x-direction, and the dual ("perpendicular") **Maxwell** honeycomb-like 2-form

$$\boldsymbol{{}^*F} = \frac{1}{2} \, {}^* F_{\mu\nu} \, \boldsymbol{dx}^{\mu} \wedge \boldsymbol{dx}^{\nu} = {}^* F_{23} \, \boldsymbol{dx}^2 \wedge \boldsymbol{dx}^3 = F^{01} \, \boldsymbol{dx}^2 \wedge \boldsymbol{dx}^3 = F_{10} \, \boldsymbol{dx}^2 \wedge \boldsymbol{dx}^3 = E_x \, \boldsymbol{dy} \wedge \boldsymbol{dz}.$$

Represent in geometric form the field of a point-charge of strength e at rest at the origin. Operate in flat space with spherical polar coordinates:

$$\begin{aligned} ds^2 &= -d\tau^2 = g_{\mu\nu} \, dx^{\mu} \, dx^{\nu} \\ &= -dt^2 + dr^2 + r^2 \, d\theta^2 + r^2 \sin^2\theta \, d\varphi^2. \end{aligned} \tag{4.15}$$

The electric field in the r-direction being $E_r = e/r^2$, it follows that the 2-form **F** or **Faraday** is

$$\boldsymbol{F} = \frac{1}{2} F_{\mu\nu} \, \boldsymbol{dx}^{\mu} \wedge \boldsymbol{dx}^{\nu} = -E_r \, \boldsymbol{dt} \wedge \boldsymbol{dr} = -\frac{e}{r^2} \, \boldsymbol{dt} \wedge \boldsymbol{dr}. \tag{4.16}$$

Its dual, according to the prescription in exercise 3.14, is **Maxwell**:

$$\boldsymbol{Maxwell} = \boldsymbol{{}^*F} = e \sin\theta \, \boldsymbol{d\theta} \wedge \boldsymbol{d\varphi}, \tag{4.17}$$

as illustrated in Figure 4.5.

Pattern of tubes in dual structure **Maxwell** for point-charge at rest

Take a tour in the positive sense around a region of the surface of the sphere illustrated in Figure 4.5. The number of tubes of ***F** encompassed in the route will be precisely

$$\binom{\text{number}}{\text{of tubes}} = e \binom{\text{solid}}{\text{angle}}.$$

The whole number of tubes of ***F** emergent over the entire sphere will be $4\pi e$, in conformity with Faraday's picture of tubes of force.

Box 4.3 DUALITY OF 2-FORMS IN SPACETIME

Given a general 2-form (containing six exterior or wedge products)

$$\boldsymbol{F} = E_x \, \boldsymbol{dx} \wedge \boldsymbol{dt} + E_y \, \boldsymbol{dy} \wedge \boldsymbol{dt} + \cdots + B_z \, \boldsymbol{dx} \wedge \boldsymbol{dy},$$

one gets to its dual ("perpendicular") by the prescription

$$*\boldsymbol{F} = -B_x \, \boldsymbol{dx} \wedge \boldsymbol{dt} - \cdots + E_y \, \boldsymbol{dz} \wedge \boldsymbol{dx} + E_z \, \boldsymbol{dx} \wedge \boldsymbol{dy}.$$

Duality Rotations

Note that the dual of the dual is the negative of the original 2-form; thus

$$**\boldsymbol{F} = -E_x \, \boldsymbol{dx} \wedge \boldsymbol{dt} - \cdots - B_z \, \boldsymbol{dx} \wedge \boldsymbol{dy} = -\boldsymbol{F}.$$

In this sense $*$ has the same property as the imaginary number i: $** = ii = -1$. Thus one can write

$$e^{*\alpha} = \cos \alpha + *\sin \alpha.$$

This operation, applied to \boldsymbol{F}, carries attention from the generic 2-form in its simplest representation (see exercise 4.1)

$$\boldsymbol{F} = E_x \, \boldsymbol{dx} \wedge \boldsymbol{dt} + B_x \, \boldsymbol{dy} \wedge \boldsymbol{dz}$$

to another "duality rotated electromagnetic field"

$$e^{*\alpha}\boldsymbol{F} = (E_x \cos \alpha - B_x \sin \alpha) \, \boldsymbol{dx} \wedge \boldsymbol{dt} + (B_x \cos \alpha + E_x \sin \alpha) \, \boldsymbol{dy} \wedge \boldsymbol{dz}.$$

If the original field satisfied Maxwell's empty-space field equations, so does the new field. With suitable choice of the "complexion" α, one can annul one of the two wedge products at any chosen point in spacetime and have for the other

$$(B_x^{\ 2} + E_x^{\ 2})^{1/2} \, \boldsymbol{dy} \wedge \boldsymbol{dz}.$$

Field of a point-charge in motion

How can one determine the structure of tubes associated with a charged particle moving at a uniform velocity? First express $*\boldsymbol{F}$ in rectangular coordinates moving with the particle (barred coordinates in this comoving "rocket" frame of reference; unbarred coordinates will be used later for a laboratory frame of reference). The relevant steps can be listed:

(a)

$$*\boldsymbol{F} = e \sin \overline{\theta} \, \boldsymbol{d\overline{\theta}} \wedge \boldsymbol{d\overline{\varphi}} = -e(\boldsymbol{d} \cos \overline{\theta}) \wedge \boldsymbol{d\overline{\varphi}};$$

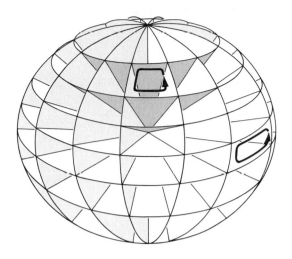

Figure 4.5.
The field of 2-forms **Maxwell** = ***F** = $e \sin \theta \, d\theta \wedge d\phi$ that describes the electromagnetic field of a charge e at rest at the origin. This picture is actually the intersection of ***F** with a 3-surface of constant time t; i.e., the time direction is suppressed from the picture.

(b)

$$\bar{\varphi} = \arctan \frac{\bar{y}}{\bar{x}}; \qquad d\bar{\varphi} = \frac{\bar{x} \, d\bar{y} - \bar{y} \, d\bar{x}}{\bar{x}^2 + \bar{y}^2};$$

(c)

$$\cos \bar{\theta} = \frac{\bar{z}}{\bar{r}}; \qquad -d(\cos \bar{\theta}) = \frac{-d\bar{z}}{\bar{r}} + \frac{\bar{z}}{\bar{r}^3}(\bar{x} \, d\bar{x} + \bar{y} \, d\bar{y} + \bar{z} \, d\bar{z});$$

(d) combine to find

$$*F = (e/\bar{r}^3)(\bar{x} \, d\bar{y} \wedge d\bar{z} + \bar{y} \, d\bar{z} \wedge d\bar{x} + \bar{z} \, d\bar{x} \wedge d\bar{y}) \qquad (4.18)$$

(electromagnetic field of point charge in a comoving Cartesian system; spherically symmetric). Now transform to laboratory coordinates:

velocity parameter α

velocity $\beta = \tanh \alpha$

$$\frac{1}{\sqrt{1 - \beta^2}} = \cosh \alpha, \qquad \frac{\beta}{\sqrt{1 - \beta^2}} = \sinh \alpha$$

(a)
$$\begin{cases} \bar{t} = t \cosh \alpha - x \sinh \alpha, \\ \bar{x} = -t \sinh \alpha + x \cosh \alpha, \\ \bar{y} = y \qquad \bar{z} = z; \end{cases}$$

(b)
$$\bar{r} = [(x \cosh \alpha - t \sinh \alpha)^2 + y^2 + z^2]^{1/2};$$

(c)
$$\begin{aligned} *F = (e/\bar{r}^3)[&(x \cosh \alpha - t \sinh \alpha) \, dy \wedge dz + y \, dz \wedge \\ &(\cosh \alpha \, dx - \sinh \alpha \, dt) + z(\cosh \alpha \, dx - \sinh \alpha \, dt) \wedge dy]; \quad (4.19) \end{aligned}$$

(d) compare with the general dual 2-form,

$$*F = E_x \, dy \wedge dz + E_y \, dz \wedge dx + E_z \, dx \wedge dy$$
$$+ \, B_x \, dt \wedge dx + B_y \, dt \wedge dy + B_z \, dt \wedge dz;$$

and get the desired individual field components

(e)
$$\begin{cases} E_x = (e/\overline{r}^3)(x \cosh \alpha - t \sinh \alpha), & B_x = 0, \\ E_y = (e/\overline{r}^3)y \cosh \alpha, & B_y = -(e/\overline{r}^3)z \sinh \alpha, \\ E_z = (e/\overline{r}^3)z \cosh \alpha, & B_z = (e/\overline{r}^3)y \sinh \alpha. \end{cases} \qquad (4.20)$$

One can verify that the invariants

$$B^2 - E^2 = \frac{1}{2} F_{\alpha\beta} F^{\alpha\beta}, \qquad (4.21)$$

$$E \cdot B = \frac{1}{4} F_{\alpha\beta} {}^*F^{\alpha\beta} \qquad (4.22)$$

have the same value in the laboratory frame as in the rocket frame, as required. Note that the honeycomb structure of the differential form is not changed when one goes from the rocket frame to the laboratory frame. What changes is only the mathematical formula that describes it.

§4.4. RADIATION FIELDS

The **Maxwell** structure of tubes associated with a charge in uniform motion is more remarkable than it may seem at first sight, and not only because of the Lorentz contraction of the tubes in the direction of motion. The tubes arbitrarily far away move on in military step with the charge on which they center, despite the fact that there is no time for information "emitted" from the charge "right now" to get to the faraway tube "right now." The structure of the faraway tubes "right now" must therefore derive from the charge at an earlier moment on its uniform-motion, straight-line trajectory. This circumstance shows up nowhere more clearly than in

How an acceleration causes radiation

what happens to the field in consequence of a sudden change, in a short time $\Delta\tau$, from one uniform velocity to another uniform velocity (Figure 4.6). The tubes have the standard patterns for the two states of motion, one pattern within a sphere of radius r, the other outside that sphere, where r is equal to the lapse of time ("cm of light-travel time") since the acceleration took place. The necessity for the two patterns to fit together in the intervening zone, of thickness $\Delta r = \Delta\tau$, forces the field there to be multiplied up by a "stretching factor," proportional to r. This factor is responsible for the well-known fact that radiative forces fall off inversely only as the first power of the distance (Figure 4.6).

When the charge continuously changes its state of motion, the structure of the electromagnetic field, though based on the same simple principles as those illustrated in Figure 4.6, nevertheless looks more complex. The following is the **Faraday** 2-form

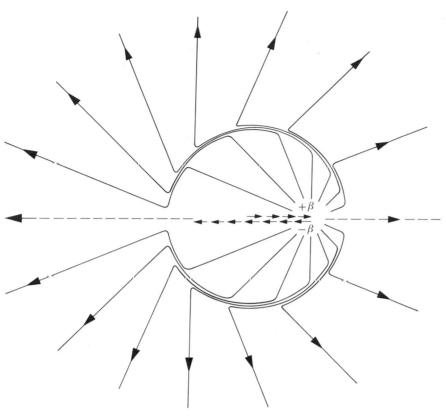

Figure 4.6.

Mechanism of radiation. J. J. Thomson's way to understand why the strength of an electromagnetic wave falls only as the inverse first power of distance r and why the amplitude of the wave varies (for low velocities) as $\sin\theta$ (maximum in the plane perpendicular to the line of acceleration). The charge was moving to the left at uniform velocity. Far away from it, the lines of force continue to move as if this uniform velocity were going to continue forever (Coulomb field of point-charge in slow motion). However, closer up the field is that of a point-change moving to the right with uniform velocity ($1/r^2$ dependence of strength upon distance). The change from the one field pattern to another is confined to a shell of thickness $\Delta\tau$ located at a distance r from the point of acceleration (amplification of field by "stretching factor" $r\sin\theta\,\Delta\beta/\Delta\tau$; see text). We thank C. Teitelboim for the construction of this diagram.

for the field of an electric dipole of magnitude p_1 oscillating up and down parallel to the z-axis:

Field of an oscillating dipole

$$\boldsymbol{F} = E_x\,\boldsymbol{dx}\wedge\boldsymbol{dt} + \cdots + B_x\,\boldsymbol{dy}\wedge\boldsymbol{dz} + \cdots = \text{real part of }\{p_1 e^{i\omega r - i\omega t}$$

$$\underbrace{\left[2\cos\theta\left(\frac{1}{r^3}-\frac{i\omega}{r^2}\right)\boldsymbol{dr}\wedge\boldsymbol{dt}\right.}_{\text{gives }E_r} + \underbrace{\sin\theta\left(\frac{1}{r^3}-\frac{i\omega}{r^2}-\frac{\omega^2}{r}\right)r\,\boldsymbol{d\theta}\wedge\boldsymbol{dt}}_{\text{gives }E_\theta}$$

$$\underbrace{+ \sin\theta\left(\frac{-i\omega}{r^2}-\frac{\omega^2}{r}\right)\boldsymbol{dr}\wedge r\,\boldsymbol{d\theta}]\}}_{\text{gives }B_\phi} \qquad\qquad (4.23)$$

and the dual 2-form **Maxwell** = *__*F__* is

$$*\textbf{\textit{F}} = -B_x \, \textbf{\textit{dx}} \wedge \textbf{\textit{dt}} - \cdots + E_x \, \textbf{\textit{dy}} \wedge \textbf{\textit{dz}} + \cdots = \text{real part of } \{ p_1 e^{i\omega r - i\omega t}$$

$$\underbrace{[\sin\theta \left(\frac{-i\omega}{r^2} - \frac{\omega^2}{r} \right) \textbf{\textit{dt}} \wedge r \sin\theta \, \textbf{\textit{d}}\phi}_{\text{gives } B_\phi}$$

$$\underbrace{+ 2\cos\theta \left(\frac{1}{r^3} - \frac{i\omega}{r^2} \right) r \, \textbf{\textit{d}}\theta \wedge r \sin\theta \, \textbf{\textit{d}}\phi}_{\text{gives } E_r}$$

$$\underbrace{+ \sin\theta \left(\frac{1}{r^3} - \frac{i\omega}{r^2} - \frac{\omega^2}{r} \right) r \sin\theta \, \textbf{\textit{d}}\phi \wedge \textbf{\textit{dr}}]\}.}_{\text{gives } E_\theta} \tag{4.24}$$

§4.5. MAXWELL'S EQUATIONS

The general 2-form **_F_** is written as a superposition of wedge products with a factor $\frac{1}{2}$,

$$\textbf{\textit{F}} = \frac{1}{2} F_{\mu\nu} \, \textbf{\textit{dx}}^\mu \wedge \textbf{\textit{dx}}^\nu, \tag{4.25}$$

because the typical term appears twice, once as $F_{xy} \, \textbf{\textit{dx}} \wedge \textbf{\textit{dy}}$ and the second time as $F_{yx} \, \textbf{\textit{dy}} \wedge \textbf{\textit{dx}}$, with $F_{yx} = -F_{xy}$ and $\textbf{\textit{dy}} \wedge \textbf{\textit{dx}} = -\textbf{\textit{dx}} \wedge \textbf{\textit{dy}}$.

If differentiation ("taking the gradient"; the operator **_d_**) produced out of a scalar a 1-form, it is also true that differentiation (again the operator **_d_**, but now generally known under Cartan's name of "exterior differentiation") produces a 2-form out of the general 1-form; and applied to a 2-form produces a 3-form; and applied to a 3-form produces a 4-form, the form of the highest order that spacetime will accommodate. Write the general *f*-form as

$$\boldsymbol{\phi} = \frac{1}{f!} \phi_{\alpha_1 \alpha_2 \ldots \alpha_f} \, \textbf{\textit{dx}}^{\alpha_1} \wedge \textbf{\textit{dx}}^{\alpha_2} \wedge \cdots \wedge \textbf{\textit{dx}}^{\alpha_f} \tag{4.26}$$

where the coefficient $\phi_{\alpha_1 \alpha_2 \ldots \alpha_f}$, like the wedge product that follows it, is antisymmetric under interchange of any two indices. Then the exterior derivative of **_ϕ_** is

Taking exterior derivative

$$\textbf{\textit{d}}\boldsymbol{\phi} \equiv \frac{1}{f!} \frac{\partial \phi_{\alpha_1 \alpha_2 \ldots \alpha_f}}{\partial x^{\alpha_0}} \, \textbf{\textit{dx}}^{\alpha_0} \wedge \textbf{\textit{dx}}^{\alpha_1} \wedge \textbf{\textit{dx}}^{\alpha_2} \wedge \cdots \wedge \textbf{\textit{dx}}^{\alpha_f}. \tag{4.27}$$

Take the exterior derivative of **_Faraday_** according to this rule and find that it vanishes, not only for the special case of the dipole oscillator, but also for a general electromagnetic field. Thus, in the coordinates appropriate for a local Lorentz frame, one has

$$\boldsymbol{dF} = \boldsymbol{d}(E_x\,\boldsymbol{dx} \wedge \boldsymbol{dt} + \cdots + B_x\,\boldsymbol{dy} \wedge \boldsymbol{dz} + \cdots)$$

$$= \left(\frac{\partial E_x}{\partial t}\,\boldsymbol{dt} + \frac{\partial E_x}{\partial x}\,\boldsymbol{dx} + \frac{\partial E_x}{\partial y}\,\boldsymbol{dy} + \frac{\partial E_x}{\partial z}\,\boldsymbol{dz}\right) \wedge \boldsymbol{dx} \wedge \boldsymbol{dt}$$

$$+ \cdots \text{(5 more such sets of 4 terms each)}\ldots. \tag{4.28}$$

Note that such a term as $\boldsymbol{dy} \wedge \boldsymbol{dy} \wedge \boldsymbol{dz}$ is automatically zero ("collapse of egg-crate cell when stamped on"). Collect the terms that do not vanish and find

$$\boldsymbol{dF} = \left(\frac{\partial B_x}{\partial x} + \frac{\partial B_y}{\partial y} + \frac{\partial B_z}{\partial z}\right)\boldsymbol{dx} \wedge \boldsymbol{dy} \wedge \boldsymbol{dz}$$

$$+ \left(\frac{\partial B_x}{\partial t} + \frac{\partial E_z}{\partial y} - \frac{\partial E_y}{\partial z}\right)\boldsymbol{dt} \wedge \boldsymbol{dy} \wedge \boldsymbol{dz}$$

$$+ \left(\frac{\partial B_y}{\partial t} + \frac{\partial E_x}{\partial z} - \frac{\partial E_z}{\partial x}\right)\boldsymbol{dt} \wedge \boldsymbol{dz} \wedge \boldsymbol{dx}$$

$$+ \left(\frac{\partial B_z}{\partial t} + \frac{\partial E_y}{\partial x} - \frac{\partial E_x}{\partial y}\right)\boldsymbol{dt} \wedge \boldsymbol{dx} \wedge \boldsymbol{dy}. \tag{4.29}$$

Each term in this expression is familiar from Maxwell's equations

$$\text{div } \boldsymbol{B} = \nabla \cdot \boldsymbol{B} = 0$$

and

$$\text{curl } \boldsymbol{E} = \nabla \times \boldsymbol{E} = -\dot{\boldsymbol{B}}.$$

Each vanishes, and with their vanishing *Faraday* itself is seen to have zero exterior derivative:

$$\boldsymbol{dF} = 0. \tag{4.30}$$

In other words, "*Faraday* is a closed 2-form"; "the tubes of \boldsymbol{F} nowhere come to an end."

Faraday structure: tubes nowhere end

A similar calculation gives for the exterior derivative of the dual 2-form *Maxwell* the result

$$\boldsymbol{d}*\boldsymbol{F} = \boldsymbol{d}(-B_x\,\boldsymbol{dx} \wedge \boldsymbol{dt} - \cdots + E_x\,\boldsymbol{dy} \wedge \boldsymbol{dz} + \cdots)$$

$$= \left(\frac{\partial E_x}{\partial x} + \frac{\partial E_y}{\partial y} + \frac{\partial E_z}{\partial z}\right)\boldsymbol{dx} \wedge \boldsymbol{dy} \wedge \boldsymbol{dz}$$

$$+ \left(\frac{\partial E_x}{\partial t} - \frac{\partial B_z}{\partial y} + \frac{\partial B_y}{\partial z}\right)\boldsymbol{dt} \wedge \boldsymbol{dy} \wedge \boldsymbol{dz}$$

$$+ \cdots$$

$$= 4\pi(\rho\,\boldsymbol{dx} \wedge \boldsymbol{dy} \wedge \boldsymbol{dz}$$
$$- J_x\,\boldsymbol{dt} \wedge \boldsymbol{dy} \wedge \boldsymbol{dz}$$
$$- J_y\,\boldsymbol{dt} \wedge \boldsymbol{dz} \wedge \boldsymbol{dx}$$
$$- J_z\,\boldsymbol{dt} \wedge \boldsymbol{dx} \wedge \boldsymbol{dy}) = 4\pi\,{}^*\boldsymbol{J};$$

$$\boldsymbol{d}*\boldsymbol{F} = 4\pi\,{}^*\boldsymbol{J}. \tag{4.31}$$

Maxwell structure: density of tube endings given by charge-current 3-form

In empty space this exterior derivative, too, vanishes; there **Maxwell** is a closed 2-form; the tubes of ***F***, like the tubes of ***F***, nowhere come to an end.

In a region where charge is present, the situation changes. Tubes of **Maxwell** take their origin in such a region. The density of endings is described by the 3-form ***J* = *charge***, a "collection of eggcrate cells" collected along bundles of world lines.

The two equations

$$\boldsymbol{dF} = 0$$

and

$$\boldsymbol{d*F} = 4\pi \, \boldsymbol{*J}$$

summarize the entire content of Maxwell's equations in geometric language. The forms ***F* = *Faraday***, and ***F* = *Maxwell***, can be described in any coordinates one pleases—or in a language (honeycomb and egg-crate structures) free of any reference whatsoever to coordinates. Remarkably, neither equation makes any reference whatsoever to *metric*. As Hermann Weyl was one of the most emphatic in stressing (see also Chapters 8 and 9), the concepts of form and exterior derivative are metric-free. Metric made an appearance only in one place, in the concept of duality ("perpendicularity") that carried attention from ***F*** to the dual structure ***F***.

Duality: the only place in electromagnetism where metric must enter

§4.6. EXTERIOR DERIVATIVE AND CLOSED FORMS

The words "honeycomb" and "egg crate" may have given some feeling for the geometry that goes with electrodynamics. Now to spell out these concepts more clearly and illustrate in geometric terms, with electrodynamics as subject matter, what it means to speak of "exterior differentiation." Marching around a boundary, yes; but how and why and with what consequences? It is helpful to return to functions and 1-forms, and see them and the 2-forms **Faraday** and **Maxwell** and the 3-form ***charge*** as part of an ordered progression (see Box 4.4). Two-forms are seen in this box to be of two kinds: (1) a special 2-form, known as a "closed" 2-form, which has the property that as many tubes enter a closed 2-surface as emerge from it (exterior derivative of 2-form zero; no 3-form derivable from it other than the trivial *zero* 3-form!); and (2) a general 2-form, which sends across a closed 2-surface a non-zero net number of tubes, and therefore permits one to define a nontrivial 3-form ("exterior derivative of the 2-form"), which has precisely as many egg-crate cells in any closed 2-surface as the net number of tubes of the 2-form emerging from that same closed 2-surface (generalization of Faraday's concept of tubes of force to the world of spacetime, curved as well as flat).

Closed 2-form contrasted with general 2-form

(*continued on page 120*)

Box 4.4 THE PROGRESSION OF FORMS AND EXTERIOR DERIVATIVES

0-Form or Scalar, f

An example in the context of 3-space and Newtonian physics is temperature, $T(x, y, z)$, and in the context of spacetime, a scalar potential, $\phi(t, x, y, z)$.

From Scalar to *1*-Form

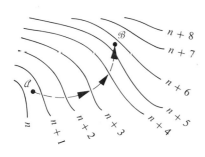

Take the gradient or "exterior derivative" of a scalar f to obtain a special 1-form, $\boldsymbol{\gamma} = \boldsymbol{d}f$. Comments: (a) Any additive constant included in f is erased in the process of differentiation; the quantity n in the diagram at the left is unknown and irrelevant. (b) The 1-form $\boldsymbol{\gamma}$ is special in the sense that surfaces in one region "mesh" with surfaces in a neighboring region ("closed 1-form"). (c) Line integral $\int_{\mathscr{A}}^{\mathscr{B}} \boldsymbol{d}f$ is independent of path for any class of paths equivalent to one another under continuous deformation. (d) The 1-form is a machine to produce a number ("bongs of bell" as each successive integral surface is crossed) out of a displacement (approximation to concept of a tangent vector).

General *1*-Form $\boldsymbol{\beta} = \beta_\alpha \, \boldsymbol{d}x^\alpha$

This is a pattern of surfaces, as illustrated in the diagram at the right; i.e., a machine to produce a number ("bongs of bell"; $\langle \boldsymbol{\beta}, \boldsymbol{u} \rangle$) out of a vector. A 1-form has a reality and position in space independent of all choice of coordinates. Surfaces do not ordinarily mesh. Integral $\int \boldsymbol{\beta}$ around indicated closed loop does not give zero ("more bongs than antibongs").

From *1*-Form to 2-Form $\boldsymbol{\xi} = \boldsymbol{d}\boldsymbol{\beta} = \dfrac{\partial \beta_\alpha}{\partial x^\mu} \, \boldsymbol{d}x^\mu \wedge \boldsymbol{d}x^\alpha$

$\boldsymbol{\xi}$ is a pattern of honeycomb-like cells, with a direction of circulation marked on each, so stationed

Box 4.4 (*continued*)

that the number of cells encompassed in the dotted closed path is identical to the net contribution (excess of bongs over antibongs) for the same path in the diagram of $\boldsymbol{\beta}$ above. The "exterior derivative" is *defined* so this shall be so; the generalized Stokes theorem codifies it. The word "exterior" comes from the fact that the path goes around the periphery of the region under analysis. Thus the 2-form is a machine to get a number (number of tubes, $\langle \boldsymbol{\xi}, \boldsymbol{u} \wedge \boldsymbol{v} \rangle$) out of a bit of surface ($\boldsymbol{u} \wedge \boldsymbol{v}$) that has a sense of circulation indicated upon it. The 2-form thus defined is special in this sense: a rubber sheet "supported around its edges" by the dotted curve or any other closed curve is crossed by the same number of tubes when: (a) it bulges up in the middle; (b) it is pushed down in the middle; (c) it experiences any other continuous deformation. The **Faraday** or 2-form **F** of electromagnetism, always expressible as $\boldsymbol{F} = \boldsymbol{dA}$ (\boldsymbol{A} = 4-potential, a 1-form), also has always this special property ("conservation of tubes").

0-Form to *1*-Form to *2*-Form? No!

Go from scalar f to 1-form $\boldsymbol{\gamma} = \boldsymbol{df}$. The next step to a 2-form $\boldsymbol{\alpha}$ is vacuous. The net contribution of the line integral $\int \boldsymbol{\gamma}$ around the dotted closed path is automatically zero. To reproduce that zero result requires a zero 2-form. Thus $\boldsymbol{\alpha} = \boldsymbol{d\gamma} = \boldsymbol{dd}f$ has to be the zero 2-form. This result is a special instance of the general result $\boldsymbol{dd} = 0$.

General *2*-Form $\sigma = \dfrac{1}{2} \sigma_{\alpha\beta} \, \boldsymbol{dx}^\alpha \wedge \boldsymbol{dx}^\beta$, with $\sigma_{\alpha\beta} = -\sigma_{\beta\alpha}$

Again, this is a honeycomb-like structure, and again a machine to get a number (number of tubes, $\langle \boldsymbol{\sigma}, \boldsymbol{u} \wedge \boldsymbol{v} \rangle$) out of a surface ($\boldsymbol{u} \wedge \boldsymbol{v}$) that has a sense of circulation indicated on it. It is general in the sense that the honeycomb structures in one region do not ordinarily mesh with those

in a neighboring region. In consequence, a closed
2-surface, such as the box-like surface indicated
by dotted lines at the right, is ordinarily crossed
by a non-zero net number of tubes. The net num-
ber of tubes emerging from such a closed surface
is, however, exactly zero when the 2-form is the
exterior derivative of a 1-form.

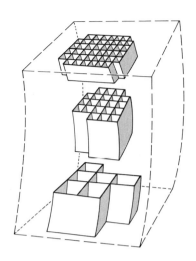

From *2*-Form to *3*-Form $\boldsymbol{\mu} = \boldsymbol{d\sigma} = \dfrac{\partial \sigma_{|\alpha\beta|}}{\partial x^{\gamma}} \, \boldsymbol{dx}^{\gamma} \wedge \boldsymbol{dx}^{\alpha} \wedge \boldsymbol{dx}^{\beta},$

where $\boldsymbol{dx}^{\gamma} \wedge \boldsymbol{dx}^{\alpha} \wedge \boldsymbol{dx}^{\beta} \equiv 3! \, \boldsymbol{dx}^{[\gamma} \otimes \boldsymbol{dx}^{\alpha} \otimes \boldsymbol{dx}^{\beta]}$

This egg-crate type of structure is a machine to
get a number (number of cells $\langle \boldsymbol{\mu}, \boldsymbol{u} \wedge \boldsymbol{v} \wedge \boldsymbol{w} \rangle$)
from a volume (volume $\boldsymbol{u} \wedge \boldsymbol{v} \wedge \boldsymbol{w}$ within which
one counts the cells). A more complete diagram
would provide each cell and the volume of inte-
gration itself with an indicator of orientation
(analogous to the arrow of circulation shown for
cells of the 2-form). The contribution of a given
cell to the count of cells is $+1$ or -1, according
as the orientation indicators have same sense or
opposite sense. The number of egg-crate cells of
$\boldsymbol{\mu} = \boldsymbol{d\sigma}$ in any given volume (such as the volume
indicated by the dotted lines) is tailored to give
precisely the same number as the net number of
tubes of the 2-form $\boldsymbol{\sigma}$ (diagram above) that emerge
from that volume (generalized Stokes theorem).
For electromagnetism, the exterior derivative of
Faraday or 2-form \boldsymbol{F} gives a null 3-form, but the
exterior derivative of ***Maxwell*** or 2-form $\boldsymbol{*F}$ gives
4π times the 3-form $\boldsymbol{*J}$ of charge:

$$\boldsymbol{*J} = \rho \, \boldsymbol{dx} \wedge \boldsymbol{dy} \wedge \boldsymbol{dz} - J_{x} \, \boldsymbol{dt} \wedge \boldsymbol{dy} \wedge \boldsymbol{dz}$$
$$- J_{y} \, \boldsymbol{dt} \wedge \boldsymbol{dz} \wedge \boldsymbol{dx} - J_{z} \, \boldsymbol{dt} \wedge \boldsymbol{dx} \wedge \boldsymbol{dy}.$$

Box 4.4 (*continued*)

From *1*-Form to *2*-Form to *3*-Form? No!

Starting with a 1-form (electromagnetic 4-potential), construct its exterior derivative, the 2-form $\boldsymbol{F} = \boldsymbol{dA}$ (*Faraday*). The tubes in this honeycomb-like structure never end. So the number of tube endings in any elementary volume, and with it the 3-form $\boldsymbol{dF} = \boldsymbol{ddA}$, is automatically zero. This is another example of the general result that $\boldsymbol{dd} = 0$.

From *2*-Form to *3*-Form to *4*-Form? No!

Starting with 2-form $\boldsymbol{{}^*F}$ (*Maxwell*), construct its exterior derivative, the 3-form $4\pi \, \boldsymbol{{}^*J}$. The cells in this egg-crate type of structure extend in a fourth dimension ("hypertube"). The number of these hypertubes that end in any elementary 4-volume, and with it the 4-form

$$\boldsymbol{d}(4\pi \, \boldsymbol{{}^*J}) = \boldsymbol{dd\,{}^*F},$$

is automatically zero, still another example of the general result that $\boldsymbol{dd} = 0$. This result says that

$$\boldsymbol{d\,{}^*J} = \left(\frac{\partial \rho}{\partial t} + \frac{\partial J_x}{\partial x} + \frac{\partial J_y}{\partial y} + \frac{\partial J_z}{\partial z} \right) \boldsymbol{dt} \wedge \boldsymbol{dx} \wedge \boldsymbol{dy} \wedge \boldsymbol{dz} = 0$$

("law of conservation of charge"). Note:

$$\boldsymbol{dx}^\alpha \wedge \boldsymbol{dx}^\beta \wedge \boldsymbol{dx}^\gamma \wedge \boldsymbol{dx}^\delta \equiv 4! \, \boldsymbol{dx}^{[\alpha} \otimes \boldsymbol{dx}^\beta \otimes \boldsymbol{dx}^\gamma \otimes \boldsymbol{dx}^{\delta]}.$$

This implies $\boldsymbol{dt} \wedge \boldsymbol{dx} \wedge \boldsymbol{dy} \wedge \boldsymbol{dz} = \boldsymbol{\varepsilon}$.

From *3*-Form to *4*-Form $\boldsymbol{\tau} = \boldsymbol{dv} = \dfrac{\partial \nu_{|\alpha\beta\gamma|}}{\partial x^\delta} \boldsymbol{dx}^\delta \wedge \boldsymbol{dx}^\alpha \wedge \boldsymbol{dx}^\beta \wedge \boldsymbol{dx}^\gamma$

This four-dimensional "super-egg-crate" type structure is a machine to get a number (number of cells, $\langle \boldsymbol{\tau}, \boldsymbol{n} \wedge \boldsymbol{u} \wedge \boldsymbol{v} \wedge \boldsymbol{w} \rangle$) from a 4-volume $\boldsymbol{n} \wedge \boldsymbol{u} \wedge \boldsymbol{v} \wedge \boldsymbol{w}$.

From *4*-Form to *5*-Form? No!

Spacetime, being four-dimensional, cannot accommodate five-dimensional egg-crate structures. At least two of the dx^μ's in

$$dx^\alpha \wedge dx^\beta \wedge dx^\gamma \wedge dx^\delta \wedge dx^\varepsilon$$

must be the same; so, by antisymmetry of "\wedge," this "basis 5-form" must vanish.

Results of Exterior Differentiation, Summarized

0-form	f				
1-form	df	A			
2-form	$ddf \equiv 0$	$F = dA$	$*F$		
3-form		$dF = ddA \equiv 0$	$4\pi\,*J = d\,*F$	ν	
4-form			$d(4\pi\,*J) = dd\,*F \equiv 0$	$\tau = d\nu$	μ
5-form?	No!			$d\tau \equiv 0$	$d\mu \equiv 0$

New Forms from Old by Taking Dual (see exercise 3.14)

Dual of scalar f is 4-form: $*f = f\,dx^0 \wedge dx^1 \wedge dx^2 \wedge dx^3 = f\varepsilon$.

Dual of 1-form J is 3-form: $*J = J^0\,dx^1 \wedge dx^2 \wedge dx^3 - J^1\,dx^2 \wedge dx^3 \wedge dx^0$
$\qquad\qquad + J^2\,dx^3 \wedge dx^0 \wedge dx^1 - J^3\,dx^0 \wedge dx^1 \wedge dx^2.$

Dual of 2-form F is 2-form: $*F = F^{|\alpha\beta|}\varepsilon_{\alpha\beta|\mu\nu|}\,dx^\mu \wedge dx^\nu$, where
$\qquad\qquad F^{\alpha\beta} = \eta^{\alpha\lambda}\eta^{\beta\delta}F_{\lambda\delta}.$

Dual of 3-form K is 1-form: $*K = K^{012}\,dx^3 - K^{123}\,dx^0 + K^{230}\,dx^1 - K^{301}\,dx^2$,
$\qquad\qquad$ where $K^{\alpha\beta\gamma} = \eta^{\alpha\mu}\eta^{\beta\nu}\eta^{\gamma\lambda}K_{\mu\nu\lambda}.$

Dual of 4-form L is a scalar: $L = L_{0123}\,dx^0 \wedge dx^1 \wedge dx^2 \wedge dx^3$;
$\qquad\qquad *L = L^{0123} = -L_{0123}.$

Note 1: This concept of duality between *one form and another* is to be distinguished from the concept of duality between the *vector basis* e_α and the *1-form basis* w^α of a given frame. The two types of duality have nothing whatsoever to do with each other!

Box 4.4 (*continued*)

Note 2: In spacetime, the operation of taking the dual, applied twice, leads back to the original form for forms of odd order, and to the negative thereof for forms of even order. In Euclidean 3-space the operation reproduces the original form, regardless of its order.

Duality Plus Exterior Differentiation

Start with scalar ϕ. Its gradient $\boldsymbol{d}\phi$ is a 1-form. Take its dual, to get the 3-form $*\boldsymbol{d}\phi$. Take its exterior derivative, to get the 4-form $\boldsymbol{d}*\boldsymbol{d}\phi$. Take its dual, to get the scalar $\Box\phi \equiv -*\boldsymbol{d}*\boldsymbol{d}\phi$. Verify by index manipulations that \Box as defined here is the wave operator; i.e., in any Lorentz frame, $\Box\phi = \phi_{,\alpha}{}^{,\alpha} = -(\partial^2\phi/\partial t^2) + \boldsymbol{\nabla}^2\phi$.

 Start with 1-form \boldsymbol{A}. Get 2-form $\boldsymbol{F} = \boldsymbol{dA}$. Take its dual $*\boldsymbol{F} = *\boldsymbol{dA}$, also a 2-form. Take its exterior derivative, obtaining the 3-form $\boldsymbol{d}*\boldsymbol{F}$ (has value $4\pi *\boldsymbol{J}$ in electromagnetism). Take its dual, obtaining the 1-form $*\boldsymbol{d}*\boldsymbol{F} = *\boldsymbol{d}*\boldsymbol{dA} = 4\pi\boldsymbol{J}$ ("Wave equation for electromagnetic 4-potential"). Reduce in index notation to

$$F_{\mu\nu}{}^{,\nu} = A_{\nu,\mu}{}^{,\nu} - A_{\mu,\nu}{}^{,\nu} = 4\pi J_\mu.$$

[More in Flanders (1963) or Misner and Wheeler (1957); see also exercise 3.17.]

§4.7. DISTANT ACTION FROM LOCAL LAW

Differential forms are a powerful tool in electromagnetic theory, but full power requires mastery of other tools as well. Action-at-a-distance techniques ("Green's functions," "propagators") are of special importance. Moreover, the passage from Maxwell field equations to electromagnetic action at a distance provides a preview of how Einstein's local equations will reproduce (approximately) Newton's $1/r^2$ law.

 In flat spacetime and in a Lorentz coordinate system, express the coordinates of particle A as a function of its proper time α, thus:

$$a^\mu = a^\mu(\alpha), \qquad \frac{da^\mu}{d\alpha} = \dot{a}^\mu(\alpha), \qquad \frac{d^2a^\mu}{d\alpha^2} = \ddot{a}^\mu(\alpha). \tag{4.32}$$

Dirac found it helpful to express the distribution of charge and current for a particle of charge e following such a motion as a superposition of charges that momentarily

flash into existence and then flash out of existence. Any such flash has a localization in space and time that can be written as the product of four Dirac delta functions [see, for example, Schwartz (1950–1951), Lighthill (1958)]:

$$\delta^4(x^\mu - a^\mu) = \delta[x^0 - a^0(\alpha)]\,\delta[x^1 - a^1(\alpha)]\,\delta[x^2 - a^2(\alpha)]\,\delta[x^3 - a^3(\alpha)]. \quad (4.33)$$

World line of charge regarded as succession of flash-on, flash-off charges

Here any single Dirac function $\delta(x)$ ("symbolic function"; "distribution"; "limit of a Gauss error function" as width is made indefinitely narrow and peak indefinitely high, with integrated value always unity) both (1) vanishes for $x \neq 0$, and (2) has the integral $\int_{-\infty}^{+\infty} \delta(x)\,dx = 1$. Described in these terms, the density-current vector for the particle has the value ("superposition of flashes")

$$J^\mu = e \int \delta^4[x^\nu - a^\nu(\alpha)]\dot{a}^\mu(\alpha)\,d\alpha. \quad (4.34)$$

The density-current (4.34) drives the electromagnetic field, \mathbf{F}. Write $\mathbf{F} = d\mathbf{A}$ to satisfy automatically half of Maxwell's equations ($d\mathbf{F} = dd\mathbf{A} \equiv 0$):

$$F_{\mu\alpha} = \frac{\partial A_\alpha}{\partial x^\mu} - \frac{\partial A_\mu}{\partial x^\alpha}. \quad (4.35)$$

In flat space, the remainder of Maxwell's equations ($d*\mathbf{F} = 4\pi\,*\mathbf{J}$) become

$$\frac{\partial F_\mu{}^\nu}{\partial x^\nu} = 4\pi J_\mu$$

or

$$\frac{\partial}{\partial x^\mu}\frac{\partial A^\nu}{\partial x^\nu} - \eta^{\nu\alpha}\frac{\partial^2 A_\mu}{\partial x^\nu\,\partial x^\alpha} = 4\pi J_\mu. \quad (4.36)$$

Make use of the freedom that exists in the choice of 4-potentials A^ν to demand

$$\frac{\partial A^\nu}{\partial x^\nu} = 0 \quad (4.37)$$

(Lorentz gauge condition; see exercise 3.17). Thus get

$$\Box A_\mu = -4\pi J_\mu. \quad (4.38)$$

The electromagnetic wave equation

The density-current being the superposition of "flashes," the effect (\mathbf{A}) of this density-current can be expressed as the superposition of the effects E of elementary flashes; thus

$$A^\mu(x) = \int E[x - a(\alpha)]\dot{a}^\mu(\alpha)\,d\alpha, \quad (4.39)$$

The solution of the wave equation

where the "elementary effect" E ("kernel"; "Green's function") satisfies the equation

$$\Box E(x) = -4\pi\,\delta^4(x). \quad (4.40)$$

One solution is the "half-advanced-plus-half-retarded potential,"

$$E(x) = \delta(\eta_{\alpha\beta}x^\alpha x^\beta). \quad (4.41)$$

It vanishes everywhere except on the backward and forward light cones, where it has equal strength. Normally more useful is the retarded solution,

$$R(x) = \begin{cases} 2E(x) & \text{if } x^0 > 0, \\ 0 & \text{if } x^0 < 0, \end{cases} \tag{4.42}$$

which is obtained by doubling (4.41) in the region of the forward light cone and nullifying it in the region of the backward light cone. All electrodynamics (Coulomb forces, Ampère's law, electromagnetic induction, radiation) follows from the simple expression (4.39) for the vector potential [see, e.g., Wheeler and Feynman (1945) and (1949), also Rohrlich (1965)].

EXERCISES

Exercise 4.1. GENERIC LOCAL ELECTROMAGNETIC FIELD EXPRESSED IN SIMPLEST FORM

In the laboratory Lorentz frame, the electric field is E, the magnetic field B. Special cases are: (1) pure electric field ($B = 0$); (2) pure magnetic field ($E = 0$); and (3) "radiation field" or "null field" (E and B equal in magnitude and perpendicular in direction). All cases other than (1), (2), and (3) are "generic." In the generic case, calculate the Poynting density of flow of energy $E \times B/4\pi$ and the density of energy $(E^2 + B^2)/8\pi$. Define the direction of a unit vector n and the magnitude of a velocity parameter α by the ratio of energy flow to energy density:

$$n \tanh 2\alpha = \frac{2E \times B}{E^2 + B^2}.$$

View the same electromagnetic field in a rocket frame moving in the direction of n with the velocity parameter α (not 2α; factor 2 comes in because energy flow and energy density are components, not of a vector, but of a tensor). By employing the formulas for a Lorentz transformation (equation 3.23), or otherwise, show that the energy flux vanishes in the rocket frame, with the consequence that \bar{E} and \bar{B} are parallel. No one can prevent the \bar{z}-axis from being put in the direction common to \bar{E} and \bar{B}. Show that with this choice of direction, **Faraday** becomes

$$F = \bar{E}_z \, d\bar{z} \wedge d\bar{t} + \bar{B}_z \, d\bar{x} \wedge d\bar{y}$$

(only two wedge products needed to represent the generic local field; "canonical representation"; valid in one frame, valid in any frame).

Exercise 4.2. FREEDOM OF CHOICE OF 1-FORMS IN CANONICAL REPRESENTATION OF GENERIC LOCAL FIELD

Deal with a region so small that the variation of the field from place to place can be neglected. Write **Faraday** in canonical representation in the form

$$F = dp_I \wedge dq^I + dp_{II} \wedge dq^{II},$$

where p_A ($A = I$ or II) and q^A are scalar functions of position in spacetime. Define a "canonical transformation" to new scalar functions of position $p_{\bar{A}}$ and $q^{\bar{A}}$ by way of the "equation of transformation"

$$p_A \, dq^A = dS + p_{\bar{A}} \, dq^{\bar{A}},$$

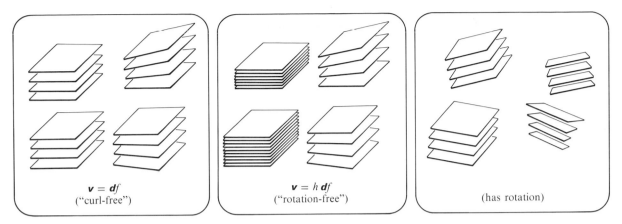

Figure 4.7.
Some simple types of 1-forms compared and contrasted.

where the "generating function" S of the transformation is an arbitrary function of the q^A and the $q^{\bar{A}}$:

$$\boldsymbol{d}S = (\partial S/\partial q^A)\,\boldsymbol{d}q^A + (\partial S/\partial q^{\bar{A}})\,\boldsymbol{d}q^{\bar{A}}.$$

(a) Derive expressions for the two p_A's and the two $p_{\bar{A}}$'s in terms of S by equating coefficients of dq^I, dq^{II}, $dq^{\bar{I}}$, $dq^{\bar{II}}$ individually on the two sides of the equation of transformation.

(b) Use these expressions for the p_A's and $p_{\bar{A}}$'s to show that $\boldsymbol{F} = \boldsymbol{d}p_A \wedge \boldsymbol{d}q^A$ and $\bar{\boldsymbol{F}} = \boldsymbol{d}p_{\bar{A}} \wedge \boldsymbol{d}q^{\bar{A}}$, ostensibly different, are actually expressions for one and the same 2-form in terms of alternative sets of 1-forms.

Exercise 4.3. A CLOSED OR CURL-FREE 1-FORM IS A GRADIENT

Given a 1-form $\boldsymbol{\sigma}$ such that $\boldsymbol{d\sigma} = 0$, show that $\boldsymbol{\sigma}$ can be expressed in the form $\boldsymbol{\sigma} = \boldsymbol{d}f$, where f is some scalar. The 1-form $\boldsymbol{\sigma}$ is said to be "curl-free," a narrower category of 1-form than the "rotation-free" 1-form of the next exercise (expressible as $\boldsymbol{\sigma} = h\,\boldsymbol{d}f$), and it in turn is narrower (see Figure 4.7) than the category of "1-forms with rotation" (not expressible in the form $\boldsymbol{\sigma} = h\,\boldsymbol{d}f$). When the 1-form $\boldsymbol{\sigma}$ is expressed in terms of basis 1-forms $\boldsymbol{d}x^\alpha$, multiplied by corresponding components σ_α, show that "curl-free" implies $\sigma_{[\alpha,\beta]} = 0$.

Exercise 4.4. CANONICAL EXPRESSION FOR A ROTATION-FREE 1-FORM

In three dimensions a rigid body turning with angular velocity ω about the z-axis has components of velocity $v_y = \omega x$, and $v_x = -\omega y$. The quantity curl $\boldsymbol{v} = \boldsymbol{\nabla} \times \boldsymbol{v}$ has z-component equal to 2ω, and all other components equal to zero. Thus the scalar product of \boldsymbol{v} and curl \boldsymbol{v} vanishes:

$$v_{[i,j}v_{k]} = 0.$$

The same concept generalizes to four dimensions,

$$v_{[\alpha,\beta}v_{\gamma]} = 0,$$

and lends itself to expression in coordinate-free language, as the requirement that a certain 3-form must vanish:

$$\boldsymbol{d}\boldsymbol{v} \wedge \boldsymbol{v} = 0.$$

Any 1-form \boldsymbol{v} satisfying this condition is said to be "rotation-free." Show that a 1-form is rotation-free if and only if it can be written in the form

$$\boldsymbol{v} = h\,\boldsymbol{df},$$

where h and f are scalar functions of position (the "Frobenius theorem").

Exercise 4.5. FORMS ENDOWED WITH POLAR SINGULARITIES

List the principal results on how such forms are representable, such as

$$\boldsymbol{\Phi}_1 = \frac{\boldsymbol{dS}}{S} \wedge \boldsymbol{\psi}_1 + \boldsymbol{\theta}_1,$$

and the conditions under which each applies [for the meaning and answer to this exercise, see Lascoux (1968)].

Exercise 4.6. THE FIELD OF THE OSCILLATING DIPOLE

Verify that the expressions given for the electromagnetic field of an oscillating dipole in equations (4.23) and (4.24) satisfy $\boldsymbol{dF} = 0$ everywhere and $\boldsymbol{d}^*\boldsymbol{F} = 0$ everywhere except at the origin.

Exercise 4.7. THE 2-FORM MACHINERY TRANSLATED
INTO TENSOR MACHINERY

This exercise is stated at the end of the legend caption of Figure 4.1.

Exercise 4.8. PANCAKING THE COULOMB FIELD

Figure 4.5 shows a spacelike slice, $t = $ const, through the **Maxwell** of a point-charge at rest. By the following pictorial steps, verify that the electric-field lines get compressed into the transverse direction when viewed from a moving Lorentz frame: (1) Draw a picture of an equatorial slice ($\theta = \pi/2$; t, r, ϕ variable) through **Maxwell** $= $ ***F***. (2) Draw various spacelike slices, corresponding to constant time in various Lorentz frames, through the resultant geometric structure. (3) Interpret the intersection of **Maxwell** $= $ ***F*** with each Lorentz slice in the manner of Figure 4.3.

Exercise 4.9. COMPUTATION OF SURFACE INTEGRALS

In Box 4.1 the definition

$$\int \boldsymbol{\alpha} = \int \cdots \int \left\langle \boldsymbol{\alpha}, \frac{\partial \mathscr{P}}{\partial \lambda^1} \wedge \cdots \wedge \frac{\partial \mathscr{P}}{\partial \lambda^p} \right\rangle d\lambda^1 \ldots d\lambda^p$$

is given for the integral of a p-form $\boldsymbol{\alpha}$ over a p-surface $\mathscr{P}(\lambda^1, \ldots, \lambda^p)$ in n-dimensional space. From this show that the following computational rule (also given in Box 4.1) works: (1) substitute the equation for the surface,

$$x^k = x^k(\lambda^1, \ldots, \lambda^p),$$

into $\boldsymbol{\alpha}$ and collect terms in the form

$$\boldsymbol{\alpha} = a(\lambda^1, \ldots, \lambda^p)\,\boldsymbol{d}\lambda^1 \wedge \cdots \wedge \boldsymbol{d}\lambda^p;$$

(2) integrate

$$\int \boldsymbol{\alpha} = \int \cdots \int a(\lambda^1, \ldots, \lambda^p)\,d\lambda^1 \ldots d\lambda^p$$

using the elementary definition of integration.

Exercise 4.10. WHITAKER'S CALUMOID, OR, THE LIFE OF A LOOP

Take a closed loop, bounding a 2-dimensional surface \mathcal{S}. It entraps a certain flux of **Faraday** $\Phi_F = \int_{\mathcal{S}} \mathbf{F}$ ("magnetic tubes") and a certain flux of **Maxwell** $\Phi_M = \int_{\mathcal{S}} {}^{*}\mathbf{F}$ ("electric tubes").

(a) Show that the fluxes Φ_F and Φ_M depend only on the choice of loop, and not on the choice of the surface \mathcal{S} bounded by the loop, if and only if $\mathbf{dF} = \mathbf{d}{}^{*}\mathbf{F} = 0$ (no magnetic charge; no electric charge). *Hint:* use generalized Stokes theorem, Boxes 4.1 and 4.6.

(b) Move the loop in space and time so that it continues to entrap the same two fluxes. Move it forward a little more here, a little less there, so that it continues to do so. In this way trace out a 2-dimensional surface ("calumoid"; see E. T. Whitaker 1904) $\mathcal{P} = \mathcal{P}(a, b)$; $x^{\mu} = x^{\mu}(a, b)$. Show that the elementary bivector in this surface, $\mathbf{\Sigma} = \partial\mathcal{P}/\partial a \wedge \partial\mathcal{P}/\partial b$ satisfies $\langle \mathbf{F}, \mathbf{\Sigma} \rangle = 0$ and $\langle {}^{*}\mathbf{F}, \mathbf{\Sigma} \rangle = 0$.

(c) Show that these differential equations for $x^{\mu}(a, b)$ can possess a solution, with given initial condition $x^{\mu} = x^{\mu}(a, 0)$ for the initial location of the loop, if $\mathbf{dF} = 0$ and $\mathbf{d}{}^{*}\mathbf{F} = 0$ (no magnetic charge, no electric charge).

(d) Consider a static, uniform electric field $\mathbf{F} = -E_x \, \mathbf{dt} \wedge \mathbf{dx}$. Solve the equations, $\langle \mathbf{F}, \mathbf{\Sigma} \rangle = 0$ and $\langle {}^{*}\mathbf{F}, \mathbf{\Sigma} \rangle = 0$ to find the equation $\mathcal{P}(a, b)$ for the most general calumoid. [*Answer:* $y = y(a)$, $z = z(a)$, $x = x(b)$, $t = t(b)$.] Exhibit two special cases: (i) a calumoid that lies entirely in a hypersurface of constant time [loop moves at infinite velocity; analogous to super light velocity of point of crossing for two blades of a pair of scissors]; (ii) a calumoid whose loop remains forever at rest in the t, x, y, z Lorentz frame.

Exercise 4.11. DIFFERENTIAL FORMS AND HAMILTONIAN MECHANICS

Consider a dynamic system endowed with two degrees of freedom. For the definition of this system as a Hamiltonian system (special case: here the Hamiltonian is independent of time), one needs (1) a definition of canonical variables (see Box 4.5) and (2) a knowledge of the Hamiltonian H as a function of the coordinates q^1, q^2 and the canonically conjugate momenta p_1, p_2. To derive the laws of mechanics, consider the five-dimensional space of p_1, p_2, q^1, q^2, and t, and a curve in this space leading from starting values of the five coordinates (subscript A) to final values (subscript B), and the value

$$I = \int_A^B p_1 \, \mathbf{dq}^1 + p_2 \, \mathbf{dq}^2 - H(p, q) \, \mathbf{dt} = \int_A^B \boldsymbol{\omega}$$

of the integral I taken along this path. The difference of the integral for two "neighboring" paths enclosing a two-dimensional region \mathcal{S}, according to the theorem of Stokes (Boxes 4.1 and 4.6), is

$$\delta I = \oint \boldsymbol{\omega} = \int_{\mathcal{S}} \mathbf{d\boldsymbol{\omega}}.$$

The principle of least action (principle of "extremal history") states that the representative point of the system must travel along a route in the five-dimensional manifold (route with tangent vector $d\mathcal{P}/dt$) such that the variation vanishes for this path; i.e.,

$$\mathbf{d\boldsymbol{\omega}}(\ldots, d\mathcal{P}/dt) = 0$$

(2-form $\mathbf{d\boldsymbol{\omega}}$ with a single vector argument supplied, and other slot left unfilled, gives the 1-form in 5-space that must vanish). This fixes only the direction of $d\mathcal{P}/dt$; its magnitude can be normalized by requiring $\langle \mathbf{dt}, d\mathcal{P}/dt \rangle = 1$.

(a) Evaluate $\mathbf{d\boldsymbol{\omega}}$ from the expression $\boldsymbol{\omega} = p_j \, \mathbf{dq}^j - H \, \mathbf{dt}$.

(b) Set $d\mathcal{P}/dt = \dot{q}^j(\partial\mathcal{P}/\partial q^j) + \dot{p}_j(\partial\mathcal{P}/\partial p_j) + \dot{t}(\partial\mathcal{P}/\partial t)$, and expand $\mathbf{d\boldsymbol{\omega}}(\ldots, d\mathcal{P}/dt) = 0$ in terms of the basis $\{\mathbf{dp}_j, \mathbf{dq}^k, \mathbf{dt}\}$.

Box 4.5 METRIC STRUCTURE AND HAMILTONIAN OR "SYMPLECTIC STRUCTURE" COMPARED AND CONTRASTED

	Metric structure	*Symplectic structure*
1. Physical application	Geometry of spacetime	Hamiltonian mechanics
2. Canonical structure	$(\ldots\cdot\ldots) = \text{``}\boldsymbol{ds}^2\text{''} = -\boldsymbol{dt} \otimes \boldsymbol{dt}$ $+ \boldsymbol{dx} \otimes \boldsymbol{dx} + \boldsymbol{dy} \otimes \boldsymbol{dy}$ $+ \boldsymbol{dz} \otimes \boldsymbol{dz}$	$\boldsymbol{\Theta} = \boldsymbol{dp}_1 \wedge \boldsymbol{dq}^1 + \boldsymbol{dp}_2 \wedge \boldsymbol{dq}^2$
3. Nature of "metric"	Symmetric	Antisymmetric
4. Name for given coordinate system and any other set of four coordinates in which metric has same form	Lorentz coordinate system	System of "canonically" (or "dynamically") conjugate coordinates
5. Field equation for this metric	$\boldsymbol{R}_{\mu\nu\alpha\beta} = 0$ (zero Riemann curvature; flat spacetime)	$\boldsymbol{d\Theta} = 0$ ("closed 2-form"; condition automatically satisfied by expression above).
6. The four-dimensional manifold	Spacetime	Phase space
7. Coordinate-free description of the structure of this manifold	$\boldsymbol{Riemann} = 0$	$\boldsymbol{d\Theta} = 0$
8. Canonical coordinates distinguished from other coordinates (allowable but less simple)	Make metric take above form (item 2)	Make metric take above form (item 2)

(c) Show that this five-dimensional equation can be written in the 4-dimensional phase space of $\{q^j, p_k\}$ as

$$\boldsymbol{\Theta}(\ldots, d\mathscr{P}/dt) = \boldsymbol{d}H,$$

where $\boldsymbol{\Theta}$ is the 2-form defined in Box 4.5.

(d) Show that the components of $\boldsymbol{\Theta}(\ldots, d\mathscr{P}/dt) = \boldsymbol{d}H$ in the $\{q^j, p_k\}$ coordinate system are the familiar Hamilton equations. Note that this conclusion depends only on the form assumed for $\boldsymbol{\Theta}$, so that one also obtains the standard Hamilton equations in any other phase-space coordinates $\{\bar{q}^j, \bar{p}_k\}$ ("canonical variables") for which

$$\boldsymbol{\Theta} = \boldsymbol{d\bar{p}}_1 \wedge \boldsymbol{d\bar{q}}^1 + \boldsymbol{d\bar{p}}_2 \wedge \boldsymbol{d\bar{q}}^2.$$

Exercise 4.12. SYMMETRY OPERATIONS AS TENSORS

We define the meaning of square and round brackets enclosing a set of indices as follows:

$$V_{(\alpha_1\ldots\alpha_p)} \equiv \frac{1}{p!} \, \Sigma \, V_{\alpha_{\pi_1}\ldots\alpha_{\pi_p}}, \qquad V_{[\alpha_1\ldots\alpha_p]} \equiv \frac{1}{p!} \, \Sigma (-1)^\pi V_{\alpha_{\pi_1}\ldots\alpha_{\pi_p}}.$$

Box 4.6 BIRTH OF STOKES' THEOREM

Central to the mathematical formulation of electromagnetism are the theorems of Gauss (taken up in Chapter 5) and Stokes. Both today appear together as one unity when expressed in the language of forms. In earlier times the unity was not evident. Everitt (1970) recalls the history of Stokes' theorem: "The Smith's Prize paper set by [G. C.] Stokes [Lucasian Professor of Mathematics] and taken by Maxwell in [February] 1854 . . .

 5. Given the centre and two points of an ellipse, and the length of the major axis, find its direction by a geometrical construction.

 6. Integrate the differential equation

$$(a^2 - x^2)\, dy^2 + 2xy\,dy\,dx + (a^2 - y^2)\, dx^2 = 0.$$

Has it a singular solution?

 7. In a double system of curves of double curvature, a tangent is always drawn at the variable point P; shew that, as P moves away from an arbitrary fixed point Q, it must begin to move along a generating line of an elliptic cone having Q for vertex in order that consecutive tangents may ultimately intersect, but that the conditions of the problem may be impossible.

 8. If X, Y, Z be functions of the rectangular co-ordinates x, y, z, dS an element of any limited surface, l, m, n the cosines of the inclinations of the normal at dS to the axes, ds an element of the bounding line, shew that

$$\iint \left\{ l\left(\frac{dZ}{dy} - \frac{dY}{dz}\right) + m\left(\frac{dX}{dz} - \frac{dZ}{dx}\right) + n\left(\frac{dY}{dx} - \frac{dX}{dy}\right) \right\} dS$$

$$= \int \left(X\frac{dx}{ds} + Y\frac{dy}{ds} + Z\frac{dz}{ds}\right) ds,$$

the differential coefficients of X, Y, Z being partial, and the single integral being taken all round the perimeter of the surface

marks the first appearance in print of the formula connecting line and surface integrals now known as Stokes' theorem. This was of great importance to Maxwell's development of electromagnetic theory. The earliest explicit proof of the theorem appears to be that given in a letter from Thomson to Stokes dated July 2, 1850." [Quoted in Campbell and Garnett (1882), pp. 186–187.]

Here the sum is taken over all permutations π of the numbers $1, 2, \ldots, p$, and $(-1)^\pi$ is $+1$ or -1 depending on whether the permutation is even or odd. The quantity V may have other indices, not shown here, besides the set of p indices $\alpha_1, \alpha_2, \ldots, \alpha_p$, but only this set of indices is affected by the operations described here. The numbers $\pi_1, \pi_2, \ldots, \pi_p$ are the numbers $1, 2, \ldots, p$ rearranged according to the permutation π. (Cases $p = 2, 3$ were treated in exercise 3.12.) We therefore have machinery to convert any rank-p tensor with components $V_{\alpha_1 \ldots \alpha_p}$ into a new tensor with components

$$[\boldsymbol{Alt(V)}]_{\mu_1 \ldots \mu_p} = V_{[\mu_1 \ldots \mu_p]}.$$

Since this machinery \boldsymbol{Alt} is linear, it can be viewed as a tensor which, given suitable arguments $\boldsymbol{u}, \boldsymbol{v}, \ldots, \boldsymbol{w}, \boldsymbol{\alpha}, \boldsymbol{\beta}, \ldots, \boldsymbol{\gamma}$ produces a number

$$u^\mu v^\nu \ldots w^\lambda \alpha_{[\mu} \beta_\nu \ldots \gamma_{\lambda]}.$$

(a) Show that the components of this tensor are

$$(\boldsymbol{Alt})_{\beta_1 \ldots \beta_p}{}^{\alpha_1 \ldots \alpha_p} = (p!)^{-1} \delta^{\alpha_1 \ldots \alpha_p}_{\beta_1 \ldots \beta_p} \qquad \text{(Note: indices of } \delta \text{ are almost never raised or lowered, so this notation leads to no confusion.)}$$

where

$$\delta^{\alpha_1 \ldots \alpha_p}_{\beta_1 \ldots \beta_p} = \begin{cases} +1 \text{ if } (\alpha_1, \ldots, \alpha_p) \text{ is an even permutation of } (\beta_1, \ldots, \beta_p), \\ -1 \text{ if } (\alpha_1, \ldots, \alpha_p) \text{ is an odd permutation of } (\beta_1, \ldots, \beta_p), \\ 0 \text{ if (i) } \text{ any two of the } \alpha\text{'s are the same,} \\ 0 \text{ if (ii) } \text{ any two of the } \beta\text{'s are the same,} \\ 0 \text{ if (iii) } \text{ the } \alpha\text{'s and } \beta\text{'s are different sets of integers.} \end{cases}$$

Note that the demonstration, and therefore these component values, are correct in any frame.

(b) Show for any "alternating" (i.e., "*completely antisymmetric*") tensor $A_{\alpha_1 \ldots \alpha_p} = A_{[\alpha_1 \ldots \alpha_p]}$ that

$$\frac{1}{p!} A_{\alpha_1 \ldots \alpha_p} \delta^{\alpha_1 \ldots \alpha_p \beta_1 \ldots \beta_q}_{\gamma_1 \ldots \gamma_p \gamma_{p+1} \ldots \gamma_{p+q}}$$

$$= \sum_{\alpha_1 < \alpha_2 < \ldots < \alpha_p} A_{\alpha_1 \ldots \alpha_p} \delta^{\alpha_1 \ldots \alpha_p \beta_1 \ldots \beta_q}_{\gamma_1 \ldots \ldots \ldots \gamma_{p+q}}$$

$$\equiv A_{|\alpha_1 \ldots \alpha_p|} \delta^{\alpha_1 \ldots \alpha_p \beta_1 \ldots \beta_q}_{\gamma_1 \ldots \ldots \ldots \gamma_{p+q}}.$$

The final line here introduces the convention that a summation over indices enclosed between vertical bars includes only terms with those indices in increasing order. Show, consequently or similarly, that

$$\delta^{\alpha_1 \ldots \alpha_p \beta_1 \ldots \beta_q}_{\gamma_1 \ldots \ldots \ldots \gamma_{p+q}} \delta^{\mu_1 \ldots \mu_q}_{|\beta_1 \ldots \beta_q|} = \delta^{\alpha_1 \ldots \alpha_p \mu_1 \ldots \mu_q}_{\gamma_1 \ldots \ldots \ldots \gamma_{p+q}}.$$

(c) *Define* the exterior ("wedge") product of any two alternating tensors by

$$(\boldsymbol{\alpha} \wedge \boldsymbol{\beta})_{\lambda_1 \ldots \lambda_{p+q}} = \delta^{\mu_1 \ldots \mu_p \nu_1 \ldots \nu_q}_{\lambda_1 \ldots \lambda_p \lambda_{p+1} \ldots \lambda_{p+q}} \alpha_{|\mu_1 \ldots \mu_p|} \beta_{|\nu_1 \ldots \nu_q|};$$

and similarly

$$(\boldsymbol{U} \wedge \boldsymbol{V})^{\lambda_1 \ldots \lambda_{p+q}} = \delta^{\lambda_1 \ldots \lambda_p \lambda_{p+1} \ldots \lambda_{p+q}}_{\mu_1 \ldots \mu_p \nu_1 \ldots \nu_q} U^{|\mu_1 \ldots \mu_p|} V^{|\nu_1 \ldots \nu_q|}.$$

Show that this implies equation (3.45b). Establish the associative law for this product rule by showing that

$$[(\boldsymbol{\alpha} \wedge \boldsymbol{\beta}) \wedge \boldsymbol{\gamma}]_{\sigma_1 \dots \sigma_{p+q+r}}$$
$$= \delta^{\lambda_1 \dots \lambda_p \mu_1 \dots \mu_q \nu_1 \dots \nu_r}_{\sigma_1 \dots \sigma_{p+q+r}} \alpha_{|\lambda_1 \dots \lambda_p|} \beta_{|\mu_1 \dots \mu_q|} \gamma_{|\nu_1 \dots \nu_r|}$$
$$= [\boldsymbol{\alpha} \wedge (\boldsymbol{\beta} \wedge \boldsymbol{\gamma})]_{\sigma_1 \dots \sigma_{p+q+r}};$$

and show that this reduces to the 3-form version of Equation (3.45c) when $\boldsymbol{\alpha}$, $\boldsymbol{\beta}$, and $\boldsymbol{\gamma}$ are all 1-forms.

(d) Derive the following formula for the components of the exterior product of p vectors

$$(\boldsymbol{u}_1 \wedge \boldsymbol{u}_2 \wedge \cdots \wedge \boldsymbol{u}_p)^{\alpha_1 \dots \alpha_p} = \delta^{\alpha_1 \dots \alpha_p}_{\mu \dots \nu}(u_1)^\mu \dots (u_p)^\nu$$
$$= p! u_1^{[\alpha_1} u_2^{\alpha_2} \dots u_p^{\alpha_p]}$$
$$= \delta^{\alpha_1 \alpha_2 \dots \alpha_p}_{1 \ 2 \ \dots \ p} \det [(u_\mu)^\lambda].$$

CHAPTER **5**

STRESS-ENERGY TENSOR
AND CONSERVATION LAWS

§5.1. TRACK-1 OVERVIEW

"Geometry tells matter how to move, and matter tells geometry how to curve." However, it will do no good to look into curvature (Part III) and Einstein's law for the production of curvature by mass-energy (Part IV) until a tool can be found to determine how much mass-energy there is in a unit volume. That tool is the stress-energy tensor. It is the focus of attention in this chapter.

The essential features of the stress-energy tensor are summarized in Box 5.1 for the benefit of readers who want to rush on into gravitation physics as quickly as possible. Such readers can proceed directly from Box 5.1 into Chapter 6—though by doing so, they close the door on several later portions of track two, which lean heavily on material treated in this chapter.

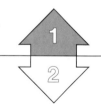

§5.2. THREE-DIMENSIONAL VOLUMES AND DEFINITION OF THE STRESS-ENERGY TENSOR

The rest of this chapter is Track 2.

It depends on no preceding Track-2 material.

It is needed as preparation for Chapter 20 (conservation laws for mass and angular momentum).

It will be extremely helpful in all applications of gravitation theory (Chapters 18–40).

Spacetime contains a flowing "river" of 4-momentum. Each particle carries its 4-momentum vector with itself along its world line. Many particles, on many world lines, viewed in a smeared-out manner (continuum approximation), produce a continuum flow—a river of 4-momentum. Electromagnetic fields, neutrino fields, meson fields: they too contribute to the river.

How can the flow of the river be quantified? By means of a linear machine: the stress-energy tensor **T**.

Choose a small, three-dimensional parallelepiped in spacetime with vectors **A**, **B**, **C** for edges (Figure 5.1). Ask how much 4-momentum crosses that volume in

Box 5.1 CHAPTER 5 SUMMARIZED

A. STRESS-ENERGY TENSOR AS A MACHINE

At each event in spacetime, there exists a stress-energy tensor. It is a machine that contains a knowledge of the energy density, momentum density, and stress as measured by any and all observers at that event. Included are energy, momentum, and stress associated with all forms of matter and all nongravitational fields.

The stress-energy tensor is a linear, symmetric machine with two slots for the insertion of two vectors: $\boldsymbol{T}(\ldots,\ldots)$. Its output, for given input, can be summarized as follows.

(1) Insert the 4-velocity \boldsymbol{u} of an observer into one of the slots; leave the other slot empty. The output is

$$\boldsymbol{T}(\boldsymbol{u},\ldots) = \boldsymbol{T}(\ldots,\boldsymbol{u}) = -\begin{pmatrix} \text{density of 4-momentum,} \\ \text{``}d\boldsymbol{p}/dV\text{,'' i.e., 4-momentum} \\ \text{per unit of three-dimensional volume,} \\ \text{as measured in observer's} \\ \text{Lorentz frame at event where} \\ \boldsymbol{T} \text{ is chosen} \end{pmatrix};$$

i.e., $T^{\alpha}{}_{\beta}u^{\beta} = T_{\beta}{}^{\alpha}u^{\beta} = -(dp^{\alpha}/dV)$ for observer with 4-velocity u^{α}.

(2) Insert 4-velocity of observer into one slot; insert an arbitrary unit vector \boldsymbol{n} into the other slot. The output is

$$\boldsymbol{T}(\boldsymbol{u},\boldsymbol{n}) = \boldsymbol{T}(\boldsymbol{n},\boldsymbol{u}) = -\begin{pmatrix} \text{component, ``}\boldsymbol{n}\cdot d\boldsymbol{p}/dV\text{'', of} \\ \text{4-momentum density along the} \\ \boldsymbol{n} \text{ direction, as measured in} \\ \text{observer's Lorentz frame} \end{pmatrix};$$

i.e., $T_{\alpha\beta}u^{\alpha}n^{\beta} = T_{\alpha\beta}n^{\alpha}u^{\beta} = -n_{\mu}dp^{\mu}/dV$.

(3) Insert 4-velocity of observer into both slots. The output is the density of mass-energy that he measures in his Lorentz frame:

$$\boldsymbol{T}(\boldsymbol{u},\boldsymbol{u}) = \begin{pmatrix} \text{mass-energy per unit volume as measured} \\ \text{in frame with 4-velocity } \boldsymbol{u} \end{pmatrix}.$$

(4) Pick an observer and choose two spacelike basis vectors, \boldsymbol{e}_{j} and \boldsymbol{e}_{k}, of his Lorentz frame. Insert \boldsymbol{e}_{j} and \boldsymbol{e}_{k} into the slots of \boldsymbol{T}. The output is the j,k component of the stress as measured by that observer:

$$T_{jk} = \boldsymbol{T}(\boldsymbol{e}_{j},\boldsymbol{e}_{k}) = T_{kj} = \boldsymbol{T}(\boldsymbol{e}_{k},\boldsymbol{e}_{j})$$

$$= \begin{pmatrix} j\text{-component of force acting} \\ \text{from side } x^{k} - \varepsilon \text{ to side } x^{k} + \varepsilon, \\ \text{across a unit surface area with} \\ \text{perpendicular direction } \boldsymbol{e}_{k} \end{pmatrix} = \begin{pmatrix} k\text{-component of force acting} \\ \text{from side } x^{j} - \varepsilon \text{ to side } x^{j} + \varepsilon, \\ \text{across a unit surface area with} \\ \text{perpendicular direction } \boldsymbol{e}_{j} \end{pmatrix}.$$

Box 5.1 (*continued*)

B. STRESS-ENERGY TENSOR FOR A PERFECT FLUID

One type of matter studied extensively later in this book is a *"perfect fluid."* A perfect fluid is a fluid or gas that (1) moves through spacetime with a 4-velocity \boldsymbol{u} which may vary from event to event, and (2) exhibits a density of mass-energy ρ and an isotropic pressure p in the rest frame of each fluid element. Shear stresses, anisotropic pressures, and viscosity must be absent, or the fluid is not perfect. The stress-energy tensor for a perfect fluid at a given event can be constructed from the metric tensor, \boldsymbol{g}, the 4-velocity, \boldsymbol{u}, and the rest-frame density and pressure, ρ and p:

$$\boldsymbol{T} = (\rho + p)\boldsymbol{u} \otimes \boldsymbol{u} + p\boldsymbol{g}, \qquad \text{or } T_{\alpha\beta} = (\rho + p)u_\alpha u_\beta + pg_{\alpha\beta}.$$

In the fluid's rest frame, the components of this stress-energy tensor have the expected form (insert into a slot of \boldsymbol{T}, as 4-velocity of observer, just the fluid's 4-velocity):

$$T^\alpha{}_\beta u^\beta = [(\rho + p)u^\alpha u_\beta + p\delta^\alpha{}_\beta]u^\beta = -(\rho + p)u^\alpha + pu^\alpha = -\rho u^\alpha;$$

i.e.,

$$T^0{}_\beta u^\beta = -\rho = -(\text{mass-energy density}) = -dp^0/dV,$$
$$T^j{}_\beta u^\beta = 0 = -(\text{momentum density}) = -dp^j/dV;$$

also

$$T_{jk} = \boldsymbol{T}(\boldsymbol{e}_j, \boldsymbol{e}_k) = p\delta_{jk} = \text{stress-tensor components.}$$

C. CONSERVATION OF ENERGY-MOMENTUM

In electrodynamics the conservation of charge can be expressed by the differential equation

$$\partial(\text{charge density})/\partial t + \boldsymbol{\nabla} \cdot (\text{current density}) = 0;$$

i.e., $J^0{}_{,0} + \boldsymbol{\nabla} \cdot \boldsymbol{J} = 0$; i.e. $J^\alpha{}_{,\alpha} = 0$; i.e., $\boldsymbol{\nabla} \cdot \boldsymbol{J} = 0$. Similarly, conservation of energy-momentum can be expressed by the fundamental geometric law

$$\boldsymbol{\nabla} \cdot \boldsymbol{T} = 0.$$

(Because \boldsymbol{T} is symmetric, it does not matter on which slot the divergence is taken.) This law plays an important role in gravitation theory.

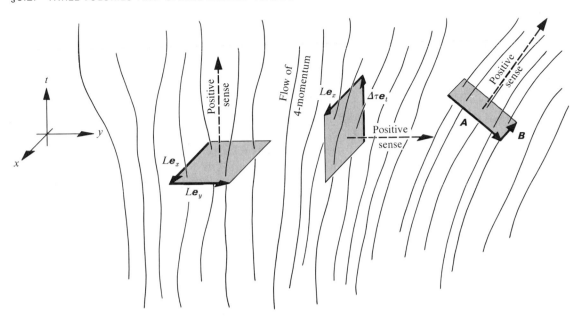

Figure 5.1.
The "river" of 4-momentum flowing through spacetime, and three different 3-volumes across which it flows. (One dimension is suppressed from the picture; so the 3-volumes look like 2-volumes.) The first 3-volume is the interior of a cubical soap box momentarily at rest in the depicted Lorentz frame. Its edges are Le_x, Le_y, Le_z; and its volume 1-form, with "positive" sense toward future ("standard orientation"), is $\Sigma = L^3\, dt = -Vu$ ($V = L^3$ = volume as measured in rest frame; $u = -dt$ = 4-velocity of box). The second 3-volume is the "world sheet" swept out in time $\Delta\tau$ by the top of a second cubical box. The box top's edges are Le_x and Le_z; and its volume 1-form, with "positive" sense away from the box's interior, in direction of increasing y, is $\Sigma = L^2\,\Delta\tau\, dy = \mathcal{Q}\,\Delta\tau\sigma$ ($\mathcal{Q} = L^2$ = area of box top; $\sigma = dy$ = unit 1-form containing world tube). The third 3-volume is an arbitrary one, with edges A, B, C and volume 1-form $\Sigma_\mu = \epsilon_{\mu\alpha\beta\gamma}A^\alpha B^\beta C^\gamma$.

its positive sense (i.e., from its "negative side" toward its "positive side"). To calculate the answer: (1) Construct the *"volume 1-form"*

Mathematical representation of 3-volumes

$$\Sigma_\mu = +\epsilon_{\mu\alpha\beta\gamma}A^\alpha B^\beta C^\gamma; \qquad (5.1)$$

the parallelepiped lies in one of the 1-form surfaces, and the positive sense across the parallelepiped is defined to be the positive sense of the 1-form Σ. (2) Insert this volume 1-form into the second slot of the stress-energy tensor T. The result is

Momentum crossing a 3-volume calculated, using stress-energy tensor

$$T(\ldots, \Sigma) = p = \left(\begin{array}{c}\text{momentum crossing from}\\ \text{negative side toward positive side}\end{array}\right). \qquad (5.2)$$

empty ↑
slot

(3) To get the projection of the 4-momentum along a vector w or 1-form α, insert the volume 1-form Σ into the second slot and w or α into the first:

$$T(w, \Sigma) = w \cdot p, \qquad T(\alpha, \Sigma) = \langle \alpha, p \rangle. \qquad (5.3)$$

This defines the stress-energy tensor.

The key features of 3-volumes and the stress-energy tensor are encapsulated by the above three-step procedure. But encapsulation is not sufficient; deep understanding is also required. To gain it, one must study special cases, both of 3-volumes and of the operation of the stress-energy machinery.

A Special Case

Interior of a soap box:

A soap box moves through spacetime. A man at an event \mathscr{P}_0 on the box's world line peers inside it, and examines all the soap, air, and electromagnetic fields it contains. He adds up all their 4-momenta to get a grand total $\boldsymbol{p}_{\text{box at }\mathscr{P}_0}$. How much is this grand total? One can calculate it by noting that the 4-momentum inside the box at \mathscr{P}_0 is precisely the 4-momentum crossing the box from past toward future there (Figure 5.1). Hence, the 4-momentum the man measures is

$$\boldsymbol{p}_{\text{box at }\mathscr{P}_0} = \boldsymbol{T}(\ldots, \boldsymbol{\Sigma}), \tag{5.4}$$

Its volume 1-form

where $\boldsymbol{\Sigma}$ is the box's volume 1-form at \mathscr{P}_0. But for such a soap box, $\boldsymbol{\Sigma}$ has a magnitude equal to the box's volume V as measured by a man in its momentary rest frame, and the box itself lies in one of the hyperplanes of $\boldsymbol{\Sigma}$; equivalently,

$$\boldsymbol{\Sigma} = -V\boldsymbol{u}, \tag{5.5}$$

Its 4-momentum content

where \boldsymbol{u} is the soap box's 4-velocity at \mathscr{P}_0 (minus sign because \boldsymbol{u}, regarded as a 1-form, has positive sense toward the past, $u_0 < 0$); see Box 5.2. Hence, the total 4-momentum inside the box is

$$\boldsymbol{p}_{\text{box at }\mathscr{P}_0} = \boldsymbol{T}(\ldots, -V\boldsymbol{u}) = -V\boldsymbol{T}(\ldots, \boldsymbol{u}), \tag{5.6}$$

or, in component notation,

$$(p^\alpha)_{\text{box at }\mathscr{P}_0} = -VT^{\alpha\beta}u_\beta. \tag{5.6'}$$

The energy in the box, as measured in its rest frame, is minus the projection of the 4-momentum on the box's 4-velocity:

$$E = -\boldsymbol{u} \cdot \boldsymbol{p}_{\text{box at }\mathscr{P}_0} = +VT^{\alpha\beta}u_\alpha u_\beta = V\boldsymbol{T}(\boldsymbol{u}, \boldsymbol{u});$$

so

Its energy density

$$\left(\begin{array}{l}\text{energy density as} \\ \text{measured in box's} \\ \text{rest frame}\end{array}\right) = \frac{E}{V} = \boldsymbol{T}(\boldsymbol{u}, \boldsymbol{u}). \tag{5.7}$$

Another Special Case

A man riding with the same soap box opens its top and pours out some soap. In a very small interval of time $\varDelta\tau$, how much total 4-momentum flows out of the box?

Box 5.2 THREE-DIMENSIONAL VOLUMES

A. General Parallelepiped

1. *Edges* of parallelepiped are three vectors **A**, **B**, **C**. One must order the edges; e.g., "**A** is followed by **B** is followed by **C**."

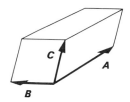

(One dimension, that orthogonal to the parallelepiped, is suppressed here.)

2. *Volume trivector* is defined to be **A** ∧ **B** ∧ **C**. It enters into the sophisticated theory of volumes (Chapter 4), but is not used much in the elementary theory.

3. *Volume 1-form* is defined by $\Sigma_\mu = \epsilon_{\mu\alpha\beta\gamma} A^\alpha B^\beta C^\gamma$. (**A**, **B**, **C** must appear here in standard order as chosen in step 1.) Note that the vector "corresponding" to **Σ** and the volume trivector are related by **Σ** = −*(**A** ∧ **B** ∧ **C**).

4. *Orientation* of the volume is defined to agree with the orientation of its 1-form **Σ**. More specifically: the edges **A**, **B**, **C** lie in a hyperplane of **Σ**(⟨**Σ**, **A**⟩ = ⟨**Σ**, **B**⟩ = ⟨**Σ**, **C**⟩ = 0; no "bongs of bell"). Thus, *the volume itself is one of **Σ**'s hyperplanes!* The positive sense moving away from the volume is defined to be the positive sense of **Σ**. *Note:* reversing the order of **A**, **B**, **C** reverses the positive sense!

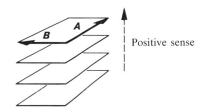

Positive sense

(One dimension, that along which **C** extends, is suppressed here.)

5. *The "standard orientation"* for a spacelike 3-volume has the positive sense of the 1-form **Σ** toward the future, corresponding to **A**, **B**, **C** forming a righthanded triad of vectors.

B. 3-Volumes of Arbitrary Shape

Can be analyzed by being broken up into union of parallelepipeds.

C. Interior of a Soap Box (Example)

1. *Analysis in soap box's rest frame.* Pick an event on the box's world line. The box's three edges there are three specific vectors **A**, **B**, **C**. In the box's rest frame they are purely spatial: $A^0 = B^0 = C^0 = 0$. Hence, the volume 1-form has components $\Sigma_j = 0$ and

Box 5.2 (*continued*)

$$\Sigma_0 = \epsilon_{0ijk}A^iB^jC^k = \det \begin{Vmatrix} A^1 & A^2 & A^3 \\ B^1 & B^2 & B^3 \\ C^1 & C^2 & C^3 \end{Vmatrix}$$

$= \boldsymbol{A} \cdot (\boldsymbol{B} \times \boldsymbol{C})$, in the standard notation of 3-dimensional vector analysis;

$= +V$ (V = volume of box) if (A, B, C) are righthand ordered (positive sense of $\boldsymbol{\Sigma}$ toward future; standard orientation);

$= -V$ (V = volume of box) if (A, B, C) are lefthand ordered (positive sense of $\boldsymbol{\Sigma}$ toward past).

2. This result reexpressed in geometric language: Let \boldsymbol{u} be the box's 4-velocity and V be its volume, as measured in its rest frame. Then either

$\boldsymbol{\Sigma} = -V\boldsymbol{u}$, in which case the "positive side" of the box's 3-surface is the future side, and its edges are ordered in a righthanded manner—the standard orientation;

or else

$\boldsymbol{\Sigma} = +V\boldsymbol{u}$, in which case the "positive side" is the past side, and the box's edges are ordered in a lefthanded manner.

D. 3-Volume Swept Out in Time $\Delta\tau$ by Two-Dimensional Top of a Soap Box (Example)

1. *Analysis in box's rest frame:* Pick an event on box's world line. There the two edges of the box top are vectors \boldsymbol{A} and \boldsymbol{B}. In the box's rest frame, orient the space axes so that \boldsymbol{A} and \boldsymbol{B} lie in the y,z-plane. During the lapse of a proper time $\Delta\tau$, the box top sweeps out a 3-volume whose third edge is $\boldsymbol{u}\,\Delta\tau$ ($\boldsymbol{u} = $ 4-velocity of box). In the box's rest-frame, with ordering "\boldsymbol{A} followed by \boldsymbol{B} followed by $\boldsymbol{u}\,\Delta\tau$," the volume 1-form has components

$\Sigma_0 = \Sigma_2 = \Sigma_3 = 0$, and
$\Sigma_1 = \epsilon_{1jk0}A^jB^k\,\Delta\tau u^0 = -\epsilon_{01jk}A^jB^k\,\Delta\tau$

$= -\mathcal{A}\,\Delta\tau$ (\mathcal{A} = area of box top) if (\boldsymbol{e}_x, A, B) are righthand ordered

$= +\mathcal{A}\,\Delta\tau$ (\mathcal{A} = area of box top) if (\boldsymbol{e}_x, A, B) are lefthand ordered.

(Note: No standard orientation can be defined in this case, because $\boldsymbol{\Sigma}$ can be carried continuously into $-\boldsymbol{\Sigma}$ by purely spatial rotations.)

2. *This result reexpressed in geometric language:* Let \mathcal{A} be the area of the box top as measured in its rest frame; and let $\boldsymbol{\sigma}$ be a unit 1-form, one of whose surfaces contains the box top and its 4-velocity (i.e., contains the box top's "world sheet"). Orient the positive sense of $\boldsymbol{\sigma}$ with the (arbitrarily chosen) positive sense of the box-top 3-volume. Then

$$\boldsymbol{\Sigma} = \mathcal{A}\,\Delta\tau\,\boldsymbol{\sigma}.$$

To answer this question, consider the three-dimensional volume swept out during $\Delta\tau$ by the box's opened two-dimensional top ("world sheet of top"). The 4-momentum asked for is the 4-momentum that crosses this world sheet in the positive sense (see Figure 5.1); hence, it is

The top of a soap box:

$$\boldsymbol{p}_{\text{flows out}} = \boldsymbol{T}(\dots, \boldsymbol{\Sigma}), \tag{5.8}$$

where $\boldsymbol{\Sigma}$ is the world sheet's volume 1-form. Let \mathscr{C} be the area of the box top, and $\boldsymbol{\sigma}$ be the outward-oriented unit 1-form, whose surfaces contain the world sheet (i.e., contain the box top and its momentary 4-velocity vector). Then

Its volume 1-form

$$\boldsymbol{\Sigma} = \mathscr{C}\,\Delta\tau\,\boldsymbol{\sigma} \tag{5.9}$$

(see Box 5.2); so the 4-momentum that flows out during $\Delta\tau$ is

Its 4-momentum that flows across

$$\boldsymbol{p}_{\text{flows out}} = \mathscr{C}\,\Delta\tau\,\boldsymbol{T}(\dots, \boldsymbol{\sigma}). \tag{5.10}$$

§5.3. COMPONENTS OF STRESS-ENERGY TENSOR

Like all other tensors, the stress-energy tensor is a machine whose definition and significance transcend coordinate systems and reference frames. But any one observer, locked as he is into some one Lorentz frame, pays more attention to the components of \boldsymbol{T} than to \boldsymbol{T} itself. To each component he ascribes a specific physical significance. Of greatest interest, perhaps, is the "time-time" component. It is the total density of mass-energy as measured in the observer's Lorentz frame:

Physical interpretation of stress-energy tensor's components:

$$T_{00} = -T_0{}^0 = T^{00} = \boldsymbol{T}(\boldsymbol{e}_0, \boldsymbol{e}_0) = \text{density of mass-energy} \tag{5.11}$$

T^{00}: energy density

(cf. equation 5.7, with the observer's 4-velocity \boldsymbol{u} replaced by the basis vector $\boldsymbol{e}_0 = \boldsymbol{u}$).

The "spacetime" components T^{j0} can be interpreted by considering the interior of a soap box at rest in the observer's frame. If its volume is V, then its volume 1-form is $\boldsymbol{\Sigma} = -V\boldsymbol{u} = +V\,\boldsymbol{dt}$; and the μ-component of 4-momentum inside it is

$$p^\mu = \langle \boldsymbol{dx}^\mu, \boldsymbol{p} \rangle = \boldsymbol{T}(\boldsymbol{dx}^\mu, \boldsymbol{\Sigma}) = V\boldsymbol{T}(\boldsymbol{dx}^\mu, \boldsymbol{dt}) = VT^{\mu 0}.$$

Thus, the 4-momentum per unit volume is

$$p^\mu/V = T^{\mu 0}, \tag{5.12a}$$

or, equivalently:

$$T^{00} = \text{density of mass-energy} \tag{5.13a}$$
$$\text{(units: g/cm}^3\text{, or erg/cm}^3\text{, or cm}^{-2}\text{);}$$

$$T^{j0} = \text{density of j-component of momentum} \tag{5.13b}$$
$$\text{(units: g (cm/sec) cm}^{-3}\text{, or cm}^{-2}\text{).}$$

T^{j0}: momentum density

The components $T^{\mu k}$ can be interpreted using a two-dimensional surface of area \mathscr{C}, at rest in the observer's frame with positive normal pointing in the k-direction.

During the lapse of time Δt, this 2-surface sweeps out a 3-volume with volume 1-form $\boldsymbol{\Sigma} = \mathcal{A} \, \Delta t \, \boldsymbol{dx}^k$ (see Box 5.2). The μ-component of 4-momentum that crosses the 2-surface in time Δt is

$$p^\mu = \boldsymbol{T}(\boldsymbol{dx}^\mu, \boldsymbol{\Sigma}) = \mathcal{A} \, \Delta t \, \boldsymbol{T}(\boldsymbol{dx}^\mu, \boldsymbol{dx}^k) = \mathcal{A} \, \Delta t \, T^{\mu k}.$$

Thus, the flux of 4-momentum (4-momentum crossing a unit surface oriented perpendicular to \boldsymbol{e}_k, in unit time) is

$$(p^\mu / \mathcal{A} \, \Delta t)_{\text{crossing surface} \perp \text{to } \boldsymbol{e}_k} = T^{\mu k}, \tag{5.12b}$$

or, equivalently:

<div style="margin-left:2em;">

T^{0k}: energy flux

T^{jk}: stress

</div>

$T^{0k} = k$-component of energy flux
 (units: erg/cm^2 sec, or cm^{-2}); $\qquad\qquad\qquad\qquad\qquad$ (5.13c)

$T^{jk} = j, k$ component of "stress"
 $\equiv k$-component of flux of j-component of momentum $\qquad\qquad$ (5.13d)
 $\equiv j$-component of force produced by fields and matter at $x^k - \epsilon$ acting
 on fields and matter at $x^k + \epsilon$ across a unit surface, the perpendicular
 to which is \boldsymbol{e}_k
 (units: dynes/cm^2, or cm^{-2}).

(Recall that "momentum transfer per second" is the same as "force.")

The stress-energy tensor is necessarily symmetric, $T^{\alpha\beta} = T^{\beta\alpha}$; but the proof of this will be delayed until several illustrations have been examined.

§5.4. STRESS-ENERGY TENSOR FOR A SWARM OF PARTICLES

Consider a swarm of particles. Choose some event \mathcal{P} inside the swarm. Divide the particles near \mathcal{P} into categories, $A = 1, 2, \ldots$, in such a way that all particles in the same category have the same properties:

$$
\begin{aligned}
m_{(A)}, &\qquad\qquad \text{rest mass;} \\[1em]
\boldsymbol{u}_{(A)}, &\qquad\qquad \text{4-velocity;} \\[1em]
\boldsymbol{p}_{(A)} = m_{(A)}\boldsymbol{u}_{(A)}, &\qquad\qquad \text{4-momentum.}
\end{aligned}
$$

Number-flux vector for swarm of particles defined

Let $N_{(A)}$ be the number of category-A particles per unit volume, as measured in the particles' own rest frame. Then the *"number-flux vector"* $\boldsymbol{S}_{(A)}$, defined by

$$\boldsymbol{S}_{(A)} \equiv N_{(A)}\boldsymbol{u}_{(A)}, \tag{5.14}$$

has components with simple physical meanings. In a frame where category-A particles have ordinary velocity $\boldsymbol{v}_{(A)}$, the meanings are:

$$S_{(A)}^0 = N_{(A)}u_{(A)}^0 = N_{(A)}\,[1 - v_{(A)}{}^2]^{-1/2} = \text{number density};\qquad (5.15a)$$

Number density in
particles' rest frame

Lorentz contraction
factor for volume

$$\boldsymbol{S}_{(A)} = N_{(A)}\boldsymbol{u}_{(A)} = S_{(A)}^0\boldsymbol{v}_{(A)} = \text{flux of particles}.\qquad (5.15b)$$

Consequently, the 4-momentum density has components

$$T_{(A)}^{\mu0} = p_{(A)}^\mu S_{(A)}^0 = m_{(A)}u_{(A)}^\mu N_{(A)}u_{(A)}^0$$
$$= m_{(A)}N_{(A)}u_{(A)}^\mu u_{(A)}^0;$$

and the flux of μ-component of momentum across a surface with perpendicular direction \boldsymbol{e}_j is

$$T_{(A)}^{\mu j} = p_{(A)}^\mu S_{(A)}^j = m_{(A)}u_{(A)}^\mu N_{(A)}u_{(A)}^j$$
$$- m_{(A)}N_{(A)}u_{(A)}^\mu u_{(A)}^j.$$

These equations are precisely the $\mu, 0$ and μ, j components of the geometric, frame-independent equation

$$\boldsymbol{T}_{(A)} = m_{(A)}N_{(A)}\boldsymbol{u}_{(A)} \otimes \boldsymbol{u}_{(A)} = \boldsymbol{p}_{(A)} \otimes \boldsymbol{S}_{(A)}.\qquad (5.16)$$

Stress-energy tensor for
swarm of particles

The total number-flux vector and stress-energy tensor for all particles in the swarm near \mathscr{P} are obtained by summing over all categories:

$$\boldsymbol{S} = \sum_A N_{(A)}\boldsymbol{u}_{(A)};\qquad (5.17)$$

$$\boldsymbol{T} = \sum_A m_{(A)}N_{(A)}\boldsymbol{u}_{(A)} \otimes \boldsymbol{u}_{(A)} = \sum_A \boldsymbol{p}_{(A)} \otimes \boldsymbol{S}_{(A)}.\qquad (5.18)$$

§5.5. STRESS-ENERGY TENSOR FOR A PERFECT FLUID

There is no simpler example of a fluid than a gas of noninteracting particles ("ideal gas") in which the velocities of the particles are distributed isotropically. In the Lorentz frame where isotropy obtains, symmetry argues equality of the diagonal space-space components of the stress-energy tensor,

Ideal gas defined

$$T_{xx} = T_{yy} = T_{zz} = \sum_A \frac{m_{(A)}v_{x(A)}}{(1 - v_{(A)}{}^2)^{1/2}}\frac{N_{(A)}v_{x(A)}}{(1 - v_{(A)}{}^2)^{1/2}},\qquad (5.19)$$

and vanishing of all the off-diagonal components. Moreover, (5.19) represents a product: the number of particles per unit volume, multiplied by velocity in the x-direction (giving flux in the x-direction) and by momentum in the x-direction,

giving the standard kinetic-theory expression for the pressure, p. Therefore, the stress-energy tensor takes the form

$$T_{\alpha\beta} = \begin{Vmatrix} \rho & 0 & 0 & 0 \\ 0 & p & 0 & 0 \\ 0 & 0 & p & 0 \\ 0 & 0 & 0 & p \end{Vmatrix} \qquad (5.20)$$

in this special Lorentz frame—the "rest frame" of the gas. Here the quantity ρ has nothing directly to do with the rest-masses of the constituent particles. It measures the density of rest-plus-kinetic energy of these particles.

Rewrite (5.20) in terms of the 4-velocity $u^{\alpha} = (1, 0, 0, 0)$ of the fluid in the gas's rest frame, and find

$$T_{\alpha\beta} = \begin{Vmatrix} \rho & 0 & 0 & 0 \\ 0 & 0 & 0 & 0 \\ 0 & 0 & 0 & 0 \\ 0 & 0 & 0 & 0 \end{Vmatrix} + \begin{Vmatrix} 0 & 0 & 0 & 0 \\ 0 & p & 0 & 0 \\ 0 & 0 & p & 0 \\ 0 & 0 & 0 & p \end{Vmatrix}$$

$$= \rho u_{\alpha} u_{\beta} + p(\eta_{\alpha\beta} + u_{\alpha} u_{\beta}),$$

or, in frame-independent, geometric language

<div style="float:left; width:30%;">

Stress-energy tensor for ideal gas or perfect fluid

Perfect fluid defined

</div>

$$T = p\boldsymbol{g} + (\rho + p)\boldsymbol{u} \otimes \boldsymbol{u}. \qquad (5.21)$$

Expression (5.21) has general application. It is exact for the "ideal gas" just considered. It is also exact for any fluid that is "perfect" in the sense that it is free of such transport processes as heat conduction and viscosity, and therefore (in the rest frame) free of shear stress (diagonal stress tensor; diagonal components identical, because if they were not identical, a rotation of the frame of reference would reveal presence of shear stress). However, for a general perfect fluid, density ρ of mass-energy as measured in the fluid's rest frame includes not only rest mass plus kinetic energy of particles, but also energy of compression, energy of nuclear binding, and all other sources of mass-energy [total density of mass-energy as it might be determined by an idealized experiment, such as that depicted in Figure 1.12, with the sample mass at the center of the sphere, and the test particle executing oscillations of small amplitude about that location, with $\omega^2 = (4\pi/3)\rho$].

§5.6. ELECTROMAGNETIC STRESS-ENERGY

Faraday, with his picture of tensions along lines of force and pressures at right angles to them (Figure 5.2), won insight into new features of electromagnetism. In addition to the tension $\boldsymbol{E}^2/8\pi$ (or $\boldsymbol{B}^2/8\pi$) along lines of force, and an equal pressure at right angles, one has the Poynting flux $(\boldsymbol{E} \times \boldsymbol{B})/4\pi$ and the Maxwell expression for the

Figure 5.2.
Faraday stresses at work. When the electromagnet is connected to an alternating current, the aluminum ring flies into the air.

energy density, $(E^2 + B^2)/8\pi$. All these quantities find their places in the Maxwell stress-energy tensor, defined by

$$4\pi T^{\mu\nu} = F^{\mu\alpha}F^{\nu}{}_{\alpha} - \frac{1}{4}\eta^{\mu\nu}F_{\alpha\beta}F^{\alpha\beta}. \qquad (5.22)$$

Stress-energy tensor for electromagnetic field

EXERCISE

Exercise 5.1.

Show that expression (5.22), evaluated in a Lorentz coordinate frame, gives

$$T^{00} = (E^2 + B^2)/8\pi, \qquad T^{0j} = T^{j0} = (E \times B)^j/4\pi,$$

$$T^{jk} = \frac{1}{4\pi}\left[-(E^j E^k + B^j B^k) + \frac{1}{2}(E^2 + B^2)\delta^{jk} \right]. \qquad (5.23)$$

Show that the stress tensor does describe a tension $(E^2 + B^2)/8\pi$ along the field lines and a pressure $(E^2 + B^2)/8\pi$ perpendicular to the field lines, as stated in the text.

§5.7. SYMMETRY OF THE STRESS-ENERGY TENSOR

All the stress-energy tensors explored above were symmetric. That they could not have been otherwise one sees as follows.

Calculate in a specific Lorentz frame. Consider first the momentum density (components T^{j0}) and the energy flux (components T^{0j}). They must be equal because energy = mass ("$E = Mc^2 = M$"):

Proof that stress-energy tensor is symmetric

$$
\begin{aligned}
T^{0j} &= \text{(energy flux)} \\
&= \text{(energy density)} \times \text{(mean velocity of energy flow)}^j \\
&= \text{(mass density)} \times \text{(mean velocity of mass flow)}^j \\
&= \text{(momentum density)} = T^{j0}.
\end{aligned} \qquad (5.24)
$$

Only the stress tensor T^{jk} remains. For it, one uses the same standard argument as in Newtonian theory. Consider a very small cube, of side L, mass-energy $T^{00}L^3$,

and moment of inertia $\sim T^{00}L^5$. With the space coordinates centered at the cube, the expression for the z-component of torque exerted on the cube by its surroundings is

$$\tau^z = \underbrace{(-T^{yx}L^2)}_{\begin{pmatrix} y\text{-component} \\ \text{of force on} \\ +x \text{ face} \end{pmatrix}} \underbrace{(L/2)}_{\begin{pmatrix} \text{lever} \\ \text{arm to} \\ +x \text{ face} \end{pmatrix}} + \underbrace{(T^{yx}L^2)}_{\begin{pmatrix} y\text{-component} \\ \text{of force on} \\ -x \text{ face} \end{pmatrix}} \underbrace{(-L/2)}_{\begin{pmatrix} \text{lever} \\ \text{arm to} \\ -x \text{ face} \end{pmatrix}}$$

$$- \underbrace{(-T^{xy}L^2)}_{\begin{pmatrix} x\text{-component} \\ \text{of force on} \\ +y \text{ face} \end{pmatrix}} \underbrace{(L/2)}_{\begin{pmatrix} \text{lever} \\ \text{arm to} \\ +y \text{ face} \end{pmatrix}} - \underbrace{(T^{xy}L^2)}_{\begin{pmatrix} x\text{-component} \\ \text{of force on} \\ -y \text{ face} \end{pmatrix}} \underbrace{(-L/2)}_{\begin{pmatrix} \text{lever} \\ \text{arm to} \\ -y \text{ face} \end{pmatrix}}$$

$$= (T^{xy} - T^{yx})L^3.$$

Since the torque decreases only as L^3 with decreasing L, while the moment of inertia decreases as L^5, the torque will set an arbitrarily small cube into arbitrarily great angular acceleration—which is absurd. To avoid this, the stresses distribute themselves so the torque vanishes:

$$T^{yx} = T^{xy}.$$

Put differently, if the stresses were not so distributed, the resultant infinite angular accelerations would instantaneously redistribute them back to equilibrium. This condition of torque balance, repeated for all other pairs of directions, is equivalent to symmetry of the stresses:

$$T^{jk} = T^{kj}. \tag{5.25}$$

§5.8. CONSERVATION OF 4-MOMENTUM: INTEGRAL FORMULATION

Energy-momentum conservation has been a cornerstone of physics for more than a century. Nowhere does its essence shine forth so clearly as in Einstein's geometric formulation of it (Figure 5.3,a). There one examines a four-dimensional region of spacetime \mathcal{V} bounded by a closed, three-dimensional surface $\partial\mathcal{V}$. As particles and fields flow into \mathcal{V} and later out, they carry 4-momentum. Inside \mathcal{V} the particles collide, break up, radiate; radiation propagates, jiggles particles, produces pairs. But at each stage in this complex maze of physical processes, total energy-momentum is conserved. The energy-momentum lost by particles goes into fields; the energy-momentum lost by fields goes into particles. So finally, when the "river" of 4-momentum exits from \mathcal{V}, it carries out precisely the same energy-momentum as it carried in.

Restate this equality by asking for the total flux of 4-momentum *outward* across $\partial\mathcal{V}$. Count inflowing 4-momentum negatively. Then "inflow equals outflow" means "total outflow vanishes":

Integral conservation law for 4-momentum:

$$\oint_{\partial\mathcal{V}} \boldsymbol{T} \cdot d^3\boldsymbol{\Sigma} = 0$$

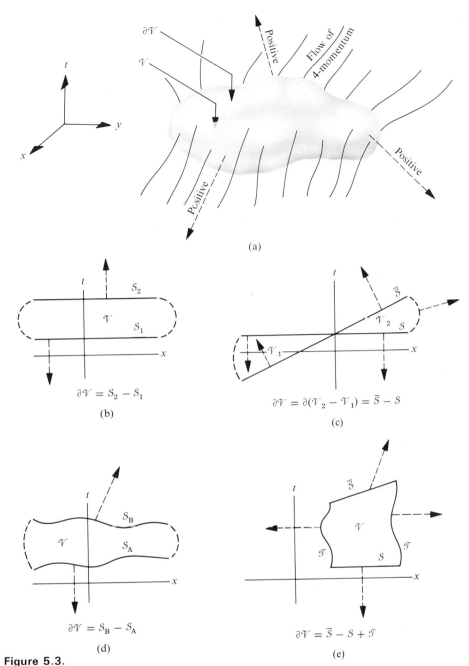

(a)

(b)

$\partial \mathcal{V} = S_2 - S_1$

(c)

$\partial \mathcal{V} = \partial(\mathcal{V}_2 - \mathcal{V}_1) = \bar{S} - S$

(d)

$\partial \mathcal{V} = S_B - S_A$

(e)

$\partial \mathcal{V} = \bar{S} - S + \mathcal{T}$

Figure 5.3.

(a) A four-dimensional region of spacetime \mathcal{V} bounded by a closed three-dimensional surface $\partial \mathcal{V}$. The positive sense of $\partial \mathcal{V}$ is defined to be everywhere outward (away from \mathcal{V}). Conservation of energy-momentum demands that every bit of 4-momentum which flows into \mathcal{V} through $\partial \mathcal{V}$ must somewhere flow back out; none can get lost inside; the interior contains no "sinks." Equivalently, the total flux of 4-momentum across $\partial \mathcal{V}$ in the positive (outward) sense must be zero:

$$\oint_{\partial \mathcal{V}} T^{\mu\alpha} \, d^3\Sigma_\alpha = 0.$$

Figures (b), (c), (d), and (e) depict examples to which the text applies this law of conservation of 4-momentum. All symbols \mathcal{V} (or S) in these figures mean spacetime volumes (or spacelike 3-volumes) with standard orientations. The dotted arrows indicate the positive sense of the closed surface $\partial \mathcal{V}$ used in the text's discussion of 4-momentum conservation. How $\partial \mathcal{V}$ is constructed from the surfaces S and \mathcal{T} is indicated by formulas below the figures. For example, in case (b), $\partial \mathcal{V} = S_2 - S_1$ means that $\partial \mathcal{V}$ is made by joining together S_2 with its standard orientation and S_1 with reversed orientation.

> *Total flux of 4-momentum outward across a closed*
> *three-dimensional surface must vanish.* (5.26)

To calculate the total outward flux in the most elementary of fashions, approximate the closed 3-surface $\partial\mathcal{V}$ by a large number of flat 3-volumes ("boiler plates") with positive direction oriented outward (away from \mathcal{V}). Then

$$\boldsymbol{P}_{\text{total out}} = \sum_{\text{boiler plates } A} \boldsymbol{T}(\dots, \boldsymbol{\Sigma}_{(A)}) = 0, \tag{5.27}$$

where $\boldsymbol{\Sigma}_{(A)}$ is the volume 1-form of boiler plate A. Equivalently, in component notation

$$p^{\mu}{}_{\text{total out}} = \sum_{A} T^{\mu\alpha} \Sigma_{(A)\alpha}. \tag{5.27'}$$

To be slightly more sophisticated about the calculation, take the limit as the number of boiler plates goes to infinity and their sizes go to zero. The result is an integral (Box 5.3, at the end of this section),

$$p^{\mu}{}_{\text{total out}} = \oint_{\partial\mathcal{V}} T^{\mu\alpha} \, d^3\Sigma_{\alpha} = 0. \tag{5.28}$$

Think of this (like all component equations) as a convenient way to express a coordinate-independent statement:

$$\boldsymbol{P}_{\text{total out}} = \oint_{\partial\mathcal{V}} \boldsymbol{T} \cdot d^3\boldsymbol{\Sigma} = 0. \tag{5.29}$$

To be more sophisticated yet (not recommended on first reading of this book) and to simplify the computations in practical cases, interpret the integrands as exterior differential forms (Box 5.4, at the end of this section).

But however one calculates it, and however one interprets the integrands, the statement of the result is simple: the total flux of 4-momentum outward across a closed 3-surface must vanish.

Several special cases of this "integral conservation law," shown in Figure 5.3, are instructive. There shown, in addition to the general case (a), are:

Special cases of integral conservation law:

Case (b)

The closed 3-surface $\partial\mathcal{V}$ is made up of two slices taken at constant time t of a specific Lorentz frame, plus timelike surfaces at "infinity" that join the two slices together. The surfaces at infinity do not contribute to $\oint_{\partial\mathcal{V}} T^{\mu\alpha} \, d^3\Sigma_{\alpha}$ if the stress-energy tensor dies out rapidly enough there. The boundary $\partial\mathcal{V}$ of the standard-oriented 4-volume \mathcal{V}, by definition, has its positive sense away from \mathcal{V}. This demands nonstandard

orientation of S_1 (positive sense toward past), as is indicated by writing $\partial \mathcal{V} = S_2 - S_1$; and it produces a sign flip in the evaluation of the hypersurface integral

$$0 = \oint_{\partial \mathcal{V}} T^{\alpha\mu} \, d^3\Sigma_\mu = -\int_{S_1} T^{\alpha 0} \, dx \, dy \, dz + \int_{S_2} T^{\alpha 0} \, dx \, dy \, dz.$$

Because $T^{\alpha 0}$ is the density of 4-momentum, this equation says

$$\begin{pmatrix} \text{total 4-momentum in} \\ \text{all of space at time } t_1 \end{pmatrix} = \int_{S_1} T^{\alpha 0} \, dx \, dy \, dz$$

Total 4-momentum conserved
in time

$$\qquad\qquad (5.30)$$

$$= \begin{pmatrix} \text{total 4-momentum in} \\ \text{all of space at time } t_2 \end{pmatrix} = \int_{S_2} T^{\alpha 0} \, dx \, dy \, dz.$$

Case (c)

Here one wants to compare hypersurface integrals over S and \bar{S}, which are slices of constant time, $t = \text{const}$ and $\bar{t} = \text{const}$ in two different Lorentz frames. To form a closed surface, one adds time-like hypersurfaces at infinity and assumes they do not contribute to the integral. The orientations fit together smoothly and give a closed surface

$$\partial \mathcal{V} = \bar{S} - S + \text{(surfaces at infinity)}$$

only if one takes $\mathcal{V} = \mathcal{V}_2 - \mathcal{V}_1$—i.e., only if one uses the nonstandard 4-volume orientation in \mathcal{V}_1. (See part A.1 of Box 5.3 for "standard" versus "non-standard" orientation.) The integral conservation law then gives

$$0 = \int_{\bar{S}} \boldsymbol{T} \cdot d^3\boldsymbol{\Sigma} - \int_{S} \boldsymbol{T} \cdot d^3\boldsymbol{\Sigma},$$

or, equivalently,

Total 4-momentum the same
in all Lorentz frames

$$\int_{\bar{S}} \boldsymbol{T} \cdot d^3\boldsymbol{\Sigma} = \text{(total 4-momentum } \boldsymbol{p} \text{ on } \bar{S})$$

$$\qquad\qquad (5.31)$$

$$= \int_{S} \boldsymbol{T} \cdot d^3\boldsymbol{\Sigma} = \text{(total 4-momentum } \boldsymbol{p} \text{ on } S).$$

This says that observers in different Lorentz frames measure the same total 4-momentum \boldsymbol{p}. It does *not* mean that they measure the same components ($p^\alpha \neq p^{\bar\alpha}$); rather, it means they measure the same geometric vector

$$\boldsymbol{p}_{\text{on } S} = p^\alpha \boldsymbol{e}_\alpha = \boldsymbol{p}_{\text{on } \bar{S}} = p^{\bar\alpha} \boldsymbol{e}_{\bar\alpha},$$

a vector whose components are connected by the usual Lorentz transformation law

$$p^\alpha = \Lambda^\alpha{}_{\bar\beta} p^{\bar\beta}. \qquad\qquad (5.32)$$

Case (d)

Here the contribution to the integral comes entirely from two arbitrary spacelike hypersurfaces, \mathcal{S}_A and \mathcal{S}_B, cutting all the way across spacetime. As in cases (a) and (b), the integral form of the conservation law says

<div style="float:left; width:30%;">Total 4-momentum independent of hypersurface where measured</div>

$$\boldsymbol{P}_{\text{on } \mathcal{S}_A} = \boldsymbol{P}_{\text{on } \mathcal{S}_B}; \tag{5.33}$$

i.e., *the total 4-momentum on a spacelike slice through spacetime is independent of the specific slice chosen*—so long as the energy-momentum flux across the "hypersurface at infinity" connecting \mathcal{S}_A and \mathcal{S}_B is zero.

Case (e)

This case concerns a box whose walls oscillate and accelerate as time passes. The three-dimensional boundary $\partial \mathcal{V}$ is made up of (1) the interior \mathcal{S} of the box, at an initial moment of time $t = $ constant in the box's initial Lorentz frame, taken with nonstandard orientation; (2) the interior $\bar{\mathcal{S}}$ of the box, at $\bar{t} = $ constant in its final Lorentz frame, with standard orientation; (3) the 3-volume \mathcal{T} swept out by the box's two-dimensional faces between the initial and final states, with positive sense oriented outward. The integral conservation law $\int_{\partial \mathcal{V}} \boldsymbol{T} \cdot d^3 \boldsymbol{\Sigma} = 0$ says

Change with time of 4-momentum in a box equals flux of 4-momentum across its faces

$$\begin{pmatrix} \text{total 4-momentum} \\ \text{in box at } \bar{\mathcal{S}} \end{pmatrix} - \begin{pmatrix} \text{total 4-momentum} \\ \text{in box at } \mathcal{S} \end{pmatrix}$$

$$= \begin{pmatrix} \text{total 4-momentum that enters box through} \\ \text{its faces between states } \mathcal{S} \text{ and } \bar{\mathcal{S}} \end{pmatrix}. \tag{5.34}$$

§5.9. CONSERVATION OF 4-MOMENTUM: DIFFERENTIAL FORMULATION

Complementary to any "integral conservation law in flat spacetime" is a "differential conservation law" with identical information content. To pass back and forth between them, one can use Gauss's theorem.

Gauss's theorem in four dimensions, applied to the law of 4-momentum conservation, converts the surface integral of $T^{\mu\alpha}$ into a volume integral of $T^{\mu\alpha}{}_{,\alpha}$:

$$0 = \oint_{\partial \mathcal{V}} T^{\mu\alpha} \, d^3 \Sigma_\alpha = \int_{\mathcal{V}} T^{\mu\alpha}{}_{,\alpha} \, dt \, dx \, dy \, dz. \tag{5.35}$$

(See Box 5.3 for elementary discussion; Box 5.4 for sophisticated discussion.) If the integral of $T^{\mu\alpha}{}_{,\alpha}$ is to vanish, as demanded, for any and every 4-volume \mathcal{V}, then $T^{\mu\alpha}{}_{,\alpha}$ must itself vanish everywhere in spacetime:

Differential conservation law for 4-momentum: $\boldsymbol{\nabla} \cdot \boldsymbol{T} = 0$

$$T^{\mu\alpha}{}_{,\alpha} = 0; \text{ i.e., } \boldsymbol{\nabla} \cdot \boldsymbol{T} = 0 \text{ everywhere.} \tag{5.36}$$

(*continued on page 152*)

Box 5.3 VOLUME INTEGRALS, SURFACE INTEGRALS, AND GAUSS'S THEOREM IN COMPONENT NOTATION

A. Volume Integrals in Spacetime

1. By analogy with three-dimensional space, the volume of a "hyperparallele-piped" with vector edges **A**, **B**, **C**, **D** is

$$4\text{-volume} \equiv \Omega \equiv \epsilon_{\alpha\beta\gamma\delta} A^\alpha B^\beta C^\gamma D^\delta = \det \begin{Vmatrix} A^0 & A^1 & A^2 & A^3 \\ B^0 & B^1 & B^2 & B^3 \\ C^0 & C^1 & C^2 & C^3 \\ D^0 & D^1 & D^2 & D^3 \end{Vmatrix}$$

$$= {}^*(\mathbf{A} \wedge \mathbf{B} \wedge \mathbf{C} \wedge \mathbf{D}).$$

Here, as for 3-volumes, orientation matters; interchange of any two edges reverses the sign of Ω. The *standard orientation* for any 4-volume is the one which makes Ω positive; thus, $\mathbf{e}_0 \wedge \mathbf{e}_1 \wedge \mathbf{e}_2 \wedge \mathbf{e}_3$ has standard orientation if \mathbf{e}_0 points toward the future and \mathbf{e}_1, \mathbf{e}_2, \mathbf{e}_3 are a righthanded triad.

2. The "volume element" whose edges in a specific, standard-oriented Lorentz frame are

$$A^\alpha = (\Delta t, 0, 0, 0), \; B^\alpha = (0, \Delta x, 0, 0), \; C^\alpha = (0, 0, \Delta y, 0), \; D^\alpha = (0, 0, 0, \Delta z)$$

has a 4-volume, according to the above definition, given by

$$\Delta^4\Omega = \epsilon_{0123} \, \Delta t \, \Delta x \, \Delta y \, \Delta z = \Delta t \, \Delta x \, \Delta y \, \Delta z.$$

3. Thus, the volume integral of a tensor **S** over a four-dimensional region \mathcal{V} of spacetime, defined as

$$\mathbf{M} \equiv \underset{\begin{pmatrix}\text{number of}\\\text{elementary}\\\text{volumes}\\\to\infty\end{pmatrix}}{\mathrm{Lim}} \quad \underset{\begin{pmatrix}\text{elementary}\\\text{volumes } \mathcal{A}\\\text{in } \mathcal{V}\end{pmatrix}}{\Sigma} \quad \mathbf{S}_{\text{at center of } \mathcal{A}} \, (\text{volume of } \mathcal{A}),$$

can be calculated in a Lorentz frame by

$$M^\alpha{}_{\beta\gamma} = \int_{\mathcal{V}} S^\alpha{}_{\beta\gamma} \, d^4\Omega = \int_{\mathcal{V}} S^\alpha{}_{\beta\gamma} \, dt \, dx \, dy \, dz.$$

Box 5.3 (*continued*)

B. Integrals over *3*-Surfaces in Spacetime

1. Introduce arbitrary coordinates a, b, c on the three-dimensional surface. The elementary volume bounded by coordinate surfaces

$$a_0 < a < a_0 + \Delta a, \qquad b_0 < b < b_0 + \Delta b,$$
$$c_0 < c < c_0 + \Delta c$$

has edges

$$A^\alpha = \frac{\partial x^\alpha}{\partial a} \Delta a, \; B^\beta = \frac{\partial x^\beta}{\partial b} \Delta b, \; C^\gamma = \frac{\partial x^\gamma}{\partial c} \Delta c;$$

so its volume 1-form is

$$\Delta^3 \Sigma_\mu = \epsilon_{\mu\alpha\beta\gamma} \frac{\partial x^\alpha}{\partial a} \frac{\partial x^\beta}{\partial b} \frac{\partial x^\gamma}{\partial c} \Delta a \, \Delta b \, \Delta c.$$

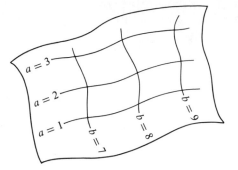

2. The integral of a tensor **S** over the 3-surface \mathcal{S} thus has components

$$N^\alpha{}_\beta = \int_{\mathcal{S}} S^\alpha{}_\beta{}^\gamma \, d^3\Sigma_\gamma = \int_{\mathcal{S}} S^\alpha{}_\beta{}^\gamma \epsilon_{\gamma\mu\nu\lambda} \frac{\partial x^\mu}{\partial a} \frac{\partial x^\nu}{\partial b} \frac{\partial x^\lambda}{\partial c} \, da \, db \, dc.$$

An equivalent formula involving a Jacobian is often used (see exercise 5.5):

$$N^\alpha{}_\beta = \int_{\mathcal{S}} S^\alpha{}_\beta{}^\gamma \frac{1}{3!} \epsilon_{\gamma\mu\nu\lambda} \frac{\partial(x^\mu, x^\nu, x^\lambda)}{\partial(a, b, c)} \, da \, db \, dc.$$

C. Gauss's Theorem Stated

1. Consider a bounded four-dimensional region of spacetime \mathcal{V} with closed boundary $\partial\mathcal{V}$. Orient the volume 1-forms on $\partial\mathcal{V}$ so that the "positive sense" is away from \mathcal{V}.
2. Choose a tensor field **S**. Integrate its divergence over \mathcal{V}, and integrate it itself over $\partial\mathcal{V}$. The results must be the same (*Gauss's theorem*):

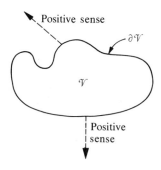

$$\int_{\mathcal{V}} S^\alpha{}_\beta{}^\gamma{}_{,\gamma} \, d^4\Omega = \oint_{\partial\mathcal{V}} S^\alpha{}_\beta{}^\gamma \, d^3\Sigma_\gamma.$$

D. Proof of Gauss's Theorem

1. The indices α and β of $S^{\alpha}{}_{\beta}{}^{\gamma}$ "go along for a free ride," so one can suppress them from the proof. Then the equation to be derived is

$$\int_{\mathcal{V}} S^{\gamma}{}_{,\gamma}\, dt\, dx\, dy\, dz = \oint_{\partial\mathcal{V}} S^{\gamma}\, d^3\Sigma_{\gamma}.$$

2. Since the integral of a derivative is just the original function, the volume integral of $S^0{}_{,0}$ is

$$\int_{\mathcal{V}} S^0{}_{,0}\, dt\, dx\, dy\, dz$$

$$= \int_{\text{"up"}} S^0\, dx\, dy\, dz - \int_{\text{"down"}} S^0\, dx\, dy\, dz.$$

3. The surface integral $\int_{\partial\mathcal{V}} S^0\, d^3\Sigma_0$ can be reduced to the same set of terms:

 a. Use x, y, z as coordinates on $\partial\mathcal{V}$. On the "up" side, $d^3\Sigma_0$ must be positive to achieve a "positive" sense pointing away from \mathcal{V}, so (see part B above)

$$d^3\Sigma_0 = \epsilon_{0\alpha\beta\gamma}\frac{\partial x^{\alpha}}{\partial x}\frac{\partial x^{\beta}}{\partial y}\frac{\partial x^{\gamma}}{\partial z}\, dx\, dy\, dz = \epsilon_{0123}\, dx\, dy\, dz = dx\, dy\, dz.$$

 b. On the "down" side, $d^3\Sigma_0$ must be negative, so

$$d^3\Sigma_0 = -dx\, dy\, dz.$$

 c. Hence,

$$\int_{\partial\mathcal{V}} S^0\, d^3\Sigma_0 = \int_{\text{"up"}} S^0\, dx\, dy\, dz - \int_{\text{"down"}} S^0\, dx\, dy\, dz.$$

4. Equality is proved for the other components in the same manner. Adding components produces the result desired:

$$\int_{\mathcal{V}} S^{\gamma}{}_{,\gamma}\, d^4\Omega = \oint_{\partial\mathcal{V}} S^{\gamma}\, d^3\Sigma_{\gamma}.$$

FOR THE READER WHO HAS STUDIED CHAPTER 4

Box 5.4 I. EVERY INTEGRAL IS THE INTEGRAL OF A FORM.
II. THE THEOREM OF GAUSS IN THE LANGUAGE OF FORMS.

I. *Every integral encountered* in Chapter 5 can be interpreted as *the integral of an exterior differential form*. This circumstance shows up in fourfold and threefold integrals, for example, in the fact that

$$d^4\Omega = \boldsymbol{\varepsilon} = {}^*1 = \varepsilon_{0123}\, \boldsymbol{dt} \wedge \boldsymbol{dx} \wedge \boldsymbol{dy} \wedge \boldsymbol{dz}$$

and

$$d^3\Sigma_\mu = \varepsilon_{\mu|\alpha\beta\gamma|}\, \boldsymbol{dx}^\alpha \wedge \boldsymbol{dx}^\beta \wedge \boldsymbol{dx}^\gamma$$

are basis 4- and 3-forms. (Recall: the indices $\alpha\beta\gamma$ between vertical bars are to be summed only over $0 \leq \alpha < \beta < \gamma \leq 3$.) A more extensive glossary of notations is found in C below.

II. *Gauss's Theorem for a tensor integral* in flat space reads

$$\int_{\mathcal{V}} (\boldsymbol{\nabla} \cdot \boldsymbol{S})\, d^4\Omega = \oint_{\partial\mathcal{V}} \boldsymbol{S} \cdot d\boldsymbol{\Sigma}$$

for any tensor, such as $\boldsymbol{S} = S^\alpha{}_\beta{}^\gamma \boldsymbol{e}_\alpha \otimes \boldsymbol{w}^\beta \otimes \boldsymbol{e}_\gamma$ (see Box 5.3 for component form). It is an application of the generalized Stokes Theorem (Box 4.1), and *depends on the fact that the basis vectors* \boldsymbol{e}_α *and* \boldsymbol{w}^β *of a global Lorentz frame are constants*, i.e., are independent of x. The definitions follow in A; the proof is in B.

A. *Tensor-valued integrals* can be defined in flat spaces because one uses constant basis vectors. Thus one defines

$$\int \boldsymbol{S} \cdot d^3\boldsymbol{\Sigma} = \boldsymbol{e}_\alpha \otimes \boldsymbol{w}^\beta \int S^\alpha{}_\beta{}^\gamma\, d^3\Sigma_\gamma$$

for a tensor of the indicated rank. One justifies pulling basis vectors and forms outside the integral sign because they are constants, independent of location in spacetime. Each of the numbers $\int S^\alpha{}_\beta{}^\gamma\, d^3\Sigma_\gamma$ (for $\alpha, \beta = 0, 1, 2, 3$) is then evaluated by substituting *any* properly oriented parametrization of the hypersurface into the 3-form $S^\alpha{}_\beta{}^\gamma\, d^3\Sigma_\gamma$ as described in Box 4.1 (arbitrary curvilinear parametrization in the part of the calculation not involving the "free indices" α and β). In other words, $\boldsymbol{S} \cdot d^3\boldsymbol{\Sigma} = \boldsymbol{e}_\alpha \otimes \boldsymbol{w}^\beta \otimes S^\alpha{}_\beta{}^\gamma\, d^3\Sigma_\gamma$ is considered a "tensor-valued 3-form." Under an integral sign, it is contracted with the hyperplane element tangent to the 3-surface $\mathcal{P}(\lambda^1, \lambda^2, \lambda^3)$ of integration to form the integral

$$\int \boldsymbol{S} \cdot d^3\boldsymbol{\Sigma} = \int \left\langle \boldsymbol{S} \cdot d^3\boldsymbol{\Sigma}, \frac{\partial\mathcal{P}}{\partial\lambda^1} \wedge \frac{\partial\mathcal{P}}{\partial\lambda^2} \wedge \frac{\partial\mathcal{P}}{\partial\lambda^3} \right\rangle d\lambda^1\, d\lambda^2\, d\lambda^3$$

$$= \boldsymbol{e}_\alpha \otimes \boldsymbol{w}^\beta \int S^\alpha{}_\beta{}^\gamma \varepsilon_{\gamma|\lambda\mu\nu|} \underbrace{\frac{\partial(x^\lambda, x^\mu, x^\nu)}{\partial(\lambda^1, \lambda^2, \lambda^3)}}_{\substack{\text{Jacobian} \\ \text{determinant}}} d\lambda^1\, d\lambda^2\, d\lambda^3.$$

Although constant basis vectors $\boldsymbol{e}_\alpha, \boldsymbol{w}^\beta$ derived from rectangular coordinates are essential here, *a completely general parametrization of the hypersurface may be used*.

B. *The proof of Gauss's Theorem* is a computation:

$$\oint_{\partial \mathcal{V}} \boldsymbol{S} \cdot d^3\boldsymbol{\Sigma} = \boldsymbol{e}_\alpha \otimes \boldsymbol{\omega}^\beta \oint_{\partial \mathcal{V}} S^\alpha{}_\beta{}^\gamma \, d^3\Sigma_\gamma \qquad (\boldsymbol{e}_\alpha, \boldsymbol{\omega}^\beta \text{ are constant})$$

$$= \boldsymbol{e}_\alpha \otimes \boldsymbol{\omega}^\beta \int_{\mathcal{V}} \boldsymbol{d}(S^\alpha{}_\beta{}^\gamma \, d^3\Sigma_\gamma) \qquad \text{(Stokes Theorem)}$$

$$= \boldsymbol{e}_\alpha \otimes \boldsymbol{\omega}^\beta \int_{\mathcal{V}} S^\alpha{}_\beta{}^\gamma{}_{,\gamma} \, {*}1 \qquad \text{(see below)}$$

$$= \int_{\mathcal{V}} (\boldsymbol{\nabla} \cdot \boldsymbol{S}) \, d^4\Omega. \qquad \text{(merely notation)}$$

The missing computational step above is

$$\boldsymbol{d}(S^\alpha{}_\beta{}^\gamma \, d^3\Sigma_\gamma) = (\partial S^\alpha{}_\beta{}^\gamma / \partial x^\rho) \, \boldsymbol{d}x^\rho \wedge d^3\Sigma_\gamma$$
$$= (\partial S^\alpha{}_\beta{}^\gamma / \partial x^\gamma) \, {*}1.$$

Here the first step uses $\boldsymbol{d}(d^3\Sigma_\gamma) = 0$ (which follows from $\varepsilon_{\mu\alpha\beta\gamma} = \text{const}$ in flat spacetime). The second step uses

$$\boldsymbol{d}x^\rho \wedge d^3\Sigma_\gamma = \delta^\rho_\gamma \, {*}1.$$

[Write the lefthand side of this identity as $\varepsilon_{\gamma|\mu\nu\lambda|} \, \boldsymbol{d}x^\rho \wedge \boldsymbol{d}x^\mu \wedge \boldsymbol{d}x^\nu \wedge \boldsymbol{d}x^\lambda$. The only possible non-zero term in the sum over $\mu\nu\lambda$ is the one with $\mu < \nu < \lambda$ all different from ρ. The righthand side is the value of this term.]

C. *Glossary of notations.*

Charge density 3-form:

$$\begin{aligned}{*}\boldsymbol{J} &= J^\mu \, d^3\Sigma_\mu = \boldsymbol{J} \cdot d^3\boldsymbol{\Sigma} \\ &= \underbrace{J^\mu \varepsilon_{\mu\alpha\beta\gamma}}_{({*}J)_{\alpha\beta\gamma}} \underbrace{\boldsymbol{d}x^\alpha \wedge \boldsymbol{d}x^\beta \wedge \boldsymbol{d}x^\gamma /3!}_{d^3\Sigma_\mu}.\end{aligned}$$

Maxwell and Faraday 2-forms:

$${*}\boldsymbol{F} = \frac{1}{2} F^{\mu\nu} \, d^2 S_{\mu\nu};$$

$$\boldsymbol{F} = \frac{1}{2} F_{\mu\nu} \, \boldsymbol{d}x^\mu \wedge \boldsymbol{d}x^\nu.$$

Basis 2-forms:

$$\boldsymbol{d}x^\alpha \wedge \boldsymbol{d}x^\beta; \qquad \text{(one way to label)}$$

$$d^2 S_{\mu\nu} = \varepsilon_{\mu\nu|\alpha\beta|} \, \boldsymbol{d}x^\alpha \wedge \boldsymbol{d}x^\beta. \qquad \text{(dual way to label)}$$

Energy-momentum density 3-form:

$$\underbrace{\boldsymbol{T} \cdot d^3\boldsymbol{\Sigma} \equiv \boldsymbol{e}_\mu T^{\mu\nu} \, d^3\Sigma_\nu \equiv {*}\boldsymbol{T};}_{\text{dual on last index, } ({*}T)^\mu{}_{\alpha\beta\gamma} = T^{\mu\nu} \varepsilon_{\nu\alpha\beta\gamma}.}$$

Angular momentum density 3-form:

$$\boldsymbol{\mathcal{J}} \cdot d^3\boldsymbol{\Sigma} \equiv \frac{1}{2} \boldsymbol{e}_\mu \wedge \boldsymbol{e}_\nu \mathcal{J}^{\mu\nu\alpha} \, d^3\Sigma_\alpha \equiv {*}\boldsymbol{\mathcal{J}};$$

$$({*}\mathcal{J})^{\mu\nu}{}_{\alpha\beta\gamma} = \mathcal{J}^{\mu\nu\lambda} \varepsilon_{\lambda\alpha\beta\gamma}.$$

(In the frame-independent equation $\boldsymbol{\nabla} \cdot \boldsymbol{T} = 0$, one need not worry about which slot of \boldsymbol{T} to take the divergence on; the slots are symmetric, so either can be used.)

The equation $\boldsymbol{\nabla} \cdot \boldsymbol{T} = 0$ is *the differential formulation of the law of 4-momentum conservation.* It is also called the *equation of motion for stress-energy,* because it places constraints on the dynamic evolution of the stress-energy tensor. To examine these constraints for simple systems is to realize the beauty and power of the equation $\boldsymbol{\nabla} \cdot \boldsymbol{T} = 0.$

§5.10. SAMPLE APPLICATIONS OF $\boldsymbol{\nabla} \cdot \boldsymbol{T} = 0$

Newtonian fluid characterized by $|v^j| \ll 1$, $p \ll \rho$

The equation of motion $\boldsymbol{\nabla} \cdot \boldsymbol{T} = 0$ makes contact with the classical (Newtonian) equations of hydrodynamics, when applied to a nearly Newtonian fluid. Such a fluid has low velocities relative to the Lorentz frame used, $|v^j| \ll 1$; and in its rest frame its pressure is small compared to its density of mass-energy, $p/\rho = p/\rho c^2 \ll 1$. For example, the air in a hurricane has

$$|v^j| \sim 100 \text{ km/hour} \sim 3{,}000 \text{ cm/sec} \sim 10^{-7} c = 10^{-7} \ll 1,$$

$$\frac{p}{\rho} \sim \frac{1 \text{ atmosphere}}{10^{-3} \text{ g/cm}^3} \sim \frac{10^6 \text{ dynes/cm}^2}{10^{-3} \text{ g/cm}^3} = 10^9 \frac{\text{cm}^2}{\text{sec}^2} \sim 10^{-12} c^2 = 10^{-12} \ll 1.$$

Stress-energy tensor and equation of motion for a Newtonian fluid

The stress-energy tensor for such a fluid has components

$$T^{00} = (\rho + p)u^0 u^0 - p \approx \rho, \tag{5.37a}$$

$$T^{0j} = T^{j0} = (\rho + p)u^0 u^j \approx \rho v^j, \tag{5.37b}$$

$$T^{jk} = (\rho + p)u^j u^k + p\,\delta^{jk} \approx \rho v^j v^k + p\,\delta^{jk}; \tag{5.37c}$$

and the equation of motion $\boldsymbol{\nabla} \cdot \boldsymbol{T} = 0$ has components

$$T^{00}{}_{,0} + T^{0j}{}_{,j} = \partial\rho/\partial t + \boldsymbol{\nabla} \cdot (\rho \boldsymbol{v}) = 0$$

$$\text{(``equation of continuity'');} \tag{5.38a}$$

and

$$T^{j0}{}_{,0} + T^{jk}{}_{,k} = \partial(\rho v^j)/\partial t + \partial(\rho v^j v^k)/\partial x^k + \partial p/\partial x^j = 0,$$

or, equivalently (by combining with the equation of continuity),

$$\frac{\partial \boldsymbol{v}}{\partial t} + (\boldsymbol{v} \cdot \boldsymbol{\nabla})\boldsymbol{v} = -\frac{1}{\rho}\boldsymbol{\nabla}p \qquad \text{(``Euler's equation'').} \tag{5.38b}$$

Box 5.5 derives and discusses these results from the Newtonian viewpoint.

Application of $\boldsymbol{\nabla} \cdot \boldsymbol{T} = 0$ to an electrically charged, vibrating rubber block

As a second application of $\boldsymbol{\nabla} \cdot \boldsymbol{T} = 0$, consider a composite system: a block of rubber with electrically charged beads imbedded in it, interacting with an electromagnetic field. The block of rubber vibrates, and its accelerating beads radiate electromagnetic waves; at the same time, incoming electromagnetic waves push on the beads, altering the pattern of vibration of the block of rubber. The interactions shove 4-momentum back and forth between beaded block and electromagnetic field.

Box 5.5 NEWTONIAN HYDRODYNAMICS REVIEWED

Consider a classical, nonrelativistic, perfect fluid. Apply Newton's law $\mathbf{F} = m\mathbf{a}$ to a "fluid particle"; that is, to a small fixed mass of fluid followed in its progress through space:

$$\frac{d}{dt} \text{(momentum per unit mass)} = \text{(force per unit mass)}$$

$$= \frac{\text{(force per unit volume)}}{\text{(density)}} = \frac{-\text{(gradient of pressure)}}{\text{(density)}}$$

or

$$\frac{d\mathbf{v}}{dt} = -\frac{1}{\rho} \mathbf{\nabla} p. \tag{1}$$

Translate from time-rate of change following the fluid to time-rate of change as measured at a fixed location, finding

$$\begin{pmatrix} \text{rate of change} \\ \text{with time} \\ \text{following fluid} \end{pmatrix} = \begin{pmatrix} \text{rate of change} \\ \text{with time at} \\ \text{fixed location} \end{pmatrix} + \begin{pmatrix} \text{velocity} \\ \text{of fluid} \end{pmatrix} \cdot \begin{pmatrix} \text{rate of change} \\ \text{with position} \end{pmatrix}$$

or

$$\frac{\partial \mathbf{v}}{\partial t} + (\mathbf{v} \cdot \mathbf{\nabla})\mathbf{v} = -\frac{1}{\rho} \mathbf{\nabla} p \tag{2}$$

or

$$\frac{\partial v_i}{\partial t} + v_{i,k} v_k = -\frac{1}{\rho} p_{,i}.$$

(Latin indices run from 1 to 3; summation convention; upper and lower indices used indifferently for space dimensions in flat space!) This is *Euler's fundamental equation* for the hydrodynamics of a perfect fluid.

Two further equations are needed to complete the description of a perfect fluid. One states the absence of heat transfer by requiring that the specific entropy (entropy per unit mass) be constant for each fluid "particle":

$$\frac{ds}{dt} = 0, \quad \text{or} \quad \frac{\partial s}{\partial t} + (\mathbf{v} \cdot \mathbf{\nabla})s = 0. \tag{3}$$

The final equation expresses the conservation of mass:

$$\frac{\partial \rho}{\partial t} + \mathbf{\nabla} \cdot (\rho \mathbf{v}) = 0, \tag{4}$$

or

$$\frac{\partial \rho}{\partial t} + (\rho v_k)_{,k} = 0;$$

Box 5.5 (*continued*)

it is analogous in every way to the equation that expresses conservation of charge in electrodynamics and that bears the same name, *"equation of continuity."*

The Newtonian stress-energy tensor, like its relativistic counterpart, is linked to conservation of momentum and mass. Therefore examine the time-rate of change of the density of fluid momentum, ρv_i, contained in a unit volume; thus,

$$\partial(\rho v_i)/\partial t = -(\rho v_i v_k)_{,k} - p_{,i}. \tag{5}$$

Momentum flows into the little volume element on the left ("force equals time-rate of change of momentum") and out on the right; similarly at the other faces. Therefore the righthand side of (5) must represent the divergence of this momentum flux:

$$\partial(\rho v_i)/\partial t = -T_{ik,k}. \tag{6}$$

Consequently, we take for the momentum flux itself

$$T^{ik} = T_{ik} = \underbrace{\rho v_i v_k}_{\text{"convection"}} + \underbrace{\delta_{ik} p}_{\text{"push"}}. \tag{7}$$

For the momentum density, the Newtonian value is

$$T^{0i} = T^{i0} = \rho v_i. \tag{8}$$

With this notation, the equation for the time-rate of change of momentum becomes

$$\partial T^{i\mu}/\partial x^\mu = 0; \tag{9}$$

and with $T^{00} = \rho$, the equation of continuity reads

$$\partial T^{0\mu}/\partial x^\mu = 0. \tag{10}$$

In conclusion, these Newtonian considerations give a reasonable approximation to the relativistic stress-energy tensor:

$$\left\| \begin{array}{c:c} \rho & \rho v^j \\ \hdashline \rho v^i & \rho v^i v^j + \delta^{ij} p \end{array} \right\| \simeq \left\| \begin{array}{c:c} (p + \rho)u^0 u^0 - p & (p + \rho)u^0 u^j \\ \hdashline (p + \rho)u^0 u^i & (p + \rho)u^i u^j + \delta^{ij} p \end{array} \right\| \tag{11}$$

The 4-momentum of neither block nor field is conserved; neither $\mathbf{\nabla} \cdot \boldsymbol{T}_{(\text{block})}$ nor $\mathbf{\nabla} \cdot \boldsymbol{T}_{(\text{em field})}$ vanishes. But total 4-momentum must be conserved, so

$$\mathbf{\nabla} \cdot (\boldsymbol{T}_{(\text{block})} + \boldsymbol{T}_{(\text{em field})}) \text{ must vanish.} \tag{5.39}$$

For a general electromagnetic field interacting with any source, $\mathbf{\nabla} \cdot \boldsymbol{T}_{(\text{em field})}$ has the form

$$T^{\mu\nu}_{(\text{em field}),\nu} = -F^{\mu\alpha}J_\alpha. \tag{5.40}$$

(This was derived in exercise 3.18 by combining $T^{\mu\nu}_{,\nu} = 0$ with expression 5.22 for the electromagnetic stress-energy tensor, and with Maxwell's equations.) For our beaded block, \boldsymbol{J} is the 4-current associated with the vibrating, charged beads, and \boldsymbol{F} is the electromagnetic field tensor. The time component of equation (5.40) reads

$$
\begin{aligned}
T^{0\nu}_{(\text{em field}),\nu} &= -F^{0k}J_k = -\boldsymbol{E} \cdot \boldsymbol{J} \\
&= -\begin{pmatrix} \text{rate at which electric field } \boldsymbol{E} \text{ does work} \\ \text{on a unit volume of charged beads} \end{pmatrix}.
\end{aligned}
\tag{5.41}
$$

For comparison, $T^{00}_{(\text{block}),0}$ is the rate at which the block's energy density changes with time, $-T^{0j}_{(\text{block}),j}$ is the contribution of the block's energy flux to this rate of change of energy density, and consequently their difference $T^{0\nu}_{(\text{block}),\nu}$ has the meaning

$$
T^{0\nu}_{(\text{block}),\nu} = \begin{pmatrix} \text{rate at which mass-energy of block per} \\ \text{unit volume increases due to actions} \\ \text{other than internal mechanical forces} \\ \text{between one part of block and another} \end{pmatrix}.
\tag{5.42}
$$

Hence, the conservation law

$$(T^{0\nu}_{(\text{em field})} + T^{0\nu}_{(\text{block})}),_\nu = 0$$

says that the mass-energy of the block increases at precisely the same rate as the electric field does work on the beads. A similar result holds for momentum:

$$
\begin{aligned}
T^{k\nu}_{(\text{em field}),\nu}\boldsymbol{e}_k &= -F^{k\nu}J_\nu\boldsymbol{e}_k = -(J^0\boldsymbol{E} + \boldsymbol{J} \times \boldsymbol{B}) \\
&= -\begin{pmatrix} \text{Lorentz force per unit volume} \\ \text{acting on beads} \end{pmatrix},
\end{aligned}
\tag{5.43}
$$

$$
T^{k\nu}_{(\text{block}),\nu}\boldsymbol{e}_k = \begin{pmatrix} \text{rate at which momentum per unit volume} \\ \text{of block increases due to actions} \\ \text{other than its own stresses} \end{pmatrix};
\tag{5.44}
$$

so the conservation law

$$(T^{k\nu}_{(\text{em field})} + T^{k\nu}_{(\text{block})}),_\nu = 0$$

says that the rate of change of the momentum of the block equals the force of the electromagnetic field on its beads.

§5.11. ANGULAR MOMENTUM

Angular momentum defined
and its integral conservation
law derived

The symmetry, $T^{\mu\nu} = T^{\nu\mu}$, of the stress-energy tensor enables one to define a conserved angular momentum $J^{\alpha\beta}$, analogous to the linear momentum p^{α}. The angular momentum is defined relative to a specific but arbitrary origin—an event \mathcal{A} with coordinates, in a particular Lorentz frame,

$$x^{\alpha}(\mathcal{A}) = a^{\alpha}. \qquad (5.45)$$

The angular momentum about \mathcal{A} is defined using the tensor

$$\mathcal{J}^{\alpha\beta\gamma} = (x^{\alpha} - a^{\alpha})T^{\beta\gamma} - (x^{\beta} - a^{\beta})T^{\alpha\gamma}. \qquad (5.46)$$

(Note that $x^{\alpha} - a^{\alpha}$ is the vector separation of the "field point" x^{α} from the "origin" \mathcal{A}; $T^{\alpha\gamma}$ is here evaluated at the "field point".) Because of the symmetry of \mathbf{T}, $\mathcal{J}^{\alpha\beta\gamma}$ has vanishing divergence:

$$\mathcal{J}^{\alpha\beta\gamma}{}_{,\gamma} = \delta^{\alpha}{}_{\gamma}T^{\beta\gamma} + (x^{\alpha} - a^{\alpha})\underbrace{T^{\beta\gamma}{}_{,\gamma}}_{0} - \delta^{\beta}{}_{\gamma}T^{\alpha\gamma} - (x^{\beta} - a^{\beta})\underbrace{T^{\alpha\gamma}{}_{,\gamma}}_{0}$$

$$= T^{\beta\alpha} - T^{\alpha\beta} = 0. \qquad (5.47)$$

Consequently, its integral over any closed 3-surface vanishes

$$\oint_{\partial\mathcal{V}} \mathcal{J}^{\alpha\beta\gamma}\, d^3\Sigma_{\gamma} = 0 \qquad (5.48)$$

("*integral form of the law of conservation of angular momentum*").
The integral over a spacelike surface of constant time t is

$$J^{\alpha\beta} = \int \mathcal{J}^{\alpha\beta 0}\, dx\, dy\, dz = \int [(x^{\alpha} - a^{\alpha})T^{\beta 0} - (x^{\beta} - a^{\beta})T^{\alpha 0}]\, dx\, dy\, dz. \quad (5.49)$$

Recalling that $T^{\beta 0}$ is momentum density, one sees that (5.49) has the same form as the equation "$\mathbf{J} = \mathbf{r} \times \mathbf{p}$" of Newtonian theory. Hence the name "total angular momentum" for $J^{\alpha\beta}$. Various aspects of this conserved angular momentum, including the tie to its Newtonian cousin, are explored in Box 5.6.

EXERCISES

Exercise 5.2. CHARGE CONSERVATION
Exercise 3.16 revealed that the charge-current 4-vector \mathbf{J} satisfies the differential conservation law $\boldsymbol{\nabla} \cdot \mathbf{J} = 0$. Write down the corresponding integral conservation law, and interpret it for the four closed surfaces of Fig. 5.3.

Exercise 5.3. PARTICLE PRODUCTION
Inside highly evolved, massive stars, the temperature is so high that electron-positron pairs are continually produced and destroyed. Let \mathbf{S} be the number-flux vector for electrons and positrons, and denote its divergence by

$$\epsilon \equiv \boldsymbol{\nabla} \cdot \mathbf{S}. \qquad (5.50)$$

Box 5.6 ANGULAR MOMENTUM

A. Definition of Angular Momentum

(a) Pick an arbitrary spacelike hypersurface S and an arbitrary event \mathcal{A} with coordinates $x^\alpha(\mathcal{A}) \equiv a^\alpha$. (Use globally inertial coordinates throughout.)

(b) Define *"total angular momentum on S about \mathcal{A}"* to be

$$J^{\mu\nu} \equiv \int_S \mathcal{J}^{\mu\nu\alpha}\, d^3\Sigma_\alpha,$$

$$\mathcal{J}^{\mu\nu\alpha} \equiv (x^\mu - a^\mu)T^{\nu\alpha} - (x^\nu - a^\nu)T^{\mu\alpha}.$$

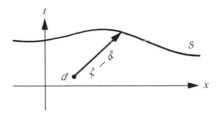

(c) If S is a hypersurface of constant time t, this becomes

$$J^{\mu\nu} = \int \mathcal{J}^{\mu\nu 0}\, dx\, dy\, dz.$$

B. Conservation of Angular Momentum

(a) $T^{\mu\nu}{}_{,\nu} = 0$ implies $\mathcal{J}^{\mu\nu\alpha}{}_{,\alpha} = 0$.

(b) This means that $J^{\mu\nu}$ is independent of the hypersurface S on which it is calculated (Gauss's theorem):

$$J^{\mu\nu}(S_A) - J^{\mu\nu}(S_B)$$

$$= \int_{\partial\mathcal{V}} \mathcal{J}^{\mu\nu\alpha}\, d^3\Sigma_\alpha$$

$$= \int_{\mathcal{V}} \mathcal{J}^{\mu\nu\alpha}{}_{,\alpha}\, d^4x = 0.$$

(Note: $\partial\mathcal{V} \equiv$ (boundary of \mathcal{V}) includes S_A, S_B, and timelike surfaces at spatial infinity; contribution of latter dropped—localized source.)

C. Change of Point About Which Angular Momentum is Calculated

Let b^α be vector from \mathcal{A}_0 to \mathcal{A}_1: $b^\alpha = a_1{}^\alpha - a_0{}^\alpha$. Then

$$J^{\mu\nu}(\text{about } \mathcal{A}_1) - J^{\mu\nu}(\text{about } \mathcal{A}_0)$$

$$= -b^\mu \int_S T^{\nu\alpha}\, d^3\Sigma_\alpha + b^\nu \int_S T^{\mu\alpha}\, d^3\Sigma_\alpha$$

$$= -b^\mu P^\nu + b^\nu P^\mu,$$

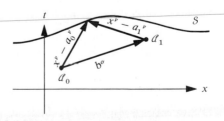

where P^μ is total 4-momentum.

Box 5.6 (*continued*)

D. Intrinsic Angular Momentum

(a) Work, for a moment, in the system's rest frame, where

$$P^0 = M, \qquad P^j = 0, \qquad x_{CM}{}^j = \frac{1}{M}\int x^j T^{00}\, d^3x = \text{location of center of mass.}$$

Intrinsic angular momentum is defined as angular momentum about any event $(a^0, x_{CM}{}^j)$ on center of mass's world line. Its components are denoted $S^{\mu\nu}$ and work out to be

$$S^{0j} = 0, \; S^{jk} = \epsilon^{jk\ell} S^\ell,$$

where

$$\boldsymbol{S} \equiv \int (\boldsymbol{x} - \boldsymbol{x}_{CM}) \times (\textbf{momentum density})\, d^3x$$
$$\equiv \text{``intrinsic angular momentum vector.''}$$

(b) Define "intrinsic angular momentum 4-vector" S^μ to be that 4-vector whose components in the rest frame are $(0, \boldsymbol{S})$; then the above equations say

$$S^{\mu\nu} = U_\alpha S_\beta \epsilon^{\alpha\beta\mu\nu},$$
$$U_\beta \equiv P_\beta/M = \text{4-velocity of center of mass,}$$
$$U_\beta S^\beta = 0.$$

E. Decomposition of Angular Momentum into Intrinsic and Orbital Parts

(a) Pick an arbitrary event \mathcal{A}, whose perpendicular displacement from center-of-mass world line is $-Y^\alpha$, so

$$U_\beta Y^\beta = 0.$$

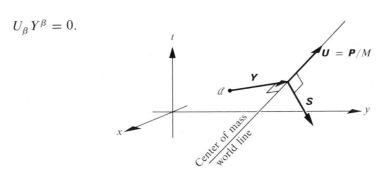

(b) Then, by Part C, the angular momentum about \mathcal{Q} is

$$J^{\mu\nu} = \underbrace{U_\alpha S_\beta \epsilon^{\alpha\beta\mu\nu}}_{S^{\mu\nu}\ \text{(intrinsic)}} + \underbrace{Y^\mu P^\nu - Y^\nu P^\mu}_{L^{\mu\nu}\ \text{(orbital)}}.$$

(c) Knowing the angular momentum about \mathcal{Q}, and the 4-momentum (and hence 4-velocity), one can calculate the vector from \mathcal{Q} to the center-of-mass world line,

$$Y^\mu = -J^{\mu\nu}P_\nu/M^2,$$

and the intrinsic angular momentum

$$S_\rho = \frac{1}{2}U^\sigma J^{\mu\nu}\epsilon_{\sigma\mu\nu\rho}.$$

Use Gauss's theorem to show that ϵ is the number of particles created (minus the number destroyed) in a unit four-dimensional volume of spacetime.

Exercise 5.4. INERTIAL MASS PER UNIT VOLUME

Consider a stressed medium in motion with ordinary velocity $|v| \ll 1$ with respect to a specific Lorentz frame.

(a) Show by Lorentz transformations that the spatial components of the momentum density are

$$T^{0j} = \sum_k m^{jk} v^k, \tag{5.51}$$

where

$$m^{jk} = T^{\bar{0}\bar{0}}\delta^{jk} + T^{\bar{j}\bar{k}} \tag{5.52}$$

and $T^{\bar{\mu}\bar{\nu}}$ are the components of the stress-energy tensor in the rest frame of the medium. Throughout the solar system $T^{\bar{0}\bar{0}} \gg |T^{\bar{j}\bar{k}}|$ (see, e.g., discussion of hurricane in §5.10), so one is accustomed to write $T^{0j} = T^{\bar{0}\bar{0}}v^j$, i.e., "(momentum density) = (rest-mass density) \times (velocity)". But inside a neutron star $T^{\bar{0}\bar{0}}$ may be of the same order of magnitude as $T^{\bar{j}\bar{k}}$, so one must replace "(momentum density) = (rest-mass density) \times (velocity)" by equations (5.51) and (5.52), at low velocities.

(b) Derive equations (5.51) and (5.52) from Newtonian considerations plus the equivalence of mass and energy. (*Hint:* the total mass-energy carried past the observer by a volume V of the medium includes both the rest mass $T^{\bar{0}\bar{0}}V$ and the work done by forces acting across the volume's faces as they "push" the volume through a distance.)

(c) As a result of relation (5.51), the force per unit volume required to produce an acceleration dv^k/dt in a stressed medium, which is at rest with respect to the man who applies the force, is

$$F^j = dT^{0j}/dt = \sum_k m^{jk}\,dv^k/dt. \tag{5.53}$$

This equation suggests that one call m^{jk} the "inertial mass per unit volume" of a stressed medium at rest. In general m^{jk} is a symmetric 3-tensor. What does it become for the special case of a perfect fluid?

(d) Consider an isolated, stressed body at rest and in equilibrium ($T^{\alpha\beta}{}_{,0} = 0$) in the laboratory frame. Show that its total inertial mass, defined by

$$M^{ij} = \int_{\substack{\text{stressed}\\\text{body}}} m^{ij}\, dx\, dy\, dz, \tag{5.54}$$

is isotropic and equals the rest mass of the body

$$M^{ij} = \delta^{ij} \int T^{00}\, dx\, dy\, dz. \tag{5.55}$$

Exercise 5.5. DETERMINANTS AND JACOBIANS

(a) Write out explicitly the sum defining d^2S_{01} in

$$d^2S_{\mu\nu} \equiv \epsilon_{\mu\nu\alpha\beta} \frac{\partial x^\alpha}{\partial a} \frac{\partial x^\beta}{\partial b}\, da\, db.$$

Thereby establish the formula

$$d^2S_{\mu\nu} = \epsilon_{\mu\nu|\alpha\beta|} \frac{\partial(x^\alpha, x^\beta)}{\partial(a, b)}\, da\, db = \frac{1}{2!}\, \epsilon_{\mu\nu\alpha\beta} \frac{\partial(x^\alpha, x^\beta)}{\partial(a, b)}\, da\, db.$$

(Expressions such as these should occur only under integral signs. In this exercise one may either supply an $\int \ldots$ wherever necessary, or else interpret the differentials in terms of the exterior calculus, $da\, db \longrightarrow da \wedge db$; see Box 5.4.) The notation used here for Jacobian determinants is

$$\frac{\partial(f, g)}{\partial(a, b)} = \begin{vmatrix} \dfrac{\partial f}{\partial a} & \dfrac{\partial f}{\partial b} \\[2mm] \dfrac{\partial g}{\partial a} & \dfrac{\partial g}{\partial b} \end{vmatrix}.$$

(b) By a similar inspection of a specific case, show that

$$d^3\Sigma_\mu \equiv \epsilon_{\mu\alpha\beta\gamma} \frac{\partial x^\alpha}{\partial a} \frac{\partial x^\beta}{\partial b} \frac{\partial x^\gamma}{\partial c}\, da\, db\, dc = \frac{1}{3!}\, \epsilon_{\mu\alpha\beta\gamma} \frac{\partial(x^\alpha, x^\beta, x^\gamma)}{\partial(a, b, c)}\, da\, db\, dc.$$

(c) Cite a precise definition of the value of a determinant as a sum of terms (with suitably alternating signs), with each term a product containing one factor from each row and simultaneously one factor from each column. Show that this definition can be stated (in the 4×4 case, with the $p \times p$ case an obvious extension) as

$$\det A \equiv \det \|A^\lambda{}_\rho\| = \epsilon_{\alpha\beta\gamma\delta} A^\alpha{}_0 A^\beta{}_1 A^\gamma{}_2 A^\delta{}_3.$$

(d) Show that

$$\det A = \frac{1}{4!}\, \delta^{\mu\nu\rho\sigma}_{\alpha\beta\gamma\delta} A^\alpha{}_\mu A^\beta{}_\nu A^\gamma{}_\rho A^\delta{}_\sigma$$

(for a definition of $\delta^{\mu\nu\rho\sigma}_{\alpha\beta\gamma\delta}$, see exercises 3.13 and 4.12).

(e) Use properties of the δ-symbol to show that the matrix A^{-1} inverse to A has entries $(A^{-1})^{\mu}{}_{\alpha}$ given by

$$(A^{-1})^{\mu}{}_{\alpha}(\det A) = \frac{1}{3!}\,\delta^{\mu\nu\rho\sigma}_{\alpha\beta\gamma\delta}A^{\beta}{}_{\nu}A^{\gamma}{}_{\rho}A^{\delta}{}_{\sigma}.$$

(f) By an "index-mechanics" computation, from the formula for det A in part (d) derive the following expression for the derivative of the logarithm of the determinant

$$\boldsymbol{d}\ln|\det A| = \text{trace}(A^{-1}\,\boldsymbol{d}A).$$

Here $\boldsymbol{d}A$ is the matrix $\|\boldsymbol{d}A^{\alpha}{}_{\mu}\|$ whose entries are 1-forms.

Exercise 5.6. CENTROIDS AND SIZES

Consider an isolated system with stress-energy tensor $T^{\mu\nu}$, total 4-momentum P^{α}, magnitude of 4-momentum $M = (-\boldsymbol{P}\cdot\boldsymbol{P})^{1/2}$, intrinsic angular momentum tensor $S^{\alpha\beta}$, and intrinsic angular momentum vector S^{α}. (See Box 5.6.) An observer with 4-velocity u^{α} defines the *centroid* of the system, at his Lorentz time $x^0 = t$ and in his own Lorentz frame, by

$$X^{j}_{\boldsymbol{u}}(t) = (1/P^0)\int_{x^0=t} x^{j}T^{00}\,d^3x \qquad \text{in Lorentz frame where } \boldsymbol{u} = \partial\mathscr{P}/\partial x^0. \qquad (5.56)$$

This centroid depends on (i) the particular system being studied, (ii) the 4-velocity \boldsymbol{u} of the observer, and (iii) the time t at which the system is observed.

(a) Show that the centroid moves with a uniform velocity

$$dX^{j}_{\boldsymbol{u}}/dt = P^{j}/P^{0}, \qquad (5.57)$$

corresponding to the 4-velocity

$$\boldsymbol{U} = \boldsymbol{P}/M. \qquad (5.57')$$

Note that this "4-velocity of centroid" is independent of the 4-velocity \boldsymbol{u} used in defining the centroid.

(b) The centroid associated with the rest frame of the system (i.e., the centroid defined with $\boldsymbol{u} = \boldsymbol{U}$) is called the *center of mass;* see Box 5.6. Let $\boldsymbol{\xi}_{\boldsymbol{u}}$ be a vector reaching from any event on the center-of-mass world line to any event on the world line of the centroid associated with 4-velocity \boldsymbol{u}; thus the components of $\boldsymbol{\xi}_{\boldsymbol{u}}$ in any coordinate system are

$$\xi^{\alpha}_{\boldsymbol{u}} = X^{\alpha}_{\boldsymbol{u}} - X^{\alpha}_{\boldsymbol{U}}. \qquad (5.58)$$

Show that $\boldsymbol{\xi}_{\boldsymbol{u}}$ satisfies the equation

$$[(\xi^{\alpha}_{\boldsymbol{u}}P^{\beta} - P^{\alpha}\xi^{\beta}_{\boldsymbol{u}}) - S^{\alpha\beta}]u_{\beta} = 0. \qquad (5.59)$$

[*Hint:* perform the calculation in a Lorentz frame where $\boldsymbol{u} = \partial\mathscr{P}/\partial x^0$.]

(c) Show that, as seen in the rest-frame of the system at any given moment of time, the above equation reduces to the three-dimensional Euclidean equation

$$\boldsymbol{\xi}_{\boldsymbol{u}} = -(\boldsymbol{v}\times\boldsymbol{S})/M, \qquad (5.59')$$

where $\boldsymbol{v} = \boldsymbol{u}/u^0$ is the ordinary velocity of the frame associated with the centroid.

(d) Assume that the energy density measured by any observer anywhere in spacetime is

non-negative ($\boldsymbol{u} \cdot \boldsymbol{T} \cdot \boldsymbol{u} \geq 0$ for all timelike \boldsymbol{u}). In the rest frame of the system, construct the smallest possible cylinder that is parallel to \boldsymbol{S} and that contains the entire system ($T_{\alpha\beta} = 0$ everywhere outside the cylinder). Show that the radius r_0 of this cylinder is limited by

$$r_0 \geq |\boldsymbol{S}|/M. \tag{5.60}$$

Thus, a system with given intrinsic angular momentum \boldsymbol{S} and given mass M has a minimum possible size $r_{0\,\mathrm{min}} = |\boldsymbol{S}|/M$ as measured in its rest frame.

CHAPTER **6**

ACCELERATED OBSERVERS

The objective world simply is; *it does not* happen. *Only to the gaze of my consciousness, crawling upward along the life line* [*world line*] *of my body, does a section of this world come to life as a fleeting image in space which continuously changes in time.*

HERMAN WEYL (1949, p. 116)

§6.1. ACCELERATED OBSERVERS CAN BE ANALYZED USING SPECIAL RELATIVITY

It helps in analyzing gravitation to consider a situation where gravity is mocked up by acceleration. Focus attention on a region so far from any attracting matter, and so free of disturbance, that (to some proposed degree of precision) spacetime there can be considered to be flat and to have Lorentz geometry. Let the observer acquire the feeling that he is subject to gravity, either because of jet rockets strapped to his legs or because he is in a rocket-driven spaceship. How does physics look to him?

Dare one answer this question? At this early stage in the book, is one not too ignorant of gravitation physics to predict what physical effects will be measured by an observer who thinks he is in a gravitational field, although he is really in an accelerated spaceship? Quite the contrary; special relativity was developed precisely to predict the physics of accelerated objects—e.g., the radiation from an accelerated charge. Even the fantastic accelerations

Accelerated motion and accelerated observers can be analyzed using special relativity

$$a_{\text{nuclear}} \sim v^2/r \sim 10^{31} \text{ cm/sec}^2 \sim 10^{28} \text{ "earth gravities"}$$

suffered by a neutron bound in a nucleus, and the even greater accelerations met in high-energy particle-scattering events, are routinely and accurately treated within

Box 6.1 GENERAL RELATIVITY IS BUILT ON SPECIAL RELATIVITY

A tourist in a powered interplanetary rocket feels "gravity." Can a physicist by local effects convince him that this "gravity" is bogus? Never, says Einstein's principle of the local equivalence of gravity and accelerations. But then the physicist will make no errors if he deludes himself into treating true gravity as a local illusion caused by acceleration. Under this delusion, he barges ahead and solves gravitational problems by using special relativity: if he is clever enough to divide every problem into a network of local questions, each solvable under such a delusion, then he can work out all influ-ences of any gravitational field. Only three basic principles are invoked: special-relativity physics, the equivalence principle, and the local nature of physics. They are simple and clear. To apply them, however, imposes a double task: (1) take space-time apart into locally flat pieces (where the principles are valid), and (2) put these pieces together again into a comprehensible picture. To undertake this dissection and reconstitution, to see curved dynamic spacetime inescapably take form, and to see the consequences for physics: that is general relativity.

the framework of special relativity. The theoretician who confidently applies special relativity to antiproton annihilations and strange-particle resonances is not about to be frightened off by the mere illusions of a rocket passenger who gullibly believed the travel brochures advertising "earth gravity all the way." When spacetime is flat, move however one will, special relativity can handle the job. (It can handle bigger jobs too; see Box 6.1.) The essential features of *how* special relativity handles the job are summarized in Box 6.2 for the benefit of the Track-1 reader, who can skip the rest of the chapter, and also for the benefit of the Track-2 reader, who will find it useful background for the rest of the chapter.

Box 6.2 ACCELERATED OBSERVERS IN BRIEF

An accelerated observer can carry clocks and measuring rods with him, and can use them to set up a reference frame (coordinate system) in his neighborhood.

His clocks, if carefully chosen so their structures are affected negligibly by acceleration (e.g., atomic clocks), will tick at the same rate as unaccelerated clocks moving momentarily along with him:

$$\Delta\tau \equiv \begin{pmatrix} \text{time interval ticked off} \\ \text{by observer's clocks as he} \\ \text{moves a vector displacement} \\ \boldsymbol{\xi} \text{ along his world line} \end{pmatrix} = [-\boldsymbol{g}(\boldsymbol{\xi}, \boldsymbol{\xi})]^{1/2}.$$

And his rods, if chosen to be sufficiently rigid, will measure the same lengths as

momentarily comoving, unaccelerated rods do. (For further discussion, see §16.4, and Boxes 16.2 to 16.4.)

Let the observer's coordinate system be a Cartesian latticework of rods and clocks, with the origin of the lattice always on his world line. He must keep his latticework small:

$$l \equiv \begin{pmatrix} \text{spatial dimensions} \\ \text{of lattice} \end{pmatrix} \ll \begin{pmatrix} \text{the acceleration measured} \\ \text{by accelerometers he carries} \end{pmatrix}^{-1} \equiv \frac{1}{g}.$$

At distances l away from his world line, strange things of dimensionless magnitude gl happen to his lattice—e.g., the acceleration measured by accelerometers differs from g by a fractional amount $\sim gl$ (exercise 6.7); also, clocks initially synchronized with the clock on his world line get out of step (tick at different rates) by a fractional amount $\sim gl$ (exercise 6.6). (Note that an acceleration of one "earth gravity" corresponds to

$$g^{-1} \sim 10^{-3} \text{ sec}^2/\text{cm} \sim 10^{18} \text{ cm} \sim 1 \text{ light-year},$$

so the restriction $l \ll 1/g$ is normally not severe.)

To deduce the results of experiments and observations performed by an accelerated observer, one can analyze them in coordinate-independent, geometric terms, and then project the results onto the basis vectors of his accelerated frame. Alternatively, one can analyze the experiments and observations in a Lorentz frame, and then transform to the accelerated frame.

As deduced in this manner, the results of experiments performed locally (at $l \ll 1/g$) by an accelerated observer differ from the results of the same experiments performed in a Lorentz frame in only three ways:

(1) There are complicated fractional differences of order $gl \ll 1$ mentioned above, that can be made negligible by making the accelerated frame small enough.

(2) There are Coriolis forces of precisely the same type as are encountered in Newtonian theory (exercise 6.8). These the observer can get rid of by carefully preventing his latticework from rotating—e.g., by tying it to gyroscopes that he accelerates with himself by means of forces applied to their centers of mass (no torque!). Such a nonrotating latticework has "Fermi-Walker transported" basis vectors (§6.5),

$$\frac{d\boldsymbol{e}_{\alpha'}}{d\tau} = \boldsymbol{u}(\boldsymbol{a} \cdot \boldsymbol{e}_{\alpha'}) - \boldsymbol{a}(\boldsymbol{u} \cdot \boldsymbol{e}_{\alpha'}), \tag{1}$$

where \boldsymbol{u} = 4-velocity, and $\boldsymbol{a} = d\boldsymbol{u}/d\tau$ = 4-acceleration.

(3) There are inertial forces of precisely the same type as are encountered in Newtonian theory (exercise 6.8). These are due to the observer's acceleration, and he cannot get rid of them except by stopping his accelerating.

The rest of this chapter is Track 2.

It depends on no preceding Track-2 material.

It is needed as preparation for

(1) the mathematical analysis of gyroscopes in curved spacetime (exercise 19.2, §40.7), and

(2) the mathematical theory of the proper reference frame of an accelerated observer (§13.6).

It will be helpful in many applications of gravitation theory (Chapters 18–40).

§6.2. HYPERBOLIC MOTION

Study a rocket passenger who feels "gravity" because he is being accelerated in flat spacetime. Begin by describing his motion relative to an inertial reference frame. His 4-velocity satisfies the condition $\boldsymbol{u}^2 = -1$. To say that it is fixed in magnitude is to say that the 4-acceleration,

$$\boldsymbol{a} = d\boldsymbol{u}/d\tau, \tag{6.1}$$

is orthogonal to the 4-velocity:

$$0 = (d/d\tau)(-1/2) = (d/d\tau)\left(\frac{1}{2}\boldsymbol{u}\cdot\boldsymbol{u}\right) = \boldsymbol{a}\cdot\boldsymbol{u}. \tag{6.2}$$

This equation implies that $a^0 = 0$ in the rest frame of the passenger (that Lorentz frame, where, at the instant in question, $\boldsymbol{u} = \boldsymbol{e}_0$); in this frame the space components of a^μ reduce to the ordinary definition of acceleration, $a^i = d^2x^i/dt^2$. From the components $a^\mu = (0; a^i)$ in the rest frame, then, one sees that the magnitude of the acceleration in the rest frame can be computed as the simple invariant

$$\boldsymbol{a}^2 = a^\mu a_\mu = (d^2\boldsymbol{x}/dt^2)^2_{\text{as measured in rest frame}}.$$

Consider, for simplicity, an observer who feels always a constant acceleration g. Take the acceleration to be in the x^1 direction of some inertial frame, and take $x^2 = x^3 = 0$. The equations for the motion of the observer in that inertial frame become

$$\frac{dt}{d\tau} = u^0, \qquad \frac{dx}{d\tau} = u^1; \qquad \frac{du^0}{d\tau} = a^0, \qquad \frac{du^1}{d\tau} = a^1. \tag{6.3}$$

Write out the three algebraic equations

$$u^\mu u_\mu = -1,$$
$$u^\mu a_\mu = -u^0 a^0 + u^1 a^1 = 0,$$

and

$$a^\mu a_\mu = g^2.$$

Solve for the acceleration, finding

$$a^0 = \frac{du^0}{d\tau} = gu^1, \qquad a^1 = \frac{du^1}{d\tau} = gu^0. \tag{6.4}$$

These linear differential equations can be solved immediately. The solution, with a suitable choice of the origin, reads

$$t = g^{-1}\sinh g\tau, \qquad x = g^{-1}\cosh g\tau. \tag{6.5}$$

Uniformly accelerated observer moves on hyperbola in spacetime diagram

Note that $x^2 - t^2 = g^{-2}$. The world line is a hyperbola in a spacetime diagram ("hyperbolic motion"; Figure 6.1). Several interesting aspects of this motion are

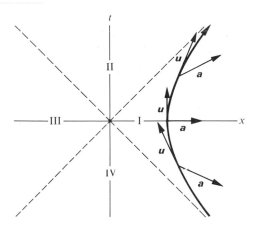

Figure 6.1.
Hyperbolic motion. World line of an object that (or an observer who) experiences always a fixed acceleration g with respect to an inertial frame that is instantaneously comoving (different inertial frames at different instants!). The 4-acceleration \mathbf{a} is everywhere orthogonal (Lorentz geometry!) to the 4-velocity \mathbf{u}.

treated in the exercises. Let the magnitude of the constant acceleration g be the acceleration of gravity, $g = 980 \text{ cm/sec}^2$ experienced on earth: $g \simeq (10^3 \text{ cm/sec}^2)/(3 \times 10^{10} \text{ cm/sec})^2 = (3 \times 10^7 \text{ sec} \cdot 3 \times 10^{10} \text{ cm/sec})^{-1} = (1 \text{ light-year})^{-1}$. Thus the observer will attain relativistic velocities after maintaining this acceleration for something like one year of his own proper time. He can outrun a photon if he has a head start on it of one light-year or more.

Exercise 6.1. A TRIP TO THE GALACTIC NUCLEUS

Compute the proper time required for the occupants of a rocket ship to travel the ~30,000 light-years from the Earth to the center of the Galaxy. Assume that they maintain an acceleration of one "earth gravity" (10^3 cm/sec^2) for half the trip, and then decelerate at one earth gravity for the remaining half.

Exercise 6.2. ROCKET PAYLOAD

What fraction of the initial mass of the rocket can be payload for the journey considered in exercise 6.1? Assume an ideal rocket that converts rest mass into radiation and ejects all the radiation out the back of the rocket with 100 per cent efficiency and perfect collimation.

Exercise 6.3. TWIN PARADOX

(a) Show that, of all timelike world lines connecting two events \mathscr{A} and \mathscr{B}, the one with the *longest* lapse of proper time is the unaccelerated one. (*Hint:* perform the calculation in the inertial frame of the unaccelerated world line.)

(b) One twin chooses to move from \mathscr{A} to \mathscr{B} along the unaccelerated world line. Show that the other twin, by an appropriate choice of accelerations, can get from \mathscr{A} to \mathscr{B} in arbitrarily small proper time.

(c) If the second twin prefers to ride in comfort, with the acceleration he feels never exceeding one earth gravity, g, what is the shortest proper time-lapse he can achieve between \mathscr{A} and \mathscr{B}? Express the answer in terms of g and the proper time-lapse $\Delta\tau$ measured by the unaccelerated twin.

(d) Evaluate the answer numerically for several interesting trips.

Exercise 6.4. RADAR SPEED INDICATOR

A radar set measures velocity by emitting a signal at a standard frequency and comparing it with the frequency of the signal reflected back by another object. This redshift measurement is then converted, using the standard special-relativistic formula, into the corresponding velocity, and the radar reads out this velocity. How useful is this radar set as a velocity-measuring instrument for a uniformly accelerated observer?

(a) Consider this problem first for the special case where the object and the radar set are at rest with respect to each other at the instant the radar pulse is reflected. Compute the redshift $1 + z = \omega_e/\omega_0$ that the radar set measures in this case, and the resulting (incorrect) velocity it infers. Simplify by making use of the symmetries of the situation.

(b) Now consider the situation where the object has a non-zero velocity in the momentary rest frame of the observer at the instant it reflects the radar pulse. Compute the ratio of the actual relative velocity to the velocity read out by the radar set.

Exercise 6.5. RADAR DISTANCE INDICATOR

Use radar as a distance-measuring device. The radar set measures its proper time τ between the instant at which it emits a pulse and the later instant when it receives the reflected pulse. It then performs the simple computation $L_0 = \tau/2$ and supplies as output the "distance" L_0. How accurate is the output reading of the radar set for measuring the actual distance L to the object, when used by a uniformly accelerated observer? (L is defined as the distance in the momentary rest frame of the observer at the instant the pulse is reflected, which is at the observer's proper time halfway between emitting and receiving the pulse.) Give a correct formula relating $L_0 \equiv \tau/2$ to the actual distance L. Show that the reading L_0 becomes infinite as L approaches g^{-1}, where g is the observer's acceleration, as measured by an accelerometer he carries.

§6.3. CONSTRAINTS ON SIZE OF AN ACCELERATED FRAME

Difficulties in constructing "the coordinate system of an accelerated observer":

It is very easy to put together the words "the coordinate system of an accelerated observer," but it is much harder to find a concept these words might refer to. The most useful first remark one can make about these words is that, if taken seriously, they are self-contradictory. The definite article "the" in this phrase suggests that one is thinking of some unique coordinate system naturally associated with some specified accelerated observer, such as one whose world line is given in equation (6.5). If the coordinate system is indeed natural, one would expect that the coordinates of any event could be determined by a sufficiently ingenious observer by sending and receiving light signals. But from Figure 6.1 it is clear that the events composing one quarter of all spacetime (Zone III) can neither send light signals to, nor receive light signals from, the specified observer. Another half of spacetime suffers lesser disabilities in this respect: Zone II cannot send to the observer, Zone IV cannot receive from him. It is hard to see how the observer could define in any natural way a coordinate system covering events with which he has no causal relationship, which he cannot see, and from which he cannot be seen!

Breakdown in communication between observer and events at distance
$l > $ (acceleration)$^{-1}$

Difficulties also occur when one considers an observer who begins at rest in one frame, is accelerated for a time, and maintains thereafter a constant velocity, at rest in some other inertial coordinate system. Do his motions define in any natural way

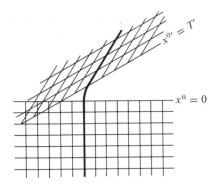

Figure 6.2.
World line of an observer who has undergone a brief period of acceleration. In each phase of motion at constant velocity, an inertial coordinate system can be set up. However, there is no way to reconcile these discordant coordinates in the region of overlap (beginning at distance g^{-1} to the left of the region of acceleration).

a coordinate system? Then this coordinate system (1) should be the inertial frame x^μ in which he was at rest for times x^0 less than 0, and (2) should be the other inertial frame $x^{\mu'}$ for times $x^{0'} > T'$ in which he was at rest in that other frame. Evidently some further thinking would be required to decide how to define the coordinates in the regions not determined by these two conditions (Figure 6.2). More serious, however, is the fact that these two conditions are inconsistent for a region of spacetime that satisfies simultaneously $x^0 < 0$ and $x^{0'} > T'$. In both examples of accelerated motion (Figures 6.1 and 6.2), the serious difficulties about defining a coordinate system begin only at a finite distance g^{-1} from the world line of the accelerated observer. The problem evidently has no solution for distances from the world line greater than g^{-1}. It does possess a natural solution in the immediate vicinity of the observer. This solution goes under the name of "Fermi-Walker transported orthonormal tetrad." The essential idea lends itself to simple illustration for hyperbolic motion, as follows.

Natural coordinates inconsistent at distance $l >$ (acceleration)$^{-1}$

§6.4. THE TETRAD CARRIED BY A UNIFORMLY ACCELERATED OBSERVER

An infinitesimal version of a coordinate system is supplied by a "tetrad," or "moving frame" (Cartan's "repère mobile"), or set of basis vectors $e_{0'}$, $e_{1'}$, $e_{2'}$, $e_{3'}$ (subscript tells which vector, not which component of one vector!) Let the time axis be the time axis of a comoving inertial frame in which the observer is momentarily at rest. Thus the zeroth basis vector is identical with his 4-velocity: $e_{0'} = u$. The space axes e_2 and e_3 are not affected by Lorentz transformations in the 1-direction. Therefore take $e_{2'}$ and $e_{3'}$ to be the unit basis vectors of the all-encompassing Lorentz frame relative to which the hyperbolic motion of the observer has already been described in equations (6.5): $e_{2'} = e_2$; $e_{3'} = e_3$. The remaining basis vector, $e_{1'}$, orthogonal to the other three, is parallel to the acceleration vector, $e_{1'} = g^{-1}a$ [see equation (6.4)]. There is a more satisfactory way to characterize this moving frame: the time axis $e_{0'}$ is the observer's 4-velocity, so he is always at rest in this frame; and the

Orthonormal tetrad of basis vectors carried by uniformly accelerated observer

other three vectors $\boldsymbol{e}_{1'}$ are chosen in such a way as to be (1) orthogonal and (2) nonrotating. These basis vectors are:

$$(e_{0'})^\mu = (\cosh g\tau;\ \sinh g\tau,\ 0,\ 0);$$

$$(e_{1'})^\mu = (\sinh g\tau;\ \cosh g\tau,\ 0,\ 0);$$

$$(e_{2'})^\mu = (0;\ 0,\ 1,\ 0);$$

$$(e_{3'})^\mu = (0;\ 0,\ 0,\ 1). \tag{6.6}$$

There is a simple prescription to obtain these four basis vectors. Take the four basis vectors \boldsymbol{e}_0, \boldsymbol{e}_1, \boldsymbol{e}_2, \boldsymbol{e}_3 of the original global Lorentz reference frame, and apply to them a simple boost in the 1-direction, of such a magnitude that $\boldsymbol{e}_{0'}$ comes into coincidence with the 4-velocity of the observer. The fact that these vectors are all orthogonal to each other and of unit magnitude is formally stated by the equation

$$\boldsymbol{e}_{\mu'} \cdot \boldsymbol{e}_{\nu'} = \eta_{\mu'\nu'}. \tag{6.7}$$

§6.5. THE TETRAD FERMI-WALKER TRANSPORTED BY AN OBSERVER WITH ARBITRARY ACCELERATION

Turn now from an observer, or an object, in hyperbolic motion to one whose acceleration, always finite, varies arbitrarily with time. Here also we impose three criteria on the moving, infinitesimal reference frame, or tetrad: (1) the basis vectors $\boldsymbol{e}_{\mu'}$ of the tetrad must remain orthonormal [equation (6.7)]; (2) the basis vectors must form a rest frame for the observer at each instant ($\boldsymbol{e}_{0'} = \boldsymbol{u}$); and (3) the tetrad should be "nonrotating."

This last criterion requires discussion. The basis vectors of the tetrad at any proper time τ must be related to the basis vectors \boldsymbol{e}_0, \boldsymbol{e}_1, \boldsymbol{e}_2, \boldsymbol{e}_3 of some given inertial frame by a Lorentz transformation $\boldsymbol{e}_{\mu'}(\tau) = \Lambda^\nu{}_{\mu'}(\tau)\boldsymbol{e}_\nu$. Therefore the basis vectors at two successive instants must also be related to each other by a Lorentz transformation. But a Lorentz transformation can be thought of as a "rotation" in spacetime. The 4-velocity \boldsymbol{u}, always of unit magnitude, changes in direction. The very concept of acceleration therefore implies "rotation" of velocity 4-vector. How then is the requirement of "no rotation" to be interpreted? Demand that the tetrad $\boldsymbol{e}_{\mu'}(\tau)$ change from instant to instant by precisely that amount implied by the rate of change of $\boldsymbol{u} = \boldsymbol{e}_{0'}$, and by no additional arbitrary rotation. In other words, (1) accept the inevitable pseudorotation in the timelike plane defined by the velocity 4-vector and the acceleration, but (2) rule out any ordinary rotation of the three space vectors.

Nonrelativistic physics describes the rotation of a vector (components v_i) by an instantaneous angular velocity vector (components ω_i). This angular velocity appears in the formula for the rate of change of \boldsymbol{v},

$$(dv_i/dt) = (\boldsymbol{\omega} \times \boldsymbol{v})_i = \varepsilon_{ijk}\omega_j v_k. \tag{6.8}$$

For the extension to four-dimensional spacetime, it is helpful to think of the rotation

Orthonormal tetrad of arbitrarily accelerated observer: should be "nonrotating"

"Nonrotating" means rotation only in timelike plane of 4-velocity and 4-acceleration

Mathematics of rotation in 3-space

as occurring in the plane perpendicular to the angular velocity vector $\boldsymbol{\omega}$. Thus rewrite (6.8) as

$$dv_i/dt = -\Omega_{ik}v_k, \tag{6.9}$$

where

$$\Omega_{jk} = -\Omega_{kj} = \omega_i\varepsilon_{ijk} \tag{6.10}$$

has non-zero components only in the plane of the rotation. In other words, to speak of "a rotation in the $(1, 2)$-plane" is more useful than to speak of a rotation about the 3-axis. The concept of "plane of rotation" carries over to four dimensions. There a rotation in the $(1, 2)$-plane will leave constant not only the v_3 but also the v_0 component of the velocity. The four-dimensional definition of a rotation is

<div style="text-align:right">Mathematics of rotation in spacetime</div>

$$\frac{dv^\mu}{d\tau} = -\Omega^{\mu\nu}v_\nu, \quad \text{with} \quad \Omega^{\mu\nu} = -\Omega^{\nu\mu}. \tag{6.11}$$

To test the appropriateness of this definition of a generalized rotation or infinitesimal Lorentz transformation, verify that it leaves invariant the length of the 4-vector:

$$d(v_\mu v^\mu)/d\tau = 2v_\mu(dv^\mu/d\tau) = -2\Omega^{\mu\nu}v_\mu v_\nu = 0. \tag{6.12}$$

The last expression vanishes because $\Omega^{\mu\nu}$ is antisymmetric, whereas $v_\mu v_\nu$ is symmetric. Note also that the antisymmetric tensor $\Omega^{\mu\nu}$ ("rotation matrix"; "infinitesimal Lorentz transformation") has $4 \times 3/2 = 6$ independent components. This number agrees with the number of components in a finite Lorentz transformation (three parameters for rotations, plus three parameters for the components of a boost). The "infinitesimal Lorentz transformation" here must (1) generate the appropriate Lorentz transformation in the timelike plane spanned by the 4-velocity and the 4-acceleration, and (2) exclude a rotation in any other plane, in particular, in any spacelike plane. The unique answer to these requirements is

$$\Omega^{\mu\nu} = a^\mu u^\nu - a^\nu u^\mu; \quad \text{i.e.,} \quad \boldsymbol{\Omega} = \boldsymbol{a} \wedge \boldsymbol{u}. \tag{6.13}$$

Apply this rotation to a spacelike vector \boldsymbol{w} orthogonal to \boldsymbol{u} and \boldsymbol{a}, ($\boldsymbol{u} \cdot \boldsymbol{w} = 0$ and $\boldsymbol{a} \cdot \boldsymbol{w} = 0$). Immediately compute $\Omega^{\mu\nu}w_\nu = 0$. Thus verify the absence of any space rotation. Now check the over-all normalization of $\Omega^{\mu\nu}$ in equation (6.13). Apply the infinitesimal Lorentz transformation to the velocity 4-vector \boldsymbol{u} of the observer. Thus insert $v^\mu = u^\mu$ in (6.11). It then reads

$$du^\mu/d\tau \equiv a^\mu = u^\mu(a^\nu u_\nu) - a^\mu(u^\nu u_\nu) = a^\mu.$$

This result is an identity, since $\boldsymbol{u} \cdot \boldsymbol{u} = -1$ and $\boldsymbol{u} \cdot \boldsymbol{a} = 0$.

A vector \boldsymbol{v} that undergoes the indicated infinitesimal Lorentz transformation,

<div style="text-align:right">Fermi-Walker law of transport for "nonrotating" tetrad of basis vectors carried by an accelerated observer</div>

$$dv^\mu/d\tau = (u^\mu a^\nu - u^\nu a^\mu)v_\nu, \tag{6.14}$$

is said to experience "Fermi-Walker transport" along the world line of the observer.

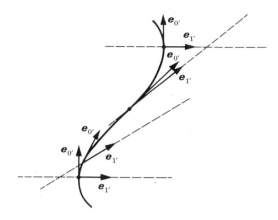

Figure 6.3.
Construction of spacelike hyperplanes (dashed) orthogonal to
the world line (heavy line) of an accelerated particle at selected
moments along that world line. Note crossing of hyperplanes
at distance $g^{-1}(\tau)$ (time-dependent acceleration!) from the
world line.

*The natural moving frame associated with an accelerated observer consists of four
orthonormal vectors, each of which is Fermi-Walker transported along the world line
and one of which is $\boldsymbol{e}_{0'} = \boldsymbol{u}$* (the 4-velocity of the observer). Fermi-Walker transport
of the space basis vectors $\boldsymbol{e}_{j'}$ can be achieved in practice by attaching them to
gyroscopes (see Box 6.2 and exercise 6.9).

§6.6. THE LOCAL COORDINATE SYSTEM OF
AN ACCELERATED OBSERVER

Extend this moving frame or "infinitesimal coordinate system" to a "local coordinate
system" covering a finite domain. Such local coordinates can escape none of the
problems encountered in "hyperbolic motion" (Figure 6.1) and "briefly accelerated
motion" (Figure 6.2). Therefore the local coordinate system has to be restricted to
a region within a distance g^{-1} of the observer, where these problems do not arise.
Figure 6.3 illustrates the construction of the local coordinates $\xi^{\mu'}$. At any given proper
time τ the observer sits at a specific event $\mathcal{P}(\tau)$ along his world line. Let the displace-
ment vector, from the origin of the original inertial frame to his position $\mathcal{P}(\tau)$, be
$\boldsymbol{z}(\tau)$. At $\mathcal{P}(\tau)$ the observer has three spacelike basis vectors $\boldsymbol{e}_{1'}(\tau)$, $\boldsymbol{e}_{2'}(\tau)$, $\boldsymbol{e}_{3'}(\tau)$. The
point $\mathcal{P}(\tau)$ plus those basis vectors define a spacelike hyperplane. The typical point
of this hyperplane can be represented in the form

$$\boldsymbol{x} = \xi^{1'}\boldsymbol{e}_{1'}(\tau) + \xi^{2'}\boldsymbol{e}_{2'}(\tau) + \xi^{3'}\boldsymbol{e}_{3'}(\tau) + \boldsymbol{z}(\tau)$$

$$= \text{(separation vector from origin of original inertial frame)}. \tag{6.15}$$

Here the three numbers $\xi^{k'}$ play the role of Euclidean coordinates in the hyperplane.
This hyperplane advances as proper time unrolls. Eventually the hyperplane cuts
through the event \mathcal{P}_0 to which it is desired to assign coordinates. *Assign to this event
as coordinates the numbers* $\xi^{0'} = \tau$, $\xi^{k'}$ *given by* (6.15). *Call these four numbers*

Tetrad used to construct
"local coordinate system of
accelerated observer"

"coordinates relative to the accelerated observer." In detail, the prescription for the determination of these four coordinates consists of the four equations

$$x^\mu = \xi^{k'}[e_{k'}(\tau)]^\mu + z^\mu(\tau),\tag{6.16}$$

in which the x^μ are considered as known, and the coordinates τ, $\xi^{k'}$ are considered unknowns.

At a certain distance from the accelerated world line, successive spacelike hyperplanes, instead of advancing with increasing τ, will be retrogressing. At this distance, and at greater distances, the concept of "coordinates relative to the accelerated observer" becomes ambiguous and has to be abandoned. To evaluate this distance, note that any sufficiently short section of the world line can be approximated by a hyperbola ("hyperbolic motion with acceleration g"), where the time-dependent acceleration $g(\tau)$ is given by the equation $g^2 = a^\mu a_\mu$.

Apply the above general prescription to hyperbolic motion, arriving at the equations

Local coordinate system for uniformly accelerated observer

$$
\begin{aligned}
x^0 &= (g^{-1} + \xi^{1'})\sinh(g\xi^{0'}),\\
x^1 &= (g^{-1} + \xi^{1'})\cosh(g\xi^{0'}),\\
x^2 &= \xi^{2'},\\
x^3 &= \xi^{3'}.
\end{aligned}\tag{6.17}
$$

The surfaces of constant $\xi^{0'}$ are the hyperplanes with $x^0/x^1 = \tanh g\xi^{0'}$ sketched in Figure 6.4. Substitute expressions (6.17) into the Minkowski formula for the line element to find

$$
\begin{aligned}
ds^2 &= \eta_{\mu\nu}\,dx^\mu\,dx^\nu\\
&= -(1 + g\xi^{1'})^2(d\xi^{0'})^2 + (d\xi^{1'})^2 + (d\xi^{2'})^2 + (d\xi^{3'})^2.
\end{aligned}\tag{6.18}
$$

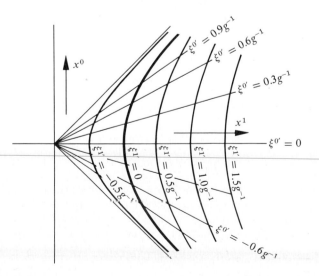

Figure 6.4.
Local coordinate system associated with an observer in hyperbolic motion (heavy black world line). The local coordinate system fails for $\xi^{1'}$ less than $-g^{-1}$.

The coefficients of $d\xi^{\mu'} d\xi^{\nu'}$ in this expansion are not the standard Lorentz metric components. The reason is clear. The $\xi^{\mu'}$ do not form an inertial coordinate system. However, at the position of the observer, $\xi^{1'} = 0$, the coefficients reduce to the standard form. Therefore these "local coordinates" approximate a Lorentz coordinate system in the immediate neighborhood of the observer.

EXERCISES

Exercise 6.6. CLOCK RATES VERSUS COORDINATE TIME IN ACCELERATED COORDINATES

Let a clock be attached to each grid point, $(\xi^{1'}, \xi^{2'}, \xi^{3'}) = $ constant, of the local coordinate system of an accelerated observer. Assume for simplicity that the observer is in hyperbolic motion. Use equation (6.18) to show that proper time as measured by a lattice clock differs from coordinate time at its lattice point:

$$d\tau/d\xi^{0'} = 1 + g\xi^{1'}.$$

(Of course, very near the observer, at $\xi^{1'} \ll g^{-1}$, the discrepancy is negligible.)

Exercise 6.7. ACCELERATION OF LATTICE POINTS IN ACCELERATED COORDINATES

Let an accelerometer be attached to each grid point of the local coordinates of an observer in hyperbolic motion. Calculate the magnitude of the acceleration measured by the accelerometer at $(\xi^{1'}, \xi^{2'}, \xi^{3'})$.

Exercise 6.8. OBSERVER WITH ROTATING TETRAD

An observer moving along an arbitrarily accelerated world line chooses *not* to Fermi-Walker transport his orthonormal tetrad. Instead, he allows it to rotate. The antisymmetric rotation tensor $\boldsymbol{\Omega}$ that enters into his transport law

$$d\boldsymbol{e}_{\alpha'}/d\tau = -\boldsymbol{\Omega} \cdot \boldsymbol{e}_{\alpha'} \tag{6.19}$$

splits into a Fermi-Walker part plus a spatial rotation part:

$$\Omega^{\mu\nu} = \underbrace{a^{\mu}u^{\nu} - a^{\nu}u^{\mu}}_{\Omega^{\mu\nu}_{(FW)}} + \underbrace{u_{\alpha}\omega_{\beta}\epsilon^{\alpha\beta\mu\nu}}_{\Omega^{\mu\nu}_{(SR)}} \tag{6.20}$$

$\boldsymbol{\omega}$ = a vector orthogonal to 4-velocity \boldsymbol{u}.

(a) The observer chooses his time basis vector to be $\boldsymbol{e}_{0'} = \boldsymbol{u}$. Show that this choice is permitted by his transport law (6.19), (6.20).

(b) Show that $\Omega^{\mu\nu}_{(SR)}$ produces a rotation in the plane perpendicular to \boldsymbol{u} and $\boldsymbol{\omega}$—i.e., that

$$\boldsymbol{\Omega}_{(SR)} \cdot \boldsymbol{u} = 0, \qquad \boldsymbol{\Omega}_{(SR)} \cdot \boldsymbol{\omega} = 0. \tag{6.21}$$

(c) Suppose the accelerated observer Fermi-Walker transports a second orthonormal tetrad $\boldsymbol{e}_{\alpha''}$. Show that the space vectors of his first tetrad rotate relative to those of his second tetrad with angular velocity vector equal to $\boldsymbol{\omega}$. *Hint:* At a moment when the tetrads coincide, show that (in three-dimensional notation, referring to the 3-space orthogonal to the observer's world line):

$$d(\boldsymbol{e}_{j'} - \boldsymbol{e}_{j''})/d\tau = \boldsymbol{\omega} \times \boldsymbol{e}_{j'}. \tag{6.22}$$

(d) The observer uses the same prescription [equation (6.16)] to set up local coordinates based on his rotating tetrad as for his Fermi-Walker tetrad. Pick an event \mathcal{Q} on the observer's world line, set $\tau = 0$ there, and choose the original inertial frame of prescription (6.16) so (1) it comoves with the accelerated observer at \mathcal{Q}, (2) its origin is at \mathcal{Q}, and (3) its axes coincide with the accelerated axes at \mathcal{Q}. Show that these conditions translate into

$$z^\mu(0) = 0, \qquad \boldsymbol{e}_{\alpha'}(0) = \boldsymbol{e}_\alpha. \tag{6.23}$$

(e) Show that near \mathcal{Q}, equations (6.16) for the rotating, accelerated coordinates reduce to:

$$x^0 = \xi^{0'} + a_{k'}\xi^{k'}\xi^{0'} + O([\xi^{\alpha'}]^3);$$
$$x^j = \xi^{j'} + \frac{1}{2}a^j\xi^{0'2} + \epsilon^{jk\ell}\omega^k\xi^{\ell'}\xi^{0'} + O([\xi^{\alpha'}]^3). \tag{6.24}$$

(f) A freely moving particle passes through the event \mathcal{Q} with ordinary velocity \boldsymbol{v} as measured in the inertial frame. By transforming its straight world line $x^j = v^j x^0$ to the accelerated, rotating coordinates, show that its coordinate velocity and acceleration there are:

$$(d\xi^{j'}/d\xi^{0'})_{\text{at } \mathcal{Q}} = v^j;$$

$$(d^2\xi^{j'}/d\xi^{0'2})_{\text{at } \mathcal{Q}} = \underbrace{-a^j}_{} - \underbrace{2\epsilon^{jk\ell}\omega^k v^\ell}_{} + \underbrace{2v^j a^k v^k}_{}. \tag{6.25}$$

inertial acceleration Coriolis acceleration relativisitc correction to inertial acceleration

Exercise 6.9. THOMAS PRECESSION

Consider a spinning body (gyroscope, electron, ...) that accelerates because forces act at its center of mass. Such forces produce no torque; so they leave the body's intrinsic angular-momentum vector \boldsymbol{S} unchanged, except for the unique rotation in the $\boldsymbol{u} \wedge \boldsymbol{a}$ plane required to keep \boldsymbol{S} orthogonal to the 4-velocity \boldsymbol{u}. Mathematically speaking, the body Fermi-Walker transports its angular momentum (no rotation in planes other than $\boldsymbol{u} \wedge \boldsymbol{a}$):

$$d\boldsymbol{S}/d\tau = (\boldsymbol{u} \wedge \boldsymbol{a}) \cdot \boldsymbol{S}. \tag{6.26}$$

This transport law applies to a spinning electron that moves in a circular orbit of radius r around an atomic nucleus. As seen in the laboratory frame, the electron moves in the x, y-plane with constant angular velocity, ω. At time $t = 0$, the electron is at $x = r$, $y = 0$; and its spin (as treated classically) has components

$$S^0 = 0, \qquad S^x = \frac{1}{\sqrt{2}}\hbar, \qquad S^y = 0, \qquad S^z = \frac{1}{2}\hbar.$$

Calculate the subsequent behavior of the spin as a function of laboratory time, $S^\mu(t)$. Answer:

$$S^x = \frac{1}{\sqrt{2}}\hbar\,(\cos\omega t\cos\omega\gamma t + \gamma\sin\omega t\sin\omega\gamma t);$$

$$S^y = \frac{1}{\sqrt{2}}\hbar\,(\sin\omega t\cos\omega\gamma t - \gamma\cos\omega t\sin\omega\gamma t);$$

$$S^z = \frac{1}{2}\hbar; \qquad S^0 = -\frac{1}{\sqrt{2}}\hbar\,v\gamma\sin\omega\gamma t; \tag{6.27}$$

$$v = \omega r; \qquad \gamma = (1 - v^2)^{-1/2}.$$

Rewrite the time-dependent spatial part of this as

$$S^x + iS^y = \frac{\hbar}{\sqrt{2}} [e^{-i(\gamma - 1)\omega t} + i(1 - \gamma)\sin(\omega\gamma t)e^{i\omega t}]. \tag{6.28}$$

The first term rotates steadily in a retrograde direction with angular velocity

$$\omega_{\text{Thomas}} = (\gamma - 1)\omega$$
$$\approx \frac{1}{2}v^2\omega \ll \omega \text{ if } v \ll 1. \tag{6.29}$$

It is called the Thomas precession. The second term rotates in a righthanded manner for part of an orbit ($0 < \omega\gamma t < \pi$) and in a lefthanded manner for the rest ($\pi < \omega\gamma t < 2\pi$). Averaged in time, it does nothing. Moreover, in an atom it is very small ($\gamma - 1 \ll 1$). It must be present, superimposed on the Thomas precession, in order to keep

$$\boldsymbol{S} \cdot \boldsymbol{u} = \boldsymbol{S} \cdot \boldsymbol{u} - S^0 u^0 = 0, \tag{6.30}$$

and

$$\boldsymbol{S}^2 = \boldsymbol{S}^2 - (S^0)^2 = 3\hbar^2/4 = \text{constant}. \tag{6.31}$$

It comes into play with righthanded rotation when $\boldsymbol{S} \cdot \boldsymbol{u}$ is negative; it goes out of play when $\boldsymbol{S} \cdot \boldsymbol{u} = 0$; and it returns with lefthanded rotation when $\boldsymbol{S} \cdot \boldsymbol{u}$ turns positive.

The Thomas precession can be understood, alternatively, as a spatial rotation that results from the combination of successive boosts in slightly different directions. [See, e.g., exercise 103 of Taylor and Wheeler (1966).] For an alternative derivation of the Thomas precession (6.29) from "spinor formalism," see §41.4.

CHAPTER **7**

INCOMPATIBILITY OF GRAVITY
AND SPECIAL RELATIVITY

§7.1. ATTEMPTS TO INCORPORATE GRAVITY INTO SPECIAL RELATIVITY

The discussion of special relativity so far has consistently assumed an absence of gravitational fields. Why must gravity be ignored in special relativity? This chapter describes the difficulties that gravitational fields cause in the foundations of special relativity. After meeting these difficulties, one can appreciate fully the curved-space-time methods that Einstein introduced to overcome them.

Start, then, with what one already knows about gravity, Newton's formulation of its laws:

$$d^2x^i/dt^2 = -\partial\Phi/\partial x^i, \tag{7.1}$$

$$\nabla^2\Phi = 4\pi G\rho. \tag{7.2}$$

These equations cannot be incorporated as they stand into special relativity. The equation of motion (7.1) for a particle is in three-dimensional rather than four-dimensional form; it requires modification into a four-dimensional vector equation for $d^2x^\mu/d\tau^2$. Likewise, the field equation (7.2) is not Lorentz-invariant, since the appearance of a three-dimensional Laplacian operator instead of a four-dimensional d'Alembertian operator means that the potential Φ responds instantaneously to changes in the density ρ at arbitrarily large distances away. In brief, Newtonian gravitational fields propagate with infinite velocity.

One's first reaction to these problems might be to think that they are relatively straightforward to correct. Exercises at the end of this section study some relatively straightforward generalizations of these equations, in which the gravitational potential Φ is taken to be first a scalar, then a vector, and finally a symmetric tensor field. Each of these theories has significant shortcomings, and all fail to agree with observations. The best of them is the tensor theory (exercise 7.3, Box 7.1), which, however,

This chapter is entirely Track 2.

It depends on no preceding Track-2 material.

It is not needed as preparation for any later chapter, but will be helpful in Chapter 18 (weak gravitational fields), and in Chapters 38 and 39 (experimental tests and other theories of gravity).

Newton's gravitational laws must be modified into four-dimensional, geometric form

All straightforward modifications are unsatisfactory

Best modification (tensor theory in flat spacetime) is internally inconsistent; when repaired, it becomes general relativity.

is internally inconsistent and admits no exact solutions. This difficulty has been attacked in recent times by Gupta (1954, 1957, 1962), Kraichnan (1955), Thirring (1961), Feynman (1963), Weinberg (1965), Deser (1970). They show how the flat-space tensor theory may be modified within the spirit of present-day relativistic field theory to overcome these inconsistencies. By this field-theory route (part 5 of Box 17.2), they arrive uniquely at standard 1915 general relativity. Only at this end point does one finally recognize, from the mathematical form of the equations, that what ostensibly started out as a flat-space theory of gravity is really Einstein's theory, with gravitation being a manifestation of the curvature of spacetime. This book follows Einstein's line of reasoning because it keeps the physics to the fore.

EXERCISES

EXERCISES ON FLAT-SPACETIME THEORIES OF GRAVITY

The following three exercises provide a solid challenge. Happily, all three require similar techniques, and a solution to the most difficult one (exercise 7.3) is presented in Box 7.1. Therefore, it is reasonable to proceed as follows. (a) Work either exercise 7.1 (scalar theory of gravity) or 7.2 (vector theory of gravity), skimming exercise 7.3 and Box 7.1 (tensor theory of gravity) for outline and method, not for detail, whenever difficulties arise. (b) Become familiar with the results of the other exercise (7.2 or 7.1) by discussing it with someone who has worked it in detail. (c) Read in detail the solution to exercise 7.3 as presented in Box 7.1, and compare with the computed results for the other two theories. (d) Develop computational power by checking some detailed computations from Box 7.1.

Exercise 7.1. SCALAR GRAVITATIONAL FIELD, Φ

A. Consider the variational principle $\delta I = 0$, where

$$I = -m \int e^{\Phi} \left(-\eta_{\alpha\beta} \frac{dz^{\alpha}}{d\lambda} \frac{dz^{\beta}}{d\lambda} \right)^{1/2} d\lambda, \tag{7.3}$$

Here $m = $ (rest mass) and $z^{\alpha}(\lambda) = $ (parametrized world line) for a test particle in the scalar gravitational field Φ. By varying the particle's world line, derive differential equations governing the particle's motion. Write them using the particle's proper time as the path parameter,

$$d\tau = \left(-\eta_{\alpha\beta} \frac{dz^{\alpha}}{d\lambda} \frac{dz^{\beta}}{d\lambda} \right)^{1/2} d\lambda,$$

so that $u^{\alpha} = dz^{\alpha}/d\tau$ satisfies $u^{\alpha} u^{\beta} \eta_{\alpha\beta} = -1$.

B. Obtain the field equation for $\Phi(\boldsymbol{x})$ implied by the variational principle $\delta I = 0$, where $I = \int \mathcal{L} \, d^4x$ and

$$\mathcal{L} = -\frac{1}{8\pi G} \eta^{\alpha\beta} \frac{\partial \Phi}{\partial x^{\alpha}} \frac{\partial \Phi}{\partial x^{\beta}} - \int m e^{\Phi} \delta^4[\boldsymbol{x} - \boldsymbol{z}(\tau)] \, d\tau. \tag{7.4}$$

Show that the second term here gives the same integral as that studied in part A (equation 7.3).

Discussion: The field equations obtained describe how a single particle of mass m generates the scalar field. If many particles are present, one includes in \mathcal{L} a term $-\int m e^{\Phi} \delta^4[\boldsymbol{x} - \boldsymbol{z}(\tau)] \, d\tau$ for each particle.

C. Solve the field equation of part B, assuming a single source particle at rest. Also assume that $e^{\Phi} = 1$ is an adequate approximation in the neighborhood of the particle. Then check this assumption from your solution; i.e., what value does it assign to e^{Φ} at the surface of the earth? (Units with $c = 1$ are used throughout; one may also set $G = 1$, if one wishes.)

D. Now treat the static, spherically symmetric field Φ from part C as the field of the sun acting as a given external field in the variational principle of part A, and study the motion of a planet determined by this variational principle. Constants of motion are available from the spherical symmetry and time-independence of the integrand. Use spherical coordinates and assume motion in a plane. Derive a formula for the perihelion precession of a planet.

E. Pass to the limit of a zero rest-mass particle in the equations of motion of part A. Do this by using a parameter λ different from proper time, so chosen that $k^\mu = dx^\mu/d\lambda$ is the energy-momentum vector, and by taking the limit $m \longrightarrow 0$ with $k^0 = \gamma m = E$ remaining finite (so $u^0 = \gamma \longrightarrow \infty$). Use these equations to show that the quantities $q^\mu = k^\mu e^\Phi$ are constants of motion, and from this deduce that there is no bending of light by the sun in this scalar theory.

Exercise 7.2. VECTOR GRAVITATIONAL FIELD, Φ_μ

A. Verify that the variational principle $\delta I = 0$ gives Maxwell's equations by varying A_μ, and the Lorentz force law by varying $z^\mu(\tau)$, when

$$I = \frac{-1}{16\pi} \int F_{\mu\nu} F^{\mu\nu} \, d^4x + \frac{1}{2} m \int \frac{dz^\mu}{d\tau} \frac{dz_\mu}{d\tau} \, d\tau + e \int \frac{dz^\mu}{d\tau} A_\mu(z) \, d\tau. \tag{7.5}$$

Here $F_{\mu\nu}$ is an abbreviation for $A_{\nu,\mu} - A_{\mu,\nu}$. Hint: to vary $A_\mu(\boldsymbol{x})$, rewrite the last term as a spacetime integral by introducing a delta function $\delta^4[\boldsymbol{x} - \boldsymbol{z}(\tau)]$ as in exercise 7.1, parts A and B.

B. Define, by analogy to the above, a vector gravitational field Φ_μ with $G_{\mu\nu} \equiv \Phi_{\nu,\mu} - \Phi_{\mu,\nu}$ using a variational principle with

$$I = + \frac{1}{16\pi G} \int G_{\mu\nu} G^{\mu\nu} \, d^4x + \frac{1}{2} m \int \frac{dz^\mu}{d\tau} \frac{dz_\mu}{d\tau} \, d\tau + m \int \Phi_\mu \frac{dz^\mu}{d\tau} \, d\tau. \tag{7.6}$$

(Note: if many particles are present, one must augment I by terms $\frac{1}{2}m\int(dz^\mu/d\tau)(dz_\mu/d\tau)\,d\tau + m\int\Phi_\mu(dz^\mu/d\tau)\,d\tau$ for each particle.) Find the "Coulomb" law in this theory, and verify that the coefficients of the terms in the variational principle have been chosen reasonably.

C. Compute the perihelion precession in this theory.

D. Compute the bending of light in this theory (i.e., scattering of a highly relativistic particle $u^0 = \gamma \longrightarrow \infty$), as it passes by the sun, because of the sun's Φ_μ field.

E. Obtain a formula for the total field energy corresponding to the Lagrangian implicit in part B. Use the standard method of Hamiltonian mechanics, with

$$I_{\text{field}} = \frac{1}{16\pi G} \int G_{\mu\nu} G^{\mu\nu} \, d^4x \equiv \int \mathcal{L} \, d^4x;$$

\mathcal{L} is the Lagrangian density and $L \equiv \int \mathcal{L} \, d^3x$ is the Lagrangian. The corresponding Hamiltonian density (\equiv energy density) is

$$\mathcal{H} = \sum_\mu \Phi_{\mu,0} \frac{\partial \mathcal{L}}{\partial \Phi_{\mu,0}} - \mathcal{L}.$$

Show that vector gravitational waves carry negative energy.

Exercise 7.3. SYMMETRIC TENSOR GRAVITATIONAL FIELD, $h_{\mu\nu} = h_{\nu\mu}$

Here the action principle is, as for the vector field, $\delta I = 0$, with $I = I_{\text{field}} + I_{\text{particle}} + I_{\text{interaction}}$. I_{particle} is the same as for the vector field:

$$I_{\text{particle}} = \frac{1}{2} m \int \frac{dz^\mu}{d\tau} \frac{dz_\mu}{d\tau} \, d\tau. \tag{7.7}$$

However, I_{field} and $I_{\text{interaction}}$ are different:

$$I_{\text{field}} = \int \mathcal{L}_f \, d^4x, \tag{7.8a}$$

$$\mathcal{L}_f = \frac{-1}{32\pi G}\left(\frac{1}{2}\, h_{\nu\beta,\alpha}\bar{h}^{\nu\beta,\alpha} - \bar{h}_{\mu\alpha}{}^{,\alpha}\bar{h}^{\mu\beta}{}_{,\beta}\right) \qquad \begin{bmatrix}\text{Note that}\\ \text{one } h \text{ here}\\ \text{is not an } \bar{h}\end{bmatrix}, \tag{7.8b}$$

with

$$\bar{h}_{\mu\nu} \equiv h_{\mu\nu} - \frac{1}{2}\eta_{\mu\nu}h^\sigma{}_\sigma; \tag{7.8c}$$

$$I_{\text{interaction}} = \frac{1}{2}\int h_{\mu\nu}T^{\mu\nu}\, d^4x. \tag{7.9}$$

Here $T^{\mu\nu}$ is the stress-energy tensor for all nongravitational fields and matter present. For a system of point particles (used throughout this exercise),

$$T^{\mu\nu}(\boldsymbol{x}) = \int m\,\frac{dz^\mu}{d\tau}\frac{dz^\nu}{d\tau}\,\delta^4[\boldsymbol{x} - \boldsymbol{z}(\tau)]\, d\tau. \tag{7.10}$$

A. Obtain the equations of motion of a particle by varying $z^\mu(\tau)$ in $\delta(I_{\text{particle}} + I_{\text{interaction}}) = 0$. Express your result in terms of the "gravitational force field"

$$\Gamma_{\nu\alpha\beta} = \frac{1}{2}\left(h_{\nu\alpha,\beta} + h_{\nu\beta,\alpha} - h_{\alpha\beta,\nu}\right) \tag{7.11}$$

derived from the tensor gravitational potentials $h_{\mu\nu} = h_{\nu\mu}$.

B. Obtain the field equations from $\delta(I_{\text{field}} + I_{\text{interaction}}) = 0$; express them in terms of

$$-H^{\mu\alpha\nu\beta} \equiv \bar{h}^{\mu\nu}\eta^{\alpha\beta} + \bar{h}^{\alpha\beta}\eta^{\mu\nu} - \bar{h}^{\alpha\nu}\eta^{\mu\beta} - \bar{h}^{\mu\beta}\eta^{\alpha\nu}. \tag{7.12}$$

Discuss gauge invariance, and the condition $\bar{h}^{\mu\alpha}{}_{,\alpha} = 0$.

C. Find the tensor gravitational potentials $h_{\mu\nu}$ due to the sun (treated as a point mass).

D. Compute the perihelion precession.

E. Compute the bending of light.

F. Consider a gravitational wave

$$\bar{h}^{\mu\nu} = A^{\mu\nu}\exp(ik_\alpha x^\alpha). \tag{7.13}$$

What conditions are imposed by the field equations? By the gauge condition

$$\bar{h}^{\mu\alpha}{}_{,\alpha} = 0? \tag{7.14}$$

Show that, by further gauge transformations

$$h_{\mu\nu} \longrightarrow h_{\mu\nu} + \xi_{\mu,\nu} + \xi_{\nu,\mu} \tag{7.15}$$

that preserve the $\bar{h}^{\mu\alpha}{}_{,\alpha} = 0$ restrictions, further conditions

$$u_\alpha\bar{h}^{\alpha\mu} = 0, \qquad \bar{h}^\alpha{}_\alpha = 0 \tag{7.16}$$

can be imposed, where u^α is a fixed, timelike vector. It is sufficient to consider the case, obtained by a suitable choice of reference frame, where $u^\alpha = (1; 0, 0, 0)$ and $k^\alpha = (\omega; 0, 0, \omega)$.

G. From the Hamiltonian density

$$\mathcal{H} \equiv \dot{h}_{\mu\nu}(\partial\mathcal{L}/\partial\dot{h}_{\mu\nu}) - \mathcal{L} \tag{7.17}$$

for the field, show that the energy density of the waves considered in part F is positive.

H. Compute $T^{\mu\nu}{}_{,\nu}$ for the stress-energy tensor of particles $T^{\mu\nu}$ that appears in the action integral I. Does $T^{\mu\nu}{}_{,\nu}$ vanish (e.g., for the earth in orbit around the sun)? Why? Show that the coupled equations for fields and particles obtained from $\delta I = 0$ have no solutions.

(continued on page 187)

Box 7.1 AN ATTEMPT TO DESCRIBE GRAVITY BY A SYMMETRIC TENSOR FIELD IN FLAT SPACETIME [Solution to exercise 7.3]

Attempts to describe gravity within the framework of special relativity would naturally begin by considering the gravitational field to be a scalar (exercise 7.1) as it is in Newtonian theory, or a vector (exercise 7.2) by analogy to electromagnetism. Only after these are found to be deficient (e.g., no bending of light in either theory; negative-energy waves in the vector theory) would one face the computational complexities of a symmetric tensor gravitational potential, $h_{\mu\nu} = h_{\nu\mu}$, which has more indices.

 The foundations of the most satisfactory of all tensor theories of gravity in flat spacetime are laid out at the beginning of exercise 7.3. The choice of the Lagrangian made there (equations 7.8) is dictated by the demand that $h_{\mu\nu}$ be a "Lorentz covariant, massless, spin-two field." The meaning of this demand, and the techniques of special relativity required to translate it into a set of field equations, are customarily found in books on elementary particle physics or quantum field theory; see, e.g., Wentzel (1949), Feynman (1963), or Gasiorowicz (1966). Fierz and Pauli (1939) were the first to write down this Lagrangian and investigate the resulting theory. The conclusions of the theory are spelled out here in the form of a solution to exercise 7.3.

A. Equation of Motion for a Test Particle (exercise 7.3A)

Carry out the integration in equation (7.9), using the particle stress-energy tensor of equation (7.10), to find

$$I_{p+i} \equiv I_{\text{particle}} + I_{\text{interaction}} = \frac{1}{2} m \int (\eta_{\mu\nu} + h_{\mu\nu})\dot{z}^{\mu}\dot{z}^{\nu} \, d\tau, \qquad (1)$$

where

$$\dot{z}^{\mu} \equiv dz^{\mu}/d\tau.$$

Then compute δI_{p+i}, and find that the coefficient of the arbitrary variation in path δz^{μ} vanishes if and only if

$$(d/d\tau)[(\eta_{\mu\nu} + h_{\mu\nu})\dot{z}^{\nu}] - \frac{1}{2} h_{\alpha\beta,\mu}\dot{z}^{\alpha}\dot{z}^{\beta} = 0.$$

Rewrite this equation of motion in the form

$$(\eta_{\mu\nu} + h_{\mu\nu})\ddot{z}^{\nu} + \Gamma_{\mu\alpha\beta}\dot{z}^{\alpha}\dot{z}^{\beta} = 0, \qquad (2)$$

where $\Gamma_{\mu\alpha\beta}$ is defined in equation (7.11).

B$_1$. Field Equations (exercise 7.3B)

Use I_{field} and $I_{\text{interaction}}$ in the forms given in equations (7.8) and (7.9); but for the quickest and least messy derivation, do *not* use the standard Euler-Lagrange equations. Instead, compute directly the first-order change $\delta\mathcal{L}_f$ produced by a small

Box 7.1 (*continued*)

variation $\delta h_{\alpha\beta}$ of the field. For the second term of \mathcal{L}_f, it is clear (by relabeling dummy indices as needed) that varying each factor gives the same result, so the two terms from the product rule combine:

$$\delta(\bar{h}_{\mu\alpha}{}^{,\alpha}\,\bar{h}^{\mu\beta}{}_{,\beta}) = 2\bar{h}^{\mu\beta}{}_{,\beta}\,\delta\bar{h}_{\mu\alpha}{}^{,\alpha}.$$

A similar result holds for the first term of \mathcal{L}_f, in view of the identity $a_{\mu\nu}\bar{b}^{\mu\nu} = \bar{a}_{\mu\nu}b^{\mu\nu}$, which holds for the "bar" operation of equations (7.8); each side here is just $a_{\mu\nu}b^{\mu\nu} - \frac{1}{2}a^\mu{}_\mu b^\nu{}_\nu$. Consequently,

$$-(32\pi G)\,\delta\mathcal{L}_f = \bar{h}^{\nu\beta,\alpha}\,\delta h_{\nu\beta,\alpha} - 2\bar{h}^{\mu\beta}{}_{,\beta}\,\delta\bar{h}_{\mu\alpha}{}^{,\alpha}. \tag{3}$$

Next use this expression in δI_{field}; and, by an integration by parts, remove the derivatives from $\delta h_{\mu\nu}$, giving

$$\delta I_{\text{field}} = (32\pi G)^{-1} \int [\bar{h}^{\nu\beta,\alpha}{}_{,\alpha}\,\delta h_{\nu\beta} - 2\bar{h}^{\mu\beta}{}_{,\beta}{}^{,\alpha}\,\delta\bar{h}_{\mu\alpha}]\,d^4x.$$

To find the coefficient of $\delta h_{\mu\nu}$ in this expression, write (from equation 7.8c)

$$\delta\bar{h}_{\alpha\beta} = (\delta^\mu{}_\alpha\,\delta_{\nu\beta} - \frac{1}{2}\eta_{\alpha\beta}\eta^{\mu\nu})\,\delta h_{\mu\nu};$$

and then rearrange and relabel dummy (summation) indices to obtain

$$\delta I_{\text{field}} = (32\pi G)^{-1} \int [\bar{h}^{\mu\beta,\alpha}{}_{,\alpha}\,\delta h_{\nu\beta} - 2\bar{h}^{\mu\beta}{}_{,\beta}{}^{,\alpha}\,\delta\bar{h}_{\mu\alpha}]\,d^4x.$$

By combining this with $\delta I_{\text{interaction}} = \frac{1}{2}T^{\mu\nu}\,\delta h_{\mu\nu}\,d^4x$, and by using the symmetry $\delta h_{\mu\nu} = \delta h_{\nu\mu}$, obtain

$$-\bar{h}^{\mu\nu}{}_{,\alpha}{}^{,\alpha} - \eta^{\mu\nu}\bar{h}^{\alpha\beta}{}_{,\alpha\beta} + \bar{h}^{\mu\alpha}{}_{,\alpha}{}^{,\nu} + \bar{h}^{\nu\alpha}{}_{,\alpha}{}^{,\mu} = 16\pi G T^{\mu\nu}. \tag{4}$$

The definition made in equation (7.12) allows this to be rewritten as

$$H^{\mu\alpha\nu\beta}{}_{,\alpha\beta} = 16\pi G T^{\mu\nu}. \tag{4'}$$

B$_2$. Gauge Invariance (exercise 7.3B, continued)

The symmetries,

$$H^{\mu\alpha\nu\beta} = H^{[\mu\alpha][\nu\beta]} = H^{\nu\beta\mu\alpha},$$

of $H^{\mu\alpha\nu\beta}$ imply an identity

$$H^{\mu\alpha\nu\beta}{}_{,\alpha\beta\nu} = H^{\mu\alpha[\nu\beta]}{}_{,\alpha(\beta\nu)} \equiv 0$$

analogous to $F^{\mu\nu}{}_{,\nu\mu} \equiv 0$ in electromagnetism.

Thus $T^{\mu\nu}{}_{,\nu} = 0$ is required of the sources, just as is $J^\mu{}_{,\mu} = 0$ in electromagnetism (exercise 3.16). These identities make the field equations (4') too weak to fix $h_{\mu\nu}$

completely. In particular, by direct substitution in equations (4), one verifies that to any solution one can add a gauge field

$$h_{\mu\nu}{}^{(\text{gauge})} = \xi_{\mu,\nu} + \xi_{\nu,\mu},$$
$$\bar{h}_{\mu\nu}{}^{(\text{gauge})} = \xi_{\mu,\nu} + \xi_{\nu,\mu} - \eta_{\mu\nu}\xi^{\alpha}{}_{,\alpha},$$

(5)

without changing $T^{\mu\nu}$.

Let ξ_{μ} vanish outside some finite spacetime volume, but be otherwise arbitrary. Then $h_{\mu\nu}$ and $h_{\mu\nu} = h_{\mu\nu} + h_{\mu\nu}{}^{(\text{gauge})}$ both satisfy the source equation (4) for the same source $T^{\mu\nu}$ and the same boundary conditions at infinity. We therefore expect them to be physically equivalent.

By a specialization of the gauge analogous to the "Lorentz" specialization $A^{\alpha}{}_{,\alpha} = 0$ of electromagnetism (equation 3.58a; exercise 3.17), one imposes the condition

$$\bar{h}^{\mu\alpha}{}_{,\alpha} = 0.$$

(6)

This reduces the field equations (4) to the simple d'Alembertian form

$$\Box\bar{h}^{\mu\nu} \equiv \bar{h}^{\mu\nu}{}_{,\alpha}{}^{\alpha} = -16\pi T^{\mu\nu}$$

(7)

(see exercise 18.2). Here and henceforth we set $G = 1$ ("geometrized units").

C. Field of a Point Mass (exercise 7.3C)

For a static source, the wave equation (7) reduces to a Laplace equation

$$\nabla^2\bar{h}_{\mu\nu} = -16\pi T_{\mu\nu}.$$

The stress-energy tensor for a static point mass (equation 7.10) is $T^{00} = M\delta^3(x)$ and $T^{\mu k} = 0$. Put this into the Laplace equation, solve for $\bar{h}_{\mu\nu}$, and use equation (7.8c) to obtain $h_{\mu\nu}$. The result is:

$$h_{00} = 2M/r; \qquad h_{0k} = 0; \qquad h_{ik} = \delta_{ik}(2M/r)$$

(8)

(see equation 18.15a).

D. Perihelion Precession (exercise 7.3D)

Direct substitution of the potential (8) into the equations of motion (2) is tedious and not very instructive. Variational principles are popular in mechanics because they simplify such calculations. Return to the basic variational principle $\delta I_{p+i} = 0$ (equation 1), and insert the potential (8) for the sun. Convert to spherical coordinates so oriented that the orbit lies in the equatorial ($\theta = \pi/2$) plane:

$$I_{p+i} = \int L \, d\tau;$$

(9)

$$L = \frac{1}{2}m[-(1 - 2Mr^{-1})\dot{t}^2 + (1 + 2Mr^{-1})(\dot{r}^2 + r^2\dot{\phi}^2)].$$

(10)

Box 7.1 (continued)

From the absence of explicit t-, ϕ-, and τ-dependence in L, infer three constants of motion: the canonical momenta

$$P_t \equiv -m\gamma = \partial L/\partial \dot{t}$$

(this defines γ) and

$$P_\phi \equiv m\alpha = \partial L/\partial \dot{\phi}$$

(this defines α); and the Hamiltonian

$$H = \dot{x}^\mu(\partial L/\partial \dot{x}^\mu) - L,$$

which can be set equal to $-\frac{1}{2}m$ by appropriate normalization of the path parameter τ. From these constants of the motion, derive an orbit equation as follows: (1) calculate $H = -\frac{1}{2}m$ in terms of r, \dot{r}, $\dot{\phi}$, and \dot{t}; (2) eliminate \dot{t} and $\dot{\phi}$ in favor of the constants γ and α; (3) as in Newtonian orbit problems, define $u = M/r$, and write

$$\frac{du}{d\phi} = \frac{\dot{u}}{\dot{\phi}} = -\frac{M\dot{r}}{r^2\dot{\phi}} = -\frac{M}{\alpha}(1 + 2u)\dot{r};$$

(4) in H, eliminate \dot{r} in favor of $du/d\phi$ via the above equation, and eliminate r in favor of u; (5) solve for $du/d\phi$. The result is

$$\left(\frac{du}{d\phi}\right)^2 + u^2 = (\gamma^2 - 1 + 2u)\frac{M^2}{\alpha^2}\left[\frac{1 + 2u}{1 - 2u}\right]. \tag{11}$$

Neglecting cubic and higher powers of $u = GM/c^2r \sim (1 - \gamma^2)$ in this equation, derive the perihelion shift. (For details of method, see exercise 40.4, with the γ and α of this box renamed \widetilde{E} and \widetilde{L}, and with the γ and β of that exercise set equal to 1 and 0.) The resulting shift per orbit is

$$\Delta\phi = 8\pi M/r_0 + O([M/r_0]^2). \tag{12}$$

This is $\frac{4}{3}$ the prediction of general relativity; and it disagrees with the observations on Mercury (see Box 40.3).

E. Bending of Light (exercise 7.3E)

The deflection angle for light passing the sun is, on dimensional grounds, a small quantity, $\Delta\phi \sim M_\odot/R_\odot \sim 10^{-6}$; from the outset, one makes approximations based on this smallness. A diagram of the photon trajectory, set in the x, z-plane, shows that, for initial motion parallel to the z-axis, the deflection angle can be expressed in terms of the final momentum as $\Delta\phi = p_x/p_z$. Compute the final p_x by an integral along the trajectory,

$$p_x = \int_{-\infty}^{+\infty} (dp_x/dz)\, dz,$$

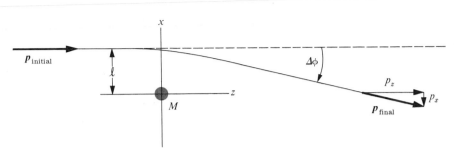

treating p_z as essentially constant. This computation requires generalization of the equation of motion (2) to the case of zero rest mass. To handle the limit $m \longrightarrow 0$, introduce a new parameter $\lambda = \tau/m$; then $p^\mu = m(dz^\mu/d\tau) = dz^\mu/d\lambda$. Also define $p_\mu = (\eta_{\mu\nu} + h_{\mu\nu})p^\nu$, since this quantity appears simply in equation (2) and agrees with p_μ in the limit $r \longrightarrow \infty$, where one will need to evaluate it. Then equation (2) reads, for any m, including $m = 0$,

$$\frac{dP_\mu}{d\lambda} = \frac{1}{2} h_{\alpha\beta,\mu} p^\alpha p^\beta .$$

On the righthand side here, since $h_{\alpha\beta,\mu}$ is small, a crude approximation to p^μ suffices; $p^1 = p^2 = 0$, $p^0 = p^3 = dz/d\lambda = \omega = $ constant. Thus,

$$\frac{dP_1}{d\lambda} = \frac{1}{2}(h_{00} + 2h_{03} + h_{33})_{,1}\omega^2$$

and

$$\frac{1}{p_3}\frac{dP_1}{dz} = \frac{1}{2}(h_{00} + 2h_{03} + h_{33})_{,1}.$$

For the sun, $h_{00} = h_{33} = 2M/r$, and $h_{03} = 0$ (equation 8), so

$$\Delta\phi = -\left(\frac{P_1}{P_3}\right)_{\text{final}} = -\left(\frac{P_1}{P_3}\right)_{\text{final}} = \int_{-\infty}^{\infty} \frac{2Ml\,dz}{(l^2 + z^2)^{3/2}} = \frac{2M}{l}\int_{-\infty}^{\infty} \frac{d\zeta}{(1 + \zeta^2)^{3/2}} = \frac{4M}{l}. \quad (13)$$

For light grazing the sun, $l = R_\odot$, this gives $\Delta\phi = 4M_\odot/R_\odot$ radians $= 1''.75$, which is also the prediction of general relativity, and is consistent with the observations (see Box 40.1).

F. Gravitational Waves (exercise 7.3F)

The field equations (4) and gauge properties (5) of the present flat-spacetime theory are identical to those of Einstein's "linearized theory." Thus, the treatment of gravitational waves using linearized theory, which is presented in §§18.2, 35.3, and 35.4, applies here.

G. Positive Energy of the Waves (exercise 7.3G)

Computing a general formula for \mathcal{K} from equation (7.17) is tedious, but it is sufficient to consider only the special case of a plane wave (equation 7.13)—or better still,

Box 7.1 (*continued*)

a plane wave with only $h_{12} = h_{21} = f(z - t)$. Any gravitational wave can be constructed as a superposition of such plane waves. First compute the Langrangian for this case. According to equation (7.8), it reads

$$\mathcal{L}_f = (32\pi)^{-1}[(h_{12,0})^2 - (h_{12,3})^2].$$

Now the full content of the formula (7.17) defining \mathcal{K} is precisely the following: start from the Lagrangian; keep all terms that are quadratic in time derivatives; omit all terms that are linear in time derivatives; and reverse the sign of terms that contain no time derivatives. The result is

$$\mathcal{K} = (32\pi)^{-1}[(h_{12,0})^2 + (h_{12,3})^2], \tag{14}$$

which is positive.

H. Self-Inconsistency of the Theory (exercise 7.3H)

From equation (7.10), find

$$T^{\mu\nu}{}_{,\nu} = m \int \frac{dz^\mu}{d\tau} \frac{dz^\nu}{d\tau} \frac{\partial}{\partial x^\nu} \delta^4[\mathbf{x} - \mathbf{z}(\tau)] \, d\tau.$$

But $\delta^4(\mathbf{x} - \mathbf{z})$ depends only on the difference $x^\mu - z^\mu$, so $-\partial/\partial z^\nu$ can replace $\partial/\partial x^\nu$ when acting on the δ-function. Noting that

$$\frac{dz^\nu}{d\tau} \frac{\partial}{\partial z^\nu} \delta^4[\mathbf{x} - \mathbf{z}(\tau)] = \frac{d}{d\tau} \delta^4[\mathbf{x} - \mathbf{z}(\tau)],$$

rewrite $T^{\mu\nu}{}_{,\nu}$ as

$$T^{\mu\nu}{}_{,\nu} = -m \int \dot{z}^\mu (d/d\tau) \, \delta^4[\mathbf{x} - \mathbf{z}(\tau)] \, d\tau = +m \int \ddot{z}^\mu \delta^4[\mathbf{x} - \mathbf{z}(\tau)] \, d\tau.$$

(The last step is obtained by an integration by parts.) Thus $T^{\mu\nu}{}_{,\nu} = 0$ holds *if and only if* $\ddot{z}^\mu = 0$. But $\ddot{z}^\mu = 0$ means that the gravitational fields have no effect on the motion of the particle. But this contradicts the equation of motion (2), which follows from the theory's variational principle. Thus, this tensor theory of gravity is inconsistent. [Stated briefly, equation (4) requires $T^{\mu\nu}{}_{,\nu} = 0$, while equation (2) excludes it.]

The fact that, in this theory, gravitating bodies cannot be affected by gravity, also holds for bodies made of arbitrary stress-energy (e.g., rubber balls or the Earth). Since all bodies gravitate, since the field equations imply $T^{\mu\nu}{}_{,\nu} = 0$, and since this "equation of motion for stress-energy" implies conservation of a body's total 4-momentum $P^\mu = \int T^{\mu 0} \, d^3x$, no body can be accelerated by gravity. The Earth cannot be attracted by the sun; it must fly off into interstellar space!

Straightforward steps to repair this inconsistency in the theory lead inexorably to general relativity (see Box 17.2 part 5). Having adopted general relativity as *the* theory of gravity, one can then use the present flat-spacetime theory as an approximation to it ("Linearized general relativity"; treated in Chapters 18, 19, and 35; see especially discussion at end of §18.3).

§7.2. GRAVITATIONAL RED SHIFT DERIVED
FROM ENERGY CONSERVATION

Einstein argued against the existence of any ideal, straight-line reference system such as is assumed in Newtonian theory. He emphasized that nothing in a natural state of motion, not even a photon, could ever give evidence for the existence or location of such ideal straight lines.

That a photon must be affected by a gravitational field Einstein (1911) showed from the law of conservation of energy, applied in the context of Newtonian gravitation theory. Let a particle of rest mass m start from rest in a gravitational field g at point \mathcal{C} and fall freely for a distance h to point \mathcal{B}. It gains kinetic energy mgh. Its total energy, including rest mass, becomes

Gravitational redshift derived from energy considerations

$$m + mgh. \qquad (7.18)$$

Now let the particle undergo an annihilation at \mathcal{B}, converting its total rest mass plus kinetic energy into a photon of the same total energy. Let this photon travel upward in the gravitational field to \mathcal{C}. If it does not interact with gravity, it will have its original energy on arrival at \mathcal{C}. At this point it could be converted by a suitable apparatus into another particle of rest mass m (which could then repeat the whole process) plus an excess energy mgh that costs nothing to produce. To avoid this contradiction of the principal of conservation of energy, which can also be stated in purely classical terms, Einstein saw that the photon must suffer a red shift. The energy of the photon must decrease just as that of a particle does when it climbs out of the gravitational field. The photon energy at the top and the bottom of its path through the gravitational field must therefore be related by

$$E_{\text{bottom}} = E_{\text{top}}(1 + gh) = E_{\text{top}}(1 + g_{\text{conv}}h/c^2). \qquad (7.19)$$

The drop in energy because of work done against gravitation implies a drop in frequency and an increase in wavelength (red shift; traditionally stated in terms of a red shift parameter, $z = \Delta\lambda/\lambda$); thus,

$$1 + z = \frac{\lambda_{\text{top}}}{\lambda_{\text{bottom}}} = \frac{h\nu_{\text{bottom}}}{h\nu_{\text{top}}} = \frac{E_{\text{bottom}}}{E_{\text{top}}} = 1 + gh. \qquad (7.20)$$

The redshift predicted by this formula has been verified to 1 percent by Pound and Snider (1964, 1965), refining an experiment by Pound and Rebka (1960).

§7.3. GRAVITATIONAL REDSHIFT IMPLIES
SPACETIME IS CURVED

An argument by Schild (1960, 1962, 1967) yields an important conclusion: the existence of the gravitational redshift shows that a consistent theory of gravity cannot be constructed within the framework of special relativity.

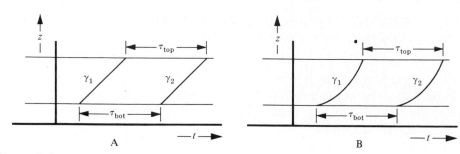

Figure 7.1.

Successive pulses of light rising from height z_1, to height $z_2 = z_1 + h$ against the gravitational field of the earth. The paths γ_1, and γ_2 must be exactly congruent, whether sloped at 45° (left) or having variable slope (right).

Assume gravity is described by an (unspecified) field in flat spacetime . . .

Whereas Einstein's argument (last section) was formulated in Newtonian theory, Schild's is formulated in special relativity. It analyzes gravitational redshift experiments in the field of the Earth, using a global Lorentz frame tied to the Earth's center. It makes no demand that free particles initially at rest remain at rest in this global Lorentz frame (except far from the Earth, where gravity is negligible). On the contrary, it demands that free particles be accelerated relative to the Lorentz frame by the Earth's gravitational field. It is indifferent to the mathematical nature of that field (scalar, vector, tensor,...), but it docs insist that the gravitational accelerations agree with experiment. And, of course, it demands that proper lengths and times be governed by the metric of special relativity.

Schild's argument proceeds as follows. Consider one observer at rest on the Earth's surface at height z_1, and a second above the Earth's surface at height $z_2 = z_1 + h$ (Figure 7.1). The observers may verify that they are at rest relative to each other and relative to the Earth's Lorentz frame by, for instance, radar ranging to free particles that are at rest in the Earth's frame far outside its gravitational field. The bottom experimenter then emits an electromagnetic signal of a fixed standard frequency ω_b which is received by the observer on top. For definiteness, let the signal be a pulse exactly N cycles long. Then the interval of time* $\delta\tau_{bot}$ required for the emission of the pulse is given by $2\pi N = \omega_b\,\delta\tau_{bot}$. The observer at the top is then to receive these same N cycles of the electromagnetic wave pulse and measure the time interval* $\delta\tau_{top}$ required. By the definition of "frequency," it satisfies $2\pi N = \omega_t\,\delta\tau_{top}$. The redshift effect, established by experiment (for us) or by energy conservation (for Einstein), shows $\omega_t < \omega_b$; consequently the time intervals are different, $\delta\tau_{top} > \delta\tau_{bot}$. Transfer this information to the special-relativity spacetime diagram of the experiment (Figure 7.1). The waves are light rays; so one might show them as traveling along 45° null lines in the spacetime diagram (Figure 7.1,A). In this

*Proper time equals Lorentz coordinate time for both observers, since they are at rest in the Earth's Lorentz frame.

simplified but slightly inadequate form of the argument, one reaches a contradiction by noticing that here one has drawn a *parallelogram* in Minkowski spacetime in which two of the sides are unequal, $\tau_{top} > \tau_{bot}$, whereas a parallelogram in flat Minkowski spacetime cannot have opposite sides unequal. One concludes that *special relativity cannot be valid* over any sufficiently extended region. Globally, spacetime, as probed by the tracks of light rays and test particles, departs from flatness ("curvature"; Parts III and IV of this book), despite the fine fit that Lorentz-Minkowski flatness gives to physics locally.

This assumption is incompatible with gravitational redshift

Figure 7.1,B, repairs an oversimplification in this argument by recognizing that the propagation of light will be influenced by the gravitational field. Therefore photons might not follow straight lines in the diagram. Consequently, the world lines γ_1 and γ_2 of successive pulses are curves. However, the gravitational field is static and the experimenters do not move. Therefore nothing in the experimental setup changes with time. Whatever the path γ_1, the path γ_2 must be a *congruent* path of exactly the same shape, merely translated in time. On the basis of this congruence and the fact that the observers are moving on parallel world lines, one would again conclude, if flat Minkowski geometry were valid, that $\tau_{bot} = \tau_{top}$, thus contradicting the observed redshift experiment. The experimenters do not need to understand the propagation of light in a gravitational field. They need only use their radar apparatus to verify the fact that they are at rest relative to each other and relative to the source of the gravitational field. They know that, whatever influence the gravitational field has on their radar apparatus, it will not be a time-dependent influence. Moreover, they do not have to know how to compute their separation in order to verify that the separation remains constant. They only need to verify that the round-trip time required for radar pulses to go out to each other and back is the same every time they measure it.

Schild's redshift argument does not reveal what kind of curvature must exist, or whether the curvature exists in the neighborhood of the observational equipment or some distance away from it. It does say, however, quite unambiguously, that the flat spacetime of special relativity is inadequate to describe the situation, and it should therefore motivate one to undertake the mathematical analysis of curvature in Part III.

Conclusion: spacetime is curved

§7.4. GRAVITATIONAL REDSHIFT AS EVIDENCE FOR THE PRINCIPLE OF EQUIVALENCE

Einstein (1908, 1911) elevated the idea of the universality of gravitational interactions to the status of a fundamental *principle of equivalence, that all effects of a uniform gravitational field are identical to the effects of a uniform acceleration of the coordinate system.* This principle generalized a result of Newtonian gravitation theory, in which a uniform acceleration of the coordinate system in equation (7.1) gives rises to a

Principle of equivalence: a uniform gravitational field is indistinguishable from a uniform acceleration of a reference frame

supplementary uniform gravitational field. However, the Newtonian theory only gives this result for particle mechanics. Einstein's principle of equivalence asserts that a similar correspondence will hold for all the laws of physics, including Maxwell's equations for the electromagnetic field.

The rules of the game—the "scientific method"—require that experimental support be sought for any new theory or principle, and Einstein could treat the gravitational redshift as the equivalent of experimental confirmation of his principle of equivalence. There are two steps in such a confirmation: first, the theory or principle must predict an effect (the next paragraph describes how the equivalence principle implies the redshift); second, the predicted effect must be observed. With the Pound-Rebka-Snider experiments, one is in much better shape today than Einstein was for this second step. Einstein himself had to rely on the experiments supporting the general concept of energy conservation, plus the necessity of a redshift to preserve energy conservation, as a substitute for direct experimental confirmation.

Gravitational redshift derived from equivalence principle

The existence of the gravitational redshift can be deduced from the equivalence principle by considering two experimenters in a rocket ship that maintains a constant acceleration g. Let the distance between the two observers be h in the direction of the acceleration. Suppose for definiteness that the rocket ship was at rest in some inertial coordinate system when the bottom observer sent off a photon. It will require time $t = h$ for the photon to reach the upper observer. In that time the top observer acquires a velocity $v = gt = gh$. He will therefore detect the photon and observe a Doppler redshift $z = v = gh$. The results here are therefore identical to equation (7.20). The principle of equivalence of course requires that, if this redshift is observed in an experiment performed under conditions of uniform acceleration in the absence of gravitational fields, then the same redshift must be observed by an experiment performed under conditions where there is a uniform gravitational field, but no acceleration. Consequently, by the principle of equivalence, one can derive equation (7.20) as applied to the gravitational situation.

§7.5. LOCAL FLATNESS, GLOBAL CURVATURE

The equivalence principle helps one to discern the nature of the spacetime curvature, whose existence was inferred from Schild's argument. Physics is the same in an accelerated frame as it is in a laboratory tied to the Earth's surface. Thus, an Earth-bound lab can be regarded as accelerating upward, with acceleration g, relative to the Lorentz frames in its neighborhood.* Equivalently, relative to the lab and the Earth's surface, all Lorentz frames must accelerate downward. But the downward (radial) direction is different at different latitudes and longitudes. Hence, local Lorentz frames, initially at rest with respect to each other but on opposite sides of the Earth, subsequently fall toward the center and go flying through each other. Clearly they cannot be meshed to form the single global Lorentz frame, tied to the

Equivalence principle implies nonmeshing of local Lorentz frames near Earth (spacetime curvature!)

*This upward acceleration of the laboratory, plus equation (6.18) for the line element in an accelerated coordinate system, explains the nonequality of the bottom and top edges of the parallelograms in Figure 7.1.

Earth, that was assumed in Schild's argument. This nonmeshing of local Lorentz frames, like the nonmeshing of local Cartesian coordinates on a curved 2-surface, is a clear manifestation of spacetime curvature.

Geographers have similar problems when mapping the surface of the earth. Over small areas, a township or a county, it is easy to use a standard rectangular coordinate system. However, when two fairly large regions are mapped, each with one coordinate axis pointing north, then one finds that the edges of the maps overlap each other best if placed at a slight angle (spacetime analog: relative velocity of two local Lorentz frames meeting at center of Earth). It is much easier to start from a picture of a spherical globe, and then talk about how small flat maps might be used as good approximations to parts of it, than to start with a huge collection of small maps and try to piece them together to build up a picture of the globe. The exposition of the geometry of spacetime in this book will therefore take the first approach. Now that one recognizes that the problem is to fit together local, flat spacetime descriptions of physics into an over-all view of the universe, one should be happy to jump, in the next chapter, into a broadscale study of geometry. From this more advantageous viewpoint, one can then face the problem of discussing the relationship between the local inertial coordinate systems appropriate to two nearby regions that have slightly different gravitational fields.

Nonmeshing of local Lorentz frames motivates study of geometry

There are actually two distinguishable ways in which geometry enters the theory of general relativity. One is the geometry of lengths and angles in four-dimensional spacetime, which is inherited from the metric structure ds^2 of special relativity. Schild's argument already shows (without a direct appeal to the equivalence principle) that the special-relativistic ideas of length and angle must be modified. The modified ideas of metric structure lead to Riemannian geometry, which will be treated in Chapters 8 and 13. However, geometry also enters general relativity because of the equivalence principle. An equivalence principle can already be stated within Newtonian gravitational theory, in which no concepts of a *spacetime* metric enter, but only the Euclidean metric structure of three-dimensional *space*. The equivalence-principle view of Newtonian theory again insists that the local standard of reference be the freely falling particles. This requirement leads to the study of a *spacetime* geometry in which the curved world lines of freely falling particles are defined to be locally straight. They play the role in a curved spacetime geometry that straight lines play in flat spacetime. This "affine geometry" will be studied in Chapters 10–12. It leads to a quantitative formulation of the ideas of "covariant derivative" and "curvature" and even "curvature of Newtonian spacetime"!

Two types of geometry relevant to spacetime:

Riemannian geometry (lengths and angles)

Affine geometry ("straight lines" and curvature)

THE MATHEMATICS OF CURVED SPACETIME

*Wherein the reader is exposed to the charms of a new temptress—
Modern Differential Geometry—and makes a decision:
to embrace her for eight full chapters; or,
having drunk his fill, to escape after one.*

CHAPTER **8**

DIFFERENTIAL GEOMETRY: AN OVERVIEW

I am coming more and more to the conviction that the necessity of our geometry cannot be demonstrated, at least neither by, nor for, the human intellect. . . . geometry should be ranked, not with arithmetic, which is purely aprioristic, but with mechanics.
(1817)

We must confess in all humility that, while number is a product of our mind alone, space has a reality beyond the mind whose rules we cannot completely prescribe. (1830)

CARL FRIEDRICH GAUSS

§8.1. AN OVERVIEW OF PART III

Gravitation is a manifestation of spacetime curvature, and that curvature shows up in the deviation of one geodesic from a nearby geodesic ("relative acceleration of test particles"). The central issue of this part of the book is clear: *How can one quantify the "separation," and the "rate of change" of "separation," of two "geodesics" in "curved" spacetime?* A clear, precise answer requires new concepts.

Concepts to be developed in Part III:

"Separation" between geodesics will mean "vector." But the concept of vector as employed in flat Lorentz spacetime (a bilocal object: point for head and point for tail) must be sharpened up into the local concept of *tangent vector,* when one passes to curved spacetime. Chapter 9 does the sharpening. It also reveals how the passage to curved spacetime affects 1-forms and tensors.

Tangent vector

It takes one tool (vectors in curved geometry, Chapter 9) to define "separation" clearly as a vector; it takes another tool (parallel transport in curved spacetime, Chapter 10) to compare separation vectors at neighboring points and to define the "rate of change of separation." No transport, no comparison; no comparison, no meaning to the term "rate of change"! The notion of parallel transport founds itself

Geodesic

Covariant derivative

Geodesic deviation

Spacetime curvature

This chapter: a Track-1 overview of differential geometry

on the idea of *"geodesic,"* the world line of a freely falling particle. The special mathematical properties of a geodesic are explored in Chapter 10. That chapter uses geodesics to define parallel transport, uses parallel transport to define *covariant derivative,* and—completing the circle—uses covariant derivative to describe geodesics.

Chapter 11 faces up to the central issue: *geodesic deviation* ("rate of change of separation vector between two geodesics"), and its use in defining the *curvature* of spacetime.

But to define curvature is not enough. The man who would understand gravity deeply must also see curvature at work, producing relative accelerations of particles in Newtonian spacetime (Chapter 12); he must learn how, in Einstein spacetime, distances (metric) determine completely the curvature and the law of parallel transport (Chapter 13); he must be the master of powerful tools for computing curvature (Chapter 14); and he must grasp the geometric significance of the algebraic and differential symmetries of curvature (Chapter 15).

Unfortunately, such deep understanding requires time—far more time than one can afford in a ten-week or fifteen-week course, far more than a lone reader may wish to spend on first passage through the book. For the man who must rush on rapidly, this chapter contains a "Track-1" overview of the essential mathematical tools (§§8.4–8.7). From it one can gain an adequate, but not deep, understanding of spacetime curvature, of tidal gravitational forces, and of the mathematics of curved spacetime. This overview is also intended for the Track-2 reader; it will give him a taste of what is to come. The ambitious reader may also wish to consult other introductions to differential geometry (see Box 8.1).

Box 8.1 BOOKS ON DIFFERENTIAL GEOMETRY

There are several mathematics texts that may be consulted for a more detailed and extensive discussion of modern differential geometry along the line taken here. Bishop and Goldberg (1968) is the no. 1 reference. Hicks (1965) could be chosen as a current standard graduate-level text, with O'Neill (1966) at the undergraduate level introducing many of the same topics without presuming that the reader finds easy and obvious the current style in which pure mathematicians think and write. Auslander and MacKenzie (1963) at a somewhat more advanced level also allow for the reader to whom differential equations are more

familiar than homomorphisms. Willmore (1959) is easy to read but presents no challenge, and leads to little progress in adapting to the style of current mathematics. Trautman (1965) and Misner (1964a, 1969a) are introductions somewhat similar to ours, except for deemphasis of pictures; like ours, they are aimed at the student of relativity. Flanders (1963) is easy and useful as an introduction to exterior differential forms; it also gives examples of their application to a wide variety of topics in physics and engineering.

§8.2. TRACK 1 VERSUS TRACK 2: DIFFERENCE IN OUTLOOK AND POWER

Nothing is more wonderful about the relation between Einstein's theory of gravity and Newton's theory than this, as discovered by Élie Cartan (1923, 1924): that both theories lend themselves to description in terms of curvature; that in both this curvature is governed by the density of mass-energy; and that this curvature allows itself to be defined and measured without any use of or reference to any concept of metric. The difference between the two theories shows itself up in this: Einstein's theory in the end (or in the beginning, depending upon how one presents it!) does define an interval between every event and every nearby event; Newton's theory not only does not, but even says that any attempt to talk of *spacetime* intervals violates Newton's laws. This being the case, Track 2 will forego for a time (Chapters 9–12) any use of a spacetime metric ("Einstein interval"). It will extract everything possible from a metric-free description of spacetime curvature (all of Newton's theory; important parts of Einstein's theory).

Preview of Track-2 differential geometry

Geodesic deviation is a measurer and definer of curvature, but the onlooker is forbidden to reduce a vector description of separation to a numerical measure of distance (no metric at this stage of the analysis): what an impossible situation! Nevertheless, that is exactly the situation with which Chapters 9–12 will concern themselves: how to do geometry without a metric. Speaking physically, one will overlook at this stage the fact that the geometry of the physical world is always and everywhere locally Lorentz, and endowed with a light cone, but one will exploit to the fullest the Galileo-Einstein principle of equivalence: in any given locality one can find a frame of reference in which every neutral test particle, whatever its velocity, is free of acceleration. The tracks of these neutral test particles define the geodesics of the geometry. These geodesics provide tools with which one can do much: define parallel transport (Chapter 10), define covariant derivative (Chapter 10), quantify geodesic deviation (Chapter 11), define spacetime curvature (Chapter 11), and explore Newtonian gravity (Chapter 12). Only after this full exploitation of metric-free geodesics will Track 2 admit the Einstein metric back into the scene (Chapters 13–15).

But to forego use of the metric is a luxury which Track 1 can ill afford; too little time would be left for relativistic stars, cosmology, black holes, gravitational waves, experimental tests, and the dynamics of geometry. Therefore, the Track-1 overview in this chapter keeps the Einstein metric throughout. But in doing so, it pays a heavy price: (1) no possibility of seeing curvature at work in Newtonian spacetime (Chapter 12); (2) no possibility of comparing and contrasting the geometric structures of Newtonian spacetime (Chapter 12) and Einstein spacetime (Chapter 13), and hence no possibility of grasping fully the Newtonian-based motivation for the Einstein field equations (Chapter 17); (3) no possibility of understanding *fully* the mathematical interrelationships of "geodesic," "parallel transport," "covariant derivative," "curvature," and "metric" (Chapters 9, 10, 11, 13); (4) no possibility of introducing the mathematical subjects *"differential topology"* (geometry without metric or covariant

What the Track-1 reader will miss

derivative, Chapter 9) and *"affine geometry"* (geometry with covariant derivative but no metric, Chapters 10 and 11), subjects which find major application in modern analytical mechanics [see, e.g., Arnold and Avez (1968); also exercise 4.11 of this book], in Lie group theory with its deep implications for elementary particle physics [see, e.g., Hermann (1966); also exercises 9.12, 9.13, 10.16, and 11.12 of this book], in the theory and solution of partial differential equations [see, e.g., Sternberg (1969)], and, of course, in gravitation theory.

§8.3. THREE ASPECTS OF GEOMETRY: PICTORIAL, ABSTRACT, COMPONENT

Geometry from three viewpoints: pictorial, abstract, component

Gain the power in §8.4 and Chapter 9 to discuss tangent vectors, 1-forms, tensors in curved spacetime; gain the power in §8.5 and Chapter 10 to parallel-transport vectors, to differentiate them, to discuss geodesics; use this power in §8.7 and Chapter 11 to discuss geodesic deviation, to define curvature; But full power this will be only if it can be exercised in three ways: in pictures, in abstract notation, and in component notation (Box 8.3). Élie Cartan (Box 8.2) gave new insight into both

Box 8.2 ÉLIE CARTAN, 1869–1951

Élie Cartan is a most remarkable figure in recent mathematical history. One learns from his obituary [Chern and Chevalley (1952)] that he was born a blacksmith's son in southern France, and proved the value of government scholarship aid by rising through the system to a professorship at the Sorbonne in 1912 when he was 43. At the age of 32

he invented the exterior derivative [Cartan (1901)], which he used then mostly in differential equations and the theory of Lie groups, where he had already made significant contributions. He was about fifty when he began applying it to geometry, and sixty before Riemannian geometry specifically was the object of this research, including his text [Cartan (1928)], which is still reprinted and worth studying. Although universally recognized, his work did not find responsive readers until he neared retirement around 1940, when the "Bourbaki" generation of French mathematicians began to provide a conceptual framework for (among other things) Cartan's insights and methods. This made Cartan communicable and teachable as his own writings never were, so that by the time of his death at 82 in 1951 his influence was obviously dominating the revolutions then in full swing in all the fields (Lie groups, differential equations, and differential geometry) in which he had primarily worked.

The modern, abstract, coordinate-free approach to geometry, which is used extensively in this book, is due largely to Élie Cartan. He also discovered the geometric approach to Newtonian gravity that is developed and exploited in Chapter 12.

Box 8.3 THREE LEVELS OF DIFFERENTIAL GEOMETRY

(1) Purely *pictorial* treatment of geometry:

tangent vector is conceived in terms of the separation of two points in the limit in which the points are indefinitely close;

vectors are added and subtracted locally as in flat space;

vectors at distinct points are compared by parallel transport from one point to another;

this parallel transport is accomplished by a "Schild's ladder construction" of geodesics (Box 10.2);

diagrams, yes; algebra, no;

it is tied conceptually as closely as possible to the world of test particles and measurements.

(2) *Abstract* differential geometry:

treats a tangent vector as existing in its own right, without necessity to give its breakdown into components,

$$\boldsymbol{A} = A^0\boldsymbol{e}_0 + A^1\boldsymbol{e}_1 + A^2\boldsymbol{e}_2 + A^3\boldsymbol{e}_3,$$

just as one is accustomed nowadays in electromagnetism to treat the electric vector \boldsymbol{E}, without having to write out its components;

uses a similar approach to differentiation (compare gradient operator ∇ of elementary vector analysis, as distinguished from coordinate-dependent pieces of such an operator, such as $\partial/\partial x$, $\partial/\partial y$, etc.);

is the quickest, simplest mathematical scheme one knows to derive general results in differential geometry.

(3) Differential geometry as expressed in the language of *components:*

is indispensible in programming large parts of general relativity for a computer;

is convenient or necessary or both when one is dealing even at the level of elementary algebra with the most simple applications of relativity, from the expansion of the Friedmann universe to the curvature around a static center of attraction.

Newtonian gravity (Chapter 12) and the central geometric simplicity of Einstein's field equations (Chapter 15), because he had full command of all three methods of doing differential geometry. Today, no one has full power to communicate with others about the subject who cannot express himself in all three languages. Hence the interplay between the three forms of expression in what follows.

It is not new to go back and forth between the three languages, as witnesses the textbook treatment of the velocity and acceleration of a planet in Kepler motion around the sun. The velocity is written

Planetary orbit as example of three viewpoints

$$\boldsymbol{v} = v^{\hat{r}}\boldsymbol{e}_{\hat{r}} + v^{\hat{\phi}}\boldsymbol{e}_{\hat{\phi}}. \tag{8.1}$$

(The hats "ˆ" on $\boldsymbol{e}_{\hat{r}}$ and $\boldsymbol{e}_{\hat{\phi}}$ signify that these are unit vectors.) The acceleration is

$$\boldsymbol{a} = \frac{d\boldsymbol{v}}{dt} = \frac{dv^{\hat{r}}}{dt}\boldsymbol{e}_{\hat{r}} + \frac{dv^{\hat{\phi}}}{dt}\boldsymbol{e}_{\hat{\phi}} + v^{\hat{r}}\frac{d\boldsymbol{e}_{\hat{r}}}{dt} + v^{\hat{\phi}}\frac{d\boldsymbol{e}_{\hat{\phi}}}{dt}. \tag{8.2}$$

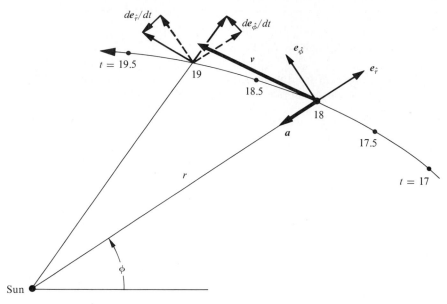

Figure 8.1.

A Keplerian orbit in the sun's gravitational field, as treated using the standard Euclidean-space version of Newtonian gravity. The basis vectors themselves change from point to point along the orbit [equations (8.3)]. This figure illustrates the pictorial aspect of differential geometry. Later (exercise 8.5) it will illustrate the concepts of "covariant derivative" and "connection coefficients."

The unit vectors are turning (Figure 8.1) with the angular velocity $\omega = d\phi/dt$; so

$$\frac{d\boldsymbol{e}_{\hat{r}}}{dt} = \omega \boldsymbol{e}_{\hat{\phi}} = \frac{d\phi}{dt} \boldsymbol{e}_{\hat{\phi}},$$

$$\frac{d\boldsymbol{e}_{\hat{\phi}}}{dt} = -\omega \boldsymbol{e}_{\hat{r}} = -\frac{d\phi}{dt} \boldsymbol{e}_{\hat{r}}.$$

(8.3)

Thus the components of the acceleration have the values

$$a^{\hat{r}} = \frac{dv^{\hat{r}}}{dt} - v^{\hat{\phi}} \frac{d\phi}{dt} = \frac{d^2r}{dt^2} - r\left(\frac{d\phi}{dt}\right)^2$$

(8.4a)

and

$$a^{\hat{\phi}} = \frac{dv^{\hat{\phi}}}{dt} + v^{\hat{r}} \frac{d\phi}{dt} = \frac{d}{dt}\left(r\frac{d\phi}{dt}\right) + \frac{dr}{dt}\frac{d\phi}{dt}.$$

(8.4b)

Here is the acceleration in the language of components; \boldsymbol{a} was the acceleration in abstract language; and Figure 8.1 shows the acceleration as an arrow. Each of these three languages will receive its natural generalization in the coming sections and chapters from two-dimensional flat space (with curvilinear coordinates) to four-dimensional curved spacetime, and from spacetime to more general manifolds (see §9.7 on manifolds).

Turn now to the Track-1 overview of differential geometry.

§8.4. TENSOR ALGEBRA IN CURVED SPACETIME

To see spacetime curvature at work, examine tidal gravitational forces (geodesic deviation); and to probe these forces, make measurements in a finite-sized laboratory. Squeeze the laboratory to infinitesimal size; all effects of spacetime curvature become infinitesimal; the physicist cannot tell whether he is in flat spacetime or curved spacetime. Neither can the mathematician, in the limit as his domain of attention squeezes down to a single event, \mathscr{P}_o.

At the event \mathscr{P}_o (in the infinitesimal laboratory) both physicist and mathematician can talk of vectors, of 1-forms, of tensors; and no amount of spacetime curvature can force the discussion to change from its flat-space form. A particle at \mathscr{P}_o has a 4-momentum \boldsymbol{p}, with squared length

$$\boldsymbol{p}^2 = \boldsymbol{p} \cdot \boldsymbol{p} = \boldsymbol{g}(\boldsymbol{p}, \boldsymbol{p}) = -m^2.$$

The squared length, as always, is calculated by inserting \boldsymbol{p} into both slots of a linear machine, the metric \boldsymbol{g} at \mathscr{P}_o. The particle also has a 4-acceleration \boldsymbol{a} at \mathscr{P}_o; and, if the particle is charged and freely moving, then \boldsymbol{a} is produced by the electromagnetic field tensor \boldsymbol{F}:

$$m\boldsymbol{a} = e\boldsymbol{F}(\ldots, \boldsymbol{u}).$$

In no way can curvature affect such local, coordinate-free, geometric relations. And in no way can it prevent one from introducing a local Lorentz frame at \mathscr{P}_o, and from performing standard, flat-space index manipulations in it:

$$\boldsymbol{p} = p^\alpha \boldsymbol{e}_\alpha, \qquad \boldsymbol{p}^2 = p^\alpha p^\beta \eta_{\alpha\beta} = p^\alpha p_\alpha, \qquad m a^\alpha = e F^{\alpha\beta} u_\beta.$$

But local Lorentz frames are not enough for the man who would calculate in curved spacetime. Non-Lorentz frames (nonorthonormal basis vectors $\{\boldsymbol{e}_\alpha\}$) often simplify calculations. Fortunately, no effort at all is required to master the rules of "index mechanics" in an arbitrary basis at a fixed event \mathscr{P}_o. The rules are identical to those in flat spacetime, except that (1) the covariant Lorentz components $\eta_{\alpha\beta}$ of the metric are replaced by

$$g_{\alpha\beta} \equiv \boldsymbol{e}_\alpha \cdot \boldsymbol{e}_\beta \equiv \boldsymbol{g}(\boldsymbol{e}_\alpha, \boldsymbol{e}_\beta); \tag{8.5}$$

(2) the contravariant components $\eta^{\alpha\beta}$ are replaced by $g^{\alpha\beta}$, where

$$\|g^{\alpha\beta}\| \equiv \|g_{\alpha\beta}\|^{-1} \text{ (matrix inverse)}; \tag{8.6}$$

i.e.,

$$g_{\alpha\beta} g^{\beta\gamma} = \delta_\alpha{}^\gamma; \tag{8.6'}$$

(3) the Lorentz transformation matrix $\|\Lambda^{\alpha'}{}_\beta\|$ and its inverse $\|\Lambda^\beta{}_{\alpha'}\|$ are replaced by an arbitrary but nonsingular transformation matrix $\|L^{\alpha'}{}_\beta\|$ and its inverse $\|L^\beta{}_{\alpha'}\|$:

$$\boldsymbol{e}_\beta = \boldsymbol{e}_{\alpha'} L^{\alpha'}{}_\beta, \qquad p^\beta = L^\beta{}_{\alpha'} p^{\alpha'}, \tag{8.7}$$

$$\|L^\beta{}_{\alpha'}\| = \|L^{\alpha'}{}_\beta\|^{-1}; \tag{8.8}$$

Tensor algebra:

(1) occurs in infinitesimal neighborhood of an event

(2) is same in curved spacetime as in flat

(3) rules for component manipulation change slightly when using nonorthonormal basis

Components of metric

Transformation of basis

(4) in the special case of "coordinate bases," $\boldsymbol{e}_\alpha = \partial\mathscr{P}/\partial x^\alpha$, $\boldsymbol{e}_{\beta'} = \partial\mathscr{P}/\partial x^{\beta'}$,

$$L^{\alpha'}{}_\beta = \partial x^{\alpha'}/\partial x^\beta, \qquad L^\beta{}_{\alpha'} = \partial x^\beta/\partial x^{\alpha'}; \tag{8.9}$$

Components of Levi-Civita tensor

and (5) the Levi-Civita tensor $\boldsymbol{\varepsilon}$, like the metric tensor, has components that depend on how nonorthonormal the basis vectors are (see exercise 8.3): if \boldsymbol{e}_0 points toward the future and \boldsymbol{e}_1, \boldsymbol{e}_2, \boldsymbol{e}_3 are righthanded, then

$$\varepsilon_{\alpha\beta\gamma\delta} = (-g)^{1/2}[\alpha\beta\gamma\delta],$$
$$\varepsilon^{\alpha\beta\gamma\delta} = g^{-1}\varepsilon_{\alpha\beta\gamma\delta} = -(-g)^{-1/2}[\alpha\beta\gamma\delta], \tag{8.10a}$$

where $[\alpha\beta\gamma\delta]$ is the completely antisymmetric symbol

$$[\alpha\beta\gamma\delta] \equiv \begin{cases} +1 & \text{if } \alpha\beta\gamma\delta \text{ is an even permutation of } 0123, \\ -1 & \text{if } \alpha\beta\gamma\delta \text{ is an odd permutation of } 0123, \\ 0 & \text{if } \alpha\beta\gamma\delta \text{ are not all different}, \end{cases} \tag{8.10b}$$

and where g is the determinant of the matrix $\|g_{\alpha\beta}\|$

$$g \equiv \det\|g_{\alpha\beta}\| = \det\|\boldsymbol{e}_\alpha \cdot \boldsymbol{e}_\beta\|. \tag{8.11}$$

Read Box 8.4 for full discussion and proofs; work exercise 8.1 below for fuller understanding and mastery.

Several dangers are glossed over in this discussion. In flat spacetime one often does not bother to say where a vector, 1-form, or tensor is located. One freely moves geometric objects from event to event without even thinking. Of course, the unwritten rule of transport is: hold all lengths and directions fixed while moving; i.e., hold all Lorentz-frame components fixed; i.e., "parallel-transport" the object. But in

Box 8.4 TENSOR ALGEBRA AT A FIXED EVENT IN AN *ARBITRARY* BASIS

A. Bases

Tangent-vector basis: Pick \boldsymbol{e}_0, \boldsymbol{e}_1, \boldsymbol{e}_2, \boldsymbol{e}_3 at \mathscr{P}_0 arbitrarily—but insist they be linearly independent.

"Dual basis" for 1-forms: The basis $\{\boldsymbol{e}_\alpha\}$ determines a 1-form basis $\{\boldsymbol{\omega}^\alpha\}$ (its "dual basis") by

$$\langle\boldsymbol{\omega}^\alpha, \boldsymbol{e}_\beta\rangle = \delta^\alpha{}_\beta$$

[see equation (2.19)].

Geometric interpretation (Figure 9.2): \boldsymbol{e}_2, \boldsymbol{e}_3, \boldsymbol{e}_0 lie parallel to surfaces of $\boldsymbol{\omega}^1$; \boldsymbol{e}_1 pierces precisely one surface of $\boldsymbol{\omega}^1$.

Function interpretation: $\langle\boldsymbol{\omega}^\alpha, \boldsymbol{e}_\beta\rangle = \delta^\alpha{}_\beta$ determines the value of $\boldsymbol{\omega}^\alpha$ on any vector $\boldsymbol{u} = u^\beta\boldsymbol{e}_\beta$ (number of "bongs of bell" as \boldsymbol{u} pierces $\boldsymbol{\omega}^\alpha$):

$$\langle \boldsymbol{w}^{\alpha}, \boldsymbol{u} \rangle = \langle \boldsymbol{w}^{\alpha}, u^{\beta}\boldsymbol{e}_{\beta} \rangle = u^{\beta}\langle \boldsymbol{w}^{\alpha}, \boldsymbol{e}_{\beta} \rangle = u^{\beta}\delta^{\alpha}{}_{\beta} = u^{\alpha}.$$

Special case: *coordinate bases*. Choose an arbitrary coordinate system $\{x^{\alpha}(\mathscr{P})\}$. At \mathscr{P}_0 choose $\boldsymbol{e}_{\alpha} = \partial\mathscr{P}/\partial x^{\alpha}$ as basis vectors. Then the dual basis is $\boldsymbol{w}^{\alpha} = \boldsymbol{d}x^{\alpha}$. *Proof:* the general coordinate-free relation $\langle \boldsymbol{df}, \boldsymbol{v} \rangle = \partial_{\boldsymbol{v}} f$ [equation (2.17)], with $f = x^{\alpha}$ and $\boldsymbol{v} = \partial\mathscr{P}/\partial x^{\beta}$, reads

$$\langle \boldsymbol{d}x^{\alpha}, \partial\mathscr{P}/\partial x^{\beta} \rangle = (\partial/\partial x^{\beta})x^{\alpha} = \delta^{\alpha}{}_{\beta}.$$

B. Algebra of Tangent Vectors and 1-Forms

The Lorentz-frame discussion of equations (2.19) to (2.22) is completely unchanged when one switches to an arbitrary basis. Its conclusions:

expansion, $\boldsymbol{u} = \boldsymbol{e}_{\alpha}u^{\alpha}$, $\boldsymbol{\sigma} = \sigma_{\alpha}\boldsymbol{w}^{\alpha}$;

calculation of components, $u^{\alpha} = \langle \boldsymbol{w}^{\alpha}, \boldsymbol{u} \rangle$, $\sigma_{\alpha} = \langle \boldsymbol{\sigma}, \boldsymbol{e}_{\alpha} \rangle$;

value of form on vector, $\langle \boldsymbol{\sigma}, \boldsymbol{u} \rangle = \sigma_{\alpha}u^{\alpha}$.

Application to gradients of functions:

expansion, $\boldsymbol{df} = f_{,\alpha}\boldsymbol{w}^{\alpha}$ [defines $f_{,\alpha}$];

calculation of components, $f_{,\alpha} = \langle \boldsymbol{df}, \boldsymbol{e}_{\alpha} \rangle = \partial_{\boldsymbol{e}_{\alpha}} f$ [see equation (2.17)].

Raising and lowering of indices is accomplished with $g_{\alpha\beta}$ and $g^{\alpha\beta}$ [equations (8.5) and (8.6)]. Proof:

$\tilde{\boldsymbol{u}}$, the 1-form corresponding to \boldsymbol{u}, is defined by $\langle \tilde{\boldsymbol{u}}, \boldsymbol{v} \rangle = \boldsymbol{u} \cdot \boldsymbol{v}$ for all \boldsymbol{v};

thus, $u_{\alpha} \equiv \langle \tilde{\boldsymbol{u}}, \boldsymbol{e}_{\alpha} \rangle = \boldsymbol{u} \cdot \boldsymbol{e}_{\alpha} = u^{\beta}\boldsymbol{e}_{\beta} \cdot \boldsymbol{e}_{\alpha} = u^{\beta}g_{\beta\alpha}$;

inverting this equation yields $u^{\beta} = g^{\beta\alpha}u_{\alpha}$.

C. Change of Basis

The discussion of Lorentz transformations in equations (2.39) to (2.43) is applicable to general changes of basis if one replaces $\|\Lambda^{\alpha'}{}_{\beta}\|$ by an arbitrary but nonsingular matrix $\|L^{\alpha'}{}_{\beta}\|$ [equations (8.7), (8.8)]. Conclusions:

$$
\begin{aligned}
\boldsymbol{e}_{\alpha'} &= \boldsymbol{e}_{\beta}L^{\beta}{}_{\alpha'}, & \boldsymbol{e}_{\beta} &= \boldsymbol{e}_{\alpha'}L^{\alpha'}{}_{\beta}; \\
\boldsymbol{w}^{\alpha'} &= L^{\alpha'}{}_{\beta}\boldsymbol{w}^{\beta}, & \boldsymbol{w}^{\beta} &= L^{\beta}{}_{\alpha'}\boldsymbol{w}^{\alpha'}; \\
v^{\alpha'} &= L^{\alpha'}{}_{\beta}v^{\beta}, & v^{\beta} &= L^{\beta}{}_{\alpha'}v^{\alpha'}; \\
\sigma_{\alpha'} &= \sigma_{\beta}L^{\beta}{}_{\alpha'}, & \sigma_{\beta} &= \sigma_{\alpha'}L^{\alpha'}{}_{\beta}.
\end{aligned}
$$

When both bases are coordinate bases, then $L^{\beta}{}_{\alpha'} = \partial x^{\beta}/\partial x^{\alpha'}$, $L^{\alpha'}{}_{\beta} = \partial x^{\alpha'}/\partial x^{\beta}$. *Proof:*

$$\boldsymbol{e}_{\alpha'} = \frac{\partial}{\partial x^{\alpha'}} = \frac{\partial x^{\beta}}{\partial x^{\alpha'}}\frac{\partial}{\partial x^{\beta}} = \frac{\partial x^{\beta}}{\partial x^{\alpha'}}\boldsymbol{e}_{\beta}; \quad \text{similarly } \boldsymbol{e}_{\beta} = \frac{\partial x^{\alpha'}}{\partial x^{\beta}}\boldsymbol{e}_{\alpha'}.$$

Box 8.4 (*continued*)

D. Algebra of Tensors

The discussions of tensor algebra given in §3.2 [equations (3.8) to (3.22)] and in §3.5 (excluding gradient and divergence) are unchanged, except that

$$\eta_{\alpha\beta} \longrightarrow g_{\alpha\beta} \qquad \eta^{\alpha\beta} \longrightarrow g^{\alpha\beta}, \qquad \Lambda^{\alpha'}{}_{\beta} \longrightarrow L^{\alpha'}{}_{\beta}, \qquad \Lambda^{\beta}{}_{\alpha'} \longrightarrow L^{\beta}{}_{\alpha'};$$

and the components of the Levi-Civita tensor are changed from (3.50) to (8.10) [see exercise 8.3[.

Chief conclusions:

 expansion, $\boldsymbol{S} = S^{\alpha}{}_{\beta\gamma}\boldsymbol{e}_{\alpha} \otimes \boldsymbol{\omega}^{\beta} \otimes \boldsymbol{\omega}^{\gamma}$;

 components, $S^{\alpha}{}_{\beta\gamma} = \boldsymbol{S}(\boldsymbol{\omega}^{\alpha}, \boldsymbol{e}_{\beta}, \boldsymbol{e}_{\gamma})$;

 raising and lowering indices, $S_{\mu\beta}{}^{\nu} = g_{\mu\alpha}g^{\nu\gamma}S^{\alpha}{}_{\beta\gamma}$;

 change of basis, $S^{\lambda'}{}_{\mu'\nu'} = L^{\lambda'}{}_{\alpha}L^{\beta}{}_{\mu'}L^{\gamma}{}_{\nu'}S^{\alpha}{}_{\beta\gamma}$;

 machine operation, $\boldsymbol{S}(\boldsymbol{\sigma}, \boldsymbol{u}, \boldsymbol{v}) = S^{\alpha}{}_{\beta\gamma}\sigma_{\alpha}u^{\beta}v^{\gamma}$;

 tensor product, $\boldsymbol{T} = \boldsymbol{u} \otimes \boldsymbol{v} \Longleftrightarrow T^{\alpha\beta} = u^{\alpha}v^{\beta}$;

 contraction, "\boldsymbol{M} = contraction of \boldsymbol{R} on slots 1 and 3" $\Longleftrightarrow M_{\mu\nu} = R^{\alpha}{}_{\mu\alpha\nu}$;

 wedge product, $\boldsymbol{\alpha} \wedge \boldsymbol{\beta}$ has components $\alpha^{\mu}\beta^{\nu} - \beta^{\mu}\alpha^{\nu}$;

 Dual, $*J_{\alpha\beta\gamma} = J^{\mu}\varepsilon_{\mu\alpha\beta\gamma}$, $*F_{\alpha\beta} = \frac{1}{2}F^{\mu\nu}\varepsilon_{\mu\nu\alpha\beta}$, $*B_{\alpha} = \frac{1}{6}B^{\lambda\mu\nu}\varepsilon_{\lambda\mu\nu\alpha}$.

E. Commutators (exercise 8.2; §9.6; Box 9.2)

If \boldsymbol{u} and \boldsymbol{v} are tangent vector fields, one takes the view that $\boldsymbol{u} = \partial_{\boldsymbol{u}}$ and $\boldsymbol{v} = \partial_{\boldsymbol{v}}$, and one defines

$$[\boldsymbol{u}, \boldsymbol{v}] \equiv [\partial_{\boldsymbol{u}}, \partial_{\boldsymbol{v}}] \equiv \partial_{\boldsymbol{u}}\,\partial_{\boldsymbol{v}} - \partial_{\boldsymbol{v}}\,\partial_{\boldsymbol{u}}.$$

This commutator is itself a tangent vector field.

Components in a coordinate basis:

$$[\boldsymbol{u}, \boldsymbol{v}] = (u^{\beta}v^{\alpha}{}_{,\beta} - v^{\beta}u^{\alpha}{}_{,\beta})(\partial/\partial x^{\alpha}).$$
$$\llcorner [= \boldsymbol{e}_{\alpha}]$$

Commutation coefficients of a basis:

$$[\boldsymbol{e}_{\alpha}, \boldsymbol{e}_{\beta}] \equiv c_{\alpha\beta}{}^{\gamma}\boldsymbol{e}_{\gamma}, \qquad c_{\alpha\beta\mu} \equiv c_{\alpha\beta}{}^{\gamma}g_{\gamma\mu}.$$

Coordinate basis ("holonomic") $c_{\alpha\beta}{}^{\gamma} = 0$;

Noncoordinate basis ("anholonomic") some $c_{\alpha\beta}{}^{\gamma} \neq 0$ (see exercise 9.9).

curved spacetime there is no global Lorentz coordinate system in which to hold components fixed; and objects initially parallel, after "parallel transport" along different curves cease to be parallel ("geodesic deviation"; Earth's meridians, parallel at equator, cross at north and south poles). Thus, in curved spacetime one must not blithely move a geometric object from point to point, without carefully specifying how it is to be moved and by what route. Each local geometric object has its own official place of residence (event \mathcal{P}_0); it can interact with other objects residing there (tensor algebra); but it cannot interact with any object at another event \mathcal{Q}, until it has been carefully transported from \mathcal{P}_0 to \mathcal{Q}.

Vectors and tensors must not be moved blithely from point to point

This line of reasoning, pursued further, leads one to speak of the "*tangent space*" at each event, in which that event's vectors (arrows) and 1-forms (families of surfaces) lie, and in which its tensors (linear machines) operate. One even draws heuristic pictures of the tangent space, as in Figure 9.1 (p. 231).

Tangent space defined

Another danger in curved spacetime is the temptation to regard vectors as arrows linking two events ("point for head and point for tail")—i.e., to regard the tangent space of Figure 9.1 as lying in spacetime itself. This practice can be useful for heuristic purposes, but it is incompatible with complete mathematical precision. (How is the tangent space to be molded into a warped surface?) Four definitions of a vector were given in Figure 2.1 (page 49): three definitions relying on "point for head and point for tail"; one, "$d\mathcal{P}/d\lambda$", purely local. Only the local definition is wholly viable in curved spacetime, and even it can be improved upon, in the eyes of mathematicians, as follows.

Definitions of vector in curved spacetime:

(1) as $d\mathcal{P}/d\lambda$

There is a one-to-one correspondence (complete "isomorphism") between vectors \boldsymbol{u} and directional derivative operators $\partial_{\boldsymbol{u}}$. The concept of vector is a bit fuzzy, but "directional derivative" is perfectly well-defined. To get rid of all fuzziness, exploit the isomorphism to the full: *define* the tangent vector \boldsymbol{u} to be equal to the corresponding directional derivative

(2) as directional derivative

$$\boldsymbol{u} \equiv \partial_{\boldsymbol{u}}. \tag{8.12}$$

(This practice, unfamiliar as it may be to a physicist at first, has mathematical power; so this book will use it frequently. For a fuller discussion, see §9.2.)

Exercise 8.1. PRACTICE WITH TENSOR ALGEBRA

EXERCISES

Let t, x, y, z be Lorentz coordinates in flat spacetime, and let

$$r = (x^2 + y^2 + z^2)^{1/2}, \qquad \theta = \cos^{-1}(z/r), \qquad \phi = \tan^{-1}(y/x)$$

be the corresponding spherical coordinates. Then

$$\boldsymbol{e}_0 = \partial\mathcal{P}/\partial t, \qquad \boldsymbol{e}_r = \partial\mathcal{P}/\partial r, \qquad \boldsymbol{e}_\theta = \partial\mathcal{P}/\partial\theta, \qquad \boldsymbol{e}_\phi = \partial\mathcal{P}/\partial\phi,$$

is a coordinate basis, and

$$\boldsymbol{e}_{\hat{0}} = \frac{\partial\mathcal{P}}{\partial t}, \qquad \boldsymbol{e}_{\hat{r}} = \frac{\partial\mathcal{P}}{\partial r}, \qquad \boldsymbol{e}_{\hat{\theta}} = \frac{1}{r}\frac{\partial\mathcal{P}}{\partial\theta}, \qquad \boldsymbol{e}_{\hat{\phi}} = \frac{1}{r\sin\theta}\frac{\partial\mathcal{P}}{\partial\phi}$$

is a noncoordinate basis.

(a) Draw a picture of e_θ, e_ϕ, $e_{\hat\theta}$, and $e_{\hat\phi}$ at several different points on a sphere of constant t, r. [Answer for e_θ, e_ϕ should resemble Figure 9.1.]

(b) What are the 1-form bases $\{\boldsymbol{\omega}^\alpha\}$ and $\{\boldsymbol{\omega}^{\hat\alpha}\}$ dual to these tangent-vector bases? [Answer: $\boldsymbol{\omega}^0 = \boldsymbol{dt}$, $\boldsymbol{\omega}^r = \boldsymbol{dr}$, $\boldsymbol{\omega}^\theta = \boldsymbol{d\theta}$, $\boldsymbol{\omega}^\phi = \boldsymbol{d\phi}$; $\boldsymbol{\omega}^{\hat0} = \boldsymbol{dt}$, $\boldsymbol{\omega}^{\hat r} = \boldsymbol{dr}$, $\boldsymbol{\omega}^{\hat\theta} = r\,\boldsymbol{d\theta}$, $\boldsymbol{\omega}^{\hat\phi} = r\sin\theta\,\boldsymbol{d\phi}$.]

(c) What is the transformation matrix linking the original Lorentz frame to the spherical coordinate frame $\{e_\alpha\}$? [Answer: nonzero components are

$$L^t{}_0 = 1, \qquad L^z{}_r = \frac{\partial z}{\partial r} = \cos\theta, \qquad L^z{}_\theta = \frac{\partial z}{\partial\theta} = -r\sin\theta,$$

$$L^x{}_r = \sin\theta\cos\phi, \qquad L^x{}_\theta = r\cos\theta\cos\phi, \qquad L^x{}_\phi = -r\sin\theta\sin\phi,$$

$$L^y{}_r = \sin\theta\sin\phi, \qquad L^y{}_\theta = r\cos\theta\sin\phi, \qquad L^y{}_\phi = r\sin\theta\cos\phi.]$$

(d) Use this transformation matrix to calculate the metric components $g_{\alpha\beta}$ in the spherical coordinate basis, and invert the resulting matrix to get $g^{\alpha\beta}$. [Answer:

$$g_{00} = -1, \quad g_{rr} = 1, \quad g_{\theta\theta} = r^2, \quad g_{\phi\phi} = r^2\sin^2\theta, \quad \text{all other } g_{\alpha\beta} = 0.$$
$$g^{00} = -1, \quad g^{rr} = 1, \quad g^{\theta\theta} = r^{-2}, \quad g^{\phi\phi} = r^{-2}\sin^{-2}\theta, \quad \text{all other } g^{\alpha\beta} = 0.]$$

(e) Show that the noncoordinate basis $\{e_{\hat\alpha}\}$ is orthonormal everywhere; i.e., that $g_{\hat\alpha\hat\beta} = \eta_{\alpha\beta}$; i.e. that

$$g = -\boldsymbol{\omega}^{\hat0} \otimes \boldsymbol{\omega}^{\hat0} + \boldsymbol{\omega}^{\hat r} \otimes \boldsymbol{\omega}^{\hat r} + \boldsymbol{\omega}^{\hat\theta} \otimes \boldsymbol{\omega}^{\hat\theta} + \boldsymbol{\omega}^{\hat\phi} \otimes \boldsymbol{\omega}^{\hat\phi}.$$

(f) Write the gradient of a function f in terms of the spherical coordinate and noncoordinate bases. [Answer:

$$\boldsymbol{df} = \frac{\partial f}{\partial t}\boldsymbol{dt} + \frac{\partial f}{\partial r}\boldsymbol{dr} + \frac{\partial f}{\partial\theta}\boldsymbol{d\theta} + \frac{\partial f}{\partial\phi}\boldsymbol{d\phi}$$
$$- \frac{\partial f}{\partial t}\boldsymbol{\omega}^{\hat0} + \frac{\partial f}{\partial r}\boldsymbol{\omega}^{\hat r} + \frac{1}{r}\frac{\partial f}{\partial\theta}\boldsymbol{\omega}^{\hat\theta} + \frac{1}{r\sin\theta}\frac{\partial f}{\partial\phi}\boldsymbol{\omega}^{\hat\phi}.]$$

(g) What are the components of the Levi-Civita tensor in the spherical coordinate and noncoordinate bases? [Answer for coordinate basis:

$$\varepsilon_{0r\theta\phi} = -\varepsilon_{r0\theta\phi} = +\varepsilon_{r\theta0\phi} = \cdots = r^2\sin\theta,$$
$$\varepsilon^{0r\theta\phi} = -\varepsilon^{r0\theta\phi} = +\varepsilon^{r\theta\phi0} = \cdots = -r^{-2}\sin^{-1}\theta.]$$

Exercise 8.2. COMMUTATORS

Take the mathematician's viewpoint that tangent vectors and directional derivatives are the same thing, $\boldsymbol{u} \equiv \partial_{\boldsymbol{u}}$. Let \boldsymbol{u} and \boldsymbol{v} be two vector fields, and define their commutator in the manner familiar from quantum mechanics

$$[\boldsymbol{u}, \boldsymbol{v}] \equiv [\partial_{\boldsymbol{u}}, \partial_{\boldsymbol{v}}] \equiv \partial_{\boldsymbol{u}}\partial_{\boldsymbol{v}} - \partial_{\boldsymbol{v}}\partial_{\boldsymbol{u}}. \tag{8.13a}$$

(a) Derive the following expression for $[\boldsymbol{u}, \boldsymbol{v}]$, valid in any *coordinate* basis,

$$[\boldsymbol{u}, \boldsymbol{v}] = (u^\beta v^\alpha{}_{,\beta} - v^\beta u^\alpha{}_{,\beta})\boldsymbol{e}_\alpha. \tag{8.13b}$$

Thus, despite the fact that it looks like a second-order differential operator, $[\boldsymbol{u}, \boldsymbol{v}]$ is actually of first order—i.e., it is a tangent vector.

(b) For any basis $\{\boldsymbol{e}_\alpha\}$, one defines the "*commutation coefficients*" $c_{\beta\gamma}{}^\alpha$ and $c_{\beta\gamma\alpha}$ by

$$[\boldsymbol{e}_\beta, \boldsymbol{e}_\gamma] \equiv c_{\beta\gamma}{}^\alpha \boldsymbol{e}_\alpha; \qquad c_{\beta\gamma\alpha} \equiv g_{\alpha\mu} c_{\beta\gamma}{}^\mu. \tag{8.14}$$

Show that $c_{\beta\gamma}{}^\alpha = c_{\beta\gamma\alpha} = 0$ for any coordinate basis.

(c) Calculate $c_{\hat\beta\hat\gamma}{}^{\hat\alpha}$ for the spherical noncoordinate basis of exercise 8.1. [*Answer:* All vanish except

$$c_{\hat r\hat\theta}{}^{\hat\theta} = -c_{\hat\theta\hat r}{}^{\hat\theta} = -1/r,$$

$$c_{\hat r\hat\phi}{}^{\hat\phi} = -c_{\hat\phi\hat r}{}^{\hat\phi} = -1/r,$$

$$c_{\hat\theta\hat\phi}{}^{\hat\phi} = -c_{\hat\phi\hat\theta}{}^{\hat\phi} = -\cot\theta/r.]$$

Exercise 8.3. COMPONENTS OF LEVI-CIVITA TENSOR IN NONORTHONORMAL FRAME

(a) Show that expressions (8.10) are the components of $\boldsymbol\varepsilon$ in an arbitrary basis, with $\boldsymbol e_0$ pointing toward the future and $\boldsymbol e_1$, $\boldsymbol e_2$, $\boldsymbol e_3$ right-handed. [Hints: (1) Review the discussion of $\boldsymbol\varepsilon$ in Lorentz frames, given in exercise 3.13. (2) Calculate $\varepsilon_{\alpha\beta\gamma\delta}$ and $\varepsilon^{\alpha\beta\gamma\delta}$ by transforming from a local Lorentz frame $\{\boldsymbol e_{\hat\mu}\}$, e.g.,

$$\varepsilon_{\alpha\beta\gamma\delta} = L^{\hat\mu}{}_\alpha L^{\hat\nu}{}_\beta L^{\hat\lambda}{}_\gamma L^{\hat\rho}{}_\delta \varepsilon_{\hat\mu\hat\nu\hat\lambda\hat\rho}.$$

(3) Show that these expressions reduce to

$$\varepsilon_{\alpha\beta\gamma\delta} = \det\|L^{\hat\mu}{}_\nu\| \varepsilon_{\hat\alpha\hat\beta\hat\gamma\hat\delta}, \qquad \varepsilon^{\alpha\beta\gamma\delta} = \det\|L^\nu{}_{\hat\mu}\| \varepsilon^{\hat\alpha\hat\beta\hat\gamma\hat\delta}.$$

(4) Show, from the transformation law for the metric components, that

$$(\det\|L^\nu{}_{\hat\mu}\|)^2 \det\|g_{\alpha\beta}\| = -1.$$

(5) Combine these results to obtain expressions (8.10).]

 (b) Show that the components of the permutation tensors [defined by equations (3.50h)–(3.50j)] have the same values [equations (3.50k)–(3.50m)] in arbitrary frames as in Lorentz frames.

 Additional exercises on tensor algebra: exercises 9.3 and 9.4 (page 234).

§8.5. PARALLEL TRANSPORT, COVARIANT DERIVATIVE, CONNECTION COEFFICIENTS, GEODESICS

The vehicle that carries one from classical mechanics to quantum mechanics is the correspondence principle. Similarly, the vehicle between flat spacetime and curved spacetime is the equivalence principle: "The laws of physics are the same in any local Lorentz frame of curved spacetime as in a global Lorentz frame of flat spacetime." But to apply the equivalence principle, one must first have a mathematical representation of a local Lorentz frame. The obvious choice is this: *A local Lorentz frame at a given event \mathscr{P}_0 is the closest thing there is to a global Lorentz frame at that event; i.e., it is a coordinate system in which*

Equivalence principle as vehicle between flat spacetime and curved

Local Lorentz frame: mathematical representation

$$g_{\mu\nu}(\mathscr{P}_0) = \eta_{\mu\nu}, \tag{8.15a}$$

and in which $g_{\mu\nu}$ holds as tightly as possible to $\eta_{\mu\nu}$ in the neighborhood of \mathscr{P}_0:

$$g_{\mu\nu,\alpha}(\mathscr{P}_0) = 0. \tag{8.15b}$$

More tightly than this it cannot hold in general [$g_{\mu\nu,\alpha\beta}(\mathscr{P}_0)$ cannot be set to zero]; spacetime curvature forces it to change. [Combine §11.5 with equations (8.24) and (8.44).]

Parallel transport defined

An observer in a local Lorentz frame in curved spacetime can compare vectors and tensors at neighboring events, just as he would in flat spacetime. But to make the comparison, he must parallel-transport them to a common event. For him the act of parallel transport is simple: he keeps all Lorentz-frame components fixed, just as if he were in flat spacetime. But for a man without a local Lorentz frame—perhaps with no coordinate system or basis vectors at all—parallel transport is less trivial. He must either ask his Lorentz-based friend the result, or he must use a more sophisticated technique. One technique he can use—a "Schild's ladder" construction that requires no coordinates or basis vectors whatsoever—is described in §10.2 and Box 10.2. But the Track-1 reader need not master Schild's ladder. He can always ask a local Lorentz observer what the result of any given parallel transport is, or he can use general formulas worked out below.

Comparison by parallel transport is the foundation on which rests the gradient of a tensor field, ∇T. No mention of parallel transport was made in §3.5, where the gradient was first defined, but parallel transport occurred implicitly: one defined ∇T in such a way that its components were $T^{\alpha}{}_{\beta,\gamma} = \partial T^{\alpha}{}_{\beta}/\partial x^{\gamma}$ [for T a $\binom{1}{1}$ tensor]; i.e., one asked ∇T to measure how much the Lorentz-frame components of T change from point to point. But "no change in Lorentz components" would have meant "parallel transport," so one was implicitly asking for the change in T relative to what T would have been after pure parallel transport.

Covariant derivative defined

To codify in abstract notation this concept of differentiation, proceed as follows. First define the "covariant derivative" $\nabla_u T$ of T along a curve $\mathcal{P}(\lambda)$, whose tangent vector is $u = d\mathcal{P}/d\lambda$:

$$(\nabla_u T)_{\text{at } \mathcal{P}(0)} = \lim_{\varepsilon \to 0} \left\{ \frac{T[\mathcal{P}(\varepsilon)]_{\text{parallel-transported to } \mathcal{P}(0)} - T[\mathcal{P}(0)]}{\varepsilon} \right\}. \qquad (8.16)$$

Gradient defined

(See Figure 8.2 for the special case where T is a vector field v.) Then define ∇T to be the linear machine, that gives $\nabla_u T$ when u is inserted into its last slot:

$$\nabla T(\ldots, \ldots, u) \equiv \nabla_u T. \qquad (8.17)$$

The result is the same animal ("gradient") as was defined in §3.5 (for proof see exercise 8.8). But this alternative definition makes clear the relationship to parallel transport, including the fact that

$$\nabla_u T = 0 \iff T \text{ is parallel-transported along } u = d\mathcal{P}/d\lambda. \qquad (8.18)$$

In a local Lorentz frame, the components of ∇T are directional derivatives of the components of T: $T^{\beta}{}_{\alpha,\gamma}$. Not so in a general basis. If $\{e_{\beta}(\mathcal{P})\}$ is a basis that varies arbitrarily but smoothly from point to point, and $\{\omega^{\alpha}(\mathcal{P})\}$ is its dual basis, then $\nabla T = \nabla(T^{\beta}{}_{\alpha} e_{\beta} \otimes \omega^{\alpha})$ will contain contributions from ∇e_{β} and $\nabla \omega^{\alpha}$, as well as from $\nabla T^{\beta}{}_{\alpha} \equiv dT^{\beta}{}_{\alpha} = T^{\beta}{}_{\alpha,\gamma}\omega^{\gamma}$.

Connection coefficients defined

To quantify the contributions from ∇e_{β} and $\nabla \omega^{\alpha}$, i.e., to quantify the twisting, turning, expansion, and contraction of the basis vectors and 1-forms, one defines "connection coefficients":

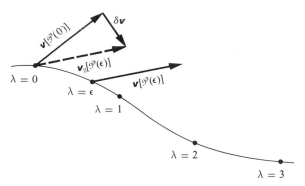

Figure 8.2.
Definition of the covariant derivative "$\mathbf{\nabla}_{\boldsymbol{u}}\boldsymbol{v}$" of a vector field \boldsymbol{v} along a curve $\mathcal{P}(\lambda)$, with tangent vector $\boldsymbol{u} \equiv d\mathcal{P}/d\lambda$: (1) choose a point $\mathcal{P}(0)$ on the curve, at which to evaluate $\mathbf{\nabla}_{\boldsymbol{u}}\boldsymbol{v}$. (2) Choose a nearby point $\mathcal{P}(\varepsilon)$ on the curve. (3) Parallel-transport $\boldsymbol{v}[\mathcal{P}(\varepsilon)]$ along the curve back to $\mathcal{P}(0)$, getting the vector $\boldsymbol{v}_{\parallel}[\mathcal{P}(\varepsilon)]$. (4) Take the difference $\delta\boldsymbol{v} \equiv \boldsymbol{v}_{\parallel}[\mathcal{P}(\varepsilon)] - \boldsymbol{v}[\mathcal{P}(0)]$. (5) Then $\mathbf{\nabla}_{\boldsymbol{u}}\boldsymbol{v}$ is defined by

$$\mathbf{\nabla}_{\boldsymbol{u}}\boldsymbol{v} \equiv \lim_{\varepsilon \to 0} \frac{\delta\boldsymbol{v}}{\varepsilon} = \lim_{\varepsilon \to 0} \left\{ \frac{\boldsymbol{v}_{\parallel}[\mathcal{P}(\varepsilon)] - \boldsymbol{v}[\mathcal{P}(0)]}{\varepsilon} \right\}.$$

$$\Gamma^{\alpha}{}_{\beta\gamma} \equiv \langle \boldsymbol{\omega}^{\alpha}, \mathbf{\nabla}_{\gamma}\boldsymbol{e}_{\beta} \rangle \qquad \left[\begin{array}{l} \text{Note reversal of } \beta \text{ and } \gamma \text{ to make the} \\ \text{differentiating index come last on } \Gamma \end{array} \right] \quad (8.19a)$$

$$[\equiv \mathbf{\nabla}_{\boldsymbol{e}_{\gamma}}]$$

$$= \begin{pmatrix} \alpha \text{ component of change in } \boldsymbol{e}_{\beta}, \text{ relative} \\ \text{to parallel transport, along } \boldsymbol{e}_{\gamma} \end{pmatrix},$$

and one proves (exercise 8.12) that

$$\langle \mathbf{\nabla}_{\gamma}\boldsymbol{\omega}^{\alpha}, \boldsymbol{e}_{\beta} \rangle = -\Gamma^{\alpha}{}_{\beta\gamma}. \qquad (8.19b)$$

In terms of these coefficients and

$$T^{\beta}{}_{\alpha,\gamma} \equiv \mathbf{\nabla}_{\gamma}T^{\beta}{}_{\alpha} \equiv \partial_{\boldsymbol{e}_{\gamma}}T^{\beta}{}_{\alpha} \equiv \partial_{\gamma}T^{\beta}{}_{\alpha}, \qquad (8.20)$$

the components of the gradient, denoted $T^{\beta}{}_{\alpha;\gamma}$, are

$$T^{\beta}{}_{\alpha;\gamma} = T^{\beta}{}_{\alpha,\gamma} + \Gamma^{\beta}{}_{\mu\gamma}T^{\mu}{}_{\alpha} - \Gamma^{\mu}{}_{\alpha\gamma}T^{\beta}{}_{\mu} \qquad (8.21)$$

Components of gradient in arbitrary frame

(see exercise 8.13). If the basis at the event where $\mathbf{\nabla}\boldsymbol{T}$ is calculated were a local Lorentz frame, the components of $\mathbf{\nabla}\boldsymbol{T}$ would just be $T^{\beta}{}_{\alpha,\gamma}$. Because it is not, one must correct this "Lorentz-frame" value for the twisting, turning, expansion, and contraction of the basis vectors and 1-forms. The "ΓT" terms in equation (8.21) are the necessary corrections—one for each index of \boldsymbol{T}. The pattern of these correction terms is easy to remember: (1) "+" sign if index being corrected is up, "−" sign if it is down; (2) differentiation index (γ in above case) always at end of Γ; (3) index being corrected (β in first term, α in second) shifts from T onto Γ and gets replaced on T by a dummy summation index (μ).

Knowing the components (8.21) of the gradient, one can calculate the components of the covariant derivative $\nabla_{\boldsymbol{u}} \boldsymbol{T}$ by a simple contraction into u^γ [see equation (8.17)]:

$$\nabla_{\boldsymbol{u}} \boldsymbol{T} = (T^\beta{}_{\alpha;\gamma} u^\gamma) \boldsymbol{e}_\beta \otimes \boldsymbol{w}^\alpha. \tag{8.22}$$

Components of covariant derivative

When \boldsymbol{u} is the tangent vector to a curve $\mathscr{P}(\lambda)$, $\boldsymbol{u} = d\mathscr{P}/d\lambda$, one uses the notation $DT^\beta{}_\alpha/d\lambda$ for the components of $\nabla_{\boldsymbol{u}} \boldsymbol{T}$:

$$\frac{DT^\beta{}_\alpha}{d\lambda} \equiv T^\beta{}_{\alpha;\gamma} u^\gamma \overset{\text{[if basis is a coordinate basis so } u^\gamma = dx^\gamma/d\lambda]}{=} T^\beta{}_{\alpha;\gamma} \frac{dx^\gamma}{d\lambda}$$

$$= (T^\beta{}_{\alpha,\gamma} + \text{``}\Gamma T\text{''} \text{ corrections}) \, dx^\gamma/d\lambda \tag{8.23}$$

$$= \frac{dT^\beta{}_\alpha}{d\lambda} + (\Gamma^\beta{}_{\mu\gamma} T^\mu{}_\alpha - \Gamma^\mu{}_{\alpha\gamma} T^\beta{}_\mu) \frac{dx^\gamma}{d\lambda}.$$

The ";" in $T^\beta{}_{\alpha;\gamma}$ reminds one to correct $T^\beta{}_{\alpha,\gamma}$ with "ΓT" terms; similarly, the "D" in $DT^\beta{}_\alpha/d\lambda$ reminds one to correct $dT^\beta{}_\alpha/d\lambda$ with "ΓT" terms.

Calculation of connection coefficients from metric and commutators

This is all well and good, but how does one find out the connection coefficients $\Gamma^\alpha{}_{\beta\gamma}$ for a given basis? The answer is derived in exercise 8.15. It says: (1) take the metric coefficients in the given basis; (2) calculate their directional derivatives along the basis directions

$$g_{\beta\gamma,\mu} \equiv \partial_\mu g_{\beta\gamma} \quad = \partial g_{\beta\gamma}/\partial x^\mu;$$
$$\overset{[\equiv \partial_{\boldsymbol{e}_\mu}]}{\uparrow} \qquad \overset{[\text{if a coordinate basis, } \boldsymbol{e}_\mu = \partial\mathscr{P}/\partial x^\mu, \text{ is being used}]}{\uparrow} \tag{8.24a}$$

(3) calculate the commutation coefficients of the basis [equations (8.14) in general; $c_{\mu\beta\gamma} = 0$ in special case of coordinate basis]; (4) calculate the "covariant connection coefficients"

$$\Gamma_{\mu\beta\gamma} = \frac{1}{2} (g_{\mu\beta,\gamma} + g_{\mu\gamma,\beta} - g_{\beta\gamma,\mu} + \underline{c_{\mu\beta\gamma} + c_{\mu\gamma\beta} - c_{\beta\gamma\mu}}); \tag{8.24b}$$

$$\begin{bmatrix} \text{these terms are 0 for} \\ \text{coordinate basis} \end{bmatrix}$$

(5) raise an index to get the connection coefficients:

$$\Gamma^\alpha{}_{\beta\gamma} = g^{\alpha\mu} \Gamma_{\mu\beta\gamma}. \tag{8.24c}$$

[*Note on terminology:* a coordinate basis always has $c_{\alpha\beta\gamma} = 0$, and is sometimes called *holonomic;* a noncoordinate basis always has some of its $c_{\alpha\beta\gamma}$ nonzero, and is sometimes called *anholonomic.* In the holonomic case, the connection coefficients are sometimes called *Christoffel symbols.*]

The component notation, with its semicolons, commas, D's, connection coefficients, etc., looks rather formidable at first. But it bears great computational power, one discovers as one proceeds deep into gravitation theory; and its rules of manipulation

are simple enough to be learned easily. By contrast, the abstract notation ($\mathbf{\nabla}\boldsymbol{T}$, $\mathbf{\nabla}_u\boldsymbol{T}$, etc.) is poorly suited to complex calculations; but it possesses great conceptual power.

This contrast shows clearly in the way the two notations handle the concept of *geodesic*. A geodesic of spacetime is a curve that is straight and uniformly parametrized, as measured in each local Lorentz frame along its way. If the geodesic is timelike, then it is a possible world line for a freely falling particle, and its uniformly ticking parameter λ (called *"affine parameter"*) is a multiple of the particle's proper time, $\lambda = a\tau + b$. (Principle of equivalence: test particles move on straight lines in local Lorentz frames, and each particle's clock ticks at a uniform rate as measured by any Lorentz observer.) This definition of geodesic is readily translated into abstract, coordinate-free language: a geodesic is a curve $\mathcal{P}(\lambda)$ that parallel-transports its tangent vector $\boldsymbol{u} = d\mathcal{P}/d\lambda$ along itself—

<div style="text-align:right">Geodesic and affine parameter defined</div>

$$\mathbf{\nabla}_u\boldsymbol{u} = 0. \tag{8.25}$$

(See Figure 10.1.) What could be simpler conceptually? But to compute the geodesic, given an initial event \mathcal{P}_0 and initial tangent vector $\boldsymbol{u}(\mathcal{P}_0)$ there, one must use the component formalism. Introduce a coordinate system $x^\alpha(\mathcal{P})$, in which $u^\alpha = dx^\alpha/d\lambda$, and write the component version of equation (8.25) as

$$0 = \frac{D(dx^\alpha/d\lambda)}{d\lambda} = \frac{d(dx^\alpha/d\lambda)}{d\lambda} + \left(\Gamma^\alpha{}_{\mu\gamma}\frac{dx^\mu}{d\lambda}\right)\frac{dx^\gamma}{d\lambda}$$

[see equation (8.23), with one less index on T]; i.e.,

$$\frac{d^2x^\alpha}{d\lambda^2} + \Gamma^\alpha{}_{\mu\gamma}\frac{dx^\mu}{d\lambda}\frac{dx^\gamma}{d\lambda} = 0. \tag{8.26}$$

<div style="text-align:right">Geodesic equation</div>

This *geodesic equation* can be solved (in principle) for the coordinates of the geodesic, $x^\alpha(\lambda)$, when initial data [x^α and $dx^\alpha/d\lambda$ at $\lambda = \lambda_0$] have been specified.

The geodesics of the Earth's surface (great circles) are a foil against which one can visualize connection coefficients; see Figure 8.3.

The material of this section is presented more deeply and from a different viewpoint in Chapters 10 and 13. The Track-2 reader who plans to study those chapters is advised to ignore the following exercises. The Track-1 reader who intends to skip Chapters 9–15 will gain necessary experience with the component formalism by working exercises 8.4–8.7. Less important to him, but valuable nonetheless, are exercises 8.8–8.15, which develop the formalism of covariant derivatives and connection coefficients in a systematic manner. The most important results of these exercises will be summarized in Box 8.6 (pages 223 and 224).

Exercise 8.4. PRACTICE IN WRITING COMPONENTS OF GRADIENT

<div style="text-align:right">**EXERCISES**</div>

Rewrite the following quantities in terms of ordinary derivatives ($f_{,\gamma} \equiv \partial_{e_\gamma}f \equiv \mathbf{\nabla}_\gamma f$) and "$\Gamma T$" correction terms: (a) $T_{;\gamma}$ where T is a function. (b) $T^\alpha{}_{;\gamma}$ where \boldsymbol{T} is a vector. (c) $T_{\alpha;\gamma}$ where \boldsymbol{T} is a 1-form. (d) $T^\alpha{}_{\beta\delta}{}^\varepsilon{}_{;\gamma}$. [*Answer:*

(a) $T_{;\gamma} = T_{,\gamma}$. (b) $T^\alpha{}_{;\gamma} = T^\alpha{}_{,\gamma} + \Gamma^\alpha{}_{\mu\gamma}T^\mu$. (c) $T_{\alpha;\gamma} = T_{\alpha,\gamma} - \Gamma^\mu{}_{\alpha\gamma}T_\mu$.

(d) $T^\alpha{}_{\beta\delta}{}^\varepsilon{}_{;\gamma} = T^\alpha{}_{\beta\delta}{}^\varepsilon{}_{,\gamma} + \Gamma^\alpha{}_{\mu\gamma}T^\mu{}_{\beta\delta}{}^\varepsilon - \Gamma^\mu{}_{\beta\gamma}T^\alpha{}_{\mu\delta}{}^\varepsilon - \Gamma^\mu{}_{\delta\gamma}T^\alpha{}_{\beta\mu}{}^\varepsilon + \Gamma^\varepsilon{}_{\mu\gamma}T^\alpha{}_{\beta\delta}{}^\mu.$]

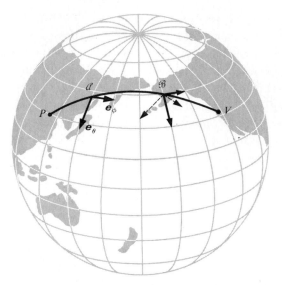

Figure 8.3.

The why of connection coefficients, schematically portrayed. The aviator pursuing his great circle route from Peking to Vancouver finds himself early going north, but later going south, although he is navigating the straightest route that is at all open to him (geodesic). The apparent change in direction indicates a turning, not in his route, but in the system of coordinates with respect to which his route is described. The vector \boldsymbol{v} of his velocity (a vector defined not on spacetime but rather on the Earth's two-dimensional surface), carried forward by parallel transport from an earlier moment to a later moment, finds itself in agreement with the velocity that he is then pursuing; or, in the abstract language of coordinate-free differential geometry, the covariant derivative $\boldsymbol{\nabla_v v}$ vanishes along the route ("equation of a geodesic"). Though \boldsymbol{v} is in this sense constant, the individual pieces of which the navigator considers this vector to be built, $\boldsymbol{v} = v^\theta \boldsymbol{e}_\theta + v^\phi \boldsymbol{e}_\phi$, are not constant.

In the language of components, the quantities v^θ and v^ϕ are changing along the route at a rate that annuls the covariant derivative of \boldsymbol{v}; thus

$$\boldsymbol{\nabla_v v} = \boldsymbol{a} = a^\phi \boldsymbol{e}_\phi + a^\theta \boldsymbol{e}_\theta = 0,$$

or

$$0 = a^\theta = \frac{dv^\theta}{dt} + \Gamma^\theta{}_{mn} v^m v^n,$$

$$0 = a^\phi = \frac{dv^\phi}{dt} + \Gamma^\phi{}_{mn} v^m v^n.$$

In this sense the connection coefficients $\Gamma^j{}_{mn}$ serve as "turning coefficients" to tell how fast to "turn" the components of a vector in order to keep that vector constant (against the turning influence of the base vectors).

Alternatively, the navigator can use an "automatic pilot system" which parallel-transports its own base vectors along the plane's route:

$$\boldsymbol{\nabla_v e}_{\theta'} = \boldsymbol{\nabla_v e}_{\phi'} = 0;$$

solid vectors at \mathscr{A} become dotted vectors at \mathscr{B}. Then the components of \boldsymbol{v} must be kept fixed to achieve a great-circle route,

$$\frac{dv^{\theta'}}{dt} = \frac{dv^{\phi'}}{dt} = 0;$$

and the turning coefficients are used to describe the turning of the lines of latitude and longitude relative to this parallel-transported basis:

$$\boldsymbol{\nabla_v e}_\theta = \boldsymbol{e}_m \Gamma^m{}_{\theta n} v^n,$$

$$\boldsymbol{\nabla_v e}_\phi = \boldsymbol{e}_m \Gamma^m{}_{\phi n} v^n.$$

The same turning coefficients enter into both viewpoints. The only difference is in how these coefficients are used.

Exercise 8.5. A SHEET OF PAPER IN POLAR COORDINATES

The two-dimensional metric for a flat sheet of paper in polar coordinates (r, θ) is $ds^2 = dr^2 + r^2\, d\phi^2$—or, in modern notation, $\boldsymbol{g} = \boldsymbol{dr} \otimes \boldsymbol{dr} + r^2\, \boldsymbol{d\phi} \otimes \boldsymbol{d\phi}$.

(a) Calculate the connection coefficients using equations (8.24). [*Answer:* $\Gamma^r{}_{\phi\phi} = -r$; $\Gamma^\phi{}_{r\phi} = \Gamma^\phi{}_{\phi r} = 1/r$; all others vanish.]

(b) Write down the geodesic equation in (r, ϕ) coordinates. [*Answer:* $d^2 r/d\lambda^2 - r(d\phi/d\lambda)^2 = 0$; $d^2\phi/d\lambda^2 + (2/r)(dr/d\lambda)(d\phi/d\lambda) = 0$.]

(c) Solve this geodesic equation for $r(\lambda)$ and $\phi(\lambda)$, and show that the solution is a uniformly parametrized straight line ($x \equiv r\cos\phi = a\lambda + b$ for some a and b; $y \equiv r\sin\phi = j\lambda + k$ for some j and k).

(d) Verify that the noncoordinate basis $\boldsymbol{e}_{\hat{r}} \equiv \boldsymbol{e}_r = \partial\mathscr{P}/\partial r$, $\boldsymbol{e}_{\hat{\phi}} \equiv r^{-1}\boldsymbol{e}_\phi$ $r^{-1}\,\partial\mathscr{P}/\partial\phi$, $\boldsymbol{w}^r = \boldsymbol{dr}$, $\boldsymbol{w}^{\hat{\phi}} = r\,\boldsymbol{d\phi}$ is orthonormal, and that $\langle \boldsymbol{w}^\alpha, \boldsymbol{e}_\beta \rangle = \delta^{\hat{\alpha}}{}_\beta$. Then calculate the connection coefficients of this basis from a knowledge [part (a)] of the connection of the coordinate basis. [*Answer:*

$$\begin{aligned}
\Gamma^{\hat{\phi}}{}_{\hat{\phi}\hat{r}} &= \langle \boldsymbol{w}^{\hat{\phi}}, \boldsymbol{\nabla}_{\hat{r}}\boldsymbol{e}_{\hat{\phi}} \rangle = \langle r\,\boldsymbol{d\phi}, \boldsymbol{\nabla}_r(r^{-1}\boldsymbol{e}_\phi) \\
&= r\langle \boldsymbol{d\phi}, (\boldsymbol{\nabla}_r r^{-1})\boldsymbol{e}_\phi + r^{-1}(\boldsymbol{\nabla}_r \boldsymbol{e}_\phi)\rangle = r\langle \boldsymbol{d\phi}, -r^{-2}\boldsymbol{e}_\phi \rangle + \langle \boldsymbol{d\phi}, \boldsymbol{\nabla}_r\boldsymbol{e}_\phi \rangle \\
&= -r^{-1} + \Gamma^\phi{}_{\phi r} = -r^{-1} + r^{-1} = 0;
\end{aligned}$$

similarly, $\Gamma^{\hat{\phi}}{}_{\hat{r}\hat{\phi}} = +1/r$, $\Gamma^{\hat{r}}{}_{\hat{\phi}\hat{\phi}} = -1/r$; all others vanish.]

(e) Consider the Keplerian orbit of Figure 8.1 and §8.3 as a nongeodesic curve in the sun's two-dimensional, Euclidean, equatorial plane. In place of the old notation $d\boldsymbol{v}/dt$, $d\boldsymbol{e}_{\hat{r}}/dt$, etc., use the new notation $\boldsymbol{\nabla}_v\boldsymbol{v}$, $\boldsymbol{\nabla}_v\boldsymbol{e}_{\hat{r}}$, etc. Then $\boldsymbol{v} = d\mathscr{P}/dt$ is the tangent to the orbit, and $\boldsymbol{a} = \boldsymbol{\nabla}_v\boldsymbol{v}$ is the acceleration. Derive equations (8.4) for $a^{\hat{r}}$ and $a^{\hat{\phi}}$ using component manipulations and connection coefficients in the orthonormal basis.

Exercise 8.6. SPHERICAL COORDINATES IN FLAT SPACETIME

The spherical noncoordinate basis $\{\boldsymbol{e}_{\hat{\alpha}}\}$ of Exercise 8.1 was orthonormal, $g_{\hat{\alpha}\hat{\beta}} = \eta_{\hat{\alpha}\hat{\beta}}$, but had nonvanishing commutation coefficients [part (c) of Exercise 8.2].

(a) Calculate the connection coefficients for this basis, using equations (8.24). [*Answer:*

$$\Gamma^{\hat{\theta}}{}_{\hat{r}\hat{\theta}} = \Gamma^{\hat{\phi}}{}_{\hat{r}\hat{\phi}} = -\Gamma^{\hat{r}}{}_{\hat{\theta}\hat{\theta}} = -\Gamma^{\hat{r}}{}_{\hat{\phi}\hat{\phi}} = 1/r;$$

$$\Gamma^{\hat{\phi}}{}_{\hat{\theta}\hat{\phi}} = -\Gamma^{\hat{\theta}}{}_{\hat{\phi}\hat{\phi}} = \cot\theta/r;$$

all others vanish.]

(b) Write down expressions for $\boldsymbol{\nabla}_{\hat{\alpha}}\boldsymbol{e}_{\hat{\beta}}$ in terms of $\boldsymbol{e}_{\hat{\gamma}}$, and verify the correctness of these expressions by drawing sketches of the basis vectors on a sphere of constant t and r. [*Answer:*

$$\boldsymbol{\nabla}_{\hat{\theta}}\boldsymbol{e}_{\hat{r}} = r^{-1}\boldsymbol{e}_{\hat{\theta}}, \qquad \boldsymbol{\nabla}_{\hat{\theta}}\boldsymbol{e}_{\hat{\theta}} = -r^{-1}\boldsymbol{e}_{\hat{r}}, \qquad \boldsymbol{\nabla}_{\hat{\phi}}\boldsymbol{e}_{\hat{r}} = r^{-1}\boldsymbol{e}_{\hat{\phi}},$$

$$\boldsymbol{\nabla}_{\hat{\phi}}\boldsymbol{e}_{\hat{\theta}} = (\cot\theta/r)\boldsymbol{e}_{\hat{\phi}}, \qquad \boldsymbol{\nabla}_{\hat{\phi}}\boldsymbol{e}_{\hat{\phi}} = -r^{-1}\boldsymbol{e}_{\hat{r}} - (\cot\theta/r)\boldsymbol{e}_{\hat{\theta}}.$$

All others vanish.]

(c) Calculate the divergence of a vector, $\boldsymbol{\nabla} \cdot \boldsymbol{A} = A^{\hat{\alpha}}{}_{;\hat{\alpha}}$, in this basis. [*Answer:*

$$\begin{aligned}
\boldsymbol{\nabla} \cdot \boldsymbol{A} &= A^{\hat{t}}{}_{,\hat{t}} + r^{-2}(r^2 A^{\hat{r}})_{,\hat{r}} + (\sin\theta)^{-1}(\sin\theta A^{\hat{\theta}})_{,\hat{\theta}} + A^{\hat{\phi}}{}_{,\hat{\phi}} \\
&= \frac{\partial A^{\hat{t}}}{\partial t} + \frac{1}{r^2}\frac{\partial(r^2 A^{\hat{r}})}{\partial r} + \frac{1}{r\sin\theta}\frac{\partial(\sin\theta A^{\hat{\theta}})}{\partial\theta} + \frac{1}{r\sin\theta}\frac{\partial A^{\hat{\phi}}}{\partial\phi}.
\end{aligned}$$

This answer should be familiar from flat-space vector analysis.]

Exercise 8.7. SYMMETRIES OF CONNECTION COEFFICIENTS

From equation (8.24b), the symmetry of the metric, and the antisymmetry $(c_{\beta\gamma\mu} = -c_{\gamma\beta\mu})$

of the commutation coefficients, show that: $\Gamma_{\alpha[\beta\gamma]} = 0$ (last two indices are symmetric) in a coordinate basis; $\Gamma_{(\hat\alpha\hat\beta)\hat\gamma} = 0$ (first two indices are antisymmetric) in a globally orthonormal basis, $g_{\hat\alpha\hat\beta} = \eta_{\alpha\beta}$.

SYSTEMATIC DERIVATION OF RESULTS IN §8.5

Exercise 8.8. NEW DEFINITION OF ▼*T* COMPARED WITH OLD DEFINITION

The new definition of $\boldsymbol{\nabla T}$ is given by equations (8.16) and (8.17). Use the fact that parallel transport keeps local-Lorentz components fixed to derive, from (8.16), the Lorentz-frame equation $\boldsymbol{\nabla_u T} = T^\beta{}_{\alpha,\gamma} u^\gamma \boldsymbol{e}_\beta \otimes \boldsymbol{w}^\alpha$. From this and equation (8.17), infer that the Lorentz-frame components of $\boldsymbol{\nabla T}$ are $T^\beta{}_{\alpha,\gamma}$—which accords with the old definition of $\boldsymbol{\nabla T}$.

Exercise 8.9. CHAIN RULE FOR ▼*u**T*

(a) Use calculations in a local Lorentz frame to show that "$\boldsymbol{\nabla_u}$" obeys the standard chain rule for derivatives:

$$\boldsymbol{\nabla_u}(f\boldsymbol{A} \otimes \boldsymbol{B}) = (\boldsymbol{\nabla_u} f)\boldsymbol{A} \otimes \boldsymbol{B} + f(\boldsymbol{\nabla_u A}) \otimes \boldsymbol{B} + f\boldsymbol{A} \otimes (\boldsymbol{\nabla_u B}). \qquad (8.27)$$

Here \boldsymbol{A} and \boldsymbol{B} are arbitrary vectors, 1-forms, or tensors; and f is an arbitrary function. [*Hint:* assume for concreteness that \boldsymbol{A} is a $\binom{1}{1}$ tensor and \boldsymbol{B} is a vector. Then this equation reads, in Lorentz-frame component notation,

$$(fA^\alpha{}_\beta B^\gamma)_{,\delta} u^\delta = (f_{,\delta} u^\delta)A^\alpha{}_\beta B^\gamma + f(A^\alpha{}_{\beta,\delta} u^\delta)B^\gamma + fA^\alpha{}_\beta(B^\gamma{}_{,\delta} u^\delta).] \qquad (8.27')$$

(b) Rewrite equation (8.27) in component notation in an arbitrary basis. [*Answer:* same as (8.27'), except "," is replaced everywhere by ";". But *note* that $f_{;\delta} u^\delta = f_{,\delta} u^\delta$, because the function f "has no components to correct".]

Exercise 8.10. COVARIANT DERIVATIVE COMMUTES WITH CONTRACTION

(a) Let \boldsymbol{S} be a $\binom{1}{2}$ tensor. Using components in a local Lorentz frame show that

$$\boldsymbol{\nabla_u}\ (\text{contraction on slots 1 and 2 of } \boldsymbol{S}) = (\text{contraction on slots 1 and 2 of } \boldsymbol{\nabla_u S}). \qquad (8.28)$$

[Hint: in a local Lorentz frame this equation makes the trivial statement

$$\left(\sum_\alpha S^\alpha{}_{\alpha\beta}\right)_{,\gamma} u^\gamma = \sum_\alpha (S^\alpha{}_{\alpha\beta,\gamma} u^\gamma).]$$

Exercise 8.11. ALGEBRAIC PROPERTIES OF ▼

Use calculations in a local Lorentz frame to show that

$$\boldsymbol{\nabla_{au+bv} S} = a\boldsymbol{\nabla_u S} + b\boldsymbol{\nabla_v S} \qquad (8.29)$$

for all tangent vectors $\boldsymbol{u}, \boldsymbol{v}$ and numbers a, b; also that

$$\boldsymbol{\nabla_u}(\boldsymbol{S} + \boldsymbol{M}) = \boldsymbol{\nabla_u S} + \boldsymbol{\nabla_u M} \qquad (8.30)$$

for any two tensor fields \boldsymbol{S} and \boldsymbol{M} of the same rank; also that

$$\nabla_u w - \nabla_w u = [u, w],$$

$$\underbrace{\qquad}_{\begin{bmatrix}\text{commutator of } u \text{ and } w; \\ \text{discussed in exercise 8.2}\end{bmatrix}}$$

$$(8.31)$$

for any two vector fields u and w.

Exercise 8.12. CONNECTION COEFFICIENTS FOR 1-FORM BASIS

Show that the same connection coefficients $\Gamma^\alpha{}_{\beta\gamma}$ that describe the changes in $\{e_\beta\}$ from point to point [definition (8.19a)] also describe the changes in $\{\omega^\alpha\}$, except for a change in sign [equation (8.19b)]. {*Answer:* (1) $\langle \omega^\alpha, e_\beta \rangle = \delta^\alpha{}_\beta$ is a constant function (0 or 1, depending on whether $\alpha = \beta$). (2) Thus, $\nabla_\gamma \langle \omega^\alpha, e_\beta \rangle = \partial_{e_\gamma} \langle \omega^\alpha, e_\beta \rangle = 0$. (3) But $\langle \omega^\alpha, e_\beta \rangle$ is the contraction of $\omega^\alpha \otimes e_\beta$, so equation (8.28) implies $0 = \nabla_\gamma(\text{contraction of } \omega^\alpha \otimes e_\beta)$ = contraction of $[\nabla_\gamma(\omega^\alpha \otimes e_\beta)]$. (4) Apply the chain rule (8.27) to conclude $0 = $ contraction of $[(\nabla_\gamma \omega^\alpha) \otimes e_\beta + \omega^\alpha \otimes (\nabla_\gamma e_\beta)] = \langle \nabla_\gamma \omega^\alpha, e_\beta \rangle + \langle \omega^\alpha, \nabla_\gamma e_\beta \rangle$. (5) Finally, use definition (8.19a) to arrive at the desired result, (8.19b).}

Exercise 8.13. "ΓT" CORRECTION TERMS FOR $T^\beta{}_{\alpha;\gamma}$

Derive equation (8.21) for $T^\beta{}_{\alpha;\gamma}$ in an arbitrary basis by first calculating the components of $\nabla_u T$ for arbitrary u, and by then using equation (8.17) to infer the components of ∇T. [*Answer:* (1) Use the chain rule (8.27) to get

$$\begin{aligned}\nabla_u T &= \nabla_u(T^\beta{}_\alpha e_\beta \otimes \omega^\alpha) \\ &= (\nabla_u T^\beta{}_\alpha) e_\beta \otimes \omega^\alpha + T^\beta{}_\alpha (\nabla_u e_\beta) \otimes \omega^\alpha + T^\beta{}_\alpha e_\beta \otimes (\nabla_u \omega^\alpha).\end{aligned}$$

(2) Write u in terms of its components, $u = u^\gamma e_\gamma$; use linearity of ∇_u in u from equation (8.29), to get $\nabla_u = u^\gamma \nabla_\gamma$; and use this in $\nabla_u T$:

$$\nabla_u T = u^\gamma \{ T^\beta{}_{\alpha,\gamma} e_\beta \otimes \omega^\alpha + T^\beta{}_\alpha (\nabla_\gamma e_\beta) \otimes \omega^\alpha + T^\beta{}_\alpha e_\beta \otimes (\nabla_\gamma \omega^\alpha) \}.$$

(3) Use equations (8.19a,b), rewritten as

$$\nabla_\gamma e_\beta = \Gamma^\mu{}_{\beta\gamma} e_\mu, \qquad \nabla_\gamma \omega^\alpha = -\Gamma^\alpha{}_{\mu\gamma} \omega^\mu,$$

$$(8.32)$$

to put $\nabla_u T$ in the form

$$\nabla_u T = u^\gamma \{ T^\beta{}_{\alpha,\gamma} e_\beta \otimes \omega^\alpha + \Gamma^\mu{}_{\beta\gamma} T^\beta{}_\alpha e_\mu \otimes \omega^\alpha - \Gamma^\alpha{}_{\mu\gamma} T^\beta{}_\alpha e_\beta \otimes \omega^\mu \}.$$

(4) Rename dummy indices so that the basis tensor $e_\beta \otimes \omega^\alpha$ can be factored out:

$$\nabla_u T = u^\gamma \{ T^\beta{}_{\alpha,\gamma} + \Gamma^\beta{}_{\mu\gamma} T^\mu{}_\alpha - \Gamma^\mu{}_{\alpha\gamma} T^\beta{}_\mu \} e_\beta \otimes \omega^\alpha.$$

(5) By comparison with

$$\nabla_u T = \nabla T(\ldots, \ldots, u) = (T^\beta{}_{\alpha;\gamma} u^\gamma) e_\beta \otimes \omega^\alpha,$$

read off the value of $T^\beta{}_{\alpha;\gamma}$.]

Exercise 8.14. METRIC IS COVARIANTLY CONSTANT

Show on physical grounds (using properties of local Lorentz frames) that

$$\nabla g = 0$$

$$(8.33)$$

or, equivalently, that $\nabla_u g = 0$ for any vector u. Then deduce as a mathematical consequence the obviously desirable product rule

$$\nabla_u(A \cdot B) = (\nabla_u A) \cdot B + A \cdot (\nabla_u B).$$

[Answer: (1) As discussed following equation (8.18), the components of ∇g in a local Lorentz frame are $g_{\mu\nu,\alpha}$. Just use g for T in that discussion. But these components all vanish by equation (8.15b). Therefore equation (8.33) holds in this frame, and—as a tensor equation—in all frames. (2) The product rule is also a tensor equation, true immediately via components in a local Lorentz frame. (3) Prove the product rule also the hard way, to see where equation (8.33) enters. Use the chain rule of exercise 8.9 to write

$$\nabla_u(g \otimes A \otimes B) = (\nabla_u g) \otimes A \otimes B + g \otimes (\nabla_u A) \otimes B$$
$$+ g \otimes A \otimes (\nabla_u B).$$

Use equation (8.33) to drop one term, then contract, forming

$$A \cdot B = \text{contraction}(g \otimes A \otimes B)$$

and the other inner products. Exercise 8.10 is used to justify commuting the contraction with ∇_u on the lefthand side.]

Exercise 8.15. CONNECTION COEFFICIENTS IN TERMS OF METRIC

Use the fact that the metric is covariantly constant [equation (8.33)] to derive equation (8.24b) for the connection coefficients. Treat equation (8.24c) as a definition of $\Gamma_{\mu\beta\gamma}$ in terms of $\Gamma^\alpha{}_{\beta\gamma}$. [*Answer:* (1) Calculate the components of ∇g in an arbitrary frame:

$$g_{\alpha\beta;\gamma} = 0 = g_{\alpha\beta,\gamma} - \Gamma^\mu{}_{\alpha\gamma}g_{\mu\beta} - \Gamma^\mu{}_{\beta\gamma}g_{\mu\alpha}$$
$$\equiv g_{\alpha\beta,\gamma} - \Gamma_{\beta\alpha\gamma} - \Gamma_{\alpha\beta\gamma};$$

thereby conclude that $g_{\alpha\beta,\gamma} = 2\Gamma_{(\alpha\beta)\gamma}$. (Round brackets denote symmetric part.) (2) Construct the metric terms in the claimed answer for $\Gamma_{\mu\beta\gamma}$:

$$\frac{1}{2}(g_{\mu\beta,\gamma} + g_{\mu\gamma,\beta} - g_{\beta\gamma,\mu}) = \Gamma_{(\mu\beta)\gamma} + \Gamma_{(\mu\gamma)\beta} - \Gamma_{(\beta\gamma)\mu}$$
$$= \frac{1}{2}[\Gamma_{\mu\beta\gamma} + \Gamma_{\beta\mu\gamma} + \Gamma_{\mu\gamma\beta} + \Gamma_{\gamma\mu\beta} - \Gamma_{\beta\gamma\mu} - \Gamma_{\gamma\beta\mu}]$$
$$= \Gamma_{\mu\beta\gamma} + (-\Gamma_{\mu[\beta\gamma]} + \Gamma_{\beta[\mu\gamma]} + \Gamma_{\gamma[\mu\beta]}).$$

(3) Infer from equation (8.31), with u and w chosen as two basis vectors ($u = e_\mu$, $w = e_\nu$) that

$$c_{\mu\nu}{}^\rho e_\rho \equiv [e_\mu, e_\nu] = \nabla_\mu e_\nu - \nabla_\nu e_\mu = (\Gamma^\rho{}_{\nu\mu} - \Gamma^\rho{}_{\mu\nu})e_\rho = 2\Gamma^\rho{}_{[\nu\mu]}e_\rho;$$

i.e.,

$$\Gamma^\rho{}_{[\mu\nu]} = -\frac{1}{2}c_{\mu\nu}{}^\rho; \qquad \Gamma_{\rho[\mu\nu]} = -\frac{1}{2}c_{\mu\nu\rho}. \tag{8.34}$$

(4) This, combined with step (2) yields the desired formula for $\Gamma_{\mu\beta\gamma}$.]

§8.6. LOCAL LORENTZ FRAMES: MATHEMATICAL DISCUSSION

An observer falling freely in curved spacetime makes measurements in his local Lorentz frame. What he discovers has been discussed extensively in Parts I and II of this book. Try now to derive his basic discoveries from the formalism of the last section.

Pick an event \mathscr{P}_0 on the observer's world line. His local Lorentz frame there is a coordinate system $x^\alpha(\mathscr{P})$ in which

Local Lorentz frame:

$$g_{\alpha\beta} \equiv \boldsymbol{e}_\alpha \cdot \boldsymbol{e}_\beta \equiv \frac{\partial \mathscr{P}}{\partial x^\alpha} \cdot \frac{\partial \mathscr{P}}{\partial x^\beta} = \eta_{\alpha\beta} \text{ at } \mathscr{P}_0 \qquad (8.35a)$$

(Lorentz metric at \mathscr{P}_0), and in which

$$\partial g_{\alpha\beta}/\partial x^\mu = 0 \text{ at } \mathscr{P}_0 \qquad (8.35b)$$

(metric as Lorentz as possible near \mathscr{P}_0). [See equation (8.15).] In addition, by virtue of equations (8.24),

$$\Gamma^\alpha{}_{\beta\gamma} = 0 \text{ at } \mathscr{P}_0 \qquad (8.36)$$

(no "correction terms" in covariant derivatives). Of course, the observer must be at rest in his local Lorentz frame; i.e., his world line must be

$$x^j = x^j(\mathscr{P}_0) = \text{constant}; \qquad x^0 \text{ varying.} \qquad (8.37)$$

Query: Equations (8.35) to (8.37) guarantee that the observer is at rest in a local Lorentz frame. Do they imply that he is freely falling? (They should!) *Answer:* Calculate the observer's 4-acceleration $\boldsymbol{a} = d\boldsymbol{u}/d\tau$ (notation of chapter 6) $= \boldsymbol{\nabla}_{\boldsymbol{u}}\boldsymbol{u}$ (notation of this chapter). His 4-velocity, calculated from equation (8.37) is

Origin falls freely along a geodesic

$$\boldsymbol{u} = (dx^\alpha/d\tau)\boldsymbol{e}_\alpha = (dx^0/d\tau)\boldsymbol{e}_0 = \boldsymbol{e}_0; \qquad (8.38)$$
$$\begin{bmatrix} \text{because } \boldsymbol{u} \text{ and } \boldsymbol{e}_0 \text{ both} \\ \text{have unit length} \end{bmatrix}$$

so his 4-acceleration is

$$\boldsymbol{a} = \boldsymbol{\nabla}_{\boldsymbol{u}}\boldsymbol{u} = \boldsymbol{\nabla}_0\boldsymbol{e}_0 = \Gamma^\alpha{}_{00}\boldsymbol{e}_\alpha$$
$$= 0 \text{ at } \mathscr{P}_0. \qquad (8.39)$$

Thus, he is indeed freely falling ($\boldsymbol{a} = 0$); and he moves along a geodesic ($\boldsymbol{\nabla}_{\boldsymbol{u}}\boldsymbol{u} = 0$).

Query: Do freely falling particles move along straight lines ($d^2x^\alpha/d\tau^2 = 0$) in the observer's local Lorentz frame at \mathscr{P}_0? (They should!) *Answer:* A freely falling particle experiences zero 4-acceleration

Freely falling particles move on straight lines

$$\boldsymbol{a}_{\text{particle}} = \boldsymbol{\nabla}_{\boldsymbol{u}_{\text{particle}}}\boldsymbol{u}_{\text{particle}} = 0;$$

i.e., it parallel-transports its 4-velocity; i.e., it moves along a geodesic of spacetime

with affine parameter equal to its proper time. The geodesic equation for its world line, in local Lorentz coordinates, says

$$\frac{d^2x^\alpha}{d\tau^2} = -\Gamma^\alpha{}_{\mu\nu}\frac{dx^\mu}{d\tau}\frac{dx^\nu}{d\tau}$$

$$= 0 \text{ at } \mathcal{P}_o.$$

The particle's world line is, indeed, straight at \mathcal{P}_o.

Query: Does the freely falling observer Fermi-Walker-transport his spatial basis vectors \boldsymbol{e}_j; i.e., can he attach them to gyroscopes that he carries? (He should be able to!) *Answer:* Fermi-Walker transport (Box 6.2) would say

$$\underbrace{\frac{d\boldsymbol{e}_j}{d\tau}}_{\substack{\uparrow \\ \text{old} \\ \text{notation}}} \equiv \underbrace{\boldsymbol{\nabla}_{\!\boldsymbol{u}}\boldsymbol{e}_j}_{\substack{\uparrow \\ \text{new} \\ \text{notation}}} = \boldsymbol{u}(\boldsymbol{a}\cdot\boldsymbol{e}_j) - \boldsymbol{a}(\boldsymbol{u}\cdot\boldsymbol{e}_j).$$

But $\boldsymbol{u} = \boldsymbol{e}_0$, $\boldsymbol{e}_0\cdot\boldsymbol{e}_j = 0$, and $\boldsymbol{a} = 0$ for the observer; so Fermi-Walker transport in this case reduces to parallel transport along \boldsymbol{e}_0: thus $\boldsymbol{\nabla}_0\boldsymbol{e}_j = 0$. This is, indeed, how \boldsymbol{e}_j is transported through \mathcal{P}_o, because

$$\boldsymbol{\nabla}_0\boldsymbol{e}_j = \Gamma^\alpha{}_{j0}\boldsymbol{e}_\alpha = 0 \text{ at } \mathcal{P}_o.$$

§8.7. GEODESIC DEVIATION AND THE RIEMANN CURVATURE TENSOR

"Gravitation is a manifestation of spacetime curvature, and that curvature shows up in the deviation of one geodesic from a nearby geodesic (relative acceleration of test particles)." To make this statement precise, first quantify the "deviation" or "relative acceleration" of neighboring geodesics.

Focus attention on a family of geodesics $\mathcal{P}(\lambda, n)$; see Figure 8.4. The smoothly varying parameter n ("selector parameter") distinguishes one geodesic from the next. For fixed n, $\mathcal{P}(\lambda, n)$ is a geodesic with affine parameter λ and with tangent vector

$$\boldsymbol{u} = \partial\mathcal{P}/\partial\lambda; \tag{8.40}$$

thus $\boldsymbol{\nabla}_{\!\boldsymbol{u}}\boldsymbol{u} = 0$ (geodesic equation). The vector

$$\boldsymbol{n} \equiv \partial\mathcal{P}/\partial n \tag{8.41}$$

measures the separation between points with the same value of λ on neighboring geodesics.

An observer falling freely along the "fiducial geodesic" $n = 0$ watches a test particle fall along the "test geodesic" $n = 1$. The velocity of the test particle relative

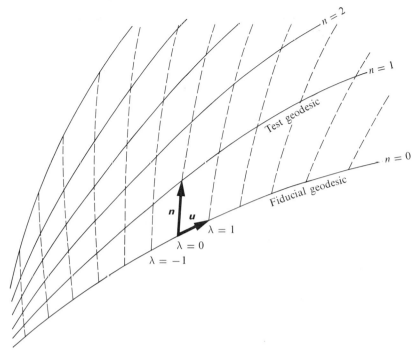

Figure 8.4.

A family of geodesics $\mathcal{P}(\lambda, n)$. The selector parameter n tells "which" geodesic; the affine parameter λ tells "where" on a given geodesic. The separation vector $\boldsymbol{n} \equiv \partial\mathcal{P}/\partial n$ at a point $\mathcal{P}(\lambda, 0)$ along the fiducial geodesic, $n = 0$, reaches (approximately) to the point $\mathcal{P}(\lambda, 1)$ with the same value of λ on the test geodesic, $n = 1$.

to him he quantifies by $\boldsymbol{\nabla}_{\boldsymbol{u}}\boldsymbol{n}$. This relative velocity, like the separation vector \boldsymbol{n}, is an arbitrary "initial condition." Not arbitrary, however, is the "relative acceleration," $\boldsymbol{\nabla}_{\boldsymbol{u}}\boldsymbol{\nabla}_{\boldsymbol{u}}\boldsymbol{n}$ of the test particle relative to the observer (see Boxes 11.2 and 11.3). It would be zero in flat spacetime. In curved spacetime, it is given by

$$\boldsymbol{\nabla}_{\boldsymbol{u}}\boldsymbol{\nabla}_{\boldsymbol{u}}\boldsymbol{n} + \boldsymbol{Riemann}\,(\ldots, \boldsymbol{u}, \boldsymbol{n}, \boldsymbol{u}) = 0, \tag{8.42}$$

Riemann curvature tensor defined by relative acceleration of geodesics

or, in component notation,

$$\frac{D^2 n^\alpha}{d\lambda^2} + R^\alpha{}_{\beta\gamma\delta}u^\beta n^\gamma u^\delta = 0. \tag{8.43}$$

This equation serves as a definition of the *"Riemann curvature tensor;"* and it can also be used to derive the following expressions for the components of *Riemann* in a coordinate basis:

Components of **Riemann**

$$\begin{aligned} R^\alpha{}_{\beta\gamma\delta} &= \langle \boldsymbol{dx}^\alpha, [\boldsymbol{\nabla}_\gamma, \boldsymbol{\nabla}_\delta]\boldsymbol{e}_\beta \rangle \\ &= \frac{\partial \Gamma^\alpha{}_{\beta\delta}}{\partial x^\gamma} - \frac{\partial \Gamma^\alpha{}_{\beta\gamma}}{\partial x^\delta} + \Gamma^\alpha{}_{\mu\gamma}\Gamma^\mu{}_{\beta\delta} - \Gamma^\alpha{}_{\mu\delta}\Gamma^\mu{}_{\beta\gamma}. \end{aligned} \tag{8.44}$$

(For proof, read Box 11.4, Box 11.5, and exercise 11.3, in that order.) For a glimpse of the man who first analyzed the curvature of spaces with three and more dimensions, see Box 8.5.

Effects of curvature

Spacetime curvature causes not only geodesic deviation, but also route dependence in parallel transport (parallel transport around a closed curve changes a vector or tensor—Box 11.7); it causes covariant derivatives to fail to commute [equation (8.44)]; and it prevents the existence of a global Lorentz coordinate system (§11.5).

At first sight one might think **Riemann** has $4 \times 4 \times 4 \times 4 = 256$ independent components. But closer examination (§13.5) reveals a variety of symmetries

Symmetries of Riemann

$$R_{\alpha\beta\gamma\delta} = R_{[\alpha\beta][\gamma\delta]} = R_{[\gamma\delta][\alpha\beta]}, \qquad R_{[\alpha\beta\gamma\delta]} = 0, \qquad R_{\alpha[\beta\gamma\delta]} = 0 \qquad (8.45)$$

Box 8.5 GEORG FRIEDRICH BERNHARD RIEMANN

September 17, 1826, Breselenz, Hanover—July 20, 1866, Selasca, Lake Maggiore

With his famous doctoral thesis of 1851, "Foundations for a general theory of functions of a single complex variable," Riemann founded one branch of modern mathematics (the theory of Riemann surfaces); and with his famous lecture of three years later founded another (Riemannian geometry). These and other writings will be found in his collected works, edited by H. Weber (1953).

"The properties which distinguish space from other conceivable triply-extended magnitudes are only to be deduced from experience. . . . At every point the three-directional measure of curvature can have an arbitrary value if only the effective curvature of every measurable region of space does not differ noticeably from zero." [G. F. B. Riemann, "On the hypotheses that lie at the foundations of geometry," *Habilitationsvorlesung* of June 10, 1854, on entry into the philosophical faculty of the University of Göttingen.]

Dying of tuberculosis twelve years later, occu-

(antisymmetry on first two indices; antisymmetry on last two; symmetry under exchange of first pair with last pair; vanishing of completely antisymmetric parts). These reduce **Riemann** (in four dimensions) from 256 to 20 independent components.

Besides these algebraic symmetries, **Riemann** possesses differential symmetries called *"Bianchi identities,"*

$$R^\alpha{}_{\beta[\lambda\mu;\nu]} = 0, \tag{8.46}$$

Bianchi identities

which have deep geometric significance (Chapter 15).

From **Riemann** one can form several other curvature tensors by contraction. The easiest to form are the *"Ricci curvature tensor,"*

pied with an attempt at a unified explanation of gravity and electromagnetism, Riemann communicated to Betti his system of characterization of multiply-connected topologies (which opened the door to the view of electric charge as "lines of force trapped in the topology of space"), making use of numbers that today are named after Betti but that are identified with a symbol, R_n, that honors Riemann.

"A more detailed scrutiny of a surface might disclose that what we had considered an elementary piece in reality has tiny handles attached to it which change the connectivity character of the piece, and that a microscope of ever greater magnification would reveal ever new topological complications of this type, *ad infinitum*. The Riemann point of view allows, also for real space, topological conditions entirely different from those realized by Euclidean space. I believe that only on the basis of the freer and more general conception of geometry which had been brought out by the development of mathematics during the last century, and with an open mind for the imaginative possibilities which it has revealed, can a philosophically fruitful attack upon the space problem be undertaken." H. Weyl (1949, p. 91).

"But ... physicists were still far removed from such a way of thinking; space was still, for them, a rigid, homogeneous something, susceptible of no change or conditions. Only the genius of Riemann, solitary and uncomprehended, had already won its way by the middle of the last century to a new conception of space, in which space was deprived of its rigidity, and in which its power to take part in physical events was recognized as possible." A. Einstein (1934, p. 68).

Riemann formulated the first known model for superspace (for which see Chapter 43), a superspace built, however, not of the totality of all 3-geometries with positive definite Riemannian metric (the dynamic arena of Einstein's general relativity), but of all conformally equivalent closed Riemannian 2-geometries of the same topology, a type of superspace known today as Teichmüller space, for more on Riemann's contributions to which and the subsequent development of which, see the chapters by L. Bers and J. A. Wheeler in Gilbert and Newton (1970).

Ricci curvature tensor

$$R_{\mu\nu} \equiv R^{\alpha}{}_{\mu\alpha\nu} = \Gamma^{\alpha}{}_{\mu\nu,\alpha} - \Gamma^{\alpha}{}_{\mu\alpha,\nu} + \Gamma^{\alpha}{}_{\beta\alpha}\Gamma^{\beta}{}_{\mu\nu} - \Gamma^{\alpha}{}_{\beta\nu}\Gamma^{\beta}{}_{\mu\alpha},$$

$$\text{[in a coordinate frame]} \tag{8.47}$$

and the *"scalar curvature,"*

Scalar curvature

$$R \equiv R^{\mu}{}_{\mu}. \tag{8.48}$$

But of much greater geometric significance is the *"Einstein curvature tensor"*

Einstein curvature tensor

$$G^{\mu}{}_{\nu} \equiv \frac{1}{2}\varepsilon^{\mu\alpha\beta\gamma}R_{\beta\gamma}{}^{\rho\sigma}\frac{1}{2}\varepsilon_{\nu\alpha\rho\sigma} = R^{\mu}{}_{\nu} - \frac{1}{2}\delta^{\mu}{}_{\nu}R. \tag{8.49}$$

Of all second-rank curvature tensors one can form by contracting **Riemann**, only **Einstein = G** retains part of the Bianchi identities (8.46): it satisfies

Contracted Bianchi identities

$$G^{\mu\nu}{}_{;\nu} = 0. \tag{8.50}$$

For the beautiful geometric meaning of these *"contracted Bianchi identities"* ("the boundary of a boundary is zero"), see Chapter 15.

Box 8.6 summarizes the above equations describing curvature, as well as the fundamental equations for covariant derivatives.

EXERCISE

[The following exercises from Track 2 are appropriate for the Track-1 reader who wishes to solidify his understanding of curvature: 11.6, 11.9, 11.10, 13.7–11, and 14.3.]

Exercise 8.16. SOME USEFUL FORMULAS IN COORDINATE FRAMES

In any coordinate frame, define g to be the determinant of the matrix $g_{\alpha\beta}$ [equation 8.11]. Derive the following relations, valid in any coordinate frame.

(a) Contraction of connection coefficients:

$$\Gamma^{\alpha}{}_{\beta\alpha} = (\ln\sqrt{-g})_{,\beta}. \tag{8.51a}$$

[*Hint:* Use the results of exercise 5.5.]

(b) Components of Ricci tensor:

$$R_{\alpha\beta} = \frac{1}{\sqrt{-g}}(\sqrt{-g}\,\Gamma^{\mu}{}_{\alpha\beta})_{,\mu} - (\ln\sqrt{-g})_{,\alpha\beta} - \Gamma^{\mu}{}_{\nu\alpha}\Gamma^{\nu}{}_{\beta\mu}. \tag{8.51b}$$

(c) Divergence of a vector A^{α} or *antisymmetric* tensor $F^{\alpha\beta}$:

$$A^{\alpha}{}_{;\alpha} = \frac{1}{\sqrt{-g}}(\sqrt{-g}\,A^{\alpha})_{,\alpha}, \qquad F^{\alpha\beta}{}_{;\beta} = \frac{1}{\sqrt{-g}}(\sqrt{-g}\,F^{\alpha\beta})_{,\beta}. \tag{8.51c}$$

(d) Integral of a scalar field Ψ over the proper volume of a 4-dimensional region \mathcal{V}:

$$\int_{\mathcal{V}}\Psi\,d(\text{proper volume}) = \int_{\mathcal{V}}\Psi\sqrt{-g}\,dt\,dx\,dy\,dz. \tag{8.51d}$$

[*Hint:* In a local Lorentz frame, $d(\text{proper volume}) = d\hat{t}\,d\hat{x}\,d\hat{y}\,d\hat{z}$. Use a Jacobian to transform this volume element to the given coordinate frame, and prove from the transformation law

$$g_{\alpha\beta} = \frac{\partial x^{\hat{\mu}}}{\partial x^{\alpha}}\frac{\partial x^{\hat{\nu}}}{\partial x^{\beta}}\eta_{\mu\nu}$$

that the Jacobian is equal to $\sqrt{-g}$.]

Box 8.6 COVARIANT DERIVATIVE AND CURVATURE: FUNDAMENTAL EQUATIONS

Entity	Abstract notation	Component notation
Covariant Derivative	$\boldsymbol{\nabla_u T} - \boldsymbol{\nabla T}(\ldots,\ldots,\boldsymbol{u})$	$T^{\beta}{}_{\alpha;\gamma}u^{\gamma} = DT^{\beta}{}_{\alpha}/d\lambda \quad (\boldsymbol{u} = d\mathcal{P}/d\lambda)$ $$= \frac{dT^{\beta}{}_{\alpha}}{d\lambda} + (\Gamma^{\beta}{}_{\nu\mu}T^{\nu}{}_{\alpha} - \Gamma^{\nu}{}_{\alpha\mu}T^{\beta}{}_{\nu})u^{\mu}$$
	$\boldsymbol{\nabla_u}f = \partial_{\boldsymbol{u}}f = \langle \boldsymbol{df},\boldsymbol{u}\rangle$	$f_{;\alpha}u^{\alpha} = f_{,\alpha}u^{\alpha}$
algebraic properties (Exercise 8.11)	$\boldsymbol{\nabla}_{a\boldsymbol{u}+b\boldsymbol{v}}\boldsymbol{T} = a\boldsymbol{\nabla_u T} + b\boldsymbol{\nabla_v T}$ $\boldsymbol{\nabla_u}(\boldsymbol{S}+\boldsymbol{M}) = \boldsymbol{\nabla_u S} + \boldsymbol{\nabla_u M}$ $\boldsymbol{\nabla_u w} - \boldsymbol{\nabla_w u} = [\boldsymbol{u},\boldsymbol{w}]$ for $\boldsymbol{u},\boldsymbol{w}$ both vector fields	$T^{\beta}{}_{\alpha;\gamma}(au^{\gamma}+bv^{\gamma}) = aT^{\beta}{}_{\alpha;\gamma}u^{\gamma} + bT^{\beta}{}_{\alpha;\gamma}v^{\gamma}$ $(S^{\beta}{}_{\alpha}+M^{\beta}{}_{\alpha})_{;\gamma}u^{\gamma} = S^{\beta}{}_{\alpha;\gamma}u^{\gamma} + M^{\beta}{}_{\alpha;\gamma}u^{\gamma}$ $\Gamma^{\rho}{}_{[\mu\nu]} = -\frac{1}{2}c_{\mu\nu}{}^{\rho}$ [equation (8.34)]
chain rule	$\boldsymbol{\nabla_u}(\boldsymbol{A}\otimes\boldsymbol{B}) = (\boldsymbol{\nabla_u A})\otimes\boldsymbol{B} + \boldsymbol{A}\otimes(\boldsymbol{\nabla_u B})$ $\boldsymbol{\nabla_u}(f\boldsymbol{A}) = (\boldsymbol{\nabla_u}f)\boldsymbol{A} + f\boldsymbol{\nabla_u A}$	$(A^{\alpha}{}_{\beta}B_{\gamma})_{;\mu}u^{\mu} = A^{\alpha}{}_{\beta;\mu}B_{\gamma}u^{\mu} + A^{\alpha}{}_{\beta}B_{\gamma;\mu}u^{\mu}$ $(fA^{\alpha}{}_{\beta})_{;\mu}u^{\mu} = f_{,\mu}A^{\alpha}{}_{\beta}u^{\mu} + fA^{\alpha}{}_{\beta;\mu}u^{\mu}$
$\boldsymbol{\nabla_u}$ and contraction commute	$\boldsymbol{\nabla_u}(\text{contraction of }\boldsymbol{S})$ $= (\text{contraction of }\boldsymbol{\nabla_u S})$	$\left(\sum_{\alpha}S^{\alpha}{}_{\alpha\gamma}\right)_{;\mu}u^{\mu} = \sum_{\alpha}(S^{\alpha}{}_{\alpha\gamma;\mu}u^{\mu})$
*metric covariantly constant	$\boldsymbol{\nabla_u g} = 0$	$g_{\alpha\beta;\gamma}u^{\gamma} = 0$
Gradient	$\boldsymbol{\nabla T}$	$T^{\beta}{}_{\alpha;\gamma} = T^{\beta}{}_{\alpha,\gamma} + \Gamma^{\beta}{}_{\mu\gamma}T^{\mu}{}_{\alpha} - \Gamma^{\mu}{}_{\alpha\gamma}T^{\beta}{}_{\mu}$
Connection Coefficients	$\Gamma^{\alpha}{}_{\beta\gamma} = \langle \boldsymbol{\omega}^{\alpha},\boldsymbol{\nabla}_{\gamma}\boldsymbol{e}_{\beta}\rangle$ $= -\langle \boldsymbol{\nabla}_{\gamma}\boldsymbol{\omega}^{\alpha},\boldsymbol{e}_{\beta}\rangle$	$\Gamma^{\alpha}{}_{\beta\gamma} = g^{\alpha\mu}\Gamma_{\mu\beta\gamma},^{*}$ $\Gamma_{\mu\beta\gamma} = \frac{1}{2}(g_{\mu\beta,\gamma} + g_{\mu\gamma,\beta} - g_{\beta\gamma,\mu}$ $+ c_{\mu\beta\gamma} + c_{\mu\gamma\beta} - c_{\beta\gamma\mu})^{*}$ $c_{\beta\gamma\mu} = g_{\mu\alpha}c_{\beta\gamma}{}^{\alpha}$ $= g_{\mu\alpha}\langle \boldsymbol{\omega}^{\alpha},[\boldsymbol{e}_{\beta},\boldsymbol{e}_{\gamma}]\rangle^{*}$
*Local Lorentz frame at \mathcal{P}_0		Coordinate system with $g_{\mu\nu}(\mathcal{P}_0) = \eta_{\mu\nu}, \qquad \Gamma^{\alpha}{}_{\beta\gamma}(\mathcal{P}_0) = 0$
Parallel transport	$\boldsymbol{\nabla_u S} = 0$	$S^{\alpha}{}_{\beta,\gamma}u^{\gamma} = 0$

Box 8.6 (continued)

Entity	Abstract notation	Component notation
Geodesic Equation	$\nabla_u u = 0$	$d^2 x^\alpha / d\lambda^2 + \Gamma^\alpha{}_{\mu\nu}(dx^\mu/d\lambda)(dx^\nu/d\lambda) = 0$ in a coordinate basis
Riemann Curvature Tensor	$\textbf{Riemann}(\sigma, \boldsymbol{C}, \boldsymbol{A}, \boldsymbol{B}) \equiv \langle \sigma, \mathscr{R}(\boldsymbol{A}, \boldsymbol{B})\boldsymbol{C}\rangle$ $\mathscr{R}(\boldsymbol{A}, \boldsymbol{B}) \equiv [\nabla_{\boldsymbol{A}}, \nabla_{\boldsymbol{B}}] - \nabla_{[\boldsymbol{A},\boldsymbol{B}]}$ (not track-one formulas; see Chapter 11)	$R^\alpha{}_{\beta\gamma\delta} = \dfrac{\partial \Gamma^\alpha{}_{\beta\delta}}{\partial x^\gamma} - \dfrac{\partial \Gamma^\alpha{}_{\beta\gamma}}{\partial x^\delta}$ $\quad + \Gamma^\alpha{}_{\mu\gamma}\Gamma^\mu{}_{\beta\delta} - \Gamma^\alpha{}_{\mu\delta}\Gamma^\mu{}_{\beta\gamma}$ in coordinate frame [see equation (11.13) for formula in non-coordinate frame]
Ricci Curvature Tensor	$\textbf{Ricci} = $ contraction on slots 1 and 3 of $\textbf{Riemann}$	$R_{\mu\nu} = R^\alpha{}_{\mu\alpha\nu} = \Gamma^\alpha{}_{\mu\nu,\alpha} - \Gamma^\alpha{}_{\mu\alpha,\nu} + \Gamma^\alpha{}_{\beta\alpha}\Gamma^\beta{}_{\mu\nu}$ $\qquad\qquad - \Gamma^\alpha{}_{\beta\nu}\Gamma^\beta{}_{\mu\alpha}$ in coordinate frame
*Curvature Scalar	$R = $ (contraction of \textbf{Ricci})	$R = R^\alpha{}_\alpha$
*Einstein Curvature Tensor	$\boldsymbol{G} = \textbf{Ricci} - \dfrac{1}{2}\boldsymbol{g}R$	$G_{\alpha\beta} = R_{\alpha\beta} - \dfrac{1}{2} g_{\alpha\beta}R$ Useful formulas for computing $G^\alpha{}_\beta$ (derived in §14.2): $G^0{}_0 = -(R^{12}{}_{12} + R^{23}{}_{23} + R^{31}{}_{31})$, $G^0{}_1 = R^{02}{}_{12} + R^{03}{}_{13}$, $G^1{}_1 = -(R^{02}{}_{02} + R^{03}{}_{03} + R^{23}{}_{23})$, $G^1{}_2 = R^{10}{}_{20} + R^{13}{}_{23}$, etc.
*Symmetries of Curvature Tensors		$R_{\alpha\beta\gamma\delta} = R_{[\alpha\beta][\gamma\delta]} = R_{[\gamma\delta][\alpha\beta]}, \; R_{[\alpha\beta\gamma\delta]} = 0, \; R_{\alpha[\beta\gamma\delta]} = 0$ $R_{\alpha\beta} = R_{(\alpha\beta)}, \; G_{\alpha\beta} = G_{(\alpha\beta)}$
Bianchi Identities		$R^\alpha{}_{\beta[\mu\nu;\lambda]} = 0$
*Contracted Bianchi Identities		$G^{\alpha\beta}{}_{;\beta} = 0$
Geodesic Deviation	$\nabla_u \nabla_u n + \textbf{Riemann}(\dots, \boldsymbol{u}, \boldsymbol{n}, \boldsymbol{u}) = 0$	$\dfrac{D^2 n^\alpha}{d\lambda^2} + R^\alpha{}_{\beta\gamma\delta} u^\beta n^\gamma u^\delta = 0$
Parallel Transport around closed curve (§11.4)	$\delta \boldsymbol{A} + \textbf{Riemann}(\dots \boldsymbol{A}, \boldsymbol{u}, \boldsymbol{v}) = 0$ if $\boldsymbol{u}, \boldsymbol{v}$ are edges of curve	$\delta A^\alpha + R^\alpha{}_{\beta\gamma\delta} A^\beta u^\gamma v^\delta = 0$

*If metric is absent, these starred formulas cannot be formulated. All other formulas are valid in absence of metric.

CHAPTER **9**

DIFFERENTIAL TOPOLOGY

*In analytic geometry, many relations which are independent of
any frame must be expressed with respect to some particular
frame. It is therefore preferable to devise new methods—
methods which lead directly to intrinsic properties without any
mention of coordinates. The development of the topology of
general spaces and of the objects which occur in them, as well
as the development of the geometry of general metric spaces,
are steps in this direction.*

KARL MENGER, in Schilpp (1949), p. 467.

§9.1. GEOMETRIC OBJECTS IN METRIC-FREE, GEODESIC-FREE SPACETIME

Curved spacetime without metric or geodesics or parallel transport, i.e., "differential
topology," is the subject of this easy chapter. It is easy because all the necessary
geometric objects (event, curve, vector, 1-form, tensor) are already familiar from
flat spacetime. Yet it is also necessary, because one's viewpoint must be refined when
one abandons the Lorentz metric of flat spacetime.

Events

The primitive concept of an event \mathcal{P} (Figure 1.2) needs no refinement. The essential
property here is identifiability, which is not dependent on the Lorentz metric struc-
ture of spacetime.

> This chapter is entirely Track 2.
> It depends on no preceding
> Track-2 material.
> It is needed as preparation
> for
> (1) Chapters 10–13
> (differential geometry;
> Newtonian gravity),
> and
> (2) Box 30.1 (mixmaster
> cosmology).
> It will be helpful in
> (1) Chapter 14 (calculation
> of curvature) and in
> (2) Chapter 15 (Bianchi
> identities).

Metric is abandoned

Geometric concepts must be refined

Curves

Again no refinement. A "curve" $\mathscr{P}(\lambda)$ is also too primitive to care whether spacetime has a metric—*except* that, with metric gone, there is no concept of "proper length" along the curve. This is in accord with Newton's theory of gravity, where one talks of the lengths of curves in "space," but never in "spacetime."

Vectors

Here refinement is needed. In special relativity one could dress primitive ("identifiable") events in enough algebraic plumage to talk of vectors as differences $\mathscr{P} - \mathscr{Q}$ between "algebraic" events. Now the plumage is gone, and the old bilocal ("point for head and point for tail") version of a vector must be replaced by a purely local version (§9.2). Also vectors cannot be moved around; each vector must be attached to a specific event (§§9.2 and 9.3).

1-Forms

Almost no refinement needed, except that, with metric gone, there is no way to tell which 1-form corresponds to a given vector (no way to raise and lower indices), and each 1-form must be attached to a specific event (§9.4).

Tensors

Again almost no refinement, except that each slot of a tensor is specific: if it accepts vectors, then it cannot accommodate 1-forms, and conversely (no raising and lowering of indices); also, each tensor must be attached to a specific event (§9.5).

§9.2. "VECTOR" AND "DIRECTIONAL DERIVATIVE" REFINED INTO TANGENT VECTOR

Flat spacetime can accommodate several equivalent definitions of a vector (§2.3): a vector is an arrow reaching from an event \mathscr{P}_0 to an event \mathscr{Q}_0; it is the parameterized straight line, $\mathscr{P}(\lambda) = \mathscr{P}_0 + \lambda(\mathscr{Q}_0 - \mathscr{P}_0)$ extending from \mathscr{P}_0 at $\lambda = 0$ to \mathscr{Q}_0 at $\lambda = 1$; it is the rate of change of the point $\mathscr{P}(\lambda)$ with increasing λ, $d\mathscr{P}/d\lambda$.

Old definitions of vector break down when metric is abandoned

With Lorentz metric gone, the "arrow" definition and the "parametrized-straight line" definition must break down. By what route is the arrow or line to be laid out between \mathscr{P}_0 and \mathscr{Q}_0? There is no concept of straightness; all routes are equally straight or bent.

Such fuzziness forces one to focus on the "rate-of-change-of-point-along-curve"

Box 9.1 TANGENT VECTORS AND TANGENT SPACE

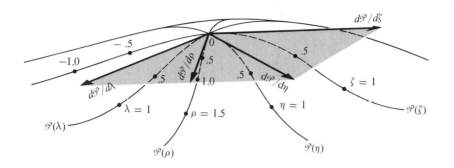

A tangent vector $d\mathcal{P}/d\lambda$ is defined to be "the limit, when $N \longrightarrow \infty$, of N times the displacement of \mathcal{P} as λ ranges from 0 to $1/N$." One cannot think of this final displacement $d\mathcal{P}/d\lambda$ as lying in spacetime; fuzziness forbids (no concept of straightness). Instead, one visualizes $d\mathcal{P}/d\lambda$ as lying in a "tangent plane" or "tangent space," which makes contact with spacetime only at $\mathcal{P}(0)$, the event where $d\mathcal{P}/d\lambda$ is evaluated. All other tangent vectors at $\mathcal{P}(0)$—e.g., $d\mathcal{P}/d\rho$, $d\mathcal{P}/d\eta$, $d\mathcal{P}/d\zeta$—lie in this same tangent space.

To make precise these concepts of tangent space and tangent vector, one may regard spacetime as embedded in a flat space of more than four di-mensions. One can then perform the limiting process that leads to $d\mathcal{P}/d\lambda$, using straight arrows in the flat embedding space. The result is a higher-dimensional analog of the figure shown above.

But such a treatment is dangerous. It suggests, falsely, that the tangent vector $d\mathcal{P}/d\lambda$ and the tangent space at \mathcal{P}_0 depend on how the embedding is done, or depend for their existence on the em-bedding process. They do not. And to make clear that they do not is one motivation for defining the directional derivative operator "$d/d\lambda$" to be the tangent vector, rather than using Cartan's more pictorial concept "$d\mathcal{P}/d\lambda$".

definition, $d\mathcal{P}/d\lambda$. It, under the new name *"tangent vector,"* is explored briefly in Box 9.1, and in greater depth in the following paragraphs.

Even "$d\mathcal{P}/d\lambda$" is a fuzzy definition of tangent vector, most mathematicians would argue. More acceptable, they suggest, is this definition: *the tangent vector \boldsymbol{u} to a curve $\mathcal{P}(\lambda)$ is the directional derivative operator along that curve*

$$\boldsymbol{u} = \partial_{\boldsymbol{u}} = (d/d\lambda)_{\text{along curve}}. \tag{9.1}$$

Best new definition: "tangent vector equals directional derivative operator"
$\boldsymbol{u} = d/d\lambda$

Tangent vector equals directional derivative operator? Preposterous! A vector started out as a happy, irresponsible trip from \mathcal{P}_0 to \mathcal{Q}_0. It ended up loaded with the social responsibility to tell how something else changes at \mathcal{P}_0. At what point did the vector get saddled with this unexpected load? And did it really change its character all that much, as it seems to have done? For an answer, go back and try

to redo the "rate-of-change-of-point" definition, $d\mathscr{P}/d\lambda$, in the form of a limiting process:

0. Choose a curve $\mathscr{P}(\lambda)$ whose tangent vector $d\mathscr{P}/d\lambda$ at $\lambda = 0$ is desired.
1. Take the displacement of \mathscr{P} as λ ranges from 0 to 1; that is *not* $d\mathscr{P}/d\lambda$.
2. Take twice the displacement of \mathscr{P} as λ ranges from 0 to $\frac{1}{2}$; that is *not* $d\mathscr{P}/d\lambda$.
N. Take N times the displacement of \mathscr{P} as λ ranges from 0 to $1/N$; that is *not* $d\mathscr{P}/d\lambda$.
∞. Take the limit of such displacements as $N \longrightarrow \infty$; that *is* $d\mathscr{P}/d\lambda$.

This definition has the virtue that $d\mathscr{P}/d\lambda$ describes the properties of the curve $\mathscr{P}(\lambda)$, not over the huge range from $\lambda = 0$ to $\lambda = 1$, where the curve might be doing wild things, but only in an infinitesimal neighborhood of the point $\mathscr{P}_0 = \mathscr{P}(0)$.

The deficiency in this definition is that no meaning is assigned to steps $1, 2, \ldots,$ $N, \ldots,$ so there is nothing, yet, to take the limit of. To make each "displacement of \mathscr{P}" a definite mathematical object in a space where "limit" has a meaning, one can imagine the original manifold to be a low-dimensional surface in some much higher-dimensional *flat* space. Then $\mathscr{P}(1/N) - \mathscr{P}(0)$ is just a straight arrow connecting two points, i.e. a segment of a straight line, which, in general, will not lie in the surface itself—see Box 9.1. The resulting mental picture of a tangent vector makes its essential properties beautifully clear, but at the cost of some artifacts. The picture relies on a specific but arbitrary way of embedding the manifold of interest (metric-free spacetime) in an extraneous flat space. In using this picture, one must ignore everything that depends on the peculiarities of the embedding. One must think like the chemist, who uses tinkertoy molecular models to visualize many essential properties of a molecule clearly, but easily ignores artifacts of the model (colors of the atoms, diameters of the pegs, its tendency to collapse) that do not mimic quantum-mechanical reality.

Élie Cartan's approach to differential geometry, including the $d\mathscr{P}/d\lambda$ idea of a tangent vector, suggests that he always thought of manifolds as embedded in flat spaces this way, and relied on insights that he did not always formalize to separate the essential geometry of these pictures from their embedding-dependent details. Acceptance of his methods of calculation came late. Mathematicians, who mistrusted their own ability to distinguish fact from artifact, exacted this price for acceptance: stop talking about the movement of the point itself, and start dealing only with concrete measurable changes that take place within the manifold, changes in any or all scalar functions f as the point moves. The limiting process then reads:

0. Choose a curve $\mathscr{P}(\lambda)$ whose tangent vector at $\lambda = 0$ is desired.
1. Compute the number $f[\mathscr{P}(1)] - f[\mathscr{P}(0)]$, which measures the change in f as the point $\mathscr{P}(\lambda)$ moves from $\mathscr{P}_0 = \mathscr{P}(0)$ to $\mathscr{Q}_0 = \mathscr{P}(1)$.
2. Compute $2\{ f[\mathscr{P}(\frac{1}{2})] - f[\mathscr{P}(0)]\}$, which is twice the change in f as the point goes from $\mathscr{P}(0)$ to $\mathscr{P}(\frac{1}{2})$.
N. Compute $N\{ f[\mathscr{P}(1/N)] - f[\mathscr{P}(0)]\}$, which is N times the change in f as the point goes from $\mathscr{P}(0)$ to $\mathscr{P}(1/N)$.

Alternative definition, $\boldsymbol{u} = d\mathscr{P}/d\lambda$, requires embedding in flat space of higher dimensionality

Refinement of $d\mathscr{P}/d\lambda$ into $d/d\lambda$

∞. Same in the limit as $N \longrightarrow \infty$: (change in f) $= df/d\lambda$.

0. The vector is not itself the change in f. It is instead the operation $d/d\lambda$, which, when applied to f, gives the change $df/d\lambda$. Thus

$$\text{tangent vector} = d/d\lambda$$

[cf. definition (9.1)].

The operation $d/d\lambda$ clearly involves nothing but the last steps $N \longrightarrow \infty$ in this limiting process, and only those aspects of these steps that are independent of f. But this means it involves the infinitesimal displacements of the point \mathcal{P} *and nothing more*.

One who wishes both to stay in touch with the present and to not abandon Cartan's deep geometric insight (Box 9.1) can seek to keep alive a distinction between:

(A) the tangent vector itself in the sense of Cartan, the displacement $d\mathcal{P}/d\lambda$ of a point; and

(B) the "tangent vector operator," or "directional derivative operator," telling what happens to a function in this displacement: (tangent vector operator) $= d/d\lambda$.

However, present practice drops (or, if one will, "slurs") the word "operator" in (B), and uses the phrase "tangent vector" itself for the operator, as will be the practice here from now on. The ideas (A) and (B) should also slur or coalesce in one's mind, so that when one visualizes an embedding diagram with arrows drawn tangent to the surface, one always realizes that the arrow characterizes an infinitesimal motion of a point $d\mathcal{P}/d\lambda$ that takes place purely within the surface, and when one thinks of a derivative operator $d/d\lambda$, one always visualizes this same infinitesimal motion of a point in the manifold, a motion that must occur in constructing any derivative $df(\mathcal{P})/d\lambda$. In this sense, one should regard a vector $d\mathcal{P}/d\lambda \equiv d/d\lambda$ as both "a displacement that carries attention from one point to another" and "a purely geometric object built on points and nothing but points."

The hard-nosed physicist may still be inclined to say "Tangent vector equals directional derivative operator? Preposterous!" Perhaps he will be put at ease by another argument. He is asked to pick an event \mathcal{P}_0. At that event he chooses any set of four noncoplanar vectors (vectors defined in whatever way seems reasonable to him); he names them $\boldsymbol{e}_0, \boldsymbol{e}_1, \boldsymbol{e}_2, \boldsymbol{e}_3$; and he uses them as a basis on which to expand all other vectors at \mathcal{P}_0:

$$\boldsymbol{u} = u^\alpha \boldsymbol{e}_\alpha, \qquad \boldsymbol{v} = v^\alpha \boldsymbol{e}_\alpha. \qquad (9.2)$$

He is asked to construct the four directional derivative operators $\partial_\alpha \equiv \partial_{\boldsymbol{e}_\alpha}$ along his four basis vectors. As in flat spacetime, so also here; the same expansion coefficients that appear in $\boldsymbol{u} = u^\alpha \boldsymbol{e}_\alpha$ also appear in the expansion for the directional derivative:

Isomorphism between directional derivatives and vectors

$$\partial_{\boldsymbol{u}} = u^\alpha \partial_\alpha, \qquad \partial_{\boldsymbol{v}} = v^\alpha \partial_\alpha. \qquad (9.3)$$

Hence, every relation between specific vectors at \mathscr{P}_0 induces an identical relation between their differential operators:

$$
\begin{aligned}
\boldsymbol{u} = a\boldsymbol{w} + b\boldsymbol{v} &\Longleftrightarrow u^\alpha = aw^\alpha + bv^\alpha \\
&\Longleftrightarrow \partial_{\boldsymbol{u}} = a\,\partial_{\boldsymbol{w}} + b\,\partial_{\boldsymbol{v}}.
\end{aligned} \tag{9.4}
$$

There is a complete "isomorphism" between the vectors and the corresponding directional derivatives. So how can the hard-nosed physicist deny the hard-nosed mathematician the right to identify completely each tangent vector with its directional derivative? No harm is done; no answer to any computation can be affected.

This isomorphism extends to the concept "*tangent space.*" Because linear relations (such as $\partial_{\boldsymbol{u}} = a\,\partial_{\boldsymbol{w}} + b\,\partial_{\boldsymbol{v}}$) among directional derivatives evaluated *at one and the same point* \mathscr{P}_0 are meaningful and obey the usual addition and multiplication rules, these derivative operators form an abstract (but finite-dimensional) vector space called the tangent space at \mathscr{P}_0. In an embedding picture (Box 9.1) one uses these derivatives (as operators in the flat embedding space) to construct tangent vectors $\boldsymbol{u} = \partial_{\boldsymbol{u}}\mathscr{P},\ \boldsymbol{v} = \partial_{\boldsymbol{v}}\mathscr{P}$, in the form of straight arrows. Thereby one identifies the abstract tangent space with the geometrically visualized tangent space.

Tangent space defined

§9.3. BASES, COMPONENTS, AND TRANSFORMATION LAWS FOR VECTORS

Coordinate-induced basis defined

An especially useful basis in the tangent space at an event \mathscr{P}_0 is induced by any coordinate system [four functions, $x^0(\mathscr{P}),\ x^1(\mathscr{P}),\ x^2(\mathscr{P}),\ x^3(\mathscr{P})$]:

$$
\boldsymbol{e}_0 \equiv \left(\frac{\partial}{\partial x^0}\right)_{x^1,x^2,x^3} = \left(\begin{array}{l}\text{directional derivative along the}\\ \text{curve with constant } (x^1,x^2,x^3)\\ \text{and with parameter } \lambda = x^0\end{array}\right)_{\text{at }\mathscr{P}_0}, \tag{9.5}
$$

$$
\boldsymbol{e}_1 = \frac{\partial}{\partial x^1}, \qquad \boldsymbol{e}_2 = \frac{\partial}{\partial x^2}, \qquad \boldsymbol{e}_3 = \frac{\partial}{\partial x^3}.
$$

(See Figure 9.1.)

A transformation from one basis to another in the tangent space at \mathscr{P}_0, like any change of basis in any vector space, is produced by a nonsingular matrix,

Changes of basis: transformation matrices defined

$$
\boldsymbol{e}_{\alpha'} = \boldsymbol{e}_\beta L^\beta{}_{\alpha'}; \tag{9.6}
$$

and, as always (including the Lorentz frames of flat spacetime), the components of a vector must transform by the inverse matrix

$$
u^{\alpha'} = L^{\alpha'}{}_\beta u^\beta; \tag{9.7}
$$

$$
\|L^{\alpha'}{}_\beta\| = \|L^\beta{}_{\gamma'}\|^{-1}, \text{ i.e.,} \begin{cases} L^{\alpha'}{}_\beta L^\beta{}_{\gamma'} = \delta^{\alpha'}{}_{\gamma'}, \\ L^\delta{}_{\alpha'} L^{\alpha'}{}_\beta = \delta^\delta{}_\beta. \end{cases} \tag{9.8}
$$

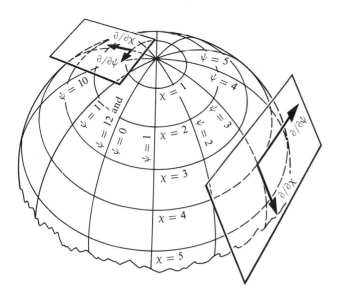

Figure 9.1.

The basis vectors induced, by a coordinate system, into the tangent space at each event. Here a truncated, two-dimensional spacetime is shown (two other dimensions suppressed), with coordinates $\chi(\mathcal{P})$ and $\psi(\mathcal{P})$, and with corresponding basis vectors $\partial/\partial\chi$ and $\partial/\partial\psi$.

This "inverse" transformation law guarantees compatibility between the expansions $\boldsymbol{u} = \boldsymbol{e}_{\alpha'}u^{\alpha'}$ and $\boldsymbol{u} = \boldsymbol{e}_{\beta}u^{\beta}$:

$$\boldsymbol{u} = \boldsymbol{e}_{\alpha'}u^{\alpha'} = (\boldsymbol{e}_{\gamma}L^{\gamma}{}_{\alpha'})(L^{\alpha'}{}_{\beta}u^{\beta}) = \boldsymbol{e}_{\gamma}\delta^{\gamma}{}_{\beta}u^{\beta}$$
$$= \boldsymbol{e}_{\beta}u^{\beta}.$$

In the special case of transformations between coordinate-induced bases, the transformation matrix has a simple form:

$$\frac{\partial}{\partial x^{\alpha'}} = \frac{\partial x^{\beta}}{\partial x^{\alpha'}}\frac{\partial}{\partial x^{\beta}} \text{ (by usual rules of calculus)},$$

so

$$L^{\beta}{}_{\alpha'} = (\partial x^{\beta}/\partial x^{\alpha'})_{\text{at event } \mathcal{P}_0 \text{ where tangent space lies}}. \tag{9.9}$$

(*Note:* this generalizes the Lorentz-transformation law $x^{\beta} = \Lambda^{\beta}{}_{\alpha'}x^{\alpha'}$, which has the differential form $\Lambda^{\beta}{}_{\alpha'} = \partial x^{\beta}/\partial x^{\alpha'}$; also, it provides a good way to remember the signs in the Λ matrices.)

§9.4. 1-FORMS

When the Lorentz metric is removed from spacetime, one must sharpen up the concept of a 1-form $\boldsymbol{\sigma}$ by insisting that it, like any tangent vector \boldsymbol{u}, be attached to a specific event \mathcal{P}_0 in spacetime. The family of surfaces representing $\boldsymbol{\sigma}$ resides in the tangent space at \mathcal{P}_0, not in spacetime itself. The piercing of surfaces of $\boldsymbol{\sigma}$ by an arrow \boldsymbol{u} to produce the number $\langle\boldsymbol{\sigma}, \boldsymbol{u}\rangle$ ("bongs of bell") occurs entirely in the tangent space.

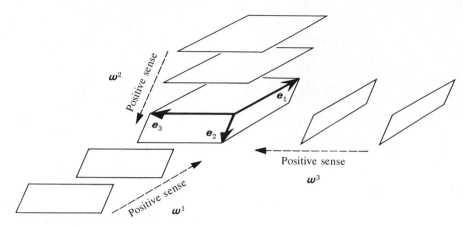

Figure 9.2.

The basis vectors \boldsymbol{e}_α and dual basis 1-forms \boldsymbol{w}^β in the tangent space of an event \mathcal{P}_0. The condition

$$\langle \boldsymbol{w}^\beta, \boldsymbol{e}_\alpha \rangle = \delta^\beta{}_\alpha$$

dictates that the vectors \boldsymbol{e}_2 and \boldsymbol{e}_3 lie parallel to the surfaces of \boldsymbol{w}^1, and that \boldsymbol{e}_1 extend from one surface of \boldsymbol{w}^1 to the next (precisely 1.00 surfaces pierced).

Notice that this picture could fit perfectly well into a book on X-rays and crystallography. There the vectors $\boldsymbol{e}_1, \boldsymbol{e}_2, \boldsymbol{e}_3$ would be the edges of a unit cell of the crystal; and the surfaces of $\boldsymbol{w}^1, \boldsymbol{w}^2, \boldsymbol{w}^3$ would be the surfaces of unit cells. Also, for an X-ray diffraction experiment, with wavelength of radiation and orientation of crystal appropriately adjusted, the successive surfaces of \boldsymbol{w}^1 would produce Bragg reflection. For other choices of wavelength and orientation, the surfaces of \boldsymbol{w}^2 or \boldsymbol{w}^3 would produce Bragg reflection.

Dual basis of 1-forms defined

Given any set of basis vectors $\{\boldsymbol{e}_0, \boldsymbol{e}_1, \boldsymbol{e}_2, \boldsymbol{e}_3\}$ at an event \mathcal{P}_0, one constructs the "*dual basis*" of 1-forms $\{\boldsymbol{w}^0, \boldsymbol{w}^1, \boldsymbol{w}^2, \boldsymbol{w}^3\}$ by choosing the surfaces of \boldsymbol{w}^β such that that

$$\langle \boldsymbol{w}^\beta, \boldsymbol{e}_\alpha \rangle = \delta^\beta{}_\alpha. \tag{9.10}$$

See Figure 9.2. A marvelously simple formalism for calculating and manipulating components of tangent vectors and 1-forms then results:

Component-manipulation formulas

$$\boldsymbol{u} = \boldsymbol{e}_\alpha u^\alpha \qquad \text{(definition of components of } \boldsymbol{u}\text{)}, \tag{9.11a}$$

$$\boldsymbol{\sigma} = \sigma_\beta \boldsymbol{w}^\beta \qquad \text{(definition of components of } \boldsymbol{\sigma}\text{)}, \tag{9.11b}$$

$$u^\alpha = \langle \boldsymbol{w}^\alpha, \boldsymbol{u} \rangle \qquad \text{(way to calculate components of } \boldsymbol{u}\text{)}, \tag{9.11c}$$

$$\sigma_\beta = \langle \boldsymbol{\sigma}, \boldsymbol{e}_\beta \rangle \qquad \text{(way to calculate components of } \boldsymbol{\sigma}\text{)}, \tag{9.11d}$$

$$\langle \boldsymbol{\sigma}, \boldsymbol{u} \rangle = \sigma_\alpha u^\alpha \qquad \text{(way to calculate } \langle \boldsymbol{\sigma}, \boldsymbol{u} \rangle \text{ using} \atop \text{components)}, \tag{9.11e}$$

$$\boldsymbol{w}^{\alpha'} = L^{\alpha'}{}_\beta \boldsymbol{w}^\beta \qquad \text{(transformation law for 1-form} \atop \text{basis, corresponding to equation 9.6)}, \tag{9.11f}$$

$$\sigma_{\alpha'} = \sigma_\beta L^\beta{}_{\alpha'} \qquad \text{(transformation law for 1-form} \atop \text{components)}. \tag{9.11g}$$

(Exercise 9.1 below justifies these equations.)

In the absence of a metric, there is no way to pick a specific 1-form \tilde{u} at an event \mathscr{P}_0 and say that it corresponds to a specific tangent vector u at \mathscr{P}_0. The correspondence set up in flat spacetime,

Correspondence between vectors and 1-forms rubbed out

$$\langle \tilde{u}, v \rangle = u \cdot v \qquad \text{for all } v,$$

was rubbed out when "\cdot" was rubbed out. Restated in component language: the raising of an index, $u^\alpha = \eta^{\alpha\beta} u_\beta$, is impossible because the $\eta^{\alpha\beta}$ do not exist; similarly, lowering of an index, $u_\beta = \eta_{\beta\alpha} u^\alpha$, is impossible.

The 1-form gradient df was introduced in §2.6 with absolutely no reference to metric. Consequently, it and its mathematical formalism are the same here, without metric, as there with metric, except that, like all other 1-forms, df now resides in the tangent space rather than in spacetime itself. For example, there is no change in the fundamental equation relating the projection of the gradient to the directional derivative:

Gradient of a function

$$\langle df, u \rangle = \partial_u f = u[f]. \qquad (9.12)$$

$$\begin{bmatrix} \text{old notation for} \\ \text{directional derivative} \end{bmatrix} \uparrow \qquad \uparrow \begin{bmatrix} \text{new notation;} \\ \text{recall } u = \partial_u. \end{bmatrix}$$

Similarly, there are no changes in the component equations,

$$df = f_{,\alpha} \omega^\alpha \qquad \text{(expansion of } df \text{ in arbitrary basis)}, \qquad (9.13a)$$

$$f_{,\alpha} = \partial_\alpha f = e_\alpha[f] \qquad \text{(way to calculate components of } df \text{)}, \qquad (9.13b)$$

$$f_{,\alpha} = \partial f / \partial x^\alpha \qquad \text{if } \{e_\alpha\} \text{ is a coordinate basis,}$$

except that they work in arbitrary bases, not just in Lorentz bases. And, as in Lorentz frames, so also in general: the one-form basis $\{dx^\alpha\}$ and the tangent-vector basis $\{\partial/\partial x^\alpha\}$, which are induced into a tangent space by the same coordinate system, are the duals of each other,

$$\langle dx^\alpha, \partial/\partial x^\beta \rangle = \delta^\alpha{}_\beta. \qquad (9.14)$$

(See exercise 9.2 for proofs.) Also, most aspects of Cartan's "Exterior Calculus" (parts A, B, C of Box 4.1) are left unaffected by the removal of metric.

§9.5. TENSORS

A tensor S, in the absence of Lorentz metric, differs from the tensors of flat, Lorentz spacetime in two ways. (1) S must reside at a specific event \mathscr{P}_0, just as any vector or 1-form must. (2) Each slot of S is specific; it will accept either vectors or 1-forms, but not both, because it has no way to convert a 1-form \tilde{u} into a "corresponding

Specificity of tensor slots

vector" \boldsymbol{u} as it sends $\tilde{\boldsymbol{u}}$ through its linear machinery. Thus, if \boldsymbol{S} is a $\binom{1}{2}$ tensor

$$\boldsymbol{S}(\ldots, \ldots, \ldots), \tag{9.15}$$

insert 1-form here \longrightarrow \uparrow \llcorner insert vector here

insert vector here

then it *cannot* be converted alternatively to a $\binom{2}{1}$ tensor, or a $\binom{3}{0}$ tensor, or a $\binom{0}{3}$ tensor by the procedure of §3.2. In component language, the indices of \boldsymbol{S} cannot be raised and lowered.

Except for these two restrictions (attachment to a specific event; specificity of slots), a tensor \boldsymbol{S} is the same linear machine as ever. And the algebra of component manipulations is the same:

$$S^{\alpha}{}_{\beta\gamma} = \boldsymbol{S}(\boldsymbol{\omega}^{\alpha}, \boldsymbol{e}_{\beta}, \boldsymbol{e}_{\gamma}) \qquad (\boldsymbol{S}, \boldsymbol{\omega}^{\alpha}, \boldsymbol{e}_{\beta} \text{ must all reside at same event}) \tag{9.16}$$

$$\boldsymbol{S} = S^{\alpha}{}_{\beta\gamma} \boldsymbol{e}_{\alpha} \otimes \boldsymbol{\omega}^{\beta} \otimes \boldsymbol{\omega}^{\gamma}, \tag{9.17}$$

$$\boldsymbol{S}(\boldsymbol{\sigma}, \boldsymbol{u}, \boldsymbol{v}) = S^{\alpha}{}_{\beta\gamma} \sigma_{\alpha} u^{\beta} v^{\gamma}. \tag{9.18}$$

EXERCISES

Exercise 9.1. COMPONENT MANIPULATIONS

Derive equations (9.11c) through (9.11g) from (9.10), (9.11a, b), (9.6), (9.7), and (9.8).

Exercise 9.2. COMPONENTS OF GRADIENT, AND DUALITY OF COORDINATE BASES

In an arbitrary basis, define $f_{,\alpha}$ by the expansion (9.13a). Then combine equations (9.11d) and (9.12) to obtain the method (9.13b) of computing $f_{,\alpha}$. Finally, combine equations (9.12) and (9.13b) to show that the bases $\{\boldsymbol{d}x^{\alpha}\}$ and $\{\partial/\partial x^{\beta}\}$ are the duals of each other.

Exercise 9.3. PRACTICE MANIPULATING TANGENT VECTORS

Let \mathscr{P}_0 be the point with coordinates $(x = 0, y = 1, z = 0)$ in a three-dimensional space; and define three curves through \mathscr{P}_0 by

$$\mathscr{P}(\lambda) = (\lambda, 1, \lambda),$$

$$\mathscr{P}(\zeta) = (\sin\zeta, \cos\zeta, \zeta),$$

$$\mathscr{P}(\rho) = (\sinh\rho, \cosh\rho, \rho + \rho^3).$$

(a) Compute $(d/d\lambda)f$, $(d/d\zeta)f$, and $(d/d\rho)f$ for the function $f = x^2 - y^2 + z^2$ at the point \mathscr{P}_0. (b) Calculate the components of the tangent vectors $d/d\lambda$, $d/d\zeta$, and $d/d\rho$ at \mathscr{P}_0, using the basis $\{\partial/\partial x, \partial/\partial y, \partial/\partial z\}$.

Exercise 9.4. MORE PRACTICE WITH TANGENT VECTORS

In a three-dimensional space with coordinates (x, y, z), introduce the vector field $\boldsymbol{v} = y^2 \partial/\partial x - x \partial/\partial z$, and the functions $f = xy$, $g = z^3$. Compute

(a) $\boldsymbol{v}[f]$ (c) $\boldsymbol{v}[fg]$ (e) $\boldsymbol{v}[f^2 + g^2]$

(b) $\boldsymbol{v}[g]$ (d) $f\boldsymbol{v}[g] - g\boldsymbol{v}[f]$ (f) $\boldsymbol{v}\{\boldsymbol{v}[f]\}$

Exercise 9.5. PICTURE OF BASIS 1-FORMS INDUCED BY COORDINATES
In the tangent space of Figure 9.1, draw the basis 1-forms $\boldsymbol{d\psi}$ and $\boldsymbol{d\chi}$ induced by the ψ, χ-coordinate system.

Exercise 9.6. PRACTICE WITH DUAL BASES
In a three-dimensional space with spherical coordinates r, θ, ϕ, one often likes to use, instead of the basis $\partial/\partial r, \partial/\partial \theta, \partial/\partial \phi$, the basis

$$\boldsymbol{e}_{\hat{r}} = \frac{\partial}{\partial r}, \qquad \boldsymbol{e}_{\hat{\theta}} = \frac{1}{r}\frac{\partial}{\partial \theta}, \qquad \boldsymbol{e}_{\hat{\phi}} = \frac{1}{r \sin \theta}\frac{\partial}{\partial \phi}.$$

(a) What is the 1-form basis $\{\boldsymbol{\omega}^{\hat{r}}, \boldsymbol{\omega}^{\hat{\theta}}, \boldsymbol{\omega}^{\hat{\phi}}\}$ dual to this tangent-vector basis? (b) On the sphere $r = 1$, draw pictures of the bases $\{\partial/\partial r, \partial/\partial \theta, \partial/\partial \phi\}$, $\{\boldsymbol{e}_{\hat{r}}, \boldsymbol{e}_{\hat{\theta}}, \boldsymbol{e}_{\hat{\phi}}\}$, $\{\boldsymbol{dr}, \boldsymbol{d\theta}, \boldsymbol{d\phi}\}$, and $\{\boldsymbol{\omega}^{\hat{r}}, \boldsymbol{\omega}^{\hat{\theta}}, \boldsymbol{\omega}^{\hat{\phi}}\}$.

§9.6. COMMUTATORS AND PICTORIAL TECHNIQUES

A vector \boldsymbol{u}_0 given only at one point \mathscr{P}_0 suffices to compute the derivative $\boldsymbol{u}_0[f] \equiv \partial_{\boldsymbol{u}_0} f$, which is simply a number associated with the point \mathscr{P}_0. In contrast, a vector field \boldsymbol{u} provides a vector $\boldsymbol{u}(\mathscr{P})$—which is a differential operator $\partial_{\boldsymbol{u}(\mathscr{P})}$—at each point \mathscr{P} in some region of spacetime. This vector field operates on a function f to produce not just a number, but another function $\boldsymbol{u}[f] \equiv \partial_{\boldsymbol{u}} f$. A second vector field \boldsymbol{v} can perfectly well operate on this new function, to produce yet another function

$$\boldsymbol{v}\{\boldsymbol{u}[f]\} = \partial_{\boldsymbol{v}}(\partial_{\boldsymbol{u}} f).$$

Does this function agree with the result of applying \boldsymbol{v} first and then \boldsymbol{u}? Equivalently, does the "commutator"

$$[\boldsymbol{u}, \boldsymbol{v}][f] \equiv \boldsymbol{u}\{\boldsymbol{v}[f]\} - \boldsymbol{v}\{\boldsymbol{u}[f]\} \tag{9.19}$$

Commutator defined

vanish?

The simplest special case is when \boldsymbol{u} and \boldsymbol{v} are basis vectors of a coordinate system, $\boldsymbol{u} = \partial/\partial x^{\alpha}$, $\boldsymbol{v} = \partial/\partial x^{\beta}$. Then the commutator does vanish, because partial derivatives always commute:

$$[\partial/\partial x^{\alpha}, \partial/\partial x^{\beta}][f] = \partial^2 f/\partial x^{\beta} \partial x^{\alpha} - \partial^2 f/\partial x^{\alpha} \partial x^{\beta} = 0.$$

But in general the commutator is nonzero, as one sees from a coordinate-based calculation:

$$[\boldsymbol{u}, \boldsymbol{v}]f = u^{\alpha}\frac{\partial}{\partial x^{\alpha}}\left(v^{\beta}\frac{\partial f}{\partial x^{\beta}}\right) - v^{\alpha}\frac{\partial}{\partial x^{\alpha}}\left(u^{\beta}\frac{\partial f}{\partial x^{\beta}}\right)$$

$$= \left[(u^{\alpha}v^{\beta}{}_{,\alpha} - v^{\alpha}u^{\beta}{}_{,\alpha})\frac{\partial}{\partial x^{\beta}}\right]f.$$

Commutator of two vector fields is a vector field

Notice however, that the commutator $[\boldsymbol{u}, \boldsymbol{v}]$, like \boldsymbol{u} and \boldsymbol{v} themselves, is a vector field, i.e., a linear differential operator at each event:

$$[\boldsymbol{u}, \boldsymbol{v}] = (\boldsymbol{u}[v^\beta] - \boldsymbol{v}[u^\beta]) \frac{\partial}{\partial x^\beta} = (u^\alpha v^\beta{}_{,\alpha} - v^\alpha u^\beta{}_{,\alpha}) \frac{\partial}{\partial x^\beta}. \qquad (9.20)$$

Such results should be familiar from quantum theory's formalism for angular momentum operators (exercise 9.8).

The three levels of geometry—pictorial, abstract, and component—yield three different insights into the commutator. (1) The abstract expression $[\boldsymbol{u}, \boldsymbol{v}]$ suggests the close connection to quantum theory, and brings to mind the many tools developed there for handling operators. But *recall* that the operators of quantum theory need not be first-order differential operators. The kinetic energy is second order and the potential is zeroth order in the familiar Schrödinger equation. Only first-order operators are vectors. (2) The component expression $u^\alpha v^\beta{}_{,\alpha} - v^\alpha u^\beta{}_{,\alpha}$, valid in any coordinate basis, brings the commutator into the reaches of the powerful tools of index mechanics. (3) The pictorial representation of $[\boldsymbol{u}, \boldsymbol{v}]$ (Box 9.2) reveals its fundamental role as a "closer of curves"—a role that will be important in Chapter 11's analysis of curvature.

Commutator as a "closer of curves"

Commutators find application in the distinction between a coordinate-induced basis, $\{\boldsymbol{e}_\alpha\} = \{\partial/\partial x^\alpha\}$, and a noncoordinate basis. Because partial derivatives always commute,

$$[\boldsymbol{e}_\alpha, \boldsymbol{e}_\beta] = [\partial/\partial x^\alpha, \partial/\partial x^\beta] = 0 \text{ in any coordinate basis.} \qquad (9.21)$$

Box 9.2 THE COMMUTATOR AS A CLOSER OF QUADRILATERALS

A. Pictorial Representation in Flat Spacetime

1. For ease of visualization, consider flat spacetime, so the two vector fields $\boldsymbol{u}(\mathcal{P})$ and $\boldsymbol{v}(\mathcal{P})$ can be laid out in spacetime itself.
2. Choose an event \mathcal{P}_0 where the commutator $[\boldsymbol{u}, \boldsymbol{v}]$ is to be calculated.
3. Give the names $\mathcal{P}_1, \mathcal{P}_2, \mathcal{P}_3, \mathcal{P}_4$ to the events pictured in the diagram.
4. Then the vector $\mathcal{P}_4 - \mathcal{P}_3$, which measures how much the four-legged curve fails to close, can be expressed in a coordinate basis as

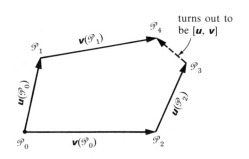

$$\begin{aligned}
\mathcal{P}_4 - \mathcal{P}_3 &= [\boldsymbol{u}(\mathcal{P}_0) + \boldsymbol{v}(\mathcal{P}_1)] - [\boldsymbol{u}(\mathcal{P}_2) + \boldsymbol{v}(\mathcal{P}_0)] \\
&= [\boldsymbol{v}(\mathcal{P}_1) - \boldsymbol{v}(\mathcal{P}_0)] - [\boldsymbol{u}(\mathcal{P}_2) - \boldsymbol{u}(\mathcal{P}_0)] \\
&= (v^\beta{}_{,\alpha} u^\alpha \boldsymbol{e}_\beta)_{\mathcal{P}_0} - (u^\beta{}_{,\alpha} v^\alpha \boldsymbol{e}_\beta)_{\mathcal{P}_0} + \text{errors} \\
&= [\boldsymbol{u}, \boldsymbol{v}]_{\mathcal{P}_0} + \text{errors}.
\end{aligned}$$

\llcorner[terms such as $v^\beta{}_{,\mu\nu} u^\mu u^\nu \boldsymbol{e}_\beta$]

5. Notice that if \boldsymbol{u} and \boldsymbol{v} are halved everywhere, then $[\boldsymbol{u}, \boldsymbol{v}]$ is cut down by a factor of 4, while the error terms in the above go down by a factor of 8. Thus, $[\boldsymbol{u}, \boldsymbol{v}]$ represents accurately the gap in the four-legged curve (*"quadrilateral"*) in the limit where \boldsymbol{u} and \boldsymbol{v} are sufficiently short; i.e., $[\boldsymbol{u}, \boldsymbol{v}]$ *"closes the quadrilateral"* whose edges are the vectors fields \boldsymbol{u} and \boldsymbol{v}.

B. Pictorial Representation in Absence of Metric, or in Curved Spacetime with a Metric

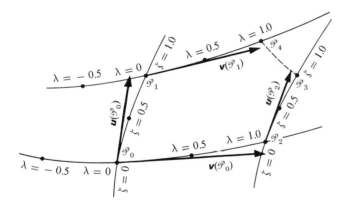

1. The same picture must work, but now one dares not (at least initially) lay out the vector fields in spacetime itself. Instead one lays out two families of curves: the curves for which $\boldsymbol{u}(\mathcal{P})$ is the tangent vector; and the curves for which $\boldsymbol{v}(\mathcal{P})$ is the tangent vector.

2. The gap "$\mathcal{P}_4 - \mathcal{P}_3$" in the four-legged curve can be characterized by the difference $f(\mathcal{P}_4) - f(\mathcal{P}_3)$ in the values of an arbitrary function at \mathcal{P}_4 and \mathcal{P}_3. That difference is, in a coordinate basis,

Box 9.2 (*continued*)

$$f(\mathcal{P}_4) - f(\mathcal{P}_3) = \underbrace{[f(\mathcal{P}_4) - f(\mathcal{P}_1)]}_{\left(f_{,\alpha}v^\alpha + \frac{1}{2}f_{,\alpha\beta}v^\alpha v^\beta\right)_{\mathcal{P}_1}} + \underbrace{[f(\mathcal{P}_1) - f(\mathcal{P}_0)]}_{\left(f_{,\alpha}u^\alpha + \frac{1}{2}f_{,\alpha\beta}u^\alpha u^\beta\right)_{\mathcal{P}_0}}$$

$$-\underbrace{[f(\mathcal{P}_2) - f(\mathcal{P}_0)]}_{\left(f_{,\alpha}v^\alpha + \frac{1}{2}f_{,\alpha\beta}v^\alpha v^\beta\right)_{\mathcal{P}_0}} - \underbrace{[f(\mathcal{P}_3) - f(\mathcal{P}_2)]}_{\left(f_{,\alpha}u^\alpha + \frac{1}{2}f_{,\alpha\beta}u^\alpha u^\beta\right)_{\mathcal{P}_2}}$$

$$= [(f_{,\alpha}v^\alpha)_{,\beta}u^\beta - (f_{,\alpha}u^\alpha)_{,\beta}v^\beta]_{\mathcal{P}_0} + \text{``cubic errors''}$$
$$= [(u^\beta v^\alpha{}_{,\beta} - v^\beta u^\alpha{}_{,\beta})\partial f/\partial x^\alpha]_{\mathcal{P}_0} + \text{``cubic errors''}$$
$$= \{[\boldsymbol{u}, \boldsymbol{v}][f]\}_{\mathcal{P}_0} + \text{``cubic errors.''}$$

Here "cubic errors" are cut down by a factor of 8, while $[\boldsymbol{u}, \boldsymbol{v}]f$ is cut down by one of 4, whenever \boldsymbol{u} and \boldsymbol{v} are cut in half.

3. The result

$$f(\mathcal{P}_4) - f(\mathcal{P}_3) = \{[\boldsymbol{u}, \boldsymbol{v}][f]\}_{\mathcal{P}_0} + \text{``cubic errors''}$$

says that $[\boldsymbol{u}, \boldsymbol{v}]$ is a tangent vector at \mathcal{P}_0 that describes the separation between the points \mathcal{P}_3 and \mathcal{P}_4. Its description gets arbitrarily accurate when \boldsymbol{u} and \boldsymbol{v} get arbitrarily short. Thus, $[\boldsymbol{u}, \boldsymbol{v}]$ closes the quadrilateral whose edges are the projections of \boldsymbol{u} and \boldsymbol{v} into spacetime.

C. Philosophy of Pictures

1. Pictures are no substitute for computation. Rather, they are useful for (a) suggesting geometric relationships that were previously unsuspected and that one verifies subsequently by computation; (b) interpreting newly learned geometric results.
2. This usual noncomputational role of pictures permits one to be sloppy in drawing them. No essential new insight was gained in part B over part A, when one carefully moved the tangent vectors into their respective tangent spaces, and permitted only curves to lie in spacetime. Moreover, the original picture (part A) was clearer because of its greater simplicity.
3. This motivates one to draw "sloppy" pictures, with tangent vectors lying in spacetime itself—so long as one keeps those tangent vectors short and occasionally checks the scaling of errors when the lengths of the vectors are halved.

Conversely, if one is given a field of basis vectors ("frame field") $\{e_\alpha(\mathcal{P})\}$, but one does not know whether a coordinate system $\{x^\alpha(\mathcal{P})\}$ exists in which $\{e_\alpha\} = \{\partial/\partial x^\alpha\}$, one can find out by a simple test: calculate all $(4 \times 3)/2 = 6$ commutators $[e_\alpha, e_\beta]$; if they all vanish, then there exists such a coordinate system. If not, there doesn't. Stated more briefly, $\{e_\alpha(\mathcal{P})\}$ *is a coordinate-induced basis if and only if* $[e_\alpha, e_\beta] = 0$ *for all* e_α *and* e_β. (See exercise 9.9 for proof; see §11.5 for an important application.) Coordinate-induced bases are sometimes called *"holonomic."* In an *"anholonomic basis"* (noncoordinate basis), one defines the commutation coefficients $c_{\mu\nu}{}^\alpha$ by

Vanishing commutator: a test for coordinate bases

Commutation coefficients defined

$$[e_\mu, e_\nu] = c_{\mu\nu}{}^\alpha e_\alpha. \tag{9.22}$$

They enter into the component formula for the commutator of arbitrary vector fields \boldsymbol{u} and \boldsymbol{v}:

$$[\boldsymbol{u}, \boldsymbol{v}] = (\boldsymbol{u}[v^\beta] - \boldsymbol{v}[u^\beta] + u^\mu v^\nu c_{\mu\nu}{}^\beta) e_\beta \tag{9.23}$$

(see exercise 9.10).

[*Warning!* In notation for functions and fields, mathematicians and physicists often use the same symbols to mean contradictory things. The physicist may write l when considering the length of some critical component in an instrument he is designing, then switch to $l(T)$ when he begins to analyze its response to temperature changes. Thus l is a number, whereas $l(T)$ is a function. The mathematician, in contrast, will write f for a function that he may be considering as an element in some infinite-dimensional function space. Once the function is supplied with an argument, he then contemplates $f(x)$, which is merely a number: the value of f at the point x. Caught between these antithetical rituals of the physics and mathematics sects, the authors have adopted a clear policy: vacillation. Usually physics-sect statements, like "On a curve $\mathcal{P}(\lambda) \dots$," are used; and the reader can translate them himself into mathematically precise language: "Consider a curve \mathcal{C} on which a typical point is $\mathcal{P} = \mathcal{C}(\lambda)$; on this curve \dots." But on occasion the reader will encounter a pedantic-sounding paragraph written in mathematics-sect jargon (Example: Box 23.3). Such paragraphs deal with concepts and relationships so complex that standard physics usage would lead to extreme confusion. They also should prevent the reader from becoming so conditioned to physics usage that he is allergic to the mathematical literature, where great advantages of clarity and economy of thought are achieved by consistent reliance on wholly unambiguous notation.]

Physicists' notation vs. mathematicians' notation

Exercise 9.7. PRACTICE WITH COMMUTATORS

Compute the commutator $[e_{\hat\theta}, e_{\hat\phi}]$ of the vector fields

$$e_{\hat\theta} = \frac{1}{r} \frac{\partial}{\partial \theta}, \qquad e_{\hat\phi} = \frac{1}{r \sin\theta} \frac{\partial}{\partial \phi}.$$

Express your result as a linear combination of $e_{\hat\theta}$ and $e_{\hat\phi}$.

EXERCISES

Exercise 9.8. ANGULAR MOMENTUM OPERATORS

In Cartesian coordinates of three-dimensional Euclidean space, one defines three *"angular-momentum operators"* (vector fields) \boldsymbol{L}_j by

$$\boldsymbol{L}_j \equiv \epsilon_{jkl} x^k (\partial/\partial x^l).$$

Draw a picture of these three vector fields. Calculate their commutators both pictorially and analytically.

Exercise 9.9. COMMUTATORS AND COORDINATE-INDUCED BASES

Let \boldsymbol{u} and \boldsymbol{v} be vector fields in spacetime. Show that in some neighborhood of any given point there exists a coordinate system for which

$$\boldsymbol{u} = \partial/\partial x^1, \qquad \boldsymbol{v} = \partial/\partial x^2,$$

if and only if \boldsymbol{u} and \boldsymbol{v} are linearly independent and commute:

$$[\boldsymbol{u}, \boldsymbol{v}] = 0.$$

First make this result plausible from the second figure in Box 9.2; then prove it mathematically. *Note:* this result can be generalized to four arbitrary vector fields $\boldsymbol{e}_0, \boldsymbol{e}_1, \boldsymbol{e}_2, \boldsymbol{e}_3$. There exists a coordinate system in which $\boldsymbol{e}_\alpha = \partial/\partial x^\alpha$ if and only if $\boldsymbol{e}_0, \boldsymbol{e}_1, \boldsymbol{e}_2, \boldsymbol{e}_3$ are linearly independent and $[\boldsymbol{e}_\mu, \boldsymbol{e}_\nu] = 0$ for all pairs $\boldsymbol{e}_\mu, \boldsymbol{e}_\nu$.

Exercise 9.10. COMPONENTS OF COMMUTATOR IN NON-COORDINATE BASIS

Derive equation (9.23).

Exercise 9.11. LIE DERIVATIVE

The "Lie derivative" of a vector field $\boldsymbol{v}(\mathcal{P})$ along a vector field $\boldsymbol{u}(\mathcal{P})$ is defined by

$$\pounds_{\boldsymbol{u}} \boldsymbol{v} \equiv [\boldsymbol{u}, \boldsymbol{v}]. \tag{9.24}$$

Draw a space-filling family of curves (a *"congruence"*) on a sheet of paper. Draw an arbitrary vector \boldsymbol{v} at an arbitrary point \mathcal{P}_0 on the sheet. Transport that vector along the curve through \mathcal{P}_0 by means of the *"Lie transport law"* $\pounds_{\boldsymbol{u}} \boldsymbol{v} = 0$, where $\boldsymbol{u} = d/dt$ is the tangent to the curve. Draw the resulting vector \boldsymbol{v} at various points $\mathcal{P}(t)$ along the curve.

Exercise 9.12. A CHIP OFF THE OLD BLOCK

(a) Prove the Jacobi identity

$$[\boldsymbol{u}, [\boldsymbol{v}, \boldsymbol{w}]] + [\boldsymbol{v}, [\boldsymbol{w}, \boldsymbol{u}]] + [\boldsymbol{w}, [\boldsymbol{u}, \boldsymbol{v}]] = 0 \tag{9.25}$$

by picking out all terms of the form $\partial_{\boldsymbol{u}} \partial_{\boldsymbol{v}} \partial_{\boldsymbol{w}}$, showing that they add to zero, and arguing from symmetry that all other terms, e.g., $\partial_{\boldsymbol{w}} \partial_{\boldsymbol{u}} \partial_{\boldsymbol{v}}$ terms, must similarly cancel.

(b) State this identity in index form.

(c) Draw a picture corresponding to this identity (see Box 9.2).

§9.7. MANIFOLDS AND DIFFERENTIAL TOPOLOGY

Spacetime is not the only arena in which the ideas of this chapter can be applied. Points, curves, vectors, 1-forms, and tensors exist in any "differentiable manifold."

Their use to study differentiable manifolds constitutes a branch of mathematics called *"differential topology"*—hence the title of this chapter.

The mathematician usually begins his development of differential topology by introducing some very primitive concepts, such as sets and topologies of sets, by building a fairly elaborate framework out of them, and by then using that framework to define the concept of a differentiable manifold. But most physicists are satisfied with a more fuzzy, intuitive definition of manifold: roughly speaking, an *n*-dimensional *differentiable manifold* is a set of "points" tied together continuously and differentiably, so that the points in any sufficiently small region can be put into a one-to-one correspondence with an open set of points of R^n. [R^n is the number space of n dimensions, i.e., the space of ordered n-tuples (x^1, x^2, \ldots, x^n).] That correspondence furnishes a coordinate system for the neighborhood.

A few examples will convey the concept better than this definition. Elementary examples (Euclidean 3-spaces, the surface of a sphere) bring to mind too many geometric ideas from richer levels of geometry; so one is forced to contemplate something more complicated. Let R^3 be a three-dimensional number space with the usual advanced-calculus ideas of continuity and differentiability. Points $\boldsymbol{\xi}$ of R^3 are triples, $\boldsymbol{\xi} = (\xi_1, \xi_2, \xi_3)$, of real numbers. Let a *ray* \mathscr{P} in R^3 be any half-line from the origin consisting of all $\boldsymbol{\xi}$ of the form $\boldsymbol{\xi} = \lambda\boldsymbol{\eta}$ for some fixed $\boldsymbol{\eta} \neq 0$ and for all positive real numbers $\lambda > 0$. (See Figure 9.3.) A good example of a differentiable manifold then is the set S^2 of all distinct rays. If f is a real-valued function with a specific value $f(\mathscr{P})$ for any ray \mathscr{P} [so one writes $f: S^2 \longrightarrow R: \mathscr{P} \longrightarrow f(\mathscr{P})$], it should be intuitively (or even demonstrably) clear that we can define what we mean by saying f is continuous or differentiable. In this sense S^2 itself is continuous and differentiable. Thus S^2 is a manifold, and the rays \mathscr{P} are the points of S^2. There are many other manifolds that differential topology finds indistinguishable from S^2. The simplest is the two-dimensional spherical surface (2-sphere), which is the standard representation of S^2; it is the set of points $\boldsymbol{\xi}$ of R^3 satisfying $(\xi_1)^2 + (\xi_2)^2 + (\xi_3)^2 = 1$. Clearly a different point \mathscr{P} of S^2 (one ray in R^3) intersects each point of this standard 2-sphere surface, and the correspondence is continuous and differentiable in either direction (ray to point; point to ray). The same is true for any ellipsoidal surface in R^3 enclosing the origin, and for any other surface enclosing the origin that has

Differentiable manifold "defined"

Examples of differentiable manifolds

The manifold S^2

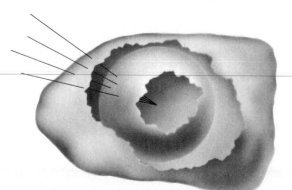

Figure 9.3.
Three different representations of the differentiable manifold S^2. The first is the set of all rays emanating from the origin; the second is the sphere they intersect; the third is an odd-shaped, closed surface that each ray intersects precisely once.

The manifold T^2

The manifold $SO(3)$ (rotation group)

Affine geometry and Riemannian geometry defined

a different ray through each point of itself. They each embody the same global continuity and differentiability concepts, and represent the same abstract differentiable manifold S^2, the 2-sphere. They, and the bundle of rays we started with, all have the same geometric properties at this rudimentary level of geometry. A two-dimensional manifold that has a different geometric structure at this level (a different "differentiable structure") is the torus T^2, the surface of a donut. There is no way to imbed this surface smoothly in R^3 so that a distinct ray $\mathscr{P} \in S^2$ intersects each of its points; there is no invertible and differentiable correspondence between T^2 and S^2.

Another example of a manifold is the rotation group $SO(3)$, whose points \mathscr{P} are all the 3×3 orthogonal matrices of unit determinant, so $\mathscr{P} = \|P_{ij}\|$ with $\mathscr{P}^T \mathscr{P} = 1$ and det $\mathscr{P} = 1$. This is a three-dimensional space (one often uses the three Euler-angle parameters in computations), where differential ideas (e.g., angular velocity) are employed; hence, it is a manifold. So is the Lorentz group.

The differentiability of a manifold (i.e., the possibility of defining differentiable functions on it) permits one to introduce coordinate systems locally, if not globally, and also curves, tangent spaces, tangent vectors, 1-forms, and tensors, just as is done for spacetime. *But* the mere fact that a manifold is differentiable does not mean that such concepts as geodesics, parallel transport, curvature, metric, or length exist in it. These are additional layers of structure possessed by some manifolds, but not by all. Roughly speaking, every manifold has smoothness properties and topology, but without additional structure it is shapeless and sizeless.

That branch of mathematics which adds geodesics, parallel transport, and curvature (shape) to a manifold is called *affine geometry;* that branch which adds a metric is called *Riemannian geometry.* They will be studied in the next few chapters.

EXERCISES

EXERCISES ON THE ROTATION GROUP

As the exposition of differential geometry becomes more and more sophisticated in the following chapters, the exercises will return time and again to the rotation group as an example of a manifold. Then, in Box 30.1, the results developed in these exercises will be used to analyze the "Mixmaster universe," which is a particularly important cosmological solution to Einstein's field equation.

Before working these exercises, the reader may wish to review the Euler-angle parametrization for rotation matrices, as treated, e.g., on pp. 107–109 of Goldstein (1959).

Exercise 9.13. ROTATION GROUP: GENERATORS

Let \mathscr{K}_l be three 3×3 matrices whose components are $(K_l)_{mn} = \epsilon_{lmn}$.
(a) Display the matrices \mathscr{K}_1, $(\mathscr{K}_1)^2$, $(\mathscr{K}_1)^3$, and $(\mathscr{K}_1)^4$.
(b) Sum the series

$$\mathscr{R}_x(\theta) \equiv \exp(\mathscr{K}_1 \theta) = \sum_{n=0}^{\infty} \frac{\theta^n}{n!} (\mathscr{K}_1)^n. \tag{9.26}$$

Show that $\mathscr{R}_x(\theta)$ is a rotation matrix and that it produces a rotation through an angle θ about the x-axis.

(c) Show similarly that $\mathcal{R}_z(\phi) = \exp(\mathcal{K}_3\phi)$ and $\mathcal{R}_y(\chi) = \exp(\mathcal{K}_2\chi)$ are rotation matrices, and that they produce rotations through angles ϕ and χ about the z- and y-axes, respectively.

(d) Explain why $\mathcal{P} = \mathcal{R}_z(\psi)\mathcal{R}_x(\theta)\mathcal{R}_z(\phi)$ defines the Euler-angle coordinates, ψ, θ, ϕ for the generic element $\mathcal{P} \in SO(3)$ of the rotation group.

(e) Let \mathcal{C} be the curve $\mathcal{P} = \mathcal{R}_z(t)$ through the identity matrix, $\mathcal{C}(0) = \mathcal{I} \in SO(3)$. Show that its tangent, $(d\mathcal{C}/dt)(0) \equiv \dot{\mathcal{C}}(0)$ does not vanish by computing $\dot{\mathcal{C}}(0)f_{12}$, where f_{12} is the function $f_{12}(\mathcal{P}) = P_{12}$, whose value is the 12 matrix element of \mathcal{P}.

(f) Define a vector field \boldsymbol{e}_3 on $SO(3)$ by letting $\boldsymbol{e}_3(\mathcal{P})$ be the tangent (at $t = 0$) to the curve $\mathcal{C}(t) = \mathcal{R}_z(t)\mathcal{P}$ through \mathcal{P}. Show that $\boldsymbol{e}_3(\mathcal{P})$ is nowhere zero. *Note:* $\boldsymbol{e}_3(\mathcal{P})$ is called the "generator of rotations about the z-axis," because it points from \mathcal{P} toward neighboring rotations, $\mathcal{R}_z(t)\mathcal{P}$, which differ from \mathcal{P} by a rotation about the z-axis.

(g) Show that $\boldsymbol{e}_3 = (\partial/\partial\psi)_{\theta\phi}$.

(h) Derive the following formulas, valid for $t \ll 1$:

$$\mathcal{R}_x(t)\mathcal{R}_z(\psi)\mathcal{R}_x(\theta)\mathcal{R}_z(\phi) = \mathcal{R}_z(\psi - t\sin\psi\cot\theta)\mathcal{R}_x(\theta + t\cos\psi)\mathcal{R}_z(\phi + t\sin\psi/\sin\theta);$$
$$\mathcal{R}_y(t)\mathcal{R}_z(\psi)\mathcal{R}_x(\theta)\mathcal{R}_z(\phi) = \mathcal{R}_z(\psi + t\cos\psi\cot\theta)\mathcal{R}_x(\theta + t\sin\psi)\mathcal{R}_z(\phi - t\cos\psi/\sin\theta).$$

(i) Define $\boldsymbol{e}_1(\mathcal{P})$ and $\boldsymbol{e}_2(\mathcal{P})$ to be the tangent vectors (at $t = 0$) to the curves $\mathcal{C}(t) = \mathcal{R}_x(t)\mathcal{P}$ and $\mathcal{C}(t) = \mathcal{R}_y(t)\mathcal{P}$, respectively. Show that

$$\boldsymbol{e}_1 = \cos\psi\frac{\partial}{\partial\theta} - \sin\psi\left(\cot\theta\frac{\partial}{\partial\psi} - \frac{1}{\sin\theta}\frac{\partial}{\partial\phi}\right),$$

$$\boldsymbol{e}_2 = \sin\psi\frac{\partial}{\partial\theta} + \cos\psi\left(\cot\theta\frac{\partial}{\partial\psi} - \frac{1}{\sin\theta}\frac{\partial}{\partial\phi}\right).$$

\boldsymbol{e}_1 and \boldsymbol{e}_2 are the "generators of rotations about the x- and y-axes."

Exercise 9.14. ROTATION GROUP: STRUCTURE CONSTANTS

Use the three vector fields constructed in the last exercise,

$$\boldsymbol{e}_1 = \cos\psi\frac{\partial}{\partial\theta} - \sin\psi\left(\cot\theta\frac{\partial}{\partial\psi} - \frac{1}{\sin\theta}\frac{\partial}{\partial\phi}\right),$$

$$\boldsymbol{e}_2 = \sin\psi\frac{\partial}{\partial\theta} + \cos\psi\left(\cot\theta\frac{\partial}{\partial\psi} - \frac{1}{\sin\theta}\frac{\partial}{\partial\phi}\right), \tag{9.27}$$

$$\boldsymbol{e}_3 = \frac{\partial}{\partial\psi},$$

as basis vectors for the manifold of the rotation group. The above equations express this "basis of generators" in terms of the Euler-angle basis. Show that the commutation coefficients for this basis are

$$c_{\alpha\beta}{}^\gamma = -\epsilon_{\alpha\beta\gamma}, \tag{9.28}$$

independently of location \mathcal{P} in the rotation group. These coefficients are also called the *structure constants* of the rotation group.

CHAPTER **10**

AFFINE GEOMETRY: GEODESICS, PARALLEL TRANSPORT, AND COVARIANT DERIVATIVE

Galilei's Principle of Inertia is sufficient in itself to prove conclusively that the world is affine in character.

HERMANN WEYL

§10.1. GEODESICS AND THE EQUIVALENCE PRINCIPLE

Freely falling particles and their clocks

Free fall is the "natural state of motion," so natural, in fact, that *the path through spacetime of a freely falling, neutral test body is independent of its structure and composition* (the "weak equivalence principle" of Einstein, Eötvös, Dicke; see Box 1.2 and §38.3).

Picture spacetime as filled with free-fall trajectories. Pick an event. Pick a velocity there. They determine a unique trajectory.

Be more precise. Ask for the maximum amount of information tied up in each trajectory. Is it merely the sequence of points along which the test body falls? No; there is more. Each test body can carry a clock with itself (same kind of clock—"good" clock in sense of Figure 1.9—regardless of structure or composition of test body). The clock ticks as the body moves, labeling each event on its trajectory with a number: the time λ the body was there. Result: the free-fall trajectory is not just a sequence of points; it is a parametrized sequence, a *"curve"* $\mathcal{P}(\lambda)$.

But is the parametrization unique? Not entirely. Quite arbitrary are (1) the choice of time origin, $\mathcal{P}(0)$; and (2) the units (centimeters, seconds, furlongs, . . .) in which clock time λ is measured. Hence, λ is unique only up to linear transformations

$$\lambda_{\text{new}} = a\lambda_{\text{old}} + b; \tag{10.1}$$

Figure 10.1.

A geodesic viewed as a rule for "straight-on parallel transport." Pick an event \mathscr{P}_0 and a tangent vector $\boldsymbol{u} = d/d\lambda$ there. Construct the unique geodesic $\mathscr{P}(\lambda)$ that (1) passes through \mathscr{P}_0: $\mathscr{P}(0) = \mathscr{P}_0$; and (2) has \boldsymbol{u} as its tangent vector there: $(d\mathscr{P}/d\lambda)_{\lambda=0} = \boldsymbol{u}$. This geodesic can be viewed as a rule for picking up \boldsymbol{u} from $\mathscr{P}(0)$ and laying it down again at its tip, $\mathscr{P}(1)$, in as straight a manner as possible,

$$\boldsymbol{u}_{\lambda=1} = (d\mathscr{P}/d\lambda)_{\lambda=1};$$

and for then picking it up and laying it down as straight as possible again at $\mathscr{P}(2)$,

$$\boldsymbol{u}_{\lambda=2} = (d\mathscr{P}/d\lambda)_{\lambda=2};$$

etc. This sequence of "straight as possible," "tail-on-tip" transports gives meaning to the idea that $(d\mathscr{P}/d\lambda)_{\lambda=17}$ and $\boldsymbol{u} = (d\mathscr{P}/d\lambda)_{\lambda=0}$ are "the same vector" at different points along the geodesic; or, equivalently, that one has been obtained from the other by "*straight-on parallel transport.*"

b ("new origin of clock time") is a number independent of location on this specific free-fall trajectory, and a ("ratio of new units to old") is also.

In the curved spacetime of Einstein (and in that of Cartan-Newton, Chapter 12), these parametrized free-fall trajectories are the straightest of all possible curves. Consequently, one gives these trajectories the same name, "*geodesics,*" that mathematicians use for the straight lines of a curved manifold; and like the mathematicians, one uses the name "*affine parameter*" for the parameter λ along a free-fall geodesic. Equation (10.1) then says "the affine parameter of a geodesic is unique up to linear transformations." — Geodesic defined as a free-fall trajectory

The affine parameter ("clock time") along a geodesic has nothing to do, à priori, with any metric. It exists even in the absence of metric (e.g., in Cartan-Newtonian spacetime). It gives one a method for comparing the separation between events on a geodesic (\mathscr{B} and \mathscr{A} are "twice as far apart" as \mathscr{R} and \mathscr{Q} if $[\lambda_{\mathscr{B}} - \lambda_{\mathscr{A}}] = 2[\lambda_{\mathscr{R}} - \lambda_{\mathscr{Q}}]$). But the affine parameter measures relative separations only along its own geodesic. It has no means of reaching off the geodesic. — Affine parameter defined as clock time along free-fall trajectory

The above features of geodesics, and others, are summarized in Figure 10.1 and Box 10.1.

§10.2. PARALLEL TRANSPORT AND COVARIANT DERIVATIVE: PICTORIAL APPROACH

Two test bodies, initially falling through spacetime on parallel, neighboring geodesics, get pushed toward each other or apart by tidal gravitational forces (spacetime curvature). To quantify this statement, one must quantify the concepts of "parallel" and "rate of acceleration away from each other." Begin with parallelism.

Box 10.1 GEODESICS

Geodesic in brief

Give point, give tangent vector; get unique, affine-parametrized curve ("geodesic").

Geodesic: in context of gravitation physics

World line of a neutral test particle ("Einstein's geometric theory of gravity"; also "Cartan's translation into geometric terms of Newton's theory of gravity"):

(1) "given point": some event on this world line;

(2) "given vector": vector ("displacement per unit increase of parameter") tangent to world line at instant defined by that event;

(3) "unique curve": every neutral test particle with a specified initial position and a specified initial velocity follows the same world line, regardless of its composition and regardless of its mass (small; test mass!; "weak equivalence principle of Einstein-Eötvös-Dicke");

(4) "affine parameter": in Cartan-Newton theory, Newton's "universal time" (which is measured by "good" clocks); in the real physical world, "proper time" (as measured by a "good" clock) along a timelike geodesic;

(5) "parametrized curve": (a) affine parameter unique up to a transformation of the form $\lambda \longrightarrow a\lambda + b$, where a and b are constants (no arbitrariness along a given geodesic other than zero of parameter and unit of parameter); or equivalently (b) given any three events \mathcal{A}, \mathcal{B}, \mathcal{C} on the geodesic, one can find by well-determined physical construction ("clocking") a unique fourth event \mathcal{D} on the geodesic such that $(\lambda_{\mathcal{D}} - \lambda_{\mathcal{C}})$ is equal to $(\lambda_{\mathcal{B}} - \lambda_{\mathcal{A}})$; or equivalently (c) [differential version] given a tangent vector with components $(dx^{\alpha}/d\lambda)_{\mathcal{A}}$ at point \mathcal{A}, one can find by physical construction (again "clocking") "the same tangent vector" at point \mathcal{C} with uniquely determined components $(dx^{\alpha}/d\lambda)_{\mathcal{C}}$ (vector "equal"; components ordinarily not equal because of twisting and turning of arbitrary base vectors between \mathcal{A} and \mathcal{C}).

Comparison of vectors at different events by parallel transport

Consider two neighboring events \mathcal{A} and \mathcal{B} connected by a curve $\mathcal{P}(\lambda)$. A vector $\boldsymbol{v}_{\mathcal{A}}$ lies in the tangent space at \mathcal{A}, and a vector $\boldsymbol{v}_{\mathcal{B}}$ lies in the tangent space at \mathcal{B}. How can one say whether $\boldsymbol{v}_{\mathcal{A}}$ and $\boldsymbol{v}_{\mathcal{B}}$ are parallel, and how can one compare their lengths? The equivalence principle gives an answer: an observer travels (using rocket power as necessary) through spacetime along the world line $\mathcal{P}(\lambda)$. He carries the vector $\boldsymbol{v}_{\mathcal{A}}$ with himself as he moves, and he uses flat-space Newtonian or Minkowskian standards to keep it always unchanging (flat-space physics is valid locally

according to the equivalence principle!). On reaching event \mathcal{B} the observer compares his "parallel-transported vector" $\boldsymbol{v}_{\mathcal{A}}$ with the vector $\boldsymbol{v}_{\mathcal{B}}$. If they are identical, then the original vector $\boldsymbol{v}_{\mathcal{A}}$ was (by definition) parallel to $\boldsymbol{v}_{\mathcal{B}}$, and they had the same length. (No metric means no way to quantify length; nevertheless, parallel transport gives a way to compare length!)

The equivalence principle entered this discussion in a perhaps unfamiliar way, applied to an observer who may be accelerated, rather than to one who is freely falling. But one cannot evade a basic principle by merely confronting it with an intricate application. (Ingenious perpetual-motion machines are as impossible as simpleminded ones!) The equivalence principle states that no local measurement that is insensitive to gravitational tidal forces can detect any difference whatsoever between flat and curved spacetime. The spaceship navigator has an inertial guidance system (accelerometers, gyroscopes, computers) capable of preserving an inertial reference frame in flat spacetime; and in flat spacetime it can compute the attitude and velocity of any object in the spaceship relative to a given inertial frame. The purchaser may specify whether he wants a guidance computer programmed with the laws of zero-gravity Newtonian mechanics, or with those of special-relativity physics. Use this same guidance system—including the same computer program—in curved spacetime. A vector is being parallel transported if the guidance system's computer says it is not changing.

Parallel transport defined using inertial guidance systems and equivalence principle

Will the result of transport in this way be independent of the curve used to link \mathcal{A} and \mathcal{B}? Clearly yes, in gravity-free spacetime, since this is a principal performance criterion that the purchaser of an inertial guidance system can demand of the manufacturer. But in a curved spacetime, the answer is "NO!" If $\boldsymbol{v}_{\mathcal{A}}$ agrees with $\boldsymbol{v}_{\mathcal{B}}$ after parallel transport along one curve, it need not agree with $\boldsymbol{v}_{\mathcal{B}}$ after parallel transport along another. Spacetime curvature produces discrepancies. But one is not ready to study and quantify those discrepancies (Chapter 11), until one has developed the mathematical formalism of parallel transport, which, in turn, cannot be done until one has made precise the "flat-space standards for keeping the vector $\boldsymbol{v}_{\mathcal{A}}$ always unchanging" as it is transported along a curve.

Result of parallel transport depends on route

The flat-space standards are made precise in Box 10.2. They lead to (1) a "Schild's ladder" construction for performing parallel transport; (2) the concept "covariant derivative," $\boldsymbol{\nabla}_{\boldsymbol{u}}\boldsymbol{v}$, of a vector field \boldsymbol{v} along a curve with tangent \boldsymbol{u}; (3) the "equation of motion" $\boldsymbol{\nabla}_{\boldsymbol{u}}\boldsymbol{u} = 0$ for a geodesic, which states that "a geodesic parallel transports its own tangent vector along itself;" and (4) a link between the tangent spaces at adjacent events (Figure 10.2).

Schild's ladder for performing parallel transport; its consequences

§10.3. PARALLEL TRANSPORT AND COVARIANT DERIVATIVE: ABSTRACT APPROACH

From the "Schild's ladder" construction of Box 10.2, one learns the following properties of spacetime's covariant derivative:

(continued on page 252)

**Box 10.2　FROM GEODESICS TO PARALLEL TRANSPORT TO COVARIANT
DIFFERENTIATION TO GEODESICS TO . . .**

"Parallel transport" as
defined by geodesics

A. Transport any sufficiently short stretch of a
curve $\mathcal{A}\mathcal{X}$ (i.e., any tangent vector) parallel to
itself along curve $\mathcal{A}\mathcal{B}$ to point \mathcal{B} as follows:

1. Take some point \mathcal{M} along $\mathcal{A}\mathcal{B}$ close to \mathcal{A}.
Take geodesic $\mathcal{X}\mathcal{M}$ through \mathcal{X} and \mathcal{M}. Take
any affine parametrization λ of $\mathcal{X}\mathcal{M}$ and
define a unique point \mathcal{N} by the condition
$\lambda_{\mathcal{N}} = \frac{1}{2}(\lambda_{\mathcal{X}} + \lambda_{\mathcal{M}})$ ("equal stretches of time
in $\mathcal{X}\mathcal{N}$ and $\mathcal{N}\mathcal{M}$").

2. Take geodesic that starts at \mathcal{A} and passes
through \mathcal{N}, and extend it by an equal pa-
rameter increment to point \mathcal{P}.

3. Curve $\mathcal{M}\mathcal{P}$ gives vector $\mathcal{A}\mathcal{X}$ as propagated
parallel to itself from \mathcal{A} to \mathcal{M} (for suffi-
ciently short $\mathcal{A}\mathcal{X}$ and $\mathcal{A}\mathcal{M}$). This construc-
tion certainly yields parallel transport in flat
spacetime (Newtonian or Einsteinian).
Moreover, it is local (vectors $\mathcal{A}\mathcal{X}$, $\mathcal{A}\mathcal{M}$, etc.,
very short). Therefore, it must work even in
curved spacetime. (It embodies the equival-
ence principle.)

4. Repeat process over and over, and eventually end up with $\mathcal{A}\mathcal{X}$ propagated parallel to itself from \mathcal{A} to \mathcal{B}. Call this construction "Schild's Ladder," from Schild's (1970) similar construction. [See also Ehlers, Pirani, and Schild (1972).] Note that curve $\mathcal{A}\mathcal{B}$ need not be a geodesic. There is no requirement that $\mathcal{M}\mathcal{Q}$ be the straight-on continuation of $\mathcal{A}\mathcal{M}$ similar to the geodesic requirement in the "cross-brace" that $\mathcal{N}\mathcal{P}$ be the straight-on continuation of $\mathcal{A}\mathcal{N}$.

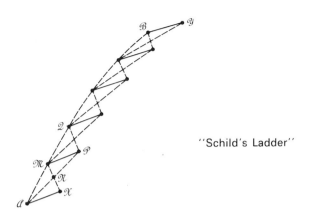

"Schild's Ladder"

5. Result of propagating $\mathcal{A}\mathcal{X}$ parallel to itself from \mathcal{A} to \mathcal{B} depends on choice of world line $\mathcal{A}\mathcal{B}$ ("evidence of curvature of spacetime").

B. Ask how rapidly a vector field \boldsymbol{v} is changing along a curve with tangent vector $\boldsymbol{u} = d/d\lambda$. The answer, $d\boldsymbol{v}/d\lambda \equiv \boldsymbol{\nabla}_{\boldsymbol{u}}\boldsymbol{v} \equiv$ "rate of change of \boldsymbol{v} with respect to λ" \equiv "covariant derivative of \boldsymbol{v} along \boldsymbol{u}," is constructed by the following obvious procedure: (1) Take \boldsymbol{v} at $\lambda = \lambda_0 + \varepsilon$. (2) Parallel transport it back to $\lambda = \lambda_0$. (3) Calculate how much it differs from \boldsymbol{v} there. (4) Divide by ε (and take limit as $\varepsilon \longrightarrow 0$):

"Covariant differentiation" as defined by parallel transport

$$\boldsymbol{\nabla}_{\boldsymbol{u}}\boldsymbol{v} = \lim_{\varepsilon \to 0}\left\{\frac{[\boldsymbol{v}(\lambda_0 + \varepsilon)]_{\text{parallel transported to }\lambda_0} - \boldsymbol{v}(\lambda_0)}{\varepsilon}\right\}.$$

If $\boldsymbol{u} = d/d\lambda$ is short compared to scale of inhomogeneities in the vector field \boldsymbol{v}, then $\boldsymbol{\nabla}_{\boldsymbol{u}}\boldsymbol{v}$ can be read directly off drawing I, or, equally well, off drawing II.

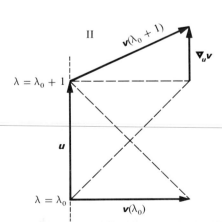

Box 10.2 (*continued*)

"Symmetry" of covariant
differentiation

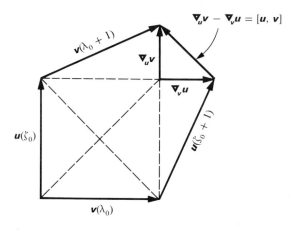

C. Take two vector fields. Combine into one the two diagrams for $\nabla_u v$ and $\nabla_v u$. Thereby discover that $\nabla_u v - \nabla_v u$ is the vector by which the *v-u-v-u* quadrilateral fails to close—i.e. (see Box 9.2), it is the commutator $[u, v]$: $\nabla_u v - \nabla_v u = [u, v]$.

Terminology: ∇ is said to be a *"symmetric"* or *"torsion-free"* covariant derivative when $\nabla_u v - \nabla_v u = [u, v]$. Other types of covariant derivatives, as studied by mathematicians, have no relevance for any gravitation theory based on the equivalence principle.

Chain rule for covariant
differentiation

D. The "take-the-difference" and "take-the-limit" process used to define $\nabla_u v$ guarantees that it obeys the usual rule for differentiating products:

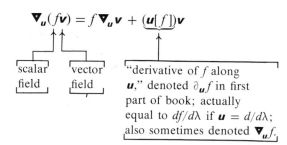

$$\nabla_u(fv) = f\nabla_u v + (u[f])v$$

scalar field vector field "derivative of f along u," denoted $\partial_u f$ in first part of book; actually equal to $df/d\lambda$ if $u = d/d\lambda$; also sometimes denoted $\nabla_u f$.

(for proof, see exercise 10.2.)

Additivity for covariant
differentiation

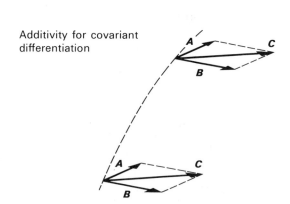

E. In the real physical world, be it Newtonian or relativistic, parallel transport of a triangle cannot break its legs apart: (1) A, B, C initially such that $A + B = C$; (2) A, B, C each parallel transported with himself by freely falling (inertial) observer; (3) then $A + B = C$ always. Any other result would violate the equivalence principle!

1. Consequence of this (as seen by following through definition of covariant derivative, and by noting that any vector u can be regarded as the tangent vector to a freely falling world line):

$$\nabla_u(v + w) = \nabla_u v + \nabla_u w$$

for any vector u and vector fields v and w.

2. Consequence of this, combined with symmetry of covariant derivative, and with additivity of the "closer of quadrilaterals" $[u, v]$:

$$\nabla_{u+n} v = \nabla_u v + \nabla_n v.$$

(See exercise 10.1.) This can be inferred, alternatively, from the equivalence principle: in a local inertial frame, as in special relativity or Newtonian theory, the change in v along $u + n$ should equal the sum of the changes along u and along n.

3. Consequence of above: choose n to be a multiple of u; thereby conclude

$$\nabla_{au} v = a \nabla_u v.$$

F. The "Schild's ladder" construction process for parallel transport (beginning of this box), applied to the tangent vector of a geodesic (exercise 10.6) guarantees: *a geodesic parallel transports its own tangent vector along itself.* Translated into covariant-derivative language:

$$\begin{pmatrix} u = d/d\lambda \text{ is a tangent} \\ \text{vector to a curve, and} \\ \nabla_u u = 0 \end{pmatrix} \Rightarrow \begin{pmatrix} \text{the curve is} \\ \text{a geodesic} \end{pmatrix}.$$

Geodesics as defined by parallel transport or covariant differentiation

Thus closes the circle: geodesic to parallel transport to covariant derivative to geodesic.

Covariant derivative: basic
properties

Symmetry: $\nabla_u v - \nabla_v u = [u, v]$ for any vector fields u and v; (10.2a)

Chain rule: $\nabla_u(fv) = f\nabla_u v + v\,\partial_u f$ for any function f, (10.2b)
vector field v, and vector u;

Additivity: $\nabla_u(v + w) = \nabla_u v + \nabla_u w$ for any vector (10.2c)
fields v and w, and vector u;

$\nabla_{au+bn} v = a\,\nabla_u v + b\,\nabla_n v$ for any vector (10.2d)
field v, vectors or vector fields u and n,
and numbers or functions a and b.

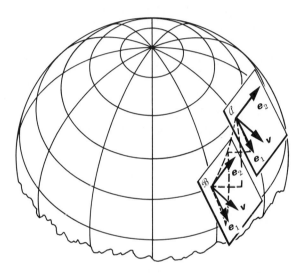

Figure 10.2.
The link between the tangent spaces at neighboring points, made possible by a parallel-transport law. Choose basis vectors e_1 and e_2 at the event \mathcal{Q}. Parallel transport them to a neighboring event \mathcal{B}. (Schild's ladder for transport of e_1 is shown in the figure.) Then any other vector v that is parallel transported from \mathcal{Q} to \mathcal{B} will have the same components at the two events (parallel transport cannot break the legs of a triangle; see Box 10.2):

$$v = v^1 e_1 + v^2 e_2 \text{ at } \mathcal{Q} \Rightarrow v = v^1 e_1 + v^2 e_2 \text{ at } \mathcal{B}.$$

[same numerically as at \mathcal{Q}]

[parallel transported from \mathcal{Q} to \mathcal{B}]

Thus, parallel transport provides a unique and complete link between the tangent space at \mathcal{Q} and the tangent space at \mathcal{B}. It identifies a unique vector at \mathcal{B} with each vector at \mathcal{Q} in a way that preserves all algebraic relations. Similarly (see §10.3), it identifies a unique 1-form at \mathcal{B} with each 1-form at \mathcal{Q}, and a unique tensor at \mathcal{B} with each tensor at \mathcal{Q}, preserving all algebraic relations such as $\langle \sigma, v \rangle = 19.9$ and $\boldsymbol{S}(\sigma, v, w) = 37\,1$.

Actually, all this is true only in the limit when \mathcal{Q} and \mathcal{B} are arbitrarily close to each other. When \mathcal{Q} and \mathcal{B} are close but not arbitrarily close, the result of parallel transport is slightly different for different paths; so the link between the tangent spaces is slightly nonunique. But the differences decrease by a factor of 4 each time the affine-parameter distance between \mathcal{Q} and \mathcal{B} is cut in half; see Chapter 11.

Any "rule" ▼, for producing new vector fields from old, that satisfies these four conditions, is called by differential geometers a *"symmetric covariant derivative."* Such a rule is not inherent in the more primitive concepts (Chapter 9) of curves, vectors, tensors, etc. In the arena of a spacetime laboratory, there are as many ways of defining a covariant derivative rule ▼ as there are of rearranging sources of the gravitational field. Different free-fall trajectories (geodesics) result from different distributions of masses.

Given the geodesics of spacetime, or of any other manifold, one can construct a unique corresponding covariant derivative by the Schild's ladder procedure of Box 10.2. Given any covariant derivative, one can discuss parallel transport via the equation

$$d\mathbf{v}/d\lambda \equiv \nabla_{\mathbf{u}}\mathbf{v} = 0 \iff \text{the vector field } \mathbf{v} \text{ is parallel transported} \qquad (10.3)$$
$$\text{along the vector } \mathbf{u} = d/d\lambda;$$

Equation for parallel transport

and one can test whether any curve is a geodesic via

$$\nabla_{\mathbf{u}}\mathbf{u} = 0 \iff \text{the curve } \mathscr{P}(\lambda) \text{ with tangent vector } \mathbf{u} = d/d\lambda$$
$$\text{parallel transports its own tangent vector } \mathbf{u} \qquad (10.4)$$
$$\iff \mathscr{P}(\lambda) \text{ is a geodesic.}$$

Thus a knowledge of all geodesics is completely equivalent to a knowledge of the covariant derivative.

Knowledge of all geodesics is equivalent to knowledge of covariant derivative

The covariant derivative ▼ generalizes to curved spacetime the flat-space gradient ▼. Like its flat-space cousin, it can be viewed as a machine for producing a number $\langle \boldsymbol{\sigma}, \nabla_{\mathbf{u}}\mathbf{v} \rangle$ out of a 1-form $\boldsymbol{\sigma}$, a vector \mathbf{u}, and a vector field \mathbf{v}. This machine viewpoint is explored in Box 10.3. Note there an important fact: despite its machine nature, ▼ is *not* a tensor; it is a nontensorial geometric object.

Covariant derivative generalizes flat-space gradient

In curved as in flat spacetime, ▼ can be applied not only to vector fields, but also to functions, 1-form fields, and tensor fields. Its action on functions is defined in the obvious manner:

Action of covariant derivative on functions, 1-forms, and tensors

$$\nabla f \equiv df; \qquad \nabla_{\mathbf{u}} f \equiv \partial_{\mathbf{u}} f \equiv \mathbf{u}[f] \equiv \langle df, \mathbf{u} \rangle. \qquad (10.5)$$

Its action on 1-form fields and tensor fields is defined by the curved-space generalization of equation (3.39): ▼\mathbf{S} is a linear machine for calculating the change in output of \mathbf{S}, from point to point, when "constant" (i.e., parallel transported) vectors are inserted into its slots. Example: the gradient of a $\binom{0}{1}$ tensor, i.e., of a 1-form field $\boldsymbol{\sigma}$. Pick an event \mathscr{P}_0; pick two vectors \mathbf{u} and \mathbf{v} in the tangent space at \mathscr{P}_0; construct from \mathbf{v} a "constant" vector field $\mathbf{v}(\mathscr{P})$ by parallel transport along the direction of \mathbf{u}, $\nabla_{\mathbf{u}}\mathbf{v} = 0$. Then ▼$\boldsymbol{\sigma}$ is a $\binom{0}{2}$ tensor, and $\nabla_{\mathbf{u}}\boldsymbol{\sigma}$ is a $\binom{0}{1}$ tensor defined at \mathscr{P}_0 by

$$\nabla\boldsymbol{\sigma}(\mathbf{v}, \mathbf{u}) \equiv \langle \nabla_{\mathbf{u}}\boldsymbol{\sigma}, \mathbf{v} \rangle \equiv \nabla_{\mathbf{u}}(\langle \boldsymbol{\sigma}, \mathbf{v} \rangle) \equiv \frac{d}{d\lambda}\langle \boldsymbol{\sigma}, \mathbf{v} \rangle, \qquad (10.6)$$

where $\mathbf{u} = d/d\lambda$. This defines ▼$\boldsymbol{\sigma}$ and $\nabla_{\mathbf{u}}\boldsymbol{\sigma}$, because it states their output for any

(continued on page 257)

Box 10.3 COVARIANT DERIVATIVE VIEWED AS A MACHINE; CONNECTION COEFFICIENTS AS ITS COMPONENTS

A. The Machine View

1. The covariant derivative operator $\boldsymbol{\nabla}$, like most other geometric objects, can be regarded as a machine with slots. There is one such machine at each event \mathscr{P}_0 in spacetime. In brief, the machine interpretation of $\boldsymbol{\nabla}$ at \mathscr{P}_0 says

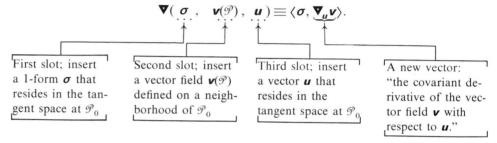

$$\boldsymbol{\nabla}(\ \underset{\cdots}{\boldsymbol{\sigma}}\ ,\quad \underset{\cdots}{\boldsymbol{v}(\mathscr{P})},\quad \underset{\cdots}{\boldsymbol{u}}\) \equiv \langle \boldsymbol{\sigma}, \underbrace{\boldsymbol{\nabla}_{\boldsymbol{u}}\boldsymbol{v}}\rangle.$$

First slot; insert a 1-form $\boldsymbol{\sigma}$ that resides in the tangent space at \mathscr{P}_0

Second slot; insert a vector field $\boldsymbol{v}(\mathscr{P})$ defined on a neighborhood of \mathscr{P}_0

Third slot; insert a vector \boldsymbol{u} that resides in the tangent space at \mathscr{P}_0

A new vector: "the covariant derivative of the vector field \boldsymbol{v} with respect to \boldsymbol{u}."

[Note: this slot notation for $\boldsymbol{\nabla}$ serves no useful purpose except to emphasize the "machine"-nature of $\boldsymbol{\nabla}$. This box is the only place it will be used.]

2. Geometrically, the output of the machine, $\langle \boldsymbol{\sigma}, \boldsymbol{\nabla}_{\boldsymbol{u}}\boldsymbol{v} \rangle$, is obtained as follows:

(a) Calculate the rate of change of \boldsymbol{v}, $\boldsymbol{\nabla}_{\boldsymbol{u}}\boldsymbol{v}$, along the vector \boldsymbol{u}; when \boldsymbol{u} and \boldsymbol{v} are infinitesimally small, the calculation can be represented pictorially:

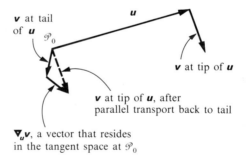

\boldsymbol{v} at tail of \boldsymbol{u}

\boldsymbol{u}

\boldsymbol{v} at tip of \boldsymbol{u}

\boldsymbol{v} at tip of \boldsymbol{u}, after parallel transport back to tail

$\boldsymbol{\nabla}_{\boldsymbol{u}}\boldsymbol{v}$, a vector that resides in the tangent space at \mathscr{P}_0

(b) Count how many surfaces of the 1-form $\boldsymbol{\sigma}$ are pierced by the vector $\boldsymbol{\nabla}_{\boldsymbol{u}}\boldsymbol{v}$ (piercing occurs in tangent space at \mathscr{P}_0)

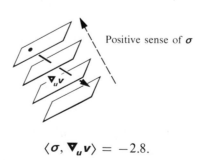

Positive sense of $\boldsymbol{\sigma}$

$\boldsymbol{\nabla}_{\boldsymbol{u}}\boldsymbol{v}$

$$\langle \boldsymbol{\sigma}, \boldsymbol{\nabla}_{\boldsymbol{u}}\boldsymbol{v} \rangle = -2.8.$$

This number is the output of the machine $\boldsymbol{\nabla}$, when $\boldsymbol{\sigma}$, $\boldsymbol{v}(\mathscr{P})$ and \boldsymbol{u} are inserted into its slots.

3. Another, equivalent, statement of covariant derivative as a machine. Leave first slot empty (no mention of any 1-form $\boldsymbol{\sigma}$); get a new vector field from original vector field \boldsymbol{v}:

$$\boldsymbol{\nabla}(\underbrace{\quad}_{\text{empty}}, \boldsymbol{v}(\mathscr{P}), \boldsymbol{u}) \equiv \boldsymbol{\nabla}_{\boldsymbol{u}}\boldsymbol{v}$$

empty

$$= \text{"covariant derivative of vector field } \boldsymbol{v} \text{ along vector } \boldsymbol{u}."$$

4. A third machine operation. Leave first and third slots empty (no mention of any 1-form $\boldsymbol{\sigma}$; no mention of any vector \boldsymbol{u} along which to differentiate); get a $\binom{1}{1}$ tensor field from original vector field \boldsymbol{v}:

$$\boldsymbol{\nabla}(\underbrace{\quad}_{\text{empty}}, \boldsymbol{v}(\mathscr{P}), \underbrace{\quad}_{\text{empty}}) \equiv \boldsymbol{\nabla}\boldsymbol{v}$$

empty empty

$$= \text{"covariant derivative" or "gradient" of vector field } \boldsymbol{v}.$$

This tensor field, $\boldsymbol{\nabla}\boldsymbol{v}$, is the curved-space generalization of the flat-space $\boldsymbol{\nabla}\boldsymbol{v}$ studied in §3.5. It has two slots (the two left empty in its definition). Its output for given input is

$$\boldsymbol{\nabla}\boldsymbol{v}(\underbrace{\quad}_{\text{empty}}, \boldsymbol{u}) = \boldsymbol{\nabla}_{\boldsymbol{u}}\boldsymbol{v}$$

empty

$$\boldsymbol{\nabla}\boldsymbol{v}(\boldsymbol{\sigma}, \boldsymbol{u}) = \langle \boldsymbol{\sigma}, \boldsymbol{\nabla}_{\boldsymbol{u}}\boldsymbol{v} \rangle.$$

5. Summary of the quantities defined above:
 (a) $\boldsymbol{\nabla}$ is a *covariant derivative operator;* to get a number from it, insert $\boldsymbol{\sigma}$, $\boldsymbol{v}(\mathscr{P})$, and \boldsymbol{u}; the result is $\langle \boldsymbol{\sigma}, \boldsymbol{\nabla}_{\boldsymbol{u}}\boldsymbol{v} \rangle$.
 (b) $\boldsymbol{\nabla}\boldsymbol{v}$ is the *gradient of* \boldsymbol{v}; to get a number from it, insert $\boldsymbol{\sigma}$ and \boldsymbol{u}; the result is $\langle \boldsymbol{\sigma}, \boldsymbol{\nabla}_{\boldsymbol{u}}\boldsymbol{v} \rangle$ [same as in (a)].
 (c) $\boldsymbol{\nabla}_{\boldsymbol{u}}\boldsymbol{v}$ is the *covariant derivative of* \boldsymbol{v} *along* \boldsymbol{u}; to get a number from it, insert $\boldsymbol{\sigma}$; the result is $\langle \boldsymbol{\sigma}, \boldsymbol{\nabla}_{\boldsymbol{u}}\boldsymbol{v} \rangle$ [same as in (a) and (b)].

B. How $\boldsymbol{\nabla}$ Differs from a Tensor

The machine $\boldsymbol{\nabla}$ differs from a tensor in two ways. (1) The middle slot of $\boldsymbol{\nabla}$ will not accept a vector; it demands a vector field—the vector field that is to be differentiated. (2) $\boldsymbol{\nabla}$ is not a linear machine (whereas a tensor must be linear!):

Box 10.3 (*continued*)

$$\boldsymbol{\nabla}(a\boldsymbol{\sigma}, f(\mathcal{P})\mathbf{v}(\mathcal{P}), b\mathbf{u}) \equiv \langle a\boldsymbol{\sigma}, \boldsymbol{\nabla}_{b\mathbf{u}} f\mathbf{v}\rangle$$
$$= abf\langle\boldsymbol{\sigma}, \boldsymbol{\nabla}_{\mathbf{u}}\mathbf{v}\rangle + \underbrace{ab\langle\boldsymbol{\sigma}, \mathbf{v}\rangle\,\boldsymbol{\nabla}_{\mathbf{u}} f}$$

$$\begin{bmatrix}\text{this would be absent if } \boldsymbol{\nabla} \text{ were a} \\ \text{linear machine.}\end{bmatrix}$$

C. The "Connection Coefficients" as Components of $\boldsymbol{\nabla}$

Given a tensor \boldsymbol{S} of rank $\binom{1}{2}$, a basis of tangent vectors $\{\boldsymbol{e}_\alpha\}$ at the event \mathcal{P}_0 where \boldsymbol{S} resides, and the dual basis of 1-forms $\{\boldsymbol{\omega}^\alpha\}$, one defines the components of \boldsymbol{S} by

$$S^\alpha{}_{\beta\gamma} \equiv \boldsymbol{S}(\boldsymbol{\omega}^\alpha, \boldsymbol{e}_\beta, \boldsymbol{e}_\gamma).$$

One defines the components of $\boldsymbol{\nabla}$ similarly, except that for $\boldsymbol{\nabla}$ one needs not only a basis $\{\boldsymbol{e}_\alpha\}$ at the event \mathcal{P}_0, but also a basis $\{\boldsymbol{e}_\alpha(\mathcal{P})\}$ at each event \mathcal{P} in its neighborhood:

$$\Gamma^\alpha{}_{\beta\gamma} \equiv \text{components of } \boldsymbol{\nabla} = \boldsymbol{\nabla}(\boldsymbol{\omega}^\alpha, \boldsymbol{e}_\beta(\mathcal{P}), \boldsymbol{e}_\gamma)$$
$$\equiv \langle\boldsymbol{\omega}^\alpha, \boldsymbol{\nabla}_{\boldsymbol{e}_\gamma}\boldsymbol{e}_\beta\rangle$$
$$\simeq \begin{pmatrix}\text{"α-component of change in basis vector } \boldsymbol{e}_\beta, \text{ when} \\ \text{in evaluating } \boldsymbol{e}_\beta \text{ one moves from tail to tip of } \boldsymbol{e}_\gamma\end{pmatrix}.$$

These components of $\boldsymbol{\nabla}$ are called the "*connection coefficients*" of the basis $\{\boldsymbol{e}_\alpha\}$. They are the "coordinate representation" of the covariant derivative operator $\boldsymbol{\nabla}$.

The covariant derivative operator $\boldsymbol{\nabla}$ and the connection coefficients $\Gamma^\alpha{}_{\mu\nu}$ provide different mathematical representations of the same geometric animal? Preposterous! The one animal runs from place to place and barks, or at least bites (takes difference, for example, between vector fields at one place and at a nearby place). The other animal, endowed with forty faces (see exercise 10.9) sits quietly at one spot. It would be difficult for two animals to look more different. Yet they do the same jobs in any world compatible with the equivalence principle: (1) they summarize the properties of all geodesics that go through the point in question; and, so doing, (2) they provide a physical means (parallel transport) to compare the values of vector fields and tensor fields at two neighboring events.

given input vectors \mathbf{v} and \mathbf{u}. If $\mathbf{v}(\mathcal{P})$ is not constrained to be "constant" along $\mathbf{u} = d/d\lambda$, then $(d/d\lambda) \langle \boldsymbol{\sigma}, \mathbf{v} \rangle$ has contributions from both the change in \mathbf{v} and the change in $\boldsymbol{\sigma}$:

$$\frac{d}{d\lambda} \langle \boldsymbol{\sigma}, \mathbf{v} \rangle \equiv \boldsymbol{\nabla}_{\mathbf{u}} \langle \boldsymbol{\sigma}, \mathbf{v} \rangle = \langle \boldsymbol{\nabla}_{\mathbf{u}} \boldsymbol{\sigma}, \mathbf{v} \rangle + \langle \boldsymbol{\sigma}, \boldsymbol{\nabla}_{\mathbf{u}} \mathbf{v} \rangle \tag{10.7}$$

(see exercise 10.3).

Similarly, if \mathbf{S} is a $\binom{1}{2}$ tensor field, then its gradient $\boldsymbol{\nabla}\mathbf{S}$ is a $\binom{1}{3}$ tensor field defined as follows. Pick an event \mathcal{P}_0; pick three vectors $\mathbf{u}, \mathbf{v}, \mathbf{w}$, and a 1-form $\boldsymbol{\sigma}$ in the tangent space at \mathcal{P}_0; turn \mathbf{v}, \mathbf{w}, and $\boldsymbol{\sigma}$ into "constant" vector fields and a "constant" 1-form field near \mathcal{P}_0 by means of parallel transport ($\boldsymbol{\nabla}_{\mathbf{u}} \mathbf{v} = \boldsymbol{\nabla}_{\mathbf{u}} \mathbf{w} = \boldsymbol{\nabla}_{\mathbf{u}} \boldsymbol{\sigma} = 0$ at \mathcal{P}_0); then define

$$\begin{aligned}\boldsymbol{\nabla}\mathbf{S}(\boldsymbol{\sigma}, \mathbf{v}, \mathbf{w}, \mathbf{u}) &\equiv (\boldsymbol{\nabla}_{\mathbf{u}} \mathbf{S})(\boldsymbol{\sigma}, \mathbf{v}, \mathbf{w}) \equiv \boldsymbol{\nabla}_{\mathbf{u}} [\mathbf{S}(\boldsymbol{\sigma}, \mathbf{v}, \mathbf{w})] \\ &= \partial_{\mathbf{u}} [\mathbf{S}(\boldsymbol{\sigma}, \mathbf{v}, \mathbf{w})].\end{aligned} \tag{10.8}$$

Exercise 10.1. ADDITIVITY OF COVARIANT DIFFERENTIATION

Show that the commutator ("closer of quadrilaterals") is additive:

$$[\mathbf{u}, \mathbf{v} + \mathbf{w}] = [\mathbf{u}, \mathbf{v}] + [\mathbf{u}, \mathbf{w}]; \qquad [\mathbf{u} + \mathbf{n}, \mathbf{v}] = [\mathbf{u}, \mathbf{v}] + [\mathbf{n}, \mathbf{v}].$$

Use this result, the additivity condition $\boldsymbol{\nabla}_{\mathbf{u}}(\mathbf{v} + \mathbf{w}) = \boldsymbol{\nabla}_{\mathbf{u}} \mathbf{v} + \boldsymbol{\nabla}_{\mathbf{u}} \mathbf{w}$, and symmetry of the covariant derivative, $\boldsymbol{\nabla}_{\mathbf{u}} \mathbf{v} - \boldsymbol{\nabla}_{\mathbf{v}} \mathbf{u} = [\mathbf{u}, \mathbf{v}]$, to prove that

$$\boldsymbol{\nabla}_{\mathbf{u} + \mathbf{n}} \mathbf{v} = \boldsymbol{\nabla}_{\mathbf{u}} \mathbf{v} + \boldsymbol{\nabla}_{\mathbf{n}} \mathbf{v}.$$

Exercise 10.2. CHAIN RULE FOR COVARIANT DIFFERENTIATION

Use pictures, and the "take-the-difference-and-take-the-limit" definition of $\boldsymbol{\nabla}_{\mathbf{u}} \mathbf{v}$ (Box 10.2) to show that

$$\boldsymbol{\nabla}_{\mathbf{u}}(f\mathbf{v}) = f\boldsymbol{\nabla}_{\mathbf{u}} \mathbf{v} + \mathbf{v} \, \partial_{\mathbf{u}}[f]. \tag{10.9}$$

Exercise 10.3. ANOTHER CHAIN RULE

Derive equation (10.7), using the "take-the-difference-and-take-the-limit" definitions of derivatives. *Hint:* Before taking the differences, parallel transport $\boldsymbol{\sigma}[\mathcal{P}(\lambda)]$ and $\mathbf{v}[\mathcal{P}(\lambda)]$ back from $\mathcal{P}(\lambda)$ to $\mathcal{P}(0)$.

Exercise 10.4. STILL ANOTHER CHAIN RULE

Show that, as in flat spacetime, so also in curved spacetime,

$$\boldsymbol{\nabla}_{\mathbf{u}}(\mathbf{v} \otimes \mathbf{w}) = (\boldsymbol{\nabla}_{\mathbf{u}} \mathbf{v}) \otimes \mathbf{w} + \mathbf{v} \otimes (\boldsymbol{\nabla}_{\mathbf{u}} \mathbf{w}). \tag{10.10}$$

Write down the more familiar component version of this equation in flat spacetime.

Solution to first part of exercise: Choose 1-forms $\boldsymbol{\sigma}$ and $\boldsymbol{\rho}$ at the event \mathcal{P}_0 in question, and extend them along the vector $\mathbf{u} = d/d\lambda$ by parallel transport, $\boldsymbol{\nabla}_{\mathbf{u}} \boldsymbol{\rho} = \boldsymbol{\nabla}_{\mathbf{u}} \boldsymbol{\sigma} = 0$. Then

$$[\mathbf{\nabla}_u(v \otimes w)](\rho, \sigma) = \frac{d}{d\lambda}[(v \otimes w)(\rho, \sigma)] \qquad \text{(def of } \mathbf{\nabla}_u \text{ on a tensor)}$$

$$= \frac{d}{d\lambda}[\langle \rho, v\rangle\langle \sigma, w\rangle] \qquad \text{(def of tensor product "}\otimes\text{")}$$

$$= \frac{d\langle \rho, v\rangle}{d\lambda}\langle \sigma, w\rangle + \langle \rho, v\rangle\frac{d\langle \sigma, w\rangle}{d\lambda} \qquad \text{(chain rule for derivatives)}$$

$$= \langle \rho, \mathbf{\nabla}_u v\rangle\langle \sigma, w\rangle + \langle \rho, v\rangle\langle \sigma, \mathbf{\nabla}_u w\rangle$$

$$\text{(by equation 10.7 with } \rho, \sigma \text{ const)}$$

$$= [(\mathbf{\nabla}_u v) \otimes w](\rho, \sigma) + [v \otimes (\mathbf{\nabla}_u w)](\rho, \sigma)$$

$$\text{(def of tensor product "}\otimes\text{")}.$$

Exercise 10.5. ONE MORE CHAIN RULE

Show, using techniques similar to those in exercise 10.4, that

$$\mathbf{\nabla}_u(\sigma \otimes \rho \otimes v) = (\mathbf{\nabla}_u \sigma) \otimes \rho \otimes v + \sigma \otimes (\mathbf{\nabla}_u \rho) \otimes v + \sigma \otimes \rho \otimes (\mathbf{\nabla}_u v). \quad (10.11)$$

Exercise 10.6. GEODESIC EQUATION

Use the "Schild's ladder" construction process for parallel transport (beginning of Box 10.2) to show that a geodesic parallel transports its own tangent vector along itself (end of Box 10.2).

§10.4. PARALLEL TRANSPORT AND COVARIANT DERIVATIVE: COMPONENT APPROACH

The pictorial approach motivates the mathematics; the abstract approach makes the pictorial ideas precise; but usually one must use the component approach in order to actually do complex calculations.

To work with components, one needs a set of basis vectors $\{e_\alpha\}$ and the dual set of basis 1-forms $\{\boldsymbol{\omega}^\alpha\}$. In flat spacetime a single such basis suffices; all events can use the same Lorentz basis. Not so in curved spacetime! There each event has its own tangent space, and each tangent space requires a basis of its own. As one travels from event to event, comparing their bases via parallel transport, one sees the bases twist and turn. They must do so. In no other way can they accommodate themselves to the curvature of spacetime. Bases at points \mathscr{P}_0 and \mathscr{P}_1, which are the same when compared by parallel transport along one curve, must differ when compared along another curve (see "Curvature"; Chapter 11).

To quantify the twisting and turning of a "field" of basis vectors $\{e_\alpha(\mathscr{P})\}$ and forms $\{\boldsymbol{\omega}^\alpha(\mathscr{P})\}$, use the covariant derivative. Examine the changes in vector fields along a basis vector e_β, abbreviating

$$\mathbf{\nabla}_{e_\beta} \equiv \mathbf{\nabla}_\beta \qquad \text{(def of } \mathbf{\nabla}_\beta\text{)}; \qquad (10.12)$$

and especially examine the rate of change of some basis vector: $\mathbf{\nabla}_\beta e_\alpha$. This rate of change is itself a vector, so it can be expanded in terms of the basis:

$$\nabla_{\beta} \boldsymbol{e}_{\alpha} = \boldsymbol{e}_{\mu} \Gamma^{\mu}{}_{\alpha\beta} \qquad \text{(def of } \Gamma^{\mu}{}_{\alpha\beta}\text{);} \qquad\qquad (10.13)$$

note reversal of order of α and β!

Connection coefficients
defined

and the resultant *"connection coefficients"* $\Gamma^{\mu}{}_{\alpha\beta}$ can be calculated by projection on the basis 1-forms:

$$\langle \boldsymbol{\omega}^{\mu}, \nabla_{\beta} \boldsymbol{e}_{\alpha} \rangle = \Gamma^{\mu}{}_{\alpha\beta}. \qquad\qquad (10.14)$$

(See exercise 10.7; also Box 10.3.) Because the basis 1-forms are "locked into" the basis vectors ($\langle \boldsymbol{\omega}^{\nu}, \boldsymbol{e}_{\alpha} \rangle = \delta^{\nu}{}_{\alpha}$), these same connection coefficients $\Gamma^{\nu}{}_{\alpha\beta}$ tell how the 1-form basis changes from point to point:

$$\nabla_{\beta} \boldsymbol{\omega}^{\nu} = -\Gamma^{\nu}{}_{\alpha\beta} \boldsymbol{\omega}^{\alpha}, \qquad\qquad (10.15)$$

$$\langle \nabla_{\beta} \boldsymbol{\omega}^{\nu}, \boldsymbol{e}_{\alpha} \rangle = -\Gamma^{\nu}{}_{\alpha\beta}. \qquad\qquad (10.16)$$

(See exercise 10.8.)

The connection coefficients do even more. They allow one to calculate the components of the gradient of an arbitrary tensor \boldsymbol{S}. In a Lorentz frame of flat spacetime, the components of $\nabla\boldsymbol{S}$ are obtained by letting the basis vectors $\boldsymbol{e}_{\alpha} = \partial \mathcal{P}/\partial x^{\alpha} = \partial/\partial x^{\alpha}$ act on the components of \boldsymbol{S}. Thus for a $\binom{1}{2}$ tensor field \boldsymbol{S} one finds that

$$\nabla\boldsymbol{S} \text{ has components } S^{\alpha}{}_{\beta\gamma,\delta} = \frac{\partial}{\partial x^{\delta}} [S^{\alpha}{}_{\beta\gamma}].$$

Not so in curved spacetime, or even in a non-Lorentz basis in flat spacetime. There the basis vectors turn, twist, expand, and contract, so even if \boldsymbol{S} were constant ($\nabla\boldsymbol{S} = 0$), its components on the twisting basis vectors would vary. The connection coefficients, properly applied, will compensate for this twisting and turning. As one learns in exercise 10.10, the components of $\nabla\boldsymbol{S}$, called $S^{\alpha}{}_{\beta\gamma;\delta}$ so that

Components of the gradient
of a tensor field

$$\nabla\boldsymbol{S} = S^{\alpha}{}_{\beta\gamma;\delta} \boldsymbol{e}_{\alpha} \otimes \boldsymbol{\omega}^{\beta} \otimes \boldsymbol{\omega}^{\gamma} \otimes \boldsymbol{\omega}^{\delta}, \qquad\qquad (10.17)$$

can be calculated from those of \boldsymbol{S} by the usual flat-space method, plus a correction applied to each index (i.e., to each basis vector):

["+" when correcting "up" indices] [interchange and sum on index being corrected]

[differentiating index]

$$S^{\alpha}{}_{\beta\gamma;\delta} = S^{\alpha}{}_{\beta\gamma,\delta} + S^{\mu}{}_{\beta\gamma}\Gamma^{\alpha}{}_{\mu\delta} - S^{\alpha}{}_{\mu\gamma}\Gamma^{\mu}{}_{\beta\delta} - S^{\alpha}{}_{\beta\mu}\Gamma^{\mu}{}_{\gamma\delta}. \qquad (10.18)$$

["−" when correcting "down" indices]

[interchange and sum on index being corrected]

[differentiating index]

Here

$$S^{\alpha}{}_{\beta\gamma,\delta} \equiv \boldsymbol{e}_{\delta}[S^{\alpha}{}_{\beta\gamma}] \equiv \partial_{\boldsymbol{e}_{\delta}} S^{\alpha}{}_{\beta\gamma}. \qquad\qquad (10.19)$$

Equation (10.18) looks complicated; but it is really very simple, once the pattern has been grasped.

Just as one uses special notation, $S^\alpha{}_{\beta\gamma;\delta}$, for the components of $\boldsymbol{\nabla}S$, so one introduces special notation, $DS^\alpha{}_{\beta\gamma}/d\lambda$, for components of the covariant derivative $\boldsymbol{\nabla}_u S$ along $\boldsymbol{u} = d/d\lambda$:

Components of the covariant derivative of a tensor field

$$\boldsymbol{\nabla}_u S = (DS^\alpha{}_{\beta\gamma}/d\lambda)\boldsymbol{e}_\alpha \otimes \boldsymbol{w}^\beta \otimes \boldsymbol{w}^\gamma; \tag{10.20}$$

$$\frac{DS^\alpha{}_{\beta\gamma}}{d\lambda} = S^\alpha{}_{\beta\gamma;\delta}u^\delta = (S^\alpha{}_{\beta\gamma,\delta} + \text{correction terms})u^\delta.$$

Since for any f

$$f_{,\delta}u^\delta = \partial_u f = df/d\lambda$$

this reduces to

$$\frac{DS^\alpha{}_{\beta\gamma}}{d\lambda} = \frac{dS^\alpha{}_{\beta\gamma}}{d\lambda} + S^\mu{}_{\beta\gamma}\Gamma^\alpha{}_{\mu\delta}u^\delta - S^\alpha{}_{\mu\gamma}\Gamma^\mu{}_{\beta\delta}u^\delta - S^\alpha{}_{\beta\mu}\Gamma^\mu{}_{\gamma\delta}u^\delta. \tag{10.21}$$

The power of the component approach shows up clearly when one discusses chain rules for covariant derivatives. The multitude of abstract-approach chain rules (equations 10.2b, 10.7, 10.10, 10.11) all boil down into a single rule for components: *The gradient operation ";" obeys the standard partial-differentiation chain rule of ordinary calculus.* Example:

Chain rule for gradient

$$(fv^\alpha)_{;\mu} = f_{;\mu}v^\alpha + fv^\alpha{}_{;\mu} \tag{10.22a}$$
$$[= f_{,\mu} \text{ because } f \text{ has no indices to correct}]$$

(contract this with u^μ to get chain rule 10.2b). Another example:

$$(\sigma_\alpha v^\alpha)_{;\mu} = \sigma_{\alpha;\mu}v^\alpha + \sigma_\alpha v^\alpha{}_{;\mu} \tag{10.22b}$$
$$[= (\sigma_\alpha v^\alpha)_{,\mu} \text{ because } \sigma_\alpha v^\alpha \text{ has no free indices to correct}]$$

(contract this with u^μ to get chain rule 10.7). Another example:

$$(\sigma_\alpha \rho_\beta v^\gamma)_{;\mu} = \sigma_{\alpha;\mu}\rho_\beta v^\gamma + \sigma_\alpha \rho_{\beta;\mu}v^\gamma + \sigma_\alpha \rho_\beta v^\gamma{}_{;\mu} \tag{10.22c}$$

(contract this with u^μ to get chain rule 10.11). Another example: see Exercise (10.12) below.

EXERCISES

Exercise 10.7. COMPUTATION OF CONNECTION COEFFICIENTS
Derive equation (10.14) for $\Gamma^\mu{}_{\alpha\beta}$ from equation (10.13).

Exercise 10.8. CONNECTION FOR 1-FORM BASIS
Derive equations (10.15) and (10.16), which relate $\boldsymbol{\nabla}_\beta \boldsymbol{w}^\nu$ to $\Gamma^\nu{}_{\alpha\beta}$, from equation (10.14). *Hint:* use equation (10.7).

Exercise 10.9. SYMMETRY OF CONNECTION COEFFICIENTS

Show that the symmetry of spacetime's covariant derivative (equation 10.2a) is equivalent to the following symmetry condition on the connection coefficients:

$$(\text{antisymmetric part of } \Gamma^\mu{}_{\alpha\beta}) \equiv \frac{1}{2}(\Gamma^\mu{}_{\alpha\beta} - \Gamma^\mu{}_{\beta\alpha})$$

$$\equiv \Gamma^\mu{}_{[\alpha\beta]} = -\frac{1}{2}\langle \boldsymbol{\omega}^\mu, \underbrace{[\boldsymbol{e}_\alpha, \boldsymbol{e}_\beta]}\rangle \equiv -\frac{1}{2}c_{\alpha\beta}{}^\mu. \qquad (10.23)$$

[commutator of basis vectors]

As a special case, $\Gamma^\mu{}_{\alpha\beta}$ *is symmetric in α and β when a coordinate basis* ($\boldsymbol{e}_\alpha = \partial/\partial x^\alpha$) *is used.* Show that in a coordinate basis this symmetry reduces the number of independent connection coefficients at each event from $4 \times 4 \times 4 = 64$ to $4 \times 10 = 40$.

Exercise 10.10. COMPONENTS OF GRADIENT

Derive equation (10.18) for the components of the gradient, $S^\alpha{}_{\beta\gamma;\delta}$. *Hint:* Expand \boldsymbol{S} in terms of the given basis, and then evaluate the righthand side of

$$\boldsymbol{\nabla}_{\boldsymbol{u}}\boldsymbol{S} = \boldsymbol{\nabla}_{\boldsymbol{u}}(S^\alpha{}_{\beta\gamma}\boldsymbol{e}_\alpha \otimes \boldsymbol{\omega}^\beta \otimes \boldsymbol{\omega}^\gamma),$$

for an arbitrary vector \boldsymbol{u}. Use the chain rules (10.2b) and (10.11). By comparing the result with

$$\boldsymbol{\nabla}_{\boldsymbol{u}}\boldsymbol{S} = S^\alpha{}_{\beta\gamma;\delta}u^\delta\boldsymbol{e}_\alpha \otimes \boldsymbol{\omega}^\beta \otimes \boldsymbol{\omega}^\gamma,$$

read off the components $S^\alpha{}_{\beta\gamma;\delta}$.

Exercise 10.11. DIVERGENCE

Let \boldsymbol{T} be a $\binom{2}{0}$ tensor field, and define the divergence on its second slot by the same process as in flat spacetime: $\boldsymbol{\nabla} \cdot \boldsymbol{T} = $ contraction of $\boldsymbol{\nabla}\boldsymbol{T}$; i.e.,

$$(\boldsymbol{\nabla} \cdot \boldsymbol{T})^\alpha = T^{\alpha\beta}{}_{;\beta}. \qquad (10.24)$$

Write the components $T^{\alpha\beta}{}_{;\beta}$ in terms of $T^{\alpha\beta}{}_{,\beta}$ plus correction terms for each of the two indices of \boldsymbol{T}.
[*Answer:*

$$T^{\alpha\beta}{}_{;\beta} = T^{\alpha\beta}{}_{,\beta} + \Gamma^\alpha{}_{\mu\beta}T^{\mu\beta} + \Gamma^\beta{}_{\mu\beta}T^{\alpha\mu}.]$$

Exercise 10.12. VERIFICATION OF CHAIN RULE

Let $S^{\alpha\beta}{}_\gamma$ be components of a $\binom{2}{1}$ tensor field, and $M_\beta{}^\gamma$ be components of a $\binom{1}{1}$ tensor field. By contracting these tensor fields, one obtains a vector field $S^{\alpha\beta}{}_\gamma M_\beta{}^\gamma$. The chain rule for the divergence of this vector field reads

$$(S^{\alpha\beta}{}_\gamma M_\beta{}^\gamma)_{;\alpha} = S^{\alpha\beta}{}_{\gamma;\alpha}M_\beta{}^\gamma + S^{\alpha\beta}{}_\gamma M_\beta{}^\gamma{}_{;\alpha}.$$

Verify the validity of this chain rule by expressing both sides of the equation in terms of directional derivatives ($,_{\boldsymbol{e}_\alpha}$) plus connection-coefficient corrections. *Hint:* the left side becomes

$$(S^{\alpha\beta}{}_\gamma M_\beta{}^\gamma)_{;\alpha} = \underbrace{(S^{\alpha\beta}{}_\gamma M_\beta{}^\gamma)_{,\alpha}} + \Gamma^\alpha{}_{\mu\alpha}(S^{\mu\beta}{}_\gamma M_\beta{}^\gamma).$$

$$\begin{bmatrix} S^{\alpha\beta}{}_{\gamma,\alpha}M_\beta{}^\gamma + S^{\alpha\beta}{}_\gamma M_\beta{}^\gamma{}_{,\alpha} \\ \text{by chain rule for directional derivative} \end{bmatrix}$$

The right side has many more correction terms (three on $S^{\alpha\beta}{}_{\gamma;\alpha}$; two on $M_\beta{}^\gamma{}_{;\alpha}$), but they must cancel against each other, leaving only one.

Exercise 10.13. TRANSFORMATION LAW FOR CONNECTION COEFFICIENTS

Let $\{e_\alpha\}$ and $\{e_{\mu'}\}$ be two different fields of basis vectors related by the transformation law

$$e_{\mu'}(\mathscr{P}) = L^\alpha{}_{\mu'}(\mathscr{P})e_\alpha(\mathscr{P}). \tag{10.25}$$

Show that the corresponding connection coefficients are related by

$$\Gamma^{\alpha'}{}_{\beta'\gamma'} = \underbrace{L^{\alpha'}{}_\rho L^\mu{}_{\beta'} L^\nu{}_{\gamma'} \Gamma^\rho{}_{\mu\nu}}_{\substack{\text{standard transformation law} \\ \text{for components of a tensor}}} + L^{\alpha'}{}_\mu L^\mu{}_{\beta',\gamma'} \tag{10.26}$$

Exercise 10.14. POLAR COORDINATES IN FLAT 2-DIMENSIONAL SPACE

On a sheet of paper draw an (r, ϕ) polar coordinate system. At neighboring points, draw the basis vectors $e_{\hat{r}} = \partial/\partial r$ and $e_{\hat{\phi}} \equiv r^{-1}\,\partial/\partial\phi$. (a) Use this picture, and Euclid's version of parallel transport, to justify the relations

$$\nabla_{\hat{r}}e_{\hat{r}} = 0, \qquad \nabla_{\hat{r}}e_{\hat{\phi}} = 0, \qquad \nabla_{\hat{\phi}}e_{\hat{r}} = r^{-1}e_{\hat{\phi}}, \qquad \nabla_{\hat{\phi}}e_{\hat{\phi}} = -r^{-1}e_{\hat{r}}.$$

(b) From these relations write down the connection coefficients. (c) Let $\boldsymbol{A} = A^{\hat{r}}e_{\hat{r}} + A^{\hat{\phi}}e_{\hat{\phi}}$ be a vector field. Show that its divergence, $\nabla \cdot \boldsymbol{A} = A^{\hat{\alpha}}{}_{;\hat{\alpha}} = A^{\hat{\alpha}}{}_{,\hat{\alpha}} + \Gamma^{\hat{\alpha}}{}_{\hat{\mu}\hat{\alpha}}A^{\hat{\mu}}$, can be calculated using the formula

$$\nabla \cdot \boldsymbol{A} = \frac{1}{r}\frac{\partial A^{\hat{\phi}}}{\partial\phi} + \frac{1}{r}\frac{\partial(rA^{\hat{r}})}{\partial r}$$

(which should be familiar to most readers).

§10.5. GEODESIC EQUATION

Geodesics—the parametrized paths of freely falling particles—were the starting point of this chapter. From them parallel transport was constructed (Schild's ladder; Box 10.2); and parallel transport in turn produced the covariant derivative and its connection coefficients. Given the covariant derivative, one recovered the geodesics: they were the curves whose tangent vectors, $\boldsymbol{u} = d\mathscr{P}/d\lambda$, satisfy $\nabla_{\boldsymbol{u}}\boldsymbol{u} = 0$ (\boldsymbol{u} is parallel transported along itself).

Geodesic equation: abstract version

Let a coordinate system $\{x^\alpha(\mathscr{P})\}$ be given. Let it induce basis vectors $e_\alpha = \partial/\partial x^\alpha$ into the tangent space at each event. Let the connection coefficients $\Gamma^\alpha{}_{\beta\gamma}$ for this "coordinate basis" be given. Then the component version of the "geodesic equation" $\nabla_{\boldsymbol{u}}\boldsymbol{u} = 0$ becomes a differential equation for the geodesic $x^\alpha(\lambda)$:

$$(1) \qquad \boldsymbol{u} = \frac{d}{d\lambda} = \frac{dx^\alpha}{d\lambda}\frac{\partial}{\partial x^\alpha} \qquad \Longrightarrow \qquad \text{components of } \boldsymbol{u} \text{ are } u^\alpha = \frac{dx^\alpha}{d\lambda};$$

(2) then components of $\boldsymbol{\nabla}_{\boldsymbol{u}}\boldsymbol{u} = 0$ are

$$0 = u^{\alpha}{}_{;\beta}u^{\beta} = (u^{\alpha}{}_{,\beta} + \Gamma^{\alpha}{}_{\gamma\beta}u^{\gamma})u^{\beta}$$
$$= \frac{\partial}{\partial x^{\beta}}\left(\frac{dx^{\alpha}}{d\lambda}\right)\frac{dx^{\beta}}{d\lambda} + \Gamma^{\alpha}{}_{\gamma\beta}\frac{dx^{\gamma}}{d\lambda}\frac{dx^{\beta}}{d\lambda},$$

which reduces to the differential equation

$$\frac{d^{2}x^{\alpha}}{d\lambda^{2}} + \Gamma^{\alpha}{}_{\gamma\beta}\frac{dx^{\gamma}}{d\lambda}\frac{dx^{\beta}}{d\lambda} = 0. \qquad (10.27)$$

Component version

This component version of the geodesic equation gives an analytic method ("translation" of Schild's ladder) for constructing the parallel transport law from a knowledge of the geodesics. Pick an event \mathscr{P}_0 and set up a coordinate system in its neighborhood. Watch many clock-carrying particles pass through (or arbitrarily close to) \mathscr{P}_0. For each particle read off the values of $d^2x^{\alpha}/d\lambda^2$ and $dx^{\alpha}/d\lambda$ at \mathscr{P}_0. Insert all the data for many particles into equation (10.27), and solve for the connection coefficients. Do not be disturbed that only the symmetric part of $\Gamma^{\alpha}{}_{\gamma\beta}$ is obtained thereby; the antisymmetric part, $\Gamma^{\alpha}{}_{[\gamma\beta]}$, vanishes identically in any coordinate frame! (See exercise 10.9.) Knowing $\Gamma^{\alpha}{}_{\gamma\beta}$, use them to parallel transport any desired vector along any desired curve through \mathscr{P}_0:

How to construct parallel transport law from knowledge of geodesics

$$\boldsymbol{\nabla}_{\boldsymbol{u}}\boldsymbol{v} = 0 \quad\Longleftrightarrow\quad \frac{dv^{\alpha}}{d\lambda} + \Gamma^{\alpha}{}_{\gamma\beta}v^{\gamma}\frac{dx^{\beta}}{d\lambda} = 0. \qquad (10.28)$$

Exercise 10.15. COMPONENTS OF PARALLEL-TRANSPORT LAW

EXERCISES

Show that equation (10.28) is the component version of the law for parallel transporting a vector \boldsymbol{v} along the curve $\mathscr{P}(\lambda)$ with tangent vector $\boldsymbol{u} = d\mathscr{P}/d\lambda$.

Exercise 10.16. GEODESICS IN POLAR COORDINATES

In rectangular coordinates on a flat sheet of paper, Euclid's straight lines (geodesics) satisfy $d^2x/d\lambda^2 = d^2y/d\lambda^2 = 0$. Transform this geodesic equation into polar coordinates ($x = r \cos \phi$, $y = r \sin \phi$); and read off the resulting connection coefficients by comparison with equation (10.27). These are the connection coefficients for the coordinate basis ($\partial/\partial r$, $\partial/\partial\phi$). From them calculate the connection coefficients for the basis

$$\boldsymbol{e}_{\hat{r}} = \frac{\partial}{\partial r}, \qquad \boldsymbol{e}_{\hat{\phi}} = \frac{1}{r}\frac{\partial}{\partial\phi}.$$

The answer should agree with the answer to part (b) of Exercise 10.14. *Hint:* Use such relations as

$$\boldsymbol{\nabla}_{\boldsymbol{e}_{\hat{\phi}}}\boldsymbol{e}_{\hat{r}} = \boldsymbol{\nabla}_{(1/r)(\partial/\partial\phi)}(\partial/\partial r) = \frac{1}{r}\boldsymbol{\nabla}_{(\partial/\partial\phi)}(\partial/\partial r).$$

Exercise 10.17. ROTATION GROUP: GEODESICS AND CONNECTION COEFFICIENTS

[Continuation of exercises 9.13 and 9.14.] In discussing the rotation group, one must make a clear distinction between the *Euclidean space* (coordinates x, y, z; basis vectors $\partial/\partial x$, $\partial/\partial y$, $\partial/\partial z$) in which the rotation matrices act, and the *group manifold SO(3)* (coordinates ψ, θ, ϕ; coordinate basis $\partial/\partial\psi$, $\partial/\partial\theta$, $\partial/\partial\phi$; basis of "generators" \boldsymbol{e}_1, \boldsymbol{e}_2, \boldsymbol{e}_3), whose points \mathscr{P} are rotation matrices.

(a) Pick a vector

$$\boldsymbol{n} = n^x\,\partial/\partial x + n^y\,\partial/\partial y + n^z\,\partial/\partial z$$

in Euclidean space. Show that

$$\mathscr{R}_{\boldsymbol{n}}(t) \equiv \exp[(n^x\mathscr{K}_1 + n^y\mathscr{K}_2 + n^z\mathscr{K}_3)t] \tag{10.29}$$

is a rotation matrix that rotates the axes of Euclidean space by an angle

$$t|\boldsymbol{n}| = t[(n^x)^2 + (n^y)^2 + (n^z)^2]^{1/2}$$

about the direction \boldsymbol{n}. (\mathscr{K}_j are matrices defined in exercise 9.13.)

(b) In the group manifold $SO(3)$, pick a point (rotation matrix) \mathscr{P}, and pick a tangent vector $\boldsymbol{u} = u^\alpha\boldsymbol{e}_\alpha$ at \mathscr{P}. Let \boldsymbol{u} be a vector in Euclidean space with the same components as \boldsymbol{u} has in $SO(3)$:

$$\boldsymbol{u} = u^1\boldsymbol{e}_1 + u^2\boldsymbol{e}_2 + u^3\boldsymbol{e}_3; \qquad \boldsymbol{u} = u^1\,\partial/\partial x + u^2\,\partial/\partial y + u^3\,\partial/\partial z. \tag{10.30}$$

Show that \boldsymbol{u} is the tangent vector (at $t = 0$) to the curve

$$\mathscr{C}(t) = \mathscr{R}_{\boldsymbol{u}}(t)\mathscr{P}. \tag{10.31}$$

The curve $\mathscr{C}(t)$ through the arbitrary point \mathscr{P} with arbitrary tangent vector $\boldsymbol{u} = (d\mathscr{C}/dt)_{t=0}$ is a very special curve: every point on it differs from \mathscr{P} by a rotation $\mathscr{R}_{\boldsymbol{u}}(t)$ about one and the same direction \boldsymbol{u}. No other curve in $SO(3)$ with "starting conditions" $\{\mathscr{P}, \boldsymbol{u}\}$ has such beautiful simplicity. Hence it is natural to decree that each such $\mathscr{C}(t)$ is a geodesic of the group manifold $SO(3)$. This decree adds new geometric structure to $SO(3)$; it converts $SO(3)$ from a differentiable manifold into something more special: an *affine manifold*.

One has no guarantee that an arbitrarily chosen family of curves in an arbitrary manifold *can* be decreed to be geodesics. Most families of curves simply do not possess the right geometric properties to function as geodesics. Most will lead to covariant derivatives that violate one or more of the fundamental conditions (10.2). To learn whether a given choice of geodesics is possible, one can try to derive connection coefficients $\Gamma^\alpha{}_{\beta\gamma}$ (for some given basis) corresponding to the chosen geodesics. If the derivation is successful, the choice of geodesics was a possible one. If the derivation produces inconsistencies, the chosen family of curves have the wrong geometric properties to function as geodesics.

(c) For the basis of generators $\{\boldsymbol{e}_\alpha\}$ derive connection coefficients corresponding to the chosen geodesics, $\mathscr{C}(t) = \mathscr{R}_{\boldsymbol{u}}(t)\mathscr{P}$, of $SO(3)$. *Hint:* show that the components $u^\alpha = \langle\boldsymbol{w}^\alpha, \boldsymbol{u}\rangle$ of the tangent $\boldsymbol{u} = d\mathscr{C}/dt$ to a given geodesic are independent of position $\mathscr{C}(t)$ along the geodesic. Then use the geodesic equation $\boldsymbol{\nabla}_{\boldsymbol{u}}\boldsymbol{u} = 0$, expanded in the basis $\{\boldsymbol{e}_\alpha\}$, to calculate the symmetric part of the connection $\Gamma^\alpha{}_{(\beta\gamma)}$. Finally use equation (10.23) to calculate $\Gamma^\alpha{}_{[\beta\gamma]}$. [*Answer:*

$$\Gamma^\alpha{}_{\beta\gamma} = \frac{1}{2}\,\epsilon_{\alpha\beta\gamma}, \tag{10.32}$$

where $\epsilon_{\alpha\beta\gamma}$ is the completely antisymmetric symbol with $\epsilon_{123} = +1$. This answer is independent of location \mathscr{P} in $SO(3)$!]

CHAPTER **11**

GEODESIC DEVIATION AND SPACETIME CURVATURE

§11.1. CURVATURE, AT LAST!

Spacetime curvature manifests itself as gravitation, by means of the deviation of one geodesic from a nearby geodesic (relative acceleration of test particles).

Let the geodesics of spacetime be known. Then the covariant derivative ∇ and its connection coefficients $\Gamma^\alpha{}_{\beta\gamma}$ are also known. How, from this information, does one define, calculate, and understand geodesic deviation and spacetime curvature? The answer unfolds in this chapter, and is summarized in Box 11.1. To disclose the answer one must (1) define the "relative acceleration vector" $\nabla_u \nabla_u n$, which measures the deviation of one geodesic from another (§11.2); (2) derive an expression in terms of ∇ or $\Gamma^\alpha{}_{\beta\gamma}$ for the "Riemann curvature tensor," which produces the geodesic deviation (§11.3); (3) see Riemann curvature at work, producing changes in vectors that are parallel transported around closed circuits (§11.4); (4) see Riemann curvature test whether spacetime is flat (§11.5); and (5) construct a special coordinate system, "Riemann normal coordinates," which is tied in a special way to the Riemann curvature tensor (§11.6).

> This chapter is entirely Track 2. Chapters 9 and 10 are necessary preparation for it. It will be needed as preparation for
> (1) Chapters 12 and 13 (Newtonian gravity; Riemannian geometry),
> (2) the second half of Chapter 14 (calculation of curvature), and
> (3) the details, but not the message, of Chapter 15 (Bianchi identities).

Overview of chapter

§11.2. THE RELATIVE ACCELERATION OF NEIGHBORING GEODESICS

Focus attention on a family of geodesics (Figure 11.1). Let one geodesic be distinguished from another by the value of a "selector parameter" n. The family includes not only geodesics $n = 0, 1, 2, \ldots$ but also geodesics for all intervening values of

Geometry of a family of geodesics:

Selector parameter

Box 11.1 GEODESIC DEVIATION AND RIEMANN CURVATURE IN BRIEF

"Geodesic separation" \boldsymbol{n} is displacement (tangent vector) from point on fiducial geodesic to point on nearby geodesic characterized by same value of affine parameter λ.

Geodesic separation changes with respect to λ (i.e., changes along the tangent vector $\boldsymbol{u} = d/d\lambda$) at a rate given by the *equation of geodesic deviation*

$$\boldsymbol{\nabla_u}\,\boldsymbol{\nabla_u}\boldsymbol{n} + \boldsymbol{Riemann}\,(\ldots,\boldsymbol{u},\boldsymbol{n},\boldsymbol{u}) = 0 \quad (1)$$

(second-order equation; see §§1.6 and 1.7; Figures 1.10, 1.11, 1.12).

In terms of components of the Riemann tensor the driving force ("tidal graviational force") is

$$\boldsymbol{Riemann}\,(\ldots,\boldsymbol{u},\boldsymbol{n},\boldsymbol{u}) = \boldsymbol{e}_\alpha R^\alpha{}_{\beta\gamma\delta}u^\beta n^\gamma u^\delta. \quad (2)$$

The components of the Riemann curvature tensor in a coordinate frame are given in terms of the connection coefficients by the formula

$$R^\alpha{}_{\beta\gamma\delta} = \frac{\partial \Gamma^\alpha{}_{\beta\delta}}{\partial x^\gamma} - \frac{\partial \Gamma^\alpha{}_{\beta\gamma}}{\partial x^\delta}$$
$$+ \Gamma^\alpha{}_{\mu\gamma}\Gamma^\mu{}_{\beta\delta} - \Gamma^\alpha{}_{\mu\delta}\Gamma^\mu{}_{\beta\gamma}. \quad (3)$$

This curvature tensor not only quantifies the concept of "tidal gravitational force," but also enters into Einstein's law, by which "matter tells spacetime how to curve." That law, to be studied in later chapters, takes the following operational-computational form in a given coordinate system:

(a) Write down trial formula for dynamic evolution of metric coefficients $g_{\mu\nu}$ with time.

(b) Calculate the connection coefficients from

$$\Gamma^\alpha{}_{\mu\nu} = g^{\alpha\beta}\Gamma_{\beta\mu\nu}; \quad (4)$$
$$\Gamma_{\beta\mu\nu} = \frac{1}{2}\left(\frac{\partial g_{\beta\nu}}{\partial x^\mu} + \frac{\partial g_{\beta\mu}}{\partial x^\nu} - \frac{\partial g_{\mu\nu}}{\partial x^\beta}\right) \quad (5)$$

(derived in Chapter 13).

(c) Calculate Riemann curvature tensor from equation (3).

(d) Calculate Einstein curvature tensor from

$$G_{\mu\nu} = R^\alpha{}_{\mu\alpha\nu} - \frac{1}{2}g_{\mu\nu}g^{\sigma\tau}R^\alpha{}_{\sigma\alpha\tau} \quad (6)$$

(geometric significance in Chapter 15).

(e) Insert into Einstein's equations (Chapter 17):

$$G_{\mu\nu} = 0 \qquad \text{(empty space),}$$
$$G_{\mu\nu} = 8\pi T_{\mu\nu} \qquad \text{(when mass-energy is present).}$$

(f) Test whether the trial formula for the dynamic evolution of the geometry was correct, and, if not, change it so it is.

Affine parameter

n. The typical point \mathscr{P} on the typical geodesic will be a continuous, doubly differentiable function of the selector parameter n and the affine parameter λ; thus

$$\mathscr{P} = \mathscr{P}(\lambda, n). \quad (11.1)$$

Tangent vector

The tangent vector

$$\boldsymbol{u} = \frac{\partial \mathscr{P}}{\partial \lambda} \qquad \text{(Cartan notation)}$$

or

$$\boldsymbol{u} = \frac{\partial}{\partial \lambda} \qquad \text{(notation of this book)} \quad (11.2)$$

is constant along any given geodesic in this sense: the vector \boldsymbol{u} at any point, trans-

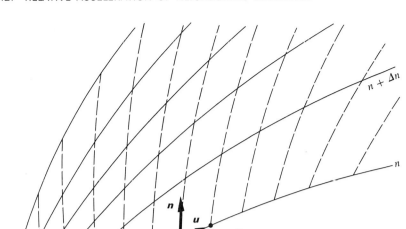

Figure 11.1.
One-parameter family of geodesics. The "selector parameter" n tells which geodesic. The affine parameter λ tells where on a given geodesic. The two tangent vectors indicated in the diagram are $\boldsymbol{u} = \partial/\partial\lambda$ (Cartan: $\partial\mathcal{P}/\partial\lambda$) and $\boldsymbol{n} = \partial/\partial n$ (Cartan: $\partial\mathcal{P}/\partial n$).

ported parallel to itself along the geodesic, arrives at a second point coincident in direction and length with the \boldsymbol{u} already existing at that point.

The *"separation vector"*

Separation vector

$$\boldsymbol{n} = \frac{\partial\mathcal{P}}{\partial n} \qquad \text{(Cartan notation)}$$

or

$$\boldsymbol{n} = \frac{\partial}{\partial n} \qquad \text{(notation of this book)} \tag{11.3}$$

measures the separation between the geodesic n, regarded as the fiducial geodesic, and the typical nearby geodesic, $n + \Delta n$ (for small Δn), in the sense that

$$(\Delta n)\boldsymbol{n} = \begin{Bmatrix} \Delta n\, \dfrac{\partial\mathcal{P}}{\partial n} \\[2mm] \Delta n\, \dfrac{\partial}{\partial n} \end{Bmatrix} \begin{Bmatrix} \text{measures the} \\ \text{change in} \end{Bmatrix} \begin{Bmatrix} \text{position} \\ \text{any} \\ \text{function} \end{Bmatrix} \tag{11.4}$$

brought about by transfer of attention from the one geodesic to the other at a fixed value of the affine parameter λ. This vector is represented by the arrow $\mathcal{M}2$ in the first diagram in Box 11.2.

(*continued on page 270*)

Box 11.2 GEODESIC DEVIATION REPRESENTED AS AN ARROW

"Fiducial geodesic" n. Separation vector $\boldsymbol{n}\,\Delta n = \mathcal{M}\mathcal{Q}$ leads from point \mathcal{M} on it, to point \mathcal{Q} with same value of affine parameter λ (timelike quantity) on neighboring "test geodesic" $n + \Delta n$.

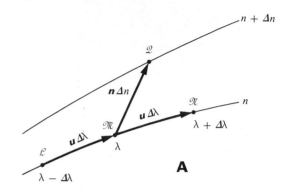

Parallel transport of $\mathcal{M}\mathcal{Q}$ by "Schild's ladder construction" (Box 10.2) to $\mathcal{N}\mathcal{B}$ and $\mathcal{L}\mathcal{A}$. If the test geodesic $n + \Delta n$ had kept a constant separation from the fiducial geodesic n, its tracer point would have arrived at \mathcal{A} at the value $(\lambda - \Delta\lambda)$ of the affine parameter, and at \mathcal{B} at $(\lambda + \Delta\lambda)$.

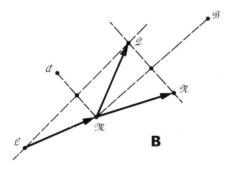

Actual location of tracer point of test geodesic at values of the timelike affine parameter $(\lambda - \Delta\lambda)$, λ, and $(\lambda + \Delta\lambda)$.

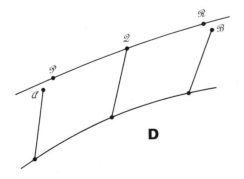

Confrontation between actual course of tracer point on test geodesic and "canonical course": course it would have had to take to keep constant separation from the tracer point moving along the fiducial geodesic.

Test geodesic same as before, except for uniform stretchout in scale of affine parameter. Any measure of departure of the actual course of geodesic from the canonical course ($\mathcal{A}\mathcal{Q}\mathcal{B}$), to be useful, should be independent of this stretchout. Hence, take as measure of geodesic deviation, not the vector $\mathcal{B}\mathcal{R}$ alone, nor the vector $\mathcal{A}\mathcal{P}$, but the stretch-independent combination $\boldsymbol{\delta}_2 = (\mathcal{B}\mathcal{R}) + (\mathcal{A}\mathcal{P})$. Here the sign of addition implies that the two vectors have been transported parallel to themselves, before addition, to a common point (\mathcal{Q} in the diagram; \mathcal{M} in the differential calculus limit $\Delta n \longrightarrow 0$, $\Delta\lambda \longrightarrow 0$).

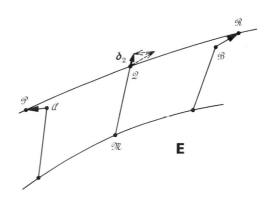

Alternative courses that the test geodesic of **D** could have taken through \mathcal{Q} (families of geodesics characterized by different degrees of divergence from the left or convergence towards the right). Tilt changes values of $\mathcal{A}\mathcal{P}$ (to $\mathcal{A}\overline{\mathcal{P}}$) and $\mathcal{B}\mathcal{R}$ (to $\mathcal{B}\overline{\mathcal{R}}$) individually, but not value of the sum $\boldsymbol{\delta}_2 = (\mathcal{B}\mathcal{R}) + (\mathcal{A}\mathcal{P})$ ("lever principle").

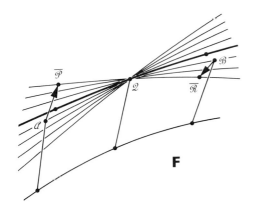

Note that arrow $\mathcal{B}\mathcal{R}$ is of first order in $\Delta\lambda$ and of first order in Δn; similarly for $\mathcal{A}\mathcal{P}$; hence the combination $\boldsymbol{\delta}_2$ is of second order in $\Delta\lambda$ and first order in Δn. Conclude that *the arrow $\boldsymbol{\delta}_2/(\Delta\lambda)^2(\Delta n)$ is the desired measure of geodesic deviation* in the sense that:

size of mesh (ultimately to go to zero) cancels out;
parameterization of test geodesic cancels out;
slope of test geodesic cancels out.

Give this arrow the name *"relative-acceleration vector"*; and by examining it more closely (Box 11.3), discover the formula

$$\boldsymbol{\delta}_2/(\Delta\lambda)^2(\Delta n) = \boldsymbol{\nabla}_u \boldsymbol{\nabla}_u \boldsymbol{n}$$

for it.

Box 11.2 illustrates what it means to speak of geodesic deviation. One transports the separation $\boldsymbol{n}\,\Delta n = \mathcal{MZ}$ parallel to itself along the fiducial geodesic. The tip of this vector traces out the canonical course that the nearby tracer point would have to pursue if it were to maintain constant separation from the fiducial tracer point. The actual course of the test geodesic deviates from this "canonical" course. The deviation, a vector (\mathcal{AP} of Box 11.2), changes with the affine parameter (\mathcal{AP} at \mathcal{A}, 0 at \mathcal{Z}, \mathcal{BR} at \mathcal{B}). The first derivative of this vector with respect to the affine parameter is sensitive to the scale of parameterization along the test geodesic, and to its slope (Box 11.2, **F**). Not so the second derivative. It depends only on the tangent vector \boldsymbol{u} of the fiducial geodesic, and on the separation vector $\boldsymbol{n}\,\Delta n$. Divide this second derivative of the deviation by Δn and give it a name: the *"relative-acceleration vector"*. Discover (Box 11.3) a simple formula for it

Relative-acceleration vector

$$\text{(relative-acceleration vector)} = \boldsymbol{\nabla_u}\,\boldsymbol{\nabla_u}\,\boldsymbol{n}. \tag{11.5}$$

§11.3. TIDAL GRAVITATIONAL FORCES AND RIEMANN CURVATURE TENSOR

With "relative acceleration" now defined, turn to the "tidal gravitational force" (i.e., "spacetime curvature") that produces it. Use a Newtonian analysis of tidal forces

Box 11.3 GEODESIC DEVIATION: ARROW CORRELATED WITH SECOND COVARIANT DERIVATIVE

The arrow $\boldsymbol{\delta}_2$ in Box 11.2 measures, not the rate of change of the separation of the test geodesic $n + \Delta n$ from the "canonical course" \mathcal{AZB} as baseline, but the second derivative:

$$\left(\text{first derivative at } \lambda + \tfrac{1}{2}\,\Delta\lambda\right) = \boldsymbol{\nabla_u}\boldsymbol{n} = \frac{\mathcal{NR} - \mathcal{NB}}{\Delta\lambda\,\Delta n} = \frac{\mathcal{BR}}{\Delta\lambda\,\Delta n};$$

$$\left(\text{first derivative at } \lambda - \tfrac{1}{2}\,\Delta\lambda\right) = \boldsymbol{\nabla_u}\boldsymbol{n} = \frac{\mathcal{LA} - \mathcal{LP}}{\Delta\lambda\,\Delta n} = \frac{-\mathcal{AP}}{\Delta\lambda\,\Delta n}.$$

Transpose to common location λ, take difference, and divide it by $\Delta\lambda$ to obtain the second covariant derivative with respect to the vector \boldsymbol{u}; thus

$$\boldsymbol{\nabla_u}\,\boldsymbol{\nabla_u}\boldsymbol{n} = \frac{(\boldsymbol{\nabla_u}\boldsymbol{n})_{\lambda + \frac{1}{2}\Delta\lambda} - (\boldsymbol{\nabla_u}\boldsymbol{n})_{\lambda - \frac{1}{2}\Delta\lambda}}{\Delta\lambda}$$

$$= \frac{(\mathcal{BR} + \mathcal{AP})_{\substack{\text{vectors transported to} \\ \text{common location}}}}{(\Delta\lambda)^2\,\Delta n} = \frac{\boldsymbol{\delta}_2}{(\Delta\lambda)^2\,\Delta n}$$

$$= \text{"relative acceleration vector" for neighboring geodesics.}$$

(left half of Box 11.4) to motivate the geometric analysis (right half of same box). Thereby arrive at the remarkable equation

$$\underset{\substack{\uparrow \\ \text{"relative} \\ \text{acceleration"}}}{\nabla_u \nabla_u n} + \underset{\substack{\uparrow \\ \text{"tide-producing} \\ \text{gravitational forces"}}}{[\nabla_n, \nabla_u]u} = 0. \qquad (11.6)$$

Tide-producing gravitational forces expressed in terms of a commutator

This equation is remarkable, because at first sight it seems crazy. The term $[\nabla_n, \nabla_u]u$ involves second derivatives of u, and a first derivative of ∇_n:

$$[\nabla_n, \nabla_u]u \equiv \nabla_n \nabla_u u - \nabla_u \nabla_n u. \qquad (11.7)$$

It thus must depend on how u and n vary from point to point. But the relative acceleration it produces, $\nabla_u \nabla_u n$, is known to depend only on the values of u and n at the fiducial point, not on how u and n vary (see Box 11.2, **F**). How is this possible?

Somehow all derivatives must drop out of the tidal-force quantity $[\nabla_n, \nabla_u]u$. One must be able to regard $[\nabla_{...}, \nabla_{...}]...$ as a purely local, algebraic machine with three slots, whose output is a vector. If it is purely local and not differential, then it is even linear (as one sees from the additivity properties of ∇), so it must be a tensor. Give this tensor the name **Riemann**, and give it a fourth slot for inputting a 1-form:

$$\textbf{Riemann}\,(\ldots, \textbf{C}, \textbf{A}, \textbf{B}) \equiv [\nabla_A, \nabla_B]\textbf{C};$$

$$\textbf{Riemann}\,(\sigma, \textbf{C}, \textbf{A}, \textbf{B}) \equiv \langle \sigma, [\nabla_A, \nabla_B]\textbf{C}\rangle.$$

This is only a tentative definition of **Riemann**. Before accepting it, one should verify that it is, indeed, a tensor. Does it *really* depend on only the values of \textbf{A}, \textbf{B}, \textbf{C} at the point of evaluation, and not on how they are changing there? The answer (derived in Box 11.5) is "almost." It fails the test, but with a slight modification it will pass. The modification is to replace the commutator $[\nabla_A, \nabla_B]$ by the "*curvature operator*"

Curvature operator defined

$$\mathscr{R}(\textbf{A}, \textbf{B}) \equiv [\nabla_A, \nabla_B] - \nabla_{[A,B]}, \qquad (11.8)$$

where $\nabla_{[A,B]}$ is the derivative along the vector $[\textbf{A}, \textbf{B}]$ (commutator of \textbf{A} and \textbf{B}). ($\mathscr{R}(\textbf{A}, \textbf{B}) \equiv [\nabla_A, \nabla_B]$ for the fields $\textbf{A} = \textbf{n}$ and $\textbf{B} = \textbf{u}$ of the geodesic-deviation problem, because $[\textbf{n}, \textbf{u}] = 0$.) Then the modified and acceptable *definition of the Riemann curvature tensor is*

Riemann curvature tensor defined

$$\textbf{Riemann}\,(\ldots, \textbf{C}, \textbf{A}, \textbf{B}) \equiv \mathscr{R}(\textbf{A}, \textbf{B})\textbf{C};$$
$$\textbf{Riemann}\,(\sigma, \textbf{C}, \textbf{A}, \textbf{B}) \equiv \langle \sigma, \mathscr{R}(\textbf{A}, \textbf{B})\textbf{C}\rangle. \qquad (11.9)$$

To define **Riemann** thus, and to verify its tensorial character (exercise 11.2), does not by any means teach one what curvature is all about. To understand curvature, one must scrutinize **Riemann** from all viewpoints. That is the task of the rest of this chapter.

(*continued on page 275*)

Box 11.4 RELATIVE ACCELERATION OF TEST PARTICLES—
GEOMETRIC ANALYSIS PATTERNED ON NEWTONIAN ANALYSIS

Newtonian Analysis

1. Consider a family of test-particle trajectories $x^j(t, n)$ in ordinary, three-dimensional space: "t" is time measured by particle's clock, or any clock; "n" is "selector parameter."

2. Equation of motion for each trajectory:

$$\left(\frac{\partial^2 x^j}{\partial t^2}\right)_n + \frac{\partial \Phi}{\partial x^j} = 0,$$

where Φ is Newtonian potential.

3. Take difference between equations of motion for neighboring trajectories, n and $n + \Delta n$, and take limit as $\Delta n \longrightarrow 0$—i.e., take derivative

$$\left(\frac{\partial}{\partial n}\right)_t \left[\left(\frac{\partial^2 x^j}{\partial t^2}\right)_n + \frac{\partial \Phi}{\partial x^j}\right] = 0.$$

4. When $\partial/\partial n$ acts on second term, rewrite it as

$$\left(\frac{\partial}{\partial n}\right)_t = \left(\frac{\partial x^k}{\partial n}\right)_t \frac{\partial}{\partial x^k} = n^k \frac{\partial}{\partial x^k};$$

Thereby obtain

$$\left(\frac{\partial}{\partial n}\right)_t \left(\frac{\partial}{\partial t}\right)_n \left(\frac{\partial x^j}{\partial t}\right)_n + \frac{\partial^2 \Phi}{\partial x^j \partial x^k} n^k = 0.$$

Geometric Analysis

1. Consider a family of test-particle trajectories (geodesics), $\mathcal{P}(\lambda, n)$, in spacetime: "λ" is affine-parameter, i.e., time measured by particle's clock; "n" is "selector parameter."

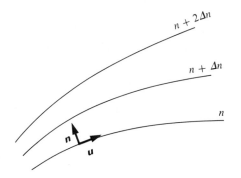

2. Geodesic equation for each trajectory:

$$\boldsymbol{\nabla}_u \boldsymbol{u} = 0.$$

[Looks like first-order equation; is actually second-order because the "\boldsymbol{u}" being differentiated is itself a derivative, $\boldsymbol{u} = (\partial \mathcal{P}/\partial \lambda)_n$.]

3. Take difference between geodesic equations for neighboring geodesics n and $n + \Delta n$, and take limit as $\Delta n \longrightarrow 0$—i.e., take covariant derivative

$$\boldsymbol{\nabla}_n [\boldsymbol{\nabla}_u \boldsymbol{u}] = 0.$$

4. There is no second term, so leave equation in form

$$\boldsymbol{\nabla}_n [\boldsymbol{\nabla}_u \boldsymbol{u}] = 0.$$

5. To obtain equation for relative acceleration, move $(\partial/\partial n)_t$ through both of the $(\partial/\partial t)_n$ terms (permissible because partial derivatives commute!):

$$\left(\frac{\partial}{\partial t}\right)_n \left(\frac{\partial}{\partial t}\right)_n \left(\frac{\partial x^j}{\partial n}\right)_t + \frac{\partial^2 \Phi}{\partial x^j \, \partial x^k}\, n^k = 0.$$

This is equivalent to

$$\left(\frac{\partial^2 n^j}{\partial t^2}\right) + \frac{\partial^2 \Phi}{\partial x^j \, \partial x^k}\, n^k = 0.$$

"relative acceleration" ⟶ ⟵ "tide-producing gravitational forces"

5. To obtain equation for relative acceleration, $\nabla_u \nabla_u n$, move ∇_n through ∇_u and through the $\partial/\partial\lambda$ of $u = \partial\mathcal{P}/\partial\lambda$:

a. *First step:* In $\nabla_n \nabla_u u = 0$, move ∇_n through ∇_u. The result:

$$(\nabla_u \nabla_n + [\nabla_n, \nabla_u])u = 0.$$

commutator; must be included as protection against possibility that $\nabla_u \nabla_n \neq \nabla_n \nabla_u$.

b. *Second step:* Move ∇_n through $\partial/\partial\lambda$ of $u = \partial\mathcal{P}/\partial\lambda$; i.e., write

$$\nabla_n \frac{\partial\mathcal{P}}{\partial\lambda} = \nabla_n u = \nabla_u n = \nabla_u \frac{\partial\mathcal{P}}{\partial n}$$

[def. of u] [def. of n]

Why? Because symmetry of covariant derivative says $\nabla_n u - \nabla_u n = [n, u]$

$$= \left[\frac{\partial}{\partial n}, \frac{\partial}{\partial\lambda}\right] = \frac{\partial^2}{\partial n \, \partial\lambda} - \frac{\partial^2}{\partial\lambda \, \partial n} = 0.$$

c. Result:

$$\nabla_u \nabla_u n + [\nabla_n, \nabla_u]u = 0$$

"relative acceleration" "tide-producing gravitational forces"; i.e., "spacetime curvature"

Box 11.5 RIEMANN CURVATURE TENSOR

A. Definition of *Riemann* Motivated by Tidal Gravitational Forces:

1. Tidal forces (spacetime curvature) produce relative acceleration of test particles (geodesics) given by

$$\nabla_u \nabla_u n + [\nabla_n, \nabla_u]u = 0. \qquad (1)$$

Box 11.5 (*continued*)

2. This motivates the definition

$$\textbf{\textit{Riemann}} \, (.\,.\,.\,, \textbf{\textit{C}}, \textbf{\textit{A}}, \textbf{\textit{B}}) = [\boldsymbol{\nabla}_{\textbf{\textit{A}}}, \boldsymbol{\nabla}_{\textbf{\textit{B}}}]\textbf{\textit{C}}. \qquad (2)$$

[empty slot for inserting a one-form]

B. Failure of this Definition

1. Definition acceptable only if **_Riemann_** $(\ldots, \textbf{\textit{C}}, \textbf{\textit{A}}, \textbf{\textit{B}})$ is a linear machine, independent of how **_A_**, **_B_**, **_C_** vary from point to point.
2. Check, in part: change variations of **_C_**, but not **_C_** itself, at event \mathscr{P}_0:

$$\textbf{\textit{C}}_{\text{NEW}}(\mathscr{P}) = f(\mathscr{P})\textbf{\textit{C}}_{\text{OLD}}(\mathscr{P}).$$

[arbitrary function except $f(\mathscr{P}_0) = 1$]

3. Does this change $[\boldsymbol{\nabla}_{\textbf{\textit{A}}}, \boldsymbol{\nabla}_{\textbf{\textit{B}}}]\textbf{\textit{C}}$? *Yes!* Exercise 11.1 shows

$$\{[\boldsymbol{\nabla}_{\textbf{\textit{A}}}, \boldsymbol{\nabla}_{\textbf{\textit{B}}}]\textbf{\textit{C}}_{\text{NEW}}\}_{\text{at}\,\mathscr{P}_0} - \{[\boldsymbol{\nabla}_{\textbf{\textit{A}}}, \boldsymbol{\nabla}_{\textbf{\textit{B}}}]\textbf{\textit{C}}_{\text{OLD}}\}_{\mathscr{P}_0} = \textbf{\textit{C}}_{\text{OLD}}\,\boldsymbol{\nabla}_{[\textbf{\textit{A}},\textbf{\textit{B}}]}f.$$

C. Modified Definition of **_Riemann_**:

1. The term causing trouble, $\textbf{\textit{C}}_{\text{OLD}}\,\boldsymbol{\nabla}_{[\textbf{\textit{A}},\textbf{\textit{B}}]}f$, can be disposed of by subtracting a "correction term" resembling it from **_Riemann_**—i.e., by redefining

$$\textbf{\textit{Riemann}} \, (\ldots, \textbf{\textit{C}}, \textbf{\textit{A}}, \textbf{\textit{B}}) \equiv \mathscr{R}(\textbf{\textit{A}}, \textbf{\textit{B}})\textbf{\textit{C}}, \qquad (3)$$

$$\mathscr{R}(\textbf{\textit{A}}, \textbf{\textit{B}}) \equiv [\boldsymbol{\nabla}_{\textbf{\textit{A}}}, \boldsymbol{\nabla}_{\textbf{\textit{B}}}] - \boldsymbol{\nabla}_{[\textbf{\textit{A}},\textbf{\textit{B}}]} \qquad (4)$$

2. The above calculation then gives a result independent of the "modifying function" f:

$$\{\mathscr{R}(\textbf{\textit{A}}, \textbf{\textit{B}})\textbf{\textit{C}}_{\text{NEW}}\}_{\text{at}\,\mathscr{P}_0} = \{\mathscr{R}(\textbf{\textit{A}}, \textbf{\textit{B}})\textbf{\textit{C}}_{\text{OLD}}\}_{\text{at}\,\mathscr{P}_0}.$$

D. Is Modified Definition Compatible with Equation for Tidal Gravitational Forces?

1. One would like to write $\boldsymbol{\nabla}_{\textbf{\textit{u}}}\boldsymbol{\nabla}_{\textbf{\textit{u}}}\textbf{\textit{n}} + \textbf{\textit{Riemann}} \, (\ldots, \textbf{\textit{u}}, \textbf{\textit{n}}, \textbf{\textit{u}}) = 0$.
2. This works just as well for modified definition of **_Riemann_** as for original definition, because

$$\mathscr{R}(\boldsymbol{n}, \boldsymbol{u}) = [\boldsymbol{\nabla}_n, \boldsymbol{\nabla}_u] - \boldsymbol{\nabla}_{[n,u]} = [\boldsymbol{\nabla}_n, \boldsymbol{\nabla}_u].$$

$$\left[\begin{array}{l} = 0 \text{ because } \boldsymbol{n} = (\partial/\partial n)_\lambda \text{ and} \\ \boldsymbol{u} = (\partial/\partial\lambda)_n \text{ commute} \end{array} \right]$$

Geodesic deviation and tidal forces cannot tell the difference between $\mathscr{R}(\boldsymbol{n}, \boldsymbol{u})$ and $[\boldsymbol{\nabla}_n, \boldsymbol{\nabla}_u]$, nor consequently between old and new definitions of **Riemann**.

E. Is Modified Definition Acceptable?

I.e., is **Riemann** $(\ldots, \boldsymbol{C}, \boldsymbol{A}, \boldsymbol{B}) \equiv \mathscr{R}(\boldsymbol{A}, \boldsymbol{B})\boldsymbol{C}$ a linear machine with output independent of how \boldsymbol{A}, \boldsymbol{B}, \boldsymbol{C} vary near point of evaluation? YES! (See exercise 11.2.)

Take stock, first, of what one knows already about the Riemann curvature tensor. (1) **Riemann** *is* a tensor; despite the appearance of $\boldsymbol{\nabla}$ in its definition (11.9), no derivatives actually act on the input vectors \boldsymbol{A}, \boldsymbol{B}, and \boldsymbol{C}. (2) **Riemann** is a $\binom{1}{3}$ tensor; its first slot accepts a 1-form; the others, vectors. (3) **Riemann** is determined entirely by $\boldsymbol{\nabla}$, or equivalently by the geodesics of spacetime, or equivalently by spacetime's parallel transport law; nothing but $\boldsymbol{\nabla}$ and the input vectors and 1-form are required to fix **Riemann**'s output. (4) **Riemann** produces the tidal gravitational forces that pry geodesics (test-particle trajectories) apart or push them together; i.e., it characterizes the "curvature of spacetime":

Tide-producing gravitational forces expressed in terms of **Riemann**

$$\boldsymbol{\nabla}_u \boldsymbol{\nabla}_u \boldsymbol{n} + \textbf{Riemann} (\ldots, \boldsymbol{u}, \boldsymbol{n}, \boldsymbol{u}) = 0. \tag{11.10}$$

(This *"equation of geodesic deviation"* follows from equations 11.6, 11.8, and 11.9, and the relation $[\boldsymbol{n}, \boldsymbol{u}] = 0$.)

All these facets of **Riemann** are *pictorial* (e.g., geodesic deviation; see Boxes 11.2 and 11.3) or *abstract* (e.g., equations 11.8 and 11.9 for **Riemann** in terms of $\boldsymbol{\nabla}$). **Riemann**'s *component* facet,

Components of **Riemann** expressed in terms of connection coefficients

$$R^\alpha{}_{\beta\gamma\delta} \equiv \textbf{Riemann} (\boldsymbol{\omega}^\alpha, \boldsymbol{e}_\beta, \boldsymbol{e}_\gamma, \boldsymbol{e}_\delta) \equiv \langle \boldsymbol{\omega}^\alpha, \mathscr{R}(\boldsymbol{e}_\gamma, \boldsymbol{e}_\delta)\boldsymbol{e}_\beta \rangle, \tag{11.11}$$

is related to the component facet of $\boldsymbol{\nabla}$ by the following equation, valid in any coordinate basis $\{\boldsymbol{e}_\alpha\} = \{\partial/\partial x^\alpha\}$:

$$R^\alpha{}_{\beta\gamma\delta} = \frac{\partial \Gamma^\alpha{}_{\beta\delta}}{\partial x^\gamma} - \frac{\partial \Gamma^\alpha{}_{\beta\gamma}}{\partial x^\delta} + \Gamma^\alpha{}_{\mu\gamma}\Gamma^\mu{}_{\beta\delta} - \Gamma^\alpha{}_{\mu\delta}\Gamma^\mu{}_{\beta\gamma}. \tag{11.12}$$

(See exercise 11.3 for derivation, and exercise 11.4 for the extension to noncoordinate bases.) These components of **Riemann**, with no sign of any derivative operator anywhere, may leave one with a better feeling in one's stomach than the definition (11.8) with its nondifferentiating derivatives!

EXERCISES

Exercise 11.1. $[\nabla_A, \nabla_B]C$ DEPENDS ON DERIVATIVES OF C

(Based on Box 11.5.) Let C_{NEW} and C_{OLD} be vector fields related by

$$C_{\text{NEW}}(\mathscr{P}) = f(\mathscr{P})C_{\text{OLD}}(\mathscr{P}).$$

$$\text{[arbitrary function, except } f(\mathscr{P}_0) = 1]$$

Show that

$$\{[\nabla_A, \nabla_B]C_{\text{NEW}}\}_{\text{at }\mathscr{P}_0} - \{[\nabla_A, \nabla_B]C_{\text{OLD}}\}_{\text{at }\mathscr{P}_0} = C_{\text{OLD}}\,\nabla_{[A,B]}f.$$

Exercise 11.2. PROOF THAT *Riemann* IS A TENSOR

Show from its definition (11.8, 11.9) that **Riemann** is a tensor. Hint: Use the following procedure.

(a) If $f(\mathscr{P})$ is an arbitrary function, show that

$$\mathscr{R}(A, B)fC = f\mathscr{R}(A, B)C.$$

(b) Similarly show that

$$\mathscr{R}(fA, B)C = f\mathscr{R}(A, B)C \quad\text{and}\quad \mathscr{R}(A, fB)C = f\mathscr{R}(A, B)C.$$

(c) Show that $\mathscr{R}(A, B)C$ is linear; i.e.,

$$\mathscr{R}(A + a, B)C = \mathscr{R}(A, B)C + \mathscr{R}(a, B)C;$$

$$\mathscr{R}(A, B + b)C = \mathscr{R}(A, B)C + \mathscr{R}(A, b)C;$$

$$\mathscr{R}(A, B)(C + c) = \mathscr{R}(A, B)C + \mathscr{R}(A, B)c.$$

(d) Now use the above properties to prove the most crucial feature of $\mathscr{R}(A, B)C$: Modify the variations (gradients) of A, B, and C in an arbitrary manner, but leave A, B, C unchanged at \mathscr{P}_0:

$$\left.\begin{array}{l} A \longrightarrow A + a^\alpha e_\alpha \\ B \longrightarrow B + b^\alpha e_\alpha \\ C \longrightarrow C + c^\alpha e_\alpha \end{array}\right\} \quad \begin{array}{l} a^\alpha(\mathscr{P}),\ b^\alpha(\mathscr{P}),\ c^\alpha(\mathscr{P}) \text{ arbitrary except} \\ \text{they all vanish at } \mathscr{P} = \mathscr{P}_0. \end{array}$$

Show that this modification leaves $\mathscr{R}(A, B)C$ unchanged at \mathscr{P}_0.
(e) From these facts, conclude that **Riemann** is a tensor.

Exercise 11.3. COMPONENTS OF *Riemann* IN COORDINATE BASIS

Derive equation (11.12) for the components of the Riemann tensor in a coordinate basis. [*Solution:*

$$R^\alpha{}_{\beta\gamma\delta} = \textbf{\textit{Riemann}}\,(\omega^\alpha, e_\beta, e_\gamma, e_\delta) \qquad \begin{bmatrix} \text{standard way to} \\ \text{calculate components} \end{bmatrix}$$

$$= \langle \omega^\alpha, \mathscr{R}(e_\gamma, e_\delta)e_\beta \rangle \qquad \text{[by definition (11.9)]}$$

$$= \langle \omega^\alpha, (\nabla_\gamma \nabla_\delta - \nabla_\delta \nabla_\gamma)e_\beta \rangle \qquad \begin{bmatrix} \text{by definition (11.8) plus} \\ [e_\gamma, e_\delta] = 0 \text{ in coord. basis} \end{bmatrix}$$

$$= \langle \omega^\alpha, e_\mu \Gamma^\mu{}_{\beta\delta,\gamma} + (e_\nu \Gamma^\nu{}_{\mu\gamma})\Gamma^\mu{}_{\beta\delta} - e_\mu \Gamma^\mu{}_{\beta\gamma,\delta} - (e_\nu \Gamma^\nu{}_{\mu\delta})\Gamma^\mu{}_{\beta\gamma} \rangle$$

$$= (\Gamma^\mu{}_{\beta\delta,\gamma} - \Gamma^\mu{}_{\beta\gamma,\delta})\langle \omega^\alpha, e_\mu \rangle + (\Gamma^\nu{}_{\mu\gamma}\Gamma^\mu{}_{\beta\delta} - \Gamma^\nu{}_{\mu\delta}\Gamma^\mu{}_{\beta\gamma})\langle \omega^\alpha, e_\nu \rangle,$$

which reduces (upon using $\langle \omega^\alpha, e_\mu \rangle = \delta^\alpha{}_\mu$) to (11.12).]

**Exercise 11.4. COMPONENTS OF *RIEMANN*
IN NONCOORDINATE BASIS**

In a noncoordinate basis with commutation coefficients $c_{\alpha\beta}{}^{\gamma}$ defined by equation (9.22), derive the following equation for the components of *Riemann*:

$$R^{\alpha}{}_{\beta\gamma\delta} = \Gamma^{\alpha}{}_{\beta\delta,\gamma} - \Gamma^{\alpha}{}_{\beta\gamma,\delta} + \Gamma^{\alpha}{}_{\mu\gamma}\Gamma^{\mu}{}_{\beta\delta} - \Gamma^{\alpha}{}_{\mu\delta}\Gamma^{\mu}{}_{\beta\gamma} - \Gamma^{\alpha}{}_{\beta\mu}c_{\gamma\delta}{}^{\mu}. \qquad (11.13)$$

§11.4. PARALLEL TRANSPORT AROUND A CLOSED CURVE

What are the effects of spacetime curvature, and how can one quantify them? One effect is geodesic deviation (relative acceleration of test bodies), quantified by equation (11.10). Another effect, almost as important, is the change in a vector caused by parallel transport around a closed curve. This effect shows up most clearly in the same problem, geodesic deviation, that motivated curvature in the first place. The relative acceleration vector $\boldsymbol{\nabla}_u\boldsymbol{\nabla}_u\boldsymbol{n}$ is also the change $\delta\boldsymbol{u}$ in the vector \boldsymbol{u} caused by parallel transport around the curve whose legs are the vectors \boldsymbol{n} and \boldsymbol{u}:

Change in a vector due to parallel transport around a closed curve:

Related to geodesic deviation

$$\boldsymbol{\nabla}_u\boldsymbol{\nabla}_u\boldsymbol{n} = \delta\boldsymbol{u}.$$

(See Box 11.6 for proof.) Hence, in this special case one can write

$$\delta\boldsymbol{u} + \boldsymbol{Riemann}\,(\ldots,\boldsymbol{u},\boldsymbol{n},\boldsymbol{u}) = 0.$$

The expected generalization is obvious: pick a closed quadrilateral with legs $\boldsymbol{u}\,\Delta a$ and $\boldsymbol{v}\,\Delta b$ (Figure 11.2; Δa and Δb are small parameters, to go to zero at end of discussion). Parallel transport the vector \boldsymbol{A} around this quadrilateral. The resultant change in \boldsymbol{A} should satisfy the equation

$$\delta\boldsymbol{A} + \boldsymbol{Riemann}\,(\ldots,\boldsymbol{A},\boldsymbol{u}\,\Delta a,\boldsymbol{v}\,\Delta b) = 0; \qquad (11.14)$$

Equation for change

or, equivalently,

$$\delta\boldsymbol{A} + \Delta a\,\Delta b\,\mathcal{R}(\boldsymbol{u},\boldsymbol{v})\boldsymbol{A} = 0; \qquad (11.14')$$

or, more precisely,

$$\lim_{\substack{\Delta a\to 0 \\ \Delta b\to 0}}\left(\frac{\delta\boldsymbol{A}}{\Delta a\,\Delta b}\right) + \boldsymbol{Riemann}\,(\ldots,\boldsymbol{A},\boldsymbol{u},\boldsymbol{v}) = 0. \qquad (11.14'')$$

The proof is enlightening, for it reveals the geometric origin of the correction term $\boldsymbol{\nabla}_{[u,v]}$ in the curvature operator.

The circuit of transport (Figure 11.2) is to be made from two arbitrary vector fields $\boldsymbol{u}\,\Delta a$ and $\boldsymbol{v}\,\Delta b$. However, a circuit made only of these fields has a gap in it, for a simple reason. The magnitude of \boldsymbol{u} varies the wrong way from place to place. The displacement $\boldsymbol{u}\,\Delta a$ that reaches across at the bottom of the quadrilateral from

Derivation of equation for change

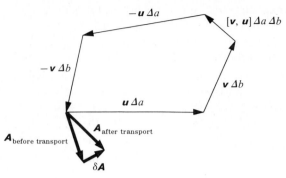

Figure 11.2.

The change $\delta\mathbf{A}$ in a vector \mathbf{A} as a result of parallel transport around a closed curve. The edges of the curve are the vector fields $\mathbf{u}\,\Delta a$ and $\mathbf{v}\,\Delta b$, plus the "closer of the quadrilateral" $[\mathbf{v}\,\Delta b, \mathbf{u}\,\Delta a] = [\mathbf{v}, \mathbf{u}]\,\Delta a\,\Delta b$ (see Box 9.2).

one line of \mathbf{v}'s to another cannot make the connection at the top of the quadrilateral. Similarly the \mathbf{v}'s vary the wrong way from place to place to connect the \mathbf{u}'s. To close the gap and complete the circuit, insert the "closer of quadrilaterals" $[\mathbf{v}\,\Delta b, \mathbf{u}\,\Delta a] = [\mathbf{v}, \mathbf{u}]\,\Delta a\,\Delta b$. (See Box 9.2 for why this vector closes the gap.)

With the route now specified, the vector \mathbf{A} is to be transported around it. One way to do this, "geometrical construction" by the method of Schild's ladder applied over and over, is the foundation for planning a possible experiment. For planning an abstract and coordinate-free calculation (the present line of action), introduce a "fiducial field," only to take it away at the end of the calculation. *Plan:* Conceive of \mathbf{A}, not as a localized vector defined solely at the start of the trip, but as a vector field (defined throughout the trip). *Purpose:* To provide a standard of reference (comparison of \mathbf{A} transported from the origin with \mathbf{A} at the place in question). *Principle:* The standard of reference will cancel out in the end. *Procedure:*

$$-\delta\mathbf{A} = -\begin{pmatrix}\text{Net change made in taking the vector } \mathbf{A}, \text{ originally localized at the}\\ \text{start of the circuit, and transporting it parallel to itself ("mobile } \mathbf{A}\text{")}\\ \text{around the closed circuit. This quantity cannot be evaluated until}\\ \text{completion of circuit because there is no preexisting standard of}\\ \text{reference along the way.}\end{pmatrix}$$

$$= +\begin{pmatrix}\text{A quantity subject to analysis for each leg of circuit individually. This}\\ \text{new quantity is defined by introducing throughout the whole region}\\ \text{a vector field } \mathbf{A}^{\text{(field)}}, \text{ smoothly varying, and in agreement at starting}\\ \text{point with the original localized } \mathbf{A}, \text{ but otherwise arbitrary. This new}\\ \text{quantity is then given by } \mathbf{A}^{\text{(field)}} \text{ at starting point (same as } \mathbf{A}^{\text{(localized)}}\\ \text{at starting point) minus } \mathbf{A}^{\text{(mobile)}} \text{ at finish point (after transit).}\end{pmatrix}$$

$$= \sum_{\substack{\text{legs of}\\\text{circuit}}}\begin{pmatrix}\text{Change in } \mathbf{A}^{\text{(field)}} \text{ relative to } \mathbf{A}^{\text{(mobile)}} \text{ in the course of transport along}\\ \text{specified leg. Value for any one leg depends on the arbitrary choice}\\ \text{of } \mathbf{A}^{\text{(field)}}, \text{ but this arbitrariness cancels out in end because of closure}\\ \text{of circuit.}\end{pmatrix}$$

$$\left(\begin{array}{l}\text{Change in } \textbf{A}^{\text{(field)}} \text{ relative to the parallel-transported } \textbf{A}^{\text{(mobile)}} \text{ as standard} \\ \text{of reference, made up of contributions along following legs of Figure} \\ \text{11.2:}\end{array}\right.$$

$$= \left(\begin{array}{ll}\textbf{\textit{v}} \, \Delta b, \text{ giving} & \boldsymbol{\nabla}_{\!v}\textbf{A}^{\text{(field)}} \, \Delta b \text{ (on line displaced } \textbf{\textit{u}} \, \Delta a \text{ from start)} \\ -\textbf{\textit{v}} \, \Delta b, \text{ giving} - \boldsymbol{\nabla}_{\!v}\textbf{A}^{\text{(field)}} \, \Delta b \text{ (on line through starting point)} \\ -\textbf{\textit{u}} \, \Delta a, \text{ giving} - \boldsymbol{\nabla}_{\!u}\textbf{A}^{\text{(field)}} \, \Delta a \text{ (on line displaced } \textbf{\textit{v}} \, \Delta b \text{ from start)} \\ +\textbf{\textit{u}} \, \Delta a, \text{ giving} & \boldsymbol{\nabla}_{\!u}\textbf{A}^{\text{(field)}} \, \Delta a \text{ (on line through starting point)} \\ +[\textbf{\textit{v}}, \textbf{\textit{u}}] \, \Delta a \, \Delta b, \text{ giving} \, \boldsymbol{\nabla}_{\![v,u]}\textbf{A}^{\text{(field)}} \, \Delta a \, \Delta b \end{array}\right)$$

$$= \{\boldsymbol{\nabla}_{\!u}\boldsymbol{\nabla}_{\!v} - \boldsymbol{\nabla}_{\!v}\boldsymbol{\nabla}_{\!u} + \boldsymbol{\nabla}_{\![v,u]}\}\textbf{A}^{\text{(field)}} \, \Delta a \, \Delta b$$

$$= \textbf{\textit{Riemann}} \, (\ldots, \textbf{A}^{\text{(field)}}, \textbf{\textit{u}}, \textbf{\textit{v}}) \, \Delta a \, \Delta b = \mathscr{R}\,(\textbf{\textit{u}}, \textbf{\textit{v}})\textbf{A}^{\text{(field)}} \, \Delta a \, \Delta b. \qquad (11.15)$$

Profit: The curvature operator

$$\textbf{\textit{Riemann}} \, (\ldots, \ldots, \textbf{\textit{u}}, \textbf{\textit{v}}) = \mathscr{R}\,(\textbf{\textit{u}}, \textbf{\textit{v}}) = [\boldsymbol{\nabla}_{\!u}, \boldsymbol{\nabla}_{\!v}] - \boldsymbol{\nabla}_{\![u,v]},$$

Box 11.6 GEODESIC DEVIATION AND PARALLEL TRANSPORT AROUND CLOSED CURVE: TWO ASPECTS OF SAME CONSTRUCTION

Geodesic Deviation

$$\boldsymbol{\nabla}_{\!u}\boldsymbol{\nabla}_{\!u}\textbf{\textit{n}} = \operatorname*{Lim}_{\substack{\Delta\lambda \to 0 \\ \Delta n \to 0}} \left\{\frac{\mathscr{AP} + \mathscr{BR}}{(\Delta\lambda)^2 \, \Delta n}\right\}.$$

(See Boxes 11.2 and 11.3)

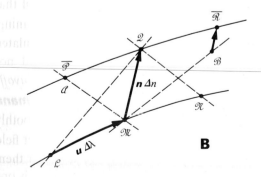

Geodesic Deviation

Same result; different construction. To simplify the connection with closed-curve transport, change the tilt and dilate the parametrization of geodesic $\mathscr{P2R}$ in **A**. The result: **B**, where $\overline{\mathscr{P}}$ and \mathscr{Q} coincide. From **F** of Box 11.2 one knows $\mathscr{AP} + \mathscr{BR} = \mathscr{A}\overline{\mathscr{P}} + \mathscr{B}\overline{\mathscr{R}}$—i.e. $\boldsymbol{\nabla}_{\!u}\boldsymbol{\nabla}_{\!u}\textbf{\textit{n}}$ is the

Box 11.6 (*continued*)

same for this family of geodesics as for the original family

$$\boldsymbol{\nabla_u}\,\boldsymbol{\nabla_u n} = \operatorname*{Lim}_{\substack{\Delta\lambda \to 0 \\ \Delta n \to 0}} \left\{ \frac{\mathcal{B}\bar{\mathcal{R}}}{(\Delta\lambda)^2\,\Delta n} \right\}.$$

Also, to simplify discussion set $\Delta n = \Delta\lambda = 1$, and assume \boldsymbol{n} and \boldsymbol{u} are small enough that one can evaluate $\boldsymbol{\nabla_u}\,\boldsymbol{\nabla_u n}$ without taking the limit:

$$\boldsymbol{\nabla_u}\,\boldsymbol{\nabla_u n} = \mathcal{B}\bar{\mathcal{R}}.$$

Parallel Transport Around Closed Curve, Performed by Same Construction

Plan: Parallel transport the vector $\boldsymbol{u}\,\Delta\lambda = \mathcal{2}\bar{\mathcal{R}}$ counterclockwise around the curve $\mathcal{2} \longrightarrow \bar{\mathcal{P}} \longrightarrow \mathcal{L} \longrightarrow \mathcal{M} \longrightarrow \mathcal{2}$. *Execution:* (1) Call transported vector $\boldsymbol{u}^{(\mathrm{m})}$ ("m" for "mobile"). (2) At $\mathcal{2}$, $\boldsymbol{u}^{(\mathrm{m})} = \mathcal{2}\bar{\mathcal{R}}$. (3) At $\bar{\mathcal{P}}$, $\boldsymbol{u}^{(\mathrm{m})} = \bar{\mathcal{P}}\mathcal{2}$ because $\bar{\mathcal{P}}\mathcal{2}\bar{\mathcal{R}}$ is a geodesic and $\boldsymbol{u}^{(\mathrm{m})}$ is its tangent vector. (4) At \mathcal{L}, $\boldsymbol{u}^{(\mathrm{m})} = \mathcal{L}\mathcal{M}$ according to Schild's ladder of the picture. (5) At \mathcal{M}, $\boldsymbol{u}^{(\mathrm{m})} = \mathcal{M}\mathcal{N}$ because $\mathcal{L}\mathcal{M}\mathcal{N}$ is a geodesic and $\boldsymbol{u}^{(\mathrm{m})}$ is now its tangent vector. (6) At $\mathcal{2}$, $\boldsymbol{u}^{(\mathrm{m})} = \mathcal{2}\mathcal{B}$ according to Schild's ladder. Result: The change in $\boldsymbol{u}^{(\mathrm{m})}$ is $-\mathcal{B}\bar{\mathcal{R}}$. Had the curve been circuited in opposite direction ($\mathcal{L} \longrightarrow \bar{\mathcal{P}} \longrightarrow \mathcal{2} \longrightarrow \mathcal{M} \longrightarrow \mathcal{L}$), the change would have been $+\mathcal{B}\bar{\mathcal{R}}$:

$$(\delta\boldsymbol{u})_{\substack{\text{due to parallel transport up } \boldsymbol{n}, \text{ out } \boldsymbol{u}, \text{ down } -\boldsymbol{n}, \text{ and} \\ \text{back along } -\boldsymbol{u} \text{ to starting point}}} = \mathcal{B}\bar{\mathcal{R}} = \boldsymbol{\nabla_u}\,\boldsymbol{\nabla_u n}.$$

applied to the vector field $\boldsymbol{A}^{(\mathrm{field})}$, gives the negative of the change in the localized vector $\boldsymbol{A}^{(\mathrm{localized})}$ (called $\boldsymbol{A}^{(\mathrm{mobile})}$ during the phase of travel) on parallel transport around the closed circuit. It does not give the change in $\boldsymbol{A}^{(\mathrm{field})}$ on traversal of that circuit, for $\boldsymbol{A}^{(\mathrm{field})}$ has the same value at the end of the journey as at the beginning. Equation (11.14′) expresses that change in terms of the conveniently calculated differential operator, $\mathcal{R}(\boldsymbol{u}, \boldsymbol{v}) = [\boldsymbol{\nabla_u}, \boldsymbol{\nabla_v}] - \boldsymbol{\nabla_{[u,v]}}$. *Paradox:* Neither wanted nor evaluated is the change in the quantity $\boldsymbol{A}^{(\mathrm{field})}$ acted on by this operator. *Payoff:* Ostensibly differential in the character of its action on \boldsymbol{A}, the operator **Riemann** $(\dots,\dots,\boldsymbol{u}, \boldsymbol{v}) = \mathcal{R}(\boldsymbol{u}, \boldsymbol{v})$ is actually local. Thus, replace the proposed smoothly varying vector field $\boldsymbol{A}^{(\mathrm{field})}$ by a quite different but also smoothly varying vector field $\boldsymbol{A}^{(\mathrm{field, new})}$. Then the two fields need agree only at the one point in question for them to give the same output **Riemann** $(\dots, \boldsymbol{A}, \boldsymbol{u}, \boldsymbol{v}) = \mathcal{R}(\boldsymbol{u}, \boldsymbol{v})\boldsymbol{A}$ at that point. This one

knows from the fact that $\delta \boldsymbol{A}$, the quantity calculated, has an existence and value independent of the choice of $\boldsymbol{A}^{(\text{field})}$. This one can also verify by detailed calculation (exercise 11.2). *Power:* Although they cancel out in their response to any change of \boldsymbol{A} with location, the several differentiations in the curvature operator respond directly to the "rate of change of geometry with location" ("geodesic deviation"). *Prolongation:* The closed curve need not be a quadrilateral. The curvature operator tells how a vector changes on parallel transport about small curves of arbitrary shape (Box 11.7).

Exercise 11.5. COPLANARITY OF CLOSED CURVES **EXERCISE**

Let \boldsymbol{f}_1 and \boldsymbol{f}_2 be the bivectors (see Box 11.7) for two small closed curves at the same event. Show that the curves are coplanar if and only if $\boldsymbol{f}_1 = a\boldsymbol{f}_2$ for some number a.

Box 11.7 THE LAW FOR PARALLEL TRANSPORT ABOUT A CLOSED CURVE

A. Special Case

Curve is closed quadrilateral formed by vector fields \boldsymbol{u} and \boldsymbol{v}.

1. Law says (in component form)

$$\delta A^\alpha + R^\alpha{}_{\beta\gamma\delta} A^\beta u^\gamma v^\delta = 0. \tag{1}$$

2. On what characteristics of the closed curve does this depend?
 a. Notice that $R^\alpha{}_{\beta\gamma\delta} = -R^\alpha{}_{\beta\delta\gamma}$ (antisymmetry in last two indices; obvious in equation 11.12 for components; also obvious because reversing the direction the curve is traversed—i.e., interchanging \boldsymbol{u} and \boldsymbol{v}—should reverse sign of $\delta \boldsymbol{A}$).
 b. Equation (1) contracts $\boldsymbol{u} \otimes \boldsymbol{v}$ into these antisymmetric, last two indices. The symmetric part of $\boldsymbol{u} \otimes \boldsymbol{v}$ must give zero. Only the antisymmetric part, $\boldsymbol{u} \wedge \boldsymbol{v} = \boldsymbol{u} \otimes \boldsymbol{v} - \boldsymbol{v} \otimes \boldsymbol{u}$ can contribute:

$$\delta A^\alpha + \frac{1}{2} R^\alpha{}_{\beta\gamma\delta} A^\beta (\boldsymbol{u} \wedge \boldsymbol{v})^{\gamma\delta} = 0. \tag{2}$$

3. This antisymmetric part is a *"bivector."* It is independent of the curve's shape; it depends only on (a) the plane the curve lies in, and (b) the area enclosed by the curve. [Although without metric "area" is meaningless, "relative areas at an event in a given plane" have just as much meaning as "relative lengths at an

Box 11.7 (*continued*)

event along a given direction." Two vectors at the same event lie on the same line if they are multiples of each other; their relative length in that case is their ratio. Similarly, two small closed curves at the same event lie in the same plane if their bivectors are multiples of each other (exercise 11.5); their relative area in that case is the ratio of their bivectors.]

B. General Case

Arbitrary but small closed curve.

1. Break the curve down into a number of quadrilaterals, all lying in the same plane as the curve.
2. Traverse each quadrilateral once in the same sense as the curve is to be traversed. Result: all interior edges get traversed twice in opposite directions (no net traversal); the outer edge (the curve itself) gets traversed once.
3. Thus, $\delta\boldsymbol{A}$ due to traversing curve is the sum of the $\delta\boldsymbol{A}$'s from traversal of each quadrilateral:

$$\delta A^\alpha = -\frac{1}{2} \sum_{\text{quadrilaterals}} R^\alpha{}_{\beta\gamma\delta} A^\beta (\boldsymbol{u} \wedge \boldsymbol{v}_{\text{for given quadrilateral}})^{\gamma\delta}.$$

Define the bivector \boldsymbol{f} for the curve as the sum of the bivectors for its component quadrilaterals:

$$\boldsymbol{f} \equiv \sum_{\text{quadrilaterals}} (\mathbf{u} \wedge \boldsymbol{v})_{\text{quadrilateral}}$$

(add "areas"; keep plane the same).

4. Then

$$\delta A^\alpha + \frac{1}{2} R^\alpha{}_{\beta\gamma\delta} A^\beta f^{\gamma\delta} \equiv \delta A^\alpha + R^\alpha{}_{\beta|\gamma\delta|} A^\beta f^{\gamma\delta} = 0.$$

C. Warning

This is valid only for closed curves of small compass: $\delta\boldsymbol{A}$ doubles when the area doubles; but the error increases by a factor $\sim 2^{3/2}$ [$\delta\boldsymbol{A} \propto \Delta a\, \Delta b$ in calculation of §11.4; but error $\propto (\Delta a)^2\, \Delta b$ or $\Delta a\, (\Delta b)^2$].

§11.5. FLATNESS IS EQUIVALENT TO ZERO RIEMANN CURVATURE

To say that space or spacetime or any other manifold is flat is to say that there exists a coordinate system $\{x^\alpha(\mathcal{P})\}$ in which all geodesics appear straight:

Flatness of a manifold defined

$$x^\alpha(\lambda) = a^\alpha + b^\alpha \lambda. \tag{11.16}$$

(Example: Lorentz spacetime of special relativity, where test bodies move on such straight lines.) They can appear so if and only if the connection coefficients in the geodesic equation

$$\frac{d^2 x^\beta}{d\lambda^2} + \Gamma^\beta{}_{\mu\nu} \frac{dx^\mu}{d\lambda} \frac{dx^\nu}{d\lambda} = 0, \tag{11.17}$$

expressed in the same coordinate system, all vanish:

$$\Gamma^\beta{}_{\mu\nu} = 0. \tag{11.18}$$

From the vanishing of these connection coefficients, it follows immediately (equation 11.12) that all the components of the curvature tensor are zero:

Flatness implies Riemann = 0

$$R^\beta{}_{\gamma\mu\nu} = 0. \tag{11.19}$$

[Geometric restatement of (11.16) \longrightarrow (11.18) \longrightarrow (11.19): For all geodesics to be straight in a given coordinate system means that initially parallel geodesics preserve their separation; the geodesic deviation is zero; and therefore the curvature vanishes.]

Is the converse true? Does zero Riemann curvature imply the existence of a coordinate system in which all geodesics appear straight? Yes, as one sees by the following construction.

Transport a vector parallel to itself from \mathcal{P}_0 to \mathcal{Q}, and then back from \mathcal{Q} to \mathcal{P}_0 along a slightly different route. It returns to its starting point with no alteration in magnitude or direction, *because Riemann* everywhere vanishes. Therefore parallel transport of a base vector \boldsymbol{e}_μ from \mathcal{P}_0 to \mathcal{Q} yields at \mathcal{Q} a base vector \boldsymbol{e}_μ that is independent, both in magnitude and in direction, of the route of transportation (for routes obtainable one from the other by any continuous sequence of deformations). As for \mathcal{Q}, so for all points of the manifold; and as for the one base vector \boldsymbol{e}_μ, so for a complete set of base vectors ($\mu = 0, 1, 2, 3$): Parallel transport of a basis $\{\boldsymbol{e}_\alpha(\mathcal{P}_0)\}$ yields everywhere a field of frames ("frame field"), each base vector of which suffers zero change (relative to the frame field) on parallel transport from any point to any nearby point; thus,

Proof that Riemann = 0 implies flatness

$$\boldsymbol{\nabla}\boldsymbol{e}_\mu = 0; \tag{11.20}$$

or

$$\boldsymbol{\nabla}_\nu\boldsymbol{e}_\mu \, (\equiv \boldsymbol{\nabla}_{\boldsymbol{e}_\nu}\boldsymbol{e}_\mu) = 0. \tag{11.21}$$

With the vanishing of these individual derivatives, there also vanishes the commutator of any two basis-vector fields:

$$[\boldsymbol{e}_\mu, \boldsymbol{e}_\nu] = \boldsymbol{\nabla}_\mu\boldsymbol{e}_\nu - \boldsymbol{\nabla}_\nu\boldsymbol{e}_\mu = 0 - 0 = 0. \tag{11.22}$$

The gap in the quadrilateral of Figure 11.2 (there read "\boldsymbol{e}_μ" for "\boldsymbol{u}," "\boldsymbol{e}_ν" for "\boldsymbol{v}") closes up completely. Thereupon one can introduce coordinates x^μ, each of which increases with a motion in the direction of the corresponding vector field; and with appropriate scaling of these coordinates, one can write

$$\boldsymbol{e}_\mu = \frac{\partial}{\partial x^\mu} \tag{11.23}$$

(see exercise 9.9). With this coordinate basis in hand, one can employ the formula

$$\boldsymbol{\nabla}_\alpha \boldsymbol{e}_\beta = \boldsymbol{e}_\mu \Gamma^\mu{}_{\beta\alpha} \tag{11.24}$$

to calculate the connection coefficients. From the vanishing of the quantities on the left, one concludes that all the connection coefficients on the right ("bending of geodesics") must be zero; so spacetime is indeed flat.

Summary: Spacetime is flat—i.e., there exist "flat coordinates" in which $\Gamma^\mu{}_{\alpha\beta} = 0$ everywhere and geodesics are straight lines, $x^\alpha(\lambda) = a^\alpha + b^\alpha\lambda$—*if and only if* **Riemann** $= 0$.

Note: In the spacetime of Einstein, which has a metric, one can choose $\{\boldsymbol{e}_\mu(\mathscr{P}_0)\}$ in the above argument to be orthonormal, $\boldsymbol{e}_\mu \cdot \boldsymbol{e}_\nu = \eta_{\mu\nu}$ at \mathscr{P}_0. The resulting field of frames will then be orthonormal everywhere, and the resulting coordinate system will be Lorentz. Thus, in Einsteinian gravity the above summary can be rewritten: *spacetime is flat (there exists a Lorentz coordinate system) if and only if* **Riemann** $= 0$.

Warning: Flatness does not necessarily imply Euclidean topology. Take a sheet of paper. It is flat. Roll it up into a cylinder. It is still flat, intrinsically. The tracks of geodesics over it have not changed. Distances between neighboring points have not changed. Only the topology has changed, so far as an observer confined forever to the sheet is concerned. (The "extrinsic geometry"—the way the sheet is embedded in the surrounding three-dimensional space—has also changed; but an observer on the sheet knows nothing of this, and it is not the subject of the present chapter. See, instead §21.5.)

Take this cylinder. Bend it around and glue its two ends together, without changing its flat intrinsic geometry. Doing so is impossible if the cylinder remains embedded in flat, three-dimensional Euclidean space; perfectly possible if it is embedded in a Euclidean space of 4 dimensions. However, embedding is unimportant to observers confined to the cylinder, since all they ever measure is intrinsic geometry; so all that matters to them is the *topological identification* of the two ends of the cylinder with each other. The result is topologically a torus; but the tracks of geodesics are still unchanged; the intrinsic geometry is flat; **Riemann** vanishes.

By analogy, take flat Minkowskii spacetime. Pick some Lorentz frame, and in it pick a cube 10^{10} light years on each side ($0 < x < 10^{10}$ light years; similarly for y and z). Identify opposite faces of the cube so that a geodesic exiting across one face enters across the other. The result is topologically a three-torus: a "closed universe" with finite volume, with flat, Minkowskii geometry, and with a form that changes not at all as Lorentz time t passes (no expansion, no contraction).

Lorentz coordinates exist if and only if **Riemann** $= 0$

Flatness does not imply Euclidean topology

§11.6. RIEMANN NORMAL COORDINATES

In curved spacetime one can never find a coordinate system with $\Gamma^{\alpha}{}_{\beta\gamma} = 0$ every-
where. But one can always construct local inertial frames at a given event \mathcal{P}_0; and
as viewed in such frames, free particles must move along straight lines, at least
locally—which means $\Gamma^{\alpha}{}_{\beta\gamma}$ must vanish, at least locally.

A very special and useful realization of such a local inertial frame is a *Riemann-
normal coordinate system*. Pick an event \mathcal{P}_0 and a set of basis vectors $\{\boldsymbol{e}_{\alpha}(\mathcal{P}_0)\}$ to
be used there by an inertial observer. Fill spacetime, near \mathcal{P}_0, with geodesics radiating
out from \mathcal{P}_0 like the quills of a hedgehog or porcupine. Each geodesic is determined
by its tangent vector \boldsymbol{v} at \mathcal{P}_0; and the general point on it can be denoted

$$\mathcal{P} = \mathcal{G}(\lambda; \boldsymbol{v}). \qquad (11.25)$$

$$\begin{bmatrix} \text{affine parameter;} \\ \text{tells "where" on geodesic} \end{bmatrix} \qquad \begin{Bmatrix} \text{tangent vector at } \mathcal{P}_0; \\ \text{tells "which geodesic"} \end{Bmatrix}$$

*Riemann normal coordinates:
a realization of local inertial
frames*

*Geometric construction of
Riemann normal coordinates*

Actually, this gives more geodesics than are needed. One reaches the same point
after parameter length $\frac{1}{2}\lambda$ if the initial tangent vector is $2\boldsymbol{v}$, as one reaches after
λ if the tangent vector is \boldsymbol{v}:

$$\mathcal{G}(\lambda; \boldsymbol{v}) = \mathcal{G}\left(\frac{1}{2}\lambda; 2\boldsymbol{v}\right) = \mathcal{G}(1; \lambda\boldsymbol{v}).$$

Thus, by fixing $\lambda = 1$ and varying \boldsymbol{v} in all possible ways, one can reach every point
in some neighborhood of \mathcal{P}_0. This is the foundation for constructing Riemann normal
coordinates. Choose an event \mathcal{P}. Find that tangent vector \boldsymbol{v} at \mathcal{P}_0 for which $\mathcal{P} =
\mathcal{G}(1; \boldsymbol{v})$. Expand that \boldsymbol{v} in terms of the chosen basis and give its components the
names x^{α}:

$$\mathcal{P} = \mathcal{G}(1; x^{\alpha}\boldsymbol{e}_{\alpha}). \qquad (11.26)$$

The point \mathcal{P} determines x^{α} uniquely (if \mathcal{P} is near enough to \mathcal{P}_0 that spacetime
curvature has not caused geodesics to cross each other). Similarly, x^{α} determines
\mathcal{P} uniquely. Hence, x^{α} can be chosen as the coordinates of \mathcal{P}—its "Riemann-normal
coordinates, based on the event \mathcal{P}_0 and basis $\{\boldsymbol{e}_{\alpha}(\mathcal{P}_0)\}$."

Equation (11.26) summarizes Riemann-normal coordinates concisely. Other
equations, derived in exercise 11.9, summarize their powerful properties:

*Mathematical properties of
Riemann normal coordinates*

$$\boldsymbol{e}_{\alpha}(\mathcal{P}_0) = (\partial/\partial x^{\alpha})_{\mathcal{P}_0}; \qquad (11.27)$$

$$\Gamma^{\alpha}{}_{\beta\gamma}(\mathcal{P}_0) = 0; \qquad (11.28)$$

$$\Gamma^{\alpha}{}_{\beta\gamma,\mu}(\mathcal{P}_0) = -\frac{1}{3}(R^{\alpha}{}_{\beta\gamma\mu} + R^{\alpha}{}_{\gamma\beta\mu}). \qquad (11.29)$$

If spacetime has a metric (as it does in actuality), and if the observer's frame at
\mathcal{P}_0 has been chosen orthonormal ($\boldsymbol{e}_{\alpha} \cdot \boldsymbol{e}_{\beta} = \eta_{\alpha\beta}$), then

$$g_{\alpha\beta}(\mathcal{P}_0) = \eta_{\alpha\beta}, \tag{11.30}$$

$$g_{\alpha\beta,\mu}(\mathcal{P}_0) = 0, \tag{11.31}$$

$$g_{\alpha\beta,\mu\nu}(\mathcal{P}_0) = -\frac{1}{3}(R_{\alpha\mu\beta\nu} + R_{\alpha\nu\beta\mu}) \tag{11.32}$$

$$= -\frac{2}{3}J_{\alpha\beta\mu\nu},$$

$$R_{\alpha\beta\gamma\delta}(\mathcal{P}_0) = g_{\alpha\delta,\beta\gamma}(\mathcal{P}_0) - g_{\alpha\gamma,\beta\delta}(\mathcal{P}_0). \tag{11.32'}$$

Here $J_{\alpha\beta\mu\nu}$ are components of the Jacobi curvature tensor (see exercise 11.7).

Is this the only coordinate system that is locally inertial at \mathcal{P}_0 (i.e., has $\Gamma^{\alpha}{}_{\beta\gamma} = 0$ there) and is tied to the basis vectors \boldsymbol{e}_α there (i.e., has $\partial/\partial x^\alpha = \boldsymbol{e}_\alpha$ there)? No. But all such coordinate systems (called *"normal coordinates"*) will be the same to second order:

Other mathematical realizations of a local inertial frame

$$x^\alpha_{\text{NEW}}(\mathcal{P}) = x^\alpha_{\text{OLD}}(\mathcal{P}) + \text{corrections of order } (x^\alpha_{\text{OLD}})^3.$$

Moreover, only those the same to third order,

$$x^\alpha_{\text{NEW}}(\mathcal{P}) = x^\alpha_{\text{OLD}}(\mathcal{P}) + \text{corrections of order } (x^\alpha_{\text{OLD}})^4,$$

will preserve the beautiful ties (11.29) and (11.32) to the Riemann curvature tensor.

EXERCISES

Exercise 11.6. SYMMETRIES OF *Riemann*

(To be discussed in Chapter 13). Show that ***Riemann*** has the following symmetries:

$$R^{\alpha}{}_{\beta\gamma\delta} = R^{\alpha}{}_{\beta[\gamma\delta]} \qquad \text{(antisymmetric on last 2 indices)} \tag{11.33a}$$

$$R^{\alpha}{}_{[\beta\gamma\delta]} = 0 \qquad \text{(vanishing of completely antisymmetric part)} \tag{11.33b}$$

**Exercise 11.7. GEODESIC DEVIATION MEASURES ALL
 CURVATURE COMPONENTS**

The equation of geodesic deviation, written up to now in the form

$$\boldsymbol{\nabla_u}\,\boldsymbol{\nabla_u}\boldsymbol{n} + \boldsymbol{Riemann}\,(\ldots,\boldsymbol{u},\boldsymbol{n},\boldsymbol{u}) = 0$$

or

$$\boldsymbol{\nabla_u}\,\boldsymbol{\nabla_u}\boldsymbol{n} + \mathcal{R}(\boldsymbol{n},\boldsymbol{u})\boldsymbol{u} = 0,$$

also lets itself be written in the Jacobi form $\boldsymbol{\nabla_u}\,\boldsymbol{\nabla_u}\boldsymbol{n} + \mathcal{J}(\boldsymbol{u},\boldsymbol{u})\boldsymbol{n} = 0$. Here $\mathcal{J}(\boldsymbol{u},\boldsymbol{v})$, the *"Jacobi curvature operator,"* is defined by

$$\mathcal{J}(\boldsymbol{u},\boldsymbol{v})\boldsymbol{n} \equiv \frac{1}{2}[\mathcal{R}(\boldsymbol{n},\boldsymbol{u})\boldsymbol{v} + \mathcal{R}(\boldsymbol{n},\boldsymbol{v})\boldsymbol{u}], \tag{11.34}$$

and is related to the "Jacobi curvature tensor" by

$$\boldsymbol{Jacobi}\,(\ldots,\boldsymbol{n},\boldsymbol{u},\boldsymbol{v}) \equiv \mathcal{J}(\boldsymbol{u},\boldsymbol{v})\boldsymbol{n}, \tag{11.35}$$

which implies

$$J^\mu{}_{\nu\alpha\beta} = J^\mu{}_{\nu\beta\alpha} = \frac{1}{2}(R^\mu{}_{\alpha\nu\beta} + R^\mu{}_{\beta\nu\alpha}). \tag{11.36}$$

(a) Show that $J^\mu{}_{(\alpha\beta\gamma)} = 0$ follows from $R^\mu{}_{\alpha\beta\gamma} = R^\mu{}_{\alpha[\beta\gamma]}$.

(b) Show that by studying geodesic deviation (allowing arbitrary \boldsymbol{u} and \boldsymbol{n} in $\boldsymbol{\nabla_u}\boldsymbol{\nabla_u}\boldsymbol{n} + \mathcal{J}(\boldsymbol{u},\boldsymbol{u})\boldsymbol{n} = 0$) one can measure *all* components of **Jacobi**.

(c) Show that **Jacobi** contains precisely the same information as **Riemann**. [Hint: show that

$$R^\mu{}_{\alpha\nu\beta} = \frac{2}{3}(J^\mu{}_{\nu\alpha\beta} - J^\mu{}_{\beta\alpha\nu}); \tag{11.37}$$

this plus equation (11.36) for $J^\mu{}_{\nu\alpha\beta}$ proves "same information content".] Hence, by studying geodesic deviation one can also measure all the components of **Riemann**.

(d) Show that the symmetry of $R^\mu{}_{[\nu\alpha\beta]} = 0$ is essential in the equivalence between **Jacobi** and **Riemann** by exhibiting proposed values for $R^\mu{}_{\nu\alpha\beta} = -R^\mu{}_{\nu\beta\alpha}$ for which $R^\mu{}_{[\nu\alpha\beta]} \neq 0$, and from which one would find $J^\mu{}_{\nu\alpha\beta} = 0$.

Exercise 11.8. GEODESIC DEVIATION IN GORY DETAIL

Write out the equation of geodesic deviation in component form in a coordinate system. Expand all covariant derivatives (semicolon notation) in terms of ordinary (comma) derivatives and in terms of Γ's to show all Γ and ∂ terms explicitly.

Exercise 11.9. RIEMANN NORMAL COORDINATES IN GENERAL

Derive properties (11.27), (11.28), (11.29), (11.31), (11.32), and (11.32′) of Riemann normal coordinates. *Hint:* Proceed as follows.

(a) From definition (11.26), derive $(\partial\mathcal{P}/\partial x^\alpha)_{\mathcal{P}_0} = \boldsymbol{e}_\alpha$.

(b) Similarly, from definition (11.26), show that each of the curves $x^\alpha = v^\alpha\lambda$ (where the v^α are constants) is a geodesic through \mathcal{P}_0, with affine parameter λ.

(c) Show that $\Gamma^\alpha{}_{\beta\gamma}(\mathcal{P}_0) = 0$ by substituting $x^\alpha = v^\alpha\lambda$ into the geodesic equation.

(d) Since the curves $x^\alpha = v^\alpha\lambda$ are geodesics for every choice of the parameters v^α, they provide not only a geodesic tangent $\boldsymbol{u} \equiv (\partial/\partial\lambda)_{v^\alpha}$, but also several deviation vectors $\boldsymbol{N}_{(\alpha)} \equiv (\partial/\partial v^\alpha)_\lambda$. Compute the components of these vectors in the Riemann normal coordinate system, and substitute into the geodesic deviation equation as written in exercise 11.8.

(e) Equate to zero the coefficients of the zeroth and first powers of λ in the geodesic deviation equation of part (d), using

$$\Gamma^\alpha{}_{\beta\gamma}\Big|_{x^\mu = v^\mu\lambda} = \lambda v^\mu \Gamma^\alpha{}_{\beta\gamma,\mu}(\mathcal{P}_0) + O(\lambda^2),$$

which is a Taylor series for Γ. In this way arrive at equation (11.29) for $\Gamma^\alpha{}_{\beta\gamma,\mu}$ in terms of the Riemann tensor.

(f) From equations (11.28), (11.29), and (8.24) for the connection coefficients in terms of the metric, derive equations (11.31), (11.32), and (11.32′).

Exercise 11.10. BIANCHI IDENTITIES

Show that the Riemann curvature tensor satisfies the following "Bianchi identities"

$$R^\alpha{}_{\beta[\gamma\delta;\epsilon]} = 0. \tag{11.38}$$

The geometric meaning of these identities will be discussed in Chapter 15. [*Hint:* Perform the calculation at the origin of a Riemann normal coordinate system.]

Exercise 11.11. CURVATURE OPERATOR ACTS ON 1-FORMS

Let $\mathcal{R}(\boldsymbol{u}, \boldsymbol{v})$ be the operator $\mathcal{R}(\boldsymbol{u}, \boldsymbol{v}) = [\boldsymbol{\nabla_u}, \boldsymbol{\nabla_v}] - \boldsymbol{\nabla}_{[\boldsymbol{u}, \boldsymbol{v}]}$ when acting on 1-forms $\boldsymbol{\sigma}$ (or other tensors) as well as on tangent vectors. Show that

$$\langle \mathcal{R}(\boldsymbol{u}, \boldsymbol{v})\boldsymbol{\sigma}, \boldsymbol{w} \rangle = -\langle \boldsymbol{\sigma}, \mathcal{R}(\boldsymbol{u}, \boldsymbol{v})\boldsymbol{w} \rangle.$$

Exercise 11.12. ROTATION GROUP: RIEMANN CURVATURE

[Continuation of exercises 9.13, 9.14, and 10.17.] Calculate the components of the Riemann curvature tensor for the rotation group's manifold $SO(3)$; use the basis of generators $\{\boldsymbol{e}_\alpha\}$. [*Answer:*

$$R^\alpha{}_{\beta\gamma\delta} = \frac{1}{2}\delta^{\alpha\beta}_{\gamma\delta}, \tag{11.39}$$

where $\delta^{\alpha\beta}_{\gamma\delta}$ is the permutation symbol defined in equation (3.50l):

$$\delta^{\alpha\beta}_{\gamma\delta} \equiv (\delta^\alpha{}_\gamma \delta^\beta{}_\delta - \delta^\alpha{}_\delta \delta^\beta{}_\gamma).$$

Note that this answer is independent of location \mathcal{P} in the group manifold.]

CHAPTER **12**

NEWTONIAN GRAVITY IN THE LANGUAGE OF CURVED SPACETIME

The longest period of time for which a modern painting has hung upside down in a public gallery unnoticed is 47 days. This occurred to Le Bateau *by Matisse in the Museum of Modern Art, New York City. In this time 116,000 people had passed through the gallery.*

McWHIRTER AND McWHIRTER (1971)

§12.1. NEWTONIAN GRAVITY IN BRIEF

The equivalence principle is not unique to Einstein's description of the facts of gravity. What is unique to Einstein is the combination of the equivalence principle and local Lorentz geometry. To return to the world of Newton, forget everything discovered in the last century about special relativity, light cones, the limiting speed of light, and proper time. Return to the "universal time" t of earlier centuries. In terms of that universal time, and of rectangular, "Galilean" space coordinates, Newtonian theory gives for the trajectories of neutral test particles

$$\frac{d^2x^j}{dt^2} + \frac{\partial \Phi}{\partial x^j} = 0; \tag{12.1}$$

$$\Phi \text{ (sometimes denoted } -U) = \text{Newtonian potential.} \tag{12.2}$$

Customarily one interprets these equations as describing the "curved paths" $x^j(t)$ along which test particles fall in Euclidean space (*not* spacetime). These curved paths include circular orbits about the Earth and the parabolic trajectory of a baseball. Cartan (1923, 1924) asks one to abandon this viewpoint. Instead, he says, regard these trajectories as geodesics $[t(\lambda), x^j(\lambda)]$ in curved spacetime. (This change of viewpoint was embodied in Figures B and C of Box 1.6.) Since the "affinely ticking"

This chapter is entirely Track 2. Chapters 9–11 are necessary preparation for it. It is not needed for any later chapter, but it will be helpful in
(1) Chapter 17 (Einstein field equations) and
(2) Chapters 38 and 39 (experimental tests and other theories of gravity).

Newtonian gravity: original formulation

Newtonian gravity: translation into language of curved spacetime

Newtonian clocks carried by test particles read universal time (or some multiple, $\lambda = at + b$, thereof), the equation of motion (12.1) can be rewritten

$$\frac{d^2t}{d\lambda^2} = 0, \qquad \frac{d^2x^j}{d\lambda^2} + \frac{\partial\Phi}{\partial x^j}\left(\frac{dt}{d\lambda}\right)^2 = 0. \tag{12.3}$$

By comparing with the geodesic equation

$$d^2x^\alpha/d\lambda^2 + \Gamma^\alpha{}_{\beta\gamma}(dx^\beta/d\lambda)(dx^\gamma/d\lambda) = 0,$$

one can read off the values of the connection coefficients:

$$\Gamma^j{}_{00} = \partial\Phi/\partial x^j; \qquad \text{all other } \Gamma^\alpha{}_{\beta\gamma} \text{ vanish.} \tag{12.4}$$

And by inserting these into the standard equation (11.12) for the components of the Riemann tensor, one learns (exercise 12.1)

$$R^j{}_{0k0} = -R^j{}_{00k} = \frac{\partial^2\Phi}{\partial x^j \, \partial x^k}; \qquad \text{all other } R^\alpha{}_{\beta\gamma\delta} \text{ vanish.} \tag{12.5}$$

Finally, the source equation for the Newtonian potential

$$\nabla^2\Phi \equiv \sum_j \Phi_{,jj} = 4\pi\rho \tag{12.6}$$

one can rewrite with the help of the *"Ricci curvature tensor"*

$$R_{\alpha\beta} \equiv R^\mu{}_{\alpha\mu\beta} \text{ (contraction of } \boldsymbol{Riemann}) \tag{12.7}$$

in the geometric form (exercise 12.2)

$$R_{00} = 4\pi\rho; \qquad \text{all other } R_{\alpha\beta} \text{ vanish.} \tag{12.8}$$

Equation (12.4) for $\Gamma^\alpha{}_{\beta\gamma}$, equation (12.5) for $R^\alpha{}_{\beta\gamma\delta}$, equation (12.8) for $R_{\alpha\beta}$, plus the law of geodesic motion are the full content of Newtonian gravity, rewritten in geometric language.

It is one thing to pass quickly through these component manipulations. It is quite another to understand fully, in abstract and pictorial terms, the meanings of these equations and the structure of Newtonian spacetime. To produce such understanding, and to compare Newtonian spacetime with Einsteinian spacetime, are the goals of this chapter, which is based on the work of Cartan (1923, 1924), Trautman (1965), and Misner (1969a).

EXERCISES

Exercise 12.1. RIEMANN CURVATURE OF NEWTONIAN SPACETIME
Derive equation (12.5) for $R^\alpha{}_{\beta\gamma\delta}$ from equation (12.4) for $\Gamma^\alpha{}_{\beta\gamma}$.

Exercise 12.2. NEWTONIAN FIELD EQUATION
Derive the geometric form (12.8) of the Newtonian field equation from (12.5) through (12.7).

§12.2. STRATIFICATION OF NEWTONIAN SPACETIME

Galileo and Newton spoke of a flat, Euclidean "absolute space" and of an "absolute time," two concepts distinct and unlinked. In absolute space Newtonian physics took place; and as it took place, absolute time marched on. No hint was there that space and time might be two aspects of a single entity, a curved "*spacetime*"—until Einstein made the unification in relativity physics, and Cartan (1923) followed suit in Newtonian physics in order to provide clearer insight into Einstein's ideas.

How do the absolute space of Galileo and Newton, and their absolute time, fit into Cartan's "*Newtonian spacetime*"? The key to the fit is *stratification;* stratification produced by the universal time coordinate *t*.

Regard *t* as a function (scalar field) defined once and for all in Newtonian spacetime

The geometry of Newtonian spacetime:

$$t = t(\mathscr{P}). \tag{12.9}$$

"Universal time" as a scalar field

Without it, spacetime could not be Newtonian, for "*t*" is every bit as intrinsic to Newtonian spacetime as the metric "\boldsymbol{g}" is to Lorentz spacetime. The layers of spacetime are the slices of constant *t*—the "*space slices*"—each of which has an identical geometric structure: the old "absolute space."

Adopting Cartan's viewpoint, ask what kind of geometry is induced onto each space slice by the surrounding geometry of spacetime. A given space slice is endowed, by the Galilean coordinates of §12.1, with basis vectors $\boldsymbol{e}_j = \partial/\partial x^j$; and this basis has vanishing connection coefficients, $\Gamma^j{}_{kl} = 0$ [cf. equation (12.4)]. Consequently, *the geometry of each space slice is completely flat.*

"Absolute space" is Euclidean in its geometry, according to the old viewpoint, and the Galilean coordinates are Cartesian. Translated into Cartan's language, this says: not only is each space slice (*t* = constant) flat, and not only do its Galilean coordinates have vanishing connection coefficients, but also *each space slice is endowed with a three-dimensional metric, and its Galilean coordinate basis is orthonormal,*

Space slices with Euclidean geometry

$$\boldsymbol{e}_i \cdot \boldsymbol{e}_j = (\partial/\partial x^i) \cdot (\partial/\partial x^j) = \delta_{ij}. \tag{12.10}$$

If the space slices are really so flat, where do curvature and geodesic deviation enter in? They are properties of *spacetime*. Parallel transport a vector around a closed curve lying entirely in a space slice; it will return to its starting point unchanged. But transport it forward in time by Δt, northerly in space by Δx^k, back in time by $-\Delta t$, and southerly by $-\Delta x^k$ to its starting point; it will return changed by

Curvature acts in spacetime, not in space slices

$$\delta \boldsymbol{A} = -\mathscr{R}\left(\Delta t \frac{\partial}{\partial t}, \Delta x^k \frac{\partial}{\partial x^k}\right)\boldsymbol{A};$$

i.e.,

$$\delta A^0 = 0, \qquad \delta A^j = -R^j{}_{00k} A^0 (\Delta t)(\Delta x^k) = \frac{\partial^2 \Phi}{\partial x^j \, \partial x^k} A^0 (\Delta t)(\Delta x^k). \tag{12.11}$$

Geodesics of a space slice (Euclid's straight lines) that are initially parallel remain

always parallel. But geodesics of spacetime (trajectories of freely falling particles) initially parallel get pried apart or pushed together by spacetime curvature,

$$\mathbf{\nabla}_u \mathbf{\nabla}_u \mathbf{n} + \mathcal{R}(\mathbf{n}, \mathbf{u})\mathbf{u} = 0,$$

or equivalently in Galilean coordinates:

$$n^0 = dn^0/dt = 0 \text{ initially} \implies n^0 = 0 \text{ always;} \tag{12.12a}$$

$$\frac{d^2 n^j}{dt^2} + \frac{\partial^2 \Phi}{\partial x^j \, \partial x^k} \, n^k = 0 \tag{12.12b}$$

(see Box 12.1 and exercise 12.3).

EXERCISE

Exercise 12.3. GEODESIC DEVIATION DERIVED

Produce a third column for Box 11.4, one that carries out the "geometric analysis" in component notation using the Galilean connection coefficients (12.4) of Newtonian spacetime. Thereby achieve a deeper understanding of how the geometric analysis parallels the old Newtonian analysis.

§12.3. GALILEAN COORDINATE SYSTEMS

The Lorentz spacetime of special relativity has an existence and structure completely independent of any coordinate system. But a special property of its geometry (zero curvature) allows the introduction of a special class of coordinates (Lorentz coordinates), which cling to spacetime in a special way

$$(\partial/\partial x^\alpha) \cdot (\partial/\partial x^\beta) = \eta_{\alpha\beta} \text{ everywhere.}$$

By studying these special coordinate systems and the relationships between them (Lorentz transformations), one learns much about the structure of spacetime itself (breakdown in simultaneity; Lorentz contraction; time dilatation; ...).

Galilean coordinates defined

Similarly for Newtonian spacetime. Special properties of its geometry (explored in abstract later; Box 12.4) permit the introduction of special coordinates (Galilean coordinates), which cling to spacetime in a special way

$$x^0(\mathcal{P}) = t(\mathcal{P});$$

$$(\partial/\partial x^j) \cdot (\partial/\partial x^k) = \delta_{jk};$$

$$\Gamma^j{}_{00} = \Phi_{,j} \text{ for some scalar field } \Phi, \text{ and all other } \Gamma^\alpha{}_{\beta\gamma} \text{ vanish.}$$

To understand Newtonian spacetime more deeply, study the relations between these Galilean coordinate systems.

Box 12.1 GEODESIC DEVIATION IN NEWTONIAN SPACETIME

Coordinate system for calculation: Galilean space coordinates x^j and universal time coordinate t. General component form of equation:

$$\frac{D^2 n^\alpha}{d\lambda^2} + R^\alpha{}_{\beta\gamma\delta} \frac{dx^\beta}{d\lambda} n^\gamma \frac{dx^\delta}{d\lambda} = 0.$$

Special conditions for this calculation: let the particles' clocks (affine parameters) all be normalized to read universal time, $\lambda = t$. This means that the separation vector

$$n^\alpha = (\partial x^\alpha / \partial n)_\lambda$$

between geodesics has zero time component, $n^0 = 0$; i.e., in abstract language,

$$\langle \boldsymbol{d}t, \boldsymbol{n} \rangle = t_{,\alpha} n^\alpha = n^0 = 0;$$

i.e., in geometric language, \boldsymbol{n} lies in a space slice (surface of constant t).

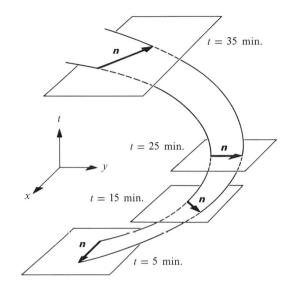

Evaluation of covariant derivative:

$$\frac{Dn^\alpha}{D\lambda} = \frac{dn^\alpha}{d\lambda} + \underbrace{\Gamma^\alpha{}_{\beta\mu} n^\beta \frac{dx^\mu}{d\lambda}}_{} = \frac{dn^\alpha}{d\lambda},$$

[0 unless $\beta = 0$] [0 unless β is a space index $(n^0 = 0)$]

$$\frac{D^2 n^\alpha}{d\lambda^2} = \frac{d(Dn^\alpha/d\lambda)}{d\lambda} + \underbrace{\Gamma^\alpha{}_{\beta\mu} \frac{dn^\beta}{d\lambda} \frac{dx^\mu}{d\lambda}}_{} = \frac{d(Dn^\alpha/d\lambda)}{d\lambda} = \frac{d^2 n^\alpha}{d\lambda^2} \equiv \frac{d^2 n^\alpha}{dt^2}.$$

[since $\lambda = t$]

[0 unless $\beta = 0$] [0 unless β is space index]

Evaluation of tidal accelerations:

$$R^0{}_{\beta\gamma\delta} \frac{dx^\beta}{d\lambda} n^\gamma \frac{dx^\delta}{d\lambda} = 0 \quad \text{since } R^j{}_{0k0} \text{ and } R^j{}_{00k} \text{ are only nonzero components.}$$

$$\underbrace{R^j{}_{\beta\gamma\delta} \frac{dx^\beta}{d\lambda} n^\gamma \frac{dx^\delta}{d\lambda}}_{} = R^j{}_{0k0} \frac{dt}{d\lambda} n^k \frac{dt}{d\lambda} = R^j{}_{0k0} n^k = \frac{\partial^2 \Phi}{\partial x^j \, \partial x^k} n^k.$$

[0 unless γ is space index]

[for γ a space index: 0 unless $\beta = \delta = 0$]

Resultant equation of geodesic deviation:

$$\frac{d^2 n^0}{dt^2} = 0 \qquad \left(\begin{array}{l} \text{agrees with result } n^0 = 0 \text{ always, which} \\ \text{followed from choice } \lambda = t \text{ for all particles} \end{array} \right)$$

$$\frac{d^2 n^j}{dt^2} + \frac{\partial^2 \Phi}{\partial x^j \, \partial x^k} n^k = 0 \qquad \left(\begin{array}{l} \text{agrees with Newton-type calculation} \\ \text{in Box 11.4; see also exercise 12.3} \end{array} \right).$$

Point of principle: how can one write down the laws of gravity and properties of spacetime in Galilean coordinates first (§12.1), and only afterward (here) come to grip with the nature of the coordinate system and its nonuniqueness? Answer: (a quotation from §3.1, slightly modified): "Here and elsewhere in science, as emphasized not least by Henri Poincaré, that view is out of date which used to say 'Define your terms before you proceed.' All the laws and theories of physics, including Newton's laws of gravity, have this deep and subtle character, that they both define the concepts they use (here Galilean coordinates) and make statements about these concepts."

The Newtonian laws of gravity, written in a Galilean coordinate system

$$x^0 = t, \qquad (\partial/\partial x^j) \cdot (\partial/\partial x^k) = \delta_{jk}$$

make the statement "$\Gamma^j{}_{00} = \Phi_{,j}$ and all other $\Gamma^\alpha{}_{\beta\gamma} = 0$" about the geometry of spacetime. This statement in turn gives information about the relationships between different Galilean systems. Let one Galilean system $\{x^\alpha(\mathcal{P})\}$ be given, and seek the most general coordinate transformation leading to another, $\{x^{\alpha'}(\mathcal{P})\}$. The following constraints exist: (1) $x^{0'} = x^0 = t$ (both time coordinates must be universal time); (2) at fixed t (i.e., in a fixed space slice) both sets of space coordinates must be Euclidean, so they must be related by a rotation and a translation:

$$x^{j'} = A_{j'k} x^k + a^{j'} \tag{12.13a}$$

[translation]

[rotation matrix, i.e., $A_{j'\ell} A_{k'\ell} = \delta_{j'k'}$]

$$x^k = A_{j'k} x^{j'} - a^k, \text{ with } a^k \equiv A_{j'k} a^{j'}. \tag{12.13b}$$

The rotation and translation might, *a priori,* be different on different slices, $A_{j'k} = A_{j'k}(t)$ and $a^j = a^j(t)$; but (3) they must be constrained by the required special form of the connection coefficients. Calculate the connection coefficients in the new coordinate system, given their form in the old. The result (exercise 12.4) is:

$$\Gamma^{j'}{}_{0'k'} = \Gamma^{j'}{}_{k'0'} = A_{j'\ell} \dot{A}_{k'\ell} \qquad \text{(produces "Coriolis forces");}$$

$$\Gamma^{j'}{}_{0'0'} = \frac{\partial \Phi}{\partial x^{j'}} + A_{j'k}(\ddot{A}_{\ell'k} x^{\ell'} - \ddot{a}^k); \tag{12.14}$$

["centrifugal forces"] ["inertial forces"]

all other $\Gamma^{\alpha'}{}_{\beta'\gamma'}$ vanish

("Euclidean" index conventions; repeated space indices to be summed even if both are down; dot denotes time derivative). These have the standard Galilean form (12.4) if and only if

$$\dot{A}_{j'k} = 0, \qquad \Phi' = \Phi - \ddot{a}^k x^k + \text{constant}. \tag{12.15}$$

$\begin{bmatrix} \text{Newtonian potential in} \\ \text{new coordinate system} \end{bmatrix}$ $\begin{bmatrix} \text{Newtonian potential in} \\ \text{old coordinate system} \end{bmatrix}$

These results can be restated in words: any two Galilean coordinate systems are related by (1) a time-independent rotation of the space grid (*same* rotation on each space slice), and (2) a time-dependent translation of the space grid (translation possibly *different* on different slices)

$$x^{j'} = A_{j'k}x^k + a^{j'}(t). \qquad (12.16)$$

[constant] ↑ ↑[time-dependent]

<div style="float:right">Transformations linking Galilean coordinate systems</div>

The Newtonian potential is not a function defined in spacetime with existence independent of all coordinate systems. (There is no coordinate-free way to measure it.) Rather, it depends for its existence on a particular choice of Galilean coordinates; and if the choice is changed via equation (12.16), then Φ is changed:

$$\Phi' = \Phi - \ddot{a}^k x^k. \qquad (12.17)$$

<div style="float:right">Newtonian potential depends on choice of Galilean coordinate system</div>

(By contrast, an existence independent of all coordinates is granted to the universal time $t(\mathscr{P})$ and the covariant derivative ▼.)

Were all the matter in the universe concentrated in a finite region of space and surrounded by emptiness ("island universe"), then one could impose the global boundary condition

$$\Phi \longrightarrow 0 \text{ as } r \equiv (x^k x^k)^{1/2} \longrightarrow \infty. \qquad (12.18)$$

<div style="float:right">Absolute Galilean coordinates defined</div>

This would single out a subclass of Galilean coordinates ("*absolute*" Galilean coordinates), with a unique, common Newtonian potential. The transformation from one absolute Galilean coordinate system to any other would be

$$x^{j'} = A_{j'k}x^k + a^{j'} + v^{j'}t \qquad (12.19)$$

[constant rotation] ↑ ↑ ↑ [constant velocity]

[constant displacement]

<div style="float:right">Transformations linking absolute Galilean coordinate systems</div>

("Galilean transformation"). *But,* (1) by no local measurements could one ever distinguish these absolute Galilean coordinate systems from the broader class of Galilean systems (to distinguish, one must integrate the locally measurable quantity $\Phi_{,j} = \Gamma^j{}_{00}$ out to infinity); and (2) astronomical data deny that the real universe is an island of matter surrounded by emptiness.

It is instructive to compare Galilean coordinates and Newtonian spacetime as described above with Lorentz coordinates and the Minkowskii spacetime of special relativity, and with the general coordinates and Einstein spacetime of general relativity; see Boxes 12.2 and 12.3.

(*continued on page 298*)

Box 12.2 NEWTONIAN SPACETIME, MINKOWSKIIAN SPACETIME, AND EINSTEINIAN SPACETIME: COMPARISON AND CONTRAST

Query	Newtonian spacetime	Minkowskiian spacetime (special relativity)	Einsteinian spacetime (general relativity)
What *a priori* geometric structures does spacetime possess?	(1) Universal time function *t* (2) Covariant derivative ▽ (3) Spatial metric "·"; but spacetime metric can *not* be defined here (exercise 12.10)	A spacetime metric that is flat (vanishing Riemann curvature)	A spacetime metric
What preferred coordinate systems are present?	(1) Galilean coordinates in general (2) Absolute Galilean coordinates in an island universe (this case not considered here)	Lorentz coordinates	In general, every coordinate system is equally preferred (though in special cases with symmetry there are special preferred coordinates)
What is required to select out a particular preferred coordinate system?	(1) A single spatial orientation, the same throughout all spacetime (three Euler angles) (2) The arbitrary world line of the origin of space coordinates (three functions of time)	(1) A single spatial orientation, the same throughout all spacetime (three Euler angles) (2) The location of the origin of coordinates (four numbers) (3) The velocity of the origin of space coordinates (three numbers)	All four functions of position $x^\alpha(\mathcal{P})$
Under what conditions is "\mathcal{P} and \mathcal{Q} are simultaneous" well-defined?	In general; it is a coordinate-free geometric concept	Only after a choice of Lorentz frame has been made; "simultaneity" depends on the frame's velocity	Only after arbitrary choice of time coordinate has been made
Under what conditions is "\mathcal{P} and \mathcal{Q} occur at same point in space" well-defined?	Only after choice of Galilean coordinates has been made	Only after choice of Lorentz coordinates has been made	Only after arbitrary choice of space coordinates has been made
Under what conditions is "u and v, at different events, point in same direction" well-defined?	Only if u and v are both spatial vectors ($\langle dt, u \rangle = \langle dt, v \rangle = 0$); or if they lie in the same space slice and are arbitrary vectors; or if there exists a preferred route connecting their locations, along which to compare them by parallel transport	Always	Only if u and v lie at events infinitesimally close together; or if there exists a preferred route (e.g., a unique geodesic) connecting their locations, along which to compare them by parallel transport
Under what conditions is "the invariant distance between \mathcal{P} and \mathcal{Q}" well-defined?	Only if \mathcal{P} and \mathcal{Q} lie in the same space slice	Always	Only if \mathcal{P} and \mathcal{Q} are sufficiently close together; or if there exists a unique preferred world line (e.g., a geodesic) linking them, along which to measure the distance

Box 12.3 NEWTONIAN GRAVITY Á LA CARTAN, AND EINSTEINIAN GRAVITY: COMPARISON AND CONTRAST

Property	Newton-Cartan	Einstein
Idea in brief (formulations of the equivalence principle of very different scope)	Laws of motion of free particles in a local, freely falling, nonrotating frame are identical to Newton's laws of motion as expressed in a gravity-free Galilean frame	Laws of physics in a local, freely falling, nonrotating frame are identical with the laws of physics as formulated in special relativity in a Lorentz frame
Idea even more briefly stated	Point mechanics simple in a local inertial frame	Everything simple in a local inertial frame
Consequence (tested to one part in 10^{11} by Roll-Krotkov-Dicke experiment)	Test particles of diverse composition started with same initial position and same initial velocity follow the same world line ("definition of geodesic")	Test particles of diverse composition started with same initial position and same initial velocity follow the same world line ("definition of geodesic")
Another consequence	In every local region, there exists a local frame ("freely falling frame") in which all geodesics appear straight (all $\Gamma^\alpha{}_{\mu\nu} = 0$)	In every local region there exists a local frame ("freely falling frame") in which all geodesics appear straight (all $\Gamma^\alpha{}_{\mu\nu} = 0$)
Consequence of way light rays travel in real physical world?	Disregarded or evaded. All light rays have same velocity? Speed depend on motion of source? Speed depend on motion of observer? Possible to move fast enough to catch up with a light ray? No satisfactory position on any of these issues	Spacetime always and everywhere has local Lorentz character
Summary of spacetime structure	Stratified into spacelike slices; geometry in each slice Euclidean; each slice characterized by value of universal time (geodesic parameter); displacement of one slice with respect to another not specified; no such thing as a spacetime interval	No stratification. Well-defined interval between every event and every nearby event; spacetime has everywhere local Lorentz character, with one local frame (specific space and time axes) as good as another (other space and time axes); "homogeneous" rather than stratified
This structure expressed in mathematical language	$\Gamma^\alpha{}_{\mu\nu}$'s, yes; spacetime metric $g_{\mu\nu}$, no;	$\Gamma^\alpha{}_{\mu\nu}$'s have no independent existence; all derived from

$$\Gamma^i{}_{00} = \frac{\partial \Phi}{\partial x^i} \ (i = 1, 2, 3);$$

all other $\Gamma^\alpha{}_{\mu\nu}$ vanish

$$\Gamma^\alpha{}_{\mu\nu} = g^{\alpha\beta} \frac{1}{2} \left(\frac{\partial g_{\beta\nu}}{\partial x^\mu} + \frac{\partial g_{\beta\mu}}{\partial x^\nu} - \frac{\partial g_{\mu\nu}}{\partial x^\beta} \right)$$

("metric theory of gravity")

EXERCISES

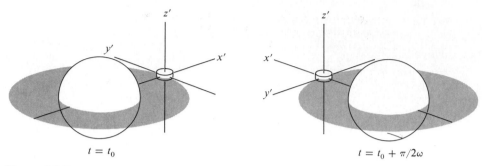

Figure 12.1.
The coordinate system carried by an orbital laboratory
as it moves in a circular orbit about the Earth.

Exercise 12.4. CONNECTION COEFFICIENTS FOR ROTATING, ACCELERATING COORDINATES

Beginning with equation (12.4) for the connection coefficients of a Galilean coordinate system $\{x^\alpha(\mathscr{P})\}$, derive the connection coefficients (12.14) of the coordinate system $\{x^{\alpha'}(\mathscr{P})\}$ of equations (12.13). From this, verify that (12.15) are necessary and sufficient for $\{x^{\alpha'}(\mathscr{P})\}$ to be Galilean.

Exercise 12.5. EINSTEIN'S ELEVATOR

Use the formalism of this chapter to discuss "Einstein's elevator"—i.e., the equivalence of "gravity" to an acceleration of one's reference frame. Which aspects of "gravity" are equivalent to an acceleration, and which are not?

Exercise 12.6. GEODESIC DEVIATION ABOVE THE EARTH

A manned orbital laboratory is put into a circular orbit about the Earth [radius of orbit = r_0, angular velocity = $\omega = (M/r_0^3)^{1/2}$—why?]. An astronaut jetisons a bag of garbage and watches it move along its geodesic path. He observes its motion relative to (non-Galilean) space coordinates $\{x^{j'}(\mathscr{P})\}$ which—see Figure 12.1—(1) are Euclidean at each moment of universal time [$(\partial/\partial x^{j'}) \cdot (\partial/\partial x^{k'}) = \delta_{jk}$], (2) have origin at the laboratory's center, (3) have $\partial/\partial x'$ pointing away from the Earth, (4) have $\partial/\partial x'$ and $\partial/\partial y'$ in the plane of orbit. Use the equation of geodesic deviation to calculate the motion of the garbage bag in this coordinate system. Verify the answer by examining the Keplerian orbits of laboratory and garbage. *Hints:* (1) Calculate $R^{\alpha'}{}_{\beta'\gamma'\delta'}$ in this coordinate system by a trivial transformation of tensorial components. (2) Use equation (12.14) to calculate $\Gamma^{\alpha'}{}_{\beta'\gamma'}$ at the center of the laboratory (i.e., on the fiducial geodesic).

§12.4. GEOMETRIC, COORDINATE-FREE FORMULATION OF NEWTONIAN GRAVITY

To restate Newton's theory of gravity in coordinate-independent, geometric language is the principal goal of this chapter. It has been achieved, thus far, with extensive assistance from a special class of coordinate systems, the Galilean coordinates. To

climb out of Galilean coordinates and into completely coordinate-free language is straightforward in principle. One merely passes from index notation to abstract notation.

Example: Restate in coordinate-free language the condition $\Gamma^0{}_{\alpha\beta} = 0$ of Galilean coordinates.

Solution: Write $\Gamma^0{}_{\alpha\beta} = -\langle \boldsymbol{\nabla}_\beta\, dt, \boldsymbol{e}_\alpha \rangle$; the vanishing of this for all α means $\boldsymbol{\nabla}_\beta\, dt = 0$ for all β, which in turn means $\boldsymbol{\nabla}_{\boldsymbol{u}}\, dt = 0$ for all \boldsymbol{u}. In words: *the gradient of universal time is covariantly constant.*

By this process one can construct a set of coordinate-free statements about Newtonian spacetime (Box 12.4) that are completely equivalent to the standard, nongeometric version of Newton's gravitation theory. From standard Newtonian theory, one can deduce these geometric statements (exercise 12.7); from these geometric statements, regarded as axioms, one can deduce standard Newtonian theory (exercise 12.8).

Coordinate-free, geometric axioms for Newton's theory of gravity

Exercise 12.7. FROM NEWTON TO CARTAN

From the standard axioms of Newtonian theory (last part of Box 12.4) derive the geometric axioms (first part of Box 12.4). *Suggested procedure:* Verify each of the geometric axioms by a calculation in the Galilean coordinate system. Make free use of the calculations and results in §12.1.

Exercise 12.8. FROM CARTAN TO NEWTON

From the geometric axioms of Newtonian theory (first part of Box 12.4) derive the standard axioms (last part of Box 12.4). *Suggested procedure:* (1) Pick three orthonormal, spatial basis vectors (\boldsymbol{e}_j with $\boldsymbol{e}_j \cdot \boldsymbol{e}_k = \delta_{jk}$) at some event \mathscr{P}_0. Parallel transport each of them by arbitrary routes to all other events in spacetime.

(2) Use the condition $\mathscr{R}(\boldsymbol{u}, \boldsymbol{n})\boldsymbol{e}_j = 0$ for all \boldsymbol{u} and \boldsymbol{n} [axiom (3)] and an argument like that in §11.5 to conclude: (a) the resultant vector fields \boldsymbol{e}_j are independent of the arbitrary transport routes, (b) $\boldsymbol{\nabla}\boldsymbol{e}_j = 0$ for the resultant fields, and (c) $[\boldsymbol{e}_j, \boldsymbol{e}_k] = 0$.

(3) Pick an arbitrary "time line", which passes through each space slice (slice of constant t) once and only once. Parametrize it by t and select its tangent vector as the basis vector \boldsymbol{e}_0 at each event along it. Parallel transport each of these \boldsymbol{e}_0's throughout its respective space slice by arbitrary routes.

(4) From axiom (4) conclude that the resultant field is independent of the transport routes; also show that the above construction process guarantees $\boldsymbol{\nabla}_j\boldsymbol{e}_0 = \boldsymbol{\nabla}_0\boldsymbol{e}_j = 0$.

(5) Show that $[\boldsymbol{e}_\alpha, \boldsymbol{e}_\beta] = 0$ for all pairs of the four basis-vector fields, and conclude from this that there exists a coordinate system ("Galilean coordinates") in which $\boldsymbol{e}_\alpha = \partial/\partial x^\alpha$ (see §11.5 and exercise 9.9).

(6) Show that in this coordinate system $\boldsymbol{e}_j \cdot \boldsymbol{e}_k = \delta_{jk}$ everywhere (space coordinates are Euclidean), and the only nonzero components of the connection coefficient are $\Gamma^j{}_{00}$; here axioms (6) and (2) will be helpful.

(7) From the self-adjoint property of the Jacobi curvature operator (axiom 7) show that $R^j{}_{0k0} = R^k{}_{0j0}$; show that in terms of the connection coefficients this reads $\Gamma^j{}_{00,k} = \Gamma^k{}_{00,j}$; and from this conclude that there exists a potential Φ such that $\Gamma^j{}_{00} = \Phi_{,j}$.

(8) Show that the geometric field equation (axiom 5) reduces to Poisson's equation $\nabla^2\Phi = 4\pi\rho$.

(9) Show that the geodesic equation for free fall (axiom 8) reduces to the Newtonian equation of motion $d^2x^j/dt^2 + \Phi_{,j} = 0$.

EXERCISES

(continued on page 302)

**Box 12.4 NEWTONIAN GRAVITY: GEOMETRIC FORMULATION
CONTRASTED WITH STANDARD FORMULATION**

Geometric Formulation

Newton's theory of gravity and the properties of Newtonian spacetime can be derived from the following axioms. (For derivation see exercise 12.8.)

(1) There exists a function t called "universal time", and a symmetric covariant derivative $\mathbf{\nabla}$ (with associated geodesics, parallel transport law, curvature operator, etc.).

(2) The 1-form \boldsymbol{dt} is covariantly constant; i.e.,

$$\mathbf{\nabla_u}\, \boldsymbol{dt} = 0 \text{ for all } \boldsymbol{u}.$$

[Consequence: if \boldsymbol{w} is a spatial vector field (i.e., \boldsymbol{w} lies everywhere in a surface of constant t; i.e. $\langle \boldsymbol{dt}, \boldsymbol{w} \rangle = 0$ everywhere), then $\mathbf{\nabla_u w}$ is also spatial for every \boldsymbol{u},

$$\langle \boldsymbol{dt}, \mathbf{\nabla_u w} \rangle = \mathbf{\nabla_u}\, \underbrace{\langle \boldsymbol{dt}, \boldsymbol{w} \rangle}_{\text{[0 always]}} - \langle \underbrace{\mathbf{\nabla_u}\, \boldsymbol{dt}}_{\text{[0 always]}}, \boldsymbol{w} \rangle = 0.]$$

(3) Spatial vectors arc unchanged by parallel transport around infinitesimal closed curves; i.e.,

$$\mathscr{R}(\boldsymbol{u}, \boldsymbol{n})\boldsymbol{w} = 0 \text{ if } \boldsymbol{w} \text{ is spatial, for every } \boldsymbol{u} \text{ and } \boldsymbol{n}.$$

(4) All vectors are unchanged by parallel transport around infinitesimal, spatial, closed curves; i.e.,

$$\mathscr{R}(\boldsymbol{v}, \boldsymbol{w}) = 0 \text{ for every spatial } \boldsymbol{v} \text{ and } \boldsymbol{w}.$$

(5) The Ricci curvature tensor, $R_{\alpha\beta} \equiv R^{\mu}{}_{\alpha\mu\beta}$, has the form

$$\boldsymbol{Ricci} = 4\pi\rho\, \boldsymbol{dt} \otimes \boldsymbol{dt},$$

where ρ is the density of mass.

(6) There exists a metric "\cdot" defined on spatial vectors only, which is compatible with the covariant derivative in this sense: for any spatial \boldsymbol{w} and \boldsymbol{v}, and for any \boldsymbol{u} whatsoever,

$$\mathbf{\nabla_u}(\boldsymbol{w} \cdot \boldsymbol{v}) = (\mathbf{\nabla_u w}) \cdot \boldsymbol{v} + \boldsymbol{w} \cdot (\mathbf{\nabla_u v}).$$

[Note: axioms (1), (2), and (3) guarantee that such a spatial metric can exist; see exercise 12.9.]

(7) The Jacobi curvature operator $\mathcal{J}(\boldsymbol{u}, \boldsymbol{n})$, defined for any vectors $\boldsymbol{u}, \boldsymbol{n}, \boldsymbol{p}$ by

$$\mathcal{J}(\boldsymbol{u}, \boldsymbol{n})\boldsymbol{p} = \frac{1}{2}[\mathcal{R}(\boldsymbol{p}, \boldsymbol{n})\boldsymbol{u} + \mathcal{R}(\boldsymbol{p}, \boldsymbol{u})\boldsymbol{n}],$$

is *"self-adjoint"* when operating on spatial vectors; i.e.,

$$\boldsymbol{v} \cdot [\mathcal{J}(\boldsymbol{u}, \boldsymbol{n})\boldsymbol{w}] = \boldsymbol{w} \cdot [\mathcal{J}(\boldsymbol{u}, \boldsymbol{n})\boldsymbol{v}] \text{ for all spatial } \boldsymbol{v}, \boldsymbol{w};$$
$$\text{and for any } \boldsymbol{u}, \boldsymbol{n}.$$

(8) "Ideal rods" measure the lengths that are calculated with the spatial metric; "ideal clocks" measure universal time t (or some multiple thereof); and "freely falling particles" move along geodesics of $\boldsymbol{\nabla}$. [Note: this can be regarded as a definition of "ideal rods," "ideal clocks," and "freely falling particles." A more complete theory (e.g., general relativity; see §16.4) would predict in advance whether a given physical rod or clock is ideal, and whether a given real particle is freely falling.]

Note: For an alternative but equivalent set of axioms, see pp. 106–107 of Trautman (1965).

Standard Formulation

The following standard axioms are equivalent to the above.

(1) There exist a universal time t, a set of Cartesian space coordinates x^j (called "Galilean coordinates"), and a Newtonian gravitational potential Φ.

(2) The density of mass ρ generates the Newtonian potential by Poisson's equation,

$$\boldsymbol{\nabla}^2\Phi \equiv \frac{\partial^2\Phi}{\partial x^j \partial x^j} = 4\pi\rho.$$

(3) The equation of motion for a freely falling particle is

$$\frac{d^2x^j}{dt^2} + \frac{\partial\Phi}{\partial x^j} = 0.$$

(4) "Ideal rods" measure the Galilean coordinate lengths; "ideal clocks" measure universal time.

Exercise 12.9. SPATIAL METRIC ALLOWED BY OTHER AXIOMS

Show that the geometric axioms (1), (2), and (3) of Box 12.4 permit one to introduce a spatial metric satisfying axiom (6). *Hint:* Pick an arbitrary spatial basis $\{\boldsymbol{e}_j\}$ at some event. *Define* it to be orthonormal, $\boldsymbol{e}_j \cdot \boldsymbol{e}_k \equiv \delta_{jk}$. Extend this basis through all spacetime by the method used in (1) of exercise 12.8. Define $\boldsymbol{e}_j \cdot \boldsymbol{e}_k \equiv \delta_{jk}$ everywhere in spacetime for this basis. Then prove that the resulting metric satisfies the compatibility condition of axiom (6).

Exercise 12.10. SPACETIME METRIC FORBIDDEN BY OTHER AXIOMS

Show that in Newtonian spacetime it is impossible to construct a nondegenerate spacetime metric \boldsymbol{g}, defined on *all* vectors, that is compatible with the covariant derivative in the sense that

$$\boldsymbol{\nabla}_u g(n, p) = g(\boldsymbol{\nabla}_u n, p) + g(n, \boldsymbol{\nabla}_u p). \tag{12.20}$$

Note: to prove this requires mastery of the material in Chapter 8 or 13; so study either 8 or 13 before tackling it. *Hint:* Assume that such a \boldsymbol{g} exists. Show, by the methods of exercise 12.8, that in a Galilean coordinate system the spatial components g_{jk} are independent of position in spacetime. Then use this and the form of $R^{\alpha}{}_{\beta\gamma\delta}$ in Galilean coordinates to prove R_{i0k0} and $-R_{0ik0}$ are not identical, a result that conflicts with the symmetries of the Riemann tensor [eq. (8.45)] in a manifold with compatible metric and covariant derivative.

§12.5. THE GEOMETRIC VIEW OF PHYSICS: A CRITIQUE

An important digression is in order.

"Every physical quantity must be describable by a (coordinate-free) geometric object, and the laws of physics must all be expressible as geometric relationships between these geometric objects." This view of physics, sometimes known as the *"principle of general covariance,"* pervades twentieth-century thinking. But does it have any forcible content? No, not at all, according to one viewpoint that dates back to Kretschmann (1917). Any physical theory originally written in a special coordinate system can be recast in geometric, coordinate-free language. Newtonian theory is a good example, with its equivalent geometric and standard formulations (Box 12.4). Hence, as a sieve for separating viable theories from nonviable theories, the principle of general covariance is useless.

But another viewpoint is cogent. It constructs a powerful sieve in the form of a slightly altered and slightly more nebulous principle: "Nature likes theories that are simple when stated in coordinate-free, geometric language."* According to this principle, Nature must love general relativity, and it must hate Newtonian theory. Of all theories ever conceived by physicists, general relativity has the simplest, most elegant geometric foundation [three axioms: (1) there is a metric; (2) the metric is governed by the Einstein field equation $\boldsymbol{G} = 8\pi\boldsymbol{T}$; (3) all special relativistic laws of physics are valid in local Lorentz frames of metric]. By contrast, what diabolically

The principle of general covariance has no forcible content

Twentieth-century viewpoint judges a theory by simplicity of its geometric formulation

Einstein's theory of gravity is simple; Newton's is complex

*Admittedly, this principle is anthropomorphic: twentieth-century physicists like such theories and even find them effective in correlating observational data. Therefore, Nature must like them too!

clever physicist would ever foist on man a theory with such a complicated geometric foundation as Newtonian theory?

Of course, from the nineteenth-century viewpoint, the roles are reversed. It judged simplicity of theories by examining their coordinate formulations. In Galilean coordinates, Newtonian theory is beautifully simple. Expressed as differential equations for the metric coefficients in a specific coordinate system, Einstein's field equations (ten of them now!) are horrendously complex.

The geometric, twentieth-century view prevails because it accords best with experimental data (see Chapters 38–40). In Chapter 17 it will be applied ruthlessly to make Einstein's field equation seem compelling.

CHAPTER **13**

RIEMANNIAN GEOMETRY: METRIC AS FOUNDATION OF ALL

Philosophy is written in this great book (by which I mean the universe) which stands always open to our view, but it cannot be understood unless one first learns how to comprehend the language and interpret the symbols in which it is written, and its symbols are triangles, circles, and other geometric figures, without which it is not humanly possible to comprehend even one word of it; without these one wanders in a dark labyrinth.

GALILEO GALILEI (1623)

§13.1. NEW FEATURES IMPOSED ON GEOMETRY BY LOCAL VALIDITY OF SPECIAL RELATIVITY

This chapter is entirely Track 2. Chapters 9–11 are necessary preparation for it. It will be needed as preparation for
(1) the second half of Chapter 14 (calculation of curvature), and
(2) the details, but not the message, of Chapter 15 (Bianchi identities).
§13.6 (proper reference frame) will be useful throughout the applications of gravitation theory (Chapters 18–40).

Constraints imposed on spacetime by special relativity

Freely falling particles (geodesics) define and probe the structure of spacetime. This spacetime is curved. Gravitation is a manifestation of its curvature. So far, so good, in terms of Newton's theory of gravity as translated into geometric language by Cartan. What is absolutely unacceptable, however, is the further consequence of the Cartan-Newton viewpoint (Chapter 12): stratification of spacetime into slidable slices, with no meaning for the spacetime separation between an event in one slice and an event in another.

Of all the foundations of physics, none is more firmly established than special relativity; and of all the lessons of special relativity none stand out with greater force than these. (1) Spacetime, far from being stratified, is homogeneous and isotropic throughout any region small enough ("local region") that gravitational tide-producing effects ("spacetime curvatures") are negligible. (2) No local experiment whatsoever can distinguish one local inertial frame from another. (3) The speed of light is the same in every local inertial frame. (4) It is not possible to give frame-independent meaning to the separation in time ("no Newtonian stratifica-

tion"). (5) Between every event and every nearby event there exists a frame-independent, coordinate-independent spacetime interval ("Riemannian geometry"). (6) Spacetime is always and everywhere locally Lorentz in character ("local Lorentz character of this Riemannian geometry").

What mathematics gives all these physical properties? A metric; a metric that is locally Lorentz (§§13.2 and 13.6). All else then follows. In particular, the metric destroys the stratified structure of Newtonian spacetime, as well as its gravitational potential and universal time coordinate. But not destroyed are the deepest features of Newtonian gravity: (1) the equivalence principle (as embodied in geodesic description of free-fall motion, §§13.3 and 13.4); and (2) spacetime curvature (as measured by tidal effects, §13.5).

Metric: the instrument which imposes the constraints

The skyscraper of vectors, forms, tensors (Chapter 9), geodesics, parallel transport, covariant derivative (Chapter 10), and curvature (Chapter 11) has rested on crumbling foundations—Newtonian physics and a geodesic law based on Newtonian physics. But with metric now on the scene, the whole skyscraper can be transferred to new foundations without a crack appearing. Only one change is necessary: the geodesic law must be selected in a new, relativistic way; a way based on metric (§§13.3 and 13.4). Resting on metric foundations, spacetime curvature acquires additional and stronger properties (the skyscraper is redecorated and extended), which are studied in §13.5 and in Chapters 14 and 15, and which lead almost inexorably to Einstein's field equation.

§13.2. METRIC

A spacetime metric; a curved spacetime metric; a locally Lorentz, curved spacetime metric. This is the foundation of spacetime geometry in the real, physical world. Therefore take a moment to recall what "metric" is in three contrasting languages.

Metric described in three languages

In the language of elementary geometry, "metric" is a table giving the interval between every event and every other event (Box 13.1 and Figure 13.1). In the language of coordinates, "metric" is a set of ten functions of position, $g_{\mu\nu}(x^\alpha)$, such that the expression

$$\Delta s^2 = -\Delta\tau^2 = g_{\mu\nu}(x^\alpha)\,\Delta x^\mu\,\Delta x^\nu \tag{13.1}$$

gives the interval between any event x^α and any nearby event $x^\alpha + \Delta x^\alpha$. In the language of abstract differential geometry, metric is a bilinear machine, $\boldsymbol{g} \equiv (\ldots \cdot \ldots)$, to produce a number ["scalar product $\boldsymbol{g}(\boldsymbol{u}, \boldsymbol{v}) \equiv (\boldsymbol{u} \cdot \boldsymbol{v})$"] out of two tangent vectors, \boldsymbol{u} and \boldsymbol{v}.

The link between the abstract, machine viewpoint and the concrete coordinate viewpoint is readily exhibited. Let the tangent vector

$$\boldsymbol{\xi} \equiv \Delta x^\alpha \boldsymbol{e}_\alpha = \Delta x^\alpha (\partial/\partial x^\alpha)$$

represent the displacement between two neighboring events. The abstract viewpoint gives

$$\Delta s^2 \equiv \boldsymbol{\xi} \cdot \boldsymbol{\xi} \equiv \boldsymbol{g}(\Delta x^\mu \boldsymbol{e}_\mu, \Delta x^\nu \boldsymbol{e}_\nu) = \Delta x^\mu \, \Delta x^\nu \boldsymbol{g}(\boldsymbol{e}_\mu, \boldsymbol{e}_\nu)$$

(continued on page 310)

Box 13.1 METRIC DISTILLED FROM DISTANCES

Raw Data on Distances

Let the shape of the earth be described as in Figure 13.1, by giving distances between some of the principal identifiable points: buoys, ships, icebergs, lighthouses, peaks, and flags: points to a total of $n = 2 \times 10^7$. The total number of distances to be given is $n(n - 1)/2 = 2 \times 10^{14}$. With 200 distances per page of printout, this means

First point	Second point	Distance (Nautical miles)	First point	Seco... p...
9,316,434	14,117,103	1410.316	9,316,434	
9,316,434	14,117,104	1812.717	9,316,434	
9,316,434	14,117,105	1629.291	9.3...	

10^{12} pages weighing 6 g each, or 6×10^6 metric tons of data. With 6 tons per truck this means 10^6 truckloads of data; or with one truck passing by every 5 seconds, 5×10^6 seconds or 2 months of night and day traffic to get in the data.

First Distillation: Distances to Nearby Points Only

Get distances between faraway points by adding distances covered on the elementary short legs of the trip. Boil down the table of distances to give only the distance between each point and the hundred nearest points. Now have $100\, n = 2 \times 10^9$ distances, or $2 \times 10^9/200 = 10^7$ pages of data, or 60 tons of records, or 10 truckloads.

Second Distillation: Distances Between Nearby Points in Terms of Metric

Idealize the surface of the earth as smooth. Then in any sufficiently limited region the geometry is Euclidean. This circumstance has a happy consequence. It is enough to know a few distances between the nearby points to be able to determine all the

distances between the nearby points. Locate point 2 so that (102) is a right triangle; thus, $(12)^2 = (10)^2 + (20)^2$. Consider a point 3 close to 0. Define

$$x(3) = (13) - (10)$$

and

$$y(3) = (23) - (20).$$

Then the distance (03) does not have to be supplied independently; it can be calculated from the formula*

$$(03)^2 = [x(3)]^2 + [y(3)]^2.$$

Similarly for a point 4 and its distance (04) from the local origin 0. Similarly for the distance (mn) between any two points m and n that are close to 0:

$$(mn)^2 = [x(m) - x(n)]^2 + [y(m) - y(n)]^2.$$

Thus it is only needful to have the distance $(1m)$ (from point 1) and $(2m)$ (from point 2) for each point m close to 0 $(m = 3, 4, \ldots, N + 2)$ to be able to work out

*If the distance (03) is given arbitrarily, the resulting four-vertex figure will burst out of the plane. Regarded as a tetrahedron in a three-dimensional Euclidean space, it has a volume given by the formula of Niccolo Fontana Tartaglia (1500–1557), generalized today (Blumenthal 1953) to

$$\begin{pmatrix} \text{volume of} \\ n\text{-dimensional} \\ \text{simplex} \\ \text{spanned by} \\ (n+1) \text{ points} \end{pmatrix} = \left(\frac{(-1)^{n+1}}{2^n}\right)^{1/2} \frac{1}{n!} \begin{vmatrix} 0 & 1 & 1 & 1 & \ldots & 1 \\ 1 & 0 & (01)^2 & (02)^2 & \ldots & (0n)^2 \\ 1 & (10)^2 & 0 & (12)^2 & \ldots & (1n)^2 \\ \cdots & \cdots & \cdots & \cdots & \cdots & \cdots \\ 1 & (n0)^2 & (n1)^2 & (n2)^2 & \ldots & 0 \end{vmatrix}^{1/2},$$

which reduces for three points to the standard textbook formula of Hero of Alexandria (A.D. 62 to A.D. 150).

$$\text{area} = \{s[s - (01)][s - (02)][s - (12)]\}^{1/2},$$
$$2s = (01) + (02) + (12),$$

for the area of a triangle. Conversely, if the four points are to remain in two-dimensional Euclidean space, the tetrahedron must collapse to zero volume. This requirement supplies one condition on the one distance (03). It simplifies the discussion of this condition to take (03) small and (102) to be a right triangle, as above. However, the general principle is independent of such approximations, and follows directly from the extended Hero-Tartaglia formula. It is enough in a locally Euclidean or Lorentz space of n dimensions to have laid down $(n + 1)$ fiducial points 0, 1, 2, \ldots, n, and to know the distance of every other point j, k, \ldots from these fiducial points, in order to be able to calculate the distance of these points j, k, \ldots from one another ("distances between nearby points in terms of coordinates"; metric as distillation of distance data).

Box 13.1 (continued)

its distance from every point n close to 0. The prescription to determine the $N(N-1)/2$ distances between these N nearby points can be reexpressed to advantage in these words: (1) each point has two coordinates, x and y; and (2) the distance is given in terms of these coordinates by the standard Euclidean metric; thus

$$(\Delta s)^2 = (\Delta x)^2 + (\Delta y)^2.$$

Having gone this far on the basis of "distance geometry" (for more on which, see Robb 1914 and 1936), one can generalize from a small region (Euclidean) to a large region (not Euclidean). Introduce any arbitrary smooth pair of everywhere-independent curvilinear coordinates x^k, and express distance, not only in the immediate neighborhood of the point 0, but also in the immediate neighborhood of every point of the surface (except places where one has to go to another coordinate patch; at least two patches needed for 2-sphere) in terms of the formula

$$ds^2 = g_{jk}\, dx^j\, dx^k.$$

Thus out of the table of distances between nearby points one has distilled now five numbers per point (two coordinates, x^1, x^2, and three metric coefficients, g_{11}, $g_{12} = g_{21}$, and g_{22}), down by a factor of $100/5 = 20$ from what one had before (now 3 tons of data, or half a truckload).

Third Distillation: Metric Coefficients Expressed as Analytical Functions of Coordinates

Instead of giving the three metric coefficients at each of the 2×10^7 points of the surface, give them as functions of the two coordinates x^1, x^2, in terms of a power series or an expansion in spherical harmonics or otherwise with some modest number, say 100, of adjustable coefficients. Then the information about the geometry itself (as distinct from the coordinates of the 2×10^7 points located on that geometry) is caught up in these three hundred coefficients, a single page of printout. Goodbye to any truck! In brief, metric provides a shorthand way of giving the distance between every point and every other point—but its role, its justification and its meaning lies in these distances and only in these many distances.

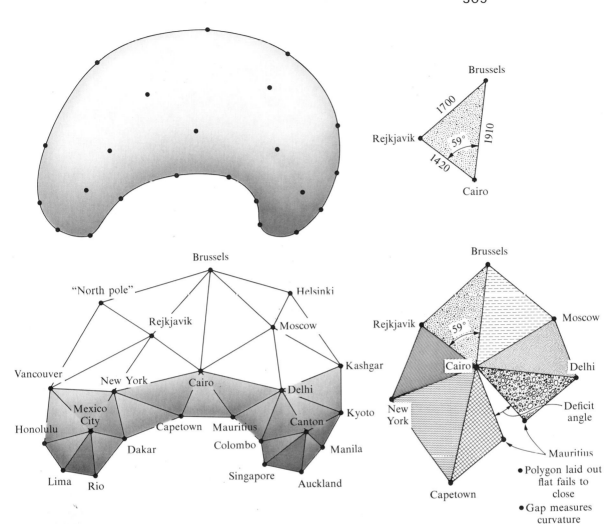

Figure 13.1.

Distances determine geometry. Upper left: Sufficiently great tidal forces, applied to the earth with tailored timing, have deformed it to the shape of a tear drop. Lower left: This tear drop is approximated by a polyhedron built out of triangles ("skeleton geometry"). The approximation can be made arbitrarily good by making the number of triangles sufficiently great and the size of each sufficiently small. Upper right: The geometry in each triangle is Euclidean: giving the three edge lengths fixes all the features of the figure, including the indicated angle. Lower right: The triangles that belong to a given vertex, laid out on a flat surface, fail to meet. The deficit angle measures the amount of curvature concentrated at that vertex on the tear-drop earth. The sum of these deficit angles for all vertices of the tear drop equals 4π. This "Gauss-Bonnet theorem" is valid for any figure with the topology of the 2-sphere; for the simplest figure of all, a tetrahedron, four vertices with a deficit angle at each of 180° are needed—3 triangles × 60° per triangle available = 180° deficit. In brief, the shape of the tear drop, in the given skeleton-geometry approximation, is determined by its 50 visible edge lengths plus, say, 32 more edge lengths hidden behind the figure, or a total of 82 edge lengths, and by nothing more ("distances determine geometry"). "Metric" tells the distance between every point and every nearby point. If volcanic action raises Rejkjavik, the distances between that Icelandic capital and nearby points increase accordingly; distances again reveal shape. Conversely, that there is not a great bump on the earth in the vicinity of Iceland, and that the earth does not now have a tear-drop shape, can be unambiguously established by analyzing the pattern of distances from point to point in a sufficiently well-distributed network of points, with no call for any observations other than measurements of distance.

for the interval between those events; comparison with the coordinate viewpoint [equation (13.1)] reveals

Covariant components of metric

$$g_{\mu\nu} = \boldsymbol{g}(\boldsymbol{e}_\mu, \boldsymbol{e}_\nu) = \boldsymbol{e}_\mu \cdot \boldsymbol{e}_\nu \qquad (13.2)$$

(standard equation for calculating components of a tensor).

"Line element" compared with "metric as bilinear machine"

Just as modern differential geometry replaces the old style "differential" df by the "differential form" $\boldsymbol{d}f$ (Box 2.3, page 63), so it also replaces the old-style "line element"

$$ds^2 = g_{\mu\nu}\,dx^\mu\,dx^\nu = (\text{"interval between } x^\alpha \text{ and } x^\alpha + dx^\alpha\text{"}) \qquad (13.3)$$

by the bilinear machine ("metric tensor")

$$\boldsymbol{g} \equiv \boldsymbol{ds}^2 \equiv g_{\mu\nu}\,\boldsymbol{dx}^\mu \otimes \boldsymbol{dx}^\nu. \qquad (13.4)$$

The output $\boldsymbol{g}(\boldsymbol{\xi}, \boldsymbol{\xi})$ of this machine, for given displacement-vector input, is identical to the old-style interval. Hence, $\boldsymbol{ds}^2 = g_{\mu\nu}\,\boldsymbol{dx}^\mu \otimes \boldsymbol{dx}^\nu$ represents the interval of an unspecified displacement; and the act of inserting $\boldsymbol{\xi}$ into the slots of \boldsymbol{ds}^2 is the act of making explicit the interval $\boldsymbol{g}(\boldsymbol{\xi}, \boldsymbol{\xi}) = g_{\mu\nu}\,\Delta x^\mu\,\Delta x^\nu$ of an explicit displacement.

Metric produces a correspondence between 1-forms and tangent vectors

In curved spacetime with metric, just as in flat spacetime with metric (§2.5), a particular 1-form $\tilde{\boldsymbol{u}}$ corresponds to any given tangent vector \boldsymbol{u}:

$$\tilde{\boldsymbol{u}} \text{ is defined by "}\langle\tilde{\boldsymbol{u}}, \boldsymbol{v}\rangle \equiv \boldsymbol{g}(\boldsymbol{u}, \boldsymbol{v}) \text{ for all } \boldsymbol{v}\text{"} \qquad (13.5)$$

("representation of the same physical quantity in the two alternative versions of vector and 1-form"; "corresponding representations" as $\binom{1}{0}$-tensor and as $\binom{0}{1}$-tensor). Example: the 1-form $\tilde{\boldsymbol{u}}$ corresponding to a basis vector $\boldsymbol{u} = \boldsymbol{e}_\alpha$ has components

$$u_\beta = \langle\tilde{\boldsymbol{u}}, \boldsymbol{e}_\beta\rangle \equiv \boldsymbol{g}(\boldsymbol{u}, \boldsymbol{e}_\beta) = \boldsymbol{g}(\boldsymbol{e}_\alpha, \boldsymbol{e}_\beta) = g_{\alpha\beta};$$

$$\begin{bmatrix}\text{standard way} \\ \text{to compute } u_\beta\end{bmatrix} \qquad \text{[equation (13.2)]}$$

$$\text{[by } \boldsymbol{u} = \boldsymbol{e}_\alpha\text{]}$$

$$\text{[definition (13.5)]}$$

thus

$$g_{\alpha\beta}\boldsymbol{\omega}^\beta \text{ is the 1-form } \tilde{\boldsymbol{e}}_\alpha \text{ corresponding to } \boldsymbol{e}_\alpha. \qquad (13.6)$$

Also as in flat spacetime (§3.2), a tensor can accept either a vector or a 1-form into any given slot

$$\boldsymbol{S}(\tilde{\boldsymbol{u}}, \boldsymbol{\sigma}, \boldsymbol{v}) \equiv \boldsymbol{S}(\boldsymbol{u}, \boldsymbol{\sigma}, \boldsymbol{v}). \qquad (13.7)$$

Lowering indices

Equivalently, in component language, the indices of a tensor can be lowered with the covariant components of the metric

$$S_\alpha{}^\beta{}_\gamma = \boldsymbol{S}(\boldsymbol{e}_\alpha, \boldsymbol{\omega}^\beta, \boldsymbol{e}_\gamma) = \boldsymbol{S}(\tilde{\boldsymbol{e}}_\alpha, \boldsymbol{\omega}^\beta, \boldsymbol{e}_\gamma) = \boldsymbol{S}(g_{\alpha\mu}\boldsymbol{\omega}^\mu, \boldsymbol{\omega}^\beta, \boldsymbol{e}_\gamma) = g_{\alpha\mu}S^{\mu\beta}{}_\gamma. \qquad (13.8)$$

$$\text{[definition of } S_\alpha{}^\beta{}_\gamma\text{]} \qquad \text{[by equation (13.6)]}$$

The basis vectors $\{\boldsymbol{e}_\alpha\}$ can be chosen arbitrarily at each event. Therefore the corresponding components $g_{\alpha\beta}$ of the metric are quite arbitrary (though symmetric: $g_{\alpha\beta} = g_{\beta\alpha}$). But the mixed components $g^\alpha{}_\beta$ are not arbitrary. In particular, equations (13.5) and (13.7) imply

$$\boldsymbol{g}(\tilde{\boldsymbol{u}}, \boldsymbol{v}) \equiv \boldsymbol{g}(\boldsymbol{u}, \boldsymbol{v}) \equiv \langle \tilde{\boldsymbol{u}}, \boldsymbol{v} \rangle. \tag{13.9}$$

Therefore one concludes that the metric tensor in mixed representation is identical with the unit matrix:

Mixed and contravariant components of metric

$$g^\alpha{}_\beta \equiv \boldsymbol{g}(\boldsymbol{w}^\alpha, \boldsymbol{e}_\beta) \equiv \langle \boldsymbol{w}^\alpha, \boldsymbol{e}_\beta \rangle = \delta^\alpha{}_\beta. \tag{13.10}$$

This feature of the metric in turn fixes the contravariant components of the metric:

$$g^{\alpha\mu} g_{\mu\beta} = g^\alpha{}_\beta = \delta^\alpha{}_\beta; \tag{13.11}$$
$$\text{[“lowering an index” of } g^{\alpha\mu}]$$

i.e.,

$$\|g^{\alpha\beta}\| \text{ is the matrix inverse of } \|g_{\alpha\beta}\|. \tag{13.12}$$

This reciprocity enables one to undo the lowering of tensor indices (i.e., raise indices) with $g^{\alpha\beta}$:

Raising indices

$$S^{\mu\beta}{}_\gamma = \delta^\mu{}_\alpha S^{\alpha\beta}{}_\gamma = g^{\mu\nu} g_{\nu\alpha} S^{\alpha\beta}{}_\gamma = g^{\mu\nu} S_\nu{}^\beta{}_\gamma. \tag{13.13}$$

The last two paragraphs may be summarized in brief:

(1) $g^\alpha{}_\beta = \delta^\alpha{}_\beta$;
(2) $\|g^{\alpha\beta}\| = \|g_{\alpha\beta}\|^{-1}$;
(3) tensor indices are lowered with $g_{\alpha\beta}$;
(4) tensor indices are raised with $g^{\alpha\beta}$.

In this formalism of metric and index shuffling, a big question demands attention: how can one tell whether the metric is locally Lorentz rather than locally Euclidean or locally something else? Of course, one criterion (necessary; not sufficient!) is dimensionality—a locally Lorentz spacetime must have four dimensions. (Recall the method of §1.2 to determine dimensionality.) Confine attention, then, to four-dimensional manifolds. What else must one demand? One must demand that at every event \mathscr{P} there exist an orthonormal frame (orthonormal set of basis vectors $\{\boldsymbol{e}_{\hat{\alpha}}\}$) in which the components of the metric have their flat-spacetime form

Metric must be locally Lorentz

$$g_{\hat{\alpha}\hat{\beta}} \equiv \boldsymbol{e}_{\hat{\alpha}} \cdot \boldsymbol{e}_{\hat{\beta}} = \eta_{\alpha\beta} \equiv \text{diagonal} \, (-1, 1, 1, 1). \tag{13.14}$$

To test for this is straightforward (exercise 13.1). (1) Search for a timelike vector \boldsymbol{u} ($\boldsymbol{u} \cdot \boldsymbol{u} < 0$). If none exist, spacetime is not locally Lorentz. If one is found, then (2) examine all non-zero vectors \boldsymbol{v} perpendicular to \boldsymbol{u}. If they are all spacelike ($\boldsymbol{v} \cdot \boldsymbol{v} > 0$), then spacetime is locally Lorentz. Otherwise it is not.

EXERCISES

Exercise 13.1. TEST WHETHER SPACETIME IS LOCAL LORENTZ

Prove that the above two-step procedure for testing whether spacetime is locally Lorentz is valid: i.e., prove that if the procedure says "yes," then there exists an orthonormal basis with $g_{\hat\alpha\hat\beta} = \eta_{\alpha\beta}$ at the event in question; if it says "no," then no such basis exists.

Exercise 13.2. PRACTICE WITH METRIC

A four-dimensional manifold with coordinates v, r, θ, ϕ has line element (old-style notation)

$$ds^2 = -(1 - 2M/r)\, dv^2 + 2\, dv\, dr + r^2(d\theta^2 + \sin^2\theta\, d\phi^2),$$

corresponding to metric (new-style notation)

$$\boldsymbol{ds}^2 = -(1 - 2M/r)\, \boldsymbol{dv} \otimes \boldsymbol{dv} + \boldsymbol{dv} \otimes \boldsymbol{dr} + \boldsymbol{dr} \otimes \boldsymbol{dv} + r^2(\boldsymbol{d\theta} \otimes \boldsymbol{d\theta} + \sin^2\theta\, \boldsymbol{d\phi} \otimes \boldsymbol{d\phi}),$$

where M is a constant.

(a) Find the *"covariant"* components $g_{\alpha\beta}$ and *"contravariant"* components $g^{\alpha\beta}$ of the metric in this coordinate system. [Answer: $g_{vv} = -(1 - 2M/r)$, $g_{vr} = g_{rv} = 1$, $g_{\theta\theta} = r^2$, $g_{\phi\phi} = r^2 \sin^2\theta$; all other $g_{\alpha\beta}$ vanish; $g^{vr} = g^{rv} = 1$, $g^{rr} = (1 - 2M/r)$, $g^{\theta\theta} = r^{-2}$, $g^{\phi\phi} = r^{-2}\sin^{-2}\theta$, all other $g^{\alpha\beta}$ vanish.]

(b) Define a scalar field t by

$$t \equiv v - r - 2M \ln[(r/2M) - 1].$$

What are the covariant and contravariant components (u_α and u^α) of the 1-form $\tilde{\boldsymbol{u}} \equiv \boldsymbol{dt}$? What is the squared length $\boldsymbol{u}^2 \equiv \boldsymbol{u} \cdot \boldsymbol{u}$, of the corresponding vector? Show that \boldsymbol{u} is timelike in the region $r > 2M$. [Answer: $u_v = 1$, $u_r = -1/(1 - 2M/r)$, $u_\theta = u_\phi = 0$; $u^v = -1/(1 - 2M/r)$, $u^r = 0$, $u^\theta = u^\phi = 0$; $\boldsymbol{u}^2 = -1/(1 - 2M/r)$.]

(c) Find the most general non-zero vector \boldsymbol{w} orthogonal to \boldsymbol{u} in the region $r > 2M$, and show that it is spacelike. Thereby conclude that spacetime is locally Lorentz in the region $r > 2M$. [Answer: Since $\boldsymbol{w} \cdot \boldsymbol{u} = w_\alpha u^\alpha = -w_v/(1 - 2M/r)$, w_v must vanish, but w_r, w_θ, w_ϕ are arbitrary, and $\boldsymbol{w}^2 = (1 - 2M/r)w_r{}^2 + r^{-2}w_\theta{}^2 + r^{-2}\sin^{-2}\theta w_\phi{}^2 > 0$.]

(d) Let t, r, θ, ϕ be new coordinates for spacetime. Find the line element in this coordinate system. [Answer: This is the "Schwarzschild" line element

$$ds^2 = -\left(1 - \frac{2M}{r}\right)dt^2 + \frac{dr^2}{1 - 2M/r} + r^2\, d\theta^2 + r^2 \sin^2\theta\, d\phi^2.]$$

(e) Find an orthonormal basis, for which $g_{\hat\alpha\hat\beta} = \eta_{\alpha\beta}$ in the region $r > 2M$. [Answer: $\boldsymbol{e}_{\hat{0}} \equiv (1 - 2M/r)^{-1/2}\, \partial/\partial t$, $\boldsymbol{e}_{\hat{r}} \equiv (1 - 2M/r)^{1/2}\, \partial/\partial r$, $\boldsymbol{e}_{\hat{\theta}} \equiv r^{-1}\, \partial/\partial\theta$, $\boldsymbol{e}_{\hat{\phi}} = (r\sin\theta)^{-1}\, \partial/\partial\phi$.]

§13.3. CONCORD BETWEEN GEODESICS OF CURVED SPACETIME GEOMETRY AND STRAIGHT LINES OF LOCAL LORENTZ GEOMETRY

More could be said about the mathematical machinery and physical implications of "metric," but an issue of greater urgency presses for attention. What has metric (or spacetime interval) to do with geodesic (or world line of test particle)? Answer:

Two mathematical objects ("straight line in a local Lorentz frame" and "geodesic of the over-all global curved spacetime geometry") equal to the same physical object ("world line of test particle") must be equal to each other ("condition of consistency"). As a first method to spell out this consistency requirement, examine the two mathematical representations of the world line of a test particle in the neighborhood of a given event \mathscr{P}_0. The local-Lorentz representation says:

"Pick a local Lorentz frame at \mathscr{P}_0. [As spelled out in exercise 13.3, such a local Lorentz frame is the closest thing there is to a global Lorentz frame at \mathscr{P}_0; i.e., it is a coordinate system in which

Local-Lorentz description of straight lines

$$g_{\alpha\beta}(\mathscr{P}_0) = \eta_{\alpha\beta} \text{ (flat-spacetime metric)}, \tag{13.15a}$$

$$g_{\alpha\beta,\gamma}(\mathscr{P}_0) = 0, \tag{13.15b}$$

$$g_{\alpha\beta,\gamma\delta}(\mathscr{P}_0) \neq 0 \text{ except in special cases, such as flat space.]} \tag{13.15c}$$

The world line in that frame has zero acceleration,

$$d^2x^\mu/d\tau^2 = 0 \text{ at } \mathscr{P}_0 \text{ ("straight-line equation")}, \tag{13.16}$$

where τ is proper time as measured by the particle's clock."

The geodesic representation says

"In the local Lorentz frame, as in any coordinate frame, the world line satisfies the geodesic equation

Geodesic description of straight lines

$$d^2x^\alpha/d\tau^2 + \Gamma^\alpha{}_{\beta\gamma}(dx^\beta/d\tau)(dx^\gamma/d\tau) = 0 \tag{13.17}$$

(τ is an affine parameter because it is time as measured by the test particle's clock)." Consistency of the two representations for any and every choice of test particle (any and every choice of $dx^\alpha/d\tau$ at \mathscr{P}_0) demands

Condition of consistency: $\Gamma^\alpha{}_{\beta\gamma} = 0$ in local Lorentz frame

$$\Gamma^\alpha{}_{\beta\gamma}(\mathscr{P}_0) = 0 \text{ in any local Lorentz frame [coordinate system satisfying equations (13.15) at } \mathscr{P}_0]; \tag{13.18}$$

i.e., it demands that *every local Lorentz frame is a local inertial frame*. (On local inertial frames see §11.6.) In such a frame, all local effects of "gravitation" disappear. That is the physical shorthand for (13.18).

One does not have to speak in the language of a specific coordinate system when one demands identity between the geodesic (derived from the $\Gamma^\alpha{}_{\beta\gamma}$) and the straight line of the local Lorentz geometry ($g_{\mu\nu}$). The local Lorentz specialization of coordinates may be the most immediate way to see the physics ("no local effects of gravitation"), but it is not the right way to formulate the basic mathematical requirement in its full generality and power. The right way is to demand

Consistency reformulated: $\boldsymbol{\nabla}\boldsymbol{g} = 0$.

$$\boldsymbol{\nabla}\boldsymbol{g} = 0 \text{ ("compatibility of } \boldsymbol{g} \text{ and } \boldsymbol{\nabla} \text{")}. \tag{13.19}$$

Stated in the language of an arbitrary coordinate system, this requirement reads

$$g_{\alpha\beta;\gamma} \equiv \frac{\partial g_{\alpha\beta}}{\partial x^\gamma} - \Gamma^\mu{}_{\alpha\gamma}g_{\mu\beta} - \Gamma^\mu{}_{\beta\gamma}g_{\alpha\mu} = 0. \tag{13.19'}$$

That this covariant requirement is fulfilled in every coordinate system follows from its validity in one coordinate system: a local Lorentz frame. (The first term in this equation, and the last two terms, are separately required to vanish in the local Lorentz frame at point \mathscr{P}_0—and required to vanish by the *physics*.) From $\boldsymbol{\nabla g} = 0$, one can derive both the abstract chain rule

$$\boldsymbol{\nabla}_u\,(\boldsymbol{v}\cdot\boldsymbol{w}) = (\boldsymbol{\nabla}_u\boldsymbol{v})\cdot\boldsymbol{w} + \boldsymbol{v}\cdot(\boldsymbol{\nabla}_u\boldsymbol{w}) \tag{13.20}$$

$\Gamma^\alpha{}_{\beta\gamma}$ expressed in terms of metric

(Exercise 13.4) and the following equations for the connection coefficients in *any* frame in terms of (1) the metric coefficients, $g_{\alpha\beta} = \boldsymbol{e}_\alpha\cdot\boldsymbol{e}_\beta$, and (2) the covariant commutation coefficients

$$c_{\alpha\beta\gamma} \equiv c_{\alpha\beta}{}^\mu g_{\mu\gamma} \equiv \langle\boldsymbol{w}^\mu, [\boldsymbol{e}_\alpha, \boldsymbol{e}_\beta]\rangle g_{\mu\gamma} \tag{13.21}$$

of that frame:

$$\Gamma^\alpha{}_{\beta\gamma} = g^{\alpha\mu}\Gamma_{\mu\beta\gamma} \qquad \text{(definition of } \Gamma_{\mu\beta\gamma}\text{)}, \tag{13.22}$$

$$\Gamma_{\mu\beta\gamma} = \frac{1}{2}\left(g_{\mu\beta,\gamma} + c_{\mu\beta\gamma} + g_{\mu\gamma,\beta} + c_{\mu\gamma\beta} - g_{\beta\gamma,\mu} - c_{\beta\gamma\mu}\right)$$
$$= \frac{1}{2}\left(g_{\mu\beta,\gamma} + g_{\mu\gamma,\beta} - g_{\beta\gamma,\mu}\right) \text{ in any coordinate frame.} \tag{13.23}$$

(See Exercise 13.4).

Equations (13.23) are the connection coefficients required to make the geodesics of curved spacetime coincide with the straight lines of the local Lorentz geometry. And they are fixed uniquely; no other choice of connection coefficients will do the job!

Summary: in curved spacetime with a local Lorentz metric, the following seemingly different statements are actually equivalent: (1) the geodesics of curved spacetime coincide with the straight lines of the local Lorentz geometry; (2) every local Lorentz frame [coordinates with $g_{\alpha\beta}(\mathscr{P}_0) = \eta_{\alpha\beta}$, $g_{\alpha\beta,\gamma}(\mathscr{P}_0) = 0$] is a local inertial frame $[\Gamma^\alpha{}_{\beta\gamma}(\mathscr{P}_0) = 0]$; (3) the metric and covariant derivative satisfy the compatibility condition $\boldsymbol{\nabla g} = 0$; (4) the covariant derivative obeys the chain rule (13.20); (5) the connection coefficients are determined by the metric in the manner of equations (13.23). A sixth equivalent statement, derived in the next section, says (6) the geodesics of curved spacetime coincide with world lines of extremal proper time.

EXERCISES

Exercise 13.3. MATHEMATICAL REPRESENTATION OF LOCAL LORENTZ FRAME

By definition, a local Lorentz frame at a given event \mathscr{P}_0 is the closest thing there to a global Lorentz frame. Thus, it should be a coordinate system with $g_{\mu\nu}(\mathscr{P}_0) = \eta_{\mu\nu}$, and with as many derivatives of $g_{\mu\nu}$ as possible vanishing at \mathscr{P}_0. Prove that there exist coordinates in which $g_{\mu\nu}(\mathscr{P}_0) = \eta_{\mu\nu}$ and $g_{\mu\nu,\rho}(\mathscr{P}_0) = 0$, but that $g_{\mu\nu,\rho\sigma}(\mathscr{P}_0)$ cannot vanish in general. Hence, such coordinates are the mathematical representation of a local Lorentz frame. [Hint: Let $\{x^{\alpha'}(\mathscr{P})\}$ be an arbitrary but specific coordinate system, and $\{x^\mu(\mathscr{P})\}$ be a local Lorentz frame, both

with origins at \mathcal{P}_0. Expand the coordinate transformation between the two in powers of x^μ

$$x^{\alpha'} = M^{\alpha}{}_{\mu} x^\mu + \frac{1}{2} N^{\alpha}{}_{\mu\nu} x^\mu x^\nu + \frac{1}{6} P^{\alpha}{}_{\mu\nu\rho} x^\mu x^\nu x^\rho + \dots ;$$

and use the transformation matrix $L^{\alpha'}{}_{\mu} \equiv \partial x^{\alpha'}/\partial x^\mu$ to get $g_{\mu\nu}(\mathcal{P}_0)$, $g_{\mu\nu,\rho}(\mathcal{P}_0)$, and $g_{\mu\nu,\rho\sigma}(\mathcal{P}_0)$ in terms of $g_{\alpha'\beta'}$ and its derivatives and the constants $M^{\alpha}{}_{\mu}$, $N^{\alpha}{}_{\mu\nu}$, $P^{\alpha}{}_{\mu\nu\rho}$. Show that whatever $g_{\alpha'\beta'}$ may be (so long as it is nonsingular, so $g^{\alpha'\beta'}$ exists!), one can choose the 16 constants $M^{\alpha}{}_{\mu}$ to make $g_{\mu\nu} = \eta_{\mu\nu}$ (ten conditions); one can choose the $4 \times 10 = 40$ constants $N^{\gamma}{}_{\mu\nu}$ to make the $10 \times 4 = 40$ $g_{\mu\nu,\rho}(\mathcal{P}_0)$ vanish; but one cannot in general choose the $4 \times 20 = 80$ $P^{\alpha}{}_{\mu\nu\rho}$ to make the $10 \times 10 = 100$ $g_{\mu\nu,\rho\sigma}$ vanish.]

Exercise 13.4. CONSEQUENCES OF COMPATIBILITY BETWEEN g AND ∇

(a) From the condition of compatibility $\nabla g = 0$, derive the chain rule (13.20).

(b) From the condition of compatibility $\nabla g = 0$ and definitions (13.21) and (13.22), derive equation (13.23) for the connection coefficients. [Answer: See exercise 8.15, p. 216.]

§13.4. GEODESICS AS WORLD LINES OF EXTREMAL PROPER TIME

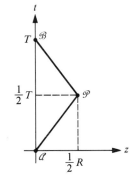

In a local Lorentz frame, it is easy to distinguish a world line that is straight from one that is not. Position the Lorentz frame and so orient it that the starting point of the world line, \mathcal{Q}, lies at the origin and the end point, \mathcal{B}, lies at $x = 0$, $y = 0$, $z = 0$, $t = T$. As an example of a nonstraight world line, consider passage at uniform velocity from \mathcal{Q} to point \mathcal{P} with coordinates $(\frac{1}{2}T; 0, 0, \frac{1}{2}R)$ and from there again with uniform velocity to point \mathcal{B}. The lapse of proper time from start to finish ("length of world line") is

$$\tau = (T^2 - R^2)^{1/2}.$$

Thus the lapse of proper time is diminished from its straight-line value, and diminished moreover for any choice of R whatsoever, except for the zero or straight-line value $R = 0$. As for this simple nonstraight curve, so also for any other nonstraight curve: the lapse of proper time between \mathcal{Q} and \mathcal{B} is less than the straight-line lapse (Exercise 6.3). Thus, in flat spacetime, extremal length of world line is an indicator of straightness.

In flat spacetime, straight lines have extremal length

Any local region of the curved spacetime of the real, physical world is Lorentz in character. In this local Lorentz geometry, it is easy to set up Lorentz coordinates and carry out the extremal-length analysis just sketched to distinguish between a straight line and a nonstraight line:

Extremal length in curved spacetime

$$\tau = \int_{\mathcal{Q}}^{\mathcal{B}} d\tau = \int_{\mathcal{Q}}^{\mathcal{B}} (-\eta_{\mu\nu}\, dx^\mu\, dx^\nu)^{1/2}$$

$$= \left(\begin{array}{l} \text{a maximum for straight line} \\ \text{as compared to any variant of} \\ \text{the straight line} \end{array} \right). \qquad (13.24)$$

Such a test for straightness can be carried out separately in each local Lorentz region along the world line, or, with greater efficiency, it can be carried out over many local Lorentz regions simultaneously, i.e., over a region with endpoints \mathcal{A} and \mathcal{B} so widely separated that no single Lorentz frame can possibly contain them both. To carry out the analysis, one must abandon local Lorentz coordinates. Therefore introduce a general curvilinear coordinate system and find

$$\tau = \int_{\mathcal{A}}^{\mathcal{B}} d\tau = \int_{\mathcal{A}}^{\mathcal{B}} (-g_{\mu\nu}\, dx^\mu\, dx^\nu)^{1/2}$$

$$= \left(\begin{array}{l} \text{an extremum for timelike world line that} \\ \text{is straight in each local Lorentz frame} \\ \text{along its path, as compared to any "nearby"} \\ \text{variant of this world line} \end{array}\right). \tag{13.25}$$

In the real world, the path of extremal τ, being straight in every local Lorentz frame, must be a geodesic of spacetime.

Notice that the word "maximum" in equation (13.24) has been replaced by "extremum" in the statement (13.25). When \mathcal{A} and \mathcal{B} are widely separated, they may be connected by several different geodesics with differing lapses of proper time (Figure 13.2). Each timelike geodesic extremizes τ with respect to nearby deformations of itself, but the extremum need not be a maximum. When several distinct geodesics connect two events, the typical one is not a local maximum ("mountain peak") but a saddle point ("mountain pass") in such a diagram as Figure 13.2 or 13.3.

Concord between locally straight lines (lines of extremal τ) and geodesics of curved spacetime demands that timelike geodesics have extremal proper length. If so, then any curve $x^\mu(\lambda)$ between \mathcal{A} (where $\lambda = 0$) and \mathcal{B} (where $\lambda = 1$) that extremizes τ should satisfy the geodesic equation. To test for an extremal by comparing times, pick a curve suspected to be a geodesic, and deform it slightly but arbitrarily:

Proof that curves of extremal length are geodesics

$$\text{original curve, } x^\mu = a^\mu(\lambda);$$
$$\text{deformed curve, } x^\mu = a^\mu(\lambda) + \delta a^\mu(\lambda). \tag{13.26}$$

Along either curve the lapse of proper time is

$$\tau = \int_{\mathcal{A}}^{\mathcal{B}} d\tau = \int_0^1 \left(-g_{\mu\nu} \frac{dx^\mu}{d\lambda} \frac{dx^\nu}{d\lambda}\right)^{1/2} d\lambda. \tag{13.27}$$

At fixed λ the metric coefficient $g_{\mu\nu}[x^\alpha(\lambda)]$ differs from one curve to the other by

$$\delta g_{\mu\nu} \equiv g_{\mu\nu}[a^\alpha(\lambda) + \delta a^\alpha(\lambda)] - g_{\mu\nu}[a^\alpha(\lambda)] = \frac{\partial g_{\mu\nu}}{\partial x^\sigma} \delta a^\sigma(\lambda); \tag{13.28}$$

and the components $dx^\nu/d\lambda$ of the tangent vector differ by

$$\delta\left(\frac{dx^\nu}{d\lambda}\right) \equiv \frac{d(a^\nu + \delta a^\nu)}{d\lambda} - \frac{da^\nu}{d\lambda} = \frac{d}{d\lambda}(\delta a^\nu). \tag{13.29}$$

These changes in $g_{\mu\nu}$ and $dx^\nu/d\lambda$, at fixed λ, produce corresponding changes in the lapse of proper time in equation (13.27):

$$\delta\tau = \int_0^1 \left\{ \frac{-g_{\mu\nu}(da^\mu/d\lambda)d(\delta a^\nu)/d\lambda - \frac{1}{2}(g_{\mu\nu,\sigma}\delta a^\sigma)(da^\mu/d\lambda)(da^\nu/d\lambda)}{[-g_{\gamma\delta}(da^\gamma/d\lambda)(da^\delta/d\lambda)]^{1/2}} \right\} d\lambda.$$

Integrate the first term by parts. Strike out the end-point terms, because both paths must pass through \mathcal{A} and \mathcal{B} ($\delta a^\mu = 0$ at $\lambda = 0$ and $\lambda = 1$). Thus find

$$\delta\tau = \int_{\lambda=0}^{\lambda=1} f_\sigma(\lambda)\, \delta a^\sigma \left[-g_{\gamma\delta} \frac{da^\gamma}{d\lambda} \frac{da^\delta}{d\lambda} \right]^{1/2} d\lambda. \tag{13.30}$$

Here the f_σ ("force terms") in the integrand are abbreviations for the four expressions

$$f_\sigma(\lambda) = \frac{1}{\left[-g_{\gamma\delta} \dfrac{da^\gamma}{d\lambda} \dfrac{da^\delta}{d\lambda} \right]^{1/2}} \frac{d}{d\lambda} \frac{g_{\sigma\nu} \dfrac{da^\nu}{d\lambda}}{\left[-g_{\gamma\delta} \dfrac{da^\gamma}{d\lambda} \dfrac{da^\delta}{d\lambda} \right]^{1/2}} - \frac{\frac{1}{2} \dfrac{\partial g_{\mu\nu}}{\partial x^\sigma} \dfrac{da^\mu}{d\lambda} \dfrac{da^\nu}{d\lambda}}{\left[-g_{\gamma\delta} \dfrac{da^\gamma}{d\lambda} \dfrac{da^\delta}{d\lambda} \right]}. \tag{13.31}$$

An extremum is achieved, and the first-order change $\delta\tau$ vanishes for every first-order deformation $\delta a^\sigma(\lambda)$ from an optimal path $x^\sigma = a^\sigma(\lambda)$, when the four quantities f_σ that multiply the δa^σ all vanish. Thus one arrives at the four conditions

$$f_\sigma(\lambda) = 0 \tag{13.32}$$

for the determination of an extremal world line. (An alternative viewpoint on the extremization is spelled out in Figure 13.3.)

 Sufficient these four equations are, but *independent* they are not, by reason of a "bead argument" (automatic vanishing of $\delta\tau$ for any set of changes that merely slide points, like beads, along an existing world line). The operation of mere "sliding of beads" implies the trivial change

$$\delta a^\sigma(\lambda) = h(\lambda) \frac{da^\sigma}{d\lambda}, \tag{13.33}$$

where $h(\lambda)$ is an arbitrary function of position along the world line ("more sliding here than there"). Already knowing that this operation cannot change τ, one is guaranteed that the integrand in (13.30) must vanish when one inserts (13.33) for δa^σ; and must vanish, moreover, whatever choice is made for the arbitrary "magnitude of slide" factor $h(\lambda)$. This requirement implies and demands that the scalar product $f_\sigma\, da^\sigma/d\lambda$ must automatically vanish; or, otherwise stated,

$$f_\sigma \frac{da^\sigma}{d\tau} = 0. \tag{13.34}$$

The argument applies, and this equation holds, whether one is or is not dealing with an optimal world line. An equation of this type, valid whether or not the world line is an allowable track for a free test particle (track of extremal lapse of proper

time), is known as an *identity*. Equation (13.34), an important identity in the realm of spacetime geodesics, is an appropriate forerunner for the Bianchi identities of Chapter 15: the most important identities in the realm of spacetime curvature.

The freedom that exists to "slide λ-values along the world line" can be exploited to replace the arbitrary parameter λ by the physically more interesting parameter of proper time itself,

$$d\tau = \left[-g_{\gamma\delta} \frac{da^\gamma}{d\lambda} \frac{da^\delta}{d\lambda} \right]^{1/2} d\lambda. \tag{13.35}$$

Figure 13.2.

Star oscillating back and forth through the plane of a disc galaxy, as an example of a situation where two events \mathscr{A} and \mathscr{B} can be connected by more than one geodesic. Upper left: The galaxy seen edge-on, showing (dashed line) the path of the star in question, referred to a local frame partaking of and comoving with the general revolution of the nearby "disc stars." Upper right: The effective potential sensed by the star, according to Newtonian gravitation theory, is like that experienced by a ball which rolls down one inclined plane and up another ("free fall toward galactic plane" with acceleration $g = \frac{1}{2}$ in the units used here). The three central frames: Possible and impossible world lines for the star connecting two given events \mathscr{A} (plane of galaxy at $t = 0$) and \mathscr{B} (plane of galaxy at $t = 2$). Right: Throw star up from the galactic plane with enough velocity so that it just gets back to the plane at $t = 2$. Left: Throw it up with half the velocity and it will come back in half the time (very contrary to behavior of a simple harmonic oscillation, but in accord with galaxy's v-shaped potential!), thus being able to make two excursions in the allotted time between \mathscr{A} and \mathscr{B}. Center: A conceivable world line (conceivable with rocket propulsion!) but not a geodesic. Bottom: Comparison of these and any other paths that allow themselves to be approximated in the form

$$z = a_1 \sin (\pi t/2) + a_2 \sin (2\pi t/2).$$

Here the two adjustable parameters, a_1 and a_2, provide the coordinates in a two-dimensional "function space" (approximation to the infinite-dimensional function space required to depict all conceivable world lines connecting \mathscr{A} and \mathscr{B}; note comparison in right center frame between one-term Fourier approximation and exact, parabolic law of free fall; similarly in left center frame, where the two curves agree too closely to be shown separate on the diagram). Details: In the context of general relativity, take an arbitrary world line that connects \mathscr{A} and \mathscr{B}, evaluate lapse of proper time, repeat for other world lines, and say that a given world line represents a possible motion ("geodesic") when for it the proper time is an extremum with respect to all nearby world lines. In the Newtonian approximation, the difference between the lapse of proper time and the lapse ($t_{\mathscr{B}} - t_{\mathscr{A}}$) of coordinate time is all that comes to attention, in the form of the "action integral" (on a "per-unit-mass basis")

$$I = \int_{\mathscr{A}}^{\mathscr{B}} \left[\begin{pmatrix} \text{kinetic} \\ \text{energy} \end{pmatrix} - \begin{pmatrix} \text{potential} \\ \text{energy} \end{pmatrix} \right] dt$$

$$= \int \left[\frac{1}{2} \left(\frac{dz}{dt} \right)^2 - |z| \right] dt$$

(maximum, or other extremum, in the proper time implies minimum, or corresponding other extremum, in the action I). The integration gives

$$I = (\pi^2 a_1^2/8) - (4|a_1|/\pi) + (\pi^2 a_2^2/2)$$

for $|a_2| < \frac{1}{2}|a_1|$ (one-excursion motions), and for $|a_2| > \frac{1}{2}|a_1|$ (two-excursion motions),

$$I = (\pi^2 a_1^2/8) + (\pi^2 a_2^2/2) - (4|a_2|/\pi) - (a_1^2/\pi|a_2|).$$

The one-excursion motion minimizes the action (maximizes the lapse of proper time). The two-excursion motion extremizes the action but does not minimize it ("saddle point"; "mountain pass" in the topography). Choquard (1955) gives other examples of problems of mechanics where there is more than one extremum. Morse (1934) and Morse and Cairns (1969) give a theorem connecting the number of saddles of various types with the numbers of maxima and minima ("critical-point theorem of the calculus of variations in the large").

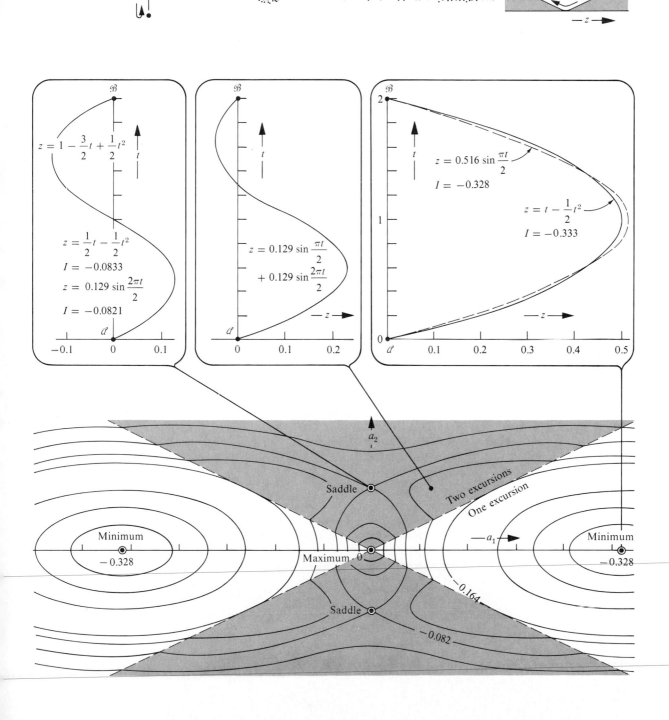

Star • ↕ z

Galaxy

Potential

$z = 1 - \dfrac{3}{2}t + \dfrac{1}{2}t^2$

$z = \dfrac{1}{2}t - \dfrac{1}{2}t^2$
$I = -0.0833$
$z = 0.129 \sin \dfrac{2\pi t}{2}$
$I = -0.0821$

$z = 0.129 \sin \dfrac{\pi t}{2}$
$+ 0.129 \sin \dfrac{2\pi t}{2}$

$z = 0.516 \sin \dfrac{\pi t}{2}$
$I = -0.328$

$z = t - \dfrac{1}{2}t^2$
$I = -0.333$

Two excursions

One excursion

Saddle

Minimum
−0.328

Maximum 0

a_1

Minimum
−0.328

−0.164

Saddle

−0.082

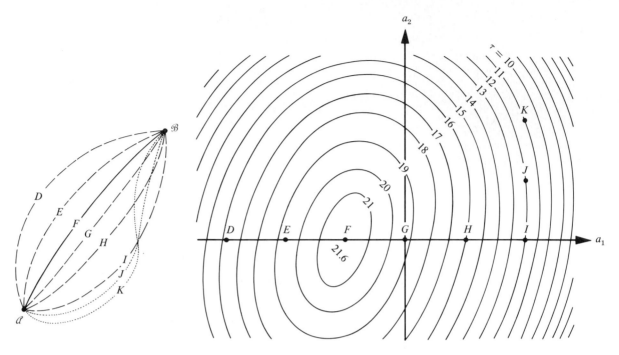

Figure 13.3.
Extremizing lapse of proper time by suitable choice of world line. Left: Spacetime; and world line F that extremizes the lapse of proper time τ from \mathscr{A} to \mathscr{B} compared to other world lines. The specific world lines depicted in the diagram happen to be distinguished from fiducial world line G by two "Fourier amplitudes" a_1 and a_2:

$$\delta a^\mu(\lambda) = a_1 \sin(\pi\lambda) + a_2 \sin(2\pi\lambda),$$

where the arbitrary scaling of λ, and its zero, are so adjusted that $\lambda(\mathscr{A}) = 0$, $\lambda(\mathscr{B}) = 1$.

Right: "Path space." The coordinates in this space are the Fourier amplitudes a_1 and a_2. Only these two amplitudes ("two dimensions") are shown out of what in principle are infinitely many amplitudes ("infinite-dimensional path space") required to represent the general timelike world line connecting \mathscr{A} and \mathscr{B}. Any given contour curve runs through all those points (in path space) for which the corresponding world lines (in spacetime) rack up the indicated lapse of proper time τ. Foregoing description is classical; according to quantum mechanics, all the timelike world lines connecting \mathscr{A} and \mathscr{B} occur with the same probability amplitude ("principle of democracy of histories") with the only difference from one to another being the phase of this complex probability amplitude $\exp(-im\tau/\hbar)$ (m = mass of particle, \hbar = quantum of angular momentum). In the sum over these probability amplitudes, however, destructive interference wipes out the contributions from all those histories which differ too much from the optimal or classical history ("Fresnel wave zone"; "Feynman's principle of sum over histories"; see Feynman and Hibbs, 1965). Capitalizing on this wave-mechanical background to show how the machinery of the physical world works, Box 25.3 spells out the Hamilton-Jacobi method ("short-wavelength limit of quantum mechanics") for determining geodesics, a method considerably more convenient for most applications than the usual "second-order differential equations for geodesics" (equation 10.27).

Focus on a specific world line, $x^\mu = a^\mu(\lambda)$, with all deformations of it gone from view; one may replace $a^\mu(\lambda)$ by $x^\mu(\lambda)$ everywhere. Then the differential equations (13.32) for an extremal world line reduce to

$$g_{\sigma\nu}\frac{d^2 x^\nu}{d\tau^2} + \frac{1}{2}\left(\frac{\partial g_{\sigma\nu}}{\partial x^\mu} + \frac{\partial g_{\sigma\mu}}{\partial x^\nu} - \frac{\partial g_{\mu\nu}}{\partial x^\sigma}\right)\frac{dx^\mu}{d\tau}\frac{dx^\nu}{d\tau} = 0. \qquad (13.36)$$

As an aside, note that the identity (13.34) now follows by one differentiation (with respect to τ) of the equation

$$g_{\sigma\nu} \frac{dx^\sigma}{d\tau} \frac{dx^\nu}{d\tau} + 1 = 0. \tag{13.37}$$

Thus the identity is to be interpreted as saying that 4-velocity and 4-acceleration are orthogonal for any world line, extremal or not. Now return to (13.36), raise an index with $g^{\beta\sigma}$, and thereby bring the equation for a straight line of local Lorentz geometry into the form

$$\frac{d^2x^\beta}{d\tau^2} + g^{\beta\sigma} \frac{1}{2} \left(\frac{\partial g_{\sigma\nu}}{\partial x^\mu} + \frac{\partial g_{\sigma\mu}}{\partial x^\nu} - \frac{\partial g_{\mu\nu}}{\partial x^\sigma} \right) \frac{dx^\mu}{d\tau} \frac{dx^\nu}{d\tau} = 0. \tag{13.38}$$

Compare with the standard form of the equation for a geodesic in "premetric geometry,"

$$\frac{d^2x^\beta}{d\lambda^2} + \Gamma^\beta{}_{\mu\nu} \frac{dx^\mu}{d\lambda} \frac{dx^\nu}{d\lambda} = 0. \tag{13.39}$$

Conclude that the geodesics of the premetric geometry will agree with the straight lines of the local Lorentz geometry if and only if two conditions are satisfied: (1) the 40 connection coefficients $\Gamma^\beta{}_{\mu\nu}$ that define geodesics, covariant derivatives, and parallel transport must be given in terms of the 10 metric coefficients $g_{\mu\nu}$ ("Einstein gravitation potentials") by the equations (13.22) and (13.23) previously derived; and (2) the geodesic parameter λ must agree with the proper time τ up to an arbitrary normalization of zero point and an arbitrary but constant scale factor; thus

$$\lambda = a\tau + b.$$

(Nothing in the formalism has any resemblance whatsoever to the universal time t of Newton "flowing everywhere uniformly"; rather, there is a separate proper time τ for each geodesic). See Box 13.3 for another variational principle, which gives in one step both the extremal world line and the right parametrization on that line.

With this step, one has completed the transfer of the ideas of curved-space geometry from a foundation based on geodesics to a foundation based on metric. The resulting geometry always and everywhere anchors itself to the principle of "local Lorentz character," as the geometry of Newton-Cartan never did and never could.

Exercise 13.5. ONCE TIMELIKE, ALWAYS TIMELIKE **EXERCISES**

Show that a geodesic of spacetime which is timelike at one event is everywhere timelike. Similarly, show that a geodesic initially spacelike is everywhere spacelike, and a geodesic initially null is everywhere null. [*Hint:* This is the easiest exercise in the book!]

(*continued on page 324*)

Box 13.2 "GEODESIC" VERSUS "EXTREMAL WORLD LINE"

Once the connection coefficients $\Gamma^\alpha{}_{\mu\nu}$ have been expressed in terms of Einstein's gravitational potentials $g_{\mu\nu}$ by the equations (13.22) and (13.23), as they are now and hereafter will be in this book ("Riemannian or metric geometry"), it is permissible and appropriate to subsume under the one word "geodesic" two previously distinct ideas: (1) a parametrized world line that satisfies the geodesic equation

$$\frac{d^2x^\alpha}{d\lambda^2} + \Gamma^\alpha{}_{\mu\nu}\frac{dx^\mu}{d\lambda}\frac{dx^\nu}{d\lambda} = 0;$$

and (2) a world line that extremizes the proper time (or, if spacelike, a curve that extremizes the proper distance) between two events \mathcal{A} and \mathcal{B}. The one possible source of confusion is this, that (1) presupposes a properly parametrized curve (as was essential, for example, in the Schild's ladder construction employed for parallel transport in Chapter 10), whereas (2) cares only about the course of the world line through spacetime, being indifferent to what parametrization is used or whether any parametrization at all is introduced. This is not to deny the possibility of "marking in afterward" along the extremal curve the most natural and easily evaluated of all parameters, the proper time itself, whereupon the extremal curve of (2) satisfies the geodesic equation of (1). Ambiguity is avoided by insisting on proper parametrization: henceforth the word "curve" means a parametrized curve, the word "geodesic" means a properly parametrized geodesic.

Box 13.3 "DYNAMIC" VARIATIONAL PRINCIPLE FOR GEODESICS

If the principle of extremal length

$$\tau = \int_{\mathcal{A}}^{\mathcal{B}}\left[-g_{\mu\nu}\frac{dx^\mu}{d\lambda}\frac{dx^\nu}{d\lambda} \right]^{1/2} d\lambda = \text{extremum} \tag{1}$$

is indifferent to choice of parametrization ["$d\lambda$" canceling out in (1)] and if the geodesic equation finds the proper parametrization a matter of concern, it is appropriate to search for another extremal principle that yields in one package both the right curve and the right parameter. By analogy with elementary mechanics, one expects that an equation of motion [the geodesic equation

$$d^2x^\mu/d\lambda^2 + \Gamma^\mu{}_{\alpha\beta}(dx^\alpha/d\lambda)(dx^\beta/d\lambda) = 0]$$

whose leading term has the form "\ddot{x}" can be derived from a Lagrangian with leading term "$\frac{1}{2}\dot{x}^2$" ("kinetic energy"; "dynamic" term). The simplest coordinate invariant generalization of $\frac{1}{2}\dot{x}^2$ is

$$\frac{1}{2}g_{\mu\nu}(dx^\mu/d\lambda)(dx^\nu/d\lambda).$$

Thus one is led to try, in place of the "geometric" principle of extremal length, a new "dynamic" extremal principle:

$$
I = \frac{1}{2} \int_{\mathscr{A}}^{\mathscr{B}} g_{\mu\nu} \frac{dx^\mu}{d\lambda} \frac{dx^\nu}{d\lambda} \, d\lambda
$$

$$
= \int_{\mathscr{A}}^{\mathscr{B}} L\left(x^\sigma, \frac{dx^\sigma}{d\lambda}\right) d\lambda = \text{extremum}
\tag{2}
$$

(replacement of square root in previous variational principle by first power). The condition for an extremum, here as before [equations (13.30) to (13.32)] is annulment of the so-called Euler-Lagrange "functional derivative"

$$
0 = \frac{\delta I}{\delta x^\sigma} \equiv \begin{pmatrix} \text{coefficient of } \delta x^\sigma \text{ in} \\ \text{the integrand of } \delta I \end{pmatrix}
$$

$$
= \frac{\partial L}{\partial x^\sigma} - \frac{d}{d\lambda} \frac{\partial L}{\partial \left(\dfrac{dx^\sigma}{d\lambda}\right)};
\tag{3}
$$

or, written out in full detail,

$$
g_{\sigma\nu} \frac{d^2 x^\nu}{d\lambda^2} + \frac{1}{2}\left(\frac{\partial g_{\sigma\nu}}{\partial x^\mu} + \frac{\partial g_{\sigma\mu}}{\partial x^\nu} - \frac{\partial g_{\mu\nu}}{\partial x^\sigma}\right) \frac{dx^\mu}{d\lambda} \frac{dx^\nu}{d\lambda} = 0;
\tag{4}
$$

or, after multiplication by the reciprocal metric,

$$
\frac{d^2 x^\alpha}{d\lambda^2} + g^{\alpha\sigma} \frac{1}{2}\left(\frac{\partial g_{\sigma\nu}}{\partial x^\mu} + \frac{\partial g_{\sigma\mu}}{\partial x^\nu} - \frac{\partial g_{\mu\nu}}{\partial x^\sigma}\right) \frac{dx^\mu}{d\lambda} \frac{dx^\nu}{d\lambda} = 0;
\tag{5}
$$

which translates into the geodesic equation

$$
\frac{d^2 x^\alpha}{d\lambda^2} + \Gamma^\alpha{}_{\mu\nu} \frac{dx^\mu}{d\lambda} \frac{dx^\nu}{d\lambda} = 0.
\tag{6}
$$

Thus, the new "dynamic" expression (2) is indeed extremal for geodesic curves—and, by contrast with proper length, (1), it is extremal when and only when the geodesic is affinely parametrized. [Its "Euler-Lagrange equations" (6) remain satisfied only under parameter changes $\lambda_{\text{new}} = a\lambda_{\text{old}} + b$, which keep the parameter affine; by contrast, the Euler-Lagrange equations (13.31) and (13.32) for the "principle of extremal length" (1) remain satisfied for any change of parameter whatsoever.]

Exercise 13.6. SPACELIKE GEODESICS HAVE EXTREMAL LENGTH

Show that any spacelike curve linking two events \mathcal{Q} and \mathcal{B} is a geodesic if and only if it extremizes the proper length

$$s = \int_{\mathcal{Q}}^{\mathcal{B}} (g_{\mu\nu}\, dx^\mu\, dx^\nu)^{1/2}.$$

[*Hint:* This is almost as easy as exercise 13.5 if one has already proved the analogous theorem for timelike geodesics.]

Exercise 13.7. METRIC TENSOR MEASURED BY LIGHT SIGNALS AND
** FREE PARTICLES [Kuchař]**

(a) Instead of parametrizing a timelike geodesic by the proper time τ, parametrize it by an arbitrary parameter μ,

$$\tau = F(\mu).$$

Write the geodesic equation in the μ-parametrization.

(b) Use now the coordinate time t as a parameter. Throw out a cloud of free particles with different "velocities" $v^i = dx^i/dt$ and observe their "accelerations" $a^i = d^2x^i/dt^2$. Discuss what combinations of the components of the affine connection $\Gamma^\iota{}_{\kappa\lambda}$ one can measure in this way. (Assume that no standard clocks measuring τ are available!)

(c) Show that one can measure the conformal metric $\bar{g}_{\iota\kappa}$, i.e., the ratios of the components of the metric tensor $g_{\iota\kappa}$ to a given component (say, g_{00})

$$\bar{g}_{\iota\kappa} = \Lambda g_{\iota\kappa}, \qquad \Lambda \equiv (-g_{00})^{-1},$$

using only the light signals moving along the null geodesics $g_{\iota\kappa}\, dx^\iota\, dx^\kappa = 0$.

(d) Combine now the results of (b) and (c). Assume that $\Gamma^\iota{}_{\kappa\lambda}$ is generated by the metric tensor by (13.22), (13.23), in the coordinate frame x^ι. Show that one can determine Λ everywhere, if one prescribes it at one event (equivalent to fixing the unit of time).

§13.5. METRIC-INDUCED PROPERTIES OF *RIEMANN*

Symmetries of *Riemann* in absence of metric

In Newtonian spacetime, in the real, physical spacetime of Einstein—indeed, in any manifold with covariant derivative—the Riemann curvature tensor has these symmetries (exercise 11.6):

$$R^\alpha{}_{\beta\gamma\delta} \equiv R^\alpha{}_{\beta[\gamma\delta]} \qquad \text{(antisymmetry on last two indices)} \qquad (13.40)$$

$$R^\alpha{}_{[\beta\gamma\delta]} \equiv 0 \qquad \text{(vanishing of completely antisymmetric part).} \quad (13.41)$$

In addition, it satisfies a differential identity (exercise 11.10):

$$R^\alpha{}_{\beta[\gamma\delta;\epsilon]} \equiv 0 \qquad \text{("Bianchi identity")} \qquad (13.42)$$

(see Chapter 15 for geometric significance).

New symmetries imposed by metric

When metric is brought onto the scene, whether in Einstein spacetime or elsewhere, it impresses on *Riemann* the additional symmetry (exercise 13.8)

$$R_{\alpha\beta\gamma\delta} \equiv R_{[\alpha\beta]\gamma\delta} \quad \text{(antisymmetry on first two indices).} \tag{13.43}$$

This, together with (13.40) and (13.41), forms a complete set of symmetries for **Riemann**; other symmetries that follow from these (exercise 13.10) are

$$R_{\alpha\beta\gamma\delta} = R_{\gamma\delta\alpha\beta} \qquad \text{(symmetry under pair exchange),} \tag{13.44}$$

and

$$R_{[\alpha\beta\gamma\delta]} = 0 \qquad \text{(vanishing of completely antisymmetric part).} \tag{13.45}$$

These symmetries reduce the number of independent components of **Riemann** from $4 \times 4 \times 4 \times 4 = 256$ to 20 (exercise 13.9).

With metric present, one can construct a variety of new curvature tensors from **Riemann**. Some that will play important roles later are as follows.

(1) The *double dual of* **Riemann**, $G \equiv$ ***Riemann*** (analog of **Maxwell** \equiv ***Faraday**), which has components

The curvature tensor **G**

$$G^{\alpha\beta}{}_{\gamma\delta} \equiv \frac{1}{2}\epsilon^{\alpha\beta\mu\nu}R_{\mu\nu}{}^{\rho\sigma}\frac{1}{2}\epsilon_{\rho\sigma\gamma\delta} = -\frac{1}{4}\delta^{\alpha\beta\mu\nu}_{\rho\sigma\gamma\delta}R_{\mu\nu}{}^{\rho\sigma} \tag{13.46}$$

(exercise 13.11).

(2) The *Einstein curvature tensor,* which is symmetric (exercise 13.11)

Einstein tensor

$$G^{\beta}{}_{\delta} \equiv G^{\mu\beta}{}_{\mu\delta}; \quad G_{\beta\delta} = G_{\delta\beta}. \tag{13.47}$$

(3) The *Ricci curvature tensor,* which is symmetric, and the *curvature scalar*

Ricci tensor

Curvature scalar

$$R^{\beta}{}_{\delta} = R^{\mu\beta}{}_{\mu\delta}, \qquad R_{\beta\delta} = R_{\delta\beta}; \qquad R \equiv R^{\beta}{}_{\beta}; \tag{13.48}$$

which are related to the Einstein tensor by (exercise 13.12)

$$R^{\beta}{}_{\delta} = G^{\beta}{}_{\delta} + \frac{1}{2}R\delta^{\beta}{}_{\delta}. \tag{13.49}$$

(4) *The Weyl conformal tensor* (exercise 13.13)

Weyl conformal tensor

$$C^{\alpha\beta}{}_{\gamma\delta} = R^{\alpha\beta}{}_{\gamma\delta} - 2\delta^{[\alpha}{}_{[\gamma}R^{\beta]}{}_{\delta]} + \frac{1}{3}\delta^{[\alpha}{}_{[\gamma}\delta^{\beta]}{}_{\delta]}R. \tag{13.50}$$

The Bianchi identity (13.42) takes a particularly simple form when rewritten in terms of the double dual **G**:

Bianchi identities

$$G_{\alpha\beta\gamma}{}^{\delta}{}_{;\delta} \equiv 0 \qquad \text{(``Bianchi identity'')} \tag{13.51}$$

(exercise 13.11); and it has the obvious consequence

$$G_{\alpha}{}^{\beta}{}_{;\beta} \equiv 0 \qquad \text{(``contracted Bianchi identity'').} \tag{13.52}$$

Chapter 15 will be devoted to the deep geometric significance of these Bianchi identities.

EXERCISES

**Exercise 13.8. *RIEMANN* ANTISYMMETRIC IN FIRST
TWO INDICES**

(a) Derive the antisymmetry condition (13.43). [Hint: Prove by abstract calculations that any vector fields *s*, *u*, *v*, *w* satisfy $0 = \mathcal{R}(\boldsymbol{u}, \boldsymbol{v})(\boldsymbol{s} \cdot \boldsymbol{w}) = \boldsymbol{s} \cdot [\mathcal{R}(\boldsymbol{u}, \boldsymbol{v})\boldsymbol{w}] + \boldsymbol{w} \cdot [\mathcal{R}(\boldsymbol{u}, \boldsymbol{v})\boldsymbol{s}]$. Then from this infer (13.43).]

(b) Explain in geometric terms the meaning of this antisymmetry.

**Exercise 13.9. NUMBER OF INDEPENDENT COMPONENTS
OF *RIEMANN***

(a) In the absence of metric, a complete set of symmetry conditions for ***Riemann*** is $R^{\alpha}{}_{\beta\gamma\delta} = R^{\alpha}{}_{\beta[\gamma\delta]}$ and $R^{\alpha}{}_{[\beta\gamma\delta]} = 0$. Show that in four-dimensional spacetime these reduce the number of independent components from $4 \times 4 \times 4 \times 4 = 256$ to $4 \times 4 \times 6 - 4 \times 4 = 96 - 16 = 80$.

(b) Show that in a manifold of n dimensions without metric, the number of independent components is

$$\frac{n^3(n-1)}{2} - \frac{n^2(n-1)(n-2)}{6} = \frac{n^2(n^2-1)}{3}. \tag{13.53}$$

(c) In the presence of metric, a complete set of symmetries is $R_{\alpha\beta\gamma\delta} = R_{[\alpha\beta][\gamma\delta]}$, and $R_{\alpha[\beta\gamma\delta]} = 0$. Show that in four-dimensional spacetime, these reduce the number of independent components to $6 \times 6 - 4 \times 4 = 36 - 16 = 20$.

(d) Show that in a manifold of n dimensions with metric, the number of independent components is

$$\left[\frac{n(n-1)}{2}\right]^2 - \frac{n^2(n-1)(n-2)}{6} = \frac{n^2(n^2-1)}{12}. \tag{13.54}$$

**Exercise 13.10. *RIEMANN* SYMMETRIC IN EXCHANGE OF PAIRS;
COMPLETELY ANTISYMMETRIC PART VANISHES**

From the complete set of symmetries in the presence of a metric, $R_{\alpha\beta\gamma\delta} = R_{[\alpha\beta][\gamma\delta]}$ and $R_{\alpha[\beta\gamma\delta]} = 0$, derive: (a) symmetry under pair exchange, $R_{\alpha\beta\gamma\delta} = R_{\gamma\delta\alpha\beta}$, and (b) vanishing of completely antisymmetric part, $R_{[\alpha\beta\gamma\delta]} = 0$. Then (c) show that the following form a complete set of symmetries:

$$R_{\alpha\beta\gamma\delta} = R_{[\alpha\beta][\gamma\delta]} = R_{\gamma\delta\alpha\beta}, \qquad R_{[\alpha\beta\gamma\delta]} = 0. \tag{13.55}$$

Exercise 13.11. DOUBLE DUAL OF *RIEMANN*; *EINSTEIN*

(a) Show that $\boldsymbol{G} \equiv {}^*\boldsymbol{Riemann}^*$ contains precisely the same amount of information as ***Riemann***, and satisfies precisely the same set of symmetries [(13.40), (13.41), (13.43) to (13.45)].

(b) From the symmetries of \boldsymbol{G}, show that ***Einstein*** [defined in (13.47)] is symmetric ($G_{[\beta\delta]} = 0$).

(c) Show that the Bianchi identities (13.42), when written in terms of \boldsymbol{G}, take the form (13.51) ("vanishing divergence," $\boldsymbol{\nabla} \cdot \boldsymbol{G} = 0$).

(d) By contracting the Bianchi identities $\boldsymbol{\nabla} \cdot \boldsymbol{G} = 0$, show that $\boldsymbol{G} \equiv \boldsymbol{Einstein}$ has vanishing divergence [equation (13.52)].

Exercise 13.12. *RICCI* AND *EINSTEIN* RELATED

(a) From the symmetries of ***Riemann***, show that ***Ricci*** is symmetric ($R_{[\beta\delta]} = 0$).

(b) Show that ***Ricci*** is related to ***Einstein*** by equation (13.49).

Exercise 13.13. THE WEYL CONFORMAL TENSOR

(a) Show that the Weyl conformal tensor (13.50) possesses the same symmetries [(13.40), (13.41), (13.43) to (13.45)] as the Riemann tensor.

(b) Show that the Weyl tensor is completely "trace-free"; i.e., that

$$\text{contraction of } C_{\alpha\beta\gamma\delta} \text{ on any pair of slots vanishes.} \qquad (13.56)$$

Thus, $C_{\alpha\beta\gamma\delta}$ can be regarded as the trace-free part of **Riemann**, and $R_{\alpha\beta}$ can be regarded as the trace of **Riemann**. **Riemann** is determined entirely by its trace-free part $C_{\alpha\beta\gamma\delta}$ and its trace $R_{\alpha\beta}$ [see equation (13.50), and recall $R = R^{\alpha}{}_{\alpha}$].

(c) Show that in spacetime the Weyl tensor has 10 independent components.

(d) Show that in an n-dimensional manifold the number of independent components of **Weyl** [defined by a modification of (13.50) that maintains (13.56)] is

$$\frac{n^2(n^2 - 1)}{12} - \frac{n(n + 1)}{2} \text{ for } n \geq 3, \qquad 0 \text{ for } n \leq 3. \qquad (13.57)$$

Thus, in manifolds of 1, 2, or 3 dimensions, the Weyl tensor is identically zero, and the Ricci tensor completely determines the Riemann tensor.

§13.6. THE PROPER REFERENCE FRAME OF AN ACCELERATED OBSERVER

A physicist performing an experiment in a jet airplane (e.g., an infrared astronomy experiment) may use several different coordinate systems at once. But a coordinate system of special utility is one at rest relative to all the apparatus bolted into the floor and walls of the airplane cabin. This "proper reference frame" has a rectangular "\hat{x}, \hat{y}, \hat{z}" grid attached to the walls of the cabin, and one or more clocks at rest in the grid. That this proper reference frame is accelerated relative to the local Lorentz frames, the physicist knows from his own failure to float freely in the cabin, or, with greater precision, from accelerometer measurements. That his proper reference frame is rotating relative to local Lorentz frames he knows from the Coriolis forces he feels, or, with greater precision, from the rotation of inertial-guidance gyroscopes relative to the cabin walls.

Exercise 6.8 gave a mathematical treatment of such an accelerated, rotating, but locally orthonormal reference frame in flat spacetime. This section does the same in curved spacetime. In the immediate vicinity of the spatial grid's origin $x^{\hat{j}} = 0$ (region of spatial extent so small that curvature effects are negligible), no aspect of the coordinate system can possibly reveal whether spacetime is curved or flat. Hence, all the details of exercise 6.8 must remain valid in curved spacetime. Nevertheless, it is instructive to rediscuss those details, and some new ones, using the powerful mathematics of the last few chapters.

Begin by making more precise the coordinate grid to be used. The following is perhaps the most natural way to set up the grid.

(1) Let τ be proper time as measured by the accelerated observer's clock (clock at center of airplane cabin in above example). Let $\mathcal{P} = \mathcal{P}_0(\tau)$ be the observer's world line, as shown in Figure 13.4,a.

Proper reference frame described physically

Six-step construction of coordinate grid for proper frame

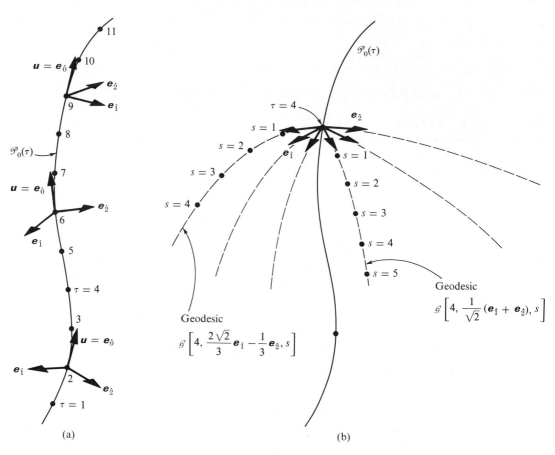

(a) (b)

Figure 13:4.

The proper reference frame of an accelerated observer. Diagram (a) shows the observer's orthonormal tetrad $\{e_{\hat{\alpha}}\}$ being transported along his world line $\mathscr{P}_0(\tau)$ [transport law (13.60)]. Diagram (b) shows geodesics bristling out perpendicularly from an arbitrary event $\mathscr{P}_0(4)$ on the observer's world line. Each geodesic is specified uniquely by (1) the proper time τ at which it originates, and (2) the direction (unit tangent vector $\boldsymbol{n} = d/ds = n^{\hat{j}}e_{\hat{j}}$ along which it emanates). A given event on the geodesic is specified by τ, \boldsymbol{n}, and proper distance s from the geodesic's emanation point; hence the notation

$$\mathscr{P} = \mathscr{G}[\tau, \boldsymbol{n}, s]$$

for the given event. The observer's proper reference frame attributes to this given event the coordinates

$$x^{\hat{0}}(\mathscr{G}[\tau, \boldsymbol{n}, s]) = \tau,$$
$$x^{\hat{j}}(\mathscr{G}[\tau, \boldsymbol{n}, s]) = sn^{\hat{j}}.$$

(2) The observer carries with himself an orthonormal tetrad $\{e_{\hat{\alpha}}\}$ (Figure 13.4,a), with

$$e_{\hat{0}} = \boldsymbol{u} = d\mathscr{P}_0/d\tau = \text{(4-velocity of observer)} \qquad (13.58)$$

($e_{\hat{0}}$ points along observer's "time direction"), and with

$$e_{\hat{\alpha}} \cdot e_{\hat{\beta}} = \eta_{\alpha\beta} \qquad (13.59)$$

(orthonormality).

(3) The tetrad changes from point to point along the observer's world line, relative to parallel transport:

Transport law for observer's tetrad

$$\boldsymbol{\nabla}_{\boldsymbol{u}}\boldsymbol{e}_{\hat{\alpha}} = -\boldsymbol{\Omega}\cdot\boldsymbol{e}_{\hat{\alpha}}, \tag{13.60}$$

$$\Omega^{\mu\nu} = a^{\mu}u^{\nu} - u^{\mu}a^{\nu} + u_{\alpha}\omega_{\beta}\epsilon^{\alpha\beta\mu\nu} \tag{13.61}$$
$$= \text{"generator of infinitesimal Lorentz transformation."}$$

This transport law has the same form in curved spacetime as in flat (§6.5 and exercise 6.8) because curvature can only be felt over finite distances, not over the infinitesimal distance involved in the "first time-rate of change of a vector" (equivalence principle). As in exercise 6.8,

$$\boldsymbol{a} = \boldsymbol{\nabla}_{\boldsymbol{u}}\boldsymbol{u} = \text{(4-acceleration of observer)},$$

$$\boldsymbol{w} = \begin{pmatrix} \text{angular velocity of rotation of spatial} \\ \text{basis vectors } \boldsymbol{e}_{\hat{j}} \text{ relative to Fermi-} \\ \text{Walker-transported vectors, i.e.,} \\ \text{relative to inertial-guidance gyroscopes} \end{pmatrix}, \tag{13.62}$$

$$\boldsymbol{u}\cdot\boldsymbol{a} = \boldsymbol{u}\cdot\boldsymbol{w} = 0.$$

If \boldsymbol{w} were zero, the observer would be Fermi-Walker-transporting his tetrad (gyroscope-type transport). If both \boldsymbol{a} and \boldsymbol{w} were zero, he would be freely falling (geodesic motion) and would be parallel-transporting his tetrad, $\boldsymbol{\nabla}_{\boldsymbol{u}}\boldsymbol{e}_{\hat{\alpha}} = 0$.

(4) The observer constructs his proper reference frame (local coordinate system) in a manner analogous to the Riemann-normal construction of §11.6. From each event $\mathscr{P}_0(\tau)$ on his world line, he sends out purely spatial geodesics (geodesics orthogonal to $\boldsymbol{u} = d\mathscr{P}_0/d\tau$), with affine parameter equal to proper length,

$$\mathscr{P} = \mathscr{G}[\tau, \boldsymbol{n}, s]. \tag{13.63}$$

$$\begin{bmatrix} \text{proper time; tells} \\ \text{"starting point" of} \\ \text{geodesic} \end{bmatrix} \quad \begin{bmatrix} \text{tangent vector to} \\ \text{geodesic at starting} \\ \text{point; tells "which"} \\ \text{geodesic} \end{bmatrix} \quad \begin{bmatrix} \text{proper length along} \\ \text{geodesic from starting} \\ \text{point; tells "where"} \\ \text{on geodesic} \end{bmatrix}$$

(See Figure 13.4,b.) The tangent vector has unit length, because the chosen affine parameter is proper length:

$$\boldsymbol{n} = (\partial\mathscr{G}/\partial s)_{s=0}; \quad n^{\mu} = (dx^{\mu}/ds) \text{ along geodesic,} \tag{13.64}$$
$$\boldsymbol{n}\cdot\boldsymbol{n} = g_{\mu\nu}\left(\frac{dx^{\mu}}{ds}\right)\left(\frac{dx^{\nu}}{ds}\right) = \frac{ds^2}{ds^2} = 1.$$

(5) Each event near the observer's world line is intersected by precisely one of the geodesics $\mathscr{G}[\tau, \boldsymbol{n}, s]$. [Far away, this is not true; the geodesics may cross, either because of the observer's acceleration, as in Figure 6.3, or because of the curvature of spacetime ("geodesic deviation").]

(6) Pick an event \mathcal{P} near the observer's world line. The geodesic through it originated on the observer's world line at a specific time τ, had original direction $\boldsymbol{n} = n^{\hat{j}}\boldsymbol{e}_{\hat{j}}$, and needed to extend a distance s before reaching \mathcal{P}. Hence, the four numbers

$$(x^{\hat{0}}, x^{\hat{1}}, x^{\hat{2}}, x^{\hat{3}}) \equiv (\tau, sn^{\hat{1}}, sn^{\hat{2}}, sn^{\hat{3}}) \tag{13.65}$$

are a natural way of identifying the event \mathcal{P}. These are the coordinates of \mathcal{P} in the observer's proper reference frame.

(7) Restated more abstractly,

$$\begin{aligned} x^{\hat{0}}(\mathcal{G}[\tau, \boldsymbol{n}, s]) &= \tau, \\ x^{\hat{j}}(\mathcal{G}[\tau, \boldsymbol{n}, s]) &= sn^{\hat{j}} = sn_{\hat{j}} = s\boldsymbol{n} \cdot \boldsymbol{e}_{\hat{j}}. \end{aligned} \tag{13.65'}$$

In flat spacetime this construction process and the resulting coordinates $x^{\hat{\alpha}}(\mathcal{P})$ are identical to the process and resulting coordinates $\xi^{\alpha'}(\mathcal{P})$ of exercise 6.8.

For use in calculations one wants not only the coordinate system, but also its metric coefficients and connection coefficients. Fortunately, $g_{\hat{\alpha}\hat{\beta}}$ and $\Gamma^{\hat{\alpha}}{}_{\hat{\beta}\hat{\gamma}}$ are needed only along the observer's world line, where they are especially simple. Only a foolish observer would try to use his own proper reference frame far from his world line, where its grid ceases to be orthonormal and its geodesic grid lines may even cross! (See §6.3.)

All along the observer's world line $\mathcal{P}_0(\tau)$, the basis vectors of his coordinate grid are identical (by construction) to his orthonormal tetrad

$$\partial/\partial x^{\hat{\alpha}} = \boldsymbol{e}_{\hat{\alpha}}, \tag{13.66}$$

and therefore its metric coefficients are

$$g_{\hat{\alpha}\hat{\beta}} = \boldsymbol{e}_{\hat{\alpha}} \cdot \boldsymbol{e}_{\hat{\beta}} = \eta_{\alpha\beta} \text{ all along } \mathcal{P}_0(\tau). \tag{13.67}$$

Connection coefficients along observer's world line

Some of the connection coefficients are determined by the transport law (13.60) for the observer's orthonormal tetrad:

$$\begin{aligned} \boldsymbol{\nabla}_{\boldsymbol{u}}\boldsymbol{e}_{\hat{\alpha}} = \boldsymbol{\nabla}_{\hat{0}}\boldsymbol{e}_{\hat{\alpha}} &= \boldsymbol{e}_{\beta}\Gamma^{\hat{\beta}}{}_{\hat{\alpha}\hat{0}} \\ &= -\boldsymbol{\Omega} \cdot \boldsymbol{e}_{\hat{\alpha}} = -\boldsymbol{e}_{\beta}\Omega^{\hat{\beta}}{}_{\hat{\alpha}}. \end{aligned}$$

Thus

$$\Gamma^{\hat{\beta}}{}_{\hat{\alpha}\hat{0}} = -\Omega^{\hat{\beta}}{}_{\hat{\alpha}} \text{ all along } \mathcal{P}_0(\tau). \tag{13.68}$$

Since $\boldsymbol{\Omega}$ has the form (13.61) and the observer's 4-velocity and 4-acceleration have components $u_{\hat{0}} = -1$, $u_{\hat{j}} = 0$, $a_{\hat{0}} = 0$ in the observer's own proper frame, these connection coefficients are

$$\left. \begin{aligned} \Gamma^{\hat{0}}{}_{\hat{0}\hat{0}} &= \Gamma_{\hat{0}\hat{0}\hat{0}} = 0, \\ \Gamma^{\hat{0}}{}_{\hat{j}\hat{0}} &= -\Gamma_{\hat{0}\hat{j}\hat{0}} = +\Gamma_{\hat{j}\hat{0}\hat{0}} = +\Gamma^{\hat{j}}{}_{\hat{0}\hat{0}} = a^{\hat{j}}, \\ \Gamma^{\hat{j}}{}_{\hat{k}\hat{0}} &= \Gamma_{\hat{j}\hat{k}\hat{0}} = -\omega^{\hat{i}}\epsilon_{\hat{0}\hat{i}\hat{j}\hat{k}}, \end{aligned} \right\} \text{ all along } \mathcal{P}_0(\tau). \tag{13.69a}$$

The remaining connection coefficients can be read from the geodesic equation for the geodesics $\mathcal{G}[\tau, \boldsymbol{n}, s]$ that emanate from the observer's world line. According to equation (13.65), the coordinate representation of each such geodesic is

$$x^{\hat{0}}(s) = \tau = \text{constant}, \qquad x^{\hat{j}}(s) = n^{\hat{j}} s;$$

hence, $d^2 x^{\hat{\alpha}}/ds^2 = 0$ all along the geodesic, and the geodesic equation reads

$$0 = \frac{d^2 x^{\hat{\alpha}}}{ds^2} + \Gamma^{\hat{\alpha}}{}_{\hat{\beta}\hat{\gamma}} \frac{dx^{\hat{\beta}}}{ds} \frac{dx^{\hat{\gamma}}}{ds} = \Gamma^{\hat{\alpha}}{}_{\hat{j}\hat{k}} n^{\hat{j}} n^{\hat{k}}.$$

This equation is satisfied on the observer's world line for all spatial geodesics (all $n^{\hat{j}}$) if and only if

$$\Gamma^{\hat{\alpha}}{}_{\hat{j}\hat{k}} = \Gamma_{\hat{\alpha}\hat{j}\hat{k}} = 0 \text{ all along } \mathcal{P}_0(\tau). \qquad (13.69b)$$

The values (13.69) of the connection coefficients determine uniquely the partial derivatives of the metric coefficients [see equation (13.19′)]:

$$\left.\begin{array}{ll} g_{\hat{\alpha}\hat{\beta},\hat{0}} = 0, & g_{\hat{j}\hat{k},\hat{l}} = 0, \\[2mm] g_{\hat{0}\hat{0},\hat{j}} = -2a_{\hat{j}}, & g_{\hat{0}\hat{j},\hat{k}} = -\epsilon_{\hat{0}\hat{j}\hat{k}\hat{l}}\omega^{\hat{l}} \end{array}\right\} \text{ all along } \mathcal{P}_0(\tau); \qquad (13.70)$$

and these, plus the orthonormality condition $g_{\hat{\alpha}\hat{\beta}}[\mathcal{P}_0(\tau)] = \eta_{\alpha\beta}$, imply that the line element near the observer's world line is

Metric of proper reference frame, and its physical interpretation

$$\begin{aligned} ds^2 = &-(1 + 2a_{\hat{j}}x^{\hat{j}}) \, dx^{\hat{0}2} - 2(\epsilon_{\hat{j}\hat{k}\hat{l}}x^{\hat{k}}\omega^{\hat{l}}) \, dx^{\hat{0}} \, dx^{\hat{j}} \\ &+ \delta_{\hat{j}\hat{k}} \, dx^{\hat{j}} \, dx^{\hat{k}} + O(|x^{\hat{j}}|^2) \, dx^{\hat{\alpha}} \, dx^{\hat{\beta}}. \end{aligned} \qquad (13.71)$$

Several features of this line element deserve notice, as follows.

(1) On the observer's world line $\mathcal{P}_0(\tau)$—i.e., $x^{\hat{j}} = 0$—$ds^2 = \eta_{\alpha\beta} \, dx^{\hat{\alpha}} \, dx^{\hat{\beta}}$.

(2) The observer's acceleration shows up in a correction term to $g_{\hat{0}\hat{0}}$,

$$\delta g_{\hat{0}\hat{0}} = -2\boldsymbol{a} \cdot \boldsymbol{x}, \qquad (13.72a)$$

which is proportional to distance along the acceleration direction. For the flat-space-time derivation of this correction term, see §6.6.

(3) The observer's rotation relative to inertial-guidance gyroscopes shows up in a correction term to $g_{\hat{0}\hat{j}}$, which can be rewritten in 3-vector notation

$$\delta g_{\hat{0}\hat{j}}\boldsymbol{e}_{\hat{j}} = -\boldsymbol{x} \times \boldsymbol{\omega} = +\boldsymbol{\omega} \times \boldsymbol{x}. \qquad (13.72b)$$

(4) These first-order corrections to the line element are unaffected by spacetime curvature and contain no information about curvature. Only at second order, $O(|x^{\hat{j}}|^2)$, will curvature begin to show up.

(5) *In the special case of zero acceleration and zero rotation ($\boldsymbol{a} = \boldsymbol{\omega} = 0$), the observer's proper reference frame reduces to a local Lorentz frame ($g_{\hat{\alpha}\hat{\beta}} = \eta_{\alpha\beta}$, $\Gamma^{\hat{\alpha}}{}_{\hat{\beta}\hat{\gamma}} = 0$) all along his geodesic world line!* By contrast, the local Lorentz coordinate

systems constructed earlier in the book ("general" local Lorentz coordinates of §8.6, "Riemann normal coordinates" of §11.6) are local Lorentz only at a single event.

In the case of zero rotation and zero acceleration, one can derive the following expression for the metric, accurate to second order in $|x^{\hat{j}}|$:

$$ds^2 = (-1 - R_{\hat{0}\hat{l}\hat{0}\hat{m}} x^{\hat{l}} x^{\hat{m}}) \, dt^2 - \left(\frac{4}{3} R_{\hat{0}\hat{l}\hat{j}\hat{m}} x^{\hat{l}} x^{\hat{m}}\right) dt \, dx^{\hat{j}}$$

$$+ \left(\delta_{\hat{i}\hat{j}} - \frac{1}{3} R_{\hat{i}\hat{l}\hat{j}\hat{m}} x^{\hat{l}} x^{\hat{m}}\right) dx^{\hat{i}} \, dx^{\hat{j}} + O(|x^{\hat{j}}|^3) \, dx^{\hat{\alpha}} \, dx^{\hat{\beta}} \tag{13.73}$$

[see, e.g., Manasse and Misner (1963)]. Here $R_{\hat{\alpha}\hat{\beta}\hat{\gamma}\hat{\delta}}$ are the components of the Riemann tensor along the world line $x^{\hat{j}} = 0$. Such coordinates are called "Fermi Normal Coordinates."

EXERCISES

Exercise 13.14. INERTIAL AND CORIOLIS FORCES

An accelerated observer studies the path of a freely falling particle as it passes through the origin of his proper reference frame. If

$$v \equiv (dx^{\hat{j}}/dx^{\hat{0}}) e_{\hat{j}} \tag{13.74}$$

is the particle's ordinary velocity, show that its ordinary acceleration relative to the observer's proper reference frame is

$$\frac{d^2 x^{\hat{j}}}{dx^{\hat{0}2}} e_{\hat{j}} = -a - 2\omega \times v + 2(a \cdot v)v. \tag{13.75}$$

[inertial acceleration] [Coriolis acceleration] $\begin{bmatrix}\text{relativistic correction} \\ \text{to inertial acceleration}\end{bmatrix}$

Here a is the observer's own 4-acceleration, and ω is the angular velocity with which his spatial basis vectors $e_{\hat{j}}$ are rotating [see equations (13.62)]. [*Hint:* Use the geodesic equation at the point $x^{\hat{j}} = 0$ of the particle's trajectory. *Note:* This result was derived in flat spacetime in exercise 6.8 using a different method.]

Exercise 13.15. ROTATION GROUP: METRIC

(Continuation of exercises 9.13, 9.14, 10.17 and 11.12). Show that for the manifold $SO(3)$ of the rotation group, there exists a metric g that is compatible with the covariant derivative ∇. Prove existence by exhibiting the metric components explicitly in the noncoordinate basis of generators $\{e_\alpha\}$. [*Answer:*

$$g_{\alpha\beta} = \delta_{\alpha\beta}. \tag{13.76}$$

Restated in words: If one postulates that: (1) the manifold of the rotation group is locally Euclidean; (2) the generators of infinitesimal rotations $\{e_\alpha\}$ are orthonormal, $e_\alpha \cdot e_\beta = \delta_{\alpha\beta}$; and (3) $\{e_\alpha\}$ obey the standard rotation-group commutation relations

$$[e_\alpha, e_\beta] = -\epsilon_{\alpha\beta\gamma} e_\gamma; \tag{13.77}$$

then the resulting geodesics of $SO(3)$ agree with the geodesics chosen in exercise 10.17.]

CHAPTER **14**

CALCULATION OF CURVATURE

§14.1. CURVATURE AS A TOOL FOR UNDERSTANDING PHYSICS

Elementary physics sometimes allows one to shortcircuit any systematized calculation of curvature (frequency of oscillation of test particle; tide-producing acceleration near a center of attraction; curvature of a closed 3-sphere model universe; effect of parallel transport on gyroscope or vector; see Figures 1.1, 1.10, and 1.12, and Boxes 1.6 and 1.7); but on other occasions a calculation of curvature is the quickest way into the physics. This chapter is designed for such occasions. It describes three ways to calculate curvature and gives the components of the Einstein curvature tensor for a plane gravitational wave (Box 14.4, equation 5), for the Friedmann geometry of the universe (Box 14.5), and for Schwarzschild geometry, both static (exercise 14.13) and dynamic (exercise 14.16). These and other calculations of curvature elsewhere are indexed under "curvature tensors."

It is enough to look at an expression for a 4-geometry as complicated as

> This chapter is entirely Track 2.
> Chapter 4 (differential forms) and Chapter 10, 11, and 13 (differential geometry) are necessary preparation for §§14.5–14.6.
> This chapter is needed as preparation for Chapter 15 (Bianchi identities).
> It will be helpful in many applications of gravitation theory (Chapters 23–40).

$$ds^2 = -(x/3^{1/2}L + y^2/12L^2)^{-3^{1/2}}\left(\int \frac{v\,dz}{z}\right)^{-1}(-z/L)^{3^{-1/2}}\,dt^2$$

$$+(x/3^{1/2}L + y^2/12L^2)^{1+3^{1/2}}\left(\int \frac{v\,dz}{z}\right)^{1+2/3^{1/2}}(-z/L)^{-1+3^{-1/2}}\,dx^2$$

$$+(x/3^{1/2}L + y^2/12L^2)^{2+3^{1/2}}\left(\int \frac{v\,dz}{z}\right)^{1+2/3^{1/2}}(-z/L)^{-3^{-1/2}}\,dy^2$$

$$+(x/3^{1/2}L + y^2/12L^2)^{3+3^{1/2}}\left(\int \frac{v\,dz}{z}\right)^{1+2/3^{1/2}}(-z/L)^{-2-3^{-1/2}} \times$$

$$\times \left(\frac{v^2-1}{-1-z/L}\right)dz^2 \quad (14.1)$$

[Harrison (1959)] to realize that one might understand the physical situation better if one knew what the curvature is; similarly with any other complicated expressions for metrics that arise from solving Einstein's equations or that appear undigested in the literature. In any such case, the appropriate method often is: curvature first, understanding second.

Curvature is the simplest local measure of geometric properties (see Box 14.1). Curvature is therefore a good first step toward a more comprehensive picture of the spacetime in question.

One sometimes has an expression for a spacetime metric first, and then makes calculations of curvature to understand it. But more often one makes calculations of curvature, subject to specified conditions of symmetry in space and time, as an aid in arriving at an expression for a physically interesting metric (stars, Chapters 23 to 26; model cosmologies, Chapters 27 to 30; collapse and black holes, Chapters 31 to 34; and gravitational waves, Chapters 35 to 37).

The basic "standard procedure for computing curvature" is illustrated in Box 14.2. Two formulas in Box 14.2, derived previously, are used in succession. The first (equations 1 and 2) has the form $\Gamma \sim g \, \partial g$ and provides the $\Gamma^\mu{}_{\alpha\beta}$. The other (equation 3) has the form $R \sim \partial\Gamma + \Gamma^2$ and gives the curvature components $R^\mu{}_{\nu\alpha\beta}$.

After the curvature components have been computed, there are helpful ways to present the results. (1) Form the Ricci tensor $R_{\mu\nu} = R^\alpha{}_{\mu\alpha\nu}$ and the scalar curvature $R = R^\mu{}_\mu$. (2) Form other invariants such as $R^{\mu\nu}{}_{\alpha\beta} R^{\alpha\beta}{}_{\mu\nu}$. (3) Form components $R^{\hat\mu\hat\nu}{}_{\hat\alpha\hat\beta}$ in a judiciously chosen orthonormal frame $\boldsymbol{w}^{\hat\alpha} = L^{\hat\alpha}{}_\beta \, \boldsymbol{d}x^\beta$, and (4) display $R^{[\hat\mu\hat\nu]}{}_{[\hat\alpha\hat\beta]}$ as a 6×6 matrix (in four dimensions; a 3×3 matrix in three dimensions) where $[\hat\mu\hat\nu] = [\hat0\hat1], [\hat0\hat2], [\hat0\hat3], [\hat2\hat3], [\hat3\hat1], [\hat1\hat2]$ labels the rows and $[\hat\alpha\hat\beta]$ labels the columns (exercises 14.14 and 14.15). (5) Last, but by far the most important for general relativity, form the Einstein tensor $G^{\hat\mu}{}_{\hat\nu}$ as described in §14.2.

The method of computation outlined above and described in more detail in Box 14.2 is used wherever it is quicker to employ a standard method than to learn or invent a better method. The standard method is always preferable for the student in a short course where physical insight has higher priority than technical facility. It is, however, a dull method, better suited to computers than to people. Even the algebra can be handled by a computer (see Box 14.3).

EXERCISES

Exercise 14.1. CURVATURE OF A TWO-DIMENSIONAL HYPERBOLOID

Compute the curvature of the hyperboloid $t^2 - x^2 - y^2 = T^2 = $ const in $2 + 1$ Minkowski spacetime with $ds_3^2 = -dt^2 + dx^2 + dy^2$. First show that intervals within this two-dimensional surface can be expressed in the form $ds^2 = T^2(d\alpha^2 + \sinh^2\alpha \, d\phi^2)$ by a suitable choice of coordinates α, ϕ, on the hyperboloid.

Exercise 14.2. RIEMANNIAN CURVATURE EXPRESSIBLE IN TERMS OF
RICCI CURVATURE IN TWO AND THREE DIMENSIONS

In two dimensions, there is only one independent curvature component, R_{1212}. Evidently the single scalar quantity R must carry the same information. The two-dimensional identity $R_{\mu\nu\alpha\beta} = \frac{1}{2}R(g_{\mu\alpha}g_{\nu\beta} - g_{\mu\beta}g_{\nu\alpha})$ is established by noting that it is the only tensor formula giving

(continued on page 343)

Box 14.1 PERSPECTIVES ON CURVATURE

1. Historical point of departure: a curved line on a plane. There is no way to define the curvature of a line by measurements confined to ("intrinsic to") the line itself. One needs, for example, the azimuthal bearing θ of the tangent vector relative to a fixed direction in the plane, as a function of proper distance s measured along the curve; thus, $\theta = \theta(s)$. Then curvature κ and its reciprocal, the radius of curvature ρ, are given by $\kappa(s) = 1/\rho(s) = d\theta(s)/ds$. Alternatively, one can examine departure, y, measured normally off from the tangent line as a function of distance x measured along that tangent line; then $\kappa = 1/\rho = d^2y/dx^2$.

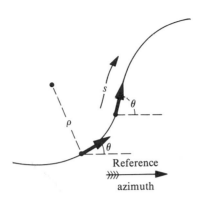

2. This concept was later extended to a curved surface embedded in flat (Euclidean) 3-space. Departure, z, of the smooth curved surface from the flat surface tangent to it at a given point is described in the neighborhood of that point by the quadratic expression

$$z = \frac{1}{2}ax^2 + bxy + \frac{1}{2}cy^2.$$

Rotation of the axes by an appropriate angle α,

$$x = \xi \cos\alpha + \eta \sin\alpha,$$

$$y = -\xi \sin\alpha + \eta \cos\alpha,$$

reduces this expression to

$$z = \frac{1}{2}\kappa_1\xi^2 + \frac{1}{2}\kappa_2\eta^2,$$

with

$$\kappa_1 = 1/\rho_1$$

and

$$\kappa_2 = 1/\rho_2$$

representing the two "principal curvatures" of the surface.

3. Gauss (1827) conceived the idea of defining curvature by measurements confined entirely to the surface ("society of ants"). From a given point \mathscr{P} on the surface, proceed on a geodesic on the surface for a proper distance ϵ measured entirely within the surface. Repeat, starting at the original point but proceeding in other directions.

Box 14.1 (continued)

Obtain an infinity of points. They define a "circle". Determine its proper circumference, again by measurements confined entirely to the surface. Using the metric corresponding to the embedding viewpoint

$$ds^2 = dz^2 + d\xi^2 + d\eta^2 \qquad \text{(Euclidean 3-space)}$$

$$= [(\kappa_1 \xi\, d\xi + \kappa_2 \eta\, d\eta)^2 + (d\xi^2 + d\eta^2)] \qquad \begin{pmatrix} \text{metric intrinsic} \\ \text{to the curved} \\ \text{2-geometry} \end{pmatrix},$$

one can calculate the result of such an "intrinsic measurement." One calculates that the circumference differs from the Euclidean value, $2\pi\epsilon$, by a fractional correction that is proportional to the square of ϵ; specifically,

$$\lim_{\epsilon \to 0} \frac{6}{\epsilon^2} \left(1 - \frac{\text{circumference}}{2\pi\epsilon} \right) = \kappa_1 \kappa_2 = \frac{1}{\rho_1 \rho_2} = \det \begin{pmatrix} a & b \\ b & c \end{pmatrix}.$$

Note especially the first equality sign. Gauss did not conceal the elation he felt on discovering that something defined by measurements entirely within the surface agrees with the product of two quantities, κ_1 and κ_2, that individually demand for their definition measurements extrinsic to the surface.

4. The contrast between "extrinsic" and "intrinsic" curvature is summarized in the terms,

$$(\text{extrinsic curvature}) = \kappa = (\kappa_1 + \kappa_2)(\text{cm}^{-1}),$$

$$\begin{pmatrix} \text{intrinsic or Gaussian} \\ \text{curvature} \end{pmatrix} = \kappa_1 \kappa_2 (\text{cm}^{-2})$$

(the latter being identical with half the scalar curvature invariant, R, of the 2-geometry). Draw a $3:4:5$ triangle on a flat piece of paper; then curl up the paper. The Euclidean 2-geometry intrinsic to the piece of paper is preserved by this bending. The Gaussian curvature intrinsic to the surface remains unaltered; it keeps the Euclidean value of zero (κ_2, non-zero; κ_1, zero; product, $\kappa_1 \kappa_2 = $ zero). However, the extrinsic curvature is changed from $\kappa_1 + \kappa_2 = 0$ to a non-zero value, $\kappa_1 + \kappa_2 \neq 0$.

5. The curvature dealt with in this chapter is curvature intrinsic to spacetime; that is, curvature defined without any use of, and repelling every thought of, any embedding in any hypothetical higher-dimensional flat manifold (concept of Riemann,

Clifford, and Einstein that geometry is a dynamic participant in physics, not some God-given perfection above the battles of matter and energy).

6. The curvature of the geometry of spacetime imposes curvature on any spacelike slice (3-geometry; "initial-value hypersurface") through that spacetime (see "relations of Gauss and Codazzi" in Chapter 21, on the initial-value problem of geometrodynamics).

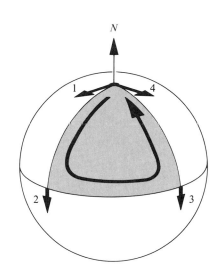

7. Rotation of a vector transported parallel to itself around a closed loop provides a definition of curvature as useful in four and three as in two dimensions. (In a curved two-dimensional geometry, at a point there is only one plane. Consequently only one number is required to describe the Gaussian curvature there. In three and four dimensions, there are more independent planes through a point and therefore more numbers are required to describe the curvature.) In the diagram, start with a vector at position 1 (North Pole). Transport it parallel to itself (position $2, 3, \ldots$) around a $90°\text{–}90°\text{–}90°$ spherical triangle. It arrives back at the starting point (position 4) turned through $90°$:

$$\begin{pmatrix} \text{Gaussian} \\ \text{curvature} \end{pmatrix} = \frac{\begin{pmatrix} \text{angle turned} \\ \text{through} \end{pmatrix}}{\begin{pmatrix} \text{area circum-} \\ \text{navigated} \end{pmatrix}} = \frac{(\pi/2)}{(1/8)(4\pi a^2)} = \frac{1}{a^2}$$

(positive; sense of rotation same as sense of circumnavigation).

8. Still staying for simplicity with a curved two-dimensional manifold, describe the curvature of the 2-surface as a 2-form ("box-like structure") defined over the entire surface. The number of boxes enclosed by any given route gives immediately the angle in radians (or tenths or hundredths of a radian, etc., depending on chosen fineness of subdivision) turned through by a vector carried parallel to itself around that route. The contribution of a given box is counted as positive or negative depending on whether the sense of the arrow marked on it (see magnified view) agrees or disagrees with the sense of circumnavigation of the route.

Box 14.1 (*continued*)

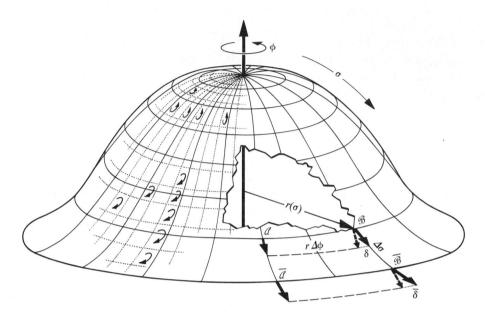

9. Curvature 2-form for the illustrated surface of rotational symmetry ("pith helmet") with metric $ds^2 = d\sigma^2 + r^2(\sigma)\,d\phi^2$ is

$$\boldsymbol{curvature} = -\frac{1}{r}\frac{d^2r}{d\sigma^2}\,\boldsymbol{d\sigma} \wedge r\,\boldsymbol{d\phi} \tag{1}$$

(positive on crown of helmet, negative around brim, as indicated by sense of arrows in the "boxes of the 2-form" shown at left). "Meaning" of r is illustrated by imbedding the surface in Euclidean 3-space, a convenience for visualization; but more important is the idea of a 2-geometry defined by measurements intrinsic to it, with no embedding.

10. How lengths ("metric") determine curvature in quantitative detail is shown nowhere more clearly than in this two-dimensional example, a model for "what is going on behind the scene" in the mathematical calculations done in this chapter with 1-forms and 2-forms in four-dimensional spacetime.

 a. Net rotation in going around element of surface $\mathcal{A}\mathcal{B}\bar{\mathcal{B}}\bar{\mathcal{A}}\mathcal{A}$ is $\delta - \bar{\delta}$ (no turn of vector to left or to right in its transport along a meridian $\mathcal{A}\bar{\mathcal{A}}$ or $\mathcal{B}\bar{\mathcal{B}}$).

 b. Rotation of vector in going from \mathcal{A} to \mathcal{B}, relative to coordinate system (directions of meridians), is

$$(\text{angle } \delta) = \frac{\text{arc}}{\text{length}} = \frac{r(\sigma + d\sigma)\,\Delta\phi - r(\sigma)\,\Delta\phi}{d\sigma} = \left(\frac{dr}{d\sigma}\right)_\sigma \Delta\phi.$$

c. Rotation of vector in going from \bar{a} to $\bar{\mathscr{B}}$ is similarly

$$(\text{angle } \bar{\delta}) = \left(\frac{dr}{d\sigma}\right)_{\sigma + \Delta\sigma} \Delta\phi.$$

d. Thus net rotation is:

$$\delta - \bar{\delta} = -\left(\frac{d^2r}{d\sigma^2}\right)_\sigma \Delta\sigma \, \Delta\phi.$$

e. Expressed as a form, this gives immediately equation (1).

f. Ideas and calculations are more complicated in four dimensions, primarily because one has to deal with different choices for the orientation of the surface to be studied at the point in question.

11. Translation of these geometric ideas into the language of forms is most immediate when one stays with this example of two dimensions. A sample vector $A^i = (A^1, A^2)$ carried around the boundary of an element of surface comes back to its starting point slightly changed in direction:

$$-\binom{\text{change}}{\text{in } A^i} = \mathscr{R}^i_{\ j}A^j. \tag{2}$$

a. To be more specific, it is convenient to adopt as the basis 1-forms $\boldsymbol{\omega}^{\hat{1}} = \boldsymbol{d\sigma}$ and $\boldsymbol{\omega}^{\hat{2}} = r\,\boldsymbol{d\phi}$, and have $A^{\hat{1}}$ as the component of \boldsymbol{A} along the direction of increasing σ, $A^{\hat{2}}$ as the component of \boldsymbol{A} along the direction of increasing ϕ. The matrix $\mathscr{R}^{\hat{i}}_{\ \hat{j}}$ is a rotation matrix, which produces a change in direction but no change in length (zero diagonal components); thus here

$$\|\mathscr{R}^{\hat{i}}_{\ \hat{j}}\| = \left\|\begin{array}{cc} 0 & \mathscr{R}^{\hat{1}}_{\ \hat{2}} \\ -\mathscr{R}^{\hat{1}}_{\ \hat{2}} & 0 \end{array}\right\|. \tag{3}$$

In this example, $\mathscr{R}^{\hat{1}}_{\ \hat{2}}$ evidently represents the angle through which the vector \boldsymbol{A} turns on transport parallel to itself around the element of surface.

b. So far the rotation is "indefinite" because the size of the element of surface has not yet been specified. It is most conveniently conceived as an elementary parallelogram, defined by two vectors ("bivector"). Thus $\mathscr{R}^{\hat{i}}_{\ \hat{j}}$, or, specifically, the one element that counts, $\mathscr{R}^{\hat{1}}_{\ \hat{2}}$ (the "angle of rotation"), has to be envisaged as a mathematical object ("2-form") endowed with two slots, into which these two vectors are inserted to get a definite number (angle in radians). In the example of the pith helmet, one has, from equation (1)

$$\mathscr{R}^{\hat{1}}_{\ \hat{2}} = -\frac{1}{r}\frac{d^2r}{d\sigma^2}\,\boldsymbol{\omega}^{\hat{1}} \wedge \boldsymbol{\omega}^{\hat{2}}. \tag{4}$$

Thus the $\mathscr{R}^\mu_{\ \nu}$ in the text are called "curvature 2-forms."

Box 14.1 (*continued*)

c. The text tells one how to read out of such expressions the components of the Riemann curvature tensor; for example here,

$$R^{\hat{1}}{}_{\hat{2}\hat{1}\hat{2}} = -R^{\hat{1}}{}_{\hat{2}\hat{2}\hat{1}} = (-1/r)(d^2r/d\sigma^2) \text{ (coefficients of } \boldsymbol{\omega}^{\hat{1}} \wedge \boldsymbol{\omega}^{\hat{2}} \text{ or } \boldsymbol{\omega}^{\hat{2}} \wedge \boldsymbol{\omega}^{\hat{1}}).$$

d. Generalizing to four dimensions, one understands by $R^{\alpha}{}_{\beta\mu\nu}$ the factor that one has to multiply by three numbers to obtain a fourth. The number obtained is the change (with reversed sign) that takes place in the αth component of a vector when that vector is transported parallel to itself around a closed path, defined, for example, by a parallelogram built from two vectors \boldsymbol{u} and \boldsymbol{v}. The factors that multiply $R^{\alpha}{}_{\beta\mu\nu}$ are (1) the component of the vector \boldsymbol{A} in the βth direction and (2, 3) the $\mu\nu$ component of the extension of the parallelogram, $(u^{\mu}v^{\nu} - u^{\nu}v^{\mu})$. Thus

$$\delta A^{\alpha} = -R^{\alpha}{}_{\beta|\mu\nu|}A^{\beta}(u^{\mu}v^{\nu} - u^{\nu}v^{\mu}).$$

Box 14.2 STRAIGHTFORWARD CURVATURE COMPUTATION
(Illustrated for a Globe)

The elementary and universally applicable method for computing the components $R^{\mu}{}_{\nu\alpha\beta}$ of the Riemann curvature tensor starts from the metric components $g_{\mu\nu}$ in a coordinate basis, and proceeds by the following scheme:

$$g_{\mu\nu} \xrightarrow{\Gamma \,\dot{\sim}\, \partial g} \Gamma_{\mu\alpha\beta} \longrightarrow \Gamma^{\mu}{}_{\alpha\beta} \xrightarrow{R \,\sim\, \partial\Gamma + \Gamma\Gamma} R^{\mu}{}_{\nu\alpha\beta}.$$

The formulas required for these three steps are

$$\Gamma_{\mu\alpha\beta} = \frac{1}{2}\left(\frac{\partial g_{\mu\alpha}}{\partial x^{\beta}} + \frac{\partial g_{\mu\beta}}{\partial x^{\alpha}} - \frac{\partial g_{\alpha\beta}}{\partial x^{\mu}}\right), \tag{1}$$

$$\Gamma^{\mu}{}_{\alpha\beta} = g^{\mu\nu}\Gamma_{\nu\alpha\beta}, \tag{2}$$

and

$$R^{\mu}{}_{\nu\alpha\beta} = \frac{\partial \Gamma^{\mu}{}_{\nu\beta}}{\partial x^{\alpha}} - \frac{\partial \Gamma^{\mu}{}_{\nu\alpha}}{\partial x^{\beta}} + \Gamma^{\mu}{}_{\rho\alpha}\Gamma^{\rho}{}_{\nu\beta} - \Gamma^{\mu}{}_{\rho\beta}\Gamma^{\rho}{}_{\nu\alpha}. \tag{3}$$

The metric of the two-dimensional surface of a sphere of radius a is

$$ds^2 = a^2(d\theta^2 + \sin^2\theta\, d\phi^2). \tag{4}$$

To compute the curvature by the standard method, use the formula for ds^2 as a table of g_{kl} values. It shows that $g_{\theta\theta} = a^2$, $g_{\theta\phi} = 0$, $g_{\phi\phi} = a^2\sin^2\theta$. Compute the six possible different $\Gamma_{jkl} = \Gamma_{jlk}$ (there will be 40 in four dimensions) from formula

(1). Thus

$$\Gamma_{\theta\phi\phi} = -a^2\sin\theta\cos\theta = -\Gamma_{\phi\phi\theta},$$

$$\Gamma_{\theta\theta\theta} = \Gamma_{\phi\phi\phi} = 0, \tag{5}$$

$$\Gamma_{\theta\theta\phi} = \Gamma_{\phi\theta\theta} = 0.$$

Raise the first index:

$$\Gamma^{\theta}{}_{\phi\phi} = -\sin\theta\cos\theta,$$

$$\Gamma^{\phi}{}_{\phi\theta} = \cot\theta, \tag{6}$$

$$\Gamma^{\theta}{}_{\theta\theta} = \Gamma^{\theta}{}_{\theta\phi} = 0 = \Gamma^{\phi}{}_{\theta\theta} = \Gamma^{\phi}{}_{\phi\phi}.$$

Choose a suitable curvature component (one that is not automatically zero by reason of the elementary symmetry $R_{\mu\nu\alpha\beta} = R_{[\mu\nu][\alpha\beta]}$, nor previously computed in another form, as by $R_{\mu\nu\alpha\beta} = R_{\alpha\beta\mu\nu}$). In this two-dimensional example, there is only one choice (compared to 21 such computations in four dimensions); it is

$$R^{\theta}{}_{\phi\theta\phi} = \frac{\partial\Gamma^{\theta}{}_{\phi\phi}}{\partial\theta} - \frac{\partial\Gamma^{\theta}{}_{\phi\theta}}{\partial\phi} + \Gamma^{\theta}{}_{k\theta}\Gamma^{k}{}_{\phi\phi} - \Gamma^{\theta}{}_{k\phi}\Gamma^{k}{}_{\phi\theta}$$

$$= \frac{\partial\Gamma^{\theta}{}_{\phi\phi}}{\partial\theta} - 0 + 0 - \Gamma^{\theta}{}_{\phi\phi}\Gamma^{\phi}{}_{\phi\theta}$$

$$= \sin^2\theta - \cos^2\theta + \sin\theta\cos\theta\cot\theta;$$

so

$$R^{\theta}{}_{\phi\theta\phi} = \sin^2\theta \tag{7}$$

or

$$R^{\theta\phi}{}_{\theta\phi} = \frac{1}{a^2}. \tag{8}$$

Contraction gives the components of the Ricci tensor,

$$R^{\theta}{}_{\theta} = R^{\phi}{}_{\phi} = \frac{1}{a^2}, \qquad R^{\theta}{}_{\phi} = 0, \tag{9}$$

and further contraction gives the curvature scalar

$$R = 2/a^2. \tag{10}$$

A convenient orthonormal frame in this manifold is

$$\boldsymbol{\omega}^{\hat{\theta}} = a\,\boldsymbol{d}\theta, \qquad \boldsymbol{\omega}^{\hat{\phi}} = a\sin\theta\,\boldsymbol{d}\phi. \tag{11}$$

More generally one writes $\boldsymbol{\omega}^{\hat{\alpha}} = L^{\hat{\alpha}}{}_{\beta}\,\boldsymbol{d}x^{\beta}$. To transform the curvature tensor to orthonormal components in this simple but illuminating example of a diagonal metric requires a single normalization factor for each index on a tensor. Thus $v^{\hat{\theta}} = av^{\theta}$, $v^{\hat{\phi}} = a\sin\theta\,v^{\phi}$, $v_{\hat{\theta}} = a^{-1}v_{\theta}$, $v_{\hat{\phi}} = (a\sin\theta)^{-1}v_{\phi}$. Similarly, from $R^{\theta}{}_{\phi\theta\phi} = \sin^2\theta$ one finds the components of the curvature tensor,

$$R^{\hat{\theta}}{}_{\hat{\phi}\hat{\theta}\hat{\phi}} = \frac{1}{a^2} = R^{\hat{\theta}\hat{\phi}}{}_{\hat{\theta}\hat{\phi}}, \tag{12}$$

in the orthonormal frame.

Box 14.3 ANALYTICAL CALCULATIONS ON A COMPUTER

Research in gravitation physics and general relativity is sometimes beset by long calculations, requiring meticulous care, of such quantities as the Einstein and Riemann curvature tensors for a given metric, or the divergence of a given stress-energy tensor, or the Newman-Penrose tetrad equations under given algebraic assumptions. Such calculations are sufficiently straightforward and deductive in logical structure that they can be handled by a computer. Since 1966, computers have been generally taking over such tasks.

There are several computer languages in which the investigator can program his analytic calculations. The computer expert may wish to work in a machine-oriented language such as LISP [see, e.g., the work of Fletcher (1966) and of Hearn (1970)]. However, most appliers of relativity will prefer user-oriented languages such as REDUCE [created by Hearn (1970) and available for the IBM 360 and 370, and the PDP 10 computers], ALAM [created by D'Inverno (1969) and available on Atlas computers], CAMAL [created by Barton, Bourne, and Fitch (1970) and available on Atlas computers], and FORMAC [created by Tobey *et al.* (1967) and available on IBM 7090, 7094, 360, and 370]. For a review of activity in this area, see Barton and Fitch (1971). Here we describe only FORMAC. It is the most widely available and widely used of the languages; but it is probably *not* the most powerful [see, e.g., D'Inverno (1969)]. FORMAC is to analytic work what the earliest and most primitive versions of FORTRAN were to numerical work.

FORMAC manipulates algebraic expressions involving: numerical constants, such as $1/3$; symbolic constants, such as x or u; specific elementary functions, such as $\sin(u)$ or $\exp(x)$; and symbolic functions of several variables, such as $f(x, u)$ or $g(u)$. For example, it can add $ax + bx^2$ to $2x + (3 + b)x^2$ and get $(a + 2)x + (3 + 2b)x^2$; it can take the partial derivative of $x^2 uf(x, u) + \cos(x)$ with respect to x and get

$$2xuf(x, u) + x^2u\, \partial f(x, u)/\partial x - \sin(x).$$

It can do any algebraic or differential-calculus

computation that a human can do—but without making mistakes! Unfortunately, it cannot integrate analytically; integration requires inductive logic rather than deductive logic.

PL/1 is a language that can be used simultaneously with FORMAC or independently of it. PL/1 manipulates strings of characters—e.g., "Z$/1\times29-+/$." It knows symbolic logic; it can tell whether two strings are identical; it can insert new characters into a string or remove old ones; but it does not know the rules of algebra or differential calculus. Thus, its primary use is as an adjunct to FORMAC (though from the viewpoint of the computer system FORMAC is an adjunct of PL/1).

FORMAC programs for evaluating Einstein's tensor in terms of given metric components and for doing other calculations are available from many past users [see, e.g., Fletcher, Clemens, Matzner, Thorne, and Zimmerman (1967); Ernst (1968); Harrison (1970)]. However, programming in FORMAC is sufficiently simple that one ordinarily does not have difficulty creating one's own program to do a given task. If a difficulty does arise, it may be because the analytic computation exhausts the core of the computer. It is easy to create an expression too large to fit in the core of any existing computer by several differentiations of an expression half a page long!

Users of FORMAC, confronted by core-exhaustion, have devised several ways to solve their problems. One is to remove unneeded parts of the program and of the FORMAC system from the core. Routines called PURGE and KILL have been developed for this purpose by Clemens and Matzner (1967). Another is to create the answer to a given calculation in manageable-sized pieces and output those pieces from the computer's core onto its disk. One must then add all the pieces together—a task that is impossible using FORMAC alone, or even FORMAC plus PL/1, but a task that James Hartle has solved [see Hartle and Thorne (1974)] by using a combination of FORMAC, PL/1, and IBM data-manipulation routines called SORT.

$R_{\mu\nu\alpha\beta}$ as a linear function of R, constructed from R and the metric alone, and with the correct contracted value $R^{\mu\nu}{}_{\mu\nu} = R$. Establish a corresponding three-dimensional identity expressing R_{ijkl} in terms of the Ricci tensor R_{jk} and the metric.

Exercise 14.3. CURVATURE OF 3-SPHERE IN ORTHONORMAL FRAME
Compute the curvature tensor for a 3-sphere

$$ds^2 = a^2[d\chi^2 + \sin^2\chi(d\theta^2 + \sin^2\theta \, d\phi^2)] \tag{14.2}$$

or for a 3-hyperboloid

$$ds^2 = a^2[d\chi^2 + \sinh^2\chi(d\theta^2 + \sin^2\theta \, d\phi^2)]. \tag{14.3}$$

Convert the coordinate-based components $R^i{}_{jkl}$ to a corresponding orthonormal basis, $R^{\hat{i}}{}_{\hat{j}\hat{k}\hat{l}}$. Display $R^{\hat{i}\hat{j}}{}_{\hat{k}\hat{l}} = R^{[\hat{i}\hat{j}]}{}_{[\hat{k}\hat{l}]}$ as a 3×3 matrix with appropriately labeled rows and columns.

§14.2. FORMING THE EINSTEIN TENSOR

The distribution of matter in space does not immediately tell all details of the local curvature of space, according to Einstein. The stress-energy tensor provides information only about a certain combination of components of the Riemann curvature tensor, the combination that makes up the Einstein tensor. Chapter 13 described two equivalent ways to calculate the Einstein tensor: (1) by successive contractions of the Riemann tensor

Three ways to compute the Einstein tensor from the Riemann tensor

$$R_{\mu\nu} = R^\alpha{}_{\mu\alpha\nu}, \qquad R = g^{\mu\nu}R_{\mu\nu},$$
$$G_{\mu\nu} = R_{\mu\nu} - \frac{1}{2}g_{\mu\nu}R \tag{14.4}$$

[equations (13.48) and (13.49)]; (2) by forming the dual of the Riemann tensor and then contracting:

$$\mathcal{G}_{\alpha\beta}{}^{\gamma\delta} \equiv (*R*)_{\alpha\beta}{}^{\gamma\delta} = \epsilon_{\alpha\beta\mu\nu}R^{|\mu\nu|}{}_{|\rho\sigma|}\epsilon^{\rho\sigma\gamma\delta}$$
$$= -\delta^{\rho\sigma\gamma\delta}{}_{\alpha\beta\mu\nu}R^{|\mu\nu|}{}_{|\rho\sigma|}, \tag{14.5a}$$

$$G_\beta{}^\delta = \mathcal{G}_{\alpha\beta}{}^{\alpha\delta} \tag{14.5b}$$

[equations (13.46) and (13.47)]. A third method, usually superior to either of these, is discovered by combining equations (14.5a,b):

$$G_\beta{}^\delta = G^\delta{}_\beta = -\delta^{\delta\rho\sigma}{}_{\beta\mu\nu}R^{|\mu\nu|}{}_{|\rho\sigma|}. \tag{14.6}$$

[Note: in any frame, orthonormal or not, the permutation tensor $\delta^{\delta\rho\sigma}{}_{\beta\mu\nu}$ has components

$$\delta^{\delta\rho\sigma}{}_{\beta\mu\nu} = \delta_{\beta\mu\nu}{}^{\delta\rho\sigma} = \begin{cases} +1 \text{ if } \delta\rho\sigma \text{ is an even permutation of } \beta\mu\nu, \\ -1 \text{ if } \delta\rho\sigma \text{ is an odd permutation of } \beta\mu\nu, \\ \;\; 0 \text{ otherwise;} \end{cases}$$

to see this, simply evaluate $\delta^{\delta\rho\sigma}{}_{\beta\mu\nu}$ using definition (3.50h) and using the components (8.10) of $\epsilon_{\alpha\beta\mu\nu}$ and $\epsilon^{\rho\sigma\gamma\delta}$.] Equation (14.6) for the Einstein tensor, written out explicitly, reads

$$G^0{}_0 = -(R^{12}{}_{12} + R^{23}{}_{23} + R^{31}{}_{31}),$$

$$G^1{}_1 = -(R^{02}{}_{02} + R^{03}{}_{03} + R^{23}{}_{23}),$$

$$G^0{}_1 = R^{02}{}_{12} + R^{03}{}_{13},$$

$$G^1{}_2 = R^{10}{}_{20} + R^{13}{}_{23},$$

(14.7)

and every other component is given by a similar formula, obtainable by obvious permutations of indices.

§14.3. MORE EFFICIENT COMPUTATION

Standard method of computing curvature is wasteful

Ways to avoid "waste":

(1) geodesic Lagrangian method

(2) method of curvature 2-forms

If the answer to a problem or the result of a computation is not simple, then there is no simple way to obtain it. But when a long computation gives a short answer, *then* one looks for a better method. Many of the best-known applications of general relativity present one with metric forms in which many of the components $g_{\mu\nu}$, $\Gamma^\mu{}_{\alpha\beta}$, and $R^\mu{}_{\nu\alpha\beta}$ are zero; for them the standard computation of the curvature (Box 14.2) involves much "wasted" effort. One computes many $\Gamma^\mu{}_{\alpha\beta}$ that turn out to be zero. One checks off many terms in a sum like $-\Gamma^\mu{}_{\rho\beta}\Gamma^\rho{}_{\alpha\mu}$ that are zero, or cancel with others to give zero. Two alternative procedures are available to eliminate some of this "waste." The "geodesic Lagrangian" method provides an economical way to tabulate the $\Gamma^\mu{}_{\alpha\beta}$. The method of "curvature 2-forms" reorganizes the description from beginning to end, and computes both the connection and the curvature.

The geodesic Lagrangian method is only a moderate improvement over the standard method, but it also demands only a modest investment in the calculus of variations, an investment that pays off in any case in other contexts in the world of mathematics and physics. In contrast, the method of curvature 2-forms is efficient, but demands a heavier investment in the mathematics of 1-forms and 2-forms than anyone would normally find needful for any introductory survey of relativity. Anyone facing several days' work at computing curvatures, however, would do well to learn the algorithm of the curvature 2-forms.

§14.4. THE GEODESIC LAGRANGIAN METHOD

One normally thinks that the connection coefficients $\Gamma^\mu{}_{\alpha\beta}$ must be known before one can write the geodesic equation

$$\ddot{x}^\mu + \Gamma^\mu{}_{\alpha\beta}\dot{x}^\alpha\dot{x}^\beta = 0.$$

(14.8)

(Here and below dots denote derivative with respect to the affine parameter, λ.) However, the argument can be reversed. Once the geodesic equations have been

written down, the connection coefficients can be read out of them. For instance, on the 2-sphere as treated in Box 14.2, the geodesic equations are

$$\ddot{\theta} - \sin\theta \cos\theta \, \dot{\phi}^2 = 0, \tag{14.9\theta}$$

$$\ddot{\phi} + 2\cot\theta \, \dot{\phi}\dot{\theta} = 0. \tag{14.9\phi}$$

The first equation here shows that $\Gamma^{\theta}_{\phi\phi} = -\sin\theta \cos\theta$; the second equation shows that $\Gamma^{\phi}_{\phi\theta} = \Gamma^{\phi}_{\theta\phi} = \cot\theta$; and the absence of any further terms shows that all other Γ^{i}_{jk} are zero.

The first essential principle is thus clear: an explicit writing out of the geodesic equation is equivalent to a tabulation of all the connection coefficients $\Gamma^{\mu}_{\alpha\beta}$.

The second principle says more: one can write out the geodesic equation without ever having computed the $\Gamma^{\mu}_{\alpha\beta}$. In order to arrive at the equations for a geodesic (see Box 13.3), one need only recall that a geodesic is a parametrized curve that extremizes the integral

$$I = \frac{1}{2} \int g_{\mu\nu} \dot{x}^{\mu} \dot{x}^{\nu} \, d\lambda \tag{14.10}$$

in the sense

$$\delta I = 0.$$

In practical applications of this variational principle, *the first step is to rewrite equation (14.10) in the simplest possible form, inserting the specific values of $g_{\mu\nu}$ for the problem at hand.* If one's interest attaches to the geodesics themselves, one can recognize many constants of motion even without carrying out any variations (see Chapter 25 on geodesic motion in Schwarzschild geometry, especially §25.2 on conservation laws and constants of motion). For the purpose of computing the $\Gamma^{\mu}_{\alpha\beta}$, one proceeds to vary each coordinate in turn, obtaining four equations. Next these equations are rearranged so that their leading terms are \ddot{x}^{μ}. In this form they must be precisely the geodesic equations (14.8). Consequently, the $\Gamma^{\mu}_{\alpha\beta}$ are immediately available as the coefficients in these four equations. For the final step in computing curvature by this method, one returns to the standard method and to formulas of the type $R \sim \partial\Gamma + \Gamma\Gamma$, treated in the standard way (Box 14.2); and as the need arises for each Γ in turn, one scans the geodesic equation to find it. The procedure is best understood by following an example: Box 14.4 provides one.

Geodesic Lagrangian method in 4 steps:

(1) write I in simple form

(2) vary I to get geodesic equation

(3) read off $\Gamma^{\alpha}_{\beta\gamma}$

(4) compute $R^{\alpha}_{\beta\gamma\delta}$ etc. by standard method

Exercise 14.4. EINSTEIN EQUATIONS FOR THE CLOSED FRIEDMANN UNIVERSE CALCULATED BY USING THE GEODESIC LAGRANGIAN METHOD

EXERCISE

The line element of interest here is (see Chapter 27)

$$ds^2 = -dt^2 + a^2(t)\,[d\chi^2 + \sin^2\chi\,(d\theta^2 + \sin^2\theta\,d\phi^2)].$$

(continued on page 348)

Box 14.4 GEODESIC LAGRANGIAN METHOD SHORTENS SOME CURVATURE COMPUTATIONS

Aim: Compute the curvature for the line element

$$ds^2 = L^2(e^{2\beta} dx^2 + e^{-2\beta} dy^2) - 2\, du\, dv \tag{1}$$

where L and β are functions of u only. [This metric is discussed as an example of a gravitational wave in §§35.9–35.12.]

Method: Obtain the $\Gamma^\mu{}_{\alpha\beta}$ from the geodesic equations as inferred from the variational principle (14.10), then compute $R^\mu{}_{\nu\alpha\beta} \sim \partial\Gamma + \Gamma^2$ as in Box 14.2.

Step 1. State the variational integral. For the metric under consideration, equation (14.10) requires $\delta I = 0$ for

$$I = \int \left[\frac{1}{2} L^2(e^{2\beta}\dot{x}^2 + e^{-2\beta}\dot{y}^2) - \dot{u}\dot{v}\right] d\lambda. \tag{2}$$

A world line that extremizes this integral is a geodesic.

Step 2: Vary the coordinates of the world line, one at a time, in their dependence on λ. First vary $x(\lambda)$, keeping fixed the functions $y(\lambda)$, $u(\lambda)$, and $v(\lambda)$. Then

$$\delta I = \int (L^2 e^{2\beta}\dot{x})\,\delta\dot{x}\, d\lambda = -\int (L^2 e^{2\beta}\dot{x})^{\cdot}\,\delta x\, d\lambda.$$

The requirement that $\delta I = 0$ for this variation (among others) gives

$$0 = (L^2 e^{2\beta}\dot{x})^{\cdot} = L^2 e^{2\beta}\ddot{x} + \dot{x}\dot{u}\frac{\partial}{\partial u}(L^2 e^{2\beta}).$$

Varying y, u, v, in the same way gives

$$0 = (L^2 e^{-2\beta}\dot{y})^{\cdot} = L^2 e^{-2\beta}\ddot{y} + \dot{y}\dot{u}\frac{\partial}{\partial u}(L^2 e^{-2\beta}),$$

$$0 = \ddot{v} + \frac{1}{2}\dot{x}^2\frac{\partial}{\partial u}(L^2 e^{2\beta}) + \frac{1}{2}\dot{y}^2\frac{\partial}{\partial u}(L^2 e^{-2\beta}),$$

$$0 = \ddot{u}.$$

Step 3: Rearrange to get \ddot{x}^μ **leading terms.** If this step is not straightforward, this method will not save time, and the technique of either Box 14.2 or Box 14.5 will be more suitable. In the example here, one quickly writes, using a prime for $\partial/\partial u$,

$$0 = \ddot{x} + 2(L^{-1}L' + \beta')\dot{x}\dot{u}, \tag{3x}$$

$$0 = \ddot{y} + 2(L^{-1}L' - \beta')\dot{y}\dot{u}, \tag{3y}$$

$$0 = \ddot{v} + (L^2 e^{2\beta})(L^{-1}L' + \beta')\dot{x}^2 + (L^2 e^{-2\beta})(L^{-1}L' - \beta')\dot{y}^2, \tag{3v}$$

$$0 = \ddot{u}. \tag{3u}$$

Step 3′: Interpret these equations as a tabulation of $\Gamma^\mu{}_{\alpha\beta}$. Equations (3) are the standard equations for a geodesic,

$$\ddot{x}^\mu + \Gamma^\mu{}_{\alpha\beta}\dot{x}^\alpha\dot{x}^\beta = 0.$$

Therefore it is enough to scan them to find the value of any desired Γ. For instance $\Gamma^x{}_{yu}$ must appear in the coefficient $(\Gamma^x{}_{yu} + \Gamma^x{}_{uy}) = 2\Gamma^x{}_{yu}$ of the $\dot{y}\dot{u}$ term in the equation for \ddot{x}. But no $\dot{y}\dot{u}$ term appears in equation (3x). Therefore $\Gamma^x{}_{yu}$ is zero in this example. Note that equations (3) are simple, in the sense that they contain few terms; therefore most of the $\Gamma^\mu{}_{\alpha\beta}$ must be zero. For instance, it follows from equation (3u) that all ten $\Gamma^u{}_{\alpha\beta}$ are zero. The only non-zero Γ's are $\Gamma^x{}_{xu} = \Gamma^x{}_{ux} = (L^{-1}L' + \beta')$ from equation (3x), $\Gamma^y{}_{yu} = \Gamma^y{}_{uy} = (L^{-1}L' - \beta')$ from equation (3y), and $\Gamma^v{}_{xx}$ and $\Gamma^v{}_{yy}$ from equation (3v).

Step 4: Compute each $R^\mu{}_{\nu\alpha\beta}$, **etc.** There is little relief from routine in systematically applying equation (3) from Box 14.2. One must list 21 components $R^\mu{}_{\nu\alpha\beta}$ that are not related by any of the symmetries $R_{\mu\nu\alpha\beta} = R_{\alpha\beta\mu\nu} = -R_{\mu\nu\beta\alpha}$, and compute each. In the example here, one notes that $\Gamma^u{}_{\alpha\beta} = 0$ implies $R^u{}_{\alpha\beta\gamma} = -R_{v\alpha\beta\gamma} = 0$. Therefore 15 of the list of 21 vanish at one swat. The list then is:

$$R_{v\alpha\beta\gamma} = -R^u{}_{\alpha\beta\gamma} = 0,$$

$$R_{uxux} = -R^v{}_{xux} = -(\Gamma^v{}_{xx})' + \Gamma^v{}_{xx}\Gamma^x{}_{xu}$$
$$= -(L^2 e^{2\beta})\left(\frac{L''}{L} + \beta'' + 2\frac{L'}{L}\beta' + \beta'^2\right),$$

$$R_{uxxy} = -R^v{}_{xxy} = 0,$$

$$R_{uxyu} = -R^v{}_{xyu} = 0, \tag{4}$$

$$R_{uyuy} = -R^v{}_{yuy} = -(\Gamma^v{}_{yy})' + \Gamma^v{}_{yy}\Gamma^y{}_{yu}$$
$$= -(L^2 e^{-2\beta})\left(\frac{L''}{L} - \beta'' - 2\frac{L'}{L}\beta' + \beta'^2\right),$$

$$R_{uyxy} = -R^v{}_{yxy} = 0,$$

$$R_{xyxy} = (L^2 e^{2\beta})R^x{}_{yxy} = 0.$$

One can now calculate the Einstein tensor via equation (14.7). In the example here, however, it is equally simple to form first the Ricci tensor by the straightforward contraction $R^\mu{}_{\alpha\mu\beta}$. Only $\mu = x$ and $\mu = y$ give any contribution, because no superscript index can be a u, and no subscript a v. Thus one finds

$$R_{uu} = -2[L^{-1}L'' + \beta'^2], \tag{5}$$
$$\text{all other } R_{\alpha\beta} = 0,$$

and

$$R = 0. \tag{6}$$

From this last result, it follows that here the desired Einstein tensor is identical with the Ricci tensor.

(a) Set up the variational integral (14.10) for a geodesic in this metric, then successively vary t, χ, θ, and ϕ to obtain, after some rearrangement, four equations $0 = \ddot{t} + \cdots$, $0 = \ddot{\chi} + \cdots$, etc. displaying the Γ's in the form of equation (14.8).

(b) Use this display as a table of Γ's to compute $R^t{}_{\chi\mu\nu}$ and $R^\chi{}_{\theta\mu\nu}$, of which only $R^t{}_{\chi t\chi}$ and $R^\chi{}_{\theta\chi\theta}$ are non-zero (consequence of the complete equivalence of all directions tangent to the $\chi\theta\phi$ sphere).

(c) Convert to an orthonormal frame with $\boldsymbol{w}^{\hat{t}} = \boldsymbol{d}t$, $\boldsymbol{w}^{\hat{\chi}} = a\,\boldsymbol{d}\chi$, $\boldsymbol{w}^{\hat{\theta}} = ?$, $\boldsymbol{w}^{\hat{\phi}} = ?$, and list $R^{\hat{t}\hat{\chi}}{}_{\hat{t}\hat{\chi}}$ and $R^{\hat{\chi}\hat{\theta}}{}_{\hat{\chi}\hat{\theta}}$. Explain why all other components are known by symmetry in terms of these two.

(d) Calculate, using equations (14.7), all independent components of the Einstein tensor $G^{\hat{\mu}}{}_{\hat{\nu}}$. [*Answer:* See Box 14.5.]

§14.5. CURVATURE 2-FORMS

In electrodynamics the abstract notation

$$\boldsymbol{F} = \boldsymbol{d}\boldsymbol{A}$$

saves space compared to the explicit notation

$$F_{31} = \frac{\partial A_1}{\partial x^3} - \frac{\partial A_3}{\partial x^1},$$

$$F_{12} = \frac{\partial A_2}{\partial x^1} - \frac{\partial A_1}{\partial x^2},$$

$$\ldots, \text{etc. (six equations)};$$

Concepts needed for method of curvature 2-forms

there is no reason to shun similar economies in dealing with the dynamics of geometry. Cartan introduced the decisive ideas, seen above, of differential forms (where a simple object replaces a listing of four components; thus, $\boldsymbol{\sigma} = \sigma_\mu\,\boldsymbol{d}x^\mu$), and of the exterior derivative \boldsymbol{d}. He went on (1928, 1946) to package the 21 components $R_{\mu\nu\alpha\beta}$ of the curvature tensor into six curvature 2-forms,

$$\mathscr{R}^{\mu\nu} = -\mathscr{R}^{\nu\mu}.$$

Regarded purely as notation, these 2-forms automatically produce a profit. They cut down the weight of paper work required to list one's answer after one has it. They also provide a route into deeper insight on "curvature as a geometric object," although that is not the objective of immediate concern in this chapter.

Cartan's exterior derivative \boldsymbol{d} automatically effects many cancelations in the calculation of curvature. It often cancels terms before they ever need to be evaluated.

Extension of Cartan's calculus from electromagnetism and other applications (Chapter 4) to the analysis of curvature (this chapter) requires two minor additions to the armament of forms and exterior derivative: (1) the idea of a vector-valued (or tensor-valued) exterior differential form; and (2) a corresponding generalization

of the exterior derivative \boldsymbol{d}. This section uses both these tools in deriving the key formulas (14.18), (14.25), (14.31), and (14.32). Once derived, however, these formulas demand no more than the standard exterior derivative for all applications and for all calculations of curvature (§14.6 and Box 14.5).

The extended exterior derivative leads to nothing new in the first two contexts to which one applies it: a scalar function ("0-form") and a vector field ("vector-valued 0-form"). Thus, take any function f. Its derivative in an unspecified direction is a 1-form; or, to make a new distinction that will soon become meaningful, a "scalar-valued 1-form." Specify the direction in which differentiation is to occur ("fill in the slot in the 1-form"). Thereby obtain the ordinary derivative as it applies to a function

Extended exterior derivative:

(1) acting on a scalar

$$\langle \boldsymbol{df}, \boldsymbol{u} \rangle = \partial_{\boldsymbol{u}} f. \tag{14.11}$$

Next, take any vector field \boldsymbol{v}. Its covariant derivative in an unspecified direction is a "vector-valued 1-form." Specify the direction \boldsymbol{u} in which differentiation is to occur ("fill in the slot in the 1-form"). Thereby obtain the covariant derivative

(2) acting on a vector

$$\langle \boldsymbol{dv}, \boldsymbol{u} \rangle \equiv \boldsymbol{\nabla}_{\boldsymbol{u}} \boldsymbol{v}. \tag{14.12a}$$

This object too is not new; it is the covariant derivative of the vector \boldsymbol{v} taken in the direction of the vector \boldsymbol{u}. When one abstracts away from any special choice of the direction of differentiation \boldsymbol{u}, one finds an expression that one has encountered before, though not under its new name of "vector-valued 1-form." This expression measures the covariant derivative of the vector \boldsymbol{v} in an unspecified direction ("slot for direction not yet filled in"). From a look at (14.12a), one sees that this extended exterior derivative is applied to \boldsymbol{v}, without reference to \boldsymbol{u}, is

$$\boldsymbol{dv} = \boldsymbol{\nabla} \boldsymbol{v}. \tag{14.12b}$$

Similarly, for any "tensor-valued 0-form" [i.e. $\binom{n}{0}$ tensor] \boldsymbol{S}, $\boldsymbol{dS} \equiv \boldsymbol{\nabla S}$.

Before proceeding further with the exterior (soon to be marked as "antisymmetric") differentiation of tensors, write down a formula (see exercise 14.5) for the exterior (antisymmetric) derivative of a product of forms:

$$\boldsymbol{d}(\boldsymbol{\alpha} \wedge \boldsymbol{\beta}) = (\boldsymbol{d\alpha}) \wedge \boldsymbol{\beta} + (-1)^p \boldsymbol{\alpha} \wedge \boldsymbol{d\beta}, \tag{14.13a}$$

where $\boldsymbol{\alpha}$ is a p-form and $\boldsymbol{\beta}$ is a q-form.

Now extend the exterior derivative from elementary forms to the exterior product of a tensor-valued p-form \boldsymbol{S} with any ordinary q-form, $\boldsymbol{\beta}$; thus,

(3) defined in general

$$\boldsymbol{d}(\boldsymbol{S} \wedge \boldsymbol{\beta}) = \boldsymbol{dS} \wedge \boldsymbol{\beta} + (-1)^p \boldsymbol{S} \wedge \boldsymbol{d\beta}. \tag{14.13b}$$

This equation can be regarded as a general definition of the extended exterior derivative. For example, if \boldsymbol{S} is a tensor-valued 2-form, $\boldsymbol{S} = S^{\alpha\beta}{}_{|\gamma\delta|} \boldsymbol{e}_\alpha \boldsymbol{e}_\beta \, dx^\gamma \wedge dx^\delta$, then equation (14.13b) says

$$\boldsymbol{dS} = \boldsymbol{d}[(\boldsymbol{e}_\alpha \boldsymbol{e}_\beta S^{\alpha\beta}{}_{|\gamma\delta|})(dx^\gamma \wedge dx^\delta)] = \boldsymbol{d}(\boldsymbol{e}_\alpha \boldsymbol{e}_\beta S^{\alpha\beta}{}_{|\gamma\delta|}) \wedge (dx^\gamma \wedge dx^\delta).$$

As another example, use (14.13b) to calculate $d(u\sigma)$, where u is a vector-valued 0-form (vector) and σ is a scalar-valued 1-form (1-form):

$$d(u\sigma) = (du) \wedge \sigma + u\, d\sigma.$$

If one were following the practice of earlier chapters, one would have written $u \otimes \sigma$ where $u\sigma$ appears here, $u \otimes d\sigma$ instead of $u\, d\sigma$, and $e_\alpha \otimes e_\beta$ instead of $e_\alpha e_\beta$. However, to avoid overcomplication in the notation, all such tensor product symbols are omitted here and hereafter.

Equations (14.12) and (14.13) do more than define the (extended) exterior derivative d and provide a way to use it in computations. They also allow one to define and calculate the antisymmetrized second derivatives, e.g., d^2v. The relation

$$d^2v = \mathscr{R}v$$

where v is a vector will then introduce the "operator-valued" or "$\binom{1}{1}$-tensor valued" curvature 2-form \mathscr{R}. The notation of the extended exterior derivative puts a new look on the old apparatus of base vectors and parallel transport, and opens a way to calculate the curvature 2-form \mathscr{R}.

Let the vector field v be expanded in terms of some field of basis vectors e_μ; thus

$$v = e_\mu v^\mu.$$

Then the exterior derivative of this vector is

$$dv = de_\mu v^\mu + e_\mu\, dv^\mu.$$

Expand the typical vector-valued 1-form de_μ in the form

Definition of $\boldsymbol{\omega}^\mu{}_\nu$

$$de_\mu = e_\nu \omega^\nu{}_\mu. \tag{14.14}$$

Here the "components" $\omega^\nu{}_\mu$ in the expansion of de_μ are 1-forms. Recall from equation (10.13) that the typical $\omega^\nu{}_\mu$ is related to the connection coefficients by

$$\omega^\nu{}_\mu = \Gamma^\nu{}_{\mu\lambda}\omega^\lambda. \tag{14.15}$$

Therefore the expansion of the "vector" (really, "vector-valued 1-form") is

$$dv = e_\mu(dv^\mu + \omega^\mu{}_\nu v^\nu). \tag{14.16}$$

Now differentiate once again to find

$$
\begin{aligned}
d^2v &= de_\alpha \wedge (dv^\alpha + \omega^\alpha{}_\nu v^\nu) \\
&\quad + e_\mu(d^2v^\mu + d\omega^\mu{}_\nu v^\nu - \omega^\mu{}_\nu \wedge dv^\nu) \\
&= e_\mu(\omega^\mu{}_\alpha \wedge dv^\alpha + \omega^\mu{}_\alpha \wedge \omega^\alpha{}_\nu v^\nu \\
&\quad + d^2v^\mu + d\omega^\mu{}_\nu v^\nu - \omega^\mu{}_\alpha \wedge dv^\alpha).
\end{aligned}
$$

The simplifications made here use (1) the equation (14.14), for a second time; and (2) the product rule (14.13a), which introduced the minus sign in the last term, ready

to cancel the first term. Now consider the term d^2v^μ. Recall that any given component, for example, v^3, is an ordinary scalar function of position (as contrasted to \mathbf{v} or \mathbf{e}_3 or $\mathbf{e}_3 v^3$). Therefore the standard exterior derivative (Chapter 4) as applied to a scalar function is all that d can mean in $d^2 v^\mu$. But for the standard exterior derivative applied twice, one has automatically $d^2 v^\mu = 0$ (Box 4.1, B; Box 4.4). This circumstance reduces the expansion for $d^2\mathbf{v}$ to the form

$$d^2\mathbf{v} = \mathbf{e}_\mu \mathcal{R}^\mu{}_\nu v^\nu, \tag{14.17}$$

where the $\mathcal{R}^\mu{}_\nu$ are abbreviations for *the curvature 2-forms*

$$\mathcal{R}^\mu{}_\nu \equiv d\boldsymbol{\omega}^\mu{}_\nu + \boldsymbol{\omega}^\mu{}_\alpha \wedge \boldsymbol{\omega}^\alpha{}_\nu. \tag{14.18}$$

Curvature 2-forms $\mathcal{R}^\mu{}_\nu$:

(1) in terms of $d^2\mathbf{v}$

(2) in terms of $\boldsymbol{\omega}^\mu{}_\nu$

Ordinarily, equation (14.18) surpasses in efficiency every other known method for calculating the curvature 2-forms.

The remarkable form of equation (14.17) deserves comment. On the left appear two d's, reminders that one has twice differentiated the vector field \mathbf{v}. But on the right, as the result of the differentiation, one has only the vector field \mathbf{v} at the point in question, undifferentiated. How \mathbf{v} varies from place to place enters not one whit in the answer. All that matters is how the geometry varies from place to place. Here is curvature coming into evidence. It comes into evidence free of any special features of the vector field \mathbf{v}, because the operation d^2 is an antisymmetrized covariant derivative [compare equation (11.8) for this antisymmetrized covariant derivative in the previously developed abstract language, and see Boxes 11.2 and 11.6 for what is going on behind the scene expressed in the form of pictures]. In brief, the result of operating on \mathbf{v} twice with d is an algebraic linear operation on \mathbf{v}; thus,

$$d^2\mathbf{v} = \mathcal{R}\mathbf{v}. \tag{14.19}$$

Here \mathcal{R} is an abbreviation for the "$\binom{1}{1}$-tensor valued 2-form,"

$$\mathcal{R} = \mathbf{e}_\mu \otimes \boldsymbol{\omega}^\nu \mathcal{R}^\mu{}_\nu. \tag{14.20}$$

Tensor-valued curvature 2-form \mathcal{R}

If d is a derivative with a "slot in it" in which to insert the vector saying in what direction the differentiation is to proceed, then the $d^2\mathbf{w}$ of $d^2\mathbf{w} = \mathcal{R}\mathbf{w}$ has two slots and calls for two vectors, say, \mathbf{u} and \mathbf{v}. These two vectors define the plane in which the antisymmetrized exterior derivative of (14.19) is to be evaluated (change in \mathbf{w} upon going around the elementary route defined by \mathbf{u} and \mathbf{v} and coming back to its starting point; Boxes 11.6 and 11.7). To spell out explicitly this insertion of vectors into slots, return first to a simpler context, and see the exterior derivative of a 1-form (itself a 2-form) "evaluated" for a bivector $\mathbf{u} \wedge \mathbf{v}$ ("count of honeycomblike cells of the 2-form over the parallelogram-shaped domain defined by the two vectors \mathbf{u} and \mathbf{v}"), and see the result of the evaluation (exercise 14.6) expressed as a commutator,

$$\langle d\boldsymbol{\alpha}, \mathbf{u} \wedge \mathbf{v} \rangle = \partial_\mathbf{u} \langle \boldsymbol{\alpha}, \mathbf{v} \rangle - \partial_\mathbf{v} \langle \boldsymbol{\alpha}, \mathbf{u} \rangle - \langle \boldsymbol{\alpha}, [\mathbf{u}, \mathbf{v}] \rangle. \tag{14.21}$$

This result generalizes itself to a tensor-valued 1-form \boldsymbol{S} of any rank in an obvious way; thus,

$$\langle d\boldsymbol{S}, \boldsymbol{u} \wedge \boldsymbol{v} \rangle = \boldsymbol{\nabla}_{\boldsymbol{u}} \langle \boldsymbol{S}, \boldsymbol{v} \rangle - \boldsymbol{\nabla}_{\boldsymbol{v}} \langle \boldsymbol{S}, \boldsymbol{u} \rangle - \langle \boldsymbol{S}, [\boldsymbol{u}, \boldsymbol{v}] \rangle. \tag{14.22}$$

Apply this result to the vector-valued 1-form $\boldsymbol{S} = d\boldsymbol{w}$. Recall the expression for a directional derivative, $\langle d\boldsymbol{w}, \boldsymbol{u} \rangle = \boldsymbol{\nabla}_{\boldsymbol{u}} \boldsymbol{w}$. Thus find the result

$$\begin{aligned}\langle \boldsymbol{d}^2 \boldsymbol{w}, \boldsymbol{u} \wedge \boldsymbol{v} \rangle &= \boldsymbol{\nabla}_{\boldsymbol{u}} \boldsymbol{\nabla}_{\boldsymbol{v}} \boldsymbol{w} - \boldsymbol{\nabla}_{\boldsymbol{v}} \boldsymbol{\nabla}_{\boldsymbol{u}} \boldsymbol{w} - \boldsymbol{\nabla}_{[\boldsymbol{u},\boldsymbol{v}]} \boldsymbol{w} \\ &= \mathscr{R}(\boldsymbol{u}, \boldsymbol{v}) \boldsymbol{w},\end{aligned} \tag{14.23}$$

Relation of curvature 2-form \mathscr{R} to curvature operator \mathscr{R}

where $\mathscr{R}(\boldsymbol{u}, \boldsymbol{v})$ is the curvature operator defined already in Chapter 11 [equation (11.8)]. The conclusion is simple: the (1_1)-tensor-valued 2-form \mathscr{R} of (14.19), evaluated on the bivector ("parallelogram") $\boldsymbol{u} \wedge \boldsymbol{v}$, is identical with the curvature operator $\mathscr{R}(\boldsymbol{u}, \boldsymbol{v})$ introduced previously; thus

$$\langle \mathscr{R}, \boldsymbol{u} \wedge \boldsymbol{v} \rangle = \mathscr{R}(\boldsymbol{u}, \boldsymbol{v}). \tag{14.24}$$

Now go from the language of abstract operators to a language that begins to make components show up. Substitute on the left the expression (14.20) and on the right the value of the curvature operator from (11.11); and rewrite (14.24) in the form

$$\boldsymbol{e}_\mu \otimes \boldsymbol{w}^\nu \langle \mathscr{R}^\mu{}_\nu, \boldsymbol{u} \wedge \boldsymbol{v} \rangle = \boldsymbol{e}_\mu \otimes \boldsymbol{w}^\nu R^\mu{}_{\nu\alpha\beta} u^\alpha v^\beta.$$

Compare and conclude that the typical individual curvature 2-form is given by the formula

Relation of \mathscr{R} to components of *Riemann*

$$\mathscr{R}^\mu{}_\nu = R^\mu{}_{\nu|\alpha\beta|} \boldsymbol{w}^\alpha \wedge \boldsymbol{w}^\beta \tag{14.25}$$

(sum over α, β, restricted to $\alpha < \beta$; so each index pair occurs only once).

Equation (14.25) provides the promised packaging of 21 curvature components into six curvature 2-forms; and equation (14.18) provides the quick means to calculate these curvature 2-forms. It is not necessary to take the key calculational equations (14.18) on faith, or to master the extended exterior derivative to prove or use them. Not one mention of any \boldsymbol{d} do they make except the standard exterior \boldsymbol{d} of Chapter 4. These key equations, moreover, can be verified in detail (exercise 14.8) by working in a coordinate frame. One adopts basis 1-forms $\boldsymbol{w}^\alpha = \boldsymbol{d}x^\alpha$. One goes on to use $\boldsymbol{w}^\mu{}_\nu = \Gamma^\mu{}_{\nu\lambda} \boldsymbol{d}x^\lambda$ from equation (14.15). In this way one obtains the "standard formula for the curvature" [equation (11.12) and equation (3) of Box 14.2] by standard methods.

In summary, the calculus of forms and exterior derivatives reduces the

$$\Gamma^\mu{}_{\alpha\beta} \longrightarrow R^\mu{}_{\nu\alpha\beta}$$

calculation to the

$$\boldsymbol{w}^\mu{}_\nu \longrightarrow \mathscr{R}^\mu{}_\nu$$

computation. Now look at the other link in the chain that leads from metric to curvature. It used to be

$$g_{\mu\nu} \longrightarrow \Gamma^\mu{}_{\alpha\beta}.$$

It now reduces to the calculation of "connection 1-forms"; thus

$$g_{\mu\nu} \longrightarrow \boldsymbol{\omega}^{\mu}{}_{\nu}.$$

Two principles master this first step in the curvature computation: (1) the symmetry of the covariant derivative; and (2) its compatibility with the metric. Condition (1), symmetry, appears in hidden guise in the principle

$$\boldsymbol{d}^{2}\mathscr{P} = 0. \tag{14.26}$$

Symmetry of covariant derivative:

(1) expressed as $\boldsymbol{d}^{2}\mathscr{P} = 0$

Here the notation "\mathscr{P} for point" comes straight out of Cartan. He thought of a vector as defined by the movement of one point to another point infinitesimally close to it. To write $\boldsymbol{d}\mathscr{P}$ was therefore to take the "derivative of a point" [make a construction with a "point deleted" (tail of vector) and "point reinserted nearby" (tip of vector)]. The direction of the derivative \boldsymbol{d} in $\boldsymbol{d}\mathscr{P}$ is indefinite. In other words, $\boldsymbol{d}\mathscr{P}$ contains a "slot." Only when one inserts into this slot a definite vector \boldsymbol{v} does $\boldsymbol{d}\mathscr{P}$ give a definite answer for Cartan's vector. What is that vector that $\boldsymbol{d}\mathscr{P}$ then gives? It is \boldsymbol{v} itself. "The movement that is \boldsymbol{v} tells the point \mathscr{P} to reproduce the movement that is \boldsymbol{v}"; or in concrete notation,

$$\langle \boldsymbol{d}\mathscr{P}, \boldsymbol{v} \rangle = \boldsymbol{v}. \tag{14.27}$$

Put the content of this equation into more formalistic terms. The quantity $\boldsymbol{d}\mathscr{P}$ is a $\binom{1}{1}$-tensor

$$\boldsymbol{d}\mathscr{P} = \boldsymbol{e}_{\mu}\boldsymbol{\omega}^{\mu}. \tag{14.28}$$

It is distinguished from the generic $\binom{1}{1}$-tensor

$$\boldsymbol{T} = \boldsymbol{e}_{\mu}T^{\mu}{}_{\nu}\boldsymbol{\omega}^{\nu}$$

by the special value of its components

$$T^{\mu}{}_{\nu} = \delta^{\mu}{}_{\nu}.$$

In this sense it deserves the name of "unit tensor." Insert this tensor in place of \boldsymbol{S} into equation (14.22) and obtain the result

$$\langle \boldsymbol{d}^{2}\mathscr{P}, \boldsymbol{u} \wedge \boldsymbol{v} \rangle = \boldsymbol{\nabla}_{\boldsymbol{u}}\boldsymbol{v} - \boldsymbol{\nabla}_{\boldsymbol{v}}\boldsymbol{u} - [\boldsymbol{u}, \boldsymbol{v}] = 0. \tag{14.29}$$

The zero on the right is a restatement of equation (10.2a) or of "the closing of the vector diagram" in the picture called "symmetry of covariant differentiation" in Box 10.2. The vanishing of the righthand side for arbitrary \boldsymbol{u} and \boldsymbol{v} demands the vanishing of $\boldsymbol{d}^{2}\mathscr{P}$ on the left; and conversely, the vanishing of $\boldsymbol{d}^{2}\mathscr{P}$ demands the symmetry of the covariant derivative. The other principle basic to the forthcoming computations is "compatibility of covariant derivative with metric," as expressed in the form

$$\boldsymbol{d}(\boldsymbol{u} \cdot \boldsymbol{v}) = (\boldsymbol{d}\boldsymbol{u}) \cdot \boldsymbol{v} + \boldsymbol{u} \cdot (\boldsymbol{d}\boldsymbol{v}). \tag{14.30}$$

It is essential here to ascribe to the metric (the "dot") a vanishing covariant derivative; thus

$$\boldsymbol{d}(\cdot) = 0.$$

Capitalize on the symmetry and compatibility of the covariant derivative by using basis vectors (and where appropriate the basis 1-forms dual to these basis vectors) in equations (14.26) and (14.30). Thus from

$$d\mathscr{P} = \boldsymbol{e}_\mu \boldsymbol{\omega}^\mu$$

compute

$$\begin{aligned}0 = \boldsymbol{d}^2\mathscr{P} &= \boldsymbol{d}\boldsymbol{e}_\mu \wedge \boldsymbol{\omega}^\mu + \boldsymbol{e}_\mu \boldsymbol{d}\boldsymbol{\omega}^\mu \\ &= \boldsymbol{e}_\mu(\boldsymbol{\omega}^\mu{}_\nu \wedge \boldsymbol{\omega}^\nu + \boldsymbol{d}\boldsymbol{\omega}^\mu),\end{aligned}$$

and conclude that the coefficient of \boldsymbol{e}_μ must vanish; or

$$0 = \boldsymbol{d}\boldsymbol{\omega}^\mu + \boldsymbol{\omega}^\mu{}_\nu \wedge \boldsymbol{\omega}^\nu \qquad \text{(``symmetry'')}. \qquad (14.31a)$$

Next, into (14.30) in place of the general \boldsymbol{u} and \boldsymbol{v} insert the specific \boldsymbol{e}_μ and \boldsymbol{e}_ν, respectively, and find

$$\boldsymbol{d}g_{\mu\nu} = \boldsymbol{\omega}_{\mu\nu} + \boldsymbol{\omega}_{\nu\mu} \qquad \text{(``compatibility'')}, \qquad (14.31b)$$

where

$$\boldsymbol{\omega}_{\mu\nu} \equiv g_{\mu\alpha}\boldsymbol{\omega}^\alpha{}_\nu = \Gamma_{\mu\nu\alpha}\boldsymbol{\omega}^\alpha. \qquad (14.31c)$$

In equations (14.31) one has the connection between metric and connection forms expressed in the most compact way.

§14.6. COMPUTATION OF CURVATURE USING EXTERIOR DIFFERENTIAL FORMS

The use of differential forms for the computation of curvature is illustrated in Box 14.5. This section outlines the method. There are three main steps: compute $\boldsymbol{\omega}^\mu{}_\nu$; compute $\mathscr{R}^\mu{}_\nu$; and compute $G^\mu{}_\nu$. More particularly, first select a metric and a frame. Thereby fix the basis forms $\boldsymbol{\omega}^\mu = L^\mu{}_{\alpha'}\,\boldsymbol{d}x^{\alpha'}$ and the metric components $g_{\mu\nu}$ in $\boldsymbol{d}s^2 = g_{\mu\nu}\boldsymbol{\omega}^\mu \otimes \boldsymbol{\omega}^\nu$. Then determine the connection forms $\boldsymbol{\omega}^\mu{}_\nu$, and determine them uniquely, as solutions of the equations

$$0 = \boldsymbol{d}\boldsymbol{\omega}^\mu + \boldsymbol{\omega}^\mu{}_\nu \wedge \boldsymbol{\omega}^\nu, \qquad (14.31a)$$

$$\boldsymbol{d}g_{\mu\nu} = \boldsymbol{\omega}_{\mu\nu} + \boldsymbol{\omega}_{\nu\mu}. \qquad (14.31b)$$

The "guess and check" method of finding a solution to these equations (described and illustrated in Box 14.5) is often quick and easy. [Exercise (14.7) shows that a solution always exists by showing that the Christoffel formula (14.36) is the unique solution in coordinate frames.] It is usually most convenient to use an orthonormal frame with $g_{\mu\nu} = \eta_{\mu\nu}$ (or some other simple frame where $g_{\mu\nu} = $ const, e.g., a null frame). Then $\boldsymbol{d}g_{\mu\nu} = 0$ and equation (14.31b) shows that $\boldsymbol{\omega}_{\mu\nu} = -\boldsymbol{\omega}_{\nu\mu}$. Therefore there are only six $\boldsymbol{\omega}_{\mu\nu}$ for which to solve in four dimensions.

(continued on page 358)

(Margin notes:)

(2) expressed as
$$\boldsymbol{d}\boldsymbol{\omega}^\mu + \boldsymbol{\omega}^\mu{}_\nu \wedge \boldsymbol{\omega}^\nu = 0$$

Compatibility of \boldsymbol{g} and ∇ expressed as
$$\boldsymbol{d}g_{\mu\nu} = \boldsymbol{\omega}_{\mu\nu} + \boldsymbol{\omega}_{\nu\mu}$$

Method of curvature 2-forms in 4 steps:

(1) select metric and frame

(2) calculate connection 1-forms $\boldsymbol{\omega}^\mu{}_\nu$

Box 14.5 CURVATURE COMPUTED USING EXTERIOR DIFFERENTIAL FORMS (METRIC FOR FRIEDMANN COSMOLOGY)

The Friedmann metric

$$ds^2 = -dt^2 + a^2(t)[d\chi^2 + \sin^2\chi(d\theta^2 + \sin^2\theta \, d\phi^2)]$$

(Box 27.1) represents a spacetime where each constant-t hypersurface is a three-dimensional hypersphere of proper circumference $2\pi a(t)$. An orthonormal basis is easily found in this spacetime; thus,

$$\boldsymbol{ds}^2 = -(\boldsymbol{\omega}^{\hat{t}})^2 + (\boldsymbol{\omega}^{\hat{\chi}})^2 + (\boldsymbol{\omega}^{\hat{\theta}})^2 + (\boldsymbol{\omega}^{\hat{\phi}})^2,$$

where

$$\boldsymbol{\omega}^{\hat{t}} = \boldsymbol{dt},$$
$$\boldsymbol{\omega}^{\hat{\chi}} = a \, \boldsymbol{d\chi},$$
$$\boldsymbol{\omega}^{\hat{\theta}} = a \sin \chi \, \boldsymbol{d\theta}, \tag{1}$$
$$\boldsymbol{\omega}^{\hat{\phi}} = a \sin \chi \sin \theta \, \boldsymbol{d\phi}.$$

A. Connection Computation

Equation (14.31b) gives, since $\boldsymbol{dg}_{\mu\nu} = \boldsymbol{d\eta}_{\mu\nu} = 0$, just

$$\boldsymbol{\omega}_{\mu\nu} = -\boldsymbol{\omega}_{\nu\mu}; \tag{2}$$

so there are only six 1-forms $\boldsymbol{\omega}_{\mu\nu}$ to be found. Turn to the second basic equation (14.31a). The game now is to *guess* a solution (because this is so often quicker than using systematic methods) to the equations $0 = \boldsymbol{d\omega}^\mu + \boldsymbol{\omega}^\mu{}_\nu \wedge \boldsymbol{\omega}^\nu$ in which the $\boldsymbol{\omega}^\nu$ and thus also $\boldsymbol{d\omega}^\mu$ are known, and $\boldsymbol{\omega}^\mu{}_\nu$ are unknown. The solution $\boldsymbol{\omega}^\mu{}_\nu$ is known to be unique; so guessing (if it leads to any answer) can only give the right answer.

Proceed from the simplest such equation. From $\boldsymbol{\omega}^{\hat{t}} = \boldsymbol{dt}$, compute

$$\boldsymbol{d\omega}^{\hat{t}} = 0.$$

Compare this with $\boldsymbol{d\omega}^{\hat{t}} = -\boldsymbol{\omega}^{\hat{t}}{}_\mu \wedge \boldsymbol{\omega}^\mu$ or (since $\boldsymbol{\omega}^{\hat{t}}{}_{\hat{t}} = -\boldsymbol{\omega}_{\hat{t}\hat{t}} = 0$, by $\boldsymbol{\omega}_{\mu\nu} = -\boldsymbol{\omega}_{\nu\mu}$)

$$\boldsymbol{d\omega}^{\hat{t}} = -\boldsymbol{\omega}^{\hat{t}}{}_k \wedge \boldsymbol{\omega}^k = 0.$$

This equation could be satisfied by having $\boldsymbol{\omega}^{\hat{t}}{}_k \propto \boldsymbol{\omega}^k$, or in more complicated ways with cancelations among different terms, or more simply by $\boldsymbol{\omega}^{\hat{t}}{}_k = 0$. Proceed, not

Box 14.5 (continued)

looking for trouble, until some non-zero $\boldsymbol{w}^\mu{}_\nu$ is required. From $\boldsymbol{w}^{\hat\chi} = a\,\boldsymbol{d}\chi$, find

$$\boldsymbol{dw}^{\hat\chi} = \dot{a}\,\boldsymbol{dt} \wedge \boldsymbol{d}\chi$$
$$= (\dot{a}/a)\boldsymbol{w}^{\hat{t}} \wedge \boldsymbol{w}^{\hat\chi} = -(\dot{a}/a)\boldsymbol{w}^{\hat\chi} \wedge \boldsymbol{w}^{\hat{t}}.$$

Compare this with

$$\boldsymbol{dw}^{\hat\chi} = -\boldsymbol{w}^{\hat\chi}{}_\mu \wedge \boldsymbol{w}^\mu$$
$$= -\boldsymbol{w}^{\hat\chi}{}_{\hat{t}} \wedge \boldsymbol{w}^{\hat{t}} - \boldsymbol{w}^{\hat\chi}{}_{\hat\theta} \wedge \boldsymbol{w}^{\hat\theta} - \boldsymbol{w}^{\hat\chi}{}_{\hat\phi} \wedge \boldsymbol{w}^{\hat\phi}.$$

Guess that $\boldsymbol{w}^{\hat\chi}{}_{\hat{t}} = (\dot{a}/a)\boldsymbol{w}^{\hat\chi}$ from the first term; and hope the other terms vanish. (Note that this allows $\boldsymbol{w}^{\hat{t}}{}_{\hat\chi} \wedge \boldsymbol{w}^{\hat\chi} = -\boldsymbol{w}_{\hat{t}\hat\chi} \wedge \boldsymbol{w}^{\hat\chi} = \boldsymbol{w}_{\hat\chi\hat{t}} \wedge \boldsymbol{w}^{\hat\chi} = 0$ in the $\boldsymbol{dw}^{\hat{t}}$ equation.) Look at $\boldsymbol{w}^{\hat\theta} = a \sin\chi\,\boldsymbol{d}\theta$, and write

$$\boldsymbol{dw}^{\hat\theta} = (\dot{a}/a)\boldsymbol{w}^{\hat{t}} \wedge \boldsymbol{w}^{\hat\theta} + a^{-1}\cot\chi\,\boldsymbol{w}^{\hat\chi} \wedge \boldsymbol{w}^{\hat\theta}$$
$$= -\boldsymbol{w}^{\hat\theta}{}_{\hat{t}} \wedge \boldsymbol{w}^{\hat{t}} - \boldsymbol{w}^{\hat\theta}{}_{\hat\chi} \wedge \boldsymbol{w}^{\hat\chi} - \boldsymbol{w}^{\hat\theta}{}_{\hat\phi} \wedge \boldsymbol{w}^{\hat\phi}.$$

Guess, consistent with previously written equations, that

$$\boldsymbol{w}^{\hat\theta}{}_{\hat{t}} = \boldsymbol{w}^{\hat{t}}{}_{\hat\theta} = (\dot{a}/a)\boldsymbol{w}^{\hat\theta},$$
$$\boldsymbol{w}^{\hat\theta}{}_{\hat\chi} = -\boldsymbol{w}^{\hat\chi}{}_{\hat\theta} = a^{-1}\cot\chi\,\boldsymbol{w}^{\hat\theta}.$$

Finally from

$$\boldsymbol{dw}^{\hat\phi} = (\dot{a}/a)\boldsymbol{w}^{\hat{t}} \wedge \boldsymbol{w}^{\hat\phi} + a^{-1}\cot\chi\,\boldsymbol{w}^{\hat\chi} \wedge \boldsymbol{w}^{\hat\phi}$$
$$+ (a\sin\chi)^{-1}\cot\theta\,\boldsymbol{w}^{\hat\theta} \wedge \boldsymbol{w}^{\hat\phi}$$
$$= -\boldsymbol{w}^{\hat\phi}{}_{\hat{t}} \wedge \boldsymbol{w}^{\hat{t}} - \boldsymbol{w}^{\hat\phi}{}_{\hat\chi} \wedge \boldsymbol{w}^{\hat\chi} - \boldsymbol{w}^{\hat\phi}{}_{\hat\theta} \wedge \boldsymbol{w}^{\hat\theta},$$

deduce values of $\boldsymbol{w}^{\hat\phi}{}_{\hat{t}}$, $\boldsymbol{w}^{\hat\phi}{}_{\hat\chi}$, and $\boldsymbol{w}^{\hat\phi}{}_{\hat\theta}$. These are not inconsistent with previous assumptions that terms like $\boldsymbol{w}^{\hat\theta}{}_{\hat\phi} \wedge \boldsymbol{w}^{\hat\phi}$ vanish (in the $\boldsymbol{dw}^{\hat\theta}$ equation); so one has in fact solved $\boldsymbol{dw}^\mu = -\boldsymbol{w}^\mu{}_\nu \wedge \boldsymbol{w}^\nu$ for a set of connection forms $\boldsymbol{w}^\mu{}_\nu$, as follows:

$$\boldsymbol{w}^k{}_{\hat{t}} = \boldsymbol{w}^{\hat{t}}{}_k = (\dot{a}/a)\boldsymbol{w}^k,$$

$$\boldsymbol{w}^{\hat\theta}{}_{\hat\chi} = -\boldsymbol{w}^{\hat\chi}{}_{\hat\theta} = a^{-1}\cot\chi\,\boldsymbol{w}^{\hat\theta}$$
$$= \cos\chi\,\boldsymbol{d}\theta,$$

$$\boldsymbol{w}^{\hat\phi}{}_{\hat\chi} = -\boldsymbol{w}^{\hat\chi}{}_{\hat\phi} = a^{-1}\cot\chi\,\boldsymbol{w}^{\hat\phi} \qquad\qquad (3)$$
$$= \cos\chi\sin\theta\,\boldsymbol{d}\phi,$$

$$\boldsymbol{w}^{\hat\phi}{}_{\hat\theta} = -\boldsymbol{w}^{\hat\theta}{}_{\hat\phi} = (a\sin\chi)^{-1}\cot\theta\,\boldsymbol{w}^{\hat\phi}$$
$$= \cos\theta\,\boldsymbol{d}\phi.$$

Of course, if these hit-or-miss methods of finding $\boldsymbol{w}^\mu{}_\nu$ do not work easily in some problem, one may simply use equations (14.32) and (14.33).

B. Curvature Computation

The curvature computation is a straightforward substitution of $\boldsymbol{\omega}^{\mu}{}_{\nu}$ from equations (3) above into equation (14.34), which is

$$\mathscr{R}^{\mu}{}_{\nu} = \boldsymbol{d\omega}^{\mu}{}_{\nu} + \boldsymbol{\omega}^{\mu}{}_{\alpha} \wedge \boldsymbol{\omega}^{\alpha}{}_{\nu}.$$

This equation is short enough that one can write out the sum

$$\mathscr{R}^{\hat{t}}{}_{\hat{\chi}} = \boldsymbol{d\omega}^{\hat{t}}{}_{\hat{\chi}} + \boldsymbol{\omega}^{\hat{t}}{}_{\hat{\theta}} \wedge \boldsymbol{\omega}^{\hat{\theta}}{}_{\hat{\chi}} + \boldsymbol{\omega}^{\hat{t}}{}_{\hat{\phi}} \wedge \boldsymbol{\omega}^{\hat{\phi}}{}_{\hat{\chi}}$$

in contrast to the ten terms in the corresponding $R = \partial \Gamma + \Gamma^2$ equation [equation (3) of Box 14.2]. *Warning!:* From $\boldsymbol{\omega}^{\hat{t}}{}_{\hat{\chi}} = (\dot{a}/a)\boldsymbol{\omega}^{\hat{\chi}}$, do not compute $\boldsymbol{d\omega}^{\hat{t}}{}_{\hat{\chi}} = (\dot{a}/a)^{\cdot}\,\boldsymbol{\omega}^{\hat{t}} \wedge \boldsymbol{\omega}^{\hat{\chi}}$. Missing is the term $(\dot{a}/a)\,\boldsymbol{d\omega}^{\hat{\chi}}$. Instead write $\boldsymbol{\omega}^{\hat{t}}{}_{\hat{\chi}} = (\dot{a}/a)\boldsymbol{\omega}^{\hat{\chi}} = \dot{a}\,\boldsymbol{d\chi}$, and then find $\boldsymbol{d\omega}^{\hat{t}}{}_{\hat{\chi}} = \ddot{a}\,\boldsymbol{dt} \wedge \boldsymbol{d\chi} = (\ddot{a}/a)\boldsymbol{\omega}^{\hat{t}} \wedge \boldsymbol{\omega}^{\hat{\chi}}$. With elementary care, then, in correctly substituting from (3) for the $\boldsymbol{\omega}^{\mu}{}_{\nu}$ in the formula for $\mathscr{R}^{\mu}{}_{\nu}$, one finds

$$\mathscr{R}^{\hat{t}}{}_{\hat{\chi}} = (\ddot{a}/a)\boldsymbol{\omega}^{\hat{t}} \wedge \boldsymbol{\omega}^{\hat{\chi}},$$

and

$$\mathscr{R}^{\hat{\chi}}{}_{\hat{\theta}} = (1 + \dot{a}^2)a^{-2}\boldsymbol{\omega}^{\hat{\chi}} \wedge \boldsymbol{\omega}^{\hat{\theta}}.$$

This completes the computation of the $R^{\mu}{}_{\nu\alpha\beta}$, since in this isotopic model universe, all space directions in the orthonormal frame $\boldsymbol{\omega}^{\mu}$ are algebraically equivalent. One can therefore write

$$\mathscr{R}^{\hat{t}}{}_{k} = (\ddot{a}/a)\boldsymbol{\omega}^{\hat{t}} \wedge \boldsymbol{\omega}^{k},$$

$$\mathscr{R}^{k}{}_{\ell} = a^{-2}(1 + \dot{a}^2)\boldsymbol{\omega}^{k} \wedge \boldsymbol{\omega}^{\ell}, \tag{4}$$

for the complete list of $\mathscr{R}^{\mu}{}_{\nu}$. Specific components, such as

$$R^{\hat{t}}{}_{\hat{\chi}\hat{t}\hat{\chi}} = \ddot{a}/a, \qquad R^{\hat{t}}{}_{\hat{\chi}\hat{t}\hat{\theta}} = 0, \text{ etc.,}$$

or

$$R^{\hat{t}}{}_{\hat{\theta}\hat{\chi}\hat{\phi}} = 0, \qquad R^{\hat{\theta}}{}_{\hat{\phi}\hat{\theta}\hat{\phi}} = a^{-2}(1 + \dot{a}^2),$$

are easily read out of this display of $\mathscr{R}^{\mu}{}_{\nu}$.

C. Contraction

From equations (14.7), find

$$G^{\hat{t}\hat{t}} = +3a^{-2}(1 + \dot{a}^2), \tag{5a}$$

$$G^{\hat{t}\hat{\chi}} = G^{\hat{t}\hat{\theta}} = G^{\hat{t}\hat{\phi}} = 0 = G^{\hat{\chi}\hat{\theta}} = G^{\hat{\theta}\hat{\phi}} = G^{\hat{\phi}\hat{\chi}}, \tag{5b}$$

$$G^{\hat{\chi}\hat{\chi}} = G^{\hat{\theta}\hat{\theta}} = G^{\hat{\phi}\hat{\phi}} = -[2a^{-1}\ddot{a} + a^{-2}(1 + \dot{a}^2)], \tag{5c}$$

and

$$R = -G^{\mu}{}_{\mu} = 6[a^{-1}\ddot{a} + a^{-2}(1 + \dot{a}^2)]. \tag{6}$$

If guessing is not easy, there is a systematic way to solve equations (14.31) in an orthonormal frame or in any other frame in which $dg_{\mu\nu} = 0$. Compute the $d\boldsymbol{w}^{\mu}$ and arrange them in the format

$$d\boldsymbol{w}^{\alpha} = -c_{|\mu\nu|}{}^{\alpha}\boldsymbol{w}^{\mu} \wedge \boldsymbol{w}^{\nu}. \tag{14.32}$$

In this way display the 24 "commutation coefficients" $c_{\mu\nu}{}^{\alpha}$. These quantities enter into the formula

$$\boldsymbol{w}_{\mu\nu} = \frac{1}{2}(c_{\mu\nu\alpha} + c_{\mu\alpha\nu} - c_{\nu\alpha\mu})\boldsymbol{w}^{\alpha} \tag{14.33}$$

to provide the six $\boldsymbol{w}^{\mu}{}_{\nu}$ (exercise 14.12).

(3) calculate curvature 2-forms $\mathscr{R}^{\mu}{}_{\nu}$

Once the $\boldsymbol{w}_{\mu\nu}$ are known, one computes the curvature forms $\mathscr{R}^{\mu}{}_{\nu}$ (again only six in four dimensions, since $\mathscr{R}^{\mu\nu} = -\mathscr{R}^{\nu\mu}$) by use of the formula

$$\mathscr{R}^{\mu}{}_{\nu} = d\boldsymbol{w}^{\mu}{}_{\nu} + \boldsymbol{w}^{\mu}{}_{\alpha} \wedge \boldsymbol{w}^{\alpha}{}_{\nu}. \tag{14.34}$$

(4) calculate components of curvature tensors

Out of this tabulation, one reads the individual components of the curvature tensor by using the identification scheme

$$\mathscr{R}^{\mu\nu} = R^{\mu\nu}{}_{|\alpha\beta|}\boldsymbol{w}^{\alpha} \wedge \boldsymbol{w}^{\beta}. \tag{14.35}$$

The Einstein tensor $G^{\mu}{}_{\nu}$ is computed by scanning the $\mathscr{R}^{\mu\nu}$ display to find the appropriate $R^{\mu\nu}{}_{\alpha\beta}$ components for use in formulas (14.7).

EXERCISES

Exercise 14.5. EXTERIOR DERIVATIVE OF A PRODUCT OF FORMS

Establish equation (14.13a) by working up recursively from forms of lower order to forms of higher order. [*Hints*: Recall from equation (4.27) that for a p-form

$$\boldsymbol{\alpha} = \alpha_{|\mu_1 \ldots \mu_p|}\boldsymbol{dx}^{\mu_1} \wedge \cdots \wedge \boldsymbol{dx}^{\mu_p},$$

the exterior derivative is defined by

$$\boldsymbol{d\alpha} = \frac{\partial \alpha_{|\mu_1 \ldots \mu_p|}}{\partial x^{\mu_0}}\boldsymbol{dx}^{\mu_0} \wedge \boldsymbol{dx}^{\mu_1} \wedge \cdots \wedge \boldsymbol{dx}^{\mu_p}.$$

Applied to the product $\boldsymbol{\alpha} \wedge \boldsymbol{\beta}$ of two 1-forms, this formula gives

$$\begin{aligned}
\boldsymbol{d}(\boldsymbol{\alpha} \wedge \boldsymbol{\beta}) &= \boldsymbol{d}[(\alpha_\lambda \boldsymbol{dx}^\lambda) \wedge (\beta_\mu \boldsymbol{dx}^\mu)] \\
&= \boldsymbol{d}[(\alpha_\lambda \beta_\mu)(\boldsymbol{dx}^\lambda \wedge \boldsymbol{dx}^\mu)] \\
&= \frac{\partial(\alpha_\lambda \beta_\mu)}{\partial x^\kappa}\boldsymbol{dx}^\kappa \wedge \boldsymbol{dx}^\lambda \wedge \boldsymbol{dx}^\mu \\
&= \left(\frac{\partial\alpha_\lambda}{\partial x^\kappa}\boldsymbol{dx}^\kappa \wedge \boldsymbol{dx}^\lambda\right) \wedge \beta_\mu \boldsymbol{dx}^\mu - (\alpha_\lambda \boldsymbol{dx}^\lambda) \wedge \left(\frac{\partial\beta_\mu}{\partial x^\kappa}\boldsymbol{dx}^\kappa \wedge \boldsymbol{dx}^\mu\right) \\
&= (\boldsymbol{d\alpha}) \wedge \boldsymbol{\beta} - \boldsymbol{\alpha} \wedge \boldsymbol{d\beta}.
\end{aligned}$$

Extend the reasoning to forms of higher order.]

**Exercise 14.6. RELATIONSHIP BETWEEN EXTERIOR DERIVATIVE
AND COMMUTATOR**

Establish formula (14.21) by showing (a) that the righthand side is an *algebraic* linear function of \boldsymbol{u} and an algebraic linear function of \boldsymbol{v}, and (b) that the equation holds when \boldsymbol{u} and \boldsymbol{v} are coordinate basis vectors $\boldsymbol{u} = \partial/\partial x^k$, $\boldsymbol{v} = \partial/\partial x^l$.

Exercise 14.7. CHRISTOFFEL FORMULA DERIVED FROM CONNECTION FORMS

In a coordinate frame $\boldsymbol{w}^\mu = \boldsymbol{d}x^\mu$, show that equation (14.31a) requires $\Gamma^\mu{}_{\alpha\beta} = \Gamma^\mu{}_{\beta\alpha}$, and that, with this symmetry established, equation (14.31b) gives an expression for $\partial g_{\mu\nu}/\partial x^\alpha$ which can be solved to give the Christoffel formula

$$\Gamma^\mu{}_{\alpha\beta} = \frac{1}{2} g^{\mu\nu}\left(\frac{\partial g_{\nu\alpha}}{\partial x^\beta} + \frac{\partial g_{\nu\beta}}{\partial x^\alpha} - \frac{\partial g_{\alpha\beta}}{\partial x^\nu}\right). \tag{14.36}$$

**Exercise 14.8. RIEMANN-CHRISTOFFEL CURVATURE FORMULA
RELATED TO CURVATURE FORMS**

Substitute $\boldsymbol{w}^\mu{}_\nu = \Gamma^\mu{}_{\nu\lambda}\,\boldsymbol{d}x^\lambda$ into equation (14.18), and from the result read out, according to equation (14.25), the classical formula (3) of Box 14.2 for the components $R^\mu{}_{\nu\alpha\beta}$.

**Exercise 14.9. MATRIX NOTATION FOR REVIEW OF CARTAN
STRUCTURE EQUATIONS**

Let $e \equiv (\boldsymbol{e}_1, \ldots, \boldsymbol{e}_n)$ be a row matrix whose entries are the basis vectors, and let ω be a column of basis 1-forms \boldsymbol{w}^μ. Similarly let $\Omega = \|\boldsymbol{w}^\mu{}_\nu\|$ and $\mathscr{R} = \|\mathscr{R}^\mu{}_\nu\|$ be square matrices with 1-form and 2-form entries. This gives a compact notation in which $\boldsymbol{d}e_\mu = \boldsymbol{e}_\nu \boldsymbol{w}^\nu{}_\mu$ and $\boldsymbol{d}\mathscr{P} = \boldsymbol{e}_\mu \boldsymbol{w}^\mu$ read

$$\boldsymbol{d}e = e\Omega \quad \text{and} \quad \boldsymbol{d}\mathscr{P} = e\omega, \tag{14.37}$$

respectively.

(a) From equations (14.37) and $\boldsymbol{d}^2\mathscr{P} = 0$, derive equation (14.31a) in the form

$$0 = \boldsymbol{d}\omega + \Omega \wedge \omega. \tag{14.38}$$

[Solution: $\boldsymbol{d}^2\mathscr{P} = \boldsymbol{d}e \wedge \omega + e\,\boldsymbol{d}\omega = e(\Omega \wedge \omega + \boldsymbol{d}\omega).$]

(b) Compute $\boldsymbol{d}^2 e$ as motivation for definition (14.18), which reads

$$\mathscr{R} = \boldsymbol{d}\Omega + \Omega \wedge \Omega. \tag{14.39}$$

(c) From $\boldsymbol{d}^2\omega = 0$, deduce $\mathscr{R} \wedge \omega = 0$ and then decompress the notation to get the antisymmetry relation $R^\mu{}_{[\alpha\beta\gamma]} = 0$.

(d) Compute $\boldsymbol{d}\mathscr{R}$ from equation (14.39), and relate it to the Bianchi identity $R^\mu{}_{\nu[\alpha\beta;\gamma]} = 0$.

(e) Let $v = \{v^\mu\}$ be a column of functions; so $\boldsymbol{v} = ev = \boldsymbol{e}_\mu v^\mu$ is a vector-field. Compute, in compact notation, $\boldsymbol{d}v$ and $\boldsymbol{d}^2 v$ to show $\boldsymbol{d}^2 v = e\mathscr{R}v$ (which is equation 14.17).

**Exercise 14.10. TRANSFORMATION RULES FOR CONNECTION FORMS
IN COMPACT NOTATION**

Using the notation of the previous exercise, write $e' = eA$ in place of $\boldsymbol{e}_{\mu'} = \boldsymbol{e}_\nu A^\nu{}_{\mu'}$, and similarly $\omega' = A^{-1}\omega$, to represent a change of frame. Show that $\boldsymbol{d}\mathscr{P} \equiv e\omega = e'\omega'$. Substitute $e' = eA$ in $\boldsymbol{d}e' = e'\Omega'$ to deduce the transformation law

$$\Omega' = A^{-1}\Omega A + A^{-1}\,\boldsymbol{d}A. \tag{14.40}$$

Rewrite this in decompressed notation for coordinate frames with $A^\nu{}_{\mu'} = \partial x^\nu/\partial x^{\mu'}$ as a formula of the form $\Gamma^{\mu'}{}_{\alpha'\beta'} = (?)$.

Exercise 14.11. SPACE IS FLAT IF THE CURVATURE VANISHES (see §11.5)

If coordinates exist in which all straight lines ($d^2x^\mu/d\lambda^2 = 0$) are geodesics, then one says the space is flat. Evidently all $\Gamma^\mu{}_{\alpha\beta}$ and $R^\mu{}_{\nu\alpha\beta}$ vanish in this case, by equation (14.8) and equation (3) in Box 14.2. Show conversely that, if $\mathscr{R} = 0$, then such coordinates exist. Use the results of the previous problem to find differential equations for a transformation A to a basis e' where $\Omega' = 0$. What are the conditions for complete integrability of these equations? [Note that $df_K = F_K(x, f)$ is completely integrable if $d^2f_K = 0$ modulo the original equations.] Why will the basis forms $\boldsymbol{\omega}^{\mu'}$ in this new frame be coordinate differentials $\boldsymbol{\omega}^{\mu'} = \boldsymbol{d}x^{\mu'}$?

Exercise 14.12. SYSTEMATIC COMPUTATION OF CONNECTION FORMS IN ORTHONORMAL FRAMES

Deduce equation (14.32) by applying equation (14.21) to basis vectors, using equations (8.14) to define $c_{\mu\nu}{}^\alpha$. Then show that, in an orthonormal frame (or any frames with $g_{\mu\nu} = $ const), equation (14.33) provides a solution of equations (14.31), which define $\boldsymbol{\omega}^\mu{}_\nu$. [Compare also equation (8.24b).]

Exercise 14.13. SCHWARZSCHILD CURVATURE FORMS

Use the obvious orthonormal frame $\boldsymbol{\omega}^{\hat{t}} = e^\Phi \, \boldsymbol{d}t$, $\boldsymbol{\omega}^{\hat{r}} = e^\Lambda \, \boldsymbol{d}r$, $\boldsymbol{\omega}^{\hat{\theta}} = r \, \boldsymbol{d}\theta$, $\boldsymbol{\omega}^{\hat{\phi}} = r \sin\theta \, \boldsymbol{d}\phi$ for the Schwarzschild metric

$$ds^2 = -e^{2\Phi} \, dt^2 + e^{2\Lambda} \, dr^2 + r^2(d\theta^2 + \sin^2\theta \, d\phi^2), \tag{14.41}$$

in which Φ and Λ are functions of r only; and compute the curvature forms $\mathscr{R}^{\hat{\mu}}{}_{\hat{\nu}}$ and the Einstein tensor $G^{\hat{\mu}}{}_{\hat{\nu}}$ by the methods of Box 14.5. [*Answer:* $\mathscr{R}^{\hat{t}\hat{r}} = E\boldsymbol{\omega}^{\hat{t}} \wedge \boldsymbol{\omega}^{\hat{r}}$, $\mathscr{R}^{\hat{t}\hat{\theta}} = \bar{E}\boldsymbol{\omega}^{\hat{t}} \wedge \boldsymbol{\omega}^{\hat{\theta}}$, $\mathscr{R}^{\hat{t}\hat{\phi}} = \bar{E}\boldsymbol{\omega}^{\hat{t}} \wedge \boldsymbol{\omega}^{\hat{\phi}}$, $\mathscr{R}^{\hat{\theta}\hat{\phi}} = F\boldsymbol{\omega}^{\hat{\theta}} \wedge \boldsymbol{\omega}^{\hat{\phi}}$, $\mathscr{R}^{\hat{\phi}\hat{r}} = \bar{F}\boldsymbol{\omega}^{\hat{\phi}} \wedge \boldsymbol{\omega}^{\hat{r}}$, $\mathscr{R}^{\hat{r}\hat{\theta}} = \bar{F}\boldsymbol{\omega}^{\hat{r}} \wedge \boldsymbol{\omega}^{\hat{\theta}}$, with

$$E = -e^{-2\Lambda}(\Phi'' + \Phi'^2 - \Phi'\Lambda'),$$

$$\bar{E} = -\frac{1}{r} e^{-2\Lambda}\Phi',$$

$$F = \frac{1}{r^2}(1 - e^{-2\Lambda}), \tag{14.42}$$

$$\bar{F} = \frac{1}{r} e^{-2\Lambda}\Lambda';$$

and then

$$\begin{aligned} G^{\hat{t}}{}_{\hat{t}} &= -(F + 2\bar{F}), \\ G^{\hat{r}}{}_{\hat{r}} &= -(F + 2\bar{E}), \\ G^{\hat{\theta}}{}_{\hat{\theta}} &= G^{\hat{\phi}}{}_{\hat{\phi}} = -(E + \bar{E} + \bar{F}), \\ G^{\hat{t}}{}_{\hat{r}} &= G^{\hat{t}}{}_{\hat{\theta}} = G^{\hat{t}}{}_{\hat{\phi}} = 0 = G^{\hat{r}}{}_{\hat{\theta}} = G^{\hat{r}}{}_{\hat{\phi}} = G^{\hat{\theta}}{}_{\hat{\phi}}.] \end{aligned} \tag{14.43}$$

Exercise 14.14. MATRIX DISPLAY OF THE RIEMANN-TENSOR COMPONENTS

Use the symmetries of the Riemann tensor to justify displaying its components in an orthonormal frame in the form

$$R^{\hat{\mu}\hat{\nu}}{}_{\alpha\beta} = \begin{matrix} 01 \\ 02 \\ 03 \\ 23 \\ 31 \\ 12 \end{matrix} \left(\begin{array}{c|c} E & H \\ \hline -H^T & F \end{array} \right), \tag{14.44}$$

where the rows are labeled by index pairs $\hat{\mu}\hat{\nu} = 01, 02$, etc., as shown; and the columns $\hat{\alpha}\hat{\beta}$, similarly. Here E, F, and H are each 3×3 matrices with (why?)

$$E = E^T, \qquad F = F^T, \qquad \text{trace } H = 0, \tag{14.45}$$

where E^T means the transpose of E.

Exercise 14.15. RIEMANN MATRIX WITH VANISHING EINSTEIN TENSOR

Show that the empty-space Einstein equations $G^{\hat{\mu}}_{\phantom{\hat{\mu}}\hat{\nu}} = 0$ allow the matrix in equation (14.44) to be simplified to the form

$$R^{\hat{\mu}\hat{\nu}}_{\phantom{\hat{\mu}\hat{\nu}}\hat{\alpha}\hat{\beta}} = \left(\begin{array}{c|c} E & H \\ \hline -H & E \end{array} \right), \tag{14.46}$$

where now, in addition to the equality $E = F$ that this form implies, the further conditions

$$\text{trace } E = 0, \quad H = H^T \tag{14.47}$$

hold.

Exercise 14.16. COMPUTATION OF CURVATURE FOR A PULSATING OR COLLAPSING STAR

Spherically symmetric motions of self-gravitating bodies are discussed in Chapters 26 and 32. A metric form often adopted in this situation is

$$ds^2 = -e^{2\Phi}\, dT^2 + e^{2\Lambda}\, dR^2 + r^2(d\theta^2 + \sin^2\theta\, d\phi^2), \tag{14.48}$$

where now Φ, Λ, and r are each functions of the two coordinates R and T. Compute the curvature 2-forms and the Einstein tensor for this metric, using the methods of Box 14.5. In the guessing of the $\boldsymbol{\omega}^{\mu}_{\nu}$, most of the terms will already be evident from the corresponding calculation in exercise 14.13. [*Answer*, in the obvious orthonormal frame $\boldsymbol{\omega}^{\hat{T}} = e^{\Phi}\, \boldsymbol{dT}$, $\boldsymbol{\omega}^{\hat{R}} = e^{\Lambda}\, \boldsymbol{dR}$, $\boldsymbol{\omega}^{\hat{\theta}} = r\, \boldsymbol{d\theta}$, $\boldsymbol{\omega}^{\hat{\phi}} = r \sin\theta\, \boldsymbol{d\phi}$:

$$
\begin{aligned}
\mathscr{R}^{\hat{T}}_{\phantom{\hat{T}}\hat{R}} &= E \boldsymbol{\omega}^{\hat{T}} \wedge \boldsymbol{\omega}^{\hat{R}}, \\
\mathscr{R}^{\hat{T}}_{\phantom{\hat{T}}\hat{\theta}} &= \bar{E} \boldsymbol{\omega}^{\hat{T}} \wedge \boldsymbol{\omega}^{\hat{\theta}} + H \boldsymbol{\omega}^{\hat{R}} \wedge \boldsymbol{\omega}^{\hat{\theta}}, \\
\mathscr{R}^{\hat{T}}_{\phantom{\hat{T}}\hat{\phi}} &= \bar{E} \boldsymbol{\omega}^{\hat{T}} \wedge \boldsymbol{\omega}^{\hat{\phi}} + H \boldsymbol{\omega}^{\hat{R}} \wedge \boldsymbol{\omega}^{\hat{\phi}}, \\
\mathscr{R}^{\hat{\theta}}_{\phantom{\hat{\theta}}\hat{\phi}} &= F \boldsymbol{\omega}^{\hat{\theta}} \wedge \boldsymbol{\omega}^{\hat{\phi}}, \\
\mathscr{R}^{\hat{R}}_{\phantom{\hat{R}}\hat{\theta}} &= \bar{F} \boldsymbol{\omega}^{\hat{R}} \wedge \boldsymbol{\omega}^{\hat{\theta}} - H \boldsymbol{\omega}^{\hat{T}} \wedge \boldsymbol{\omega}^{\hat{\theta}}, \\
\mathscr{R}^{\hat{R}}_{\phantom{\hat{R}}\hat{\phi}} &= \bar{F} \boldsymbol{\omega}^{\hat{R}} \wedge \boldsymbol{\omega}^{\hat{\phi}} - H \boldsymbol{\omega}^{\hat{T}} \wedge \boldsymbol{\omega}^{\hat{\phi}},
\end{aligned}
\tag{14.49}
$$

which, in the matrix display of exercise 14.14, gives

$$
R^{\mu\nu}_{\alpha\beta} = \left(
\begin{array}{ccc|ccc}
E & \cdot & \cdot & \cdot & \cdot & \cdot \\
\cdot & \bar{E} & \cdot & \cdot & \cdot & H \\
\cdot & \cdot & \bar{E} & \cdot & -H & \cdot \\
\hline
\cdot & \cdot & \cdot & F & \cdot & \cdot \\
\cdot & \cdot & H & \cdot & \bar{F} & \cdot \\
\cdot & -H & \cdot & \cdot & \cdot & \bar{F}
\end{array}
\right)
\begin{array}{l}
\hat{T}\hat{R} \\
\hat{T}\hat{\theta} \\
\hat{T}\hat{\phi} \\
\hat{\theta}\hat{\phi} \\
\hat{\phi}\hat{R} \\
\hat{R}\hat{\theta}
\end{array} . \tag{14.50}
$$

Here

$$E = e^{-2\Phi}(\ddot{\Lambda} + \dot{\Lambda}^2 - \dot{\Lambda}\dot{\Phi}) - e^{-2\Lambda}(\Phi'' + \Phi'^2 - \Phi'\Lambda'),$$

$$\bar{E} = \frac{1}{r}e^{-2\Phi}(\ddot{r} - \dot{r}\dot{\Phi}) - \frac{1}{r}e^{-2\Lambda}r'\Phi',$$

$$H = \frac{1}{r}e^{-\Phi-\Lambda}(\dot{r}' - \dot{r}\Phi' - r'\dot{\Lambda}), \qquad\qquad (14.51)$$

$$F = \frac{1}{r^2}(1 - r'^2 e^{-2\Lambda} + \dot{r}^2 e^{-2\Phi}),$$

$$\bar{F} = \frac{1}{r}e^{-2\Phi}\dot{r}\dot{\Lambda} + \frac{1}{r}e^{-2\Lambda}(r'\Lambda' - r'').$$

The Einstein tensor is

$$
\begin{aligned}
G^{\hat{T}\hat{T}} &= -G^{\hat{T}}{}_{\hat{T}} = F + 2\bar{F}, \\
G^{\hat{T}\hat{R}} &= G^{\hat{T}}{}_{\hat{R}} = 2H, \\
G^{\hat{T}}{}_{\hat{\theta}} &= G^{\hat{T}}{}_{\hat{\phi}} = 0, \\
&\quad G^{\hat{R}}{}_{\hat{R}} = -(2\bar{E} + F), \\
G^{\hat{\theta}}{}_{\hat{\theta}} &= G^{\hat{\phi}}{}_{\hat{\phi}} = -(E + \bar{E} + \bar{F}), \\
G^{\hat{R}}{}_{\hat{\theta}} &= G^{\hat{R}}{}_{\hat{\phi}} = G^{\hat{\theta}}{}_{\hat{\phi}} = 0.]
\end{aligned}
\qquad (14.52)
$$

Exercise 14.17. BIANCHI IDENTITY IN $d\mathcal{R} = 0$ FORM

Define the Riemann tensor as a bivector-valued 2-form,

$$\mathcal{R} = \frac{1}{2}\boldsymbol{e}_\mu \wedge \boldsymbol{e}_\nu \mathcal{R}^{\mu\nu}, \qquad\qquad (14.53)$$

and evaluate $\boldsymbol{d}\mathcal{R}$ to make it manifest that $\boldsymbol{d}\mathcal{R} = 0$. Use

$$\mathcal{R}^{\mu\nu} = \boldsymbol{d}\boldsymbol{\omega}^{\mu\nu} - \boldsymbol{\omega}^\mu{}_\alpha \wedge \boldsymbol{\omega}^{\nu\alpha}, \qquad\qquad (14.54)$$

which is derived easily in an orthonormal frame (adequate for proving $\boldsymbol{d}\mathcal{R} = 0$), or (as a test of skill) in a general frame where $\mathcal{R}^{\mu\nu} = \mathcal{R}^\mu{}_\alpha g^{\alpha\nu}$ and (why?) $\boldsymbol{d}g^{\mu\nu} = -g^{\mu\alpha}(\boldsymbol{d}g_{\alpha\beta})g^{\beta\nu}$. [*Note:* only wedge products between forms (not those between vectors) count in fixing signs in the product rule (14.13) for \boldsymbol{d}.]

Exercise 14.18. LOCAL CONSERVATION OF ENERGY AND MOMENTUM: $\boldsymbol{d}*\boldsymbol{T} = 0$ MEANS $\boldsymbol{\nabla}\cdot\boldsymbol{T} = 0$

Let the duality operator $*$, as defined for exterior differential forms in Box 4.1, act on the forms, *but not on the contravariant vectors,* which appear when the stress-energy tensor \boldsymbol{T} or the Einstein tensor \boldsymbol{G} is written as a mixed $\binom{1}{1}$ tensor:

$$\boldsymbol{T} = \boldsymbol{e}_\mu T^\mu{}_\nu \boldsymbol{\omega}^\nu$$

or

$$\boldsymbol{G} = \boldsymbol{e}_\mu G^\mu{}_\nu \boldsymbol{\omega}^\nu.$$

 (a) Give an expression for $*\boldsymbol{T}$ (or $*\boldsymbol{G}$) expanded in terms of basis vectors and forms.

(b) Show that

$$^*\boldsymbol{T} = \boldsymbol{e}_\mu T^{\mu\nu}\, d^3\boldsymbol{\Sigma}_\nu,$$

where $d^3\boldsymbol{\Sigma}_\nu = \epsilon_{\nu|\alpha\beta\gamma|}\boldsymbol{\omega}^\alpha \wedge \boldsymbol{\omega}^\beta \wedge \boldsymbol{\omega}^\gamma$ [see Box 5.4 and equations (8.10)].

(c) Compute $\boldsymbol{d}\,^*\boldsymbol{T}$ using the generalized exterior derivative \boldsymbol{d}; find that

$$\boldsymbol{d}\,^*\boldsymbol{T} = \boldsymbol{e}_\mu T^{\mu\nu}{}_{;\nu}\sqrt{|g|}\,\boldsymbol{\omega}^0 \wedge \boldsymbol{\omega}^1 \wedge \boldsymbol{\omega}^2 \wedge \boldsymbol{\omega}^3.$$

CHAPTER **15**

BIANCHI IDENTITIES AND
THE BOUNDARY OF A BOUNDARY

This chapter is entirely Track 2.
As preparation, one needs to have covered (1) Chapter 4 (differential forms) and (2) Chapter 14 (computation of curvature).
In reading it, one will be helped by Chapters 9–11 and 13.
It is not needed as preparation for any later chapter, but it will be helpful in Chapter 17 (Einstein field equations).

§15.1. BIANCHI IDENTITIES IN BRIEF

Geometry gives instructions to matter, but how does matter manage to give instructions to geometry? Geometry conveys its instructions to matter by a simple handle: "pursue a world line of extremal lapse of proper time (geodesic)." What is the handle by which matter can act back on geometry? How can one identify the right handle when the metric geometry of Riemann and Einstein has scores of interesting features? Physics tells one what to look for: *a machinery of coupling between gravitation (spacetime curvature) and source* (matter; stress-energy tensor T) *that will guarantee the automatic conservation of the source* ($\mathbf{\nabla} \cdot \mathbf{T} = 0$). Physics therefore asks mathematics: "What tensor-like feature of the geometry is automatically conserved?" Mathematics comes back with the answer: "The Einstein tensor." Physics queries, "How does this conservation come about?" Mathematics, in the person of Élie Cartan, replies, "Through the principle that 'the boundary of a boundary is zero'" (Box 15.1).

Actually, two features of the curvature are automatically conserved; or, otherwise stated, the curvature satisfies two Bianchi identities, the subject of this chapter. Both features of the curvature, both "geometric objects," lend themselves to representation in diagrams, moreover, diagrams that show in action the principle that "the boundary of a boundary is zero." In this respect, the geometry of spacetime shows a striking analogy to the field of Maxwell electrodynamics.

Identities and conservation of the source: electromagnetism and gravitation compared:

In electrodynamics there are four potentials that are united in the 1-form $\mathbf{A} \equiv A_\mu \, \mathbf{d}x^\mu$. Out of this quantity by differentiation follows the ***Faraday***, $\mathbf{F} = \mathbf{d}A$. This

field satisfies the identity $dF = 0$ (identity, yes; identity lending itself to the definition $dF \equiv 0$
of a conserved source, no).

In gravitation there are ten potentials (metric coefficients $g_{\mu\nu}$) that are united in
the metric tensor $\boldsymbol{g} = g_{\mu\nu}\, \boldsymbol{d}x^\mu \otimes \boldsymbol{d}x^\nu$. Out of this quantity by two differentiations
follows the curvature operator

$$\mathcal{R} = \frac{1}{4}\, \boldsymbol{e}_\mu \wedge \boldsymbol{e}_\nu\, R^{\mu\nu}{}_{\alpha\beta}\, \boldsymbol{d}x^\alpha \wedge \boldsymbol{d}x^\beta.$$

This curvature operator satisfies the Bianchi identity $d\mathcal{R} = 0$, where now "\boldsymbol{d}" is a $d\mathcal{R} \equiv 0$
generalization of Cartan's exterior derivative, described more fully in Chapter 14
(again an identity, but again one that does not lend itself to the definition of a
conserved source).

In electromagnetism, one has to go to the dual, $*F$, to have any feature of the $dd\,^*\!F \equiv 0$ plus Maxwell
field that offers a handle to the source, $\boldsymbol{d}^*\boldsymbol{F} = 4\pi\,^*\boldsymbol{J}$. The conservation of the equations $\Longrightarrow \boldsymbol{d}^*\boldsymbol{J} = 0$
source, $\boldsymbol{d}^*\boldsymbol{J} = 0$, appears as a consequence of the identity $dd\,^*\!F = 0$; or, by a
rewording of the reasoning (Box 15.1), as a consequence of the vanishing of the
boundary of a boundary.

(*continued on page 370*)

Box 15.1 THE BOUNDARY OF A BOUNDARY IS ZERO

A. The Idea in Its 1-2-3-Dimensional Form

Begin with an oriented cube or approximation to
a cube (3-dimensional).

 Its boundary is composed of six oriented faces,
each two-dimensional. Orientation of each face is
indicated by an arrow.

 Boundary of any one oriented face consists of
four oriented edges or arrows, each one-dimen-
sional.

 Every edge unites one face with another. No
edge stands by itself in isolation.

 "Sum" over all these edges, with due regard to
sign. Find that any given edge is counted twice,
once going one way, once going the other.

 Conclude that the one-dimensional boundary of
the two-dimensional boundary of the three-di-
mensional cube is identically zero.

Box 15.1 (*continued*)

B. The Idea in Its 2-3-4-Dimensional Form

Begin with an oriented four-dimensional cube or approximation thereto. The coordinates of the typical corner of the four-cube may be taken to be $(t_0 \pm \frac{1}{2}\Delta t, \; x_0 \pm \frac{1}{2}\Delta x, \; y_0 \pm \frac{1}{2}\Delta y, \; z_0 \pm \frac{1}{2}\Delta z)$; and, accordingly, a sample corner itself, in an obvious abbreviation, is conveniently abbreviated $+ - - +$. There are 16 of these corners. Less complicated in appearance than the 4-cube itself are

its three-dimensional faces, which are "exploded off of it" into the surrounding area of the diagram, where they can be inspected in detail.

The boundary of the 4-cube is composed of eight oriented hyperfaces, each of them three-dimensional (top hyperface with extension $\Delta x \, \Delta y \, \Delta z$, for example; a "front" hyperface with extension $\Delta t \, \Delta y \, \Delta z$; etc.)

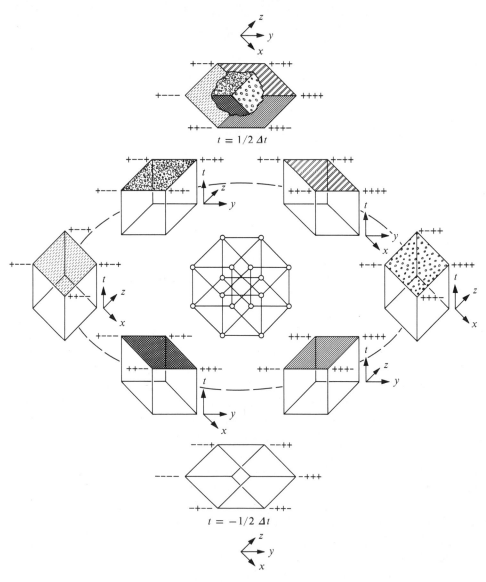

Boundary of any one hyperface ("cube") consists of six oriented faces, each two-dimensional.

Every face (for example, the hatched face $\Delta x\,\Delta y$ in the lower lefthand corner) unites one hypersurface with another (the "3-cube side face" $\Delta t\,\Delta x\,\Delta y$ in the lower lefthand corner with the "3-cube top face" $\Delta x\,\Delta y\,\Delta z$, in this example). No face stands by itself in isolation. The three-dimensional boundary of the 4-cube exposes no 2-surface to the outside world. It is *faceless*.

"Sum" over all these faces, with due regard to orientation. Find any given face is counted twice, once with one orientation, once with the opposite orientation.

Conclude that the two-dimensional boundary of the three-dimensional boundary of the four-dimensional cube is identically zero.

C. The Idea in Its General Abstract Form

$\partial\partial = 0$ (the boundary of a boundary is zero).

D. Idea Behind Application to Gravitation and Electromagnetism

The one central point is a law of conservation (conservation of charge; conservation of momentum-energy).

The other central point is "automatic fulfillment" of this conservation law.

"Automatic conservation" requires that source not be an agent free to vary arbitrarily from place to place and instant to instant.

Source needs a tie to something that, while having degrees of freedom of its own, will cut down the otherwise arbitrary degrees of freedom of the source sufficiently to guarantee that the source automatically fulfills the conservation law. Give the name "field" to this something.

Define this field and "wire it up" to the source in such a way that the conservation of the source shall be an automatic consequence of the "zero boundary of a boundary." Or, more explicitly: Conservation demands no creation or destruction of source inside the four-dimensional cube shown in the diagram. Equivalently, integral of "creation events" (integral of $d*J$ for electric charge; integral of $d*T$ for energy-momentum) over this four-dimensional region is required to be zero.

Integral of creation over this four-dimensional region translates into integral of source density-current ($*J$ or $*T$) over three-dimensional boundary of this region. This boundary consists of eight hyperfaces, each taken with due regard to orientation. Integral over upper hyperface ("$\Delta x\,\Delta y\,\Delta z$)" gives amount of source present at later moment; over lower hyperface gives amount of souce present at earlier moment; over such hyperfaces as "$\Delta t\,\Delta x\,\Delta y$" gives outflow of source over intervening period of time. Conservation demands that sum of these eight three-dimensional integrals shall be zero (details in Chapter 5).

Box 15.1 (*continued*)

Vanishing of this sum of three-dimensional integrals states the conservation requirement, but does not provide the machinery for "automatically" (or, in mathematical terms, "identically") meeting this requirement. For that, turn to principle that "boundary of a boundary is zero."

Demand that integral of source density-current over any oriented hyperface \mathcal{V} (three-dimensional region; "cube") shall equal integral of field over faces of this "cube" (each face being taken with the appropriate orientation and the cube being infinitesimal):

$$4\pi \int_{\mathcal{V}} {}^{*}\boldsymbol{J} = \int_{\partial\mathcal{V}} {}^{*}\boldsymbol{F}; \qquad 8\pi \int_{\mathcal{V}} {}^{*}\boldsymbol{T} = \int_{\partial\mathcal{V}} \begin{pmatrix} \text{moment of} \\ \text{rotation} \end{pmatrix}.$$

Sum over the six faces of this cube and continue summing until the faces of all eight cubes are covered. Find that any given face (as, for example, the hatched face in the diagram) is counted twice, once with one orientation, once with the other ("boundary of a boundary is zero"). Thus is guaranteed the conservation of source: integral of source density-current over three-dimensional boundary of four-dimensional region is automatically zero, making integral of creation over interior of that four-dimensional region also identically zero.

Repeat calculation with boundary of that four-dimensional region slightly displaced in one locality [the "bubble differentiation" of Tomonaga (1946) and Schwinger (1948)], and conclude that conservation is guaranteed, not only in the four-dimensional region as a whole, but at every point within it, and, by extension, everywhere in spacetime.

E. Relation of Source to Field

One view: Source is primary. Field may have other duties, but its prime duty is to serve as "slave" of source. Conservation of source comes first; field has to adjust itself accordingly.

Alternative view: Field is primary. Field takes the responsibility of seeing to it that the source obeys the conservation law. Source would not know what to do in absence of the field, and would not even exist. Source is "built" from field. Conservation of source is consequence of this construction.

One model illustrating this view in an elementary context: Concept of "classical" electric charge as nothing but "electric lines of force trapped in the topology of a multiply connected space" [Weyl (1924b); Wheeler (1955); Misner and Wheeler (1957)].

On any view: Integral of source density-current over any three-dimensional region (a "cube" in simplified analysis above) equals integral of field over boundary of this region (the six faces of the cube above). No one has ever found any other way to understand the correlation between field law and conservation law.

F. Electromagnetism as a Model: How to "Wire Up" Source to Field to Give Automatic Conservation of Source Via "$\partial\partial = 0$" in Its 2-3-4-Dimensional Form

Conservation means zero creation of charge (zero creation in four-dimensional region Ω).

Conservation therefore demands zero value for integral of charge density-current over three-dimensional boundary of this volume; thus,

$$0 = \int_\Omega \frac{\partial J^\mu}{\partial x^\mu}\, d^4\Omega = \int_{\partial\Omega} J^\mu\, d^3\Sigma_\mu$$

in the Track-1 language of Chapters 3 and 5. Equivalently, in the coordinate-free abstract language of §§4.3-4.6, one has

$$0 = \int_\Omega \boldsymbol{d}{*}\boldsymbol{J} = \int_{\partial\Omega} {*}\boldsymbol{J},$$

where

$$\begin{aligned}{*}\boldsymbol{J} = {}&{*}J_{123}\, \boldsymbol{d}x^1 \wedge \boldsymbol{d}x^2 \wedge \boldsymbol{d}x^3 + {*}J_{023}\, \boldsymbol{d}x^0 \wedge \boldsymbol{d}x^2 \wedge \boldsymbol{d}x^3 \\ &+ {*}J_{031}\, \boldsymbol{d}x^0 \wedge \boldsymbol{d}x^3 \wedge \boldsymbol{d}x^1 + {*}J_{012}\, \boldsymbol{d}x^0 \wedge \boldsymbol{d}x^1 \wedge \boldsymbol{d}x^2\end{aligned}$$

("eggcrate-like structure" of the 3-form of charge-density and current-density).

Fulfill this conservation requirement automatically ("identically") through the principle that "the boundary of a boundary is zero" by writing $4\pi\, {*}\boldsymbol{J} = \boldsymbol{d}{*}\boldsymbol{F}$; thus,

$$4\pi \int_{\partial\Omega} {*}\boldsymbol{J} = \int_{\partial\Omega} \boldsymbol{d}{*}\boldsymbol{F} = \int_{\partial\partial\Omega\text{(zero!)}} {*}\boldsymbol{F} \equiv 0.$$

or, in Track-1 language, write $4\pi J^\mu = F^{\mu\nu}{}_{;\nu}$, and have

$$4\pi \int_{\partial\Omega} J^\mu\, d^3\Sigma_\mu = \int_{\partial\Omega} F^{\mu\nu}{}_{;\nu}\, d^3\Sigma_\mu = \int_{\partial\partial\Omega\text{(zero!)}} F^{\mu\alpha}\, d^2\Sigma_{\mu\alpha} \equiv 0.$$

In other words, half of Maxwell's equations in their familiar flat-space form,

$$\text{div } \boldsymbol{E} = \boldsymbol{\nabla} \cdot \boldsymbol{E} = 4\pi\rho, \qquad \text{curl } \boldsymbol{B} = \boldsymbol{\nabla} \times \boldsymbol{B} = \dot{\boldsymbol{E}} + 4\pi\boldsymbol{J},$$

"wire up" the source to the field in such a way that the law of conservation of source follows directly from "$\partial\partial\Omega = 0$."

G. Electromagnetism Also Employs "$\partial\partial = 0$" in its 1-2-3-Dimensional Form ("No Magnetic Charge")

Magnetic charge is linked with field via $4\pi\boldsymbol{J}_{\text{mag}} = \boldsymbol{d}\boldsymbol{F}$ (see point **F** above for translation of this compact Track-2 language into equivalent Track-1 terms). Absence of

Box 15.1 (*continued*)

any magnetic charge says that integral of $\boldsymbol{J}_{\mathrm{mag}}$ over any 3-volume \mathcal{V} is necessarily zero; or ("integration by parts," generalized Stokes theorem)

$$0 = \int_{\mathcal{V}} d\boldsymbol{F} = \int_{\partial\mathcal{V}} \boldsymbol{F} = \begin{pmatrix} \text{total magnetic flux} \\ \text{exiting through } \partial\mathcal{V} \end{pmatrix}.$$

In order to satisfy this requirement "automatically," via principle that "the boundary of a boundary is zero," write $\boldsymbol{F} = d\boldsymbol{A}$ ("expression of field in terms of 4-potential"), and have

$$\int_{\partial\mathcal{V}} \boldsymbol{F} = \int_{\partial\mathcal{V}} d\boldsymbol{A} = \int_{\partial\partial\mathcal{V}\,(\text{zero!})} \boldsymbol{A} \equiv 0.$$

H. Structure of Electrodynamics in Outline Form

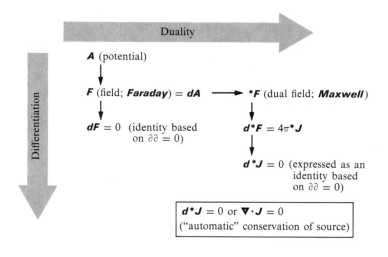

In gravitation physics, one has to go to the "double dual" (two pairs of alternating indices, two places to take the dual) $\boldsymbol{G} = {}^{*}\boldsymbol{R}{}^{*}$ of $\boldsymbol{Riemann}$ to have a feature of the field that offers a handle to the source:

$$\boldsymbol{G} = \mathrm{Tr}\boldsymbol{G} = \boldsymbol{Einstein} = 8\pi\boldsymbol{T} = 8\pi \times (\text{density of energy-momentum}).$$

The conservation of the source $\boldsymbol{T} \equiv \boldsymbol{e}_{\mu} T^{\mu}{}_{\nu} \boldsymbol{\omega}^{\nu}$ can be stated $\boldsymbol{\nabla} \cdot \boldsymbol{T} = 0$. But better suited for the present purpose is the form (see Chapter 14 and exercise 14.18)

I. Structure of Geometrodynamics in Outline Form

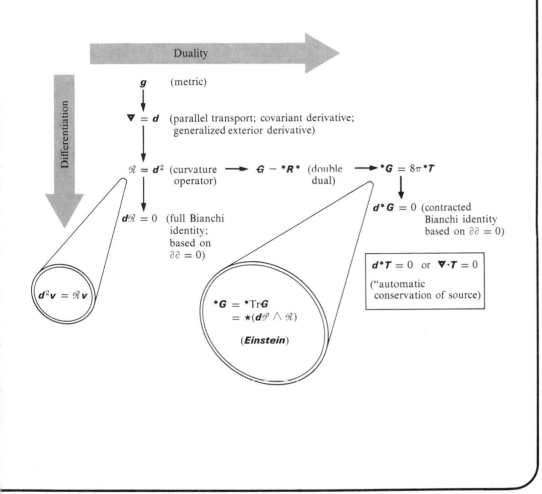

$$d*T = 0,$$

$$d*G \equiv 0 \text{ plus Einstein field equation} \implies d*T = 0$$

where

$$*T \equiv e_\mu T^\mu{}_\nu(*\omega^\nu) = e_\mu T^{\mu\nu} \, d^3\Sigma_\nu.$$

This conservation law arises as a consequence of the "contracted Bianchi identity", $d*G = 0$, again interpretable in terms of the vanishing of the boundary of a boundary.

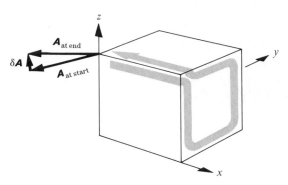

Figure 15.1.

Combine rotations associated with each of the six faces of the illustrated 3-volume and end up with zero net rotation ("full Bianchi identity"). Reason: Contribution of any face is measured by change in a test vector **A** carried in parallel transport around the perimeter of that face. Combine contributions of all faces and end up with each edge traversed twice, once in one direction, once in the other direction [boundary (here one-dimensional) of boundary (two-dimensional) of indicated three-dimensional figure is zero]. Detail: The vector **A**, residing at the indicated site, is transported parallel to itself over to the indicated face, then carried around the perimeter of that face by parallel transport, experiencing in the process a rotation measured by the spacetime curvature associated with that face, then transported parallel to itself back to the original site. To the lowest relevant order of small quantities one can write

$$(\text{change in } \boldsymbol{A}) = -\Delta y\, \Delta z\, \mathcal{R}(\boldsymbol{e}_y, \boldsymbol{e}_z)\, \boldsymbol{A}$$

in operator notation; or in coordinate language,

$$-\delta A^\alpha = R^\alpha{}_{\beta yz}(\text{at } x + \Delta x)A^\beta\, \Delta y\, \Delta z.$$

§15.2. BIANCHI IDENTITY $d\mathcal{R} = 0$ AS A MANIFESTATION OF ''BOUNDARY OF BOUNDARY = 0''

Bianchi identity, $\boldsymbol{d}\mathcal{R} \equiv 0$, interpreted in terms of parallel transport around the six faces of a cube.

Such is the story of the two Bianchi identities in outline form; it is now appropriate to fill in the details. Figure 15.1 illustrates the full Bianchi identity, $\boldsymbol{d}\mathcal{R} = 0$ (see exercise 14.17), saying in brief, "The sum of the curvature-induced rotations associated with the six faces of any elementary cube is zero." The change in a vector **A** associated with transport around the perimeter of the indicated face evaluated to the lowest relevant order of small quantities is given by

$$-\delta A^\alpha = R^\alpha{}_{\beta yz} \text{ (at } x + \Delta x)A^\beta\, \Delta y\, \Delta z. \tag{15.1}$$

The opposite face gives a similar contribution, except that now the sign is reversed and the evaluation takes place at x rather than at $x + \Delta x$. The combination of the contributions from the two faces gives

$$\frac{\partial R^\alpha{}_{\beta yz}}{\partial x} A^\beta\, \Delta x\, \Delta y\, \Delta z, \tag{15.2}$$

when Riemann normal coordinates are in use. In such coordinates, the vanishing of the total $-\delta A^\alpha$ contributed by all six faces implies

$$R^\alpha{}_{\beta yz;x} + R^\alpha{}_{\beta zx;y} + R^\alpha{}_{\beta xy;z} = 0. \tag{15.3}$$

Here semicolons (covariant derivatives) can be and have been inserted instead of commas (ordinary derivatives), because the two are identical in the context of Riemann normal coordinates; and the covariant version (15.3) generalizes itself to arbitrary curvilinear coordinates. Turn from an xyz cube to a cube defined by any set of coordinate axes, and write Bianchi's identity in the form

$$R^\alpha{}_{\beta[\lambda\mu;\nu]} = 0. \tag{15.4}$$

(See exercise 14.17 for one reexpression of this identity in the abstract coordinate-independent form, $d\mathcal{R} = 0$, and §15.3 for another.) This identity occupies much the same place in gravitation physics as that occupied by the identity $dF = ddA \equiv 0$ in electromagnetism:

$$F_{[\lambda\mu,\nu]} = F_{[\lambda\mu;\nu]} = 0. \tag{15.5}$$

§15.3. MOMENT OF ROTATION: KEY TO CONTRACTED BIANCHI IDENTITY

The contracted Bianchi identity, the identity that offers a "handle to couple to the source," was shown by Élie Cartan to deal with "moments of rotation" [Cartan (1928); Wheeler (1964b); Misner and Wheeler (1972)]. Moments are familiar in elementary mechanics. A rigid body will not remain at rest unless all the forces acting on it sum to zero:

$$\sum_i \mathbf{F}^{(i)} = 0. \tag{15.6}$$

Although necessary, this condition is not sufficient. The sum of the moments of these forces about some point \mathcal{P} must also be zero:

$$\sum_i (\mathcal{P}^{(i)} - \mathcal{P}) \wedge \mathbf{F}^{(i)} = 0. \tag{15.7}$$

Exactly what point these moments are taken about happily does not matter, and this for a simple reason. The arbitrary point in the vector product (15.7) has for coefficient the quantity $\Sigma_i \mathbf{F}^{(i)}$, which already has been required to vanish. The situation is similar in the elementary cube of Figure 15.1. Here the rotation associated with a given face is the analog of the force $\mathbf{F}^{(i)}$ in mechanics. That the sum of these rotations vanishes when extended over all six faces of the cube is the analog of the vanishing of the sum of the forces $\mathbf{F}^{(i)}$.

What is the analog for curvature of the moment of the force that one encounters in mechanics? It is the *moment of the rotation associated with a given face of the*

Net moment of rotation over all six faces of a cube:

(1) described

(2) equated to integral of source, $\int *T$, over interior of cube

cube. The value of any individual moment depends on the reference point \mathcal{P}. However, the sum of these moments taken over all six faces of the cube will have a value independent of the reference point \mathcal{P}, for the same reason as in mechanics. Therefore \mathcal{P} can be taken where one pleases, inside the elementary cube or outside it. Moreover, the cube may be viewed as a bit of a hypersurface sliced through spacetime. Therefore \mathcal{P} can as well be off the slice as on it. It is only required that all distances involved be short enough that one obtains the required precision by calculating the moments and the sum of moments in a local Riemann-normal coordinate system. One thus arrives at a \mathcal{P}-independent totalized moment of rotation (not necessarily zero; gravitation is not mechanics!) associated with the cube in question.

Now comes the magic of "the boundary of the boundary is zero." Identify this net moment of rotation of the cube, evaluated by summing individual moments of rotation associated with individual faces, with the integral of the source density-current (energy-momentum tensor $*T$) over the interior of the 3-cube. Make this identification not only for the one 3-cube, but for all eight 3-cubes (hyperfaces) that bound the four-dimensional cube in Box 15.1. Sum the integrated source density-current $*T$ not only for the one hyperface of the 4-cube, but for all eight hyperfaces. Thus have

$$\int_{\substack{\text{4-cube}}} \begin{pmatrix} \text{source} \\ \text{creation} \\ d*T \end{pmatrix} = \int_{\substack{\text{3-boundary} \\ \text{of this 4-cube}}} \begin{pmatrix} \text{source current-} \\ \text{density, } *T \end{pmatrix}$$

$$= \sum_{\substack{\text{these eight} \\ \text{bounding} \\ \text{3-cubes}}} \begin{pmatrix} \text{net moment of rotation} \\ \text{associated with speci-} \\ \text{fied cube} \end{pmatrix}$$

$$= \sum_{\substack{\text{eight} \\ \text{bounding} \\ \text{3-cubes}}} \underbrace{\sum_{\substack{\text{six faces} \\ \text{bounding} \\ \text{given 3-cube}}} \begin{pmatrix} \text{moment of rotation} \\ \text{associated with specified} \\ \text{face of specified cube} \end{pmatrix}}. \quad (15.8)$$

$$\text{(zero!)}$$

(3) conserved

Let the moments of rotation, not only for the six faces of one cube, but for all the faces of all the cubes, be taken with respect to one and the same point \mathcal{P}. Recall (Box 15.1) that any given face joins two cubes or hyperfaces. It therefore appears twice in the count of faces, once with one orientation ("sense of circumnavigation in parallel transport to evaluate rotation") and once with the opposite orientation. Therefore the double sum vanishes identically (boundary of a boundary is zero!) This identity establishes existence of a new geometric object, a feature of the curvature, that is conserved, and therefore provides a handle to which to couple a source. The desired result has been achieved. Now to translate it into standard mathematics!

§15.4. CALCULATION OF THE MOMENT OF ROTATION

It remains to find the tensorial character and value of this conserved Cartan moment of rotation that appertains to any elementary 3-volume. The rotation associated with the front face $\Delta y\, \Delta z\, \boldsymbol{e}_y \wedge \boldsymbol{e}_z$ of the cube in Figure 15.1 will be represented by the bivector

(4) evaluated

$$\begin{pmatrix} \text{rotation associated} \\ \text{with front } \Delta y\, \Delta z \text{ face} \end{pmatrix} = \boldsymbol{e}_\lambda \wedge \boldsymbol{e}_\mu R^{|\lambda\mu|}{}_{yz}\, \Delta y\, \Delta z \qquad (15.9)$$

located at $\mathscr{P}_{\text{front}} = (t - \tfrac{1}{2}\Delta t,\ x + \Delta x,\ y + \tfrac{1}{2}\Delta y,\ z + \tfrac{1}{2}\Delta z)$. This equation uses Riemann normal coordinates; indices enclosed by strokes, as in $|\lambda\mu|$, are summed with the restriction $\lambda < \mu$. The moment of this rotation with respect to the point \mathscr{P} will be represented by the trivector

$$\begin{pmatrix} \text{moment of rotation} \\ \text{associated with} \\ \text{front } \Delta y\, \Delta z \text{ face} \end{pmatrix} = (\mathscr{P}_{\substack{\text{center} \\ \text{of front} \\ \text{face}}} - \mathscr{P}) \wedge \boldsymbol{e}_\lambda \wedge \boldsymbol{e}_\mu R^{|\lambda\mu|}{}_{yz}\, \Delta y\, \Delta z. \qquad (15.10)$$

Here neither $\mathscr{P}_{\text{center front}}$ nor \mathscr{P} has any well-defined meaning whatsoever as a vector, but their difference is a vector in the limit of infinitesimal separation, $\Delta\mathscr{P} = \mathscr{P}_{\text{center front}} - \mathscr{P}$. With the back face a similar moment of rotation is associated, with the opposite sign, and with $\mathscr{P}_{\text{center front}}$ replaced by $\mathscr{P}_{\text{center back}}$. In the difference between the two terms, the factor \mathscr{P} is of no interest, because one is already assured it will cancel out [Bianchi identity (15.4); analog of $\Sigma F^{(i)} = 0$ in mechanics]. The difference $\mathscr{P}_{\text{center front}} - \mathscr{P}_{\text{center back}}$ has the value $\Delta x\, \boldsymbol{e}_x$. Summing over all six faces, one has

$$\begin{pmatrix} \text{net moment of} \\ \text{rotation associated} \\ \text{with cube or hyper-} \\ \text{face } \Delta x\, \Delta y\, \Delta z \end{pmatrix} =$$

$$\begin{aligned} & \boldsymbol{e}_x \wedge \boldsymbol{e}_\lambda \wedge \boldsymbol{e}_\mu R^{|\lambda\mu|}{}_{yz}\, \Delta x\, \Delta y\, \Delta z \ \text{(front and back)} \\ + & \boldsymbol{e}_y \wedge \boldsymbol{e}_\lambda \wedge \boldsymbol{e}_\mu R^{|\lambda\mu|}{}_{zx}\, \Delta y\, \Delta z\, \Delta x \ \text{(sides)} \\ + & \boldsymbol{e}_z \wedge \boldsymbol{e}_\lambda \wedge \boldsymbol{e}_\mu R^{|\lambda\mu|}{}_{xy}\, \Delta z\, \Delta x\, \Delta y \ \text{(top and bottom)}. \quad (15.11) \end{aligned}$$

This sum one recognizes as the value (on the volume element $\boldsymbol{e}_x \wedge \boldsymbol{e}_y \wedge \boldsymbol{e}_z\, \Delta x\, \Delta y\, \Delta z$) of the 3-form

$$\boldsymbol{e}_\nu \wedge \boldsymbol{e}_\lambda \wedge \boldsymbol{e}_\mu R^{|\lambda\mu|}{}_{|\alpha\beta|}\, \boldsymbol{dx}^\nu \wedge \boldsymbol{dx}^\alpha \wedge \boldsymbol{dx}^\beta.$$

Moreover this 3-form is defined, and precisely defined, at a point, whereas (15.11), applying as it does to an extended region, does not lend itself to an analysis that is at the same time brief and precise. Therefore forego (15.11) in favor of the 3-form. Only remember, when it comes down to interpretation, that this 3-form is to be

evaluated for the "cube" $\boldsymbol{e}_x \wedge \boldsymbol{e}_y \wedge \boldsymbol{e}_z \, \Delta x \, \Delta y \, \Delta z$. Now note that the "trivector-valued moment-of-rotation 3-form" can also be written as

(5) abstracted to give
$d\mathscr{P} \wedge \mathscr{R}$

$$\begin{pmatrix} \text{moment of} \\ \text{rotation} \end{pmatrix} = d\mathscr{P} \wedge \mathscr{R} = \boldsymbol{e}_\nu \wedge \boldsymbol{e}_\lambda \wedge \boldsymbol{e}_\mu R^{|\lambda\mu|}{}_{|\alpha\beta|} \, \boldsymbol{dx}^\nu \wedge \boldsymbol{dx}^\alpha \wedge \boldsymbol{dx}^\beta. \quad (15.12)$$

Here

$$d\mathscr{P} = \boldsymbol{e}_\sigma \, \boldsymbol{dx}^\sigma \qquad (15.13)$$

is Cartan's $\binom{1}{1}$ unit tensor. Also \mathscr{R} is the curvature operator, treated as a bivector-valued 2-form:

$$\mathscr{R} = \boldsymbol{e}_\lambda \wedge \boldsymbol{e}_\mu R^{|\lambda\mu|}{}_{|\alpha\beta|} \, \boldsymbol{dx}^\alpha \wedge \boldsymbol{dx}^\beta. \qquad (15.14)$$

Using the language of components as in (15.11), or the abstract language introduced in (15.12), one finds oneself dealing with a trivector. A trivector can be left a trivector, as, in quite another context, an element of 3-volume on a hypersurface in 4-space can be left as a trivector. However, there it is more convenient to take the dual representation, and speak of the element of volume as a vector. Denote by \star a duality operation that acts only on contravariant vectors, trivectors, etc. (but *not* on forms). Then in a Lorentz frame one has $\star(\boldsymbol{e}_1 \wedge \boldsymbol{e}_2 \wedge \boldsymbol{e}_3) = \boldsymbol{e}_0$; but $\star(\boldsymbol{dx}^3) = \boldsymbol{dx}^3$. More generally,

$$\star(\boldsymbol{e}_\nu \wedge \boldsymbol{e}_\lambda \wedge \boldsymbol{e}_\mu) = \varepsilon_{\nu\lambda\mu}{}^\sigma \boldsymbol{e}_\sigma. \qquad (15.15)$$

(6) abstracted to give
$\star(d\mathscr{P} \wedge \mathscr{R}) = \boldsymbol{e}_\sigma G^{\sigma\tau} \, d^3\Sigma_\tau$

In this notation, the "vector-valued moment-of-rotation 3-form" is

$$\begin{pmatrix} \text{moment} \\ \text{of rotation} \end{pmatrix} = \star(d\mathscr{P} \wedge \mathscr{R}) = \boldsymbol{e}_\sigma \varepsilon_{\nu\lambda\mu}{}^\sigma R^{|\lambda\mu|}{}_{|\alpha\beta|} \, \boldsymbol{dx}^\nu \wedge \boldsymbol{dx}^\alpha \wedge \boldsymbol{dx}^\beta$$
$$= \boldsymbol{e}_\sigma (*R)_\nu{}^\sigma{}_{|\alpha\beta|} \, \boldsymbol{dx}^\nu \wedge \boldsymbol{dx}^\alpha \wedge \boldsymbol{dx}^\beta,$$

or, in one more step,

$$\begin{pmatrix} \text{moment} \\ \text{of rotation} \end{pmatrix} = \star(d\mathscr{P} \wedge \mathscr{R}) = \boldsymbol{e}_\sigma (*R*)_\nu{}^{\sigma\nu\tau} d^3\Sigma_\tau. \qquad (15.16)$$

Here $d^3\Sigma_\tau$ is a notation for basis 3-forms, as in Box 5.4; thus,

$$\boldsymbol{dx}^\nu \wedge \boldsymbol{dx}^\alpha \wedge \boldsymbol{dx}^\beta = \varepsilon^{\nu\alpha\beta\tau} d^3\Sigma_\tau. \qquad (15.17)$$

(In a local Lorentz frame, $\boldsymbol{dx}^1 \wedge \boldsymbol{dx}^2 \wedge \boldsymbol{dx}^3 = d^3\Sigma_0$.)

Nothing is more central to the analysis of curvature than the formula (15.16). It starts with an element of 3-volume and ends up giving the moment of rotation in that 3-volume. The tensor that connects the starting volume with the final moment, the "contracted double-dual" of **Riemann**, is so important that it deserves and receives a name of its own, $\boldsymbol{G} \equiv \boldsymbol{Einstein}$; thus

$$(\text{Einstein})^{\sigma\tau} \equiv G^{\sigma\tau} = G_\nu{}^{\sigma\nu\tau} = (*R*)_\nu{}^{\sigma\nu\tau}. \qquad (15.18)$$

This tensor received attention in §§13.5 and 14.2, and also in the examples at the

end of Chapter 14. In terms of **Einstein**, the connection between element of 3-volume and "vector-valued moment of rotation" is

$$\begin{pmatrix} \text{moment} \\ \text{of rotation} \end{pmatrix} = \star(d\mathcal{P} \wedge \mathcal{R}) = e_\sigma G^{\sigma\tau}\, d^3\Sigma_\tau. \tag{15.19}$$

The amount of "vector-valued moment of rotation" contained in the element of 3-volume $d^3\Sigma_\mu$ is identified by general relativity with the amount of energy-momentum contained in that 3-volume. However, defer this identification for now. Concentrate instead on the conservation properties of this moment of rotation. See them once in the formulation of integral calculus, as a consequence of the principle "$\partial\partial \equiv 0$." See them then a second time, in differential formulation, as a consequence of "$dd \equiv 0$."

§15.5. CONSERVATION OF MOMENT OF ROTATION SEEN FROM "BOUNDARY OF A BOUNDARY IS ZERO"

The moment of rotation defines an automatically conserved quantity. In other words, the value of the moment of rotation for an elementary 3-volume $\Delta x\,\Delta y\,\Delta z$ after the lapse of a time Δt is equal to the value of the moment of rotation for the same 3-volume at the beginning of that time, corrected by the inflow of moment of rotation over the six faces of the 3-volume in that time interval (quantities proportional to $\Delta y\,\Delta z\,\Delta t$, etc.) Now verify this conservation of moment of rotation in the language of "the boundary of a boundary." Follow the pattern of equation (15.8), but translate the words into formulas, item by item. Evaluate the amount of moment of rotation created in the elementary 4-cube Ω, and find

Conservation of net moment of rotation:

(1) derived from "$\partial\partial = 0$"

$$\text{"creation"} \equiv \int \begin{pmatrix} \text{"creation of moment of} \\ \text{rotation" in the elementary} \\ \text{4-cube of spacetime } \Omega \end{pmatrix} = \int_\Omega d{\star}G;$$

definition definition

$$\int_\Omega d{\star}G = \int_{\partial\Omega} {\star}G = \int_{\partial\Omega} \star(d\mathcal{P} \wedge \mathcal{R}) = \sum_{\substack{\text{the eight} \\ \text{3-cubes} \\ \text{that bound }\Omega}} \star \begin{pmatrix} \text{moment of rotation} \\ \int_{\text{3-cube}} (d\mathcal{P} \wedge \mathcal{R}) \\ \text{associated with} \\ \text{specified 3-cube} \end{pmatrix} =$$

step 1 step 2 step 3 step 4

$$= \sum_{\substack{\text{eight bounding} \\ \text{3-cubes}}} \sum_{\substack{\text{six faces bounding} \\ \text{specified 3-cube}}} \star \begin{pmatrix} \text{moment of rotation} \\ \int_{\text{face}} (\mathcal{P} \wedge \mathcal{R}) \\ \text{associated with} \\ \text{specified face of} \\ \text{specified cube} \end{pmatrix} \equiv 0. \tag{15.20}$$

step 4 step 5

Here step 1 is the theorem of Stokes. Step 2 is the identification established by (15.19) between the Einstein tensor and the moment of rotation. Step 3 breaks down the integral over the entire boundary $\partial\Omega$ into integrals over the individual 3-cubes that constitute this boundary. Moreover, in all these integrals, the star \star is treated as a constant and taken outside the sign of integration. The reason for such treatment is simple: the duality operation \star involves only the metric, and the metric is locally constant throughout the infinitesimal 4-cube over the boundary of which the integration extends. Step 4 uses the formula

$$\boldsymbol{d}(\mathscr{P} \wedge \mathscr{R}) = \boldsymbol{d}\mathscr{P} \wedge \mathscr{R} + \mathscr{P} \wedge \boldsymbol{d}\mathscr{R} = \boldsymbol{d}\mathscr{P} \wedge \mathscr{R} \tag{15.21}$$

and the theorem of Stokes to express each 3-cube integral as an integral of $\mathscr{P} \wedge \mathscr{R}$ over the two-dimensional boundary of that cube. The culminating step is 5. It has nothing to do with the integrand. It depends solely on the principle $\partial\partial \equiv 0$.

In brief, the conservation of moment of rotation follows from two circumstances. (1) The moment of rotation associated with any elementary 3-cube is by definition a net value, obtained by adding the six moments of rotation associated with the six faces of that cube. (2) When one sums these net values for all eight 3-cubes in (15.20), which are the boundary of the elementary 4-cube Ω, one counts the contribution of a given 2-face twice, once with one sign and once with the opposite sign. In virtue of the principle that "the boundary of a boundary is zero," the conservation of moment of rotation is thus an identity.

§15.6. CONSERVATION OF MOMENT OF ROTATION EXPRESSED IN DIFFERENTIAL FORM

(2) derived from "$\boldsymbol{dd} = 0$"

Every conservation law stated in integral form lends itself to restatement in differential form, and conservation of moment of rotation is no exception. The calculation is brief. Evaluate the generalized exterior derivative of the moment of rotation in three steps, and find that it vanishes; thus:

$$
\begin{aligned}
\boldsymbol{d}{*}\boldsymbol{G} &= \boldsymbol{d}[\star(\boldsymbol{d}\mathscr{P} \wedge \mathscr{R})] \\
&= \star[\boldsymbol{d}(\boldsymbol{d}\mathscr{P} \wedge \mathscr{R})] \\
&= \star[\boldsymbol{d}^2\mathscr{P} \wedge \mathscr{R} - \boldsymbol{d}\mathscr{P} \wedge \boldsymbol{d}\mathscr{R}] \\
&= 0
\end{aligned}
\qquad
\begin{aligned}
&\} \text{ step 1} \\
&\} \text{ step 2} \\
&\} \text{ step 3}
\end{aligned}
$$

Step 1 uses the relation $\boldsymbol{d}\star = \star\boldsymbol{d}$. The star duality and the generalized exterior derivative commute because when \boldsymbol{d} is applied to a contravariant vector, it acts as a covariant derivative, and when \star is applied to a covariant vector or 1-form, it is without effect. Step 2 applies the standard rule for the action of \boldsymbol{d} on a product of tensor-valued forms [see equation (14.13b)]. Step 3 deals with two terms. The first term vanishes because the first factor in it vanishes; thus, $\boldsymbol{d}^2\mathscr{P} = 0$ [Cartan's equation of structure; expresses the "vanishing torsion" of the covariant derivative; see equation (14.26)]. The second term also vanishes, in this case, because the second factor in it vanishes; thus, $\boldsymbol{d}\mathscr{R} = 0$ (the full Bianchi identity). Thus briefly is conservation of moment of rotation established.

Box 15.2 THE SOURCE OF GRAVITATION AND THE MOMENT OF ROTATION: THE TWO KEY QUANTITIES AND THE MOST USEFUL MATHEMATICAL REPRESENTATIONS FOR THEM

	Energy-momentum as source of gravitation (curvature of space-time)	Moment of rotation as automatically conserved feature of the geometry
Representation as a vector-valued 3-form, a coordinate-independent geometric object	Machine to tell how much energy-momentum is contained in an elementary 3-volume: $$*\boldsymbol{T} = \boldsymbol{e}_\sigma T^{\sigma\tau}\, d^3\Sigma_\tau$$ ("dual of stress-energy tensor")	Machine to tell how much net moment of rotation—expressed as a vector—is obtained by adding the six moments of rotation associated with the six faces of the elementary 3-cube: $$\star(\boldsymbol{d\mathscr{P}} \wedge \mathscr{R}) = *\boldsymbol{G} = \boldsymbol{e}_\sigma G^{\sigma\tau}\, d^3\Sigma_\tau$$ ("dual of Einstein")
Representation as a $\binom{2}{0}$-tensor (also a coordinate independent geometric object)	Stress-energy tensor itself: $$\boldsymbol{T} = \boldsymbol{e}_\sigma T^{\sigma\tau}\boldsymbol{e}_\tau$$	Einstein itself: $$\boldsymbol{G} = \boldsymbol{e}_\sigma G^{\sigma\tau}\boldsymbol{e}_\tau$$
Representation in language of components (values depend on choice of coordinate system)	$T^{\sigma\tau}$	$G^{\sigma\tau}$
Conservation law in language of components	$T^{\sigma\tau}{}_{;\tau} = 0$	$G^{\sigma\tau}{}_{;\tau} \equiv 0$
Conservation in abstract language, for the $\binom{2}{0}$-tensor	$\boldsymbol{\nabla} \cdot \boldsymbol{T} = 0$	$\boldsymbol{\nabla} \cdot \boldsymbol{G} \equiv 0$
Conservation in abstract language, as translated into exterior derivative of the dual tensor (vector-valued 3-form)	$\boldsymbol{d}*\boldsymbol{T} = 0$	$\boldsymbol{d}*\boldsymbol{G} \equiv 0$ or $\boldsymbol{d}\star(\boldsymbol{d\mathscr{P}} \wedge \mathscr{R}) \equiv 0$
Same conservation law expressed in integral form for an element of 4-volume Ω	$\displaystyle\int_{\partial\Omega} *\boldsymbol{T} = 0$	$\displaystyle\int_{\partial\Omega} *\boldsymbol{G} \equiv 0$ or $\displaystyle\star\int_{\partial\Omega} (\boldsymbol{d\mathscr{P}} \wedge \mathscr{R}) \equiv 0$ or $\displaystyle\star\int_{\partial\partial\Omega} (\mathscr{P} \wedge \mathscr{R}) \equiv 0$

§15.7. FROM CONSERVATION OF MOMENT OF ROTATION TO EINSTEIN'S GEOMETRODYNAMICS: A PREVIEW

Mass, or mass-energy, is the source of gravitation. Mass-energy is one component of the energy-momentum 4-vector. Energy and momentum are conserved. The amount of energy-momentum in the element of 3-volume $d^3\boldsymbol{\Sigma}$ is

Einstein field equation "derived" from demand that (conservation of net moment of rotation) \Rightarrow (conservation of source)

$$*\boldsymbol{T} = \boldsymbol{e}_\sigma T^{\sigma\tau}\, d^3\Sigma_\tau \tag{15.22}$$

(see Box 15.2). Conservation of energy-momentum for an elementary 4-cube Ω expresses itself in the form

$$\int_{\partial\Omega} *\boldsymbol{T} = 0. \tag{15.23}$$

This conservation is not an accident. According to Einstein and Cartan, it is "automatic"; and automatic, moreover, as a consequence of exact equality between energy-momentum and an automatically conserved feature of the geometry. What is this feature? It is the moment of rotation, which satisfies the law of automatic conservation,

$$\int_{\partial\Omega} {}^{*}\boldsymbol{G} = 0. \tag{15.24}$$

In other words, the conservation of momentum-energy is to be made geometric in character and automatic in action by the following prescription: *Identify the stress-energy tensor* (up to a factor 8π, or $8\pi G/c^4$, or other factor that depends on choice of units) *with the moment of rotation;* thus,

$$\star(\boldsymbol{d\mathscr{P}} \wedge \mathscr{R}) = {}^{*}\boldsymbol{G} = 8\pi\, {}^{*}\boldsymbol{T}; \tag{15.25}$$

or equivalently (still in the language of vector-valued 3-forms)

$$\begin{pmatrix}\text{moment of}\\ \text{rotation}\end{pmatrix} = \star(\boldsymbol{d\mathscr{P}} \wedge \mathscr{R}) = \boldsymbol{e}_\sigma G^{\sigma\tau}\, d^3\Sigma_\tau = 8\pi\boldsymbol{e}_\sigma T^{\sigma\tau}\, d^3\Sigma_\tau; \tag{15.26}$$

or, in the language of tensors,

$$\boldsymbol{G} = \boldsymbol{e}_\sigma G^{\sigma\tau}\boldsymbol{e}_\tau = 8\pi\boldsymbol{e}_\sigma T^{\sigma\tau}\boldsymbol{e}_\tau = 8\pi\boldsymbol{T}; \tag{15.27}$$

or, in the language of components,

$$G^{\sigma\tau} = 8\pi T^{\sigma\tau} \tag{15.28}$$

(Einstein's field equation; more detail, and more on the question of uniqueness, will be found in Chapter 17; see also Box 15.3). Thus simply is all of general relativity tied to the principle that the boundary of a boundary is zero. No one has ever discovered a more compelling foundation for the principle of conservation of momentum and energy. No one has ever seen more deeply into that action of matter on space, and space on matter, which one calls gravitation.

In summary, *the Einstein theory realizes the conservation of energy-momentum as the identity, "the boundary of a boundary is zero."*

EXERCISES

Exercise 15.1. THE BOUNDARY OF THE BOUNDARY OF A 4-SIMPLEX

In the analysis of the development in time of a geometry lacking all symmetry, when one is compelled to resort to a computer, one can, as one option, break up the 4-geometry into simplexes [four-dimensional analog of two-dimensional triangle, three-dimensional tetrahedron; vertices of "central simplex" conveniently considered to be at $(t, x, y, z) = (0, 1, 1, 1)$, $(0, 1, -1, -1)$, $(0, -1, 1, -1)$, $(0, -1, -1, 1)$, $(5^{1/2}, 0, 0, 0)$, for example], sufficiently numerous, and each sufficiently small, that the geometry inside each can be idealized as flat (Lorentzian), with all the curvature concentrated at the join between simplices (see discussion of dynamics of geometry via Regge calculus in Chapter 42). Determine ("give a mathematical

Box 15.3 OTHER IDENTITIES SATISFIED BY THE CURVATURE

(1) The source of gravitation is energy-momentum. (2) Energy-momentum is expressed by stress-energy tensor (or by its dual) as a vector-valued 3-form ("energy-momentum per unit 3-volume"). (3) This source is conserved (no creation in an elementary spacetime 4-cube).

These principles form the background for the probe in this chapter of the Bianchi identities. That is why two otherwise most interesting identities [Allendoerfer and Weil (1943); Chern (1955, 1962)] are dropped from attention. One deals with the 4-form

$$II = \frac{1}{24\pi^2} g^{\alpha\gamma} g^{\beta\delta} \mathscr{R}_{\alpha\beta} \wedge \mathscr{R}_{\gamma\delta}, \qquad (1)$$

and the other with the 4-form

$$\Gamma = \frac{1}{8\pi^2 |\det g_{\mu\nu}|^{1/2}} (\mathscr{R}_{12} \wedge \mathscr{R}_{30} + \mathscr{R}_{13} \wedge \mathscr{R}_{02} + \mathscr{R}_{10} \wedge \mathscr{R}_{23}). \qquad (2)$$

Both quantities are built from the tensorial "curvature 2-forms"

$$\mathscr{R}_{\alpha\gamma} = \frac{1}{2} R_{\alpha\gamma\beta\delta} \, \boldsymbol{dx}^{\beta} \wedge \boldsymbol{dx}^{\delta}. \qquad (3)$$

The four-dimensional integral of either quantity over a four-dimensional region Ω has a value that (1) is a scalar, (2) is not identically equal to zero, (3) depends on the boundary of the region of spacetime over which the integral is extended, but (4) is independent of any changes made in the spacetime geometry interior to that surface (provided that these changes neither abandon the continuity nor change the connectivity of the 4-geometry in that region). Property (1) kills any possibility of identifying the integral, a scalar, with energy-momentum, a 4-vector. Property (2) kills it for the purpose of a conservation law, because it implies a non-zero creation in Ω.

Also omitted here is the Bel-Robinson tensor (see exercise 15.2), built bilinearly out of the curvature tensor, and other tensors for which see, e.g., Synge (1962).

One or all of these quantities may be found someday to have important physical content.

The integral of the 4-form Γ of equation (2) over the entire manifold gives a number, an integer, the so-called Euler-Poincaré characteristic of the manifold, whenever the integral and the integer are well-defined. This result is the four-dimensional generalization of the Gauss-Bonnet integral, widely known in the context of two-dimensional geometry:

$$\int \left(\begin{array}{l} \text{Riemannian scalar curvature} \\ \text{invariant (value } 2/a^2 \\ \text{for a sphere of radius } a) \end{array} \right) g^{1/2} \, d^2x.$$

This integral has the value 8π for any closed, oriented, two-dimensional manifold with the topology of a 2-sphere, no matter how badly distorted; and the value 0 for any 2-torus, again no matter how rippled and twisted; and other equally specific values for other topologies.

description of") the boundary (three-dimensional) of such a simplex. Take one piece of this boundary and determine its boundary (two-dimensional). For one piece of this two-dimensional boundary, verify that there is at exactly one other place, and no more, in the bookkeeping on the boundary of a boundary, another two-dimensional piece that cancels it ("facelessness" of the 3-boundary of the simplex).

Exercise 15.2. THE BEL-ROBINSON TENSOR [Bel (1958, 1959, 1962),
Robinson (1959b), Sejnowski (1973); see also Pirani (1957)
and Lichnerowicz (1962)].

Define the Bel-Robinson tensor by

$$T_{\alpha\beta\gamma\delta} = R_{\alpha\rho\gamma\sigma} R_{\beta}{}^{\rho}{}_{\delta}{}^{\sigma} + {}^{*}R_{\alpha\rho\gamma\sigma} {}^{*}R_{\beta}{}^{\rho}{}_{\delta}{}^{\sigma}. \qquad (15.29)$$

Show that in empty spacetime this tensor can be rewritten as

$$T_{\alpha\beta\gamma\delta} = R_{\alpha\rho\gamma\sigma} R_{\beta}{}^{\rho}{}_{\delta}{}^{\sigma} + R_{\alpha\rho\delta\sigma} R_{\beta}{}^{\rho}{}_{\gamma}{}^{\sigma} - \frac{1}{8} g_{\alpha\beta} g_{\gamma\delta} R_{\rho\sigma\lambda\mu} R^{\rho\sigma\lambda\mu}. \qquad (15.30a)$$

Show also that in empty spacetime

$$T^{\alpha}{}_{\beta\gamma\delta;\alpha} = 0, \qquad (15.30b)$$

$$T_{\alpha\beta\gamma\delta} \text{ is symmetric and traceless on all pairs of indices.} \qquad (15.30c)$$

Discussion: It turns out that Einstein's "canonical energy-momentum pseudotensor" (§20.3) for the gravitational field in empty spacetime has a second derivative which, in a Riemann-normal coordinate system, is

$$t_{\mathrm{E}\alpha\beta,\gamma\delta} = -\frac{4}{9} \left(T_{\alpha\beta\gamma\delta} - \frac{1}{4} S_{\alpha\beta\gamma\delta} \right). \qquad (15.31a)$$

Here $T_{\alpha\beta\gamma\delta}$ is the completely symmetric Bel-Robinson tensor, and $S_{\alpha\beta\gamma\delta}$ is defined by

$$S_{\alpha\beta\gamma\delta} \equiv R_{\alpha\delta\rho\sigma} R_{\beta\gamma}{}^{\rho\sigma} + R_{\alpha\gamma\rho\sigma} R_{\beta\delta}{}^{\rho\sigma} + \frac{1}{4} g_{\alpha\beta} g_{\gamma\delta} R_{\mu\nu\rho\sigma} R^{\mu\nu\rho\sigma}. \qquad (15.31b)$$

$S_{\alpha\beta\gamma\delta}$ appears in the empty-space covariant wave equation

$$\Delta R_{\alpha\beta\gamma\delta} \equiv -R_{\alpha\beta\gamma\delta;\mu}{}^{;\mu} + R_{\alpha\beta\rho\sigma} R_{\gamma\delta}{}^{\rho\sigma} + 2(R_{\alpha\rho\gamma\sigma} R_{\beta}{}^{\rho}{}_{\delta}{}^{\sigma} - R_{\alpha\rho\delta\sigma} R_{\beta}{}^{\rho}{}_{\gamma}{}^{\sigma}) = 0, \quad (15.31c)$$

where Δ is a variant of the Lichnerowicz-de Rham wave operator [Lichnerowicz (1964)], when one rewrites this wave equation as

$$\Box R_{\alpha\beta}{}^{\gamma\delta} \equiv R_{\alpha\beta}{}^{\gamma\delta}{}_{;\mu}{}^{;\mu} = 2 S_{[\alpha}{}^{[\gamma}{}_{\beta]}{}^{\delta]}. \qquad (15.31d)$$

EINSTEIN'S GEOMETRIC
THEORY OF GRAVITY

*Wherein the reader is seduced into marriage with the most elegant
temptress of all—Geometrodynamics—and learns from her
the magic potions and incantations that control the universe.*

CHAPTER **16**

EQUIVALENCE PRINCIPLE AND MEASUREMENT OF THE "GRAVITATIONAL FIELD"

Rather than have one global frame with gravitational forces we have many local frames without gravitational forces.

STEPHEN SCHUTZ (1966)

§16.1. OVERVIEW

With the mathematics of curved spacetime now firmly in hand, one is tempted to rush headlong into a detailed study of Einstein's field equations. But such temptation must be resisted for a short time more. To grasp the field equations fully, one must first understand how the classical laws of physics change, or do not change, in the transition from flat spacetime to curved (§§16.2 and 16.3); and one must understand how the "gravitational field" (metric; covariant derivative; spacetime curvature; . . .) can be "measured" (§§16.4 and 16.5).

Purpose of this chapter

§16.2. THE LAWS OF PHYSICS IN CURVED SPACETIME

Wherever one is and whenever one probes, one finds that then and there one can introduce a local inertial frame in which all test particles move along straight lines. Moreover, this local inertial frame is also locally Lorentz: in it the velocity of light has its standard value, and light rays, like world lines of test particles, are straight. But physics is more, and the analysis of physics demands more than an account solely of the motions of test particles and light rays. What happens to Maxwell's equations, the laws of hydrodynamics, the principles of atomic structure, and all the rest of physics under the influence of "powerful gravitational fields"?

Einstein's equivalence
principle

The answer is simple: *in any and every local Lorentz frame, anywhere and anytime in the universe, all the (nongravitational) laws of physics must take on their familiar special-relativistic forms.* Equivalently: there is no way, by experiments confined to infinitesimally small regions of spacetime, to distinguish one local Lorentz frame in one region of spacetime from any other local Lorentz frame in the same or any other region. This is Einstein's principle of equivalence in its strongest form—a principle that is compelling both philosophically and experimentally. (For the relevant experimental tests, see §38.6.)

Equivalence principle as tool
to mesh nongravitational
laws with gravity

The principle of equivalence has great power. With it one can generalize all the special relativistic laws of physics to curved spacetime. And the curvature need not be small. It may be as large as that in the center of a neutron star; as large as that at the edge of a black hole; arbitrarily large, in fact—or almost so. Only at the endpoint of gravitational collapse and in the initial instant of the "big bang," i.e., only at "singularities of spacetime," will there be a breakdown in the conditions needed for direct application of the equivalence principle (see §§28.3, 34.6, 43.3, 43.4, and chapter 44). Everywhere else the equivalence principle acts as a tool to mesh all the nongravitational laws of physics with gravity.

Example: Mesh the "law of local energy-momentum conservation," $\boldsymbol{\nabla} \cdot \boldsymbol{T} = 0$, with gravity. *Solution:*

(1) The law in flat spacetime, written in abstract geometric form, reads

$$\boldsymbol{\nabla} \cdot \boldsymbol{T} = 0. \tag{16.1a}$$

(2) Rewritten in a global Lorentz frame of flat spacetime, it reads

$$T^{\mu\nu}{}_{,\nu} = 0. \tag{16.1b}$$

(3) Application of equivalence principle gives same equation in local Lorentz frame of curved spacetime:

$$T^{\hat{\mu}\hat{\nu}}{}_{,\hat{\nu}} = 0 \text{ at origin of local Lorentz frame.} \tag{16.1c}$$

Because the connection coefficients vanish at the origin of the local Lorentz frame, this can be rewritten as

$$T^{\hat{\mu}\hat{\nu}}{}_{;\hat{\nu}} = 0 \text{ at origin of local Lorentz frame.} \tag{16.1d}$$

(4) The geometric law in curved spacetime, of which these are the local-Lorentz components, is

$$\boldsymbol{\nabla} \cdot \boldsymbol{T} = 0; \tag{16.1e}$$

and its component formulation in any reference frame reads

$$T^{\mu\nu}{}_{;\nu} = 0. \tag{16.1f}$$

Compare the abstract geometric law (16.1e) in curved spacetime with the corresponding law (16.1a) in flat spacetime. They are identical! That this is not an accident one can readily see by tracing out the above four-step argument for any other law

of physics (e.g., Maxwell's equation $\boldsymbol{\nabla} \cdot \boldsymbol{F} = 4\pi \boldsymbol{J}$). *The laws of physics, written in abstract geometric form, differ in no way whatsoever between curved spacetime and flat spacetime;* this is guaranteed by, and in fact is a mere rewording of, the equivalence principle.

Compare the component version of the law $\boldsymbol{\nabla} \cdot \boldsymbol{T} = 0$, as written in an arbitrary frame in curved spacetime [equation (16.1f)], with the component version in a global Lorentz frame of flat spacetime [equation (16.1b)]. They differ in only one way: the comma (partial derivative; flat-spacetime gradient) is replaced by a semicolon (covariant derivative; curved-spacetime gradient). This procedure for rewriting the equations has universal application. *The laws of physics, written in component form, change on passage from flat spacetime to curved spacetime by a mere replacement of all commas by semicolons* (no change at all physically or geometrically; change due only to switch in reference frame from Lorentz to non-Lorentz!). This statement, like the nonchanging of abstract geometric laws, is nothing but a rephrased version of the equivalence principle.

"Comma-goes-to-semicolon" rule

The transition in formalism from flat spacetime to curved spacetime is a trivial process when performed as outlined above. But it is nontrivial in its implications. It meshes gravity with all the laws of physics. Gravity enters in an essential way through the covariant derivative of curved spacetime, as one sees clearly in the following exercise.

Exercise 16.1. HYDRODYNAMICS IN A WEAK GRAVITATIONAL FIELD

EXERCISES

(a) In §18.4 it will be shown that for a nearly Newtonian system, analyzed in an appropriate nearly global Lorentz coordinate system, the metric has the form

$$ds^2 = -(1 + 2\varPhi)\,dt^2 + (1 - 2\varPhi)(dx^2 + dy^2 + dz^2) \qquad (16.2a)$$

where \varPhi is the Newtonian potential ($-1 \ll \varPhi < 0$). Consider a nearly Newtonian perfect fluid [stress-energy tensor

$$T^{\alpha\beta} = (\rho + p)u^\alpha u^\beta + pg^{\alpha\beta}, \qquad p \ll \rho; \qquad (16.2b)$$

see Box 5.1 and §5.10] moving in such a spacetime with ordinary velocity

$$v^j \equiv dx^j/dt \ll 1. \qquad (16.2c)$$

Show that the equations $T^{\mu\nu}{}_{;\nu} = 0$ for this system reduce to the familiar Newtonian law of mass conservation, and the Newtonian equation of motion for a fluid in a gravitational field:

$$\frac{d\rho}{dt} = -\rho\,\frac{\partial v^j}{\partial x^j}, \qquad \rho\,\frac{dv^j}{dt} = -\rho\,\frac{\partial \varPhi}{\partial x^j} - \frac{\partial p}{\partial x^j}, \qquad (16.3a)$$

where d/dt is the time derivative comoving with the matter

$$\frac{d}{dt} \equiv \frac{\partial}{\partial t} + v^j \frac{\partial}{\partial x^j}. \qquad (16.3b)$$

(b) Use these equations to calculate the pressure gradient in the Earth's atmosphere as a function of temperature and pressure. In the calculation, use the nonrelativistic relation $\rho = n_M \mu_M$, where n_M is the number density of molecules and μ_M is the mean rest mass per molecule; use the ideal-gas equation of state

$$p = n_M k T \qquad (k = \text{Boltzmann's constant});$$

and use the spherically symmetric form, $\Phi = -M/r$, for the Earth's Newtonian potential. If the pressure at sea level is 1.01×10^6 dynes/cm^2, what, approximately, is the pressure on top of Mount Everest (altitude 8,840 meters)? (Make a reasonable assumption about the temperature distribution of the atmosphere.)

Exercise 16.2. WORLD LINES OF PHOTONS

Show that in flat spacetime the conservation law for the 4-momentum of a freely moving photon can be written

$$\nabla_{\boldsymbol{p}} \boldsymbol{p} = 0. \qquad (16.4a)$$

According to the equivalence principle, this equation must be true also in curved spacetime. Show that this means photons move along null geodesics of curved spacetime with affine parameter λ related to 4-momentum by

$$\boldsymbol{p} = d/d\lambda \qquad (16.4b)$$

In exercise 18.6 this result will be used to calculate the deflection of light by the sun.

§16.3. FACTOR-ORDERING PROBLEMS IN THE EQUIVALENCE PRINCIPLE

Factor-ordering problems and coupling to curvature

On occasion in applying the equivalence principle to get from physics in flat space-time to physics in curved spacetime one encounters "factor-ordering problems" analogous to those that beset the transition from classical mechanics to quantum mechanics.* *Example:* How is the equation (3.56) for the vector potential of electrodynamics to be translated into curved spacetime? If the flat-spacetime equation is written

$$-A^{\alpha,\mu}{}_{\mu} + A^{\mu}{}_{,\mu}{}^{\alpha} = 4\pi J^{\alpha},$$

then its transition ("comma goes to semicolon") reads

$$-A^{\alpha;\mu}{}_{\mu} + A^{\mu}{}_{;\mu}{}^{\alpha} = 4\pi J^{\alpha}. \qquad (16.5)$$

However, if the flat-spacetime equation is written with two of its partial derivatives interchanged

$$-A^{\alpha,\mu}{}_{\mu} + A^{\mu,\alpha}{}_{\mu} = 4\pi J^{\alpha},$$

*For a discussion of quantum-mechanical factor-ordering problems, see, e.g., Merzbacher (1961), pp. 138–39 and 334–35; also Pauli (1934).

then its translation reads

$$-A^{\alpha;\mu}{}_\mu + A^{\mu;\alpha}{}_\mu = 4\pi J^\alpha,$$

which can be rewritten

$$-A^{\alpha;\mu}{}_\mu + A^\mu{}_{;\mu}{}^\alpha + R^\alpha{}_\mu A^\mu = 4\pi J^\alpha. \tag{16.5'}$$

(Ricci tensor appears as result of interchanging covariant derivatives; see exercise 16.3.) Which equation is correct—(16.5) or (16.5')? This question is nontrivial, just as the analogous factor-ordering problems of quantum theory are nontrivial. For rules-of-thumb that resolve this and most factor-ordering problems, see Box 16.1. These rules tell one that (16.5') is correct and (16.5) is wrong (see Box 16.1 and §22.4).

Exercise 16.3. NONCOMMUTATION OF COVARIANT DERIVATIVES

Let \mathbf{B} be a vector field and \mathbf{S} be a second-rank tensor field. Show that

$$B^\mu{}_{;\alpha\beta} = B^\mu{}_{;\beta\alpha} + R^\mu{}_{\nu\beta\alpha}B^\nu \tag{16.6a}$$

$$S^{\mu\nu}{}_{;\alpha\beta} = S^{\mu\nu}{}_{;\beta\alpha} + R^\mu{}_{\rho\beta\alpha}S^{\rho\nu} + R^\nu{}_{\rho\beta\alpha}S^{\mu\rho}. \tag{16.6b}$$

From equation (16.6a), show that

$$B^{\mu;\alpha}{}_\mu = B^\mu{}_{;\mu}{}^\alpha + R^\alpha{}_\mu B^\mu. \tag{16.6c}$$

[*Hint* for Track-1 calculation: Work in a local Lorentz frame, where $\Gamma^\alpha{}_{\beta\gamma} = 0$ but $\Gamma^\alpha{}_{\beta\gamma,\delta} \ne 0$; expand the lefthand side in terms of Christoffel symbols and partial derivatives; and use equation (8.44) for the Riemann tensor. An alternative Track-2 calculation notices that $\nabla_\beta \nabla_\alpha \mathbf{B}$ is *not* linear in \mathbf{e}_α, and that $B^\mu{}_{;\alpha\beta}$ are *not* its components; but, rather, that

$$B^\mu{}_{;\alpha\beta} \equiv \nabla\nabla\mathbf{B}(\boldsymbol{\omega}^\mu, \mathbf{e}_\alpha, \mathbf{e}_\beta). \tag{16.7}$$

$$\uparrow$$
$$\underset{}{\llcorner}\text{[Third-rank tensor]}$$

The calculation then proceeds as follows:

$$
\begin{aligned}
\langle \boldsymbol{\omega}^\mu, \nabla_\beta \nabla_\alpha \mathbf{B} \rangle &= \langle \boldsymbol{\omega}^\mu, \nabla_\beta (\mathbf{e}_\alpha \cdot \nabla\mathbf{B}) \rangle \\
&= \langle \boldsymbol{\omega}^\mu, (\nabla_\beta \mathbf{e}_\alpha) \cdot \nabla\mathbf{B} + \mathbf{e}_\alpha \cdot (\nabla_\beta \nabla\mathbf{B}) \rangle \\
&= \langle \boldsymbol{\omega}^\mu, \Gamma^\nu{}_{\alpha\beta} \mathbf{e}_\nu \cdot \nabla\mathbf{B} + \nabla\nabla\mathbf{B}(\dots, \mathbf{e}_\alpha, \mathbf{e}_\beta) \rangle \\
&= B^\mu{}_{;\nu} \Gamma^\nu{}_{\alpha\beta} + B^\mu{}_{;\alpha\beta}.
\end{aligned}
$$

Consequently

$$
\begin{aligned}
B^\mu{}_{;\alpha\beta} - B^\mu{}_{;\beta\alpha} &= \langle \boldsymbol{\omega}^\mu, [\nabla_\beta, \nabla_\alpha]\mathbf{B} \rangle - B^\mu{}_{;\nu}(\Gamma^\nu{}_{\alpha\beta} - \Gamma^\nu{}_{\beta\alpha}) \\
&= \langle \boldsymbol{\omega}^\mu, [\nabla_\beta, \nabla_\alpha]\mathbf{B} \rangle - \langle \boldsymbol{\omega}^\mu, \nabla_{(\nabla_\beta \mathbf{e}_\alpha - \nabla_\alpha \mathbf{e}_\beta)}\mathbf{B} \rangle \\
&= \langle \boldsymbol{\omega}^\mu, ([\nabla_\beta, \nabla_\alpha] - \nabla_{[\mathbf{e}_\beta, \mathbf{e}_\alpha]})\mathbf{B} \rangle = \langle \boldsymbol{\omega}^\mu, \mathscr{R}(\mathbf{e}_\beta, \mathbf{e}_\alpha)\mathbf{B} \rangle \\
&= R^\mu{}_{\nu\beta\alpha}B^\nu,
\end{aligned}
$$

in agreement with (16.6a). Note: because of slight ambiguity in the abstract notation, one must think carefully about each step in the above calculation. Component notation, by contrast, is completely unambiguous.]

(*continued on page 392*)

Box 16.1 FACTOR ORDERING AND CURVATURE COUPLING IN APPLICATIONS OF THE EQUIVALENCE PRINCIPLE

The Problem

In what order should derivatives be written when applying the "comma-goes-to-semicolon rule"? Interchanging derivatives makes no difference in flat spacetime, but in curved spacetime it produces terms that couple to curvature, e.g., $2B^\alpha{}_{;[\gamma\beta]} \equiv B^\alpha{}_{;\gamma\beta} - B^\alpha{}_{;\beta\gamma} = R^\alpha{}_{\mu\beta\gamma}B^\mu$ for any vector field (see exercise 16.3). Hence, the problem can be restated: *When must the comma-goes-to-semicolon rule be augmented by terms that couple to curvature?*

The Solution

There is no solution in general, but in most cases the following types of mathematical and physical reasoning resolve the problem unambiguously.

A. *Mathematically, curvature terms almost always arise from the noncommutation of covariant derivatives.* Consequently, one needs to worry about curvature terms in any equation that contains a double covariant derivative (e.g., $-A^{\alpha,\mu}{}_\mu + A^\mu{}_{,\mu}{}^\alpha = 4\pi J^\alpha$); or in any equation whose derivation from more fundamental laws involves double covariant derivatives (e.g. $\boldsymbol{\nabla}_u\boldsymbol{S} = 0$ in Example B(3) below). But one can ignore curvature coupling everywhere else (e.g., in Maxwell's first-order equations).

B. *Coupling to curvature can surely not occur without some physical reason.* Therefore, if one applies the comma-goes-to-semicolon rule only to physically measurable quantities (e.g., to the electromagnetic field, but not to the vector potential), one can "intuit" whether coupling to curvature is likely. *Examples:*

(1) *Local energy-momentum conservation.* A coupling to curvature in the equations $T^{\alpha\beta}{}_{;\beta} = 0$—e.g., replacing them by $T^{\alpha\beta}{}_{;\beta} = R^\alpha{}_{\beta\gamma\delta}T^{\beta\gamma}u^\delta$—would not make sense at all. In a local inertial frame such terms as $R^\alpha{}_{\beta\gamma\delta}T^{\beta\gamma}u^\delta$ would be interpreted as forces produced at a single point by curvature. But it should not be possible to feel curvature except over finite regions (geodesic deviation, etc.)! Put differently, the second derivatives of the gravitational potential (metric) can hardly produce net forces at a point; they should only produce tidal forces!

(2) *Maxwell's equations* for the electromagnetic field tensor. Here it would also be unnatural to introduce curvature terms. They would cause a breakdown in charge conservation, in the sense of termination of electric and magnetic field lines at points where there is curvature but no charge. To maintain charge conservation, one omits curvature coupling when one translates Maxwell's equations (3.32) and (3.36) into curved spacetime:

$$F^{\alpha\beta}{}_{;\beta} = 4\pi J^{\alpha}, \qquad F_{\alpha\beta;\gamma} + F_{\beta\gamma;\alpha} + F_{\gamma\alpha;\beta} = 0.$$

Moreover, one continues to regard $F_{\mu\nu}$ as arising from a vector potential by the curved-spacetime translation of (3.54′)

$$F_{\mu\nu} = A_{\nu;\mu} - A_{\mu;\nu}.$$

These points granted, one can verify that the second of Maxwell's equations is automatically satisfied, and verify also that the first is satisfied if and only if

$$-A^{\alpha;\mu}{}_{\mu} + A^{\mu}{}_{;\mu}{}^{\alpha} + R^{\alpha}{}_{\mu}A^{\mu} = 4\pi J^{\alpha}.$$

(See §22.4 for fuller discussion and derivation.)

(3) *Transport law for Earth's angular-momentum vector*. If the Earth were in flat spacetime, like any other isolated body it would parallel-transport its angular-momentum vector **S** along the straight world line of its center of mass, $\boldsymbol{\nabla}_u\boldsymbol{S} = 0$ ("conservation of angular momentum"). When translating this transport law into curved spacetime (where the Earth actually resides!), can one ignore curvature coupling? No! Spacetime curvatures due to the moon and sun produce tidal gravitational forces in the Earth; and because the Earth has an equatorial bulge, the tidal forces produce a nonzero net torque about the Earth's center of mass. (In Newtonian language: the piece of bulge nearest the Moon gets pulled with greater force, and hence greater torque, than the piece of bulge farthest from the Moon.) Thus, in curved spacetime one expects a transport law of the form

$$\boldsymbol{\nabla}_u\boldsymbol{S} = (\text{Riemann tensor}) \times (\text{Earth's quadrupole moment}).$$

This curvature-coupling torque produces a precession of the Earth's rotation axis through a full circle in the plane of the ecliptic once every 26,000 years ("general precession"; "precession of the equinoxes"; discovered by Hipparchus about 150 B.C.). The precise form of the curvature-coupling term is derived in exercise 16.4.

Exercise 16.4. PRECESSION OF THE EQUINOXES

(a) Show that the transport law for the Earth's intrinsic angular momentum vector S^α in curved spacetime is

$$\frac{DS^\alpha}{D\tau} = \epsilon^{\alpha\beta\gamma\delta} \mathit{f}_{\beta\mu} R^\mu{}_{\nu\gamma\zeta} u_\delta u^\nu u^\zeta. \tag{16.8}$$

Here $d/d\tau = \boldsymbol{u}$ is 4-velocity along the Earth's world line; $\mathit{f}_{\beta\mu}$ is the Earth's "reduced quadrupole moment" (trace-free part of second moment of mass distribution), defined in the Earth's local Lorentz frame by

$$\mathit{f}_{\hat{0}\hat{0}} = \mathit{f}_{\hat{0}\hat{j}} = 0, \qquad \mathit{f}_{\hat{j}\hat{k}} = \int \rho \left(x^{\hat{j}} x^{\hat{k}} - \frac{1}{3} \hat{r}^2 \delta_{jk} \right) d^3\hat{x}; \tag{16.9}$$

and $R^\mu{}_{\nu\gamma\zeta}$ is the Riemann curvature produced at the Earth's location by the moon, sun, and planets. [*Hint:* Derive this result in the Earth's local Lorentz frame, ignoring the spacetime curvature due to the Earth. (In this essentially Newtonian situation, curvature components $R^{\hat{j}}{}_{\hat{0}\hat{k}\hat{0}}$ due to the Earth, sun, moon, and planets superpose linearly; "gravity too weak to be nonlinear"). Integrate up the torque produced about the Earth's center of mass by tidal gravitational forces ("geodesic deviation"):

$$\left(\begin{array}{l} \text{acceleration at } x^{\hat{j}}, \text{ relative to center of mass } (x^{\hat{j}} = 0), \\ \text{produced by tidal gravitational forces but counterbalanced} \\ \text{in part by Earth's internal stresses} \end{array} \right)$$

$$= \left(\frac{d^2 x^{\hat{k}}}{d\hat{t}^2} \right)_{\substack{\text{geodesic} \\ \text{deviation}}} = -R^{\hat{k}}{}_{\hat{0}\hat{i}\hat{0}} x^{\hat{i}} \quad [\text{see equation (1.8')}];$$

$$\left(\begin{array}{l} \text{force per unit volume due to this} \\ \text{acceleration, relative to center} \\ \text{of mass} \end{array} \right)^{\hat{k}} = \overset{\text{mass density}}{\underset{\downarrow}{\rho}} \frac{d^2 x^{\hat{k}}}{d\hat{t}^2} = -\rho R^{\hat{k}}{}_{\hat{0}\hat{i}\hat{0}} x^{\hat{i}};$$

$$\left(\begin{array}{l} \text{torque per unit volume relative} \\ \text{to center of mass} \end{array} \right)_{\hat{i}} = \epsilon_{\hat{0}\hat{i}\hat{j}\hat{k}} x^{\hat{j}} (-\rho R^{\hat{k}}{}_{\hat{0}\hat{i}\hat{0}} x^{\hat{i}});$$

$$\left(\begin{array}{l} \text{total torque about center} \\ \text{of mass} \end{array} \right)_{\hat{i}} = \int [\epsilon_{\hat{0}\hat{i}\hat{j}\hat{k}} x^{\hat{j}} (-\rho R^{\hat{k}}{}_{\hat{0}\hat{i}\hat{0}} x^{\hat{i}})] d^3\hat{x}.$$

Put this expression into a form involving $\mathit{f}_{\hat{j}\hat{i}}$, equate it to $dS_{\hat{i}}/d\tau$, and then reexpress it in frame-independent, component notation. The result should be equation (16.8).]

(b) Rewrite equation (16.8) in the Earth's local Lorentz frame, using the equation

$$R^{\hat{j}}{}_{\hat{0}\hat{k}\hat{0}} = \partial^2 \Phi / \partial x^{\hat{j}} \, \partial x^{\hat{k}}$$

for the components of **Riemann** in terms of the Newtonian gravitational potential. (Newtonian approximation to Einstein theory. Track-2 readers have met this equation in Chapter 12; track-one readers will meet it in §17.4.)

(c) Calculate $dS^{\hat{j}}/d\hat{t}$ using Newton's theory of gravity from the beginning. The answer should be identical to that obtained in part (b) using Einstein's theory.

(d) Idealizing the moon and sun as point masses, calculate the long-term effect of the spacetime curvatures that they produce upon the Earth's rotation axis. Use the result of part (b), together with moderately accurate numerical values for the relevant solar-system parameters. [*Answer:* The Earth's rotation axis precesses relative to the axes of its local Lorentz frame ("precession of the equinoxes"; "general precession"); the precession period is 26,000 years. The details of the calculation will be found in any textbook on celestial mechanics.]

§16.4. THE RODS AND CLOCKS USED TO MEASURE SPACE AND TIME INTERVALS

Turn attention now from the laws of physics in the presence of gravity to the nature of the rods and clocks that must be used for measuring the length and time intervals appearing in those laws.

One need not—and indeed must not!—postulate that proper length s is measured by a certain type of rod (e.g., platinum meter stick), or that proper time τ is measured by a certain type of clock (e.g., hydrogen-maser clock). Rather, one must ask the laws of physics themselves what types of rods and clocks will do the job. Put differently, one *defines an "ideal" rod or clock* to be one which measures proper length as given by $ds = (g_{\alpha\beta}\,dx^\alpha\,dx^\beta)^{1/2}$ or proper time as given by $d\tau = (-g_{\alpha\beta}\,dx^\alpha\,dx^\beta)^{1/2}$ (the kind of clock to which one was led by physical arguments in §1.5). One must then determine the accuracy to which a given rod or clock is ideal under given circumstances by using the laws of physics to analyze its behavior.

"Ideal" rods and clocks defined

How ideal are real clocks?

As an obvious example, consider a pendulum clock. If it is placed at rest on the Earth's surface, if it is tiny enough that redshift effects from one end to the other and time dilation effects due to its swinging velocity are negligible, and if the accuracy one demands is small enough that time variations in the local gravitational acceleration due to Earth tides can be ignored, then the laws of physics report (Box 16.2) that the pendulum clock is "ideal." However, in any other context (e.g., on a rocket journey to the moon), a pendulum clock should be far from ideal. Wildly changing accelerations, or no acceleration at all, will make it worthless!

(1) pendulum clocks

Of greater interest are atomic and nuclear clocks of various sorts. Such a clock is analyzed most easily if it is freely falling. One can then study it in its local Lorentz rest frame, using the standard equations of quantum theory; and, of course, one will find that it measures proper time to within the precision ($\Delta t/t \sim 10^{-9}$ to 10^{-14}) of the technology used in its construction. However, one rarely permits his atomic clock to fall freely. (The impact with the Earth's surface can be expensive!) Nevertheless, even when accelerated at "1 g" = 980 cm/sec^2 on the Earth's surface, and even when accelerated at "2 g" in an airliner trying to avoid a midair collision (Box 16.3), an atomic clock—if built solidly—will still measure proper time $d\tau = (-g_{\alpha\beta}\,dx^\alpha\,dx^\beta)^{1/2}$ along its world line to nearly the same accuracy as if it were freely falling. To discover this one can perform an experiment. Alternatively, one can analyze the clock in its own "proper reference frame" (§13.6), with Fermi-Walker-transported basis vectors, using the standard local Lorentz laws of quantum mechanics as adapted to accelerated frames (local Lorentz laws plus an "inertial force," which can be treated as due to a potential with a uniform gradient.

(2) atomic clocks

Of course, any clock has a "breaking point," beyond which it will cease to function properly (Box 16.3). But that breaking point depends entirely on the construction of the clock—and not at all on any "universal influence of acceleration on the march of time." Velocity produces a universal time dilation; acceleration does not.

The aging of the human body is governed by the same electromagnetic and quantum-mechanical laws as govern the periodicities and level transitions in atoms and molecules. Consequently, aging, like atomic processes, is tied to proper time

(3) human clocks

(continued on page 396)

**Box 16.2 PROOF THAT A PENDULUM CLOCK AT REST
ON THE EARTH'S SURFACE IS IDEAL**

That is, a proof that it measures the interval $d\tau = (-g_{\alpha\beta}\,dx^\alpha\,dx^\beta)^{1/2}$.

A. Constraint on the Pendulum

It must be so small that it cannot couple to the spacetime curvature—i.e., so small that the Earth's gravitational field looks uniform in its neighborhood—and that the velocity of its ball is totally negligible compared to the speed of light.

B. Coordinate System and Metric

(1) General coordinate system: because the Earth's field is nearly Newtonian, one can introduce the coordinates of "linearized theory" (§18.4; one must take this on faith until one reaches that point) in which

$$ds^2 = -(1 + 2\Phi)\,dt'^2 + (1 - 2\Phi)(dx'^2 + dy'^2 + dz'^2),$$

where Φ is the Newtonian potential.

(2) Put the origin of coordinates at the pendulum's equilibrium position, and orient the x',z'-plane so the pendulum swings in it.

(3) Renormalize the coordinates so they measure proper length and proper time at the equilibrium position

$$t = [1 + 2\Phi(0)]^{1/2}t', \qquad x^j = [1 - 2\Phi(0)]^{1/2}x^{j'}.$$

Then near the pendulum (inhomogeneities in the field neglected!)

$$\Phi = \Phi(0) + gz, \qquad g = \text{``acceleration of gravity,''} \tag{1}$$

$$ds^2 = -(1 + 2gz)\,dt^2 + (1 - 2gz)(dx^2 + dy^2 + dz^2). \tag{2}$$

C. Analysis of Pendulum Motion

(1) Put the total mass m of the pendulum in its ball (negligible mass in its rod). Let its rod have proper length l.

(2) Calculate the 4-acceleration $\boldsymbol{a} = \boldsymbol{\nabla_u u}$ of the pendulum's ball in terms of $d^2 x^\alpha / dt^2$, using the velocity condition $v <\!\!<\!\!< 1$ and $dt/d\tau \approx 1$:

$$
\begin{aligned}
a^x &= d^2 x/d\tau^2 + \Gamma^x{}_{00}(dt/d\tau)^2 = d^2 x/dt^2 + \Gamma^x{}_{00} = d^2 x/dt^2, \\
a^z &= d^2 z/d\tau^2 + \Gamma^z{}_{00}(dt/d\tau)^2 = d^2 z/dt^2 + \Gamma^z{}_{00} = d^2 z/dt^2 + g.
\end{aligned}
\tag{3}
$$

(3) This 4-acceleration must be produced by the forces in the rod, and must be directed up the rod so that (for $x \ll l$ so $g \gg d^2 z/dt^2$)

$$d^2 x/dt^2 = a^x = -(x/l)a^z = -(g/l)x. \tag{4}$$

(4) Solve this differential equation to obtain

$$x = x_0 \cos (t \sqrt{g/l}). \tag{5}$$

(5) Thus conclude that the pendulum is periodic in t, which is proper time at the ball's equilibrium position (see equation 2). This means that *the pendulum is an ideal clock when it is at rest on the Earth's surface.*

Note: The above analysis ignores the Earth's rotation; for an alternative analysis including rotation, one can perform a similar calculation at the origin of the pendulum's "proper reference frame" [§13.6; line element (13.71)]. The answer is the same; but now "g" is a superposition of the "gravitational acceleration," and the "centrifugal acceleration produced by Earth's rotation."

Box 16.3 RESPONSE OF CLOCKS TO ACCELERATION AND TO TIDAL GRAVITATIONAL FORCES

Consider an atomic clock with frequency stabilized by some atomic or molecular process—for example, fixed by the "umbrella vibrations" of ammonia molecules [see Feynman *et. al.* (1964)]. When subjected to sufficiently strong accelerations or tidal forces, such a clock will cease to measure proper time with its normal precision. Two types of effects could lead to such departures from "ideality":

A. Influence of the acceleration or tidal force on the atomic process that provides the frequency stability. Example: If tidal forces are significant over distances of a few angstroms (e.g., near a spacetime "singularity" terminating gravitational collapse), then they can and will deform an ammonia molecule and destroy the regularity of its umbrella vibrations, thereby making useless *any* ammonia atomic clock, no matter how constructed. Similarly, if an ammonia molecule is subjected to accelerations of magnitude comparable to its internal atomic accelerations $(a \sim 10^{12}\text{"}g\text{"} \sim 10^{15} \text{ cm/sec}^2)$, which change in times of the order of the "umbrella" vibration period, then it must cease to vibrate regularly, and any clock based on its vibrations must fail. Such limits of principle on the ideality of a clock will vary from one atomic process to another. However, they are far from being a limiting factor on clock construction in 1973. Much more important today is:

B. Influence of the acceleration or tidal force on the macroscopic structure of the clock—a structure dictated by current technology. The crystal oscillator, which produces the periodic signal output, must be locked to the regulating atomic process in some way. The lock will be disturbed by moderate accelerations. The toughest task for the manufacturer of aircraft clocks is to guarantee that precise locking will be maintained, even when the aircraft is maneuvering desperately to avoid collision with another aircraft or with a missile. In 1972 a solidly built rubidium clock will maintain its lock, with no apparent degradation of stability

$$[\Delta t/t \sim 10^{-12}(1 \text{ sec}/t)^{1/2} \text{ for } 1 \text{ sec} \lesssim t \lesssim 10^3 \text{ sec}]$$

under steady-state accelerations up to 50 "g" or more. But, because of the finite bandwidth of the lock loop (typically $\Delta\nu \sim 20$ to 50 Hz), sudden changes in acceleration will temporarily break the lock, degrading the clock stability to that of the unlocked crystal oscillator—for which an acceleration a produces a change in frequency of about $(a/1 \text{ "}g\text{"}) \times 10^{-9}$. But the lock to the rubidium standard is restored quickly $(\delta t \sim 1/\Delta\nu)$, bringing the clock back to its normal highly stable performance.*

Tidal forces are so small in the solar system that the clock manufacturer can ignore them. However, a 1973 atomic clock, subjected to the tidal accelerations near a spacetime singularity, should break the "lock" to its atomic process long before the tidal forces become strong enough to influence the atomic process itself.

*For this information on the response of rubidium clocks to acceleration, we thank H. P. Stratemeyer of General Radio Company, Concord, Massachusetts.

as governed by the metric—though, of course, it is also tied to other things, such as cigarette smoking.

In principle, one can build ideal rods and clocks from the geodesic world lines of freely falling test particles and photons. (See Box 16.4.) In other words, spacetime has its own rods and clocks built into itself, even when matter and nongravitational fields are absent!

Ideal rods and clocks constructed from geodesic world lines

Box 16.4 IDEAL RODS AND CLOCKS BUILT FROM GEODESIC WORLD LINES*

The Standard Interval. A specific timelike interval—the interval between two particular neighboring events \mathcal{A} and \mathcal{B}—is chosen as the standard interval, and is assigned unit length. It is used to calibrate a huge set of geodesic clocks that pass through \mathcal{A}.

Each *geodesic clock* is constructed and calibrated as follows:

(1) A timelike geodesic $\mathcal{A}\mathcal{C}$ (path of freely falling particle) passes through \mathcal{A}.

(2) A neighboring world line, everywhere parallel to $\mathcal{A}\mathcal{C}$ (and thus not a geodesic), is constructed by the method of Schild's ladder (Box 10.2), which relies only on geodesics.

(3) Light rays (null geodesics) bounce back and forth between these parallel world lines; each round trip constitutes one "tick."

(4) The proper time lapse, τ_0, between ticks is related to the interval $\mathcal{A}\mathcal{B}$ by

$$-1 \equiv (\mathcal{A}\mathcal{B})^2 = -(N_1\tau_0)(N_2\tau_0),$$

where N_1 and N_2 are the number of ticks between the events shown in the diagrams. [*Proof:* see diagram at right.]

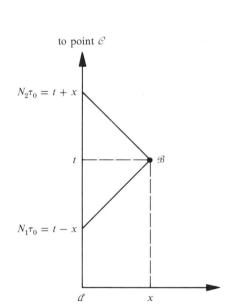

In local Lorentz rest frame of geodesic clock:

$$(N_1\tau_0)(N_2\tau_0) = (t - x)(t + x)$$
$$= t^2 - x^2 = -(\mathcal{A}\mathcal{B})^2$$

Spacetime is filled with such geodesic clocks. Those that pass through \mathcal{A} are calibrated as above against the standard interval $\mathcal{A}\mathcal{B}$, and are used subsequently to calibrate all other clocks they meet.

*Based on Marzke and Wheeler (1964).

Box 16.4 (*continued*)

Any interval $\mathscr{P}\mathscr{Q}$ along the world line of a geodesic clock can be measured by the same method as was used in calibration. The interval $\mathscr{P}\mathscr{Q}$ can be time-like, spacelike, or null; its squared length in all three cases will be

$$(\mathscr{P}\mathscr{Q})^2 = -(N_3\tau_0)(N_4\tau_0)$$

To achieve a precision of measurement good to one part in N, where N is some large number, take two precautions:

(1) Demand that the intervals $\mathscr{A}\mathscr{B}$ and $\mathscr{P}\mathscr{Q}$ be sufficiently small compared to the scale of curvature of spacetime; or specifically,

$$R^{(AB)}(\mathscr{A}\mathscr{B})^2 \ll 1/N$$

and

$$R^{(PQ)}(\mathscr{P}\mathscr{Q})^2 \ll 1/N,$$

where $R^{(AB)}$ and $R^{(PQ)}$ are the largest relevant components of the curvature tensor in the two regions in question.

(2) Demand that the time scale, τ_0, of the geodesic clocks employed be small compared to $\mathscr{A}\mathscr{B}$ and $\mathscr{P}\mathscr{Q}$ individually; thus,

$$\tau_0 \ll \mathscr{A}\mathscr{B}/N,$$
$$\tau_0 \ll \mathscr{P}\mathscr{Q}/N.$$

The Einstein principle that spacetime is described by Riemannian geometry exposes itself to destruction by a "thousand" tests. Thus, from the fiducial interval, $\mathscr{A}\mathscr{B}$, to the interval under measurement, $\mathscr{P}\mathscr{Q}$, there are a "score" of routes of intercomparison, all of which must give the same value for the ratio $\mathscr{P}\mathscr{Q}/\mathscr{A}\mathscr{B}$. Moreover, one can easily select out "fifty" intervals $\mathscr{P}\mathscr{Q}$ to which the same kind of test can be applied. Such tests are not all items for the future.

Some 5×10^9 years ago, electrons arrived by different routes at a common location, a given atom of iron in the core of the earth. This iron atom does not collapse. The Pauli principle of

exclusion keeps the electrons from all falling into the K-orbit. The Pauli principle would not apply if the electrons were not identical or nearly so. From this circumstance it would appear possible to draw an important conclusion (Marzke and Wheeler). With each electron is associated a standard length, its Compton wavelength, \hbar/mc. If these lengths had started different, or changed by different amounts along the different routes, and if the resulting difference in properties were as great as one part in

$$\sim(5 \times 10^9 \text{ yr}) \times (3 \times 10^7 \text{ sec/yr})$$
$$\times (5 \times 10^{18} \text{ rev/sec}) \sim 10^{36},$$

by now this difference would have shown up, the varied electrons would have fallen into the K-orbit, and the earth would have collapsed, contrary to observation.

The Marzke-Wheeler construction expresses an arbitrary small interval $\mathcal{P}\mathcal{Q}$, anywhere in spacetime, in terms of the fiducial interval $\mathcal{A}\mathcal{B}$, an interval which itself may be taken for definiteness to be the "geometrodynamic standard centimeter" of §1.5. This construction thus gives a vivid meaning to the idea of Riemannian geometry.

The M-W construction makes no appeal what-soever to rods and clocks of atomic constitution. This circumstance is significant for the following reasons. The length of the usual platinum meter stick is some multiple, $N_1(\hbar^2/me^2)$, of the Bohr atomic radius. Similarly, the wavelength of the Kr^{86} line is some multiple, $N_2(\hbar c/e^2)(\hbar^2/me^2)$, of a second basic length that depends on the atomic constants in quite a different way. Thus, if there is any change with time in the dimensionless ratio $\hbar c/e^2 = 137.038$, one or the other or both of these atomic standards of length must get out of kilter with the geometrodynamic standard centimeter. In this case, general relativity says, "Stick to the geometrodynamic standard centimeter."

Hermann Weyl at first thought that one could carry out the comparison of lengths by light rays alone, but H. A. Lorentz pointed out that one can dispense with the geodesics neither of test particles nor of light rays in the measurement process, the construction for which, however, neither Weyl nor Lorentz supplied [literature in Marzke and Wheeler (1964)]. Ehlers, Pirani, and Schild (1972) have given a deeper analysis of the separate parts played in the measurement process by the affine connection, by the conformal part of the metric, and by the full metric.

§16.5. THE MEASUREMENT OF THE GRAVITATIONAL FIELD

"I know how to measure the electromagnetic field using test charges; what is the analogous procedure for measuring the gravitational field?" This question has, at the same time, many answers and none.

It has no answers because nowhere has a precise definition of the term "gravitational field" been given—nor will one be given. Many different mathematical entities are associated with gravitation: the metric, the Riemann curvature tensor, the Ricci curvature tensor, the curvature scalar, the covariant derivative, the connection coefficients, etc. Each of these plays an important role in gravitation theory, and none is so much more central than the others that it deserves the name "gravitational field." Thus it is that throughout this book the terms "gravitational field" and "gravity" refer in a vague, collective sort of way to all of these entities. Another, equivalent term used for them is the "geometry of spacetime."

The many faces of gravity, and how one measures them

To "measure the gravitational field," then, means to "explore experimentally various properties of the spacetime geometry." One makes different kinds of measurements, depending on which geometric property of spacetime one is interested in. However, all such measurements must involve a scrutiny of the effects of the spacetime geometry (i.e., of gravity) on particles, on matter, or on nongravitational fields.

For example, to "measure" the metric near a given event, one typically lays out a latticework of rods and clocks (local orthonormal frame, small enough that curvature effects are negligible), and uses it to determine the interval between neighboring events. To measure the Riemann curvature tensor near an event, one typically studies the geodesic deviation (relative accelerations) that curvature produces between the world lines of a variety of neighboring test particles; alternatively, one makes measurements with a "gravity gradiometer" (Box 16.5) if the curvature is static or slowly varying; or with a gravitational wave antenna (Chapter 37) if the curvature fluctuates rapidly. To study the large-scale curvature of spacetime, one examines large-scale effects of gravity, such as the orbits of planets and satellites, or the bending of light by the sun's gravitational field.

But whatever aspect of gravity one measures, and however one measures it, one is studying the geometry of spacetime.

EXERCISE

Exercise 16.5. GRAVITY GRADIOMETER

The gravity gradiometer of Box 16.5 moves through curved spacetime along an accelerated world line. Calculate the amplitude and phase of oscillation of one arm of the gradiometer relative to the other. [Hint: Perform the calculation in the gradiometer's "proper reference frame" (§13.6), with Fermi-Walker-transported basis vectors. Use, as the equation for the relative angular acceleration of the two arms,

$$2m\ell^2(\ddot{\alpha} + \dot{\alpha}/\tau_0 + \omega_0^2\alpha) = \begin{pmatrix} \text{Driving torque produced by} \\ \text{Riemann curvature} \end{pmatrix},$$

where

$$2m\ell^2 = \text{(moment of inertia of one arm)},$$
$$\alpha = \text{(angular displacement of one arm from equilibrium)},$$
$$\frac{\pi}{2} + 2\alpha = \text{(angular separation of the two arms)},$$
$$2m\ell^2\omega_0^2 = \text{(torsional spring constant)},$$
$$\omega_0 = \text{(angular frequency of free vibrations)},$$
$$\tau_0 = \text{(decay time for free vibrations to damp out due to}$$
$$\text{internal frictional forces).}$$

If $\boldsymbol{\xi}$ is the vector from the center of mass of the gradiometer to mass 1, then one has

$$\begin{pmatrix} \text{curvature-produced} \\ \text{acceleration of mass 1} \\ \text{relative to center of} \\ \text{gradiometer} \end{pmatrix}_{\hat{k}} = \left(\frac{D^2\xi_{\hat{k}}}{d\tau^2}\right)_{\substack{\text{geodesic} \\ \text{deviation}}} = -R_{\hat{k}\hat{0}\hat{l}\hat{0}}\xi_{\hat{l}};$$

(continued on page 403)

Box 16.5 GRAVITY GRADIOMETER FOR MEASURING THE RIEMANN CURVATURE OF SPACETIME

This gravity gradiometer was designed and built by Robert M. Forward and his colleagues at Hughes Research Laboratories, Malibu, California. It measures the Riemann curvature of spacetime produced by nearby masses. By flying a more advanced version of such a gradiometer in an airplane above the Earth's surface, one should be able to measure subsurface mass variations due to varying geological structure. In an Earth-orbiting satellite, such a gradiometer could measure the gravitational multipole moments of the Earth. Technical details of the gradiometer are spelled out in the papers of Forward (1972), and Bell, Forward, and Williams (1970). The principles of its operation are outlined below.

The gradiometer consists of two orthogonal arms with masses m on their ends, connected at their centers by a torsional spring. When the arms are twisted out of orthogonal alignment, they oscillate. A piezoelectric strain transducer is used to measure the oscillation amplitude.

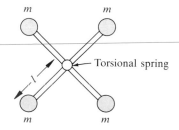

Box 16.5 (continued)

When placed near an external mass, M, the gradiometer experiences a torque: because of the gradient in the gravitational field of M (i.e., because of the spacetime curvature produced by M), the Newtonian forces F_1 and F_2 are greater than F_3 and F_4; so a net torque pulls masses 1 and 2 toward each other, and 3 and 4 toward each other. [*Note:* the forces F_1, F_2, F_3, F_4 depend on whether the gradiometer is in free fall (geodesic motion; $\nabla_u u = 0$) or is moving on an accelerated world line. But the net torque is unaffected by acceleration; acceleration produces equal Newtonian forces on all four masses, with zero net torque.]

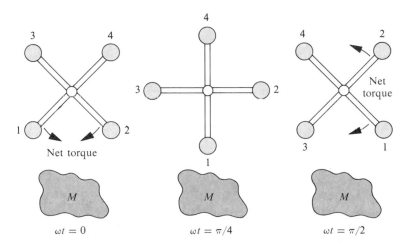

$\omega t = 0$ $\omega t = \pi/4$ $\omega t = \pi/2$

When in operation the gradiometer rotates with angular velocity ω about its center. As it rotates, the torques on its arms oscillate:

at $\omega t = 0$ net torque pushes 1 and 2 toward each other;

at $\omega t = \pi/4$ net torque is zero;

at $\omega t = \pi/2$ net torque pushes 1 and 2 away from each other.

The angular frequency of the oscillating torque is 2ω. If 2ω is set equal to $\omega_0 \equiv$ (natural oscillation frequency of the arms), the oscillating torque drives the arms into resonant oscillation. The resulting oscillation amplitude, in the 1970 prototype of the gradiometer, was easily detectable for gravity gradients (Riemann curvatures) of magnitude

$$\gtrsim 0.0002 \left[\frac{2(\text{mass of earth})}{(\text{radius of earth})^3} \right]$$

$$\sim 1 \times 10^{-30} \text{ cm}^{-2} \sim .01 \text{ g/cm}^3$$

$$\sim \begin{bmatrix} \text{Riemann curvature produced by a two-kilometer} \\ \text{high mountain, idealized as a two-kilometer high} \\ \text{cube, at a distance of 15 kilometers. (Neglected} \\ \text{in this idealization are isostacy and any lowering} \\ \text{of density of Earth's crust in regions of mountain} \\ \text{uplift.)} \end{bmatrix}$$

For a mathematical analysis of the gradiometer, see exercise 16.5.

$$\begin{pmatrix} \text{torque acting on mass 1} \\ \text{relative to center of} \\ \text{gradiometer} \end{pmatrix}_{\hat{i}} = \epsilon_{ijk}\xi_j(-mR_{\hat{k}\hat{0}\hat{l}\hat{0}}\xi_l).$$

The torque on mass 4 is identical to this (replace $\boldsymbol{\xi}$ by $-\boldsymbol{\xi}$), so the total torque on arm 1-4 is twice this. The components $R_{\hat{k}\hat{0}\hat{l}\hat{0}}$ of **Riemann** can be regarded as components of a 3×3 symmetric matrix. By appropriate orientation of the reference frame's spatial axes (orientation along "principal axes" of $R_{\hat{k}\hat{0}\hat{l}\hat{0}}$), one can make $R_{\hat{k}\hat{0}\hat{l}\hat{0}}$ diagonal at some initial moment of time

$$R_{\hat{x}\hat{0}\hat{x}\hat{0}} \neq 0,\ R_{\hat{y}\hat{0}\hat{y}\hat{0}} \neq 0,\ R_{\hat{z}\hat{0}\hat{z}\hat{0}} \neq 0,\ \text{all others vanish.}$$

Assume that **Riemann** changes sufficiently slowly along the gradiometer's world line that throughout the experiment $R_{\hat{j}\hat{0}\hat{k}\hat{0}}$ remains diagonal and constant. For simplicity, place the gradiometer in the \hat{x}, \hat{y}-plane, so it rotates about the \hat{z} axis with angular velocity $\omega \approx \tfrac{1}{2}\omega_0$:

$$\begin{pmatrix} \text{Angle of arm 1-4} \\ \text{relative to } \hat{x} \text{ axis} \end{pmatrix} = \omega t.$$

Show that the resultant equation of oscillation is

$$\ddot{\alpha} + \dot{\alpha}/\tau_0 + \omega_0^2\alpha = \frac{1}{2}(R_{\hat{x}\hat{0}\hat{x}\hat{0}} - R_{\hat{y}\hat{0}\hat{y}\hat{0}})\sin 2\omega t;$$

and that the steady-state oscillations are

$$\alpha = Im\left\{ \frac{\dfrac{1}{2}(R_{\hat{x}\hat{0}\hat{x}\hat{0}} - R_{\hat{y}\hat{0}\hat{y}\hat{0}})}{2\omega_0(\omega_0 - 2\omega + i/2\tau_0)} e^{i2\omega t} \right\}.$$

Thus, for fixed ω (e.g., $2\omega = \omega_0$), by measuring the amplitude and phase of the oscillations, one can learn the magnitude and sign of $R_{\hat{x}\hat{0}\hat{x}\hat{0}} - R_{\hat{y}\hat{0}\hat{y}\hat{0}}$. The other differences, $R_{\hat{y}\hat{0}\hat{y}\hat{0}} - R_{\hat{z}\hat{0}\hat{z}\hat{0}}$ and $R_{\hat{z}\hat{0}\hat{z}\hat{0}} - R_{\hat{x}\hat{0}\hat{x}\hat{0}}$ can be measured by placing the gradiometer's rotation axis along the \hat{x} and \hat{y} axes, respectively.]

CHAPTER **17**

HOW MASS-ENERGY
GENERATES CURVATURE

The physical world is represented as a four-dimensional continuum. If in this I adopt a Riemannian metric, and look for the simplest laws which such a metric can satisfy, I arrive at the relativistic gravitation theory of empty space. If I adopt in this space a vector field, or the antisymmetrical tensor field derived from it, and if I look for the simplest laws which such a field can satisfy, I arrive at the Maxwell equations for free space. . . . at any given moment, out of all conceivable constructions, a single one has always proved itself absolutely superior to all the rest . . .

ALBERT EINSTEIN (1934, p. 18)

§17.1. AUTOMATIC CONSERVATION OF THE SOURCE AS THE CENTRAL IDEA IN THE FORMULATION OF THE FIELD EQUATION

This section derives the "Einstein field equation"

Turn now from the response of matter to geometry (motion of a neutral test particle on a geodesic; "comma-goes-to-semicolon rule" for the dynamics of matter and fields), and analyze the response of geometry to matter.

Mass is the source of gravity. The density of mass-energy as measured by any observer with 4-velocity \boldsymbol{u} is

$$\rho = \boldsymbol{u} \cdot \boldsymbol{T} \cdot \boldsymbol{u} = u^\alpha T_{\alpha\beta} u^\beta. \tag{17.1}$$

Therefore the stress-energy tensor \boldsymbol{T} is the frame-independent "geometric object" that must act as the source of gravity.

This source, this geometric object, is not an arbitrary symmetric tensor. It must have zero divergence

$$\boldsymbol{\nabla} \cdot \boldsymbol{T} = 0,$$ (17.2)

because only so can the law of conservation of momentum-energy be upheld.

Place this source, \boldsymbol{T}, on the righthand side of the equation for the generation of gravity. On the lefthand side will stand a geometric object that characterizes gravity. That object, like \boldsymbol{T}, must be a symmetric, divergence-free tensor; and if it is to characterize gravity, it must be built out of the geometry of spacetime and nothing but that geometry. Give this object the name "Einstein tensor" and denote it by \boldsymbol{G}, so that the equation for the generation of gravity reads

Equation describing how matter generates gravity must have form $\boldsymbol{G} = \kappa \boldsymbol{T}$, where \boldsymbol{T} is stress-energy tensor

$$\boldsymbol{G} = \kappa \boldsymbol{T}.$$ (17.3)

$$\left[\begin{array}{l}\text{proportionality factor;} \\ \text{to be evaluated later}\end{array}\right]$$

(Do not assume that \boldsymbol{G} is the same Einstein tensor as was encountered in Chapters 8, 13, 14, and 15; that will be proved below!)

The vanishing of the divergence $\boldsymbol{\nabla} \cdot \boldsymbol{G}$ is not to be regarded as a consequence of $\boldsymbol{\nabla} \cdot \boldsymbol{T} = 0$. Rather, the obedience of all matter and fields to the conservation law $\boldsymbol{\nabla} \cdot \boldsymbol{T} = 0$ is to be regarded (1) as a consequence of the way [equation (17.3)] they are wired into the geometry of spacetime, and therefore (2) as required and enforced by an *automatic* conservation law, or *identity*, that holds for any smooth Riemannian spacetime whatsoever, physical or not: $\boldsymbol{\nabla} \cdot \boldsymbol{G} \equiv 0$. (See Chapter 15 for a fuller discussion and §17.2 below for a fuller justification.) Accordingly, look for a symmetric tensor \boldsymbol{G} that is an "automatically conserved measure of the curvature of spacetime" in the following sense:

Properties that the tensor \boldsymbol{G} must have

(1) \boldsymbol{G} vanishes when spacetime is flat.
(2) \boldsymbol{G} is constructed from the Riemann curvature tensor and the metric, and from nothing else.
(3) \boldsymbol{G} is distinguished from other tensors which can be built from *Riemann* and *g* by the demands (i) that it be linear in *Riemann*, as befits any natural measure of curvature; (ii) that, like \boldsymbol{T}, it be symmetric and of second rank; and (iii) that it have an automatically vanishing divergence,

$$\boldsymbol{\nabla} \cdot \boldsymbol{G} \equiv 0.$$ (17.4)

Apart from a multiplicative constant, there is only one tensor (exercise 17.1) that satisfies these requirements of being an automatically conserved, second-rank tensor, linear in the curvature, and of vanishing when spacetime is flat. It is the Einstein curvature tensor, \boldsymbol{G}, expressed in Chapter 8 in terms of the Ricci curvature tensor:

Proof that \boldsymbol{G} must be the Einstein curvature tensor of Chapter 8

$$R_{\mu\nu} = R^{\alpha}{}_{\mu\alpha\nu},$$
$$G_{\mu\nu} = R_{\mu\nu} - \frac{1}{2}g_{\mu\nu}R.$$ (17.5)

This quantity was given vivid meaning in Chapter 15 as the "moment of rotation of the curvature" or, more simply, the "moment of rotation," constructed by taking the double-dual

$$\mathbf{G} = {}^{*}\mathbf{Riemann}^{*} \tag{17.6a}$$

of the Riemann curvature tensor, and then contracting this double dual,

$$G_{\mu\nu} = \mathcal{G}^{\alpha}{}_{\mu\alpha\nu}. \tag{17.6b}$$

In Chapter 15 the vanishing of $\boldsymbol{\nabla} \cdot \mathbf{G}$ was shown to follow as a consequence of the elementary principle of topology that "the boundary of a boundary is zero."

Evaluation of κ (in $\mathbf{G} = \kappa\mathbf{T}$) by comparing with Newtonian theory of gravity

To evaluate the proportionality constant κ in the "Einstein field equation" $\mathbf{G} = \kappa\mathbf{T}$, one can compare with the well-tested Newtonian theory of gravity. To facilitate the comparison, examine the relative acceleration (geodesic deviation) of particles that fall down a pipe inserted into an idealized Earth of uniform density ρ (Figure 1.12). According to Newton, the relative acceleration is governed by the density; according to Einstein, it is governed by the Riemann curvature of spacetime. Direct comparison of the Newtonian and Einstein predictions using Newtonian coordinates (where $g_{\mu\nu} \approx \eta_{\mu\nu}$) reveals the relation

$$R_{00} \equiv R^{\alpha}{}_{0\alpha0} = 4\pi\rho. \tag{17.7}$$

(See §1.7 for details of the derivation; see Chapter 12 for extensive discussion of Newtonian gravity using this equation.) When applied to the Earth's interior, the Einstein field equation $\mathbf{G} = \kappa\mathbf{T}$ must thus reduce to $R_{00} = 4\pi\rho$. In component form, the Einstein field equation reads

$$G_{\mu\nu} = R_{\mu\nu} - \frac{1}{2} g_{\mu\nu} R = \kappa T_{\mu\nu}.$$

Its trace reads

$$-R = R - 2R = \kappa T.$$

In consequence, it predicts

$$R_{00} = \frac{1}{2} g_{00} R + \kappa T_{00} = \frac{1}{2}\kappa(2 T_{00} - \underbrace{g_{00}}_{-1} T)$$

$$= \frac{1}{2} \kappa[2 T_{00} + (T^{0}{}_{0} + T^{j}{}_{j})]$$

$$= \frac{1}{2}\kappa(T_{00} + T^{j}{}_{j}),$$

which reduces to

$$R_{00} = \frac{1}{2}\kappa\rho \tag{17.8}$$

when one recalls that for the Earth—as for any nearly Newtonian system—the stresses T_{jk} are very small compared to the density of mass-energy $T_{00} = \rho$:

$$\frac{|T_{jk}|}{T_{00}} \sim \frac{\text{pressure}}{\text{density}} \sim \frac{dp}{d\rho} \sim (\text{velocity of sound})^2 \ll 1.$$

The equation $R_{00} = 4\pi\rho$ (derived by comparing relative accelerations in the Newton and Einstein theories) and the equation $R_{00} = \frac{1}{2}\kappa\rho$ (derived directly from the Einstein field equation) can agree only if the proportionality constant κ is 8π.

Thus, the Einstein field equation, describing the generation of curvature by mass-energy, must read

$$\boldsymbol{G} = 8\pi\boldsymbol{T}. \tag{17.9}$$

<div style="float:right">Result: "Einstein field equation" $\boldsymbol{G} = 8\pi\boldsymbol{T}$</div>

The lefthand side ("curvature") has units cm^{-2}, since a curvature tensor is a linear machine into which one inserts a displacement (units: cm) and from which one gets a relative acceleration (units: $\text{cm/sec}^2 \sim \text{cm/cm}^2 \sim \text{cm}^{-1}$). The right-hand side also has dimensions cm^{-2}, since it is a linear machine into which one inserts 4-velocity (dimensionless) and from which one gets mass density [units: $\text{g/cm}^3 \sim \text{cm/cm}^3 \sim \text{cm}^{-2}$; recall from equation (1.12) and Box 1.8 that $1\text{g} = (1\text{g}) \times (G/c^2) = (1\text{g}) \times (0.742 \times 10^{-28}\ \text{cm/g}) = 0.742 \times 10^{-28}\ \text{cm}$].

This concludes the simplest derivation of Einstein's field equation that has come to hand, and establishes its correspondence with the Newtonian theory of gravity under Newtonian conditions. That correspondence had to be worked out to determine the factor $\kappa = 8\pi$ on the righthand side of (17.9). Apart from this factor, the central point in the derivation was the demand for, and the existence of, a unique tensorial measure of curvature \boldsymbol{G} with an identically vanishing divergence.

EXERCISES

Exercise 17.1. UNIQUENESS OF THE EINSTEIN TENSOR

(a) Show that the most general second-rank, symmetric tensor constructable from **Riemann** and **g**, and linear in **Riemann**, is

$$aR_{\alpha\beta} + bRg_{\alpha\beta} + \Lambda g_{\alpha\beta}$$
$$= aR^{\mu}{}_{\alpha\mu\beta} + bR^{\mu\nu}{}_{\mu\nu}g_{\alpha\beta} + \Lambda g_{\alpha\beta}, \tag{17.10}$$

where a, b, and Λ are constants.

(b) Show that this tensor has an automatically vanishing divergence if and only if $b = -\frac{1}{2}a$.

(c) Show that, in addition, this tensor vanishes in flat spacetime, if and only if $\Lambda = 0$—i.e., if and only if it is a multiple of the Einstein tensor $G_{\alpha\beta} = R_{\alpha\beta} - \frac{1}{2}Rg_{\alpha\beta}$. (Do not bother to prove that $\boldsymbol{\nabla} \cdot \boldsymbol{G} \equiv 0$; assume it as a result from Chapter 13.)

Exercise 17.2. NO TENSOR CONSTRUCTABLE FROM FIRST DERIVATIVES OF METRIC

Show that there exists *no* tensor with components constructable from the ten metric coefficients $g_{\alpha\beta}$ and their 40 first derivatives $g_{\alpha\beta,\mu}$—except the metric tensor \boldsymbol{g}, and products of it with itself; e.g., $\boldsymbol{g} \otimes \boldsymbol{g}$. [*Hint:* Assume there exists some other such tensor, and examine its hypothesized components in a local inertial frame.]

Exercise 17.3. *RIEMANN* **AS THE ONLY TENSOR CONSTRUCTABLE**
FROM, AND LINEAR IN SECOND DERIVATIVES OF METRIC

Show that (1) *Riemann*, (2) *g*, and (3) tensors (e.g., *Ricci*) formed from *Riemann* and *g* but linear in *Riemann*, are the only tensors that (a) are constructable from the ten $g_{\alpha\beta}$, the 40 $g_{\alpha\beta,\mu}$, and the 100 $g_{\alpha\beta,\mu\nu}$, and (b) are linear in the $g_{\alpha\beta,\mu\nu}$. [*Hint:* Assume there exists some other such tensor, and examine its hypothesized components in an orthonormal, Riemann-normal coordinate system. Use equations (11.30) to (11.32).]

Exercise 17.4. **UNIQUENESS OF THE EINSTEIN TENSOR**

(a) Show that the Einstein tensor, $G_{\alpha\beta} = R_{\alpha\beta} - \frac{1}{2}Rg_{\alpha\beta}$, is the only second-rank, symmetric tensor that (1) has components constructable solely from $g_{\alpha\beta}, g_{\alpha\beta,\mu}, g_{\alpha\beta,\mu\nu}$; (2) has components linear in $g_{\alpha\beta,\mu\nu}$; (3) has an automatically vanishing divergence, $\mathbf{\nabla} \cdot \mathbf{G} = 0$; and (4) vanishes in flat spacetime. This provides added motivation for choosing the Einstein tensor as the left side of the field equation $\mathbf{G} = 8\pi\mathbf{T}$.

(b) Show that, when condition (4) is dropped, the most general tensor is $\mathbf{G} + \Lambda\mathbf{g}$, where Λ is a constant. (See §17.3 for the significance of this.)

§17.2. AUTOMATIC CONSERVATION OF THE SOURCE: A DYNAMIC NECESSITY

The answer $\mathbf{G} = 8\pi\mathbf{T}$ is now on hand; but what is the question? An equation has been derived that connects the Einstein-Cartan "moment of rotation" \mathbf{G} with the stress-energy tensor \mathbf{T}, but what is the purpose for which one wants this equation in the first place? If geometry tells matter how to move, and matter tells geometry how to curve, does one not have in one's hands a Gordian knot? And how then can one ever untie it?

The story is no different in character for the dynamics of geometry than it is for other branches of dynamics. To predict the future, one must first specify, on an "initial" hypersurface of "simultaneity," the position and velocity of every particle, and the amplitude and time-rate of change of every field that obeys a second-order wave equation. One can then evolve the particles and fields forward in time by means of their dynamic equations. Similarly, one must give information about the geometry and its first time-rate of change on the "initial" hypersurface if the Einstein field equation is to be able to predict completely and deterministically the future time-development of the entire system, particles plus fields plus geometry. (See Chapter 21 for details.)

If a prediction is to be made of the geometry, how much information has to be supplied for this purpose? The geometry of spacetime is described by the metric

$$ds^2 = g_{\alpha\beta}(\mathscr{P})\, dx^\alpha\, dx^\beta;$$

that is, by the ten functions $g_{\alpha\beta}$ of location \mathscr{P} in spacetime. It might then seem that ten functions must be predicted; and, if so, that one would need for the task ten

Einstein field equation governs the evolution of spacetime geometry

equations. Not so. Introduce a new set of coordinates $x^{\bar{\mu}}$ by way of the coordinate transformations

$$x^\alpha = x^\alpha(x^{\bar{\mu}}),$$

and find the same spacetime geometry, with all the same bumps, rills, and waves, described by an entirely new set of metric coefficients $g_{\bar{\alpha}\bar{\beta}}(\mathscr{P})$.

It would transgress the power as well as the duty of Einstein's "geometrodynamic law" $\boldsymbol{G} = 8\pi\boldsymbol{T}$ if, out of the appropriate data on the "initial-value hypersurface," it were to provide a way to calculate, on out into the future, values for all ten functions $g_{\alpha\beta}(\mathscr{P})$. To predict all ten functions would presuppose a choice of the coordinates; and to make a choice among coordinate systems is exactly what the geometrodynamic law cannot and must not have the power to do. That choice resides of necessity in the man who studies the geometry, not in the Nature that makes the geometry. The geometry in and by itself, like an automobile fender in and by itself, is free of coordinates. The coordinates are the work of man.

It follows that the ten components $G_{\alpha\beta} = 8\pi T_{\alpha\beta}$ of the field equation must not determine completely and uniquely all ten components $g_{\mu\nu}$ of the metric. On the contrary, $G_{\alpha\beta} = 8\pi T_{\alpha\beta}$ must place only six independent constraints on the ten $g_{\mu\nu}(\mathscr{P})$, leaving four arbitrary functions to be adjusted by man's specialization of the four coordinate functions $x^\alpha(\mathscr{P})$.

How can this be so? How can the ten equations $G_{\alpha\beta} = 8\pi T_{\alpha\beta}$ be in reality only six? Answer: by virtue of the "automatic conservation of the source." More specifically, the identity $G^{\alpha\beta}{}_{;\beta} \equiv 0$ guarantees that the ten equations $G_{\alpha\beta} = 8\pi T_{\alpha\beta}$ contain the four "conservation laws" $T^{\alpha\beta}{}_{;\beta} = 0$. These four conservation laws—along with other equations—govern the evolution of the source. They do not constrain in any way the evolution of the geometry. The geometry is constrained only by the six remaining, independent equations in $G_{\alpha\beta} = 8\pi T_{\alpha\beta}$.

When viewed in this way, the "automatic conservation of the source" is not merely a philosophically attractive principle. It is, in fact, an absolute dynamic necessity. Without "automatic conservation of the source," the ten $G_{\alpha\beta} = 8\pi T_{\alpha\beta}$ would place ten constraints on the ten $g_{\alpha\beta}$, thus fixing the coordinate system as well as the geometry. With "automatic conservation," the ten $G_{\alpha\beta} = 8\pi T_{\alpha\beta}$ place four constraints (local conservation of energy and momentum) on the source, and six constraints on the ten $g_{\alpha\beta}$, leaving four of the $g_{\alpha\beta}$ to be adjusted by adjustment of the coordinate system.

$\boldsymbol{G} = 8\pi\boldsymbol{T}$ must determine only six metric components; the other four are adjustable by changes of coordinates

$\boldsymbol{G} = 8\pi\boldsymbol{T}$ leaves four components of metric free because it satisfies four identities
$$0 \equiv \boldsymbol{\nabla}\cdot\boldsymbol{G} = 8\pi\boldsymbol{\nabla}\cdot\boldsymbol{T}$$
("automatic conservation of source")

§17.3. COSMOLOGICAL CONSTANT

In 1915, when Einstein developed his general relativity theory, the permanence of the universe was a fixed item of belief in Western philosophy. "The heavens endure from everlasting to everlasting." Thus, it disturbed Einstein greatly to discover (Chapter 27) that his geometrodynamic law $\boldsymbol{G} = 8\pi\boldsymbol{T}$ predicts a *non*permanent universe; a dynamic universe; a universe that originated in a "big-bang" explosion,

Einstein's motivation for introducing a cosmological constant

or will be destroyed eventually by contraction to infinite density, or both. Faced with this contradiction between his theory and the firm philosophical beliefs of the day, Einstein weakened; he modified his theory.

The only conceivable modification that does not alter vastly the structure of the theory is to change the lefthand side of the geometrodynamic law $\boldsymbol{G} = 8\pi\boldsymbol{T}$. Recall that the lefthand side is *forced* to be the Einstein tensor, $G_{\alpha\beta} = R_{\alpha\beta} - \frac{1}{2}Rg_{\alpha\beta}$, by three assumptions:

(1) \boldsymbol{G} vanishes when spacetime is flat.
(2) \boldsymbol{G} is constructed from the Riemann curvature tensor and the metric and nothing else.
(3) \boldsymbol{G} is distinguished from other tensors that can be built from ***Riemann*** and ***g*** by the demands (1) that it be linear in ***Riemann***, as befits any natural measure of curvature; (2) that, like ***T***, it be symmetric and of second rank; and (3) that it have an automatically vanishing divergence, $\boldsymbol{\nabla} \cdot \boldsymbol{G} \equiv 0$.

Denote a new, modified lefthand side by "\boldsymbol{G}", with quotation marks to avoid confusion with the standard Einstein tensor. To abandon $\boldsymbol{\nabla} \cdot$ "\boldsymbol{G}" $\equiv 0$ is impossible on dynamic grounds (see §17.2). To change the symmetry or rank of "\boldsymbol{G}" is impossible on mathematical grounds, since "\boldsymbol{G}" must be equated to \boldsymbol{T}. To let "\boldsymbol{G}" be nonlinear in ***Riemann*** would vastly complicate the theory. To construct "\boldsymbol{G}" from anything except ***Riemann*** and ***g*** would make "\boldsymbol{G}" no longer a measure of spacetime geometry and would thus violate the spirit of the theory. After much anguish, one concludes that the assumption which one might drop with least damage to the beauty and spirit of the theory is assumption (1), that "\boldsymbol{G}" vanish when spacetime is flat. But even dropping this assumption is painful: (1) although "\boldsymbol{G}" might still be in some sense a measure of geometry, it can no longer be a measure of curvature; and (2) flat, empty spacetime will no longer be compatible with the geometrodynamic law ($\boldsymbol{G} \neq 0$ in flat, empty space, where $\boldsymbol{T} = 0$). Nevertheless, these consequences were less painful to Einstein than a dynamic universe.

The only tensor that satisfies conditions (2) and (3) [with (1) abandoned] is the Einstein tensor plus a multiple of the metric:

$$\text{"}G_{\alpha\beta}\text{"} = R_{\alpha\beta} - \frac{1}{2}g_{\alpha\beta}R + \Lambda g_{\alpha\beta} = G_{\alpha\beta} + \Lambda g_{\alpha\beta}$$

(exercise 17.1; see also exercise 17.4). Thus was Einstein (1917) led to his modified field equation

Einstein's field equation with the cosmological constant

$$\boldsymbol{G} + \Lambda\boldsymbol{g} = 8\pi\boldsymbol{T}. \tag{17.11}$$

The constant Λ he called the "cosmological constant"; it has dimensions cm^{-2}.

The modified field equation, by contrast with the original, admits a static, unchanging universe as one particular solution (see Box 27.5). For this reason, Einstein in 1917 was inclined to place his faith in the modified equation. But thirteen years later Hubble discovered the expansion of the universe. No longer was the cosmological constant necessary. Einstein, calling the cosmological constant "the biggest

Why Einstein abandoned the cosmological constant

blunder of my life," abandoned it and returned to his original geometrodynamic law, $G = 8\pi T$ [Einstein (1970)].

A great mistake Λ was indeed!—not least because, had Einstein stuck by his original equation, he could have claimed the expansion of the universe as the most triumphant prediction of his theory of gravity.

A mischievous genie, once let out of a bottle, is not easily reconfined. Many workers in cosmology are unwilling to abandon the cosmological constant. They insist that it be abandoned only after cosmological observations reveal it to be negligibly small. As a modern-day motivation for retaining the cosmological constant, one sometimes rewrites the modified field equation in the form

$$G = 8\pi[T + T^{(\text{VAC})}], \tag{17.12a}$$

$$T^{(\text{VAC})} \equiv -(\Lambda/8\pi)g \tag{17.12b}$$

A modern-day motivation for the cosmological constant: vacuum polarization

and interprets $T^{(\text{VAC})}$ as a stress-energy tensor associated with the vacuum. This viewpoint speculates [Zel'dovich (1967)] that the vacuum polarization of quantum field theory endows the vacuum with the nonzero stress-energy tensor (17.12b), which is completely unobservable except by its gravitational effects. Unfortunately, today's quantum field theory is too primitive to allow a calculation of $T^{(\text{VAC})}$ from first principles. (See, however, exercise 17.5.)

The mass-energy density that the cosmological constant attributes to the vacuum is

$$\rho^{(\text{VAC})} = T^{(\text{VAC})}_{\hat{0}\hat{0}} = +\Lambda/8\pi. \tag{17.13}$$

If $\Lambda \neq 0$, it must at least be so small that $\rho^{(\text{VAC})}$ has negligible gravitational effects $[|\rho^{(\text{VAC})}| < \rho^{(\text{MATTER})}]$ wherever Newton's theory of gravity gives a successful account of observations. The systems of lowest density to which one applies Newtonian theory with some (though not great) success are small clusters of galaxies. Hence, one can place the limit

Observational limit on the cosmological constant

$$|\rho^{(\text{VAC})}| = |\Lambda|/8\pi \lesssim \rho^{(\text{CLUSTER})} \sim 10^{-29} \text{ g/cm}^3 \sim 10^{-57} \text{ cm}^{-2} \tag{17.14}$$

on the value of the cosmological constant. Evidently, even if $\Lambda \neq 0$, Λ is so small that it is totally unimportant on the scale of a galaxy or a star or a planet or a man or an atom. Consequently it is reasonable to stick with Einstein's original geometrodynamic law ($G = 8\pi T$; $\Lambda = 0$) everywhere, except occasionally when discussing cosmology (Chapters 27–30).

Why one ignores the cosmological constant everywhere except in cosmology

Exercise 17.5. MAGNITUDE OF COSMOLOGICAL CONSTANT

(a) What is the order of magnitude of the influence of the cosmological constant on the celestial mechanics of the solar system if $\Lambda \sim 10^{-57}$ cm^{-2}?

EXERCISE

(b) Show that the mass-energy density of the vacuum $\rho^{\text{(VAC)}} = \Lambda/8\pi \sim 10^{-29}\,\text{g/cm}^3$, corresponding to the maximum possible value of Λ, agrees in very rough magnitude with

$$\frac{\text{rest mass of an elementary particle}}{(\text{Compton wavelength of particle})^3} \times (\text{gravitational fine-structure constant})$$

$$= \frac{m}{(\hbar/m)^3}\,\frac{m^2}{\hbar} = \frac{m^6}{\hbar^4}$$

[Zel'dovich (1967, 1968)]. This numerology is suggestive, but has not led to any believable derivation of a stress-energy tensor for the vacuum.

§17.4. THE NEWTONIAN LIMIT

Just as quantum mechanics reduces to classical mechanics in the "correspondence limit" of large actions, $I \gg \hbar$, so general relativity reduces to Newtonian theory in the "correspondence limit" of weak gravity and low velocities. (On "correspondence limits," see Box 17.1.) This section elucidates, in some mathematical detail, the correspondence between general relativity and Newtonian theory. It begins with "passive" aspects of gravitation (response of matter to gravity) and then turns to "active" aspects (generation of gravity by matter).

Consider an isolated system—e.g., the solar system—in which Newtonian theory is highly accurate. In order that special relativistic effects not be noticeable, all

Box 17.1 CORRESPONDENCE PRINCIPLES

A. General Remarks and Specific Examples

1. As physics develops and expands, its unity is maintained by a network of correspondence principles, through which simpler theories maintain their vitality by links to more sophisticated but more accurate ones.
 a. Physical optics, with all the new diffraction and interference phenomena for which it accounted, nevertheless also had to account, and did account, for the old, elementary, geometric optics of mirrors and lenses. Geometric optics is recovered from physical optics in the mathematical "correspondence principle limit" in which the wavelength is made indefinitely small in comparison with all other relevant dimensions of the physical system.
 b. Newtonian mechanics is recovered from the mechanics of special relativity in the mathematical "correspondence principle limit" in which all relevant velocities are negligibly small compared to the speed of light.
 c. Thermodynamics is recovered from its successor theory, statistical mechanics, in the mathematical "correspondence principle limit" in which so many particles are taken into account that fluctuations in pressure,

particle number, and other physical quantities are negligible compared to the average values of these parameters of the system.

d. Classical mechanics is recovered from quantum mechanics in the "correspondence principle limit" in which the quantum numbers of the quantum states in question are so large, or the quantities of action that come into play are so great compared to \hbar, that wave and diffraction phenomena make negligible changes in the predictions of standard deterministic classical mechanics. Niels Bohr formulated and took advantage of this correspondence principle even before any proper quantum theory existed. He used it to predict approximate values of atomic energy levels and of intensities of spectral lines. He also expounded it as a guide to all physicists, first in searching for a proper version of the quantum theory, and then in elucidating the content of this theory after it was found.

2. In all these examples and others, the newer, more sophisticated theory is "better" than its predecessor because it gives a good description of a more extended domain of physics, or a more accurate description of the same domain, or both.

3. The correspondence between the newer theory and its predecessor (a) gives one the power to recover the older theory from the newer; (b) can be exhibited by straightforward mathematics; and (c), according to the historical record, often guided the development of the newer theory.

B. Correspondence Structure of General Relativity

1. Einstein's theory of gravity has as distinct limiting cases (a) special relativity; (b) the "linear-ized theory of gravity"; (c) Newton's theory of gravity; and (d) the post-Newtonian theory of gravity. Thus, it has a particularly rich correspondence structure.

a. *Correspondence with special relativity:* General relativity has two distinct kinds of correspondence with special relativity. The first is the limit of vanishing gravitational field everywhere (vanishing curvature); in this limit one can introduce a global inertial frame, set $g_{\mu\nu} = \eta_{\mu\nu}$, and recover completely and precisely the theory of special relativity. The second is local rather than global; it is the demand ("correspondence principle"; "equivalence principle") that in a local inertial frame all the laws of physics take on their special relativistic forms. As was seen in Chapter 16, this puts no restrictions on the metric (except that $g_{\mu\nu} = \eta_{\mu\nu}$ and $g_{\mu\nu,\alpha} = 0$ in local inertial frames); but it places severe constraints on the behavior of matter and fields in the presence of gravity.

b. *Correspondence with Newtonian theory:* In the limit of weak gravitational fields, low velocities, and small pressures, general relativity reduces to Newton's theory of gravity. The correspondence structure is explored mathematically in the text of §17.4.

c. *Correspondence with post-Newtonian theory:* When Newtonian theory is nearly valid, but "first-order relativistic corrections" might be important, one often uses the "post-Newtonian theory of gravity." Chapter 39 expounds the post-Newtonian theory and its correspondence with both general relativity and Newtonian theory.

d. *Correspondence with linearized theory:* In the limit of weak gravitational fields, but possibly large velocities and pressures ($v \sim 1$, $T_{jk} \sim T_{00}$) general relativity reduces to the "linearized theory of gravity". This correspondence is explored in Chapter 18.

Conditions which a system
must satisfy for Newton's
theory of gravity to be
accurate

velocities in the system, relative to its center of mass and also relative to the Newtonian coordinates, must be small compared to the speed of light

$$v \ll 1. \tag{17.15a}$$

As a particle falls from the outer region of the system to the inner region, gravity accelerates it to a kinetic energy $\frac{1}{2}mv^2 \sim |m\Phi|_{\max}$. [Here $\Phi < 0$ is Newton's gravitational potential, so normalized that $\Phi(\infty) = 0$.] The resulting velocity will be small only if

$$|\Phi| \ll 1. \tag{17.15b}$$

Internal stresses in the system also produce motion—e.g., sound waves. Such waves have characteristic velocities of the order of $|T^{ij}/T^{00}|^{1/2}$—for example, the speed of sound in a perfect fluid is

$$v = (dp/d\rho)^{1/2} \sim (p/\rho)^{1/2} \sim |T^{ij}/T^{00}|^{1/2}.$$

In order that these velocities be small compared to the speed of light, all stresses must be small compared to the density of mass-energy

$$|T^{ij}|/T^{00} = |T^{ij}|/\rho \ll 1. \tag{17.15c}$$

When, and only when conditions (17.15) hold, one can expect Newtonian theory to describe accurately the system being studied. Correspondence of general relativity with Newtonian theory for gravity in a passive role then demands that the geodesic world lines of freely falling particles reduce to the Newtonian world lines

$$d^2x^i/dt^2 = -\partial\Phi/\partial x^i. \tag{17.16}$$

Moreover, they must reduce to this form in *any* relativistic coordinate system where the source and test particles have low velocities $v \ll 1$, and where coordinate lengths and times agree very nearly with the lengths and times of the Newtonian coordinates—which in turn are proper lengths and times as measured by rods and clocks. Thus, the relevant coordinates (called "*Galilean*" *or* "*Newtonian*" *coordinates*) are ones in which

"Newtonian coordinates"
defined

$$g_{\mu\nu} = \eta_{\mu\nu} + h_{\mu\nu}, \qquad |h_{\mu\nu}| \ll 1, \qquad |v^j| = |dx^j/dt| \ll 1 \tag{17.17}$$

(weak gravitational field; nearly inertial coordinates; low velocities). In such a coordinate system, the geodesic world lines of test particles have the form

$$\frac{d^2x^i}{dt^2} = \frac{d^2x^i}{d\tau^2} \qquad \text{(since } dt/d\tau \approx 1 \text{ when } |h_{\mu\nu}| \ll 1 \text{ and } |v^j| \ll 1)$$

$$= -\Gamma^i{}_{\alpha\beta}\frac{dx^\alpha}{d\tau}\frac{dx^\beta}{d\tau} \qquad \text{(geodesic equation)}$$

$$= -\Gamma^i{}_{00} \qquad \text{(since } dt/d\tau \approx 1 \text{ and } |dx^j/d\tau| \ll 1)$$

$$= -\Gamma_{i00} \qquad \text{(since } g_{\mu\nu} \approx \eta_{\mu\nu})$$

$$= \frac{1}{2}h_{00,i} - h_{0i,0} \qquad \text{(equation for } \Gamma_{\alpha\beta\gamma} \text{ in terms of } g_{\alpha\beta,\gamma})$$

$$= \frac{1}{2}h_{00,i} \qquad \left(\begin{array}{l} \text{all velocities small compared to } c \text{ implies time} \\ \text{derivatives small compared to space derivatives} \\ \text{—i.e., } h_{\alpha\beta,0} \sim vh_{\alpha\beta,i} \end{array} \right).$$

These geodesic world lines do, indeed, reduce to those of Newtonian theory [equation (17.16)] if one makes the identification

$$\Gamma^i{}_{00} = -\frac{1}{2} h_{00,i} = \Phi_{,i}. \tag{17.18}$$

Together with the boundary conditions $\Phi(r = \infty) = 0$ and $h_{\mu\nu}(r = \infty) = 0$ (coordinates Lorentz far from the source), this identification implies $h_{00} = -2\Phi$; i.e.,

$$g_{00} = -1 - 2\Phi \text{ for nearly Newtonian systems in Newtonian coordinates.} \tag{17.19}$$

Einstein gravity reduces to Newton gravity only if, in Newtonian coordinates, $g_{00} = -1 - 2\Phi$

Note that the correspondence tells one the form of h_{00} for nearly Newtonian systems, but not the forms of the other components of the metric perturbation. In fact, the other $h_{\mu\nu}$ could perfectly well be of the same order of magnitude as $h_{00} \sim \Phi$, without influencing the world lines of slowly moving particles, because they always enter the geodesic equation multiplied by the small numbers v or v^2, or differentiated by t rather than by x^i. The forms of the other $h_{\mu\nu}$ and their small corrections to the Newtonian motion will be explored in Chapters 18, 39, and 40.

The relation $g_{00} = -1 - 2\Phi$ is the mathematical embodiment of the correspondence between general relativity theory and Newtonian theory for passive aspects of gravity. Together with the "validity conditions" (17.15, 17.17), it is a foundation from which one can derive all other aspects of the correspondence for "passive gravity," including the relation

The correspondence between Einstein theory and Newton theory for all "passive" aspects of gravity

$$R^i{}_{0j0} = \partial^2\Phi / \partial x^i \, \partial x^j \tag{17.20}$$

(exercise 17.6). Alternatively, all other aspects of this correspondence can be derived by direct comparison of Newton's predictions with Einstein's. For example, to derive equation (17.20), examine the relative acceleration of two test particles, one at $x^i + \xi^i$ and the other at x^i. According to Newton

$$\frac{d^2\xi^i}{dt^2} = \frac{d^2(x^i + \xi^i)}{dt^2} - \frac{d^2 x^i}{dt^2}$$

$$= -\frac{\partial\Phi}{\partial x^i}\bigg|_{\text{at } x^j + \xi^j} + \frac{\partial\Phi}{\partial x^i}\bigg|_{\text{at } x^j} = \frac{-\partial^2\Phi}{\partial x^i \, \partial x^j}\xi^j.$$

For comparison, Einstein predicts (equation of geodesic deviation)

$$\frac{D^2\xi^i}{d\tau^2} = \frac{d^2\xi^i}{dt^2} = -R^i{}_{0j0}\xi^j.$$

$$\uparrow \text{—[by conditions (17.15) and (17.17)]}$$

Direct comparison gives relation (17.20).

Turn now from correspondence for passive aspects of gravity to correspondence for active aspects. According to Einstein, mass generates gravity (spacetime curvature) by the geometrodynamic law $\mathbf{G} = 8\pi\mathbf{T}$. Apply this law to a nearly Newtonian system, and by the chain of reasoning that preceeds equation (17.8) derive the relation

The Newtonian limit of the Einstein field equation is $\nabla^2\Phi = 4\pi\rho$

$$R_{00} = 4\pi\rho. \tag{17.21}$$

Combine with the contraction of (17.20),

$$R_{00} = R^i{}_{0i0} + R^0{}_{000} = \partial^2 \Phi / \partial x^i \, \partial x^i = \nabla^2 \Phi,$$
$$\uparrow$$
$$0$$

and thereby obtain Newton's equation for the generation of gravity by mass

$$\nabla^2 \Phi = 4\pi\rho. \tag{17.22}$$

Thus, Einstein's field equation reduces to Newton's field equation in the Newtonian limit.

The correspondence between Newton and Einstein, although clear and straightforward as outlined above, is even more clear and straightforward when Newton's theory of gravity is rewritten in Einstein's language of curved spacetime (Chapter 12; exercise 17.7).

EXERCISES

Exercise 17.6. RAMIFICATIONS OF CORRESPONDENCE FOR GRAVITY IN A PASSIVE ROLE

From the correspondence relation $g_{00} = -1 - 2\Phi$, and from conditions (17.15) and (17.17) for Newtonian physics, derive the correspondence relations

$$\Gamma^i{}_{00} = \partial\Phi / \partial x^i, \qquad R^i{}_{0j0} = \partial^2\Phi / \partial x^i \, \partial x^j.$$

Exercise 17.7. CORRESPONDENCE IN THE LANGUAGE OF CURVED SPACETIME [Track 2]

Exhibit the correspondence between the Einstein theory and Cartan's curved-spacetime formulation of Newtonian theory (Chapter 12).

§17.5. AXIOMATIZE EINSTEIN'S THEORY?

Find the most compact and reasonable axiomatic structure one can for general relativity? Then from the axioms derive Einstein's field equation,

$$\boldsymbol{G} = 8\pi\boldsymbol{T}?$$

There are many ways (Box 17.2) to derive the Einstein field equation

That approach would follow tradition. However, it may be out of date today. More than half a century has gone by since November 25, 1915. For all that time the equation has stood unchanged, if one ignores Einstein's temporary "aberration" of adding the cosmological constant. In contrast the derivations have evolved and become more numerous and more varied. In the beginning axioms told what equation is acceptable. By now the equation tells what axioms are acceptable. Box 17.2 sketches a variety of sets of axioms, and the resulting derivations of Einstein's equation.

(continued on page 429)

Box 17.2 SIX ROUTES TO EINSTEIN'S GEOMETRODYNAMIC LAW OF THE EQUALITY OF CURVATURE AND ENERGY DENSITY ("EINSTEIN'S FIELD EQUATION")

[Recommended to the attention of Track-1 readers are only route 1 (automatic conservation of the source, plus correspondence with Newtonian theory) and route 2 (Hilbert's variational principle); and even Track-2 readers are advised to finish the rest of this chapter before they study route 3 (physics on a spacelike slice), route 4 (going from superspace to Einstein's equation), route 5 (field of spin 2 in an "unobservable flat spacetime" background), and route 6 (gravitation as an elasticity of space that arises from particle physics).]

1. Model geometrodynamics after electrodynamics and treat "automatic conservation of the source" and correspondence with the Newtonian theory of gravity as the central considerations.
 a. Particle responds in electrodynamics to field; in general relativity, to geometry.
 b. The potential for the electromagnetic field is the 4-vector \boldsymbol{A} (components A_μ). The potential for the geometry is the metric tensor \boldsymbol{g} (components $g_{\mu\nu}$).
 c. The electromagnetic potential satisfies a wave equation with source term (4-current) on the right,

 $$\left(\frac{\partial A_\nu}{\partial x^\mu} - \frac{\partial A_\mu}{\partial x^\nu}\right)^{;\nu} = 4\pi j_\mu, \tag{1}$$

 so constructed that conservation of the source, $j_\mu{}^{;\mu} = 0$, is automatic (consequence of an identity fulfilled by the lefthand side). By analogy, the geometrodynamic potential must also satisfy a wave equation with source term (stress-energy tensor) on the right,

 $$G_{\mu\nu} = 8\pi T_{\mu\nu}, \tag{2}$$

 so constructed that conservation of the source, $T_{\mu\nu}{}^{;\nu} = 0$ (Chapter 16) is "automatic." This conservation is automatic here because the lefthand side of the equation is a tensor (the Einstein tensor; see Box 8.6 or Chapter 15), built from the metric components and their second derivatives, that fulfills the identity $G_{\mu\nu}{}^{;\nu} \equiv 0$.
 d. No other tensor which (1) is linear in the second derivatives of the metric components, (2) is free of higher derivatives, and (3) vanishes in flat spacetime, satisfies such an identity.
 e. The constant of proportionality (8π) is fixed by the choice of units [here geometric; see Box 1.8] and by the requirement ("correspondence with Newtonian theory") that a test particle shall oscillate back and forth through a collection of matter of density ρ, or revolve in circular orbit around that collection of matter, at a circular frequency given by $\omega^2 = (4\pi/3)\rho$ (Figure

Box 17.2 (continued)

1.12). The foregoing oversimplifies, and omits Einstein's temporary false turns, but otherwise summarizes the reasoning he pursued in arriving at his field equation. This reasoning is spelled out in more detail in the text of Chapter 17.

2. Take variational principle as central.
 a. Construct out of the metric components the only scalar that exists that (1) is linear in the second derivatives of the metric tensor, (2) contains no higher derivatives, and (3) vanishes in flat spacetime: namely, the Riemann scalar curvature invariant, R.
 b. Construct the invariant integral,

 $$I = \frac{1}{16\pi} \int_\Omega R(-g)^{1/2}\, d^4x. \tag{3}$$

 c. Make small variations, $\delta g^{\mu\nu}$, in the metric coefficients $g^{\mu\nu}$ in the interior of the four-dimensional region Ω, and find that this integral changes by the amount

 $$\delta I = \frac{1}{16\pi} \int_\Omega G_{\mu\nu}\, \delta g^{\mu\nu}(-g)^{1/2}\, d^4x. \tag{4}$$

 d. Demand that I should be an extremum with respect to the choice of geometry in the region interior to Ω ($\delta I = 0$ for arbitrary $\delta g^{\mu\nu}$; "principle of extremal action").
 e. Thus arrive at the Einstein field equation for empty space,

 $$G_{\mu\nu} = 0. \tag{5}$$

 f. The continuation of the reasoning leads to the identity

 $$G_{\mu\nu}{}^{;\nu} = 0.$$

 Chapter 21, on the variational principle, gives more detail and takes up the additional term that appears on the righthand side of (5) when matter or fields or both are present.
 g. This approach goes back to David Hilbert (1915). No route to the field equations is quicker. Moreover, it connects immediately (see the following section here, 2′) with the quantum principle of the "democracy of all histories" [Feynman (1942); Feynman and Hibbs (1965)]. The variational principle is spelled out in more detail in Chapter 21.

2′. An aside on the meaning of the classical action integral for the real world of quantum physics.
 a. A "history of geometry," H, is a spacetime, that is to say, a four-dimensional manifold with four-dimensional $-+++$ Riemann metric that (1) reduces on one spacelike hypersurface ("hypersurface of simultaneity") to a specified "initial value 3-geometry," A, with positive definite metric and (2) reduces on

another spacelike hypersurface to a specified "final value 3-geometry," B, also with positive definite metric.

b. The classical variational principle of Hilbert, as reformulated by Arnowitt, Deser, and Misner, provides a prescription for the dynamical path length, I_H, of any conceivable history H, classically allowed or not, that connects A and B (see Chapter 21 for a fuller statement for what can and must be specified on the initial hypersurface of simultaneity, and on the final one, and for alternative choices of the integrand in the action principle).

c. Classical physics says that a history H is allowed only if it extremizes the dynamic path length I as compared to all nearby histories. Quantum physics says that all histories occur with equal probability amplitude, in the following sense. The probability amplitude for "the dynamic geometry of space to transit from A to B" by way of the history H with action integral I_H, and by way of histories that lie within a specified infinitesimal range, $\mathcal{D}H$, of the history H, is given by the expression

$$\left(\begin{array}{l} \text{probability amplitude} \\ \text{to transit from } A \text{ to} \\ B \text{ by way of history } H \\ \text{and histories lying} \\ \text{within the range } \mathcal{D}H \\ \text{about } H \end{array} \right) \sim \exp\left(iI_H/\hbar\right)N\mathcal{D}H. \tag{6}$$

Here the normalization factor, N, is the same for all conceivable histories H, allowed or not, that lead from A to B ("principle of democracy of histories"). The quantum of angular momentum, $\hbar = h/2\pi$, expressed in geometric units, has the value

$$\hbar = \hbar_{\text{conv}}G/c^3 = (L^*)^2, \tag{7}$$

where L^* is the Planck length, $L^* = 1.6 \times 10^{-33}$ cm.

d. The classically allowed history receives "preference without preference." That history, and histories H that differ from it so little that $\delta I = I_H - I_{class}$ is only of the order \hbar and less, give contributions to the probability amplitude that interfere constructively. In contrast, destructive interference effectively wipes out the contribution (to the probability amplitude for a transition) that comes from histories that differ more from the classically allowed history. Thus there are quantum fluctuations in the geometry, but they are fluctuations of limited magnitude. The smallness of \hbar ensures that the scale of these fluctuations is unnoticeable at everyday distances (see the further discussion in Chapters 43 and 44). In this sense classical geometrodynamics is a good approximation to the geometrodynamics of the real world of quantum physics.

3. "Physics on a spacelike slice or hypersurface of simultaneity," again with electromagnetism as the model.

a. Say over and over "lines of magnetic force never end" and come out with half of Maxwell's equations. Say over and over "lines of electric force end

The rest of this chapter is Track 2. No previous track-2 material is needed as preparation for it, nor is it necessary preparation for any later chapter, but it will be helpful in Chapter 21 (initial-value equations and variational principle) and in Chapter 39 (other theories of gravity).

Box 17.2 (continued)

only on charge" and arrive at the other half of Maxwell's equations. Similarly, say over and over

$$
\begin{pmatrix} \text{intrinsic} \\ \text{curvature} \\ \text{scalar} \end{pmatrix} + \begin{pmatrix} \text{extrinsic} \\ \text{curvature} \\ \text{scalar} \end{pmatrix} = 16\pi \begin{pmatrix} \text{local density} \\ \text{of mass-} \\ \text{energy} \end{pmatrix} \tag{8}
$$

and end up with all ten components of Einstein's equation. To "say over and over" is an abbreviation for demanding that the stated principles hold on every spacelike slice through every event of spacetime.

b. Spell out explicitly this "spacelike-slice formulation" of the equations of Maxwell and Einstein. Consider an arbitrary point of spacetime, \mathcal{P} ("event"), and an arbitrary "simultaneity" S through \mathcal{P} (hypersurface of simultaneity; spacelike slice through spacetime). Magnetic lines of force run about throughout S, but nowhere is even a single one of them permitted to end. Recall (§3.4) that the demand "lines of magnetic force never end," when imposed on *all* reference frames at \mathcal{P} (for all choices of the "simultaneity" S), guarantees not only $\nabla \cdot \boldsymbol{B} = 0$, but also $\nabla \times \boldsymbol{E} + \partial\boldsymbol{B}/\partial t = 0$. Similarly (§3.4) the demand that "electric lines of force never end except on electric charge," $\nabla \cdot \boldsymbol{E} = 4\pi J^0$, when imposed on all "simultaneities" through \mathcal{P}, guarantees the remaining Maxwell equation $\nabla \times \boldsymbol{B} = \partial\boldsymbol{E}/\partial t + 4\pi\boldsymbol{J}$.

c. Each simultaneity S through \mathcal{P} has its own slope and curvature. The possibility of different slopes (different local Lorentz frames at \mathcal{P}) is essential for deriving all of Maxwell's equations from the requirements of conservation of flux. Relevant though the slope thus is, the curvature of the hypersurface S never matters for the analysis of electromagnetism. It does matter, however, for any analysis of gravitation modeled on the foregoing treatment of electromagnetism.

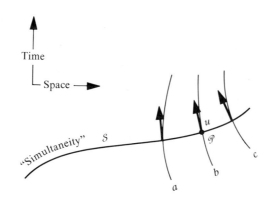

"Simultaneity" S (spacelike hypersurface or "slice through spacetime") that cuts through event \mathcal{P}. The "simultaneity" may be considered to be defined by a set of "observers" a,b,c, Their world lines cross the simultaneity orthogonally, and their clocks all read the same proper time at the instant of crossing. Another simultaneity through \mathcal{P} may have at \mathcal{P} a different curvature or a different slope or both; and it is defined by a different band of observers, with other wrist watches.

d. "Mass-energy curves space" is the central principle of gravitation. To spell out this principle requires one to examine in succession the terms "space" and "curvature of space" and "density of mass-energy in a given region of space." "Space" means spacelike hypersurface; or, more specifically, a hypersurface of simultaneity S that includes the point \mathscr{P} where the physics is under examination.

e. Denote by \boldsymbol{u} the 4-vector normal to S at \mathscr{P}. Then the density of mass-energy in the spacelike hypersurface S at \mathscr{P} is

$$\rho = u^\alpha T_{\alpha\beta} u^\beta, \tag{9}$$

in accordance with the definition of the stress-energy tensor given in Chapter 5.

f. This density is a single number, dependent on the inclination of the slice one cuts through spacetime, but independent of how curved one cuts this slice. If it is to be equated to "curvature of space," that curvature must also be independent of how curved one cuts the slice.

g. Conclude that the geometric quantity, "curvature of space," must (1) be a single number (a scalar) that (2) depends on the inclination \boldsymbol{u} of the cut one makes through spacetime at \mathscr{P} in constructing the hypersurface S, but (3) must be unaffected by how one curves his cut. The demand made here appears paradoxical. One seems to be asking for a measure of curvature that is independent of curvature!

h. A closer look discloses that three distinct ideas come into consideration here. One is the scalar curvature invariant $^{(3)}R$ of the 3-geometry intrinsic to the hypersurface S at \mathscr{P}: "intrinsic" in the sense that it is defined by, and depends exclusively on, measurements of distance made within the hypersurface. The second is the "extrinsic curvature" of this 3-geometry relative to the 4-geometry of the enveloping spacetime ("how curved one cuts his slice"; see Box 14.1 for more on the distinction between extrinsic and intrinsic curvature). The third is the curvature of the four-dimensional spacetime itself, "normal to \boldsymbol{u}," in some sense yet to be more closely defined. This is the quantity that is independent of how curved one cuts his slice. It is the quantity that is to be identified, up to a factor that depends on the choice of units, with the density of mass-energy.

Box 17.2 (*continued*)

i. These three quantities are related in the following way:

$$
\left(
\begin{array}{l}
\text{scalar curvature invariant,} \\
{}^{(3)}R, \text{ of the 3-geometry} \\
\text{intrinsic to the spacelike} \\
\text{hypersurface } \mathcal{S}, \text{ a quantity} \\
\text{dependent on ``how curved} \\
\text{one cuts the slice''}
\end{array}
\right)
+
\left(
\begin{array}{l}
\text{a correction term that (a)} \\
\text{depends only on the ``ex-} \\
\text{trinsic curvature'' } K_{\alpha\beta} \text{ (Box} \\
\text{14.1 and Chapter 21) of the} \\
\text{hypersurface relative to} \\
\text{the four-dimensional geometry} \\
\text{in which it is imbedded, and} \\
\text{(b) is so calculated (a} \\
\text{uniquely determinate calcu-} \\
\text{lation) that the sum of this} \\
\text{correction term and } {}^{(3)}R \text{ is} \\
\text{independent of ``how curved} \\
\text{one cuts his slice,'' and (c)} \\
\text{has the precise value} \\
(\operatorname{Tr} \boldsymbol{K})^2 - \operatorname{Tr} \boldsymbol{K}^2 \equiv (K_\alpha{}^\alpha)^2 - K_{\alpha\beta}K^{\alpha\beta}
\end{array}
\right)
$$

$$
=
\left(
\begin{array}{l}
\text{a measure of the curvature} \\
\text{of spacetime that depends} \\
\text{on the 4-geometry of} \\
\text{the spacetime and on} \\
\text{the inclination } \boldsymbol{u} \text{ of the} \\
\text{spacelike slice } \mathcal{S} \text{ cut} \\
\text{through spacetime, but is} \\
\text{independent, by construc-} \\
\text{tion, of ``how curved one} \\
\text{cuts the slice''}
\end{array}
\right)
=
\left(
\begin{array}{l}
\text{a scalar quantity that} \\
\text{(a) is completely defined} \\
\text{by what has just been} \\
\text{said and (b) can there-} \\
\text{fore be calculated in all} \\
\text{completeness by standard} \\
\text{differential geometry} \\
\text{(details in Chapter 21)}
\end{array}
\right)
$$

$$
=
\left(
\begin{array}{l}
2u^\alpha G_{\alpha\beta} u^\beta, \text{ where } G_{\alpha\beta} \text{ is} \\
\text{the Einstein curvature} \\
\text{tensor of equation 8.49} \\
\text{and Box 8.6}
\end{array}
\right)
= 2
\left(
\begin{array}{l}
\text{a quantity interpreted in} \\
\text{Track 2, Chapter 15, as the} \\
\text{``moment of rotation'' asso-} \\
\text{ciated with a unit element} \\
\text{of 3-volume located at } \mathcal{P} \\
\text{in the hypersurface orth-} \\
\text{ogonal to } \boldsymbol{u}
\end{array}
\right)
$$

$$
\tag{10}
$$

j. Conclude that the central principle, "mass-energy curves space," translates to the formula

$$
{}^{(3)}R + (\operatorname{Tr} \boldsymbol{K})^2 - \operatorname{Tr} \boldsymbol{K}^2 = 16\pi\rho, \tag{11}
$$

or, in shorthand form,

$$\begin{pmatrix} \text{moment of} \\ \text{rotation} \end{pmatrix} = \begin{pmatrix} \text{intrinsic} \\ \text{curvature} \end{pmatrix} + \begin{pmatrix} \text{extrinsic} \\ \text{curvature} \end{pmatrix} = \begin{pmatrix} \text{density of} \\ \text{mass-energy} \end{pmatrix}, \quad (12)$$

valid for every spacelike slice through spacetime at any arbitrary point \mathscr{P}.

k. All of Einstein's geometrodynamics is contained in this statement as truly as all of Maxwell's electrodynamics is contained in the statement that the number of lines of force that end in an element of volume is equal to 4π times the amount of charge in that element of volume. The factor 16π is appropriate for the geometric system of units in use in this book (density ρ in cm^{-2} given by $G/c^2 = 0.742 \times 10^{-28}$ cm/g multiplied by the density ρ_{conv} expressed in the conventional units of g/cm^3).

l. Reexpress the principle that "mass-energy curves space" in the form

$$2u^\alpha G_{\alpha\beta} u^\beta = 16\pi u^\alpha T_{\alpha\beta} u^\beta. \tag{13}$$

Demand that this equation should hold for every simultaneity that cuts through \mathscr{P}, whatever its "inclination" \boldsymbol{u}.

m. Conclude that the coefficients on the two sides of (13) must agree; thus,

$$G_{\alpha\beta} = 8\pi T_{\alpha\beta}, \tag{14}$$

Einstein's equation in the language of components; or, in the language of abstract geometric quantities,

$$\boldsymbol{G} = 8\pi \boldsymbol{T}. \tag{15}$$

4. Going from superspace to Einstein's equation rather than from Einstein's equation to superspace.

a. A fourth route to Einstein's equation starts with the advanced view of geometrodynamics that is spelled out in Chapter 43. One notes there that the dynamics of geometry unfolds in superspace. Superspace has an infinite number of dimensions. Any one point in superspace describes a complete 3-geometry, $^{(3)}\mathscr{G}$, with all its bumps and curvatures. The dynamics of geometry leads from point to point in superspace.

b. Like the dynamics of a particle, the dynamics of geometry lends itself to distinct but equivalent mathematical formulations, associated with the names of Lagrange, of Hamilton, and of Hamilton and Jacobi. Of these the most convenient for the present analysis is the last ("H-J").

c. In the problem of one particle moving in one dimension under the influence of a potential $V(x)$, the H-J equation reads

$$-\underbrace{\frac{\partial S}{\partial t}}_{\substack{\text{total} \\ \text{energy}}} = \underbrace{\frac{1}{2m}\left(\frac{\partial S}{\partial x}\right)^2}_{\substack{\text{kinetic} \\ \text{energy}}} + V(x). \tag{16}$$

Box 17.2 (continued)

It has the solution

$$S_E(x,t) = -Et + \int^x [2m(E - V)]^{1/2}\, dx. \tag{17}$$

Out of this solution one reads the motion by applying the "condition of constructive interference,"

$$\frac{\partial S_E(x,t)}{\partial E} = 0 \tag{18}$$

(one equation connecting the two quantities x and t; for more on the condition of constructive interference and the H-J method in general, see Boxes 25.3 and 25.4).

 d. In the corresponding equation for the dynamics of geometry, one deals with a function $S = S(^{(3)}\mathcal{G})$ of the 3-geometry. It depends on the 3-geometry itself, and not on the vagaries of one's choice of coordinates or on the corresponding vagaries in the metric coefficients of the 3-geometry,

$$ds^2 = {}^{(3)}g_{mn}\, dx^m\, dx^n \tag{19}$$

($^{(3)}$ to indicate 3-geometry omitted hereafter for simplicity). This function obeys the H-J equation [the analog of (16)]

$$-(16\pi)^2 \frac{1}{2g}(g_{im}g_{jn} + g_{in}g_{jm} - g_{ij}g_{mn})\frac{\delta S}{\delta g_{ij}}\frac{\delta S}{\delta g_{mn}} + {}^{(3)}R = 16\pi\rho. \tag{20}$$

 e. Out of this equation for the dynamics of geometry in superspace one can deduce the Einstein field equation by reasoning similar to that employed in going from (17) to (18) (Gerlach 1969).

 f. It would appear that one must break new ground, and establish new foundations, if one is to find out how to regard the "Einstein-Hamilton-Jacobi equation" (20) as more basic than the Einstein field equation that one derives from it. [Since done, by Hojman, Kuchař, and Teitelboim (1973 preprint).]

5. Einstein's geometrodynamics viewed as the standard field theory for a field of spin 2 in an "unobservable flat spacetime" background.

 a. This approach to Einstein's field equation has a long history, references to which will be found in §7.1 and §18.1. (Further discussion of this approach will be found in those two sections and in Box 7.1, exercise 7.3, and Box 18.1).

 b. The following summary is quoted from Deser (1970): "We wish to give a simple physical derivation of the nonlinearity . . . , using a now familiar argument . . . leading from the linear, massless, spin-2 field to the full Einstein equations

c. "The Einstein equations may be derived nongeometrically by noting that the free, massless, spin-2 field equations,

$$R^L{}_{\mu\nu}(\phi) - \frac{1}{2} R^L{}_{\alpha\alpha}(\phi)\eta_{\mu\nu} \equiv G^L{}_{\mu\nu}(\phi) \equiv [(\eta_{\mu\alpha}\eta_{\nu\beta} - \eta_{\mu\nu}\eta_{\alpha\beta})\Box$$

$$+ \eta_{\mu\nu}\partial_\alpha\partial_\beta + \eta_{\alpha\beta}\partial_\mu\partial_\nu - \eta_{\mu\alpha}\partial_\nu\partial_\beta - \eta_{\nu\beta}\partial_\mu\partial_\alpha]\phi_{\alpha\beta} = 0, \quad (21)$$

whose source is the matter stress-tensor $T_{\mu\nu}$, must actually be coupled to the *total* stress-tensor, including that of the ϕ-field itself. That is, while the free-field equations (21) are of course quite consistent as they stand, [they are not] when there is a dynamic system's $T_{\mu\nu}$ as a source. For then the left side, which is identically divergenceless, is inconsistent with the right, since the coupling implies that $T^{\mu\nu}{}_{,\nu}$, as computed from the matter equations of motion, is no longer conserved.

d. "To remedy this [violation of the principle of conservation of momentum and energy], the stress tensor $^{(2)}\theta_{\mu\nu}$ arising from the quadratic Lagrangian $^{(2)}L$ responsible for equation (21) is then inserted on the right.

e. "But the Lagrangian $^{(3)}L$ leading to these modified equations is then cubic, and itself contributes a cubic $^{(3)}\theta_{\mu\nu}$.

f. "This series continues indefinitely, and sums (if properly derived!) to the full nonlinear Einstein equations, $G_{\mu\nu}$ ([calculated from] $\eta_{\alpha\beta} + \phi_{\alpha\beta}) = -\kappa T_{\mu\nu}$ [$+ 8\pi T_{\mu\nu}$ in the geometric units and sign conventions of this book], which are an infinite series in the deviation $\phi_{\mu\nu}$ of the metric $g_{\mu\nu}$ from its Minkowskian value $\eta_{\mu\nu}$.

g. Once the iteration is begun (whether or not a $T_{\mu\nu}$ is actually present), it must be continued to all orders, since conservation only holds for the full series

$$\sum_{n=2}^{\infty} {}^{(n)}\theta_{\mu\nu}.$$ Thus, the theory is either left in its (physically irrelevant) free linear

form (21), or it *must* be an infinite series."

h. For details, see Deser (1970); the paper goes on (1) to take advantage of a well-chosen formalism (2) to rearrange the calculation, and thus (3) to "derive the full Einstein equations, on the basis of the same self-coupling requirement, but with the advantages that the full theory emerges in closed form with just one added (cubic) term, rather than as an infinite series."

i. Deser summarizes the analysis at the end thus: "Consistency has therefore led us to universal coupling, which implies the equivalence principle. It is at this point that the geometric interpretation of general relativity arises, since *all matter* now moves in an effective Riemann space of metric $\mathfrak{g}^{\mu\nu} \equiv \eta^{\mu\nu} + h^{\mu\nu}$. . . . [The] initial flat 'background' space is no longer observable." In other words, this approach to Einstein's field equation can be summarized as "curvature without curvature" or—equally well—as "flat spacetime without flat spacetime"!

Box 17.2 (continued)

6. Sakharov's view of gravitation as an elasticity of space that arises from particle physics.

a. The resistance of a homogeneous isotropic solid to deformation is described by two elastic constants, Young's modulus and Poisson's ratio.

b. The resistance of space to deformation is described by one elastic constant, the Newtonian constant of gravity. It makes its appearance in the action principle of Hilbert

$$I = \frac{1}{16\pi G}\int {}^{(4)}R(-g)^{1/2}\,d^4x$$
$$+ \int (L_{\text{matter}} + L_{\text{fields}})(-g)^{1/2}\,d^4x = \text{extremum.} \tag{22}$$

c. According to the historical records, it was first learned how many elastic constants it takes to describe a solid from microscopic molecular models of matter (Newton, Laplace, Navier, Cauchy, Poisson, Voigt, Kelvin, Born), not from macroscopic considerations of symmetry and invariance. Thus, count the energy stored up in molecular bonds that are deformed from natural length or natural angle or both. Arrive at an expression for the energy of deformation per unit volume of the elastic material of the form

$$e = A(\text{Tr } \mathbf{s})^2 + B\,\text{Tr}(\mathbf{s}^2). \tag{23}$$

Here the strain tensor

$$s_{mn} = \frac{1}{2}\left(\frac{\partial \xi_m}{\partial x^n} + \frac{\partial \xi_n}{\partial x^m}\right) \tag{24}$$

measures the strain produced in the elastic medium by motion of the typical point that was at the location x^m to the location $x^m + \xi^m(x)$. The constants A and B are derived out of microscopic physics. They fix the values of the two elastic constants of the macroscopic theory of elasticity.

d. Andrei Sakharov (1967) (*the* Andrei Sakharov) has proposed a similar microscopic foundation for gravitation or, as he calls it, the "metric elasticity of space." He identifies the action term of Einstein's geometrodynamics [the first term in (22)] "with the change in the action of quantum fluctuations of the vacuum [associated with the physics of particles and fields and brought about] when space is curved."

e. Sakharov notes that present-day quantum field theory "gets rid by a renormalization process" of an energy density in the vacuum that would formally be infinite if not removed by this renormalization. Thus, in the standard analysis of the degrees of freedom of the electromagnetic field in flat space, one counts the number of modes of vibration per unit volume in the range

of circular wave numbers from k to $k + dk$ as $(2 \cdot 4\pi/8\pi^3)k^2\, dk$. Each mode of oscillation, even at the absolute zero of temperature, has an absolute irreducible minimum of "zero-point energy of oscillation," $\frac{1}{2}h\nu = \frac{1}{2}\hbar ck$ [the fluctuating electric field associated with which is among the most firmly established of all physical effects. It acts on the electron in the hydrogen atom in supplement to the electric field caused by the proton alone, and thereby produces most of the famous Lamb-Retherford shift in the energy levels of the hydrogen atom, as made especially clear by Welton (1948) and Dyson (1954)]. The totalized density of zero-point energy of the electromagnetic field per unit volume of spacetime (units: cm⁴) formally diverges as

$$(\hbar/2\pi^2)\int_0^\infty k^3\, dk. \tag{25}$$

Equally formally this divergence is "removed" by "renormalization" [for more on renormalization see, for example, Hepp (1969)].

f. Similar divergences appear when one counts up formally the energy associated with other fields and with vacuum fluctuations in number of pairs of electrons, μ-mesons, and other particles in the limit of quantum energies large in comparison with the rest mass of any of these particles. Again these divergences in formal calculations are "removed by renormalization."

g. Removed by renormalization is a contribution not only to the energy density, and therefore to the stress-energy tensor, but also to the total Lagrange function \mathcal{L} of the variational principle for all these fields and particles,

$$I = \int \mathcal{L}\, d^4x = \text{extremum.} \tag{26}$$

h. Curving spacetime alters all these energies, Sakharov points out, extending an argument of Zel'dovich (1967). Therefore the process of "renormalization" or "subtraction" no longer gives zero. Instead, the contribution of zero-point energies to the Lagrangian, expanded as a power series in powers of the curvature, with numerical coefficients A, B, \ldots of the order of magnitude of unity, takes a form simplified by Ruzmaikina and Ruzmaikin (1969) to the following:

$$\mathcal{L}(R) = A\hbar \int k^3\, dk + B\hbar^{(4)}R \int k\, dk$$
$$+ \hbar[C({}^{(4)}R)^2 + DR^{\alpha\beta}R_{\alpha\beta}]\int k^{-1}\, dk$$
$$+ \text{(higher-order terms)}. \tag{27}$$

[For the alteration in the number of standing waves per unit frequency in a curved manifold, see also Berger (1966), Sakharov (1967), Hill in De Witt (1967c), Polievktov-Nikoladze (1969), and Berger, Gauduchon, and Mazet (1971).]

i. Renormalization physics argues that the first term in (27) is to be dropped. The second term, Sakharov notes, is identical in form to the Hilbert action

Box 17.2 (*continued*)

principle, equation (3) above, with the exception that there the constant that multiplies the Riemann scalar curvature invariant is $-c^3/16\pi G$ (in conventional units), whereas here it is $B\hbar\int k\,dk$ (in the same conventional units). The higher order terms in (27) lead to what Sakharov calls "corrections . . . to Einstein's equations."

j. Overlooking these corrections, one evidently obtains the action principle of Einstein's theory when one insists on the equality

$$G = \left(\begin{matrix}\text{Newtonian}\\ \text{constant of gravity}\end{matrix}\right) = \frac{c^3}{16\pi B\hbar\int k\,dk}. \tag{28}$$

With B a dimensionless numerical factor of the order of unity, it follows, Sakharov argues, that the effective upper limit or "cutoff" in the formally divergent integral in (28) is to be taken to be of the order of magnitude of the reciprocal Planck length [see equation (7)],

$$k_{\text{cut off}} \sim (c^3/\hbar G)^{1/2} = 1/L^* = 1/1.6 \times 10^{-33}\ \text{cm}. \tag{29}$$

In effect Sakharov is saying (1) that field physics suffers a sea change into something new and strange for wavelengths less than the Planck length, and for quantum energies of the order of $\hbar c k_{\text{cutoff}} \sim 10^{28}$ eV or 10^{-5} g or more; (2) that in consequence the integral $\int k\,dk$ is cut off; and (3) that the value of this cutoff, arising purely out of the physics of fields and particles, governs the value of the Newtonian constant of gravity, G.

k. In this sense, Sakharov's analysis suggests that gravitation is to particle physics as elasticity is to chemical physics: merely a statistical measure of residual energies. In the one case, molecular bindings depend on departures of molecule-molecule bond lengths from standard values. In the other case, particle energies are affected by curvatures of the geometry.

l. Elasticity, which looks simple, gets its explanation from molecular bindings, which are complicated; but molecular bindings, which are complicated, receive their explanation in terms of Schrödinger's wave equation and Coulomb's law of force between charged point-masses, which are even simpler than elasticity.

m. Einstein's geometrodynamics, which looks simple, is interpreted by Sakharov as a correction term in particle physics, which is complicated. Is particle physics, which is complicated, destined some day in its turn to unravel into something simple—something far deeper and far simpler than geometry ("pregeometry"; Chapter 44)?

§17.6. "NO PRIOR GEOMETRY": A FEATURE DISTINGUISHING EINSTEIN'S THEORY FROM OTHER THEORIES OF GRAVITY

Whereas Einstein's theory of gravity is exceedingly compelling, one can readily construct less compelling and less elegant alternative theories. The physics literature is replete with examples [see Ni (1972), and Thorne, Ni, and Will (1971) for reviews]. However, when placed among its competitors, Einstein's theory stands out sharp and clear: it agrees with experiment; most of its competitors do not (Chapters 38–40). It describes gravity entirely in terms of geometry; most of its competitors do not. It is free of any "prior geometry"; most of its competitors are not.

Einstein's theory compared with other theories of gravity

Set aside, until Chapter 38, the issue of agreement with experiment. Einstein's theory remains unique. Every other theory either introduces auxiliary gravitational fields [e.g., the scalar field of Brans and Dicke (1961)], or involves "prior geometry," or both. Thus, every other theory is more complicated conceptually than Einstein's theory. Every other theory contains elements of complexity for which there is no experimental motivation.

All other theories introduce auxiliary gravitational fields or prior geometry

The concept of "prior geometry" requires elucidation, not least because the rejection of prior geometry played a key role in the reasoning that originally led Einstein to his geometrodynamic equation $\mathbf{G} = 8\pi\mathbf{T}$. By "prior geometry" one means *any aspect of the geometry of spacetime that is fixed immutably, i.e., that cannot be changed by changing the distribution of gravitating sources*. Thus, prior geometry is not generated by or affected by matter; it is not dynamic. Example: Nordstrøm (1913) formulated a theory in which the physical metric of spacetime \mathbf{g} (the metric that enters into the equivalence principle) is generated by a "background" flat-spacetime metric $\boldsymbol{\eta}$, and by a scalar gravitational field ϕ:

"Prior geometry" defined

Nordstrom's theory as an illustration of prior geometry

$$\eta^{\alpha\beta}\phi_{,\alpha\beta} = -4\pi\phi\eta^{\alpha\beta}T_{\alpha\beta} \quad \left(\begin{array}{c}\text{generation of } \phi \text{ by}\\ \text{stress-energy}\end{array}\right), \qquad (17.23\text{a})$$

$$g_{\alpha\beta} = \phi^2\eta_{\alpha\beta} \quad \left(\begin{array}{c}\text{construction of } \mathbf{g}\\ \text{from } \phi \text{ and } \boldsymbol{\eta}\end{array}\right). \qquad (17.23\text{b})$$

In this theory, the physical metric \mathbf{g} (governor of rods and clocks and of test-particle motion) has but one changeable degree of freedom—the freedom in ϕ. The rest of \mathbf{g} is fixed by the flat spacetime metric ("prior geometry") $\boldsymbol{\eta}$. One does not remove the prior geometry by rewriting Nordstrøm's equations (17.23) in a form

$$\underbrace{R}_{\substack{\text{[curvature scalar}\\ \text{constructed from } \mathbf{g}]}} = 24\,\pi\underbrace{T}_{[g^{\alpha\beta}T_{\alpha\beta}]}, \qquad \underbrace{C^{\alpha\beta}{}_{\mu\nu}}_{\substack{\text{[Weyl tensor}\\ \text{constructed from } \mathbf{g}]}} = 0 \qquad (17.24)$$

devoid of reference to $\boldsymbol{\eta}$ and ϕ [Einstein and Fokker (1914); exercise 17.8]. Mass can still influence only one degree of freedom in the spacetime geometry. The other degrees of freedom are fixed *à priori*—they are *prior geometry*. And this prior geometry can perfectly well (in principle) be detected by physical experiments that make no reference to any equations (Box 17.3).

Box 17.3 AN EXPERIMENT TO DETECT OR EXCLUDE CERTAIN TYPES OF PRIOR GEOMETRY
(Based on December 1970 discussions between Alfred Schild and Charles W. Misner)

Choose a momentarily static universe populated with a large supply of suitable pulsars. The pulsars should be absolutely regular, periodically emitting characteristic pulses of both gravitational and electromagnetic waves.

Two fleets of spaceships containing receivers are sent out "on station" to collect the experimental data. Admiral Weber's fleet carries gravitational-wave receivers; Admiral Hertz's fleet, electromagnetic receivers. The captain of each spaceship holds himself "on station" by monitoring three suitably chosen pulsars (of identical frequency) and maneuvering so that their pulses always arrive in coincidence. The experimental data he collects consist of the pulses received from all other pulsars, which he is not using for station keeping, each registered as coincident with or interlaced among the reference (stationary) pulses. [For display purposes, the pattern produced by any single pulsar can be converted to acoustic form. The reference pulses can be played acoustically (by the data-processing computer) on one drum at a fixed rate, and the pulses from other pulsars can be played on a second drum. A pattern of rythmic beats will result.]

When the data fleet is checked out and tuned up, each captain reports stationary patterns. Now the experiment begins. One or more massive stars are towed in among the fleet. The fleet reacts to stay on station, and reports changes in the data patterns. The spaceships on the outside edges of the fleet verify that no detectable changes occur at their stations; so the incident radiation from the distant pulsars can be regarded as unaffected by the newly placed stars. Data stations nearer the movable stars report the interesting data.

What are the results?

In a universe governed by the laws of special relativity (spacetime always flat), no patterns change. (Weber's fleet was unable to get checked out in the first place, as no gravitational waves were ever detected from the pulsars). Neither stars, nor anything else, can produce gravitational fields. All aspects of the spacetime geometry are fixed *a priori* (complete prior geometry!). There is no gravity; and no light deflection takes place to make Hertz's captains adjust their positions.

In a universe governed by Nordstrøm's theory of gravity (see text) both fleets get checked out—i.e., both see waves. But neither fleet sees any changes in the rhythmic pattern of beats. The stars being towed about have no influence on either gravitational waves or electromagnetic waves. The prior geometry (η) present in the theory precludes any light deflection or any gravitational-wave deflection.

In a universe governed by Whitehead's (1922) theory of gravity [see Will (1971b) and references cited therein], radio waves propagate along geodesics of the "physical metric" g, and get deflected by the gravitational fields of the stars. But gravitational waves propagate along geodesics of a *flat* background metric η, and are thus unaffected by the stars. Consequently, Hertz's captains must maneuver to keep on station; and they hear a changing beat pattern between the reference pulsars and the other pulsars. But Weber's fleet remains on station and records no changes in the beat pattern. The prior geometry (η) shows itself clearly in the experimental result.

In a universe governed by Einstein's theory, both fleets see effects (no sign of prior geometry because Einstein's theory has no prior geometry). Moreover, if the fleets were originally paired, one Weber ship and one Hertz at each station, they remain paired. No differences exist between the propagation of high-frequency light waves and high-frequency gravitational waves. Both propagate along geodesics of g.

Mathematics was not sufficiently refined in 1917 to cleave apart the demands for "no prior geometry" and for a "geometric, coordinate-independent formulation of physics." Einstein described both demands by a single phrase, "general covariance." The "no-prior-geometry" demand actually fathered general relativity, but by doing so anonymously, disguised as "general covariance," it also fathered half a century of confusion. [See, e.g., Kretschmann (1917).]

A systematic treatment of the distinction between prior geometry ("absolute objects") and dynamic fields ("dynamic objects") is a notable feature of Anderson's (1967) relativity text.

"No prior geometry" as a part of Einstein's principle of "general covariance"

Exercise 17.8. EINSTEIN-FOKKER REDUCES TO NORDSTRØM

The vanishing of the Weyl tensor [equation (13.50)] for a spacetime metric g guarantees that the metric is conformally flat—i.e., that there exists a scalar field ϕ such that $g = \phi^2 \eta$, where η is a flat-spacetime metric. [See, e.g., Schouten (1954) for proof.] Thus, the Einstein-Fokker equation (17.24), $C^{\alpha\beta}{}_{\mu\nu} = 0$, is equivalent to the Nordstrøm equation (17.23b). With this fact in hand, show that the Einstein-Fokker field equation $R = 24\pi T$ reduces to the Nordstrøm field equation (17.23a).

EXERCISE

§17.7. A TASTE OF THE HISTORY OF EINSTEIN'S EQUATION

Nothing shows better what an idea is and means today than the battles and changes it has undergone on its way to its present form. A complete history of general relativity would demand a book. Here let a few key quotes from a few of the great papers give a little taste of what a proper history might encompass.

Einstein (1908): "We . . . will therefore in the following assume the complete physical equivalence of a gravitational field and of a corresponding acceleration of the reference system. . . . the clock at a point P for an observer anywhere in space runs $(1 + \Phi/c^2)$ times faster than the clock at the coordinate origin. . . . it follows that light rays are curved by the gravitational field. . . . an amount of energy E has a mass E/c^2."

Einstein and Grossmann (1913): "The theory described here originates from the conviction that the proportionality between the inertial and the gravitational mass of a body is an exact law of nature that must be expressed as a foundation principle of theoretical physics. . . . An observer enclosed in an elevator has no way to decide whether the elevator is at rest in a static gravitational field or whether the elevator is located in gravitation-free space in an accelerated motion that is maintained by forces acting on the elevator (equivalence hypothesis). . . . In the decay of radium, for example, that decrease [of mass] amounts to 1/10,000 of the total mass. If those changes in inertial mass did not correspond to changes in gravitational mass, then deviations of inertial from gravitational masses would arise that are far larger than the Eötvös experiments allow. It must therefore be considered as very probable that the identity of gravitational and inertial mass is exact.

"The sought for generalization will surely be of the form

$$\Gamma_{\mu\nu} = \kappa T_{\mu\nu},$$

where κ is a constant and $\Gamma_{\mu\nu}$ is a contravariant tensor of the second rank that arises out of the fundamental tensor $g_{\mu\nu}$ through differential operations. . . . it proved impossible to find a differential expression for $\Gamma_{\mu\nu}$ that is a generalization of [Poisson's] $\Delta\phi$, and that is a tensor with respect to arbitrary transformations. . . . It seems most natural to demand that the system of equations should be covariant against arbitrary transformations. That stands in conflict with the result that the equations of the gravitational field do not possess this property."

Einstein and Grossman (1914): "In a 1913 treatment . . . we could not show general covariance for these gravitational equations. [Origin of their difficulty: part of the two-index curvature tensor was put on the left, to constitute the second-order part of the field equation, and part was put on the right with $T_{\mu\nu}$ and was called gravitational stress-energy. It was asked that lefthand and righthand sides transform as tensors, which they cannot do under general coordinate transformations.]

Einstein (1915a): "In recent years I had been trying to found a general theory of relativity on the assumption of the relativity even of nonuniform motions. I believed in fact that I had found the only law of gravitation that corresponds to a reasonably formulated postulate of general relativity, and I sought to establish the necessity of exactly this solution in a paper that appeared last year in these proceedings.

"A renewed analysis showed me that that necessity absolutely was not shown in the approach adopted there; that it nevertheless appeared to be shown rested on an error.

"For these reasons, I lost all confidence in the field equations I had set up, and I sought for an approach that would limit the possibilities in a natural way. In this way I was led back to the demand for the general covariance of the field equations, from which I had departed three years ago, while working with my friend Grossmann, only with a heavy heart. In fact we had already at that time come quite near to the solution of the problem that is given in what follows.

"According to what has been said, it is natural to postulate the field equations of gravitation in the form

$$R_{\mu\nu} = -\kappa T_{\mu\nu},$$

since we already know that these equations are covariant with respect to arbitrary transformations of determinant 1. In fact, these equations satisfy all conditions that we have to impose on them. [Here $R_{\mu\nu}$ is a piece of the Ricci tensor that Einstein regarded as covariant.] . . .

"Equations (22a) give in the first approximation

$$\frac{\partial^2 g^{\alpha\beta}}{\partial x^\alpha \partial x^\beta} = 0.$$

By this [condition] the coordinate system is still not determined, in the sense that for this determination four equations are necessary." (Session of Nov. 4, 1915, published Nov. 11.)

Einstein (1915b): "In a recently published investigation, I have shown how a theory of the gravitational field can be founded on Riemann's covariant theory of many-di-

mensional manifolds. Here it will now be proved that, by introducing a surely bold additional hypothesis on the structure of matter, a still tighter logical structure of the theory can be achieved. . . . it may very well be possible that in the matter to which the given expression refers, gravitational fields play an essential part. Then $T^\mu{}_\mu$ can appear to be positive for the entire structure, although in reality only $T^\mu{}_\mu + t^\mu{}_\mu$ is positive, and $T^\mu{}_\mu$ vanishes everywhere. We assume in the following that in fact the condition $T^\mu{}_\mu = 0$ is fulfilled [quite] generally.

"Whoever does not from the beginning reject the hypothesis that molecular [small-scale] gravitational fields constitute an essential part of matter will see in the following a strong support for this point of view.

"Our hypothesis makes it possible . . . to give the field equations of gravitation in a generally covariant form . . .

$$G_{\mu\nu} = -\kappa T_{\mu\nu}$$

[where $G_{\mu\nu}$ is the Ricci tensor]." (Session of Nov. 11, 1915; published Nov. 18.)

Einstein (1915c): "I have shown that no objection of principle stands in the way of this hypothesis [the field equations], by which space and time are deprived of the last trace of objective reality. In the present work I find an important confirmation of this most radical theory of relativity: it turns out that it explains qualitatively and quantitatively the secular precession of the orbit of Mercury in the direction of the orbital motion, as discovered by Leverrier, which amounts to about 45″ per century, without calling on any special hypothesis whatsoever."

Einstein (1915d; session of Nov. 25, 1915; published Dec. 2): "More recently I have found that one can proceed without hypotheses about the energy tensor of matter when one introduces the energy tensor of matter in a somewhat different way than was done in my two earlier communications. The field equations for the motion of the perihelion of Mercury are undisturbed by this modification. . . .

"Let us put

$$G_{im} = -\kappa \left(T_{im} - \frac{1}{2} g_{im} T \right)$$

[where G_{im} is the Ricci tensor]." . . .

. . . these equations, in contrast to (9), contain no new condition, so that no other assumption has to be made about the energy tensor of matter than obedience to the energy-momentum [conservation] laws.

"With this step, general relativity is finally completed as a logical structure. The postulate of relativity in its most general formulation, which makes the spacetime coordinates into physically meaningless parameters, leads compellingly to a completely determinate theory of gravitation that explains the perihelion motion of Mercury. In contrast, the general-relativity postulate is able to open up to us nothing about the nature of the other processes of nature that special relativity has not already taught. The opinion on this point that I recently expressed in these proceedings was erroneous. Every physical theory compatible with special relativity can be aligned into the system of general relativity by means of the absolute differential calculus, without [general relativity] supplying any criterion for the acceptability of that theory."

Hilbert (1915): "Axiom I [notation changed to conform to usage in this book]. The

law of physical events is determined through a world function [Mie's terminology; better known today as "Lagrangian"] L, that contains the following arguments:

$$g_{\mu\nu}, \frac{\partial g_{\mu\nu}}{\partial x^\alpha}, \frac{\partial^2 g_{\mu\nu}}{\partial x^\alpha \partial x^\beta},$$

$$A_\sigma, \frac{\partial A_\sigma}{\partial x^\tau},$$

and specifically the variation of the integral

$$\int L(-g)^{1/2} \, d^4x$$

must vanish for [changes in] every one of the 14 potentials $g_{\sigma\nu}$, A_σ. . . .

"Axiom II (axiom of general invariance). The world function L is invariant with respect to arbitrary transformations of the world parameters [coordinates] x^α. . . .

"For the world function L, still further axioms are needed to make its choice unambiguous. If the gravitation equations are to contain only second derivatives of the potentials $g^{\sigma\nu}$, then L must have the form

$$L = R + L_{\text{elec}},$$

where R is the invariant built from the Riemann tensor (curvature of the four-dimensional manifold)." (Session of Nov. 20, 1915.)

Einstein (1916c): "Recently H. A. Lorentz and D. Hilbert have succeeded in giving general relativity an especially transparent form in deriving its equations from a single variation principle. This will be done also in the following treatment. There it is my aim to present the basic relations as transparently as possible and in a way as general as general relativity allows."

Einstein (1916b): "From this it follows, first of all, that gravitational fields spread out with the speed of light. . . . [plane] waves transport energy. . . . One thus gets . . . the radiation of the system per unit time. . . .

$$\frac{G}{24\pi} \sum_{\alpha,\beta} \left(\frac{\partial^3 J_{\alpha\beta}}{\partial t^3}\right)^2."$$

Hilbert (1917): "As for the principle of causality, the physical quantities and their time-rates of change may be known at the present time in any given coordinate system; a prediction will then have a physical meaning only when it is invariant with respect to all those transformations for which exactly those coordinates used for the present time remain unchanged. I declare that predictions of this kind for the future are all uniquely determined; that is, that the causality principle holds in this formulation:

"From the knowledge of the 14 physical potentials $g_{\mu\nu}$, A_σ, in the present, all predictions about the same quantities in the future follow necessarily and uniquely insofar as they have physical meaning."

CHAPTER **18**

WEAK GRAVITATIONAL FIELDS

The way that can be walked on is not the perfect way.
The word that can be said is not the perfect word.

LAO-TZU (~3rd century B.C.)

§18.1. THE LINEARIZED THEORY OF GRAVITY

Because of the geometric language and abbreviations used in writing them, Einstein's field equations, $G_{\mu\nu} = 8\pi T_{\mu\nu}$, hardly seem to be differential equations at all, much less ones with many familiar properties. The best way to see that they are is to apply them to weak-field situations

$$g_{\mu\nu} = \eta_{\mu\nu} + h_{\mu\nu}, \qquad |h_{\mu\nu}| \ll 1, \tag{18.1}$$

e.g., to the solar system, where $|h_{\mu\nu}| \sim |\Phi| \lesssim M_\odot/R_\odot \sim 10^{-6}$; or to a weak gravitational wave propagating through interstellar space.

In a weak-field situation, one can expand the field equations in powers of $h_{\mu\nu}$, using a coordinate frame where (18.1) holds; and without much loss of accuracy, one can keep only linear terms. The resulting formalism is often called "*the linearized theory of gravity,*" because it is an important theory in its own right. In fact, it is precisely this "linearized theory" that one obtains when one asks for the classical field corresponding to quantum-mechanical particles of (1) zero rest mass and (2) spin two in (3) flat spacetime [see Fierz and Pauli (1939)]. Track-2 readers have already explored linearized theory somewhat in §7.1, exercise 7.3, and Box 7.1. There it went under the alternative name, "tensor-field theory of gravity in flat spacetime."

"Linearized theory of gravity":

(1) as weak-field limit of general relativity

(2) as standard "field-theory" description of gravity in "flat spacetime"

(3) as a foundation for "deriving" general relativity

Just as one can "descend" from general relativity to linearized theory by linearizing about flat spacetime (see below), so one can "bootstrap" one's way back up from linearized theory to general relativity by imposing consistency between the linearized field equations and the equations of motion. or, equivalently, by asking about: (1) the stress-energy carried by the linearized gravitational field $h_{\mu\nu}$; (2) the influence of this stress-energy acting as a source for corrections $h^{(1)}{}_{\mu\nu}$ to the field; (3) the stress-energy carried by the corrections $h^{(1)}{}_{\mu\nu}$; (4) the influence of this stress-energy acting as a source for corrections $h^{(2)}{}_{\mu\nu}$ to the corrections $h^{(1)}{}_{\mu\nu}$; (5) the stress-energy carried by the corrections to the corrections; and so on. This alternative way to derive general relativity has been developed and explored by Gupta (1954, 1957, 1962), Kraichnan (1955), Thirring (1961), Feynman (1963a), Weinberg (1965), and Deser (1970). But because the outlook is far from geometric (see Box 18.1), the details of the derivation are not presented here. (But see part 5 of Box 17.2.)

Details of linearized theory:

Here attention focuses on deriving linearized theory from general relativity. Adopt the form (18.1) for the metric components. The resulting connection coefficients [equations (8.24b)], when linearized in the metric perturbation $h_{\mu\nu}$, read

(1) connection coefficients

$$\Gamma^{\mu}{}_{\alpha\beta} = \frac{1}{2}\eta^{\mu\nu}(h_{\alpha\nu,\beta} + h_{\beta\nu,\alpha} - h_{\alpha\beta,\nu})$$

$$\equiv \frac{1}{2}(h_{\alpha}{}^{\mu}{}_{,\beta} + h_{\beta}{}^{\mu}{}_{,\alpha} - h_{\alpha\beta}{}^{,\mu}). \tag{18.2}$$

The second line here introduces the convention, used routinely whenever one expands in powers of $h_{\mu\nu}$, that indices of $h_{\mu\nu}$ are raised and lowered using $\eta^{\mu\nu}$ and $\eta_{\mu\nu}$, *not* $g^{\mu\nu}$ and $g_{\mu\nu}$. A similar linearization of the Ricci tensor [equation (8.47)] yields

$$R_{\mu\nu} = \Gamma^{\alpha}{}_{\mu\nu,\alpha} - \Gamma^{\alpha}{}_{\mu\alpha,\nu}$$

$$= \frac{1}{2}(h_{\mu}{}^{\alpha}{}_{,\nu\alpha} + h_{\nu}{}^{\alpha}{}_{,\mu\alpha} - h_{\mu\nu,\alpha}{}^{\alpha} - h_{,\mu\nu}), \tag{18.3}$$

where

$$h \equiv h^{\alpha}{}_{\alpha} = \eta^{\alpha\beta}h_{\alpha\beta}. \tag{18.4}$$

After a further contraction to form $R \equiv g^{\mu\nu}R_{\mu\nu} \approx \eta^{\mu\nu}R_{\mu\nu}$, one finds that the Einstein equations, $2G_{\mu\nu} = 16\pi T_{\mu\nu}$, read

$$h_{\mu\alpha,\nu}{}^{\alpha} + h_{\nu\alpha,\mu}{}^{\alpha} - h_{\mu\nu,\alpha}{}^{\alpha} - h_{,\mu\nu}$$

$$- \eta_{\mu\nu}(h_{\alpha\beta}{}^{,\alpha\beta} - h_{,\beta}{}^{\beta}) = 16\pi T_{\mu\nu}. \tag{18.5}$$

The number of terms has increased in passing from $R_{\mu\nu}$ (18.3) to $G_{\mu\nu} = R_{\mu\nu} - \frac{1}{2}g_{\mu\nu}R$ (18.5), but this annoyance can be counteracted by defining

(2) "gravitational potentials" $\bar{h}_{\mu\nu}$

$$\bar{h}_{\mu\nu} \equiv h_{\mu\nu} - \frac{1}{2}\eta_{\mu\nu}h \tag{18.6}$$

and using a bar to imply a corresponding operation on any other symmetric tensor.

Box 18.1 **DERIVATIONS OF GENERAL RELATIVITY FROM GEOMETRIC VIEWPOINT AND FROM SPIN-TWO VIEWPOINT, COMPARED AND CONTRASTED**

	Einstein derivation	*Spin-2 derivation*
Nature of primordial spacetime geometry?	Not primordial; geometry is a dynamic participant in physics	"God-given" flat Lorentz spacetime manifold
Topology (multiple connectedness) of spacetime?	Laws of physics are local; they do not specify the topology	Simply connected Euclidean topology
Vision of physics?	Dynamic geometry is the "master field" of physics	This field, that field, and the other field all execute their dynamics in a flat-spacetime manifold
Starting points for this derivation of general relativity	1. Equivalence principle (world lines of photons and test particles are geodesics of the spacetime geometry) 2. That tensorial conserved quantity which is derived from the curvature (Cartan's moment of rotation) is to be identified with the tensor of stress-momentum-energy (see Chapter 15).	1. Begin with field of spin two and zero rest mass in flat spacetime. 2. Stress-energy tensor built from this field serves as a source for this field.
Resulting equations	Einstein's field equations	Einstein's field equations
Resulting assessment of the spacetime geometry from which derivation started	Fundamental dynamic participant in physics	None. Resulting theory eradicates original flat geometry from all equations, showing it to be unobservable
View about the greatest single crisis of physics to emerge from these equations: complete gravitational collapse	Central to understanding the nature of matter and the universe	Unimportant or at most peripheral

Thus $G_{\mu\nu} = \bar{R}_{\mu\nu}$ to first order in the $h_{\mu\nu}$, and $\bar{\bar{h}}_{\mu\nu} = h_{\mu\nu}$; i.e., $h_{\mu\nu} = \bar{h}_{\mu\nu} - \frac{1}{2}\eta_{\mu\nu}\bar{h}$. With this notation *the linearized field equations become*

$$-\bar{h}_{\mu\nu,\alpha}{}^{\alpha} - \eta_{\mu\nu}\bar{h}_{\alpha\beta,}{}^{\alpha\beta} + \bar{h}_{\mu\alpha,}{}^{\alpha}{}_{\nu} + \bar{h}_{\nu\alpha,}{}^{\alpha}{}_{\mu} = 16\pi T_{\mu\nu}. \qquad (18.7)$$

(3) linearized field equations

The first term in these linearized equations is the usual flat-space d'Alembertian, and the other terms serve merely to keep the equations "gauge-invariant" (see Box

18.2). In Box 18.2 it is shown that, without loss of generality, one can impose the "gauge conditions"

(4) gauge conditions

$$\bar{h}^{\mu\alpha}{}_{,\alpha} = 0. \tag{18.8a}$$

These gauge conditions are the tensor analog of the Lorentz gauge $A^{\alpha}{}_{,\alpha} = 0$ of electromagnetic theory. The field equations (18.7) then become

(5) field equations and metric in Lorentz gauge

$$-\bar{h}_{\mu\nu,\alpha}{}^{\alpha} = 16\pi T_{\mu\nu}. \tag{18.8b}$$

The gauge conditions (18.8a), the field equations (18.8b), and the definition of the metric

$$g_{\mu\nu} = \eta_{\mu\nu} + h_{\mu\nu} = \eta_{\mu\nu} + \bar{h}_{\mu\nu} - \frac{1}{2}\eta_{\mu\nu}\bar{h} \tag{18.8c}$$

are the fundamental equations of the linearized theory of gravity written in Lorentz gauge.

EXERCISES

Exercise 18.1. GAUGE INVARIANCE OF THE RIEMANN CURVATURE

Show that in linearized theory the components of the Riemann tensor are

$$R_{\alpha\mu\beta\nu} = \frac{1}{2}(h_{\alpha\nu,\mu\beta} + h_{\mu\beta,\nu\alpha} - h_{\mu\nu,\alpha\beta} - h_{\alpha\beta,\mu\nu}). \tag{18.9}$$

Then show that these components are left unchanged by a gauge transformation of the form discussed in Box 18.2 [equation (4b)]. Since the Einstein tensor is a contraction of the Riemann tensor, this shows that it is also gauge-invariant.

Exercise 18.2. JUSTIFICATION OF LORENTZ GAUGE

Let a particular solution to the field equations (18.7) of linearized theory be given, in an arbitrary gauge. Show that there necessarily exist four generating functions $\xi_{\mu}(t, x^j)$ whose gauge transformation [Box 18.2, eq. (4b)] makes

$$\bar{h}^{\text{new}\,\mu\alpha}{}_{,\alpha} = 0 \qquad \text{(Lorentz gauge).}$$

Also show that a subsequent gauge transformation leaves this Lorentz gauge condition unaffected if and only if its generating functions satisfy the sourceless wave equation

$$\xi^{\alpha,\beta}{}_{\beta} = 0.$$

Exercise 18.3. EXTERNAL FIELD OF A STATIC, SPHERICAL BODY

Consider the external gravitational field of a static spherical body, as described in the body's (nearly) Lorentz frame—i.e., in a nearly rectangular coordinate system $|h_{\mu\nu}| \ll 1$, in which the body is located at $x = y = z = 0$ for all t. By fiat, adopt Lorentz gauge.

(a) Show that the field equations (18.8b) and gauge conditions (18.8a) imply

$$\bar{h}_{00} = 4M/(x^2 + y^2 + z^2)^{1/2}, \qquad\qquad \bar{h}_{0j} = \bar{h}_{jk} = 0,$$

$$h_{00} = h_{xx} = h_{yy} = h_{zz} = 2M/(x^2 + y^2 + z^2)^{1/2}, \qquad h_{\alpha\beta} = 0 \text{ if } \alpha \neq \beta,$$

where M is a constant (the mass of the body; see §19.3).

Box 18.2 GAUGE TRANSFORMATIONS AND COORDINATE TRANSFORMATIONS IN LINEARIZED THEORY

A. The Basic Equations of Linearized Theory, written in any coordinate system that is nearly globally Lorentz, are (18.1) and (18.7):

$$g_{\mu\nu} = \eta_{\mu\nu} + h_{\mu\nu}, \qquad |h_{\mu\nu}| \ll 1; \tag{1}$$

$$-\bar{h}_{\mu\nu,\alpha}{}^{\alpha} - \eta_{\mu\nu}\bar{h}_{\alpha\beta}{}^{,\alpha\beta} + \bar{h}_{\mu\alpha}{}^{,\alpha}{}_{\nu} + \bar{h}_{\nu\alpha}{}^{,\alpha}{}_{\mu} = 16\pi T_{\mu\nu}. \tag{2}$$

Two different types of coordinate transformations connect nearly globally Lorentz systems to each other: global Lorentz transformations, and infinitesimal coordinate transformations.

1. *Global Lorentz Transformations:*

$$x^{\mu} = \Lambda^{\mu}{}_{\alpha'} x^{\alpha'}, \qquad \Lambda^{\mu}{}_{\alpha'}\Lambda^{\nu}{}_{\beta'}\eta_{\mu\nu} = \eta_{\alpha'\beta'}. \tag{3a}$$

These transform the metric coefficients via

$$\eta_{\alpha'\beta'} + h_{\alpha'\beta'} = g_{\alpha'\beta'} = \frac{\partial x^{\mu}}{\partial x^{\alpha'}}\frac{\partial x^{\nu}}{\partial x^{\beta'}} g_{\mu\nu} = \Lambda^{\mu}{}_{\alpha'}\Lambda^{\nu}{}_{\beta'}(\eta_{\mu\nu} + h_{\mu\nu})$$

$$= \eta_{\alpha'\beta'} + \Lambda^{\mu}{}_{\alpha'}\Lambda^{\nu}{}_{\beta'}h_{\mu\nu}.$$

Thus, $h_{\mu\nu}$—and likewise $\bar{h}_{\mu\nu}$—transform like components of a tensor in flat spacetime

$$h_{\alpha'\beta'} = \Lambda^{\mu}{}_{\alpha'}\Lambda^{\nu}{}_{\beta'}h_{\mu\nu}. \tag{3b}$$

2. *Infinitesimal Coordinate Transformations* (creation of "ripples" in the coordinate system):

$$x^{\mu'}(\mathcal{P}) = x^{\mu}(\mathcal{P}) + \xi^{\mu}(\mathcal{P}), \tag{4a}$$

where $\xi^{\mu}(\mathcal{P})$ are four arbitrary functions small enough to leave $|h_{\mu'\nu'}| \ll 1$. Infinitesimal transformations of this sort make tiny changes in the functional forms of all scalar, vector, and tensor fields. *Example:* the temperature T is a unique function of position, $T(\mathcal{P})$; so when written as a function of coordinates it changes

$$T(x^{\mu'} = a^{\mu}) = T(x^{\mu} + \xi^{\mu} = a^{\mu}) = T(x^{\mu} = a^{\mu} - \xi^{\mu})$$
$$= T(x^{\mu} = a^{\mu}) - T_{,\mu}\xi^{\mu};$$

i.e., if $\xi^{0} = 0.001 \sin(x^{1})$, and if $T = \cos^{2}(x^{0})$, then

$$T = \cos^{2}(x^{0'}) + 0.002 \sin(x^{1'}) \cos(x^{0'}) \sin(x^{0'}).$$

Box 18.2 (*continued*)

These tiny changes can be ignored in all quantities except the metric, where tiny deviations from $\eta_{\mu\nu}$ contain all the information about gravity. The usual tensor transformation law for the metric

$$g_{\rho'\sigma'}[x^{\alpha'}(\mathcal{P})] = g_{\mu\nu}[x^{\alpha}(\mathcal{P})]\frac{\partial x^{\mu}}{\partial x^{\rho'}}\frac{\partial x^{\nu}}{\partial x^{\rho'}},$$

when combined with the transformation law (4a) and with

$$g_{\mu\nu}[x^{\alpha}(\mathcal{P})] = \eta_{\mu\nu} + h_{\mu\nu}[x^{\alpha}(\mathcal{P})],$$

reveals that

$$g_{\rho'\sigma'}(x^{\alpha'} = a^{\alpha}) = \eta_{\rho\sigma} + h_{\rho\sigma}(x^{\alpha} = a^{\alpha}) - \xi_{\rho,\sigma} - \xi_{\sigma,\rho}$$
$$+ \text{ negligible corrections} \sim h_{\rho\sigma,\alpha}\xi^{\alpha} \text{ and} \sim h_{\rho\alpha}\xi^{\alpha}{}_{,\sigma}.$$

Hence, *the metric perturbation functions in the new* ($x^{\mu'}$) *and old* (x^{μ}) *coordinate systems are related by*

$$h_{\mu\nu}^{\text{new}} = h_{\mu\nu}^{\text{old}} - \xi_{\mu,\nu} - \xi_{\nu,\mu}, \tag{4b}$$

whereas the functional forms of all other scalars, vectors, and tensors are unaltered, to within the precision of linearized theory.

B. Gauge Transformations and Gauge Invariance. In linearized theory one usually regards equation (4b) as gauge transformations, analogous to those

$$A_{\mu}^{\text{new}} = A_{\mu}^{\text{old}} + \Psi_{,\mu} \tag{5a}$$

of electromagnetic theory. The fact that gravitational gauge transformations do not affect the functional forms of scalars, vectors, or tensors (i.e., observables) is called "gauge invariance." Just as a straightforward calculation reveals the gauge invariance of the electromagnetic field,

$$F_{\mu\nu}^{\text{new}} = A_{\nu,\mu}^{\text{new}} - A_{\mu,\nu}^{\text{new}} = A_{\nu,\mu}^{\text{old}} + \Psi_{,\nu\mu} - A_{\mu,\nu}^{\text{old}} - \Psi_{,\mu\nu} = F_{\mu\nu}^{\text{old}}, \tag{5b}$$

so a straightforward calculation (exercise 18.1) reveals the gauge invariance of the Riemann tensor

$$R_{\mu\nu\alpha\beta}^{\text{new}} = R_{\mu\nu\alpha\beta}^{\text{old}}. \tag{6}$$

Such gauge invariance was already guaranteed by the fact that $R_{\mu\nu\alpha\beta}$ are the components of a tensor, and are thus essentially the same whether calculated in an orthonormal frame $g_{\hat{\mu}\hat{\nu}} = \eta_{\mu\nu}$, in the old coordinates where $g_{\mu\nu} = \eta_{\mu\nu} + h_{\mu\nu}^{\text{old}}$, or in the new coordinates where $g_{\mu\nu} = \eta_{\mu\nu} + h_{\mu\nu}^{\text{new}}$.

Like the Riemann tensor, the Einstein tensor and the stress-energy tensor are unaffected by gauge transformations. Hence, *if one knows a specific solution $\bar{h}_{\mu\nu}$ to the linearized field equations (2) for a given $T^{\mu\nu}$, one can obtain another solution that describes precisely the same physical situation (all observables unchanged) by the change of gauge (4), in which ξ_{μ} are four arbitrary but small functions.*

C. Lorentz Gauge. One can show (exercise 18.2) that for any physical situation, one can specialize the gauge (i.e., the coordinates) so that $\bar{h}^{\mu\alpha}{}_{,\alpha} = 0$. This is the Lorentz gauge introduced in §18.1. The Lorentz gauge is not fixed uniquely. The gauge condition $\bar{h}^{\mu\alpha}{}_{,\alpha} = 0$ is left unaffected by any gauge transformation for which

$$\xi^{\alpha,\beta}{}_{\beta} = 0.$$

(See exercise 18.2.)

D. Curvilinear Coordinate Systems. Once the gauge has been fixed by fiat for a given system (e.g., the solar system), one can regard $h_{\mu\nu}$ and $\bar{h}_{\mu\nu}$ as components of tensors in flat spacetime; and one can regard the field equations (2) and the chosen gauge conditions as geometric, coordinate-independent equations in flat spacetime. This viewpoint allows one to use curvilinear coordinates (e.g., spherical coordinates centered on the sun), if one wishes. But in doing so, one must everywhere replace the Lorentz components of the metric, $\eta_{\mu\nu}$, by the metric's components $g_{\mu\nu\,\text{flat}}$ in the flat-spacetime curvilinear coordinate system; and one must replace all ordinary derivatives ("commas") in the field equations and gauge conditions by covariant derivatives whose connection coefficients come from $g_{\mu\nu\,\text{flat}}$. See exercise 18.3 for an example.

(b) Adopt spherical polar coordinates,

$$x = r\sin\theta\cos\phi, \qquad y = r\sin\theta\sin\phi, \qquad z = r\cos\theta.$$

By regarding $h_{\mu\nu}$ and $\bar{h}_{\mu\nu}$ as components of tensors in flat spacetime (see end of Box 18.2), and by using the usual tensor transformation laws, put the solution found in (a) into the form

$$\bar{h}_{00} = 4M/r, \qquad \bar{h}_{0j} = \bar{h}_{jk} = 0,$$

$$h_{00} = \frac{2M}{r}, \qquad h_{0j} = 0, \qquad h_{jk} = \frac{2M}{r}g_{jk\,\text{flat}}$$

where $g_{\alpha\beta\,\text{flat}}$ are the components of the flat-spacetime metric in the spherical coordinate system

$$g_{00\,\text{flat}} = -1, \qquad g_{rr\,\text{flat}} = 1, \qquad g_{\theta\theta\,\text{flat}} = r^2,$$

$$g_{\phi\phi\,\text{flat}} = r^2\sin^2\theta, \qquad g_{\alpha\beta\,\text{flat}} = 0 \text{ when } \alpha \neq \beta.$$

Thereby conclude that the general relativistic line element, accurate to linearized order, is

$$ds^2 = -(1 - 2M/r)\,dt^2 + (1 + 2M/r)(dr^2 + r^2\,d\theta^2 + r^2\sin^2\!\theta\,d\phi^2).$$

(c) Derive this general, static, spherically symmetric, Lorentz-gauge, vacuum solution to the linearized field equations from scratch, working entirely in spherical coordinates. [*Hint:* As discussed at the end of Box 18.2, $\eta_{\mu\nu}$ in equation (18.8c) must be replaced by $g_{\mu\nu\,\text{flat}}$; and in the field equations and gauge conditions (18.8a, b), all commas (partial derivatives) must be replaced by covariant derivatives, whose connection coefficients come from $g_{\mu\nu\,\text{flat}}$.] (d) Calculate the Riemann curvature tensor for this gravitational field. The answer should agree with equation (1.14).

§18.2. GRAVITATIONAL WAVES

Linearized theory and electromagnetic theory compared

The gauge conditions and field equations (18.8a, b) of linearized theory bear a close resemblance to the equations of electromagnetic theory in Lorentz gauge and flat spacetime,

$$A^\alpha{}_{,\alpha} = 0, \tag{18.10a}$$

$$-A^\mu{}_{,\alpha}{}^\alpha = 4\pi J^\mu. \tag{18.10b}$$

They differ only in the added index ($h^{\mu\nu}$ versus A^μ, $T^{\mu\nu}$ versus J^μ). Consequently, from past experience with electromagnetic theory, one can infer much about linearized gravitation theory.

For example, the field equations (18.8b) must have gravitational-wave solutions. The analog of the electromagnetic plane wave

$$A^x = A^x(t - z), \qquad A^y = A^y(t - z), \qquad A^z = 0, \qquad A^0 = 0,$$

Plane gravitational waves

will be the gravitational plane wave

$$\bar{h}^{xx} = \bar{h}^{xx}(t - z), \qquad \bar{h}^{xy} = \bar{h}^{xy}(t - z), \qquad \bar{h}^{yy} = \bar{h}^{yy}(t - z),$$

$$\bar{h}^{\mu 0} = \bar{h}^{\mu z} = 0 \text{ for all } \mu. \tag{18.11}$$

Although a detailed study of such waves will be delayed until Chapters 35–37, some properties of these waves are explored in the exercises at the end of the next section.

§18.3. EFFECT OF GRAVITY ON MATTER

How to analyze effects of weak gravity on matter

The effects of weak gravitational fields on matter can be computed by using the linearized metric (18.1) and Christoffel symbols (18.2) in the appropriate equations of motion—i.e., in the geodesic equation (for the motion of particles or light rays), in the hydrodynamic equations (for fluid matter), in Maxwell's equations (for electromagnetic waves), or in the equation $\boldsymbol{\nabla} \cdot \boldsymbol{T} = 0$ for the total stress-energy tensor

of whatever fields and matter may be present. Exercises 18.5, 18.6 and 18.7 provide examples, as do the Newtonian-limit calculations in exercises 16.1 and 16.4, and in §17.4. If, however, the lowest-order (linearized) gravitational "forces" (Christoffel-symbol terms) have a significant influence on the motion of the sources of the gravitational field, one finds that the linearized field equation (18.7) is inadequate, and better approximations to Einstein's equations must be considered. [Thus emission of gravitational waves by a mechanically or electrically driven oscillator falls within the scope of linearized theory, but emission by a double-star system, or by stellar oscillations that gravitational forces maintain, will require discussion of nonlinear terms (gravitational "stress-energy") in the Einstein equations; see §§36.9 to 36.11.]

The above conclusions follow from a consideration of conservation laws associated with the linearized field equation. Just as the electromagnetic equations (18.10a, b) guarantee charge conservation

$$J^{\mu}{}_{,\mu} = 0, \qquad \int_{\text{all space}} J^{0}(t, \mathbf{x})\, dx\, dy\, dz \equiv Q = \text{const},$$

so the gravitational equations (18.8a, b) guarantee conservation of the total 4-momentum and angular momentum of any body bounded by vacuum:

<div style="float:right; font-style:italic;">Conservation of 4-momentum and angular momentum in linearized theory</div>

$$T^{\mu\nu}{}_{,\nu} = 0, \tag{18.12a}$$

$$\int_{\text{body}} T^{\mu 0}(t, \mathbf{x})\, dx\, dy\, dz \equiv P^{\mu} = \text{const}; \tag{18.12b}$$

$$(x^{\alpha} T^{\beta\mu} - x^{\beta} T^{\alpha\mu})_{,\mu} = 0, \tag{18.13a}$$

$$\int_{\text{body}} (x^{\alpha} T^{\beta 0} - x^{\beta} T^{\alpha 0})\, dx\, dy\, dz \equiv J^{\alpha\beta} = \text{const}. \tag{18.13b}$$

(See §5.11 for the basic properties of angular momentum in special relativity. The angular momentum here is calculated relative to the origin of the coordinate system.) Now it is important that the stress-energy components $T^{\mu\nu}$, which appear in the linearized field equations (18.7) and in these conservation laws, are precisely the components one would calculate using special relativity (with $g_{\mu\nu} = \eta_{\mu\nu}$). As a result, the energy-momentum conservation formulated here contains no contributions or effects of gravity! From this one sees that linearized theory assumes that gravitational forces do no significant work. For example, energy losses due to gravitational radiation-damping forces are neglected by linearized theory. Similarly, conservation of 4-momentum P^{μ} for each of the bodies acting as sources of $h_{\mu\nu}$ means that each body moves along a geodesic of $\eta_{\mu\nu}$ (straight lines in the nearly Lorentz coordinate system) rather than along a geodesic of $g_{\mu\nu} = \eta_{\mu\nu} + h_{\mu\nu}$. Thus, linearized theory can be used to calculate the motion of test particles and fields, using $g_{\mu\nu} = \eta_{\mu\nu} + h_{\mu\nu}$; but to include gravitational corrections to the motion of the sources themselves—to allow them to satisfy $T^{\mu\nu}{}_{;\nu} = 0$ rather than $T^{\mu\nu}{}_{,\nu} = 0$—one must reinsert into the field equations the nonlinear terms that linearized theory discards. (See, e.g., Chapter 20 on conservation laws; §§36.9–36.11 on the generation of gravitational waves and radiation reaction; and Chapter 39 on the post-Newtonian approximation.)

<div style="float:right; font-style:italic;">Limit on validity of linearized theory: gravity must not affect motions of sources significantly</div>

The energy, momentum, and angular momentum radiated by gravitational waves in linearized theory can be calculated by special-relativistic methods analogous to those used in electromagnetic theory for electromagnetic waves [Fiertz and Pauli (1939)], but it will be more informative and powerful to use a fully gravitational approach (Chapters 35 and 36).

EXERCISES

**Exercise 18.4. SPACETIME CURVATURE FOR A PLANE
 GRAVITATIONAL WAVE**

Calculate the components of the Riemann curvature tensor [equations (18.9)] for the gravitational plane wave (18.11). [*Answer:*

$$R_{x0x0} = -R_{y0y0} = -R_{x0xz} = +R_{y0yz} = +R_{xzxz} = -R_{yzyz} = -\frac{1}{4}(\bar{h}_{xx} - \bar{h}_{yy})_{,tt};$$

$$R_{x0y0} = -R_{x0yz} = +R_{xzyz} = -R_{xzy0} = -\frac{1}{2}\bar{h}_{xy,tt};$$

all other components vanish except those obtainable from the above by the symmetries $R_{\alpha\beta\gamma\delta} = R_{[\alpha\beta][\gamma\delta]} = R_{\gamma\delta\alpha\beta}$.

**Exercise 18.5. A PRIMITIVE GRAVITATIONAL-WAVE DETECTOR
 (see Figure 18.1)**

Two beads slide almost freely on a smooth stick; only slight friction impedes their sliding. The stick falls freely through spacetime, with its center moving along a geodesic and its ends attached to gyroscopes, so they do not rotate. The beads are positioned equidistant (distance $\frac{1}{2}\ell$) from the stick's center. Plane gravitational waves [equation (18.11) and exercise 18.4], impinging on the stick, push the beads back and forth ("geodesic deviation"; "tidal gravitational forces"). The resultant friction of beads on stick heats the stick; and the passage of the waves is detected by measuring the rise in stick temperature.* (Of course, this is not the best of all conceivable designs!) Neglecting the effect of friction on the beads' motion, calculate the proper distance separating them as a function of time. [*Hints:* Let $\boldsymbol{\xi}$ be the separation between the beads; and let $\boldsymbol{n} = \boldsymbol{\xi}/|\boldsymbol{\xi}|$ be a unit vector that points along the stick in the stick's own rest frame. Then their separation has magnitude $\ell = \boldsymbol{\xi} \cdot \boldsymbol{n}$. The fact that the stick is nonrotating is embodied in a parallel-transport law for \boldsymbol{n}, $\boldsymbol{\nabla}_{\boldsymbol{u}}\boldsymbol{n} = 0$. ("Fermi-Walker transport" of §§6.5, 6.6, and 13.6 reduces to parallel transport, because the stick moves along a geodesic with $\boldsymbol{a} = \boldsymbol{\nabla}_{\boldsymbol{u}}\boldsymbol{u} = 0$.) Thus,

$$d\ell/d\tau = \boldsymbol{\nabla}_{\boldsymbol{u}}(\boldsymbol{\xi} \cdot \boldsymbol{n}) = (\boldsymbol{\nabla}_{\boldsymbol{u}}\boldsymbol{\xi}) \cdot \boldsymbol{n},$$

$$d^2\ell/d\tau^2 = \boldsymbol{\nabla}_{\boldsymbol{u}}\boldsymbol{\nabla}_{\boldsymbol{u}}(\boldsymbol{\xi} \cdot \boldsymbol{n}) = (\boldsymbol{\nabla}_{\boldsymbol{u}}\boldsymbol{\nabla}_{\boldsymbol{u}}\boldsymbol{\xi}) \cdot \boldsymbol{n},$$

where τ is the stick's proper time. But $\boldsymbol{\nabla}_{\boldsymbol{u}}\boldsymbol{\nabla}_{\boldsymbol{u}}\boldsymbol{\xi}$ is produced by the Riemann curvature of the wave (geodesic deviation):

$$\boldsymbol{\nabla}_{\boldsymbol{u}}\boldsymbol{\nabla}_{\boldsymbol{u}}\boldsymbol{\xi} = \text{projection along } \boldsymbol{n} \text{ of } [-\boldsymbol{Riemann}\,(\ldots, \boldsymbol{u}, \boldsymbol{\xi}, \boldsymbol{u})].$$

(The geodesic-deviation forces perpendicular to the stick, i.e., perpendicular to \boldsymbol{n}, are coun-

*This thought experiment was devised by Bondi [1957, 1965; Bondi and McCrea (1960)] as a means for convincing skeptics of the reality of gravitational waves.

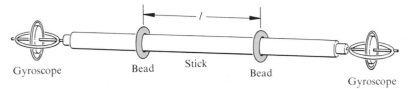

Figure 18.1.
A primitive detector for gravitational waves, consisting of a beaded stick with gyroscopes on its ends [Bondi (1957)]. See exercise 18.5 for discussion.

terbalanced by the stick's pushing back on the beads to stop them from passing through it—no penetration of matter by matter!) Thus,

$$d^2\mathit{l}/d\tau^2 = -\textbf{\textit{Riemann}}\,(\dots, \textbf{\textit{u}}, \textbf{\textit{\xi}}, \textbf{\textit{u}}) \cdot \textbf{\textit{n}} = -\textbf{\textit{Riemann}}\,(\textbf{\textit{n}}, \textbf{\textit{u}}, \textbf{\textit{\xi}}, \textbf{\textit{u}}).$$

Evaluate this acceleration in the stick's local Lorentz frame. Orient the coordinates so the waves propagate in the z-direction and the stick's direction has components $n^z = \cos\theta$, $n^x = \sin\theta\cos\phi$, $n^y = \sin\theta\sin\phi$. Solve the resulting differential equation for $\mathit{l}(\tau)$.] [*Answer:*

$$\mathit{l} = \mathit{l}_0\left[1 + \frac{1}{4}(\bar{h}_{xx} - h_{yy})\sin^2\theta\cos 2\phi + \frac{1}{2}\bar{h}_{xy}\sin^2\theta\sin 2\phi\right],$$

where \bar{h}_{jk} are evaluated on the stick's world line ($x = y = z = 0$). Notice that, if the stick is oriented along the direction of wave propagation (if $\theta = 0$), the beads do not move. In this sense, the effect of the waves (geodesic deviation) is purely transverse. For further discussion, see §§35.4 to 35.6.]

§18.4. NEARLY NEWTONIAN GRAVITATIONAL FIELDS

The general solution to the linearized field equations in Lorentz gauge [equations (18.8a, b)] lends itself to expression as a retarded integral of the form familiar from electromagnetic theory:

Retarded-integral solution of linearized field equation

$$\bar{h}_{\mu\nu}(t, \textbf{\textit{x}}) = \int \frac{4T_{\mu\nu}(t - |\textbf{\textit{x}} - \textbf{\textit{x}}'|, \textbf{\textit{x}}')}{|\textbf{\textit{x}} - \textbf{\textit{x}}'|}\,d^3x'. \qquad (18.14)$$

The gravitational-wave aspects of this solution will be studied in Chapter 36. Here focus attention on a nearly Newtonian source: $T_{00} \gg |T_{0j}|$, $T_{00} \gg |T_{jk}|$, and velocities slow enough that retardation is negligible. In this case, (18.14) reduces to

Newtonian gravity as a limit of linearized theory

$$\bar{h}_{00} = -4\Phi, \qquad \bar{h}_{0j} = \bar{h}_{jk} = 0, \qquad (18.15a)$$

$$\Phi(t, \textbf{\textit{x}}) = -\int \frac{T_{00}(t, \textbf{\textit{x}}')}{|\textbf{\textit{x}} - \textbf{\textit{x}}'|}\,d^3x' = \text{Newtonian potential.} \qquad (18.15b)$$

The corresponding metric (18.8c) is

$$ds^2 = -(1 + 2\Phi)\,dt^2 + (1 - 2\Phi)(dx^2 + dy^2 + dz^2) \qquad (18.15c)$$
$$\approx -(1 - 2M/r)\,dt^2 + (1 + 2M/r)(dx^2 + dy^2 + dz^2) \text{ far from source.}$$

The errors in this metric are: (1) missing corrections of order Φ^2 due to nonlinearities of which linearized theory is oblivious; (2) missing corrections due to setting $\bar{h}_{0j} = 0$ (these are of order $\bar{h}_{0j} \sim \Phi v$, where $v \sim |T_{0j}|/T_{00}$ is a typical velocity in the source); (3) missing corrections due to setting $\bar{h}_{jk} = 0$ [these are of order $\bar{h}_{jk} \sim \Phi(|T_{jk}|/T_{00})$]. In the solar system all these errors are $\sim 10^{-12}$, whereas $\Phi \sim 10^{-6}$.

Passive correspondence with Newtonian theory demanded only that $g_{00} = -(1 + 2\Phi)$; see equation (17.19). However, linearized theory determines all the metric coefficients, up to errors of $\sim \Phi v$, $\sim \Phi^2$, and $\sim \Phi(|T_{jk}|/T_{00})$. This is sufficient accuracy to predict correctly (fractional errors $\sim 10^{-6}$) the bending of light and the gravitational redshift in the solar system, but not perihelion shifts.

Bending of light and gravitational redshift predicted by linearized theory

EXERCISES

Exercise 18.6. BENDING OF LIGHT BY THE SUN

To high precision, the sun is static and spherical, so its external line element is (18.15c) with $\Phi = -M/r$; i.e.,

$$ds^2 = -(1 - 2M/r)\,dt^2 + (1 + 2M/r)(dx^2 + dy^2 + dz^2) \text{ everywhere outside sun.} \quad (18.16)$$

A photon moving in the equatorial plane ($z = 0$) of this curved spacetime gets deflected very slightly from the world line

$$x = t, \qquad y = b \equiv \text{"impact parameter,"} \qquad z = 0. \quad (18.17)$$

Calculate the amount of deflection as follows.

(a) Write down the geodesic equation (16.4a) for the photon's world line,

$$\frac{dp^\alpha}{d\lambda^*} + \Gamma^\alpha{}_{\beta\gamma} p^\beta p^\gamma = 0. \quad (18.18)$$

[Here $\boldsymbol{p} = d/d\lambda^* =$ (4-momentum of photon) = (tangent vector to photon's null geodesic).]

(b) By evaluating the connection coefficients in the equatorial plane, and by using the approximate values, $|p^y| \ll p^0 \approx p^x$, of the 4-momentum components corresponding to the approximate world line (18.17), show that

$$\frac{dp^y}{d\lambda^*} = \frac{-2Mb}{(x^2 + b^2)^{3/2}} p^x \frac{dx}{d\lambda^*}, \qquad p^x = p^0\left[1 + O\left(\frac{M}{b}\right)\right] = \text{const}\left[1 + O\left(\frac{M}{b}\right)\right].$$

(c) Integrate this equation for p^y, assuming $p^y = 0$ at $x = -\infty$ (photon moving precisely in x-direction initially); thereby obtain

$$p^y(x = +\infty) = -\frac{4M}{b} p^x.$$

(d) Show that this corresponds to deflection of light through the angle

$$\Delta\phi = 4M/b = 1''.75 \, (R_\odot/b), \quad (18.19)$$

where R_\odot is the radius of the sun. For a comparison of this prediction with experiment, see Box 40.1.

Exercise 18.7. GRAVITATIONAL REDSHIFT

(a) Use the geodesic equation for a photon, written in the form

$$dp_\mu/d\lambda^* - \Gamma^\alpha{}_{\mu\beta} p_\alpha p^\beta = 0,$$

to prove that any photon moving freely in the sun's gravitational field [line element (18.16)] has $dp_0/d\lambda^* = 0$; i.e.,

$$p_0 = \text{constant along photon's world line.} \qquad (18.20)$$

(b) An atom at rest on the sun's surface emits a photon of wavelength λ_e, as seen in its orthonormal frame. [Note:

$$h\nu_e = h/\lambda_e = (\text{energy atom measures}) = -\boldsymbol{p} \cdot \boldsymbol{u}_e, \qquad (18.21)$$

where \boldsymbol{p} is the photon's 4-momentum and \boldsymbol{u}_e is the emitter's 4-velocity.] An atom at rest far from the sun receives the photon, and measures its wavelength to be λ_r. [Note: $h/\lambda_r = -\boldsymbol{p} \cdot \boldsymbol{u}_r$.] Show that the photon is redshifted by the amount

$$z \equiv \frac{\lambda_r - \lambda_e}{\lambda_e} = \frac{M_\odot}{R_\odot} = 2 \times 10^{-6}. \qquad (18.22)$$

[*Hint:* $\boldsymbol{u}_r = \partial/\partial t$; $\boldsymbol{u}_e = (1 - 2M/r)^{-1/2} \partial/\partial t$. Why?] For further discussion of the gravitational redshift and experimental results, see §§7.4 and 38.5; also Figures 38.1 and 38.2.

CHAPTER **19**

MASS AND ANGULAR MOMENTUM OF A GRAVITATING SYSTEM

§19.1. EXTERNAL FIELD OF A WEAKLY GRAVITATING SOURCE

Metric far from a weakly gravitating system, as a power series in $1/r$:

Consider an isolated system with gravity so weak that in calculating its structure and motion one can completely ignore self-gravitational effects. (This is true of an asteroid, and of a nebula with high-energy electrons and protons spiraling in a magnetic field; it is not true of the Earth or the sun.) Assume nothing else about the system—for example, by contrast with Newtonian theory, allow velocities to be arbitrarily close to the speed of light, and allow stresses T^{jk} and momentum densities T^{0j} to be comparable to the mass-energy density T^{00}.

(1) derivation

Calculate the weak gravitational field,

$$g_{\mu\nu} = \eta_{\mu\nu} + h_{\mu\nu}, \tag{19.1}$$

$$\bar{\bar{h}}_{\mu\nu} \equiv h_{\mu\nu} = \int \frac{4\,\bar{T}_{\mu\nu}(t - |\boldsymbol{x} - \boldsymbol{x}'|, \boldsymbol{x}')}{|\boldsymbol{x} - \boldsymbol{x}'|}\, d^3x', \tag{19.2}$$

produced by such a system [see "barred" version of equation (18.14)]. Restrict attention to the spacetime region far outside the system, and expand $h_{\mu\nu}$ in powers of $\boldsymbol{x}'/r \equiv \boldsymbol{x}'/|\boldsymbol{x}|$, using the relations

$$\bar{T}_{\mu\nu}(t - |\boldsymbol{x} - \boldsymbol{x}'|, \boldsymbol{x}') = \sum_{n=0}^{\infty} \frac{1}{n!}\left[\frac{\partial^n}{\partial t^n}\,\bar{T}_{\mu\nu}(t - r, \boldsymbol{x}')\right](r - |\boldsymbol{x} - \boldsymbol{x}'|)^n, \tag{19.3a}$$

$$r - |\boldsymbol{x} - \boldsymbol{x}'| = x^j\left(\frac{x^{j'}}{r}\right) + \frac{1}{2}\frac{x^j x^k}{r}\left(\frac{x^{j'}x^{k'} - r'^2\,\delta_{jk}}{r^2}\right) + \cdots, \tag{19.3b}$$

$$\frac{1}{|\boldsymbol{x} - \boldsymbol{x}'|} = \frac{1}{r} + \frac{x^j}{r^2}\frac{x^{j'}}{r} + \frac{1}{2}\frac{x^j x^k}{r^3}\frac{(3x^{j'}x^{k'} - r'^2\,\delta_{jk})}{r^2} + \cdots. \tag{19.3c}$$

Perform the calculation in the system's rest frame, where

$$P^j \equiv \int T^{0j}\, d^3x = 0, \tag{19.4a}$$

with origin of coordinates at the system's center of mass

$$\int x^j T^{00}\, d^3x = 0. \tag{19.4b}$$

The result, after a change of gauge to simplify h_{00} and h_{0j}, is

$$ds^2 = -\left[1 - \frac{2M}{r} + 0\!\left(\frac{1}{r^3}\right)\right] dt^2 - \left[4\epsilon_{jk\ell}S^k \frac{x^\ell}{r^3} + 0\!\left(\frac{1}{r^3}\right)\right] dt\, dx^j \tag{19.5}$$
$$+ \left[\left(1 + \frac{2M}{r}\right)\delta_{jk} + \binom{\text{gravitational radiation terms}}{\text{that die out as } 0(1/r)}\right] dx^j\, dx^k.$$

(2) result

(see exercise 19.1 for derivation.) Here M and S^k are the body's mass and intrinsic angular momentum.

$$M = \int T^{00}\, d^3x, \tag{19.6a}$$
$$S_k = \int \epsilon_{k\ell m} x^\ell T^{m0}\, d^3x. \tag{19.6b}$$

The corresponding Newtonian potential is

$$\Phi = -\frac{1}{2}(g_{00} - \eta_{00}) = -\frac{M}{r} + 0\!\left(\frac{1}{r^3}\right). \tag{19.6c}$$

Conclusion: With an appropriate choice of gauge, Φ and g_{00} far from any weak source are time-independent and are determined uniquely by the source's mass M; g_{0j} is time-independent and is fixed by the source's intrinsic angular momentum S^j; but g_{jk} has time-dependent terms (gravitational waves!) of $0(1/r)$.

How metric depends on system's mass M and angular momentum S

The rest of this chapter focuses on the "imprints" of the mass and angular momentum in the gravitational field; the gravitational waves will be ignored, or almost so, until Chapter 35.

Exercise 19.1. DERIVATION OF METRIC FAR OUTSIDE A WEAKLY GRAVITATING BODY

EXERCISE

(a) Derive equation (19.5). [*Hints:* (1) Follow the procedure outlined in the text. (2) When calculating h_{00}, write out explicitly the $n = 0$ and $n = 1$ terms of (19.2), to precision $0(1/r^2)$, and simplify the $n = 0$ term using the identities

$$T^{jk} = \frac{1}{2}(T^{00}x^j x^k)_{,00} + (T^{ij}x^k + T^{ik}x^j)_{,\ell} - \frac{1}{2}(T^{\ell m}x^j x^k)_{,\ell m}, \tag{19.7a}$$

$$T^{\ell\ell}x^m = \left(T^{0\ell}x^\ell x^m - \frac{1}{2}T^{0m}r^2\right)_{,0} + \left(T^{\ell k}x^k x^m - \frac{1}{2}T^{\ell m}r^2\right)_{,\ell}. \tag{19.7b}$$

(Verify that these identities follow from $T^{\alpha\beta}{}_{,\beta} = 0$.) (3) When calculating h_{0m}, write out explicitly the $n = 0$ term of (19.2), to precision $0(1/r^2)$, and simplify it using the identity

$$T^{0k}x^j + T^{0j}x^k = (T^{00}x^jx^k)_{,0} + (T^{0l}x^jx^k)_{,l}. \tag{19.7c}$$

(Verify that this follows from $T^{\alpha\beta}{}_{,\beta} = 0$.) (4) Simplify h_{00} and h_{0m} by the gauge transformation generated by

$$\xi_0 = \frac{1}{2r}\frac{\partial}{\partial t}\int T^{00'}r'^2\,d^3x' + \frac{x^j}{r^3}\int\left(T^{0k'}x^{k'}x^{j'} - \frac{1}{2}T^{0j'}r'^2\right)d^3x'$$

$$+ \int(T_{00'} + T_{ll'})\left[\frac{x^jx^{j'}}{r^2} + \frac{(3x^{j'}x^{k'} - r'^2\delta_{jk})x^jx^k}{2r^4}\right]d^3x'$$

$$+ \sum_{n=2}^{\infty}\frac{1}{n!}\frac{\partial^{n-1}}{\partial t^{n-1}}\int(T_{00'} + T_{kk'})\frac{(r - |x - x'|)^n}{|x - x'|}\,d^3x',$$

$$\xi_m = -\frac{2x^j}{r^3}\int T_{00'}x^{j'}x^{m'}\,d^3x' + 4\sum_{n=1}^{\infty}\frac{1}{n!}\frac{\partial^{n-1}}{\partial t^{n-1}}\int T_{0m'}\frac{(r - |x - x'|)^n}{|x - x'|}\,d^3x'$$

$$+ \frac{x^m}{r}\xi_0 - \frac{1}{2}\left(\frac{1}{r}\right)_{,m}\int T_{00'}r'^2\,d^3x' - \left(\frac{x^k}{r^2}\right)_{,m}\int\left(T^{0j'}x^{j'}x^{k'} - \frac{1}{2}T^{0k'}r'^2\right)d^3x'$$

$$- \sum_{n=2}^{\infty}\frac{1}{n!}\frac{\partial^{n-2}}{\partial t^{n-2}}\int(T_{00'} + T_{kk'})\left[\frac{(r - |x - x'|)^n}{|x - x'|}\right]_{,m}d^3x'$$

Here $T_{\mu\nu}{}'$ denotes $T_{\mu\nu}(t - r, x')$.]

(b) Prove that the system's mass and angular momentum are conserved. [*Note:* Because $T^{\alpha\beta}{}_{,\beta} = 0$ (self-gravity has negligible influence), the proof is no different here than in flat spacetime (Chapter 5).]

§19.2. MEASUREMENT OF THE MASS AND ANGULAR MOMENTUM

For a weakly gravitating system:

The values of a system's mass and angular momentum can be measured by probing the imprint they leave in its external gravitational field. Of all tools one might use to probe, the simplest is a test particle in a gravitationally bound orbit. If the particle is sufficiently far from the source, its motion is affected hardly at all by the source's angular momentum or by the gravitational waves; only the spherical, Newtonian part of the gravitational field has a significant influence. Hence, the particle moves in an elliptical Keplerian orbit. To determine the source's mass M, one need only apply Kepler's third law (perhaps better called "Kepler's 1-2-3 law"):

(1) total mass M can be measured by applying Kepler's "1-2-3" law to orbiting particles

$$M = \left(\frac{2\pi}{\text{orbital period}}\right)^2\left(\begin{matrix}\text{Semi-major axis}\\\text{of ellipse}\end{matrix}\right)^3; \quad \text{i.e., } M^1 = \omega^2a^3. \tag{19.8}$$

The source's angular momentum is not measured quite so easily. One must use a probe that is insensitive to Newtonian gravitational effects, but "feels" the off-diagonal term,

$$g_{0j} = -2\epsilon^{jkl}S^kx^l/r^3, \tag{19.9}$$

in the metric (19.5). One such probe is the precession of the perihelion of a corevolving satellite, relative to the precession for a counterrevolving satellite. A gyroscope is another such probe. Place a gyroscope at rest in the source's gravitational field. By a force applied to its center of mass, prevent it from falling. As time passes, the g_{0j} term in the metric will force the gyroscope to precess relative to the basis vectors $\partial/\partial x^j$; and since these basis vectors are "tied" to the coordinate system, which in turn is tied to the Lorentz frames at infinity, which in turn are tied to the "fixed stars" (cf. §39.12), the precession is relative to the "fixed stars." The angular velocity of precession, as derived in exercise 19.2, is

$$\mathbf{\Omega} = \frac{1}{r^3}\left[-\mathbf{S} + \frac{3(\mathbf{S}\cdot\mathbf{x})\mathbf{x}}{r^2}\right]. \tag{19.10}$$

One sometimes says that the source's rotation "drags the inertial frames near the source," thereby forcing the gyroscope to precess. For further discussion, see §§21.12, 40.7, and 33.4.

Exercise 19.2. GYROSCOPE PRECESSION

EXERCISE

Derive equation (19.10) for the angular velocity of gyroscope precession. [*Hints:* Place an orthonormal tetrad at the gyroscope's center of mass. Tie the tetrad rigidly to the coordinate system, and hence to the "fixed stars"; more particularly, choose the tetrad to be that basis $\{e_{\hat{\alpha}}\}$ which is dual to the following 1-form basis:

$$\boldsymbol{\omega}^{\hat{t}} = [1 - (2M/r)]^{1/2}\,\boldsymbol{dt} + 2\epsilon_{jkl}S^k(x^l/r^3)\,\boldsymbol{dx}^j, \qquad \boldsymbol{\omega}^{\hat{j}} = [1 + (2M/r)]^{1/2}\,\boldsymbol{dx}^j. \tag{19.11}$$

The spatial legs of the tetrad, $e_{\hat{j}}$, rotate relative to the gyroscope with an angular velocity $\boldsymbol{\omega}$ given by [see equation (13.69)]

$$-\epsilon_{\hat{i}\hat{j}\hat{k}}\omega^{\hat{k}} = \Gamma_{\hat{i}\hat{j}\hat{0}}.$$

Consequently, the gyroscope's angular momentum vector \mathbf{L} precesses relative to the tetrad with angular velocity $\mathbf{\Omega} = -\boldsymbol{\omega}$:

$$\frac{dL^{\hat{j}}}{d\hat{t}} = \epsilon_{\hat{j}\hat{k}\hat{l}}\Omega^{\hat{k}}L^{\hat{l}}, \qquad \epsilon_{\hat{i}\hat{j}\hat{k}}\Omega^{\hat{k}} = \Gamma_{\hat{i}\hat{j}\hat{0}}. \tag{19.12}$$

Calculate $\Gamma_{\hat{i}\hat{j}\hat{0}}$ for the given orthonormal frame, and thereby obtain equation (19.10) for $\mathbf{\Omega}$.]

§19.3. MASS AND ANGULAR MOMENTUM OF FULLY RELATIVISTIC SOURCES

Abandon, now, the restriction to weakly gravitating sources. Consider an isolated, gravitating system inside which spacetime may or may not be highly curved—a black hole, a neutron star, the Sun, ... But refuse, for now, to analyze the system's interior or the "strong-field region" near the system. Instead, restrict attention to the weak

gravitational field far from the source, and analyze it using linearized theory in vacuum. Expand $h_{\mu\nu}$ in multipole moments and powers of $1/r$; and adjust the gauge, the Lorentz frame, and the origin of coordinates to simplify the resulting metric. The outcome of such a calculation is a gravitational field identical to that for a weak source [equation (19.5)]! (Details of the calculation are not spelled out here because of their length; but see exercise 19.3.)

But before accepting this as the distant field of an arbitrary source, one should examine the nonlinear effects in the vacuum field equations. Two types of nonlinearities turn out to be important far from the source: (1) nonlinearities in the static, Newtonian part of the metric, which generate metric corrections

$$\delta g_{00} = -2M^2/r^2, \qquad \delta g_{jk} = \frac{3}{2}(M^2/r^2)\,\delta_{jk},$$

(see exercise 19.3 and §39.8), thereby putting the metric into the form

<div style="float:left; width:25%;">

Metric far from *any* gravitating system, as a power series in $1/r$

</div>

$$ds^2 = -\left[1 - \frac{2M}{r} + \frac{2M^2}{r^2} + 0\left(\frac{1}{r^3}\right)\right]dt^2 - \left[4\epsilon_{jkl}S^k\frac{x^l}{r^3} + 0\left(\frac{1}{r^3}\right)\right]dt\,dx^j$$
$$+ \left[\left(1 + \frac{2M}{r} + \frac{3M^2}{2r^2}\right)\delta_{jk} + \left(\begin{array}{l}\text{gravitational radiation terms}\\\text{that die out as } 0(1/r)\end{array}\right)\right]dx^j\,dx^k; \qquad (19.13)$$

(2) a gradual decrease in the source's mass, gradual changes in its angular momentum, and gradual changes in its "rest frame" to compensate for the mass, angular momentum, and linear momentum carried off by gravitational waves (see Box 19.1, which is best read only after finishing this section).

By measuring the distant spacetime geometry (19.13) of a given source, one cannot discover whether that source has strong internal gravity, or weak. But when one expresses the constants M and S_j, which determine g_{00} and g_{0j}, as integrals over the interior of the source, one discovers a crucial difference: if the internal gravity is weak, then linearized theory is valid throughout the source, and

<div style="float:left; width:25%;">

Failure of volume integrals for M and S when source has strong internal gravity

</div>

$$M = \int T_{00}\,d^3x, \qquad S_j = \int \epsilon_{jkl}x^k T^{l0}\,d^3x; \qquad (19.14)$$

but if the gravity is strong, these formulas fail. Does this failure prevent one, for strong gravity, from identifying the constants M and S_j of the metric (19.13) as the source's mass and angular momentum? Not at all, according to the following argument.

Consider, first, the mass of the sun. For the sun one expects Newtonian theory to be highly accurate (fractional errors $\sim M_\odot/R_\odot \sim 10^{-6}$); so one can assert that the constant M appearing in the line element (19.13) is, indeed

$$M = \int \rho\,d^3x = \int T_{00}\,d^3x = \text{total mass}.$$

But might this assertion be wrong? To gain greater confidence and insight, adopt the viewpoint of "*controlled ignorance*"; i.e., do not pretend to know more than what is needed. (This style of physical argument goes back to Newton's famous "Hypotheses non fingo," i.e. "I do not feign hypotheses.") In evaluating the volume integral of T_{00} (usual Newtonian definition of M), one needs a theory of the internal structure

of the sun. For example, one must know that the visible surface layers of the sun do not hide a massive central core, so dense and large that relativistic gravitational fields $|\varPhi| \sim 1$ exist there. If one makes use in the analysis of a fluid-type stress-energy tensor $T^{\mu\nu}$, one needs to know equations of state, opacities, and theories of energy generation and transport. One needs to justify the fluid description as an adequate approximation to the atomic constitution of matter. One needs to assume that an ultimate theory of matter explaining the rest masses of protons and electrons will not assign an important fraction of this mass to strong (nonlinear) gravitational fields on a submicroscopic scale. It is plausible that one could do all this, but it is also obvious that this is not the way the mass of the sun is, in fact, determined by astronomers! Theories of stellar structure are adjusted to give the observed mass; they are not constructed to let one deduce the mass from nongravitational observations. The mass of the sun is measured in practice by studying the orbits of planets in its external gravitational field, a procedure equivalent to reading the mass M off the line element (19.13), rather than evaluating the volume integral $\int T^{00}\, d^3x$.

To avoid all the above uncertainties, and to make theory correspond as closely as possible to experiment, *one defines the "total mass-energy" M of the sun or any other body to be the constant that appears in the line element (19.13) for its distant external spacetime geometry. Similarly, one defines the body's intrinsic angular momentum as the constant 3-vector S appearing in its line element (19.13)*. Operationally, the total mass-energy M is measured via Kepler's third law; the angular momentum S is measured via its influence on the precession of a gyroscope or a planetary orbit. This is as true when the body is a black hole or a neutron star as when it is the sun.

What kind of a geometric object is the intrinsic angular momentum S? It is defined by measurements made far from the source, where, with receding distance, spacetime is becoming flatter and flatter (asymptotically flat). Thus, it can be regarded as a 3-vector in the *"asymptotically flat spacetime"* that surrounds the source. But in what Lorentz frame is S a 3-vector? Clearly, in the asymptotic Lorentz frame where the line element (19.13) is valid; i.e., in the asymptotic Lorentz frame where the source's distant "coulomb" ("M/r") field is static; i.e., in the *"asymptotic rest frame"* of the source. Alternatively, one can regard S as a 4-vector, \mathbf{S}, which is purely spatial ($S^0 = 0$) in the asymptotic rest frame. If one denotes the 4-velocity of the asymptotic rest frame by \mathbf{U}, then the fact that \mathbf{S} is purely spatial can be restated geometrically as $\mathbf{S} \cdot \mathbf{U} = 0$, or

$$\mathbf{S} \cdot \mathbf{P} = 0, \qquad\qquad (19.15)$$

where

$$\mathbf{P} \equiv M\mathbf{U} \equiv \text{``total 4-momentum of source''} \qquad (19.16)$$

is still another vector residing in the asymptotically flat region of spacetime.

The total 4-momentum \mathbf{P} and intrinsic angular momentum \mathbf{S} satisfy conservation laws that are summarized in Box 19.1. These conservation laws are valuable tools in gravitation theory and relativistic astrophysics, but the derivation of these laws (Chapter 20) does not compare in priority to topics such as neutron stars and basic cosmology; so most readers will wish to skip it on a first reading of this book.

(continued on page 456)

Definition of "total mass-energy" M and "angular momentum" S in terms of external gravitational field

S as a geometric object in an asymptotically flat region far outside source

"Asymptotic rest frame" and "total 4-momentum"

Conservation laws for total 4-momentum and angular momentum

Box 19.1 TOTAL MASS-ENERGY, 4-MOMENTUM, AND ANGULAR MOMENTUM OF AN ISOLATED SYSTEM

A. Spacetime is divided into (1) the source's interior; which is surrounded by (2) a strong-field vacuum region; which in turn is surrounded by (3) a weak-field, asymptotically flat, near-zone region; which in turn is surrounded by (4) a weak-field, asymptotically flat, radiation-zone region. This box and this chapter treat only the asymptotically flat regions. The interior and strong-field regions are treated in the next chapter.

Interior

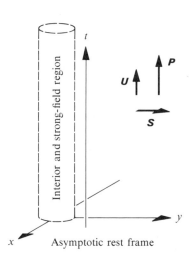

Asymptotic rest frame

B. The asymptotic rest frame of the source is that global, asymptotically Lorentz frame (coordinates t, x, y, z) in which the distant, "coulomb" part of the source's field is at rest (see diagram). The asymptotic rest frame does not extend into the strong-field region; any such extension of it would necessarily be forced by the curvature into a highly non-Lorentz, curvilinear form. The spatial origin of the asymptotic rest frame is so adjusted that the source is centered on it—i.e., that the distant Newtonian potential is $\Phi = -M/(x^2 + y^2 + z^2)^{1/2} + 0(1/r^3)$; i.e., that Φ has no dipole term, $\boldsymbol{D} \cdot \boldsymbol{x}/r^3$, such as would originate from an offset of the coordinates.

C. To the source one can attribute a total mass-energy M, **a 4-velocity** \boldsymbol{U}, **a total 4-momentum** \boldsymbol{P}, **and an intrinsic angular momentum vector,** \boldsymbol{S}. The 4-vectors \boldsymbol{U}, \boldsymbol{P}, and \boldsymbol{S} reside in the asymptotically flat region of spacetime and can be moved about freely there (negligible curvature \Rightarrow parallel transport around closed curves does not change \boldsymbol{U}, \boldsymbol{P}, or \boldsymbol{S}). The source's 4-velocity \boldsymbol{U} is defined to equal the 4-velocity of the asymptotic rest frame ($U^0 = 1$, $\boldsymbol{U} = 0$ in rest frame). The total mass-energy M is measured via Kepler's third ("1-2-3") law [equation (19.8)]. The total 4-momentum is defined by $\boldsymbol{P} \equiv M\boldsymbol{U}$. The intrinsic angular momentum \boldsymbol{S} is orthogonal to the 4-velocity \boldsymbol{U}, $\boldsymbol{S} \cdot \boldsymbol{U} = 0$ (so $S^0 = 0$; $\boldsymbol{S} \neq 0$ in general in asymptotic rest frame); \boldsymbol{S} is measured via gyroscope precession or differential perihelion precession (§19.2).

In the asymptotic rest frame, with an appropriate choice of gauge (i.e., of ripples in the coordinates), *the slight deviations from flat-spacetime geometry are described by the line element*

$$ds^2 = -\left[1 - \frac{2M}{r} + \frac{2M^2}{r^2} + 0\left(\frac{1}{r^3}\right)\right] dt^2 - \left[4\epsilon_{jk\ell}S^k \frac{x^\ell}{r^3} + 0\left(\frac{1}{r^3}\right)\right] dt\,dx^j$$
$$+ \left[\left(1 + \frac{2M}{r} + \frac{3M^2}{2r^2}\right)\delta_{jk} + \text{(gravitational radiation terms)}\right] dx^j\,dx^k. \tag{1}$$

D. Conservation of 4-momentum and angular momentum: Suppose that particles fall into a source or are ejected from it; suppose that electromagnetic waves flow in and out; suppose the source emits gravitational waves. All such processes break the source's isolation and can change its total 4-momentum **P**, its intrinsic angular momentum **S**, and its asymptotic rest frame. Surround the source with a spherical shell \mathcal{S}, which is far enough out to be in the asymptotically flat region. Keep this shell always at rest in the source's momentary asymptotic rest frame. By probing the source's gravitational field near \mathcal{S}, measure its 4-momentum **P** and intrinsic angular momentum **S** as functions of the shell's proper time τ. An analysis given in the next chapter reveals that the 4-momentum is conserved, in the sense that

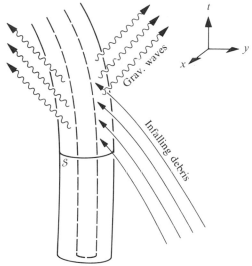

Interstellar debris falls into a black hole, and gravitational waves emerge.

$$\frac{dP^\alpha}{d\tau} = -\int_S T^{\alpha j}n_j\,d(\text{area}) = \begin{pmatrix}\text{rate at which 4-momentum}\\ \text{flows inward through shell}\end{pmatrix}, \tag{2}$$

where **n** is the unit outward normal to \mathcal{S} and the integral is performed in the shell's momentary rest frame. In words: *the rate at which 4-momentum flows through the shell, as measured in the standard special relativistic manner, equals the rate of change of the source's gravitationally measured 4-momentum. Similarly, the angular momentum is conserved in the sense that*

$$\frac{dS_i}{d\tau} = -\int_S (\epsilon_{ijk}x^j T^{k\ell})n_\ell\,d(\text{area}) = \begin{pmatrix}\text{rate at which angular}\\ \text{momentum flows inward}\\ \text{through the shell}\end{pmatrix}, \tag{3a}$$

$$\frac{dS_0}{d\tau} = -\frac{dU^\alpha}{d\tau}S_\alpha = \begin{pmatrix}\text{change required to keep } \textbf{S} \text{ orthogonal to } \textbf{U};\\ \text{``Fermi-Walker-transport law''; cf. §§6.5, 13.6}\end{pmatrix}. \tag{3b}$$

In these conservation laws $T^{\alpha\beta}$ is the total stress-energy tensor at the shell, including contributions from matter, electromagnetic fields, and gravitational waves. The gravitational-wave contribution, called $T^{(GW)\alpha\beta}$, is treated in Chapter 35.

Note: The conservation laws in the form stated above contain fractional errors of order M/r (contributions from "gravitational potential energy" of infalling material), but such errors go to zero in the limit of a very large shell ($r \longrightarrow \infty$).

Note: The formulation of these conservation laws given in the next chapter is more precise and more rigorous, but less physically enlightening than the one here.

EXERCISE

Exercise 19.3. GRAVITATIONAL FIELD FAR FROM A STATIONARY, FULLY RELATIVISTIC SOURCE

Derive the line element (19.13) for the special case of a source that is time-independent ($g_{\mu\nu,t} = 0$). This can be a difficult problem, if one does not proceed carefully along the following outlined route. (1) Initially ignore all nonlinearities in the Einstein field equations. The field is weak far from the source. These nonlinearities will be absent from the dominant terms. (2) Calculate the dominant terms using linearized theory in the Lorentz gauge [equations (18.8)]. (3) In particular, write the general solution to the vacuum, time-independent wave equation (18.8b) in the following form involving $n^j \equiv x^j/r \equiv$ (unit vector in radial direction):

$$\bar{h}_{00} = \frac{A^0}{r} + \frac{B^j n^j}{r^2} + 0\left(\frac{1}{r^3}\right),$$

$$\bar{h}_{0j} = \frac{A^j}{r} + \frac{B^{jk} n^k}{r^2} + 0\left(\frac{1}{r^3}\right),$$

$$\bar{h}_{jk} = \frac{A^{jk}}{r} + \frac{B^{jkl} n^l}{r^2} + 0\left(\frac{1}{r^3}\right),$$

$$A^{jk} = A^{(jk)}, \qquad B^{jkl} = B^{(jk)l}.$$

(19.17)

(Round brackets denote symmetrization.) (4) Then impose the Lorentz gauge conditions $\bar{h}_\alpha{}^\beta{}_{,\beta} = 0$ on this general solution, thereby learning

$$A^j = 0, \qquad A^{jk} = 0,$$

$$B^{jk}(\delta^{jk} - 3n^j n^k) = 0,$$

$$B^{jkl}(\delta^{kl} - 3n^k n^l) = 0.$$

(19.18)

(5) Write B^{jk} as the sum of its trace $3B$, its traceless symmetric part S^{jk}, and its traceless antisymmetric part (these are its "irreducible parts"):

$$B^{jk} = B\delta^{jk} + S^{jk} + \epsilon^{jkl} F^l, \qquad S^{jj} = 0.$$

(19.19)

Show that any tensor B^{jk} can be put into such a form. Then show that the gauge conditions (19.18) imply $S^{jk} = 0$. (6) Similarly show that any tensor B^{jkl} that is symmetric on its first two indices can be put into the form

$$B^{jkl} = \delta^{jk} A^l + C^{(j}\delta^{k)l} + \epsilon^{ml(j} E^{k)m} + S^{jkl},$$

E^{km} symmetric and traceless, i.e., $\quad E^{km} = E^{(km)}, \qquad E^{kk} = 0,$

(19.20)

S^{jkl} symmetric and traceless, i.e., $\qquad S^{jkl} = S^{(jkl)},$

$$S^{jjl} = S^{jkk} = S^{jkj} = 0.$$

Then show that the gauge conditions (19.18) imply $C^j = -2A^j$ and $E^{km} = S^{jkl} = 0$. (7) Combining all these results, conclude that

$$\bar{h}_{00} = \frac{A^0}{r} + \frac{B^j n^j}{r^2} + 0\left(\frac{1}{r^3}\right),$$

$$\bar{h}_{0j} = \frac{\epsilon^{jkl} n^k F^l}{r^2} + \frac{B n^j}{r^2} + 0\left(\frac{1}{r^3}\right),$$

(19.21)

$$\bar{h}_{jk} = \frac{\delta^{jk} A^l n^l - A^j n^k - A^k n^j}{r^2} + 0\left(\frac{1}{r^3}\right).$$

Then use gauge transformations, which stay within Lorentz gauge, to eliminate B and A^j from \bar{h}_{0j} and \bar{h}_{jk}; so

$$\bar{h}_{00} = \frac{A^0}{r} + \frac{(B^j + A^j)n^j}{r^2} + 0\left(\frac{1}{r^3}\right),$$

$$\bar{h}_{0j} = \frac{\epsilon^{jkl}n^k F^l}{r^2} + 0\left(\frac{1}{r^3}\right), \qquad (19.22)$$

$$\bar{h}_{jk} = 0\left(\frac{1}{r^3}\right).$$

(8) Translate the origin of coordinates so $x^j{}_{\text{new}} = x^j{}_{\text{old}} - (B^j + A^j)/A^0$; in the new coordinate system $\bar{h}_{\alpha\beta}$ has the same form as (19.22), but with $B^j + A^j$ removed. From the resultant $\bar{h}_{\alpha\beta}$, construct the metric and redefine the constants A^0 and F^l to agree with equation (19.13). (9) All linear terms in the metric are now accounted for. The dominant nonlinear terms must be proportional to the square, $(M/r)^2$, of the dominant linear term. The easiest way to get the proportionality constant is to take the Schwarzschild geometry for a fully relativistic, static, spherical source [equation (31.1)], by a change of coordinates put it in the form

$$ds^2 = -\left(\frac{1 - M/2r}{1 + M/2r}\right)^2 dt^2 + \left(1 + \frac{M}{2r}\right)^4 (dx^2 + dy^2 + dz^2) \qquad (19.23)$$

(exercise 25.8), and expand it in powers of M/r.

§19.4. MASS AND ANGULAR MOMENTUM OF A CLOSED UNIVERSE

"There are no snakes in Ireland."

Statement of St Patrick
after driving the snakes
out of Ireland (legend*)

There is no such thing as "the energy (or angular momentum, or charge) of a closed universe," according to general relativity, and this for a simple reason. To weigh something one needs a platform on which to stand to do the weighing.

To weigh the sun, one measures the periods and semimajor axes of planetary orbits, and applies Kepler's "1-2-3" law, $M = \omega^2 a^3$. To measure the angular momentum, S, of the sun (a task for space technology in the 1970's or 1980's!), one measures the precession of a gyroscope in a near orbit about the sun, or one examines some other aspect of the "dragging of inertial frames." To determine the electric charge

For a closed universe the total mass-energy M and angular momentum S are undefined and undefinable

*Stokes (1887) and other standard references deny this legend. In part I of Stokes the basic manuscript references are listed, including especially codex manuscript Rawlinson B.512 in 154 folios, in double columns, written by various hands in the fourteenth and fifteenth centuries (*cf. Catalogi codicum manuscriptorum Bibliothecae Bodleianae Partis Quintae Fasciculus Primus*, Oxford, 1862, col. 728–732). In this manuscript, folio 97b.1, line 14, reads in the translation of Stokes, Part I, p. xxx: "as Paradise is without beasts, without a snake, without a lion, without a dragon, without a scorpion, without a mouse, without a frog, so is Ireland in the same manner without any harmful animal, save only the wolf. . ."

of a body, one surrounds it by a large sphere, evaluates the electric field normal to the surface at each point on this sphere, integrates over the sphere, and applies the theorem of Gauss. But within any closed model universe with the topology of a 3-sphere, a Gaussian 2-sphere that is expanded widely enough from one point finds itself collapsing to nothingness at the antipodal point. Also collapsed to nothingness is the attempt to acquire useful information about the "charge of the universe": the charge is trivially zero. By the same token, every "surface integral" (see details in Chapter 20) to determine mass-energy or angular momentum collapses to nothingness. To make the same point in another way: around a closed universe there is no place to put a test object or gyroscope into Keplerian orbit to determine either any so-called "total mass" or "rest frame" or "4-momentum" or "angular momentum" of the system. These terms are undefined and undefinable. Words, yes; meaning, no.

Not having a defined 4-momentum for a closed universe may seem at first sight disturbing; but it would be far more disturbing to be given four numbers and to be told authoritatively that they represent the components of some purported "total energy-momentum 4-vector of the universe." Components with respect to what local Lorentz frame? At what point? And what about the change in this vector on parallel transport around a closed path leading back to that strangely preferred point? It is a happy salvation from these embarrassments that the issue does not and cannot arise!

Imagine a fantastically precise measurement of the energy of a γ-ray. The experimenter wishes to know how much this γ-ray contributes to the total mass-energy of the universe. Having measured its energy in the laboratory, he then corrects it for the negative gravitational energy by which it is bound to the Earth. The result,

$$E_{\text{corrected}} = h\nu(1 - M_\oplus/R_\oplus),$$

is the energy the photon will have after it climbs out of the Earth's gravitational field. But this is only the first in a long chain of corrections for energy losses (redshifts) as the photon climbs out of the gravitational fields of the solar system, the galaxy, the local cluster of galaxies, the supercluster, and then what? These corrections show no sign of converging, unless to $E_{\text{corrected}} = 0$.

Asymptotic flatness as the key to the definability of M and S

Quite in contrast to the charge-energy-angular-momentum facelessness of a closed universe are the attractive possibilities of defining and measuring all three quantities in any space that is asymptotically flat. One does not have to revolutionize present-day views of cosmology to talk of asymptotically flat space. It is enough to note how small is the departure from flatness, as measured by the departure of $(-g_{00})^{1/2}$ from unity, in cases of astronomical or astrophysical interest (Box 19.2). Surrounding a region where any dynamics, however complicated, is going on, whenever the geometry is asymptotically flat to some specified degree of precision, then to that degree of precision it makes sense to speak of the total energy-momentum 4-vector of the dynamic region, \boldsymbol{P}, and its total intrinsic angular momentum, \boldsymbol{S}. Parallel transport of either around any closed curve in the flat region brings it back to its

Box 19.2 METRIC CORRECTION TERM NEAR SELECTED HEAVENLY BODIES

	m	m	r	$\dfrac{m}{r} = 1 - (-g_{00})^{1/2}$
At shoulder of Venus de Milo	2×10^5 g	$= 1.5 \times 10^{-23}$ cm	30 cm	5×10^{-25}
At surface of Earth	6×10^{27} g	$= 4 \times 10^{-1}$ cm	6.4×10^8 cm	6×10^{-10}
At Earth's distance from sun	2×10^{33} g	$= 1.5 \times 10^5$ cm	1.5×10^{13} cm	1×10^{-8}
At sun's distance from center of galaxy	2×10^{44} g	$= 1.5 \times 10^{16}$ cm	2.5×10^{22} cm	6×10^{-7}
At distance of galaxy from center of Virgo cluster of galaxies	6×10^{47} g	$= 4 \times 10^{19}$ cm	3×10^{25} cm	1×10^{-6}

starting point unchanged. Moreover, it makes no difference how enormous are the departures from flatness in the dynamic region (black holes, collapsing stars, intense gravitational waves, etc.); far away the curvature will be weak, and the 4-momentum and angular momentum will reveal themselves by their imprints on the spacetime geometry.

CHAPTER **20**

CONSERVATION LAWS FOR
4-MOMENTUM AND
ANGULAR MOMENTUM

*We denote as energy of a material system in a certain state
the contribution of all effects (measured in mechanical units of
work) produced outside the system when it passes in an arbitrary
manner from its state to a reference state which has been
defined ad hoc.*

WILLIAM THOMPSON (later Lord Kelvin),
as quoted by Max von Laue in Schilpp (1949), p. 514.

All forms of energy possess inertia.

ALBERT EINSTEIN, conclusion
from his paper of September 26, 1905,
as summarized by von Laue in Schilpp (1949), p. 523.

§20.1. OVERVIEW

Chapter 5 (stress-energy
tensor) is needed as
preparation for this chapter,
which in turn is needed as
preparation for the Track-2
portion of Chapter 36
(generation of gravitational
waves) and will be useful in
understanding Chapter 35
(propagation of gravitational
waves).

Chapter 19 expounded the key features of total 4-momentum P and total angular momentum S for an arbitrary, gravitating system. But one crucial feature was left unproved: the conservation laws for P and S (Box 19.1). To prove those conservation laws is the chief purpose of this chapter. But other interesting, if less important, aspects of P and S will be encountered along the route to the proof—Gaussian flux integrals for 4-momentum and angular momentum; a stress-energy "pseudotensor" for the gravitational field, which is a tool in constructing volume integrals for P and S; and the nonlocalizability of the energy of the gravitational field.

§20.2. GAUSSIAN FLUX INTEGRALS FOR 4-MOMENTUM AND ANGULAR MOMENTUM

In electromagnetic theory one can determine the conserved total charge of a source by adding up the number of electric field lines emanating from it—i.e., by performing a Gaussian flux integral over a closed 2-surface surrounding it:

Gaussian flux integrals for charge and Newtonian mass

$$Q = \frac{1}{4\pi} \oint E^j \, d^2S_j = \frac{1}{4\pi} \oint F^{0j} \, d^2S_j. \tag{20.1}$$

Similarly, in Newtonian theory one can determine the mass of a source by evaluating the Gaussian flux integral

$$M = \frac{1}{4\pi} \oint \Phi_{,j} \, d^2S_j. \tag{20.2}$$

These flux integrals work because the charge and mass of a source place indelible imprints on the electromagnetic and gravitational fields that envelop it.

The external gravitational field (spacetime geometry) in general relativity possesses similar imprints, imprints not only of the source's total mass-energy M, but also of its total 4-momentum \boldsymbol{P} and its intrinsic angular momentum \boldsymbol{S} (see Box 19.1). Hence, it is reasonable to search for Gaussian flux integrals that represent the 4-momentum and angular momentum of the source.

To simplify the search, carry it out initially in linearized theory, and use Maxwell electrodynamics as a guide. In electrodynamics the Gaussian flux integral for charge follows from Maxwell's equations $F^{\mu\nu}{}_{,\nu} = 4\pi J^\mu$, plus the crucial fact that $F^{\mu\nu}$ is antisymmetric, so that $F^{0\mu}{}_{,\mu} = F^{0j}{}_{,j}$:

$$Q = \int J^0 \, d^3x = \frac{1}{4\pi} \int F^{0\nu}{}_{,\nu} \, d^3x = \frac{1}{4\pi} \int F^{0j}{}_{,j} \, d^3x = \frac{1}{4\pi} \oint F^{0j} \, d^2S_j.$$

$$\text{[Gauss's theorem]}$$

To find analogous flux integrals in linearized theory, rewrite the linearized field equations (18.7) in an analogous form involving an entity with analogous crucial symmetries. The entity needed turns out to be

$$H^{\mu\alpha\nu\beta} \equiv -(\bar{h}^{\mu\nu}\eta^{\alpha\beta} + \eta^{\mu\nu}\bar{h}^{\alpha\beta} - \bar{h}^{\alpha\nu}\eta^{\mu\beta} - \bar{h}^{\mu\beta}\eta^{\alpha\nu}). \tag{20.3}$$

$H^{\mu\alpha\nu\beta}$ defined

As one readily verifies from this expression, it has the same symmetries as the Riemann tensor

$$H^{\mu\alpha\nu\beta} = H^{\nu\beta\mu\alpha} = H^{[\mu\alpha][\nu\beta]},$$
$$H^{\mu[\alpha\nu\beta]} = 0. \tag{20.4}$$

This entity, like $\bar{h}^{\mu\nu}$, transforms as a tensor under the Lorentz transformations of linearized theory; but it is not gauge-invariant, so it is not a tensor in the general relativistic sense.

Linearized field equations in
terms of $H^{\mu\alpha\nu\beta}$

In terms of $H^{\mu\alpha\nu\beta}$, the linearized field equations (18.7) take on the much simplified form

$$2G^{\mu\nu} = H^{\mu\alpha\nu\beta}{}_{,\alpha\beta} = 16\pi T^{\mu\nu}; \tag{20.5}$$

and from these, by antisymmetry of $H^{\mu\alpha\nu\beta}$ in ν and β, follow the source conservation laws of linearized theory,

$$T^{\mu\nu}{}_{,\nu} = \frac{1}{16\pi} H^{\mu\alpha\nu\beta}{}_{,\alpha\beta\nu} = 0,$$

which were discussed back in §18.3. The same antisymmetry as yields these equations of motion also produces a Gaussian flux integral for the source's total 4-momentum:

Gaussian flux integrals in
linearized theory: (1) for
4-momentum

$$P^\mu = \int T^{\mu 0}\, d^3x = \frac{1}{16\pi}\int H^{\mu\alpha 0\beta}{}_{,\alpha\beta}\, d^3x = \frac{1}{16\pi}\int H^{\mu\alpha 0j}{}_{,\alpha j}\, d^3x$$

$$= \frac{1}{16\pi}\oint_S H^{\mu\alpha 0j}{}_{,\alpha}\, d^2S_j.$$

[Gauss's theorem] $\tag{20.6}$

Here the closed 2-surface of integration \mathcal{S} must completely surround the source and must lie in a 3-surface of constant time x^0. The integral (20.6) for the source's energy P^0, which is used more frequently than the integrals for P^j, reduces to an especially simple form in terms of $g_{\alpha\beta} = \eta_{\alpha\beta} + h_{\alpha\beta}$:

$$P^0 = \frac{1}{16\pi}\int_S (g_{jk,k} - g_{kk,j})\, d^2S_j \tag{20.7}$$

(see exercise 20.1).

(2) for angular momentum

A calculation similar to (20.6), but more lengthy (exercise 20.2), yields a flux integral for total angular momentum about the origin of coordinates:

$$J^{\mu\nu} = \int (x^\mu T^{\nu 0} - x^\nu T^{\mu 0})\, d^3x$$

$$= \frac{1}{16\pi}\oint_S (x^\mu H^{\nu\alpha 0j}{}_{,\alpha} - x^\nu H^{\mu\alpha 0j}{}_{,\alpha} + H^{\mu j 0\nu} - H^{\nu j 0\mu})\, d^2S_j. \tag{20.8}$$

Generalization of Gaussian
flux integrals to full general
relativity

To evaluate the flux integrals in (20.6) to (20.8) (by contrast with the volume integrals), one need utilize only the gravitational field far outside the source. Since that gravitational field has the same form in full general relativity for strong sources as in linearized theory for weak sources, the flux integrals can be used to calculate P^μ and $J^{\mu\nu}$ for any isolated source whatsoever, weak or strong:

$$P^\mu = \frac{1}{16\pi}\oint_S H^{\mu\alpha 0j}{}_{,\alpha}\, d^2S_j,$$

$$P^0 = \frac{1}{16\pi}\oint_S (g_{jk,k} - g_{kk,j})\, d^2S_j,$$

$$J^{\mu\nu} = \frac{1}{16\pi}\oint_S (x^\mu H^{\nu\alpha 0j}{}_{,\alpha} - x^\nu H^{\mu\alpha 0j}{}_{,\alpha}$$
$$+ H^{\mu j 0\nu} - H^{\nu j 0\mu})\, d^2S_j.$$

in full general relativity theory, for any isolated source, when the closed surface of integration \mathcal{S} is in the asymptotically flat region surrounding the source, and when asymptotically Minkowskian coordinates are used. $\tag{20.9}$

Knowing P^μ and $J^{\mu\nu}$, one can calculate the source's total mass-energy M and intrinsic angular momentum S^μ by the standard procedure of Box 5.6:

Total mass-energy, center of mass, and intrinsic angular momentum

$$M = (-P^\mu P_\mu)^{1/2}, \tag{20.10}$$

$$Y^\mu = -J^{\mu\nu}P_\nu/M^2 = \begin{pmatrix} \text{vector by which the source's asymptotic,} \\ \text{``}M/r\text{'', spherical field is displaced from} \\ \text{being centered on the origin of coordinates} \end{pmatrix} \tag{20.11}$$

$$S_\rho = \frac{1}{2}\epsilon_{\mu\nu\sigma\rho}(J^{\mu\nu} - Y^\mu P^\nu + Y^\nu P^\mu)P^\sigma/M. \tag{20.12}$$

Note especially that *the integrands of the flux integrals (20.9) are not gauge-invariant*. In any local inertial frame at an event $\mathcal{P}_0 [g_{\mu\nu}(\mathcal{P}_0) = \eta_{\mu\nu}, g_{\mu\nu,\alpha}(\mathcal{P}_0) = 0]$ they vanish, since

$$g_{\mu\nu,\alpha} = h_{\mu\nu,\alpha} = 0 \Rightarrow H^{\mu\nu\alpha\beta}{}_{,\alpha} = 0; \qquad g_{\mu\nu} = \eta_{\mu\nu} \Rightarrow H^{\mu\nu\alpha\beta} = 0.$$

This is reasonable behavior; their Newtonian analog, the integrand $\Phi_{,j}$ — (gravitational acceleration) of the Newtonian flux integral (20.2), similarly vanishes in local inertial frames.

Although the integrands of the flux integrals are not gauge-invariant, the total integrals P^μ (4-momentum) and $J^{\mu\nu}$ (angular momentum) most assuredly are! They have meaning and significance independent of any coordinate system and gauge. They are tensors in the asymptotically flat region surrounding the source.

The spacetime must be asymptotically flat if there is to be any possibility of defining energy and angular momentum. Only then can linearized theory be applied; and only on the principle that linearized theory applies far away can one justify using the flux integrals (20.9) in the full nonlinear theory. Nobody can compel a physicist to move in close to define energy and angular momentum. He has no need to move in close; and he may have compelling motives not to: the internal structure of the sources may be inaccessible, incomprehensible, uninteresting, dangerous, expensively distant, or frightening. This requirement for far-away flatness is a remarkable feature of the flux integrals (20.9); it is also a decisive feature. Even the coordinates must be asymptotically Minkowskian; otherwise most formulas in this chapter fail or require modification. In particular, *when evaluating the 4-momentum and angular momentum of a localized system, one must apply the flux integrals (20.9) only in asymptotically Minkowskian coordinates. If such coordinates do not exist (spacetime not flat at infinity), one must completely abandon the flux integrals, and the quantities that rely on them for definition: the total mass, momentum, and angular momentum of the gravitating source.* In this connection, recall the discussion of §19.4.

Gaussian flux integrals valid only in asymptotically flat region of spacetime and in asymptotically Minkowskian coordinates

It described, in physical terms, why "total mass-energy" is a limited concept, useful only when one adopts a limited viewpoint that ignores cosmology. (Compare "light ray" or "particle," concepts of enormous value, but concepts that break down when wave optics or wave mechanics enter significantly.)

Summary: Attempts to use formulas (20.9) in ways that lose sight of the Minkowski boundary conditions (and especially simply adopting them unmodified in curvilinear coordinates) easily and unavoidably produce nonsense.

EXERCISES

Exercise 20.1. FLUX INTEGRAL FOR TOTAL MASS-ENERGY IN LINEARIZED THEORY

Show that the flux integral (20.6) for P^0 reduces to (20.7). Then show that, when applied to a nearly Newtonian source [line element (18.15c)], it reduces further to the familiar Newtonian flux integral (20.2).

Exercise 20.2. FLUX INTEGRAL FOR ANGULAR MOMENTUM IN LINEARIZED THEORY

Derive the Gaussian flux integral (20.8) for $J^{\mu\nu}$. [*Hint:* use the field equations (20.5) to show

$$16\pi x^\mu T^{\nu 0} = (x^\mu H^{\nu\alpha 0k}{}_{,\alpha})_{,k} - H^{\nu j0\mu}{}_{,j} - H^{\nu 00\mu}{}_{,0}; \tag{20.13}$$

and then use Gauss's theorem to evaluate the volume integral of equation (20.8)].

Exercise 20.3. FLUX INTEGRALS FOR AN ARBITRARY STATIONARY SOURCE

(a) Use the flux integrals (20.9) to calculate P^μ and $J^{\mu\nu}$ for an arbitrary stationary source. For the asymptotically flat metric around the source, use (19.13), with the gravitational radiation terms set to zero.

(b) Verify that the "auxiliary equations" (20.10) to (20.12) give the correct answer for this source's total mass-energy M and intrinsic angular momentum S^μ.

§20.3. VOLUME INTEGRALS FOR 4-MOMENTUM AND ANGULAR MOMENTUM

It is easy, in linearized theory, to convert the surface integrals for P^μ and $J^{\mu\nu}$ into volume integrals over the source; one can simply trace backward the steps that led to the surface integrals in the first place [equation (20.6); exercise 20.2]. How, in full general relativity, can one similarly convert from the surface integrals to volume integrals? The answer is rather easy, if one thinks in the right direction. One need only put the full Einstein field equations into the form

The full Einstein field equations in terms of $H^{\mu\alpha\nu\beta}$

$$H^{\mu\alpha\nu\beta}{}_{,\alpha\beta} = 16\pi T^{\mu\nu}_{\text{eff}} \tag{20.14}$$

analogous to equations (20.5) of linearized theory. Here $H^{\mu\alpha\nu\beta}$ is to be defined in terms of $h_{\mu\nu} \equiv g_{\mu\nu} - \eta_{\mu\nu}$ by equation (20.3), even deep inside the source where $|h_{\mu\nu}|$ might be $\gtrsim 1$. This form of the Einstein equations then permits a conversion of the Gaussian flux integrals into volume integrals, just as in linearized theory:

Volume integrals for 4-momentum and angular momentum in full general relativity

$$P^\mu = \frac{1}{16\pi} \oint H^{\mu\alpha 0j}{}_{,\alpha} \, d^2S_j = \frac{1}{16\pi} \int H^{\mu\alpha 0j}{}_{,\alpha j} \, d^3x = \frac{1}{16\pi} \int H^{\mu\alpha 0\beta}{}_{,\alpha\beta} \, d^3x$$
$$= \int T^{\mu 0}_{\text{eff}} \, d^3x. \tag{20.15}$$

Similarly,

$$J^{\mu\nu} = \int (x^\mu T^{\nu 0}_{\text{eff}} - x^\nu T^{\mu 0}_{\text{eff}}) \, d^3x. \tag{20.16}$$

[Crucial to the conversion is the use of partial derivatives rather than covariant derivatives in equations (20.14).] In these volume integrals, as throughout the preceeding discussion, the coordinates must become asymptotically Lorentz ($g_{\mu\nu} \to \eta_{\mu\nu}$) far from the source.

The form of $T_{\text{eff}}^{\mu\nu}$ can be calculated by recalling that $H^{\mu\alpha\nu\beta}{}_{,\alpha\beta}$ is a linearized approximation to the Einstein curvature tensor (20.5). Define the nonlinear corrections by

$$16\pi t^{\mu\nu} \equiv H^{\mu\alpha\nu\beta}{}_{,\alpha\beta} - 2G^{\mu\nu}. \qquad (20.17)$$

$t^{\mu\nu}$ ("stress-energy pseudotensor") defined

(To calculate them in terms of $g_{\mu\nu}$ or $h_{\mu\nu} = g_{\mu\nu} - \eta_{\mu\nu}$ is straightforward but lengthy. The precise form of these corrections will never be needed in this book.) Then Einstein's equations read

$$H^{\mu\alpha\nu\beta}{}_{,\alpha\beta} = 16\pi t^{\mu\nu} + 2G^{\mu\nu} = 16\pi(t^{\mu\nu} + T^{\mu\nu}),$$

so that

$$T_{\text{eff}}^{\mu\nu} = T^{\mu\nu} + t^{\mu\nu}. \qquad (20.18)$$

$T_{\text{eff}}^{\mu\nu}$ defined

The quantity $t^{\mu\nu}$ is sometimes called a "stress-energy pseudotensor for the gravitational field." The Einstein field equations (20.14) imply, because $H^{\mu\alpha\nu\beta}{}_{,\alpha\beta}$ is antisymmetric in ν and β, that

$$T_{\text{eff},\nu}^{\mu\nu} = (T^{\mu\nu} + t^{\mu\nu})_{,\nu} = 0. \qquad (20.19)$$

Conservation law for $T_{\text{eff}}^{\mu\nu}$

These equations are equivalent to $T^{\mu\nu}{}_{;\nu} = 0$, but they are written with partial derivatives rather than covariant derivatives—a fact that permits conversions back and forth between volume integrals and surface integrals.

All the quantities $H^{\mu\alpha\nu\beta}$, $T_{\text{eff}}^{\mu\nu}$, and $t^{\mu\nu}$ depend for their definition and existence on the choice of coordinates; they have no existence independent of coordinates; they are not components of tensors or of any other geometric object. Correspondingly, the equations (20.14) to (20.19) involving $T_{\text{eff}}^{\mu\nu}$ and $t^{\mu\nu}$ have no geometric, coordinate-free significance; they are not "covariant tensor equations." There is, nevertheless, adequate invariance under general coordinate transformations to give the values P^μ and $J^{\mu\nu}$ of the volume integrals (20.15) and (20.16) geometric, coordinate-free significance in the asymptotically flat region far outside the source. Although this invariance is hard to see in the volume integrals themselves, it is clear from the surface-integral forms (20.9) that no coordinate transformation which changes the coordinates only inside some spatially bounded region can influence the values of the integrals. For coordinate changes in the distant, asymptotically flat regions, linearized theory guarantees that under Lorentz transformations the integrals for P^μ and $J^{\mu\nu}$ will transform like special relativistic tensors, and that under infinitesimal coordinate transformations (gauge changes) they will be invariant.

$H^{\mu\alpha\nu\beta}$, $t^{\mu\nu}$, and $T_{\text{eff}}^{\mu\nu}$ are coordinate-dependent objects

Because $t^{\mu\nu}$ are not tensor components, they can vanish at a point in one coordinate system but not in another. The resultant ambiguity in defining a localized energy density t^{00} for the gravitational field has a counterpart in ambiguities that exist in

Other, equally good versions of $H^{\mu\alpha\nu\beta}$, $t^{\mu\nu}$, $T^{\mu\nu}_{\text{eff}}$:

(1) $H^{\mu\alpha\nu\beta}_{\text{L-L}}$

(2) $t^{\alpha\beta}_{\text{L-L}}$

(3) $T^{\mu\nu}_{\text{L-L eff}}$

the formal definition of $t^{\mu\nu}$. It is clear that any quantities $H^{\mu\alpha\nu\beta}_{\text{new}}$ which agree with the original $H^{\mu\alpha\nu\beta}$ in the asymptotic weak-field region will give the same values as $H^{\mu\alpha\nu\beta}$ does for the P^μ and $J^{\mu\nu}$ surface integrals (20.9). One especially convenient choice has been given by Landau and Lifshitz (1962; §100), who define

$$H^{\mu\alpha\nu\beta}_{\text{L-L}} = \mathfrak{g}^{\mu\nu}\mathfrak{g}^{\alpha\beta} - \mathfrak{g}^{\alpha\nu}\mathfrak{g}^{\mu\beta}, \tag{20.20}$$

where $\mathfrak{g}^{\mu\nu} \equiv (-g)^{1/2}g^{\mu\nu}$. Landau and Lifshitz show that Einstein's equations can be written in the form

$$H^{\mu\alpha\nu\beta}_{\text{L-L},\alpha\beta} = 16\pi(-g)(T^{\mu\nu} + t^{\mu\nu}_{\text{L-L}}), \tag{20.21}$$

where the Landau-Lifshitz pseudotensor components

$$
\begin{aligned}
(-g)\,t^{\alpha\beta}_{\text{L-L}} = \frac{1}{16\pi}\Big\{ &\mathfrak{g}^{\alpha\beta}{}_{,\lambda}\mathfrak{g}^{\lambda\mu}{}_{,\mu} - \mathfrak{g}^{\alpha\lambda}{}_{,\lambda}\mathfrak{g}^{\beta\mu}{}_{,\mu} + \tfrac{1}{2}g^{\alpha\beta}g_{\lambda\mu}\mathfrak{g}^{\lambda\nu}{}_{,\rho}\mathfrak{g}^{\rho\mu}{}_{,\nu} \\
&- (g^{\alpha\lambda}g_{\mu\nu}\mathfrak{g}^{\beta\nu}{}_{,\rho}\mathfrak{g}^{\mu\rho}{}_{,\lambda} + g^{\beta\lambda}g_{\mu\nu}\mathfrak{g}^{\alpha\nu}{}_{,\rho}\mathfrak{g}^{\mu\rho}{}_{,\lambda}) + g_{\lambda\mu}g^{\nu\rho}\mathfrak{g}^{\alpha\lambda}{}_{,\nu}\mathfrak{g}^{\beta\mu}{}_{,\rho} \\
&+ \tfrac{1}{8}(2g^{\alpha\lambda}g^{\beta\mu} - g^{\alpha\beta}g^{\lambda\mu})(2g_{\nu\rho}g_{\sigma\tau} - g_{\rho\sigma}g_{\nu\tau})\mathfrak{g}^{\nu\tau}{}_{,\lambda}\mathfrak{g}^{\rho\sigma}{}_{,\mu} \Big\}
\end{aligned}
\tag{20.22}
$$

are precisely quadratic in the first derivatives of the metric. (Einstein also gave a pseudotensor $t_E{}^\mu{}_\nu$ with this property, but it was not symmetric and so did not lead to an integral for $J^{\mu\nu}$.) Because $H^{\mu\alpha\nu\beta}_{\text{L-L}}$ has the same symmetries as $H^{\mu\alpha\nu\beta}$ and equals $H^{\mu\alpha\nu\beta}$ far from the source (exercise 20.4), and because the field equations (20.21) in terms of $H^{\mu\alpha\nu\beta}_{\text{L-L}}$ have the same form as in terms of $H^{\mu\alpha\nu\beta}$, it follows that

$$T^{\mu\nu}_{\text{L-L eff}} \equiv (-g)(T^{\mu\nu} + t^{\mu\nu}_{\text{L-L}}) \tag{20.23a}$$

has all the properties of the $T^{\mu\nu}_{\text{eff}}$ introduced earlier in this section:

$$T^{\mu\nu}_{\text{L-L eff},\nu} = 0, \tag{20.23b}$$

$$P^\mu = \int T^{\mu 0}_{\text{L-L eff}}\, d^3x, \tag{20.23c}$$

$$J^{\mu\nu} = \int (x^\mu T^{\nu 0}_{\text{L-L eff}} - x^\nu T^{\mu 0}_{\text{L-L eff}})\, d^3x. \tag{20.23d}$$

EXERCISE

Exercise 20.4. FORM OF $H^{\mu\alpha\nu\beta}_{\text{L-L}}$ FAR FROM SOURCE

Show that the entities $H^{\mu\alpha\nu\beta}_{\text{L-L}}$ of equations (20.20) reduce to $H^{\mu\alpha\nu\beta}$ (20.3) in the weak-field region far outside the source.

§20.4. WHY THE ENERGY OF THE GRAVITATIONAL FIELD CANNOT BE LOCALIZED

Consider an element of 3-volume $d\Sigma_\nu$ and evaluate the contribution of the "gravitational field" in that element of 3-volume to the energy-momentum 4-vector, using

in the calculation either the pseudotensor $t^{\mu\nu}$ or the pseudotensor $t^{\mu\nu}_{\text{L-L}}$ discussed in the last section. Thereby obtain

$$\boldsymbol{p} = \boldsymbol{e}_\mu t^{\mu\nu} \, d\Sigma_\nu$$

or

$$\boldsymbol{p} = \boldsymbol{e}_\mu t^{\mu\nu}_{\text{L-L}} \, d\Sigma_\nu.$$

Right? No, the question is wrong. The motivation is wrong. The result is wrong. The idea is wrong.

To ask for the amount of electromagnetic energy and momentum in an element of 3-volume makes sense. First, there is one and only one formula for this quantity. Second, and more important, this energy-momentum in principle "has weight." It curves space. It serves as a source term on the righthand side of Einstein's field equations. It produces a relative geodesic deviation of two nearby world lines that pass through the region of space in question. It is observable. Not one of these properties does "local gravitational energy-momentum" possess. There is no unique formula for it, but a multitude of quite distinct formulas. The two cited are only two among an infinity. Moreover, "local gravitational energy-momentum" has no weight. It does not curve space. It does not serve as a source term on the righthand side of Einstein's field equations. It does not produce any relative geodesic deviation of two nearby world lines that pass through the region of space in question. It is not observable.

Anybody who looks for a magic formula for "local gravitational energy-momentum" is looking for the right answer to the wrong question. Unhappily, enormous time and effort were devoted in the past to trying to "answer this question" before investigators realized the futility of the enterprise. Toward the end, above all mathematical arguments, one came to appreciate the quiet but rock-like strength of Einstein's equivalence principle. One can always find in any given locality a frame of reference in which all local "gravitational fields" (all Christoffel symbols; all $\Gamma^\alpha{}_{\mu\nu}$) disappear. No Γ's means no "gravitational field" and no local gravitational field means no "local gravitational energy-momentum."

Nobody can deny or wants to deny that gravitational forces make a contribution to the mass-energy of a gravitationally interacting system. The mass-energy of the Earth-moon system is less than the mass-energy that the system would have if the two objects were at infinite separation. The mass-energy of a neutron star is less than the mass-energy of the same number of baryons at infinite separation. Surrounding a region of empty space where there is a concentration of gravitational waves, there is a net attraction, betokening a positive net mass-energy in that region of space (see Chapter 35). At issue is not the existence of gravitational energy, but the localizability of gravitational energy. It is not localizable. The equivalence principle forbids.

Look at an old-fashioned potato, replete with warts and bumps. With an orange marking pen, mark on it a "North Pole" and an "equator". The length of the equator is very far from being equal to 2π times the distance from the North Pole to the

Why one cannot define a localized energy-momentum for the gravitational field

equator. The explanation, "curvature," is simple, just as the explanation, "gravitation", for the deficit in mass of the earth-moon system (or deficit for the neutron star, or surplus for the region of space occupied by the gravitational waves) is simple. Yet it is not possible to ascribe the deficit in the length of the equator in the one case, or in mass in the other case, in any uniquely right way to different elements of the manifold (2-dimensional in the one case, 3-dimensional in the other). Look at a small region on the surface of the potato. The geometry there is locally flat. Look at any small region of space in any of the three gravitating systems. In an appropriate coordinate system it is free of gravitational field. The over-all effect one is looking at is a global effect, not a local effect. That is what the mathematics cries out. That is the lesson of the nonuniqueness of the $t^{\mu\nu}$!

§20.5. CONSERVATION LAWS FOR TOTAL 4-MOMENTUM AND ANGULAR MOMENTUM

Consider a system such as our galaxy or the solar system, which is made up of many gravitating bodies. Some of the bodies may be highly relativistic (black holes; neutron stars), while others are not. However, insist that in the regions between the bodies spacetime be nearly flat (gravity be weak)—so flat, in fact, that one can cover the entire system with coordinates which are (almost) globally inertial, except in a small neighborhood of each body where gravity may be strong. Such coordinates can exist only if the Newtonian gravitational potential, $\Phi \approx \frac{1}{2}(\eta_{00} - g_{00})$, in the interbody region is small:

$$\Phi_{\text{interbody}} \sim (\text{Mass of system})/(\text{radius of system}) \ll 1.$$

The solar system certainly satisfies this condition ($\Phi_{\text{interbody}} \sim 10^{-7}$), as does the Galaxy ($\Phi_{\text{interbody}} \sim 10^{-6}$), as do clusters of galaxies ($\Phi_{\text{interbody}} \sim 10^{-6}$); *but the universe as a whole does not* ($\Phi_{\text{interbody}} \sim 1$)!

In evaluating volume integrals for the system's total 4-momentum, split its volume into a region containing each body (denoted "A") plus an interbody region; and neglect the pseudotensor contribution from the almost-flat interbody region:

Total 4-momentum and angular momentum for a system of gravitating bodies

$$
\begin{aligned}
P^{\mu}_{\text{system}} &= \sum_A \int_A T^{\mu 0}_{\text{eff}} \, d^3x + \int_{\substack{\text{interbody} \\ \text{region}}} T^{\mu 0}_{\text{eff}} \, d^3x \\
&= \sum_A P^{\mu}_A + \int_{\substack{\text{interbody} \\ \text{region}}} T^{\mu 0} \, d^3x.
\end{aligned}
\tag{20.24a}
$$

Because spacetime is asymptotically flat around each body, $P_A{}^\mu$ is the 4-momentum of body A as measured gravitationally by an experimenter near it. The integral of $T^{\mu 0}$ over the interbody region is the contribution of any gas, particles, or magnetic

fields out there to the total 4-momentum. A similar breakup of the angular momentum reads

$$J^{\mu\nu}_{\text{system}} = \sum_A J^{\mu\nu}_A + \int_{\substack{\text{interbody} \\ \text{region}}} (x^\mu T^{\nu 0} - x^\nu T^{\mu 0})\, d^3x. \qquad (20.24b)$$

In operational terms, these breakups show that *the total 4-momentum and angular momentum of the system, as measured gravitationally by an experimenter outside it, are sums of P^μ and $J^{\mu\nu}$ for each individual body, as measured gravitationally by an experimenter near it, plus contributions of the usual special-relativistic type from the interbody matter and fields.* This is true even if some of the bodies are hurtling through the system with speeds near that of light; their gravitationally measured P^μ and $J^{\mu\nu}$ contribute, on an equal footing with anyone else's, to the system's total P^μ and $J^{\mu\nu}$!

Surround this asymptotically flat system by a two-dimensional surface \mathcal{S} that is at rest in some asymptotic Lorentz frame. Then the 4-momentum and angular momentum inside \mathcal{S} change at a rate (as measured in \mathcal{S}'s rest frame) given by

<div style="text-align:right">Rates of change of total 4-momentum and angular momentum:</div>

$$\frac{dP^\mu}{dt} = \frac{d}{dt}\int T^{\mu 0}_{\text{eff}}\, d^3x = \int T^{\mu 0}_{\text{eff},0}\, d^3x = -\int T^{\mu j}_{\text{eff},j}\, d^3x$$
$$= -\oint T^{\mu j}_{\text{eff}}\, d^2S_j, \qquad (20.25)$$

<div style="text-align:right">(1) expressed as flux integrals of $T^{\mu\nu}_{\text{eff}}$</div>

and similarly

$$\frac{dJ^{\mu\nu}}{dt} = -\oint_{S_2} (x^\mu T^{\nu j}_{\text{eff}} - x^\nu T^{\mu j}_{\text{eff}})\, d^2S_j. \qquad (20.26)$$

Although the pseudotensor $t^{\mu\nu}$, in the interbody region and outside the system, contributes negligibly to the total 4-momentum and angular momentum (by assumption), its contribution via gravitational waves to the time derivatives dP^μ/dt and $dJ^{\mu\nu}/dt$ can be important when added up over astronomical periods of time. Thus, one must not ignore it in the flux integrals (20.25), (20.26).

In evaluating these flux integrals, it is especially convenient to use the Landau-Lifshitz form of $T^{\mu\nu}_{\text{eff}}$, since that form contains no second derivatives of the metric. Thus set

$$T^{\mu\nu}_{\text{eff}} = (-g)(T^{\mu\nu} + t^{\mu\nu}_{\text{L-L}}) \approx (T^{\mu\nu} + t^{\mu\nu}_{\text{L-L}}),$$

where $t^{\mu\nu}_{\text{L-L}}$ are given by equations (20.22). Only those portions of $t^{\mu\nu}_{\text{L-L}}$ that die out as $1/r^2$ or $1/r^3$ at large r can contribute to the flux integrals (20.25), (20.26). For static solutions [$g_{\mu\nu} \sim$ const. $+ O(1/r)$], $t^{\mu\nu}_{\text{L-L}}$ dies out as $1/r^4$. Hence, the only contributions come from dynamic parts of the metric, which, at these large distances, are entirely in the form of gravitational waves. The study of gravitational waves in Chapter 35 will reveal that when $t^{\mu\nu}_{\text{L-L}}$ is averaged over several wavelengths, it becomes a stress-energy tensor $T^{(\text{GW})\mu\nu}$ for the waves, which has all the properties one ever requires of any stress-energy tensor. (For example, via Einstein's equations

$G^{(B)\mu\nu} = 8\pi T^{(GW)\mu\nu}$, it contributes to the "background" curvature of the spacetime through which the waves propagate.) Moreover, averaging $t^{\mu\nu}_{L\text{-}L}$ over several wavelengths before evaluating the flux integrals (20.25), (20.26) cannot affect the values of the integrals. Therefore, one can freely make in these integrals the replacement

$$T^{\mu\nu}_{\text{eff}} = T^{\mu\nu} + T^{(GW)\mu\nu},$$

thereby obtaining

(2) expressed as flux integrals
of $T^{\mu\nu} + T^{(GW)\mu\nu}$

$$-\frac{dP^\mu}{dt} = \oint_S (T^{\mu j} + T^{(GW)\mu j})\, d^2 S_j, \tag{20.27}$$

$$-\frac{dJ^{\mu\nu}}{dt} = \oint_S [x^\mu(T^{\nu j} + T^{(GW)\nu j}) - x^\nu(T^{\mu j} + T^{(GW)\mu j})]\, d^2 S_j. \tag{20.28}$$

These are tensor equations in the asymptotically flat spacetime surrounding the system. All reference to pseudotensors and other nontensorial entities has disappeared.

Equations (20.27) and (20.28) say that *the rate of loss of 4-momentum and angular momentum from the system, as measured gravitationally, is precisely equal to the rate at which matter, fields, and gravitational waves carry off 4-momentum and angular momentum.*

This theorem is extremely useful in thought experiments where one imagines changing the 4-momentum or angular momentum of a highly relativistic body (e.g., a rotating neutron star) by throwing particles onto it from far away [see, e.g., Hartle (1970)].

EXERCISE

Exercise 20.5. TOTAL MASS-ENERGY IN NEWTONIAN LIMIT

(a) Calculate $t^{\alpha\beta}_{L\text{-}L}$ for the nearly Newtonian metric

$$ds^2 = -(1 + 2\Phi)\, dt^2 + (1 - 2\Phi)\, \delta_{jk}\, dx^j\, dx^k$$

(see §18.4). Assume the source is slowly changing, so that time derivatives of Φ can be neglected compared to space derivatives. [*Answer:*

$$t^{00}_{L\text{-}L} = -\frac{7}{8\pi}\Phi_{,j}\Phi_{,j},$$

$$t^{0j}_{L\text{-}L} = 0,$$

$$t^{jk}_{L\text{-}L} = \frac{1}{4\pi}(\Phi_{,j}\Phi_{,k} - \tfrac{1}{2}\,\delta_{jk}\Phi_{,l}\Phi_{,l}).] \tag{20.29}$$

(*Note:* $t^{jk}_{L\text{-}L}$ as given here is the "stress tensor for a Newtonian gravitational field"; *cf.* exercises 39.5 and 39.6.)

(b) Let the source of the gravitational field be a perfect fluid with

$$T^{\mu\nu} = (\rho + p)u^\mu u^\nu + pg^{\mu\nu}, \qquad p/\rho \sim v^2 \equiv (dx/dt)^2 \sim |\Phi|.$$

Let the Newtonian potential satisfy the source equation

$$\Phi_{,jj} = 4\pi\rho.$$

Show that the energy of the source is

$$P^0 = \int (T^{00} + t^{00})(-g)\, d^3x$$
$$= \int [\underbrace{\rho/(1 - v^2)^{1/2}}_{} + \underbrace{\tfrac{1}{2}\rho v^2}_{} + \underbrace{\tfrac{1}{2}\rho\Phi}_{}](\underbrace{g_{xx}g_{yy}g_{zz})^{1/2}\, dx\, dy\, dz}_{} \qquad (20.30)$$

$$\begin{bmatrix}\text{Lorentz} \\ \text{contraction} \\ \text{factor}\end{bmatrix} \quad \begin{bmatrix}\text{kinetic} \\ \text{energy}\end{bmatrix} \quad \begin{bmatrix}\text{potential} \\ \text{energy}\end{bmatrix} \quad \begin{bmatrix}\text{proper} \\ \text{volume}\end{bmatrix}$$

$$+ \text{ higher-order corrections.}$$

(c) Show that the "equations of motion" $T^{\mu\nu}_{\text{L-L eff},\nu} = 0$ reduce to the standard equations (16.3) of Newtonian hydrodynamics.

§20.6. EQUATIONS OF MOTION DERIVED FROM THE FIELD EQUATION

Consider the Einstein field equation

$$\boldsymbol{G} = 8\pi\boldsymbol{T} \qquad (20.31)$$

under conditions where space is empty of everything except a source-free electromagnetic field:

$$T^{\mu\nu} = \frac{1}{4\pi}\left(F^{\mu\alpha}g_{\alpha\beta}F^{\nu\beta} - \frac{1}{4}g^{\mu\nu}F_{\sigma\tau}F^{\sigma\tau}\right) \qquad (20.32)$$

(*cf.* the expression for stress-energy tensor of the electromagnetic field in §5.6). To predict from (20.31) how the geometry changes with time, one has to know how the electromagnetic field changes with time. The field is expressed as the "exterior derivative" of the 4-potential,

$$\boldsymbol{F} = \boldsymbol{dA} \text{ (language of forms)}$$

or

$$F_{\mu\nu} = \frac{\partial A_\nu}{\partial x^\mu} - \frac{\partial A_\mu}{\partial x^\nu} \text{ (language of components)}, \qquad (20.33)$$

and the time rate of change of the field is governed by the Maxwell equation

$$\boldsymbol{d*F} = 0$$

or

$$F^{\mu\nu}{}_{;\nu} = 0. \qquad (20.34)$$

Vacuum Maxwell equations
derived from Einstein field
equation

If it seems a fair division of labor for the Maxwell equation to predict the development in time of the Maxwell field and the Einstein equation to do the same for the Einstein field, then it may come as a fresh surprise to discover that the Einstein equation (20.31), plus expression (20.32) for the Maxwell stress-energy, can do both jobs. One does not have to be given the Maxwell "equation of motion" (20.34). One can derive it fresh from (20.31) plus (20.32). The proof proceeds in five steps (see also exercise 3.18 and §5.10). Step one: The Bianchi identity $\nabla \cdot \boldsymbol{G} \equiv 0$ implies conservation of energy-momentum $\nabla \cdot \boldsymbol{T} = 0$. Step two: Conservation expresses itself in the language of components in the form

$$0 = 8\pi T^{\mu\nu}{}_{;\nu} = 2F^{\mu\alpha}{}_{;\nu}g_{\alpha\beta}F^{\nu\beta} + 2F^{\mu\alpha}g_{\alpha\beta}F^{\nu\beta}{}_{;\nu}$$
$$- g^{\mu\nu}F_{\sigma\tau;\nu}F^{\sigma\tau}. \tag{20.35}$$

Step three: Leaving the middle term unchanged, rearrange the first term so that, like the last term, it carries a factor $F^{\sigma\tau}$. Thus in that first term let the indices $\nu\beta$ of $F^{\nu\beta}$ be replaced in turn by $\sigma\tau$ and by $\tau\sigma$, to subdivide that term into

$$F^{\mu\alpha}{}_{;\sigma}g_{\alpha\tau}F^{\sigma\tau} + F^{\mu\alpha}{}_{;\tau}g_{\alpha\sigma}F^{\tau\sigma}$$
$$= (F^{\mu}{}_{\tau;\sigma} - F^{\mu}{}_{\sigma;\tau})F^{\sigma\tau}$$
$$= g^{\mu\nu}(F_{\nu\tau;\sigma} + F_{\sigma\nu;\tau})F^{\sigma\tau}. \tag{20.36}$$

Step four: Combine the first and the last terms in (20.35) to give

$$g^{\mu\nu}(F_{\nu\tau;\sigma} + F_{\sigma\nu;\tau} + F_{\tau\sigma;\nu})F^{\sigma\tau}. \tag{20.37}$$

The indices on the derivatives of the field quantities stand in cyclic order. This circumstance annuls all the terms in the connection coefficients $\Gamma^{\alpha}{}_{\beta\gamma}$ when one writes out the covariant derivatives explicitly. Thus one can replace the covariant derivatives by ordinary derivatives. Moreover, these three derivatives annul one another identically when one substitutes for the fields their expressions (20.33) in terms of the potentials. Consequently, nothing remains in the conservation law (20.35) except the middle term, giving rise to four statements ($\mu = 0, 1, 2, 3$)

$$F^{\mu}{}_{\beta}F^{\beta\nu}{}_{;\nu} = 0 \tag{20.38}$$

about the four quantities ($\beta = 0, 1, 2, 3$)

$$F^{\beta\nu}{}_{;\nu}. \tag{20.39}$$

Step five: The determinant of the coefficients in the four equations (20.38) for the four unknowns (20.39) has the value

$$\begin{vmatrix} F^0{}_0 F^0{}_1 F^0{}_2 F^0{}_3 \\ \cdots \cdots \cdots \cdots \\ \cdots \cdots \cdots \cdots \\ F^3{}_0 F^3{}_1 F^3{}_2 F^3{}_3 \end{vmatrix} = -(\boldsymbol{E} \cdot \boldsymbol{B})^2 \tag{20.40}$$

(see exercise 20.6, part i). In the generic case, this one function of the four variables (t, x, y, z) vanishes on one or more hypersurfaces; but off any such hypersurface (i.e., at "normal points" in spacetime) it differs from zero. At all normal points, the solution of the four linear equations (20.38) with their nonvanishing determinant gives identically zero for the four unknowns (20.39); that is to say, Maxwell's "equations of motion"

$$F^{\beta\nu}{}_{;\nu} = 0$$

are fulfilled and must be fulfilled as a straight consequence of Einstein's field equation (20.31)—plus expression 20.32 for the stress-energy tensor. Special cases admit counterexamples (see exercise 20.8); but in the generic case one need not invoke Maxwell's equations of motion; one can deduce them from the Einstein field equation.

Turn from the dynamics of the Maxwell field itself to the dynamics of a charged particle moving under the influence of the Maxwell field. Make no more appeal to outside providence for the Lorentz equation of motion than for the Maxwell equation of motion. Instead, to generate the Lorentz equation call once more on the Einstein field equation or, more directly, on its consequence, the principle of the local conservation of energy-momentum.

Lorentz force equation derived from the Einstein field equation

Keep track of the world line of the particle from $t = t$ to $t = t + \Delta t$ (Figure 20.1). Generate a "world tube" around this world line. Thus, at each value of the time coordinate t, take the location of the particle as center; construct a sphere of radius ϵ around this center; and note how the successive spheres sweep out the desired world tube. Construct "caps" on this tube at times t and $t + \Delta t$. The two caps, together with the world tube proper, bound a region of spacetime in which energy and momentum can be neither created nor destroyed ("no creation of moment of rotation," in the language of the Bianchi identities, Chapter 15). Therefore the energy-momentum emerging out of the "top" cap has to equal the energy-momentum entering the "bottom" cap, supplemented by the amount of energy-momentum carried in across the world-tube by the Maxwell field. Out of such an analysis, as performed in flat spacetime, one ends up with the Lorentz equation of motion in its elementary form (see Chapters 3 and 4),

Figure 20.1.
"World tube." The change in the 4-momentum of the particle is governed by the flow of 4-momentum across the boundary of the world tube.

$$dp/d\tau = e\langle F, u\rangle \qquad \text{(language of forms)}$$

or in curved spacetime, the Lorentz equation of motion in covariant form,

$$\nabla_u p = m\,\nabla_u u = e\langle F, u\rangle \qquad \text{(form language)}$$

or

$$m\left[\frac{d^2x^\alpha}{d\tau^2} + \Gamma^\alpha{}_{\mu\nu}\frac{dx^\mu}{d\tau}\frac{dx^\nu}{d\tau}\right] = eF^\alpha{}_\beta\frac{dx^\beta}{d\tau} \quad \text{(component language).} \quad (20.41)$$

"One ends up with the Lorentz equation of motion"—but only after hurdling problems of principle along the way. One would understand what a particle is if one understood how to do the calculation of balance of energy-momentum with all rigor! Few calculations in all of physics have been done in so many ways by so many leading investigators, from Lorentz and his predecessors to Dirac and Rohrlich [see Teitelboim (1970, 1971) for still further insights]. Among the issues that develop are two that never cease to compel attention. (1) The particle responds according to the Lorentz force law (20.41) to a field. This field is the sum of a contribution from external sources and from the particle itself. How is the field exerted by the particle on itself to be calculated? Insofar as it is not already included in its effects in the "experimental mass" m in (20.41), this force is to be calculated as half the difference between the retarded field and the advanced field caused by that particle (see §36.11 for a more detailed discussion of the corresponding point for an emitter of gravitational radiation). This difference is singularity-free. On the world line, it has the following simple value [valid in general for point particles; valid for finite-sized particles when and only when the particle changes its velocity negligibly compared to the speed of light during the light-travel time across itself—see, e.g., Burke (1970)]

A particle acted on by its own electromagnetic field ("radiation damping")

$$\frac{1}{2}(F_{\text{ret}} - F_{\text{adv}})^{\mu\nu} = \frac{2e}{3}\left(\frac{dx^\mu}{d\tau}\frac{d^3x^\nu}{d\tau^3} - \frac{d^3x^\mu}{d\tau^3}\frac{dx^\nu}{d\tau}\right). \qquad (20.42)$$

Every acceptable line of reasoning has always led to expression (20.42). It also represents the field required to reproduce the long-known and thoroughly tested law of radiation damping. (2) "Infinite self-energy." Around a particle at rest, or close to a particle in an arbitrary state of motion, the field is e/r^2 and the field energy is

Infinite self-energy of a point particle

$$(1/8\pi)\int_{r_{\min}}^{\epsilon} (e/r^2)^2 4\pi r^2\, dr = (e^2/2)(r_{\min}^{-1} - \epsilon^{-1}). \qquad (20.43)$$

This expression diverges as r_{\min} is allowed to go to zero. To hurdle this difficulty, one arranges the calculation of energy balance in such a way that there always appears the sum of this "self-energy" and the "bare mass." The two terms individually are envisaged as "going to infinity" as r_{\min} goes to zero; but the sum is identified with the "experimental mass" and is required to remain finite. Of course, no particle is a classical object. A proper calculation of the energy has to be conducted at the quantum level. There it is easier to hide from sight the separate infinities—but they

are still present, and promise to remain until the structure of a particle is understood.

Before one turns from the Maxwell and Lorentz equations of motion to a final example (deriving the geodesic equations of motion for an uncharged particle), is it not time to object to the whole program of "deriving an equation of motion from Einstein's field equation"? First, is it not a pretensious parade of pomposity to say it comes "from Einstein's field equation" (and even more, "from Einstein's field equations") when it really comes from a principle so elementary and long established as the law of conservation of 4-momentum? It cannot be contested that this conservation principle, in historical fact, came before geometrodynamics, just as it came before electrodynamics and before the theories of all other established fields. However, in no theory but Einstein's is this principle incorporated as an identity. Only here does the conservation of energy-momentum appear as a fully automatic consequence of the inner working of the machinery of the world (energy density tied to moment of rotation, and moment of rotation automatically conserved; see Chapter 17). Out of Einstein's theory one can derive the equation of motion of a particle. Out of Maxwell's one cannot. Thus, nothing prevents one from acting on a charge with an "external" force, over and above the Lorentz force, nor from tailoring this force in such a way that the charge follows some prescribed world line ("engine-driven source"). It makes no difficulties whatsoever for Maxwell's equations that one has shifted attention from a world line that follows the Lorentz equation of motion to one that does not. Quite the contrary is true in general relativity. To shift from right world line (geodesic) to wrong world line makes the difference between satisfying Einstein's field equation in the vicinity of that world line and being unable to satisfy Einstein's field equation.

The Maxwell field equations are so constructed that they automatically fulfill and demand the conservation of charge; but not everything has charge. The Einstein field equation is so constructed that it automatically fulfills and demands the conservation of momentum-energy; and everything does have energy. The Maxwell field equations are indifferent to the interposition of an "external" force, because that force in no way threatens the principle of conservation of charge. The Einstein field equation cares about every force, because every force is a medium for the exchange of energy.

Electromagnetism has the motto, "I count all the electric charge that's here." All that bears no charge escapes its gaze.

"I weigh all that's here" is the motto of spacetime curvature. No physical entity escapes this surveillance.

Why, then, is the derivation of the geodesic equation of motion of an object said to be based on "Einstein's geometrodynamic field equation" rather than on "the principle of conservation of 4-momentum"? Because geometry responds by its curvature to mass-energy in every form. Most of all, because geometry outside tells about mass-energy inside, free of all concern about issues of internal structure (violent motions, unknown forces, tortuously curved and even multiply-connected geometry).

If one objection to the plan to derive the equation of motion of a particle "from the field equation" has been disposed of, then the moment has come to deal with

Why one is justified to regard equations of motion as consequences of the Einstein field equation

How one can avoid
complexities of particle
structure when deriving
equations of motion: the
"external viewpoint"

the other natural objection: Is there not an inner contradiction in trying to apply
to a "particle" (implying idealization to a point) a field equation that deals with
the continuum? Answer: There *is* a contradiction in dealing with a point. Therefore
do not deal with a point. Do not deal with internal structure at all. Analyze the
motion by looking at the geometry outside the object. That geometry provides all
the handle one needs to follow the motion.

Already here one sees the difference from the derivation of the Lorentz equation
of motion as sketched out above. There (1) no advantage was taken of geometry
outside as indicator of motion inside; (2) a detailed bookkeeping was envisaged of
the localization in space of the electromagnetic energy; and (3) this bookkeeping
brought up the issue of the internal structure of the particle, which could not be
satisfactorily resolved.

Derivation of geodesic
motion from Einstein field
equation:

Now begin the analysis in the new geometrodynamic spirit. Surrounding "the
Schwarzschild zone of influence" of the object, mark out a "buffer zone" (Figure
20.2) that extends out to the region where the "background geometry" begins to
depart substantially from flatness. Idealize the geometry in the buffer zone as that
of an unchanging source merging asymptotically ("boundary \mathscr{B} of buffer zone") into
flat space. It suffices to recall the properties of the spacetime geometry far outside

(1) derivation in brief

an unchanging (i.e., nonradiating) source (exercise 19.3) to draw the key conclusion:
relative to this flat spacetime and regardless of its internal structure, the object
remains at rest, or continues to move in a straight line at uniform velocity (conserva-
tion of total 4-momentum; §20.5). In other words, it obeys the geodesic equation
of motion. If this is the result in a flash, then it is appropriate to go back a step
to review it, to find out what it means and what it demands.

(2) derivation with care

When the object is absent and the background geometry alone has to be consid-
ered, then the geodesic is a well-defined mathematical construct. Moreover, Fermi-
Walker transport along this geodesic gives a well-defined way to construct a comoving
local inertial frame (see §13.6). Relative to this frame, the representative point of
the geodesic remains for all time at rest at the origin.

Coupling of curvature to
particle moments produces
deviations from geodesic
motion

In what way does the presence of the object change this picture? The object
possesses an angular momentum, mass quadrupole moments, and higher multipole
moments. They interact with the tide-producing accelerations (Riemann curvature)
of the background geometry. Depending on the orientation in space of these mo-
ments, the interactions drive the object off its geodesic course in one direction or
another (see §40.9). These anomalies in the motion go hand in hand with anomalies
in the geometry. On and near the ideal mathematical geodesic the metric is Min-
kowskian. At a point removed from this geodesic by a displacement with Riemann
normal coordinates ξ^1, ξ^2, ξ^3 (see §11.6), the metric components differ from their
canonical values $(-1, 1, 1, 1)$ by amounts proportional (1) to the squares and
products of the ξ^m and (2) to the components of the Riemann curvature tensor
(tide-producing acceleration) of the background geometry. These second-order terms
produce departures from ideality in the buffer zone, departures that may be described
symbolically as of order

$$\delta(\text{metric}) \sim r^2 \cdot R \cdot (\text{spherical harmonic of order two}). \qquad (20.44)$$

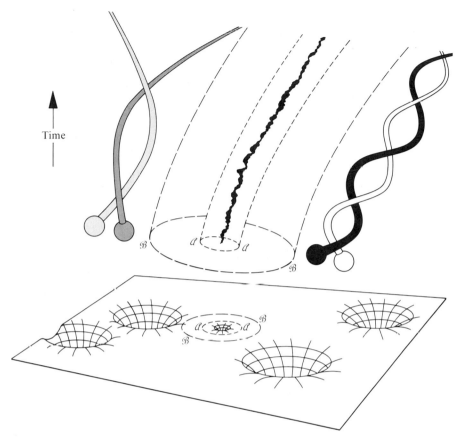

Time

Figure 20.2.
"Buffer zone": the shell of space between \mathcal{A} and \mathcal{B}, where the geometry is appropriately idealized as the spherically symmetric "Schwarzschild geometry" of a localized center of attraction (the object under study) in an asymptotically flat space. Inside \mathcal{A}: the "zone of influence" of the object. In the general case where this object lacks all symmetry, the metric is found to depart more and more from ideal "Schwarzschild character" as the exploration of the geometry is carried inward from \mathcal{A} (effect of angular momentum of the object on the metric; effect of quadrupole moment; effect of higher moments). Outside \mathcal{B}: the "background geometry." As this geometry is explored at greater and greater distances outside \mathcal{B}, it is found to depart more and more from flatness (effect of concentrations of mass, gravitational waves, and other geometrodynamics).

Here r is the distance from the geodesic and R is the magnitude of the significant components of the curvature tensor. The object produces not only the standard "Schwarzschild" departure from flatness,

$$\delta(\text{metric}) \sim m/r, \tag{20.45}$$

which by itself (in a flat background) would bring about no departure from geodesic motion, but also correction terms which may be symbolized as

$$\delta(\text{metric}) \sim (S/r^2) \text{ (spherical harmonic of order one)} \qquad (20.46)$$

and

$$\delta(\text{metric}) \sim (f/r^3) \text{ (spherical harmonic of order two)} \qquad (20.47)$$

and higher-order terms. Here $S(\text{cm}^2)$ is a typical component of the angular momentum vector or "spin"; $f(\text{cm}^3)$ is a representative component of the moment of inertia or quadrupole tensor (see Chapter 36 for details), and higher terms have higher-order coefficients.

Coupling of spin to curvature

The tide-producing acceleration generated by the surroundings of the object ("background geometry") acts on the spin of the object with a force of order RS and pulls it away from geodesic motion with an acceleration of the order

$$\text{acceleration (cm}^{-1}) \sim \frac{R(\text{cm}^{-2}) \, S(\text{cm}^2)}{m(\text{cm})} \qquad (20.48)$$

(see exercise 40.8). Otherwise stated, the surrounding and the spin both put warps in the geometry, and these warps conspire together to push the object off track.

The sum of the relevant two perturbations in the metric is qualitatively of the form

$$\delta g \sim r^2 R + S/r^2. \qquad (20.49)$$

The sum is least where r has a value of the order

$$r \sim (S/R)^{1/4}, \qquad (20.50)$$

and there it has the magnitude

$$\delta g \sim (SR)^{1/2}. \qquad (20.51)$$

To "derive the geodesic equation of motion with some preassigned accuracy ϵ" may be defined to mean that the metric in the buffer zone is Minkowskian within the latitude ϵ. In the illustrative example, this means that $(SR)^{1/2}$ is required to be of the order of ϵ or less. Nothing can be done about the value of R because the background curvature R is a feature of the background geometry. One can meet the requirement only by imposing limits on the mass and moments of the object. In the example, where the dominating moment is the angular momentum, one must require that this parameter of the object be less in order of magnitude than the limit

$$S \sim \epsilon^2/R. \qquad (20.52)$$

Evidently this and similar conditions on the higher moments are most easily satisfied by demanding that the object have spherical symmetry ($S = 0$, $f = 0$, higher

moments $= 0$). Then the perturbation in the metric, again disregarding angle factors and indices, is qualitatively of the form

$$\delta g \sim r^2 R + m/r, \tag{20.53}$$

and the buffer zone is best designed to bracket the minimizing value of r,

$$r_{\mathscr{A}} \leq [r \sim (m/R)^{1/3}] \leq r_{\mathscr{B}}. \tag{20.54}$$

The departure of the metric from Minkowskian perfection in the buffer zone is of the order

$$\delta g \sim (m^2 R)^{1/3}. \tag{20.55}$$

To achieve any preassigned accuracy ϵ for δg, one must demand that the mass be less than a limit of the order

$$m \sim \epsilon^{3/2}/R^{1/2}. \tag{20.56}$$

No object of finite mass moving under the influence of a complex background will admit a buffer zone where the geometry approaches Minkowskian values with arbitrary precision. Therefore it is incorrect to say that such an object follows a geodesic world line. It is meaningless to say that an object of finite rest mass follows a geodesic world line. World line of what? If the object is a black hole, there is no point inside its "horizon" (capture surface; one-way membrane; see Chapters 33 and 34) that is relevant to the physics going on outside. Geodesic world line within what background geometry? It has no sense to speak of a geometry that "lies behind" or is "background to" a black hole.

Turn from one motion of one object in one spacetime to a continuous one-parameter family of spacetimes, with the mass m of the object being the parameter that distinguishes one of these solutions of Einstein's field equation from another. Go to the limit $m = 0$. Then the size of the buffer zone shrinks to zero and the departure of the metric from Minkowskian perfection in the buffer zone also goes to zero. In this limit ("test particle"), it makes sense to say that the object moves in a straight line with uniform velocity in the local inertial frame or, otherwise stated, it pursues a geodesic in the background geometry. Moreover, this background geometry is well-defined: it is the limit of the spacetime geometry as the parameter m goes to zero [see Infeld and Schild (1949)]. In this sense, the geodesic equation of motion follows as an inescapable consequence of Einstein's field equation.

The concept of "background" as limit of a one-parameter family of spacetimes extends itself to the case where the object bears charge as well as mass, and where the surrounding space is endowed with an electromagnetic field. This time the one-parameter family consists of solutions of the combined Einstein-Maxwell equations. The charge-to-mass ratio e/m is fixed. The mass m is again the adjustable parameter. In the limit when m goes to zero, one is left with (1) a background geometry, (2) a background electromagnetic field, and (3) a world line that obeys

The sense in which *no* body can move on a geodesic of spacetime

The sense in which test particles *do* move on geodesics of a background geometry

Motion of a charged test particle in curved spacetime

the general-relativity version of the Lorentz equation of motion in this background as a consequence of the field equations [Chase (1954)]. In contrast, a so-called "unified field theory of gravitation and electromagnetism" that Einstein tentatively put forward at one stage of his thinking, as a conceivable alternative to the combination of his standard 1915 geometrodynamics with Maxwell's standard electrodynamics, has been shown [Callaway (1953)] to lead to the wrong equation of motion for a charged particle. It moves as if uncharged no matter how much charge is piled on its back. If that theory were correct, no cyclotron could operate, no atom could exist, and life itself would be impossible.

Thus the ability to yield the correct equation of motion of a particle has today become an added ace in the hand of general relativity. The idea for such a treatment dates back to Einstein and Grommer (1927). Corrections to the geodesic equation of motion arising from interaction between the spin of the object (when it has finite dimensions) and the curvature of the background geometry are treated by Papapetrou (1951) and more completely by Pirani (1956) (see exercise 40.8). A book on the subject exists [Infeld and Plebanski (1960)]. Section 40.9 describes how corrections to geodesic motion affect lunar and planetary orbits. Some of the problems that arise when the object under study fragments or emits a directional stream of radiation, and unresolved issues of principle, are discussed by Wheeler (1961).

When one turns from the limit of infinitesimal mass to an object of finite mass, no simpler situation presents itself then a system of uncharged black holes (Chapter 33). Everything about the motion of these objects follows from an application of the source-free Einstein equation $\mathbf{G} = 0$ to the region of spacetime outside the horizons (see Chapter 34) of the several objects. The theory of motion is then geometrodynamics and nothing but geometrodynamics.

It has to be emphasized that all the considerations on motion in this section are carried out in the context of classical theory. In the real world of quantum physics, the geometry everywhere experiences unavoidable, natural, zero-point fluctuations (Chapter 43). The calculated local curvatures associated with these fluctuations at the Planck scale of distances $[L = (\hbar G/c^3)^{1/2} = 1.6 \times 10^{-33}$ cm] are enormous $[R \sim 1/L^2 \sim 0.4 \times 10^{66}$ cm$^{-2}]$ compared to the curvature produced on much larger scales by any familiar object (electron or star). No detailed analysis of the interaction of these two curvatures has ever been made. Such an analysis would define a smoothed-out average of the geometry over regions larger than the local quantum fluctuations. With respect to this average geometry, the object will follow geodesic motion: this is the expectation that no one has ever seen any reason to question—but that no one has proved.

References on derivation of equations of motion from Einstein field equation

Quantum mechanical limitations on the derivation

EXERCISES

**Exercise 20.6. SIMPLE FEATURES OF THE ELECTROMAGNETIC FIELD
 AND ITS STRESS-ENERGY TENSOR**

(a) Show that the "scalar" $- 1/2 F_{\alpha\beta}F^{\alpha\beta}$ (invariant with respect to coordinate transformations) and the "pseudoscalar" $1/4 F_{\alpha\beta}{}^*F^{\alpha\beta}$ (reproduces itself under a coordinate transformation up to a \pm sign, according as the sign of the Jacobian of the transformation is positive

or negative) have in any local inertial frame the values $E^2 - B^2$ and $E \cdot B$, respectively ("the two Lorentz invariants" of the electromagnetic field).

(b) Show that the Poynting flux $(E \times B)/4\pi$ is less in magnitude than the energy density $(E^2 + B^2)/8\pi$, save for the exceptional case where both Lorentz invariants of the field vanish (case where the field is locally "null").

(c) A charged pith ball is located a small distance from the North Pole of a bar magnet. Draw the pattern of electric and magnetic lines of force, indicating where the electromagnetic field is "null" in character. Is it legitimate to say that a "null field" is a "radiation field"?

(d) A plane wave is traveling in the z-direction. Show that the corresponding electromagnetic field is everywhere null.

(e) Show that the superposition of two monochromatic plane waves traveling in different directions is null on at most a set of points of measure zero.

(f) In the "generic case" where the field (E, B) at the point of interest is not null, show that the Poynting flux is reduced to zero by viewing the field from a local inertial frame that is traveling in the direction of $E \times B$ with a velocity

$$v = \tanh \alpha, \tag{20.57}$$

where the velocity parameter α is given by the formula

$$\tanh 2\alpha = \frac{(\text{Poynting flux})}{(\text{energy density})} = \frac{2|E \times B|}{E^2 + B^2}. \tag{20.58}$$

(g) Show that all components of the electric and magnetic field in this new frame can be taken to be zero except E_x and B_x.

(h) Show that the 4×4 determinant built out of the components of the field in mixed representation, $F_\alpha{}^\beta$, is invariant with respect to general coordinate transformations. (Hint: Use the theorem that the determinant of the product of three matrices is equal to the product of the determinants of those three matrices.)

(i) Show that this determinant has the value $-(E \cdot B)^2$ by evaluating it in the special local inertial frame of (f).

(j) Show that in this special frame the Maxwell stress-energy tensor has the form

$$\|T^\mu{}_\nu\| = \frac{E_x{}^2 + B_x{}^2}{8\pi} \begin{Vmatrix} -1 & 0 & 0 & 0 \\ 0 & -1 & 0 & 0 \\ 0 & 0 & +1 & 0 \\ 0 & 0 & 0 & +1 \end{Vmatrix} \tag{20.59}$$

(Faraday tension along the lines of force; Faraday pressure at right angles to the lines of force).

(k) In the other case, where the field is locally null, show that one can always find a local inertial frame in which the field has the form $E = (0, F, 0)$, $B = (0, 0, F)$ and the stress-energy tensor has the value

$$\|T^\mu{}_\nu\| = \frac{F^2}{4\pi} \begin{Vmatrix} -1 & 1 & 0 & 0 \\ -1 & 1 & 0 & 0 \\ 0 & 0 & 0 & 0 \\ 0 & 0 & 0 & 0 \end{Vmatrix} \quad (\mu \text{ for row, } \nu \text{ for column}). \tag{20.60}$$

(l) Regardless of whether the electromagnetic field is or is not null, show that the Maxwell stress-energy tensor has zero trace, $T^\mu{}_\mu = 0$, and that its square is a multiple of the unit tensor,

$$T^\mu{}_\alpha T^\alpha{}_\nu = \frac{\delta^\mu{}_\nu}{(8\pi)^2} [(E^2 - B^2)^2 + (2E \cdot B)^2]$$

$$= \frac{\delta^\mu{}_\nu}{(8\pi)^2} [(E^2 + B^2)^2 - (2E \times B)^2]. \tag{20.61}$$

Exercise 20.7. THE STRESS-ENERGY TENSOR DETERMINES THE ELECTROMAGNETIC FIELD EXCEPT FOR ITS COMPLEXION

(a) Given a non-zero symmetric 4×4 tensor $T^{\mu\nu}$ which has zero trace $T^{\mu}{}_{\mu} = 0$ and whose square is a multiple, $M^4/(8\pi)^2$, of the unit matrix, show that, according as this multiple is zero ("null case") or positive, the tensor can be transformed to the form (20.60) or (20.59) by a suitable rotation in 3-space or by a suitable choice of local inertial frame, respectively.

(b) In the generic (non-null) case in the frame in question, show that $T^{\mu\nu}$ is the Maxwell tensor of the "extremal electromagnetic field" $\xi_{\mu\nu}$ with components

$$\boldsymbol{E}^{(\text{extremal})} = (M, 0, 0),$$
$$\boldsymbol{B}^{(\text{extremal})} = (0, 0, 0). \tag{20.62}$$

Show that it is also the Maxwell tensor of the "dual extremal field" $*\xi_{\mu\nu}$ with components

$$*\boldsymbol{E}^{(\text{extremal})} = (0, 0, 0),$$
$$*\boldsymbol{B}^{(\text{extremal})} = (M, 0, 0). \tag{20.63}$$

(c) Recalling that the duality operation $*$ applied twice to an antisymmetric second-rank tensor (2-form) in four-dimensional space leads back to the negative of that tensor, show that the operator $e^{*\alpha}$ ("duality rotation") has the value

$$e^{*\alpha} = (\cos \alpha) + (\sin \alpha)*. \tag{20.64}$$

(d) Show that the most general electromagnetic field which will reproduce the non-null tensor $T^{\mu\nu}$ in the frame in question, and therefore in any coordinate system, is

$$F_{\mu\nu} = e^{*\alpha}\xi_{\mu\nu}. \tag{20.65}$$

(e) Derive a corresponding result for the null case. [The field $F_{\mu\nu}$ defined in the one frame and therefore in every coordinate system by (d) and (e) is known as the "Maxwell square root" of $T^{\mu\nu}$; $\xi_{\mu\nu}$ is known as the "extremal Maxwell square root" of $T^{\mu\nu}$; and the angle α is called the "complexion of the electromagnetic field." See Misner and Wheeler (1957); see also Boxes 20.1 and 20.2, adapted from that paper.]

Box 20.1 CONTRAST BETWEEN PROPER LORENTZ TRANSFORMATION AND DUALITY ROTATION

Quantity	General proper Lorentz transformation	Duality rotation
Components of the Maxwell stress-energy tensor or the "Maxwell square" of the field \boldsymbol{F}	Transformed	Unchanged
The invariants $E^2 - B^2$ and $(\boldsymbol{E} \cdot \boldsymbol{B})^2$	Unchanged	Transformed
The combination $[(E^2 - B^2)^2 + (2\boldsymbol{E} \cdot \boldsymbol{B})^2] = [(E^2 + B^2)^2 - (2\boldsymbol{E} \times \boldsymbol{B})^2]$	Unchanged	Unchanged

Box 20.2 TRANSFORMATION OF THE GENERIC (NON-NULL) ELECTROMAGNETIC FIELD TENSOR $F = (E, B)$ IN A LOCAL INERTIAL FRAME

Field values	At start	After simplifying duality rotation
At start	E, B	E and B perpendicular, and E greater than B
After simplifying Lorentz transformation	E and B parallel to each other and parallel to x-axis	E parallel to x-axis and $B = 0$

Exercise 20.8. THE MAXWELL EQUATIONS CANNOT BE DERIVED FROM THE LAW OF CONSERVATION OF STRESS-ENERGY WHEN $(E \cdot B) = 0$ OVER AN EXTENDED REGION

Supply a counter-example to the idea that the Maxwell equations,

$$F^{\mu\nu}{}_{;\nu} = 0,$$

follow from the Einstein equation; or, more precisely, show that (1) the condition that the Maxwell stress-energy tensor should have a vanishing divergence plus (2) the condition that this Maxwell field is the curl of a 4-potential A_μ can both be satisfied, while yet the stated Maxwell equations are violated. [Hint: It simplifies the analysis without obscuring the main point to consider the problem in the context of flat spacetime. Refer to the paper of Teitelboim (1970) for the decomposition of the retarded field of an arbitrarily accelerated charge into two parts, of which the second, there called $F^{\mu\nu}{}_{II}$, meets the stated requirements, and has everywhere off the worldline $(E \cdot B) = 0$, but does not satisfy the cited Maxwell equations.]

Exercise 20.9. EQUATION OF MOTION OF A SCALAR FIELD AS CONSEQUENCE OF THE EINSTEIN FIELD EQUATION

The stress-energy tensor of a massless scalar field is taken to be

$$T_{\mu\nu} = (1/4\pi)(\phi_{,\mu}\phi_{,\nu} - 1/2 g_{\mu\nu}\phi_{,\alpha}\phi^{,\alpha}). \tag{20.66}$$

Derive the equation of motion of this scalar field from Einstein's field equation.

CHAPTER **21**

VARIATIONAL PRINCIPLE
AND INITIAL-VALUE DATA

*Whenever any action occurs in nature, the quantity of action
employed by this change is the least possible.*

PIERRE MOREAU DE MAUPERTUIS (1746)

*In the theory of gravitation, as in all other branches of
theoretical physics, a mathematically correct statement of a
problem must be determinate to the extent allowed by the nature
of the problem; if possible, it must ensure the
uniqueness of its solution.*

VLADIMIR ALEXANDROVITCH FOCK (1959)

Things are as they are because they were as they were.

THOMAS GOLD (1972)

Calculemus

G. W. LEIBNIZ

This chapter is entirely Track 2.
No earlier Track-2 material is
needed as preparation for it,
but Chapters 9–11 and 13–15
will be helpful. It is needed as
preparation for Box 30.1
(mixmaster universe) and for
Chapters 42 and 43.

§21.1. DYNAMICS REQUIRES INITIAL-VALUE DATA

No plan for predicting the dynamics of geometry could be at the same time more
mistaken and more right than this: "Give the distribution of mass-energy; then solve
Einstein's second-order equation,

$$\boldsymbol{G} = 8\pi \boldsymbol{T}, \tag{21.1}$$

for the geometry." Give the distribution of mass-energy in spacetime and solve for
the spacetime geometry? No. Give the fields that generate mass-energy, and their

To Karel Kuchař, Claudio Teitelboim, and James York go warm thanks for their collaboration in
the preparation of this chapter, and for permission to draw on the lecture notes of K. K. and to quote
results of K. K. [especially exercise 21.10] and of J. Y. [especially equations (21.87), (21.88), and (21.152)]
prior to publication elsewhere.

time-rates of change, and give 3-geometry of space and its time-rate of change, all at one time, and solve for the 4-geometry of spacetime at that one time? Yes. And only then let one's equations for geometrodynamics and field dynamics go on to predict for all time, in and by themselves, needing no further prescriptions from outside (needing only work!), both the spacetime geometry and the flow of mass-energy throughout this spacetime. This, in brief, is the built-in "plan" of geometrodynamics, the plan spelled out in more detail in this chapter.

Give initial data, predict geometry

Contest the plan. Point out that the art of solving any coupled set of equations lies in separating the unknowns from what is known or to be prescribed. Insist that this separation is already made in (21.1). On the right already stands the source of curvature. On the left already stands the receptacle of curvature in the form of what one wants to know, the metric coefficients, twice differentiated. Claim therefore that one has nothing to do except to go ahead and solve these equations for the metric coefficients. However, in analyzing the structure of the equations to greater depth [see Cartan (1922a) for the rationale of analyzing a coupled set of partial differential equations], one discovers that one can only make the split between "the source and the receptacle" in the right way when one has first recognized the still more important split between "the initial-value data and the future." Thus—to summarize the results before doing the analysis—four of the ten components of Einstein's law connect the curvature of space here and now with the distribution of mass-energy here and now, and the other six equations tell how the geometry as thus determined then proceeds to evolve.

Four of ten components of Einstein equation are conditions on initial-value data

In determining what are appropriate initial-value data to give, one discovers no guide more useful than the Hilbert variational principle,

$$I = \int \mathcal{L}\, d^4x = \int L(-g)^{1/2}\, d^4x = \int L\, d(\text{proper 4-volume}) = \text{extremum} \quad (21.2)$$

[exercise 8.16]

or the Arnowitt-Deser-Misner ("ADM") variant of it (§21.6) and generalizations thereof by Kuchař (§21.9). Out of this principle one can recognize most directly what one must hold fixed at the limits (on an initial spacelike hypersurface and on a final spacelike hypersurface) as one varies the geometry (§21.2) throughout the spacetime "filling of this sandwich," if one is to have a well-defined extremum problem.

The Lagrange function L (scalar function) or the Lagrangian density $\mathcal{L} = (-g)^{1/2}L$ (quantity to be integrated over coordinate volume) is built of geometry alone, when one deals with curved empty space, but normally fields are present as well, and contribute also to the Lagrangian; thus,

$$\mathcal{L} = \mathcal{L}_{\text{geom}} + \mathcal{L}_{\text{field}} = (-g)^{1/2}L;$$
$$L = L_{\text{geom}} + L_{\text{field}}. \quad (21.3)$$

The variation of the field Lagrangian with respect to the typical metric coefficient proves to be, of all ways, the one most convenient for generating (that is, for calculating) the corresponding component of the symmetric stress-energy tensor of the field (§21.3).

New view of stress-energy tensor

A computer, allowing for the effect of this field on the geometry and computing ahead from instant to instant the evolution of the metric with time, imposes its own ordering on the events of spacetime. In effect, it slices spacetime into a great number of spacelike slices. It finds it most convenient (§21.4) to do separate bookkeeping on (1) the 3-geometry of the individual slices and (2) the relation between one such slice and the next, as expressed in a "lapse function" N and a 3-vector "shift function" N_i.

The 3-geometry internal to the individual slice or "simultaneity" defines in and by itself the three-dimensional Riemannian curvature intrinsic to this hypersurface; but for a complete account of this hypersurface one must know also the extrinsic curvature (§21.5) telling how this hypersurface is curved with respect to the enveloping four-dimensional spacetime manifold.

In terms of the space-plus-time split of the 4-geometry, the action principle of Hilbert takes a simple and useful form (§21.6).

In the most elementary example of the application of an action principle in mechanics, where one writes

$$I = \int_{x',t'}^{x,t} L(dx/dt, x, t) \, dt \tag{21.4}$$

and extremizes the integral, one already knows that the resultant "dynamic path length" or "dynamic phase" or "action,"

$$S(x, t) = I_{\text{extremum}}, \tag{21.5}$$

is an important quantity, not least because it gives (up to a factor \hbar) the phase of the quantum-mechanical wave function. Moreover, the rate of change of this action function with position is what one calls momentum,

$$p = \partial S(x, t)/\partial x; \tag{21.6}$$

and the (negative of the) rate of change with time gives energy (Figure 21.1),

$$E = -\partial S(x, t)/\partial t; \tag{21.7}$$

and the relation between these two features of a system of wave crests,

Hamiltonian as a dispersion relation

$$E = H(p, x), \tag{21.8}$$

call it "dispersion relation" or call it what one will, is the central topic of mechanics.

When dealing with the dynamics of geometry in the Arnowitt-Deser-Misner formulation,* one finds it convenient to think of the specified quantities as being

Historical remark. No one knew until recently what coordinate-free geometric-physical quantity *really* is fixed at limits in the Hilbert-Palatini variational principle. In his pioneering work on the Hamiltonian formulation of general relativity, Dirac paid no particular attention to any variational principle. He had to generalize the Hamiltonian formalism to accommodate it to general relativity, introducing "first- and second-class constraints" and generalizations of the Poisson brackets of classical mechanics. The work of Arnowitt, Deser, and Misner, by contrast, took the variational principle as the foundation for the whole treatment, even though they too did not ask what it is that is fixed at limits in the sense of

Figure 21.1.

Momentum and (the negative of the) energy viewed as rate of change of "dynamic phase" or "action,"

$$S(x, t) = I_{\text{extremum}}(x, t) = \binom{\text{extremum}}{\text{value of}} \int_{x',t'}^{x,t} L(x, \dot{x}, t)\, dt, \quad (1)$$

with respect to position and time; thus,

$$\delta S = p\, \delta x - E\, \delta t. \quad (2)$$

The variation of the integral I with respect to changes of the history along the way, $\delta x(t)$, is already zero by reason of the optimization of the history; so the only change that takes place is

$$\delta S = \delta I_{\text{extremum}} = L(x, \dot{x}, t)\, \delta t + \int_{x',t'}^{x+\Delta x, t} \delta I.\, dt$$

$$= L\, \delta t + \int_{x',t'}^{x+\Delta x, t} \left(\frac{\partial L}{\partial \dot{x}} \delta \dot{x} + \frac{\partial L}{\partial x} \delta x \right) dt$$

$$= L\, \delta t + \frac{\partial L}{\partial \dot{x}} \Delta x + \int_{x',t'}^{x+\Delta x, t} \underbrace{\left(\frac{\partial L}{\partial x} - \frac{d}{dt} \frac{\partial L}{\partial \dot{x}} \right)}_{\begin{bmatrix}\text{zero by reason}\\\text{of extremization}\end{bmatrix}} \delta x\, dt. \quad (3)$$

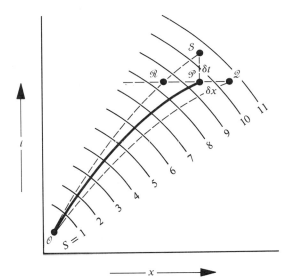

When one contemplates only a change δx in the coordinates (x, t) of the end point (change of history from \mathcal{OP} to $\mathcal{O2}$), one has $\Delta x = \delta x$. When one makes only a change δt in the end point (change of history from \mathcal{OP} to \mathcal{OS}), one has $\Delta x = $ (indicator of change from \mathcal{P} to \mathcal{R}) = $-\dot{x}\, \delta t$. For the general variation of the final point, one thus has $\Delta x = \delta x - \dot{x}\, \delta t$ and

$$\delta S = \frac{\partial L}{\partial \dot{x}} \delta x - \left(\dot{x} \frac{\partial L}{\partial \dot{x}} - L \right) \delta t. \quad (4)$$

One concludes that the "dispersion relation" is obtained by taking the relations [compare (2) and (4)]

$$\begin{pmatrix}\text{rate of change of}\\\text{dynamic phase}\\\text{with position}\end{pmatrix} = (\text{momentum}) = p = \frac{\partial L(x, \dot{x}, t)}{\partial \dot{x}} \quad (5)$$

and

$$-\begin{pmatrix}\text{rate of change of}\\\text{dynamic phase}\\\text{with time}\end{pmatrix} = (\text{energy}) = E = \dot{x} \frac{\partial L}{\partial \dot{x}} - L, \quad (6)$$

and eliminating \dot{x} from them [solve (5) for \dot{x} and substitute that value of \dot{x} into (6)]; thus

$$E = H(p, x, t) \quad (7)$$

or

$$-\frac{\partial S}{\partial t} = H\left(\frac{\partial S}{\partial x}, x, t \right). \quad (8)$$

Every feature of this elementary analysis has its analog in geometrodynamics.

a coordinate-free geometric-physical quantity. The great payoff of this work was recognition of the lapse and shift functions of equation (21.40) as Lagrange multipliers, the coefficients of which gave directly and simply Dirac's constraints. They did not succeed in arriving at a natural and simple time-coordinate, but that goal has in the meantime been achieved in the "extrinsic time" of Kuchař and York (§21.11). However, the Arnowitt-Deser Misner approach opened the door to the "intrinsic time" of Sharp, Baierlein, and Wheeler, where 3-geometry is fixed at limits, and 3-geometry is the carrier of information about time; and this led directly to Wheeler's "superspace version" of the treatment of Arnowitt, Deser, and Misner.

the 3-geometry $^{(3)}\mathcal{G}$ of the initial spacelike hypersurface and the 3-geometry $^{(3)}\mathcal{G}$ of the final spacelike hypersurface. One envisages the action integral as extremized with respect to the choice of the spacetime that fills the "sandwich" between these two faces. If one has thus determined the spacetime, one has automatically by that very act determined the separation in proper time of the two hypersurfaces. There is no additional time-variable to be brought in or considered. The one concept $^{(3)}\mathcal{G}$ thus takes the place in geometrodynamics of the two quantities x, t of particle dynamics. The action S that there depended on x and t here depends on the 3-geometry of the face of the sandwich; thus,

Action viewed as dependent on 3-geometry

$$S = S(^{(3)}\mathcal{G}). \tag{21.9}$$

A change in the 3-geometry changes the action. The amount of the change in action per elementary change in 3-geometry defines the "field momentum" π^{ij}_{true} conjugate to the geometrodynamic field coordinate g_{ij}, according to the formula

$$\delta S = \int \pi^{ij}_{\text{true}} \, \delta g_{ij} \, d^3x. \tag{21.10}$$

Comparing this equation out of the Arnowitt, Deser, and Misner (ADM) canonical formulation of geometrodynamics (§21.7) with the expression for change of action with change of endpoint in elementary mechanics,

$$\delta S = p \, \delta x - E \, \delta t, \tag{21.11}$$

one might at first think that something is awry, there being no obvious reference to time in (21.10). However, the 3-geometry is itself automatically the carrier of information about time; and (21.10) is complete. Moreover, with no "time" variable other than the information that $^{(3)}\mathcal{G}$ itself already carries about time, there is also no "energy." Thus the "dispersion relation" that connects the rates of change of action with respect to the several changes that one can make in the "field coordinates" or 3-geometry takes the form

$$\mathcal{H}(\pi^{ij}, g_{mn}) = 0, \tag{21.12}$$

with the E-term of (21.8) equal to zero (details in §21.7). All the content of Einstein's general relativity can be extracted from this one Hamiltonian, or "super-Hamiltonian," to give it a more appropriate name [see DeWitt (1967a), pp. 1113–1118, for an account of the contributions of Dirac, of Arnowitt, Deser, and Misner, and of others to the Hamiltonian formulation of geometrodynamics; and see §21.7 and subsequent sections of this chapter for the meaning and payoffs of this formulation].

Hamiltonian versus super-Hamiltonian

The difference between a Hamiltonian and a super-Hamiltonian [see, for example, Kramers (1957)] shows nowhere more clearly than in the problem of a charged particle moving in flat space under the influence of the field derived from the electromagnetic 4-potential, $A_\mu(x^\alpha)$. The Hamiltonian treatment derives the equation of motion from the action principle,

$$0 = \delta I = \delta \int \left[p_i \frac{dx^i}{dt} - H(p_j, x^k, t) \right] dt$$

with

$$H = -\frac{e}{c}\phi + \left[m^2 + \eta^{ij}\left(p_i + \frac{e}{c}A_i \right)\left(p_j + \frac{e}{c}A_j \right) \right]^{1/2}.$$

The super-Hamiltonian analysis gets the equations of motion from the action principle

$$0 = \delta I' = \delta \int \left[p_\mu \frac{dx^\mu}{d\lambda} - \mathcal{H}(p_\alpha, x^\beta) \right] d\lambda.$$

Here the super-Hamiltonian is given by the expression

$$\mathcal{H}(p_\alpha, x^\beta) = \frac{1}{2}\left[m^2 + \eta^{\mu\nu}\left(p_\mu + \frac{e}{c}A_\mu \right)\left(p_\nu + \frac{e}{c}A_\nu \right) \right].$$

The variational principle gives Hamilton's equations for the rates of change

$$dx^\alpha/d\lambda = \partial \mathcal{H}/\partial p_\alpha$$

and

$$dp_\beta/d\lambda = -\partial \mathcal{H}/\partial x^\beta.$$

From these equations, one discovers that \mathcal{H} itself must be a constant, independent of the time-like parameter λ. The value of this constant has to be imposed as an initial condition, $\mathcal{H} = 0$ ("specification of particle mass"), thereafter maintained by the Hamiltonian equations themselves. This vanishing of \mathcal{H} in no way kills the partial derivatives,

$$\partial \mathcal{H}/\partial p_\alpha \quad \text{and} \quad -\partial \mathcal{H}/\partial x^\beta,$$

that enter Hamilton's equations for the rates of change,

$$dx^\alpha/d\lambda \quad \text{and} \quad dp_\beta/d\lambda.$$

Whether derived in the one formalism or the other, the equations of motion are equivalent, but the covariance shows more clearly in the formalism of the super-Hamiltonian, and similarly in general relativity.

Granted values of the "field coordinates" $g_{ij}(x, y, z)$ $^{(3)}\mathcal{G}$ and field momenta $\pi^{ij}_{\text{true}}(x, y, z) = \delta S/\delta g_{ij}$ compatible with (21.12), one has what are called "compatible initial-value data on an initial spacelike hypersurface." One can proceed as described in §21.8 to integrate ahead in time step by step from one spacelike hypersurface to another and another, and construct the whole 4-geometry. Here one is dealing with what in mathematical terminology are hyperbolic differential equations that have the character of a wave equation.

Dynamic evolution of geometry

In contrast, one deals with elliptic differential equations that have the character of a Poisson potential equation when one undertakes in the first place to construct the needed initial-value data (§21.9). In the analysis of these elliptic equations, it

proves helpful to distinguish in the 3-geometry between (1) the part of the metric that determines relative lengths at a point, which is to say angles ("the conformal part of the metric") and (2) the common multiplicative factor that enters all the components of the g_{ij} at a point to determine the absolute scale of lengths at that point. This breakdown of the 3-geometry into two parts provides a particularly simple way to deal with two special initial-value problems known as the time-symmetric and time-antisymmetric initial-value problems (§21.10).

The ADM formalism is today in course of development as summarized in §21.11. In Wheeler's (1968a) "superspace" form, the ADM treatment takes the 3-geometry to be fixed on each of the bounding spacelike hypersurfaces. In contrast, York (§21.11) goes back to the original Hilbert action principle, and discovers what *it* takes to be fixed on each of the bounding spacelike hypersurfaces. The appropriate data turn out to be the "conformal part of the 3-geometry" plus something closely related to what Kuchař (1971a and 1972) calls the "extrinsic time." The contrast between Wheeler's approach and the Kuchař-York approach shows particularly clearly when one (1) deals with a flat spacetime manifold, (2) takes a flat spacelike section through this spacetime, and then (3) introduces a slight bump on this slice, of height ϵ. The 3-geometry intrinsic to this deformed slice differs from Euclidean geometry only to the second order in ϵ. Therefore to read back from the full 3-geometry to the time ("the forward advance of the bump") requires in this case an operation something like extracting a square root. In contrast, the Kuchař-York treatment deals with the "extrinsic curvature" of the slice, something proportional to the first power of ϵ, and therefore provides what is in some ways a more convenient measure of time [see especially Kuchař (1971) for the construction of "extrinsic time" for arbitrarily strong cylindrical gravitational waves; see also Box 30.1 on "time" as variously defined in "mixmaster cosmology"]. York shows that the time-variable is most conveniently identified with the variable "dynamically conjugate to the conformal factor in the 3-geometry."

The initial-value problem of geometrodynamics can be formulated either in the language of Wheeler or in the language of Kuchař and York. In either formulation (§21.9 or §21.11) it throws light on what one ought properly today to understand by Mach's principle (§21.12). That principle meant to Mach that the "acceleration" dealt with in Newtonian mechanics could have a meaning only if it was acceleration with respect to the fixed stars or to something equally well-defined. It guided Einstein to general relativity. Today it is summarized in the principle that "mass-energy there governs inertia here," and is given mathematical expression in the initial-value equations.

The analysis of the initial-value problem connected past and future across a spacelike hypersurface. In contrast, one encounters a hypersurface that accommodates a timelike vector when one deals (§21.13) with the junction conditions between one solution of Einstein's field equation (say, the Friedmann geometry interior to a spherical cloud of dust of uniform density) and another (say, the Schwarzschild geometry exterior to this cloud of dust). Section 21.13, and the chapter, terminate with notes on gravitational shock waves and the characteristic initial-value problem (the statement of initial-value data on a light cone, for example).

Another choice of what to fix at boundary hypersurface: conformal part of 3-geometry plus extrinsic time

Mach updated: mass-energy there governs inertia here

§21.2. THE HILBERT ACTION PRINCIPLE AND THE PALATINI METHOD OF VARIATION

Five days before Einstein presented his geometrodynamic law in its final and now standard form, Hilbert, animated by Einstein's earlier work, independently discovered (1915a) how to formulate this law as the consequence of the simplest action principle of the form (21.2–21.3) that one can imagine:

Variational principle the simplest route to Einstein's equation

$$L_{\text{geom}} = (1/16\pi)^{(4)}R. \tag{21.13}$$

(Replace $1/16\pi$ by $c^3/16\pi G$ when going from the present geometric units to conventional units; or divide by $\hbar \sim L^{*2}$ to convert from dynamic phase, with the units of action, to actual phase of a wave function, with the units of radians). Here $^{(4)}R$ is the four-dimensional scalar curvature invariant, as spelled out in Box 8.4.

This action principle contains second derivatives of the metric coefficients. In contrast, the action principle for mechanics contains only first derivatives of the dynamic variables; and similarly only derivatives of the type $\partial A_\alpha/\partial x^\beta$ appear in the action principle for electrodynamics. Therefore one might also have expected only first derivatives, of the form $\partial g_{\mu\nu}/\partial x^\gamma$, in the action principle here. However, no scalar invariant lets itself be constructed out of these first derivatives. Thus, to be an invariant, L_{geom} has to have a value independent of the choice of coordinate system. But in the neighborhood of a point, one can always so choose a coordinate system that all first derivatives of the $g_{\mu\nu}$ vanish. Apart from a constant, there is no scalar invariant that can be built homogeneously out of the metric coefficients and their first derivatives.

When one turns from first derivatives to second derivatives, one has all twenty distinct components of the curvature tensor to work with. Expressed in a local inertial frame, these twenty components are arbitrary to the extent of the six parameters of a local Lorentz transformation. There are thus $20 - 6 = 14$ independent local features of the curvature ("curvature invariants") that are coordinate-independent, any one of which one could imagine employing in the action principle. However, $^{(4)}R$ is the only one of these 14 quantities that is linear in the second derivatives of the metric coefficients. Any choice of invariant other than Hilbert's complicates the geometrodynamic law, and destroys the simple correspondence with the Newtonian theory of gravity (Chapter 17).

Scalar curvature invariant the only natural choice

Hilbert originally conceived of the independently adjustable functions of x, y, z, t in the variational principle as being the ten distinct components of the metric tensor in contravariant representation, $g^{\mu\nu}$. Later Palatini (1919) discovered a simpler and more instructive listing of the independently adjustable functions: not the ten $g^{\mu\nu}$ alone, but the ten $g^{\mu\nu}$ plus the forty $\Gamma^\alpha_{\mu\nu}$ of the affine connection.

To give up the standard formula for the connection Γ in terms of the metric g and let Γ "flap in the breeze" is not a new kind of enterprise in mathematical physics. Even in the simplest problem of mechanics, one can give up the standard formula for the momentum p in terms of a time-derivative of the coordinate x and also let

Idea of varying coordinate
and momentum
independently

p "flap in the breeze." Then $x(t)$ and $p(t)$ become two independently adjustable functions in a new variational principle,

$$I = \int_{x',t'}^{x,t} \left[p(t) \frac{dx(t)}{dt} - H(p(t), x(t), t) \right] dt = \text{extremum}. \tag{21.14}$$

Happily, out of the extremization with respect to choice of the function $p(t)$, one recovers the standard formula for the momentum in terms of the velocity. The extremization with respect to choice of the other function, $x(t)$, gives the equation of motion just as does the more elementary variational analysis of Euler and Lagrange, where $x(t)$ is the sole adjustable function. A further analysis of this equivalence between the two kinds of variational principles in particle mechanics appears in Box 21.1. In that box, one also sees the two kinds of variational principle as applied to electrodynamics.

To express the Hilbert variational principle in terms of the $\Gamma^{\lambda}_{\mu\nu}$ and $g^{\alpha\beta}$ regarded as the primordial functions of t, x, y, z, note that the Lagrangian density is

$$L_{\text{geom}}(-g)^{1/2} = (1/16\pi)^{(4)}R(-g)^{1/2} = (1/16\pi)g^{\alpha\beta} R_{\alpha\beta}(-g)^{1/2}. \tag{21.15}$$

Here, as in any spacetime manifold with an affine connection, one has (Chapter 14)

$$R_{\alpha\beta} = R^{\lambda}{}_{\alpha\lambda\beta}, \tag{21.16}$$

where

$$R^{\lambda}{}_{\alpha\mu\beta} = \partial\Gamma^{\lambda}{}_{\alpha\beta}/\partial x^{\mu} - \partial\Gamma^{\lambda}{}_{\alpha\mu}/\partial x^{\beta} + \Gamma^{\lambda}{}_{\sigma\mu}\Gamma^{\sigma}{}_{\alpha\beta} - \Gamma^{\lambda}{}_{\sigma\beta}\Gamma^{\sigma}{}_{\alpha\mu}, \tag{21.17}$$

and every Γ is given in advance (in a coordinate frame) as symmetric in its two lower indices. In order that the integral I of (21.2–21.3) should be an extremum, one requires that the variation in I caused by changes both in the $g^{\mu\nu}$ and in the Γ's should vanish; thus,

$$0 = \delta I = (1/16\pi) \int \delta[g^{\alpha\beta} R_{\alpha\beta}(-g)^{1/2}] \, d^4x + \int \delta[L_{\text{field}}(-g)^{1/2}] \, d^4x. \tag{21.18}$$

Consider now the variations of the individual factors in the first and second integrals in (21.18). The variation of the first factor is trivial, $\delta g^{\alpha\beta}$. In the variation of the second factor, $R_{\alpha\beta}$, changes in the $g^{\alpha\beta}$ play no part; only changes in the Γ's appear. Moreover, the variation $\delta\Gamma^{\lambda}_{\alpha\beta}$ is a tensor even though $\Gamma^{\lambda}_{\alpha\beta}$ itself is not. Thus in the transformation formula

Variation of connection is a
tensor

$$\Gamma^{\bar{\gamma}}{}_{\bar{\alpha}\bar{\beta}} = \left[\Gamma^{\lambda}{}_{\sigma\tau} \frac{\partial x^{\sigma}}{\partial x^{\bar{\alpha}}} \frac{\partial x^{\tau}}{\partial x^{\bar{\beta}}} + \frac{\partial^2 x^{\lambda}}{\partial x^{\bar{\alpha}} \partial x^{\bar{\beta}}} \right] \frac{\partial x^{\bar{\gamma}}}{\partial x^{\lambda}}, \tag{21.19}$$

the last term destroys the tensor character of any set of $\Gamma^{\lambda}_{\sigma\tau}$ individually, but subtracts out in the difference $\delta\Gamma^{\lambda}_{\sigma\tau}$ between two alternative sets of Γ's. Note that the variation $\delta R^{\lambda}{}_{\alpha\mu\beta}$ of the typical component of the curvature tensor consists of two terms of

(continued on page 500)

**Box 21.1 RATE OF CHANGE OF ACTION WITH DYNAMIC COORDINATE
(= "MOMENTUM") AND WITH TIME, AND THE DISPERSION
RELATION (= "HAMILTONIAN") THAT CONNECTS THEM
IN PARTICLE MECHANICS AND IN ELECTRODYNAMICS**

A. PROLOG ON THE PARTICLE-MECHANICS ANALOG OF THE PALATINI METHOD

In particle mechanics, one considers the history $x = x(t)$ to be adjustable be-
tween the end points (x', t') and (x, t) and varies it to extremize the integral $I = \int L(x, \dot{x}, t)\, dt$ taken between these two limits.

Expressed in terms of coordinates and momenta
(see Figure 21.1), the integral has the form

$$I = \int [p\dot{x} - H(p, x, t)]\, dt, \qquad (1)$$

where $x(t)$ is again the function to be varied and
p is only an abbreviation for a certain function of
x and \dot{x}; thus, $p = \partial L(x, \dot{x}, t)/\partial \dot{x}$. Viewed in this
way, the variation, $\delta p(t)$, of the momentum is gov-
erned by, and is only a reflection of, the variation
$\delta x(t)$.

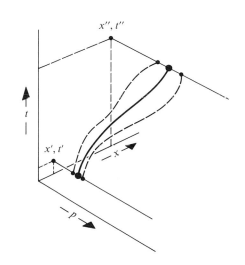

1. Momentum Treated as Independently Variable

There miraculously exists, however, quite another
way to view the problem (see inset). One can re-
gard $x(t)$ and $p(t)$ as two quite uncorrelated and
independently adjustable functions. One abandons
the formula $p = \partial L(x, \dot{x}, t)/\partial \dot{x}$, only to recover it,
or the equivalent of it, from the new "independ-
ent-coordinate-and-momentum version" of the
variation principle.

The variation of (1), as defined and calculated in this new way, becomes

$$\delta I = p\, \delta x \, \Big|_{x', t'}^{x'', t''} + \int_{x', t'}^{x'', t''} \left[\left(\dot{x} - \frac{\partial H}{\partial p} \right) \delta p + \left(-\dot{p} - \frac{\partial H}{\partial x} \right) \delta x \right] dt. \qquad (2)$$

Demand that the coefficient of δp vanish and have the sought-for new version,

$$\dot{x} = \frac{\partial H(p, x, t)}{\partial p}$$

Box 21.1 (continued)

of the old relation, $p = \partial L(x, \dot{x}, t)/\partial \dot{x}$, between momentum and velocity. The vanishing of the coefficient of δx gives the other Hamilton equation,

$$\dot{p} = -\frac{\partial H(p, x, t)}{\partial x}, \tag{3}$$

equivalent in content to the original Lagrange equation of motion,

$$\frac{d}{dt}\frac{\partial L}{\partial \dot{x}} - \frac{\partial L}{\partial x} = 0. \tag{4}$$

That $p(t)$ in this double variable conception is—before the extremization!—a function of time quite separate from and independent of the function $x(t)$ shows nowhere more clearly than in the circumstance that $p(t)$ has no end point conditions imposed on it, whereas x' and x'' are specified. Thus not only is the shape of the history subject to adjustment in x, p, t space in the course of achieving the extremum, but even the end points are subject to being slid along the two indicated lines in the inset, like beads on a wire.

2. Action as Tool for Finding Dispersion Relation

Denote by $S(x, t)$ the "action," or extremal value of I, for the classical history that starts with (x', t') and ends at (x, t) ($= \hbar$ times phase of de Broglie wave). To change the end points to $(x + \delta x, t)$ makes the change in action

$$\delta S = p\,\delta x. \tag{5}$$

Thus momentum is "rate of change of action with dynamic coordinate."

To change the end point to

$$(x + \delta x, t + \delta t) = ([x + \dot{x}\,\delta t] + [\delta x - \dot{x}\,\delta t], t + \delta t) \tag{6}$$

makes the change in action

$$\delta S = p[\delta x - \dot{x}\,\delta t] + L\,\delta t = p\,\delta x - H\,\delta t. \tag{7}$$

Thus the Hamiltonian is the negative of "the rate of change of action with time."

In terms of the Hamiltonian $H = H(p, x)$, the "dispersion relation" for de Broglie waves becomes

$$-\frac{\partial S}{\partial t} = H\left(\frac{\partial S}{\partial x}, x\right). \tag{8}$$

In the derivation of this dispersion relation, one can profitably short-cut all talk of $p(t)$ and $x(t)$ as independently variable quantities, and derive the result in hardly

more than one step from the definition $I = \int L(x, \dot{x}, t)\, dt$. Similarly in electrody-namics.

The remainder of this box best follows a first perusal of Chapter 21.

B. ANALOG OF THE PALATINI METHOD IN ELECTRODYNAMICS

In source-free electrodynamics, one considers as given two spacelike hypersurfaces S' and S'', and the magnetic fields-as-a-function-of-position in each, B' and B'' (this second field will later be written without the $''$ superscript to simplify the notation). To be varied is an integral extended over the region of spacetime between the two hypersurfaces,

$$I_{\text{Maxwell}} \equiv \int \mathcal{L}_{\text{Maxwell}}\, d^4x = -\frac{1}{16\pi} \int F^{\mu\nu} F_{\mu\nu} (-g)^{1/2}\, d^4x. \tag{9}$$

1. Variation of Field on Hypersurface and Variation of Location of Hypersurface are Cleanly Separated Concepts in Electromagnetism

The electromagnetic field \mathbf{F} is the physically relevant quantity in electromagnetism (compare the 3-geometry in geometrodynamics). By contrast, the 4-potential \mathbf{A} has no direct physical significance. A change of gauge in the potentials,

$$A_\mu = A_{\mu_{\text{new}}} + \partial\lambda/\partial x^\mu$$

leaves unchanged the field components

$$F_{\mu\nu} = \partial A_\nu/\partial x^\mu - \partial A_\mu/\partial x^\nu$$

(compare the coordinate transformation that changes the $g_{\mu\nu}$ while leaving un-changed the $^{(3)}\mathcal{G}$). The variation of the fields within the body of the sandwich is nevertheless expressed most conveniently in terms of the effect of changes δA_μ in the potentials.

One also wants to see how the action integral is influenced by changes in the location of the upper spacelike hypersurface ("many-fingered time"). Think of the point of the hypersurface that is presently endowed with coordinates $x, y, z, t(x, y, z)$ as being displaced to $x, y, z, t + \delta t(x, y, z)$. Now renounce this use of a privi-leged coordinate system. Describe the displacement of the simultaneity in terms of a 4-vector $\delta\mathbf{n}$ (not a unit 4-vector) normal to the hypersurface Σ. The ele-ment of 4-volume $\delta\Omega$ included between the original upper face of the sandwich and the new upper face, that had in the privileged coordinate system the form $(-g)^{1/2}\,\delta t(x, y, z)\, d^3x$, in the notation of Chapter 20 becomes

$$\delta\Omega = \delta n^\mu\, d^3\Sigma_\mu = (\delta\mathbf{n} \cdot d^3\mathbf{\Sigma}), \tag{10}$$

where the element of surface $d^3\Sigma_\mu$ already includes the previously listed factor $(-g)^{1/2}$.

Box 21.1 (continued)

Counting together the influence of changes in the field values on the upper hypersurface and changes in the location of that hypersurface, one has

$$\delta S = \delta I_{\text{extremal}} = -(1/16\pi) \int_{\text{upper } \Sigma} F^{\mu\nu} F_{\mu\nu}(\delta \boldsymbol{n} \cdot d^3\boldsymbol{\Sigma})$$

$$+ (1/4\pi) \int_{\text{upper } \Sigma} F^{\mu\nu} \underbrace{\Delta A_\mu} d^3\Sigma_\nu \qquad (11)$$

replace by
its equivalent
$$(\delta A_\mu - \delta n^\alpha A_{\mu;\alpha})$$

$$+ (1/4\pi) \int_{\text{4-volume}} \underbrace{F^{\mu\nu}{}_{;\nu}} \delta A_\mu (-g)^{1/2} \, d^4x.$$

has to vanish
because integral has
been extremized

Simplify this expression by arranging the coordinates so that the hypersurface shall be a hypersurface of constant t, and so that lines of constant x, y, z shall be normal to this hypersurface. Then it follows that the element of volume on that hypersurface contains a single nonvanishing component, $d^3\Sigma_0 = (-g)^{1/2} \, d^3x$. The antisymmetry of the field quantity $F^{0\nu}$ in its two indices requires that ν be a spacelike label, $i = 1, 2, 3$. The variation of the action becomes

$$\delta S = \int \left[\frac{(-g)^{1/2}F^{i0}}{4\pi} \delta A_i - \left\{ \underbrace{\frac{(-g)^{1/2}F^{i0}}{4\pi} A_{i;0}} - \mathcal{L}_{\text{Maxwell}} \right\} \delta t \right] d^3x.$$

$$\underbrace{\qquad\qquad\qquad\qquad\qquad\qquad}_{\text{add and subtract } \left\{ \frac{(-g)^{1/2}F^{i0}}{4\pi} A_0 \, \delta t \right\}_{,i}} \qquad (12)$$

2. Meaning of Field "Momentum" in Electrodynamics

Identify this expression with the quantity

$$\delta S = \int \pi^i_{EM} \, \delta A_i \, d^3x - \int \mathcal{H} \, \delta\Omega, \qquad (13)$$

where

$$\pi^i_{EM} = \frac{\delta S}{\delta A_i} = \begin{pmatrix} \text{``density of electromagnetic} \\ \text{momentum dynamically canon-} \\ \text{ically conjugate to } A_i\text{''} \end{pmatrix} = \frac{(-g)^{1/2}F^{i0}}{4\pi} = -\frac{\mathcal{E}^i}{4\pi} \quad (14)$$

is a simple multiple of the electric field and where

$$\mathcal{H} = -\frac{\delta S}{\delta \Omega} = \begin{pmatrix} \text{``density of} \\ \text{electromagnetic} \\ \text{Hamiltonian''} \end{pmatrix} = (1/16\pi)[F^{\mu\nu}F_{\mu\nu} + 4F^{i0}(A_{i;0} - A_{0;i})] \quad (15)$$

$$= (1/8\pi)(\boldsymbol{E}^2 + \boldsymbol{B}^2).$$

The concept of dynamic Hamiltonian density agrees with the usual concept of density of electromagnetic energy, despite the very different context in which the two quantities are derived and used. However, the canonical momentum π^i_{EM} has nothing directly whatsoever to do with the density of electromagnetic momentum as defined, for example, by the Poynting vector, despite the confusing similarity in the standard names for the two quantities. Note that there is no term δA_0 in (13); that is, $\pi^0_{EM} \equiv 0$.

3. Bubble Differentiation

The "bubble differentiation" with respect to "many-fingered time" that appears in (15) was first introduced by Tomonaga (1946). One thinks of a spacelike hypersurface Σ_1, a magnetic field \boldsymbol{B} defined as a function of position on this hypersurface (by an observer on a world line normal to this hypersurface), and a prescription S that carries one from this information to a single number, the action. (Divided by \hbar, this action gives the phase of the "wave function" or "probability amplitude" for the occurrence of this particular distribution of field values over this particular hypersurface.) One goes to a second hypersurface Σ_2 (see inset), which is identical with Σ_1, except in the immediate vicinity of a given point. Take a distribution of field values over Σ_2 that is identical with the original distribution over Σ_1, "identity of location" being defined by means of the normal. Evaluate the difference, δS, in the value of the dynamic phase or action in the two cases. Divide this difference by the amount of proper 4-volume $\delta \Omega = \int (\delta \boldsymbol{n} \cdot d^3\boldsymbol{\Sigma})$ contained in the "bubble" between the two hypersurfaces. Take the quotient, evaluate it in the limit in which the size of the bubble goes to zero, and in this way get the "bubble-time derivative," $\delta S/\delta \Omega$, of the action.

Box 21.1 (*continued*)

What does it mean to say that the action, S, besides depending on the hypersurface, Σ, depends also on the distribution of the magnetic field, B, over that hypersurface? The action depends on the physical quantity, $B = \nabla \times A$, not on the prephysical quantity, A. Thus a change in gauge $\delta A_i = \partial\lambda/\partial x^i$, cannot make any change in S. On the other hand, the calculated value of the change in S for this alteration in A is

$$
\begin{aligned}
\delta(\text{action}) = \delta S &= \int \frac{\delta S}{\delta A_i} \delta A_i \, d^3x \\
&= \int \frac{\delta S}{\delta A_i} \frac{\partial \lambda}{\partial x^i} \, d^3x = - \int \left(\frac{\delta S}{\delta A_i} \right)_{,i} \lambda(x, y, z) \, d^3x.
\end{aligned}
\tag{16}
$$

In order that there shall be no dependence of action on gauge, it follows that this expression must vanish for arbitrary $\lambda(x, y, z)$, a result only possible if $S(\Sigma, B) = S(\text{hypersurface, field on hypersurface})$ satisfies the identity

$$
\left(\frac{\delta S}{\delta A_i} \right)_{,i} = \pi^i_{EM,i} = -(1/4\pi)\mathcal{E}^i{}_{,i} = 0.
\tag{17}
$$

4. Hamilton-Jacobi ''Propagation Law'' for Electrodynamics

The "dispersion relation" or "Hamilton-Jacobi equation" for electromagnetism relates (1) the changes of the "dynamic phase" or "action" brought about by alterations in the dynamic variables A_i (the generalization of the x of particle dynamics) with (2) the changes brought about by alterations in many-fingered time (the generalization of the single time t of particle dynamics); thus (15) translates into

$$
-\frac{\delta S}{\delta \Omega} = \frac{(4\pi)^2}{8\pi} \left(\frac{\delta S}{\delta A} \right)^2 + \frac{1}{(8\pi)} (\nabla \times A)^2
\tag{18}
$$

C. DISPERSION RELATIONS FOR GEOMETRODYNAMICS AND ELECTRODYNAMICS COMPARED AND CONTRASTED

Geometrodynamics possesses a direct analog of equation (17) ("action depends on no information carried by the vector potential A except the magnetic field $B = \nabla \times A$"), in an equation that says the action depends on no information carried by the metric g_{ij} on the "upper face of the sandwich" except the 3-geometry there, $^{(3)}\mathcal{G}$. It also possesses a direct analog of equation (18) ("dynamic equation for the propagation of the action") with this one difference: in electrodynamics the field variable B and the many-fingered time are distinct in character, whereas in geometrodynamics the "field" and the "many-fingered time" can be regarded as two aspects of one and the same $^{(3)}\mathcal{G}$:

D. ACTION PRINCIPLE AND DISPERSION RELATION ARE ROOTED IN THE QUANTUM PRINCIPLE; FEYNMAN'S PRINCIPLE OF THE DEMOCRATIC EQUALITY OF ALL HISTORIES

For more on action principles in physics, see for example Mercier (1953), Lanczos (1970), and Yourgrau and Mandelstam (1968).

Newton (1687) in the first page of the preface to the first edition of his *Principia* notes that "The description of right lines..., upon which geometry is founded, belongs to mechanics. Geometry does not teach us to draw these lines, but requires them to be drawn."

Newton's remark is also a question. Mechanics moves a particle along a straight line, but what is the machinery by which mechanics accomplishes this miracle? The quantum principle gives the answer. The particle moves along the straight line only by not moving along the straight line. In effect it "feels out" every conceivable world line that leads from the start, (x', t'), to the point of detection, (x'', t''), "compares" one with another, and takes the extremal world line. How does it accomplish this miracle?

The particle is governed by a "probability amplitude to transit from (x', t') to (x'', t'')." This amplitude or "propagator," $\langle x'', t'' | x', t' \rangle$, is the democratic sum with equal weight of contributions from every world line that leads from start to finish; thus,

$$\langle x'', t'' | x', t' \rangle = N \int e^{i I_H / \hbar} \, \mathcal{D}x. \tag{15}$$

Here N is a normalization factor, the same for all histories.

$\mathcal{D}x$ is the "volume element" for the sum over histories. For a "skeleton history" defined by giving x_n at $t_n = t_0 + n \, \Delta t$, one has $\mathcal{D}x$ equal, up to a multiplicative constant, to $dx_1 \, dx_2 \ldots dx_N$. When the history is defined by the Fourier coefficients in such an expression as

$$x(t) = \frac{x'(t'' - t) + x''(t - t')}{(t'' - t')} + \sum_n a_n \sin n\pi \frac{(t - t')}{(t'' - t')}, \tag{16}$$

the volume element, again up to a multiplicative factor, is $da_1 \, da_2 \ldots$.

Destructive interference in effect wipes out the contribution to the transition probability from histories that differ significantly from the "extremal history" or "classical history." Histories that are near that extremal history, on the other hand, contribute constructively, and for a simple reason: a small departure of the first order from the classical history brings about a change in phase which is only of the second order in the departure.

In this elementary example, one sees illustrated why it is that extremal principles play such a large part in classical dynamics. They remind one that all classical physics rests on a foundation of quantum physics. The central ideas are (1) the principle

Box 21.1 (continued)

of superposition of probability amplitudes, (2) constructive and destructive interference, (3) the "democracy of all histories," and (4) the probability amplitude associated with a history H is $e^{iI_H/\hbar}$, apart from a normalizing factor that is a multiplicative constant.

For more on the democracy of histories and the sum over histories see Feynman (1942, 1948, 1949, 1951, and 1955), and the book of Feynman and Hibbs (1965); also Hibbs (1951), Morette (1951), Choquard (1955), Polkinghorne (1955), Fujiwara (1962), and the survey and literature references in Kursunoglu (1962); also reports of Dempster (1963) and Symanzik (1963). This outlook has been applied by many workers to discuss the quantum formulation of geometrodynamics, the first being Misner (1957) and one of the latest being Faddeev (1971).

the form $\delta\Gamma^\lambda{}_{\alpha\beta,\mu}$ and four terms of the form $\Gamma\,\delta\Gamma$ (indices being dropped for simplicity). One coordinate system is as good as another in dealing with a tensor. Therefore pick a coordinate system in which all the Γ's vanish at the point under study. The terms $\Gamma\,\delta\Gamma$ drop out. In this coordinate system, the variation of the curvature is expressed in terms of first derivatives of quantities like $\delta\Gamma^\lambda_{\alpha\beta}$. One then need only replace the ordinary derivatives by covariant derivatives to obtain a formula correct in any coordinate system,

$$\delta R^\lambda_{\alpha\mu\beta} = \delta\Gamma^\lambda{}_{\alpha\beta\,;\mu} - \delta\Gamma^\lambda{}_{\alpha\mu\,;\beta}, \tag{21.20}$$

along with its contraction,

$$\delta R_{\alpha\beta} = \delta\Gamma^\lambda{}_{\alpha\beta\,;\lambda} - \delta\Gamma^\lambda{}_{\alpha\lambda\,;\beta}. \tag{21.21}$$

The third factor that appears in the variation principle is $(-g)^{1/2}$. Its variation (exercise 21.1) is

$$\delta(-g)^{1/2} = -\frac{1}{2}(-g)^{1/2}g_{\mu\nu}\,\delta g^{\mu\nu}. \tag{21.22}$$

The other integrand, the Lagrange density L_{field}, will depend on the fields present and their derivatives, but will be assumed to contain the metric only as $g^{\mu\nu}$ itself, never in the form of any derivatives of $g^{\mu\nu}$.

In order for an extremum to exist, the following expression has to vanish:

$$(1/16\pi)\int\left[\left(R_{\alpha\beta} - \frac{1}{2}g_{\alpha\beta}R\right)\delta g^{\alpha\beta} + g^{\alpha\beta}(\delta\Gamma^\lambda{}_{\alpha\beta\,;\lambda} - \delta\Gamma^\lambda{}_{\alpha\lambda\,;\beta})\right](-g)^{1/2}\,d^4x$$
$$+ \int\left(\frac{\delta L_{\text{field}}}{\delta g^{\alpha\beta}} - \frac{1}{2}g_{\alpha\beta}L_{\text{field}}\right)\delta g^{\alpha\beta}(-g)^{1/2}\,d^4x = 0 \tag{21.23}$$

Focus attention on the term in (21.23) that contains the variations of Γ,

$$(1/16\pi) \int g^{\alpha\beta}(\delta\Gamma^{\lambda}_{\alpha\beta;\lambda} - \delta\Gamma^{\lambda}_{\alpha\lambda;\beta})(-g)^{1/2}\,d^4x,$$

and integrate by parts to eliminate the derivatives of the $\delta\Gamma$. To prepare the way for this integration, introduce the concept of *tensor density,* a notational device widely applied in general relativity. The concept of tensor density aims at economy. Without this concept, one will treat the tensor

Concept of tensor density

$$\epsilon_{\mu\alpha\beta\gamma} = (-g)^{1/2}[\mu\alpha\beta\gamma]$$

(see exercise 3.13) as having $4^4 = 256$ components, and its covariant derivative as having $4^5 = 1{,}024$ components, of which one is

$$\begin{aligned}
\epsilon_{0123;\rho} &= \partial(-g)^{1/2}/\partial x^{\rho}\,\epsilon_{[0123]} - \Gamma^{\sigma}_{0\rho}\epsilon_{\sigma123} - \Gamma^{\sigma}_{1\rho}\epsilon_{0\sigma23} \\
&\quad - \Gamma^{\sigma}_{2\rho}\epsilon_{01\sigma3} - \Gamma^{\sigma}_{3\rho}\epsilon_{012\sigma} \\
&= [(-g)^{1/2}{}_{,\rho} - \Gamma^{\sigma}_{\sigma\rho}(-g)^{1/2}][0123].
\end{aligned}$$

The symbol $[\alpha\beta\gamma\delta]$, with values $(0, -1, +1)$, introduces what is largely excess baggage, doing mere bookkeeping on alternating indices. Drop this unhandiness. Introduce instead the non-tensor $(-g)^{1/2}$ and *define* for it the law of covariant differentiation,

$$(-g)^{1/2}{}_{;\rho} = (-g)^{1/2}{}_{,\rho} - \Gamma^{\sigma}_{\sigma\rho}(-g)^{1/2}. \tag{21.24}$$

These four components take the place of the 1,024 components and communicate all the important information that was in them.

Associated with the vector j_{μ} is the vector density

$$\mathfrak{j}_{\mu} = (-g)^{1/2}j_{\mu};$$

with the tensor $T_{\mu\nu}$, the tensor density

$$\mathfrak{T}_{\mu\nu} = (-g)^{1/2}T_{\mu\nu};$$

and so on; the German gothic letter is a standard indicator for the presence of the factor $(-g)^{1/2}$. On some occasions (see, for example, §21.11) it is convenient to multiply the components of a tensor with a power of $(-g)^{1/2}$ other than 1. According to the value of the exponent, the resulting assemblage of components is then called a tensor density of this or that *weight.*

The law of differentiation of an ordinary or standard tensor density formed from a tensor of arbitrary order,

$$\mathfrak{A}^{\cdots}_{\cdots} = (-g)^{1/2}A^{\cdots}_{\cdots},$$

is

$$(\mathfrak{A}^{\cdots}_{\cdots})_{;\rho} = (\mathfrak{A}^{\cdots}_{\cdots})_{,\rho} + (\text{standard } \Gamma^{\cdots}_{\cdots} \text{ terms of a standard covariant} \\ \text{derivative multiplied into } \mathfrak{A}^{\cdots}_{\cdots}) - (\mathfrak{A}^{\cdots}_{\cdots})\Gamma^{\sigma}_{\sigma\rho}.$$

The covariant derivative of a product is the sum of two terms: the covariant deriva-

tive of the first, times the second, plus the first times the covariant derivative of the second.

Now return to the integral to be evaluated. Combine the factors $g^{\alpha\beta}$ and $(-g)^{1/2}$ into the tensor density $\mathfrak{g}^{\alpha\beta}$. Integrate covariantly by parts, as justified by the rule for the covariant derivative of a product. Get a "term at limits," plus the integral

$$-(1/16\pi) \int (\mathfrak{g}^{\alpha\beta}{}_{;\lambda} - \delta^\beta_\lambda \mathfrak{g}^{\alpha\gamma}{}_{;\gamma}) \, \delta\Gamma^\lambda_{\alpha\beta} \, d^4x.$$

This integral is the only term in the action integral that contains the variations of the Γ's at the "interior points" of interest here. For the integral to be an extremum, the symmetrized coefficient of $\delta\Gamma^\lambda_{\alpha\beta}$ must vanish,

$$\mathfrak{g}^{\alpha\beta}{}_{;\lambda} - \frac{1}{2}\delta^\alpha_\lambda \mathfrak{g}^{\beta\gamma}{}_{;\gamma} - \frac{1}{2}\delta^\beta_\lambda \mathfrak{g}^{\alpha\gamma}{}_{;\gamma} = 0.$$

This set of forty equations for the forty covariant derivative $\mathfrak{g}^{\alpha\beta}{}_{;\lambda}$ has only the zero solution,

$$\mathfrak{g}^{\alpha\beta}{}_{;\lambda} = 0. \tag{21.25}$$

Thus the "density formed from the reciprocal metric tensor" is covariantly constant.

This simple result (1) brings many simple results in its train: the covariant constancy of (2) $(-g)^{1/2}$, (3) $g^{\alpha\beta}$, (4) $g_{\alpha\beta}$, and (5) $\mathfrak{g}_{\alpha\beta}$. Of these, (4) is of special interest here, and (2) is needed in proving it, as follows. Take definition (21.24) for the covariant derivative of $(-g)^{1/2}$, and calculate the ordinary derivative that appears in the first term from exercise 21.1. One encounters in this calculation terms of the form $\partial\mathfrak{g}^{\alpha\beta}/\partial x^\lambda$. Use (21.25) to evaluate them, and end up with the result

$$(-g)^{1/2}{}_{;\lambda} = 0.$$

From this result it follows that the covariant derivative of the $\binom{1}{1}$-tensor density $(-g)^{1/2}\delta^\alpha_\gamma$ is also zero. But this tensor density is the product of the tensor density $\mathfrak{g}^{\alpha\beta}$ by the ordinary metric tensor $g_{\beta\gamma}$. In the covariant derivative of this product by x^λ, one already knows that the derivative of the first factor is zero. Therefore the first factor times the derivative of the second must be zero,

$$\mathfrak{g}^{\alpha\beta} g_{\beta\gamma;\lambda} = 0,$$

and from this it follows that

$$g_{\beta\gamma;\lambda} = 0, \tag{21.26}$$

as was to be proven; or, explicitly,

$$\frac{\partial g_{\beta\gamma}}{\partial x^\lambda} - g_{\gamma\sigma}\Gamma^\sigma_{\beta\lambda} - g_{\beta\sigma}\Gamma^\sigma_{\gamma\lambda} = 0.$$

Solve these equations for the Γ's, which up to now have been independent of the $g_{\beta\gamma}$, and end up with the standard equation for the connection coefficients,

$$\Gamma^\rho_{\mu\nu} = \frac{1}{2}g^{\rho\sigma}(g_{\mu\sigma,\nu} + g_{\sigma\nu,\mu} - g_{\mu\nu,\sigma}), \tag{21.27}$$

as required for Riemannian geometry.

Similarly, equate to zero the coefficient of $\delta g^{\alpha\beta}$ in the variation (21.23), and find all ten components of Einstein's field equation, in the form

$$
G_{\alpha\beta} = 8\pi \underbrace{\left(g_{\alpha\beta} L_{\text{field}} - 2 \frac{\delta L_{\text{field}}}{\delta g^{\alpha\beta}} \right).}_{\uparrow \begin{bmatrix} \text{identified in §21.3 with} \\ \text{the stress-energy tensor } T_{\alpha\beta} \end{bmatrix}}
$$

(21.28)

Among variations of the metric, one of the simplest is the change

$$
g_{\text{new } \mu\nu} = g_{\mu\nu} + \delta g_{\mu\nu} = g_{\mu\nu} + \xi_{\mu;\nu} + \xi_{\nu;\mu}
$$

(21.29)

brought about by the infinitesimal coordinate transformation

$$
x^{\mu}_{\text{new}} = x^{\mu} - \xi^{\mu}.
$$

(21.30)

Although the metric changes, the 3-geometry does not. It does not matter whether the spacetime geometry that one is dealing with extremizes the action principle or not, whether it is a solution of Einstein's equations or not; the action integral I is a scalar invariant, a number, the value of which depends on the physics but not at all on the system of coordinates in which that physics is expressed. This invariance even obtains for both parts of the action principle individually (I_{geom} and I_{fields}). Therefore neither part will be affected in value by the variation (21.29). In other words, the quantity

Action unaffected by mere change in coordinatization

$$
\begin{aligned}
\delta I_{\text{geom}} &= (1/16\pi) \int G_{\alpha\beta}(\xi^{\alpha;\beta} + \xi^{\beta;\alpha})(-g)^{1/2} \, d^4x \\
&= -(1/8\pi) \int G_{\alpha\beta}{}^{;\beta}\xi^{\alpha}(-g)^{1/2} \, d^4x
\end{aligned}
$$

(21.31)

$\underset{\uparrow}{\hspace{1em}}$["covariant integration by parts"]

must vanish whatever the 4-geometry and whatever the change ξ^{α}. In this way, one sees from a new angle the contracted Bianchi identities of Chapter 15,

$$
G_{\alpha\beta}{}^{;\beta} = 0.
$$

(21.32)

The "neutrality" of the action principle with respect to a mere coordinate transformation such as (21.29) shows once again that the variational principle—and with it Einstein's equation—cannot determine the coordinates or the metric, but only the 4-geometry itself.

Exercise 21.1. VARIATION OF THE DETERMINANT OF THE METRIC TENSOR **EXERCISE**

Recalling that the change in the value of any determinant is given by multiplying the change in each element of that determinant by its cofactor and adding the resulting products (exercise 5.5) prove that

$$
\delta(-g)^{1/2} = \frac{1}{2}(-g)^{1/2}g^{\mu\nu} \, \delta g_{\mu\nu} \quad \text{and} \quad \delta(-g)^{1/2} = -\frac{1}{2}(-g)^{1/2}g_{\mu\nu} \, \delta g^{\mu\nu}.
$$

Also show that

$$
g = \det \|\mathfrak{g}^{\mu\nu}\| \quad \text{and} \quad \delta(-g)^{1/2} = +\frac{1}{2}g_{\mu\nu} \, \delta\mathfrak{g}^{\mu\nu}.
$$

§21.3. MATTER LAGRANGIAN AND STRESS-ENERGY TENSOR

The derivation of Einstein's geometrodynamic law from Hilbert's action principle puts on the righthand side a source term that is derived from the field Lagrangian. In contrast, the derivation of Chapter 17 identified the source term with the stress-energy tensor of the field. For the two derivations to be compatible, the stress-energy tensor must be given by the expression

$$T_{\alpha\beta} = -2 \frac{\delta L_{\text{field}}}{\delta g^{\alpha\beta}} + g_{\alpha\beta} L_{\text{field}}, \qquad (21.33a)$$

or

$$(-g)^{1/2} T^{\alpha\beta} \equiv \mathfrak{T}^{\alpha\beta} = 2 \frac{\delta \mathcal{L}_{\text{field}}}{\delta g_{\alpha\beta}}. \qquad (21.33b)$$

What are the consequences of this identification?

By the term "Lagrange function of the field" as employed here, one means the Lagrange function of the classical theory as formulated in flat spacetime, with the flat-spacetime metric replaced wherever it appears by the actual metric, and with the "comma-goes-to-semicolon rule" of Chapter 16 applied to all derivatives.

Were one dealing with a general tensorial field, the comma-goes-to-semicolon rule would introduce, in addition to the derivative of the tensorial field with all its indices, a number of Γ's equal to the number of indices. The presence of these Γ's in the field Lagrangian would have unhappy consequences for the Palatini variational procedure described in §21.2. No longer would the Γ's end up given in terms of the metric coefficients by the standard formula (21.27). No longer would the geometry, as derived from the Hilbert-Palatini variation principle, be Riemannian. Then what?

These troublesome issues do not arise in two well-known simple cases, a scalar field and an electromagnetic field. In the one case, the field Lagrangian becomes

$$L_{\text{field}} = (1/8\pi)[-g^{\alpha\beta}(\partial\phi/\partial x^{\alpha})(\partial\phi/\partial x^{\beta}) - m^2\phi^2]. \qquad (21.34)$$

No connection coefficient comes in; the quantity being differentiated is a scalar. In the other case, the field Lagrangian is built on first derivatives of the 4-potential A_{μ}. Therefore Γ's should appear, according to the standard rules for covariant differentiation (Box 8.4). However, the derivatives of the A's appear, never alone, but always in an antisymmetric combination where the Γ's cancel, making covariant derivatives equivalent to ordinary derivatives:

$$F_{\mu\nu} = A_{\nu;\mu} - A_{\mu;\nu} = A_{\nu,\mu} - A_{\mu,\nu}. \qquad (21.35)$$

In both cases, the differentiations of (21.33) to generate the stress-energy tensor are easily carried out (exercises 21.2 and 21.3) and give the standard expressions already seen [(5.22) and (5.23)] for $T_{\mu\nu}$ in one of these two cases in an earlier chapter.

Field theory provides a quite other method to generate a so-called canonical expression for the stress-energy tensor of a field [see, for example, Wentzel (1949)].

By the very manner of construction, such an expression is guaranteed also to satisfy the law of conservation of momentum and energy, and by this circumstance it too becomes useful in certain contexts. However, the canonical tensor is often not symmetric in its two indices, and in such cases violates the law of conservation of angular momentum (see discussion in §5.7). Even when symmetric, it may give a quite different localization of stress and energy than that given by (21.33). Field theory in and by itself is unable to decide between these different pictures of where the field energy is localized. However, direct measurements of the pull of gravitation provide in principle [see, for example, Feynman (1964)] a means to distinguish between alternative prescriptions for the localization of stress-energy, because gravitation responds directly to density of mass-energy and momentum. It is therefore a happy circumstance that the theory of gravity in the variational formulation gives a unique prescription for fixing the stress-energy tensor, a prescription that, besides being symmetric, also automatically satisfies the laws of conservation of momentum and energy (exercises 21.2 and 21.3). [For an early discussion of the symmetrization of the stress-energy tensor, see Rosenfeld (1940) and Belinfante (1940). A more extensive discussion is given by Corson (1953) and Davis (1970), along with extensive references to the literature.]

When one deals with a spinor field, one finds it convenient to take as the quantities to be varied, not the metric coefficients themselves, but the components of a tetrad of orthonormal vectors defined as a tetrad field over all space [see Davis (1970) for discussion and references].

Exercise 21.2. STRESS-ENERGY TENSOR FOR A SCALAR FIELD

Given the Lagrange function (21.34) of a scalar field, derive the stress-energy tensor for this field. Also write down the field equation for the scalar field that one derives from this Lagrange function (in the general case where the field executes its dynamics within the arena of a curved spacetime). Show that as a consequence of this field equation, the stress-energy tensor satisfies the conservation law, $T_{\alpha\beta}{}^{;\beta} = 0$.

Exercise 21.3. FARADAY-MAXWELL STRESS-ENERGY TENSOR

Given the Lagrangian density $-F_{\mu\nu}F^{\mu\nu}/16\pi$, reexpress it in terms of the variables A_μ and $g^{\mu\nu}$, and by use of (21.33) derive the stress-energy tensor as discussed in §5.6. Also derive from the Lagrange variation principle the field equation $F_{\alpha\beta}{}^{;\beta} = 0$ (curved spacetime, but—for simplicity—a charge-free region of space). As a consequence of this field equation, show that the Faraday-Maxwell stress-energy tensor satisfies the conservation law, $T_{\alpha\beta}{}^{;\beta} = 0$. For a more ambitious project, show that any stress-energy tensor derived from a field Lagrangian by the prescription of equation (21.33) will automatically satisfy the conservation law $T_{\alpha\beta}{}^{;\beta} = 0$.

EXERCISES

§21.4. SPLITTING SPACETIME INTO SPACE AND TIME

There are many ways to "push forward" many-fingered time and explore spacetime faster here and slower there, or faster there and slower here. However, a computer is most efficiently programmed only when it follows one definite prescription. The

Figure 21.2.
Building two 3-geometries into a thin sandwich 4-geometry, by interposing perpendicular connectors between the two, with preassigned lengths and shifts. What would otherwise be flexible thereupon becomes rigid. The flagged point illustrates equation (21.40).

Slice spacetime to compute spacetime

successive hypersurfaces on which it gives the geometry are most conveniently described by successive values of a time-parameter t. One treats on a different footing the 3-geometries of these hypersurfaces and the 4-geometry that fills in between these laminations.

The slicing of spacetime into a one-parameter family of spacelike hypersurfaces is called for, not only by the analysis of the dynamics along the way, but also by the boundary conditions as they pose themselves in any action principle of the form, "Give the 3-geometries on the two faces of a sandwich of spacetime, and adjust the 4-geometry in between to extremize the action."

Thin sandwich 4-geometry

There is no simpler sandwich to consider than one of infinitesimal thickness (Figure 21.2). Choosing coordinates adapted to the $(3 + 1)$-space-time split, designate the "lower" (earlier) hypersurface in the diagram as $t =$ constant and the "upper" (later) one as $t + dt =$ constant (names, only names; no direct measure whatsoever of proper time). Compare the two hypersurfaces with two ribbons of steel out of which one wants to construct a rigid structure. To give the geometry on the two ribbons by no means fixes this structure; for that purpose, one needs cross-connectors between the one ribbon and the other. It is not even enough (1) to specify that these connectors are to be welded on perpendicular to the lower ribbon; (2) to specify where each is to be welded; and (3) to give its length. One must in addition tell where each connector joins the upper surface. If the proper distances between tops of the connectors are everywhere shorter than the distances between the bases of the connectors, the double ribbon will have the curve of the cable of a suspension bridge; if everywhere longer, the curve of the arch of a masonry bridge. The data necessary for the construction of the sandwich are thus (1) the metric of the 3-geometry of the lower hypersurface,

$$g_{ij}(t, x, y, z)\, dx^i\, dx^j,\qquad(21.36)$$

telling the (distance)2 between one point in that hypersurface and another; (2) the metric on the upper hypersurface,

$$g_{ij}(t + dt, x, y, z) \, dx^i \, dx^j; \tag{21.37}$$

(3) a formula for the proper length,

$$\begin{pmatrix} \text{lapse of} \\ \text{proper time} \\ \text{between lower} \\ \text{and upper} \\ \text{hypersurface} \end{pmatrix} = \begin{pmatrix} \text{``lapse} \\ \text{function''} \end{pmatrix} dt = N(t, x, y, z) \, dt, \tag{21.38}$$

of the connector that is based on the point (x, y, z) of the lower hypersurface; and (4) a formula for the place on the upper hypersurface,

$$x^i_{\text{upper}}(x^m) = x^i - N^i(t, x, y, z) \, dt, \tag{21.39}$$

where this connector is to be welded. Omit part of this information, and find the structure deprived of rigidity.

The rigidity of the structure of the thin sandwich is most immediately revealed in the definiteness of the 4-geometry of the spacetime filling of the sandwich. Ask for the proper interval ds or $d\tau$ between $x^\alpha = (t, x^i)$ and $x^\alpha + dx^\alpha = (t + dt, x^i + dx^i)$. The Pythagorean theorem in its 4-dimensional form

<div style="float:right">Metric of 4-geometry depends on lapse and shift of connectors of the two 3-geometries</div>

$$ds^2 = \begin{pmatrix} \text{proper distance} \\ \text{in base 3-geometry} \end{pmatrix}^2 - \begin{pmatrix} \text{proper time from} \\ \text{lower to upper 3-geometry} \end{pmatrix}^2$$

yields the result (see Figure 21.2).

$$ds^2 = g_{ij}(dx^i + N^i \, dt)(dx^j + N^j \, dt) - (N \, dt)^2 \tag{21.40}$$

Here as in (21.36) the g_{ij} are the metric coefficients of the 3-geometry, distinguished by their Latin labels from the Greek-indexed components of the 4-metric,

$$ds^2 = {}^{(4)}g_{\alpha\beta} \, dx^\alpha \, dx^\beta, \tag{21.41}$$

labeled here with a suffix $^{(4)}$ to reduce the possibility of confusion. Comparing (21.41) and (21.40), one arrives at the following construction of the 4-metric out of the 3-metric and the lapse and shift functions [Arnowitt, Deser, and Misner (1962)]:

<div style="float:right">Details of the 4-geometry</div>

$$\left\| \begin{matrix} {}^{(4)}g_{00} & {}^{(4)}g_{0k} \\ {}^{(4)}g_{i0} & {}^{(4)}g_{ik} \end{matrix} \right\| = \left\| \begin{matrix} (N_s N^s - N^2) & N_k \\ N_i & g_{ik} \end{matrix} \right\|. \tag{21.42}$$

The welded connectors do the job!

In (21.42), the quantities N^m are the components of the shift in its original primordial contravariant form, whereas the $N_i = g_{im}N^m$ are the covariant components, as calculated within the 3-geometry with the 3-metric. To invert this relation,

$$N^m = g^{ms}N_s \tag{21.43}$$

is to deal with the reciprocal 3-metric, a quantity that has to be distinguished sharply from the reciprocal 4-metric. Thus, the reciprocal 4-metric is

$$\left\| \begin{matrix} {}^{(4)}g^{00} & {}^{(4)}g^{0m} \\ {}^{(4)}g^{k0} & {}^{(4)}g^{km} \end{matrix} \right\| = \left\| \begin{matrix} -(1/N^2) & (N^m/N^2) \\ (N^k/N^2) & (g^{km} - N^k N^m/N^2) \end{matrix} \right\|, \qquad (21.44)$$

a result that one checks by calculating out the product

$$ {}^{(4)}g_{\alpha\beta} {}^{(4)}g^{\beta\gamma} = {}^{(4)}\delta_\alpha{}^\gamma $$

according to the standard rules for matrix multiplication.

The volume element has the form

$$(-{}^{(4)}g)^{1/2} \, dx^0 \, dx^1 \, dx^2 \, dx^3 = N g^{1/2} \, dt \, dx^1 \, dx^2 \, dx^3. \qquad (21.45)$$

Welding the connectors to the two steel ribbons, or adding the lapse and shift functions to the 3-metric, by rigidifying the 4-metric, also automatically determines the components of the unit timelike normal vector \boldsymbol{n}. The condition of normalization on this 4-vector is most easily formulated by saying that there exists a 1-form, also called \boldsymbol{n} for the sake of convenience, dual to \boldsymbol{n}, and such that the product of this vector by this 1-form has the value

$$\langle \boldsymbol{n}, \boldsymbol{n} \rangle = -1. \qquad (21.46)$$

This 1-form has the value

$$\boldsymbol{n} = n_\beta \, \boldsymbol{dx}^\beta = -N \, \boldsymbol{dt} + 0 + 0 + 0. \qquad (21.47)$$

Only so can this 1-form, this structure of layered surfaces, automatically yield a value of unity, one bong of the bell, when pierced as in Figure 2.4 by a vector that represents an advance of one unit in proper time, regardless of what x, y, and z displacements it also has. Thus the unit timelike normal vector in covariant 1-form representation necessarily has the components

<div style="float:left; width:30%">The components of the unit normal</div>

$$n_\beta = (-N, 0, 0, 0) \qquad (21.48)$$

Raise the indices via (21.44) to obtain the contravariant components of the same normal, represented as a tangent vector; thus,

$$n^\alpha = [(1/N), -(N^m/N)]. \qquad (21.49)$$

This result receives a simple interpretation on inspection of Figure 21.2. Thus the typical "perpendicular connector" in the diagram can be said to have the components

$$(dt, -N^m \, dt)$$

and to have the proper length $d\tau = N \, dt$; so, ratioed down to a vector \boldsymbol{n} of unit proper length, the components are precisely those given by (21.49).

§21.5. INTRINSIC AND EXTRINSIC CURVATURE

The central concept in Einstein's account of gravity is curvature, so it is appropriate to analyze curvature in the language of the $(3 + 1)$-space-time split. The curvature intrinsic to the 3-geometry of a spacelike hypersurface may be defined and calculated by the same methods described and employed in the calculation of four-dimensional curvature in Chapter 14. Of all measures of the intrinsic curvature, one of the simplest is the Riemann scalar curvature invariant $^{(3)}R$ (written for simplicity of notation in what follows without the prefix, as R); and of all ways to define this invariant (see Chapter 14), one of the most compact uses the limit (see exercise 21.4)

$$R \begin{pmatrix} \text{at point} \\ \text{under study} \end{pmatrix} = \underset{\varepsilon \to 0}{\text{Lim}} \ 18 \ \frac{4\pi\varepsilon^2 - \begin{pmatrix} \text{proper area of a surface (approximately} \\ \text{a 2-sphere) defined as the locus of the} \\ \text{points at a proper distance } \varepsilon \end{pmatrix}}{4\pi \ \varepsilon^4}$$

<div style="text-align:right">Scalar curvature as measure of area deficit</div>

$$(21.50)$$

For a more detailed description of the curvature intrinsic to the 3-geometry, capitalize on differential geometry as already developed in Chapters 8 through 14, amending it only as required to distinguish what is three-dimensional from what is four-dimensional. Begin by considering a displacement

$$d\mathscr{P} = \boldsymbol{e}_i \, \boldsymbol{dx}^i \qquad (21.51)$$

within the hypersurface. Here the \boldsymbol{e}_i are the basis tangent vectors $\boldsymbol{e}_i = \partial/\partial x^i$ (in one notation) or $\boldsymbol{e}_i = \partial\mathscr{P}/\partial x^i$ (in another notation) dual to the three coordinate 1-forms \boldsymbol{dx}^i. Any field of tangent vectors \boldsymbol{A} that happens to lie in the hypersurface lets itself be expressed in terms of the same basis vectors:

$$\boldsymbol{A} = \boldsymbol{e}_i A^i. \qquad (21.52)$$

The scalar product of this vector with the base vector \boldsymbol{e}_j is

$$(\boldsymbol{A} \cdot \boldsymbol{e}_j) = A^i(\boldsymbol{e}_i \cdot \boldsymbol{e}_j) = A^i g_{ij} = A_j. \qquad (21.53)$$

Now turn attention from a vector at one point to the parallel transport of the vector to a nearby point.

A vector lying on the equator of the Earth and pointing toward the North Star, transported parallel to itself along a meridian to a point still on the Earth's surface, but 1,000 km to the north, will no longer lie in the 2-geometry of the surface of the Earth. A telescope located in the northern hemisphere has to raise its tube to see the North Star! The generalization to a three-dimensional hypersurface imbedded in a 4-geometry is immediate. Take vector \boldsymbol{A}, lying in the hypersurface, and transport it along an elementary route lying in the hypersurface, and in the course of this transport displace it at each stage parallel to itself, where "parallel" means parallel with respect to the geometry of the enveloping 4-manifold. Then \boldsymbol{A} will ordinarily

end up no longer lying in the hypersurface. Thus the "covariant derivative" of \boldsymbol{A} in the direction of the i-th coordinate direction in the geometry of the enveloping spacetime (that is, the \boldsymbol{A} at the new point diminished by the transported \boldsymbol{A}) has the form (see §10.4)

$$^{(4)}\boldsymbol{\nabla}_{\boldsymbol{e}_i}\boldsymbol{A} = {}^{(4)}\boldsymbol{\nabla}_i\boldsymbol{A} = {}^{(4)}\boldsymbol{\nabla}_i(\boldsymbol{e}_j A^j) = \boldsymbol{e}_j \frac{\partial A^j}{\partial x^i} + ({}^{(4)}\Gamma^\mu_{ji}\boldsymbol{e}_\mu)A^j. \tag{21.54}$$

A special instance of this formula is the equation for the covariantly measured change of the base vector \boldsymbol{e}_m itself,

$$^{(4)}\boldsymbol{\nabla}_i\boldsymbol{e}_m = {}^{(4)}\Gamma^\mu_{mi}\boldsymbol{e}_\mu. \tag{21.55}$$

In both (21.54) and (21.55) the presence of the "out-of-the-hypersurface component"

$$(A^{j(4)}\Gamma^0_{ji})(\boldsymbol{e}_0 \cdot \boldsymbol{n}) \tag{21.56}$$

From parallel transport in 4-geometry to parallel transport in 3-geometry

is quite evident. Now kill this component. Project $^{(4)}\boldsymbol{\nabla}\boldsymbol{A}$ orthogonally onto the hypersurface. In this way arrive at a parallel transport and a covariant derivative that are intrinsic to the 3-geometry of the hypersurface. By rights this covariant derivative should be written $^{(3)}\boldsymbol{\nabla}$; but for simplicity of notation it will be written as $\boldsymbol{\nabla}$ in the rest of this chapter, except where ambiguity might arise. To get the value of the new covariant derivative, one has only to rewrite (21.54) with the suffix $^{(4)}$ replaced everywhere by a $^{(3)}$, or, better, dropped altogether and with the "dummy index" of summation $\mu = (0, 1, 2, 3)$ replaced by $m = (1, 2, 3)$. However, it is more convenient, following Israel (1966), to turn from an expression dealing with contravariant components A^i of \boldsymbol{A} to one dealing with covariant components $A_i = (\boldsymbol{A} \cdot \boldsymbol{e}_i)$. Thus the covariant derivative of \boldsymbol{A} in the direction of the i-th coordinate direction in the hypersurface, calculated with respect to the 3-geometry intrinsic to the hypersurface itself, has for its h-th covariant component the quantity [see equation (10.18)]

A new covariant derivative, taken with respect to the 3-geometry

$$A_{h|i} = \boldsymbol{e}_h \cdot {}^{(3)}\boldsymbol{\nabla}_{\boldsymbol{e}_i}\boldsymbol{A} \equiv \boldsymbol{e}_h \cdot \boldsymbol{\nabla}_i\boldsymbol{A} = \frac{\partial A_h}{\partial x^i} - A^m\Gamma_{mhi} \;(= A_{h;i} \text{ for } \boldsymbol{A} \text{ in } \Sigma). \tag{21.57}$$

Here the notation of the vertical stroke distinguishes this covariant derivative from the covariant derivative taken with respect to the 4-geometry, as, for example, in equations (10.17ff). The connection coefficients here for three dimensions, like those dealt with earlier for four dimensions [see the equations leading from (14.14) through (14.15)], allow themselves to be expressed in terms of the metric coefficients and their first derivatives, and have the interpretation

$$^{(3)}\Gamma_{mhi} \equiv \Gamma_{mhi} = \boldsymbol{e}_m \cdot \boldsymbol{\nabla}_i\boldsymbol{e}_h. \tag{21.58}$$

From the connection coefficients in turn, one calculates as in Chapter 14 the full Riemann curvature tensor $^{(3)}R^i{}_{jmn}$ of the 3-geometry intrinsic to the hypersurface.

Over and above the curvature intrinsic to the simultaneity, one now encounters a concept not covered in previous chapters (except fleetingly in Box 14.1), the *extrinsic curvature* of the 3-geometry. This idea has no meaning for a 3-geometry

$$\delta \boldsymbol{n} = -\boldsymbol{K}(\delta \mathscr{P})$$

Figure 21.3.

Extrinsic curvature measures the fractional shrinkage and deformation of a figure lying in the spacelike hypersurface Σ that takes place when each point in the figure is carried forward a unit interval of proper time "normal" to the hypersurface out into the enveloping spacetime. (No enveloping spacetime? No extrinsic curvature!) The extrinsic curvature tensor is a positive multiple of the unit tensor when elementary displacements $\delta \mathscr{P}$, in whatever direction within the surface they point, all experience the same fractional shrinkage. Thus the extrinsic curvature of the hypersurface illustrated in the figure is positive. The dashed arrow represents the normal vector \boldsymbol{n} at the fiducial point \mathscr{P} after parallel transport to the nearby point $\mathscr{P} + \delta \mathscr{P}$.

conceived in and by itself. It depends for its existence on this 3-geometry's being imbedded as a well-defined slice in a well-defined enveloping spacetime. It measures the curvature of this slice relative to that enveloping 4-geometry (Figure 21.3).

Take the normal that now stands at the point \mathscr{P} and, "keeping its base in the hypersurface" Σ, transport it parallel to itself as a "fiducial vector" to the point $\mathscr{P} + \delta \mathscr{P}$, and there subtract it from the normal vector that already stands at that point. The difference, $\delta \boldsymbol{n}$, may be regarded in the appropriate approximation as a "vector," the value of which is governed by and depends linearly on the "vector" of displacement $\delta \mathscr{P}$.

To obviate any appeal to the notion of approximation, go from the finite displacement $\delta \mathscr{P}$ to the limiting concept of the vector-valued "displacement 1-form" $\boldsymbol{d}\mathscr{P}$ [see equation 15.13]. Also replace the finite but not rigorously defined vector $\delta \boldsymbol{n}$ by the limiting concept of a vector-valued 1-form \boldsymbol{dn}. This quantity, regarded as a vector, being the change in a vector \boldsymbol{n} that does not change in length, must represent a change in direction and thus stand perpendicular to \boldsymbol{n}. Therefore it can be regarded as lying in the hypersurface Σ. Depending linearly on $\boldsymbol{d}\mathscr{P}$, it can be represented in the form

$$\boldsymbol{dn} = -\boldsymbol{K}(\boldsymbol{d}\mathscr{P}). \qquad (21.59)$$

Extrinsic curvature as an operator

Here the linear operator \boldsymbol{K} is the extrinsic curvature presented as an abstract coordinate-independent geometric object. The sign of \boldsymbol{K} as defined here is positive when the tips of the normals in Figure 21.3 are closer than their bases, as they are, for example, during the recontraction of a model universe, in agreement with the conventions employed by Eisenhart (1926), Schouten (1954), and Arnowitt, Deser and Misner (1962), but opposite to the convention of Israel (1966).

Into the slots in the 1-forms that appear on the lefthand and righthand sides of (21.59), insert in place of the general tangent vector [which is to describe the general

local displacement, so far left open, as in the discussion following (2.12a)] a very special tangent vector, the basis vector \boldsymbol{e}_i, for a displacement in the i-th coordinate direction. Thus find (21.59) reading

$$^{(4)}\boldsymbol{\nabla}_i \boldsymbol{n} = -\boldsymbol{K}(\boldsymbol{e}_i) = -K_i{}^j \boldsymbol{e}_j, \tag{21.60}$$

where the $K_i{}^j$ are the components of the linear operator \boldsymbol{K} in a coordinate representation. Take the scalar product of both sides of (21.60) with the basis vector \boldsymbol{e}_m. Recall $(\boldsymbol{e}_m \cdot \boldsymbol{n}) = 0$. Thus establish the symmetry of the tensor K_{im}, covariantly presented, in its two indices:

$$\begin{aligned} K_{im} &= K_i{}^j g_{jm} = K_i{}^j(\boldsymbol{e}_j \cdot \boldsymbol{e}_m) = -\boldsymbol{e}_m \cdot {}^{(4)}\boldsymbol{\nabla}_i \boldsymbol{n} = \boldsymbol{n} \cdot {}^{(4)}\boldsymbol{\nabla}_i \boldsymbol{e}_m \\ &= (\boldsymbol{n} \cdot \boldsymbol{e}_0)^{(4)}\Gamma^0_{mi} = \boldsymbol{n} \cdot {}^{(4)}\boldsymbol{\nabla}_m \boldsymbol{e}_i = K_{mi}. \end{aligned} \tag{21.61}$$

$$\text{[see (21.55)]}$$

A knowledge of the tensor K_{ij} of extrinsic curvature assists in revealing the changes of the four vectors $\boldsymbol{n}, \boldsymbol{e}_1, \boldsymbol{e}_2, \boldsymbol{e}_3$ under parallel transport. Equation (21.60) already tells how \boldsymbol{n} changes under parallel transport. The change of \boldsymbol{e}_m is to be read off from (21.55) as a vector. It is adequate identification of this vector to know its scalar product with each of four independent vectors: with the basis vectors $\boldsymbol{e}_1, \boldsymbol{e}_2$, and \boldsymbol{e}_3, or, more briefly, with the \boldsymbol{e}_s, in (21.58); and with the normal vector \boldsymbol{n} in (21.61). Thus one arrives, following Israel (1966), at what are known as the equations of Gauss and Weingarten, in happy oversight of all change of notation in the intervening century:

Gauss-Weingarten equation for 4-transport in terms of extrinsic curvature

$$^{(4)}\boldsymbol{\nabla}_i \boldsymbol{e}_j = K_{ij} \frac{\boldsymbol{n}}{\boldsymbol{n} \cdot \boldsymbol{n}} + {}^{(3)}\Gamma^h_{ji} \boldsymbol{e}_h. \tag{21.62}$$

Knowing from this equation how each basis vector in Σ changes, one also knows how to rewrite (21.54) for the change in any vector field \boldsymbol{A} that lies in Σ. The change in both cases is expressed relative to a fiducial vector transported from a fiducial nearby point. By the term "parallel transport" one now means "parallel with respect to the geometry of the enveloping spacetime":

$$^{(4)}\boldsymbol{\nabla}_i \boldsymbol{A} = A^j{}_{|i} \boldsymbol{e}_j + K_{ij} A^j \frac{\boldsymbol{n}}{(\boldsymbol{n} \cdot \boldsymbol{n})}. \tag{21.63}$$

Of special importance is the evaluation of extrinsic curvature when spacetime is sliced up into spacelike slices according to the plan of Arnowitt, Deser, and Misner as described in §21.4. The 4-geometry of the thin sandwich illustrated in Figure 21.2, rudimentary though it is, is fully defined by the 3-metric on the two faces of the sandwich and by the lapse and shift functions N and N^i. The normal in covariant representation according to (21.47) has the components

$$(n_0, n_1, n_2, n_3) = (-N, 0, 0, 0). \tag{21.64}$$

The change in \boldsymbol{n} relative to "\boldsymbol{n} transported parallel to itself in the enveloping 4-geometry," according to the definition of parallel transport, is

$$(\boldsymbol{dn})_i = n_{i;k}\, \boldsymbol{dx}^k$$

$$= \left[\frac{\partial n_i}{\partial x^k} - {}^{(4)}\Gamma^{\sigma}_{ik}n_{\sigma}\right]\boldsymbol{dx}^k \tag{21.65}$$

$$= N\,{}^{(4)}\Gamma^{0}_{ik}\, \boldsymbol{dx}^k$$

Compare to the same change as expressed in terms of the extrinsic curvature tensor,

$$(\boldsymbol{dn})_i = -K_{ik}\, \boldsymbol{dx}^k. \tag{21.66}$$

Conclude that this tensor has the value

$$K_{ik} = -n_{i;k} = -N\,{}^{(4)}\Gamma^{0}_{ik} = -N[{}^{(4)}g^{00\,(4)}\Gamma_{0ik} + {}^{(4)}g^{0p\,(4)}\Gamma_{pik}],$$

or, with the help of equations (21.42) and (21.44),

$$K_{ik} = (1/N)[{}^{(4)}\Gamma_{0ik} - N^{p\,(3)}\Gamma_{pik}]$$

$$= \frac{1}{2N}\left[\frac{\partial N_i}{\partial x^k} + \frac{\partial N_k}{\partial x^i} - \frac{\partial g_{ik}}{\partial t} - 2\Gamma_{pik}N^p\right] \tag{21.67}$$

$$= \frac{1}{2N}\left[N_{i|k} + N_{k|i} - \frac{\partial g_{ik}}{\partial t}\right].$$

Extrinsic curvature in terms of shift and change of 3-metric

This is the extrinsic curvature expressed in terms of the ADM lapse and shift functions [Arnowitt, Deser, and Misner (1962)].

As an example, let Σ have the geometry of a 3-sphere

Extrinsic curvature of expanding 3-sphere

$$ds^2 = a^2[d\chi^2 + \sin^2\chi(d\theta^2 + \sin^2\theta\, d\phi^2)]. \tag{21.68}$$

Let the nearby spacelike slice in the one-parameter family of slices, the slice with the label $t + dt$ (only a label!) have a 3-metric given by the same formula with the radius a replaced by $a + da$. The 4-geometry of the thin sandwich between these two slices is completely undetermined until one gives the lapse and shift functions. For simplicity, take the shift vector N^i (see Figure 21.2) to be everywhere zero and the lapse function at every point on Σ to have the same value N. The separation in proper time between the two spheres is thus $d\tau = N\, dt$. Any geometric figure located in Σ expands with time. The fractional increase of any length in this figure per unit of proper time is the same in whatever direction that length is oriented, and has the value

$$\begin{pmatrix}\text{fractional increase}\\ \text{of length per unit}\\ \text{of proper time}\end{pmatrix} = \frac{1}{a}\frac{da}{d\tau} = \frac{1}{2N}\frac{1}{a^2}\frac{d(a^2)}{dt}. \tag{21.69}$$

The negative of this quantity, multiplied by the $\binom{1}{1}$ unit tensor, $\boldsymbol{1} = \boldsymbol{d}\mathscr{P}$, gives the extrinsic curvature tensor in $\binom{1}{1}$ representation,

$$\boldsymbol{K} = -\frac{1}{2N}\frac{1}{a^2}\frac{d(a^2)}{dt}\,\boldsymbol{1}. \tag{21.70}$$

One confirms this result (exercise 21.5) by direct calculation of the components K_i^j using the ADM formula (21.67) as the starting point.

The Riemann curvature $R^a_{\ bcd} = {}^{(3)}R^a_{\ bcd}$ intrinsic to the hypersurface Σ, together with the extrinsic curvature K_{ij}, give one information on the Riemann and Einstein curvatures of the 4-geometry. In the calculation, it is not convenient to use the coordinate basis,

$$
\begin{array}{cc}
\text{basis vectors,} & \text{basis 1-forms} \\
\boldsymbol{e}_0 = \partial_t, & \boldsymbol{d}t, \\
\boldsymbol{e}_i = \partial_i, & \boldsymbol{d}x^i,
\end{array}
$$

because ordinarily the basis vector \boldsymbol{e}_0 does not stand perpendicular to the hypersurface (see Figure 21.2). Adopt a different basis but one that is still self-dual,

<div style="float:left; width:30%;">

Basic forms for calculating 4-curvature

</div>

$$
\begin{array}{cc}
\text{basis vectors,} & \text{basis 1-forms,} \\
\boldsymbol{e}_n \equiv \boldsymbol{n} = N^{-1}(\partial_t - N^m\,\partial_m), & \boldsymbol{\omega}^n = N\,\boldsymbol{d}t = (\boldsymbol{n}\cdot\boldsymbol{n})\boldsymbol{n} \\
\boldsymbol{e}_i = \partial_i, & \boldsymbol{\omega}^i \equiv \boldsymbol{d}x^i + N^i\,\boldsymbol{d}t.
\end{array}
\tag{21.71}
$$

Also use Greek labels $\bar{\alpha} = n, 1, 2, 3$, instead of Greek labels $\alpha = 0, 1, 2, 3$, to list components.

Recall that curvature is measured by the change in a vector on transport around a closed route; or, from equation (14.23),

$$
\mathscr{R}(\boldsymbol{u}, \boldsymbol{v})\boldsymbol{w} = \boldsymbol{\nabla}_{\boldsymbol{u}}\boldsymbol{\nabla}_{\boldsymbol{v}}\boldsymbol{w} - \boldsymbol{\nabla}_{\boldsymbol{v}}\boldsymbol{\nabla}_{\boldsymbol{u}}\boldsymbol{w} - \boldsymbol{\nabla}_{[\boldsymbol{u},\boldsymbol{v}]}\boldsymbol{w}.
\tag{21.72}
$$

Let the vector transported be \boldsymbol{e}_i and let the route be defined by \boldsymbol{e}_j and \boldsymbol{e}_k. The latter two vectors belong to a coordinate basis. Therefore the "route closes automatically", $[\boldsymbol{e}_j, \boldsymbol{e}_k] = 0$, and the final term in (21.72) drops out of consideration. Call on (21.62) and (21.60) to find

$$
\begin{aligned}
{}^{(4)}\boldsymbol{\nabla}_{\boldsymbol{e}_j}{}^{(4)}\boldsymbol{\nabla}_{\boldsymbol{e}_k}\boldsymbol{e}_i &= {}^{(4)}\boldsymbol{\nabla}_{\boldsymbol{e}_j}\left[K_{ik}\frac{\boldsymbol{n}}{(\boldsymbol{n}\cdot\boldsymbol{n})} + {}^{(3)}\Gamma^m_{ik}\boldsymbol{e}_m \right] \\
&= K_{ik,j}\frac{\boldsymbol{n}}{(\boldsymbol{n}\cdot\boldsymbol{n})} - K_{ik}K_j^{\ m}\boldsymbol{e}_m\frac{1}{(\boldsymbol{n}\cdot\boldsymbol{n})} + {}^{(3)}\Gamma^m_{ik,j}\boldsymbol{e}_m \\
&\quad + {}^{(3)}\Gamma^m_{ik}\left[K_{mj}\frac{\boldsymbol{n}}{(\boldsymbol{n}\cdot\boldsymbol{n})} + {}^{(3)}\Gamma^s_{mj}\boldsymbol{e}_s \right].
\end{aligned}
\tag{21.73}
$$

Evaluate similarly the term with indices j and k reversed, subtract from (21.73), simplify, and find

<div style="float:left; width:30%;">

Gauss-Codazzi: 4-curvature in terms of intrinsic 3-geometry and extrinsic curvature

</div>

$$
\begin{aligned}
\mathscr{R}(\boldsymbol{e}_j, \boldsymbol{e}_k)\boldsymbol{e}_i &= (K_{ik|j} - K_{ij|k})\frac{\boldsymbol{n}}{(\boldsymbol{n}\cdot\boldsymbol{n})} \\
&\quad + [(\boldsymbol{n}\cdot\boldsymbol{n})^{-1}(K_{ij}K_k^{\ m} - K_{ik}K_j^{\ m}) + {}^{(3)}R^m_{\ ijk}]\boldsymbol{e}_m.
\end{aligned}
\tag{21.74}
$$

The coefficients give directly the desired components of the curvature tensor

$$
{}^{(4)}R^m_{\ ijk} = {}^{(3)}R^m_{\ ijk} + (\boldsymbol{n}\cdot\boldsymbol{n})^{-1}(K_{ij}K_k^{\ m} - K_{ik}K_j^{\ m})
\tag{21.75}
$$

and

$$
{}^{(4)}R^n_{\ ijk} = (\boldsymbol{n}\cdot\boldsymbol{n})^{-1\,(4)}R_{nijk} = -(\boldsymbol{n}\cdot\boldsymbol{n})^{-1}(K_{ij|k} - K_{ik|j}).
\tag{21.76}
$$

Equations (21.75) and (21.76) are known as the equations of Gauss and Codazzi [for literature, see Eisenhart (1926)]. It follows from (21.75) that the components of the curvature of the 3-geometry will normally only then agree with the corresponding components of the curvature of the 4-geometry when the imbedding happens to be accomplished at the point under study with a hypersurface free of extrinsic curvature. The directly opposite situation is illustrated by the familiar example of a 2-sphere imbedded in a flat 3-space, where the lefthand side of (21.75) (with dimensions lowered by one unit throughout!) is zero, and the extrinsic and intrinsic curvature on the right exactly cancel.

Important components of the Einstein curvature let themselves be evaluated from the Gauss-Codazzi results. In doing the calculation, it is simplest to think of \boldsymbol{e}_i, \boldsymbol{e}_j and \boldsymbol{e}_k as being an orthonormal tetrad, \boldsymbol{n} being itself already normalized and orthogonal to every vector in the hypersurface. Then, employing (14.7) and (21.75), one finds

<div style="float:right; width:30%; text-align:left;">

Einstein curvature in terms of extrinsic curvature

</div>

$$
\begin{aligned}
-G_0^0 &= {}^{(4)}R^{12}{}_{12} + {}^{(4)}R^{23}{}_{23} + {}^{(4)}R^{31}{}_{31} \\
&= {}^{(3)}R^{12}{}_{12} + {}^{(3)}R^{23}{}_{23} + {}^{(3)}R^{31}{}_{31} \\
&\quad + (\boldsymbol{n}\cdot\boldsymbol{n})^{-1}[(K_1^2 K_2^1 - K_2^2 K_1^1) + (K_2^3 K_3^2 - K_3^3 K_2^2) \\
&\quad + (K_3^1 K_1^3 - K_1^1 K_3^3)] \\
&= \tfrac{1}{2} R - \tfrac{1}{2}(\boldsymbol{n}\cdot\boldsymbol{n})^{-1}[(\operatorname{Tr}\boldsymbol{K})^2 - \operatorname{Tr}(\boldsymbol{K}^2)].
\end{aligned}
\tag{21.77}
$$

Here R is the 3-dimensional scalar curvature invariant and Tr stands for "trace of"; thus,

$$
\operatorname{Tr}\boldsymbol{K} = g^{ij}K_{ij} = g_{ij}K^{ij} = K_j^j
\tag{21.78}
$$

and

$$
\operatorname{Tr}\boldsymbol{K}^2 = (K^2)_j^j = K_j^m K_m^{\ j} = g_{js}K^{sm}g_{mi}K^{ij}.
\tag{21.79}
$$

The result, though obtained in an orthonormal tetrad, plainly is covariant with respect to general coordinate transformations within the spacelike hypersurface; and it makes no explicit reference whatever to any time coordinate, in this respect providing a coordinate-free description of the Einstein curvature.

The Einstein field equation equates (21.77) to $8\pi\rho$, where ρ is the density of mass-energy. Expression (21.77) is the "measure of curvature that is independent of how curved one cuts a spacelike slice." This measure of curvature is central to the derivation of Einstein's field equation that is sketched in Box 17.2, item 3, "Physics on a Spacelike Slice."

<div style="float:right; width:30%; text-align:left;">

Equation (21.77) is the central Einstein equation, "mass-energy fixes curvature"

</div>

The other component of the Einstein curvature tensor that is easily evaluated by (14.7) from the results at hand has the form

$$
\begin{aligned}
G_1^n &= {}^{(4)}R^{n2}{}_{12} + {}^{(4)}R^{n3}{}_{13} \\
&= -(\boldsymbol{n}\cdot\boldsymbol{n})^{-1}(K_{1|2}^2 - K_{2|1}^2 + K_{1|3}^3 - K_{3|1}^3),
\end{aligned}
\tag{21.80}
$$

when referred to an orthonormal frame. One immediately translates to a form valid for any frame \boldsymbol{e}_1, \boldsymbol{e}_2, \boldsymbol{e}_3 in the hypersurface, orthonormal or not,

$$
G_i^n = -(\boldsymbol{n}\cdot\boldsymbol{n})^{-1}[K_{i|m}^m - (\operatorname{Tr}\boldsymbol{K})_{|i}].
\tag{21.81}
$$

<div style="float:right; width:30%; text-align:left;">

The other initial-value equation

</div>

The Einstein field equation equates this quantity to 8π times the i-th covariant component of the density of momentum carried by matter and fields other than gravity.

The four components of the Einstein field equation so far written down will have a central place in what follows as "initial-value equations" of general relativity. The other six components will not be written out: (1) the dynamics lets itself be analyzed more simply by Hamiltonian methods; and (2) the calculation takes work. It demands that one evaluate the remaining type of object, $\mathscr{R}(\boldsymbol{e}_j, \boldsymbol{n})\boldsymbol{e}_i$. One step towards that calculation will be found in exercise 21.7. Sachs does the calculation (1964, equation 10) but only after specializing to Gaussian normal coordinates. These coordinates presuppose a very special slicing of spacetime: (1) geodesics issuing normally from the spacelike hypersurface $n = 0$ cut all subsequent simultaneities $n =$ constant normally; and (2) the n coordinate directly measures lapse of proper time, or proper length, whichever is appropriate,* along these geodesics. In coordinates so special it is not surprising that the answer looks simple:

$$^{(4)}R^n{}_{ink} = (\boldsymbol{n}\cdot\boldsymbol{n})^{-1}\left(\frac{\partial K_{ik}}{\partial n} + K_{im}K^m{}_k\right). \qquad \begin{pmatrix}\text{Gaussian normal}\\ \text{coordinates}\end{pmatrix} \quad (21.82)$$

Additional terms come into (21.82) when one uses, instead of the Gaussian normal coordinate system, the coordinate system of Arnowitt, Deser, and Misner. The ADM coordinates are employed here because they allow one to analyze the dynamics as one *wants* to analyze the dynamics, with freedom to push the spacelike hypersurface ahead in time at different rates in different places ("many-fingered time"). Fischer (1971) shows how to evaluate and understand the geometric content of such formulas in a coordinate-free way by using the concept of Lie derivative of a tensor field, an introduction to which is provided by exercise 21.8.

*Here Sachs' equation (10) is generalized to the case where the unit normal \boldsymbol{n} is not necessarily timelike. Sachs used $\boldsymbol{n} = \partial/\partial t$.

EXERCISES

Exercise 21.4.　SCALAR CURVATURE INVARIANT IN TERMS OF AREA DEFICIT

It being 10,000 km from North Pole to equator, one would have 62,832 km for the length of the "equator" if the earth were flat, as contrasted to the actual \sim40,000 km, a difference reflecting the fact that the surface is curved up into closure. Turn from this "pre-problem" to the actual problem, a 3-sphere

$$ds^2 = a^2[d\chi^2 + \sin^2\chi(d\theta^2 + \sin^2\theta\, d\phi^2)].$$

Measure off from $\chi = 0$ a 2-sphere of proper radius $\varepsilon = a\chi$. Determine the proper area of this 2-sphere as a function of χ. Verify that relation (21.50) on the area deficit gives in the limit $\varepsilon \longrightarrow 0$ the correct result $R = 6/a^2$. For a more ambitious exercise: (1) take a general (smooth) 3-geometry; (2) express the metric near any chosen point in terms of Riemann's normal coordinates as given in §11.6; (3) determine the locus of the set of points at the proper distance ε to the lowest interesting power of ε in terms of the spherical polar angles θ and ϕ (direction of start of geodesic of length ε); (4) determine to the lowest interesting power of ε the proper area of the figure defined by these points; and thereby establish (21.50) [for more on this topic see, for example, Cartan (1946), pp. 252–256].

Exercise 21.5. EXTRINSIC CURVATURE TENSOR FOR SLICE OF FRIEDMANN GEOMETRY

Confirm the result (21.70) for the extrinsic curvature by direct calculation from formula (21.67).

Exercise 21.6. EVALUATION OF $\mathcal{R}(e_j, e_k)n$

Evaluate this quantity along the model of (21.74) or otherwise. How can it be foreseen that the coefficient of n in the result must vanish identically? Comparing coefficients of e_m, find $^{(4)}R^m{}_{njk}$ and test for equivalence to equation (21.76).

Exercise 21.7. EVALUATION OF THE COMMUTATOR $[e_j, n]$

The evaluation of this commutator is a first step toward the calculation of a quantity like $\mathcal{R}(e_j, n)e_i$. Expressing e_j as the differential operator $\partial/\partial x^j$, use (21.49) to represent n also as a differential operator. In this way, show that the commutator in question has the value $-(N_{,j}/N)n - (N^m{}_{,j}/N)e_m$.

Exercise 21.8. LIE DERIVATIVE OF A TENSOR (exercise provided by J. W. York, Jr.)

Define the Lie derivative of a tensor field and explore some of its properties. The Lie derivative along a vector field n is a differential operator that operates on tensor fields T of type $\binom{r}{s}$, converting them into tensors $\mathcal{L}_n T$, also of type $\binom{r}{s}$. The Lie differentiation process obeys the usual chain rule and has additivity properties [compare equations (10.2b, 10.2c, 10.2d) for the covariant derivative]. For scalar functions f, one has $\mathcal{L}_n f \equiv n[f] = f_{,\mu} n^\mu$. The Lie derivative of a vector field u along a vector field v was defined in exercise 9.11 by

$$\mathcal{L}_u v \equiv [u, v].$$

If the action of \mathcal{L}_n on 1-forms is defined, the extension to tensors of general type will be simple, because the latter can always be decomposed into a sum of tensor products of vectors and 1-forms. If σ is a 1-form and v is a vector, then one defines $\mathcal{L}_n \sigma$ to be that 1-form satisfying

$$\langle \mathcal{L}_n \sigma, v \rangle = n[\langle \sigma, v \rangle] - \langle \sigma, [n, v] \rangle$$

for arbitrary v.

(a) Show that in a coordinate basis

$$\mathcal{L}_n \sigma = (\sigma_{\alpha,\beta} n^\beta + \sigma_\beta n^\beta{}_{,\alpha}) dx^\alpha.$$

(b) Show that in a coordinate basis

$$\mathcal{L}_n T = (T_{\alpha\beta,\mu} n^\mu + T_{\mu\beta} n^\mu{}_{,\alpha} + T_{\alpha\mu} n^\mu{}_{,\beta}) dx^\alpha \otimes dx^\beta$$

where T is of type $\binom{0}{2}$.

(c) Show that in (a) and (b), all partial derivatives can be replaced by covariant derivatives. [Observe that Lie differentiation is defined independently of the existence of an affine connection. For more information, see, for example, Bishop and Goldberg (1968) and Schouten (1954)].

Exercise 21.9. EXPRESSION FOR DYNAMIC COMPONENTS OF THE CURVATURE TENSOR (exercise provided by J. W. York, Jr.)

The Gauss-Codazzi equations can be viewed as giving 14 of the 20 algebraically independent components of the spacetime curvature tensor in terms of the intrinsic and extrinsic geometry of three-dimensional (non-null) hypersurfaces. In order to accomplish a space-plus-time splitting of the Hilbert Lagrangian $\sqrt{-g}\,{}^{(4)}R$, one must express, in addition, the remaining

6 components of the curvature tensor in an analogous manner. It is convenient for this purpose to express all tensors as spacetime tensors, and to use Lie derivation in the direction of the timelike unit normal field of the spacelike hypersurfaces as a generalized notion of time differentiation. A number of preliminary results must be proven:

(a)
$$\mathcal{L}_{\boldsymbol{u}} g_{\mu\nu} = u_{\mu;\nu} + u_{\nu;\mu},$$

(b)
$$\mathcal{L}_{\boldsymbol{u}}(g_{\mu\nu} + u_{\mu} u_{\nu}) \equiv \mathcal{L}_{\boldsymbol{u}}(\gamma_{\mu\nu})$$
$$= u_{\mu;\nu} + u_{\nu;\mu} + u_{\mu} a_{\nu} + a_{\mu} u_{\nu},$$

where $\gamma_{\mu\nu}$ is the metric of the spacelike hypersurface, expressed in the spacetime coordinate basis, and $a^{\mu} \equiv u^{\lambda} \nabla_{\lambda} u^{\mu}$ is the curvature vector (4-acceleration) of the timelike normal curves whose tangent field is u^{μ}. (Recall that $u_{\mu} a^{\mu} = 0$.)

(c) Prove that the extrinsic curvature tensor is given by

$$K_{\mu\nu} = -\frac{1}{2} \mathcal{L}_{\boldsymbol{u}} \gamma_{\mu\nu}.$$

(d) The unit tensor of projection into the hypersurface is defined by

$$\perp^{\mu}_{\nu} \equiv \delta^{\mu}_{\nu} + u^{\mu} u_{\nu}.$$

In terms of \perp show that one can write

$$u_{\alpha;\beta} \equiv -K_{\alpha\beta} - \omega_{\alpha\beta} - a_{\alpha} u_{\beta},$$

where

$$K_{\alpha\beta} = -\perp^{\mu}_{\alpha} \perp^{\nu}_{\beta} u_{(\mu;\nu)}$$

and

$$\omega_{\alpha\beta} = -\perp^{\mu}_{\alpha} \perp^{\nu}_{\beta} u_{[\mu;\nu]}.$$

(e) From the fact that u^{μ} is the unit normal field for a family of spacelike hypersurfaces, show that $\omega_{\alpha\beta} = 0$.

(f) The needed tools are now on hand. To obtain the result:

(i) Write down $\mathcal{L}_{\boldsymbol{u}} K_{\mu\nu}$ (see exercise 21.8);

(ii) Insert this expression into the Ricci identity in the form

$$u^{\sigma} \nabla_{\sigma} \nabla_{\mu} u_{\nu} = u^{\sigma} \nabla_{\mu} \nabla_{\sigma} u_{\nu} + {}^{(4)}R_{\rho\nu\mu\sigma} u^{\sigma} u^{\rho};$$

(iii) Project the two remaining free indices into the hypersurface using \perp, and show that one obtains

$$\perp^{\mu}_{\alpha} \perp^{\rho}_{\beta} {}^{(4)}R_{\mu\nu\rho\sigma} u^{\nu} u^{\sigma} = \mathcal{L}_{\boldsymbol{u}} K_{\alpha\beta} + K_{\alpha\gamma} K^{\gamma}_{\beta}$$
$$+ {}^{(3)}\nabla_{(\alpha} a_{\beta)} + a_{\alpha} a_{\beta},$$

where ${}^{(3)}\nabla_{\alpha} a_{\beta} \equiv \perp^{\mu}_{\alpha} \perp^{\nu}_{\beta} \nabla_{\mu} a_{\nu}$ can be shown to be the three-dimensional covariant derivative of a_{β}. In Gaussian normal coordinates, show that one obtains from this result

$$R_{0i0j} = \frac{\partial}{\partial t} K_{ij} + K_{ik} K^{k}_{j}.$$

(g) Finally, in the construction of ${}^{(4)}R$, one needs to show that

$$\gamma^{\mu\nu}[{}^{(3)}\nabla_{(\mu} a_{\nu)} + a_{\mu} a_{\nu}] = g^{\mu\nu}[{}^{(3)}\nabla_{(\mu} a_{\nu)} + a_{\mu} a_{\nu}] = a^{\lambda}_{;\lambda}.$$

**Exercise 21.10. EXPRESSION OF $^{(4)}R^i{}_{nin}$ IN TERMS OF EXTRINSIC
CURVATURE, PLUS A COVARIANT DIVERGENCE
(exercise provided by K. Kuchař)**

Let α' be an arbitrary smooth set of four coordinates, not necessarily coordinated in any
way with the choice of the 1-parameter family of hypersurfaces.

(a) Show that

$$^{(4)}R^i{}_{nin} = g^{\alpha'\gamma'}n^{\beta'}(n_{\alpha';\beta'\gamma'} - n_{\alpha';\gamma'\beta'}).$$

(b) Show that the covariant divergences

$$(n^{\beta'}n^{\gamma'}{}_{;\beta'})_{;\gamma'}$$

and

$$-(n^{\beta'}n^{\gamma'}{}_{;\gamma'})_{;\beta'}$$

can be removed from this expression in such a way that what is left behind contains only
first derivatives of the unit normal vector \boldsymbol{n}.

(c) Noting that the basis vectors \boldsymbol{e}_i and \boldsymbol{n} form a complete set, justify the formula

$$g^{\beta'\mu'} = e_i{}^{\beta'}\omega^{i\mu'} + (\boldsymbol{n}\cdot\boldsymbol{n})^{-1}n^{\beta'}n^{\mu'},$$

where ω^i is the 1-form dual to \boldsymbol{e}_i.

(d) Noting that $n_{\alpha';\beta'}n^{\alpha'} = 0$ and

$$K_{ij} = -e_{i\alpha'}n^{\alpha'}{}_{;\beta'}e_j{}^{\beta'},$$

show that

$$^{(4)}R^i{}_{nin} = (\mathrm{Tr}\boldsymbol{K})^2 - \mathrm{Tr}\boldsymbol{K}^2 \text{ plus a covariant divergence.}$$

§21.6. THE HILBERT ACTION PRINCIPLE AND
THE ARNOWITT-DESER-MISNER MODIFICATION
THEREOF IN THE SPACE-PLUS-TIME SPLIT

For analyzing the dynamics, it happily proves unnecessary to possess the missing
formula for $^{(4)}R^n{}_{ink}$. It is essential, however, to have the Lagrangian density,

$$16\pi\mathscr{L}_{\text{geom}} = (-^{(4)}g)^{1/2\,(4)}R, \tag{21.83}$$

in the Hilbert action principle as the heart of all the dynamic analysis. In the present
ADM (1962) notation, this density has the form

$$
\begin{aligned}
(-^{(4)}g)^{1/2\,(4)}R &= (-^{(4)}g)^{1/2}[^{(4)}R^{ij}{}_{ij} + 2\,^{(4)}R^{in}{}_{in}] \\
&= (-^{(4)}g)^{1/2}[R + (\boldsymbol{n}\cdot\boldsymbol{n})(\mathrm{Tr}\,\boldsymbol{K}^2 - (\mathrm{Tr}\,\boldsymbol{K})^2) + 2(\boldsymbol{n}\cdot\boldsymbol{n})\,^{(4)}R^i{}_{nin}].
\end{aligned}
\tag{21.84}
$$

Kuchař (1971b; see also exercise 21.10) shows how to calculate a sufficient part of
this quantity without calculating all of it. The difference between the "sufficient part"
and the "whole" is a time derivative plus a divergence, a quantity of the form

Drop a complete derivative
from the Hilbert action
principle to get the ADM
principle

$$[(-^{(4)}g)^{1/2}A^\alpha]_{,\alpha} = (-^{(4)}g)^{1/2}A^\alpha{}_{;\alpha}. \tag{21.85}$$

When one multiplies (21.83) by $dt\, dx^1\, dx^2\, dx^3$ and integrates to obtain the action integral, the term (21.85) integrates out to a surface term. Variations of the geometry interior to this surface make no difference in the value of this surface term. Therefore it has no influence on the equations of motion to drop the term (21.85). The result of the calculation (exercise 21.10) is simple: what is left over after dropping the divergence merely changes the sign of the terms in $\text{Tr}\, \boldsymbol{K}^2$ and $(\text{Tr}\, \boldsymbol{K})^2$ in (21.84). Thus the variation principle becomes

$$(\text{extremum}) = I_{\text{modified}} = \int \mathcal{L}_{\text{modified}}\, d^4x$$
$$= (1/16\pi) \int [R + (\boldsymbol{n} \cdot \boldsymbol{n})((\text{Tr}\, \boldsymbol{K})^2 - \text{Tr}\, \boldsymbol{K}^2)]N g^{1/2}\, dt\, d^3x + \int \mathcal{L}_{\text{fields}}\, d^4x. \tag{21.86}$$

This expression, rephrased, is the starting point for Arnowitt, Deser, and Misner's analysis of the dynamics of geometry.

Two supplements from a paper of York (1972b; see also exercise 21.9) enlarge one's geometric insight into what is going on in the foregoing analysis. First, the tensor of extrinsic curvature lets itself be defined [see also Fischer (1971)] most naturally in the form

$$\boldsymbol{K} = -\frac{1}{2}\, \pounds_n \boldsymbol{g}, \tag{21.87}$$

where \boldsymbol{g} is the metric tensor of the 3-geometry, \boldsymbol{n} is the timelike unit normal field, and \pounds is the Lie derivative as defined in exercise 21.8. Second, the divergence (21.85), which has to be added to the Lagrangian of (21.86) to obtain the full Hilbert Lagrangian, is

$$-2[(-{}^{(4)}g)^{1/2}(n^{\alpha'}\text{Tr}\, \boldsymbol{K} + a^{\alpha'})]_{,\alpha'}, \tag{21.88}$$

where the coordinates are general (see exercise 21.10), and

$$a^{\alpha'} = n^{\alpha'}{}_{;\beta'}n^{\beta'} \tag{21.89}$$

is the 4-acceleration of an observer traveling along the timelike normal \boldsymbol{n} to the successive slices.

§21.7. THE ARNOWITT, DESER, AND MISNER FORMULATION OF THE DYNAMICS OF GEOMETRY

Dirac (1959, 1964, and earlier references cited therein) formulated the dynamics of geometry in a (3 + 1)-dimensional form, using generalizations of Poisson brackets and of Hamilton equations. Arnowitt, Deser, and Misner instead made the Hilbert-Palatini variational principle the foundation for this dynamics. Because of its simplicity, this ADM (1962) approach is followed here. The gravitational part of the integrand in the Hilbert-Palatini action principle is rewritten in the condensed but standard form (after inserting a 16π that ADM avoid by other units) as

$$16\pi\mathcal{L}_{\text{geom true}} = \mathcal{L}_{\text{geom ADM}} = -g_{ij}\, \partial \pi^{ij}/\partial t - N\mathcal{H} - N_i \mathcal{H}^i$$
$$- 2\left[\pi^{ij}N_j - \frac{1}{2}N^i\, \text{Tr}\, \boldsymbol{n} + N^{|i}(g)^{1/2}\right]_{,i}. \tag{21.90}$$

Here each item of abbreviation has its special meaning and will play its special part, a part foreshadowed by the name now given it:

$$\pi^{ij}_{\text{true}} = \frac{\delta(\text{action})}{\delta g_{ij}} = \begin{pmatrix} \text{``geometrodynamic} \\ \text{field momentum'' dyn-} \\ \text{amically conjugate to} \\ \text{the ``geometrodynamic} \\ \text{field coordinate''} \ g_{ij} \end{pmatrix} = \frac{\pi^{ij}}{16\pi}; \ \pi^{ij} = g^{1/2}(g^{ij}\text{Tr}\,\boldsymbol{K} - K^{ij})$$

Momenta conjugate to the dynamic g_{ij}

(21.91)

(here the π^{ij} of ADM is usually more convenient than π^{ij}_{true}); and

$$\mathcal{H}_{\text{true}} = \mathcal{H}(\pi^{ij}_{\text{true}}, g_{ij}) = (\text{``super-Hamiltonian''}) = \mathcal{H}/16\pi;$$

$$\mathcal{H}(\pi^{ij}, g_{ij}) = g^{-1/2}\left(\text{Tr}\,\boldsymbol{n}^2 - \frac{1}{2}(\text{Tr}\,\boldsymbol{n})^2\right) - g^{1/2}R;$$

(21.92)

and

$$16\pi\mathcal{H}^i_{\text{true}} = \mathcal{H}^i = \mathcal{H}^i(\pi^{ij}, g_{ij}) = (\text{``supermomentum''}) = -2\pi^{ik}{}_{|k}. \quad (21.93)$$

Here the covariant derivative is formed treating π^{ik} as a tensor density, as its definition in (21.91) shows it to be (see §21.2). The quantities to be varied to extremize the action are the coefficients in the metric of the 4-geometry, as follows: the six g_{ij} and the lapse function N and shift function N_i; and also the six "geometrodynamic momenta," π^{ij}. To vary these momenta as well as the metric is (1) to follow the pattern of elementary Hamiltonian dynamics (Box 21.1), where, by taking the momentum p to be as independently variable as the coordinate x, one arrives at two Hamilton equations of the first order instead of one Lagrange equation of the second order, and (2) to follow in some measure the lead of the Palatini variation principle of §21.2. There, however, one had 40 connection coefficients to vary, whereas here one has come down to only six π^{ij}. To know these momenta and the 3-metric is to know the extrinsic curvature. Before carrying out the variation, drop the divergence $-2[\ \]_{,i}$ from (21.90), since it gives rise only to surface integrals and therefore in no way affects the equations of motion that will come out of the variational principle. Also rewrite the first term in (21.90) in the form

$$-(\partial/\partial t)(g_{ij}\pi^{ij}) + \pi^{ij}\,\partial g_{ij}/\partial t, \quad (21.94)$$

and drop the complete time-derivative from the variation principle, again because it is irrelevant to the resulting equations of motion. The action principle now takes the form

$$\text{extremum} = I_{\text{true}} = I_{\text{ADM}}/16\pi$$
$$= (1/16\pi)\int[\pi^{ij}\,\partial g_{ij}/\partial t - N\mathcal{H}(\pi^{ij}, g_{ij}) - N_i\mathcal{H}^i(\pi^{ij}, g_{ij})]\,d^4x$$
$$+ \int \mathcal{L}_{\text{field}}\,d^4x. \quad (21.95)$$

The action principle itself, here as always, tells one what must be fixed to make the action take on a well-defined value (if and when the action possesses an extremum). Apart from appropriate potentials having to do with fields other than geom-

Action principle says, fix
3-geometry on each face of
sandwich

etry, the only quantities that have to be fixed appear at first sight to be the values
of the six g_{ij} on the initial and final spacelike hypersurfaces. However, the ADM
action principle is invariant with respect to any change of coordinates x^1, x^2, x^3
$\longrightarrow x^{\bar{1}}$, $x^{\bar{2}}$, $x^{\bar{3}}$ within the successive spacelike slices. Therefore the quantities that
really have to be fixed on the two faces of the sandwich are the 3-geometries $^{(3)}\mathcal{G}'$
(on the initial hypersurface) and $^{(3)}\mathcal{G}$ (on the final hypersurface) and nothing more.

What a 3-geometry is

In mathematical terms, a 3-geometry $^{(3)}\mathcal{G}$ is the "equivalence class" of a set of
differentiable manifolds that are isometrically equivalent to each other under diffeo-
morphisms. In the terms of the everyday physicist, a 3-geometry is the equivalence
class of 3-metrics $g_{ij}(x, y, z)$ that are equivalent to one another under coordinate
transformations. In more homely terms, two automobile fenders have one and the
same 2-geometry if they have the same shape, regardless of how much the coordinate
rulings painted on the one may differ from the coordinate rulings painted on the
other.

Electromagnetism gives
example of momentum
conjugate to "field
coordinate"

To have in equation (21.95) an example of a field Lagrangian that is at the same
time physically relevant and free of avoidable complications, take the case of a
source-free electromagnetic field. It would be possible to take the field Lagrangian
to have the standard Maxwell value,

$$(1/8\pi)(\boldsymbol{E}^2 - \boldsymbol{B}^2) \longrightarrow -(1/16\pi)F_{\mu\nu}F^{\mu\nu}, \tag{21.96}$$

with

$$F_{\mu\nu} = \partial A_\nu/\partial x^\mu - \partial A_\mu/\partial x^\nu. \tag{21.97}$$

The variation of the Lagrangian with respect to the independent dynamic variables
of the field, the four potentials A_α, would then immediately give the four second-order
partial differential wave equations for these four potentials. However, to have instead
a larger number of first-order equations is as convenient for electrodynamics as it
is for geometrodynamics. One seeks for the analog of the Hamiltonian equations
of particle dynamics,

$$
\begin{aligned}
dx/dt &= \partial H(x, p)/\partial p, \\
dp/dt &= -\partial H(x, p)/\partial x.
\end{aligned}
\tag{21.98}
$$

One gets those equations by replacing the Lagrange integral $\int L(x, \dot{x})\, dt$ by the
Hamilton integral $\int [p\dot{x} - H(x, p)]\, dt$. Likewise, here one replaces the action inte-
grand of (21.96) by what in flat spacetime would be

$$(1/4\pi)\left[A_{\mu, \nu}F^{\mu\nu} + \frac{1}{4}F_{\mu\nu}F^{\mu\nu} \right]. \tag{21.99}$$

In actuality, spacetime is to be regarded as not only curved but also sliced up into
spacelike hypersurfaces. This $(3 + 1)$ split of the geometry made it desirable to split
the ten geometrodynamic potentials into the six g_{ij} and the four lapse and shift
functions. Here one similarly splits the four A_μ into the three components A_i of the
vector potential and the scalar potential $A_0 = -\phi$ (with the sign so chosen that,
in flat spacetime in a Minkowski coordinate system, $\phi = A^0$). In this notation, the

Lagrange density function, including the standard density factor $(-^{(4)}g)^{1/2}$ but dropping a complete time integral $(\partial/\partial t)(A_i \mathscr{E}^i)$ that has no influence on the equations of motion, is given by the formula

$$4\pi \mathscr{L}_{\text{field}} = -\mathscr{E}^i \, \partial A_i/\partial t + \phi \mathscr{E}^i{}_{,i}$$
$$-\frac{1}{2} Ng^{-1/2}g_{ij}(\mathscr{E}^i \mathscr{E}^j + \mathscr{B}^i \mathscr{B}^j) + N^i[ijk]\mathscr{E}^j \mathscr{B}^k. \qquad (21.100)$$

Lagrange density for
electromagnetism

Here use is made of the alternating symbol $[ijk]$, defined as changing sign on the interchange of any two labels, and normalized so that $[123] = 1$. Note that the 3-tensor ε^{ijk} and the alternating symbol $[ijk]$ are related much as are the corresponding four-dimensional objects in equation (8.10), so that one can write

$$\mathscr{B}^i = \frac{1}{2}[ijk](A_{k,j} - A_{j,k}). \qquad (21.101)$$

The quantities \mathscr{B}^i are the components of the magnetic field in the spacelike slice. They are not regarded as independently variable. They are treated as fully fixed by the choice of the three potentials A_i. The converse is the case for the components \mathscr{E}^i of the electric field: they are treated like momenta, and as independently variable.

Extremizing the action with respect to the \mathscr{E}^i (exercise 21.11) gives the analog of the equation $dx/dt = p/m$ in particle mechanics, and the analog of the equation

$$E_i = -\partial A_i/\partial t - \partial \phi/\partial x^i \qquad (21.102)$$

of flat-spacetime electrodynamics; namely,

$$-\partial A_i/\partial t - \phi_{,i} - Ng^{-1/2}g_{ij}\mathscr{E}^j - [ijk]N^j\mathscr{B}^k = 0. \qquad (21.103)$$

The initial-value equation of
electromagnetism

Here the last term containing the shift functions N^j, arises from the obliquity of the coordinate system. ADM give the following additional but equivalent ways to state the result (21.103):

$$\mathscr{E}^i = \frac{1}{2}[ijk] * F_{jk}$$
$$= \frac{1}{2}[ijk]\left\{\frac{1}{2}[jk\mu\nu](-^{(4)}g)^{1/2\,(4)}g^{\mu\alpha\,(4)}g^{\nu\beta}F_{\alpha\beta}\right\}. \qquad (21.104)$$

They note that \mathscr{E}^j and \mathscr{B}^j are not directly the contravariant components of the fields in the simultaneity Σ,

$$\boldsymbol{E} = E^j \boldsymbol{e}_j, \; \boldsymbol{B} = B^j \boldsymbol{e}_j, \qquad (21.105)$$

but the contravariant densities,

$$\mathscr{E}^j = g^{1/2}E^j, \; \mathscr{B}^j = g^{1/2}B^j. \qquad (21.106)$$

Extremizing the action with respect to the three A_i (exercise 21.12) gives the curved-spacetime analog of the Maxwell equations,

$$\partial \boldsymbol{E}/\partial t = \boldsymbol{\nabla} \times \boldsymbol{B}. \qquad (21.107)$$

The remaining potential, ϕ, enters the action principle at only one point. Extremizing with respect to it gives immediately the divergence relation of source-free electromagnetism,

$$\mathscr{E}^i{}_{,i} = 0. \tag{21.108}$$

If an action principle tells in and by itself what quantities are to be fixed at the limits, what lessons does (21.100) give on this score? One can go back to the example of particle mechanics in Hamiltonian form, as in Box 21.1, and note that there the momentum p could "flap in the breeze." Only the coordinate x had to be fixed at the limits. Thus the variation of the action was

$$\delta I = \delta \int [p\dot{x} - H(x, p)]\, dt \tag{21.109}$$
$$= \int \{[\dot{x} - \partial H/\partial p]\,\delta p + (d/dt)(p\,\delta x) + [-\dot{p} - \partial H/\partial x]\,\delta x\}\, dt.$$

To arrive at a well-defined extremum of the action integral I, it was not enough to annul the coefficients, in square brackets, of δp and δx; that is, to impose Hamilton's equations of motion. It was necessary in addition to annul the quantities at limits, $p\,\delta x$; that is, to specify x at the start and at the end of the motion. Similarly
here. The quantities ϕ and \mathscr{E}^i flap in the breeze, but the magnetic field has to be specified on the two faces of the sandwich to allow one to speak of a well-defined extremum of the action principle. Why the magnetic field, or the three quantities

$$\partial A_j/\partial x^i - \partial A_i/\partial x^j; \tag{21.110}$$

why not the three A_i themselves? When one varies (21.100) with respect to the A_i, and integrates the variation of the first term by parts, as one must to arrive at the dynamic equations, one obtains a term at limits

$$\int_{\Sigma_{\text{initial}}} \mathscr{E}^i\,\delta A_i\, d^3x - \int_{\Sigma_{\text{final}}} \mathscr{E}^i\,\delta A_i\, d^3x. \tag{21.111}$$

One demands that both these terms at limits must vanish in order to have a well-defined variational problem. Go from the given vector potential to another vector potential, $A_{i_{\text{new}}}$, by the gauge transformation

$$A_{i_{\text{new}}} = A_i + \delta A_i = A_i + \partial\lambda/\partial x^i. \tag{21.112}$$

The magnetic-field components given by the three $A_{i_{\text{new}}}$ differ in no way from those listed in (21.110). Moreover the "variation at limits,"

$$\int \mathscr{E}^i\,\delta A_i\, d^3x = \int \mathscr{E}^i\,\partial\lambda/\partial x^i\, d^3x = -\int \lambda \mathscr{E}^i{}_{,i}\, d^3x, \tag{21.113}$$

is automatically zero by virtue of the divergence condition (21.108), for any arbitrary choice of λ. Therefore the quantities fixed at limits are not the three A_i themselves (mere potentials) but the physically significant quantities (21.110), the components of the magnetic field. Moreover, the divergence condition $\mathscr{E}^i{}_{,i} = 0$ now becomes *the initial-value equation for the determination of the potential* ϕ.

In the preceding paragraph one need only replace "the three A_i" by "the six g_{ij}" and "the components of the magnetic field" by "the 3-geometry $^{(3)}\mathcal{G}$" and "the potential ϕ" by "the lapse and shift functions N and N^i" to pass from electrodynamics to geometrodynamics.

With this parallelism in view, turn back to the variational principle (21.95) of general relativity in the ADM formulation. With the 3-geometry fixed on the two faces of the sandwich, vary conditions in between to extremize the action, varying in turn the π^{ij}, the g_{ij}, and the lapse and shift functions. The geometrodynamic momenta appear everywhere only algebraically in the action principle, except in the term $-2N_i\pi^{ij}{}_{|j}$. Variation and integration by parts gives $2N_{i|j}\,\delta\pi^{ij}$. Collecting coefficients of $\delta\pi^{ij}$ and annuling the sum of these coefficients, one arrives at one of the several conditions required for an extremum,

ADM principle reproduces formula for geometrodynamic momentum

$$\partial g_{ij}/\partial t = 2Ng^{-1/2}\left(\pi_{ij} - \frac{1}{2}g_{ij}\mathrm{Tr}\boldsymbol{\pi}\right) + N_{i|j} + N_{j|i}. \qquad (21.114)$$

This result agrees with what one gets from equations (21.91) defining geometrodynamic momentum in terms of extrinsic curvature, together with expression (21.67) for extrinsic curvature in terms of lapse and shift. The result (21.114) here is no less useful than the result

$$dx/dt = \partial H(x,p)/\partial p = p/m$$

in the most elementary problem in mechanics: it marks the first step in splitting a second-order equation or equations into twice as many first-order equations.

Now vary the action with respect to the g_{ij} and again, after appropriate integration by parts and rearrangement, find the remaining first-order dynamic equations of general relativity [simplified by use of equations (21.116) and (21.117)],

Dynamic and initial-value equations out of ADM formalism

$$\partial\pi^{ij}/\partial t = -Ng^{1/2}\left(R^{ij} - \frac{1}{2}g^{ij}R\right) + \frac{1}{2}Ng^{-1/2}g^{ij}\left(\mathrm{Tr}\,\boldsymbol{\pi}^2 - \frac{1}{2}(\mathrm{Tr}\,\pi)^2\right)$$

$$- 2Ng^{-1/2}\left(\pi^{im}\pi_m{}^j - \frac{1}{2}\pi^{ij}\mathrm{Tr}\,\boldsymbol{\pi}\right)$$

$$+ g^{1/2}(N^{|ij} - g^{ij}N^{|m}{}_{|m}) + (\pi^{ij}N^m)_{|m} \qquad (21.115)$$

$$- N^i{}_{|m}\pi^{mj} - N^j{}_{|m}\pi^{mi} + \begin{bmatrix}\text{source terms arising from fields} \\ \text{other than geometry, omitted here for} \\ \text{simplicity, but discussed by ADM (1962)}\end{bmatrix}^{ij}.$$

Finally extremize the action (21.95) with respect to the lapse function N and the shift functions N_i, and find the four so-called initial-value equations of general relativity, equivalent to (21.77) and (21.81) or to $G^\alpha_n = 8\pi T^\alpha_n$; thus,

$$-(1/16\pi)\mathcal{H}(\pi^{ij}, g_{ij}) = (1/8\pi)Ng^{-1/2}g_{ij}(\mathcal{E}^i\mathcal{E}^j + \mathcal{B}^i\mathcal{B}^j), \qquad (21.116)$$

$$-(1/16\pi)\mathcal{H}^i(\pi^{ij}, g_{ij}) = -(1/4\pi)[ijk]\mathcal{E}^j\mathcal{B}^k. \qquad (21.117)$$

EXERCISES

Exercise 21.11. FIRST EXPLOITATION OF THE ADM VARIATIONAL PRINCIPLE FOR THE ELECTROMAGNETIC FIELD

Extremize the action principle (21.100) with respect to the \mathscr{E}^i and derive the result (21.103).

Exercise 21.12. SECOND EXPLOITATION OF THE ADM VARIATIONAL PRINCIPLE FOR THE ELECTROMAGNETIC FIELD

Extremize (21.100) with respect to the A_i, and verify that the resulting equations in any Minkowski-flat region are equivalent to (21.107).

Exercise 21.13. FARADAY-MAXWELL SOURCE TERM IN THE DYNAMIC EQUATIONS OF GENERAL RELATIVITY

Evaluate the final indicated source terms in (21.115) from the Lagrangian (21.100) of Maxwell electrodynamics, regarded as a function of the A_i and the g_{ij}.

Exercise 21.14. THE CHOICE OF ϕ DOESN'T MATTER

Prove the statement in the text that the dynamic development of the electric and magnetic fields themselves is independent of the choice made for the scalar potential $\phi(t, x, y, z)$ in the analysis (a) in flat spacetime in Minkowski coordinates and (b) in general relativity, according to equations (21.103), and (21.107) as generalized in exercise 21.12.

Exercise 21.15. THE CHOICE OF SLICING OF SPACETIME DOESN'T MATTER

Given a metric $^{(3)}g_{ij}(x, y, z)$ and an extrinsic curvature $K^{ij}(x, y, z)$ on a spacelike hypersurface $\mathbf{\Sigma}$, and given that these quantities satisfy the initial-value equations (21.116) and (21.117), and given two alternative choices for the lapse and shift functions (N, N_i) and $(N + \delta N, N_i + \delta N_i)$, show that the curvature itself (as distinguished from its components in these two distinct coordinate systems), as calculated at a point \mathscr{P} a "little way" (first order of small quantities) off the hypersurface, by way of the dynamic equations (21.114) and (21.115), is independent of this choice of lapse and shift.

§21.8. INTEGRATING FORWARD IN TIME

In the Hamiltonian formalism of Arnowitt, Deser, and Misner [see also the many papers by many workers on the quantization of general relativity—primarily putting Einstein's theory into Hamiltonian form—cited, for example, in references 1 and 2 of Wheeler (1968)], the dynamics of geometry takes a form quite similar to the Hamiltonian dynamics of geometry. There one gives x and p at a starting time and integrates two first-order equations for dx/dt and dp/dt ahead in time to find these dynamically conjugate variables at all future times. Here one gives appropriate values of g_{ij} and π^{ij} over an initial spacelike hypersurface and integrates the two first-order equations (21.114) and (21.115) ahead in time to find the geometry at future times. For example, one can rewrite the differential equations as difference equations according to the practice by now familiar in modern hydrodynamics, and then carry out the integration on an electronic digital computer of substantial memory capacity.

Time in general relativity has a many-fingered quality very different from the one-parameter nature of time in nonrelativistic particle mechanics [see, however, Dirac, Fock, and Podolsky (1932) for a many-time formalism for treating the relativistic dynamics of a system of many interacting particles]. He who is studying the geometry is free to push ahead the spacelike hypersurface faster at one place than another, so long as he keeps it spacelike. This freedom expresses itself in the lapse function $N(t, x, y, z)$ at each stage, t, of the integration. Equations (21.114) and (21.115) are not a conduit to feed out information on N to the analyst. They are a conduit for the analyst to feed in information on N. The choice of N is to be made, not by nature, but by man. The dynamic equations cannot begin to fulfill their purpose until this choice is made. The "time parameter" t is only a label to distinguish one spacelike hypersurface from another in a one-parameter family of hypersurface; but N thus tells the spacing in proper time, as it varies from place to place, between the successive slices on which one chooses to record the time-evolution of the geometry. A cinema camera can record what happens only one frame at a time, but the operator can make a great difference in what that camera sees by his choice of angle for the filming of the scene. So here, with the choice of slicing.

Lapse and shift chosen to push forward the integration in time as one finds most convenient

Another choice is of concern to the analyst, especially one doing his analysis on a digital computer. He is in the course of determining, via (21.114–21.115) written as difference equations, what happens on a lattice work of points, typified by $x = \dots, 73, 74, 75, 76, 77, \dots$, etc. He finds that the curvatures are developing most strongly in a localized region in the range around $x = 83$ to $x = 89$. He wants to increase the density of coverage of his tracer points in this region. He does so by causing points at lesser and greater x values to drift into this region moment by moment as t increases: $t = \dots, 122, 123, 124, \dots$. He makes the tracer points at lesser x-values start to move to the right (N_1 positive) and points at greater x-values move to the left (N_1 negative). In other words, the choice of the three shift functions $N_i(t, x, y, z)$ is just as much the responsibility of the analyst as is the choice of the lapse function N. The equations will never tell him what to pick. He has to tell the equations.

These options, far from complicating dynamic equations (21.114–21.115), make them flexible and responsive to the wishes of the analyst in following the course of whatever geometrodynamic process is in his hands for study.

The freedom that exists in general relativity in the choice of the four functions N, N_i, is illuminated from another side by comparing it with the freedom one has in electrodynamics to pick the one function $\phi(t, x, y, z)$, the scalar potential. In no way do the dynamic Maxwell equations (21.103) and (21.107), as generalized in exercise 21.12 determine ϕ. Instead they demand that it be determined (by the analyst) as the price for predicting the time-development of the vector potential A_i. An altered choice of $\phi(t, x, y, z)$ in its dependence on position and time means altered results from the dynamic equations for the development of the three A_i in time and space. However, the physically significant quantities, the electric and magnetic fields themselves on successive hypersurfaces, come out the same (exercise 21.14) regardless of this choice of ϕ. Similarly in geometrodynamics: an altered choice for the four

Same 4-geometry regardless of lapse and shift options

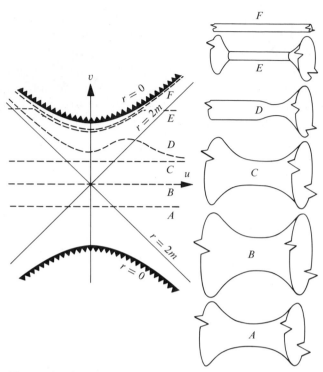

Figure 21.4.
Some of the many ways to make distinct spacelike slices through
one and the same $^{(4)}\mathcal{G}$, the complete Schwarzschild 4-geometry.

functions N, N_i, means (a) an altered laying down of coordinates in spacetime, and
therefore (b) altered results for the intrinsic metric $^{(3)}g_{ij}$ and extrinsic curvature K^{ij}
of successive spacelike hypersurfaces, but yields the same 4-geometry $^{(4)}\mathcal{G}$ (Figure
21.4) regardless of this choice of coordinatization (exercise 21.15).

§21.9. THE INITIAL-VALUE PROBLEM IN
THE THIN-SANDWICH FORMULATION

Initial-value data: what is
freely disposable? and what
is thereby fixed?

Given appropriate initial-value data, one can integrate the dynamic equations ahead
in time and determine the evolution of the geometry; but what are "appropriate
initial-value data"? They are six functions $^{(3)}g_{ij}(x, y, z)$ plus six more functions
$\pi^{ij}(x, y, z)$ or $K^{ij}(x, y, z)$ that together satisfy the four initial-value equations (21.116)
and (21.117). To be required to give coordinates and momenta accords with the
familiar plan of Hamiltonian mechanics; but to have consistency conditions or
"constraints" imposed on such data is less familiar. A particle moving in two-dimen-
sional space is catalogued by coordinates x, y, and coordinates p_x, p_y; but a particle
forced to remain on the circle $x^2 + y^2 = a^2$ satisfies the constraint $xp_x + yp_y = 0$.
Thus the existence of a "constraint" is a signal that the system possesses fewer degrees

of freedom than one would otherwise suppose. Fully to analyze the four "initial-value" or "constraint" conditions (21.116) and (21.117) is thus to determine (1) how many dynamic degrees of freedom the geometry possesses and (2) what these degrees of freedom are; that is to say, precisely what "handles" one can freely adjust to govern completely the geometry and its evolution with time. The counting one can do today, with the conclusion that the geometry possesses the same count of true degrees of freedom as the electromagnetic field. The identification of the "handles," or freely adjustable features of the dynamics, is less advanced for geometry than it is for electromagnetism (Box 21.2), but most instructive so far as it goes.

By rights the identification of the degrees of freedom of the field, whether that of Einstein or that of Faraday and Maxwell, requires nothing more than knowing what must be fixed on initial and final spacelike hypersurfaces to make the appropriate variation principle well-defined. One then has the option whether (1) to give that quantity on both hypersurfaces or (2) to give that quantity and its dynamic conjugate on one hypersurface or (3) to give the quantity on both hypersurfaces, as in (1), but go to the limit of an infinitely thin sandwich, so that one ends up specifying the quantity and its time rate of change on one hypersurface. This third "thin sandwich" procedure is simplest for a quick analysis of the initial-value problem in both electrodynamics and geometrodynamics. Take electrodynamics first, as an illustration.

Give the divergence-free magnetic field and its time-rate of change: on an arbitrary smooth spacelike hypersurface in curved spacetime in the general case; on the hypersurface $t = 0$ in Minkowski spacetime in the present illustrative treatment,

In electromagnetism, give magnetic field and its rate of change as initial data

$$\mathcal{B}^i(0, x, y, z) \text{ given}, \tag{21.118}$$

$$\dot{\mathcal{B}}^i(0, x, y, z) = \left(\frac{\partial \mathcal{B}^i}{\partial t}\right) \text{ also given}. \tag{21.119}$$

These quantities together contain four and only four independent data per space point. How is one now to obtain the momenta $\pi^i \sim -\mathcal{E}^i$ so that one can start integrating the dynamic equations (21.103) and (21.107) forward in time? (1) Find a set of three functions $A_i(0, x, y, z)$ such that their curl gives the three specified \mathcal{B}^i. That this can be done at all is guaranteed by the vanishing of the divergence $\mathcal{B}^i{}_{,i}$. However, the choice of the A_i is not unique. The new set of potentials $A_{i\text{new}} = A_i + \partial\lambda/\partial x^i$ with arbitrary smooth λ, provide just as good a solution as the original A_i. No matter. Pick one solution and stick to it. (2) Similarly, find a set of three $\dot{A}_i(0, x, y, z)$ such that their curl gives the specified $\dot{\mathcal{B}}^i(0, x, y, z)$, and resolve all arbitrariness of choice by *fiat*. (3) Recall that the electric field (negative of the field momentum) is given by

$$\mathcal{E}_i = -\dot{A}_i - \partial\phi/\partial x^i \tag{21.120}$$

(formula valid without amendment only in flat space). The initial-value or constraint equation $\mathcal{E}^i{}_{,i} = 0$ translates to the form

$$\nabla^2\phi = -\eta^{ij}\dot{A}_{i,j}. \tag{21.121}$$

Box 21.2 COUNTING THE DEGREES OF FREEDOM OF THE ELECTROMAGNETIC FIELD

A. First Approach: Number of "Field Coordinates" per Spacepoint

Superficial tally of the degrees of freedom of the source-free electromagnetic field gives three field coordinates $A_i(x, y, z)$ per spacepoint on the initial simultaneity Σ, plus three field momenta $\pi^i_{\text{true}} = \pi^i/4\pi$ [with $\pi^i = -\mathcal{E}^i(x, y, z)$] per spacepoint.

Closer inspection reveals that the number of coordinate degrees of freedom per spacepoint is not three but two. Thus the change in vector potential $A_i \longrightarrow A_i + \partial\lambda/\partial x^i$ makes no change in the actual physics, the magnetic field components,

$$B^i = \frac{1}{2}[ijk](\partial A_k/\partial x^j - \partial A_j/\partial x^k).$$

Moreover, though those components are three in number, they satisfy one condition per spacepoint, $\mathcal{B}^i{}_{,i} = 0$, thus reducing the effective net number of coordinate degrees of freedom per spacepoint to two.

The momentum degrees of freedom per spacepoint are likewise reduced from three to two by the one condition per spacepoint $\mathcal{E}^i{}_{,i} = 0$.

B. Alternative Approach: Count Fourier Coefficients

In textbooks on field theory [see, for example, Wentzel (1949)], attention focuses on flat spacetime. The electromagnetic field is decomposed by Fourier analysis into individual running waves. Instead of counting degrees of freedom per point in coordinate space, one does the equivalent: counts up degrees of freedom per point in wavenumber space. Thus for each (k_x, k_y, k_z), there are two independent states of polarization. Each state of polarization requires for its description an amplitude ("coordinate") and time-rate of change of amplitude ("momentum") at the initial time, t'_0. Thus the number of degrees of freedom per point in wave-number space is two for coordinates and two for momenta, in accord with what one gets by carrying out the count in coordinate space.

In curved spacetime, Fourier analysis is a less convenient way of identifying the degrees of freedom of the electromagnetic field [for such a Fourier analysis, see Misner and Wheeler (1957), especially their Table X and following text] than direct analysis in space, as above.

C. Another Alternative: Analyze "Deformation of Structure"

Still a third way to get a handle on the degrees of freedom of a divergence-free field, whether \mathcal{E} or \mathcal{B}, rests on the idea of deformation of structure [diagram from Wheeler (1964)]. Represent the

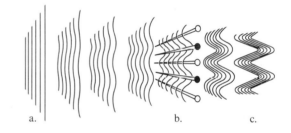

a. b. c.

magnetic field by Faraday's picture of lines of force (a) continuing through space without ever ending, automatic guarantee that $\mathcal{B}^i{}_{,i}$ is everywhere zero. Insert "knitting needles" (b) into the spaghetti-like structure of the lines of force and move these needles as one will. Sliding the "knitting needles" along a line of force causes no movement of the line of force. (c) With the help of two knitting needles perpendicular to each other and to the line of force, one can give any given line of force any small displacement one pleases perpendicular to its length: again two degrees of freedom per spacepoint. Granted any non-zero field to begin with, no matter how small, one can build it up by a sequence of such small deformations to agree with any arbitrary field pattern of zero divergence, no matter what its complexity and strength may be.

Solve for ϕ. Then (4) equation (21.120) gives the initial-value electric field, or electrodynamic field momentum $\pi^i \sim -\mathscr{E}^i$, required (along with the field coordinate A_i) for starting the integration of the dynamic equations (21.103) and (21.107). [Misner and Wheeler (1957) deal with the additional features that come in when the space is multiply connected. Each wormhole or handle of the geometry is able to trap electric lines of force. The flux trapped in any one wormhole defines the classical electric charge q_w associated with that wormhole. One has to specify all these charges once and for all in addition to the data (21.118) and (21.119) in order to determine fully the dynamic evolution of the electromagnetic field. There is no geometrodynamic analog to electric charge, according to Unruh (1971).] (5) In this integration, the scalar potential ϕ at each subsequent time step is not to be calculated; it is to be chosen. Only when one has made this free choice definite do the dynamic equations come out with definite results for the A_i and the π^i or \mathscr{E}^i at these successive steps.

In the thin-sandwich formulation of the initial-value problem of electrodynamics, to summarize, one gives \mathscr{B}^i and $\dot{\mathscr{B}}^i$ (equivalent to \mathcal{B} on two nearby hypersurfaces). One chooses the A_i and \dot{A}_i with much arbitrariness to represent these initial-value data. The arbitrariness having been seized on to give the initial A_i and A_i, there is no arbitrariness left in the initial ϕ. However, at all subsequent times the situation is just the other way around. All the arbitrariness is sopped up in the choice of the ϕ, leaving no arbitrariness whatever in the three A_i (as given by the integration of the dynamic equation).

The situation is quite similar in geometrodynamics. One gives the beginnings of a 1-parameter family of spacelike hypersurfaces; namely,

$$^{(3)}\mathscr{G}(0) \text{ given,} \tag{21.122}$$

$$^{(3)}\dot{\mathscr{G}}(0) = \frac{\partial\,^{(3)}\mathscr{G}}{\partial t} \text{ given,} \tag{21.123}$$

Then (1) one picks a definite set of coordinates $x^i = (x, y, z)$ and in terms of those coordinates finds the unique metric coefficients $g_{ij}(x, y, z)$ that describe that 3-geometry. The existence of a solution is guaranteed by the circumstance that $^{(3)}\mathscr{G}$ is a Riemannian geometry. However, one could have started with different coordinates and ended up with different metric coefficients for the description of the same 3-geometry. No matter. Pick one set of coordinates, take the resulting metric coefficients, and stick to them as giving half the required initial-value data. (2) Similarly, to describe the 3-geometry $^{(3)}\mathscr{G} + {}^{(3)}\dot{\mathscr{G}}\,dt$ at the value of the parameter $t + dt$, make use of coordinates $x^i + \dot{x}^i\,dt$ and arrive at the metric coefficients $g_{ij} + \dot{g}_{ij}\,dt$. The arbitrariness in the x^i having thus been resolved by *fiat*, and the $^{(3)}\mathscr{G}$ being given as definite initial physical data, the g_{ij} are thereby completely fixed. (3) Recall that the components of the extrinsic curvature K_{ij} or the momenta π^{ij} are given in terms of the g_{ij} and \dot{g}_{ij} and the lapse and shift functions N and N_i by (21.67) or by (21.67) plus (21.91) or by (21.114). The four initial-value or "constraint" equations (21.116) and (21.117) thus become four conditions for finding the four

Scalar potential: fixed at start; freely disposable later

In ADM treatment, give 3-geometry and its time-rate of change

quantities N, N_i. One can shorten the writing of these conditions by introducing the abbreviations

$$\gamma_{ij} = \frac{1}{2}[N_{i|j} + N_{j|i} - \partial g_{ij}/\partial t] \tag{21.124}$$

and

$$\gamma_2 = \begin{pmatrix} \text{"shift} \\ \text{anomaly"} \end{pmatrix} = (\text{Tr } \boldsymbol{\gamma})^2 - \text{Tr } \boldsymbol{\gamma}^2 \tag{21.125}$$

(both for functions of x, y, z on the initial simultaneity). Then one has

$$^{(3)}R + \gamma_2/N^2 = 16\pi T_{nn} = 16\pi T^{nn} \tag{21.126}$$

for the one initial-value equation; and for the other three,

$$\left[\frac{\gamma_i{}^k - \delta_i{}^k \text{Tr } \gamma}{N} \right]_{|k} = -8\pi T_i{}^n. \tag{21.127}$$

In summary, one chooses the g_{ij} and \dot{g}_{ij} with much arbitrariness (because of the arbitrariness in the coordinates, not by reason of any arbitrariness in the physics) to represent the given initial-value data, $^{(3)}\mathcal{G}$ and $^{(3)}\dot{\mathcal{G}}$. The arbitrariness at the initial time all having been soaked up in this way, one expects no arbitrariness to be left in the initial N and N_i as obtained by solving (21.126) and (21.127). However, on all later spacelike slices, the award of the arbitrariness is reversed. The lapse and shift functions are freely disposable, but, with them once chosen, there is no arbitrariness whatever in the six g_{ij} (and the six K^{ij} or π^{ij}) as given by the integration of the dynamic equations (21.114) and (21.115). The analogy with electrodynamics is clear. There the one "gauge-controlled" function ϕ was fixed at the start by the elliptic equation (21.121), but was thereafter free. Here the four lapse and shift functions are fixed at the start by the four equations (21.126) and (21.127), but are thereafter free.

Lapse and shift initially determinate; thereafter freely disposable

Exercise 21.16 applies the initial-value equation (21.126) to analyze the whole evolution in time of any Friedmann universe in which one knows the equation $p = p(\rho)$ connecting pressure with density. Exercise 21.17 looks for a variation principle on the spacelike hypersurface Σ equivalent in content to the elliptic initial-value equation (21.121) for the scalar potential ϕ. Exercises 21.18 and 21.19 look for similar variation principles to determine the lapse and shift functions.

How many degrees of freedom, or how many "handles," are there in the specification of the 4-geometry that one will obtain? The metric coefficients of the initial 3-geometry provided six numbers per space point. However, they were arbitrary to the extent of a coordinate transformation, specified by three functions of position,

Counting initial-value data

$$x = x(x', y', z'),$$
$$y = y(x', y', z'),$$
$$z = z(x', y', z').$$

The net number of quantities per space point with any physical information was therefore $6 - 3 = 3$. One can visualize these three functions as the three diagonal components of the metric in a coordinate system in which g_{ij} has been transformed to diagonal form. Ordinarily it is not useful to go further and actually spell out the analysis in any such narrowly circumscribed coordinate system.

Now think of the $^{(3)}\mathcal{G}$ in question as imbedded in the $^{(4)}\mathcal{G}$ that comes out of the integrations. Moreover, think of that $^{(4)}\mathcal{G}$ as endowed with the lumps, bumps, wiggles, and waves that distinguish it from other generic 4-geometries and that make Minkowski geometry and special cosmologies so unrepresentative. The $^{(3)}\mathcal{G}$ is a slice in that $^{(4)}\mathcal{G}$. It partakes of the lumps, bumps, wiggles, and waves present in all those regions of the $^{(4)}\mathcal{G}$ that it intersects. To the extent that the $^{(4)}\mathcal{G}$ is generic, it does not allow the $^{(3)}\mathcal{G}$ to be moved to another location without becoming a different $^{(3)}\mathcal{G}$. If one tries to push the $^{(3)}\mathcal{G}$ "forward in time" a little in a certain locality, leaving it unchanged in location elsewhere, one necessarily changes the $^{(3)}\mathcal{G}$. By this circumstance, one sees that the $^{(3)}\mathcal{G}$ "carries information about time" [Sharp (1960); Baierlein, Sharp, and Wheeler (1962)]. Moreover, this "forward motion in time" demands for its description one number per space point. It is possible to think of this number in concrete terms by imagining an arbitrary coordinate system $\bar{t}, \bar{x}, \bar{y}, \bar{z}$ laid down in the $^{(4)}\mathcal{G}$. Then the hypersurface can be conceived as defined by the value $\bar{t} = \bar{t}(\bar{x}, \bar{y}, \bar{z})$ at which it cuts the typical line $\bar{x}, \bar{y}, \bar{z}$. A forward movement carries it to $\bar{t}(\bar{x}, \bar{y}, \bar{z}) + \delta\bar{t}(\bar{x}, \bar{y}, \bar{z})$, and changes shape and metric coefficients on $^{(3)}\mathcal{G}$ accordingly. It is usually better not to tie one's thinking down to such a concrete model, but rather to recognize as a general point of principle (1) that the location of the $^{(3)}\mathcal{G}$ in spacetime demands for its specification one datum per spacepoint, and (2) that this datum is already willy-nilly present in the three data per spacepoint that mark any $^{(3)}\mathcal{G}$.

In conclusion, there are only two data per spacepoint in a $^{(3)}\mathcal{G}$ that really tell anything about the $^{(4)}\mathcal{G}$ in which it is imbedded, or to be imbedded (as distinguished from where the $^{(3)}\mathcal{G}$ slices through that $^{(4)}\mathcal{G}$). Similarly for the other $^{(3)}\mathcal{G}$ that defines the other "face of the sandwich," whether thick or thin. Thus one concludes that the specification of $^{(3)}\mathcal{G}$ and $^{(3)}\dot{\mathcal{G}}$ actually gives four net pieces of dynamic information per spacepoint about the $^{(4)}\mathcal{G}$ (all the rest of the information being "many-fingered time," telling where the 3-geometries are located in that $^{(4)}\mathcal{G}$). According to this line of reasoning, geometrodynamics has the same number of dynamic degrees of freedom as electrodynamics. One arrives at the same conclusion in quite another way through the weak-field analysis (§35.3) of gravitational waves on a flat spacetime background: the same ranges of possible wave numbers as for Maxwell waves; and for each wave number two states of polarization; and for each polarization one amplitude and one phase (the equivalent of one coordinate and one momentum).

In electrodynamics in a prescribed spacetime manifold, one has a clean separation between the one time-datum per spacepoint (when one deals with electromagnetism in the context of many-fingered time) and the two dynamic variables per spacepoint; but not so in the superspace formulation of geometrodynamics. There the two kinds of quantities are inextricably mixed together in the one concept of 3-geometry.

Four pieces of geometrodynamic information per space point on initial simultaneity

Turn from initial- and final-value data to the action integral that is determined by (1) these data and (2) the principle that the action be an extremum,

$$I = I_{\text{extremum}} = S.$$

The action depends on the variables on the final hypersurface, according to the formula

$$S = S(\Sigma, \boldsymbol{B}) \tag{21.128}$$

in electrodynamics, but according to the formula

$$S = S({}^{(3)}\mathcal{G}) \tag{21.129}$$

in geometrodynamics. In each case, there are three numbers per spacepoint in the argument of the functional (one in Σ; two in a divergence-free magnetic field; three in ${}^{(3)}\mathcal{G}$).

This mixing of the one many-fingered time and the two dynamic variables in a 3-geometry makes it harder in general relativity than in Maxwell theory to know when one has in hand appropriate initial value data. Give Σ and give \mathcal{B} and $\dot{\mathcal{B}}$ on Σ: that was enough for electrodynamics. For geometrodynamics, to give the six $g_{ij}(x, y, z)$ and the six $\dot{g}_{ij}(x, y, z)$ is not necessarily enough. For example, let the time parameter t be a fake, so that dt, instead of leading forward from a given hypersurface Σ to a new hypersurface $\Sigma + d\Sigma$, merely recoordinatizes the present hypersurface:

$$x^i \longrightarrow x^i - \xi^i \, dt,$$

$$g_{ij} \longrightarrow g_{ij} + (\xi_{i|j} + \xi_{j|i}) \, dt. \tag{21.130}$$

A first inspection may make one think that one has adequate data in the six g_{ij} and the six

$$\dot{g}_{ij} = \xi_{i|j} + \xi_{j|i}, \tag{21.131}$$

but in the end one sees that one has not both faces of the thin sandwich, as required, but only one. Thus one must reject, as improperly posed data in the generic problem of dynamics, any set of six \dot{g}_{ij} that let themselves be expressed in the form (21.131) [Belasco and Ohanian (1969)].

Similar difficulties occur when the two faces of the thin sandwich, instead of coinciding everywhere, coincide in a limited region, be it three-dimensional, two-dimensional, or even one-dimensional ("crossover of one face from being earlier than the other to being later"). Thus it is enough to have (21.131) obtaining even on only a curved line in Σ to reject the six g_{ij} as inappropriate initial-value data.

That one can impose conditions on the g_{ij} and \dot{g}_{ij} which will guarantee existence and uniqueness of the solution $N(x, y, z)$, $N_i(x, y, z)$ of the initial-value equations (21.126) and (21.127) is known as the "thin-sandwich conjecture," a topic on which there has been much work by many investigators, but so far no decisive theorem.

Problem in assuring completeness and consistency of initial data

The "thin sandwich conjecture"

To presuppose existence and uniqueness is to make the first step in giving mathematical content to Mach's principle that the distribution of mass-energy throughout space determines inertia (§21.12).

§21.10. THE TIME-SYMMETRIC AND TIME-ANTISYMMETRIC INITIAL-VALUE PROBLEMS

Turn from the general initial-value problem to two special initial-value problems that lend themselves to detailed treatment, one known as the time-symmetric initial-value problem, the other as the time-antisymmetric problem.

A 4-geometry is said to be time-symmetric when there exists a spacelike hypersurface Σ at all points of which the extrinsic curvature vanishes. In this case the three initial value equations (21.127) are automatically satisfied, and the fourth reduces to a simple requirement on the three-dimensional scalar curvature invariant,

$$R = 16\pi\rho. \tag{21.132}$$

Still further simplifications result when one limits attention to empty space. Simplest of all is the case of spherical symmetry in which (21.132) yields at once the full Schwarzschild geometry at the moment of time symmetry (two asymptotically flat spaces connected by a throat), as developed in exercise 21.20.

Consider a 3-geometry with metric

$$ds_1{}^2 = g_{(1)ik}\, dx^i\, dx^k. \tag{21.133}$$

Call it a "base metric." Consider another 3-geometry with metric

$$ds_2{}^2 = \psi^4(x^i)\, ds_1{}^2. \tag{21.134}$$

Angles are identical in the two geometries. On this account they are said to be conformally equivalent. The scalar curvature invariants of the two 3-geometries are related by the formula [Eisenhart (1926)]

$$R_2 = -8\psi^{-5}\,\boldsymbol{\nabla}_1{}^2\psi + \psi^{-4}R_1, \tag{21.135}$$

where

$$\boldsymbol{\nabla}_1{}^2\psi = \psi_{|i}{}^{|i} = g_1{}^{-1/2}(\partial/\partial x^i)[g_1{}^{1/2}g^{ik}(\partial\psi/\partial x^k)] \tag{21.136}$$

Demand that the scalar curvature invariant R_2 vanish, and arrive [Brill (1959)] at the "wave equation"

$$\boldsymbol{\nabla}_1{}^2\psi - (R_1/8)\psi = 0 \tag{21.137}$$

for the conformal correction factor ψ. Brill takes the base metric to have the form suggested by Bondi,

$$ds_1{}^2 = e^{2Aq_1(\rho,z)}(dz^2 + d\rho^2) + \rho^2\, d\phi^2, \tag{21.138}$$

and takes the conformal correction factor ψ also to possess axial symmetry. In the application:

$q_1(\rho, z)$ measures the "distribution of gravitational wave amplitude," assumed for simplicity to vanish outside $r = (\rho^2 + z^2)^{1/2} = a$;

A measures the "amplitude of the distribution of gravitational wave amplitude";

$\psi(\rho, z)$ is the conformal correction factor, which varies with position at large distances as $1 + (m/2r)$. The quantity m(cm) is uniquely determined by the condition that the geometry be asymtotically flat. It measures the mass-energy of the distribution of gravitational radiation.

Wave amplitude determines
mass-energy: $m = m(A)$

The mass m of the gravitational radiation is proportional to A^2 for small values of the amplitude A. It is inversely proportional to the reduced wavelength $\lambda = $ (effective wavelength$/2\pi$) that measures the scale of rapid variations in the gravitational wave amplitude $q_1(\rho, z)$ in the "active zone." Thus the metric is dominated by wiggles, proportional in amplitude to A, in the active zone, and at larger distances dominated by something close to a Schwarzschild $(1 + 2m/r)$ factor in the metric. When the amplitude A is increased, a critical value is attained, $A = A_{\text{crit}}$, at which m goes to infinity and the geometry curves up into closure ("universe closed by its own content of gravitational-wave energy"). Further analysis and examples will be found in Wheeler (1964a), pp. 399–451, also in Wheeler (1964c).

"Time-antisymmetric"
initial-value data

Brill has carried out a similar analysis [Brill (1961)] for the vacuum case of what he calls time-antisymmetric initial-value conditions, sketched below as amended by York (1973). (1) The initial slice is maximal, Tr $\boldsymbol{K} = 0$. (2) This slice is conformally flat,

$$g_{ij} = \psi^4 \, \delta_{ij}. \tag{21.139}$$

(3) Work in the "base space" with metric δ_{ij} and afterwards transform to the geometry (21.139). Three of the initial-value equations become

$$K^{ij}_{\text{base}, j} = 0. \tag{21.140}$$

To solve these equations, (1) take any localized trace-free symmetric tensor B_{km}; (2) solve the flat-space Laplace equation $\nabla^2 A = (3/2) \, \partial^2 B_{km}/\partial x^k \, \partial x^m$ for A; (3) define the six potentials $A_{km} = B_{km} + \tfrac{1}{3} A \, \delta_{km}$; and (4) calculate

$$K^{ij}_{\text{base}} = [ik\ell][jmn] \, \partial^2 A_{km}/\partial x^\ell \, \partial x^n, \tag{21.141}$$

that automatically satisfy (21.140) and give Tr $\boldsymbol{K}_{\text{base}} = 0$. Then $K^{ij} = \psi^{-10} K^{ij}_{\text{base}}$ *also* automatically satisfies these conditions, but now in the *curved* geometry (21.139). The final initial-value equation becomes a quasilinear elliptic equation, in the flat base space, for the conformal factor ψ,

$$8\nabla^2_{\text{base}} \psi + \psi^{-7} \sum_{i,j} (K_{\text{base } ij})^2 = 0. \tag{21.142}$$

The asymptotic form of ψ reveals that the mass of the wave is positive.

In addition to the time-symmetric and time-antisymmetric cases, there are at least two further cases where the initial-value problem possess special simplicity. One is the case of a geometry endowed with a symmetry, as, for example, for the Friedmann universe of Chapter 27 or the mixmaster universe of Chapter 30 or cylindrical gravitational waves in the treatment of Kuchař (1971a). One starts with a spacelike slice on which the g_{ij} and π^{ij} have a special symmetry, and makes all future spacelike slices in a way that preserves this symmetry. The geometry on any one of these simultaneities, though almost entirely governed by these symmetry considerations, still typically demands some countable number of parameters for its complete determination, such as the radius of the Friedmann universe, or the three principal radii of curvature of the mixmaster universe. These parameters and the momenta conjugate to them define a miniphase space. In this miniphase space, the dynamics runs its course as for any other problem of classical dynamics [see, for example, Box 30.1 and Misner (1969) for the mixmaster universe; Kuchař (1971a) and (1972) for waves endowed with cylindrical symmetry; and Gowdy (1973) for waves with spherical symmetry]. Even the evidence for the existence of many-fingered time, most characteristic feature of general relativity, is suppressed as the price for never having to give attention to any spacelike slice that departs from the prescribed symmetry.

Finite dimensional dynamics for geometries endowed with high symmetry

EXERCISES

Exercise 21.16. POOR MAN'S WAY TO DO COSMOLOGY

Consider a spacetime with the metric

$$ds^2 = -dt^2 + a^2(t)[d\chi^2 + \sin^2\chi(d\theta^2 + \sin^2\theta\, d\phi^2)],$$

corresponding to a 3-geometry with the form of a sphere of radius $a(t)$ changing with time. Show that the tensor of extrinsic curvature as expressed in a local Euclidean frame of reference is

$$\boldsymbol{K} = -a^{-1}(da/dt)\boldsymbol{1},$$

where $\boldsymbol{1}$ is the unit tensor. Show that the initial value equation (21.77) reduces to

$$(6/a^2)(da/dt)^2 + (6/a^2) = 16\pi\rho(a)$$

[for the value of the second term on the left, see exercise 14.3 and Boxes 14.2 and 14.5], and explain why it is appropriate to write the term on the right as $6a_0/a^3$ for a "dust-filled model universe." More generally, given any equation of state, $p = p(\rho)$, explain how one can find $\rho = \rho(a)$ from

$$d(\rho a^3) = -p\, d(a^3);$$

and how one can thus forecast the history of expansion and recontraction, $a = a(t)$.

Exercise 21.17. THIN-SANDWICH VARIATIONAL PRINCIPLE FOR THE SCALAR POTENTIAL IN ELECTRODYNAMICS

(a) Choose the unknown U^m in the expression

$$\frac{1}{8\pi} g^{mn} \frac{\partial\phi}{\partial x^m} \frac{\partial\phi}{\partial x^n} + U^m \frac{\partial\phi}{\partial x^n}$$

in such a way that this expression, multiplied by the volume element $g^{1/2} d^3x$, and integrated over the simultaneity Σ, is extremized by a ϕ, and only by a ϕ, that satisfies the initial-value equation (21.108) of electrodynamics.

(b) Show that the resulting variational principle, instead of having to be invented "out of the blue," is none other than what follows directly from the action principle build on the Lagrangian density (21.100) of electrodynamics (independent variation of ϕ and the three A_i everywhere between the two faces of a sandwich to extremize I, subject only to the prior specification of the A_i on the two faces of the sandwich, in the limit where the thickness of the sandwich goes to zero).

Exercise 21.18. THIN-SANDWICH VARIATIONAL PRINCIPLE FOR THE LAPSE AND SHIFT FUNCTIONS IN GEOMETRODYNAMICS

(a) Extremize the action integral

$$I_3 = \int \{[R - (\mathrm{Tr}\boldsymbol{K})^2 + \mathrm{Tr}\boldsymbol{K}^2 - 2T^*_{nn}]N \\ - 2T^{*k}_n N_k\} g^{1/2} d^3x$$

with respect to the lapse and shift functions, and show that one arrives in this way at the four initial-value equations of geometrodynamics. It is understood that one has given the six g_{ij} and the six $\partial g_{ij}/\partial t$ on the simultaneity where the analysis is being done. The extrinsic curvature is considered to be expressed as in (21.67) in terms of these quantities and the lapse and shift. The energy density and energy flow are referred to a unit normal vector \boldsymbol{n} and three arbitrary coordinate basis vectors \boldsymbol{e}_i within the simultaneity, as earlier in this chapter, and the asterisk is an abbreviation for an omitted factor of 8π.

(b) Derive this variational principle from the ADM variational principle by going to the limit of an infinitesimally thin sandwich [see derivation in Wheeler (1964)].

Exercise 21.19. CONDENSED THIN-SANDWICH VARIATIONAL PRINCIPLE

(a) Extremize the action I_3 of the preceding exercise with respect to the lapse function N.

(b) What is the relation between the result and the principle that "3-geometry is a carrier of information about time"?

(c) By elimination of N, arrive at a "condensed thin-sandwich variational principle" in which the only quantities to be varied are the three shift functions N_i.

Exercise 21.20. POOR MAN'S WAY TO SCHWARZSCHILD GEOMETRY

On curved empty space evolving deterministically in time, impose the conditions (1) that it possess a moment of time-symmetry, a spacelike hypersurface, the extrinsic curvature of which, with respect to the enveloping spacetime, is everywhere zero, and (2) that this spacelike hypersurface be endowed with spherical symmetry. Write the metric of the 3-geometry in the form

$$ds^2 = \psi^4(\bar{r})(d\bar{r}^2 + \bar{r}^2 d\theta^2 + \bar{r}^2 \sin^2\theta \, d\phi^2).$$

From the initial-value equation (21.127), show that the conformal factor ψ up to a multiplicative factor must have the form $\psi = (1 + m/2\bar{r})$. Show that the proper circumference $2\pi\bar{r}\psi^2(\bar{r})$ assumes a minimum value at a certain value of \bar{r}, thus defining the *throat* of the 3-geometry. Show that the 3-geometry is mirror-symmetric with respect to reflection in this throat in the sense that the metric is unchanged in form under the substitution $r' = m^2/4\bar{r}$. Find the transformation from the conformal coordinate \bar{r} to the Schwarzschild coordinate r.

§21.11. YORK'S "HANDLES" TO SPECIFY A 4-GEOMETRY

On a simultaneity—or on *the* simultaneity—of extremal proper volume, give the conformal part of the 3-geometry and give the two inequivalent components of the dynamically conjugate momentum in order (1) to have freely specifiable, but also complete, initial-value data and thus (2) to determine completely the whole generic four-dimensional spacetime manifold. This in brief is York's extension (1971, 1972b) to the generic case of what Brill did for special cases (see the preceding section). York and Brill acknowledge earlier considerations of Lichnerowicz (1944) and Bruhat (1962 and earlier papers cited there on conformal geometry and the initial-value problem). But why conformal geometry, and why pick such a special spacelike hypersurface on which to give the four dynamic data per spacepoint?

The degrees of freedom of the geometry in brief

Few solutions of Maxwell's equations are simpler than an infinite plane monochromatic wave in Minkowski's flat spacetime, and few look more complex when examined on a spacelike slice cut through that spacetime in an arbitrary way, with local wiggles and waves, larger-scale lumps and bumps, and still larger-scale general curvatures. No one who wants to explore electrodynamics in its evolution with many-fingered time can avoid these complexities; and no one will accept these complexities of many-fingered time who wants to see the degrees of freedom of the electromagnetic field in and by themselves exhibited in their neatest form. He will pick the simplest kind of timelike slice he can find. On that simultaneity, there are two and only two field coordinates, and two and only two field momenta per spacepoint. Similarly in geometrodynamics.

When one wants to untangle the degrees of freedom of the geometry, as distinct from analyzing the dynamics of the geometry, one therefore retreats from the three items of information per spacepoint that are contained in a 3-geometry [or in any other way of analyzing the geometrodynamics, as especially seen in the "extrinsic time" formulation of Kuchař (1971b and 1972)] and following York (1) picks the simultaneity to have maximal proper volume and (2) on this simultaneity specifies the two "coordinate degrees of freedom per spacepoint" that are contained in the conformal part of the 3-geometry.

Pick hypersurface of extremal proper volume

An element of proper volume $g^{1/2}\, d^3x$ on the spacelike hypersurface Σ undergoes, in the next unit interval of proper time as measured normal to the hypersurface, a fractional increase of proper volume [see Figure 21.3 and equations 21.59 and 21.66] given by

$$-\operatorname{Tr} \boldsymbol{K} = -\frac{1}{2} g^{-1/2} \operatorname{Tr} \boldsymbol{\pi}. \tag{21.143}$$

For the volume to be extremal this quantity must vanish at every point of Σ. This condition is satisfied in a Friedmann universe (Chapter 27) and in a Taub universe (Chapter 30) at that value of the natural time-coordinate t at which the universe switches over from expansion to recontraction. It is remarkable that the same condition on the choice of simultaneity, Σ, lets itself be formulated in the same natural way,

$$\operatorname{Tr} \boldsymbol{K} = 0 \ \text{or} \ \operatorname{Tr} \boldsymbol{\pi} = 0, \tag{21.144}$$

for a closed universe altogether deprived of any symmetry whatsoever. Alternatively, one can deal with a spacetime that is topologically the product of an open 3-space by the real line (time). Then it is natural to think of specifying the location in it of a bounding spacelike 2-geometry S with the topology of a 2-sphere. Then one has many ways to fill in the interior of S with a spacelike 3-geometry Σ; but of all these Σ's, only the one that is extremal, or only the ones that are extremal, satisfy (21.144).

Who is going to specify this 2-geometry with the topology of a 2-sphere? The choice of that 2-geometry is not a matter of indifference. In a given 4-geometry, distinct choices for the bounding 2-geometry will ordinarily give distinct results for the extremizing 3-geometry, and therefore different choices for the "initial-value simultaneity," Σ. No consideration immediately thrusts itself forward that would give preference to one choice of 2-geometry over another. However, no such infinity of options presents itself when one limits attention to a closed 3-geometry. Therefore it will give concreteness to the following analysis to consider it applied to a closed universe, even though the analysis surely lets itself be made well-defined in an open region by appropriate specification of boundary values on the closed 2-geometry that bounds that open region. In brief, by limiting attention to a closed 3-geometry, one lets the obvious condition of closure take the place of boundary conditions that are not obvious.

York's analysis remains simple when his extrinsic time

$$\tau = \frac{2}{3} g^{-1/2} \operatorname{Tr} \boldsymbol{\pi} = \frac{4}{3} \operatorname{Tr} \boldsymbol{K}$$

has any constant value on the hypersurface, not only the value $\tau = 0$ appropriate for the hypersurface of extremal proper volume.

On the simultaneity Σ specified by the condition of constant extrinsic time, $\tau = $ constant, begin by giving the conformal 3-geometry,

$$< \; = \; {}^{(3)}< \; = \; \left(\begin{array}{l} \text{the equivalence class of all those positive definite} \\ \text{Riemannian three-dimensional metrics that are} \\ \text{equivalent to each other under (1) diffeomorphism} \\ \text{(smooth sliding of the points over the mainfold to} \\ \text{new locations) or (2) changes of scale that vary} \\ \text{smoothly from point to point, leaving fixed all} \\ \text{local angles (ratios of local distances), but} \\ \text{changing local distances themselves or (3) both.} \end{array} \right) \quad (21.145)$$

The conformal 3-geometry is a geometric object that lends itself to definition and interpretation quite apart from the specific choice of coordinate system and even without need to use any coordinates at all. The conformal 3-geometry (*on the hypersurface Σ where $\tau = $ constant*) may be regarded much as one regards the magnetic field in electromagnetism. The case of conformally flat 3-geometry,

$$ds^2 = \psi^4(x, y, z) \, ds_{\text{base}}^2 \quad (21.146)$$

(with $g_{ij\,\text{base}} = \delta_{ij}$), is analogous to those initial-value situations in electromagnetism where the magnetic field is everywhere zero (the time-antisymmetric initial-value problem of Brill); but now we consider the case of general ds^2_{base}.

The six metric coefficients g_{ij} of the conformal 3-geometry, subject to being changed by change of the three coordinates x^i, and undetermined at any one point up to a common position-dependent multiplicative factor, carry $6 - 3 - 1 = 2$ pieces of information per spacepoint. In this respect, they are like the components of the divergenceless magnetic field \mathscr{B}. The corresponding field momentum $\pi^i_{EM} \propto \mathscr{E}^i$ (Box 21.1, page 496) has its divergence specified by the charge density, and so also carries

> two pieces of information (in addition to the prescribed information about the density of charge) per spacepoint. (21.147)

The comparison is a little faulty between the components of \mathscr{B} and the metric coefficients. They are more like potentials than like components of the physically relevant field.

The appropriate measure of the "field" in geometrodynamics is the curvature tensor; but how can one possibly define a curvature tensor for a geometry that is as rudimentary as a conformal 3-geometry? York (1971) has raised and answered this question. The Weyl conformal-curvature tensor [equation (13.50) and exercise 13.13] is independent [in the proper $\binom{2}{2}$ representation], in spaces of higher dimensionality, of the position-dependent factor ψ^4 with which one multiplies the metric coefficients, but vanishes identically in three-dimensional space (exercise 21.21). One arrives at a non-zero conformally invariant measure of the curvature only when one goes to one higher derivative (exercise 21.22). In this way, one comes to *York's curvature* $\widetilde{\beta}^{ab}$, here called Y^{ab}, a tensor density with these properties:

York's curvature tensor

$$Y^{ab} = Y^{ba} \text{ (symmetric)};$$

$$Y^a_a = 0 \text{ (traceless)};$$

$$Y^{ab}{}_{|b} = 0 \text{ (transverse)};$$

Y^{ab} invariant with respect to position-dependent changes in the conformal scale factor;

$Y^{ab} = 0$ when and only when the 3-geometry is conformally flat. (21.148)

Y^{ab} provides what York calls the pure spin-two representation of the 3-geometry intrinsic to Σ. It is the analog of the field \mathscr{B} of electrodynamics on the spacelike initial-value simultaneity. It directly carries physical information about the conformal 3-geometry.

In addition to the conformal geometry $^{(3)}\!<$, specified by the "potentials" $g_{ij}/g^{1/3}$, and measured by the "field components" Y^{ij}, one must also specify on Σ the corresponding conjugate momenta:

$$\widetilde{\pi}^{ab} = \widetilde{\pi}^{ab} \text{ (symmetric)}; \qquad \widetilde{\pi}^a_a = 0 \text{ (traceless)};$$

$$\widetilde{\pi}^{ab}_{|b} = 0 \text{ (transverse) in case there is no flow of energy in space; otherwise}$$

$$\widetilde{\pi}^{ab}_{|b} = 8\pi \text{ (density of flow of energy)}^a;$$

two pieces of information (in addition to the prescribed information about the flow of energy) per spacepoint. (21.149)

It might appear to be essential to specify with respect to which of the 3-geometries, distinguished from one another by different values of the conformal factor one calculates the covariant derivatives of tensor densities of weight 5/3 (see §21.2) in (21.148) and (21.149). However, York has shown that the conditions (21.149) do not in any way depend on the value of the conformal factor ψ^4.

These equations (21.149) for what York calls the "momentum density of weight 5/3,"

$$\widetilde{\pi}^{ab} = g^{1/3}\left(\pi^{ab} - \frac{1}{3}g^{ab}\,\text{Tr}\,\boldsymbol{n}\right), \qquad (21.150)$$

are linear, and therefore lend themselves to analysis by standard methods. It is a great help in this enterprise that York (1973a,b) has provided a "conformally invariant orthogonal decomposition of symmetric tensors on Riemannian manifolds" that allows one to generate solutions of these requirements ("transverse traceless," "conformal Killing," and "trace" parts, respectively, measure deformation of conformal part of geometry, mere recoordinatization, and change of scale). It is a further assistance, as York notes, that one has the same $\widetilde{\pi}^{ab}$ for an entire conformal equivalence class of metrics; that is, for a given

$$\widetilde{g}_{ab} = g^{-1/3}g_{ab}, \qquad (21.151)$$

no matter how different the g_{ab} and ψ themselves may be.

The conformal 3-geometry and the "momentum density of weight 5/3" once picked, the remaining initial-value equation (21.116) then becomes the "scale" equation,

$$8\boldsymbol{\nabla}^2\psi - {}^{(3)}R\psi + M\psi^{-7} + Q\psi^{-3} - \frac{3}{8}\tau^2\psi^5 = 0, \qquad (21.152)$$

for the determination of the conformal factor ψ. Here $\boldsymbol{\nabla}^2$ stands for the Laplacian

$$\boldsymbol{\nabla}^2\psi \equiv g^{-1/2}(\partial/\partial x^a)g^{1/2}g^{ab}(\partial\psi/\partial x^b). \qquad (21.153)$$

It, like ${}^{(3)}R$, M, and Q, refers to the base space. It is interesting that

$$\boldsymbol{\nabla}^2 - \frac{1}{8}{}^{(3)}R$$

is a conformally invariant wave operator, whereas $\boldsymbol{\nabla}^2$ itself is not. The quantity M in York's analysis is an abbreviation for

$$M \equiv g^{-5/3}g_{ac}g_{bd}\widetilde{\pi}^{ab}\widetilde{\pi}^{cd}, \qquad (21.154a)$$

and

$$Q \equiv 16\pi\rho_{\text{base}}(= 16\pi\psi^8\rho = 16\pi\psi^8\rho_{\text{in final 3-geometry}}). \qquad (21.154b)$$

One seeks a solution ψ that is continuous over the closed manifold and everywhere real and positive. When does such a solution ψ of the elliptic equation (21.152) exist? When is it unique? *Always* (when $M > 0$ and $\tau \neq 0$), is the result of O'Murchadha and York (1973); see also earlier investigations of Choquet-Bruhat (1972). Some of the physical considerations that come into this kind of problem have been discussed by Wheeler (1964a, pp. 370–381).

§21.12. MACH'S PRINCIPLE AND THE ORIGIN OF INERTIA

> *In my opinion the general theory of relativity can only solve this problem [of inertia] satisfactorily if it regards the world as spatially self-enclosed.*
>
> ALBERT EINSTEIN (1934), p. 52.

On June 25, 1913, two years before he had discovered the geometrodynamic law that bears his name, Einstein (1913b) wrote to Ernst Mach (Figure 21.5) to express his appreciation for the inspiration that he had derived for his endeavors from Mach's ideas. In his great book, *The Science of Mechanics,* Mach [(1912), Chapter 2, section 6] had reasoned that it could not make sense to speak of the acceleration of a mass relative to absolute space. Anyone trying to clear physics of mystical ideas would do better, he reasoned, to speak of acceleration relative to the distant stars. But how can a star at a distance of 10^9 light-years contribute to inertia in the here and the now? To make a long story short, one can say at once that Einstein's theory (1) identifies gravitation as the mechanism by which matter there influences inertia here; (2) says that this coupling takes place on a spacelike hypersurface [in what one, without a closer examination, might mistakenly think to be a violation of the principle of causality; see Fermi (1932) for a discussion and clarification of the similar apparent paradox in electrodynamics; see also Einstein (1934), p. 84: "Moreover I believed that I could show on general considerations a law of gravitation invariant in relation to any transformation of coordinates whatever was inconsistent with the principle of causation. These were errors of thought which cost me two years of excessively hard work, until I finally recognized them as such at the end of 1915"]; (3) supplies in the initial-value equations of geometrodynamics a mathematical tool to describe this coupling; (4) demands closure of the geometry in space [one conjectures; see Wheeler (1959, 1964c) and Hönl (1962)], as a boundary condition on the initial-value equations if they are to yield a well-determined [and, we know now, a unique] 4-geometry; and (5) identifies the collection of local Lorentz frames near any point in this resulting spacetime as what one means quantitatively by speaking of inertia at that point. This is how one ends up with inertia here determined by density and flow of mass-energy there.

No violation of causality, despite appearances

There are many scores of papers in the literature on Mach's principle, including many—even one by Lenin (English translation, 1927)—one could call anti-Machian; and many of them make interesting points [see especially the delightful dialog by Weyl (1924a) on "inertia and the cosmos," and the article (1957) and book (1961) of Sciama]. However, most of them were written before one had anything like the understanding of the initial-value problem that one possesses today. Therefore no

An enormous literature

(*continued on page 546*)

544

Figure 21.5.
Einstein's appreciation of Mach, written to Ernst Mach June 25, 1913, while Einstein was working hard at arriving at the final November 1915 formulation of standard general relativity. Regarding confirmation at a forthcoming eclipse: "If so, then your happy investigations on the foundations of mechanics, Planck's unjustified criticism notwithstanding, will receive brilliant confirmation. For it necessarily turns out that inertia originates in a kind of interaction between bodies, quite in the sense of your considerations on Newton's pail experiment. The first consequence is on p. 6 of my paper. The following additional points emerge: (1) If one accelerates a heavy shell of matter *S*, then a mass enclosed by that shell experiences an accelerative force. (2) If one rotates the shell relative to the fixed stars about an axis going through its center, a Coriolis force arises in the interior of the shell; that is, the plane of a Foucault pendulum is dragged around (with a practically unmeasurably small angular velocity)." Following the death of Mach, Einstein (1916a) wrote a tribute to the man and his work. Reprinted with the kind permission of the estate of Albert Einstein, Helen Dukas and Otto Nathan, executors.

glänzende Bestätigung. Denn es ergibt sich mit Notwendigkeit, dass die Trägheit in einer Art Wechselwirkung der Körper ihren Ursprung hat, ganz im Sinne Ihrer Überlegungen zum Newton'schen Eimer-Versuch.

Eine erste Konsequenz in diesem Sinne finden Sie oben auf Seite 6 der Arbeit Es hat sich ferner folgendes ergeben:

1) Beschleunigt man eine träge Kugelschale S, so erfährt nach der Theorie ein von ihr eingeschlossener Körper eine beschleunigende Kraft

2) Rotiert die Schale Ihnen eine durch ihren Mittelpunkt gehende Achse (relativ zum System der Fixsterne ("Bezugsystem"), so entsteht im Inneren der Schale ein Coriolis-Feld, d. h. (die Ebene des) das Foucault-Pendels wird (mit jener allerdings nicht merklich messbaren "Geschwindigkeit) mitgenommen

Es ist mir eine grosse Freude, Ihnen dies mitteilen zu können, zumal jene Kritik Plancks mir schon immer höchst ungerecht fertigt erscheinen war.

Mit grössten Hochachtung grüsst Sie herzlich

Ihr ergebener A. Einstein.

Ich danke Ihnen herzlich für die Übersendung Ihres Buches

attempt will be made to summarize or analyze the literature, which would demand a book in itself. Moreover, Mach's principle as presented here is more sharply formulated than Einstein ever put it in the literature [except for his considerations arguing that the universe must be closed; see Einstein's book (1950), pp. 107–108]; and Mach would surely have disowned it, for he could never bring himself to accept general (or even special) relativity. Nevertheless, it is a fact that Mach's principle— that matter there governs inertia here—and Riemann's idea—that the geometry of space responds to physics and participates in physics—were the two great currents of thought which Einstein, by means of his powerful equivalence principle, brought together into the present-day geometric description of gravitation and motion.

Mach's principle updated and spelled out

"Specify everywhere the distribution and flow of mass-energy and thereby determine the inertial properties of every test particle everywhere and at all times". Spelled out, this prescription demands (1) a way of speaking about "everywhere": a spacelike hypersurface Σ. Let one insist—in conformity with Einstein—(2) that it be a closed 3-geometry, and for convenience, not out of necessity, (3) that τ be independent of position on Σ. (4) Specify this 3-geometry to the extent of giving the conformal metric; without the specification of at least this much 3-geometry, there would be no evident way to say "where" the mass-energy is to be located. (5) Give density ρ_{base} as a function of position in this conformal 3-geometry. (6) Recognize that giving the mass-energy only of fields other than gravity is an inadequate way to specify the distribution of mass-energy throughout space. Formalistically, to be sure, the gravitational fields does not and cannot make any contribution to the source term that stands on the righthand side of Einstein's field equation. However, the analysis of gravitational waves (Chapters 18 and 35) shows that perturbations in the geometry of scale small compared to the scale of observation have to be regarded as carrying an effective content of mass-energy. Moreover, one has in a geon [Wheeler (1955); Brill and Hartle (1964); for more on gravitational-wave energy, see §35.14] an object built out of gravitational waves (or electromagnetic waves, or neutrinos, or any combination of the three) that holds itself together for a time that is long in comparison to the characteristic period of vibration of the waves. It looks from a distance like any other mass, even though nowhere in its interior can one put a finger and say "here is mass." Therefore it, like any other mass, must have "its influence on inertia." But to specify this mass, one must give enough information to characterize completely the gravitational waves on the simultaneity Σ. For this, it is not enough merely to have given the two "wave-coordinates" per spacepoint that one possesses in $^{(3)}<$. One must give in addition (7) the two "wave-momenta" per spacepoint that appear in York's "momentum density of weight 5/3," $\widetilde{\pi}^{ab}$; and at the same time, as an inextricable part of this operation, one must (8) specify the density of flow of field energy. (9) Solve for the conformal factor ψ. (10) Then one has complete initial-value data that satisfy the initial-value equations of general relativity. (11) These data now known, the remaining, dynamic, components of the field equation determine the 4-geometry into the past and the future. (12) In this way, the inertial properties of every test particle are determined everywhere and at all times, giving concrete realization to Mach's principle.

Much must still be done to spell out the physics behind these equations and to

see this physics in action. Some significant progress had already been made in this direction before the present stage in one's understanding of the initial-value equations. Especially interesting are results of Thirring (1918) and (1921) and of Thirring and Lense (1918), discussed by Einstein (1950) in the third edition of his book, *The Meaning of Relativity*.

The Foucault pendulum

Consider a bit of solid ground near the geographic pole, and a support erected there, and from it hanging a pendulum. Though the sky is cloudy, the observer watches the track of the Foucault pendulum as it slowly turns through 360°. Then the sky clears and, miracle of miracles, the pendulum is found to be swinging all the time on an arc fixed relative to the far-away stars. If "mass there governs inertia here," as envisaged by Mach, how can this be?

The dragging of the inertial frame

Enlarge the question. By the democratic principle that equal masses are created equal, the mass of the earth must come into the bookkeeping of the Foucault pendulum. Its plane of rotation must be dragged around with a slight angular velocity, ω_{drag}, relative to the so-called "fixed stars." How much is ω_{drag}? And how much would ω_{drag} be if the pendulum were surrounded by a rapidly spinning spherical shell of mass M and radius R_{shell}, turning at angular velocity ω_{shell}?

Einstein's theory says that inertia is a manifestation of the geometry of spacetime. It also says that geometry is affected by the presence of matter to an extent proportional to the factor $G/c^2 = 0.742 \times 10^{-28}$ cm/g. Simple dimensional considerations leave no room except to say that the rate of drag is proportional to a expression of the form

$$\omega_{\text{drag}} = k \frac{G}{c^2} \frac{m_{\text{shell, conv}}}{R_{\text{shell}}} \omega_{\text{shell}} = k \frac{m_{\text{shell}}}{R_{\text{shell}}} \omega_{\text{shell}}. \tag{21.155}$$

Here k is a numerical factor to be found only by detailed calculation. Lense and Thirring [(1918) and (1921)], starting with a flat background spacetime manifold, calculated in the weak-field approximation of Chapter 18 the effect of the moving current of mass on the metric. Expressed in polar coordinates, the metric acquires a non-zero coefficient $g_{\phi t}$. Inserted into the equation of geodesic motion, this off-diagonal metric coefficient gives rise to a precession. This precession (defined here about an axis parallel to the axis of rotation, not about the local vertical) is given by an expression of the form (21.155), where the precession factor k has the value

$$k = 4/3. \tag{21.156}$$

There is a close parallelism between the magnetic component of the Maxwell field and the precession component of the Einstein field. In neither field does a source at rest produce the new kind of effect when acting on a test particle that is also at rest. One designs a circular current of charge to produce a magnetic field; and a test charge, in order to respond to this magnetic field, must also be in motion. Similarly here: no pendulum vibration means no pendulum precession. Moreover, the direction of the precession depends on where the pendulum is, relative to the rotating shell of mass. The precession factor k has the following values:

$k = 4/3$ for pendulum anywhere inside rotating
 shell of mass;

$k = 4/3$ for pendulum at North or South pole; (21.157)

$k = -2/3$ for pendulum just outside the rotating
 shell at its equator.

This position-dependence of the drag, ω_{drag}, makes still more apparent the analogy with magnetism, where the field of a rotating charged sphere points North at the center of the sphere, and North at both poles, but South at the equator.

Whether the Foucault pendulum is located in imagination at the center of the earth or in actuality at the North pole, the order of magnitude of the expected drag is

$$\omega_{\text{drag}} \sim \frac{m_{\text{earth}}}{R_{\text{earth}}} \omega_{\text{earth}} \sim \frac{0.44 \text{ cm}}{6 \times 10^8 \text{ cm}} \frac{1 \text{ radian}}{13700 \text{ sec}} \qquad (21.158)$$
$$\sim 5 \times 10^{-14} \text{ rad/sec},$$

too small to allow detection, let alone actual measurement, by any device so far built—but perhaps measurable by gyroscopes now under construction (§40.7). By contrast, near a rapidly spinning neutron star or near a black hole endowed with substantial angular momentum, the calculated drag effect is not merely detectable; it is even important (see Chapter 33 on the physics of a rotating black hole).

The distant stars must influence the natural plane of vibration of the Foucault pendulum as the nearby rotating shell of matter does, provided that the stars are not so far away ($r \sim$ radius of universe) that the curvature of space begins to introduce substantial corrections into the calculation of Thirring and Lense. In other words, no reason is apparent why all masses should not be treated on the same footing, so that (21.158) more appropriately, if also somewhat symbolically, reads

The "sum for inertia"

$$\omega_{\substack{\text{plane of}\\\text{vibration}\\\text{of Foucault}\\\text{pendulum}}} \sim \frac{m_{\text{shell}}}{R_{\text{shell}}} \omega_{\text{shell}} + \sum_{\substack{\text{far-away}\\\text{"stars"}}} \frac{m_{\text{"star"}}}{r_{\text{"star"}}} \omega_{\text{"star"}}. \qquad (21.159)$$

Moreover, when there is no nearby shell of matter, or when it has negligible effects, the plane of vibration of the pendulum, if experience is any guide, cannot turn with respect to the frame defined by the far-away "stars." In this event ω_{Foucault} must be identical with ω_{stars}; or the "sum for inertia,"

$$\sum_{\substack{\text{far-away}\\\text{"stars"}}} \frac{m_{\text{"star"}}}{r_{\text{"star"}}} \sim \frac{m_{\text{universe}}}{r_{\text{universe}}}, \qquad (21.160)$$

must be of the order of unity. Just such a relation of approximate identity between the mass content of the universe and its radius at the phase of maximum expansion is a characteristic feature of the Friedman model and other simple models of a closed universe (Chapters 27 and 30). In this respect, Einstein's theory of Mach's principle exhibits a satisfying degree of self-consistency.

At phases of the dynamics of the universe other than the stage of maximum expansion, $r_{universe}$ can become arbitrarily small compared to $m_{universe}$. Then the ratio (21.160) can depart by powers of ten from unity. Regardless of this circumstance, one has no option but to understand that the *effective* value of the "sum for inertia" is still unity after all corrections have been made for the dynamics of contraction or expansion, for retardation, etc. Only so can $\omega_{Foucault}$ retain its inescapable identity with $\omega_{far\text{-}away\,stars}$. Fortunately, one does not have to pursue the theology of the "sum for inertia" to the uttermost of these sophistications to have a proper account of inertia. Mach's idea that mass there determines inertia here has its complete mathematical account in Einstein's geometrodynamic law, as already spelled out. For the first strong-field analysis of the dragging of the inertial reference system in the context of relativistic cosmology, see Brill and Cohen (1966) and Cohen and Brill (1967); see also §33.4 for dragging by a rotating black hole.

Still another clarification is required of what Mach's principle means and how it is used. The inertial properties of a test particle are perfectly well-determined when that particle is moving in ideal Minkowski space. "Point out, please," the anti-Machian critic says, "the masses that are responsible for this inertia." In answer, recall that Einstein's theory includes not only the geometrodynamic law, but also, in Einstein's view, the boundary condition that the universe be closed. Thus the section of spacetime that is flat is to be viewed, not as infinite, but as part of a closed universe. (For a two-dimensional analog, fill a rubber balloon with water and set it on a glass tabletop and look at it from underneath). The part of the universe that is curved acquires its curvature by reason of its actual content of mass-energy or—if animated only by gravitational waves—by reason of its effective content of mass-energy. This mass-energy, real or effective, is to be viewed as responsible for the inertial properties of the test particle that at first sight looked all alone in the universe.

Minkowski geometry as limit of a closed 3-geometry

It in no way changes the qualitative character of the result to turn attention to a model universe where the region of Minkowski flatness, and all the other linear dimensions of the universe, have been augmented tenfold ("ten times larger balloon; ten times larger face"). The curvature and density of the curved part of the model universe are down by a factor of 100, the volume is up by a factor of 1,000, the mass is up by a factor of 10; but the ratio of mass to radius, or the "sum for inertia" (the poor man's substitute for a complete initial-value calculation) is unchanged.

Einstein acknowledged a debt of parentage for his theory to Mach's principle (Figure 21.5). It is therefore only justice that Mach's principle should in return today owe its elucidation to Einstein's theory.

**Exercise 21.21. WHY THE WEYL CONFORMAL CURVATURE
 TENSOR VANISHES**

EXERCISES

How many independent components does the Riemann curvature tensor have in three-dimensional space? How many does the Ricci curvature tensor have? Show that the two tensors are related by the formula

$$R^d{}_{abc} = \delta^d_b R_{ac} - \delta^d_c R_{ab} + g_{ac} R^d{}_b - g_{ab} R^d_c$$
$$+ \frac{1}{2} R(\delta^d_c g_{ab} - \delta^d_b g_{ac})$$

with no need of any Weyl conformal-curvature tensor to specify (as in higher dimensions) the further details of the Riemann tensor. Show that the Weyl tensor, from an n-dimensional modification of equation (13.50) as in exercise 13.13, vanishes for $n = 2$.

Exercise 21.22. YORK'S CURVATURE

[York (1971)]. (a) Define the tensor [Eisenhart (1926)]

$$R_{abc} = R_{ab|c} - R_{ac|b} + \frac{1}{4}(g_{ac} R_{|b} - g_{ab} R_{|c}).$$

(b) Show that a 3-geometry is conformally flat when and only when $R_{abc} = 0$.

(c) Show that the following identities hold and reduce to five the number of independent components of R_{abc}:

$$R^a{}_{ac} = g^{ab} R_{bac} = 0;$$
$$R_{abc} + R_{acb} = 0;$$
$$R_{abc} + R_{cab} + R_{bca} = 0.$$

(d) Show that Yorks' curvature

$$Y^{ab} = g^{1/3}[aef]\left(R_f{}^b - \frac{1}{4}\delta_f{}^b R\right)_{|e}$$
$$= -\frac{1}{2} g^{1/3}[aef] g^{bm} R_{mef}$$

is conformally invariant and has the properties listed in equations (21.148).

Exercise 21.23. PULLING THE POYNTING FLUX VECTOR "OUT OF THE AIR"

From the condition that the Hamilton-Jacobi functional $S(g_{ij}, A_m)$ (extremal of the action integral) for the combined Einstein and Maxwell fields, ostensibly dependent on the six metric coefficients $g_{ij}(x, y, z)$ and the three potentials $A_m(x, y, z)$, shall actually depend only on the 3-geometry of the spacelike hypersurface and the distribution of magnetic field strength on this hypersurface, show that the geometrodynamic field momentum $\pi^{ij} = \delta S/\delta g_{ij}$ satisfies a condition of the form

$$\pi^{ij}{}_{|j} = c[imn]\mathcal{E}_m \mathcal{B}_n,$$

and evaluate the coefficient c in this equation [Wheeler (1968b)]. Hint: Note that the transformation

$$x^i \longrightarrow x^i - \xi^i, \quad g_{ij} \longrightarrow g_{ij} + \xi_{i|j} + \xi_{j|i}$$

in no way changes the 3-geometry itself, and therefore the corresponding induced change in S,

$$\delta S = \int \left[\frac{\delta S}{\delta g_{ij}} \delta g_{ij} + \frac{\delta S}{\delta A_m} \delta A_m\right] d^3x$$

must vanish identically for arbitrary choice of the $\xi^i(x, y, z)$, which measure the equivalent of the sliding of a ruled transparent rubber sheet over an automobile fender.

Exercise 21.24. **THE EXTREMAL ACTION ASSOCIATED WITH THE HILBERT ACTION PRINCIPLE DEPENDS ON CONFORMAL 3-GEOMETRY AND EXTRINSIC TIME [K. Kuchař (1972) and J. York (1972)]**

Show that the data demanded by the Hilbert action principle $\delta \int {}^{(4)}R(-{}^{(4)}g)^{1/2}\, d^4x = 0$ on each of the two bounding spacelike hypersurfaces consist of (1) the conformal 3-geometry ${}^{(3)}<$ of the hypersurface plus (2) the extrinsic time variable defined by

$$\tau = \frac{2}{3} g^{-1/2} \operatorname{Tr} \boldsymbol{n} = \frac{4}{3} \operatorname{Tr} \boldsymbol{K},$$

conveniently represented by the pictogram ⊁, measured by one number per spacepoint, and independent of the conformal factor in the metric of the 3-geometry. This done, explain in a few words why in this formulation of geometrodynamics the Hamilton-Jacobi function (\hbar times the phase of the wave function in the semiclassical or JWKB approximation) is appropriately expressed in the form

$$S = S({}^{(3)}<, \text{⊁}).$$

§21.13. JUNCTION CONDITIONS

The intrinsic and extrinsic curvatures of a hypersurface, which played such fundamental roles in the initial-value formalism, are also powerful tools in the analysis of "junction conditions."

Recall the junction conditions of electrodynamics: across any surface (e.g., a capacitor plate), the tangential part of the electric field, E_\parallel, and the normal part of the magnetic field, B_\perp, must be continuous; thus,

Junction conditions for electrodynamics

$$[E_\parallel] \equiv (\text{discontinuity in } E_\parallel)$$
$$\equiv (E_\parallel \text{ on "+" side of surface}) - (E_\parallel \text{ on "−" side of surface})$$
$$\equiv E_\parallel^+ - E_\parallel^- = 0, \tag{21.161a}$$

$$[B_\perp] \equiv B_\perp^+ - B_\perp^- = 0; \tag{21.161b}$$

while the "jump" in the parts E_\perp and B_\parallel must be related to the charge density (charge per unit area) σ, the current density (current per unit area) j, and the unit normal to the surface n by the formulas

$$[E_\perp] = E_\perp^+ - E_\perp^- = 4\pi\sigma n, \tag{21.161c}$$

$$[B_\parallel] = B_\parallel^+ - B_\parallel^- = 4\pi j \times n. \tag{21.161d}$$

Recall also that one derives these junction conditions by integrating Maxwell's equations over a "pill box" that is centered on the surface.

Similar junction conditions, derivable in a similar manner, apply to the gravitational field (spacetime curvature), and to the stress-energy that generates it.* Focus

*The original formulation of gravitational junction conditions stemmed from Lanczos (1922, 1924). The formulation given here, in terms of intrinsic and extrinsic curvature, was developed by Darmois (1927), Misner and Sharp (1964), and Israel (1966). For further references to the extensive literature, see Israel.

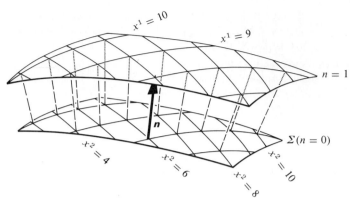

Figure 21.6.

Gaussian normal coordinates in the neighborhood of a 3-surface Σ. The metric in Gaussian normal coordinates has the form

$$ds^2 = (\boldsymbol{n} \cdot \boldsymbol{n})^{-1} \, dn^2 + g_{ij} \, dx^i \, dx^j$$

with $\boldsymbol{n} = \partial/\partial n$, $(\boldsymbol{n} \cdot \boldsymbol{n}) = -1$ if the surface is spacelike, and $(\boldsymbol{n} \cdot \boldsymbol{n}) = 1$ if it is timelike. (See exercise 27.2.) The extrinsic curvature of the surfaces $n = $ constant is $K_{ij} = -\frac{1}{2} \partial g_{ij}/\partial n$, and the Einstein field equations written in "3 + 1" form are (21.162).

attention on a specific three-dimensional slice through spacetime—the 3-surface Σ on Figure 21.6. Let the surface be either spacelike [unit normal \boldsymbol{n} timelike; $(\boldsymbol{n} \cdot \boldsymbol{n}) = -1$] or timelike [$\boldsymbol{n}$ spacelike; $(\boldsymbol{n} \cdot \boldsymbol{n}) = +1$]. The null case will be discussed later. As an aid in deriving junction conditions, introduce Gaussian normal coordinates in the neighborhood of Σ [see the paragraph preceding equation (21.82)]. In terms of the intrinsic and extrinsic curvatures of Σ and of neighboring 3-surfaces

Einstein equation in "3 + 1" form

$n = $ constant, the Einstein tensor and Einstein field equation have components

$$G^n{}_n = -\frac{1}{2}{}^{(3)}R + \frac{1}{2}(\boldsymbol{n} \cdot \boldsymbol{n})^{-1}\{(\operatorname{Tr} \boldsymbol{K})^2 - \operatorname{Tr}(\boldsymbol{K}^2)\} = 8\pi T^n{}_n, \quad (21.162\mathrm{a})$$

$$G^n{}_i = -(\boldsymbol{n} \cdot \boldsymbol{n})^{-1}\{K_i{}^m{}_{|m} - (\operatorname{Tr} \boldsymbol{K})_{|i}\} = 8\pi T^n{}_i, \quad\quad (21.162\mathrm{b})$$

$$G^i{}_j = {}^{(3)}G^i{}_j + (\boldsymbol{n} \cdot \boldsymbol{n})^{-1}\{(K^i{}_j - \delta^i{}_j \operatorname{Tr} \boldsymbol{K})_{,n}$$
$$- (\operatorname{Tr} \boldsymbol{K})K^i{}_j + \frac{1}{2}\delta^i{}_j(\operatorname{Tr} \boldsymbol{K})^2 + \frac{1}{2}\delta^i{}_j \operatorname{Tr}(\boldsymbol{K}^2)\} = 8\pi T^i{}_j. \quad (21.162\mathrm{c})$$

[See equations (21.77), (21.81), (21.76), and (21.82).]

Suppose that the stress-energy tensor $T^\alpha{}_\beta$ contains a "delta-function singularity" at Σ—i.e., suppose that Σ is the "world tube" of a two-dimensional surface with finite 4-momentum per unit area (analog of surface charge and surface current in electrodynamics). Then define the *surface stress-energy tensor on Σ* to be the integral

Surface stress-energy tensor

of $T^\alpha{}_\beta$ with respect to proper distance (n), measured perpendicularly through Σ:

$$S^\alpha{}_\beta = \operatorname*{Lim}_{\varepsilon \to 0} \left[\int_{-\varepsilon}^{+\varepsilon} T^\alpha{}_\beta \, dn \right]. \quad\quad (21.163)$$

To discover the effect of this surface layer on the spacetime geometry, perform a "pill-box integration" of the Einstein field equation (21.162)

Derivation of junction conditions

$$\lim_{\varepsilon \to 0} \left[\int_{-\varepsilon}^{+\varepsilon} G^\alpha{}_\beta \, dn \right] = 8\pi S^\alpha{}_\beta. \qquad (21.164)$$

Examine the integral of $G^\alpha{}_\beta$. If the 3-metric g_{ij} were to contain a delta function or a discontinuity at Σ, then Σ would not have any well-defined 3-geometry—a physically inadmissible situation, even in the presence of surface layers. Absence of delta functions, $\delta(n)$, in g_{ij} means absence of delta functions in $^{(3)}R$; absence of discontinuities in g_{ij} means absence of delta functions in $K_{ij} = -\frac{1}{2}g_{ij,n}$. Thus, equations (21.162) when integrated say

$$\int G^n{}_n \, dn = 0 = 8\pi S^n{}_n, \qquad (21.165a)$$

$$\int G^n{}_i \, dn = 0 = 8\pi S^n{}_i, \qquad (21.165b)$$

$$\int G^i{}_j \cdot dn = (\boldsymbol{n} \cdot \boldsymbol{n})(\gamma^i{}_j - \delta^i{}_j \operatorname{Tr} \boldsymbol{\gamma}) = 8\pi S^i{}_j, \qquad (21.165c)$$

where $\gamma^i{}_j$ is the "jump" in the components of the extrinsic curvature

$$\boldsymbol{\gamma} \equiv [\boldsymbol{K}] \equiv (\boldsymbol{K} \text{ on "}n = +\varepsilon \text{ side" of } \Sigma) - (\boldsymbol{K} \text{ on "}n = -\varepsilon \text{ side" of } \Sigma) \qquad (21.166)$$
$$\equiv \boldsymbol{K}^+ - \boldsymbol{K}^-.$$

In the absence of a delta-function surface layer, the above junction conditions say, simply, that $\boldsymbol{\gamma} \equiv [\boldsymbol{K}] = 0$. In words: if one examines how Σ is embedded in the spacetime above its "upper" face, and how it is embedded in the spacetime below its "lower" face, one must discover identical embeddings—i.e., identical extrinsic curvatures \boldsymbol{K}. Of course, the intrinsic curvature of Σ must also be the same, whether viewed from above or below. More briefly:

Junction conditions in absence of surface layers

$$\text{(absence of surface layers)} \rightleftarrows \text{("continuity" of } g_{ij} \text{ and } K_{ij}). \qquad (21.167)$$

If a surface layer is present, then Σ must be the world tube of a two-dimensional layer of matter, and the normal to Σ must be spacelike, $(\boldsymbol{n} \cdot \boldsymbol{n}) = +1$. The junction conditions (21.165a,b) then have the simple physical meaning

Junction conditions for a surface layer

$$\boldsymbol{S}(\boldsymbol{n}, \ldots) = 0 \xleftrightarrow{} \begin{pmatrix} \text{the momentum flow is entirely in } \Sigma; \\ \text{i.e., no momentum associated with the} \\ \text{surface layer flows out of } \Sigma; \text{ i.e., } \Sigma \\ \text{is the world tube of the surface layer} \end{pmatrix}, \qquad (21.168a)$$

which tells one nothing new. The junction condition (21.165c) says that the surface stress-energy generates a discontinuity in the extrinsic curvature (different embedding in spacetime "above" Σ than "below" Σ), given by

$$\gamma^i{}_j - \delta^i{}_j \operatorname{Tr} \boldsymbol{\gamma} = 8\pi S^i{}_j. \qquad (21.168b)$$

Of course, the intrinsic geometry of Σ must be the same as seen from above and below,

$$g_{ij} \text{ continuous across } \Sigma. \qquad (21.169)$$

In analyzing surface layers, one uses not only the junction conditions (21.168a) to (21.169), but also the four-dimensional Einstein field equation applied on each side of the surface Σ separately, and also an equation of motion for the surface stress-energy. The equation of motion is derived by examining the jump in the field equation $G^n{}_i = 8\pi T^n{}_i$ (equation 21.162b); thus $[G^n{}_i] = 8\pi[T^n{}_i]$ says

$$(\gamma_i{}^m - \delta_i{}^m \operatorname{Tr} \boldsymbol{\gamma})_{|m} = -8\pi[T^n{}_i];$$

and when reexpressed in terms of $S_i{}^m$ by means of the junction condition (21.168b), it says

Equation of motion for a surface layer

$$S^{im}{}_{|m} + [T^{in}] = 0. \tag{21.170}$$

[For intuition into this equation of motion, see Exercises 21.25 and 21.26. For applications of the "surface-layer formalism" see exercise 21.27; also Israel (1966), Kuchař (1968), Papapetrou and Hamoui (1968).]

Gravitational-wave shock fronts

When one turns attention to junction conditions across a *null* surface Σ, one finds results rather different from those in the spacelike and timelike cases. A "pill-box" integration of the field equations reveals that even in vacuum the extrinsic curvature may be discontinuous. A discontinuity in K_{ij} across a null surface, without any stress-energy to produce it, is the geometric manifestation of a *gravitational-wave shock front* (analog of a shock-front in hydrodynamics). For quantitative details see, e.g., Pirani (1957), Papapetrou and Treder (1959, 1962), Treder (1962), and especially Choquet-Bruhat (1968b).

That a discontinuity in the curvature tensor can propagate with the speed of light is a reminder that all gravitational effects, like all electromagnetic effects, obey a causal law. The initial-value data on a spacelike initial-value hypersurface uniquely determine the resulting spacetime geometry [see the work of Cartan, Stellmacher, Lichnerowicz, and Bruhat (also under the names Fourès-Bruhat and Choquet-Bruhat) and others cited and summarized in the article of Bruhat (1962)] but determine it in a way consistent with causality. Thus a change in these data throughout a limited region of the initial value 3-geometry makes itself felt on a slightly later hypersurface solely in a region that is also limited, and only a little larger than the original region.

When one turns from classical dynamics to quantum dynamics, one sees new reason to focus attention on a spacelike initial-value hypersurface: the observables at different points on such a hypersurface commute with one another; i.e., are in principle simultaneously observable.

Not every four-dimensional manifold admits a global singularity-free spacelike hypersurface. Those manifolds that do admit such a hypersurface have more to do with physics, it is possible to believe, than those that do not.

Even in a manifold that does admit a spacelike hypersurface, attention has been given sometimes, in the context of classical theory, to initial-value data on a hypersurface that is not spacelike but "characteristic," in the sense that it accomodates null geodesics [see, for example, Sachs (1964) and references cited there]. It is typical in such situations that one can predict the future but not the past, or predict the past but not the future.

Children of light and children of darkness is the vision of physics that emerges from this chapter, as from other branches of physics. The children of light are the differential equations that predict the future from the present. The children of darkness are the factors that fix these initial conditions.

Exercise 21.25. EQUATION OF MOTION FOR A SURFACE LAYER

(a) Let \boldsymbol{u} be the "mean 4-velocity" of the matter in a surface layer—so defined that an observer moving with 4-velocity \boldsymbol{u} sees zero energy flux. Let σ be the total mass-energy per unit proper surface area, as measured by such a "comoving observer." Show that the surface stress-energy tensor can be expressed in the form

$$\boldsymbol{S} = \sigma\boldsymbol{u} \otimes \boldsymbol{u} + \boldsymbol{t}, \text{ where } (\boldsymbol{t} \cdot \boldsymbol{u}) = 0, \tag{21.171}$$

and where \boldsymbol{t} is a symmetric stress tensor.

(b) Show that the component along \boldsymbol{u} of the equation of motion (21.170) is

$$d\sigma/d\tau = -\sigma u^j{}_{|j} + u_j t^{jk}{}_{|k} + u_j[T^{jn}], \tag{21.172}$$

where $d/d\tau = \boldsymbol{u}$. Give a physical interpretation for each term.

(c) Let a_j be that part of the 4-acceleration of the comoving observer which lies in the surface layer Σ. By projecting the equation of motion (21.170) perpendicular to \boldsymbol{u}, show that

$$\sigma a_j = -P_{ja}\{t^{ab}{}_{|b} + [T^{an}]\}, \tag{21.173}$$

where P_{ja} is the projection operator

$$P_{ja} = g_{ja} + u_j u_a. \tag{21.174}$$

Give a physical interpretation for each term of equation (21.182).

Exercise 21.26. THIN SHELLS OF DUST

For a thin shell of dust surrounded by vacuum ($[T^{jn}] = 0$, $\boldsymbol{t} = 0$), derive the following equations

$$d\sigma/d\tau = -\sigma u^b{}_{|b}, \tag{21.175a}$$

$$\boldsymbol{a}^+ + \boldsymbol{a}^- = 0, \tag{21.175b}$$

$$\boldsymbol{a}^+ - \boldsymbol{a}^- = (4\pi\sigma)\boldsymbol{n} \tag{21.175c}$$

$$\boldsymbol{\gamma} = 8\pi\sigma\left(\boldsymbol{u} \otimes \boldsymbol{u} + \frac{1}{2}\boldsymbol{g}\right). \tag{21.175d}$$

Here \boldsymbol{a}^+ and \boldsymbol{a}^- are the 4-accelerations as measured by accelerometers that are fastened onto the outer and inner sides of the shell, and \boldsymbol{g} is the 3-metric of the shell. Show that the first of these equations is the law of "conservation of rest mass."

Exercise 21.27. SPHERICAL SHELL OF DUST

Apply the formalism of exercise 21.25 to a collapsing spherical shell of dust [Israel (1967b)]. For the metric inside and outside the shell, take the flat-spacetime and vacuum Schwarzschild expressions (Chapter 23),

$$ds^2 = -dt^2 + dr^2 + r^2(d\theta^2 + \sin^2\theta \, d\phi^2) \text{ inside,} \tag{21.176a}$$

$$ds^2 = -\left(1 - \frac{2M}{r}\right)dt^2 + \frac{dr^2}{1 - 2M/r} + r^2(d\theta^2 + \sin^2\theta \, d\phi^2) \text{ outside.} \tag{21.176b}$$

Let the "radius" of the shell, as a function of proper time measured on the shell, be

$$R \equiv \frac{1}{2\pi} \times \text{(proper circumference of shell)} = R(\tau). \tag{21.176c}$$

Show that the shell's mass density varies with time as

$$\sigma(\tau) = \mu/4\pi R^2(\tau), \qquad \mu = \text{constant} = \text{"total rest mass"}; \tag{21.176d}$$

and derive and solve the equation of motion

$$M = \mu\left\{1 + \left(\frac{dR}{d\tau}\right)^2\right\}^{1/2} - \frac{\mu}{2R}. \tag{21.176e}$$

CHAPTER **22**

THERMODYNAMICS, HYDRODYNAMICS, ELECTRODYNAMICS, GEOMETRIC OPTICS, AND KINETIC THEORY

§22.1. THE WHY OF THIS CHAPTER

Astrophysical applications of gravitation theory are the focus of the rest of this book, except for Chapters 41–44. Each application—stars, star clusters, cosmology, collapse, black holes, gravitational waves, solar-system experiments—can be pursued by itself at an elementary level, without reference to the material in this chapter. But deep understanding of the applications requires a prior grasp of thermodynamics, hydrodynamics, electrodynamics, geometric optics, and kinetic theory, all in the context of curved spacetime. Hence, most Track-2 readers will want to probe these subjects at this point.

This chapter is entirely Track 2. No earlier Track-2 material is needed as preparation for it, but Chapter 5 (stress-energy tensor) will be helpful.

§22.5 (geometric optics) is needed as preparation for Chapter 34 (singularities and global methods). The rest of the chapter is not needed as preparation for any later chapter; but it will be extremely helpful in most applications of gravitation theory (Chapters 23–40).

§22.2. THERMODYNAMICS IN CURVED SPACETIME*

Consider, for concreteness and simplicity, the equilibrium thermodynamics of a perfect fluid with fixed chemical composition ("simple perfect fluid")—for example, the gaseous interior of a collapsing supermassive star. The thermodynamic state of a fluid element, as it passes through an event \mathscr{P}_0, can be characterized by various thermodynamic potentials, such as n, ρ, p, T, s, μ. The numerical value of each potential at \mathscr{P}_0 is measured in the proper reference frame (§13.6) of an observer who moves with the fluid element—i.e., in the fluid element's "rest frame." Despite

Thermodynamic potentials are defined in rest frame of fluid

*For more detailed treatments of this subject see, e.g., Stueckelberg and Wanders (1953), Kluitenberg and de Groot (1954), Meixner and Reik (1959), and references cited therein; see also the references on hydrodynamics cited at the beginning of §22.3, and the references on kinetic theory cited at the beginning of §22.6.

this use of rest frame to measure the potentials, the potentials are frame-independent functions (scalar fields). At the chosen event \mathscr{P}_0, a given potential (e.g., n) has a unique value $n(\mathscr{P}_0)$; so n is a perfectly good frame-independent function.

The values of n, ρ, p, T, s, μ measure the following quantities in the rest frame of the fluid element:

Definitions of thermodynamic potentials

n, baryon number density; i.e., number of baryons per unit three-dimensional volume of rest frame, with antibaryons (if any) counted negatively.

ρ, density of total mass-energy; i.e., total mass-energy (including rest mass, thermal energy, compressional energy, etc.) contained in a unit three-dimensional volume of the rest frame.

p, isotropic pressure in rest frame.

T, temperature in rest frame.

s, entropy per baryon in rest frame. (The entropy per unit volume is ns.)

μ, chemical potential of baryons in rest frame [see equation (22.8) below].

Definition of "simple fluid"

The chemical composition of the fluid (number density of hydrogen molecules, number density of hydrogen atoms, number density of free protons and electrons, number density of photons, number density of ^{238}U nuclei, number density of Λ hyperons . . .) is assumed to be fixed uniquely by two thermodynamic variables—e.g., by the total number density of baryons n and the entropy per baryon s. In this sense the fluid is a "simple fluid." Simple fluids occur whenever the chemical abundances are "frozen" (reaction rates too slow to be important on the time scales of interest; for example, in a supermassive star except during explosive burning and except at temperatures high enough for $e^- - e^+$ pair production). Simple fluids also occur in the opposite extreme of complete chemical equilibrium (reaction rates fast enough to maintain equilibrium despite changing density and entropy; for example, in neutron stars, where high pressures speed up all reactions). When one examines nuclear burning in a nonconvecting star, or explosive nuclear burning, or pair production and neutrino energy losses at high temperatures, one must usually treat the fluid as "multicomponent." Then one introduces a number density n_J and a chemical potential μ_J for each chemical species with abundance not fixed by n and s. For further details see, e.g., Zel'dovich and Novikov (1971).

Law of baryon conservation

The most fundamental law of thermodynamics—even more fundamental than the "first" and "second" laws—is *baryon conservation*. Consider a fluid element whose moving walls are attached to the fluid so that no baryons flow in or out. As the fluid element moves through spacetime, deforming along the way, its volume V changes. But the number of baryons in it must remain fixed, so

$$\frac{d}{d\tau}(nV) = 0. \tag{22.1}$$

The changes in volume are produced by the flow of neighboring bits of fluid away from or toward each other—explicitly (exercise 22.1)

$$dV/d\tau = (\boldsymbol{\nabla} \cdot \boldsymbol{u})V, \tag{22.2}$$

where $u = d/d\tau$ is the 4-velocity of the fluid. Consequently, baryon conservation [equation (22.1)] can be reexpressed as

$$0 = \frac{dn}{d\tau} + \frac{n}{V}\frac{dV}{d\tau} = \boldsymbol{\nabla}_u n + n(\boldsymbol{\nabla} \cdot \boldsymbol{u}) = \boldsymbol{u} \cdot \boldsymbol{\nabla} n + n(\boldsymbol{\nabla} \cdot \boldsymbol{u}) = \boldsymbol{\nabla} \cdot (n\boldsymbol{u});$$

i.e.,

$$\boldsymbol{\nabla} \cdot \boldsymbol{S} = 0, \tag{22.3}$$

$$\boldsymbol{S} = n\boldsymbol{u} = \text{baryon number-flux vector} \tag{22.4}$$

(see §5.4 and exercise 5.3.) Moreover, this abstract geometric version of the law must be just as valid in curved spacetime as in flat (equivalence principle).

Note the analogy with the law of charge conservation, $\boldsymbol{\nabla} \cdot \boldsymbol{J} = 0$, in electrodynamics (exercise 3.16) and with the local law of energy-momentum conservation, $\boldsymbol{\nabla} \cdot \boldsymbol{T} = 0$ (§§5.9 and 16.2). In a very deep sense, the forms of these three laws are dictated by the theorem of Gauss (§5.9, and Boxes 5.3, 5.4).

The second law of thermodynamics states that, in flat spacetime or in curved, entropy can be generated but not destroyed. Apply this law to a fluid element of volume V containing a fixed number of baryons N. The entropy it contains is

Second law of thermodynamics

$$S = Ns = nsV.$$

Entropy may flow in and out across the faces of the fluid element ("heat flow" between neighboring fluid elements); but for simplicity assume it does not; or if it does, assume that it flows too slowly to have any significance for the problem at hand. Then the entropy in the fluid element can only increase:

$$d(nsV)/d\tau \geq 0 \quad \text{when negligible entropy is exchanged between neighboring fluid elements;}$$

i.e. [combine with equation (22.1)]

$$ds/d\tau \geq 0 \text{ (no entropy exchange).} \tag{22.5}$$

So long as the fluid element remains in thermodynamic equilibrium, its entropy will actually be conserved ["=" in equation (22.5)]; but at a shock wave, where equilibrium is momentarily broken, the entropy will increase (conversion of "relative kinetic energy" of neighboring fluid elements into heat). [For discussions of heat flow in special and general relativity, see Exercise 22.7. For discussion of shock waves, see Taub (1948), de Hoffman and Teller (1950), Israel (1960), May and White (1967), Zel'dovich and Rayzer (1967), Lichnerowicz (1967, 1971), and Thorne (1973a).]

Shock waves and heat flow

The first law of thermodynamics, in the proper reference frame of a fluid element, is identical to the first law in flat spacetime ("principle of equivalence"); and in flat spacetime the first law is merely the law of energy conservation:

First law of thermodynamics

$$d\begin{pmatrix}\text{energy in a volume element containing} \\ \text{a fixed number, } A, \text{ of baryons}\end{pmatrix} = -p\, d(\text{volume}) + T\, d(\text{entropy});$$

i.e.,

$$d(\rho A/n) = -p\, d(A/n) + T\, d(As);$$

i.e.,

$$d\rho = \frac{\rho + p}{n}\, dn + nT\, ds.$$

Query: what kind of a "d" appears here? For a simple fluid, the values of two potentials, e.g., n and s, fix all the others uniquely; so *any* change in ρ must be determined uniquely by the changes in n and s. It matters not whether the changes are measured along the world line of a given fluid element, or in some other direction. Thus, the "d" in the first law can be interpreted as an exterior derivative

$$\boldsymbol{d}\rho = \frac{\rho + p}{n}\, \boldsymbol{d}n + nT\, \boldsymbol{d}s; \tag{22.6}$$

and the changes along a given direction in the fluid (along a given tangent vector \boldsymbol{v}) can be written

$$\boldsymbol{\nabla}_{\boldsymbol{v}}\rho \equiv \langle \boldsymbol{d}\rho, \boldsymbol{v} \rangle = \frac{\rho + p}{n} \langle \boldsymbol{d}n, \boldsymbol{v} \rangle + nT\langle \boldsymbol{d}s, \boldsymbol{v} \rangle$$

$$= \frac{\rho + p}{n}\, \boldsymbol{\nabla}_{\boldsymbol{v}}n + nT\, \boldsymbol{\nabla}_{\boldsymbol{v}}s.$$

Pressure and temperature calculated from $\rho(n, s)$

Equation (22.6) lends itself to interpretation in two opposite senses: as a way to deduce the density of mass-energy of the medium from information about pressure (as a function of n and s) and temperature (as a function of n and s); and conversely, as a way to deduce the two functions $p(n, s)$ and $T(n, s)$ from the one function $\rho(n, s)$. It is natural to look at the second approach first; who does not like a strategy that makes an intellectual profit? Regarding ρ as a known (or calculable) function of n and s, one deduces from (22.6)

$$\frac{\rho + p}{n} = \left(\frac{\partial \rho}{\partial n}\right)_{s},$$

$$nT = \left(\frac{\partial \rho}{\partial s}\right)_{n},$$

and thence pressure and temperature individually,

$$p(n, s) = n\left(\frac{\partial \rho}{\partial n}\right)_{s} - \rho, \tag{22.7a}$$

$$T(n, s) = \frac{1}{n}\left(\frac{\partial \rho}{\partial s}\right)_{n} \tag{22.7b}$$

("two equations of state from one"). The analysis simplifies still further when the fluid, already assumed to be everywhere of the same composition, is also everywhere

endowed with the same entropy per baryon, s, and is in a state of adiabatic flow (no shocks or heat conduction). Then the density $\rho = \rho(n, s)$ reduces to a function of one variable out of which one derives everything (ρ, p, μ) needed for the hydrodynamics and the gravitation physics of the system (next chapter). Other choices of the "primary thermodynamic potential" are appropriate under other circumstances (see Box 22.1).

If differentiation leads from $\rho(n, s)$ to $p(n, s)$ and $T(n, s)$, it does not follow that one can take any two functions $p(n, s)$ and $T(n, s)$ and proceed "backwards" (by integration) to the "primary function", $\rho(n, s)$. To be compatible with the first law of thermodynamics (22.6), the two functions must satisfy the consistency requirement ["Maxwell relation"; equality of second partial derivatives of ρ]

Maxwell relation

$$(\partial p / \partial s)_n = n^2 (\partial T / \partial n)_s. \tag{22.7c}$$

Box 22.1 PRINCIPAL ALTERNATIVES FOR "PRIMARY THERMODYNAMIC POTENTIAL" TO DESCRIBE A FLUID

Primary thermodynamic potential and quantities on which it is most appropriately envisaged to depend	"Secondary" thermodynamic quantities obtained by differentiation of primary with or without use of $$d\left(\frac{\rho}{n}\right) + pd\left(\frac{1}{n}\right) - T\,ds = 0$$	Conditions under which convenient, appropriate, and relevant
"Density"; total amount of mass-energy (rest + thermal + \cdots) per unit volume $$\rho = \rho(n, s)$$	$p(n, s) = n\left(\dfrac{\partial \rho}{\partial n}\right)_s - \rho$ $T(n, s) = \dfrac{1}{n}\left(\dfrac{\partial \rho}{\partial s}\right)_n$ $\mu(n, s) = \dfrac{p + \rho}{n} = \left(\dfrac{\partial \rho}{\partial n}\right)_s$	Conditions of adiabatic flow (no shocks or heat conduction), so that s stays constant along streamline
"Physical free energy" $$a(n, T) = \dfrac{\rho}{n} - Ts$$	$p(n, T) = n^2\left(\dfrac{\partial a}{\partial n}\right)_T$ $s(n, T) = -\left(\dfrac{\partial a}{\partial T}\right)_n$ $\rho(n, T) = -nT^2\left[\dfrac{\partial (a/T)}{\partial T}\right]_n$	Know or can calculate a (or the "sum over states" of statistical mechanics) for conditions of specified volume per baryon and temperature
"Chemical free energy" $$f(p, T) = \dfrac{\rho + p}{n} - Ts$$	$1/n(p, T) = (\partial f / \partial p)_T$ $s(p, T) = -(\partial f / \partial T)_p$ $\rho(p, T) = \dfrac{f - T(\partial f / \partial T)_p}{(\partial f / \partial p)_T} - p$	Relevant for determining equilibrium when pressure and temperature are specified
"Chemical potential" ("energy to inject" expressed on a "per baryon" basis) $$\mu(p, s) = \dfrac{p + \rho}{n}$$	$1/n(p, s) = (\partial \mu / \partial p)_s$ $T(p, s) = (\partial \mu / \partial s)_p$ $\rho(p, s) = \dfrac{\mu}{(\partial \mu / \partial p)_s} - p$	When injection energy [= Fermi energy for an ideal Fermi gas, relativistic or not; see exercise 22.3] is the center of attention

Chemical potential equals
"injection energy" at
fixed entropy per baryon and
total volume

The chemical potential μ is also a unique function of n and s. It is defined as follows. (1) Take a sample of the simple fluid in a fixed thermodynamic state (fixed n and s). (2) Take, separately, a much smaller sample of the same fluid, containing δA baryons in the same thermodynamic state as the large sample (same n and s). (3) Inject the smaller sample into the larger one, holding the volume of the large sample fixed during the injection process. (4) The total mass-energy injected,

$$\delta M_{\text{injected}} = \rho \times (\text{volume of injected fluid}) = \rho(\delta A/n),$$

plus the work required to perform the injection

$$\delta W_{\text{injection}} = \begin{pmatrix} \text{work done against pressure of large sample} \\ \text{to open up space in it for the injected fluid} \end{pmatrix}$$
$$= p(\text{volume of injected fluid}) = p(\delta A/n),$$

is equal to $\mu \, \delta A$:

$$\mu \, \delta A = \delta M_{\text{injected}} + \delta W_{\text{injection}} = \frac{\rho + p}{n} \delta A.$$

Stated more briefly:

$$\mu = \begin{pmatrix} \text{total mass-energy required, per baryon, to "create" and} \\ \text{inject a small additional amount of fluid into a given} \\ \text{sample, without changing } s \text{ or volume of the sample} \end{pmatrix}$$
$$= \frac{\rho + p}{n} = \left(\frac{\partial \rho}{\partial n}\right)_s.$$
$$\text{[by first law of thermodynamics (22.6)]}$$

(22.8)

All the above laws and equations of thermodynamics are the same in curved spacetime as in flat spacetime; and the same in (relativistic) flat spacetime as in classical nonrelativistic thermodynamics—except for the inclusion of rest mass, together with all other forms of mass-energy, in ρ and μ. The reason is simple: the laws are all formulated as scalar equations linking thermodynamic variables that one measures in the rest frame of the fluid.

§22.3.　HYDRODYNAMICS IN CURVED SPACETIME*

Laws of hydrodynamics for
simple fluid without heat flow
or viscosity:

A simple perfect fluid flows through spacetime. It might be the Earth's atmosphere circulating in the Earth's gravitational field. It might be the gaseous interior of the Sun at rest in its own gravitational field. It might be interstellar gas accreting on-to a black hole. But whatever and wherever the fluid may be, its motion will be governed by the curved-spacetime laws of thermodynamics (§22.2) plus the local

*For more detailed treatments of this subject see, e.g., Ehlers (1961), Taub (1971), Ellis (1971), Lichnerowicz (1967), Cattaneo (1971), and references cited therein; see also the references on kinetic theory cited at the beginning of §22.6.

law of energy-momentum conservation, $\boldsymbol{\nabla} \cdot \boldsymbol{T} = 0$. The chief objective of this section is to reduce the equation $\boldsymbol{\nabla} \cdot \boldsymbol{T} = 0$ to usable form. The reduction will be performed in the text using abstract notation; the reader is encouraged to repeat the reduction using index notation. (1) Origins of laws

The stress-energy tensor for a perfect fluid, in curved spacetime as in flat (equivalence principle!), is

$$\boldsymbol{T} = (\rho + p)\boldsymbol{u} \otimes \boldsymbol{u} + p\boldsymbol{g}. \tag{22.9}$$

(See §5.5.) Its divergence is readily calculated using the chain rule; using the "compatibility relation between \boldsymbol{g} and $\boldsymbol{\nabla}$," $\boldsymbol{\nabla g} = 0$; using the identity $(\boldsymbol{\nabla}p) \cdot \boldsymbol{g} = \boldsymbol{\nabla}p$ (which one readily verifies in index notation); and using

$$0 = \boldsymbol{\nabla} \cdot \boldsymbol{T} = [\boldsymbol{\nabla}(\rho + p) \cdot \boldsymbol{u}]\boldsymbol{u} + [(\rho + p)\boldsymbol{\nabla} \cdot \boldsymbol{u}]\boldsymbol{u} + [(\rho + p)\boldsymbol{u}] \cdot \boldsymbol{\nabla}\boldsymbol{u} + (\boldsymbol{\nabla}p) \cdot \boldsymbol{g}$$

\lfloor[divergence on first slot]

$$= [\boldsymbol{\nabla}_{\boldsymbol{u}}\rho + \boldsymbol{\nabla}_{\boldsymbol{u}}p + (\rho + p)\boldsymbol{\nabla} \cdot \boldsymbol{u}]\boldsymbol{u} + (\rho + p)\boldsymbol{\nabla}_{\boldsymbol{u}}\boldsymbol{u} + \boldsymbol{\nabla}p. \tag{22.10}$$

The component of this equation along the 4-velocity is especially simple (recall that $\boldsymbol{u} \cdot \boldsymbol{\nabla}_{\boldsymbol{u}}\boldsymbol{u} = \frac{1}{2}\boldsymbol{\nabla}_{\boldsymbol{u}}\boldsymbol{u}^2 = 0$ because $\boldsymbol{u}^2 \equiv -1$):

$$0 = \boldsymbol{u} \cdot (\boldsymbol{\nabla} \cdot \boldsymbol{T}) = -[\boldsymbol{\nabla}_{\boldsymbol{u}}\rho + \boldsymbol{\nabla}_{\boldsymbol{u}}p + (\rho + p)\boldsymbol{\nabla} \cdot \boldsymbol{u}] + \boldsymbol{\nabla}_{\boldsymbol{u}}p$$

$$= -\boldsymbol{\nabla}_{\boldsymbol{u}}\rho - (\rho + p)\boldsymbol{\nabla} \cdot \boldsymbol{u}.$$

Combine this with the equation of baryon conservation (22.3); the result is

$$\frac{d\rho}{d\tau} = \frac{(\rho + p)}{n}\frac{dn}{d\tau}. \tag{22.11a}$$

(2) Local energy conservation: adiabaticity of flow

Notice that this is identical to the first law of thermodynamics (22.6) applied along a flow line, plus the assumption that the entropy per baryon is conserved along a flow line

$$ds/d\tau = 0. \tag{22.11b}$$

There is no reason for surprise at this result. To insist on thermodynamic equilibrium and to demand that the entropy remain constant is to require zero exchange of heat between one element of the fluid and another. But the stress-energy tensor (22.9) recognizes that heat exchange is absent. Any heat exchange would show up as an energy flux term in \boldsymbol{T} (Ex. 22.7); but no such term is present. Consequently, when one studies local energy conservation by evaluating $\boldsymbol{u} \cdot (\boldsymbol{\nabla} \cdot \boldsymbol{T}) = 0$, the stress-energy tensor reports that no heat flow is occurring—i.e. that $ds/d\tau = 0$.

Three components of $\boldsymbol{\nabla} \cdot \boldsymbol{T} = 0$ remain: the components orthogonal to the fluid's 4-velocity. One can pluck them out of $\boldsymbol{\nabla} \cdot \boldsymbol{T} = 0$, leaving behind the component along \boldsymbol{u}, by use of the "projection tensor"

$$\boldsymbol{P} \equiv \boldsymbol{g} + \boldsymbol{u} \otimes \boldsymbol{u}. \tag{22.12}$$

Box 22.2 THERMODYNAMICS AND HYDRODYNAMICS FOR A SIMPLE PERFECT FLUID IN CURVED SPACETIME

A. Ten Quantities Characterize the Fluid

Thermodynamic potentials all measured in rest frame

n, baryon number density
ρ, density of total mass-energy
p, pressure
T, temperature
s, entropy per baryon
μ, chemical potential per baryon

Four components of the fluid 4-velocity

B. Ten Equations Govern the Fluid's Motion

Two *equations of state*

$$p = p(n, s), \qquad T = T(n, s) \qquad (1), (2)$$

subject to the compatibility constraint ("Maxwell relation," which follows from first law of thermodynamics)

$$(\partial p/\partial s)_n = n^2(\partial T/\partial n)_s.$$

First law of thermodynamics

$$\boldsymbol{d}\rho = \frac{\rho + p}{n}\,\boldsymbol{d}n + nT\,\boldsymbol{d}s, \qquad (3)$$

which can be integrated to give $\rho(n, s)$.

Equation for chemical potential

$$\mu = (\rho + p)/n, \qquad (4)$$

which can be combined with $\rho(n, s)$ and $p(n, s)$ to give $\mu(n, s)$.

Law of baryon conservation

$$dn/d\tau \equiv \boldsymbol{\nabla}_{\boldsymbol{u}}n = -n\,\boldsymbol{\nabla}\cdot\boldsymbol{u}. \qquad (5)$$

Conservation of energy along flow lines, which (assuming no energy exchange between adjacent fluid elements) means "*adiabatic flow*"

$ds/d\tau = 0$ except in shock waves, where
$$ds/d\tau > 0. \qquad (6)$$

[Shock waves are not treated in this book; see Taub (1948), de Hoffman and Teller (1950), Israel (1960), May and White (1967), Zel'dovich and Rayzer (1967); Lichnerowicz (1967, 1971); and Thorne (1973a).]

Euler equations

$$(\rho + p)\,\boldsymbol{\nabla}_{\boldsymbol{u}}\boldsymbol{u} = -(\boldsymbol{g} + \boldsymbol{u}\otimes\boldsymbol{u})\cdot\boldsymbol{\nabla}p,$$
$$(7), (8), (9)$$

which determine the flow lines to which \boldsymbol{u} is tangent.

Normalization of 4-velocity

$$\boldsymbol{u}\cdot\boldsymbol{u} = -1. \qquad (10)$$

(See exercise 22.4.) Contracting \boldsymbol{P} with $\boldsymbol{\nabla}\cdot\boldsymbol{T} = 0$ [equation (22.10)] gives

(3) Euler equation

$$(\rho + p)\,\boldsymbol{\nabla}_{\boldsymbol{u}}\boldsymbol{u} = -\boldsymbol{P}\cdot(\boldsymbol{\nabla}p) \equiv -[\boldsymbol{\nabla}p + (\boldsymbol{\nabla}_{\boldsymbol{u}}p)\boldsymbol{u}]. \qquad (22.13)$$

This is the "*Euler equation*" of relativistic hydrodynamics. It has precisely the same form as the corresponding flat-spacetime Euler equation:

$$\begin{pmatrix} \text{inertial mass} \\ \text{per unit volume} \\ \text{[exercise 5.4]} \end{pmatrix} \times \begin{pmatrix} \text{4-acceleration} \\ \text{of fluid} \end{pmatrix} = - \begin{pmatrix} \text{pressure gradient} \\ \text{in the 3-surface} \\ \text{orthogonal to 4-velocity} \end{pmatrix}. \quad (22.13')$$

The pressure gradient, not "gravity," is responsible for all deviation of flow lines from geodesics.

Box 22.2 reorganizes and summarizes the above laws of thermodynamics and hydrodynamics.

Exercise 22.1. DIVERGENCE OF FLOW LINES PRODUCES VOLUME CHANGES **EXERCISES**

Derive the equation $dV/d\tau = (\boldsymbol{\nabla} \cdot \boldsymbol{u})V$ [equation (22.2)] for the rate of change of volume of a fluid element. [*Hint:* Pick an event \mathscr{P}_0, and calculate in a local Lorentz frame at \mathscr{P}_0 which momentarily moves with the fluid ("rest frame at \mathscr{P}_0").] [*Solution:* At events near \mathscr{P}_0 the fluid has a very small ordinary velocity $v^j = dx^j/dt$. Consequently a cube of fluid at \mathscr{P}_0 with edges $\Delta x = \Delta y = \Delta z = L$ changes its edges, after time δt, by the amounts

$$\delta(\Delta x) = [(dx/dt)\,\delta t]_{\text{at "front face"}} - [(dx/dt)\,\delta t]_{\text{at "back face"}}$$
$$= (\partial v^x/\partial x)L\,\delta t,$$

$$\delta(\Delta y) = (\partial v^y/\partial y)L\,\delta t,$$

$$\delta(\Delta z) = (\partial v^z/\partial z)L\,\delta t.$$

The corresponding change in volume is

$$\delta(\Delta x\,\Delta y\,\Delta z) = (\partial v^j/\partial x^j)L^3\,\delta t;$$

so the rate of change of volume is

$$\partial V/\partial t = V(\partial v^j/\partial x^j).$$

But in the local Lorentz rest frame at and near \mathscr{P}_0 (where $x^\alpha = 0$), the metric coefficients are $g_{\mu\nu} = \eta_{\mu\nu} + 0(|x^\alpha|^2)$, and the ordinary velocity is $v^j = 0(|x^\alpha|)$; so

$$u^0 = \frac{dt}{d\tau} = \frac{dt}{(-g_{\mu\nu}\,dx^\mu\,dx^\nu)^{1/2}} = 1 + 0(|x^\alpha|^2),$$

$$u^j = \frac{dx^j}{d\tau} = v^j + 0(|x^\alpha|^3).$$

Thus, the derivatives $\partial V/\partial t$ and $V(\partial v^j/\partial x^j)$ at \mathscr{P}_0 are

$$\partial V/\partial t = u^\alpha \partial V/\partial x^\alpha = u^\alpha V_{,\alpha} = dV/d\tau$$
$$= V(\partial v^j/\partial x^j) = V(\partial u^\alpha/\partial x^\alpha) = Vu^\alpha{}_{;\alpha} = V(\boldsymbol{\nabla} \cdot \boldsymbol{u}). \quad \text{Q.E.D.}]$$

[Note that by working in flat spacetime, one could have inferred more easily that $\partial V/\partial t = dV/d\tau$ and $\partial v^j/\partial x^j = \boldsymbol{\nabla} \cdot \boldsymbol{u}$; one would then have concluded $dV/d\tau = (\boldsymbol{\nabla} \cdot \boldsymbol{u})V$; and one could have invoked the equivalence principle to move this law into curved spacetime.]

Exercise 22.2. EQUATION OF CONTINUITY

Show that in the nonrelativistic limit in flat spacetime the equation of baryon conservation (22.3) becomes the "equation of continuity"

$$\frac{\partial n}{\partial t} + \frac{\partial}{\partial x^j}(nv^j) = 0.$$

Exercise 22.3. CHEMICAL POTENTIAL FOR IDEAL FERMI GAS

Show that the chemical potential of an ideal Fermi gas, nonrelativistic or relativistic, is (at zero temperature) equal to the Fermi energy (energy of highest occupied momentum state) of that gas.

Exercise 22.4. PROJECTION TENSORS

Show that contraction of a tangent vector \boldsymbol{B} with the "projection tensor" $\boldsymbol{P} \equiv \boldsymbol{g} + \boldsymbol{u} \otimes \boldsymbol{u}$ projects \boldsymbol{B} into the 3-surface orthogonal to the 4-velocity vector \boldsymbol{u}. [Hint: perform the

calculation in an orthonormal frame with $\boldsymbol{e}_{\hat{0}} = \boldsymbol{u}$, and write $\boldsymbol{B} = B^{\alpha}\boldsymbol{e}_{\hat{\alpha}}$; then show that $\boldsymbol{P}\cdot\boldsymbol{B} = B^{\hat{j}}\boldsymbol{e}_{\hat{j}}$.] If \boldsymbol{n} is a unit spacelike vector, show that $\boldsymbol{P} \equiv \boldsymbol{g} - \boldsymbol{n} \otimes \boldsymbol{n}$ is the corresponding projection operator. Note: There is no *unique* concept of "the projection orthogonal to a null vector." Why? [Hint: draw pictures in flat spacetime suppressing one spatial dimension.]

Exercise 22.5. PRESSURE GRADIENT IN STATIONARY GRAVITATIONAL FIELD

A perfect fluid is at rest (flow lines have $x^j = $ constant) in a stationary gravitational field (metric coefficients are independent of x^0). Show that the pressure gradient required to "support the fluid against gravity" (i.e., to make its flow lines be $x^j = $ constant instead of geodesics) is

$$\frac{\partial p}{\partial x^0} = 0, \qquad \frac{\partial p}{\partial x^j} = -(\rho + p)\,\frac{\partial \ln \sqrt{-g_{00}}}{\partial x^j}. \tag{22.14}$$

Evaluate this pressure gradient in the Newtonian limit, using the coordinate system and metric coefficients of equation (18.15c).

Exercise 22.6. EXPANSION, ROTATION, AND SHEAR

Let a field of fluid 4-velocities $\boldsymbol{u}(\mathscr{P})$ be given.

(a) Show that $\boldsymbol{\nabla u}$ can be decomposed in the following manner:

$$u_{\alpha;\beta} = \omega_{\alpha\beta} + \sigma_{\alpha\beta} + \tfrac{1}{3}\theta P_{\alpha\beta} - a_{\alpha}u_{\beta}, \tag{22.15a}$$

where \boldsymbol{a} is the *4-acceleration* of the fluid

$$a_{\alpha} \equiv u_{\alpha;\beta}u^{\beta}, \tag{22.15b}$$

θ is the *"expansion"* of the fluid world lines

$$\theta \equiv \boldsymbol{\nabla}\cdot\boldsymbol{u} = u^{\alpha}{}_{;\alpha}, \tag{22.15c}$$

$P_{\alpha\beta}$ is the *projection tensor*

$$P_{\alpha\beta} \equiv g_{\alpha\beta} + u_{\alpha}u_{\beta}, \tag{22.15d}$$

$\sigma_{\alpha\beta}$ is the *shear tensor* of the fluid

$$\sigma_{\alpha\beta} \equiv \tfrac{1}{2}(u_{\alpha;\mu}P^{\mu}{}_{\beta} + u_{\beta;\mu}P^{\mu}{}_{\alpha}) - \tfrac{1}{3}\theta P_{\alpha\beta}, \tag{22.15e}$$

and $\omega_{\alpha\beta}$ is the *rotation 2-form* of the fluid

$$\omega_{\alpha\beta} \equiv \tfrac{1}{2}(u_{\alpha;\mu}P^{\mu}{}_{\beta} - u_{\beta;\mu}P^{\mu}{}_{\alpha}). \tag{22.15f}$$

(b) Each of the component parts of this decomposition has a simple physical interpretation in the local rest frames of the fluid. The interpretation of the 4-acceleration \boldsymbol{a} in terms of accelerometer readings should be familiar. Exercise 22.1 showed that the expansion $\theta = \boldsymbol{\nabla}\cdot\boldsymbol{u}$ describes the rate of increase of the volume of a fluid element,

$$\theta = (1/V)(dV/d\tau). \tag{22.15g}$$

Exercise 22.4 explored the meaning and use of the projection tensor \boldsymbol{P}. Verify that in a local Lorentz frame ($g_{\hat{\alpha}\hat{\beta}} = \eta_{\hat{\alpha}\hat{\beta}}$, $\Gamma^{\hat{\alpha}}{}_{\hat{\beta}\hat{\gamma}} = 0$) momentarily moving with the fluid ($u^{\hat{\alpha}} = \delta^{\alpha}{}_{0}$), $\sigma_{\hat{\alpha}\hat{\beta}}$ and $\omega_{\hat{\alpha}\hat{\beta}}$ reduce to the classical (nonrelativistic) shear and rotation of the fluid. [See, e.g., §§2.4 and 2.5 of Ellis (1971) for both classical and relativistic descriptions of shear and rotation.]

Exercise 22.7. HYDRODYNAMICS WITH VISCOSITY AND HEAT FLOW.*

(a) In §15 of Landau and Lifshitz (1959), one finds an analysis of viscous stresses for a classical (nonrelativistic) fluid. By carrying that analysis over directly to the local Lorentz rest frame of a relativistic fluid, and by then generalizing to frame-independent language, show that the contribution of viscosity to the stress-energy tensor is

$$\mathbf{T}^{(\text{visc})} = -2\eta\boldsymbol{\sigma} - \zeta\theta\mathbf{P}, \tag{22.16a}$$

where $\eta \geq 0$ is the *"coefficient of dynamic viscosity"*; $\zeta \geq 0$ is the *"coefficient of bulk viscosity"*; and $\boldsymbol{\sigma}$, θ, \mathbf{P} are the shear, expansion, and projection tensor of the fluid.

(b) An idealized description of heat flow in a fluid introduces the *heat-flux 4-vector* \mathbf{q} with components in the local rest-frame of the fluid,

$$q^{\hat{0}} = 0, \qquad q^{\hat{j}} = \begin{pmatrix} \text{energy per unit time crossing unit} \\ \text{surface perpendicular to } \mathbf{e}_{\hat{j}} \end{pmatrix}. \tag{22.16b}$$

By generalizing from the fluid rest frame to frame-independent language, show that the contribution of heat flux to the stress-energy tensor is

$$\mathbf{T}^{(\text{heat})} = \mathbf{u} \otimes \mathbf{q} + \mathbf{q} \otimes \mathbf{u}. \tag{22.16c}$$

Thereby conclude that, in this idealized picture, the stress-energy tensor for a viscous fluid with heat conduction is

$$T^{\alpha\beta} = \rho u^{\alpha}u^{\beta} + (p - \zeta\theta)P^{\alpha\beta} - 2\eta\sigma^{\alpha\beta} + q^{\alpha}u^{\beta} + u^{\alpha}q^{\beta}. \tag{22.16d}$$

(c) Define the entropy 4-vector \mathbf{s} by

$$\mathbf{s} \equiv ns\mathbf{u} + \mathbf{q}/T. \tag{22.16e}$$

By calculations in the local rest-frame of the fluid, show that

$$\boldsymbol{\nabla} \cdot \mathbf{s} = \begin{pmatrix} \text{rate of increase of entropy} \\ \text{in a unit volume} \end{pmatrix} - \begin{pmatrix} \text{rate at which heat and fluid} \\ \text{carry entropy into a unit volume} \end{pmatrix}$$

$$= \begin{pmatrix} \text{rate at which entropy is being} \\ \text{generated in a unit volume} \end{pmatrix}. \tag{22.16f}$$

Thereby arrive at the following form of the *second law of thermodynamics:*

$$\boldsymbol{\nabla} \cdot \mathbf{s} \geq 0. \tag{22.16g}$$

(d) Calculate the law of local energy conservation, $\mathbf{u} \cdot \boldsymbol{\nabla} \cdot \mathbf{T} = 0$, for a viscous fluid with heat flow. Combine with the first law of thermodynamics and with the law of baryon conservation to obtain

$$T\boldsymbol{\nabla} \cdot \mathbf{s} = \zeta\theta^2 + 2\eta\sigma_{\alpha\beta}\sigma^{\alpha\beta} - q^{\alpha}(T_{,\alpha}/T + a_{\alpha}). \tag{22.16h}$$

Interpret each term of this equation as a contribution to entropy generation (*example:* $2\eta\sigma_{\alpha\beta}\sigma^{\alpha\beta}$ describes entropy generation by viscous heating). [*Note:* The term $q^{\alpha}a_{\alpha}$ is relativistic in origin. It is associated with the inertia of the flowing heat.]

(e) When one takes account of the inertia of the flowing heat, one obtains the following generalization of the classical law of heat conduction:

$$q^{\alpha} = -\kappa P^{\alpha\beta}(T_{,\beta} + Ta_{\beta}) \tag{22.16i}$$

*Exercise supplied by John M. Stewart.

(Eckart 1940). Here κ is the *coefficient of thermal conductivity*. Use this equation to show that, for a fluid at rest in a stationary gravitational field (Exercise 22.5),

$$q_0 = 0, \qquad q_j = -\frac{\kappa}{\sqrt{-g_{00}}}(T\sqrt{-g_{00}})_{,j}. \tag{22.16j}$$

[Thus, thermal equilibrium corresponds not to constant temperature, but to the redshifted temperature distribution $T\sqrt{-g_{00}} = $ constant; Tolman (1934a), p. 313.] Also, use the idealized law of heat conduction (22.16i) to reexpress the rate of entropy generation as

$$T\boldsymbol{\nabla}\cdot\boldsymbol{s} = \zeta\theta^2 + 2\eta\sigma_{\alpha\beta}\sigma^{\alpha\beta} + (\kappa/T)P^{\alpha\beta}(T_{,\alpha} + Ta_\alpha)(T_{,\beta} + Ta_\beta)$$
$$\geq 0. \tag{22.16k}$$

[For further details about heat flow and for discussions of the limitations of the above idealized description, see e.g., §4.18 of Ehlers (1971); also Marle (1969), Anderson (1970), Stewart (1971), and papers cited therein.]

§22.4. ELECTRODYNAMICS IN CURVED SPACETIME

Electric and magnetic fields

In a local Lorentz frame in the presence of gravity, an observer can measure the electric and magnetic fields \boldsymbol{E} and \boldsymbol{B} using the usual Lorentz force law for charged particles. As in special relativity, he can regard \boldsymbol{E} and \boldsymbol{B} as components of an electromagnetic field tensor,

$$F^{\hat{0}\hat{j}} = -F^{\hat{j}\hat{0}} = E^{\hat{j}}, \qquad F^{\hat{j}\hat{k}} = \epsilon^{\hat{j}\hat{k}\hat{l}}B^{\hat{l}};$$

he can regard the charge and current densities as components of a 4-vector $J^{\hat{\alpha}}$, and he can write Maxwell's equations and the Lorentz force equation in the special relativistic form,

$$F^{\hat{\alpha}\hat{\beta}}{}_{,\hat{\beta}} = 4\pi J^{\hat{\alpha}}, \qquad F_{\hat{\alpha}\hat{\beta},\hat{\gamma}} + F_{\hat{\beta}\hat{\gamma},\hat{\alpha}} + F_{\hat{\gamma}\hat{\alpha},\hat{\beta}} = 0,$$

$$ma^{\hat{\alpha}} = F^{\hat{\alpha}\hat{\beta}}qu_{\hat{\beta}} \qquad \begin{pmatrix} m = \text{mass of particle, } q = \text{charge,} \\ u^{\hat{\alpha}} = \text{4-velocity, } a^{\hat{\alpha}} = \text{4-acceleration} \end{pmatrix}.$$

In any other frame these equations will have the same form, but with commas replaced by semicolons

Maxwell equations and Lorentz force law

$$F^{\alpha\beta}{}_{;\beta} = 4\pi J^\alpha, \tag{22.17a}$$

$$F_{\alpha\beta;\gamma} + F_{\beta\gamma;\alpha} + F_{\gamma\alpha;\beta} = 0, \tag{22.17b}$$

$$ma^\alpha = F^{\alpha\beta}qu_\beta. \tag{22.17c}$$

These are the basic equations of electrodynamics in the presence of gravity. From them follows everything else. For example, as in special relativity, so also here (exercise 22.9), they imply the equation of charge conservation

Charge conservation

$$J^\alpha{}_{;\alpha} = 0; \tag{22.18a}$$

and for an electromagnetic field interacting with charged matter (exercise 22.10) they imply vanishing divergence for the sum of the stress-energy tensors

$$(T^{(\mathrm{EM})\alpha\beta} + T^{(\mathrm{MATTER})\alpha\beta})_{;\beta} = 0. \tag{22.18b}$$

<div style="float:right">Local conservation of energy-momentum</div>

As in special relativity, so also here, one can introduce a vector potential A^{μ}. Replacing commas by semicolons in the usual special-relativistic expression for $F^{\mu\nu}$ in terms of A^{μ}, one obtains

<div style="float:right">Vector potential</div>

$$F_{\mu\nu} = A_{\nu;\mu} - A_{\mu;\nu}. \tag{22.19a}$$

If all is well, this equation should guarantee (as in special relativity) that the Maxwell equations (22.17b) are satisfied. Indeed, it does, as one sees in exercise 22.8. To derive the wave equation that governs the vector potential, insert expression (22.19a) into the remaining Maxwell equations (22.17a), obtaining

$$-A^{\alpha;\beta}{}_{\beta} + A^{\beta;\alpha}{}_{\beta} = 4\pi J^{\alpha}; \tag{22.19b}$$

then commute covariant derivatives in the first term using the identity (16.6c), to obtain

$$-A^{\alpha;\mu}{}_{\mu} + A^{\mu}{}_{;\mu}{}^{;\alpha} + R^{\alpha}{}_{\mu}A^{\mu} = 4\pi J^{\alpha}. \tag{22.19b'}$$

Finally, adopting the standard approach of special relativity, impose the Lorentz gauge condition

$$A^{\mu}{}_{;\mu} = 0, \tag{22.19c}$$

<div style="float:right">Lorentz gauge condition</div>

thereby bringing the wave equation (22.19b') into the form

$$(\Delta_{dR}A)^{\alpha} \equiv -A^{\alpha;\beta}{}_{\beta} + R^{\alpha}{}_{\beta}A^{\beta} = 4\pi J^{\alpha}. \tag{22.19d}$$

<div style="float:right">Wave equation for vector potential</div>

The "de Rham vector wave operator" Δ which appears here is, apart from sign, a generalized d'Alambertian for vectors in curved spacetime. Mathematically it is more powerful than $-A^{\alpha;\beta}{}_{;\beta}$, and than any other operator that reduces to (minus) the d'Alambertian in special relativity. [For a discussion, see de Rham (1955).]

Although the electrodynamic equations (22.17a)–(22.19b) are all obtained from special relativity by the comma-goes-to-semicolon rule, the wave equation (22.19d) for the vector potential is not ("curvature coupling"; see Box 16.1). Nevertheless, when spacetime is flat (so $R^{\alpha}{}_{\beta} = 0$), (22.19d) does reduce to the usual wave equation of special relativity.

Exercise 22.8. THE VECTOR POTENTIAL FOR ELECTRODYNAMICS

<div style="float:right">**EXERCISES**</div>

Show that in any coordinate frame the connection coefficients cancel out of both equations (22.19a) and (22.17b), so they can be written

$$F_{\mu\nu} = A_{\nu,\mu} - A_{\mu,\nu}, \tag{22.20a}$$

$$F_{\alpha\beta,\gamma} + F_{\beta\gamma,\alpha} + F_{\gamma\alpha,\beta} = 0. \tag{22.20b}$$

(In the language of differential forms these equations are $\boldsymbol{F} = \boldsymbol{dA}$, $\boldsymbol{dF} = 0$.) Then use this form of the equations to show that equation (22.19a) implies equation (22.17b), as asserted in the text.

Exercise 22.9. CHARGE CONSERVATION IN THE PRESENCE OF GRAVITY

Show that Maxwell's equations (22.17a,b) imply the equation of charge conservation (22.18a) when gravity is present, just as they do in special relativity theory. [Hints: Use the antisymmetry of $F^{\alpha\beta}$; and beware of the noncommutation of the covariant derivatives, which must be handled using equations (16.6). Alternatively, show that in coordinate frames, equation (22.17a) can be written as

$$\frac{1}{\sqrt{|g|}} \frac{\partial}{\partial x^\beta} (\sqrt{|g|} F^{\alpha\beta}) = 4\pi J^\alpha \qquad (22.17a')$$

and (22.18a) as

$$J^\alpha{}_{;\alpha} \equiv \frac{1}{\sqrt{|g|}} \frac{\partial}{\partial x^\alpha} (\sqrt{|g|} J^\alpha) = 0, \qquad (22.18a')$$

and carry out the demonstration in a coordinate frame.]

Exercise 22.10. INTERACTING ELECTROMAGNETIC FIELD AND CHARGED MATTER

As in special relativity, so also in the presence of gravity ("equivalence principle"), the stress-energy tensor for an electromagnetic field is

$$T^{(\mathrm{EM})}{}_{\alpha\beta} = \frac{1}{4\pi} \left(F_{\alpha\mu} F_\beta{}^\mu - \frac{1}{4} F_{\mu\nu} F^{\mu\nu} g_{\alpha\beta} \right). \qquad (22.21)$$

Use Maxwell's equations (22.17a,b) in the presence of gravity to show that

$$T^{(\mathrm{EM})\alpha\beta}{}_{;\beta} = -F^{\alpha\beta} J_\beta. \qquad (22.22)$$

But $F^{\alpha\beta} J_\beta$ is just the Lorentz 4-force per unit volume with which the electromagnetic field acts on the charged matter [see the Lorentz force equation (22.17c); also equation (5.43)]; i.e., it is $T^{(\mathrm{MATTER})\alpha\beta}{}_{;\beta}$. Consequently, the above equation can be rewritten in the form (22.18b) cited in the text.

§22.5. GEOMETRIC OPTICS IN CURVED SPACETIME*

Radio waves from the quasar 3C279 pass near the sun and get deflected by its gravitational field. Light rays emitted by newborn galaxies long ago and far away propagate through the cosmologically curved spacetime of the universe, and get focused (and redshifted) producing curvature-enlarged (but dim) images of the galaxies on the Earth's sky.

*Based in part on notes prepared by William L. Burke at Caltech in 1968. For more detailed treatments of geometric optics in curved spacetime, see, e.g., Sachs (1961), Jordan, Ehlers, and Sachs (1961), and Robinson (1961); also references discussed and listed in §41.11.

These and most other instances of the propagation of light and radio waves are subject to the laws of geometric optics. This section derives those laws, in curved spacetime, from Maxwell's equations.

The fundamental laws of geometric optics are: (1) light rays are null geodesics; (2) the polarization vector is perpendicular to the rays and is parallel-propagated along the rays; and (3) the amplitude is governed by an adiabatic invariant which, in quantum language, states that the number of photons is conserved.

Overview of geometric optics

The conditions under which these laws hold are defined by conditions on three lengths: (1) the typical reduced wavelength of the waves,

Conditions for validity of geometric optics

$$\lambda\!\!\!\!- \equiv \frac{\lambda}{2\pi} = \begin{pmatrix} \text{``classical distance of closest approach for} \\ \text{a photon with one unit of angular momentum''} \end{pmatrix}, \quad (22.23a)$$

as measured in a typical local Lorentz frame (e.g., a frame at rest relative to nearby galaxies); (2) the typical length \mathcal{L} over which the amplitude, polarization, and wavelength of the waves vary, e.g., the radius of curvature of a wave front, or the length of a wave packet produced by a sudden outburst in a quasar; (3) the typical radius of curvature \mathcal{R} of the spacetime through which the waves propagate,

$$\mathcal{R} \equiv \begin{vmatrix} \text{typical component of } \textbf{\textit{Riemann}} \text{ as measured} \\ \text{in typical local Lorentz frame} \end{vmatrix}^{-1/2}. \quad (22.23b)$$

Geometric optics is valid whenever the reduced wavelength is very short compared to each of the other scales present,

$$\lambda\!\!\!\!- \ll \mathcal{L} \quad \text{and} \quad \lambda\!\!\!\!- \ll \mathcal{R}, \quad (22.23c)$$

so that the waves can be regarded *locally* as plane waves propagating through spacetime of negligible curvature.

Mathematically one exploits the geometric-optics assumption, $\lambda\!\!\!\!- \ll \mathcal{L}$ and $\lambda\!\!\!\!- \ll \mathcal{R}$, as follows. Focus attention on waves that are highly monochromatic over regions $\lesssim \mathcal{L}$. (More complex spectra can be analyzed by superposition, i.e., by Fourier analysis.) Split the vector potential of electromagnetic theory into a rapidly changing, real phase,

The "two-length-scale" expansion underlying geometric optics

$$\theta \sim (\text{distance propagated})/\lambda\!\!\!\!-,$$

and a slowly changing, complex amplitude (i.e. one with real and imaginary parts),

$$\textbf{\textit{A}} = \text{Real part of}\{\text{amplitude} \times e^{i\theta}\} \equiv \Re\{\text{amplitude} \times e^{i\theta}\}.$$

Imagine holding fixed the scale of the amplitude variation, \mathcal{L}, and the scale of the spacetime curvature, \mathcal{R}, while making the reduced wavelength, $\lambda\!\!\!\!-$, shorter and shorter. The phase will get larger and larger ($\theta \propto 1/\lambda\!\!\!\!-$) at any fixed event in spacetime, but the amplitude as a function of location in spacetime can remain virtually unchanged,

$$\text{Amplitude} = \begin{bmatrix} \text{dominant part,} \\ \text{independent of } \lambda\!\!\!\!- \end{bmatrix} + \begin{bmatrix} \text{small corrections (deviations from} \\ \text{geometric optics) due to finite wavelength} \end{bmatrix}.$$

This circumstance allows one to expand the amplitude in powers of λbar:*

$$\text{Amplitude} = \boldsymbol{a} + \boldsymbol{b} + \boldsymbol{c} + \ldots.$$

$$\begin{bmatrix}\text{independent}\\ \text{of } \lambdabar\end{bmatrix} \quad [\propto \lambdabar] \quad [\propto \lambdabar^2]$$

[Actually, the expansion proceeds in powers of the dimensionless number

$$\lambdabar/(\text{minimum of } \mathcal{L} \text{ and } \mathcal{R}) \equiv \lambdabar/L. \tag{22.24}$$

Applied mathematicians call this a "two-length-scale expansion"; see, e.g., Cole (1968). The basic short-wavelength approximation here has a long history; see, e.g., Liouville (1837), Rayleigh (1912). Following a suggestion of Debye, it was applied to Maxwell's equations by Sommerfeld and Runge (1911). It is familiar as the WKB approximation in quantum mechanics, and has many other applications as indicated by the bibliography in Keller, Lewis, and Seckler (1956). The contribution of higher order terms is considered by Kline (1954) and Lewis (1958). See especially the book of Fröman and Fröman (1965).]

It is useful to introduce a parameter ε that keeps track of how rapidly various terms approach zero (or infinity) as \lambdabar/L approaches zero:

The vector potential in geometric optics

$$A_\mu = \mathfrak{R}\{(a_\mu + \varepsilon b_\mu + \varepsilon^2 c_\mu + \cdots)e^{i\theta/\varepsilon}\}. \tag{22.25}$$

Any term with a factor ε^n in front of it varies as $(\lambdabar/L)^n$ in the limit of very small wavelengths [$\theta \propto (\lambdabar/L)^{-1}$; $c_\mu \propto (\lambdabar/L)^2$; etc.]. By convention, ε is a dummy expansion parameter with eventual value unity; so it can be dropped from the calculations when it ceases to be useful. And by convention, all "post-geometric-optics corrections" are put into the amplitude terms $\boldsymbol{b}, \boldsymbol{c}, \ldots$; none are put into θ.

Note that, while the phase θ is a real function of position in spacetime, the amplitude and hence the vectors $\boldsymbol{a}, \boldsymbol{b}, \boldsymbol{c}, \ldots$ are complex. For example, to describe monochromatic waves with righthand circular polarization, propagating in the z direction, one could set $\theta = \omega(z - t)$ and $\boldsymbol{a} = 1/\sqrt{2}a(\boldsymbol{e}_x + i\boldsymbol{e}_y)$ with a real; so

$$\boldsymbol{A} = \mathfrak{R}\left\{\frac{1}{\sqrt{2}} a(\boldsymbol{e}_x + i\boldsymbol{e}_y)e^{i\omega(z-t)}\right\} = \frac{1}{\sqrt{2}} a\{\cos[\omega(z - t)]\boldsymbol{e}_x - \sin[\omega(z - t)]\boldsymbol{e}_y\}$$

Basic concepts of geometric optics:

The assumed form (22.25) for the vector potential is the mathematical foundation of geometric optics. All the key equations of geometric optics result from inserting this vector potential into the source-free wave equation $\boldsymbol{\Delta A} = 0$ [equation (22.19d)] and into the Lorentz gauge condition $\boldsymbol{\nabla \cdot A} = 0$ [equation (22.19c)]. The resulting equations (derived below) take their simplest form only when expressed in terms of the following:

*The equations for \boldsymbol{A} are linear. Therefore the analysis would proceed equally well assuming, instead of an amplitude independent of λbar, a dominant term $\boldsymbol{a} \propto \lambdabar^n$, with $\boldsymbol{b} \propto \lambdabar^{n+1}$, $\boldsymbol{c} \propto \lambdabar^{n+2}$, etc. The results are independent of n. Choosing $n = 1$ would give field strengths $F_{\mu\nu}$ and energy densities $T_{\mu\nu} \propto F^2 \propto A^2/\lambdabar^2 \propto$ constant as $\lambdabar \longrightarrow 0$.

"wave vector," $\boldsymbol{k} \equiv \boldsymbol{\nabla}\theta$; (22.26a)

"scalar amplitude," $a \equiv (\boldsymbol{a} \cdot \bar{\boldsymbol{a}})^{1/2} = (a^\mu \bar{a}_\mu)^{1/2}$; (22.26b)

"polarization vector," $\boldsymbol{f} \equiv \boldsymbol{a}/a =$ "unit complex vector along \boldsymbol{a}". (22.26c)

(1) wave vector

(2) scalar amplitude

(3) polarization vector

(Here $\bar{\boldsymbol{a}}$ is the complex conjugate of \boldsymbol{a}.) *Light rays* are defined to be the curves $\mathscr{P}(\lambda)$ normal to surfaces of constant phase θ. Since $\boldsymbol{k} \equiv \boldsymbol{\nabla}\theta$ is the normal to these surfaces, the differential equation for a light ray is

(4) light rays

$$\frac{dx^\mu}{d\lambda} = k^\mu(x) = g^{\mu\nu}(x)\theta_{,\nu}(x).$$ (22.26d)

Box 22.3, appropriate for study at this point, shows the polarization vector, wave vector, surfaces of constant phase, and light rays for a propagating wave; the scalar amplitude, not shown there, merely tells the length of the vector amplitude \boldsymbol{a}. Insight into the complex polarization vector, if not familiar from electrodynamics, can be developed later in Exercise 22.12.

So much for the foundations. Now for the calculations. First insert the geometric-optics vector potential (22.25) into the Lorentz gauge condition:

Derivation of laws of geometric optics

$$0 = A^\mu{}_{;\mu} = \Re\left\{\left[\frac{i}{\varepsilon}k_\mu(a^\mu + \varepsilon b^\mu + \cdots) + (a^\mu + \varepsilon b^\mu + \cdots)_{;\mu}\right]e^{i\theta/\varepsilon}\right\}.$$ (22.27)

The leading term (order $1/\varepsilon$) says

$$\boldsymbol{k} \cdot \boldsymbol{a} = 0 \quad \textit{(amplitude is perpendicular to wave vector)};$$ (22.28)

or, equivalently

$$\boldsymbol{k} \cdot \boldsymbol{f} = 0 \quad \textit{(polarization is perpendicular to wave vector)}.$$ (22.28')

The post-geometric-optics breakdown in this orthogonality condition is governed by the higher-order terms $[0(1), 0(\varepsilon), 0(\varepsilon^2), \ldots]$ in the gauge condition (22.27); for example, the $0(1)$ terms say

$$\boldsymbol{k} \cdot \boldsymbol{b} = i\boldsymbol{\nabla} \cdot \boldsymbol{a}.$$

Next insert the vector potential (22.25) into the source-free wave equation (22.19d):

$$0 = (\varDelta_{dR}\boldsymbol{A})^\alpha = -A^{\alpha;\beta}{}_\beta + R^\alpha{}_\beta A^\beta$$

$$= \Re\left\{\left[\frac{1}{\varepsilon^2}k^\beta k_\beta(a^\alpha + \varepsilon b^\alpha + \varepsilon^2 c^\alpha + \cdots) - 2\frac{i}{\varepsilon}k^\beta(a^\alpha + \varepsilon b^\alpha + \cdots)_{;\beta}\right.\right.$$

$$\left.\left. - \frac{i}{\varepsilon}k^\beta{}_{;\beta}(a^\alpha + \varepsilon b^\alpha + \cdots) - (a^\alpha + \cdots)^{;\beta}{}_\beta + R^\alpha{}_\beta(a^\beta + \cdots)\right]e^{i\theta/\varepsilon}\right\}.$$ (22.29)

Collect terms of order $1/\varepsilon^2$ and $1/\varepsilon$ (terms of order higher than $1/\varepsilon$ govern post-geometric-optics corrections):

(continued on page 576)

Box 22.3 GEOMETRY OF AN ELECTROMAGNETIC WAVE TRAIN

direction of propagation as
seen in local Lorentz frame

one wavelength, $2\pi\lambda$,
as seen in local Lorentz frame

The drawing shows surfaces of constant phase, $\theta = $ constant, emerging through the "surface of simultaneity", $t = 0$, of a local Lorentz frame. The surfaces shown are alternately "crests" ($\theta = 1764\pi$, $\theta = 1766\pi$, . . .) and "troughs" ($\theta = 1765\pi$, $\theta = 1767\pi, . . .$) of the wave train. These surfaces make up a 1-form, $\widetilde{k} = d\theta$. The "corresponding vector" $k = \boldsymbol{\nabla}\theta$ is the "*wave vector.*" The wave vector is null, $k \cdot k = 0$, according to Maxwell's equations [equation (22.30)]. Therefore it lies in a surface of constant phase:

$$\begin{pmatrix}\text{number of surfaces}\\ \text{pierced by } \boldsymbol{k}\end{pmatrix} = \langle \boldsymbol{d\theta}, \boldsymbol{k} \rangle = \langle \widetilde{\boldsymbol{k}}, \boldsymbol{k} \rangle = \boldsymbol{k} \cdot \boldsymbol{k} = 0.$$

But not only does it lie in a surface of constant phase; it is also perpendicular to that surface! Any vector \boldsymbol{v} in that surface must satisfy $\boldsymbol{k} \cdot \boldsymbol{v} = \langle \widetilde{\boldsymbol{k}}, \boldsymbol{v} \rangle = \langle \boldsymbol{d\theta}, \boldsymbol{v} \rangle = 0$ because it pierces no surfaces.

Geometric optics assumes that the reduced wavelength λ, as measured in a typical local Lorentz frame, is small compared to the scale \mathcal{L} of inhomogeneities in the wave train and small compared to the radius of curvature of spacetime, \mathcal{R}. Thus, over regions much larger than λ but smaller than \mathcal{L} or \mathcal{R}, the waves are plane-fronted

and monochromatic, and there exist Lorentz reference frames (Riemann normal coordinates). In one of these "extended" local Lorentz frames, the phase must be

$$\theta = k_\alpha x^\alpha + \text{constant};$$

no other expression will yield $\boldsymbol{\nabla}\theta = \boldsymbol{k}$. The corresponding vector potential [equation (22.25)] will be

$$A^\mu = \Re\{a^\mu \exp[i(\boldsymbol{k}\cdot\boldsymbol{x} - k^0 t)]\} + (\text{"post-geometric-optics corrections"});$$

hence,

$$k^0 = 2\pi/(\text{period of wave}) = 2\pi\nu = \omega \equiv (\text{angular frequency}),$$
$$|\boldsymbol{k}| = 2\pi/(\text{wavelength of wave}) = 1/\lambdabar = \omega,$$
$$\boldsymbol{k} \text{ points along direction of propagation of wave.}$$

At each event in spacetime there is a wave vector; and these wave vectors, tacked end-on-end, form a family of curves—the "light rays" or simply "rays"—whose tangent vector is \boldsymbol{k}. The rays, like their tangent vector, lie both in and perpendicular to the surfaces of constant phase.

The affine parameter λ of a ray (not to be confused with wavelength $= 2\pi\lambdabar$) satisfies $\boldsymbol{k} = d/d\lambda$; therefore it is given by

$$\lambda = t/k^0 + \text{constant} = t/\omega + \text{constant},$$

where t is proper time along the ray as measured, not by the ray itself (its proper time is zero!), but by the local Lorentz observer who sees angular frequency ω. Thus, while ω is a frame-dependent quantity and t is also a frame-dependent quantity, their quotient t/ω when measured along the ray (*not off the ray*) is the frame-independent affine parameter. For a particle it is possible and natural to identify the affine parameter λ with proper time τ. For a light ray this identification is unnatural and impossible. The lapse of proper time along the ray is identically zero. The springing up of λ to take the place of the vanished τ gives one a tool to do what one might not have suspected to be possible. Given a light ray shot out at event \mathscr{A} and passing through event \mathscr{B}, one can give a third event \mathscr{C} along the same null world line that is twice as "far" from \mathscr{A} as \mathscr{B} is "far," in a new sense of "far" that has nothing whatever directly to do with proper time (zero!), but is defined by equal increments of the affine parameter ($\lambda_\mathscr{C} - \lambda_\mathscr{B} = \lambda_\mathscr{B} - \lambda_\mathscr{A}$). The "affine parameter" has a meaning for any null geodesic analyzed even in isolation. In this respect, it is to be distinguished from the so-called "luminosity distance" which is sometimes introduced in dealing with the propagation of radiation through curved spacetime, and which is defined by the spreading apart of two or more light rays coming from a common source.

Maxwell's equations as explored in the text [equation (22.28')] guarantee that the complex polarization vector \boldsymbol{f} is perpendicular to the wave vector \boldsymbol{k} and that, therefore, it lies in a surface of constant phase (see drawing). Intuition into the polarization vector is developed in exercise 22.12.

$$0\left(\frac{1}{\varepsilon^2}\right): \qquad\qquad k^\beta k_\beta a^\alpha = 0$$

$$\implies \boldsymbol{k}\cdot\boldsymbol{k} = 0 \ (\textit{wave vector is null}); \qquad (22.30)$$

$$0\left(\frac{1}{\varepsilon}\right): \qquad \underbrace{k^\beta k_\beta b^\alpha}_{[=\,0]} - 2i\left(k^\beta a^\alpha{}_{;\beta} + \frac{1}{2}k^\beta{}_{;\beta}a^\alpha\right) = 0$$

$$\implies \boldsymbol{\nabla}_{\!\boldsymbol{k}}\boldsymbol{a} = -\frac{1}{2}(\boldsymbol{\nabla}\cdot\boldsymbol{k})\boldsymbol{a} \ (\textit{propagation equation for vector amplitude}). \quad (22.31)$$

These equations (22.30, 22.31) together with equation (22.28) are the basis from which all subsequent results will follow. As a first consequence, one can obtain the geodesic law from equation (22.30). Form the gradient of $\boldsymbol{k}\cdot\boldsymbol{k} = 0$,

$$0 = (k^\beta k_\beta)_{;\alpha} = 2k^\beta k_{\beta;\alpha},$$

and use the fact that $k_\beta \equiv \theta_{,\beta}$ is the gradient of a *scalar* to interchange indices, $\theta_{;\beta\alpha} = \theta_{;\alpha\beta}$ or

$$0 = k^\beta k_{\beta;\alpha} = k^\beta k_{\alpha;\beta}.$$

<div style="float:left">The main laws of geometric optics:</div>

The result is

$$\boldsymbol{\nabla}_{\!\boldsymbol{k}}\boldsymbol{k} = 0 \ (\textit{propagation equation for wave vector}). \qquad (22.32)$$

Notice that this is the geodesic equation! Combined with equation (22.30), it is the statement, derived from Maxwell's equations in curved spacetime, that *light rays are null geodesics,* the first main result of geometric optics.

<div style="float:left">(1) Light rays are null geodesics</div>

Turn now from the propagation vector $\boldsymbol{k} = \boldsymbol{\nabla}\theta$ to the wave amplitude $\boldsymbol{a} = a\boldsymbol{f}$, and obtain separate equations for the magnitude a and polarization \boldsymbol{f}. Use equation (22.31) to compute

$$2a\,\partial_{\boldsymbol{k}}a = 2a\,\boldsymbol{\nabla}_{\!\boldsymbol{k}}a = \boldsymbol{\nabla}_{\!\boldsymbol{k}}a^2 = \boldsymbol{\nabla}_{\!\boldsymbol{k}}(\boldsymbol{a}\cdot\bar{\boldsymbol{a}}) = \bar{\boldsymbol{a}}\cdot\boldsymbol{\nabla}_{\!\boldsymbol{k}}\boldsymbol{a} + \boldsymbol{a}\cdot\boldsymbol{\nabla}_{\!\boldsymbol{k}}\bar{\boldsymbol{a}}$$

$$= -\frac{1}{2}(\boldsymbol{\nabla}\cdot\boldsymbol{k})(\bar{\boldsymbol{a}}\cdot\boldsymbol{a} + \boldsymbol{a}\cdot\bar{\boldsymbol{a}}) = -a^2\,\boldsymbol{\nabla}\cdot\boldsymbol{k};$$

so

$$\partial_{\boldsymbol{k}}a = -\frac{1}{2}(\boldsymbol{\nabla}\cdot\boldsymbol{k})a \ (\textit{propagation equation for scalar amplitude}). \quad (22.33)$$

Next write $\boldsymbol{a} = a\boldsymbol{f}$ in equation (22.31) to obtain

$$0 = \boldsymbol{\nabla}_{\!\boldsymbol{k}}(a\boldsymbol{f}) + \frac{1}{2}(\boldsymbol{\nabla}\cdot\boldsymbol{k})a\boldsymbol{f} = a\,\boldsymbol{\nabla}_{\!\boldsymbol{k}}\boldsymbol{f} + \boldsymbol{f}\left[\boldsymbol{\nabla}_{\!\boldsymbol{k}}a + \frac{1}{2}(\boldsymbol{\nabla}\cdot\boldsymbol{k})a\right] = a\,\boldsymbol{\nabla}_{\!\boldsymbol{k}}\boldsymbol{f}$$

or

$$\boldsymbol{\nabla}_{\boldsymbol{k}} \boldsymbol{f} = 0 \quad (\textit{propagation equation for polarization vector}). \qquad (22.34)$$

This together with equation (22.28′), constitutes the second main result of geometric optics, that *the polarization vector is perpendicular to the rays and is parallel-propagated along the rays*. It is now possible to see that these results, derived from equations (22.30) and (22.31) are consistent with the gauge condition (22.28). The vectors \boldsymbol{k} and \boldsymbol{f}, specified at one point, are fixed along the entire ray by their propagation equations. But because both propagation equations are parallel-transport laws, the conditions $\boldsymbol{k} \cdot \boldsymbol{k} = 0$, $\boldsymbol{f} \cdot \bar{\boldsymbol{f}} = 1$, and $\boldsymbol{k} \cdot \boldsymbol{f} = 0$, once imposed on the vectors at one point, will be satisfied along the entire ray.

The equation (22.33) for the scalar amplitude can be reformulated as a conservation law. Since $\partial_{\boldsymbol{k}} \equiv (\boldsymbol{k} \cdot \boldsymbol{\nabla})$, one rewrites the equation as $(\boldsymbol{k} \cdot \boldsymbol{\nabla})a^2 + a^2 \boldsymbol{\nabla} \cdot \boldsymbol{k} = 0$, or

$$\boldsymbol{\nabla} \cdot (a^2 \boldsymbol{k}) = 0. \qquad (22.35)$$

Consequently the vector $a^2 \boldsymbol{k}$ is a "conserved current," and the integral $\int a^2 k^\mu \, d^3\Sigma_\mu$ has a fixed, unchanging value for each 3-volume cutting a given tube formed of light rays. (The tube must be so formed of rays that an integral of $a^2\boldsymbol{k}$ over the walls of the tube will give zero.) What is conserved? To remain purely classical, one could say it is the "number of light rays" and call $a^2 k^0$ the "density of light rays" on an $x^0 = $ constant hypersurface. But the proper correspondence and more concrete physical interpretation make one prefer to call equation (22.35) *the law of conservation of photon number*. It is the third main result of geometric optics. Photon number, of course, is not always conserved; it is an adiabatic invariant, a quantity that is not changed by influences (e.g., spacetime curvature, $\sim 1/\mathscr{R}^2$) which change slowly ($\mathscr{R} \gg \lambda$) compared to the photon frequency.

Box 22.4 summarizes the above equations of geometric optics, along with others derived in the exercises.

<div style="text-align: right">(2) polarization vector is perpendicular to ray and is parallel propagated along ray</div>

<div style="text-align: right">(3) conservation of "photon number"</div>

Exercise 22.11. ELECTROMAGNETIC FIELD AND STRESS ENERGY
Derive the equations given in part D of Box 22.4 for \boldsymbol{F}, \boldsymbol{E}, \boldsymbol{B}, and \boldsymbol{T}.

Exercise 22.12. POLARIZATION
At an event \mathscr{P}_0 through which geometric-optics waves are passing, introduce a local Lorentz frame with z-axis along the direction of propagation. Then $\boldsymbol{k} = \omega(\boldsymbol{e}_0 + \boldsymbol{e}_z)$. Since the polarization vector is orthogonal to \boldsymbol{k}, it is $\boldsymbol{f} = f^0(\boldsymbol{e}_0 + \boldsymbol{e}_z) + f^1\boldsymbol{e}_x + f^2\boldsymbol{e}_y$; and since $\boldsymbol{f} \cdot \bar{\boldsymbol{f}} = 1$, it has $|f^1|^2 + |f^2|^2 = 1$.

(a) Show that the component f^0 of the polarization vector has no influence on the electric and magnetic fields measured in the given frame; i.e., show that one can add a multiple of \boldsymbol{k} to \boldsymbol{f} without affecting any physical measurements.

<div style="text-align: right">**EXERCISES**</div>

(continued on page 581)

Box 22.4 GEOMETRIC OPTICS IN CURVED SPACETIME
(Summary of Results Derived in Text and Exercises)

A. Geometric Optics Assumption

Electromagnetic waves propagating in a source-free region of spacetime are locally plane-fronted and monochromatic (reduced wavelength $\lambda \ll$ scale \mathcal{L} over which amplitude, wavelength, or polarization vary; and $\lambda \ll \mathcal{R} =$ mean radius of curvature of spacetime).

B. Rays, Phase, and Wave Vector (see Box 22.3)

Everything (amplitude, polarization, energy, etc.) is transported along *rays;* and the quantities on one ray do not influence the quantities on any other ray.

The rays are null geodesics of curved spacetime, with tangent vectors ("*wave vectors*")\boldsymbol{k}:

$$\boldsymbol{\nabla}_{\boldsymbol{k}}\boldsymbol{k} = 0.$$

The rays both lie in and are perpendicular to surfaces of constant phase, $\theta =$ const.; and their tangent vectors are the gradient of θ:

$$\boldsymbol{k} = \boldsymbol{\nabla}\theta.$$

In a local Lorentz frame, k^0 is the "angular frequency" and $k^0/2\pi$ is the ordinary frequency of the waves, and

$$\boldsymbol{n} = \boldsymbol{k}/k^0$$

is a unit 3-vector pointing along their direction of propagation.

C. Amplitude and Polarization Vector

The waves are characterized by a real amplitude a and a complex polarization vector \boldsymbol{f} of unit length, $\boldsymbol{f} \cdot \bar{\boldsymbol{f}} = 1$. (Of the fundamental quantities θ, \boldsymbol{k}, a, \boldsymbol{f}, all are real except \boldsymbol{f}. See exercise 22.12 for deeper understanding of \boldsymbol{f}.)

The polarization vector is everywhere orthogonal to the rays, $\boldsymbol{k} \cdot \boldsymbol{f} = 0$; and is parallel-transported along them, $\boldsymbol{\nabla}_{\boldsymbol{k}}\boldsymbol{f} = 0$.

The propagation law for the amplitude is

$$\partial_{\boldsymbol{k}} a = -\frac{1}{2}(\boldsymbol{\nabla} \cdot \boldsymbol{k})a.$$

This propagation law is equivalent to a *law of conservation of photons* (classically: of rays); $a^2\boldsymbol{k}$ is the "conserved current" satisfying $\boldsymbol{\nabla}\cdot(a^2\boldsymbol{k})=0$; and $(8\pi\hbar)^{-1}\int a^2 k^0\sqrt{|g|}\,d^3x$ is the number of photons (rays) in the 3-volume of integration on any $x^0=\text{constant}$ hypersurface, and is constant as this volume is carried along the rays.

The propagation law holds separately on each hypersurface of constant phase. There it can be interpreted as conservation of a $a^2\mathcal{C}$, where \mathcal{C} is a two-dimensional cross-sectional area of a pulse of photons or rays. See exercise 22.13.

D. Vector Potential, Electromagnetic Field, and Stress-Energy-Momentum

At any event the vector potential in Lorentz gauge is

$$\boldsymbol{A}=\Re\{ae^{i\theta}\boldsymbol{f}\},$$

where \Re denotes the real part.

The electromagnetic field tensor is orthogonal to the rays, $\boldsymbol{F}\cdot\boldsymbol{k}=0$, and is given by

$$\boldsymbol{F}=\Re\{iae^{i\theta}\boldsymbol{k}\wedge\boldsymbol{f}\}.$$

The corresponding electric and magnetic fields in any local Lorentz frame are

$$\boldsymbol{E}=\Re\{iak^0 e^{i\theta}(\text{projection of }\boldsymbol{f}\text{ perpendicular to }\boldsymbol{k})\},$$

$$\boldsymbol{B}=\boldsymbol{n}\times\boldsymbol{E},\text{ where }\boldsymbol{n}\equiv\boldsymbol{k}/k^0.$$

The stress-energy tensor, averaged over a wavelength, is

$$\boldsymbol{T}=(1/8\pi)a^2\boldsymbol{k}\otimes\boldsymbol{k},$$

corresponding to an energy density in a local Lorentz frame of

$$T^{00}=(1/8\pi)(ak^0)^2$$

and an energy flux of

$$T^{0j}=T^{00}n^j,$$

so that energy flows along the rays (in $\boldsymbol{n}=\boldsymbol{k}/k^0$ direction) with the speed of light. This is identical with the stress-energy tensor that would be produced by a beam of photons with 4-momenta $\boldsymbol{p}=\hbar\boldsymbol{k}$.

Conservation of energy-momentum $\boldsymbol{\nabla}\cdot\boldsymbol{T}=0$ follows from the ray conservation law $\boldsymbol{\nabla}\cdot(a^2\boldsymbol{k})=0$ and the geodesic law $\boldsymbol{\nabla}_{\boldsymbol{k}}\boldsymbol{k}\equiv(\boldsymbol{k}\cdot\boldsymbol{\nabla})\boldsymbol{k}=0$:

$$8\pi\,\boldsymbol{\nabla}\cdot\boldsymbol{T}=\boldsymbol{\nabla}\cdot(a^2\boldsymbol{k}\otimes\boldsymbol{k})=[\boldsymbol{\nabla}\cdot(a^2\boldsymbol{k})]\boldsymbol{k}+a^2(\boldsymbol{k}\cdot\boldsymbol{\nabla})\boldsymbol{k}=0.$$

Box 22.4 (*continued*)

The adiabatic (geometric optics) invariant "ray number" a^2k^0 or "photon number" $(8\pi\hbar)^{-1}a^2k^0$ in a unit volume is proportional to the energy, $(8\pi)^{-1}a^2(k^0)^2$, divided by the frequency, k^0—corresponding exactly to the harmonic oscillator adiabatic invariant E/ω [Einstein (1912), Ehrenfest (1916), Landau and Lifshitz (1960)].

E. Photon Reinterpretation of Geometric Optics

The laws of geometric optics can be reinterpreted as follows. This reinterpretation becomes a foundation of the standard quantum theory of the electromagnetic field (see, e.g., Chapters 1 and 13 of Baym (1969)]; and the classical limit of that quantum theory is standard Maxwell electrodynamics.

Photons are particles of zero rest mass that move along null geodesics of spacetime (the null rays).

The 4-momentum of a photon is related to the tangent vector of the null ray (wave vector) by $\boldsymbol{p} = \hbar\boldsymbol{k}$. A renormalization of the affine parameter,

$$\text{(new parameter)} = (1/\hbar) \times \text{(old parameter)},$$

makes \boldsymbol{p} the tangent vector to the ray.

Each photon possesses a polarization vector, \boldsymbol{f}, which is orthogonal to its 4-momentum ($\boldsymbol{p} \cdot \boldsymbol{f} = 0$), and which it parallel-transports along its geodesic world line ($\boldsymbol{\nabla}_{\boldsymbol{p}}\boldsymbol{f} = 0$).

A swarm of photons, all with nearly the same 4-momentum \boldsymbol{p} and polarization vector \boldsymbol{f} (as compared by parallel transport), make up a classical electromagnetic wave. The scalar amplitude a of the wave is determined by equating the stress-energy tensor of the wave

$$\boldsymbol{T} = \frac{1}{8\pi}a^2\boldsymbol{k} \otimes \boldsymbol{k} = \frac{1}{8\pi}\left(\frac{a}{\hbar}\right)^2\boldsymbol{p} \otimes \boldsymbol{p}$$

to the stress-energy tensor of a swarm of photons with number-flux vector \boldsymbol{S},

$$\boldsymbol{T} = \boldsymbol{p} \otimes \boldsymbol{S}$$

[see equation (5.18)]. The result:

$$\boldsymbol{S} = \frac{1}{8\pi}\left(\frac{a}{\hbar}\right)^2\boldsymbol{p} = \frac{1}{8\pi\hbar}a^2\boldsymbol{k}$$

or, in any local Lorentz frame,

$$a = (8\pi\hbar^2 S^0/p^0)^{1/2} = (8\pi)^{1/2}\hbar\left(\frac{\text{number density of photons}}{\text{energy of one photon}}\right)^{1/2}.$$

(b) Show that the following polarization vectors correspond to the types of polarization listed:

$f = e_x$, linear polarization in x direction;

$f = e_y$, linear polarization in y direction;

$f = \dfrac{1}{\sqrt{2}}(e_x + ie_y)$, righthand circular polarization;

$f = \dfrac{1}{\sqrt{2}}(e_x - ie_y)$, lefthand circular polarization;

$f = \alpha e_x + i(1 - \alpha^2)^{1/2}e_y$, righthand elliptical polarization.

(c) Show that the type of polarization (linear; circular; elliptical with given eccentricity of ellipse) is the same as viewed in any local Lorentz frame at any event along a given ray. [*Hint:* Use pictures and abstract calculations rather than Lorentz transformations and component calculations.]

Exercise 22.13. THE AREA OF A BUNDLE OF RAYS

Write equation (22.31) in a coordinate system in which one of the coordinates is chosen to be $x^0 = \theta$, the phase (a retarded time coordinate).

(a) Show that $g^{00} = 0$ and that no derivatives $\partial/\partial\theta$ appear in equation (22.33); so propagation of a can be described within a single $\theta = $ constant hypersurface.

(b) Perform the following construction (see Figure 22.1). Pick a ray \mathcal{C}_0 along which a is to be propagated. Pick a bundle of rays, with two-dimensional cross section, that (i) all lie in the same constant-phase surface as \mathcal{C}_0, and (ii) surround \mathcal{C}_0. (The surface is three-di-

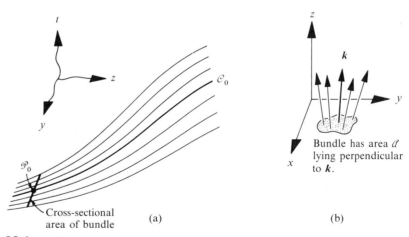

Figure 22.1.
Geometric optics for a bundle of rays with two-dimensional cross section, all lying in a surface of constant phase, $\theta = $ const. Sketch (a) shows the bundle, surrounding a central ray \mathcal{C}_0, in a spacetime diagram with one spatial dimension suppressed. Sketch (b) shows the bundle as viewed on a slice of simultaneity in a local Lorentz frame at the event \mathcal{P}_0. Slicing the bundle turns each ray into a "photon"; so the bundle becomes a two-dimensional surface filled with photons. The area \mathcal{A} of this photon-filled surface obeys the following laws (see exercises 22.13 and 22.14): (1) \mathcal{A} is independent of the choice of Lorentz frame; it depends only on location \mathcal{P}_0 along the ray \mathcal{C}_0. (2) The amplitude a of the waves satisfies

$$\mathcal{A}a^2 = \text{constant all along the ray } \mathcal{C}_0$$

("*conservation of photon flux*"). (3) \mathcal{A} obeys the "*propagation equation*" (22.36).

mensional, so any bundle filling it has a two-dimensional cross section.) At any event \mathcal{P}_0, in any local Lorentz frame there, on a "slice of simultaneity" $x^0 = $ constant, measure the cross-sectional area \mathcal{A} of the bundle. (Note: the area being measured is perpendicular to \boldsymbol{k} in the three-dimensional Euclidean sense; it can be thought of as the region occupied momentarily by a group of photons propagating along, side by side, in the \boldsymbol{k} direction.) Show that the area \mathcal{A} is the same, at a given event \mathcal{P}_0, *regardless of what Lorentz frame is used to measure it; but the area changes from point to point along the ray \mathcal{C}_0 as a result of the rays' divergence away from each other or convergence toward each other:*

$$\partial_{\boldsymbol{k}}\mathcal{A} = (\boldsymbol{\nabla} \cdot \boldsymbol{k})\mathcal{A}. \tag{22.36}$$

Then show that $\mathcal{A}a^2$ is a constant everywhere along the ray \mathcal{C}_0 ("conservation of photon flux"). [Hints: (i) Any vector $\boldsymbol{\xi}$ connecting adjacent rays in the bundle is perpendicular to \boldsymbol{k}, because $\boldsymbol{\xi}$ lies in a surface of constant θ and $\boldsymbol{k} \cdot \boldsymbol{\xi} = \langle \widetilde{\boldsymbol{k}}, \boldsymbol{\xi} \rangle = \langle \boldsymbol{d}\theta, \boldsymbol{\xi} \rangle = $ (change in θ along $\boldsymbol{\xi}$) $= 0$. (ii) Consider, for simplicity, a bundle with rectangular cross section as seen in a specific local Lorentz frame at a specific event \mathcal{P}_0 [edge vectors \boldsymbol{v} and \boldsymbol{w} with $\boldsymbol{v} \cdot \boldsymbol{w} = 0$ (edges perpendicular) and $\boldsymbol{v} \cdot \boldsymbol{e}_0 = \boldsymbol{w} \cdot \boldsymbol{e}_0 = 0$ (edges in surface of constant time) and $\boldsymbol{v} \cdot \boldsymbol{k} = \boldsymbol{w} \cdot \boldsymbol{k} = 0$ (since edge vectors connect adjacent rays of the bundle)]. Show pictorially that in any other Lorentz frame at \mathcal{P}_0, the edge vectors are $\boldsymbol{v}' = \boldsymbol{v} + \alpha\boldsymbol{k}$ and $\boldsymbol{w}' = \boldsymbol{w} + \beta\boldsymbol{k}$ for some α and β. Conclude that in all Lorentz frames at \mathcal{P}_0 the cross section has identical shape and identical area, and is spatially perpendicular to the direction of propagation ($\boldsymbol{k} \cdot \boldsymbol{v} = \boldsymbol{k} \cdot \boldsymbol{w} = 0$). (iii) By a calculation in a local Lorentz frame show that $\partial_{\boldsymbol{k}}\mathcal{A} = (\boldsymbol{\nabla} \cdot \boldsymbol{k})\mathcal{A}$. (iv) Conclude from $\partial_{\boldsymbol{k}}a = -\frac{1}{2}(\boldsymbol{\nabla} \cdot \boldsymbol{k})a$ that $\partial_{\boldsymbol{k}}(\mathcal{A}a^2) = 0$.]

Exercise 22.14. FOCUSING THEOREM

The cross-sectional area \mathcal{A} of a bundle of rays all lying in the same surface of constant phase changes along the central ray of the bundle at the rate (22.36) (see Figure 22.1).

(a) Derive the following equation ("*focusing equation*") for the second derivative of $\mathcal{A}^{1/2}$:

$$\frac{d^2\mathcal{A}^{1/2}}{d\lambda^2} = -\left(|\sigma|^2 + \frac{1}{2}R_{\alpha\beta}k^\alpha k^\beta\right)\mathcal{A}^{1/2}, \tag{22.37}$$

where λ is affine parameter along the central ray ($\boldsymbol{k} = d/d\lambda$), and the "magnitude of the shear of the rays", $|\sigma|$, is defined by the equation

$$|\sigma|^2 \equiv \frac{1}{2}k_{\alpha;\beta}k^{\alpha;\beta} - \frac{1}{4}(k^\mu{}_{;\mu})^2. \tag{22.38}$$

[Hint: This is a vigorous exercise in index manipulations. The key equations needed in the manipulations are $\mathcal{A}_{,\alpha}k^\alpha = (k^\alpha{}_{;\alpha})\mathcal{A}$ [equation (22.36)]; $k^\alpha{}_{;\beta}k^\beta = 0$ [geodesic equation (22.32) for rays]; $k_{\alpha;\beta} = k_{\beta;\alpha}$ [which follows from $k_\alpha \equiv \theta_{,\alpha}$]; and the rule (16.6c) for interchanging covariant derivatives of a vector.]

(b) Show that, in a local Lorentz frame where $\boldsymbol{k} = \omega(\boldsymbol{e}_t + \boldsymbol{e}_z)$ at the origin,

$$|\sigma|^2 = \frac{1}{4}(k_{x,x} - k_{y,y})^2 + (k_{x,y})^2. \tag{22.39}$$

Thus, $|\sigma|^2$ is nonnegative, which justifies the use of the absolute value sign.

(c) *Discussion:* The quantity $|\sigma|$ is called the *shear* of the bundle of rays because it measures the extent to which neighboring rays are sliding past each other [see, e.g., Sachs (1964)]. Hence, the focusing equation (22.37) says that shear focuses a bundle of rays (makes $d^2\mathcal{A}^{1/2}/d\lambda^2 < 0$); and spacetime curvature also focuses it if $R_{\alpha\beta}k^\alpha k^\beta > 0$, but defocuses it if $R_{\alpha\beta}k^\alpha k^\beta < 0$. (When a bundle of toothpicks, originally circular in cross section, is squeezed into an elliptic cross section, it is sheared.)

(d) Assume that the energy density $T_{\hat{0}\hat{0}}$, as measured by any observer anywhere in space-time, is nonnegative. By combining the focusing equation (22.37) with the Einstein field equation, conclude that

$$\frac{d^2 \mathcal{A}^{1/2}}{d\lambda^2} \leq 0 \quad \left(\begin{array}{l}\text{for any bundle of rays, all in the same} \\ \text{surface of constant phase, anywhere in} \\ \text{spacetime}\end{array}\right) \qquad (22.40)$$

(*focusing theorem*). This theorem plays a crucial role in black-hole physics (§34.5) and in the theory of singularities (§34.6).

§22.6. KINETIC THEORY IN CURVED SPACETIME*

The stars in a galaxy wander through spacetime, each on its own geodesic world line, each helping to produce the spacetime curvature felt by all the others. Photons, left over from the hot phases of the big bang, bathe the Earth, bringing with themselves data on the homogeneity and isotropy of the universe. Theoretical analyses of these and many other problems are unmanageable, if they attempt to keep track of the motion of every single star or photon. But a statistical description gives accurate results and is powerful. Moreover, for most problems in astrophysics and cosmology, the simplest of statistical descriptions—one ignoring collisions—is adequate. Usually collisions are unimportant for the large-scale behavior of a system (e.g.; a galaxy), or they are so important that a fluid description is possible (e.g., in a stellar interior).

Consider, then, a swarm of particles (stars, or photons, or black holes, or . . .) that move through spacetime on geodesic world lines, without colliding. Assume, for simplicity, that the particles all have the same rest mass. Then all information of a statistical nature about the particles can be incorporated into a single function, the "*distribution function*" or "*number density in phase space*", \mathcal{N}.

Define \mathcal{N} in terms of measurements made by a specific local Lorentz observer at a specific event \mathcal{P}_0 in curved spacetime. Give the observer a box with 3-volume \mathcal{V}_x (and with imaginary walls). Ask the observer to count how many particles, N, are inside the box *and* have local-Lorentz momentum components p^j in the range

Volume in phase space for a group of identical particles

$$P^j - \frac{1}{2}\Delta p^j < p^j < P^j + \frac{1}{2}\Delta p^j.$$

(He can ignore the particle energies p^0; since all particles have the same rest mass m, energy

$$p^0 = (m^2 + \boldsymbol{p}^2)^{1/2}$$

*For more detailed and sophisticated treatments of this topic, see, e.g., Tauber and Weinberg (1961), and Lindquist (1966), Marle (1969), Ehlers (1971), Stewart (1971), Israel (1972), and references cited therein. Ehlers (1971) is a particularly good introductory review article.

is fixed uniquely by momentum.) The volume in momentum space occupied by the N particles is $\mathcal{V}_p = \Delta p^x \, \Delta p^y \, \Delta p^z$; and *the volume in phase space is*

$$\mathcal{V} \equiv \mathcal{V}_x \mathcal{V}_p. \tag{22.41}$$

Other observers at \mathcal{P}_0, moving relative to the first, will disagree on how much spatial volume \mathcal{V}_x and how much momentum volume \mathcal{V}_p these same N particles occupy:

$$\mathcal{V}_x \text{ and } \mathcal{V}_p \text{ depend on the choice of Lorentz frame.} \tag{22.42}$$

However, all observers will agree on the value of the product $\mathcal{V} \equiv \mathcal{V}_x \mathcal{V}_p$ ("volume in phase space"):

> The phase-space volume \mathcal{V} occupied by a given set of N identical particles at a given event in spacetime is independent of the local Lorentz frame in which it is measured. (22.43)

(See Box 22.5 for proof.) Moreover, as the same N particles move through spacetime along their geodesic world lines (and through momentum space), the volume \mathcal{V} they span in phase space remains constant:

> The \mathcal{V} occupied by a given swarm of N particles is independent of location along the world line of the swarm ("*Liouville's theorem in curved spacetime*"). (22.44)

(See Box 22.6 for proof.)

More convenient for applications than the volume \mathcal{V} in phase space occupied by a given set of N particles is the "*number density in phase space*" ("*distribution function*") in the neighborhood of one of these particles:

$$\mathcal{N} \equiv N/\mathcal{V}. \tag{22.45}$$

On what does this number density depend? It depends on the location in spacetime, \mathcal{P}, at which the measurements are made. It also depends on the 4-momentum \boldsymbol{p} of the particle in whose neighborhood the measurements are made. But because the particles all have the same rest mass, \boldsymbol{p} cannot take on any and every value in the tangent space at \mathcal{P}. Rather, \boldsymbol{p} is confined to the "forward mass hyperboloid" at \mathcal{P}:

$$\boldsymbol{p}^2 = m^2; \qquad \boldsymbol{p} \text{ lies inside future light cone.}$$

Thus,

$$\mathcal{N} = \mathcal{N} \left[\begin{pmatrix} \text{location, } \mathcal{P}, \\ \text{in spacetime} \end{pmatrix}, \begin{pmatrix} \text{4-momentum } \boldsymbol{p}, \text{ which must lie} \\ \text{on the forward mass hyperboloid} \\ \text{of the tangent space at } \mathcal{P} \end{pmatrix} \right]. \tag{22.46}$$

Pick some one particle in the swarm, with geodesic world line $\mathcal{P}(\lambda)$ [$\lambda =$ (affine parameter) = (proper time, if particle has finite rest mass)], and with 4-momentum

Box 22.5 VOLUME IN PHASE SPACE

A. For Swarm of Identical Particles with Nonzero Rest Mass

Pick an event \mathscr{P}_0, through which passes a particle named "John" with a 4-momentum named "**P**". In John's local Lorentz rest frame at \mathscr{P}_0 ("barred frame", \bar{S}), select a small 3-volume, $\mathcal{V}_{\bar{x}} \equiv \Delta\bar{x}\,\Delta\bar{y}\,\Delta\bar{z}$, containing him. Also select a small "3-volume in momentum space," $\mathcal{V}_{\bar{p}} \equiv \Delta p^{\bar{x}}\,\Delta p^{\bar{y}}\,\Delta p^{\bar{z}}$ centered on John's momentum, which is $P^{\bar{x}} = P^{\bar{y}} = P^{\bar{z}} = 0$. Focus attention on all particles whose world lines pass through $\mathcal{V}_{\bar{x}}$ and which have momenta $p^{\bar{j}}$ in the range $\mathcal{V}_{\bar{p}}$ surrounding $P^{\bar{j}} = 0$.

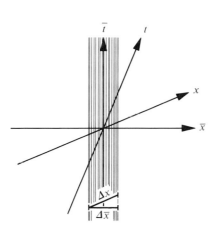

Examine this bundle in another local Lorentz frame ("unbarred frame", S) at \mathscr{P}_0, which moves with speed β relative to the rest frame. Orient axes so the relative motion of the frames is in the x and \bar{x} directions. Then the space volume \mathcal{V}_x occupied in the new frame has $\Delta y = \Delta\bar{y}$, $\Delta z = \Delta\bar{z}$ (no effect of motion on transverse directions), and $\Delta x = (1 - \beta^2)^{1/2}\,\Delta\bar{x}$ (Lorentz contraction in longitudinal direction). Hence $\mathcal{V}_x = (1 - \beta^2)^{1/2}\mathcal{V}_{\bar{x}}$ ("transformation law for space volumes") or, equivalently [since $P^0 = m/(1 - \beta^2)^{1/2}$]:

$$P^0\mathcal{V}_x = m\mathcal{V}_{\bar{x}} = \begin{pmatrix} \text{constant, independent} \\ \text{of Lorentz frame} \end{pmatrix}.$$

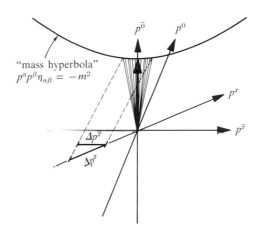

A momentum-space diagram, analogous to the spacetime diagram, depicts the momentum spread for particles in the bundle, and shows that $\Delta p^x = \Delta p^{\bar{x}}/(1 - \beta^2)^{1/2}$. The Lorentz transformation from \bar{S} to S leaves transverse components of momenta unaffected; so $\Delta p^y = \Delta p^{\bar{y}}$, $\Delta p^z = \Delta p^{\bar{z}}$. Hence $\mathcal{V}_p = \mathcal{V}_{\bar{p}}/(1 - \beta^2)^{1/2}$ ("transformation law for momentum volumes"); or, equivalently

$$\frac{\mathcal{V}_p}{P^0} = \frac{\mathcal{V}_{\bar{p}}}{m} = \begin{pmatrix} \text{constant, independent} \\ \text{of Lorentz frame} \end{pmatrix}.$$

Although the spatial 3-volumes \mathcal{V}_x and $\mathcal{V}_{\bar{x}}$ differ from one frame to another, and the momentum 3-volumes \mathcal{V}_p and $\mathcal{V}_{\bar{p}}$ differ, the volume in six-dimensional phase space is Lorentz-invariant:

$$\mathcal{V} \equiv \mathcal{V}_{\bar{x}}\mathcal{V}_{\bar{p}} = \mathcal{V}_x\mathcal{V}_p.$$

It is a frame-independent, geometric object!

B. For Swarm of Identical Particles with Zero Rest Mass

Examine a sequence of systems, each with particles of smaller rest mass and of higher velocity relative to a laboratory. For every bundle of particles in each system, $P^0\mathcal{V}_x$, \mathcal{V}_p/P^0, and $\mathcal{V}_x\mathcal{V}_p$ are Lorentz-invariant. Hence, in the limit as $m \longrightarrow 0$, as $\beta \longrightarrow 1$, and as $P^0 = m/(1 - \beta^2)^{1/2} \longrightarrow$ finite value (particles of zero rest mass moving with speed of light), $P^0\mathcal{V}_x$ and \mathcal{V}_p/P^0 and $\mathcal{V}_x\mathcal{V}_p$ are still Lorentz-invariant, geometric quantities.

Box 22.6 CONSERVATION OF VOLUME IN PHASE SPACE

Examine a very small bundle of identical particles that move through curved spacetime on neighboring geodesics. Measure the bundle's volume in phase space, \mathcal{V} ($\mathcal{V} = \mathcal{V}_x \mathcal{V}_p$ in any local Lorentz frame), as a function of affine parameter λ along the central geodesic of the bundle. The following calculation shows that

$$d\mathcal{V}/d\lambda = 0 \qquad \left(\begin{array}{l}\text{"Liouville theorem in}\\\text{curved spacetime"}\end{array}\right).$$

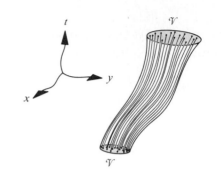

Proof for particles of finite rest mass: Examine particle motion during time interval $\delta\tau$, using local Lorentz rest frame of central particle. All velocities are small in this frame, so

$$p^{\bar{j}} = m\,dx^{\bar{j}}/d\bar{t}.$$

Hence (see pictures) the spreads in momentum and position conserve $\Delta\bar{x}\,\Delta p^{\bar{x}}$, $\Delta\bar{y}\,\Delta p^{\bar{y}}$, and $\Delta\bar{z}\,\Delta p^{\bar{z}}$; i.e.,

$$\frac{d\mathcal{V}}{d\tau} = \frac{\delta(\Delta\bar{x}\,\Delta\bar{y}\,\Delta\bar{z}\,\Delta p^{\bar{x}}\,\Delta p^{\bar{y}}\,\Delta p^{\bar{z}})}{\delta\bar{t}} = 0.$$

But $\tau = a\lambda + b$ for some arbitrary constants a and b; so $d\mathcal{V}/d\lambda = 0$.

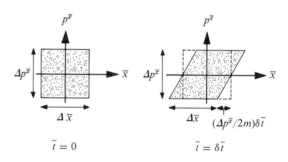

Each particle moves with speed $d\bar{x}/d\bar{t}$ proportional to height in diagram

$$d\bar{x}/d\bar{t} = p^{\bar{x}}/m,$$

and conserves its momentum, $dp^{\bar{x}}/d\bar{t} = 0$. Hence the region occupied by particles deforms, but maintains its area. Same is true for $(y - p^y)$ and $(z - p^z)$.

Proof for particles of zero rest mass. Examine particle motion in local Lorentz frame where central particle has $\boldsymbol{P} = P^0(\boldsymbol{e}_0 + \boldsymbol{e}_x)$. In this frame, all particles have $p^y \ll p^0$, $p^z \ll p^0$, $p^x = p^0 + O([p^y]^2/P^0) \approx P^0$. Since $p^\alpha = dx^\alpha/d\lambda$ for appropriate normalization of affine parameters (see Box 22.4), one can write $dx^j/dt = p^j/p^0$; i.e.,

$$\frac{dx}{dt} = 1 + O([p^y/P^0]^2 + [p^z/P^0]^2)$$

$$\approx 1,$$

$$\frac{dy}{dt} = \frac{p^y}{P^0}, \qquad \frac{dz}{dt} = \frac{p^z}{P^0}.$$

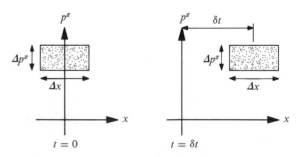

Each particle ("photon") moves with $dx/dt = 1$ and $dp^x/dt = 0$ in the local Lorentz frame. Area and shape of occupied region are preserved.

Hence (see pictures) $\Delta x\,\Delta p^x$, $\Delta y\,\Delta p^y$, and $\Delta z\,\Delta p^z$ are all conserved; and

$$\frac{d\mathcal{V}}{dt} = \frac{\delta(\Delta x\,\Delta y\,\Delta z\,\Delta p^x\,\Delta p^y\,\Delta p^z)}{\delta t} = 0.$$

But t and the affine parameter λ of central particle are related by $t = P^0\lambda$ [cf. equation (16.4)]; thus

$$d\mathcal{V}/d\lambda = 0.$$

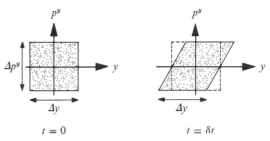

Particle ("photon") speeds are proportional to height in diagram

$$dy/dt = p^y/P^0,$$

and $dp^y/dt = 0$. Hence, occupied region deforms but maintains its area. Same is true of $z - p^z$.

$\boldsymbol{p}(\lambda)$. Examine the density in phase space in this particle's neighborhood at each point along its world line:

$$\mathcal{N} = \mathcal{N}[\mathcal{P}(\lambda), \boldsymbol{p}(\lambda)].$$

Calculate $\mathcal{N}(\lambda)$ as follows: (1) Pick an initial event $\mathcal{P}(0)$ on the world line, and a phase-space volume \mathcal{V} containing the particle. (2) Cover with red paint all the particles contained in \mathcal{V} at $\mathcal{P}(0)$. (3) Watch the red particles move through spacetime alongside the initial particle. (4) As they move, the phase-space region they occupy changes shape extensively; but its volume \mathcal{V} remains fixed (Liouville's theorem). Moreover, no particles can enter or leave that phase-space region (once in, always in; once out, always out; boundaries of phase-space region are attached to and move with the particles). (5) Hence, at any λ along the initial particle's world line, the particle is in a phase-space region of unchanged volume \mathcal{V}, unchanged number of particles N, and unchanged ratio $\mathcal{N} = N/\mathcal{V}$:

$$\frac{d\mathcal{N}[\mathcal{P}(\lambda), \boldsymbol{p}(\lambda)]}{d\lambda} = 0. \tag{22.47}$$

Collisionless Boltzmann equation (kinetic equation)

This equation for the conservation of \mathcal{N} along a particle's trajectory in phase space is called the "collisionless Boltzmann equation," or the "kinetic equation."

Photons provide an important application of the Boltzmann equation. But when discussing photons one usually does not think in terms of the number density in phase space. Rather, one speaks of the "specific intensity" I_ν of radiation at a given frequency ν, flowing in a given direction, \boldsymbol{n}, as measured in a specified local Lorentz frame:

$$I_\nu \equiv \frac{d(\text{energy})}{d(\text{time})\,d(\text{area})\,d(\text{frequency})\,d(\text{solid angle})}. \tag{22.48}$$

Distribution function for
photons expressed in terms
of specific intensity, I_ν

Invariance and conservation
of I_ν/ν^3

(See Figure 22.2). A simple calculation in the local Lorentz frame reveals that

$$\mathscr{N} = h^{-4}(I_\nu/\nu^3),\qquad(22.49)$$

where h is Planck's constant (see Figure 22.2). Thus, if two different observers at the same or different events in spacetime look at the same photon (and neighboring photons) as it passes them, they will see different frequencies ν ("doppler shift," "cosmological red shift," "gravitational redshift"), and different specific intensities I_ν; but they will obtain identical values for the ratio I_ν/ν^3. Thus I_ν/ν^3, like \mathscr{N}, is invariant from observer to observer and from event to event along a given photon's world line.

EXERCISES

Exercise 22.15. INVERSE SQUARE LAW FOR FLUX

The *specific flux* of radiation entering a telescope from a given source is defined by

$$F_\nu = \int I_\nu \, d\Omega,\qquad(22.50)$$

where integration is over the total solid angle (assumed $\ll 4\pi$) subtended by the source on the observer's sky. Use the Boltzmann equation (conservation of I_ν/ν^3) to show that $F_\nu \propto$ (distance from source)$^{-2}$ for observers who are all at rest relative to each other in flat spacetime.

Exercise 22.16. BRIGHTNESS OF THE SUN

Does the surface of the sun look any brighter to an astronaut standing on Mercury than to a student standing on Earth?

Exercise 22.17. BLACK BODY RADIATION

An "optically thick" source of black-body radiation (e.g., the surface of a star, or the hot matter filling the universe shortly after the big bang) emits photons isotropically with a specific intensity, as seen by an observer at rest near the source, given (Planck radiation law) by

$$I_\nu = \frac{2h\nu^3}{e^{h\nu/kT} - 1}.\qquad(22.51)$$

Here T is the temperature of the source. Show that any observer, in any local Lorentz frame, anywhere in the universe, who examines this radiation as it flows past him, will also see a black-body spectrum. Show, further, that if he calculates a temperature by measuring the specific intensity I_ν at any one frequency, and if he calculates a temperature from the shape of the spectrum, those temperatures will agree. (Radiation remains black body rather than being "diluted" into "grey-body.") Finally, show that the temperature he measures is redshifted by precisely the same factor as the frequency of any given photon is redshifted,

$$\frac{T_{\text{observed}}}{T_{\text{emitted}}} = \left(\frac{\nu_{\text{observed}}}{\nu_{\text{emitted}}}\right) \text{ for a given photon.}\qquad(22.52)$$

[Note that the redshifts can be "Doppler" in origin, "cosmological" in origin, "gravitational" in origin, or some inseparable mixture. All that matters is the fact that the parallel-transport law for a photon's 4-momentum, $\boldsymbol{\nabla}_{\boldsymbol{p}}\boldsymbol{p} = 0$, guarantees that the redshift $\nu_{\text{observed}}/\nu_{\text{emitted}}$ is independent of frequency emitted.]

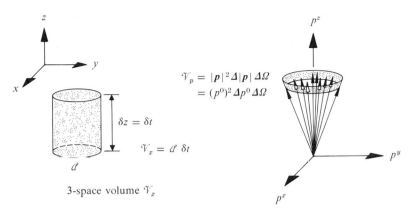

$$\mathcal{V}_p = |\boldsymbol{p}|^2 \Delta |\boldsymbol{p}| \Delta\Omega$$
$$= (p^0)^2 \Delta p^0 \Delta\Omega$$

$$\delta z = \delta t$$

$$\mathcal{V}_x = \mathcal{Q} \, \delta t$$

3-space volume \mathcal{V}_x

3-momentum volume, with direction
of momentum vectors reversed for
ease of visualization (telescope as
an emitter, not a receiver!)

Figure 22.2.
Number density in phase space for photons, interpreted in terms of the specific intensity I_ν. An astronomer has a telescope with filter that admits only photons arriving from within a small solid angle $\Delta\Omega$ about the z-direction, and having energies between p^0 and $p^0 + \Delta p^0$. The collecting area, \mathcal{Q}, of his telescope lies in the x, y-plane (perpendicular to the incoming photon beam). Let δN be the number of photons that cross the area \mathcal{Q} in a time interval δt. [All energies, areas, times, and lengths are measured in the orthonormal frame ("proper reference frame; §13.6) which the astronomer Fermi-Walker transports with himself along his (possibly accelerated) world line—or, equivalently, in a local Lorentz frame momentarily at rest with respect to the astronomer.] The δN photons, just before the time interval δt begins, lie in the cylinder of area \mathcal{Q} and height $\delta z = \delta t$ shown above. Their spatial 3-volume is thus $\mathcal{V}_x = \mathcal{Q} \, \delta t$. Their momentum 3-volume is $\mathcal{V}_p = (p^0)^2 \, \Delta p^0 \, \Delta\Omega$ (see drawing). Hence, their number density in phase space is

$$\mathcal{N} = \frac{\delta N}{\mathcal{V}_x \mathcal{V}_p} = \frac{\delta N}{\mathcal{Q} \, \delta t (p^0)^2 (\Delta p^0) \, \Delta\Omega} = \frac{\delta N}{h^3 \, \mathcal{Q} \, \delta t \, \nu^2 \, \Delta\nu \, \Delta\Omega}$$

where ν is the photon frequency measured by the telescope ($p^0 = h\nu$).

The specific intensity of the photons, I_ν (a standard concept in astronomy), is the energy per unit area per unit time per unit frequency per unit solid angle crossing a surface perpendicular to the beam: i.e.,

$$I_\nu = \frac{h\nu \, \delta N}{\mathcal{Q} \, \delta t \, \Delta\nu \, \Delta\Omega}.$$

Direct comparison reveals $\mathcal{N} = h^{-4}(I_\nu/\nu^3)$.

Thus, conservation of \mathcal{N} along a photon's world line implies conservation of I_ν/ν^3. This conservation law finds important applications in cosmology (e.g., Box 29.2 and Ex. 29.5) and in the gravitational lens effect (Refsdal 1964); see also exercises 22.15–22.17.

Exercise 22.18. STRESS-ENERGY TENSOR

(a) Show that the stress-energy tensor for a swarm of identical particles at an event \mathscr{P}_0 can be written as an integral over the mass hyperboloid of the momentum space at \mathscr{P}_0:

$$\boldsymbol{T} = \int (\mathcal{N}\boldsymbol{p} \otimes \boldsymbol{p})(d\mathcal{V}_p/p^0), \tag{22.53}$$

$$\frac{d\mathcal{V}_p}{p^0} \equiv \frac{dp^x \, dp^y \, dp^z}{p^0} \text{ in a local Lorentz frame.} \tag{22.54}$$

(Notice from Box 22.5 that $d\mathcal{V}_p/p^0$ is a Lorentz-invariant volume element for any segment of the mass hyperboloid.)

(b) Verify that the Boltzmann equation, $d\mathcal{N}/d\lambda = 0$, implies $\boldsymbol{\nabla} \cdot \boldsymbol{T} = 0$ for any swarm of identical particles. [*Hint:* Calculate $\boldsymbol{\nabla} \cdot \boldsymbol{T}$ in a local Lorentz frame, using the above expression for \boldsymbol{T}, and using the geodesic equation in the form $Dp^\mu/d\lambda = 0$.]

Exercise 22.19. KINETIC THEORY FOR NONIDENTICAL PARTICLES

For a swarm of particles with a wide distribution of rest masses, define

$$\mathcal{N} = \frac{\Delta N}{\mathcal{V}_x \mathcal{V}_p \, \Delta m}, \tag{22.55}$$

where \mathcal{V}_x and \mathcal{V}_p are spatial and momentum 3-volumes, and ΔN is the number of particles in the region $\mathcal{V}_x \mathcal{V}_p$ with rest masses between $m - \Delta m/2$ and $m + \Delta m/2$. Show the following.

(a) $\mathcal{V}_x \mathcal{V}_p \, \Delta m$ is independent of Lorentz frame and independent of location on the world tube of a bundle of particles.

(b) \mathcal{N} can be regarded as a function of location \mathcal{P} in spacetime and 4-momentum \boldsymbol{p} inside the future light cone of the tangent space at \mathcal{P}:

$$\mathcal{N} = \mathcal{N}(\mathcal{P}, \boldsymbol{p}). \tag{22.56}$$

(c) \mathcal{N} satisfies the collisionless Boltzmann equation (kinetic equation)

$$\frac{d\mathcal{N}[\mathcal{P}(\lambda), \boldsymbol{p}(\lambda)]}{d\lambda} = 0 \qquad \text{along geodesic trajectory of any particle.} \tag{22.57}$$

(d) \mathcal{N} can be rewritten in a local Lorentz frame as

$$\mathcal{N} = \frac{\Delta N}{[(p^0/m) \, \Delta x \, \Delta y \, \Delta z][\Delta p^0 \, \Delta p^x \, \Delta p^y \, \Delta p^z]}. \tag{22.58}$$

(e) The stress-energy tensor at an event \mathcal{P} can be written as an integral over the interior of the future light cone of momentum space

$$T^{\mu\nu} = \int (\mathcal{N}p^\mu p^\nu)m^{-1} \, dp^0 \, dp^1 \, dp^2 \, dp^3 \tag{22.59}$$

in a local Lorentz frame (Track-1 notation for integral; see Box 5.3);

$$\boldsymbol{T} = \int (\mathcal{N}\boldsymbol{p} \otimes \boldsymbol{p})m^{-1} {}^*1 \qquad \text{in frame-independent notation}$$

$$= \int (\mathcal{N}\boldsymbol{p} \otimes \boldsymbol{p})m^{-1} \, \boldsymbol{dp}^0 \wedge \boldsymbol{dp}^1 \wedge \boldsymbol{dp}^2 \wedge \boldsymbol{dp}^3 \tag{22.59'}$$

in a local Lorentz frame (Track-2 notation; see Box 5.4).

RELATIVISTIC STARS

*Wherein the reader, armed
with the magic potions and powers
of Geometrodynamics, conquers the stars.*

CHAPTER **23**

SPHERICAL STARS

§23.1. PROLOG

Beautiful though gravitation theory may be, it is a sterile subject until it touches the real physical world. Only the hard reality of experiments and of astronomical observations can bring gravitation theory to life. And only by building theoretical models of stars (Part V), of the universe (Part VI), of stellar collapse and black holes (Part VII), of gravitational waves and their sources (Part VIII), and of gravitational experiments (Part IX), can one understand clearly the contacts between gravitation theory and reality.

Preview of the rest of this book

The model-building in this book will follow the tradition of theoretical physics. Each Part (stars, universe, collapse, . . .) will begin with the most oversimplified model conceivable, and will subsequently add only those additional touches of realism necessary to make contact with the least complex of actual physical systems. The result will be a tested intellectual framework, ready to support and organize the additional complexities demanded by greater realism. Greater realism will not be attempted in this book. But the reader seeking it could start in no better place than the two-volume treatise on *Relativistic Astrophysics* by Zel'dovich and Novikov (1971, 1974).

Begin, now, with models for relativistic stars. As a major simplification, insist (initially) that all stars studied be static. Thereby exclude not only exploding and pulsating stars, but even quiescent ones with stationary rotational motions. From the static assumption, plus a demand that the star be made of "perfect fluid" (no shear stresses allowed!), plus Einstein's field equations, it probably follows that the star is spherically symmetric. However, nobody has yet given a proof. [For proofs under more restricted assumptions, see Avez (1964) and Kunzle (1971).] In the absence of a proof, assume the result: insist that all stars studied be spherical as well as static.

Static stars must be spherical

§23.2. COORDINATES AND METRIC FOR A STATIC, SPHERICAL SYSTEM

Metric for any static, spherical system:

To deduce the gravitational field for a static spherical star—or for any other static, spherical system—begin with the metric of special relativity (no gravity) in the spherically symmetric form

$$ds^2 = -dt^2 + dr^2 + r^2 \, d\Omega^2, \tag{23.1}$$

where

$$d\Omega^2 = d\theta^2 + \sin^2 \theta \, d\phi^2. \tag{23.2}$$

(1) generalized from flat spacetime

Try to modify this metric to allow for curvature due to the gravitational influence of the star, while preserving spherical symmetry. The simplest and most obvious guess is to allow those metric components that are already non-zero in equation (23.1) to assume different values:

$$ds^2 = -e^{2\Phi} \, dt^2 + e^{2\Lambda} \, dr^2 + R^2 \, d\Omega^2, \tag{23.3}$$

where Φ, Λ, and R are functions of r only. (The static assumption demands $\partial g_{\mu\nu}/\partial t = 0$.) To verify that this guess is good, use it in constructing stellar models, and check that the resulting models have the same generality (same set of quantities freely specifiable) as in Newtonian theory and as expected from general physical considerations. An apparently more general metric

$$ds^2 = -a^2 \, dt^2 - 2ab \, dr \, dt + c^2 \, dr^2 + R^2 \, d\Omega^2 \tag{23.4}$$

actually is not more general in any physical sense. One can perform a coordinate transformation to a new time coordinate t' defined by

$$e^\Phi \, dt' = a \, dt + b \, dr. \tag{23.5}$$

By inserting this in equation (23.4), and by defining $e^{2\Lambda} \equiv b^2 + c^2$, one obtains the postulated line element (23.3), apart from a prime on the t.[*]

The necessity to allow for arbitrary coordinates in general relativity may appear burdensome when one is formulating the theory; but it gives an added flexibility, something one should always try to turn to one's advantage when formulating and solving problems. The $g_{rt} = 0$ simplification (called a *coordinate condition*) in equation (23.3) results from an advantageous choice of the t coordinate. The r coordinate, however, is also at one's disposal (as long as one chooses it in a way that respects spherical symmetry; thus not $r' = r + \cos\theta$). One can turn this freedom to advantage by introducing a new coordinate $r'(r)$ defined by

(2) specialized to "Schwarzschild form"

$$r' = R(r). \tag{23.6}$$

[*] Of course, equation (23.5) only succeeds in defining a new time coordinate t' if it is integrable as a differential equation for t'. By choosing the integrating factor e^Φ to be just $e^\Phi = a(r)$, one sees that $t' = t + \int [b(r)/a(r)] \, dr$ is the integral of (23.5); thus the required t' coordinate always exists, no matter what the functions $a(r)$, $b(r)$, $c(r)$, and $R(r)$ in equation (23.4) may be.

With this choice of the radial coordinate, and with the primes dropped, equation (23.3) reduces to

$$ds^2 = -e^{2\Phi} dt^2 + e^{2\Lambda} dr^2 + r^2 d\Omega^2, \tag{23.7}$$

a line element with just two unknown functions, $\Phi(r)$ and $\Lambda(r)$. This coordinate system and metric have been used in most theoretical models for relativistic stars since the pioneering work of Schwarzschild (1916b), Tolman (1939), and Oppenheimer and Volkoff (1939). These particular coordinates are sometimes called "curvature coordinates" and sometimes "Schwarzschild coordinates." The central idea of these coordinates, in a nutshell, is (Schwarzschild r-coordinate) = (proper circumference)/2π.

For a more rigorous proof that in any static spherical system Schwarzschild coordinates can be introduced, bringing the metric into the simple form (23.7), see Box 23.3 at the end of this chapter.

(3) derived more rigorously

Exercise 23.1. ISOTROPIC COORDINATES AND NEWTONIAN LIMIT

EXERCISE

An alternative set of coordinates sometimes used for static, spherical systems is the "isotropic coordinate system" $(t, \bar{r}, \theta, \phi)$. The metric in isotropic coordinates has the form

$$ds^2 = -e^{2\Phi} dt^2 + e^{2\mu}[d\bar{r}^2 + \bar{r}^2 d\Omega^2], \tag{23.8}$$

with Φ and μ being functions of \bar{r}.

(a) Exhibit the coordinate transformation connecting the Schwarzschild coordinates (23.7) to the isotropic coordinates (23.8).

(b) From equation (16.2a) [or equivalently (18.15c)], show that, in the Newtonian limit, the metric coefficient Φ of the isotropic line element becomes the Newtonian potential; and μ becomes equal to $-\Phi$. By combining with part (a), discover that $\Lambda = r \, d\Phi/dr$ in the Newtonian limit.

§23.3. PHYSICAL INTERPRETATION OF SCHWARZSCHILD COORDINATES

In general relativity, because the use of arbitrary coordinates is permitted, the physical significance of statements about tensor or vector components and other quantities is not always obvious. There are, however, some situations where the interpretation is almost as straightforward as in special relativity. The most obvious example is the center point of a local inertial coordinate system, where the principle of equivalence allows one to treat all local quantities (quantities not involving spacetime curvature) exactly as in special relativity. Schwarzschild coordinates for a spherical system turn out to be a second example.

One's first reaction when meeting a new metric should be to examine it, not in order to learn about the gravitational field, for which the curvature tensor is more

The form of any metric can reveal the nature of the coordinates being used

directly informative, but to learn about the coordinates. (Are they, for instance, locally inertial at some point?)

The names given to the coordinates have no intrinsic significance. A coordinate transformation $t' = \theta$, $r' = \phi$, $\theta' = r$, $\phi' = t$ is perfectly permissible, and has no influence on the physics or the mathematics of a relativistic problem. The only thing it affects is easy communication between the investigator who adopts it and his colleagues. Thus the names $tr\theta\phi$ for the Schwarzschild coordinates (23.7) provide a mnemonic device pointing out the geometric content of the coordinates.* In particular, the names θ, ϕ are justified by the fact that on each two-dimensional surface of constant r and t, the distance between two nearby events is given by $ds^2 = r^2\, d\Omega^2$, as befits standard θ, ϕ coordinates on a sphere of radius r. The area of this two-dimensional sphere is clearly

Geometric significance of the Schwarzschild coordinates:

(1) θ, ϕ are angles on sphere

$$A = \int (r\, d\theta)(r \sin\theta\, d\phi) = 4\pi r^2; \tag{23.9}$$

hence, the metric (23.7) tells how to measure the r coordinate that it employs. One can merely measure (in proper length units) the area A of the sphere, composed of all points rotationally equivalent to the point \mathcal{P} for which the value $r(\mathcal{P})$ is desired; and one can then calculate

(2) r measures surface area of sphere

$$r(\mathcal{P}) = \left(\frac{\text{proper area of sphere}}{\text{through point } \mathcal{P}}\bigg/ 4\pi\right)^{1/2}. \tag{23.9'}$$

The Schwarzschild coordinates have been picked for convenience, and not for the ease with which one could build a coordinate-measuring machine. This makes it more difficult to design a machine to measure t than machines to measure r, θ, ϕ.

The geometric properties of t on which a measuring device can be based are: (1) the time-independent distances $(\partial g_{\alpha\beta}/\partial t = 0)$ between world lines of constant r, θ, ϕ; (2) the orthogonality $(g_{tr} = g_{t\theta} = g_{t\phi} = 0)$ of these world lines to the $t = $ constant hypersurfaces; and (3) a labeling of these hypersurfaces by Minkowski (special relativistic) coordinate time at spatial infinity, where spacetime becomes flat. This labeling produces a constraint

(3) t has 3 special geometric properties

$$\Phi(\infty) = 0 \tag{23.10}$$

in the metric (23.7). [Mathematically, this constraint is imposed by a simple rescaling transformation $t' = e^{\Phi(\infty)}t$, and by then dropping the prime.]

One "machine" design which constructs (mentally) such a t coordinate, and in the process measures it, is the following. Observers using radar sets arrange to move along the coordinate lines $r, \theta, \phi = $ const. They do this by adjusting their velocities until each finds that the radar echos from his neighbors, or from "benchmark" reference points in the asymptotically flat space, require the same round-trip time at each repetition. Equivalently, each returning echo must show zero doppler shift;

(4) description of a "machine" to measure t

*For an example of misleading names, consider those in the equation

$$ds^2 = -e^{2\Phi(\theta')}\, d\phi'^2 + e^{2\Lambda(\theta')}\, d\theta'^2 + \theta'^2\, (dt'^2 + \sin^2 t'\, dr'^2),$$

which is equivalent to equation (23.7), but employs the coordinates $t' = \theta$, $r' = \phi$, $\theta' = r$, $\phi' = t$.

it must return with the same frequency at which it was sent out. Next a master clock is set up near spatial infinity (far from the star). It is constructed to measure proper time—which, for it, is Minkowski time "at infinity"—and to emit a standard one-Hertz signal. Each observer adjusts the rate of his "coordinate clock" to beat in time with the signals he receives from the master clock. To set the zero of his "coordinate clock," now that its rate is correct, he synchronizes with the master clock, taking account of the coordinate time Δt required for radar signals to travel from the master to him. [To compute the transit time, he assumes that for radar signals $(t_{\text{reflection}} - t_{\text{emission}}) = (t_{\text{return}} - t_{\text{reflection}}) = \Delta t$, so that the echo is obtained by time-inversion about the reflection event. This time-reversal invariance distinguishes the time t in the metric (23.7) from the more general t coordinates allowed by equation (23.4).] Each observer moving along a coordinate line $(r, \theta, \phi = \text{const.})$ now has a clock that measures coordinate time t in his neighborhood.

The above discussion identifies the Schwarzschild coordinates of equation (23.7) by their intrinsic geometric properties. Not only are r and t radial and time variables, respectively (in that $\partial/\partial r$ and $\partial/\partial t$ are spacelike and timelike, respectively, and are orthogonal also to the spheres defined by rotational symmetry), but they have particular properties [$4\pi r^2 =$ surface area; $\partial g_{\mu\nu}/\partial t = 0$; $\partial/\partial r \cdot \partial/\partial t = g_{rt} = 0$; $\partial/\partial t \cdot \partial/\partial t = g_{tt} = -1$ at $r = \infty$] that distinguish them from other possible coordinate choices [$r' = f(r)$, $t' = t + F(r)$]. No claim is made that these are the only coordinates that might reasonably be called r and t; for an alternative choice ("isotropic coordinates"), see exercise 23.1. However, they provide a choice that is reasonable, unambiguous, useful, and often used.

Other coordinates are possible, but Schwarzschild are particularly simple

§23.4. DESCRIPTION OF THE MATTER INSIDE A STAR

To high precision, the matter inside any star is a perfect fluid. (Shear stresses are negligible, and energy transport is negligible on a "hydrodynamic time scale.") Thus, it is reasonable in model building to describe the matter by perfect-fluid parameters:

Material inside star to be idealized as perfect fluid

$\rho = \rho(r) =$ density of mass-energy in rest-frame of fluid;

$p = p(r) =$ isotropic pressure in rest-frame of fluid;

$n = n(r) =$ number density of baryons in rest-frame of fluid; (23.11)

$u^\mu = u^\mu(r) =$ 4-velocity of fluid;

$T^{\mu\nu} = (\rho + p)u^\mu u^\nu + p g^{\mu\nu} =$ stress-energy tensor of fluid. (23.12)

Parameters describing perfect fluid:
(1) ρ, p, n

(For Track-1 discussion, see Box 5.1; for greater Track-2 detail, see §§22.2 and 22.3.) In order that the star be static, each element of fluid must remain always at rest in the static coordinate system; i.e., each element must move along a world line of constant r, θ, ϕ; i.e., each element must have 4-velocity components

(2) u

$$u^r = dr/d\tau = 0, \qquad u^\theta = d\theta/d\tau = 0, \qquad u^\phi = d\phi/d\tau = 0. \quad (23.13a)$$

The normalization of 4-velocity,

$$-1 = \boldsymbol{u} \cdot \boldsymbol{u} = g_{\mu\nu}u^{\mu}u^{\nu} = g_{tt}u^t u^t = -e^{2\Phi}u^t u^t,$$

then determines u^t,

$$u^t = dt/d\tau = e^{-\Phi}, \qquad \boldsymbol{u} = e^{-\Phi}\,\partial/\partial t; \qquad (23.13b)$$

(3) **T**

and this, together with the general form (23.12) of the stress-energy tensor and the form (23.7) of the metric, determines $T^{\mu\nu}$:

$$T^{00} = \rho e^{-2\Phi}, \qquad T^{rr} = pe^{-2\Lambda}, \qquad T^{\theta\theta} = pr^{-2}, \qquad T^{\phi\phi} = pr^{-2}\sin^{-2}\theta,$$

$$T^{\alpha\beta} = 0 \text{ if } \alpha \neq \beta. \qquad (23.14)$$

Although these components of the stress-energy tensor in Schwarzschild coordinates are useful for calculations, the normalization factors $e^{-2\Phi}$, $e^{-2\Lambda}$, r^{-2}, $r^{-2}\sin^{-2}\theta$ make them inconvenient for physical interpretations. More convenient are components on orthonormal tetrads carried by the fluid elements ("proper reference frames"; see §13.6):

Proper reference frame of fluid

$$\boldsymbol{e}_{\hat{t}} \equiv \frac{d}{d\tau} = \frac{1}{e^{\Phi}}\frac{\partial}{\partial t}, \qquad \boldsymbol{e}_{\hat{r}} = \frac{1}{e^{\Lambda}}\frac{\partial}{\partial r}, \qquad \boldsymbol{e}_{\hat{\theta}} = \frac{1}{r}\frac{\partial}{\partial\theta}, \qquad \boldsymbol{e}_{\hat{\phi}} = \frac{1}{r\sin\theta}\frac{\partial}{\partial\phi}; \quad (23.15a)$$

$$\boldsymbol{w}^{\hat{t}} = e^{\Phi}\,\boldsymbol{d}t, \qquad \boldsymbol{w}^{\hat{r}} = e^{\Lambda}\,\boldsymbol{d}r, \qquad \boldsymbol{w}^{\hat{\theta}} = r\,\boldsymbol{d}\theta, \qquad \boldsymbol{w}^{\hat{\phi}} = r\sin\theta\,\boldsymbol{d}\phi; \quad (23.15b)$$

Components of **u** and **T** in proper reference frame

$$\boldsymbol{u} = \boldsymbol{e}_{\hat{t}}; \qquad u^{\hat{t}} = 1, \qquad u^{\hat{r}} = u^{\hat{\theta}} = u^{\hat{\phi}} = 0; \qquad (23.15c)$$

$$T_{\hat{t}\hat{t}} \equiv T_{\hat{0}\hat{0}} = \rho, \qquad T_{\hat{r}\hat{r}} = T_{\hat{\theta}\hat{\theta}} = T_{\hat{\phi}\hat{\phi}} = p, \qquad T_{\hat{\alpha}\hat{\beta}} = 0 \text{ if } \alpha \neq \beta. \quad (23.15d)$$

See exercise 23.2 below.

The structure of a star—i.e., the set of functions $\Phi(r)$, $\Lambda(r)$, $\rho(r)$, $p(r)$, $n(r)$—is determined in part by the Einstein field equations, $G^{\mu\nu} = 8\pi T^{\mu\nu}$, and in part by the law of local conservation of energy-momentum in the fluid, $T^{\mu\nu}{}_{;\nu} = 0$. However, these are not sufficient to fix the structure uniquely. Also necessary is the functional dependence of pressure p and density ρ on number density of baryons n:

$$p = p(n), \qquad \rho = \rho(n). \qquad (23.16)$$

Equation of state:
(1) in general

Normally one cannot deduce p and ρ from a knowledge solely of n. One must know, in addition, the temperature T or the entropy per baryon s; then the laws of thermodynamics plus equations of state will determine all remaining thermodynamic variables:

$$p = p(n, s), \qquad \rho = \rho(n, s), \dots .$$

(2) idealized to "one-parameter form" $p = p(n), \ \rho = \rho(n)$

(See §22.2 and Box 22.1 for full Track-2 discussions.) To pass from the given thermodynamic knowledge, $p(n, s)$ and $\rho(n, s)$, to the desired knowledge, $p(n)$ and $\rho(n)$, one needs information about the star's thermal properties, and especially about the way in which energy generation plus heat flow have conspired to distribute the entropy, $s = s(n)$:

$$p(n) = p[n, s(n)], \qquad \rho(n) = \rho[n, s(n)].$$

There exist three important applications of the theory of relativistic stars: neutron stars, white dwarfs, and supermassive stars (stars with $M \gtrsim 10^3 M_\odot$, which may exist according to theory, but the existence of which has never yet been confirmed by observation). In all three cases, happily, the passage from $p = p(n, s)$, $\rho(n, s)$, to $p = p(n)$, $\rho = \rho(n)$, is trivial.

Justification for idealized equation of state:

Consider first a neutron star. Though hot by ordinary standards, a neutron star is so cold by any nuclear-matter scale of temperatures that essentially all its thermal degrees of freedom are frozen out ("degenerate gas"; "quantum fluid"). It is not important that a detailed treatment of the substance of a neutron star is beyond the capability of present theory (allowance for the interaction between baryon and baryon; production at sufficiently high pressures of hyperons and mesons). The simple fact is that one is dealing with matter at densities comparable to the density of matter in an atomic nucleus (2×10^{14} g/cm³) and higher. Everything one knows about nuclear matter [see, for example, Bohr and Mottelson (1969)] tells one that it is degenerate, and that one can estimate in order of magnitude its degeneracy temperature by treating it as though it were an ideal Fermi neutron gas. (In a normal atomic nucleus, a little more than 50 per cent of all baryons are neutrons, the rest are protons; in a neutron star, as many as 99 per cent are neutrons.) When approximating the neutron-star matter as an ideal Fermi neutron gas, one considers the neutrons to occupy free-particle quantum states, with two particles of opposite spin in each occupied state, and a sharp drop from 100 per cent occupancy of quantum states to empty states when the particle energy rises to the level of the "Fermi energy" [for more on such an ideal Fermi gas, see Kittel, Section 19 (1958); or at an introductory level, see Sears, Section 16-5 (1953)]. In matter at nuclear density, the Fermi energy is of the order

$$E_{\text{Fermi}} \sim 30 \text{ MeV or } 3 \times 10^{11} \text{ K};$$

and at higher density the temperature required to unfreeze the degeneracy is even greater. In other words, for matter at and above nuclear densities, already at zero temperature the kinetic energy of the particles (governed by the Pauli exclusion principle and by their Fermi energy) is a primary source of pressure. Nuclear forces make a large correction to this pressure, but for $T \lesssim 30 \text{ MeV} = 3 \times 10^{11}$ K, energies of thermal agitation do not.

A star, in collapsing from a normal state to a neutron-star state (see Chapter 24), emits a huge flux of neutrinos at temperatures $\geq 10^{10}$ K, and thereby cools to $T \ll 3 \times 10^{11}$ K within a few seconds after formation. Consequently, in all neutron stars older than a few seconds one can neglect thermal contributions to the pressure and density; i.e., one can set

$$p(n, s) = p(n, s = 0) = p(n), \qquad \rho(n, s) = \rho(n, s = 0) = \rho(n).$$

A white dwarf is similar, except that here electrons rather than neutrons are the source of Fermi gas pressure and degeneracy. Typical white-dwarf temperatures satisfy

$$kT \ll E_{\text{Fermi electrons}};$$

the Fermi kinetic energy (Pauli exclusion principle), and not random kT energy, is primarily responsible for the pressure and energy density; and one can set

$$p(n, s) = p(n, s = 0) = p(n), \qquad \rho(n, s) = \rho(n, s = 0) = \rho(n).$$

(3) in supermassive stars

In a supermassive star (see Chapter 24), the situation is quite different. There temperature and entropy are almost the whole story, so far as pressure and energy density are concerned. However, convection keeps the star stirred up and produces a uniform entropy distribution

$$s = \text{const. independent of radius;}$$

so one can write

$$p(n, s) = p_s(n), \qquad\qquad\qquad \rho(n, s) = \rho_s(n).$$

$$\uparrow \begin{bmatrix} \text{functions depending on} \\ \text{uniform entropy per baryon,} \\ s, \text{ in the star} \end{bmatrix} \uparrow$$

In all three cases—neutron stars, white dwarfs, supermassive stars—one regards the relations $p(n)$ and $\rho(n)$ as "equations of state"; and having specified them, one can calculate the star's structure without further reference to its thermal properties.

EXERCISE

Exercise 23.2. PROPER REFERENCE FRAMES OF FLUID ELEMENTS

(a) Verify that equations (23.15a,b) define an orthonormal tetrad and its dual basis of 1-forms, at each event in spacetime.

(b) Verify that the components of the fluid 4-velocity relative to these tetrads are given by equations (23.15c). Why do these components guarantee that the tetrads form "proper reference frames" for the fluid elements?

(c) Verify equations (23.15d) for the components of the stress-energy tensor.

§23.5. EQUATIONS OF STRUCTURE

Five equations needed to determine 5 stellar-structure functions: Φ, Λ, p, ρ, n

The structure of a relativistic star is determined by five functions of radius r: the metric functions $\Phi(r)$, $\Lambda(r)$, the pressure $p(r)$, the density of mass-energy $\rho(r)$, and the number density of baryons, $n(r)$. Hence, to determine the structure uniquely, one needs five equations of structure, plus boundary conditions. Two equations of structure, the equations of state $p(n)$ and $\rho(n)$, are already in hand. The remaining three must be the essential content of the Einstein field equations and of the law of local energy-momentum conservation, $T^{\mu\nu}{}_{;\nu} = 0$.

One knows that the law of local energy-momentum conservation for the fluid follows as an identity from the Einstein field equations. Without loss of information,

one can therefore impose all ten field equations and ignore local energy-momentum conservation. But that is an inefficient way to proceed. Almost always the equations $T^{\mu\nu}{}_{;\nu} = 0$ can be reduced to usable form more easily than can the field equations. Hence, the most efficient procedure is to: (1) evaluate the four equations $T^{\mu\nu}{}_{;\nu} = 0$; (2) evaluate enough field equations (six) to obtain a complete set $(6 + 4 = 10)$; and (3) evaluate the remaining four field equations as checks of the results of (1) and (2).

The most efficient procedure for solving Einstein equations

The Track-2 reader has learned (§22.3) that the equations $T^{\mu\nu}{}_{;\nu} = 0$ for a perfect fluid take on an especially simple form when projected (1) on the 4-velocity \boldsymbol{u} of the fluid itself, and (2) orthogonal to \boldsymbol{u}. Projection along \boldsymbol{u} $(u_\mu T^{\mu\nu}{}_{;\nu} = 0)$ gives the local law of energy conservation (22.11a),

$$\frac{d\rho}{d\tau} = -(\rho + p)\boldsymbol{\nabla} \cdot \boldsymbol{u} = \frac{\rho + p}{n}\frac{dn}{d\tau},$$

where $\boldsymbol{u} = d/d\tau$; i.e., τ is proper time along the world line of any chosen element of the fluid. For a static star, or for any other static system, both sides of this equation must vanish identically (no fluid element ever sees any change in its own density).

Projection of $T^{\mu\nu}{}_{;\nu} = 0$ orthogonal to \boldsymbol{u} gives the reasonable equation

Equation of hydrostatic equilibrium derived

$$\begin{pmatrix}\text{inertial mass} \\ \text{per unit volume}\end{pmatrix} \times (\text{4-acceleration}) = -\begin{pmatrix}\text{pressure gradient, projected} \\ \text{perpendicular to } \boldsymbol{u}\end{pmatrix}$$

i.e.,

$$(\rho + p)\boldsymbol{\nabla}_{\boldsymbol{u}}\boldsymbol{u} = -[\boldsymbol{\nabla}p + (\boldsymbol{\nabla}_{\boldsymbol{u}}p)\boldsymbol{u}].$$

[see equation (22.13)]. When applied to a static star, this equation tells how much pressure gradient is needed to prevent a fluid element from falling. Only the radial component of this equation has content, since the pressure depends only on r. The radial component in the Schwarzschild coordinate system says [see the line element (23.7) and the 4-velocity components (23.13)],

$$\begin{aligned}(\rho + p)u_{r;\nu}u^\nu &= -(\rho + p)\Gamma^\alpha{}_{r\nu}u_\alpha u^\nu = -(\rho + p)\Gamma^0{}_{r0}u_0 u^0 \\ &= (\rho + p)\Phi_{,r} = -p_{,r}.\end{aligned} \tag{23.17}$$

(Track-1 readers can derive this from scratch at the end of the section, exercise 23.3.) In the Newtonian limit, Φ becomes the Newtonian potential (since $g_{00} = -e^{2\Phi} \approx -1 - 2\Phi$), and the pressure becomes much smaller than the mass-energy density; consequently equation (23.17) becomes

$$\rho\Phi_{,r} = -p_{,r}. \tag{23.17N}$$

This is the Newtonian version of the equation describing the balance between gravitational force and pressure gradient.

The pressure gradient that prevents a fluid element from falling appears in Einstein's theory as the source of an acceleration. This acceleration, by keeping the fluid element at a fixed r value, causes it to depart from geodesic motion (from "fiducial world line"; from motion of free fall into the center of the star). Newtonian

Comparison of Newton and Einstein views of hydrostatic equilibrium

theory, on the other hand, views as the fiducial world line the one that stays at a fixed r value. It regards the "gravitational force" as trying (without success, because balanced by the pressure gradient) to pull a particle from a fixed-r world line onto a geodesic world line. In the two theories the magnitudes of the acceleration, whether "actually taking place" (Einstein theory) or "trying to take place" (Newtonian theory), are the same to lowest order (but opposite in direction); so it is no surprise that (23.17) and (23.17N) differ only in detail.

Turn next to the Einstein field equation. Here, as is often the case, the components of the field equation in the fluid's orthonormal frame [equations (23.15a,b)] are simpler than the components in the coordinate basis. One already knows the stress-energy tensor $T_{\hat{\alpha}\hat{\beta}}$ in the orthonormal frame [equation (23.15d)]; and Track-2 readers have already calculated the Einstein tensor $G_{\hat{\alpha}\hat{\beta}}$ (exercise 14.13; Track-1 readers will face the task at the end of this section, exercise 23.4). All that remains is to equate $G_{\hat{\alpha}\hat{\beta}}$ to $8\pi T_{\hat{\alpha}\hat{\beta}}$. Examine first the $\hat{0}\hat{0}$ component of the field equations:

Equation for Λ derived

$$
\begin{aligned}
G_{\hat{0}\hat{0}} &= r^{-2} - r^{-2}e^{-2\Lambda} - r^{-1}(d/dr)(e^{-2\Lambda}) \\
&= r^{-2}(d/dr)[r(1 - e^{-2\Lambda})] = 8\pi T_{\hat{0}\hat{0}} = 8\pi\rho.
\end{aligned}
$$

This equation becomes easy to solve as soon as one notices that it is a differential equation *linear* in the quantity $e^{-2\Lambda}$; a bit of tidying up then focuses attention on the quantity $r(1 - e^{-2\Lambda})$. Give this quantity the name $2m(r)$ (so far only a name!); thus,

$$
2m \equiv r(1 - e^{-2\Lambda}); \qquad e^{2\Lambda} = (1 - 2m/r)^{-1}. \tag{23.18}
$$

In this notation the $\hat{0}\hat{0}$ component of the Einstein tensor becomes

$$
G_{\hat{0}\hat{0}} = \frac{2}{r^2} \frac{dm(r)}{dr} = 8\pi\rho.
$$

Integrate and find

$$
m(r) = \int_0^r 4\pi r^2 \rho \, dr + m(0). \tag{23.19}
$$

For the constant of integration $m(0)$, a zero value means a space geometry smooth at the origin (physically acceptable); a non-zero value means a geometry with a singularity at the origin (physically unacceptable: no local Lorentz frame at $r = 0$):

$$
\begin{aligned}
ds^2 &= [1 - 2m(0)/r]^{-1} dr^2 + r^2(d\theta^2 + \sin^2\theta \, d\phi^2) \\
&\approx -[r/2m(0)] dr^2 + r^2(d\theta^2 + \sin^2\theta \, d\phi^2) \qquad \text{at } r \approx 0 \text{ if } m(0) \neq 0; \\
ds^2 &= [1 - (8\pi/3)\rho_c r^2]^{-1} dr^2 + r^2(d\theta^2 + \sin^2\theta \, d\phi^2) \\
&\approx dr^2 + r^2(d\theta^2 + \sin^2\theta \, d\phi^2) \qquad\qquad \text{at } r \approx 0 \text{ if } m(0) = 0.
\end{aligned} \tag{23.20}
$$

"Mass-energy inside radius r," $m(r)$, defined

The quantity $m(r)$, defined by equation (23.18) and calculated from equation (23.19) with $m(0) = 0$, is a relativistic analog of the "mass-energy inside radius r." Box 23.1 spells out the analogy in detail.

Box 23.1 MASS-ENERGY INSIDE RADIUS r

The total mass-energy M of an isolated star is well-defined (Chapter 19). But not well-defined, in general, is the distribution of that mass-energy from point to point inside the star and in its gravitational field (no unique "gravitational stress-energy tensor"). This was the crucial message of §20.4 (Track 2).

The message is true in general. But for the case of a spherical star—and only for that case—the message loses its bite. Spherical symmetry allows one to select a distribution of the total mass-energy that is physically reasonable. In Schwarzschild coordinates, it is defined by

$$\text{``total mass-energy inside radius } r\text{''} \equiv m(r) = \int_0^r 4\pi r^2 \rho \, dr. \tag{1}$$

The fully convincing argument for this definition is found only by considering a generalization of it to time-dependent spherically symmetric stars (pulsating, collapsing, or exploding stars; see Chapters 26 and 32, and especially exercise 32.7). For them one finds that the mass-energy m associated with a given ball of matter (fixed baryon number) can change in time only to the extent that locally measurable energy fluxes can be detected at the boundary of the ball. [Such energy fluxes could be the power expended by pressure forces against the moving boundary surface, or heat fluxes, or radiation (photon or neutrino) fluxes. But since spherically symmetric gravitational waves do not exist (Chapters 35 and 36), neither physical intuition nor Einstein's equations require that problems of localizing gravitational-wave energy be faced.] Thus the energy m is localized, not by a mathematical convention, but by the circumstance that transfer of energy (with this definition of m) is detectable by local measurements. [For the mathematical details of $m(r, t)$ in the time-dependent case, see Misner and Sharp (1964), Misner (1965), and exercise 32.7.]

In addition to the critical "local energy flux" property of $m(r)$ described above, there are three further properties that verify its identification as mass-energy. They are: (1) Everywhere outside the star

$$m(r) = M \equiv \begin{pmatrix} \text{total mass-energy of star as measured from} \\ \text{Kepler's third law for distant planets} \end{pmatrix}; \tag{2}$$

see §23.6 for proof. (2) For a Newtonian star, where "mass inside radius r" has a unique meaning, $m(r)$ is that mass. (3) For a relativistic star, $m(r)$ splits nicely into "rest mass-energy" $m_0(r)$ plus "internal energy" $U(r)$ plus "gravitational potential energy" $\Omega(r)$.

To recognize and appreciate the split

$$m(r) = m_0(r) + U(r) + \Omega(r), \tag{3}$$

proceed as follows. First split the total density of mass-energy, ρ, into a part $\mu_0 n$ due to rest mass—where μ_0 is the average rest mass of the baryonic species pres-

Box 23.1 (continued)

ent—and a part $\rho - \mu_0 n$ due to internal thermal energy, compressional energy, etc. Next notice that the proper volume of a shell of thickness dr is

$$d\mathcal{V} = 4\pi r^2 (e^\Lambda \, dr) = 4\pi r^2 (1 - 2m/r)^{-1/2} \, dr, \tag{4}$$

not $4\pi r^2 \, dr$. Consequently, the total rest mass inside radius r is

$$m_0 = \int_0^r \mu_0 n \, d\mathcal{V} = \int_0^r 4\pi r^2 (1 - 2m/r)^{-1/2} \mu_0 n \, dr, \tag{5}$$

and the total internal energy is

$$U = \int_0^r (\rho - \mu_0 n) \, d\mathcal{V} = \int_0^r 4\pi r^2 (1 - 2m/r)^{-1/2} (\rho - \mu_0 n) \, dr. \tag{6}$$

Subtract these from the total mass-energy, m; the quantity that is left must be the gravitational potential energy,

$$\Omega = -\int_0^r \rho[(1 - 2m/r)^{-1/2} - 1] 4\pi r^2 \, dr$$

$$\approx -\int_0^r (\rho m/r) 4\pi r^2 \, dr. \tag{7}$$

$\underset{\uparrow}{\quad}$[Newtonian limit, $m/r \ll 1$]

(See exercise 23.7.)

Equation for Φ derived

Turn next to the $\hat{r}\hat{r}$ component of the field equations:

$$G_{\hat{r}\hat{r}} = -r^{-2} + r^{-2} e^{-2\Lambda} + 2r^{-1} e^{-2\Lambda} \, d\Phi/dr$$
$$= 8\pi T_{\hat{r}\hat{r}} = 8\pi p.$$

Solving this equation for the derivative of Φ, and replacing $e^{-2\Lambda}$ by $1 - 2m/r$, one obtains an expression for the gradient of the potential Φ:

$$\frac{d\Phi}{dr} = \frac{m + 4\pi r^3 p}{r(r - 2m)}. \tag{23.21}$$

This expression reduces to the familiar formula

$$d\Phi/dr = m/r^2 \tag{23.21N}$$

in the Newtonian limit.

In most studies of stellar structure, one replaces equation (23.17) by the equivalent equation obtained with the help of (23.21),

$$\frac{dp}{dr} = -\frac{(\rho + p)(m + 4\pi r^3 p)}{r(r - 2m)}.$$ (23.22)

Equation of hydrostatic
equilibrium rewritten in
"OV" form

This is called the Oppenheimer-Volkoff (OV) equation of hydrostatic equilibrium. Its Newtonian limit,

$$dp/dr = -\rho m/r^2,$$ (23.22N)

is familiar.

Compare two stellar models, one relativistic and the other Newtonian. Suppose that at a given radius r [determined in both cases by (proper area) $= 4\pi r^2$], the two configurations have the same values of ρ, p, and m. Then in the relativistic model the pressure gradient is

$$\frac{dp}{d(\text{proper radial distance})} = \frac{dp}{e^\Lambda \, dr}$$

$$= -\frac{(\rho + p)(m + 4\pi r^3 p)}{r^2(1 - 2m/r)^{1/2}}.$$ (23.23)

In contrast, Newtonian theory gives for the pressure gradient

$$\frac{dp}{d(\text{proper radial distance})} = \frac{dp}{dr} = -\frac{\rho m}{r^2}.$$ (23.23N)

The relativistic expression for the gradient is larger than the Newtonian expression (1) because the numerator is larger (added pressure term in both factors) and (2) because the denominator is smaller [shrinkage factor $(1 - 2m/r)^{1/2}$]. Therefore, as one proceeds deeper into the star, one finds pressure rising faster than Newtonian gravitation theory would predict. Moreover, this rise in pressure is in a certain sense "self-regenerative." The more the pressure goes up, the larger the pressure-correction terms become in the numerator of (23.23); and the larger these terms become, the faster is the further rise of the pressure as one probes still deeper into the star. The geometric factor $[1 - 2m(r)/r]^{1/2}$ in the denominator of (23.23) further augments this regenerative rise of pressure towards the center. It is appropriate to summarize the situation in short-hand terms by saying that general relativity predicts stronger gravitational forces in a stationary body than does Newtonian theory. These forces, among their other important effects, can pull certain white-dwarf stars and super-massive stars into gravitational collapse under circumstances (see Chapter 24) where Newtonian theory would have predicted stable hydrostatic equilibrium. As the most elementary indication that a new factor has surfaced in the analysis of stability, note that no star in hydrostatic equilibrium can ever have $2m(r)/r \geq 1$ (see Box 23.2 for one illustration and §23.8 for discussion), a phenomenon alien to Newtonian theory.

Comparison of pressure
gradients in Newtonian and
relativistic stars

Now in hand are five equations of structure [two equations of state (23.16); equation (23.19), expressing $m(r) = \frac{1}{2}r(1 - e^{-2\Lambda})$ as a volume integral of ρ; the source

Equations of stellar structure
summarized

equation (23.21) for Φ; and the OV equation of hydrostatic equilibrium (23.22)] for the five structure functions ρ, p, n, Φ, Λ. If the theory of relativistic stars as outlined above is well posed, then each of the remaining eight Einstein field equations $G_{\hat{\alpha}\hat{\beta}} = 8\pi T_{\hat{\alpha}\hat{\beta}}$ must be either vacuous ("$0 = 0$"), or must be a consequence of the five equations of structure. This is, indeed, the case, as one can verify by straightforward but tedious computations.

To construct a stellar model, one needs boundary conditions as well as structure equations. To facilitate the presentation of boundary conditions, the next section will examine the star's external gravitational field.

EXERCISES

Exercise 23.3. LAW OF LOCAL ENERGY-MOMENTUM CONSERVATION
(for readers who have not studied Chapter 22)

Evaluate the four components of the equation $T^{\alpha\beta}{}_{;\beta} = 0$ for the stress-energy tensor (23.14) in the Schwarzschild coordinate system of equation (23.7). [Answer: only $T^{r\beta}{}_{;\beta} = 0$ gives a nonvacuous result; it gives equation (23.17).]

Exercise 23.4. EINSTEIN CURVATURE TENSOR
(for readers who have not studied Chapter 14)

Calculate the components of the Einstein curvature tensor, $G_{\alpha\beta}$, in Schwarzschild coordinates. Then perform a transformation to obtain $G_{\hat{\alpha}\hat{\beta}}$, the components in the orthonormal frame of equations (23.15a,b). [See Box 8.6, or Box 14.2 and equation (14.7).]

Exercise 23.5. TOTAL NUMBER OF BARYONS IN A STAR

Show that, if $r = R$ is the location of the surface of a static star, then the total number of baryons inside the star is

$$A = \int_0^R 4\pi r^2 n e^{\Lambda}\, dr. \tag{23.24}$$

[Hint: See the discussion of m_0 in Box 23.1.]

Exercise 23.6. BUOYANT FORCE IN A STAR

An observer at rest at some point inside a relativistic star measures the radial pressure-buoyant force, F_{buoy}, on a small fluid element of volume V. Let him use the usual laboratory techniques. Do not confuse him by telling him he is in a relativistic star. What value will he find for F_{buoy}, in terms of ρ, p, m, V, and dp/dr? If he equates this buoyant force to an equal and opposite gravitational force, F_{grav}, what will F_{grav} be in terms of ρ, p, m, V, and r? (Use equation 23.22.) How do these results differ from the corresponding Newtonian results?

Exercise 23.7. GRAVITATIONAL ENERGY OF A NEWTONIAN STAR

Calculate in Newtonian theory the energy one would gain from gravity if one were to construct a star by adding one spherical shell of matter on top of another, working from the inside outward. Use Laplace's equation $(r^2\Phi_{,r})_{,r} = 4\pi r^2\rho$ and the equation of hydrostatic equilibrium $p_{,r} = -\rho\Phi_{,r}$ to put the answer in the following equivalent forms:

(energy gained from gravity) ≡ −(gravitational potential energy)

$$= \int_0^R (\rho r \Phi_{,r}) 4\pi r^2 \, dr = \int_0^R (\rho m/r) 4\pi r^2 \, dr$$

$$= -\frac{1}{2} \int_0^R (\rho \Phi) 4\pi r^2 \, dr = \frac{1}{8\pi} \int_0^\infty (\Phi_{,r})^2 4\pi r^2 \, dr$$

$$= 3 \int_0^R 4\pi r^2 p \, dr.$$

§23.6. EXTERNAL GRAVITATIONAL FIELD

Outside a star the density and pressure vanish, so only the metric parameters Φ and $\Lambda = -\frac{1}{2} \ln (1 - 2m/r)$ need be considered. From equation (23.19) one sees that "the mass inside radius r," $m(r)$, stays constant for values of r greater than R (outside the star). Its constant value is denoted by M:

$$m(r) = M \qquad \text{for } r > R \text{ (i.e., outside the star).} \qquad (23.25)$$

By integrating equation (23.21) with $p = 0$ and $m = M$, and by imposing the boundary condition (23.10) on Φ at $r = \infty$ ("normalization of scale of time at $r = \infty$"), one finds

$$\Phi(r) = \frac{1}{2} \ln (1 - 2M/r) \qquad \text{for } r > R. \qquad (23.26)$$

Consequently, outside the star the spacetime geometry (23.7) becomes

$$ds^2 = -\left(1 - \frac{2M}{r}\right) dt^2 + \frac{dr^2}{(1 - 2M/r)} + r^2(d\theta^2 + \sin^2\theta \, d\phi^2). \qquad (23.27)$$

Spacetime outside star possesses "Schwarzschild" geometry

This is called the "Schwarzschild geometry" or "Schwarzschild gratitational field" or "Schwarzschild line element," because Karl Schwarzschild (1916a) discovered it as an exact solution to Einstein's field equations a few months after Einstein formulated general relativity theory.

In that region of spacetime, $r \gg 2M$, where the geometry is nearly flat, Newton's theory of gravity is valid, and the Newtonian potential is

$$\Phi = -M/r \qquad \text{for } r > R, \, r \gg 2M. \qquad (23.26\text{N})$$

Consequently, M is the mass that governs the Keplerian motions of planets in the distant, Newtonian gravitational field—i.e., it is the star's "total mass-energy" (see Chapters 19 and 20). Since the metric (23.27) far outside the star is precisely diagonal ($g_{tj} \equiv 0$), the star's total angular momentum must vanish. This result accords with the absence of internal fluid motions.

Total mass-energy of star

§23.7. HOW TO CONSTRUCT A STELLAR MODEL

Equations of stellar structure collected together

The equations of stellar structure (23.16), (23.19), (23.21), (23.22), and associated boundary conditions (to be discussed below), all gathered together along with the line element, read as follows.

Line Element

$$ds^2 = -e^{2\Phi}\, dt^2 + \frac{dr^2}{1 - 2m/r} + r^2(d\theta^2 + \sin^2\theta\, d\phi^2) \tag{23.27'}$$

$$= -\left(1 - \frac{2M}{r}\right) dt^2 + \frac{dr^2}{1 - 2M/r} + r^2(d\theta^2 + \sin^2\theta\, d\phi^2) \quad \text{for } r > R.$$

Mass Equation

$$m = \int_0^r 4\pi r^2\rho\, dr, \text{ with } m(r = 0) = 0. \tag{23.28a}$$

OV Equation of Hydrostatic Equilibrium

$$\frac{dp}{dr} = -\frac{(\rho + p)(m + 4\pi r^3 p)}{r(r - 2m)}, \text{ with } p(r = 0) = p_c = \text{central pressure.} \tag{23.28b}$$

Equations of State

$$p = p(n), \tag{23.28c}$$

$$\rho = \rho(n). \tag{23.28d}$$

Source Equation for Φ

$$\frac{d\Phi}{dr} = \frac{(m + 4\pi r^3 p)}{r(r - 2m)}, \quad \text{with } \Phi(r = R) = \frac{1}{2}\ln(1 - 2M/R). \tag{23.28e}$$

How to solve the equations of stellar structure

To construct a stellar model one can proceed as follows. First specify the equations of state (23.28c,d) and a value of the central pressure, p_c. Also specify an arbitrary (later to be renormalized) value, Φ_0, for $\Phi(r = 0)$. The boundary conditions $p(r = 0) = p_c$, $\Phi(r = 0) = \Phi_0$, $m(r = 0) = 0$ are sufficient to determine uniquely the solution to the coupled equations (23.28). Integrate these coupled equations outward from $r = 0$ until the pressure vanishes. [The OV equation, (23.28b), guarantees that the pressure will decrease monotonically so long as the equations of state obey the

reasonable restriction $\rho \geq 0$ for all $p \geq 0$.] The point at which the pressure reaches zero is the star's surface; the value of r there is the star's radius, R; and the value of m there is the star's total mass-energy, M. Having reached the surface, renormalize Φ by adding a constant to it everywhere, so that it obeys the boundary condition (23.28e). The result is a relativistic stellar model whose structure functions Φ, m, ρ, p, n satisfy the equations of structure.

Notice that for any fixed choice of the equations of state $p = p(n)$, $\rho = \rho(n)$, the stellar models form a one-parameter sequence (parameter p_c). Once the central pressure has been specified, the model is determined uniquely.

The next chapter describes a variety of realistic stellar models constructed numerically by the above prescription. For an idealized stellar model constructed analytically, see Box 23.2.

Exercise 23.8. NEWTONIAN STARS OF UNIFORM DENSITY **EXERCISE**

Calculate the structures of uniform-density configurations in Newtonian theory. Show that the relativistic configurations of Box 23.2 become identical to the Newtonian configurations in the weak-gravity limit. Also show that there are no mass or radius limits in Newtonian theory.

(*continued on page 612*)

Box 23.2 RELATIVISTIC MODEL STAR OF UNIFORM DENSITY

For realistic equations of state (see next chapter), the equations of stellar structure (23.28) cannot be integrated analytically; numerical integration is necessary. However, analytic solutions exist for various idealized and *ad hoc* equations of state. One of the most useful analytic solutions [Karl Schwarzschild (1916b)] describes a star of uniform density,

$$\rho = \rho_0 = \text{constant for all } p. \tag{1}$$

It is not necessary to indulge in the fiction of "an incompressible fluid" to accept this model as interesting. Incompressibility would imply a speed of sound, $v = (dp/d\rho)^{1/2}$, of unlimited magnitude, therefore in excess of the speed of light, and therefore in contradiction with a central principle of special relativity ("principle of causality") that no physical effect can be propagated at a speed $v > 1$. (If a source could cause an effect so quickly in one local Lorentz frame, then there would exist another local Lorentz frame in which the effect would occur before the source had acted!) However, that the part of the fluid in the region of high pressure has the same density as the part of the fluid in the region of low pressure is an idea easy to admit, if only one thinks of the fluid having a composition that varies from one

Box 23.2 (*continued*)

r value to another ("hand-tailored"). Whether one thinks along this line, or simply has in mind a globe of water limited in size to a small fraction of the dimensions of the earth, one has in Schwarzschild's model an instructive example of hydrostatics done in the framework of Einstein's theory.

The mass equation (23.28a) gives immediately

$$m = \begin{cases} (4\pi/3)\rho_0 r^3 & \text{for } r < R \\ M = (4\pi/3)\rho_0 R^3 & \text{for } r > R \end{cases}. \tag{2}$$

from which follows the length-correction factor in the metric

$$\frac{d(\text{proper distance})}{dr} = e^\Lambda = [1 - 2m(r)/r]^{-1/2}. \tag{3}$$

When for ease of visualization the space geometry (r, ϕ) of an equatorial slice through the star is viewed as embedded in a Euclidean 3-geometry (z, r, ϕ) [see §23.8], the "lift" out of the plane $z = 0$ is

$$z(r) = \begin{cases} (R^3/2M)^{1/2}[1 - (1 - 2Mr^2/R^3)^{1/2}] & \text{for } r \leq R, \\ (R^3/2M)^{1/2}[1 - (1 - 2M/R)^{1/2}] + [8M(r - 2M)]^{1/2} - [8M(R - 2M)]^{1/2} \\ \qquad\qquad \text{for } r \geq R. \end{cases} \tag{4}$$

The knowledge of $m(r)$ from (2) allows the equation of hydrostatic equilibrium (23.28b) to be integrated to give the pressure:

$$p = \rho_0 \left\{ \frac{(1 - 2Mr^2/R^3)^{1/2} - (1 - 2M/R)^{1/2}}{3(1 - 2M/R)^{1/2} - (1 - 2Mr^2/R^3)^{1/2}} \right\} \text{ for } r < R. \tag{5}$$

The pressure in turn leads via (23.28e) to the time-correction factor in the metric.

$$\frac{d(\text{proper time})}{dt} = e^\Phi = \begin{cases} \dfrac{3}{2}\left(1 - \dfrac{2M}{R}\right)^{1/2} - \dfrac{1}{2}\left(1 - \dfrac{2Mr^2}{R^3}\right)^{1/2} & \text{for } r < R \\ (1 - 2M/r)^{1/2} & \text{for } r > R \end{cases}. \tag{6}$$

Several features of these uniform-density configurations are noteworthy. (1) For fixed energy density, ρ_0, the central pressure

$$p_c = \rho_0 \left\{ \frac{1 - (1 - 2M/R)^{1/2}}{3(1 - 2M/R)^{1/2} - 1} \right\}, \tag{7}$$

increases monotonically as the radius, R, increases—and, hence, also as the mass, $M = (4\pi/3)\rho_0 R^3$, and the ratio ("strength of gravity")

$$2M/R = (8\pi/3)\rho_0 R^2 \tag{8}$$

increase. This is natural, since, as more and more matter is added to the star, a greater and greater pressure is required to support it. (2) The central pressure becomes infinite when M, R, and $2M/R$ reach the limiting values

$$R_{\text{lim}} = (9/4)M_{\text{lim}} = (3\pi\rho_0)^{-1/2}, \tag{9}$$

$$(2M/R)_{\text{lim}} = 8/9. \tag{10}$$

No star of uniform density can have a mass and radius exceeding these limits. These limits are purely relativistic phenomena; no such limits occur in Newtonian theory. (3) Inside the star the space geometry (geometry of a hypersurface $t = constant$) is that of a three-dimensional spherical surface with radius of curvature

$$a = (3/8\pi\rho_0)^{1/2}. \tag{11}$$

[See equation (4), above.] Outside the star the (Schwarzschild) space geometry is that of a three-dimensional paraboloid of revolution. The interior and exterior geometries join together smoothly. All these details are shown in the following three diagrams. There all quantities are given in the following geometric units (to convert mass in g or density in g/cm^3 into mass in cm or density in cm^{-2}, multiply by 0.742×10^{-28} cm/g): lengths, in units $(3/8\pi\rho_0)^{1/2}$; pressure, in units ρ_0; mass, in units $(3/32\pi\rho_0)^{1/2}$.

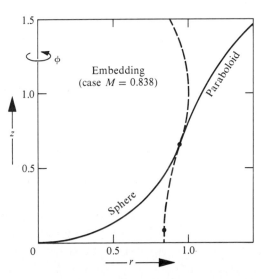

Box 23.2 (continued)

The mass "after assembly" is what is called M. The mass of the same fluid, dispersed in droplets at infinite separation, is called M_{before} in the following table.

M_{before}	small	0.0882	0.894	1.0913	1.374
M	small	0.0828	0.636	0.729	0.838 (critical)
Difference (binding):	$\frac{3}{10}M^{5/3}$	0.0054	0.258	0.362	0.536

§23.8. THE SPACETIME GEOMETRY FOR A STATIC STAR

Surface area of spheres, $4\pi r^2$:

(1) increases monotonically from center of star outward

For a highly relativistic star, the spacetime geometry departs strongly from Euclid-Lorentz flatness. Consequently, there is no *a priori* reason to expect that the surface area $4\pi r^2$, and hence also the radial coordinate r, will increase monotonically as one moves from the center of the star outward. Fortunately, *the equations of stellar structure guarantee that* r *will increase monotonically from* 0 *at the star's center to* ∞ *at an infinite distance away from the star,* so long as $\rho \geq 0$ and so long as the star is static (equilibrium).

The monotonicity of r can be seen as follows. Introduce as a new radial coordinate proper distance, ℓ, from the center of the star. By virtue of expression (23.27′) for the metric, ℓ and r are related by

$$dr = \pm(1 - 2m/r)^{1/2}\,d\ell. \qquad (23.29)$$

Note that r is zero at the center of the star (where $m \propto r^3$), and note that r is always nonnegative by definition. Therefore r must at first increase with ℓ as one moves outward from $\ell = 0$; $r(\ell)$ can later reach a maximum and start decreasing only at a point where $2m/r$ becomes unity [see equation (23.29)]. Such a behavior can and does happen in a closed model universe, a 3-sphere of uniform density and radius a, where

$$r(\ell) = a\sin(\ell/a)$$

[see Chapter 27; especially the embedding diagram of Box 27.2(A)]. However, the field equations demand that such a system be dynamic. Here, on the contrary, attention is limited to a system where conditions are static. In such a system, the condition of hydrostatic equilibrium (23.28b) applies. Then the pressure gradient is given by an expression with the factor $[1 - 2m(r)/r]$ in its denominator. If $2m/r$ approaches unity with increasing ℓ in some region of the star, the pressure gradient

there becomes so large that one comes to the point $p = 0$ (surface of the star) before one comes to any point where $2m(r)/r$ might attain unit value. Moreover, after the surface of the star is passed, m remains constant, $m(r) = M$, and $2m(r)/r$ decreases. Consequently, *2m/r is always less than unity;* and $r(\ell)$ cannot have a maximum, Q.E.D. (Details of the proof are left to the reader as exercise 23.9.)

Although the radii of curvature, r, and corresponding spherical surface areas, $4\pi r^2$, increase monotonically from the center of a star outward, they do not increase at the same rate as they would in flat spacetime. In flat spacetime the rate of increase is given by dr/d(proper radial distance) $- dr/d\ell = 1$. In a star it is given by $dr/d\ell = (1 - 2m/r)^{1/2} < 1$. Consequently, if one were to climb a long ladder outward from the center of a relativistic star, measuring for each successive spherical shell its Schwarzschild r-value ("proper circumference"$/2\pi$), one would find these r-values to increase surprisingly slowly.

(2) but increases more slowly than in flat spacetime

This strange behavior is most easily visualized by means of an "embedding diagram." It would be too much for any easy visualization if one were to attempt to embed the whole curved four-dimensional manifold in some higher-dimensional flat space. [See, however, Fronsdal (1959) and Clarke (1970) for a global embedding in $5 + 1$ dimensions, and Kasner (1921b) for a local embedding in $4 + 2$ dimensions. One can never embed a non-flat, vacuum metric ($G_{\mu\nu} = 0$) in a flat space of 5 dimensions (Kasner, 1921c).] Therefore seek a simpler picture (Flamm 1916). Space at one time in the context of a static system has the same 3-geometry as space at another time. Therefore, depict 3-space only as it is at one time, $t = constant$. Moreover, at any one time the space itself has spherical symmetry. Consequently, one slice through the center, $r = 0$, that divides the space symmetrically into two halves (for example, the equatorial slice, $\theta = \pi/2$) has the same 2-geometry as any other such slice (any selected angle of tilt, at any azimuth) through the center. Therefore limit attention to the 2-geometry of the equatorial slice. The geometry on this slice is described by the line element

Embedding of spacetime in a flat space of higher dimensionality

$$ds^2 = [1 - 2m(r)/r]^{-1} dr^2 + r^2 d\phi^2. \qquad (23.30)$$

Now one may embed this two-dimensional curved-space geometry in the flat geometry of a Euclidean three-dimensional manifold.

If the curvature of the two-dimensional slice is zero or negligible, the embedding is trivial. In this event, identify the 2-geometry with the slice $z = 0$ of the Euclidean 3-space. Moreover, introduce into that 3-space the familiar cylindrical coordinates z, r, ϕ, that one employs for any problem with axial symmetry (see Fig. 23.1 and Box 23.2 for more detail). Then one recognizes the flat two-dimensional slice as the set of points of the Euclidean space with $z = 0$, with ϕ running from 0 to 2π, and r from 0 to ∞. One has identified the r and ϕ of the slice with the r and ϕ of the Euclidean 3-space.

Construction of "embedding diagram" for equatorial slice through star

If the 2-geometry is curved, as it is when the equatorial section is taken through a real star, then maintain the identification between the r, ϕ, of the slice and the r, ϕ, of the Euclidean 3-geometry, but bend up the slice out of the plane $z = 0$ (except at the origin, $r = 0$). At the same time, insist that the bending be axially symmetric. In other words, require that the amount of the "lift" above the plane $z = 0$ shall

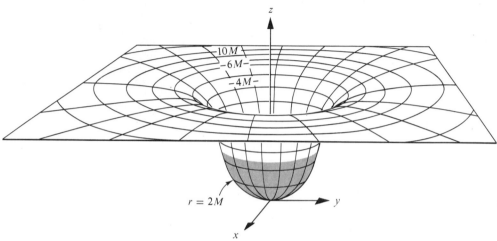

Figure 23.1.
Geometry within (grey) and around (white) a star of radius $R = 2.66M$, schematically displayed. The star is in hydrostatic equilibrium and has zero angular momentum (spherical symmetry). The two-dimensional geometry

$$ds^2 = [1 - 2m(r)/r]^{-1} dr^2 + r^2 d\phi^2$$

of an equatorial slice through the star ($\theta = \pi/2$, $t = $ constant) is represented as embedded in Euclidean 3-space, in such a way that distances between any two nearby points (r, ϕ) and $(r + dr, \phi + d\phi)$ are correctly reproduced. Distances measured off the curved surface have no physical meaning; points off that surface have no physical meaning; and the Euclidean 3-space itself has no physical meaning. Only the curved 2-geometry has meaning. A circle of Schwarzschild coordinate radius r has proper circumference $2\pi r$ (attention limited to equatorial plane of star, $\theta = \pi/2$). Replace this circle by a sphere of proper area $4\pi r^2$, similarly for all the other circles, in order to visualize the entire 3-geometry in and around the star at any chosen moment of Schwarzschild coordinate time t. The factor $[1 - 2m(r)/r]^{-1}$ develops no singularity as r decreases within $r = 2M$, because $m(r)$ decreases sufficiently fast with decreasing r.

be independent of ϕ, whatever may be its dependence on r. Thus the whole story of the embedding is summarized by the single function, the lift,

$$z = z(r) \quad \text{(“embedding formula”).}$$

The geometry on this curved two-dimensional locus in Euclidean space (a made-up 3-space; it has nothing whatever to do with the real world) is to be identical with the geometry of the two-dimensional equatorial slice through the actual star; in other words, the line elements in the two cases are to be identical. To work out this requirement in mathematical terms, write the line element in three-dimensional Euclidean space in the form

$$ds^2 = dz^2 + dr^2 + r^2 d\phi^2. \tag{23.31}$$

Restrict to the chosen locus (“lifted surface”) by writing $z = z(r)$ or $dz = (dz/dr) dr$. Thus have

$$ds^2 = \left[1 + \left(\frac{dz(r)}{dr}\right)^2\right] dr^2 + r^2 d\phi^2 \tag{23.32}$$

on the two-dimensional locus in the 3-geometry, to be identified with

$$ds^2 = [1 - 2m(r)/r]^{-1} dr^2 + r^2 d\phi^2$$

in the actual star. Compare and conclude

$$\left(\frac{dz(r)}{dr}\right)^2 + 1 = [1 - 2m(r)/r]^{-1}. \tag{23.33}$$

This equation is information enough to find the lift as a function of r; thus,

$$z(r) = \int_0^r \frac{dr}{\left[\dfrac{r}{2m(r)} - 1\right]^{1/2}} \qquad \text{everywhere,} \tag{23.34a}$$

$$z(r) = [8M(r - 2M)]^{1/2} + \text{constant} \qquad \text{outside the star.} \tag{23.34b}$$

Outside the star this embedded surface is a segment of a paraboloid of revolution. Its form inside the star depends on how the mass, m, varies as a function of r. Recall that $m(r)$ varies as $(4\pi/3)\rho_c r^3$ near the center of the star. Conclude that the embedded surface there looks like a segment of a sphere of radius $a = (3/8\pi\rho_c)^{1/2}$; thus,

Description of embedded surface

$$[a - z(r)]^2 + r^2 = a^2 \qquad \text{for } r \ll a = (3/8\pi\rho_c)^{1/2}. \tag{23.34c}$$

In the special case of a star with uniform density (Box 23.2), the entire interior is of the spherical form (23.34c); in the general case it is not. In all cases, because $r > 2m(r)$, equation (23.34a) produces a surface with z and r as monotonically increasing functions of each other. This means that the embedded surface always opens upward and outward like a bowl; it always looks qualitatively like Figure 23.1; it never has a neck, and it never flattens out except asymptotically at $r = \infty$. At the star's surface, even though the density may drop discontinuously to zero (ρ finite inside when $p = 0$; ρ zero outside), the interior and exterior geometries will join together smoothly [dz/dr, as given by equation (23.33), is continuous].

It must be emphasized that only points lying on the embedded 2-surface have physical significance so far as the stellar geometry is concerned: the three-dimensional regions inside and outside the bowl of Figure 23.1 are physically meaningless. So is the Euclidean embedding space. It merely permits one to visualize the geometry of space around the star in a convenient manner.

Exercise 23.9. GOOD BEHAVIOR OF r

EXERCISES

Carry out explicitly the full details of the proof, at the beginning of this section, that $2m/r$ is always less than unity and r is a monotonic function of l.

Exercise 23.10. CENTER OF STAR OCCUPIED BY IDEAL FERMI GAS AT EXTREME RELATIVISTIC LIMIT

Opposite to the idealization of a star built from an incompressible fluid is the idealization in which it is built from an ideal Fermi gas [ideal neutron star; see Oppenheimer and Volkoff (1939)] at zero temperature, so highly compressed that the particles have relativistic energies,

in comparison with which any rest mass they possess is negligible. In this limit, with two particles per occupied cell of volume h^3 in phase space, one has

$$\begin{pmatrix} \text{number density} \\ \text{of fermions} \end{pmatrix} = n = (2/h^3)4\pi \int_0^{p_F} p^2 \, dp = 8\pi p_F{}^3/3h^3,$$

$$\begin{pmatrix} \text{density of} \\ \text{mass-energy} \end{pmatrix} = \rho = (2/h^3)4\pi \int_0^{p_F} cp \cdot p^2 \, dp = 2\pi c p_F{}^4/h^3,$$

and finally

$$p = -\frac{d\begin{pmatrix} \text{energy} \\ \text{per particle} \end{pmatrix}}{d\begin{pmatrix} \text{volume} \\ \text{per particle} \end{pmatrix}} = -\frac{d(\rho/n)}{d(1/n)} = 2\pi c p_F{}^4/3h^3 = \rho/3,$$

as if one were dealing with radiation instead of particles (p_F = Fermi momentum; momentum of highest occupied state).

Box 23.3 RIGOROUS DERIVATION OF THE SPHERICALLY SYMMETRIC LINE ELEMENT

Section 23.2 gave a heuristic derivation of the general spherically symmetric line element (23.7). This box attempts a more rigorous derivation, applicable to nonstatic systems, as well as static ones.

Begin with a manifold M^4 on which a metric ds^2 of Lorentz signature is defined. Assume M^4 to be spherically symmetric in the sense that to any 3×3 rotation matrix A there corresponds a mapping (rotation) of M^4, also called A ($A : M^4 \longrightarrow M^4 : \mathcal{P} \longrightarrow A\mathcal{P}$), that preserves the lengths of all curves. Further assumptions and constructions will be numbered (i), (ii), etc., so one can see what specializations are needed to get to the line element (23.7). Daggers (†) indicate assumptions that are found inapplicable to some other physically interesting situations.

For any point \mathcal{P}, form the set $s = \mathcal{S}(\mathcal{P}) = \{A\mathcal{P} \in M^4 | A \in SO(3)\}$ of all points equivalent to \mathcal{P} under rotations. Assume (i)† that s is a two-dimensional surface (except for center points, where s is zero-dimensional), and (ii) that the metric on s is that of a standard 2-sphere. Then on s one will have

$$(ds^2)_s = R^2(s) \, d\Omega^2, \qquad (1)$$

where $d\Omega^2$ is the standard metric of a unit sphere ($d\Omega^2 = d\theta^2 + \sin^2\theta \, d\phi^2$ for some θ, ϕ, defined on s), and where $2\pi R$ is the circumference of s. If M^2 is the set of all such surfaces s, then $\mathcal{S} : M^4 \longrightarrow$

$M^2 : \mathcal{P} \longrightarrow s = \mathcal{S}(\mathcal{P})$ allows one to obtain, from $R : M^2 \longrightarrow \mathcal{R} : s \longrightarrow R(s)$ [the "circumference" function on M^2 as defined by equation (1)], a corresponding function $R : M^4 \longrightarrow \mathcal{R} : \mathcal{P} \longrightarrow R(\mathcal{S}(\mathcal{P}))$ on M^4 which in some cases can eventually be used as a coordinate on M^4. (Note: \mathcal{R} denotes here the real numbers.)

Now assume (iii)† there is a spherically symmetric 4-velocity field \boldsymbol{u}, defined so that if $\mathcal{P} = \mathcal{C}(\tau)$ is one trajectory of \boldsymbol{u} with $\boldsymbol{u} = d/d\tau$, then each curve $\mathcal{P} = A\mathcal{C}(\tau)$ obtained by a rotation must also be a trajectory of \boldsymbol{u}. The orthogonal projection of \boldsymbol{u} onto any sphere s must then vanish, as there are no rotation invariant non-zero vector fields on 2-spheres. Thus \boldsymbol{u} is orthogonal to each s. Also, if two trajectories of \boldsymbol{u} start on some same sphere s, so $\mathcal{C}_1(0) = A\mathcal{C}_2(0)$, then the same rotation A will always relate them, $\mathcal{C}_1(\tau) = A\mathcal{C}_2(\tau)$, since trajectories are uniquely defined by any one point on them. Then $\mathcal{S}(\mathcal{C}_1(\tau))$ and $\mathcal{S}(\mathcal{C}_2(\tau))$ are both the same curve in M^2, whose tangent $d/d\tau$ one can call also \boldsymbol{u}; in this way one obtains a vector field \boldsymbol{u} on M^2. Give each trajectory of \boldsymbol{u} on M^2 a different label r to define a function $r(s)$ on M^2. Denote by $r = r(\mathcal{S}(\mathcal{P}))$ a corresponding function r on M^4 with $dr/d\tau = 0$. Since functions and their gradients on M^4 define corresponding quantities on M^2, inner products such as $\boldsymbol{df} \cdot \boldsymbol{dg}$ can be defined on M^2 by their values on M^4; thus, from the metric on M^4 one obtains a metric on M^2. Then by equa-

(a) Write out the relativistic equation of hydrostatic equilibrium for a substance satisfying the equation of state $p = \rho/3$.

(b) Show that there exists a well-defined analytic solution for the limiting case of infinite central density, in which $m(r)/r$ has the value $3/14$.

(c) Find $\rho(r)$, $p(r)$, and $n(r)$.

(d) Show that the number of particles out to any finite r-value is finite, despite the fact that $n(r)$ is infinite at the origin.

(e) Show that the 3-geometry has a "conical singularity" at $r = 0$.

(f) Make an "embedding diagram" for this 3-geometry ["lift" $z(r)$ as a function of r from (23.34)]. (Note that the conical singularity at $r = 0$, otherwise physically unreasonable, arises because the density of mass-energy goes to infinity at that point. Note also that the calculated mass of the system diverges to infinity as $r \longrightarrow \infty$. In actuality with decreasing density the Fermi momentum falls from relativistic to nonrelativistic values, the equation of state changes its mathematical form, and the total mass M converges to a finite value).

tion (23.5) or equivalently by drawing curves in M^2 orthogonal to the $r =$ const. lines, and giving each a different label t, one obtains coordinates with $g^{rt} = \mathbf{dr} \cdot \mathbf{dt} = 0$. Both r and t labels were assigned arbitrarily on the corresponding curves, so it is clear that transformations $t' = t'(t)$ and $r' = r'(r)$ are not excluded.

On one 2-sphere s in M^4, on the $t = 0$ hypersurface, choose a set of θ, ϕ coordinates by picking the pole ($\theta = 0$) and the prime meridian ($\phi = 0$) arbitrarily. Then extend the definition of θ, ϕ, over the $t = 0$ hypersurface by requiring θ and ϕ to be constant on curves orthogonal to each 2-sphere s, i.e., by demanding that $(\partial/\partial r)_{\theta\phi}$ be orthogonal to each s at $t = 0$. Extend the definition of θ and ϕ to $t \neq 0$ by requiring them to be constant on curves with tangent \mathbf{u}, so $(\partial/\partial t)_{r\theta\phi} \propto \mathbf{u}$. But each s is a surface of constant r and t; so $(\partial/\partial\theta)_{rt\phi}$ and $(\partial/\partial\phi)_{rt\theta}$ are tangent to s, while $\mathbf{u} \propto (\partial/\partial t)$ is orthogonal to each s. Consequently,

$$g_{t\theta} = (\partial/\partial t) \cdot (\partial/\partial\theta) = 0 \qquad (2)$$

and

$$g_{t\phi} = (\partial/\partial t) \cdot (\partial/\partial\phi) = 0 \qquad (3)$$

in the $tr\theta\phi$ coordinate system just constructed. The vector $(\partial/\partial r)_{t\theta\phi}$ does not depend on the arbitrary directions introduced in the original choice of θ, ϕ coordinates on one sphere s; it is invariant under transformations $\theta = \theta(\theta', \phi')$, $\phi = \phi(\theta', \phi')$. But nothing except θ and ϕ introduced nonrotationally invariant elements into the discussion; so $(\partial/\partial r)_{t\theta\phi}$ must be a rotationally invariant vector field (un-

like, say, $\partial/\partial\phi$); so it is, like \mathbf{u}, orthogonal to each 2-sphere s. This invariance then gives

$$g_{r\theta} = (\partial/\partial r) \cdot (\partial/\partial\theta) = 0, \qquad (4)$$
$$g_{r\phi} = (\partial/\partial r) \cdot (\partial/\partial\phi) = 0, \qquad (5)$$

which, with $g^{tr} = 0$ as previously established, gives $g_{tr} = 0$. The result is a line element of the form (23.3). Further specialization, a change of radial and time coordinates to R and T, where R is defined by (1) above and

$$\mathbf{dT} = e^\psi \left[\frac{1}{g_{rr}} \frac{\partial R}{\partial r} \mathbf{dt} - \frac{1}{g_{tt}} \frac{\partial R}{\partial t} \mathbf{dr} \right],$$

$$e^\psi = \binom{\text{integrating}}{\text{factor}},$$

followed by a change of notation, leads to Schwarzschild coordinates and the line element (23.7)—though such a transformation is possible (i.e., nonsingular) only where $\mathbf{dR} \wedge \mathbf{dT} \neq 0$:

$$(\boldsymbol{\nabla} R)^2 = \frac{(\partial R/\partial t)^2}{g_{tt}} + \frac{(\partial R/\partial r)^2}{g_{rr}} \neq 0.$$

If (iv)† spacetime is asymptotically flat, so $r \longrightarrow \infty$ is a region where the metric can take on its special relativity values, then the arbitrariness in the t coordinate, $t' = t'(t)$, can be eliminated by requiring $g_{tt} = -1$ as $r \longrightarrow \infty$. Then $(\partial/\partial t)_{r\theta\phi}$ is uniquely determined by natural requirements (independent of the arbitrary θ, ϕ, choices), and whenever it is desired to make the further physical assumption (v)† of a time-independent geometry, this can be appropriately restated as $\partial g_{\mu\nu}/\partial t = 0$.

CHAPTER **24**

PULSARS AND NEUTRON STARS; QUASARS AND SUPERMASSIVE STARS

Go, wond'rous creature, mount where Science guides,
Go, measure earth, weigh air, and state the tides;
Instruct the planets in what orbs to run,
Correct old time, and regulate the sun.

ALEXANDER POPE (1733)

§24.1. OVERVIEW

Types of stellar configurations where relativity should be important

Five kinds of stellar configurations are recognized in which relativistic effects should be significant: white dwarfs, neutron stars, black holes, supermassive stars, and relativistic star clusters. The key facts about each type of configuration are summarized in Box 24.1; and the most important details are described in the text of this chapter (white dwarfs in §24.2; neutron stars and their connection to pulsars in §§24.2 and 24.3; supermassive stars and their possible connection to quasars and galactic nuclei in §§24.4 and 24.5; and relativistic star clusters in §24.6; a detailed discussion of black holes is delayed until Chapter 33).

The book *Stars and Relativity* by Zel'dovich and Novikov (1971) presents a clear and very complete treatment of all these astrophysical applications of relativistic stellar theory. In a sense, that book can be regarded as a companion volume to this one; it picks up, with astrophysical emphasis, all the topics that this book treats with gravitational emphasis. This chapter is meant only to give the reader a brief survey of the material to be found in *Stars and Relativity*.

(continued on page 621)

Box 24.1. STELLAR CONFIGURATIONS WHERE RELATIVISTIC EFFECTS ARE IMPORTANT

[For detailed analyses and references on all these topics,
see Zel'dovich and Novikov (1971).]

A. White Dwarf Stars

Are stars of about one solar mass, with radii
about 5,000 kilometers and densities about
10^6 g/cm$^3 \sim 1$ ton/cm^3; support themselves
against gravity by the pressure of degenerate
electrons; have stopped burning nuclear fuel,
and are gradually cooling as they radiate
away their remaining store of thermal energy.
Were observed and studied astronomically long
before they were understood theoretically.
Key points in history:
 August 1926, Dirac (1926) formulated Fermi-
 Dirac statistics, following Fermi (February).
 December 1926, R. H. Fowler (1926) used
 Fermi-Dirac statistics to explain the nature
 of white dwarfs; he invoked electron de-
 generacy pressure to hold the star out
 against the inward pull of gravity.
 1930, S. Chandrasekhar (1931a,b) calculated
 white-dwarf models taking account of spe-
 cial relativistic effects in the electron-de-
 generacy equation of state; he discovered
 that *no white dwarf can be more massive
 than ~1.2 solar masses ("Chandrasekhar
 Limit")*.
 1932, L. D. Landau (1932) gave an elemen-
 tary explanation of the Chandrasekhar
 limit.
 1949, S. A. Kaplan (1949) derived the effects
 of general relativity on the mass-radius
 curve for massive white dwarfs, and de-
 duced that general relativity probably in-
 duces an instability when the radius be-
 comes smaller than 1.1×10^3 km.
Role of general relativity in white dwarfs:
 negligible influence on structure;
 significant influence on stability, on pulsation

frequencies, and on form of mass-radius
curve near the Chandrasekhar limit (i.e., in
massive white dwarfs). Electron capture also
significant. See, e.g., Zel'dovich and Novi-
kov (1971); Faulkner and Gribbin (1968).

B. Neutron Stars

Are stars of about one solar mass, with radii
about 10 km and densities about 10^{14} g/cm^3
(same as density of an atomic nucleus); are
supported against gravity by the pressure of
degenerate neutrons and by nucleon-nucleon
strong-interaction forces; are not burning nu-
clear fuel; the energy being radiated is the
energy of rotation and the remaining store of
internal thermal energy.
Theoretical calculations predicted their exist-
ence in 1934, but they were not verified to
exist observationally until 1968.
Key points in history:
 1932, neutron discovered by Chadwick (1932).
 1933-34, Baade and Zwicky (1934a,b,c) (1)
 invented the concept of neutron star; (2)
 identified a new class of astronomical ob-
 jects which they called "supernovae"; (3)
 suggested that supernovae might be created
 by the collapse of a normal star to form a
 neutron star. (See Figure 24.1.)
 1939, Oppenheimer and Volkoff (1939) per-
 formed the first detailed calculations of the
 structures of neutron stars; in the process,
 they laid the foundations of the general
 relativistic theory of stellar structure as pre-
 sented in Chapter 23. (See Figure 24.1.)
 1942, Duyvendak (1942) and Mayall and Oort
 (1942) deduced that the Crab nebula is a
 remnant of the supernova observed by Chi-

Box 24.1 (*continued*)

nese astronomers in A.D. 1054. Baade (1942) and Minkowskii (1942) identified the "south preceding star," near the center of the Crab Nebula, as probably the (collapsed) remnant of the star that exploded in 1054 (see frontispiece).

1967, Pulsars were discovered by Hewish *et al.* (1968).

1968, Gold (1968) advanced the idea that pulsars are rotating neutron stars; and subsequent observations confirmed this suggestion.

1969, Cocke, Disney, and Taylor (1969) discovered that the "south preceding star" of the Crab nebula is a pulsar, thereby clinching the connection between supernovae, neutron stars, and pulsars.

Role of general relativity in neutron stars:
 significant effects (as much as a factor of 2) on structure and vibration periods;
 gravitational radiation reaction may be the dominant force that damps nonradial vibrations.

C. Black Holes

Are objects created when a star collapses to a size smaller than twice its geometrized mass ($R < 2M \sim (M/M_\odot) \times 3$ km), thereby creating such strong spacetime curvatures that it can no longer communicate with the external universe (detailed analysis of black holes in Chapters 33 and 34).

No one who accepts general relativity has found any way to escape the prediction that black holes must exist in our galaxy. This prediction depends in no way on the complexity of the collapse that forms the black holes, or on unknown properties of matter at high density. However, the existence of black holes has not yet been verified observationally.

Key points in history:

1795, Laplace (1795) noted that, according to Newtonian gravity and Newton's corpuscular theory of light, light cannot escape from a sufficiently massive object (Figure 24.1).

1939, Oppenheimer and Snyder (1939) calculated the collapse of a homogeneous sphere of pressure-free fluid, using general relativity, and discovered that the sphere cuts itself off from communication with the rest of the universe. This was the first calculation of how a black hole can form (Figure 24.1).

1965, Beginning of an era of intensive theoretical investigation of black-hole physics.

Role of general relativity in black-hole physics:
 No sensible account of black holes possible in Newtonian theory. The physics of black holes calls on Einstein's description of gravity from beginning to end.

D. Supermassive Stars

Are stars of mass between 10^3 and 10^9 solar masses, constructed from a hot plasma of density typically less than that in normal stars; are supported primarily by the pressure of photons, which are trapped in the plasma and are in thermal equilibrium with it; burn nuclear fuel (hydrogen) at some stages in their evolution.

Theoretical calculations suggest (but *not* with complete confidence) that supermassive stars exist in the centers of galaxies and quasars, and perhaps elsewhere. Supermassive stars conceivably could be the energy sources for some quasars and galactic nuclei. However, astronomical observations have not yet yielded definitive evidence about their existence or their roles in the universe if they do exist.

Key points in history:

1963, Hoyle and Fowler (1963a,b) conceived the idea of supermassive stars, calculated their properties, and suggested that they might be associated with galactic nuclei and quasars.

1963–64, Chandrasekhar (1964a,b) and Feynman (1964) developed the general relativistic theory of stellar pulsations; and Feynman used it to show that supermassive stars, although Newtonian in structure, are subject to a general-relativistic instability.

1964 and after, calculations by many workers have elaborated on and extended the ideas of Hoyle and Fowler, but have not produced any spectacular breakthrough.

Role of general relativity in supermassive stars:

negligible influence on structure, except in the extreme case of a compact, rapidly rotating, disc-like configuration [see Bardeen and Wagoner (1971); Salpeter and Wagoner (1971)].

significant influence on stability.

E. Relativistic Star Clusters

Are clusters of stars so dense that relativistic corrections to Newtonian theory modify their structure.

Theoretical calculations suggest that relativistic star clusters might, but quite possibly do not, form in the nuclei of some galaxies and quasars; if they do try to form, they might be destroyed during formation by star-star collisions, which convert the cluster into supermassive stars or into a dense conglomerate of stars and gas. Astronomical observations have yielded no definitive evidence, as yet, about the existence of relativistic clusters.

Key points in history:

1965, Zel'dovich and Podurets (1965) conceived the idea of relativistic star clusters, developed the theory of their structure using general relativity and kinetic theory (cf. §25.7), and speculated about their stability.

1968, Ipser (1969) developed the theory of star-cluster stability and showed (in agreement with the Zel'dovich-Podurets speculations) that, when it becomes too dense, a cluster begins to collapse to form a black hole.

Role of general relativity in star clusters:

significant effect on structure when gravitational redshift from center to infinity exceeds $z_c \equiv \Delta\lambda/\lambda \sim 0.05$.

induces collapse of cluster to form black hole when central redshift reaches $z_c \approx 0.50$.

§24.2. THE ENDPOINT OF STELLAR EVOLUTION

After the normal stages of evolution, stars "die" by a variety of processes. Some stars explode, scattering themselves into the interstellar medium; others contract into a white-dwarf state; and others—according to current theory—collapse to a neutron-star state, or beyond, into a black hole. Although one knows little at present about a star's dynamic evolution into its final state, much is known about the final states themselves. The final states include dispersed nebulae, which are of no interest here; cold stellar configurations, the subject of this section; and "black holes," the subject of Part VII.

(continued on page 624)

VOLUME 45

PHYSICAL REVIEW

JANUARY 15, 1934

Proceedings
of the
American Physical Society

MINUTES OF THE STANFORD MEETING, DECEMBER 15–16, 1933

38. Supernovae and Cosmic Rays. W. BAADE, *Mt. Wilson Observatory,* AND F. ZWICKY, *California Institute of Technology.*—Supernovae flare up in every stellar system (nebula) once in several centuries. The lifetime of a supernova is about twenty days and its absolute brightness at maximum may be as high as $M_{vis} = -14^M$. The visible radiation L_v of a supernova is about 10^8 times the radiation of our sun, that is, $L_v = 3.78 \times 10^{41}$ ergs/sec. Calculations indicate that the total radiation, visible and invisible, is of the order $L_\tau = 10^7 L_v = 3.78 \times 10^{48}$ ergs/sec. The supernova therefore emits during its life a total energy $E_\tau \geq 10^5 L_\tau = 3.78 \times 10^{53}$ ergs. If supernovae initially are quite ordinary stars of mass $M < 10^{34}$ g, E_τ/c^2 is of the same order as M itself. In the *supernova* process *mass in bulk is annihilated.* In addition the hypothesis suggests itself that *cosmic rays are produced by supernovae.* Assuming that in every nebula one supernova occurs every thousand years, the intensity of the cosmic rays to be observed on the earth should be of the order $\sigma = 2 \times 10^{-3}$ erg/cm² sec. The observational values are about $\sigma = 3 \times 10^{-3}$ erg/cm² sec. (Millikan, Regener). With all reserve we advance the view that supernovae represent the transitions from ordinary stars into *neutron stars,* which in their final stages consist of extremely closely packed neutrons.

FEBRUARY 15, 1939

PHYSICAL REVIEW

VOLUME 55

On Massive Neutron Cores

J. R. OPPENHEIMER AND G. M. VOLKOFF
Department of Physics, University of California, Berkeley, California
(Received January 3, 1939)

It has been suggested that, when the pressure within stellar matter becomes high enough, a new phase consisting of neutrons will be formed. In this paper we study the gravitational equilibrium of masses of neutrons, using the equation of state for a cold Fermi gas, and general relativity. For masses under $\frac{1}{3}\odot$ only one equilibrium solution exists, which is approximately described by the nonrelativistic Fermi equation of state and Newtonian gravitational theory. For masses $\frac{1}{3}\odot < m < \frac{3}{4}\odot$ two solutions exist, one stable and quasi-Newtonian, one more condensed, and unstable. For masses greater than $\frac{3}{4}\odot$ there are no static equilibrium solutions. These results are qualitatively confirmed by comparison with suitably chosen special cases of the analytic solutions recently discovered by Tolman. A discussion of the probable effect of deviations from the Fermi equation of state suggests that actual stellar matter after the exhaustion of thermonuclear sources of energy will, if massive enough, contract indefinitely, although more and more slowly, never reaching true equilibrium.

Figure 24.1.
Two important arrivals on the astrophysical scene:
the neutron star (1933) and the black hole (1795, 1939).
No proper account of either can forego general relativity.

EXPOSITION

DU SYSTEME

DU MONDE,

Par Pierre-Simon LAPLACE,
de l'Institut National de France, et
du Bureau des Longitudes.

TOME SECOND.

A PARIS,

De l'Imprimerie du Cercle-Social, rue du
Théâtre Français, N°. 4.

L'an IV de la République Française.

aussi sensibles à la distance qui nous en sepaparee ; et combien ils doivent surpasser ceux
que nous observons à la surface du soleil ?
Tous ces corps devenus invisibles , sont à
la même place où ils ont été observés , puisqu'ils n'en ont point changé , durant leur apparition ; il existe donc dans les espaces célestes , des corps obscurs aussi considérables ,
et peut être en aussi grand nombre , que les
étoiles. Un astre lumineux de même densité
que la terre , et dont le diamètre serait deux
cents cinquante fois plus grand que celui du
soleil , ne laisserait en vertu de son attraction , parvenir aucun de ses rayons jusqu'à
nous ; il est donc possible que les plus grands
corps lumineux de l'univers , soient par cela
même , invisibles. Une étoile qui , sans être de
cette grandeur , surpasserait considérablement
le soleil ; affaiblirait sensiblement la vîtesse
de la lumière , et augmenterait ainsi l'étendue
de son aberration. Cette différence dans l'aberration des étoiles ; un catalogue de celles qui
ne font que paraître , et leur position observée
au moment de leur éclat passager ; la détermination de toutes les étoiles changeantes ,

VOLUME 56

PHYSICAL REVIEW

SEPTEMBER 1, 1939

On Continued Gravitational Contraction

J. R. Oppenheimer and H. Snyder
University of California, Berkeley, California
(Received July 10, 1939)

When all thermonuclear sources of energy are exhausted a sufficiently heavy star will
collapse. Unless fission due to rotation, the radiation of mass, or the blowing off of mass by
radiation, reduce the star's mass to the order of that of the sun, this contraction will continue
indefinitely. In the present paper we study the solutions of the gravitational field equations
which describe this process. In I, general and qualitative arguments are given on the
behavior of the metrical tensor as the contraction progresses: the radius of the star approaches asymptotically its gravitational radius; light from the surface of the star is progressively reddened, and can escape over a progressively narrower range of angles. In II, an
analytic solution of the field equations confirming these general arguments is obtained for the
case that the pressure within the star can be neglected. The total time of collapse for an observer comoving with the stellar matter is finite, and for this idealized case and typical stellar
masses, of the order of a day; an external observer sees the star asymptotically shrinking to
its gravitational radius.

What does one mean in principle by the term "the final state of stellar evolution"? Start with a star containing a given number, A, of baryons and let it evolve to the absolute, burned-out end point of thermonuclear combustion (minimum mass-energy possible for the A-baryon system). If the normal course of thermonuclear combustion is too slow, speed it up by catalysis. If an explosion occurs, collect the outgoing matter, extract its kinetic energy, and let it fall back onto the system. Repeat this operation as many times as needed to arrive at burnout (cold Fe^{56} for the part of the system under modest pressure; other nuclear species in the region closer to the center; "cold matter catalyzed to the end point of thermonuclear combustion" throughout). End up finally with the system in its absolutely lowest energy state, with all angular momentum removed and all heat extracted, so that it sits at the absolute zero of temperature and has zero angular velocity. Such a "dead" system, depending upon its mass and prior history (two distinct energy minima for certain A-values), ends up as a cold stellar configuration (neutron star, or "white" dwarf), or as a "dead" black hole.

The analysis of a cold stellar configuration demands an equation of state. The temperature is fixed at zero; the nuclear composition in principle is specified uniquely by the density; and therefore the pressure is also fixed uniquely once the density has been specified [equation of state $p(\rho)$ for "cold catalyzed matter"].

The white dwarfs and neutron stars observed by astronomers are not really built of cold catalyzed matter. However, the matter in them is sufficiently near the end point of thermonuclear evolution and sufficiently cold that it can be idealized with fair accuracy as cold and catalyzed (see §23.4).

The equation of state, $\rho(p)$, for cold catalyzed matter is shown graphically in Figure 24.2. This version of the equation of state was constructed by Harrison and Wheeler in 1958. Other versions constructed more recently [see Cameron (1970) and Baym, Bethe, and Pethick (1971) for references] are almost identical to the Harrison-Wheeler version at densities well below nuclear densities, $\rho < 3 \times 10^{13}$ g/cm^3. At nuclear and supernuclear densities, all versions differ because of differing assumptions about nucleon-nucleon interactions. Along with the equation of state, in Figure 24.2 are shown properties of the models of cold stars constructed from this equation of state by integrating numerically the equations of structure (23.28).

The equation of state can be understood by following the transformations that occur as a sample of cold catalyzed matter is compressed to higher and higher densities. At each stage in the compression, each possible thermonuclear reaction is to be catalyzed to its endpoint and the resultant thermal energy is to be removed.

When the sample is at zero pressure, it is a ball of pure, cold Fe^{56}, since Fe^{56} is the most tightly bound of all nuclei. It has the density 7.86 g/cm^3. As the sample is compressed, its internal pressure is provided at first by normal solid-state forces; but the atoms are soon squeezed so closely together that the electrons become quite oblivious of their nuclei, and begin to form a degenerate Fermi gas. By the time a density of $\rho = 10^5$ g/cm^3 has been reached, valence forces are completely negligible, the degenerate electron pressure dominates, and the compressibility index, γ (see legend for Figure 24.2), is 5/3, the value for a nonrelativistically degenerate Fermi gas. Between 10^5 and 10^7 g/cm^3, the pressure-providing electrons gradually

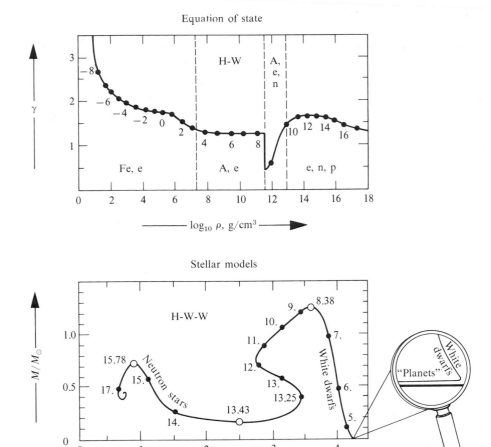

Figure 24.2.

The Harrison-Wheeler equation of state for cold matter at the absolute end point of thermonuclear evolution, and the corresponding Harrison-Wakano-Wheeler stellar models. The equation of state is exhibited in the form of a plot of "compressibility index,"

$$\gamma = \frac{\rho + p}{p} \frac{dp}{d\rho},$$

as a function of density of mass-energy, ρ. (Small γ corresponds to easy compressibility.) The curve is parameterized by the logarithm of the pressure, $\log_{10} p$, in units of g/cm³ [same units as ρ; note that $p(\text{g/cm}^3) = (1/c^2) \times p(\text{dyne/cm}^2)$]. The chemical composition of the matter as a function of density is indicated as follows: Fe, Fe^{56} nuclei; A, nuclei more neutron rich than Fe^{56}; e, electrons; n, free neutrons; p, free protons.

The first law of thermodynamics [equation (22.6)], when applied to cold matter (zero entropy) says $d\rho/(\rho + p) = dn/n$; i.e.,

$$n = \frac{\rho + p}{\mu_{Fe}/56} \exp\left(-\int_0^p \frac{dp}{\rho + p}\right).$$

Here μ_{Fe}, the rest mass of an Fe^{56} atom, is the ratio between $\rho + p \approx \rho$ and $n/56$ in the limit of zero density. From this equation and a knowledge of $\rho(p)$—see Figure—one can calculate $n(p)$.

The equilibrium configurations are represented by curves of total mass-energy, M, versus radius, R. (R is defined such that $4\pi R^2$ is the star's surface area.) The $M(R)$ curve is parameterized by the logarithm of the central density, $\log_{10} \rho_c$, measured in g/cm³. Only configurations along two branches of the curve are stable against small perturbations and can therefore exist in nature: the white dwarfs, with $\log_{10} \rho_c < 8.38$, and the neutron stars, with $13.43 < \log_{10} \rho_c < 15.78$ (see Box 26.1).

For greater detail on both the equation of state and the equilibrium configurations, see Harrison, Thorne, Wakano, and Wheeler (1965); also, for an updated table of the equation of state, see Hartle and Thorne (1968).

become relativistically degenerate, and γ approaches 4/3. Above $\rho = 1.4 \times 10^7$ g/cm³, the rest mass of 62 Fe_{26}^{56} nuclei, plus the rest mass of 44 electrons, plus the rather large Fermi kinetic energy of 44 electrons at the top of the Fermi sea, exceeds the rest mass of 56 Ni_{28}^{62} nuclei. Consequently, as the catalyzed sample of matter is compressed past $\rho = 1.4 \times 10^7$ g/cm³, the nuclear reaction

$$62 \ Fe_{26}^{56} \begin{pmatrix} \text{highly compressed} \\ \text{neutral atoms} \end{pmatrix} \longrightarrow$$

$$56 \ Ni_{28}^{62} \begin{pmatrix} \text{highly compressed} \\ \text{neutral atoms} \end{pmatrix}$$

(24.1)

goes to its end point, with a release of energy. As the compression continues beyond this point, the rising Fermi energy of the electrons induces new nuclear reactions similar to (24.1), but involving different nuclei. In these reactions more and more electrons are swallowed up to form new nuclei, which are more and more neutron-rich. When the density reaches $\rho = 3 \times 10^{11}$ g/cm³, the nuclei are so highly neutron-rich (Y_{39}^{122}) that neutrons begin to drip off them. The matter now becomes highly compressible for a short time ($3 \times 10^{11} \lesssim \rho \lesssim 4 \times 10^{11}$), since most of the remaining electrons are swallowed up very rapidly by the dripping nuclei. Above $\rho \sim 4 \times 10^{11}$ g/cm³ free neutrons become plentiful and their degeneracy pressure exceeds that of the electrons. Further compression to $\rho \sim 10^{13}$ g/cm³ completely disintegrates the remaining nuclei, leaving the sample almost pure neutrons with $\gamma = 5/3$, the value for a nonrelativistically degenerate Fermi gas. Intermixed with the neutrons are just enough degenerate electrons to prevent the neutrons from decaying, and just enough protons to maintain charge neutrality. Compression beyond $\rho \sim 10^{13}$ g/cm³ pushes the sample into the domain of nuclear densities where the physics of matter is only poorly understood. This Harrison-Wheeler version of the equation of state ignores all nucleon-nucleon interactions at and above nuclear densities; it idealizes matter as a noninteracting mixture of neutrons, protons, and electrons with neutrons dominating; and it shows a compressibility index of 5/3 while the neutrons are nonrelativistic, but 4/3 after they attain relativistic Fermi energies. Other versions of the equation of state attempt to take into account the nucleon-nucleon interactions in a variety of ways [see Cameron (1970), Baym, Bethe, and Pethick (1971), and many references cited therein].

Equilibrium configurations for cold, catalyzed matter:

(1) forms and stability

Corresponding to each value of the central density, ρ_c, there is one stellar equilibrium configuration. Equilibrium, yes; but is the equilibrium stable? Stability studies (Chapter 26, especially Box 26.1) show that many of the models are unstable against small radial perturbations, which lead to gravitational collapse. Only white-dwarf stars in the range $\log_{10} \rho_c < 8.4$ and neutron stars in the range $13.4 \lesssim \log_{10} \rho_c \lesssim 15.8$ are stable. Instability for the region of $\log_{10} \rho_c$ values between 8.4 and 13.4 is caused by a combination of (1) relativistic strengthening of the gravitational forces, and (2) high compressibility of the matter due to electron capture and neutron drip by

the atomic nuclei. Neutron stars are stable for a simple reason. Neutron-dominated matter is so difficult to compress that even the relativistically strengthened gravitational forces cannot overcome it. Above $\log_{10} \rho_c \sim 15.8$, the gravitational forces become strong enough to win out over the pressure of the nuclear matter, and the stars are all unstable. [See Gerlach (1968) for the possibility—which, however, he rates as unlikely—that there might exist a third family of stable equilibrium configurations, additional to white dwarfs and neutron stars.]

The white-dwarf stars have masses below 1.2 M_\odot and radii between ~ 3000 and $\sim 20,000$ km. They are supported almost entirely by the pressure of the degenerate electron gas. Relativistic deviations from Newtonian structure are only a fraction of a per cent, but relativistic effects on stability and pulsations are important from $\rho_c \approx 10^8$ g/cm^3 to the upper limit of the white-dwarf family at $\rho_c = 10^{8.4}$ g/cm^3 [see, e.g., Faulkner and Gribbin (1968)]. The properties of white-dwarf models are fairly independent of whose version of the equation of state is used in the calculations. \qquad (2) white-dwarf stars

The properties of neutron stars are moderately dependent on the equation of state used. However, all versions lead to upper and lower limits on the mass and central density. The correct lower limits probably lie in the range \qquad (3) neutron stars

$$13.4 \lesssim \log_{10} \rho_{c\,\min} \lesssim 14.0,$$
$$0.05\, M_\odot \lesssim M_{\min} \lesssim 0.2\, M_\odot; \qquad (24.2)$$

the correct upper limits are probably in the range

$$15.0 \lesssim \log_{10} \rho_{c\,\max} \lesssim 16.0,$$
$$0.5\, M_\odot \lesssim M_{\max} \lesssim 3\, M_\odot \qquad (24.3)$$

[see Rhoades (1971)]. Neutron stars typically have radii between ~ 6 km and ~ 100 km. Relativistic deviations from Newtonian structure are great, sometimes more than 50 per cent.

It appears certain that no cold stellar configuration can have a mass exceeding $\sim 5\, M_\odot$ [Rhoades (1971)] (1.2 M_\odot according to the Harrison-Wheeler equation of state, Figure 24.2). Any star more massive than this must reduce its mass below this limit if it is to fade away into quiet obscurity, otherwise relativistic gravitational forces will eventually pull it into catastrophic gravitational collapse past white-dwarf radii, past neutron-star radii, and into a black hole a few kilometers in size (see Part VII). \qquad (4) black holes

§24.3. PULSARS

Theory predicts that, when a star more massive than the Chandrasekhar limit of 1.2 M_\odot has exhausted the nuclear fuel in its core and has compressed its core to white-dwarf densities, an instability pushes the star into catastrophic collapse. The

Birth of a neutron star by
stellar collapse

core implodes upon itself until nucleon-nucleon repulsion halts the implosion. The result is a neutron star, unless the core's mass is so great that gravity overcomes the nucleon-nucleon repulsion and pulls the star on in to form a black hole. Not all the star's mass should become part of the neutron star or black hole. Much of it, perhaps most, can be ejected into interstellar space by the violence that accompanies the collapse—violence due to flash nuclear burning, shock waves, and energy transport by neutrinos ("stick of dynamite in center of star, ignited by collapse").

Dynamics of a newborn
neutron star

The collapsed core holds more interest for gravitation theory than the ejected envelope. That core, granted a mass small enough to avoid the black-hole fate, will initially be a hot, wildly pulsating, rapidly rotating glob of nuclear matter with a strong, embedded magnetic field (see Figure 24.3). The pulsations must die out quickly. They emit a huge flux of gravitational radiation, and radiation reaction damps them in a characteristic time of ~ 1 second [see Wheeler (1966); Thorne (1969a)]. Moreover, the pulsations push and pull elementary particle reactions back and forth by raising and lowering the Fermi energies in the core's interior; these particle reactions can convert pulsation energy into heat at about the same rate as the pulsation energy is radiated by gravity. [See Langer and Cameron (1969); also §11.5 of Zel'dovich and Novikov (1971) for details and references.]

The result, after a few seconds, is a rapidly rotating centrifugally flattened neutron star with a strong (perhaps 10^{12} gauss) magnetic field; all the pulsations are gone. If the star is deformed from axial symmetry (e.g., by centrifugal forces or by a nonsymmetric magnetic field), its rotation produces a steady outgoing stream of gravitational waves, which act back on the star to remove rotational energy. Whether or not this occurs, the rotating magnetic field itself radiates electromagnetic waves. They slow the rotation and transport energy into the surrounding, exploding gas cloud (nebula). [See Pacini (1968), Goldreich and Julian (1968), and Ostriker and Gunn (1969) for basic considerations.]

Neutron star as a pulsar

Somehow, but nobody understands in detail how, the rotating neutron star beams coherent radio waves and light out into space. Each time the beam sweeps past the Earth optical and radio telescopes see a pulse of radiation. The light is emitted synchronously with the radio waves, but the light pulses reach Earth earlier (~ 1 second for the pulsar in the crab nebula) because of the retardation of the radio waves by the plasma along the way. This is the essence of the 1973 theory of pulsars, accepted by most astrophysicists.

Pulsar radiation as a tool for
studying neutron stars

Although the mechanism of coherent emission is not understood, the pulsar radiation can nevertheless be a powerful tool in the experimental study of neutron stars. Anything that affects the stellar rotation rate, even minutely (fractional changes as small as 10^{-9}) will produce measurable irregularities in the timing of the pulses at Earth. If the star's crust and mantle are crystalline, as 1973 theory predicts, they may be subject to cracking, faulting, or slippage ("starquake") that changes the moment of inertia, and thence the rotation rate. Debris falling into the star will also change its rotation. Whichever the cause, after such a disturbance the star may rotate differentially for awhile; and how it returns to rigid rotation may depend on such phenomena as superfluidity in its deep interior. Thus, pulsar-timing data may eventually give information about the interior and crust of the neutron star, and

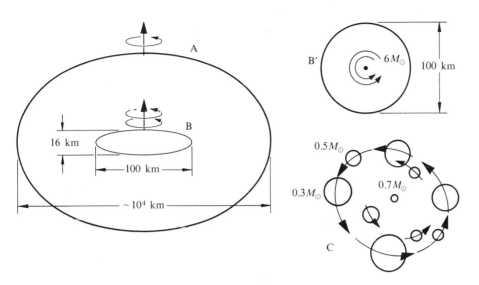

Figure 24.3.
"Collapse, pursuit, and plunge scenario" [schematic from Ruffini and Wheeler (1971b)].

- A star with white-dwarf core (A), slowly rotating,
 - evolves by straightforward astrophysics,
 - arrives at the point of gravitational instability,
 - collapses, and
 - ends up as a rapidly spinning neutron-star pancake (B,B′).
- It then fragments (C) because it has too much angular momentum to collapse into a single stable object. If the substance of the neutron-star pancake were an incompressible fluid, the fragmentation would have a close tie to well-known and often observed phenomena ("drop formation"). However, the more massive a neutron star is, the smaller it is, so one's insight into this and subsequent stages of the scenario are of necessity subject to correction or amendment. One can not today guarantee that fragmentation takes place at all; nevertheless, fragmentation will be assumed in what follows.
- The fragments dissipate energy and angular momentum via gravitational radiation.
- One by one as they revolve they coalesce ("pursuit and plunge scenario").
- In each such plunge a pulse of gravitational radiation emerges.
- Fragments of debris fall onto the coalesced objects (neutron stars or black holes, as the case may be), changing their angular momenta.
- Eventually the distinct neutron stars or black holes or both unite into one such collapsed object with a final pulse of gravitational radiation.
- The details of the complete scenario differ completely from one evolving star to another, depending on
 - the mass of its core, and
 - the angular momentum of this core.
- An entirely different kind of picture therefore has to be drawn for altered values of these two parameters.
- Even for the values of these parameters adopted in the drawing, the present picture can at best possess only qualitative validity.
- Detailed computer analysis would seem essential for any firm prediction about the course of any selected scenario.

thence (by combination with theory) about its mass and radius. These issues are discussed in detail in a review article by Ruderman (1972) as well as in Zel'dovich and Novikov (1971).

§24.4. SUPERMASSIVE STARS AND STELLAR INSTABILITIES

Theory of the stability of Newtonian stars

When a Newtonian star of mass M oscillates adiabatically in its fundamental mode, the change in its radius, δR, obeys a harmonic-oscillator equation,

$$M\,\delta\ddot{R} = -k\,\delta R, \tag{24.4}$$

with a "spring constant" k that depends on the star's mean adiabatic index $\bar{\Gamma}_1$ [recall: $\Gamma_1 \equiv (n/p)(\partial p/\partial n)_{\text{const. entropy}}$], on its gravitational potential energy Ω, on the trace $I = \int \rho r^2\, d\mathcal{V}$ of the second moment of its mass distribution, and on its mass M,

$$k = 3M(\bar{\Gamma}_1 - 4/3)|\Omega|/I \tag{24.5}$$

(See Box 24.2). If $\bar{\Gamma}_1 > 4/3$ the Newtonian star is stable and oscillates; if $\bar{\Gamma}_1 < 4/3$ the star is unstable and either collapses or explodes, depending on its initial conditions and overall energetics. This result is a famous theorem in Newtonian stellar theory—but it is relevant only for adiabatic oscillations.

Box 24.2 OSCILLATION OF A NEWTONIAN STAR

The following is a volume-averaged analysis of the lowest mode of radial oscillation. Such analyses are useful in understanding the qualitative behavior and stability of a star. [See Zel'dovich and Novikov (1971) for an extensive exploitation of them.] However, for precise quantitative results, one must perform a more detailed analysis [see, e.g., Ledoux and Walraven (1958); also Chapter 26 of this book].

1. Let $M =$ star's total mass

 $R =$ star's radius

 $\bar{\rho} =$ mean density $= (3/4\pi)M/R^3$

 $\bar{p} =$ mean pressure

 $\bar{\Gamma}_1 =$ mean adiabatic index $= (\bar{n}/\bar{p})(\partial \bar{p}/\partial \bar{n})_{\text{adiabatic}}$

 $= (\bar{\rho}/\bar{p})(\partial \bar{p}/\partial \bar{\rho})_{\text{adiabatic}}$ in Newtonian limit, where $\rho = \text{const.} \times n$.

2. Then the mean pressure-buoyancy force \bar{F}_{buoy} and the counterbalancing gravitational force \bar{F}_{grav} in the equilibrium star are

$$\bar{F}_{\text{buoy}} = \bar{p}/R$$
$$= \bar{F}_{\text{grav}} = \bar{\rho}M/R^2 = (4\pi/3)\bar{\rho}^2 R.$$

3. When the oscillating star has expanded or contracted so its radius is $R + \delta R$, then its mean density will have changed to

$$\bar{\rho} + \delta\bar{\rho} = (3/4\pi)M[R^{-3} + \delta(R^{-3})] = \bar{\rho} - 3(\bar{\rho}/R)\,\delta R,$$

and its mean pressure will be

$$\bar{p} + \delta\bar{p} = \bar{p} + (\bar{p}/\bar{\rho})\bar{\Gamma}_1\,\delta\bar{\rho} = \bar{p} - 3(\bar{\Gamma}_1\bar{p}/R)\,\delta R.$$

The corresponding changes in the forces will be

$$\delta\bar{F}_{\text{buoy}} = \frac{\delta\bar{p}}{R} - \frac{\bar{p}}{R^2}\delta R = -(3\Gamma_1 + 1)\frac{\bar{p}}{R}\frac{\delta R}{R} = -(3\bar{\Gamma}_1 + 1)\bar{F}_{\text{buoy}}\left(\frac{\delta R}{R}\right),$$

$$\delta\bar{F}_{\text{grav}} = \left(\frac{4\pi}{3}\right)(2\bar{\rho}R\,\delta\bar{\rho} + \bar{\rho}^2\,\delta R) = \left(\frac{4\pi}{3}\bar{\rho}^2R\right)\left(-5\frac{\delta R}{R}\right) = -5\bar{F}_{\text{grav}}\left(\frac{\delta R}{R}\right).$$

Consequently, the restoring force will be (recall: $\bar{F}_{\text{buoy}} = \bar{F}_{\text{grav}}$)

$$\delta\bar{F}_{\text{grav}} - \delta\bar{F}_{\text{buoy}} = 3\left(\bar{\Gamma}_1 - \frac{4}{3}\right)\bar{F}_{\text{grav}}\frac{\delta R}{R}.$$

4. This restoring force produces an acceleration,

$$\delta\bar{F}_{\text{grav}} - \delta\bar{F}_{\text{buoy}} = -\bar{\rho}\,\delta\ddot{R}.$$

Hence, the equation of motion for the oscillations is

$$\delta\ddot{R} = -3(\bar{\Gamma}_1 - 4/3)(4\pi/3)\bar{\rho}\,\delta R,$$

corresponding to a "spring constant" k and angular frequency of oscillation ω, given by $\omega^2 = 4\pi(\bar{\Gamma}_1 - 4/3)\bar{\rho}$, and $k = M\omega^2$.

5. A more nearly exact analysis (see exercise 39.7 for details, or Box 26.2 for an alternative derivation) yields the improved formula

$$\omega^2 = 3(\bar{\Gamma}_1 - 4/3)|\Omega|/I,$$

$$\Omega = \binom{\text{star's self-gravitational}}{\text{energy}} = \frac{1}{2}\int \rho\Phi\,d\mathcal{V} = -\frac{1}{2}\int \frac{\rho\rho'}{|\mathbf{x} - \mathbf{x}'|}\,d\mathcal{V}\,d\mathcal{V}',$$

$$I = \binom{\text{trace of second moment of}}{\text{star's mass distribution}} = \int \rho r^2\,d\mathcal{V},$$

for the square of the oscillation frequency.

6. Note that $\bar{\Gamma}_1 > 4/3$ corresponds to stable oscillations; $\bar{\Gamma}_1 < 4/3$ corresponds to exponentially developing collapse or explosion.

In a real star no oscillation is precisely adiabatic. The oscillations in temperature cause corresponding oscillations in the stellar opacity and in nuclear burning rates. These insert energy into or extract energy from the gas vibrations.

All main-sequence stars thus far observed and studied have masses below 60 M_\odot. For such small masses, theory predicts low enough temperatures that gas pressure dominates over radiation pressure, and the adiabatic index is nearly that of nonrelativistic gas, $\bar{\Gamma}_1 \approx 5/3$. Such stars vibrate stably. The net effect of the oscillating opacity and burning rate is usually to extract energy from the vibrations. Thus, they damp. (The vibrations of Cepheid variable stars are a notable exception.)

Stability theory predicts "engine-driven oscillations" and quick death for stars of $M > 60M_\odot$

No one has yet seen a main-sequence star with mass above about 60 M_\odot. This is explained as follows. For masses above 60 M_\odot, the temperature should be so high that radiation pressure dominates over gas pressure, and the adiabatic index $\bar{\Gamma}_1$ is only slightly above the value 4/3 for pure radiation. Consequently the "spring constant" of the star, although positive, is very small. On the inward stroke of an oscillation, the central temperature rises, and nuclear burning speeds up. (The nuclear burning rate goes as a very high power of the central temperature; for example, in a massive star HCNO burning releases energy at a rate $\varepsilon_{\mathrm{HCNO}} \propto T_c^{11}$.) Because the spring constant is so small, the inward stroke lasts for a long time, and the enhanced nuclear burning produces a significant excess of thermal energy and pressure. Hence, on the outward stroke the star expands more vigorously than it contracted ("engine"). Successive vibrations are driven to higher and higher amplitudes. Eventually, calculations suggest, the star either explodes, or it ejects enough mass by its vigorous vibrations to drop below the critical limit of $M \sim 60\ M_\odot$. Hence, stars of mass above 60 M_\odot should not live long enough that astronomers could have a reasonable probability of discovering them.

Possible existence of supermassive stars

Of course, this "engine action" does not prevent massive stars from forming, living a short time, and then disrupting themselves. Such a possibility is particularly intriguing for *supermassive stars* [M between $10^3\ M_\odot$ and $10^9\ M_\odot \sim 0.01 \times$ (mass of a galaxy)]. Although such stars may be exceedingly rare, by their huge masses and huge release of explosive energy they might play an important role in the universe. Moreover, it is conceivable that the oscillations of such stars, like those of Cepheid variables, might be sustained at large amplitudes for long times (a million years?), with nonlinear damping processes preventing their further growth.

Relativistic instabilities in a supermassive star

Theory predicts that general relativistic effects should strongly influence the oscillations of a supermassive star. The increase in "gravitational force," δF_{grav}, acting on a shell of matter on the inward stroke is greater in general relativity than in Newtonian theory, and the decrease on the outward stroke is also greater. Consequently the "effective index" $\Gamma_{1\,\mathrm{crit}}$ of gravitational forces is increased above the Newtonian value of 4/3; thus,

$$\begin{pmatrix} \text{fractional increase in} \\ \text{"pressure-like force of} \\ \text{gravity" per unit fractional} \\ \text{change in baryon-number} \\ \text{density} \end{pmatrix} \equiv \Gamma_{1\,\mathrm{crit}} = (4/3) + \alpha(M/R) + O(M^2/R^2), \quad (24.6)$$

where α is a constant of the order of unity that depends on the structure of the star (see Box 26.2). To resist gravity, one has only the elasticity of the relativistic material of the star:

$$\begin{pmatrix} \text{fractional increase in} \\ \text{``pressure-like resisting} \\ \text{force'' per unit fractional} \\ \text{change in baryon number} \\ \text{density} \end{pmatrix} = \bar{\Gamma}_1 = \left\langle \frac{p}{n}\left(\frac{\partial p}{\partial n}\right)_s \right\rangle_{\substack{\text{effective average} \\ \text{over star}}} . \qquad (24.7)$$

The effective spring constant for the vibrations of the star is governed by the delicate margin between these two indices:

$$k = \begin{pmatrix} \text{effective} \\ \text{spring constant} \end{pmatrix} = \begin{pmatrix} \text{contribution} \\ \text{of ``elastic} \\ \text{forces''} \end{pmatrix} - \begin{pmatrix} \text{contribution} \\ \text{of gravity} \end{pmatrix}$$

$$= 3M(\bar{\Gamma}_1 - \Gamma_{1\,\text{crit}})\frac{|\Omega|}{I}. \qquad (24.8)$$

(derivation in Chapter 26). The relativistic rise in the effective index of gravity above 4/3 [equation (24.6)] brings on the transition from stability (positive k; vibration) to instability (negative k; explosion or collapse) under conditions when one otherwise would have expected stability. For supermassive stars, Fowler and Hoyle (1964) show that

$$\bar{\Gamma}_1 = 4/3 + \zeta(M/M_\odot)^{-1/2},$$

where ζ is a constant of order unity. As a newly formed supermassive star contracts inward, heating up, but not yet hot enough to ignite its nuclear fuel, it approaches nearer and nearer to instability against collapse. Unless burning halts the contraction, collapse sets in at a radius R_{crit} given by

$$\bar{\Gamma}_1 = 4/3 + \zeta(M/M_\odot)^{-1/2} = \Gamma_{1\,\text{crit}} = 4/3 + \alpha M/R;$$

i.e.,

$$R = (\alpha/2\zeta)(M/M_\odot)^{1/2} \times (\text{Schwarzschild Radius})$$
$$\sim 10^4 \times (\text{Schwarzschild Radius}) \text{ if } M = 10^8\, M_\odot.$$

The relativistic instability occurs far outside the Schwarzschild radius when the star is very massive. Relativity hardly modifies the star's structure at all; but because of the delicate balance between $\delta\bar{F}_{\text{grav}}$ and $\delta\bar{F}_{\text{buoy}}$ in the Newtonian oscillations (Box 24.2), tiny relativistic corrections to these forces can completely change the stability.

 In practice, the story of a supermassive star is far more complicated than has been indicated here. Rotation can stabilize it against relativistic collapse for a while. However, after the star has lost all angular momentum in excess of the critical value

Temporary stabilization by rotation

$J_{\text{crit}} = M^2$ ("extreme Kerr limit"; see Chapter 33), and after it has contracted to near the Schwarzschild radius, rotation is helpless to stave off implosion. Depending on its mass and angular momentum, the star may ignite its fuel before or after relativistic collapse begins, and before or after implosion through the Schwarzschild radius. When the fuel is ignited, it can wreak havoc, because even if the star is not then imploding, its adiabatic index will be very near the critical one, and the burning may drive oscillations to higher and higher amplitudes. These processes are so complex that in 1973 one is far from having satisfactory analyses of them, but for reviews of what is known and has been done, the reader can consult Fowler (1966), Thorne (1967), and Zel'dovich and Novikov (1971).

The theory of stellar pulsations in general relativity is presented for Track-2 readers in Chapter 26 of this book.

§24.5. QUASARS AND EXPLOSIONS IN GALACTIC NUCLEI

Supermassive stars were first conceived by Hoyle and Fowler (1963a,b) as an explanation for explosions in the nuclei of galaxies. Shortly thereafter, when quasars were discovered, Hoyle and Fowler quite naturally appealed to their supermassive stars for an explanation of these puzzles as well. Whether galactic explosions or quasars are driven by supermassive stars remains a subject of debate in astronomical circles even as this book is being finished, in 1973. Hence, this book will avoid the issue except for the following remark.

Whatever is responsible for quasars and galactic explosions must be a machine of great mass ($M \sim 10^6$ to $10^{10}\, M_\odot$) and small radius (light-travel time across the machine, as deduced from light variations, is sometimes less than a day). The machine might be a coherent object, i.e., a supermassive star; or it might be a dense mixture of ordinary stars and much gas. Actually these two possibilities may not be distinct. Star-star collisions in a dense cluster can lead to stellar coalescence and the gradual building up of one or more supermassive stars [Sanders (1970); Spitzer (1971); Colgate (1967)]. Thus, at one stage in its life, a galactic nucleus or quasar might be driven by collisions in a dense star cluster; and at a later stage it might be driven by a supermassive star; and at a still later stage that star might collapse to leave behind a massive black hole (10^6–$10^9\, M_\odot$), but a black hole that is still "live" and active (Chapter 33).

§24.6. RELATIVISTIC STAR CLUSTERS

The normal astrophysical evolution of a galactic nucleus is estimated [Sanders (1970); Spitzer (1971)] to lead under some circumstances to a star cluster so dense that general relativity influences its structure and evolution. The theory of relativistic star clusters is closely related to that of relativistic stars, as developed in Chapter 23. A star is a swarm of gas molecules that collide frequently; a star cluster is a swarm of stars that collide rarely. But the frequency of collisions is relatively unim-

portant in a steady state. For the theory of relativistic star clusters, see: §25.7 of this book; Zel'dovich and Podurets (1965); Fackerell, Ipser, and Thorne (1969); Chapter 12 of Zel'dovich and Novikov (1971); and references cited there. A relativistic star cluster is a latent volcano. No future is evident for it except to evolve with enormous energy release to a massive black hole, either by direct collapse (possibly a star at a time) or by first coalescing into a supermassive star that later collapses.

CHAPTER **25**

THE "PIT IN THE POTENTIAL" AS THE CENTRAL NEW FEATURE OF MOTION IN SCHWARZSCHILD GEOMETRY

"Eccentric, intervolved, yet regular
Then most, when most irregular they seem;
And in their motions harmony divine"

MILTON, 1665

Overview of this chapter

§25.1. FROM KEPLER'S LAWS TO THE EFFECTIVE POTENTIAL FOR MOTION IN SCHWARZSCHILD GEOMETRY

No greater glory crowns Newton's theory of gravitation than the account it gives of the principal features of the solar system: a planet in its motion sweeps out equal areas in equal times; its orbit is an ellipse, with one focus at the sun; and the cube of the semimajor axis, a, of the ellipse, multiplied by the square of the average angular velocity of the planet in its orbit ($\omega = 2\pi/\text{period}$) gives a number with the dimensions of a length, the same number for all the planets (Box 25.1), equal to the mass of the sun:

$$M = \omega^2 a^3.$$

Exactly the same is true for the satellites of Jupiter (Figure 25.1), and of the Earth (Box 25.1), and true throughout the heavens. What more can one possibly expect of Einstein's theory of gravity when it in its turn grapples with this centuries-old theme of a test object moving under the influence of a spherically symmetric center of attraction? The principal new result can be stated in a single sentence: The particle is governed by an "effective potential" (Figure 25.2 and §§25.5, 25.6) that possesses not only (1) the long distance $-M/r$ attractive behavior and (2) the shorter distance

(angular momentum)$^2/r^2$ repulsive behavior of Newtonian gravitational theory, but also (3) at still shorter distances *a pit in the potential,* which (1) captures a particle that comes too close; (2) establishes a critical distance of closest approach for this black-hole capture process; (3) for a particle that approaches this critical point without crossing it, lengthens the turn-around time as compared to Newtonian expectations; and thereby (4) makes the period for a radial excursion longer than the period of a revolution; (5) causes an otherwise Keplerian orbit to precess; and (6) deflects a fast particle and a photon through larger angles than Newtonian theory would predict.

The *pit in the potential* being thus the central new feature of motion in Schwarzschild geometry and the source of major predictions (Box 25.2), it is appropriate to look for the most direct road into the concept of effective potential and its meaning and application. In this search no guide is closer to hand than Newtonian mechanics.

Analytic mechanics offers several ways to deal with the problem of motion in a central field of force, and among them are two of central relevance here: (1) the world-line method, which includes second-order differential equations of motion, Lagrange's equations, search for constants of integration, reduction to first-order equations, and further integration in rather different ways according as one wants the shape of the orbit, $\theta = \theta(r)$, or the time to get to a given point on the world line, $t = t(r)$; and (2) the wave-crest method, otherwise known as the "eikonal method" or "Hamilton-Jacobi method," which gives the motion by the condition of "constructive interference of wave crests," thus making a single leap from the Hamilton-Jacobi equation to the motion of the test object. Both methods are em-

(continued on page 641)

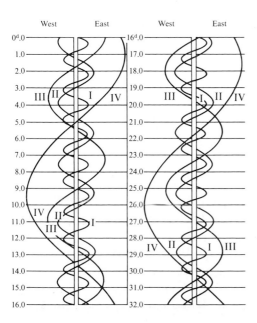

Figure 25.1.

Jupiter's satellites, as followed from night to night with field glasses or telescope, provide an opportunity to check for oneself the central ideas of gravitation physics in the Newtonian approximation (distances large compared to Schwarzschild radius). For the practically circular orbits of these satellites, Kepler's law becomes $M^1 = \omega^2 r^3$ ("1-2-3 principle") and the velocity in orbit is $\beta = \omega r$. Out of observations on any two of the quantities β, M, ω, r, one can find the other two. (In the opposite limiting case of two objects, each of mass M, going around their common center of gravity with separation r, one has $M = \omega^2 r^3/2$, $\beta = \omega r/2$). The configurations of satellites I–IV of Jupiter as given here for December 1964 (days 0.0, 1.0, 2.0, etc. in "universal time," for which see any good dictionary or encyclopedia) are taken from *The American Ephemeris and Nautical Almanac for 1964* [U.S. Government Printing Office (1962)].

Box 25.1 MASS FROM MEAN ANGULAR FREQUENCY AND
SEMIMAJOR AXIS: $M = \omega^2 a^3$

Appropriateness of Newtonian analysis shown by smallness of mass (or "half-Schwarzschild radius" or "extension of the pit in the potential") as listed in last column compared to the semimajor axis a in the next-to-last column. Basic data from compilation of Allen (1963).

Object	Period[a] (days)	$\omega(cm^{-1})$	$a(cm)$	$\omega^2 a^3(cm)$
Planets				
Mercury	87.9686	275.8×10^{-19}	0.5791×10^{13}	1.477×10^5
Venus	224.700	107.95	1.0821	1.477
Earth	365.257	66.41	1.4960	1.477
Mars	686.980	35.31	2.2794	1.477
Jupiter	4332.587	5.599	7.783	1.478
Saturn	10759.20	2.255	14.27	1.477
Uranus	30685	0.7905	28.69	1.476
Neptune	60188	0.4030	44.98	1.478
Pluto	90700	0.2674×10^{-19}	59.00×10^{13}	1.469×10^5
Major satellites of Jupiter				
Io	1.769 138	13.711×10^{-16}	0.422×10^{11}	141.3
Europa	3.551 181	6.831	0.671	141.0
Ganymede	7.154 553	3.391	1.070	140.8
Callisto	16.689 018	1.454×10^{-16}	1.883×10^{11}	141.1
Two satellites of Earth				
OSO 5[b]	95.6 min.	3.65×10^{-14}	6.916×10^8	0.442
Moon	27.32	0.888×10^{-16}	3.84×10^{10}	0.446

[a] Sidereal period: time to make one revolution relative to fixed stars.
[b] Orbiting scientific observatory launched Jan. 22, 1969, to observe x-rays and ultraviolet radiation from the sun. Perigee 531 km, apogee 560 km, above earth.

SOME TYPICAL MASSES AND TIMES IN CONVENTIONAL AND
GEOMETRIC UNITS. Conversion factor for mass,
$G/c^2 = 0.742 \times 10^{-28}$ **cm/g**

Mass	Galaxy	Sun	Jupiter	Earth
M_{conv} (g)	2.2×10^{44}	1.989×10^{33}	1.899×10^{30}	5.977×10^{27}
M (cm)	1.6×10^{16}	1.47×10^5	112	0.444

Conversion factor for time, $c = 2.998 \times 10^{10}$ cm/sec. One sidereal year = 365.256 days or 3.1558×10^7 sec.

Period	1 sec	1 min	1 hr	1 day
ω_{conv} (sec^{-1})	6.28	1.046×10^{-1}	1.75×10^{-3}	7.28×10^{-5}
ω (cm^{-1})	2.09×10^{-10}	3.48×10^{-12}	5.80×10^{-14}	2.42×10^{-15}

	1 week	1 month	1 year
	1.04×10^{-5}	2.39×10^{-6}	1.99×10^{-7}
	3.46×10^{-16}	7.95×10^{-17}	6.63×10^{-18}

Figure 25.2.

Effective potential for motion of a test particle in the Schwarzschild geometry of a concentrated mass M. Energy, in units of the rest mass μ of the particle, is denoted $\tilde{E} = E/\mu$; angular momentum, $\tilde{L} = L/\mu$. The quantity r denotes the Schwarzschild r coordinate. The effective potential (also in units of μ) is defined by equation (25.16) or, equivalently, by the equation

$$\left(\frac{dr}{d\tau}\right)^2 + \tilde{V}^2(r) = \tilde{E}^2$$

(see also §25.5) and has the value

$$\tilde{V} = [(1 - 2M/r)(1 + \tilde{L}^2/r^2)]^{1/2}.$$

It represents that value of \tilde{E} at which the radial kinetic energy of the particle, at r, reduces to zero (\tilde{E}-value that makes r into a "turning point": $\tilde{V}(r) = \tilde{E}$. Note that one could equally well regard $\tilde{V}^2(r)$ as the effective potential, and define a turning point by the condition $\tilde{V}^2 = \tilde{E}^2$. Which definition one chooses depends on convenience, on the intended application, on the tie to the archetypal differential equation $\frac{1}{2}\dot{x}^2 + V(x) = E$, and on the stress one wishes to put on correspondence with the effective potential of Newtonian theory). Stable circular orbits are possible (representative point sitting at minimum of effective potential) only for \tilde{L} values in excess of $2\sqrt{3}\,M$. For any such fixed \tilde{L} value, the motion departs slightly from circularity as the energy is raised above the potential minimum (see the two heavy horizontal lines for $\tilde{L} = 3.75\,M$). In classical physics, the motion is limited to the region of positive kinetic energy. In quantum physics, the particle can tunnel through the region where the kinetic energy, as calculated classically, is negative (dashed prolongations of heavy horizontal lines) and head for the "pit in the potential" (capture by black hole). Such tunneling is absolutely negligible when the center of attraction has any macroscopic dimension, but in principle becomes important for a black hole of mass 10^{17} g (or 10^{-11} cm) if such an object can in principle exist.

The diagram at the right gives values of the minimum and maximum of the potential as they depend on the angular momentum of the test particle. The roots of $\partial \tilde{V}/\partial r$ are given in terms of the "reduced angular momentum parameter" $L^\dagger = \tilde{L}/M = L/M\mu$ by

$$r = \frac{6M}{1 + (1 - 12/L^{\dagger 2})^{1/2}},$$

$$\tilde{E}^2 = \frac{(L^{\dagger 2} + 36) + (L^{\dagger 2} - 12)(1 - 12/L^{\dagger 2})^{1/2}}{54}$$

$$[= 8/9 \text{ for } L^\dagger = (12)^{1/2}; 1 \text{ for } L^\dagger = 4; (L^{\dagger 2}/27) + (1/3) + (1/L^{\dagger 2})$$
$$+ \cdots \text{ for } L^\dagger \longrightarrow \infty]$$

(plus root for maximum of the effective potential; minus root for minimum; see exercise 25.18).

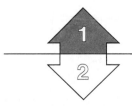

Box 25.2 MOTION IN SCHWARZSCHILD GEOMETRY REGARDED AS A CENTRAL POINT OF DEPARTURE FOR MAJOR APPLICATIONS OF EINSTEIN'S GEOMETRODYNAMICS

1. Newtonian effect of sun on planets and of earth on moon and man.
2. Bending of light by sun.
3. Red shift of light from sun.
4. Precession of the perihelion of Mercury around the sun.
5. Capture of a test object by a black hole as simple exemplar of gravitational collapse.
6. Dynamics of Friedmann universe derived from model of Schwarzschild "lattice universe." Lattice universe is constructed from 120 or some other magic number of concentrations of mass, each mass in an otherwise empty lattice cell of its own. Each lattice cell, though actually polygonal, is idealized (see Wigner-Seitz approximation of solid-state physics) as spherical. A test object at the interface between two cells falls toward the center of each [standard radial motion in Schwarzschild geometry; see discussion following equation (25.27). Therefore the two masses fall toward each other at a calculable rate. From this simple argument follows the entire dynamics of the closed 3-sphere lattice universe, in close concord with the predictions of the Friedmann model [see Lindquist and Wheeler (1957)].
7. Perturbations of Schwarzschild geometry, I. Gravitational waves are incident on, scattered by, and captured into a black hole. Waves with wavelength short compared to the Schwarzschild radius can be analyzed to good approximation by the methods of geometric optics (exercises 35.15 and 35.16), as employed in this chapter to treat the motions of particles and photons. For longer wavelengths, there occur important physical-optics corrections to this

geometric-optics idealization (see §35.8 and exercises 32.10, 32.11). Similar considerations apply to electromagnetic and de Broglie waves.

8. Lepton number for an electron in its lowest quantum state in the geometry ("gravitational field of force") of a black hole is calculated to be transcended (capture of the electron!) or not according as the mass of this black hole is large or small compared to a certain critical mass $M_{*e} = M^{*2}/m_e$ ($\sim 10^{17}$ g or 10^{-11} cm) [Hartle (1971, 1972); Wheeler (1971b,c); Teitelboim (1972b,c)]. Similarly (with another value for the critical mass) for conservation of baryon number [Bekenstein (1972a,b), Teitelboim (1972a)]. To analyze "transcendence or not" one must solve quantum-mechanical wave equations, of which the Hamilton-Jacobi equation for particle and photon orbits is a classical limit. These quantum wave equations contain effective potentials identical—aside from spin-dependent and wavelength-dependent corrections—to the effective potentials for particle and photon motion.

9. Perturbations of Schwarzschild geometry, II. Those small changes in standard Schwarzschild black-hole geometry which remain stationary in time describe the alterations in a "dead" black hole that make it into a "live" black hole, one endowed with angular momentum as well as mass (see Chapter 33). To analyze such changes in black-hole geometry, one must again solve wave equations, but wave equations which are now classical. Once more the wave equations are closely related to the Hamilton-Jacobi equation, and their effective potentials are close kin to those for particle motion.

ployed here in turn because each gives special insights. The Hamilton-Jacobi method (Box 25.3) leads quickly to the major results of interest (Box 25.4), and it has a close tie to the quantum principle. The world-line method (§§25.2, 25.3, 25.4) starts with the geodesic equations of motion themselves. It provides a more familiar way into the subject for a reader not acquainted with the Hamilton-Jacobi approach. Moreover, in attempting to solve the geodesic equations of motion, one must analyze symmetry properties of the geometry, an enterprise that continues to pay dividends when one moves from Schwarzschild geometry to Kerr-Newman geometry (Chapter 33), and from Friedmann cosmology (Chapter 27) to more general cosmologies (Chapter 30).

(*continued on page 650*)

**Box 25.3 THE HAMILTON-JACOBI DESCRIPTION OF MOTION:
NATURAL BECAUSE RATIFIED BY THE QUANTUM PRINCIPLE**

1. Purely classical (nonquantum).

2. Originated with William Rowan Hamilton out of conviction that mechanics is similar in its character to optics; that the "particle world line" of mechanics is an idealization analogous to the "light ray" of geometric optics. Localization of energy of light ray is approximate only. Its spread is governed by wavelength of light ("geometric optics"). Hamilton had glimmerings of same idea for particles: "quantum physics before quantum physics." The way that he and Jacobi developed to analyze motion through the Hamilton-Jacobi function $S(x, t)$—to take the example of a dynamic system with only one degree of freedom, x—makes the leap from classical ideas to quantum ideas as short as one knows how to make it. Moreover, the real world is a quantum world. Classical mechanics is not born out of a vacuum. It is an idealization of and approximation to quantum mechanics.

3. Key idea is idealization to a particle wavelength so short that quantum-mechanical spread or uncertainty in location of particle (or spread of configuration coordinates of more complex system) is negligible. No better way was ever discovered to unite the spirit of quantum mechanics and the precision of location of classical mechanics.

4. Call Hamiltonian $H(p, x) = p^2/2m + V(x)$. Call energy of particle E. Then there is no way whatever consistent with the quantum principle to describe the motion of the particle in space and time. The uncertainty principle forbids (sharply defined energy $\Delta E \longrightarrow 0$, in $\Delta E \, \Delta t \geq \hbar/2$, implies uncertainty $\Delta t \longrightarrow \infty$; also $\Delta p \longrightarrow 0$ in $\Delta p \, \Delta x \geq \hbar/2$ implies $\Delta x \longrightarrow \infty$). The quantum-mechanical wave function is spread out over all space. This spread shows in the so-called semi-

Box 25.3 (continued)

classical or Wentzel-Kramers-Brillouin ["WKB"; see, for example, Kemble (1937)] approximation for the probability amplitude function,

$$\psi_E(x, t) = \begin{pmatrix} \text{slowly varying} \\ \text{amplitude function} \end{pmatrix} e^{(i/\hbar)S_E(x,t)}. \tag{1}$$

5. It is of no help in localizing the probability distribution that $\hbar = 1.054 \times 10^{-27}$ g cm^2/s [or $\hbar = (1.6 \times 10^{-33}$ cm$)^2$ in geometric units] is very small compared to the "quantities of action" or "magnitudes of the Hamilton-Jacobi function, S" or "dynamic phase, S" encountered in most everyday applications.

6. It is of no help in localizing the probability distribution that this dynamic phase obeys the simple Hamilton-Jacobi law of propagation,

$$-\frac{\partial S}{\partial t} = H\left(\frac{\partial S}{\partial x}, x\right) = \frac{1}{2m}\left(\frac{\partial S}{\partial x}\right)^2 + V(x). \tag{2}$$

7. It is of no help in localizing the probability distribution that the solution of this equation for a particle of energy E is extraordinarily simple,

$$S(x, t) = -Et + \int_{x_0}^{x} \{2m[E - V(x)]\}^{1/2}\, dx + \delta_E \tag{3}$$

(with δ_E an arbitrary additive phase constant). The probability amplitude is still spread all over everywhere. There is not the slightest trace of anything like a localized world line, $x = x(t)$.

8. To localize the particle, build a probability-amplitude wave packet by superposing mono-frequency (monoenergy) terms, according to a prescription qualitatively of the form

$$\psi(x, t) = \psi_E(x, t) + \psi_{E + \Delta E}(x, t) + \cdots \tag{4}$$

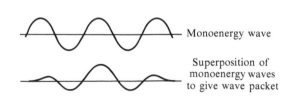

Destructive interference takes place almost everywhere. The wave packet is concentrated in the region of constructive interference. There the phases of the various waves agree; thus

$$S_E(x, t) = S_{E + \Delta E}(x, t). \qquad (5)$$

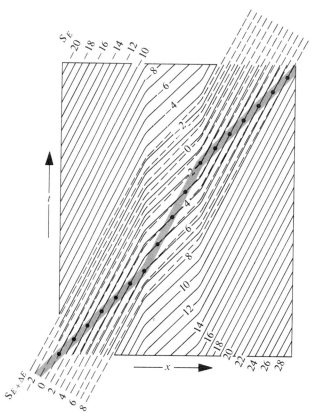

At last one has moved from a wave spread everywhere to a localized wave and thence, in the limit of indefinitely small wavelength, to a classical world line. This one equation of constructive interference ties together x and t (locus of world line in x, t, diagram). Smooth lines -20, -19, -18, etc. are wave crests of ψ_E; dashed lines, wave crests for $\psi_{E + \Delta E}$. Shaded area is region of constructive interference (wave packet). Black dots mark locus of classical world line,

$$\lim_{\Delta E \to 0} \frac{S_{E + \Delta E}(x, t) - S_E(x, t)}{\Delta E} = 0.$$

9. The Newtonian course of the world line through spacetime follows at once from this condition of constructive interference when one goes to the classical limit (\hbar negligible compared to amounts of action involved; hence wavelength negligibly short; hence spread of energies ΔE required to build well-localized wave packet also negligible); thus

$$\frac{S_{E + \Delta E}(x, t) - S_E(x, t)}{\Delta E} = 0$$

reduces to

$$\frac{\partial S_E(x, t)}{\partial E} = 0.$$

Box 25.3 (*continued*)

This condition in turn, applied to expression (3), gives the time required to travel to the point x; thus,

$$-t + \int_{x_0}^{x} \frac{dx}{\{(2/m)[E - V(x)]\}^{1/2}} + t_0 = 0,$$

where t_0 is an abbreviation for the quantity

$$t_0 = d\,\delta_E/dE$$

("difference in base value of dynamic phase per unit difference of energy").

10. Not one trace of the quantum of action comes into this final Newtonian result, for a simple reason: \hbar has been treated as negligible and the wavelength has been treated as negligible. In this limit the location of the wave "packet" reduces to the location of the wave crest. The location of the wave crest is precisely what is governed by $S_E(x, t)$; and the condition of "constructive interference" $\partial S_E(x, t)/\partial E = 0$ gives without approximation the location of the sharply defined Newtonian world line $x = x(t)$.

11. Relevance in the context of motion in a central field of force? Quickest known route to the concept of effective potential (Box 25.4).

Box 25.4 MOTION UNDER GRAVITATIONAL ATTRACTION OF A CENTRAL MASS ANALYZED BY HAMILTON-JACOBI METHOD

A. Newtonian Theory of Gravitation

Hamiltonian
$$\widetilde{H} = \frac{\widetilde{p}_r^{\,2}}{2} + \frac{\widetilde{p}_\theta^{\,2}}{2r^2} + \frac{\widetilde{p}_\phi^{\,2}}{2r^2 \sin^2\theta} - \frac{M}{r} \tag{1}$$

(tildes over energy, momentum, etc., refer to test object of unit mass; test particle of mass μ follows same motion with energy $E = \mu\widetilde{E}$, momentum $\boldsymbol{p} = \mu\widetilde{\boldsymbol{p}}$, etc.). Equation of Hamilton-Jacobi for propagation of wave crests:

$$-\frac{\partial \widetilde{S}}{\partial t} = \frac{1}{2}\left(\frac{\partial \widetilde{S}}{\partial r}\right)^2 + \frac{1}{2r^2}\left(\frac{\partial \widetilde{S}}{\partial \theta}\right)^2 + \frac{1}{2r^2 \sin^2\theta}\left(\frac{\partial \widetilde{S}}{\partial \phi}\right)^2 - \frac{M}{r}. \tag{2}$$

Box 25.4 (continued)

Solve by "method of separation of variables" with convention that $\sqrt{a^2} \equiv \pm a$,

$$\widetilde{S} = -\widetilde{E}t + \widetilde{p}_\phi\phi + \int^\theta \left(\widetilde{L}^2 - \frac{\widetilde{p}_\phi{}^2}{\sin^2\theta}\right)^{1/2} d\theta$$
$$+ \int^r \left[2\left(\widetilde{E} + \frac{M}{r} - \frac{\widetilde{L}^2}{2r^2}\right)\right]^{1/2} dr + \delta_{\widetilde{p}_\phi,\widetilde{L},\widetilde{E}}. \tag{3}$$

(Check by substituting into Hamilton-Jacobi equation. Solution as *sum* of four terms corresponding to the four independent variables goes hand in hand with expression of probability amplitude in quantum mechanics as *product* of four factors, because $iS/\hbar = i\mu\widetilde{S}/\hbar$ is exponent in approximate expression for the probability amplitude.)

Constructive interference of waves:

(1) with slightly different \widetilde{E} values (impose "condition of constructive interference" $\partial\widetilde{S}_{\widetilde{p}_\phi,\widetilde{L},\widetilde{E}}(t, r, \theta, \phi)/\partial\widetilde{E} = 0$) tells when the particle arrives at a given r (that is, gives relation between t and r);

(2) with slightly different values of the "parameter of total angular momentum per unit mass," \widetilde{L} (impose condition of constructive interference $\partial\widetilde{S}_{\widetilde{p}_\phi,\widetilde{L},\widetilde{E}}(t, r, \theta, \phi)/\partial\widetilde{L} = 0$) tells correlation between θ and r (a major feature of the shape of the orbit);

(3) with slightly different values of the "parameter of azimuthal angular momentum per unit mass," \widetilde{p}_ϕ (impose condition $\partial\widetilde{S}/\partial\widetilde{p}_\phi = 0$) gives correlation between θ and ϕ,

$$0 = \frac{\partial\widetilde{S}}{\partial\widetilde{p}_\phi} = \phi - \int^\theta \frac{(\widetilde{p}_\phi/\widetilde{L})\, d\theta}{\sin\theta\,(\sin^2\theta - \widetilde{p}_\phi{}^2/\widetilde{L}^2)^{1/2}} \tag{4}$$

Planar character of the orbit.

Puzzle out the value of this last integral with the help of a table of integrals? It is quicker and clearer to capture the content without calculation: the particle moves in a plane. The vector associated with the angular momentum \widetilde{L} stands perpendicular to this plane. The projection of this angular momentum along the z-axis is $\widetilde{p}_\phi = \widetilde{L}\cos\alpha$ (definition of orbital inclination, α). Straight line connecting origin with particle cuts unit sphere in a point \mathscr{P}. As time runs on, \mathscr{P} traces out a great circle on the unit sphere. The plane of this great circle cuts the equatorial plane in a "line of nodes," at which "hinge-line" the two planes are separated by a dihedral angle, α. The orbit of the point \mathscr{P} is described by $\hat{x} = r\cos\psi$, $\hat{y} = r\sin\psi$, $\hat{z} = 0$ in a Cartesian system of coordinates in which \hat{y} runs along the line of nodes and in which \hat{x} lies in the plane of the *orbit*.

Box 25.4 (*continued*)

In a coordinate system in which y runs along the line of nodes and x lies in the plane of the *equator,* one has:

$$r \cos \theta = z = \hat{z} \cos \alpha + \hat{x} \sin \alpha = r \cos \psi \sin \alpha;$$

$$r \sin \theta \cos \phi = x = -\hat{z} \sin \alpha + \hat{x} \cos \alpha = r \cos \psi \cos \alpha;$$

$$r \sin \theta \sin \phi = y = \hat{y} = r \sin \psi.$$

Eliminate reference to the Cartesian coordinates and, by taking ratios, also eliminate reference to r. Thus find the equation of the great circle route in parametric form,

$$\tan \phi = \tan \psi / \cos \alpha$$

and

$$\cos \theta = \cos \psi \sin \alpha.$$

Here increasing values of ψ spell out successive points on the great circle. Eliminate ψ via the relation

$$\sec^2\psi - \tan^2\psi = 1$$

to find

$$\frac{\sin^2\alpha}{\cos^2\theta} - \tan^2\phi \cos^2\alpha = 1$$

or, more briefly,

$$\sec \phi = \tan \alpha \tan \theta. \tag{5}$$

One verifies that ϕ as calculated from (5) provides an integral of (4), thus confirming the physical argument just traced out. Moreover, the arbitrary constant of integration that comes from (4), left out for the sake of simplicity from (5), is easily inserted by replacing ϕ there by $\phi - \phi_0$ (rotation of line of nodes to a new azimuth). The kind of physics just done in tracing out the relation between θ and ϕ is evidently elementary solid geometry and nothing more. The same geometric relationships also show up, with no relativistic corrections whatsoever (how could there be any?!) for motion in Schwarzschild geometry. Therefore it is appropriate to drop this complication from attention here and hereafter. Let the particle move entirely in the direction of increasing θ, not at all in the direction of increasing ϕ; that is, let it move in an orbit of zero angular momentum \tilde{p}_ϕ (total angular momentum vector \tilde{L} inclined at angle $\alpha = \pi/2$ to z-axis). Consequently the dynamic phase S (to be divided by \hbar to obtain phase of Schrödinger wave function when one turns from classical to quantum mechanics) becomes

$$\tilde{S} = -\tilde{E}t + \tilde{L}\theta + \int^r \left[2\left(\tilde{E} + \frac{M}{r} - \frac{\tilde{L}^2}{2r^2} \right) \right]^{1/2} dr + \delta_{\tilde{L},\tilde{E}}. \tag{6}$$

Shape of orbit:

$$0 = \frac{\partial \widetilde{S}}{\partial \widetilde{L}} = \theta - \int^{r} \frac{\widetilde{L}\, dr/r^2}{[2(\widetilde{E} + M/r - \widetilde{L}^2/2r^2)]^{1/2}}, \tag{7}$$

whence

$$r = \frac{\widetilde{L}^2/M}{1 + e \cos \theta}. \tag{8}$$

Here e is an abbreviation for the eccentricity of the orbit,

$$e = (1 + 2\widetilde{E}\widetilde{L}^2/M^2)^{1/2} \tag{9}$$

(greater than 1 for positive \widetilde{E}, hyperbolic orbit; equal to 1 for zero \widetilde{E}, parabolic orbit; less than 1 for negative \widetilde{E}, elliptic orbit). A constant of integration has been omitted from (8) for simplicity. To reinstall it, replace θ by $\theta - \theta_0$ (rotation of direction of principal axis in the plane of the orbit). Other features of the orbit:

$\begin{pmatrix} \text{semimajor axis of} \\ \text{orbit when elliptic} \end{pmatrix}$
$$a = \frac{\widetilde{L}^2/M}{1 - e^2} = \frac{M}{(-2\widetilde{E})}; \tag{10}$$

$\begin{pmatrix} \text{semiminor axis of} \\ \text{orbit when elliptic} \end{pmatrix}$
$$b = \frac{\widetilde{L}^2/M}{(1 - e^2)^{1/2}} = \frac{\widetilde{L}}{(-2\widetilde{E})^{1/2}}; \tag{11}$$

$\begin{pmatrix} \text{"impact parameter"} \\ \text{for hyperbolic orbit,} \\ \text{or "distance of closest} \\ \text{approach in} \\ \text{absence of deflection"} \end{pmatrix}$
$$b = \frac{(\text{angular momentum per unit mass})}{(\text{linear momentum per unit mass})}$$
$$= \frac{\widetilde{L}}{(2\widetilde{E})^{1/2}}; \tag{12}$$

$\begin{pmatrix} \text{actual distance of} \\ \text{closest approach} \end{pmatrix}$
$$r_{\min} = \frac{\widetilde{L}^2/M}{(1 + 2\widetilde{E}\widetilde{L}^2/M^2)^{1/2} + 1}; \tag{13}$$

$\begin{pmatrix} \text{angle of deflection} \\ \text{in hyperbolic orbit} \end{pmatrix}$
$$\Theta = \pi - 2 \arccos (1/e)$$
$$= 2 \arctan [M/(2\widetilde{E})^{1/2}\widetilde{L}] \tag{14}$$
$$= 2 \arctan [M/2\widetilde{E}b];$$

$\begin{pmatrix} \text{differential scattering} \\ \text{cross section} \end{pmatrix}$
$$\frac{d\sigma}{d\Omega} = \frac{2\pi b\, db}{2\pi \sin \Theta\, d\Theta}$$
$$= \frac{M^2}{(4\widetilde{E} \sin^2\Theta/2)^2} \text{ (Rutherford).} \tag{15}$$

Box 25.4 (continued)

Time as correlated with position:

$$0 = \frac{\partial \widetilde{S}}{\partial \widetilde{E}} = -t + \int^r \frac{dr}{\left[2\left(\widetilde{E} + \frac{M}{r} - \frac{\widetilde{L}^2}{2r^2}\right)\right]^{1/2}}. \tag{16}$$

Write

$$r = \frac{M}{(-2\widetilde{E})}(1 - e\cos u) \tag{17}$$

to simplify the integration. Get

$$t = \frac{M}{(-2\widetilde{E})^{3/2}}(u - e\sin u), \tag{18}$$

$$\begin{pmatrix}\text{mean circular}\\\text{frequency}\end{pmatrix} = \frac{2\pi}{(\text{period})} = \omega = \frac{(-2\widetilde{E})^{3/2}}{M} = \left(\frac{M}{a^3}\right)^{1/2}. \tag{19}$$

Here u is the so-called "mean eccentric anomaly" (Bessel's time parameter). In terms of this quantity, one has also:

$$\sin u = \frac{(1 - e^2)^{1/2}\sin\theta}{1 + e\cos\theta};$$

$$\cos u = \frac{\cos\theta + e}{1 + e\cos\theta};$$

$$\cos\theta = \frac{\cos u - e}{1 - e\cos u};$$

$$\sin\theta = \frac{(1 - e^2)^{1/2}\sin u}{1 - e\cos u};$$

$$x = r\cos\theta = \frac{M}{(-2\widetilde{E})}(\cos u - e); \tag{20}$$

$$y = r\sin\theta = \frac{\widetilde{L}}{(-2\widetilde{E})^{1/2}}\sin u. \tag{21}$$

These expressions lend themselves to Fourier analysis into harmonic functions of the time, with coefficients that are standard Bessel functions:

$$J_n(z) = \frac{1}{2\pi}\int_{-\pi}^{\pi} e^{i(z\sin u - nu)}\, du; \tag{22}$$

$$x/a = -\frac{3}{2}e + \sum_{\substack{k=-\infty\\k\neq 0}}^{+\infty} k^{-1}J_{k-1}(ke)\cos k\omega t; \tag{23}$$

$$y/a = (1 - e^2)^{1/2}\sum_{\substack{k=-\infty\\k\neq 0}}^{+\infty} k^{-1}J_{k-1}(ke)\sin k\omega t \tag{24}$$

[for these and further formulas of this type, see, for example, Wintner (1941), Siegel (1956), and Siegel and Moser (1971)]. Via such Fourier analysis one is in a position to calculate the intensity of gravitational radiation emitted at the fundamental circular frequency ω and at the overtone frequencies (see Chapter 36).

B. Einstein's Geometric Theory of Gravitation

Connection between energy and momentum for a test particle of rest mass μ traveling in curved space,

$$g^{\alpha\beta}p_\alpha p_\beta + \mu^2 = 0. \tag{25}$$

Gravitation shows up in no way other than in curvature of the geometry, in which the particle moves as free of all "real" force. Refer all quantities to basis of a test object of unit rest mass by dealing throughout with $\widetilde{p} = p/\mu$. Also write $\widetilde{p}_\alpha = \partial\widetilde{S}/\partial x^\alpha$. Thus Hamilton-Jacobi equation for propagation of wave crests in Schwarzschild geometry (external field of a star; §23.6) becomes

$$-\frac{1}{(1 - 2M/r)}\left(\frac{\partial\widetilde{S}}{\partial t}\right)^2 + (1 - 2M/r)\left(\frac{\partial\widetilde{S}}{\partial r}\right)^2 + \frac{1}{r^2}\left(\frac{\partial\widetilde{S}}{\partial\theta}\right)^2 + \frac{1}{r^2\sin^2\theta}\left(\frac{\partial\widetilde{S}}{\partial\phi}\right)^2$$
$$+ 1 = 0. \tag{26}$$

Solve Hamilton-Jacobi equation. As in Newtonian problem, simplify by eliminating all motion in direction of increasing ϕ. Thus set $0 = \widetilde{p}_\phi = \partial\widetilde{S}/\partial\phi$ (dynamic phase independent of ϕ) and have

$$\widetilde{S} = -\widetilde{E}t + \widetilde{L}\theta + \int^r [\widetilde{E}^2 - (1 - 2M/r)(1 + \widetilde{L}^2/r^2)]^{1/2}\frac{dr}{(1 - 2M/r)}. \tag{27}$$

Find shape of orbit by "principle of constructive interference"; thus,

$$0 = \frac{\partial\widetilde{S}}{\partial\widetilde{L}} = \theta - \int^r \frac{\widetilde{L}\,dr/r^2}{[\widetilde{E}^2 - (1 - 2M/r)(1 + \widetilde{L}^2/r^2)]^{1/2}}. \tag{28}$$

[See equation (25.41) and associated discussion in text; also Figure 25.6.]
Find time to get to given r by considering "interference of wave crests" belonging to slightly different \widetilde{E} values:

$$0 = \frac{\partial\widetilde{S}}{\partial\widetilde{E}} = -t + \int^r \frac{\widetilde{E}}{[\widetilde{E}^2 - (1 - 2M/r)(1 + \widetilde{L}^2/r^2)]^{1/2}}\frac{dr}{(1 - 2M/r)}. \tag{29}$$

[See equation (25.32) and associated discussion in text; also Figure 25.5 and exercise 25.15.]

§25.2. SYMMETRIES AND CONSERVATION LAWS

From symmetries to
conservation laws by:
(1) Lagrangian or
 Hamiltonian approach

(2) Killing-vector approach

Killing vector, ξ, defined

In analytic mechanics, one knows that symmetries of a Lagrangian or Hamiltonian result in conservation laws. Exercises 25.1 to 25.4 describe how these general principles are used to deduce, from the symmetries of Schwarzschild spacetime, constants of motion for the trajectories (geodesics) of freely falling particles in the gravitational field outside a star. The same constants of motion are obtained in a different language in differential geometry, where a "Killing vector" is the standard tool for the description of symmetry. This section considers the general question of metric symmetries before proceeding to Schwarzschild spacetime.

Let the metric components $g_{\mu\nu}$ relative to some coordinate basis dx^α be independent of one of the coordinates x^K, so

$$\partial g_{\mu\nu}/\partial x^\alpha = 0 \text{ for } \alpha = K. \tag{25.1}$$

This relation possesses a geometric interpretation. Any curve $x^\alpha = c^\alpha(\lambda)$ can be translated in the x^K direction by the coordinate shift $\Delta x^K = \varepsilon$ to form a "congruent" (equivalent) curve:

$$x^\alpha = c^\alpha(\lambda) \text{ for } \alpha \neq K \text{ and } x^K = c^K(\lambda) + \varepsilon.$$

Let the original curve run from $\lambda = \lambda_1$ to $\lambda = \lambda_2$ and have length

$$L = \int_{\lambda_1}^{\lambda_2} [g_{\mu\nu}(x(\lambda))(dx^\mu/d\lambda)(dx^\nu/d\lambda)]^{1/2} \, d\lambda.$$

Then the displaced curve has length

$$L(\varepsilon) = \int_{\lambda_1}^{\lambda_2} \left[\left\{ g_{\mu\nu}(x(\lambda)) + \varepsilon \frac{\partial g_{\mu\nu}}{\partial x^K} \right\} (dx^\mu/d\lambda)(dx^\nu/d\lambda) \right]^{1/2} d\lambda.$$

But the coefficient of ε in the integrand is zero. Therefore the length of the new curve is identical to the length of the original curve: $dL/d\varepsilon = 0$.

The vector

$$\xi \equiv d/d\varepsilon = (\partial/\partial x^K) \tag{25.2}$$

provides an infinitesimal description of these length-preserving "translations." It is called a "Killing vector." It satisfies Killing's equation*

Killing's equation derived

$$\xi_{\mu;\nu} + \xi_{\nu;\mu} = 0 \tag{25.3}$$

(condition on the vector field ξ necessary and sufficient to ensure that all lengths are preserved by the displacement $\varepsilon\xi$). This condition is expressed in covariant form.

*Historical note: Wilhelm K. J. Killing, born May 10, 1847, in Burbach, Westphalia, died February 11, 1923 in Münster, Westphalia; Professor of Mathematics at the University of Münster, 1892–1920. The key article that gives the name "Killing vector" to the kind of isometries considered here appeared almost a century ago [Killing (1892)].

Therefore it is enough to establish it in the preferred coordinate system of (25.1) in order to have it hold in every coordinate system. In that preferred coordinate system, the vector field, according to (25.2), has components

$$\xi^\mu = \delta^\mu{}_K.$$

Therefore the covariant derivative of this vector field has components

$$\xi_{\mu;\nu} = g_{\mu\alpha}\xi^\alpha{}_{;\nu} = g_{\mu\alpha}\left(\frac{\partial \xi^\alpha}{\partial x^\nu} + \Gamma^\alpha{}_{\nu\sigma}\xi^\sigma\right)$$

$$= g_{\mu\alpha}\Gamma^\alpha{}_{\nu K} = \Gamma_{\mu\nu K} = \frac{1}{2}\left(\frac{\partial g_{\mu K}}{\partial x^\nu} + \frac{\partial g_{\mu\nu}}{\partial x^K} - \frac{\partial g_{\nu K}}{\partial x^\mu}\right) \qquad (25.4)$$

$$= \frac{1}{2}(g_{\mu K,\nu} - g_{\nu K,\mu}).$$

One sees that $\xi_{\mu;\nu}$ is antisymmetric in the labels μ and ν, as stated in Killing's equation (25.3).

The geometric significance of a Killing vector is spelled out in Box 25.5.

From Killing's equation, $\xi_{(\mu;\nu)} = 0$, and from the geodesic equation $\boldsymbol{\nabla_p p} = 0$ for the tangent vector $\boldsymbol{p} = d/d\lambda$ to any geodesic, follows an important theorem: *In any geometry endowed with a symmetry described by a Killing vector field $\boldsymbol{\xi}$, motion along any geodesic whatsoever leaves constant the scalar product of the tangent vector with the Killing vector:*

Conservation of $\boldsymbol{p}\cdot\boldsymbol{\xi}$ for geodesic motion

$$p_K = \boldsymbol{p}\cdot\boldsymbol{\xi} = \text{constant}. \qquad (25.5)$$

In verification of this result, evaluate the rate of change of the constant p_K along the course of the typical geodesic (affine parameter λ; result therefore as applicable to light rays—with zero lapse of proper time—as to particles); thus,

$$dp_K/d\lambda = (p^\mu\xi_\mu)_{;\nu}p^\nu = (p^\mu{}_{;\nu}p^\nu)\xi_\mu + p^{(\mu}p^{\nu)}\xi_{[\mu;\nu]} = 0. \qquad (25.6)$$

Turn back from a general coordinate system to the coordinates of (25.1), where the Killing vector field of the symmetry lets itself be written $\boldsymbol{\xi} = \partial/\partial x^K$. Then the scalar product of (25.5) becomes constant $\equiv p_\alpha\xi^\alpha = p_\alpha\delta^\alpha{}_K = p_K$. *The symmetry of the geometry guarantees the conservation of the K-th covariant coordinate-based component of the momentum.*

On a timelike geodesic in spacetime, the momentum of a test particle of mass μ is

$$\boldsymbol{p} \equiv d/d\lambda = \mu\boldsymbol{u} = \mu d/d\tau. \qquad (25.7)$$

Thus the affine parameter λ most usefully employed in the above analysis, when it is concerned with a particle, is not proper time τ but rather the ratio $\lambda = \tau/\mu$.

When the metric $g_{\mu\nu}$ is independent of a coordinate x^K, that coordinate is called cyclic, and the corresponding conserved quantity, p_K, is called the "momentum conjugate to that cyclic coordinate" in a terminology borrowed from nonrelativistic mechanics.

Terminology: "cyclic coordinate," "conjugate momentum"

(*continued on page 654*)

Box 25.5 KILLING VECTORS AND ISOMETRIES (Illustrated by a Donut)

A. On a given manifold (e.g., spacetime, or the donut pictured here), in a given coordinate system, the metric components are independent of a particular coordinate x^K. Example of donut:

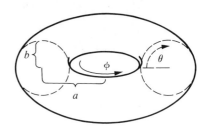

$$ds^2 = b^2 \, d\theta^2 + (a - b \cos \theta)^2 \, d\phi^2;$$

$$g_{\mu\nu} \text{ independent of } x^K = \phi.$$

B. Translate an arbitrary curve \mathcal{C} through the infinitesimal displacement

$$\varepsilon \boldsymbol{\xi} \equiv \varepsilon(\partial/\partial x^K) = \varepsilon(\partial/\partial\phi), \qquad \varepsilon \ll 1$$

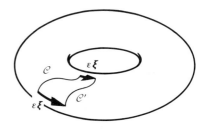

to form a new curve \mathcal{C}'. In coordinate language \mathcal{C} is $\theta = \theta(\lambda)$, $\phi = \phi(\lambda)$; while \mathcal{C}' is $\theta = \theta(\lambda)$, $\phi = \phi(\lambda) + \varepsilon$. (Translation of all points through $\Delta\phi = \varepsilon$.) Because $\partial g_{\mu\nu}/\partial\phi = 0$, the curves \mathcal{C} and \mathcal{C}' have the same length (see text).

C. Pick a set of neighboring points \mathcal{A}, \mathcal{B}, \mathcal{C}, \mathcal{D}; and translate each of them through $\varepsilon \boldsymbol{\xi}$ to obtain points \mathcal{A}', \mathcal{B}', \mathcal{C}', \mathcal{D}'. Since the length of every curve is preserved by this translation, the distances between neighboring points are also preserved:

(distance between \mathcal{A}' and \mathcal{B}') =
\qquad (distance between \mathcal{A} and \mathcal{B}).

But geometry is equivalent to a table of all infinitesimal distances (see Box 13.1). *Thus the geometry of the manifold is left completely unchanged by a translation of all points through $\varepsilon\boldsymbol{\xi}$.* [This is the coordinate-free version of the statement $\partial g_{\mu\nu}/\partial\phi = 0$.] One says that $\boldsymbol{\xi} = \partial/\partial\phi$ *is the generator of an "isometry" (or "group of motions") on the manifold.*

D. In general (see text), a vector field $\boldsymbol{\xi}(\mathcal{P})$ generates an isometry if and only if it satisfies Killing's equation $\xi_{(\alpha;\beta)} = 0$.

E. If $\boldsymbol{\xi}(\mathscr{P})$ generates an isometry (i.e. if $\boldsymbol{\xi}$ is a "Killing vector"), then the curves

$$\mathscr{P}(x^K, \underbrace{\alpha_1, \ldots, \alpha_n}_{})$$

$$\begin{bmatrix} \text{parameter} \\ \text{on curve} \end{bmatrix} \uparrow \qquad \uparrow \begin{bmatrix} \text{labels to tell} \\ \text{"which" curve} \end{bmatrix}$$

to which $\boldsymbol{\xi}$ is tangent $[\boldsymbol{\xi} = (\partial\mathscr{P}/\partial x^K)_{\alpha_1,\ldots,\alpha_n}]$ are called "trajectories of the isometry."

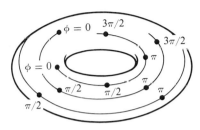

Three different trajectories on a donut

Parameter on trajectories is $x^K = \phi$

F. The geometry is invariant under a translation of all points of the manifold through the same Δx^K along these trajectories $[\mathscr{P}(x^K, \alpha_1, \ldots, \alpha_n) \longrightarrow \mathscr{P}(x^K + \Delta x^K, \alpha_1, \ldots, \alpha_n)$; "finite motion" built up from many "infinitesimal motions" $\varepsilon\boldsymbol{\xi}$.]

G. This isometry is described in physical terms as follows. Station a family of observers throughout the manifold. Have each observer report to a central computer (1) all aspects of the manifold's geometry near him, and (2) the distances and directions to all neighboring observers (directions relative to "preferred" directions that are determined by anisotropies in the local geometry). Through each observer's position passes a unique trajectory of the isometry. Move each observer through the same fixed Δx^K (e.g., $\Delta x^K = 17$) along his trajectory, leaving the manifold itself unchanged. Then have each observer report to the central computer the same geometric information as before his motion. The information received by the computer after the motion will be identical to that received before the motion. There is no way whatsoever, by geometric measurements, to discover that the motion has occurred! This is the significance of an isometry.

EXERCISES

**Exercise 25.1. CONSTANT OF MOTION OBTAINED FROM
HAMILTON'S PRINCIPLE**

Prove the above theorem of conservation of $p_K \equiv \boldsymbol{p} \cdot \boldsymbol{\xi}$ from Hamilton's principle (Box 13.3)

$$\delta \int \frac{1}{2} g_{\mu\nu}(x)(dx^\mu/d\lambda)(dx^\nu/d\lambda)\,d\lambda = 0 \tag{25.8}$$

as applied to geodesic paths. Recall: In this action principle, $g_{\mu\nu}$ is to be regarded as a known function of position, x, along the path; and the path itself, $x^\mu(\lambda)$, is to be varied.

**Exercise 25.2. SUPER-HAMILTONIAN FORMALISM
FOR GEODESIC MOTION**

Show that a set of differential equations in Hamiltonian form results from varying p_μ and x^μ independently in the variational principle $\delta I = 0$, where

$$I = \int (p_\mu\,dx^\mu - \mathcal{H}\,d\lambda) \tag{25.9}$$

and

$$\mathcal{H} \equiv \frac{1}{2} g^{\mu\nu}(x)p_\mu p_\nu. \tag{25.10}$$

Show that the "super-Hamiltonian" \mathcal{H} is a constant of motion, and that the solutions of these equations are geodesics. What do the choices $\mathcal{H} = +\frac{1}{2}$, $\mathcal{H} = 0$, $\mathcal{H} = -\frac{1}{2}\mu^2$, or $\mathcal{H} = -\frac{1}{2}$ mean for the geodesic and its parametrization?

Exercise 25.3. KILLING VECTORS IN FLAT SPACETIME

Find ten Killing vectors in flat Minkowski spacetime that are linearly independent. (Restrict attention to linear relationships with constant coefficients).

Exercise 25.4. POISSON BRACKET AS KEY TO CONSTANTS OF MOTION

If $\boldsymbol{\xi}$ is a Killing vector, show that $p_K \equiv \xi^\mu p_\mu$ commutes (has vanishing Poisson bracket) with the Hamiltonian \mathcal{H} of the previous problem, $[\mathcal{H}, p_K] = 0$, so $dp_K/d\lambda = 0$. (Hint: Use a convenient coordinate system.)

Exercise 25.5. COMMUTATOR OF KILLING VECTORS IS A KILLING VECTOR

Consider two Killing vectors, $\boldsymbol{\xi}$ and $\boldsymbol{\eta}$, which happen not to commute [as differential operators; i.e., the commutator of equations (8.13) does not vanish; consider rotations about perpendicular directions as a case in point]; thus,

$$[\boldsymbol{\xi}, \boldsymbol{\eta}] \equiv \boldsymbol{\zeta} \neq 0.$$

(a) Show that no single coordinate system can be simultaneously adapted, in the sense of equation (25.1), to both the $\boldsymbol{\xi}$ and $\boldsymbol{\eta}$ symmetries (see exercise 9.9).

(b) Let $p_\xi = p_\mu\xi^\mu$, $p_\eta = p_\mu\eta^\mu$, and $p_\zeta = p_\mu\zeta^\mu$, and derive the Poisson-bracket relationship $[p_\xi, p_\eta] = -p_\zeta$. In a geometry, the symmetries of which are related in this way, show that p_ζ is also a constant of motion.

(c) In a coordinate system where $\boldsymbol{\zeta} = (\partial/\partial x^K)$, define \mathcal{H} as in (25.10) and show from $[\mathcal{H}, p_\zeta] = 0$ that $\boldsymbol{\zeta}$ is a Killing vector.

Thus the commutator of two Killing vectors is itself a Killing vector.

Exercise 25.6. EIGENVALUE PROBLEM FOR KILLING VECTORS

Show that any Killing vector satisfies $\xi^\mu{}_{;\mu} = 0$, and is an eigenvector with eigenvalue $\kappa = 0$ of the equation

$$\xi^{\mu;\nu}{}_{;\nu} + R^\mu{}_\nu \xi^\nu = \kappa \xi^\mu. \qquad (25.11)$$

Find a variational principle (Raleigh-Ritz type) for this eigenvalue equation.

§25.3. CONSERVED QUANTITIES FOR MOTION IN SCHWARZSCHILD GEOMETRY

Consider a test particle moving in the Schwarzschild geometry, described by the line element

$$ds^2 = -(1 - 2M/r)\,dt^2 + \frac{dr^2}{(1 - 2M/r)} + r^2(d\theta^2 + \sin^2\theta\,d\phi^2). \qquad (25.12)$$

This expression for the geometry applies outside any spherically symmetric center of attraction of total mass-energy M. It makes no difference, for the motion of the particle outside, what the geometry is inside, because the particle never gets there; before it can, it collides with the surface of the star—if the center of attraction is a star, that is to say, a fluid mass in hydrostatic equilibrium. At each point throughout such an equilibrium configuration, the Schwarzschild coordinate r exceeds the local value of the quantity $2m(r)$; see §23.8. Therefore the Schwarzschild coordinate R of the surface exceeds $2M$. Consequently, expression (25.12) applies outside any equilibrium configuration, no matter how compact ($r > R > 2M$ implies that one need not face the issue of the "singularity" at $r = 2M$). The more compact the configuration, however, the more of the Schwarzschild geometry the test particle can explore. The ideal limit is not a star in hydrostatic equilibrium. It is a star that has undergone complete gravitational collapse to a black hole. Then (25.12) applies arbitrarily close to $r = 2M$. This idealization is assumed here ("black hole"), because the analysis can then cover the maximum range of possible situations.

Wherever the test particle lies, and however fast it moves, project that point and project that 3-velocity radially onto a sphere of some fixed r value, say, the unit sphere $r = 1$. The point and the vector together define a point and a vector on the surface of the unit sphere; and they in turn mark the beginning and define the totality of a great circle. As the particle continues on its way, the radial projection of its position will continue to lie on that great circle. To depart from the great circle on one side or the other would be to give preference to the one hemisphere or the other of the unit sphere, contrary to the symmetry of the situation.

Orient the coordinate system so that the radial projection of the orbit coincides with the equator, $\theta = \pi/2$, of the polar coordinates (see Box 25.4 for the spherical trigonometry of a more general orientation of the orbit, and for eventual specializa-

Why attention focuses on particle orbits around a black hole

Choice of coordinates to make particle orbit lie in "equator," $\theta = \pi/2$

tion to a polar orbit, in contrast to the equatorial orbit considered here). In polar coordinates as so oriented, the particle has at the start, and continues to have, zero momentum in the θ direction; thus,

$$p^\theta = d\theta/d\lambda = 0.$$

The expression (25.12) for the line element shows that the geometry is unaffected by the translations $t \longrightarrow t + \Delta t$, $\phi \longrightarrow \phi + \Delta\phi$. Thus the coordinates t and ϕ are "cyclic." The conjugate momenta $p_0 \equiv -E$ and $p_\phi \equiv \pm L \, (L \geq 0)$ are therefore conserved. This circumstance allows one immediately to deduce the major features of the motion, as follows.

The magnitude of the 4-vector of energy-momentum is given by the rest mass of the particle,

$$g_{\alpha\beta}p^\alpha p^\beta + \mu^2 = g^{\alpha\beta}p_\alpha p_\beta + \mu^2 = 0 \tag{25.13}$$

or

$$-\frac{E^2}{(1 - 2M/r)} + \frac{1}{(1 - 2M/r)}\left(\frac{dr}{d\lambda}\right)^2 + \frac{L^2}{r^2} + \mu^2 = 0. \tag{25.14}$$

Moreover, one knows from the equivalence principle that test particles follow the same world line regardless of mass. Therefore what is relevant for the motion of particles is not the energy and angular momentum themselves, but the ratios

$$\widetilde{E} = E/\mu = \begin{pmatrix}\text{energy per unit}\\\text{rest mass}\end{pmatrix},$$

$$\widetilde{L} = L/\mu = \begin{pmatrix}\text{angular momentum}\\\text{per unit rest mass}\end{pmatrix}. \tag{25.15}$$

Recall also

$$\lambda = \tau/\mu = \begin{pmatrix}\text{proper time per}\\\text{unit rest mass}\end{pmatrix}.$$

Then (25.14) becomes an equation for the change of r-coordinate with proper time in which the rest mass makes no appearance:

$$\left(\frac{dr}{d\tau}\right)^2 = \widetilde{E}^2 - (1 - \overline{2M}/r)(1 + \widetilde{L}^2/r^2)$$

$$= \widetilde{E}^2 - \widetilde{V}^2(r). \tag{25.16a}$$

Here

$$\widetilde{V}(r) = [(1 - 2M/r)(1 + \widetilde{L}^2/r^2)]^{1/2} \tag{25.16b}$$

is the "effective potential" mentioned in §25.1 and Figure 25.2 and to be discussed

in §25.5. For the rate of change of the other two relevant coordinates with proper time, one has, assuming a "direct" orbit ($d\phi/d\tau > 0$; $p_\phi = +L$ rather than $-L$),

$$\frac{d\phi}{d\tau} = \frac{1}{\mu}\frac{d\phi}{d\lambda} = \frac{p^\phi}{\mu} = \frac{g^{\phi\phi}L}{\mu} = \frac{\tilde{L}}{r^2} \tag{25.17}$$

and

$$\frac{dt}{d\tau} = \frac{1}{\mu}\frac{dt}{d\lambda} = \frac{p^0}{\mu} = -\frac{g^{00}E}{\mu} = \frac{\tilde{E}}{1 - 2M/r}. \tag{25.18}$$

Knowing r as a function of τ from (25.16), one can find ϕ and t in their dependence on τ from (25.17) and (25.18). Symmetry considerations have in effect reduced the four coupled second-order differential equations $p^\mu{}_{;\nu}p^\nu = 0$ of geodesic motion to the single first-order equation (25.16).

For objects of zero rest mass, it makes no sense to refer to proper time, and a slightly different treatment is appropriate (§25.6).

Before looking, in §25.5, at the motions predicted by equations (25.16) to (25.18), it is useful to analyze the physical significance of the constants p_0 and p_ϕ, and to identify other physically significant quantities whose values will be of interest in studying these orbits. One calls $E = -p_0$ the "energy at infinity"; and $L = |p_\phi|$, for equatorial orbits, the "total angular momentum." To justify these names, compare them with standard quantities measured by an observer at rest on the equator of the Schwarzschild coordinate system as the test particle flies past him in its orbit. Let

Interpretation of E as "energy at infinity" and L as "angular momentum"

$$E_{\text{local}} \equiv p^{\hat{0}} \equiv \langle \boldsymbol{\omega}^{\hat{0}}, \boldsymbol{p} \rangle \equiv \langle |g_{00}|^{1/2}\,\boldsymbol{dt}, \boldsymbol{p} \rangle = |g_{00}|^{1/2}p^0$$
$$= |g_{00}|^{1/2}\,dt/d\lambda = (1 - 2M/r)^{1/2}\,dt/d\lambda \tag{25.19}$$

be the energy he measures in his proper reference frame, and let

$$v_{\hat{\phi}} \equiv \frac{p^{\hat{\phi}}}{p^{\hat{0}}} \equiv \frac{\langle \boldsymbol{\omega}^{\hat{\phi}}, \boldsymbol{p} \rangle}{E_{\text{local}}} = \frac{\langle |g_{\phi\phi}|^{1/2}\,\boldsymbol{d\phi}, d/d\lambda \rangle}{E_{\text{local}}}$$
$$= \frac{r(d\phi/d\lambda)}{E_{\text{local}}} = \frac{p_\phi}{rE_{\text{local}}} \tag{25.20}$$

be the tangential component of the ordinary velocity he measures. [Note: $\boldsymbol{\omega}^{\hat{\alpha}}$ are the basis one-forms of the observer's proper reference frame; see equations (23.15a,b).] In terms of these locally measured quantities, the energy-at-infinity is

$$E = -p_0 = -g_{00}p^0 = |g_{00}|^{1/2}E_{\text{local}} = (1 - 2M/r)^{1/2}E_{\text{local}} = \text{constant.} \tag{25.21}$$

It therefore represents the locally measured energy E_{local}, corrected by a factor $|g_{00}|^{1/2}$. For any particle that flies freely (geodesic motion) from this observer to $r = \infty$, the correction factor reduces to unity, and E_{local} (as measured by a second observer, this time at infinity) becomes identical with E. Similarly the angular momentum from (25.20) is

$$p_\phi = E_{\text{local}}v_{\hat{\phi}}r = \text{constant.} \tag{25.22}$$

This, like $E = -p_0$, represents a quantity that is conserved, and whose interpretation for $r \longrightarrow \infty$ on any orbit is familiar. Finally, recall that the total 4-momentum of two colliding particles $\boldsymbol{p}_1 + \boldsymbol{p}_2$ or $(p_\mu)_1 + (p_\mu)_2$ is conserved in a point collision (at any r). Therefore the totals $(E)_1 + (E)_2 = (-p_0)_1 + (-p_0)_2$ and $(p_\phi)_1 + (p_\phi)_2$ are also conserved. One of the colliding particles may be on an orbit that could never reach out to $r = \infty$, but this makes no difference. This conservation principle allows and forces one to take over the terms $E = $ "energy at infinity" and $L = $ "angular momentum," valid for orbits that do reach to infinity, for an orbit that does not reach to infinity.

EXERCISES

Exercise 25.7. RADIAL VELOCITY OF A TEST PARTICLE

Obtain a formula for the radial component of velocity $v_{\hat{r}}$ that an observer at r would measure [see (25.20) for $v_{\hat{\phi}}$]. Express E_{local}, $v_{\hat{r}}$, and $v_{\hat{\phi}}$ in terms of r and the constants E, p_ϕ.

Exercise 25.8. ROTATIONAL KILLING VECTORS FOR SCHWARZSCHILD GEOMETRY

(a) Show that in the isotropic coordinates of exercise 23.1, the metric for the Schwarzschild geometry takes the form

$$ds^2 = -(1 - M/2\bar{r})^2(1 + M/2\bar{r})^{-2}\, dt^2 + (1 + M/2\bar{r})^4(d\bar{r}^2 + \bar{r}^2\, d\Omega^2). \quad (25.23)$$

(b) Exhibit a coordinate transformation that brings this into the form

$$ds^2 = -(1 - M/2\bar{r})^2(1 + M/2\bar{r})^{-2}\, dt^2 + (1 + M/2\bar{r})^4(dx^2 + dy^2 + dz^2), \quad (25.24)$$

with $\bar{r} = (x^2 + y^2 + z^2)^{1/2}$.

(c) Show that $\boldsymbol{\xi}_x = y(\partial/\partial z) - z(\partial/\partial y)$ and similar vectors $\boldsymbol{\xi}_y$ and $\boldsymbol{\xi}_z$ are each Killing vectors, by verifying (see exercise 25.5c) that the Poisson brackets $[\mathcal{H}, L_K]$ vanish for each $L_K = \boldsymbol{p} \cdot \boldsymbol{\xi}_K$, $K = x, y, z$.

(d) Show that $\boldsymbol{\xi}_z = (\partial/\partial\phi)_{t,r,\theta}$; and show that for orbits in the equatorial plane $L_z = p_\phi$, $L_x = L_y = 0$.

Exercise 25.9. CONSERVATION OF TOTAL ANGULAR MOMENTUM OF A TEST PARTICLE

Prove by a Poisson-bracket calculation that the total angular momentum squared, $L^2 = p_\theta{}^2 + (\sin\theta)^{-2}p_\phi{}^2$ is a constant of motion for any Schwarzschild geodesic.

Exercise 25.10. SELECTING EQUATION BY SELECTING WHAT IS VARIED

Write out the integral I that is varied in (25.8) for the special case of the Schwarzschild metric (25.12). What equation results from the demand $\delta I = 0$ if only $\phi(\lambda)$ is varied? If only $t(\lambda)$?

Exercise 25.11. MOTION DERIVED FROM SUPER-HAMILTONIAN

Write out the super-Hamiltonian (25.10) for the special case of the Schwarzschild metric. Deduce from its form that p_0 and p_ϕ are constants of motion. Derive (25.14), (25.17), and (25.18) from this super-Hamiltonian formalism.

§25.4. GRAVITATIONAL REDSHIFT

The conservation law $|g_{00}|^{1/2}E_{local} =$ constant (equation 25.21), which is valid in this form for any time-independent metric with $g_{0j} = 0$ and for particles with both zero and non-zero rest mass, is sometimes called the "law of energy red-shift." It describes how the locally measured energy of any particle or photon changes (is "red-shifted" or "blue-shifted") as it climbs out of or falls into a static gravitational field. For a particle of zero rest mass (photon or neutrino), the locally measured energy E_{local}, and wavelength λ_{local} (not to be confused with affine parameter!), are related by $E_{local} = h/\lambda_{local}$, where h is Planck's constant. Consequently, the law of energy red-shift can be rewritten as

Law of "energy redshift" ("gravitational redshift")

$$\lambda_{local}|g_{00}|^{-1/2} = \text{constant.} \qquad (25.25)$$

A photon emitted by an atom at rest in the gravitational field at radius r, and received by an astronomer at rest at infinity is red-shifted by the amount

$$z \equiv \Delta\lambda/\lambda = (\lambda_{received} - \lambda_{emitted})/\lambda_{emitted} = |g_{00}(r)|^{-1/2} - 1,$$

$$z = (1 - 2M/r)^{-1/2} - 1, \qquad (25.26)$$

$$z \approx M/r \text{ in Newtonian limit.} \qquad (25.26N)$$

Note that these expressions are valid whether the photon travels along a radial path or not.

Exercise 25.12. REDSHIFT BY TIMED PULSES

Derive expression (25.26) for the photon redshift by considering two pulses of light emitted successively by an atom at rest at radius r. [*Hint:* If $\Delta\tau_{em}$ is the proper time between pulses as measured by the emitting atom, and $\Delta\tau_{rec}$ is the proper time separation as measured by the observer at $r = \infty$, then one can idealize λ_{em} as $\Delta\tau_{em}$ and λ_{rec} as $\Delta\tau_{rec}$.]

EXERCISE

§25.5. ORBITS OF PARTICLES

Turn attention now from energy red-shift to the orbit of a particle in the Schwarzschild geometry. The position as a function of proper time follows upon solving (25.16a),

$$\left(\frac{dr}{d\tau}\right)^2 + \tilde{V}^2(r) = \tilde{E}^2, \qquad (25.16a)$$

where \tilde{V} is the "effective potential" defined by

$$\tilde{V}^2(r) = (1 - 2M/r)(1 + \tilde{L}^2/r^2) \qquad (25.16b)$$

and illustrated in Figure 25.2 and Box 25.6. The first diagram in Box 25.6 gives $\widetilde{V}^2(r)$ as a function of r. It is relevant even in the "domain inside the black hole" ($r < 2M$), where \widetilde{V}^2 is negative (see Chapter 31). It serves as a model for, and is closely related to, the "effective potential" $B^{-2}(r)$ used in §25.6 to analyze photon orbits. The final diagram in Box 25.6 gives $\widetilde{V}(r)$ itself as a function of r. Energy levels in this diagram or in Figure 25.2 can be interpreted as in any conventional energy-level diagram. The difference in energy between two levels represents energy, as measured at infinity, of the photon given off in the transition from the one level to the other. Whether one plots $\widetilde{V}(r)$ or $\widetilde{V}^2(r)$ as a function of r is largely a matter of convenience. The important point is this: a value of r where $\widetilde{V}(r)$ becomes equal to the available energy \widetilde{E}, or $\widetilde{V}^2(r)$ becomes equal to \widetilde{E}^2, is a *turning point*. A particle that was moving to larger r values, once arrived at a turning point, turns around and moves to smaller r values. Or when a particle moving to smaller r values comes to a turning point, it reverses its motion and proceeds to larger r values. A turning point is not a point of equilibrium. A stone thrown straight up does not sit at a point of equilibrium at the top of its flight. However, when $\widetilde{E} - \widetilde{V}(r)$, or $\widetilde{E}^2 - \widetilde{V}^2(r)$, instead of having a single root, has a double root, then one does deal with a point of equilibrium (only possible because of "centrifugal force" fighting against gravity). When this equilibrium occurs at a minimum of $\widetilde{V}(r)$, it is a stable equilibrium; at a maximum, an unstable equilibrium. Thus all the major features of the motion in the r direction can be read from a plot of the effective potential as a function of r (plot depends on value of \widetilde{L}) and from a knowledge of the \widetilde{E} value (Figure 25.2, with further details in Box 25.6).

Qualitative features of orbits diagnosed from effective-potential diagram

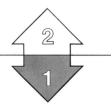

Box 25.6 QUALITATIVE FEATURES OF ORBITS OF A PARTICLE MOVING IN SCHWARZSCHILD GEOMETRY

A. Equations Governing Orbit (see text for derivation)

1. Effective-potential equation for radial part of motion:

$$(dr/d\tau)^2 + \widetilde{V}^2(\widetilde{L}, r) = \widetilde{E}^2,$$
$$\widetilde{V}^2(\widetilde{L}, r) = (1 - 2M/r)(1 + \widetilde{L}^2/r^2),$$

$\widetilde{E} =$ (energy at infinity per unit rest mass),
$\widetilde{L} =$ (angular momentum per unit rest mass).

2. Supplementary equations for angular and time motion for "direct" orbit, $d\phi/d\tau > 0$:

$$d\phi/d\tau = \widetilde{L}/r^2,$$
$$\frac{dt}{d\tau} = \frac{\widetilde{E}}{1 - 2M/r}.$$

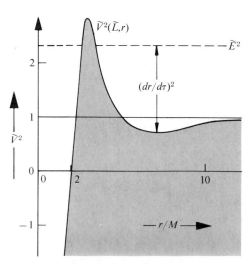

"Turning points" of orbit occur where horizontal line of height \widetilde{E}^2 crosses \widetilde{V}^2

B. Newtonian Limit, $|\widetilde{E} - 1| \ll 1$, $M/r \ll 1$, $\widetilde{L}/r \ll 1$

1. Speak not about "energy-at-infinity per unit rest mass," $\widetilde{E} = E/\mu = (1 - v_\infty^2)^{-1/2}$, but instead about the "nonrelativistic energy per unit rest mass,"

$$\varepsilon \equiv \frac{1}{2}(\widetilde{E}^2 - 1) \approx \widetilde{E} - 1 \approx \frac{1}{2}v_\infty^2.$$

2. Speak not about $\widetilde{V}^2(\widetilde{L}, r)$ but instead about the Newtonian effective potential,

$$V_N(\widetilde{L}, r) \equiv \frac{1}{2}(\widetilde{V}^2 - 1) \approx -\frac{M}{r} + \frac{\widetilde{L}^2}{2r^2}.$$

3. Rewrite effective-potential equation in the form

$$\frac{1}{2}\left(\frac{dr}{d\tau}\right)^2 + V_N(\widetilde{L}, r) = \varepsilon.$$

4. From the effective-potential diagram and the subsidiary equation $d\phi/d\tau = \widetilde{L}/r^2$, conclude that:

 a. Particles with $\varepsilon \geq 0$ ($\widetilde{E} \geq 1$) come in from $r = \infty$ along hyperbolic or parabolic orbits, are reflected off the effective potential at $\varepsilon = V_N[\widetilde{E}^2 = \widetilde{V}^2$; "turning point"; $(dr/d\tau)^2 = 0]$, and return to $r = \infty$.

 b. Particles with $\varepsilon < 0$ ($\widetilde{E} < 1$) move back and forth in an effective potential well between periastron (inner turning point of elliptic orbit) and apastron (outer turning point).

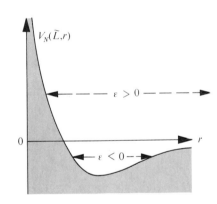

C. Relativistic Orbits

Use the effective-potential diagram of part A (reproduced here for various \widetilde{L}), in the same way one uses the Newtonian diagram of part B, to deduce the qualitative features of the orbits. The main conclusions are these.

Box 25.6 (*continued*)

1. Orbits with periastrons at $r \gg M$ are Keplerian in form, except for the periastron shift (exercise 25.16; §40.5) familiar for Mercury.

2. Orbits with periastrons at $r \lesssim 10M$ differ markedly from Keplerian orbits.

3. For $\widetilde{L}/M \leq 2\sqrt{3}$ there is no periastron; any incoming particle is necessarily pulled into $r = 2M$.

4. For $2\sqrt{3} < \widetilde{L}/M < 4$ there are bound orbits in which the particle moves in and out between periastron and apastron; but any particle coming in from $r = \infty$ (unbound; $\widetilde{E}^2 \geq 1$) necessarily gets pulled into $r = 2M$.

5. For $L^{\dagger} = \widetilde{L}/M > 4$, there are bound orbits; particles coming in from $r = \infty$ with

$$\widetilde{E}^2 < \widetilde{V}_{\max}^2 = (1 - 2u_m)(1 + L^{\dagger 2}u_m^2),$$

$$u_m \equiv \frac{1 + \sqrt{1 - 12/L^{\dagger 2}}}{6}$$

reach periastrons and then return to $r = \infty$; but particles from $r = \infty$ with $\widetilde{E}^2 > \widetilde{V}_{\max}^2$ get pulled into $r = 2M$.

6. There are stable circular orbits at the minimum of the effective potential; the minimum moves inward from $r = \infty$ for $\widetilde{L} = \infty$ to $r = 6M$ for $L^{\dagger} = \widetilde{L}/M = 2\sqrt{3}$. The most tightly bound, stable circular orbit ($\widetilde{L}/M = 2\sqrt{3}$, $r = 6M$) has a fractional binding energy of

$$\frac{\mu - E}{\mu} = 1 - \widetilde{E} = 1 - \sqrt{8/9} = 0.0572.$$

7. There are unstable circular orbits at the maximum of the effective potential; the maximum moves outward from $r = 3M$ for $\widetilde{L} = \infty$ to $r = 6M$ for $\widetilde{L}/M = 2\sqrt{3}$. A particle in such a circular orbit, if perturbed inward, will spiral into $r = 2M$. If perturbed outward, and if it has $\widetilde{E}^2 > 1$, it will escape to $r = \infty$. If perturbed out-

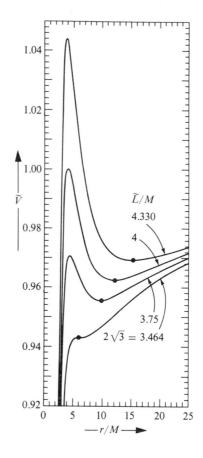

ward, and if it has $\widetilde{E}^2 < 1$, it will either reach an apastron and then enter a spiraling orbit that eventually falls into the star (e.g., if $\delta\widetilde{E} > 0$, with unchanged angular momentum); or it will move out and in between apastron and periastron, in a stable bound orbit (e.g., if $\delta\widetilde{E} < 0$, again with unchanged angular momentum).

When one turns from qualitative features to quantitative results, one finds it appropriate to write down explicitly the proper time $\Delta\tau$ required for the particle to augment its Schwarzschild coordinate by the amount Δr; thus (with the convention that square roots may be negative or positive, $\sqrt{a^2} \equiv \pm a$)

$$\tau = \int d\tau = \int \frac{dr}{[\tilde{E}^2 - (1 - 2M/r)(1 + \tilde{L}^2/r^2)]^{1/2}}. \tag{25.27}$$

The integration is especially simple for a particle falling straight in, or climbing straight out, for then the angular momentum vanishes and the integral can be written in an elementary form that applies (with the change $\tau \longrightarrow t$) even in Newtonian mechanics,

$$\tau = \int d\tau = \int \frac{dr}{[2M/r - 2M/R]^{1/2}}. \tag{25.27'}$$

Here $R \equiv 2M/(1 - \tilde{E}^2)$ is the radius at which the particle has zero velocity ("apastron"). The motion follows the same "cycloid principle" that is so useful in nonrelativistic mechanics (Figure 25.3). Thus, in parametric form, one has

$$r = \frac{R}{2}(1 + \cos\eta),$$

$$\tau = \frac{R}{2}\left(\frac{R}{2M}\right)^{1/2}(\eta + \sin\eta), \tag{25.28}$$

with the total proper time to fall from rest at $r = R$ into $r = 0$ given by the expression

$$\tau = \frac{\pi}{2}R\left(\frac{R}{2M}\right)^{1/2} \tag{25.29}$$

(shorter by a factor $1/\sqrt{2}$ than the time for fall under pull of the same mass, distributed over a sphere of radius R; see dotted curve in Figure 25.3).

What about the Schwarzschild-coordinate time taken for a given motion? Take equation (25.16a) for general motion (radial or nonradial), and where $dr/d\tau$ appears, replace it by

$$\frac{dr}{d\tau} = \frac{dr}{dt}\frac{dt}{d\tau} = \frac{dr}{dt}\frac{\tilde{E}}{1 - 2M/r} = \tilde{E}\frac{dr^*}{dt}. \tag{25.30}$$

Here r^* is an abbreviation for a new "tortoise coordinate,"

$$r^* = \int dr^* = \int \frac{dr}{1 - 2M/r} = r + 2M\ln\left(\frac{r}{2M} - 1\right), \tag{25.31}$$

which was introduced by Wheeler (1955) and popularized by Regge and Wheeler (1957). Thus find the equation

$$\left(\tilde{E}\frac{dr^*}{dt}\right)^2 + \tilde{V}^2 = \tilde{E}^2. \tag{25.32}$$

Radial orbits:

(1) "cycloidal" form of $r(\tau)$ for radial bound orbits

(2) "tortoise" radial coordinate as function of coordinate time, $r^*(t)$

Figure 25.3.

A cycloid gives the relation between proper time and Schwarschild r coordinate for a test particle falling straight in toward center of gravitational attraction of negligible dimensions. The angle of turn of the wheel as it rolls on the base line and generates the cycloid is denoted by η. In terms of this parameter, one has

$$r = \frac{R}{2}(1 + \cos\eta) \qquad \text{(Schwarzschild } r\text{-coordinate)}$$

$$\tau = \frac{R}{2}\left(\frac{R}{2M}\right)^{1/2}(\eta + \sin\eta) \qquad \text{(proper time)}$$

(note difference in scale factors in expressions for r and for τ). The total lapse of proper time to fall from $r = R$ to $r = 0$ is $\tau = (\pi/2)(R^3/2M)^{1/2}$. The same cycloid relation and the same expression for time to fall holds in Newton's nonrelativistic theory of gravitation, except that there the symbol τ is to be replaced by the symbol t (ordinary time). Were one dealing in Newtonian theory with the same attracting mass M spread uniformly over a sphere of radius R, with a pipe thrust through it to make a channel for the motion of the test particle, then that particle would execute simple harmonic oscillations (dotted curve above). The angular frequency ω of these vibrations would be identical with the angular frequency of revolution of the test particle in a circle just grazing the surface of the planet, a frequency given by Kepler's law $M = \omega^2 R^3$. In this case, the time to fall to the center would be $(\pi/2)(R^3/M)^{1/2}$, longer by a factor $2^{1/2}$ than for a concentrated center of attraction (concentrated mass: stronger acceleration and higher velocity in the later phases of the fall). The expression for the Schwarzschild-coordinate time t required to reach any point r in the fall under the influence of a concentrated center of attraction is complicated and is not shown here (see equation 25.37 and Figure 25.5).

The same cycloidal relation that connects r with time for free fall of a particle also connects the radius of the "Friedmann dust-filled universe" with time (see Box 27.1), except that there the cycloid diagram applies directly, without any difference in scale between the two key variables:

$$\binom{\text{radius of}}{\text{3-sphere}} = \frac{a}{2}(1 - \cos\eta) \simeq \frac{a}{4}\eta^2 \text{ (for small } \eta\text{)},$$

$$\begin{pmatrix}\text{coordinate time}\\ \text{identical with}\\ \text{proper time as}\\ \text{measured on dust}\\ \text{particle}\end{pmatrix} = \frac{a}{2}(\eta - \sin\eta) \simeq \frac{a}{12}\eta^3 \text{ (for small } \eta\text{)}.$$

The starting point of η is renormalized to time of start of expansion; see Lindquist and Wheeler (1957) for more on correlation between fall of particle and expansion of universe.

Here the effective potential is the same effective potential that one dealt with before,

$$\widetilde{V} = [(1 - 2M/r)(1 + \widetilde{L}^2/r^2)]^{1/2}. \qquad (25.33)$$

Moreover, the \widetilde{E} on the righthand side is the same \widetilde{E} that appeared in the earlier equation for $(dr/d\tau)^2$. Therefore the turning points and the qualitative description of the motion are both the same as before. "A turning point is a turning point is

a turning point." Right? Right about turning points; wrong about the conclusion.

(3) details of the approach to the Schwarzschild radius ($r = 2M$)

The story has it that Achilles never could pass the tortoise. Whenever he caught up with where it had been, it had moved ahead to a new location; and when he got there, it was still further ahead; and so on *ad infinitum*. Imagine the race between Achilles and the tortoise as running to the left and the expected point of passing as lying at $r = 2M$. The r-coordinate has no inhibition about passing through the value $r = 2M$. Not so r^*, the "tortoise coordinate." It can go arbitrarily far in the direction of minus infinity (corresponding to the infinitely many times when Achilles catches up with where the tortoise was) and still r remains outside $r = 2M$:

$r/2M$	1.000001	1.0001	1.01	1.278465	2	5	10	10,000
$r^*/2M$	-12.8155	-8.2102	-3.5952	0	2	6.386	12.303	10,009.210

It follows that there is a great difference between the description of the motion in terms of the proper time τ of a clock on the falling particle (r goes all the way from $r = R$ down to $r = 0$ in the finite proper time of 25.29) and a description of the motion in terms of the Schwarzschild-coordinate time t appropriate for the faraway observer (r^* goes all the way from $r^* = R^*$ down to $r^* = -\infty$; infinite t required for this; but even in infinite time, as r^* goes down to $-\infty$, r is only brought asymptotically down to $r \sim 2M$). Thus the second description of the motion leaves out, and has no alternative but to leave out, the whole range of r values from $r = 2M$ down to zero: perfectly good physics, and physics that the falling particle is going to see and explore, but physics that the faraway observer never will see and never can see. If the tortoise coordinate did not exist, it would have to be invented. It invests each factor ten of closer approach to $r = 2M$ with the same interest as the last factor ten and the next to come. It proportions itself in accord with the amount of Schwarzschild-coordinate time available to the faraway observer to study these more and more microscopic amounts of motion in more and more detail.

Figure 25.4 shows the effective potential \tilde{V} of (25.33) and of Figure 25.2 replotted as a function of the tortoise coordinate. The approach of \tilde{V} to zero at $r = 2M$ shows up as an exponential approach of \tilde{V} to zero as r^* goes to minus infinity. Thus in moving "towards the black hole" ($r = 2M$, $r^* = -\infty$), the particle, as described in coordinate time t, soon casts off any effective influence of any potential, and moves essentially freely toward decreasing r^*, in accordance with the equation

$$\left(\tilde{E} \frac{dr^*}{dt} \right)^2 \simeq \tilde{E}^2; \tag{25.34}$$

that is, "with the speed of light" ($dr^*/dt \simeq -1$). This dependence of r^* on t implies at once an asymptotic dependence of r itself on Schwarzschild-coordinate time t, of the form

$$r = 2M + (\text{constant} \times e^{-t/2M}). \tag{25.35}$$

This result is independent of the angular momentum of the particle and independent also of the energy, provided only that the energy-per-unit-mass \tilde{E} is enough to

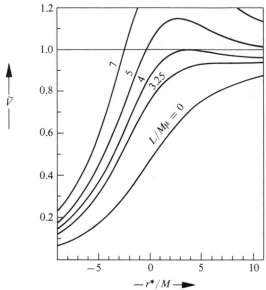

Figure 25.4.
Effective potential for motion in Schwarzschild geometry, expressed as a function of the tortoise coordinate, for selected values of the angular momentum of the test particle. The angular momentum L is expressed in units $M\mu$, where M is the mass of the black hole and μ the mass of the test particle. The effective potential (including rest mass) is expressed in units μ; thus, $\widetilde{V} = V/\mu$. The tortoise coordinate $r^* = r + 2M \ln (r/2M - 1)$ is given in units M.

surmount the barrier (Figure 25.4) of the effective potential-per-unit-mass \widetilde{V}. (More will be said on the approach to $r = 2M$ in Chapter 32, on gravitational collapse.)

To replace the asymptotic formula (25.35) by a complete formula requires one to integrate (25.32); thus,

$$
\begin{aligned}
t = \int dt &= \int \frac{\widetilde{E}\, dr^*}{[\widetilde{E}^2 - \widetilde{V}^2]^{1/2}} \\
&= \int \frac{\widetilde{E}}{[\widetilde{E}^2 - (1 - 2M/r)(1 + \widetilde{L}^2/r^2)]^{1/2}} \frac{dr}{(1 - 2M/r)}.
\end{aligned} \tag{25.36}
$$

The integration here is not easy, even for pure radial motion ($\widetilde{L} = 0$), as is seen in the complication of the resulting expression (Khuri 1957):

$$
\begin{aligned}
t = &\left[\left(\frac{R}{2} + 2M\right)\left(\frac{R}{2M} - 1\right)^{1/2}\right]\eta + \frac{R}{2}\left(\frac{R}{2M} - 1\right)^{1/2}\sin\eta \\
&+ 2M \ln \left| \frac{(R/2M - 1)^{1/2} + \tan(\eta/2)}{(R/2M - 1)^{1/2} - \tan(\eta/2)} \right|.
\end{aligned} \tag{25.37}
$$

Here η is the same cycloid parameter that appears in equation (25.28) and Figure 25.3 (see the detailed plot in Figure 25.5 of the correlation between r and t, illustrat-

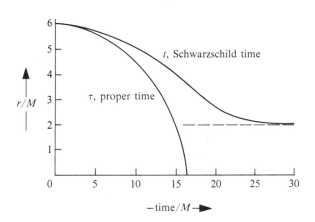

Figure 25.5.

Fall toward a Schwarzschild black hole as described (a) by a comoving observer (proper time τ) and (b) by a faraway observer (Schwarzschild-coordinate time t). In the one description, the point $r = 0$ is attained, and quickly [see equation (25.28)]. In the other description, $r = 0$ is never reached and even $r = 2M$ is attained only asymptotically [equations (25.35) and (25.37)]. The qualitative features of the motion in both cases are most easily deduced by inspection of the "effective potential-per-unit-mass" \widetilde{V} in its dependence on r (Figure 25.2) when one is interested in proper time; or the same effective potential \widetilde{V} in its dependence on the "tortoise coordinate" r^* [Figure 25.4 and equation (25.31)] when one is interested in Schwarzschild-coordinate time t.

ing the asymptotic approach to $r = 2M$). The difficulty in the integration for t, as compared to the ease of the integration for τ (25.28), has a simple origin. Only two r-values appear in (25.27a) as special points when \widetilde{L} is zero: the starting point, $r = R$, where the velocity vanishes, and the point $r = 0$, where $dr/d\tau$ becomes infinite. In contrast (25.36), rewritten as

$$t = \int dt = \int \frac{[1 - 2M/R]^{1/2}}{[2M/r - 2M/R]^{1/2}} \frac{dr}{(1 - 2M/r)}, \qquad (25.36')$$

contains three special points: $r = R$, $r = 0$, and the added point with all the new physics, $r = 2M$. To admit angular momentum is to increase the number of special points still further, and to make the integral unmanageable except numerically or qualitatively (via the potential diagram of Figure 25.4), or in terms of elliptic functions [Hagihara (1931)].

It is often convenient to abstract away from the precise value $r = R$ at the start of the collapse. In this event, one deals with the limit $R \longrightarrow \infty$. Then it is convenient to displace the zero of proper time to the instant of final catastrophe. In this limit, one has

$$\tau/2M = -(2/3)(r/2M)^{3/2},$$

$$t/2M = -(2/3)(r/2M)^{3/2} - 2(r/2M)^{1/2} + \ln \frac{(r/2M)^{1/2} + 1}{(r/2M)^{1/2} - 1}. \qquad (25.38)$$

At very large negative time, the particle is far away and approaching only very slowly. Then one can write

$$r = (9M\tau^2/2)^{1/3} \simeq (9Mt^2/2)^{1/3} \qquad (25.39a)$$

(4) free-fall from $r = \infty$

whether one refers to coordinate time or to proper time. However, the final stages of infall are again very different, when expressed in terms of proper time ($\tau \longrightarrow 0$, $r \longrightarrow 0$), from what they are as expressed in terms of Schwarzschild-coordinate time,

$$r/2M = 1 + 4e^{-8/3}e^{-t/2M}. \tag{25.39b}$$

Nonradial orbits:
(1) Fourier analysis

Turning from pure radial motion to motion endowed with angular momentum, one has a situation where one would like to express the principal quantities of the motion (components of displacement, velocity, and acceleration) in Fourier series (in Schwarzschild-coordinate time), these being so convenient in the Newtonian limit in analyzing radiation and perturbations of one orbit by another and tidal perturbations of the moving particle itself by the tide-producing action of the center of attraction. Any exact evaluation of these coefficients would appear difficult. For the time being, the values of the Fourier amplitudes would seem best developed by successive approximations starting from the Newtonian analysis (see Box 25.4 and references cited there).

In connection with any such Fourier analysis, it is appropriate to recall that the fundamental frequency alone appears, and all higher harmonics have zero amplitude, when the motion takes place in an exactly circular orbit (opposite extreme from the pure radial motion of $\widetilde{L} = 0$). Therefore it is of interest to note (exercise 25.19) that the circular frequency ω of this motion, as measured by a faraway observer, is correlated with the Schwarzschild r-value of the orbit by exactly the Keplerian formula of non-relativistic physics:

$$\omega^2 r^3 = M \qquad \text{(exact; general relativity).} \tag{25.40}$$

(2) details of angular motion

Turn now from the correlation between r and time to the correlation between r and angle of revolution (ϕ in the analysis here; θ in the Hamilton-Jacobi analysis of Box 25.4; this difference in name is irrelevant in what follows). Return to equation (25.16),

$$\left(\frac{dr}{d\tau}\right)^2 + \widetilde{V}^2(r) = \widetilde{E}^2,$$

and recall also equation (25.17)

$$\frac{d\phi}{d\tau} = \frac{\widetilde{L}}{r^2}.$$

Solve the second equation for $d\tau$, and substitute into the first to find

$$\left(\frac{\widetilde{L}\,dr}{r^2\,d\phi}\right)^2 + \widetilde{V}^2(r) = \widetilde{E}^2, \tag{25.41}$$

or equivalently, with $u = M/r$ and $L^\dagger = \widetilde{L}/M = L/M\mu$,

$$\left(\frac{du}{d\phi}\right)^2 = \frac{\widetilde{E}^2 - (1 - 2u)(1 + L^{\dagger 2}u^2)}{L^{\dagger 2}}. \tag{25.42}$$

Exercise 25.16 presents an alternative differential equation derived from this formula, and uses it to obtain the following expression for the angle swept out by the particle or planet, moving in a nearly circular orbit, between two successive points of closest approach:

$$\Delta\phi = \frac{2\pi}{(1 - 6M/r_0)^{1/2}}. \tag{25.43}$$

The radial motion turns around from ingoing to outgoing, or from outgoing to ingoing, whenever the quantity $\tilde{E}^2 - \tilde{V}^2(r)$, or $\tilde{E} - \tilde{V}(r)$, plotted as a function of r, undergoes a change of sign, and this as clearly here in the correlation between r and ϕ as in the earlier correlation between r and time. Recall again the curves of Figure 25.2 for $\tilde{V}(r)$ as a function of r for selected \tilde{L} values. From them one can read out, without any calculation at all, the principal features of typical orbits (Box 25.6) obtained by detailed numerical calculation. Characteristic features are

(1) circular orbit when \tilde{E} coincides with a minimum of the effective potential $\tilde{V}(r)$,

(2) precession when \tilde{E} is a little more than \tilde{V}_{min},

(3) temporary "orbiting" (many turns around the center of attraction) when \tilde{E} is close to a maximum \tilde{V}_{max} of the effective potential,

(4) "capture into the black hole" when \tilde{E} exceeds \tilde{V}_{max}.

A more detailed analysis appears in Box 25.6. [For explicit analytic calculation of orbits in the Schwarzschild geometry, see Hagihara (1931), Darwin (1959 and 1961), and Mielnik and Plebanski (1962).]

For orbits of positive energy, no feature of the inverse-square force is better known than the Rutherford scattering formula. It gives the "effective amount of target area" presented by the center of attraction for throwing particles into a faraway receptor that picks up everything coming off into a unit solid angle at a specified angle of deflection Θ:

$$\frac{d\sigma}{d\Omega} = \frac{M^2}{[4(\tilde{E} - 1)\sin^2\Theta/2]^2} \text{ (Rutherford; nonrelativistic)} \tag{25.44}$$

(derivation in equations 8 to 15 of Box 25.4). When one turns from the Newtonian analysis to the general-relativity treatment, one finds two striking new features of the scattering associated with the phenomenon of orbiting. (1) The particles that come off at a given angle of deflection Θ now include not only those that have really been deflected by Θ (the only contribution in Rutherford scattering), but also those that have been deflected by $\Theta + 2\pi$, $\Theta + 4\pi, \ldots$ etc. (an infinite series of contributions). (2) These supplementary contributions, while finite in amount, and even finite in amount "per unit range of Θ," are not finite in amount when expressed "per unit of solid angle $d\Omega = 2\pi \sin\Theta\,d\Theta$" in either the forward direction ($\Theta = 0$) or the backward direction ($\Theta = \pi$). This circumstance produces no spectacular change in the forward scattering, for that is already infinite in the nonrelativistic approximation (infinity in Rutherford value of $d\sigma/d\Omega$ as $\Theta = 0$ is approached, arising from

particles flying past with large impact parameters and experiencing small deflections; see exercise 25.21). In contrast, the backward scattering, which was perfectly finite in the Rutherford analysis, acquires also an infinity:

$$\left(\frac{d\sigma}{d\Omega}\right)_{\Theta \sim \pi} \sim \frac{\text{constant}}{\sin \Theta}. \tag{25.45}$$

This concentration of scattering in the backward direction is known as a "glory." The effect is most readily seen by looking at the brilliant illumination that surrounds the shadow of one's plane on clouds far below (180° scattering of light ray within waterdrop). It is also clearly seen in observations on the scattering of atoms by atoms near $\Theta = 180°$. No dwarf star, not even any neutron star, is sufficiently compact to be out of the way of a high-speed particle trying to make such a 180° turn. Only a black hole is compact enough to produce this effect.

Further interesting features of motion in Schwarzschild geometry appear in the exercises below.

EXERCISES

Exercise 25.13. QUALITATIVE FORMS OF PARTICLE ORBITS
Verify the statements about particle orbits made in part C of Box 25.6.

Exercise 25.14. IMPACT PARAMETER
For a scattering orbit (i.e., unbound orbit), show that $\widetilde{L} = \widetilde{E}v_\infty b$, where b is the impact parameter and v_∞ the asymptotic ordinary velocity; also show that

$$b = \widetilde{L}/(\widetilde{E}^2 - 1)^{1/2}. \tag{25.46}$$

Draw a picture illustrating the physical significance of the impact parameter.

Exercise 25.15. TIME TO FALL TO $r = 2M$
Show from equation (25.16) and the first picture in Box 25.6 that orbits (general \widetilde{L} value!) which approach $r = 2M$ do so in a finite proper time, but (equation 25.32) an infinite coordinate time t. For equilibrium stars, which must have radii $R > 2M$, the coordinate time t to fall to the surface is finite, of course.

Exercise 25.16. PERIASTRON SHIFT FOR NEARLY CIRCULAR ORBITS
Rewrite equation (25.42) in the form

$$(du/d\phi)^2 + (1 - 6u_0)(u - u_0)^2 - 2(u - u_0)^3 = (\widetilde{E}^2 - \widetilde{E}_0^2)/L^{\dagger 2}. \tag{25.47}$$

Express the constant $u_0 \equiv M/r_0$ in terms of \widetilde{L}/M, and express \widetilde{E}_0 in terms of u_0. Show for a nearly circular orbit of radius r_0 that the angle swept out between two successive periastra (points of closest approach to the star) is

$$\Delta\phi = 2\pi(1 - 6M/r_0)^{-1/2}. \tag{25.48}$$

Sketch the shape of the orbit for $r_0 = 8M$.

Exercise 25.17. ANGULAR MOTION DURING INFALL

From equation (25.42), deduce that the total angle $\Delta\phi$ swept out on a trajectory falling into $r = 0$ is finite. The computation is straightforward; but the interpretation, in view of the behavior of $t(\lambda)$ on the same trajectory (equation 25.32 and exercise 25.15), is not. The interpretation will be elucidated in Chapter 31.

Exercise 25.18. MAXIMUM AND MINIMUM OF EFFECTIVE POTENTIAL

Derive the expressions given in the caption of Figure 25.2 for the locations of the maximum and the minimum of the effective potential as a function of angular momentum. Determine also the limiting form of the dependence of barrier height on angular momentum in the limit in which \tilde{L} is very large compared to M.

Exercise 25.19. KEPLER LAW VALID FOR CIRCULAR ORBITS

From $d\phi/d\tau$ of (25.17) and $dt/d\tau$ of (25.18), deduce an expression for the circular frequency of revolution as seen by a faraway observer; and from the results of exercise 25.18 (or otherwise) show that it fulfills exactly the Kepler relation

$$\omega^2 r^3 = M$$

for any circular orbit of Schwarzschild r-value equal to r, whether stable (potential minimum) or unstable (potential maximum).

Exercise 25.20. HAMILTON-JACOBI FUNCTION

Construct the locus in the r, θ diagram of points of constant dynamic phase $\widetilde{S}(t, r, \theta) = 0$ for $t = 0$ and for values $\tilde{L} = 4M$, $\tilde{E} = 1$ (or for $\tilde{L} = 2\sqrt{3}M$, $\tilde{E} = (8/9)^{1/2}$, or for some other equally simple set of values for these two parameters). Show that the whole set of surfaces of constant \widetilde{S} can be obtained by rotating the foregoing locus through one angle, then another and another, and recopying or retracing. Interpret physically the principal features of the resulting pattern of curves.

Exercise 25.21. DEFLECTION BY GRAVITY CONTRASTED WITH
DEFLECTION BY ELECTRIC FORCE

A test particle of arbitrary velocity β flies past a mass M at an impact parameter b so great that the deflection is small. Show that the deflection is

$$\theta = \frac{2M}{b\beta^2}(1 + \beta^2). \tag{25.49}$$

Derive the deflection according to Newtonian mechanics for a particle moving with the speed of light. Show that (25.49) in the limit $\beta \longrightarrow 1$ is twice the Newtonian deflection. Derive also (flat-space analysis) the contrasting formula for the deflection of a fast particle of rest mass μ and charge e by a nucleus of charge Ze,

$$\theta = \frac{2Ze^2}{\mu b\beta^2}(1 - \beta^2)^{1/2}. \tag{25.50}$$

How feasible is it to rule out a "vector" theory of gravitation [see, for example, Brillouin (1970)], patterned after electromagnetism, by observations on the bending of light by the sun? [Hint: To simplify the mathematical analysis, go back to (25.42). Differentiate once with respect to ϕ to convert into a second-order equation. Rearrange to put on the left all those terms that would be there in the absence of gravity, and on the right all those that originate from the $-2u$ term (gravitation) in the factor $(1 - 2u)$. Neglect the right-hand side of the equation and solve exactly (straight-line motion). Evaluate the perturbing term

on the right as a function of ϕ by inserting in it the unperturbed expression for $u(\phi)$. Solve again and get the deflection.]

Exercise 25.22. CAPTURE BY A BLACK HOLE

Over and above any scattering of particles by a black hole, there is direct capture into the black hole. Show that the cross section for capture is πb_{crit}^2, with the critical impact parameter b_{crit} given by $L_{\text{crit}}/(E^2 - \mu^2)^{1/2}$. From the formulas in the caption of Fig. 25.2 or otherwise, show that for high-energy particles this cross section varies with energy as

$$\sigma_{\text{capt}} = 27\pi M^2 \left(1 + \frac{2}{3\widetilde{E}^2} + \cdots \right) \tag{25.51}$$

(photon limit for $\widetilde{E} \longrightarrow \infty$) and for low energies as

$$\sigma_{\text{capt}} = 16\pi M^2/\beta^2, \tag{25.52}$$

where β is the velocity relative to the velocity of light [Bogorodsky (1962)].

§25.6. ORBIT OF A PHOTON, NEUTRINO, OR GRAVITON IN SCHWARZSCHILD GEOMETRY

Orbits for particles of zero rest mass:

The concepts of "energy per unit of rest mass" and "angular momentum per unit of rest mass" make no sense for an object of zero rest mass (photon, neutrino, even the graviton of exercise 35.16). However, there is nothing about the motion of such an entity that cannot be discovered by considering the motion of a particle of finite rest mass μ and going to the limit $\mu \longrightarrow 0$. In this limit the quantities

$$\widetilde{E} = E/\mu$$

and

$$\widetilde{L} = L/\mu$$

individually go to infinity; but the ratio

(1) impact parameter defined

$$\begin{pmatrix} \text{impact para-} \\ \text{meter} \end{pmatrix} = b = \frac{\begin{pmatrix} \text{angular} \\ \text{momentum} \end{pmatrix}}{\begin{pmatrix} \text{linear} \\ \text{momentum} \end{pmatrix}} = \frac{L}{(E^2 - \mu^2)^{1/2}} = \frac{\widetilde{L}}{(\widetilde{E}^2 - 1)^{1/2}} \tag{25.53}$$

goes to the finite value

$$\operatorname*{Lim}_{\mu \to 0} \frac{\widetilde{L}}{\widetilde{E}} = b. \tag{25.54}$$

(2) shape of orbit

In this limit, equation (25.41) for the shape of the orbit reduces at once to the simple form

$$\left(\frac{1}{r^2}\frac{dr}{d\phi}\right)^2 + \frac{1 - 2M/r}{r^2} = \frac{1}{b^2}, \tag{25.55}$$

or

$$\left(\frac{1}{r^2}\frac{dr}{d\phi}\right)^2 + B^{-2}(r) = b^{-2}, \tag{25.56}$$

or

$$\left(\frac{du}{d\phi}\right)^2 + u^2(1 - 2u) = \left(\frac{M}{b}\right)^2 \equiv \frac{1}{\tilde{b}^2}. \tag{25.57}$$

Whichever way the differential equation for the orbit is written, one term in it depends on the choice of orbit (the term $1/b^2$) the other on the properties of the Schwarzschild geometry, but not on the choice of orbit. This second term defines a kind of effective potential,

$$\begin{pmatrix} \text{"effective} \\ \text{potential for} \\ \text{photon"} \end{pmatrix} \equiv B^{-2}(r) \equiv \frac{1 - 2M/r}{r^2}. \tag{25.58}$$

(3) effective potential

No attempt is made here to take the square root, as was done for a particle of finite rest mass. There one took the root in order to have a quantity that reduced to the Newtonian effective potential (plus the rest mass) in the nonrelativistic limit; but for light ($v = 1$) there is no nonrelativistic limit. Therefore the effective potential (25.58) is plotted directly in Box 25.7, and used there to analyze some of the principal features of the orbits of a photon in Schwarzschild geometry.

On occasion it has proved useful to plot as a function of r, not the "effective potential" of (25.58), but the "potential impact parameter $B(r)$" calculated from that formula [see, for example, Power and Wheeler (1957), Zel'dovich and Novikov (1971)]. This potential impact parameter has the following interpretation: A ray, in order to reach the point r, must have an impact parameter b that is equal to or less than $B(r)$:

$$b \le B(r) \text{ ("condition of accessibility").} \tag{25.59}$$

A ray with zero impact parameter (head-on impact), or any impact parameter less than $b_{\text{crit}} = \min[B(r)] = 3\sqrt{3}M$, can get to any and all r values.

(4) critical impact parameter

The beautifully simple "effective potential" defined by (25.58) is used in (25.56) to determine the shape of an orbit; that is, the azimuth ϕ that the photon has when it gets to a given r-value. In other connections, it can be equally interesting to know when, or at what Schwarzschild coordinate time, the photon gets to a given r value. More broadly, the geodesic of a photon, for which proper time has no meaning, admits of analysis from first principles by way of an affine parameter λ, as contrasted with the device of first considering a particle and then going to the limit $\mu \longrightarrow 0$.

(5) affine parameter

(*continued on page 676*)

Box 25.7 QUALITATIVE ANALYSIS OF ORBITS OF A PHOTON IN SCHWARZSCHILD GEOMETRY

A. Equations Governing Orbit

1. Effective-potential equation for radial part of motion:

$$\left(\frac{dr}{d\lambda}\right)^2 + B^{-2}(r) = b^{-2};$$

$$B^{-2}(r) = r^{-2}(1 - 2M/r);$$

$$b = \text{(impact parameter)}.$$

2. Supplementary equations to determine angular and time motion:

$$d\phi/d\lambda = 1/r^2;$$

$$dt/d\lambda = b^{-1}(1 - 2M/r)^{-1}.$$

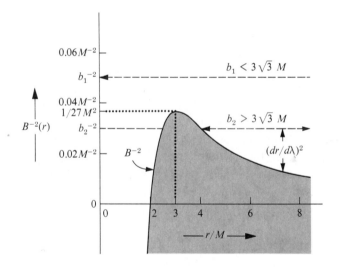

B. Qualitative Features of Orbits (deduced from effective-potential diagram)

1. A zero-mass particle with $b > 3\sqrt{3}M$, which falls in from $r = \infty$, is "reflected off the potential barrier" (periastron; $b = B$; $dr/d\lambda = 0$) and returns to infinity.
 a. For $b \gg 3\sqrt{3}M$, the orbit is a straight line, except for a slight deflection of angle $4M/b$ (exercise 25.21; §40.3).
 b. For $0 < b - 3\sqrt{3}M \ll M$, the particle circles the star many times ("unstable circular orbit") at $r \approx 3M$ before flying back to $r = \infty$.

2. A zero-mass particle with $b < 3\sqrt{3}M$, which falls in from $r = \infty$, falls into $r = 2M$ (no periastron).

3. A zero-mass particle emitted from near $r = 2M$ escapes to infinity only if it has $b < 3\sqrt{3}M$; otherwise it reaches an apastron and then gets pulled back into $r = 2M$.

C. Escape Versus Capture as a Function of Propagation Direction

An observer at rest in the Schwarzschild gravitational field measures the ordinary velocity of a zero-mass particle relative to his orthonormal frame [equations (23.15)]:

$$v_{\hat{r}} = \frac{|g_{rr}|^{1/2}\, dr/d\lambda}{|g_{00}|^{1/2}\, dt/d\lambda} = \pm(1 - b^2/B^2)^{1/2};$$

$$v_{\hat{\phi}} = \frac{|g_{\phi\phi}|^{1/2}\, d\phi/d\lambda}{|g_{00}|^{1/2}\, dt/d\lambda} = b/B;$$

$$(v_{\hat{r}})^2 + (v_{\hat{\phi}})^2 = 1;$$

$\delta \equiv$ (angle between propagation direction and radial direction)
$= \cos^{-1} v_{\hat{r}} = \sin^{-1} v_{\hat{\phi}}.$

To be able to cross over the potential barrier, the particle must have $b < 3\sqrt{3}M$, or $v_{\hat{\phi}}^2 B^2 < 27M^2$, or $\sin^2\delta < 27M^2/B^2$. This result, restated:

1. *A particle of zero rest mass at $r < 3M$ will eventually escape to infinity, rather than be captured by a black hole at $r = 2M$ if and only if v_r is positive and*

$$\sin\delta < 3\sqrt{3}MB^{-1}(r).$$

2. *A particle of zero rest mass at $r > 3M$ will eventually escape to infinity if and only if: (1) v_r is positive, or (2) v_r is negative and*

$$\sin\delta > 3\sqrt{3}MB^{-1}(r).$$

White, escape; black, to black hole;
directions in proper reference frame

$$\boldsymbol{e}_{\hat{\phi}} = r^{-1}\, \partial/\partial\phi$$

$$\boldsymbol{e}_{\hat{r}} = (1 - 2M/r)^{1/2}\, \partial/\partial r$$

$$\longrightarrow r/M \longrightarrow$$

0 2 4 6 8 10

Return to the statement of the conservation laws (25.17) and (25.18) in the form that makes reference to the affine parameter λ but no reference to the rest mass μ; thus

$$\frac{d\phi}{d\lambda} = \frac{L}{r^2} \tag{25.60}$$

and

$$\frac{dt}{d\lambda} = \frac{E}{1 - 2M/r}. \tag{25.61}$$

Recall that the course of a photon in a gravitational field is governed by its direction but not by its energy. Therefore neither E nor L individually are relevant but only their ratio, the impact parameter $b = L/E$ of (25.54) and exercise 25.14. This circumstance leads one to replace the affine parameter λ by a new affine parameter,

$$\lambda_{\text{new}} = L\lambda, \tag{25.62}$$

that is equally constant along the world line of the photon. In this notation (drop the subscript "new" hereafter), the conservation laws take the form

(6) equations for orbit

$$\frac{d\phi}{d\lambda} = \frac{1}{r^2}, \tag{25.63}$$

$$\frac{dt}{d\lambda} = \frac{1}{b(1 - 2M/r)}. \tag{25.64}$$

The statement that the world line of the photon is a line of zero lapse of proper time,

$$g_{\alpha\beta} \frac{dx^\alpha}{d\lambda} \frac{dx^\beta}{d\lambda} = 0 \tag{25.65}$$

leads to the "radial equation"

$$\left(\frac{dr}{d\lambda}\right)^2 + B^{-2}(r) = b^{-2}. \tag{25.66}$$

Here one encounters again the "effective potential" $B^{-2}(r)$ of (25.58). The present fuller set of equations for the geodesic of a photon have the advantage that they reach beyond space to a description of the world line in spacetime.

Return to space! Figure 25.6 shows typical orbits for a photon in Schwarzschild geometry. Figure 25.7 shows angle of deflection as a function of impact parameter. From the information contained in this curve, one can evaluate the contributions to the differential scattering cross section

(7) scattering cross section

$$\frac{d\sigma}{d\Omega} = \sum_{\text{"branches"}} \left| \frac{2\pi b \, db}{2\pi \sin\Theta \, d\Theta} \right| \tag{25.67}$$

from the various "branches" of the scattering curve of Figure 25.7 [one turn around the center of attraction, two turns, etc.; for more on these branches and the central

Figure 25.6.

The orbit of a photon in the "equatorial plane" of a black hole, plotted in terms of the Schwarzschild coordinates r and ϕ, for selected values of the turning point of the orbit, $r_{TP}/M = 2.99$, 3.00 (unstable circular orbit), 3.01, 3.5, 4, 5, 6, 7, 8, 9. The impact parameter is given by the formula $b = r_{TP}(1 - 2M/r_{TP})^{-1/2}$. In none of the cases shown, even for the inward plunging spiral, is the impact parameter less than $b_{crit} = (27)^{1/2} M$, nor are any of these orbits able to cross the circle $r = 3M$. That only happens for orbits with b less than b_{crit}. For such orbits there is no turning point; the photon comes in from infinity and ends up at $r = 0$: straight in for $b = 0$ (head-on impact); only after many loops near $r = 3M$, when $b/M = (27)^{1/2} - \varepsilon$, where ε is a very small quantity. Appreciation is expressed to Prof. R. H. Dicke for permission to publish these curves, which he had a digital calculator compute and plot out directly from the formula $d^2u/d\phi^2 = 3u^2 - u$, where $u = M/r$.

role of the deflection function $\Theta = \Theta(b)$ in the analysis of scattering, see, for example, Ford and Wheeler (1959a,b)]. For small angles the "Rutherford" part of the scattering predominates. The major part of the small-angle scattering, and in the limit $\Theta \longrightarrow 0$ all of it, comes from large impact parameters, for which one has

$$\Theta = \frac{4M}{b} \tag{25.68}$$

(see exercises 25.21 and 25.24). It follows that the limiting form of the cross section is

$$\frac{d\sigma}{d\Omega} = \left(\frac{4M}{\Theta^2}\right)^2 \qquad \text{(small } \Theta\text{)}. \tag{25.69}$$

Also, at $\Theta = \pi$ one has a singularity in the differential scattering cross section, with the character of a glory [see discussion following equation (25.44)]. Writing down the contributions of the several branches of the scattering function to the differential cross section, and summing them, one has, near $\Theta = \pi$,

$$\frac{d\sigma}{d\Omega} = \frac{M^2}{\pi - \Theta}(1.75 + 0.0029 + 0.0000055 + \cdots) = 1.75\frac{M^2}{\pi - \Theta}. \tag{25.70}$$

Thus, in principle, if one shines a powerful source of light onto a black hole, one gets a direct return of a few photons from it. Equation (25.70) provides a means to calculate the strength of this return. See exercise 25.26.

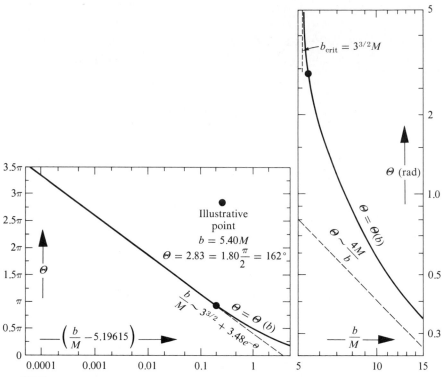

Figure 25.7.

Deflection of a photon by a Schwarzschild black hole, or by any spherically symmetric center of attraction small enough not to block the trajectory of the photon. The accurate calculations (smooth curves) are compared with formulas (dashed curves) valid asymptotically in the two limiting cases of an impact parameter, b: (1) very close to $b_{\text{crit}} = 3^{3/2}M$ (many turns around the center of attraction); and (2) very large compared to b_{crit} (small deflection). The algorithm for the accurate calculation of the deflection proceeds as follows (all distances being given, for simplicity, in units of the mass value, M). (1) Choose a value, $r = R$, for the Schwarzschild coordinate of the point of closest approach. (2) Calculate the impact parameter, b, from $b^2 = R^3/(R-2)$. (3) Calculate Q from $Q^2 = (R-2)(R+6)$. (4) Determine the modulus, k, of an "elliptic integral of the first kind" from $\sin^2\theta = k^2 = (Q-R+6)/2Q$. (5) Determine the so-called amplitude $\phi = \phi_{\min}$ of the same elliptic function from $\text{sn}^2 u_{\min} = \sin^2\phi_{\min} = (2+Q-R)/(6+Q-R)$. (6) Then the total deflection is

$$\Theta = 4(R/Q)^{1/2}[F(\pi/2,\theta) - F(\phi_{\min},\theta)] - \pi.$$

The values plotted here were kindly calculated by James A. Isenberg on the basis of the work of C. G. Darwin (1959, 1961).

(8) gravitational lens effect

When the source of illumination, instead of being on the observer's side of the black hole, is on the opposite side, then in addition to the "lens effect" experienced by photons flying by with large impact parameter [literature too vast to summarize here, but see, e.g., Refsdal (1964)], and subsumed in equation (25.68), there is a glory type of illumination (intensity $\sim 1/\sin\Theta$, with now, however, Θ close to zero) received from photons that have experienced deflections $\Theta = 2\pi, 4\pi, \ldots$. This illumination comes from "rings of brightness" located at impact parameters given by $b/M - 3^{3/2} = 0.0065, 0.000012, \ldots$. Interesting though all these optical effects are as matters of principle, they are, among all the ways to observe a black hole, the worst; see part VI, C, of Box 33.3 for a detailed discussion.

Exercise 25.23. QUALITATIVE FEATURES OF PHOTON ORBITS

Verify all the statements about orbits for particles of zero rest mass made in Box 25.7.

Exercise 25.24. LIGHT DEFLECTION

Using the dimensionless variable $u = M/r$ in place of r itself, and $u_b = M/b$ in place of the impact parameter, transform (25.55) into the first-order equation

$$\left(\frac{du}{d\phi}\right)^2 + (1 - 2u)u^2 = u_b^2 \tag{25.71}$$

and thence, by differentiation, into

$$\frac{d^2u}{d\phi^2} + u = 3u^2. \tag{25.72}$$

(a) In the large-impact-parameter or small-u approximation, in which the term on the right is neglected, show that the solution of (25.72) yields elementary rectilinear motion (zero deflection).

(b) Insert this zero-order solution into the perturbation term $3u^2$ on the righthand side of (25.72), and solve anew for u ("rectilinear motion plus first-order correction"). In this way, verify the formula for the bending of light by the sun given by putting $\beta = 1$ in equation (25.49).

Exercise 25.25. CAPTURE OF LIGHT BY A BLACK HOLE

Show that a Schwarzschild black hole presents a cross section $\sigma_{capt} = 27\pi M^2$ for capture of light.

Exercise 25.26. RETURN OF LIGHT FROM A BLACK HOLE

Show that flashing a powerful pulse of light onto a black hole leads in principle to a return from rings of brightness located at $b/M - 3^{3/2} = 0.151, 0.00028, \ldots$. How can one evaluate the difference in time delays of these distinct returns? Show that the intensity I of the return (erg/cm^2) as a function of the energy E_0(erg/steradian) of the original pulse, the mass M(cm) of the black hole, the distance R to it, and the lateral distance r from the "flashlight" to the receptor of returned radiation is

$$I = \frac{E_0}{R^3 r} \sum_{\Theta = (2N+1)\pi} \left| \frac{2b\,db}{d\Theta} \right| = \frac{E_0 M^2}{R^3 r}(1.75 + 0.0029 + 0.0000055 + \cdots)$$

under conditions where diffraction can be neglected.

§25.7. SPHERICAL STAR CLUSTERS

By combining orbit theory, as developed in this chapter, with kinetic theory in curved spacetime as developed in §22.6, one can formulate the theory of relativistic star clusters.

Consider, for simplicity, a spherically symmetric cluster of stars (e.g., a globular cluster, but one so dense that relativistic gravitational effects might be important).

Static, spherical star clusters:

Demand that the cluster be static, in the sense that the number density in phase space \mathscr{N} is independent of time. (New stars, flying along geodesic orbits, enter a fixed region in phase space at the same rate as "old" stars leave it.) Ignore collisions and close encounters between stars; i.e., treat each star's orbit as a geodesic in the spherically symmetric spacetime of the cluster as a whole.

With these idealizations accepted, one can write down a manageable set of equations for the structure of the cluster.* Since the cluster is static and spherical, so must be its gravitational field. Consequently, one can introduce the same kind of coordinate system ("Schwarzschild coordinates") as was used for a static spherical star in Chapter 23:

$$ds^2 = -e^{2\Phi}\,dt^2 + e^{2\Lambda}\,dr^2 + r^2\,d\Omega^2; \qquad \Phi = \Phi(r), \qquad \Lambda = \Lambda(r). \quad (25.73)$$

In the tangent space at each event in spacetime reside the momentum vectors of the swarming stars. For coordinates in this tangent space ("momentum space"), it is convenient to use the physical components of 4-momentum, $p^{\hat{\alpha}}$—i.e., components on the orthonormal frame

$$\boldsymbol{\omega}^{\hat{t}} = e^{\Phi}\,\boldsymbol{dt}, \qquad \boldsymbol{\omega}^{\hat{r}} = e^{\Lambda}\,\boldsymbol{dr}, \qquad \boldsymbol{\omega}^{\hat{\theta}} = r\,\boldsymbol{d\theta}, \qquad \boldsymbol{\omega}^{\hat{\phi}} = r\sin\theta\,\boldsymbol{d\phi}. \quad (25.74)$$

Then the number density of stars in phase space is a spherically symmetric, static function

$$\mathscr{N} = \mathscr{N}[r, p^{\hat{0}}, p^{\hat{r}}, (p^{\hat{\theta}2} + p^{\hat{\phi}2})^{1/2}]. \quad (25.75)$$

[\mathscr{N} is independent of t because the cluster is static; and independent of θ, ϕ, and angle $\Theta = \tan^{-1}(p^{\hat{\phi}}/p^{\hat{\theta}})$ because of spherical symmetry.]

The functions describing the structure of the cluster, Φ, Λ, and \mathscr{N}, are determined by the kinetic (also, in this context, called the Vlasoff) equation (§22.6)

$$d\mathscr{N}/d\lambda = 0, \text{i.e., } \mathscr{N} \text{ conserved along orbit} \atop \text{of each star in phase space;} \quad (25.76\text{a})$$

and by the Einstein field equations

$$G^{\hat{\alpha}\hat{\beta}} = 8\pi T^{\hat{\alpha}\hat{\beta}} = 8\pi \int (\mathscr{N}p^{\hat{\alpha}}p^{\hat{\beta}})\mu^{-1}\,dp^{\hat{0}}\,dp^{\hat{r}}\,dp^{\hat{\theta}}\,dp^{\hat{\phi}}. \quad (25.76\text{b})$$

[The Vlasoff equation for Newtonian star clusters is treated by Ogorodnikov (1965). The above expression for the stress-energy tensor of a swarm of particles (stars) was derived in exercise 22.18. Here, as in exercise 22.18, the particles (stars) are assumed *not* all to have the same rest mass. Note that rest mass is here denoted μ, but in Chapter 22 it was denoted m.]

To solve the Vlasoff equation, one need only note that \mathscr{N} is conserved along stellar orbits and therefore must be a function of the constants of the orbital motion. There is a constant of motion corresponding to each Killing vector in the cluster's static, spherical spacetime (see exercise 25.8):

(1) foundations for analysis

(2) solution of Vlasoff equation

*These equations were first derived and explored by Zel'dovich and Podurets (1965).

$$E = \text{``energy at infinity''} = -\boldsymbol{p} \cdot (\partial/\partial t) = -p_0,$$

$$L_z = \text{``}z\text{-component of angular momentum''} = p \cdot \boldsymbol{\xi}_z = p \cdot (\partial/\partial\phi) = p_\phi,$$

$$L_y = \text{``}y\text{-component of angular momentum''} = p \cdot \boldsymbol{\xi}_y, \qquad (25.77a)$$

$$L_x = \text{``}x\text{-component of angular momentum''} = p \cdot \boldsymbol{\xi}_x.$$

In addition, each star's rest mass

$$\mu = (p^{\hat{0}2} - p^{\hat{r}2} - p^{\hat{\theta}2} - p^{\hat{\phi}2})^{1/2} \qquad (25.77b)$$

is a constant of its motion. The general solution of the Vlasoff equation, then, has the form

$$\mathcal{N} = H(E, L_x, L_y, L_z, \mu).$$

But this general solution is not spherically symmetric. For example, the distribution function

$$\mathcal{N} = H(E, \mu, L_z)\, \delta(L_y)\, \delta(L_x),$$

corresponds to a cluster of stars with orbits all in the equatorial plane $\theta = \pi/2$ ($L_y = L_x = 0$ for all stars in cluster). To be spherical the cluster's distribution function must depend only on the magnitude

$$L = (L_x{}^2 + L_y{}^2 + L_z{}^2)^{1/2}$$

of the angular momentum, and not on its direction (not on the orientation of a star's orbital plane). Thus, the general spherical solution to the Vlasoff equation in a static, spherical spacetime must have the form

$$\mathcal{N} = F(E, L, \mu). \qquad (25.78)$$

To use this general solution, one must reexpress the constants of the motion E, L, μ, in terms of the agreed-on phase-space coordinates $(t, r, \theta, \phi, p^{\hat{0}}, p^{\hat{r}}, p^{\hat{\theta}}, p^{\hat{\phi}})$. The rest mass of a star is given by (25.77b). The energy-at-infinity is obtained by red-shifting the locally measured energy

$$E = -p_0 = e^{\Phi}p^{\hat{0}}. \qquad (25.79a)$$

For an orbit in the equatorial plane ($p_\theta = p^\theta = p^{\hat{\theta}} = 0$; $L_x = L_y = 0$), the total angular momentum has the form

$$L = |L_z| = |p_\phi| = |rp^{\hat{\phi}}| = r \times \text{(``tangential'' component of 4-momentum)}.$$

By symmetry, the equation $L = r \times$ ("tangential" component of \boldsymbol{p}) must hold true also for orbits in other planes; it must be perfectly general:

$$L = rp^{\hat{t}}, \qquad (25.79b)$$

$$p^{\hat{t}} \equiv \text{(tangential component of 4-momentum)} = [(p^{\hat{\theta}})^2 + (p^{\hat{\phi}})^2]^{1/2} \qquad (25.80)$$

(see exercise 25.9).

$$p^{\hat\theta} \, dp^{\hat\theta} + p^{\hat\phi}\, dp^{\hat\phi}$$

$$dp^{\hat{t}} = \frac{dp^{\hat{t}}}{dp^{\hat\theta}}\, dp^{\hat\theta} + \frac{dp^{\hat{t}}}{dp^{\hat\phi}}\, p^{\hat\phi}$$

$$= 2p^{\hat\theta} \cdot \tfrac{1}{2}(p^{\hat{t}})^{-1}\, dp^{\hat\theta} + p^{\hat\phi}(p^{\hat{t}})^{-1}\, dp\phi = \frac{p^{\hat\theta}}{p^{\hat{t}}}\, dp^{\hat\theta} + \frac{p^{\hat\phi}}{p^{\hat{t}}}\, dp^{\hat\phi}$$

(3) "smeared-out"
stress-energy tensor due
to stars

Before solving the Einstein field equations, one finds it useful to reduce the stress-energy tensor to a more explicit form than (25.76b). The off-diagonal components $T^{\hat{0}\hat{j}}$ and $T^{\hat{j}\hat{k}}$ ($j \neq k$) all vanish because their integrands are odd functions of $p^{\hat{j}}$. The integrands for the diagonal components $T^{\hat{0}\hat{0}}$, $T^{\hat{r}\hat{r}}$, and $\frac{1}{2}(T^{\hat{\theta}\hat{\theta}} + T^{\hat{\phi}\hat{\phi}})$ are independent of angle $\Theta \equiv \tan^{-1}(p^{\hat{\phi}}/p^{\hat{\theta}})$ in the tangential momentum plane; so the momentum volume element can be rewritten as

$$dp^{\hat{0}} \, dp^{\hat{r}} \, dp^{\hat{\theta}} \, dp^{\hat{\phi}} \longrightarrow 2\pi p^{\hat{T}} \, dp^{\hat{T}} \, dp^{\hat{r}} \, dp^{\hat{0}}.$$

Changing variables from $(p^{\hat{T}}, p^{\hat{r}}, p^{\hat{0}})$ to $(p^{\hat{T}}, \mu, p^{\hat{0}})$ where

$$\mu = [(p^{\hat{0}})^2 - (p^{\hat{r}})^2 - (p^{\hat{T}})^2]^{1/2},$$

and recognizing that two values of $p^{\hat{r}}$ ($\pm p^{\hat{r}}$) correspond to each value of μ, one brings the volume element into the form

$$2\pi p^{\hat{T}} \, dp^{\hat{T}} \, dp^{\hat{r}} \, dp^{\hat{0}} \longrightarrow 4\pi(p^{\hat{T}}\mu/p^{\hat{r}}) \, dp^{\hat{T}} \, dp^{\hat{0}} \, d\mu.$$

The diagonal components of \boldsymbol{T} [equation (25.76b)] then read

$$\rho \equiv T^{\hat{0}\hat{0}} = \text{(total density of mass-energy)}$$
$$= 4\pi \int F(e^{\Phi}p^{\hat{0}}, rp^{\hat{T}}, \mu)(p^{\hat{0}2}p^{\hat{T}}/p^{\hat{r}}) \, dp^{\hat{T}} \, dp^{\hat{0}} \, d\mu, \tag{25.81a}$$

$$P_T \equiv \frac{1}{2}(T^{\hat{\theta}\hat{\theta}} + T^{\hat{\phi}\hat{\phi}}) = T^{\hat{\theta}\hat{\theta}} = T^{\hat{\phi}\hat{\phi}} = \text{(tangential pressure)}$$
$$\text{[by spherical symmetry]} \tag{25.81b}$$
$$= 2\pi \int F(e^{\Phi}p^{\hat{0}}, rp^{\hat{T}}, \mu)[(p^{\hat{T}})^3/p^{\hat{r}}] \, dp^{\hat{T}} \, dp^{\hat{0}} \, d\mu,$$

$$P_r \equiv T^{\hat{r}\hat{r}} = \text{(radial pressure)}$$
$$= 4\pi \int F(e^{\Phi}p^{\hat{0}}, rp^{\hat{T}}, \mu)(p^{\hat{r}}p^{\hat{T}}) \, dp^{\hat{T}} \, dp^{\hat{0}} \, d\mu. \tag{25.81c}$$

When performing these integrals, one must express $p^{\hat{r}}$ in terms of the variables of integration,

$$p^{\hat{r}} = [(p^{\hat{0}})^2 - (p^{\hat{T}})^2 - \mu^2]^{1/2}. \tag{25.81d}$$

(4) solution of field equations

The Einstein field equations for this stress-energy tensor and the metric (25.73), after use of expressions (14.43) for $G^{\hat{\alpha}\hat{\beta}}$ and after manipulations analogous to those for a spherical star (§23.5), reduce to

$$e^{2\Lambda} = (1 - 2m/r)^{-1}, \qquad m = \int_0^r 4\pi r^2 \rho \, dr; \tag{25.82a}$$

$$\frac{d\Phi}{dr} = \frac{m + 4\pi r^3 P_r}{r(r - 2m)}. \tag{25.82b}$$

These equations, together with the assumed form $F(E, L, \mu)$ of the distribution

function and the integrals (25.81) for ρ, P_r, and P_T, determine the structure of the cluster. Box 25.8 gives an overview of these structure equations, and specializes them for an isotropic velocity distribution. Box 25.9 presents and discusses the solution to the equations for an isothermal star cluster (truncated Maxwellian velocity distribution).

EXERCISES

Exercise 25.27. ISOTROPIC STAR CLUSTER

For a cluster with distribution function independent of angular momentum, derive properties B.1 to B.6 of Box 25.8.

Exercise 25.28. SELF-SIMILAR CLUSTER [See Bisnovatyi-Kogan and Zel'dovich (1969), Bisnovatyi-Kogan and Thorne (1970).]

(a) Find a solution to the equations of structure for a spherical star of infinite central density, with the equation of state $P = \gamma\rho$, where γ is a constant ($0 < \gamma < 1/3$).

(b) Find an isotropic distribution function $F(E, \mu)$ that leads to a star cluster with the same distributions of ρ, P, m, and Φ as in the gas sphere of part (a). (See Box 25.8.) [*Answer:*

$$P = \gamma\rho = \frac{\gamma^2}{1 + 6\gamma + \gamma^2}\frac{1}{2\pi r^2},$$

$$\begin{cases} e^{2\Lambda} = (1 - 2m/r)^{-1} = (1 + 6\gamma + \gamma^2)/(1 + \gamma)^2, \\ e^{2\Phi} = Br^{4\gamma/(1+\gamma)}, \qquad B = \text{const}; \end{cases}$$

$$F = A(E/B^{1/2})^{-(1+\gamma)/\gamma}\,\delta(\mu - \mu_0) = Ar^{-2}(E_{\text{local}})^{-(1+\gamma)/\gamma}, \qquad A = \text{const}.]$$

Exercise 25.29. CLUSTER WITH CIRCULAR ORBITS

What must be the form of the distribution function to guarantee that all stars move in circular orbits? Specialize the equations of structure to this case. Analyze the stability of the orbits of individual stars in the cluster, using an effective-potential diagram. What conditions must the distribution function satisfy if all orbits are to be stable? [See Einstein (1939), Zapolsky (1968).]

Box 25.8 EQUATIONS OF STRUCTURE FOR A SPHERICAL STAR CLUSTER

A. To Build a Model for a Star Cluster, Proceed as Follows

1. Specify the distribution function $\mathscr{N} = F(E, L, \mu)$, where

$$E = \text{energy-at-infinity of a star,}$$

$$L = \text{angular momentum of a star,}$$

$$\mu = \text{rest mass of a star.}$$

2. Solve the following two integro-differential equations for the metric functions $m = \frac{1}{2}r(1 - e^{-2\Lambda})$ and Φ of the line element

Box 25.8 (continued)

$$ds^2 = -e^{2\Phi} dt^2 + e^{2\Lambda} dr^2 + r^2 d\Omega^2:$$

$$m = \int_0^r 4\pi r^2 \rho \, dr,$$

$$\frac{d\Phi}{dr} = \frac{m + 4\pi r^3 P_r}{r(r - 2m)},$$

where

$$\rho = 4\pi \int F(e^\Phi p^{\hat{0}}, rp^{\hat{T}}, \mu)[(p^{\hat{0}})^2 p^{\hat{T}}/p^{\hat{r}}] \, dp^{\hat{T}} \, dp^{\hat{0}} \, d\mu,$$

$$P_T = 2\pi \int F(e^\Phi p^{\hat{0}}, rp^{\hat{T}}, \mu)[(p^{\hat{T}})^3/p^{\hat{r}}] \, dp^{\hat{T}} \, dp^{\hat{0}} \, d\mu,$$

$$P_r = 4\pi \int F(e^\Phi p^{\hat{0}}, rp^{\hat{T}}, \mu)(p^{\hat{r}} p^{\hat{T}}) \, dp^{\hat{T}} \, dp^{\hat{0}} \, d\mu,$$

$$p^{\hat{r}} = [(p^{\hat{0}})^2 - (p^{\hat{T}})^2 - \mu^2]^{1/2}.$$

The integrations for ρ, P_T, and P_r go over all positive $p^{\hat{T}}$, $p^{\hat{0}}$, μ for which $(p^{\hat{0}})^2 - (p^{\hat{T}})^2 - \mu^2 \geq 0$.

B. If the Distribution Function is Independent of Angular Momentum, Then

1. $F = F(E, \mu)$.
2. The distribution of stellar velocities at each point in the cluster is isotropic.
3. $\rho = 4\pi \int F(e^\Phi p^{\hat{0}}, \mu)[(p^{\hat{0}})^2 - \mu^2]^{1/2}(p^{\hat{0}})^2 \, dp^{\hat{0}} \, d\mu.$
4. The pressure is isotropic:

$$P_r = P_T \equiv P \equiv \frac{4\pi}{3} \int F(e^\Phi p^{\hat{0}}, \mu)(p^{\hat{0}2} - \mu^2)^{3/2} \, dp^{\hat{0}} \, d\mu.$$

5. The total density of mass-energy ρ, the pressure P, and the metric functions Φ and $m = \frac{1}{2}r(1 - e^{-2\Lambda})$ satisfy the equations of structure for a gas sphere ("star"),

$$m = \int 4\pi r^2 \rho \, dr,$$

$$\frac{d\Phi}{dr} = \frac{m + 4\pi r^3 P}{r(r - 2m)},$$

$$\frac{dP}{dr} = -\frac{(\rho + P)(m + 4\pi r^3 P)}{r(r - 2m)}.$$

6. Thus, to every static, spherical star cluster with isotropic velocity distribution, there corresponds a unique gas sphere that has the same distributions of ρ, P, m, and Φ.
7. Conversely [see Fackerell (1968)], given a gas sphere (solution to equations of stellar structure for ρ, P, m, and Φ), one can always find a distribution function $F(E, \mu)$ that describes a cluster with the same ρ, P, m, and Φ. But for some gas spheres F is necessarily negative in part of phase space, and is thus unphysical.

Box 25.9 ISOTHERMAL STAR CLUSTERS

A. Distribution Function

1. In any relativistic star cluster, one might expect that occasional close encounters between stars would "thermalize" the stellar distribution function. This suggests that one study isotropic, spherical clusters with the Boltzmann distribution function (tacitly assumed zero for $p^{\hat{0}} = Ee^{-\Phi} < \mu_0$)

$$\mathscr{N} = F(E, L, \mu) = Ke^{-E/T} \delta(\mu - \mu_0). \tag{1}$$

Here K is a normalization constant, T is a constant "temperature," and for simplicity the stars are all assumed to have the same rest mass μ_0.

2. In such a cluster, an observer at radius r sees a star of energy-at-infinity E to have locally measured energy

$$p^{\hat{0}} = (\text{rest mass-energy}) + (\text{kinetic energy}) = \frac{\mu_0}{(1 - v^2)^{1/2}} = Ee^{-\Phi(r)}. \tag{2}$$

Consequently, the stars in his neighborhood have a Boltzmann distribution

$$\frac{dN}{d^3\hat{p}\, d^3\hat{x}\, d\mu} = \mathscr{N} = K\exp(-p^{\hat{0}}/T_{\text{loc}})\, \delta(\mu - \mu_0) \tag{3}$$

with locally measured temperature

$$T_{\text{loc}}(r) = Te^{-\Phi(r)}. \tag{4}$$

Thus, the temperature of the cluster is subject to identically the same redshift-blueshift effects as photons, particles, and stars that move about in the cluster. (For a derivation of this same temperature-redshift law for a gas in thermal equilibrium, see part (e) of exercise 22.7.)

3. Actually, the Boltzmann distribution (1) can never be achieved. Stars with $E > \mu_0$ are gravitationally unbound from the cluster and will escape. The Boltzmann distribution presumes that, as such stars go zooming off toward $r = \infty$, an equal number of stars with the same energies come zooming in from $r = \infty$ to maintain an unchanged distribution function. Such a situation is clearly unrealistic. Instead, one expects the escape of stars to truncate the distribution at some energy E_{max} slightly less than μ_0. The result, in idealized form, is the "truncated Boltzmann distribution"

$$\mathscr{N} = F(E, L, \mu) = \begin{cases} Ke^{-E/T}\delta(\mu - \mu_0), & E < E_{\text{max}}, \\ 0, & E > E_{\text{max}}. \end{cases} \tag{5}$$

Box 25.9 (*continued*)

B. Structure and Stability of Cluster Models

1. Models for star clusters with truncated Boltzmann distributions have been constructed by Zel'dovich and Podurets (1965), by Fackerell (1966), and by Ipser (1969), using the procedure of Box 25.8. Ipser has analyzed the collisionless radial vibrations of such clusters.

2. In general, these clusters form a 4-parameter family (K, T, μ_0, E_{\max}). Replace the parameter K by the total rest mass of the cluster, $M_0 = \mu_0 N$, where N is the total number of stars. Replace T by the temperature per unit rest mass, $\widetilde{T} = T/\mu_0$. Replace E_{\max} by the maximum energy per unit rest mass, $\widetilde{E}_{\max} = E_{\max}/\mu_0$. Then the clusters are parametrized by $(M_0, \widetilde{T}, \mu_0, \widetilde{E}_{\max})$. When one now doubles μ_0, holding $M_0, \widetilde{T}, \widetilde{E}_{\max}$ fixed (and thus halving the total number of stars), all macroscopic features of the cluster remain unchanged. In this sense μ_0 is a "trivial parameter" and can henceforth be ignored or changed at will. The total rest mass of the cluster M_0 can be regarded as a "scaling factor"; all dimensionless features of the cluster are independent of it. For example, if ρ_c is the central density of mass-energy [equation (25.81a), evaluated at $r = 0$], then $\rho_c M_0{}^2$ is dimensionless and is thus independent of M_0, which means that $\rho_c \propto M_0{}^{-2}$. Only two nontrivial parameters remain: \widetilde{T} and \widetilde{E}_{\max}.

3. Consider as an instructive special case [Zel'dovich and Podurets (1965)] the one-parameter sequence with $\widetilde{E}_{\max} = 1 - \frac{1}{2}\widetilde{T}$. The following figure, computed by Ipser (1969), plots for this sequence the fractional binding energy,

$$E_{\text{bind}}/M_0 \equiv (M_0 - M)/M_0 \tag{6}$$

(here M is total mass-energy); the square of the angular frequency for collisionless vibrations (vibration amplitude $\propto e^{-i\omega t}$) divided by central density of mass-energy, ω^2/ρ_c; and the redshift, z_c, of photons emitted from the center of the cluster and received at infinity. All these quantities are dimensionless, and thus depend only on the choice of $\widetilde{T} = T/\mu_0$.

4. Notice that all models beyond the point of maximum binding energy $(z_c \gtrsim 0.5)$ are unstable against collisionless radial perturbations (ω imaginary; amplitude of perturbation $\propto e^{|\omega|t}$). When perturbed slightly, such clusters must collapse to form black holes. (See Chapter 26 for an analysis of the analogous instability in stars).

5. These results suggest an idealized story of the evolution of a spherical cluster [Zel'dovich and Podurets (1965); Fackerell, Ipser, and Thorne (1969)]. The

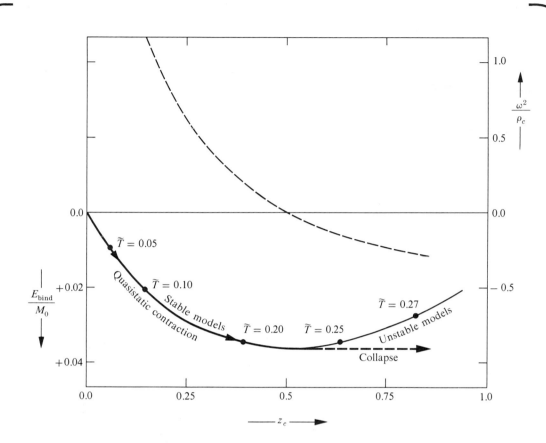

cluster would evolve quasistatically along a sequence of spherical equilibrium configurations such as those of the figure. The evolution would be driven by stellar collisions and by the evaporation of stars. When two stars collide and coalesce, they increase the cluster's rest mass and hence its fractional binding energy. When a star gains enough energy from such encounters to escape from the cluster, it carries away excess kinetic energy, leaving the cluster more tightly bound. Thus, both collisions and evaporation should drive the cluster toward states of tighter and tighter binding. When the cluster reaches the point, along its sequence, of maximum fractional binding energy, it can no longer evolve quasistatically. Relativistic gravitational collapse sets in: the stars spiral inward through the gravitational radius of the cluster toward its center, leaving behind a black hole with, perhaps, some remaining stars orbiting it.

It is tempting to speculate that violent events in the nuclei of some galaxies and in quasars might be associated with the onset of such a collapse, or with encounters between an already collapsed cluster (black hole) and surrounding stars.

CHAPTER **26**

STELLAR PULSATIONS

This chapter is entirely
Track 2, but it neither depends
on nor prepares for any other
chapter.

The *raison d'etre* of this
chapter

§26.1. MOTIVATION

In relativistic astrophysics, as elsewhere in physics, most problems of deep physical
interest are too difficult and too complex to be solved exactly. They can be solved
only by use of approximation techniques. And of all approximation techniques, the
one that has the widest range of application is perturbation theory.

Perturbation calculations are typically long, tedious, and filled with complicated
mathematical expressions. Therefore, they are not appropriate for a textbook such
as this. Nevertheless, because it is so important that aspiring astrophysicists know
how to set up and carry out perturbation calculations in general relativity, the authors
have chosen to present one example in detail.

The example chosen is an analysis of adiabatic, radial pulsations of a nonrotating,
relativistic star. Two features of this example are noteworthy: (1) it is sufficiently
complex to be instructive, but sufficiently simple to be presentable; (2) in the results
of the calculation one can discern and quantify the relativistic instability that is so
important for modern astrophysics (see Chapter 24).

The calculation presented here is patterned after that of Chandrasekhar (1964a,b),
which first revealed the existence of the relativistic instability. For an alternative
calculation, based on the concept of "extremal energy," see Appendix B of Harrison,
Thorne, Wakano, and Wheeler (1965); and for a calculation based on extremal
entropy, see Cocke (1965).

The authors are deeply indebted to Mr. Carlton M. Caves, who found and corrected many errors in
the equations of this chapter and of a dozen other chapters.

§26.2. SETTING UP THE PROBLEM

The system to be analyzed is a sphere of perfect fluid, pulsating radially with very small amplitude. To analyze the pulsations one needs (a) the exact equations governing the equilibrium configuration about which the sphere pulsates; (b) a coordinate system for the vibrating sphere that reduces, for zero pulsation amplitude, to the standard Schwarzschild coordinates of the equilibrium sphere; (c) a set of small functions describing the pulsation (radial displacement and velocity, pressure and density perturbations, first-order changes in metric coefficients), in which to linearize; and (d) a set of equations governing the evolution of these "perturbation functions."

Setting up the analysis of stellar pulsations

a. Equilibrium Configuration

The equations of structure for the equilibrium sphere are those summarized in §23.7. It will be useful to rewrite them here, with a few changes of notation (use of subscript "o" to denote "unperturbed configuration"; use of $\Lambda = -\frac{1}{2}\ln(1 - 2m/r)$ in place of m in all equations; use of a prime to denote derivatives with respect to r):

Equilibrium configuration of star

$$ds^2 = -e^{2\Phi_o}\,dt^2 + e^{2\Lambda_o}\,dr^2 + r^2(d\theta^2 + \sin^2\theta\,d\phi^2), \qquad (26.1a)$$

$$\Lambda_o' = \frac{1}{2r}(1 - e^{2\Lambda_o}) + 4\pi r\rho_o e^{2\Lambda_o}, \qquad (26.1b)$$

$$p_o' = -(\rho_o + p_o)\Phi_o', \qquad (26.1c)$$

$$\Phi_o' = -\frac{1}{2r}(1 - e^{2\Lambda_o}) + 4\pi r p_o e^{2\Lambda_o}. \qquad (26.1d)$$

b. Coordinates for Perturbed Configuration

The gas sphere pulsates in a radial, i.e., spherically symmetric, manner. Consequently, its spacetime geometry must be spherical. In Box 23.3 it is shown that for any spherical spacetime, whether dynamic or static, one can introduce Schwarzschild coordinates with a line element

$$ds^2 = -e^{2\Phi}\,dt^2 + e^{2\Lambda}\,dr^2 + r^2(d\theta^2 + \sin^2\theta\,d\phi^2), \qquad (26.2)$$

$$\Phi = \Phi(t, r), \qquad \Lambda = \Lambda(t, r).$$

Coordinates for perturbed configuration

One uses these coordinates for the pulsating sphere because they reduce to the unperturbed coordinates when the pulsations have zero amplitude.

c. Perturbation Functions

When the pulsations have very small amplitude, the metric coefficients, Φ and Λ, and the thermodynamic variables p, ρ, and n as measured in the fluid's rest frame

have very nearly their unperturbed values. Denote by $\delta\Phi$, $\delta\Lambda$, δp, $\delta\rho$, and δn the perturbations at fixed coordinate locations:

Perturbation functions

$$\Phi(t, r) = \Phi_o(r) + \delta\Phi(t, r), \qquad \Lambda(t, r) = \Lambda_o(r) + \delta\Lambda(t, r),$$

$$p(t, r) = p_o(r) + \delta p(t, r), \qquad \rho(t, r) = \rho_o(r) + \delta\rho(t, r), \qquad \text{(26.3a)}$$

$$n(t, r) = n_o(r) + \delta n(t, r).$$

Besides $\delta\Phi$, $\delta\Lambda$, δp, $\delta\rho$, and δn, one more perturbation function is needed to describe the pulsations: the radial displacement ξ of the fluid from its equilibrium position:

> A fluid element located at coordinate radius r in the unperturbed configuration is displaced to coordinate radius $r + \xi(r, t)$ at coordinate time t in the vibrating configuration. (26.3b)

To make the analysis of the pulsations tractable, all equations will be linearized in the perturbation functions ξ, $\delta\Phi$, $\delta\Lambda$, δp, $\delta\rho$, and δn.

d. Equations of Evolution

How to derive equations governing the perturbation functions

The evolution of the perturbation functions with time will be governed by the Einstein field equations, the local law of conservation of energy-momentum $\boldsymbol{\nabla} \cdot \boldsymbol{T} = 0$, and the laws of thermodynamics—all appropriately linearized. The analysis from here on is nothing but a reduction of those equations to "manageable form." Of course, the reduction will proceed most efficiently if one knows in advance what form one seeks. The goal in this calculation and in most similar calculations is simple: (1) obtain a set of *dynamic equations* for the true dynamic degrees of freedom (only the fluid displacement ξ in this case; the fluid displacement and the amplitudes of the gravitational waves in a nonspherical case, where waves are possible); and (2) obtain a set of *initial-value equations* expressing the remaining perturbation functions ($\delta\Phi$, $\delta\Lambda$, δp, $\delta\rho$, and δn in this case) in terms of the dynamic degrees of freedom (ξ).

§26.3. EULERIAN VERSUS LAGRANGIAN PERTURBATIONS

Before deriving the dynamic and initial-value equations, it is useful to introduce a new concept: the "Lagrangian perturbation" in a thermodynamic variable. The

Eulerian perturbations defined

Lagrangian perturbations defined

perturbations δp, $\delta\rho$, and δn of equations (26.3) are *Eulerian perturbations* in p, ρ, and n; i.e., they are changes measured by an observer who sits forever at a fixed point (t, r, θ, ϕ) in the coordinate grid. By contrast, the *Lagrangian perturbations*—denoted Δp, $\Delta\rho$, and Δn—are changes measured by an observer who moves with

the fluid; i.e., by an observer who would sit at radius r in the unperturbed configuration, but sits at $r + \xi(t, r)$ in the perturbed configuration:

$$\Delta p(t, r) = p[t, r + \xi(t, r)] - p_o(r) \tag{26.4a}$$
$$\approx \delta p + p_o'\xi;$$

Relation between Eulerian and Lagrangian perturbations

$$\Delta\rho(t, r) = \rho[t, r + \xi(t, r)] - \rho_o(r) \tag{26.4b}$$
$$\approx \delta\rho + \rho_o'\xi;$$

$$\Delta n(t, r) = n[t, r + \xi(t, r)] - n_o(r) \tag{26.4c}$$
$$\approx \delta n + n_o'\xi.$$

§26.4. INITIAL-VALUE EQUATIONS

a. Baryon Conservation

The law of baryon conservation, $\boldsymbol{\nabla} \cdot (n\boldsymbol{u}) = 0$ (§22.2), governs the evolution of perturbations Δn and δn in baryon number. By applying the chain rule to the divergence and using the relation $\boldsymbol{u} \cdot \boldsymbol{\nabla} n = \boldsymbol{\nabla}_{\boldsymbol{u}} n = dn/d\tau$, one can rewrite the conservation law as

Derivation of initial value equations:
(1) for baryon perturbations Δn and δn

$$dn/d\tau = -n(\boldsymbol{\nabla} \cdot \boldsymbol{u}).$$
$\quad\quad\uparrow$
$\quad\quad$———[derivative of n along fluid world line]

In terms of Δn, the perturbation measured by an observer moving with the fluid, this equation can be rewritten as

$$\frac{d\,\Delta n}{d\tau} = -n(\boldsymbol{\nabla} \cdot \boldsymbol{u}). \tag{26.5}$$

To reduce this equation further, one needs an expression for the fluid's 4-velocity. It is readily derived from

$$\frac{u^r}{u^t} = \left(\frac{dr/d\tau}{dt/d\tau}\right) = \left(\frac{dr}{dt}\right)_{\text{along world line}} = \frac{\partial\xi}{\partial t} \equiv \dot{\xi},$$
$$(u^t)^2 e^{2\Phi} - (u^r)^2 e^{2\Lambda} = 1.$$

The result to first order in ξ, $\delta\Lambda$, and $\delta\Phi$ is

$$u^t = e^{-\Phi} = e^{-\Phi_o}(1 - \delta\Phi), \quad u^r = \dot{\xi}e^{-\Phi_o}. \tag{26.6}$$

Using these components in equation (26.5), and using the relations

$$\frac{d}{d\tau} = \boldsymbol{u} = u^\alpha\frac{\partial}{\partial x^\alpha}, \quad \boldsymbol{\nabla} \cdot \boldsymbol{u} = \frac{1}{\sqrt{-g}}(\sqrt{-g}u^\alpha)_{,\alpha},$$

together with the vibrating metric (26.2), one reduces equation (26.5) to a relation whose time integral is

$$\Delta n = -n_o[r^{-2}e^{-\Lambda_o}(r^2 e^{\Lambda_o}\xi)' + \delta\Lambda]. \tag{26.7}$$

This is the initial value equation for Δn in terms of the dynamic variable ξ. The initial-value equation for δn, which will not be needed later, one obtains by combining with equation (26.4c).

b. Adiabaticity

(2) for pressure perturbations Δp and δp

For adiabatic vibrations (negligible heat transfer between neighboring fluid elements), the Lagrangian changes in number density and pressure are related by

$$\left(\frac{\partial \ln p}{\partial \ln n}\right)_s \equiv \Gamma_1 = \frac{n}{p}\frac{\Delta p}{\Delta n}. \qquad (26.8)$$

$$\left[\begin{array}{l}\text{definition of adia-}\\ \text{batic index, } \Gamma_1\end{array}\right]$$

Combining this adiabatic relation with equation (26.7) for Δn, and equation (26.4a) for δp in terms of Δp, one obtains the following *initial-value equation for δp:*

$$\delta p = -\Gamma_1 p_o[r^{-2}e^{-\Lambda_o}(r^2 e^{\Lambda_o}\xi)' + \delta\Lambda] - \xi p_o'. \qquad (26.9)$$

c. Energy Conservation

(3) for density perturbations $\Delta\rho$ and $\delta\rho$

The local law of energy conservation [first law of thermodynamics; $\boldsymbol{u} \cdot (\boldsymbol{\nabla} \cdot \boldsymbol{T}) = 0$; see §§22.2 and 22.3] says that

$$\frac{d\rho}{d\tau} = \frac{(\rho + p)}{n}\frac{dn}{d\tau}.$$

Rewritten in terms of Lagrangian perturbations (recall: $d/d\tau$ is a time derivative as measured by an observer moving with the fluid), this reads

$$\frac{d\,\Delta\rho}{d\tau} = \frac{\rho + p}{n}\frac{d\,\Delta n}{d\tau},$$

which has as its time integral (first-order analysis!)

$$\Delta\rho = \frac{\rho_o + p_o}{n_o}\,\Delta n. \qquad (26.10)$$

(The constant of integration is zero, because, when $\Delta n = 0$, $\Delta\rho$ must also vanish.) Combining this with equation (26.7) for Δn and equation (26.4b) for $\delta\rho$ in terms of $\Delta\rho$, one obtains the following *initial-value equation for $\delta\rho$:*

$$\delta\rho = -(\rho_o + p_o)[r^{-2}e^{-\Lambda_o}(r^2 e^{\Lambda_o}\xi)' + \delta\Lambda] - \xi\rho_o'. \qquad (26.11)$$

d. Einstein Field Equations

Two of the Einstein field equations, when linearized, reduce to initial-value equations for the metric perturbations $\delta\Lambda$ and $\delta\Phi$. The equations needed, expressed in an orthonormal frame

$$\boldsymbol{w}^{\hat{t}} = e^{\Phi}\,\boldsymbol{dt}, \qquad \boldsymbol{w}^{\hat{r}} = e^{\Lambda}\,\boldsymbol{dr}, \qquad \boldsymbol{w}^{\hat{\theta}} = r\,\boldsymbol{d\theta}, \qquad \boldsymbol{w}^{\hat{\phi}} = r\sin\theta\,\boldsymbol{d\phi}, \quad (26.12)$$

(4) for metric perturbations $\delta\Lambda$ and $\delta\Phi$

are $G_{\hat{r}\hat{t}} = 8\pi T_{\hat{r}\hat{t}}$, and $G_{\hat{r}\hat{r}} = 8\pi T_{\hat{r}\hat{r}}$. The components of the Einstein tensor in this orthonormal frame were evaluated in exercise 14.16:

$$\overset{\displaystyle\text{[linearized]}}{\underset{\displaystyle\downarrow}{}}$$

$$G_{\hat{r}\hat{t}} = 2(\dot{\Lambda}/r)e^{-(\Lambda + \Phi)} = 2r^{-1}e^{-(\Lambda_o + \Phi_o)}\,\delta\dot{\Lambda}; \qquad (26.13a)$$

$$G_{\hat{r}\hat{r}} = 2(\Phi'/r)e^{-2\Lambda} + r^{-2}(e^{-2\Lambda} - 1) \qquad (26.13b)$$

$$= (G_{\hat{r}\hat{r}})_o + 2r^{-1}e^{-2\Lambda_o}\,\delta\Phi' - 2e^{-2\Lambda_o}(2r^{-1}\Phi_o' + r^{-2})\,\delta\Lambda.$$

$$\overset{\displaystyle}{\underset{\displaystyle\text{[linearized]}}{\uparrow}}$$

The components of the stress-energy tensor, $T_{\hat\alpha\hat\beta} = (\rho + p)u_{\hat\alpha}u_{\hat\beta} + p\eta_{\alpha\beta}$, as calculated using the 4-velocity (26.6) [transformed into the form $u_{\hat 0} = -1$, $u_{\hat r} = \dot\xi e^{\Lambda_o - \Phi_o}$] and using expressions (26.3a) for ρ and p, reduce to

$$T_{\hat{r}\hat{t}} = -(\rho_o + p_o)e^{\Lambda_o - \Phi_o}\dot\xi, \qquad T_{\hat{r}\hat{r}} = p_o + \delta p. \qquad (26.14)$$

Consequently, the field equation $G_{\hat{r}\hat{t}} = 8\pi T_{\hat{r}\hat{t}}$—after integration with respect to time and choice of the constant of integration, so that $\delta\Lambda = 0$ when $\xi = 0$—reduces to

$$\delta\Lambda = -4\pi(\rho_o + p_o)re^{2\Lambda_o}\xi = -(\Lambda_o' + \Phi_o')\xi. \qquad (26.15)$$

This is the initial-value equation for $\delta\Lambda$. The field equation $G_{\hat{r}\hat{r}} = 8\pi T_{\hat{r}\hat{r}}$, after using (26.15) to remove $\delta\Lambda$ and (26.9) to remove δp, and (26.1c) to remove Φ_o', reduces to

$$\delta\Phi' = -4\pi\Gamma_1 p_o r^{-1}e^{2\Lambda_o + \Phi_o}(r^2 e^{-\Phi_o}\xi)' \qquad (26.16)$$
$$+ [4\pi p_o'r - 4\pi(\rho_o + p_o)]e^{2\Lambda_o}\xi.$$

This is the initial-value equation for $\delta\Phi$.

§26.5. DYNAMIC EQUATION AND BOUNDARY CONDITIONS

The dynamic evolution of the fluid displacement $\xi(t, r)$ is governed by the Euler equation (22.13):

Derivation of equation of motion for fluid displacement ξ

$$(\rho + p) \times (\text{4-acceleration}) = -(\text{projection of } \boldsymbol\nabla p \text{ orthogonal to } \boldsymbol{u}). \quad (26.17)$$

The 4-acceleration $\boldsymbol{a} = \boldsymbol\nabla_{\boldsymbol u}\boldsymbol{u}$ corresponding to the 4-velocity (26.6) in the metric (26.2) has as its only non-zero, linearized, covariant component:

$$a_r = \Phi_o' + \delta\Phi' + e^{2(\Lambda_o - \Phi_o)}\ddot\xi.$$

[The component a_t is trivial in the sense that it leads to an Euler equation that duplicates (26.1c).] Combining this with $\rho + p = \rho_o + p_o + \delta\rho + \delta p$, with the radial component $p_o' + \delta p'$ for the projection of $\boldsymbol{\nabla}p$, and with the zero-order equation of hydrostatic equilibrium (26.1c), one obtains for the Euler equation

$$(\rho_o + p_o)e^{2(\Lambda_o - \Phi_o)}\ddot{\xi} = -\delta p' - (\delta\rho + \delta p)\Phi_o' - (\rho_o + p_o)\,\delta\Phi'. \qquad (26.18)$$

This equation of motion is put into its most useful form by using the initial-value equations (26.9), (26.11), and (26.16) to reexpress δp, $\delta\rho$, and $\delta\Phi'$ in terms of ξ, and by then manipulating terms extensively with the aid of the zero-order equations of structure (26.1). The result is

$$W\ddot{\zeta} = (P\zeta')' + Q\zeta, \qquad (26.19)$$

where ζ is a "renormalized displacement function," and W, P, Q are functions of radius determined by the structure of the equilibrium star:

$$\zeta \equiv r^2 e^{-\Phi_o}\xi; \qquad (26.20)$$

$$W \equiv (\rho_o + p_o)r^{-2}e^{3\Lambda_o + \Phi_o}; \qquad (26.21a)$$

$$P \equiv \Gamma_1 p_o r^{-2}e^{\Lambda_o + 3\Phi_o}; \qquad (26.21b)$$

$$Q \equiv e^{\Lambda_o + 3\Phi_o}\left[\frac{(p_o')^2}{\rho_o + p_o}r^{-2} - 4p_o'r^{-3} - 8\pi(\rho_o + p_o)p_o r^{-2}e^{2\Lambda_o}\right]. \qquad (26.21c)$$

Equation (26.19) is the dynamic equation governing the stellar pulsations. [This equation could be written in other forms; for instance, it could be multiplied by W^{-1} or any other non-zero factor, and terms could be regrouped. The form given in equation (26.19) is preferred because it leads to a self-adjoint eigenvalue problem for the oscillation frequencies, as indicated in Box 26.1.]

Not all solutions of the dynamic equation are acceptable. To be physically acceptable, the displacement function must produce noninfinite density and pressure perturbations ($\delta\rho$ and δp) at the center of the sphere, which means

Boundary conditions on fluid displacement

$$(\xi/r) \text{ finite or zero in limit as } r \longrightarrow 0 \qquad (26.22a)$$

[see (26.9) and (26.11)]; also, it must leave the pressure equal to zero at the star's surface, which means

$$\Delta p = -\Gamma_1 p_o r^{-2}e^{\Phi_o}(r^2 e^{-\Phi_o}\xi)' \longrightarrow 0 \text{ as } r \longrightarrow R \qquad (26.22b)$$

$$\text{[surface radius]}{\longrightarrow}$$

[see (26.8), (26.7), and (26.15)].

§26.6. SUMMARY OF RESULTS

Summary of theory of stellar pulsations

If an initial displacement of the fluid, $\xi(t = 0, r)$, is specified subject to the boundary conditions (26.22), then its subsequent evolution $\xi(t, r)$ can be calculated by inte-

grating the dynamic equation (26.19); and the form of the pressure, density, and metric perturbations can be calculated from $\xi(t, r)$ using the initial-value equations (26.9), (26.11), (26.15), and (26.16).

Several important consequences of these results are explored in Boxes 26.1 and 26.2.

(*continued on page 699*)

Box 26.1 EIGENVALUE PROBLEM AND VARIATIONAL PRINCIPLE FOR NORMAL-MODE PULSATIONS OF A STAR

Assume that the renormalized displacement function (26.20) has a sinusoidal time dependence:

$$\zeta = \zeta(r)e^{-i\omega t}.$$

Then the dynamic equation (26.19) and boundary conditions (26.22) reduce to an eigenvalue problem for the angular frequency ω and amplitude $\zeta(r)$:

$$(P\zeta')' + Q\zeta + \omega^2 W\zeta = 0, \tag{1}$$

$$\zeta/r^3 \text{ finite or zero as } r \longrightarrow 0, \tag{2a}$$

$$\Gamma_1 p_0 r^{-2} e^{\Phi_0}\zeta' \longrightarrow 0 \text{ as } r \longrightarrow R. \tag{2b}$$

Methods for solving this eigenvalue problem are catalogued and discussed by Bardeen, Thorne, and Meltzer (1966). One method (but *not* the best for numerical calculations) is the variational principle:

$$\omega^2 = \text{extremal value of } \left\{ \frac{\left| \int_0^R (P\zeta'^2 - Q\zeta^2)\, dr \right|}{\int_0^R W\zeta^2\, dr} \right\}, \tag{3}$$

where ζ is varied over all functions satisfying the boundary conditions (2). [See e.g., §12.3 of Mathews and Walker (1965) for discussion of the equivalence between this variational principle and the original eigenvalue problem.]

The absolute minimum value of expression (3) is the squared frequency of the fundamental mode of pulsation. If it is negative, the star is unstable ($e^{-i\omega t}$ grows exponentially in time). If it is positive, the star is stable against adiabatic, radial perturbations. Therefore, since the denominator of expression (3) is positive definite,

$$\begin{bmatrix} \text{stability against} \\ \text{adiabatic radial} \\ \text{perturbations} \end{bmatrix} \Longleftrightarrow \begin{bmatrix} \int_0^R (P\zeta'^2 - Q\zeta^2)\, dr > 0 \text{ for all functions} \\ \zeta \text{ satisfying (2)} \end{bmatrix}. \tag{4}$$

Box 26.1 (continued)

By numerical solution of the eigenvalue equation (1), the pulsation frequencies have been calculated for a wide variety of models of neutron stars and supermassive stars. *Example:* The figure gives a plot of pulsation frequency as a function of central density for the lowest four normal modes of the Harrison-Wakano-Wheeler models at the endpoint of stellar evolution. (Make a detailed comparison with Figure 24.2.) These curves are based on calculations by Meltzer and Thorne (1966), with corrections for the fundamental mode of massive white dwarfs by Faulkner and Gribbin (1968).

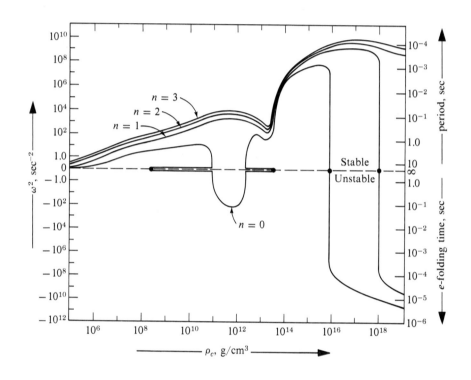

Box 26.2 THE CRITICAL ADIABATIC INDEX FOR NEARLY NEWTONIAN STARS

A. Fully Newtonian Stars

1. For a Newtonian star that pulsates sinusoidally, $\xi = \xi(r)e^{-i\omega t}$, the dynamic equation (26.19) reduces to

$$[\Gamma_1 p_o r(\xi/r)']' + 3(\Gamma_1 p_o \xi/r)' - 4p_o'\xi/r + \omega^2\rho_o\xi = 0. \tag{1}$$

2. If $\Gamma_1 = 4/3$ throughout the star, the physically acceptable solution [solution satisfying boundary conditions (26.22)] for the fundamental mode of vibration (mode with lowest value of ω^2) is

$$\omega^2 = 0, \quad \xi = \epsilon r, \quad \epsilon = \text{const.} \tag{2}$$

Thus, for $\Gamma_1 = 4/3$ the fundamental mode is "neutrally stable" and has a "homologous" displacement function—independent of the star's equation of state or structure.

3. If Γ_1 is allowed to differ slightly from 4/3 in an r-dependent way, then $\xi(r)$ will differ slightly from the homologous form:

$$\xi = \epsilon r[1 + r\text{-dependent corrections of magnitude } (\Gamma_1 - 4/3)].$$

Consequently, if one uses the homologous expression $\xi = \epsilon r$ as a trial function in the variational principle of Box 26.1, one will obtain ω^2 accurate to $O[(\Gamma_1 - 4/3)^2]$. (Recall: first-order errors in trial function produce second-order errors in value of variational expression.) The Newtonian limit of the variational expression [equation (3) of Box 26.1] becomes, with the homologous choice of trial function,

$$\omega^2 = (3\bar{\Gamma}_1 - 4)\frac{\displaystyle\int_0^R 3p_o r^2\, dr}{\displaystyle\int_0^R \rho_o r^4\, dr} + O[(3\bar{\Gamma}_1 - 4)^2], \tag{3}$$

where $\bar{\Gamma}_1$ is the pressure-averaged adiabatic index

$$\bar{\Gamma}_1 = \frac{\displaystyle\int_0^R \Gamma_1 p_o 4\pi r^2\, dr}{\displaystyle\int_0^R p_o 4\pi r^2\, dr}. \tag{4}$$

Box 26.2 (*continued*)

By use of the Newtonian virial theorem for the nonpulsating star [equation (39.21b) or exercise 23.7], one can convert equation (3) into the form

$$\omega^2 = (3\bar{\Gamma}_1 - 4)|\Omega|/I, \tag{5}$$

where Ω is the star's self-gravitational energy and $I = \int(\rho_o r^2)4\pi r^2\, dr$ is the trace of the second moment of its mass distribution (see Box 24.2 and exercise 39.6).

B. Nearly Newtonian Stars

1. When one takes into account first-order relativistic corrections (corrections of order M/R), but ignores higher-order corrections, one can rewrite the variational expression [equation (3) of Box 26.1] in the form

$$\omega^2 = \frac{\displaystyle\int_0^R p_o[\Gamma_1 r^4 \eta'^2 + (3\Gamma_1 - 4)(r^3\eta^2)'](1 + \Lambda_o + 3\Phi_o)\, dr - \int_0^R F_o\eta^2\, dr}{\displaystyle\int_0^R \rho_o r^4(1 + 3\Lambda_o + \Phi_o + p_o/\rho_o)\eta^2\, dr}, \tag{6}$$

where

$$F_o \equiv 8\pi r^4 p_o\rho_o + 8rm_o p_o + \rho_o m_o^2, \qquad \eta = \zeta/r^3 = (\xi/r)(1 - \Phi_o), \tag{7}$$

and $m_o(r)$ is the equilibrium mass inside radius r.

2. For a relativistic star with $\Gamma_1 - 4/3$ of order M/R and with $M/R \ll 1$, the homologous trial function $\xi = \epsilon r$ will still be highly accurate. Equally accurate, and easier to work with, will be $\xi = \epsilon re^{\Phi_o} \approx \epsilon r(1 + \Phi_o)$, which corresponds to $\eta = \epsilon =$ constant. Its fractional errors will be of order M/r; and the errors which it produces in ω^2 will be of order $(M/R)^2$. By inserting this trial function into the variational principle (6) and keeping only relativistic corrections of order M/R, one obtains

$$\omega^2 = 3(\bar{\Gamma}_1 - \Gamma_{1\,\mathrm{crit}})|\Omega|/I. \tag{8}$$

Here $\bar{\Gamma}_1$ is the pressure-averaged adiabatic index, and the critical value of the adiabatic index $\Gamma_{1\,\mathrm{crit}}$ is

$$\Gamma_{1\,\mathrm{crit}} = \frac{4}{3} + \alpha M/R, \tag{9}$$

with α a positive constant of order unity given by

$$\alpha = \frac{1}{3} \frac{R}{M|\Omega|} \int_0^R \left(3\rho_o \frac{m_o^2}{r^2} + 4p_o \frac{m_o}{r}\right) 4\pi r^2 \, dr. \tag{10}$$

Expressions (8) and (9) for the pulsation frequency and the adiabatic index play an important role in the theory of supermassive stars (§24.4).

3. For alternative derivations of the above result, see Chandrasekhar (1964a,b; 1965c), Fowler (1964, 1965), Wright (1964).

Exercise 26.1. DRAGGING OF INERTIAL FRAMES BY A SLOWLY ROTATING STAR

EXERCISE

A fluid sphere rotates very slowly. Analyze its rotation using perturbation theory; keep only effects and terms *linear* in the angular velocity of rotation. [Hints: (1) Centrifugal forces are second-order in angular velocity. Therefore, to first order the star is undeformed; its density and pressure distributions remain spherical and unperturbed. (2) Show, by symmetry and time-reversal arguments, that one can introduce coordinates in which

$$ds^2 = -e^{2\Phi} \, dt^2 + e^{2\Lambda} \, dr^2 + r^2[d\theta^2 + \sin^2\theta \, d\phi^2] \tag{26.23}$$
$$- 2(r^2 \sin^2\theta)\omega \, d\phi \, dt,$$

where

$$\Phi = \Phi(r), \ \Lambda = \Lambda(r), \text{ and } \omega = \omega(r, \theta). \tag{26.24}$$

Show that $\Phi = \Phi_o$ and $\Lambda = \Lambda_o$ (no perturbations!) to first-order in angular velocity. (3) Adopt the following precise definition of the angular velocity $\Omega(r, \theta)$:

$$\Omega \equiv u^\phi/u^t = (d\phi/dt)_{\text{moving with the fluid}}. \tag{26.25}$$

Assuming $u^r = u^\theta = 0$ (i.e., rotation in the ϕ direction), calculate the 4-velocity of the fluid. (4) Use the Einstein field equations to derive a differential equation for the metric perturbation ω in terms of the angular velocity Ω. (5) Solve that differential equation outside the star in terms of elementary functions, and express the solution for $\omega(r, \theta)$ in terms of the star's total angular momentum S, as measured using distant gyroscopes (see Chapter 19).] For the original analyses of this problem and of related topics, see Gurovich (1965), Doroshkevich, Zel'dovich, and Novikov (1965), Hartle and Sharp (1965), Brill and Cohen (1966), Hartle (1967), Krefetz (1967), Cohen and Brill (1968), Cohen (1968).

THE UNIVERSE

Wherein the reader, flushed with joy at his conquest of the stars, seeks to control the entire universe, and is foiled by an unfathomed mystery: the Initial Singularity.

CHAPTER **27**

IDEALIZED COSMOLOGIES

*From my point of view one cannot arrive, by way of theory, at any at least
somewhat reliable results in the field of cosmology, if one makes no use of the
principle of general relativity.*

ALBERT EINSTEIN (1949b, p. 684)

§27.1. THE HOMOGENEITY AND ISOTROPY OF THE UNIVERSE

Astronomical observations reveal that the universe is homogeneous and isotropic on scales of $\sim 10^8$ light years and larger. Taking a "fine-scale" point of view, one sees the agglomeration of matter into stars, galaxies, and clusters of galaxies in regions of size ~ 1 light year, $\sim 10^6$ light years, and $\sim 3 \times 10^7$ light years, respectively. But taking instead a "large-scale" viewpoint, one sees little difference between an elementary volume of the universe of the order of 10^8 light years on a side centered on the Earth and other elementary volumes of the same size located elsewhere.

Cosmology, summarized in its simplest form in Box 27.1, takes the large-scale viewpoint as its first approximation; and as its second approximation, it treats the fine-scale structure as a perturbation on the smooth, large-scale background. This chapter (27) treats in detail the large-scale, homogeneous approximation. Chapter 28 considers such small-scale phenomena as the primordial formation of the elements, and the condensation of galaxies out of the primeval plasma during the expansion of the universe. Chapter 29 discusses observational cosmology.

Evidence for the large-scale homogeneity and isotropy of the universe comes from several sources. (1) There is evidence in the distribution of galaxies on the sky and in the distribution of their apparent magnitudes and redshifts [see, e.g., Hubble (1934b, 1936); Sandage (1972a); Sandage, Tamman, and Hardy (1972); but note the papers claiming "hierarchic" deviations from homogeneity, which Sandage cites and attacks]. (2) There is evidence in the isotropy of the distribution of radio sources on the sky [see, e.g., Holden (1966), and Hughes and Longair (1967)]. (3) There is evidence in the remarkable isotropy of the cosmic microwave radiation [see, e.g., Boughn, Fram, and Partridge (1971)]. For a review of most of the evidence, see Chapter 2 of Peebles (1971).

The universe: fine-scale condensations contrasted with large-scale homogeneity

Evidence for large-scale homogeneity and isotropy

(continued on page 711)

Box 27.1 COSMOLOGY IN BRIEF

Uniform density. Idealize the stars and atoms as scattered like dust through the heavens with an effective average density ρ of mass-energy everywhere the same.

Geometry homogeneous and isotropic. Idealize the curvature of space to be everywhere the same.

Closure. Accept the term, "Einstein's geometric theory of gravity" as including not only his field equation $\boldsymbol{G} = 8\pi\boldsymbol{T}$, but also his boundary condition of closure imposed on any solution of this equation.*

A three-sphere satisfies the three requirements of homogeneity, isotropy, and closure, and is the natural generalization of the metric on a circle and a 2-sphere:

Spheres of selected dimensionality	Visualized as embedded in a Euclidean space of one higher dimension[a]	Transformation from Cartesian to polar coordinates	Metric on S^n expressed in terms of these polar coordinates
S^1	$x^2 + y^2 = a^2$	$x = a \cos\phi$ $y = a \sin\phi$	$ds^2 = a^2\, d\phi^2$
S^2	$x^2 + y^2 + z^2 = a^2$	$x = a \sin\theta \cos\phi$ $y = a \sin\theta \sin\phi$ $z = a \cos\theta$	$ds^2 = a^2(d\theta^2 + \sin^2\theta\, d\phi^2)$
S^3	$x^2 + y^2 + z^2 + w^2 = a^2$	$x = a \sin\chi \sin\theta \cos\phi$ $y = a \sin\chi \sin\theta \sin\phi$ $z = a \sin\chi \cos\theta$ $w = a \cos\theta$	$ds^2 = a^2[d\chi^2 + \sin^2\chi(d\theta^2 + \sin^2\theta\, d\phi^2)]$

[a]Excursion off the sphere is physically meaningless and is forbidden. The superfluous dimension is added to help the reason in reasoning, not to help the traveler in traveling. Least of all does it have anything whatsoever to do with time.

The spacetime geometry is described by the metric

$$ds^2 = -dt^2 + a^2(t)[d\chi^2 + \sin^2\chi(d\theta^2 + \sin^2\theta\, d\phi^2)]. \tag{1}$$

The dynamics of the geometry is known in full when one knows the radius a as a function of the time t.

*"Thus we may present the following arguments against the conception of a space-infinite, and for the conception of a space-bounded, universe:

"1. From the standpoint of the theory of relativity, the condition for a closed surface is very much simpler than the corresponding boundary condition at infinity of the quasi-Euclidean structure of the universe.

"2. The idea that Mach expressed, that inertia depends upon the mutual action of bodies, is contained, to a first approximation, in the equations of the theory of relativity; . . . But this idea of Mach's corresponds only to a finite universe, bounded in space, and not to a quasi-Euclidean, infinite universe" [Einstein (1950), pp. 107–108].

Many workers in cosmology are skeptical of Einstein's boundary condition of closure of the universe, and will remain so until astronomical observations confirm it.

Einstein's field equation (doubled, for convenience), $2\boldsymbol{G} = 16\pi\boldsymbol{T}$, has its whole force concentrated in its $\hat{0}\hat{0}$ (or $\hat{\imath}\hat{\imath}$) component,

$$\frac{6}{a^2}\left(\frac{da}{dt}\right)^2 + \frac{6}{a^2} = 16\pi\rho \qquad (2)$$

[equation (5a) of Box 14.5]. This component of Einstein's equation is as central as the component $\boldsymbol{\nabla} \cdot \boldsymbol{E} = 4\pi\rho$ of Maxwell's equations. It is described in the Track-2 Chapter 21 as the "initial-value equation" of geometrodynamics. There the two terms on the left receive separate names: the "second invariant" of the "extrinsic curvature" of a "spacelike slice" through the 4-geometry (tells how rapidly all linear dimensions are being stretched from instant to instant); and the "intrinsic curvature" or three-dimensional scalar curvature invariant $^{(3)}R$ of the "spacelike slice" (here a 3-sphere) at the given instant itself.

The amount of mass-energy in the universe changes from instant to instant in accordance with the work done by pressure during the expansion,

$$d\left[\left(\begin{array}{c}\text{density of}\\ \text{mass-energy}\end{array}\right) \times (\text{volume})\right] = -(\text{pressure})\, d(\text{volume}). \qquad (3)$$

Today the pressure of radiation is negligible compared to the density of mass-energy, and the righthand side of this equation ("work done") can be neglected. The same was true in the past, one estimates, back to a time when linear dimensions were about a thousand times smaller than they are today. During this "matter-dominated phase" of the expansion of the universe, the product

$$\left(\begin{array}{c}\text{density of}\\ \text{mass-energy}\end{array}\right) \times (\text{volume})$$

remained a constant,

$$\rho \cdot 2\pi^2 a^3 = M. \qquad (4)$$

Here the symbol M can look like mass in the form of matter, and can even be called mass; but one has to recall again (see §19.4) that the concept of total mass-energy of a closed universe has absolutely no well-defined meaning whatsoever, not least because there is no "platform" outside the universe on which to stand to measure its attraction via periods of Keplerian orbits or in any other way. More convenient than M, because more significant in what follows, is the quantity a_{\max} ("radius of universe at phase of maximum expansion") defined by

$$a_{\max} = 4M/3\pi. \qquad (5)$$

Box 27.1 (*continued*)

The decisive component of the Einstein field equation, in the terms of this notation, becomes

$$\frac{6}{a^2}\left(\frac{da}{dt}\right)^2 + \frac{6}{a^2} = \frac{6a_{max}}{a^3}$$

or

$$\left(\frac{da}{dt}\right)^2 - \frac{a_{max}}{a} = -1.\tag{6}$$

The first term in (6) has the qualitative character of "kinetic energy" in an elementary problem in Newtonian mechanics. The second term has the qualitative character of a "potential energy,"

$$V(a) = -\frac{a_{max}}{a}$$

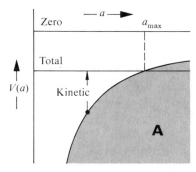

(see diagram **A**), resulting from an inverse-square Newtonian force. Pursuing the analogy, one identifies the "-1" on the righthand side with the total energy in the Newtonian problem. The qualitative character of the dynamics shows up upon an inspection of diagram **A**. Values of the radius of the universe, a, greater than a_{max} are not possible. If a were to become greater than a_{max}, the "potential energy" would exceed the total "energy" and the "kinetic energy" of expansion would have to become negative, which is impossible. Consequently the geometrodynamic system can never be in a state more expanded than $a = a_{max}$. Starting in a state of small a, ($a \ll a_{max}$) and expanding, the universe has for each a value a perfectly definite da/dt value. This velocity of expansion decreases as the expansion proceeds. It falls to zero at the turning point $a = a_{max}$. Thereafter the system recontracts.

Lack of option is the striking feature of the dynamics. Granted a specific amount of matter [specific M value in (5)], one has at his disposal no free parameter whatsoever. The value of a_{max} is uniquely specified by the amount of matter present, and by nothing more. There is no such thing as an "adjustable constant of energy," such as there would have been in a traditional problem of Newtonian dynamics. Where such an adjustable parameter might have appeared in equation (6), there appears instead the fixed number "-1." This fixity is the decisive feature of a system bound up into closure. Were one dealing with a collection of rocks out in space, one would have a choice about the amount of dynamite one placed at their center. With a low charge of explosive, one would find the rocks flying out for only a limited distance before gravity halted their flight and brought them to collapse together again. With more propellant, they would fly out with escape velocity and never return. But no such options present themselves here, exactly because Einstein's condition of closure has been imposed; and once closed, always closed. Collapse of the universe is universal. This is simple cosmology in brief.

Einstein's unhappiness at this result was great. At the time he developed general relativity, the permanence of the universe was a fixed item of belief in Western philosophy: "The heavens endure from everlasting to everlasting." Yet the reasoning that led to the fixed equation left open no natural way to change that equation or its fantastic prediction. Therefore Einstein (1917), much against his will, introduced the least unnatural change he could imagine, a so-called cosmological term (§27.11), the whole purpose of which was to avoid the expansion of the universe. A decade later, Hubble (1929) verified the predicted expansion. Thereupon Einstein abandoned the cosmological term, calling it "the biggest blunder of my life" [Einstein (1970)]. Thus ended the first great cycle of apparent contradiction to general relativity, test, and dramatic vindication. Will one ever penetrate the mystery of creation? There is no more inspiring evidence that the answer will someday be "yes" than man's power to predict, and predict correctly, and predict against all expectations, so fantastic a phenomenon as the expansion of the universe.

"Newtonian cosmology" provides an "equation of energy" similar to that of Einstein cosmology, but fails to provide any clean or decisive argument for closure or for the unique constant "-1." It considers the mass in any elementary spherical region of space of momentary radius r, and the gravitational acceleration of a test particle at the boundary of this sphere toward the center of the sphere; thus,

$$\frac{d^2r}{dt^2} = -\frac{(\text{mass})}{(\text{distance})^2} = -\frac{(4\pi/3)\rho r^3}{r^2} = -\frac{4\pi\rho}{3}r. \qquad (7)$$

Consider such imaginary spheres of varied radii drawn in the cosmological medium with the same center. Note that doubling the radius doubles the acceleration. This proportionality between acceleration and distance is compatible with a homogeneous

Box 27.1 (*continued*)

deceleration of the expansion of the universe. Therefore define an expansion param-
eter a^* as the ratio between the radius of any one of these spheres now and the
radius of the same sphere at some fiducial instant; thus, $a^* = r/r_o$ is to be considered
as independent of the particular sphere under consideration. Write $\rho = \rho_o r_o^3/r^3$,
where ρ_o is the density at the fiducial instant. Insert this expression for ρ into the
deceleration equation (7), multiply both sides of the equation through by dr/dt,
integrate, and translate the result from an equation for dr/dt to an equation for
da^*/dt, finding

$$\left(\frac{da^*}{dt}\right)^2 - \frac{(8\pi\rho_o/3)}{a^*} = \text{constant},\tag{8}$$

in agreement with equation (6), except for (1) the trivial differences that arise because
a^* is a dimensionless expansion ratio, whereas a is an absolute radius with the
dimensions of cm, and (2) the all-important difference that here the constant is
disposable, whereas in standard Einstein geometrodynamics it has the unique ca-
nonical value "-1." For more on Newtonian insights into cosmology, see especially
Bondi (1961).

Free fall of a particle towards a Newtonian center of attraction according to
Newtonian mechanics gives an equation of energy of the same form as (6), except
that the "radius of the universe," a, is replaced by distance, r, from the center of

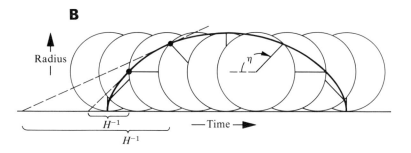

attraction. The solution of this problem of free fall is described by a cycloid (diagram
B; see also Figure 25.3 and Box 25.4), generated by rolling a circle of diameter a_{max}
on a line through an ever increasing angle η; thus,

$$a = \frac{1}{2}a_{max}(1 - \cos\eta),$$

$$t = \frac{1}{2}a_{max}(\eta - \sin\eta).\tag{9}$$

Immediately observable today is the present rate of expansion of the universe, with every distance increasing at a rate directly proportional to the magnitude of that distance:*

$$\frac{\left(\begin{array}{l}\text{velocity of recession}\\\text{of a galaxy}\end{array}\right)}{\text{(distance to a galaxy)}} = (\text{Hubble ``constant,''}\ H_0) \sim 55\ \text{km/sec megaparsec}$$

$$= \frac{1}{18 \times 10^9\ \text{yr}}\ \text{or}\ \frac{1}{1.7 \times 10^{28}\ \text{cm}}$$

$$= \frac{\left(\begin{array}{l}\text{rate of increase of the}\\\text{radius of the universe itself}\end{array}\right)}{\text{(radius of the universe)}} = \frac{da/dt}{a}. \tag{10}$$

The Hubble time, $H_0^{-1} \sim 18 \times 10^9$ yr (linearly extrapolated back to zero separation on the basis of the expansion rate observed today, as illustrated in the diagram) is predicted to be greater by a factor 1.5 or more (Box 27.3) than the actual time back to the start of the expansion as deduced from the rate of the development of stars ($\sim 10 \times 10^9$ yr). No such satisfactory concord between prediction and observation on this inequality existed in the 1940's. The scale of distances between galaxy and galaxy in use at that time was short by a factor more than five. The error arose from misidentifications of Cepheid variable stars and of HII regions, which are used as standards of intensity to judge the distance of remote galaxies. The linearly extrapolated time,

$$(\text{Hubble time}) = \frac{\text{(distance today)}}{\text{(recession velocity today)}},$$

back to the start of the expansion was correspondingly short by a factor more than five. The Hubble time came out to be only of the order of 3×10^9 yr. This number obviously violates the inequality

$$\left(\begin{array}{l}\sim 3 \times 10^9\ \text{yr Hubble}\\\text{time}\end{array}\right) \geq 1.5 \left(\begin{array}{l}\sim 10 \times 10^9\ \text{yr; actual time}\\\text{back to start of expansion}\end{array}\right).$$

It implies a curve for dimensions as a function of time not bending down, as in diagram **B**, but bending up. On some sides the proposal was made to regard the actual curve as rising exponentially. Thus began an era of "theories of continuous creation of matter," all outside the context of Einstein's standard geometrodynamics.

*H_0 is predicted to be independent of the choice of galaxy insofar as local motions are unimportant, and insofar as the difference between recession velocity now and recession velocity at the time when the light was emitted is unimportant. The latter condition is well fulfilled by galaxies close enough to admit of the necessary measurement of distance, for they have redshifts only of the order of $z \sim 0.1$ and less (little lapse of time between emission of light and its reception on earth; therefore little change in recession velocity between then and now; see §29.3 and Box 29.4 for a fuller analysis).

Box 27.1 (*continued*)

This era ended when, for the first time, the distinction between stellar populations of classes I and II was recognized and as a result Cepheid variables were correctly identified, by Baade (1952, 1956) and when Sandage (1958) discovered that Hubble had misidentified as bright stars the HII regions in distant galaxies. Then the scale of galactic distances was set straight. Thus ended the second great cycle of an apparent contradiction to general relativity, then test, and then dramatic vindication.

The mystery of the missing matter marks a third cycle of doubt and test with the final decision yet to come. It follows from equation (2) that, if Einstein's closure boundary condition is correct, then the density of mass-energy must exceed a certain lower limit given by the equation

$$\rho \geq \rho_H = \frac{3}{8\pi} H_o^2 \tag{11}$$

("critical amount of mass-energy required to curve up the geometry of the universe into closure"). A Hubble expansion rate of $H_o = 55$ km/sec Megaparsec implies a lower limit to the density of

$$\rho_H = \frac{3}{8\pi} \frac{1}{(1.7 \times 10^{28} \text{ cm})^2} \text{ or } \rho_{H,\text{conv}} = 5 \times 10^{-30} \text{ g/cm}^3 \tag{12}$$

as compared to $\rho \sim 2 \times 10^{-31}$ g/cm^3 of "luminous matter" observed in galaxies (§29.6) and more being searched for today in the space between the galaxies.

A fuller treatment of cosmology deals with conditions back in the past corresponding to redshifts of 10,000 or more and dimensions 10,000 times less than they are today, when radiation could not be neglected, and even dominated (§27.10). It also considers even earlier conditions, when anisotropy oscillations of the geometry of the universe as a whole (analogous to the transformation from a cigar to a pancake and back again) may conceivably have dominated (Chapter 30). More broadly, it takes up the evolution of the universe into its present state (Chapter 28) and the present state and future evolution of the universe (Chapter 29). The present chapter examines the basic assumptions that underlie the simple standard cosmology thus traced out, and §27.11 examines what kinds of qualitative changes would result if one or another of these assumptions were to be relaxed.

§27.2. STRESS-ENERGY CONTENT OF THE UNIVERSE— THE FLUID IDEALIZATION

By taking the large-scale viewpoint, one can treat galaxies as "particles" of a "gas" that fills the universe. These particles have internal structure (stars, globular clusters, etc.); but one ignores it. The "particles" cluster on a small scale (clusters of galaxies of size $\lesssim 3 \times 10^7$ light years); but one ignores the clustering. To simplify calculations, one even ignores the particulate nature of the "gas" [though one can take it into account, if one wishes, by adopting a kinetic-theory description; see §22.6 for kinetic theory, and Ehlers, Geren, and Sachs (1968) for its application to cosmology]. One removes the particulate structure of the gas from view by treating it in the perfect-fluid approximation. Thus, one characterizes the gas by a *4-velocity*, **u** (the 4-velocity of an observer who sees the galaxies in his neighborhood to have *no* mean motion), by a *density of mass-energy*, ρ (the smoothed-out density of mass-energy seen in the frame with 4-velocity **u**; this includes the rest mass plus kinetic energy of the galaxies in a unit volume, divided by the volume), and by a *pressure p* (the kinetic pressure of the galaxies). The stress-energy tensor for this "fluid of galaxies" is the familiar one

$$\mathbf{T} = (\rho + p)\mathbf{u} \otimes \mathbf{u} + \mathbf{g}p, \tag{27.1}$$

where **g** is the metric tensor.

Astronomical observations reveal that the rest-mass density of the galaxies is much greater than their density of kinetic energy. The typical ordinary velocities of the galaxies—and of stars in them—relative to each other are

$$\langle v \rangle \sim 200 \text{ km/sec} \sim 10^{-3}. \tag{27.2}$$

Consequently, the ratios of kinetic-energy density and of pressure to rest-mass density are

$$\varepsilon_{\text{kin}}/\rho_{\text{rm}} = \frac{1}{2} \langle v^2 \rangle \approx 10^{-6},$$

$$p/\rho_{\text{rm}} = \frac{1}{3} \langle v^2 \rangle \approx 10^{-6}. \tag{27.3}$$

At least, these are the ratios today. Very early in the life of the universe, conditions must have been quite different.

The total density of mass-energy, ρ, is thus very nearly the rest-mass density of the galaxies, ρ_{rm}. Astronomical observations yield for ρ_{rm} today

$$\rho_{\text{rm}} \gtrsim 2 \times 10^{-31} \text{ g/cm}^3 \tag{27.4}$$

(see §29.6).

The rest of this chapter, *except* for Box 27.4, is Track 2.

No earlier track-2 material is needed as preparation for it, but it is needed as preparation for Chapter 29 (Present state and future evolution of the universe).

Idealization of matter in universe as a perfect fluid ("fluid of galaxies")

Large-scale conditions in universe today:

(1) kinetic energy and pressure of stars and galaxies

(2) density of mass in galaxies

(3) cosmic-ray density

Not all the matter in the universe is tied up in galaxies; there is also matter in cosmic rays, with an averaged-out density of mass-energy

$$\rho_{\text{cr}} \lesssim 10^{-33} \text{ g/cm}^3, \tag{27.5}$$

(4) density of intergalactic gas

and, perhaps, gas in intergalactic space with

$$\rho_{\text{ig}} \lesssim 10^{-28} \text{ g/cm}^3. \tag{27.6}$$

[Delineating more sharply the value of ρ_{ig} is one of the most important goals of current cosmological research. For a review of this question as of 1971, see "The mean mass density of the universe," pp. 56–120 in Peebles (1971).] These sources of mass density, and the associated pressures, one can lump together with the galaxies into the "cosmological fluid," with stress-energy tensor (27.1).

Not all the stress-energy in the universe is in the form of matter. There are also magnetic fields, with mean energy density that almost certainly does not exceed the limit

(5) magnetic fields

$$\rho_{\text{mag}} \lesssim 10^{-35} \text{ g/cm}^3 \tag{27.7}$$

(6) radiation density

(corresponding to $B_{\text{avg}} \lesssim 10^{-6} \, G$), and radiation (electromagnetic radiation, neutrino radiation, and perhaps gravitational radiation) totaling, one estimates,

$$\rho_{\text{rad}} \approx 10^{-33} \text{ g/cm}^3. \tag{27.8}$$

The magnetic fields will be ignored in this chapter; they are unimportant for large-scale cosmology, except perhaps very near the "big-bang beginning" of the universe—if they existed then. However, the radiation cannot be ignored, for it plays a crucial role.

The cosmic microwave radiation

Most of the radiation density is in the form of "cosmic microwave radiation," which was discovered by Penzias and Wilson (1965) [see also Dicke, Peebles, Roll, and Wilkinson (1965)], and has been studied extensively since then [for a review, see Partridge (1969)]. The evidence is very strong that this cosmic microwave radiation is a remnant of the big-bang beginning of the universe. This interpretation will be accepted here.

The cosmic microwave radiation has just the form one would expect if the earth were enclosed in a box ("black-body cavity") with temperature 2.7K. The spectrum is a Planck spectrum with this temperature, and the radiation is isotropic [Boughn, Fram, and Partridge (1971)]. Consequently, its pressure and density of mass-energy are given by the formula,

$$\begin{aligned} \rho_{\text{microwave}} &= 3p_{\text{microwave}} = aT^4 \\ &= 4 \times 10^{-34} \text{ g/cm}^3. \end{aligned} \tag{27.9}$$

Thermodynamic considerations (§27.10) suggest that the universe should also be filled with neutrino radiation and perhaps gravitational radiation that have Planck spectra at approximately the same temperature (\sim3K). However, they are not detectable with today's technology.

To high accuracy ($\lesssim 300$ km/sec) the mean rest frame of the cosmic microwave radiation near Earth is the same as the mean rest frame of the galaxies in the neighborhood of Earth [Boughn, Fram and Partridge (1971)]. Consequently, the radiation can be included, along with the matter, in the idealized cosmological fluid.

Summary: From the large-scale viewpoint, the stress-energy of the universe can be idealized as a perfect fluid with 4-velocity \boldsymbol{u}, density of mass-energy ρ, pressure p, and stress-energy tensor

<div align="right">Summary of fluid idealization of matter in universe</div>

$$\boldsymbol{T} = (\rho + p)\boldsymbol{u} \otimes \boldsymbol{u} + p\boldsymbol{g}. \tag{27.10}$$

The 4-velocity \boldsymbol{u} at a given event \mathscr{P} in spacetime is the mean 4-velocity of the galaxies near \mathscr{P}; it is also the 4-velocity with which one must move in order to measure an isotropic intensity for the cosmic microwave radiation. The density ρ is made up of material density (rest mass plus negligible kinetic energy of galaxies; rest mass plus kinetic energy of cosmic rays; rest mass plus thermal energy of intergalactic gas—all "smeared out" over a unit volume), and also of radiation energy density (electromagnetic radiation, neutrino radiation, gravitational radiation). The pressure p, like the density ρ, is due to both matter and radiation. Today the pressure of the matter is much less than its mass-energy density,

$$p_{\text{matter}} \ll \rho_{\text{matter}} \text{ today}, \tag{27.11a}$$

but this strong inequality cannot have held long ago. Always the pressure of the radiation is $\frac{1}{3}$ its mass-energy density:

$$p_{\text{radiation}} = \frac{1}{3}\,\rho_{\text{radiation}} \text{ always}. \tag{27.11b}$$

§27.3. GEOMETRIC IMPLICATIONS OF HOMOGENEITY AND ISOTROPY

This chapter will idealize the universe to be *completely* homogeneous and isotropic. This idealization places tight constraints on the geometry of spacetime and on the motion of the cosmological fluid through it. In order to discover these constraints, one must first give precise mathematical meaning to the concepts of homogeneity and isotropy.

Homogeneity means, roughly speaking, that the universe is the same everywhere at a given moment of time. A given moment of what time? Whose time? This is the crucial question that the investigator asks.

In Newtonian theory there is no ambiguity about the concept "a given moment of time." In special relativity there is some ambiguity because of the nonuniversality of simultaneity, but once an inertial reference frame has been specified, the concept becomes precise. In general relativity there are no global inertial frames (unless spacetime is flat); so the concept of "a given moment of time" is completely ambiguous. However, another, more general concept replaces it: the concept of a three-dimensional spacelike hypersurface. This hypersurface may impose itself on one's

<div align="right">Spacelike hypersurface as generalization of "moment of time"</div>

attention by reason of natural symmetries in the spacetime. Or it may be selected at the whim or convenience of the investigator. He may find it more convenient to explore spacetime here and there than elsewhere, and to push the hypersurface forward accordingly ("many-fingered time"; the dramatically new conception of time that is part of general relativity). At each event on a spacelike hypersurface, there is a local Lorentz frame whose surface of simultaneity coincides locally with the hypersurface. Of course, this Lorentz frame is the one whose 4-velocity is orthogonal to the hypersurface. These Lorentz frames at various events on the hypersurface do not mesh to form a global inertial frame, but their surfaces of simultaneity do mesh to form the spacelike hypersurface itself.

The intuitive phrase "at a given moment of time" translates, in general relativity, into the precise phrase "on a given spacelike hypersurface." The investigator can go further. He can "slice up" the entire spacetime geometry by means of a "one-parameter family" of such spacelike surfaces. He can give the parameter that distinguishes one such slice from the next the name of "time." Such a one-parameter family of slices through spacetime is not required in the Regge calculus of Chapter 42. However, such a "slicing" is a necessity in most other practical methods for analyzing the dynamics of the geometry of the universe (Chapters 21, 30, and 43). The choice of slicing may dissolve away the difficulties of the dynamic analysis or may merely recognize those difficulties. The successive slices of "moments of time" may shine with simplicity or may only do a tortured legalistic bookkeeping for the dynamics. Which is the case depends on whether the typical spacelike hypersurface is distinguished by natural symmetries or, instead, is drawn arbitrarily.

Homogeneity of the universe means, then, that through each event in the universe there passes a spacelike "hypersurface of homogeneity" (physical conditions identical at every event on this hypersurface). At each event on such a hypersurface the density, ρ, and pressure, p, must be the same; and the curvature of spacetime must be the same.

The concept of isotropy must also be made precise. Clearly, the universe cannot look isotropic to all observers. For example, an observer riding on a 10^{20} eV cosmic ray will see the matter of the universe rushing toward him from one direction and receding in the opposite direction. Only an observer who is moving with the cosmo-logical fluid can possibly see things as isotropic. One considers such observers in defining isotropy:

Isotropy of the universe means that, at any event, an observer who is "moving with the cosmological fluid" cannot distinguish one of his space directions from the others by any local physical measurement.

Isotropy of the universe actually implies homogeneity; of this one can convince oneself by elementary reasoning (exercise 27.1).

Isotropy guarantees that the world lines of the cosmological fluid are orthogonal to each hypersurface of homogeneity. This one sees as follows. An observer "moving with the fluid" can discover by physical measurements on which hypersurface through a given event conditions are homogeneous. Moreover, he can measure his own ordinary velocity relative to that hypersurface. If that ordinary velocity is nonzero, it provides the observer with a way to distinguish one space direction in

"Homogeneity of universe" defined in terms of spacelike hypersurfaces

"Isotropy of universe" defined

Isotropy implies fluid world lines orthogonal to homogeneous hypersurfaces

his rest frame from all others—in violation of isotropy. Thus in an isotropic universe, where the concept of "observer moving with the fluid" makes sense, each such observer must discover that he is at rest relative to the hypersurface of homogeneity. His world line is orthogonal to that hypersurface.

Exercise 27.1. ISOTROPY IMPLIES HOMOGENEITY

Use elementary thought experiments to show that isotropy of the universe implies homogeneity.

EXERCISE

§27.4. COMOVING, SYNCHRONOUS COORDINATE SYSTEMS FOR THE UNIVERSE

The results of the last section enable one to set up special coordinate systems in the spacetime manifold of an isotropic model universe (Figure 27.1). Choose a hypersurface of homogeneity S_I. To all the events on it assign coordinate time, t_I. Lay out, in any manner desired, a grid of space coordinates (x^1, x^2, x^3) on S_I. "Propagate" these coordinates off S_I and throughout all spacetime by means of the world lines of the cosmological fluid. In particular, assign to every event on a given world line the space coordinates (x^1, x^2, x^3) at which that world line intersects S_I. This assignment has a simple consequence. The fluid is always at rest relative to the space coordinates. In other words, *the space coordinates are "comoving"*; they are merely labels for the world lines of the fluid. For the time coordinate t of a given event \mathcal{P}, use the lapse of proper time, $\int d\tau$, of \mathcal{P} from S_I, as measured along the fluid world line that passes through \mathcal{P}, plus t_I ("standard of time" on the initial hypersurface S_I); thus,

Construction of a "comoving, synchronous" coordinate system for the universe

$$t(\mathcal{P}) = t_I + \left(\int_{S_I}^{\mathcal{P}} d\tau \right)_{\text{along world line of fluid}} \tag{27.12}$$

The surfaces t = constant of such a coordinate system will coincide with the hypersurfaces of homogeneity of the universe. This one sees by focusing attention on observations made by two different observers, A and B, who move with the fluid along different world lines. At coordinate time t_I (on S_I) the universe looks the same to B as to A. Let A and B make observations again after their clocks have ticked away the same time interval $\Delta\tau$. Homogeneity of the initial hypersurface S_I, plus the deterministic nature of Einstein's field equations, guarantees that A and B will again see identical physics. (Identical initial conditions on S_I, plus identical lapses of proper time during which Einstein's equations govern the evolution of the universe near A and B, guarantee identical final conditions.) Therefore, after time lapse $\Delta\tau$, A and B are again on the same hypersurface of homogeneity—albeit a different

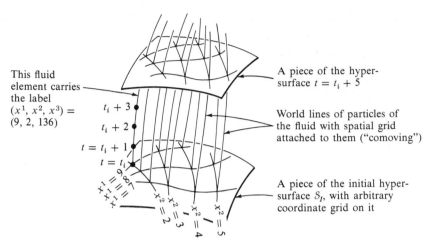

This fluid element carries the label $(x^1, x^2, x^3) = (9, 2, 136)$

$t_i + 3$
$t_i + 2$
$t = t_i + 1$
$t = t_i$

A piece of the hyper-surface $t = t_i + 5$

World lines of particles of the fluid with spatial grid attached to them ("comoving")

A piece of the initial hyper-surface \mathcal{S}_I, with arbitrary coordinate grid on it

Figure 27.1.
Comoving, synchronous coordinate system for the universe, as constructed in §27.4 of the text. Key features of such a coordinate system are as follows (see §§27.4 and 27.5). (1) The spatial coordinates move with the fluid, and the time coordinate is proper time along the fluid world lines; i.e., the coordinate description of a particular fluid world line is

$$(x^1, x^2, x^3) = \text{constant}, \quad x^0 \equiv t = \tau + \text{constant}.$$

$$\left[\begin{array}{l}\text{proper time measured}\\\text{along world line}\end{array}\right]$$

(2) Any surface of constant coordinate time is a hypersurface of homogeneity of the universe. Every such hypersurface is orthogonal to the world lines of all particles of the fluid. (3) The spatial grid on some initial hypersurface \mathcal{S}_I is completely arbitrary. (4) If $\gamma_{ij}\, dx^i\, dx^j$ is the metric on the initial hypersurface in terms of its arbitrary coordinates (with γ_{ij} a function of x^1, x^2, x^3), then the metric of spacetime in terms of the comoving, synchronous coordinate system is

$$ds^2 = -dt^2 + a^2(t)\gamma_{ij}\, dx^i\, dx^j.$$

Thus, the entire dynamics of the geometry of the universe is embodied in a single function of time, $a(t) = $ "expansion factor"; while the shape (but not size) of the hypersurfaces of homogeneity is embodied in the spatial 3-metric $\gamma_{ij}\, dx^i\, dx^j$.

one from \mathcal{S}_I, where they began. By virtue of definition (27.12) of coordinate time, the time coordinate at the intersection of B's world line with this new hypersurface is $t = t_I + \Delta\tau$; and similarly for A. Moreover, observers A and B were arbitrary. Consequently the new hypersurface of homogeneity, like \mathcal{S}_I, is a hypersurface of constant coordinate time. Q.E.D.

Because the hypersurfaces of homogeneity are given by $t = $ constant, the basis vectors $\partial/\partial x^i$ at any given event \mathcal{P} are tangent to the hypersurface of homogeneity that goes through that event. On the other hand, the time basis vector, $\partial/\partial t$, is tangent to the world line of the fluid through \mathcal{P}, since that world line has $x^i = $ constant along it. Consequently, orthogonality of the world line to the hypersurface guarantees orthogonality of $\partial/\partial t$ to $\partial/\partial x^i$:

$$(\partial/\partial t) \cdot (\partial/\partial x^i) = 0 \text{ for } i = 1, 2, 3. \tag{27.13a}$$

The time coordinate has another special property: it measures lapse of proper time along the world lines of the fluid. Because of this, and because $\partial/\partial t$ is tangent to the world lines, one can write

$$\partial/\partial t = (d/d\tau)_{\text{along fluid's world lines}} \tag{27.13b}$$
$$= \boldsymbol{u},$$

where \boldsymbol{u} is the 4-velocity of the "cosmological fluid." The 4-velocity always has unit length,

$$(\partial/\partial t) \cdot (\partial/\partial t) = -1. \tag{27.13c}$$

Conditions (27.13a,c) reveal that, in the comoving coordinate frame [where $g_{\alpha\beta} \equiv (\partial/\partial x^{\alpha}) \cdot (\partial/\partial x^{\beta})$], the line element for spacetime reads

$$ds^2 = -dt^2 + g_{ij}\, dx^i\, dx^j. \tag{27.14}$$

Form of the line element in this coordinate system

Any coordinate system in which the line element has this form is said to be "*synchronous*" (1) because the coordinate time t measures proper time along the lines of constant x^i (i.e., $g_{tt} = -1$), and (2) because the surfaces $t = \text{constant}$ are (locally) surfaces of simultaneity for the observers who move with $x^i = \text{constant}$ [i.e., $g_{ti} = (\partial/\partial t) \cdot (\partial/\partial x^i) = 0$]; it is also called a "Gaussian normal coordinate system" (*cf.* Figure 21.6).

A hypersurface of homogeneity, $t = \text{constant}$, has a spatial, three-dimensional geometry described by equation (27.14) with $dt = 0$:

$$(ds^2)_{\text{on hypersurface of homogeneity}} = d\sigma^2 \tag{27.15}$$
$$= [g_{ij}]_{t = \text{const}}\, dx^i\, dx^j.$$

To know everything about the 3-geometry on each of these hypersurfaces is to know everything about the geometry of spacetime.

Exercise 27.2. SYNCHRONOUS COORDINATES IN GENERAL **EXERCISE**

In an arbitrary spacetime manifold (not necessarily homogeneous or isotropic), pick an initial spacelike hypersurface S_I, place an arbitrary coordinate grid on it, eject geodesic world lines orthogonal to it, and give these world lines the coordinates

$$(x^1, x^2, x^3) = \text{constant}, \qquad x^0 \equiv t = t_I + \tau,$$

where τ is proper time along the world line, beginning with $\tau = 0$ on S_I. Show that in this coordinate system the metric takes on the synchronous (Gaussian normal) form (27.14).

§27.5. THE EXPANSION FACTOR

To determine the 3-geometry, $d\sigma^2 = g_{ij}(t, x^k)\, dx^i\, dx^j$, of each of the hypersurfaces of homogeneity, split the problem into two parts: (1) the nature of the 3-geometry on an arbitrary initial hypersurface (dealt with in next section); and (2) the evolution of the 3-geometry as time passes, i.e., as attention moves from the initial hypersurface to a subsequent hypersurface, and another, and another, . . . (dealt with in this section).

Assume that one knows the initial 3-geometry—i.e., the coefficients in the space part of the metric,

$$\gamma_{ij}(x^k) \equiv g_{ij}(t_I, x^k), \tag{27.16}$$

on the initial hypersurface \mathcal{S}_I—in its arbitrary but explicitly chosen coordinate system. What form will the metric coefficients $g_{ik}(t, x^k)$ have on the other hypersurfaces of homogeneity? This question is easily answered by the following argument: Consider two adjacent world lines, \mathcal{A} and \mathcal{B}, of the cosmological fluid, with coordinates (x^1, x^2, x^3) and $(x^1 + \Delta x^1, x^2 + \Delta x^2, x^3 + \Delta x^3)$. At time t_I (on surface \mathcal{S}_I) they are separated by the proper distance

$$\Delta\sigma(t_I) = (\gamma_{ij}\, \Delta x^i\, \Delta x^j)^{1/2}. \tag{27.17}$$

At some later time t (on surface \mathcal{S}), they will be separated by some other proper distance $\Delta\sigma(t)$. Isotropy of spacetime guarantees that the ratio of separations $\Delta\sigma(t)/\Delta\sigma(t_I)$ will be independent of the direction from \mathcal{A} to \mathcal{B} (no shearing motion of the fluid). For any given direction, the additivity of small separations guarantees that $\Delta\sigma(t)/\Delta\sigma(t_I)$ will be independent of $\Delta\sigma(t_I)$. Thus $\Delta\sigma(t)/\Delta\sigma(t_I)$ must be the same for all pairs of world lines near a given world line. Finally, homogeneity guarantees that this scalar ratio will be independent of position on the initial surface \mathcal{S}_I—i.e., independent of x^1, x^2, x^3. Define $a(t)$ to be this spatially constant ratio,

$$a(t) \equiv \Delta\sigma(t)/\Delta\sigma(t_I). \tag{27.18}$$

Thus, $a(t)$ is the factor by which the separations of world lines expand between time t_I and time t. In other words, the function $a(t)$ is a universal *"expansion factor,"* or *"scale factor."*

By combining equations (27.17) and (27.18), one obtains for the separation of adjacent world lines at time t

$$\Delta\sigma(t) = a(t)[\gamma_{ij}(x^k)\, \Delta x^i\, \Delta x^j]^{1/2}.$$

This corresponds to the spatial metric at time t,

$$d\sigma^2 = a^2(t)\gamma_{ij}(x^k)\, dx^i\, dx^j, \tag{27.19}$$

and to the spacetime metric,

$$ds^2 = -dt^2 + a^2(t)\gamma_{ij}(x^k)\, dx^i\, dx^j. \tag{27.20}$$

Proof that, aside from an over-all "expansion factor," all homogeneous hypersurfaces in the universe have the same 3-geometry

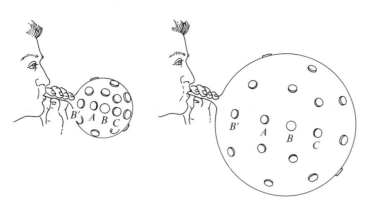

Figure 27.2.
Inflation of a balloon covered with pennies as a model for the expansion of the universe. Each penny A may well consider itself to be the center of the expansion because the distance from A to any neighbor B or C increases the more the more remote that neighbor was to begin with ("the Hubble relation"). The pennies themselves do not expand (constancy of sun-Earth distance, no expansion of a meter stick, no increase of atomic dimensions). The spacing today between galaxy and galaxy ($\sim 10^6$ lyr) is roughly ten times the typical dimension of a galaxy ($\sim 10^5$ lyr).

Notice that the coefficients $\gamma_{ij}(x^k)$ describe the shape not only of the initial hypersurface, but also of all other hypersurfaces of homogeneity. All that changes in the geometry from one hypersurface to the next is the scale of distances. All distances between spatial grid points (fluid world lines) expand by the same factor $a(t)$, leaving the shape of the hypersurface unchanged. This is a consequence of homogeneity and isotropy; and it is precisely true only if the model universe is precisely homogeneous and isotropic.

Of all the disturbing implications of "the expansion of the universe," none is more upsetting to many a student on first encounter than the nonsense of this idea. The universe expands, the distance between one cluster of galaxies and another cluster expands, the distance between sun and earth expands, the length of a meter stick expands, the atom expands? Then how can it make any sense to speak of any expansion at all? Expansion relative to what? Expansion relative to nonsense! Only later does he realize that the atom does not expand, the meter stick does not expand, the distance between sun and earth does not expand. Only distances between clusters of galaxies and greater distances are subject to the expansion. Only at this gigantic scale of averaging does the notion of homogeneity make sense. Not so at smaller distances. No model more quickly illustrates the actual situation than a rubber balloon with pennies affixed to it, each by a drop of glue. As the balloon is inflated (Figure 27.2) the pennies increase their separation one from another but not a single one of them expands! [For mathematical detail see, e.g., Noerdlinger and Petrosian (1971).]

What expands in the universe, and what does not

EXERCISE

Exercise 27.3. ARBITRARINESS IN THE EXPANSION FACTOR

How much arbitrariness is there in the definition of the expansion factor $a(t)$? Civilization A started long ago at time t_A. For it, the expansion factor is

$$\frac{\left(\begin{array}{l}\text{proper distance between}\\ \text{two particles of the ``cos-}\\ \text{mological fluid'' at time } t\end{array}\right)}{\left(\begin{array}{l}\text{proper distance between}\\ \text{same two particles}\\ \text{at time } t_A\end{array}\right)} = a_A(t).$$

Subsequently men planted civilization B at time t_B on a planet in a nearby galaxy. [At this time, the expansion factor a_A had the value $a_A(t_B)$]. Civilization B defines the expansion factor relative to the time of its own beginning:

$$\frac{\left(\begin{array}{l}\text{proper distance between}\\ \text{two particles of the ``cos-}\\ \text{mological fluid'' at time } t\end{array}\right)}{\left(\begin{array}{l}\text{proper distance between}\\ \text{the same two particles}\\ \text{at time } t_B\end{array}\right)} = a_B(t).$$

At two subsequent events, C and D, of which both civilizations are aware, they assign to the universe in their bookkeeping by no means identical expansion factors,

$$a_A(t_C) \neq a_B(t_C),$$

$$a_A(t_D) \neq a_B(t_D).$$

Show that the relative expansion of the model universe in passing from stage C to stage D in its evolution is nevertheless the same in the two systems of bookkeeping:

$$\frac{a_A(t_D)}{a_A(t_C)} = \left(\begin{array}{l}\text{relative expansion}\\ \text{from } C \text{ to } D\end{array}\right) = \frac{a_B(t_D)}{a_B(t_C)}.$$

§27.6. POSSIBLE 3-GEOMETRIES FOR A HYPERSURFACE OF HOMOGENEITY

Riemann tensor for homogeneous, isotropic hypersurfaces

Turn now to the 3-geometry $\gamma_{ij}\, dx^i\, dx^j$ for the arbitrary initial hypersurface \mathcal{S}_I. This 3-geometry must be homogeneous and isotropic. A close scrutiny of its three-dimensional Riemann curvature must yield no "handles" to distinguish one point on \mathcal{S}_I from any other, or to distinguish one direction at a given point from any other. "No handles" means that $^{(3)}\boldsymbol{Riemann}$ must be constructed algebraically from pure numbers and from the only "handle-free" tensors that exist: the 3-metric γ_{ij} and

the three-dimensional Levi-Civita tensor ε_{ijk}. (All other tensors pick out preferred directions or locations.) One possible expression for $^{(3)}\boldsymbol{Riemann}$ is

$$^{(3)}R_{ijkl} = K(\gamma_{ik}\gamma_{jl} - \gamma_{il}\gamma_{jk}); \quad K = \text{``curvature parameter''} = \text{constant.} \quad (27.21)$$

Trial and error soon convince one that this is the *only* expression that both has the correct symmetries for a curvature tensor and can be constructed solely from constants, γ_{ij}, and ε_{ijk}. Hence, this must be the 3-curvature of \mathcal{S}_I. [One says that any manifold with a curvature tensor of this form is a manifold of *"constant curvature."*]

As one might expect, the metric for \mathcal{S}_I is completely determined, up to coordinate transformations, by the form (27.21) of its curvature tensor. (See exercise 27.4 below). With an appropriate choice of coordinates, the metric reads (see exercise 27.5 below),

Metric for homogeneous, isotropic hypersurfaces: three possibilities—positive, zero, or negative spatial curvature

$$do^2 = \gamma_{ij}\,dx^i\,dx^j = K^{-1}[d\chi^2 + \sin^2\chi(d\theta^2 + \sin^2\theta\,d\phi^2)] \text{ if } K > 0,$$
$$do^2 = \gamma_{ij}\,dx^i\,dx^j = d\chi^2 + \chi^2(d\theta^2 + \sin^2\theta\,d\phi^2) \text{ if } K = 0, \quad (27.22)$$
$$do^2 = \gamma_{ij}\,dx^i\,dx^j = (-K)^{-1}[d\chi^2 + \sinh^2\chi(d\theta^2 + \sin^2\theta\,d\phi^2)] \text{ if } K < 0.$$

Absorb the factor $K^{-1/2}$ or $(-K)^{-1/2}$ into the expansion factor $a(t)$ [see exercise 27.3], and define the function

$$\Sigma \equiv \sin\chi, \qquad \text{if } k \equiv K/|K| = +1 \text{ (``positive spatial curvature''),}$$
$$\Sigma \equiv \chi, \qquad \text{if } k \equiv K = 0 \text{ (``zero spatial curvature''),} \quad (27.23)$$
$$\Sigma \equiv \sinh\chi, \qquad \text{if } k \equiv K/|K| = -1 \text{ (``negative spatial curvature'').}$$

Thus write the full spacetime geometry in the form

$$ds^2 = -dt^2 + a^2(t)\gamma_{ij}\,dx^i\,dx^j,$$
$$\gamma_{ij}\,dx^i\,dx^j = d\chi^2 + \Sigma^2(d\theta^2 + \sin^2\theta\,d\phi^2), \quad (27.24)$$

and the 3-curvatures of the homogeneous hypersurfaces in the form

$$^{(3)}R_{ijkl} = [k/a^2(t)][\gamma_{ik}\gamma_{jl} - \gamma_{il}\gamma_{jk}]. \quad (27.25a)$$

The curvature parameter K, after this renormalization, is evidently

$$K = k/a^2(t). \quad (27.25b)$$

Why is the word "renormalization" appropriate? Previously $a(t)$ was a scale factor describing expansion of linear dimensions relative to the linear dimensions as they stood at some arbitrarily chosen epoch; but the choice of that fiducial epoch was a matter of indifference. Now $a(t)$ has lost that arbitrariness. It has been normalized so that its value here and now gives the curvature of a spacelike hypersurface of homogeneity here and now. Previously the factor $a(t)$ was conceived as dimensionless. Now it has the dimensions of a length. This length is called the "radius of the model universe" when the curvature is positive. Even when the curvature is negative one sometimes speaks of $a(t)$ as a "radius." Only for zero curvature does the normaliza-

Significance of normalization of the expansion factor

tion of $a(t)$ still retain its former arbitrariness. Thus, for zero-curvature, consider two choices for $a(t)$, one of them $a(t)$, the other $\bar{a}(t) = 2a(t)$. Then with $\bar{\chi} = \frac{1}{2}\chi$, one can write proper distances in the three directions of interest with perfect indifference in either of two ways:

$$\begin{pmatrix} \text{proper distance} \\ \text{in the direction} \\ \text{of increasing } \chi \end{pmatrix} = a(t)\, d\chi = \bar{a}(t)\, d\bar{\chi},$$

$$\begin{pmatrix} \text{proper distance} \\ \text{in the direction} \\ \text{of increasing } \theta \end{pmatrix} = a(t)\chi\, d\theta = \bar{a}(t)\bar{\chi}\, d\theta,$$

$$\begin{pmatrix} \text{proper distance} \\ \text{in the direction} \\ \text{of increasing } \phi \end{pmatrix} = a(t)\chi \sin\theta\, d\phi = \bar{a}(t)\bar{\chi} \sin\theta\, d\phi,$$

No such freedom of choice is possible when the model universe is curved, because then the χ's in the last two lines are replaced by a function, $\sin\chi$ or $\sinh\chi$, that is not linear in its argument.

Despite the feasibility in principle of determining the absolute value of the "radius" $a(t)$ of a curved universe, in practice today's accuracy falls short of what is required to do so. Therefore it is appropriate in many contexts to continue to regard $a(t)$ as a factor of relative expansion, the absolute value of which one tries to keep from entering into any equation exactly because it is difficult to determine. This motivation will account for the way much of the analysis of expansion is carried out in what follows, with calculations arranged to deal with ratios of a values rather than with absolute a values.

Box 27.2 explores and elucidates the geometry of a hypersurface of homogeneity.

EXERCISES

**Exercise 27.4. UNIQUENESS OF METRIC FOR 3-SURFACE
 OF CONSTANT CURVATURE**

Let γ_{ij} and $\gamma_{i'j'}$ be two sets of metric coefficients, in coordinate systems $\{x^i\}$ and $\{x^{i'}\}$, that have Riemann curvature tensors [derived by equations (8.22) and (8.42)] of the constant-curvature type (27.21). Let it be given in addition that the curvature parameters K and K' are equal. Show that γ_{ij} and $\gamma_{i'j'}$ are related by a coordinate transformation. [For a solution, see §8.10 of Robertson and Noonan (1968), or §§10 and 27 of Eisenhart (1926).]

Exercise 27.5. METRIC FOR 3-SURFACE OF CONSTANT CURVATURE

(a) Show that the following metric has expression (27.21) as its curvature tensor

$$\gamma_{ij} = \left(1 + \frac{1}{4} K \delta_{kl} x^k x^l\right)^{-2} \delta_{ij}. \tag{27.26}*$$

*With this choice of spatial coordinates, the spacetime metric reads

$$ds^2 = -dt^2 + \frac{(dx^2 + dy^2 + dz^2)}{[1 + \frac{1}{4}K(x^2 + y^2 + z^2)]^2}.$$

This is often called the *"Robertson-Walker line element,"* because Robertson (1935, 1936) and Walker (1936) gave the first proofs that it describes the most general homogeneous and isotropic spacetime geometry.

(b) By transforming to spherical coordinates (R, θ, ϕ) and then changing to a Schwarzschild radial coordinate ($2\pi r = $ "proper circumference"), transform this metric into the form

$$d\sigma^2 = \frac{dr^2}{1 - Kr^2} + r^2(d\theta^2 + \sin^2\theta \, d\phi^2). \quad (27.27)$$

(c) Find a further change of radial coordinate that brings the metric into the form (27.22).

Exercise 27.6. PROPERTIES OF THE 3-SURFACES

Verify all statements made in Box 27.2.

Exercise 27.7. ISOTROPY IMPLIES HOMOGENEITY

Use the contracted Bianchi identity $^{(3)}G^{ik}{}_{|k} = 0$ (where the stroke indicates a covariant derivative based on the 3-geometry alone) to show (1) that $^{(3)}\nabla K = 0$ in equation (27.21), and therefore to show (2) that direction-independence of the curvature [isotropy; curvature of form (27.21)] implies and demands homogeneity (K constant in space).

(*continued on page 726*)

Box 27.2 THE 3-GEOMETRY OF HYPERSURFACES OF HOMOGENEITY

A. Universe with Positive Spatial Curvature '''Spatially Closed Universe'')

Metric of each hypersurface is

$$d\sigma^2 = a^2[d\chi^2 + \sin^2\chi(d\theta^2 + \sin^2\theta \, d\phi^2)]. \quad (1)$$

To visualize this 3-geometry, imagine embedding it in a four-dimensional Euclidean space (such embedding possible here; not possible for general three-dimensional manifold; only four freely disposable functions [w, x, y, z] of three variables [α, β, γ] are at one's disposal to try to reproduce six prescribed functions [$g_{mn}(\alpha, \beta, \gamma)$] of those same three variables).

The embedding is achieved by

$$w = a \cos \chi, \qquad z = a \sin \chi \cos \theta,$$
$$x = a \sin \chi \sin \theta \cos \phi, \quad (2)$$
$$y = a \sin \chi \sin \theta \sin \phi,$$

since it follows that

$$d\sigma^2 \equiv dw^2 + dx^2 + dy^2 + dz^2$$
$$= a^2[d\chi^2 + \sin^2\chi(d\theta^2 + \sin^2\theta \, d\phi^2)]. \quad (3)$$

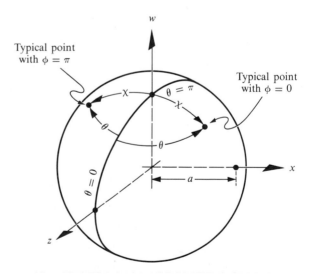

A 3-surface of positive curvature embedded in four-dimensional Euclidean space. One rotational degree of freedom is suppressed by setting $\phi = 0$ and π ("slice through pole," 3-sphere in 4-space looks like a 2-sphere in 3-space).

Box 27.2 (*continued*)

Equations (2) for the embedded surface imply that

$$w^2 + x^2 + y^2 + z^2 = a^2; \qquad (4)$$

i.e., the surface is a 3-dimensional sphere in 4-dimensional Euclidean space.

To verify homogeneity and isotropy, one need only notice that rotations in the four-dimensional embedding space can move any given point [any given (w, x, y, z) on the 3-sphere] and any given direction at that point into any other point and direction—while leaving unchanged the line element

$$d\sigma^2 = dw^2 + dx^2 + dy^2 + dz^2.$$

The above equations and the picture show that
(1) The 2-surfaces of fixed χ (which look like circles in the picture, because one rotational degree of freedom is suppressed) are actually 2-spheres of surface area $4\pi a^2 \sin^2\chi$; and (θ, ϕ) are standard spherical coordinates on these 2-spheres.
(2) As χ ranges from 0 to π, one moves outward from the "north pole" of the hypersurface, through successive 2-spheres ("shells") of area $4\pi a^2 \sin^2\chi$ (2-spheres look like circles in picture). The area of these shells increases rapidly at first and then more slowly as one approaches the "equator" of the hypersurface, $\chi = \pi/2$. Beyond the equator the area decreases slowly at first, and then more rapidly as one approaches the "south pole", ($\chi = \pi$; area $= 0$).
(3) The entire hypersurface is swept out by

$$0 \leq \chi \leq \pi,$$
$$0 \leq \theta \leq \pi,$$
$$0 \leq \phi \leq 2\pi$$

(ϕ is cyclic; $\phi = 0$ is same as $\phi = 2\pi$);

its 3-volume is

$$\mathcal{V} = \int (a\,d\chi)(a\sin\chi\,d\theta)(a\sin\chi\sin\theta\,d\phi)$$
$$= \int_0^\pi 4\pi a^2 \sin^2\chi\,(a\,d\chi) = 2\pi^2 a^3. \qquad (5)$$

B. Universe with Zero Spatial Curvature ("Spatially Flat Universe")

Metric of each hypersurface is

$$d\sigma^2 = a^2[d\chi^2 + \chi^2(d\theta^2 + \sin^2\theta\,d\phi^2)]. \qquad (6)$$

This is a perfectly flat, three-dimensional, Euclidean space described in spherical coordinates. In Cartesian coordinates

$$x = a\chi\sin\theta\cos\phi,$$
$$y = a\chi\sin\theta\sin\phi, \qquad (7)$$
$$z = a\chi\cos\theta,$$

the metric is

$$d\sigma^2 = dx^2 + dy^2 + dz^2. \qquad (8)$$

The entire hypersurface is swept out by

$$0 \leq \chi < \infty,$$
$$0 \leq \theta \leq \pi, \qquad (9)$$
$$0 \leq \phi \leq 2\pi;$$

and its volume is infinite.

C. Universe with Negative Spatial Curvature ("Spatially open Universe")

Metric of each hypersurface is

$$d\sigma^2 = a^2[d\chi^2 + \sinh^2\chi(d\theta^2 + \sin^2\theta\,d\phi^2)]. \qquad (10)$$

This 3-geometry cannot be embedded in a four-dimensional Euclidean space; but it can be embedded in a flat Minkowski space

$$d\sigma^2 = -dw^2 + dx^2 + dy^2 + dz^2. \qquad (11)$$

To achieve the embedding, set

$$w = a\cosh\chi, \qquad z = a\sinh\chi\cos\theta,$$
$$x = a\sinh\chi\sin\theta\cos\phi, \qquad (12)$$
$$y = a\sinh\chi\sin\theta\sin\phi;$$

insert this into equation (11), and thereby obtain (10).

Equations (12) for the embedded surface imply that

$$w^2 - x^2 - y^2 - z^2 = a^2; \qquad (13)$$

i.e., the surface is a three-dimensional hyperboloid in four-dimensional Minkowski space. (It has the

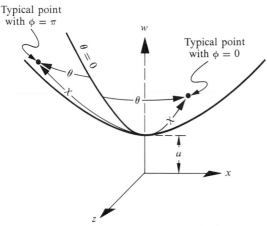

Typical point with $\phi = \pi$

Typical point with $\phi = 0$

$\theta = 0$

A 3-surface of negative curvature embedded in four-dimensional Minkowski space. One rotational degree of freedom is suppressed by setting $\phi = 0$ and π ("slice through pole"; 3-hyperboloid in 4-space looks like 2-hyperboloid in 3-space).

same form as a mass hyperboloid in momentum space; see Box 22.5.)

To verify homogeneity and isotropy, one need only notice that "Lorentz transformations" in the embedding space can move any given point on the 3-hyperboloid and any direction through that point into any other point and direction—while leaving unchanged the line element

$$d\sigma^2 = -dw^2 + dx^2 + dy^2 + dz^2.$$

The above equations and the picture show that

(1) The 2-surfaces of fixed χ (which look like circles in the picture because one rotational degree of freedom is suppressed) are actually 2-spheres of surface area $4\pi a^2 \sinh^2\chi$; and (θ, ϕ) are standard spherical coordinates on these 2-spheres.

(2) As χ ranges from 0 to ∞, one moves outward from the (arbitrarily chosen) "pole" of the hypersurface, through successive 2-spheres ("shells") of ever increasing area $4\pi a^2 \sinh^2\chi$. For large χ, surface area increases far more rapidly than it would if the hypersurface were flat

$$\frac{\text{(proper surface area)}}{4\pi \, \text{(proper distance)}^2} = \frac{A}{4\pi \, l^2}$$

$$= \frac{4\pi a^2 \sinh^2\chi}{4\pi a^2 \chi^2} \quad (14)$$

$$\approx \left(\frac{e^{l/a}}{2 \, l/a}\right)^2 \longrightarrow \infty.$$

The entire hypersurface is swept out by

$$0 \le \chi < \infty,$$
$$0 \le \theta \le \pi,$$
$$0 \le \phi \le 2\pi, \quad (15)$$

(ϕ is cyclic; $\phi = 0$ is same as $\phi = 2\pi$).
The volume of the hypersurface is infinite.

D. Nonuniqueness of Topology

Warning: Although the demand for homogeneity and isotropy determines completely the local geometric properties of a hypersurface of homogeneity up to the single disposable factor K, it leaves the global topology of the hypersurface undetermined. The above choices of topology are the most straightforward. But other choices are possible.

This arbitrariness shows most simply when the hypersurface is flat ($k = 0$). Write the full spacetime metric in Cartesian coordinates as

$$ds^2 = -dt^2 + a^2(t)[dx^2 + dy^2 + dz^2]. \quad (16)$$

Then take a cube of coordinate edge L

$$0 < x < L, \quad 0 < y < L, \quad 0 < z < L,$$

and identify opposite faces (process similar to rolling up a sheet of paper into a tube and gluing its edges together; see last three paragraphs of §11.5 for detailed discussion). The resulting geometry is still described by the line element (16), but now all three spatial coordinates are "cyclic," like the ϕ coordinate of a spherical coordinate system:

$$(t, x, y, z) \text{ is the same event as}$$
$$(t, x + L, y + L, z + L).$$

The homogeneous hypersurfaces are now "3-toruses" of finite volume

$$V = a^3 L^3,$$

analogous to the 3-toruses which one meets under the name "periodic boundary conditions" when analyzing electron waves and acoustic waves in solids and electromagnetic waves in space.

Another example: The 3-sphere described in part A above (case of "positive curvature") has the same geometry, but not the same topology, as the manifold of the rotation group, $SO(3)$ [see exercises 9.12, 9.13, 10.16, and 11.12]. For detailed discussion, see for example Weyl (1946), Coxeter (1963), and Auslander and Markus (1959).

§27.7. EQUATIONS OF MOTION FOR THE FLUID

After the above analysis of any one hypersurface of homogeneity, return to the dynamics of the universe. Examine, first, the evolution of the fluid, as governed by the law $\boldsymbol{\nabla} \cdot \boldsymbol{T} = 0$.

Recall (§22.3 and 23.5) that for a perfect fluid the equations of motion split into two parts. The component along the 4-velocity, $\boldsymbol{u} \cdot (\boldsymbol{\nabla} \cdot \boldsymbol{T}) = 0$, reproduces the first law of thermodynamics

$$(d/d\tau)(\rho V) = -p(dV/d\tau), \tag{27.28a}$$

where V is the volume of any fluid element. The part orthogonal to the 4-velocity, $(\boldsymbol{g} + \boldsymbol{u} \otimes \boldsymbol{u}) \cdot (\boldsymbol{\nabla} \cdot \boldsymbol{T}) = 0$, gives the force equation ("Euler equation")

$$(\rho + p) \times \text{(4-acceleration)} = -\text{(component of } \boldsymbol{\nabla}p \text{ orthogonal to } \boldsymbol{u}). \tag{27.28b}$$

Euler equation is vacuous for a homogeneous universe

For a static star (§23.5) the first law of thermodynamics was vacuous, but the force equation was crucial. For a homogeneous universe, the converse is true; the force equation is vacuous (no accelerations), but the first law of thermodynamics is crucial.

To see that the force equation is vacuous, notice that isotropy guarantees the vanishing of both sides of equation (27.28b). If either side were nonzero at any event \mathscr{P}, it would distinguish a direction in the homogeneous hypersurface at \mathscr{P}.

In applying the first law of thermodynamics (27.28a) to cosmology, divide the density and pressure into contributions due to matter and contributions due to radiation:

$$\rho = \rho_m + \rho_r; \qquad p = p_m + p_r. \tag{27.29}$$

"Equations of state" for matter and radiation

First discuss the density of mass-energy. Today $\rho_m (\gtrsim 10^{-31} \text{ g/cm}^3)$ dominates over $\rho_r (\sim 10^{-33} \text{ g/cm}^3)$. Matter did not always dominate. Therefore, one cannot set $\rho_r = 0$. Now discuss the pressure. During that epoch of the universe when pressure was significant cosmologically, p_r dominated over p_m. Consequently, one can neglect p_m at all times, and one can use the "equation of state" for radiation, $p_r = \frac{1}{3}\rho_r$, to write

$$\rho = \rho_m + \rho_r; \qquad p = \frac{1}{3}\rho_r. \tag{27.30}$$

When (27.30) is inserted into the first law of thermodynamics (27.28a), it yields the result

$$(d/d\tau)(\rho_m V) + (d/d\tau)(\rho_r V) = -\frac{1}{3}\rho_r \, dV/d\tau. \tag{27.31}$$

Energy exchange between matter and radiation is negligible

One cannot integrate this equation until one knows how mass-energy is fed back and forth between matter and radiation—i.e., until one knows another relationship between $\rho_m V$ and $\rho_r V$. All estimates indicate that, except in the first few seconds of the life of the universe, the energy exchanged between radiation and matter was

negligible compared to $\rho_m V$ and $\rho_r V$ individually (see §28.1). Under these conditions, equation (27.31) can be split into two parts:

$$(d/d\tau)(\rho_m V) = 0, \tag{27.32a}$$

First law of thermodynamics used to express densities of radiation and matter in terms of expansion factor

and

$$(d/d\tau)(\rho_r V) + \frac{1}{3}\rho_r \, dV/d\tau = 0. \tag{27.32b}$$

The solutions are simple:

$$\rho_m V = \text{constant (conservation of matter)} \tag{27.33a}$$

and

$$\rho_r V^{4/3} = \text{const} = \frac{\rho_r}{V^{-1/3}} \, V \, \begin{pmatrix} \text{constancy of number} \\ \text{of photons} \end{pmatrix} \tag{27.33b}$$

$$\underset{\underset{\text{to a factor of}}{\underset{\text{one photon, up}}{\text{energy } hc/\lambda \text{ of}}}}{\uparrow}$$
proportionality

Now what is V? It is the volume of any fluid element. It has the value

$$V = a^3 \Sigma^2 \sin\theta \, \Delta\chi \, \Delta\theta \, \Delta\phi$$

for a fluid element with edges $\Delta\chi$, $\Delta\theta$, $\Delta\phi$. Here χ, θ, ϕ are constant along each world line of the fluid (comoving coordinates). Therefore the element of hyperspherical solid angle $\Sigma^2 \sin\theta \, \Delta\chi \, \Delta\theta \, \Delta\phi$ (or pseudohyperspherical solid angle for the model of an open universe) is constant throughout all time for any fluid element. Therefore the volume of the fluid element grows in direct proportion to the cube of the expansion parameter a; thus,

$$V/a^3 = \text{constant}.$$

Combining this result with the constancy of $\rho_m V$ and $\rho_r V^{4/3}$, one sees that

$$\rho_m a^3 = \text{constant}, \qquad \rho_r a^4 = \text{constant}. \tag{27.34}$$

Let ρ_{mo} be the density of matter today, ρ_{ro} be the density of radiation today, and a_o be the expansion factor for the universe today. Then, at any time in the past,

$$\rho(t) = \rho_{mo} \frac{a_o^3}{a^3(t)} + \rho_{ro} \frac{a_o^4}{a^4(t)} \tag{27.35a}$$

and

$$p(t) = \frac{1}{3}\rho_{ro} \frac{a_o^4}{a^4(t)}. \tag{27.35b}$$

These results were based on two key claims, which will be justified in detail later (Chapter 28): the claim that in the epoch when pressure was important p_m was much smaller than p_r; and the claim that exchange of mass-energy between radiation and matter was always negligible (except in the first few seconds after the "creation").

§27.8. THE EINSTEIN FIELD EQUATION

Once the time evolution of the expansion factor, $a(t)$, is known, one can read off the time evolution of the density and pressure directly from equations (27.35). The density and pressure, in turn, determine how the expansion proceeds in time, via Einstein's field equations. Thus the field equations "close the logic loop" and give one a closed mathematical system from which to determine all three quantities, $a(t)$, $p(t)$ and $\rho(t)$.

One can readily calculate the components of the Einstein tensor for the model universe using the orthonormal basis one-forms,

$$\boldsymbol{w}^{\hat{t}} \equiv \boldsymbol{d}t, \qquad \boldsymbol{w}^{\hat{\chi}} \equiv a(t)\,\boldsymbol{d}\chi, \qquad \boldsymbol{w}^{\hat{\theta}} \equiv a(t)\Sigma\,\boldsymbol{d}\theta, \qquad \boldsymbol{w}^{\hat{\phi}} \equiv a(t)\Sigma\sin\theta\,\boldsymbol{d}\phi. \quad (27.36)$$

Evaluation of the Einstein field equation for a homogeneous universe:

The result [see equations (5) of Box 14.5] is

$$G_{\hat{t}\hat{t}} = 3\left(\frac{a_{,t}}{a}\right)^2 + \frac{3k}{a^2}, \qquad (27.37a)$$

$$G_{\hat{\chi}\hat{\chi}} = G_{\hat{\theta}\hat{\theta}} = G_{\hat{\phi}\hat{\phi}} = -2\frac{a_{,tt}}{a} - \left(\frac{a_{,t}}{a}\right)^2 - \frac{k}{a^2}, \qquad (27.37b)$$

$$G_{\hat{\mu}\hat{\nu}} = 0 \text{ if } \mu \neq \nu. \qquad (27.37c)$$

(With foresight, one will notice ahead of time that isotropy guarantees the equality $G_{\hat{\chi}\hat{\chi}} = G_{\hat{\theta}\hat{\theta}} = G_{\hat{\phi}\hat{\phi}}$, and similar equalities for the Riemann tensor; and one will calculate only $G_{\hat{\chi}\hat{\chi}}$, the component that is most easily calculated.)

The basis one-forms, $\boldsymbol{w}^{\hat{t}}$, $\boldsymbol{w}^{\hat{\chi}}$, $\boldsymbol{w}^{\hat{\theta}}$, $\boldsymbol{w}^{\hat{\phi}}$, are the orthonormal basis carried along by an observer who moves with the "cosmological fluid." Consequently, $T_{\hat{t}\hat{t}}$ is the mass-energy density, ρ, that he measures; $T_{\hat{j}\hat{j}}$ is the pressure, p; $T_{\hat{t}\hat{j}}$ vanishes, because he sees no energy flux (no momentum density); and $T_{\hat{i}\hat{j}}$ vanishes for $i \neq j$ because he sees no shear stresses:

$$T_{\hat{t}\hat{t}} = \rho, \qquad (27.38a)$$

$$T_{\hat{\chi}\hat{\chi}} = T_{\hat{\theta}\hat{\theta}} = T_{\hat{\phi}\hat{\phi}} = p, \qquad (27.38b)$$

$$T_{\hat{\mu}\hat{\nu}} = 0 \text{ when } \mu \neq \nu. \qquad (27.38c)$$

Equate the Einstein ("moment of rotation") tensor of equations (27.37) to the stress-energy tensor of equations (27.38). And if one insists, include the so-called "Λ-term" or "cosmological term" in the field equations [Einstein (1970): "the biggest blunder of my life"). Thus obtain two nonvacuous field equations. The first is an

"initial value equation," which relates $a_{,t}$ to a and ρ at any initial moment of time:

$$\left(\frac{a_{,t}}{a}\right)^2 = -\frac{k}{a^2} + \underbrace{\frac{\Lambda}{3}}_{\text{omit}} + \frac{8\pi}{3}\rho. \qquad (27.39a)$$

(1) initial value equation

The second is a "dynamic equation," which gives the second time-derivative of the expansion factor, and thereby governs the dynamic evolution away from the initial moment of time,

$$2\frac{a_{,tt}}{a} = -\left(\frac{a_{,t}}{a}\right)^2 - \frac{k}{a^2} + \underbrace{\Lambda}_{\text{omit}} - 8\pi p. \qquad (27.39b)$$

(2) dynamic equation

If (27.39b) is to be compared with anything in Newtonian mechanics, it is to be compared with an equation for acceleration (equation of motion), and in the same spirit (27.39a) is to be compared with a first integral of the equation of motion; that is, an equation of energy. In accordance with this comparison, note that one only has to differentiate (27.39a) and combine it with the relation satisfied by the pressure,

Why the dynamic equation is superfluous

$$(\rho a^3)_{,t} = -p(a^3)_{,t}$$

("law of conservation of energy") to get the acceleration equation (27.39b). Without any loss of information, one can therefore ignore the "acceleration equation" or "dynamic equation" (27.39b) henceforth, and work with the analog of an energy equation or what is more properly known as an "initial-value equation" (details of initial-value problem for Track-2 readers in Chapter 21).

What shows up here in the limited context of Friedmann cosmology is appropriately viewed in the wider context of general geometrodynamics. Conservation of energy plus one field equation have just been seen to reproduce the other field equations. Conversely, by accepting both field equations, one can derive the law of conservation of energy in the form just stated. Thus, the very act of writing the field equation $\mathbf{G} = 8\pi\mathbf{T}$ (or, if one insists upon the "cosmological term,"

Side remarks about initial value equations, dynamic equations, and Bianchi identities in more general contexts

$$\mathbf{G} + \underbrace{\Lambda\mathbf{g}}_{\text{omit}} = 8\pi\mathbf{T})$$

was encouraged by and founded on the automatic vanishing of the divergence $\boldsymbol{\nabla} \cdot \mathbf{G}$ (or the vanishing of the divergences of \mathbf{G} and \mathbf{g}), because one knew to begin with that energy and momentum are conserved, $\boldsymbol{\nabla} \cdot \mathbf{T} = 0$. It is not surprising, then, that there should be a redundancy between the conservation law, $\boldsymbol{\nabla} \cdot \mathbf{T} = 0$, and the field equations. Neither is it surprising in the dynamics of the Friedmann universe that one can use what is here the one and only interesting component of the conservation law, plus the one and only interesting initial value component ($G_{\hat{t}\hat{t}}$ component) of the field equations, to obtain the one and only interesting dynamic component ($G_{\hat{x}\hat{x}}$ component) of the field equations.

In a similar way, in more general problems that lack symmetry, one can always eliminate *some* of the dynamic field equations, but when gravitational radiation is present, one cannot eliminate them all. The dynamic field equations that cannot be eliminated, even in principle, govern the propagation of the gravitational waves. No gravitational waves are present in a perfectly homogeneous and isotropic cosmological model; its high degree of symmetry—in particular, its spherical (2-sphere!) symmetry about $\chi = 0$—is incompatible with gravitational waves.

Now turn back from general dynamics to Friedmann cosmology. To determine the time evolution of the expansion factor, a, insert into the initial-value equation (27.39a) the expression for the density of mass-energy given in (27.35a), and arrive at an equation ready for integration,

Differential equation for expansion factor

$$\left(\frac{a_{,t}}{a}\right)^2 = -\frac{k}{a^2} + \underbrace{\frac{\Lambda}{3}}_{\text{omit}} + \underbrace{\frac{(8\pi\rho_{mo}a_o^3/3)}{a^3} + \frac{(8\pi\rho_{ro}a_o^4/3)}{a^4}}_{(8\pi/3)\rho(a)} \tag{27.40}$$

When one has completed the integration of this equation for $a = a(t)$, one turns back to equation (27.35a,b) to get $\rho(t)$ and $p(t)$, and to expression (27.24) to get the geometry,

$$ds^2 = -dt^2 + a^2(t)[d\chi^2 + \Sigma^2(d\theta^2 + \sin^2\theta \, d\phi^2)], \tag{27.41}$$

thus completing the solution of the problem.

§27.9. TIME PARAMETERS AND THE HUBBLE CONSTANT

To the analysis of this dynamic problem, many investigators have contributed over the years, beginning with Friedmann himself in 1922. They discovered, among other results, that there are three natural choices of time variable, the one of greatest utility depending on the application that one has at hand.

Three choices of time parameter for universe:

(1) proper time, t

First is t, the original time variable. This quantity gives directly proper time elapsed since the start of the expansion. This is the time available for the formation of galaxies. It is also the time during which radioactive decay and other physical processes have been taking place.

(2) expansion factor, a

Second is $a(t)$, the expansion factor, which grows with time, which therefore serves to distinguish one phase of the expansion from another, and which consequently can be regarded as a parametric measure of time in its own right. The ratio of $a(t)$ at two times gives the ratio of the dimensions of the universe (cube root of volume) at those two times. It also gives the ratio $(1 + z)$ of wavelengths at those two times (see §29.2). A knowledge of the red shift, z, experienced in time past by radiation received today is equivalent to a knowledge of $a(t)/a_o$, where a_o is the expansion factor today. Specifically, radiation coming in with $z = 999$ is radiation coming in from a time in the history of the universe when it had 10^{-3} of its present dimensions and 10^{-9} of its present volume. During the interval of time while the expansion

parameter is increasing from a to $a + da$, the lapse of proper time, according to (27.40), is

$$dt = \frac{da}{[-k + (8\pi/3)a^2\rho(a) + \underbrace{(\Lambda/3)a^2}_{\text{omit}}]^{1/2}}. \qquad (27.42)$$

In terms of a as a new time parameter, it follows from this formula that the metric takes the form [Hughston (1969)]

$$ds^2 = \frac{-(da)^2}{-k + (8\pi/3)a^2\rho(a) + \underbrace{(\Lambda/3)a^2}_{\text{omit}}} + a^2[d\chi^2 + \Sigma^2(d\theta^2 + \sin^2\theta\, d\phi^2)]. \quad (27.43)$$

Third is $\eta(t)$, the "arc-parameter measure of time." During the interval of time dt, a photon traveling on a hypersphere of radius $a(t)$ covers an arc measured in radians equal to

(3) arc parameter, η

$$d\eta = \frac{dt}{a(t)}. \qquad (27.44)$$

When the model universe is open instead of closed, the same parameter lets itself be defined. Only the words "hypersphere" and "arc" have to be replaced by the corresponding words for a flat hypersurface of homogeneity ($k = 0$) or a hyperboloidal hypersurface ($k = -1$). In all three cases, the "arc parameter" is defined by the integral of this expression from the start of the expansion:

$$\eta = \int_0^t \frac{dt}{a(t)}; \qquad (27.45)$$

thus small values of the "arc parameter time," η, mean early times; and larger values mean later times. In terms of this "arc-parameter measure of time," the metric takes the form

$$ds^2 = a^2(\eta)[-d\eta^2 + d\chi^2 + \Sigma^2(d\theta^2 + \sin^2\theta\, d\phi^2)]. \qquad (27.46)$$

Let a photon start at the "North Pole" of the 3-sphere ($\chi = 0$; any θ and ϕ) at the "arc parameter time" $\eta = \eta_1$. Then, by the "arc parameter time" $\eta = \eta_2$, the photon has traveled to a new point on the hypersphere and encountered a new set of particles of the "cosmological fluid." They lie at the hyperpolar angle

$$\chi = \eta_2 - \eta_1.$$

When one makes a spacetime diagram on a piece of paper to show what is happening when an effect propagates from one point to another in the universe, one finds it most convenient to take (1) the space coordinate to be χ (the life histories of distinct particles of the "cosmological fluid" thus being represented by distinct vertical lines), and (2) the time coordinate to be η (so that photons are described by lines inclined at $\pm 45°$). No time parameter is more natural to use than η when one is tracing

out the course of null geodesics. For an example, see the treatment of the cosmological redshift in §29.2. It also turns out that it is simpler analytically (when Λ is taken to be zero) to give $a = a(\eta)$ and $t = t(\eta)$ than to give a directly as a function of time. Thus one gets the connection between the dimension a and the "arc-parameter time" η from the formula

$$\eta = \int d\eta = \int \frac{dt}{a(t)} = \int \frac{da}{[-ka^2 + (8\pi/3)a^4\rho(a) + \underbrace{(\Lambda/3)a^4}_{\text{omit}}]^{1/2}}. \qquad (27.47)$$

From a knowledge of the dimension a as a function of this time parameter, one immediately gets proper time itself in terms of this time parameter, from the formula

$$dt = a(\eta)\, d\eta. \qquad (27.48)$$

An equation (27.40) for the expansion factor and a choice of parameters for marking out time have now set the stage for a detailed analysis of idealized cosmology, and some of the relevant questions have even been asked: How does the characteristic dimension, a, of the geometry (radius of 3-sphere, in the case of closure) change with time? What is the spacetime geometry? How do geodesics, especially null geodesics, travel in this geometry? However, additional questions are equally important: Is the expansion of the universe decelerating and, if so, how fast? How do density and pressure of matter and radiation vary with time? And finally, for the simplest and most immediate tie between theory and observation, what is the expansion rate?

Hubble constant and Hubble time

In speaking of expansion rate, one refers to the "Hubble constant," the fractional rate of increase of distances,

$$H \equiv \frac{\dot{a}(t)}{a(t)}, \qquad (27.49)$$

which is normally evaluated today $H(\text{today}) \equiv H_o$, but is in principle defined as a function of time for every phase of the history of the universe. The reciprocal of H is the "Hubble time," H^{-1}. This quantity represents the time it would have taken for the galaxies to attain their present separations, starting from a condition of infinite compaction, if they had maintained for all time their present velocities ("time for expansion with dimensions linearly extrapolated back to the start"). For the conversion from astrophysical to geometric units and to years, take the currently accepted value, $H_o = 55$ km/sec megaparsec (Box 29.4), as an illustration:

$$H_o = \frac{55 \text{ km/sec}}{(299{,}793 \text{ km/sec})(3.0856 \times 10^{24} \text{ cm or } 3.2615 \times 10^6 \text{ yr of time})}$$

$$= 0.59 \times 10^{-28} \text{ per cm of light-travel time}$$

$$\text{or } 5.6 \times 10^{-11} \text{ fractional expansion per yr}, \qquad (27.50)$$

$$H_o^{-1} = 1.7 \times 10^{28} \text{ cm of light-travel time or } 18 \times 10^9 \text{ yr}.$$

§27.10. THE ELEMENTARY FRIEDMANN COSMOLOGY OF A CLOSED UNIVERSE

Take the simplest cosmological model, an isotropic homogeneous closed universe with $\Lambda = 0$, and trace out its features in all detail in the two limiting cases where matter dominates and where radiation dominates. The term "Friedmann universe" is used here for both cases, although the matter-dominated model is sometimes referred to as the Friedmann universe and the radiation-dominated one as the Tolman universe. In this analysis, it will be appropriate to let the variable $a(t)$ represent the radius of the universe, as measured in cm, because only by reference to this radius does one have the tool in hand to discuss all the interesting geometric effects that in principle lend themselves to observation. After this discussion, it will be enough, in dealing with other models, to summarize their principal parts and comment on their differences from this simple model, without repeating the full investigation. Any reference to an open universe or any so-called "cosmological constant" or its effects will therefore be deferred to a brief final section, §27.11. There the variable $a(t)$ will sometimes be taken to represent only a parameter of relative expansion, as is appropriate for discussions reaching out only to, say, $z = 0.1$, where global geometric issues are not taken up.

Rewrite the controlling component (27.40) of Einstein's field equation in the form

$$\left(\frac{da}{dt}\right)^2 - \frac{8\pi\rho_{mo}a_o^3/3}{a} - \frac{8\pi\rho_{ro}a_o^4/3}{a^2} = -1. \tag{27.51}$$

According as one neglects the radiation term or the matter term in this equation, the equation idealizes to

$$\left(\frac{da}{dt}\right)^2 - \frac{a_{max}}{a} = -1, \qquad \left(\begin{matrix}27.52;\ \text{matter} \\ \text{dominates}\end{matrix}\right)$$

or

$$\left(\frac{da}{dt}\right)^2 - \frac{a^{*2}}{a^2} = -1. \qquad \left(\begin{matrix}27.52;\ \text{radiation} \\ \text{dominates}\end{matrix}\right)$$

In both cases, the problem lends itself to comparison to the problem of particle motion in Newtonian mechanics with "total energy" -1 and with an "effective potential energy" of the qualitative form shown in diagram **A** of Box 27.1—apart from minor differences in shape according as the potential goes as $-1/a$ or as $-1/a^2$. The principal features of the solution are collected in Box 27.3.

It is a striking feature of the radiation-dominated era of the early Friedmann universe that the density of the radiation depends on time according to a simple universal law,

$$\rho_r = 3/32\pi t^2 \tag{27.53}$$

(final line and final column of Box 27.3). This circumstance may someday provide

Features of a closed Friedmann universe with $\Lambda = 0$:

(1) radius as function of time

(2) early era, when radiation dominates: types of radiation

(continued on page 736)

Box 27.3 SOLUTIONS FOR THE ELEMENTARY FRIEDMANN COSMOLOGY OF A CLOSED UNIVERSE IN THE TWO LIMITING CASES IN WHICH (1) MATTER DOMINATES AND RADIATION IS NEGLIGIBLE, AND (2) RADIATION DOMINATES AND MATTER IS NEGLIGIBLE

Idealization for dynamics of 3-sphere	Matter dominated	Radiation dominated
Model relevant when?	back into past to redshift $z \sim 10,000$; through today and through phase of maximum expansion, and re-contraction down to dimensions $\sim 10,000$-fold smaller than today	very early phase of expansion, for redshifts $z \sim 1,000$ and greater; and corresponding phase in late stages of recontraction; not directly relevant today.
Effective "potential" in $\left(\dfrac{da}{dt}\right)^2 + V(a) = -1$	$V(a) = -\dfrac{a_{max}}{a}$	$V(a) = -\dfrac{a^{*2}}{a^2}$
Value of constant in this "potential" in terms of conditions at some standard epoch	$a_{max} = \dfrac{8\pi}{3} a_o^3 \rho_{mo}$	$a^{*2} = \dfrac{8\pi}{3} a_o^4 \rho_{ro}$
Solution of dynamic equation expressed parametrically in terms of "arc parameter" η (radians of arc distance on 3-sphere covered by a photon travelling ever since start of expansion)	$a = \dfrac{a_{max}}{2}(1 - \cos \eta)$ $t = \dfrac{a_{max}}{2}(\eta - \sin \eta)$	$a = a^* \sin \eta$ $t = a^*(1 - \cos \eta)$
Range of η from start of expansion to end of recontraction	2π (one trip around the universe)	π (gets only as far as antipodal point of universe)
Nature of curve relating radius a to time t	cycloid	semicircle
Hubble time $H^{-1} = \dfrac{a}{(da/dt)} = \dfrac{a^2}{(da/d\eta)}$	$\dfrac{a_{max}}{2} \dfrac{(1 - \cos \eta)^2}{\sin \eta}$	$a^* \dfrac{\sin^2 \eta}{\cos \eta}$

Idealization for dynamics of 3-sphere	Matter dominated	Radiation dominated
Inequality between Hubble or "extrapolated" time and actual time back to start of expansion	$H^{-1} \geq 1.5t$	$H^{-1} \geq 2t$
Density of mass-energy	$\rho_m = \dfrac{3}{\pi a_{\max}^2 (1 - \cos \eta)^3}$	$\rho_r = \dfrac{3}{8\pi a^{*2} \sin^4 \eta}$
This density expressed in terms of Hubble expansion rate	$\rho_m = \dfrac{3H^2}{8\pi} \dfrac{2}{1 + \cos \eta}$	$\rho_r = \dfrac{3H^2}{8\pi} \dfrac{1}{\cos^2 \eta}$
Inequality satisfied by density	$\rho_m \geq \dfrac{3H^2}{8\pi}$	$\rho_r \geq \dfrac{3H^2}{8\pi}$
Analysis of magnification of distant galaxy by curvature of intervening space	§29.5 and Figure 29.2	§29.5
Limiting form of law of expansion for early times	$\begin{cases} t = \dfrac{a_{\max}}{12} \eta^3 \\[2mm] a = \dfrac{a_{\max}}{4} \eta^2 \\[2mm] a = \left(\dfrac{9 a_{\max} t^2}{4}\right)^{1/3} \end{cases}$	$\begin{cases} t = \dfrac{a^*}{2} \eta^2 \\[2mm] a = a^* \eta \\[2mm] a = (2a^* t)^{1/2} \end{cases}$
Other features of expansion at early times	$\begin{cases} H^{-1} = \dfrac{a_{\max}}{8} \eta^3 = 1.5t \\[2mm] \rho_m = \dfrac{a_{\max}}{(8\pi a^3/3)} \\[2mm] \quad = \dfrac{1}{6\pi t^2} = \dfrac{3H^2}{8\pi} \end{cases}$	$\begin{cases} H^{-1} = a^* \eta^2 = 2t \\[2mm] \rho_r = 3p_r = \dfrac{a^{*2}}{(8\pi a^4/3)} \\[2mm] \quad = \dfrac{3}{32\pi t^2} = \dfrac{3H^2}{8\pi} \end{cases}$

a tool to tell how many kinds of radiation contributed to ρ_r in the early universe; or, in other words, to learn about field physics from observational cosmology. Express the density of radiation in the form

$$\rho_r(\text{cm}^{-2}) = \frac{G}{c^4} \rho_{r,\text{conv}}(\text{erg/cm}^3) = \frac{Gf\pi^2}{c^4 120} \frac{(kT)^4}{\hbar^3 c^3}. \qquad (27.54)$$

It would be surprising if electromagnetism made the sole contribution to the radiation density, since the following additional mechanisms are available to sop up thermal energy from a violently radiating source:

electromagnetic radiation (already considered), $\qquad f_{em} = 8$;

gravitational black body radiation, $\qquad f_g = 8$;

neutrino plus antineutrino radiation of the electron-neutrino type [its contribution depends on the chemical potential of the neutrinos, on which see Brill and Wheeler (1957); a zero value is assumed here for that potential], $\qquad f_{e\nu} = 7$;

neutrino plus antineutrino radiation of the muon-neutrino type [with the same assumptions as for ν_e's], $\qquad f_{\mu\nu} = 7$;

pairs of positive and negative electrons produced out of the vacuum when temperatures are of the order of $T = mc^2/k = 0.59 \times 10^{10}$ K and higher, evaluated in the approximation in which these particles are treated as overwhelmingly more numerous than the unpaired electrons that one sees today, $\qquad f_{e^+ e^-} = 14$;

other particles such as mesons created out of the vacuum when temperatures are two orders of magnitude higher ($\sim 10^{12}$ K), and baryon-antibaryon pairs created out of the vacuum when temperatures are of the order of $\sim 10^{13}$ K and higher, $\qquad f_{\mu^+ \mu^-}, f_\pi, \dots;$

sum of these f-values, $\qquad \overline{ f. \qquad (27.55)}$

As the expansion proceeds and temperatures drop below 10^{13} K, then 10^{12}, then 10^{10}, the various particle pairs presumably annihilate and disappear [see, however, Alfvén and Klein (1962), Alfvén (1971), Klein (1971), and Omnes (1969)]. One is left with the radiations of zero rest mass, and only these radiations, contributing to the specific heat of the vacuum. At the phases of baryon-antibaryon and electron-positron annihilation, the thermal gravitational radiation present has already effectively decoupled itself from the matter, according to all current estimates. Therefore the energy set free by annihilation of matter and antimatter is expected to pour at first into the other two carriers of energy: neutrinos and electromagnetic radiation. However, the neutrinos also decouple early (after baryon-antibaryon annihilation; before full electron-positron annihilation), because the mean free path for neutrinos

rises rapidly with expansion. The energy of the subsequent annihilations goes almost exclusively into electromagnetic radiation. Thus the temperatures of the three radiations at the present time are expected to stand in the order

$$T_{\text{em}} > T_{\nu} > T_{g}. \tag{27.56}$$

T_{em} has been measured to be 2.7 K; T_{ν} is calculated to be $(4/11)^{1/3} T_{\text{em}} = 1.9$ K, and T_{g} has been calculated to be 1.5 K [Matzner (1968)] in a model where gravitons decouple during an early, quark-dominated era.

Decoupled radiation, once in a Planck spectrum, remains in a Planck spectrum (see Box 29.2). Expansion leaves constant the product $\rho_{r,\text{decoupled}}\, a^4$ or the product $T^4_{r,\text{decoupled}}\, a^4$. Compare the temperature of this particular radiation now to the temperature of the same radiation at any chosen fiducial time t_{fid} after its era of decoupling. Find

$$T_{r,\text{fid}} = \frac{a_{\text{now}}}{a_{\text{fid}}} T_{r,\text{now}} = (1 + z)T_{r,\text{now}}. \tag{27.57}$$

Here z represents the red shift of any "tracer" spectral line, given off at the fiducial time, and observed today, relative to the standard wavelength of the same transition as observed in the laboratory.

If the three radiations could be catalyzed into thermodynamic equilibrium, then all radiations could be treated on the same footing during the radiation-dominated era of cosmology. Their individual f values could be added directly to give $f = 8 + 8 + 7 + 7 = 30$. Temperature and time would then be connected by the formula

$$(T/10^{10}\ K)^2(t/1\ \text{sec}) = 1.19. \tag{27.58a}$$

This formula together with (27.57) implies the relation

$$\left[\left(\frac{T_{r,\text{now}}}{10^{10}\ K}\right)(1 + z)\right]^2\left(\frac{t_{\text{fid}}}{1\ \text{sec}}\right) = 1.19. \tag{27.58b}$$

This relation concerns two radiations: (1) the actual electromagnetic radiation with Planck spectrum (a continuum); and (2) the redshift and time of emission of a "tracer radiation" (a line spectrum). A measured departure from this relation could serve as one potential (indirect) indication that, in accordance with standard theory, neutrinos and gravitational radiation today are cooler than electromagnetic radiation.

Turn now from the radiation-dominated era of cosmology to the matter-dominated era. Numbers sometimes elicit more response from the imagination than formulas. Therefore idealize to a matter-dominated cosmology, and for the moment arbitrarily adopt 20×10^9 yr and 10×10^9 yr as Hubble time and actual time, respectively, back to the start of the expansion. It is certain that future work will show both numbers to require revision, but probably not by more than a factor 2, in the opinion of observational cosmologists. Since any judgment on the best numbers is subject

(3) later era, when matter dominates

to uncertainty, one can pick the numbers to be simple as well as reasonable. From Box 27.3, one then deduces the present value of the arc parameter time η,

$$\frac{20 \times 10^9 \text{ lyr}}{10 \times 10^9 \text{ lyr}} = \frac{H^{-1}}{t} = \frac{\dfrac{a_{\max}}{2} \dfrac{(1 - \cos \eta)^2}{\sin \eta}}{\dfrac{a_{\max}}{2}(\eta - \sin \eta)} \tag{27.59}$$

or

$$\eta = 1.975 \text{ (or } 113.2°) \tag{27.60}$$

(arc traveled by a photon on the 3-sphere from the start of the expansion to today.) This fixed, all other numbers emerge as shown in Box 27.4.

Box 27.4 A TYPICAL COSMOLOGICAL MODEL COMPATIBLE WITH ASTRONOMICAL OBSERVATIONS AND WITH EINSTEIN'S CONCEPTION OF COSMOLOGY ($\Lambda = 0$; Universe Closed)

Radius at phase of maximum expansion,	18.94×10^9 lyr;
Time from start to maximum,	29.76×10^9 yr;
Time from start to final recontraction,	59.52×10^9 yr;
Time from start to today (adopted value),	10×10^9 yr;
Radius today,	13.19×10^9 lyr;
Hubble time today (adopted value),	20×10^9 yr;
Hubble expansion rate today,	49.0 km/sec Megaparsec;
Deceleration parameter today, q_o [equation (29.1b)]	1.7
Density today $(3/8\pi a_o{}^2) + (3H_o^2/8\pi)$,	$(7.67 + 3.33) \times 10^{-58}$ cm^{-2} $= 11.00 \times 10^{-58}$ cm^{-2} or 14.8×10^{-30} g/cm^3;
Volume today, $2\pi^2 a_o{}^3$,	38.3×10^{84} cm^3;
Density at maximum $(3/8\pi a^2) + (3H^2/8\pi)$,	$(3.70 + 0.00) \times 10^{-58}$ cm^{-2} $= 5.0 \times 10^{-30}$ g/cm^3;
Volume at maximum,	114×10^{84} cm^3;
Rate of increase of radius today,	13.19×10^9 lyr/20×10^9 yr $= 0.66$ lyr/yr;
Rate of increase of volume today,	1.82×10^{68} cm^3/sec;
Amount of matter,	5.68×10^{56} g;
Equivalent number of solar masses,	2.86×10^{23};
Equivalent number of baryons,	3.39×10^{80}.
Fraction visible today	0.74

It must be emphasized that these numbers do not deserve the title of "canonical," however convenient that adjective may be for describing them; they can at most be called illustrative.

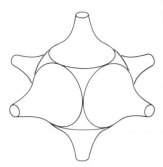

Figure 27.3.
Many Schwarzschild zones are fitted together to make a closed universe. This universe is dynamic because a test particle at the interface between two zones rises up against the gravitational attraction of each and falls back under the gravitational attraction of each. Therefore the two centers themselves have to move apart and move back together again. The same being true for all other pairs of centers, it follows that the lattice universe itself expands and recontracts, even though each Schwarzschild geometry individually is viewed as static. This diagram is taken from Lindquist and Wheeler (1957).

If every five seconds a volume of space is added to the universe, a volume equivalent to a cube 10^5 lyr ($= 0.95 \times 10^{23}$ cm) on an edge, about equal to the volume occupied by the Milky Way, where does that volume make its entry? Rather than look for an answer, one had better reexamine the question. Space is not like water. The outpouring of fresh water beneath the ocean at the Jesuit Spring off Mount Desert Island can be detected and measured by surrounding the site with flowmeters. There is no such thing as a flowmeter to tell "how fast space is streaming past." The very idea that "space flows" is mistaken. There is no way to define a flow of space, not least because there is no way to measure a flow of space. Water, yes; space, no. Life is very different for the flowmeter, according as it is stationary or moving with respect to the water. For a particle in empty space, however, physics is indistinguishable regardless of whether the particle is at rest or moves at high velocity relative to some chosen inertial frame. To try to pinpoint where those cubic kilometers of space get born is a mistaken idea, because it is a meaningless idea.

One can get a fresh perspective on what is going on in expansion and recontraction by turning from a homogeneous isotropic closed universe to a Schwarzschild lattice closed universe. [Lindquist and Wheeler (1957)]. In the former case, the mass is idealized as distributed uniformly. In the latter, the mass is concentrated into 120 identical Schwarzschild black holes. Each is located at the center of its own cell, of dodecahedral shape, bounded by 12 faces, each approximately a pentagon; and space is empty. The dynamics is easy to analyze in the approximation in which each lattice cell is idealized as spherical, a type of treatment long familiar in solid-state physics as the "Wigner-Seitz approximation" (references in Lindquist and Wheeler). In this approximation, the geometry inside each lattice cell is treated as having exactly the Schwarzschild character (Figure 27.3); a test particle placed midway between black hole A and black hole B rises against the attraction of each, and ultimately falls back toward each, according to the law developed in Chapter 25 [equation (25.28) with a shift of π in the starting point for defining η],

$$r = \frac{R}{2}(1 - \cos\eta),$$

$$\tau = \frac{R}{2}\left(\frac{R}{2M}\right)^{1/2}(\eta - \sin\eta). \tag{27.61}$$

Accordingly, the two masses in question must fall toward each other; and so it is with all the masses. One comes out in this way with the conclusion that the lattice

(4) "Where is the new space created during expansion?"—a meaningless question

universe follows the same law of expansion and recontraction as the Friedmann universe to an accuracy of better than 4 per cent [Lindquist and Wheeler; Wheeler (1964a), pp. 370–381]. Now ask again the same meaningless question about where the cubic kilometers of space pour into the universe while it is expanding, and where they pour out while it is recontracting. Receive a fuller picture why the question is meaningless. Surrounding each center of mass, the geometry is and remains the Schwarzschild geometry (until eventually the black holes come so close together that they coalesce). The situation inside each cell is therefore static. Moreover, the interface between cell and cell is defined in imagination by a sprinkling of test particles so light that they have no influence on the geometry or its dynamics. The matchup between the geometry in one cell and the next is smooth ("tangency between the two geometries"). There is nothing abnormal whatsoever in the space-time on and near the interface. One has as little right to say those cubic kilometers are "created" here as anywhere else. To speak of the "creation" of space is a bad way of speaking, and the original question is a bad question. The right way of speaking is to speak of a dynamic geometry. So much for one question!

In charting the dynamics of the geometry of a Friedmann universe, one often finds that it simplifies things to take as space coordinate the hyperpolar angle χ, measured from some chosen world line (moving with the "cosmological fluid") as standard of reference; and to take as time coordinate the arc-parameter measure of time, η, as illustrated in Figure 27.4.

(5) causal isolation of various regions of universe from each other

Inspection of the (χ, η)-diagram makes it clear that photons emitted from matter at one point cannot reach, in a limited time, any matter except that which is located in a limited fraction of the 3-sphere. In a short time t, according to Box 27.3, a photon can cover an arc distance on the 3-sphere equal only to $\eta = (2t/a^*)^{1/2}$. Moreover, what is true of photons is true of other fields, forces, pressures, energies and influences: they cannot reach beyond this limit. Evidently the 3-sphere at time t is divided into a number of "zones,"

$$
N = \begin{pmatrix} \text{number of} \\ \text{"zones"} \end{pmatrix} = \frac{\begin{pmatrix} \text{hyperspherical solid} \\ \text{angle of entire 3-sphere} \end{pmatrix}}{\begin{pmatrix} \text{hyperspherical solid} \\ \text{angle of one zone} \end{pmatrix}} = \frac{2\pi^2}{4\pi\chi^3/3} = \frac{3\pi}{2^{5/2}}\left(\frac{a^*}{t}\right)^{3/2},
$$
(27.62)

effectively decoupled one from the other. As time goes on, there are fewer separate zones, and ultimately every particle has been subjected to influences from every other particle in the model universe.

EXERCISES

Exercise 27.8.　MATTER-DOMINATED AND RADIATION-DOMINATED REGIMES OF FRIEDMANN COSMOLOGY

Derive the results listed in the last two columns of Box 27.3, except for the focusing properties of the curved space.

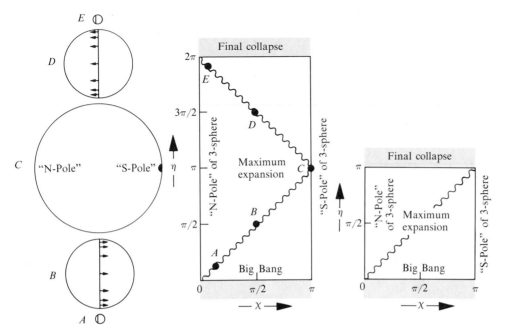

Figure 27.4.

Use of "arc parameter" η as a time coordinate and hyperpolar angle χ as a space coordinate to describe travel of a photon ($\pm 45°$ line) in a Friedmann universe that is matter-dominated (center) or radiation-dominated (right). The burst of photons is emitted from the "N-pole" of the 3-sphere at a time very little after the big bang, and the locus of the cloud of photons at subsequent stages of the expansion and recontraction is indicated by sections of the 3-sphere in the diagrams at the left. The matter-dominated Friedmann universe appears to be a reasonable model for the physical universe, except when its dimensions have fallen to the order of one ten-thousandth of those at maximum expansion or less ("radiation regime").

Exercise 27.9. TRANSITION FROM RADIATION-DOMINATED REGIME TO MATTER-DOMINATED REGIME

Including both the radiation and the matter terms in equation (27.51), restate the equation in terms of the arc parameter η (with $d\eta = dt/a$) as independent variable, and integrate to find

$$a = (a_{max}/2) - [(a_{max}/2)^2 + a^{*2}]^{1/2} \cos(\eta + \delta), \tag{27.63}$$

$$t = (a_{max}/2)\eta - [(a_{max}/2)^2 + a^{*2}]^{1/2}[\sin(\eta + \delta) - \sin \delta], \tag{27.64}$$

where

$$\delta = \arctan[a^*/(a_{max}/2)]. \tag{27.65}$$

(a) Verify that under appropriate conditions these expressions reduce at early times to a "circle" relation between radius and time and to a "cycloid" relation later.

(b) Assign to a^{*2} the value $a_o a_{max}/10{,}000$ (why?) and construct curves for the dimensionless measures of density,

$$\log_{10}\left[(8\pi a_{max}^2/3)\left\{\begin{matrix}\rho_m \\ \rho_r \\ \rho_m + \rho_r\end{matrix}\right\}\right],$$

as a function of the dimensionless measure of time,

$$\log_{10}(t/a_{max}).$$

What conclusions emerge from inspecting the logarithmic slope of these curves?

Exercise 27.10. THE EXPANDING AND RECONTRACTING SPHERICAL WAVE FRONT

An explosion takes place at the "N-pole" of the matter-dominated Friedmann model universe at the value of the "arc parameter time" $\eta = \pi/3$, when the radius of the universe has reached half its peak value. The photons from the explosion race out on a spherical wave front. Through what fraction of the "cosmological fluid" has this wave front penetrated at that instant when the wave front has its largest proper surface area?

§27.11. HOMOGENEOUS ISOTROPIC MODEL UNIVERSES THAT VIOLATE EINSTEIN'S CONCEPTION OF COSMOLOGY

Open Friedmann universe with $\Lambda = 0$:

(1) expansion factor as function of time

(2) early stage—same as for closed universe

It violates Einstein's conception of cosmology (Box 27.1)—though not the equations of his theory—to replace the closed 3-sphere of radius a by the open hyperboloidal geometry of equation (27.22) with the same scale length a. Even so, the results of Box 27.3 continue to apply in the two limiting regimes of matter-dominated and radiation-dominated dynamics when the following changes are made. (1) Change the constant -1 on the righthand side of the analog of a "Newtonian energy equation" to $+1$, thus going over from a bound system (maximum expansion) to an open system (forever expanding). (2) Replace $(1 - \cos\eta)$ by $(\cosh\eta - 1)$, $\sin\eta$ by $\sinh\eta$, $\cos\eta$ by $\cosh\eta$, and $(\eta - \sin\eta)$ by $(\sinh\eta - \eta)$. (3) The range of the "arc parameter" η now extends from 0 to ∞, and the curve relating "radius" a to time t changes from cycloid or circle to an ever-rising curve. (4) The listed inequalities on the Hubble time (as related to the actual time of expansion) and on the density (as related to $3H_o^2/8\pi$) no longer hold. (5) The formulas given in Box 27.3 for conditions at early times continue to hold, for a simple reason: at early times the curvature of spacetime "in the direction of increasing time" [the extrinsic curvature $(6/a^2)(da/dt)^2$ as it appears in Box 27.1, equation (2)] is overwhelmingly more important than the curvature within any hypersurface of homogeneity, $\pm 6/a^2$ (the intrinsic curvature); therefore it makes no detectable difference at early times whether the sign is plus or minus, whether the space is closed or open, or whether the geometry of space is spherical or hyperboloidal.

Why doesn't it make a difference? Not why mathematically, but why physically, doesn't it make a difference in early days whether the space is open or closed?

Because photons, signals, pressures, forces, and energies cannot get far enough to "smell out" the difference between closure and openness. The "zones of influence" of (27.62) are too small for any one by itself to sense or to respond significantly to any difference between a negative space curvature $-6/a^2$ and a positive space curvature $+6/a^2$. Therefore the simple power-law time-dependence of the density of the mass-energy of radiation given in Box 27.3 for a closed universe holds equally well in the earliest days of a radiation-dominated, open, isotropic model universe; thus,

$$\rho_r = 3/32\pi t^2. \tag{27.66}$$

Only at a later stage of the expansion, when the "extrinsic curvature" term [equation (2), Box 27.1], $(6/a^2)(da/dt)^2$ (initially varying as $1.5t^{-2}$, according to Box 27.3) has fallen to a value of the same order of magnitude as the "intrinsic curvature" term $\pm 6/a^2$ (initially varying as $\pm 3a^{*-1}t^{-1}$), does the sign of the intrinsic curvature begin to matter. Only then do the differences in rate of expansion begin to show up that distinguish the open model universe from the closed one.

The open model goes on expanding forever. Therefore the density of mass-energy, whether matter-dominated and proportional to a_{max}/a^3, or radiation-dominated and proportional to a^{*2}/a^4, or some combination of the two, (1) ultimately falls to a level that is negligible in comparison with the intrinsic curvature, $-6/a^2$, and (2) thereafter can be neglected. Under these circumstances, the only term left to balance the intrinsic curvature is the extrinsic curvature. The important component of the field equation (after removal from all terms of a common factor 3) now reads

(3) late stage—expansion forever

$$\frac{1}{a^2}\left(\frac{da}{dt}\right)^2 - \frac{1}{a^2} = 0. \tag{27.67}$$

For a closed universe, the two terms (one sixth the extrinsic curvature and one sixth the intrinsic curvature) have the same sign, and any equation like (27.67) leads to an impossibility. Here, however, rather than impossibility, one has the remarkably simple solution

$$a = t, \tag{27.68}$$

and the corresponding metric

$$ds^2 = -dt^2 + t^2[d\chi^2 + \sinh^2\chi(d\theta^2 + \sin^2\theta\, d\phi^2)]. \tag{27.69}$$

Write

$$r = t \sinh \chi,$$
$$t_{new} = t \cosh \chi, \tag{27.70}$$

and find that (27.69), solution as it is of Einstein's empty-space field equation, is identical with the Lorentz-Minkowski metric of flat spacetime,

$$ds^2 = -dt^2_{new} + dr^2 + r^2(d\theta^2 + \sin^2\theta\, d\phi^2) \tag{27.71}$$

(see Box 27.2C). This geometry had acquired the flavor of an expanding universe

because the cosmological fluid, too thinly spread to influence the dynamics of the geometry, and serving only to provide marker points, was flying out in all directions [for a fuller discussion of this "expanding Minkowski universe," see, for example, Chapter 16 of Robertson and Noonan (1968)]. The typical spacelike hypersurface of homogeneity looks to have a curved 3-space geometry, and does have a curved geometry (instrinsic curvature), because the slice (27.70) through flat spacetime is itself curved (extrinsic curvature).

Turn now to a second violation of Einstein's conception of cosmology: a cosmological term in the field equation (27.39),

$$\left(\frac{da/dt}{a}\right)^2 + \frac{k}{a^2} - \frac{\Lambda}{3} = \frac{8\pi}{3}\rho(a) = \frac{8\pi\rho_{mo}a_o^3/3}{a^3} + \frac{8\pi\rho_{ro}a_o^4/3}{a^4}. \qquad (27.72)$$

In analyzing the implications of this broadened equation, turn attention from the "radius" $a(t)$ itself, which was the focus of interest in the previous section, §27.10, on Friedmann cosmology. Recognize that present measurements have not yet provided a good, direct handle on the absolute dimension $a(t)$ of the universe. However, they do give good figures for the redshift z and therefore for the ratio between a at the time of emission and $a = a_o$ now,

$$a_o/a = 1 + z \qquad (27.73)$$

For any comparison with observations designed to fix limits on k (Einstein value: $k = +1$) and on Λ (expected to be zero), it is therefore appropriate to rewrite the foregoing equations so that they refer as much as possible only to ratios. Thus one rephrases (27.72) as the "generalized Friedmann equation,"

$$\left[\frac{d}{dt}\left(\frac{a(t)}{a_o}\right)\right]^2 + V(a/a_o) = -\frac{k}{a_o^2} \equiv -K_o. \qquad (27.74)$$

Here the quantity

$$V(a/a_o) \equiv -\frac{8\pi}{3}\left[\rho_{mo}\left(\frac{a_o}{a}\right) + \rho_{ro}\left(\frac{a_o}{a}\right)^2\right] - \frac{1}{3}\Lambda\left(\frac{a}{a_o}\right)^2 \qquad (27.75)$$

acts as an "effective potential" for the dynamics of the expansion. The constant term K_o represents one sixth of the intrinsic curvature of the model universe today. Its negative, $-K_o$, plays the role of an "effective energy" in the generalized Friedmann equation (Box 27.5). All the qualitative features of the cosmology can be read off from the curve for the effective potential as a function of (a/a_o) and from the value of K_o.

For a quantitative analysis, the log-log diagram of Figure 27.5 is often more useful than the straight linear plot of V against (a/a_o) of Box 27.5.

All the limiting features shown in the varied types of cosmology have been encountered before in the analysis of the elementary Friedmann cosmology (big bang out of a configuration of infinite compaction; reaching a maximum expansion at a turning point, or continued expansion to a Minkowski universe; recollapse to

infinite density) or lend themselves to simple visualization (static but unstable Einstein universe; "hesitation" model; "turnaround" model), except for the even more rapid expansion that occurs when Λ is positive and the dimension a has surpassed a certain critical value. In this expansion, a eventually increases as $\exp[(\Lambda/3)^{1/2}t]$ irrespective of the openness or closure of the universe ($k = 0, \pm1$). This expansion dominates every other feature of the cosmology. Therefore, in discussing it, it is appropriate to suppress every other feature of the cosmology, take the density of matter to be negligible, and take $k = 0$ (hypersurfaces of homogeneity endowed with flat 3-space geometry). In this limit, one has the following empty-space solution of Einstein's field equation with cosmological constant:

(3) de Sitter universe

$$ds^2 = -dt^2 + a_o^2 e^{2(\Lambda/3)^{1/2}t}[d\chi^2 + \chi^2(d\theta^2 + \sin^2\theta\, d\phi^2)]. \tag{27.76}$$

This "de Sitter universe" [de Sitter (1917a,b)] may be regarded as a four-dimensional surface,

$$-(z^0)^2 + (z^1)^2 + (z^2)^2 + (z^3)^2 + (z^4)^2 = 3/\Lambda, \tag{27.77}$$

in a five-dimensional space endowed with the metric

$$(ds)^2 = -(dz^0)^2 + (dz^1)^2 + (dz^2)^2 + (dz^3)^2 + (dz^4)^2. \tag{27.78}$$

The correctness of this description may be checked directly by making the substitutions

$$z^0 = (3/\Lambda)^{1/2} \sinh[(\Lambda/3)^{1/2}t] + \frac{1}{2}(\Lambda/3)^{1/2}e^{(\Lambda/3)^{1/2}t}a_o^2\chi^2,$$

$$z^4 = (3/\Lambda)^{1/2} \cosh[(\Lambda/3)^{1/2}t] - \frac{1}{2}(\Lambda/3)^{1/2}e^{(\Lambda/3)^{1/2}t}a_o^2\chi^2,$$

$$z^1 = a_o e^{(\Lambda/3)^{1/2}t}\chi \sin\theta \cos\phi,$$

$$z^2 = a_o e^{(\Lambda/3)^{1/2}t}\chi \sin\theta \sin\phi,$$

$$z^3 = a_o e^{(\Lambda/3)^{1/2}t}\chi \cos\theta.$$

(27.79)

Because of its beautiful group-theoretical properties and invariance with respect to $5 \times 4/2 = 10$ independent rotations, the de Sitter geometry has been the subject of scores of mathematical investigations. The physical implications of a cosmology following the de Sitter model are described for example by Robertson and Noonan (1968, especially their §16.2). The de Sitter model is the only model obeying Einstein's equations (with $\Lambda \neq 0$) which (1) continually expands and (2) looks the same to any observer who moves with the cosmological fluid, regardless of his position or his time. Any model of the universe satisfying condition (2) is said to obey the so-called "perfect cosmological principle." This phrase arose in the past in studying models that lie outside the framework of general relativity, models in which matter is envisaged as continuously being created, and to which the name of "steady-state universe" has been given. Any such model has been abandoned by most investigators today, not least because it gives no satisfactory account of the 2.7 K background radiation.

Other non-Einsteinian cosmologies:

(1) steady-state model

(continued on page 748)

**Box 27.5 EFFECT OF VALUE OF COSMOLOGICAL CONSTANT AND
OF INTRINSIC CURVATURE OF MODEL UNIVERSE "TODAY"
ON THE PREDICTED COURSE OF COSMOLOGY**

The "effective potential" V in the generalized Friedmann equation (27.74) is represented schematically here as a function of the expansion ratio a/a_o. The diagram illustrates the influence on the cosmology of (1) the cosmological constant Λ (determines the behavior of the effective potential at large values of a/a_o; see dashed curves) and (2) the value adopted for $K_o =$ (one sixth of the intrinsic curvature of 3-space at the present epoch). The value of the quantity $-K_o$ determines the "effective energy level" and is shown in the diagram as a horizontal line. The difference between this horizontal line and the effective potential determines $(a_o^{-1}\,da/dt)^2$. Regions where this difference is negative are inaccessible. From the diagram one can read off the histories of 3-space on the facing page.

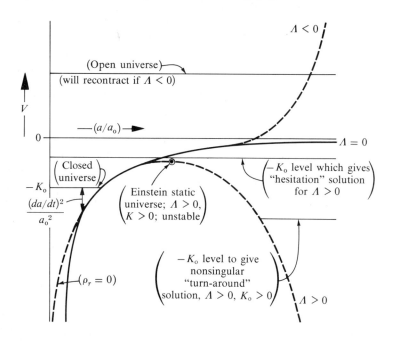

The diagram is schematic, not quantitative. Representative values might be $\Lambda_{\mathrm{conv}} = 0$ or $\pm 3 \times 10^{-28}$ g/cm³; $\rho_{mo,\mathrm{conv}} = 10^{-30}$ g/cm³ or $\rho_{mo,\mathrm{conv}} = 10^{-28}$ g/cm³; and $(a_o^{-1}\,da/dt)^2 = H_o^2 = (1/20 \times 10^9 \text{ yr})^2$ or 3.8×10^{-29} g/cm³. At small values of a/a_o the cosmological term $-(\Lambda/3)(a/a_o)^2$ is negligible. Not negligible at small values of (a/a_o) is the difference between a model universe curved only by the density of matter (the dashed curve in the diagram) and one curved also by a density of radiation (the full curve). The different dependence of "radius" and density on time at early times in these two cases of a matter-dominated cosmology and a radiation-dominated cosmology is spelled out in the last part of Box 27.3, giving in the one case $\rho = 1/6\pi t^2$ and in the other $\rho = 3/32\pi t^2$.

Intrinsic curvature of space today	Λ	Cosmology
Hyperbolic; K_o negative	negative	Universe starts in a condition of infinite density. It expands to a maximum extent (or minimum density) governed by the value of Λ. It then recontracts at an ever increasing rate to a condition of infinite density.
Hyperbolic; K_o negative	zero	Universe starts in a condition of infinite density. It expands. Ultimately the rate of expansion reaches a steady rate, $da/dt = 1$. The 4-geometry is Minkowski flat spacetime. Only the curvature of the spacelike slices taken through this flat 4-geometry gives the 3-geometry its hyperbolic character [see equation (27.70)].
Closed; K_o positive	zero	Standard Friedmann cosmology: expansion from infinite compaction to a finite radius and recontraction and collapse.
Closed; K_o positive	negative	Qualitatively same as foregoing. Quantitatively a slightly smaller radius at the phase of maximum expansion and a slightly shorter time from start to end.
Closed; K_o positive	Λ more positive than a certain critical value: $\Lambda > \Lambda_{\text{crit}}$	"Summit" of "effective potential" is reduced to a value slightly less than $-K_o$. The closed universe once again starts its expansion from a condition of infinite compaction. This expansion once again slows down as the expansion proceeds and then looks almost as if it is going to stop ("moment of hesitation"). However, the representative point slowly passes over the summit of the potential. Thereafter the expansion gathers more and more speed. It eventually follows the exponential law $$a = \text{constant} \times \exp\left[(\Lambda/3)^{1/2}t\right].$$
Closed; K_o positive	Λ positive and exactly equal to the critical value, $\Lambda = \Lambda_{\text{crit}}$, that puts the "summit of the potential" into coincidence with $-K_o$	Situation similar to that of a pencil with its tip dug into the table and provided with just enough energy to rise asymptotically in infinite time to the vertical position. Universe starts from a compact configuration and expanding approaches a certain radius ("Einstein radius", a_E) according to a law of the form $$a = a_E - \text{constant} \times \exp(-\alpha t).$$ Or (Einstein's original proposal, when he thought that the universe is static, and added the "cosmological term," against his will, to the field equation to permit a static universe) the representative point sits forever at the "summit of the effective potential" (Einstein universe). Aside from contradicting present-day evidence on expansion, this configuration has the same instability as does a pencil trying to stand on its tip. The least disturbance will cause it to "fall" either way, toward collapse or toward accelerating expansion, in the expansion case ultimately approaching the law $$a = \text{constant} \times \exp\left[(\Lambda/3)^{1/2}t\right].$$
Closed; K_o positive	Λ less positive than the critical value: $0 < \Lambda < \Lambda_{\text{crit}}(K_o)$	Motion on the large a side of the "potential barrier." Far back in the past the model universe has enormous dimensions, but is also contracting with enormous rapidity, in approximate accord with the formula $$a = \text{constant} \times \exp\left[-(\Lambda/3)^{1/2}t\right].$$ The radius a reaches a minimum value and thereafter the universe reexpands ("turn-around solution"), ultimately approaching the asymptotic law $$a = \text{constant} \times \exp\left[(\Lambda/3)^{1/2}t\right].$$

Figure 27.5. (*facing page*)

Log-log plot of the effective potential $V(a)$ of equation (27.75) and Box 27.5 as it enters the generalized Friedmann equation

$$\left[\frac{d}{dt}\left(\frac{a(t)}{a_o}\right)\right]^2 + V\left(\frac{a}{a_o}\right) = -\frac{k}{a_o^2} = -K_o.$$

Horizontally is given the expansion ratio referred to $(a/a_o)_{\text{today}} = 1$ as standard of reference. Vertically is given the value of $V(a/a_o)$ in the geometric units of cm^{-2}. The supplementary scale at the right translates to $-(c^2/G)(3/8\pi)V(a/a_o)$ as an equivalent density, expressed in g/cm^3. The contribution of radiation density to the effective potential is indicated by the wavy line in the diagram. It is normalized to a value of the radiation density today of $\rho_{ro} = 10^{-33}$ g/cm^3 and has a logarithmic slope of two. The contribution of matter density to the "effective potential" has a logarithmic slope of unity. Two choices are made for it, corresponding to a density of matter today of $\rho_{mo} = 10^{-30}$ g/cm^3 and $\rho_{mo} = 10^{-28}$ g/cm^3 (dashed lines in the diagram). The total effective potential in the two cases is also indicated in the diagram: a heavy line for the case $\rho_{mo} = 10^{-30}$ g/cm^3 (no cosmological term included) and a light line for the case $\rho_{mo} = 10^{-28}$ g/cm^3. In this second case, a cosmological term is included, with the cosmological constant given by $(3/8\pi)(\Lambda/3) = 10^{-29}$ g/cm^3. The line describing the contribution of this term has a negative slope of magnitude two (dashed line). The horizontal or "level line" is drawn for a value of the Hubble expansion rate today, H_o, equal to $1/(20 \times 10^9$ years). The vertical separation on the log plot between the potential curve and the level line gives the ratio $-V/H_o^2$. This ratio as evaluated at any time t has the value $\dot{a}^2(t)/\dot{a}_o^2 + K_o H_o^{-2}$, where $\dot{a} \equiv da/dt$. As evaluated "today" $(a/a_o = 1)$ this ratio has the value $1 + K_o H_o^{-2}$. Knowing the Hubble expansion rate H_o^2 today, and knowing (or trying a certain set of parameters for) the potential curve, one can therefore deduce from the spread between the two the value of $1 + K_o H_o^{-2}$, hence the value of $K_o H_o^{-2}$, hence the present value, K_o, of the curvature factor. As an example, consider the case of the low-density universe (heavy line) and read off "today's" value, $1 + K_o H_o^{-2} = 0.223$. From this follows $K_o = -0.777 \, H_o^2$ (open or hyperbolic universe), hence $k = -1$ and $a_o = (k/K_o)^{1/2} = (1/0.777)^{1/2} \, 20 \times 10^9$ yr $= 22.7 \times 10^9$ yr. For the high-density model universe, with $\rho_{mo} = 10^{-28}$ g/cm^3, one similarly finds $1 + K_o H_o^{-2} = 24.5$, hence $K_o = +23.5 H_o^2$, hence $k = +1$ (closed universe) and $a_o = (k/K_o)^{1/2} = (1/23.5)^{1/2} \, 20 \times 10^9$ yr $= 4.12 \times 10^9$ yr. Expansion stops, if and when it stops, at that stage when the ratio $-V/H_o^2$ between the potential and the level line, or $\dot{a}^2(t)/\dot{a}_o^2 + K_o H_o^{-2}$, falls from its "present value" of $1 + K_o H_o^{-2}$ to $0 + K_o H_o^{-2}$; that is, from 0.223 to -0.777 in the one case, and from 24.5 to 23.5 in the other case. This log-log plot should be replaced by the linear plot of Box 27.5 when $\Lambda < 0$.

(2) hierarchic model

However great a departure it is from Einstein's concept of cosmology to give any heed to a cosmological constant or an open universe, it is a still greater departure to contemplate a "hierarchic model" of the universe, in which clusters of galaxies, and clusters of clusters of galaxies, in this part of the universe are envisaged to grade off in density with distance, with space at great distances becoming asymptotically flat [Alfvén and Klein (1962), Alfvén (1971), Klein (1971), Moritz (1969), de Vaucouleurs (1971), Steigman (1971)]. The viewpoint adopted here is expressed by Oskar Klein in these words, "Einstein's cosmology was adapted to the discovery by Hubble that the observed part is expanding; the so-called cosmological postulate has been used as a kind of an axiomatic background which, when analyzed, makes it appear that this expansion is shared by a very big, but still bounded system. This implies that our expanding metagalaxy is probably just one of a type of stellar objects in different phases of evolution, some expanding and some contracting."

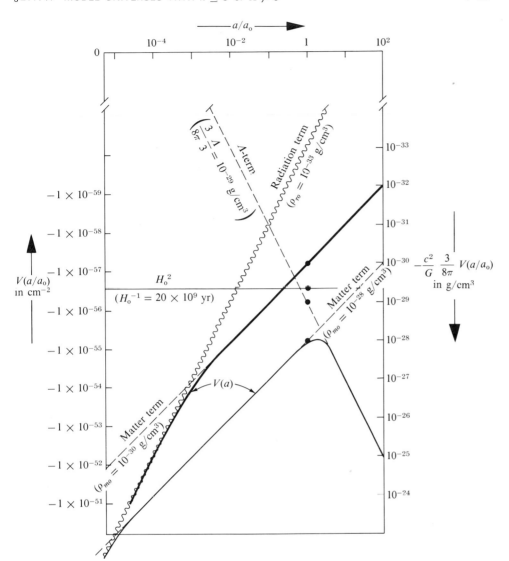

The contrast between the hierarchic cosmology and Einstein's cosmology [Einstein (1931) advocates a closed Friedmann cosmology] appears nowhere more strongly than here, that the one regards asymptotically flat spacetime as a requirement; the other, as an absurdity. "Only the genius of Riemann, solitary and uncomprehended," Einstein (1934) puts it, "had already won its way by the middle of the last century to a new conception of space, in which space was deprived of its rigidity, and in which its power to take part in physical events was recognized as possible." That statement epitomizes cosmology today.

But today's view of cosmology, as dominated by Einstein's boundary condition of closure ($k = +1$) and his belief in $\Lambda = 0$, need not be accepted on faith forever. Einstein's predictions are clear and definite. They expose themselves to destruction. Observational cosmology will ultimately confirm or destroy them, as decisively as it has already destroyed the 1920 belief in a static universe and the 1948 steady-state models (see Box 27.7 on the history of cosmology).

EXERCISES

Exercise 27.11. ON SEEING THE BACK OF ONE'S HEAD

Can a being at rest relative to the "cosmological fluid" ever see the back of his head by means of photons that travel all the way around a closed model universe that obeys the Friedmann cosmology and has a non-zero cosmological constant (see the entries in Box 27.3 for the case of a zero cosmological constant)?

Exercise 27.12. DO THE CONSERVATION LAWS FORBID THE PRODUCTION OF PARTICLE-ANTIPARTICLE PAIRS OUT OF EMPTY SPACE BY TIDAL GRAVITATION FORCES?

Find out what is wrong with the following argument: "The classical equations

$$G_{\alpha\beta} = 8\pi T_{\alpha\beta}$$

are not compatible with the production of pairs, since they lead to the identity $T_\alpha{}^\beta{}_{;\beta} \equiv 0$. Let the initial state be vacuum, and let $T_{\alpha\beta}$ and its derivative be equal to zero on the hypersurface $t = \text{const}$ or $t = -\infty$. It then follows from $T_\alpha{}^\beta{}_{;\beta} = 0$, that the vacuum is always conserved." [*Answer:* See Zel'dovich (1970, 1971, 1972). Also see §30.8.]

Exercise 27.13. TURN-AROUND UNIVERSE MODEL NEGLECTING MATTER DENSITY

If turn-around (minimum radius) occurs far to the right (large a) of the maximum of the potential $V(a)$ in equation (27.75), the matter terms will be negligible. Let $\rho_{mo} = \rho_{ro} = 0$. Then (what signs of k, Λ are needed for turn-around?), solve to show that $\Lambda = 3(a_{\min})^{-2}$, $H = (a_{\min})^{-1}\tanh(t/a_{\min})$ near turnaround ($t = 0$) and that the deceleration parameter $q \equiv -(1/H^2 a)(d^2a/dt^2)$ has the value

$$q = -a^2(a^2 - a_{\min}^2)^{-1} < -1.$$

Exercise 27.14. "HESITATION" UNIVERSE

Neglect radiation in equation (27.75) but assume K_o and Λ to be chosen so that the universe spent a very long time with $a(t)$ near a_h (a_h measures location of highest point of the barrier, or the size of the universe at which the universe is most sluggish). Choose $a_h = a_o/3$ to produce an abnormally great number of quasar redshifts near $z = 2$ [as Burbidge and Burbidge (1969a,b) believe to be the case, though their opinion is not shared by all observers]. Show (a) that the density of matter now would account for only 10 per cent of the value of $H_o^2 = (\dot{a}/a)^2_{\text{now}}$ in equation (27.75) ["missing matter", i.e., K_o and Λ terms, account for 90 per cent], (b) that $a_h \simeq 20^{1/2}H_o^{-1}$, and (c) that the deceleration parameter defined in the previous exercise, as evaluated "today", has the value $q_o = -13/10$.

Exercise 27.15. UNIVERSE OPAQUE TO BLACK-BODY RADIATION AT A NONSINGULAR PAST TURN-AROUND REQUIRES IMPOSSIBLE PARAMETERS

From a plot like that in Box 27.5, construct a model of the universe that contains 2.7 K black-body radiation at the present, but, with $k = +1$ and $\Lambda > 0$, had a past turn-around (minimum radius) at which the blackbody temperature reached 3,000 K where hydrogen would be ionized. Try to use values of H_o^{-1} and ρ_{mo} that are as little as possible smaller than presently accepted values.

Box 27.6 ALEXANDER ALEXANDROVITCH FRIEDMANN
St. Petersburg, June 17, 1888—Leningrad, September 15, 1925

Graduated from St. Petersburg University, 1909; doctorate, 1922; 1910, mathematical assistant in the Institute of Bridges and Roads; 1912, lecturer on differential calculus in the Institute of Mines; 1913, physicist in the Aerological Institute of Pavlov; dirigible ascent in preparation for observing eclipse of the sun of August 1914; volunteer in air corps on war front near Osovets, 1914; head of military air navigation service, 1916–1917; professor of mechanics at Perm University, 1918; St. Petersburg University, 1920; lectures in hydrodynamics, tensor analysis; author of books, *Experiments in the Hydromechanics of Compressible Liquids* and *The World as Space and Time,* and the path-breaking paper, *On the Curvature of Space,* 1922; a director of researches in the department of theoretical meteorology of the Main Geophysical Laboratory, Leningrad, and, from February 1925 until his death of typhoid fever seven months later, director of that Laboratory; with L. V. Keller "introduced the concept of coupling moments, i.e., mathematical expectation values for the products of pulsations of hydrodynamic elements at different points and at different instants . . . to elucidate the physical structure of turbulence" [condensed from Polubarinova-Kochina (1963), which also contains a bibliography of items by and about Friedmann].

Box 27.7 SOME STEPS IN COSMOLOGY ON THE WAY TO WIDER PERSPECTIVES AND FIRMER FOUNDATIONS [For general reference on the history of cosmology, see among others Munitz (1957), Nasr (1964), North (1965), Peebles (1971), Rindler (1969), and Sciama (1971); and especially see Peebles and Sciama for bibliographical references to modern developments listed below in abbreviated form.]

A. Before the Twentieth Century

Concepts of very early Indian cosmology [summarized by Zimmer (1946)]: "One thousand mahāyugas—4,320,000,000 years of human reckoning—constitute a single day of Brahmā, a single kalpa. . . . I have known the dreadful dissolution of the universe. I have seen all perish, again and again, at every cycle. At that terrible time, every single atom dissolves into the primal, pure waters of eternity, whence all originally arose."

 Plato, ca. 428 to ca. 348 B.C. [from the *Timaeus,* written late in his life, as translated by Cornford (1937)]: "The world [universe] has been fashioned on the model of that which is comprehensible by rational discourse and understanding and is always in the same state. . . . this world came to be . . . a living creature with soul and reason. . . . its maker did not make two worlds nor yet an indefinite number; but this Heaven has come to be and is and shall be hereafter one and unique. . . . he fashioned it complete and free from age and sickness. . . . he turned its shape rounded and spherical . . . It had no need of eyes, for nothing visible was left outside; nor of hearing, for there was nothing outside to be heard. . . . in order that Time might be brought into being, Sun and Moon and five other stars—'wanderers,' as they are called—were made to define and preserve the numbers of Time. . . . the generation of this universe was a mixed result of the combination of Necessity and Reason . . . we must also bring in the Errant Cause. . . . that which is to receive in itself all kinds [all forms] must be free from all characters [all form] For this reason, then, the mother and Receptacle of what has come to be visible and otherwise sensible must not be called earth or air or fire or water . . . but a nature invisible and characterless, all-receiving, partaking in some very puzzling way of the intelligible, and very hard to apprehend."

 Aristotle, 384–322 B.C. [from *On the Heavens,* as translated by Guthrie (1939)]: "Throughout all past time, according to the records handed down from generation to generation, we find no trace of change either in the whole of the outermost heaven or in any one of its proper parts. . . . the shape of the heaven must be spherical. . . . From these considerations [motion invariably in a straight line toward the center; regularity of rising and setting of stars; natural motion of earth toward the center of the universe] it is clear that the earth does not move, neither does it lie anywhere but at the center. . . . the earth . . . must have grown in the form of a sphere. This [shape of segments cut out of moon at time of eclipse of moon; and ability to see

in Egypt stars not visible in more northerly lands] proves both that the earth is spherical and that its periphery is not large . . . Mathematicians who try to calculate the circumference put it at 400,000 stades [1 stade = 600 Greek feet = 606 English feet; thus 24.24×10^7 ft/(6080.2 ft/nautical mile) = 39,900 nautical miles—the oldest recorded calculation of the earth's circumference, and reportedly known to Columbus—85 per cent more than the true circumference, $60 \times 360 = 21,600$ nautical miles]."

Aristotle [from the *Metaphysics,* as translated by Warrington (1956)]: "Euxodus [of Cnidos, 408–355 B.C.] supposed that the motion of the sun and moon involves, in each case, three spheres. . . . He further assumed that the motion of the planets involves, in each case, four spheres. . . . Calippus [of Cyzicus, flourished 330 B.C.] . . . considered that, in the light of observation, two more spheres should be added to the sun, two to the moon, and one more to each of the other planets."

Eratosthenes, 276–194 B.C. [a calculation attributed to him by Claudius Ptolemy, who observed at Alexandria from 127 to 141 or 151 A.D., in his *Almagest,* I, §12; see the translation by Taliaferro (1952)]:

(Maximum distance of moon from earth) = $(64\frac{1}{6})$ (radius of earth);

(Minimum distance of sun from earth) = (1,160) (radius of earth).

Abū 'Alī al-Husain ibn 'Abdallāh ibn Sīnā, otherwise known as **Avicenna,** 980–1037; physician, philosopher, codifier of Aristotle, and one of the most influential of those who preserved Greek learning and thereby made possible its transmission to mediaeval Europe [quoted in Nasr (1964), p. 225]: "Time is the measure of motion."

From the *Rasā'il,* a 51-treatise encyclopedia, sometimes known as the *Koran after the Koran,* of the **Ikhwān al-Safā'** or Brothers of Sincerity, whose main center was at Basra, Iraq, roughly A.D. 950–1000 [as quoted by Nasr (1964), p. 64; see p. 78 for a list of distances to the planets (in terms of Earth radii) taken from the *Rasā'il,* as well as sizes of planets and the motions of rotation of the various Ptolemaic carrier-spheres]: Space is "a form abstracted from matter and existing only in the consciousness."

Abū Raihān al-Bīrūnī, 973–1030, a scholar, but concerned also with experiment, observation, and measurement, who calculated the circumference of the Earth from measurements he made in India as 80,780,039 cubits (about 4 per cent larger than the value accepted today), and gave a table of distances to the planets [as quoted in Nasr (1964), pp. 120 and 130]: "Both [kinds of eclipses] do not happen together except at the time of the total collapse of the universe."

Étiene Tempier, Bishop of Paris, in 1277, to settle a controversy then dividing much of the French theological community, ruled that one could not deny the power of God to create as many universes as He pleases.

Roger Bacon, 1214–1294, in his *Opus Majus* (1268), gave the diameter of the sphere that carries the stars, on the authority of Alfargani, as 130,715,000 Roman miles

Box 27.7 (*continued*)

[mile equal to 1,000 settings down of the right foot]; the volume of the sun, 170 times that of the Earth; first-magnitude star, 107 times; sixth-magnitude, 18 times Earth.

Nicolas Cusanus, 1401–1464 [from *Of Learned Ignorance* (1440), as translated by Heron (1954)]: "Necessarily all parts of the heavens are in movement. . . . It is evident from the foregoing that the Earth is in movement . . . the world [universe], its movement and form . . . will appear as a wheel in a wheel, a sphere in a sphere without a center or circumference anywhere. . . . It is now evident that this Earth really moves, though to us it seems stationary. In fact, it is only by reference to something fixed that we detect the movement of anything. How would a person know that a ship was in movement, if . . . the banks were invisible to him and he was ignorant of the fact that water flows?"

Nicolaus Copernicus, February 19, 1473, to May 24, 1543 [from *De Revolutionibus Orbium Coelestrum* (1543), as translated by Dobson and Brodetsky (1947)]: "I was induced to think of a method of computing the motions of the spheres by nothing less than the knowledge that the mathematicians are inconsistent in these investigations. . . . they cannot explain or observe the constant length of the seasonal year. . . . some use only concentric circles, while others eccentrics and epicycles. . . . Nor have they been able thereby to discern or deduce the principal thing—namely the shape of the universe and the unchangeable symmetry of its parts. . . .

"I found first in Cicero that Nicetas had realized that the Earth moved. Afterwards I found in Plutarch [∼A.D. 46–120] . . . 'The rest hold the Earth to be stationary, but Philolaus the Pythagorean [born ∼480 B.C.] says that she moves around the (central) fire on an oblique circle like the Sun and Moon. Heraclides of Pontus [flourished in 4th century B.C.] and Ecphantus the Pythagorean also make the Earth to move, not indeed through space but by rotating round her own center as a wheel on an axle from West to East.' Taking advantage of this I too began to think of the mobility of the Earth. . . .

"Should we not be more surprised if the vast Universe revolved in twenty-four hours, than that little Earth should do so? . . . Idle therefore is the fear of Ptolemy that Earth and all thereon would be disintegrated by a natural rotation. . . . That the Earth is not the center of all revolutions is proved by the apparently irregular motions of the planets and the variations in their distances from the Earth. . . . We therefore assert that the center of the Earth, carrying the Moon's path, passes in a great orbit among the other planets in an annual revolution round the Sun; that near the Sun is the center of the Universe; and that whereas the Sun is at rest, any apparent motion of the Sun can be better explained by motion of the Earth. . . . Particularly Mars, when he shines all night, appears to rival Jupiter in magnitude, being distinguishable only by his ruddy color; otherwise he is scarce equal to a star of the second magnitude, and can be recognized only when his movements are

carefully followed. All these phenomena proceed from the same cause, namely Earth's motion. . . . That there are no such phenomena for the fixed stars proves their immeasurable distance, compared to which even the size of the Earth's orbit is negligible and the parallactic effect unnoticeable."

Thomas Digges, 1546–1595 [in *a Perfit Description of the Caelestiall Orbes according to the most aunciente doctrine of the Pythagoreans, latelye reuiued by Copernicus and by Geometricall Demonstrations approued* (1576), the principal vehicle by which Copernicus reached England, as quoted in Johnson (1937)]: "Of whiche lightes Celestiall it is to bee thoughte that we onely behoulde sutch as are in the inferioure partes of the same Orbe, and as they are hygher, so seeme they of lesse and lesser quantity, euen tyll our sighte beinge not able farder to reach or conceyue, the greatest part rest by reason of their wonderfull distance inuisible vnto vs."

Giordano Bruno, born ca. 1548, burned at the stake in the Campo dei Fiori in Rome, February 17, 1600 [from *On the Infinite Universe and Worlds*, written on a visit to England in 1583–1585, as translated by Singer (1950)]: "Thus let this surface be what it will, I must always put the question, what is beyond? If the reply is NOTHING, then I call that the VOID or emptiness. And such a Void or Emptiness hath no measure and no outer limit, though it hath an inner; and this is harder to imagine than is an infinite or immense universe. . . . There are then innumerable suns, and an infinite number of earths revolve around those suns, just as the seven we can observe revolve around this sun which is close to us."

Johann Kepler established the laws of elliptic orbits and of equal areas (1609), and established the connection between planetary periods and semimajor axes (1619).

Galileo Galilei observed the satellites of Jupiter and realized they provided support for Copernican theory, and interpreted the Milky Way as a collection of stars (1610). In 1638 he wrote:

"*Salvati*. Now what shall we do, Simplicio, with the fixed stars? Do we want to sprinkle them through the immense abyss of the universe, at various distances from any predetermined point, or place them on a spherical surface extending around a center of their own so that each of them will be at the same distance from that center?

"*Simplicio*. I had rather take a middle course, and assign to them an orb described around a definite center and included between two spherical surfaces . . ."

Isaac Newton (1687): "Gravitation toward the sun is made up out of the gravitations toward the several particles of which the body of the sun is composed, and in receding from the sun decreases accurately as the inverse square of the distances as far as the orbit of Saturn, as evidently appears from the quiescence of the aphelion of the planets."

Isaac Newton [in a letter of Dec. 10, 1692, to Richard Bentley, quoted in Munitz (1957)]: "If the matter of our sun and planets and all the matter of the universe were evenly scattered throughout all the heavens, and every particle had an innate gravity toward all the rest, and the whole space throughout which this matter was

Box 27.7 (*continued*)

scattered was but finite, the matter on the outside of this space would, by its gravity, tend toward all the matter on the inside and, by consequence, fall down into the middle of the whole space and there compose one great spherical mass. But if the matter was evenly disposed throughout an infinite space, it could never convene into one mass; but some of it would convene into one mass and some into another, so as to make an infinite number of great masses scattered at great distances from one to another throughout all that infinite space. And thus might the sun and fixed stars be formed."

Christiaan Huygens, 1629–1695 [in his posthumously published *Cosmotheoros* (1698)]: "Seeing then that the stars . . . are so many suns, if we do but suppose one of them [Sirius, the Dog-star] equal to ours, it will follow [details, including telescope directed at sun; thin plate; hole in it; comparison with Sirius] . . . that his distance to the distance of the sun from us is as 27,664 to 1. . . . Indeed it seems to me certain that the universe is infinitely extended."

Edmund Halley (1720): "If the number of the Fixt Stars were more than finite, the whole superficies of their apparent Sphere [i.e., the sky] would be luminous" [by today's reasoning the same temperature as the surface of the average star; this is known today as Olber's paradox, or the paradox of P. L. de Chéseaux (1744) and Heinrich Wilhelm Matthias Olbers (1826)].

Thomas Wright of Durham (1750): "To . . . solve the Phaenomena of the Via Lactea . . . granted . . . that the *Milky Way* is formed of an infinite number of small Stars . . . imagine a vast infinite gulph, or medium, every way extended like a plane, and inclosed between two surfaces, nearly even on both sides. . . . Now in this space let us imagine all the Stars scattered promiscuously, but at such an adjusted distance from one another, as to fill up the whole medium with a kind of regular irregularity of objects. [Considering its appearance] "to an eye situated . . . anywhere about the middle plane" . . . all the irregularity we observe in it at the Earth, I judge to be entirely owing to our Sun's position . . . and the diversity of motion . . . amongst the stars themselves, which may here and there . . . occasion a cloudy knot of stars."

Immanuel Kant, 1724–1804 (1755): "It was reserved for an Englishman, Mr. Wright of Durham, to make a happy step . . . we will try to discover the cause that has made the positions of the fixed stars come to be in relation to a common plane. . . . granted . . . that the whole host of [the fixed stars] are striving to approach each other through their mutual attraction . . . ruin is prevented by the action of the centrifugal forces . . . the same cause [centrifugal force] . . . has also so directed their orbits that they are all related to one plane. . . . [The needed motion is calculated to be] one degree [or less] in four thousand years; . . . careful observers . . . will be required for it. . . . Mr. Bradley has observed almost imperceptible displacements of the stars" [known from later work to be caused by aberration (effect of observer velocity) rather than real parallax (effect of position of observer)].

Asks for the first time how a very remote galaxy would appear: "circular if its plane is presented directly to the eye, and elliptical if it is seen from the side or obliquely. The feebleness of its light, its figure, and the apparent size of its diameter will clearly distinguish such a phenomenon when it is presented, from all the stars that are seen single. . . . this phenomenon . . . has been distinctly perceived by different observers [who] . . . have been astonished at its strangeness. . . . Analogy thus does not leave us to doubt that these systems [planets, stars, galaxies] have been formed and produced . . . out of the smallest particles of the elementary matter that filled empty space."

Goes on to consider seriously "the successive expansion of the creation [of planets, stars, galaxies] through the infinite regions of space that have the matter for it. . . . attraction is just that universal relation which unites all the parts of nature in one space. It reaches, therefore, to . . . all the distance of nature's infinitude."

Johann Heinrich Lambert, 1728–1777 (1761): "The fixed stars obeying central forces move in orbits. The Milky Way comprehends several systems of fixed stars. . . . Each system has its center, and several systems taken together have a common center. Assemblages of their assemblages likewise have theirs. In fine, there is a universal center for the whole world round which all things revolve." [First spelling out of a "hierarchical model" for the universe, later taken up by C. V. I. Charlier and by H. Alfvén and O. Klein (1962); see also O. Klein (1966 and 1971)].

Auguste Comte (1835) concluded that it is meaningless to speak of the chemical composition of distant stars because man will never be able to explore them; "the field of positive philosophy lies wholly within the limits of our solar system, the study of the universe being inaccessible in any positive sense."

The first successful determination of the parallax [1 second of parallax: 1 pc $= 3.08 \times 10^{18}$ cm $= 3.26$ lyr] of any star was made in 1838 (for α Centauri by Henderson, for α Lyrae by Struve, and for 61 Cygni by Bessel).

B. The Twentieth Century

Derivation by James Jeans in 1902 of the critical wavelength that separates short-wavelength acoustical modes of vibration of a hot primordial gas and longer wavelength modes of commencement of gravitational condensation of this gas. Application of these considerations by P. J. E. Peebles and R. H. Dicke in 1968 to explain why globular star clusters have masses of the order of $10^5 M_\odot$.

Investigations of cosmic rays from first observation by V. F. Hess and W. Kolhörster in 1911–1913 to date; determination that the energy density in interstellar space (in this galaxy) is about 1 eV/cm^3 or 10^{-12} erg/cm^3, comparable to the density of energy of starlight, to the kinetic energy of clouds of ionized interstellar gas, averaged over the galaxy, and to the energy density of the interstellar

Box 27.7 (continued)

magnetic field ($\sim 10^{-5}$ gauss). In connection with this equality, see especially E. Fermi (1949).

Discovery by Henrietta Leavitt in 1912 that there is a well-defined relation between the period of a Cepheid variable and its absolute luminosity.

First determination of the radial velocity of a galaxy by V. M. Slipher in 1912: Andromeda approaching at 200 km/sec. Thirteen galaxies investigated by him by 1915; all but two receding at roughly 300 km/sec.

Albert Einstein (1915d): Interpreted gravitation as a manifestation of geometry; gave final formulation of the law that governs the dynamic development of the geometry of space with the passage of time.

Albert Einstein (1917): Idealized the universe as a 3-sphere filled with matter at effectively uniform density; the radius of this 3-sphere could not be envisaged as static without altering his standard 1915 geometrodynamic law; for this reason Einstein introduced a so-called "cosmological term," which he later dropped as "the biggest blunder" in his life [Gamow (1970)].

Formulation by W. de Sitter in 1917 of a cosmological model in which (1) the universe is everywhere isotropic (and therefore homogeneous) and (2) the universe does not change with time, so that the mean density of mass-energy and the mean curvature of space are constant, but in which perforce (3) a cosmological term ("repulsion") of the Einstein type has to be added to balance the attraction of the matter. Observation by de Sitter that he could obtain another static model by removing all the matter from the original model, but that the Λ-term would cause test particles to accelerate away from one another.

From 1917 to 1920, debate about whether spiral nebulae are mere nebulous objects (Harlow Shapley) or are "island universes" or galaxies similar to but external to the Milky Way (H. D. Curtis).

Discovery by Harlow Shapley in 1918, by mapping distribution of about 100 globular clusters of this Galaxy (10^4 to 10^6 stars each) in space that center is in direction of Sagittarius (present value of distance from sun ~ 10 kpc).

Independent derivation of evolving homogeneous and isotropic cosmological models [also leading to the relation $v = H \cdot$ (distance)] by A. Friedmann in 1922 and G. Lemaître in 1927, with Lemaître tieing in his theoretical analysis with the then-ongoing Mt. Wilson work, to become the "father of the big-bang cosmology". (Universe, however, taken to expand smoothly away from Einstein's static $\Lambda > 0$ solution in Lemaître's original paper).

Remark by H. Weyl in 1923 that test particles in de Sitter model will separate at a rate given by a formula of the form $v = H \cdot$ (distance).

In 1924, resolution of debate about nature of spiral nebulae by Edwin P. Hubble with Mount Wilson 100-inch telescope; discovery of Cepheid variables in Andromeda and other spiral nebulae, and consequent determination of distances to these nebulae.

Determination by Jan Oort in 1927 of characteristic pattern of radial velocities of stars near sun,

$$\delta v_r = Ar \cos 2(\theta - \delta),$$

showing that: (1) axis of rotation of stars in Milky Way is perpendicular to disc; (2) sun makes a complete revolution in $\sim 10^8$ yr; and (3) the effective mass pulling on the sun required to produce a revolution with this period is of the order $\sim 10^{44}$ g or $\sim 10^{11} M_\odot$.

Age of a *uranium ore* as established from lead-uranium ratio: greatest value found up to 1927, 1.3×10^9 yr (A. Holmes and R. W. Lawson). Age of the lead in the "average" *surface rocks* of the earth as calculated from time required to produce this lead from the uranium in the same surface rocks, 2×10^9 yr to 6×10^9 yr. Age of *elemental uranium* as estimated by Rutherford from time required for U^{235} and U^{238} to decay from assumed roughly equal ratio in early days to known very unequal ratio today, $\sim 3 \times 10^9$ yr.

Establishment by Hubble in 1929 that out to 6×10^6 lyr the velocity of recession of a galaxy is proportional to its distance.

Note by A. S. Eddington in 1930 that Einstein $\Lambda > 0$ static universe is unstable against any small increase or decrease in the radius of curvature.

Recommendation from Einstein in 1931 hereafter to drop the so-called cosmological term.

Proposal by Einstein and de Sitter in 1932 that one tentatively adopt the simplest assumption that $\Lambda = 0$, that pressure is negligible, and that the reciprocal of the square of the radius of curvature of the universe is neither positive nor negative (spherical or hyperbolic universe) but zero ("cosmologically flat"), thus leading to the relation $\rho = 3H^2/8\pi$ (in geometric units).

Evidence uncovered by Grote Reber in 1934 for the existence of a discrete radio source in Cygnus; evidence for this source, Cygnus A, firmed up by J. S. Hey, S. J. Parsons, and J. W. Phillips in 1946; six other discrete radio sources, including Taurus A and Centaurus A, discovered by J. G. Bolton in 1948.

Discovery by E. A. Milne and W. H. McCrea in 1934 of close correspondence between Newtonian dynamics of a large gas cloud and Einstein theory of a dynamic universe, with the scale factor of the expansion satisfying the same equation in both theories, so long as pressure is negligible.

Demonstration by H. P. Robertson and by A. G. Walker, independently, in 1935 that the Lemaître type of line element provides the most general Riemannian geometry compatible with homogeneity and isotropy.

Classification of nebulae as spiral, barred spiral, elliptical, and irregular by Hubble in 1936.

First detailed theory of thermonuclear energy generation in the sun, H. A. Bethe, 1939.

Box 27.7 (continued)

Reasoning by George Gamow in 1946 that matter in the early universe was dense enough and hot enough to undergo rapid thermonuclear reaction, and that energy densities were radiation-dominated.

Proposal of so-called "steady-state cosmology" by H. Bondi, T. Gold, and F. Hoyle in 1948, lying outside the framework of Einstein's standard general relativity, with "continuous creation of matter" taking place throughout the universe, and the mean age of the matter present being equal to one third of the Hubble time.

Prediction by R. A. Alpher, H. A. Bethe, and G. Gamow in 1948 that the black-body radiation that originally filled the universe should today have a Planck spectrum corresponding to a temperature of 25 K. Independent conception of same idea by R. H. Dicke in 1964 and start of an experimental search for this primordial cosmic-fireball radiation. Discovery of unwanted and unexpected 7 cm background radiation in 1965 by A. A. Penzias and R. W. Wilson with a temperature of about 3.5 K; immediate identification of this radiation by Dicke, P. J. E. Peebles, P. G. Roll, and D. T. Wilkinson as the expected relict radiation.

Radio sources Taurus A, Virgo A, and Centaurus A tentatively and, as it later proved, correctly identified with the Crab Nebula and with the galaxies NGC 4486 and NGC 5128 by J. G. Bolton, G. J. Stanley, and O. B. Slee in 1949.

Analysis by Lemaître in 1950 of big-bang expansion approaching very closely the Einstein static universe ($\Lambda > 0$) and then, at first slowly, subsequently more and more rapidly, going into exponential expansion.

Discovery by Walter Baade in 1952 that there are two types of Cepheid variables with different period-luminosity relations; consequent increase in Hubble distance scale by factor of about 2.6, and a corresponding increase in the original value (roughly 2×10^9 yr) of the Hubble time, H_o^{-1}.

Identification of radio source Cygnus A by W. Baade and R. Minkowski in 1954 with the brightest member of a faint cluster of galaxies, contrary to the then widely held view that the majority of radio sources lie within the Milky Way. Determination of redshift in the optical spectrum of $\delta\lambda/\lambda = z = 0.057$ by Minkowski, implying for Cygnus A a distance of 170 Mpc and a radio luminosity of 10^{45} erg/sec, 10^7 times the radio power and ten times the optical power of a normal galaxy.

Resolution of radio source Cygnus A in 1956 into two components symmetrically located on either side of the optical galaxy, the first indication that most radio sources are double. Still unsolved is the mystery of the explosion or other mechanism that caused this and other double sources.

Calculation by G. R. Burbidge in 1956 of the kinetic energy in the electrons giving off synchrotron radiation in a radio galaxy and the energy of the magnetic field that holds these electrons in orbit; minimization of the sum of these two energies;

determination that this minimum is of the order of 10^{60} ergs (energy of annihilation of half a million suns) for Hercules A, for example.

Solar system determined to have an age of 4.55×10^9 yr or more from relative abundances of $Pb^{204,206,207}$ and $U^{235,238}$ in meteorites and oceanic sediments by C. Patterson in 1956; and by others in 1965 and 1969 from evidence on the processes $Rb^{87} \longrightarrow Sr^{87}$ and $K^{40} \longrightarrow A^{40}$ in meteorites.

Discovery by Allen Sandage in 1958 that what Hubble had identified in distant galaxies as bright stars were H II regions, clumps of hot stars surrounded by a plasma ionized by stars, and consequent upping of Hubble's distance scale by a further factor of about 2.2.

Estimation by Jan Oort in 1958, from luminosity of other galaxies, that matter in galaxies contributes to the density of mass-energy in the universe roughly 3×10^{-31} g/cm^3 [see Peebles (1971) for updated analysis], this being one or two orders of magnitude less than that called for by Einstein's concept that the universe is curved up into closure, and thereby giving rise to "the mystery of the missing matter," the focus of much present-day research.

Discovery of celestial (nonsolar) X-rays in 1962 by Giacconi, Gursky, Paolini, and Rossi. Majority of sources in plane of the Milky Way, presumably local to this galaxy, as is the Crab nebula. Extragalactic sources include the radio galaxy Virgo A and the quasar 3C273.

Revised "3C-catalog" of radio sources published in 1962 by A. S. Bennett, containing 328 sources, nearly complete in coverage between declinations $-5°$ and $+90°$ for sources brighter than 9 flux units (9×10^{-26} watt/m^2Hz) at 178 MHz.

Identification of the first quasistellar object (QSO) by Maarten Schmidt at Mt. Palomar in 1963: radio-position determination of 3C273 to better than 1 second of arc by C. Hazard, M. B. Mackey, and A. J. Shimmins in 1962, followed by Schmidt's taking an optical spectrum of the star-like source and, despite all presumptions that it was a star in this galaxy, trying to fit it, and succeeding, with a redshift of the magnitude (unprecedented for a "star") of $\delta\lambda/\lambda = z = 0.158$. Distance implied by Hubble relation, 1.5×10^9 lyr; optical brightness, 100 times brightest known galaxy. Largest redshift of any QSO known in 1972, $z = 2.88$ (4C05.34; C. R. Lynds). Such a source detectable even if it had a redshift of 3; but no QSO's known in 1972 with such redshifts. See Box 28.1.

Reasoning by Dennis Sciama in 1964 [see also Sciama (1971)] that intergalactic hydrogen can best escape observation if at a temperature between 3×10^5 K and 10^6 K. With as many as 10^{-5} protons and 10^{-5} electrons per cm^3 and a temperature lower than 3×10^5 K, the number density of neutral atoms would be great enough and the resulting absorption of Lyman α from a distant galaxy ($z = 2$) would be strong enough to show up, contrary to observation.

In 1964 J. E. Gunn and B. A. Peterson, E. J. Wampler, and others determined that, at a temperature greater than 10^6 K, the intensity of 0.25 keV or 50 Å x-rays

Box 27.7 (continued)

from intergalactic space would be too high to be compatible with the observations.

Emphasis by Wheeler (1964a) that the dynamic object in Einstein's general relativity is 3-geometry, not 4-geometry, and that this dynamics, both classical and quantum, unrolls in the arena of superspace.

Discovery by Sandage in 1965 of quasistellar galaxies (radio-quiet QSO's).

Discovery by E. M. Burbidge, G. R. Burbidge, C. R. Lynds, and A. N. Stockton in 1965 of a QSO, 3C191, with numerous absorption lines, implying the coexistence of several redshifts in one spectrum.

Fraction (by mass) of matter converted to helium in early few minutes of universe nearly independent of the relative numbers of photons and baryons, over a 10^6 range in values of this number ratio, so long as the universe at 10^{10} K is still radiation-dominated. Value of this plateau helium abundance (following earlier work of others) first accurately calculated as 27 per cent by P. J. E. Peebles in 1966 and by R. V. Wagoner, W. A. Fowler, and F. Hoyle in 1967.

Proposal by C. W. Misner in 1968 to consider as an important part of early cosmology the anisotropy vibrations of the geometry of space previously brought to attention by E. Kasner and by I. M. Khalatnikov and E. Lifshitz. [Misner's hope to account naturally in this way for the otherwise so puzzling homogeneity of the universe was later dashed.]

Proof on the basis of standard general relativity by S. W. Hawking, G. F. R. Ellis, and R. Penrose in 1968 and 1969 [see also related work of earlier investigators cited in Chapter 44] that a model universe presently expanding and filled with matter and radiation obeying a physically acceptable equation of state must have been singular in the past, however wanting in symmetry it is today.

Discovery of pulsars in 1968 by Hewish, Bell, Pilkington, Scott, and Collins, and their interpretation as spinning neutron stars (see Chapter 24).

"No poet, nor artist of any art, has his complete meaning alone. His significance, his appreciation, is the appreciation of his relation to the dead poets and artists. You cannot value him alone; you must set him, for contrast and comparison, among the dead . . . when a new work of art is created . . . something . . . happens simultaneously to all the works of art which preceded it. The existing monuments form an ideal order among themselves, which is modified by the introduction of the new (the really new) work of art among them."

T. S. ELIOT (1920).

EVOLUTION OF THE UNIVERSE INTO ITS PRESENT STATE

Cosmology . . . restrains the aberrations of the mere undisciplined imagination.

ALFRED NORTH WHITEHEAD (1929, p. 21)

§28.1. THE "STANDARD MODEL" OF THE UNIVERSE

Since the discovery of the cosmic microwave radiation in 1965, extensive theoretical research has produced a fairly detailed picture of how the universe probably evolved into its present state. This picture, called the "standard hot big-bang model" of the universe, is sketched in the present chapter, and its main features appear in Figure 28.1. Gravitation dominates the over-all expansion; but otherwise most details of the evolution are governed much less by gravitation than by the laws of thermodynamics, hydrodynamics, atomic physics, nuclear physics, and high-energy physics. This fact, and the existence of three excellent recent books on the subject [Sciama (1971); Peebles (1972); Zel'dovich and Novikov (1974)], encourage brevity here.

The past evolution of the universe is qualitatively independent of the nature of the homogeneous hypersurfaces ($k = -1, 0$, or $+1$) and qualitatively independent of the cosmological constant, since the contributions of k and Λ to the evolution are not important in early stages of the history (small a/a_o) [see equation (27.40) and Figure 27.5]. One crucial assumption underlies the standard hot big-bang model: that the universe "began" in a state of rapid expansion from a very nearly homogeneous, isotropic condition of infinite (or near infinite) density and temperature.

During the first second after the beginning, according to this analysis, the temperature of the universe was so high that there was complete thermodynamic equilib-

Evolution of universe according to "standard hot big-bang model":

(1) initial state

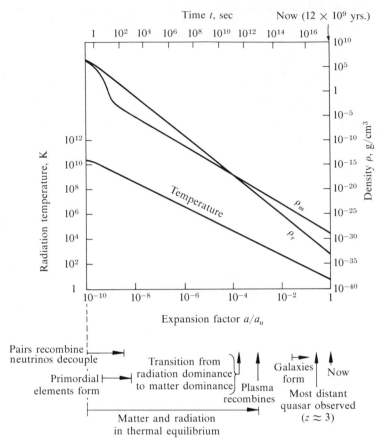

Figure 28.1.

Evolution of the universe into its present state, according to the standard hot big-bang model. The curves are drawn assuming

$$\rho_{mo} = 5 \times 10^{-30} \text{ g/cm}^3, \qquad \rho_{ro} = 10^{-33} \text{ g/cm}^3, \qquad k = 0;$$

but for other values of ρ_{mo}, ρ_{ro}, and k within the limits of observation, the curves are virtually the same (see exercise 28.1). See text and Box 28.1 for detailed discussion of the processes described at the bottom of the figure. [This figure is adapted from Dicke, Peebles, Roll, and Wilkinson (1965).]

(2) thermal equilibrium, decay of particles, recombination of pairs ($0 < t \lesssim 10$ sec.)

(3) decoupling and free propagation of gravitons and neutrinos ($t \lesssim 1$ sec.)

rium between photons, neutrinos, electrons, positrons, neutrons, protons, various hyperons and mesons, and perhaps even gravitons (gravitational waves) [see, e.g., Kundt (1971) and references cited therein]. However, by the time the universe was a few seconds old, its temperature had dropped to about 10^{10} K and its density was down to $\sim 10^5$ g/cm³; so all nucleon-antinucleon pairs had recombined, all hyperons and mesons had decayed, and all neutrinos and gravitons had decoupled from matter. The universe then consisted of freely propagating neutrinos, and perhaps gravitons, with black-body spectra at temperatures $T \sim 10^{10}$ K, plus electron-positron pairs in the process of recombining, plus electrons, neutrons, protons, and photons all in thermal equilibrium at $T \sim 10^{10}$ K.

Since that early state, the gravitons (if present) and neutrinos have continued

to propagate freely, maintaining black-body spectra; but their temperatures have been redshifted by the expansion of the universe in accordance with the law

$$T \propto 1/a \qquad (28.1)$$

(Box 29.2). Consequently, today their temperatures should be roughly 3 K, and they should still fill the universe. Unfortunately, today's technology is far from being able to detect such a "sea" of neutrinos or gravitons. However, if and when they can be detected, they will provide *direct* observational information about the first one second of the life of the universe!

As the universe continued to expand after the first few seconds, it entered a period lasting from $t \sim 2$ seconds to $t \sim 1,000$ seconds ($T \sim 10^{10}$ to $\sim 10^9$ K, $\rho \sim 10^{+5}$ to 10^{-1} g/cm^3), during which primordial element formation occurred. Before this period, there were so many high-energy protons around that they could blast apart any atomic nucleus (e.g., deuterium or tritium or He3 or He4) the moment it formed; after this period, the protons were too cold (had kinetic energies too low) to penetrate each others' coulomb barriers, and all the freely penetrating neutrons from the earlier, hotter stage had decayed into electrons plus protons. Only during the short, crucial period from $t \sim 2$ seconds to $t \sim 1,000$ seconds were conditions right for making elements. Calculations by Gamow (1948), by Alpher and Hermann (1948a,b; 1950), by Fermi and Turkevitch (1950), by Peebles (1966), and by Wagoner, Fowler, and Hoyle (1967) reveal that about 25 per cent of the baryons in the universe should have been converted into He4 (alpha particles) during this period, and about 75 per cent should have been left as H^1 (protons). Traces of deuterium, He3, and Li should have also been created, but essentially no heavy elements. All the heavy elements observed today must have been made later, in stars [see, e.g., Fowler (1967) or Clayton (1968)]. Current astronomical studies of the abundances of the elements give some support for these predictions; but the observational data are not yet very conclusive [see, e.g., Danziger (1968) and pp. 268–275 of Peebles (1971)].

After primordial element formation, the matter and radiation continued to interact thermally through frequent ionization and recombination of atoms, keeping each other at the same temperature. Were the temperatures of radiation and matter not locked together, the radiation would cool more slowly than the matter (for adiabatic expansion, $T_r \propto 1/a$, but $T_m \propto 1/a^2$). Thus thermal equilibrium was maintained only by a constant transfer of energy from radiation to matter. But the heat capacity of the radiation was far greater than that of the matter. Therefore the energy transfer had a negligible effect on ρ_r, p_r, and T_r. It held up the temperature of the matter ($T_m = T_r$) without significantly lowering the temperature of the radiation. On the other hand, the total mass-energy of matter was and is dominated by rest mass. Therefore the energy transfer had negligible influence on ρ_m. [This circumstance justifies the approximation of ignoring energy transfer when passing from equation (27.31) to (27.32).]

When the falling temperature reached a few thousand degrees ($a/a_o \sim 10^{-3}$, $\rho \sim 10^{-20}$ g/cm^3, $t \sim 10^5$ years), two things of interest happened: the universe ceased to be radiation-dominated and became matter-dominated [$\rho_m = \rho_{mo}(a_o/a)^3$ came to exceed $\rho_r = \rho_{ro}(a_o/a)^4$]; and the photons ceased to be energetic enough to keep

(4) primordial element formation (2 sec. $\lesssim t \lesssim 1,000$ sec.)

(5) thermal interaction of matter and radiation (1,000 sec. $\lesssim t \lesssim 10^5$ years)

(6) plasma recombination and transition to matter dominance ($t \sim 10^5$ yrs.)

hydrogen atoms ionized, so the electrons and protons quickly recombined. That these two events were roughly coincident is a result of the specific, nearly conserved value that the entropy per baryon has in our universe:

$$s \equiv \text{entropy per baryon} \sim \frac{\text{(number of photons in universe)}}{\text{(number of baryons in universe)}} \sim 10^8.$$

Why the universe began with this value of s, rather than some other value (e.g. unity), nobody has been able to explain.

Recombination of the plasma at $t \sim 10^5$ years was crucial, because it brought an end to the interaction and thermal equilibrium between radiation and matter ("decoupling"). Thereafter, with very few free electrons off which to scatter, and with Rayleigh scattering off atoms and molecules unimportant, the photons propagated almost freely through space. Unless energy-releasing processes reionized the intergalactic medium sometime between $a/a_o \sim 10^{-3}$ and $a/a_o \sim 0.1$, the photons have been propagating freely ever since the plasma recombined. Even if reionization occurred, the photons have been propagating freely at least since $a/a_o \sim 0.1$.

(7) subsequent propagation of photons ($t \gtrsim 10^5$ yrs.)

The expansion of the universe has redshifted the temperature of the freely propagating photons in accordance with the equation $T \propto 1/a$ (see Box 29.2). As a consequence, today they have a black-body spectrum with a temperature of 2.7 K. They are identified with the cosmic microwave radiation that was discovered in 1965, and they give one direct information about the nature of the universe at the time they last interacted with matter ($a/a_o \sim 10^{-3}$, $t \sim 10^5$ years if reionization did not occur; $a/a_o \sim 0.1$, $t \sim 5 \times 10^8$ years if reionization did occur.)

(8) condensation of stars, galaxies and clusters (10^8 yrs. $\lesssim t \lesssim 10^9$ yrs.)

Return to the history of matter. Before plasma recombination, the photon pressure ("elasticity of the cosmological fluid") prevented the uniform matter (25 per cent He^4, 75 per cent H) from condensing into stars, galaxies, or clusters of galaxies. However, after recombination, the photon pressure was gone, and condensation could begin. Small perturbations in the matter density, perhaps dating back to the beginning of expansion, then began to grow larger and larger. Somewhere between $a/a_o \sim 1/30$ and $a/a_o \sim 1/10$ (10^8 years $\lesssim t \lesssim 10^9$ years) these perturbations began developing into stars, galaxies, and clusters of galaxies. Slightly later, at $a/a_o \sim 1/4$, quasars probably "turned on," emitting light which astronomers now receive at Earth (see Box 28.1).

EXERCISE

Exercise 28.1. UNCERTAINTY IN EVOLUTION

Current observations, plus the assumption of complete homogeneity and isotropy at the beginning of expansion, plus the assumption that the excess of leptons over antileptons is less than or of the order of the excess of baryons over antibaryons, place the following limits on the cosmological parameters today:

Matter density today = ρ_{mo}, between 10^{-28} and 2×10^{-31} g/cm^3;
$k = 0$ or $+1$ or -1;
temperature of electromagnetic radiation today = 2.7 ± 0.1 K.
Total radiation density [observed photons, plus neutrinos and gravitons that presumably
 originated in big bang in thermal equilibrium with photons] = ρ_{ro}, between 0.7×10^{-33}
 and 1.2×10^{-33} g/cm^3.

(*continued on page 769*)

Box 28.1 EVOLUTION OF THE QUASAR POPULATION

That the large-scale, average properties of the universe are changing markedly with time one can infer from quasar data. In brief, there appear to have been about 50 times more quasars in the universe at a redshift $z \approx 2$ than at $z \approx 0.5$; and there may well have been fewer, or none, at redshifts $z > 3$. (On the use of redshift to characterize time since the big bang, see Box 29.3.) In greater detail, Schmidt (1972) gives the following analysis of the data:*

1. Schmidt assumes from the outset that quasar redshifts are cosmological in origin [redshift = (Hubble constant) \times (distance); §29.2]. The evidence for this is
 a. Observational: Some quasars are located in clusters of galaxies [as evidenced both by position on sky and by quasar having same redshift as galaxies in cluster; see Gunn (1971)]. Since the evidence for the cosmological distance-redshift relation for galaxies is overwhelming (Boxes 29.4 and 29.5), the redshifts of these particular quasars *must* be cosmological.
 b. Theoretical: Observed quasar redshifts of $z \sim 1$ to 3 cannot be gravitational in origin; objects with gravitational redshifts larger than $z \approx 0.5$ are unstable against collapse (see Chapters 24 and 26 and Box 25.9). Nor are the quasar redshifts likely to be Doppler; how could so massive an object be accelerated to $v \approx 1$ without complete disruption? The only remaining possibility is a cosmological redshift. For this reason, opponents of the cosmological hypothesis usually feel pressed to invoke in the quasars a breakdown of the laws of physics as one understands them today. [See, e.g., Arp (1971) and references cited therein. These references also describe evidence against the cosmological assumption, evidence that a few prominent investigators find compelling, but that most do not as of 1972.]

2. Schmidt then asks how many quasars, N, there were in the universe at a time corresponding to the redshift z, and with absolute luminosity per unit frequency, $L_\nu(2{,}500 \text{ Å})$ at the wavelength 2500 Å as measured in the quasar's local Lorentz frame.

3. The data on quasars available in 1972 are not at all sufficient to determine $N[z, L_\nu(2{,}500 \text{ Å})]$ uniquely. But they *are* sufficient to show unequivocally that:
 a. There *must* have been evolution; $N(z, L_\nu)$ cannot be independent of z.
 b. The evolution cannot have resided primarily in the luminosities: the total number of quasars,

$$N_{\text{tot}}(z) \equiv \sum_{L_\nu(2{,}500\text{Å})} N(z, L_\nu)$$

 must have changed markedly with time (with z).

*Our version of Schmidt's (1972) argument is oversimplified. The reader interested in greater precision should consult his original paper.

Box 28.1 (*continued*)

c. If the evolution was primarily in the total number, $N_{tot}(z)$, i.e., if the changes in the relative luminosity distribution at 2,500 Å

$$f(z, L_\nu) \equiv [1/N_{tot}(z)]N(z, L_\nu)$$

were negligible, and if the universe today is characterized by $\sigma_0 = q_0 = 1$ (see Chapter 29 for notation), then the data show

$$N_{tot}(z = 2) \approx 50 N_{tot}(z = 0.5).$$

This steep increase in number as one goes backward in time—and all other basic features of the observed quasar redshift and magnitude distributions for $z \lesssim 2$—can be fit in a universe with $\sigma_0 = q_0 = 1$ by either of the evolution laws

$$N_{tot}(z) \propto (1 + z)^6,$$

$$N_{tot}[z(t)] \propto 10^{5(t_0 - t)/t_0}.$$

Here t_0 is the current age of the universe and t was the age at redshift z.

d. These evolution laws, when extrapolated beyond a redshift $z \approx 2$ and when combined with the observed relative luminosity function $f(z, L_\nu)$ for quasars near apparent magnitude 18, predict that an observer on Earth should *see* the following fractions of nineteenth and twentieth-magnitude quasars to have redshifts greater than 2.5:

evolution law	fraction with $z > 2.5$	
	$m = 19$	$m = 20$
$(1 + z)^6$	29%	49%
$10^{5(t_0 - t)/t_0}$	12%	14%

In 1972 about 30 quasars fainter than $m = 18.5$ are known, and of these only 1 (3%) has $z > 2.5$. This shows, in Schmidt's words, "that the density law $(1 + z)^6$ cannot persist beyond a redshift of around 2.5." Schmidt regards the $10^{5(t_0 - t)/t_0}$ law (which becomes nearly constant at $z > 2.5$) to be also in apparent conflict with the observations, but he says that "further spectroscopic work on faint quasars is needed to confirm this suspicion."

One reason for caution is the difficult problem of removing "observational selection effects" from the data. Schmidt, Sandage, and others have independently searched for selection effects that might produce an artificial apparent decrease in the number of quasars at $z > 2.5$. None have been found. In the words of Sandage (1972d) "The apparent cutoff in quasar redshifts near $z = 2.8$ [has been] examined for selection effects that could produce it artificially. If the cutoff is real, it may be the time of the birth of the first quasars, although the suggested redshift is unexpectedly small. At $z = 3$ in a $q_0 = 1$ universe, the look-back time is 89 per cent of the Friedmann age. Assessment of the observational selection effects shows that none are positively established that could produce the cutoff artificially."

(The uncertainties taken into account in ρ_{ro} are uncertainty about whether quadrupole moments at early times were sufficient to create gravitons at the full level corresponding to thermal equilibrium, and uncertainty about the number and statistical weights of particle species in equilibrium at the time gravitons decoupled.) Use the equations in §27.10 to calculate the uncertainties in the evolutionary history (Figure 28.1) caused by these uncertainties in the present state of the universe.

§28.2. STANDARD MODEL MODIFIED FOR PRIMORDIAL CHAOS

The standard hot big-bang model is remarkably powerful and accords well with observations (primordial helium abundances; existence, temperature, and isotropy of cosmic microwave radiation; homogeneity and isotropy of universe in the large; close accord between age of universe as measured by expansion and ages of oldest stars; . . .). However, in 1972 it encounters apparent difficulty with one item: the origin of galaxies. In a universe that is initially homogeneous and isotropic it is not clear that random fluctuations will give rise (after plasma recombination) to perturbations in the density of matter of sufficient amplitude to condense into galaxies. The perturbations that eventually form galaxies might have to reside in the initial, exploding state of the universe. [See Zel'dovich and Novikov (1974) for detailed review and discussion; see also references cited in §30.1.]

Is it reasonable to assume a small amount of initial inhomogeneity? Is it not much more reasonable to assume either perfect homogeneity (one extreme) or perfect chaos (the other extreme)?

Thus, if perfect initial homogeneity turns out to be incompatible with the origin of galaxies, it is attractive to try "perfect initial chaos"—i.e., completely random initial conditions, with a full spectrum of fluctuations in density, entropy, and local expansion rate [Misner (1968, 1969b)]. It is conceivable, but far from proved, that during its subsequent evolution such a model universe will quickly smooth itself out by natural processes (Chapter 30) such as "Mixmaster oscillations," neutrino-induced viscosity [see, e.g., Matzner and Misner (1971)], and gravitational curvature-induced creation of particle pairs [Zel'dovich (1972)]. Will one be left, after a few seconds or less, with a nearly homogeneous and isotropic, Friedmann universe, containing just enough remaining perturbations to condense eventually into galaxies? Theoretical calculations have not yet been carried far enough to give a clear answer. Of course, after the initial chaos subsides, if it subsides, such a model universe will evolve in accord with the standard big-bang model of the last section.

What if the universe began chaotic?

§28.3. WHAT "PRECEDED" THE INITIAL SINGULARITY?

No problem of cosmology digs more deeply into the foundations of physics than the question of what "preceded" the "initial state" of infinite (or near infinite) density, pressure, and temperature. And, unfortunately, no problem is farther from solution in 1973.

The initial singularity and
quantum gravitational effects

General relativity predicts, inexorably, that even if the "initial state" was chaotic rather than smooth, it must have involved a spacetime "singularity" of some sort [see Hawking and Ellis (1968); also §34.6 of this book]. And general relativity is incapable of projecting backward through the singularity to say what "preceded" it. Perhaps only by coming to grip with quantum gravitational effects (marriage of quantum theory with classical geometrodynamics) will one ever reach a clear understanding of the initial state and of what, if anything, "preceded" it [see Misner (1969c), Wheeler (1971c)]. For further discussion of these deep issues, see §§34.6, 43.4, the final section of Box 30.1, and Chapter 44.

§28.4. OTHER COSMOLOGICAL THEORIES

Cosmologies that violate
general relativity

This book confines attention to the cosmology of general relativity. If one were to abandon general relativity, one would have a much wider set of possibilities, including (1) the steady-state theory [Hoyle (1948); Bondi and Gold (1948)], which has not succeeded in accounting for the cosmic microwave radiation or in explaining observed evolutionary effects in radio sources and quasars [Box 28.1]; (2) the Klein-Alfvén "hierarchic cosmology" of matter in an asymptotically flat spacetime [Alfvén and Klein (1962), Alfvén (1971), Klein (1971), Moritz (1969), de Vaucouleurs (1971)], which disagrees with cosmic-ray and gamma-ray observations [Steigman (1971)]; and the Brans-Dicke cosmologies [Dicke (1968), Greenstein (1968a,b), Morganstern (1973)], which are qualitatively the same and quantitatively almost the same as the standard hot big-bang model. However, no motivation or justification is evident for abandoning general relativity. The experimental basis of general relativity has been strengthened substantially in the past decade (Chapters 38–40); and the standard big-bang model of the universe predicted by general relativity accords remarkably well with observations—far better than any other model ever proposed!

PRESENT STATE AND FUTURE EVOLUTION OF THE UNIVERSE

§29.1. PARAMETERS THAT DETERMINE THE FATE OF THE UNIVERSE

This chapter is entirely Track 2. Chapter 27 (idealized cosmological models) is needed as preparation for it, but this chapter is not needed as preparation for any later chapter.

Will the universe continue to expand forever; or will it slow to a halt, reverse into contraction, and implode back to a state of infinite (or near infinite) density, pressure, temperature, and curvature? The answer is not yet known for certain. To discover the answer is one of the central problems of cosmology today.

The only known way to discover the answer is to measure, observationally, the present state of the universe; and then to calculate the future evolution using Einstein's field equations. The field equations have already been solved in §§27.10 and 27.11. From those solutions one reads off the following correlation between the present state of the universe and its future.

Expansion forever vs. recontraction of universe

If $\Lambda = 0$ [in accord with Einstein's firmly held principle of simplicity]:

Expansion forever \Longleftrightarrow negative or zero spatial curvature for hypersurfaces of homogeneity, i.e., $k/a_o{}^2 \leq 0$ ("open" or "flat");

Recontraction \Longleftrightarrow positive spatial curvature for homogeneous hypersurfaces, i.e., $k/a_o{}^2 > 0$ ("closed").

If $\Lambda \neq 0$:

Expansion forever $\Longleftrightarrow \Lambda \geq \Lambda_{\text{crit}} \equiv \begin{cases} 0 & \text{if } k \leq 0, \\ (4\pi\rho_{mo}a_o{}^3)^{-2} & \text{if } k > 0; \end{cases}$

Recontraction $\Longleftrightarrow \Lambda < \Lambda_{\text{crit}}$.

Evidently three parameters are required to predict the future: the cosmological constant, Λ; the curvature parameter today for the hypersurface of homogeneity, $k/a_o{}^2$; and the density of matter today, ρ_{mo}. (To extrapolate into the past, as was done in the last chapter, one needs, besides these quantities, the radiation density today, ρ_{ro}. But ρ_{ro} is so small now and is getting smaller so fast ($\rho_r \propto a^{-4}$; $\rho_m \propto a^{-3}$) that it can have no influence on the decision between the possibilities just listed.

Parameters required to predict future of universe:

(1) "relativity parameters" Λ, $k/a_o{}^2$, ρ_{mo}

The task of predicting the future, then, reduces to the task of measuring the "relativity parameters" Λ, k/a_o^2, and ρ_{mo}.

In tackling this task, observational cosmologists prefer to replace the three "relativity parameters," which have immediate significance for relativity theory, by parameters that are more directly observable. One parameter close to the observations is the Hubble expansion rate *today*, i.e., the "*Hubble constant*,"

(2) "observational parameters" H_o, q_o, σ_o

$$H_o \equiv (a_{,t}/a)_o. \tag{29.1a}$$

Another is the dimensionless "*deceleration parameter*" today, q_o, defined by

$$q_o \equiv -\frac{a_{,tt}}{a}\frac{1}{H_o^2} = -\left(\frac{aa_{,tt}}{a_{,t}^2}\right)_o. \tag{29.1b}$$

And a third is the dimensionless "*density parameter*," today,

$$\sigma_o \equiv \frac{4\pi\rho_{mo}}{3H_o^2}. \tag{29.1c}$$

(3) relationship between relativity parameters and observational parameters

The relationships between these three "observational parameters" and the three "relativity parameters" Λ, k/a_o^2, and ρ_{mo} (together making six "cosmological parameters") can be calculated by combining definitions (29.1) with the Einstein field equations (27.39), which, evaluated today, say

$$H_o^2 = -\frac{k}{a_o^2} + \frac{\Lambda}{3} + \frac{8\pi}{3}\rho_{mo},$$
$$-2q_oH_o^2 = -H_o^2 - \frac{k}{a_o^2} + \Lambda. \tag{29.2}$$

By combining these equations, one finds the relationships shown in Box 29.1, where the implications of several values of σ_o and q_o are also shown.

EXERCISE

Exercise 29.1. IMPLICATIONS OF PARAMETER VALUES
Derive the results quoted in Box 29.1.

§29.2. COSMOLOGICAL REDSHIFT

One of the key pieces of observational data used in measurements of H_o, q_o, and σ_o is the cosmological redshift: spectral lines emitted by galaxies far from Earth and received at Earth are found to be shifted in wavelength toward the red. For example, the [O II]λ3727 line, when both emitted *and* observed in an Earth-bound laboratory, has a wavelength of 3727 Å. However, when it is emitted by a star in the galaxy

Observed features of cosmological redshift

Box 29.1 OBSERVATIONAL PARAMETERS COMPARED TO RELATIVITY PARAMETERS

A. Relativity Parameters

1. Matter density today,

$$\rho_{mo}$$

2. Curvature of hypersurface of homogeneity today,

$$k/a_o{}^2$$

3. Cosmological constant,

$$\Lambda$$

4. Radiation density today, ρ_{ro} (unimportant for the present dynamics of the universe, and therefore ignored in this chapter)

B. Observational Parameters

1. Hubble constant (Hubble expansion rate today),

$$H_o \equiv (a_{,t}/a)_o$$

2. Deceleration parameter,

$$q_o \equiv -\frac{a_{,tt}}{a}\frac{1}{H_o{}^2}$$

3. Density parameter,

$$\sigma_o \equiv \frac{4\pi\rho_{mo}}{3H_o{}^2}$$

C. Observational Parameters in Terms of Relativity Parameters

$$H_o{}^2 = (8\pi/3)\rho_{mo} - k/a_o{}^2 + \Lambda/3, \tag{1}$$

$$q_o = \frac{(4\pi/3)\rho_{mo} - \Lambda/3}{(8\pi/3)\rho_{mo} - k/a_o{}^2 + \Lambda/3}, \tag{2}$$

$$\sigma_o = \frac{(4\pi/3)\rho_{mo}}{(8\pi/3)\rho_{mo} - k/a_o{}^2 + \Lambda/3}. \tag{3}$$

D. Relativity Parameters in Terms of Observational Parameters

$$\rho_{mo} = (3/4\pi)H_o{}^2\sigma_o, \tag{4}$$

$$k/a_o{}^2 = H_o{}^2(3\sigma_o - q_o - 1), \tag{5}$$

$$\Lambda = 3H_o{}^2(\sigma_o - q_o). \tag{6}$$

E. Implications of Specific Parameter Values

1. $\Lambda = 0$ (in accord with Einstein's point of view) if and only if $\sigma_o = q_o$.
2. Sign of Λ is same as sign of $\sigma_o - q_o$.

Box 29.1 (continued)

3. If $\Lambda = 0$

(a) $q_o > \dfrac{1}{2} \iff \rho_{mo} > \rho_{\text{crit}} \equiv \dfrac{3}{8\pi}H_o{}^2 \iff k > 0$ $\qquad \left(\begin{array}{l}\text{positive curvature;}\\ \text{``closed'' universe}\end{array}\right)$

$\qquad\qquad \iff$ universe will eventually recontract;

(b) $q_o = \dfrac{1}{2} \iff \rho_{mo} = \rho_{\text{crit}} \equiv \dfrac{3}{8\pi}H_o{}^2 \iff k = 0$ $\qquad \left(\begin{array}{l}\text{zero curvature;}\\ \text{``flat'' universe}\end{array}\right)$

$\qquad\qquad \implies$ universe will expand forever;

(c) $q_o < \dfrac{1}{2} \iff \rho_{mo} < \rho_{\text{crit}} \equiv \dfrac{3}{8\pi}H_o{}^2 \iff k < 0$ $\qquad \left(\begin{array}{l}\text{negative curvature;}\\ \text{``open'' universe}\end{array}\right)$

$\qquad\qquad \implies$ universe will expand forever.

4. If $\Lambda \neq 0$

(a) $\sigma_o > \dfrac{1}{3}(q_o + 1) \iff k > 0$ $\qquad\qquad\qquad \left(\begin{array}{l}\text{positive curvature;}\\ \text{``closed'' universe}\end{array}\right),$

and in this case,

$$\sigma_o - q_o \geq \dfrac{1}{\sigma_o{}^2}\left(\sigma_o - \dfrac{q_o + 1}{3}\right)^3 \iff \text{universe will expand forever,}$$

$$\sigma_o - q_o < \dfrac{1}{\sigma_o{}^2}\left(\sigma_o - \dfrac{q_o + 1}{3}\right)^3 \iff \text{universe will eventually recontract;}$$

(b) $\sigma_o = \dfrac{1}{3}(q_o + 1) \iff k = 0$ $\qquad\qquad\qquad \left(\begin{array}{l}\text{zero curvature;}\\ \text{``flat'' universe}\end{array}\right),$

and in this case,

$\sigma_o \geq q_o \iff$ universe will expand forever,

$\sigma_o < q_o \iff$ universe will eventually recontract;

(c) $\sigma_o < \dfrac{1}{3}(q_o + 1) \iff k < 0$ $\qquad\qquad\qquad \left(\begin{array}{l}\text{negative curvature;}\\ \text{``open'' universe}\end{array}\right),$

and in this case,

$\sigma_o \geq q_o \iff$ universe will expand forever,

$\sigma_o < q_o \iff$ universe will eventually recontract.

3C 295 (presumably with the same wavelength, $\lambda_{em} = 3727$ Å) and received at Earth, it is measured here to have the wavelength $\lambda_{rec} = 5447$ Å. The fractional change in wavelength is

$$z \equiv (\lambda_{rec} - \lambda_{em})/\lambda_{em} = 0.4614 \text{ for 3C 295.} \tag{29.3}$$

The cosmological redshift is observed to affect all spectral lines alike, and not only lines in the visible spectrum. Thus, the 21-cm line of hydrogen, with 400,000 times the wavelength of the central region of the visible, undergoes a redshift that agrees (within the errors of the measurements) with the redshifts of lines in the visible for recession velocities of the order of $v \sim 0.005$, according to observation of thirty objects by Dieter, Epstein, Lilley, and Roberts (1962) and further observations by Roberts (1965).

No one has ever put forward a satisfactory explanation for the cosmological redshift other than the expansion of the universe (see below). The idea has been proposed at various times by various authors that some new process is at work ("tired light") in which photons interact with atoms or electrons on their way from source to receptor, and thereby lose bits and pieces of their energy. Ya. B. Zel'dovich (1963) gives a penetrating analysis of the difficulties with any such ideas:

Why redshift cannot be due to "tired light"

> (1) "If the energy loss is caused by an interaction with the intergalactic matter, it is accompanied by a transfer of momentum; that is, there is a change of the direction of motion of the photon. There would then be a smearing out of images; a distant star would be seen as a disc, not a point, and that is not what is observed." (2) "Let us suppose that the photon decays, $\gamma \longrightarrow \gamma' + k$, giving up a small part of its energy to some particle, k. It follows from the conservation laws that k must move in the direction of the photon (this, by the way, avoids a smearing out), and must have zero rest mass. Because of the statistical nature of the process, however, some photons would lose more energy than others, and there would be a spectral broadening of the lines, which is also not observed."

(3) If there does exist any such decay process, then simple arguments of special relativity that Zel'dovich attributes to M. P. Bronshtein, and spells out in detail, demand the relationship

$$\begin{pmatrix} \text{probability per} \\ \text{second of} \\ \text{"photon decay"} \end{pmatrix} = \frac{\begin{pmatrix} \text{a universal constant with} \\ \text{the dimensions sec}^{-2} \end{pmatrix}}{(\text{frequency of photon in sec}^{-1})}.$$

> "Thus," Zel'dovich concludes, "if the decay of photons is possible at all, those in radio waves must decay especially rapidly! This would mean that the Maxwell equation for a static electric field would have to be changed There is no experimental indication of such effects: the radio-frequency radiation from distant sources is transmitted to us not a bit more poorly than visible light, and the red shift measured in different parts of the spectrum is exactly the same Thus, suggestions that there is an explanation of the red shift other than Friedmann's fail completely."

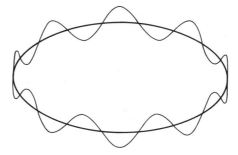

Emission:
atom excites n-node standing wave;
universe small, $a(t_e) = a_{em}$;
wavelengths small, $\lambda(t_e) = \lambda_{em}$.

Reception:
universe larger, $a(t_r) = a_{rec}$;
wavelengths larger, $\lambda(t_r) = \lambda_{rec}$;
number of nodes in standing
wave unchanged;

$$n = \text{constant} = \frac{2\pi a_{rec}}{2\pi \lambda_{rec}} = \frac{a_{em}}{\lambda_{em}}.$$

Figure 29.1.
Redshift as an effect of standing waves. The ratio of wavelengths, $\lambda_{rec}/\lambda_{em}$, is identical with the ratio of dimensions, a_{rec}/a_{em} in any closed spherically symmetrical (Friedmann) model universe. The atom excites an n-node standing wave in the universe. The number n stays constant during the expansion. Therefore wavelengths increase in the same proportion as the dimensions of the universe. One sees immediately in this way that the redshift is independent of all such details as (1) why the expansion came about (spherical symmetry, but arbitrary equation of state); (2) the rate—uniform or nonuniform—at which it came about; and (3) the distance between source and receptor at emission, at reception, or at any time in-between. The reasoning in the diagram appears to depend on the closure of the universe (standing waves; $k = +1$ rather than 0 or -1). That closure is not required for this simple result is seen from the further analysis given in the text.

Not the least among the considerations that lead one to accept the general recession of the galaxies as the explanation for the redshift is the circumstance that this general recession was predicted [Friedmann (1922)] before the redshift was observed [Hubble (1929)].

The cosmological redshift is easily understood (Figure 29.1) in terms of the standard big-bang model for the universe. A detailed analysis focuses attention on three processes: emission of the light, propagation of the light through curved spacetime from emitter to receiver, and reception of the light. Emission and reception occur in the proper reference frames (orthonormal tetrads) of the emitter and receiver; they are special-relativistic phenomena. Propagation, by contrast, is a general-relativistic process; it is governed by the law of geodesic motion in curved spacetime.

In calculating all three processes—emission, propagation, and absorption—one

Derivation of redshift
formula:

$$\lambda \propto \begin{pmatrix} \text{expansion} \\ \text{factor} \end{pmatrix}$$

needs a coordinate system. Use the coordinates (t, χ, θ, ϕ) or $(\eta, \chi, \theta, \phi)$ introduced in Chapter 27; and orient the space coordinates in such a way that the paths of the light rays through the coordinate system are simple. This is best done by putting the origin of the coordinate system $(\chi = 0)$ at the Earth. Then the emitting galaxy will lie at some "radius" χ_e and some angular position (θ_e, ϕ_e). The cosmological line element

$$ds^2 = -dt^2 + a^2(t)[d\chi^2 + \Sigma^2(d\theta^2 + \sin^2\theta\, d\phi^2)]$$
$$= a^2(\eta)[-d\eta^2 + d\chi^2 + \Sigma^2(d\theta^2 + \sin^2\theta\, d\phi^2)], \qquad (29.4a)$$

$$\Sigma = \begin{cases} \sin\chi & \text{if } k = +1, \\ \chi & \text{if } k = 0, \\ \sinh\chi & \text{if } k = -1, \end{cases} \qquad (29.4b)$$

is spherically symmetric about $\chi = 0$ (i.e., about the Earth) whether $k = -1, 0,$ or $+1$. Consequently, the geodesics (photon world lines) that pass through both Earth and the emitting galaxy must all be radial

$$\theta = \theta_e, \qquad \phi = \phi_e, \qquad \chi = \chi(t). \qquad (29.5)$$

(One who wishes to forego any appeal to symmetry can examine the geodesic equation in the (t, χ, θ, ϕ) coordinate system, and discover that if $d\theta/d\lambda = d\phi/d\lambda = 0$, then $d^2\theta/d\lambda^2 = d^2\phi/d\lambda^2 = 0$. Consequently a geodesic that is initially radial will always remain radial.)

Consider, now, emission. A galaxy at rest (moving with the "cosmological fluid") at $(\chi_e, \theta_e, \phi_e)$ emits two successive crests, A and B, of a wave train toward Earth at coordinate times t_{Ae} and t_{Be}. It has been arranged that proper time as measured on the galaxy is the same as coordinate time ($t = \tau + $ const. was part of the construction process for the coordinate system in §27.4). Consequently the period of the radiation as seen by the emitter is $P_{em} = t_{eB} - t_{eA}$; and the wavelength is the same as the period when geometrized units are used:

$$\lambda_{em} = t_{eB} - t_{eA}. \qquad (29.6)$$

Next examine propagation. Wave crests A and B propagate along null geodesics. This fact enables one to read the world lines of the wave crests, $\chi_A(t)$ and $\chi_B(t)$, directly from the line element (29.4): $ds^2 = 0$ guarantees that $a(t)\, d\chi = -dt$ ($-$, not $+$, because the light propagates toward the Earth at $\chi = 0$). Consequently, the world lines are

$$\chi_e - \chi_A(t \text{ or } \eta) = \eta - \eta_{eA} = \int_{t_{eA}}^{t} a^{-1}\, dt, \qquad (29.7)$$

$$\chi_e - \chi_B(t \text{ or } \eta) = \eta - \eta_{eB} = \int_{t_{eB}}^{t} a^{-1}\, dt.$$

Finally, examine reception. The receiver on Earth moves with the "cosmological fluid," just as does the distant emitter. (Ignore the Earth's "peculiar motion" relative

to the fluid—motion around the sun, motion around center of our Galaxy, etc.; it can be taken into account by an ordinary Doppler correction.) Thus, for receiver as for emitter, proper time is the same as coordinate time, and

$$\lambda_{\text{rec}} = t_{rB} - t_{rA},\tag{29.8}$$

where t_{rB} and t_{rA} are the times of reception of the successive wave crests.

Now combine equations (29.6), (29.7), and (29.8) to obtain the redshift. The receiver is at $\chi = 0$. Therefore equations (29.7) say

$$0 = \chi_e - \int_{t_{eA}}^{t_{rA}} a^{-1}\, dt,$$

$$0 = \chi_e - \int_{t_{eB}}^{t_{rB}} a^{-1}\, dt.\tag{29.9}$$

Subtract these equations from each other to obtain

$$0 = \int_{t_{eB}}^{t_{rB}} a^{-1}\, dt - \int_{t_{eA}}^{t_{rA}} a^{-1}\, dt$$

$$= \int_{t_{rA}}^{t_{rB}} a^{-1}\, dt - \int_{t_{eA}}^{t_{eB}} a^{-1}\, dt \approx \frac{t_{rB} - t_{rA}}{a(t_r)} - \frac{t_{eB} - t_{eA}}{a(t_e)};$$

and combine with (29.6) and (29.8) to discover

$$\frac{\lambda_{\text{rec}}}{a(t_r)} = \frac{\lambda_{\text{em}}}{a(t_e)};\tag{29.10}$$

i.e.,

$$z \equiv \Delta\lambda/\lambda = a(t_r)/a(t_e) - 1.\tag{29.11}$$

These redshift equations confirm the simple result of Figure 29.1: As the light ray propagates, its wavelength (as measured by observers moving with the "fluid") increases in direct proportion to the linear expansion of the universe. *The ratio of the wavelength to the expansion factor, λ/a, remains constant.* For important applications of this result, see Boxes 29.2 and 29.3.

EXERCISES

Exercise 29.2. ALTERNATIVE DERIVATION OF REDSHIFT

Notice that the only part of the line element that is relevant for the light ray is

$$ds^2 = -dt^2 + a^2(t)\, d\chi^2,$$

since $d\theta = d\phi = 0$ along its world line (spherical symmetry!). Regard the light ray as made

Box 29.2 COSMOLOGICAL REDSHIFT OF THE PRIMORDIAL RADIATION

As an important application of the redshift formula

$$\lambda/a = \text{constant} \qquad (1)$$

[equation (29.10)], consider the radiation emerging from the hot big bang. Because it is initially in thermal equilibrium with matter, this primordial radiation initially has a Planck black-body spectrum. Subsequent interactions with matter cannot change the spectrum, because the matter remains in thermal equilibrium with the radiation so long as interactions are occurring. The cosmological redshift can and does change the spectrum, however. It was shown in exercise 22.17, using kinetic theory, that radiation with a Planck spectrum as viewed by one observer has a Planck spectrum as viewed by all observers; but the observed temperature is redshifted in precisely the same manner as the frequency of an individual photon is redshifted. Consequently, as seen by observers at rest in the "fluid," the temperature of the primordial radiation is redshifted

$$T \propto 1/a. \qquad (2)$$

This is true after plasma recombination, when the radiation and matter are decoupled, as well as before recombination, when they are interacting. And it is true not only for the primordial photons but also for thermalized neutrinos and gravitons emerging from the hot big bang.

There is another way to derive the redshift equation (2). Combine the equation

$$\rho_r \propto T^4 \qquad (3)$$

for the energy density of black-body radiation in terms of temperature, with the equation

$$\rho_r \propto (\text{volume})^{-4/3} \propto (a^3)^{-4/3} \propto a^{-4} \qquad (4)$$

for the decrease of energy density with adiabatic expansion.

Box 29.3 USE OF REDSHIFT TO CHARACTERIZE DISTANCES AND TIME

Distance: When discussing objects within the Earth's cluster of galaxies, astronomers typically describe distances in units of lightyears or parsecs. But when dealing with more distant objects (galaxies, quasars, etc.), astronomers find it more convenient to describe distance in terms of what is actually observed: redshift. For example, the statement "the galaxy 3C 295 is at a redshift of 0.4614" means that "3C 295 is at that distance from Earth [given by equation (29.16)] which corresponds to a redshift of $z = 0.4614$."

Time: When discussing events that occurred during the last few 10^9 years, astronomers usually measure time in units of years. Example: "The solar system condensed out of interstellar gas 4.6×10^9 years ago" [see Wasserburg and Burnett (1968)]. But when dealing with events much nearer the beginning of the universe, all of which have essentially the same age, of about 12×10^9 years, astronomers find it more convenient to describe time in terms of redshift. Example: "The primordial plasma recombined at a redshift of 1,000" means that "If a photon had been emitted at the time of plasma recombination, and had propagated freely ever since, it would have experienced a total redshift between then and now of $z = 1,000$." Equivalently, since $1 + z = (a_o/a)$ [see equation (29.11)], "the plasma recombined when the universe was a factor of $1 + z \approx 1,000$ smaller than it is today." [Application: In Figure 28.1, where the past evolution of the universe is summarized, one can freely replace the horizontal scale a/a_o by $1/(1 + z)$, and thereby see that primordial element formation occurred at a redshift of $z \approx 10^9$.) The conversion from redshift units to time units is strongly dependent on the parameters ρ_{mo}, ρ_{ro}, and k/a_o^2 [see §§27.10 and 27.11; also equation (29.15)].

of photons with 4-momenta \boldsymbol{p}. From the geodesic equation (or, for the reader who has studied chapter 25, from arguments about Killing vectors), show that

$$p_\chi \equiv \boldsymbol{p} \cdot (\partial/\partial\chi)$$

is conserved along the photon's world line. Use this fact, the fact that a photon's 4-momentum is null, $\boldsymbol{p} \cdot \boldsymbol{p} = 0$, and the equation $E = -\boldsymbol{p} \cdot \boldsymbol{u}$ for the energy measured by an observer with 4-velocity \boldsymbol{u}, to derive the redshift equation (29.11).

Exercise 29.3. REDSHIFT OF PARTICLE DE BROGLIE WAVELENGTHS

A particle of finite rest mass μ moves along a geodesic world line through the expanding cosmological fluid. Let

$$p \equiv (\boldsymbol{p} \cdot \boldsymbol{p})^{1/2} \equiv \frac{\mu v}{(1 - v^2)^{1/2}}$$

be the spatial 4-momentum of the particle as measured by observers at rest in the fluid. (The ordinary velocity they measure in their proper reference frames is v.) The associated "de Broglie wavelength" of the particle is $\lambda \equiv h/p$.

(a) Show that this de Broglie wavelength is redshifted in precisely the same manner as a photon wavelength:

$$\lambda/a = \text{constant}.$$

(b) Employing this result, show that, for the molecules of an ideal gas that fills the universe, their mean kinetic energy decreases in inverse proportion to a^2 when the gas is nonrelativistic and (like photon energies) in inverse proportion to a when the gas is highly relativistic.

§29.3. THE DISTANCE-REDSHIFT RELATION; MEASUREMENT OF THE HUBBLE CONSTANT

Equation (29.11) expresses the redshift in terms of the change in expansion factor between the event of emission and the event of reception. For "nearby" emitters (emitters at distances much less than $1/H_o$, the "Hubble length") it is more convenient to express the redshift in terms of the distance between the emitter and Earth. That distance ("present distance") is defined on the hypersurface of homogeneity that passes through Earth today, since that hypersurface agrees locally with the surface of simultaneity of the receiver today, and it is also, locally, a surface of simultaneity for any observer moving today with the "cosmological fluid."

Derivation of distance-redshift relation

The distance between emitter and observer today [the distance along the spatial geodesic of constant (t, θ, ϕ) connecting $(t_r, 0, \theta_e, \phi_e)$ and $(t_r, \chi_e, \theta_e, \phi_e)$] can be read directly from the line element (29.4):

$$\ell = a(t_r)(\chi_e - \chi_r) = a(t_r)\chi_e. \tag{29.12}$$

Using expression (29.9) for χ_e, one finds

$$\ell = a(t_r) \int_{t_e}^{t_r} a^{-1} \, dt. \tag{29.12'}$$

In the recent past, $a(t)$ was given by

$$a(t) = a(t_r) + (a_{,t})_{t_r}(t - t_r) + \frac{1}{2}(a_{,tt})_{t_r}(t - t_r)^2 + \cdots$$

$$= a(t_r)[1 + H_0(t - t_r) - \frac{1}{2}q_0 H_0^2(t - t_r)^2 + \cdots], \tag{29.13}$$

where definitions (29.1) for the Hubble constant H_0 and the deceleration parameter q_0 have been used. Putting this expression into equation (29.12′) and integrating, one finds for the distance the expression

$$\ell = (t_r - t_e) + \frac{1}{2}H_0(t_r - t_e)^2 + \cdots$$

or, equivalently,

$$t_r - t_e = \ell - \frac{1}{2}H_0\ell^2 + \cdots. \tag{29.14}$$

The redshift [equation (29.11)] can be expressed as a power series in $t_r - t_e$ by using equation (29.13):

$$z = \frac{a(t_r) - a(t_e)}{a(t_e)} = \frac{a(t_r)[H_0(t_r - t_e) + \frac{1}{2}q_0 H_0^2(t_r - t_e)^2 + \cdots]}{a(t_r)[1 - H_0(t_r - t_e) + \cdots]}$$

$$= H_0(t_r - t_e) + H_0^2\left(1 + \frac{1}{2}q_0\right)(t_r - t_e)^2 + \cdots. \tag{29.15}$$

Combining this with equation (29.14) for $t_r - t_e$ in terms of ℓ, one finally obtains

$$z = H_0\ell + \frac{1}{2}(1 + q_0)(H_0\ell)^2 + O([H_0\ell]^3). \tag{29.16}$$

Result for distance-redshift relation

This is the "*distance-redshift relation*" for the standard big-bang model of the universe.

By comparing this distance-redshift relation with astronomical observations (see Box 29.4, which is best read after the next section), Allan Sandage (1972a) obtains a Hubble constant of

Measurement of Hubble constant H_0

$$H_0 = 55 \pm 7 \text{ km sec}^{-1} \text{ Mpc}^{-1}; \tag{29.17}$$

i.e.,

$$H_0^{-1} = (18 \pm 2) \times 10^9 \text{ years}. \tag{29.18}$$

(Note: 1 Mpc \equiv one Megaparsec is 3.26×10^6 light years, or 3.08×10^{24} cm.) The uncertainty of ± 7 km sec^{-1} Mpc^{-1} quoted here is the "one-sigma" statistical uncertainty associated with the distance-redshift data. Systematic errors, not now understood, might be somewhat larger; but the true value of H_0 almost certainly is within a factor 2 of Sandage's value, 55 km sec^{-1} Mpc^{-1}.

Note that, if $\Lambda = 0$, then the "critical density" marking the dividing line between a "closed" universe and an "open" universe—i.e., between eventual recontraction and expansion forever—is

Value of critical density

$$\rho_{\text{crit}} = \frac{3}{8\pi} H_o{}^2 = 5 \times 10^{-30} \text{ g/cm}^3. \tag{29.19}$$

(As described in Box 29.1, $\rho > \rho_{\text{crit}} \iff$ "closed" \iff recontraction; $\rho < \rho_{\text{crit}} \iff$ "open" \iff expansion forever.) Comparison with the actual density will be delayed until §29.6.

The distance measurements are not accurate enough to yield useful information about the deceleration parameter, q_o.

§29.4. THE MAGNITUDE-REDSHIFT RELATION; MEASUREMENT OF THE DECELERATION PARAMETER

Information about q_o is best obtained by comparing the apparent magnitudes of galaxies with their redshifts.

Apparent magnitude defined

In astronomy one defines the apparent (bolometric) magnitude, m, of an object by the formula

$$
\begin{aligned}
m &= -2.5 \log_{10}(S/2.52 \times 10^{-5} \text{ erg cm}^{-2} \text{ sec}^{-1}) \\
&= -2.5 \log_{10} S + \text{constant},
\end{aligned}
\tag{29.20}
$$

where S is the flux of energy (energy per unit time per unit area) that arrives at Earth from the object. [Of course, one cannot measure the flux over the entire wavelength range $0 < \lambda < \infty$; so one distinguishes various apparent magnitudes (m_U, m_B, m_V, . . .) corresponding to fluxes in various wavelength ranges ("U" \equiv "ultraviolet"; "B" \equiv "blue"; "V" \equiv "visual"). However, these subtleties are too far from gravitation physics to be treated here.]

Derivation of magnitude-redshift relation

Calculate the apparent magnitude for a galaxy of intrinsic luminosity L and redshift z. To simplify the calculation, put the emitter at the origin of the space coordinates ($\chi_e = 0$); and put the Earth at (χ_r, θ_r, ϕ_r). (Note the reversal of locations compared to redshift calculation of §29.2.) On Earth, place a photographic plate of area A perpendicular to the incoming light. Then at time t_r the plate is a tiny segment of a spherical two-dimensional surface ($t = t_r, \chi = \chi_r; \theta$ and ϕ vary) about the emitting galaxy. The total area of the 2-sphere surrounding the galaxy is

$$\mathcal{A} = 4\pi[a(t_r)\Sigma(\chi_r)]^2. \tag{29.21}$$

Therefore, the ratio of the area of the plate to the area of the 2-sphere is given by

$$\frac{A}{\mathcal{A}} = \frac{A}{4\pi[a(t_r)\Sigma(\chi_r)]^2}. \tag{29.22}$$

The plate catches a fraction A/\mathcal{A} of the energy that pours out through the 2-sphere.

If there were no redshift, the power crossing the entire 2-sphere at time t_r would be precisely the luminosity of the emitter at time t_e. However, the redshift modifies this result in two ways. (1) The energy of each photon that crosses the 2-sphere is smaller, as measured in the local Lorentz frame of the fluid there, than it was as measured by the emitter:

$$E_{rec}/E_{em} = \lambda_{em}/\lambda_{rec} = 1/(1 + z). \tag{29.23}$$

(2) Two photons with the same θ and ϕ, which are separated by a time Δt_r as measured by an observer stationary with respect to the "cosmological fluid" at the 2-sphere, were separated by a shorter time Δt_e as measured by the emitter:

$$\Delta t_r/\Delta t_e = \lambda_r/\lambda_e = 1 + z. \tag{29.24}$$

The luminosity, L, as measured at the source, is the sum of the energies E_{emJ} of the individual photons (labeled with the index J) emitted in a time interval Δt_e, divided by Δt_e:

$$L = (1/\Delta t_e) \sum_J E_{emJ}. \tag{29.25}$$

The power that crosses the 2-sphere a time $t_r - t_e$ later, as measured by the fluid at the 2-sphere, is

$$P = (1/\Delta t_r) \sum_J E_{recJ}, \tag{29.26}$$

where the summation runs over the same set of photons.

Combining equations (29.23) to (29.26), one sees that the power crossing the 2-sphere is

$$P = L/(1 + z)^2.$$

Of this, a fraction,

$$P_A = \frac{A}{\mathcal{A}} P = \frac{AL}{4\pi[(1 + z)a(t_r)\Sigma(\chi_r)]^2},$$

crosses the photographic plate; so the flux measured at the Earth is

$$S = \frac{P_A}{A} = \frac{L}{4\pi R^2(1 + z)^2}, \tag{29.27}$$

where R is the "radius of curvature" of the 2-sphere surrounding the emitter and passing through the receiver at the time of reception,

$$R \equiv a_o \Sigma(\chi_r - \chi_e) = \begin{cases} a_o \sinh(\chi_r - \chi_e) & \text{if } k = -1, \\ a_o[\chi_r - \chi_e] & \text{if } k = 0, \\ a_o \sin(\chi_r - \chi_e) & \text{if } k = +1 \end{cases} \tag{29.28}$$

[recall: χ_e is 0 according to the present conventions, and $a_o = a(t_r)$]. The corresponding apparent magnitude [equation (29.20)] is

$$m = +5 \log_{10}[(1 + z)R] - 2.5 \log_{10}L + \text{constant}. \tag{29.29}$$

In order to relate the apparent magnitude to the redshift of the emitter, one must express the quantity R in terms of z. From equation (29.7) for the photon propagation (with sign reversed because positions of receiver and emitter have been reversed), one knows that

$$\chi_r - \chi_e = \int_{t_e}^{t_r} a^{-1}\, dt = \int_1^{a(t_r)/a(t_e)} \left[\frac{a}{a(t_r)}\right]\left[\frac{dt}{da}\right] d\left[\frac{a(t_r)}{a}\right], \tag{29.30}$$

and from equation (29.11) one knows that

$$z = a(t_r)/a(t_e) - 1.$$

Hence

$$\chi_r - \chi_e = \int_1^{1+z} \left[\frac{a}{a(t_r)}\right]\left[\frac{dt}{da}\right] d\left[\frac{a(t_r)}{a}\right]. \tag{29.31}$$

Equations (4) to (6) of Box 29.1, and (27.40), determine the function dt/da in terms of $a/a(t_r)$ and the constants H_o, q_o, σ_o. By inserting that result into equation (29.31) and integrating, one obtains $\chi_r - \chi_e$ in terms of the redshift z and the cosmological parameters H_o, q_o, σ_o:

$$\chi_r - \chi_e = |1 + q_o - 3\sigma_o|^{1/2} \int_1^{1+z} \frac{du}{[2\sigma_o u^3 + (1 + q_o - 3\sigma_o)u^2 + \sigma_o - q_o]^{1/2}}. \tag{29.32a}$$

The 2-sphere radius of curvature R is obtained by inserting this expression into the equation

$$R = \frac{H_o^{-1}}{|1 + q_o - 3\sigma_o|^{1/2}} \Sigma(\chi_r - \chi_e) \tag{29.32b}$$

[equation (29.28), with a_o evaluated by equation (5) of Box 29.1].

Equations (29.29) and (29.32) determine the apparent magnitude, m, in terms of redshift, z.

For the case of vanishing cosmological constant ($\sigma_o = q_o$; $\Lambda = 0$), the integral (29.32a) can be expressed in terms of elementary functions, yielding

$$R = \frac{H_o^{-1}}{q_o^2(1 + z)}[-q_o + 1 + q_o z + (q_o - 1)(2q_o z + 1)^{1/2}],$$

$$\approx H_o^{-1}z\left[1 - \frac{1}{2}(1 + q_o)z + O(z^2)\right], \tag{29.33}$$

so that

Result for magnitude-redshift relation

$$m = 5 \log_{10}[1 - q_o + q_o z + (q_o - 1)(2q_o z + 1)^{1/2}] - 2.5 \log_{10} L + \text{const.}$$

$$\approx 5 \log_{10} z + 1.086(1 - q_o)z + O(z^2) - 2.5 \log_{10} L + \text{const.} \tag{29.34}$$

$$\text{for } z \ll 1.$$

(Note: the factor 1.086 is actually 2.5/ln 10.) A power-series solution for nonzero Λ (for $\sigma_o \neq q_o$) reveals a dependence on σ_o only at $O(z^2)$ and higher:

$$R \approx H_o^{-1} z \left[1 - \frac{1}{2}(1 + q_o)z + \text{(corrections of } O(z^2) \text{ depending on } \sigma_o \text{ and } q_o) \right],$$

<div align="right">(29.35a)</div>

$$m \approx 5 \log_{10} z + 1.086(1 - q_o)z + O(z^2) - 2.5 \log_{10} L + \text{const.} \quad (29.35b)$$

Sheldon (1971) gives the exact solution for $\Lambda \neq 0$ in terms of the Weierstrass elliptic function. Refsdal *et al.* (1967) tabulate and plot the exact solution.

By comparing the theoretical magnitude-redshift relation (29.35b) with observations of the brightest galaxies in 82 clusters, Allan Sandage (1972a,c,d) obtains the following value for the deceleration parameter:

<div align="right">Measurement of deceleration
parameter, q_o</div>

$$q_o = 1.0 \pm 0.5, \qquad \text{if } \sigma_o = q_o (\text{i.e. } \Lambda = 0). \quad (29.36)$$

(Note: 0.5 is the "one-sigma" uncertainty. Sandage estimates with 68 per cent confidence that $0.5 < q_o < 1.5$, and with 95 per cent confidence that $0 < q_o < 2$—providing unknown evolutionary effects are negligible.) The observations leading to this result and the uncertainties due to evolutionary effects are described in Box 29.4. Box 29.5 gives a glimpse of Edwin Hubble, the man who laid the foundations for such cosmological measurements.

(continued on page 794)

**Box 29.4 MEASUREMENT OF HUBBLE CONSTANT AND
DECELERATION PARAMETER**

I. *Hubble Constant, H_o*

 A. *Objective:* To measure the constant H_o by comparing observational data with the distance-redshift relation

 $$z = H_o \ell + \frac{1}{2}(1 + q_o)(H_o \ell)^2 + O([H_o \ell]^3).$$

 Here ℓ is distance from Earth to source today; and z is redshift of source as measured at Earth.

 B. *Key Difficulty:* This distance-redshift relation does not apply to stars in our Galaxy: the Galaxy is gravitationally bound and therefore is impervious to the universal expansion. Nor does the distance-redshift relation apply to the separations between our Galaxy and nearby galaxies (the "local group"); gravitational attraction between our Galaxy and its neighbors is so great it perturbs their motions substantially away from universal expansion. Only on

Box 29.4 (continued)

scales large enough to include many galaxies (scales where each galaxy or cluster of galaxies can be thought of as a "grain of dust," with the grains distributed roughly homogeneously)—only on such large scales should the distance-redshift relation hold with good accuracy. But it is very difficult to obtain reliable measurements of the distances ℓ to galaxies that are so far away!

C. *Procedure by which H_o has been measured* [*Sandage and Tamman, as summarized in Sandage (1972a)*]:

1. Cepheid variables are pulsating stars with pulsation periods (as measured by oscillations in light output) that are very closely correlated with their luminosities L—or, equivalently, with their absolute (bolometric) magnitudes, M:

$$M \equiv \begin{pmatrix} \text{apparent magnitude star would have were it at a} \\ \text{distance of 10 parsecs} = 32.6 \text{ light years} \end{pmatrix} \tag{1}$$

$$= -2.5 \log_{10}(L/3.0 \times 10^{35} \text{ erg sec}^{-1})$$

[see equation (29.20).] By measurements within our Galaxy, astronomers have obtained the "period-luminosity relation" for cepheid variables.

2. Cepheid variables are clearly visible in galaxies as far away as \sim4 Mpc (4 Megaparsecs $\equiv 4 \times 10^6$ parsecs). In each such galaxy one measures the periods of the cepheids; one then infers their absolute magnitudes M from the period-luminosity relation; one measures their apparent magnitudes m; and one then calculates their distances ℓ from Earth using the relation

$$m - M = 5 \log_{10}(\ell/10 \text{ pc}). \tag{2}$$

By this means one obtains the distances ℓ to all galaxies within \sim4 Mpc of our own. Unfortunately, such galaxies are too close to participate cleanly in the universal expansion. (They include only the "local group," the "M81 group," and the "south polar group.") Thus, one must push the distance scale out still farther before attempting to measure H_o.

3. Galaxies of types Sc, Sd, Sm, and Ir within \sim4 Mpc contain huge clouds of ionized hydrogen, which shine brightly in "Hα light." These clouds, called "H II regions," exhibit a very tight correlation between diameter D of the H II region and luminosity L of the galaxy (or, equivalently, between D and absolute magnitude of galaxy, M). In fact, for a given galaxy luminosity L, the fractional spread in H II diameters is $\sigma(\Delta D/D) \approx 0.12$. Using (a) the distances (\leq4 Mpc) to these galaxies as determined via cepheid variables, (b) the apparent magnitudes of the galaxies, and (c) the angular diameters of H II regions in the galaxies, one calculates

the actual H II diameters D and galaxy luminosities L, and thereby obtains the "diameter-luminosity relation" $D(L)$.

4. H II regions are large enough to be seen clearly in galaxies as far away as \sim60 Mpc. By measuring the H II angular diameters $\alpha = D/\ell$ and galaxy apparent (bolometric) magnitudes

$$m = -2.5 \, \log_{10} \left(\frac{L/4\pi\ell^2}{2.52 \times 10^{-5} \, \text{erg cm}^{-2} \, \text{sec}^{-1}} \right), \qquad (3)$$

and by combining with the diameter-luminosity relation, one obtains the distances ℓ to all galaxies of types Sc, Sd, Sm, and Ir which possess H II regions and lie within \sim60 Mpc of Earth. Unfortunately, this is *still* not far enough away for local motions to be negligible compared with the universal expansion.

5. Within \sim60 Mpc reside enough galaxies of type Sc I for one to discover that their luminosities (absolute magnitudes) are rather constant (difference in L from one Sc I galaxy to another \lesssim50 per cent). Using the distances to such Sc I galaxies, as measured via H II regions, and using measurements of their apparent magnitudes, one calculates their universal absolute magnitude (measured photographically) to be $M_{pg} = -21.2$.

The Sc I galaxy M101 at a distance $\ell \sim 3$ Mpc from Earth, as photographed with the 200-inch telescope. (Courtesy of Hale Observatories)

Box 29.4 (*continued*)

6. One then examines all known Sc I galaxies with distances greater than ~70 Mpc. For each, one measures the apparent magnitude and compares it with the universal absolute magnitude to obtain the distance ℓ from Earth. And for each, one measures the redshift $z = \Delta\lambda/\lambda$ of the spectral lines. From the resulting redshift-distance relation—and taking into account the statistical uncertainties in all steps leading up to it—Sandage and Tamman (work carried out in 1965–1972) obtain the value $H_o = dz/d\ell = 55 \pm 7$ (km/sec) Mpc^{-1} = $1/[(18 \pm 2) \times 10^9$ years]. [For a review see Sandage (1972a).] The quoted error is purely statistical. Systematic errors are surely larger—but they almost surely do not exceed a factor 2 [i.e., $30 < H_o < 110$ (km/sec) Mpc^{-1}].

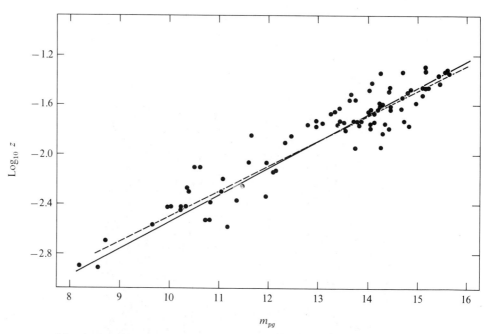

Magnitude-redshift relation for Sc I galaxies at distances $\gtrsim 70$ Mpc. Solid line is a least-squares fit to the data; dotted line has the theoretical slope of 5. [From Sandage and Tamman.]

II. *Deceleration Parameter, q_o.*

A. *Objective:* To measure the constant q_o by comparing observational data with the magnitude-redshift relation:

$$m = 5 \log_{10} z + 1.086(1 - q_o)z + O(z^2) - 2.5 \log_{10} L + \text{const.} \quad (4)$$

[Note: This relation is valid even if the cosmological constant is nonzero, i.e., even if $\sigma_o \neq q_o$. Dependence on σ_o occurs only at $O(z^2)$ and higher.]

B. *Key Difficulty:* One must use data for objects with the same absolute luminosity L ("standard candles"). But one cannot measure L at distances great enough for the effects of q_o to show up.

C. *The Search for a Standard Candle:* One obvious choice for the standard candle would be Sc I galaxies, since they were found to all have nearly the same L (see above). But they are not bright enough to be seen at distances great enough for effects of q_o to show up. An alternative choice, quasars, are bright enough to be seen at very large redshifts (z as large as \sim3). But their absolute luminosities have enormous scatter—or so one infers from the failure of quasars to fall on a straight line, even at small z, in the magnitude-redshift diagram. The best choice is the brightest type of object that has small scatter in L. Sandage (1972a,b,c) chooses the brightest galaxy in "recognized regular clusters of galaxies." Such clusters are composed predominately of E-type galaxies, and the brightest members are remarkably similar from one cluster

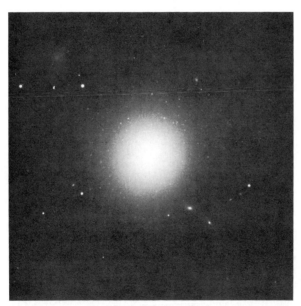

The E-type galaxy M87 at a distance $\ell \sim 11$ Mpc from Earth, as photographed with the 200-inch telescope. (Courtesy of Hale Observatories)

to another (scatter in L is \sim25 per cent). The similarity shows up in their spectra and in the very precise straight lines they give when one plots angular diameter versus redshift (next page), or apparent magnitude versus redshift (next page), or angular diameter versus apparent magnitude.

Box 29.4 (continued)

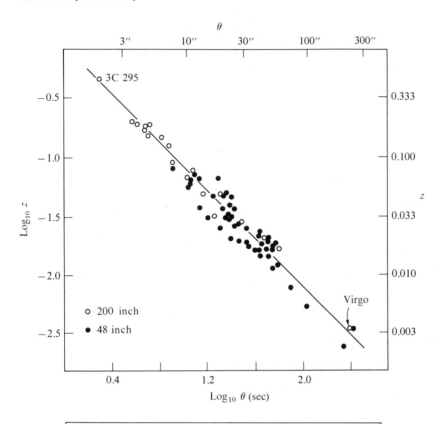

Angular diameter versus redshift for brightest galaxy in recognized regular clusters. From Sandage (1972a,b). [These data are not sufficiently precise to yield useful information about q_o and σ_o; but improvements in 1973 may bring the needed precision; see §29.5.]

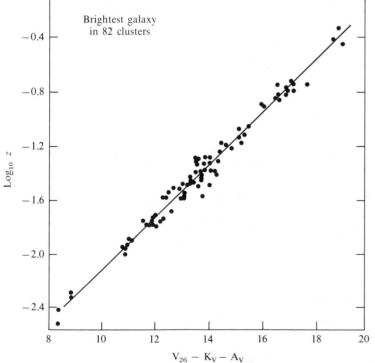

Magnitude versus redshift for brightest galaxy in recognized regular clusters. $V_{26} - K_V - A_V$ is the apparent magnitude with certain corrections taken into account. The line plotted corresponds to $\sigma_o = q_o = 1$ (straight line of slope 5). From Sandage (1972a).

D. *Procedure by which q_o has been measured* [*Sandage (1972a,c)*]:
 1. Data on magnitude versus redshift have been gathered for the brightest galaxy in 82 recognized regular clusters (see above).
 2. The data, when fitted with a straight line, show a slope of

$$dm/d \log_{10} z = 5.150 \pm 0.268 \text{ (rms)}, \tag{5}$$

 by comparison with a theoretical slope of 5.
 3. The data, when fitted to the theoretical relation

$$m = 5 \log_{10} z + 1.086(1 - q_o)z + O(z^2) + \text{const}, \tag{6}$$

 [with the correct $O(z^2)$ and higher terms included; see equations (29.29), (29.32), and (29.34)] yield

$$\left. \begin{array}{l} q_o = 1 \pm 0.5 \text{ (one-sigma)} \\ = 1 \pm 1 \quad \text{(two-sigma)} \end{array} \right\} \text{ if } \sigma_o = q_o(\Lambda = 0). \tag{7}$$

 The data are inadequate to determine σ_o and q_o simultaneously. [The $O(z^2)$ terms, which depend on σ_o, play a significant role in the fit to the data. For a graphical depiction of their theoretical effects see Figure 2 of Refsdal *et. al* (1967).]

E. *Evolutionary uncertainties*
 1. Sandage's fit of data to theory assumes that the luminosities of his "standard candles" are constant in time. If, because of evolution of old stars and formation of new ones, his galaxies were to dim by 0.09 magnitudes per 10^9 years, then galaxies 10^9 light-years away, which one sees as they were 10^9 years ago, would be 0.09 magnitudes brighter intrinsically than identical nearby galaxies. Correction for this effect would lower the most probable value of q_o from 1 to 0 [Sandage (1972c)].
 2. Knowledge of the evolution of galaxies in 1972 is too rudimentary to confirm or rule out such an effect. [See references cited by Sandage (1972c).]

792

Box 29.5 EDWIN POWELL HUBBLE
November 20, 1889, Marshfield Missouri—
September 28, 1953, Pasadena, California

Edwin Hubble, at age 24, earned a law degree from Oxford University and began practicing law in Louisville, Kentucky. After a year of practice he became fed up and, in his own words, "chucked the law for astronomy, and I knew that even if I were second-rate or third-rate it was astronomy that mattered." He chose the University of Chicago and Yerkes Observatory as the site for his astronomy education, and three years later (1917) completed a Ph.D. thesis on "Photographic Investigations of Faint Nebulae."

When Hubble entered astronomy, it was suspected that some nebulae lie outside the Galaxy, but the evidence was exceedingly weak. During the subsequent two decades, Hubble, more than anyone else, was responsible for opening to man's purview the extragalactic universe. Working with the 60-inch and 100-inch telescopes at Mount Wilson, Hubble developed irrefutable evidence of the extragalactic nature of spiral nebulae, elliptical nebulae, and irregular nebulae (now called galaxies). He devised the classification scheme for galaxies which is still in use today. He systematized the entire subject of extragalactic research: determining distance scales, luminosities, star densities, and the peculiar motion of our Galaxy; and obtaining extensive evidence that the laws of physics outside the Galaxy are the same as near Earth (in Hubble's words: "verifying the principle of the uniformity of nature"). He discovered and quantified the large-scale homogeneity of the universe. And—his greatest triumph of all!—he discovered the expansion of the universe.

The details of Hubble's pioneering work are best sketched in his own words:

"Extremely little is known of the nature of nebulae; and no classification has yet been suggested. . . . The agreement [between the velocity of escape from a spiral nebula and that from our galaxy] is such as to lend some color to the hypothesis that the spirals are stellar systems at distances to be measured often in millions of light years."

(1920; Ph.D. THESIS; PUBLICATION DELAYED 3 YEARS BY WORLD WAR I)

This box is based largely on the biography of Hubble by Mayall (1970).

"The present investigation [using Cepheid variables for the first time as an indicator of distances beyond the Magellanic clouds] identifies NGC 6822 as an isolated system of stars and nebulae of the same type as the Magellanic clouds, although somewhat smaller and much more distant. A consistent structure is thus reared on the foundation of the Cepheid criterion, in which the dimensions, luminosities, and densities, both of the system [NGC 6822] as a whole and its separate members, are of orders of magnitude which are thoroughly familiar. The distance is the only quantity of a new order. The principle of the uniformity of nature thus seems to rule undisturbed in this remote region of space."

(1925)

"Critical tests made with the 100-inch reflector, the highest resolving power available, show no difference between the photographic images of the so-called condensations in Messier 33 and the images of ordinary galactic stars. . . . The period-luminosity relation is conspicuous among the thirty-five Cepheids and indicates a distance about 8.1 times that of the Small Magellanic Cloud. Using Shapley's value for the latter, the distance of the spiral is about 263,000 parsecs."

(1926a)

"[To the present paper (1926b)] is prefaced a general classification of nebulae . . . the various types [of extragalactic nebulae] are homogeneously distributed over the sky. . . . The data are now available for deriving a value for the order of the density of space. This is accomplished by means of the formulae for the numbers of nebulae to a given limiting magnitude and for the distance in terms of the magnitude. [The result is]

$\rho = 1.5 \times 10^{-31}$ *grams per cubic centimeter.*

This must be considered as a lower limit, for loose material scattered between the systems is entirely ignored. The mean density of space can be used to determine the dimensions of the finite but boundless universe of general relativity . . .

$$R = \frac{c}{\sqrt{4\pi k}} \frac{1}{\sqrt{\rho}} = \cdots = 2.7 \times 10^{10} \text{ parsecs."}$$

(1926b)

"The data . . . indicate a linear correlation between distances and velocities [for extragalactic nebulae]. Two solutions have been made, one using the 24 nebulae individually, the other combining them into 9 groups according the proximity in direction and distance. The results are . . . 24 objects: K = 465 ± 50 km/sec per 10^6 parsecs; 9 groups: K = 513 ± 60 km/sec per 10^6 parsecs. . . . The outstanding feature, however, is the possibility that the velocity-distance relation may represent the de Sitter effect, and hence that numerical data may be introduced into discussions of the general curvature of space."

(1929)*

*Hubble's value of K (the "Hubble constant") was later revised downward by the work of Baade and Sandage; see section titled *The Hubble Time* in Box 27.1.

Box 29.5 (*continued*)

"The velocity-distance relation is re-examined with the aid of 40 new velocities. . . . The new data extend out to about eighteen times the distance available in the first formulation of the velocity-distance relation, but the form of the relation remains unchanged except for [Shapley's 10 per cent] revision of the unit of distance."

(1931), WITH M. L. HUMASON

"Many ways of producing such effects [redshifts in extragalactic nebulae] are known, but of them all, only one will produce large redshifts without introducing other effects which should be conspicuous but actually are not found. This one known permissible explanation interprets redshifts as due to actual motion away from the observer."

(1934a)

"We now have a hasty sketch of some of the general features of the observable region as a unit. The next step will be to follow the reconnaissance with a survey—to repeat carefully the explorations with an eye to accuracy and completeness. The program, with its emphasis on methods, will be a tedious series of successive approximations."

(1934b)

Most of the remainder of Hubble's career was dedicated to this "tedious series of successive approximations." Shortly before Hubble's death the 200-inch telescope went into operation at Palomar Mountain; and Hubble's student, Alan Sandage, began using it in a continuation of Hubble's quest into the true nature of the universe. (See Box 29.4).

EXERCISES

Exercise 29.4. m(z) DERIVED USING STATISTICAL PHYSICS

Derive the magnitude-redshift relation using a statistical description of the photon distribution [cf. eq. (22.49) and associated discussion].

Exercise 29.5. DOPPLER SHIFT VERSUS COSMOLOGICAL REDSHIFT

(a) Consider, in flat spacetime, a galaxy moving away from the Earth with velocity v, and emitting light that is received at Earth. Let the distance between Earth and galaxy, as measured in the Earth's Lorentz frame at some specific moment of emission, be r; and let the Doppler shift of the radiation when it is eventually received be $z = \Delta\lambda/\lambda$. Show that the flux of energy S received at the Earth is related to the galaxy's intrinsic luminosity L by

$$S = \frac{L}{4\pi r^2(1 + z)^4}.$$

(29.37)

[Track-2 readers will find it most convenient to use the statistical formalism of equation (22.49).]

(b) Compare this formula for the flux with formula (29.27), where the redshift is of cosmological origin. Why is the number of factors of $1 + z$ different for the two formulas? [Mathematical answer: equation (6.28a) of Ellis (1971).]

§29.5. SEARCH FOR "LENS EFFECT" OF THE UNIVERSE

Curved space should act as a lens of great focal length. The curving of light rays has little effect on the apparent size of nearby objects. However, distant galaxies— galaxies from a quarter of the way up to halfway around the universe—are expected to have greatly magnified angular diameters [Klauder, Wakano, Wheeler, and Willey (1959)]. To see a normal galaxy at such a distance by means of an optical telescope seems out of the question. However, radio telescopes resolve features in quasistellar sources and other radiogalaxies at redshifts of $z = 2$ or more. Moreover, paired radio telescopes at intercontinental distances (for example, Goldstone, California, and Woomera, Australia) resolve distant sources to better than $0''.001$ or 4.8×10^{-9} radians or 15 lightyears for an object at a distance of 3×10^9 lightyears (Euclidean geometry temporarily being assumed). A radio telescope in space paired with a radio telescope on earth will be able to do even better on angular resolution. Will one be able to find any fiducial distance characteristic of any one class of objects that will serve as a natural standard of length, for very great distances ($z = 2$ to $z = 3$) as well as for galaxies closer at hand? Perhaps not. However, it would seem unwise to discount this possibility, with all the advantages it would bring, in view of the demonstrated ability of skilled observers to find regularities elsewhere where one had no right to expect them in advance.

The hope for a fiducial length in distant objects

Let L denote the actual length of a fiducial element (if any be found) in a galaxy; and let $\delta\theta$ (radians!) denote the apparent length of the object, idealized as perpendicular to the line of sight, as seen by the observer. The ratio of these two quantities defines the "angle effective distance" of the source,

Angle effective distance defined

$$\text{(angle effective distance)} = r_{\text{aed}} = L/\delta\theta. \tag{29.38a}$$

In flat space and for objects with zero relative velocity, this distance is to be identified with the actual distance, r, to the source or with the actual time of flight, t, of light from source to observer. The situation is changed in an expanding universe.

To calculate the angle effective distance as a function of redshift, place the Earth (receiver) at $\chi_r = 0$; and place the object under study (emitter) at χ_e. Let the fiducial length L lie on the sphere at χ_e (perpendicular to line of sight), and let it run from θ_e to $\theta_e + \delta\theta$ [one end of fiducial element at $(\chi_e, \theta_e, \phi_e)$; other at $(\chi_e, \theta_e + \delta\theta, \phi_e)$]. Then

$$L = a(t_e)\Sigma(\chi_e)\,\delta\theta,$$

and

$$r_{\text{aed}} = a(t_e)\Sigma(\chi_e) = [a(t_r)\Sigma(\chi_e - \chi_r)]a(t_e)/a(t_r);$$

i.e. [see equation (29.28), with χ_r and χ_e reversed],

$$r_{\text{aed}} = R/(1 + z). \tag{29.38b}$$

Angle effective distance as function of redshift

Here R is given as a function of redshift of source, z, and cosmological parameters H_o, q_o, σ_o, by equations (29.32) in general, or by (29.33) if $\Lambda = 0$. [Equation (29.38b) is modified if the beam preferentially traverses regions of low mass density ("vacuum between the galaxies"); see equation (22.37) and Gunn (1967).]

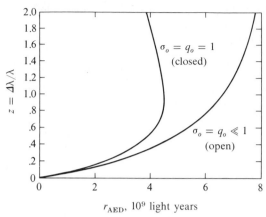

Figure 29.2.

Angle effective distance versus redshift for two typical cosmo-logical models—one open ($0 < \sigma_o = q_o \ll 1$); the other closed ($\sigma_o = q_o = 1$); both with zero cosmological constant; both with $H_o^{-1} = 18 \times 10^9$ lyr.

Angle effective distance as a tool for determining whether universe is closed

Figure 29.2 shows angle effective distance as a function of redshift for a few selected choices of the relevant parameters. It is evident that the angle effective distance has a maximum for a redshift roughly of the order $z \sim 1$, provided that the universe is closed. However, there is a big difference if the universe is open (Figure 29.2). The rapid improvements taking place in radio astronomy make increasingly attractive the possibility it provides for testing whether the universe is closed, as Einstein argued it should be [Einstein (1950), pp. 107–108]. Moreover, even with optical telescopes, in 1973 one may be on the verge of measuring q_o by studies of angle effective distance: preliminary studies [Sandage (1972b)] suggest that the optical size of the brightest E-type galaxies may be a usable fiducial length.

§29.6. DENSITY OF THE UNIVERSE TODAY

Measurements of mean mass density of universe:

(1) luminous matter in galaxies

It is exceedingly difficult to measure the mean density ρ_{mo} of the universe today. A large amount of matter may be in forms that astronomers have not yet managed to observe (intergalactic matter, black holes, etc.). Therefore, the best one can do is to add up all the luminous matter in galaxies and regard the resulting number as a lower limit on ρ_{mo}. Even adding up the luminous matter is a difficult and risky task, so difficult that even today no analysis is more definitive than the classic work of Oort (1958). [See, however, the very detailed review of the problem in Chapter 4 of Peebles (1971)]. Oort's result is

$$\rho_{\text{luminous matter}} \sim (2 \times 10^{-31} \text{ g/cm}^3)(H_o/55 \text{ km sec}^{-1} \text{ Mpc}^{-1})^2, \quad (29.39)$$

corresponding to

$$\sigma_o \gtrsim 0.02 \quad \text{(independent of the value of } H_o\text{).} \quad (29.40)$$

As an example (albeit an atypical one) of the danger inherent in any such estimate, Oort points to the Virgo cluster of galaxies. If the Virgo cluster is not gravitationally bound, then its ~2,500 galaxies will go flying apart, destroying any semblance of a cluster, in about one billion years. If it is gravitationally bound, then the mean velocity of its galaxies relative to each other, when combined with the virial theorem, yields an estimate of the cluster's total mass. That estimate is 25 times larger than the value one gets by Oort's method of adding up the luminous mass of the cluster.

Although one has no *definitive* evidence for or against large amounts of matter (enough to close the universe) in intergalactic space, one has tentative indirect limits: (1) If $\Lambda = 0$ (in accord with Einstein), then $\sigma_o = q_o$; so Sandage's value of $q_o \lesssim 1$—stretched to $q_o < 10$ under the most wild of assumptions about galaxy evolution—implies

$$\rho_{ig} < 10^{-28}\,\text{g/cm}^3 \qquad (\sigma_o = q_o < 10).$$

(2) Gott and Gunn (1971) point out that, if the density of gas in intergalactic space were $\geq 10^{-30}$ g/cm^3 (i.e., if σ_o were ≥ 0.1), one would expect gas falling into the Coma cluster of galaxies to form a shock wave, which would emit large amounts of X-rays. From the current X-ray observations, one can place a limit on the amount of such infalling matter—and therefrom a limit

$$\rho_{ig} \lesssim 10^{-30}\,\text{g/cm}^3 \qquad (\sigma_o \lesssim 0.1)$$

on the density of gas in intergalactic space. But these limits, like others obtained in other ways [see Chapter 4 of Peebles (1971) for a review] are far from definitive; they depend too much on theoretical calculations to make one feel fully comfortable.

§29.7. SUMMARY OF PRESENT KNOWLEDGE ABOUT COSMOLOGICAL PARAMETERS

The best data available in 1973 [equations (29.18), (29.36), (29.40)] reveal

$$H_o^{-1} = (18 \pm 2) \times 10^9 \text{ years,}$$

$$q_o = 1 \pm 0.5 \text{ (one-sigma)} \qquad \text{if } \sigma_o = q_o (\Lambda = 0), \qquad (29.41)$$

$$\sigma_o \geq 0.02,$$

for the observational parameters of the universe. These numbers are inadequate to reveal whether the universe is closed or open, and whether it will continue to expand forever or will eventually slow to a halt and recontract.

If one is disappointed in this lack of knowledge, one can at least be consoled by the following. (1) There is excellent agreement between theory and observation for the linear (low-z) parts of the distance-redshift, magnitude-redshift, and angular diameter-redshift relations (Box 29.4). (2) There is remarkably good agreement between (a) the age of the universe (18 billion years if $q_o = \sigma_o \ll 1$; 12 billion years if $q_o = \sigma_o = \frac{1}{2}$) as calculated from the measured value of H_o; (b) the ages of the

(2) matter in intergalactic space

Summary of observational parameters of universe

Some quantitative triumphs of cosmology

oldest stars ($\sim 10 \times 10^9$ years) as calculated by comparing the theory of stellar evolution with the properties of the observed stars; (c) the time (~ 9 billion years) since nucleosynthesis of the uranium, thorium, and plutonium atoms that one finds on Earth, as calculated from the measured relative abundances of various nucleides; and (d) the ages (4.6 billion years) of the oldest meteorites and oldest lunar rock samples, as calculated from measured relative abundances of other nucleides. For further detail see, e.g., Sandage (1968, 1970), Wasserburg *et al.* (1969), Wasserburg and Burnett (1968), and Fowler (1972). (3) Observations of the cosmic microwave radiation and measurements of helium abundance are now capable of giving direct information about physical processes in the universe at redshifts $z \gg 1$ (Chapter 28). (4) One may yet find "fiducial lengths" in radio sources, visible out to $z \gtrsim 1$, with which to measure q_o and σ_o by the angle-effective-distance method (§29.5). (5) The enigmas of the nature of quasars and of their peculiar distribution with redshift (great congregation at $z \sim 2$; absence at $z \gtrsim 3$) may yet be cracked and may yield, in the process, much new information about the origin of structure in the universe (Box 28.1). (6) The next decade may well bring as many great observational surprises, and corresponding new insights, as has the last decade.

The bright prospects for observational cosmology

EXERCISES

Exercise 29.6. SOURCE COUNTS

Suppose that one could find (which one cannot) a family of light or radio sources that (1) are all identical with intrinsic luminosities L, (2) are distributed uniformly throughout the universe, and (3) are born at the same rate as they die so that the number in a unit comoving coordinate volume is forever fixed.

(a) Show that the number of such sources $N(z)$ with redshifts less than z, as observed from Earth today, would be

$$N(z) = (\text{constant}) \cdot z^3 \left[1 - \frac{3}{2}(1 + q_o)z + O(z^2) \right]. \qquad (29.42)$$

(b) Show that the number of sources $N(S)$ with fluxes greater than S as observed at Earth today would be

$$N(S) = (\text{constant}) \cdot \left(\frac{LH_o^2}{4\pi S} \right)^{3/2} \left[1 - 3 \left(\frac{LH_o^2}{4\pi S} \right)^{1/2} + O \left(\frac{LH_o^2}{4\pi S} \right) \right] \qquad (29.43)$$

$$\underbrace{\qquad}_{= z^2 + O(z^3)} \quad \underbrace{\left[\begin{array}{l} \text{first-order correction} \\ \text{independent of } q_o \text{ and } \sigma_o \end{array} \right]}$$

[*Answer:* See §15.7 of Robertson and Noonan (1968).]

Exercise 29.7. COSMIC-RAY DENSITY (Problem devised by Maarten Schmidt)

Suppose the universe has contained the same number of galaxies indefinitely into the past. Suppose further that the cosmic rays in the universe were created in galaxies and that a negligible fraction of them have been degraded or lost since formation. Derive an expression for the average density of energy in cosmic rays in the universe today in terms of: (1) the number density of galaxies, N_o, today; and (2) the nonconstant rate, dE/dz, at which the average galaxy created cosmic-ray energy during the past history of the universe. [At redshift z in range dz, the average galaxy liberates energy $(dE/dz)\, dz$ into cosmic rays.]

Exercise 29.8. FRACTION OF SKY COVERED BY GALAXIES

Assume that the redshifts of quasars are cosmological. Let the number of galaxies per unit physical volume in the universe today be N_o, and assume that no galaxies have been created or destroyed since a redshift of ≥ 7. Let D be the average angular diameter of a galaxy. Calculate the probability that the light from a quasar at redshift z, has passed through at least one intervening galaxy during its travel to Earth. [For a detailed discussion of this problem, see Wagoner (1967).]

CHAPTER **30**

ANISOTROPIC AND INHOMOGENEOUS COSMOLOGIES

This chapter is entirely Track 2.
The main text requires no
 special preparation, although
 Chapters 27–29 would be
 helpful.
Box 30.1 contains more
 technical sections: ideal
 preparation for it would be
 Chapters 4, 9–14, 21, and
 27–29, plus §25.2; minimal
 preparation would be
 exercises 9.13, 9.14, and
 25.2, Chapter 21 through
 §21.8, and §§27.8, 27.11,
 and 29.2.
Chapter 30 is not needed as
 preparation for any later
 chapter.

Motivation for studying
inhomogeneous and
anisotropic cosmologies: Why
is universe so uniform?

§30.1. WHY IS THE UNIVERSE SO HOMOGENEOUS AND ISOTROPIC?

The last three chapters studied the Friedmann cosmological models and the relatively satisfactory picture they give of the universe and its evolution. This chapter describes less simplified cosmological models, and uses them to begin answering the question, "Why are the very simple Friedmann models satisfactory?" This question is intended to probe more deeply than the first, obvious answer—namely, that the models are satisfactory because they do not contradict observations. Accepting the agreement with observations, we want to understand *why the laws of physics should demand* (*rather than merely permit*) *a universe that is homogeneous and isotropic to high accuracy on large scales.* Because this question cannot be answered definitively in 1972, many readers will prefer to omit this chapter on the first reading and return to it only after they have surveyed the major results in other areas such as black holes (Chapter 33), gravitational waves (Chapters 35–37), and solar-system experiments (Chapter 40).

The approach described here to the question "Why is the universe so highly symmetric?" is to ask Einstein's equations to describe what would have happened if the universe had started out highly irregular.

The first step in this approach is to ask what would have happened if the universe had started a little bit irregular. This problem can be tackled by analyzing small perturbations away from the high symmetry of the Friedmann models. Such an analysis is most fruitful in its discussion of the beginnings of galaxy formation, and

in its ability to relate small upper limits on the present-day anisotropy of the microwave background radiation to limits on density and temperature irregularities that might have existed ten billion years ago, when the radiation was emitted. These studies are described so well in the book by Zel'dovich and Novikov (1974) [see also Field (1973), Peebles (1969), Peebles and Yu (1970), Jones and Peebles (1972), and references cited therein] that we omit them here.

Another approach is to allow large deviations from the symmetry of the Friedmann universes, but to put the asymmetries into only a few degrees of freedom.

§30.2. THE KASNER MODEL FOR AN ANISOTROPIC UNIVERSE

The prototype for cosmological models with great asymmetry in a few degrees of freedom is the Kasner (1921a) metric,

Kasner metric: an example of an anisotropic model universe

$$ds^2 = -dt^2 + t^{2p_1} dx^2 + t^{2p_2} dy^2 + t^{2p_3} dz^2, \tag{30.1}$$

which was first studied as a cosmological model by Schücking and Heckmann (1958). In this metric the p_i are constants satisfying

$$p_1 + p_2 + p_3 = (p_1)^2 + (p_2)^2 + (p_3)^2 = 1. \tag{30.2}$$

Each $t = $ constant hypersurface of this cosmological model is a flat three-dimensional space. The world lines of constant x, y, z are timelike geodesics along which galaxies or other matter, treated as test particles, can be imagined to move. This model represents an expanding universe, since the volume element

$$\sqrt{-g} = \sqrt{^{(3)}g} = t$$

is constantly increasing. But it is an anisotropically expanding universe. The separation between two standard (constant x, y, z) observers is $t^{p_1} \Delta x$ if only their x-coordinates differ. Thus, distances parallel to the x-axis expand at one rate, $\ell_1 \propto t^{p_1}$, while those along the y-axis can expand at a different rate, $\ell_2 \propto t^{p_2}$. Most remarkable perhaps is the fact that along one of the axes distances contract rather than expand. This contraction shows up mathematically in the fact that equations (30.2) require one of the p's, say p_1, to be nonpositive:

$$-\frac{1}{3} \le p_1 \le 0. \tag{30.3}$$

As a consequence, in a universe of this sort, if black-body radiation were emitted at one time t and never subsequently scattered, later observers would see blue shifts near one pair of antipodes on the sky and red shifts in most other directions. In terms of this example, the fundamental cosmological question is why the Friedmann metrics should be a more accurate approximation to the real universe than is this Kasner metric.

§30.3. ADIABATIC COOLING OF ANISOTROPY

Kasner model with matter
becomes isotropic in "old
age"

In seeking an answer, ask a question. Ask, in particular, what would become of a universe that starts out near $t = 0$ with a form described by the Kasner metric of equation (30.1). This metric is an exact solution of the vacuum Einstein equation $\mathbf{G} = 0$. It approximates a situation where the matter terms in the Einstein equations are negligible by comparison with typical non-zero components of the Riemann tensor. Schücking and Heckmann (1958) give solutions with matter included as a pressureless fluid. In this situation, the curvature of empty spacetime dominates both the geometry and the expansion rate at early times, $t \longrightarrow 0$; but after some characteristic time t_m the matter terms become more important, and the metric reduces asymptotically to the homogeneous, isotropic model with $k = 0$.

This example illustrates the possibility that the universe might achieve a measure of isotropy and homogeneity in old age, even if it were born in a highly irregular state. Whether the symmetry of our universe can be explained along these lines is not yet clear in 1972. The model universe just mentioned is only a hint, especially since the critical parameter t_m can be given any value whatsoever.

Anisotropy energy

The standard Einstein general-relativity physics of this model can be described in other language (Misner, 1968) by ascribing to the anisotropic motions of empty spacetime an "effective energy density" ρ_{aniso}, which enters the G_{00} component of the Einstein equation on an equal footing with the matter-energy density, and thereby helps to account for the expansion of the universe:

$$H^2 = \left(\frac{1}{3} \frac{d}{dt} \ln \sqrt{{}^{(3)}g} \right)^2 = \frac{8\pi}{3} \left(\rho_{\text{aniso}} + \rho_{\text{matter}} \right). \tag{30.4}$$

The anisotropy energy density is found to have an equation of state

$$\rho_{\text{aniso}} \propto {}^{(3)}g^{-1} = (\text{volume})^{-2},$$

while

$$\rho_{\text{matter}} \propto {}^{(3)}g^{-\gamma/2} = (\text{volume})^{-\gamma}.$$

For pressureless matter $\gamma = 1$; for a radiation fluid $\gamma = 4/3$; for a nonrelativistic ideal gas $\gamma = 5/3$.

Adiabatic cooling of
anisotropy

This arrangement of the Einstein equation allows one to think of the anisotropy motions as being adiabatically cooled by the expansion of the universe, just as the thermal motions of an ideal gas would be. Since the adiabatic index for homogeneous anisotropy is $\gamma = 2$, the anisotropy will be the dominant source of "effective energy" in a highly compressed state, whereas the matter will dominate in an expanded state.

§30.4. VISCOUS DISSIPATION OF ANISTROPY

The model universe sketched above can be further elaborated by introducing dissipative mechanisms that convert anisotropy energy into thermal energy. Suppose that

such an anisotropic universe were filled at one time with thermal radiation. If the radiation were collisionless or nearly so, the quanta moving parallel to the contracting x-axis would get blueshifted and would develop an energy distribution corresponding to a high temperature. The quanta moving parallel to the other (expanding) axes would be redshifted to an energy distribution corresponding to a low temperature. Any collisions taking place between these two systems of particles would introduce a "thermal contact" between them, and would transfer energy from the hot system to the cold one, with a corresponding large production of entropy. This provides an irreversible dissipative process, which decreases ρ_{aniso} and increases $\rho_{radiation}$ relative to the values they would have had under conditions of adiabatic expansion. [For further details, see, e.g., Matzner and Misner (1972).]

It is possible that both the adiabatic cooling of anisotropy and the dissipation of anisotropy by its action on a gas of almost collisionless quanta have played significant roles in the evolution of our universe. In particular, neutrinos above 10^{10} K may have undergone sufficient ν-e scattering to have provided strong dissipation during the first few seconds of the life of the universe.

§30.5. PARTICLE CREATION IN AN ANISOTROPIC UNIVERSE

Adiabatic cooling and viscous dissipation might not be the chief destroyers of anisotropy in an expanding universe. More powerful still might be another highly dissipative process, which might occur at still earlier times, very near the initial "singularity." This is a process of particle creation which was first treated by DeWitt (1953), then explored by Parker (1966 and 1969) for isotropic cosmologies and finally by Zel'dovich (1970) in the present context of anisotropic cosmologies. In this process one again turns to the Kasner metric for the simplest example, but now quantum-mechanical considerations enter the picture. One realizes that not only would real quanta propagating in different directions be subject to red shifts and blue shifts, but that virtual quanta must be considered as well. Vacuum fluctuations (zero-point oscillations) entail a certain minimum number of virtual quanta, which are subject to the redshifting and blueshifting action of the strong gravitational fields. Virtual quanta that are blueshifted sufficiently violently can materialize as real particles, thanks to their energy gain. In this context "sufficiently violently" means *not adiabatically*.

Creation of particles by anisotropy of expansion

In an adiabatic expansion, the number of particles does not change, although the energy of each one does. This adiabatic limit is just the geometric-optics approximation to wave equations, which was discussed in §22.5. There one saw that, if spacetime were not flat on the scale of a wavelength, then the wave equation could not be replaced by a particle description with conserved particle numbers. Thus, the adiabatic limit (geometric-optics approximation) is violated in the conditions of high curvature near the singularity at the beginning of the universe.

By studying wave equations in the Kasner background metric, Zel'dovich and Starobinsky (1971) find quantitatively the consequences of the failure of the adia-

batic approximation near the singularity. Classically, the amplitudes of waves at frequencies comparable to the Hubble constant for any given epoch increase faster than a simple blue-shift calculation would imply (amplification through parametric resonance). Quantum-mechanically, the same amplification, applied to zero-point oscillations, leads to the creation of particle-antiparticle pairs. The calculations indicate that this effect is very strong at the characteristic time $t_q = \sqrt{G\hbar/c^5} \simeq 10^{-43}$ sec. (All calculations performed thus far are inadequate when the effect becomes strong, thus for $t \lesssim t_q$).

For the creation of massless particles, it is essential that an anisotropically expanding universe be postulated (except for scalar particles, for which particle creation occurs already in the Friedmann universe, *unless* the particle satisfies the conformal-invariant wave equation). The isotropic Friedmann universes are all conformally flat, so that solutions of the wave equation for a field of zero rest mass can be given in terms of solutions for flat-space wave equations where there is no particle creation. There is some particle creation even in the isotropic Friedmann universe when the particle has finite rest mass and low energy. However, the particle-creation process normally uses anisotropy energy as the energy supply that it converts into radiation energy.

The pioneering work by Parker and Zel'dovich suggests that one should study in detail cosmological models in which the initial conditions are a singularity, and in which quantum effects near the time $t = t_q$ dissipate all anisotropies and simultaneously give rise to the matter content of the model. This program of research, which is in its infancy, seems to require extrapolating laws of physics down to the very natural looking but preposterously small dimension $\sqrt{G\hbar/c^5} \simeq 10^{-43}$ sec, or equivalently $\sqrt{G\hbar/c^3} \sim 10^{-33}$ cm.

§30.6. INHOMOGENEOUS COSMOLOGIES

The model universes considered above were all homogeneous although anisotropic. It is also crucial to study inhomogeneous cosmological models, in which the metric has a nontrivial dependence on the space coordinates. One class of such models is spherically symmetric universes, where the matter density, expansion rate, and all other locally measurable physical quantities have spherical symmetry about some preferred origin. Models of this sort were first considered by Lemaitre (1933a,b), Tolman (1934b), and Datt (1938), and were also treated by Bondi in 1947. These models provide a means for studying density perturbations of large amplitude.

Inhomogeneous cosmological models:

(1) with spherical symmetry

A recent tool is making it possible to study large-amplitude, spatially varying curvature perturbations of other symmetries; this tool is the Gowdy (1971, 1973) metrics. These metrics, which are exact solutions of the Einstein equations, represent closed universes with various topologies (S^3, $S^1 \times S^2$, T^3) containing gravitational waves. The wave form in these solutions is essentially arbitrary, but all the waves propagate along a single preferred direction and have a common polarization.

(2) with (rather symmetric) gravitational waves

A rather different approach to understanding the behavior of inhomogeneous and anisotropic solutions of the Einstein equations has been developed by Khalatnikov,

(3) near a singularity, with few or no symmetries

Lifshitz, and their colleagues. Rather than truncate the Einstein theory by limiting attention to specialized situations where exact solutions can be obtained, they have sought to study the widest possible class of solutions, but to describe their behavior only in the immediate neighborhood of the singularity. These studies give a greatly enhanced significance to some of the exact solutions, by showing that phenomena found in them are in fact typical of much broader classes of solutions.

Thus, in the first large class of solutions studied [Lifshitz and Khalatnikov (1963)], it was found that near the singularity solutions containing matter showed no features not already found in the vacuum solutions. Furthermore, space derivatives in the Einstein equations became negligible near the singularity in these solutions, with the consequence that a metric of the Kasner form [equation (30.1)] described the local behavior of spacetime near the singularity, but with a different set of p_i values possible at each point of the singular hypersurface. Subsequently, broadened studies of solutions near a singularity [Belinsky and Khalatnikov (1970)] showed that the mixmaster universe [Misner (1969b); Belinsky, Khalatnikov, and Lifshitz (1970)] is a still better homogeneous prototype for singularity behavior than the Kasner metric.

§30.7. THE MIXMASTER UNIVERSE

The simplest example of a mixmaster universe is described in Box 30.1. It shows how, near the singularity, the Kasner exponents p_i can become functions of time. The result is most simply described in terms of the Khalatnikov-Lifshitz parameter u:

Mixmaster universe:

(1) "anisotropy oscillations" explained in terms of Kasner model

$$p_1 = -u/(1 + u + u^2),$$
$$p_2 = (1 + u)/(1 + u + u^2), \qquad (30.5)$$
$$p_3 = u(1 + u)/(1 + u + u^2).$$

As one extrapolates backward in time toward the singularity, one finds that the expansion rates in the three principal directions correspond to those of the Kasner metric of equation (30.1), with p_i values corresponding to some fixed u parameter. In these mixmaster models, however, the metric is not independent of the space coordinates (the spacelike hypersurfaces can, for instance, have the same 3-sphere topology as the closed Friedmann universes).

The Kasner-like behavior at fixed u can persist through many decades of volume expansion before effects of the spatial derivatives of the metric come into play. The role then played by the space curvature is brief and decisive. The expansion is converted from a type corresponding to a parameter value $u = u_0$ to a type corresponding to the value $u = -u_0$ (which is equivalent, under a relabeling of the axes, to the value $u = u_0 - 1$). Extrapolating still farther back toward the singularity, one finds a previous period with $u = u_0 - 2$. Throughout an entire sequence $u = u_0$, $u_0 - 1$, $u_0 - 2$, $u_0 - 3, \ldots$, with $u_0 \gg 1$, nearly the entire volume expansion is due to expansion in the 3-direction, whereas the 1- and 2-directions change very little, alternating at each step between expansion and contraction. Sufficiently far in the past, however, such a sequence leads to a value of u between 0 and 1. This value

can be interpreted as the starting point for another, similar sequence, through the transformation $u \longrightarrow 1/u$, which interchanges the names of axes 2 and 3.

The extrapolation of the universe's evolution back toward the singularity at $t = 0$ therefore shows an extraordinarily complex behavior, in which similar but not precisely identical sequences of behavior are repeated infinitely many times. In terms of a time variable which is approximately $\log(\log t^{-1})$, these behaviors are quasi-periodic. In the generic example to which the Khalatnikov-Lifshitz methods lead, one has a metric whose asymptotic behavior near the singularity is at each point of the singular hypersurface described by a mixmaster-type behavior, but with the principal axes of expansion changing their directions as well as their roles (as characterized by the u parameter) at each step, and with the mixmaster parameters spatially variable. [For more details see Belinsky, Lifshitz, and Khalatnikov (1971), and Ryan (1971, 1972).]

It is not yet (1972) known whether there are important solutions or classes of solutions relevant to the cosmological problem, with asymptotic singularity behavior *not* described by the Khalatnikov-Lifshitz generic case. The difficulty in reaching a definitive assessment here is that Khalatnikov and Lifshitz use essentially local methods, confined to a single coordinate patch, whereas the desired assessment poses an essentially global question. The global approaches (described in Chapter 34) have not, however, provided any comparable description of the nature of the singularity whose necessity they prove. One attempt to bridge these differences in technique and content is the work by Eardley, Liang, and Sachs (1972).

(*continued on page 815*)

Box 30.1 THE MIXMASTER COSMOLOGY

The Mixmaster Cosmology is a valuable example. As described in §30.7, it shows a singularity behavior which illustrates most of the features of the most general examples known. In particular, it shows how properties of empty space reminiscent of an elastic solid become evident near the cosmological singularity.

The mathematical path to this example, as given in this box, also illustrates several important techniques in using the variational principles for the Einstein equations to elucidate the solution of these equations. The Mixmaster example can also be used to provide simple examples of superspace ideas and of quantum formulations of the laws of gravity [Misner (1972a)].

A Generalized Kasner Model

Two generalizations must be implemented in order to progress from the Kasner example (30.1) of a cosmological singularity to the Mixmaster example. The first is to allow a more general time-dependence while preserving some of the simplicity of the conditions (30.2) on the exponents p_i. Note that these exponents satisfy, e.g., $p_2 \equiv d \ln g_{22}/d \ln g$. Therefore one is led to parametrize the 3×3 spatial metric as

$$g_{ij} = e^{2\alpha}(e^{2\beta})_{ij} \qquad (1)$$

or equivalently, $(\ln g)_{ij} = 2\alpha \, \delta_{ij} + 2\beta_{ij}$, where β_{ij} is a traceless 3×3 symmetric matrix, and the exponential is a matrix power series, so $\det e^{2\beta} = 1$ and

$$\sqrt{g} = e^{3\alpha}. \qquad (2)$$

For the purposes of this paragraph only, define

(2) as a prototype for generic behavior near singularities

Are there any other generic types of behavior near singularities?

$p_{ij} = d(\ln g)_{ij}/d \ln \det g$. Then from equations (1) and (2), one computes

$$p_{ij} = \frac{1}{3}[\delta_{ij} + (d\beta_{ij}/d\alpha)]; \qquad (3)$$

so the one Kasner condition

$$1 = \sum_i p_i \equiv \text{trace } p_{ij} = 1 + \frac{1}{3}\text{trace }(d\beta/d\alpha)$$

is an identity in view of trace $\beta_{ij} = 0$. The second condition on the Kasner exponents is trace $(p^2) = 1$, and becomes $(d\beta_{ij}/d\alpha)^2 = 6$ by equation (3). This is not an identity, but a consequence of the Einstein equations in empty space. For the (Bianchi Type I) metric

$$ds^2 = -dt^2 + e^{2\alpha}(e^{2\beta})_{ij}\, dx^i\, dx^j, \qquad (4)$$

and in the case when β_{ij} is diagonal, the Einstein equations are,

$$\left(\frac{d\alpha}{dt}\right)^2 = \frac{8\pi}{3}\left[T^{00} + \frac{1}{16\pi}(d\beta_{ij}/dt)^2\right] \qquad (5)$$

and

$$e^{-3\alpha}\frac{d}{dt}\left(e^{3\alpha}\frac{d\beta_{ij}}{dt}\right) = 8\pi\left(T_{ij} - \frac{1}{3}\delta_{ij}T_{kk}\right), \qquad (6)$$

together with a redundant equation involving T_{kk} and the equation $T_{0k} = 0$. [The stress components here refer to an orthonormal frame with basis 1-forms $\omega^i = e^\alpha(e^\beta)_{ij}\, dx^j$.] From equation (5) one immediately derives

$$\rho_{\text{aniso(I)}} = (c^2/16\pi G)(d\beta_{ij}/dt)^2 \qquad (7)$$

as a formula for the effectiveness of Type I anisotropy in contributing to the Hubble constant $H = d\alpha/dt$ on a basis comparable to matter energy, as in equation (30.4). Similarly, for equation (6) in the case of fluid matter (isotropic pressures), the stress terms vanish, and one obtains $\rho_{\text{aniso(I)}}e^{6\alpha} =$ const., as in the equation following (30.4). The Kasner condition $\Sigma p_i^2 = 1$ or $(d\beta_{ij}/d\alpha)^2 = 6$ follows from equation (5) whenever $T^{00} \ll \rho_{\text{aniso}}$.

In the diagonal case, β_{ij} has only two independent components, and it is convenient at times to define them explicitly by the parameterization

$$\begin{aligned}
\beta_{11} &= \beta_+ + \sqrt{3}\beta_-, \\
\beta_{22} &= \beta_+ - \sqrt{3}\beta_-, \\
\beta_{33} &= -2\beta_+.
\end{aligned} \qquad (8)$$

For these the Kasner condition $(d\beta_{ij}/d\alpha)^2 = 6$ becomes

$$(d\beta_+/d\alpha)^2 + (d\beta_-/d\alpha)^2 = 1. \qquad (9)$$

The β_\pm are related to the Kasner exponents p_i or the u parameter of equations (30.5) by

$$\begin{aligned}
d\beta_+/d\alpha &= \frac{1}{2}(1 - 3p_3) \\
&= -1 + (3/2)(1 + u + u^2)^{-1} \\
d\beta_-/d\alpha &= \frac{1}{2}\sqrt{3}(p_1 - p_2) \\
&= -\frac{1}{2}\sqrt{3}(1 + 2u)(1 + u + u^2)^{-1}.
\end{aligned} \qquad (10)$$

Introducing Space Curvature

The first step in generalizing the Kasner metric has focused attention on the "velocity" $\boldsymbol{\beta}' \equiv (d\beta_+/d\alpha, d\beta_-/d\alpha)$ which is a derivative of anisotropy with respect to expansion. The effects of matter or, as will soon appear, space curvature can change the magnitude $\|\boldsymbol{\beta}'\|$ from the Kasner value of unity. The second step of generalization is to introduce space curvature. This one achieves in a simple example by retaining the metric components of equation (1), but employing them in a non-holonomic basis. Use the basis vectors introduced in exercises 9.13 and 9.14 on the rotation group $SO(3)$, whose dual 1-forms are

$$\begin{aligned}
\sigma^1 &= \cos\psi\, d\theta + \sin\psi\sin\theta\, d\phi, \\
\sigma^2 &= \sin\psi\, d\theta - \cos\psi\sin\theta\, d\phi, \\
\sigma^3 &= d\psi + \cos\theta\, d\phi,
\end{aligned} \qquad (11)$$

to form the metric

$$ds^2 = -N^2\, dt^2 + e^{2\alpha}(e^{2\beta})_{ij}\sigma^i\sigma^j, \qquad (12)$$

where N, α, and β_{ij} are functions of t only. When

Box 30.1 (continued)

$\alpha = 0 = \beta_{ij}$, the three-dimensional space metric here reduces to the one studied in exercise 13.15, which is the metric of highest symmetry on the group space $SO(3)$. The simply connected covering space has the 3-sphere topology, and is obtained by extending the range of the Euler angle ψ to give it a 4π period [$SU(2)$ or spin $\frac{1}{2}$ covering of the rotation group]. With $N = 1$, $\frac{1}{2}a = e^\alpha$, and $\beta_{ij} = 0$, one obtains from equation (12) the same metric (in different coordinates) as that treated in exercise 14.4 and in Chapter 27 in discussions of the closed Friedmann cosmological model. A non-zero value for β_{ij} allows the 3-sphere to have a different circumference on great circles in each of 3 mutually orthogonal principal directions, thus destroying its isotropy but not its homogeneity.

Let us consider only the case with β_{ij} diagonal, as in equation (8). Then the T^{00} Einstein equation becomes (with $N = 1$ as a time-coordinate condition)

$$3(\dot{\alpha}^2 - \dot{\beta}_+{}^2 - \dot{\beta}_-{}^2) + \frac{1}{2}(^3R_{IX}) = 8\pi T^{00}, \quad (13)$$

where only the term

$$^3R_{IX} = \frac{1}{2}e^{-2\alpha}\text{ trace }(2e^{-2\beta} - e^{4\beta}) \quad (14)$$

is different from equation (5). This term [see equation (21.92)] is the scalar curvature of a three-dimensional slice, $t = $ const [which has symmetry properties known as "Bianchi Type IX" for the metric of equations (11) and (12)]. If equation (13) is interpreted in terms of an anisotropy energy density contributing, with T^{00}, to the volume expansion $\dot{\alpha}^2$, then there are not only kinetic energy terms $\dot{\beta}^2$ [as in equations (5) and (7)], but also a potential energy term. This term shows that negative scalar curvature, which can be produced by anisotropy ($\beta \neq 0$), is equivalent to a positive potential (or "internal") energy, and suggests that empty space has properties with analogies to an elastic solid and resists shear strains. The more detailed analysis which follows shows that, near

the singularity, the scalar curvature is always negligible when positive.

Negative curvatures, however, arise in this closed universe from large shear (β) deformations near the singularity and become large enough to reverse one Kasner shear motion [u-value, etc.; equation (10)] and change it to another.

These conclusions and further details of the time-evolution of the "Mixmaster" metric (11, 12) require, in principle, the study of all the Einstein equations, not just equation (13) for T^{00}. As described in Chapter 21, however, this T^{00} constraint equation is central, and actually contains implicitly the full content of the Einstein equations when formulated properly.

Variational Principles

One adequate formulation, adopted here, involves treating equation (13) not as an energy equation (involving velocities), but as a Hamiltonian (involving momenta). Take the Einstein variational principle (21.15) in ADM form (21.95) and carry out the space integration, using

$$\int \sigma^1 \wedge \sigma^2 \wedge \sigma^3 = \int \sin\theta \, d\phi \wedge d\theta \wedge d\psi = (4\pi)^2,$$

to obtain the action integral in the form

$$I = (\pi) \int \left\{ \pi^{ij} \, dg_{ij} + Ne^{3\alpha}\left[^3R_{IX} + e^{-6\alpha}\left(\frac{1}{2}(\pi^\ell{}_\ell)^2 - \pi^{ik}\pi_{ik}\right)\right] dt \right\}. \quad (15)$$

When introducing the specific form (1) and (8) for g_{ij}, it is convenient also to parameterize the diagonal matrix $\pi^i{}_k$ as follows:

$$p_\alpha = (2\pi)\pi^k{}_k,$$
$$p^i{}_k = (2\pi)\left(\pi^i{}_k - \frac{1}{3}\delta^i{}_k\pi^\ell{}_\ell\right), \quad (16)$$

with

$$6p^1{}_1 = p_+ + p_-\sqrt{3},$$
$$6p^2{}_2 = p_+ - p_-\sqrt{3}, \quad (17)$$
$$6p^3{}_3 = -2p_+$$

[see equation (8)]. The result is

$$I = \int p_+ \, d\beta_+ + p_- \, d\beta_- + p_\alpha \, d\alpha$$
$$- \frac{Ne^{-3\alpha}}{24\pi} [-p_\alpha^2 + p_+^2 + p_-^2$$
$$- 24\pi^2 e^{6\alpha} (^3R_{\mathrm{IX}})] \, dt.$$

This is cleaned up for further study as follows. Write

$$^3R_{\mathrm{IX}} = \frac{3}{2} e^{-2\alpha}(1 - V), \qquad (18)$$

where

$$V = V(\beta) = \frac{1}{3} \text{ trace } (1 - 2e^{-2\beta} + e^{4\beta}) \quad (19)$$

so $V(0) = 0$; and adjust the zero of α ($\alpha \longrightarrow \alpha - \alpha_0$) so that $e^{2\alpha} \longrightarrow (6\pi)^{-1}e^{2\alpha}$. Then *the metric is*

$$ds^2 = -N^2 \, dt^2 + (6\pi)^{-1}e^{2\alpha}(e^{2\beta})_{ij}\sigma^i\sigma^j, \quad (20)$$

and the variational integral is

$$I = \int p_+ \, d\beta_+ + p_- \, d\beta_- + p_\alpha \, d\alpha$$
$$- (3\pi/2)^{1/2}Ne^{-3\alpha}\mathcal{H} \, dt, \qquad (21)$$

with

$$2\mathcal{H} \equiv -p_\alpha^2 + p_+^2 + p_-^2 + e^{4\alpha}(V - 1). \quad (22)$$

One demands $\delta I = 0$ *for arbitrary independent variations of* p_\pm, p_α, β_\pm, α, N *to obtain the Einstein equations.* From varying N, one obtains the fundamental constraint equation $\mathcal{H} = 0$ [which would reduce to the vacuum version of equation (13) when the momenta are replaced by velocities (via equations obtained by varying the p's) if the coordinate condition $N = 1$ were imposed.]

ADM Hamiltonian

The standard ADM prescription for reducing this variational principle to canonical (Hamiltonian) form is to choose one of the field variables or momenta as a time-coordinate, and solve the constraint for its conjugate Hamiltonian. Here an obvious and satisfactory choice is to set $t = \alpha$, and solve $\mathcal{H} = 0$ for

$$H_{\mathrm{ADM}} = -p_\alpha = [p_+^2 + p_-^2 + e^{4\alpha}(V - 1)]^{1/2}. \qquad (23)$$

The $\dot{\alpha}$ equation [vary p_α in equation (21)] is

$$\dot{\alpha} = -(3\pi/2)^{1/2}Ne^{-3\alpha}p_\alpha \qquad (24)$$

and shows that the choice $\alpha = t$ (so $\dot{\alpha} = 1$) requires

$$N_{\mathrm{ADM}} = (2/3\pi)^{1/2}e^{3\alpha}/H_{\mathrm{ADM}}. \qquad (25)$$

The reduced, canonical, variational principle which results when equation (23) is used to eliminate p_α reads $\delta I_{\mathrm{red}} = 0$ with

$$I_{\mathrm{red}} = \int p_+ \, d\beta_+ + p_- \, d\beta_- - H_{\mathrm{ADM}} \, d\alpha \quad (26)$$

and must be supplemented by equation (25).

Super-Hamiltonian

A more convenient approach here is one more closely related to the Dirac Hamiltonian methods than those of ADM. Note, however, that one does not remove the arbitrariness in the lapse function by taking it to be some specified function $N(t)$ of the coordinates. Instead the procedure adopted here is to eliminate N from the variational principle (21) by choosing it (coordinate condition!) to be some chosen function of the field variables and momenta, $N = N(\alpha, \beta_\pm, p_\alpha, p_\pm)$. Any such choice, inserted in equation (21), leaves a variational integral in canonical Hamiltonian form. The content of this new variational principle becomes equivalent to the original one only when supplemented by the constraint

$$\mathcal{H} = 0, \qquad (27)$$

which can no longer be derived from the variational principle. [The other Euler-Lagrange equations for these two principles differ only by terms proportional to \mathcal{H}, and thus are equivalent when

Box 30.1 (*continued*)

$\mathcal{H} = 0$ is imposed on the initial conditions.] The choice

$$N = (2/3\pi)^{1/2} e^{3\alpha} \qquad (28)$$

is obvious and convenient. It makes \mathcal{H} become a super-Hamiltonian in the resulting variational principle

$$I = \int p_+ \, d\beta_+ + p_- \, d\beta_- + p_\alpha \, d\alpha - \mathcal{H} \, d\lambda, \quad (29)$$

where $t \equiv \lambda$ has been written to label the specific time-coordinate choice that equation (28) implies.

Mixmaster Dynamics

If matter terms with no additional degrees of freedom are included, the super-Hamiltonian in equation (29) is modified simply. For an example, choose

$$T^{\hat{0}\hat{0}} = -T^0_{\ 0} = (3/4)^2 (\mu e^{-3\alpha} + \Gamma e^{-4\alpha}) \quad (30)$$

for the energy density of matter in a frame with time-axis $\boldsymbol{e}_{\hat{0}} = N^{-1}(\partial/\partial t)$. The two terms represent a nonrelativistic perfect fluid ($\rho \propto V^{-1}$) and a radiation fluid ($\rho \propto V^{-4/3}$), respectively, and lead to

$$2\mathcal{H} = -p_\alpha^2 + p_+^2 + p_-^2 + e^{4\alpha}(V - 1) \\ + \mu e^{3\alpha} + \Gamma e^{2\alpha}. \qquad (31)$$

This Hamiltonian, with its simple quadratic momentum dependence, differs in only two ways from the Hamiltonians of elementary mechanics, namely, (1) in the sign of the p_α^2 term and (2) in the detailed shape of the "potential" term as function of α and β_\pm, the study of which reduces to a study of the function $V(\beta)$. Hamilton's equations, from varying α, β_\pm, p_α, and p_\pm in equation (29), yield

$$\frac{d^2\beta_\pm}{d\lambda^2} = -\frac{\partial \mathcal{H}}{\partial \beta_\pm} = -\frac{1}{2} e^{4\alpha} \frac{\partial V}{\partial \beta_\pm} \qquad (32)$$

and

$$\frac{d^2\alpha}{d\lambda^2} = +\frac{\partial \mathcal{H}}{\partial \alpha} = 2e^{4\alpha}(V - 1) + \frac{3}{2}\mu e^{3\alpha} + \Gamma e^{2\alpha}. \qquad (33)$$

Thus the sign of the p_α^2 term causes α to accelerate toward (rather than away from) higher values of the "potential" terms $e^{4\alpha}(V - 1) + \mu e^{3\alpha} + \Gamma e^{2\alpha}$. When $|V| \ll 1$ (small anisotropy), equation (33) is identical to its form in the isotropic Friedmann model, and allows a deceleration only when α is large enough that the positive curvature term ($-e^{4\alpha}$) dominates over matter ($\mu e^{3\alpha}$) and radiation ($\Gamma e^{2\alpha}$). Near the singularity ($\alpha \longrightarrow -\infty$), the positive curvature term is always negligible compared to radiation and matter.

For studies of the singularity behavior, it is sufficient to study the simplified super-Hamiltonian

$$2\mathcal{H} \sim -p_\alpha^2 + p_+^2 + p_-^2 + e^{4\alpha}V(\beta), \quad (34)$$

since the other terms obviously vanish for $\alpha \longrightarrow -\infty$. This form retains only the V term in $^3R_{IX} = \frac{3}{2}e^{-2\alpha}(1 - V)$, which dominates when the curvature of this closed universe becomes negative, $V \gg 1$. If the term in $V(\beta)$ were also negligible, then $\mathcal{H} = -p_\alpha^2 + p_+^2 + p_-^2$ would make each p_α, p_\pm constant, giving the Kasner behavior with

$$d\beta_\pm/d\alpha = p_\pm/p_\alpha = \text{const}$$

and $|d\boldsymbol{\beta}/d\alpha|^2 = 1$ as expected (since matter and curvature have been neglected). To proceed further, a study of $V(\beta)$ is required, based on equations (19) and (8), and their immediate consequence:

$$V(\beta) = \frac{1}{3} e^{-8\beta_+} - \frac{4}{3} e^{-2\beta_+} \cosh 2\sqrt{3}\beta_- \\ + 1 + \frac{2}{3} e^{4\beta_+}(\cosh 4\sqrt{3}\beta_- - 1). \qquad (35)$$

One finds that $V(\beta)$ is a *positive definite* "potential well" which has the same symmetries as an equi-

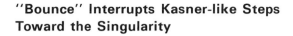

lateral triangle in the $\beta_+\beta_-$ plane. Near the origin, $\beta_\pm = 0$, the equipotentials are circles, since

$$V(\beta) = 8(\beta_+{}^2 + \beta_-{}^2) + 0(\beta^3). \quad (36)$$

For large β values, one finds

$$V(\beta) \sim \frac{1}{3}e^{-8\beta_+}, \qquad \beta_+ \longrightarrow -\infty, \quad (37)$$

and

$$V(\beta) \sim 1 + 16\beta_-{}^2 e^{4\beta_+}, \qquad \begin{array}{l} \beta_+ \longrightarrow +\infty, \\ |\beta_-| \ll 1. \end{array} \quad (38)$$

These two asymptotic forms, together with the triangular symmetry, give a complete asymptotic description of $V(\beta)$, as sketched in the figure, where on successive levels separated by $\Delta\beta = 1$, the potential V increases by a *factor* of $e^8 = 3 \times 10^3$.

"Bounce" Interrupts Kasner-like Steps Toward the Singularity

The dominant feature of the $V(\beta)$ potential is evidently its steep (exponential) triangular walls, with equation (37) representing the typical one for study. Under the influence of this potential wall, the evolution of this model universe is governed by the super-Hamiltonian

$$2\mathcal{H} \sim -p_\alpha{}^2 + p_+{}^2 + p_-{}^2 + \frac{1}{3}e^{4(\alpha - 2\beta_+)}. \quad (39)$$

If $\alpha \longrightarrow -\infty$ with $d\beta_+/d\alpha > 1/2$ [recall $d\beta_+/d\alpha$ = const., $|d\boldsymbol{\beta}/d\alpha| = 1$, when the last term in (39) is small], then the potential term grows and will eventually become large enough to influence the motion. A simple "Lorentz" transformation, suggested by the superspace metric (coefficients of the

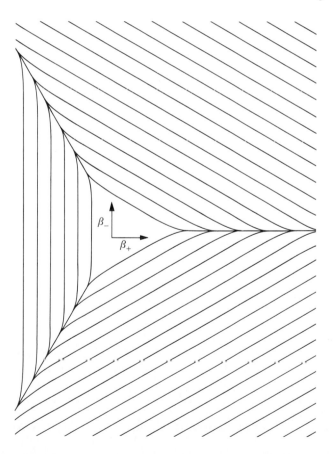

Some equipotentials, $V(\beta) = $ constant, are shown for the function defined in equation (35). Equipotentials near the origin of the β-plane are closed curves for $V < 1$ and are omitted here.

Box 30.1 (*continued*)

quadratic in the momenta) simplifies the computation further. Set

$$\bar{\beta}_+ = \left(\beta_+ - \frac{1}{2}\alpha\right)/\sqrt{3/4},$$

$$\bar{\alpha} = \left(\alpha - \frac{1}{2}\beta_+\right)/\sqrt{3/4},$$

and find

$$2\mathcal{K} = -\bar{p}_\alpha{}^2 + \bar{p}_+{}^2 + p_-{}^2 + \frac{1}{3}\exp(-4\sqrt{3}\,\bar{\beta}_+).$$

$$(40)$$

For this super-Hamiltonian both \bar{p}_α and p_- are constants of motion, whereas the $\bar{\beta}_+$-Hamiltonian, $\bar{p}_+{}^2 + \frac{1}{3}e^{-4\sqrt{3}\bar{\beta}_+}$, represents a simple bounce against a one-dimensional potential wall with the initial and final values of \bar{p}_+ different only in sign. The behavior of the anisotropy parameters β_\pm near the singularity thus consists of a simple Kasner step (where $d\beta_\pm/d\alpha = $ const., with the $d\beta_+/d\alpha \geq \frac{1}{2}$, or conditions equivalent by symmetry, satisfied relative to one of the three walls), followed by a bounce against that wall, beginning a new Kasner step with other Kasner parameters. [The most detailed description of this behavior and its relation to more general cosmological models can be found in Belinsky, Khalatnikov, and Lifshitz (1970)—see also the briefer report, Khalatnikov and Lifshitz (1970)—using quite different methods. For detailed developments by Hamiltonian methods, which supercede the partial Lagrangian methods of Misner (1969b), see Misner (1970, 1972a), and Ryan (1972a,b).]

Steady-State, Quasiperiodic Infinity of "Bounces" Approaching the Singularity

Some comprehensive features of the singularity behavior, involving many Kasner-like steps, can be exhibited by another transformation of the parameter space (superspace) of the metric field.

The transformation introduces a "radial" t-coordinate out from the origin of $\alpha\beta_\pm$ space, while respecting the metric properties of this superspace implied by the form of the super-Hamiltonian. Thus one defines (for any constant α_0)

$$\begin{aligned}\alpha_0 - \alpha &= e^t \cosh\zeta, \\ \beta_+ &= e^t \sinh\zeta\cos\phi, \\ \beta_- &= e^t \sinh\zeta\sin\phi,\end{aligned} \qquad (41)$$

and finds

$$2\mathcal{K} = e^{-2t}[(-p_t{}^2 + p_\zeta{}^2 + p_\phi{}^2\sinh^{-2}\zeta) + e^{2t}e^{4\alpha}V]. \qquad (42)$$

The advantage of this transformation is that in the limit $t \longrightarrow \infty$ ($\alpha \longrightarrow -\infty$, singularity) the potential terms become, in first approximation, independent of t. Thus equation (37) gives, for one potential wall,

$$e^{2t}e^{4\alpha}V \sim \frac{1}{3}e^{2t}\exp\Big[4\alpha_0$$
$$- 8e^t\Big(\sinh\zeta\cos\phi + \frac{1}{2}\cosh\zeta\Big)\Big]. \qquad (43)$$

For $t \longrightarrow \infty$ this expression evidently tends to either zero or infinity, depending on the sign of the expression in parentheses. Therefore define the asymptotic potential walls by

$$\tanh\zeta + \frac{1}{2}\sec\phi = 0 \qquad (44)$$

in the sector $|\phi - \pi| < \pi/3$, and equivalent formulae in which ϕ is replaced by $\phi \pm (2\pi/3)$ for the other sides of the triangle. Consequently, an asymptotic approximation to the super-Hamiltonian is

$$2\mathcal{K} = e^{-2t}[-p_t{}^2 + p_\zeta{}^2 + p_\phi{}^2\sinh^{-2}\zeta + V'(\zeta, \phi)], \qquad (45)$$

where $V'(\zeta, \phi)$ vanishes inside the asymptotic walls (44) and equals $+\infty$ outside. Because the remaining t-dependence is a common factor in (45), a simple change of independent variable $e^{-2t}\,d\lambda = d\lambda'$ in equation (29)—equivalent to the choice

$$N = (2/3\pi)^{1/2}e^{-2t}\exp[3(\alpha_0 - e^t\cosh\zeta)] \qquad (46)$$

in place of equation (28)—gives a new super-Hamiltonian $\mathcal{H}' = e^{2t}\mathcal{H}$ with the variational integral

$$I = \int p_t\, dt + p_\zeta\, d\zeta + p_\phi\, d\phi - \mathcal{H}'\, d\lambda'. \quad (47)$$

In the asymptotic approximation where

$$2\mathcal{H}' = -p_t{}^2 + p_\zeta{}^2 + p_\phi{}^2 \sinh^{-2}\zeta + V'(\zeta, \phi), \quad (48)$$

one immediately sees that p_t is a constant of motion, and that the "bouncing" of the $\zeta\phi$ values within the asymptotic potential walls is a stationary, quasi-periodic process in this time-coordinate λ' (or t, since $dt/d\lambda' = -p_t = $ const). [More detailed studies based on this asymptotic super-Hamiltonian show that the motion is even ergodic, with $\zeta\phi$ approaching arbitrarily close to any given value infinitely many times as $t \longrightarrow \infty$; see Chitre (1972a).]

Summary

One has found the singularity behavior in this Mixmaster example to be extraordinarily active. In the simple Kasner singularity, two axes collapse, but the third is stretched in a simple tidal deformation accompanied by volume compression. But in the Mixmaster example, every such collapse attempt is defeated by the high negative curvature it implies. Or rather it is diverted to another attempt as compression continues inexorably, but the tidal deformations attempt first one configuration, then another, in an infinitely recurring probing of all possible configurations.

Speculations on Time and the Singularity

The cosmological singularity (in all examples where its character is not known to be unstable) involves infinite curvature and infinite density. One's abhorrence of such a theoretical prediction is particularly heightened by the correlative prediction that these infinities occurred at a finite proper time in the past, and would—if they recur—occur again at some finite proper time in the future. The singularity prediction would be more tolerable if the infinite densities could be removed to the infinitely distant past. The universe could then, as now, find its natural state to be one of expansion, so every *finite* density will have been experienced at some suitably remote past time, but *infinite* density becomes a formal abstraction never realized in the course of evolution.

To push infinite curvature out of the finite past might be achieved in two ways. It is not known which, if either, works. One way is to change the physical laws which require the singularity, changing them perhaps only in obvious and desirable ways, such as stating the laws of gravity in a proper quantum language. Computations of quantum geometry are not yet definitive, however, and some (perhaps inadequate) approximations [Misner (1972a)] do not remove the singularity problem.

Another way to discard the singularity is to accept the mathematics of the classical Einstein equations, but reinterpret it in terms of an infinite past time. There are, of course, simple and utterly inadequate ways to do this by arbitrary coordinate transformations such as $t = \ln \tau$ which change a $\tau = 0$ singularity into one at $t = -\infty$. But an arbitrary coordinate is without significance. The problem is that the singularity occurs at a finite *proper* time in the past, and proper time is the most physically significant, most physically real time we know. It corresponds to the ticking of physical clocks and measures the natural rhythms of actual events. To reinterpret finite past time as infinite, one must attack proper time on precisely these grounds, and claim it is inadequately physical. On a local basis, where special relativity is valid, no challenge to the physical significance of proper time can succeed. It is on a more global scale that the physical primacy of proper time needs to be reviewed.

"The cosmological singularity occurred ten thousand million years ago." In this statement, take time to mean the proper time along the world line of the solar system, ephemeris time. Then the statement would have a most direct physical sig-

Box 30.1 (*continued*)

nificance if it meant that the Earth had completed 10^{10} orbits about the sun since the beginning of the universe. But proper time is not that closely tied to actual physical phenomena. The statement merely implies that those 5×10^9 orbits which the earth may have actually accomplished give a standard of time which is to be extrapolated in prescribed ways, thus giving theoretical meaning to the other 5×10^9 years which are asserted to have preceeded the formation of the solar system.

A hardier standard clock changes the details of the argument, but not its qualitative conclusion. To interpret 10^{10} years in terms of SI (Systeme Internationale) seconds assigns a past history containing some 3×10^{27} oscillations of a hyperfine transition in neutral Cesium. But again the critical early ticks of the clock (needed to locate the singularity in time by actual physical events) are missing. The time needed for stellar nucleosynthesis to produce the first Cesium disqualifies this clock on historical grounds, and the still earlier high temperatures nearer the singularity would have ionized all Cesium even if this element had predated stars.

Thus proper time near the singularity is not a direct counting of simple and actual physical phenomena, but an elaborate mathematical extrapolation. Each actual clock has its "ticks" discounted by a suitable factor—3×10^7 seconds per orbit from the Earth-sun system, 1.1×10^{-10} seconds per oscillation for the Cesium transition, etc. Since no single clock (because of its finite size and strength) is conceivable all the way back to the singularity, a statement about the proper time since the singularity involves the concept of an infinite sequence of successively smaller and sturdier clocks with their ticks then discounted and

added. "Finite proper time," then, need not imply that any finite sequence of events was possible. It may describe a necessarily infinite number of events ("ticks") in any physically conceivable history, converted by mathematics into a finite sum by the action of a non-local convergence factor, the "discount" applied to convert "ticks" into "proper time."

Here one has the conceptual inverse of Zeno's paradox. One rejects Zeno's suggestion that a single swing of a pendulum is infinitely complicated—being composed of a half period, plus a quarter period, plus 2^{-n} *ad infinitum*—because the terms in his infinite series are mathematical abstractions, not physically achieved discrete acts in a drama that must be played out. By a comparable standard, one should ignore as a mathematical abstraction the finite sum of the proper-time series for the age of the universe, if it can be proved that there must be an infinite number of discrete acts played out during its past history. In both cases, finiteness would be judged by counting the number of discrete ticks on realizable clocks, not by assessing the weight of unrealizable mathematical abstractions.

Whether the universe is infinitely old by this standard remains to be determined. The quantum influences, in particular, remain to be calculated. The decisive question is whether each present-epoch event is subject to the influence of infinitely many previous discrete events. In that case statistical assumptions (large numbers, random phases, etc.) could enter in stronger ways into theories of cosmology. The Mixmaster cosmological model does have an infinite past history in this sense, since each "bounce" from one Kasner-like motion to another is a recognizable cosmological event, of which infinitely many must be realized between any finite epoch and the singularity.

§30.8. HORIZONS AND THE ISOTROPY OF THE MICROWAVE BACKGROUND

The fundamental cosmological question—"Must a universe that is born chaotic necessarily become as homogeneous and isotropic as our universe is, and do so before life evolves?"—entails one further issue. This issue is *horizons*. As was discussed in §27.10, at any given epoch in the expansion of a Friedmann universe (e.g., the present epoch), there may be significant portions of the universe from which no light signal or other causally propagating influence will have yet reached Earth in the time available since the initial singularity. "If we should live so long," the question would arise, "will the new portions of the universe which first come into view during the next ten billion years look statistically identical to the neighboring portions which are already being seen?"

Horizons in a Friedmann universe

Fortunately, this question need not be posed only for the future. It can be asked as of some past time, and the answer then is yes. Microwave background radiation arrives at the earth from all directions in the sky with very nearly the same temperature. [The data of Boughn, Fram, and Partridge (1971) and of Conklin (1969) show $\Delta T/T \lesssim 0.004$.] The plasma that emitted the microwave radiation coming to earth from one direction in the sky had not been able, before the epoch of emission, to communicate causally with the plasma emitting the radiation that arrives from other directions. If one adopts a Friedmann model of the universe, then different sectors of the microwave sky are disjoint from each other in this sense if they are separated from each other by more than 30°, even if the microwaves were emitted as recently as $z = 7$. (The critical angle is much smaller if the microwaves were last scattered at $z = 1,000$.) From this, one concludes that the foundations for the homogeneity and isotropy of the universe were laid long before the universe became approximately Friedmann, for if statistical homogeneity and isotropy of the universe had not already been achieved at the longest wavelengths earlier, these horizon limitations would have prevented any further synchronization of conditions over large scales while the universe was in a nearly Friedmann state, and small amplitude (10%) deviations from isotropy should be observed now.

Observed isotropy of microwave radiation proves foundations for homogeneity were laid before universe became Friedmann-like

The mixmaster universe received its name from the hope that it could contribute to the solution of this problem. The very large u values that occur sporadically an infinite number of times near the singularity in a mixmaster universe give a geometry close to that of the Kasner model with $p_1 = 1$, $p_2 = p_3 = 0$. This model can be written in the form

What made the universe homogeneous and isotropic?

(1) Mixmaster oscillations?— probably not

$$ds^2 = e^{2\eta}(-d\eta^2 + dx^2) + dy^2 + dz^2, \tag{30.6}$$

where $\eta = \ln t$. If this metric is converted into a closed-universe model by interpreting x, y, z as angle coordinates each with period 4π, then one sees that a light ray can circumnavigate the universe in the x-direction in a time interval $\Delta\eta = 4\pi$, which corresponds to a volume expansion by a factor $\sqrt{-g_1}/\sqrt{-g_2} = e^{4\pi}$. Unfortunately, a quantitative analysis of the degree and frequency with which the mixmaster universe achieves this specific Kasner form suggests that the horizon breaking

(2) particle creation near singularity?

is inadequate to explain the present state of the universe [Doroshkevich, Lukash, and Novikov (1971); Chitre (1972)]. It may turn out that particle creation near the singularity can solve this horizon question, as well as provide for the dissipation of anisotropy. Hope is provided by the fact that particle creation, when described in purely classical terms, has some acausal appearances, even though it is a strictly causal process at the quantum level [Zel'dovich (1972)].

GRAVITATIONAL COLLAPSE
AND BLACK HOLES

*Wherein the reader is transported to the land of black holes, and
encounters colonies of static limits, ergospheres,
and horizons—behind whose veils are hidden gaping,
ferocious singularities.*

CHAPTER **31**

SCHWARZSCHILD GEOMETRY

§31.1. INEVITABILITY OF COLLAPSE FOR MASSIVE STARS

There is no equilibrium state at the endpoint of thermonuclear evolution for a star containing more than about twice the number of baryons in the sun ($A > A_{\max} \sim 2A_\odot$). This is one of the most surprising—and disturbing—consequences of the discussion in Chapter 24. Stated differently: A star with $A > A_{\max} \sim 2A_\odot$ must eject all but A_{\max} of its baryons—e.g., by nova or supernova explosions—before settling down into its final resting state; otherwise there will be no final resting state for it to settle down into.

What is the fate of a star that fails to eject its excess baryons before nearing the endpoint of thermonuclear evolution? For example, after a very massive supernova explosion, what will become of the collapsed degenerate-neutron core when it contains more than A_{\max} baryons? Such a supercritical mass cannot explode, since it is gravitationally bound and it has no more thermonuclear energy to release. Nor can it reach a static equilibrium state, since there exists no such state for so large a mass. There remains only one alternative; the supercritical mass must collapse through its "gravitational radius," $r = 2M$, leaving behind a gravitating "black hole" in space.

The phenomenon of collapse through the gravitational radius, as described by classical general relativity, will be the subject of the next chapter. However, before tackling it, one must understand more fully than heretofore the Schwarzschild spacetime geometry, which surrounds black holes and collapsing stars as well as static stars.

This chapter will concern itself with two topics that, at first sight, appear to be disconnected. One is the fall of a test particle in a preexisting Schwarzschild geometry, which is regarded as static, but can also be visualized as all that remains of a star that underwent collapse some time ago. The second topic is the physical

This chapter, on Schwarzschild geometry, is key preparation for understanding gravitational collapse (next chapter) and black holes (following chapter)

character of this geometry, regarded in and by itself. For the exploration of this geometry, the test particle serves as the best of all explorers. But the test particle may also be regarded in another light. It can be viewed as a rag-tag johnny-come-lately piece of the matter of the falling star. Regarded in this way, it provides the simplest of all illustrations of an asymmetry in the distribution of mass of a collapsing star. That this asymmetry irons itself out will therefore give one some preliminary insight into how more complicated asymmetries also iron themselves out. In brief, the motion of the test particle and the dynamics of the Schwarzschild geometry (for this geometry will prove to be dynamic), two apparently different problems, have the happy ability to throw light on each other.

§31.2. THE NONSINGULARITY OF THE GRAVITATIONAL RADIUS

The Schwarzschild spacetime geometry

$$ds^2 = -\left(1 - \frac{2M}{r}\right)dt^2 + \frac{dr^2}{1 - 2M/r} + r^2(d\theta^2 + \sin^2\theta\, d\phi^2) \tag{31.1}$$

The Schwarzschild line element becomes singular at $r = 2M$ ("gravitational radius")

appears to behave badly near $r = 2M$; there g_{tt} becomes zero, and g_{rr} becomes infinite. However, one cannot be sure without careful study whether this pathology in the line element is due to a pathology in the spacetime geometry itself, or merely to a pathology of the (t, r, θ, ϕ) coordinate system near $r = 2M$. (As an example of a coordinate-induced pathology, consider the neighborhood of $\theta = 0$ on one of the invariant spheres, $t = \text{const}$ and $r = \text{const}$. There $g_{\phi\phi}$ becomes zero because the coordinate system behaves badly; however, the intrinsic, coordinate-independent geometry of the sphere is well-behaved there. For another example, see Figure 1.4.

The worrisome region of the Schwarzschild geometry, $r = 2M$, is called the "gravitational radius," or the "Schwarzschild radius," or the "Schwarzschild surface," or the "Schwarzschild horizon," or the "Schwarzschild sphere." It is also called the "Schwarzschild singularity" in some of the older literature; but that is a misnomer, since, as will be shown, the spacetime geometry is not singular there.

To determine whether the spacetime geometry is singular at the gravitational radius, send an explorer in from far away to chart it. For simplicity, let him fall freely and radially into the gravitational radius, carrying his orthonormal tetrad with him as he falls. His trajectory through spacetime ["parabolic orbit"; radial geodesic of metric (31.1)] is

$$\frac{\tau}{2M} = -\frac{2}{3}\left(\frac{r}{2M}\right)^{3/2} + \text{constant},$$

$$\frac{t}{2M} = -\frac{2}{3}\left(\frac{r}{2M}\right)^{3/2} - 2\left(\frac{r}{2M}\right)^{1/2} + \ln\left|\frac{(r/2M)^{1/2} + 1}{(r/2M)^{1/2} - 1}\right| + \text{constant}. \tag{31.2}$$

[See §25.5 and especially equation (25.38) for derivation and discussion.] One obtains the r coordinate of the explorer in terms of the proper time measured on a clock

he carries, $r(\tau)$, by inverting the first equation; one finds his r coordinate in terms of coordinate time, $r(t)$, by inverting the second equation.

Of all the features of the traveler's trajectory, one stands out most clearly and disturbingly: to reach the gravitational radius, $r = 2M$, requires a finite lapse of proper time, but an infinite lapse of coordinate time:

An infalling observer reaches $r = 2M$ in finite proper time but infinite coordinate time

$$r/2M = 1 - (\tau + \text{constant})/2M \quad \text{when near } r = 2M;$$
$$r/2M = 1 + \text{constant} \times \exp(-t/2M) \quad \text{in limit as } t \longrightarrow \infty. \tag{31.3}$$

(see Fig. 25.5.) Of course, proper time is the relevant quantity for the explorer's heart-beat and health. No coordinate system has the power to prevent him from reaching $r = 2M$. Only the coordinate-independent geometry of spacetime could possibly do that; and equation (31.3) shows it does not!

Let the explorer approach and reach $r = 2M$, then. What spacetime geometry does he measure there? Is it singular or nonsingular? Restated in terms of measurements, do infinite tidal gravitational forces tear the traveler apart and crush him as he approaches $r = 2M$, or does he feel only finite tidal forces which in principle his body can withstand?

The tidal forces felt by the explorer as he passes a given radius r are measured by the components of the Riemann curvature tensor with respect to his orthonormal frame there (equation of geodesic deviation). To calculate those curvature components at r, proceed in two steps. (1) Calculate the components, not in the traveler's frame, but rather in the "static" orthonormal frame

$$\boldsymbol{\omega}^{\hat{t}} = \left(1 - \frac{2M}{r}\right)^{1/2} \boldsymbol{dt}, \ \boldsymbol{\omega}^{\hat{r}} = \frac{\boldsymbol{dr}}{(1 - 2M/r)^{1/2}}, \ \boldsymbol{\omega}^{\hat{\theta}} = r\,\boldsymbol{d\theta}, \ \boldsymbol{\omega}^{\hat{\phi}} = r\sin\theta\,\boldsymbol{d\phi} \quad (31.4a)$$

located at the event through which he is passing; the result [obtainable from equations (14.50) and (14.51) by setting $e^{2\Phi} = e^{-2\Lambda} = 1 - 2M/r$] is

$$R_{\hat{t}\hat{r}\hat{t}\hat{r}} = \frac{-2M}{r^3}, \qquad R_{\hat{t}\hat{\theta}\hat{t}\hat{\theta}} = R_{\hat{t}\hat{\phi}\hat{t}\hat{\phi}} = \frac{M}{r^3},$$
$$R_{\hat{\theta}\hat{\phi}\hat{\theta}\hat{\phi}} = \frac{2M}{r^3}, \qquad R_{\hat{r}\hat{\theta}\hat{r}\hat{\theta}} = R_{\hat{r}\hat{\phi}\hat{r}\hat{\phi}} = \frac{-M}{r^3}; \tag{31.4b}$$

all other $R_{\hat{\alpha}\hat{\beta}\hat{\gamma}\hat{\delta}}$ vanish except those obtainable from the above by symmetries of **Riemann**.

(2) Calculate the components in the explorer's frame by applying to the "static-frame" components (31.4b) the appropriate transformation—for $r > 2M$, a Lorentz boost in the $\boldsymbol{e}_{\hat{r}}$ direction with ordinary velocity $v^{\hat{r}}$; for $r < 2M$, not a "boost," but a transformation given by the standard boost formula (Box 2.4) with $v^{\hat{r}} > 1$. Here

$$v^{\hat{r}} = \frac{(g_{rr})^{1/2}\,dr}{(-g_{tt})^{1/2}\,dt} = \frac{dr/dt}{1 - 2M/r} = -\left(\frac{2M}{r}\right)^{1/2}. \tag{31.5}$$

The amazing result (a consequence of special algebraic properties of the Schwarzschild geometry, and somewhat analogous to what happens—or, rather, does not hap-

pen—to the components of the electromagnetic field, E and B, when they are both parallel to a boost) is this: all the components of **Riemann** are left completely unaffected by the boost. If $e_{\hat{\rho}}$ is the traveler's radial basis vector, and $e_{\hat{\tau}} = u$ is his time basis vector, then

$$R_{\hat{\tau}\hat{\rho}\hat{\tau}\hat{\rho}} = -2M/r^3, \qquad R_{\hat{\tau}\hat{\theta}\hat{\tau}\hat{\theta}} = R_{\hat{\tau}\hat{\phi}\hat{\tau}\hat{\phi}} = M/r^3,$$

$$R_{\hat{\theta}\hat{\phi}\hat{\theta}\hat{\phi}} = 2M/r^3, \qquad R_{\hat{\rho}\hat{\theta}\hat{\rho}\hat{\theta}} = R_{\hat{\rho}\hat{\phi}\hat{\rho}\hat{\phi}} = -M/r^3. \tag{31.6}$$

(See exercise 31.1.)

The infalling observer does not feel infinite tidal forces at $r = 2M$

The payoff of this calculation: according to equations (31.6), none of the components of **Riemann** in the explorer's orthonormal frame become infinite at the gravitational radius. The tidal forces the traveler feels as he approaches $r = 2M$ are finite; they do not tear him apart—at least not when the mass M is sufficiently great, because at $r = 2M$ the typical non-zero component $R_{\hat{\alpha}\hat{\beta}\hat{\gamma}\hat{\delta}}$ of the curvature tensor is of the order $1/M^2$. The gravitational radius is a perfectly well-behaved, nonsingular region of spacetime, and nothing there can prevent the explorer from falling on inward.

Thus, the spacetime geometry is well behaved at $r = 2M$, but the coordinate system is pathological

By contrast, deep inside the gravitational radius, at $r = 0$, the traveler must encounter infinite tidal forces, independently of the route he uses to reach there. One says that "$r = 0$ is a physical singularity of spacetime." To see this, one need only calculate from equation (31.4b) or (31.6) the "curvature invariant":

$$I \equiv R_{\hat{\alpha}\hat{\beta}\hat{\gamma}\hat{\delta}} R^{\hat{\alpha}\hat{\beta}\hat{\gamma}\hat{\delta}} = 48M^2/r^6. \tag{31.7}$$

Box 31.1 THE "SCHWARZSCHILD SINGULARITY": HISTORICAL REMARKS

Although Eddington (1924) was the first to construct a coordinate system that is nonsingular at $r = 2M$, he seems not to have recognized the significance of his result. Lemaître (1933c, especially p. 82) appears to have been the first to recognize that the so-called "Schwarzschild singularity" at $r = 2M$ is not a singularity. He wrote, "La singularité du champ de Schwarzschild est donc une singularité fictive, analogue à celle qui se présentait a l'horizon du centre dans la forme originale de l'univers de de Sitter". He also provided a coordinate system to go through $r = 2M$. However, his coordinate system, like Eddington's, covered only half of the Schwarzschild geometry:

regions I and II of Figure 31.3. Synge (1950) was the first to discover the incompleteness in the Eddington and Lemaître coordinate systems, and to provide coordinates that cover the entire geometry (regions I, II, III, IV of Figure 31.3). Fronsdal (1959), unaware of Synge's work, rediscovered the global structure of the Schwarzschild geometry by means of embedding diagrams and calculations. The coordinate system that provides maximum insight into the Schwarzschild geometry is the one generally known as the Kruskal-Szekeres coordinate system. It was constructed independently by Kruskal (1960) and by Szekeres (1960).

In every local Lorentz frame this will be a sum of products of curvature components, and it will have the same value $48M^2/r^6$. Thus, in every local Lorentz frame, including the traveler's, **Riemann** will have one or more infinite components as $r \longrightarrow 0$; i.e., tidal forces will become infinite.

At $r = 0$ the curvature is infinite

Exercise 31.1. TIDAL FORCES ON INFALLING EXPLORER

(a) Carry out the details of the derivation of the Riemann tensor components (31.6).

 (b) Calculate, roughly, the critical mass M_{crit} such that, if $M > M_{crit}$ the explorer's body (a human body made of normal flesh and bones) can withstand the tidal forces at $r = 2M$, but if $M < M_{crit}$ his body is mutilated by them. [Answer: $M_{crit} \sim 1000 M_\odot$. Evidently, if $M \sim M_\odot$ the physicist should transform himself into an ant before taking the plunge! For details see §32.6.]

EXERCISE

§31.3. BEHAVIOR OF SCHWARZSCHILD COORDINATES AT $r = 2M$

Since the spacetime geometry is well behaved at the gravitational radius, the singular behavior there of the Schwarzschild metric components, $g_{tt} = -(1 - 2M/r)$ and $g_{rr} = (1 - 2M/r)^{-1}$, must be due to a pathology there of the Schwarzschild coordinates t, r, θ, ϕ. Somehow one must find a way to get rid of that pathology—i.e., one must construct a new coordinate system from which the pathology is absent. Before doing this, it is helpful to understand better the precise nature of the pathology.

 The most obvious pathology at $r = 2M$ is the reversal there of the roles of t and r as timelike and spacelike coordinates. In the region $r > 2M$, the t direction, $\partial/\partial t$, is timelike ($g_{tt} < 0$) and the r direction, $\partial/\partial r$, is spacelike ($g_{rr} > 0$); but in the region $r < 2M$, $\partial/\partial t$ is spacelike ($g_{tt} > 0$) and $\partial/\partial r$ is timelike ($g_{rr} < 0$).

 What does it mean for r to "change in character from a spacelike coordinate to a timelike one"? The explorer in his jet-powered spaceship prior to arrival at $r = 2M$ always has the option to turn on his jets and change his motion from decreasing r (infall) to increasing r (escape). Quite the contrary is the situation when he has once allowed himself to fall inside $r = 2M$. Then the further decrease of r represents the passage of time. No command that the traveler can give to his jet engine will turn back time. That unseen power of the world which drags everyone forward willy-nilly from age twenty to forty and from forty to eighty also drags the rocket in from time coordinate $r = 2M$ to the later value of the time coordinate $r = 0$. No human act of will, no engine, no rocket, no force (see exercise 31.3) can make time stand still. As surely as cells die, as surely as the traveler's watch ticks away "the unforgiving minutes," with equal certainty, and with never one halt along the way, r drops from $2M$ to 0.

 At $r = 2M$, where r and t exchange roles as space and time coordinates, g_{tt} vanishes while g_{rr} is infinite. The vanishing of g_{tt} suggests that *the surface $r = 2M$, which*

Nature of the coordinate pathology at $r = 2M$:

(1) t and r reverse roles as timelike and spacelike coordinates

(2) the region $r = 2M$,
$-\infty < t < +\infty$ is
two-dimensional rather
than three

appears to be three-dimensional in the Schwarzschild coordinate system ($-\infty < t < +\infty$, $0 < \theta < \pi$, $0 < \phi < 2\pi$) *has zero volume and thus is actually only two-dimensional, or else is null;* thus,

$$\int_{r=2M} |g_{tt} g_{\theta\theta} g_{\phi\phi}|^{1/2} \, dt \, d\theta \, d\phi = 0;$$

$$\int_{(r=2M,\, t=\text{const})} |g_{\theta\theta} g_{\phi\phi}|^{1/2} \, d\theta \, d\phi = 4\pi(2M)^2. \tag{31.8}$$

The divergence of g_{rr} at $r = 2M$ does *not* mean that $r = 2M$ is infinitely far from all other regions of spacetime. On the contrary, the proper distance from $r = 2M$ to a point with arbitrary r is

$$\int_{2M}^{r} |g_{rr}|^{1/2} \, dr = \begin{cases} [r(r - 2M)]^{1/2} + 2M \ln |(r/2M - 1)^{1/2} + (r/2M)^{1/2}| \\ \qquad\qquad\qquad\qquad\qquad\qquad\qquad\qquad \text{when } r > 2M, \\[6pt] -2M \cot^{-1}[r^{1/2}/(2M - r)^{1/2}] - [r(2M - r)]^{1/2} \\ \qquad\qquad\qquad\qquad\qquad\qquad\qquad\qquad \text{when } r < 2M, \end{cases} \tag{31.9}$$

which is finite for all $0 < r < \infty$.

Just how the region $r < 2M$ is physically connected to the region $r > 2M$ can be discovered by examining the radial geodesics of the Schwarzschild metric. Focus attention, for concreteness, on the trajectory of a test particle that gets ejected from the singularity at $r = 0$, flies radially outward through $r = 2M$, reaches a maximum radius r_{max} ("top of orbit") at proper time $\tau = 0$ and coordinate time $t = 0$, and then falls back down through $r = 2M$ to $r = 0$. The solution of the geodesic equation for such an orbit was derived in §25.5 and described in Figure 25.3. It has the "cycloid form" (with the parameter η running from $-\pi$ to $+\pi$),

$$r = \frac{1}{2} r_{\text{max}}(1 + \cos \eta), \tag{31.10a}$$

$$\tau = (r_{\text{max}}^3/8M)^{1/2}(\eta + \sin \eta), \tag{31.10b}$$

$$t = 2M \ln \left| \frac{(r_{\text{max}}/2M - 1)^{1/2} + \tan(\eta/2)}{(r_{\text{max}}/2M - 1)^{1/2} - \tan(\eta/2)} \right|$$
$$+ 2M\left(\frac{r_{\text{max}}}{2M} - 1\right)^{1/2} \left[\eta + \left(\frac{r_{\text{max}}}{4M}\right)(\eta + \sin \eta)\right]. \tag{31.10c}$$

Figure 31.1 plots this orbit in the r, t-coordinate plane (curve F-F'-F''), along with several other types of radial geodesics.

(3) radial geodesics reveal
that the regions $r = 2M$,
$t = \pm\infty$ are "finite"
parts of spacetime

Every radial geodesic except a "set of geodesics of measure zero" crosses the gravitational radius at $t = +\infty$ (or at $t = -\infty$, or both), according to Figure 31.1 and the calculations behind that figure (exercises for the student! See Chapter 25). One therefore suspects that all the physics at $r = 2M$ is consigned to $t = \pm\infty$ by reason of some unhappiness in the choice of the Schwarzschild coordinates. A better coordinate system, one begins to believe, will take these two "points at infinity" and

Figure 31.1.

Typical radial geodesics of the Schwarzschild geometry, as charted in Schwarzschild coordinates (schematic). $FF'F''$ [see equations (31.10)] is the timelike geodesic of a test particle that starts at rest at $r = 5.2M$ and falls straight in, arriving in a finite proper time at the singularity $r = 0$ (zig-zag marking). The unhappiness of the Schwarzschild coordinate system shows in two ways: (1) in the fact that t goes to ∞ partway through the motion; and (2) in the fact that t thereafter decreases as τ (not shown) continues to increase. The course of the same trajectory prior to $t = 0$ may be constructed by reflecting the diagram in the horizontal axis ("time inversion"). The time-reversed image of F'' marks the ejection of the test particle from the singularity. $AA'A''$ is a timelike geodesic which comes in from $r = +\infty$. $BB'B''$ is the null geodesic travelled by a photon that falls straight in (no summit; never at rest!). $DD'D''$ is a spacelike radial geodesic. So is CC', but $E'E''$ is timelike. Neither of the latter two ever succeed in crossing $r = 2M$. (Unanswered questions about these geodesics will answer themselves in Figure 31.4, where the same world lines are charted in a "Kruskal-Szekeres diagram").

Described mathematically via equation (31.10), the geodesic $F''_{\text{inverse}} F'_{\text{inverse}} FF'F''$ starts with ejection at

$$r = 0 \text{ at } t = -2\pi M\left(\frac{r_{\max}}{2M} - 1\right)^{1/2}\left(\frac{r_{\max}}{4M} + 1\right), \quad \tau = -\frac{\pi}{2}\left(\frac{r_{\max}^3}{2M}\right)^{1/2};$$

it flies outward with increasing proper time τ, but decreasing coordinate time, t, until it reaches the gravitational radius

$$r = 2M \text{ at } t = -\infty, \quad \tau = -\left(\frac{r_{\max}^3}{8M}\right)^{1/2}\cos^{-1}\left(\frac{4M}{r_{\max}} - 1\right) - r_{\max}\left(1 - \frac{2M}{r_{\max}}\right)^{1/2};$$

it then continues to fly on outward, but with coordinate time now increasing from $t = -\infty$, until it reaches its maximum radius

$$r = r_{\max} \text{ at } t = 0, \quad \tau = 0 \text{ (event } F \text{ in diagram)};$$

it then falls inward, with t continuing to increase, until it crosses the gravitational radius again

$$r = 2M \text{ at } t = +\infty, \quad \tau = +\left(\frac{r_{\max}^3}{8M}\right)^{1/2}\cos^{-1}\left(\frac{4M}{r_{\max}} - 1\right) + r_{\max}\left(1 - \frac{2M}{r_{\max}}\right)^{1/2}$$
$$\text{(event } F' \text{ in diagram)};$$

and it finally falls on in with decreasing t (but, of course, still increasing τ) to

$$r = 0 \text{ at } t = +2\pi M\left(\frac{r_{\max}}{2M} - 1\right)^{1/2}\left(\frac{r_{\max}}{4M} + 1\right), \quad \tau = +\frac{\pi}{2}\left(\frac{r_{\max}^3}{2M}\right)^{1/2}$$
$$\text{(event } F'' \text{ in diagram)}.$$

spread them out into a line in a new (r_{new}, t_{new})-plane; and will squeeze the "line" $(r = 2M, t$ from $-\infty$ to $+\infty)$ into a single point in the (r_{new}, t_{new})-plane. One is the more prepared to accept this tentative conclusion and act on it because one has already seen (equation 31.8) that the region covering the (θ, ϕ) 2-sphere at $r = 2M$, and extending from $t = -\infty$ to $t = +\infty$, has zero proper volume. What timelier indication could one want that the "line" $r = 2M, -\infty < t < \infty$, is actually a point?

§31.4. SEVERAL WELL-BEHAVED COORDINATE SYSTEMS

The well-behaved coordinate system that is easiest to visualize is one in which the radially moving test particles of equations (31.10) remain always at rest ("comoving coordinates"). Such coordinates were first used by Novikov (1963). Novikov attaches a specific value of his radial coordinate, R^*, to each test particle as it emerges from the singularity of infinite tidal forces at $r = 0$, and insists that the particle carry that value of R^* throughout its "cycloidal life"—up through $r = 2M$ to $r = r_{max}$, then back down through $r = 2M$ to $r = 0$. For definiteness, Novikov expresses the R^* value for each particle in terms of the peak point on its trajectory by

$$R^* = (r_{max}/2M - 1)^{1/2}. \tag{31.11}$$

As a time coordinate, Novikov uses proper time τ of the test particles, normalized so $\tau = 0$ at the peak of the orbit. Every particle in the swarm is ejected in such a manner that it arrives at the summit of its trajectory $(r = r_{max}, \tau = 0)$ at one and the same value of the Schwarzschild coordinate time; namely, at $t = 0$.

Simple though they may be conceptually, the Novikov coordinates are related to the original Schwarzschild coordinates by a very complicated transformation: (1) combine equations (31.10b) and (31.11) to obtain $\eta(\tau, R^*)$; (2) combine $\eta(\tau, R^*)$ with (31.10a) and (31.11) to obtain $r(\tau, R^*)$; (3) combine $\eta(\tau, R^*)$ with (31.10c) and (31.11) to obtain $t(\tau, R^*)$. The resulting coordinate transformation, when applied to the Schwarzschild metric (31.1), yields the line element

$$ds^2 = -d\tau^2 + \left(\frac{R^{*2} + 1}{R^{*2}}\right)\left(\frac{\partial r}{\partial R^*}\right)^2 dR^{*2} + r^2(d\theta^2 + \sin^2\theta \, d\phi^2). \tag{31.12a}$$

("Schwarzschild geometry in Novikov coordinates".) Here r is no longer a radial coordinate; it is now a metric function $r(\tau, R^*)$ given implicitly by

$$\frac{\tau}{2M} = \pm (R^{*2} + 1)\left[\frac{r}{2M} - \frac{(r/2M)^2}{R^{*2} + 1}\right]^{1/2}$$
$$+ (R^{*2} + 1)^{3/2} \cos^{-1}\left[\left(\frac{r/2M}{R^{*2} + 1}\right)^{1/2}\right]. \tag{31.12b}$$

Figure 31.2 shows the locations of several key regions of Schwarzschild spacetime in this coordinate system. The existence of two distinct regions with $r = 0$ (singularities) and two distinct regions with $r \longrightarrow \infty$ (asymptotically flat regions; recall that $4\pi r^2 = $ surface area!) will be discussed in §31.5.

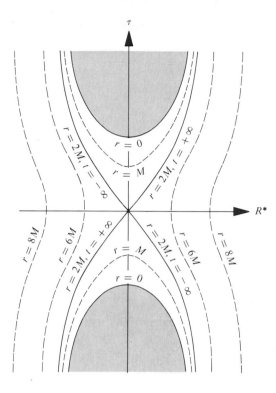

Figure 31.2.

The Novikov coordinate system for Schwarzschild spacetime (schematic). The dashed curves are curves of constant r (recall: $4\pi r^2 =$ surface area about center of symmetry). The region shaded gray is not part of spacetime; it corresponds to $r < 0$, a region that cannot be reached because of the singularity of spacetime at $r = 0$. Notice that the "line" ($r = 2M$, $-\infty < t < +\infty$) of the Schwarzschild coordinate diagram (Figure 31.1) has been compressed into a point here, in accordance with the discussion at the end of §31.3.

Although Novikov's coordinate system is very simple conceptually, the mathematical expressions for the metric components in it are rather unwieldy. Simpler, more usable expressions have been obtained in a different coordinate system ("Kruskal-Szekeres coordinates") by Kruskal (1960), and independently by Szekeres (1960).

Kruskal and Szekeres use a dimensionless radial coordinate u and a dimensionless time coordinate v related to the Schwarzschild r and t by

Kruskal-Szekeres coordinates

$$\left.\begin{array}{l} u = (r/2M - 1)^{1/2}e^{r/4M}\cosh(t/4M) \\ v = (r/2M - 1)^{1/2}e^{r/4M}\sinh(t/4M) \end{array}\right\} \text{ when } r > 2M, \qquad (31.13a)$$

$$\left.\begin{array}{l} u = (1 - r/2M)^{1/2}e^{r/4M}\sinh(t/4M) \\ v = (1 - r/2M)^{1/2}e^{r/4M}\cosh(t/4M) \end{array}\right\} \text{ when } r < 2M. \qquad (31.13b)$$

(Motivation for introducing such coordinates is given in Box 31.2.) By making this change of coordinates in the Schwarzschild metric (31.1), one obtains the following line element:

$$ds^2 = (32M^3/r)e^{-r/2M}(-dv^2 + du^2) + r^2(d\theta^2 + \sin^2\theta\, d\phi^2) \qquad (31.14a)$$

("*Schwarzschild geometry in Kruskal-Szekeres coordinates*"). Here r is to be regarded as a function of u and v defined implicitly by

$$(r/2M - 1)e^{r/2M} = u^2 - v^2 \qquad (31.14b)$$

[cf. equations (31.13)].

(*continued on page 833*)

Box 31.2 MOTIVATION FOR KRUSKAL-SZEKERES COORDINATES*

A. EDDINGTON-FINKELSTEIN COORDINATES

The motivation for the Kruskal-Szekeres system begins by introducing a different coordinate system, first devised by Eddington (1924) and rediscovered by Finkelstein (1958). Eddington and Finkelstein use as the foundation of their coordinate system, not freely falling particles as did Novikov, but freely falling photons. More particularly, they introduce coordinates \widetilde{U} and \widetilde{V}, which are labels for outgoing and ingoing, radial, null geodesics. The geodesics are given by

$$ds^2 = 0 = -(1 - 2M/r)\,dt^2 + (1 - 2M/r)^{-1}\,dr^2.$$

Equivalently, outgoing geodesics are given by $\widetilde{U} = $ const, where

$$\widetilde{U} \equiv t - r^*; \tag{1a}$$

and ingoing geodesics are given by $\widetilde{V} = $ const, where

$$\widetilde{V} \equiv t + r^*. \tag{1b}$$

Here r^* is the "tortoise coordinate" of §25.5 and Figure 25.4:

$$r^* \equiv r + 2M \ln |r/2M - 1|. \tag{1c}$$

Ingoing Eddington-Finkelstein Coordinates—Adopt r and \widetilde{V} as coordinates in place of r and t

The Schwarzschild metric becomes.

$$ds^2 = -(1 - 2M/r)\,d\widetilde{V}^2 + 2\,d\widetilde{V}\,dr + r^2\,d\Omega^2. \tag{2}$$

The radial light cone, $ds^2 = 0$, has one leg

$$d\widetilde{V}/dr = 0, \tag{3a}$$

and the other leg

$$\frac{d\widetilde{V}}{dr} = \frac{2}{1 - 2M/r}. \tag{3b}$$

From this, and this alone, one can infer all features of the drawing.

*This box is based on Misner (1969a).

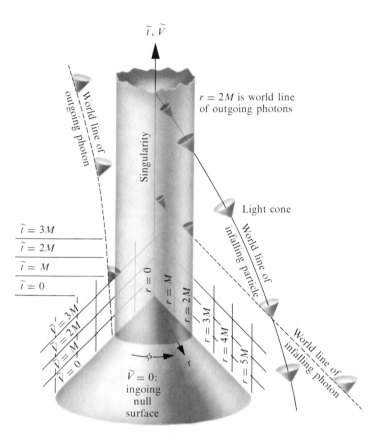

Ingoing Eddington-Finkelstein coordinates (one rotational degree of freedom is suppressed; i.e., θ is set equal to $\pi/2$). Surfaces of constant \widetilde{V}, being ingoing null surfaces, are plotted on a 45-degree slant, just as they would be in flat spacetime. Equivalently, surfaces of constant

$$\widetilde{t} \equiv \widetilde{V} - r = t + 2M \ln |r/2M - 1|$$

are plotted as horizontal surfaces.

Outgoing Eddington-Finkelstein Coordinates—Adopt r and \widetilde{U} as coordinates in place of r and t

The Schwarzschild metric becomes

$$ds^2 = -(1 - 2M/r)\, d\widetilde{U}^2 - 2\, d\widetilde{U}\, dr + r^2\, d\Omega^2. \tag{4}$$

Box 31.2 (continued)

The radial light cone, $ds^2 = 0$, has one leg

$$d\widetilde{U}/dr = 0, \tag{5a}$$

and the other leg

$$\frac{d\widetilde{U}}{dr} = -\frac{2}{1 - 2M/r}. \tag{5b}$$

From this, and this alone, one can infer all features of the drawing.

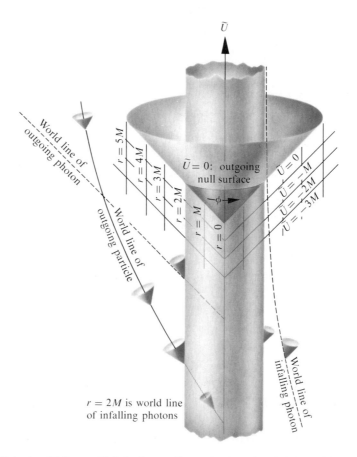

Outgoing Eddington-Finkelstein coordinates (one rotational degree of freedom is suppressed). (Surfaces of constant \widetilde{U}, being outgoing null surfaces, are plotted on a 45-degree slant, just as they would be in flat spacetime.)

Notice that both Eddington-Finkelstein coordinate systems are better behaved at the gravitational radius than is the Schwarzschild coordinate system; but they are not *fully* well-behaved. The outgoing coordinates $(\widetilde{U}, r, \theta, \phi)$ describe in a non-pathological manner the ejection of particles outward from $r = 0$ through $r = 2M$; but their description of infall through $r = 2M$ has the same pathology as the description given by Schwarzschild coordinates (Figure 31.1). Similarly, the ingoing coordinates $(\widetilde{V}, r, \theta, \phi)$ describe well the infall of a particle through $r = 2M$, but they give a pathological description of outgoing trajectories. Moreover, the contrast between the two diagrams seems paradoxical: in one the gravitational radius is made up of world lines of outgoing photons; in the other it is made up of world lines of ingoing photons! To resolve the paradox, one must seek another, better-behaved coordinate system. [But *note:* because the ingoing Eddington-Finklestein coordinates describe infall so well, they are used extensively in discussions of gravitational collapsc (Chapter 32) and black holes (Chapters 33 and 34).]

B. TRANSITION FROM EDDINGTON-FINKELSTEIN TO KRUSKAL-SZEKERES

Perhaps one would obtain a fully well-behaved coordinate system by dropping r from view and using \widetilde{U}, \widetilde{V}, as the two coordinates in the radial-time plane. The resulting coordinate system is related to Schwarzschild coordinates by [see equations (1)]

$$\widetilde{V} - \widetilde{U} = 2r^*, \tag{6a}$$

$$\widetilde{V} + \widetilde{U} = 2t; \tag{6b}$$

and the line element in terms of the new coordinates reads

$$ds^2 = -(1 - 2M/r)\, d\widetilde{U}\, d\widetilde{V} + r^2(d\theta^2 + \sin^2\theta\, d\phi^2). \tag{7}$$

Contrary to one's hopes, this coordinate system is pathological at $r = 2M$.

Second thoughts about the construction reveal the trouble: the surfaces $\widetilde{U} =$ constant (outgoing null surfaces) used in constructing it are geometrically well-defined, as are the surfaces $\widetilde{V} =$ constant (ingoing null surfaces); but the way of labeling them is not. Any relabeling, $\widetilde{u} = F(\widetilde{U})$ and $\widetilde{v} = G(\widetilde{V})$, will leave the surfaces unchanged physically. What one needs is a relabeling that will get rid of the singular factor $1 - 2M/r$ in the line element (7). A successful relabeling is suggested by the equation

$$\exp\left[(\widetilde{V} - \widetilde{U})/4M\right] = \exp\left(r^*/2M\right) = (r/2M - 1)\exp\left(r/2M\right), \tag{8}$$

Box 31.2 (continued)

which follows from equations (6a) and (1c). Experimenting with this relation quickly reveals that the relabeling

$$\widetilde{u} \equiv -e^{-\widetilde{U}/4M} = -(r/2M - 1)^{1/2}e^{r/4M}e^{-t/4M}, \tag{9a}$$

$$\widetilde{v} \equiv e^{+\widetilde{V}/4M} = (r/2M - 1)^{1/2}e^{r/4M}e^{t/4M}, \tag{9b}$$

will remove the offending $1 - 2M/r$ from the metric coefficients. In terms of these new coordinates, the line element reads

$$ds^2 = -(32M^3/r)e^{-r/2M}\,d\widetilde{v}\,d\widetilde{u} + r^2(d\theta^2 + \sin^2\theta\,d\phi^2). \tag{10a}$$

Here r is still defined by $4\pi r^2 =$ surface area, but it must be regarded as a function of \widetilde{v} and \widetilde{u}:

$$(r/2M - 1)e^{r/2M} = -\widetilde{u}\widetilde{v}. \tag{10b}$$

One can readily verify that this equation determines r uniquely (recall: $r > 0$!) in terms of the product $\widetilde{u}\widetilde{v}$ [details in Misner (1969a)].

The coordinates, $\widetilde{u}, \widetilde{v}$, which label the ingoing and outgoing null surfaces, are null coordinates; i.e.,

$$\partial/\partial\widetilde{u} \cdot \partial/\partial\widetilde{u} = g_{\widetilde{u}\widetilde{u}} = 0, \qquad \partial/\partial\widetilde{v} \cdot \partial/\partial\widetilde{v} = g_{\widetilde{v}\widetilde{v}} = 0$$

[see equation (10a)]. If one is not accustomed to working with null coordinates, it is helpful to replace \widetilde{u} and \widetilde{v} by spacelike and timelike coordinates, u and v (Kruskal-Szekeres coordinates!) defined by

$$u \equiv \frac{1}{2}(\widetilde{v} - \widetilde{u}) = (r/2M - 1)^{1/2}e^{r/4M}\cosh(t/4M), \tag{11a}$$

$$v \equiv \frac{1}{2}(\widetilde{v} + \widetilde{u}) = (r/2M - 1)^{1/2}e^{r/4M}\sinh(t/4M), \tag{11b}$$

so that

$$dv^2 - du^2 = d\widetilde{v}\,d\widetilde{u}. \tag{12}$$

In terms of these coordinates, the line element has the Kruskal form (31.14), which is fully well-behaved at the gravitational radius.

Although the Kruskal-Szekeres line element is well behaved at $r = 2M$, the transformation (11) from Schwarzschild to Kruskal-Szekeres is not; it becomes meaningless (u and v "imaginary") when one moves from $r > 2M$ to $r < 2M$. Of course, this is a manifestation of the pathologies of Schwarzschild coordinates. By trial and error, one readily finds a new transformation, to replace (11) at $r < 2M$, leading from Schwarzschild to Kruskal-Szekeres coordinates:

$$u = (1 - r/2M)^{1/2}e^{r/4M}\sinh(t/4M), \tag{11c}$$

$$v = (1 - r/2M)^{1/2}e^{r/4M}\cosh(t/4M). \tag{11d}$$

§31.5. RELATIONSHIP BETWEEN KRUSKAL-SZEKERES COORDINATES AND SCHWARZSCHILD COORDINATES

In the Kruskal-Szekeres coordinate system, the singularity $r = 0$ is located at $v^2 - u^2 = 1$. Thus there are actually *two* singularities, not one; both

$$v = +(1 + u^2)^{1/2} \text{ and } v = -(1 + u^2)^{1/2} \text{ correspond to } r = 0! \quad (31.15)$$

This is not the only surprise that lies hidden in the Kruskal-Szekeres line element (31.14). Notice also that $r \gg 2M$ (the region of spacetime far outside the gravitational radius) is given by $u^2 \gg v^2$. Thus there are actually *two* exterior regions*; both

$$u \gg +|v| \text{ and } u \ll -|v| \text{ correspond to } r \gg 2M! \quad (31.16)$$

How can this be? When the geometry is charted in Schwarzschild coordinates, it contains one singularity and one exterior region; but when expressed in Kruskal-Szekeres coordinates, it shows two of each. The answer must be that the Schwarzschild coordinates cover only part of the spacetime manifold; they must be only a local coordinate patch on the full manifold. Somehow, by means of the coordinate transformation that leads to Kruskal-Szekeres coordinates, one has analytically extended the limited Schwarzschild solution for the metric to cover all (or more nearly all) of the manifold.

To understand this covering more clearly, transform back from Kruskal-Szekeres coordinates to Schwarzschild coordinates (see Figure 31.3). The transformation equations, as written down in (31.13) were valid only for the quadrants $u > |v|$ [equation (31.13a)] and $v > |u|$ [equation (31.13b)] of Kruskal coordinates. Denote these quadrants by the numerals I and II; and denote the other quadrants by III and IV (see Figure 31.3). In the other quadrants, one can also transform the Kruskal-Szekeres line element (31.14) into the Schwarzschild line element (31.1); but slightly different transformation equations are needed. One easily verifies that the following sets of transformations work:

<div style="text-align:right">

Kruskal-Szekeres coordinates reveal that Schwarzschild spacetime has two "$r = 0$ singularities" and two "$r \longrightarrow \infty$ exterior regions"

</div>

$$\text{(I)} \begin{cases} u = (r/2M - 1)^{1/2} e^{r/4M} \cosh (t/4M) \\ v = (r/2M - 1)^{1/2} e^{r/4M} \sinh (t/4M) \end{cases}, \quad (31.17a)$$

$$\text{(II)} \begin{cases} u = (1 - r/2M)^{1/2} e^{r/4M} \sinh (t/4M) \\ v = (1 - r/2M)^{1/2} e^{r/4M} \cosh (t/4M) \end{cases}, \quad (31.17b)$$

$$\text{(III)} \begin{cases} u = -(r/2M - 1)^{1/2} e^{r/4M} \cosh (t/4M) \\ v = -(r/2M - 1)^{1/2} e^{r/4M} \sinh (t/4M) \end{cases}, \quad (31.17c)$$

$$\text{(IV)} \begin{cases} u = -(1 - r/2M)^{1/2} e^{r/4M} \sinh (t/4M) \\ v = -(1 - r/2M)^{1/2} e^{r/4M} \cosh (t/4M) \end{cases}. \quad (31.17d)$$

Transformation between Schwarzschild coordinates and Kruskal-Szekeres coordinates

*The global structure of the Schwarzschild geometry, including the existence of two singularities and two exterior regions, was first discovered by Synge (1950). See Box 31.1.

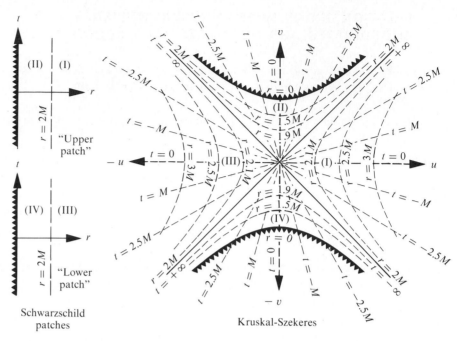

Figure 31.3.

The transformation of the Schwarzschild vacuum geometry between Schwarzschild and Kruskal-Szekeres coordinates. Two Schwarzschild coordinate patches I, II, and III, IV (illustrated in the upper and lower portions of Figure 31.5,a) are required to cover the complete Schwarzschild geometry, whereas a single Kruskal-Szekeres coordinate system suffices. The Schwarzschild geometry consists of four regions I, II, III, IV. Regions I and III represent two distinct, but identical, asymptotically flat universes in which $r > 2M$; while regions II and IV are two identical, but time-reversed, regions in which physical singularities ($r = 0$) evolve. The transformation laws that relate the Schwarzschild and Kruskal-Szekeres coordinate systems to each other are given by equations (31.17) and (31.18). In the Kruskal-Szekeres u,v-plane, curves of constant r are hyperbolae with asymptotes $u = \pm v$, while curves of constant t are straight lines through the origin.

The inverse transformations are

$$(r/2M - 1)e^{r/2M} = u^2 - v^2 \text{ in I, II, III, IV;} \tag{31.18a}$$

$$t = \begin{cases} 4M \tanh^{-1}(v/u) \text{ in I and III,} \\ 4M \tanh^{-1}(u/v) \text{ in II and IV.} \end{cases} \tag{31.18b}$$

Two Schwarzschild coordinate patches are required to cover all of spacetime

These coordinate transformations are exhibited graphically in Figure 31.3. Notice that two Schwarzschild coordinate patches, I, II, and III, IV, are required to cover the entire Schwarzschild geometry; but a single Kruskal coordinate system suffices. Schwarzschild patch I, II, is divided into two regions—region I, which is outside the gravitational radius ($r > 2M$), and region II, which is inside the gravitational radius ($r < 2M$). Similarly, Schwarzschild patch III, IV, consists of an exterior region (III) and an interior region (IV).

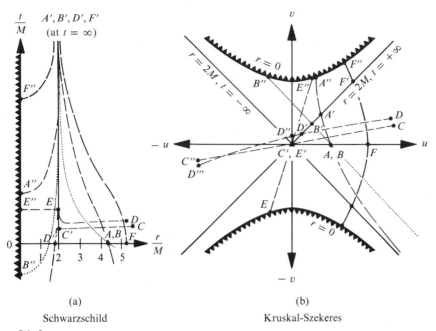

(a) (b)

Schwarzschild Kruskal-Szekeres

Figure 31.4.

(a) Typical radial timelike (*A*, *E*, *F*), lightlike (*B*), and spacelike (*C*, *D*) geodesics of the Schwarzschild geometry, as seen in the Schwarzschild coordinate system (schematic only). This is a reproduction of Figure 31.1.

(b) The same geodesics, as seen in the Kruskal-Szekeres coordinate system, and as extended either to infinite length or to the singularity of infinite curvature at *r* = 0 (schematic only).

Equations (31.18) reveal that the regions of constant *r* (constant surface area) are hyperbolae with asymptotes $u = \pm v$ in the Kruskal-Szekeres diagram, and that regions of constant *t* are straight lines through the origin.

Several radial geodesics of the complete Schwarzschild geometry are depicted in the Kruskal-Szekeres coordinate system in Figure 31.4. Notice how much more reasonable the geodesic curves look in Kruskal-Szekeres coordinates than in Schwarzschild coordinates. Notice also that *radial, lightlike geodesics (paths of radial light rays) are 45-degree lines in the Kruskal-Szekeres coordinate system*. This can be seen from the Kruskal-Szekeres line element (31.14), for which $du = \pm dv$ guarantees $ds = 0$. Because of this 45-degree property, the radial light cone in a Kruskal-Szekeres diagram has the same form as in the space-time diagram of special relativity. Any radial curve that points "generally upward" (i.e., makes an angle of less than 45 degrees with the vertical, *v*, axis) is timelike; and curves that point "generally outward" are spacelike. This property enables a Kruskal-Szekeres diagram to exhibit easily the causality relation between one event in spacetime and another (see exercises 31.2 to 31.4).

Properties of the Kruskal-Szekeres coordinate system

EXERCISES

Exercise 31.2. NONRADIAL LIGHT CONES

Show that the world line of a photon traveling nonradially makes an angle less than 45 degrees with the vertical v-axis of a Kruskal-Szekeres coordinate diagram. From this, infer that particles with finite rest mass, traveling nonradially or radially, must always move "generally upward" (angle less than 45 degrees with vertical v-axis).

Exercise 31.3. THE CRACK OF DOOM

Use a Kruskal diagram to show the following.

(a) If a man allows himself to fall through the gravitational radius $r = 2M$, there is no way whatsoever for him to avoid hitting (and being killed in) the singularity at $r = 0$.

(b) Once a man has fallen inward through $r = 2M$, there is no way whatsoever that he can send messages out to his friends at $r > 2M$, but he can still receive messages from them (e.g., by radio waves, or laser beam, or infalling "CARE packages").

Exercise 31.4. HOW LONG TO LIVE?

Show that once a man falling inward reaches the gravitational radius, no matter what he does subsequently (no matter in what directions, how long, and how hard he blasts his rocket engines), he will be pulled into the singularity and killed in a proper time of

$$\tau < \tau_{\text{max}} = \pi M = 1.54 \times 10^{-5}(M/M_\odot) \text{ seconds.} \tag{31.19}$$

[*Hint:* The trajectory of longest proper time lapse must be a geodesic. Use the mathematical tools of Chapter 25 to show that the geodesic of longest proper time lapse between $r = 2M$ and $r = 0$ is the radial geodesic (31.10a), with $r_{\text{max}} = 2M$, for which the time lapse is πM.]

Exercise 31.5. EDDINGTON-FINKELSTEIN AND KRUSKAL-SZEKERES COMPARED

Use coordinate diagrams to compare the ingoing and outgoing Eddington-Finkelstein coordinates of Box 31.2 with the Kruskal-Szekeres coordinates. Pattern the comparison after that between Schwarzschild and Kruskal-Szekeres in Figures 31.3 and 31.4.

Exercise 31.6. ANOTHER COORDINATE SYSTEM

Construct a coordinate diagram for the $\tilde{U}, \tilde{V}, \theta, \phi$ coordinate system of Box 31.2 [equations (6) and (7)]. Show such features as (1) the relationship to Schwarzschild and to Kruskal-Szekeres coordinates; (2) the location of $r = 2M$; and (3) radial geodesics.

§31.6. DYNAMICS OF THE SCHWARZSCHILD GEOMETRY

What does the Schwarzschild geometry look like? This question is most readily answered by means of embedding diagrams analogous to those for an equilibrium star (§23.8; Figure 23.1; and end of Box 23.2) and for Friedmann universes of positive and negative spatial curvature [equations (27.23) and (27.24) and Box 27.2].

Examine, first, the geometry of the spacelike hypersurface $v = 0$, which extends from $u = +\infty$ $(r = \infty)$ into $u = 0$ $(r = 2M)$ and then out to $u = -\infty$ $(r = \infty)$. In Schwarzschild coordinates this surface is a slice of constant time, $t = 0$ [see equation (31.18b)]; it is precisely the surface for which an embedding diagram was calculated in equation (23.34b). The embedded surface, with one degree of rotational freedom suppressed, is described by the paraboloid of revolution

$$\bar{r} = 2M + \bar{z}^2/8M \tag{31.20}$$

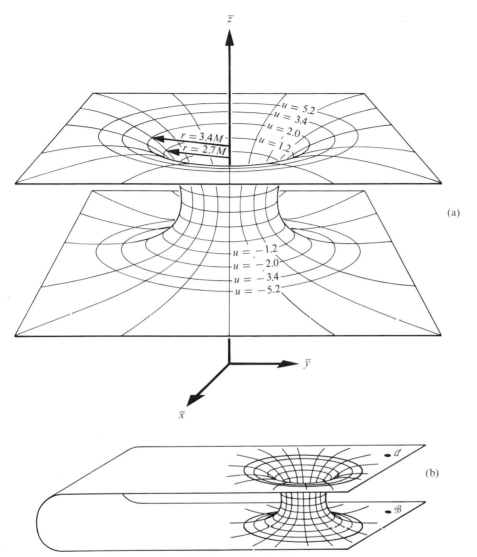

Figure 31.5.

(a) The Schwarzschild space geometry at the "moment of time" $t = v = 0$, with one degree of rotational freedom suppressed ($\theta = \pi/2$). To restore that rotational freedom and obtain the full Schwarzschild 3-geometry, one mentally replaces the circles of constant $\bar{r} = (\bar{x}^2 + \bar{y}^2)^{1/2}$ with spherical surfaces of area $4\pi\bar{r}^2$. Note that the resultant 3-geometry becomes flat (Euclidean) far from the throat of the bridge in both directions (both "universes").

(b) An embedding of the Schwarzschild space geometry at "time" $t = v = 0$, which is geometrically identical to the embedding (a), but which is topologically different. Einstein's field equations fix the local geometry of spacetime, but they do not fix its topology; see the discussion at end of Box 27.2. Here the Schwarzschild "wormhole" connects two distant regions of a single, asymptotically flat universe. For a discussion of issues of causality associated with this choice of topology, see Fuller and Wheeler (1962).

in the flat Euclidean space with metric

$$d\sigma^2 = d\bar{r}^2 + d\bar{z}^2 + \bar{r}^2 \, d\bar{\phi}^2. \tag{31.21}$$

(See Figure 31.5.)

Notice from the embedding diagram of Figure 31.5,a, that the Schwarzschild

The 3-surface $v = t = 0$ is a "wormhole" connecting two asymptotically flat universes, or two different regions of one universe

geometry on the spacelike hypersurface $t = $ const consists of a bridge or "wormhole" connecting two distinct, but identical, asymptotically flat universes. This bridge is sometimes called the "Einstein-Rosen bridge" and sometimes the "Schwarzschild throat" or the "Schwarzschild wormhole." If one so wishes, one can change the topology of the Schwarzschild geometry by connecting the two asymptotically flat universes together in a region distant from the Schwarzschild throat [Fuller and Wheeler (1962); Fig. 31.5b]. The single, unique universe then becomes multiply connected, with the Schwarzschild throat providing one spacelike path from point \mathcal{A} to point \mathcal{B}, and the nearly flat universe providing another. For concreteness, focus attention on the interpretation of the Schwarzschild geometry, not in terms of Wheeler's multiply connected single universe, but rather in terms of the Einstein-Rosen double universe of Figure 31.5,a.

Schwarzschild geometry is dynamic in regions $r < 2M$

One is usually accustomed to think of the Schwarzschild geometry as static. However, the static "time translations," $t \longrightarrow t + \Delta t$, which leave the Schwarzschild geometry unchanged, are time translations in the strict sense of the words only in regions I and III of the Schwarzschild geometry. In regions II and IV, $t \longrightarrow t + \Delta t$ is a spacelike motion, not a timelike motion (see Fig. 31.3). Consequently, a spacelike hypersurface, such as the surface $t = $ const of Figure 31.5,a, which extends from region I through $u = v = 0$ into region III, is *not* static. As this spacelike hypersurface is pushed forward in time (in the $+v$ direction of the Kruskal diagram), it enters region II, and its geometry begins to change.

In order to examine the time-development of the Schwarzschild geometry, one needs a sequence of embedding diagrams, each corresponding to the geometry of a spacelike hypersurface to the future of the preceding one. But how are the hypersurfaces to be chosen? In Newtonian theory or special relativity, one chooses hypersurfaces of constant time. But in dynamic regions of curved spacetime, no naturally preferred time coordinate exists. This situation forces one to make a totally *arbitrary* choice of hypersurfaces to use in visualizing the time-development of geometry, and to keep in mind how very arbitrary that choice was.

Time evolution of the wormhole: creation; expansion; recontraction; and pinch-off

Figure 31.6 uses two very different choices of hypersurfaces to depict the time-development of the Schwarzschild geometry. (Still other choices are shown in Figure 21.4.) Notice that the precise geometry of the evolving bridge depends on the arbitrary choice of spacelike hypersurfaces, but that the qualitative nature of the evolution is independent of the choice of hypersurfaces. Qualitatively speaking, the two asymptotically flat universes begin disconnected, with each one containing a singularity of infinite curvature ($r = 0$). As the two universes evolve in time, their singularities join each other and form a nonsingular bridge. The bridge enlarges, until it reaches a maximum radius at the throat of $r = 2M$ (maximum circumference of $4\pi M$; maximum surface area of $16\pi M^2$). It then contracts and pinches off, leaving the two universes disconnected and containing singularities ($r = 0$) once again. The

Communication through the wormhole is impossible: it pinches off too fast

formation, expansion, and collapse of the bridge occur so rapidly, that no particle or light ray can pass across the bridge from the faraway region of the one universe to the faraway region of the other without getting caught and crushed in the throat as it pinches off. (To verify this, examine the Kruskal-Szekeres diagram of Figure 31.3, where radial light rays move along 45-degree lines.)

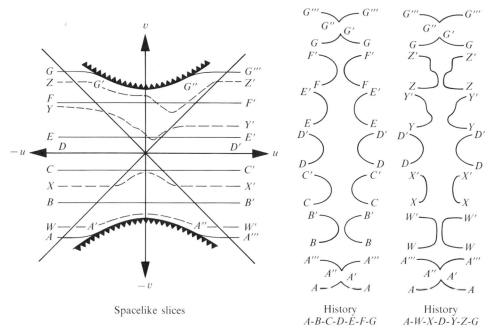

| | Spacelike slices | History
A-B-C-D-E-F-G | History
A-W-X-D-Y-Z-G |

Figure 31.6.

Dynamical evolution of the Einstein-Rosen bridge of the vacuum Schwarzschild geometry (schematic). Shown here are two sequences of embedding diagrams corresponding to two different ways of viewing the evolution of the bridge—History *A-B-C-D-E-F-G*, and History *A-W-X-D-Y-Z-G*. The embedding diagrams are skeletonized in that each diagram must be rotated about the appropriate vertical axis in order to become two-dimensional surfaces analogous to Figure 31.5,a. [Notice that the hypersurfaces of which embedding diagrams are given intersect the singularity only tangentially. Hypersurfaces that intersect the singularity at a finite angle in the u,v-plane are not shown because they cannot be embedded in a Euclidean space. Instead, a Minkowski space (indefinite metric) must be used, at least near $r = 0$. For an example of an embedding in Minkowski space, see the discussion of a universe with constant negative spatial curvature in equations (27.23) and (27.24) and Box 27.2C.] Figure 21.4 exhibits embedding diagrams for other spacelike slices in the Schwarzschild geometry.

From the Kruskal-Szekeres diagram and the 45-degree nature of its radial light rays, one sees that any particle that ever finds itself in region IV of spacetime must have been "created" in the earlier singularity; and any particle that ever falls into region II is doomed to be crushed in the later singularity. Only particles that stay forever in one of the asymptotically flat universes I or III, outside the gravitational radius ($r > 2M$), are forever safe from the singularities.

Some investigators, disturbed by the singularities at $r = 0$ or by the "double-universe" nature of the Schwarzschild geometry, have proposed modifications of its topology. One proposal is that the earlier and later singularities be identified with each other, so that a particle which falls into the singularity of region II, instead of being destroyed, will suddenly reemerge, being ejected, from the singularity of region IV. One cannot overstate the objections to this viewpoint: the region $r = 0$ is a physical singularity of infinite tidal gravitation forces and infinite Riemann curvature. Any particle that falls into that singularity must be destroyed by those

Creation and destruction in the singularities

Nonviable proposals for modifying the topology of Schwarzschild spacetime

forces. Any attempt to extrapolate its fate through the singularity using Einstein's field equations must fail; the equations lose their predictive power in the face of infinite curvature. Consequently, to postulate that the particle reemerges from the earlier singularity is to make up an *ad hoc* mathematical rule, one unrelated to physics. It is conceivable, but few believe it true, that any object of finite mass will modify the geometry of the singularity as it approaches $r = 0$ to such an extent that it can pass through and reemerge. However, whether such a speculation is correct must be answered not by *ad hoc* rules, but by concrete, difficult computations within the framework of general relativity theory (see Chapter 34).

A second proposal for modifying the topology of the Schwarzschild geometry is this: one should avoid the existence of two different asymptotically flat universes by identifying each point (v, u, θ, ϕ) with its opposite point $(-v, -u, \theta, \phi)$ in the Kruskal-Szekeres coordinate system. Two objections to this proposal are: (1) it produces a sort of "conical" singularity (absence of local Lorentz frames) at $(v, u) = (0, 0)$, i.e., at the neck of the bridge at its moment of maximum expansion; and (2) it leads to causality violations in which a man can meet himself going backward in time.

One good way for the reader to become conversant with the basic features of the Schwarzschild geometry is to reread §§31.1–31.4 carefully, reinterpreting everything said there in terms of the Kruskal-Szekeres diagram.

EXERCISES

**Exercise 31.7. SCHWARZSCHILD METRIC
 IN ISOTROPIC COORDINATES**

(a) Show that, rewritten in the isotropic coordinates of Exercise 23.1, the Schwarzschild metric reads

$$ds^2 = -\left(\frac{1 - M/2\bar{r}}{1 + M/2\bar{r}}\right)^2 dt^2 + \left(1 + \frac{M}{2\bar{r}}\right)^4 [d\bar{r}^2 + \bar{r}^2(d\theta^2 + \sin^2\theta \, d\phi^2)]; \qquad (31.22)$$

and derive the transformation

$$r = \bar{r}(1 + M/2\bar{r})^2 \qquad (31.23)$$

between the two radial coordinates.

(b) Which regions of spacetime (I, II, III, IV; see Figure 31.3) are covered by the isotropic coordinate patch, and which are not?

(c) Calculate and construct an embedding diagram for the spacelike hypersurface $t = 0$, $0 < \bar{r} < \infty$.

(d) Find a coordinate transformation that interchanges the region near $\bar{r} = 0$ with the region near $\bar{r} = \infty$, while leaving the metric coefficients in their original form.

Exercise 31.8. REISSNER-NORDSTRØM GEOMETRY

(a) Solve the Einstein field equations for a spherically symmetric, static gravitational field

$$ds^2 = -e^{2\Phi(r)} dt^2 + e^{2\Lambda(r)} dr^2 + r^2(d\theta^2 + \sin^2\theta \, d\phi^2),$$

with no matter present, but with a radial electric field $\boldsymbol{B} = 0$, $\boldsymbol{E} = f(r)\boldsymbol{e}_{\hat{r}}$ in the static orthonormal frame

$$\boldsymbol{w}^{\hat{t}} = e^{\Phi}\,\boldsymbol{dt}, \qquad \boldsymbol{w}^{\hat{r}} = e^{\Lambda}\,\boldsymbol{dr}, \qquad \boldsymbol{w}^{\hat{\theta}} = r\,\boldsymbol{d\theta}, \qquad \boldsymbol{w}^{\hat{\phi}} = r\sin\theta\,\boldsymbol{d\phi}.$$

Use as a source in the Einstein field equations the stress-energy of the electric field. [*Answer:*

$$\boldsymbol{E} = (Q/r^2)\boldsymbol{e}_{\hat{r}}, \tag{31.24a}$$

$$ds^2 = -\left(1 - \frac{2M}{r} + \frac{Q^2}{r^2}\right)dt^2 + \left(1 - \frac{2M}{r} + \frac{Q^2}{r^2}\right)^{-1}dr^2 + r^2(d\theta^2 + \sin^2\theta\,d\phi^2). \tag{31.24b}$$

This is called the "Reissner (1916)-Nordstrøm (1918) metric".]

(b) Show that the constant Q is the total charge as measured by a distant observer ($r \gg 2M$ and $r \gg Q$), who uses a Gaussian flux integral, or who studies the coulomb-force-dominated orbits of test charges with charge-to-mass ratio $e/\mu \gg M/Q$. What is the charge-to-mass ratio, in dimensionless units, for an electron? Show that the constant M is the total mass as measured by a distant observer using the Keplerian orbits of electrically neutral particles.

(c) Show that for $Q > M$, the Reissner-Nordstrøm coordinate system is well-behaved from $r = \infty$ down to $r = 0$, where there is a physical singularity and infinite tidal forces.

(d) Explore the nature of the spacetime geometry for $Q < M$, using all the techniques of this chapter (coordinate transformations, Kruskal-like coordinates, studies of particle orbits, embedding diagrams, etc.).
[*Solution:* see Graves and Brill (1960); also Fig. 34.4 of this book.]

(e) Similarly explore the spacetime geometry for $Q = M$. [*Solution:* see Carter (1966b).]

(f) For the case of a large ratio of charge to mass [$Q > M$ as in part (c)], show that the region near $r = 0$ is unphysical. More precisely, show that any spherically symmetric distribution of charged stressed matter that gives rise to the fields (31.24) outside its boundary must modify these fields for $r < r_0 = Q^2/2M$. [Hint: Study the quantity $m(r)$ defined in equations (23.18) and (32.22h), noting its values deduced from equation (31.24), on the one hand, and from the appropriate Einstein equation within the matter distribution, on the other hand. See Figure 26 of Misner (1969a) for a similar argument.]

CHAPTER **32**

GRAVITATIONAL COLLAPSE

Now, here, you see, it takes all the running you can do, to keep in the same place. If you want to get somewhere else, you must run twice as fast as that.

The Red Queen, in *Through the Looking Glass*,
LEWIS CARROLL (1871)

§32.1. RELEVANCE OF SCHWARZSCHILD GEOMETRY

The story that unfolded in the preceeding chapter was fantastic! One began with the innocuous looking Schwarzschild line element

$$ds^2 = -\left(1 - \frac{2M}{r}\right)dt^2 + \frac{dr^2}{1 - 2M/r} + r^2(d\theta^2 + \sin^2\theta\, d\phi^2), \qquad (32.1)$$

which was derived originally as the external field of a static star. One asked what happens if the star is absent; i.e., one probed the nature of the Schwarzschild geometry when no star is present to generate it. One might have expected the geometry to be that of a point mass sitting at $r = 0$. But it was not. It turned out to represent a "wormhole" connecting two asymptotically flat universes. Moreover, the wormhole was dynamic. It was created by the "joining together" of two "$r = 0$" singularities, one in each universe; it expanded to a maximum circumference of $4\pi M$; it then recontracted and pinched off, leaving the two universes disconnected once again, each with its own "$r = 0$" singularity.

The roles and relevance of the Schwarzschild geometry

As a solution to Einstein's field equations, this expanding and recontracting wormhole must be taken seriously. It is an exact solution; and it is one of the simplest of all exact solutions. But there is no reason whatsoever to believe that such wormholes exist in the real universe! They can exist only if the expanding universe, $\sim 10 \times 10^9$ years ago, was "born" with the necessary initial conditions—with "$r = 0$"

Schwarzschild singularities ready and waiting to blossom forth into wormholes. There is no reason at all to believe in such pathological initial conditions!

Why, then, was so much time and effort spent in Chapter 31 on understanding the Schwarzschild geometry? (1) Because it illustrates clearly the highly non-Euclidean character of spacetime geometry when gravity becomes strong; (2) because it illustrates many of the techniques one can use to analyze strong gravitational fields; and most importantly (3) because, when appropriately truncated, it is the spacetime geometry of a black hole and of a collapsing star—as well as of a wormhole.

This chapter explores the role of the Schwarzschild geometry in gravitational collapse; the next chapter explores its role in black-hole physics.

§32.2. BIRKHOFF'S THEOREM

That the Schwarzschild geometry is relevant to gravitational collapse follows from *Birkhoff's (1923) theorem: Let the geometry of a given region of spacetime (1) be spherically symmetric, and (2) be a solution to the Einstein field equations in vacuum. Then that geometry is necessarily a piece of the Schwarzschild geometry.* The external field of any electrically neutral, spherical star satisfies the conditions of Birkhoff's theorem, whether the star is static, vibrating, or collapsing. Therefore the external field must be a piece of the Schwarzschild geometry.

The uniqueness of the Schwarzschild geometry: Birkhoff's theorem

Birkhoff's theorem is easily understood on physical grounds. Consider an equilibrium configuration that is unstable against gravitational collapse and that, like all equilibrium configurations (see §23.6), has the Schwarzschild geometry as its external gravitational field. Perturb this equilibrium configuration in a spherically symmetric way, so that it begins to collapse radially. The perturbation and subsequent collapse cannot affect the external gravitational field so long as exact spherical symmetry is maintained. Just as Maxwell's laws prohibit monopole electromagnetic waves, so Einstein's laws prohibit monopole gravitational waves. There is no possible way for any gravitational influence of the radial collapse to propagate outward.

The physics underlying Birkhoff's theorem

Not only is Birkhoff's theorem easy to understand, but it is also fairly easy to prove. Consider a spherical region of spacetime. Spherical symmetry alone is sufficient to guarantee that conditions (i), (ii), and (iii) of Box 23.3 are satisfied, and thus to guarantee that one can introduce Schwarzschild coordinates

Proof of Birkhoff's theorem

$$ds^2 = -e^{2\Phi}\,dt^2 + e^{2\Lambda}\,dr^2 + r^2(d\theta^2 + \sin^2\theta\,d\phi^2),$$
$$\Phi = \Phi(t, r), \text{ and } \Lambda = \Lambda(t, r). \tag{32.2}$$

[See Box 23.3 for proof; and notice that: (1) for generality one must allow $g_{tt} = -e^{2\Phi}$ and $g_{rr} = e^{2\Lambda}$ to be positive or negative (no constraint on sign!); (2) at events where the gradient of the "circumference function" r is zero or null, Schwarzschild coordinates cannot be introduced. The special case $(\boldsymbol{\nabla}r)^2 = 0$ is treated in exercise 32.1.]

Impose Einstein's vacuum field equation on the metric (32.2), using the orthonormal components of the Einstein tensor as derived in exercise 14.16:

$$G_{\hat{t}\hat{t}} = r^{-2}(1 - e^{-2\Lambda}) + 2(\Lambda_{,r}/r)e^{-2\Lambda} = 0, \tag{32.3a}$$

$$G_{\hat{t}\hat{r}} = G_{\hat{r}\hat{t}} = 2(\Lambda_{,t}/r)e^{-(\Lambda + \Phi)} = 0, \tag{32.3b}$$

$$G_{\hat{r}\hat{r}} = 2(\Phi_{,r}/r)e^{-2\Lambda} + r^{-2}(e^{-2\Lambda} - 1) = 0, \tag{32.3c}$$

$$G_{\hat{\theta}\hat{\theta}} = G_{\hat{\phi}\hat{\phi}} = +(\Phi_{,rr} + \Phi_{,r}^2 - \Phi_{,r}\Lambda_{,r} + \Phi_{,r}/r - \Lambda_{,r}/r)e^{-2\Lambda}$$
$$- (\Lambda_{,tt} + \Lambda_{,t}^2 - \Lambda_{,t}\Phi_{,t})e^{-2\Phi} = 0. \tag{32.3d}$$

Equation (32.3b) guarantees that Λ is a function of r only, and equation (32.3a) then guarantees that Λ has the same form as for the Schwarzschild metric:

$$\Lambda = -\frac{1}{2}\ln|1 - 2M/r|. \tag{32.4a}$$

Equations (32.3c,d) then become two equivalent equations for $\Phi(t, r)$—equivalent by virtue of the Bianchi identity, $\boldsymbol{\nabla} \cdot \boldsymbol{G} = 0$—whose solution is

$$\Phi = \frac{1}{2}\ln|1 - 2M/r| + f(t). \tag{32.4b}$$

Here f is an arbitrary function. Put expressions (32.4) into the line element (32.2); thereby obtain

$$ds^2 = -e^{2f(t)}\left(1 - \frac{2M}{r}\right)dt^2 + \frac{dr^2}{1 - 2M/r} + r^2(d\theta^2 + \sin^2\theta\, d\phi^2).$$

Then redefine the time coordinate

$$t_{\text{new}} = \int e^{f(t)}\, dt,$$

and thereby bring the line element into the Schwarzschild form

$$ds^2 = -\left(1 - \frac{2M}{r}\right)dt^2 + \frac{dr^2}{1 - 2M/r} + r^2(d\theta^2 + \sin^2\theta\, d\phi^2).$$

Conclusion: When the spacetime surrounding any object has spherical symmetry and is free of charge, mass, and all fields other than gravity, then one can introduce coordinates in which the metric is that of Schwarzschild. Conclusion restated in coordinate-free language: the geometry of any spherically symmetric vacuum region of spacetime is a piece of the Schwarzschild geometry (Birkhoff's theorem). Q.E.D.

EXERCISE

Exercise 32.1. UNIQUENESS OF REISSNER-NORDSTRØM GEOMETRY [Track 2]

Prove the following generalization of Birkhoff's theorem. Let the geometry of a given region of spacetime (1) be spherically symmetric, and (2) be a solution to the Einstein field equations

with an electromagnetic field as source. Then that geometry is necessarily a piece of the Reissner-Nordstrøm geometry [equation (31.24b)] with electric and magnetic fields, as measured in the standard static orthonormal frames

$$E = (Q_e/r^2)e_{\hat{r}}, \qquad B = (Q_m/r^2)e_{\hat{r}}, \qquad Q = (Q_e^2 + Q_m^2)^{1/2}.$$

[*Hints:* (1) First consider regions of spacetime in which $(\nabla r)^2 \neq 0$, using the same methods as the text uses for Birkhoff's theorem. The result is the Reissner-Nordstrøm solution. (2) Any region of dimensionality less than four, in which $(\nabla r)^2 = 0$ (e.g., the Schwarzschild radius), can be treated as the join between four-dimensional regions with $(\nabla r)^2 \neq 0$. Moreover, the geometry of such a region is determined uniquely by the geometry of the adjoining four-dimensional regions ("junction conditions"; §21.13). Since the adjoining regions are necessarily Reissner-Nordstrøm (step 1), then so are such "sandwiched" regions. (3) Next consider four-dimensional regions in which $\nabla r = dr$ is null and nonzero. Show that in such regions there exist coordinate systems with

$$ds^2 = -2\Psi\,dr\,dt + r^2(d\theta^2 + \sin^2\theta\,d\phi^2),$$

where $\Psi = \Psi(r, t)$. Show further that the Ricci tensor for this line element has an orthonormalized component

$$R_{\hat{\theta}\hat{\theta}} = 1/r^2,$$

whereas the stress-energy tensor for a spherically symmetric electromagnetic field has

$$8\pi T_{\hat{\theta}\hat{\theta}} = 8\pi\left(T_{\hat{\theta}\hat{\theta}} - \frac{1}{2}g_{\hat{\theta}\hat{\theta}}T\right) = Q^2/r^4, \qquad Q = \text{const}.$$

These quantities, $R_{\hat{\theta}\hat{\theta}}$ and $8\pi T_{\hat{\theta}\hat{\theta}}$, must be equal (Einstein's field equation) but cannot be because of their different r-dependence. Thus, an electromagnetic field cannot generate regions with $dr \neq 0$, $dr \cdot dr = 0$. (4) Finally, consider four-dimensional regions in which $dr = 0$. Denote the constant value of r by a, and show that any event can be chosen as the origin of a locally well-behaved coordinate system with

$$ds^2 = a^2(-d\tilde{\tau}^2 + e^{2\lambda}\,dz^2 + d\theta^2 + \sin^2\theta\,d\phi^2),$$

$$\lambda = \lambda(\tilde{\tau}, z), \qquad \lambda(\tilde{\tau} = 0, z) = 0, \qquad \dot{\lambda}(\tilde{\tau} = 0, z) = 0.$$

[Novikov-type coordinate system; see §31.4.] Show that, in the associated orthonormal frame, spherical symmetry demands

$$E = (Q_e/a^2)e_{\hat{z}}, \qquad B = (Q_m/a^2)e_{\hat{z}}, \qquad Q \equiv (Q_e^2 + Q_m^2)^{1/2},$$

and that the Einstein field equations then require $Q = a$ and $e^\lambda = \cos\tilde{\tau}$, so that

$$ds^2 = Q^2(-d\tilde{\tau}^2 + \cos^2\tilde{\tau}\,dz^2 + d\theta^2 + \sin^2\theta\,d\phi^2).$$

(5) This solution of the field equations [sometimes called the "Bertotti (1959)-Robinson (1959a) Electromagnetic Universe," and explored in this coordinate system by Lindquist (1960)] is actually the throat of the Reissner-Nordstrøm solution for the special case $Q = M$. Verify this claim by performing the following coordinate transformation on the Reissner-Nordstrøm throat region [equation (31.24b) with $Q = M$ and $|r - Q| \ll Q$]:

$$r - Q = Qe^{-z}\cos\tilde{\tau}, \qquad t = Qe^z\tan\tilde{\tau}.$$

(6) Thus, each possible case leads either to no solution at all, or to a segment of the Reissner-Nordstrøm geometry. Q.E.D.] Note: The missing case, $(\boldsymbol{\nabla} r)^2 = 0$, in the text's proof of Birkhoff's theorem, is resolved by noting that, for $Q = 0$, steps (3) and (4) above lead to no solutions at all. We thank G. F. R. Ellis for pointing out the omission of the case $(\boldsymbol{\nabla} r)^2 = 0$ in the preliminary version of this book.

§32.3. EXTERIOR GEOMETRY OF A COLLAPSING STAR

Consider a star that is momentarily static, but will subsequently begin to collapse. Its space geometry at the initial moment of Schwarzschild coordinate time, $t = 0$, has two parts: in the exterior, vacuum region ($r > R > 2M$), it is the Schwarzschild geometry (Birkhoff's theorem!); but in the star's interior, it is some other, totally different geometry. Whatever the interior geometry may be, it has an embedding diagram at time $t = 0$ which is qualitatively like that of Figure 23.1. (For discussion and proof of this, see §23.8.) Notice that the star's space geometry is obtained by discarding the lower universe of the full Schwarzschild geometry (Figure 31.5,a), and replacing it with a smooth "bowl" on which the matter of the star is contained.

Gravitational collapse analyzed by examining the star's exterior, Schwarzschild geometry

To follow the subsequent collapse of this star in the Schwarzschild coordinate system, or in the Kruskal-Szekeres coordinate system, or in an ingoing Eddington-Finkelstein coordinate system, one can similarly discard that part of the coordinate diagram which lies inside the star's surface, and keep only the exterior Schwarzschild region. (See Figure 32.1.) In place of the discarded interior Schwarzschild region, one must introduce some other coordinate system, line element, and diagram that correctly describe the interior of the collapsing star.

From truncated coordinate diagrams (such as Figures 32.1,a,b,c), one can readily discover and understand the various peculiar features of collapse through the gravitational radius.

The gravitational radius as a point of no return, and the "crushing" at $r = 0$

(1) No matter how stiff may be the matter of which a (spherical) star is made, once its surface has collapsed within the gravitational radius, the star will continue to collapse until its surface gets crushed in the singularity at $r = 0$. This one discovers by recalling that the star's surface cannot move faster than the speed of light, so its world line must always make an angle of less than 45 degrees with the v-axis of the Kruskal-Szekeres diagram.

(2) No signal (e.g., photon) emitted from the star's surface after it collapses inside the gravitational radius can ever escape to an external observer. Rather, all signals emitted from inside the gravitational radius get caught and destroyed by the collapse of the surrounding geometry into the singularity at $r = 0$ as space "pinches off" around the star.

(3) Consequently, an external observer can never see the star after it passes the gravitational radius; and he can never see the singularity that terminates its collapse—unless he chooses to fall through the gravitational radius himself and pay the price of death for the knowledge gained.

Does this mean that the collapsing star instantaneously and completely disappears from external view as it reaches the gravitational radius? No, not according to the analysis depicted in Figure 31.1,c: Place an astrophysicist on the surface of a collapsing star, and have him send a series of uniformly spaced signals to a distant astronomer, at rest at $r \gg 2M$, to inform him of the progress of the collapse. These signals propagate along null lines in the spacetime diagram of Figure 31.1c. The signals originate on the world line of the stellar surface, and they are received by the distant astronomer when they intersect his world line, $r = $ constant $\gg M$. As the star collapses closer and closer to its gravitational radius, $R = 2M$, the signals, which are sent at equally spaced intervals according to the astrophysicist's clock, are received by the astronomer at more and more widely spaced intervals. The astronomer does not receive a signal emitted just before the gravitational radius is reached until after an infinite amount of time has elapsed; and he never receives signals emitted after the gravitational radius has been passed. Those signals, like the astrophysicist who sends them, after brief runs get caught and destroyed by the collapsing geometry in the singularity, at $r = 0$. It is not only the star that collapses. The geometry around the star collapses.

The redshift of signals emitted just before passage through the gravitational radius

Hence, to the distant astronomer, the collapsing star appears to slow down as it approaches its gravitational radius: light from the star becomes more and more red-shifted. Clocks on the star appear to run more and more slowly. It takes an infinite time for the star to reach its gravitational radius; and, as seen by the distant astronomer, the star never gets beyond there.

The optical appearance of a collapsing star was first analyzed mathematically, giving main attention to radially propagating photons, by J. R. Oppenheimer and H. Snyder (1939). More recently a number of workers have reexamined the problem [see, e.g., Podurets (1964), Ames and Thorne (1968) and Jaffe (1969)]. The most important quantitative results of these studies are as follows. In the late stages of collapse, when the distant astronomer sees the star to be very near its gravitational radius, he observes its total luminosity to decay exponentially in time

Optical appearance of the collapsing star

$$L \propto \exp\left(-\frac{2}{3\sqrt{3}} \frac{t}{2M}\right). \tag{32.5}$$

Simultaneously, photons that travel to him along radial trajectories arrive with exponentially increasing redshifts

$$z = \Delta\lambda/\lambda \propto e^{t/4M}. \tag{32.6}$$

However, the light from the star is dominated in these late stages, not by photons flying along radial trajectories from near the gravitational radius, but by photons that were deposited by the star in unstable circular orbits as its surface passed through $r = 3M$ (see §25.6 and Box 25.7). As time passes, these photons gradually leak out the diffuse spherical shell of trapped photons at $r = 3M$ and fly off to the distant observer, who measures them to have redshift $z \simeq 2$. Consequently, in the late stages of collapse the star's spectral lines are broadened enormously, but they are brightest at redshift $z \simeq 2$.

(continued on page 850)

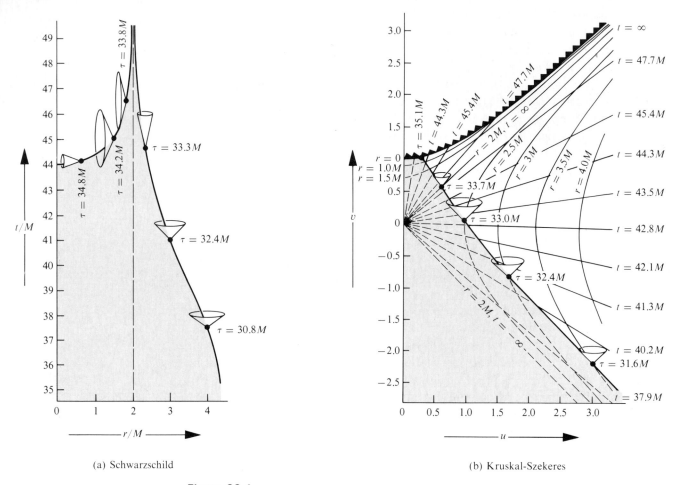

(a) Schwarzschild

(b) Kruskal-Szekeres

Figure 32.1.

The free-fall collapse of a star of initial radius $R_i = 10\,M$ as depicted alternatively in (a) Schwarzschild coordinates, (b) Kruskal-Szekeres coordinates, and (c) ingoing Eddington-Finkelstein coordinates (see Box 31.2). The region of spacetime inside the collapsing star is grey, that outside it is white. Only the geometry of the exterior region is that of Schwarzschild. The curve separating the grey and white regions is the geodesic world line of the surface of the collapsing star (equations [31.10] or [32.10] with $r_{max} = R_i = 10\,M$). This world line is parameterized by proper time, τ, as measured by an observer who sits on the surface of the star; the radial light cones, as calculated from $ds^2 = 0$, are attached to it.

Notice that, although the shapes of the light cones are not all the same relative to Schwarzschild coordinates or relative to Eddington-Finkelstein coordinates, they are all the same relative to Kruskal-Szekeres coordinates. This is because light rays travel along 45-degree lines in the u,v-plane ($dv = \pm du$), but they travel along curved paths in the r,t-plane and r,\widetilde{V}-plane.

The Kruskal-Szekeres spacetime diagram shown here is related to the Schwarzschild diagram by equations (31.13) plus a translation of Schwarzschild time: $t \longrightarrow t + 42.8\,M$. The Eddington-Finkelstein diagram is related to the Schwarzschild diagram by

$$\widetilde{V} = t + r^* = t + r + 2\,M \ln |r/2\,M - 1|$$

(see Box 31.2).

It is evident from these diagrams that the free-fall collapse is characterized by a constantly diminishing radius, which drops from $R = 10\,M$ to $R = 0$ in a finite and short comoving proper time interval, $\Delta\tau = 35.1\,M$. The point $R = 0$ and the entire region $r = 0$ outside the star make up a physical "singularity" at which infinite tidal gravitational forces—according to classical, unquantized general relativity—can and do crush matter to infinite density (see end of §31.2; also §32.6).

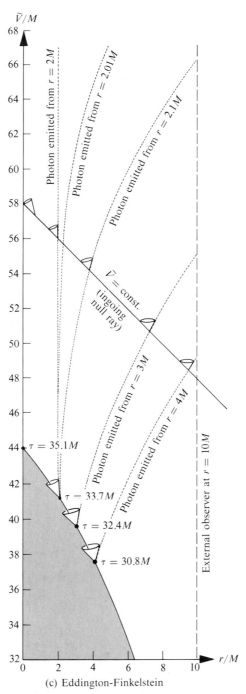

(c) Eddington-Finkelstein

The Eddington-Finkelstein diagram depicts a series of photons emitted radially from the surface of the collapsing star, and received by an observer at $r = R_{initial} = 10\,M$. The observer eventually receives all photons emitted radially from outside the gravitational radius; all photons emitted after the star passes through its gravitational radius eventually get pulled into the singularity at $r = 0$; and any photon emitted radially at the gravitational radius stays at the gravitational radius forever.

Non-free-fall collapse is similar to the collapse depicted here. When pressure gradients are present, only the detailed shape of the world line of the star's surface changes.

Notice how short is the characteristic e-folding time for the decay of luminosity and for the radial redshift:

$$\tau_{\text{char}} = 2M \approx 1 \times 10^{-5}(M/M_\odot) \text{ sec}$$
$$= \left(\begin{matrix}\text{light-travel time across a flat-space} \\ \text{distance equal to the gravitational radius}\end{matrix}\right). \tag{32.7}$$

Here M_\odot denotes one solar mass.

EXERCISE

The rest of this chapter is Track 2. No previous Track-2 material is needed as preparation for it, but it is needed as preparation for (1) the Track-2 part of Chapter 33 (black holes), and (2) Chapter 34 (singularities and global methods).

Exercise 32.2. REDSHIFTS DURING COLLAPSE

(a) Let a radio transmitter on the surface of a collapsing spherical star emit monochromatic waves of wavelength λ_e; and let a distant observer, at the same θ, ϕ, as the transmitter, receive the waves. Show that at late times the wavelength received varies as

$$\lambda_{\text{rec}}/\lambda_{\text{em}} \propto e^{t/4M} \tag{32.8a}$$

[equation (32.6)], where t is proper time as measured by the distant observer.

(b) [Track 2] Use kinetic theory for the outgoing photons (conservation of density in phase space: Liouville's theorem; §22.6) to show that the energy flux of the radiation received (ergs/cm^2 sec) varies as

$$F \propto e^{-t/M}. \tag{32.8b}$$

(c) Suppose that nuclear reactions at the center of the collapsing star generate neutrinos of energy E_e, and that these neutrinos flow freely outward (negligible absorption in star). Show that the energy of the neutrinos received by a distant observer decreases at late times as

$$E_{\text{rec}}/E_e \propto e^{-t/4M}. \tag{32.9a}$$

(d) Show that the flux of neutrino energy dies out at late times as

$$F \propto e^{-t/2M}. \tag{32.9b}$$

(e) Explain in elementary terms why the decay laws (32.8a) and (32.9a) for energy are the same, but the decay laws (32.8b) and (32.9b) for energy flux are different.

(f) Let a collapsing star emit photons from its surface at the black-body rate

$$\frac{dN}{d\tau} = \left(1.5 \times 10^{11}\frac{\text{photons}}{\text{cm}^2 \text{ sec K}^3}\right) \times \left(\begin{matrix}\text{surface area} \\ \text{of star}\end{matrix}\right) \times \left(\begin{matrix}\text{temperature} \\ \text{of surface}\end{matrix}\right)^3.$$

Let a distant observer count the photons as they pass through his sphere of radius $r \gg M$. Let him begin his count (time $t = 0$) when he sees (via photons traveling radially outward) the center of the star's surface pass through the radius $r = 3M$. Show that, in order of magnitude, the time he and his associates must wait, until the last photon that will ever get out has reached them, is

$$t = (M/M_\odot)[8 \times 10^{-4} + 5 \times 10^{-5} \log_{10}(T_{11}M/M_\odot)] \text{ seconds}, \tag{32.9c}$$

where T_{11} is the star's surface temperature in units of 10^{11} K.

§32.4. COLLAPSE OF A STAR WITH UNIFORM DENSITY AND ZERO PRESSURE

When one turns attention to the interior of a collapsing star and to the precise world line that its surface follows in the Schwarzschild geometry, one encounters rather complicated mathematics. The simplest case to treat is that of a "star" with uniform density and zero pressure; and, indeed, until recently that was the only case which had been treated in detail. The original—and very complete—analysis of the collapse of such a uniform-density "ball of dust" was given in the classic paper of Oppenheimer and Snyder (1939). More recently, other workers have discussed it from slightly different points of view and using different coordinate systems. The approach taken here was devised by Beckedorff and Misner (1962).

The collapse, from rest, of a uniform-density ball of "dust":

Because no pressure gradients are present to deflect their motion, the particles on the surface of any ball of dust must move along radial geodesics in the exterior Schwarzschild geometry. For a ball that begins at rest with finite radius, $R = R_i$, at time $t = 0$, the subsequent geodesic motion of its surface is given by equations (31.10):

$$R = (R_i/2)(1 + \cos \eta), \tag{32.10a}$$

$$t = 2M \ln \left| \frac{(R_i/2M - 1)^{1/2} + \tan (\eta/2)}{(R_i/2M - 1)^{1/2} - \tan (\eta/2)} \right|$$
$$+ 2M(R_i/2M - 1)^{1/2}[\eta + (R_i/4M)(\eta + \sin \eta)]. \tag{32.10b}$$

(1) world line of ball's surface in exterior Schwarzschild coordinates

Here R is the Schwarzschild radial coordinate (i.e., $4\pi R^2$ is the star's surface area) at Schwarzschild time t. This world line is plotted in Figure 32.1 for $R_i = 10M$, in terms of Schwarzschild coordinates, Kruskal-Szekeres coordinates, and Eddington-Finkelstein coordinates. The proper time read by a clock on the surface of the collapsing star is given by equation (31.10b):

$$\tau = (R_i^3/8M)^{1/2}(\eta + \sin \eta). \tag{32.10c}$$

Note that the collapse begins when the parameter η is zero ($R = R_i$, $t = \tau = 0$); and it terminates at the singularity ($R = 0, \eta = \pi$) after a lapse of proper time, as measured on any test particle falling with the dust, equal to

$$\Delta\tau = \pi(R_i^3/8M)^{1/2}.$$

It is interesting, though coincidental, that this is precisely the time-lapse required for free-fall collapse to infinite density in Newtonian theory [see equation (25.27′), Figure 25.3, and associated discussion].

What is the behavior of the interior of the ball of dust as it collapses? A variety of different interiors for pressureless dust can be conceived (exercise 32.8). But here attention focuses on the simplest of them: an interior that is homogeneous and isotropic everywhere, except at the surface—i.e., an interior locally identical to a dust-filled Friedmann cosmological model (Box 27.1). Is the Friedmann interior to be "open" ($k = -1$), "flat" ($k = 0$), or "closed" ($k = +1$)? Only the closed case

(2) interior of ball is identical to a portion of a closed Friedmann universe

is appropriate, since one has already demanded [equation (32.10)] that the star be at rest initially (initial rate of change of density equals zero; "moment of maximum expansion").

Using comoving hyperspherical coordinates, χ, θ, ϕ, for the star's interior, and putting the origin of coordinates at the star's center, one can write the line element in the interior in the familiar Friedmann form

$$ds^2 = -d\tau^2 + a^2(\tau)[d\chi^2 + \sin^2\chi(d\theta^2 + \sin^2\theta \, d\phi^2)]. \tag{32.11}$$

Here $a(\tau)$ is given by the familiar cycloidal relation,

$$a = \frac{1}{2}a_m(1 + \cos\eta),$$

$$\tau = \frac{1}{2}a_m(\eta + \sin\eta); \tag{32.12}$$

and the density is given by

$$\rho = (3a_m/8\pi)a^{-3} = (3/8\pi a_m{}^2)\left[\frac{1}{2}(1 + \cos\eta)\right]^{-3} \tag{32.13}$$

[see equations (1), (9), (4), and (5) of Box 27.1, with η replaced by $\eta + \pi$].

(3) the join between Friedmann interior and Schwarzschild exterior

There is one possible difficulty with this interior solution. In the cosmological case, the solution was homogeneous and isotropic everywhere. Here homogeneity and isotropy are broken at the star's surface—which lies at some radius

$$\chi = \chi_0 \tag{32.14}$$

for all τ, as measured in terms of the hyperspherical polar angle χ, a comoving coordinate (first picture in Box 27.2). At that surface (i.e., three-dimensional world tube enclosing the star's fluid) the interior Friedmann geometry must match smoothly onto the exterior Schwarzschild geometry. If the match cannot be achieved, then the Friedmann line element (32.11) cannot represent the interior of a collapsing star. An example of a case in which the matching could not be achieved is an interior of uniform and nonzero pressure, as well as uniform density. In that case there would be an infinite pressure gradient at the star's surface, which would blow off the outer layers of the star, and would send a rarefaction wave propagating inward toward its center. The uniform distribution of density and pressure would quickly be destroyed.

For the case of zero pressure, the match is possible. As a partial verification of the match, one can examine the separate and independent predictions made by the interior and exterior solutions for the star's circumference, $C = 2\pi R$, as a function of proper time τ at the star's surface. The external Schwarzschild solution predicts the cycloidal relation,

$$C = 2\pi R = 2\pi(R_i/2)(1 + \cos\eta),$$

$$\tau = (R_i{}^3/8M)^{1/2}(\eta + \sin\eta) \tag{32.15}$$

[equations (32.10)]. The interior Friedmann solution predicts a similar cycloidal relation:

$$C = 2\pi R = 2\pi a \sin \chi_0 = 2\pi\left(\frac{1}{2} a_m \sin \chi_0\right)(1 + \cos \eta),$$

$$\tau = \frac{1}{2} a_m (\eta + \sin \eta). \tag{32.16}$$

The two predictions agree perfectly for all time if and only if

$$R_i = a_m \sin \chi_0, \tag{32.17a}$$

$$M = \frac{1}{2} a_m \sin^3 \chi_0. \tag{32.17b}$$

A more complete verification of the match is given in exercise 32.4.

For further insight into this idealized model of gravitational collapse, see Box 32.1.

Exercise 32.3. EMBEDDING DIAGRAMS AND PHOTON PROPAGATION FOR COLLAPSING STAR

Verify in detail the features of homogeneous collapse described in Box 32.1.

Exercise 32.4. MATCH OF FRIEDMANN INTERIOR TO SCHWARZSCHILD EXTERIOR

The Einstein field equations are satisfied on a star's surface if and only if the intrinsic and extrinsic geometries of the surface's three-dimensional world tube are the same, whether measured on its interior or on its exterior (see §21.13 for proof and discussion). Verify that for the collapsing star discussed above, the intrinsic and extrinsic geometries match at the join between the Friedmann interior and the Schwarzschild exterior. [Hints: (a) Use η, θ, ϕ, as coordinates on the world tube of the star's surface, and show that the intrinsic geometry has the same line element

$$ds^2 = a^2(\eta)[-d\eta^2 + \sin^2\chi_0(d\theta^2 + \sin^2\theta \, d\phi^2)], \tag{32.18a}$$

whether measured in the Schwarzschild exterior or in the Friedmann interior. (b) Show that the extrinsic curvature of the world tube has the same components

$$K_{\eta\eta} = K_{\eta\theta} = K_{\eta\phi} = K_{\theta\phi} = 0,$$
$$K_{\theta\theta} = K_{\phi\phi}/\sin^2\theta = -a(\eta) \sin \chi_0 \cos \chi_0, \tag{32.18b}$$

whether measured in the Schwarzschild exterior or in the Friedmann interior.]

Exercise 32.5. STARS THAT COLLAPSE FROM INFINITY

(a) Patch together a truncated Schwarzschild geometry and the geometry of a truncated "flat" ($k = 0$), dust-filled Friedmann universe to obtain a model of a star that collapses from rest at an infinite initial radius. [*Hint:* The world line of the star's surface in the Schwarzschild geometry is given by equations (31.2).]

(b) Similarly patch together a truncated Schwarzschild geometry and the geometry of a truncated "open" ($k = -1$), dust-filled Friedmann universe to obtain a star which collapses from infinity with finite initial inward velocity.

(*continued on page 857*)

**Box 32.1 AN IDEALIZED COLLAPSING STAR
 WITH FRIEDMANN INTERIOR
 AND SCHWARZSCHILD EXTERIOR**

**(See §32.4 and exercises 32.3 and 32.4
for justification of the results
described here.)**

Initial State

(1) Take a Friedmann universe of radius $a = a_m$ at its moment of maximum expansion, $\eta = 0$; and slice off and discard the region $\chi_0 < \chi \leq \pi$, where χ_0 is some angle less than $\pi/2$. (2) Take a Schwarzschild geometry of mass $M = (a_m/2) \sin^3 \chi_0$ at the moment $t = 0$; and slice off and discard the region $r < R_i = a_m \sin \chi_0$. (3) Glue the remaining pieces of Friedmann and Schwarzschild geometry together smoothly along their cut surfaces. The resultant object will be a momentarily static star of uniform density $\rho_i = 3/(8\pi a_m{}^2)$, of mass $M = (a_m/2) \sin^3 \chi_0$, and of radius $R_i = a_m \sin \chi_0$.

Subsequent Evolution

Release this star from its intial state, and let it collapse in accord with Einstein's field equations. The interior, truncated Friedmann universe and the exterior, truncated Schwarzschild geometry will evolve just as though they had never been cut up and patched together; and this evolution will preserve the smoothness of the match between interior and exterior!

Details of the Collapse

Probe the details of the collapse using sequences of embedding diagrams (histories $ABCD$ and $AWXY$), and using photons that propagate radially outward (photons α, β, γ, δ, ϵ). The example shown here has $\chi_0 = 0.96$ and $R_i/M = 2/\sin^2 \chi_0 = 3$.

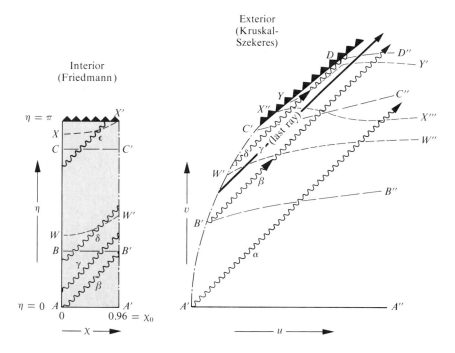

Interior
(Friedmann)

Exterior
(Kruskal-
Szekeres)

History of Collapse as Probed by Hypersurfaces ABCD:

(1) Initial configuration, $A - A' - A''$, is that constructed by cutting and sewing at times $\eta = t = 0$.

(2) Each subsequent configuration has as its interior a slice of constant Friedmann time η.

(3) The interior remains always a spherical cup of half-angle χ_0; but it contracts from radius $R = R_i = a_m \sin \chi_0$ to $R = 0$ as time increases.

(4) The matter in the star is all crushed simultaneously to infinite density when R reaches zero, and the external Schwarzschild "funnel" develops a cusp-like singularity at that point.

(5) As time increases further, this cusp pulls the region $r < 2M$ of the funnel into $r = 0$ so fast that the outward-traveling photon δ is gobbled up and crushed.

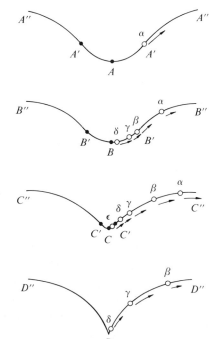

These embedding diagrams must be rotated about the vertical axes in order to become 2-dimensional surfaces analogous to Figure 23.1.

Box 32.1 (*continued*)

History as Probed by Hypersurfaces AWXY

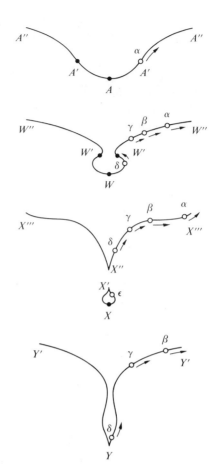

(1) Initial configuration, $A - A' - A''$, is again that constructed by cutting and sewing at $\eta = t = 0$.

(2) Subsequent hypersurfaces are very different from $\eta = $ const.

(3) As time passes, a neck develops in the geometry just outside the surface of the star.

(4) This neck becomes tighter and tighter and then pinches off, leaving the star completely isolated from the rest of the universe, and leaving a deadly cusp-like singularity in the exterior geometry where the star used to be.

(5) The isolated star, in its own little closed universe, continues to contract until it is crushed to infinite density, while the external geometry begins to develop another neck and the cusp quickly gobbles up photon δ.

The extreme difference between histories $ABCD$ and $AWXY$ dramatizes the "many-fingered time" of general relativity. The hypersurface on which one explores the geometry can be pushed ahead faster in time in one region, at the option of the party of explorers. Thus whether one region of the star collapses first, or another, or the entire star collapses simultaneously, is a function both of the spacetime geometry and of the choice of slicing. The party of explorers has this choice of slicing at their own control, and thus they themselves to this extent govern what kind of spacelike slices they will see as their exploration moves forward in time. The spacetime geometry that they slice, however, is in no way theirs to control or to change. To the extent that their masses are negligible and they serve merely as test objects, they have no influence whatsoever on the spacetime. It was fixed completely by the specification of the initial conditions for the collapse. In brief, spacetime is four-dimensional and slices are only three-dimensional (and in the pictures here look only two-dimensional or one-dimensional). Any one set of slices captures only a one-sided view of the whole story. To see the entire picture one must either examine the dynamics of the geometry as it reveals itself in varied choices of the slicing or become accustomed to visualizing the spacetime geometry as a whole.

§32.5. SPHERICALLY SYMMETRIC COLLAPSE WITH INTERNAL PRESSURE FORCES

So far as the external gravitational field is concerned, the only difference between a freely collapsing star and a collapsing, spherically symmetric star with internal pressure is this: that the surfaces of the two stars move along different world lines in the exterior Schwarzschild geometry. Because the exterior geometry is the same in both cases, *the qualitative aspects of free-fall collapse as described in the last section can be carried over directly to the case of nonnegligible internal pressure.*

An important and fascinating question to ask is this: can large internal pressures in any way prevent a collapsing star from being crushed to infinite density by infinite tidal gravitational forces? From the Kruskal-Szekeres diagram of Figure 32.1,b, it is evident that, once a star has passed inside its gravitational radius ($R < 2M$), no internal pressures, regardless of how large they may be, can prevent the star's surface from being crushed in a singularity. The surface must move along a time-like world line, and all such world lines inside $r = 2M$ hit $r = 0$. Although there is no such theorem now available, one can reasonably conjecture that, if the surface of a spherical configuration is crushed in the $r = 0$ singularity, the entire interior must also be crushed.

The details of the interior dynamics of a spherically symmetric collapsing star with pressure are not so well-understood as the exterior Schwarzschild dynamics. However, major advances in one's understanding of the interior dynamics are now being made by means of numerical computations and analytic analyses [see Misner (1969a) for a review]. In these computations and analyses, no new features (at least, no unexpected ones) have been encountered beyond those that occurred in the simple uniform-density, free-fall collapse of the last section.

Spherical collapse with pressure is qualitatively the same as without pressure

Exercise 32.6. GENERAL SPHERICAL COLLAPSE: METRIC IN COMOVING COORDINATES

EXERCISES

Consider an inhomogeneous star with pressure, undergoing spherical collapse. Spherical symmetry alone is enough to guarantee the existence of a Schwarzschild coordinate system (t, r, θ, ϕ) throughout the interior and exterior of the star [see equation (32.2) and preceding discussion]. Label each spherical shell of the star by a parameter a, which tells how many baryons are contained interior to that shell. Then $r(a, t)$ is the world line of the shell with label a. The expression for these world lines can be inverted to obtain $a(t, r)$, the number of baryons interior to radius r at time t. Show that there exists a new time coordinate $\widetilde{t}(t, r)$, such that the line element (32.2), rewritten in the coordinates $(\widetilde{t}, a, \theta, \phi)$, has the form

$$ds^2 = -e^{2\widetilde{\Phi}}\, d\widetilde{t}^2 + \left[\frac{(\partial r/\partial a)_{\widetilde{t}}\, da}{\Gamma}\right]^2 + r^2\, d\Omega^2, \tag{32.19a}$$

$$\widetilde{\Phi} = \widetilde{\Phi}(\widetilde{t}, a), \qquad r = r(\widetilde{t}, a), \qquad \Gamma = \Gamma(\widetilde{t}, a). \tag{32.19b}$$

These are "comoving, synchronous coordinates" for the stellar interior.

Exercise 32.7. ADIABATIC SPHERICAL COLLAPSE: EQUATIONS OF EVOLUTION [Misner and Sharp (1964)]

Describe the interior of a collapsing star by the comoving, synchronous metric (32.19), by the number density of baryons n, by the total density of mass-energy ρ, and by the pressure p. The 4-velocity of the star's fluid is

$$\boldsymbol{u} = e^{-\widetilde{\Phi}}\, \partial/\partial \widetilde{t}, \tag{32.20}$$

since the fluid is at rest in the coordinate system. Let a dot denote a proper time derivative as seen by the fluid—e.g.,

$$\dot{n} \equiv \boldsymbol{u}[n] = e^{-\widetilde{\Phi}}(\partial n/\partial \widetilde{t})_a;$$

and let a prime denote a partial derivative with respect to baryon number,—e.g.

$$n' \equiv (\partial n/\partial a)_{\widetilde{t}}.$$

Denote by U the rate of change of $(1/2\pi) \times$ (circumference of shell), as measured by a man riding in a given shell:

$$U \equiv \dot{r}; \tag{32.21a}$$

and denote by $m(\widetilde{t}, a)$ the "total mass-energy interior to shell a at time \widetilde{t}:

$$m(\widetilde{t}, a) \equiv \int_0^a 4\pi r^2 \rho(\widetilde{t}, a) r'\, da. \tag{32.21b}$$

(See Box 23.1 for discussion of this method of localizing mass-energy.) Assume that the collapse is adiabatic (no energy flow between adjacent shells; stress-energy tensor entirely that of a perfect fluid).

(a) Show that the equations of collapse [baryon conservation, (22.3); local energy conservation, (22.11a); Euler equation, (22.13); and Einstein field equations (ex. 14.16)] can be reduced to the following eight equations for the eight functions $\widetilde{\Phi}$, Γ, r, n, ρ, p, U, m:

$$\dot{r} = U \qquad\qquad \text{(dynamic equation for } r\text{)}; \tag{32.22a}$$

$$\frac{(nr^2)^{\boldsymbol{\cdot}}}{nr^2} = -\frac{U'}{r'} \qquad\qquad \text{(dynamic equation for } n\text{)}; \tag{32.22b}$$

$$\frac{\dot{\rho}}{\rho + p} = \frac{\dot{n}}{n}, \qquad\qquad \begin{array}{l}\text{except at a shock front, where adiabaticity}\\ \text{breaks down (dynamic equation for } \rho\text{)}; \end{array} \tag{32.22c}$$

$$\dot{U} = -\frac{\Gamma^2}{\rho + p}\frac{p'}{r'} - \frac{m + 4\pi r^3 p}{r^2} \qquad \text{(dynamic equation for } U\text{)}; \tag{32.22d}$$

$$p = p(n, \rho) \qquad\qquad \text{(equation of state)}; \tag{32.22e}$$

$$\widetilde{\Phi}' = -p'/(\rho + p),\ \widetilde{\Phi} = 0 \text{ at star's surface} \quad \text{(source equation for } \widetilde{\Phi}\text{)}; \tag{32.22f}$$

$$m' = 4\pi r^2 \rho r', \qquad m = 0 \text{ at } a = 0, \qquad \text{(source equation for } m\text{)}; \tag{32.22g}$$

$$\Gamma = \operatorname{sign}(r')(1 + U^2 - 2m/r)^{1/2} \qquad \text{(algebraic equation for } \Gamma\text{)}. \tag{32.22h}$$

(b) The preceding equations are in a form useful for numerical calculations. [For particular numerical solutions and for the handling of shocks, see May and White (1966).] For analytic work it is often useful to replace (32.22b) by

$$n = \Gamma/(4\pi r^2 r'), \tag{32.22b$'$}$$

and (32.22d) by

$$\dot{m} = -4\pi r^2 pU. \tag{32.22d'}$$

Derive these equations.

(c) Explain why equations (32.22g) and (32.22d') justify the remarks made in Box 23.1 about localizability of energy.

Exercise 32.8. ANALYTIC SOLUTIONS FOR PRESSURE-FREE COLLAPSE
[Tolman (1934b); Datt (1938)]

Show that the general solution to equations (32.22) in the case of zero pressure can be generated as follows.

(a) Specify the mass inside shell a, $m(a)$; by equation (32.22d'), with $p = 0$, it will not change with time \tilde{t}.

(b) Assume that all the dust particles have rest masses μ that depend upon radius, $\mu(a)$; so

$$\rho = \mu n. \tag{32.23a}$$

(c) Calculate Γ from the equation

$$\Gamma = m'/\mu; \tag{32.23b}$$

it will be independent of \tilde{t}.

(d) Specify an initial distribution of circumference $2\pi r$ as function of a, and solve the dynamic equation

$$\left(\frac{\partial r}{\partial \tilde{t}}\right)^2 - \frac{2m(a)}{r} = \Gamma^2(a) - 1 \tag{32.23c}$$

to obtain the subsequent evolution of $r(\tilde{t}, a)$. Notice that this equation has identically the same form as in Newtonian theory!

(e) Calculate the remaining quantities of interest from the algebraic equations

$$ds^2 = -d\tilde{t}^2 + (r' \, da/\Gamma)^2 + r^2 \, d\Omega^2, \tag{32.23d}$$

$$\rho = \mu n = m'/(4\pi r^2 r'), \tag{32.23e}$$

$$\tilde{\Phi} = 0, \qquad U = \partial r/\partial \tilde{t}. \tag{32.23f}$$

[Note: In this solution, successive "shells" may pass through each other, producing a surface of infinite density as they do ($r' \longrightarrow 0$ where $m' \neq 0$), since there is no pressure built up to stop shell crossing. When this happens, the coordinate system becomes pathological (a no longer increases monotonically outward), but spacetime remains well-behaved. The surface of infinite density (1) produces negligible tidal forces on neighboring dust particles; and (2) like the surface layers of §21.13, it is an idealization that gets smeared down to finite density by finite pressure.]

Exercise 32.9. COLLAPSE WITH UNIFORM DENSITY

Recover the Friedmann-Schwarzschild solution for collapse with uniform density and zero pressure by specifying appropriate forms for $m(a)$ and $r(a)$ in the prescription of exercise 32.8. In the interior of the star, give the dust particles nonzero rest masses, $\mu = \text{constant} \neq 0$; in the exterior give them zero rest masses, $\mu = 0$ ("imaginary dust particles" in vacuum). Reduce the resulting metric (32.23d) to that of Friedmann inside the star, and to that of Novikov for the Schwarzschild geometry outside the star [equations (31.12)].

§32.6. THE FATE OF A MAN WHO FALLS INTO THE SINGULARITY AT $r = 0$

The effect of tidal forces on the body of a man falling into the $r = 0$ singularity:

Consider the plight of an experimental astrophysicist who stands on the surface of a freely falling star as it collapses to $R = 0$.

As the collapse proceeds toward $R = 0$, the various parts of the astrophysicist's body experience different gravitational forces. His feet, which are on the surface of the star, are attracted toward the star's center by an infinitely mounting gravitational force; while his head, which is farther away, is accelerated downward by a somewhat smaller, though ever rising force. The difference between the two accelerations (tidal force) mounts higher and higher as the collapse proceeds, finally becoming infinite as R reaches zero. The astrophysicist's body, which cannot withstand such extreme forces, suffers unlimited stretching between head and foot as R drops to zero.

But this is not all. Simultaneous with this head-to-foot stretching, the astrophysicist is pulled by the gravitational field into regions of spacetime with ever-decreasing circumferential area, $4\pi r^2$. In order to accomplish this, tidal gravitational forces must compress the astrophysicist on all sides as they stretch him from head to foot. The circumferential compression is actually more extreme than the longitudinal stretching; so the astrophysicist, in the limit $R \longrightarrow 0$, is crushed to zero volume and indefinitely extended length.

The above discussion can be put on a mathematical footing as follows.

There are three stages in the killing of the astrophysicist: (1) the early stage, when his body successfully resists the tidal forces; (2) the intermediate stage, when it is gradually succumbing; and (3) the final stage, when it has been completely overwhelmed.

Stage 1: body resists deformation; stresses build up

During the early stage, one can analyze the tidal forces by means of the equation of geodesic deviation, evaluated in the astrophysicist's orthonormal frame $\boldsymbol{\omega}^{\hat{\tau}}$, $\boldsymbol{\omega}^{\hat{\rho}}$, $\boldsymbol{\omega}^{\hat{\theta}}$, $\boldsymbol{\omega}^{\hat{\phi}}$ (see §31.2). In this frame, the nonvanishing components of the Riemann curvature tensor are given by equations (31.6):

$$R_{\hat{\tau}\hat{\rho}\hat{\tau}\hat{\rho}} = -2M/r^3, \qquad R_{\hat{\tau}\hat{\theta}\hat{\tau}\hat{\theta}} = R_{\hat{\tau}\hat{\phi}\hat{\tau}\hat{\phi}} = M/r^3,$$
$$R_{\hat{\theta}\hat{\phi}\hat{\theta}\hat{\phi}} = 2M/r^3, \qquad R_{\hat{\rho}\hat{\theta}\hat{\rho}\hat{\theta}} = R_{\hat{\rho}\hat{\phi}\hat{\rho}\hat{\phi}} = -M/r^3. \tag{32.24a}$$

The equation of geodesic deviation says that two *freely moving* particles, momentarily at rest in the astrophysicist's local inertial frame, and separated by the 3-vector

$$\boldsymbol{\xi} = \xi^{\hat{j}} \boldsymbol{e}_{\hat{j}},$$

must accelerate apart with a relative acceleration given by

$$D^2 \xi^{\hat{j}} / d\tau^2 = -R^{\hat{j}}_{\hat{\tau}\hat{k}\hat{\tau}} \xi^{\hat{k}} = -R_{\hat{j}\hat{\tau}\hat{k}\hat{\tau}} \xi^{\hat{k}}$$
$$= -R_{\hat{\tau}\hat{j}\hat{\tau}\hat{k}} \xi^{\hat{k}}.$$

Using the components (32.24a) of the curvature tensor, one sees that

$$D^2\xi^{\hat{\rho}}/d\tau^2 = +(2M/r^3)\xi^{\hat{\rho}},$$

$$D^2\xi^{\hat{\theta}}/d\tau^2 = -(M/r^3)\xi^{\hat{\theta}}, \qquad (32.24b)$$

$$D^2\xi^{\hat{\phi}}/d\tau^2 = -(M/r^3)\xi^{\hat{\phi}}.$$

To apply these equations to the astrophysicist's body, idealize it (for simplicity) as a homogeneous rectangular box of mass $\mu \approx 165$ pounds ≈ 75 kg, of length $l \approx 70$ inches ≈ 1.8 m in the $\boldsymbol{e}_{\hat{\rho}}$ direction, and of width and depth $w \approx 10$ inches ≈ 0.2 m in the $\boldsymbol{e}_{\hat{\theta}}$ and $\boldsymbol{e}_{\hat{\phi}}$ directions. Then calculate the stresses that must be set up in this idealized body to prevent its particles from moving along diverging (and converging) geodesics.

From the form of equations (32.24), it is evident that the principal directions of the stress will be $\boldsymbol{e}_{\hat{\rho}}$, $\boldsymbol{e}_{\hat{\theta}}$, and $\boldsymbol{e}_{\hat{\phi}}$ (i.e., in the $\boldsymbol{e}_{\hat{\rho}}$, $\boldsymbol{e}_{\hat{\theta}}$, $\boldsymbol{e}_{\hat{\phi}}$ basis, the stress tensor will be diagonal). The longitudinal component of the stress, at the astrophysicist's center of mass, can be evaluated as follows. A volume element of his body with mass $d\mu$, located at a height h above the center of mass (distance h measured along $\boldsymbol{e}_{\hat{\rho}}$ direction) would accelerate with $a = (2M/r^3)h$ away from the center of mass, if it were allowed to move freely. To prevent this acceleration, the astrophysicist's muscles must exert a force

$$dF = a\,d\mu = (2M/r^3)h\,d\mu.$$

This force contributes to the stress across the horizontal plane ($\boldsymbol{e}_{\hat{\theta}} \wedge \boldsymbol{e}_{\hat{\phi}}$ plane) through the center of mass. The total force across that plane is the sum of the forces on all mass elements above it (which is also equal to the sum of the forces on the mass elements below it):

$$F = \int_{\substack{\text{(region above plane)}}} a\,d\mu = \int_0^{l/2} \left(\frac{2Mh}{r^3}\right)\left(\frac{\mu}{lw^2}\right)(w^2\,dh)$$

$$= \frac{1}{4}\frac{\mu M l}{r^3}.$$

The stress is this force divided by the cross-sectional area w^2, with a minus sign because it is a tension rather than a pressure:

$$T_{\hat{\rho}\hat{\rho}} = -\frac{1}{4}\frac{\mu M l}{w^2 r^3} \approx -1.1 \times 10^{15}\frac{M/M_{\odot}}{(r/1\text{ km})^3}\frac{\text{dynes}}{\text{cm}^2}. \qquad (32.25a)$$

The components of the stress in the $\boldsymbol{e}_{\hat{\theta}}$ and $\boldsymbol{e}_{\hat{\phi}}$ directions at the center of mass are, similarly,

$$T_{\hat{\theta}\hat{\theta}} = T_{\hat{\phi}\hat{\phi}} = +\frac{1}{8}\frac{\mu M}{l r^3} \approx +0.7 \times 10^{13}\frac{M/M_{\odot}}{(r/1\text{ km})^3}\frac{\text{dynes}}{\text{cm}^2}. \qquad (32.25b)$$

(Recall that one atmosphere of pressure is 1.01×10^6 dynes/cm^2.)

Stage 2: body gives way;
man dies

Stage 3: body gets crushed
and distended

The human body cannot withstand a tension or pressure of ≥ 100 atmospheres $\approx 10^8$ dynes/cm² without breaking. Consequently, an astrophysicist on a freely collapsing star of one solar mass will be killed by tidal forces when the star's radius is $R \sim 200$ km $\gg 2M \approx 3$ km.

By the time the star is much smaller than its gravitational radius, the baryons of the astrophysicist's body are moving along geodesics; his muscles and bones have completely given way. In this final stage of collapse, the timelike geodesics are curves along which the Schwarzschild "time"-coordinate, t, is almost constant [cf. the narrowing down of the light cones near $r = 0$ in Figure 32.1,a; also equation (31.2) in the limit $r \ll 2M$]. The astrophysicist's feet touch the star's surface at one value of t—say $t = t_f$—while his head moves along the curve $t = t_h > t_f$. Consequently, the length of the astrophysicist's body increases according to the formula

$$\ell_{\text{astroph}} = [g_{tt}(R)]^{1/2}[t_h - t_f] = [2M/R]^{1/2}[t_h - t_f]$$
$$\propto R^{-1/2} \propto (\tau_{\text{collapse}} - \tau)^{-1/3}. \qquad (32.26a)$$

Here $\tau = [-\int^R |g_{rr}|^{1/2}\, dr + \text{constant}]$ is proper time as it would be measured by the astrophysicist if he were still alive, and τ_{collapse} is the time at which he hits $r = 0$. The gravitational field also constrains the baryons of the astrophysicist's body to fall along world lines of constant θ and ϕ during the final stages of collapse. Consequently, his cross-sectional area decreases according to the law

$$\mathcal{Q}_{\text{astroph}} = [g_{\theta\theta}(R)g_{\phi\phi}(R)]^{1/2} \Delta\theta\, \Delta\phi \propto R^2 \propto (\tau_{\text{collapse}} - \tau)^{4/3}. \qquad (32.26b)$$

By combining equations (32.26a,b), one sees that the volume of the astrophysicist's body decreases, during the last few moments of collapse, according to the law

$$\mathcal{V}_{\text{astroph}} = \ell_{\text{astroph}}\mathcal{Q}_{\text{astroph}} \propto R^{3/2} \propto (\tau_{\text{collapse}} - \tau). \qquad (32.26c)$$

This crushing of matter to infinite density by infinitely large tidal gravitational forces can occur not only on the surface of the collapsing star, but also at any other point along the $r = 0$ singularity outside the surface of the star. Hence, any foolish rocketeer who ventures below the radius $r = 2M$ of the external gravitational field is doomed to destruction.

For further discussion of spacetime singularities, and of the possibility that quantum gravitational effects might force a reconsideration of the singularities predicted by classical gravitation theory, see Chapter 30, §34.6, and Chapter 44.

§32.7. REALISTIC GRAVITATIONAL COLLAPSE— AN OVERVIEW

Review of spherical collapse

Instability, implosion, horizon, and singularity; these are the key stages in the spherical collapse of any star. *Instability:* The star, having exhausted its nuclear fuel, and having contracted slowly inward, begins to squeeze its pressure-sustaining electrons or photons onto its atomic nuclei; this softens the equation of state, which induces an instability [see, e.g., §§10.15 and 11.4 of Zel'dovich and Novikov (1971)

for details]. *Implosion:* Within a fraction of a second the instability develops into a full-scale implosion; for realistic density distributions, the stellar core falls rapidly inward on itself, and the outer envelopes trail along behind [see, e.g., the numerical calculations of Colgate and White (1966), Arnett (1966, 1967), May and White (1966), and Ivanova, Imshennik, and Nadezhin (1969)]. *Horizon:* In the idealized spherical case, the star's surface falls through its gravitational radius ("horizon"; end of communication with the exterior; point of no return). From the star's vantage point this happens after a finite, short lapse of proper time. But from an external vantage point the star requires infinite time to reach the horizon, though it becomes black exponentially rapidly in the process [e-folding time $\sim M \sim 10^{-5}(M/M_\odot)$ sec]. The result is a "black hole", whose boundary is the horizon (gravitational radius), and whose interior can never communicate with the exterior. *Singularity:* From the star's interior vantage point, within a short proper time interval $\Delta\tau \sim M \sim 10^{-5}(M/M_\odot)$ sec after passing through the horizon, a singularity is reached (zero radius, infinite density, infinite tidal gravitational forces).

Does this basic picture—instability, implosion, horizon, singularity—have any relevance for real stars? Might complications such as rotation, nonsphericity, magnetic fields, and neutrino fluxes alter the qualitative picture? No, not for small initial perturbations from sphericity. Perturbation theory analyses described in Box 32.2 and exercise 32.10 show that *realistic, almost-spherically symmetric collapse, like idealized collapse, is characterized by instability, implosion, horizon;* and Penrose (1965b; see §34.6) proves that *some type of singularity then follows.*

Summary of 1972 knowledge about realistic, nonspherical collapse:

Highly nonspherical collapse is more poorly understood, of course. Nevertheless, a number of detailed calculations and precise theorems point with some confidence to two conclusions: (1) *horizons (probably) form when and only when a mass M gets compacted into a region whose circumference in EVERY direction is $\mathcal{C} \lesssim 4\pi M$* (Box 32.3); (2) *the external gravitational field of a horizon (black hole), after all the "dust" and gravitational waves have cleared away, is almost certainly the Kerr-Newman generalization of the Schwarzschild geometry* (Chapter 33). *If so, then the external field is determined uniquely by the mass, charge, and angular momentum that went "down the hole".* (This nearly proved theorem carries the colloquial title "A black hole has no hair.")

(1) horizon

(2) black hole

The interior of the horizon, and the endpoint (if any) of the collapse are very poorly understood today. The various possibilities will be reviewed in Chapter 34. That a singularity occurs one can state with much certainty, thanks to theorems of Penrose, Hawking, and Geroch. But whether all, only some, or none of the collapsing matter and fields ultimately encounter the singularity one does not know.

(3) singularity

Exercise 32.10. PRICE'S THEOREM FOR A SCALAR FIELD
 [See Price (1971, 1972a), also Thorne (1972),
 for more details than are presented here.]

EXERCISES

A collapsing spherical star, with an arbitrary nonspherical "scalar charge distribution," generates an external scalar field Φ. The vacuum field equation for Φ is $\square\Phi = \Phi_{;\alpha}{}^{\alpha} = 0$. Ignore the back-reaction of the field's stress-energy on the geometry of spacetime.

(continued on page 868)

Box 32.2 COLLAPSE WITH SMALL NON-SPHERICAL PERTURBATIONS
[based on detailed calculations by Richard H. Price (1971, 1972a,b)].

A. Density Perturbations

1. When star begins to collapse, it possesses a small nonspherical "lump" in its density distribution.

2. As collapse proceeds, lump grows larger and larger [instability of collapse against small perturbations—a phenomenon well known in Newtonian theory; see, e.g., Hunter (1967); Lin, Mestel, and Shu (1965)].

3. The growing lump radiates gravitational waves.

4. Waves of short wavelength ($\lambda \ll M$), emitted from near horizon ($r - 2M \lesssim M$), partly propagate to infinity and partly get backscattered by the "background" Schwarzschild curvature of spacetime. Backscattered waves propagate into horizon (surface of black hole; gravitational radius) formed by collapsing star.

5. Waves of long wavelength ($\lambda \gg M$), emitted from near horizon ($r - 2M \lesssim M$), get fully backscattered by spacetime curvature; they never reach out beyond $r \sim 3M$; they end up propagating "down the hole."

6. Is lump on star still there as star plunges through horizon, and does star thereby create a deformed (lumpy) horizon? Yes, according to calculations.

7. *But* external observers can only learn about existence of "final lump" by examining deformation (quadrupole moment) in final gravitational field. That final deformation in field does not and cannot propagate outward with infinite speed (no instantaneous "action at a distance"). It propagates with speed of light, in form of gravitational waves with near-infinite wavelength (infinite redshift from edge of horizon to any external radius). Deformation in final field, like any other wave of long wavelength, gets fully backscattered by curvature of spacetime at $r \lesssim 3M$; it cannot reach external observers. External observers can never learn of existence

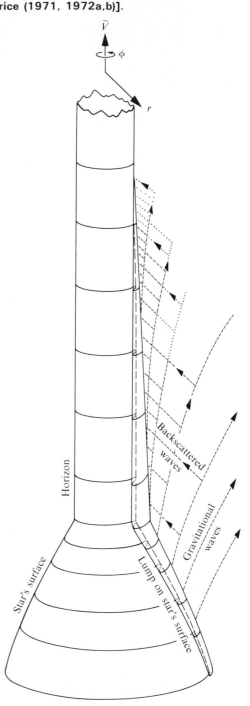

Collapse depicted in ingoing
Eddington-Finkelstein coordinates

of final lump. *Final external field is perfectly spherical, lump-free, Schwarzschild geometry!*

8. Even in region of backscatter ($2M < r \lesssim 3M$), final external field is lump-free. Backscattered waves, carrying information about existence of final lump, interfere destructively with outgoing waves carrying same information. Result is destruction of all deformation in external field and in horizon!

9. Final black hole is a Schwarzschild black hole!

B. Perturbations in Angular Momentum

1. When star begins to collapse, it possesses a small, nonzero intrinsic angular momentum ("spin") S.

2. As collapse proceeds, S is conserved (except for a tiny, negligible change due to angular momentum carried off by waves; that change is proportional to square of amplitude of waves, i.e., to square of amplitude of perturbations in star, i.e., to S^2).

3. Consequently, external field always and everywhere carries imprint of angular momentum S (on imprints, see Chapter 19). There is no need for that imprint to propagate outward from near horizon. Moreover, it could not so propagate even if it tried, because of the conservation law for S (absence of dipole gravitational waves; see §§36.1 and 36.10).

4. Hence, the final external field is that of an undeformed, slowly rotating black hole:

$$ds^2 = \underbrace{-\left(1 - \frac{2M}{r}\right)dt^2 + \frac{dr^2}{1 - 2M/r} + r^2\,d\Omega^2}_{\text{Schwarzschild geometry}}$$

$$\underbrace{-\left(\frac{4S\sin\theta}{r^2}\right)(r\sin\theta\,d\phi)\,dt.}_{\substack{\text{rotational imprint, see exercise} \\ \text{26.1; also Chapter 19.}}}$$

Here the polar axis has been oriented along S.

C. Perturbations in Electromagnetic Field

1. Star possesses a magnetic field generated by currents in its interior, and an electric field due to an arbitrary internal charge distribution; and electromagnetic radiation is emitted by its hot matter. For simplicity, S is assumed zero.

2. Evolution of external electromagnetic field is similar to evolution of perturbations in external gravitational field. Distant observer can never learn "final" values of changeable quantities (magnetic dipole moment, electric dipole moment, quadrupole moments, . . .). Final values try to propagate out from horizon, carried by electromagnetic waves of near-infinite wavelength. But they cannot get out: spacetime curvature reflects them back down the hole; and they superpose destructively with their outgoing counterparts, to produce zero net field.

3. By contrast with all other quantities, which are changeable, the electric monopole moment (total flux of electric field; equal to 4π times total electric charge) is conserved. It never changes from before star collapses, through the collapse stage, into the quiescent black-hole stage.

4. Hence, the final external electromagnetic field is a spherically symmetric coulomb field

$$\left.\begin{array}{l} \boldsymbol{E} = (Q/r^2)\boldsymbol{e}_{\hat{r}} \\ \boldsymbol{B} = 0 \end{array}\right\} \begin{array}{l} \text{as measured by static} \\ \text{observer } (r, \theta, \phi, \text{constant});\end{array}$$

and the final spacetime geometry is that of Reissner and Nordstrøm (charged black hole; see exercises 31.8 and 32.1):

$$ds^2 = -\left(1 - \frac{2M}{r} + \frac{Q^2}{r^2}\right)dt^2$$

$$+ \frac{dr^2}{(1 - 2M/r + Q^2/r^2)} + r^2\,d\Omega^2.$$

Box 32.2 (*continued*)

D. Generalization; Price's Theorem

1. Let the star generate a "zero-rest-mass, integer-spin field." ["Zero rest mass" refers to the quantized particles associated with the classical field. Classically it means the field has a Coulomb-law $(1/r)$ fall off at large distances. The spin also is a property of the quantized particles; classically it is most easily visualized as describing the symmetries of a monochromatic plane wave under rotations about the direction of propagation; see §35.6. A scalar field has spin zero; an electromagnetic field has spin one; Einstein's gravitational field has spin two; Of such fields, only gravitational $(s = 2)$ and electromagnetic $(s = 1)$ are known to exist in the real universe. See, e.g., Dirac (1936), Gårding (1945), Bargmann and Wigner (1948), Penrose (1965a), for further discussion.]

2. Let the spin-s field be sufficiently weak that its stress-energy perturbs the star's external, Schwarzschild geometry only very slightly.

3. Resolve the external field into spherical harmonics (scalar spherical harmonics for $s = 0$; vector spherical harmonics for $s = 1$; tensor spherical harmonics for $s \geq 2$); and label the spherical harmonics by the usual integer l ($l = 0$ for monopole; $l = 1$ for dipole; $l = 2$ for quadrupole; etc.).

4. All multipole fields with $l < s$ are conserved during the collapse (theorem from classical radiation theory). A scalar field $(s = 0)$ conserves nothing. The electromagnetic field $(s = 1)$ conserves only its monopole parts (electric Coulomb field, and vanishing magnetic Coulomb field). The gravitational field $(s = 2)$ conserves its monopole part (with imprint equal to mass), and its dipole parts (with imprints measuring the angular momentum, and the standard gravitational dipole moment—which vanishes if coordinate system is centered on star).

5. For $l \geq s$, and only for $l \geq s$, radiation is possible (scalar waves can have any multipolarity; electromagnetic waves must be dipole and higher; gravitational waves must be quadrupole and higher; see §36.1).

6. Price's theorem states that, as the nearly spherical star collapses to form a black hole, all things that can be radiated (all multipoles $l \geq s$) get radiated completely away—in part "off to infinity"; in part "down the hole" ("what is permitted is compulsory"). The final field is characterized completely by its conserved quantities (multipole moments with $l < s$).

7. For proof of Price's theorem in the case of a scalar field, see exercise 32.10.

E. Generalization to Nonclassical Fields

See Hartle (1971, 1972) and Teitelboim (1972b,c) for neutrino fields; Bekenstein (1972a,b) and Teitelboim (1972a) for pion fields.

Box 32.3 COLLAPSE IN ONE AND TWO DIMENSIONS

A. The Question

To produce a black hole (horizon from which nothing can emerge), must one compact matter strongly in all three spatial dimensions, to circumferences $\mathcal{C} \lesssim 4\pi M$ (quasispherical compaction); or is it sufficient to compact only in one or two dimensions?

B. The Answer for One Dimension

Consider, as an example readily generalized, the gravitational collapse of a spheroid of dust (zero pressure). Let the spheroid be highly Newtonian ($r \ggg 2M$) in its initial, momentary state of rest; and let it be slightly flattened (oblate). In Newtonian theory, any homogeneous, nonrotating spheroid of dust remains homogeneous as it collapses; but its deformations grow [see, e.g., Lin, Mestel, and Shu (1965) for details]. Hence, the spheroid of interest implodes to form a pancake of infinite density but finite mass per unit surface area. The final kinetic energy of the dust particles is roughly equal to their final potential energy:

$$\frac{1}{2}v^2 \sim \frac{M}{(\mathcal{C}/2\pi)}$$

M = mass of spheroid,
\mathcal{C} = circumference of final pancake.

Consequently, so long as $\mathcal{C}/2\pi \gg 2M$, the collapse velocities remain much smaller than light, and the gravitational energy remains much smaller than the rest mass-energy. This means that for $\mathcal{C}/2\pi \gg 2M$, the Newtonian analysis is an excellent approximation to general relativity all the way down to the pancake endpoint. Hence, *no horizon can form*, hardly any gravitational waves are emitted, and the whole story is exceedingly simple and fully Newtonian. However, since the pancake endpoint is not a singularity of spacetime (see the remarks at end of exercise 32.8), the evolution can proceed beyond it; and as \mathcal{C} contracts to $\lesssim 4\pi M$, the evolu-

tion will become very complicated and highly relativistic (see the "collapse, pursuit, and plunge scenario" of Figure 24.3).

C. The Answer for Two Dimensions

Consider, as an example *not* so readily generalized, the gravitational collapse of a homogeneous prolate spheroid of dust, initially highly Newtonian. Such a spheroid collapses to form a thin "thread" or "spindle" [see Lin, Mestel, and Shu (1965)]. Assume that the spheroid is still Newtonian when its threadlike state is reached. It then has a length ℓ, a mass per unit length $\lambda = M/\ell \ll 1$, and a rapidly contracting equatorial radius $R \ll \ell$. Subsequently, each segment of the thread collapses radially as though it were part of an infinite cylinder. [Ignore the instability of breakup into "beads"; see, e.g., Hunter (1967), Chandrasekhar (1968).] The radial collapse velocity approaches the speed of light and the gravitational energy approaches the rest mass-energy only when the thread has become exceedingly thin, $R \lesssim R_{\text{crit}} \sim \ell \exp(-1/4\lambda)$. At this stage, relativistic deviations from Newtonian collapse come into play. Thorne (1972) and Morgan and Thorne (1973) have analyzed the relativistic effects using an idealized infinite-cylinder model. The results are very different from either the spherical case or the pancake case. The collapsing cylinder emits a large flux of gravitational waves; but they are powerless to halt the collapse. *The collapse proceeds inward to a thread-like singularity, without the creation of any horizon (no black hole!).*

D. Objection to the Answer, a Reply, and a Conjecture

One can object that the collapses of both pancake and cylinder can be halted short of their endpoints, especially that of the pancake. As the thickness of

Box 32.3 (*continued*)

the pancake approaches zero, the vertical pull of gravity remains finite, but the pressure gradient caused by any finite pressure goes to infinity. Hence, pressure halts the collapse. Subsequently the rim of the pancake contracts toward the relativistic regime $\mathcal{C}/2\pi \lesssim 2M$. In the collapse of a cylinder according to Newtonian theory, with a pressure-density relation $p \propto \rho^\gamma$, the gravitational acceleration a_g and pressure-buoyancy acceleration a_p vary as

$$a_g = -2\lambda/R, \qquad a_p \sim \rho^{-1}(p/R) \propto \rho^{\gamma-1}/R.$$

Hence, for $\gamma > 1$ (the most common realistic case) pressure halts the collapse, but for $\gamma < 1$ it does

not. Whether this is true also after the relativistic domain is reached, one does not yet know.

Actually, the ability of pressure to halt the collapse is of no importance to the issue of black holes and horizons. The important thing is that *in oblate collapse with final circumference $\mathcal{C} \gg 4\pi M$, and also in prolate collapse with final thread length $\mathit{l} \gg 2M$, no horizons are created.* This, coupled with the omnipresent horizons in nearly spherical collapse (Box 32.2) suggests the following conjecture [Thorne (1972)]: *Black holes with horizons form when and only when a mass M gets compacted into a region whose circumference in EVERY direction is $\mathcal{C} \leq 4\pi M$.* (Like most conjectures, this one is sufficiently vague to leave room for many different mathematical formulations!)

(a) Resolve the external field into scalar spherical harmonics, using Schwarzschild coordinates for the external Schwarzschild geometry:

$$\Phi = \sum_{\mathit{l}} \frac{1}{r} \Psi_{\mathit{l}}(t, r) Y_{\mathit{l}m}(\theta, \phi). \tag{32.27a}$$

Show that the vacuum field equation reduces to

$$-\Psi_{\mathit{l},tt} + \Psi_{\mathit{l},r^*r^*} = \left(1 - \frac{2M}{r}\right)\left(\frac{2M}{r^3} + \frac{\mathit{l}(\mathit{l}+1)}{r^2}\right)\Psi_{\mathit{l}}, \tag{32.27b}$$

where r^* is the "tortoise coordinate" of §25.5 and Figure 25.4:

$$r^* = r + 2M \ln(r/2M - 1). \tag{32.27c}$$

Notice that (32.27b) is a flat-space, one-dimensional wave equation with effective potential

$$V_{\text{eff}}(r) = \left(1 - \frac{2M}{r}\right)\left(\frac{2M}{r^3} + \frac{\mathit{l}(\mathit{l}+1)}{r^2}\right). \tag{32.27d}$$

Part of this effective potential $[\mathit{l}(\mathit{l}+1)/r^2]$ is the "centrifugal barrier," and part $[2M/r]$ is due to the curvature of spacetime. Notice the similarity of this effective potential for scalar waves, to the effective potentials for particles and photons moving in the Schwarzschild geometry,

$$(\widetilde{V}^2)_{\text{particles}} = (1 - 2M/r)(1 + \widetilde{L}^2/r^2),$$

$$(B^{-2})_{\text{photons}} = (1 - 2M/r)r^{-2}$$

(Boxes 25.6 and 25.7). The scalar-wave potential, like the photon potential, is positive for all $r > 2M$. It rises, from 0 at $r = 2M$, to a barrier summit; then falls back to 0 at $r = \infty$.

(b) Show that *there exist no physically acceptable, static scalar-wave* perturbations of a Schwarzschild black hole. [More precisely, show that all static solutions to equation (32.27b) become infinite at either the horizon ($r = 2M$, $r^* = -\infty$) or at radial infinity.] This suggests that somehow the black hole formed by collapse must divest itself of the star's external scalar field before it can settle down into a quiescent state.

(c) The general solution to the wave equation (32.27b) can be written in terms of a Fourier transform. For waves that begin near the horizon, propagate outward, and are partially transmitted and partially reflected ("rightward-propagating waves"), show that the general solution is

$$\Psi_l(t, r^*) = \int_{-\infty}^{\infty} A(k) R_k^l(r^*) e^{-ikt} \, dk, \tag{32.28a}$$

where

$$d^2 R_k^l / dr^{*2} = [-k^2 + V_{\text{eff}}(r)] R_k^l, \tag{32.28b}$$

$$R_k^l = e^{ikr^*} + \Gamma_k^{(R)} e^{-ikr^*} \quad \text{as } r^* \longrightarrow -\infty,$$

$$R_k^l = T_k^{(R)} e^{ikr^*} \quad \text{as } r^* \longrightarrow \infty. \tag{32.28c}$$

Show that the "reflection and transmission coefficients for rightward-propagating waves," $\Gamma_k^{(R)}$ and $T_k^{(R)}$, have the following asymptotic forms for $|k| \ll 1/M$ (short wave number; long wavelength):

$$\Gamma_k^{(R)} = -1 + \alpha 2Mik, \qquad T_k^{(R)} = \frac{\beta}{(2l-1)!!} (2Mik)^{l+1} \tag{32.28d}$$

produces complete reflection and complete destructive interference in limit $k \longrightarrow 0$; see Box 32.2 for detailed discussion of consequences

no transmission in limit $k \longrightarrow 0$; see Box 32.2

where α and β are constants of order unity. Give a similar analysis for waves that impinge on a Schwarzschild black hole from outside ("leftward-propagating waves").

(d) Show that, as the star collapses into the horizon, the world line of its surface in (t, r^*) coordinates is

$$r^* = R^*(t) \equiv -t - R_0^* \exp(-t/2M) + \text{const.}, \tag{32.29a}$$

where R_0^* is related to the magnitude a of the surface's 4-acceleration ($a > 0$ for outward 4-acceleration) by

$$R_0^* = (8M/e)\left\{1 + 16Ma\left[Ma + \left(M^2a^2 + \frac{1}{8}\right)^{1/2}\right]\right\}. \tag{32.29b}$$

Thus, the world line of the surface appears to become null near the horizon ($t + r^* \equiv \tilde{V} = $ constant); of course, this is due to pathology of the coordinate system there. Show, further, that the scalar field on the star's surface ($\tilde{V} = $ constant) must vary as

$$\Psi_l = Q_{l0} + Q_{l1} e^{-\tilde{U}/4M}, \qquad \tilde{U} \equiv t - r^*. \tag{32.29c}$$

when the star is approaching the horizon ($t \longrightarrow \infty$, $r^* \longrightarrow -\infty$, $\tilde{U} \longrightarrow \infty$), in order that the rate of change of Ψ_l be finite as measured on the star's surface. Notice that Q_{l0} is the "final value" of the scalar field on the star's surface. It can be regarded as an outgoing wave with zero wave number (infinite wavelength); and, consequently, *it gets completely and*

destructively reflected by the effective potential [see equation (32.28d); also Box 32.2]. Conclusion: All multipoles of the scalar field die out at finite r^* as $t \longrightarrow \infty$. (Price's theorem for a scalar field.) For a more detailed analysis, including the rates at which the multipoles die out, see Price (1971, 1972a) or Thorne (1972).

Exercise 32.11. NEWMAN-PENROSE "CONSTANTS"
[See Press and Bardeen (1971), Bardeen and Press (1972),
and Piir (1971) for more details than are presented here.]

Wheeler (1955) showed that Maxwell's equations for an ℓ-pole electromagnetic field residing in the Schwarzschild geometry can be reduced to the wave equation

$$-\Psi_{\ell,tt} + \Psi_{\ell,r^*r^*} = \left(1 - \frac{2M}{r}\right)\frac{\ell(\ell + 1)}{r^2}\,\Psi_\ell \tag{32.30}$$

[electromagnetic analogue of (32.27b)]. After this equation has been solved, the components of the electromagnetic field can be obtained by applying certain differential operators to $\Psi_\ell(t, r^*)Y_{\ell m}(\theta, \phi)$.

(a) Show that the general solution to the electromagnetic wave equation (32.30) for dipole ($\ell = 1$) fields, with outgoing-wave boundary conditions at $r^* \longrightarrow +\infty$, has the form

$$\Psi_1 = f_0(\widetilde{U}) + \frac{f_1(\widetilde{U})}{r} + \frac{f_2(\widetilde{U})}{r^2} + \dots, \tag{32.31a}$$

where

$$\widetilde{U} = t - r^* \text{ is "retarded time", and}$$

$$f_1' = f_0, \qquad f_2' = 0, \qquad \dots, \qquad f_n' = -\frac{(n + 1)(n - 2)}{2n}f_{n-1} + (n - 2)Mf_{n-2}. \tag{32.31b}$$

When spacetime is flat ($M = 0$), this solution becomes

$$\Psi_1 = f_1'(\widetilde{U}) + f_1(\widetilde{U})/r. \tag{32.31N}$$

[The $1/r$ fall-off of the radiation field $f_1'(\widetilde{U})$ has been factored out of Ψ_1; see the scalar-wave function (32.27a).] The terms $f_2(\widetilde{U})/r^2 + \dots$, which are absent in flat spacetime, are attributable to backscatter of the outgoing waves by the curvature of spacetime. They are sometimes called the "tail" of the waves.

(b) Show that the general static dipole field has the form (32.31a) with

$$(f_0)_{\text{static}} = 0; \qquad (f_1)_{\text{static}} \equiv D = \text{dipole moment};$$

$$(f_2)_{\text{static}} = \frac{3}{2}\,MD; \dots. \tag{32.32}$$

(c) Consider a star (not a black hole!) with a dipole field that is initially static. At time $t = 0$, let the star suddenly change its dipole moment to a new static value D'. Equations (32.31b) demand that f_2 be conserved ["*Newman-Penrose (1965) constant*"]. Hence, f_2 *will always exhibit a value,* $\frac{3}{2}MD$, *corresponding to the old dipole moment; it can never change to* $\frac{3}{2}MD'$. This is a manifestation of the tail of the waves that are generated by the sudden change in dipole moment. To understand this tail effect more clearly, and to discover an important flaw in the above result, evaluate the solution (32.31) for retarded time $\widetilde{U} > 0$, using the assumptions

$$\text{(1) field has static form (32.32) for } \widetilde{U} < 0,$$

$$\text{(2) } f_1 = D' \text{ for } \widetilde{U} > 0. \tag{32.33}$$

Put the answer in the form

$$
\Psi_1 = \frac{D'}{r} + \frac{\frac{3}{2}MD}{r^2} + \sum_{n=3}^{\infty} \frac{2M(D'-D)(-1)^{n+1}(n+1)\tilde{U}^{n-2}}{(2r)^n} + O\left(\frac{M^2}{r^3}, \frac{M^2\tilde{U}}{r^4}\right). \quad (32.34)
$$

(The terms neglected are small compared to those kept for all \tilde{U}/r, so long as $r \gg M$.) Evidently, so long as the series converges the Newman-Penrose "constant" (coefficient of $1/r^2$) remembers the old D value and is conserved, as claimed above. Show, however, that the series diverges for $\tilde{U} > 2r$—i.e., it diverges inside a sphere that moves outward with asymptotically $\frac{1}{3}$ the speed of light. Thus, *the Newman-Penrose "constant" is well-defined and conserved only outside the "$\frac{1}{3}$-speed-of-light cone."*

(d) Sum the series in (32.34) to obtain a solution valid even for $\tilde{U} > 2r$:

$$
\Psi = \underbrace{\frac{D'}{r} + \frac{3}{2}\frac{MD'}{r^2}}_{\substack{\text{new static}\\\text{solution}}} - \underbrace{\frac{2M(D'-D)}{r}\frac{(\tilde{U}+3r)}{(\tilde{U}+2r)^2}}_{\text{"tail term"}} + O\left(\frac{M^2}{r^3}\right)
$$

$$
= \text{the series (32.34) for } \tilde{U} < 2r \text{ (domain of convergence of that series)}
$$

$$
= \frac{D'}{r} + \frac{3}{2}\frac{MD'}{r^2} + O\left(\frac{M}{\tilde{U}r}, \frac{M^2}{r^3}\right) \qquad \text{for } \tilde{U} \gg r \gg M.
$$

(32.35)

From this result conclude that *at fixed r and late times the electromagnetic field becomes asymptotically static, and the Newman-Penrose "constant" assumes the new value $\frac{3}{2}MD'$ appropriate to the new dipole moment.*

CHAPTER **33**

BLACK HOLES

*A luminous star, of the same density as the Earth, and whose diameter should be
two hundred and fifty times larger than that of the Sun, would not, in
consequence of its attraction, allow any of its rays to arrive at us; it is therefore
possible that the largest luminous bodies in the universe may, through this cause,
be invisible.*

P. S. LAPLACE (1798)

§33.1. WHY "BLACK HOLE"?

A dialog explaining why
black holes deserve their
name

Sagredus. What is all this talk about "black holes"? When an external observer
watches a star collapse, he sees it implode with ever-increasing speed, until
the relativistic stage is reached. Then it appears to slow down and become
"frozen," just outside its horizon (gravitational radius). However long the
observer waits, he never sees the star proceed further. How can one reasonably
give the name "black hole" to such a frozen object, which never disappears
from sight?

Salvatius. Let us take the name "black hole" apart. Consider first the blackness.
Surely nothing can be blacker than a black hole. The very redshift that makes
the collapsing star appear to freeze also makes it darken and become black.
In the continuum approximation, where one ignores the discreteness of
photons, the intensity of the radiation received by distant observers decreases
exponentially as time passes, $L \propto \exp(-t/3\sqrt{3}M)$, with an exceedingly short
e-folding time

$$\tau = 3\sqrt{3}M = (2.6 \times 10^{-5} \text{ sec})(M/M_\odot).$$

Within a fraction of a second, the star is essentially black. Discreteness of
photons makes it even blacker. The number of photons emitted before the
star crosses its horizon is finite, so the exponential decay cannot continue

For a more detailed exposition of the foundations of "black-hole physics," see DeWitt and DeWitt
(1973).

forever. Eventually—only $10^{-3}(M/M_\odot)$ seconds after the star begins to dim (see exercise 32.2)—the last photon that will ever get out reaches the distant observers. Thereafter nothing emerges. The star is not merely "essentially black"; it is *absolutely black."*

Sagredus. Agreed. But it is the word "hole" that concerns me, not "black." How can one possibly regard the name "hole" as appropriate for an object that hovers forever just outside its horizon. True, absence of light makes the object invisible. But couldn't one always see it by shining a flashlight onto its surface? And couldn't one always fly down to its surface in a rocket ship and scoop up a few of the star's baryons? After all, as seen from outside the baryons at its surface will never, never, never manage to fall into the horizon!

Salvatius. Your argument *sounds* persuasive. To test its validity, examine the collapse of a spherically symmetric system, using the ingoing Eddington-Finkelstein diagram of Figure 33.1. Let a family of external observers shine their flashlights onto the star's surface, as you have suggested. Let the surface of the star be carefully silvered so it reflects back all light that reaches it. Initially (low down in the spacetime diagram of Figure 33.1) the ingoing light beams

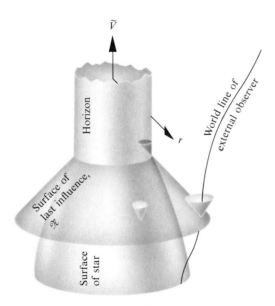

Figure 33.1.
Spherical gravitational collapse of a star to form a black hole, as viewed in ingoing Eddington-Finkelstein coordinates. The "surface of last influence," \mathcal{R}, is an ingoing null surface that intersects the horizon in coincidence with the surface of the collapsing star. After an external observer, moving forward in time, has passed through the surface of last influence, he cannot interact with and influence the star before it plunges through the horizon. Thus, one can think of the surface of last influence as the "birthpoint" of the black hole. Before passing through this surface, the external observer can say his flashlight is probing the shape of a collapsing star; afterwards, he can regard his signals as probes of a black hole. For further discussion, see text.

reach the star's surface and are reflected back to the flashlights with no difficulty. But there is a critical point—an ingoing radial null surface \mathfrak{N}—beyond which reflection is impossible. Photons emitted inward along \mathfrak{N} reach the star just as it is passing through its horizon. After reflection these photons fly "outward" along the horizon, remaining forever at $r = 2M$. Other photons, emitted inward after the flashlight has passed through \mathfrak{N}, reach the surface of the star and are reflected only after the star is inside its horizon. Such photons can never return to the shining flashlights. Once inside the horizon, they can never escape. Thus, the total number of photons returned is finite and is subject to the same blackness decay law as is the intrinsic luminosity of the star. Moreover, *if the observers do not turn on their flashlights until after they pass through the null surface \mathfrak{N}, they can never receive back any reflected photons!* Evidently, flashlights are of no help in seeing the "frozen star."

Sagredus. I cannot escape the logic of your argument. Nevertheless, seeing is not the only means of interacting with the frozen star. I have already suggested swooping down in a rocket ship and stealing a few baryons from its surface. Similarly, one might let matter fall radially inward onto the frozen star. When the matter hits the star's surface, its great kinetic energy of infall will be converted into heat and into outpouring radiation.

Salvatius. Thus it might seem at first sight. But examine again Figure 33.1. No swooping rocket ship and no infalling matter can move inward faster than a light ray. Thus, if the decision to swoop is made after the ship passes through the surface \mathfrak{N}, the rocket ship has no possibility of reaching the star before it plunges through the horizon; the rocket and pilot cannot touch the star, sweep up baryons, and return to tell their tale. Similarly, infalling matter to the future of \mathfrak{N} can never hit the star's surface before passing through the horizon. The surface \mathfrak{N} is, in effect, a "surface of last influence." Once anybody or anything has passed through \mathfrak{N}, he or it has no possibility whatever of influencing or interacting with the star in any way before it plunges through the horizon. *Thus, from a "causal" or "interaction" standpoint, the collapsing star becomes a hole in space at the surface \mathfrak{N}.* This hole is not black at first. Radiation from the collapsing star still emerges after \mathfrak{N} because of finite light-propagation times, just as radiation still reaches us today from the hot big-bang explosion of the universe. But if an observer at radius $r \gg 2M$ waits for a time $2r$ after passing through \mathfrak{N} (time for \mathfrak{N} to reach horizon, plus time for rays emitted at $R \sim 3M$ to get back to observer), then he will see the newly formed hole begin to turn black; and within a time $\Delta t \sim (10^{-3}$ seconds$)(M/M_\odot)$ thereafter, it will be completely black.

Sagredus. You have convinced me. For all practical purposes the phrase "black hole" is an excellent description. The alternative phrases "frozen star" and "collapsed star," which I find in the pre-1969 physics literature, emphasize an optical-illusion aspect of the phenomenon. Attention must be directed away from the star that created the black hole, because beyond the surface of last influence one has no means to interact with that star. The star is irrelevant

to the subsequent physics and astrophysics. Only the horizon and its external spacetime geometry are relevant for the future. Let us agree to call that horizon the "surface of a black hole," and its external geometry the "gravitational field of the black hole."

Salvatius. Agreed.

§33.2. THE GRAVITATIONAL AND ELECTROMAGNETIC FIELDS OF A BLACK HOLE

The collapse of an electrically neutral star endowed with spherical symmetry produces a spherical black hole with external gravitational field described by the Schwarzschild line element

$$ds^2 = -(1 - 2M/r)\,dt^2 + \frac{dr^2}{1 - 2M/r} + r^2(d\theta^2 + \sin^2\theta\,d\phi^2). \quad (33.1)$$

The surface of the black hole, i.e., the horizon, is located at $r = 2M = $ (gravitational radius). Only the region on and outside the black hole's surface, $r \geq 2M$, is relevant to external observers. Events inside the horizon can never influence the exterior.

The gravitational collapse of a realistic star (nonspherical, collapse with small but nonzero net charge of one sign or the other) produces a black hole somewhat different from the simple Schwarzschild hole. For collapse with small charge and small asymmetries, perturbation-theory calculations (Box 32.2) predict a final black hole with external field determined entirely by the mass M, charge Q, and intrinsic angular momentum S of the collapsing star. For fully relativistic collapse, with large asymmetries and possibly a large charge, the final black hole (if one forms) is also characterized uniquely by M, Q, and S. This is the conclusion that strongly suggests itself in 1972 from a set of powerful theorems described in Box 33.1.

Why M, Q, and S should be the complete governors of the final external field of the black hole, one can understand heuristically as follows. Of all quantities intrinsic to any isolated source of gravity and electromagnetism, only M, Q, and S possess (and are defined in terms of) *unique, conserved imprints* in the distant external fields of the source (conserved Gaussian flux integrals; see Box 19.1 and §20.2). When a star collapses to form a black hole, its distant external fields are forced to maintain unchanged the imprints of M, Q, and S. In effect, M, Q, and S provide anchors or constraints on the forms of the fields. Initially other constraints are produced by the distributions of mass, momentum, stress, charge, and current inside the star. But ultimately the star plunges through a horizon, cutting itself off causally from the external universe. (The nonpropagation of long-wavelength waves through curved spacetime plays a key role in this cutoff; see Box 32.2.) Subsequently, the only anchors remaining for the external fields are the conserved imprints of M, Q, and S. Consequently, the external fields quickly settle down into unique shapes corresponding to the given M, Q, and S. Of course, the settling down involves dynamic changes of the fields and an associated outflow of gravitational and electro-

The structure of a black hole is determined uniquely by its mass M, charge Q, and intrinsic angular momentum, S

Heuristic explanation of the M-Q-S uniqueness

Box 33.1 A BLACK HOLE HAS NO "HAIR"

The following theorems come close to proving that *the external gravitational and electromagnetic fields of a stationary black hole* (a black hole that has settled down into its "final" state) *are determined uniquely by the hole's mass M, charge Q, and intrinsic angular momentum S*—i.e., the black hole can have no "hair" (no other independent characteristics). For a detailed review, see Carter (1973).

I. Stephen Hawking (1971b, 1972a): A stationary black hole must have a horizon with spherical topology; and it must either be static (zero angular momentum), or axially symmetric, or both.

II. Werner Israel (1967a, 1968): Any *static* black hole with event horizon of spherical topology has external fields determined uniquely by its mass M and charge Q; moreover, those external fields are the Schwarzschild solution if $Q = 0$, and the Reissner-Nordstrøm solution (exercises 31.8 and 32.1) if $Q \neq 0$ (both special cases of Kerr-Newman; see §33.2).

III. Brandon Carter (1970): "All uncharged, stationary, axially symmetric black holes with event horizons of spherical topology fall into disjoint families not deformable into each other. The black holes in each family have external gravitational fields determined uniquely by two parameters: the mass M and the angular momentum S." (Note: the "Kerr solutions"—i.e., "Kerr-Newman" with $Q = 0$—form one such family; it is very likely that there are no others, but this has not been proved as of December 1972. It is also likely that Carter's theorem can be extended to the case with charge; but this has also not yet been done.)

IV. Conclusions made by combining all three theorems:
 (a) All stationary black holes are axially symmetric.
 (b) All static (nonrotating) black holes are characterized uniquely by M and Q, and have the Reissner-Nordstrøm form.
 (c) All uncharged, rotating black holes fall into distinct and disjoint families, with each black hole in a given family characterized uniquely by M and S. The Kerr solutions form one such family. There may well be no other family.

V. Remarks and Caveats:
 (a) The above statements of the theorems are all somewhat heuristic. Each theorem makes several highly technical assumptions, not stated here, about the global properties of spacetime. These assumptions seem physically reasonable and innocuous, but they might not be.
 (b) Progress in black-hole physics is so rapid that, by the time this book is published, there may well exist theorems more powerful than the above, which really prove that "a black hole has no hair."
 (c) For insight into the techniques of "global geometry" used in proving the above theorems and others like them, see Chapter 34; for greater detail see the forthcoming book by Hawking and Ellis (1973).
 (d) For analyses which show that a black hole cannot exert any weak-interaction forces caused by the leptons which have gone down it, see Hartle (1971, 1972) and Teitelboim (1972b,c). For similar analyses which show absence of strong-interaction forces from baryons that have gone down the hole, see Bekenstein (1972a,b) and Teitelboim (1972a).

magnetic waves. And, of course, the outflowing waves carry off mass and angular momentum (but not charge), thereby leaving M and S changed. And, of course, the external fields must then readjust themselves to the new M and S. But the process will quickly converge, producing a black hole with specific final values of M, Q, and S and with external fields determined uniquely by those values.

The problem of calculating the external fields for given M, Q, and S and their given imprints, is analogous to the problem of Plateau—to calculate the shape of a soap film anchored to a wire of given shape.* One calculates the shape of the soap film by seeking a surface of minimum area spanning the bent wire. The condition of minimum area leads to a differential equation describing the soap film, which must be solved subject to the constraint imposed by the shape of the wire.

To calculate the external fields of a black hole, one can extremize the "action integral" $\int (\mathcal{R} + \mathcal{L})\sqrt{-g}\, d^4x$ for interacting gravitational and electromagnetic fields (see Chapter 21) subject to the anchored-down imprints of M, Q, and S at radial infinity, and subject to the existence of a physically nonsingular horizon (no infinite curvature at horizon!). Extremizing the action is equivalent to solving the coupled Einstein-Maxwell field equations subject to the constraints imprinted by M, Q, and S, and the existence of the horizon. The derivation of the solution and the proof of its uniqueness are much too complex to be given here. (See references cited in Box 33.1.) However, the solution turns out to be the "*Kerr-Newman geometry*" and its associated electromagnetic field.†

Written in the t, r, θ, ϕ coordinates of Boyer and Lindquist (1967) (generalization of Schwarzschild coordinates), the Kerr-Newman geometry has the form

$$ds^2 = -\frac{\Delta}{\rho^2}[dt - a\sin^2\theta\, d\phi]^2 + \frac{\sin^2\theta}{\rho^2}[(r^2 + a^2)\, d\phi - a\, dt]^2$$
$$+ \frac{\rho^2}{\Delta}dr^2 + \rho^2\, d\theta^2, \tag{33.2}$$

where

$$\Delta \equiv r^2 - 2Mr + a^2 + Q^2, \tag{33.3a}$$

$$\rho^2 \equiv r^2 + a^2\cos^2\theta, \tag{33.3b}$$

$$a \equiv S/M \equiv \text{angular momentum per unit mass.} \tag{33.4}$$

The corresponding electromagnetic field tensor, written as a 2-form (recall: $\boldsymbol{d}x^\alpha \wedge \boldsymbol{d}x^\beta \equiv \boldsymbol{d}x^\alpha \otimes \boldsymbol{d}x^\beta - \boldsymbol{d}x^\beta \otimes \boldsymbol{d}x^\alpha$) is

$$\boldsymbol{F} = Q\rho^{-4}(r^2 - a^2\cos^2\theta)\,\boldsymbol{d}r \wedge [\boldsymbol{d}t - a\sin^2\theta\,\boldsymbol{d}\phi]$$
$$+ 2Q\rho^{-4}ar\cos\theta\sin\theta\,\boldsymbol{d}\theta \wedge [(r^2 + a^2)\,\boldsymbol{d}\phi - a\,\boldsymbol{d}t]. \tag{33.5}$$

Variational principle for black-hole structure

Details of black-hole structure:

(1) metric ("Kerr-Newman geometry")

(2) electromagnetic field

*On the problem of Plateau see, e.g., Courant (1937), Darboux (1941), or p. 157 of Lipman Bers (1952).

†The uncharged ($Q = 0$) version was first found as a solution to Einstein's vacuum field equations by Kerr (1963). The charged generalization was first found as a solution to the Einstein-Maxwell field equations by Newman, Couch, Chinnapared, Exton, Prakash, and Torrence (1965). Only later was the connection to black holes discovered; see Box 33.1.

Expressions (33.2) for the metric and (33.5) for the electromagnetic field are sufficiently long to be somewhat frightening. Therefore, it is helpful to develop some qualitative insight into them and into their implications before attempting detailed computations with them. Boxes 33.2, 33.3, and 33.4 develop qualitative insight by presenting, without derivation, a summary of the key features of the Kerr-Newman geometry and a summary of the physics and astrophysics of black holes. The remainder of this chapter is a Track-2 justification and derivation of some, but not all, of the results cited in Boxes 33.2–33.4.

(*continued on page 891*)

Box 33.2 KERR-NEWMAN GEOMETRY AND ELECTROMAGNETIC FIELD

I. Equations for metric and electromagnetic field

A. Parameters appearing in equations:
M = mass, Q = charge, $a \equiv S/M$ = angular momentum per unit mass, all as measured by their standard imprints on the distant fields.

B. Constraint on parameters:
The Kerr-Newman geometry has a horizon, and therefore describes a black hole, if and only if $M^2 \geq Q^2 + a^2$. It seems likely that in any collapsing body which violates this constraint, centrifugal forces and/or electrostatic repulsion will halt the collapse before a size $\sim M$ is reached; see equation (33.56).

C. Limiting cases:

$Q = 0$,	Kerr (1963) geometry;
$S = 0$,	Reissner-Nordstrøm geometry and electromagnetic field (exercises 31.8 and 32.1);
$Q = S = 0$,	Schwarzschild geometry;
$M^2 = Q^2 + a^2$	"Extreme Kerr-Newman geometry."

D. Boyer-Lindquist (1967) coordinates (t, r, θ, ϕ—generalization of Schwarzschild coordinates; black hole rotates in ϕ direction):

$$ds^2 = -(\Delta/\rho^2)[dt - a\sin^2\theta\, d\phi]^2 + (\sin^2\theta/\rho^2)[(r^2 + a^2)\, d\phi - a\, dt]^2 \qquad (1)$$
$$+ (\rho^2/\Delta)\, dr^2 + \rho^2\, d\theta^2;$$

$$\Delta \equiv r^2 - 2Mr + a^2 + Q^2, \qquad \rho^2 \equiv r^2 + a^2\cos^2\theta. \qquad (2)$$

$$\boldsymbol{F} = Q\rho^{-4}(r^2 - a^2\cos^2\theta)\, \boldsymbol{dr} \wedge [\boldsymbol{dt} - a\sin^2\theta\, \boldsymbol{d\phi}]$$
$$+ 2Q\rho^{-4}ar\cos\theta\sin\theta\, \boldsymbol{d\theta} \wedge [(r^2 + a^2)\, \boldsymbol{d\phi} - a\, \boldsymbol{dt}]. \qquad (3)$$

E. Kerr coordinates [$\widetilde{V}, r, \theta, \widetilde{\phi}$—generalization of ingoing Eddington-Finkelstein coordinates; $(\widetilde{V}, \theta, \widetilde{\phi})$ = constant is an ingoing, "radial," null geodesic; black hole rotates in $\widetilde{\phi}$ direction]:
Relationship to Boyer-Lindquist:

$$d\widetilde{V} = dt + (r^2 + a^2)(dr/\Delta),$$

$$d\widetilde{\phi} = d\phi + a(dr/\Delta). \tag{4}$$

$$ds^2 = -[1 - \rho^{-2}(2Mr - Q^2)]\,d\widetilde{V}^2 + 2\,dr\,d\widetilde{V} + \rho^2\,d\theta^2$$
$$+ \rho^{-2}[(r^2 + a^2)^2 - \Delta a^2 \sin^2\theta]\sin^2\theta\,d\widetilde{\phi}^2 - 2a\sin^2\theta\,d\widetilde{\phi}\,dr \tag{5}$$
$$- 2a\rho^{-2}(2Mr - Q^2)\sin^2\theta\,d\widetilde{\phi}\,d\widetilde{V}.$$

$$\boldsymbol{F} = Q\rho^{-4}[(r^2 - a^2\cos^2\theta)\,\boldsymbol{dr} \wedge \boldsymbol{d\widetilde{V}} - 2a^2 r\cos\theta\sin\theta\,\boldsymbol{d\theta} \wedge \boldsymbol{d\widetilde{V}} \tag{6}$$
$$- a\sin^2\theta(r^2 - a^2\cos^2\theta)\,\boldsymbol{dr} \wedge \boldsymbol{d\widetilde{\phi}} + 2ar(r^2 + a^2)\cos\theta\sin\theta\,\boldsymbol{d\theta} \wedge \boldsymbol{d\widetilde{\phi}}].$$

II. Properties of spacetime geometry

A. Symmetries (§33.4):
The metric coefficients in Boyer-Lindquist coordinates are independent of t and ψ, and in Kerr coordinates are independent of \widetilde{V} and $\widetilde{\phi}$. Thus the spacetime geometry is "time-independent" (stationary) and axially symmetric. The "Killing vectors" (§25.2) associated with these two symmetries are $(\partial/\partial t)_{r,\theta,\phi} = (\partial/\partial\widetilde{V})_{r,\theta,\widetilde{\phi}}$ and $(\partial/\partial\phi)_{t,r,\theta} = (\partial/\partial\widetilde{\phi})_{\widetilde{V},r,\theta}$.

B. Dragging of inertial frames and static limit (§33.4):

1. The "dragging of inertial frames" by the black hole's angular momentum produces a precession of gyroscopes relative to distant stars. By this precession one defines and measures the angular momentum of the black hole (see §§19.2 and 19.3).

2. The dragging becomes more and more extreme the nearer one approaches the horizon of the black hole. Before the horizon is reached, at a surface described by

$$r = r_0(\theta)$$
$$\equiv M + \sqrt{M^2 - Q^2 - a^2\cos^2\theta}, \tag{7}$$

the dragging becomes so extreme that no observer can possibly remain at rest there (i.e., be "static") relative to the distant stars. At and inside this surface (called the "static limit"), all observers with fixed r and θ must orbit the black hole in the same direction in which the hole rotates:

$$\Omega \equiv d\phi/dt$$
$$> \frac{a\sin\theta - \sqrt{\Delta}}{(r^2 + a^2)\sin\theta - \sqrt{\Delta}a\sin^2\theta}$$
$$(\geq 0 \text{ for } a = S/M > 0 \text{ and } r \leq r_0).$$

No matter how hard an observer, at fixed (r, θ) inside the static limit, blasts his rocket engines, he can never halt his angular motion relative to the distant stars.

3. The mathematical foundation for the above statement is this: world lines of the form $(r, \theta, \phi) = $ constant [tangent vector $\propto \partial/\partial t = $ "Killing vector in time direction"] change from being timelike outside the static limit to being spacelike inside it. Therefore, on and inside the static limit, no observer can remain at rest.

C. Horizon (§33.4):

1. The horizon is located at

$$r = r_+ \equiv M + \sqrt{M^2 - Q^2 - a^2}. \tag{8}$$

2. As with the Schwarzschild horizon of a nonrotating black hole, so also here, particles and photons can fall inward through the horizon; but no particle or

Box 33.2 (continued)

photon can emerge outward through it.

3. The horizon is "generated" by outgoing null geodesics (outgoing photon world lines).

D. Ergosphere (§33.4):

1. The "ergosphere" is the region of space-time between the horizon and the static limit. It plays a fundamental role in the physics of black holes (Box 33.3; §33.7).

2. The static limit and the horizon touch at the point where they are cut by the axis of rotation of the black hole ($\theta = 0, \pi$); they are well-separated elsewhere with the static limit outside the horizon, unless $a = 0$ (no rotation). When $a = 0$, the static limit and horizon coincide; there is no dragging of inertial frames; there is no ergosphere.

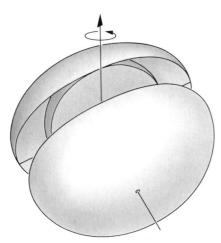

Qualitative representation of horizon, ergosphere, and static limit [adapted from Ruffini and Wheeler (1971b)].

E. Singularity in Boyer-Lindquist coordinates:

1. For a nonrotating black hole, the Schwarzschild coordinates become singular at the horizon. One manifestation

of the singularity is the infinite amount of coordinate time required for any particle or photon to fall inward through the horizon, $t \longrightarrow \infty$ as $r \longrightarrow 2M$. One way to remove the singularity (Eddington-Finkelstein way) is to replace t by a null coordinate
$$\widetilde{V} = t + r + 2M \ln |r/2M - 1|$$
attached to infalling photons [so $(\partial/\partial r)_{\widetilde{V},\theta,\phi}$ is vector tangent to photon world lines].

2. For a rotating black hole, the Boyer-Lindquist coordinates, being generalizations of the Schwarzschild coordinates, are also singular at the horizon. It requires an infinite coordinate time for any particle or photon to fall inward through the horizon, $t \longrightarrow \infty$ as $r \longrightarrow r_+$. But that is not all. The dragging of inertial frames forces particles and photons near the horizon to orbit the black hole with $\Omega \equiv d\phi/dt > 0$. Consequently, for a particle falling through the horizon ($r \longrightarrow r_+$), just as $t \longrightarrow \infty$, so also $\phi \longrightarrow \infty$ (infinite twisting of world lines around horizon).

3. To remove the coordinate singularity, one must perform an infinite compression of coordinate time, and an infinite untwisting in the neighborhood of the horizon. Kerr coordinates achieve this by replacing t with a null coordinate \widetilde{V}, and ϕ with an untwisted angular coordinate $\widetilde{\phi}$:

$$d\widetilde{V} = dt + (r^2 + a^2)(dr/\Delta),$$
$$d\widetilde{\phi} = d\phi + a(dr/\Delta).$$

Both of the new coordinates are attached to the world lines of a particular family of infalling photons; $(\partial/\partial r)_{\widetilde{V},\theta,\widetilde{\phi}}$ is the field of vectors tangent to the world lines of this family of photons (ingoing principal null congruence; §33.6).

F. Spacetime diagram:

1. A spacetime diagram in Kerr coordi-
 nates looks much like an Eddington-
 Finkelstein diagram for the Schwarz-
 schild geometry. In both cases, one plots
 the surfaces of constant \widetilde{V} not as hori-
 zontal planes, but as "backward light
 cones" ("45-degree surfaces"), because
 they are generated by the world lines
 of ingoing photons. Equivalently, one
 plots surfaces of constant $\widetilde{t} \equiv \widetilde{V} - r$ as
 horizontal planes.

2. The key differences between a Kerr
 diagram and an Eddington-Finkelstein
 diagram are: (a) Because the Kerr-
 Newman geometry is not spherical, a
 Kerr diagram with one rotational de-
 gree of freedom suppressed loses infor-
 mation about the geometry. Kerr dia-
 grams are usually made for the
 equatorial "plane," $\theta = \pi/2$. (b) Just as
 the horizon pulls the light cones inward,
 so the dragging of inertial frames tilts
 the light cones in the direction of in-
 creasing $\widetilde{\phi}$, for $a > 0$ and $r = $ constant.
 (c) The *ingoing* edge of a light cone
 $(dr/d\widetilde{V} = -\infty)$ does *not* tilt toward in-
 creasing $\widetilde{\phi}$; the transformation from
 Boyer-Lindquist coordinates to Kerr
 coordinates untwists the tilt with de-
 creasing r, which would otherwise be
 produced by "frame dragging."

3. The shapes of the light cones reveal the
 special features of the static limit and
 horizon. At the static limit, a vertical
 world line $[r, \theta, \widetilde{\phi}$ constant; $(\partial/\partial\widetilde{V})_{r,\theta,\widetilde{\phi}}$
 $= (\partial/\partial t)_{r,\theta,\phi} = $ tangent vector] lies on
 the light cone. At the horizon the light
 cones tilts fully inward, except for a
 single line of tangency to the horizon.
 Notice that the line of tangency has
 $d\widetilde{\phi}/d\widetilde{V} = a/(r_{+}{}^2 + a^2) \neq 0$. Equiva-
 lently, the outgoing null geodesics,
 which generate the horizon, twist about
 it ("barber-pole-twist")—yet another
 manifestation of the dragging of inertial
 frames.

Kerr diagram for equatorial slice
($\theta = \pi/2$) through the spacetime of an
"extreme Kerr" black hole ($Q = 0, a = M$).

View from above showing the shapes of
the light cones as a function of radius

Box 33.2 (continued)

4. The Kerr diagram, like the Eddington-Finkelstein diagram, describes infall through the horizon in a faithful, non-singular way.

5. [The term "Kerr diagram" is a misnomer. Kerr has not published such diagrams himself, though nowadays others construct such diagrams using his coordinate system. Penrose is the originator and greatest exploiter of such diagrams (see, e.g., Penrose, 1969). But several other types of diagrams bear Penrose's name, so it would be confusing to name them all after him.]

G. Maximal analytic extension of Kerr-Newman geometry:

1. When one abstracts the Schwarzschild geometry away from all sources (Chapter 31), one discovers that it describes an expanding and recontracting bridge, connecting two different universes. But in the context of black holes, only half of the Schwarzschild geometry (regions I and II) is relevant. The other half (regions III and IV) gets fully replaced by the interior of the star that collapsed to form the black hole. Because only a

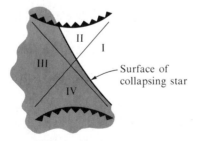

Surface of collapsing star

part of the Schwarzschild geometry comes into play, ingoing Eddington-Finkelstein coordinates—which describe I and II well, but III and IV badly—are well-suited to black-hole physics.

2. Similarly, when one abstracts the Kerr-Newman geometry away from all sources, one discovers that it describes a much larger, and more complex space-time manifold than one might ever have suspected. This "maximum analytic extension" of the Kerr-Newman geometry has been analyzed in detail by Boyer and Lindquist (1967) and by Carter (1966a, 1968a). But it is totally irrelevant to the subject of black holes, for two reasons. First, as with Schwarzschild, the star that collapsed to form the black hole replaces most of the inward extension of the Kerr-Newman manifold. Second, even outside the star, the Kerr-Newman geometry does not properly represent the true geometry at early times. At early times the star has not got far down the road to collapse. Gravitational moments of the star arise from mountains or prominences or turbulence or other particularities that have not yet gone into the meat grinder. The geometry departs from flatness (1) by a term that varies for large distances as mass divided by distance, and (2) by another term that varies as angular momentum divided by the square of the distance and multiplied by a spherical harmonic of order one, but also (3) by higher-order terms proportional to higher-order mass moments multiplied by higher spherical harmonics. These higher-order terms normally will deviate at early times from the corresponding terms in the mathematical analysis of the Kerr-Newman geometry—though the deviations will die out as time passes. For a system endowed with spherical symmetry, no such higher-order terms do occur or can occur. Therefore the geometry outside is

Schwarzschild in character at all stages of the collapse. However, when the system lacks spherical symmetry, the geometry outside initially departs from Kerr-Newman character. *Only well after the collapse occurs* (asymptotic future), *and in the region at and outside the horizon, is the Kerr-Newman geometry a faithful descriptor of a black hole.* This region is described in a nonsingular manner by Kerr coordinates and Kerr diagrams; and it is the only region that this book will explore.

H. Test-particle orbits
 See §§33.5–33.8 and Box 33.5.

III. Properties of electromagnetic field (§33.3):

A. Far from the black hole, where spacetime is nearly flat, in the usual spherical orthonormal frame ($\boldsymbol{\omega}^{\hat{t}} = \boldsymbol{d}t$, $\boldsymbol{\omega}^{\hat{r}} = \boldsymbol{d}r$, $\boldsymbol{\omega}^{\hat{\theta}} = r\,\boldsymbol{d}\theta$, $\boldsymbol{\omega}^{\hat{\phi}} = r\sin\theta\,\boldsymbol{d}\phi$), the electric and magnetic fields have dominant components

$$E_{\hat{r}} = \frac{Q}{r^2};$$

$$B_{\hat{r}} = \frac{2Qa}{r^3}\cos\theta, \quad B_{\hat{\theta}} = \frac{Qa}{r^3}\sin\theta.$$

These reveal that

Q = charge of black hole,

$\mathfrak{M} \equiv Qa$ = magnetic dipole moment of black hole.

B. Notice that the gyromagnetic ratio, $\gamma \equiv$ (magnetic moment)/(angular momentum), is equal to Q/M = (charge/mass), just as for an electron!

C. Notice that the value of the magnetic moment, like all other features of the black hole, is determined uniquely by the hole's mass, charge, and angular momentum: $\mathfrak{M} = QS/M$. This illustrates the theorem (Box 33.1) that a black hole has no "hair."

D. Other electric and magnetic moments are nonzero, but are determined uniquely by M, S, and Q.

E. Near the black hole, the curvature of spacetime deforms the electric and magnetic fields produced by the charged, rotating black hole. For a mathematical description of this deformed field, see Cohen and Wald (1971); for a diagrammatic representation, Hanni and Ruffini (1973).

Box 33.3 THE ASTROPHYSICS OF BLACK HOLES

Black holes in nature should participate in astrophysical processes that are as varied as those for stars. By searching for observable phenomena associated with these processes, astronomers have a good chance of discovering the first black hole sometime during the 1970's. This box lists some possible astrophysical processes, and a few relevant references.

I. Mechanisms of Formation

A. "Direct, in isolation": A massive star ($M \gtrsim 3M_\odot$) collapses, almost spherically, producing a collapsed neutron-star core that is too massive to support itself against gravity. Gravity pulls the core on inward, producing a horizon and black hole. [May

Box 33.3 (*continued*)

and White (1966, 1967); Chapter 32 of this book.]

B. "Indirect, in isolation": "Collapse, pursuit, and plunge scenario" depicted in Figure 24.3 [Ruffini and Wheeler (1971b).]

C. "In the thick of things": Stars collected into a dense cluster (e.g., the nucleus of a galaxy) exchange energy. Some acquire energy and move out into a halo. Others lose energy and make a more compact cluster. This process of segregation continues. The cluster becomes so compact that collisions ensue and gas is driven off. The gas moves toward the center of the gravitational potential well. Out of it new stars form. The process continues. Eventually star-star collisions may become sufficiently energetic and inelastic that the centers of the colliding stars coalesce. In this way supermassive objects may be built up and may evolve. Ultimately (1) many "small" stars may collapse to form "small" black holes ($M \sim M_\odot$); (2) one or more supermassive stars may collapse to form huge black holes ($M \sim 10^4 M_\odot$ to $10^9 M_\odot$); (3) the entire conglomerate of stars and gas and holes may become so dense that it collapses to form a single gigantic hole. [Sanders (1970), Spitzer (1971), Lynden-Bell (1967, 1969), Colgate (1967), §§24.5, 24.6, 25.7 of this book.]

D. "Primordially": Perturbations in the initial density distribution of the expanding universe may produce collapse, resulting in "primordial black holes." Those holes would subsequently grow by accretion of radiation and matter. By today all such holes might have grown into enormous objects [$M \sim 10^{17} M_\odot$; Zel'dovich and Novikov (1966)]; but some of them might have avoided such growth and might be as small as 10^{-5} grams [Hawking (1971a)].

II. How many black holes are there in our galaxy today?

Peebles (1972) has given an excellent review of this issue and of prospects for finding black holes in the near future. He says "a good fraction of the mass of the disc of our galaxy was deposited [long ago] in stars capable of collapsing to black holes. ... The indication is that the galaxy's disk may contain on the order of 10^9 black holes."

III. "Live" black holes versus "dead" black holes

A. A Schwarzschild black hole is "dead" in the sense that one can never extract from it any of its mass-energy. One aspect of this "deadness"—the fact that a Schwarzschild black hole is stable against small perturbations—is essential (1) to the identification of a black hole with the ultimate "ground state" of a large mass, and (2) to any assertion that general relativity theory predicts the possible existence of black holes. [For a proof of stability see Vishveshwara (1970). The problem was formulated, and most of the necessary techniques developed, by Regge and Wheeler (1957), with essential contributions also by Zerilli (1970a).] Thus a small pulse of gravitational (or other) radiation impinging on a Schwarzschild black hole does not initiate a transition of the black hole into a very different object or state.

B. A Kerr-Newman black hole—which is rotating or charged or both—is not dead. The rotational and electromagnetic contributions to the mass-energy *can* be extracted. (See §§33.7 and 33.8 for mathematical details.) Thus, such black holes are "live"; they can inject energy into their surroundings. By a suitable arrangement of external apparatus, one can trigger an exponentially growing energy release [Press and Teukolsky (1972).] But for a perturbed

black hole in isolation, the release is always "controlled" and damped; i.e., Kerr black holes are stable in any classical context [Press and Teukolsky (1973)].

C. Most objects (massive stars; galactic nuclei; . . .) that can collapse to form black holes have so much angular momentum that the holes they produce should be "very live" (a nearly equal to M; S nearly equal to M^2). [Bardeen (1970a).]

D. By contrast, it is quite probable (but far from certain) that no black hole in the universe has substantial charge—i.e., that all black holes have $Q \ll M$. A black hole with $Q \sim M$ (say, $Q > 0$ for concreteness) would exert attractive electrostatic forces on electrons, and repulsive electrostatic forces on protons, that are larger than the hole's gravitational pull by the factor

$$\frac{\text{(electrostatic force)}}{\text{(gravitational force)}} = \frac{eQ}{\mu M} \sim \frac{e}{\mu} \sim 10^{20}.$$

Here e is the electron charge and μ is the electron (or proton) mass. Such huge differential forces are likely to pull in enough charge from outside the hole to neutralize it.

E. But one has learned from the "unipolar induction process" for neutron stars [Goldreich and Julian (1968)] that charge neutralization can sometimes be circumvented. Whether any black-hole process can possibly prevent neutralization one does not know in 1972.

IV. Interaction of a black hole with its environment

A. Gravitational pull: A black hole exerts a gravitational pull on surrounding matter and stars. The pull is indistinguishable, at radii $r \gg M$, from the pull of a star with the same mass.

B. Accretion and emission of x-rays and γ-rays: Gas surrounding a black hole gets pulled inward and is heated by adiabatic compression, by shock waves, by turbulence, by viscosity, etc. Before it reaches the horizon, the gas may become so hot that it emits a large flux of x-rays and perhaps even γ-rays. Thus, accreting matter can convert a black hole into a glowing "white" body [for a review of the literature, see Novikov and Thorne (1973)]. Accretion from a nonrotating gas cloud tends to decrease the angular momentum of a black hole [preferential accretion of particles with "negative" angular momentum; Doroshkevich (1966), Godfrey (1970a)]. But the gas surrounding a hole is likely to be rotating in the same direction as the hole itself, and to maintain $S \sim M^2$ [more precisely, $S \approx 0.998 M^2$; Thorne (1973b)].

C. A lump of matter (an "asteroid" or a "planet" or a star) falling into a black hole should emit a burst of gravitational waves as it falls. The total energy radiated is $E \sim 0.01\mu(\mu/M)$, where μ is the mass of the object. [Zerilli (1970b); Davis, Ruffini, Press, and Price (1971); Figure 36.2 of this book.]

D. An object in a stable orbit around a black hole should spiral slowly inward because of loss of energy through gravitational radiation, until it reaches the most tightly bound, stable circular orbit. It should then fall quickly into the hole, emitting a "last-gasp burst" of waves. The total energy radiated during the slow inward spiral is equal to the binding energy of the last stable circular orbit:

$$E_{\text{radiated}} = \mu - E_{\text{last orbit}}$$
$$= \begin{cases} 0.0572\mu \text{ for Schwarzschild hole,} \\ 0.4235\mu \text{ for Kerr hole with} \\ \qquad\qquad S = M^2, Q = 0. \end{cases}$$

Here μ is the rest mass of the captured object. [Box 33.5.] The total energy in the last-gasp burst is $E \sim 0.01\mu(\mu/M)$ if $\mu \ll M$. [Fig. 36.2.]

Box 33.3 (*continued*)

E. When matter falls down a black hole, it can excite the hole's external spacetime geometry into vibration. The vibrations are gradually converted into gravitational waves, some of which escape, others go down the hole. [Press (1971, Goebel (1972).] These vibrations are analogous to an "incipient gravitational geon" [Wheeler (1962); Christodoulou (1971)]—except that for a vibrating black hole the background Kerr geometry holds the vibration energy together (prevents it from propagating away immediately), whereas in a geon it is curvature produced by the "vibration energy" itself that prevents disruption.

F. By a non-Newtonian, induction-zone (i.e., nonradiative) gravitational interaction, a black hole gradually transfers its angular momentum to any non-axially-symmetric, nearby distribution of matter or fields. [Hawking (1972a); Ipser (1971), Press (1972), Hawking and Hartle (1972).]

G. A star or planet falling into a large black hole will get torn apart by tidal gravitational forces. If the tearing occurs near but outside the horizon, it may eject a blob of stellar matter that goes out with relativistic velocity ("tube-of-toothpaste effect"). Moreover, the outgoing jet may extract a substantial amount of rotational energy from the hole's ergosphere—i.e., the hole might throw it off with a rest mass plus kinetic energy in excess of the rest mass of the original infalling object. [Wheeler (1971d); §§33.7 and 33.8.]

H. The magnetic field lines of a charged black hole may be anchored to surrounding plasma, may get wound up as the hole rotates, and may shake, twitch, and excite the plasma.

V. Collisions between black holes

A. Two black holes can collide and coalesce; but there is no way to blast a black hole apart into several black holes [Hawking (1972a); exercise 34.4].

B. When two black holes collide and coalesce, the surface area of the final black hole must exceed the sum of the surface areas of the two initial black holes ("second law of black-hole dynamics"; Hawking (1971a,b); Box 33.4; §34.5). This constraint places an upper limit on the amount of gravitational radiation emitted in the collision. For example, if all three holes are of the Schwarzschild variety and the two initial holes have equal masses $M/2$, then

$$4\pi(2M_{\text{final}})^2 \geq 4\pi[2(M/2)]^2 + 4\pi[2(M/2)]^2,$$
$$M_{\text{final}} \geq M/\sqrt{2},$$

so the energy radiated is

$$E_{\text{radiated}} \leq M - M/\sqrt{2} = 0.293M.$$

VI. Where and how to search for a black hole [For a detailed review, see Peebles (1971)]:

A. When it forms, by the burst or bursts of gravitational radiation given off during formation [Figure 24.3].

B. In a binary star system: black-hole component optically invisible, but may emit x-rays and γ-rays due to accretion; visible component shows telltale Doppler shifts [Hoyle, Fowler, Burbidge, and Burbidge (1964); Zel'dovich and Guseynov (1965); Trimble and Thorne (1969); Pringle and Rees (1972); Shakura and Sunyaev (1973)]. The velocity of the visible component and the period give information on the mass of the invisible component. If

mass of this invisible component is four solar masses or more, it cannot be an ordinary star, because an ordinary star of that mass would have $(4)^3 = 64$ times the luminosity of the sun. Neither can it be a white dwarf or a neutron star because either object, so heavy, would instantly collapse to a black hole. Therefore, it is attractive—though not necessarily compelling [see Trimble and Thorne (1969)]—to identify the invisible object as a black hole.

C. [*But* one must not expect to see any noticeable gravitational lens action from a black hole in a binary system: if it taxed the abilities of astronomers for decades to see the black disc of Mercury, 4,800 km in diameter, swim across the great face of the sun, little hope there is to see a black hole with an effective radius of only \sim3 km, enormously more remote, occult a companion star. Significant lens action requires that the lens (black hole) be separated by a normal interstellar distance from the star it focuses; whence the impact parameter of the focused rays is more than a stellar radius, so the lens action is not more than that of a normal star. Moreover,

even with 10^9 black holes in the galaxy, only one per year would pass directly between the Earth and a more distant star, and produce significant lens action (Refsdal, 1964). Chance of watching the right spot on the sky at the right time with a sufficiently strong telescope: nil!]

D. At the center of a globular cluster, where a black hole may settle down, attract normal stars to its vicinity, and thereby produce a cusp in the distribution of light from the cluster. [Cameron and Truran (1971), Peebles (1971).]

E. In the nucleus of a galaxy, including even the Milky Way, where a single huge black hole ($M \sim 10^4$ to $10^8 M_\odot$) might sit as an end-product of earlier activity of the galactic nucleus. Such a hole will emit gravitational waves, light, and radio waves as it accretes matter. Much of the light may be converted into infrared radiation by surrounding dust. The black hole may also produce jets and other nuclear activity. [Lynden-Bell (1969), Lynden-Bell and Rees (1971), Wheeler (1971d), Peebles (1971).]

Box 33.4 THE LAWS OF BLACK-HOLE DYNAMICS

The black-hole processes described in Box 33.3 are governed by the standard laws of physics: general relativity, plus Maxwell electrodynamics, plus hydrodynamic, quantum mechanical, and other laws for the physics of matter and radiation. From these standard laws of physics, one can derive certain "rules" or "constraints," which all black-hole processes must satisfy. Those rules have a power, elegance, and simplicity that rival and resemble the power, elegance, and simplicity of the laws of thermodynamics. Therefore, they have been given the analogous name "the laws of black-hole dynamics" (Israel 1971). This box states two of the laws of black-hole

Box 33.4 (*continued*)

dynamics and some of their ramifications. Two additional laws, not discussed here, have been formulated by Bardeen, Carter, and Hawking (1973).

I. The First and Second Laws of Black-Hole Dynamics.

 A. *The first law.*

 1. Like the first law of thermodynamics, the first law of black-hole dynamics is the standard law of conservation of total energy, supplemented by the laws of conservation of total momentum, angular momentum, and charge. For detailed discussions of these conservation laws, see Box 19.1 and Chapter 20.

 2. Specialized to the case where matter falls down a black hole and gravitational waves pour out, the first law takes the form depicted and discussed near the end of Box 19.1.

 3. Specialized to the case of infalling electric charge, the first law says that the total charge Q of a black hole, as measured by the electric flux emerging from it, changes by an amount equal to the total charge that falls down the hole,

 $$\Delta Q = q_{\text{that falls in}}.$$

 4. Specialized to the case where two black holes collide and coalesce (example given in Box 33.3), the first law says: (a) Let \boldsymbol{P}_1 and \boldsymbol{P}_2 be the 4-momenta of the two black holes as measured gravitationally, when they are so well-separated that they have negligible influence on each other. (\boldsymbol{P}_1 and \boldsymbol{P}_2 are 4-vectors in the surrounding asymptotically flat spacetime.) Similarly, let \boldsymbol{J}_1 and \boldsymbol{J}_2 be their total angular-momentum tensors (not intrinsic angular-momentum vectors!) relative to some arbitrarily chosen origin of coordinates, \mathscr{P}_0, in the surrounding asymptotically flat spacetime (\boldsymbol{J}_1 and \boldsymbol{J}_2 contain orbital angular momentum, as well as intrinsic angular momentum; see Box 5.6.). (b) Let \boldsymbol{P}_3 and \boldsymbol{J}_3 be the similar total 4-momentum and angular momentum of the final black hole. (c) Let \boldsymbol{P}_r and \boldsymbol{J}_r be the total 4-momentum and angular momentum radiated as gravitational waves during the collision and coalescence. *Then*

 $$\boldsymbol{P}_3 = \boldsymbol{P}_1 + \boldsymbol{P}_2 - \boldsymbol{P}_r, \qquad \boldsymbol{J}_3 = \boldsymbol{J}_1 + \boldsymbol{J}_2 - \boldsymbol{J}_r.$$

 [Note: to calculate the mass and intrinsic angular momentum of the final black hole from a knowledge of \boldsymbol{P}_3 and \boldsymbol{J}_3, follow the prescription of Box 5.6. In that prescription, the world line of the final black hole is that world line, in the distant asymptotically Lorentz coordinates, on which the hole's distant spherical field is centered.

B. *The second law* [expounded and applied by Hawking (1971b, 1972a)].
 When anything falls down a black hole, or when several black holes collide
 and coalesce or collide and scatter, or in any other process whatsoever involv-
 ing black holes, *the sum of the surface areas* (or squares of "irreducible
 masses"—see equation 3 below) *of all black holes involved can never decrease.*
 (See §34.5 for proof.) This is the second law of black-hole dynamics.

II. Reversible and Irreducible Transformations; Irreducible Mass
 [Christodoulou (1970); Christodoulou and Ruffini (1971)—results derived inde-
 pendently of and simultaneously with Hawking's discovery of the second law.]

A. Consider a single Kerr-Newman black hole interacting with surrounding
 matter and fields. Its surface area, at any moment of time, is given in terms
 of its momentary mass M, charge Q, and intrinsic angular momentum per
 unit mass $a \equiv S/M$ by

$$A = 4\pi[r_+^2 + a^2] = 4\pi[(M + \sqrt{M^2 - Q^2 - a^2})^2 + a^2] \qquad (1)$$

 (exercise 33.12). Interaction with matter and fields may change M, Q, and
 a in various ways; M can even be decreased—i.e., energy can be extracted
 from the black hole! [Penrose (1969); §33.7.] But whatever may be the
 changes, they can never reduce the surface area A. Moreover, if any change
 in M, Q, and a ever increases the surface area, no future process can ever
 reduce it back to its initial value.

B. Thus, one can classify black-hole processes into two groups.
 1. *Reversible transformations* change M, Q, or a or any set thereof, while
 leaving the surface area fixed. They can be reversed, bringing the black
 hole back to its original state.
 2. *Irreversible transformations* change M, Q, or a or any set thereof, and
 increase the surface area in the process. Such a transformation can never
 be reversed. The black hole can never be brought back to its original state
 after an irreversible transformation.

C. Examples of reversible transformations and of irreversible transformations
 induced by infalling particles are presented in §§33.7 and 33.8.

D. The reversible extraction of charge and angular momentum from a black hole
 (decrease in Q and a holding A fixed) necessarily reduces the black hole's
 mass (energy extraction!). By the time all charge and angular momentum
 have been removed, the mass has dropped to a final "irreducible value" of

$$M_{\text{ir}} = (A/16\pi)^{1/2} = \begin{pmatrix} \text{mass of Schwarzschild} \\ \text{black hole of surface area } A \end{pmatrix}. \qquad (2)$$

Box 33.4 (*continued*)

E. Expressed in terms of this final, irreducible mass, the initial mass-energy of the black hole (with charge Q and intrinsic angular momentum S) is

$$M^2 = \left(M_{\text{ir}} + \frac{Q^2}{4M_{\text{ir}}} \right)^2 + \frac{S^2}{4M_{\text{ir}}{}^2} \qquad (3)$$

[This formula, derived by Christodoulou and Ruffini, may be obtained by combining equations (1), (2), and $S = Ma$].

F. Thus, *one can regard the total mass-energy of a black hole as made up of an irreducible mass, an electromagnetic mass-energy, and a rotational energy. But one must resist the temptation to think of these contributions as adding linearly. On the contrary, they combine in a way* [equation (3)] *analogous to the way rest mass and linear momentum combine to give energy,* $E^2 = m^2 + p^2$.

G. Contours of constant M/M_{ir} are depicted below in the "charge-angular momentum plane." Black holes can exist only in the interior of the region depicted ($Q^2 + a^2 \leq M^2$). [Diagram adapted from Christodoulou (1971).]

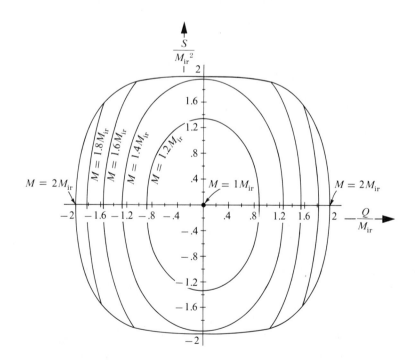

H. Since a black hole's irreducible mass is proportional to the square root of its surface area, one can restate the second law of black-hole dynamics as follows:

> *In black-hole processes the sum of the squares of the irreducible masses of all black holes involved can never decrease.*

§33.3. MASS, ANGULAR MOMENTUM, CHARGE, AND MAGNETIC MOMENT

It is instructive to verify that the constants M, Q, and a, which appear in equations (33.2)–(33.5) for the Kerr-Newman geometry and electromagnetic field, are actually the black hole's mass, charge, and angular momentum per unit mass, as claimed above.

Mass and angular momentum are defined by their imprints on the spacetime geometry far from the black hole. Therefore, to calculate the mass and angular momentum, one can expand the line element (33.2) in powers of $1/r$ and examine the leading terms:

$$ds^2 = -\left[1 - \frac{2M}{r} + O\left(\frac{1}{r^2}\right)\right]dt^2 - \left[\frac{4aM}{r}\sin^2\theta + O\left(\frac{1}{r^2}\right)\right]dt\,d\phi$$
$$+ \left[1 + O\left(\frac{1}{r}\right)\right][dr^2 + r^2(d\theta^2 + \sin^2\theta\,d\phi^2)].$$

(33.6)

The examination is facilitated by transforming to asymptotically Lorentz coordinates—$x = r\sin\theta\cos\phi$, $y = r\sin\theta\sin\phi$, $z = r\cos\theta$:

$$ds^2 = -\left[1 - \frac{2M}{r} + O\left(\frac{1}{r^2}\right)\right]dt^2 - \left[\frac{4aM}{r^3} + O\left(\frac{1}{r^4}\right)\right][x\,dy - y\,dx]$$
$$+ \left[1 + O\left(\frac{1}{r}\right)\right][dx^2 + dy^2 + dz^2].$$

(33.6')

Direct comparison with the "standard form" [equation (19.13)] of the metric far from a stationary rotating source reveals that (1) the parameter M is, indeed, the mass of the black hole; and (2) the intrinsic angular momentum vector of the black hole is

$$\boldsymbol{S} = (aM)\,\partial/\partial z = (aM)\cdot\binom{\text{unit vector pointing along polar axis}}{\text{of Boyer-Lindquist coordinates}}. \quad (33.7)$$

The charge is defined for the black hole, as for any source, by a Gaussian flux integral of its electric field over a closed surface surrounding the hole. The electric

The rest of this chapter is Track 2. To be prepared for it, one needs to have covered the Track-2 part of Chapter 32 (gravitational collapse). In reading it, one will be helped greatly by Chapter 25 (orbits in Schwarzschild geometry). The rest of this chapter is needed as preparation for Chapter 34 (singularities and global methods).

The metric far outside a black hole: imprints of mass and angular momentum

The electromagnetic field far outside a black hole:

(1) electric field

field in the asymptotic rest frame of the black hole has as its orthonormal components

$$E_{\hat{r}} = E_r = F_{rt} = Q/r^2 + O(1/r^3),$$
$$E_{\hat{\theta}} = E_\theta/r = F_{\theta t}/r = O(1/r^4), \tag{33.8}$$
$$E_{\hat{\phi}} = E_\phi/r \sin\theta = F_{\phi t}/r \sin\theta = 0.$$

Hence, the electric field is purely radial with a Gaussian flux integral of $4\pi Q$, which reveals Q to be the black hole's charge.

A similar calculation of the dominant components of the magnetic field reveals

(2) magnetic field

$$B_{\hat{r}} = F_{\hat{\theta}\hat{\phi}} = \frac{F_{\theta\phi}}{r^2 \sin\theta} = 2\frac{Qa}{r^3}\cos\theta + O\left(\frac{1}{r^4}\right),$$

$$B_{\hat{\theta}} = F_{\hat{\phi}\hat{r}} = \frac{F_{\phi r}}{r \sin\theta} = \frac{Qa}{r^3}\sin\theta + O\left(\frac{1}{r^4}\right), \tag{33.9}$$

$$B_{\hat{\phi}} = F_{\hat{r}\hat{\theta}} = \frac{F_{r\theta}}{r} = 0.$$

This is a dipole magnetic field, and from it one immediately reads off the value

(3) magnetic dipole moment

$$\mathfrak{M} = Qa = (Q/M)S = \underbrace{(\text{charge}/\text{mass})}_{\text{["gyromagnetic ratio"]}} \times (\text{angular momentum}) \tag{33.10}$$

for the magnetic moment of the black hole.

Nonspherical shape of hole's geometry

Just as the rotation of the black hole produces a magnetic field, so it also produces nonspherical deformations in the gravitational field of the black hole [see Hernandez (1967) for quantitative discussion]. But those deformations, like the magnetic moment, are *not* freely specifiable. They are determined uniquely by the mass, charge, and angular momentum of the black hole.

§33.4. SYMMETRIES AND FRAME DRAGGING

The metric components (33.2) of a Kerr-Newman black hole are independent of the Boyer-Lindquist time coordinate t and angular coordinate ϕ. This means (see §25.2) that

Killing vectors for the Kerr-Newman geometry

$$\boldsymbol{\xi}_{(t)} \equiv (\partial/\partial t)_{r,\theta,\phi} \quad \text{and} \quad \boldsymbol{\xi}_{(\phi)} \equiv (\partial/\partial\phi)_{t,r,\theta} \tag{33.11}$$

are Killing vectors associated with the stationarity (time-translation invariance) and axial symmetry of the black hole. The scalar products of these Killing vectors with themselves and each other are

$$\boldsymbol{\xi}_{(t)} \cdot \boldsymbol{\xi}_{(t)} = g_{tt} = -\left(\frac{\varDelta - a^2 \sin^2\theta}{\rho^2}\right) = -\left(1 - \frac{2Mr - Q^2}{\rho^2}\right), \quad (33.12a)$$

$$\boldsymbol{\xi}_{(t)} \cdot \boldsymbol{\xi}_{(\phi)} = g_{t\phi} = \frac{a \sin^2\theta \, (\varDelta - r^2 - a^2)}{\rho^2} = -\frac{(2Mr - Q^2)a \sin^2\theta}{\rho^2}, \quad (33.12b)$$

$$\boldsymbol{\xi}_{(\phi)} \cdot \boldsymbol{\xi}_{(\phi)} = g_{\phi\phi} = \frac{[(r^2 + a^2)^2 - \varDelta a^2 \sin^2\theta] \sin^2\theta}{\rho^2}. \quad (33.12c)$$

Since Killing vectors are geometric properties of spacetime, with existence independent of any and all coordinate systems, their scalar products also have coordinate-free meaning. It so happens (not by chance, but by careful choice of coordinates!) that the Boyer-Lindquist metric components g_{tt}, $g_{t\phi}$, and $g_{\phi\phi}$ are equal to these coordinate-independent scalar products. Thus g_{tt}, $g_{t\phi}$, and $g_{\phi\phi}$ can be thought of as three scalar fields which embody information about the symmetries of spacetime. By contrast, the metric coefficients $g_{rr} = \rho^2/\varDelta$ and $g_{\theta\theta} = \rho^2$ carry no information at all about the symmetries.* They depend, for their existence and values, on the specific Boyer-Lindquist choice of coordinates.

Any observer who moves along a world line of constant (r, θ) with uniform angular velocity sees an unchanging spacetime geometry in his neighborhood. Hence, such an observer can be thought of as *"stationary"* relative to the local geometry. If and only if his angular velocity is zero, that is, if and only if he moves along a world line of constant (r, θ, ϕ), will he also be *"static"* relative to the black hole's asymptotic Lorentz frame (i.e., relative to the "distant stars").

The precise definition of "angular velocity relative to the asymptotic rest frame"—or simply "angular velocity"—is

Stationary observers

Static observers

Angular velocity and 4-velocity of a stationary observer

$$\varOmega \equiv \frac{d\phi}{dt} = \frac{d\phi/d\tau}{dt/d\tau} = \frac{u^\phi}{u^t} \quad (33.13a)$$

(see exercise 33.2). In terms of \varOmega, the Killing vectors, and the scalar products of Killing vectors, *the 4-velocity of a stationary observer is*

$$\begin{aligned} \boldsymbol{u} &= u^t(\partial/\partial t + \varOmega \partial/\partial\phi) = \frac{\boldsymbol{\xi}_{(t)} + \varOmega \boldsymbol{\xi}_{(\phi)}}{|\boldsymbol{\xi}_{(t)} + \varOmega \boldsymbol{\xi}_{(\phi)}|} \\ &= \frac{\boldsymbol{\xi}_{(t)} + \varOmega \boldsymbol{\xi}_{(\phi)}}{(-g_{tt} - 2\varOmega g_{t\phi} - \varOmega^2 g_{\phi\phi})^{1/2}}. \end{aligned} \quad (33.13b)$$

A stationary observer is static if and only if \varOmega vanishes.

The stationary observers at given r, θ cannot have any and every angular velocity. Only those values of \varOmega are allowed for which the 4-velocity \boldsymbol{u} lies inside the future light cone—i.e., for which

$$(\boldsymbol{\xi}_{(t)} + \varOmega \boldsymbol{\xi}_{(\phi)})^2 = g_{tt} + 2\varOmega g_{t\phi} + \varOmega^2 g_{\phi\phi} < 0.$$

*This is not quite true. Kerr-Newman spacetime possesses, in addition to its two Killing vectors, also a "Killing tensor" which is closely linked to the Boyer-Lindquist coordinates r and θ. See Walker and Penrose (1970); also §33.5.

Frame dragging, static limit, and ergosphere

Thus, *the angular velocities of stationary observers are constrained by*

$$\Omega_{\min} < \Omega < \Omega_{\max}, \tag{33.14}$$

where

$$\Omega_{\min} = \omega - \sqrt{\omega^2 - g_{tt}/g_{\phi\phi}}, \tag{33.15a}$$

$$\Omega_{\max} = \omega + \sqrt{\omega^2 - g_{tt}/g_{\phi\phi}}, \tag{33.15b}$$

$$\omega \equiv \frac{1}{2}(\Omega_{\min} + \Omega_{\max}) = -\frac{g_{\phi t}}{g_{\phi\phi}} = \frac{(2Mr - Q^2)a}{(r^2 + a^2)^2 - \Delta a^2 \sin^2\theta}, \tag{33.16}$$

and it is assumed that $S/M = a > 0$. The following features of these limits are noteworthy. (1) Far from the black hole, one has $r\Omega_{\min} = -1$ and $r\Omega_{\max} = +1$, corresponding to the standard limits imposed by the speed of light in flat spacetime. (2) With decreasing radius, Ω_{\min} increases ("dragging of inertial frames"). Finally, when g_{tt} reaches zero, i.e., at

$$r = r_0(\theta) \equiv M + \sqrt{M^2 - Q^2 - a^2 \cos^2\theta}, \tag{33.17}$$

Ω_{\min} becomes zero. At and inside this surface, all stationary observers must orbit the black hole with positive angular velocity. Thus, *static observers exist outside and only outside* $r = r_0(\theta)$. For this reason $r = r_0(\theta)$ is called the "static limit"; see Box 33.2. (3) As one moves through the static limit into the "ergosphere," one sees the allowed range of angular velocities become ever more positive (ever more "frame dragging"). At the same time, one sees the allowed range narrow down, until finally, at the horizon

$$r = r_+ \equiv M + \sqrt{M^2 - Q^2 - a^2}, \tag{33.18}$$

the limits Ω_{\min} and Ω_{\max} coalesce ($\omega^2 = g_{tt}/g_{\phi\phi}$). Thus, at the horizon there are no stationary observers. All timelike world lines point inward. There is no escape from the black hole's "pull."

Further features of stationary observers and "frame dragging" are explored in the exercises.

EXERCISES

Exercise 33.1. KERR DESCRIPTION OF KILLING VECTORS

(a) Use the transformation law from Boyer-Lindquist coordinates to Kerr coordinates [equation (4) of Box 33.2] to show that

$$\boldsymbol{\xi}_{(t)} \equiv (\partial/\partial t)_{r,\theta,\phi} = (\partial/\partial \widetilde{V})_{r,\theta,\widetilde{\phi}}, \tag{33.19a}$$

$$\boldsymbol{\xi}_{(\phi)} \equiv (\partial/\partial \phi)_{t,r,\theta} = (\partial/\partial \widetilde{\phi})_{\widetilde{V},r,\theta}. \tag{33.19b}$$

Verify explicitly by examining metric components that

$$g_{\widetilde{V}\widetilde{V}} = g_{tt}, \qquad g_{\widetilde{V}\widetilde{\phi}} = g_{t\phi}, \qquad g_{\widetilde{\phi}\widetilde{\phi}} = g_{\phi\phi}, \tag{33.19c}$$

in accordance with equations (33.19a,b).

(b) Show that for a stationary observer (world line of constant r, θ), the angular velocity expressed in terms of Kerr coordinates is

$$\Omega \equiv d\phi/dt = d\tilde{\phi}/d\tilde{V} = u^{\tilde{\phi}}/u^{\tilde{V}},$$

so that the entire discussion of stationary observers in terms of Kerr coordinates is identical to the discussion in terms of Boyer-Lindquist coordinates. Differences between the coordinate systems show up only when one moves along world lines of changing r. Reconcile this fact with the fact that both coordinate systems use the *same* coordinates (r, θ) but different time and azimuthal coordinates (t, ϕ versus $\tilde{V}, \tilde{\phi}$).

Exercise 33.2. OBSERVATIONS OF ANGULAR VELOCITY

An observer, far from a black hole and at rest in the hole's asymptotic Lorentz frame, watches (with his eyes) as a particle moves along a stationary (nongeodesic) orbit near the black hole. Let $\Omega = d\phi/dt$ be the particle's angular velocity, as defined and discussed above. The distant observer uses his stopwatch to measure the time required for the particle to make one complete circuit around the black hole (one complete circuit relative to the distant observer himself; i.e., relative to the hole's asymptotic Lorentz frame).

(a) Show that the circuit time measured is $2\pi/\Omega$. Thus, *Ω can be regarded as the particle's "angular velocity as measured from infinity."*

(b) Let the observer moving with the particle measure its circuit time relative to the asymptotic Lorentz frame, using his eyes and a stopwatch he carries. Show that his answer for the circuit time must be

$$\Delta\tau = \frac{2\pi}{\Omega} \underbrace{(-g_{tt} - 2\Omega g_{t\phi} - \Omega^2 g_{\phi\phi})^{1/2}}. \qquad (33.20)$$

$$\uparrow\text{["redshift factor"]}$$

Exercise 33.3. LOCALLY NONROTATING OBSERVERS
(Bardeen 1970b)

(a) Place a rigid, circular mirror ("ring mirror") at fixed (r, θ) around a black hole. Let an observer at (r, θ) with angular velocity Ω emit a flash of light. Some of the photons will get caught by the mirror and will skim along its surface, circumnavigating the black hole in the positive-ϕ direction. Others will get caught and will skim along in the negative-ϕ direction. Show that the observer will receive back the photons from both directions simultaneously only if his angular velocity is

$$\Omega = \omega(r, \theta) = \text{expression (33.16)}.$$

Thus in this case, and only in this case, can the observer regard the $+\phi$ and $-\phi$ directions as equivalent in terms of local geometry. Put differently, in this case and only in this case is the observer "nonrotating relative to the local spacetime geometry." Thus, it is appropriate to use the name *"locally nonrotating observer"* for an observer who moves with the angular velocity $\Omega = \omega(r, \theta)$.

(b) Associated with the axial symmetry of a black hole is a conserved quantity, $p_\phi \equiv \boldsymbol{p} \cdot \boldsymbol{\xi}_{(\phi)}$, for geodesic motion. This quantity for any particle—whether it is moving along a geodesic or not—is called the "component of angular momentum along the black hole's spin axis," or simply the particle's "angular momentum." (See §33.5 below.) Show that of all stationary observers at fixed (r, θ), only the "locally nonrotating observer" has zero angular momentum. [*Note:* Bardeen, Press, and Teukolsky (1972) have shown that the "locally nonrotating observer" can be a powerful tool in the analysis of physical processes near a black hole.]

Exercise 33.4. ORTHONORMAL FRAMES OF LOCALLY NONROTATING OBSERVERS

(a) Let spacetime be filled with world lines of locally nonrotating observers, and let each such observer carry an orthonormal frame with himself. Show that the spatial orientations of these frames can be so chosen that their basis 1-forms are

$$\boldsymbol{w}^{\hat{t}} = |g_{tt} - \omega^2 g_{\phi\phi}|^{1/2} \, \boldsymbol{dt}, \qquad \boldsymbol{w}^{\hat{\phi}} = (g_{\phi\phi})^{1/2}(\boldsymbol{d\phi} - \omega \, \boldsymbol{dt}),$$

$$\boldsymbol{w}^{\hat{r}} = (\rho/\Delta^{1/2}) \, \boldsymbol{dr}, \qquad \boldsymbol{w}^{\hat{\theta}} = \rho \, \boldsymbol{d\theta}. \tag{33.21}$$

More specifically, show that these 1-forms are orthonormal and that the dual basis has

$$\partial/\partial\hat{t} = \boldsymbol{u} \equiv \text{4-velocity of locally nonrotating observer.} \tag{33.22}$$

Show that $\boldsymbol{u} = -\boldsymbol{w}^{\hat{t}}$ is a rotation-free field of 1-forms [$\boldsymbol{dw}^{\hat{t}} \wedge \boldsymbol{w}^{\hat{t}} = 0$; exercise 4.4].

(b) One sometimes meets the mistaken notion that a "locally nonrotating observer" is in some sense locally inertial. To destroy this false impression, verify that: (i) such an observer has nonzero 4-acceleration,

$$\boldsymbol{a} = \Gamma_{\hat{j}\hat{t}\hat{t}}\boldsymbol{e}_{\hat{j}} = \frac{1}{2} \boldsymbol{\nabla} \ln |g_{tt} - \omega^2 g_{\phi\phi}|; \tag{33.23}$$

(ii) if such an observer carries gyroscopes with himself, applying the necessary accelerations at the gyroscope centers of mass, he sees the gyroscopes precess relative to his orthonormal frame (33.21) with angular velocity

$$\boldsymbol{\Omega}^{\text{(precess)}} = \Gamma_{\hat{\theta}\hat{\phi}\hat{t}}\boldsymbol{e}_{\hat{r}} + \Gamma_{\hat{\phi}\hat{r}\hat{t}}\boldsymbol{e}_{\hat{\theta}}$$

$$= \frac{1}{2} \frac{g_{\phi\phi}^{1/2}}{|g_{tt} - \omega^2 g_{\phi\phi}|^{1/2}} \left[\frac{\omega_{,\theta}}{\rho} \boldsymbol{e}_{\hat{r}} - \frac{\Delta^{1/2}\omega_{,r}}{\rho} \boldsymbol{e}_{\hat{\theta}} \right]. \tag{33.24}$$

[Hints: See exercise 19.2, equation (13.69), and associated discussions. The calculation of the connection coefficients is performed most easily using the methods of differential forms; see §14.6.]

Exercise 33.5. LOCAL LIGHT CONES

Calculate the shapes of the light cones depicted in the Kerr diagram for an uncharged ($Q = 0$) Kerr black hole (part II.F of Box 33.2). In particular, introduce a new time coordinate

$$\tilde{t} \equiv \tilde{V} - r \tag{33.25}$$

for which the slices of constant \tilde{t} are horizontal surfaces in the Kerr diagram. Then the Kerr diagram plots \tilde{t} vertically, r radially, and $\tilde{\phi}$ azimuthally, while holding $\theta = \pi/2$ ("equatorial slice through black hole").

(a) Show that the light cone emanating from given $\tilde{t}, r, \tilde{\phi}$ has the form

$$\frac{dr}{d\tilde{t}} = a\left(\frac{d\tilde{\phi}}{d\tilde{t}}\right) - \frac{2M/r}{1 + 2M/r} \pm \sqrt{\frac{1}{(1 + 2M/r)^2} - \frac{r^2(d\tilde{\phi}/d\tilde{t})^2}{1 + 2M/r}}.$$

(b) Show that the light cone slices through the surface of constant radius along the curves

$$dr/d\tilde{t} = 0, \quad d\tilde{\phi}/d\tilde{t} = \Omega_{\min} \text{ and } \Omega_{\max}, \tag{33.26b}$$

where Ω_{\min} and Ω_{\max} are given by expressions (33.15a,b) (minimum and maximum allowed angular velocities for stationary observers).

(c) Show that at the static limit, $r = r_0(\pi/2)$, the light cone is tangent to a curve of constant $r, \theta, \tilde{\phi}$.

(d) Show that the light cone slices the surface of constant $\widetilde{\phi}$ along the curves

$$\frac{d\widetilde{\phi}}{d\widetilde{t}} = 0, \qquad \frac{dr}{d\widetilde{t}} = -1 \text{ and } \frac{1-2M/r}{1+2M/r}. \tag{33.26c}$$

(e) Show that the light cone is tangent to the horizon.

(f) Make pictures of the shapes of the light cone as a function of radius.

(g) Describe qualitatively how the light cone must look near the horizon in Boyer-Lindquist coordinates. (*Note:* it will look "crazy" because the coordinates are singular at the horizon.)

§33.5. EQUATIONS OF MOTION FOR TEST PARTICLES
[Carter (1968a)]

Let a test particle with electric charge e and rest mass μ move in the external fields of a black hole. Were there no charge down the black hole, the test particle would move along a geodesic (zero 4-acceleration). But the charge produces an electromagnetic field, which in turn produces a Lorentz force on the particle: $\mu\boldsymbol{a} = e\boldsymbol{F}\cdot\boldsymbol{u}$. (Here \boldsymbol{u} is the particle's 4-velocity, and $\boldsymbol{a} \equiv \boldsymbol{\nabla_u u}$ is its 4-acceleration.)

The geodesic equation, $\boldsymbol{a} = 0$, for the uncharged case is equivalent to Hamilton's equations

$$dx^{\mu}/d\lambda = \partial\mathcal{H}/\partial p_{\mu}, \qquad dp_{\mu}/d\lambda = -\partial\mathcal{H}/\partial x^{\mu}, \tag{33.27a}$$

where λ is an affine parameter so normalized that

$$d/d\lambda = \boldsymbol{p} = \text{4-momentum}, \tag{33.27b}$$

and where

$$\mathcal{H} \equiv \text{"super-Hamiltonian"} = \frac{1}{2} g^{\mu\nu} p_{\mu} p_{\nu} \tag{33.27c}$$

(see exercise 25.2). Similarly (see exercise 33.6) the Lorentz-force equation, $\mu\boldsymbol{a} = e\boldsymbol{F}\cdot\boldsymbol{u}$, for the charged case is equivalent to Hamilton's equations written in terms of position x^{μ} and "generalized momentum" π_{μ}:

$$dx^{\mu}/d\lambda = \partial\mathcal{H}/\partial\pi_{\mu}, \qquad d\pi_{\mu}/d\lambda = -\partial\mathcal{H}/\partial x^{\mu}. \tag{33.28a}$$

The form of the superhamiltonian \mathcal{H}, in terms of the metric coefficients at the particle's location, $g^{\mu\nu}(x^{\alpha})$, and the particle's charge e and generalized momentum π_{μ}, is

$$\mathcal{H} = \frac{1}{2} g^{\mu\nu}(\pi_{\mu} - eA_{\mu})(\pi_{\nu} - eA_{\nu}). \tag{33.28b}$$

Superhamiltonian for a charged test particle in any electromagnetic field in curved spacetime

[See §7.3 of Goldstein (1959) for the analogous superhamiltonian in flat spacetime.]

The first of Hamilton's equations for this superhamiltonian reduces to

$$p^\mu \equiv (\text{4-momentum}) \equiv dx^\mu/d\lambda = \pi^\mu - eA^\mu \tag{33.29a}$$

(value of π^μ in terms of p^μ, e, and A^μ); the second, when combined with the first, reduces to the Lorentz-force equation

$$dp^\mu/d\lambda + \Gamma^\mu{}_{\alpha\beta} p^\alpha p^\beta = eF^{\mu\nu} p_\nu. \tag{33.29b}$$

$$\left[\begin{array}{c} p_\nu, \text{ not } u_\nu \text{ because} \\ \lambda = \tau/\mu \end{array} \right]$$

For a Kerr-Newman black hole, the vector potential in Boyer-Lindquist coordinates can be put in the form

Vector potential for a charged
black hole

$$\boldsymbol{A} = -\frac{Qr}{\rho^2}(\boldsymbol{dt} - a\sin^2\theta\,\boldsymbol{d\phi}), \tag{33.30}$$

as one verifies by checking that

$$\boldsymbol{dA} = \frac{1}{2}(A_{\beta,\alpha} - A_{\alpha,\beta})\,\boldsymbol{dx}^\alpha \wedge \boldsymbol{dx}^\beta$$

reduces to the Faraday 2-form of equation (33.5).

There is good reason for going through all this formalism, rather than tackling head-on the Lorentz-force equation in its most elementary coordinate version,

$$\frac{d^2x^\alpha}{d\lambda^2} + \Gamma^\alpha{}_{\beta\gamma}\frac{dx^\beta}{d\lambda}\frac{dx^\gamma}{d\lambda} = eF^\alpha{}_\beta\frac{dx^\beta}{d\lambda}.$$

"Constants of motion" for a charged test particle moving around a charged black hole:

The Hamiltonian formalism enables one to discover immediately two constants of the motion; the elementary Lorentz-force equation does not. The key point is that the components A_μ of \boldsymbol{A} [equation (33.30)] and the components $g^{\mu\nu}$ of the metric [inverse of $g_{\mu\nu}$ of equation (33.2); see (33.35)] are independent of t and ϕ (stationarity and axial symmetry of both the electromagnetic field and the spacetime geometry). Consequently, the superhamiltonian is also independent of t and ϕ; and therefore Hamilton's equation

$$d\pi_\alpha/d\lambda = -\partial\mathcal{K}/\partial x^\alpha$$

(1) "energy at infinity" E
(2) "axial component of angular momentum" L_z

guarantees that π_t and π_ϕ are constants of the motion.

Far from the black hole, where the vector potential vanishes and the metric becomes

$$ds^2 = -dt^2 + dr^2 + r^2(d\theta^2 + \sin^2\theta\,d\phi^2),$$

the constants of the motion become

$$\pi_t = p_t = -p^t = -\text{energy},$$

$$\pi_\phi = p_\phi = rp^{\hat{\phi}} = \left(\begin{array}{l}\text{projection of angular momentum}\\\text{along black hole's rotation axis}\end{array}\right).$$

Thus it is appropriate to adopt the names and notation

$$E \equiv (\text{``energy at infinity''}) \equiv -\pi_t = -(p_t + eA_t), \qquad (33.31a)$$

$$L_z \equiv \begin{pmatrix} \text{``axial component of angular} \\ \text{momentum'', or simply} \\ \text{``angular momentum''} \end{pmatrix} \equiv \pi_\phi = p_\phi + eA_\phi \qquad (33.31b)$$

for the constants of the motion $-\pi_t$ and π_ϕ.

A third constant of the motion is the particle's rest mass

$$\mu = |\boldsymbol{p}| = (-g^{\alpha\beta} p_\alpha p_\beta)^{1/2}. \qquad (33.31c)$$

In general, four constants of the motion are needed to determine uniquely the orbit of a particle through four-dimensional spacetime. If the black hole were to possess an additional symmetry—e.g., if it were spherical, rather than merely axially symmetric—then automatically there would be a fourth constant of the motion. But in general, black holes are not spherical; so test-particle motion around a black hole possesses only three *obvious* constants. It is rather remarkable, then, that a constant turns out to exist. It was discovered by Carter (1968a), using Hamilton-Jacobi methods. As of 1973, nobody has given a cogent geometric explanation of why this fourth constant should exist—although hints of an explanation may be found in Carter (1968c) and Walker and Penrose (1970).

Carter's "fourth constant" of the motion, as derived in exercise 33.7, is

$$\mathscr{Q} = p_\theta{}^2 + \cos^2\theta[a^2(\mu^2 - E^2) + \sin^{-2}\theta L_z{}^2]. \qquad (33.31d)$$

The constant of the motion

$$\mathscr{K} \equiv \mathscr{Q} + (L_z - aE)^2, \qquad (33.31e)$$

obtained by combining \mathscr{Q}, L_z, and E, is often used in place of \mathscr{Q}. Whereas \mathscr{Q} can be negative, \mathscr{K} is always nonnegative:

$$\mathscr{K} = p_\theta{}^2 + (L_z - aE\sin^2\theta)^2/\sin^2\theta + a^2\mu^2\cos^2\theta$$

≥ 0 everywhere

$= 0$ only for case of photon ($\mu = 0$) moving along polar axis ($\theta = 0, \pi$).

The contravariant components of the test particle's 4-momentum, $p^\alpha = dx^\alpha/d\lambda$, are readily expressed in terms of the constants E, L_z, μ, \mathscr{Q}, by combining equations (33.31) with the metric coefficients (33.2) and the components of the vector potential (33.30). The result is

$$\rho^2 \, d\theta/d\lambda = \sqrt{\Theta}, \qquad (33.32a)$$

$$\rho^2 \, dr/d\lambda = \sqrt{R}, \qquad (33.32b)$$

$$\rho^2 \, d\phi/d\lambda = -(aE - L_z/\sin^2\theta) + (a/\Delta)P, \qquad (33.32c)$$

$$\rho^2 \, dt/d\lambda = -a(aE\sin^2\theta - L_z) + (r^2 + a^2)\Delta^{-1}P. \qquad (33.32d)$$

(3) rest mass μ

(4) "\mathscr{Q}"

Equations of motion for charged test particles

Here $\rho^2 = r^2 + a^2 \cos^2\theta$ as defined in equation (33.3b), and the functions Θ, R, P are defined by

$$\Theta = \mathcal{Q} - \cos^2\theta \, [a^2(\mu^2 - E^2) + L_z^{\,2}/\sin^2\theta], \qquad (33.33a)$$

$$P = E(r^2 + a^2) - L_z a - eQr, \qquad (33.33b)$$

$$R = P^2 - \Delta[\mu^2 r^2 + (L_z - aE)^2 + \mathcal{Q}]. \qquad (33.33c)$$

When working in Kerr coordinates (to avoid the coordinate singularity at the horizon), one must replace equations (33.32c) and (33.32d) by

$$\rho^2 \, d\widetilde{V}/d\lambda = -a(aE \sin^2\theta - L_z) + (r^2 + a^2)\Delta^{-1}(\sqrt{R} + P), \qquad (33.32c')$$

$$\rho^2 \, d\widetilde{\phi}/d\lambda = -(aE - L_z/\sin^2\theta) + a\Delta^{-1}(\sqrt{R} + P). \qquad (33.32d')$$

[These follow from (33.32) and the transformation between the two coordinate systems—see equations (4) of Box 33.2.] In the above equations, the signs of \sqrt{R} and $\sqrt{\Theta}$ can be chosen independently; but once chosen, they must be used consistently everywhere.

Applications of these equations of motion will play a key role in the rest of this chapter.

EXERCISES

Exercise 33.6. SUPERHAMILTONIAN FOR CHARGED-PARTICLE MOTION

Show that Hamilton's equations (33.28a) for the Hamiltonian (33.28b) reduce to equation (33.29a) for the value of the generalized momentum, and to the Lorentz force equation (33.29b). [*Hint:* Use the relation $(g^{\alpha\beta}g_{\beta\gamma})_{,\mu} = 0$.]

Exercise 33.7. HAMILTON-JACOBI DERIVATION OF EQUATIONS OF MOTION
[Based on Carter (1968a)]

Derive the first-order equations of motion (33.32) for a charged particle moving in the external fields of a Kerr-Newman black hole. Use the Hamilton-Jacobi method [Boxes 25.3 and 25.4 of this book; also Chapter 9 of Goldstein (1959)], as follows.

(a) Throughout the superhamiltonian \mathcal{H} of equation (33.28b), replace the generalized momentum π_α by the gradient $\partial S/\partial x^\alpha$ of the Hamilton-Jacobi function.

(b) Write down the Hamilton-Jacobi equation [generalization of equation (2) of Box 25.4] in the form

$$-\frac{\partial S}{\partial \lambda} = \mathcal{H}\left[x^\alpha, \frac{\partial S}{\partial x^\beta}\right] = \frac{1}{2} g^{\alpha\beta}\left(\frac{\partial S}{\partial x^\alpha} - eA_\alpha\right)\left(\frac{\partial S}{\partial x^\beta} - eA_\beta\right). \qquad (33.34a)$$

(c) Show that the metric components $g^{\alpha\beta}$ for a Kerr-Newman black hole in Boyer-Lindquist coordinates are given by

$$\boldsymbol{g} \equiv g^{\alpha\beta}\frac{\partial}{\partial x^\alpha}\frac{\partial}{\partial x^\beta} = -\frac{1}{\Delta\rho^2}\left[(r^2 + a^2)\frac{\partial}{\partial t} + a\frac{\partial}{\partial \phi}\right]^2 + \frac{1}{\rho^2 \sin^2\theta}\left[\frac{\partial}{\partial \phi} + a\sin^2\theta\frac{\partial}{\partial t}\right]^2$$
$$+ \frac{\Delta}{\rho^2}\left(\frac{\partial}{\partial r}\right)^2 + \frac{1}{\rho^2}\left(\frac{\partial}{\partial \theta}\right)^2. \qquad (33.35)$$

(d) Use these metric components and the components (33.30) of the vector potential to bring the Hamilton-Jacobi equation (33.33) into the concrete form

$$
-\frac{\partial S}{\partial \lambda} = -\frac{1}{2}\frac{1}{\Delta \rho^2}\left[(r^2 + a^2)\frac{\partial S}{\partial t} + a\frac{\partial S}{\partial \phi} - eQr\right]^2
$$
$$
+ \frac{1}{2}\frac{1}{\rho^2 \sin^2\theta}\left[\frac{\partial S}{\partial \phi} + a\sin^2\theta \frac{\partial S}{\partial t}\right]^2 + \frac{1}{2}\frac{\Delta}{\rho^2}\left(\frac{\partial S}{\partial r}\right)^2 + \frac{1}{2}\frac{1}{\rho^2}\left(\frac{\partial S}{\partial \theta}\right)^2. \tag{33.34b}
$$

(e) Solve this Hamilton-Jacobi equation by separation of variables. [*Hint:* Because the equation has no explicit dependence on λ, ϕ, or t, the solution must take the form

$$
S = \frac{1}{2}\mu^2\lambda - Et + L_z\phi + S_r(r) + S_\theta(\theta), \tag{33.36a}
$$

where the values of the "integration constants" follow from $\partial S/\partial \lambda = -\mathcal{H}$, $\partial S/\partial t = \pi_t$, $\partial S/\partial \phi = \pi_\phi$. Insert this assumed form into (33.35) and solve for $S_r(r)$ and $S_\theta(\theta)$ to obtain

$$
S_r = \int \Delta^{-1}\sqrt{R}\,dr, \qquad S_\theta = \int \sqrt{\Theta}\,d\theta, \tag{33.36b}
$$

where $R(r)$ and $\Theta(\theta)$ are the functions defined in equation (33.33). Notice that the constant \mathcal{Q} arises naturally as a "separation-of-variables constant" in this procedure. It was in this way that Carter originally discovered \mathcal{Q}, following Misner's suggestion that he seek analogies to a constant in Newtonian dipole fields (Corben and Stehle, 1960, p. 209).]

(f) By successively setting $\partial S/\partial[\mathcal{Q} + (L_z - aE)^2]$, $\partial S/\partial \mu^2$, $\partial S/\partial E$, and $\partial S/\partial L_z$ to zero, obtain the following equations describing the test-particle orbits:

$$
\int^\theta \frac{d\theta}{\sqrt{\Theta}} = \int^r \frac{dr}{\sqrt{R}}, \tag{33.37a}
$$

$$
\lambda = \int^\theta \frac{a^2 \cos^2\theta}{\sqrt{\Theta}}\,d\theta + \int^r \frac{r^2}{\sqrt{R}}\,dr, \tag{33.37b}
$$

$$
t = \int^\theta \frac{-a(aE\sin^2\theta - L_z)}{\sqrt{\Theta}}\,d\theta + \int^r \frac{(r^2 + a^2)P}{\Delta\sqrt{R}}\,dr, \tag{33.37c}
$$

$$
\phi = \int \frac{-(aE\sin^2\theta - L_z)}{\sin^2\theta \sqrt{\Theta}}\,d\theta + \int \frac{aP}{\Delta\sqrt{R}}\,dr. \tag{33.37d}
$$

(g) By differentiating these equations and combining them, obtain the equations of motion (33.32) cited in the text.

(h) Derive equations (33.31) for E, L_z, μ, and \mathcal{Q} by setting $\partial S/\partial x^\alpha = \pi_\alpha = p_\alpha + eA_\alpha$.

§33.6. PRINCIPAL NULL CONGRUENCES

Two special families of *photon* trajectories "mold themselves into" the Kerr-Newman geometry in an especially harmonious way. They are called the "*principal null congruences*" of the geometry. ("Congruence" is an elegant word that means "space-

Principal null congruences for the spacetime geometry of a black hole

filling family of curves.") These congruences are the solutions to the test-particle equations of motion (33.32) with

$$\mu = 0 \text{ (zero rest mass; photon)}, \tag{33.38a}$$

$$e = 0 \text{ (zero charge on photon)}, \tag{33.38b}$$

$$L_z = aE \sin^2\theta \quad \begin{pmatrix} \text{a permissible value for } L_z \\ \text{only because } d\theta/d\lambda \text{ turns} \\ \text{out to be zero} \end{pmatrix}, \tag{33.38c}$$

$$\mathscr{Q} = -(L_z - aE)^2 = -a^2 E^2 \cos^4\theta. \tag{33.38d}$$

For these values of the constants of motion, the equations of motion (33.32) reduce to

$$k^\theta \equiv d\theta/d\lambda = 0, \tag{33.39a}$$

$$k^r \equiv dr/d\lambda = \pm E \quad \begin{array}{l} \text{("+" for outgoing photons,} \\ \text{"$-$" for ingoing)}, \end{array} \tag{33.39b}$$

$$k^\phi \equiv d\phi/d\lambda = aE/\Delta, \tag{33.39c}$$

$$k^t \equiv dt/d\lambda = (r^2 + a^2)E/\Delta. \tag{33.39d}$$

Significance of the principal null congruences

In what sense are these photon trajectories more interesting than others? (1) They mold themselves to the spacetime curvature in such a way that, if $C_{\alpha\beta\gamma\delta}$ is the Weyl conformal tensor (§13.5), and $^*C_{\alpha\beta\gamma\delta} = \epsilon_{\alpha\beta\mu\nu}C^{|\mu\nu|}{}_{\gamma\delta}$ is its dual, then

$$C_{\alpha\beta\gamma[\delta}k_{\epsilon]}k^\beta k^\gamma = 0, \qquad {}^*C_{\alpha\beta\gamma[\delta}k_{\epsilon]}k^\beta k^\gamma = 0. \tag{33.40}$$

[This relationship implies that the Kerr-Newman geometry is of "Petrov-Pirani type D" and that these photon trajectories are "doubly degenerate, principal null congruences." For details of the meanings and implications of these terms see, e.g., §8 of Sachs (1964), or Ehlers and Kundt (1962), or the original papers by Petrov (1954, 1969) and Pirani (1957).] (2) By suitable changes of coordinates (exercise 33.8), one can bring the Kerr-Newman metric into the form

$$ds^2 = (\eta_{\alpha\beta} + 2Hk_\alpha k_\beta)\, dx^\alpha \, dx^\beta, \tag{33.41}$$

where H is a scalar field and k_α are the components of the wave vector for one of the principal null congruences (either one; but not both!). [This was the property of the Kerr-Newman metric that led to its original discovery (Kerr, 1963). For further detail on metrics of this form, see Kerr and Schild (1965).] (3) In Kerr coordinates (Box 33.2), the ingoing principal null congruence is

$$r = -E\lambda, \qquad \theta = \text{const}, \qquad \widetilde{\phi} = \text{const}, \qquad \widetilde{V} = \text{const}. \tag{33.42a}$$

$$\begin{bmatrix} \text{arbitrary normalization} \\ \text{factor; can be removed} \\ \text{by redefinition of } \lambda \end{bmatrix}$$

These ingoing photon world lines are the generators of the conical surface \widetilde{V} = const. in the Kerr diagram of Box 33.2. (4) The only kind of particle that can remain forever at the horizon is a photon with world line in the outgoing principal null congruence (exercise 33.9). Such photon world lines are "generators" of the horizon (dotted curves with a "barber-pole twist" in Kerr diagram of Box 33.2). They have angular velocity

$$\Omega = \frac{d\phi}{dt} = \frac{d\widetilde{\phi}}{d\widetilde{V}} = \frac{a}{r_+^2 + a^2} = \frac{a}{2M^2 - Q^2 + 2M(M^2 - a^2 - Q^2)^{1/2}}. \quad (33.42b)$$

Exercise 33.8. KERR-SCHILD COORDINATES

(a) Show that in Kerr coordinates the ingoing null congruence (33.39) has the form (33.42a). Also show that the covariant components of the wave vector—after changing to a new affine parameter $\lambda_{\text{new}} = \lambda_{\text{old}}E$—are

$$k_r^{(\text{in})} = 0, \qquad k_\theta^{(\text{in})} = 0, \qquad k_{\widetilde{\phi}}^{(\text{in})} = a \sin^2\theta, \qquad k_{\widetilde{V}}^{(\text{in})} = -1. \quad (33.43)$$

(b) Introduce new coordinates \widetilde{t}, x, y, z, defined by

$$x + iy = (r + ia)e^{i\widetilde{\phi}} \sin\theta, \qquad z = r \cos\theta, \qquad \widetilde{t} = \widetilde{V} - r; \quad (33.44a)$$

and show that in this "Kerr-Schild coordinate system" the metric takes the form

$$ds^2 = (\eta_{\alpha\beta} + 2H k_\alpha^{(\text{in})} k_\beta^{(\text{in})}) \, dx^\alpha \, dx^\beta, \quad (33.44b)$$

where

$$H = \frac{Mr - \frac{1}{2}Q^2}{r^2 + a^2(z/r)^2}, \quad (33.44c)$$

$$k_\alpha^{(\text{in})} \, dx^\alpha = -\frac{r(x \, dx + y \, dy) - a(x \, dy - y \, dx)}{r^2 + a^2} - \frac{z \, dz}{r} - d\widetilde{t}. \quad (33.44d)$$

For the transformation to analogous coordinates in which

$$ds^2 = (\eta_{\alpha\beta} + 2H k_\alpha^{(\text{out})} k_\beta^{(\text{out})}) \, dx^\alpha \, dx^\beta.$$

see, e.g., Boyer and Lindquist (1967).

Exercise 33.9. NULL GENERATORS OF HORIZON

(a) Show that in Kerr coordinates the outgoing principle null congruence is described by the tangent vector

$$\frac{d\theta}{d\lambda} = 0, \qquad \frac{dr}{d\lambda} = E, \qquad \frac{d\widetilde{\phi}}{d\lambda} = 2a\frac{E}{\Delta}, \qquad \frac{d\widetilde{V}}{d\lambda} = 2(r^2 + a^2)\frac{E}{\Delta}. \quad (33.45)$$

(b) These components of the wave vector become singular at the horizon ($\Delta = 0$), not because of a singularity in the coordinate system—the coordinates are well-behaved!—but because of poor normalization of the affine parameter. For each outgoing geodesic, let Δ_0

be a constant, defined as the value of Δ at the event where the geodesic slices the hypersurface $\widetilde{V} = 0$. Then renormalize the affine parameter for each geodesic

$$\lambda_{\text{new}} = (E/\Delta_0)\lambda_{\text{old}}. \tag{33.46}$$

Show that the resulting wave vectors

$$\frac{d\theta}{d\lambda} = 0, \qquad \frac{dr}{d\lambda} = \Delta_0, \qquad \frac{d\widetilde{\phi}}{d\lambda} = 2a\frac{\Delta_0}{\Delta}, \qquad \frac{d\widetilde{V}}{d\lambda} = 2(r^2 + a^2)\frac{\Delta_0}{\Delta} \tag{33.45'}$$

are well-behaved as one approaches the horizon; and show that the geodesics on the horizon have the form

$$\theta = \text{const.}, \qquad r = r_+ = \text{const.}, \qquad \widetilde{\phi} = 2a\lambda, \qquad \widetilde{V} = 2(r_+{}^2 + a^2)\lambda. \tag{33.47}$$

(c) Show that these are the only test-particle trajectories that remain forever on the horizon. [*Hint:* Examine the light cone.]

§33.7. STORAGE AND REMOVAL OF ENERGY FROM BLACK HOLES [Penrose (1969)]

When an object falls into a black hole, it changes the hole's mass, charge, and intrinsic angular momentum (first law of black-hole dynamics; Box 33.4). If the infalling object is large, its fall produces much gravitational and electromagnetic radiation. To calculate the radiation emitted, and the energy and angular momentum it carries away—which are prerequisites to any calculation of the final state of the black hole—is an enormously difficult task. But if the object is very small (size of object ≪ size of horizon; mass of object ≪ mass of hole), and has sufficiently small charge, the radiation it emits in each circuit around the hole is negligible. For example, for gravitational radiation

When a small object falls down a large hole:

(1) energy radiated is negligible compared to object's rest mass

$$\frac{\text{(energy emitted per circuit)}}{\text{(rest mass of object)}} \sim \frac{\text{(rest mass of object)}}{\text{(mass of hole)}} \tag{33.48}$$

[see §36.5; also Bardeen, Press, and Teukolsky (1972)]. Because the energy emitted is negligible, radiation reaction is also negligible, and the object moves very nearly along a test-particle trajectory. In this case, application of the first law of black-hole dynamics is simple and straightforward.

Consider, initially, a small object that falls directly into the black hole from far away. According to the first law, it produces the following changes in the mass, charge, and angular momentum of the black hole:

(2) hole's mass, charge, and angular momentum change by $\Delta M = E$, $\Delta Q = e$, $\Delta S = L_z$

$$\Delta M = E = (\text{"energy at infinity" of infalling object}), \tag{33.49a}$$

$$\Delta Q = e = (\text{charge of infalling object}), \tag{33.49b}$$

$$\Delta S \equiv \Delta|\boldsymbol{S}| = L_z = \begin{pmatrix} \text{component of object's angular momentum} \\ \text{on black hole's rotation axis} \end{pmatrix}. \tag{33.49c}$$

The infalling object will also change the direction of **S**. In the black hole's original asymptotic Lorentz frame, its initial angular momentum vector points in the z-direction,

$$(S_z)_{\text{initial}} = S, \qquad (S_x)_{\text{initial}} = 0, \qquad (S_y)_{\text{initial}} = 0.$$

Consequently, only the z-component of angular momentum of the infalling object can produce any significant change in the magnitude of **S**. But the x- and y-components, L_x and L_y, can change the direction of **S**. If the object has negligible speed at infinity, then it produces the changes (exercise 33.10):

$$\Delta S_x = L_x = -(\sin \phi_\infty)\sqrt{\Theta_\infty} - (\cot \theta_\infty \cos \phi_\infty)L_z, \tag{33.49d}$$

$$\Delta S_y = L_y = (\cos \phi_\infty)\sqrt{\Theta_\infty} - (\cot \theta_\infty \sin \phi_\infty)L_z, \tag{33.49e}$$

$$\Delta(S_x^2 + S_y^2)^{1/2} = \sqrt{\mathcal{Q}} = (L^2 - L_z^2)^{1/2}. \tag{33.49f}$$

Here a subscript "∞" means the value of a quantity at a point on the orbit far from the black hole (at "infinity").

Consider, next, a more complicated process, first conceived of by Penrose (1969): (1) Shoot a small object A into the black hole from outside with energy-at-infinity E_A, charge e_A, and axial component of angular momentum L_{zA}. (2) When the object is deep down near the horizon, let it explode into two parts, B and C, each of which subsequently moves along a new test-particle trajectory, with new constants of the motion e_B and e_C, E_B and E_C, L_{zB} and L_{zC}. (3) So design the explosion that object B falls down the hole and gets captured, but object C escapes back to radial infinity. What will be the changes in mass, charge, and angular momentum of the black hole? According to the first law of black-hole dynamics,

$$\Delta M = \begin{pmatrix} \text{total energy that distant observers see} \\ \text{fall inward past themselves minus} \\ \text{total energy that they see reemerge} \end{pmatrix}$$

$$= E_A - E_C.$$

Similarly, $\Delta Q = e_A - e_C$ and $\Delta S = L_{zA} - L_{zC}$. Not unexpectedly, these changes can be written more simply in terms of the constants of motion for object B, which went down the hole. View the explosion "$A \longrightarrow B + C$" in a local Lorentz frame down near the hole, which is centered on the explosive event. As viewed in that frame, the explosion must satisfy the special relativistic laws of physics (equivalence principle!). In particular, it must obey charge conservation

$$e_A = e_B + e_C \tag{33.50a}$$

and conservation of total 4-momentum

$$(\boldsymbol{p}_A)_{\text{immediately before explosion}} = (\boldsymbol{p}_B + \boldsymbol{p}_C)_{\text{immediately after explosion}}.$$

Moreover, conservation of 4-momentum \boldsymbol{p} and charge e implies also conservation of generalized momentum $\boldsymbol{\pi} \equiv \boldsymbol{p} - e\boldsymbol{A}$,

$$\boldsymbol{\pi}_A = \boldsymbol{p}_A - e_A\boldsymbol{A} = \boldsymbol{p}_B + \boldsymbol{p}_C - (e_B + e_C)\boldsymbol{A} = \boldsymbol{\pi}_B + \boldsymbol{\pi}_C;$$

and hence also conservation of the components of generalized momentum along the vectors $\partial/\partial t$ and $\partial/\partial\phi$,

$$E_A \equiv -\pi_{tA} = -\pi_{tB} - \pi_{tC} = E_B + E_C, \tag{33.50b}$$

$$L_{zA} \equiv \pi_{\phi B} + \pi_{\phi C} = L_{zB} + L_{zC}, \tag{33.50c}$$

(conservation of "energy-at-infinity" and "axial component of angular momentum" in explosion). Combining these conservation laws with the expressions

$$\Delta M = E_A - E_C, \qquad \Delta Q = e_A - e_C, \qquad \text{and } \Delta S = L_{zA} - L_{zC},$$

one obtains

$$\Delta M = E_B, \qquad \Delta Q = e_B, \qquad \Delta S = L_{zB}. \tag{33.51}$$

This result restated in words: the changes in mass, charge, and angular momentum are equal to the "energy-at-infinity," charge, and "axial component of angular momentum" that object B carries inward across the horizon, *even though B* may have ended up on a test-particle orbit that does not extend back to radial infinity!

Straightforward extensions of the above thought experiment produce this generalization: *In any complicated black-hole process that involves infalling, colliding, and exploding pieces of matter that emit negligible gravitational radiation, the total changes in mass, charge, and angular momentum of the black hole are*

Changes in *M, Q, S* for any nonradiative black-hole process

$$\Delta M = \begin{pmatrix} \text{sum of values of energy-at-infinity, } E, \\ \text{for all objects which cross the horizon—with} \\ E \text{ evaluated for each object at event of crossing} \end{pmatrix}, \tag{33.52a}$$

$$\Delta Q = \begin{pmatrix} \text{similar sum, of charges, } e, \text{ for} \\ \text{all objects crossing horizon} \end{pmatrix}, \tag{33.52b}$$

$$\Delta S = \begin{pmatrix} \text{similar sum of axial components of angular} \\ \text{momentum, } L_z, \text{ for all objects crossing horizon} \end{pmatrix}. \tag{33.52c}$$

This result is not at all surprising. It is precisely what one might expect from the most naive of viewpoints. Not so expected, however, is the following consequence [Penrose (1969)]: *By injecting matter into a black hole in a carefully chosen way, one can decrease the total mass-energy of the black hole—i.e., one can extract energy from the hole.*

Extraction of energy from a black hole by processes in the ergosphere

For *uncharged* infalling objects, the key to energy extraction is the ergosphere [hence its name, coined by Ruffini and Wheeler (1971a) from the Greek word "εργον" for "work"]. Outside the ergosphere, the Killing vector $\boldsymbol{\xi}_{(t)} \equiv \partial/\partial t$ is timelike, as is the 4-momentum \boldsymbol{p} of every test particle; and therefore $E = -\boldsymbol{p} \cdot \boldsymbol{\xi}_{(t)}$ is necessarily positive. But inside the ergosphere (between the horizon and the static limit), $\boldsymbol{\xi}_{(t)}$

is spacelike, so for certain choices of timelike momentum vector (certain orbits of uncharged test particles), $E = -\boldsymbol{p} \cdot \boldsymbol{\xi}_{(t)}$ is negative, whereas for others it is positive. The orbits of negative E are confined entirely to the ergosphere. Thus, to inject an uncharged object with negative E into the black hole—and thereby to extract energy from the hole—one must always change its E from positive to negative and therefore also change its orbit, after it penetrates into the ergosphere. Of course, this is not difficult in principle—and perhaps not even in practice; see Figure 33.2.

For a charged object, electromagnetic forces alter the region where there exist orbits of negative energy-at-infinity. If the charges of object and hole have opposite sign, then the hole's electromagnetic field pulls inward on the object, giving it more kinetic energy when near the hole than one would otherwise expect. Thus, $-\boldsymbol{p} \cdot \boldsymbol{\xi}_{(t)}$ becomes an overestimate of E,

$$E = -(\boldsymbol{p} - e\boldsymbol{A}) \cdot \boldsymbol{\xi}_{(t)} = -\boldsymbol{p} \cdot \boldsymbol{\xi}_{(t)} + \underbrace{eQr/\rho^2}_{[< 0 \text{ if } eQ < 0]}; \tag{33.53}$$

and orbits with $E < 0$ exist in a region somewhat larger than the ergosphere. If, on the other hand, e and Q have the same sign, then orbits with $E < 0$ are confined to a region smaller than the ergosphere. For given values e, Q, and rest mass μ, the region where there exist orbits with $E < 0$ is called the "*effective ergosphere*."

The "effective ergosphere" for charged-particle processes

Exercise 33.10. ANGULAR MOMENTUM VECTOR FOR INFALLING PARTICLE
Derive equations (33.49d,e,f) for the components L_x and L_y of the orbital angular momentum of a particle falling into a black hole. Assume negligible initial speed, $E^2 - \mu^2 \approx 0$.

§33.8. REVERSIBLE AND IRREVERSIBLE TRANSFORMATIONS
[Christodoulou (1970), Christodoulou and Ruffini (1971)]

Take a black hole of given mass M, charge Q, and angular momentum S. By injection of small objects, make a variety of changes in M, Q, and S. Can one pick an arbitrary desired change, ΔM, ΔQ, and ΔS, and devise a process that achieves it? Or are there limitations?

The second law of black-hole dynamics (nondecreasing surface area of black hole; Box 33.4; proof in §34.5 of next chapter) provides a strict limitation.

Then can all values within that limitation be achieved—and can that limitation be discovered by a direct examination of test-particle orbits?

The answer is yes; and, in fact, the limitation was discovered by Christodoulou (1970) and Christodoulou and Ruffini (1971) from an examination of test-particle orbits, independently of and simultaneously with Hawking's (1971) discovery of the second law of black-hole dynamics.

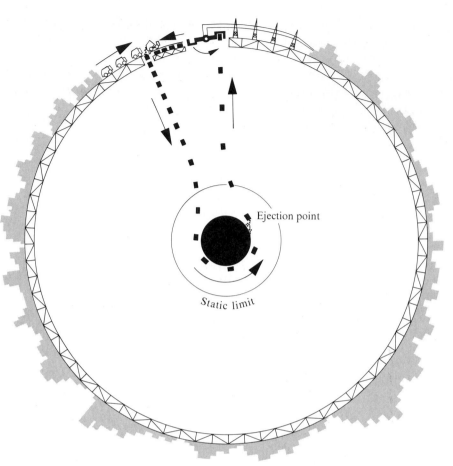

Figure 33.2.
An advanced civilization has constructed a rigid framework around a black hole, and has built a huge city on that framework. Each day trucks carry one million tons of garbage out of the city to the garbage dump. At the dump the garbage is shoveled into shuttle vehicles which are then, one after another, dropped toward the center of the black hole. Dragging of inertial frames whips each shuttle vehicle into a circling, inward-spiraling orbit near the horizon. When it reaches a certain "ejection point," the vehicle ejects its load of garbage into an orbit of negative energy-at-infinity, $E_{\text{garbage}} < 0$. As the garbage flies down the hole, changing the hole's total mass-energy by $\Delta M = E_{\text{garbage ejected}} < 0$, the shuttle vehicle recoils from the ejection and goes flying back out with more energy-at-infinity than it took down

$$E_{\text{vehicle out}} = E_{\text{vehicle + garbage down}} - E_{\text{garbage ejected}}$$
$$> E_{\text{vehicle + garbage down}}.$$

The vehicle deposits its huge kinetic energy in a giant flywheel adjacent to the garbage dump; and the flywheel turns a generator, producing electricity for the city, while the shuttle vehicle goes back for another load of garbage. The total electrical energy generated with each round trip of the shuttle vehicle is

$$\text{(Energy per trip)} = E_{\text{vehicle out}} - \text{(rest mass of vehicle)}$$
$$= (E_{\text{vehicle + garbage down}} - E_{\text{garbage ejected}}) - \text{(rest mass of vehicle)}$$
$$= \text{(rest mass of vehicle + rest mass of garbage} - \Delta M) - \text{(rest mass of vehicle)}$$
$$= \text{(rest mass of garbage)} + \text{(amount,} - \Delta M, \text{ by which hole's mass decreases)}.$$

Thus, not only can the inhabitants of the city use the black hole to convert the entire rest mass of their garbage into kinetic energy of the vehicle, and thence into electrical power, but they can also convert some of the mass of the black hole into electrical power!

To derive the limitation of nondecreasing surface area from properties of test-particle orbits, one must examine what values of energy-at-infinity, E, are allowed at a given location (r, θ) outside a black hole. Equations (33.32a,b), when combined, yield the value of E in terms of a test particle's location (r, θ), rest mass μ, charge e, axial component of angular momentum L_z, and momenta $p^r = dr/d\lambda$, $p^\theta = d\theta/d\lambda$ in the r and θ directions:

Properties of test-particle orbits:

$$\alpha E^2 - 2\beta E + \gamma = 0; \qquad E = \frac{\beta + \sqrt{\beta^2 - \alpha\gamma}}{\alpha}, \tag{33.54a}$$

(1) E as function of μ, e, L_z, r, θ, p^r

where

$$\alpha = (r^2 + a^2)^2 - \Delta a^2 \sin^2\theta > 0 \text{ everywhere outside horizon,} \tag{33.54b}$$

$$\beta = (L_z a + eQr)(r^2 + a^2) - L_z a \, \Delta, \tag{33.54c}$$

$$\gamma = (L_z a + eQr)^2 - \Delta(L_z/\sin\theta)^2 - \mu^2 \Delta\rho^2 - \rho^4[(p^r)^2 + \Delta(p^\theta)^2]. \tag{33.54d}$$

(One must take the positive square root, $+\sqrt{\beta^2 - \alpha\gamma}$, rather than the negative square root; positive square root corresponds to 4-momentum pointing toward future; while negative square root corresponds to past-pointing 4-momentum; see Figure 33.3.)

Several features of the energy equation (33.54) are noteworthy. (1) For orbits in the equatorial "plane," $\theta = \pi/2$ and $p^\theta \equiv 0$, the energy equation yields an effective potential for radial motion (Box 33.5). (2) Orbits of negative E must have $\beta < 0$ and $\gamma > 0$—which can be achieved only if $L_z a < 0$ and/or $eQ < 0$. Thus, *one cannot decrease the mass of a black hole without simultaneously decreasing the magnitude of its charge or angular momentum or both.* (3) For an orbit at given (r, θ), with given e and L_z, E is a minimum if $p^r = p^\theta = \mu = 0$. *Put differently, the rest mass and the r- and θ-components of momentum always contribute positively to E.*

(2) effective potential

(3) negative E requires $L_z a < 0$ and/or $eQ < 0$

By injecting an object into a black hole, produce small changes

$$\delta M = E, \qquad \delta Q = e, \qquad \delta S = L_z,$$

in its mass, charge, and angular momentum. For given changes in Q and S, what range of changes in M is possible? Clearly δM can be made as large as one wishes by making the rest mass μ sufficiently large. But there will be a lower limit on δM. That limit corresponds to the minimum value of E for given e and L_z. The orbit of minimum E crosses the horizon (otherwise no changes in M, Q, S would occur!), so one can evaluate E there. At the horizon, as anywhere, a minimum for E is achieved if $\mu = p^r = p^\theta = 0$. Inserting these values and $r = r_+$ (so $\Delta = 0$) into equations (33.54), one finds

Changes in black-hole properties due to injection of particles:

$$E_{\min} = \frac{L_z a + eQr_+}{r_+^2 + a^2}, \tag{33.55}$$

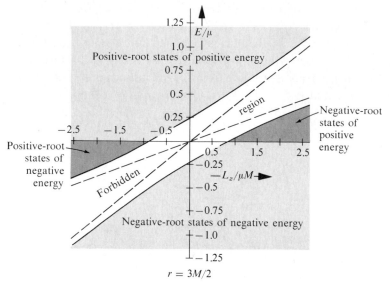

Figure 33.3.

Energy-at-infinity E allowed for a particle of angular momentum L_z and rest mass μ, which is (1) in the "equatorial plane" $\theta = \pi/2$, (2) at radius $r = 3M/2$, (3) of an uncharged ($Q = 0$) extreme-Kerr ($S = M^2$) black hole. E is here plotted against L_z. "Seas" of "positive and negative root" states are shown. The positive root states have energies-at-infinity given by equations (33.54)

$$E = \frac{\beta + \sqrt{\beta^2 - \alpha\gamma}}{\alpha}$$

and have 4-momentum vectors pointing into the future light cone. The negative root states (states of Dirac's "negative energy sea") have energies at infinity given by

$$E = \frac{\beta - \sqrt{\beta^2 - \alpha\gamma}}{\alpha},$$

and have 4-momentum vectors pointing into the past light cone. In the gap between the "seas" no orbits exist (forbidden region). The gap vanishes at the horizon $r = M$ (infinite redshift of *local* energy gap, 2μ, gives zero gap in energy-at-infinity). [Figure adapted from Christodoulou (1971).]

corresponding to changes in the black-hole properties of

(1) limit on δM for given δQ and δS

$$\delta M \geq \frac{a\,\delta S + r_+ Q\,\delta Q}{r_+{}^2 + a^2} \qquad \left(\begin{array}{l}\text{absolute minimum value of}\\ \delta M \text{ for given } \delta S \text{ and } \delta Q\end{array}\right). \qquad (33.56)$$

Notice an important consequence [Bardeen (1970a)]: if the black hole is initially of the "extreme Kerr-Newman" variety, with $M^2 = a^2 + Q^2$, so that one might fear a change which makes $M^2 < a^2 + Q^2$ and thereby destroys the horizon, one's fears are unfounded. Equation (33.56) then demands (since $r_+ = M$ and $S = Ma$)

$$M\,\delta M \geq a\,\delta a + Q\,\delta Q;$$

(2) preservation of the horizon

so M^2 remains greater than or equal to $a^2 + Q^2$, and the horizon is preserved.

Box 33.5 ORBITS OF TEST PARTICLE IN "EQUATORIAL PLANE" OF KERR-NEWMAN BLACK HOLE

Radial motion is governed by energy equation (33.54) with $\theta = p^\theta = 0$:

$$\alpha E^2 - 2\beta E + \gamma_0 - r^4(p^r)^2 = 0; \qquad E = \frac{\beta + \sqrt{\beta^2 - \alpha\gamma_0 + \alpha r^4(p^r)^2}}{\alpha}; \quad (1)$$

α, β, γ_0 arc functions of r and of constants of motion,

$$\alpha = (r^2 + a^2)^2 - \Delta a^2 > 0, \tag{2a}$$

$$\beta = (L_z a + eQr)(r^2 + a^2) - L_z a\Delta, \tag{2b}$$

$$\gamma_0 = (L_z a + eQr)^2 - \Delta L_z^2 - \mu^2 r^2 \Delta; \tag{2c}$$

$p^r =$ (radial momentum) is

$$p^r = dr/d\lambda. \tag{3}$$

Thus, equation (1) is an ordinary differential equation for $dr/d\lambda$.

Qualitative features of the radial motion can be read off an effective-potential diagram. The effective potential $V(r)$ is the minimum allowed value of E at radius r:

$$V(r) = \frac{\beta + \sqrt{\beta^2 - \alpha\gamma_0}}{\alpha}.$$

As in the Schwarzschild case (Figure 25.2), the allowed regions for a particle of energy-at-infinity E are the regions with $V(r) \leq E$; and the turning points ($p^r = dr/d\lambda = 0$) occur where $V(r) = E$.

Stable circular orbits occur at the minima of $V(r)$. By examining $V(r)$ closely, one finds that for uncharged black holes the innermost stable circular orbit (most tightly bound orbit) has the characteristics here tabulated [table adapted from Ruffini and Wheeler (1971b)].

Charac-teristic	Newtonian (Figure 25.2)	Schwarzschild ($a = Q = 0$) (Figure 25.2)	Extreme Kerr ($a^2 = M^2$, $Q = 0$) (see figure) [Bardeen (1970a)] if $L_z a > 0$	if $L_z a < 0$		
r/M	0	6	1	9		
E/μ	$-\infty$	$2\sqrt{2}/3$	$1/\sqrt{3}$	$5/(3\sqrt{3})$		
$(\mu - E)/\mu =$ "fractional binding"	$+\infty$	0.0572	0.4226	0.0377		
$	L_z	/\mu M$	0	$2\sqrt{3}$	$2/\sqrt{3}$	$22/(3\sqrt{3})$

Box 33.5 (*continued*)

For a charged extreme Kerr-Newman black hole ($M^2 = Q^2 + a^2$, $Q \neq 0$ and $a \neq 0$) stable circular orbits with 100 per cent binding ($E = 0$) are achieved in the limit

$$\frac{e}{\mu} \longrightarrow -\infty, \qquad \frac{Q}{M} \longrightarrow 0 \text{ (so } a \longrightarrow M), \qquad \text{and } \left(\frac{e}{\mu}\right) \cdot \left(\frac{Q}{M}\right) \longrightarrow -\infty.$$

[Christodoulou and Ruffini (1971)].

The effective potential for an uncharged, extreme Kerr black hole ($a = M$) is shown in the figure [figure adapted from Ruffini and Wheeler (1971b)]. For detailed diagrams of orbits in the equatorial plane, see de Felice (1968). For many interesting properties of orbits that are not confined to the equatorial plane, see Wilkins (1972).

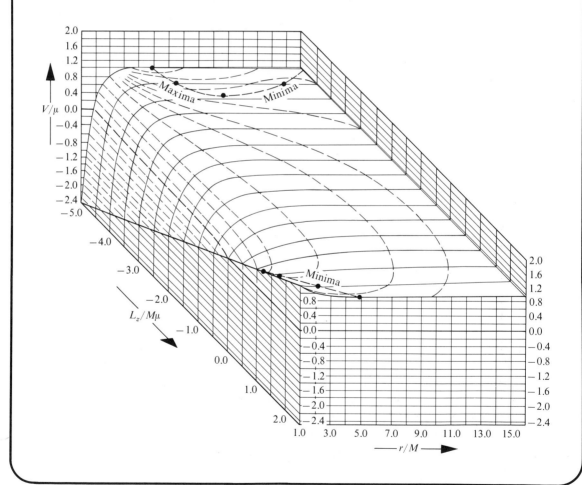

The general limit (33.56) on the change in mass can be rewritten in an alternative form [Christodoulou (1970), Christodoulou and Ruffini (1971)]:

$$\delta M_{ir} \geq 0, \tag{33.57}$$

where

$$M_{ir} \equiv \frac{1}{2} \sqrt{r_+^2 + a^2} = \frac{1}{2}[(M + \sqrt{M^2 - Q^2 - a^2})^2 + a^2]^{1/2} \tag{33.58}$$

(3) irreducible mass

is the "irreducible mass" of the black hole. *Equation (33.57) states that no black-hole transformation produced by the injection of small lumps of matter can ever reduce the irreducible mass of a black hole.* This result is actually a special case of the second law of black-hole dynamics, since the surface area of a black hole is

$$A = 16\pi M_{ir}^2 \tag{33.59}$$

(Exercise 33.12).

Equation (33.58) can be combined with $a = S/M$ and inverted to yield

$$M^2 = \left(M_{ir} + \frac{Q^2}{4 M_{ir}} \right)^2 + \frac{S^2}{4 M_{ir}^2}. \tag{33.60}$$

$$\underbrace{\begin{bmatrix}\text{irreducible con-}\\\text{tribution to mass}\end{bmatrix}}_{} \quad \underbrace{\begin{bmatrix}\text{electromagnetic con-}\\\text{tribution to mass}\end{bmatrix}}_{} \quad \underbrace{\begin{bmatrix}\text{rotational con-}\\\text{tribution to mass}\end{bmatrix}}_{}$$

A black-hole transformation that holds fixed the irreducible mass is *reversible;* one that increases it is *irreversible.* The derivation of equation (33.56) revealed that the only injection processes that actually achieve the minimum possible value for δM (and thus make $\delta M_{ir} = 0$) are those with $\mu = p^r = p^\theta = 0$ at the horizon, $r = r_+$. Restated in words: To produce a reversible transformation by injecting an object into a black hole, one must (1) give the object a rest mass μ extremely small compared to its charge ϵ or axial component of angular momentum L_z,

(4) reversible and irreversible transformations

$$\mu/e \ll 1 \text{ and/or } \mu^2/L_z^2 \ll 1;$$

and (2) set the object down "extremely gently" ($p^r = p^\theta = 0$), extremely close to the horizon ($r = r_+$). This does not sound too difficult until one recalls that objects with $p^r = p^\theta = 0$ at the horizon must be moving outward with the speed of light, and that the nearer one approaches the horizon as one sets down the object, the greater one's danger of "slipping" and getting swallowed!

Clearly, any actual injection process will depart somewhat from irreversibility. Reversibility is an idealized limit, approachable but not attainable.

EXERCISES Exercise 33.11. IRREDUCIBLE MASS IS IRREDUCIBLE

Show that condition (33.56) is equivalent to $\delta M_{\mathrm{ir}} \geq 0$.

Exercise 33.12. SURFACE AREA OF A BLACK HOLE

Show that the surface area of the horizon of the Kerr-Newman geometry [area of surface $r = r_+$ and $t = \mathrm{const}$ (Boyer-Lindquist coordinates) or $\widetilde{V} = \mathrm{const}$ (Kerr coordinates)] is $16\pi M_{\mathrm{ir}}^2$.

Exercise 33.13. ANGULAR VELOCITY OF A BLACK HOLE

A general theorem [Hartle (1970) for relativistic case; Ostriker and Gunn (1969) for nonrelativistic case] says that, if one injects angular momentum into a rotating star while holding fixed all other contributions to its total mass-energy (contributions from entropy and from baryonic rest mass), then the injection produces a change in total mass-energy given by

$$\delta(\text{mass-energy}) = \begin{pmatrix} \text{angular velocity of star} \\ \text{at point of injection} \end{pmatrix} \delta(\text{angular momentum}). \qquad (33.61)$$

By analogy, if one injects an angular momentum δS into a rotating black hole while holding fixed all other contributions to its total mass-energy (contributions from irreducible mass and from charge), one identifies the coefficient Ω_h in the equation

$$\delta M = \Omega_h\, \delta S$$

as the angular velocity of the hole:

$$\Omega_h = (\partial M/\partial S)_{Q,M_{\mathrm{ir}}}. \qquad (33.62)$$

(a) Show that the angular velocity of a black hole is equal to

$$\Omega_h = \frac{a}{r_+^{\,2} + a^2}. \qquad (33.63)$$

Notice that this is precisely the angular velocity of photons that live forever on the horizon [equation (33.42b); "barber-pole twist" of null generators of horizon].

(b) Show that *any* object falling into a black hole acquires an angular velocity (relative to Boyer-Lindquist coordinates) of $\Omega = d\phi/dt = \Omega_h$ in the late stages, as it approaches the horizon. (Recall that the horizon is a singularity of the Boyer-Lindquist coordinates. This is the reason that every object, regardless of its L_z, E, e, μ, \mathcal{Q}, can approach and does approach $\Omega = \Omega_h$.)

Exercise 33.14. SEPARATION OF VARIABLES FOR WAVE EQUATIONS

This chapter has studied extensively the motion of small objects in the external fields of black holes. Of almost equal importance, but not so well-understood yet because of its complexity, is the evolution of weak electromagnetic and gravitational perturbations ("waves") in the Kerr-Newman geometry. Just as one had no *à priori* reason to expect a "fourth constant" for test-particle motion in the Kerr-Newman geometry, so one had no reason to expect separability for Maxwell's equations, or for the wave equations describing gravitational perturbations—or even for the scalar wave equation $\Box \psi \equiv -\psi_{,\alpha}^{\ \alpha} = 0$. Thus it came as a great surprise when Carter (1968c) proved separability for the scalar wave equation, and later when Teukolsky (1972, 1973) separated both Maxwell's equations and the wave equations for gravitational perturbations.

Show that separation of variables for the scalar-wave equation in the (uncharged) Kerr geometry yields solutions of the form

$$\psi = (r^2 + a^2)^{-1/2} u_{\ell m}(r) S_{m\ell}(-i\omega a, \cos\theta) e^{i(m\phi - \omega t)}, \tag{33.64a}$$

where m and ℓ are integers with $0 \le |m| \le \ell$; $S_{m\ell}$ is a spheroidal harmonic [see Meixner and Schärfke (1954)]; and $u_{\ell m}$ satisfies the differential equation

$$-d^2u/dr^{*2} + Vu = 0. \tag{33.64b}$$

In order to put the equation in this form, define a Regge-Wheeler (1957) "tortoise"-type radial coordinate r^* by

$$dr^* = \Delta^{-1}(r^2 + a^2)\, dr, \tag{33.64c}$$

and find an effective potential $V(r^*)$ given by

$$\begin{aligned} V = &-\left(\omega - \frac{ma}{r^2 + a^2}\right)^2 + [(m - \omega a)^2 + \mathcal{Q}](r^2 + a^2)^{-2}\,\Delta \\ &+ 2(Mr - a^2)(r^2 + a^2)^{-3}\,\Delta + 3a^2(r^2 + a^2)^{-4}\,\Delta^2. \end{aligned} \tag{33.64d}$$

In this radial equation \mathcal{Q} is a constant (analog of Carter's constant for particle motion), given in terms of m and ℓ by

$$\mathcal{Q} \equiv \lambda_{m\ell} - m^2; \quad \lambda_{m\ell} = \begin{bmatrix} \text{eigenfunction of spheroidal harmonic;} \\ \text{see Meixner and Schärfke (1954)} \end{bmatrix}. \tag{33.64e}$$

[These details of the separated solution were derived by Brill *et al.* (1972). For studies of the interaction between fields and Kerr black holes—studies performed using the above solution, and using analogous solutions to the electromagnetic and gravitational wave equations—see Bardeen, Press, and Teukolsky (1972), Misner (1972b), Teukolsky (1972), Ipser (1971), Press and Teukolsky (1973), and Chrzanowski and Misner (1973).]

CHAPTER **34**

GLOBAL TECHNIQUES, HORIZONS, AND SINGULARITY THEOREMS

This chapter is entirely Track 2. §22.5 (geometric optics) and the Track-2 portions of Chapters 32 and 33 (collapse and black holes) are necessary preparation for it. It is not needed as preparation for any later chapter.

Local techniques of analyzing spacetime physics contrasted with global techniques

§34.1. GLOBAL TECHNIQUES VERSUS LOCAL TECHNIQUES

Until the 1960's, computations in gravitation theory used *local* techniques almost exclusively: the Einstein field equation describes how the stress-energy tensor \boldsymbol{T} at a given event generates curvature \boldsymbol{G} at that same event (*local* physics). When reduced to differential equations for the metric coefficients, $\boldsymbol{G} = 8\pi\boldsymbol{T}$ relates $g_{\alpha\beta}$, $\partial g_{\alpha\beta}/\partial x^{\mu}$, and $\partial^2 g_{\alpha\beta}/\partial x^{\mu}\partial x^{\nu}$ at each given event to $T_{\gamma\delta}$ at that same event (*local* equation). The solution of these differential equations is effected, on a computer or in any initial-value-type analysis, by integrating forward in time from event to event to event (*local* integration). The nongravitational laws of physics are obtained by invoking the equivalence principle in a *local* Lorentz frame at each individual event in spacetime. To build up an understanding of the global structure of spacetime, one performs *local* computations near each event, and then patches the local results together to form a global picture. Why this great reliance on local analyses? Because the laws of gravitation physics take on particularly simple forms when stated locally.

That gravitation physics is also subject to powerful and simple *global* laws, physicists did not realize until the mid 1960's. But since 1963, studies of black holes and of singularities have revealed global laws and global properties of spacetime that rival in their simplicity and elegance even the (local) equivalence principle. An example is the second law of black-hole dynamics: "In an isolated system, the sum of the surface areas of all black holes can never decrease." As a result, there has developed a powerful body of knowledge and techniques for analyzing *directly* the global properties of spacetime.

To give a full treatment of global techniques would require many chapters. Fortunately, a full treatment is being published, almost simultaneously with this book, by Hawking and Ellis (1973). Because Hawking and Ellis are much better qualified to write on this subject than are we (Misner, Thorne, and Wheeler), we have chosen to not write a "competitive" treatment. Instead, we give in this chapter only a brief taste of the subject—enough of a taste to make the reader acquainted with the types of techniques involved and several of the most important results, but

not enough to give him a working knowledge of the subject. The topics we have chosen to treat are those that contact most closely the rest of this book: properties of "infinity" in an asymptotically flat spacetime (§34.2); causality and horizons (§§34.3 and 34.4); a proof of the second law of black-hole dynamics (§34.5); and theorems about the evolution of singularities in spacetime (§34.6). For greater detail on global techniques, one can consult not only the book of Hawking and Ellis (1973), but also review articles by Geroch (1971), by Penrose (1968a, 1972), and by Hawking (1973), the thesis of Godfrey (1970b), and the more specialized papers cited in the body of this chapter.

References on global techniques

§34.2. "INFINITY" IN ASYMPTOTICALLY FLAT SPACETIMES

When performing calculations in asymptotically flat spacetime, one often must examine the asymptotic forms of fields (e.g., the metric, or the curvature tensor, or the electromagnetic field) "at infinity." For example, the mass and angular momentum of an isolated system are determined by the asymptotic form of the metric (Chapter 19). It is rarely sufficient to examine asymptotic forms near "spatial infinity." For example, if one wishes to learn how much mass was carried away by gravitational and electromagnetic waves during a supernova explosion, one must examine the asymptotic form of the metric not just at "spatial infinity," but at "future null infinity" (see Figure 34.1).

Motivation for studying properties of spacetime near infinity

Penrose (1964, 1965a) has developed a powerful body of mathematical technique for studying asymptotic properties of spacetime near "infinity." The key to his technique is a "conformal transformation" of spacetime, which brings "infinity" in to a finite radius and thereby converts asymptotic calculations into calculations at "finite points." Penrose's technique also provides rigorous definitions of several types of "infinity" that one encounters in asymptotically flat spacetimes.

The details of Penrose's technique are not of importance to the rest of this chapter. However, this chapter will refer frequently to the various types of "infinity" defined by Penrose. In heuristic terms, they are as follows (see Figure 34.2a).

Specific regions of infinity: I^+, I^0, I^-, \mathscr{I}^+, \mathscr{I}^-

$I^+ \equiv$ "future timelike infinity":
> the region $t \longrightarrow +\infty$ at finite radius r
> (region toward which timelike lines extend).

$I^- \equiv$ "past timelike infinity":
> the region $t \longrightarrow -\infty$ at finite radius r
> (region from which timelike lines come).

$I^0 \equiv$ "spacelike infinity":
> the region $r \longrightarrow \infty$ at finite time t
> (region toward which spacelike slices extend).

$\mathscr{I}^+ \equiv$ "future null infinity":
> the region $t + r \longrightarrow \infty$ at finite $t - r$
> (region toward which outgoing null lines extend).

$\mathscr{I}^- \equiv$ "past null infinity":
> the region $t - r \longrightarrow -\infty$ at finite $t + r$
> (region from which ingoing null lines come).

Note: \mathscr{I} is a script I, and is sometimes given the name "scri."

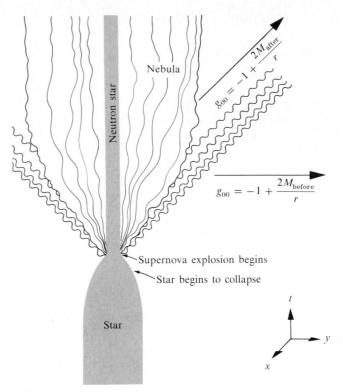

Figure 34.1.

Measurement of the mass-energy radiated as gravitational and electromagnetic waves by a supernova explosion in asymptotically flat spacetime. The mass-energy radiated equals the mass (M_{before}) of the presupernova star, minus the mass (M_{after}) of the neutron star and nebula after the explosion:

$$M_{\text{radiated}} = M_{\text{before}} - M_{\text{after}}.$$

To measure M_{before}, one can examine the asymptotic form (in suitable coordinates) of g_{00} at spatial infinity

$$g_{00} = -1 + \frac{2M_{\text{before}}}{r} + O\left(\frac{1}{r^2}\right) \qquad \text{as } r \longrightarrow \infty, \, t = \text{constant}.$$

But to measure M_{after} in the same way, one must wait, at any fixed r, until the radiation has flowed entirely past that point:

$$g_{00} = -1 + \frac{2M_{\text{after}}}{r} + O\left(\frac{1}{r^2}\right)$$

$$\text{as } r \longrightarrow \infty \text{ with } t - r = \begin{pmatrix} \text{constant value sufficiently large} \\ \text{to be inside the burst of waves} \end{pmatrix}.$$

Put differently, to measure M_{after} one must examine the asymptotic form of g_{00} *not* at "spatial infinity," but rather at "future null infinity."

Coordinate diagrams for exhibiting structure of infinity

It is often useful, in visualizing the asymptotic structure of spacetime, to introduce coordinates that attribute finite coordinate values to infinity. For example, in flat spacetime one can transform from the usual spherical coordinates t, r, θ, ϕ, with

$$ds^2 = -dt^2 + dr^2 + r^2(d\theta^2 + \sin^2\theta \, d\phi^2), \tag{34.1}$$

to new spherical coordinates ψ, ξ, θ, ϕ, with

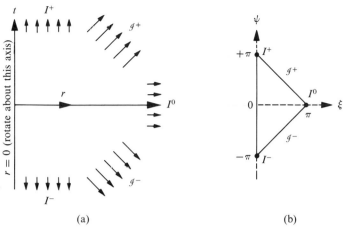

Figure 34.2.
Flat, "Minkowski" spacetime as depicted (a) in the usual spherical coordinates t, r, θ, ϕ of a global Lorentz frame, and (b) in the spherical coordinates of equations (34.2). The five regions of infinity—I^+, I^-, I^0, \mathcal{I}^+, \mathcal{I}^-—are shown in each coordinate diagram. In both coordinate systems, radial null lines make angles of 45° with the vertical axis, and nonradial null lines make angles less than 45° [see equations (34.1) and (34.2c)]. See exercise 34.1 for further detail.

$$t + r = \tan\frac{1}{2}(\psi + \xi), \tag{34.2a}$$

$$t - r = \tan\frac{1}{2}(\psi - \xi), \tag{34.2b}$$

$$ds^2 = \frac{-d\psi^2 + d\xi^2}{4\cos^2\dfrac{1}{2}(\psi + \xi)\cos^2\dfrac{1}{2}(\psi - \xi)} + r^2(d\theta^2 + \sin^2\theta\, d\phi^2). \tag{34.2c}$$

The resulting ψ, ξ coordinate diagram (Figure 34.2b) depicts I^+, I^-, I^0, \mathcal{I}^+, \mathcal{I}^- more clearly than does the usual t, r, coordinate diagram.

As another example, replace the Kruskal-Szekeres coordinates v, u, θ, ϕ for Schwarzschild spacetime by new coordinates ψ, ξ, θ, ϕ:

$$v + u = \tan\frac{1}{2}(\psi + \xi), \tag{34.3a}$$

$$v - u = \tan\frac{1}{2}(\psi - \xi), \tag{34.3b}$$

$$(1 - r/2M)e^{r/2M} = v^2 - u^2 = \tan\frac{1}{2}(\psi + \xi)\tan\frac{1}{2}(\psi - \xi), \tag{34.3c}$$

$$ds^2 = \frac{32M^3}{r}\frac{e^{-r/2M}(-d\psi^2 + d\xi^2)}{4\cos^2\dfrac{1}{2}(\psi + \xi)\cos^2\dfrac{1}{2}(\psi - \xi)} + r^2(d\theta^2 + \sin^2\theta\, d\phi^2). \tag{34.3d}$$

The resulting coordinate diagram (Figure 34.3) depicts clearly the causal connections between the horizons, the singularities, and the various regions of infinity.

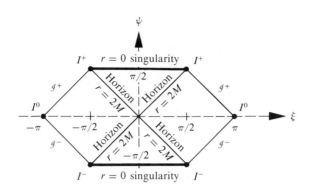

Figure 34.3.
Schwarzschild spacetime as depicted in the ψ, ξ, θ, ϕ coordinates of equations (34.3). This coordinate diagram should be compared with the Kruskal-Szekeres coordinate diagram (Figure 31.3). In both diagrams, radial null geodesics are 45° lines. Each of the asymptotically flat regions (one on each side of the "wormhole" of Figure 31.5a) has its own set of infinities I^+, I^-, I^0, \mathcal{I}^+, and \mathcal{I}^-. See exercise 34.2 for justification of this diagram.

EXERCISES

Exercise 34.1. FLAT SPACETIME IN ψ, ξ, θ, ϕ COORDINATES

(a) Derive equation (34.2c) from (34.1) and (34.2a,b).

(b) Show that the regions I^+, I^-, I^0, \mathcal{I}^+, and \mathcal{I}^- of flat spacetime are located at

$$I^+: \psi = \pi, \xi = 0,$$

$$I^-: \psi = -\pi, \xi = 0,$$

$$I^0: \psi = 0, \xi = \pi, \tag{34.4}$$

$$\mathcal{I}^+: \psi + \xi = \pi, \ -\pi < \psi - \xi < \pi,$$

$$\mathcal{I}^-: \psi - \xi = -\pi, \ -\pi < \psi + \xi < \pi.$$

[see equations (34.2)]. These are the regions depicted in Figure 34.2.

(c) Show that in flat spacetime, in a ψ, ξ coordinate diagram (Figure 34.2), radial null lines make angles of 45° with the vertical axis, and nonradial null lines make angles of less than 45°.

Exercise 34.2. SCHWARZSCHILD SPACETIME IN ψ, ξ, θ, ϕ COORDINATES

(a) Derive equations (34.3c,d) from (34.3a,b) and the Kruskal-Szekeres equations (31.14).

(b) Use equations (34.3) to justify the precise form of the coordinate diagram in Figure 34.3.

Exercise 34.3. REISSNER-NORDSTROM SPACETIME

(a) Show that there exists a coordinate system in which the Reissner-Nordstrøm geometry with $0 < |Q| < M$ (exercises 31.8 and 32.1) has the form

$$ds^2 = F^2(-d\psi^2 + d\xi^2) + r^2(d\theta^2 + \sin^2\theta \, d\phi^2),$$

$$F = F(\psi, \xi), \qquad r = r(\psi, \xi), \tag{34.5}$$

and in which the horizons and infinities are as shown in Figure 34.4. [*Note:* This is a very difficult exercise unless one has in hand the solution to exercise 31.8(d). For solution, see Carter (1966b).]

(b) Use Figure 34.4 to deduce that the Reissner-Nordstrøm geometry describes a "wormhole" or bridge, connecting two asymptotically flat spacetimes, which: (i) expands to a state of maximum circumference; (ii) recontracts toward a state of minimum circumference, and in the process disconnects its outer regions from the two I^0's (spatial infinity) and reconnects them to a pair of $r = 0$ singularities; (iii) bounces; (iv) reexpands, and in the process

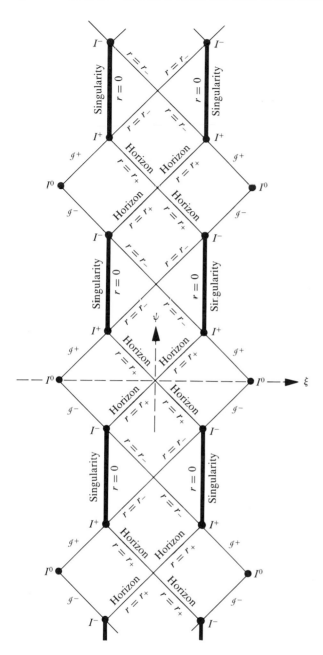

Figure 34.4.

Reissner-Nordstrøm spacetime

$$ds^2 = -\left(1 - \frac{2M}{r} + \frac{Q^2}{r^2}\right)dt^2$$

$$+ \frac{dr^2}{1 - 2M/r + Q^2/r^2} + r^2(d\theta^2 + \sin^2\theta \, d\phi^2)$$

with $0 < |Q| < M$, as depicted in a new (ψ, ξ, θ, ϕ) coordinate system where the line element has the form

$$ds^2 = F^2(-d\psi^2 + d\xi^2) + r^2(d\theta^2 + \sin^2\theta \, d\phi^2).$$

(see exercise 34.3.) This coordinate diagram reveals the global structure of the geometry, including its singularities at $r = 0$, its horizons at

$$r = r_+ = M + \sqrt{M^2 - Q^2}$$

(which limit communication with \mathscr{I}^+ and \mathscr{I}^-), the null surfaces at

$$r = r_- = M - \sqrt{M^2 - Q^2}$$

(which limit communication with the singularities), and the various asymptotically flat infinities, I^+, I^-, I^0, \mathscr{I}^+, and \mathscr{I}^-. From this diagram one can read off the "causal structure" of the geometry—i.e., the abilities of various regions to communicate with each other. For detailed discussion of the geometry, see Graves and Brill (1960) and Carter (1966b). For discussions of collapsing charged stars, for which this geometry is the external gravitational field, see Novikov (1966a,b), de la Cruz and Israel (1967), and Bardeen (1968).

disconnects its "outer regions" from the two singularities and reconnects them to a pair of I^0's in two *new* asymptotically flat universes; (v) slows its expansion to a halt; (vi) recontracts toward a state of minimum circumference, and in the process disconnects its outer regions from the two I^0's and reconnects them to a new pair of $r = 0$ singularities; etc. *ad infinitum.*

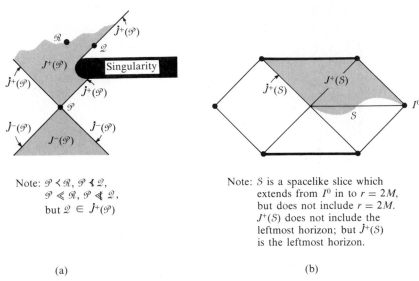

Note: $\mathscr{P} < \mathscr{R}$, $\mathscr{P} \nleq \mathscr{Q}$,
 $\mathscr{P} \ll \mathscr{R}$, $\mathscr{P} \not\ll \mathscr{Q}$,
 but $\mathscr{Q} \in \dot{J}^+(\mathscr{P})$

Note: S is a spacelike slice which
 extends from I^0 in to $r = 2M$,
 but does not include $r = 2M$.
 $J^+(S)$ does not include the
 leftmost horizon; but $\dot{J}^+(S)$
 is the leftmost horizon.

(a) (b)

Figure 34.5.
Spacetime diagrams illustrating various causal relationships. Diagram (a) is a hypothetical spacetime; diagram (b) is Schwarzschild spacetime (see Figure 34.3). In both diagrams, null lines have slopes of 45°.

§34.3. CAUSALITY AND HORIZONS

Turn now to global* techniques for analyzing black holes. The goals of the discussion will be (1) to define the concept of horizon (this section), (2) to deduce global geometric properties of horizons (next section), and (3) to prove the second law of black-hole dynamics (following section). The entire discussion will be confined to spacetime manifolds that (1) contain at least one asymptotically flat region ("the external universe"; region "outside black holes"), and (2) are "time-oriented." By *"time-oriented"* one means that at each event in spacetime a distinct choice has been made as to which light cone is the future cone and which is the past, and moreover that this choice is continuous from event to event throughout spacetime.

The discussion begins with definitions of a variety of causal relationships between events and regions of spacetime (see Figure 34.5).

Restriction of discussion to asymptotically flat, time-oriented manifolds

Definitions of several causality concepts

Definition: $\mathscr{P} \ll \mathscr{Q}$ or equivalently $\mathscr{Q} \gg \mathscr{P}$ ("the event \mathscr{P} *precedes* the event \mathscr{Q}"; "the event \mathscr{Q} *follows* the event \mathscr{P}") means that there exists at least one smooth, future-directed timelike curve that extends from \mathscr{P} to \mathscr{Q}.

Definition: A *causal curve* $\mathcal{C}(\lambda)$ is any smooth curve that is nowhere spacelike—i.e., that is timelike or null or "zero" [$\mathcal{C}(\lambda) =$ some fixed \mathscr{P}, for all λ] or some admixture thereof.

Definition: $\mathscr{P} < \mathscr{Q}$ or equivalently $\mathscr{Q} > \mathscr{P}$ ("the event \mathscr{P} *causally precedes* the event

*Global, but not fully global; the "universe" of §§34.3–34.5 is asymptotically flat; no account is taken here of possible closure or collapse of the universe or of their consequences.

\mathscr{Q}"; the event \mathscr{Q} *causally follows* the event \mathscr{P}") means that there exists at least one future-directed causal curve that extends from \mathscr{P} to \mathscr{Q}.

Definition: $J^-(\mathscr{P})$, called *the causal past of* \mathscr{P}, is the set of all events that causally precede \mathscr{P}—i.e., $J^-(\mathscr{P}) = \{\mathscr{Q} \,|\, \mathscr{Q} < \mathscr{P}\}$.

Definition: $J^+(\mathscr{P})$, called *the causal future of* \mathscr{P}, is the set of all events that causally follow \mathscr{P}—i.e., $J^+(\mathscr{P}) = \{\mathscr{Q} \,|\, \mathscr{Q} > \mathscr{P}\}$.

Definition: If S is a region of spacetime—e.g., a segment of a spacelike hypersurface—then $J^-(S)$ (*the causal past of* S) is the set of all events that causally precede at least one event in S—i.e.,

$$J^-(S) = \{\mathscr{Q} \,|\, \mathscr{Q} < \mathscr{P} \text{ for at least one } \mathscr{P} \in S\}.$$

Definition: Similarly, $J^+(S)$ (*the causal future of* S) is the set of all events that causally follow at least one event in S—i.e.,

$$J^+(S) = \{\mathscr{Q} \,|\, \mathscr{Q} > \mathscr{P} \text{ for at least one } \mathscr{P} \in S\}.$$

Definition: $\dot{J}^+(S)$ is the *boundary of* $J^+(S)$,
$\dot{J}^-(S)$ is the *boundary of* $J^-(S)$.

Definition: One defines *the future of* \mathscr{P}, $I^+(\mathscr{P})$; *the past of* \mathscr{P}, $I^-(\mathscr{P})$; *the future of* S, $I^+(S)$; *the past of* S, $I^-(S)$; *the boundary of the future of* S, $\dot{I}^+(S)$; *and the boundary of the past of* S, $\dot{I}^-(S)$ in precisely the same manner as above, except that the phrase "causally precede" is replaced by "precede," and "causally follow" is replaced by "follow." Example:

$$I^+(S) = \{\mathscr{Q} \,|\, \mathscr{Q} \gg \mathscr{P} \text{ for at least one } \mathscr{P} \in S\}.$$

Not all these definitions are needed in the following discussion; but the literature on global methods uses these concepts so extensively that the reader should be familiar with them.

Focus attention on a specific spacetime manifold, and in that manifold select out a specific asymptotically flat region. [In the external field of a star, there is but one asymptotically flat region. In the vacuum Schwarzschild geometry without source (Figure 34.3), there are two. In the Reissner-Nordstrøm geometry without source (Figure 34.4), there are infinitely many different asymptotically flat regions.] The selected asymptotically flat region ("external universe") has one future timelike infinity I^+, one past timelike infinity I^-, one spacelike infinity I^0, one future null infinity \mathscr{I}^+, and one past null infinity \mathscr{I}^-. It may also possess black holes, which form by stellar collapse, and which collide, coalesce, accrete matter, and generally wreak havoc in their immediate vicinities. The surfaces of all black holes ("future horizons") separate the external universe, which can send signals out to \mathscr{I}^+, from the black-hole interiors, which cannot. One thus has the definition:

Definition: The *totality* (*or* "*union*") *of all future horizons* (surfaces of all black holes) is the region $\dot{J}^-(\mathscr{I}^+)$—i.e., it is the boundary of the domain $J^-(\mathscr{I}^+)$ that can send future-directed causal curves out to future null infinity.

Definition: surfaces of black holes; future horizons—$\dot{J}^-(\mathscr{I}^+)$

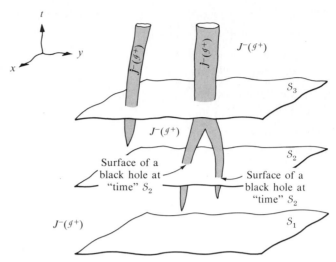

Figure 34.6.

Black holes in an asymptotically flat spacetime (schematic spacetime diagram). $J^-(\mathcal{I}^+)$ is the "external universe"—i.e., the region which can send causal curves to future null infinity. $\dot{J}^-(\mathcal{I}^+)$, the greyish region, is the boundary of the "external universe"—i.e., it is the union of all future horizons. At the "time" of spacelike slice S_1, there are no black holes in the universe. Between S_1 and S_2 two stars collapse to form black holes. The two closed 2-surfaces, in which S_2 intersects $\dot{J}^-(\mathcal{I}^+)$ are the horizons of those black holes at "time" S_2. Between S_2 and S_3, the two original black holes collide and coalesce, while a third black hole is being formed by stellar collapse.

[Similarly, one can define the totality of all past horizons to be $\dot{J}^+(\mathcal{I}^-)$. But past horizons are of little interest for astrophysics. Whereas gravitational collapse produces future horizons in a quite natural manner, past horizons must be primordial in origin—i.e., they must be postulated as initial structure in the origin of the universe [Novikov (1964), Ne'eman (1965)]. There is no good reason to believe that the universe began with or should have begun with such strange initial structure.]

Any given spacelike slice S through spacetime will intersect $\dot{J}^-(\mathcal{I}^+)$ in a number of disjoint, closed, two-dimensional surfaces. Each such 2-surface is the horizon of a single black hole at the "moment of time" S. See Figure 34.6.

§34.4. GLOBAL STRUCTURE OF HORIZONS

The union of all future horizons, $\dot{J}^-(\mathcal{I}^+)$, has an especially simple global geometric structure, as follows.

Penrose's theorem on the structure of $J^-(\mathcal{I}^+)$ (future horizons)

THEOREM [Penrose (1968a)]: $\dot{J}^-(\mathcal{I}^+)$ *is generated by null geodesics that have no future end points.* Stated more precisely (see Figure 34.7):

(1) Definition: The "*generators*" of $\dot{J}^-(\mathcal{I}^+)$ are null geodesics which (at least for some finite lapse of affine parameter) lie in $\dot{J}^-(\mathcal{I}^+)$.

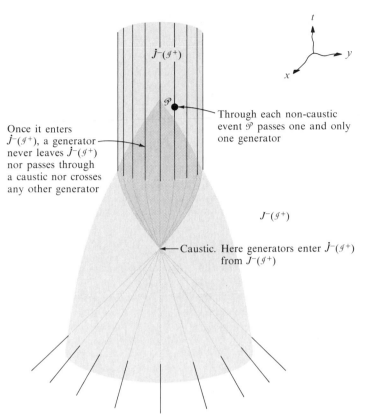

Figure 34.7.
The future horizon $\dot{J}^-(\mathcal{I}^+)$ produced by the spherical gravitational collapse of a star. This horizon illustrates the global geometric structure of $\dot{J}^-(\mathcal{I}^+)$ as spelled out in Penrose's theorem (§34.4 of text). In this special case, there is only one caustic. In general there will be many.

(2) Theorem: When followed into the past, a generator may (but does not have to!) leave $\dot{J}^-(\mathcal{I}^+)$. Each event at which a generator leaves is called a "*caustic*" of $\dot{J}^-(\mathcal{I}^+)$. When a generator leaves, it goes into $J^-(\mathcal{I}^+)$.

(3) Once a generator, being followed into the future, enters $\dot{J}^-(\mathcal{I}^+)$ from $J^-(\mathcal{I}^+)$ at a caustic, it can never thereafter leave $\dot{J}^-(\mathcal{I}^+)$, nor can it ever intersect another generator. [Generators can intersect only at the "caustics," where they enter $\dot{J}^-(\mathcal{I}^+)$.]

(4) Through each noncaustic event of $\dot{J}^-(\mathcal{I}^+)$ there passes one and (aside from normalization of affine parameter) only one generator.

This theorem is proved in Box 34.1.

For a Schwarzschild black hole, the generators of $\dot{J}^-(\mathcal{I}^+)$ are the world lines of radially outgoing photons at the gravitational radius [$r = 2M$, θ and ϕ constant,

$u = +v$; dotted line on horizon in Figure 32.1(c)]. For a Kerr-Newman black hole, the generators of $j^-(\mathcal{I}^+)$ are the "barber-pole-twist" null geodesics of Box 33.2(F)—i.e., they are those members of the outgoing principal null congruence that lie on the horizon, $r = r_+$ (§33.6; exercise 33.9). But the theorem is more general. It refers to any black hole—dynamic or static; accreting matter, or coalescing with a neighboring black hole, or existing alone in isolation—in any time-oriented, asymptotically flat spacetime.

(continued on page 931)

Box 34.1 HORIZONS ARE GENERATED BY NONTERMINATING NULL GEODESICS (Penrose 1968a)

A. *Lemma:* If (1) $\mathcal{C}_1(\lambda)$ is a causal, future-directed curve from event \mathcal{P} to event \mathcal{Q}, (2) $\mathcal{C}_2(\lambda)$ is a causal, future-directed curve from event \mathcal{Q} to event \mathcal{R}, and (3) $\mathcal{P} \not\ll \mathcal{R}$ (\mathcal{P} is not in the past of \mathcal{R}), then \mathcal{C}_1 and \mathcal{C}_2 are null geodesics, and their tangent vectors coincide (aside from normalization) at event \mathcal{Q}.

Proof:*

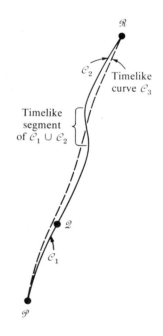

1. Suppose that \mathcal{C}_1 were not a null geodesic, or \mathcal{C}_2 were not a null geodesic, or both. Then somewhere along $\mathcal{C}_1 \cup \mathcal{C}_2$ there would be a timelike segment, or a nongeodesic null segment, or both.
 a. If $\mathcal{C}_1 \cup \mathcal{C}_2$ contained a timelike segment, then a slight deformation† of $\mathcal{C}_1 \cup \mathcal{C}_2$ would produce a smooth curve \mathcal{C}_3 from \mathcal{P} to \mathcal{R} which is everywhere timelike‡—contradicting the assumption $\mathcal{P} \not\ll \mathcal{R}$.

*The proof utilizes some elementary concepts of point-set topology; see, e.g., Wallace (1963) or Kelley (1955).

†One can always deform any curve in any spacetime manifold by a small amount in any direction one wishes, without running into singularities or into other boundaries of the manifold. This is possible because a manifold by definition is *open*. In physical terms, spacetime is open because each event in spacetime must possess a local Lorentz neighborhood which also lies in spacetime.

‡One can convince oneself of this, and of similar claims made later in the proof, by arguments using local Lorentz frames. In the literature on global geometry, claims such as this are rarely substantiated—though each author is always convinced that he could do so if forced to by a skeptic. Unfortunately, to substantiate such claims with rigorous arguments would lengthen and complicate the discussion enormously and would tend to obscure the simplicity of the underlying ideas.

b. If $\mathcal{C}_1 \cup \mathcal{C}_2$ contained a nongeodesic null segment \mathcal{C} reaching from event \mathcal{Q} to event \mathcal{B}, then, when compared to neighboring curves between \mathcal{Q} and \mathcal{B}, \mathcal{C} would not have stationary length. This means that some curves from \mathcal{Q} to \mathcal{B} would have larger squared length—i.e., would be spacelike—while others would have smaller squared length—i.e., would be timelike. Thus, a slight deformation of \mathcal{C} would produce a timelike segment from \mathcal{Q} to \mathcal{B}. Then a further deformation of $\mathcal{C}_1 \cup \mathcal{C}_2$, as described in (a) above, would produce a smooth timelike curve from \mathcal{P} to \mathcal{R}, contradicting $\mathcal{P} \not\ll \mathcal{R}$.

Thus, the supposition is wrong; i.e., both \mathcal{C}_1 and \mathcal{C}_2 must be null geodesics.

2. Suppose that the tangent vectors of \mathcal{C}_1 and \mathcal{C}_2 did not agree at their join point, \mathcal{Q}. Then one could "round off the corner" at \mathcal{Q}, producing a timelike segment there. One could then further deform $\mathcal{C}_1 \cup \mathcal{C}_2$ as in (1a) above, to produce a smooth timelike curve from \mathcal{P} to \mathcal{R}—contradicting $\mathcal{P} \not\ll \mathcal{R}$. Thus, the supposition is wrong; i.e., the tangent vectors must agree at \mathcal{Q}. Q.E.D.

Box 34.1 (continued)

B. *Lemma:* If $\mathcal{A} \in \dot{J}^-(\mathcal{I}^+)$ and $\mathcal{B} \in \dot{J}^-(\mathcal{I}^+)$, then $\mathcal{A} \not\ll \mathcal{B}$.

Proof: Assume $\mathcal{A} \ll \mathcal{B}$.

1. Then there exists a timelike curve from \mathcal{A} to \mathcal{B}.

2. A slight deformation of that curve which keeps it still timelike will make it link an arbitrary event \mathcal{Q} in some sufficiently small neighborhood $\mathcal{N}[\mathcal{A}]$ to an arbitrary event \mathcal{R} in some small $\mathcal{N}[\mathcal{B}]$.

3. Pick \mathcal{R} to lie in $J^-(\mathcal{I}^+)$. Then join the timelike curve from \mathcal{Q} to \mathcal{R} onto a causal curve from \mathcal{R} to \mathcal{I}^+. The resulting curve, when smoothed in a neighborhood of the join, becomes a causal curve from any arbitrary $\mathcal{Q} \in \mathcal{N}[\mathcal{A}]$ to \mathcal{I}^+.

4. The existence of such curves implies that $\mathcal{N}[\mathcal{A}] \subset J^-(\mathcal{I}^+)$, and hence that $\mathcal{A} \notin \dot{J}^-(\mathcal{I}^+)$—in contradiction to the original hypotheses.

Conclusion: $\mathcal{A} \not\ll \mathcal{B}$. Q.E.D.

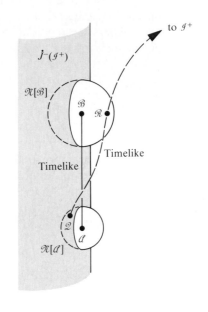

C. *Lemma:* Let $\mathcal{C}(\lambda)$ be a causal curve that intersects $\dot{J}^-(\mathcal{I}^+)$ at some event \mathcal{B}. Then when followed into the past from \mathcal{B}, $\mathcal{C}(\lambda)$ forever lies in $\dot{J}^-(\mathcal{I}^+) \cup J^-(\mathcal{I}^+)$.

Proof:

1. Pick an arbitrary event \mathcal{A} on $\mathcal{C}(\lambda)$, with $\mathcal{A} < \mathcal{B}$.

2. Construct an arbitrarily small neighborhood $\mathcal{N}[\mathcal{A}]$.

3. A small deformation of \mathcal{C}, between \mathcal{A} and \mathcal{B}, produces a timelike curve \mathcal{D} from some event $\mathcal{P} \in \mathcal{N}[\mathcal{A}]$ to \mathcal{B}.

4. Since $\mathcal{B} \in \dot{J}^-(\mathcal{I}^+)$, a slight deformation of \mathcal{D}, keeping it still timelike, produces a curve \mathcal{E} from \mathcal{P} to some event $\mathcal{Q} \in J^-(\mathcal{I}^+)$. \mathcal{E} can then be prolonged, remaining causal, until it reaches \mathcal{I}^+. The result is a causal curve from \mathcal{P} to \mathcal{I}^+. Hence, $\mathcal{P} \in J^-(\mathcal{I}^+)$.

5. But \mathcal{P} was in an arbitrarily small neighborhood $\mathcal{N}[\mathcal{A}]$. Hence, \mathcal{A} must also be in $J^-(\mathcal{I}^+)$ or else in its boundary, $\dot{J}^-(\mathcal{I}^+)$. Q.E.D.

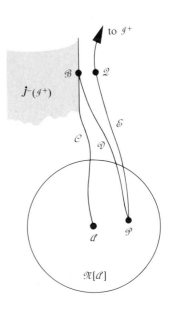

D. *Theorem* [Penrose (1968a)]: $\dot{J}^-(\mathcal{I}^+)$ is generated by null geodesics which have no future endpoints. [See text of §34.4 for more detailed statement of theorem.]

Proof:

1. Pick an arbitrary event \mathcal{P} in $\dot{J}^-(\mathcal{I}^+)$. Prove as follows that through \mathcal{P} there passes a future-directed null geodesic which lies in $\dot{J}^-(\mathcal{I}^+)$:

 a. Construct an arbitrary neighborhood $\mathcal{N}[\mathcal{P}]$. [If $\dot{J}^-(\mathcal{I}^+)$ happens somewhere to encounter a singularity of spacetime, then $\mathcal{N}[\mathcal{P}]$ must be chosen small enough to keep the singularity outside it.]

 b. In $\mathcal{N}[\mathcal{P}] \cap J^-(\mathcal{I}^+)$, construct a sequence of events $\{\mathcal{P}_i\}$ which converges to the event \mathcal{P}.

 c. For each i, construct a causal curve \mathcal{C}_i extending from \mathcal{P}_i to \mathcal{I}^+.

 d. Let \mathcal{Q}_i be the intersection of \mathcal{C}_i with $\dot{\mathcal{N}}[\mathcal{P}]$, the boundary of $\mathcal{N}[\mathcal{P}]$. Since $\dot{\mathcal{N}}[\mathcal{P}]$ is a compact set, the sequence \mathcal{Q}_i must have a limit point, \mathcal{Q}.

 e. Because there exist causal curves from events \mathcal{P}_i arbitrarily near \mathcal{P} to events \mathcal{Q}_i arbitrarily near \mathcal{Q}, there must be a causal curve from \mathcal{P} to \mathcal{Q}. Call that curve \mathcal{C}.

 f. Since \mathcal{Q} is a limit point of a sequence of events in $J^-(\mathcal{I}^+)$, \mathcal{Q} either lies in $J^-(\mathcal{I}^+)$, or else lies on its boundary $\dot{J}^-(\mathcal{I}^+)$, or both. Suppose $\mathcal{Q} \notin \dot{J}^-(\mathcal{I}^+)$.

 i. Then some small neighborhood $\mathcal{N}[\mathcal{Q}]$ is contained entirely in $J^-(\mathcal{I}^+)$.

 ii. Construct a causal curve from \mathcal{P} to \mathcal{I}^+ by going from \mathcal{P} to \mathcal{Q} along the causal curve \mathcal{C}, then from \mathcal{Q} along a timelike curve to some event $\mathcal{R} \in \mathcal{N}[\mathcal{Q}]$, and then from \mathcal{R} to \mathcal{I}^+ along a causal curve—and by smoothing at the join points \mathcal{Q} and \mathcal{R}.

 iii. Since this curve from \mathcal{P} to \mathcal{I}^+ has a timelike segment, it can be deformed smoothly, while being kept causal, so that it reaches any desired event \mathcal{S} in some small neighborhood $\mathcal{N}'[\mathcal{P}]$. But this means that $\mathcal{N}'[\mathcal{P}] \subset J^-(\mathcal{I}^+)$, hence that $\mathcal{P} \notin \dot{J}^-(\mathcal{I}^+)$—which contradicts the original definition of \mathcal{P}.

 Conclusion: $\mathcal{Q} \in \dot{J}^-(\mathcal{I}^+)$.

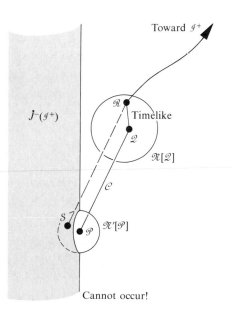

Cannot occur!

Box 34.1 (*continued*)

g. By Lemma B, since $\mathscr{P} \in \dot{J}^-(\mathscr{I}^+)$ and $\mathscr{Q} \in \dot{J}^-(\mathscr{I}^+)$, then $\mathscr{P} \ll \mathscr{Q}$. But \mathscr{C} is a future-directed causal curve from \mathscr{P} to \mathscr{Q}. Consequently, by Lemma A, \mathscr{C} is a null geodesic.

h. Since the curve \mathscr{C} intersects $\dot{J}^-(\mathscr{I}^+)$ at \mathscr{Q}, between \mathscr{P} and \mathscr{Q} it must everywhere lie in $\dot{J}^-(\mathscr{I}^+) \cup J^-(\mathscr{I}^+)$ [Lemma C]. Apply the reasoning of (f) above, with \mathscr{Q} replaced by an arbitrary point on \mathscr{C} between \mathscr{P} and \mathscr{Q}. Thereby conclude that, everywhere between \mathscr{P} and \mathscr{Q}, \mathscr{C} lies in $\dot{J}^-(\mathscr{I}^+)$.

Summary: Through every event $\mathscr{P} \in \dot{J}^-(\mathscr{I}^+)$ there passes a null geodesic \mathscr{C} which, when followed into the future from \mathscr{P}, lies in $\dot{J}^-(\mathscr{I}^+)$. This null geodesic is called a "generator" of $\dot{J}^-(\mathscr{I}^+)$.

2. Follow the generator \mathscr{C} from \mathscr{P} to \mathscr{Q} and then onward still further. Can it *ever* leave $\dot{J}^-(\mathscr{I}^+)$? No! For suppose it did leave, at some event $\mathscr{P}' \in \dot{J}^-(\mathscr{I}^+)$.

 a. Repeat the entire construction of step 1, with \mathscr{P}' replacing \mathscr{P}, to conclude that there is a null geodesic $\mathscr{C}' \subset \dot{J}^-(\mathscr{I}^+)$ extending into the causal future from \mathscr{P}' to some event \mathscr{Q}'.

 b. By Lemma B, since $\mathscr{P} \in \dot{J}^-(\mathscr{I}^+)$ and $\mathscr{Q}' \in \dot{J}^-(\mathscr{I}^+)$, $\mathscr{P} \ll \mathscr{Q}'$.

 c. Then by Lemma A the null geodesic \mathscr{C} from \mathscr{P} to \mathscr{P}' and the null geodesic \mathscr{C}' from \mathscr{P}' to \mathscr{Q}' have tangents that coincide at \mathscr{P}' (aside from normalization). Thus, with a renormalization of affine parameter, \mathscr{C}' becomes the prolongation of \mathscr{C}—which means that \mathscr{C} does not leave $\dot{J}^-(\mathscr{I}^+)$ at \mathscr{P}'.

\mathscr{C} cannot do this

Conclusion: Once a generator, being followed into the future, enters $\dot{J}^-(\mathscr{I}^+)$, it can never thereafter leave $\dot{J}^-(\mathscr{I}^+)$.

3. Figure 34.7 provides an example of how a null geodesic, being following into the future, can enter $\dot{J}^-(\mathscr{I}^+)$ and become a generator. Lemma C guarantees that, *when a null geodesic enters $\dot{J}^-(\mathscr{I}^+)$, it enters from $J^-(\mathscr{I}^+)$.*

4. As indicated by the example of Fig. 34.7, at a "caustic" [entry point of generators onto $\dot{J}^-(\mathscr{I}^+)$] generators can cross each other. Follow a generator \mathcal{C} to the causal future from its entry point onto $\dot{J}^-(\mathscr{I}^+)$. Can it ever again cross another generator? No. For suppose that at an event \mathscr{P} the generator \mathcal{C} were to cross another generator \mathscr{D}.

 a. To the causal future of \mathscr{P}, both generators always lie in $\dot{J}^-(\mathscr{I}^+)$. Thus, events \mathscr{A} and \mathscr{B} of the picture are in $\dot{J}^-(\mathscr{I}^+)$.
 b. Since \mathscr{P} is to the causal future of the caustic where \mathcal{C} enters $\dot{J}^-(\mathscr{I}^+)$, there exists an event $\mathscr{Q} \in \dot{J}^-(\mathscr{I}^+) \cap \mathcal{C}$ to the causal past of \mathscr{P}.
 c. Since $\mathscr{Q} \in \dot{J}^-(\mathscr{I}^+)$ and $\mathscr{B} \in \dot{J}^-(\mathscr{I}^+)$, $\mathscr{Q} \not\ll \mathscr{B}$ [Lemma B].
 d. Lemma A, applied to the curves \mathcal{C} from \mathscr{Q} to \mathscr{P}, and \mathscr{D} from \mathscr{P} to \mathscr{B}, then guarantees that the tangent vectors $\boldsymbol{u}_\mathcal{C}$ and $\boldsymbol{u}_\mathscr{D}$ coincide at \mathscr{P} (aside from normalization), and that therefore (aside from normalization) the geodesics \mathcal{C} and \mathscr{D} are identical. This contradicts the supposition that \mathcal{C} and \mathscr{D} are different generators which cross at \mathscr{P}.

Conclusion: Once a generator has entered $\dot{J}^-(\mathscr{I}^+)$, it can never thereafter cross any other generator.

§34.5. PROOF OF SECOND LAW OF BLACK-HOLE DYNAMICS
[Hawking (1971b, 1972a, 1973)]

All the tools are now in hand for a proof of the second law of black-hole dynamics.

Consider the union of all future horizons, $\dot{J}^-(\mathscr{I}^+)$, in an asymptotically flat spacetime, as depicted in Figure 34.8. Divide up the null-geodesic generators of $\dot{J}^-(\mathscr{I}^+)$ into a large number of infinitesimal bundles, and give each bundle an identifying number, K. As one moves from past toward future along $\dot{J}^-(\mathscr{I}^+)$, one occasionally sees new bundles of generators created in "caustics" of the 3-surface $\dot{J}^-(\mathscr{I}^+)$. The caustic sources of new generators are created by such processes as the infall of matter through the horizon (example: bundle #42 in Figure 34.8), and the collision and coalescence of two black holes (example: bundle #29). But each bundle, once created, can never be destroyed (no termination of null generators as one moves from past toward future).

Proof of second law of black-hole dynamics:

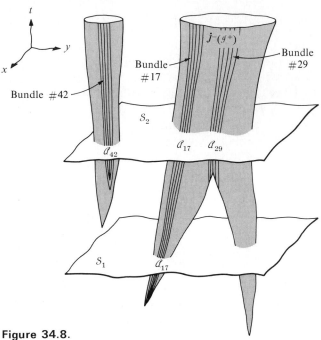

Figure 34.8.
Schematic spacetime diagram used in proving the second law of black-hole dynamics. See text for details of the proof, and see Figure 34.6 for physical interpretation of the diagram.

Focus attention on a specific bundle of generators—bundle $\#K$. At a specific event \mathscr{P} along that bundle, let various observers, moving with various velocities, measure its (two-dimensional) cross-sectional area $\mathscr{a}_K(\mathscr{P})$. As shown in Figure 22.1, exercise 22.13, and exercise 22.14: (1) the cross-sectional area $\mathscr{a}_K(\mathscr{P})$ is independent of the velocity of the observer who measures it—i.e., $\mathscr{a}_K(\mathscr{P})$ depends only on location \mathscr{P} along the bundle; and (2) \mathscr{a}_K changes from event to event along the bundle in a manner governed by the "focusing theorem"

$$\frac{d^2\mathscr{a}_K^{1/2}}{d\lambda_K^2} \leq 0 \qquad \begin{array}{l}\text{if the energy density } T_{\hat{0}\hat{0}}, \text{ as measured} \\ \text{by all observers along the bundle, is} \\ \text{nonnegative.}\end{array} \qquad (34.6)$$

Here λ_K is affine parameter along the bundle. *Assume*—in accord with all physical experience and the best assessments of modern physics—*that energy density $T_{\hat{0}\hat{0}}$ can never be negative. (This assumption underlies the second law of black-hole dynamics. If it were ever found to be invalid, then one would have to abandon the second law.)*

Suppose that $d\mathscr{a}_K^{1/2}/d\lambda_K$ were negative at some event \mathscr{P} along the bundle. Then, according to the focusing theorem, $d\mathscr{a}_K^{1/2}/d\lambda_K$ would always remain at least as

Proof assumes nonnegative energy density

negative as its value at \mathscr{P}—and, hence, after a lapse of affine parameter given by

$$\Delta\lambda_K \leq \left(\frac{\mathscr{A}_K^{1/2}}{-d\mathscr{A}_K^{1/2}/d\lambda_K} \right)_{\text{at }\mathscr{P}}, \qquad (34.7)$$

$\mathscr{A}_K^{1/2}$ would go to zero. At the point where $\mathscr{A}_K^{1/2}$ reaches zero, adjacent null geodesics in the bundle cross each other, giving rise to events in $\dot{J}^-(\mathscr{I}^+)$ through which pass more than one null geodesic generator. But this violates Penrose's theorem on the global structure of horizons (§34.4).

Thus either the supposition of negative $d\mathscr{A}_K^{1/2}/d\lambda_K$ is wrong; or else $d\mathscr{A}_K^{1/2}/d\lambda_K$ goes negative, but then, before the generators get a chance to cross [before the finite lapse (34.7) of affine parameter], the generators hit a singularity and cease to exist. *To prove the second law of black-hole dynamics, one must assume that no singularity is hit by the horizon, and thereby conclude that $d\mathscr{A}_K^{1/2}/d\lambda_K$ never goes negative. Hawking (1971b, 1972a) makes an alternative assumption which implies $d\mathscr{A}_K^{1/2}/d\lambda_K \geq 0$: Hawking assumes that spacetime is "future asymptotically predictable." In essence this means that spacetime possesses no "naked singularities"—i.e., no singularities visible from \mathscr{I}^+.* (Naked singularities could influence the evolution of the external universe; and, therefore, unless one knew the laws of physics governing singularities— which one does not—they would prevent one from predicting the future in the external universe.)

Under either assumption (no naked singularities; or horizon never hits a singularity), one concludes that

$$d\mathscr{A}_K^{1/2}/d\lambda_K \text{ is nonnegative everywhere along bundle } K. \qquad (34.8)$$

This result says that the cross-sectional area \mathscr{A}_K of each bundle can never decrease as one moves toward the future along $\dot{J}^-(\mathscr{I}^+)$. Since new bundles can be created, but old ones can never be destroyed as one moves toward the future, *the total cross-sectional area of $\dot{J}^-(\mathscr{I}^+)$ cannot decrease toward the future.* Equivalently, (see Figure 34.8), if \mathcal{S}_1 and \mathcal{S}_2 are spacelike hypersurfaces with \mathcal{S}_2 everywhere to the future of \mathcal{S}_1, then the cross-sectional area of $\dot{J}^-(\mathscr{I}^+)$ at its intersection with \mathcal{S}_2, $\mathscr{A}(\mathcal{S}_2)$, cannot be less than the cross-sectional area at \mathcal{S}_1, $\mathscr{A}(\mathcal{S}_1)$. This is the second law of black-hole dynamics, reformulated in more precise language than that of Chapter 33, and finally proved.

Proof assumes that horizon never hits a singularity (no naked singularities)

Precise formulation of second law

Exercise 34.4. A BLACK HOLE CAN NEVER BIFURCATE [Hawking (1972a)]

EXERCISE

Make plausible the theorem that no matter how hard one "zaps" a black hole, and no matter what one "zaps" it with, one can never make it bifurcate into two black holes. [*Hint:* By drawing pictures, make it plausible that, at any bifurcation point, some null geodesic generators of $\dot{J}^-(\mathscr{I}^+)$ must leave $\dot{J}^-(\mathscr{I}^+)$ as one follows them into the *future*—in violation of Penrose's theorem (§34.4). Assume that the surface of each hole is topologically a 2-sphere. *Note:* The same argument, time-reversed, shows that if two black holes coalesce, generators enter $\dot{J}^-(\mathscr{I}^+)$ from $J^-(\mathscr{I}^+)$ at the coalescence point; and the surface area of the horizon increases.]

§34.6. THEOREMS ON SINGULARITIES, AND THE "ISSUE OF THE FINAL STATE"

Overview of theorems on singularities

Just as global techniques are powerful tools in the analysis of horizons, so they also are powerful in the analysis of spacetime singularities. In fact, it was the proof of Penrose's (1965b) pioneering theorem on singularities that gave birth to global techniques for studying spacetime.

For a detailed introduction to the global analysis of singularities, one can read the book of Hawking and Ellis (1973). Now that the reader has had a taste of global techniques, attention here will focus on a qualitative description of results:

How does gravitational collapse terminate? Is the singularity at the end point of spherical collapse typical, or can asymmetries remove it? That singularities are very general phenomena, and cannot be wished away, has been known since 1965, thanks to theorems on singularities proved by Penrose, Hawking, and Geroch. [For a full list of references, see Hawking and Penrose (1969) or Hawking and Ellis (1973).]

Singularity defined

Before examining the theorems on singularities, one must make precise the concept of a *singularity*. This is not easy, as Geroch (1968) has emphasized in a long treatise on the wide variety of pathologies that can occur in spacetime manifolds. However, after vigorous efforts by many people, Schmidt (1970) finally produced a definition that appears to be satisfactory. Put in heuristic terms, Schmidt's highly technical definition goes something like this. In a spacetime manifold, consider all spacelike geodesics (paths of "tachyons"), all null geodesics (paths of photons), all timelike geodesics (paths of freely falling observers), and all timelike curves with bounded acceleration (paths along which observers are able, in principle, to move). Suppose that one of these curves terminates after the lapse of finite proper length (or finite affine parameter in the null-geodesic case). Suppose, further, that it is impossible to extend the spacetime manifold beyond that termination point—e.g., because of infinite curvature there. Then that termination point, together with all adjacent termination points, is called a "singularity." (What could be more singular than the cessation of existence for the poor tachyon, photon, or observer who moves along the terminated curve?)

Trapped surface defined

Another concept needed in the singularity theorems is that of a *trapped surface*. This concept, devised by Penrose (1965b), is motivated by a close examination of the two-dimensional, spherical surfaces $(r, t) = $ const. inside the horizon of the Schwarzschild geometry. These surfaces signal the nearness of a singularity ($r = 0$) by this property: light rays emitted from one of these surfaces in the perpendicular outward direction (i.e., outgoing, orthogonal, null geodesics) converge toward each other as they propagate; and inward light rays perpendicular to the 2-surface also converge. Penrose gives the name "trapped surface" to any closed 2-surface, spherical or not, that has this property. In Schwarzschild spacetime, the convergence of light rays, both outgoing and ingoing, can be attributed to the "intense pull of gravity," which sucks the photons into the singularity. That this might also be true in asymmetric spacetimes is suggested by the Hawking-Penrose (1969) theorem [the most powerful of a wide class; see Hawking and Penrose (1969) for references to others; and see Boxes 34.2 and 34.3 for introductions to Hawking and Penrose]:

The Hawking-Penrose
theorem on singularities

A spacetime M necessarily contains incomplete, inextendable timelike or null geodesics (and is thus singular in the Schmidt sense) if, in addition to Einstein's equations, the following four conditions hold: (1) *M* contains no closed timelike curves (reasonable causality condition); (2) at each event in *M* and for each unit timelike vector **u**, the stress-energy tensor satisfies

$$\left(T_{\alpha\beta} - \frac{1}{2} g_{\alpha\beta} T \right) u^\alpha u^\beta \geq 0$$

(reasonable energy condition); (3) the manifold is "general" (i.e., not too highly symmetric) in the sense that every timelike or null geodesic with unit tangent **u** passes through at least one event where the curvature is not lined up with it in a specific way:

$$u_{[\alpha} R_{\beta]\gamma\delta[\epsilon} u_{\rho]} u^\gamma u^\delta \neq 0 \text{ at some point on the geodesic.}$$

(4) the manifold contains a trapped surface.

All these conditions, except the trapped surface, seem eminently reasonable for any physically realistic spacetime! Note, especially, that the energy condition can be violated only if, as measured by some local observer in his proper frame, the total energy density *E* is negative or the principal pressures (eigenvalues of stress tensor) P_i are so negative that

$$\sum_i P_i < -E.$$

Relevance of the
Hawking-Penrose theorem for
gravitational collapse

The relevance of the Hawking-Penrose theorem for collapse follows from the general expectation that, in the real universe, trapped surfaces will always exist just below all future horizons, $\dot{J}^-(\mathscr{I}^+)$. (Exceptions, such as the Kerr metric with $a = M$, are probably a "set of measure zero.") Since horizons and accompanying trapped surfaces are necessarily produced by slightly nonspherical collapse (Box 32.2), and since they probably also result from moderately deformed collapse (§32.7), such collapse presumably produces singularities—or a violation of causality, which is also a rather singular occurrence!

If the singularities are really such a general feature of collapse, then the exact nature of the singularity is of life-and-death importance to anyone who falls through a horizon! Here one is on very shaky ground. Although the main results and conjectures described up to now in this section will probably survive all future research, opinions about the nature of the singularities are likely to change several times more before the whole story is in. Hence, it is safe only to describe the possibilities, not to attempt to judge them.

The nature of the singularity
at the endpoint of realistic
collapse: 4 possibilities

Possibility 1

The singularity at the endpoint of a realistic collapse is a region of infinite tidal gravitational forces (infinite curvature), which crushes the collapsing matter to infinite density. Examples: the very special, homogeneous crushing of the Oppenheimer-Snyder (1939) spherical collapse (§32.4); also the very special inhomogeneous but spherical crushing described by Podurets (1966); also the special inhomogeneous,

(continued on page 940)

Box 34.2 ROGER PENROSE: Born August 8, 1931, Colchester, Essex, England

Roger Penrose started out as an algebraic geometer. However, while at Cambridge from 1952–55 and again from 1957–60, his interest in general relativity was aroused by Hermann Bondi and Dennis Sciama. Because of his pure mathematical background, his approach to the subject was different from those which had been adopted hitherto. He was particularly interested in the global light-cone structure of spacetime and in the equations of zero rest-mass fields, both of which are preserved under conformal transformations. He exploited this conformal invariance to give an elegant and powerful treatment of gravitational radiation in terms of a null surface \mathscr{I}^+ at infinity. More recently this interest has led him to develop the theory of twistors, which are the spinors corresponding to the conformal group of Minkowski space. These offer a new and very promising approach to the quantization of spacetime.

His interest in conformal geometry also led him to study the properties of the causality relationships between points of spacetime. These in turn led him to the theorems on the occurrence of singularities in spacetime, which are probably the most important predictions of general relativity, since they seem to imply that spacetime has a beginning or an end.

"If spacetime is considered from the point of view of its conformal structure only, points at infinity can be treated on the same basis as finite points"

[PENROSE, IN INFELD (1964)]

"The argument will be to show that the existence of a trapped surface implies— irrespective of symmetry—that singularities necessarily develop"

[PENROSE (1965b)]

"While the quantum effects of gravitation are normally thought to be significant only when curvatures approach 10^{33} cm^{-1}, all our local physics is based on the Poincare group being a good approximation of a local symmetry group at dimensions greater than 10^{-13} cm. Thus, if curvatures ever even approach 10^{13} cm^{-1}, there can be little doubt but that extraordinary local effects are likely to take place"

[HAWKING AND PENROSE (1969)]

"We are thus presented with what is perhaps the most fundamental unanswered question of general-relativistic collapse theory, namely: does there exist a "cosmic censor" who forbids the appearance of naked singularities, clothing each one in an absolute event horizon?"

[PENROSE (1969)]

"Under normal circumstances, general relativity can, for practical purposes, remain remarkably apart—almost aloof—from the rest of physics. At a space-time singularity, the very reverse must surely be the case!"

"I do not believe that a real understanding of the nature of elementary particles can ever be achieved without a simultaneous deeper understanding of the nature of spacetime itself. But if we are concerned with a level of phenomena for which such an understanding is not necessary—and this will cover almost all of present-day physics—then the smooth manifold picture presents an (unreasonably!) excellent framework for the discussion of phenomena."

"The most important single lesson of relativity theory is, perhaps, that space and time are not concepts that can be considered independently of one another but must be combined together to give a four-dimensional picture of phenomena: the description in terms of spacetime"

[PENROSE (1968a)]

"If a formalism enables one to treat myriads of non-existent types of universe, then (effectively) it contains 'arbitrary parameters,' only special values of which will correspond to the world as it actually is. In the ordinary approach to spacetime as a pseudo-Riemannian differentiable manifold, the dimension of the manifold and the signature of the metric are two such arbitrary parameters."

"As we localize the position of a particle, it jumps essentially along the null cone. Other particles are produced, which leap backward and forward essentially along null directions, without apparent regard for continuity, heeding only the positions of the null cones themselves and "topology" only in the respect in which this term is applied to the structure of graphs"

[PENROSE (1966)]

"My own view is that ultimately physical laws should find their most natural expression in terms of essentially combinatorial principles, that is to say, in terms of finite processes such as counting or other basically simple manipulative procedures. Thus, in accordance with such a view, should emerge some form of discrete or combinatorial spacetime"

[PENROSE, IN KLAUDER (1972)]

"Complex numbers are . . . a very important constituent of the structure of physical laws. The twistor theory carries this further in suggesting that complex numbers may also be very basically involved in defining the nature of spacetime itself."

[PENROSE AND MACCALLUM (1973)]

"It is thus very tempting to believe that a link between spacetime curvature and quantum processes may be supplied by the use of twistors. Then, roughly speaking, it is the continual slight 'shifting' of the interpretations of the quantum (twistor) operators which results in the curvature of spacetime"

[PENROSE (1968b)]

Box 34.3 STEPHEN W. HAWKING: Born January 8, 1942, Oxford, England

As a research student of Dennis Sciama's in Cambridge, Stephen Hawking's early interest in relativity theory centered mainly on the question of spacetime singularities. With Ellis, he showed that a large class of homogeneous cosmological models must be singular. Then, encouraged by work of Penrose on the singularities arising in gravitational collapse, he developed new techniques which, in a series of papers in the Royal Society of London during 1966–67, established the important result that any plausible general-relativistic cosmology must be singular.

The major portion of his later research has been concerned with black holes. He devised a series of arguments of great ingenuity which, together with the work of Israel and Carter, established to all intents and purposes the result that (vacuum) black holes in general relativity are described by Kerr metrics, that topologies other than spherical cannot occur, and that a certain limit on the energy emitted when two black holes congeal into one must be satisfied.

Some of this work has had substantial pure mathematical interest (e.g., singularity theorems), some of it is concerned with astrophysics (e.g., work with Taylor on helium production in the big bang), some with observations (work with Gibbons on the possibility of black holes in binary star

systems) and even experimental developments (with Gibbons on gravitational-wave detectors). In such scope is exhibited not only a considerable insight, depth, and versatility, but also the gift of an extraordinary determination to overcome severe physical handicaps, to seek out and comprehend the truth.

"The observed isotropy of the microwave background indicates that the universe is rotating very little if at all. . . . This could possibly be regarded as an experimental verification of Mach's Principle"

[HAWKING (1969)]

"Undoubtedly, the most important results are the theorems . . . on the occurrence of singularities. These seem to imply either that the general theory of relativity breaks down or that there could be particles whose histories did not exist before (or after) a certain time. The author's own opinion is that the theory probably does break down but only when quantum gravitational effects become important."

"Although we have omitted the singular points from the definition of spacetime, we can still recognize the 'holes' left where they have been cut out by the existence of incomplete geodesics."

"A good physical theory should not only correctly describe the currently experimental knowledge, but should also predict new results which can be tested by experiment, the further the predictions from the original experiments, the greater the credit to the theory if they are found to be correct. Thus observations of whether or not singularities actually occurred, would provide a powerful test of the general theory of relativity in strong fields"
[HAWKING (1966a)]

"The construction of gravitational radiation detectors may open up a whole new field of 'gravitational astronomy' which could be as fruitful as radio astronomy has been in the last two decades. . . . Black hole collisions . . . would be much more effective in converting rest-mass energy into radiation than nuclear reactions, which can release only about 1 per cent of the rest-mass energy. In addition, black holes formed by collisions of smaller black holes can undergo further collisions, releasing more energy, whereas matter that has been fully processed by nuclear reactions cannot yield any more energy by the same means. . . . we are witnessing something really cataclysmic at the centre of our galaxy"
[HAWKING (1972b)]

"One might suggest that prior to the present expansion there was a collapsing phase. In this, local inhomogeneities grew large and isolated singularities occurred. Most of the matter avoided the singularities and reexpanded to give the present observed universe."

"It seems that we should draw a surface around regions where the radius of curvature is less than, say, 10^{-16} cm. On our side of this surface, a manifold picture of spacetime would be appropriate, but we have no idea what structure spacetime would have on the other side"
[HAWKING AND ELLIS (1968)]

"Presumably it would be necessary to consider quantum effects in very strong fields. However, these would not become important until the radius of curvature became of the order of 10^{-14} cm, which for practical purposes is pretty singular."

"The view has been expressed that singularities are so objectionable that if the Einstein equations were to predict their occurrence, this would be a compulsive reason for modifying them. However, the real test of a physical theory is not whether its predicted results are aesthetically attractive but whether they agree with observation. So far there are no observations which would show that singularities do not occur"
[HAWKING (1966b)]

"It is shown that a stationary black hole must have a topologically spherical boundary and must be axisymmetric if it is rotating. These results, together with those of Israel and Carter, go most of the way toward establishing the conjecture that any stationary black hole is a Kerr solution"
[HAWKING (1972a)]

"The fact that we have observed the universe to be isotropic is only a consequence of our existence."
[COLLINS AND HAWKING (1973)]

"Kasner-like" crushing of Lifschitz and Khalatnikov (1963a,b); also, most importantly, the very general "mixmaster" crushing (Chapter 30), discovered in the homogeneous case by Misner (1969b) and by Belinsky and Khalatnikov (1969a), and analyzed in the inhomogeneous case by Belinsky and Khalatnikov (1969b, 1970) and by Khalatnikov and Lifschitz (1970). The mixmaster singularities—and only they among all explicitly known singularities—appear to be generic in this sense: if one perturbs slightly but arbitrarily the initial conditions of a spacetime that evolves a mixmaster singularity, then the resultant perturbed spacetime will also evolve a mixmaster singularity. Because of this, the prevalent opinion today (1973) is that realistic collapse probably produces, inside the horizon, a mixmaster singularity. But that opinion might change tomorrow.

Possibility 2

The singularity is a region of spacetime in which timelike or null geodesics terminate, not because of infinite tidal gravitational forces or infinite crushing, but because of other, more subtle pathologies. Example: "Taub-NUT space" [see Misner and Taub (1968)]. For other examples created specially to exhibit various pathologies, see Geroch (1968).

Possibility 3

The singularity may be sufficiently limited in "size" and influence that all or most of the collapsing matter successfully avoids it. The matter cannot then explode back outward through the horizon that it went down; the horizon is a one-way membrane and will not let anything back out. Instead, the matter may reach a stage of maximum but finite contraction, and then reexplode into some other region of spacetime (multiply connected spacetime topology; "wormhole"). Analytical solutions for collapsing, charged spheres do reexplode in this manner [Novikov (1966); de la Cruz and Israel (1967); Bardeen (1968); see Figure 34.4]. Such a process requires that the "exploding" end of the wormhole be built into the initial conditions of the universe, with mass and angular momentum (as measured by Keplerian orbits and frame dragging) precisely equal to those that go down the black-hole end. This seems physically implausible. So does the "explosion."

Other Possibilities

Various combinations of the above.

Will quantization of spacetime save the universe from singularities?

If, as one suspects today, the singularities are of a very physical, infinite-curvature type, then one must face up to John Wheeler's (1964a) "issue of the final state" in its most raw and disturbing form. Wheeler, when faced with the issue, argues that infinite-curvature singularities signal a breakdown in classical general relativity—a breakdown forced by the onset of quantum gravitational phenomena (see Chapter 44). Whether quantization of gravity will actually save spacetime from such singularities one cannot know until the "fiery marriage of general relativity with quantum physics has been consummated" [Wheeler (1964a); see also Misner (1969c), and the last section of Box 30.1].

GRAVITATIONAL WAVES

*Wherein the reader voyages on stormy seas of curvature ripples,
searching for the ripple-generating storm gods, and battles
through an electromagnetic and thermal fog that allows only
uncertain visibility upon those seas.*

CHAPTER **35**

PROPAGATION OF
GRAVITATIONAL WAVES

Born: "I should like to put to Herr Einstein a question, namely, how quickly the action of gravitation is propagated in your theory. That it happens with the speed of light does not elucidate it to me. There must be a very complicated connection between these ideas."
Einstein: "It is extremely simple to write down the equations for the case when the perturbations that one introduces in the field are infinitely small. Then the g's differ only infinitesimally from those that would be present without the perturbation. The perturbations then propagate with the same velocity as light."
Born: "But for great perturbations things are surely very complicated?"
Einstein: "Yes, it is a mathematically complicated problem. It is especially difficult to find exact solutions of the equations, as the equations are nonlinear."

Excerpts from discussion after Einstein's Fall 1913 lecture in Vienna on "The present position of the problem of gravitation," already two years before he had the final field equations [EINSTEIN, 1913a]

§35.1. VIEWPOINTS

Study one idealization after another. Build a catalog of idealizations, of their properties, of techniques for analyzing them. This is the only way to come to grips with so complicated a subject as general relativity!

Spherical symmetry is the idealization that has dominated most of the last 12 chapters. Together with the idealization of matter as a perfect fluid, and of the universe as homogeneous, it has yielded insight into stars, into cosmology, into gravitational collapse.

Turn attention now to an idealization of an entirely different type, one independent of any symmetry considerations at all: the idealization of a "gravitational wave."

Just as one identifies as "water waves" small ripples rolling across the ocean, so one gives the name "gravitational waves" to small ripples rolling across spacetime.

Gravitational waves compared to water waves on ocean:

We are deeply indebted to Mr. James M. Nester, who found and corrected many errors in the equations of this chapter and of a dozen others throughout the book.

(1) approximate nature of a
 wave

Ripples of what? Ripples in the shape of the ocean's surface; ripples in the shape (i.e., curvature) of spacetime. Both types of waves are idealizations. One cannot, with infinite accuracy, delineate at any moment which drops of water are in the waves and which are in the underlying ocean: Similarly, one cannot delineate precisely which parts of the spacetime curvature are in the ripples and which are in the cosmological background. But one can almost do so; otherwise one would not speak of "waves"!

(2) local viewpoint vs.
 large-scale viewpoint

Look at the ocean from a rowboat. Waves dominate the seascape. Changes in angle and level of the surface occur every 30 feet or less. These changes propagate. They obey a simple wave equation

$$\left(\frac{1}{g^2} \frac{\partial^4}{\partial t^4} + \frac{\partial^2}{\partial y^2} + \frac{\partial^2}{\partial x^2} \right) (\text{height of surface}) = 0.$$

Now get more sophisticated. Notice from a spaceship the large-scale curvature of the ocean's surface—curvature because the Earth is round, curvature because the sun and the moon pull on the water. As waves propagate long distances, this curvature bends their fronts and changes slightly their simple wave equation. Also important over large distance are nonlinear interactions between waves, interaction with the wind, Coriolis forces due to the Earth's rotation, etc.

Spacetime is similar. Propagating through the universe, according to Einstein's theory, must be a complex pattern of small-scale ripples in the spacetime curvature, ripples produced by binary stars, by supernovae, by gravitational collapse, by explosions in galactic nuclei. Locally ("rowboat viewpoint") one can ignore the interaction of these ripples with the large-scale curvature of spacetime and their nonlinear interaction with each other. One can pretend the waves propagate in flat spacetime; and one can write down a simple wave equation for them. But globally one cannot. The large-scale curvature due to quiescent stars and galaxies will produce redshifts and will deform wave fronts; and the energy carried by the waves themselves will help to produce the large-scale curvature. This chapter treats both viewpoints, the local (§§35.2–6), and the global (§§35.7–15).

§35.2. REVIEW OF "LINEARIZED THEORY" IN VACUUM

Linearized theory of
gravitational waves:

Idealize, for awhile, the gravitational waves of our universe as propagating through flat, empty spacetime (local viewpoint). Then they can be analyzed using the "linearized theory of gravity," which was introduced in Chapter 18.

Linearized theory, one recalls, is a weak-field approximation to general relativity. The equations of linearized theory are written and solved as though spacetime were flat (special-relativity viewpoint); but the connection to experiment is made through the curved-space formalism of general relativity.

More specifically, linearized theory describes gravity by a symmetric, second-rank tensor field $\bar{h}_{\mu\nu}$. In the standard ("Lorentz," or Hilbert) gauge, this field satisfies the "gauge" or "subsidiary" conditions (coordinate conditions)

(1) Lorentz gauge condition

$$\bar{h}^{\mu\alpha}{}_{,\alpha} = 0. \tag{35.1a}$$

(Here, and throughout linearized theory, indices of $\overline{h}_{\mu\nu}$ are raised and lowered with the Minkowski metric $\eta_{\alpha\beta}$.) In this gauge the *propagation equations* for vacuum gravitational fields are the familiar wave equations

$$\square\overline{h}_{\mu\nu} \equiv \overline{h}_{\mu\nu,\alpha}{}^{\alpha} = 0. \tag{35.1b}$$

(2) propagation equation

Spacetime is really curved in linearized theory, although equations (35.1) are written and solved as though it were not. The global inertial frames of equations (35.1) are only *almost* inertial. In them the metric components are actually

$$g_{\mu\nu} = \eta_{\mu\nu} + h_{\mu\nu} + O([h_{\mu\nu}]^2); \tag{35.2a}*$$

(3) metric

and the "metric perturbation" $h_{\mu\nu}$ is related to the "gravitational field" $\overline{h}_{\mu\nu}$ by

$$h_{\mu\nu} = \overline{h}_{\mu\nu} - \frac{1}{2}\overline{h}\eta_{\mu\nu}, \qquad \overline{h}_{\mu\nu} = h_{\mu\nu} - \frac{1}{2}h\eta_{\mu\nu},$$

$$h \equiv h_{\alpha}{}^{\alpha} = -\overline{h} = -\overline{h}_{\alpha}{}^{\alpha}. \tag{35.2b}$$

The metric (35.2a) governs the motion of test particles, the propagation of light, etc., in the usual general-relativistic manner.

Recall the origin of the equations (35.1) that govern $\overline{h}_{\mu\nu}$. The subsidiary conditions $\overline{h}_{\mu}{}^{\alpha}{}_{,\alpha} = 0$ were imposed by specializing the coordinate system; and the Einstein field equations in vacuum then reduced to $\square\overline{h}_{\mu\nu} = 0$.

Actually, as was shown in Box 18.2, the coordinates of linearized theory are not fully fixed by the conditions $\overline{h}_{\mu}{}^{\alpha}{}_{,\alpha} = 0$. There remains an ambiguity embodied in further "gauge changes" (infinitesimal coordinate transformations), ξ_{μ}, which satisfy a restrictive condition

(4) residual gauge freedom

$$\xi_{\mu,\alpha}{}^{\alpha} = 0 \tag{35.3a}$$

in order to preserve conditions (35.1a). Then

$$x^{\mu}{}_{\text{new}} = x^{\mu}{}_{\text{old}} + \xi^{\mu} \tag{35.3b}$$

is the coordinate transformation and

$$\overline{h}_{\mu\nu\,\text{new}} = \overline{h}_{\mu\nu\,\text{old}} - \xi_{\mu,\nu} - \xi_{\nu,\mu} + \eta_{\mu\nu}\xi^{\alpha}{}_{,\alpha} \tag{35.3c}$$

is the gauge change. All this was derived and discussed in Chapter 18.

§35.3. PLANE-WAVE SOLUTIONS IN LINEARIZED THEORY

The simplest of all solutions to the linearized equations $\overline{h}_{\mu\nu,\alpha}{}^{\alpha} = \overline{h}_{\mu}{}^{\alpha}{}_{,\alpha} = 0$ is the monochromatic, plane-wave solution,

Monochromatic, plane wave

$$\overline{h}_{\mu\nu} = \Re[A_{\mu\nu}\exp(ik_{\alpha}x^{\alpha})]. \tag{35.4a}$$

*A more nearly rigorous treatment defines $h_{\mu\nu} \equiv g_{\mu\nu} - \eta_{\mu\nu}$, and puts the small corrections $O([h_{\mu\nu}]^2)$ into the field equations:

$$\overline{h}_{\mu}{}^{\alpha}{}_{,\alpha} = O([h_{\mu\nu}]^2{}_{,\alpha}), \qquad \overline{h}_{\mu\nu,\alpha}{}^{\alpha} = O([h_{\mu\nu}]^2{}_{,\alpha\beta}).$$

Here $\Re[.\,.\,.\,.\,.]$ means that one must take the real part of the expression in brackets; while $A_{\mu\nu}$ (*amplitude*) and k_{μ} (*wave vector*) are constants satisfying

$$k_{\alpha}k^{\alpha} = 0 \qquad (\boldsymbol{k} \text{ a null vector}), \tag{35.4b}$$

$$A_{\mu\alpha}k^{\alpha} = 0 \qquad (\boldsymbol{A} \text{ orthogonal to } \boldsymbol{k}) \tag{35.4c}$$

[consequences of $\bar{h}_{\mu\nu,\alpha}{}^{\alpha} = 0$ and $\bar{h}_{\mu,\alpha}{}^{\alpha} = 0$, respectively; see (35.10) below for the true physics associated with this wave, the curvature tensor]. Clearly, this solution describes a wave with frequency

$$\omega \equiv k^{0} = (k_{x}{}^{2} + k_{y}{}^{2} + k_{z}{}^{2})^{1/2}, \tag{35.5}$$

which propagates with the speed of light in the direction $(1/k^{0})(k_{x}, k_{y}, k_{z})$.

At first sight the amplitude $A_{\mu\nu}$ of this plane wave appears to have six independent components (ten, less the four orthogonality constraints $A_{\mu\alpha}k^{\alpha} = 0$). But this cannot be right! As Track-2 readers have learned in Chapter 21, the gravitational field in general relativity has two dynamic degrees of freedom, not six. Where has the analysis gone astray?

One went astray by forgetting the arbitrariness tied up in the gauge freedom (35.3). The plane-wave vector

$$\xi^{\mu} \equiv -iC^{\mu} \exp{(ik_{\alpha}x^{\alpha})}, \tag{35.6}$$

with four arbitrary constants C^{μ}, generates a gauge transformation that can change arbitrarily four of the six independent components of $A_{\mu\nu}$. One gets rid of this arbitrariness by choosing a specific gauge.

§35.4. THE TRANSVERSE TRACELESS (TT) GAUGE

Select a 4-velocity \boldsymbol{u}—not at just one event, but the same \boldsymbol{u} throughout all of spacetime (special-relativistic viewpoint!). By a specific gauge transformation (exercise 35.1), impose the conditions

$$A_{\mu\nu}u^{\nu} = 0. \tag{35.7a}$$

These are only three constraints on $A_{\mu\nu}$, not four, because one of them—$k^{\mu}(A_{\mu\nu}u^{\nu}) = 0$—is already satisfied (35.4c). As a fourth constraint, use a gauge transformation (exercise 35.1) to set

$$A^{\mu}{}_{\mu} = 0. \tag{35.7b}$$

One now has eight constraints in all, $A_{\mu\alpha}u^{\alpha} = A_{\mu\alpha}k^{\alpha} = A_{\alpha}{}^{\alpha} = 0$, on the ten components of the amplitude; and the coordinate system (gauge) is now fixed rigidly. Thus, the two remaining free components of $A_{\mu\nu}$ represent the two degrees of freedom (two polarizations) in the plane gravitational wave.

Plane wave has two degrees of freedom in amplitude (two polarizations)

Transverse-traceless (TT) gauge:

(1) for plane wave

It is useful to restate the eight constraints $A_{\mu\alpha}u^{\alpha} = A_{\mu\alpha}k^{\alpha} = A^{\mu}{}_{\mu} = 0$ in a Lorentz frame where $u^0 = 1$, $u^j = 0$, and in a form where k^{α} does not appear explicitly:

$$h_{\mu 0} = 0, \qquad \text{i.e., only the spatial components}$$
$$h_{jk} \text{ are nonzero;} \tag{35.8a}$$

$$h_{kj,j} = 0, \qquad \text{i.e., the spatial components are}$$
$$\text{divergence-free;} \tag{35.8b}$$

$$h_{kk} = 0, \qquad \text{i.e., the spatial components are}$$
$$\text{trace-free.} \tag{35.8c}$$

(Here and henceforth repeated spatial indices are to be summed, even if both are down; e.g., $h_{kk} \equiv \Sigma^3_{k=1}h_{kk}$.) Notice that, since $h = h_{\mu}{}^{\mu} = h_{kk} = 0$, *there is no distinction between $h_{\mu\nu}$ and $\bar{h}_{\mu\nu}$ in this gauge.*

Turn attention now away from plane waves to arbitrary gravitational waves in linearized theory. Any electromagnetic wave can be resolved into a superposition of plane waves, and so can any gravitational wave. For each plane wave in the superposition, introduce the special gauge (35.8). Note that the gauge conditions are all linear in $h_{\mu\nu}$. Therefore the arbitrary wave will also satisfy conditions (35.8). Thus arises the theorem: *Pick a specific global Lorentz frame of linearized theory (i.e., pick a specific 4-velocity \boldsymbol{u}). In that frame (where $u^{\alpha} = \delta^{\alpha}{}_0$), examine a specific gravitational wave of arbitrary form. One can always find a gauge in which $h_{\mu\nu}$ satisfies the constraints (35.8). Moreover, in this gauge only the h_{jk} are nonzero.* Therefore, one need only impose the six wave equations

(2) for any wave

$$\Box h_{jk} = h_{jk,\alpha}{}^{,\alpha} = 0. \tag{35.9}$$

Any symmetric tensor satisfying constraints (35.8) [but not necessarily the wave equations (35.9)] is called a *transverse-traceless (TT) tensor*—transverse because it is purely spatial ($h_{0\mu} = 0$) and, if thought of as a wave, is transverse to its own direction of propagation ($h_{ij,j} = h_{ij}k_j = 0$); traceless because $h_{kk} = 0$. The most general purely spatial tensor S_{ij} can be decomposed [see Arnowitt, Deser, and Misner (1962) or Box 35.1] into a part S^{TT}_{ij}, which is "transverse and traceless"; a part $S^{T}_{ij} = \frac{1}{2}(\delta_{ij}f_{,kk} - f_{,ij})$, which is "transverse" ($S^T_{ij,j} = 0$) but is determined entirely by one function f giving the trace of S ($S^T_{kk} = \nabla^2 f$); and a part $S^L_{ij} = S^L_{i,j} + S^L_{j,i}$, which is "longitudinal" and is determined by the vector field S^L_i. In linearized theory h^L_{ij} is a purely gauge part of $h_{\mu\nu}$, whereas h^T_{ij} and h^{TT}_{ij} are gauge-invariant parts of $h_{\mu\nu}$. The special gauge in which $h_{\mu\nu}$ reduces to its transverse-traceless part is called the *TT or transverse-traceless gauge*. The conditions (35.8) defining this gauge can be summarized as

Decomposition of spatial tensors

$$h_{\mu\nu} = h^{TT}_{\mu\nu}. \tag{35.8d}$$

As exercise 35.2 illustrates, only pure waves (and not more general solutions of the linearized field equations with source, $\Box h_{\mu\nu} = -16\pi T_{\mu\nu}$) can be reduced to *TT* gauge.

Curvature tensor in TT gauge In the *TT* gauge, the time-space components

$$R_{j0k0} = R_{0j0k} = -R_{j00k} = -R_{0jk0}$$

of the Riemann curvature tensor have an especially simple form [see equation (18.9) and exercise 18.4]:

$$R_{j0k0} = -\frac{1}{2} h_{jk,00}^{TT}. \tag{35.10}$$

Recall that the curvature tensor is gauge-invariant (exercise 18.1). It follows that $h_{\mu\nu}$ cannot be reduced to still fewer components than it has in the *TT* gauge.

Box 35.1 describes methods to calculate $h_{\mu\nu}^{TT}$ from a knowledge of $h_{\mu\nu}$ in some other gauge.

Box 35.1 METHODS TO CALCULATE "TRANSVERSE-TRACELESS PART" OF A WAVE

Problem: Let a gravitational wave $h_{\mu\nu}(t, x^j)$ in an arbitrary gauge of linearized theory be known. How can one calculate the metric perturbation $h_{\mu\nu}^{TT}(t, x^j)$ for this wave in the transverse-traceless gauge?

Solution 1 (valid only for waves; i.e., when $\Box \bar{h}_{\mu\nu} = 0$). Calculate the components R_{j0k0} of **Riemann** in the initial gauge; then integrate equation (35.10)

$$h_{jk,00}^{TT} = -2R_{j0k0} \tag{1}$$

to obtain h_{jk}^{TT}. When the wave is monochromatic, $h_{\mu\nu} = h_{\mu\nu}(x^l)e^{-i\omega t}$; then the solution of (1) has the simple form

$$h_{jk}^{TT} = 2\omega^{-2}R_{j0k0}. \tag{2}$$

Solution 2 (valid only for plane waves). "Project out" the *TT* components in an algebraic manner using the operator

$$P_{jk} = \delta_{jk} - n_j n_k. \tag{3}$$

Here

$$n_k = k_k/|\mathbf{k}|$$

is the unit vector in the direction of propagation. Verify that P_{jk} is a projection operator onto the transverse plane:

$$P_{jl}P_{lk} = P_{jk}, \qquad P_{jk}n_k = 0, \qquad P_{kk} = 2.$$

Then the transverse part of h_{jk} is $P_{jl}h_{lm}P_{mk}$ (or in matrix notation, PhP); and the *TT* part is this quantity diminished by its trace:

$$h_{jk}^{TT} = P_{jl}P_{mk}h_{lm} - \frac{1}{2}P_{jk}(P_{ml}h_{lm}) \tag{4}$$

(index notation),

$$h^{TT} = PhP - \frac{1}{2}P\,\mathrm{Tr}\,(Ph)$$

(matrix notation). (4')

The sequence of operations that gives h_{ij}^{TT} cuts two parts out of h_{ij}. The first part cut out is

$$h_{jk}^{T} = \frac{1}{2}P_{jk}(P_{lm}h_{lm}), \tag{5}$$

which is transverse but is built from its own trace,

$$h^{T} \equiv \mathrm{Tr}\,(PhP) = \mathrm{Tr}\,(Ph) = P_{lm}h_{ml}.$$

Exercise 35.1. TRANSFORMATION OF PLANE WAVE TO TT GAUGE

Let a plane wave of the form (35.4) be given, in some arbitrary gauge of linearized theory. Exhibit explicitly the transformation that puts it into the *TT* gauge. [*Hint:* Work in a Lorentz frame where the 4-velocity u^μ of the *TT* gauge is $u^0 = 1, u^j = 0$. Solve for the four constants C^μ of the generating function (35.6) by demanding that $\bar{h}_{\mu\nu}$ satisfy the *TT* constraints (35.7).]

EXERCISES

Exercise 35.2. LIMITATION ON EXISTENCE OF TT GAUGE

Although the metric perturbation $h_{\mu\nu}$ for any *gravitational wave* in linearized theory can be put into the *TT* form (35.8), nonradiative $h_{\mu\nu}$'s cannot. Consider, for example, the external field of a rotating, spherical star, which cannot be written as a superposition of plane waves:

The second part cut out of h_{ij} is the longitudinal part

$$h^L{}_{jk} = h_{jk} - P_{jl}P_{mk}h_{lm}$$
$$= n_l n_k h_{jl} + n_j n_l h_{lk} - n_j n_k (n_l n_m h_{lm}); \quad (6)$$

or

$$h^L = h - PhP \quad (6')$$

Solution 3 (general case). Fourier analyze any symmetric array $h_{ij} = \int h_{ij}(k, t) \exp (ik_m x^m) \, d^3k$, and apply the formulas (4) from solution 2 to each Fourier component individually. But note that in this case one can write the projection operator in the direction-independent form

$$P_{jk} = \delta_{jk} - \frac{1}{\nabla^2} \partial_j \partial_k \quad (7)$$

or

$$n_l n_m = \frac{1}{\nabla^2} \partial_l \partial_m \quad (8)$$

(provided the formulas are written with all h's standing on the right), since $\partial_l = ik_l$ under the Fourier integral. Of course the operation $1/\nabla^2$ can be evaluated by other methods, e.g., by Green's functions, as well as by Fourier analysis. [The

quantity $\psi \equiv \nabla^{-2} f$ stands for the solution ψ of the Poisson equation $\nabla^2\psi = f$.] The advantage of this method is its power in certain analytic computations (see, e.g., below).

Gauge Transformations. The change in $h_{\mu\nu}$ due to a gauge transformation is

$$\delta h_{\mu\nu} = -(\partial_\nu \xi_\mu + \partial_\mu \xi_\nu). \quad (9)$$

The transverse part of this change is

$$P_{jl} P_{km}(\delta h_{lm}) = -P_{jl}P_{km}(\partial_l \xi_m + \partial_m \xi_l) = 0. \quad (10)$$

To verify this formula for a plane wave (solution 2), note that $\partial_l = i|k|n_l$ and $P_{jl}n_l = 0$. To verify the same result in general, use equation (7) to give the result

$$P_{jl}\partial_l = 0. \quad (11)$$

Thus both h_{ij}^{TT} of equation (4), and h_{ij}^T of equation (5) are gauge-invariant:

$$\delta h_{ij}^{TT} = \delta h_{ij}^T = 0. \quad (12)$$

In empty space ($T_{\mu\nu} = 0$), both h_{ij}^T and another gauge-invariant quantity \bar{h}_{0k} (discussed in exercise 35.4) vanish, by virtue of the field equations.

$$h_{00} = \frac{2M}{r}, \qquad h_{jk} = \frac{2M}{r}\delta_{jk}, \qquad h_{0k} = -2\epsilon_{klm}\frac{S^l x^m}{r^3},$$

$$r = (x^2 + y^2 + z^2)^{1/2}$$

[see equation (19.5)]. Here M is the star's mass and S is its angular momentum. Show that this *cannot* be put into TT gauge. [Hint: Calculate R_{j0k0} and from it, by means of (35.10), infer h_{jk}^{TT}. Then calculate R_{0xyz} in both the original gauge and the new gauge, and discover that they disagree—not only by virtue of the mass term, but also by virtue of the angular-momentum term.]

Exercise 35.3. A CYLINDRICAL GRAVITATIONAL WAVE

To restore one's faith, which may have been shaken by exercise 35.2, one can consider the radiative solution whose only nonvanishing component $h_{\mu\nu}$ is

$$\bar{h}_{zz} = 4A \cos(\omega t) J_0(\omega\sqrt{x^2 + y^2}),$$

where J_0 is the Bessel function. This solution represents a superposition of ingoing and outgoing cylindrical gravitational waves. For this gravitational field calculate R_{j0k0}, and from it infer h_{jk}^{TT}. Then calculate several other components of $R_{\alpha\beta\gamma\delta}$ (e.g., R_{xyxy}) in the original gauge and in TT gauge, and verify that the answers are the same.

Exercise 35.4. NON-TT PARTS OF METRIC PERTURBATION [Track 2]

From Box 35.1 establish the formula $h^T = \nabla^{-2}(h_{kk,ll} - h_{kl,kl})$; then verify the gauge invariance of h^T directly, by showing that $h_{kk,ll} - h_{kl,kl}$ is gauge-invariant. Use $\delta h_{ij} = \xi_{i,j} + \xi_{j,i}$. Show similarly that the quantities \widetilde{h}_{0k} defined by

$$\widetilde{h}_{0k} = \bar{h}_{0k} - \nabla^{-2}(\bar{h}_{0,\mu k}^\mu + \bar{h}_{kl,l0})$$

are gauge-invariant. Show from the gauge-invariant linearized field equations (18.5) that

$$\nabla^2 h^T = -16\pi T^{00},$$

$$\nabla^2 \widetilde{h}_{0k} = -16\pi T_{0k},$$

so h^T and \widetilde{h}_{0k} must vanish for waves in empty space.

§35.5. GEODESIC DEVIATION IN A LINEARIZED GRAVITATIONAL WAVE

Action of a gravitational wave on separation of two test particles

The oscillating curvature tensor of a gravitational wave produces oscillations in the separation between two neighboring test particles, A and B. Examine the oscillations from the viewpoint of A. Use a coordinate system ("*proper reference frame of A*"), with spatial origin $x^{\hat{j}} = 0$, attached to A's world line (comoving coordinates); with coordinate time equal to A's proper time ($x^{\hat{0}} = \tau$ on world line $x^{\hat{j}} = 0$); and with orthonormal spatial axes attached to gyroscopes carried by A ("nonrotating frame"). This coordinate system, appropriately specialized, is a local Lorentz frame not just at one event \mathscr{P}_0 on A's geodesic world line, but all along A's world line:

$$ds^2 = -dx^{\hat{0}2} + \delta_{\hat{j}\hat{k}}\, dx^{\hat{j}}\, dx^{\hat{k}} + O(|x^{\hat{j}}|^2)\, dx^{\hat{\alpha}}\, dx^{\hat{\beta}}. \tag{35.11}$$

[Proof: such a "proper reference frame" was set up for accelerated particles in Track 2's §13.6. The line element (13.71) derived there, when specialized to particle A ($a_{\hat{j}} = 0$ because A falls freely; $\omega^{\hat{l}} = 0$ because the spatial axes are attached to gyroscopes) reduces to the above form, as in equation (13.73).]

As the gravitational wave passes, it produces an oscillating curvature tensor, which wiggles the separation vector \boldsymbol{n} reaching from particle A to particle B:

$$D^2 n^{\hat{j}}/d\tau^2 = -R^{\hat{j}}{}_{\hat{0}\hat{k}\hat{0}} n^{\hat{k}} = -R_{\hat{j}\hat{0}\hat{k}\hat{0}} n^{\hat{k}}. \tag{35.12}$$

The components of the separation vector are nothing but the coordinates of particle B, since particle A is at the origin of its own proper reference frame; thus,

$$n^{\hat{j}} = x_B{}^{\hat{j}} - x_A{}^{\hat{j}} = x_B{}^{\hat{j}}.$$

Moreover, at $x^{\hat{j}} = 0$ [where the calculation (35.12) is being performed], the $\Gamma^{\hat{\mu}}{}_{\hat{\alpha}\hat{\beta}}$ vanish for all $x^{\hat{0}}$; so $d\Gamma^{\hat{\mu}}{}_{\hat{\alpha}\hat{\beta}}/d\tau$ also vanish. This eliminates all Christoffel-symbol corrections in $D^2 n^{\hat{j}}/D\tau^2$. Hence, equation (35.12) reduces to

$$d^2 x_B{}^{\hat{j}}/d\tau^2 = -R_{\hat{j}\hat{0}\hat{k}\hat{0}} x_B{}^{\hat{k}}. \tag{35.13}$$

There is a *TT* coordinate system that, to first order in the metric perturbation h_{jk}^{TT}, moves with particle A and with its proper reference frame. To first order in h_{jk}^{TT}, the *TT* coordinate time t is the same as proper time τ, and $R_{j0k0}^{TT} = R_{\hat{j}\hat{0}\hat{k}\hat{0}}$. Hence, equation (35.13) can be rewritten

$$d^2 x_B{}^{\hat{j}}/dt^2 = -R_{j0k0}^{TT} x_B{}^{\hat{k}} = \frac{1}{2}(\partial^2 h_{jk}^{TT}/\partial t^2) x_B{}^{\hat{k}}. \tag{35.14}$$

Suppose, for concreteness, that the particles are at rest relative to each other before the wave arrives ($x_B{}^{\hat{j}} = x_{B(0)}{}^{\hat{j}}$ when $h_{jk}^{TT} = 0$). Then the equation of motion (35.14) can be integrated to yield

$$x_B{}^{\hat{j}}(\tau) = x_{B(0)}{}^{\hat{k}}\left[\delta_{jk} + \frac{1}{2} h_{jk}^{TT}\right]_{\text{at position of } A} \tag{35.15}$$

This equation describes the wave-induced oscillations of B's location, as measured in the proper reference frame of A.

Turn to the special case of a plane wave. Suppose the test-particle separation lies in the direction of propagation of the wave. Then the wave cannot affect the separation; there is no oscillation:

Transverse character of relative accelerations

$$h_{jk}^{TT} x_{B(0)}{}^{\hat{k}} \propto h_{jk}^{TT} k_k = 0.$$

Only separations in the transverse direction oscillate; *the wave is transverse not only in its mathematical description (h_{jk}^{TT}), but also in its physical effects (geodesic deviation)*!

Exercise 35.5. ALTERNATIVE CALCULATION OF RELATIVE OSCILLATIONS

Introduce a *TT* coordinate system in which, at time $t = 0$, the two particles are both at rest. Use the geodesic equation to *show* that subsequently they both always remain at rest in the *TT* coordinates, despite the action of the wave. This means that the contravariant components of the separation vector are always constant in the *TT* coordinate frame:

$$n^j = x_B{}^j - x_A{}^j = \text{const.}$$

Call this constant $x_{B(0)}{}^j$. *Transform* these components to the comoving orthonormal frame; the answer should be equation (35.15).

§35.6. POLARIZATION OF A PLANE WAVE

Polarization of gravitational waves:

(1) States of linear polarization, "+" and "×"

Geodesic deviation in the transverse direction provides a means for studying and characterizing the polarizations of plane waves.

Consider a plane, monochromatic wave propagating in the z direction. In the *TT* gauge the constraints $h_{0\mu}^{TT} = 0$, $h_{ij,j}^{TT} \equiv ik_j h_{ij}^{TT} = 0$, and $h_{kk}^{TT} = 0$ reveal that the only nonvanishing components of $h_{\mu\nu}^{TT}$ are

$$h_{xx}^{TT} = -h_{yy}^{TT} = \Re\{A_+ e^{-i\omega(t-z)}\},$$
$$h_{xy}^{TT} = h_{yx}^{TT} = \Re\{A_\times e^{-i\omega(t-z)}\}. \tag{35.16}$$

The amplitudes A_+ and A_\times represent two independent modes of polarization.

As for electromagnetic plane waves (Figure 35.1), so also for gravitational plane waves (Figure 35.2), one can resolve a given wave into two linearly polarized components, or, alternatively, into two circularly polarized components.

$\omega(t-z)$	Displacement, δx, for polarization			
	e_x	e_y	e_R	e_L
$2n\pi$	•	•	↑•	•↓
$\left(2n + \dfrac{1}{2}\right)\pi$	←•	•↓	←•	←•
$(2n + 1)\pi$	•	•	•↓	↑•
$\left(2n + \dfrac{3}{2}\right)\pi$	•→	↑•	•→	•→

Figure 35.1.
Plane Electromagnetic Waves.
Polarization vector: e_P
Vector Potential

$$A = \Re[A_0 e^{-i\omega(t-z)}e_P]$$

Acceleration of a test charge:

$$a = (q/m)E = (q/m)(-\partial A/\partial t)$$
$$= \Re[i\omega(q/m)A_0 e^{-i\omega(t-z)}e_P]$$

Displacement of charge relative to inertial frame:

$$\delta x = \Re\left[\frac{q/m}{i\omega}A_0 e^{-i\omega(t-z)}e_P\right]$$

For *linearly polarized waves,* the unit polarization vectors of electromagnetic theory are \boldsymbol{e}_x and \boldsymbol{e}_y. A test charge hit by a plane wave with polarization vector \boldsymbol{e}_x oscillates in the x-direction relative to an inertial frame; and similarly for \boldsymbol{e}_y. By analogy, the *unit linear-polarization tensors* for gravitational waves are

$$\boldsymbol{e}_+ \equiv \boldsymbol{e}_x \otimes \boldsymbol{e}_x - \boldsymbol{e}_y \otimes \boldsymbol{e}_y, \tag{35.17a}$$

$$\boldsymbol{e}_\times \equiv \boldsymbol{e}_x \otimes \boldsymbol{e}_y + \boldsymbol{e}_y \otimes \boldsymbol{e}_x. \tag{35.17b}$$

The plane wave (35.16), when $A_\times = 0$, has polarization \boldsymbol{e}_+ and can be rewritten

$$h_{jk} = \Re\{A_+ e^{-i\omega(t-z)} e_{+jk}\}. \tag{35.18}$$

Its effect in altering the geodesic separation between two test particles depends on the direction of their separation. To see the effect in all directions at once, consider a circular ring of test particles in the transverse (x, y) plane, surrounding a central particle (Figure 35.2). As the plane wave (35.18) (polarization \boldsymbol{e}_+) passes, it deforms what was a ring as measured in the proper reference frame of the central particle into an ellipse with axes in the x and y directions that pulsate in and out:

$$\text{O} \circ \circ \circ \circ \text{O} \ 0 \ \text{O} \circ \circ \circ \circ \circ \text{O} \ 0 \ \text{O}$$

etc. By contrast (Figure 35.2), a wave of polarization \boldsymbol{e}_\times deforms the ring at a 45-degree angle to the x and y directions: $\text{O} \ 0 \ 0 \ 0 \ \text{O} \ \text{O} \ \text{O} \ \text{O} \ \text{O}$ etc.

For circularly polarized waves, the unit polarization vectors of electromagnetic theory are

(2) States of circular polarization

$$e_R = \frac{1}{\sqrt{2}}(e_x + ie_y), \qquad e_L = \frac{1}{\sqrt{2}}(e_x - ie_y) \tag{35.19}$$

$\omega(t-z)$	Deformation of a ring of test particles			
	\boldsymbol{e}_+	\boldsymbol{e}_\times	\boldsymbol{e}_R	\boldsymbol{e}_L
$2n\pi$				
$\left(2n + \frac{1}{2}\right)\pi$				
$(2n + 1)\pi$				
$\left(2n + \frac{3}{2}\right)\pi$				

Figure 35.2.

Plane Gravitational Waves. Polarization tensor:

$$\boldsymbol{e}_P$$

Metric perturbation:

$$h_{jk} = \Re[A_0 e^{-i\omega(t-z)} e_{Pjk}]$$

Tidal acceleration between two test particles:

$$\frac{D^2 n_{\hat{j}}}{D\tau^2} = -R_{\hat{j}0\hat{k}0} n_{\hat{k}} = \frac{1}{2}\frac{\partial^2 h_{jk}}{\partial t^2} n_k$$

$$= \Re\left[-\frac{1}{2}\omega^2 A_0 e^{-i\omega(t-z)} e_{Pjk} n_{\hat{k}}\right]$$

Separation between two test particles:

$$n_{\hat{j}} = n_{\hat{j}}^{(0)} + \Re\left[\frac{1}{2}A_0 e^{-i\omega(t-z)} e_{Pjk} n_{\hat{k}}^{(0)}\right]$$

Position of test particle B in proper reference frame of test particle A. (In drawing, A is the central particle and B is any peripheral particle):

$$x_B^{\hat{j}} = x_{B(0)}^{\hat{j}} + \Re\left[\frac{1}{2}A_0 e^{-i\omega(t-z)} e_{Pjk} x_{B(0)}^{\hat{k}}\right]$$

Similarly, the *unit circular polarization tensors* of gravitation theory are

$$\boldsymbol{e}_R = \frac{1}{\sqrt{2}}(\boldsymbol{e}_+ + i\boldsymbol{e}_\times), \qquad \boldsymbol{e}_L = \frac{1}{\sqrt{2}}(\boldsymbol{e}_+ - i\boldsymbol{e}_\times). \tag{35.20}$$

A test charge hit by an electromagnetic wave of polarization \boldsymbol{e}_R moves around and around in a circle in the righthanded direction (counterclockwise for a wave propagating toward the reader); for \boldsymbol{e}_L it circles in the lefthanded (clockwise) direction (see Figure 35.1). Similarly (Figure 35.2), a gravitational wave of polarization \boldsymbol{e}_R rotates the deformation of a test-particle ring in the righthanded direction,

$$\circ\,\oslash\,\cdot\,\mathcal{O}\,\,\bigcirc\,\,\bigcirc\,\,\oslash\,\oslash\,\oslash$$

while a wave of \boldsymbol{e}_L rotates it in the lefthanded direction. The individual test particles in the ring rotate in small circles relative to the central particle. However, just as the drops in an ocean wave do not move along with the wave, so the particles on the ring do not move *around* the central particle with the rotating ellipse.

Spin-2 character of
gravitational field and its
relation to symmetries of
waves

Notice from Figure 35.2 that, at any moment of time, a gravitational wave is invariant under a rotation of $180°$ about its direction of propagation. The analogous angle for electromagnetic waves (Figure 35.1) is $360°$, and for neutrino waves it is $720°$. This behavior is intimately related to the spin of the zero-mass particles associated with the quantum-mechanical versions of these waves: gravitons have spin 2, photons spin 1, and neutrinos spin 1/2. The classical radiation field of a spin-S particle is always invariant under a rotation of $360°/S$ about its propagation direction.

A radiation field of any spin S has precisely two orthogonal states of linear polarization. They are inclined to each other at an angle of $90°/S$; thus, for a neutrino field, with $S = \frac{1}{2}$, the two states are distinguished as $|\!\uparrow\rangle$ and $|\!\downarrow\rangle$ (spin up and spin down; $180°$ angle). For an electromagnetic wave $S = 1$ and two orthogonal states of polarization are \boldsymbol{e}_x and \boldsymbol{e}_y ($90°$ angle). For a gravitational wave $S = 2$, and two orthogonal states are \boldsymbol{e}_+ and \boldsymbol{e}_\times ($45°$ angle).

EXERCISES

Exercise 35.6. ROTATIONAL TRANSFORMATIONS FOR POLARIZATION STATES

Consider two Lorentz coordinate systems, one rotated by an angle θ about the z direction relative to the other:

$$t' = t, \qquad x' = x\cos\theta + y\sin\theta, \qquad y' = y\cos\theta - x\sin\theta, \qquad z' = z. \tag{35.21}$$

Let $|\!\uparrow\rangle$ and $|\!\downarrow\rangle$ be quantum-mechanical states of a neutrino with spin-up and spin-down relative to the x direction; and similarly for $|\!\uparrow'\rangle$ and $|\!\downarrow'\rangle$. Let \boldsymbol{e}_x, \boldsymbol{e}_y, $\boldsymbol{e}_{x'}$, $\boldsymbol{e}_{y'}$ be the unit polarization vectors in the two coordinate systems for an electromagnetic wave traveling in the z-direction; and similarly \boldsymbol{e}_+, \boldsymbol{e}_\times, $\boldsymbol{e}_{+'}$, $\boldsymbol{e}_{\times'}$ for a gravitational wave in linearized theory. Derive the following transformation laws:

$$|\!\uparrow'\rangle = |\!\uparrow\rangle\cos\tfrac{1}{2}\theta + |\!\downarrow\rangle\sin\tfrac{1}{2}\theta; \qquad |\!\downarrow'\rangle = -|\!\uparrow\rangle\sin\tfrac{1}{2}\theta + |\!\downarrow\rangle\cos\tfrac{1}{2}\theta;$$

$$\boldsymbol{e}_{x'} = \boldsymbol{e}_x\cos\theta + \boldsymbol{e}_y\sin\theta; \qquad \boldsymbol{e}_{y'} = -\boldsymbol{e}_x\sin\theta + \boldsymbol{e}_y\cos\theta; \tag{35.22}$$

$$\boldsymbol{e}_{+'} = \boldsymbol{e}_+\cos 2\theta + \boldsymbol{e}_\times\sin 2\theta; \qquad \boldsymbol{e}_{\times'} = -\boldsymbol{e}_+\sin 2\theta + \boldsymbol{e}_\times\cos 2\theta.$$

What is the generalization to the linear-polarization basis states for a radiation field of arbitrary spin S?

Exercise 35.7. ELLIPTICAL POLARIZATION
Discuss elliptically polarized gravitational waves in a manner analogous to the discussion of linearly and circularly polarized waves in Figure 35.2.

§35.7. THE STRESS-ENERGY CARRIED BY A GRAVITATIONAL WAVE

Exercise 18.5 showed that in principle one can build detectors which extract energy from gravitational waves. Hence, it is clear that the waves must carry energy.

Unfortunately, to derive and justify an expression for their energy requires a somewhat more sophisticated viewpoint than linearized theory. Such a viewpoint will be developed later in this chapter (§§35.13 and 35.15). But for the benefit of Track-1 readers, the key result is stated here.

In accordance with the discussions in §§19.4 and 20.4, the stress-energy carried by gravitational waves cannot be localized inside a wavelength. One cannot say whether the energy is carried by the crest of a wave, by its trough, or by its "walls." However, one *can* say that a certain amount of stress-energy is contained in a given "macroscopic" region (region of several wavelengths' size), and one can thus talk about a tensor for an *effective* smeared-out stress-energy of gravitational waves, $T_{\mu\nu}^{(\text{GW})}$. In a (nearly) inertial frame of linearized theory, $T_{\mu\nu}^{(\text{GW})}$ is given by

> Approximate localization of energy in a gravitational wave

> Effective stress-energy tensor for gravitational waves:

$$T_{\mu\nu}^{(\text{GW})} = \frac{1}{32\pi}\langle h_{jk,\mu}^{TT} h_{jk,\nu}^{TT}\rangle, \tag{35.23}$$

> (1) expressed in terms of metric perturbations

where $\langle\ \rangle$ denotes an average over several wavelengths and h_{jk}^{TT} means the (gauge-invariant) transverse-traceless part of $h_{\mu\nu}$, which is simply h_{jk} in the TT gauge. Another formula for $T_{\mu\nu}^{(\text{GW})}$, valid in any arbitrary gauge, with $\bar{h}\neq0$, $\bar{h}_{\mu}{}^{\alpha}{}_{,\alpha}\neq0$, and $\bar{h}_{0\mu}\neq0$ is

$$T_{\mu\nu}^{(\text{GW})} = \frac{1}{32\pi}\left\langle \bar{h}_{\alpha\beta,\mu}\bar{h}^{\alpha\beta}{}_{,\nu} - \frac{1}{2}\bar{h}_{,\mu}\bar{h}_{,\nu} - \bar{h}^{\alpha\beta}{}_{,\beta}\bar{h}_{\alpha\mu,\nu} - \bar{h}^{\alpha\beta}{}_{,\beta}\bar{h}_{\alpha\nu,\mu}\right\rangle \tag{35.23'}$$

This stress-energy tensor, like any other, is divergence-free in vacuum

$$T^{(\text{GW})}{}_{\mu}{}^{\nu}{}_{,\nu} = 0; \tag{35.24}$$

> (2) subject to conservation law

and it contributes to the large-scale background curvature (which linearized theory ignores) just as any other stress-energy does:

$$G_{\mu\nu}^{(\text{B})} = 8\pi(T_{\mu\nu}^{(\text{GW})} + T_{\mu\nu}^{(\text{matter})} + T_{\mu\nu}^{(\text{other fields})}). \tag{35.25}$$

> (3) role as source of background curvature

In writing here the term $T_{\mu\nu}^{(\text{GW})}$ for the effective smeared-out energy density of the gravitational wave, one is foregoing any further insertion of the gravitational wave into the Einstein equation. Otherwise one might end up counting twice over the

contribution of the same wave to the background curvature of space, even though expressed in very different formalisms.

According to equation (35.23), the stress-energy tensor for the plane wave,

(4) for a plane, monochromatic wave

$$h_{\mu\nu} = \Re\{(A_+ e_{+\mu\nu} + A_\times e_{\times\mu\nu})e^{-i\omega(t-z)}\}, \tag{35.26}$$

is

$$T_{tt}^{(\mathrm{GW})} = T_{zz}^{(\mathrm{GW})} = -T_{tz}^{(\mathrm{GW})} = \frac{1}{32\pi}\omega^2(|A_+|^2 + |A_\times|^2). \tag{35.27}$$

Notice that the background radius of curvature \mathcal{R} (ignored by linearized theory), and the mean reduced wavelength λbar (= wavelength/2π) and amplitude \mathcal{A} of the gravitational waves satisfy

$$\mathcal{R}^{-2} \sim \text{typical magnitude of components of } R_{\alpha\beta\gamma\delta}^{(\mathrm{B})}$$
$$\sim T_{\mu\nu}^{(\mathrm{GW})} \sim \mathcal{A}^2/\lambdabar^2 \text{ if } T_{\mu\nu}^{(\mathrm{GW})} \text{ is chief source of background curvature,}$$
$$\gg T_{\mu\nu}^{(\mathrm{GW})} \sim \mathcal{A}^2/\lambdabar^2 \text{ if } T_{\mu\nu}^{(\mathrm{GW})} \text{ is not chief source.}$$

Consequently, the dimensionless numbers \mathcal{A} and \lambdabar/\mathcal{R} are related by

$$\mathcal{A} \lesssim \lambdabar/\mathcal{R}. \tag{35.28}$$

Conditions for validity of gravitational-wave formalism

Thus, *the whole concept of a small-scale ripple propagating in a background of large-scale curvature breaks down, and the whole formalism of this chapter becomes meaningless, if the dimensionless amplitude of the wave approaches unity. One must always have $\mathcal{A} \ll 1$ as well as $\lambdabar \ll \mathcal{R}$ if the concept of a gravitational wave is to make any sense!*

§35.8.　GRAVITATIONAL WAVES IN THE FULL THEORY OF GENERAL RELATIVITY

Nonlinear effects in gravitational waves:

Curving up of the background spacetime by the energy of the waves is but one of many new effects that enter, when one passes from linearized theory to the full, nonlinear general theory of relativity.

In linearized theory one can consider a localized source of gravitational waves (e.g., a vibrating bar) in steady oscillation, radiating a strictly periodic wave. But the exact theory insists that the energy of the source decrease secularly, to counterbalance the energy carried off by the radiation (energy conservation; gravitational radiation damping; see §§36.8 and 36.11). This makes an exactly periodic wave impossible, though a very nearly periodic one can certainly be emitted [Papapetrou (1958); Arnowitt, Deser, and Misner as reported by Misner (1964b)].

(1) radiation damping

In the real universe there are spacetime curvatures due not only to the energy of gravitational waves, but also, and more importantly, to the material content of the universe (planets, stars, gas, galaxies). As a gravitational wave propagates through these curvatures, its wave fronts change shape ("refraction"), its wavelength changes

(2) refraction

(gravitational redshift), and it backscatters off the curvatures to some extent. If the wave is a pulse, the backscatter will cause its shape and polarization to keep changing and will produce "tails" that spread out behind the moving pulse, traveling slower than light [see exercise 32.10; also Riesz (1949), DeWitt and Brehme (1960), DeWitt and DeWitt (1964a), Kundt and Newman (1968), Couch *et. al.* (1968)]. However, so long as $\mathcal{C} \ll 1$ and $\lambda/\mathcal{R} \ll 1$, these effects will be extremely small locally. They can only build up over distances of the order of \mathcal{R}—and sometimes not even then. Thus, locally, linearized theory will remain highly accurate.

(3) redshift
(4) backscatter
(5) tails

Even in an idealized universe containing nothing but gravitational waves, backscatter and tails are produced by the interaction of the waves with the background curvature that they themselves produce.

If the reduced wavelength $\lambda = \lambda/2\pi$ and the mass-energy m of a pulse of waves satisfy $\lambda \ll m$, it is possible (in principle) to focus the pulse into a region of size $r < m$, whereupon a part of the energy of the pulse will undergo gravitational collapse to a singularity, leaving behind a black hole [see Ruffini and Wheeler (1970), and pp. 7–24 of Christodoulou (1971)]. Short of a certain critical strength, no part of the pulse undergoes such a collapse. But it does undergo a time delay before reexpanding. This time delay is definable and measurable in the asymptotically flat space, far from the domain where the energy a little earlier underwent temporary focusing into dimensions of order λ.

(6) self-gravitational attraction

All these effects can be analyzed in general relativity theory using approximation schemes which, in first order, are similar to or identical to linearized theory. Later in this chapter (§§35.13–35.15), one such approximation scheme will be developed. But first it is helpful to study an exact solution that exhibits some of these effects.

§35.9. AN EXACT PLANE-WAVE SOLUTION

Any exact gravitational-wave solution that can be given in closed mathematical form must be highly idealized; otherwise it could not begin to cope with the complexities outlined above. Consequently, mathematically exact solutions are useful for pedagogical purposes only. However, pedogogy should not be condemned: it is needed not only by students, but also by veteran workers in the field of relativity, who even today are only beginning to develop intuition into the nonlinear regime of geometrodynamics!

The rest of this chapter is Track 2. No earlier Track-2 material is needed as preparation for it, but Chapter 20 (conservation laws) and §22.5 (geometric optics) will be found to be helpful. It is not needed as preparation for any later chapter.

From the extensive literature on exact solutions, we have chosen, as a compromise between realism and complexity, the following plane wave [Bondi *et. al.* (1959), Ehlers and Kundt (1962)]:

Exact plane-wave solution of vacuum field equation:

$$ds^2 = L^2(e^{2\beta} dx^2 + e^{-2\beta} dy^2) + dz^2 - dt^2$$
$$= L^2(e^{2\beta} dx^2 + e^{-2\beta} dy^2) - du\,dv. \qquad (35.29a)$$

(1) form of metric

Here

$$u = t - z, \qquad v = t + z, \qquad L = L(u), \qquad \beta = \beta(u). \qquad (35.29b)$$

The forms that the functions $L(u)$ ("background factor") and $\beta(u)$ ("wave factor") can take are determined by the vacuum field equations. In the null coordinate system u, v, x, y, the only component of the Ricci tensor that does not vanish identically is (see Box 14.4, allowing for the difference in coordinates, $2v_{\text{there}} = v_{\text{here}}$)

$$R_{uu} = -2L^{-1}[L'' + (\beta')^2 L], \tag{35.30}$$

where the prime denotes d/du. Thus, Einstein's equations in vacuum read

$$L'' + (\beta')^2 L = 0. \tag{35.31}$$

("effect of wave factor on background factor")

 The linearized version of this equation is $L'' = 0$, since $(\beta')^2$ is a second-order quantity. Therefore the solution corresponding to linearized theory is

$$L = 1, \qquad \beta(u) \text{ arbitrary but small.}$$

The corresponding metric is

$$ds^2 = (1 + 2\beta)\,dx^2 + (1 - 2\beta)\,dy^2 + dz^2 - dt^2, \qquad \beta = \beta(t - z). \tag{35.32}$$

Notice that this is a plane wave of polarization \boldsymbol{e}_+ propagating in the z-direction. (See exercise 35.10 at end of §35.12 for the extension to a wave possessing both polarizations, \boldsymbol{e}_+ and \boldsymbol{e}_\times.)

 Return attention to the exact plane wave, and focus on the case where the "wave factor" $\beta(u)$ is a pulse of duration $2T$, and $|\beta'| \ll 1/T$ throughout the pulse. Then the exact solution (Figure 35.3) is: (1) for $u < -T$ (flat spacetime; pulse has not yet arrived),

$$\beta = 0, \qquad L = 1; \tag{35.33a}$$

(2) for $-T < u < +T$ (interior of pulse),

$$\beta = \beta(u) \text{ is arbitrary, except that } |\beta'| \ll 1/T,$$

$$L(u) = 1 - \int_{-T}^{u} \left\{ \int_{-T}^{\bar{u}} [\beta'(\bar{\bar{u}})]^2 \, d\bar{\bar{u}} \right\} d\bar{u} + O([\beta'T]^4); \tag{35.33b}$$

(3) for $u > T$ (after the pulse has passed),

$$\beta = 0, \qquad L = 1 - \frac{u}{a}, \qquad a \equiv \frac{1}{\int_{-T}^{T} (\beta')^2 \, du} + \frac{O([\beta'T]^2)}{\int_{-T}^{T} (\beta')^2 \, du}. \tag{35.33c}$$

 Before discussing the physical interpretation of this exact solution, one must come to grips with the singularity in the metric coefficients at $u = a \gg T$. (There $L = 0$, so $g_{xx} = g_{yy} = 0$.) Is this a physical singularity like the region $r = 0$ of the Schwarzschild geometry, or is it merely a coordinate singularity as $r = 2M$ is in Schwarzschild coordinates (Chapters 31, 32, and 33)? The only nonzero components of the Riemann tensor for the metric (35.29) are (see Box 14.4)

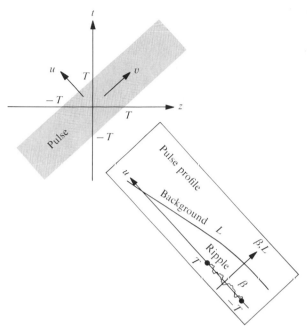

Figure 35.3.

Spacetime diagram and pulse profile for an exact plane-wave solution to Einstein's vacuum field equations. The metric has the form

$$ds^2 = L^2(e^{2\beta}\, dx^2 + e^{-2\beta}\, dy^2) + dz^2 - dt^2.$$

The "wave factor" $\beta(u) \equiv \beta(t - z)$ (short-scale ripples) and the "background factor" $L(u) \equiv L(t - z)$ (large-scale bending of the background geometry by the effective mass-energy of the "ripply" gravitational wave) are shown in the drawing and are given analytically by equations (35.33).

$$R^x{}_{uxu} = \frac{1}{2}R_{uu} - \beta'' - 2(L'/L)\beta', \qquad (35.34)$$

$$R^y{}_{uyu} = \frac{1}{2}R_{uu} + \beta'' + 2(L'/L)\beta'.$$

Moreover, these components both vanish in any extended region where $\beta = 0$. Thus, *spacetime is completely flat in regions where the "wave factor" vanishes—which is everywhere outside the pulse!* In particular, spacetime is flat near $u = a$, so the singularity there must be a coordinate singularity, not a physical singularity. To eliminate this singularity, one can perform the coordinate transformation

(5) spacetime is flat outside the pulse

$$x = \frac{X}{1 - U/a}, \qquad y = \frac{Y}{1 - U/a}, \qquad u = U, \qquad v = V + \frac{X^2 + Y^2}{a - U} \quad (35.35)$$

throughout the region to the future of the pulse ($u > T$), where

$$ds^2 = (1 - u/a)^2(dx^2 + dy^2) - du\, dv. \qquad (35.36a)$$

In the new X, Y, U, V, coordinates the metric has the explicitly flat form

$$ds^2 = dX^2 + dY^2 - dU\,dV \qquad \text{for } U = u > T. \qquad (35.36b)$$

EXERCISES

Exercise 35.8. GLOBALLY WELL-BEHAVED COORDINATES FOR PLANE WAVE
[based on Ehlers and Kundt (1962)]

Find a coordinate transformation similar to (35.35), which puts the exact plane-wave solution (35.29a), (35.31), into the form

$$ds^2 = dX^2 + dY^2 - dU\,dV + (X^2 - Y^2)F\,dU^2, \qquad (35.37)$$

$$F = F(U) \text{ completely arbitrary.}$$

This coordinate system has the advantage of no coordinate singularities anywhere; while the original coordinate system has the advantages of an easy transition to linearized theory, and easy interpretation of the action of the wave on test particles.

Exercise 35.9. GEODESIC COMPLETENESS FOR PLANE-WAVE MANIFOLD
[based on Ehlers and Kundt (1962)]

Prove that the coordinate system (X, Y, U, V) of exercise 35.8 completely covers its spacetime manifold. More specifically, show that every geodesic can be extended in both directions for an arbitrarily large affine-parameter length without leaving the X, Y, U, V coordinate system. (This property is called *geodesic completeness*.) [*Hint:* Choose an arbitrary event and an arbitrary tangent vector $d/d\lambda$ there. They determine an arbitrary geodesic. Perform a coordinate transformation that leaves the form of the metric unchanged and puts $d/d\lambda$ either in the $(\widetilde{U}, \widetilde{V}) = $ constant 2-surface, or in the $\widetilde{X}, \widetilde{Y}) = $ constant 2-surface. Verify that the two coordinate systems cover the same region of spacetime. Then analyze completeness of $d/d\lambda$'s geodesic in $(\widetilde{X}, \widetilde{Y}, \widetilde{U}, \widetilde{V})$ coordinates.]

§35.10. PHYSICAL PROPERTIES OF THE EXACT PLANE WAVE

Spacetime is completely flat both before the arrival of the plane-wave pulse $(u < -T)$ and after it has passed $(u > T)$. This is the message of the last paragraph.

Complete flatness outside the pulse is very atypical for gravitational waves in the full, nonlinear general theory of relativity. It occurs in this example only because the wave fronts (surfaces of constant u and v, i.e., constant z and t) are perfectly flat 2-surfaces. If the wave fronts were bent (e.g., spherical), the energy carried by the pulse would produce spacetime curvature outside it.

To see nonlinear effects in action, turn from the flat geometry outside the pulse to the action of the pulse on freely falling test particles. Consider a family of particles that are all at rest in the original t, x, y, z coordinate system (world lines: $[x, y, z]$ = constant) before the pulse arrives. Then even while the pulse is passing, and after it has gone, the particles remain at rest in the coordinate system. (This is true for any metric, such as (35.29a), with $g_{0\mu} = -\delta^0{}_\mu$, for then $\Gamma^\mu{}_{00} = 0$, so $x^\mu = \delta^\mu{}_0\tau + $ const. satisfies the geodesic equation.)

Flatness outside gravitational-wave pulses is unusual

Action of exact gravitational-wave pulse on test particles:

Two particles whose separation is in the direction of propagation of the pulse (z-direction) have not only constant coordinate separation, $\Delta x = \Delta y = 0$ and $\Delta z \neq 0$; they also have constant proper separation, $\Delta s = g_{zz}^{1/2} \Delta z = \Delta z$. Hence, the exact plane wave is completely transverse, like a plane wave of linearized theory.

(1) transverse character of relative accelerations

Neighboring particles transverse to the propagation direction, ($\Delta x \neq 0$, $\Delta y \neq 0$, $\Delta z = 0$) have a proper separation that wiggles as the pulse passes:

$$\Delta s = L(t - z)[e^{2\beta(t-z)}(\Delta x)^2 + e^{-2\beta(t-z)}(\Delta y)^2]^{1/2}$$
$$\approx L[(1 + 2\beta)(\Delta x)^2 + (1 - 2\beta)(\Delta y)^2]^{1/2}. \tag{35.38}$$

Superimposed on the usual linearized-theory type of wiggling, due to the "wave factor" β, is a very small net acceleration of the particles toward each other, due to the "background factor" L [note the form of $L(u)$ in Figure 35.3]. This is an acceleration of almost Newtonian type, caused by the gravitational attraction of the energy that the gravitational wave carries between the two particles. The total effect of all the energy that passes is to convert the particles from an initial state of relative rest, to a final state of relative motion with speed

(2) gravitational attraction due to energy in pulse

$$v_{\text{final}} = d\,\Delta s/dt = d(L\,\Delta s_i)/dt = -\Delta s_i/a, \tag{35.39}$$

where

$$\Delta s_i = [(\Delta x)^2 + (\Delta y)^2]^{1/2} = \text{(initial particle separation)}.$$

[Recall: $L_{\text{initial}} = 1$, $L_{\text{final}} = 1 - u/a = 1 - (t - z)/a$; equation (35.33).]

Precisely the same effect can be produced by a pulse of electromagnetic waves (§35.11).

§35.11. COMPARISON OF AN EXACT ELECTROMAGNETIC PLANE WAVE WITH THE GRAVITATIONAL PLANE WAVE

Consider the metric

$$ds^2 = L^2(u)(dx^2 + dy^2) - du\,dv, \quad \begin{Bmatrix} u = t - z \\ v = t + z \end{Bmatrix}, \tag{35.40}$$

which is always flat if it satisfies the vacuum Einstein equations ($R_{\mu\nu} = 0$ or $L'' = 0$), and therefore cannot represent a gravitational wave. In this metric the electromagnetic potential

An electromagnetic plane-wave pulse

$$\boldsymbol{A} = A_\mu\,\boldsymbol{dx}^\mu = \mathcal{A}(u)\,\boldsymbol{dx} \tag{35.41}$$

satisfies Maxwell's equations for arbitrary $\mathcal{A}(u)$. It represents an electromagnetic plane wave analogous to the gravitational plane wave of the last few sections. The only nonzero field components of this wave are

$$F_{ux} = \mathcal{A}', \text{ i.e., } F_{tx} = -F_{zx} = \mathcal{A}'; \tag{35.42}$$

so the electric vector oscillates back and forth in the x-direction, the magnetic vector oscillates in the y-direction, and the wave propagates in the z-direction. The stress-energy tensor in x, y, u, v, coordinates has only

$$T_{uu} = (4\pi L^2)^{-1}(\mathcal{A}')^2 \qquad (35.43)$$

nonzero.

The Maxwell equations are already satisfied by the potential (35.41) in the background metric (35.40), as the reader can verify. In order to make that metric itself equally acceptable, one need only impose the Einstein equations $G_{\mu\nu} = 8\pi T_{\mu\nu}$. They read [see equation (35.30) with $\beta = 0$]

$$L'' + (4\pi T_{uu})L = 0. \qquad (35.44)$$

This has exactly the form of the equation $L'' + (\beta')^2 L = 0$ for the gravitational plane wave. Consequently, the relative motions of uncharged test particles produced by the "background factor" $L(u)$ are the same whether $L(u) \neq 1$ is produced by the stress-energy of an electromagnetic wave, or by a corresponding gravitational wave with

Electromagnetic plane wave and gravitational plane wave produce same gravitational attractions	

$$[(\beta')^2/4\pi]_{\text{grav wave}} = [T_{uu}]_{\text{em wave}} = (\mathcal{A}')^2/4\pi L^2. \qquad (35.45)$$

The analogy can be made even closer. Decrease the wavelength of the waves, while holding $(\beta')^2/4\pi$ and $(\mathcal{A}')^2/4\pi L^2$ fixed:

$$\langle(\beta')^2/4\pi\rangle = \langle(\mathcal{A}')^2/4\pi L^2\rangle = \text{const}; \qquad \lambdabar \longrightarrow 0.$$

In the limit of very small wavelength (i.e., from a viewpoint whose characteristic length is $\gg \lambdabar$), the two solutions are completely indistinguishable. Their metrics are identical ($\lambdabar \longrightarrow 0$ and $\langle(\beta')^2\rangle = \text{const.}$ imply $\beta \longrightarrow 0$), and their jigglings of test particles are too small to be seen. Only their curving up of spacetime ($L \neq 1$) and the associated gravitational pull of their energy are detectable.

§35.12. A NEW VIEWPOINT ON THE EXACT PLANE WAVE

The above comparison suggests a viewpoint that was sketched briefly in the introduction to this chapter and in §35.8. Think of the exact gravitational plane-wave solution [Figure 35.3; equations (35.29) and (35.33)] as ripples in the spacetime curvature, described by $\beta(u)$, propagating on a very slightly curved background spacetime, characterized by $L(u)$. The most striking difference between the background and the ripples is not in the magnitude of their spacetime curvatures, but in their characteristic lengths. The ripples have characteristic length

Exact gravitational plane waves reexamined in the language of "short-wave approximation":	
(1) ripples vs. background	

$$\lambdabar \equiv \text{(typical reduced wavelength, } \lambda/2\pi, \text{ of waves)}; \qquad (35.46)$$

the background has characteristic length ("radius of curvature of background geometry")

$$\mathcal{R} \sim |L/L''|^{1/2}{}_{\text{inside wave}} \sim 1/|\beta'|. \tag{35.47}$$

Recall that λ is somewhat smaller than the pulse length, $2T$. Recall also that $|\beta'T| \ll 1$. Conclude that the characteristic lengths of the "wave factor" and the "background factor" differ greatly:

$$\lambda \ll \mathcal{R}. \tag{35.48}$$

This difference in scales enables one to separate out the background from the ripples.

The ripples are very much smaller in scale ($\lambda \ll \mathcal{R}$) than the background. Nevertheless the local curvature in a ripple is correspondingly larger than the background curvature [equations (35.30), (35.34)]; thus,

$$(R^x{}_{uxu})_{\text{background}} = (R^y{}_{uyu})_{\text{background}} = -L''/L \sim 1/\mathcal{R}^2,$$
$$(R^x{}_{uxu})_{\text{waves}} = -(R^y{}_{uyu})_{\text{waves}} = -\beta'' \sim |\beta'|/\lambda \sim 1/(\lambda\mathcal{R}) \tag{35.49}$$
$$\sim (\mathcal{R}/\lambda)(R^x{}_{uxu})_{\text{background}}.$$

One is reminded of the mottled surface of an orange!

The metric for the background of the gravitational plane wave is the same as for the electromagnetic one [equation (35.40)]:

$$ds^2 = g^{(B)}_{\mu\nu} dx^\mu dx^\nu = L^2(dx^2 + dy^2) - du\, dv. \tag{35.50}$$

By comparison with equation (35.29a), one sees that the metric for the full spacetime (background plus ripple) is

$$ds^2 = (g^{(B)}_{\mu\nu} + h_{\mu\nu}) dx^\mu dx^\nu, \tag{35.51}$$

$$h_{xx} = -h_{yy} = 2\beta, \text{ all other } h_{\mu\nu} = 0. \tag{35.52}$$

(Recall, in the region where $\beta \neq 0$, L is very nearly 1.) One can think of the ripples as a transverse, traceless, symmetric tensor field $h_{\mu\nu}$ analogous to the electromagnetic field A_μ, propagating in the background geometry. Just as the electromagnetic field produces the background curvature via

(2) propagation of ripples in background

$$G_{uu} = -2L''/L = 8\pi T_{uu},$$

so the gravitational-wave ripples $h_{\mu\nu}$ produce the background curvature via equation (35.31), which one can rewrite as

$$G^{(B)}_{uu} = -2L''/L = 8\pi T^{(\text{EFF})}_{uu}. \tag{35.53}$$

Here

$$T^{(\text{EFF})}_{uu} \equiv \frac{1}{4\pi}(\beta')^2 = \frac{1}{32\pi} h_{jk,u} h_{jk,u} \tag{35.54}$$

(3) effective stress-energy tensor for ripples

is the "effective stress-energy tensor" for the gravitational waves. Notice that it agrees, except for averaging, with the expression (35.23) that was written down without justification in §35.7.

EXERCISE

Exercise 35.10. PLANE WAVE WITH TWO POLARIZATIONS PRESENT

The exact plane-wave solution (35.29) has polarization \boldsymbol{e}_+. Construct a similar solution, containing two arbitrary amplitudes, $\beta(u)$ and $\gamma(u)$, for polarizations \boldsymbol{e}_+ and \boldsymbol{e}_\times. Extend the discussions of §§35.9–35.12 to this solution.

§35.13. THE SHORTWAVE APPROXIMATION

The remainder of this chapter extends the above viewpoint in a rigorous manner to very general gravitational-wave solutions. This extension is called the "shortwave formalism"; it was largely devised by Isaacson (1968a,b), though it was built on foundations laid by Wheeler (1964a) and by Brill and Hartle (1964). Versions that are even more rigorous have been given in the W.K.B. or geometric-optics limit by Choquet-Bruhat (1969), and by MacCallum and Taub (1973).

Consider gravitational waves propagating through a *vacuum* background spacetime. As in §35.7, let \mathcal{R} be the typical radius of curvature of the background; let λ and \mathcal{C} be the typical reduced wavelength ($\lambda/2\pi$) and amplitude of the waves; and demand both $\mathcal{C} \ll 1$ and $\lambda/\mathcal{R} \ll 1$. The background curvature might be due entirely to the waves, or partly to waves and partly to nearby matter and nongravitational fields.

The analysis uses a coordinate system closely "tuned" to spacetime in the sense that the metric coefficients can be split into "background" coefficients plus perturbations

$$g_{\mu\nu} = g_{\mu\nu}^{(B)} + h_{\mu\nu} \tag{35.55}$$

with these properties: (1) the amplitude of the perturbation is \mathcal{C}

$$h_{\mu\nu} \lesssim (\text{typical value of } g_{\mu\nu}^{(B)}) \cdot \mathcal{C}; \tag{35.56a}$$

(2) the scale on which $g_{\mu\nu}^{(B)}$ varies is $\geq \mathcal{R}$

$$g_{\mu\nu,\alpha}^{(B)} \lesssim (\text{typical value of } g_{\mu\nu}^{(B)})/\mathcal{R}; \tag{35.56b}$$

(3) the scale on which $h_{\mu\nu}$ varies is $\sim \lambda$

$$h_{\mu\nu,\alpha} \sim (\text{typical value of } h_{\mu\nu})/\lambda. \tag{35.56c}$$

Such coordinates are called "*steady*."

A rather long computation (exercise 35.11) shows that the Ricci tensor for an expanded metric of the form (35.55) is

$$R_{\mu\nu} = R_{\mu\nu}^{(B)} + R_{\mu\nu}^{(1)}(h) + R_{\mu\nu}^{(2)}(h) + \text{error}. \tag{35.57}$$

$$\qquad\qquad ? \qquad \mathcal{C}/\lambda^2 \qquad \mathcal{C}^2/\lambda^2 \qquad \mathcal{C}^3/\lambda^2$$

Margin notes:

Foundations for shortwave formalism:

(1) \mathcal{R}, λ, and \mathcal{C} defined

(2) demand that $\mathcal{C} \ll 1$ and $\lambda/\mathcal{R} \ll 1$

(3) split of metric into background plus perturbation; "steady coordinates"

(4) Split of Ricci curvature tensor

Here a marker (\mathcal{Q}/λ^2, etc.) has been placed under each term to show its typical order of magnitude; $R_{\mu\nu}^{(B)}$ is the Ricci tensor for the background metric $g_{\mu\nu}^{(B)}$; and $R_{\mu\nu}^{(1)}$ and $R_{\mu\nu}^{(2)}$ are expressions defined by

$$R_{\mu\nu}^{(1)}(h) \equiv \frac{1}{2}(-h_{|\mu\nu} - h_{\mu\nu|\alpha}{}^\alpha + h_{\alpha\mu|\nu}{}^\alpha + h_{\alpha\nu|\mu}{}^\alpha), \qquad (35.58a)$$

$$R_{\mu\nu}^{(2)}(h) \equiv \frac{1}{2}\left[\frac{1}{2}h_{\alpha\beta|\mu}h^{\alpha\beta}{}_{|\nu} + h^{\alpha\beta}(h_{\alpha\beta|\mu\nu} + h_{\mu\nu|\alpha\beta} - h_{\alpha\mu|\nu\beta} - h_{\alpha\nu|\mu\beta})\right.$$
$$\left. + h_\nu{}^{\alpha|\beta}(h_{\alpha\mu|\beta} - h_{\beta\mu|\alpha}) - \left(h^{\alpha\beta}{}_{|\beta} - \frac{1}{2}h^{|\alpha}\right)(h_{\alpha\mu|\nu} + h_{\alpha\nu|\mu} - h_{\mu\nu|\alpha})\right].$$
$$(35.58b)$$

In these expressions and everywhere below, indices are raised and lowered with $g_{\mu\nu}^{(B)}$, and an upright line denotes a covariant derivative with respect to $g_{\mu\nu}^{(B)}$ (whereas in Chapter 21 it denoted covariant derivative with respect to 3-geometry).

At the heart of the shortwave formalism is its method for solving the vacuum field equations $R_{\mu\nu} = 0$. One begins by selecting out of expression (35.57) the part linear in the amplitude of the wave \mathcal{Q}, and setting it equal to zero. The action of the waves to curve up the background is a nonlinear phenomenon (linearized theory shows no sign of it); so $R_{\mu\nu}^{(B)}$ cannot be linear in \mathcal{Q}. Hence, in expression (35.57), $R_{\mu\nu}^{(1)}(h)$ is the only linear term, and it must vanish by itself

$$R_{\mu\nu}^{(1)}(h) = 0. \qquad (35.59a)$$

[Of course $h_{\mu\nu}$ may contain nonlinear correction terms—call them $j_{\mu\nu}$—of order \mathcal{Q}^2, which must not be constrained by this linear equation. They will be determined by (35.59c), below.]

One next splits the remainder of $R_{\mu\nu}$ into a part that is free of ripples—i.e., that varies only on scales far larger than λ ("coarse-grain viewpoint"), and a second part that contains the fluctuations. This split can be accomplished by averaging over several wavelengths (see exercise 35.14 for a precise treatment of the averaging process, also see Choquet-Bruhat (1969) for a class of solutions where such averaging is not required):

$$\underset{?}{R_{\mu\nu}^{(B)}} + \underset{\mathcal{Q}^2/\lambda^2}{\langle R_{\mu\nu}^{(2)}(h)\rangle} + \underset{\mathcal{Q}^3/\lambda^2}{\text{error}} = 0 \quad \begin{bmatrix}\text{smooth}\\\text{part}\end{bmatrix} \qquad (35.59b)$$

$$\underset{\mathcal{Q}^2/\lambda^2}{R_{\mu\nu}^{(1)}(j)} + \underset{\mathcal{Q}^2/\lambda^2}{R_{\mu\nu}^{(2)}(h)} - \underset{\mathcal{Q}^2/\lambda^2}{\langle R_{\mu\nu}^{(2)}(h)\rangle} + \underset{\mathcal{Q}^3/\lambda^2}{\text{error}} = 0 \quad \begin{bmatrix}\text{fluctuating}\\\text{part}\end{bmatrix} \qquad (35.59c)$$

$$\begin{bmatrix}\text{nonlinear cor-}\\\text{rection to }h\end{bmatrix}$$

That's all there is to it!—except for reducing the equations to manageable form, and a fuller interpretation of the physics.

Begin with the interpretation.

Split of vacuum field equations into "wave part" ($\propto \mathcal{Q}$) plus "coarse-grain part" ($\propto \mathcal{Q}^2$ and smooth on scale λ) plus "fluctuational corrections" ($\propto \mathcal{Q}^2$ and ripply on scale λ)

Physical interpretation of the three parts of field equations:

(1) propagation of waves

(2) production of background curvature by energy of waves; $T_{\mu\nu}^{(GW)}$ defined

Equation (35.59a) is an equation for the propagation of the gravitational waves $h_{\mu\nu}$.

Equation (35.59b) shows how the stress-energy in the waves creates the background curvature. It can be rewritten in the more suggestive form

$$G_{\mu\nu}^{(B)} \equiv R_{\mu\nu}^{(B)} - \frac{1}{2} R^{(B)} g_{\mu\nu}^{(B)} = 8\pi T_{\mu\nu}^{(GW)} \text{ in vacuum,} \tag{35.60}$$

where

$$T_{\mu\nu}^{(GW)} \equiv -\frac{1}{8\pi} \left\{ \langle R_{\mu\nu}^{(2)}(h) \rangle - \frac{1}{2} g_{\mu\nu}^{(B)} \langle R^{(2)}(h) \rangle \right\} \tag{35.61}$$

is the stress-energy tensor for the gravitational waves. Now one sees the origin of the statement in §35.7, that the stress-energy of gravitational waves is well-defined only in a smeared-out sense.

(3) nonlinear self-interaction of waves

Finally, equation (35.59c) shows how the gravitational waves h generate nonlinear corrections j to themselves (wave-wave scattering, harmonics of the fundamental frequency, etc.). These higher-order effects will not be investigated in this chapter.

EXERCISE

Exercise 35.11. CONNECTION COEFFICIENTS AND CURVATURE TENSORS FOR A PERTURBED METRIC

In a specific coordinate frame of an arbitrary spacetime, write the metric coefficients in covariant representation in the form

$$g_{\mu\nu} = g_{\mu\nu}^{(B)} + h_{\mu\nu}. \tag{35.62a}$$

(At the end of the calculation, one can split $h_{\mu\nu}$ into two parts, $h_{\mu\nu} \longrightarrow h_{\mu\nu} + j_{\mu\nu}$; and out of this split obtain the formulas used in the text.) Assume that the typical components of $h_{\mu\nu}$ are much less than those of $g_{\mu\nu}^{(B)}$; so one can expand Christoffel symbols and curvature tensors in $h_{\mu\nu}$. Raise and lower indices of $h_{\mu\nu}$ with $g_{\mu\nu}^{(B)}$; and denote by a "$|$" covariant derivatives relative to $g_{\mu\nu}^{(B)}$ and by a "$;$" covariant derivatives relative to $g_{\mu\nu}$.

(a) Here $g_{\mu\nu}$ and $g_{\mu\nu}^{(B)}$ can be thought of as two different metrics coexisting in the spacetime manifold. Show that the difference between the corresponding covariant derivatives, $\mathbf{\nabla} - \mathbf{\nabla}^{(B)} \equiv \mathbf{S}$—indeed, the difference between any two covariant derivatives!—is a tensor with components

$$S^{\mu}{}_{\beta\gamma} = \Gamma^{\mu}{}_{\beta\gamma} - \Gamma^{(B)\mu}{}_{\beta\gamma} \tag{35.62b}$$

[Hint: See part B of Box 10.3.]

(b) Show that

$$g^{\mu\nu} = g^{(B)\mu\nu} - h^{\mu\nu} + h^{\mu\alpha}h_{\alpha}{}^{\nu} - h^{\mu\alpha}h_{\alpha}{}^{\beta}h_{\beta}{}^{\nu} + \cdots, \tag{35.62c}$$

and also that

$$g^{\mu\nu} = g^{(B)\mu\nu} - h^{\mu\nu} + h^{\mu\alpha}h_{\alpha}{}^{\nu} - h^{\mu\alpha}h_{\alpha}{}^{\beta}h_{\beta\gamma}g^{\gamma\nu}. \tag{35.62c'}$$

(c) By calculating in a local Lorentz frame of $g^{(B)}_{\mu\nu}$ and then transforming back to the original frame, show that

$$S^{\mu}{}_{\beta\gamma} = \frac{1}{2}g^{\mu\alpha}(h_{\alpha\beta|\gamma} + h_{\alpha\gamma|\beta} - h_{\beta\gamma|\alpha}), \tag{35.62d}$$

$$R^{\alpha}{}_{\beta\gamma\delta} - R^{(B)\alpha}{}_{\beta\gamma\delta} = S^{\alpha}{}_{\beta\delta|\gamma} - S^{\alpha}{}_{\beta\gamma|\delta} + S^{\alpha}{}_{\mu\gamma}S^{\mu}{}_{\beta\delta} - S^{\alpha}{}_{\mu\delta}S^{\mu}{}_{\beta\gamma}, \tag{35.62e}$$

$$R_{\beta\delta} - R^{(B)}{}_{\beta\delta} = S^{\alpha}{}_{\beta\delta|\alpha} - S^{\alpha}{}_{\beta\alpha|\delta} + S^{\alpha}{}_{\mu\alpha}S^{\mu}{}_{\beta\delta} - S^{\alpha}{}_{\mu\delta}S^{\mu}{}_{\beta\alpha}. \tag{35.62f}$$

(d) Show that expression (35.62f) reduces to

$$R_{\beta\delta} = R^{(B)}{}_{\beta\delta} + R^{(1)}_{\beta\delta}(h) + R^{(2)}_{\beta\delta}(h) + \cdots \tag{35.62g}$$

where $R^{(1)}$ and $R^{(2)}$ are defined by equations (35.58).

§35.14. EFFECT OF BACKGROUND CURVATURE ON WAVE PROPAGATION

Focus attention on the propagation equation $R^{(1)}_{\mu\nu}(h) = 0$. As in linearized theory, so also here, the propagation is studied more simply in terms of

$$\bar{h}_{\mu\nu} \equiv h_{\mu\nu} - \frac{1}{2}hg^{(B)}_{\mu\nu}, \tag{35.63}$$

$\bar{h}_{\mu\nu}$ defined

than in terms of $h_{\mu\nu}$. Rewritten in terms of $\bar{h}_{\mu\nu}$, $R^{(1)}_{\mu\nu}(h) = 0$ says

$$\bar{h}_{\mu\nu|\alpha}{}^{\alpha} + g^{(B)}_{\mu\nu}\bar{h}^{\alpha\beta}{}_{|\beta\alpha} - 2\bar{h}_{\alpha(\mu|}{}^{\alpha}{}_{\nu)} + 2R^{(B)}_{\alpha\mu\beta\nu}\bar{h}^{\alpha\beta} - 2R^{(B)}_{\alpha(\mu}\bar{h}_{\nu)}{}^{\alpha} = 0. \tag{35.64}$$

Propagation equation for waves on curved background

[To obtain this, invert equation (35.63) obtaining $h_{\mu\nu} = \bar{h}_{\mu\nu} - \frac{1}{2}g^{(B)}_{\mu\nu}\bar{h}$; insert this into (35.58a) and equate to zero; then commute covariant derivatives using the identity (16.6b); finally contract to obtain an expression for $\bar{h}_{|\alpha}{}^{\alpha}$ and substitute that back in.]

The propagation equation (35.64) can be simplified by a special choice of gauge. An infinitesimal coordinate transformation

Specialization to "Lorentz gauge"

$$x^{\mu}{}_{\text{new}}(\mathscr{P}) = x^{\mu}{}_{\text{old}}(\mathscr{P}) + \xi^{\mu}(\mathscr{P}) \tag{35.65a}$$

induces a first-order change in the functional forms of the metric coefficients given by

$$h_{\mu\nu\text{new}}(x^{\alpha}{}_{\text{new}}) = h_{\mu\nu\text{old}}(x^{\alpha}{}_{\text{new}}) - 2\xi_{(\mu|\nu)} \tag{35.65b}$$

[analog of the gauge transformation of linearized theory, equation (35.3c); see exercise 35.12]. By an appropriate choice of the four functions ξ^{μ}, one can enforce the four "Lorentz gauge conditions"

$$\bar{h}_{\mu}{}^{\alpha}{}_{|\alpha} = 0 \tag{35.66}$$

in the new coordinate system (exercise 35.13). This choice of gauge is analogous to that of linearized theory. It makes the second and third terms in the propagation equation vanish. (For additional gauge conditions of the "*TT*" type, see exercise 35.13.)

Coupling of waves to Ricci
tensor can be ignored

The last term of the propagation equation, $-2R^{(B)}_{\alpha(\mu}\bar{h}_{\nu)}{}^{\alpha}$, vanishes to within the precision of the analysis, for this reason: attention has been confined to vacuum; so the only source of a nonvanishing Ricci tensor is the stress-energy carried by the gravitational waves themselves [equation (35.60)]; hence $R^{(B)}_{\alpha\beta} \sim a^2/\lambdabar^2$ and

$$R^{(B)}_{\alpha(\mu}\bar{h}_{\nu)}{}^{\alpha} \sim a^3/\lambdabar^2. \tag{35.67}$$

This is of the same order as $R^{(3)}_{\mu\nu}(h)$, the third-order correction to the Ricci tensor, which is far below the precision of the analysis. For consistency in the analysis it will therefore be neglected.

Summary of this section thus far: by choosing a gauge where $\bar{h}_{\mu}{}^{\alpha}{}_{|\alpha} = 0$, and by discarding terms of higher order than the precision of the analysis, one obtains the vacuum propagation equation

Propagation equation in
Lorentz gauge and its realm
of validity

$$\bar{h}_{\mu\nu|\alpha}{}^{\alpha} + 2R^{(B)}_{\alpha\mu\beta\nu}\bar{h}^{\alpha\beta} = 0, \tag{35.68}$$

subject to the Lorentz gauge condition

$$\bar{h}_{\mu\alpha}{}^{|\alpha} = 0.$$

Equation (35.68) is accurate to first order in the amplitude [corrections $\propto a^2$ are embodied in equation (35.59c)]; and its accuracy is independent of the ratio \lambdabar/\mathscr{R}, as one sees from equations (35.59). Thus, *it can be applied whenever the waves are weak, even if the wavelength is large!*

Lists of effects absent from
and contained in propagation
equation

All nonlinear interactions of the wave with itself are neglected in this first-order propagation equation. Absent is the mechanism for waves to scatter off each other and off the background curvature that they themselves produce. Also absent are any hints of a change in shape of pulse due to self-interaction as a pulse of waves propagates. There are no signs of the gravitational collapse that one knows must occur when a mass-energy m of gravitational waves gets compressed into a region of size $\lesssim m$. To see all these effects, one must turn to corrections of second order in a and higher [e.g., equations (35.59c) and (35.60)].

Actually contained in the propagation equation are all effects due to the linear action of the background curvature on the propagating wave. These effects are explored, for short wavelengths ($\lambdabar/\mathscr{R} \ll 1$) and for nearly flat wave fronts, in exercises 35.15–35.17 at the end of the chapter. The effects considered include a gravitational redshift of gravitational radiation and gravitational deflection of the direction of propagation of gravitational radiation, identical to those for light; and also a rotation of the polarization tensor. When the wavelength is not small (\lambdabar/\mathscr{R} not $\ll 1$), the propagation equation includes a back-scatter of the gravitational waves off the background curvature and a resultant pattern of wave "tails" analogous to that explored in exercise 32.10 [see, e.g., Couch *et al.* (1968), Price (1971a), Bardeen and Press (1972), Unt and Keres (1972)].

Exercise 35.12. GAUGE TRANSFORMATIONS IN A CURVED BACKGROUND

(a) Show that the infinitesimal coordinate transformation (35.65a) induces the change (35.65b) in the functional form of the metric perturbation.

(b) Discuss the relationship between this gauge transformation and the concept of a Killing vector (§25.2).

**Exercise 35.13. TRANSVERSE-TRACELESS GAUGE FOR GRAVITATIONAL
WAVES PROPAGATING IN A CURVED BACKGROUND**

(a) Show that, in vacuum in a curved background spacetime, the gauge condition $\bar{h}_{\mu}{}^{\alpha}{}_{|\alpha} = 0$ is preserved by transformations whose generator satisfies the wave equation $\xi_{\mu}{}^{|\alpha}{}_{\alpha} = 0$.

(b) Locally (over distances much smaller than \mathcal{R}) linearized theory is applicable; so there exists such a transformation which makes [see equations (35.7b) and (35.8a)]

$$\bar{h} = 0 + \text{error}, \qquad \bar{h}_{\mu\alpha}u^{\alpha} = 0 + \text{error}. \tag{35.69}$$

Here u^{α} is a vector field that is as nearly covariantly constant as possible ($u^{\alpha}{}_{|\beta} = 0$); i.e., it is a constant vector in the inertial coordinates of linearized theory; and the errors are small over distances much less than \mathcal{R}. Show that $\bar{h} = 0$ *can be imposed globally along with* $\bar{h}_{\mu\alpha}{}^{|\alpha} = 0$; i.e., show that, if it is imposed on an initial hypersurface, the propagation equation (35.68) preserves it.

(c) Show that in general, the background curvature prevents any vector field from being covariantly constant ($u^{\hat{a}}{}_{|\beta} \sim u^{\hat{a}}/\mathcal{R}$ at best); and from this *show that* $\bar{h}_{\mu\alpha}u^{\alpha} = 0$ *cannot be imposed globally along with* $\bar{h}_{\mu}{}^{\alpha}{}_{|\alpha} = 0$.

§35.15. STRESS-ENERGY TENSOR FOR GRAVITATIONAL WAVES

Turn now to an evaluation of the effective stress-energy tensor $T^{(\text{GW})}_{\mu\nu}$ of equation (35.61). The evaluation requires averaging various quantities over several wavelengths. Useful rules for manipulating quantities inside the averaging brackets $\langle\ \rangle$ are the following (see exercise 35.14 for justification).

The averaging process involved in "coarse-grain" viewpoint

(1) Covariant derivatives commute; e.g., $\langle h\ h_{\mu\nu|\alpha\beta}\rangle = \langle h\ h_{\mu\nu|\beta\alpha}\rangle$. The fractional errors made by freely commuting are $\sim(\lambda/\mathcal{R})^2$, well below the inaccuracy of the computation.

(2) Gradients average out to zero; e.g., $\langle(h_{|\alpha}h_{\mu\nu})_{|\beta}\rangle = 0$. Fractional errors made here are $\lesssim \lambda/\mathcal{R}$.

(3) As a corollary, one can freely integrate by parts, flipping derivatives from one h to the other; e.g., $\langle h\ h_{\mu\nu|\alpha\beta}\rangle = \langle -h_{|\beta}h_{\mu\nu|\alpha}\rangle$.

A straightforward but long calculation using these rules, using equation (35.58b) for $R^{(2)}_{\mu\nu}(h)$, using definition (35.63) of $\bar{h}_{\mu\nu}$, using the propagation equation (35.64), and using the definition (35.61) of $T^{(\text{GW})}_{\mu\nu}$, yields $\langle R^{(2)}(h)\rangle = 0$, and

Evaluation of effective stress-energy tensor for gravitational waves, $T^{(\text{GW})}_{\mu\nu}$

$$T^{(\text{GW})}_{\mu\nu} = \frac{1}{32\pi}\langle \bar{h}_{\alpha\beta|\mu}\bar{h}^{\alpha\beta}{}_{|\nu} - \frac{1}{2}\bar{h}_{|\mu}\bar{h}_{|\nu} - 2\bar{h}^{\alpha\beta}{}_{|\beta}\bar{h}_{\alpha(\mu|\nu)}\rangle. \tag{35.70}$$

This is the result quoted in equation (35.23′), except that there one used an inertial

coordinate system of linearized theory, where covariant derivatives and ordinary derivatives are the same. In a gauge where $\bar{h}_{\mu}{}^{\alpha}{}_{|\alpha} = 0$, the last term vanishes. When, in addition, $\bar{h}_{\mu\nu}$ is traceless (see exercise 35.13), the second term vanishes; and there remains only

$$T_{\mu\nu}^{(\mathrm{GW})} = \frac{1}{32\pi} \langle \bar{h}_{\alpha\beta|\mu} \bar{h}^{\alpha\beta}{}_{|\nu} \rangle \qquad \text{if } \bar{h}_{\mu}{}^{\alpha}{}_{|\alpha} = \bar{h} = 0. \qquad (35.70')$$

Accuracy of expression for $T_{\mu\nu}^{(\mathrm{GW})}$

These expressions for the effective stress-energy of a gravitational wave have fractional errors of order \mathcal{A}, due to the neglect of second-order corrections to $h_{\mu\nu}$; they also have fractional errors of order λ/\mathcal{R}, due to the averaging process, which makes no sense when λ approaches \mathcal{R} in magnitude. Since $\mathcal{A} \lesssim \lambda/\mathcal{R}$ (35.28), the dominant errors in $T_{\mu\nu}^{(\mathrm{GW})}$ are $\sim \lambda/\mathcal{R}$.

Properties of $T_{\mu\nu}^{(\mathrm{GW})}$

To this accuracy, the stress-energy tensor for gravitational waves is on an equal footing with any other stress-energy tensor. It plays the same role in producing background curvature; and it enters into conservation laws in the same way. For example, one can show, either by direct calculation or from the identity $G^{(\mathrm{B})\mu\nu}{}_{|\nu} = 0$, that

$$T^{(\mathrm{GW})\mu\nu}{}_{|\nu} = 0 + \text{error}, \qquad (35.71)$$

where the error $\sim (\lambda/\mathcal{R})(T^{(\mathrm{GW})\mu\nu}/\mathcal{R})$ is negligible in the shortwave approximation.

Some of the properties of $T_{\mu\nu}^{(\mathrm{GW})}$ have already been explored in §35.7. Further properties are explored in exercises 35.18 and 35.19.

EXERCISES

Exercise 35.14.　BRILL-HARTLE AVERAGE

Isaacson (1968b) introduces the following averaging scheme, which he names "Brill-Hartle averaging."

(a) In the small region, of size several times λ, where the averaging occurs, there will be a unique geodesic of $g_{\mu\nu}^{(\mathrm{B})}$ connecting any two points \mathscr{P}' and \mathscr{P}; so given a tensor $\boldsymbol{E}(\mathscr{P}')$ at \mathscr{P}', one can parallel transport it along this geodesic to \mathscr{P}, getting there a tensor $\boldsymbol{E}(\mathscr{P}')_{\to\mathscr{P}}$.

(b) Let $f(\mathscr{P}', \mathscr{P})$ be a weighting function that falls smoothly to zero when \mathscr{P}' and \mathscr{P} are separated by many wavelengths, and such that

$$\int f(\mathscr{P}', \mathscr{P}) \sqrt{-g^{(\mathrm{B})}(\mathscr{P}')} \, d^4x' = 1. \qquad (35.72)$$

(c) Then the average of the tensor field $\boldsymbol{E}(\mathscr{P}')$ over several wavelengths about the point \mathscr{P} is

$$\langle \boldsymbol{E} \rangle_{\mathscr{P}} \equiv \int \boldsymbol{E}(\mathscr{P}')_{\to\mathscr{P}} f(\mathscr{P}', \mathscr{P}) \sqrt{-g^{(\mathrm{B})}(\mathscr{P}')} \, d^4x'. \qquad (35.73)$$

(i) Show that there exists an entity $g_\mu^{(\mathrm{B})\alpha'}(\mathscr{P}, \mathscr{P}')$, whose primed index transforms as a tensor at \mathscr{P}' and whose unprimed index transforms as a tensor at \mathscr{P}, such that (for \boldsymbol{E} second rank)

$$E_{\alpha\beta}(\mathscr{P}')_{\to\mathscr{P}} = g_\alpha^{(\mathrm{B})\mu'} g_\beta^{(\mathrm{B})\nu'} E_{\mu'\nu'}(\mathscr{P}'). \qquad (35.74)$$

This entity is called the "bivector of geodesic parallel displacement"; see DeWitt and Brehme (1960) or Synge (1960a).

(ii) Rewriting expression (35.73) in coordinate language as

$$\langle E_{\alpha\beta}(x)\rangle = \int g_{\alpha}^{(B)\mu'}(x,x')g_{\beta}^{(B)\nu'}(x,x')E_{\mu'\nu'}(x')f(x,x')\sqrt{-g^{(B)}(x')}\,d^4x', \qquad (35.73')$$

derive the three averaging rules cited at the beginning of §35.15. [For solution, see Appendix of Isaacson (1968b).]

Exercise 35.15. GEOMETRIC OPTICS

Develop geometric optics for gravitational waves of small amplitude propagating in a curved background. Pattern the analysis after geometric optics for electromagnetic waves (§22.5). In particular, let $\bar{h}_{\mu\nu}$ have an amplitude that varies slowly (on a scale $\ell \lesssim \mathscr{R}$) and a phase θ that varies rapidly ($\theta_{,\bar{\alpha}} \sim 1/\lambda$). Expand the amplitude in powers of λ/ℓ, so that

$$\bar{h}_{\mu\nu} = \Re\{A_{\mu\nu} + \epsilon B_{\mu\nu} + \epsilon^2 C_{\mu\nu} + \cdots\}e^{i\theta/\epsilon}. \qquad (35.75)$$

Here ϵ is a formal expansion parameter, actually equal to unity, which reminds one that the terms attached to ϵ^n are proportional to $(\lambda/\mathscr{R})^n$. Define the following quantities (with $A_{\mu\nu}^*$ denoting the complex conjugate of $A_{\mu\nu}$):

$$\text{``wave vector'': } k_\alpha \equiv \theta_{,\alpha} \qquad (35.76a)$$

$$\text{``scalar amplitude'': } \mathcal{A} \equiv \left(\frac{1}{2}A_{\mu\nu}^*A^{\mu\nu}\right)^{1/2} \qquad (35.76b)$$

$$\text{``polarization'': } e_{\mu\nu} \equiv A_{\mu\nu}/\mathcal{A}. \qquad (35.76c)$$

By inserting expression (35.75) into the gauge condition (35.66) and the propagation equation (35.68), derive the fundamental equations of geometrical optics as follows.

(a) The rays (curves perpendicular to surfaces of constant phase) are null geodesics; i.e.

$$k_\alpha k^\alpha = 0, \qquad (35.77a)$$

$$k_{\alpha|\beta}k^\beta = 0. \qquad (35.77b)$$

(b) The polarization is orthogonal to the rays and is parallel transported along them;

$$e_{\mu\alpha}k^\alpha = 0, \qquad (35.77c)$$

$$e_{\mu\nu|\alpha}k^\alpha = 0. \qquad (35.77d)$$

(c) The scalar amplitude decreases as the rays diverge away from each other in accordance with

$$\mathcal{A}_{,\alpha}k^\alpha = -\frac{1}{2}k^\alpha_{\ |\alpha}\mathcal{A}. \qquad (35.77e)$$

i.e.,

$$(\mathcal{A}^2 k^\alpha)_{|\alpha} = 0 \ \text{(``conservation of gravitons'')}. \qquad (35.77f)$$

(d) The correction $B_{\mu\nu}$ to the amplitude obeys

$$B_{\mu\alpha}k^\alpha = iA_{\mu\alpha}^{\ |\alpha}, \qquad (35.77g)$$

$$B_{\mu\nu|\alpha}k^\alpha = -\frac{1}{2}k^\alpha_{\ |\alpha}B_{\mu\nu} + \frac{1}{2}iA_{\mu\nu|\alpha}^{\ \ \ \alpha} + iR_{\alpha\mu\beta\nu}^{(B)}A^{\alpha\beta}. \qquad (35.77h)$$

In accordance with exercise 35.13, specialize the gauge so that $\bar{h} = 0$, i.e.,

$$e_\alpha^{\ \alpha} = 0. \qquad (35.77i)$$

Then show that the stress-energy tensor (35.70′) for the waves is

$$T_{\mu\nu}^{\text{(GW)}} = \frac{1}{32\pi}\,\mathcal{A}^2 k_\mu k_\nu. \tag{35.77j}$$

This has the same form as the stress-energy tensor for a beam of particles with zero rest mass (see §5.4). Show explicitly that $T^{\text{(GW)}\mu\nu}{}_{|\nu} = 0$.

Exercise 35.16. GRAVITONS

Show that geometric optics, as developed in the preceding exercise, is equivalent to the following: "A graviton is postulated to be a particle of zero rest mass and 4-momentum \boldsymbol{p}, which moves along a null geodesic ($\boldsymbol{\nabla_p p} = 0$). It parallel transports with itself ($\boldsymbol{\nabla_p e} = 0$) a transverse ($\boldsymbol{e \cdot p} = 0$) traceless ($e_\alpha{}^\alpha = 0$) polarization tensor \boldsymbol{e}. Geometric optics is the theory of a stream of such gravitons moving through spacetime." Exhibit the relationship between the quantities in this version of geometric optics and the quantities in the preceding version (e.g., $\boldsymbol{p} = \hbar\boldsymbol{k}$, where \hbar is Planck's reduced constant $h/2\pi$).

**Exercise 35.17. GRAVITATIONAL DEFLECTION
 OF GRAVITATIONAL WAVES**

Show that gravitational waves of short wavelength passing through the solar system experience the same redshift and gravitational deflection as does light. (One should be able to infer this directly from exercise 35.15.)

Exercise 35.18. GAUGE INVARIANCE OF $T_{\mu\nu}^{\text{(GW)}}$

Show that the stress-energy tensor $T_{\mu\nu}^{\text{(GW)}}$ of equation (35.70) is invariant under gauge transformations of the form (35.65).

**Exercise 35.19. $T_{\mu\nu}^{\text{(GW)}}$ EXPRESSED AS THE AVERAGE OF
 A STRESS-ENERGY PSEUDOTENSOR**

Calculate the average over several wavelengths of the Landau-Lifshitz stress-energy pseudotensor [equation (20.22)] for gravitational waves with $\lambda/\mathcal{R} \ll 1$. The result should be equal to $T_{\mu\nu}^{\text{(GW)}}$. [Hint: Work in a gauge where $\bar{h}_\mu{}^\alpha{}_{|\alpha} = \bar{h} = 0$, to simplify the calculation.]

**Exercise 35.20. SHORTWAVE APPROXIMATION FROM
 A VARIATIONAL VIEWPOINT**

Readers who have studied the variational approach to gravitation theory in Chapter 21 may find attractive the following derivation of the basic equations of the shortwave approximation. It was devised, independently, by Sándor Kovács and Bernard Schutz, and by Bryce DeWitt (unpublished, 1971). MacCallum and Taub (1973) give a "non-Palatini" version.

(a) Define

$$g_{\mu\nu} \equiv g_{\mu\nu}^{\text{(B)}} + h_{\mu\nu}, \qquad \bar{h}_{\mu\nu} \equiv h_{\mu\nu} - \frac{1}{2}g_{\mu\nu}^{\text{(B)}}h, \tag{35.78a}$$

$$W^\mu{}_{\beta\gamma} \equiv \frac{1}{2}g_{\text{(B)}}^{\mu\alpha}(h_{\alpha\beta|\gamma} + h_{\alpha\gamma|\beta} - h_{\beta\gamma|\alpha}). \tag{35.78b}$$

Raise and lower indices on $h_{\mu\nu}$ and $W^\mu{}_{\beta\gamma}$ with the background metric. Using the results of exercise 35.11, derive the following expression for the Lagrangian of the gravitational field:

$$\mathcal{L} \equiv \frac{1}{16\pi}(-g)^{1/2}R = \mathcal{L}' + \begin{pmatrix}\text{perfect divergence} \\ \text{of form } \partial\mathcal{Q}^\alpha/\partial x^\alpha\end{pmatrix} + \begin{pmatrix}\text{corrections of order} \\ \mathcal{C}^3/\lambda^2,\ R_{\mu\nu}^{\text{(B)}}\mathcal{C},\ \text{and smaller}\end{pmatrix}, \tag{35.78c}$$

where

$$\mathcal{L}' \equiv \frac{1}{16\pi}(-g^{(B)})^{1/2}[R^{(B)} - \bar{h}^{\mu\nu}(W^{\alpha}{}_{\mu\nu|\alpha} - W^{\alpha}{}_{\mu\alpha|\nu})$$
$$+ g^{(B)\mu\nu}(W^{\alpha}{}_{\beta\alpha}W^{\beta}{}_{\mu\nu} - W^{\alpha}{}_{\beta\nu}W^{\beta}{}_{\mu\alpha})]. \tag{35.78d}$$

[*Hint:* recall that

$$(-g^{(B)})^{1/2}B^{\alpha}{}_{|\alpha} = \partial[(-g^{(B)})^{1/2}B^{\alpha}]/\partial x^{\alpha}$$

for any B^{α}.] Drop the corrections of order \mathcal{C}^3/λ^2 from \mathcal{L}; and, knowing in advance that the field equations will demand $R^{(B)}_{\mu\nu} \sim \mathcal{C}^2/\lambda^2$, drop also the corrections of order $R^{(B)}_{\mu\nu}\mathcal{C}$. Knowing that a perfect divergence contributes nothing in an extremization calculation, drop the divergence term from \mathcal{L}. Then \mathcal{L}' is the only remaining part of \mathcal{L}.

(b) Extremize $I \equiv \int \mathcal{L}' \, d^4x$ by the Palatini method (§21.2); i.e., abandon (temporarily) definition (35.78b) of $W^{\mu}{}_{\beta\gamma}$, and extremize I with respect to independent variations of $W^{\mu}{}_{\beta\gamma} = W^{\mu}{}_{\gamma\beta}$, $\bar{h}^{\mu\nu} = \bar{h}^{\nu\mu}$, and $g^{\mu\nu}_{(B)} = g^{\nu\mu}_{(B)}$. Show that extremization with respect to $W^{\mu}{}_{\beta\gamma}$ leads back to equation (35.78b) for $W^{\mu}{}_{\beta\gamma}$ in terms of $h_{\mu\nu}$. Show that extremization with respect to $\bar{h}^{\mu\nu}$, when combined with equations (35.78a,b), leads to the propagation equation for gravitational waves (35.64). Show that extremization with respect to $g^{(B)\mu\nu}$, when combined with equations (35.78a,b) and with the propagation equation (35.64), and when averaged over several wavelengths, leads to

$$G^{(B)}_{\mu\nu} = 8\pi T^{(GW)}_{\mu\nu},$$

where $T^{(GW)}_{\mu\nu}$ is given by equation (35.70). [*Warning:* The amount of algebra in this exercise is enormous, unless one chooses to impose the gauge conditions $\bar{h} = \bar{h}_{\alpha}{}^{\beta}{}_{|\beta} = 0$ from the outset.]

CHAPTER **36**

CHAPTER **36**

GENERATION OF GRAVITATIONAL WAVES

Matter is represented by curvature, but not every curvature does represent matter; there may be curvature "in vacuo."

G. LEMAITRE in Schilpp (1949), p. 440

§36.1. THE QUADRUPOLE NATURE OF GRAVITATIONAL WAVES

Masses in an isolated, nearly Newtonian system move about each other. How much gravitational radiation do they emit?

Generation of gravitational waves analyzed by electromagnetic analog

For an order-of-magnitude estimate, one can apply the familiar radiation formulas of electromagnetic theory, with the replacement $e^2 \longrightarrow -m^2$, which converts the static coulomb law into Newton's law of attraction. This procedure treats gravity as though it were a spin-one (vector) field, rather than a spin-two (tensor) field; consequently, it introduces moderate errors in numerical factors and changes angular distributions. But it gives an adequate estimate of the total power radiated.

In electromagnetic theory, electric-dipole radiation dominates, with a power output or "luminosity," L, given (see §4.4 and Figure 4.6) by

$$L_{\text{electric dipole}} = (2/3)e^2 a^2$$

for a single particle with acceleration \boldsymbol{a} and dipole moment changing as $\ddot{\boldsymbol{d}} = e\ddot{\boldsymbol{x}} = e\boldsymbol{a}$;

$$L_{\text{electric dipole}} = (2/3)\ddot{\boldsymbol{d}}^2$$

for a general system with dipole moment \boldsymbol{d}. [Geometric units: luminosity in cm of mass-energy per cm of light travel time; charge in cm, $e = (G^{1/2}/c^2)e_{\text{conv}} = (2.87 \times 10^{-25}\,\text{cm/esu}) \times (4.8 \times 10^{-10}\,\text{esu}) = 1.38 \times 10^{-34}\,\text{cm}$, acceleration in cm of distance per cm of time per cm of time. For conventional units, with e in esu or $(\text{g cm}^3/\text{sec}^2)^{1/2}$,

insert a factor c^{-3} on the right and get L in erg/sec]. The gravitational analog of the electric dipole moment is the mass dipole moment

Why gravitational waves cannot be dipolar

$$d = \sum_{\text{particles } A} m_A x_A.$$

Its first time-rate of change is the total momentum of the system,

$$\dot{d} = \sum_{\text{particles } A}' m_A \dot{x}_A = p.$$

The second time-rate of change of the mass dipole moment has to vanish because of the law of conservation of momentum, $\ddot{d} = \dot{p} = 0$. Therefore *there can be no mass dipole radiation in gravitation physics*.

The next strongest types of electromagnetic radiation are magnetic-dipole and electric-quadrupole. Magnetic-dipole radiation is generated by the second time-derivative of the magnetic moment, $\ddot{\mu}$. Here again the gravitational analog is a constant of the motion, the angular momentum,

$$\mu = \sum_A (\text{position of } A) \times (\text{current due to } A) = \sum_A r_A \times (m v_A) = J;$$

so it cannot radiate. *Thus, there can be no gravitational dipole radiation of any sort.*

When one turns to quadrupole radiation, one finally gets a nonzero result (see Figure 36.1). The power output predicted by electromagnetic theory,

$$L_{\text{electric quadrupole}} = \frac{1}{20} \dddot{Q}^2 \equiv \frac{1}{20} \dddot{Q}_{jk} \dddot{Q}_{jk},$$

$$Q_{jk} \equiv \sum_A e_A \left(x_{Aj} x_{Ak} - \frac{1}{3} \delta_{jk} r_A^2 \right)$$

($Q_{jk\,\text{here}} = Q_{jk\,\text{in much other literature}}$), has as its gravitational counterpart

$$L_{\text{mass quadrupole}} = \frac{1}{5} \langle \dddot{I}^2 \rangle \equiv \frac{1}{5} \langle \dddot{I}_{jk} \dddot{I}_{jk} \rangle, \tag{36.1}$$

Gravitational-wave power output expressed in terms of "reduced quadrupole moment" of source

$$I_{jk} \equiv \sum_A m_A \left(x_{Aj} x_{Ak} - \frac{1}{3} \delta_{jk} r_A^2 \right) = \int \rho \left(x_j x_k - \frac{1}{3} \delta_{jk} r^2 \right) d^3x. \tag{36.2}$$

Formula (36.1) contains the correct factor of 1/5, which comes from tensor calculations (see §36.10), instead of the incorrect factor 1/20 suggested by the electromagnetic analog; and the righthand side of (36.1) has been averaged ("$\langle \; \rangle$") over several characteristic periods of the source to accord with one's inability to localize the energy of gravitational radiation inside a wavelength.

Source Receptor

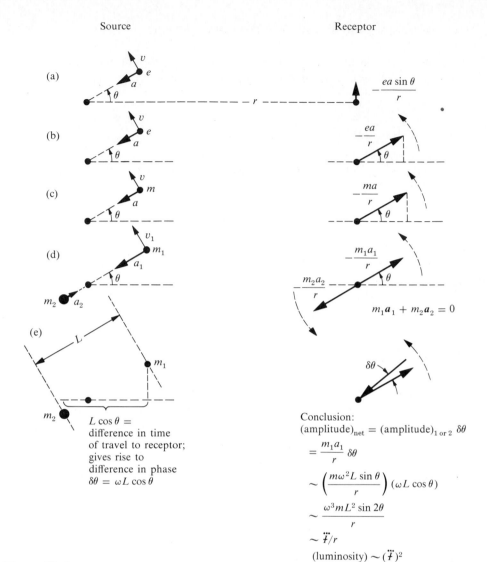

Figure 36.1.

Why gravitational radiation is ordinarily weak. In brief, contributions to the amplitude of the outgoing wave from the mass dipole moments of the separate masses cancel, $(m_1\mathbf{a}_1 + m_2\mathbf{a}_2)/r = 0$ (principle that action equals reaction).

(a) Radiation from an accelerated charge (see §4.4 and Figure 4.6).

(b) Representation of the field at the great distance r in terms of the typical rotating-vector diagram of electrical engineering; however, here, for ease of visualization, the *vertical* projection of the rotating vector gives the observed field (usual dipole-radiation field produced by a charge in circular orbit).

(c) Corresponding rotating-vector diagram for gravitational radiation, based on the simplified model of the gravitation field as a spin-one or vector field (to be contrasted with its true tensor character; hence details of angular distribution and total radiation as given by this simple diagram are not correct; but order of magnitude of luminosity is correct).

(d) The two masses m_1 and m_2 that hold each other in orbit give equal and opposite contributions to the amplitude of the outgoing wave because of the principle that action equals reaction. (In electromagnetic radiation from a hydrogen atom, the corresponding radiation amplitudes do not cancel: $e_{\text{elec}}\mathbf{a}_{\text{elec}} + e_{\text{prot}}\mathbf{a}_{\text{prot}} \sim e_{\text{elec}}\mathbf{a}_{\text{elec}} \neq 0$).

(e) In a better approximation, one has to allow for the difference in time of arrival at the receptor of the effects from the two masses. The two vectors that formerly opposed each other exactly are now drawn inclined, at the phase angle $\delta\theta$. The amplitude of the resulting field goes as $\dddot{\textit{Ｉ}}$, where $\textit{Ｉ}$ is the reduced quadrupole moment; and the luminosity is proportional to $\dddot{\textit{Ｉ}}^2$.

Notation: There is no ambiguity about the definition of the *"second moment of the mass distribution"* as it appears throughout the physics and mathematics literature

$$I_{jk} \equiv \int \rho x_j x_k \, d^3x.$$

Nor is there any ambiguity about how one constructs the moment of inertia tensor \mathcal{I}_{jk} from this second moment of the mass distribution:

$$\mathcal{I}_{jk} = \delta_{jk} \, \text{trace} \, (I_{ab}) - I_{jk} = \int \rho(r^2 \, \delta_{jk} - x_j x_k) \, d^3x.$$

The moments that characterize a source radiating quadrupole gravitational radiation are here taken, equally unambiguously, to be the *"trace-free part of the second moment of the mass distribution"*:

$$\mathcal{I}_{jk} = I_{jk} - \frac{1}{3} \delta_{jk} \, \text{trace} \, (I_{ab}) = I_{jk} - \frac{1}{3} \delta_{jk} I - \int \rho \left(x_j x_k - \frac{1}{3} \delta_{jk} r^2 \right) d^3x. \quad (36.3)$$

This notation is adopted because it simplifies formulas, it simplifies calculations, it meshes well with much of the literature of gravitational-wave theory [e.g. Peters (1964), Peres and Rosen (1964)], and it is easy to remember. Another name for the quantities \mathcal{I}_{jk} is *reduced quadrupole moment*. This terminology makes clear the distinction between the quantities used here and the three-times-larger quantities that are called quadrupole moments in the standard text of Landau and Lifshitz (1962) and in the literature on nuclear quadrupole moments, and the 3/2-times-larger quantities used in the theory of spherical harmonics:

$$Q_{zz} \begin{pmatrix} \text{Landau and Lifshitz; also} \\ \text{nuclear quadrupole moments} \end{pmatrix} = \int \rho(3z^2 - r^2) \, d^3x,$$

$$Q_{zz} \, (\text{theory of spherical harmonics}) = \int \rho \left(\frac{3}{2} z^2 - \frac{1}{2} r^2 \right) d^3x,$$

$$\mathcal{I}_{zz} \begin{pmatrix} \text{reduced quadrupole moment;} \\ \text{unambiguous measure of} \\ \text{source strength adopted here} \end{pmatrix} = \int \rho \left(z^2 - \frac{1}{3} r^2 \right) d^3x.$$

Thus the \mathcal{I}_{jk} notation has the merit of circumventing the existing ambiguity in the literature.

That electromagnetic radiation is predominantly dipolar (spherical-harmonic index $l = 1$), and gravitational radiation is quadrupolar ($l = 2$) are consequences of a general theorem. Consider a classical radiation field, whose associated quantum mechanical particles have integer spin S, and zero rest mass. Resolve that radiation field into spherical harmonics—i.e., into multipole moments. All components with $l < S$ will vanish; in general those with $l \geq S$ will not; and this is independent of the nature of the source! [See, e.g., Couch and Newman (1972).] Since the lowest nonvanishing multipoles generally dominate for a slowly moving source (speeds $\ll c$), electromagnetic radiation ($S = 1$) is ordinarily dipolar ($l = S = 1$), while gravita-

Why gravitational waves are ordinarily quadrupolar

tional radiation ($S = 2$) is ordinarily quadrupolar ($l = S = 2$). Closely connected with this theorem is the "topological fixed-point theorem" [e.g., Lifshitz (1949)], which distinguishes between scalar, vector, and tensor fields. For a scalar disturbance, such as a pressure wave, there is no difficulty in having a spherically symmetric source. Thus, over a sphere of a great radius r, there is no difficulty in having a pressure field that everywhere, at any one time, takes on the same value p. In contrast, there is no way to lay down on the surface of a 2-sphere a continuous vector field, the magnitude of which is non-zero and everywhere the same ("no way to comb smooth the hair on the surface of a billiard ball"). Likewise, there is no way to lay down on the surface of a 2-sphere a continuous non-zero transverse-traceless 2×2 matrix field that differs from one point to another at most by a rotation. Topology thus excludes the possibility of any spherically symmetric source of gravitational radiation whatsoever.

§36.2. POWER RADIATED IN TERMS OF INTERNAL POWER FLOW

Expression (36.1) for the power output can be rewritten in a form that is easier to use in order-of-magnitude estimates. Notice that the reduced quadrupole moment is

$$\dddot{\mathcal{I}}_{jk} \sim \frac{\left(\begin{array}{c}\text{mass of that part of}\\\text{system which moves}\end{array}\right) \times \left(\begin{array}{c}\text{size of}\\\text{system}\end{array}\right)^2}{\left(\begin{array}{c}\text{time for masses to move from}\\\text{one side of system to other}\end{array}\right)^3} = \frac{MR^2}{T^3}$$

$$\sim \frac{M(R/T)^2}{T} \sim \frac{\left(\begin{array}{c}\text{nonspherical part}\\\text{of kinetic energy}\end{array}\right)}{T};$$

$$\dddot{\mathcal{I}}_{jk} \sim L_{\text{internal}} \equiv \left(\begin{array}{c}\text{power flowing from one}\\\text{side of system to other}\end{array}\right): \qquad (36.4)$$

Gravitational-wave power output in terms of internal power flow of source

Consequently, equation (36.1) says that *the power output in gravitational waves* ("*luminosity*") *is roughly the square of the internal power flow*

$$L_{\text{GW}} \sim (L_{\text{internal}})^2. \qquad (36.5)$$

If this equation seems crazy (who but a fool would equate a power to the square of a power?), recall that in geometrized units power is dimensionless. The conversion factor to conventional units is

$$L_o \equiv c^5/G = 3.63 \times 10^{59} \text{ erg/sec} = 2.03 \times 10^5 M_\odot c^2/\text{sec}. \qquad (36.6)$$

One may freely insert this factor of $L_o = 1$ wherever one wishes in order to feel more comfortable with the appearance of the equations. For example, one can rewrite equation (36.5) in the form

$$L_{\text{GW}}/L_{\text{internal}} \sim L_{\text{internal}}/L_o. \qquad (36.7)$$

In applying the equation $L_{GW} \sim (L_{internal})^2$, one must be careful to ignore those internal power flows that cannot radiate at all, i.e., those that do not accompany a time-changing quadrupole moment. For example, in a star the internal power flows associated with spherical pulsation and axially symmetric rotation must be ignored.

Conservation of energy guarantees that radiation reaction forces will pull down the internal energy of the system at the same rate as gravitational waves carry energy away (see Box 19.1). The characteristic time-scale for radiation reaction to change the system markedly is

Characteristic time-scale for radiation-reaction effects

$$\tau_{react} \sim [1/(\text{rate at which energy is lost})] \times [\text{energy in motions that radiate}]$$
$$\sim [1/L_{GW}] \times [(L_{internal}) \times (\text{characteristic period } T \text{ of internal motions})]$$
$$\sim (L_{internal}/L_{GW})T \sim (L_0/L_{internal})T. \tag{36.8}$$

Consequently, *radiation reaction is important in one characteristic period only if the system achieves the enormous internal power flow*

$$L_{internal} \geq L_o = 3.63 \times 10^{59} \text{ ergs/sec} = 1!$$

§36.3. LABORATORY GENERATORS OF GRAVITATIONAL WAVES

As a laboratory generator of gravitational waves, consider a massive steel beam of radius $r = 1$ meter, length $l = 20$ meters, density $\rho = 7.8$ g/cm^3, mass $M = 4.9 \times 10^8$ g (490 tons), and tensile strength t = 40,000 pounds per square inch or 3×10^9 dyne/cm^2. Let the beam rotate about its middle (so it rotates end over end), with an angular velocity ω limited by the balance between centrifugal force and tensile strength

Power output from a rotating steel beam

$$\omega = (8t/\rho l^2)^{1/2} = 28 \text{ radians/sec.}$$

The internal power flow is

$$L_{internal} = \left(\frac{1}{2} I\omega^2\right)\omega = \frac{1}{24} Ml^2\omega^3$$
$$\approx 2 \times 10^{18} \text{ erg/sec} \approx 10^{-41} L_o.$$

The order of magnitude of the power radiated is

$$L_{GW} \sim (10^{-41})^2 L_o \sim 10^{-23} \text{ erg/sec.} \tag{36.9}$$

(An exact calculation using equation (36.1) gives 2.2×10^{-22} erg/sec; see Exercise 36.1.) Evidently the construction of a laboratory generator of gravitational radiation is an unattractive enterprise in the absence of new engineering or a new idea or both.

To rely on an astrophysical source and to build a laboratory or solar-system detector is a more natural policy to consider. Detection will be discussed in the next chapter. Here attention focuses on astrophysical sources.

EXERCISE

Exercise 36.1. GRAVITATIONAL WAVES FROM ROTATING BEAM

A long steel beam of length l and mass M rotates end over end with angular velocity ω. Show that the power it radiates as gravitational waves is

$$L_{GW} = \frac{2}{45} M^2 l^4 \omega^6. \tag{36.10}$$

Use this formula to verify that the rod described in the text radiates 2.2×10^{-22} ergs/sec.

§36.4. ASTROPHYSICAL SOURCES OF GRAVITATIONAL WAVES: GENERAL DISCUSSION

Consider a highly dynamic astrophysical system (a star pulsating and rotating wildly, or a collapsing star, or an exploding star, or a chaotic system of many stars). If its mass is M and its size is R, then according to the virial theorem (exercise 39.6) its kinetic energy is $\sim M^2/R$. The characteristic time-scale for mass to move from one side of the system to the other, T, is

$$T \sim \frac{R}{\text{(mean velocity)}} \sim \frac{R}{(M/R)^{1/2}} = \left(\frac{R^3}{M}\right)^{1/2} \tag{36.11a}$$

(\sim time of free fall; \sim time to turn one radian in Kepler orbit; see Chapter 25). Consequently, the internal power flow is

$$L_{\text{internal}} \sim \frac{\text{(kinetic energy)}}{T} \sim \left(\frac{M^2}{R}\right)\left(\frac{M}{R^3}\right)^{1/2} \sim \left(\frac{M}{R}\right)^{5/2}. \tag{36.11b}$$

The gravitational-wave output or "luminosity" is the square of this quantity, or

Power output from violent astrophysical sources, in terms of mass and radius

$$L_{GW} \sim (M/R)^5 L_o. \tag{36.11c}$$

(If the system is rather symmetric, or if only a small portion of its mass is in motion, then its quadrupole moment does not change much, and the estimate of L_{GW} must be reduced accordingly. The wave amplitude goes down in proportion to the fraction of the mass in motion, and the power is reduced in proportion to the square of that fraction.)

Clearly, *the maximum power output occurs when the system is near its gravitational radius;* and because nothing, not even gravitational waves, can escape from inside the gravitational radius, *the maximum value of the output is* $\sim L_o = 3.63 \times 10^{59}$ *ergs/second, regardless of the nature of the system!*

Upper limit on power output

Actually, the above derivation of this limit and of equation (36.11c) uses approximations to general relativity that break down near the gravitational radius. [Velocities small compared to light are required in deriving the standard formula (36.1) for

L_{GW} (see §36.7); nearly Newtonian fields are required for the virial theorem arguments of (36.11a), as well as for the L_{GW} formula.] Nevertheless, in rough order of magnitude, equation (36.11c) is valid to quite near the Schwarzschild radius, say, $R \sim 3M$; and inside that point gravity is so strong that no system can resist collapse for an effective length of time much longer than $T \sim M$.

The time required for radiation-reaction forces to affect a system substantially [equation (36.8)] is of the order

Radiation reaction in
astrophysical sources

$$\tau_{\text{react}} \sim (L_o/L_{\text{internal}})T \sim (R/M)^{5/2}T, \qquad (36.11\text{d})$$

where T is the characteristic time (36.11a) of rotation or free fall. (Note how one inserts and removes the factor $L_o = 1$ at will!) Consequently, *the effect of radiation reaction, as integrated over one period, is unimportant except when the system is near its gravitational radius.*

When a system such as a pulsating star is settling down into an equilibrium state, the radiation reaction will damp its internal motions. On the other hand, when the system, like a binary star system, is far from any state of equilibrium, then loss of energy (and angular momentum) to radiation under certain circumstances may speed up the angular velocity or speed up the internal motions and augment the radiation.

§36.5. GRAVITATIONAL COLLAPSE, BLACK HOLES, SUPERNOVAE, AND PULSARS AS SOURCES

Since $L_{GW} \sim (M/R)^5 L_o$, the most intense gravitational waves reaching Earth must come from a dynamic, deformed system near its gravitational radius (L_{GW} drops by a factor 100,000 with each increase by 10 of R!). The scenario of Figure 24.3 gives an impression of some of the dynamic processes that not only may happen but probably must happen. The sequence of events sketched out there includes pulses of gravitational radiation interspersed with intervals of continuous radiation of gradually increasing frequency. Pulse number one comes at the time of the original collapse of a star with white-dwarf core to a pancake-shaped neutron star. The details of what then goes on will differ enormously depending on the original mass and angular momentum of this "pancake." In the illustration, this pancake fragments into a constellation of corevolving neutron stars, which then one by one undergo "pursuit and plunge."

Gravitational waves from:

(1) stellar collapse and
 formation of a black hole

Whether in this kind of scenario or otherwise, perhaps the most favorable source of gravitational radiation is a star (the original very temporary "pancake" or one of the fragments therefrom) collapsing through its gravitational radius in a highly nonspherical manner. Such a star should terminate life with a last blast of gravitational waves, which carry off a sizeable fraction of its rest mass. Thus an order-of-magnitude estimate gives

$$\text{(energy radiated)} = \int L_{GW}\, dt \sim L_o \cdot \begin{pmatrix}\text{time during which} \\ \text{peak luminosity occurs}\end{pmatrix} \qquad (36.12)$$
$$\sim L_o M = M.$$

(Whether the energy radiated is $0.9M$, or $0.1M$, or $0.01M$ is not known for certain today; but it must lie in this range of orders of magnitude.) The radiation should be weak at low frequencies; it should rise to a peak at a frequency a little smaller than $1/M$; and it should cut off sharply for circular frequencies above $\omega \sim 1/M$.

(2) the fall of debris into a black hole

Matter ("debris"; see Figure 24.3) falling into a black hole can also be a significant source of gravitational waves. The infalling matter will radiate only weakly when it is far from the gravitational radius; but as it falls through the gravitational radius (between $r \sim 4M$ and $r = 2M$), it should emit a strong burst. If m is the mass of an infalling lump of matter and M is the total mass of the black hole, then the total energy in the final burst is

$$E_{\text{radiated}} \sim m^2/M, \tag{36.13}$$

and it comes off in a time $\sim M$ with a power output of $L_{\text{GW}} \sim (m/M)^2 L_o$. (See exercise 36.2.) Actually, this is an extremely rough estimate of the energy output. In the limit where the infalling lump is small in both size and mass [(size of lump) \ll (gravitational radius of black hole); $m \ll M$; "delta-function lump"], one can perform an exact calculation of the spectrum and energy radiated by treating the lump and the waves as small perturbations on the Schwarzschild geometry of the black hole. The foundations for such a treatment were given by Zerilli (1970b). Zerilli's formula was corrected and applied to the case of head-on impact by Davis, Ruffini, Press, and Price (1971). They predict the spectrum of Figure 36.2 and the total energy output

$$E_{\text{radiated}} = 0.0104 m^2/M \tag{36.14}$$

for $m \ll M$ and (size of lump) $\ll M$.

(3) collisions of black holes

A collision between black holes should also produce a strong burst of gravitational waves—through such collisions are probably very rare!

(4) supernova explosions

Not quite so rare, but still not common, are supernova explosions (about one per galaxy per 100 years). According to current theory as verified by pulsar observations, a supernova is triggered by the collapse of the core of a highly evolved star (see §24.3). The collapse itself and the subsequent wild gyrations of the collapsed core (neutron star) should produce a short, powerful burst of gravitational waves. The characteristics of the burst, as estimated with formulas (36.11), and assuming large departures from sphericity, are

$$\text{(energy radiated)} \sim \text{(neutron-star binding energy)}$$
$$\sim M^2/R \sim 0.1M \sim 10^{53} \text{ ergs,}$$

$$\text{(mean frequency)} \sim 1/T \sim (M/R^3)^{1/2} \sim 0.03M^{-1} \sim 3000 \text{ Hz,} \tag{36.15}$$

$$\text{(power output)} \sim (M/R)^5 L_o \sim 10^{-5} L_o \sim 3 \times 10^{54} \text{ ergs/sec,}$$

$$\begin{pmatrix} \text{time for gravitational} \\ \text{radiation to damp the} \\ \text{motion if turbulence,} \\ \text{heat conduction, and other} \\ \text{effects do not damp it} \\ \text{sooner} \end{pmatrix} = \tau \sim M(M/R)^{-4} \sim 0.1 \text{ sec} \sim 300 \text{ periods.}$$

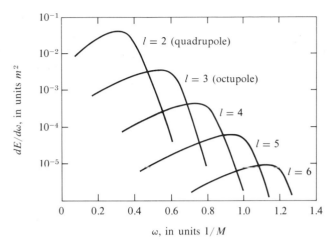

Figure 36.2.

Spectrum of the gravitational waves emitted by a "delta-function" lump of matter of mass m, falling head-on into a nonrotating (Schwarzschild) black hole of mass $M \gg m$. The total energy radiated is distributed among multipoles according to the empirical law

$$(\text{energy in } l\text{-pole waves}) \approx (0.44\ m^2/M)e^{-2l},$$

and the total spectrum peaks at angular frequency

$$\omega_{\max} = 0.32/M.$$

These results were calculated by treating the infalling lump and the gravitational waves as small perturbations on the Schwarzschild geometry of the black hole. The relevant perturbation-theory equations were derived by Zerilli (1970), and were solved numerically to give these results by Davis, Ruffini, Press, and Price (1971).

In the last stages of the stellar pulsations, when the amplitude $\xi = \delta r$ has dropped to $\delta r/r \ll 1$, one can calculate the pulsation frequencies and damping times exactly by treating the fluid motions and gravitational waves as small perturbations of an equilibrium stellar model. The results of such a calculation, which are in good agreement with the above rough estimates, are shown in Box 36.1.

Long after the pulsations of the neutron star have been damped out by gravitational radiation reaction and by other forces, the star will continue to rotate; and as it rotates, carrying along with its rotation an off-axis-pointing magnetic moment, it will beam out the radio waves, light, and x-rays that astronomers identify as "pulsar radiation." In this pulsar phase, gravitational radiation is important only if the star is somewhat deformed from axial symmetry (axial symmetry \Longrightarrow constant quadrupole moment \Longrightarrow no gravitational waves). According to estimates in exercise 36.3, a deformation that contains only 0.001 of the star's mass could radiate 10^{38} ergs per second for the youngest known pulsar (Crab nebula); and the accompanying radiation reaction could be a significant source of the pulsar's slowdown. However, it is not at all clear today (1973)—indeed, it seems unlikely—that the neutron star could support even so small a deformation.

(5) young pulsars

(*continued on page 986*)

Box 36.1 GRAVITATIONAL WAVES FROM PULSATING NEUTRON STARS

The table given here, taken from Thorne (1969a), shows various characteristics of the quadrupole oscillations of several typical neutron-star models. Note that the gravitational waves emitted by the most massive models (1) have frequencies $\nu = 1/T_n \sim 3{,}000$ Hz, (2) last for a time of $\sim\frac{1}{3}$ second, (3) damp out the stellar vibrations after only $\sim 1{,}000$ oscillations, and (4) carry off a total energy of $\sim (10^{54}$ ergs$) \times (\delta R/R)^2$, where $\delta R/R$ is the initial fractional amplitude of vibration of the star's surface.

These results are *not* based on the nearly Newtonian slow-motion formalism of this chapter [equation (36.1), §§36.7 and 36.8], because that formalism is invalid here: the reduced wavelength of the radiation, $\lambdabar \sim 15$ km for waves from the most massive star, is not large compared to the star's gravitational radius, $2M \sim 6$ km; and the star's internal gravitational field is not weak (M/R as large as 0.29). Consequently, these results were derived using an alternative technique, which is valid for rapid motions and strong internal fields, but which assumes small perturbations away from the equilibrium stellar model. See Thorne (1969a) and papers cited therein for details.

QUADRUPOLE PULSATIONS OF NEUTRON STARS

Equation of state	ρ_c (g cm^{-3})	M/M_\odot	$2M/R$	n	T_n (msec)	τ_n (sec)	τ_n/T_n	$\dfrac{E_M}{\langle(\delta R/R)^2\rangle}$ (ergs)	$\dfrac{Power}{\langle(\delta R/R)^2\rangle}$ (ergs sec^{-1})	$\dfrac{(\delta R/R)}{(\delta r/r)_c}$	$\delta\theta_s/\delta\theta_c$
H-W	3×10^{14}	0.405	0.0574	0	1.197	13.	11000	7.8×10^{50}	1.2×10^{50}	$+\ 7.4$	$+\ 3.1$
H-W	6×10^{15}	0.682	0.240	0	0.3109	0.19	610	2.8×10^{52}	2.9×10^{53}	$+\ 5.2$	$+\ 3.7$
				1	0.1713	0.28	1600	3.6×10^{51}	2.6×10^{52}	$+\ 14.$	$-\ 3.3$
				2	0.1179	1.3	11000	2.6×10^{50}	3.9×10^{50}	$+\ 55.$	$+\ 5.9$
				3	0.0938	24.	250000	8.9×10^{48}	$7.\times10^{47}$	$-350.$	$-24.$
V_γ	5.15×10^{14}	0.677	0.159	0	0.6991	1.7	2400	5.7×10^{52}	$7.\times10^{52}$	$+\ 1.4$	$+\ 1.3$
				1	0.2358	11.	47000	6.0×10^{50}	1.1×10^{50}	$-\ 38.$	$-\ 4.7$
V_γ	3×10^{15}	1.954	0.580	0	0.3777	0.22	600	1.7×10^{54}	1.6×10^{55}	$+\ 1.9$	$+\ 3.1$
				1	0.1556	1.6	10000	1.5×10^{54}	1.9×10^{54}	$-\ 2.1$	$-\ 0.66$
				2	0.1026	2.6	25000	5.2×10^{53}	4.0×10^{53}	$+\ 2.9$	$+\ 0.40$

The columns in the table have the following meanings.

Equation of state: the equation of state $p(\rho)$ used in constructing the equilibrium stellar model and in calculating the adiabatic index from $\gamma = [(\rho + p)/p]\, dp/d\rho$; H-W is the Harrison-Wheeler equation of state in the tabular form given by Hartle and Thorne (1968), Table 1; V_γ is the Levinger-Simmons-Tsuruta-Cameron V_γ equation of state in the tabular form given by Hartle and Thorne (1968), Table 2.

ρ_c: central density of total mass-energy for the equilibrium stellar model.

M/M_\odot: total mass-energy of the equilibrium model (i.e., the mass that governs distant Keplerian orbits), in units of the sun's mass.

$2M/R = 2GM/Rc^2$: ratio of the gravitational radius of the equilibrium model to its actual radius (radii are defined by $4\pi R^2 = $ surface area).

n: the "order" of the pulsational normal mode under study (for all models given here, n is also the number of nodes in the radial relative eigenfunction, $\delta r/r$.). Note: $n = 0$ is the fundamental (quadrupole) mode.

$T_n = 2\pi/\omega_n$: the pulsation period of the quasinormal mode measured in milliseconds.

τ_n: the damping time for the amplitude of the normal mode measured in seconds.

$\tau_n/T_n = \omega_n \tau_n/2\pi$: the number of pulsation periods required for the amplitude to drop by a factor of $1/e$.

$E_{\mathscr{R}}/\langle(\delta R/R)^2\rangle$: energy of pulsation of the star, divided by the square of the relative amplitude of radial motion of the star's surface averaged over its surface.

Power$/\langle(\delta R/R)^2\rangle$: the power radiated as gravitational waves, divided by the averaged square of the relative amplitude at the star's surface.

$(\delta R/R)(\delta r/r)_c^{-1}$: relative amplitude of radial motion at the star's surface divided by relative amplitude at the star's center.

$\delta\theta_s/\delta\theta_c = \delta\phi_s/\delta\phi_c$: amplitude of the angular displacement of the star's fluid at its surface divided by the same amplitude at its center.

Of the sources discussed in this section, most are "impulsive" rather than continuous (star collapsing through gravitational radius; debris falling into a black hole; collision between black holes; supernova explosion). They give rise to bursts of gravitational waves. An order-of-magnitude method of analyzing such bursts is spelled out in Box 36.2.

It is difficult and risky to pass from the above description of processes that should generate gravitational waves to an estimate of the characteristics of the waves that actually bathe the earth. For such an estimate, made in 1972 and subject to extensive revision as one's understanding of the universe improves, see Press and Thorne (1972).

EXERCISES

Exercise 36.2. GRAVITATIONAL WAVES FROM MATTER FALLING INTO A BLACK HOLE

A lump of matter with mass m falls into a black hole of mass M. Show that a burst of gravitational waves is emitted with duration $\sim M$ and power $L_{GW} \sim (m/M)^2 L_o$, so that the total energy radiated is given in crude order of magnitude by equation (36.13).

Exercise 36.3. GRAVITATIONAL WAVES FROM A VIBRATING NEUTRON STAR

Idealize a neutron star as a sphere of incompressible fluid of mass M and radius R, with structure governed by Newton's laws of gravity. Let the star pulsate in its fundamental quadrupole mode. Using Newtonian theory, calculate: the angular frequency of pulsation, ω; the energy of pulsation E_{puls}; the quantity $\frac{1}{5}\langle \dddot{I}^2 \rangle$, which, according to equation (36.1), is the power radiated in gravitational waves, L_{GW}; and the e-folding time, $\tau = E_{puls}/L_{GW}$, for radiating away the energy of the pulsations. Compare the answers with equations (36.15)—which are based on a much cruder approximation—and with the results in Box 36.1, which are based on much better approximations. [For solution, see Table 13 of Wheeler (1966).]

Exercise 36.4. PULSAR SLOWDOWN

The pulsar NPO532 in the Crab Nebula has a period of 0.033 seconds and is slowing down at the rate $dP/dt = 1.35 \times 10^{-5}$ sec/yr. Assuming the pulsar is a typical neutron star, calculate the rate at which it is losing rotational energy. If this energy loss is due primarily to gravitational radiation reaction, what is the magnitude of the star's nonaxial deformation? [For solution, see Ferrari and Ruffini (1969); for a rigorous strong-field analysis, see Ipser (1970).]

§36.6. BINARY STARS AS SOURCES

Binary stars as sources of gravitational waves:

The most numerous sources of weak gravitational waves are binary star systems. Moreover, roughly half of all stars are in binary or multiple systems [see, for example, the compilation of Allen (1962)]. According to Kepler's laws, two stars of masses m_1 and m_2 that circle each other have angular frequency ω and separation a coupled to each other by the formula

$$\omega^2 a^3 = m_1 + m_2 \equiv M.$$

Box 36.2 ANALYSIS OF BURSTS OF RADIATION FROM IMPULSE EVENTS*

	Electromagnetism	Gravitation
Typical moment relevant for radiation	$d_x(t)$	$\textit{Ŧ}_{xx}(t)$
Its Fourier transform	$(2\pi)^{-1/2}\int d_x \exp[i\omega t]\,dt$	$(2\pi)^{-1/2}\int \textit{Ŧ}_{xx} \exp[i\omega t]\,dt$
Name for this quantity	$d_x(\omega)$	$\textit{Ŧ}_{xx}(\omega)$
Time decomposition of total radiative energy loss ΔE	$c^{-3}\int \dddot{d}^2(t)\,dt$	$Gc^{-5}\int \dddot{\textit{Ŧ}}^2(t)\,dt$
Decomposition of ΔE according to circular frequency	$c^{-3}\int \dddot{d}^2(\omega)\,d\omega$	$Gc^{-5}\int \dddot{\textit{Ŧ}}^2(\omega)\,d\omega$
Integrand nearly constant with respect to ω from $\omega = 0$ up to a critical value of ω, beyond which radiation falls off very fast	$\omega_{\text{crit}} \sim 1/\Delta t$	$\omega_{\text{crit}} \sim 1/\Delta t$
$-d\,\Delta E/d\omega$ for $\omega < \omega_{\text{crit}}$	$\sim c^{-3}\dddot{d}^2(0)$	$\sim Gc^{-5}\dddot{\textit{Ŧ}}^2(0)$
Zero frequency moment that enters this formula	$\sim(e_1\,\Delta v_{x1} + e_2\,\Delta v_{x2})$	$\Delta(\text{«Kinetic Energy»})_{xx}$
Rewrite of $-d\,\Delta E/d\omega$	$\sim(e\,\Delta v)^2/c^3$	$\sim G[\Delta(\text{«K.E.»})_{xx}]^2/c^5$
Total energy of pulse	$\sim\text{This}/\Delta t$	$\sim\text{This}/\Delta t$

*Box adapted from pp. 113 and 114 of Wheeler (1962).

As sample applications of this analysis, Wheeler (1962) cites the following:

Parameter	One atomic-nucleus fission of 180 MeV	Fission bomb yield 17 kilotons at 10% efficiency	Meteorite striking earth at escape velocity	Explosion of star when 10^{-4} of mass is released
Mass	4×10^{-22} g	10^4 g	10^9 g	2×10^{33} g
Velocity	1.2×10^9 cm/s	4×10^8 cm/s	11×10^5 cm/s	4×10^8 cm/s
Energy	2.9×10^{-4} erg	7×10^{20} erg	6×10^{20} erg	1.8×10^{50} erg
Fraction assumed relevant to radiative moment	1	0.1	1	0.1
Time integral of this moment $= $ «K.E.»$_{xx}$	2.9×10^{-4} erg	7×10^{19} erg	6×10^{20} erg	1.8×10^{49} erg
«K.E.»$_{xx}/c^2$	3.2×10^{-25} g	0.08 g	0.67 g	2×10^{28} g
$\dfrac{dE}{d\omega} \sim \dfrac{G}{c}\left(\dfrac{\text{«K.E.»}_{xx}}{c^2}\right)^2$	$2.3 \times 10^{-67}\dfrac{\text{erg}}{\text{rad/s}}$	$1.4 \times 10^{-20}\dfrac{\text{erg}}{\text{rad/s}}$	$1.0 \times 10^{-18}\dfrac{\text{erg}}{\text{rad/s}}$	$9 \times 10^{38}\dfrac{\text{erg}}{\text{rad/s}}$
Δt	10^{-21} s	10^{-8} s	10^{-3} s	10^4 s
$\Delta\omega \sim 1/\Delta t$	10^{21} rad/s	10^8 rad/s	10^3 rad/s	10^{-4} rad/s
$\Delta E_{\text{radiated}}$	10^{-46} erg	10^{-12} erg	10^{-15} erg	10^{35} erg
Assumed distance to detector	10^3 cm	10^3 cm	10^9 cm	10^{23} cm
$\Delta E/4\pi r^2$	10^{-53} erg/cm^2	10^{-19} erg/cm^2	10^{-34} erg/cm^2	10^{-12} erg/cm^2

The reader might find it informative to extend this table to the bursts of waves emitted by (1) debris falling into a black hole, (2) collisions between two black holes, and (3) a supernova explosion in which a star of two solar masses collapses to nuclear densities, ejecting half its mass in the process.

In this motion the kinetic energy is

$$(\text{kinetic energy}) = -\frac{1}{2}(\text{potential energy}) = \frac{1}{2}\frac{m_1 m_2}{a}.$$

The power that they radiate as gravitational waves can be estimated roughly as the square of the circulating power, $L \sim \omega \times (\text{kinetic energy})$; thus,

$$L_{\text{GW}} \sim \frac{\mu^2 M^3}{4a^5} L_o,$$

where $\mu = m_1 m_2 / M$ is the familiar reduced mass, and $M = m_1 + m_2$ is the total mass of this binary system.

An *exact* calculation based on equation (36.1) gives a result larger than this by a factor ~ 30: for a binary system of semimajor axis a and eccentricity ε, the power output averaged over an orbital period is

(1) power output

$$L_{\text{GW}} = \frac{32}{5}\frac{\mu^2 M^3}{a^5} f(\varepsilon)L_o, \qquad (36.16a)$$

where $f(\varepsilon)$ is the dimensionless "correction function,"

$$f(\varepsilon) = \left[1 + \frac{73}{24}\varepsilon^2 + \frac{37}{96}\varepsilon^4\right][1 - \varepsilon^2]^{-7/2}. \qquad (36.16b)$$

[See exercise 36.6 at end of §36.8; also Peters and Mathews (1963).]

(2) effects of radiation reaction

As the binary system loses energy by gravitational radiation, the stars spiral in toward each other (decrease of energy; tightening of gravitational binding). For circular orbits the energy, $E = -\frac{1}{2}m_1 m_2/a = -\frac{1}{2}\mu M/a$, decreases as

$$dE/dt = 1/2(\mu M/a^2)(da/dt)$$

$$= -L_{\text{GW}} = -\frac{32}{5}\frac{\mu^2 M^3}{a^5}.$$

Consequently, the evolution of the orbital radius is given by the formula

$$a = a_o(1 - t/\tau_o)^{1/4}, \qquad (36.17a)$$

where $a_o = a_{\text{today}}$ and

$$\tau_o = \frac{1}{4}\left(\frac{-E}{L_{\text{GW}}}\right)_{\text{today}} = \frac{5}{256}\frac{a_o^4}{\mu M^2}. \qquad (36.17b)$$

Thus, unless nongravitational forces intervene, the two stars will spiral together in a time τ_o (*spiral time*). For an elliptical orbit, the eccentricity also evolves. Radiation is emitted primarily at periastron. Therefore the braking forces of radiation reaction act there with greatest force. This effect deprives the stars of some of the kinetic energy of the excursions in their separation ("radial kinetic energy"). In consequence, the orbit becomes more nearly circular. [See Peters and Mathews (1963) for detailed calculations.]

The calculated power output, flux at Earth, and damping times are shown in Box 36.3 for several known binary stars and several interesting hypothetical cases. Notice that in the most favorable known cases the period is a few hours; the damping time is the age of the universe (could the absence of better cases be due to radiation reaction's having destroyed them?); the output of power in the form of gravitational waves is $\sim 10^{30}$ to 10^{32} ergs/sec (approaching the light output of the sun, 3.9×10^{33} ergs/sec); and the calculated flux at the Earth is $\sim 10^{-10}$ to 10^{-12} ergs/sec (too small to detect in 1973, but perhaps not too small several decades hence; see Chapter 37).

The hypothetical cases in Box 36.3 illustrate the general relations for astrophysical systems that were derived in §36.4—namely, that only as the system approaches its gravitational radius can L_{GW} approach L_o, and only then can damping remove nearly the whole energy in a single period.

<div style="text-align: right">(3) particular binaries
observed by astronomers</div>

§36.7. FORMULAS FOR RADIATION FROM NEARLY NEWTONIAN SLOW-MOTION SOURCES

Turn now from illustrative astrophysical sources to rigorous formulas valid for a wide variety of sources. One such formula has already been written down,

$$L_{\mathrm{GW}} = \frac{1}{5} \langle \dddot{I}_{jk} \dddot{I}_{jk} \rangle, \tag{36.1}$$

but it has not yet been derived, nor has its realm of validity been discussed.

This formula for the power output is actually valid for any "nearly Newtonian, slow-motion source"—more particularly, for any source in which

$$\text{(size of source)/(reduced wavelength of waves)} \ll 1, \tag{36.18a}$$

$$|\text{Newtonian potential}| \ll \text{(size of source)/(reduced wavelength)}, \tag{36.18b}$$

$$\frac{|\text{typical stresses}|}{\text{(mass density)}} \ll \frac{\text{(size of source)}}{\text{(reduced wavelength)}}. \tag{36.18c}$$

The "nearly Newtonian, slow-motion approximation" for analyzing sources of gravitational waves

It is not valid, except perhaps approximately, for fast-motion or strong-field sources. Moreover, there is no formalism available today which can handle effectively and *in general* the fast-motion case or the strong-field case.

The rest of this chapter is devoted to a detailed analysis of gravitational waves from nearly Newtonian, slow-motion sources. But the analysis (Track 2; §§36.9–36.11) will be preceded by a Track-1 summary in this section and the next.

For any source of size R and mean internal velocity v, the characteristic reduced wavelength ($\lambdabar = \lambda/2\pi$) of the radiation emitted is $\lambdabar \sim$ (amplitude of motions)$/v \lesssim R/v$. Consequently the demand (36.18a) that R/\lambdabar be $\ll 1$ [i.e., that the source be confined to a small region deep inside the near (nonradiation) zone] enforces the slow-motion constraint

$$v \ll 1.$$

Box 36.3 GRAVITATIONAL RADIATION FROM SEVERAL BINARY STAR SYSTEMS[a]

Type of system	Name	Period	$\frac{m_1}{M_\odot}$	$\frac{m_2}{M_\odot}$	Distance from earth (pc)	Spiral time[b]	L_{GW} (ergs/sec)	Flux at earth (erg/sec cm²)
Solar System (Sun + Jupiter)	Solar System	11.86 yr.	1.0	9.56×10^{-4}	Earth is in near zone	2.5×10^{23} yr	5.2×10^{10}	—
Typical resolved binaries from compilation of Van de Kamp (1958)	η Cas	480 yr.	0.94	0.58	5.9	9.5×10^{24}	5.6×10^{10}	1.4×10^{-29}
	ξ Boo	149.95 yr.	0.85	0.75	6.7	3.8×10^{23}	3.6×10^{12}	6.7×10^{-28}
	Sirius	49.94 yr.	2.28	0.98	2.6	7.2×10^{21}	1.1×10^{15}	1.3×10^{-24}
	Fu 46	13.12 yr.	0.31	0.25	6.5	3.2×10^{21}	3.6×10^{14}	7.1×10^{-26}
Typical eclipsing binaries from compilation of Gaposhkin (1958)	β Lyr	12.925 day	19.48	9.74	330	7.0×10^{11}	0.057×10^{30}	0.0004×10^{-11}
	UWCMa	4.395 day	40.0	31.0	1470	8.2×10^{9}	$49. \times 10^{30}$	0.019×10^{-11}
	β Per	2.867 day	4.70	0.94	30	3.2×10^{11}	0.014×10^{30}	0.013×10^{-11}
	WUMa	0.33 day	0.76	0.57	110	6.2×10^{9}	0.47×10^{30}	0.032×10^{-11}
Favorable cases from compilation of Braginsky (1965)	UV Leo	0.6 day	1.36	1.25	68	1.0×10^{10}	0.63×10^{30}	0.012×10^{-11}
	V Pup	1.45 day	16.6	9.8	390	2.3×10^{9}	$65. \times 10^{30}$	0.36×10^{-11}
	i Boo	0.268 day	1.35	0.68	12	2.0×10^{9}	3.2×10^{30}	$18. \times 10^{-11}$
	YY Eri	0.321 day	0.76	0.50	42	6.6×10^{9}	0.42×10^{30}	0.20×10^{-11}
	SW Lac	0.321 day	0.97	0.83	75	3.5×10^{9}	1.5×10^{30}	0.21×10^{-11}
	WZ Sge	81 min	0.6	0.03	100	1.1×10^{9} yr	0.5×10^{30}	0.04×10^{-11}
Hypothetical binaries (neutron stars or black holes)	10^4 km	12.2 sec	1.0	1.0	1000	3.2 yr	3.25×10^{41}	2.7×10^{-3}
	10^3 km	0.39 sec	1.0	1.0	1000	2.8 hr	3.25×10^{46}	2.7×10^{2}
	10^2 km	12.2 msec	1.0	1.0	1000	1.0 sec	3.25×10^{51}	2.7×10^{7}
	10 km	0.39 msec	1.0	1.0	1000	0.10 msec	3.25×10^{56}	2.7×10^{12}

[a] Based on tables by Braginsky (1965) and by Ruffini and Wheeler (1971b).
[b] The spiral time, τ_0, as given by equation (36.17b) is the time for the two stars to spiral into each other if no nongravitational forces intervene.

These related conditions, $v \ll 1$ and $R \ll \lambda$, are satisfied by all presently conceived laboratory generators of gravitational waves. No one has seen how to bring a macroscopic mass up to a speed $v \sim 1$. These conditions are also satisfied by every gravitationally bound, nearly Newtonian system. Thus, for such a system of mass M, the condition for gravitational binding, $\frac{1}{2}Mv^2 \lesssim M^2/R$ guarantees that $v \lesssim (M/R)^{1/2} \ll 1$.

The conditions $M/R \ll R/\lambda$ and $|T^{jk}|/T^{00} \ll R/\lambda$ are satisfied by all nearly Newtonian sources of conceivable interest. Typical sources (e.g. binary stars) have

$$\frac{M}{R} \sim \frac{|T^{jk}|}{T^{00}} \sim \left(\frac{R}{\lambda}\right)^2 \ll \frac{R}{\lambda}$$

(virial theorem). In those rare cases where $(M/R$ or $|T^{jk}|/T^{00}) \gtrsim R/\lambda$ (e.g., a marginally stable, slowly vibrating star), the motion is so very slow that the radiation will be too weak to be interesting.

For any nearly Newtonian slow-motion system, there is a spacetime region deep inside the near zone ($r \ll \lambda$), but outside the boundary of the source ($r > R$), in which vacuum Newtonian gravitation theory is nearly valid. An observer in this Newtonian region can measure the Newtonian potential Φ and can expand it in powers of $1/r$:

Definitions of mass, dipole moment, and reduced quadrupole moment for a slow-motion source

$$\Phi = -\left(\frac{M}{r} + \frac{d_j n^j}{r^2} + \frac{3 \mathcal{I}_{jk} n^j n^k}{2r^3} + \cdots\right), \quad \text{where } n^j = x^j/r. \quad (36.19a)$$

He can then give names to the coefficients in this expansion:

$M \equiv$ "total mass-energy" = "active gravitational mass";

$d_j \equiv$ "dipole moment" [if he chooses the origin of coordinates (36.19b)
 carefully, he can make $d_j = 0$];

$\mathcal{I}_{jk} \equiv$ "reduced quadrupole moment" [because the system is nearly
 Newtonian, \mathcal{I}_{jk} is given
 by expression (36.3)].

As this Newtonian potential reaches out into the radiation zone, the static portions of it ($-M/r - d_j n^j/r^2$) maintain their Newtonian form, unchanged. But the dynamic part ($-\frac{3}{2}\mathcal{I}_{jk} n^j n^k/r^3$) ceases to be describable in Newtonian terms. As retardation effects become noticeable (at increasing r values), it gradually changes over into outgoing gravitational waves, which must be described in the full general theory of relativity, or in linearized theory, or in the "shortwave" approximation of §35.13.

If one chooses to use linearized theory in the radiation zone, and if one imposes the transverse-traceless gauge there ($h_{0\mu}^{TT} = 0$, $h_{jj}^{TT} = 0$, $h_{jk,k}^{TT} = 0$), then the gravitational waves take the form [derived later as equation (36.47)]

Properties of gravitational waves in terms of reduced quadrupole moment:

$$h_{jk}^{TT} = \frac{2}{r} \ddot{\mathcal{I}}_{jk}^{TT}(t - r) + \text{corrections of order} \left[\frac{1}{r^2} \dot{\mathcal{I}}_{jk}^{TT}(t - r)\right]. \quad (36.20)$$

(1) the wave field h_{jk}^{TT}

Here \dddot{I}_{jk}^{TT} is the second time-derivative of the transverse-traceless part of the quadrupole moment (transverse to the radial direction; see §35.4); thus,

$$\ddot{I}_{jk}^{TT} = P_{ja}\ddot{I}_{ab}P_{bk} - \frac{1}{2}P_{jk}P_{ab}\ddot{I}_{ab},$$

$$P_{ab} = (\delta_{ab} - n_a n_b) \qquad \text{(projection operator)}, \tag{36.21}$$

$$n_a = x^a/r \qquad \text{(unit radial vector)}.$$

The effective stress-energy tensor for these outgoing waves (§35.7) has the same form as for a swarm of zero-mass particles traveling radially outward with the speed of light; at large distances its components of lowest nonvanishing order are

<div style="margin-left:2em">(2) effective stress-energy tensor</div>

$$T_{00}^{(GW)} = -T_{0r}^{(GW)} = T_{rr}^{(GW)} = \frac{1}{32\pi}\langle h_{jk,0}^{TT}h_{jk,0}^{TT}\rangle = \frac{1}{8\pi r^2}\langle \ddot{I}_{jk}^{TT}\ddot{I}_{jk}^{TT}\rangle$$

$$= \frac{1}{8\pi r^2}\left\langle \ddot{I}_{jk}\ddot{I}_{jk} - 2n_i\ddot{I}_{ij}\ddot{I}_{jk}n_k + \frac{1}{2}(n_j\ddot{I}_{jk}n_k)^2 \right\rangle, \tag{36.22}$$

where $\langle\ \rangle$ denotes an average over several wavelengths. (Recall that one cannot localize the energy more closely than a wavelength!) The total power crossing a sphere of radius r at time t is

<div style="margin-left:2em">(3) total power radiated</div>

$$L_{GW}(t, r) = \int T^{(GW)0r}r^2\,d\Omega = \frac{1}{5}\langle \dddot{I}_{jk}(t-r)\dddot{I}_{jk}(t-r)\rangle. \tag{36.23}$$

(See exercise 36.9.) This is the formula with which this chapter began: equation (36.1).

The wave fronts are not precisely spherical. For example, for a binary star system the wave fronts in the equatorial plane must be spirals. This means that there is a tiny nonradial component of the momentum flux, which decreases in strength as $1/r^3$. Associated with this nonradial momentum is an angular momentum density (angular momentum relative to the system's center, $r = 0$), which drops off as $1/r^2$ [Peters (1964), as corrected by DeWitt (1971), p. 286]:

<div style="margin-left:2em">(4) density of angular momentum</div>

$$\mathcal{J}^i = \frac{1}{8\pi r^2}\epsilon^{ijk}\langle -6n_j\ddot{I}_{km}\ddot{I}_{mp}n_p + 9n_j\ddot{I}_{km}n_m n_p\ddot{I}_{pq}n_q\rangle. \tag{36.24}$$

The integral of this quantity over a sphere is the total angular momentum being transported outward per unit time,

<div style="margin-left:2em">(5) total angular momentum radiated</div>

$$-dJ_j/dt = \int \mathcal{J}_j r^2\,d\Omega = \frac{2}{5}\epsilon^{jkl}\langle \ddot{I}_{ka}\dddot{I}_{al}\rangle. \tag{36.25}$$

(See exercise 36.9.)

§36.8. RADIATION REACTION IN SLOW-MOTION SOURCES*

The conservation laws discussed in Box 19.1 and derived in §20.5 guarantee that the source must lose energy and angular momentum at the same rate as the gravitational waves carry them off. The agent that produces these losses is a tiny component of the spacetime curvature inside the source, which reverses sign if one changes from a (realistic) outgoing-wave boundary condition at infinity to the opposite (unrealistic) ingoing-wave condition. These "radiation-reaction" pieces of the curvature can be described in Newtonian language when the source obeys the nearly Newtonian, slow-motion conditions (36.18).

Outgoing-wave boundary condition gives rise to a Newtonian-type radiation-reaction potential

The dynamical part of the Newtonian potential, in its "standard form"

$$\Phi = -\frac{3}{2}\,\mathcal{I}_{jk}(t)n_j n_k/r^3 + O(1/r^4); \text{ equation (36.18)},$$

has no retardation in it. (Newtonian theory demands action at a distance!) Consequently, there is no way whatsoever for the standard potential to decide, at large radii, whether to join onto outgoing waves or onto ingoing waves. Being undecided, it takes the middle track of joining onto standing waves (half outgoing, plus half ingoing). But this is not what one wants. It turns out (see §36.11) that the join can be made to purely outgoing waves if and only if Φ is augmented by a tiny "radiation-reaction" potential

$$\Phi = \Phi_{\text{standard Newtonian theory}} + \Phi^{(\text{react})}, \tag{36.26a}$$

$$\Phi^{(\text{react})} = \frac{1}{5}\frac{d^5\mathcal{I}_{jk}}{dt^5}\,x^j x^k. \tag{36.26b}$$

Form and magnitude of the radiation-reaction potential

If, instead, one sets $\Phi = \Phi_{\text{standard}} - \Phi^{(\text{react})}$, the potential will join onto purely ingoing waves.

In order of magnitude, the radiation-reaction potential is

$$\Phi^{(\text{react})} \sim \frac{1}{\lambda^5}\,(MR^2)r^2 \sim \frac{MR^2}{r^3}\left(\frac{r}{\lambda}\right)^5. \tag{36.27}$$

Consequently, near the source it is tiny compared to the standard Newtonian potential [a factor $(R/\lambda)^5 \sim v^5$ smaller!]. However, at the inner boundary of the radiation zone ($r \sim \lambda$), it is of the same order of magnitude as the dynamic, quadrupole part of the standard potential.

The radiation-reaction part of the Newtonian potential plays the same role as a producer of accelerations that any other part of the Newtonian potential does. Any particle in the Newtonian region experiences a gravitational acceleration given by

Effects of the potential:

$$a_j = -\Phi_{,j} = -\Phi_{\text{standard},j} - \Phi^{(\text{react})}_{,j}. \tag{36.28}$$

(1) radiation-reaction accelerations

*The ideas and formalism described in this section were devised by Burke (1970), Thorne (1969b), and Chandrasekhar and Esposito (1970). Among the forerunners of these ideas were the papers of Peters (1964), and Peres and Rosen (1964).

Inside the source, this acceleration leads to energy and angular momentum losses given by

$$dE/dt = \int \rho a_j v_j \, d^3x \qquad (36.29a)$$

and

$$dJ_j/dt = \int \varepsilon_{jkl} x_k \rho a_l \, d^3x. \qquad (36.29b)$$

(Here ρ is the density, v_j is the velocity, and a_j as above is the acceleration of the matter in the source.) Standard Newtonian theory conserves the energy and angular momentum. Therefore only the reaction part of the potential can produce losses:

(2) loss of energy and angular momentum

$$dE/dt = -\int \rho \Phi^{(\mathrm{react})}_{,j} v_j \, d^3x,$$

$$dJ_j/dt = -\int \varepsilon_{jkl} x_k \rho \Phi^{(\mathrm{react})}_{,l} \, d^3x. \qquad (36.30)$$

A straightforward calculation (exercise 36.5) using expression (36.26b) for the reaction potential yields, for the time-averaged losses,

$$dE/dt = -\frac{1}{5} \langle \dddot{\mathcal{I}}_{jk} \dddot{\mathcal{I}}_{jk} \rangle,$$

$$dJ_j/dt = -\frac{2}{5} \varepsilon_{jkl} \langle \ddot{\mathcal{I}}_{ka} \dddot{\mathcal{I}}_{al} \rangle. \qquad (36.31)$$

Notice that these results agree with the energy and angular momentum carried off by the radiation as given by equations (36.1) and (36.25). The agreement is an absolute imperative. The laws of conservation of total energy and angular momentum demand it.

A slow-motion electromagnetic system emitting electric dipole radiation has a radiation-reaction potential

Radiation-reaction potential for electromagnetic waves

$$A^{(\mathrm{react})}_j = 0, \qquad A^{(\mathrm{react})}_0 = -\Phi^{(\mathrm{react})} = \frac{2}{3} \ddot{d}_j x^j, \qquad (36.32)$$

which is completely analogous to $\Phi^{(\mathrm{react})}$ of gravitation theory [see, e.g., Burke (1971)]. However, attention does not usually focus on this potential and the reaction forces it produces. Instead, it focuses on the reaction force in a special case: that of an isolated charge being accelerated by nonelectromagnetic forces. For such a charge, the reaction force is

$$\boldsymbol{F}^{(\mathrm{React})} = \frac{2}{3} e^2 \dddot{\boldsymbol{x}}. \qquad (36.33)$$

No such formula is relevant to gravitation theory, because there is no such thing as a gravitationally isolated, radiating particle (i.e., one accelerated by forces that have no coupling to gravity).

Exercise 36.5. ENERGY AND ANGULAR MOMENTUM LOSSES DUE TO RADIATION REACTION

Derive equations (36.31) for the rate at which gravitational radiation damping saps energy and angular momentum from a slow-motion source. Base the derivation on equations (36.26b) and (36.30).

Exercise 36.6. GRAVITATIONAL WAVES FROM BINARY STAR SYSTEMS

Apply the full formalism of §§36.7 and 36.8 to a binary star system with circular orbits. Calculate the angular distribution of the gravitational waves; the total power radiated; the total angular momentum radiated; the radiation-reaction forces; and the loss of energy and angular momentum due to radiation reaction. Compare the answers with the results quoted in §36.6. [For further details of the solution, see Peters and Mathews (1963).]

§36.9. FOUNDATIONS FOR DERIVATION OF RADIATION FORMULAS

Turn now from the formulas for radiation from a nearly Newtonian system in slow motion to a derivation of these formulas. Initially (this section) work in the full general theory of relativity without any approximations—not even that of slow motion. Impose only the constraint that the source be isolated, and that spacetime become asymptotically flat far away from it.

Use a coordinate system that becomes asymptotically Lorentz as rapidly as spacetime curvature permits, when one moves radially outward from the source toward infinity. Everywhere in this coordinate system, even inside the source, which may be relativistic, define

$$h_{\mu\nu} \equiv g_{\mu\nu} - \eta_{\mu\nu}. \tag{36.34}$$

The $h_{\mu\nu}$ are clearly not the components of a tensor. Neither is $\eta_{\mu\nu}$ the true metric tensor. Nevertheless, one is free to raise and lower indices on $h_{\mu\nu}$ with $\eta_{\mu\nu}$ and to *define*

$$\bar{h}_{\mu\nu} \equiv h_{\mu\nu} - \frac{1}{2}\eta_{\mu\nu}h; \qquad h = h_\alpha{}^\alpha = h_{\alpha\beta}\eta^{\alpha\beta}. \tag{36.35}$$

Moreover, one can always specialize the coordinates so that the four conditions

$$\bar{h}_\mu{}^\alpha{}_{,\alpha} = 0 \tag{36.36}$$

are exactly satisfied everywhere, including the interior of the source.

With these definitions and conventions, $\bar{h}_{\mu\nu}$ becomes the gravitational field of linearized theory far from the source, and also inside the source if gravity is weak there. But if the interior gravity is strong ($|\bar{h}_{\mu\nu}|$ not $\ll 1$), $\bar{h}_{\mu\nu}$ in the interior has no connection whatsoever to linearized theory.

The rest of this chapter is Track 2. Chapter 20 (conservation laws) is needed as preparation for it. It will be helpful in Chapter 39 (post-Newtonian formalism), but is not needed as preparation for any other chapters.

Derivation of formula for the gravitational-wave field produced by a slow-motion source:

(1) definition of $\bar{h}_{\mu\nu}$

The exact Einstein field equations can be written in terms of $\bar{h}^{\mu\nu}$ as [cf. §20.3; in particular, combine equations (20.14), (20.18), and (20.3); and impose the coordinate condition (36.36)]

$$\bar{h}^{\mu\nu}{}_{,\alpha\beta}\eta^{\alpha\beta} = -16\pi(T^{\mu\nu} + t^{\mu\nu}), \tag{36.37}$$

where $T^{\mu\nu}$ are the components of stress-energy tensor, and $t^{\mu\nu}$ are quantities (components of the "stress-energy pseudotensor for the gravitational field") that are of quadratic order and higher in $\bar{h}^{\mu\nu}$. Recall the "philosophy of controlled ignorance" expounded in §19.3. One is so ignorant that nowhere does one ever write down an explicit expression for $t^{\mu\nu}$ in terms of $\bar{h}^{\alpha\beta}$; and this ignorance is so controlled that one will never need such an expression in the calculations to follow! More specifically, the strength of the outgoing wave is proportional to the integral of a complicated expression over the interior of a system where "gravitational stresses" may be comparable to material stresses, $|t^{jk}| \sim |T^{jk}|$. No matter. All that will count for the radiation is the quadrupole part of the field. Moreover, that quadrupole moment is empirically definable by purely Newtonian measurements in the Newtonian region (1) well inside the wave zone, but (2) well outside the surface of the source. One does not have to know the inner workings of a star to define its mass (influence on Kepler orbits outside) nor does one have to know those inner workings to define its quadrupole moment *as sensed externally*.

Einstein's equations (36.37), augmented by an outgoing-wave boundary condition, are equivalent to the integral equations

$$\bar{h}^{\mu\nu}(t, x^j) = 4 \int_{\text{all space}} \frac{[T^{\mu\nu} + t^{\mu\nu}]_{\text{ret}}}{|\boldsymbol{x} - \boldsymbol{x}'|} d^3x', \tag{36.38}$$

where

$$|\boldsymbol{x} - \boldsymbol{x}'| \equiv \left[\sum_j (x^j - x^{j'})^2\right]^{1/2}, \qquad d^3x' \equiv dx^{1'}\, dx^{2'}\, dx^{3'},$$

and the subscript "ret" means the quantity is to be evaluated at the retarded spacetime point

$$(t' = t - |\boldsymbol{x} - \boldsymbol{x}'|, x^{j'}).$$

These are integral equations because the unknowns, $\bar{h}^{\mu\nu}$, appear both outside and inside the integral (inside they are contained in $t^{\mu\nu}$). Notice that in passing from the wave equations (36.37) to the integral equations (36.38), one has cavalierly behaved as though $\bar{h}^{\mu\nu}$ were fields in flat spacetime. This is certainly not true; but the mathematical manipulations are valid nevertheless!—and the integral equations (36.38) are valid for any field point (t, x^j), even inside the source.

§36.10. EVALUATION OF THE RADIATION FIELD IN THE SLOW-MOTION APPROXIMATION

Thus far the analysis has been exact. Now it is necessary to introduce the slow-motion assumption of §36.7: $R \ll \lambda$.

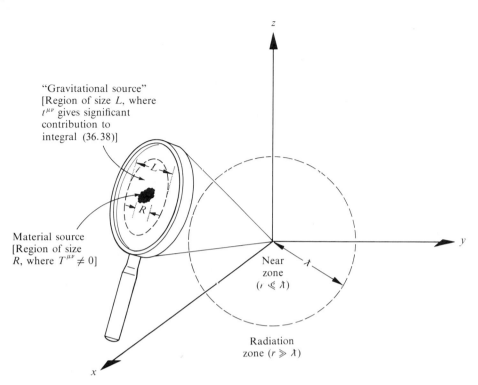

"Gravitational source"
[Region of size L, where $t^{\mu\nu}$ gives significant contribution to integral (36.38)]

Material source
[Region of size R, where $T^{\mu\nu} \neq 0$]

Near zone $(r \ll \lambda)$

Radiation zone $(r \gg \lambda)$

Figure 36.3.
A slow-motion source radiating gravitational waves. The origin of spatial coordinates is located inside the source. The size of the source, R, is very small compared to a reduced wavelength, $R \ll \lambda$. Significant contributions to the retarded integral (36.38) for $\bar{h}^{\mu\nu}$ come only from a region of size $L \sim R \ll \lambda$ surrounding the source, because outside the source—but in the near zone $(R \ll r \ll \lambda)$—the "stress-energy pseudotensor" $t^{\mu\nu}$ dies out as $1/r^4$ (see exercise 36.7).

In the radiation zone, $t^{\mu\nu}$ ceases to die out as $1/r^4$, and begins to die out as $1/r^2$; it is trying to describe (but cannot, really, without appropriate averaging) the stress-energy carried by the gravitational waves. If the source has been emitting waves long enough, contributions from the radiation zone to the retarded integral (36.38) may be nonnegligible:

$$[t^{\mu\nu}]_{\text{ret}} \sim \frac{1}{r'^2} \implies \int [t^{\mu\nu}]_{\text{ret}} \, d^3x' \sim \int \frac{1}{r'^2} r'^2 \, d\Omega' \, dr'.$$

[for $r > \lambda$] \qquad [may have significant contributions from large r']

Such contributions are ignored in the text, in calculations of the radiated waves, because they have nothing whatsoever to do with the emission process itself. Rather, they are part of the propagation process treated in the last chapter. They include the background curvature produced by the stress-energy of the waves, scattering of waves off the background curvature, wave-wave scattering, etc.; and they are totally negligible in the neighborhood of the source itself ($r \lesssim 1,000\ \lambda$, for example) because a slow-motion source radiates so very weakly.

Place the origin of spatial coordinates inside the source, as shown in Figure 36.3. For slow-motion systems, the only significant contributions to the retarded integrals (36.38) come from deep inside the near zone (from a region of size $L \sim R \ll \lambda$; see Figure 36.3). Confine attention to "field points" (points of observation) x^j far outside this "source region,"

$$|x| \equiv r \gg L \geq |x'|, \tag{36.39a}$$

and expand the retarded integral (36.38) in powers of x'/r—in just the same manner as was done in §19.1. (Such an expansion is justified by and requires the slow-motion assumption, $\lambdabar/R \sim \lambdabar/L \ll 1$.) The result is

$$\bar{h}^{\mu\nu}(t, x) = \frac{4}{r} \int [T^{\mu\nu}(x', t - r) + t^{\mu\nu}(x', t - r)]\, d^3x'$$
$$+ O\left\{ \frac{x^j}{r^2 \lambdabar} \int x^{j'}[T^{\mu\nu}(x', t - r) + t^{\mu\nu}(x', t - r)]\, d^3x' \right\}. \tag{36.40}$$

(6) calculation of \bar{h}^{jk} in
radiation zone

Of the ten components of $\bar{h}^{\mu\nu}$, only the six spatial ones, \bar{h}^{jk}, are of interest, since only they are needed in projecting out the transverse-traceless radiation field \bar{h}^{TT}_{jk}. The spatial components are expressed by equations (36.40) in terms of integrals over the "stress distribution" $T^{jk} + t^{jk}$. It will be convenient, in making comparisons with Newtonian theory, to reexpress \bar{h}^{jk} in terms of integrals over the "energy distribution" $T^{00} + t^{00}$. One can make the conversion with the help of the exact equations of motion $T^{\mu\nu}{}_{;\nu} = 0$, which have the special form

$$(T^{\mu\nu} + t^{\mu\nu})_{,\nu} = 0 \tag{36.41}$$

in the coordinate system being used [see equations (36.36) and (36.37); also the discussion in §20.3]. Applying these relations twice in succession, one obtains the identity

$$(T^{00} + t^{00})_{,00} = -(T^{0\ell} + t^{0\ell})_{,\ell 0} = -(T^{\ell 0} + t^{\ell 0})_{,0\ell}$$
$$= +(T^{\ell m} + t^{\ell m})_{,m\ell}.$$

From this and the elementary chain rule for differentiation, it follows that

$$[(T^{00} + t^{00})x^j x^k]_{,00} = (T^{\ell m} + t^{\ell m})_{,m\ell} x^j x^k$$
$$= [(T^{\ell m} + t^{\ell m})x^j x^k]_{,m\ell} - 2[(T^{\ell j} + t^{\ell j})x^k + (T^{\ell k} + t^{\ell k})x^j]_{,\ell}$$
$$+ 2(T^{jk} + t^{jk}),$$

whence

$$\int (T^{jk} + t^{jk})\, d^3x = \frac{1}{2} (d^2 I_{jk}/dt^2), \tag{36.42a}$$

where

$$I_{jk}(t) \equiv \int [T^{00}(t, x) + t^{00}(t, x)]x^j x^k\, d^3x. \tag{36.42b}$$

(7) specialization to nearly
Newtonian sources

Now introduce the nearly Newtonian assumption. It guarantees that gravitation contributes only a small fraction of the total energy:

$$t^{00} \sim (\Phi_{,j})^2 \sim M^2/R^4 \sim (M/R)T^{00} \ll T^{00};$$

hence

$$I_{jk}(t) = \int T^{00}(t, x)x^j x^k\, d^3x. \tag{36.42b'}$$

The quantity I_{jk} thus represents the second moment of the mass distribution.

By combining equations (36.42) and (36.40), and by noting that inside the source $|t^{jk}| \sim |\Phi_{,j}\Phi_{,k}| \sim T^{00}|\Phi|$, one obtains

$$\bar{h}^{jk}(t, \boldsymbol{x}) = \frac{2}{r}\frac{d^2 I_{jk}(t-r)}{dt^2} + O\left[\frac{1}{r}\left(\frac{|T^{jk}|}{T^{00}} + |\Phi|\right)\frac{R}{\lambda}M\right]$$

$$= \frac{2}{r}\frac{d^2 I_{jk}(t-r)}{dt^2}\left\{1 + \underbrace{O\left[\frac{|T^{jk}|}{T^{00}} + \frac{M}{R}\right]\frac{\lambda}{R}}_{}\right\}. \tag{36.43}$$

[negligible by assumptions (36.18)]

Actually, what one wants are h_{jk}^{TT}, not \bar{h}^{jk}. They can be obtained by first lowering indices, using $\eta_{lm} = \delta_{lm}$, and then projecting out the TT part using the projection operator for radially traveling waves:

$$P_{lm} = \delta_{lm} - n_l n_m; \qquad n_l = x^l/r \tag{36.44}$$

(8) conversion, by projection, to h_{jk}^{TT}

(see Box 35.1). (Because \bar{h}_{jk} and h_{jk} differ only in the trace, they have the same TT parts). The result is

$$h_{jk}^{TT}(t, \boldsymbol{x}) = \frac{2}{r}\frac{d^2 I_{jk}^{TT}(t-r)}{dt^2}, \tag{36.45a}$$

where

$$I_{jk}^{TT} = P_{jl}I_{lm}P_{mk} - \frac{1}{2}P_{jk}(P_{lm}I_{ml}). \tag{36.45b}$$

This is not the best form in which to write the answer, because an external observer cannot measure directly the second moment of the mass distribution, I_{jk}. Fortunately, one can replace I_{jk} by the reduced quadrupole moment,

$$\mathcal{I}_{jk} \equiv I_{jk} - \frac{1}{3}\delta_{jk}I = \int (T^{00} + t^{00})\left(x^j x^k - \frac{1}{3}\delta_{jk}r^2\right)d^3x, \tag{36.46}$$

(9) reexpression of h_{jk}^{TT} in terms of reduced quadrupole moment

and write

$$h_{jk}^{TT}(t, \boldsymbol{x}) = \frac{2}{r}\frac{d^2 \mathcal{I}_{jk}^{TT}(t-r)}{dt^2}. \tag{36.47}$$

This is allowed because the TT parts of I_{jk} and \mathcal{I}_{jk} are identical (exercise 36.8).

The reduced quadrupole moment \mathcal{I}_{jk} has a well-defined, elementary physical significance for an observer confined to the exterior of the source. In the near zone ($r \ll \lambda$), but outside the source so that vacuum Newtonian theory is very nearly valid, the Newtonian potential is

$$\Phi = -\frac{1}{2}h_{00} = -\frac{1}{2}h^{00} = -\frac{1}{2}\left(\bar{h}^{00} + \frac{1}{2}\bar{h}\right) = -\frac{1}{4}(\bar{h}^{00} + \bar{h}^{jj})$$

$$= -\int_{\text{all space}}\frac{[T^{00} + t^{00} + T^{jj} + t^{jj}]_{\text{ret}}}{|\boldsymbol{x} - \boldsymbol{x}'|}d^3x'$$

[see equation (36.38)]. Any nearly Newtonian, slow-motion source satisfies

$$|t^{00} + T^{jj} + t^{jj}| \ll T^{00}$$

[recall: $t^{\alpha\beta} \sim (\Phi_{,j})^2 \sim T^{00}|\Phi|$]. Hence, one can write

$$\Phi(\boldsymbol{x}, t) = -\int \frac{[T^{00}(\boldsymbol{x}', t)]}{|\boldsymbol{x} - \boldsymbol{x}'|} d^3x'. \tag{36.48}$$

Expanding $|\boldsymbol{x} - \boldsymbol{x}'|^{-1}$ in powers of $1/r$, one obtains

$$\Phi = -\left(\frac{M}{r} + \frac{d_j x^j}{r^3} + \frac{3 f_{jk} x^j x^k}{2r^5} + \cdots\right) \text{ for } \begin{pmatrix} r \ll \lambda, \text{ but } r \text{ neverthe-} \\ \text{less large enough that} \\ \text{vacuum Newtonian} \\ \text{theory is valid} \end{pmatrix}, \tag{36.49a}$$

where

$$M = \text{(total mass-energy of source)} = \int T^{00} d^3x,$$

$$d_j \equiv \text{(dipole moment of source)} = \int T^{00} x^j dx^3, \tag{36.49b}$$

$$f_{jk} \equiv \text{(reduced quadrupole moment of source)} = \text{expression (36.46)}.$$

Thus, *the quantities f_{jk}, whose second time-derivatives determine the radiation field by equation (36.47), are precisely the components of the star's reduced quadrupole moment, as measured by an observer who explores its Newtonian potential Φ deep inside the near zone ($r \ll \lambda$)* ("empirical quadrupole moment.").

The final answer (36.47) for the radiation field in terms of f_{jk}^{TT} was quoted in the summary of results given in §36.7. Also quoted there were expressions for the effective stress-energy tensor of the radiation and for the energy and angular momentum radiated [equations (36.22) to (36.25)]. Those expressions can be derived using the formalism of the shortwave approximation. (See exercise 36.9.)

EXERCISES

Exercise 36.7. MAGNITUDE OF $t^{\mu\nu}$

Consider a slow-motion source of gravitational waves. Show that far from the source, but in the near zone ($R \ll r \ll \lambda$) the components of the "stress-energy pseudotensor" $t^{\mu\nu}$ die out as $1/r^4$, but in the radiation zone ($r \gg \lambda$) they die out only as $1/r^2$.

Exercise 36.8. PROOF THAT THE TRANVERSE TRACELESS PARTS OF f_{jk} AND I_{jk} ARE IDENTICAL

Prove by direct computation that the TT parts of I_{jk} (36.42b) and f_{jk} (36.46) are identical, no matter where the observer is who does the TT projection (i.e., no matter what the unit vector \boldsymbol{n} in the projection operator may be).

Exercise 36.9. ENERGY AND ANGULAR MOMENTUM RADIATED

(a) For the gravitational waves in asymptotically flat spacetime described by equation (36.47), calculate the smeared-out stress-energy tensor $T_{\mu\nu}^{(\text{GW})}$ of equation (35.23). [*Answer:* equation (36.22).]

(b) Perform the integrals of equations (36.23) and (36.25) to obtain the total power and angular momentum radiated. [*Hint:* Derive and use the following averages over a sphere

$$\frac{1}{4\pi}\int n_i \, d\Omega = 0, \qquad \frac{1}{4\pi}\int n_i n_j \, d\Omega = \frac{1}{3}\delta_{ij}, \qquad \frac{1}{4\pi}\int n_i n_j n_k \, d\Omega = 0,$$

$$\frac{1}{4\pi}\int n_i n_j n_k n_l \, d\Omega = \frac{1}{15}(\delta_{ij}\delta_{kl} + \delta_{ik}\delta_{jl} + \delta_{il}\delta_{jk}).$$

Here $n \equiv x/|x|$ is the unit radial vector.]

§36.11. DERIVATION OF THE RADIATION-REACTION POTENTIAL

Turn, finally, to a derivation of the radiation-reaction results quoted in §36.8. The analysis starts with the solution (36.43) for the spatial part of the radiation field in the original (i.e., not TT) gauge:

$$\bar{h}^{jk}(t, x) = \frac{2}{r}\ddot{I}_{jk}(t - r). \qquad (36.50)$$

Derivation of formula for the radiation-reaction potential.

Although this solution was originally derived by discarding all terms that die out faster than $1/r$, it is in fact an exact solution to the vacuum field equations $\bar{h}^{jk}{}_{,\alpha}{}^{\alpha} = 0$ of linearized theory. This means that it is valid in the intermediate and near zones ($r \leq \lambda$, but $r > R$) as well as in the radiation zone.

Were one to replace the outgoing-wave condition by an ingoing-wave condition at infinity, the exact solution (36.50) for \bar{h}^{jk} would get replaced by

$$\bar{h}^{jk}(t, x) = \frac{2}{r}\ddot{I}_{jk}(t + r).$$

Thus, in order to delineate the effects of the outgoing-wave boundary condition, one can write the exact solution in the form

$$\bar{h}_{jk}(t, x) = \bar{h}^{jk}(t, x) = \frac{2}{r}\ddot{I}_{jk}(t - \epsilon r), \qquad \epsilon = \pm 1, \qquad (36.51)$$

(1) formula for \bar{h}_{jk} anywhere outside source, with either outgoing or ingoing waves

and then focus attention on the effects of the sign of ϵ.

In the near zone ($r \ll \lambda$), but outside the nearly Newtonian source, this solution for \bar{h}_{jk}, as expanded in powers of r, becomes

$$\bar{h}_{jk} = 2\left[\frac{I_{jk}^{(2)}}{r} - \epsilon I_{jk}^{(3)} + \frac{I_{jk}^{(4)}r}{2!} - \epsilon\frac{I_{jk}^{(5)}r^2}{3!} + \cdots\right], \qquad (36.52a)$$

(2) \bar{h}_{jk} specialized to near zone

where

$$I_{jk}^{(n)} \equiv d^n I_{jk}(t)/dt^n.$$

The corresponding forms of \bar{h}_{0j} and \bar{h}_{00} can be generated from this by the gauge conditions $\bar{h}_\alpha{}^\beta{}_{,\beta} = 0$; i.e., by $\bar{h}_{j0,0} = \bar{h}_{jk,k}$ and $\bar{h}_{00,0} = \bar{h}_{0j,j}$. The results are:

$$\bar{h}_{0j} = 2\left[-\frac{I^{(1)}_{jk}x^k}{r^3} + \frac{I^{(3)}_{jk}x^k}{2!r} - \epsilon\frac{2I^{(4)}_{jk}x^k}{3!} + \frac{3I^{(5)}_{jk}x^k r}{4!} - \epsilon\frac{4I^{(6)}_{jk}x^k r^2}{5!} \right]$$

$$+ \text{(static terms not associated with radiation)}; \tag{36.52b}$$

$$\bar{h}_{00} = 2\left[\frac{(3x^j x^k - r^2\delta^{jk})}{r^5}I_{jk} - \frac{(x^j x^k - r^2\delta^{jk})}{2!r^3}I^{(2)}_{jk} - \epsilon\frac{2}{3!}I^{(3)}_{jj} \right.$$

$$\left. + \frac{3(x^j x^k + r^2\delta^{jk})}{4!r}I^{(4)}_{jk} - \epsilon\frac{4(2x^j x^k + r^2\delta^{jk})}{5!}I^{(5)}_{jk} + \cdots \right] \tag{36.52c}$$

$$+ \text{(static and time-linear terms not associated with radiation)}.$$

The leading term in these expressions rises as $1/r^3$ when one approaches the source:

$$\bar{h}_{00} \approx \frac{2(3x^j x^k - r^2\delta^{jk})}{r^5}I_{jk} = \frac{6 \mathcal{I}_{jk}n^j n^k}{r^3}.$$

It is precisely the leading term in the dynamic, quadrupole part of the Newtonian potential, $\Phi = -\frac{1}{2}h_{00} = -\frac{1}{4}\bar{h}_{00}$. All other terms without ϵ's in front of them are corrections to the Newtonian potential. They produce effects like the perihelion shift of Mercury that in no way deplete the energy and angular momentum of the system.

The terms with ϵ's are associated with radiation reaction. Pluck the leading ones out and call them "reaction potentials":

$$\bar{h}^{(\text{react})}_{jk} = -2I^{(3)}_{jk} - \frac{1}{3}I^{(5)}_{jk}r^2,$$

$$\bar{h}^{(\text{react})}_{0j} = -\frac{2}{3}I^{(4)}_{jk}x^k - \frac{1}{15}I^{(6)}_{jk}x^k r^2, \tag{36.53}$$

$$\bar{h}^{(\text{react})}_{00} = -\frac{2}{3}I^{(3)}_{jj} - \frac{1}{15}(2x^j x^k + r^2\delta^{jk})I^{(5)}_{jk}.$$

The corresponding metric perturbations $h_{\alpha\beta} = \bar{h}_{\alpha\beta} - \frac{1}{2}\bar{h}\eta_{\alpha\beta}$ are

$$h^{(\text{react})}_{jk} = -2I^{(3)}_{jk} + \frac{2}{3}I^{(3)}_{\ell\ell}\delta_{jk} + 0(I^{(5)}_{jk}r^2),$$

$$h^{(\text{react})}_{0j} = -\frac{2}{3}I^{(4)}_{jk}x^k + 0(I^{(6)}_{jk}r^3), \tag{36.54}$$

$$h^{(\text{react})}_{00} = -\frac{4}{3}I^{(3)}_{\ell\ell} - \frac{1}{15}(x^j x^k + 3r^2\delta_{jk})I^{(5)}_{jk}.$$

These reaction potentials in the near zone are understood most clearly by a change of gauge that brings them into Newtonian form. Set

$$x^{\mu}{}_{\text{new}} = x^{\mu}{}_{\text{old}} + \xi^{\mu}(x), \qquad h_{\mu\nu_{\text{new}}} = h_{\mu\nu_{\text{old}}} - \xi_{\mu,\nu} - \xi_{\nu,\mu}$$

(5) conversion of radiation-reaction potentials to Newtonian gauge

with

$$\xi_j = -I^{(3)}_{jk}x^k + \frac{1}{3}I^{(3)}_{\ell\ell}x^j,$$

$$\xi_0 = -\frac{2}{3}I^{(2)}_{\ell\ell} + \frac{1}{6}I^{(4)}_{jk}x^jx^k - \frac{1}{6}I^{(4)}_{\ell\ell}r^2. \tag{36.55}$$

Then in the new gauge

$$h^{(\text{react})}_{jk} = O(I^{(5)}_{jk}r^2), \qquad h^{(\text{react})}_{0j} = O(I^{(6)}_{jk}r^3),$$

$$h^{(\text{react})}_{00} = -\frac{2}{5}I^{(5)}_{jk}x^jx^k. \tag{36.56}$$

This gauge is ideally suited to a Newtonian interpretation, since in it the geodesic equation for slowly moving particles has the form

$$d^2x^j/dt^2 = -\Phi^{(\text{react})}_{,j} + \begin{pmatrix} \text{terms not sensitive to} \\ \text{outgoing-wave condition} \end{pmatrix}, \tag{36.57}$$

with

$$\Phi^{(\text{react})} = -\frac{1}{2}h^{(\text{react})}_{00} = \frac{1}{5}I^{(5)}_{jk}x^jx^k. \tag{36.58}$$

Thus, the leading radiation-reaction effects (with fractional errors $\sim [\lambda/r]^2$) can be described in the near zone of a nearly Newtonian source by appending the term $\frac{1}{5}I^{(5)}_{jk}x^jx^k$ to the Newtonian potential. The resulting formalism and a qualitative version of the above derivation were presented in §36.8.

CHAPTER **37**

DETECTION OF GRAVITATIONAL WAVES

I often say that when you can measure what you are speaking about, and express it in numbers, you know something about it; but when you cannot measure it, when you cannot express it in numbers, your knowledge is of a meagre and unsatisfactory kind: it may be the beginning of knowledge, but you have scarcely, in your thoughts, advanced to the stage of science, whatever the matter may be.

WILLIAM THOMSON, LORD KELVIN [(1889), p. 73]

§37.1. COORDINATE SYSTEMS AND IMPINGING WAVES

The detector is even easier to analyze than the generator or the transmission when one deals with gravitational waves within the framework of general relativity. Man's potential detectors all lie in the solar system, where gravity is so weak and spacetime so nearly flat that a plane gravitational wave coming in remains for all practical purposes a plane gravitational wave. (Angle of deflection of wave front passing limb of sun is only 1″.75.) Moreover, the nearest source of significant waves is so far away that, for all practical purposes, one can consider the waves as plane-fronted when they reach the Earth. Consequently, as they propagate in the z-direction past a detector, they can be described to high accuracy by the following transverse-traceless linearized expressions

Linearized description of gravitational waves propagating past Earth

Metric perturbation: $h_{xx}^{TT} = -h_{yy}^{TT} = A_+(t - z), \quad h_{xy}^{TT} = h_{yx}^{TT} = A_\times(t - z),$ (37.1a)

Riemann tensor: $R_{x0x0} = -R_{y0y0} = -\dfrac{1}{2}\ddot{A}_+(t - z),$

$$R_{x0y0} = R_{y0x0} = -\frac{1}{2}\ddot{A}_\times(t - z).$$ (37.1b)

Figure 37.1.

The proper reference frame of a vibrating-bar detector. The bar hangs by a wire from a cross beam, which is supported by vertical posts (not shown) that are embedded in the Earth. Consequently, the bar experiences a 4-acceleration given, at the moment when this diagram is drawn, by $\boldsymbol{a} = g(\partial/\partial\hat{z})$, where g is the "local acceleration of gravity" ($g \sim 980$ cm/sec^2). Later, the spatial axes will have rotated relative to the bar ("Foucoult-pendulum effect" produced by Earth's rotation), so the components of \boldsymbol{a} but not its magnitude will have changed.

The proper reference frame relies on an imaginary clock and three imaginary gyroscopes located at the bar's center of mass (and shown above in a cut-away view). Coordinate time is equal to proper time as measured by the clock, and the directions of the spatial axes $\partial/\partial x^{\hat{j}}$ are attached to the gyroscopes. The forces that prevent the gyroscopes from falling in the Earth's field must be applied at the centers of mass of the individual gyroscopes (no torque!).

$$\text{Stress-energy:} \quad T^{(GW)}_{00} = T^{(GW)}_{zz} = -T^{(GW)}_{0z} = \frac{1}{16\pi}\langle \dot{A}^2_+ + \dot{A}^2_\times \rangle_{\text{time avg.}} \quad (37.1c)$$

(See exercise 37.1.)

To analyze most easily the response of the detector to these impinging waves, use not the *TT* coordinate system $\{x^\alpha\}$ (which is specially "tuned" to the waves), but rather use coordinates $\{x^{\hat{\alpha}}\}$ specially "tuned" to the experimenter and his detector. The detector might be a vibrating bar, or the vibrating Earth, or a loop of tubing filled with fluid (see Figures 37.1 and 37.2). But whatever it is, it will have a center of mass. Attach the spatial origin, $x^{\hat{j}} = 0$, to this center of mass; and attach *orthonormal* spatial axes, $\partial/\partial x^{\hat{j}}$, to (possibly imaginary) gyroscopes located at this spatial origin (Figure 37.1). If the detector is accelerating (i.e., not falling freely), make the gyroscopes accelerate with it by applying the necessary forces at their centers of mass (no torque!). Use, as time coordinate, the proper time $x^{\hat{0}} = \tau$ measured by a clock at the spatial origin. Finally, extend these locally defined coordinates $x^{\hat{\alpha}}$ throughout all spacetime in the "straightest" manner possible. (See

Proper reference frame of a detector

Track 2's §13.6 for full details.) The metric in this *"proper reference frame of the detector"* will have the following form

$$ds^2 = -(1 + 2a_{\hat{j}}x^{\hat{j}})(dx^{\hat{0}})^2 + \delta_{\hat{j}\hat{k}}\, dx^{\hat{j}}\, dx^{\hat{k}} + O(|x^{\hat{j}}|^2)\, dx^{\hat{\alpha}}\, dx^{\hat{\beta}}. \qquad (37.2)$$

[equation (13.71) with $\omega^{\hat{i}} = 0$]. Here $a_{\hat{j}}$ are the spatial components of the detector's 4-acceleration. (Since \boldsymbol{a} must be orthogonal to the detector's 4-velocity, $a_{\hat{0}}$ vanishes.) Notice that, except for the acceleration term in $g_{\hat{0}\hat{0}}$ ("gravitational redshift term"; see §38.5 and exercise 6.6), this reference frame is locally Lorentz.

EXERCISES

Exercise 37.1. GENERAL PLANE WAVE IN TT GAUGE

Show that the most general linearized plane wave can be described in the transverse-traceless gauge of linearized theory by expressions (37.1). [*Hint:* Express the plane wave as a superposition (Fourier integral) of monochromatic plane waves, and describe each monochromatic plane wave by expressions (35.16). Use equations (35.10) and (35.23) to calculate $R_{\alpha\beta\gamma\delta}$ and $T^{(\mathrm{GW})}_{\mu\nu}$.]

Exercise 37.2. TEST-PARTICLE MOTION IN PROPER REFERENCE FRAME

Show that a slowly moving test particle, falling freely through the proper reference frame of equation (37.2), obeys the equation of motion (geodesic equation)

$$d^2x^{\hat{j}}/d\hat{t}^2 = -a_{\hat{j}} + O(|x^{\hat{k}}|).$$

Thus, one can interpret $-a_{\hat{j}}$ as the "local acceleration of gravity" (see caption of Figure 37.1).

§37.2. ACCELERATIONS IN MECHANICAL DETECTORS

The proper reference frame of equation (37.2) is the closest thing that exists to the reference frame a Newtonian physicist would use in analyzing the detector. In fact, it is so nearly Newtonian that (according to the analysis of Box 37.1) *the equations of motion for a mechanical detector, when written in this proper reference frame, take their standard Newtonian form and can be viewed and dealt with in a fully Newtonian manner, with one exception: the gravitational waves produce a driving force of non-Newtonian origin, given by the familiar expression for geodesic deviation*

Equations of motion for a mechanical detector

$$\left(\begin{array}{l}\text{force per unit mass (i.e., acceleration)}\\ \text{of a particle at } x^{\hat{j}} \text{ relative to detector's}\\ \text{center of mass at } x^{\hat{j}} = 0\end{array}\right) = \left(\frac{d^2x^{\hat{j}}}{d\hat{t}^2}\right)_{\text{due to waves}} \qquad (37.3)$$

$$= -(R^{\hat{j}}{}_{\hat{0}\hat{k}\hat{0}})_{\text{due to waves}}\, x^{\hat{k}}.$$

To use this equation, and to calculate detector cross sections later, one must know the components of the curvature tensor $R^{\hat{\alpha}}{}_{\hat{\beta}\hat{\gamma}\hat{\delta}}$, and of the waves' stress-energy tensor, $T^{(\mathrm{GW})}_{\hat{\mu}\hat{\nu}}$, in the detector's proper reference frame. One cannot calculate $R^{\hat{\alpha}}{}_{\hat{\beta}\hat{\gamma}\hat{\delta}}$ directly

**Box 37.1 DERIVATION OF EQUATIONS OF MOTION FOR
 A MECHANICAL DETECTOR**

Consider a "mass element" in a mechanical detector (e.g., a cube of aluminum one millimeter on each edge if the detector is the bar of Figure 37.1; or an element of fluid with volume 1 mm^3 if the detector is the tube filled with fluid shown in part h of Figure 37.2). This mass element gets pushed and pulled by adjacent matter and electromagnetic fields, as the medium of the detector vibrates or flows or does whatever it is supposed to do. Let

$$\boldsymbol{f} \equiv \begin{pmatrix} \text{4-force per unit mass exerted on mass-element} \\ \text{by adjacent matter and by electromagnetic fields} \end{pmatrix}. \qquad (1)$$

This 4-force per unit mass gives the mass element a 4-acceleration $\boldsymbol{\nabla_{\!u}u} = \boldsymbol{f}$; or, in terms of components in the detector's proper reference frame, $f^{\hat{\jmath}} = Du^{\hat{\jmath}}/d\tau$. Assume that the mass element has a very small velocity ($v \ll 1$) in the detector's proper reference frame (i.e., relative to the detector's center of mass). Then, ignoring terms of $O(v^2)$, $O(|x^{\hat{\jmath}}|^2)$, and $O(|x^{\hat{\jmath}}|v)$, one has [see equation (37.2)]

$$d\hat{t}/d\tau = u^{\hat{0}} = 1 - a_{\hat{\jmath}}x^{\hat{\jmath}} \equiv 1 - \boldsymbol{a} \cdot \boldsymbol{x}, \qquad (2)$$

and

$$f^{\hat{\jmath}} = d^2 x^{\hat{\jmath}}/d\tau^2 + \Gamma^{\hat{\jmath}}_{\hat{\alpha}\hat{\beta}}u^{\hat{\alpha}}u^{\hat{\beta}} = (d^2 x^{\hat{\jmath}}/d\hat{t}^2 + \Gamma^{\hat{\jmath}}_{\hat{0}\hat{0}})(1 - 2\boldsymbol{a} \cdot \boldsymbol{x}). \qquad (3)$$

Exercise 37.3 calculates $\Gamma^{\hat{\jmath}}_{\hat{0}\hat{0}}$ to precision of $O(|x^{\hat{\jmath}}|)$. Inserting its result and rearranging terms, one finds that

$$d^2 x^{\hat{\jmath}}/d\hat{t}^2 = (1 + 2\boldsymbol{a} \cdot \boldsymbol{x})f^{\hat{\jmath}} - a^{\hat{\jmath}}(1 + \boldsymbol{a} \cdot \boldsymbol{x}) - R^{\hat{\jmath}}_{\hat{0}\hat{k}\hat{0}}x^{\hat{k}} \qquad (4)$$

("*equation of motion for mass element*").

 Examine this equation, first from the viewpoint of an Einsteinian physicist, and then from the viewpoint of a Newtonian physicist.

 The Einsteinian physicist recognizes $d^2 x^{\hat{\jmath}}/d\hat{t}^2$ as the "coordinate acceleration" of the mass element—but he keeps in mind that, to precision of $O(|x^{\hat{\jmath}}|^2)$, coordinate lengths and proper lengths are the same [see equation (37.2)]. The coordinate acceleration $d^2 x^{\hat{\jmath}}/d\hat{t}^2$ has three causes: (1) *the externally applied force,*

$$(1 + 2\boldsymbol{a} \cdot \boldsymbol{x})f^{\hat{\jmath}} = (d^2 x^{\hat{\jmath}}/d\hat{t}^2)_{\text{external force}} \qquad (5a)$$
$$= (1 + 2\boldsymbol{a} \cdot \boldsymbol{x})(d^2 x^{\hat{\jmath}}/d\tau^2)_{\text{external force}}$$

(the origin of the $\boldsymbol{a} \cdot \boldsymbol{x}$ correction is simply the conversion between coordinate time

Box 37.1 (continued)

and proper time); (2) the "inertial force" due to the acceleration of the reference frame,

$$-a^{\hat{j}}(1 + \boldsymbol{a} \cdot \boldsymbol{x}) = (d^2 x^{\hat{j}}/d\hat{t}^2)_{\text{inertial force}} \tag{5b}$$

(see exercise 37.4 for explanation of the $\boldsymbol{a} \cdot \boldsymbol{x}$ correction); and (3) a "*Riemann curvature force*," which will include Riemann curvature due to local, Newtonian gravitational fields (fields of Earth, sun, moon, etc.), plus Riemann curvature due to the impinging gravitational waves,

$$-(R^{\hat{j}}{}_{\hat{0}\hat{k}\hat{0}})_{\text{waves}} x^{\hat{k}} - (R^{\hat{j}}{}_{\hat{0}\hat{k}\hat{0}})_{\text{Newton fields}} x^{\hat{k}} = (d^2 x^{\hat{j}}/d\hat{t}^2)_{\text{curvature}} \tag{5c}$$

(linear superposition because all gravitational fields in the solar system are so weak). This "Riemann curvature force" is not, of course, "felt" by the mass element; it does not produce any 4-acceleration. Rather, like the inertial force, it originates in the choice of reference frame: The spatial coordinates $x^{\hat{j}}$ measure proper distance and direction away from the detector's center of mass; and Riemann curvature tries to change this proper distance and direction ("relative acceleration;" "geodesic deviation").

A Newtonian physicist views the equation of motion (4) in a rather different manner. Having been told that the spatial coordinates $x^{\hat{j}}$ measure proper distance and direction away from the detector's center of mass, he thinks of them as the standard Euclidean spatial coordinates of Newtonian theory. He then rewrites equation (4) in the form

$$d^2 x^{\hat{j}}/d\hat{t}^2 = F^{\hat{j}} - (R^{\hat{j}}{}_{\hat{0}\hat{k}\hat{0}})_{\text{waves}} x^{\hat{k}}, \tag{6}$$

where

$$F^{\hat{j}} \equiv \begin{pmatrix} \text{Newtonian force per unit mass} \\ \text{acting on mass element} \end{pmatrix} \tag{7}$$

$$= (1 + 2\boldsymbol{a} \cdot \boldsymbol{x}) f^{\hat{j}} - a^{\hat{j}}(1 + \boldsymbol{a} \cdot \boldsymbol{x}) - (R^{\hat{j}}{}_{\hat{0}\hat{k}\hat{0}})_{\text{Newton fields}} x^{\hat{k}}.$$

The Newtonian physicist is free to express $F^{\hat{j}}$ in a form more familiar than this. He can ignore the subtleties of the $\boldsymbol{a} \cdot \boldsymbol{x}$ "redshift effects" because (1) they are small

$$|a^{\hat{j}}(\boldsymbol{a} \cdot \boldsymbol{x})| \sim |f^{\hat{j}}(\boldsymbol{a} \cdot \boldsymbol{x})| \lesssim |(R^{\hat{j}}{}_{\hat{0}\hat{k}\hat{0}})_{\text{waves}} x^{\hat{k}}|; \tag{8}$$

and (2) they are steady in time, and therefore—by contrast with the equally small wave-induced forces—they cannot excite resonant motions of the detector. Also, he

can separate the "inertial acceleration," $-a^{\hat{j}}$, into a contribution from the local acceleration of gravity at the detector's center of mass, $-(\partial\Phi/\partial x^{\hat{j}})_{x^{\hat{j}}=0}$, plus a contribution $-a^{\hat{j}}_{\text{absolute}}$ due to acceleration of the detector relative to the "absolute space" of Newtonian theory. Finally, he can rewrite the Riemann curvature due to Newtonian gravity in the familiar form $R^{\hat{j}}{}_{\hat{0}\hat{k}\hat{0}} = \partial^2\Phi/\partial x^{\hat{j}}\,\partial x^{\hat{k}}$. The net result is

$$F^{\hat{j}} = \qquad \left[\begin{array}{l}\text{total Newtonian force per unit}\\ \text{mass acting on mass element}\end{array}\right]$$

$$+ f^{\hat{j}} \qquad \left[\begin{array}{l}\text{Newtonian force per unit mass exerted by}\\ \text{adjacent matter and by electromagnetic fields}\end{array}\right]$$

$$- a^{\hat{j}}_{\text{absolute}} \qquad \left[\begin{array}{l}\text{inertial force per unit mass due to acceleration}\\ \text{of detector relative to Newtonian absolute space}\end{array}\right]$$

$$-\left(\frac{\partial\Phi}{\partial x^{\hat{j}}}\right)_{\text{at mass element}} \left[\begin{array}{l}= -(\partial\Phi/\partial x^{\hat{j}})_{x^{\hat{j}}=0} - (\partial^2\Phi/\partial x^{\hat{j}}\,\partial x^{\hat{k}})x^{\hat{k}}\\ = \text{Newtonian gravitational acceleration}\end{array}\right]. \qquad (9)$$

Conclusion: The equation of motion for a mass element of a mechanical detector, when written in the detector's proper reference frame, has the standard Newtonian form (6), with standard Newtonian driving forces (9), plus a driving force due to the gravitational waves given by

$$(d^2 x^{\hat{j}}/d\hat{t}^2)_{\text{due to waves}} = -(R^{\hat{j}}{}_{\hat{0}\hat{k}\hat{0}})_{\text{waves}} x^{\hat{k}}. \qquad (10)$$

from the metric coefficients $g_{\hat{\alpha}\hat{\beta}}$ of expression (37.2); to do so one would need the unknown corrections of $O(|x^{\hat{j}}|^2)$. However, one can easily obtain $R^{\hat{\alpha}}{}_{\hat{\beta}\hat{\gamma}\hat{\delta}}$ and $T^{(\text{GW})}_{\hat{\mu}\hat{\nu}}$ from the corresponding components in the *TT* coordinate frame [equations (37.1)] by applying the transformation matrix $\|\partial x^{\alpha}/\partial x^{\hat{\mu}}\|$. To make the transformation trivial, orient the *TT* coordinate frame so that, to a precision of $O(|h_{\mu\nu}|) \ll 1$, it coincides with the detector's proper reference frame near the detector's center of mass at the moment of interest, $t = \hat{t} = 0$. Then the transformation matrix will be

$$\partial x^{\alpha}/\partial x^{\hat{\mu}} = \delta^{\alpha}_{\mu} + O(h_{\mu\nu}) + O(a_{\hat{j}} x^{\hat{j}}) + O(|a|\hat{t}). \qquad (37.4)$$

$$\left[\begin{array}{l}\text{corrections due to}\\ \text{ripples in spacetime}\\ \text{caused by waves}\end{array}\right] \nearrow \left[\begin{array}{l}\text{redshift}\\ \text{corrections}\end{array}\right] \nwarrow \uparrow \left[\begin{array}{l}\text{corrections due to relative}\\ \text{velocity of frames resulting}\\ \text{from detector's acceleration}\end{array}\right]$$

The acceleration the detector experiences is typically

$$|a| = \text{one "Earth gravity"} = 980 \text{ cm/sec}^2 \sim 1/(\text{light-year}).$$

Description of waves in frame of detector

Therefore to enormous precision $\|\partial x^\alpha / \partial x^{\hat\mu}\| = \|\delta^\alpha_\mu\|$, and components of tensors are the same in the two reference frames:

$$R_{\hat x \hat 0 \hat x \hat 0} = -R_{\hat y \hat 0 \hat y \hat 0} = -\frac{1}{2}\ddot A_+, \qquad R_{\hat x \hat 0 \hat y \hat 0} = R_{\hat y \hat 0 \hat x \hat 0} = -\frac{1}{2}\ddot A_\times,$$

$$T^{(\mathrm{GW})}_{\hat 0 \hat 0} = T^{(\mathrm{GW})}_{\hat z \hat z} = -T^{(\mathrm{GW})}_{\hat 0 \hat z} = \frac{1}{16\pi}\langle \dot A_+^2 + \dot A_\times^2 \rangle_{\text{time avg.}} \qquad (37.5)$$

[see equation (37.1)].

Combining equations (37.3) and (37.5), one obtains for the wave-induced accelerations relative to the center of mass of the detector

Explicit form of accelerations due to waves

$$\left(\frac{d^2 \hat x}{d\hat t^2}\right)_{\text{due to waves}} = -R_{\hat x \hat 0 \hat x \hat 0}\hat x - R_{\hat x \hat 0 \hat y \hat 0}\hat y = \frac{1}{2}(\ddot A_+ \hat x + \ddot A_\times \hat y),$$

$$\left(\frac{d^2 \hat y}{d\hat t^2}\right)_{\text{due to waves}} = -R_{\hat y \hat 0 \hat y \hat 0}\hat y - R_{\hat y \hat 0 \hat x \hat 0}\hat x = \frac{1}{2}(-\ddot A_+ \hat y + \ddot A_\times \hat x), \qquad (37.6)$$

$$\left(\frac{d^2 \hat z}{d\hat t^2}\right)_{\text{due to waves}} = 0.$$

This analysis is valid only for "small" detectors ($L \ll \lambda$)

These expressions, like the equation of geodesic deviation, are valid only over regions small compared to one wavelength. Second derivatives of the metric (i.e., the components of the Riemann tensor) give a poor measure of geodesic deviation and of wave-induced forces over regions of size $L \gtrsim \lambda$. Thus, to analyze large detectors ($L \gtrsim \lambda$), one must abandon the "local mathematics" of the curvature tensor and replace it by "global mathematics"—e.g., an analysis in the *TT* coordinate frame using the metric components $h_{\mu\nu}$. For an example, see exercise 37.6.

All detectors of high sensitivity that have been designed up until now (1973) are small compared to a wavelength, and therefore can be analyzed using the techniques of Newtonian physics and the driving forces of equations (37.6).

It is useful to develop physical intuition for the driving forces, $-R^{\hat j}{}_{\hat 0 \hat k \hat 0}x^{\hat k}$, produced by waves of various polarizations. Figure 35.2 is one aid to such intuition; Box 37.2 is another. [The reader may find it interesting to examine, compare, and reconcile them!]

EXERCISES

Exercise 37.3. CONNECTION COEFFICIENTS IN PROPER REFERENCE FRAME

(a) Calculate $\Gamma^{\hat\alpha}{}_{\hat\beta\hat\gamma}$ for the metric (37.2), ignoring corrections of $O(|x^{\hat j}|)$. [Answer: Equations (13.69) with $\omega^{\hat i} = 0$.]

(b) Calculate $R^{\hat j}{}_{\hat 0 \hat k \hat 0}$ using the standard formula (8.44), and leaving spatial derivatives of the connection coefficients unevaluated because of the unknown corrections of $O(|x^{\hat j}|)$ in $\Gamma^{\hat\alpha}{}_{\hat\beta\hat\gamma}$. [Answer: $R^{\hat j}{}_{\hat 0 \hat k \hat 0} = \Gamma^{\hat j}{}_{\hat 0 \hat 0, \hat k} - a^{\hat j}a^{\hat k}$.]

(c) Use the answer to part (b) to evaluate the $O(|x^{\hat j}|)$ corrections to $\Gamma^{\hat j}{}_{\hat 0 \hat 0}$. [Answer:

$$\Gamma^{\hat j}{}_{\hat 0 \hat 0} = a^{\hat j}(1 + a_{\hat k}x^{\hat k}) + R^{\hat j}{}_{\hat 0 \hat k \hat 0}x^{\hat k} + O(|x^{\hat k}|^2).] \qquad (37.7)$$

Box 37.2 LINES OF FORCE FOR GRAVITATIONAL-WAVE ACCELERATIONS

A. Basic Idea

Consider a plane wave propagating in the \hat{z} direction. Discuss it entirely in the proper reference frame of a detector. The relative accelerations due to the wave are entirely transverse. Relative to the center of mass of the detector (origin of spatial coordinates) they are

$$d^2\hat{x}/d\hat{t}^2 = \frac{1}{2}(\ddot{A}_+\hat{x} + \ddot{A}_\times\hat{y}),$$

$$d^2\hat{y}/d\hat{t}^2 = \frac{1}{2}(-\ddot{A}_+\hat{y} + \ddot{A}_\times\hat{x}), \qquad (1)$$

$$d^2\hat{z}/d\hat{t}^2 = 0.$$

Notice that these accelerations are divergence-free. Consequently they can be represented by "lines of force," analogous to those of a vacuum electric field. At a value of $\hat{t} - \hat{z}$ where $\ddot{A}_\times = 0$ (so polarization is entirely \boldsymbol{e}_+), the lines of force are the hyperbolas shown here [sketch (a)]. The direction of the acceleration at any point is the direction of the arrow there; the magnitude of the acceleration is the density of force lines. Since acceleration is proportional to distance from center of mass, the force lines get twice as close together when one moves twice as far away from the origin in a given direction. When \ddot{A}_+ is positive, the arrows on the force lines are as shown in (a); when it is negative, they are reversed. As $|\ddot{A}_+|$ increases, the force lines move in toward the origin so their density goes up; as $|\ddot{A}_+|$ decreases, they move out toward infinity so their density goes down.

For polarization \boldsymbol{e}_\times the force lines are rotated by 45° from the above diagram. For intermediate polarization (values of $\hat{t} - \hat{z}$ where \ddot{A}_+ and \ddot{A}_\times are both nonzero), the diagram is rotated by an intermediate angle [sketch (b)]

$$\phi_0 = \frac{1}{2}\operatorname{arc\,tan}(\ddot{A}_\times/\ddot{A}_+). \qquad (2)$$

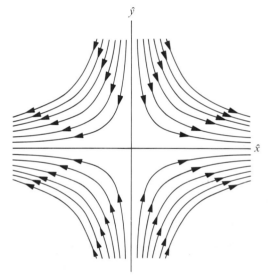

(a) Force lines for $\ddot{A}_\times = 0$, $\ddot{A}_+ > 0$

(b)

Box 37.2 (*continued*)

B. Three-Dimensional Diagram

At each value of $\hat{t} - \hat{z}$, the wave-produced accelerations have a specific polarization [orientation angle ϕ_0 of sketch (b)] and a specific amplitude (density of lines of force). Draw the lines of force in a three-dimensional $(\hat{x}, \hat{y}, \hat{z})$ diagram for fixed \hat{t}. Then as time passes the over-all diagram will remain unchanged in form, but will propagate with the speed of light in the \hat{z} direction.

Sketch (c) shows such a diagram for righthand circularly polarized waves of unchanging amplitude. *Note:* The authors are not aware of diagrams such as these [(a), (b), (c) above] and their use in analyzing detector response prior to William H. Press (1970).

(c)

Exercise 37.4. WHY THE $a \cdot x$?

Explain the origin of the $a \cdot x$ correction in equation (5b) of Box 37.1. [*Hint:* Take the viewpoint of an observer at rest at the spatial origin who watches two freely falling particles respond to the inertial force. At time $\hat{t} = 0$, put one particle at the origin and the other at $x^{\hat{j}}$. As time passes, the separation of the particles in their common Lorentz frame remains fixed; so there develops a Lorentz contraction from the viewpoint of the observer at $x^{\hat{j}} = 0$.]

Exercise 37.5. ORIENTATION OF POLARIZATION DIAGRAM

Derive equation (2) of Box 37.2.

§37.3. TYPES OF MECHANICAL DETECTORS

Eight types of mechanical
detectors:

Figure 37.2 shows eight different types of mechanical detectors for gravitational waves. (By "mechanical detector" is meant a detector that relies on the relative

(a) Oscillations in Earth-moon separation (see exercise 37.7)

(b) Normal-mode vibrations of earth and moon [see Weber (1968)]

(c) Oscillations in Earth's crust [see Dyson (1969)]

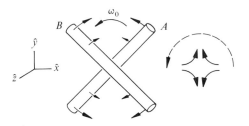

(d) Normal-mode vibrations of an elastic bar [see Weber (1969) and references cited therein]

(e) Normal-mode vibrations of an elastic square, or hoop, or tuning fork [see Douglass (1971)]

(f) Angular accelerations of rotating bars ["Heterodyne detector"; see Braginsky, Zel'dovich, and Rudenko (1969)]

(g) Angular accelerations of driven oscillators [Sakharov (1969)]

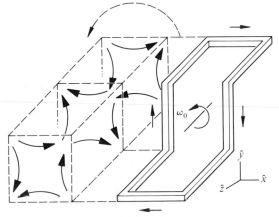

(h) Pumping of fluid in a rotating loop of pipe [Press (1970)]. The pipe rotates with the same angular velocity as the waves; so the position of the pipe in the righthand polarized lines of force remains forever fixed

Figure 37.2.
Various types of gravitational-wave detectors.

motions of matter. Nonmechanical detectors are described in §37.9, near end of this chapter.) These eight detectors, and others, can be analyzed easily using the force-line diagrams of Box 37.2. A *qualitative* discussion of each of the eight detectors is given below. (A full quantitative analysis for each one would entail experimental technicalities for which general relativity is irrelevant, and which are beyond the scope of this book. However, some quantitative details are spelled out in §§37.5–37.8.)

1. The Relative Motions of Two Freely Falling Bodies

(1) freely falling bodies

As a gravitational wave passes two freely falling bodies, their proper separation oscillates (Figure 37.3). This produces corresponding oscillations in the redshift and round-trip travel times for electromagnetic signals propagating back and forth between the two bodies. Either effect, oscillating redshift or oscillating travel time, could be used in principle to detect the passage of the waves. Examples of such detectors are the Earth-Moon separation, as monitored by laser ranging [Fig. 37.2(a)]; Earth-spacecraft separations as monitored by radio ranging; and the separation between two test masses in an Earth-orbiting laboratory, as monitored by redshift measurements or by laser interferometry. Several features of such detectors are explored in exercises 37.6 and 37.7. As shown in exercise 37.7, such detectors have so low a sensitivity that they are of little experimental interest.

Figure 37.3.

Time of round-trip travel between two geodesics responds to oscillations in the curvature of spacetime (diagram is schematic only; symbolic of a laser pulse sent from the Earth to a corner reflector on the Moon and back at a time when a very powerful, long-wavelength gravitational wave passes by; the wave would have to be powerful because a direct measure of distance to better than 10 cm is difficult, and such precision produces a much less sensitive indicator of waves than the vibrations in length [10^{-14} cm or less] of a Weber bar; see exercise 37.7). The geodesics are curved toward each other in regions where the relevant component of the Riemann curvature tensor, call it $R_{\hat{x}\hat{0}\hat{x}\hat{0}}$, has one sign, and curved away from each other in regions where it has the opposite sign. The diagram allows one to see at a glance the answer to an often expressed puzzlement: Is not any change in round-trip travel time mere trumpery flummery? The metric perturbation, $\delta h_{\mu\nu}$, of the wave changes the scale of distances slightly but also correspondingly changes the scale of time. Therefore does not every possibility of any really meaningful and measurable effect cancel out? Answer: (1) The widened separation between the geodesics is not a local effect but a cumulative one. It does not arise from the local value of $\delta h_{\mu\nu}$ directly or even from the local value of the curvature. It arises from an accumulation of the bending process over an entire half-period of the gravitational wave. (2) The change in separation of the geodesics is a true change in proper distance, and shows up in a true change in proper time (see "ticks" on the world line of one of the particles). See exercise 37.6. Note: When one investigates the separation between the geodesics, not over a single period, as here, but over a large number of periods, he finds a cumulative, systematic, net slow bending of the rapidly wiggling geodesics toward each other. This small, attractive acceleration is evidence in gravitation physics of the effective mass-energy carried by the gravitational waves (see Chapter 35).

2. Normal-Mode Vibrations of the Earth and Moon

A gravitational wave sweeping over the Earth will excite its quadrupole modes of vibration, since the driving forces in the wave have quadrupole spatial distributions [see Fig. 37.2(b)]. The fundamental quadrupole mode of the Earth has a period of 54 minutes, while that of the moon has a period of 15 minutes. Thus, the Earth and Moon should selectively pick out the 54-minute and 15-minute components of any passing wave train. Section 37.7 will analyze quantitatively the interaction between the wave and solid-body vibrations. By comparing that analysis with seismometer studies of the Earth's vibrations, Weber (1967) put the first observational limit ever on the cosmic flux of gravitational waves:

(2) Earth and Moon

$$I_\nu \equiv \frac{d\ \text{flux}}{d\ \text{frequency}} < 3 \times 10^7 \text{ erg cm}^{-2} \text{ sec}^{-1} \text{ Hz}^{-1} \text{ at } \nu = 3.1 \times 10^{-4} \text{ Hz.} \quad (37.8)$$

3. Oscillations in the Earth's Crust

If the neutron star in a pulsar is slightly deformed from axial symmetry, its rotation will produce gravitational waves. The period of the waves is half the period of the pulsar (rotation of star through 180° produces one period of waves)—i.e., it should range from 0.017 sec for NP0532 (Crab Pulsar) to 1.87 sec. for NP0527. Such a wave train cannot excite the 54-minute quadrupole vibration or any of the other normal, low-frequency modes of vibration of the Earth. The kind of vibrations it *can* excite allow themselves in principle to be described in the language of normal modes. However, they are clearly and more conveniently envisaged as vibrations of localized regions of the Earth; or, more particularly, vibrations of the Earth's crust.

(3) Earth's crust

Dyson (1969) has analyzed the response of an elastic solid, such as the Earth, to an incident, off-resonance gravitational wave. He shows that the response depends on irregularities in the elastic modulus for shear waves, and that it is strongest at a free surface [Figure 37.2(c)]. For the fraction of gravitational-wave energy crossing a flat surface that is converted into energy of elastic motion of the solid, he finds the expression

$$(\text{fraction}) = (8\pi G\rho/\omega^2)(s/c)^3 \times \sin^2\theta |\cos\theta|^{-1}[1 + \cos^2\theta + (s/v)\sin^2\theta]. \quad (37.9)$$

Here s and v are the velocities of shear waves and compressive waves, respectively, and θ is the angle between the direction of propagation of the waves and the normal to the surface. Considering a flux of 2×10^{-5} erg/cm^2 sec (an optimistic but conceivable value for waves from a pulsar) incident horizontally ($\theta = \pi/2$; "divergent" factor $|\cos\theta|^{-1}$ cancels out in calculation!), and taking s to be 4.5×10^5 cm/sec and ω to be 6 rad/sec, he calculates that the 1-Hz horizontal displacement produced in the surface has an amplitude of $\xi_0 \sim 2 \times 10^{-17}$ cm, too small by a factor of the order of 10^5 to be detected against background seismic noise. He points to the possibilities of improvements, especially via resonance (elastic waves reflected back and forth between two surfaces; Antarctic ice sheet).

4. Normal-Mode Vibrations of an Elastic Bar

As of 1972, the most often-discussed type of detector is the aluminum bar invented by Joseph Weber (1960, 1961) [see Figures 37.1 and 37.2(d)]. Weber's bars are cylindrical in shape, with length 153 cm, diameter 66 cm, and weight 1.4×10^6 g. Each bar is suspended by a wire in vacuum and is mechanically decoupled from its surroundings. Around its middle are attached piezoelectric strain transducers, which couple into electronic circuits that are sensitive to the bar's fundamental end-to-end mode of oscillation (frequency $\nu = 1{,}660$ Hz). When a gravitational wave hits the bar broadside, as shown in Figure 37.2(d), the relative accelerations carried by the wave will excite the fundamental mode of the bar. As of 1972, Weber observes sudden, simultaneous excitations in two such bars, one at the University of Maryland, near Washington, D.C.; the other at Argonne National Laboratory, near Chicago [see Weber (1969, 1970a,b)]. No one has yet come forward with a workable explanation for Weber's coincidences other than gravitational waves from outer space. However, the history of physics is rich with instances where supposedly new effects had to be attributed in the end to long familiar phenomena. Therefore it would seem difficult to rate the observed events as "battle-tested." To achieve that confidence rating would seem to require confirmation with different equipment, or under different circumstances, or both; experiments to provide that confirmation are now (1972) underway. If one makes this tentative assessment, one can be excused for expressing at the same time the greatest admiration for the experimental ingenuity, energy, and magnificent persistence that Joseph Weber has shown in his more than decade-long search for the most elusive radiation on the books of physics.

Mechanical detectors of the above four types represent systems on which measurements have been made; so practical difficulties and realizable noise levels can be estimated properly. In the continuing search for improved methods, more elaborate detectors are being studied, and in 1972 one can list a number of interesting proposals, as below. For these it is hard to know how much development would be required in order to achieve the desired performance.

5. Normal-Mode Vibrations of Elastic Bodies of Other Shapes

The "bar" of a Weber detector need not be cylindrical in shape. For a discussion of a detector with the shape of a hollow square, a hoop, or a tuning fork, see Douglass (1971); such a detector might allow its fundamental frequency to be adjusted for the most favorable response, with given mass, or given maximum dimension, or both. Sections 37.4 and 37.7 and exercises 37.9 to 37.12 analyze in detail the operation of a "vibrating-bar" detector of arbitrary shape. See also Douglass and Tyson (1971).

6. Angular Accelerations of Rotating Bars

All the potential detectors described thus far respond in the most obvious of manners to the tidal accelerations of a gravitational wave: relative distances oscillate in and

out. But the tidal accelerations contain, in addition to a length-changing component, also a tangential, rotation-producing component. In picture (a) of Box 37.2, the length-changing component dominates near the \hat{x} and \hat{y} axes, whereas the rotation-producing component dominates half-way between the axes. Vladimir B. Braginsky was the first to propose a detector that responds to the rotation-producing accelerations [see Braginsky, Zel'dovich, and Rudenko (1969); Braginsky and Nazarenko (1971)]. It consists of two metal rods, oriented perpendicular to each other, and rotating freely with angular velocity ω_0 in their common plane [see Fig. 37.2(f)]. (The rotation is relative to the gyroscopes of the proper reference frame of the detector; equivalently, it is relative to the Lorentz frame local to the detector.) Let monochromatic gravitational waves of angular frequency $\omega = 2\omega_0$ (change of phase per unit of time equals twice the angular velocity at which the pattern of lines of force turns) impinge broadside on the rotating rods. The righthand circularly polarized component of the waves will then rotate with the rods; so their orientation in its lines-of-force diagram will remain forever fixed. With the orientation of Fig. 37.2(f), rod A will undergo angular acceleration, while rod B will decelerate. The experimenter can search for the constant relative angular acceleration of the two rods (constant so long as the angle between them does not depart significantly from $90°$). Better yet, the experimenter can (all too easily) adjust the rotation rate ω_0 so it does not quite match the waves' frequency ω. Then for $\frac{1}{2}\omega_0/|\omega - 2\omega_0|$ rotations, rod A will accelerate and B will decelerate; then will follow $\frac{1}{2}\omega_0/|\omega - 2\omega_0|$ rotations in which A decelerates and B accelerates, and so on (frequency beating). The experimenter can search for oscillations in the relative orientation of the rods. One need not worry about the lefthand polarized waves marring the experiment. Since they do not rotate with the rods, their angular accelerations average out over one cycle.

Such a device is called a "heterodyne detector" by Braginsky. He envisages that such detectors might be placed in free-fall orbits about the Earth late in the 1970's. Heterodyne detectors would work most efficiently for long monochromatic wave trains such as those from pulsars; but even for short bursts of waves they may be more sensitive than vibrating bars [see Braginsky and Nazarenko (1971)].

7. Angular Accelerations of Driven Oscillators

Andrei D. Sakharov (1969) has proposed a different type of detector for the angular accelerations of a gravitational wave. Instead of two rotating bars, it consists of two identical, driven oscillators, initially parallel and nonrotating, but oscillating out of phase with each other. Each oscillator experiences angular accelerations in one direction at one phase of a passing wave, and in the opposite direction at the next phase, but the torques do not cancel out. When the oscillator is maximally distended, it experiences a greater torque (acceleration \propto length; torque \propto length2) than when it is maximally contracted. Consequently, if the driven oscillations have the same angular frequency as a passing, monochromatic wave, and if the phases are as shown in Figure 37.2(g), then oscillator A will receive an angular acceleration in the righthand direction, while B receives an angular acceleration in the lefthand direction.

(7) rotation of driven oscillators

8. Pumping of Fluid in a Rotating Loop of Pipe

A third type of detector that responds to angular accelerations has been described by William Press (1970). This detector would presumably be far less sensitive than others, and therefore not worth constructing; but it is intriguing in its novel design; and it illustrates features of gravitational waves ignored by other detectors. Press's detector consists of a loop of rotating pipe, containing a superfluid. The shape of the pipe and its constant rotation rate are chosen so that the gravitational waves will pump the fluid around inside the pipe. One conceivable pipe design (a bad one to build in practice, but an easy one to analyze) is shown in Fig. 37.2(h). Note that use is made of the variation in tidal acceleration along the direction of propagation of the wave as well as perpendicular to that direction. To analyze the response of the fluid to a righthand circularly polarized wave, one can mentally place the rotating pipe in the three-dimensional line-of-force diagram of Box 37.2(c).

EXERCISES

**Exercise 37.6. RELATIVE MOTION OF FREELY FALLING BODIES AS A
DETECTOR OF GRAVITATIONAL WAVES
[see Figures 37.2(a) and 37.3.]**

Consider two test bodies initially at rest with respect to each other in flat, empty spacetime. (The case where other, gravitating bodies are nearby can be treated without too much more difficulty; but this exercise concerns only the simplest example!) A plane, nearly monochromatic gravitational wave, with angular frequency ω and polarization e_+, impinges on the bodies, coming from the $-z$ direction. As shown in exercise 35.5, the bodies remain forever at rest in those TT coordinates that constituted the bodies' global inertial frame before the wave arrived. Calculate, for arbitrary separations ($\Delta x, \Delta y, \Delta z$) of the test bodies, the redshift and the round-trip travel time of photons going back and forth between them. Compare the answer, for large $\Delta x, \Delta y, \Delta z$, with the answer one would have obtained by using (without justification!) the equation of geodesic deviation. Physically, why does the correct answer *oscillate* with increasing separation? Discuss the feasibility and the potential sensitivity of such a detector using current technology.

**Exercise 37.7. EARTH-MOON SEPARATION AS A
GRAVITATIONAL-WAVE DETECTOR**

In the early 1970's one can monitor the Earth-moon separation using laser ranging to a precision of 10 cm, with successive observations separated by at least one round-trip travel time. Suppose that no oscillations in round-trip travel time are observed except those (of rather long periods) to be expected from the Earth-moon-sun-planets gravitational interaction. What limits can one then place on the energy flux of gravitational waves that pass the Earth? The mathematical formula for the answer should yield numerically

$$\text{Flux} \lesssim 10^{18} \text{ erg/cm}^2 \text{ sec for } 0.3 \text{ cycle/sec} \lesssim \nu \lesssim 1 \text{ cycle/day,} \qquad (37.10a)$$

corresponding to a limit on the mass density in gravitational waves of

$$\text{Density} \lesssim 10^{-13} \text{ g/cm}^3. \qquad (37.10b)$$

Why is this an uninteresting limit?

§37.4. VIBRATING, MECHANICAL DETECTORS: INTRODUCTORY REMARKS

The remainder of this chapter (except for §37.9) gives a detailed analysis of vibrating, mechanical detectors (Earth; Weber bar; "bars" with complex shapes; and so on).

The details of the analysis and its applications depend in a crucial way on the values of two dimensionless numbers: (1) the ratio τ_{GW}/τ_0, where

$$\tau_{GW} = \begin{pmatrix} \text{characteristic time scale for changes in} \\ \text{gravitational-wave amplitude and spectrum} \end{pmatrix}, \qquad (37.11a)$$

$$\tau_0 \equiv \begin{pmatrix} e\text{-folding time for detector vibrations (in} \\ \text{normal mode of interest) to die out as} \\ \text{a result of internal damping} \end{pmatrix}; \qquad (37.11b)$$

and (2) the ratio $\bar{E}_{vibration}/kT$, where

$$\bar{E}_{vibration} \equiv \begin{pmatrix} \text{mean value of detector's vibration energy (in} \\ \text{normal mode of interest) while waves are} \\ \text{passing and driving detector} \end{pmatrix}, \qquad (37.12a)$$

$$kT \equiv \begin{pmatrix} \text{Boltzman's} \\ \text{constant} \end{pmatrix} \times \begin{pmatrix} \text{detector's} \\ \text{temperature} \end{pmatrix}$$

$$= \begin{pmatrix} \text{Mean energy in normal mode} \\ \text{of interest when gravitational} \\ \text{waves are not exciting it} \end{pmatrix}. \qquad (37.12b)$$

When $\tau_{GW} \gg \tau_0$, the detector views the radiation as having a "*steady flux*," and it responds with steady-state vibrations; when $\tau_{GW} \ll \tau_0$ (short burst of waves), the waves deal a "*hammer blow*" to the detector. When $\bar{E}_{vibration} \gg kT$, the driving force of the waves dominates over the detector's random, internal, Brownian-noise forces ("*wave-dominated detector*"); when $\bar{E}_{vibration} \lesssim kT$, the driving force of the waves must compete with the detector's random, internal, Brownian-noise forces ("*noisy detector*").

Sections 37.5 to 37.7 deal with wave-dominated detectors ($\bar{E}_{vibration} \gg kT$). The key results of those sections are summarized in Box 37.3, which appears here as a quick preview (though it may not be fully understandable in advance). Section 37.8 treats noisy detectors.

Warning: Throughout the rest of this chapter prime attention focuses on the concept of cross section. This is fine for a first introduction to the theory of detectors. But cross section is not the entire story, especially when one wishes to study the detailed wave-form of the radiation. And sometimes (e.g., for the detector of Figure 37.2a), it is *none* of the story. A first-rate experimenter designing a new detector will not deal primarily in cross sections any more than a radio engineer will in designing a new radio telescope. Attention will also focus heavily on the bandwidth

(continued on page 1022)

The rest of this chapter is Track 2. No earlier track-2 material is needed as preparation for it, nor is it needed for any later chapter.

Definitions: "steady flux," "hammer-blow waves," "wave-dominated detector," "noisy detector"

Design of detectors requires much more than the concept of cross section

Box 37.3 WAYS TO USE CROSS SECTION FOR WAVE-DOMINATED DETECTORS

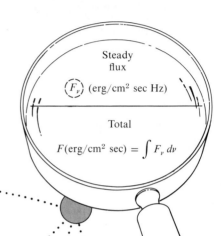

A. **To Calculate *Rate* at which
 Detector Extracts Energy from
 a *Steady* Flux of Radiation**

$(\tau_{GW} \gg \tau_0)$

Steady
flux

$\boxed{F_\nu}$ (erg/cm² sec Hz)

Total

$F(\text{erg/cm}^2 \text{ sec}) = \int F_\nu \, d\nu$

1. Frequency distribution of radiation arbitrary:

$$\binom{\text{steady rate at which detector extracts}}{\text{energy from gravitational waves}}$$

$$= \int \underbrace{\boxed{F_\nu(\nu)}}_{\text{erg/cm}^2 \text{ sec Hz}} \cdot \underbrace{\sigma(\nu)}_{\text{cm}^2} \underbrace{d\nu}_{\text{Hz}}$$

2. Frequency spread of radiation small
 compared to line width of detector:

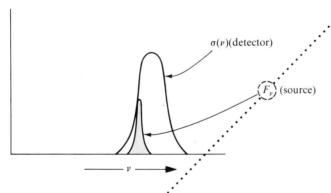

$\sigma(\nu)$(detector)

$\boxed{F_\nu}$ (source)

$\nu \longrightarrow$

$$\binom{\text{Steady rate at which}}{\text{detector extracts energy}}_{\text{from gravitational waves}}$$

$$= \sigma(\nu_{\text{source}}) \int \boxed{F_\nu} \, d\nu = \sigma F$$

$\boxed{F_\nu}$ (source)

3. Frequency spread of radiation large
 compared to line width of detector:

$$\binom{\text{steady rate at which detector extracts}}{\text{energy from gravitational waves}}$$

$$= \underbrace{\boxed{F_\nu}(\nu_{\text{detector}})}_{\text{erg/cm}^2 \text{ sec Hz}} \underbrace{\int \sigma(\nu) \, d\nu}_{\substack{\text{"resonance} \\ \text{integral", cm}^2 \text{ Hz}}}$$

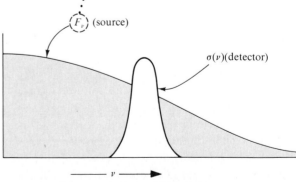

$\sigma(\nu)$(detector)

$\nu \longrightarrow$

B. To Calculate *Total* Energy Deposited in Detector by *any* Passing Wave train

1. If frequency distribution of radiation is arbitrary:

$$\begin{pmatrix}\text{total energy} \\ \text{deposited}\end{pmatrix} = \int \underbrace{\mathscr{F}_\nu(\nu)}_{\text{erg/cm}^2\,\text{Hz}} \cdot \underbrace{\sigma(\nu)}_{\text{cm}^2} \cdot \underbrace{d\nu}_{\text{Hz}}$$

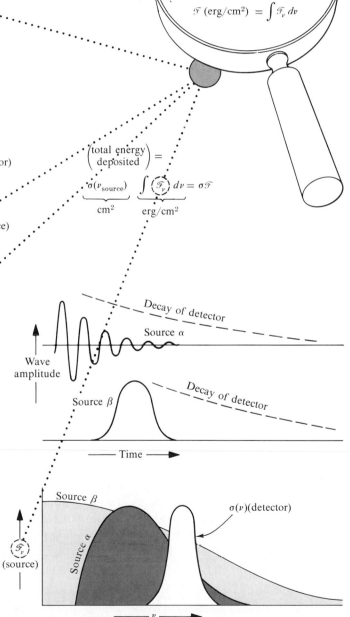

Any flux

\mathscr{F}_ν (erg/cm² Hz)

Total

\mathscr{F} (erg/cm²) $= \int \mathscr{F}_\nu \, d\nu$

2. If frequency spread of radiation is small compared to line width of detector ("monochromatic waves"):

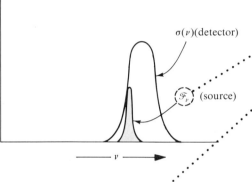

$\sigma(\nu)$(detector)

\mathscr{F}_ν (source)

$\longrightarrow \nu \longrightarrow$

$$\begin{pmatrix}\text{total energy} \\ \text{deposited}\end{pmatrix} = \underbrace{\sigma(\nu_{\text{source}})}_{\text{cm}^2} \underbrace{\int \mathscr{F}_\nu \, d\nu}_{\text{erg/cm}^2} = \sigma\mathscr{F}$$

3. If frequency spread of radiation is large compared to line width of detector (as it *must* be for hammer-blow radiation, where

$\Delta\nu_{\text{source}} \gtrsim 1/4\pi\tau_{\text{GW}} \gg 1/4\pi\tau_0 = \Delta\nu_{\text{detector}}$):

(total energy deposited) =

$$\underbrace{\mathscr{F}_\nu(\nu_{\text{detector}})}_{\text{erg/cm}^2\,\text{Hz}} \underbrace{\int \sigma(\nu) \, d\nu}_{\substack{\text{cm}^2\,\text{Hz,} \\ \text{"resonance} \\ \text{integral"}}}$$

Decay of detector

Source α

Wave amplitude

Source β

Decay of detector

\longrightarrow Time \longrightarrow

Source β

Source α

$\sigma(\nu)$(detector)

\mathscr{F}_ν (source)

$\longrightarrow \nu \longrightarrow$

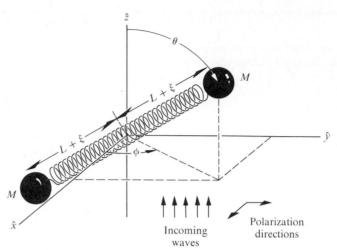

Figure 37.4.
An idealized detector (vibrator) responding to linearly polarized gravitational waves.

of the antenna, and on other, more detailed characteristics of its response, on coupling of the antenna to the displacement sensor, on response characteristics of the sensor, on antenna noise, on sensor noise, and so on. For an overview of these issues, and for discussions of detectors for which the concept of cross section is useless, see, e.g., Press and Thorne (1972).

§37.5. IDEALIZED WAVE-DOMINATED DETECTOR, EXCITED BY STEADY FLUX OF MONOCHROMATIC WAVES

Idealized detector: oscillator driven by a steady flux of monochromatic waves:

(1) derivation of equation of motion

Begin with the case of a *wave-dominated detector* ($\bar{E}_{\text{vibration}} \gg kT$) being driven by a steady flux of radiation ($\tau_{\text{GW}} \gg \tau_0$). Deal at first, not with a solid bar of arbitrary shape, but rather with the idealized detector of Figure 37.4: an oscillator made of two masses M on the ends of a spring of equilibrium length $2L$. Let the detector have a natural frequency of vibration ω_0 and a damping time $\tau_0 \gg 1/\omega_0$, so that its equation of motion (in the detector's proper reference frame) is

$$\ddot{\xi} + \dot{\xi}/\tau_0 + \omega_0^2 \xi = \text{driving acceleration.} \tag{37.13}$$

Let gravitational waves of polarization \boldsymbol{e}_+ and angular frequency ω impinge on the detector from the $-\hat{z}$ direction; and let the polar angles of the detector relative to the wave-determined $\hat{x}, \hat{y}, \hat{z}$-axes be θ and ϕ.

The incoming waves are described by equations (37.1) with the amplitude

$$A_\times = 0, \qquad A_+ = \mathcal{C}_+ e^{-i\omega(t-z)}. \tag{37.14}$$

(Here and throughout one must take the real part of all complex expressions.)

Assume that the detector is much smaller than a wavelength, so that one can set $z \approx \hat{z} = 0$ throughout it. Then the tidal accelerations produced by the wave

$$\left(\frac{d^2\hat{x}}{d\hat{t}^2}\right)_{\text{due to wave}} = -R_{\hat{x}\hat{0}\hat{j}\hat{0}}x^{\hat{j}} = -\frac{1}{2}\omega^2 a_+ e^{-i\omega\hat{t}}\hat{x},$$

$$\left(\frac{d^2\hat{y}}{d\hat{t}^2}\right)_{\text{due to wave}} = -R_{\hat{y}\hat{0}\hat{j}\hat{0}}x^{\hat{j}} = +\frac{1}{2}\omega^2 a_+ e^{-i\omega\hat{t}}\hat{y},$$

have as their component along the oscillator

$$\frac{d^2\xi}{dt^2} = \frac{\hat{x}}{L}\frac{d^2\hat{x}}{dt^2} + \frac{\hat{y}}{L}\frac{d^2\hat{y}}{dt^2} + \frac{\hat{z}}{L}\frac{d^2\hat{z}}{dt^2} = -\frac{1}{2}\omega^2 a_+ L e^{-i\omega\hat{t}}\frac{\hat{x}^2 - \hat{y}^2}{L^2}$$

$$= -\frac{1}{2}\omega^2 a_+ L e^{-i\omega\hat{t}}\sin^2\theta\,\cos 2\phi.$$

Consequently, the equation of motion for the oscillator is

$$\ddot{\xi} + \dot{\xi}/\tau_0 + \omega_0^2\xi = -\frac{1}{2}\omega^2 a_+ L e^{-i\omega\hat{t}}\sin^2\theta\,\cos 2\phi. \tag{37.15}$$

The driving force varies as $\cos 2\phi$ because of the "spin-2" nature of gravitational waves: a rotation through 180° in the transverse plane leaves the waves unchanged; a rotation through 90° reverses the phase. The $\sin^2\theta$ term results from the transverse nature of the waves [one factor of $\sin\theta$ to account for projection onto the detector's direction], plus their tidal-force nature [another factor of $\sin\theta$ to account for (relative force) \propto (distance *in transverse plane*)].

The straightforward steady-state solution of the equation of motion (37.15) is

$$\xi = \frac{\frac{1}{2}\omega^2 a_+ L\,\sin^2\theta\,\cos 2\phi}{\omega^2 - \omega_0^2 + i\omega/\tau_0}e^{-i\omega\hat{t}}. \tag{37.16}$$

(2) oscillator amplitude as function of frequency and orientation

When the incoming waves are near resonance with the detector, $|\omega \pm \omega_0| \lesssim 1/\tau_0$, the oscillator is excited to large amplitude. Otherwise the excitation is small. Focus attention henceforth on near-resonance excitations; then equation (37.16) can be simplified (*note:* ω_0 is positive, but ω may be negative or positive):

$$\xi = \frac{\frac{1}{4}\omega_0 a_+ L\,\sin^2\theta\,\cos 2\phi}{|\omega| - \omega_0 + \frac{1}{2}\text{sgn}(\omega)i/\tau_0}e^{-i\omega\hat{t}}. \tag{37.16'}$$

One measure of the detector's usefulness is its cross section for absorbing gravitational-wave energy. The steady-state vibrational energy in a detector with the above amplitude and with 2 masses M is

$$E_{\text{vibration}} = 2\cdot\frac{1}{2}\cdot M\cdot(\dot{\xi}^2)_{\text{max}} = \frac{\frac{1}{16}ML^2\omega_0^4 a_+^2\sin^4\theta\,\cos^2 2\phi}{(|\omega| - \omega_0)^2 + (1/2\tau_0)^2}. \tag{37.17}$$

This energy is being dissipated internally at a rate $E_{\text{vibration}}/\tau_0$. If one ignores reradiation of energy as gravitational waves (a negligible process!), one can equate the dissipation rate to the rate at which the detector absorbs energy from the incoming waves—which in turn equals the "cross section" σ times the incoming flux:

$$E_{\text{vibration}}/\tau_0 = -dE_{\text{waves}}/dt = \sigma T^{0z(\text{GW})} = \frac{1}{32\pi}\sigma\omega^2\mathscr{A}_+{}^2.$$

(3) cross sections for
polarized radiation

Consequently, *near resonance* ($|\omega \pm \omega_0| \ll \omega_0$), *the cross section* for interception of gravitational-wave energy is

$$\sigma = \frac{2\pi ML^2(\omega_0{}^2/\tau_0)\sin^4\theta\,\cos^22\phi}{(|\omega| - \omega_0)^2 + (1/2\tau_0)^2}, \qquad \text{for polarized radiation.} \qquad (37.18)$$

This expression applies to monochromatic radiation. However, experience with many other kinds of waves has taught that one often has to deal with a broad continuum of frequencies, with the "bandwidth" of the incident radiation far greater than the width of the detector resonance (see Box 37.3). Under these conditions, the relevant quantity is not the cross section itself, but the "resonance integral" of the cross section,

$$\int_{\text{resonance}} \sigma\,d\nu = \int \sigma(d\omega/2\pi) = 2\pi ML^2\omega_0{}^2\sin^4\theta\,\cos^22\phi,$$
$$\text{for polarized radiation.} \qquad (37.19)$$

Before examining the magnitude of this cross section, scrutinize its directionality (the "antenna-beam pattern"). The factor of $\sin^4\theta\,\cos^22\phi$ refers to linearly polarized, \boldsymbol{e}_+ radiation (see Figure 37.4). For the orthogonal mode of polarization, \boldsymbol{e}_\times, $\cos^22\phi$ is to be replaced by $\sin^22\phi$; and for unpolarized (incoherent mixture) radiation or circularly polarized radiation, the cross section is the average of these two expressions; thus

(4) cross sections for
unpolarized radiation

$$\sigma = \frac{\pi ML^2(\omega_0{}^2/\tau_0)\sin^4\theta}{(|\omega| - \omega_0)^2 + (1/2\tau_0)^2} \qquad \text{for unpolarized radiation.} \qquad (37.20)$$

Notice that this unpolarized cross section is peaked, with half-width 33°, about the equatorial plane of the detector. Averaged over all possible directions of incoming waves, the cross section is

$$\langle\sigma\rangle_{\text{all directions}} = \frac{1}{2}\int_0^\pi \sigma\sin\theta\,d\theta = \frac{8}{15}\sigma_{\text{max}}$$
$$= \frac{(8\pi/15)ML^2(\omega_0{}^2/\tau_0)}{(|\omega| - \omega_0)^2 + (1/2\tau_0)^2} \quad \text{for unpolarized radiation.} \qquad (37.21)$$

One can rewrite the above cross sections in several suggestive forms. For example, on resonance, the cross section (37.21) reads

$$\langle\sigma\rangle_{\text{all directions}} = \frac{4\pi^2}{15}\frac{4M}{2\pi/\omega_0}(\omega_0\tau_0)(2L)^2.$$

Recall that $\omega_0\tau_0$ defines the "Q" of a detector, $1/Q \equiv$ (fraction of *energy* dissipated per radian of oscillation). Note that $2\pi/\omega_0$ is the wavelength λ_0 of resonant radiation. Finally, denote by $r_g = 4M$ the gravitational radius of the detector. In terms of these three familiar quantities, find for the cross section the formula

$$\frac{\langle\sigma\rangle_{\text{all directions}}}{(2L)^2} = \frac{\text{(cross section for absorbing waves on resonance)}}{\text{("geometric" cross section of detector)}}$$

$$= (4\pi^2/15)(r_g/\lambda_0)Q \quad \begin{array}{l}\text{for unpolarized radiation} \\ \text{on resonance.}\end{array} \quad (37.22)$$

Magnitude of cross sections for any resonant detector

This relation holds in order of magnitude for any resonant detector. It shows starkly that gravitational-wave astronomy must be a difficult enterprise. How large could *you* make the factor r_g/λ_0, given a reasonable budget? Weber's 1970 detectors have $2L_{\text{effective}} \approx 1$ meter, $r_g \approx (0.74 \times 10^{-28}\ \text{cm/g}) \times (10^6\ \text{g}) \approx 10^{-22}\ \text{cm}$, $\nu_0 = \omega_0/2\pi = 1{,}660$ Hz, $\lambda_0 \approx 200$ km, $r_g/\lambda_0 \approx \frac{1}{2} \times 10^{-29}$, $\tau_0 \approx 20$ sec, $Q \approx 2 \times 10^5$; so that

$$\sigma_{\text{Weber}} \approx 3 \times 10^{-20}\ \text{cm}^2 \text{ on resonance.} \quad (37.23)$$

What flux of gravitational-wave energy would have to be incident to excite a cold detector ($\sim 0°$ K) into roughly steady-state vibrations with a vibration energy of (Boltzmann's constant) \times (room temperature) $\sim 4 \times 10^{-14}$ erg? The vibrator, if cooled enough to be wave-dominated, dissipates its energy at the rate $E_{\text{vibration}}/\tau_0 \sim 2 \times 10^{-15}$ erg/sec. The incident flux has to make up this loss, at the rate

Flux required to excite a Weber-type detector

$$T_{00}^{(\text{GW})}\sigma \sim 2 \times 10^{-15}\ \text{erg/sec}, \quad (37.24a)$$

implying an incident flux of the order of $2 \times 10^{-15}/3 \times 10^{-20} \sim 10^5$ erg/cm^2 sec. Moreover, this flux has to be concentrated in the narrow range of resonance

$$\nu \approx \nu_0 \pm 1/4\pi\tau_0 = (1660 \pm 0.004)\ \text{Hz}. \quad (37.24b)$$

By anybody's standards, this is a very high flux of gravitational radiation for such a small bandwidth ($\sim 10^7$ erg/cm^2 sec Hz, as compared to the flux of blackbody gravitational radiation, $8\pi\nu^2 kT/c^2 = 3 \times 10^{-27}$ erg/cm^2 sec Hz, that would correspond to Planck equilibrium at the same temperature; the large factor of difference is a direct reflection of the difference in rate of damping of the oscillator by friction and by gravitational radiation).

Equation (37.22) makes it seem that an optimal detector must have a large Q. This is not necessarily so. Recall that the bandwidth, $\Delta\omega \approx \omega_0/Q$, over which the cross section is large, decreases with increasing Q. When an incoming steady flux of waves of bandwidth $\Delta\omega \gg \omega_0/Q \equiv 1/\tau_0$ and of specific flux

A large Q is not necessarily optimal

$$F_\nu(\text{erg/cm}^2\ \text{sec Hz})$$

drives the detector, it deposits energy at the rate

$$\begin{pmatrix} \text{rate of deposit} \\ \text{of energy} \end{pmatrix} = \frac{dE}{dt} = \int_{\text{resonance}} F_\nu \sigma \, d\nu = F_\nu(\nu_0) \int_{\text{resonance}} \sigma \, d\nu.$$

$$\left[\begin{array}{l} \text{for radiation with} \\ \text{bandwidth } \Delta\nu \gg 1/\tau_0 \end{array} \right]$$

Consequently, the relevant measure of detector effectiveness will be the integral of the cross section over the resonance, $\int \sigma \, d\nu$ (37.19). (See next section.) This frequency-integrated cross section is independent of the detector's Q, so one must use more sophisticated reasoning (e.g., signal-to-noise theory) in deciding whether a large Q is desirable. (See §37.8.)

§37.6. IDEALIZED, WAVE-DOMINATED DETECTOR, EXCITED BY ARBITRARY FLUX OF RADIATION

Response of idealized detector to an arbitrary, non-monochromatic flux:

(1) derivation

Let plane-polarized waves of polarization e_+ but *arbitrary* spectrum [equation (37.1) with $A_\times = 0$] impinge on the idealized detector of Figure 37.4. Then the equation of motion for the detector is the same as for monochromatic waves [equation (37.15)], but with $-\omega^2 a_+ e^{-i\omega t}$ replaced by \ddot{A}_+:

$$\ddot{\xi} + \dot{\xi}/\tau_0 + \omega_0^2 \xi = \frac{1}{2} \ddot{A}_+ L \sin^2\theta \cos 2\phi. \tag{37.26}$$

[By now one is fully accustomed to the fact that all analyses of detectors (when the detector is much smaller than the wavelength of the waves) are performed in the proper reference frame, with coordinates $\hat{t}, \hat{x}, \hat{y}, \hat{z}$. *Henceforth, for ease of eyesight, abandon the "hats" on these "proper coordinates," and denote them as merely t, x, y, z.*]

Fourier-analyze the waves and the detector displacement,

$$A_+(t) = (2\pi)^{-1/2} \int_{-\infty}^{+\infty} \widetilde{A}_+(\omega) e^{-i\omega t}, \tag{37.27a}$$

$$\xi(t) = (2\pi)^{-1/2} \int_{-\infty}^{+\infty} \widetilde{\xi}(\omega) e^{-i\omega t}; \tag{37.27b}$$

and conclude from equation (37.26) that

$$\widetilde{\xi} = \frac{\frac{1}{2} \omega^2 \widetilde{A}_+ L \sin^2\theta \cos 2\phi}{\omega^2 - \omega_0^2 + i\omega/\tau_0}.$$

This Fourier amplitude is negligible unless $|\omega \pm \omega_0| \ll \omega_0$; consequently, without loss of accuracy, one can rewrite it as

$$\widetilde{\xi} = \frac{\frac{1}{4}\omega_0\widetilde{A}_+ L\,\sin^2\theta\,\cos 2\phi}{|\omega| - \omega_0 + \frac{1}{2}\,\mathrm{sgn}(\omega)i/\tau_0}. \tag{37.28}$$

[Compare with the steady-state amplitude (37.16′).]

Ask how much total energy is deposited in the detector by the gravitational waves. Do *not* seek an answer by examining the amplitude of the vibrations, $\xi(t)$, directly; since that amplitude is governed by *both* internal damping and the driving force of the waves, it does not reflect directly the energy deposited. To get the total energy deposited, integrate over time the force acting on each mass multiplied by its velocity:

$$\begin{pmatrix}\text{total energy}\\\text{deposited}\end{pmatrix} = \int_{-\infty}^{+\infty} 2\left(\frac{1}{2}M\ddot{A}_+ L\,\sin^2\theta\,\cos 2\phi\right)\dot{\xi}\,dt.$$

[2 masses] [force on each mass] [velocity of each mass]

Use Parseval's theorem (one of the most powerful tools of mathematical physics!) to replace the time integral by a frequency integral

$$\begin{pmatrix}\text{total energy}\\\text{deposited}\end{pmatrix} = \Re \int_{-\infty}^{+\infty} (ML\,\sin^2\theta\,\cos 2\phi)(-\omega^2\widetilde{A}_+{}^*)(-i\omega\widetilde{\xi})\,d\omega.$$

Then use equation (37.28) to rewrite this entirely in terms of the wave amplitude

$$\begin{pmatrix}\text{total energy}\\\text{deposited}\end{pmatrix} = \int_{-\infty}^{+\infty}\left[\frac{2\pi(\omega_0{}^2/\tau_0)ML^2\,\sin^4\theta\,\cos^2 2\phi}{(|\omega| - \omega_0)^2 + (1/2\tau_0)^2}\right]\left[\frac{\omega^2|\widetilde{A}_+|^2}{16\pi}\right]d\omega. \tag{37.29}$$

The first term in this expression is precisely the cross section for monochromatic waves, derived in the last section (37.18). The second term has an equally simple interpretation: the total energy that the gravitational waves carry past a unit surface area of detector is

$$\begin{aligned}\mathscr{F}(\text{ergs/cm}^2) &= \int T_{00}^{(\text{GW})}\,dt = \int \frac{1}{16\pi}\dot{A}_+{}^2\,dt\\ &= \int \frac{\omega^2|\widetilde{A}_+|^2}{16\pi}\,d\omega = \int \frac{\omega^2|\widetilde{A}_+|^2}{8}\,d\nu\end{aligned} \tag{37.30}$$

(Parseval's theorem again!). Consequently, the energy per unit frequency interval, per unit area carried by the waves is

$$\mathscr{F}_\nu(\text{ergs/cm}^2\,\text{Hz}) = \frac{1}{8}\omega^2|\widetilde{A}_+|^2 \tag{37.31}$$

[for $-\infty < \nu < +\infty$; double this for $0 < \nu < +\infty$, a convention we use for the rest of this chapter]. This is 2π times the second term in (37.29).

Combining equations (37.18), (37.29), and (37.31), then, one finds

(2) answer—

$$\begin{pmatrix}\text{total energy}\\\text{deposited}\end{pmatrix} = \int \sigma(\nu)\mathscr{F}_\nu(\nu)\,d\nu. \tag{37.32}$$
$$\begin{pmatrix}\text{energy}\\\text{deposited}\end{pmatrix} = \int \sigma\mathscr{F}_\nu\,d\nu$$

This is the total energy deposited, regardless of the spectrum of the waves, and regardless of whether they come in a steady flux for a long time, or in a short burst, or in any other form. It is perfectly general—so long as the detector is wave-dominated ($E_{\text{vibration}} \gg kT$) while the waves are driving it.

How can an experimenter measure the total energy deposited? He cannot measure it directly, in general, but he can measure a quantity equal to it: the total energy that goes into internal damping, i.e., into "friction." Energy is removed by "friction" at a rate $E_{\text{vibration}}/\tau_0$, when the vibration energy is much greater than kT (during period of wave-dominance). Therefore, the experimenter can measure

$$\begin{pmatrix} \text{total energy} \\ \text{deposited} \end{pmatrix} = \frac{1}{\tau_0} \int E_{\text{vibration}} \, dt, \qquad \text{in general.} \tag{37.33}$$

$$\overset{\displaystyle\longmapsto}{}\ [\text{integrate over the period that } E_{\text{vibration}} \gg kT]$$

In the special case of "hammer-blow waves" ($\tau_{\text{GW}} = $ duration of waves $\ll \tau_0$), the vibration energy is driven "instantaneously" from $\sim kT$ to a peak value, $E_{\text{vibration}}^{\text{peak}} \gg kT$, and then decays exponentially back to $\sim kT$; thus

$$\begin{pmatrix} \text{total energy} \\ \text{deposited} \end{pmatrix} = \frac{1}{\tau_0} \int_0^\infty E_{\text{vibration}}^{\text{peak}} e^{-t/\tau_0} \, dt = E_{\text{vibration}}^{\text{peak}} \tag{37.34}$$

for hammer-blow waves.

When the waves are steady for a long period of time ($\tau_{\text{GW}} \gg \tau_0$), with specific flux

$$F_\nu = \mathcal{F}_\nu/\tau_{\text{GW}} \qquad (\text{ergs/cm}^2 \text{ sec Hz}),$$

then the energy will be deposited at a constant rate

$$(dE/dt) = (\text{total energy deposited})/\tau_{\text{GW}};$$

and equation (37.32) can be rewritten

$$\begin{pmatrix} \text{rate of deposit} \\ \text{of energy} \end{pmatrix} = \int \sigma(\nu) F_\nu \, d\nu, \qquad \text{for steady waves } (\tau_{\text{GW}} \gg \tau_0). \tag{37.35}$$

Equations (37.32) and (37.35) are the key equations for application of the concept of cross section to realistic situations. They are applicable not only to polarized radiation, but also to unpolarized radiation and to radiation coming in from all directions, if one merely makes sure to use the appropriate cross section [equation (37.20) or (37.21) instead of (37.18)]. For examples of their application, see Box 37.3.

§37.7. GENERAL WAVE-DOMINATED DETECTOR, EXCITED BY ARBITRARY FLUX OF RADIATION

The cross sections of the idealized spring-plus-mass detector can be put into a form more elegant than equations (37.18) to (37.21)—a form that makes contact with many

branches of physics, and is valid for *any* vibrating resonant detector whatsoever.

Introduce the "Einstein A-coefficients," which describe the rate at which a unit amount of detector energy is lost to internal damping and to reradiation of gravitational waves:

Cross sections reexpressed in terms of "Einstein A-coefficients"

$$A_{\text{diss}} \equiv \left(\frac{\text{rate at which energy is dissipated internally}}{\text{energy in oscillations of detector}} \right) = \frac{1}{\tau_0}, \quad (37.36a)$$

$$A_{\text{GW}} \equiv \left(\frac{\text{rate at which energy is reradiated}}{\text{energy in oscillations}} \right). \quad (37.36b)$$

For the idealized detector of Figure 37.4, the standard formula (36.1) for the emission of gravitational waves yields

$$(\text{power reradiated}) = \frac{32}{15} \omega^6 M^2 L^2 \langle \xi^2 \rangle_{\text{time avg.}} \quad (37.37)$$

(see exercise 37.8). Consequently

$$A_{\text{GW}} = \frac{16}{15} M L^2 \omega^4. \quad (37.38)$$

One can use these relations to rewrite the detector cross sections in terms of A_{diss}, A_{GW}, and the reduced wavelength

$$\lambda \equiv 1/\omega \quad (37.39)$$

of the radiation. For example, the cross section (37.21)—now with $\omega \geq 0$—is

$$\langle \sigma \rangle_{\text{all directions}} = \frac{1}{2} \pi \lambda^2 \frac{A_{\text{GW}} A_{\text{diss}}}{(\omega - \omega_0)^2 + (A_{\text{diss}}/2)^2} \quad \begin{array}{l} \text{for unpolarized} \\ \text{radiation} \end{array} \quad (37.40)$$

(*recall the assumption $|\omega - \omega_0| \ll \omega_0$ in all cross-section formulas*) and the corresponding integral over the resonance is

$$\int \langle \sigma \rangle_{\text{all directions}} \, d\nu = \frac{1}{2} \pi \lambda_0^2 A_{\text{GW}} \quad \text{for polarized radiation.} \quad (37.41)$$

These expressions for the cross section are comprehensive in their application. They apply to any vibrating, resonant, gravitational-wave detector whatsoever, as one sees from the "detailed balance" calculation of exercise 37.9, and from the dynamic calculations of exercise 37.10. They also apply, with obvious changes in statistical factors and notation, to compound-nucleus reactions in nuclear physics ("Breit-Wigner formula"; see Blatt and Weisskopf, pp. 392–94, 408–10, 555–59), to the absorption of photons by atoms and molecules, to reception of electromagnetic waves by a television antenna, etc. Equation (37.41) says in effect, "Calculate the rate at which the oscillator is damped by emission of gravitational radiation; multiply that rate by the geometric factor familiar in all work with antennas, $\frac{1}{2}\pi\lambda_0^2$, and immediately obtain the resonance integral of the cross section. The result is expressed in geometric

Generality of the A-coefficient formalism

units (cm). To get the resonance integral in conventional units, multiply by the conversion factor $c = 3 \times 10^{10}$ cm Hz.

The 'dynamic analysis' of the idealized masses-on-spring detector, as developed in the last section, is readily extended to a vibrating detector of arbitrary shape (Earth; Weber's bar; an automobile fender; and so on). The extension is carried out in exercise 37.10 and its main results are summarized in Box 37.4.

Scattering of radiation by detector

Part of the energy that goes into a detector is reradiated as scattered gravitational radiation. For any detector of laboratory dimensions with laboratory damping coefficients, this fraction is fantastically small. However, in principle one can envisage a larger system and conditions where the reradiation is not at all negligible. In such an instance one is dealing with scattering. No attempt is made here to analyze such scattering processes. For a simple order-of-magnitude treatment, one can use the same type of Breit-Wigner scattering formula that one employs to calculate the scattering of neutrons at a nuclear resonance or photons at an optical resonance. A still more detailed account will analyze the correlation between the polarization of the scattered radiation and the polarization of the incident radiation. The kind of formalism useful here for gravitational radiation with its tensor character will be very much like that now used to treat polarization of radiation with a spin-1 character. Here notice especially the "Madison Convention" [Barschall and Haeberli (1971)] developed by the collaborative efforts of many workers after experience during many years with a variety of conflicting notations. Considering the way in which the best notation that is available today for spin-1 radiation was evolved, one can only feel that it is too early to canonize any one notation for describing the scattering parameters for an object that is scattering gravitational radiation.

EXERCISES

Exercise 37.8. POWER RERADIATED

The idealized gravitational wave detector of Figure 37.4 vibrates with angular frequency ω. Show that the power it radiates as gravitational waves is given by equation (37.37).

Exercise 37.9. CROSS SECTIONS CALCULATED BY DETAILED BALANCE

Use the principle of detailed balance to derive the cross sections (37.41) for a vibrating, resonant detector of any size, shape, or mass (e.g., for the vibrating Earth, or Weber's vibrating cylinder, or the idealized detector of Figure 37.4). [*Hints:* Let the detector be in thermal equilibrium with a bath of blackbody gravitational waves. Then it must be losing energy by reradiation as rapidly as it is absorbing it from the waves. (Internal damping can be ignored because, in true thermal equilibrium, energy loss by internal damping will match energy gain from random internal Brownian forces.) In detail, the balance of energy in and out reads [with I_ν = "specific intensity," equation (22.48)]

$$[4\pi I_\nu(\nu = \nu_0)]_{\text{blackbody}} \times \int \langle \sigma \rangle_{\text{all directions}} \, d\nu$$

$$= A_{\text{GW}} \times (\text{Energy in normal mode of detector}).$$

Solve for $\int \langle \sigma \rangle \, d\nu$, using the familiar form of the Planck spectrum and the fact that gravitational waves have two independent states of polarization.] *Note:* Because detailed balance

Box 37.4 VIBRATING, RESONANT DETECTOR OF ARBITRARY SHAPE

A. Physical Characteristics of Detector

1. Detector is a solid object (Earth, Weber bar, automobile fender, . . .) with density distribution $\rho(x)$ and total mass $M = \int \rho \, d^3x$.
2. Detector has normal modes of vibration. The nth normal mode is characterized by:

$$\omega_n = \text{angular frequency};$$

$$\tau_n = \begin{pmatrix} \text{e-folding time for vibration energy} \\ \text{to decay as result of internal damping} \end{pmatrix} \gg 1/\omega_n; \tag{1}$$

$$u_n(x) = \text{eigenfunction (defined here to be dimensionless and real).}$$

The eigenfunctions u_n are orthonormalized, so that

$$\int \rho u_n \cdot u_m \, d^3x = M \delta_{nm}. \tag{2}$$

3. During a normal-mode vibration with $E_{\text{vibration}} \gg kT$, a mass element originally at \hat{x} receives the displacement

$$\delta x = \xi = u_n(x) \mathcal{B}_n e^{-i\omega_n t - t/\tau_n}, \tag{3a}$$
$$\underset{\text{[constant amplitude]}}{\uparrow}$$

the density at fixed x changes by

$$\delta \rho = \quad \nabla \cdot (\rho u_n) \mathcal{B}_n e^{-i\omega_n t - t/\tau_n}, \tag{3b}$$

and the moment of inertia tensor oscillates

$$\delta I_{jk} = I_{(n)jk} \mathcal{B}_n e^{-i\omega_n t - t/\tau_n}. \tag{3c}$$

Here $I_{(n)jk}$ is the "*moment of inertia factor for the nth normal mode*":

$$I_{(n)jk} \equiv \int - (\rho u_n{}^l)_{,l} x^j x^k \, d^3x$$

$$= \int \rho (u_n{}^j x^k + u_n{}^k x^j) \, d^3x \tag{4}$$

[dimensions: mass \times length, multiply by \mathcal{B}_n (length) to get I_{jk}].

The corresponding "*reduced quadrupole factor for the nth normal mode*" is

$$\mathcal{I}_{(n)jk} \equiv I_{(n)jk} - \frac{1}{3} I_{(n)ll} \delta_{jk}. \tag{5}$$

Box 37.4 (*continued*)

B. Cross Sections for Detector (exercise 37.10)

1. For *polarized radiation* with propagation direction \boldsymbol{n} and polarization tensor \boldsymbol{e}:

$$h_{jk} = A(t - \boldsymbol{n} \cdot \boldsymbol{x})e_{jk},$$

$$e_{jk}n_k = 0, \qquad e_{jj} = 0, \qquad e_{jk}e_{jk} = 2; \tag{6}$$

$$\sigma_n(\nu) = \sigma_n(\omega/2\pi) = \frac{\pi}{4}\frac{|\mathcal{I}_{(n)jk}e_{jk}|^2}{M}\frac{\omega_n^2/\tau_n}{(|\omega| - \omega_n)^2 + (1/2\tau_n)^2}, \tag{7a}$$

$$\int_{\text{resonance}} \sigma_n \, d\nu = \frac{\pi}{4}\frac{|\mathcal{I}_{(n)jk}e_{jk}|^2}{M}\omega_n^2. \tag{7b}$$

2. For *unpolarized radiation* (random mixture of polarizations) with propagation direction \boldsymbol{n}, cross sections are

$$\sigma_n(\nu) = \sigma_n(\omega/2\pi) = \frac{\pi}{4}\frac{(\mathcal{I}_{(n)jk}^{\text{TT}})^2}{M}\frac{\omega_n^2/\tau_n}{(|\omega| - \omega_n)^2 + (1/2\tau_n)^2}, \tag{8a}$$

$$\int_{\text{resonance}} \sigma_n \, d\nu = \frac{\pi}{4}\frac{(\mathcal{I}_{(n)jk}^{\text{TT}})^2}{M}\omega_n^2. \tag{8b}$$

Here $\mathcal{I}_{(n)jk}^{\text{TT}}$ is the transverse-traceless part of $\mathcal{I}_{(n)jk}$ (transverse and traceless relative to the propagation direction \boldsymbol{n}):

$$\mathcal{I}_{(n)}^{\text{TT}} = P\mathcal{I}_{(n)}P - \frac{1}{2}P \text{ trace }(P\mathcal{I}_{(n)}), \qquad P_{jk} \equiv \delta_{jk} - n_j n_k. \tag{9}$$

(See Box 35.1)

3. Cross sections for *unpolarized radiation, averaged over all directions,* are

$$\langle\sigma_n(\nu)\rangle_{\text{all directions}} = \frac{1}{2}\pi\lambda^2\frac{A_{\text{GW}}A_{\text{diss}}}{(|\omega| - \omega_n)^2 + (A_{\text{diss}}/2)^2}, \tag{10a}$$

$$\int_{\text{resonance}} \langle\sigma_n\rangle_{\text{all directions}} \, d\nu = \frac{1}{2}\pi\lambda^2 A_{\text{GW}}, \tag{10b}$$

where the Einstein A coefficients are

$$A_{\text{diss}} = 1/\tau_n, \tag{11}$$

$$A_{\text{GW}} = \frac{1}{5}\frac{(\mathcal{I}_{(n)jk})^2}{M}\omega_n^4. \tag{12}$$

C. Spectrum Radiated by an Aperiodic Source (exercise 37.11)

It is instructive to compare these formulas with expressions for the radiation emitted by an aperiodic source.

1. Fourier-analyze the reduced quadrupole factor of the source

$$\mathcal{I}_{jk}(t) = (2\pi)^{-1/2} \int_{-\infty}^{+\infty} \tilde{\mathcal{I}}_{jk}(\omega)e^{-i\omega t}\, d\omega.$$

2. Then the total energy per unit frequency ($\nu \geq 0$) radiated over all time, into a unit solid angle about the direction \boldsymbol{n}, and with polarization tensor \boldsymbol{e}, is

$$\frac{dE}{d\nu\, d\Omega} = \frac{1}{8} \sum_{\omega = \pm 2\pi\nu} |\tilde{\mathcal{I}}_{jk}e_{jk}|^2 \omega^6 \tag{13a}$$

[compare with equations (7)]. Summed over polarizations, this is

$$\frac{dE}{d\nu\, d\Omega} = \frac{1}{2} \sum_{j,k} |\tilde{\mathcal{I}}_{jk}{}^{\mathrm{TT}}|^2 \omega^6 \tag{13b}$$

[compare with equations (8)]. Here $\nu \geq 0$.

3. The total energy radiated per unit frequency, integrated over all directions, still with $\nu \geq 0$, is

$$dE/d\nu = \frac{4\pi}{5} \sum_{j,k} |\tilde{\mathcal{I}}_{jk}|^2 \omega^6 \tag{14}$$

[compare with equations (10)–(12)].

can be applied to any kind of resonant system in interaction with any kind of thermal bath of radiation or particles, equations (37.40) and (37.41), with appropriate changes of statistical factors, have wide generality.

Exercise 37.10. NORMAL-MODE ANALYSIS OF VIBRATING,
RESONANT DETECTORS

Derive all the results for vibrating, resonant detectors quoted in Box 37.4. Pattern the derivation after the treatment of the idealized detector in §37.6. [*Guidelines:* (a) Let the detector be driven by the polarized waves of equation (6), Box 37.4; and let it be wave-dominated ($E_{\mathrm{vibration}} \gg kT$). Show that the displacements $\delta\boldsymbol{x} = \boldsymbol{\xi}(\boldsymbol{x}, t)$ of its mass elements are described by

$$\boldsymbol{\xi} = \sum_n B_n(t)\boldsymbol{u}_n(\boldsymbol{x}), \tag{37.42a}$$

where the time-dependent amplitude for the nth mode satisfies the driven-oscillator equation

$$\ddot{B}_n + \dot{B}_n/\tau_n + \omega_n{}^2 B_n = R_n(t), \tag{37.42b}$$

and where the curvature-induced driving term is

$$
\begin{aligned}
R_n(t) &= -R_{\hat{j}\hat{0}\hat{k}\hat{0}} \int (\rho/M) u_n{}^j x^k \, d^3x \\
&= \frac{1}{4} \ddot{A}(\bar{t}_{(n)jk} e_{jk})/M.
\end{aligned}
\tag{37.42c}
$$

(See Box 37.4 for notation.)

(b) Fourier-analyze the amplitudes of the detector and waves,

$$B_n(t) = (2\pi)^{-1/2} \int_{-\infty}^{+\infty} \tilde{B}_n(\omega) e^{-i\omega t} \, d\omega, \tag{37.42d}$$

$$A(t) = (2\pi)^{-1/2} \int_{-\infty}^{+\infty} \tilde{A}(\omega) e^{-i\omega t} \, d\omega, \tag{37.42e}$$

and solve the equation of motion (37.42b,c) to obtain, in the neighborhood of resonance,

$$\tilde{B}_n = \frac{\frac{1}{8} \omega_n \tilde{A}(\bar{t}_{(n)jk} e_{jk})/M}{|\omega| - \omega_n + \frac{1}{2} i/\tau_n} \qquad \text{for } |\omega \pm \omega_n| \ll \omega_n. \tag{37.42f}$$

(c) Calculate the total energy deposited in the detector by integrating

$$\begin{pmatrix} \text{energy} \\ \text{deposited} \end{pmatrix} = \int (\textbf{Force per unit volume}) \cdot (\textbf{Velocity}) \, d^3x \, dt.$$

Thereby obtain

$$\begin{pmatrix} \text{energy deposited in} \\ n\text{th normal mode} \end{pmatrix} = \frac{1}{4} (I_{(n)jk} e_{jk}) \int \ddot{A} \dot{B}_n \, dt.$$

(d) Apply Parseval's theorem and combine with expression (37.42f) to obtain

$$\begin{pmatrix} \text{energy deposited in} \\ n\text{th normal mode} \end{pmatrix} = \int \sigma_n(\nu) \mathcal{F}_\nu(\nu) \, d\nu, \tag{37.43}$$

where σ_n is given by equation (7a) of Box 37.4, and (for $-\infty < \omega < +\infty$)

$$\mathcal{F}_\nu(\nu) = \mathcal{F}_\nu(\omega/2\pi) = \frac{1}{8} \omega^2 |\tilde{A}|^2. \tag{37.44}$$

(e) Show that $\mathcal{F}_\nu(\nu)$ is the total energy per unit area per unit frequency carried by the waves past the detector.

(f) Obtain all the remaining cross sections quoted in Box 37.4 by appropriate manipulations of this cross section. Use the mathematical tools for projecting out and integrating "transverse-traceless parts," which were developed in Box 35.1 and exercise 36.9.

Exercise 37.11. SPECTRUM OF ENERGY RADIATED BY A SOURCE

Derive the results quoted in the last section of Box 37.4.

Exercise 37.12. PATTERNS OF EMISSION AND ABSORPTION

The elementary dumbbell oscillator of Figure 37.4, initially unexcited, has a cross section for absorption of unpolarized gravitational radiation proportional to $\sin^4\theta$, and when excited radiates with an intensity also proportional to $\sin^4\theta$ (Chapter 36). The patterns of emission and absorption are identical. Any other dumbbell oscillator gives the same pattern, apart from a possible difference of orientation. Consider a nonrotating oscillator of general shape undergoing free vibrations in a single nondegenerate (and therefore nonrotatory) mode, or excited from outside by *unpolarized* radiation.

(a) Show that its pattern of emission is identical with its pattern of absorption. [*Hint:* Make the comparisons suggested in the last few parts of Box 37.4.]

(b) Show that this emission pattern (\equiv absorption pattern), apart from three Euler angles that describe the orientation of this pattern in space, and apart from a fourth parameter that determines total intensity, is uniquely fixed by a single ("fifth") parameter.

(c) Construct diagrams for the pattern of intensity for the two extreme values of this parameter and for a natural choice of parameter intermediate between these two extremes.

(d) Define the parameter in question in terms of a certain dimensionless combination of the principal moments of the reduced quadrupole tensor.

Exercise 37.13. MULTIMODE DETECTOR

Consider a cylindrical bar of length very long compared to its diameter. Designate the fundamental mode of end-to-end vibration of the bar as "$n = 1$," and call the mode with $n - 1$ nodes in its eigenfunction the "nth mode." Show that the cross section for the interception of unpolarized gravitational waves at the nth resonance, integrated over that resonance, and averaged over direction, is given by the formula [Ruffini and Wheeler (1971b)]

$$\int_{\substack{n\text{th} \\ \text{resonance;} \\ \text{random}}} \sigma(\nu)d\nu = \frac{32}{15\pi}\frac{v^2}{c^2}\frac{M}{n^2} \quad \text{for } n \text{ odd (zero for even } n\text{),} \tag{37.45}$$

where v is the speed of sound in the bar expressed in the same units as the speed of light, c; and M is the mass of the bar (geometric units; multiply the righthand side by the factor $G/c = 2.22 \times 10^{-18}$ cm^2 Hz/g when employing conventional units). Show that this expression gives $\int \sigma\, d\nu = 1.0 \times 10^{-21}$ cm^2 Hz for the lowest mode of Weber's bar. Multimode detectors are (1973) under construction by William Fairbank and William Hamilton, and by David Douglass and John A. Tyson.

Exercise 37.14. CROSS SECTION OF IDEALIZED MODEL OF EARTH FOR ABSORPTION OF GRAVITATIONAL RADIATION

The observed period of quadrupole vibration of the earth is 54 minutes [see, e.g., Bolt (1964) or Press (1965) for survey and bibliography]. To analyze that mode of vibration, with all due allowance for elasticity and the variation of density in the earth, is a major enterprise. Therefore, for a first estimate of the cross section of the earth for the absorption of quadrupole radiation, treat it as a globe of fluid of uniform density held in the shape of a sphere by gravitational forces alone (zero rigidity). Let the surface be displaced from $r = a$ to

$$r = a + a\alpha P_2(\cos\theta), \tag{37.46a}$$

where θ is polar angle measured from the North Pole and α is the fractional elongation of the principal axis. The motion of lowest energy compatible with this change of shape is described by the velocity field

$$\xi^x = -\frac{1}{2}\alpha x, \qquad \xi^y = -\frac{1}{2}\alpha y, \qquad \xi^z = \alpha z \tag{37.46b}$$

(zero divergence, zero curl).

(a) Show that the sum of the kinetic energy and the gravitational potential energy is

$$E = -(3/5)(M^2/a)(1 - \alpha^2/5) + (3/20)Ma^2\dot{\alpha}^2. \qquad (37.46c)$$

(b) Show that the angular frequency of the free quadrupole vibration is

$$\omega = (16\pi/15)^{1/2}\rho^{1/2}. \qquad (37.46d)$$

(c) Show that the reduced quadrupole moments are

$$\textit{f}_{xx} = \textit{f}_{yy} = -Ma^2\alpha/5, \qquad \textit{f}_{zz} = 2Ma^2\alpha/5. \qquad (37.46e)$$

(d) Show that the rate of emission of vibrational energy, averaged over a period, is

$$-\langle dE/dt \rangle = (3/125)M^2a^4\omega^6\alpha_{\text{peak}}{}^2. \qquad (37.46f)$$

(e) Show that the exponential rate of decay of energy by reason of gravitational wave damping, or "gravitational radiation line broadening," is

$$A_{\text{GW}} = (4/25)Ma^2\omega^4. \qquad (37.46g)$$

(f) Show that the resonance integral of the absorption cross section for radiation incident from random directions with random polarization is

$$\int_{\text{resonance}} \langle \sigma(\nu) \rangle \, d\nu = (\pi/2)\lambda^2 A_{\text{GW}} = (2\pi/25)Ma^2/\lambda^2. \qquad (37.46h)$$

(g) Evaluate this resonance integral. Note: This model of a globe of fluid of uniform density would imply for the earth, with average density 5.517 g/cm³, a quadrupole vibration period of 94 min, as compared to the observed 54 min; and a moment of inertia $(2/5)Ma^2$ as compared to the observed $0.33Ma^2$. Ruffini and Wheeler (1971b) have estimated correction factors for both effects and give for the final resonance integral ~ 5 cm² Hz.

§37.8. NOISY DETECTORS

When the bandwidth of the incoming waves is large compared to the resonance width of the detector, the waves deposit a total energy in the detector given by

$$(\text{total energy deposited}) = \int \sigma\mathcal{F}_\nu \, d\nu = \mathcal{F}_\nu(\nu = \nu_0) \int \sigma \, d\nu.$$
$$[\text{ergs}]\!\uparrow \qquad\qquad [\text{erg cm}^{-2}\text{Hz}^{-1}]\!\uparrow \quad [\text{cm}^2\text{Hz}]\!\uparrow$$

At least, this is so if the detector is wave-dominated (i.e., if $E_{\text{vibration}} \gg kT$ while waves act; i.e., if initial amplitude of oscillation, produced by Brownian forces, is too small to interfere constructively or destructively with the amplitude due to waves).

Extraction of small signal from large noise—general remarks

Unfortunately, all experiments today (1973) are faced with noisy detectors. Nobody has yet found waves so strong, or constructed a detector so sensitive, that the detector is wave-dominated. Consequently, a key experimental task today is to pick a small signal out of large noise. Many techniques for doing this have been developed and used in a variety of fields of physics, as well as in astronomy, psychology and engineering [see, e.g., Davenport and Root (1958), Blackman and Tukey (1959), and

Figure 37.5.

Detection of hammer-blow gravitational waves with a noisy detector. Detection of even a weak pulse is possible if the time of the pulse is short enough. The amplitude \mathcal{B}_n of the detector's vibrations changes by an amount $\sim \mathcal{B}_n^{\text{rms}} (\Delta t/\tau_n)^{1/2}$ during a time interval Δt, due to thermal fluctuations (random-walk, Brownian-noise forces). Depicted in the inset is a change in amplitude produced by a burst of waves of duration τ_{GW} arriving out of phase with the detector's thermal motions (energy extracted by waves!). The waves are detectable because

$$\Delta|\mathcal{B}_n|_{\text{due to waves}} \gg \mathcal{B}_n^{\text{rms}} (\tau_{\text{GW}}/\tau_n)^{1/2},$$

even though $\Delta|\mathcal{B}_n| \ll \mathcal{B}_n^{\text{rms}}$.

references given there]. The key point is always to find some feature of the signal that is statistically more prominent than the same feature of the noise, plus a correlation to show that it arises from the expected signal source and not from elsewhere ("protection from systematic error"). Thus to detect steady gravitational waves from a pulsar, one might seek to define very precisely two numbers $\langle N^2 \rangle$ and $\langle (N + S)^2 \rangle = \langle N^2 \rangle + \langle S^2 \rangle$, where N and S are the noise and signal amplitudes respectively. A long series of observations (with the pulsar out of the antenna beam) gives one value of $\langle N^2 \rangle$. Another equally long series of observations, interspersed with the first series, will be expected in zeroth approximation to give the same value of $\langle N^2 \rangle$. In the next approximation one recognizes and calculates the influence of normal statistical fluctuations. In an illustrative example, theory, confirmed by statistical tests of other parameters drawn from the same data, guarantees that the fluctuations are less than $10^{-5}\langle N^2 \rangle$ with 95 per cent confidence (only 5 per cent chance of exceeding $10^{-5}\langle N^2 \rangle$; this limit is set by time and money, not by absolute limitations of physics). Let the second series of observations be carried out only at times when the pulsar is in the antenna beam. Let it give

$$\{\langle N^2 \rangle + \langle S^2 \rangle\}_{\text{2d series}} = (1 + 7.3 \times 10^{-5})\{\langle N^2 \rangle\}_{\text{1st series}}.$$

Then in first approximation one can say that $\langle S^2 \rangle$ lies with 95 per cent confidence in the limits $(7.3 \pm 1.0) \times 10^{-5}\langle N^2 \rangle$.

Many conceivable sources of gravitational radiation produce bursts rather than a steady signal strength (Figure 37.5). Thus one is led to ask in what features "hammer-blow radiation" ($\tau_{\text{GW}} \ll \tau_0$) differs from noise. The "Brownian motion" noise in the detector may be thought of as arising from large numbers of small

Rate-of-change of detector amplitude as a tool for extracting burst signals from thermal noise

(molecular) energy exchanges with a heat bath. The calculations below estimate the typical rate of change of amplitude that a series of such molecular "knocks" can produce in a detector, and compare it with the rapid amplitude change produced by a "hammer-blow" pulse of radiation. The calculations show that sudden thermally induced changes, even of very small amplitude, are rare. Thus sudden changes are a suitable feature for the observations to focus on. The actual detection of pulses requires a more extended analysis, however, which goes beyond the estimates made below. Such an analysis would calculate the probabilities that rare events (sudden changes in amplitude) occur by chance (i.e., due to thermal fluctuations) in specified periods of time, the still smaller probabilities that they occur in coincidence between two or more detectors, and the correlations with postulated sources.

Consider a realistic detector of the type described in Box 37.4. But examine it at a time when it is *not* radiation-dominated. Then its motions are being driven by internal Brownian forces (thermal fluctuations), and perhaps also by an occasional burst of gravitational waves. Focus attention on a particular normal mode (mode "n"), and describe that mode's contribution to the vibration of the detector by the vector field

$$\delta x = \xi = \mathcal{B}_n(t)e^{-i\omega_n t}u_n(x). \tag{37.47}$$

Since u_n is dimensionless with mean value unity ($\int \rho u_n^2 d^3x = M$), the complex number $\mathcal{B}_n(t)$ is the mass-weighted average of the amplitudes of motion of the detector's mass elements. This amplitude changes slowly with time (rate $\ll \omega_n$) as a result of driving by Brownian forces; but averaged over time it has a magnitude corresponding to a vibration energy of kT:

<div style="float:left; margin-right:1em;">Description of thermal noise in resonant detector</div>

$$\langle E_{\text{vibration}} \rangle = 2\left\langle \frac{1}{2}\int \rho \dot{\xi}^2 d^3x \right\rangle = \frac{1}{2}M\omega_n^2 \langle |\mathcal{B}_n|^2 \rangle = kT; \tag{37.48}$$

i.e.,

$$\mathcal{B}_n^{\text{rms}} \equiv \langle |\mathcal{B}_n|^2 \rangle^{1/2} = (2kT/M\omega_n^2)^{1/2}. \tag{37.49}$$

Example: for Weber's detector ($M \sim 10^3$ kg, $\omega_0 \sim 10^4$/sec), the fundamental mode at room temperature has

$$\mathcal{B}_0^{\text{rms}} = \left(\frac{2 \times 1.38 \times 10^{-16} \times 300 \text{ erg}}{10^6 \text{ g} \times 10^8 \text{ sec}^{-2}} \right)^{1/2} = 3 \times 10^{-14} \text{ cm}. \tag{37.50}$$

One's hope for detecting weak hammer-blow radiation lies not in an examination of the detector's vibration amplitude (or energy), but in an examination of its rate of change (Figure 37.5). The time-scale for large Brownian fluctuations in amplitude ($|\Delta\mathcal{B}_n| \sim \mathcal{B}_n^{\text{rms}}$), when the detector is noisy, is the same as the time scale τ_n for internal forces to damp the detector, when it is driven to $E_{\text{vibration}} \gg kT$. Thus, *the amplitude \mathcal{B}_n does a "random walk" under the influence of Brownian forces, with the mean time for "large walks"* ($|\Delta\mathcal{B}_n| \sim \mathcal{B}_n^{\text{rms}}$) *being $\Delta t \approx \tau_n$.* The change in \mathcal{B}_n over shorter times Δt is smaller by the "$1/\sqrt{N}$ factor," which always enters into random-walk processes:

$$\sqrt{N} = \left(\frac{\text{number of vibration cycles in time } \tau_n}{\text{number of vibration cycles in time } \Delta t}\right)^{1/2} = \left(\frac{\tau_n}{\Delta t}\right)^{1/2}; \quad (37.51)$$

$$\langle|\Delta\mathcal{B}_n^{(\text{thermal})}|\rangle \approx \mathcal{B}_0^{\text{rms}}\left(\frac{\Delta t}{\tau_n}\right)^{1/2} = \left(\frac{2kT}{M\omega_n^2}\right)^{1/2}\left(\frac{\Delta t}{\tau_n}\right)^{1/2} \begin{array}{l}\text{during} \\ \text{time } \Delta t\end{array}. \quad (37.52)^*$$

Now suppose that "hammer-blow" radiation (burst of duration $\Delta t = \tau_{\text{GW}} \ll \tau_n$) strikes the detector, producing a change $\Delta\mathcal{B}_n^{(\text{GW})}$ in the detector's amplitude. This change in amplitude, because it comes so quickly, (1) superposes linearly on any change in amplitude produced in the same time interval by the action of Brownian-motion forces; and (2) is therefore independent in value of the presence or absence of Brownian-motion forces, i.e., independent of all thermal agitation. Therefore $\Delta\mathcal{B}_n^{(\text{GW})}$ (a quantity with *both* magnitude and phase!) is identical to what it would have been if the detector were at zero temperature:

<div style="text-align: right; font-style: italic;">Effect of a burst of waves on a noisy, resonant detector</div>

$$\underbrace{\frac{1}{2}M\omega_n^2|\Delta B_n^{(\text{GW})}|^2}_{\substack{\text{energy that would} \\ \text{be deposited if} \\ \text{detector were at} \\ \text{zero temperature}}} = \int \sigma_n(\nu)\mathcal{F}_\nu(\nu)\,d\nu = \underbrace{\mathcal{F}_\nu(\omega_n/2\pi)\int \sigma_n(\nu)\,d\nu}_{\substack{\text{For hammer-blow radiation, bandwidth of} \\ \text{radiation is always} \gg \text{bandwidth of detector;} \\ \text{see Box 37.4}}};$$

i.e.,

$$|\Delta\mathcal{B}_n^{(\text{GW})}| = \left(\frac{2\mathcal{F}_\nu(\omega_n/2\pi)\int\sigma_n\,d\nu}{M\omega_n^2}\right)^{1/2}. \quad (37.53)$$

This wave-induced change in amplitude will be distinguishable from thermal changes only if it is significantly bigger than the thermal changes (37.52) expected during the same length of time τ_{GW}:

$$\left.\begin{array}{l}|\Delta\mathcal{B}_n^{(\text{GW})}| \gg \langle|\Delta\mathcal{B}_n^{(\text{thermal})}|\rangle \text{ during time } \tau_{\text{GW}} \\[2mm] \text{equivalently: } F_\nu(\omega_n/2\pi) \gg \left(\frac{kT}{\int\sigma_n\,d\nu}\right)\left(\frac{\tau_{\text{GW}}}{\tau_n}\right)\end{array}\right\} \begin{array}{l}\text{criteria for} \\ \text{detectability}\end{array} \quad (37.54)$$

<div style="text-align: right; font-style: italic;">Criteria for detectability of burst</div>

Of course, if one is equipped only to measure the magnitude of the detector's amplitude or energy, and not its phase, these criteria for detectability are not quite sufficient. The wave-induced change in squared amplitude (proportional to change in energy) will depend on the relative phases of the initial amplitude and amplitude change

*For a fuller derivation and discussion of this formula, see, e.g., Braginsky (1970). Two key points covered there are: (1) a statistical version of the formula, which describes the probability that in time Δt the amplitude will change by a given amount, from a given initial value; and (2) quantum-mechanical corrections, which come into play in the limit as $\tau_n \longrightarrow \infty$, but which are unimportant for detectors of the early 1970's.

$$\Delta|\mathcal{B}_n|^2 = |\mathcal{B}_n^{(\text{initial})} + \Delta\mathcal{B}_n^{(\text{GW})}|^2 - |\mathcal{B}_n^{(\text{initial})}|^2$$
$$\approx 2|\mathcal{B}_n^{(\text{initial})}||\Delta\mathcal{B}_n^{(\text{GW})}| \qquad \text{if in phase}$$
$$\approx 0 \qquad\qquad\qquad\qquad\quad \text{if phase difference is } \pm\pi/2 \qquad (37.55)$$
$$\approx -2|\mathcal{B}_n^{(\text{initial})}||\Delta\mathcal{B}_n^{(\text{GW})}| \qquad \text{if phase difference is } \pi.$$

Thus, only a burst that arrives in phase with the initial motion of the detector or with reversed phase will be measurable. But for such a burst, the criteria (37.54) are sufficient.

Ways to improve sensitivity of detector

Equations (37.54) make it clear that *there are three ways to improve the sensitivity of vibratory detectors to hammer-blow radiation: (1) increase the detector's integrated cross-section* [*which can be done only by increasing the rate* A_{GW} *at which it reradiates gravitational waves; see equations (10b) and (11b) of* Box 37.4]; (2) *cool the detector;* (3) *increase the detector's damping time.*

Box 37.5 applies the above detectability criteria to some detectors that seem feasible in the 1970's, and to some bursts of waves predicted by theory. The conclusions of that comparison give one hope!

To be complete, the above discussion should have analyzed not only noise in the detector, but also the noise in the sensor which one uses to measure the amplitude of the detector's displacements. However, the theory of displacement sensors is beyond the scope of this book. For a brief discussion and for references, see Press and Thorne (1972).

§37.9. NON-MECHANICAL DETECTORS

Non-mechanical detectors

When gravitational waves flow through matter, they excite it into motion. Such excitations are the basis for all detectors described thus far. But gravitational waves interact not only with matter; they also interact with electromagnetic fields; and those interactions can also be exploited in detectors. One of the most promising detectors that may be built in the future, one designed by Braginsky and Menskii (1971), relies on a resonant interaction between gravitational waves and electromagnetic waves. It is described in Box 37.6.

§37.10. LOOKING TOWARD THE FUTURE

The future of gravitational-wave astronomy

As this book is being written, it is not at all clear whether the experimental results of Joseph Weber constitute a genuine detection of gravitational waves. (See §37.4, part 4.) But whether they do or not, gravitational-wave astronomy has begun, and seems to have a bright future. The technology of 1973 appears sufficient for the construction of detectors that will register waves from a star that collapses to form a black hole anywhere in our galaxy (Box 37.5); and detectors of the late 1970's and early 1980's may well register waves from pulsars and from supernovae in other galaxies. The technical difficulties to be surmounted in constructing such detectors are enormous. But physicists are ingenious; and with the impetus provided by Joseph Weber's pioneering work, and with the support of a broad lay public sincerely interested in pioneering in science, all obstacles will surely be overcome.

Box 37.5 DETECTABILITY OF HAMMER-BLOW WAVES
FROM ASTROPHYSICAL SOURCES: TWO EXAMPLES
(The following calculations are accurate only to
within an order of magnitude or so)

A. Waves from a Star of Ten Solar Masses Collapsing to Form
a Black Hole; 1972 Detector with 1975 (?) Sensor

1. Predicted characteristics of radiation:

$$(\text{intensity at Earth}) = \mathcal{F}_\nu \sim \frac{M_\odot}{4\pi(\text{distance})^2\nu}$$

$$\sim (2 \times 10^5 \text{ ergs/cm}^2 \text{ Hz})\,[(\text{distance to center of galaxy})/(\text{distance})]^2,$$

$$(\text{frequency of waves}) = \nu \sim 10^3 \text{ Hz},$$

$$(\text{bandwidth of waves}) = \Delta\nu \sim 10^3 \text{ Hz},$$

$$(\text{duration of burst}) = \tau_{\text{GW}} \sim 10^{-3} \text{ sec to } 10^{-1} \text{ sec}.$$

2. Detector properties: A Weber bar, vibrating in its fundamental mode, with

$$M = 10^6 \text{ g}, \qquad \int \sigma\, d\nu = 10^{-21} \text{ cm}^2 \text{ Hz (exercise 37.13)},$$

$$\nu_0 = \omega_0/2\pi = 1{,}660 \text{ Hz}, \quad T = 3 \text{ K (liquid Helium temperature)},$$

$$\tau_0 = 20 \text{ seconds},$$

$$\mathcal{B}_0^{\text{rms}} = \left(\frac{2 \times 1.37 \times 10^{-16} \times 3 \text{ erg}}{10^6 \text{ g} \times 10^8 \text{ sec}^{-2}}\right)^{1/2} = 3 \times 10^{-15} \text{ cm},$$

$$|\Delta\mathcal{B}_0^{(\text{thermal})}| = (3 \times 10^{-15} \text{ cm})(10^{-3}/20)^{1/2} = 2 \times 10^{-17} \text{ cm},$$
$$\text{during } \Delta t = 10^{-3} \text{ sec},$$

$$|\Delta\mathcal{B}_0^{(\text{thermal})}| = 2 \times 10^{-16} \text{ cm, during } \Delta t = 0.1 \text{ sec}.$$

3. Effect of waves [equation (37.53)]:

$$|\Delta\mathcal{B}_0^{(\text{GW})}| = \left(\frac{2 \times 2 \times 10^5 \times 10^{-21} \text{ ergs}}{10^6 \times 10^8 \text{ sec}^{-2}}\right)^{1/2} \left(\frac{\text{center of Galaxy}}{\text{distance}}\right)$$

$$= 2 \times 10^{-15} \text{ cm} \left(\frac{\begin{array}{c}\text{distance to}\\ \text{center of Galaxy}\end{array}}{\text{distance}}\right).$$

4. Conclusion: Gravitational waves from a massive star collapsing to form a
 black hole anywhere in our galaxy are readily detectable, *if* one can construct
 a "*sensor*" to measure changes in vibration amplitudes of magnitude
 $\lesssim 10^{-15}$ cm on time scales < 0.1 seconds. This does appear to be feasible
 with 1972 technology; see Press and Thorne (1972).

Box 37.5 (*continued*)

B. Waves from a Supernova Explosion in the Virgo Cluster of Galaxies; a Detector that might be constructable by late 1970's or early 1980's

1. Predicted characteristics of radiation:

$$(\text{intensity at Earth}) = F_\nu \sim \frac{0.03\,M_\odot}{4\pi(11\text{ megaparsecs})^2\nu}$$

$$\sim 4 \times 10^{-3}\text{ ergs/cm}^2\text{ Hz},$$

$$(\text{frequency of waves}) = \nu \sim 10^3\text{ Hz},$$

$$(\text{bandwidth of waves}) \sim \nu \sim 10^3\text{ Hz},$$

$$(\text{duration of burst}) = \tau_{GW} \sim 0.3\text{ sec, or }\tau_{GW} \sim 2 \times 10^{-3}\text{ sec.*}$$

2. Detector: A Weber-type bar made not of metal, but of a 1,000-kg monocrystal of quartz, cooled to a temperature of 3×10^{-3} K. (For such a monocrystal, it is thought that the damping time would increase in inverse proportion to temperature, $\tau_0 \propto 1/T$.) Estimated properties of such a detector:

$$M \sim 10^6\text{ g}, \qquad \int \sigma\,d\nu = 10^{-21}\text{ cm}^2\text{ Hz (same as for Weber bar)},$$

$$\nu_0 = \omega_0/2\pi \sim 1,500\text{ Hz}, \qquad T = 3 \times 10^{-3}\text{ K},$$

$$\tau_0 \sim 10^6\text{ sec},$$

$$\mathscr{B}_0^{rms} = \left(\frac{2 \times 1.37 \times 10^{-16} \times 3 \times 10^{-3}\text{ erg}}{10^6\text{ g} \times 10^8\text{ sec}^{-2}}\right)^{1/2} = 1 \times 10^{-16}\text{ cm},$$

$$|\Delta\mathscr{B}_0^{(\text{thermal})}| = (1 \times 10^{-16}\text{ cm})\left(\frac{0.3\text{ or }2 \times 10^{-3}}{10^6}\right)^{1/2} = \begin{cases} 6 \times 10^{-20}\text{ cm}, \\ \text{or} \\ 5 \times 10^{-21}\text{ cm}. \end{cases}$$

3. Effect of waves [equation (37.53)]:

$$|\Delta\mathscr{B}_0^{(GW)}| = \left(\frac{2 \times 4 \times 10^{-3} \times 10^{-21}\text{ ergs}}{10^6 \times 10^8\text{ sec}^{-2}}\right)^{1/2} = 3 \times 10^{-19}\text{ cm}.$$

4. Conclusion: Gravitational waves are detectable from a supernova in the Virgo cluster, if one can construct a sensor to measure changes in vibration amplitudes of magnitude $\leq 10^{-19}$ cm on time scales of ≤ 0.1 seconds; and if one can construct a detector with the above characteristics.

*For the duration of waves from a supernova explosion, two time scales appear to be relevant: (1) the time required for the final stages of the collapse of the white-dwarf core to a neutron star or a neutron-star pancake, $\tau \sim$ (dimensions of neutron star)/(speed of sound in nuclear matter) $\sim 2 \times 10^{-3}$ sec ("pulse of gravitational radiation"); and (2) the time required for a vibrating neutron star to lose its energy of vibration by gravitational radiation ("damped train of waves"), $\tau \sim 0.3$ sec.

Box 37.6 A NONMECHANICAL DETECTOR OF GRAVITATIONAL WAVES
[Braginsky and Menskii (1971)]

The Idea in Brief
(see diagram at right)

A toroidal waveguide contains a monochromatic train of electromagnetic waves, traveling around and around it. Gravitational waves propagate perpendicular to the plane of the torus. If the circuit time for the EM waves is twice the period of the gravitational waves, then one circularly polarized component of the gravitational waves will stay always in phase with the traveling EM waves. Result: a resonance develops. In one region of the EM wave train, gravitational tidal forces always "push" the waves forward (*blue shift!*) in another region the tidal forces "push" backward (*red shift!*). An EM frequency difference builds up linearly with time; a phase difference builds up quadratically.

Outline of Quantitative Analysis

1. Let waveguide fall freely in an Earth orbit. Orient axes of waveguide's proper reference frame (\equiv local Lorentz frame) so (1) waveguide lies in \hat{x}, \hat{y}-plane, and (2) gravitational waves propagate in \hat{z} direction.

2. Let gravitational waves have amplitudes

$$A_+ - iA_\times = \mathcal{Q}e^{-i\omega(t-z)} \qquad (1)$$

[Recall: $\hat{t} \approx t, \hat{z} \approx z$; i.e., proper frame and TT coordinates almost agree.] Then in plane of waveguide ($z = 0$),

$$R_{\hat{x}\hat{0}\hat{x}\hat{0}} = -R_{\hat{y}\hat{0}\hat{y}\hat{0}} = \frac{1}{2}\omega^2\mathcal{Q}\cos(\omega t)$$
$$R_{\hat{x}\hat{0}\hat{y}\hat{0}} = R_{\hat{y}\hat{0}\hat{x}\hat{0}} = \frac{1}{2}\omega^2\mathcal{Q}\sin(\omega t) \qquad (2)$$

3. Consider two neighboring parts of the EM wave, one at $\phi = \alpha + \frac{1}{2}\omega t$; the other at $\phi = \alpha + \delta\alpha + \frac{1}{2}\omega t$. Treat them as photons. Each moves along a null geodesic, except for

ω = (angular frequency of gravitational waves) = (rate of change of phase of waves with time) = (two times angular velocity with which pattern of "lines of force" rotates)

r = (radius of torus), is adjusted so the speed of propagation of EM waves in waveguide is $v = \frac{1}{2}\omega r$.

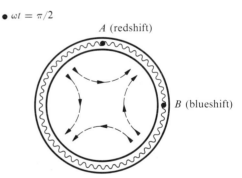

[EM waves propagate counterclockwise; gravitational line-of-force diagram rotates counterclockwise; they stay in phase.]

Box 37.6 (continued)

the deflective guidance of the wave guide. Thus, their wave vectors **k** satisfy

$$\boldsymbol{\nabla}_k \boldsymbol{k} = \begin{pmatrix} \text{deflective "acceleration"} \\ \text{of waveguide} \end{pmatrix}; \quad (3)$$

and the difference $\delta \boldsymbol{k} = \boldsymbol{\nabla}_n \boldsymbol{k}$ between the wave vectors of the two parts of the wave (difference measured via parallel transport) satisfies the equation

$$\boldsymbol{\nabla}_k \delta \boldsymbol{k} = \boldsymbol{\nabla}_k \boldsymbol{\nabla}_n \boldsymbol{k} = [\boldsymbol{\nabla}_k, \boldsymbol{\nabla}_n] \boldsymbol{k} + \boldsymbol{\nabla}_n \boldsymbol{\nabla}_k \boldsymbol{k} \quad (4)$$

$$= \boldsymbol{Riemann}\,(\ldots, \boldsymbol{k}, \boldsymbol{k}, \boldsymbol{n}) + \boldsymbol{\nabla}_n \underbrace{\boldsymbol{\nabla}_k \boldsymbol{k}}$$

[deflective acceleration of wave guide]

The waveguide influences the direction of propagation of the waves, but not their frequency. Thus only **Riemann** enters into the 0 component of the above equation:

$$k^{\hat{\alpha}} \delta k^{\hat{0}}{}_{,\hat{\alpha}} = R^{\hat{0}}{}_{\hat{\alpha}\hat{\beta}\hat{\gamma}} k^{\hat{\alpha}} k^{\hat{\beta}} n^{\hat{\gamma}}. \quad (5)$$

4. Let $k^{\hat{0}} = \omega_e$ be the angular frequency of the electromagnetic wave. The direction of the space component **k** of the propagation 4-vector is along the purely spatial vector **n**; so

$$k^{\hat{0}} = \omega_e, \qquad \boldsymbol{k} = (v\omega_e/r\delta\alpha)\boldsymbol{n}, \qquad n^{\hat{0}} = 0. \quad (6)$$

Use these relations to rewrite equation (5) as

$$(d\,\delta\omega_e/d\hat{t})_{\text{moving with photons}}$$
$$= (v\omega_e/r\delta\alpha) R^{\hat{0}}{}_{\hat{i}\hat{0}\hat{j}} n^{\hat{i}} n^{\hat{j}}. \quad (7)$$

5. Combine the expression for **n** in the spacetime diagram with equations (2) and (7), and with the world line $\phi = \alpha + \frac{1}{2}\omega t$ for the photons, to obtain

$$(d\,\delta\omega_e/d\hat{t})_{\text{moving with photons}}$$
$$= -\frac{1}{2} v\omega_e \omega^2 \mathcal{C} r\,(\cos 2\alpha)\,\delta\alpha. \quad (8)$$

6. Integrate over time and over α to obtain

$$\omega_e = \omega_{eo}\left[1 - \frac{1}{4}\mathcal{C}v\,(\sin 2\alpha)(\omega r)(\omega \hat{t})\right]. \quad (9)$$

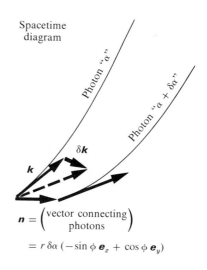

Spacetime diagram

Photon "α"

Photon "$\alpha + \delta\alpha$"

$\delta \boldsymbol{k}$

\boldsymbol{k}

$$\boldsymbol{n} = \begin{pmatrix} \text{vector connecting} \\ \text{photons} \end{pmatrix}$$
$$= r\,\delta\alpha\,(-\sin\phi\,\boldsymbol{e}_x + \cos\phi\,\boldsymbol{e}_y)$$

EXPERIMENTAL TESTS
OF GENERAL RELATIVITY

Wherein the reader is tempted by a harem of charming gravitation theories (and some not so charming), is saved from his foolish passions by an army of experiments, cleaves unto his faithful spouse, Geometrodynamics, vows to lead an honest life hereafter, and becomes a True Believer.

CHAPTER **38**

TESTING THE FOUNDATIONS
OF RELATIVITY

Provando e riprovando
(*Verify the one and disprove the other*)
GALILEO

§38.1. TESTING IS EASIER IN THE SOLAR SYSTEM
THAN IN REMOTE SPACE

For the first half-century of its life, general relativity was a theorist's paradise, but an experimentalist's hell. No theory was thought more beautiful, and none was more difficult to test.

The situation has changed. In the last few years general relativity has become one of the most exciting and fruitful branches of experimental physics. A half-century late, the march of technology has finally caught up with Einstein's genius—not only on the astronomical front, but also in laboratory experiments.

On the astronomical front, observers search for phenomena in which relativity is important, and study them: cosmology, pulsars, quasars, gravitational waves, black holes. Unfortunately, in pulsars and quasars, and in the sources of cosmological radiation and gravity waves, gravitational effects are tightly interwoven with the local hydrodynamics and local plasma physics. There is little hope of separating the several effects sufficiently sharply to get *clean* tests of the nature of gravity. Instead, astrophysicists must put the laws of gravity into their calculations along with all the other laws of physics and the observational data; and they must then seek, as output, information about the doings of matter and fields "way out there."

Thus, for clean tests of general relativity one turns to the laboratory—but to a laboratory that is much larger today than formerly: a laboratory that includes the entire solar system.

Clean tests of general relativity are currently confined to solar system

Capabilities of technology in 1970's

In the solar system all relativistic effects are tiny. Nonetheless, some of them are measurable with a precision, in the 1970's, of one part in 1,000 of their whole magnitude or better (see Box 38.1).

§38.2. THEORETICAL FRAMEWORKS FOR ANALYZING TESTS OF GENERAL RELATIVITY

There are now possible many experiments for testing general relativity. But most of them are expensive; very expensive. They involve atomic clocks flown on space-

Box 38.1 TECHNOLOGY OF THE 1970's CONFRONTED WITH RELATIVISTIC PHENOMENA

Quantity to be measured	Magnitude of relativistic effects	Precision of a one-day measurement in the early 1970's
Angular separation of two sources on the sky	Solar deflection of starlight (1) if light ray grazes edge of Sun, 1″.75 (2) if light ray comes in perpendicular to Earth-sun line, 0″.004	(a) With optical telescope, ~1″ (b) Angular separation of two quasars with radio telescope (differential measurement from day to day, not absolute measurement) in 1970, ~0″.1 in mid 1970's, ~0″.001
Distance between two bodies in solar system	(a) Perihelion shift per Earth year (1) for Mercury, 120 km (2) for Mars, 15 km (b) Relativistic time delay for radio waves from Earth, past limb of sun, to Venus (one way), 1×10^{-4} sec = 30 km (c) Periodic relativistic effects in Earth-moon separation (1) in general relativity, 100 cm (2) in Jordan-Brans-Dicke theory, 100 cm; (840 cm)/(2 + ω)	(a) Separation of another planet (Mercury, Venus, Mars) from Earth, by bouncing radar signals off it, ~0.3 km (b) Separation of a radio transponder (on another planet or in a space craft) from Earth, by measuring round-trip radio travel time, $\sim 3 \times 10^{-8}$ sec = 10 m = 0.01 km (c) Earth-moon separation by laser ranging, ~10 cm
Difference in lapse of proper time between two world lines in solar system	(a) Clock on Earth vs. clock in synchronous Earth orbit, $\Delta t/t \sim 6 \times 10^{-10}$ (b) Clock on Earth vs. clock in orbit about sun, $\Delta t/t \sim 10^{-8}$	Stability of a hydrogen maser clock, $\Delta t/t \sim 10^{-13}$ for t up to one year

craft; radar signals bounced off planets; radio beacons and transponders landed on planets or orbited about them; etc. Because of the expense, it is crucial to have as good a theoretical framework as possible for comparing the relative values of various experiments—and for proposing new ones, which might have been overlooked.

Such a framework must lie outside general relativity. It must scrutinize the foundations of Einstein's theory. It must compare Einstein's theory with other viable theories of gravity to see which experiments can distinguish between them. It must be a "theory of theories."

At present, in 1973, there are two different frameworks in broad use. One, devised largely by Dicke (1964b),* assumes almost nothing about the nature of gravity. It is used to design and discuss experiments for testing, at a very fundamental level, the nature of spacetime and gravity. Within it, one asks such questions as: Do all bodies respond to gravity with the same acceleration? Is space locally isotropic in its intrinsic properties? What are the theoretical implications of local isotropy? What types of fields, if any, are associated with gravity: scalar fields, vector fields, tensor fields, affine fields? Although some of the experiments that tackle these questions will be discussed below, this book will not attempt a detailed exposition of the Dicke framework.

The second framework in broad use is the "parametrized post-Newtonian (PPN) formalism." It has been developed to higher and higher levels of sophistication by Eddington (1922), Robertson (1962), Schiff (1962, 1967), Nordtvedt (1968b, 1969), Will (1971c), and Will and Nordtvedt (1972).

The PPN formalism is an approximation to general relativity, and also to a variety of other contemporary theories of gravity, called "metric theories." It is a good approximation whenever, as in the solar system, the sources of the field gravitate weakly ($|\Phi|/c^2 \ll 1$) and move slowly ($v^2/c^2 \ll 1$). The PPN formalism contains a set of ten parameters whose values differ from one theory to another. Solar-system experiments (measurements of perihelion shift, light deflection, etc.) can be regarded as attempts to measure some of these PPN parameters, and thereby to determine which metric theory of gravity is correct—general relativity, Brans-Dicke (1961)-Jordan (1959) theory, one of Bergmann's (1968) scalar-tensor theories, one of Nordstrøm's theories, Whitehead's (1922) theory, or something else. [For reviews of Nordstrøm and Whitehead, see Whitrow and Morduch (1965), Will (1971b), and Ni (1972). For a significant nonmetric theory, see Cartan (1920) and Trautman (1972).]

Chapter 39 will discuss the concept of a metric theory of gravity and will construct the PPN formalism; and then Chapter 40 will use the PPN formalism to analyze the systematics of the solar system, and to discuss a variety of past and future experiments that distinguish between various metric theories of gravity. But first, as a prelude to those topics, this chapter will examine experiments that test the foundations of general relativity—foundations on which most other metric theories also rest. For a more detailed discussion of most of these experiments, see Dicke (1964b).

Theoretical frameworks for analyzing gravitation experiments:

(1) Dicke framework

(2) PPN framework

*See Thorne and Will (1971), or Will (1972), for expositions of both frameworks and a comparison of them.

The rest of this chapter is Track 2.
No earlier Track-2 material is needed as preparation for it, but Chapter 7 (incompatibility of gravity and special relativity) will be helpful.
This chapter is not needed as preparation for any later chapter, but it will be helpful in Chapters 39 and 40 (other theories; PPN formalism; experimental tests).

Eötvös-Dicke experiment to test uniqueness of free fall

§38.3. TESTS OF THE PRINCIPLE OF THE UNIQUENESS OF FREE FALL: EÖTVÖS-DICKE EXPERIMENT

One fundamental building block common to Einstein's theory of gravity and to almost all other modern theories is the principle of *"uniqueness of free fall"*:* *"The world line of a freely falling test body is independent of its composition or structure."* By *"test body"* is meant an electrically neutral body, small enough that (1) its self-gravitational energy, as calculated using standard Newtonian theory, can be neglected compared to its rest mass ($M/R \ll 1$), and (2) the coupling of its multipole moments to inhomogeneities of the gravitational field can be neglected.†

The uniqueness of free fall permits one to regard spacetime as filled with a set of curves, the test-body trajectories, which are unique aside from parametrization. Through each event, along each timelike or null direction in spacetime, there passes one and only one test-body trajectory. Describe these trajectories mathematically: that is a central imperative of any theory of gravity.

When translated into Newtonian language, the uniqueness of free fall states that any two test bodies must fall with the same acceleration in a given external gravitational field. Experimental tests of this principle search for differences in acceleration from one body to another. The most precise experiments to date are of a type devised by Baron Lorand von Eötvös (Box 38.2), redesigned and pushed to much higher precision by the Princeton group of Robert H. Dicke (Box 38.3), and extended with modifications by the Moscow group of Vladimir B. Braginsky. (See Figure 1.6 and Box 1.2 for experimental details.)

These Eötvös-Dicke experiments are "null experiments." They balance the acceleration of one body against the acceleration of another, and look for tiny departures from equilibrium. The reason is simple. Null experiments typically have much higher precision than experiments measuring the value of a nonzero quantity.

Eötvös, Pekar, and Fekete (1922) checked to an accuracy of 5 parts in 10^9 that the Earth imparts the same acceleration to wood, platinum, copper, asbestos, water, magnalium (90% Al, 10% Mg), copper sulphate, and tallow. Renner (1935) checked, to 7 parts in 10^{10}, the Earth's acceleration of platinum, copper, bizmuth, brass, glass, ammonium fluoride, and an alloy of 30% Mg, 70% Cu. Dicke, and later Braginsky, chose to use the sun's gravitational acceleration rather than the Earth's, since the alternation in the direction of the sun's pull every 12 hours lends itself to amplification by resonance. (See Figure 1.6.) Roll, Krotkov, and Dicke (1964) reported an

*R. H. Dicke calls this principle "The weak equivalence principle." We prefer to avoid confusion with the equivalence principle (Chapter 16).

†In general relativity, one often uses an alternative definition of test body, which places no constraint on the self-gravitational energy [abandon condition (1) while retaining (2)]. Such a definition is preferable, in principle, because the theory of matter has not been developed sufficiently to decide whether (and no objective test has ever been proposed to decide whether), gravitational energy at the subnuclear scale is a small fraction, a large fraction, or the entirety of the rest mass. But for present purposes a definition constraining test bodies to have $M/R \ll 1$ is preferable for two reasons. First, most theories of gravity that currently "compete" with Einstein's (a) agree with the principle of uniqueness of free fall when the macroscopic, Newtonian, self-gravitational energy is neglected ($M/R \ll 1$), but (b) violate that principle when macroscopic, Newtonian self-gravitational energy is taken into account. See §40.9 for details. Second, the test bodies used in the Eötvös-Dicke experiment have M/R so small that their macroscopic, Newtonian, self-gravitational energies are, in fact, negligible ($M/R \sim E_{\text{grav}}/M \sim 10^{-27}$).

agreement of 1 part in 10^{11} between the sun's acceleration of aluminum and gold, while Braginsky and Panov (1971) reported agreement to 1 part in 10^{12} for aluminum and platinum.

From this agreement, one can infer the response of neutrons, protons, electrons, virtual electron-positron pairs, nuclear binding energy, and electrostatic energy to the sun's gravity. Gold is 60% neutrons, while aluminum is only 50% neutrons. Therefore even from the 1964 results one could conclude that neutrons and protons must have the same acceleration to within $[0.6 - 0.5 = 0.1]^{-1}$ parts in $10^{11} = 1$ part in 10^{10}. Similarly, electrons must accelerate the same as nucleons to 2 parts in 10^{7}; virtual pairs (being more abundant in gold than in aluminum) must accelerate the same to 1 part in 10^{4}; nuclear binding energy, to 1 part in 10^{7}; and electrostatic energy to 3 parts in 10^{9}.

This accuracy of testing gives one confidence in the principle of the *uniqueness of free fall*.

Theoretical implications of Eötvös-Dicke experiment

(*continued on page 1054*)

Box 38.2 BARON LORAND VON EÖTVÖS
Budapest, July 27, 1848—Budapest, April 8, 1919

Eötvös (pronounced ut'vûsh) studied at Heidelberg with Kirchhoff, Helmholtz, and Bunsen and at Königsberg with Neumann and Richelot. His 1870 Heidelberg Ph.D. thesis dealt with an issue of relativity: can the motion of a light source relative to an "ether" be detected by comparing the light intensities in the direction of the motion and in the opposite direction?

Studies of his at the same time resulted in the Eötvös law of capillarity, (surface tension) $\approx 2.12 \, (T_{\mathrm{crit}} - T)/$(specific volume)$^{2/3}$. Eötvös, made professor of physics at Budapest in 1872, concentrated on gravity from 1886 onward. He developed and extended the original Michell-Cavendish torsion balance, which measured not only $\Phi_{,xx}$ and $\Phi_{,xy}$ (where Φ is the gravitational potential) but also $\Phi_{,xz}$ and $\Phi_{,yz}$, all to a precision destined to be unexcelled for decades. He showed that the so-called "ratio between gravitational mass and inertial mass" cannot vary from material to material by more than 5 parts in 10^{9}. He investigated the paleomagnetism of bricks and other ceramic objects, and studied the shape of the earth. He served (June 1894—January 1895) as minister of public instruction and religious affairs (a cabinet position held in earlier years by his father). He founded a school which trained high-school teachers, to whose leavening influence one can give some of the credit for such outstanding scientists as von Karman, von Neuman, Teller, and Wigner. He served one year as rector of the University of Budapest.

"I can never forget the moment when my train rushed into the railroad station of Heidelberg along the banks of the Neckar. . . I cannot forget my happiness that now I could breathe the same air as those men of science whose fame attracted me there."

[EÖTVÖS IN 1887, AS QUOTED IN FEJÉR AND MIKOLA (1918), P. 259.]

Box 38.2 (*continued*)

Photograph by A. Szekely 1913

"Insofar as it is permitted on the basis of a few experiments, we can therefore declare that μ, that is, the weakening of the Earth's attraction through the intervening compensator quadrants, is less than one part in 5×10^{10}. . . . the absorption (of gravity) by the entire earth along a diameter is less than about one part in 800.
"We have carried out a series of observations which surpassed all previous ones in precision, but in no case could we discover any detectable deviation from the law of proportionality of gravitation and inertia."

[EÖTVÖS, PEKÁR, AND FEKETE (1922).]

"Science shall never find that formula by which its necessary character could be proved. Actually science itself might cease if we were to find the clue to the secret."

[EÖTVÖS, PRESIDENTIAL ADDRESS TO THE HUNGARIAN ACADEMY OF SCIENCES, 1890, AS QUOTED IN FEJÉR AND MIKOLA (1918), P. 280.]

"We should consider it as one of the most astonishing errors of the present age that so many people listen to the words of pseudoprophets who, in place of the dogmas of religion, offer scientific dogmas with medieval impatience but without historical justification."

[EÖTVÖS, 1877, AS QUOTED IN FEJÉR AND MIKOLA (1918), P. 280.]

Box 38.3 ROBERT HENRY DICKE May 6, 1916, St. Louis, Missouri
Cyrus Fogg Brackett Professor of Physics at Princeton University

During 1941–1946, Dicke was a leader in replacing the outmoded concept of lumped circuit elements by a new microwave analysis based on symmetry considerations, conservation laws, reciprocity relations, and the scattering matrix—concepts that led, among others, to the lock-in amplifier and the microwave radiometer. Searching for a means to reduce the Doppler width of spectral lines for precision measurements, Dicke discovered recoilless radiation in atomic systems held in a box or in a buffer gas. This development led to (1) the discovery of the basic idea of the gas-cell atomic clock and (2) a much more precise measurement of the gyromagnetic ratio of electrons in the 1s and 2s levels of hydrogen and of the hyperfine structure of atomic hydrogen.

A fundamental paper by Dicke in 1954 set forth the theory of coherent radiation processes and of the superradiant state, and laid the foundation for the future development of the laser and the maser, to which he also contributed. His patent no. 2,851,652 (filed May 21, 1956) was the first disclosure of a device for the generation of infrared radiation by a coherent process, and supplied the first suggestion for combining the use of an etalon resonator with an amplifying gas.

Beginning in the 1960's, Dicke brought his talent for precision measurement to the service of experimental cosmology, and with his collaborators: (1)

checked the equivalence principle with the up-to-then unprecedented accuracy of 1 part in 10^{11}; (2) determined the solar oblateness; and (3) suggested that the primordial cosmic-fireball radiation, a tool for seeing deeper into the past history of the universe than has ever before been possible, should be observable, and therefore should be hunted down and found.

"For want of a better term, a gas which is radiating strongly because of coherence will be called 'superradiant.' . . . As the system radiates it passes to states of lower m *with* r *unchanged—to the 'superradiant' region* m \sim *0"*

(1954)

"Possibilities are examined for the excitation of optical 'superradiant' states of gas"

(1957)

Box 38.3 (*continued*)

"A 'gravitational oblateness' of [the sun of] 5 × 10⁻⁵ would require the abandonment of Einstein's purely geometrical theory of gravitation. . . . Such a flattening [of the sun] could be understood as the effect of a rather rapidly rotating interior. . . . The answer appears to be that in the past, and to this day, the solar corona with its magnetic field has acted as a brake on the surface of the sun"

(1964a)

"New measurements of the solar oblateness have given a value for the fractional difference of equatorial and polar radii of (5.0 ± 0.7) × 10⁻⁵"

[DICKE AND GOLDENBERG (1967)]

"[The universe must] have aged sufficiently for there to exist elements other than hydrogen. It is well-known that carbon is required to make physicists"

(1961)

"The question of the constancy of such dimensionless numbers is to be settled not by definition but by measurements"

[BRANS AND DICKE (1961)]

"The geophysical data lead to an upper limit of 3 parts in 10¹³ per year on the rate of change of the fine-structure constant"

[DICKE AND PEEBLES (1962)]

§38.4. TESTS FOR THE EXISTENCE OF A METRIC GOVERNING LENGTH AND TIME MEASUREMENTS, AND PARTICLE KINEMATICS

Special relativity, general relativity, and all other metric theories of gravity assume the existence of a metric field and predict that this field determines the rates of ticking of atomic clocks and the lengths of laboratory rods by the familiar relation $-d\tau^2 = ds^2 = g_{\alpha\beta} dx^\alpha dx^\beta$.

Experimental evidence for existence of a metric

The experimental evidence for a metric comes largely from elementary particle physics. It is of two types: *first,* experiments that measure time intervals directly, e.g., measurements of the time dilation of the decay times of unstable particles;* *second,* experiments that reveal the fundamental role played by the Lorentz group in particle kinematics and elsewhere in particle physics.† To cast out the metric tensor entirely would leave one with no theoretical framework adequate for interpreting such experiments.

*For a 2 per cent test of time dilation with muons of $(1 - v^2)^{-1/2} \sim 12$ in a storage ring, see Farley, Bailey, Brown, Giesch, Jöstlein, van der Meer, Picasso, and Tannenbaum (1966). For earlier time-dilation experiments see Frisch and Smith (1963); Durbin, Loar, and Havens (1952); and Rossi and Hall (1941).

†See p. 18 of Lichtenberg (1965) for a discussion of Lorentz invariance, spin and statistics, the TCP theorem, and relevant experiments.

Notice what particle-physics experiments *do* and *do not* tell one about the metric tensor, **g**. *First,* they *do not* guarantee that there exist global Lorentz frames, i.e., coordinate systems extending throughout all of spacetime, in which $g_{\alpha\beta} = \eta_{\alpha\beta}$. However, they *do* suggest that at each event \mathcal{P} there exist orthonormal frames with $e_{\hat{\alpha}}(\mathcal{P}) \cdot e_{\hat{\beta}}(\mathcal{P}) = \eta_{\alpha\beta}$, which are related to each other by Lorentz transformations. These orthonormal frames provide one with a definition of the inner product between any two vectors at a given event—and, thereby, they define the metric field.

Second, particle experiments *do not* guarantee that freely falling particles move along geodesics of the metric field, i.e., along straight lines in local Lorentz frames. (Here, in §§38.4 and 38.5, the phrase "local Lorentz frame" means a "normal" coordinate system at an event \mathcal{P}, in which $g_{\alpha\beta}(\mathcal{P}) = \eta_{\alpha\beta}$ *and* $g_{\alpha\beta,\gamma}(\mathcal{P}) = 0$. The term "inertial frame" is avoided because no assertions are made, yet, about test-body motion.) In particular, one does not know from elementary-particle experiments whether the local Lorentz frames in the laboratory are freely falling (so they fly up from the center of the earth and then fall back with Newtonian acceleration $g = 980 \text{ cm/sec}^2$), whether they are forever at rest relative to the laboratory walls, or whether they undergo some other type of motion. All one is led to believe is that a metric determines the nature of the spacetime intervals ($d\tau^2 = -g_{\mu\nu}\,dx^\mu\,dx^\nu$) measured by atomic clocks, that the various local Lorentz frames in the laboratory therefore move with uniform velocity relative to each other (they are connected by Lorentz transformations), and that electric and magnetic fields and the energies and momenta of particles undergo Lorentz transformations in the passage from one local Lorentz frame to another.

Third, elementary particle experiments *do* suggest that the times measured by atomic clocks depend only on velocity, not on acceleration. The measured squared interval is $ds^2 = g_{\alpha\beta}\,dx^\alpha\,dx^\beta$, independently of acceleration (until the acceleration becomes so great it disturbs the structure of the clock; see §16.4 and Box 16.3). Equivalently, but more physically, the time interval measured by a clock moving with velocity v^j relative to a local Lorentz frame is

$$d\tau = (-\eta_{\alpha\beta}\,dx^\alpha\,dx^\beta)^{1/2} = [1 - (v^x)^2 - (v^y)^2 - (v^z)^2]^{1/2}\,dt, \qquad (38.1)$$

independently of the clock's acceleration d^2x^j/dt^2. If this were not so, then particles moving in circular orbits in strong magnetic fields would exhibit different decay rates than freely moving particles—which they do not [Farley *et al.* (1966)].*

Particle experiments do *not* guarantee existence of global Lorentz frames, or geodesic motion for test particles

Particle experiments *do* suggest proper time is independent of acceleration

§38.5. TESTS OF GEODESIC MOTION: GRAVITATIONAL REDSHIFT EXPERIMENTS

The uniqueness of free fall, as tested by the Dicke-Eötvös experiments, implies that spacetime is filled with a family of preferred curves, the test-body trajectories. There

*The experiment of Farley *et al.* is a 2 percent check of acceleration-independence of the muon decay rate for energies $E/m = (1 - v^2)^{-1/2} \sim 12$ and for accelerations, as measured in the muon rest frame, of $a = 5 \times 10^{20} \text{ cm/sec}^2 = 0.6 \text{ cm}^{-1}$.

is also another family of preferred curves, the *geodesics* of the metric **g**. It is tempting to identify these geodesics with the test-body trajectories. Einstein's geometric theory of gravity makes this identification ("equivalence principle"). One might conceive of theories that reject this identification. What is the experimental evidence on this point?

In order to see what kinds of experiments are relevant, it is helpful to elucidate the physical significance of the geodesics.

Physical meaning of a comparison between test-body trajectories and geodesics of metric

A geodesic of **g** is most readily identified locally by the fact that it is a straight line in the local Lorentz frames. Put differently, a body's motion is unaccelerated as measured in a local Lorentz frame if and only if the body moves along a geodesic of **g**. Hence, to determine whether test-body trajectories are geodesics, one must compare experimentally the motion of the spatial origin of a local Lorentz frame (as defined by atomic-clock readings) with the motion of a test body (material particle).

It is easy to study experimentally the motions of test bodies; relative to an earth-bound laboratory, they accelerate downward with $g = 980$ cm/sec^2; and this acceleration can be measured at a given location on the Earth to a precision of 1 part in. 10^6.

Unfortunately, it is much more difficult to measure the motion of a local Lorentz frame, once again as defined by atomic-clock readings. The only direct experimental handle one has on this today, with sufficient precision to be interesting, is gravitational redshift experiments. (See §§7.2–7.5 and §25.4 for theoretical discussions of the gravitational redshift in the framework of general relativity.)

Pound-Rebka-Snider redshift experiment as a test of geodesic motion

The redshift experiment of highest precision is that of Pound and Rebka (1960), as improved by Pound and Snider (1965); see Figure 38.1. It used the Mossbauer effect to measure the redshift of 14.4 keV gamma rays from Fe57. The emitter and absorber of the gamma rays were placed at rest at the bottom and top of a tower at Harvard University, separated by a height $h = 74$ feet $= 22.5$ meters. The measured redshift agreed, to 1 percent precision, with the general relativistic prediction of

$$\Delta\lambda/\lambda = gh = 2.5 \times 10^{-15}. \tag{38.2}$$

This result tells one that the local Lorentz frames are not at rest relative to the Earth's surface; rather, they are accelerating downward with the same acceleration, g, as acts on a free particle (to within 1 percent precision). To arrive at this conclusion, one analyzes the experiment in the laboratory reference frame, where everything (the experimental apparatus, the Earth, the Earth's gravitational field) is static. Relative to the laboratory a local Lorentz frame, momentarily at rest, accelerates downward (horizontal accelerations being ruled out by symmetry) with some unknown acceleration a. Equivalently, the laboratory accelerates upward (in $+z$ direction) with acceleration a relative to the local Lorentz frame. Consequently, the spacetime metric in the laboratory frame has the standard form

$$ds^2 = -(1 + 2az)\,dt^2 + dx^2 + dy^2 + dz^2 + O(|x^j|^2)\,dx^\alpha\,dx^\beta, \tag{38.3}$$

Figure 38.1.

The experiment of Pound and Rebka (1959) and Pound and Snider (1965) on the gravitational redshift of photons rising 22.5 meters against gravity through a helium-filled tube in a shaft in the Jefferson Physical Laboratory of Harvard University. The source of Co^{57} had an initial strength greater than a curie. The 14.4 keV gamma rays had to pass in through an absorber enriched in Fe^{57} to reach the large-window proportional counters. Both source and absorber were placed in temperature-regulated ovens. The velocity of the source consisted of two parts: one steady (v_M), to put the center of the emission line on the part of the transmission curve that is nearly straight; and the other alternating between $+v_J$ and $-v_J$, to sweep the transmission curve in this straight region; similarly when the steady velocity was $-v_M$. The departure from symmetry between the two cases $+v_M$ and $-v_M$ allows one to determine the offset v_D (effect of gravitational redshift) from the zero-gravity case of stationary emitter and stationary absorber. The final result for the redshift was (0.9990 ± 0.0076) times the value 4.905×10^{-15} of $2gh/c^2$ predicted from the principle of equivalence (difference between "up" experiment and "down" experiment). Diagrams adapted from Pound and Snider (1965).

which Track-2 readers have met in §§6.6 and 13.6; and Track-1 readers have met and used in Box 16.2. Moreover, in the laboratory frame the metric is static, gravity is static, and the experimental apparatus is static. Therefore the crest of each electromagnetic wave that climbs upward must follow a world line $t(z)$ identical in form to the world lines of the crests before and after it; thus,

$$\text{wave crest } \#0: \ t = t_0(z),$$
$$\text{wave crest } \#1: \ t = t_0(z) + \Delta t,$$
$$\vdots$$
$$\text{wave crest } \#n: \ t = t_0(z) + n \, \Delta t.$$

[Here, as in Schild's argument (§7.3) that redshift implies spacetime curvature, no assumption is made about the form of the wave-crest world lines $t_0(z)$; see Figure 7.1.] Hence, expressed in *coordinate* time, the interval between reception of successive wave crests is the same as the interval between emission. Both are Δt. But the atomic clocks of the experiment (Fe^{57} nuclei) are assumed to measure proper time $\Delta\tau \equiv (-g_{\alpha\beta} \, \Delta x^\alpha \, \Delta x^\beta)^{1/2}$, not coordinate time. Thus

$$\frac{\lambda_{\text{received}}}{\lambda_{\text{emitted}}} = \frac{\Delta\tau_{\text{received}}}{\Delta\tau_{\text{emitted}}} = \frac{(1 + az_{\text{received}}) \, \Delta t}{(1 + az_{\text{emitted}}) \, \Delta t}$$
$$= 1 + a(z_{\text{received}} - z_{\text{emitted}});$$

i.e.,

$$\frac{\Delta\lambda}{\lambda} = ah \quad \begin{bmatrix} \text{theoretical prediction based on assumptions} \\ (i) \ \text{that atomic clocks measure } \Delta\tau = (-g_{\alpha\beta}\Delta x^\alpha \, \Delta x^\beta)^{1/2}; \\ (ii) \ \text{that electromagnetic radiation has the form of a} \\ \quad \text{wave train;} \\ (iii) \ \text{that local Lorentz frames accelerate downward} \\ \quad \text{with acceleration } a \text{ relative to the laboratory.} \end{bmatrix} \quad (38.4)$$

Direct comparison with the experimental result (38.2) reveals that *local Lorentz frames in an Earthbound laboratory accelerate downward with the same acceleration g as acts on a test particle (to within 1 per cent precision).*

[The above discussion is basically a reworked version of Schild's proof (§7.2) that the redshift experiment implies spacetime is curved. After all, how could spacetime possibly be flat if Lorentz frames in Washington, Moscow, and Peking all accelerate toward the Earth's center with $g = 980 \text{ cm/sec}^2$?]

Other redshift experiments

Of all redshift experiments, the Pound-Rebka-Snider experiment is the easiest to interpret theoretically, because it was performed in a uniform gravitational field. Complementary to it is the experiment by Brault (1962), which measured the redshift of the sodium D_1 line emitted on the surface of the sun and received at Earth (Figure 38.2). To a precision of 5 per cent, he found a redshift of $GM_\odot/R_\odot c^2$, where M_\odot and R_\odot are the mass and radius of the sun. This is just the redshift to be expected if

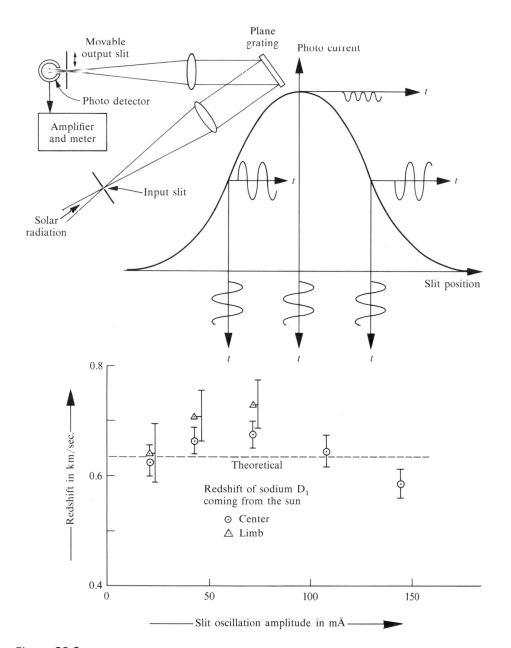

Figure 38.2.

The measurement by Brault (1962) of the redshift of the D_1 line of sodium gives 1.05 ± 0.05 of the gravitational redshift predicted by general relativity. This strong line, in contrast to the weak lines used by earlier investigators (1) is emitted high in the sun's atmosphere, above the regions strongly disturbed by the pressure and convective shifts, and yet lower than the chromosphere, and (2) comes closer to standing up cleanly above the background than any other line in the visible spectrum. Brault built a new photoelectric spectrometer (upper diagram), with its slit vibrated mechanically back and forth across a narrow region of the spectrum, to define the position of the line peak (1) electronically, (2) independently of subjective judgment, and (3) with a precision greater by a factor of the order of ten than that afforded by conventional visual methods. The slit is considered set on a line when its mean position is such that the photomultiplier current contains no signal at the frequency of the modulation. The redshift measured in this way is corrected for orbital motion and for rotation of the sun and the Earth to give the points in circles and triangles in the lower diagram. Extrapolation to zero vibration of the slit gives the cited number for the redshift. Figure adapted from thesis of Brault (1962).

the local Lorentz frames, at each point along the photon trajectory, fall in step with freely falling test bodies.*

In summary, redshift experiments reveal that, to a precision of several percent, the local Lorentz frames at the Earth's surface and near the sun are unaccelerated relative to freely falling test bodies. Equivalently, test bodies move along straight lines in the local Lorentz frames. Equivalently, *the test-body trajectories are geodesics of the metric g.*

§38.6. TESTS OF THE EQUIVALENCE PRINCIPLE

Tests of the equivalence principle:

Of all the principles at work in gravitation, none is more central than the equivalence principle. As enunciated in §16.2, it states: "*In any and every local Lorentz frame, anywhere and anytime in the universe, all the (nongravitational) laws of physics must take on their familiar special-relativistic forms.*"

(1) geodesic motion

That test bodies move along straight lines in local Lorentz frames (geodesic motion) is one aspect of the equivalence principle. Other aspects are the universality of Maxwell's equations

$$F^{\alpha\beta}{}_{,\beta} = 4\pi J^{\alpha} \text{ and } F_{\alpha\beta,\gamma} + F_{\beta\gamma,\alpha} + F_{\gamma\alpha,\beta} = 0 \tag{38.5}$$

in all local Lorentz frames; the universality of the law of local energy-momentum conservation

$$T^{\alpha\beta}{}_{,\beta} = 0; \tag{38.6}$$

and the universality of the values of the dimensionless constants that enter into the local laws of physics:

$$\alpha_e \equiv \frac{e^2}{\hbar c} = \frac{1}{137.0360\ldots} = \begin{pmatrix} \text{electromagnetic fine-} \\ \text{structure constant} \end{pmatrix};$$

$$\frac{m_{\text{neutron}}}{m_{\text{proton}}} = 1.00138\ldots, \qquad \frac{m_{\text{electron}}}{m_{\text{proton}}} = \frac{1}{1836.12\ldots}, \qquad \text{etc.} \tag{38.7}$$

(Attention here is confined to dimensionless constants, since only they are independent of one's arbitrary choice of units of measure.)

(2) physical laws are locally Lorentz-invariant

If one focuses attention on a given event and asks about invariance of the form of the physical laws [equations (38.5), (38.6), etc.] from one Lorentz frame to another, one is then in the province of special relativity. Here a multitude of experiments verify the equivalence principle (see §38.4).

(3) laws do not vary from event to event

If one asks about variations in the form of the laws from one event to another, one opens up a Pandora's box of possibilities that one hardly dares to contemplate. However, no experimental evidence has ever given the slightest warrant to consider any such "departure from democracy" in the action of the laws of physics. Moreover, astronomical observations provide strong evidence that the laws of physics are the

*For a review of other, less-precise redshift experiments, see Bertotti, Brill, and Krotkov (1962).

same in distant stellar systems as in the solar system, and the same in distant galaxies as in our own Galaxy. (See, in Box 29.5, Edwin Hubble's expressions of joy upon discovering this.)

Constancy of the dimensionless "constants" from event to event can be tested to high precision, if one assumes constancy of the physical laws. Dirac (1937, 1938), Teller (1948), Jordan (1955, 1959), Gamow (1967), and others have proposed that the fine-structure "constant" α_e might be a slowly varying scalar field, perhaps governed by a cosmological equation. However, rather stringent limits on such variations follow from data on the fine-structure splitting of the spectral lines of quasars and radio galaxies. For the quasar 3C 191 with redshift $z = 1.95$, Bahcall, Sargent, and Schmidt (1967) find $\alpha_e(3C\ 191)/\alpha_e(\text{Earth}) = 0.97 \pm 0.5$. With a cosmological interpretation of the quasar redshift, this corresponds to a limit $(1/\alpha_e)(d\alpha_e/dt) \lesssim 1/10^{11}$ years. An even tighter limit has been obtained from radio-galaxy data, where there is no question about the interpretation of the redshift. Bahcall and Schmidt (1967) measured fine-structure splitting in five radio galaxies with $z \approx 0.20$, corresponding to an emission of light 2×10^9 years ago. They obtained $\alpha_e(z = 0.20)/\alpha_e(\text{Earth}) = 1.001 \pm 0.002$, which yields the limit $|(1/\alpha_e)(d\alpha_e/dt)| \lesssim 1/10^{12}$ years.

Dyson (1972) points out that comparison of the rate of beta decay of Re^{187} in times past (via osmium-rhenium abundance ratios in old ores) with the rate of beta-decay today provides a means to check on any possible variation of α_e with time more sensitive than redshift data and more sensitive than any changes in rates of alpha decay and fission between early times and now. He summarizes the available data on Re^{187} and arrives at the limit

(4) fundamental constants do not vary from event to event

$$|(1/\alpha_e)(d\alpha_e/dt)| \lesssim 5/10^{15}\ \text{years.}$$

For further evidence of the constancy of the fundamental constants see Minkowski and Wilson (1956), Dicke (1959a,b), Dicke and Peebles (1962b).

Spatial variations of α_e, $m_{\text{neutron}}/m_{\text{proton}}$, and other "constants" in the solar system can be sought by means of Eötvös-type experiments. The reasoning [by Dicke (1969)] leading from such experiments to limits on any spatial variation of the constants is indirect. It recalls the reasoning used in standard treatises on polar molecules to deduce the acceleration of a polarizable molecule pulled on by an inhomogeneous electric field. It proceeds as follows.

Suppose one of the dimensionless "constants," "α," depends on position. This will lead to a position-dependence of the total mass-energy of a laboratory test body. For example, if α_e depends on position, then the coulomb energy of an atomic nucleus will also ($E_{\text{coul}} \propto e^4 \propto \alpha_e{}^2$; $\delta M/E_{\text{coul}} = 2\,\delta\alpha_e/\alpha_e$). One can calculate the change in a test body's mass-energy when it is moved from x^μ to $x^\mu + \delta x^\mu$ by assuming no change at all in the body's structure during its displacement:

Eötvös-type experiments as tests for spatial variation of fundamental constants

$$\delta M = (\partial M/\partial\alpha)_{\text{fixed structure}}(\partial\alpha/\partial x^\mu)\,\delta x^\mu. \tag{38.8}$$

After the displacement, a weakening of internal forces (due, e.g., to a decrease of α)

may cause a change in structure, but that change will be accompanied by a conversion of internal potential energy into internal kinetic energy, which conserves M.

Now consider the following thought experiment [an elaboration of the argument by which Einstein first derived the gravitational redshift (§7.2)]: Take n particles, each with mass-energy μ. Make the particles with a structure such that a negligible fraction of μ is associated with the "constant" of interest, α:

$$(1/\mu)(\partial\mu/\partial\alpha) = 0. \tag{38.9}$$

Place these particles at a height h in a (locally) uniform Newtonian field. Combine them together there, releasing binding energy $E_B(h)$, to form a composite body of mass

$$M = n\mu - E_B(h) \tag{38.10}$$

which depends in a significant manner on the "constant" α,

$$(1/M)(\partial M/\partial\alpha) \neq 0. \tag{38.11}$$

Lower this body, *and* the released binding energy tied up in a little bag, a distance δh. The total force acting is (in Newtonian language)

$$F = Ma + E_B(h)g. \tag{38.12}$$

Here g is acceleration experienced by the type of mass-energy that is independent of α when it is in free fall. In contrast, "free" fall of the assembled body M is not really free fall, because of the supplementary "polarization force" pulling on this object. Hence the assembled body in "free" fall experiences an acceleration, a, a little different from g. However, the mass that is accelerated is precisely M, and therefore the force required to produce this acceleration is given by the product Ma. The energy gained in lowering the body and the bag is

$$E(\text{down}) = F\,\delta h = Ma\,\delta h + E_B(h)g\,\delta h.$$

Put this energy in the bag.

At $h - \delta h$ use some of the energy from the bag to pull the body apart into its component particles. The energy required is

$$E_B(h - \delta h) = n\mu - M(h - \delta h) = n\mu - M(h) + \frac{\partial M}{\partial\alpha}\frac{d\alpha}{dh}\,\delta h$$

$$= E_B(h) + \frac{\partial M}{\partial\alpha}\frac{\partial\alpha}{\partial h}\,\delta h;$$

so an energy

$$E_{\text{bag}} = E_B(h) + E(\text{down}) - E_B(h - \delta h)$$
$$= \left[Ma + E_B(h)g - \frac{\partial M}{\partial\alpha}\frac{d\alpha}{dh}\right]\delta h \tag{38.13}$$

is left in the bag. Use this energy to raise the n particles and the bag back up to

height h. *Assume total energy conservation,* so that there will be no extra energy and no deficit when the n particles and bag have returned to the original state back at height h. This means that E_{bag} must be precisely the right amount of energy to do the raising:

$$E_{\text{bag}} = n\mu g\,\delta h = [M + E_B(h)]g\,\delta h. \tag{38.14}$$

Combining expressions (38.13) and (38.14) for E_{bag}, discover that

$$a - g = \frac{1}{M}\frac{\partial M}{\partial \alpha}\frac{d\alpha}{dh}. \tag{38.15}$$

Thus, under the assumption of total energy conservation (no perpetual-motion machines!), a spatial dependence of a physical "constant" α will lead to the anomaly (38.15) in the acceleration of a body whose mass depends on α.

Coulomb energy, which is proportional to α_e^2, amounts in a gold nucleus to 0.4 per cent of the mass, and to 0.1 per cent in an aluminum nucleus. Hence, a spatial variation in α_e should lead to a fractional difference in the gravitational accelerations of these two nuclei equal to

$$\left|\frac{a_{Au} - a_{Al}}{g}\right| \approx \frac{1}{g}\,2\,\frac{0.003}{\alpha_e}\frac{d\alpha_e}{dh} \lesssim 1\times 10^{-11};$$

i.e.,

$$\frac{1}{\alpha_e}\left|\frac{d\alpha_e}{dh}\right| \lesssim 1\times 10^{-9}\,g \approx 1\times 10^{-9}\text{ cm/sec}^2 = 1\times 10^{-30}\text{/cm} \tag{38.16}$$

$$\text{at the Earth due to the sun.}$$

Here use is made of the limit (1×10^{-11}) from Dicke's experiment (§38.3), and the acceleration $g = 0.6$ cm/sec^2 due to the sun at Earth.

Notice that this says the gradient of $\ln\alpha_e$ is less than 1×10^{-9} the gradient of the Newtonian potential!

§38.7. TESTS FOR THE EXISTENCE OF UNKNOWN LONG-RANGE FIELDS

Whether or not one accepts the assumption that test bodies move on geodesics of the metric, it remains conceivable that previously unknown long-range fields (fields with "$1/r$" fall-off at large distances) are somehow associated with gravity.

Possible existence of new long-range fields associated with gravity

If "new" long-range fields (not metric, not electromagnetic) do exist, waiting to be discovered, then there are two ways by which they could influence matter. First, they could *couple directly to matter,* producing, for example, slight deviations from geodesic motion (deviations smaller than the limits of §38.5), or slight dependences of masses of particles on position (dependences smaller than the limits of §38.6). Second (and harder to detect), they could *couple indirectly to matter* by being mere

Direct vs. indirect coupling

participants in field equations that determine the geometry of spacetime. This section will describe tests for direct-coupling effects. Theories with fields that couple indirectly will be described in Box 39.1, and tests for such fields will be discussed in Chapter 40.

Dicke (1964b), using his framework for analyzing tests of gravitation theories (§38.2), has shown that several null experiments place stringent limits on unknown, direct-coupling, long-range fields.

One of these experiments is the "Hughes-Drever Experiment" [Hughes, Robinson, and Beltran-Lopez (1960); Drever (1961)]. It can be thought of as a search for a symmetric second-rank tensor field $h_{\alpha\beta}$ that produces slight deviations of test-body trajectories from geodesics of the metric $g_{\alpha\beta}$. Unless one's experiments happen to be made in a region of spacetime where $h_{\alpha\beta}$ is a constant multiple of $g_{\alpha\beta}$ ("mere rescaling of all lengths and times by a constant factor"), this tensor field must produce anisotropies in the properties of spacetime—which, in turn, will cause anisotropies in the inertial mass of a nucleon, and in turn will cause in an atomic nucleus relative shifts of degenerate energy levels with different magnetic quantum numbers. The Hughes-Drever experiment places stringent limits on such shifts, and thereby on a possible tensor field $h_{\alpha\beta}$. To quote Dicke (1964, p. 186), "If two [tensor] fields are present with the one strongly anisotropic in a coordinate system chosen to make the other isotropic, the strength of [direct] coupling to one must be only of the order of 10^{-22} that of the other.... [Moreover], on the moving Earth with ever-changing velocity, anisotropy would be expected at some season." From the experiments of Hughes and Drever, then, one concludes that there is not the slightest evidence for the presence of a second tensor field. For further details see Dicke and Peebles (1962a).

Another series of experiments, called *"ether-drift experiments,"* places stringent limits on any unknown, long-range vector field that couples directly to mass-energy. One can imagine such a field of cosmological origin. Being cosmological, the 4-vector would most naturally be expected to point in the same direction as the 4-vector \boldsymbol{u} of the "cosmological fluid" (identical with the time direction \boldsymbol{e}_0 of a frame in which the cosmic microwave radiation is isotropic). The 4-vector of the new field would then have spatial components in any other frame. In principle an observer could use them to discern his direction of motion and speed relative to the mean rest frame of the universe. The ether-drift experiments search for effects of such a field.

For example, the experiment of Turner and Hill (1964) searches for a dependence of clock rates on such a vector field, by examining the transverse Doppler shift as a function of direction for an emitter on the rim of a centrifuge and a receiver at its center (Figure 38.3). If there is any effect, it would most naturally be expected to have the form

$$\frac{\left(\begin{matrix}\text{rate of clock moving relative}\\\text{to universe with speed }\boldsymbol{\beta}\end{matrix}\right)}{\left(\begin{matrix}\text{rate of clock at rest}\\\text{relative to universe}\end{matrix}\right)} = 1 + \gamma\beta^2, \qquad \gamma \text{ a small constant.} \quad (38.17)$$

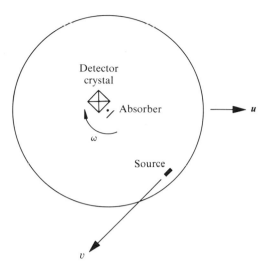

Figure 38.3.

The experiment of Turner and Hill (1964) looks for a dependence of proper clock rate (the clock being a Co^{57} source placed near the rim of the centrifuge) on velocity relative to the distant matter of the universe; or, in operational terms, relative to a "new local field" described by a 4-vector. The 14.4 keV gamma rays from the Co^{57} already experience a second-order Doppler shift of 1.3 parts in 10^{13}. One searches for an additional shift $\gamma\beta^2$ where $\boldsymbol{\beta} = \boldsymbol{u} + v(\boldsymbol{e}_x \cos \omega t + \boldsymbol{e}_y \sin \omega t)$ is the velocity relative to the frame in which the scalar field is purely timelike. The transmission of the gamma rays through the Fe^{57} absorber will drop linearly with any such additional shift, and will be noted as a drop in the counting rate of the NaI crystal. The source was 10 cm from the axis of rotation and the centrifuge turned at 15,000 rpm. The value of γ deduced from the experiment was $(1 \pm 4) \times 10^{-5}$.

A clock at the center of the centrifuge has $\boldsymbol{\beta} = \boldsymbol{u} = u\boldsymbol{e}_x$, whereas one on the rim has $\boldsymbol{\beta} = \boldsymbol{u} + v(\boldsymbol{e}_x \cos \omega t + \boldsymbol{e}_y \sin \omega t)$. Thus, the shift between rim and disk should vary with position

$$\Delta\lambda/\lambda = -\Delta\nu/\nu = -2\gamma uv \cos \omega t + \text{usual transverse shift}.$$

The data of Turner and Hill, using the Mössbauer effect, show that

$$|\gamma| < 4 \times 10^{-5}. \tag{38.18}$$

Hence, a cosmological vector field, if present, has only a weak direct coupling to matter.

For further discussion of these experiments and references on others like them, see Dicke (1964b).

CHAPTER **39**

OTHER THEORIES OF GRAVITY AND THE POST-NEWTONIAN APPROXIMATION

§39.1. OTHER THEORIES

Among all bodies of physical law none has ever been found that is simpler or more beautiful than Einstein's geometric theory of gravity (Chapters 16 and 17); nor has any theory of gravity ever been discovered that is more compelling.

As experiment after experiment has been performed, and one theory of gravity after another has fallen by the wayside a victim of the observations, Einstein's theory has stood firm. No purported inconsistency between experiment and Einstein's laws of gravity has ever surmounted the test of time.

Query: Why then bother to examine alternative theories of gravity? *Reply:* To have "foils" against which to test Einstein's theory.

To say that Einstein's geometrodynamics is "battle-tested" is to say it has won every time it has been tried against a theory that makes a different prediction. How then does one select new antagonists for decisive new trials by combat?

Not all theories of gravity are created equal. Very few, among the multitude in the literature, are sufficiently viable to be worth comparison with general relativity or with future experiments. The "worthy" theories are those which satisfy *three criteria for viability: self-consistency, completeness, and agreement with past experiment.*

Self-consistency is best illustrated by describing several theories that fail this test. The classic example of an internally inconsistent theory is the spin-two field theory of gravity [Fierz and Pauli (1939); Box 7.1 here], which is equivalent to linearized general relativity (Chapter 18). The field equations of the spin-two theory imply that all gravitating bodies move along straight lines in global Lorentz reference frames, whereas the equations of motion of the theory insist that gravity deflects

Role of alternative gravitation theories as foils for experimental tests

Criteria for viability of a theory:

(1) self-consistency

bodies away from straight-line motion. (When one tries to remedy this inconsistency, one finds oneself being "bootstrapped" up to general relativity; see route 5 of Box 17.2.) Another self-inconsistent theory is that of Kustaanheimo (1966). It predicts zero gravitational redshift when the wave version of light (Maxwell theory) is used, and nonzero redshift when the particle version (photon) is used.

Completeness: To be complete a theory of gravity must be capable of analyzing from "first principles" the outcome of every experiment of interest. It must therefore mesh with and incorporate a consistent set of laws for electromagnetism, quantum mechanics, and all other physics. No theory is complete if it *postulates* that atomic clocks measure the "interval" $d\tau = (-g_{\alpha\beta} \, dx^\alpha \, dx^\beta)^{1/2}$ constructed from a particular metric. Atomic clocks are complex systems whose behavior must be calculated from the fundamental laws of quantum theory and electromagnetism. No theory is complete if it *postulates* that planets move on geodesics. Planets are complex systems whose motion must be calculated from fundamental laws for the response of stressed matter to gravity. For further discussion see §§16.4, 20.6, and 40.9.

Agreement with past experiment: The necessity that a theory agree, to within several standard deviations, with the "four standard tests" (gravitational redshift, perihelion shift, electromagnetic-wave deflection, and radar time-delay) is obvious. Equally obvious but often forgotten is the need to agree with the expansion of the universe (historically the ace among all aces of general relativity) and with observations at the more everyday, Newtonian level. Example: Birkhoff's (1943) theory predicts the same redshift, perihelion shift, deflection, and time-delay as general relativity. But it requires that the pressure inside gravitating bodies equal the total density of mass-energy, $p = \rho$; and, as a consequence, it demands that sound waves travel with the speed of light. Of course, this prediction disagrees violently with experiment. Therefore, Birkhoff's theory is not viable. Another example: Whitehead's (1922) theory of gravity was long considered a viable alternative to Einstein's theory, because it makes exactly the same prediction as Einstein for the "four standard tests." Not until the work of Will (1971b) was it realized that Whitehead's theory predicts a time-dependence for the ebb and flow of ocean tides that is completely contradicted by everyday experience (see §40.8).

(in margin:) (2) completeness

(in margin:) (3) agreement with past experiment

§39.2. METRIC THEORIES OF GRAVITY

Two lines of argument narrow attention to a restricted class of gravitation theories, called *metric theories.*

(in margin:) Why attention focuses on metric theories of gravity

The first line of argument constitutes the theme of the preceding chapter. It examined experiment after experiment, and reached two conclusions: (1) *spacetime possesses a metric; and* (2) *that metric satisfies the equivalence principle* (the standard special relativistic laws of physics are valid in each local Lorentz frame). *Theories of gravity that incorporate these two principles are called metric theories.** In brief, Chapter 38 says, "For any adequate description of gravity, look to a metric theory."

*For a slightly narrower definition of metric theories, see Thorne and Will (1971).

Exception: Cartan's (1922b, 1923) theory ["general relativity plus torsion"; see Trautman (1972)] is nonmetric, but agrees with experiment and is experimentally indistinguishable from general relativity with the technology of the 1970's.

The second line of argument pointing to metric theories begins with the issue of completeness (preceding section). To be complete, a theory must incorporate a self-consistent version of all the nongravitational laws of physics. No one has found a way to incorporate the rest of physics with ease except to introduce a metric, and then invoke the principle of equivalence. Other approaches lead to dismaying complexity, and usually to failure of the theory on one of the three counts of self-consistency, completeness, and agreement with past experiment. *All the theories known to be viable in 1973 are metric,* except Cartan's. [See Ni (1972b); Will (1972).]

How metric theories differ

In only one significant way do metric theories of gravity differ from each other: their laws for the generation of the metric. In general relativity theory, the metric is generated directly by the stress-energy of matter and of nongravitational fields. In Dicke-Brans-Jordan theory (Box 39.1, p. 1070), matter and nongravitational fields generate a scalar field ϕ; then ϕ acts together with the matter and other fields to generate the metric. Expressed in the language of §38.7, ϕ is a "new long-range field" that couples indirectly to matter. As another example, a theory devised by Ni (1970, 1972) (Box 39.1) possesses a flat-space metric $\boldsymbol{\eta}$ and a universal time coordinate t ("prior geometry"; see §17.6); $\boldsymbol{\eta}$ acts together with matter and nongravitational fields to generate a scalar field ϕ; and then $\boldsymbol{\eta}$, t, and ϕ combine to create the physical metric \boldsymbol{g} that enters into the equivalence principle.

All three of the above theories—Einstein, Dicke-Brans-Jordan, Ni—were viable in the summer of 1971, when this section was written. But in autumn 1971 Ni's theory, and many other theories that had been regarded as viable, were proved by Nordtvedt and Will (1972) to disagree with experiment. This is an example of the rapidity of current progress in experimental tests of gravitation theory!

Henceforth, in this chapter and the next, attention will be confined to metric theories of gravity and their comparison with experiment.

§39.3. POST-NEWTONIAN LIMIT AND PPN FORMALISM

Weak-field, slow-motion
expansion of a metric theory

The solar system, where experiments to distinguish between metric theories are performed, has weak gravity,

$$|\boldsymbol{\Phi}| = |\text{Newtonian potential}| \lesssim 10^{-6}; \tag{39.1a}$$

moreover, the matter that generates solar-system gravity moves slowly

$$v^2 = (\text{velocity relative to solar-system center of mass})^2 \lesssim 10^{-7} \tag{39.1b}$$

and has small stress and internal energies

$$|T_{jk}|/\rho_o = (\text{stress divided by baryon "mass" density}) \lesssim 10^{-6}, \tag{39.1c}$$

$$\Pi = (\rho - \rho_o)/\rho_o = \binom{\text{internal energy density per}}{\text{unit baryon "mass" density}} \lesssim 10^{-6}. \tag{39.1d}$$

[Here the *baryon "mass" density* ρ_o, despite its name, and despite the fact it is sometimes even more misleadingly called "density of rest mass-energy," is actually a measure of the number density of baryons n, and nothing more. It is defined as the product of n with some standard figure for the mass per baryon, μ_0, in some well-defined standard state; thus,

$$\rho_o \equiv n\mu_0.]$$

(39.1e)

Consequently, the analysis of solar-system experiments using any metric theory of gravity can be simplified, without significant loss of accuracy, by a simultaneous expansion in the small parameters $|\Phi|$, v^2, $|T_{jk}|/\rho_o$, and Π. Such a "weak-field, slow-motion expansion" gives: (1) flat, empty spacetime in "zero order"; (2) the Newtonian treatment of the solar system in "first order"; and (3) post-Newtonian corrections to the Newtonian treatment in "second order".

The formalism of Newtonian theory plus post-Newtonian corrections is called the *"post-Newtonian approximation."* Each metric theory has its own post-Newtonian approximation. Despite the great differences between metric theories themselves, their post-Newtonian approximations are very similar. They are so similar, in fact, that one can construct a single post-Newtonian theory of gravity, devoid of any reference to indirectly coupling fields (ϕ in Dicke-Brans-Jordan; $\boldsymbol{\eta}$, t, and ϕ in Ni; see Box 39.1), that contains the post-Newtonian approximation of every conceivable metric theory as a special case. This all-inclusive post-Newtonian theory is called the *"Parametrized Post-Newtonian (PPN) Formalism."* It contains a set of parameters (called "*PPN parameters*") that can be specified arbitrarily. One set of values for these parameters makes the PPN formalism identical to the post-Newtonian limit of general relativity; another set of values makes it the post-Newtonian limit of Dicke-Brans-Jordan theory, etc.

Subsequent sections of this chapter present a version of the PPN formalism devised by Clifford M. Will and Kenneth Nordtvedt, Jr. (1972). [See also Will (1972).] This version, containing ten PPN parameters, encompasses as special cases nearly every metric theory of gravity known to the authors. The few exceptions [Whitehead (1922) and theories reviewed by Will (1973)] all disagree with experiment. One can include them in the PPN formalism by adding additional terms and parameters.

The ten parameters are described heuristically in Box 39.2, for the convenience of readers who would skip the full details of the formalism (§§39.4–39.12).

How accurate is the PPN formalism? Or, stated more precisely, how accurately does the post-Newtonian approximation agree with the metric theory from which it comes? In the solar system, where $|\Phi|$, v^2, $|T_{jk}|/\rho_o$, and Π are all $\lesssim 10^{-6}$, the post-Newtonian approximation makes fractional errors of $\lesssim 10^{-6}$ in quantities of post-Newtonian order, and fractional errors of $\lesssim 10^{-12}$ in quantities of Newtonian order. For example, it misrepresents the deflection of light by $\lesssim 10^{-6} \times$ (post-Newtonian deflection) $\sim 10^{-6}$ seconds of arc. And it ignores relativistic deformations of the Earth's orbit of magnitude $< 10^{-12} \times$ (one astronomical unit) ~ 10 centimeters. Clearly, there is no need in the 1970's to use higher-order corrections to the post-Newtonian approximation; and hence no need to construct a "parametrized post-post-Newtonian framework." However, in the words of Shapiro (1971b): "If one projects from the achievements in the last decade, it is not unreasonable to predict

Post-Newtonian approximation

PPN formalism

Accuracy of PPN formalism in solar system

(continued on page 1072)

Box 39.1 THE THEORIES OF DICKE-BRANS-JORDAN AND OF NI

A. Dicke-Brans-Jordan

References: Brans and Dicke (1961); Jordan (1959). [*Notes:* This is the special case $\eta = -1$ of Jordan's theory. An alternative mathematical representation of the theory is given by Dicke (1962).]

Fields associated with gravity:

ϕ, a long-range scalar field;

g, the metric of spacetime (from which are constructed the covariant derivative ▼ and the curvature tensors, in the usual manner).

Equations by which these fields are determined:

The trace of the stress-energy tensor generates ϕ via the curved-spacetime wave equation

$$\Box\phi = \phi^{,\alpha}{}_{;\alpha} = \frac{8\pi}{3 + 2\omega}T,$$

where ω is the dimensionless "Dicke coupling constant."

The stress-energy tensor and ϕ together generate the metric (i.e., the spacetime curvature) via the field equations

$$G_{\alpha\beta} = \frac{8\pi}{\phi}T_{\alpha\beta} + \frac{\omega}{\phi^2}\left(\phi_{,\alpha}\phi_{,\beta} - \frac{1}{2}g_{\alpha\beta}\phi_{,\mu}\phi^{,\mu}\right) + \frac{1}{\phi}(\phi_{;\alpha\beta} - g_{\alpha\beta}\Box\phi),$$

where $G_{\alpha\beta}$ is the Einstein tensor.

Variational principle for these equations:

$$\delta\int[\phi R - \omega(\phi_{,\alpha}\phi^{,\alpha}/\phi) + 16\pi L](-g)^{1/2}\,d^4x = 0,$$

where R is the scalar curvature and L is the matter Lagrangian.

Equivalence principle is satisfied:

The special-relativistic laws of physics are valid, without change, in the local Lorentz frames of the metric **g**.

Consequence: the scalar field does not exert any direct influence on matter; its only role is that of participant in the field equations that determine the geometry of spacetime. It is an "indirectly coupling field" in the sense of §38.7.

This theory is self-consistent, complete, and for $\omega > 5$ in "reasonable" accord (two standard deviations or better) with all pre-1973 experiments.

B. Ni

References: Ni (1970, 1972).

Fields associated with gravity:

η, a flat "background metric" ("prior geometry" in sense of §17.6). There exist,

by assumption, coordinate systems ("background Lorentz frames") in which everywhere at once $\eta_{00} = -1$, $\eta_{0j} = 0$, and $\eta_{jk} = \delta_{jk}$.

t, a scalar field called the "universal time coordinate" ("prior geometry" in sense of §17.6), which is so "tuned" to the background metric that

$$t_{|\alpha\beta} = 0, \qquad t_{,\alpha}t_{,\beta}\eta^{\alpha\beta} = -1,$$

where "$_|$" denotes covariant derivative with respect to $\boldsymbol{\eta}$.

This means there exists a background Lorentz frame (the "rest frame of the universe") in which $x^0 = t$.

ϕ, a scalar field called the "scalar gravitational field".

\boldsymbol{g}, the metric of spacetime (from which are constructed the covariant derivative $\boldsymbol{\nabla}$ and the curvature tensors, in the usual manner).

Equations by which these fields are determined:

The stress-energy of spacetime generates the scalar gravitational field ϕ via the wave equation

$$\Box\phi \equiv \phi^{,\alpha}{}_{;\alpha} = -2\pi T^{\alpha\beta}\,\partial g_{\alpha\beta}/\partial\phi$$
$$= 4\pi T^{\alpha\beta}[\eta_{\alpha\beta}e^{-2\phi} + (e^{2\phi} + e^{-2\phi})t_{,\alpha}t_{,\beta}].$$

ϕ, $\boldsymbol{\eta}$, and t together determine the metric of spacetime through the algebraic relation

$$\boldsymbol{g} = e^{-2\phi}\boldsymbol{\eta} + (e^{-2\phi} - e^{2\phi})\,\boldsymbol{dt} \otimes \boldsymbol{dt}.$$

Note: In the "rest frame of the universe" that is presupposed in this theory, this metric reduces to

$$ds^2 = g_{\alpha\beta}\,dx^\alpha\,dx^\beta = -e^{2\phi}\,dt^2 + e^{-2\phi}(dx^2 + dy^2 + dz^2).$$

Variational principle for the field equation for ϕ:

$$\delta \int (-2\phi^{,\alpha}\phi_{,\alpha} + 16\pi L)(-g)^{1/2}\,d^4x = 0,$$

where L is the matter Lagrangian.

Equivalence principle is satisfied:

The special-relativistic laws of physics are valid, without change, in the local Lorentz frames of the spacetime metric \boldsymbol{g}.

Consequence: ϕ, $\boldsymbol{\eta}$, and t do not exert any direct influence on matter; they are "indirectly coupling fields" in the sense of §38.7.

This theory is self-consistent and complete. If the solar system were at rest in the "rest frame of the universe", the theory would agree with all experiments to date—except, possibly, the expansion of the universe. But the motion of the solar system through the universe leads to serious disagreement with experiment (Will and Nordtvedt 1972; §40.8).

Box 39.2 HEURISTIC DESCRIPTION OF THE TEN PPN PARAMETERS

Para-meter	What it measures, relative to general relativity[a]	Value in General Relativity	Value in Dicke-Brans-Jordan Theory[b]	Value in Ni's Theory[b]
γ	How much space curvature (g_{jk}) is produced by unit rest mass?	1	$\dfrac{1+\omega}{2+\omega}$	1
β	How much nonlinearity is there in the superposition law for gravity (g_{00})?	1	1	1
β_1	How much gravity (g_{00}) is produced by unit kinetic energy ($\frac{1}{2}\rho_o v^2$)?	1	$\dfrac{3+2\omega}{4+2\omega}$	1
β_2	How much gravity (g_{00}) is produced by unit gravitational potential energy ($\rho_o U$)?	1	$\dfrac{1+2\omega}{4+2\omega}$	1
β_3	How much gravity (g_{00}) is produced by unit internal energy ($\rho_o \Pi$)?	1	1	1
β_4	How much gravity (g_{00}) is produced by unit pressure (p)?	1	$\dfrac{1+\omega}{2+\omega}$	1
ζ	How much *more* gravity (g_{00}) is produced by radial kinetic energy [$\frac{1}{2}\rho_o(\mathbf{v}\cdot\hat{\mathbf{r}})^2$]—i.e., kinetic energy of motion toward observer—than by transverse kinetic energy?	0	0	0
η	How much *more* gravity (g_{00}) is produced by radial stress [$\hat{\mathbf{r}}\cdot\mathbf{t}\cdot\hat{\mathbf{r}}$] than by transverse stress?	0	0	0
Δ_1	How much dragging of inertial frames (g_{0j}) is produced by unit momentum ($\rho_o v$)?	1	$\dfrac{10+7\omega}{14+7\omega}$	$-\dfrac{1}{7}$
Δ_2	How much easier is it for momentum ($\rho_o v$) to drag inertial frames radially (toward the observer) than in a transverse direction?	1	1	1

[a] These heuristic descriptions are based on equations (39.23).
[b] For expositions of these theories see Box 39.1. For derivation of their PPN values and of PPN values for other theories, see Ni (1972).

that in the 1980's techniques will be available to detect second-order effects of general relativity. At that point the ratio of theoretical to experimental relativists may take a sharp turn downwards."

Actually, there are a few exceptions to the claim that the post-Newtonian approximation suffices for the 1970's. These exceptions occur where the external universe impinges on and influences the solar system. For example, gravitational waves propagating into the solar system from distant sources (Chapters 35–37) are ignored by every post-Newtonian approximation and by the PPN framework. They must be treated using a full metric theory or a weak-field, "fast-motion" approximation

to such a theory. Similarly, time-dependence of the "gravitational constant" (§40.8), induced in some theories by expansion of the universe, is beyond the scope of the PPN formalism, as is the expansion itself.

The PPN formalism is used not only in interpreting experimental tests of gravitation theories, but also as a powerful tool in theoretical astrophysics. By specializing all the PPN parameters to unity, except $\zeta = \eta = 0$, one obtains the post-Newtonian approximation to Einstein's theory of gravity. This post-Newtonian approximation can then be used (and has been used extensively) to calculate general relativistic corrections to such phenomena as the structure and stability of stars.*

Applications of PPN formalism to astrophysics

Historical and Notational Notes

The earliest parametrizations of the post-Newtonian approximation were performed, and used in interpreting solar system experiments, by Eddington (1922), Robertson (1962), and Schiff (1962, 1967). However, they dealt solely with the vacuum gravitational field outside an isolated, spherical body (the sun). Nordtvedt (1968b, 1969) devised the first full PPN formalism, capable of treating all aspects of the solar system; he treated the sun, planets, and moon as made from "gases" of point-particles (atoms) that interact gravitationally and electromagnetically. Will (1971c) later used techniques devised by Chandrasekhar (1965a) to modify Nordtvedt's formalism, so that it employs a stressed, continuous-matter description of celestial bodies. The version of the formalism presented here, devised by Will and Nordtvedt (1972), generalizes all previous versions to acquire "post-Galilean invariance" [see Chandrasekhar and Contopolous (1967)]. The most detailed and up-to-date review article on the PPN formalism is Will (1972).

History and notation of PPN formalism

In the literature of post-Newtonian physics and the PPN formalism, the Newtonian potential is described traditionally not by Φ, but by

$$U \equiv -\Phi \equiv + \int \frac{\rho_o(\boldsymbol{x}')\, d^3x'}{|\boldsymbol{x} - \boldsymbol{x}'|}. \tag{39.2}$$

To avoid confusion, this chapter and the next will use U, although the rest of the book uses Φ.

Turn now to a detailed, Track-2 exposition of the PPN formalism.

§39.4. PPN COORDINATE SYSTEM

The PPN formalism covers the solar system (or whatever system is being analyzed) with coordinates $(t, x_j) \equiv (t, x^j)$ that are as nearly globally Lorentz as possible:

$$g_{\alpha\beta} = \eta_{\alpha\beta} + h_{\alpha\beta}, \qquad |h_{\alpha\beta}| \lesssim M_\odot/R_\odot \sim 10^{-6}. \tag{39.3}$$

EXPOSITION OF PPN FORMALISM:

Coordinate system

*See, e.g., a long series of papers by Chandrasekhar and his associates in the Astrophysical Journal, beginning with Chandrasekhar (1965a,b,c).

The rest of this chapter is Track 2. No earlier Track-2 material is needed as preparation for it, but the following will be helpful:
(1) Chapter 7 (incompatibility of gravity and special relativity)
(2) §17.6 (no prior geometry);
(3) §§36.9–36.11 (generation of gravitational waves); and
(4) Chapter 38 (tests of foundations).
This chapter is not needed as preparation for any later chapter, but it will be helpful in Chapter 40 (solar-system tests)

(In this sense the PPN formalism is like linearized theory; see Chapter 18.) The velocity of the coordinate system (i.e., 4-velocity of its spatial origin) is so chosen that the solar system is approximately at rest in these coordinates. (Whether the center of mass of the solar system is precisely at rest, or is moving with some low velocity $v \lesssim (M_\odot/R_\odot)^{1/2} \sim 10^{-3} \sim 300$ km/sec, is a matter for the user of the formalism to decide. For more on the options, see §§39.9 and 39.12.)

The PPN coordinates provide one with a natural "3 + 1" split of spacetime into space plus time. That split is conveniently treated using the notation of three-dimensional, flat-space vector analysis—even though spacetime and the three-dimensional hypersurfaces $x^0 = $ constant are both curved. The resultant three-dimensional formalism will look more like Newtonian theory than like general relativity—as, indeed, one wishes it to; after all, one's goal is to study small relativistic corrections to Newtonian theory!

§39.5. DESCRIPTION OF THE MATTER IN THE SOLAR SYSTEM

Description of matter

Relative to the PPN coordinates, the matter of the solar system (idealized as a stressed medium) has a coordinate-velocity field

$$v_j \equiv dx_j/dx^0. \tag{39.4}$$

Choose an event \mathcal{P}, and in its neighborhood transform to an orthonormal frame that moves with the matter there. Orient the spatial axes $\boldsymbol{e}_{\hat{j}}$ of this comoving frame so that they coincide as accurately as possible with the PPN coordinate axes. (This requirement will be made more precise in §39.10.) In the orthonormal comoving frame, define the following quantities, which describe the state of the matter:

(density of total mass-energy) $\equiv \rho;$ \hfill (39.5a)

(baryon "mass" density) $\equiv \rho_o$
$$\equiv \binom{\text{number density}}{\text{of baryons, } n} \times \binom{\text{standard rest mass per baryon, } \mu_0,}{\text{for matter in some standard state}}; \tag{39.5b}$$

(specific internal energy density) $\equiv \Pi \equiv (\rho - \rho_o)/\rho_o;$ \hfill (39.5c)

(components of stress tensor) $\equiv t_{\hat{i}\hat{j}} \equiv \boldsymbol{e}_{\hat{i}} \cdot \boldsymbol{T} \cdot \boldsymbol{e}_{\hat{j}};$ \hfill (39.5d)

(pressure) $\equiv p \equiv \dfrac{1}{3}(t_{\hat{x}\hat{x}} + t_{\hat{y}\hat{y}} + t_{\hat{z}\hat{z}})$ \hfill (39.5e)

\equiv (average of stress over all directions).

Anisotropies (i.e., shears) in the stress are important only in planets such as the Earth; and even there they are dominated by the isotropic pressure:

$$t_{\hat{i}\hat{j}} = p\,\delta_{ij} + p \times [\text{corrections} \ll 1]. \tag{39.6}$$

For many purposes, especially inside the sun, one can ignore the anisotropies, thereby approximating the solar-system matter as a perfect fluid.*

The isotropic part of the radiation field gives a significant contribution to the pressure, p, and the density of internal energy, $\rho_o \Pi$, inside the sun. However, the anisotropic radiation flux is ignored in the stress-energy tensor. This approximation is allowable because in the sun the outward energy flux carried by radiation is less than 10^{-15} of the internal energy density $\rho_o \Pi$; in planets it is even less.

§39.6. NATURE OF THE POST-NEWTONIAN EXPANSION

For any gravitationally bound configuration such as the solar system, the Newtonian approximation imposes limits on the sizes of various dimensionless physical quantities (see exercise 39.1):

$$\epsilon^2 \equiv \text{maximum value of Newtonian potential } U \tag{39.7}$$
$$\geq \text{ values anywhere of } U,\ v^2,\ p/\rho_o,\ |t_{\hat{\imath}\hat{\jmath}}|/\rho_o,\ \Pi.$$

Relative magnitudes of expansion parameters

(The Newtonian potential at the center of the sun is $\epsilon^2 \sim 10^{-5}$. The values of p/ρ_o, t_{ij}/ρ_o, and Π there are also $\sim 10^{-5}$, and they are much smaller elsewhere. The orbital velocities of the planets are all less than $100\ \mathrm{km/sec} = 3 \times 10^{-4}$, so $v^2 < 10^{-7}$.) Moreover, changes with time of all quantities at fixed x_j are due primarily to the motion of the matter. As a result, time derivatives are small by $O(\epsilon)$ compared to space derivatives,

$$\left| \frac{\partial A/\partial t}{\partial A/\partial x_j} \right| \sim |v_j| \lesssim \epsilon \text{ for any quantity } A, \tag{39.8}$$

although *not* in the radiation zone, where outgoing gravitational waves flow (distance \gtrsim one light year from Sun). Consequently, the radiation zone must be excluded from the analysis when one makes a post-Newtonian expansion. To treat it requires different techniques, e.g., those of Chapter 36.

Conditions 39.7 and 39.8 suggest that one expand the metric coefficients in powers of the small parameter ϵ, treating U, v^2, p/ρ_o, t_{ij}/ρ_o, and Π as though they were all of $O(\epsilon^2)$ (often they are smaller!), and treating time derivatives as $O(\epsilon)$ smaller than space derivatives.

Rules of the expansion

In this "post-Newtonian" expansion, terms odd in ϵ (i.e., terms such as

$$\int \frac{\rho_o(\mathbf{x}', t) v_j(\mathbf{x}', t)}{|\mathbf{x} - \mathbf{x}'|} d^3x' \sim \frac{M}{R} v \sim \epsilon^3 \tag{39.9}$$

whose total number of v's and $(\partial/\partial t)$'s is odd) change sign under time reversal,

*In the solar system, post-Newtonian corrections due to anisotropic stresses are so much smaller than other post-Newtonian corrections that there is no hope of measuring them in the 1970's. For this reason, elsewhere in the literature (but not in this book) the PPN formalism treats all stresses at the post-Newtonian level as isotropic pressures, thereby setting to zero the PPN parameter η of §§39.8–39.11.

whereas terms even in ϵ do not. Time reversal ($x^{\bar{0}} = -x^0$) also changes the sign of $g_{0j}(g_{\bar{0}j} = -g_{0j})$, but leaves g_{00} and g_{jk} unchanged. Therefore, g_{0j} *must contain only terms odd in* ϵ*; whereas* g_{00} *and* g_{jk} *must contain only even terms.* (Actually, this ceases to be the case when radiation damping enters the picture. In the real world one always insists on outgoing-wave boundary conditions. But time reversal converts outgoing waves to ingoing waves; so an extra sign change is required to convert back to out. Therefore, radiation damping terms in the near-zone metric are even in ϵ for g_{0j}, but odd for g_{00} and g_{jk}. However, radiation damping does not come into play until order ϵ^5 beyond Newtonian theory—see Chapter 36—so it will be ignored here.)

Expanded form of metric

The form of the expansion is already known through Newtonian order (see §17.4, with Φ replaced by $-U$): Newtonian gravity is only obtained when one demands that

$$g_{00} = -1 + 2U + [\text{terms} \lesssim \epsilon^4],$$

$$g_{0j} = [\text{terms} \lesssim \epsilon^3], \tag{39.10}$$

$$g_{ij} = \delta_{ij} + [\text{terms} \lesssim \epsilon^2].$$

The stated limits on the higher-order corrections are dictated by demanding that the space components of the geodesic equation agree with the Newtonian equation of motion:

$$\frac{d^2 x_j}{dt^2} \approx \frac{d^2 x_j}{d\tau^2} = -\Gamma^j{}_{\alpha\beta} \frac{dx^\alpha}{d\tau} \frac{dx^\beta}{d\tau} \approx -\Gamma^j{}_{\alpha\beta} \frac{dx^\alpha}{dt} \frac{dx^\beta}{dt} \tag{39.11}$$

$$= -\Gamma^j{}_{00} - 2\Gamma^j{}_{0k} v_k - \Gamma^j{}_{k\ell} v_k v_\ell$$

$$= U_{,j} + \text{terms of order } \{\epsilon g_{0k,j}; \epsilon^2 g_{k\ell,j}\}.$$

One would get the wrong Newtonian limit if g_{0k} were $O(\epsilon)$ or greater, and if $g_{k\ell} - \delta_{k\ell}$ were $O(1)$ or greater.

The above pattern continues on to all orders in the expansion. Thus in the geodesic equation, and also in the law of local conservation of energy-momentum $T^{\alpha\beta}{}_{;\beta} = 0$, g_{00} always goes hand-in-hand with ϵg_{0k} and $\epsilon^2 g_{jk}$ (see exercise 39.2). Therefore, the post-Newtonian expansion has the form summarized in Box 39.3.

EXERCISES

Exercise 39.1. ORDERS OF MAGNITUDE IN GRAVITATIONALLY BOUND SYSTEMS

Use Newtonian theory to derive conditions (39.7) for any gravitationally bound system. [*Hint:* Such concepts as orbital velocities, the speeds of sound and shear waves, the virial theorem, and hydrostatic equilibrium are relevant.]

Exercise 39.2. PATTERN OF TERMS IN POST-NEWTONIAN EXPANSION

Verify the statements in the paragraph following equation (39.11). In particular, suppose that one wishes to evaluate the coordinate acceleration, $d^2 x_j / dt^2$, to accuracy $\epsilon^{2N} U_{,j}$ for some

integer N. Show that this undertaking requires a knowledge of g_{00} to accuracy ϵ^{2N+2}, of g_{0k} to ϵ^{2N+1}, and of g_{jk} to ϵ^{2N}. Also suppose that one knows T^{00} to accuracy $\rho_o\epsilon^{2N}$, T^{0j} to $\rho_o\epsilon^{2N+1}$, and T^{jk} to $\rho_o\epsilon^{2N+2}$ [see, e.g., equations (39.13) for $N = 0$ and (39.42) for $N = 2$]. Show that to calculate $T^{0\alpha}{}_{;\alpha}$ with accuracy $\epsilon^{2N+1}\rho_{o,j}$ and $T^{j\alpha}{}_{;\alpha}$ with accuracy $\epsilon^{2N+2}\rho_{o,j}$, one must know g_{00} to ϵ^{2N+2}, g_{0k} to ϵ^{2N+1}, and g_{jk} to ϵ^{2N}. This dictates the pattern of Box 39.3.

§39.7. NEWTONIAN APPROXIMATION

At Newtonian order the metric has the form (39.10); and the 4-velocity and stress-energy tensor have components, relative to the PPN coordinate system,

Newtonian approximation

$$u^0 = +1 + O(\epsilon^2), \qquad u^j = v_j + O(\epsilon^3); \tag{39.12}$$

$$T^{00} = \rho_o + O(\rho_o\epsilon^2), \qquad T^{0j} = \rho_o v_j + O(\rho_o\epsilon^3),$$

$$T^{jk} = t_{\hat{j}\hat{k}} + \rho_o v_j v_k + O(\rho_o\epsilon^4) \tag{39.13}$$

(see exercise 39.3). Two sets of equations govern the structure and evolution of the solar system. (1) The Einstein field equations. As was shown in §18.4, and also in §17.4, in the Newtonian limit Einstein's equations reduce to Laplace's equation

$$U_{,jj} = -4\pi\rho_o, \tag{39.14a}$$

which has the "action-at-a-distance" solution

$$U(\boldsymbol{x}, t) = \int \frac{\rho_o(\boldsymbol{x}', t)}{|\boldsymbol{x} - \boldsymbol{x}'|} \, d^3x'. \tag{39.14b}$$

Box 39.3 POST-NEWTONIAN EXPANSION OF THE METRIC COEFFICIENTS

Level of approximation (and papers expanding general relativity to this level)	Order or value of terms		
	g_{00}	g_{0j}	g_{jk}
flat, empty spacetime	-1	0	δ_{jk}
Newtonian approximation	$2U$	0	0
post-Newtonian approximation [(Fock (1959); Chandrasekhar (1965a)]	$+$ terms $\sim \epsilon^4$	$+$ terms $\sim \epsilon^3$	$+$ terms $\sim \epsilon^2$
post-post-Newtonian approximation [Chandrasekhar and Nutku (1969)]	$+$ terms $\sim \epsilon^6$	$+$ terms $\sim \epsilon^5$	$+$ terms $\sim \epsilon^4$
radiation damping [Chandrasekhar and Esposito (1970)]	$+$ terms $\sim \epsilon^7$	$+$ terms $\sim \epsilon^6$	$+$ terms $\sim \epsilon^5$

(2) The law of local energy-momentum conservation, $T^{\alpha\beta}{}_{;\beta} = 0$. The time component of this law reduces to the conservation of rest mass

$$\partial\rho_o/\partial t + \partial(\rho_o v_j)/\partial x_j = 0 + \text{fractional errors of } O(\epsilon^2); \qquad (39.15a)$$

and the space components reduce to Newton's second law of motion, "$F = ma$":

$$\rho_o \, dv_j/dt = \rho_o(\partial U/\partial x_j) - \partial t_{\hat{j}\hat{k}}/\partial x_k + \text{fractional errors of } O(\epsilon^2), \quad (39.15b)$$

$$d/dt \equiv (\text{time derivative following the matter}) \equiv \partial/\partial t + v_k \, \partial/\partial x_k \quad (39.16)$$

(see exercise 39.3).

Equations (39.14)–(39.16), together with equations of state describing the planetary and solar matter, are the foundations for all Newtonian calculations of the structure and motion of the sun and planets. Notice that the internal energy density $\rho_o \Pi$ nowhere enters into these equations. It is of no importance to Newtonian hydrodynamics. It matters for the sun's thermal-energy balance; but that is irrelevant here.

EXERCISES

Exercise 39.3. NEWTONIAN APPROXIMATION

(a) Derive equations (39.13) for the components of the stress-energy tensor in the PPN coordinate frame. [*Hint:* In the rest frame of the matter ("comoving orthonormal frame") $T_{\hat{0}\hat{0}} = \rho = \rho_o + O(\epsilon^2)$, $T_{\hat{0}\hat{j}} = 0$, $T_{\hat{j}\hat{k}} = t_{\hat{j}\hat{k}}$; see equations (39.5). Lorentz-transform these components by a pure boost with ordinary velocity $-v_j$ to obtain $T_{\alpha\beta}$.]

(b) Show that, in the PPN coordinate frame, $T^{0\alpha}{}_{;\alpha} = 0$ reduces to equation (39.15a), and $T^{j\alpha}{}_{;\alpha} = 0$, when combined with (39.15a), reduces to equation (39.15b).]

Exercise 39.4. A USEFUL FORMULA

Derive from equations (39.15) the following useful formula, valid for any function $f(\mathbf{x}, t)$:

$$\frac{d}{dt} \int \rho_o(\mathbf{x}, t) f(\mathbf{x}, t) \, d^3x = \int \rho_o(\mathbf{x}, t) \frac{df(\mathbf{x}, t)}{dt} \, d^3x \qquad (39.17)$$

$$+ \text{ fractional errors of } O(\epsilon^2).$$

Here both integrals are extended over all of space; and df/dt is the derivative following the matter (39.16).

Exercise 39.5. STRESS TENSOR FOR NEWTONIAN GRAVITATIONAL FIELD

Define a "stress tensor for the Newtonian gravitational field U" as follows:

$$t_{jk} \equiv \frac{1}{4\pi}\left(U_{,j}U_{,k} - \frac{1}{2}\delta_{jk}U_{,l}U_{,l}\right). \qquad (39.18)$$

Show that the equations of motion for the matter (39.15b) can be rewritten in the forms

$$\rho_o \frac{dv_j}{dt} = -\frac{\partial}{\partial x^k}(t_{jk} + t_{\hat{j}\hat{k}}) + \text{fractional errors of } O(\epsilon^2), \qquad (39.19)$$

$$(\rho_o v_j)_{,t} + (t_{jk} + t_{\hat{j}\hat{k}} + \rho_o v_j v_k)_{,k} = 0 + \text{fractional errors of } O(\epsilon^2). \qquad (39.19')$$

Exercise 39.6. NEWTONIAN VIRIAL THEOREMS

(a) From equation (39.19') show that

$$d^2 I_{jk}/dt^2 = 2 \int (\mathfrak{t}_{jk} + \mathfrak{t}_{\hat{j}k} + \rho_o v_j v_k) \, d^3x + \text{fractional errors of } O(\epsilon^2), \quad (39.20a)$$

where I_{jk} is the second moment of the system's mass distribution,

$$I_{jk} = \int \rho_o x_j x_k \, d^3x.$$

This is called the "*time-dependent tensor virial theorem.*"

 (b) From this infer that, if $\langle \quad \rangle_{\text{long time}}$ denotes an average over a long period of time, then

$$\left\langle \int (\mathfrak{t}_{jk} + \mathfrak{t}_{\hat{j}k} + \rho_o v_j v_k) \, d^3x \right\rangle_{\text{long time}} = O\left(\int \rho_o \epsilon^4 \, d^3x \right). \quad (39.20b)$$

This is called the "*tensor virial theorem.*"

 (c) By contraction of indices and use of equations (39.18), (39.14a), and (39.5e), derive the (ordinary) *virial theorems*:

$$\frac{1}{2} d^2 I/dt^2 = \int \rho_o v^2 \, d^3x - \int \frac{1}{2} \rho_o U \, d^3x + 3 \int p \, d^3x + \text{fractional errors of } O(\epsilon^2), \quad (39.21a)$$

where I is the trace of the second moment of the mass distribution

$$I = I_{jj} = \int \rho_o r^2 \, d^3x;$$

and

$$\left\langle \underbrace{\int \rho_o v^2 \, d^3x}_{2 \times \binom{\text{kinetic}}{\text{energy}}} - \underbrace{\frac{1}{2} \int \rho_o U \, d^3x}_{+\binom{\text{gravitational}}{\text{energy}}} + \underbrace{3 \int p \, d^3x}_{+3 \times \binom{\text{pressure}}{\text{integral}}} \right\rangle_{\text{long time}} = O\left(\int \rho_o \epsilon^4 \, d^3x \right). \quad (39.21b)$$

Exercise 39.7. PULSATION FREQUENCY FOR NEWTONIAN STAR

Use the ordinary, time-dependent virial theorem (39.21a) to derive the following equation for the fundamental angular frequency of pulsation of a nonrotating, Newtonian star:

$$\omega^2 = (3\bar{\Gamma}_1 - 4) \frac{|\text{star's self-gravitational energy}|}{(\text{trace of second moment of star's mass distribution})}; \quad (39.22a)$$

$$\bar{\Gamma}_1 = \binom{\text{pressure-weighted average}}{\text{of adiabatic index}} = \frac{\int \Gamma_1 p \, d^3x}{\int p \, d^3x}. \quad (39.22b)$$

In the derivation assume that the pulsations are "homologous"—i.e., that a fluid element with equilibrium position x^j (relative to center of mass $x^j = 0$) gets displaced to $x^j + \xi^j(x, t)$, where

$$\xi^j = (\text{small constant}) x^j e^{-i\omega t}.$$

Assume nothing else. *Notes:* (1) The result (39.22) was derived differently in Box 26.2 and used in §24.4. (2) The assumption of homologous pulsation is fully justified if $|\Gamma_1 - 4/3| = \text{constant} \ll 1$; see Box 26.2. (3) The result (39.22) is readily generalized to slowly

rotating Newtonian stars; see, e.g., Chandrasekhar and Lebovitz (1968). It can also be generalized to nonrotating post-Newtonian stars using general relativity (Box 26.2), or using the PPN formalism for any metric theory [Ni (1973)]. And it can be generalized to slowly rotating, post-Newtonian stars [see, e.g., Chandrasekhar and Lebovitz (1968)].

§39.8. PPN METRIC COEFFICIENTS

Post-Newtonian corrections to metric:

The post-Newtonian corrections $k_{\alpha\beta}$ to the metric coefficients $g_{\alpha\beta}$ are calculated, in any metric theory of gravity, by lengthy manipulations of the field equations. (See, e.g., exercise 39.14 near the end of this chapter for general relativity.) But without ever picking some one theory, and without ever writing down any set of field equations, one can infer the *forms* of the post-Newtonian corrections $k_{\alpha\beta}$. Their

(1) rules governing forms

forms are fixed by the following constraints: (1) They must be of post-Newtonian order ($k_{00} \sim \epsilon^4$, $k_{0j} \sim \epsilon^3$, $k_{ij} \sim \epsilon^2$). (2) They must be dimensionless. (3) k_{00} must be a scalar under rotations, k_{0j} must be components of a 3-vector, and k_{jk} must be components of a 3-tensor. (4) The corrections must die out at least as fast as $1/r$ far from the solar system, so that the coordinates become globally Lorentz and spacetime becomes flat at $r = \infty$. (5) For simplicity, one can assume that the metric components are generated only by ρ_o, $\rho_o\Pi$, t_{ij}, p, products of these with the velocity v_j, and time-derivatives of such quantities;* but not by their spatial gradients. [This assumption of simplicity is satisfied by all metric theories examined up to 1973, except Whitehead (1922) and theories reviewed by Will (1973)—which disagree with experiment.] Note the further justification for this assumption in exercise 39.8.

(2) construction of corrections

Begin with the corrections to the spatial components, $k_{ij} \sim \epsilon^2$. There are only two functionals of ρ_o, p, Π, t_{jk}, v_j, that die out at least as fast as $1/r$, are dimensionless, are $O(\epsilon^2)$, and are second-rank, symmetric 3-tensors; they are

$$\delta_{ij}U(\boldsymbol{x}, t); \quad U_{ij}(\boldsymbol{x}, t) = \int \frac{\rho_o(\boldsymbol{x}', t)(x_i - x_i')(x_j - x_j')}{|\boldsymbol{x} - \boldsymbol{x}'|^3}\, d^3x'. \tag{39.23a}$$

Thus, k_{ij} must be $k_{ij} = 2\gamma\,\delta_{ij}U + 2\Gamma U_{ij}$, for some constant "PPN parameters" γ and Γ. By an infinitesimal coordinate transformation [$x_{i\,\text{NEW}} = x_{i\,\text{OLD}} + \Gamma\,\partial\chi/\partial x_i$, with $\chi(\boldsymbol{x}, t) = -\int\rho_o(\boldsymbol{x}', t)|\boldsymbol{x} - \boldsymbol{x}'|\, d^3x'$] one can set $\Gamma = 0$, thereby obtaining

$$g_{ij} = \delta_{ij} + k_{ij} = \delta_{ij}(1 + 2\gamma U) + O(\epsilon^4). \tag{39.23b}$$

*One allows for time derivatives because retarded integrals contain such terms when expanded to post-Newtonian order; thus,

$$\int \frac{\rho_o(\boldsymbol{x}', t - |\boldsymbol{x} - \boldsymbol{x}'|)}{|\boldsymbol{x} - \boldsymbol{x}'|}\, d^3x' = \int \left[\frac{\rho_o(\boldsymbol{x}', t)}{|\boldsymbol{x} - \boldsymbol{x}'|} - \frac{\partial\rho_o(\boldsymbol{x}', t)}{\partial t} + \cdots\right] d^3x'.$$

However, it turns out that, with a suitable choice of coordinates ("gauge"), all time-odd retarded terms [e.g., $\int(\partial\rho_o/\partial t)\, d^3x$] vanish, except at "the post$^{5/2}$-Newtonian order" and at higher orders of approximation; there they lead to radiation damping (see Box 39.3). For example, $\int(\partial\rho_o/\partial t)\, d^3x = (d/dt)\int\rho_o\, d^3x$ vanishes by virtue of the conservation of baryon number.

Next consider $k_{0j} \sim \epsilon^3$. Trial and error yield only two vector functionals that die out as $1/r$ or faster, are dimensionless, and are $O(\epsilon^3)$. They are

$$V_j(x, t) = \int \frac{\rho_o(x', t)v_j(x', t)}{|x - x'|} d^3x', \tag{39.23c}$$

$$W_j(x, t) = \int \frac{\rho_o(x', t)[(x - x') \cdot v(x', t)](x_j - x_j') d^3x'}{|x - x'|^3}. \tag{39.23d}$$

Thus, k_{0j} must be a linear combination of these, involving unknown constants (PPN parameters) Δ_1 and Δ_2:

$$g_{0j} = k_{0j} = -\frac{7}{2}\Delta_1 V_j - \frac{1}{2}\Delta_2 W_j + O(\epsilon^5). \tag{39.23e}$$

Finally consider $k_{00} \sim \epsilon^4$. Trial and error yields a variety of terms, which can all be combined together with the Newtonian part of g_{00} to give

$$g_{00} = -1 + 2U + k_{00} = -1 + 2U - 2\beta U^2 + 4\Psi - \zeta\mathcal{C} - \eta\mathcal{D}, \tag{39.23f}$$

where

$$\Psi(x, t) = \int \frac{\rho_o(x', t)\psi(x', t)}{|x - x'|} d^3x',$$

$$\psi = \beta_1 v^2 + \beta_2 U + \frac{1}{2}\beta_3 \Pi + \frac{3}{2}\beta_4 p/\rho_o, \tag{39.23g}*$$

$$\mathcal{C}(x, t) = \int \frac{\rho_o(x', t)[(x - x') \cdot v(x', t)]^2}{|x - x'|^3} d^3x', \tag{39.23h}$$

$$\mathcal{D}(x, t) = \int \frac{\left[t_{jk}(x', t) - \frac{1}{3}\delta_{jk} t_{ll}(x', t)\right](x_j - x_j')(x_k - x_k')}{|x - x'|^3} d^3x'. \tag{39.23i}$$

Also, β, β_1, β_2, β_3, β_4, ζ, η are unknown constants (PPN parameters). Elsewhere in the literature the term $-\eta\mathcal{D}$ in g_{00} is ignored (see footnote on p. 1075).

Yet another term is possible: one could have set

$$g_{00} = [\text{value in equation (39.23f)}]$$

$$- \Sigma \int\int \frac{\rho_o(x', t)\rho_o(x'', t)[(x - x') \cdot (x' - x'')]}{|x - x'||x' - x''|^3} d^3x' d^3x'', \tag{39.24}$$

where Σ is another PPN parameter. [It can be shown, using the Newtonian equations (39.14)–(39.16), that this expression dies out as $1/r$ far from the solar system.] If

*WARNING: *Throughout* the literature the notation Φ is used where we use Ψ for the functional (39.23g), and ϕ is used for our ψ. We are forced to violate the standard notation to avoid confusion with the Newtonian potential $\Phi = -U$. However, we urge that nobody else violate the standard notation!

such a Σ term had been included, then one could have removed it by making the infinitesimal coordinate transformation

$$x^0{}_{new} = x^0{}_{old} - \frac{1}{2}\,\Sigma \int \frac{\rho_o(\mathbf{x}',t)[(\mathbf{x}-\mathbf{x}')\cdot\mathbf{v}(\mathbf{x}',t)]}{|\mathbf{x}-\mathbf{x}'|}\,d^3x' \qquad (39.25)$$

(see exercise 39.9). Thus, there is no necessity to include the Σ term.

Rigidity of coordinate system

The absence of the Σ term from g_{00} means that the time coordinate has been fixed rigidly up through post-Newtonian order:

$$x^0 \text{ has uncertainties only of } O(R_\odot\epsilon^5) \sim 10^{-14} \text{ seconds.} \qquad (39.26a)$$

The space coordinates are also fixed rigidly through post-Newtonian order:

$$x^j \text{ has uncertainties only of } O(R_\odot\epsilon^4) \sim 0.1 \text{ cm,} \qquad (39.26b)$$

because any transformation of the form

$$x^j{}_{new} = x^j{}_{old} + \text{position-dependent terms of } O(\epsilon^2 R_\odot)$$

would destroy the form (39.23b) of the space part of the metric.

Summary of PPN metric and parameters

In summary, for almost every metric theory of gravity yet invented, accurate through post-Newtonian order the metric coefficients have the form (39.23). One theory is distinguished from another by the values of its ten "post-Newtonian parameters" β, β_1, β_2, β_3, β_4, γ, ζ, η, Δ_1 and Δ_2. These are determined by comparing the field equations of the given theory with the form (39.23) of the post-Newtonian metric. The parameter values for general relativity and for several other theories are given in Box 39.2, along with a heuristic description of each parameter.

EXERCISES

Exercise 39.8. ABSENCE OF "METRIC-GENERATES-METRIC" TERMS IN POST-NEWTONIAN LIMIT

In writing down the post-Newtonian metric corrections, one might be tempted to include terms that are generated by the Newtonian potential acting alone, without any *direct* aid from the matter. After all, general relativity and other metric theories are nonlinear; so the two-step process (matter) $\longrightarrow U \longrightarrow$ (post-Newtonian metric corrections) seems quite natural. *Show* that such terms are not needed, because the equations (39.14)–(39.16) of the Newtonian approximation enable one to reexpress them in terms of direct integrals over the matter distribution. In particular, show that

$$\int \frac{\partial^2 U(\mathbf{x}',t)/\partial x_j'\,\partial t}{|\mathbf{x}-\mathbf{x}'|}\,d^3x' = 2\pi[V_j(\mathbf{x},t) - W_j(\mathbf{x},t)] \qquad (39.27)$$

where V_j and W_j are defined by equations (39.23c,d); also show that

$$\int \frac{[\partial U(\mathbf{x}',t)/\partial x_j'][\partial U(\mathbf{x}',t)/\partial x_j']}{|\mathbf{x}-\mathbf{x}'|}\,d^3x'$$

$$= -2\pi[U(\mathbf{x},t)]^2 + 4\pi \int \frac{\rho_o(\mathbf{x}',t)U(\mathbf{x}',t)}{|\mathbf{x}-\mathbf{x}'|}\,d^3x'. \quad (39.28)$$

Note that the terms on the righthand sides of (39.27) and (39.28) are already included in the expressions (39.23e,f) for g_{0j} and g_{00}.

Exercise 39.9. REMOVAL OF Σ TERM FROM g_{00}

Show that the coordinate transformation (39.25) removes the Σ term from the metric coefficient g_{00} of equation (39.24), as claimed in the text.

Exercise 39.10. VERIFICATION OF FORMS OF POST-NEWTONIAN CORRECTIONS

Verify the claims in the text immediately preceding equations (39.23a,b,c,f).

§39.9. VELOCITY OF PPN COORDINATES RELATIVE TO "UNIVERSAL REST FRAME"

Thus far it has been assumed tacitly that the center of mass of the solar system is at rest in the PPN coordinate system. Is this really a permissible assumption? Put differently, can one always so adjust the PPN coordinate system that its origin moves with any desired velocity (e.g., that of the solar system); or is the PPN coordinate system rigidly and irrevocably attached to some "universal rest frame"?

In general relativity, the geometry of curved spacetime picks out no preferred coordinate frames (except in cases with special symmetry). Therefore, one expects the velocity of the PPN coordinate frame to be freely specifiable. Put differently, one expects the entire PPN formalism, for general relativity, to be invariant under Lorentz transformations of the PPN coordinates [combined, perhaps, with "infinitesimal coordinate transformations" to maintain the gauge conditions that the "Σ" and "U_{jk}" terms of (39.24) and (39.23a) be absent]. By contrast, in Ni's theory of gravity (Box 39.1) the geometry of spacetime *always* picks out a preferred coordinate frame: the "rest frame of the universe." One would not be surprised, in this case, to find the PPN coordinate frame rigidly attached to the universal rest frame.

Preferred-frame theories of gravity

The above intuition is correct, according to calculations by Will (1971d) and by Will and Nordtvedt (1972). When dealing with general relativity and other theories with little or no "prior geometry," one can freely specify the velocity of the PPN coordinate system (at some initial instant of time). But for theories like Ni's, with a preferred "universal rest frame" ("preferred-frame theories"), only in the preferred frame can the post-Newtonian metric assume the form derived in the last section [equations (39.23)]. This restriction on the PPN metric does not mean that one is confined, in preferred-frame theories, to perform all calculations in the universal rest frame. Rather, it means that for such theories the PPN metric requires generalization to take account of coordinate-frame motion relative to the universal rest frame.

The required generalization can be achieved by subjecting the PPN metric (39.23) to (1) a Lorentz boost from the preferred frame $\{x_{\text{OLD}}^{\alpha}\}$ to a new PPN frame $\{x_{\text{NEW}}^{\alpha}\}$, which moves with velocity w, plus (2) a change of gauge designed to keep the metric coefficients as simple as possible. The boost-plus-gauge-change is [Will and Nordtvedt (1972)]

Generalization of PPN metric to moving frames

$$x_{\text{OLD}} = x_{\text{NEW}} + \frac{1}{2}(x_{\text{NEW}} \cdot w)w + \left(1 + \frac{1}{2}w^2\right)wt_{\text{NEW}} \tag{39.29a}$$
$$+ O(\epsilon^5 t_{\text{NEW}} + \epsilon^4 x_{\text{NEW}}),$$

$$t_{\text{OLD}} = t_{\text{NEW}}\left(1 + \frac{1}{2}w^2 + \frac{3}{8}w^4\right) + \left(1 + \frac{1}{2}w^2\right)x_{\text{NEW}} \cdot w$$
$$+ \underbrace{\left(\frac{1}{2}\Delta_2 + \zeta - 1\right)w_j \frac{\partial \chi}{\partial x_{\text{NEW}}^j}}_{\text{[gauge change]}} + O(\epsilon^6 t_{\text{NEW}} + \epsilon^5 x_{\text{NEW}}), \tag{39.29b}$$

$$\chi(t_{\text{NEW}}, x_{\text{NEW}}) \equiv -\int \rho_o(t_{\text{NEW}}, x'_{\text{NEW}})|x_{\text{NEW}} - x'_{\text{NEW}}|\, d^3x'_{\text{NEW}}. \tag{39.29c}$$

[*Note:* One insists, in the spirit of the post-Newtonian approximation, that the velocity w of the new PPN frame relative to the universal rest frame be no larger than the characteristic internal velocities of the system:

$$|w| \lesssim \epsilon.] \tag{39.30}$$

This change of coordinates produces corresponding changes in the velocity of the matter

$$v_{\text{OLD}} = \frac{dx_{\text{OLD}}}{dt_{\text{OLD}}} = v_{\text{NEW}}\left(1 - w \cdot v_{\text{NEW}} - \frac{1}{2}w^2\right)$$
$$+ w\left(1 - \frac{1}{2}w \cdot v_{\text{NEW}}\right) + O(\epsilon^5). \tag{39.31}$$

A long but straightforward calculation (exercise 39.11) yields the following components for the metric in the new PPN coordinates. [Note: The subscripts NEW are here and hereafter dropped from the notation.]

Final form of metric

$$g_{jk} = \delta_{jk}(1 + 2\gamma U) + O(\epsilon^4), \tag{39.32a}$$

$$g_{0j} = -\frac{7}{2}\Delta_1 V_j - \frac{1}{2}\Delta_2 W_j + \left(\alpha_2 - \frac{1}{2}\alpha_1\right)w_j U - \alpha_2 w_k U_{kj} + O(\epsilon^5), \tag{39.32b}$$

$$g_{00} = -1 + 2U - 2\beta U^2 + 4\Psi - \zeta \mathcal{A} - \eta \mathcal{D}$$
$$+ (\alpha_2 + \alpha_3 - \alpha_1)w^2 U + (2\alpha_3 - \alpha_1)w_j V_j - \alpha_2 w_j w_k U_{jk} + O(\epsilon^6). \tag{39.32c}$$

Here α_1, α_2, and α_3 are certain combinations of PPN parameters

$$\alpha_1 = 7\Delta_1 + \Delta_2 - 4\gamma - 4, \tag{39.33a}$$

$$\alpha_2 = \Delta_2 + \zeta - 1, \tag{39.33b}$$

$$\alpha_3 = 4\beta_1 - 2\gamma - 2 - \zeta. \tag{39.33c}$$

The "gravitational potentials" U, V_j, W_j, Ψ, \mathcal{A}, and \mathcal{D} appearing here are to be calculated in the new, "moving" PPN coordinate system by the same prescriptions

as one used in the universal rest frame. Thus, their functional forms are the same as previously, but their values at any given event are different (see exercise 39.11):

$$U(x, t) = \int \frac{\rho_o(x', t)}{|x - x'|} d^3x';$$ (39.34a)

$$V_j(x, t) = \int \frac{\rho_o(x', t)v_j(x', t)}{|x - x'|} d^3x';$$ (39.34b)

$$W_j(x, t) = \int \frac{\rho_o(x', t)[(x - x') \cdot v(x', t)](x_j - x_j') \, d^3x'}{|x - x'|^3};$$ (39.34c)

$$\Psi(x, t) = \int \frac{\rho_o(x', t)\psi(x', t)}{|x - x'|} d^3x',$$

$$\psi = \beta_1 v^2 + \beta_2 U + \frac{1}{2}\beta_3 \Pi + \frac{3}{2}\beta_4 p/\rho_o;$$ (39.34d)

$$\mathcal{A}(x, t) = \int \frac{\rho_o(x', t)[(x - x') \cdot v(x', t)]^2}{|x - x'|^3} d^3x';$$ (39.34e)

$$\mathcal{D}(x, t) = \int \frac{\left[t_{jk}(x', t) - \frac{1}{3}\delta_{jk}t_{ll}(x', t)\right](x_j - x_j')(x_k - x_k')}{|x - x'|^3} d^3x'.$$ (39.34f)

The quantity U_{jk} is the gravitational potential defined in equation (39.23a):

$$U_{jk}(x, t) = \int \frac{\rho_o(x', t)(x_j - x_j')(x_k - x_k')}{|x - x'|^3} d^3x'.$$ (39.34g)

Notice that the velocity w of the PPN coordinate system relative to the universal rest frame appears explicitly in the PPN metric only if one or more of the coefficients $\alpha_1, \alpha_2, \alpha_3$, is nonzero. Thus, theories with $\alpha_1 = \alpha_2 = \alpha_3 = 0$ (e.g., general relativity) possess *no* preferred universal rest frame in the post-Newtonian limit; all their PPN frames are "created equal." By contrast, theories with at least one of $\alpha_1, \alpha_2, \alpha_3$, nonzero (e.g., Ni's theory) *do* possess a preferred frame.

The generalized form (39.32) of the PPN metric, by virtue of the process used to construct it, is invariant under a Lorentz boost plus a gauge adjustment ["Post-Galilean transformation"; see Chandrasekhar and Contopolous (1967)]:

Post-Galilean invariance

$$x_{\text{OLD}} = x_{\text{NEW}} + \frac{1}{2}(x_{\text{NEW}} \cdot \beta)\beta + \left(1 + \frac{1}{2}\beta^2\right)\beta t_{\text{NEW}}$$
$$+ O(\epsilon^5 t_{\text{NEW}} + \epsilon^4 x_{\text{NEW}}),$$ (39.35)

$$t_{\text{OLD}} = \left(1 + \frac{1}{2}\beta^2 + \frac{3}{8}\beta^4\right)t_{\text{NEW}} + \left(1 + \frac{1}{2}\beta^2\right)x_{\text{NEW}} \cdot \beta$$
$$+ \left(\frac{1}{2}\Delta_2 + \zeta_1 - 1\right)\beta \cdot \nabla_{\text{NEW}}\chi + O(\epsilon^6 t_{\text{NEW}} + \epsilon^5 x_{\text{NEW}}).$$

Of course, it is also invariant under spatial rotations.

Exercise 39.11. TRANSFORMATION TO MOVING FRAME

Show that the change of coordinates (39.29) changes the PPN metric coefficients from the form (39.23) to the form (39.32). [*Hints:* (1) Keep firmly in mind the fact that the potentials U, V_j, W_j, \mathcal{C}, and \mathcal{D} are not scalar fields. Each coordinate system possesses its own potentials. For example, by using equations (39.29) in the integral for U_{OLD}, one finds

$$U_{\text{OLD}}(x_{\text{OLD}}, t_{\text{OLD}}) = \int \frac{\rho_o(x'_{\text{OLD}}, t_{\text{OLD}})}{|x_{\text{OLD}} - x'_{\text{OLD}}|}\, d^3x'_{\text{OLD}} \tag{39.36}$$

$$= \left[U_{\text{NEW}} - w_j(V_{j\,\text{NEW}} - W_{j\,\text{NEW}}) + \frac{1}{2} w_j w_k X_{,jk} \right]_{x_{\text{NEW}}, t_{\text{NEW}}} + O(\epsilon^6).$$

(2) The law of baryon conservation (39.44) may be useful.]

§39.10. PPN STRESS-ENERGY TENSOR

The motion of the solar system is governed by the equations $T^{\alpha\beta}{}_{;\beta} = 0$. Before studying them, one must calculate the post-Newtonian corrections to the stress-energy tensor in the PPN coordinate frame. This requires a transformation from the comoving, orthonormal frame $\boldsymbol{w}^{\hat{\alpha}}$, where

$$T^{\hat{0}\hat{0}} = \rho_o(1 + \Pi), \qquad T^{\hat{0}\hat{j}} = 0, \qquad T^{\hat{j}\hat{k}} = t_{\hat{j}\hat{k}}, \tag{39.37}$$

to the coordinate frame. One can effect this transformation in two stages: stage 2 is a transformation

$$\boldsymbol{w}^{\tilde{0}} \equiv [1 - U + O(\epsilon^4)]\, \boldsymbol{dt} \tag{39.38a}$$

$$+ \left[\frac{7}{2} \Delta_1 V_j + \frac{1}{2} \Delta_2 W_j + \left(\frac{1}{2}\alpha_1 - \alpha_2 \right) w_j U + \alpha_2 w_k U_{kj} + O(\epsilon^5) \right] \boldsymbol{dx}^j,$$

$$\boldsymbol{w}^{\tilde{j}} \equiv [(1 + \gamma U)\, \delta_{jk} + O(\epsilon^4)]\, \boldsymbol{dx}^k + O(\epsilon^5)\, \boldsymbol{dt}, \tag{39.38b}$$

between the coordinate frame and an orthonormal frame attached to it; stage 1 is a pure Lorentz transformation (boost) between the two orthonormal frames $\boldsymbol{w}^{\tilde{\alpha}}$ and $\boldsymbol{w}^{\hat{\alpha}}$. The 4-velocity of this boost is minus the 4-velocity of the matter, which has components

$$u^j = v_j u^0, \qquad u^0 = 1 + \frac{1}{2} v^2 + U + O(\epsilon^4) \qquad \text{in coord. frame;} \tag{39.39}$$

$$\left.\begin{array}{c} u^{\tilde{j}} = v_{\tilde{j}} u^{\tilde{0}}, \qquad u^{\tilde{0}} = 1 + \frac{1}{2} v^2 + O(\epsilon^4), \\[2mm] v_{\tilde{j}} = v_j[1 + (1 + \gamma)U] \end{array}\right\} \text{ in } \boldsymbol{w}^{\tilde{\alpha}} \text{ frame.} \tag{39.40}$$

Combining the boost, which has ordinary velocity $\beta_{\tilde{j}} = -v_{\tilde{j}}$, with the transformation (39.38), and then inverting, one obtains the result (exercise 39.12)

$$\boldsymbol{dx}^\alpha = A^\alpha{}_\beta \boldsymbol{\omega}^{\hat{\beta}}, \quad \begin{cases} \boldsymbol{\omega}^{\hat{\beta}} = \text{orthonormal comoving basis,} \\ \boldsymbol{dx}^\alpha = \text{PPN coordinate basis;} \end{cases}$$

Transformation from rest frame of matter to PPN coordinate frame

$$A^0{}_{\hat{0}} = 1 + \frac{1}{2}v^2 + U + O(\epsilon^4),$$

$$A^0{}_{\hat{j}} = v_j\left[1 + \frac{1}{2}v^2 + (2+\gamma)U\right] - \frac{7}{2}\Delta_1 V_j - \frac{1}{2}\Delta_2 W_j$$
$$+ \left(\alpha_2 - \frac{1}{2}\alpha_1\right)w_j U - \alpha_2 w_k U_{kj} + O(\epsilon^5), \quad (39.41)$$

$$A^j{}_{\hat{0}} = v_j\left[1 + \frac{1}{2}v^2 + U\right] + O(\epsilon^5),$$

$$A^j{}_{\hat{k}} = (1 - \gamma U)\delta_{jk} + \frac{1}{2}v_j v_k + O(\epsilon^4).$$

This transformation, when applied to the stress-energy tensor (39.37) yields, in the PPN coordinate frame,

Stress-energy tensor in coordinate frame

$$T^{00} = \rho_o(1 + \Pi + v^2 + 2U) + O(\rho_o\epsilon^4), \quad (39.42a)$$

$$T^{0j} = \rho_o(1 + \Pi + v^2 + 2U)v_j + t_{\hat{j}\hat{m}}v_m + O(\rho_o\epsilon^5), \quad (39.42b)$$

$$T^{jk} = t_{\hat{j}\hat{k}}(1 - 2\gamma U) + \rho_o(1 + \Pi + v^2 + 2U)v_j v_k$$
$$+ \frac{1}{2}(v_j t_{\hat{k}\hat{m}}v_m + v_k t_{\hat{j}\hat{m}}v_m) + O(\rho_o\epsilon^6). \quad (39.42c)$$

Exercise 39.12. THE TRANSFORMATION BETWEEN COMOVING FRAME AND PPN FRAME

EXERCISE

Carry out the details of the derivation of the transformation matrix (39.41); and in the process calculate the correction of $O(\epsilon^4)$ to $A^0{}_{\hat{0}}$.

§39.11. PPN EQUATIONS OF MOTION

The post-Newtonian corrections to the Newtonian equations of motion (39.15) and (39.16) are derived from the law of conservation of baryon number $(\rho_o u^\alpha)_{;\alpha} = 0$, and from the law of conservation of local energy-momentum, $T^{\alpha\beta}{}_{;\beta} = 0$. The simplest of the equations of motion is the conservation of baryon number. Its exact expression is $(\rho_o u^\alpha)_{;\alpha} = (1/\sqrt{-g})(\sqrt{-g}\rho_o u^\alpha)_{,\alpha} = 0$. Define a new quantity

$$\rho^* \equiv \rho_o\left(1 + \frac{1}{2}v^2 + 3\gamma U\right) \quad (39.43)$$
$$= \rho_o u^0 \sqrt{-g} + O(\rho_o\epsilon^4)$$

[see (39.39) for u^0, and (39.32) for the metric]. Then rest-mass conservation takes on the same form as at the Newtonian order (39.15a), except now it is more accurate:

Law of baryon conservation

$$\rho^*_{,t} + (\rho^* v_j)_{,j} = 0 + \text{errors of } O(\rho_{o,j}\epsilon^5). \tag{39.44}$$

The next simplest equation of motion is $T^{0\alpha}_{\ \ ;\alpha} = 0$. Straightforward evaluation, using the metric of equations (39.32) and the stress-energy tensor of equations (39.42), yields

$$[\rho_o(1 + \Pi + v^2 + 2U)]_{,t} + [\rho_o(1 + \Pi + v^2 + 2U)v_j + t_{\hat{j}\hat{m}}v_m]_{,j}$$
$$+ (3\gamma - 2)\rho_o U_{,t} + (3\gamma - 3)\rho_o v_k U_{,k} = O(\rho_{o,j}\epsilon^5). \tag{39.45}$$

By subtracting equation (39.44) from this, and using the Newtonian equations of motion (39.15) and (39.16) to simplify several terms where the Newtonian approximation is adequate, one obtains

Law of energy conservation

$$\rho_o \, d\Pi/dt + t_{\hat{j}\hat{k}}v_{j,k} = 0 + \text{errors of } O(\rho_{o,j}\epsilon^5). \tag{39.46}$$

Notice that this is nothing but the first law of thermodynamics (local energy conservation) with energy flow through the matter being neglected. (Neglecting energy flow was justified in §39.5.) This first law of thermodynamics is actually a post-Newtonian equation in the context of hydrodynamics, rather than a Newtonian equation, because Π does not affect the hydrodynamic motion at Newtonian order (see §39.7).

The last of the equations of motion, $T^{j\alpha}_{\ \ ;\alpha} = 0$, reduces to the post-Newtonian Euler equation

Post-Newtonian Euler equation

$$\rho^* \frac{dv_j}{dt} - \rho^* U_{,j} + [t_{\hat{j}\hat{k}}(1 + 3\gamma U)]_{,k} - t_{\hat{j}k,k}\left(\frac{1}{2}v^2 + \Pi\right) - \frac{t_{\hat{j}\hat{k}}t_{\hat{k}\hat{l},l}}{\rho^*}$$
$$+ \rho^* \frac{d}{dt}\left[(2\gamma + 2)Uv_j - \frac{1}{2}(7\,\Delta_1 + \Delta_2)V_j - \frac{1}{2}\alpha_1 U w_j\right] - v_j\rho^* U_{,t} + v_k t_{\hat{k}\hat{j},t}$$
$$+ \frac{1}{2}\Delta_2\rho^*(V_j - W_j)_{,t} + \frac{1}{2}\rho^*[(7\,\Delta_1 + \Delta_2)v_k + (\alpha_1 - 2\alpha_3)w_k]V_{k,j}$$
$$- \rho^*\left[2\Psi - \frac{1}{2}\zeta\mathcal{A} - \frac{1}{2}\eta\mathcal{D} - \frac{1}{2}\alpha_2 w_i w_k U_{ik} + \alpha_2 w_i(V_i - W_i)\right]_{,j} \tag{39.47}$$
$$- \rho^* U_{,j}\left[\gamma v^2 - \frac{1}{2}\alpha_1 \mathbf{w}\cdot\mathbf{v} + \frac{1}{2}(\alpha_2 + \alpha_3 - \alpha_1)w^2 - (2\beta - 2)U + 3\gamma p/\rho^*\right]$$
$$+ \frac{1}{2}(v_{j,k}t_{\hat{k}\hat{m}}v_m - t_{\hat{j}\hat{m}}v_{m,k}v_k) + \frac{1}{2}[v_m(t_{\hat{m}\hat{j}}v_k)_{,k} - v_j(t_{\hat{k}\hat{l}}v_k)_{,l}] = 0.$$

Partial derivatives are denoted by commas; d/dt is the time-derivative following the matter [equation (39.16)].

Equations (39.44), (39.46), and (39.47) are a complete set of equations of motion at the post-Newtonian order.

Exercise 39.13. EQUATIONS OF MOTION

Carry out the details of the derivation of the equations of motion (39.44), (39.46), and (39.47). As part of the derivation, calculate the following values of the Christoffel symbols in the PPN coordinate frame:

$$\Gamma^0{}_{00} = -U_{,t} + O(U_{,j}\epsilon^3), \qquad \Gamma^0{}_{0j} = -U_{,j} + O(U_{,j}\epsilon^2),$$

$$\Gamma^0{}_{jk} = \gamma U_{,t}\,\delta_{jk} + \frac{7}{2}\Delta_1 V_{(j,k)} + \frac{1}{2}\Delta_2 W_{(j,k)} + \left(\frac{1}{2}\alpha_1 - \alpha_2\right) w_{(j}U_{,k)}$$

$$\qquad + \alpha_2 w_i U_{i(j,k)} + O(U_{,j}\epsilon^3).$$

$$\Gamma^j{}_{00} = -U_{,j} + \left[(\beta + \gamma)U^2 - 2\Psi + \frac{1}{2}\zeta\mathscr{A} + \frac{1}{2}\eta\mathscr{D} + \frac{1}{2}(\alpha_1 - \alpha_2 - \alpha_3)w^2 U\right.$$

$$\qquad \left. + \frac{1}{2}(\alpha_1 - 2\alpha_3)w_i V_i + \frac{1}{2}\alpha_2 w_i w_k U_{ik}\right]_{,j} - \frac{7}{2}\Delta_1 V_{j,t} - \frac{1}{2}\Delta_2 W_{j,t} \qquad (39.48)$$

$$\qquad + \left(\alpha_2 - \frac{1}{2}\alpha_1\right) w_j U_{,t} - \alpha_2 w_i U_{ij,t} + O(U_{,j}\epsilon^4),$$

$$\Gamma^j{}_{0k} = \gamma U_{,t}\,\delta_{jk} - \left(\frac{7}{2}\Delta_1 + \frac{1}{2}\Delta_2\right) V_{[j,k]} - \frac{1}{2}\alpha_1 w_{[j}U_{,k]} + O(U_{,j}\epsilon^3),$$

$$\Gamma^j{}_{kl} = -\gamma(U_{,j}\,\delta_{kl} - 2U_{,(k}\,\delta_{l)j}) + O(U_{,j}\epsilon^2).$$

Here square brackets on tensor indices denote antisymmetrization, and round brackets denote symmetrization. As part of the derivation, it may be useful to prove and use the relations

$$\chi(t, \mathbf{x}) = -\int \rho_o(t, \mathbf{x}')|\mathbf{x} - \mathbf{x}'|\, d^3x', \qquad (39.49a)$$

$$\chi_{,jk} = -\delta_{jk}U + U_{jk}, \qquad (39.49b)$$

$$\chi_{,it} = V_i - W_i + O(\epsilon^5), \qquad (39.49c)$$

$$W_{[k,j]} = V_{[k,j]}. \qquad (39.50)$$

Here χ is the function originally defined in equation (39.29c).

Exercise 39.14. POST-NEWTONIAN APPROXIMATION TO GENERAL RELATIVITY

Perform a post-Newtonian expansion of Einstein's field equations, thereby obtaining the values cited in Box 39.2 for the PPN parameters of general relativity. The calculations might best follow the approach of Chandrasekhar (1965a): Set $g_{\alpha\beta} = \eta_{\alpha\beta} + h_{\alpha\beta}$, and assume

$$h_{00} = O(\epsilon^2) + O(\epsilon^4), \qquad h_{0j} = O(\epsilon^3), \qquad h_{jk} = O(\epsilon^2). \qquad (39.51)$$

Choose the space and time coordinates so that the four "gauge conditions"

$$\left.\begin{array}{l} h_{jk,k} - \dfrac{1}{2}h_{,j} = O(\epsilon^4/R_\odot) \\[2mm] h_{0k,k} - \dfrac{1}{2}h_{kk,0} = O(\epsilon^5/R_\odot) \end{array}\right\} \text{ with } h = h_{\alpha\beta}\eta^{\alpha\beta} = -h_{00} + h_{ll} \qquad (39.52)$$

are satisfied.

 (a) Show that the spatial gauge conditions are the post-Newtonian approximations to those (35.1a) used in the study of weak gravitational waves, but that the temporal gauge condition is not.

(b) Use these gauge conditions and the post-Newtonian limit in equations (8.24) and (8.47) to obtain for the Ricci tensor, accurate to linearized order,

$$R_{00} = -\frac{1}{2} h_{00,mm} + O(\epsilon^4/R_\odot^2), \qquad R_{jk} = -\frac{1}{2} h_{jk,mm} + O(\epsilon^4/R_\odot^2), \qquad (39.53a)$$

$$R_{0j} = -\frac{1}{2} h_{0j,mm} - \frac{1}{4} h_{00,0j} + O(\epsilon^5/R_\odot^2). \qquad (39.53b)$$

(c) Combine these with the Newtonian form (39.13) of the stress-energy tensor, and with equation (39.27), to obtain the following metric coefficients, accurate to linearized order:

$$h_{00} = 2U + k_{00} + O(\epsilon^6), \qquad h_{0j} = -\frac{7}{2} V_j - \frac{1}{2} W_j + O(\epsilon^5),$$

[unknown post-Newtonian correction] (39.54)

$$h_{jk} = 2U \delta_{jk} + O(\epsilon^4).$$

Here U, V_j, and W_j are to be regarded as defined by equations (39.34a,b,c). By comparing these metric coefficients with equations (39.32), discover that

$$\gamma = 1, \qquad \varDelta_1 = 1, \qquad \varDelta_2 = 1 \qquad (39.55)$$

for general relativity.

(d) With this knowledge of the metric in linearized order, one can carry out the analysis of §39.10 (using $\gamma = \varDelta_1 = \varDelta_2 = 1$ throughout), to obtain the post-Newtonian corrections to the stress-energy tensor [equation (39.42) with $\gamma = 1$].

(e) Calculate, similarly, the post-Newtonian corrections to the Ricci tensor component R_{00}, using $g_{\alpha\beta} = \eta_{\alpha\beta} + h_{\alpha\beta}$, using $h_{\alpha\beta}$ as given in equations (39.54), and using the gauge conditions (39.52). The answer should be

$$R_{00} = \left(-U - \frac{1}{2} k_{00} - U^2\right)_{,mm} + 4UU_{,mm} + O(\epsilon^6/R_\odot^2). \qquad (39.56)$$

(f) Evaluate the Einstein equation $R_{00} = 8\pi(T_{00} - \frac{1}{2} g_{00} T)$, accurate to post-Newtonian order, and solve it to obtain the post-Newtonian metric correction

$$k_{00} = -2U^2 + 4\Psi, \qquad (39.57)$$

where Ψ is given by equation (39.43d) with $\beta_1 = \beta_2 = \beta_3 = \beta_4 = 1$. By comparing with equations (39.32c) and (39.34d), discover that

$$\beta = \beta_1 = \beta_2 = \beta_3 = \beta_4 = 1, \qquad \zeta = \eta = 0 \qquad (39.58)$$

for general relativity.

(g) Knowing the full post-Newtonian metric, and the full post-Newtonian stress-energy tensor, one can carry out the calculations of §39.11 (using $\gamma = \beta = \beta_1 = \beta_2 = \beta_3 = \beta_4 = \varDelta_1 = \varDelta_2 = 1, \zeta = \eta = 0$) to obtain the post-Newtonian equations of motion for the matter [equations (39.44), (39.46), and (39.47)].

§39.12. RELATION OF PPN COORDINATES TO SURROUNDING UNIVERSE

One crucial issue remains to be clarified: What is the orientation of the PPN coordinate system relative to the surrounding universe? More particularly: Does the PPN coordinate system rotate relative to the "fixed stars on the sky;" or is it "rigidly attached" to them, in some sense? In order to answer this question, imagine using the PPN formalism to analyze the solar system. Make no assumptions about the solar system's velocity through the PPN coordinate frame. Then, as one moves outward from the Sun, past the Earth's orbit, past Pluto's orbit, and on out toward interstellar space, one sees the PPN coordinate frame become more and more Lorentz in its global properties [$g_{\alpha\beta} = \eta_{\alpha\beta} + O(M_\odot/r)$]. Thus, *far from the solar system the PPN coordinates become a "Lorentz frame moving through the galaxy."* This means, of course, that the spatial axes of the PPN coordinate frame behave as though they were attached to gyroscopes far outside the solar system. Equivalently: The PPN coordinate system Fermi-Walker-transports its spatial axes through the spacetime geometry of the galaxy and universe.

Solar system's PPN coordinate frame is attached to a local Lorentz frame of Galaxy

§39.13. SUMMARY OF PPN FORMALISM

The PPN formalism, as constructed in this chapter, is summarized in Box 39.4. Much of the recent literature uses a different set of PPN parameters than are used in this book; for a translation from one parameter set to the other, see Box 39.5.

Exercise 39.15. MANY-BODY SYSTEM IN POST-NEWTONIAN LIMIT OF GENERAL RELATIVITY

EXERCISE

Consider, in the post-Newtonian limit of general relativity, a system made up of many gravitationally interacting bodies with separations large compared to their sizes (*example:* the solar system). Idealize each body to be spherically symmetric, to be free of internal motions, and to have isotropic internal stresses, $t_{\bar{j}k} = \delta_{jk}p$. Let the world line of the center of body A, in some chosen PPN coordinate frame, be $x_A(t)$; and let the (coordinate) velocity of the center of body A be

$$v_A(t) = dx_A/dt. \qquad (39.59a)$$

The total mass-energy of body A as measured in its neighborhood (rest mass-energy plus internal energy plus self-gravitational energy) is given by

$$M_A = \int_{\mathcal{V}_A} \left(1 + \Pi - \frac{1}{2} U_{\text{self}}\right) d(\text{rest mass}) + O(M_A\epsilon^4), \qquad (39.59b)$$

where U_{self} is the body's own Newtonian potential (no contributions from other bodies), and \mathcal{V}_A is the interior of the body.

(continued on page 1094)

Box 39.4 SUMMARY OF THE PPN FORMALISM

I. Variables

$\rho_o(\boldsymbol{x}, t)$: baryon "mass" density (§39.3), as measured in rest frame

$\Pi(\boldsymbol{x}, t)$: specific internal energy (dimensionless; §39.3), as measured in rest frame

$t_{\hat{j}\hat{k}}(\boldsymbol{x}, t)$: components of stress referred to orthonormal axes of rest frame

$v_j(\boldsymbol{x}, t)$: coordinate velocity of matter (i.e., rest frame) relative to PPN coordinates

$U(\boldsymbol{x}, t), \Psi(\boldsymbol{x}, t), \mathscr{A}(\boldsymbol{x}, t), \mathscr{D}(\boldsymbol{x}, t), V_j(\boldsymbol{x}, t), W_j(\boldsymbol{x}, t), U_{jk}(\boldsymbol{x}, t)$: gravitational potentials

$\gamma, \beta, \beta_1, \beta_2, \beta_3, \beta_4, \Delta_1, \Delta_2, \zeta, \eta$: parameters whose values distinguish one theory from another (see Box 39.2)

w: velocity of PPN coordinate frame relative to "universal rest frame" [relevant only for theories with nonzero α_1, α_2, or α_3; see eq. (39.33)].

II. Equations governing evolution of these variables

ρ_o: conservation of rest mass, equation (39.44)

Π: first law of thermodynamics, equation (39.46)

$t_{\hat{j}\hat{k}}$: determined in terms of ρ_o, Π, and other material variables (chemical composition, strains, etc.) by equations of state and the usual theory of a stressed medium—which is not discussed here

v_j: equations of motion ("$\boldsymbol{F} = m\boldsymbol{a}$"), equations (39.47)

$U, \Psi, \mathscr{A}, \mathscr{D}, V_j, W_j, U_{jk}$: source equations (39.34)

III. Quantities to be calculated from these variables

$g_{00}(\boldsymbol{x}, t), g_{0j}(\boldsymbol{x}, t), g_{jk}(\boldsymbol{x}, t)$: these components of metric in PPN coordinate frame are expressed in terms of gravitational potentials by equations (39.32)

$u^0(\boldsymbol{x}, t), u^j(\boldsymbol{x}, t)$: these components of matter 4-velocity in PPN coordinate frame are given by equations (39.39)

$T^{00}(\boldsymbol{x}, t), T^{0j}(\boldsymbol{x}, t), T^{jk}(\boldsymbol{x}, t)$: these components of stress-energy tensor in PPN coordinate frame are given by equations (39.42)

IV. Relation between rest frame, PPN coordinates, and the universe

1. Orthonormal basis $\boldsymbol{w}^{\hat{\alpha}}$ of rest frame, where $t_{\hat{j}\hat{k}}$ are defined, is related to PPN coordinate basis $\boldsymbol{d}x^\alpha$ by equations (39.41)

2. Far from the sun, the PPN coordinates become asymptotically Lorentz; i.e., they form an inertial frame moving through the spacetime geometry of the galaxy and the universe.

3. Gives no account of expansion of universe or of cosmic gravitational waves impinging on solar system.

Box 39.5 PPN PARAMETERS USED IN LITERATURE: A TRANSLATOR'S GUIDE

The original "point-particle version" of the PPN formalism [Nordtvedt (1968b)], and the original "perfect-fluid version" [Will (1971c)] used different sets of PPN parameters. This book has adopted Will's set, and has added the parameter η characterizing effects of anisotropic stresses. More recently, Will and Nordtvedt have jointly adopted a revised set of parameters, described below.

A. Translation Table

Will-Nordtvedt revised parameters[a]	Revised parameters in notation of this book[b]	Revised parameters in notation of Nordtvedt (1968b)[c]
γ	γ	γ
β	β	β
α_1	$7\Delta_1 + \Delta_2 - 4\gamma - 4$	$8\Delta - 4\gamma - 4$
α_2	$\Delta_2 + \zeta - 1$	$\alpha''' - 1$
α_3	$4\beta_1 - 2\gamma - 2 - \zeta$	$4\alpha'' - \alpha''' - 2\gamma - 1$
ζ_1	ζ	$\alpha''' - \chi$
ζ_2	$2\beta + 2\beta_2 - 3\gamma - 1$	$2\beta - \alpha' - 1$
ζ_3	$\beta_3 - 1$	absent
ζ_4	$\beta_4 - \gamma$	absent

[a] Revised parameters are used by Will and Nordtvedt (1972), Nordtvedt and Will (1972), Will (1972), and Ni (1973).
[b] Notation of this book is used by Will (1971a,b,c,d), Ni (1972), and Thorne, Ni, and Will (1971).
[c] Nordtvedt's original "point-particle" parameters were used by Nordtvedt (1968b, 1970, 1971a,b).

B. Significance of Revised Parameters

α_1, α_2, α_3 measure the extent of and nature of "preferred-frame effects"; see §39.9. Any theory of gravity with at least one α nonzero is called a *preferred-frame theory*.

ζ_1, ζ_2, ζ_3, ζ_4, α_3 measure the extent of and nature of breakdowns in global conservation laws. A theory of gravity possesses, at the post-Newtonian level, all 10 global conservation laws (4 for energy-momentum, 6 for angular momentum; see Chapters 19 and 20) if and only if $\zeta_1 = \zeta_2 = \zeta_3 = \zeta_4 = \alpha_3 = 0$. See Will (1971d), Will and Nordtvedt (1972), Will (1972), for proofs and discussion. Any theory with $\zeta_1 = \zeta_2 = \zeta_3 = \zeta_4 = \alpha_3 = 0$ is called a *conservative theory*.

In general relativity and the Dicke-Brans-Jordan theory, all α's and ζ's vanish. Thus, general relativity and Dicke-Brans-Jordan are conservative theories with no preferred-frame effects.

(a) Show that, when written in the chosen PPN coordinate frame, this expression for M_A becomes

$$M_A = \int_{\mathcal{V}_A} \rho_o \left(1 + \Pi + \frac{1}{2} v_A{}^2 + 3U - \frac{1}{2} U_{\text{self}} \right) d^3x + O(M_A \epsilon^4). \tag{39.59c}$$

Use equations (39.43), (39.44), and (39.46) to show that M_A is conserved as the bodies move about, $dM_A/dt = 0$.

(b) Pick an event (t, \boldsymbol{x}) outside all the bodies, and at time t denote

$$\boldsymbol{r}_A \equiv \boldsymbol{x}_A - \boldsymbol{x}, \qquad \boldsymbol{r}_{AB} \equiv \boldsymbol{x}_A - \boldsymbol{x}_B, \qquad r_A \equiv |\boldsymbol{r}_A|, \qquad r_{AB} \equiv |\boldsymbol{r}_{AB}|. \tag{39.59d}$$

Show that the general-relativistic, post-Newtonian metric (39.32) at the chosen event has the form

$$g_{ik} = \delta_{jk} \left(1 + 2 \sum_A \frac{M_A}{r_A} \right) + O(\epsilon^4), \tag{39.60a}$$

$$g_{0j} = -\sum_A \frac{M_A}{r_A} \left[\frac{7}{2} v_{Aj} + \frac{1}{2} \frac{(\boldsymbol{v}_A \cdot \boldsymbol{r}_A) r_{Aj}}{r_A{}^2} \right] + O(\epsilon^5), \tag{39.60b}$$

$$g_{00} = -1 + 2 \sum_A \frac{M_A}{r_A} - 2 \left(\sum_A \frac{M_A}{r_A} \right)^2 + 3 \sum_A \frac{M_A v_A{}^2}{r_A}$$

$$-2 \sum_A \sum_{B \neq A} \frac{M_A M_B}{r_A r_{AB}} + O(\epsilon^6). \tag{39.60c}$$

[*Hint:* From the Newtonian virial theorem (39.21a), applied to body A by itself in its own rest frame, conclude that

$$\int_{\mathcal{V}_A} \left(3p - \frac{1}{2} \rho_o U_{\text{self}} \right) d^3x = O(M_A \epsilon^4), \tag{39.61}$$

where the integral is performed in the PPN frame.]

(c) Perform an infinitesimal coordinate transformation,

$$t_{\text{OLD}} = t_{\text{NEW}} - \frac{1}{2} \sum_A \frac{M_A (\boldsymbol{r}_A \cdot \boldsymbol{v}_A)}{r_A}, \qquad \boldsymbol{x}_{\text{OLD}} = \boldsymbol{x}_{\text{NEW}}, \tag{39.62}$$

to bring the metric (39.60) into the standard form originally devised by Einstein, Infeld, and Hoffman (1938), and by Eddington and Clark (1938):

$$g_{jk} = \delta_{jk} \left(1 + 2 \sum_A \frac{M_A}{r_A} \right) + O(\epsilon^4), \tag{39.63a}$$

$$g_{0j} = -4 \sum_A \frac{M_A}{r_A} v_{Aj} + O(\epsilon^5), \tag{39.63b}$$

$$g_{00} = -1 + 2 \sum_A \frac{M_A}{r_A} - 2 \left(\sum_A \frac{M_A}{r_A} \right)^2 + 3 \sum_A \frac{M_A v_A{}^2}{r_A}$$

$$-2 \sum_A \sum_{B \neq A} \frac{M_A M_B}{r_A r_{AB}} - \frac{\partial^2 \chi}{\partial t^2} + O(\epsilon^6), \tag{39.63c}$$

where χ [equation (39.49a)] is given by

$$\chi = -\sum_A M_A r_A.$$

(d) The equations of motion for the bodies can be obtained in either of two ways: by performing a volume integral of the Euler equation (39.48) over the interior of each body; or by invoking the general arguments of §20.6. The latter way is the easier. Use it to conclude that any chosen body K moves along a geodesic of the metric obtained by omitting the terms $A = K$ from the sums in (39.63). Show that the geodesic equation for body K reduces to

$$
\frac{d^2 \mathbf{x}_K}{dt^2} \equiv \frac{d\mathbf{v}_K}{dt} = \sum_{A \neq K} \mathbf{r}_{AK} \frac{M_A}{r_{AK}{}^3} \left[1 - 4 \sum_{B \neq K} \frac{M_B}{r_{BK}} - \sum_{C \neq A} \frac{M_C}{r_{CA}} \left(1 - \frac{\mathbf{r}_{AK} \cdot \mathbf{r}_{CA}}{2 r_{CA}{}^2} \right) \right.
$$

$$
\left. + v_K{}^2 + 2 v_A{}^2 - 4 \mathbf{v}_A \cdot \mathbf{v}_K - \frac{3}{2} \left(\frac{\mathbf{v}_A \cdot \mathbf{r}_{AK}}{r_{AK}} \right)^2 \right]
$$

$$
- \sum_{A \neq K} (\mathbf{v}_A - \mathbf{v}_K) \frac{M_A \mathbf{r}_{AK} \cdot (3 \mathbf{v}_A - 4 \mathbf{v}_K)}{r_{AK}{}^3} \qquad (39.64)
$$

$$
+ \frac{7}{2} \sum_{A \neq K} \sum_{C \neq A} \mathbf{r}_{CA} \frac{M_A M_C}{r_{AK} r_{CA}{}^3}.
$$

Equations (39.63) and (39.64) are called the Einstein-Infeld-Hoffman ("EIH") equations for the geometry and evolution of a many-body system. They are used widely in analyses of planetary orbits in the solar system. For example, the Caltech Jet Propulsion Laboratory uses them, in modified form, to calculate ephemerides for high-precision tracking of planets and spacecraft. The above method of deriving the EIH equations and metric was devised by Fock (1959). For a similar calculation in the Dicke-Brans-Jordan theory, see Estabrook (1969); and for a derivation of the analogous many-body equations in the full PPN formalism, see Will (1972).

CHAPTER **40**

SOLAR-SYSTEM EXPERIMENTS

§40.1. MANY EXPERIMENTS OPEN TO DISTINGUISH GENERAL RELATIVITY FROM PROPOSED METRIC THEORIES OF GRAVITY

No audience will show up for a fight if in everyone's eyes the challenger has zero chance to win. No battle-hungry promoter desperately trying to finance the fight can afford to put into the ring against the champion any but the best contender that he can find. Against Einstein's metric theory of gravity, the judgment of the day (as §39.2 showed) leaves one no option except to put up another theory of gravity that is also metric (or metric plus torsion).

To put on a contest, then, is to design and perform an experiment that distinguishes general relativity from some not completely implausible metric theory of gravity. This chapter describes such experiments—some already performed; some to be performed in the future—and analyses their significance using the PPN formalism of Chapter 39.

In most of the experiments to be described, one investigates the motion of the moon, planets, spacecraft, light rays, or gyroscopes through the spacetime geometry of the solar system. That spacetime geometry is very complicated. It includes the spherical fields of the sun and all the planets, nonspherical fields due to their quadrupolar and higher-order deformations, and fields due to their momentum and angular momentum. Moreover, the spacetime geometry results—or at least in the post-Newtonian formalism it is viewed as resulting—from a *nonlinear* superposition of all these fields.*

This chapter analyzes experiments using PPN formalism

Complexity of solar system's spacetime geometry

*Of course, from the point of view of Einstein's full general relativity theory, all that legalistically counts is the one and only curved-spacetime geometry of the real physical world. All these "individual fields" are mere bookkeepers' discourse, and they are best abandoned (they cease to be useful) when one passes from the post-Newtonian limit to the full Einstein theory.

Fortunately for this discussion, several of the most important experiments are free of almost all these complications. The effects they measure are associated entirely with the spherical part of the sun's gravitational field. A description of these experiments will come first (§§40.2–40.5), and then attention will turn to experiments that are more complex in principle.

To discuss central-field experiments, one needs an expression for the external gravitational field of an idealized, isolated, static, spherical sun. In general relativity, such a gravitational field is described by the Schwarzschild line element,

$$ds^2 = -\left(1 - \frac{2M_\odot}{r}\right)dt^2 + \frac{dr^2}{1 - 2M_\odot/r} + r^2(d\theta^2 + \sin^2\phi\, d\phi^2).$$

But this line element is not what one wants, for two reasons: (1) it is "too accurate"; (2) it is written in the "wrong" coordinate system.

Why too accurate? Because it is simple only when unperturbed and unmodified; whereas some modified theories show up new effects that are so complex they are tractable only in the post-Newtonian approximation. Why wrong coordinate system? Because physicists, astronomers, and other celestial mechanics have adopted the fairly standard convention of using "isotropic coordinates" rather than "Schwarzschild coordinates" when analyzing the solar system. Example: post-Newtonian expansions, including the PPN formalism of Chapter 39, almost always use isotropic coordinates. Another example: the relativistic ephemeris for the solar system, prepared by the Caltech Jet Propulsion Laboratory [Ohandley *et al.* (1969); Anderson (1973)] and used extensively throughout the world, employs isotropic coordinates.

Modify the Schwarzschild line element, then. First transform to isotropic coordinates (Exercise 31.7); then expand the metric coefficients in powers of M_\odot/r, to post-Newtonian accuracy. Thereby obtain

$$\begin{aligned} ds^2 &= -\left[1 - 2\frac{M_\odot}{r} + 2\left(\frac{M_\odot}{r}\right)^2\right]dt^2 + \left[1 + 2\frac{M_\odot}{r}\right][dr^2 + r^2(d\theta^2 + \sin^2\theta\, d\phi^2)] \\ &= -\left[1 - 2\frac{M_\odot}{r} + 2\left(\frac{M_\odot}{r}\right)^2\right]dt^2 + \left[1 + 2\frac{M_\odot}{r}\right][dx^2 + dy^2 + dz^2]. \end{aligned} \tag{40.1}$$

Here r, θ, ϕ are related to x, y, z in the usual manner:

$$r = (x^2 + y^2 + z^2)^{1/2}, \qquad \theta = \tan^{-1}[z/(x^2 + y^2)^{1/2}], \qquad \phi = \tan^{-1}(y/x); \tag{40.2}$$

and r is the new, "isotropic" radial coordinate, not to be confused with the Schwarzschild r. (The reader who has not studied §39.6 will discover in the next section why one keeps terms of order M_\odot^2/r^2 in g_{00} but not in g_{jk}.) *Note:* this post-Newtonian expression for the metric is a special case of the result derived in exercise 19.3.

If one calculates the gravitational field of the same source (the sun) in the same post-Newtonian approximation in other metric theories of gravity, one obtains a very similar result:

Idealization of geometry to that of isolated, static, spherical sun:

(1) in Schwarzschild coordinates

(2) in isotropic coordinates

(3) in PPN formalism

$$ds^2 = -\left[1 - 2\frac{M_\odot}{r} + 2\beta\left(\frac{M_\odot}{r}\right)^2\right]dt^2 + \left[1 + 2\gamma\frac{M_\odot}{r}\right][dr^2 + r^2(d\theta^2 + \sin^2\theta\, d\phi^2)]$$

$$= -\left[1 - 2\frac{M_\odot}{r} + 2\beta\left(\frac{M_\odot}{r}\right)^2\right]dt^2 + \left[1 + 2\gamma\frac{M_\odot}{r}\right][dx^2 + dy^2 + dz^2]$$

(40.3)

(see exercise 40.1). Here γ and β are two of the ten PPN parameters described in Box 39.2. Recall from that box that γ measures "the amount of space curvature produced by unit rest mass," while β measures "the amount of nonlinearity in the superposition law for g_{00}." These heuristic descriptions find their mathematical counterparts in the above form for the idealized metric surrounding a spherically symmetric center of attraction.

By measuring the parameter γ to high precision, one can distinguish between general relativity ($\gamma = 1$) and the Dicke-Brans-Jordan theory [$\gamma = (1 + \omega)/(2 + \omega)$, where ω is the "Dicke coupling constant"]; see Box 39.2. But general relativity and Dicke-Brans-Jordan predict the same value for β ($\beta = 1$). This identity does not mean that β is unworthy of measurement. A value $\beta \neq 1$ is predicted by other theories [see Ni (1972)]; so measurements of β are useful in distinguishing such theories from general relativity.

Actually, the above form (40.3) for the sun's metric is not fully general. In any theory with a preferred "universal rest frame" (e.g., Ni's theory; Box 39.1), there are additional terms in the metric due to motion of the sun relative to that preferred frame (exercise 40.1):

(4) including preferred-frame
effects

$$ds^2 = (\text{expression } 40.3) + (\alpha_2 + \alpha_3 - \alpha_1)\frac{M_\odot}{r}w^2\, dt^2 + 2\left(\alpha_2 - \frac{1}{2}\alpha_1\right)\frac{M_\odot}{r}w_j\, dx^j\, dt$$

$$- \alpha_2\left[\frac{M_\odot}{r^3}x^j x^k - \frac{I_\odot}{r^5}\left(x^j x^k - \frac{1}{3}r^2\,\delta_{jk}\right)\right]w_j\, dt(2\, dx^k + w_k\, dt).$$

(40.3')

In these "preferred-frame terms" $I_\odot \equiv I_{jj} = \int\rho r^2\, d^3x$ is the trace of the second moment of the sun's mass distribution;

$$\alpha_1 = 7\Delta_1 + \Delta_2 - 4\gamma - 4,$$

$$\alpha_2 = \Delta_2 + \zeta - 1,$$

$$\alpha_3 = 4\beta_1 - 2\gamma - 2 - \zeta$$

are combinations of PPN parameters; and w is the sun's velocity (\equiv velocity of coordinate system) relative to the preferred frame. (Theories such as general relativity and Dicke-Brans-Jordan, which possess no preferred frame, have $\alpha_1 = \alpha_2 = \alpha_3 = 0$, and therefore have no preferred-frame terms in the metric.) For ease of exposition, all equations and calculations in this chapter will ignore the preferred-frame terms; but the consequences of those terms will be discussed and references analyzing them will be cited.

Exercise 40.1. PPN METRIC FOR IDEALIZED SUN [Track 2]
Show that for an isolated, static, spherical sun at rest at the origin of the PPN coordinate
system, the PPN metric (39.32) reduces to expressions (40.3), (40.3′). As part of the reduction,
show that the sun's total mass-energy is given by

$$M_\odot = \int_0^{R_\odot} \rho_0 (1 + 2\beta_2 U + \beta_3 \Pi + 3\beta_4 p/\rho_0) 4\pi r^2 \, dr. \qquad (40.4)$$

[*Warning:* One must not look at this formula and immediately think: "The contribution of
rest mass is $\int \rho_0 4\pi r^2 \, dr$, the contribution of gravitational energy is $\int 2\beta_2 \rho_0 U 4\pi r^2 \, dr$, etc."
Rather, in making any such interpretation one must remember that (1) spacetime is curved,
so $4\pi r^2 \, dr$ is not *proper volume* as measured by physical meter sticks; also (2) virial theorems
(exercise 39.6) and other integral theorems can be used to change the form of the integrand.
For further discussion see exercises 40.9 and 40.10 below.]

§40.2. THE USE OF LIGHT RAYS AND RADIO WAVES TO TEST GRAVITY

In the Newtonian limit, planetary and spacecraft orbits are strongly influenced by
gravity; but light propagation and radio-wave propagation (at "infinite" velocity)
are not influenced at all. For this reason, experimental studies of orbits are beset
by the problem of separating the relativistic effects from much larger standard
Newtonian effects. By contrast, experimental studies of light and radio-wave propa-
gation do not contend with any such overpowering Newtonian background. Not
surprisingly, they are to date (1973) the clearest and most definitive of the solar-
system experiments.

Light rays and radio waves
give "clean" tests of
relativity

Mathematically, the parameter that distinguishes a light ray from a planet is its
high speed. In the geodesic equation, the magnitude of the velocity determines which
metric coefficients can influence the motion. Consider, for example, a weak, static
field $g_{\alpha\beta} = \eta_{\alpha\beta} + h_{\alpha\beta}$, and a particle at $(x, y, z) = (r, 0, 0)$ moving with velocity
$(v_x, v_y, v_z) = (0, v, 0)$; see Figure 40.1. Here the effect of gravity on the trajectory
of the particle can be characterized by the quantity

$$\begin{pmatrix} \text{curvature of trajectory in 3-dimensional,} \\ \text{nearly Euclidean, space} \end{pmatrix} = \begin{pmatrix} \text{radius of curvature} \\ \text{of trajectory} \end{pmatrix}^{-1}$$

$$= \frac{d^2x}{dy^2} = \frac{d\tau}{dy} \frac{d}{d\tau}\left(\frac{d\tau}{dy} \frac{dx}{d\tau}\right) = \frac{1}{u^y} \frac{d}{d\tau}\left(\frac{u^x}{u^y}\right) = \frac{1}{(u^y)^2} \frac{du^x}{d\tau}$$

$$= -\frac{(1 - v^2)}{v^2} \Gamma^x{}_{\alpha\beta} \frac{dx^\alpha}{d\tau} \frac{dx^\beta}{d\tau} = -\frac{1}{v^2} \Gamma^x{}_{\alpha\beta} \frac{dx^\alpha}{dt} \frac{dx^\beta}{dt}$$

$$= -\Gamma^x{}_{00} v^{-2} - 2\Gamma^x{}_{0y} v^{-1} - \Gamma^x{}_{yy}$$

$$= \frac{1}{2} h_{00,x} v^{-2} + (h_{0y,x} - h_{0x,y}) v^{-1} + \left(\frac{1}{2} h_{yy,x} - h_{xy,y}\right).$$

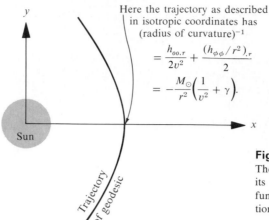

Here the trajectory as described in isotropic coordinates has (radius of curvature)$^{-1}$

$$= \frac{h_{oo,r}}{2v^2} + \frac{(h_{\phi\phi}/r^2)_{,r}}{2}$$

$$= -\frac{M_\odot}{r^2}\left(\frac{1}{v^2} + \gamma\right).$$

Figure 40.1.
The bending of the trajectory of a test body at its point of closest approach to the sun, as a function of its 3-velocity. (See text for computation and discussion.)

Reexpressed in spherical coordinates, in the terminology of the idealized solar line element (40.3), this formula says

$$\begin{pmatrix} \text{curvature of trajectory} \\ \text{in 3-space} \end{pmatrix} = \frac{1}{2}h_{00,r}v^{-2} + \frac{1}{2}(h_{\phi\phi}/r^2)_{,r}$$

$$\approx -(M_\odot/r^2)(v^{-2} + \gamma) \tag{40.5}$$

for a particle at its point of closest approach to the sun. (Compare with exercise 25.21.) Note that here γ is a PPN parameter; it is *not* $(1 - v^2)^{-1/2}$.

Notice what happens as one boosts the velocity of the particle. For slow velocities $[v^2 \sim \text{(post-Newtonian expansion parameter } \epsilon^2) \approx M_\odot/R_\odot]$, the Newtonian part of h_{00} dominates completely; and the tiny post-Newtonian corrections come equally from the ϵ^4 part of h_{00}, the ϵ^3 part of h_{0j}, and the ϵ^2 part of h_{jk}. [This was the justification for expanding h_{00} to $O(\epsilon^4)$, h_{0j} to $O(\epsilon^3)$, and h_{jk} to $O(\epsilon^2)$ in the post-Newtonian limit; see §39.6.] But as v increases, the ordering of the terms changes. In the high-v regime ($v \sim 1 \gg \epsilon^2$), the bending of the trajectory has become almost imperceptible because of the high forward momentum of the particle and the short time it receives transverse momentum from the sun. What bending is left is due to the ϵ^2 (Newtonian) part of h_{00}, and the ϵ^2 (post-Newtonian) part of h_{jk}. Nothing else can have a significant influence. Notice, moreover, that—even when one allows for "preferred-frame" effects—these dominant terms,

$$h_{00} = 2U = 2M_\odot/r \quad \text{and} \quad h_{jk} = 2\gamma U\delta_{jk} = 2\gamma(M_\odot/r)\,\delta_{jk},$$

depend only on the Newtonian potential $U \equiv -\Phi$ and the PPN parameter γ.

Light rays are governed solely by Newtonian potential and PPN parameter γ

This is a special case of a more general result: *Aside from fractional corrections of $\epsilon^2 \lesssim 10^{-6}$, relativistic effects on light and radio-wave propagation are governed entirely by the Newtonian potential U and the PPN parameter γ.* These relativistic effects include the gravitational redshift (discussed in the last chapter; independent

of γ), the gravitational deflection of light and radio waves (discussed below; dependent on γ), and the "relativistic time-delay" (discussed below; dependent on γ).

§40.3. "LIGHT" DEFLECTION

Consider a light or radio ray coming into a telescope on Earth from a distant star or quasar. Do not assume, as in the usual discussion (exercises 18.6 and 25.24), that the ray passes near the sun. The deflection by the sun's gravitational field will probably be measurable, in the middle or late 1970's, even when the ray passes far from the sun! [The calculation that follows is due to Ward (1970), but Shapiro (1967) first derived the answer.]

Orient the PPN spherical coordinates of equation (40.3) so that the ray lies in the "plane" $\theta = \pi/2$. By symmetry, if it starts out in this plane far from the Earth, it must lie in this plane always. Let the incoming ray enter the solar system along the line $\phi = 0$; and let the Earth be located at $r = r_E$, $\phi = \phi_E$ when the ray reaches it. (See Figure 40.2.) One wishes to calculate the angle α between the incoming light ray and the center of the sun, as measured in the orthonormal frame $(e_{\hat{r}}, e_{\hat{\phi}})$ of an observer on Earth. If the sun had zero mass (flat, Euclidean space), α would be $\pi - \phi_E$ (see Figure 40.2). However, the sun produces a deflection: $\alpha = \pi - \phi_E + \delta\alpha$. The deflection angle $\delta\alpha$ is the true objective of the calculation.

In the calculation, ignore the Earth's orbital and rotational motions. They lead to aberration, for which correction can be made by the usual formula of special relativity (Lorentz transformation in the neighborhood of the telescope.) Also ignore deflection of the light ray due to the Earth's gravitational field (deflection angle $\sim 2M_E/R_E \sim 0''.0003$), which might be detectable in the late 1970's.

Light deflection:

(1) derivation

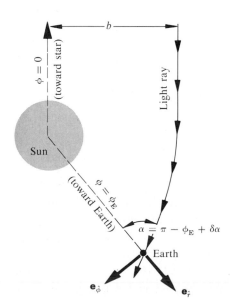

Figure 40.2.
Coordinates used in the text for calculating the deflection of light. Notice that in this diagram ϕ increases in the clockwise direction.

As the first step in calculating the deflection angle, determine the trajectory of the ray in the r, ϕ-plane. This can be calculated either using the geodesic equation, or using the eikonal method of geometric optics (Hamilton-Jacobi method; §22.5 and Box 25.4). The result of such a calculation (exercise 40.2) is an equation connecting r with ϕ; thus,

$$\frac{b}{r} = \sin\phi + \frac{(1+\gamma)M_{\odot}}{b}(1 - \cos\phi). \qquad (40.6)$$

Notice that b has a simple geometric interpretation: far from the sun, the ray trajectory is $\phi = b/r + O(M_{\odot}b/r^2)$. Consequently, b is the impact parameter in the usual sense of classical scattering theory (see Figure 40.2). The ray makes its closest approach to the sun (assuming it is not intercepted by the Earth first) at the PPN coordinate radius

$$r_{\min} = b\left[1 - \frac{(1+\gamma)M_{\odot}}{b}\right] \approx b. \qquad (40.7)$$

Thus, b can also be thought of as the radius of the ray's "perihelion."

Notice that the ray returns to $r = \infty$, not at an angle $\phi = \pi$, but rather at

$$\phi(r = \infty) = \pi + 2(1+\gamma)M_{\odot}/b. \qquad (40.8a)$$

Thus, the total deflection angle is

$$\text{(angle of total deflection)} = 2(1+\gamma)M_{\odot}/b$$
$$= \frac{1}{2}(1+\gamma)1''.75 \text{ for a ray that} \qquad (40.8b)$$
$$\text{just grazes the sun.}$$

But this is not the quantity of primary interest. Rather, one seeks the position of the star as seen by an astronomer on Earth. The angle $\alpha = \pi - \phi_E + \delta\alpha$ between the sun and the star as measured by the astronomer is given by (see Figure 40.2)

$$\tan(\pi - \phi_E + \delta\alpha) = -\tan\phi_E + \delta\alpha/\cos^2\phi_E$$
$$= \frac{u^{\hat{\phi}}}{u^{\hat{r}}} = \left[\frac{(1+\gamma M_{\odot}/r)r\,d\phi/d\lambda}{(1+\gamma M_{\odot}/r)\,dr/d\lambda}\right]_E = \left[\frac{r\,d\phi}{dr}\right]_E \qquad (40.9)$$
$$= -\left[\frac{(b/r)\,d\phi}{d(b/r)}\right]_E,$$

where $u^{\beta} = dx^{\beta}/d\lambda$ are the components of a tangent to the ray at the Earth. By inserting into this equation expression (40.6) for the trajectory of the ray, one obtains

$$\tan\phi_E - \frac{\delta\alpha}{\cos^2\phi_E} = \frac{\sin\phi_E + [(1+\gamma)M_{\odot}/b](1 - \cos\phi_E)}{\cos\phi_E + [(1+\gamma)M_{\odot}/b]\sin\phi_E} \qquad (40.10)$$
$$= \tan\phi_E - [(1+\gamma)M_{\odot}/b](1 - \cos\phi_E)/\cos^2\phi_E.$$

Thus, the deflection angle measured at the Earth is

$$\delta\alpha = \frac{(1 + \gamma)M_\odot}{b}(1 + \cos\alpha) = \frac{(1 + \gamma)M_\odot}{r_E}\left(\frac{1 + \cos\alpha}{1 - \cos\alpha}\right)^{1/2}. \qquad (40.11)$$

(2) formula for deflection angle

It ranges from zero when the ray comes in opposite to the sun's direction ($\alpha = \pi$), through the value

$$(1 + \gamma)M_\odot/r_E = \frac{1}{2}(1 + \gamma)0''.0041 \qquad (40.12)$$

when the ray comes in perpendicular to the Earth-Sun line ($\alpha = \pi/2$), to the "classical value" of $\frac{1}{2}(1 + \gamma) \times 1''.75$ when the ray comes in grazing the sun's limb.

All experiments to date (1972) have examined the case of grazing passage. The experimental results are stated and discussed in Box 40.1. They show that the PPN parameter γ has its general relativistic value of 1 to within an uncertainty of about 20 percent.

Experimental measurements of deflection

By the middle or late 1970's, measurements of the deflection of radio waves from quasars should determine γ to much better than 1 percent. Also, by that time radio astronomers may be making progress toward setting up high-precision coordinates on the sky using very long baseline interferometry. If so, they will have to use equation (40.11) to compensate for the "warping" of the coordinates caused by the sun's deflection of radio waves in all regions of the sky, not just near the solar limb.

Exercise 40.2. TRAJECTORY OF LIGHT RAY IN SUN'S GRAVITATIONAL FIELD **EXERCISE**

Derive equation (40.6) for the path of a light ray in isotropic coordinates (40.3) in the sun's "equatorial plane." Use one or more of three alternative approaches: (1) direct integration of the geodesic equation (the hardest approach!); (2) computation based on the three integrals of the motion

$$\boldsymbol{k} \cdot \boldsymbol{k} = 0, \qquad \boldsymbol{k} \cdot (\partial/\partial t) = k_0, \qquad \boldsymbol{k} \cdot (\partial/\partial\phi) = k_\phi = -bk_0;$$

$$\boldsymbol{k} \equiv d/d\lambda = \text{tangent vector to geodesic}$$

(see §§25.2 and 25.3); (3) computation based on the Hamilton-Jacobi method (Box 25.4), which for photons (zero rest mass) reduces to the "eikonal method" of geometric optics (see §22.5).

§40.4. TIME-DELAY IN RADAR PROPAGATION

Another effect of spacetime curvature on electromagnetic waves is a relativistic delay in the round-trip travel time for radar signals. It was first pointed out by Shapiro (1964); see also Muhleman and Reichley (1964, 1965).

Radar time delay:

(continued on page 1106)

Box 40.1 DEFLECTION OF LIGHT AND RADIO WAVES BY SUN: EXPERIMENTAL RESULTS

Eclipse Measurements

Until 1968 every experiment measured the deflection of star-light during total eclipse of the sun. The measurements were beset by difficulties such as poor weather, optical distortions due to temperature changes, and the strange propensity of eclipses to attain maximum time of totality in jungles, in the middles of oceans, in deserts, and in arctic tundras. Lists of all the results and references are given by Bertotti, Brill and Krotkov (1962), and by Klüber (1960). Dicke (1964b) summarizes the results as follows:

"The analyses [of the experimental data] scatter from a deflection at the limb of the sun of 1.43 seconds of arc to 2.7 seconds [compared to a general relativistic value of 1.75 seconds]. The scatter would not be too bad if one could believe that the technique was free of systematic errors. It appears that one must consider this observation uncertain to at least 10 percent, and perhaps as much as 20 percent." This result corresponds to an uncertainty in γ of 20 to 40 percent.

Measurements on the Deflection of Radio Waves

Each October 8 the sun, as seen from the Earth, passes in front of the quasar 3C279. By monitoring the angular separation between 3C279 and a nearby quasar 3C273, radio astronomers can measure the deflection by the sun of the 3C279 radio waves. The monitoring uses radio interferometers. [See references cited in table for discussion of the technique.] Technology of the early 1970's should permit measurements to a precision 0.001 seconds of arc or better, if the two ends of the interferometer are separated by several thousand kilometers ("transcontinental" or "transworld" baseline). But as of 1971 the only successful experiments were less ambitious: they used baselines of less than 10 kilometers. A summary of these pre-1971, short-baseline results is shown in the table.

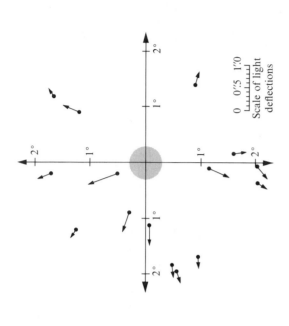

Observed light deflections (mean of two instruments) of the 15 best measured stars within 2°.5 of the sun's center in the total solar eclipse of September 21, 1922 at Wallal, Western Australia, as determined by Campbell and Trumpler (1928). The arrows represent in size and direction the observed light deflections relative to the reference stars (5° to 10° from the sun's center). (See Box 1.6 for Einstein's description of the deflection in terms of the curvature of geometry near the sun).

The 90-foot (background) and 130-foot (foreground) radio interferometer system at Caltech's Owens Valley Radio Observatory. These were used by Seielstadt, Sramek, and Weiler (1970) in their pioneering measurement of the deflection of quasar radio waves by the sun. During the experiment the two antennas were separated by 1.07 kilometers. (Photo by Alan Moffet.)

Dates of observation	Observatory	Experimenters and reference	Number of telescopes and separations	Wave lengths	Experimental results[a]		
					$\frac{1}{2}(1 + \gamma) = \left(\dfrac{\text{Observed deflection}}{\text{Einstein prediction}}\right)$	Formal standard error	One-sigma error
Sept. 30–Oct. 15 1969	Owens Valley (Caltech)	Seielstadt, Sramek, Weiler (1970)	2, 1.07 km	3.1 cm	1.01	±0.12	±0.12
Oct. 2–Oct. 10 1969	Goldstone (Caltech-JPL)	Muhleman, Ekers, Fomalont (1970)	2, 21.56 km	12.5 cm	1.04	±0.05	+0.15 −0.10
Oct. 2–Oct. 12 1970	National Radio Astronomy Observatory (USA)	Sramek (1971)	3, 0.80 km, 1.90 km, 2.70 km	11.1 cm, 3.7 cm	0.90	±0.05	±0.05
Sept. 30–Oct. 15 1970	Mullard Radio Astronomy Observatory (Cambridge Univ.)	Hill (1971)	3, 0.66 km, 1.41 km	11.1 cm, 6.0 cm	1.07	±0.17	±0.17

[a] Here (observed deflection)/(Einstein prediction) is the number $\frac{1}{2}(1 + \gamma)$ obtained by fitting the observational data to the ≥PN prediction (40.11). [For these experiments the ray passes near the solar limb; so (40.11) reduces to $\delta\alpha = \frac{1}{2}(1 + \gamma)(M_\odot/b)$.] The "formal standard error" is obtained from the data by standard statistical techniques. However, it is not usually a good measure of the certainty of the result, because it fails to take account of systematic errors. The quoted "one-sigma error" is the experimenters' best estimate of the combined statistical and systematic uncertainties. The experimenters estimate a probability of 68 percent that the true result is within "1σ" of their measured value; a probability of 95 percent that it is within "2σ"; etc.

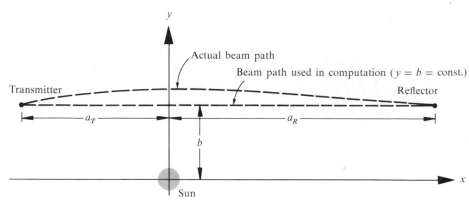

Figure 40.3.
Diagram, in the PPN coordinate system, for
the calculation of the relativistic time delay.

(1) foundations for calculation; Fermat's principle

Let a radar transmitter on Earth send a radio wave out to a reflector elsewhere in the solar system, and let the reflector return the wave to Earth. Calculate the round-trip travel time, as measured by a clock on Earth. For simplicity of calculation, idealize both Earth and reflector as nonrotating and as at rest in the static, spherical gravitational field of the Sun. At the end of the calculation, the effects of rotation and motion will be discussed separately. Also ignore time dilation of the transmitter's clock due to the Earth's gravitational field; it is easily corrected for, and it is so small that it will not come into play in these radar experiments before the middle or late 1970's. The gravitational effects of the other planets on the radio waves are too small to be discernible in the foreseeable future, unless the beam grazes the limb of one of them. However, the effects of dispersion in the solar wind and corona are discernible and must be corrected for. These corrections will not be discussed here, since they are free of any general-relativistic influence.

The calculation of the round-trip travel time can be simplified by using a general-relativistic version of Fermat's principle: *In any static field* ($g_{0j} = 0$, $g_{\alpha\beta,0} = 0$) *consider all null curves between two points in space, $x^j = a^j$ and $x^j = b^j$. Each such null curve, $x^j(t)$, requires a particular coordinate time interval Δt to get from a^j to b^j. The curves of extremal Δt are the null geodesics of spacetime.* The proof of this theorem is outlined in exercise 40.3.

Because of Fermat's principle, the lapse of coordinate time between transmission of the radar beam and reflection at the reflector, t_{TR}, is the same for a straight path in the PPN coordinates as for the slightly curved path which the beam actually follows. (The two differ by a fractional amount $\Delta t_{TR}/t_{TR} \sim$ (angle of deflection)$^2 \lesssim 10^{-12}$, which is far from discernable.) Hence, in the computation one can ignore the gravitational bending of the beam.

(2) details of calculation

Adopt Cartesian PPN coordinates with the sun at the origin; the transmitter, sun, and reflector in the $z = 0$ "plane"; and the transmitter-reflector line along the x direction (see Figure 40.3). The transmitter is at $(x, y) = (-a_T, b)$ in the PPN coordinates, and the reflector is at $(x, y) = (a_R, b)$. Recall that for a null ray $ds^2 = 0 =$

$g_{00} \, dt^2 - g_{xx} \, dx^2$. It follows that the lapse of coordinate time between transmission and reflection is

$$t_{\text{TR}} = \int_{-a_T}^{a_R} \left(\frac{g_{xx}}{-g_{00}} \right)^{1/2} dx = \int_{-a_T}^{a_R} \left[1 + \frac{(1 + \gamma)M_\odot}{\sqrt{x^2 + b^2}} \right] dx.$$

Integration yields

$$t_{\text{TR}} = a_R + a_T + (1 + \gamma)M_\odot \ln \left[\frac{(a_R + \sqrt{a_R^2 + b^2})(a_T + \sqrt{a_T^2 + b^2})}{b^2} \right]. \quad (40.13)$$

The lapse of coordinate time in round-trip travel has twice this magnitude. The lapse of proper time measured by an Earth-based clock is

$$\Delta\tau = |g_{00}|_{\text{Earth}}^{1/2} 2 t_{\text{TR}},$$

<div style="float:right">(3) formula for delay</div>

$$\begin{aligned}
\Delta\tau = 2(a_R + a_T)&\left(1 - \frac{M_\odot}{\sqrt{a_T^2 + b^2}} \right) \\
&+ 2(1 + \gamma)M_\odot \ln \left[\frac{(a_R + \sqrt{a_R^2 + b^2})(a_T + \sqrt{a_T^2 + b^2})}{b^2} \right]
\end{aligned} \quad (40.14)$$

This is the lapse of time on an Earth-based clock, aside from corrections for the orbital and rotational motion of the Earth, for the orbital motion of the reflector, for dispersion of radiation traversing the solar wind and corona, and for time dilation in the Earth's gravitational field.

Any reader is reasonable who objects to the form (40.14) in which the time-delay has been written. The quantities a_R, a_T, and b are coordinate positions in the PPN coordinate system, rather than numbers the astronomer can measure directly. They differ from coordinate positions in other, equally good coordinate systems by amounts of the order of $M_\odot \sim 1.5$ km. The objection is not mathematical in its origin. The quantities a_R, a_T, and b are perfectly well-defined [with post-post-Newtonian uncertainties of order $b(M_\odot/b)^2 \lesssim 10^{-6}$ km], because the PPN coordinate system is perfectly well-defined. But they are not quantities which the experimenter can measure directly, with precision anywhere near that required to test the relativistic terms in the time-delay formula (40.14).

<div style="float:right">(4) comparison with experiment</div>

In practice, fortunately, the experimenter does not need to measure a_R, a_T, or b with high precision. Instead, he checks the time-delay formula by measuring the changes in $\Delta\tau$ as the Earth and reflector move in their orbits about the Sun; i.e., he measures $\Delta\tau$ as a function of Earth-based time τ. Notice that when the beam is passing near the sun ($b \ll a_R$, $b \ll a_T$; but $db/d\tau \gg da_R/d\tau$ and $db/d\tau \gg da_T/d\tau$ because the Earth's and reflector's orbits are nearly circular), the change of b in the ln term of (40.14) dominates all other relativistic corrections to the Newtonian delay; consequently (using $db/d\tau \sim 10$ km/sec for typical experiments)

$$\frac{d\Delta\tau}{d\tau} - \left(\begin{array}{l} \text{Constant Newtonian} \\ \text{part} \end{array} \right) \approx -4(1 + \gamma) \frac{M_\odot}{b} \frac{db}{d\tau} \quad (40.15)$$

$$\sim 4(1 + \gamma)\left(\frac{1.5 \text{ km}}{10^6 \text{ km}} \right)\left(\frac{10 \text{ km}}{\text{sec}} \right) \sim \frac{30 \ \mu\text{sec}}{\text{day}}.$$

Such differential shifts in round-trip travel time—which rise as the Earth-reflector line moves toward the Sun and falls as it moves away—are readily observable.

In practice, in order to obtain precisions better than about 20 percent in the determination of the parameter γ by time-delay measurements, one must carefully collect and analyze data for a large fraction of a year—from a time when the beam is far from the sun ($b \sim a_T \sim 10^8$ km), to the time of superior conjunction ($b \sim R_\odot \sim 10^6$ km), and on around to a time of distant beam again. Such a long "arc" of data is needed to determine the reflector's orbit with high precision, and to take full advantage of the slow, logarithmic falloff of $\Delta\tau$ with b (40.14). When the beam is far from the sun ($b \gg R_\odot$), the simplifying assumptions behind equation (40.15) are not valid; and the relativistic time-delay gets intertwined with the orbital motions of the Earth and the reflector. The analysis then remains straightforward, but its details are so complex that one resorts to numerical integrations on a computer to carry it out. Because the orbital motions enter, the time-delay data then contain information about other metric parameters (β is the dominant one) in addition to γ.

The experimental results as of 1971 are described in Box 40.2. They yield a value for the PPN parameter γ that is more accurate than the value from light and radio-wave deflection experiments:

(5) experimental result for γ

$$\gamma = 1.02 \pm 0.08. \tag{40.16}$$

Future experiments using spacecraft may improve the precision of γ to ± 0.001 or better.

EXERCISE

Exercise 40.3. FERMAT'S PRINCIPLE

Prove Fermat's principle for a static gravitational field. [*Hint:* The proof might proceed as follows. Write down the geodesic equation in four-dimensional spacetime using an affine parameter λ. Convert from the parameter λ to coordinate time t, and use $ds^2 = 0$ to obtain

$$g_{jk}\frac{d^2x^k}{dt^2} + \Gamma_{jk\ell}\frac{dx^k}{dt}\frac{dx^\ell}{dt} - \Gamma_{j00}\frac{g_{k\ell}}{g_{00}}\frac{dx^k}{dt}\frac{dx^\ell}{dt} + \frac{d^2t/d\lambda^2}{(dt/d\lambda)^2}g_{jk}\frac{dx^k}{dt} = 0.$$

Combine with the time part of the geodesic equation

$$\frac{d^2t/d\lambda^2}{(dt/d\lambda)^2} = -2\Gamma_{0k0}\frac{dx^k/dt}{g_{00}}$$

and use the expression for the Christoffel symbols in terms of the metric to obtain

$$\gamma_{jk}\frac{d^2x^k}{dt^2} + \frac{1}{2}(\gamma_{jk,\ell} + \gamma_{j\ell,k} - \gamma_{k\ell,j})\frac{dx^k}{dt}\frac{dx^\ell}{dt} = 0, \qquad \gamma_{jk} \equiv -\frac{g_{jk}}{g_{00}}.$$

Then notice that this is a geodesic equation with affine parameter t in a three-dimensional manifold with metric γ_{jk}. The familiar extremum principle for this geodesic is

$$\delta\int_{a^j}^{b^j}(\gamma_{jk}\,dx^j\,dx^k)^{1/2} = \delta\int_{a^j}^{b^j}dt = 0,$$

which is precisely Fermat's principle!]

**Box 40.2 RADAR TIME DELAY IN THE SOLAR SYSTEM:
EXPERIMENTAL RESULTS**

Two types of experiments have been performed to measure the relativistic effects [proportional to $\frac{1}{2}(1 + \gamma)$; equation (40.14)] in the round-trip radar travel time in the solar system. In one type (*"passive"* experiment) the reflector is the surface of the planet Venus or the planet Mercury. In the other type (*"active"* experiment) the "reflector" is electronic equipment on board a spacecraft that receives the signal and transmits it back to Earth ("transponder"). Passive experiments suffer from noise due to topography of the reflecting planet (earlier radar return from mountain tops than from valley floors), and they suffer from weakness of the returned signal. Active experiments suffer from buffeting of the spacecraft by solar wind, buffeting by fluctuations in solar radiation pressure, and buffeting by leakage from gas jets ("outgassing"). Experiments of the future will solve these problems by placing a transponder on the surface of a planet or on a "drag-free" (buffeting-free) spacecraft. But experiments of the present and future must both contend with fluctuating time delays due to dispersion in the fluctuating solar wind and corona. Fortunately, these are smaller than the relativistic effects, except when

The Mariner VI spacecraft (mock-up), which was the reflector in a 1970 measurement of $\frac{1}{2}(1 + \gamma)$ by radar time delay [photo courtesy the Caltech Jet Propulsion Laboratory].

the beam passes within 2 or 3 solar radii of the sun.

The results of experiments performed before 1972 are listed in the table.

Dates of observation	Radar telescopes	Reflector	Experimenters and reference	Wave length	$\frac{1}{2}(1 + \gamma) = \left(\dfrac{Observed\ delay}{Einstein\ prediction}\right)$	Formal standard error	One-sigma error
November 1966 to August 1967	Haystack (MIT)	Venus and Mercury	Shapiro (1968)	3.8 cm	0.9		±0.2
1967 through 1970	Haystack (MIT), and Arecibo (Cornell)	Venus and Mercury	Shapiro, Ash, *et al.* (1971)	3.8 cm, and 70 cm.	1.015	±0.02	±0.05
October 1969 to January 1971	Deep Space Network (NASA)	Mariner VI and VII spacecraft	Anderson, *et al.* (1971)	14 cm.	1.00	±0.014	±0.04

The heading *Experimental result*[a] spans the last three columns.

[a] Here (observed delay)/(Einstein prediction) is the value of $\frac{1}{2}(1 + \gamma)$ obtained by fitting the observational data, $\Delta\tau(\tau)$, to a more sophisticated version of the PPN prediction (40.14). This more sophisticated version includes the gravitational influences of all the planets on the orbits of reflector and Earth; also the effect of the moon on the Earth's orbit and the effect of the Earth's rotation on the travel time; also, to as good an extent as possible, the delay due to dispersion in the solar corona and wind. "Formal standard error" and "one-sigma error" are defined in the table in Box 40.1.

§40.5. PERIHELION SHIFT AND PERIODIC PERTURBATIONS IN GEODESIC ORBITS

Perihelion shift for geodesic orbits around spherical sun, ignoring preferred-frame effects

The light-deflection and time-delay experiments both measured γ. To measure other PPN parameters, one must examine the effects of gravity on slowly moving bodies; this was the message of §40.2.

Begin with the simplest of cases: the geodesic orbit of a test body in the sun's spherical gravitational field, ignoring all gravitational effects of the planets, of solar oblateness, and of motion relative to any preferred frame. The PPN metric then has the form (40.3):

$$ds^2 = -\left[1 - 2\frac{M_\odot}{r} + 2\beta\frac{M_\odot^2}{r^2}\right]dt^2$$
$$+ \left[1 + 2\gamma\frac{M_\odot}{r}\right][dr^2 + r^2(d\theta^2 + \sin^2\theta \, d\phi^2)]. \tag{40.3}$$

Orient the coordinates so the test body moves in the equatorial "plane" $\theta = \pi/2$; and calculate the shape $r(\phi)$ of its nearly Keplerian, nearly elliptical geodesic orbit. The result, accurate to order M_\odot/r beyond Newtonian theory, is

$$r = \frac{(1 - e^2)a}{1 + e\cos[(1 - \delta\phi_0/2\pi)\phi]}, \tag{40.17}$$

where a and e are constants of integration, and $\delta\phi_0$ is defined by

$$\delta\phi_0 = \frac{(2 - \beta + 2\gamma)}{3}\frac{6\pi M_\odot}{a(1 - e^2)}$$
$$= \frac{6\pi M_\odot}{a(1 - e^2)} \text{ in general relativity.} \tag{40.18}$$

(For derivation, see exercise 40.4.)

Notice that, if $\delta\phi_0$ were zero—as it is in the Newtonian limit—then the orbit (40.18) would be an ellipse with semimajor axis a and eccentricity e (see Box 25.4). The constant $\delta\phi_0$ merely makes the ellipse precess: for r to go through a complete circuit, from perihelion to aphelion to perihelion again, $(1 - \delta\phi_0/2\pi)\phi$ must change by 2π; so ϕ must change by $2\pi + \delta\phi_0$. Thus, *the perihelion shifts forward by an angle $\delta\phi_0$ with each circuit around the ellipse*.

Relative to what does the perihelion shift? (1) Relative to the PPN coordinate system; (2) relative to inertial frames at the outskirts of the solar system (since the PPN coordinates are tied to those frames; see §39.12); (3) relative to a frame determined by the "fixed stars in the sky" (since the inertial frames at the outskirts of the solar system, inertial frames elsewhere in our galaxy, and inertial frames in our cluster of galaxies should not rotate significantly relative to each other); (4) relative to the perihelia of (other) planets, which themselves are shifting at calculable rates that decrease as one moves outward in the solar system from Mercury to Venus to Earth to. . . .

The perihelion shift is not the only relativistic effect contained in the orbital motion for a test body. There are other effects, but they are all periodic rather than cumulative with time; so, with the limited technology of the pre-space-age era, it was impossible to detect them. But the technology of the 1970's is bringing them within reach. Moreover, many space-age experiments are necessarily of short duration (\lesssim one orbit)—particularly those involving spacecraft and transponders landed on planets. For these, the periodic perturbations in an orbit are of almost as much experimental value as the cumulative perihelion shift. The periodic effects are not obvious in the PPN orbital equation (40.17); it looks like the simplest of precessing ellipses. But the quantities the observer measures directly are not a, e, and $\delta\phi_0$. Instead, he measures the time evolution of round-trip radar travel times, $\Delta\tau(\tau)$, and of angular positions on the sky $[\theta_0(\tau), \phi_0(\tau)]$. To compute these quantities is perfectly straightforward in principle, but in practice is a very complex task. The calculations predict relativistic effects periodic with the frequency of the orbit and all its harmonics. The amplitudes of these effects, for the lower harmonics, must obviously be of the order of $M_\odot \sim 1$ km ~ 10 μsec $\sim 0''.01$ arc on the sky. (The distance $M_\odot = 1.48$ km is the characteristic length for all relativistic effects in the sun's spherical field!)

<div style="text-align: right">Periodic perturbations in geodesic orbits</div>

The most favorable orbits for experimental tests of the perihelion shift and of periodic effects are those that go nearest the sun and have the highest eccentricity [see equation (40.18)]—the orbits of Mercury, Venus, Earth, Mars, and the asteroid Icarus. But how does one know that these orbits are geodesics? After all, planets are not "test bodies"; they themselves produce nonnegligible curvature in spacetime. It turns out (see §40.9 for full discussion) that there should exist tiny deviations from geodesic motion, but they are too small to compete with the perihelion shift or with the periodic effects discussed above, at least for these five bodies.

<div style="text-align: right">Comparison of theory with planetary orbits</div>

Extensive astronomical observations of planetary orbits, dating back to the mid-nineteenth century and aided by radar since 1966, have yielded accurate values for planetary perihelion shifts (accurate to ±0.4 seconds of arc per century for Mercury). From the data, which are summarized and discussed in Box 40.3, one obtains the value

$$\frac{1}{3}(2 - \beta + 2\gamma) = 1.00 \begin{cases} +0.01 \\ -0.10 \end{cases} \tag{40.19a}$$

for the ratio of observed relativistic shift to general relativistic prediction. Combining this result with the radar-delay value for γ (40.16), one obtains a value

$$\beta = 1.0 \begin{cases} +0.4 \\ -0.2 \end{cases} \tag{40.19b}$$

<div style="text-align: right">Experimental result for β</div>

for the PPN parameter β. (Recall: β measures the "amount of nonlinearity in the superposition law for g_{00}.")

The periodic effects in the planetary orbits have not yet (1973) been studied experimentally.

The above discussion and Box 40.3 have ignored the motion of the solar system relative to the preferred frame (if one exists); i.e., they have ignored the terms (40.3')

Box 40.3 PERIHELION SHIFTS; EXPERIMENTAL RESULTS

Relativistic corrections to Newtonian theory are not the only cause of shift in the perihelion of a planetary orbit. Any departure of the Newtonian gravitational field from its idealized, spherical, inverse-square-law form also produces a shift. Such nonsphericities and resulting shifts are brought about by (1) the gravitational pulls of other planets, and (2) deformation of the sun ("solar oblateness"; "quadrupole moment"). In addition, when the primary data are optical positions of planets on the sky (right ascension and declination as functions of time), there is an apparent perihelion shift caused by the precession of the Earth's axis ("general precession"; observer not on a "stable platform"; see exercise 16.4).

The perihelion shifts due to a general precession and to the gravitational pulls of other planets can be calculated with high precision. But in 1973 there is no fully reliable way to determine the solar quadrupole moment. It is conventional to quantify the sun's quadrupole moment by a dimensionless parameter J_2, which appears in the following expression for the Newtonian potential,

$$U = \frac{M_\odot}{r}\left[1 - J_2\,\frac{R_\odot{}^2}{r^2}\left(\frac{3\cos^2\theta - 1}{2}\right)\right].$$

If the sun were rotating near breakup velocity, J_2 would be near 1. Very careful measurements of the optical shape of the sun [Dicke and Goldenberg (1967)] show a flattening, which suggests J_2 may be near 3×10^{-5}.

The total perihelion shift produced by relativity plus solar quadrupole moment is (see exercise 40.5)

$$\delta\phi = \frac{6\pi M_\odot}{a(1 - e^2)}\,\lambda_p,$$

$$\lambda_p \equiv \frac{2 - \beta + 2\gamma}{3} + J_2\,\frac{R_\odot{}^2/M_\odot}{2a(1 - e^2)}.$$

The Haystack radar antenna, which Irwin Shapiro and his group have used to collect extensive data on the systematics of the inner part of the solar system. Those data are rapidly becoming the most important source of information about perihelion shifts. (Picture courtesy of Lincoln Laboratories, MIT.)

Note that relativistic and quadrupole shifts have different dependences on the semimajor axis a and ecentricity e of the orbit. This difference in dependence allows one to obtain values for both the quadrupole moment parameter J_2, and the PPN parameter $\frac{1}{3}(2 - \beta + 2\gamma)$ by combining measurements of $\delta\phi$ for more than one planet.

The experimental results, as of 1972, are as follows.

I. Data for Mercury from optical studies [Clemence (1943, 1947)]*
(general relativity with no solar oblateness predicts 43.03″/century)

Quantity	Value
(a) Total observed shift per century	5599″.74 ± 0″.41
(b) Contribution to shift caused by observer not being in an inertial frame far from the sun ("general precession" as evaluated in 1947)	5025″.645 ± 0″.50
(c) Shift per century produced by Newtonian gravitation of other planets	531″.54 ± 0″.68
(d) Residual shift per century to be attributed to general relativity plus solar oblateness	42″.56 ± 0″.94
(e) Residual shift if one uses the 1973 value for the "general precession"	41″.4 ± 0″.90
(f) Corresponding value of λ_p (see above)	$\lambda_p = 0.96 \pm 0.02$

II. 1970 Results of Shapiro (1970, 1971a,b), Shapiro *et al.* (1972)

(a) Values of λ_p obtained by reanalyzing all the world's collection of optical data, and combining it with radar data	$\begin{cases} (\lambda_p)_{\text{Mercury}} = 1.00 \pm 0.01 \\ (\lambda_p)_{\text{Mars}} = 1.07 \pm 0.10 \end{cases}$
(b) Value of J_2 obtained by comparing the observed shifts for Mercury and Mars	$J_2 \lesssim 3 \times 10^{-5}$

III. Theoretical implications of Shapiro's results

(a) Value of $(2 - \beta + 2\gamma)/3$	$1.00 \begin{cases} +0.01 \\ -0.10 \end{cases}$
(b) Value of β obtained by combining with γ from time delay experiments [equation (40.16)]	$1.0 \begin{cases} +0.4 \\ -0.2 \end{cases}$

*Clemence (1947) notes, "The observations cannot be made in a Newtonian frame of reference. They are referred to the moving equinox, that is, they are affected by the precession of the equinoxes, and the determination of the precessional motion is one of the most difficult problems of observational astronomy, if not the most difficult. In the light of all these hazards, it is not surprising that a difference of opinion could exist regarding the closeness of agreement between the observed and theoretical motions."

in the sun's metric. When one takes account of these terms, one finds an additional contribution to the perihelion shift, given for small eccentricities $e \ll 1$ by

Perihelion shift due to preferred-frame forces

$$\delta\phi_0 = -\alpha_1 \frac{\pi}{2e}\left(\frac{M_\odot}{a}\right)^{1/2} w \cdot Q - \alpha_2 \frac{\pi}{4}[(w \cdot P)^2 - (w \cdot Q)^2]$$
$$+ \alpha_3 \frac{\pi}{e}\left(\frac{|\Omega_\odot|}{M_\odot}\right)\left(\frac{\omega_\odot a^2}{M_\odot}\right) w \cdot Q$$

(40.20)

[see Nordtvedt and Will (1972)]. Here M_\odot, Ω_\odot, and ω_\odot are the sun's mass, self-gravitational energy, and rotational angular velocity; w is the sun's velocity relative to the preferred frame; a and e are the semimajor axis and eccentricity of the orbit; P is the unit vector pointing from the sun to the perihelion; and Q is a unit vector orthogonal to P and lying in the orbital plane. Comparison with observations for

Mercury—and combination with limits on α_1 and α_2 discussed below [equations (40.46b) and (40.48)]—yields the limit

Experimental result for α_3

$$\left| \alpha_3 \frac{w \cdot Q}{200 \text{ km/sec}} \right| \lesssim 2 \times 10^{-5}. \tag{40.21a}$$

Since the velocity of the sun around the Galaxy is \sim200 km/sec, and the peculiar motion of the Galaxy relative to other nearby galaxies is \sim200 km/sec, a value $w \sim 200$ km/sec is reasonable. Moreover, there is no reason to believe that w and Q are orthogonal, so one is fairly safe in concluding

$$|\alpha_3| = |4\beta_1 - 2\gamma - 2 - \zeta| \lesssim 2 \times 10^{-5} \tag{40.21b}$$

This is a stringent limit on theories that possess universal rest frames. For example, with great certainty it rules out a theory devised by Coleman (1971), which has $\beta = \gamma = 1$, but $\alpha_3 = -4$; see Ni (1972).

The future of orbital experiments

 Looking toward the future, one cannot expect data on orbits of spacecraft to give decisive tests of general relativity, despite the high precision (\sim10 meters in 1972) with which spacecraft can be tracked. Spacecraft are buffeted by the solar wind. They respond to fluctuations in this wind and in the pressure of solar radiation, and respond also to "outgassing" from leaky jets. Unless one can develop excellent "drag-free" or "conscience-guided" spacecraft, one must therefore continue to rely on planets as the source of data on geodesics. However, planetary data themselves can be greatly improved in the future by placing radar transponders on the surfaces of planets or in orbit about them, by improvements in radar technology, and by the continued accumulation of more and more observations.

EXERCISES

Exercise 40.4. DERIVATION OF PERIHELION SHIFT IN PPN FORMALISM

[See exercise 25.16 for a derivation in general relativity, accurate when gravity is strong ($2M/r$ as large as $\frac{1}{3}$) but the orbital eccentricity is small. The present exercise applies to any "metric theory" and to any eccentricity, but it assumes gravity is weak ($2M/r \ll 1$) and ignores motion relative to any universal rest frame.] Derive equation (40.17) for the shape of any bound orbit of a test particle moving in the equatorial plane of the PPN gravitational field (40.3). Keep only "first-order" corrections beyond Newtonian theory (first order in powers of M_\odot/r). [*Sketch of solution* using Hamilton-Jacobi theory (Box 25.4): (1) Hamilton-Jacobi equation, referred to a test body of unit mass, is

$$-1 = g^{\alpha\beta} \widetilde{S}_{,\alpha} \widetilde{S}_{,\beta}$$
$$= -\left[1 + 2\frac{M_\odot}{r} + (4 - 2\beta)\left(\frac{M_\odot}{r}\right)^2\right]\left(\frac{\partial \widetilde{S}}{\partial t}\right)^2 + \left[1 - 2\gamma \frac{M_\odot}{r}\right]\left[\left(\frac{\partial \widetilde{S}}{\partial r}\right)^2 + \frac{1}{r^2}\left(\frac{\partial \widetilde{S}}{\partial \phi}\right)^2\right].$$

(2) Solution to Hamilton-Jacobi equation is

$$\widetilde{S} = -\widetilde{E}t + \widetilde{L}\phi \pm \int^r \left\{ -(1 - \widetilde{E}^2) + \frac{2M_\odot}{r}[1 - (1 + \gamma)(1 - \widetilde{E}^2)] \right.$$
$$\left. - \frac{\widetilde{L}^2}{r^2}\left[1 - \frac{2M_\odot{}^2}{\widetilde{L}^2}(2 - \beta + 2\gamma)\right] \right\}^{1/2} dr, \tag{40.22}$$

where post-post-Newtonian corrections have been discarded. In discarding post-post-New-tonian corrections, recall that \tilde{E} is the conserved energy per unit rest mass and \tilde{L} is the angular momentum per unit rest mass (see Box 25.4). Consequently one has the order-of-magnitude relations

$$1 - \tilde{E}^2 \sim (\text{velocity of test body})^2 \sim M_\odot/r$$

and

$$(M_\odot/\tilde{L})^2 \sim (M_\odot/rv)^2 \sim M_\odot/r.$$

(3) The shape of the orbit is determined by the "condition of constructive interference," $\partial\tilde{S}/\partial\tilde{L} = 0$:

$$\phi = \pm \int \left\{ -\frac{1 - \tilde{E}^2}{\tilde{L}^2} + \frac{2M_\odot}{\tilde{L}^2 r}[1 - (1 + \gamma)(1 - \tilde{E}^2)] \right.$$
$$\left. - \frac{1}{r^2}\left[1 - \frac{2M_\odot^2}{\tilde{L}^2}(2 - \beta + 2\gamma)\right] \right\}^{-1/2} d(1/r).$$

(4) This integral is readily evaluated in terms of trigonometric functions. For a bound orbit ($\tilde{E} < 1$), the integral is

$$\phi = \left(1 + \frac{\delta\phi_0}{2\pi}\right)\cos^{-1}\left[\frac{(1 - e^2)a}{er} - \frac{1}{e}\right]$$

where

$$a \equiv \frac{M_\odot}{1 - \tilde{E}^2}[1 - (1 + \gamma)(1 - \tilde{E}^2)],$$

$$1 - e^2 \equiv \left(\frac{\tilde{L}}{M_\odot}\right)^2(1 - \tilde{E}^2)\left[1 + 2(1 + \gamma)(1 - \tilde{E}^2) - 2\left(\frac{M_\odot}{\tilde{L}}\right)^2(2 - \beta + 2\gamma)\right], \qquad (40.23)$$

$$\delta\phi_0 \equiv \frac{1}{3}(2 - \beta + 2\gamma)6\pi(M_\odot/\tilde{L})^2.$$

(5) Straightforward manipulations bring this result into the form of equations (40.17) and (40.18).]

Exercise 40.5. PERIHELION SHIFT FOR OBLATE SUN

(a) The Newtonian potential for an oblate sun has the form

$$U = \frac{M_\odot}{r}\left(1 - J_2\frac{R_\odot^2}{r^2}\frac{3\cos^2\theta - 1}{2}\right), \qquad (40.24)$$

where J_2 is the "quadrupole-moment parameter." One knows that $J_2 \lesssim 3 \times 10^{-5}$. Show that if an oblate sun is at rest at the origin of the PPN coordinate system, the metric of the surrounding spacetime [equations (39.32)] can be put into the form

$$ds^2 = -\left[1 - 2\frac{M_\odot}{r} - 2J_2\left(\frac{M_\odot R_\odot^2}{r^3}\right)\left(\frac{3\cos^2\theta - 1}{2}\right) + 2\beta\left(\frac{M_\odot}{r}\right)^2\right]dt^2$$

$$+ \left[1 + 2\gamma\frac{M_\odot}{r}\right][dr^2 + r^2(d\theta^2 + \sin^2\theta\,d\phi^2)] \qquad (40.25)$$

$$+ \text{corrections of post-post-Newtonian magnitude.}$$

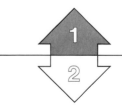

(b) Let a test particle move in a bound orbit in the equatorial plane. Use Hamilton-Jacobi theory to show that its orbit is a precessing ellipse [equation (40.17)] with a precession per orbit given by

$$\delta\phi_0 = \frac{2 - \beta + 2\gamma}{3} \frac{6\pi M_\odot}{a(1 - e^2)} + J_2 \frac{3\pi R_\odot{}^2}{a^2(1 - e^2)^2}. \tag{40.26}$$

For the significance of this result, see Box 40.3.

The rest of this chapter is Track 2. No earlier Track-2 material is needed as preparation for it, but the following will be helpful:
(1) Chapter 6 (accelerated observers);
(2) §17.6 (no prior geometry); and
(3) Chapters 38 and 39 (tests of foundations; other theories; PPN formalism).
It is not needed as preparation for any later chapter.

3-body effects in lunar orbit:

(1) theory

(2) prospects for measurement

§40.6. THREE-BODY EFFECTS IN THE LUNAR ORBIT

The relativistic effects discussed thus far all involve the spherical part of the sun's external gravitational field, and thus they can probe only the PPN parameters β and γ plus the "preferred-frame" parameters α_1, α_2, and α_3. Attempts to measure other PPN parameters can focus on three-body interactions (discussed here), the dragging of inertial frames by a rotating body (§40.7), anomalies in the locally measured gravitational constant (§40.8), and deviations of planetary and lunar orbits from geodesics (§40.9).

There is no better place to study three-body interactions than the Earth-moon orbit. The pulls of the Earth, the moon, and the sun all contribute. Perturbations in the motion of Earth and moon about their common center of gravity can be measured with high precision using laser ranging (earth-moon separation measured to ~10 cm in early 1970's) and using a radio beacon on the moon's surface (angular position on sky potentially measurable to better than 0".001 of arc).

Over and above any Newtonian three-body interactions, the Earth and the sun, acting together in a nonlinear manner, should produce relativistic perturbations in the lunar orbit that are barely within the range of this technology. These effects depend on the familiar parameters γ (measuring space curvature) and β [measuring amount of nonlinear superposition, $(U_{\text{Earth}} + U_{\text{sun}})^2$, in g_{00}]. In addition, they depend on β_2, which regulates the extent to which the sun's potential, U_{sun}, acting inside the Earth, affects the strength of the Earth's gravitational pull, causing it to vary as the Earth moves nearer and farther from the sun. These effects are expected to depend also on ζ, Δ_1, and Δ_2, which regulate the extent to which the Earth's orbital momentum and anisotropies in kinetic energy (caused by the sun) gravitate.

Bromberg (1958), Baierlein (1967), and Krogh and Baierlein (1968) have calculated the three dominant three-body effects in the Earth-moon orbit using general relativity and the Dicke-Brans-Jordan theory. These effects are noncumulative and have amplitudes of ~100 cm, ~20 cm, and ~10 cm. The 100-cm effect [which was originally discovered by de Sitter (1916)] is known to depend only on γ. The precise dependence of the other effects on the PPN parameters is not known.

The prospects for measuring these effects in the 1970's are dim; they are masked by peculiarities in the orbit of the moon that have nothing to do with relativity.

§40.7. THE DRAGGING OF INERTIAL FRAMES

Gyroscope precession:

The experiments discussed thus far study the motion of electromagnetic waves, spacecraft, planets, and asteroids through the solar system. An entirely different type of experiment measures changes in the orientation of a gyroscope moving in the gravitational field of the Earth. This experiment is particularly important because it can measure directly the "dragging of inertial frames" by the angular momentum of the Earth.

It is useful, before specializing to a rotating Earth, to derive a general expression for the precession of a gyroscope in the post-Newtonian limit. (Track-1 readers, and others who have not studied Chapters 6 and 39, may have difficulty following the derivation. No matter. It is the answer that counts!)

(1) general analysis

Let S^α be the spin of the gyroscope (i.e., its angular momentum vector), and let u^α be its 4-velocity. The spin is always orthogonal to the 4-velocity, $S^\alpha u_\alpha = 0$ (see Box 5.6). *Assume that any nongravitational forces acting on the gyroscope are applied at its center of mass,* so that there is no torque in its proper reference frame. Then the gyroscope will "Fermi-Walker transport" its spin along its world line (see §6.5):

$$\boldsymbol{\nabla}_{\boldsymbol{u}}\boldsymbol{S} = \boldsymbol{u}(\boldsymbol{a} \cdot \boldsymbol{S}), \qquad \boldsymbol{a} \equiv \boldsymbol{\nabla}_{\boldsymbol{u}}\boldsymbol{u} = \text{4-acceleration.} \qquad (40.27)$$

The objective of the calculation is to write down and analyze this transport equation in the post-Newtonian limit.

The gyroscope moves relative to the PPN coordinate grid with a velocity $v_j \equiv dx^j/dt \equiv dx_j/dt$. Assume that $v_j \lesssim \epsilon$, where ϵ is the post-Newtonian expansion parameter ($\epsilon^2 \approx M_\odot/R_\odot$). As the gyroscope moves, it carries with itself an orthonormal frame $\boldsymbol{e}_{\hat{\alpha}}$, which is related to the PPN coordinate frame by a pure Lorentz boost, plus a renormalization of the lengths of the basis vectors [transformation (39.41)]. The spin is a purely spatial vector ($S^0 = 0$) in this comoving frame; its length $(S_{\hat{j}}S_{\hat{j}})^{1/2}$ remains fixed (conservation of angular momentum); and its direction is regulated by the Fermi-Walker transport law.

The basis vectors $\boldsymbol{e}_{\hat{\alpha}}$ of the comoving frame are *not* Fermi-Walker transported, by contrast with the spin. Rather, they are tied by a pure boost (no rotation!) to the PPN coordinate grid, which in turn is tied to an inertial frame far from the solar system, which in turn one expects to be fixed relative to the "distant stars." Thus, by calculating the precession of the spin relative to the comoving frame,

$$dS_{\hat{j}}/d\tau \equiv \varepsilon_{\hat{j}\hat{k}\hat{l}}\Omega_{\hat{k}}S_{\hat{l}}, \qquad (40.28)$$

one is in effect evaluating the spin's angular velocity of precession, $\Omega_{\hat{j}}$, relative to a frame fixed on the sky by the distant stars.

Calculate $dS_{\hat{j}}/d\tau$:

$$\frac{dS_{\hat{j}}}{d\tau} = \boldsymbol{\nabla}_{\boldsymbol{u}}(\boldsymbol{S} \cdot \boldsymbol{e}_{\hat{j}}) = (\boldsymbol{\nabla}_{\boldsymbol{u}}\boldsymbol{S}) \cdot \boldsymbol{e}_{\hat{j}} + \boldsymbol{S} \cdot (\boldsymbol{\nabla}_{\boldsymbol{u}}\boldsymbol{e}_{\hat{j}}) = \boldsymbol{S} \cdot \boldsymbol{\nabla}_{\boldsymbol{u}}\boldsymbol{e}_{\hat{j}}. \qquad (40.29)$$

Here use is made of the fact that $\boldsymbol{\nabla}_{\boldsymbol{u}}\boldsymbol{S}$ is in the \boldsymbol{u} direction [equation (40.27)] and

is thus orthogonal to $\boldsymbol{e}_{\hat{\jmath}}$. The quantity $\boldsymbol{S} \cdot \boldsymbol{\nabla}_{u} \boldsymbol{e}_{\hat{\jmath}}$ is readily evaluated in the PPN coordinate frame. In the evaluation, one uses as metric coefficients [equations (39.32)] the expressions

$$g_{00} = -1 + 2U + O(\epsilon^4), \qquad g_{jk} = \delta_{jk}(1 + 2\gamma U) + O(\epsilon^4),$$

$$g_{0j} = -\frac{7}{2}\Delta_1 V_j - \frac{1}{2}\Delta_2 W_j + \left(\begin{array}{c}\text{``preferred-}\\\text{frame terms''}\end{array}\right) + O(\epsilon^5); \tag{40.30}$$

one takes as the components of $\boldsymbol{e}_{\hat{\jmath}}$ and \boldsymbol{S} [obtained via the transformation (39.41)] the expressions

$$e_{\hat{\jmath}}^{\;0} = v_j + O(\epsilon^3), \qquad e_{\hat{\jmath}}^{\;k} = (1 - \gamma U)\,\delta_{jk} + \frac{1}{2}\,v_k v_j + O(\epsilon^4),$$

$$S^0 = v_j S_{\hat{\jmath}} + O(\epsilon^3 S_j), \tag{40.31}$$

$$S^k = (1 - \gamma U)S_{\hat{k}} + \frac{1}{2}\,v_k v_j S_{\hat{\jmath}} + O(\epsilon^4 S_j);$$

and one uses the relation

$$dv_j/d\tau = a_j + U_{,j} + O(\epsilon^2 U_{,j}) \tag{40.32}$$

where a_j (assumed $\lesssim U_{,j}$) are the components of the 4-acceleration. One finds (see exercise 40.6) for the precession of the spin the result

$$dS_{\hat{\jmath}}/d\tau = \boldsymbol{S} \cdot \boldsymbol{\nabla}_{u} \boldsymbol{e}_{\hat{\jmath}} = S_{\hat{k}}[v_{[j}a_{k]} + g_{0[k,j]} - (2\gamma + 1)v_{[j}U_{,k]}].$$

Rewritten in *three-dimensional vector form* this result becomes

(2) general PPN formula for
precession

$$dS/d\tau = \boldsymbol{\Omega} \times \boldsymbol{S}, \tag{40.33a}$$

$$\boldsymbol{\Omega} \equiv -\frac{1}{2}\boldsymbol{v} \times \boldsymbol{a} - \frac{1}{2}\boldsymbol{\nabla} \times \boldsymbol{g} + \left(\gamma + \frac{1}{2}\right)\boldsymbol{v} \times \boldsymbol{\nabla}U, \tag{40.33b}$$

$$\boldsymbol{g} \equiv g_{0j}\boldsymbol{e}_{\hat{\jmath}}. \tag{40.33c}$$

In this final answer it does not matter whether the 3-vectors entering into $\boldsymbol{\Omega}$ are evaluated in the coordinate frame or in the comoving orthonormal frame, since $\boldsymbol{e}_{\hat{\jmath}}$ and $\partial/\partial x_j$ differ only by corrections of order ϵ^2.

Equations (40.33) describe in complete generality at the post-Newtonian level of approximation the precession of the gyroscope spin \boldsymbol{S} relative to the comoving orthonormal frame that is rotationally tied to the distant stars.

(3) specialization: Thomas
precession

For an electron with spin \boldsymbol{S} in orbit around a proton, only the first term, $-\frac{1}{2}\boldsymbol{v} \times \boldsymbol{a}$, is present (no gravity). This term leads to the Thomas precession, which plays an important role in the fine structure of atomic spectra [see, e.g., Ruark and Urey (1930)]. For other ways of deriving the Thomas precession, see exercise 6.9 and §41.4.

The Thomas precession comes into play for a gyroscope on the surface of the Earth (\boldsymbol{a} = Newtonian acceleration of gravity), but not for a gyroscope in a freely moving satellite.

If one ignores the rotation of the Earth and preferred-frame effects, and puts the PPN coordinate frame at rest relative to the center of the Earth, then g_{0j} vanishes and Ω is given by

$$\Omega = \mathbf{v} \times \left[-\frac{1}{2}\mathbf{a} + \left(\gamma + \frac{1}{2}\right)\boldsymbol{\nabla} U \right]$$

$$= \gamma \mathbf{v} \times \boldsymbol{\nabla} U \text{ for gyroscope on Earth's surface} \qquad (40.34)$$

$$= \left(\gamma + \frac{1}{2}\right)\mathbf{v} \times \boldsymbol{\nabla} U \text{ for gyroscope in orbit.}$$

(4) specialization: precessions due to acceleration and Earth's Newtonian potential

The general-relativistic term $(\gamma + \frac{1}{2})\mathbf{v} \times \boldsymbol{\nabla} U$ is caused by the motion of the gyroscope through the Earth's curved, static spacetime geometry. Notice that it depends solely on the same parameter γ as is tested by electromagnetic-wave experiments. In order of magnitude, for a gyroscope in a near-Earth, polar orbit,

$$\Omega \approx \frac{3}{2}\left(\frac{M_E}{R_E}\right)^{1/2}\left(\frac{M_E}{R_E^2}\right) \approx 8 \text{ seconds of arc per year.} \qquad (40.35)$$

The general-relativistic precession $\frac{3}{2}\mathbf{v} \times \boldsymbol{\nabla} U$ was derived by W. de Sitter (1916) for the "Earth-moon gyroscope" orbiting the sun. Eleven years later L. H. Thomas (1927) derived the special relativistic precession $-\frac{1}{2}\mathbf{v} \times \mathbf{a}$ for application to atomic physics.

The Earth's rotation produces off-diagonal terms, g_{0j}, in the PPN metric (exercise 40.7):

(5) specialization: precession due to Earth's rotation

$$\mathbf{g} = g_{0j}\mathbf{e}_{\hat{j}} = -\left(\frac{7}{4}\Delta_1 + \frac{1}{4}\Delta_2\right)\frac{\mathbf{J} \times \mathbf{r}}{r^3}. \qquad (40.36)$$

Here \mathbf{J} is the Earth's angular momentum. These off-diagonal terms contribute an amount

$$\Omega = -\frac{1}{2}\boldsymbol{\nabla} \times \mathbf{g} = \left(\frac{7}{8}\Delta_1 + \frac{1}{8}\Delta_2\right)\frac{1}{r^3}\left[-\mathbf{J} + \frac{3(\mathbf{J} \cdot \mathbf{r})\mathbf{r}}{r^2}\right] \qquad (40.37)$$

to the precession of the gyroscope. Notice that this contribution, unlike the others, is independent of the linear velocity of the gyroscope. One can think of it in the following way.

The gyroscope is rotationally at rest relative to the inertial frames in its neighborhood. It and the local inertial frames rotate relative to the distant galaxies with the angular velocity Ω because the Earth's rotation "drags" the local inertial frames along with it. Notice that near the north and south poles the local inertial frames rotate in the same direction as the Earth does (Ω parallel to \mathbf{J}), but near the equator they rotate in the opposite direction (Ω antiparallel to \mathbf{J}; compare Ω with the magnetic field of the Earth!). Although this might seem paradoxical at first, an analogy devised by Schiff makes it seem more reasonable.* Consider a rotating, solid sphere immersed in a viscous fluid. As it rotates, the sphere will drag the fluid along with it. At various points in the fluid, set down little rods, and watch how the fluid

*This analogy can be made mathematically rigorous; see footnote on p. 255 of Thorne (1971); see also, §21.12 on Mach's principle.

rotates them as it flows past. Near the poles the fluid will clearly rotate the rods in the same direction as the star rotates. But near the equator, because the fluid is dragged more rapidly at small radii than at large, the end of a rod closest to the sphere is dragged by the fluid more rapidly than the far end of the rod. Consequently, the rod rotates in the direction opposite to the rotation of the sphere.

In order of magnitude, the precessional angular velocity caused by the Earth's rotation is

$$\Omega \sim J_E/R_E^3 \sim 0.1 \text{ seconds of arc per year.} \tag{40.37'}$$

(6) prospects for measuring precession

Both this precession, and the larger one [equation (40.35)] due to motion through the Earth's static field, may be detectable in the 1970's. Equipment aimed at detecting them via a satellite experiment is now (1973) under construction at Stanford University; see Everitt, Fairbank, and Hamilton (1970); also O'Connell (1972).*

The gyroscope precession produced by motion of the Earth relative to the preferred frame (if any) is too small to be of much interest.

*The dragging of inertial frames by a rotating body plays important roles elsewhere in gravitation physics, e.g., in the definition of angular momentum for a gravitating body (§19.2), and in black-hole physics (Chapter 33). The effect was first discussed and calculated by Thirring and Lense (1918). More recent calculations by Brill and Cohen (1966) of idealized situations where the effect may be large give insight into the mechanism of the effect. See also the discussion of Mach's principle in §21.12.

EXERCISES

Exercise 40.6. PRECESSIONAL ANGULAR VELOCITY
Derive equations (40.33) for the precession of a gyroscope in the post-Newtonian limit. Base the derivation on equations (40.29)–(40.32).

Exercise 40.7. OFF-DIAGONAL TERMS IN METRIC ABOUT THE EARTH
Idealize the Earth as an isolated, rigidly rotating sphere with angular momentum \boldsymbol{J}. Use equations (39.34b,c) and (39.27) to show that (in three-dimensional vector notation)

$$\boldsymbol{V} \equiv V_j \boldsymbol{e}_{\hat{j}} = \boldsymbol{W} \equiv W_j \boldsymbol{e}_{\hat{j}} = \frac{1}{2} \boldsymbol{J} \times \boldsymbol{r}/r^3 \tag{40.38}$$

outside the Earth, in the Earth's PPN rest frame. From this, infer equation (40.36).

Exercise 40.8. SPIN-CURVATURE COUPLING
Consider a spinning body (e.g., the Earth or a gyroscope or an electron) moving through curved spacetime. Tidal gravitational forces produced by the curvature of spacetime act on the elementary pieces of the spinning body. These forces should depend not only on the positions of the pieces relative to the center of the object, but also on their relative velocities. Moreover, the spin of the body,

$$\boldsymbol{S} \equiv \int (\rho \boldsymbol{r} \times \boldsymbol{v}) \, d(\text{volume}) \qquad \text{in comoving orthonormal frame,}$$

is a measure of the relative positions and velocities of its pieces. Therefore one expects the spin to couple to the tidal gravitational forces—i.e., to the curvature of spacetime—producing

deviations from geodesic motion. Careful solution of the PPN equations of Chapter 39 for general relativity reveals [Papapetrou (1951), Pirani (1956)] that such coupling occurs and causes a deviation of the worldline from the course that it would otherwise take; thus,

$$m\frac{Du^\alpha}{d\tau} = -S_\mu u_\nu \frac{D^2 u_\beta}{d\tau^2}\epsilon^{\alpha\mu\nu\beta} + \frac{1}{2}(\epsilon^{\lambda\mu\rho\tau}R^{\alpha\nu}{}_{\lambda\mu})u_\nu S_\rho u_\tau. \tag{40.39}$$

Evaluate, in order of magnitude, the effects of the supplementary term on planetary orbits in the solar system.

[*Answer:* They are much too small to be detected. However, there are two other material places to look for the effect: (1) when a rapidly spinning neutron star, or a black hole endowed with substantial angular momentum enters the powerful tidal field of another neutron star or black hole; and (2) when an individual electron, or the totality of electrons in the "Dirac sea of negative energy states," enter a still more powerful tidal field (late phase of gravitational collapse). Such a tidal field, or curvature, pulls oppositely on electrons with the two opposite directions of spin [Pirani (1956); DeWitt (1962), p. 338; Schwinger (1963a,b)] just as an electric field pulls oppositely on electrons with the two opposite signs of charge ["vacuum polarization"; see especially Heisenberg and Euler (1936)]. In principle, the tidal field pulling on the spin of an electron need not be due to "background" spacetime curvature; it might be due to a nearby massive spinning object, such as a "live" black hole (chapter 33) ["gravitational spin-spin coupling"; O'Connell (1972)].

§40.8. IS THE GRAVITATIONAL CONSTANT CONSTANT?

The title and subject of this section are likely to arouse confusion. Throughout this book one has used geometrized units, in which $G = c = 1$. Therefore, one has locked oneself into a viewpoint that forbids asking whether the gravitational constant changes from event to event.

False! One can perfectly well ask the question in the context of $G = c = 1$, if one makes clear what is meant by the question.

In §§1.5 and 1.6, c was *defined* to be a certain conversion factor between centimeters and seconds; and G/c^2 was defined to be a certain conversion factor between grams and centimeters. These definitions by fiat do not guarantee, however, that a Cavendish experiment* to measure the attraction between two bodies will yield

$$\text{Force} = -Gm_1 m_2/r^2 = -m_1 m_2/r^2.$$

If general relativity correctly describes classical gravity, and if the values of the conversion factors G and c have been chosen precisely right, then any Cavendish experiment, anywhere in the universe, *will* yield "Force = $-m_1 m_2/r^2$". But if the

*See any standard textbook for a description of Cavendish experiments. By his original version of the experiment, with two separated spheres suspended by fine wires, Henry Cavendish (1798) inferred the mass and hence the density of the Earth. He reported: "By a mean of the experiments made with the wire first used, the density of the Earth comes out 5.48 times greater than that of water; and by a mean of those made with the latter wire it comes out the same; and ... the extreme results do not differ from the mean more than 0.38, or 1/14 of the whole." The most precise method of measuring G today [Rose *et al.* (1969)] gives $G_C = (6.674 \pm .004) \times 10^{-8}$ cm^3/g sec^2 (one standard deviation).

Dicke-Brans-Jordan theory, or almost any other metric theory gives the correct description of gravity, the force in the Cavendish experiment will depend on where and when the experiment is performed, as well as on m_1, m_2, and r. To discuss Cavendish experiments as tests of gravitation theory, then, one must introduce a new proportionality factor

"Cavendish gravitational constant," G_C, defined

$$G_C \equiv G_{\text{Cavendish}} \equiv (\text{"Cavendish gravitational constant"}), \qquad (40.40)$$

which enters into the Newtonian force law

$$\text{Force} = -G_C m_1 m_2 / r^2. \qquad (40.41)$$

This Cavendish constant will be unity in general relativity, but in most other metric theories it will vary from event to event in spacetime.

Changes of G_C with time

In some theories, such as Dicke-Brans-Jordan, the Cavendish constant is determined by the distribution of matter in the universe. As a result, the expansion of the universe changes its value:

$$\frac{1}{G_C}\frac{dG_C}{dt} \sim -\left(\frac{0.1 \text{ to } 1}{\text{age of universe}}\right) \sim \frac{-1}{10^{10} \text{ or } 10^{11} \text{ years}}$$

[see, e.g. Brans and Dicke (1961)]. A variety of observations place limits on such time variations. Big time changes in G_C during the last 4.6 billion years would have produced marked effects on the Earth, the sun, and the entire solar system. The expected geophysical effects have been summarized and compared with observations by Dicke and Peebles (1965). It is hard to draw firm conclusions because of the complexity of the geophysics involved, but a fairly certain limit is

$$(1/G_C)(dG_C/dt) \lesssim 1/10^{10} \text{ years} \qquad \text{(geophysical)}. \qquad (40.42a)$$

Eventually, high-precision measurements of the orbital motions of planets will yield a better limit. For the present, planetary observations show

$$(1/G_C)(dG_C/dt) \lesssim 4/10^{10} \text{ years} \qquad \text{(planetary orbits)} \qquad (40.42b)$$

[Shapiro, Smith, et al. (1971)]. These limits are tight enough to begin to be interesting, but not yet tight enough to disprove any otherwise viable theories of gravity.

Spatial variations in G_C

If G_C is determined by the distribution of matter in the universe, then it should depend on where in the universe one is, as well as when. In particular, as one moves from point to point in the solar system, closer to the Sun and then farther away, one should see G_C change. Indeed this is the case in most metric theories of gravity, though not in general relativity. Analyses of Cavendish experiments using the PPN formalism reveal spatial variation in G_C given by

$$\Delta G_C = -2G_C(\beta + \gamma - \beta_2 - 1)U \qquad (40.43)$$

[Nordtvedt (1970, 1971a); Will (1971b)].

The amplitude of these variations along the Earth's elliptical orbit is $\Delta G_C/G_C \sim 10^{-10}$, if $\beta + \gamma - \beta_2 - 1 \sim 1$. This is far too small to measure directly in the

1970's. Despite great ingenuity and effort, the most accurate experiments measuring the value of G_C have precisions in 1972 no better than 1 part in 10^4 [see Beams (1971)]. Experiments to search for yearly variations in G_C on Earth without measuring the actual value ("null-type experiments") can surely be performed with better precision than 1 in 10^4—but not with precisions approaching 1 in 10^{10}. On the other hand, *indirect* consequences of a spatial variation of G_C in the solar system are almost certainly measurable (see §40.9 below).

In Ni's theory of gravity (Box 39.1), and other two-tensor or vector-tensor theories like it, where the prior geometry picks out a preferred "universal rest frame," the Cavendish constant G_C can depend on velocity relative to the preferred frame. For Cavendish experiments with two equal masses separated by distances large compared to their sizes, G_C varies as

Dependence of G_C on velocity

$$\Delta G_C = G_C\left[\frac{1}{2}(\alpha_2 + \alpha_3 - \alpha_1)v^2 - \frac{1}{2}\alpha_2(v \cdot n)^2\right] \qquad (40.44)$$

[Will (1971b)]. Here v is the velocity of the Cavendish apparatus relative to the preferred frame, and n is the unit vector between the two masses. For experiments where one body is a massive sphere (e.g., the Earth), and the other is a small object on the sphere's surface, G_C varies as

$$\Delta G_C/G_C = \frac{1}{2}[(\alpha_3 - \alpha_1) + \alpha_2(1 - I/MR^2)]v^2$$
$$- \frac{1}{2}\alpha_2(1 - 3I/MR^2)(v \cdot n)^2 \qquad (40.44')$$

[Nordtvedt and Will (1972)]. Here M and R are the mass and radius of the sphere, and

$$I = \int (\rho r^2)4\pi r^2\, dr$$

is the trace of the second moment of its mass distribution. Consequences of these effects for planetary orbits have not yet been spelled out, but consequences for Earthbound experiments have.

Think of a Cavendish experiment in which one mass is the Earth, and the other is a gravimeter on the Earth's surface. The gravimeter gives a reading for the "local acceleration of gravity,"

Anomalies in Earth tides due to anisotropies in G_C:

$$g = G_C m_{\text{Earth}}/r_{\text{Earth}}^2. \qquad (40.45)$$

As the Earth turns, so the unit vector n between its center and the gravimeter rotates, G_C and hence g will fluctuate with a period of 12 sidereal hours and an amplitude

$$(\Delta g/g)_{\text{amplitude}} = \frac{1}{4}\alpha_2 v^2 \cos^2\theta_m.$$

Here θ_m is the minimum, as the Earth rotates, of the angle between v (constant vector) and n (rotating vector). (Note: we have used the value $I/MR^2 \simeq 0.5$ for the Earth.) These fluctuations will produce tides in the Earth of the same type as are

produced by the moon and sun. As of 1972, gravimeter measurements near the Earth's equator show no sign of any anomalous 12-sidereal-hour effects down to an amplitude of $\sim 10^{-9}$ [Will (1971b)]. Consequently,

(1) experimental value of α_2

$$|\alpha_2|^{1/2} v \cos \theta_m = |\Delta_2 + \zeta - 1|^{1/2} v \cos \theta_m \lesssim 6 \times 10^{-5} \sim 20 \text{ km/sec.} \quad (40.46a)$$

Using a rough estimate of $v \sim 200$ km/sec for the Earth's velocity relative to the universal rest frame, and $\theta_m \lesssim 60°$ for the angle between v and the Earth's equatorial plane, one obtains the rough limit

$$|\alpha_2| = |\Delta_2 + \zeta - 1| \lesssim 0.03. \quad (40.46b)$$

[This limit does not affect the three theories in Box 39.1; of them, only Ni's theory has prior geometry and a universal rest frame; and it predicts isotropic effects in $\Delta G_C/G_C$ [equation (40.44)], but no anisotropic effects. However, other theories with universal rest frames—e.g. Papapetrou's (1954a,b,c) theory—are ruled out by this limit; see Ni (1972), Nordtvedt and Will (1972).]

(2) experimental disproof of Whitehead theory

Whitehead's theory of gravity (which is a two-tensor theory with a rather different type of prior geometry from Ni's) predicts that the galaxy should produce velocity-independent anisotropies in G_C. These, in turn, would produce Earth tides with periods of 12 sidereal hours and amplitudes of

$$\Delta g/g \sim 2 \times 10^{-7} \sim 100 \times \begin{pmatrix} \text{experimental limit on} \\ \text{such amplitudes} \end{pmatrix}$$

[Will (1971b)]. The absence of such tides proves Whitehead's theory to be incorrect—a feat of disproof beyond the power of all redshift, light-deflection, time-delay, and perihelion-shift measurements. (For all these "standard experiments," the predictions of Whitehead and Einstein are identical!)

Anomalies in Earth rotation rate due to dependence of G_C on velocity

Equation (40.44′) predicts a periodic annual variation of the Cavendish constant on Earth, as the Earth moves around the sun:

$$v = \begin{pmatrix} \text{velocity of Earth} \\ \text{relative to sun} \end{pmatrix} + \begin{pmatrix} \text{velocity of sun relative} \\ \text{to preferred frame} \end{pmatrix} \equiv v_E + w;$$

$$(\Delta G_C/G_C)_{\substack{\text{averaged over} \\ \text{Surface of Earth}}} = \frac{1}{2}\left(\frac{2}{3}\alpha_2 + \alpha_3 - \alpha_1\right)\left(w^2 + v_E{}^2 + 2w \cdot v_E\right). \quad (40.47)$$

[varies sinusoidally with period of one year]

This annual variation, assuming all PPN parameters are of order unity, is 1,000 times larger than the one produced by the Earth's motion in and out through the sun's gravitational potential [equation (40.43)]. In response to this changing Cavendish constant, the Earth's self-gravitational pull should change, and the Earth should "breathe" inward (greater pull) and outward (relaxed pull). The resulting annual variations in the Earth's moment of inertia should produce annual changes in its rotation rate ω (changes in "length of day" as measured by atomic clocks):

$$\delta\omega/\omega \sim 0.1\left(\frac{2}{3}\alpha_2 + \alpha_3 - \alpha_1\right)w \cdot v_E$$

[Nordtvedt and Will (1972)]. Comparison with the measured annual variations of rotation rate (all of which geophysicists attribute to seasonal changes in the Earth's atmosphere) yields the following limit

$$\left| \frac{2}{3}\alpha_2 + \alpha_3 - \alpha_1 \right| = \lesssim 0.2.$$ (40.48)

Experimental value of $\frac{2}{3}\alpha_2 + \alpha_3 - \alpha_1$

[See Nordtvedt and Will (1972)]. This limit rules out several preferred-frame theories of gravity, including that of Ni (Boxes 39.1 and 39.2).

The experimental results (40.21), (40.46), and (40.48), when combined, place the following very rough limits on any theory that possesses a Universal rest frame:

$$|\alpha_1| = |7\,\Delta_1 + \Delta_2 - 4\gamma - 4| \lesssim 0.2,$$
$$|\alpha_2| = |\Delta_2 + \zeta - 1| \lesssim 0.03,$$ (40.49)
$$|\alpha_3| = |4\beta_1 - 2\gamma - 2 - \zeta| \lesssim 2 \times 10^{-5}.$$

These limits completely disprove all theories with preferred frames that have been examined to date except one devised by Will and Nordtvedt [see Ni (1972); Nordtvedt and Will (1972)].

In some theories of gravity, the result of a Cavendish experiment depends on the chemical composition and internal structure of the test bodies (exercises 40.9 and 40.10). Kruezer (1968) has performed the most accurate search for such effects to date. He finds that G_C is the same for fluorine and bromine to a precision of

Dependence of G_C on chemical composition

$$\left| \frac{G_C(\text{bromine}) - G_C(\text{fluorine})}{G_C} \right| \lesssim 5 \times 10^{-5}.$$ (40.50)

Exercise 40.9. CAVENDISH CONSTANT FOR IDEALIZED SUN

EXERCISES

Idealize the sun as a static sphere of perfect fluid at rest at the origin of the PPN coordinates. Then its external gravitational field has the form (40.3), with M_\odot given by (40.4). Consequently, a test body of mass m, located far away at radius r, is accelerated by a gravitational force

$$\text{Force} = -mM_\odot/r^2.$$ (40.51a)

(a) Calculate the mass of the sun, M, in the sense of the amount of energy required to construct it by adding one spherical shell of matter on top of another, working from the inside outward. [*Answer:*

$$M = \underbrace{\int_0^{R_\odot} \rho_0(1 + \Pi + 3\gamma U)4\pi r^2\,dr}_{\text{rest mass + internal energy}} - \underbrace{\frac{1}{2}\int_0^{R_\odot} \rho_0 U 4\pi r^2\,dr}_{\text{gravitational potential energy}}$$

$$= \int_0^{R_\odot} \rho_0\left[1 + \Pi + \left(3\gamma - \frac{1}{2}\right)U\right]4\pi r^2\,dr.]$$ (40.51b)

(b) Use the virial theorem [equation (39.21b)] to rewrite equation (40.4) in the form

$$M_\odot = \int_0^{R_\odot} \rho_0\left[1 + \beta_3\Pi + \left(2\beta_2 + \frac{1}{2}\beta_4\right)U\right]4\pi r^2\,dr.$$ (40.51c)

(c) Combine the above equations with the definition

$$\text{Force} = -G_C mM/r^2 \tag{40.51d}$$

of the Cavendish constant for r far outside the sun, to obtain

$$G_C = \frac{\left(\begin{array}{l}\text{mass of sun as defined by its effect in}\\\text{bending world line of a faraway test particle}\end{array}\right)}{\left(\begin{array}{l}\text{mass-energy as defined by applying law of}\\\text{conservation of energy to the steps in the}\\\text{construction of the sun}\end{array}\right)} \tag{40.52}$$

$$= 1 + \int (\rho_0/M_0)[(\beta_3 - 1)\Pi + \tfrac{1}{2}(4\beta_2 + \beta_4 - 6\gamma + 1)U]4\pi r^2\, dr.$$

Unless $\beta_3 = 1$, and $4\beta_2 + \beta_4 - 6\gamma + 1 = 0$ (as they are, of course, in Einstein's theory), G_C will depend on the sun's internal structure! Specialize equation (40.52) to "conservative theories of gravity (Box 39.5), and explain why the result is what one would expect from equation (40.43).

Exercise 40.10. CAVENDISH CONSTANT FOR ANY BODY

Extend the analysis of exercise 40.9 to a source that is arbitrarily stressed and has arbitrary shape and internal velocities (subject to the constraints $v^2 \ll 1$, $|t_{jk}|/\rho_0 \ll 1$, $U \ll 1$, $\Pi \ll 1$, of the post-Newtonian approximation). Assume that the body is at rest relative to the universal rest frame. Show that G_C depends on the internal structure of the source unless

$$2\beta_1 - \beta_4 = 1, \qquad 4\beta_2 + \beta_4 - 6\gamma = -1, \qquad \beta_3 = 1, \qquad \zeta = 0, \qquad \eta = 0. \tag{40.53}$$

Of course, these PPN constraints are all satisfied by Einstein's theory.

§40.9. DO PLANETS AND THE SUN MOVE ON GEODESICS?

Crucial to solar-system experiments is the question of whether the sun and the planets move on geodesics of spacetime. This question is complicated by the contributions to the spacetime curvature made by the moving body itself.

The sense in which general relativity predicts geodesic motion for planets and sun

To elucidate the question—and to obtain an answer in the framework of general relativity—consider an "Einstein elevator" type of argument. The astronomical object under consideration has an outer boundary, and each point on this boundary describes a world line. These world lines define a world tube. Some distance outside of this world tube construct a "buffer zone" as in §20.6. Tailor its inner and outer dimensions, according to the mass and moments of the object and the curvature of the enveloping space ("strength of the tide-producing force of the external gravitational field"), in such a way that the departure ϵ (cf. §20.6) of the metric from flatness in this buffer zone takes on values equal at most to twice the extremal achievable value ϵ_{extrem} (a minimum with respect to variations in r, a maximum

with respect to variations in direction; in other words, a minimax). Then, apart from errors of order ϵ_{extrem}, the object can be regarded as moving in an asymptotically flat space. The law of conservation of total 4-momentum applies. It assures one that the object moves in a (locally) straight line with uniform velocity. Consider, next, a "background geometry" that agrees just outside the buffer zone with the actual geometry to accuracy ϵ_{extrem} or better, but that inside is a source-free solution of Einstein's field equation. Then, to an accuracy governed by the magnitude of ϵ_{extrem}, the locally straight line along which the astronomical object moves will be a geodesic of this background geometry.

Insofar as one can give any well-defined meaning to the departure of the actual motion from this geodesic (a task complicated by the fact that the background geometry does not actually exist!), one can calculate this departure by making use of the PPN formalism or some other approximation scheme [see, e.g., Taub (1965)]. This deviation springs ordinarily in substantial measure, and sometimes almost wholly, from a coupling between the Riemann curvature tensor of the external field and the multipole moments of the astronomical object (angular momentum associated with rotation; quadrupole and higher moments associated with deformation; see, e.g., exercises 40.8 and 16.4). This coupling is important for the Earth-moon system, but one need not use relativity to calculate it; Newtonian theory does the job to far greater accuracy than needed—or would, if one understood the interiors of the Earth and the moon well enough! For the planets and sun, the effect is negligible. (Exercise: use Newtonian theory to prove so!).

Thus, in general relativity as applied to the solar system, one can approximate the orbit of the sun, the Earth-moon mass center, and each other planet, as a geodesic of that "background spacetime geometry" which would exist if its own curvature effects were absent. This is the approach used to analyze the perihelion shift for planets in §40.5 in the context of general relativity, and to derive in exercise 39.15 the post-Newtonian "many-body equations of motion."

In most other metric theories of gravity, including the Dicke-Brans-Jordan theory, there are substantial departures from geodesic motion. The "Einstein elevator" argument fails in these theories because spacetime is endowed not only with a metric, but also with a long-range field that couples indirectly (cf. §§38.7 and 39.2) to massive, gravitating bodies.

This phenomenon is best understood in terms of Dicke's argument about the influence of spatial variations of the fundamental constants on experiments of the Eötvös-Dicke type (see §38.6). In a theory where the Cavendish gravitational constant G_C depends on position (as it does not and cannot in general relativity), a body with significant self-gravitational energy E_{grav} must fall, in a perfectly uniform external Newtonian gravitational field, with an anomalous acceleration:

Deviations from geodesic motion:

(1) due to curvature coupling

(2) due to spatial dependence of gravitational constant (Nordtvedt effect)

$$\begin{pmatrix}\text{acceleration of} \\ \text{massive body}\end{pmatrix} - \begin{pmatrix}\text{acceleration of} \\ \text{test body}\end{pmatrix} = \frac{1}{M}\left(\frac{\partial E_{\text{grav}}}{\partial G_C}\right)\boldsymbol{\nabla} G_C$$

$$= \frac{E_{\text{grav}}}{M G_C}\boldsymbol{\nabla} G_C \tag{40.54}$$

[see equation (38.15)]. In Dicke-Brans-Jordan theory, G_C is essentially the reciprocal of the scalar field; and it contains a small part that is proportional to the Newtonian potential, U [equation (40.43) with the appropriate values of the parameters from Box 39.2]. As a result, the sun falls with an acceleration smaller by one part in 10^6 than the acceleration of a test body; Jupiter falls with an acceleration one part in 10^9 smaller; and the Earth, one part in 10^{10} smaller. Translated into relativistic language: the scalar field, by influencing the gravitational self-energy of a massive body, produces deviations from geodesic motion.

One can use the full PPN formalism of Chapter 29 to calculate the motion of massive bodies in any metric theory of gravity. Nordtvedt (1968b) and Will (1971a) have done this. They find that a massive body at rest in a uniform external field experiences a (Newtonian-type) PPN coordinate acceleration given by

$$\frac{d^2 x_j}{dt^2} = E_{jk} \frac{\partial U}{\partial x_k},$$

where E_{jk} is a quantity depending on the body's structure:

$$E_{jk} = \delta_{jk} \left\{ 1 - (7\Delta_1 - 3\gamma - 4\beta) \frac{E_{\text{grav}}}{m} \right\} - (2\beta + 2\beta_2 - 3\gamma + \Delta_2 - 2) \frac{\Omega_{jk}}{m},$$

$$\tag{40.55}$$

$$\Omega_{jk} = -\frac{1}{2} \int \frac{\rho_o \rho_o' (x_j - x_j')(x_k - x_k')}{|x - x'|^3} \, d^3x \, d^3x', \qquad E_{\text{grav}} = \sum \Omega_{jj}.$$

Here m is the body's total mass-energy, Ω_{jk} is the "Chandrasekhar potential-energy tensor," and E_{grav} is the body's self-gravitational energy. [Note: Dicke's method of calculating the anomalous acceleration (40.54) breaks down in theories that are not "conservative" (Box 39.5).]

In general relativity, the combinations of PPN coefficients appearing in E_{jk} vanish; so $E_{jk} = \delta_{jk}$, and the body falls with the usual acceleration—i.e., it moves along a geodesic. But in most other theories of gravity $E_{jk} \neq \delta_{jk}$; the body does *not* move on a geodesic; and its acceleration may even be in a different direction than the gradient of the Newtonian potential!

This predicted departure from geodesic motion is called the *"Nordtvedt effect."* The possibility of such an effect was first noticed in passing by Dicke (1964c), but was discovered independently and explored in great detail by Nordtvedt (1968a,b). The Nordtvedt effect in a theory other than general relativity produces a number of phenomena in the solar system that are potentially observable. [See Nordtvedt (1971b) for an enumeration and references.] The effect most suitable for a test is a "polarization" of the Earth-moon orbit due to the fact that the moon should fall toward the sun with a greater acceleration than does the Earth. This "polarization" results in an eccentricity in the orbit that points always toward the Sun and has the amplitude

Nordtvedt effect in
Earth-moon orbit

$$\delta r = 840 \left[3\gamma + 4\beta - 7\,\Delta_1 - \frac{1}{3}(2\beta + 2\beta_2 - 3\gamma + \Delta_2 - 2) \right] \text{cm} \qquad (40.56)$$

$$= 67 \text{ meters} \qquad \text{in Ni's theory (Boxes 39.1 and 39.2)}$$

$$= \frac{8.4}{2 + \omega} \text{ meters} \qquad \text{in Dicke-Brans-Jordan theory (Boxes 39.1 and 39.2)}$$

$$= 0 \qquad \text{in Einstein's theory.}$$

Box 40.4 CATALOG OF EXPERIMENTS

Type of experiment	Description of experiment	Where discussed
I. Tests of foundations of general relativity	1. Tests of uniqueness of free fall (Eötvös-Dicke-Braginsky experiments)	§38.3; Figure 1.6, Box 1.1
	2. Tests for existence of metric (time dilation of particle decays; role of Lorentz group in particle kinematics; etc.)	§38.4
	3. Searches for new, direct-coupling, long-range fields (Hughes-Drever experiment; ether-drift experiments)	§38.7; Figure 38.3
	4. Gravitational redshift experiments	§38.5; Figures 38.1 and 38.2; §§7.2, 7.3, and 7.4
	5. Constancy, in space and time, of the nongravitational physical constants	§38.6
II. Post-Newtonian ("solar-system") experiments	1. Deflection of light and radio waves by Sun	§40.3; Box 40.1
	2. Relativistic delay in round-trip travel time for radar beams passing near Sun	§40.4; Box 40.2
	3. Perihelion shifts and periodic perturbations in planetary orbits	§40.5; Box 40.3
	4. Three-body effects in the Lunar orbit	§40.6
	5. Precession of gyroscopes ("geodetic precession" and precession due to dragging of inertial frames by Earth's rotation)	§40.7
	6. Spatial variation of the Cavendish gravitational constant in the solar system	§§40.8 and 40.9
	7. Dependence of the Cavendish gravitational constant on the chemical composition of the gravitating body	§40.8
	8. Earth tides with sidereal periods	§40.8
	9. Annual variations in Earth rotation rate	§40.8
	10. Periodicities in Earth-Moon separation due to breakdown of geodesic motion	§40.9
III. Cosmological observations	1. Change of Cavendish gravitational constant with time in solar system	§40.8
	2. Large-scale features of universe (expansion, isotropy, homogeneity; existence and properties of cosmic microwave radiation; . . .)	Chapters 27–30; especially Chapter 29
	3. Agreement of various measures of age of universe (age from expansion; ages of oldest stars; age of solar system)	§29.7
IV. Gravitational-Wave experiments	Existence of waves; propagation speed; polarization properties; . . .	Chapters 35–37; especially Chapter 37

Figure 40.4. (*facing page*)

Measuring the separation between earth and moon by determining the time-delay (about 2.5 sec) between the emission of light from a laser on the earth and the return of this light to the earth. A key element in the program is a corner reflector, the first of which was landed on the moon July 20, 1969, by the Apollo 11 flight crew. In November 1971, there were three such reflectors on the moon: two American, and one French-built and Soviet-landed. A pulsed ruby laser projects a beam out of the 107-inch reflecting telescope of the McDonald Observatory of the University of Texas, on Mount Locke, 119 miles east of El Paso. This beam makes a spot of light on the moon's surface about 3.2 km in diameter. Laser light is bounced straight back to the earth by the "laser ranging retroreflectors" (LR³). Each consists of an aluminum panel of 46 cm by 46 cm with 100 fused silica corner cubes each 3.8 cm in diameter. The first reflector ever set up appears in the first inset, near the lunar landing module. It is tilted with respect to the landscape of the moon. The photograph was made shortly before astronauts Neil A. Armstrong and Edwin E. Aldrin, Jr., took off for the earth. The second inset is a photograph made by D. G. Currie of the field of view in the guiding eyepiece of the McDonald 107-inch telescope in an interval when the laser was not firing at the Apollo 11 site. One guides the telescope to Tranquility Base (small circle) by aligning fiducial marks on more visible moonscape features. In November 1971, the LR³ experiment and continuing time-of-flight measurements were the responsibility of the National Aeronautics and Space Administration and a Lunar Retroreflecting Ranging Team of representatives from several centers of research. One of the members of this team, Carroll Alley, of the University of Maryland, is hereby thanked for his kindness in providing the photographs used in this montage. Thanks to this NASA work, the distance between the laser source on the earth and the reflectors on the moon is known with an accuracy now better than half a meter. The astronauts left behind on the moon not only LR³ and a seismometer and other equipment, but also a plaque: "We came in peace for all mankind."

By the mid 1970's, lunar laser-ranging data will probably be able to determine the amplitudes of this polarization to a precision of one meter or better [see Bender *et al.* (1971); also Figure 40.4].

§40.10. SUMMARY OF EXPERIMENTAL TESTS OF GENERAL RELATIVITY

No longer is general relativity "a theorist's Paradise, but an experimentalist's Hell." It is now a Paradise for all—as one can see quickly by perusing the catalog of experiments given in Box 40.4 on page 1129. Moreover, general relativity has emerged from each of its tests unscathed—a remarkable 1973 tribute to the 1915 genius of Albert Einstein.

PART **X**

FRONTIERS

Wherein the reader—who, during a life of continued variety for forty chapters (besides the Preface), was eight chapters a mathematician, four times enticed (once by an old friend), four chapters a cosmologist, and four chapters a transported astrophysicist in the land of black holes, and who at last inherited a wealth of experiments, lived honest, and became a True Believer—now ventures forth in search of new frontiers to conquer.

CHAPTER **41**

SPINORS

§41.1. REFLECTIONS, ROTATIONS, AND
THE COMBINATION OF ROTATIONS

Spinors and their applications in relativity grew out of the analysis of "rotations," first in space, then in spacetime. Take a cube (Figure 41.1). Rotate it about one axis through 90°. Then pick another axis at right angles to the first. About it rotate the cube again through 90°. In this way the cube is carried from the orientation marked "Initial" to that marked "Final." How can one make this net transformation in a single step, with a single rotation? In other words, what is the law for the combination of rotations?

Were rotations described by vectors, then one could apply the law of combination of vectors. The resultant of two vectors of the same magnitude (90°) separated by a right angle, is a single vector that (1) lies in the same plane and (2) has the magnitude $2^{1/2} \times 90° = 127.28°$. Both predictions are wrong. To turn the cube from initial to final orientation in a single turn, (1) take an axis running from the center through the vertex A and (2) rotate through 120°.

What computational algorithm can ever reproduce a law of combination of rotations apparently so strange? On the evening of October 16, 1843, William Rowan Hamilton was walking with his wife along the Royal Canal in Dublin when the answer leaped to his mind, the fruit of years of reflection. With his knife he then and there carved on a stone on Brougham Bridge the formulas*

$$i^2 = j^2 = k^2 = ijk = -1,$$

> This chapter is entirely Track 2. No earlier Track-2 material is needed as preparation for it, nor is it needed as preparation for any later chapter.

The problem of combining rotations

*In the same city on June 21, 1972 President Eamon de Valera told one of the authors that, while in jail one evening in 1916, scheduled to be shot the next morning, he wrote down the formula of which he was so fond, $i^2 = j^2 = k^2 = ijk = -1$.

Initial Final

Figure 41.1.
Rotation about the vertical axis through 90°, followed by rotation about the horizontal axis through 90°, gives a net change in orientation that can be achieved by a single rotation through 120° about an axis emergent from the center through the corner A.

which in today's notation,

$$\sigma_x = \begin{Vmatrix} 0 & 1 \\ 1 & 0 \end{Vmatrix} = i\boldsymbol{i}, \qquad \sigma_y = \begin{Vmatrix} 0 & -i \\ i & 0 \end{Vmatrix} = i\boldsymbol{j}, \qquad \sigma_z = \begin{Vmatrix} 1 & 0 \\ 0 & -1 \end{Vmatrix} = i\boldsymbol{k}, \quad (41.1)$$

take the form

$$\sigma_x{}^2 = \sigma_y{}^2 = \sigma_z{}^2 = 1,$$
$$\sigma_x\sigma_y = -\sigma_y\sigma_x = i\sigma_z \text{ (and cyclic permutations).}$$
$$(41.2)$$

Rotation operators:
(1) defined

To any rotation is associated a quantity (Hamilton's "quaternion;" today's "spin matrix" or "spinor transformation" or "rotation operator")

$$R = \cos(\theta/2) - i\sin(\theta/2)(\sigma_x \cos\alpha + \sigma_y \cos\beta + \sigma_z \cos\gamma), \qquad (41.3)$$

where θ is the angle of rotation and α, β, γ are the angles between the axis of rotation and the coordinate axes. A rotation described by R_1 followed by a rotation described by R_2 gives a net change in orientation described by the single rotation

(2) as tools in combining rotations

$$R_3 = R_2 R_1. \qquad (41.4)$$

This is Hamilton's formula for the combination of two rotations (steps toward it by Euler in 1776; obtained by Gauss in 1819 but never published by him).

In the example in Figure 41.1,

$$R_1(\text{rotation by } \theta = 90° \text{ about } z\text{-axis}) = (1 - i\sigma_z)/2^{1/2},$$
$$R_2(\text{rotation by } \theta = 90° \text{ about } x\text{-axis}) = (1 - i\sigma_x)/2^{1/2},$$

and the product of the two is

$$R_2 R_1 = (1 - i\sigma_x + i\sigma_y - i\sigma_z)/2$$
$$= \cos 60° - i \sin 60°(\sigma_x/3^{1/2} - \sigma_y/3^{1/2} + \sigma_z/3^{1/2}).$$

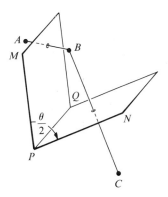

Figure 41.2.
Reflection in the plane *MPQ* carries *A* to *B*. Reflection in the plane *NPQ* carries *B* to *C*. The combination of the two reflections in the two planes separated by the angle $\theta/2$ produces the same end result (transformation from *A* to *C*) as rotation through the angle θ about the line *PQ*.

According to Hamilton's rule (41.3), this result implies a net rotation through 120° about a line that makes equal angles with the *x*-axis, the *y*-axis, and the *z*-axis, in conformity with what one already saw in Figure 41.1 (axis of rotation running from center of cube through the corner *A*).

What one has just done in the special example one can do in the general case: obtain the parameters θ_3, α_3, β_3, γ_3 of the net rotation (four unknowns!) by identifying the four coefficients of the four Hamilton units 1, $-i\sigma_x$, $-i\sigma_y$, $-i\sigma_z$ on both sides of the equation $R_3 = R_2 R_1$. In this way one arrives at the four prequaternion formulas of Olinde Rodrigues (1840) for the combination of the two rotations.

Why do half-angles put in an appearance? And what is behind the law of combination of rotations? The answer to both questions is the same: a rotation through the angle θ about a given axis may be visualized as the consequence of successive reflections in two planes that meet along that axis at the angle $\theta/2$ (Figure 41.2). Two rotations therefore imply four reflections. However, it can be arranged that reflections no. 2 and no. 3 take place in the same plane, the plane that includes the two axes of rotation. Then reflection no. 3 exactly undoes reflection no. 2. By now there remain only reflections no. 1 and no. 4, which together constitute one rotation: the net rotation that was desired (Figures 41.3 and 41.4).

The rotation

$$R = \cos(\theta/2) - i\sin(\theta/2)(\sigma_x \cos\alpha + \sigma_y \cos\beta + \sigma_z \cos\gamma) \tag{41.3}$$

is undone by the inverse rotation

$$R^{-1} = \cos(\theta/2) + i\sin(\theta/2)(\sigma_x \cos\alpha + \sigma_y \cos\beta + \sigma_z \cos\gamma). \tag{41.3'}$$

Thus the product of the two rotation operators

$$RR^{-1} = R^{-1}R = 1 \tag{41.5}$$

is an operator, the unit operator, that leaves unchanged everything that it acts on. The reciprocal R^{-1} of the combination $R = R_2 R_1$ of two rotations is

$$R^{-1} = R_1^{-1} R_2^{-1} \tag{41.5'}$$

(reverse order of factors!), as one verifies by substitution into (41.5).

Geometric reason that half angles appear in rotation operators

Algebraic properties of rotation operators

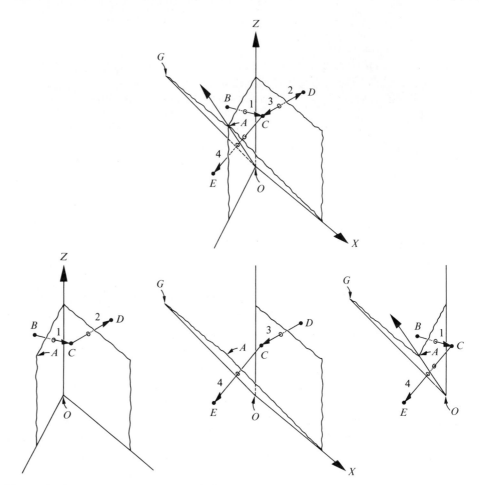

Figure 41.3.
Composition of two rotations seen in terms of reflections. The first rotation (for instance, 90° about *OZ* in the example of Figure 41.1.) is represented in terms of reflection 1 followed by reflection 2 (the planes of the two reflections being separated by 90°/2 = 45° in the example). The second reflection appears as the resultant of reflections 3 and 4. But the reflections 2 and 3 take place in the common plane *ZOX*. Therefore one reflection undoes the other. Thus the sequence of four operations 1234 collapses to the two reflections 1 and 4. Their place in turn is taken by a single rotation about the axis *OA*.

The conjugate transpose, M^*, of a matrix M is obtained by taking the conjugate complex of every element in the matrix and then interchanging rows and columns. By direct inspection of matrix expressions (41.1) one sees that $\sigma_x^* = \sigma_x$, $\sigma_y^* = \sigma_y$, $\sigma_z^* = \sigma_z$. Such matrices are said to be Hermitian. The conjugate transpose of the product $M = PQ$ of two matrices is the product $M^* = Q^*P^*$ of the individual conjugate transposed matrices taken in the reverse order. For the rotation matrix written down above, note that $R^* = R^{-1}$. Such a matrix is said to be unitary. The

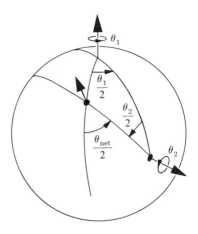

Figure 41.4.
Law of composition of rotations epitomized by a spherical triangle in which each of the three important angles represents *half* an angle of rotation.

determinant of a unitary matrix may be seen to have absolute value unity from the following line of argument:

$$
\begin{aligned}
1 &= \det\,(\text{unit matrix}) = \det\,(RR^{-1}) \\
&= \det\,(RR^*) = \det R \det R^* \\
&= |\det R|^2.
\end{aligned}
\tag{41.6}
$$

In actuality the determinant of the rotation spin matrix is necessarily unity ("unimodular matrix") as shown in the following exercises

Exercise 41.1. ELEMENTARY FEATURES OF THE ROTATION MATRIX

Write equation (41.3) in the form

$$
R(\theta) = \cos\,(\theta/2) - i\sin\,(\theta/2)(\boldsymbol{\sigma}\cdot\boldsymbol{n}),
$$

and establish the following properties:

(a) $\qquad\qquad\qquad (\boldsymbol{\sigma}\cdot\boldsymbol{n})^2 = 1 \equiv$ unit matrix;

(b) $\qquad\quad \mathrm{tr}\,(\boldsymbol{\sigma}\cdot\boldsymbol{n}) = 0$ (tr means "trace," i.e., sum of diagonal elements);

(c) $\qquad\quad \underbrace{[R,\,(\boldsymbol{\sigma}\cdot\boldsymbol{n})]}_{\uparrow\qquad\text{[commutator]}} \equiv R(\boldsymbol{\sigma}\cdot\boldsymbol{n}) - (\boldsymbol{\sigma}\cdot\boldsymbol{n})R = 0;$

(d) $\qquad\qquad\qquad \dfrac{dR}{d\theta} = -\dfrac{i}{2}(\boldsymbol{\sigma}\cdot\boldsymbol{n})R. \qquad\qquad (41.7)$

[Note that if one thinks of θ as increasing with angular velocity ω, so $d\theta/dt = \omega =$ constant, then this last equation reads

$$
\frac{dR}{dt} = -\frac{i}{2}(\boldsymbol{\sigma}\cdot\boldsymbol{\omega})R
\tag{41.7'}
$$

where $\boldsymbol{\omega} = \omega\boldsymbol{n}$.]

Exercise 41.2. ROTATION MATRIX HAS UNIT DETERMINANT

Recall from exercise 5.5 that for any matrix M one has

$$d[\ln (\det M)] = \text{tr}\,(M^{-1}\, dM)$$

and use this to show that $\det R$ in (41.7) is constant, and therefore equal to $(\det R)_{\theta=0} = 1$.

§41.2. INFINITESIMAL ROTATIONS

Infinitesimal rotations

A given rotation can be obtained by performing in turn two rotations of half the magnitude, or four rotations of a fourth the magnitude, or eight of an eighth the magnitude, and so on. Thus one arrives in the limit at the concept of an infinitesimal rotation described by the spin matrix

$$R = 1 - (i/2)(\sigma_x\, d\theta_{yz} + \sigma_y\, d\theta_{zx} + \sigma_z\, d\theta_{xy})$$

or

$$R = 1 - (i\, d\theta/2)(\boldsymbol{\sigma}\cdot\boldsymbol{n}). \tag{41.8}$$

Here the quantities

$$\begin{aligned}
d\theta_{yz} &= -d\theta_{zy} = n^x\, d\theta = \cos\alpha\, d\theta, \\
d\theta_{zx} &= -d\theta_{xz} = n^y\, d\theta = \cos\beta\, d\theta, \\
d\theta_{xy} &= -d\theta_{yx} = n^z\, d\theta = \cos\gamma\, d\theta,
\end{aligned} \tag{41.9}$$

are the components of the infinitesimal rotation in the three indicated planes. An infinitesimal rotation in the (x, y)-plane through the angle $d\theta_{xy}$ transforms the vector $\boldsymbol{x} = (x, y, z)$ into a new vector with changed components x' and y' but with unchanged component $z' = z$. More generally, the infinitesimal rotation (41.8) considered in this same "active" sense* produces the transformation

$$\boldsymbol{x} \longrightarrow \boldsymbol{x}',$$

with

$$\begin{aligned}
x' &= x - (d\theta_{xy})y - (d\theta_{xz})z, \\
y' &= -(d\theta_{yx})x + y - (d\theta_{yz})z, \\
z' &= -(d\theta_{zx})x - (d\theta_{zy})y + z.
\end{aligned} \tag{41.10}$$

Representation of a 3-vector as a spin matrix

Spinor calculus provides an alternative (and shorthand!) means to calculate the foregoing effect of a rotation on a vector. Associate with the vector \boldsymbol{x} the spin matrix

$$X = x\sigma_x + y\sigma_y + z\sigma_z = (\boldsymbol{x}\cdot\boldsymbol{\sigma}), \tag{41.11}$$

*An "*active*" transformation changes one vector into another, while leaving unchanged the underlying reference frame (if there is one). By contrast, a "*passive*" transformation leaves all vectors unchanged, but alters the reference frame. All transformations in previous chapters of this book were passive.

and with the vector x' a corresponding spin matrix or quaternion X'. Then the effect of the rotation is summarized in the formula

$$X \longrightarrow X' = RXR^*. \tag{41.12}$$

Rotation of a 3-vector described in spin-matrix language

Test this formula for the general infinitesimal rotation (41.10). It reads

$$(x' \cdot \sigma) = [1 - (i\,d\theta/2)(\sigma \cdot n)](x \cdot \sigma)[1 + (i\,d\theta/2)(\sigma \cdot n)]$$

or, to the first order in the quantity $d\theta$,

$$(x' \cdot \sigma) = (x \cdot \sigma) + (i\,d\theta/2)[(x \cdot \sigma)(\sigma \cdot n) - (\sigma \cdot n)(x \cdot \sigma)]. \tag{41.13}$$

The product of spin matrices $A = (a \cdot \sigma)$ and $B = (b \cdot \sigma)$ built from two distinct vectors a and b is

$$AB = (a \cdot \sigma)(b \cdot \sigma) = a^x b^x \sigma_x{}^2 + a^x b^y \sigma_x \sigma_y + \dots,$$

or, according to (41.2),

$$AB = (a \cdot b) + i(a \times b) \cdot \sigma. \tag{41.14}$$

Employ this formula to evaluate the right-hand side of (41.13). In the square brackets, the terms in $(x \cdot n)$ have opposite signs and cancel. In contrast, the terms in $(n \times x)$ have the same sign. They combine to cancel the factor 2 in $(d\theta/2)$. End up with

$$(x' \cdot \sigma) = (x \cdot \sigma) + d\theta(n \times x) \cdot \sigma$$

or

$$x' = [1 + (d\theta)n \times] \, x \tag{41.15}$$

in agreement with (41.10), as was to be shown.

A finite rotation about a given axis can be considered as the composition of infinitesimal rotations about that axis. To see this composition in simplest form, rewrite the spin matrix (41.8) associated with the general infinitesimal rotation as

Composition of finite rotation from infinitesimal rotations

$$R(d\theta) = e^{-(i\,d\theta/2)(\sigma \cdot n)} \tag{41.16}$$

(exponential function defined by its power-series expansion). Note that $(\sigma \cdot n)$ commutes (a) with unity and (b) with itself, and in addition (c) has a unit square. Therefore the calculation of the exponential function proceeds no differently here, for spin matrices, than for everyday algebra. The composition of the spin matrices for infinitesimal rotations about an unchanging axis proceeds by adding exponents, to give

$$R(\theta) = e^{-i(\theta/2)(\sigma \cdot n)}, \tag{41.17}$$

which can also be obtained immediately from equation (41.7). This expression can be put in another form by developing the power series; thus,

$$R(\theta) = \sum_{p=0}^{\infty} (1/p!)(-i\theta\boldsymbol{\sigma}\cdot\boldsymbol{n}/2)^p$$

$$= \sum_{\text{even }p} (1/p!)(-i\theta/2)^p + (\boldsymbol{\sigma}\cdot\boldsymbol{n}) \sum_{\text{odd }p} (1/p!)(-i\theta/2)^p \tag{41.18}$$

$$= \cos(\theta/2) - i\sin(\theta/2)(\boldsymbol{\sigma}\cdot\boldsymbol{n})$$

in agreement with the expression (41.3) originally given for a spinor transformation. The effect of one infinitesimal rotation after another after another . . . on a vector is given by

$$X' = R(d\theta) \ldots R(d\theta)XR^*(d\theta) \ldots R^*(d\theta),$$

with the consequence that even for a finite rotation $R = R(\theta)$ one is correct in employing the formula

$$X' = RXR^*. \tag{41.19}$$

EXERCISE

Exercise 41.3. MORE PROPERTIES OF THE ROTATION MATRIX

Show that for $X = \boldsymbol{x}\cdot\boldsymbol{\sigma}$ one has the commutation relation

$$[(\boldsymbol{\sigma}\cdot\boldsymbol{n}), X] = 2i(\boldsymbol{n}\times\boldsymbol{x})\cdot\boldsymbol{\sigma}.$$

Use this to obtain, from equation (41.19) in the form $X = RX_0R^*$ [where X_0 is constant, while $R(\theta)$ is given by equation (41.17)], the formula

$$\frac{d}{d\theta}(\boldsymbol{x}\cdot\boldsymbol{\sigma}) = (\boldsymbol{n}\times\boldsymbol{x})\cdot\boldsymbol{\sigma}.$$

Why is this equivalent to the standard definition

$$\frac{d\boldsymbol{x}}{dt} = \boldsymbol{\omega}\times\boldsymbol{x}$$

for the angular velocity? Reverse the argument to show that equation (41.7') correctly defines the rotation $R(t)$ resulting from a time-dependent angular velocity $\boldsymbol{\omega}(t)$, even though the simple solution $R = \exp[-\frac{1}{2}it(\boldsymbol{\sigma}\cdot\boldsymbol{\omega})]$ of this equation can no longer be written when $\boldsymbol{\omega}$ is not constant.

§41.3. LORENTZ TRANSFORMATION VIA SPINOR ALGEBRA

4-vectors and Lorentz transformations in spin-matrix language

Generate a rotation by two reflections in space? Then why not generate a Lorentz transformation by two reflections in spacetime? If for this purpose one has to turn from a real half-angle between the two planes of reflection to a complex half-angle, that development will come as no surprise; nor will it be a surprise that one can

still represent the effect of the Lorentz transformation by a matrix multiplication of the form

$$X \longrightarrow X' = LXL^*. \tag{41.20}$$

Here the "Lorentz spin transformation matrix" L is a generalization of the rotation matrix, R. Also the "coordinate-generating spin matrix" X is now generalized from (41.11) to

$$X = t + (\mathbf{x} \cdot \boldsymbol{\sigma}) \tag{41.21}$$

or

$$X = \begin{Vmatrix} t + z & x - iy \\ x + iy & t - z \end{Vmatrix}. \tag{41.22}$$

It is demanded that this matrix be Hermitian

$$X = X^*. \tag{41.23}$$

Then and only then are the coordinates (t, x, y, z) real. The conjugate transpose of the transformed spin matrix must also be Hermitian—and is:

$$\begin{aligned} (X')^* &= (LXL^*)^* \\ &= (L^*)^*(X)^*(L)^* = LXL^* = X'. \end{aligned} \tag{41.24}$$

Therefore the new coordinates (t', x', y', z') are guaranteed to be real, as desired. This reality requirement is a rationale for the form of the spin-matrix transformation (41.20), with L appearing on one side of X and L^* on the other.

A Lorentz transformation is defined by the circumstance that it leaves the interval invariant:

$$t'^2 - x'^2 - y'^2 - z'^2 = t^2 - x^2 - y^2 - z^2. \tag{41.25}$$

Note that the determinant of the matrix X as written out above has the value

$$\det X = t^2 - x^2 - y^2 - z^2. \tag{41.26}$$

Consequently the requirement for the preservation of the interval may be put in the form

$$\det X' = \det X \tag{41.27}$$

or

$$(\det L)(\det X)(\det L^*) = \det X. \tag{41.28}$$

This requirement is fulfilled by demanding

$$\det L = 1 \tag{41.29}$$

[it is not a useful generalization to multiply every element of L here by a common phase factor $e^{i\delta}$, and therefore multiply $\det L$ by $e^{2i\delta}$, because the net effect of this phase factor is nil in the formula $X' = LXL^*$].

The spin matrix associated with a rotation, whether finite or infinitesimal, already satisfied the condition det $L = 1$ [proved in exercise (41.2)]. This condition, being algebraic, will continue to hold when the real angles $d\theta_{yz}$, $d\theta_{zx}$, $d\theta_{xy}$, are replaced by complex angles, $d\theta_{yz} + i\,d\alpha_x$, $d\theta_{zx} + i\,d\alpha_y$, $d\theta_{xy} + i\,d\alpha_z$. The spin-transformation matrix acquires in this way a total of six parameters, as needed to describe the general infinitesimal Lorentz transformation. Thus the spin matrix for the general infinitesimal Lorentz transformation can be put in the form

$$\begin{aligned}
L &= 1 - (i/2)(\sigma_x\,d\theta_{yz} + \sigma_y\,d\theta_{zx} + \sigma_z\,d\theta_{xy}) \\
&\quad + (1/2)(\sigma_x\,d\alpha_x + \sigma_y\,d\alpha_y + \sigma_z\,d\alpha_z) \\
&= 1 - (i\,d\theta/2)(\boldsymbol{\sigma}\cdot\boldsymbol{n}) + (\boldsymbol{\sigma}\cdot d\boldsymbol{\alpha}/2).
\end{aligned} \tag{41.30}$$

The effect of this transformation upon the coordinates is to be read out from the formula

$$X \longrightarrow X' = LXL^*$$

or

$$\begin{aligned}
t' + (\boldsymbol{\sigma}\cdot\boldsymbol{x}') &= [1 - (i\,d\theta/2)(\boldsymbol{\sigma}\cdot\boldsymbol{n}) + (\boldsymbol{\sigma}\cdot d\boldsymbol{\alpha}/2)] \\
&\quad \times [t + (\boldsymbol{\sigma}\cdot\boldsymbol{x})][1 + (i\,d\theta/2)(\boldsymbol{\sigma}\cdot\boldsymbol{n}) + (\boldsymbol{\sigma}\cdot d\boldsymbol{\alpha}/2)]
\end{aligned} \tag{41.31}$$

Employ equation (41.14) for $(\boldsymbol{\sigma}\cdot\boldsymbol{A})(\boldsymbol{\sigma}\cdot\boldsymbol{B})$ to reduce the right side to the form

$$t + (\boldsymbol{\sigma}\cdot\boldsymbol{x}) + (\boldsymbol{\sigma}\cdot d\boldsymbol{\alpha})t + d\theta(\boldsymbol{n}\times\boldsymbol{x})\cdot\boldsymbol{\sigma} + (\boldsymbol{x}\cdot d\boldsymbol{\alpha}).$$

Now compare coefficients of 1, σ_x, σ_y and σ_z, respectively, on both sides of the equation, and find

$$\begin{aligned}
t' &= t + (\boldsymbol{x}\cdot d\boldsymbol{\alpha}) \\
\boldsymbol{x}' &= \boldsymbol{x} + t\,d\boldsymbol{\alpha} + d\theta(\boldsymbol{n}\times\boldsymbol{x}),
\end{aligned} \tag{41.32}$$

in agreement with the conventional expression for an infinitesimal Lorentz transformation or "boost" of velocity $d\boldsymbol{\alpha}$, in active form, as was to be shown.

The composition of such infinitesimal Lorentz transformations gives a finite Lorentz transformation. The result, however, can be calculated easily only when all infinitesimal transformations commute. Thus assume that $d\theta$ and $d\boldsymbol{\alpha}$ are in a fixed ratio, so

$$\boldsymbol{\omega} \equiv \boldsymbol{n}\frac{d\theta}{d\tau} \quad \text{and} \quad \boldsymbol{a} \equiv \frac{d\boldsymbol{\alpha}}{d\tau}$$

are constants, with τ a parameter. Then integration with respect to τ (composition of infinitesimal transformations) gives a finite transformation $L = \exp[-\frac{1}{2}i\tau\boldsymbol{\sigma}\cdot(\boldsymbol{\omega} + i\boldsymbol{a})]$. For $\tau = 1$, so $\theta\boldsymbol{n} = \boldsymbol{\omega}\tau$, $\boldsymbol{\alpha} = \boldsymbol{a}\tau$, this reads

$$L = \exp[(\boldsymbol{\alpha} - i\theta\boldsymbol{n})\cdot\boldsymbol{\sigma}/2]. \tag{41.33}$$

In the special case of a pure boost (no rotation; $\theta = 0$), the exponential function is evaluated along the lines indicated in (41.18), with the result

$$L = \cosh(\alpha/2) + (\boldsymbol{n}_\alpha\cdot\boldsymbol{\sigma})\sinh(\alpha/2). \tag{41.34}$$

Here $\boldsymbol{n}_\alpha = \boldsymbol{\alpha}/\alpha$ is a unit vector in the direction of the boost. The corresponding Lorentz transformation itself is evaluated from the formula

$$X' = LXL^*$$

or

$$t' + (\boldsymbol{x}' \cdot \boldsymbol{\sigma}) = [\cosh \alpha/2 + (\boldsymbol{n}_\alpha \cdot \boldsymbol{\sigma}) \sinh \alpha/2][t + (\boldsymbol{x} \cdot \boldsymbol{\sigma})]$$
$$\times [\cosh \alpha/2 + (\boldsymbol{n}_\alpha \cdot \boldsymbol{\sigma}) \sinh \alpha/2]. \tag{41.35}$$

Simplify with the help of the relations

$$\cosh^2(\alpha/2) + \sinh^2(\alpha/2) = \cosh \alpha,$$

$$2 \sinh (\alpha/2) \cosh (\alpha/2) = \sinh \alpha,$$

and

$$(\boldsymbol{n}_\alpha \cdot \boldsymbol{\sigma})(\boldsymbol{x} \cdot \boldsymbol{\sigma})(\boldsymbol{n}_\alpha \cdot \boldsymbol{\sigma}) = 2(\boldsymbol{n}_\alpha \cdot \boldsymbol{x})(\boldsymbol{n}_\alpha \cdot \boldsymbol{\sigma}) - (\boldsymbol{n}_\alpha \cdot \boldsymbol{n}_\alpha)(\boldsymbol{x} \cdot \boldsymbol{\sigma}),$$

and on both sides of the equation compare coefficients of 1 and $\boldsymbol{\sigma}$, to find

$$t' = (\cosh \alpha)t + (\sinh \alpha)(\boldsymbol{n}_\alpha \cdot \boldsymbol{x}),$$

$$\boldsymbol{x}' = [(\sinh \alpha)\boldsymbol{n}_\alpha t + (\cosh \alpha)(\boldsymbol{n}_\alpha \cdot \boldsymbol{x})\boldsymbol{n}_\alpha] \text{ ("in-line part of transformation")}$$
$$+ [\boldsymbol{x} - (\boldsymbol{x} \cdot \boldsymbol{n}_\alpha)\boldsymbol{n}_\alpha] \text{ ("perpendicular part of } \boldsymbol{x} \text{ unchanged")}. \tag{41.36}$$

In this way one verifies that the quantity α is the usual "velocity parameter," connected with the velocity itself by the relations

$$(1 - \beta^2)^{-1/2} = \cosh \alpha,$$

$$\beta(1 - \beta^2)^{-1/2} = \sinh \alpha, \tag{41.37}$$

$$\beta = \tanh \alpha.$$

That velocity parameters add for successive boosts in the same direction shows nowhere more clearly than in the representation (41.33) of the spin-transformation matrix:

$$L(\alpha_2)L(\alpha_1) = \exp [\alpha_2(\boldsymbol{n}_\alpha \cdot \boldsymbol{\sigma})/2] \exp [\alpha_1(\boldsymbol{n}_\alpha \cdot \boldsymbol{\sigma})/2] = \exp [(\alpha_2 + \alpha_1)(\boldsymbol{n}_\alpha \cdot \sigma)/2]$$
$$= L(\alpha_2 + \alpha_1). \tag{41.38}$$

Turn from this special case, and ask how to get the resultant of two arbitrary Lorentz transformations, each of which is a mixture of a rotation and a boost. No simpler method offers itself to answer this question than to use formula (41.33) together with the equation

$$L(\text{resultant}) = L_2 L_1. \tag{41.39}$$

§41.4. THOMAS PRECESSION VIA SPINOR ALGEBRA

A spinning object, free of all torque, but undergoing acceleration, changes its direction as this direction is recorded in an inertial frame of reference. This is the

Thomas precession [see exercise 6.9 and first term in equation (40.33b)]. This precession accounts for a factor two in the effective energy of coupling of spin and orbital angular momentum of an atomic electron. In a nucleus it contributes a little to the coupling of the spin and orbit of a nucleon. The evaluation of the Thomas precession affords an illustration of spin-matrix methods in action.

The precession in question can be discussed quite without reference either to angular momentum or to mass in motion. It is enough to consider a sequence of inertial frames of reference $S(t)$ with these two features. (1) To whatever point the motion has taken the mass at time t, at that point is located the origin of the frame $S(t)$. (2) The inertial frame $S(t + dt)$ at the next succeeding moment has undergone no rotation with respect to the inertial frame $S(t)$, as rotation is conceived by an observer in that inertial frame. However, it has undergone a rotation ("Thomas precession") as rotation is conceived and defined in the laboratory frame of reference.

Origin of Thomas precession: composition of two boosts is not a pure boost

How is it possible for "no rotation" to appear as "rotation"? The answer is this: one pure boost, followed by another pure boost in another direction, does not have as net result a third pure boost; instead, the net result is a boost plus a rotation. This idea is not new in kind. Figure 41.1 illustrated how a rotation about the z-axis followed by a rotation about the x-axis had as resultant a rotation about an axis with not only an x-component and a z-component but also a y-component. What is true of rotations is true of boosts: they defy the law for the addition of vectors.

Derivation of Thomas precession using spin matrices

Let the frame S_0 coincide with the laboratory frame, and let the origin of this laboratory frame be where the moving frame is at time t. Let $S(t)$ be a Lorentz frame moving with this point at time t. Let one pure boost raise its velocity relative to the laboratory from $\boldsymbol{\beta}$ to $\boldsymbol{\beta} + d\boldsymbol{\beta}$. The resulting final configuration cannot be reached from S_0 by a pure boost. Instead, first turn S_0 relative to the laboratory frame ("rotation R associated with the Thomas precession") and then send it by a simple boost to the final configuration. Only one choice of this rotation will be right to produce match-up. Thus, distinguishing the spin matrices for pure boosts and pure rotations by the letters B and R, one has the relation

$$B(\boldsymbol{\beta} + d\boldsymbol{\beta})R(\boldsymbol{\omega}\, dt) = \text{``}B(d\boldsymbol{\beta})\text{''}B(\boldsymbol{\beta}) \tag{41.40}$$

out of which to find the angular velocity $\boldsymbol{\omega}$ of the Thomas precession. The quotation marks in "$B(d\boldsymbol{\beta})$" carry a double warning: (1) the velocity of transformation that boosts $S(t)$ to $S(t + dt)$ is not $(\boldsymbol{\beta} + d\boldsymbol{\beta}) - \boldsymbol{\beta} = d\boldsymbol{\beta}$ (law of vector addition—or subtraction—not applicable to velocity), and (2) "$B(d\boldsymbol{\beta})$" does not appear as a pure boost in the laboratory frame. It appears as a pure boost only in the comoving frame.

Take care of the second difficulty first. It is only a difficulty because the formalism for combination of transformations, $R_3 = R_2 R_1$, as developed in §41.1 presupposes all operations $R_1, R_2, \ldots,$ to be defined and carried out in the laboratory reference frame. In contrast, the quantity "$B(d\boldsymbol{\beta})$" is understood to imply a pure boost as defined and carried out in the comoving frame. Such an operation can be fitted into the formalism as follows. (1) Undo any velocity that the object already has. In other words apply the operator $B^{-1}(\boldsymbol{\beta})$. Then the object is at rest in the laboratory frame. Then apply the necessary small pure boost, $B(\boldsymbol{a}_{\text{comoving}}\, d\tau)$, where $\boldsymbol{a}_{\text{comoving}}$

is the acceleration as it will be sensed by the object and $d\tau$ is the lapse of proper time as it will be sensed by the object. At the commencement of this brief acceleration the object is at rest relative to the laboratory. What is a pure boost to it is a pure boost relative to the laboratory. It is also a pure boost in the spin-matrix formalism. Then transform back from laboratory to moving frame. Thus have the relation

$$\text{``}B(d\boldsymbol{\beta})\text{''} = B(\boldsymbol{\beta})B(\boldsymbol{a}_{\text{comoving}} \, d\tau)B^{-1}(\boldsymbol{\beta}). \tag{41.41}$$

The equation for the determination of the Thomas precession now reads

$$B(\boldsymbol{\beta} + d\boldsymbol{\beta})R(\boldsymbol{\omega} \, dt) = B(\boldsymbol{\beta})B(\boldsymbol{a}_{\text{comoving}} \, d\tau) \tag{41.42}$$

or, with all unknowns put on the left,

$$R(\boldsymbol{\omega} \, dt)B^{-1}(\boldsymbol{a}_{\text{comoving}} \, d\tau) = B^{-1}(\boldsymbol{\beta} + d\boldsymbol{\beta})B(\boldsymbol{\beta}). \tag{41.43}$$

The first task, to replace the erroneous value of the velocity change $(d\boldsymbol{\beta})$ by a correct value $(\boldsymbol{a}_{\text{comoving}} \, d\tau)$, is now made part of the problem along with the evaluation of the Thomas precession itself.

Principles settled, the calculation proceeds by inserting the appropriate expressions for all four factors in (41.43), and evaluating both sides of the equation to the first order of small quantities, as follows:

$$1 - (i \, dt\boldsymbol{\omega} + d\tau\boldsymbol{a}) \cdot \boldsymbol{\sigma}/2 = [\cosh(\alpha'/2) - (\boldsymbol{n}_{\alpha'} \cdot \boldsymbol{\sigma}) \sinh(\alpha'/2)]$$
$$\times [\cosh(\alpha/2) + (\boldsymbol{n}_\alpha \cdot \boldsymbol{\sigma}) \sinh(\alpha/2)]. \tag{41.44}$$

Here α and \boldsymbol{n}_α are the velocity parameter and unit vector that go with the velocity $\boldsymbol{\beta}$; $\alpha' = \alpha + d\alpha$, and $\boldsymbol{n}_{\alpha'} = \boldsymbol{n}_\alpha + d\boldsymbol{n}_\alpha$, go with $\boldsymbol{\beta} + d\boldsymbol{\beta}$. Develop the righthand side of (41.44) by the methods of calculus, writing $\alpha' = \alpha + d\alpha$ and $\boldsymbol{n}_{\alpha'} = \boldsymbol{n}_\alpha + d\boldsymbol{n}_\alpha$, and applying the rule for the differentiation of a product. Equate coefficients of $-\boldsymbol{\sigma}/2$ and $-i\boldsymbol{\sigma}/2$ on both sides of the equation. Thus find

$$\boldsymbol{a}_{\text{comoving}} \, d\tau = (d\alpha)\boldsymbol{n}_\alpha + (\sinh \alpha) \, d\boldsymbol{n}_\alpha \tag{41.45}$$

and

$$\boldsymbol{\omega} \, dt = [2 \sinh^2(\alpha/2)] \, d\boldsymbol{n}_\alpha \times \boldsymbol{n}_\alpha. \tag{41.46}$$

Angular velocity of Thomas precession

The one expression gives the change of velocity as seen in a comoving inertial frame. The other gives the precession as seen in the laboratory frame. For low velocities the expression for the Thomas precession reduces to

$$\boldsymbol{\omega} = \boldsymbol{a} \times \boldsymbol{\beta}/2. \tag{41.47}$$

Here \boldsymbol{a} is the acceleration. Only the component perpendicular to the velocity $\boldsymbol{\beta}$ is relevant for the precession.

For an elementary account of the importance of the Thomas precession in atomic physics, see, for example, Ruark and Urey (1930).

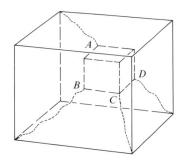

Figure 41.5.
"Orientation-entanglement relation" between a cube and the walls of a room. A 360° rotation of the cube entangles the threads. A 720° rotation might be thought to entangle them still more—but instead makes it possible completely to disentangle them.

§41.5. SPINORS

Orientation-entanglement
relation

Paint each face of a cube a different color. Then connect each corner of the cube to the corresponding corner of the room with an elastic thread (Figure 41.5). Now rotate the cube through $2\pi = 360°$. The threads become tangled. Nothing one can do will untangle them. It is impossible for every thread to proceed on its way in a straight line. Now rotate the cube about the same axis by a further 2π. The threads become still more tangled. However, a little work now completely straightens out the tangle (Figure 41.6). Every thread runs as it did in the beginning in a straight line from its corner of the cube to the corresponding corner of the room. More generally, rotations by $0, \pm 4\pi, \pm 8\pi, \ldots$, leave the cube in its standard "orientation-entanglement relation" with its surroundings, whereas rotations by $\pm 2\pi, \pm 6\pi, \pm 10\pi, \ldots$, restore to the cube only its orientation, not its orientation-entanglement relation with its surroundings. Evidently there is something about the geometry of orientation that is not fully taken into account in the usual concept of orientation; hence the concept of "orientation-entanglement relation" or (briefer term!) "version" (Latin *versor,* turn). Whether there is also a detectable difference in the physics (contact potential between a metallic object and its metallic surroundings, for example) for two inequivalent versions of an object is not known [Aharonov and Susskind (1967)].

In keeping with the distinction between the two inequivalent versions of an object, the spin matrix associated with a rotation,

$$R = \cos(\theta/2) - i(\boldsymbol{n} \cdot \boldsymbol{\sigma})\sin(\theta/2), \tag{41.48}$$

reverses sign on a rotation through an odd multiple of 2π. This sign change never shows up in the law of transformation of a vector, as summarized in the formula

$$X \longrightarrow X' = RXR^* \tag{41.49}$$

(two factors R; sign change in each!). The sign change does show up when one turns

Spinor defined

from a vector to a 2-component quantity that transforms according to the law

$$\xi \longrightarrow \xi' = R\xi. \tag{41.50}$$

Figure 41.6.
An object is connected to its surroundings by elastic threads as in Figure 41.5. (Eight are shown here; any number could be used.) Rotating the object through 720° and then following the procedure outlined (Edward McDonald) in frames 2–8 (with the object remaining fixed), one finds that the connecting threads are left disentangled, as in frame 9 (lower right).

Such a quantity is known as a spinor. A spinor reverses sign on a 360° rotation. It therefore provides a reasonable means to keep track of the difference between the two inequivalent versions of the cube. More generally, with each orientation-entanglement relation between the cube and its surroundings one can associate a different value of the spinor ξ. Moreover, there is nothing that limits the usefulness of the spinor concept to rotations. Also, for the general combination of boost and rotation, one can write

$$\xi \longrightarrow \xi' = L\xi. \qquad (41.51)$$

Lorentz transformation of a spinor

When the boost and rotation are both infinitesimal, the explicit form of this transformation is simple:

$$\xi' = [1 - (i\, d\theta/2)(\boldsymbol{n} \cdot \boldsymbol{\sigma}) + (d\boldsymbol{\beta}/2) \cdot \boldsymbol{\sigma}]\xi,$$

or, according to (41.1),

$$\begin{pmatrix} \xi'^1 \\ \xi'^2 \end{pmatrix} = \begin{Vmatrix} 1 + \dfrac{1}{2}(-i\theta_{xy} + \beta_z) & \dfrac{1}{2}(-i\theta_{yz} - \theta_{zx} + \beta_x - i\beta_y) \\ \dfrac{1}{2}(-i\theta_{yz} + \theta_{zx} + \beta_x + i\beta_y) & 1 + \dfrac{1}{2}(i\theta_{xy} - \beta_z) \end{Vmatrix} \begin{pmatrix} \xi^1 \\ \xi^2 \end{pmatrix} \tag{41.52}$$

For any combination of a boost in the z-direction of any magnitude and a finite rotation about the z-axis, one has

$$\begin{pmatrix} \xi'^1 \\ \xi'^2 \end{pmatrix} = \begin{Vmatrix} e^{-1/2 i\theta_{xy} + 1/2\beta_z} & 0 \\ 0 & e^{1/2 i\theta_{xy} - 1/2\beta_z} \end{Vmatrix} \begin{pmatrix} \xi^1 \\ \xi^2 \end{pmatrix}. \tag{41.53}$$

To keep track of the two components of the spinor, it is convenient and customary to introduce a label (capital Roman letter near beginning of alphabet) that takes on the values 1 and 2; thus (41.51) becomes

$$\xi'^A = L^A{}_B \xi^B. \tag{41.54}$$

The spinor has acquired a significance of its own through one's having pulled out half of the transformation formula

$$X' = LXL^*. \tag{41.55}$$

Second type of spinor

To be able to recover this formula, one requires the other half as well. It contains the conjugate complex of the Lorentz transformation. Therefore introduce another spinor η that transforms according to the law

$$\eta'^{\dot{U}} = \bar{L}^{\dot{U}}{}_{\dot{V}} \eta^{\dot{V}} \tag{41.56}$$

[$\dot{U} = \dot{1}, \dot{2}$; $\dot{V} = \dot{1}, \dot{2}$; dots and capital letters near the end of the alphabet are used to distinguish components that transform according to the conjugate complex (no transpose!) of the Lorentz spin matrix].

§41.6. CORRESPONDENCE BETWEEN VECTORS AND SPINORS

Vector regarded as a Hermitian second-rank spinor

To go back from spinors to vectors, note that the spin matrix X in (41.55) has the form

$$X = t + (\boldsymbol{x} \cdot \boldsymbol{\sigma}) = \begin{Vmatrix} (t + z) & (x - iy) \\ (x + iy) & (t - z) \end{Vmatrix} = \begin{Vmatrix} X^{1\dot{1}} & X^{1\dot{2}} \\ X^{2\dot{1}} & X^{2\dot{2}} \end{Vmatrix}, \tag{41.57}$$

where the labels receive dots or no dots according as they are coupled in (41.55) to L^* or to L. That equation of transformation becomes

$$X'^{A\dot{U}} = L^A{}_B \bar{L}^{\dot{U}}{}_{\dot{V}} X^{B\dot{V}} \tag{41.58}$$

(transpose obtained automatically by ordering of indices; thus $\bar{L}^{\dot{U}}{}_{\dot{V}}$, not $L^{*\dot{U}}{}_{\dot{V}}$). The coefficients in this transformation are identical with the coefficients in the law for the transformation of a "second-rank spinor with one index undotted and the other dotted:"

$$\xi'^A \eta'^{\dot{U}} = L^A{}_B \bar{L}^{\dot{U}}{}_{\dot{V}} \xi^\beta \eta^{\dot{V}}. \tag{41.59}$$

In this sense one can say that "a 4-vector transforms like a second-rank spinor." To be completely explicit about this connection between a 4-vector and a second-rank spinor, note from (41.57) the relations

$$
\begin{aligned}
X^{1\dot{1}} &= x^0 + x^3, \\
X^{1\dot{2}} &= x^1 - ix^2, \\
X^{2\dot{1}} &= x^1 + ix^2, \\
X^{2\dot{2}} &= x^0 - x^3.
\end{aligned}
\tag{41.60}
$$

In a more compact form, one has

$$X^{A\dot{U}} = [t + (\mathbf{x} \cdot \boldsymbol{\sigma})]^{A\dot{U}} = x^\mu \sigma_\mu{}^{A\dot{U}} \tag{41.61}$$

where σ_0 is the unit matrix. This equation tells immediately how to go from the components of a 4-vector, or "1-index tensor," to the components of the corresponding "1,1-spinor" (one undotted and one dotted index).

With each real 4-vector x^α is associated a 1,1-spinor that is *Hermitian* in the sense that

$$X^{A\dot{U}} = \overline{X^{U\dot{A}}}. \tag{41.62}$$

An example of a Hermitian 1,1-spinor is provided by (41.61). The concept of Hermiticity can be stated in other words, and more generally. Associated with any N,N-spinor $\boldsymbol{\Phi}$ with components $\Phi^{A_1 \cdots A_N \dot{U}_1 \cdots \dot{U}_N}$ is the *conjugate complex spinor* $\bar{\boldsymbol{\Phi}}$ with

$$(\bar{\Phi})^{A_1 \cdots A_N \dot{U}_1 \cdots \dot{U}_N} = \overline{(\Phi^{U_1 \cdots U_N \dot{A}_1 \cdots \dot{A}_N})} \tag{41.63}$$

An N,N-spinor is said to be Hermitian when it is equal to its conjugate complex.

N,N-spinors and Hermiticity

§41.7. SPINOR ALGEBRA

Equation (41.53) showed the component ξ'^1 of a spinor rising exponentially with a boost in proportion to the factor $e^{1/2\beta_z}$, and the other component, ξ'^2 falling exponentially. If from two spinors ξ and ζ, there is to be any quantity constructed which is unaffected in value by the boost, it must be formed out of such products

Spinor algebra:

(1) ε^{AB}, ε_{AB} defined

as $\xi^1\zeta^2$ and $\xi^2\zeta^1$. One can restate this product prescription in other language. Introduce the alternating symbols ε^{AB} and ε_{AB} such that $\varepsilon^{12} = \varepsilon_{12} = 1$ and

$$\varepsilon^{AB} = -\varepsilon^{BA}, \qquad \varepsilon_{AB} = -\varepsilon_{BA}, \tag{41.64}$$

the only other nonvanishing components being $\varepsilon^{21} = \varepsilon_{21} = -1$. Define the lower-label spinor ξ_A in terms of the upper-label spinor ξ^A by the equation

(2) raising and lowering
 spinor indices

$$\xi_A = \xi^B\varepsilon_{BA}, \tag{41.65}$$

with the inverse

$$\xi^B = \varepsilon^{BC}\xi_C. \tag{41.66}$$

Then the scalar product of one spinor by another is defined to be

(3) scalar products of spinors

$$\xi_A\zeta^A. \tag{41.67}$$

The value of this scalar product is unaffected by any boost or rotation or combination thereof:

$$\begin{aligned}
\xi'_A\zeta'^A &= \xi'^B\varepsilon_{BA}\zeta'^A \\
&= (L^B{}_D\xi^D)\varepsilon_{BA}(L^A{}_C\zeta^C) \\
&= (\det L)\xi^D\varepsilon_{DC}\zeta^C \\
&= \xi_C\zeta^C.
\end{aligned} \tag{41.68}$$

The proof uses the fact that the expression $L^B{}_D\varepsilon_{BA}L^A{}_C$ (1) vanishes when $D = C$, and (2) reduces to the determinant of L (unity!) or its negative when $D = 1$, $C = 2$, or $D = 2$, $C = 1$. Note that the scalar product $\xi^A\zeta_A$ is the negative of the scalar product $\xi_A\zeta^A$. The value of the scalar product of a spinor with itself is automatically zero ("built-in null character of a spinor").

(4) the mapping between
 vectors and 1,1-spinors

The components of a vector with upper index have been expressed in terms of the components of a 1,1-spinor with upper indices

$$X^{A\dot{U}} = x^\mu\sigma_\mu{}^{A\dot{U}}, \tag{41.69}$$

and a similar correlation holds between vector and 1,1-spinor with lower indices; thus,

$$X_{A\dot{U}} = x_\mu\sigma^\mu{}_{A\dot{U}}. \tag{41.70}$$

(5) σ^μ defined and related
 to σ_μ

Here the "associated basic spin matrices" have the components

$$\sigma^\mu{}_{A\dot{U}} = \eta^{\mu\nu}\sigma_\nu{}^{B\dot{V}}\varepsilon_{BA}\varepsilon_{\dot{V}\dot{U}}, \tag{41.71}$$

or, explicitly,

$$\left\| \begin{array}{cc} \sigma^{\mu}{}_{1\dot{1}} & \sigma^{\mu}{}_{1\dot{2}} \\ \sigma^{\mu}{}_{2\dot{1}} & \sigma^{\mu}{}_{2\dot{2}} \end{array} \right\| = \begin{cases} -\left\| \begin{array}{cc} 1 & 0 \\ 0 & 1 \end{array} \right\| \text{ for } \mu = 0, \\[1em] -\left\| \begin{array}{cc} 0 & 1 \\ 1 & 0 \end{array} \right\| \text{ for } \mu = 1, \\[1em] +\left\| \begin{array}{cc} 0 & -i \\ i & 0 \end{array} \right\| \text{ for } \mu = 2, \\[1em] -\left\| \begin{array}{cc} 1 & 0 \\ 0 & -1 \end{array} \right\| \text{ for } \mu = 3. \end{cases} \tag{41.72}$$

The same type of multiplication law holds for these matrices, $(\sigma^x)^2 = (\sigma^y)^2 = (\sigma^z)^2 = 1$, $\sigma^x\sigma^y = -\sigma^y\sigma^x = i\sigma^z$, etc., as for the matrices σ_x, σ_y, σ_z of (41.2). Between the "basic spin matrices," σ_μ, and the "associated basic spin matrices," σ^μ, the following orthogonality and normalization relations obtain:

$$\sigma_\mu{}^{A\dot{U}}\sigma^\mu{}_{B\dot{V}} = -2\,\delta^A_B\,\delta^{\dot{U}}_{\dot{V}} \tag{41.73}$$

and

$$\sigma_\mu{}^{A\dot{U}}\sigma^\nu{}_{A\dot{U}} = -2\,\delta^\nu_\mu. \tag{41.74}$$

One can use these relations to "go back from a quantity expressed as a 1,1-spinor ('spinor equivalent of a vector') to the same quantity expressed directly as a vector (first-rank tensor)." Thus, multiply through (41.61) on both sides by $-\frac{1}{2}\sigma^\nu{}_{A\dot{U}}$, sum over the spinor indices, and employ (41.74) to find the contravariant components of the vector,

$$x^\nu = -\frac{1}{2}\sigma^\nu{}_{A\dot{U}}X^{A\dot{U}}. \tag{41.75}$$

Similarly from (41.70) and (41.73) one finds the covariant components,

$$x_\nu = -\frac{1}{2}\sigma_\nu{}^{A\dot{U}}X_{A\dot{U}}. \tag{41.76}$$

An N-index tensor T lets itself be expressed in spinor language ("spinor equivalent of the tensor") by a generalization of (41.61) or (41.70); thus, for a mixed tensor of third order, one has

(6) the mapping between rank-N tensors and N,N-spinors

$$T_{A\dot{U}}{}^{B\dot{V}C\dot{W}} = \sigma^\alpha{}_{A\dot{U}}\sigma_\beta{}^{B\dot{V}}\sigma_\gamma{}^{C\dot{W}}T_\alpha{}^{\beta\gamma} \tag{41.77}$$

and the converse relation

$$T_\alpha{}^{\beta\gamma} = \left(-\frac{1}{2}\right)^3 \sigma_\alpha{}^{A\dot{U}}\sigma^\beta{}_{B\dot{V}}\sigma^\gamma{}_{C\dot{W}}T_{A\dot{U}}{}^{B\dot{V}C\dot{W}}. \tag{41.78}$$

Box 41.1 gives the spinor representation of several simple tensors.

Box 41.1 **SPINOR REPRESENTATION OF CERTAIN SIMPLE TENSORS IN THE CONTEXT OF A LOCAL LORENTZ FRAME**

Quantity	Tensor language	Spinor language
General 4-vector	x^α (four complex numbers)	$X^{A\dot{U}}$ (4 complex numbers)
Real 4-vector (example: 4-momentum)	$x^\alpha = \bar{x}^\alpha$ (four real numbers)	$X^{A\dot{U}} = \overline{(X^{U\dot{A}})}$ (2 real components, 1 distinct complex component)
Null 4-vector	$\eta_{\alpha\beta} x^\alpha x^\beta = 0$	$\det X^{A\dot{U}} = 0$ [see (41.57)]; hence there exist two spinors ξ^A and η^U such that $X^{A\dot{U}} = \xi^A \eta^{\dot{U}}$.
Future-pointing real null 4-vector (such as 4-momentum of a photon)	$x^\alpha = \bar{x}^\alpha$ $\eta_{\alpha\beta} x^\alpha x^\beta = 0$ $x^0 > 0$	There exists a spinor ξ^A (two complex numbers, unique up to a common multiplicative phase factor $e^{i\delta}$) such that $X^{A\dot{U}} = \xi^A (\bar{\xi})^{\dot{U}}$
Past-pointing real null 4-vector	$x^0 < 0$	$X^{A\dot{U}} = -\xi^A (\bar{\xi})^{\dot{U}}$.
Real bivector or 2-form (such as Maxwell field)	$F_{[\alpha\beta]}$ (subscript implying $F_{\alpha\beta} = -F_{\beta\alpha}$; six distinct real components)	There exists a symmetric spinor ϕ_{AB} (three distinct complex components $\phi_{11}, \phi_{12}, \phi_{22}$) such that $F_{A\dot{U}B\dot{V}} = \phi_{AB}\varepsilon_{\dot{U}\dot{V}} + \varepsilon_{AB}(\bar{\phi})_{\dot{U}\dot{V}}$
Real 2-form dual to foregoing real 2-form	$^*F_{\alpha\beta} = \frac{1}{2}\varepsilon_{\alpha\beta\gamma\delta}F^{\gamma\delta}$	$^*F_{A\dot{U}B\dot{V}} = -i\phi_{AB}\varepsilon_{\dot{U}\dot{V}} + i\varepsilon_{AB}(\bar{\phi})_{\dot{U}\dot{V}}$ (duality for 2-form corresponds to multiplication of spinor ϕ_{AB} by $-i$)
Real fourth-order tensor with symmetries of Weyl conformal curvature tensor; that is, with symmetries of Riemann curvature tensor and with additional requirement of vanishing Ricci tensor ("empty space;" "vacuum Riemann tensor")	$C_{\alpha\beta\gamma\delta} = C_{([\alpha\beta][\gamma\delta])}$ (antisymmetric in first two indices; antisymmetric in last two indices; symmetric against interchange of first pair with second pair) $C^\alpha{}_{[\beta\gamma\delta]} = 0$ (20 algebraically distinct components, as for the Riemann tensor, reduced to 10 by the further vacuum condition:) $C^\alpha{}_{\beta\alpha\delta} = 0$	There exists a completely symmetric spinor ψ_{ABCD} with five distinct complex components, ψ_{1111} ψ_{1112} ψ_{1122} ψ_{1222} ψ_{2222} such that $C_{A\dot{U}B\dot{V}C\dot{W}D\dot{X}} = \psi_{ABCD}\varepsilon_{\dot{U}\dot{V}}\varepsilon_{\dot{W}\dot{X}} + \varepsilon_{AB}\varepsilon_{CD}\overline{\psi}_{\dot{U}\dot{V}\dot{W}\dot{X}}$.

In some treatises on spinor analysis, the factor $(-\frac{1}{2})^N$ in equations like (41.78) is eliminated by the following double prescription: (1) insert into the matrices σ_μ and σ^μ a factor $1/\sqrt{2}$ not included above; and (2) use for the standard metric not diag $\eta_{\mu\nu} = (-1, 1, 1, 1)$ as above, but $(1, -1, -1, -1)$. This prescription was not adopted here (1) because the introduction of $1/\sqrt{2}$ in the matrices σ_x, σ_y, σ_z would put them out of line with the Pauli matrices as used for many years throughout

Quantity	Tensor language	Spinor language
Fully developed Riemann curvature tensor (space where matter is present)	$R_{\alpha\beta\gamma\delta} = R_{[\alpha\beta]\gamma\delta} = R_{\alpha\beta[\gamma\delta]} = R_{([\alpha\beta][\gamma\delta])}$ $R_{\alpha[\beta\gamma\delta]} = 0$ (20 algebraically distinct components)	There exists a completely symmetric spinor ψ_{ABCD} ("Weyl" or "conformal" part of curvature, or part of nonlocal origin) and a scalar Λ (measure of trace of part of curvature of local origin) and a spinor $\Phi_{AB\dot{U}\dot{V}} = \Phi_{(AB)(\dot{U}\dot{V})} = (\bar{\Phi})_{AB\dot{U}\dot{V}}$ (measure of trace-free part of curvature of local origin; last of the three irreducible parts of the curvature tensor) such that $$R_{A\dot{U}B\dot{V}C\dot{W}D\dot{X}} = \psi_{ABCD}\varepsilon_{\dot{U}\dot{V}}\varepsilon_{\dot{W}\dot{X}}$$ $$+ \varepsilon_{AB}\varepsilon_{CD}(\bar{\psi})_{\dot{U}\dot{V}\dot{W}\dot{X}}$$ $$+ 2\Lambda(\varepsilon_{AC}\varepsilon_{BD}\varepsilon_{\dot{U}\dot{V}}\varepsilon_{\dot{W}\dot{X}}$$ $$+ \varepsilon_{AB}\varepsilon_{CD}\varepsilon_{\dot{U}\dot{X}}\varepsilon_{\dot{V}\dot{W}})$$ $$+ \varepsilon_{AB}\Phi_{CD\dot{U}\dot{V}}\varepsilon_{\dot{W}\dot{X}}$$ $$+ \varepsilon_{CD}\Phi_{AB\dot{W}\dot{X}}\varepsilon_{\dot{U}\dot{V}}.$$
Each physical quantity is described by a geometric object. Every local physical quantity is described by a mathematical quantity that transforms under a proper local Lorentz transformation as an "irreducible representation of the group $L\uparrow_+$ of proper Lorentz transformations."	Each local physical quantity is described by a tensor with its own rank and specific symmetry properties.	In order to provide the required finite irreducible representation of $L\uparrow_+$ to represent a local physical quantity, the associated spinor must be completely symmetric in all of its undotted indices, and also completely symmetric in all its dotted indices [Gel'fand (1963)].

atomic and nuclear physics, and (2) because a positive definite metric within a spacelike hypersurface has the advantage of naturalness for the analysis of the initial-value problem of geometrodynamics and for the definition of what one means by a 3-geometry. The price of the factor $(-\frac{1}{2})^N$ is paid here for these advantages. Conventions that avoid this price are preferable for extensive spinor computations; see, e.g., Pirani (1965) or Penrose (1968a).

§41.8. SPIN SPACE AND ITS BASIS SPINORS

The "space" of elementary spinors is two-dimensional. Therefore it is spanned by any two linearly independent spinors λ_A and μ_A. Moreover, it is easy to diagnose a pair of spinors for possible linear dependence, that is, for existence of a relation of the form $\mu_A = \text{const}\,\lambda_A$. In this event, the scalar product of μ_A with λ^A, like the scalar product of λ_A with λ^A (41.67) automatically vanishes. Therefore a nonvanishing scalar product

$$\lambda_A \mu^A \neq 0 \tag{41.79}$$

is a necessary and sufficient condition for the linear independence of two spinors.

The general 4-vector lets itself be represented as a linear combination of four basis vectors. Similarly the general spinor lets itself be represented as a linear combination of two basis spinors:

$$\omega^A = \lambda \xi^A + \mu \eta^A. \tag{41.80}$$

Here it is understood that the term "basis spinor" implies that ξ^A and η^A satisfy the normalization condition

$$\xi_A \eta^A = 1 \; (= -\eta_A \xi^A). \tag{41.81}$$

From this condition one derives simple expressions for the expansion coefficients in (41.80):

$$\lambda = -\eta_A \omega^A \; (= \omega_B \eta^B),$$
$$\mu = \xi_A \omega^A \; (= -\omega_B \xi^B). \tag{41.82}$$

Inserting these expansion coefficients back into (41.80) will reproduce any arbitrarily chosen spinor ω^A. In other words, the following equation has to be an identity in the components of ω_B:

$$\omega^A = \varepsilon^{AB} \omega_B \equiv (\xi^A \eta^B - \eta^A \xi^B) \omega_B. \tag{41.83}$$

From this circumstance, it follows that the components of the two basic spinors are linked by the equations

$$\xi^A \eta^B - \eta^A \xi^B = \varepsilon^{AB}. \tag{41.84}$$

Given two basis spinors ξ^A and η^A, one can get two equally good basis spinors by writing

$$\xi^A{}_{\text{new}} = \xi^A,$$
$$\eta^A{}_{\text{new}} = \eta^A + \alpha \xi^A, \tag{41.85}$$

with α any real or complex constant, as one checks at once by substitution into (41.81) or (41.84). The most general "spinor mate" to a given spinor ξ^A, satisfying the normalization condition (41.81), has this form (41.85).

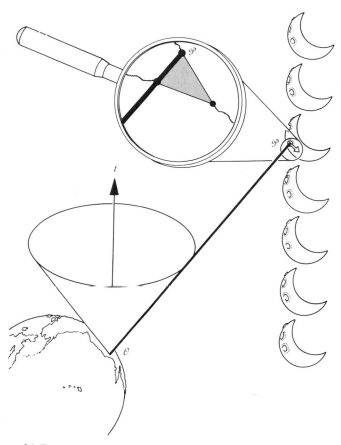

Figure 41.7.

Spinor represented by (1) "flagpole" [Penrose terminology; track of pulse of light; null vector \mathcal{OP}] plus (2) "flag" [arrow ($\mathcal{P} \longrightarrow$) flashed onto surface of moon by laser pulse from earth or, in expanded view in the inset above, a flag itself, substituted for the arrow] plus (3) the orientation-entanglement relation between the flag and its surroundings [symbolized by strings drawn from corners of flag to surroundings]. When the spinor itself is multiplied by a factor $\rho e^{i\sigma}$, the components of the null vector (flagpole) are multiplied by the factor ρ^2 and the flag is rotated through the angle 2σ about the flagpole.

§41.9. SPINOR VIEWED AS FLAGPOLE PLUS FLAG PLUS ORIENTATION-ENTANGLEMENT RELATION

How can one visualize a spinor? Aim the laser, pull the trigger, and send a megajoule pulse from the here and now (event \mathcal{O}) to the there and then (event \mathcal{P}: center of the crater Aristarchus, 400,000 km from \mathcal{O} in space, and 400,000 km from \mathcal{O} in light-travel time). The laser has been designed to produce, not a mere spot of light, but an illuminated arrow. Following Roger Penrose, speak of the null vector \mathcal{OP} as a "flagpole," and of the illuminated arrow as a "flag." A spinor (Figure 41.7) consists of this combination of (1) null flagpole plus (2) flag plus (3) the orientation-

Geometric representation of a spinor:

entanglement relation between the flag and its surroundings. "Rotate the flag" by repeatedly firing the laser, with a bit of rotation of the laser about its axis between one firing and the next. When the flag has turned through 360° and has come back to its original direction, the spinor has reversed sign. A rotation of the flag about the flagpole through any even multiple of 2π restores the spinor to its original value.

One goes from a spinor ξ, a mathematical object with two complex components ξ^1 and ξ^2, to the geometric object of "flagpole plus flag plus orientation-entanglement relation" in two steps: first the pole, then the flag. Thus, go from the spinor ξ^A to the real null 4-vector of the "pole" by way of the formula

(1) null vector (flagpole), plus

$$x^\alpha \longrightarrow X^{A\dot{U}} = \xi^A(\bar{\xi})^{\dot{U}} \tag{41.86}$$

or

$$\left\| \begin{matrix} (t+z) & (x-iy) \\ (x+iy) & (t-z) \end{matrix} \right\| = \left\| \begin{matrix} \xi^1\bar{\xi}^{\dot{1}} & \xi^1\bar{\xi}^{\dot{2}} \\ \xi^2\bar{\xi}^{\dot{1}} & \xi^2\bar{\xi}^{\dot{2}} \end{matrix} \right\|. \tag{41.87}$$

The matrix on the right has its first row identical up to a factor ξ^1/ξ^2 with its second row. Therefore the determinant of the matrix on the right vanishes. So also for the left. Therefore the 4-vector $\mathcal{OP} = (t, x, y, z)$ is a null vector. One "stretches" this vector by a factor ρ^2 when one multiplies the spinor ξ^A by the nonzero complex number $\lambda = \rho e^{i\sigma}$ (ρ, σ real); however, the vector is unchanged in direction. The 4-vector is also unaffected by the choice of the angle σ. In other words, this null 4-vector is uniquely fixed by the spinor; but the spinor is not fixed with all uniqueness by the 4-vector. To a given 4-vector corresponds a whole family of spinors. They differ from one another by a multiplicative phase factor of the form $e^{i\sigma}$ ("flag factor").

(2) bivector (flag) and its orientation-entanglement relation

Looking further to see the influence of the flag factor showing up, turn from a real vector (four components) generated out of the spinor ξ^A to a real bivector (six components) generated out of the same spinor:

$$F^{\mu\nu} \longrightarrow F^{AB\dot{U}\dot{V}} = \xi^A\xi^B\epsilon^{\dot{U}\dot{V}} + \epsilon^{AB}(\bar{\xi})^{\dot{U}}(\bar{\xi})^{\dot{V}},$$
$$\mu \longrightarrow A\dot{U}; \; \nu \longrightarrow B\dot{V}. \tag{41.88}$$

That this quantity has no more than six distinct components ($F^{\mu\nu} = -F^{\nu\mu}$) follows from interchanging A with B and \dot{U} with \dot{V}, and noting the resultant change in sign on the righthand side of (41.88). To unfold the meaning of this bivector, look in (41.88) for every appearance of the alternating factor ε^{AB}. Wherever such a factor appears, insert the expression (41.84) for this factor in terms of the starting spinor ξ^A and insert the additional spinor η^A that is needed, along with ξ^A, to supply a basis for all spinors. In this way, find

$$F^{\mu\nu} \longrightarrow F^{AB\dot{U}\dot{V}} = \xi^A\xi^B(\bar{\xi}^{\dot{U}}\bar{\eta}^{\dot{V}} - \bar{\eta}^{\dot{U}}\bar{\xi}^{\dot{V}}) + (\xi^A\eta^B - \eta^A\xi^B)\bar{\xi}^{\dot{U}}\bar{\xi}^{\dot{V}}$$

$$= \xi^A\bar{\xi}^{\dot{U}}(\xi^B\bar{\eta}^{\dot{V}} + \eta^B\bar{\xi}^{\dot{V}}) - (\xi^A\bar{\eta}^{\dot{U}} + \eta^A\bar{\xi}^{\dot{U}})\xi^B\bar{\xi}^{\dot{V}} \tag{41.89}$$

$$= X^{A\dot{U}}Y^{B\dot{V}} - Y^{A\dot{U}}X^{B\dot{V}} \longrightarrow x^\mu y^\nu - y^\mu x^\nu.$$

Thus the 2,2-spinor built from ξ^A represents a bivector constructed out of the two 4-vectors \boldsymbol{x} and \boldsymbol{y}. Of these, the first is the "real null vector of the flagpole," already seen to be determined uniquely by the spinor ξ^A. The second vector,

$$y^\alpha \longrightarrow Y^{A\dot{U}} = \xi^A \overline{\eta}^{\dot{U}} + \eta^A \overline{\xi}^{\dot{U}}, \qquad (41.90)$$

is also determined by ξ^A, but not uniquely, because the "spinor mate," η^A, to ξ^A is not unique. Go from one choice of mate, η^A, to a new choice of mate (equation 41.85),

$$\eta^A{}_{\text{new}} = \eta^A + \alpha \xi^A. \qquad (41.91)$$

Then the real 4-vector y^μ goes to the new real 4-vector

$$y^\mu{}_{\text{new}} = y^\mu + (\alpha + \overline{\alpha})x^\mu. \qquad (41.92)$$

Were the 4-vector \boldsymbol{y} unique, there would project out from the flagpole, not a flag but an arrow. The range of values open for the real constant $\alpha + \overline{\alpha}$ makes one arrow into many arrows, all coplanar; hence the flag of Penrose. Otherwise stated, the choice of a spinor ξ^A fixes no individual arrow, but does fix the totality of the collection of arrows, and thus uniquely specifies the flag.

The 4-vector \boldsymbol{y} (and with it $\boldsymbol{y}_{\text{new}}$) is orthogonal to the null 4-vector \boldsymbol{x},

$$\begin{aligned}
\boldsymbol{x} \cdot \boldsymbol{y} = x_\beta y^\beta &= -\frac{1}{2} X_{A\dot{U}} Y^{A\dot{U}} \\
&= -\frac{1}{2} \xi_A \overline{\xi}_{\dot{U}} (\xi^A \overline{\eta}^{\dot{U}} + \eta^A \overline{\xi}^{\dot{U}}) = 0,
\end{aligned} \qquad (41.93)$$

and spacelike,

$$\begin{aligned}
\boldsymbol{y} \cdot \boldsymbol{y} &= -\frac{1}{2} (\xi_A \overline{\eta}_{\dot{U}} + \eta_A \overline{\xi}_{\dot{U}})(\xi^A \overline{\eta}^{\dot{U}} + \eta^A \overline{\xi}^{\dot{U}}) \\
&= -\frac{1}{2} (\xi_A \eta^A)(\overline{\eta}_{\dot{U}} \overline{\xi}^{\dot{U}}) - \frac{1}{2} (\eta_A \xi^A)(\overline{\xi}_{\dot{U}} \overline{\eta}^{\dot{U}}) = 1
\end{aligned} \qquad (41.94)$$

("unit length of flag").

Multiplication of the spinor ξ^A by the "flag factor" $e^{i\sigma}$ rotates the flag about the flagpole by the angle 2σ, because the spinor mate, η^A, of ξ^A is multiplied by the factor $e^{-i\sigma}$ [see the normalization condition (41.81)]. These changes alter the vector \boldsymbol{y} to a rotated vector $\boldsymbol{y}_{\text{rot}}$, with

$$\begin{aligned}
y^\alpha_{\text{rot}} \longrightarrow Y^{A\dot{U}}_{\text{rot}} &= e^{2i\sigma} \xi^A \overline{\eta}^{\dot{U}} + e^{-2i\sigma} \eta^A \overline{\xi}^{\dot{U}} \\
&= \cos 2\sigma (\xi^A \overline{\eta}^{\dot{U}} + \eta^A \overline{\xi}^{\dot{U}}) + \sin 2\sigma (i\xi^A \overline{\eta}^{\dot{U}} - i\eta^A \overline{\xi}^{\dot{U}}) \\
&\longrightarrow y^\alpha \cos 2\sigma + z^\alpha \sin 2\sigma.
\end{aligned} \qquad (41.95)$$

Rotation of flag about flagpole

Here the 4-vector \boldsymbol{z} shares with the vector \boldsymbol{y} these properties: it is (1) real, (2) spacelike, (3) of unit magnitude, (4) orthogonal to the null 4-vector \boldsymbol{x} of the flagpole, and (5) uniquely specified by the original spinor ξ^A up to the additive real

multiple $(\alpha + \bar{\alpha})$ of \boldsymbol{x}. In addition, \boldsymbol{z} and \boldsymbol{y} are orthogonal. Thus \boldsymbol{y} and \boldsymbol{z} provide basis vectors in the two-dimensional space in which—to overpictorialize—the "tip of the flag" undergoes its rotation.

Recapitulate by returning to the laser pulse. Two numbers, such as the familiar polar angles θ (angle with the z-axis) and ϕ (azimuth around z-axis from x-axis) tell the direction of its flight. A third number, r, gives the distance to the moon and also the travel time for light to reach the moon. A fourth number, an angle ψ, tells the azimuth of the illuminated arrow shot onto the surface of the moon, this azimuth to be measured from the \boldsymbol{e}_θ direction (where $\psi = 0$), around the flagpole in a righthanded sense. Then the spinor associated with the flagpole plus flag (rotated arrow) is

Equations relating spinor, flagpole, and flag

$$\begin{pmatrix} \xi^1 \\ \xi^2 \end{pmatrix} = (2r)^{1/2} \begin{pmatrix} \cos(\theta/2)e^{-i\phi/2 + i\psi/2} \\ \sin(\theta/2)e^{i\phi/2 + i\psi/2} \end{pmatrix} \tag{41.96}$$

according to the conventions adopted here [see (41.87)]. The mate η_A to this spinor, unique up to an additive multiple of ξ^A, is

$$\begin{pmatrix} \eta^1 \\ \eta^2 \end{pmatrix} = (2r)^{-1/2} \begin{pmatrix} -\sin(\theta/2)e^{-i\phi/2 - i\psi/2} \\ \cos(\theta/2)e^{i\phi/2 - i\psi/2} \end{pmatrix}. \tag{41.97}$$

The 4-vector of the flagpole determined by ξ^A is found from (41.87):

$$\begin{pmatrix} x^0 \\ x^1 \\ x^2 \\ x^3 \end{pmatrix} = \begin{pmatrix} r \\ r\sin\theta\cos\phi \\ r\sin\theta\sin\phi \\ r\cos\theta \end{pmatrix}. \tag{41.98}$$

To determine the flag itself, one requires, in addition to x^α, the unit spacelike 4-vector y^α, normal to x^α, and unique up to an additive real multiple of x^α. This vector is evaluated by use of (41.90) and has the form

$$\begin{pmatrix} y^0 \\ y^1 \\ y^2 \\ y^3 \end{pmatrix} = \begin{pmatrix} 0 \\ \cos\theta\cos\phi\cos\psi + \sin\phi\sin\psi \\ \cos\theta\sin\phi\cos\psi - \cos\phi\sin\psi \\ -\sin\theta\cos\psi \end{pmatrix}. \tag{41.99}$$

From these expressions for x^μ and y^μ, one calculates the components of the bivector ("flag") $F^{\mu\nu} = x^\mu y^\nu - y^\mu x^\nu$ by simple arithmetic.

§41.10. APPEARANCE OF THE NIGHT SKY: AN APPLICATION OF SPINORS

Attention has gone here to extracting all four pieces of information contained in a spinor: separation in time (equal to separation in space), direction in space, and

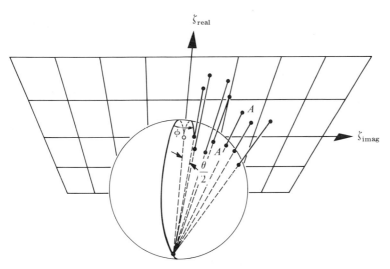

Figure 41.8.
Representation of a direction in space (one of the stars of the Big Dipper, regarded as a point on the celestial sphere) as a point in the complex ζ plane (ζ = ratio ξ^2/ξ^1 of spinor components) by stereographic projection from the South Pole.

rotation about that direction. Turn now to an application where not all that information is needed. Look at the night sky and ask (1) how to describe its appearance and (2) how to change that appearance. As one way to describe its appearance, give the direction of each star. Abandon any concern about the distance of the star, and any concern about any rotation ψ about the flagpole. In other words, the complex factor

$$(2r)^{1/2}e^{i\psi/2}$$

common to ξ^1 and ξ^2 drops from attention. All that is left as significant is the ratio ζ of these spinor components:

$$\zeta = \xi^2/\xi^1 = \tan(\theta/2)e^{i\phi}. \qquad (41.100)$$

To give the one complex number ζ ("stereographic coordinate;" Figure 41.8) for each star in the sky is to catalog the pattern of the stars.

Let the observer change his stance. The celestial sphere appears to rotate. Or let him rocket past his present location in the direction of the North Star with some substantial fraction of the velocity of light. To him all that portion of the celestial sphere is opened out, as if by a magnifying glass. To compensate, the remaining stars are packed into a smaller angular compass. Any such rotation or boost or combination of rotation and boost being described in spinor language by a transformation of the form

$$\xi^A \longrightarrow \xi^A_{\text{new}} = L^A{}_B \xi^B, \qquad (41.101)$$

Spinors used to analyze "Lorentz transformations" of appearance of night sky

implies a transformation of the complex stereographic coordinate of any given star of the form

$$\zeta \longrightarrow \zeta_{\text{new}} = \frac{\xi^2_{\text{new}}}{\xi^1_{\text{new}}} = \frac{L^2{}_2\zeta + L^2{}_1}{L^1{}_2\zeta + L^1{}_1}. \tag{41.102}$$

In the special case of a boost in the z-direction with velocity parameter α (velocity $\beta = \tanh \alpha$), the off-diagonal components $L^1{}_2$ and $L^2{}_1$ vanish. The magnification of the overhead sky then expresses itself in the simple formula

$$\zeta_{\text{new}} = e^\alpha \zeta$$

or

$$\phi_{\text{new}} = \phi,$$
$$\tan(\theta_{\text{new}}/2) = e^\alpha \tan(\theta/2). \tag{41.103}$$

Contrary to this prediction and false expectation, no magnification at all is achieved of the regions around the North Star by moving with high velocity in that direction. On the contrary, any photon coming in from a star a little off that direction, with a little transverse momentum, keeps that transverse momentum in the new frame; but its longitudinal momentum against the observer is augmented by his motion. Thus the ratio of the momenta is decreased, and the observed angle relative to the North Star is also decreased. The consequence is not magnification, but diminution ("looking through the wrong end of a telescope"). The correct formula is not (41.103) but

$$\tan(\theta_{\text{new}}/2) = e^{-\alpha} \tan(\theta/2) \tag{41.104}$$

(reversal of the sign of α). The reason for this correction is not far to seek. The spinor analysis so far had dealt with an outgoing light pulse, and a 4-vector with positive time component. That feature was built into the formula adopted to tie the spinor to the 4-vector,

$$r\mathbf{1} + (\mathbf{r} \cdot \boldsymbol{\sigma}) \equiv X = \|\xi^A \bar{\xi}^{\dot{U}}\|. \tag{41.105}$$

In contrast the 4-vector that reaches back to the origin of an incoming photon has a time component that is negative (or, alternatively, sign-reversed space components)! For any null 4-vector with negative time component, one employs instead of (41.105) the formula

$$X = -\|\xi^A \bar{\xi}^{\dot{U}}\|. \tag{41.106}$$

It is enough to mention here this point of principle without going through the details that give the altered sign for α in (41.104). From now on, to preserve the previous arithmetic, change the problem. Deal, not with incoming photons, but with outgoing photons. Replace the receiving telescope by the projector of a planetarium. It projects out into space a separate beam of light for each star of the Big Dipper and also one for the North Star itself. Let an observer move in the positive z-direction with velocity parameter α. In his frame of reference the beams actually will be widened out in full accord with (41.103).

"The magnification process changes the size of the Big Dipper but not its shape."

This statement is at the same time true and false. It is true of the Dipper and of any other constellation to the extent that the angular dimensions of that constellation can be idealized to be small compared to the entire compass of the sky. It is false in the sense that any well-rounded projected constellation, however small it may appear to an observer at rest with respect to the earth, can always be so "opened out" by the observer putting on any sufficiently high velocity, the observer still being near the earth, that the constellation encompasses a major fraction of the sky.

That the "Lorentz-transformation-induced magnification" of a small object does not change its shape can be seen in three ways. (1) Stereographic projection (Figure 41.8) and "fractional linear transformation" (41.102) are both known to leave all angles unchanged ["conformal invariance;" see for example Penrose (1959)] and known even to turn every old circle into a new circle. (2) Consider a given star, M, in the constellation and immediate neighbors, L and N, just below it and just above it in the count of the members of that constellation. Consider the flagpole pointed at M and the flag pointed first from M to L, then from M to N. The flag has turned about the flagpole through an angle ψ. The two corresponding spinors therefore differ by a phase factor $e^{i\psi/2}$. They differ in no other way. After an arbitrary Lorentz transformation they still differ by the phase factor $e^{i\psi/2}$, and in no other way. The angle between the arcs ML and MN on the celestial sphere therefore remains at its original value ψ after the Lorentz transformation (again conformal invariance of patterns on the celestial sphere!). (3) An even more elementary calculation shows that infinitesimal arc lengths on the unit celestial sphere in the direction of increasing θ and arc lengths in the direction of increasing ϕ are magnified in the same proportion, thus leaving unchanged the angle between arc and arc (conformal invariance). Thus, consider a photon shot out from the planetarium projector to a point on the celestial sphere ("planetarium version of a Big-Dipper star") with inclination θ to the z-axis, as seen by an observer at rest relative to the earth. From the standard laws of transformation of angles in a Lorentz transformation ("aberration"; Box 2.4), one has for the sine of the transformed angle

$$\sin\theta_{\text{new}} = \frac{(1-\beta^2)^{1/2}}{1-\beta\cos\theta}\sin\theta \tag{41.107}$$

and (by differentiating the expression for the cosine of the transformed angle)

$$d\theta_{\text{new}} = \frac{(1-\beta^2)^{1/2}}{1-\beta\cos\theta}d\theta. \tag{41.108}$$

From these expressions it follows at once that the inclination, relative to a meridian line, on the transformed celestial sphere is identical to the direction, relative to the same meridian line, on the original celestial sphere:

$$\tan\left(\begin{matrix}\text{new}\\\text{inclination}\end{matrix}\right) = \frac{\sin\theta_{\text{new}}\,d\phi_{\text{new}}}{d\theta_{\text{new}}} = \frac{\sin\theta\,d\phi}{d\theta}$$
$$= \tan\left(\begin{matrix}\text{original}\\\text{inclination}\end{matrix}\right) \tag{41.109}$$

(again conformal invariance!).

So much for the elementary spinor and what it has to do with a null vector, with a "flagpole" pointed to the celestial sphere, and with rotation of a "flag" about such a flagpole.

§41.11. SPINORS AS A POWERFUL TOOL IN GRAVITATION THEORY

Spinor formalism in curved spacetime

Just as vectors, tensors, and differential forms are easily generalized from flat space-time to curved, so are spinors.

Each event \mathcal{P} in curved spacetime possesses a tangent space. In that tangent space reside and operate all the vectors, tensors, and forms located at \mathcal{P}. The geometry of the tangent space is Lorentzian ("local Lorentz geometry at \mathcal{P}"), since the scalar product of any two vectors \boldsymbol{u} and \boldsymbol{v} at \mathcal{P}, expressed in an orthonormal frame at \mathcal{P}, is

$$\boldsymbol{u} \cdot \boldsymbol{v} = \boldsymbol{g}(\boldsymbol{u}, \boldsymbol{v}) = \eta_{\hat{\alpha}\hat{\beta}} u^{\hat{\alpha}} v^{\hat{\beta}}.$$

Thus, there is no mathematical difference between the tangent space at \mathcal{P} on the one hand, and flat spacetime on the other. Whatever mathematical can be done in the one can also be done in the other. In particular, *the entire formalism of spinors, developed originally in flat spacetime, can be carried over without change to the tangent space at the arbitrary event \mathcal{P} in curved spacetime.*

Let it be done. Now spinors reside at every event in curved spacetime; and at each event one can translate back and forth between spinor language and tensor language, using the equations (valid in orthonormal frames) of §§41.6 and 41.7.

Spinors needed when analyzing fermions in gravitational fields

Spinors in curved spacetime are an indispensible mathematical tool, when one wishes to study the influence of gravity on quantized particles of half-integral spin (neutrinos, electrons, protons, . . .). Consider, for example, Hartle's (1971) proof that a black hole cannot exert any long-range, weak-interaction forces on external matter (i.e., that a black hole has no "weak-interaction hair"). His proof could not function without a spinor description of neutrino fields in curved spacetime. Similarly for Wheeler's (1971b) analysis of the quasibound states of an electron in the gravitational field of a small black hole (gravitational radius $\sim 10^{-13}$ cm): it requires solving the Dirac equation for a spin-$\frac{1}{2}$ particle in the curved spacetime geometry of Schwarzschild. For a detailed discussion of the Dirac equation in curved spacetime see, e.g., Brill and Wheeler (1957).

Equivalence of spinor and tensor formalisms

To use the mathematics of spinors, one need not be dealing with quantum theory or with particles of half-integral spin. The spinor formalism is perfectly applicable in situations where only integral-spin entities (scalars, vectors, tensors) are in view, and where in fact, the spinor formalism is fully equivalent to the tensor formalism that pervades earlier chapters of this book. Equations (41.77) and (41.78) provide the translation from one formalism to the other, once an orthonormal frame has been chosen at each event in spacetime.

Certain types of problems in gravitation theory are far more tractable in the language of spinors than in the language of tensors. Examples are as follows.

(1) Geometric Optics
(the theory of "null congruences of geodesics")

Here spinors make almost trivial the lengthy tensor algebra needed in derivations of the "focusing theorem" [equation (22.37)]; and they yield an elegant, simple formalism for discussing and calculating how, with increasing affine parameter, a bundle of rays alters its size ("expansion"), its shape ("shear"), and its orientation ("rotation"). See, e.g., Sachs (1964), Pirani (1965), or Penrose (1968a) for a review and the original references.

Applications of spinor formalism in classical gravitation theory

(2) Radiation Theory in Curved Spacetime
(both gravitational and electromagnetic)

Spinors provide the most powerful of all formalisms for decomposing radiation fields into spherical harmonics and for manipulating their decomposed components. See, for example, Price's (1972a,b) analysis of how a perturbed Schwarzschild black hole radiates away all its radiatable perturbations, be they electromagnetic perturbations, gravitational perturbations, or perturbations in a fictitious field of spin 17; see, similarly, the analysis by Fackerell and Ipser (1972) and by Ipser (1971) of electromagnetic perturbations of a Kerr black hole, and the analysis by Teukolsky (1972) of gravitational perturbations of a Kerr hole. Spinors also yield an elegant and powerful analysis of the "$1/r$" expansion of a radiation field flowing out from a source into asymptotically flat space. Among its results is a *peeling theorem,* which describes the algebraic properties of the coefficients in a $1/r$ expansion of the Riemann tensor. See, e.g., Sachs (1964) or Pirani (1965) for reviews and original references.

(3) Algebraic Properties of Curvature Tensors

The spinor formalism is a more powerful method than any other for deriving the "Petrov-Pirani algebraic classification of the conformal curvature tensor," and for proving theorems about algebraic properties of curvature tensors. See, e.g., Sachs (1964) or Pirani (1965) or Penrose (1968a) for reviews and references.

Of course, the spinor formalism, like any formalism, has its limitations. For example, many of the elementary problems of gravitation theory, and a large fraction of the most difficult ones, would be more difficult in the language of spinors than in the language of tensors! But for certain classes of problems, especially those where null vectors play a central role, spinors are a most valuable tool.

Cartan gave spinors to the world's physics and mathematics. His text (American edition, 1966) is an important reference to the subject.

CHAPTER **42**

REGGE CALCULUS

§42.1. WHY THE REGGE CALCULUS?

The need for Regge calculus as a computational tool

Gravitation theory is entering an era when situations of greater and greater complexity must be analyzed. Before about 1965 the problems of central interest could mostly be handled by idealizations of special symmetry or special simplicity or both. The Schwarzschild geometry and its generalizations, the Friedmann cosmology and its generalizations, the joining together of the Schwarzschild geometry and the Friedmann geometry to describe the collapse of a bounded collection of matter, the vibrations of relativistic stars, weak gravitational waves propagating in an otherwise flat space: all these problems and others were solved by elementary means.

But today one is pressed to understand situations devoid of symmetry and not amenable to perturbation theory: How do two black holes alter in shape, and how much gravitational radiation do they emit when they collide and coalesce? What are the structures and properties of the singularities at the endpoint of gravitational collapse, predicted by the theorems of Penrose, Hawking, and Geroch? Can a Universe that begins completely chaotic smooth itself out quickly by processes such as inhomogeneous mixmaster oscillations?

To solve such problems, one needs new kinds of mathematical tools—and in response to this need, new tools are being developed. The "global methods" of Chapter 34 provide one set of tools. The Regge Calculus provides another [Regge (1961); see also pp. 467–500 of Wheeler (1964a)].

§42.2. REGGE CALCULUS IN BRIEF

Approximation of smooth geometries by skeleton structures

Consider the geodesic dome that covers a great auditorium, made of a multitude of flat triangles joined edge to edge and vertex to vertex. Similarly envisage spacetime, in the Regge calculus, as made of flat-space "simplexes" (four-dimensional

item in this progression: two dimensions, triangle; three dimensions, tetrahedron; four dimensions, simplex) joined face to face, edge to edge, and vertex to vertex. To specify the lengths of the edges is to give the engineer all he needs in order to know the shape of the roof, and the scientist all he needs in order to know the geometry of the spacetime under consideration. A smooth auditorium roof can be approximated arbitrarily closely by a geodesic dome constructed of sufficiently small triangles. A smooth spacetime manifold can be approximated arbitrarily closely by a locked-together assembly of sufficiently small simplexes. Thus the Regge calculus, reaching beyond ordinary algebraic expressions for the metric, provides a way to analyze physical situations deprived, as so many situations are, of spherical symmetry, and systems even altogether lacking in symmetry.

If the designer can give the roof any shape he pleases, he has more freedom than the analyst who is charting out the geometry of spacetime. Given the geometry of spacetime up to some spacelike slice that, for want of a better name, one may call "now," one has no freedom at all in the geometry from that instant on. Einstein's geometrodynamic law is fully deterministic. Translated into the language of the Regge calculus, it provides a means to calculate the edge lengths of new simplexes from the dimensions of the simplexes that have gone before. Though the geometry is deterministically specified, how it will be approximated is not. The original spacelike hypersurface ("now") is approximated as a collection of tetrahedrons joined together face to face; but how many tetrahedrons there will be and where their vertices will be placed is the option of the analyst. He can endow the skeleton more densely with bones in a region of high curvature than in a region of low curvature to get the most "accuracy profit" from a specified number of points. Some of this freedom of choice for the lengths of the bones remains as one applies the geometrodynamic law in the form given by Regge (1961) to calculate the future from the past. This freedom would be disastrous to any computer program that one tried to write, unless the programmer removed all indefiniteness by adding supplementary conditions of his own choice, either tailored to give good "accuracy profit," or otherwise fixed.

Having determined the lengths of all the bones in the portion of skeletonized spacetime of interest, one can examine any chosen local cluster of bones in and by themselves. In this way one can find out all there is to be learned about the geometry in that region. Of course, the accuracy of one's findings will depend on the fineness with which the skeletonization has been carried out. But in principle that is no limit to the fineness, or therefore to the accuracy, so long as one is working in the context of classical physics. Thus one ends up with a catalog of all the bones, showing the lengths of each. Then one can examine the geometry of whatever spacelike surface one pleases, and look into many other questions besides. For this purpose one has only to pick out the relevant bones and see how they fit together.

Role of Einstein field equation in fixing the skeleton structure

§42.3. SIMPLEXES AND DEFICIT ANGLES

Figure 42.1 recalls how a smoothly curved surface can be approximated by flat triangles. All the curvature is concentrated at the vertices. No curvature resides at

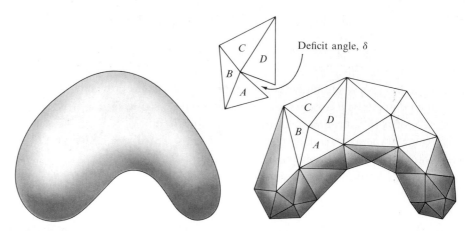

Figure 42.1.
A 2-geometry with continuously varying curvature can be approximated arbitrarily closely by a polyhedron built of triangles, provided only that the number of triangles is made sufficiently great and the size of each sufficiently small. The geometry in each triangle is Euclidean. The curvature of the surface shows up in the amount of deficit angle at each vertex (portion *ABCD* of polyhedron laid out above on a flat surface).

Deficit angle as a skeletonized measure of curvature:

(1) in two dimensions

the edge between one triangle and the next, despite one's first impression. A vector carried by parallel transport from *A* through *B* and *C* to *D*, and then carried back by another route through *C* and *B* to *A* returns to its starting point unchanged in direction, as one sees most easily by laying out this complex of triangles on a flat surface. Only if the route is allowed to encircle the vertex common to *A*, *B*, *C*, and *D* does the vector experience a net rotation. The magnitude of the rotation is equal to the indicated deficit angle, δ, at the vertex. The sum of the deficit angles over all the vertices has the same value, 4π, as does the half-integral of the continuously distributed scalar curvature ($^{(2)}R = 2/a^2$ for a sphere of radius a) taken over the entirety of the original smooth figure,

$$\sum_{\substack{\text{skeleton} \\ \text{geometry}}} \delta_i = \frac{1}{2} \int_{\substack{\text{actual smooth} \\ \text{geometry}}} {}^{(2)}R \, d(\text{surface}) = 4\pi. \tag{42.1}$$

(2) in *n* (or four) dimensions

Generalizing from the example of a 2-geometry, Regge calculus approximates a smoothly curved *n*-dimensional Riemannian manifold as a collection of *n*-dimensional blocks, each free of any curvature at all, joined by ($n - 2$)-dimensional regions in which all the curvature is concentrated (Box 42.1). For the four-dimensional spacetime of general relativity, the "hinge" at which the curvature is concentrated has the shape of a triangle, as indicated schematically in the bottom row of Figure 42.2. In the example illustrated there, ten tetrahedrons have that triangle in common. Between one of these tetrahedrons and the next fits a four-dimensional simplex. Every feature of this simplex is determined by the lengths of its ten edges. One of the features is the angle α between one of the indicated tetrahedrons or "faces" of the simplex and the next. Thus α represents the angle subtended by this simplex

Box 42.1 THE HINGES WHERE THE CURVATURE IS CONCENTRATED IN THE "ANGLE OF RATTLE" BETWEEN BUILDING BLOCKS IN A SKELETON MANIFOLD

Dimensionality of manifold	2	3	4
Elementary flat-space building block:	triangle	tetrahedron	simplex
Edge lengths to define it:	3	4	5
Hinge where cycle of such blocks meet with a deficit angle or "angle of rattle" δ:	vertex	edge	triangle
Dimensionality of hinge:	0	1	2
"Content" of such a hinge:	1	length l	area A
Contribution from all hinges within a given small region to curvature of manifold:	$\sum\limits_{\substack{\text{that}\\\text{region}}} \delta_i$	$\sum\limits_{\substack{\text{that}\\\text{region}}} l_i\,\delta_i$	$\sum\limits_{\substack{\text{that}\\\text{region}}} A_i\,\delta_i$
Continuum limit of this quantity expressed as an integral over the same small region:	$\frac{1}{2}\int {}^{(2)}R ({}^{(2)}g)^{1/2}\,d^2x$	$\frac{1}{2}\int {}^{(3)}R ({}^{(3)}g)^{1/2}\,d^3x$	$\frac{1}{2}\int {}^{(4)}R (-{}^{(4)}g)^{1/2}\,d^4x$

at the hinge. Summing the angles α for all the simplexes that meet on the given hinge \mathscr{PQR}, and subtracting from 2π, one gets the deficit angle associated with that hinge. And by then summing the deficit angles in a given small n-volume with appropriate weighting (Box 42.1), one obtains a number equal to the volume integral of the scalar curvature of the original smooth n-geometry. See Box 42.2.

§42.4. SKELETON FORM OF FIELD EQUATIONS

Rather than translate Einstein's field equations directly into the language of the skeleton calculus, Regge turns to a standard variational principle from which Einstein's law lets itself be derived. It says (see §§21.2 and 43.3) adjust the 4-geometry throughout an extended region of spacetime, subject to certain specified conditions on the boundary, so that the dimensionless integral (action in units of \hbar!),

Einstein-Hilbert variational principle reduced to skeleton form

$$I = (c^3/16\pi\hbar G) \int R(-g)^{1/2}\,d^4x, \qquad (42.2)$$

is an extremum. This statement applies when space is free of matter and electromag-

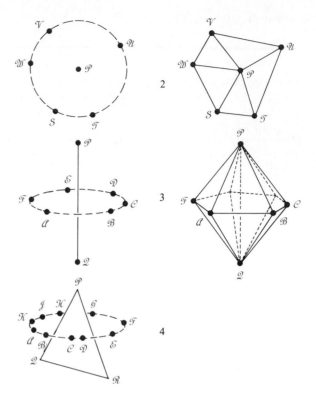

Figure 42.2.

Cycle of building blocks associated with a single hinge. Top row, two dimensions: left, schematic association of vertices \mathcal{S}, \mathcal{T}, \mathcal{U}, \mathcal{V}, \mathcal{W} with "hinge" at the vertex \mathcal{P}; right, same, but with elementary triangles indicated in full. Middle row, three dimensions: left, schematic; right, perspective representation of the six tetrahedrons that meet on the "hinge" \mathcal{PQ}. Bottom row, four dimensions; shown only schematically. The five vertices \mathcal{PQRCD} belong to one simplex, a four-dimensional region throughout the interior of which space is flat. The five vertices \mathcal{PQRDE} belong to the next simplex; and so on around the cycle of simplexes. The two simplexes just named interface at the tetrahedron \mathcal{PQRD}, inside which the geometry is also flat. Between that tetrahedron and the next, \mathcal{PQRE}, there is a certain hyperdihedral angle α subtended at the "hinge" \mathcal{PQR}. The value of this angle is completely fixed by the ten edge lengths of the intervening simplex \mathcal{PQRDE}. This dihedral angle, plus the corresponding dihedral angles subtended at the hinge \mathcal{PQR} by the other simplexes of the cycle, do not in general add up to 2π. The deficit, the "angle of rattle" or deficit angle δ, gives the amount of curvature concentrated at the hinge \mathcal{PQR}. There is no actual rattle or looseness of fit, unless one tries to imbed the cycle into an over-all flat four-dimensional space (analog of "stamping on" the collection of triangles, and seeing them open out by the amount of the deficit angle, as indicated in inset in Figure 42.1).

netic fields, a simplification that will be made in the subsequent discussion to keep it from becoming too extended. When in addition all lengths are expressed in units of the Planck length

$$L^* = (\hbar G/c^3)^{1/2} = 1.6 \times 10^{-33} \text{ cm},\qquad(42.3)$$

and the curvature integral is approximated by its expression in terms of deficit angles, Regge shows that the statement $\delta I = 0$ (condition for an extremum!) becomes

$$(1/8\pi)\,\delta \sum_{\substack{\text{hinges}\\h=1}}^{H} A_h\,\delta_h = 0.\qquad(42.4)$$

Box 42.2 FLOW DIAGRAMS FOR REGGE CALCULUS

A skeleton 4-geometry is completely determined by all its edge lengths. From the edge lengths one gets the integrated curvature by pursuing, for each hinge in the 4-geometry, the following flow diagram:

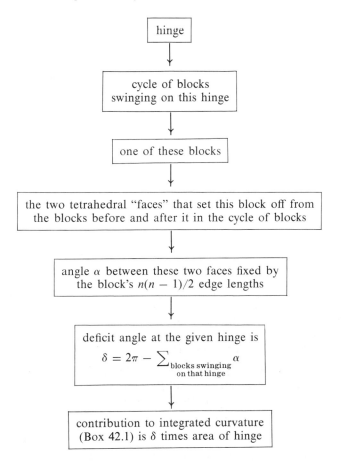

One finds it natural to apply this analysis in either of two ways. First, one can probe a given 4-geometry (given set of edge lengths!) in the sense

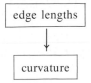

Box 42.2 (*continued*)

Second—and this is the rationale of Regge calculus—one can use the skeleton calculus to deduce a previously unknown 4-geometry from Einstein's geometrodynamic law, proceeding in the direction

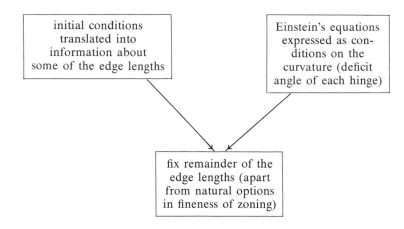

In the changes contemplated in this variational principle, certain edge lengths are thought of as being fixed. They have to do with the conditions specified at the boundaries of the region of spacetime under study. It is not necessary here to enter into the precise formulation of these boundary conditions, fortunately, since some questions of principle still remain to be clarified about the precise formulation of boundary conditions in general relativity (see §21.12). Rather, what is important is the effect of changes in the lengths of the edges of the blocks in the interior of the region being analyzed, as they augment or decrease the deficit angles at the various hinges. In his basic paper on the subject, Regge (1961) notes that the typical deficit angle δ_h depends in a complicated trigonometric way on the values of numerous edge lengths ℓ_p. However, he proves (Appendix of his paper) that "quite remarkably, we can carry out the variation as if the δ_h were constants," thus reducing the variational principle to the form

$$(1/8\pi) \sum_{\substack{\text{hinges} \\ h=1}}^{H} \delta_h \, \delta A_h = 0. \tag{42.5}$$

Here the change in area of the h-th triangle-shaped hinge, according to elementary trigonometry, is

$$\delta A_h = \frac{1}{2} \sum_p \ell_p \, \delta \ell_p \cot \theta_{ph}. \tag{42.6}$$

In this equation θ_{ph} is the angle opposite to the p-th edge in the triangle. Conse-
quently, Einstein's equations in empty space reduce in skeleton geometry to the form

$$\sum_{\substack{\text{hinges that} \\ \text{have the} \\ \text{given edge} \\ p \text{ in common}}} \delta_h \cot \theta_{ph} = 0, \quad (p = 1, 2, \ldots), \tag{42.7}$$

Einstein field equation
reduced to skeleton form

one equation for each edge length in the interior of the region of spacetime being
analyzed.

§42.5. THE CHOICE OF LATTICE STRUCTURE

Two questions arise in the actual application of Regge calculus, and it is not clear
that either has yet received the resolution which is most convenient for practical
applications of this skeleton analysis: What kind of lattice to use? How best to
capitalize on the freedom that exists in the choice of edge lengths? The first question
is discussed in this section, the second in the next section.

It might seem most natural to use a lattice made of small, nearly rectangular
blocks, the departure of each from rectangularity being conditioned by the amount
and directionality of the local curvature. However, such building blocks are "floppy."
One could give them rigidity by specifying certain angles as well as the edge lengths.
But then one would lose the cleanness of Regge's prescription: give edge lengths,
and give only edge lengths, and give each edge length freely and independently,
in order to define a geometry. In addition one would have to rederive the Regge
equations, including new equations for the determination of the new angles. There-
fore one discards the quasirectangle in favor of the simplex with its $5 \cdot 4/2 = 10$ edge
lengths. This decided, one also concludes that even in flat spacetime the simplexes
cannot all have identical edge lengths. Two-dimensional flat space can be filled with
identical equilateral triangles, but already at three dimensions it ceases to be possible
to fill out the manifold with identical equilateral tetrahedrons. One knows that a
given carbon atom in diamond is joined to its nearest neighbors with tetrahedral
bonds, but a little reflection shows that the cell assignable to the given atom is far
from having the shape of an equilateral tetrahedron.

Synthesis would appear to be a natural way to put together the building blocks:
first make one-dimensional structures; assemble these into two-dimensional struc-
tures; these, into three-dimensional ones; and these, into the final four-dimensional
structure. The one-dimensional structure is made of points, $1, 2, 3, \ldots$, alternating
with line segments, $12, 23, 34, \ldots$. To start building a two-dimensional structure,
pick up a second one-dimensional structure. It might seem natural to label its points
$1', 2', 3', \ldots$, etc. However, that labeling would imply a cross-connection between
1 and $1'$, between 2 and $2'$, etc., after the fashion of a ladder. Then the elementary
cells would be quasirectangles. They would have the "floppiness" that is to be
excluded. Therefore relabel the points of the second one-dimensional structure as
$1\frac{1}{2}', 2\frac{1}{2}', 3\frac{1}{2}'$, etc. The implication is that one cross-connects $2\frac{1}{2}'$ with points 2 and 3
of the original one-dimensional structure, etc. One ends up with something like the

The choice of lattice structure:

(1) avoiding floppiness

(2) necessity for unequal
 edge lengths

(3) construction of two-
 dimensional structures

girder structure of a bridge, fully rigid in the context of two dimensions, as desired. The same construction, extended, fills out the plane with triangles. One now has a simple, standard two-dimensional structure. One might mistakenly conclude that one is ready to go ahead to build up a three-dimensional structure: the mistake lies in the tacit assumption that the flat-space topology is necessarily correct.

Let it be the problem, for example, to determine the development in time of a 3-geometry that has the topology of a 3-sphere. This 3-sphere is perhaps strongly deformed from ideality by long-wavelength gravitational waves. A right arrangement of the points is the immediate desideratum. Therefore put aside for the present any consideration of the deformation of the geometry by the waves (alteration of edge lengths from ideality). Ask how to divide a perfect 3-sphere into two-dimensional sheets. Here each sheet is understood to be separated from the next by a certain distance. At this point two alternative approaches suggest themselves that one can call for brevity "blocks" and "spheres."

(4) 3-D structures built from 2-D structures by "method of blocks"

(1) *Blocks.* Note that a 3-sphere lets itself be decomposed into 5 identical, tetrahedron-like solid blocks (5 vertices; 5 ways to leave out any one of these vertices!) Fix on one of these "tetrahedrons." Select one vertex as summit and the face through the other three vertices as base. Give that base the two-dimensional lattice structure already described. Introduce a multitude of additional sheets piled above it as evenly spaced layers reaching to the summit. Each layer has fewer points than the layer before. The decomposition of the 3-geometry inside one "tetrahedron" is thereby accomplished. However, an unresolved question remains; not merely how to join on this layered structure in a regular way to the corresponding structure in the adjacent "tetrahedrons"; but even whether such a regular joinup is at all possible. The same question can be asked about the other two ways to break up the 3-sphere into identical "tetrahedrons" [Coxeter (1948), esp. pp. 292–293: 16 tetrahedrons defined by a total of 8 vertices or 600 tetrahedrons defined by a total of 120 vertices]. One can eliminate the question of joinup of structure in a simple way, but at the price of putting a ceiling on the accuracy attainable: take the stated number of vertices (5 or 8 or 120) as the total number of points that will be employed in the skeletonization of the 3-geometry (no further subdivision required or admitted). Considering the boundedness of the memory capacity of any computer, it is hardly ridiculous to contemplate a limitation to 120 tracer points in exploratory calculations!

(5) 3-D structures from 2-D by "method of spheres"

(2) *Spheres.* An alternative approach to the "atomization" of the 3-sphere begins by introducing on the 3-sphere a North Pole and a South Pole and the hyperspherical angle χ ($\chi = 0$ at the first pole, $\chi = \pi$ at the second, $\chi = \pi/2$ at the equator; see Box 27.2). Let each two-dimensional layer lie on a surface of constant χ (χ equal to some integer times some interval $\Delta\chi$). The structure of this 2-sphere is already to be regarded as skeletonized into elementary triangles ("fully complete Buckminster Fuller geodesic dome"). Therefore the number of "faces" or triangles F, the number of edge lengths E, and the number of vertices V must be connected by the relation of Euler:

$$F - E + V = \begin{pmatrix} \text{a topology-dependent} \\ \text{number or, "Euler character"} \end{pmatrix} = \begin{cases} 2 \text{ for 2-sphere,} \\ 0 \text{ for 2-torus.} \end{cases} \quad (42.8)$$

It follows from this relation that it is impossible for each vertex to sit at the center

of a hexagon (each vertex the point of convergence of 6 triangles). This being the case, one is not astonished that a close inspection of the pattern of a geodesic dome shows several vertices where only 5 triangles meet. It is enough to have 12 such 5-triangle vertices among what are otherwise all 6-triangle vertices in order to meet the requirements of the Euler relation:

$$n \quad \text{5-triangle vertices}$$

$$V - n \quad \text{6-triangle vertices}$$

$$F = (V - n)(6/3) + n(5/3) \text{ triangles}$$

$$E = (V - n)(6/2) + n(5/2) \text{ edges} \tag{42.9}$$

$$V = (V - n)(6/6) + n \text{ vertices}$$

$$2 = F - E + V = n/6 \quad \text{Euler characteristic}$$

$$n = 12$$

Among all figures with triangular faces, the icosahedron is the one with the smallest number of faces that meets this condition (5-triangle vertices exclusively!)

If each 2-surface has the pattern of vertices of a geodesic dome, how is one dome to be joined to the next to make a rigid skeleton 3-geometry? Were the domes imbedded in a flat 3-geometry, rigidity would be no issue. Each dome would already be rigid in and by itself. However, the 3-geometry is not given to be flat. Only by a completely deterministic skeletonization of the space between the two 2-spheres will they be given rigidity in the context of curved space geometry. (1) Not by running a single connector from each vertex in one surface to the corresponding vertex in the next ("floppy structure"!) (2) Not by displacing one surface so each of its vertices comes above, or nearly above, the center of a triangle in the surface "below." First, the numbers of vertices and triangles ordinarily will not agree. Second, even when they do, it will not give the structure the necessary rigidity to connect the vertex of the surface above to the three vertices of the triangle below. The space between will contain some tetrahedrons, but it will not be throughout decomposed into tetrahedrons. (3) A natural and workable approach to the skeletonization of the 3-geometry is to run a connector from each vertex in the one surface to the corresponding vertex in the next, but to flesh out this connection with additional structure that will give rigidity to the 3-geometry: intervening vertices and connectors as illustrated in Box 42.3.

In working up from the skeletonization of a 3-geometry to the skeletonization of a 4-geometry, it is natural to proceed similarly. (1) Use identical patterns of points in the two 3-geometries. (2) Tie corresponding points together by single connectors. (3) Halfway, or approximately half way between the two 3-geometries insert a whole additional pattern of vertices. Each of these supplementary vertices is "dual" to and lies nearly "below" the center of a tetrahedron in the 3-geometry immediately above. (4) Connect each supplementary vertex to the vertices of the tetrahedron immediately above, to the vertices of the tetrahedron immediately below, and to those other supplementary vertices that are its immediate neighbors. (5) In this way get the edge lengths needed to divide the 4-geometry into simplexes, each of rigidly defined dimensions.

(6) 4-D structures built from 3-D structures

Box 42.3 SYNTHESIS OF HIGHER-DIMENSIONAL SKELETON GEOMETRIES OUT OF LOWER-DIMENSIONAL SKELETON GEOMETRIES

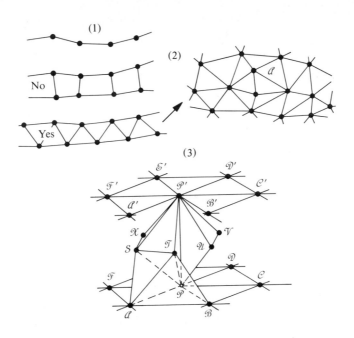

(1) One-dimensional structure as alternation of points and line segments. (2) Two-dimensional structure (a) "floppy" (unacceptable) and (b) rigidified (angles of triangles fully determined by edge lengths). When this structure is extended, as at right, the "normal" vertex has six triangles hinging on it. However, at least twelve 5-triangle vertices of the type indicated at \mathscr{A} are to be interpolated if the 2-geometry is to be able to close up into a 2-sphere. (3) Skeleton 3-geometry obtained by filling in between the skeleton 2-geometry... $\mathscr{A}\mathscr{B}...\mathscr{F}\mathscr{P}\mathscr{C}...\mathscr{E}\mathscr{D}...$ and the similar structure $...\mathscr{A}'\mathscr{B}'...\mathscr{F}'\mathscr{P}'\mathscr{C}'...\mathscr{E}'\mathscr{D}'...$ as follows. (a) Insert direct connectors such as $\mathscr{P}\mathscr{P}'$ between corresponding points in the two 2-geometries. (b) Insert an intermediate layer of "supplementary vertices" such as $\mathscr{S}\mathscr{T}\mathscr{U}\mathscr{V}\mathscr{W}\mathscr{X}....$ Each of these supplementary vertices lies roughly halfway between the center of the triangle "above" it and the center of the corresponding triangle "below" it. (c) Connect each such "supplementary vertex" with its immediate neighbors above, below, and in the same plane. (d) Give all edge lengths. (e) Then the skeleton 3-geometry between the two 2-geometries is rigidly specified. It is made up of five types of tetrahedrons, as follows. (1) "Right-through blocks," such as $\mathscr{P}\mathscr{P}'\mathscr{S}\mathscr{T}$ (six of these hinge on $\mathscr{P}\mathscr{P}'$ when \mathscr{P} is a normal vertex; five, when it is a 5-fold vertex, such as indicated by \mathscr{A} at the upper right). (2) "Lower-facing blocks," such as $\mathscr{A}\mathscr{B}\mathscr{P}\mathscr{T}$. (3) "Lower-packing blocks," such as $\mathscr{A}\mathscr{P}\mathscr{S}\mathscr{T}$. (4, 5) Corresponding "upper-facing blocks" and "upper-packing blocks" (not shown). The number of blocks of each kind is appropriately listed here for the two extreme cases of a 2-geometry that consists (a) of a normal hexagonal lattice extending indefinitely in a plane and (b) of a lattice consisting of the minimum number of 5-fold vertices ("type \mathscr{A} vertices") that will permit close-up into a 2-sphere.

2-geometry of upper (or lower) face	Hexagonal pattern of triangles	Icosahedron made of triangles
Its topology	Infinite 2-plane	2-sphere
Vertices on upper face	V	12
Nature of these vertices	6-fold	5-fold
Edge lengths on upper face	$3V$	$\frac{5}{2}V = 30$
Triangles on upper face	$2V$	20
Number of "supplementary vertices"	$2V$	20
Outer facing blocks	$2V$	20
Outer packing blocks	$3V$	30
Right through blocks	$6V$	60
Inner packing blocks	$3V$	30
Inner facing blocks	$2V$	20

§42.6. THE CHOICE OF EDGE LENGTHS

So much for the lattice structure of the 4-geometry; now for the other issue, the freedom that exists in the choice of edge lengths. Why not make the simplest choice and let all edges be light rays? Because the 4-geometry would not then be fully determined. The geometry $g_{\alpha\beta}(x^\mu)$ differs from the geometry $\lambda(x^\mu)\, g_{\alpha\beta}(x^\mu)$, even though the same points that are connected by light rays in the one geometry are also connected by light rays in the other geometry.

The choice of edge lengths:

If none of the edges is null, it is nevertheless natural to take some of the edge lengths to be spacelike and some to be timelike. In consequence the area A of the triangle in some cases will be real, in other cases imaginary. In 3-space the parallelogram (double triangle) spanned by two vectors \boldsymbol{B} and \boldsymbol{C} is described by a vector

(1) choose some timelike, others spacelike

$$2\boldsymbol{A} = \boldsymbol{B} \times \boldsymbol{C}$$

perpendicular to the two vectors. One obtains the magnitude of A from the formula

$$4A^2 = B^2C^2 - (\boldsymbol{B}\cdot\boldsymbol{C})^2.$$

In 4-space, let \boldsymbol{B} and \boldsymbol{C} be two edges of the triangle. Then, as in three dimensions, $2\boldsymbol{A}$ is *dual* to the bivector built from \boldsymbol{B} and \boldsymbol{C}. In other words, if \boldsymbol{B} goes in the t direction and \boldsymbol{C} in the z direction, then \boldsymbol{A} is a bivector lying in the (x, y) plane. Consequently its magnitude A is to be thought of as a *real* quantity. Therefore the appropriate formula for the area A is (Tullio Regge)

$$4A^2 = (\boldsymbol{B}\cdot\boldsymbol{C})^2 - B^2C^2. \tag{42.10}$$

The quantity A is real when the deficit angle δ is real. Thus the geometrically important product $A\delta$ is also real.

When the hinge lies in the (x, y) plane, on the other hand, the quantity A is purely imaginary. In that instance a test vector taken around the cycle of simplexes that swing on this hinge has undergone change only in its z and t components; that is, it has experienced a Lorentz boost; that is, the deficit angle δ is also purely imaginary. So again the product $A\delta$ is a purely real quantity.

Turn now from character of edge lengths to magnitude of edge lengths. It is desirable that the elementary building blocks sample the curvatures of space in different directions on a roughly equal basis. In other words, it is desirable not to have long needle-shaped building blocks nor pancake-shaped tetrahedrons and simplexes. This natural requirement means that the step forward in time should be comparable to the steps "sidewise" in space. The very fact that one should have to state such a requirement brings out one circumstance that should have been obvious before: the "hinge equations"

$$\sum_{\substack{\text{hinges } h \text{ that} \\ \text{have edge } p \\ \text{in common}}} \delta_h \cot \theta_{ph} = 0 \qquad (p = 1, 2, \ldots), \qquad (42.7)$$

though they are as numerous as the edges, cannot be regarded as adequate to determine all edge lengths. There are necessarily relations between these equations that keep them from being independent. The equations cannot determine all the details of the necessarily largely arbitrary skeletonization process. They cannot do so any more than the field equations of general relativity can determine the coordinate system. With a given pattern of vertices (four-dimensional generalization of drawings in Box 42.3), one still has (a) the option how close together one will take successive layers of the structure and (b) how one will distribute a given number of points in space on a given layer to achieve the maximum payoff in accuracy (greater density of points in regions of greater curvature). To prepare a practical computer program founded on Regge calculus, one has to supply the machine not only with the hinge equations and initial conditions, but also with definite algorithms to remove all the arbitrariness that resides in options (a) and (b).

Formulas from solid geometry and four-dimensional geometry, out of which to determine the necessary hyperdihedral angles α and the deficit angles δ in terms of edge lengths and nothing but edge lengths, are summarized by Wheeler (1964a, pp. 469, 470, and 490) and by C. Y. Wong (1971). Regge (1961) also gives a formula for the Riemann curvature tensor itself in terms of deficit angles and number of edges running in a given direction [see also Wheeler (1964a, p. 471)].

§42.7. PAST APPLICATIONS OF REGGE CALCULUS

Wong (1971) has applied Regge calculus to a problem where no time development shows itself, where the geometry can therefore be treated as static, and where in addition it is spherically symmetric. He determined the Schwarzschild and Reissner-Nordstrøm geometries by the method of skeletonization. Consider successive spheres

surrounding the center of attraction. Wong approximates each as an icosahedron. The condition

$$^{(3)}R = 16\pi \begin{pmatrix} \text{energy density} \\ \text{on the 3-space} \end{pmatrix}$$

(§21.5) gives a recursion relation that determines the dimension of each icosahedron in terms of the two preceding icosahedra. Errors in the skeleton representation of the exact geometry range from roughly 10 percent to less than 1 percent, depending on the method of analysis, the quantity under analysis, and the fineness of the subdivision.

Skeletonization of geometry is to be distinguished from mere rewriting of partial differential equations as difference equations. One has by now three illustrations that one can capitalize on skeletonization without fragmenting spacetime all the way to the level of individual simplexes. The first illustration is the first part of Wong's work, where the time dimension never explicitly makes an appearance, so that the building blocks are three-dimensional only. The second is an alternative treatment, also given by Wong, that goes beyond the symmetry in t to take account of the symmetry in θ and ϕ. It divides space into spherical shells, in each of which the geometry is "pseudo-flat" in much the same sense that the geometry of a paper cone is flat. The third is the numerical solution for the gravitational collapse of a spherical star by May and White (1966), in which there is symmetry in θ and ϕ, but not in r or t. This zoning takes place exclusively in the r, t-plane. Each zone is a spherical shell. The difference as compared to Regge calculus (flat geometry within each building block) is the adjustable "conicity" given to each shell. The examples show that the decision about skeletonizing the geometry in a calculation is ordinarily not "whether" but "how much."

Partial skeletonization

§42.8. THE FUTURE OF REGGE CALCULUS

In summary, Regge's skeleton calculus puts within the reach of computation problems that in practical terms are beyond the power of normal analytical methods. It affords any desired level of accuracy by sufficiently fine subdivision of the space-time region under consideration. By way of its numbered building blocks, it also offers a practical way to display the results of such calculations. Finally, one can hope that Regge's truly geometric way of formulating general relativity will someday make the content of the Einstein field equations (Cartan's "moment of rotation"; see Chapter 15) stand out sharp and clear, and unveil the geometric significance of the so-called "geometrodynamic field momentum" (analysis of the boundary-value problem associated with the variational problem of general relativity in Regge calculus; see §21.12).

Hopes for the future

CHAPTER **43**

SUPERSPACE: ARENA FOR THE DYNAMICS OF GEOMETRY

Traveler, there are no paths.
Paths are made by walking.

ANTONIO MACHADO (1940)

This chapter is entirely Track 2. Chapter 21 (initial-value formalism) is needed as preparation for it. In reading it, one will be helped by Chapter 42 (Regge calculus). It is not needed as preparation for any later chapter, but it will be helpful in Chapter 44 (vision of the future).

Superspace is the arena for geometrodynamics

§43.1. SPACE, SUPERSPACE, AND SPACETIME DISTINGUISHED

Superspace [Wheeler (1964a), pp. 459 ff] is the arena of geometrodynamics. The dynamics of Einstein's curved space geometry runs its course in superspace as the dynamics of a particle unfolds in spacetime. This chapter gives one simple version of superspace, and a little impression of alternative versions of superspace that also have their uses. It describes the classical dynamics of geometry in superspace in terms of the Hamilton-Jacobi principle of Boxes 25.3 and 25.4. No version of mechanics makes any shorter the leap from classical dynamics to quantum. Thus it provides a principle ("Einstein-Hamilton-Jacobi or EHJ equation") for the propagation of wave crests in superspace, and for finding where those wave crests give one the classical equivalent of constructive interference ("envelope formation"). In this way one finds the track of development of 3-geometry with time expressed as a sharp, thin "leaf of history" that slices through superspace. The quantum principle replaces this deterministic account with a fuzzed-out leaf of history of finite thickness. In consequence, quantum fluctuations take place in the geometry of space that dominate the scene at distances of the order of the Planck length, $L^* = (\hbar G/c^3)^{1/2} = 1.6 \times 10^{-33}$ cm, and less. The present survey simplifies by considering only the dynamics of curved empty space. When sources are present and are to be taken into account, supplementary terms are to be added, some of the literature on which is listed.

 In all the difficult investigations that led in the course of half a century to some understanding of the dynamics of geometry, both classical and quantum, the most

Box 43.1 GEOMETRODYNAMICS COMPARED WITH PARTICLE DYNAMICS

Concept	Particle dynamics	Geometrodynamics
Dynamic entity	Particle	Space
Descriptors of momentary configuration	x, t ("event")	$^{(3)}\mathcal{G}$ ("3-geometry")
Classical history	$x = x(t)$	$^{(4)}\mathcal{G}$ ("4-geometry")
History is a stockpile of configurations?	Yes. Every point on world line gives a momentary configuration of particle	Yes. Every spacelike slice through $^{(4)}\mathcal{G}$ gives a momentary configuration of space
Dynamic arena	Spacetime (totality of all points x, t)	Superspace (totality of all $^{(3)}\mathcal{G}$'s)

difficult point was also the simplest: The dynamic object is not spacetime. It is space. The geometric configuration of space changes with time. But it is space, three-dimensional space, that does the changing (see Box 43.1).

3-geometry is the dynamic object

Space will be treated here as "closed" or, in mathematical language, "compact," either because physics adds to Einstein's second-order differential equations the requirement of closure as a necessary and appropriate boundary condition [Einstein (1934, p. 52; 1950); Wheeler (1959; 1964c). Hönl (1962); see also §21.12] or because that requirement simplifies the mathematical analysis, or for both reasons together.

One can approximate a smooth, closed 3-geometry by a skeleton 3-geometry built out of tetrahedrons, as indicated schematically in Figure 43.1 (see Chapter 42 on the Regge calculus). Specify the 98 edge-lengths in this example to fix all the features of the geometry; and fix these 98 edge-lengths by giving the location of a single point in a space of 98 dimensions. This 98-dimensional manifold, this "truncated superspace," goes over into superspace [Wheeler (1964a), pp. 453, 459, 463, 495] in the idealization in which the tracer points increase in density of coverage without limit. Accounts of superspace with more mathematical detail are given by DeWitt (1967a,b), Wheeler (1970), and Fischer (1970).

Finite-dimensional "truncated superspace"

Let the representative point move from one location to a nearby location, either in truncated superspace or in full superspace. Then all edge-lengths alter, and the 3-geometry of Figure 43.1 moves as if alive. No better illustration can one easily supply of what it means to speak of the "dynamics of space."

The term "3-geometry" makes sense as well in quantum geometrodynamics as in classical theory. So does superspace. But spacetime does not. Give a 3-geometry, and give its time rate of change. That is enough, under typical circumstances (see Chapter 21) to fix the whole time-evolution of the geometry; enough in other words, to determine the entire four-dimensional spacetime geometry, provided one is

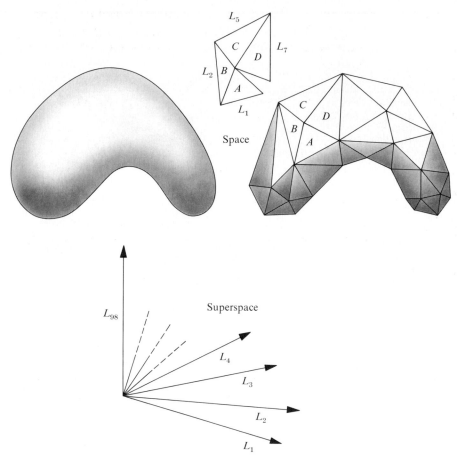

Figure 43.1.

Superspace in the simplicial approximation. Upper left, space (depicted as two-dimensional but actually three-dimensional). Upper right, simplical approximation to space. The approximation can be made arbitrarily good by going to the limit of an arbitrarily fine decomposition. The curvature at a typical location is measured by a deficit angle. This angle is completely determined by the edge lengths (L_1, L_2, ... L_8 in the figure) of the simplexes that meet at that location. When there are 98 edge lengths altogether in the simplicial representation of the geometry, then this geometry is completely specified by a single point in a 98-dimensional space (lower diagram; "superspace").

The concept of spacetime is incompatible with the quantum principle

considering the problem in the context of classical physics. In the real world of quantum physics, however, one cannot give both a dynamic variable and its time-rate of change. The principle of complementarity forbids. Given the precise 3-geometry at one instant, one cannot also know at that instant the time-rate of change of the 3-geometry. In other words, given the geometrodynamic field coordinate, one cannot know the geometrodynamic field momentum. If one assigns the intrinsic 3-geometry, one cannot also specify the extrinsic curvature.

The uncertainty principle thus deprives one of any way whatsoever to predict, or even to give meaning to, "the deterministic classical history of space evolving

in time." *No prediction of spacetime, therefore no meaning for spacetime*, is the verdict of the quantum principle. That object which is central to all of classical general relativity, the four-dimensional spacetime geometry, simply does not exist, except in a classical approximation.

These considerations reveal that the concepts of spacetime and time are not primary but secondary ideas in the structure of physical theory. These concepts are valid in the classical approximation. However, they have neither meaning nor application under circumstances where quantum geometrodynamic effects become important. Then one has to forego that view of nature in which every event, past, present, or future, occupies its preordained position in a grand catalog called "spacetime," with the Einstein interval from each event to its neighbor eternally established. There is no spacetime, there is no time, there is no before, there is no after. The question of what happens "next" is without meaning.

That spacetime is not the right way does not mean there is *no* right way to describe the dynamics of geometry consistent with the quantum principle. Superspace is the key to *one* right way to describe the dynamics (see Figure 43.2).

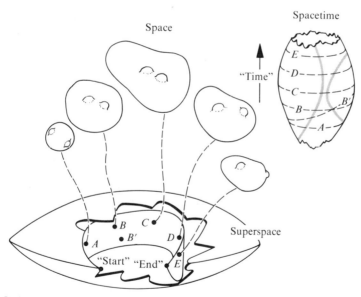

Figure 43.2.
Space, spacetime, and superspace. Upper left: Five sample configurations, *A, B, C, D, E*, attained by space in the course of its expansion and recontraction. Below: Superspace and these five sample configurations, each represented by a point in superspace. Upper right: Spacetime. A spacelike cut, like *A*, through spacetime gives a momentary configuration of space. There is no compulsion to limit attention to a one-parameter family of slices, *A, B, C, D, E* through spacetime. The phrase "many-fingered time" is a slogan telling one not to so limit one's slices, and *B'* is an example of this freedom in action. The 3-geometries *B'* and *A, B, C, D, E*, like all 3-geometries obtained by all spacelike slices whatsoever through the given classical spacetime, lie on a single bent leaf of history, indicated in the diagram, and cutting its thin slice through superspace. A different spacetime, in other words, a different solution of Einstein's field equation, means a different leaf of history (not indicated) slicing through superspace.

§43.2. THE DYNAMICS OF GEOMETRY DESCRIBED IN THE LANGUAGE OF THE SUPERSPACE OF THE $^{(3)}\mathcal{G}$'s

Spacetime is a classical leaf of history slicing through superspace

Given a spacetime, one can construct the corresponding leaf of history slicing through superspace. Conversely, given the leaf of history, one can reconstruct the spacetime.

Consider the child's toy commonly known as "Chinese boxes." One opens the outermost box only to reveal another box; when the second box is opened, there is another box, and so on, until eventually there are dozens of boxes scattered over the floor. Or conversely the boxes can be put back together, nested one inside the other, to reconstitute the original package. The packaging of $^{(3)}\mathcal{G}$'s into a $^{(4)}\mathcal{G}$ is much more sophisticated. Nature provides no monotonic ordering of the $^{(3)}\mathcal{G}$'s. Two of the dynamically allowed $^{(3)}\mathcal{G}$'s, taken at random, will often cross each other one or more times. When one shakes the $^{(4)}\mathcal{G}$ apart, one therefore gets enormously more $^{(3)}\mathcal{G}$'s "spread out over the floor" than might have been imagined. Conversely, when one puts back together all of the $^{(3)}\mathcal{G}$'s lying on the leaf of history, one gets a structure with a rigidity that might not otherwise have been foreseen. This rigidity arises from the infinitely rich interleaving and intercrossing of clear-cut, well-defined $^{(3)}\mathcal{G}$'s one with another.

In summary: (1) Classical geometrodynamics in principle constitutes a device, an algorithm, a rule for calculating and constructing a leaf of history that slices through superspace. (2) The $^{(3)}\mathcal{G}$'s that lie on this leaf of history are YES 3-geometries; the vastly more numerous $^{(3)}\mathcal{G}$'s that do not are NO 3-geometries. (3) The YES $^{(3)}\mathcal{G}$'s are the building blocks of the $^{(4)}\mathcal{G}$ that is classical spacetime. (4) The interweaving and interconnections of these building blocks give the $^{(4)}\mathcal{G}$ its existence, its dimensionality, and its structure. (5) In this structure every $^{(3)}\mathcal{G}$ has a rigidly fixed location of its own. (6) In this sense one can say that the "many-fingered time" of each 3-geometry is specified by the very interlocking construction itself. Baierlein, Sharp and Wheeler (1962) say a little more on this concept of "3-geometry as carrier of information about time."

How different from the textbook concept of spacetime! There the geometry of spacetime is conceived as constructed out of elementary objects, or points, known as "events." Here, by contrast, the primary concept is 3-geometry, *in abstracto,* and out of it is derived the idea of event. Thus, (1) the event lies at the intersection of such and such $^{(3)}\mathcal{G}$'s; and (2) it has a timelike relation to (earlier or later than, or synchronous with) some other $^{(3)}\mathcal{G}$, which in turn (3) derives from the intercrossings of all the other $^{(3)}\mathcal{G}$'s.

When one turns from classical theory to quantum theory, one gives up the concept of spacetime, except in the semiclassical approximation. Therefore, one gives up any immediate possibility whatsoever of defining the concept, normally regarded as so elemental, of an "event." The theory itself, however, here as always [Bohr and Rosenfeld (1933)], defines in and by itself, in its own natural way, the procedures-in-principle for measuring all those quantities that are in principle measurable.

Quantum theory upsets the sharp distinction between YES 3-geometries and NO

3-geometries. It assigns to each 3-geometry not a YES or a NO, but a probability amplitude,

Probability amplitude for a 3-geometry

$$\psi = \psi(^{(3)}\mathcal{G}). \tag{43.1}$$

This probability amplitude is highest near the classically forecast leaf of history and falls off steeply outside a zone of finite thickness extending a little way on either side of the leaf.

Were one to take, instead of a physically relevant probability amplitude function, a *typical* solution of the relevant wave equation, one would have to expect to see not one trace of anything like classical geometrodynamics. The typical probability amplitude function is spread all over superspace. No surprise! Already in classical theory one has to reckon with a Hamilton-Jacobi function,

$$S = S(^{(3)}\mathcal{G}), \tag{43.2}$$

spread out over superspace. Moreover, this "dynamic phase function" of classical geometrodynamics gives at once the phase of ψ, according to the formula

$$\psi(^{(3)}\mathcal{G}) = \begin{pmatrix} \text{slowly varying} \\ \text{amplitude function} \end{pmatrix} e^{(i/\hbar)S(^{(3)}\mathcal{G})}, \tag{43.3}$$

indication enough that ψ and S are both unlocalized.

Dynamics first clearly becomes recognizable when sufficiently many such spread-out probability amplitude functions are superposed to build up a localized wave packet, as in the elementary examples of Boxes 25.3 and 25.4; thus,

Wave packet recovers classical geometrodynamics

$$\psi = c_1\psi_1 + c_2\psi_2 + \cdots. \tag{43.4}$$

Constructive interference occurs where the phases of the several individual waves agree:

$$S_1(^{(3)}\mathcal{G}) = S_2(^{(3)}\mathcal{G}) = \cdots. \tag{43.5}$$

This is the condition that distinguishes YES 3-geometries from NO 3-geometries. It is the tool for constructing a leaf of history in superspace. It is the key to the dynamics of geometry. Moreover, it is an equation that says not one word about the quantum principle. It is not surprising that the equation of constructive interference in (43.5) makes the leap from classical theory to quantum theory the shortest.

§43.3. THE EINSTEIN-HAMILTON-JACOBI EQUATION

Should one write down a differential equation for the Hamilton-Jacobi function $S(^{(3)}\mathcal{G})$, solve it, and then analyze the properties of the solution? The exact opposite is simpler: look at the properties of the solution, and from that inspection find out what equation the dynamic phase or action S must satisfy.

Hilberts' principle of least action reads

$$I_{\text{Hilbert}} = (1/16\pi) \int {}^{(4)}R(-g)^{1/2}\, d^4x = \text{extremum}. \tag{43.6}$$

After one separates off complete derivatives in the integrand, what is left [see equations (21.13) and (21.95)] becomes

$$(1/16\pi)I_{\text{ADM}} = I_{\text{true}} = (1/16\pi) \int \left\{ \pi^{ij}\,\partial g_{ij}/\partial t + Ng^{1/2}R \right.$$
$$\left. + Ng^{-1/2}\left[\frac{1}{2}(\text{Tr } \boldsymbol{n})^2 - \text{Tr}(\boldsymbol{n}^2)\right] + 2N_i\pi^{ij}{}_{|j} \right\} d^4x. \tag{43.7}$$

In (43.7), but not in (43.6), g stands for the determinant of the three-dimensional metric tensor, g_{ij}, and R for the scalar curvature invariant of the 3-geometry; the suffix $^{(3)}$ is omitted for simplicity. The integral is extended from (1) a spacelike hypersurface on which a 3-geometry is given with metric $g_{ij}{}'(x, y, z)$ to (2) a spacelike hypersurface on which a 3-geometry is given with metric $g_{ij}{}''(x, y, z)$. Whatever is adjustable in the chunk of spacetime between is now to be considered as having been adjusted to extremize the integral. Therefore the value of the integral I_{ADM} becomes a functional of the metrics on the two hypersurfaces and nothing more.

Next, holding fixed the metric $g'_{ij}(x, y, z)$ on the earlier hypersurface, change slightly or even more than slightly the metric on the later hypersurface. Solve the new variation problem and get a new value of I_{ADM}. Proceeding further in this way, for each new $g_{ij}{}''$ one gets a new value of I_{ADM}. Call the functional I_{ADM} of the metric defined in this way "Hamilton's principal function," or the "action" or the "dynamic path length,*" $S(g_{ij}(x, y, z))$ of the "history of geometry" that connects the two given 3-geometries. The double prime suffix is dropped from $g_{ij}{}''$ here and hereafter to simplify the notation. One knows from other branches of mechanics that the quantity defined in this way, $S(g_{ij})$, when it exists, even though it is a special solution, nevertheless is *always* a solution of the Hamilton-Jacobi equation. Jacobi could look for more general solutions, but Hamilton already had one!

For (43.7) to be an extremal with respect to variations of the lapse N and the shift components N_i, it was necessary (see Chapter 21) that the coefficients of these four quantities should vanish; thus,

$$g^{-1/2}\left[\frac{1}{2}(\text{Tr } \boldsymbol{n})^2 - \text{Tr } \boldsymbol{n}^2\right] + g^{1/2}R = 0 \tag{43.8}$$

and

$$\pi^{ij}{}_{|j} = 0. \tag{43.9}$$

In the expression for the extremal value of the action, only one term, the first, is left:

$$S(g(x, y, z)) = I_{\text{ADM, extremal}} = \int_{g'_{ij}}^{g_{ij}} \{\pi^{ij}\,\partial g_{ij}/\partial t\}\, d^4x. \tag{43.10}$$

*Actually $S \equiv S_{\text{ADM}} \equiv 16\pi S_{\text{true}} = 16\pi$ (true dynamic path length).

The effect of a slight change, δg_{ij}, in the 3-metric at the upper limit is therefore easy to read off:

$$\delta S = \int \pi^{ij}(x, y, z)\, \delta g_{ij}(x, y, z)\, d^3x. \tag{43.11}$$

The language of "functional derivative" [see, for example, Bogoliubov and Shirkov (1959)] allows one to speak in terms of a derivative rather than an integral:

$$\frac{\delta S}{\delta g_{ij}} = \pi^{ij}. \tag{43.12}$$

Geometrodynamic momentum as rate of change of dynamic path length with respect to 3-geometry of terminal hypersurface

The "field momenta" acquire a simple meaning: they give the rate of change of the action with respect to the continuous infinitude of "field coordinates," $g_{ij}(x, y, z)$. (Here the x, y, z, as well as the i and j, serve as mere labels.)

 Although the phase function S appears to depend on all six metric coefficients g_{ij} individually, it depends in actuality only on that combination of the g_{ij} which is locked to the 3-geometry. To verify this point, express a particular 3-geometry $^{(3)}\mathcal{G}$ throughout one local coordinate patch in terms of one set of coordinates x^p by one set of metric coefficients g_{pq}. Reexpress the same 3-geometry in terms of coordinates \bar{x}^p shifted by the small amount ξ^p,

Action depends on 3-geometry, not on metric coefficients individually

$$\bar{x}^p = x^p - \xi^p. \tag{43.13}$$

To keep the 3-geometry the same, that is, to keep unchanged the distance ds from one coordinate-independent point to another, the metric coefficients have to change:

$$\bar{g}_{pq} = g_{pq} + \xi_{p|q} + \xi_{q|p}. \tag{43.14}$$

Let the phase function S (or in quantum mechanics, let the probability amplitude ψ) be considered to be expressed as a functional of the metric coefficients $g_{11}(x)$, $g_{12}(x), \ldots, g_{33}(x)$. Changes $\delta g_{pq}(x)$ in these coefficients alter the H-J phase function and the probability amplitude by the amounts

$$\delta S = \int (\delta S/\delta g_{pq})\, \delta g_{pq}\, d^3x,$$
$$\delta \psi = \int (\delta \psi/\delta g_{pq})\, \delta g_{pq}\, d^3x, \tag{43.15}$$

according to the standard definition of functional derivative. Therefore the coordinate change produces an ostensible change in the dynamic path length or phase S given by

$$\delta S = \int (\delta S/\delta g_{pq})(\xi_{p|q} + \xi_{q|p})\, d^3x$$
$$= -2 \int (\delta S/\delta g_{pq})_{|q}\, \xi_p\, d^3x. \tag{43.16}$$

This change must vanish if S is to depend on the 3-geometry alone, and not on

the coordinates in terms of which that 3-geometry is expressed; and must vanish, moreover, for arbitrary choice of the ξ_p. From this condition, one concludes

$$\left(\frac{\delta S}{\delta g_{pq}}\right)_{|q} = 0. \tag{43.17}$$

Likewise, one finds three equations on the wave function ψ itself, as distinguished from its phase S/\hbar; thus,

$$\left(\frac{\delta \psi}{\delta g_{pq}}\right)_{|q} = 0. \tag{43.18}$$

But (43.17), by virtue of (43.12), is identical with (43.9). In this sense (43.9) merely verifies what one already knew had to be true: the classical Hamilton-Jacobi function S (like the probability amplitude function ψ of quantum theory) depends on 3-geometry, not on individual metric coefficients, and not on choice of coordinates.

All the dynamic content of geometrodynamics is summarized in the sole remaining equation (43.8), which takes the form

Law of propagation of wave crests in superspace

$$g^{-1/2}\left[\frac{1}{2}g_{pq}g_{rs} - g_{pr}g_{qs}\right]\frac{\delta S}{\delta g_{pq}}\frac{\delta S}{\delta g_{rs}} + g^{1/2}R = 0. \tag{43.19}$$

This is the Einstein-Hamilton-Jacobi equation, first given explicitly in the literature by Peres (1962) on the foundation of earlier work by himself and others on the Hamiltonian formulation of geometrodynamics. This equation tells how fronts of constant S ("wave crests") propagate in superspace.

That the one EHJ equation (43.19) contains as much information as all ten components of Einstein's field equation has been demonstrated by Gerlach (1969). The central point in the analysis is the principle of constructive interference, and the main requirement for a proper treatment of this point is the concept of a completely parametrized solution of the EHJ equation.

The problem of a particle moving in two-dimensional space, as treated by the Hamilton-Jacobi method in Boxes 25.3 and 25.4, required for complete analysis a solution containing two distinct and independently adjustable parameters, the energy per unit mass, \widetilde{E}, and angular momentum per unit mass, \widetilde{L}; thus

$$S(r, \theta, t; \widetilde{E}, \widetilde{L}) = -\widetilde{E}t + \widetilde{L}\theta$$

$$+ \int^r [\widetilde{E}^2 - (1 - 2M/r)(1 + \widetilde{L}^2/r^2)]^{1/2}\frac{dr}{(1 - 2M/r)} + \delta(\widetilde{E}, \widetilde{L}). \tag{43.20}$$

Here the additive phase $\delta(\widetilde{E}, \widetilde{L})$ is required if one is to be able to arrange for the particle to arrive at a given r-value at a specified t value and at a specified value of θ. One thinks of superposing four probability amplitudes, as in (43.4), with dynamic phases, S, given by (43.20) and the parameters taking on, respectively, the following four sets of values: $(\widetilde{E}, \widetilde{L})$; $(\widetilde{E} + \Delta\widetilde{E}, \widetilde{L})$; $(\widetilde{E}, \widetilde{L} + \Delta\widetilde{L})$; and $(\widetilde{E} + \Delta\widetilde{E}, \widetilde{L} + \Delta\widetilde{L})$. The principle of constructive interference leads to the conditions

$$\partial S / \partial \widetilde{E} = 0,$$
$$\partial S / \partial \widetilde{L} = 0. \qquad (43.21)$$

The points in the spacetime (r, θ, t) that satisfy these conditions are the YES points; they lie on the world line. The ones that don't are the NO points.

The desired solution of the EHJ equation (43.19) contains not two parameters (plus an additive phase, δ, depending on these two parameters), but an infinity of parameters, and even a continuous infinity of parameters. Thus the parameters are not to be designated as $\alpha_1, \alpha_2, \ldots; \beta_1, \beta_2, \ldots$ (parameters labeled by a discrete index), but as

$$\alpha(u, v, w)$$

and

$$\beta(u, v, w)$$

(two parameters "labeled" by three continuous indices u, v, w). Accidentally omit one of this infinitude of parameters? How could one ever hope to know that what purported to be a complete solution of the EHJ equation was not in actuality complete? Happily Gerlach provides a procedure to test the parameters for completeness.

Granted completeness, Gerlach goes on to show that the "leaf of history in superspace" or collection of 3-geometries that satisfy the conditions of constructive interference,

$$\frac{\delta S(^{(3)}\mathcal{G}; \alpha(u, v, w), \beta(u, v, w))}{\delta \alpha} = 0,$$
$$\frac{\delta S(^{(3)}\mathcal{G}; \alpha(u, v, w), \beta(u, v, w))}{\delta \beta} = 0, \qquad (43.22)$$

Condition of constructive interference gives classical "leaf of history" or spacetime

is identical with the leaf of history, or equivalent 4-geometry, given by the ten components of Einstein's geometrodynamic law.

From the Hamilton-Jacobi equation for a problem in elementary mechanics, it is a short step to the corresponding Schroedinger equation; similarly in geometrodynamics. No one has done more than Bryce DeWitt to explore the meaning and consequences of this "Einstein-Schroedinger equation" [DeWitt (1967a,b)]. One of the most interesting consequences is the existence of a conserved current in superspace, analogous to the conserved current

$$j_\mu = \frac{\hbar}{2im}\left(\psi^* \frac{\partial \psi}{\partial x^\mu} - \psi \frac{\partial \psi^*}{\partial x^\mu}\right)$$

that one encounters in the Klein-Gordon wave equation for a particle of spin zero.

It is an unhappy feature of this Einstein-Schroedinger wave equation that it contains second derivatives; so one has to specify both the probability amplitude, and the normal derivative of the probability amplitude, on the appropriate "super-

hypersurface" in superspace, in order to be able to predict the evolution of this state function elsewhere in superspace. One suggested way out of this situation—it is at least an inconvenience, possibly a real difficulty—has been proposed by Leutwyler (1968): impose a natural boundary condition that reduces the number of independent solutions from the number characteristic of a second-order equation to the number characteristic of a first-order equation. Another way out is to formulate the dynamics quite differently, in the way proposed by Kuchař (see Chapter 21), in which the resulting equation is of first order in the variable analogous to time.

The exploration of quantum geometrodynamics is simplified when one treats most of the degrees of freedom of the geometry as frozen out, by imposition of a high degree of symmetry. Then one is left with one, two, or three degrees of freedom (see Chapter 30, on mixmaster cosmology), or even infinitely many, and is led to manageable problems of quantum mechanics [Misner (1972a, 1973)].

§43.4. FLUCTUATIONS IN GEOMETRY

Of all the remarkable developments of physics since World War II, none is more impressive than the prediction and verification of the effects of the vacuum fluctuations in the electromagnetic field on the motion of the electron in the hydrogen atom (Figure 43.3). That development made it impossible to overlook the effects of such fluctuations throughout all physics and, not least, in the geometry of spacetime itself.

Figure 43.3.

Symbolic representation of motion of electron in hydrogen atom as affected by fluctuations in electric field in vacuum ("vacuum" or "ground state" or "zero-point" fluctuations). The electric field associated with the fluctuation, $E_x(t) = \int E_x(\omega)e^{-i\omega t}\,d\omega$, adds to the static electric field provided by the nucleus itself. The additional field brings about in the most elementary approximation the displacement $\Delta x = \int (e/m\omega^2)E_x(\omega)e^{-i\omega t}\,d\omega$. The average vanishes but the root mean square $\langle(\Delta x)^2\rangle$ does not. In consequence the electron feels an effective atomic potential altered from the expected value $V(x, y, z)$ by the amount

$$\Delta V(x, y, z) = \frac{1}{2} \langle(\Delta x)^2\rangle \, \nabla^2 V(x, y, z).$$

The average of this perturbation over the unperturbed motion accounts for the major part of the observed Lamb-Retherford shift $\Delta E = \langle\Delta V(x, y, z)\rangle$ in the energy level. Conversely, the observation of the expected shift makes the reality of the vacuum fluctuations inescapably evident.

From the zero-point fluctuations of a single oscillator to the fluctuations of the electromagnetic field to geometrodynamic fluctuations is a natural order of progression.

A harmonic oscillator in its ground state has a probability amplitude of

Fluctuations for oscillator and for electromagnetic field

$$\psi(x) = \left(\frac{m\omega}{\pi\hbar}\right)^{1/4} e^{-(m\omega/2\hbar)x^2} \tag{43.23}$$

to be displaced by the distance x from its natural classical position of equilibrium. In this sense, it may be said to "resonate" or "fluctuate" between locations in space ranging over a region of extent

$$\Delta x \sim (\hbar/m\omega)^{1/2}. \tag{43.24}$$

The electromagnetic field can be treated as an infinite collection of independent "field oscillators," with amplitudes ξ_1, ξ_2, \ldots. When the Maxwell field is in its state of lowest energy, the probability amplitude—for the first oscillator to have amplitude ξ_1, and simultaneously the second oscillator to have amplitude ξ_2, the third ξ_3, and so on—is the product of functions of the form (43.23), one for each oscillator. When the scale of amplitudes for each oscillator is suitably normalized, the resulting infinite product takes the form

$$\psi(\xi_1, \xi_2, \ldots) = N \exp\left[-(\xi_1^2 + \xi_2^2 + \cdots)\right]. \tag{43.25}$$

This expression gives the probability amplitude ψ for a configuration $\boldsymbol{B}(x, y, z)$ of the magnetic field that is described by the Fourier coefficients ξ_1, ξ_2, \ldots. One can forgo any mention of these Fourier coefficients if one desires, however, and rewrite (43.25) directly in terms of the magnetic field configuration itself [Wheeler (1962)]:

$$\psi(\boldsymbol{B}(x, y, z)) = \mathscr{N} \exp\left(-\int\int \frac{\boldsymbol{B}(x_1)\cdot\boldsymbol{B}(x_2)}{16\pi^3\hbar c r_{12}^2} d^3x_1\, d^3x_2\right). \tag{43.26}$$

One no longer speaks of "the" magnetic field, but instead of the probability of this, that, or the other configuration of the magnetic field, even under circumstances, as here, where the electromagnetic field is in its ground state. [See Kuchař (1970) for a similar expression for the "ground state" functional of the linearized gravitational field.]

It is reasonable enough under these circumstances that the configuration of greatest probability is $\boldsymbol{B}(x, y, z) = 0$. Consider for comparison a configuration where the magnetic field is again everywhere zero except in a region of dimension L. There let the field, subject as always to the condition div $\boldsymbol{B} = 0$, be of the order of magnitude ΔB. The probability amplitude for this configuration will be reduced relative to the nil configuration by a factor $\exp(-I)$. Here the quantity I in the exponent is of the order $(\Delta B)^2 L^4/\hbar c$. Configurations for which I is large compared to 1 occur with negligible probability. Configurations for which I is small compared to 1 occur with practically the same probability as the nil configuration. In this sense, one can

say that the fluctuations in the magnetic field in a region of extension L are of the order of magnitude

$$\Delta B \sim \frac{(\hbar c)^{1/2}}{L^2}. \tag{43.27}$$

In other words, the field "resonates" between one configuration and another with the range of configurations of significance given by (43.27). Moreover, the smaller is the region of space under consideration, the larger are the field magnitudes that occur with appreciable probability.

Still another familiar way of speaking about electromagnetic field fluctuations gives additional insight relevant to geometrodynamics. One considers a measuring device responsive in comparable measure to the magnetic field at all points in a region of dimension L. One asks for the effect on this device of electromagnetic disturbances of various wavelengths. A disturbance of wavelength short compared to L will cause forces to act one way in some parts of the detector, and will give rise to nearly compensating forces in other parts of it. In contrast, a disturbance of a long wave-length λ produces forces everywhere in the same direction, but of a magnitude too low to have much effect. Thus the field, estimated from the equation

$$\begin{pmatrix} \text{energy of electromagnetic} \\ \text{wave of wavelength } \lambda \text{ in a} \\ \text{domain of volume } \lambda^3 \end{pmatrix} \sim \begin{pmatrix} \text{energy of one quantum} \\ \text{of wavelength } \lambda \end{pmatrix}$$

or

$$B^2 \lambda^3 \sim \frac{\hbar c}{\lambda}$$

or

$$B \sim \frac{(\hbar c)^{1/2}}{\lambda^2} \tag{43.28}$$

is very small if λ is large compared to the domain size L. The biggest effect is caused by a disturbance of wavelength λ comparable to L itself. This line of reasoning leads directly from (43.28) to the standard fluctuation formula (43.27).

Similar considerations apply in geometrodynamics. Quantum fluctuations in the geometry are superposed on and coexist with the large-scale, slowly varying curvature predicted by classical deterministic general relativity. Thus, in a region of dimension L, where in a local Lorentz frame the normal values of the metric coefficients will be $-1, 1, 1, 1$, there will occur fluctuations in these coefficients of the order

Fluctuations in geometry dominate at the Planck scale of distances

$$\Delta g \sim \frac{L^*}{L}, \tag{43.29}$$

fluctuations in the first derivatives of the g_{ik}'s of the order

$$\Delta \Gamma \sim \frac{\Delta g}{L} \sim \frac{L^*}{L^2}, \tag{43.30}$$

and fluctuations in the curvature of space of the order

$$\Delta R \sim \frac{\Delta g}{L^2} \sim \frac{L^*}{L^3}.$$ (43.31)

Here

$$L^* = \left(\frac{\hbar G}{c^3}\right)^{1/2} = 1.6 \times 10^{-33} \text{ cm}$$ (43.32)

is the so-called Planck length [Planck (1899)].

It is appropriate to look at orders of magnitude. The curvature of space within and near the earth, according to classical Einstein theory, is of the order

$$R \sim \left(\frac{G}{c^2}\right)\rho \sim (0.7 \times 10^{-28} \text{ cm/g})(5 \text{ g/cm}^3)$$
$$\sim 4 \times 10^{-28} \text{ cm}^{-2}.$$ (43.33)

This quantity has a very direct physical significance. It measures the "tide-producing component of the gravitational field" as sensed, for example, in a freely falling elevator or in a spaceship in free orbit around the earth. By comparison, the quantum fluctuations in the curvature of space are only

$$\Delta R \sim 10^{-33} \text{ cm}^{-2},$$ (43.34)

even in a domain of observation as small as 1 cm in extent. Thus the quantum fluctuations in the geometry of space are completely negligible under everyday circumstances.

Even in atomic and nuclear physics the fluctuations in the metric,

$$\Delta g \sim \frac{10^{-33} \text{ cm}}{10^{-8} \text{ cm}} \sim 10^{-25}$$

and

$$\Delta g \sim \frac{10^{-33} \text{ cm}}{10^{-13} \text{ cm}} \sim 10^{-20},$$ (43.35)

are so small that it is completely in order to idealize the physics as taking place in a flat Lorentzian spacetime manifold.

The quantum fluctuations in the geometry are nevertheless inescapable, if one is to believe the quantum principle and Einstein's theory. They coexist with the geometrodynamic development predicted by classical general relativity. The fluctuations widen the narrow swathe cut through superspace by the classical history of the geometry. In other words, the geometry is not deterministic, even though it looks so at the everyday scale of observation. Instead, at a submicroscopic scale it "resonates" between one configuration and another and another. This terminology means no more and no less than the following: (1) Each configuration ${}^{(3)}\mathcal{G}$ has its own probability amplitude $\psi = \psi({}^{(3)}\mathcal{G})$. (2) These probability amplitudes have comparable magnitudes for a whole range of 3-geometries included within the limits (43.29) on

either side of the classical swathe through superspace. (3) This range of 3-geometries is far too variegated on the submicroscopic sale to fit into any one 4-geometry, or any one classical geometrodynamic history. (4) Only when one overlooks these small-scale fluctuations ($\sim 10^{-33}$ cm) and examines the larger-scale features of the 3-geometries do they appear to fit into a single space-time manifold, such as comports with the classical field equations.

These small-scale fluctuations tell one that something like gravitational collapse is taking place everywhere in space and all the time; that gravitational collapse is in effect perpetually being done and undone; that in addition to the gravitational collapse of the universe, and of a star, one has also to deal with a third and, because it is constantly being undone, most significant level of gravitational collapse at the Planck scale of distances.

EXERCISES

Exercise 43.1. THE ACTION PRINCIPLE FOR A FREE PARTICLE IN NONRELATIVISTIC MECHANICS

Taking as action principle $I = \int L \, dt$ = extremum, with specified x', t' and x'', t'' at the two limits, and with $L = \frac{1}{2}m(dx/dt)^2$, find (1) the extremizing history $x = x(t)$ and (2) the dynamical path length or action $S(x'', t''; x', t') = I_{\text{extremum}}$ in its dependence on the end points. Also (3) write down the Hamilton-Jacobi equation for this problem, and (4) verify that $S(x, t; x', t')$ satisfies this equation. Finally, imagining the Hamilton-Jacobi equation not to be known, (5) derive it from the already known properties of the function S itself.

Exercise 43.2. THE ACTION FOR THE HARMONIC OSCILLATOR

The kinetic energy is $\frac{1}{2}m(dx/dt)^2$ and the potential energy is $\frac{1}{2}m\omega^2 x^2$. Carry through the analysis of parts (1), (2), (3), (4) of the preceding exercise. Partial answer:

$$S = \frac{m\omega}{2} \frac{(x^2 + x'^2) \cos \omega(t - t') - 2xx'}{\sin \omega(t - t')}.$$

Verify that $\partial S/\partial x$ gives momentum and $-\partial S/\partial t$ gives energy.

Exercise 43.3. QUANTUM PROPAGATOR FOR HARMONIC OSCILLATOR

Show that the probability amplitude for a simple harmonic oscillator to transit from x', t' to x'', t'' is

$$\langle x'', t''; x', t' \rangle$$
$$= \left(\frac{m\omega}{2\pi i\hbar \sin \omega(t'' - t')} \right)^{1/2} \times \exp \frac{im\omega[(x''^2 + x'^2) \cos \omega(t'' - t') - 2x''x']}{2\hbar \sin \omega(t'' - t')},$$

and that it reduces for the case of a free particle to

$$\langle x'', t''; x', t' \rangle = \left(\frac{m}{2\pi i\hbar(t'' - t')} \right)^{1/2} \exp \frac{im(x'' - x')^2}{2\hbar(t'' - t')}.$$

Note that one can derive all the harmonic-oscillator wave functions from the solution by use of the formula

$$\langle x'', t''; x', t' \rangle = \sum_n u_n(x'')u_n^*(x') \exp iE_n(t' - t'')/\hbar.$$

Exercise 43.4. QUANTUM PROPAGATOR FOR FREE ELECTROMAGNETIC FIELD

In flat spacetime, one is given on the spacelike hypersurface $t = t'$ the divergence-free magnetic field $B'(x, y, z)$ and on the spacelike hypersurface $t = t''$ the divergence-free magnetic field $B''(x, y, z)$. By Fourier analysis (reducing this problem to the preceding problem) or otherwise, find the probability amplitude to transit from B' at t' to B'' at t''.

Exercise 43.5. HAMILTON-JACOBI FORMULATION OF MAXWELL ELECTRODYNAMICS

Regard the four components A_μ of the electromagnetic 4-potential as the primary quantities; split them into a space part A_i and a scalar potential ψ. (1) Derive from the action principle (in flat spacetime)

$$I = (1/8\pi) \int (E^2 - B^2)\, d^4x,$$

by splitting off an appropriate divergence, an expression qualitatively similar in character to (43.7). (2) Show that the appropriate quantity to be fixed on the initial and final spacelike hypersurface is not really A_i itself, but the magnetic field, defined by $B = \text{curl}\, A$. (3) Derive the Hamilton-Jacobi equation for the dynamic phase or action $S(B, \mathcal{S})$ in its dependence on the choice of hypersurface \mathcal{S}, and the choice of magnetic field B on this hypersurface,

$$-\frac{\delta S}{\delta \Omega} = \frac{1}{8\pi} B^2 + \frac{(4\pi)^2}{8\pi}\left(\frac{\delta S}{\delta A}\right)^2.$$

The quantity on the left is Tomonaga's "bubble time" derivative [Tomonaga (1946); see also Box 21.1].

CHAPTER **44**

BEYOND THE END OF TIME

"Heaven wheels above you
Displaying to you her eternal glories
And still your eyes are on the ground"
DANTE

The world "stands before us as a great, eternal riddle"
EINSTEIN (1949a)

§44.1. GRAVITATIONAL COLLAPSE AS THE GREATEST CRISIS IN PHYSICS OF ALL TIME

This chapter is entirely Track 2. No previous Track-2 material is needed as preparation for it, but Chapter 43 will be helpful.

The universe starts with a big bang, expands to a maximum dimension, then recontracts and collapses: no more awe-inspiring prediction was ever made. It is preposterous. Einstein himself could not believe his own prediction. It took Hubble's observations to force him and the scientific community to abandon the concept of a universe that endures from everlasting to everlasting.

Later work of Tolman (1934a), Avez (1960), Geroch (1967), and Hawking and Penrose (1969) generalizes the conclusion. A model universe that is closed, that obeys Einstein's geometrodynamic law, and that contains a nowhere negative density of mass-energy, inevitably develops a singularity. No one sees any escape from the density of mass-energy rising without limit. A computing machine calculating ahead step by step the dynamical evolution of the geometry comes to the point where it can not go on. Smoke, figuratively speaking, starts to pour out of the computer. Yet physics surely continues to go on if for no other reason than this: Physics is by definition that which does go on its eternal way despite all the shadowy changes in the surface appearance of reality.

The Marchon lecture given by J. A. W. at the University of Newcastle-upon-Tyne, May 18, 1971, and the Nuffield lecture given at Cambridge University July 19, 1971, were based on the material presented in this chapter.

Some day a door will surely open and expose the glittering central mechanism of the world in its beauty and simplicity. Toward the arrival of that day, no development holds out more hope than the paradox of gravitational collapse. Why paradox? Because Einstein's equation says "this is the end" and physics says "there is no end." Why hope? Because among all paradigms for probing a puzzle, physics proffers none with more promise than a paradox.

No previous period of physics brought a greater paradox than 1911 (Box 44.1). Rutherford had just been forced to conclude that matter is made up of localized positive and negative charges. Matter as so constituted should undergo electric collapse in $\sim 10^{-17}$ sec, according to theory. Observation equally clearly proclaimed that matter is stable. No one took the paradox more seriously than Bohr. No one worked around and around the central mystery with more energy wherever work was possible. No one brought to bear a more judicious combination of daring and conservativeness, nor a deeper feel for the harmony of physics. The direct opposite

The paradox of collapse: physics stops, but physics must go on

The 1911 crisis of electric collapse

Box 44.1 COLLAPSE OF UNIVERSE PREDICTED BY CLASSICAL THEORY, COMPARED AND CONTRASTED WITH CLASSICALLY PREDICTED COLLAPSE OF ATOM

System	Atom (1911)	Universe (1970's)
Dynamic entity	System of electrons	Geometry of space
Nature of classically predicted collapse	Electron headed toward point-center of attraction is driven in a finite time to infinite energy	Not only matter but space itself arrives in a finite proper time at a condition of infinite compaction
One rejected "way out"	Give up Coulomb law of force	Give up Einstein's field equation
Another proposal for a "cheap way out" that has to be rejected	"Accelerated charge need not radiate"	"Matter cannot be compressed beyond a certain density by any pressure, however high"
How this proposal violates principle of causality	Coulomb field of point-charge cannot readjust itself with infinite speed out to indefinitely great distances to sudden changes in velocity of charge	Speed of sound cannot exceed speed of light; pressure cannot exceed density of mass-energy
A major new consideration introduced by recognizing quantum principle as overarching organizing principle of physics	Uncertainty principle; binding too close to center of attraction makes zero-point kinetic energy outbalance potential energy; consequent existence of a lowest quantum state; can't radiate because no lower state available to drop to	Uncertainty principle; propagation of representative wave packet in superspace does not lead deterministically to a singular configuration for the geometry of space; expect rather a probability distribution of outcomes, each outcome describing a universe with a different size, a different set of particle masses, a different number of particles, and a different length of time required for its expansion and recontraction.

of harmony, cacophony, is the impression that comes from sampling the literature of the 'teens on the structure of the atom. (1) Change the Coulomb law of force between electric charges? (2) Give up the principle that an accelerated charge must radiate? There was little inhibition against this and still wilder abandonings of the well-established. In contrast, Bohr held fast to these two principles. At the same time he insisted on the importance of a third principle, firmly established by Planck in quite another domain of physics, the quantum principle. With that key he opened the door to the world of the atom.

Great as was the crisis of 1911, today gravitational collapse confronts physics with its greatest crisis ever. At issue is the fate, not of matter alone, but of the universe itself. The dynamics of collapse, or rather of its reverse, expansion, is evidenced not by theory alone, but also by observation; and not by one observation, but by observations many in number and carried out by astronomers of unsurpassed ability and integrity. Collapse, moreover, is not unique to the large-scale dynamics of the universe. A white dwarf star or a neutron star of more than critical mass is predicted to undergo gravitational collapse to a black hole (Chapters 32 and 33). Sufficiently many stars falling sufficiently close together at the center of the nucleus of a galaxy are expected to collapse to a black hole many powers of ten more massive than the sun. An active search is under way for evidence for a black hole in this Galaxy (Box 33.3). The process that makes a black hole is predicted to provide an experimental model for the gravitational collapse of the universe itself, with one difference. For collapse to a black hole, the observer has his choice whether (1) to observe from a safe distance, in which case his observations will reveal nothing of what goes on inside the horizon; or (2) to follow the falling matter on in, in which case he sees the final stages of the collapse, not only of the matter itself, but of the geometry surrounding the matter, to indefinitely high compaction, but only at the cost of his own early demise. For the gravitational collapse of a closed model universe, no such choice is available to the observer. His fate is sealed. So too is the fate of matter and elementary particles, driven up to indefinitely high densities. The stakes in the crisis of collapse are hard to match: the dynamics of the largest object, space, and the smallest object, an elementary particle—and how both began.

§44.2. ASSESSMENT OF THE THEORY THAT PREDICTS COLLAPSE

No one reflecting on the paradox of collapse ("collapse ends physics"; "collapse cannot end physics") can fail to ask, "What are the limits of validity of Einstein's geometric theory of gravity?" A similar question posed itself in the crisis of 1911. The Coulomb law for the force acting between two charges had been tested at distances of meters and millimeters, but what warrant was there to believe that it holds down to the 10^{-8} cm of atomic dimensions? Of course, in the end it proves to hold not only at the level of the atom, and at the 10^{-13} cm level of the nucleus, but even down to 5×10^{-15} cm [Barber, Gittelman, O'Neill, and Richter, and Bailey *et al.* (1968), as reviewed by Farley (1969) and Brodsky and Drell (1970)], a striking

example of what Wigner (1960) calls the "unreasonable effectiveness of mathematics in the natural sciences."

No theory more resembles Maxwell's electrodynamics in its simplicity, beauty, and scope than Einstein's geometrodynamics. Few principles in physics are more firmly established than those on which it rests: the local validity of special relativity (Chapters 2–7), the equivalence principle (Chapter 16), the conservation of momentum and energy (Chapters 5, 15 and 16), and the prevalence of second-order field equations throughout physics (Chapter 17). Those principles and the demand for no "extraneous fields" (e.g., Dicke's scalar field) and "no prior geometry" (§17.6) lead to the conclusion that the geometry of spacetime must be Riemannian and the geometrodynamic law must be Einstein's.

To say that the geometry is Riemannian is to say that the interval between any two nearby events C and D, anywhere in spacetime, stated in terms of the interval AB between two nearby fiducial events, at quite another point in spacetime, has a value CD/AB independent of the route of intercomparison (Chapter 13 and Box 16.4). There are a thousand routes. By this hydraheaded prediction, Einstein's theory thus exposes itself to destruction in a thousand ways (Box 16.4).

Geometrodynamics lends itself to being disproven in other ways as well. The geometry has no option about the control it exerts on the dynamics of particles and fields (Chapter 20). The theory makes predictions about the equilibrium configurations and pulsations of compact stars (Chapters 23–26). It gives formulas (Chapters 27–29) for the deceleration of the expansion of the universe, for the density of mass-energy, and for the magnifying power of the curvature of space, the tests of which are not far off. It predicts gravitational collapse, and the existence of black holes, and a wealth of physics associated with these objects (Chapters 31–34). It predicts gravitational waves (Chapters 35–37). In the appropriate approximation, it encompasses all the well-tested predictions of the Newtonian theory of gravity for the dynamics of the solar system, and predicts testable post-Newtonian corrections besides, including several already verified effects (Chapters 38–40).

No inconsistency of principle has ever been found in Einstein's geometric theory of gravity. No purported observational evidence against the theory has ever stood the test of time. No other acceptable account of physics of comparable simplicity and scope has ever been put forward.

Continue this assessment of general relativity a little further before returning to the central issue, the limits of validity of the theory and their bearing on the issue of gravitational collapse. What has Einstein's geometrodynamics contributed to the understanding of physics?

First, it has dethroned spacetime from a post of preordained perfection high above the battles of matter and energy, and marked it as a new dynamic entity participating actively in this combat.

Second, by tying energy and momentum to the curvature of spacetime, Einstein's theory has recognized the law of conservation of momentum and energy as an automatic consequence of the geometric identity that the boundary of a boundary is zero (Chapters 15 and 17).

Third, it has recognized gravitation as a manifestation of the curvature of the

Battle-tested theory of gravitation

New view of nature flowing from Einstein's geometrodynamics

geometry of spacetime rather than as something foreign and "physical" imbedded in spacetime.

Fourth, general relativity has reinforced the view that "physics is local"; that the analysis of physics becomes simple when it connects quantities at a given event with quantities at immediately adjacent events.

Fifth, obedient to the quantum principle, it recognizes that spacetime and time itself are ideas valid only at the classical level of approximation; that the proper arena for the Einstein dynamics of geometry is not spacetime, but superspace; and that this dynamics is described in accordance with the quantum principle by the propagation of a probability amplitude through superspace (Chapter 43). In consequence, the geometry of space is subject to quantum fluctuations in metric coefficients of the order

$$\delta g \sim \frac{(\text{Planck length, } L^* = (\hbar G/c^3)^{1/2} = 1.6 \times 10^{-33} \text{ cm})}{(\text{linear extension of region under study})}.$$

Electric charge as lines of force trapped in the topology of space

Sixth, standard Einstein geometrodynamics is partial as little to Euclidean topology as to Euclidean geometry. A multiply connected topology provides a natural description for electric charge as electric lines of force trapped in the topology of a multiply connected space (Figure 44.1). Any other description of electricity postulates a breakdown in Maxwell's field equations for the vacuum at a site where charge

Figure 44.1.
Electric charge viewed as electric lines of force trapped in the topology of a multiply connected space [for the history of this concept see reference 36 of Wheeler (1968a)]. The wormhole or handle is envisaged as connecting two very different regions in the same space. One of the wormhole mouths, viewed by an observer with poor resolving power, appears to be the seat of an electric charge. Out of this region of 3-space he finds lines of force emerging over the whole 4π solid angle. He may construct a boundary around this charge, determine the flux through this boundary, incorrectly apply the theorem of Gauss and "prove" that there is a charge "inside the boundary." It isn't a boundary. Someone caught within it—to speak figuratively—can go into that mouth of the wormhole, through the throat, out the other mouth, and return by way of the surrounding space to look at his "prison" from the outside. Lines of force nowhere end. Maxwell's equations nowhere fail. Nowhere can one place a finger and say, "Here there is some charge." This classical type of electric charge has no direct relation whatsoever to quantized electric charge. There is a freedom of choice about the flux through the wormhole, and a specificity about the connection between one charge and another, which is quite foreign to the charges of elementary particle physics. For ease of visualization the number of space dimensions in the above diagram has been reduced from three to two. The third dimension, measured off the surface, has no physical meaning—it only provides an extra dimension in which to imbed the surface for more convenient diagrammatic representation. [For more detail see Misner and Wheeler (1957), reprinted in Wheeler (1962)].

is located, or postulates the existence of some foreign and "physical" electric jelly imbedded in space, or both. No one has ever found a way to describe electricity free of these unhappy features except to say that the quantum fluctuations in the geometry of space are so great at small distances that even the topology fluctuates, makes "wormholes," and traps lines of force. These fluctuations have to be viewed, not as tied to particles, and endowed with the scale of distances associated with particle physics ($\sim 10^{-13}$ cm) but as pervading all space ("foam-like structure of geometry") and characterized by the Planck distance ($\sim 10^{-33}$ cm). Thus a third type of gravitational collapse forces itself on one's attention, a collapse continually being done and being undone everywhere in space: surely a guide to the outcome of collapse at the level of a star and at the level of the universe (Box 44.2).

Box 44.2 THREE LEVELS OF GRAVITATIONAL COLLAPSE

1. Universe
2. Black hole
3. Fluctuations at the Planck scale of distances

Recontraction and collapse of the universe is a kind of mirror image of the "big bang," on which one already has so much evidence.

Collapse of matter to form a black hole is most natural at two distinct levels: (a) collapse of the dense white-dwarf core of an individual star (when that core exceeds the critical mass, $\sim 1 M_\odot$ or $2 M_\odot$, at which a neutron star is no longer a possible stable end-point for collapse) and (b) coalescence one by one of the stars in a galactic nucleus to make a black hole of mass up to $10^6 M_\odot$ or even $10^9 M_\odot$.

In either case, no feature of principle about matter falling into the black hole is more interesting than the option that the observer has (symbolized by the branching arrow in the inset). He can go along with the infalling matter, in which case he sees the final stages of collapse, but only at the cost of his own demise. Or he can stay safely outside, in which case even after indefinitely long time he sees only the first part of the collapse, with the infalling matter creeping up more and more slowly to the horizon.

In the final stages of the collapse of a closed model universe, all black holes present are caught up and driven together, amalgamating one by one. No one has any way to look at the event from safely outside; one is inevitably caught up in it oneself.

Collapse at the Planck scale of distances is taking place everywhere and all the time in quantum fluctuations in the geometry and, one believes, the topology of space. In this sense, collapse is continually being done and undone, modeling the undoing of the collapse of the universe itself, summarized in the term, "the reprocessing of the universe" (see text).

§44.3. VACUUM FLUCTUATIONS: THEIR PREVALENCE AND FINAL DOMINANCE

If Einstein's theory thus throws light on the rest of physics, the rest of physics also throws light on geometrodynamics. No point is more central than this, that empty space is not empty. It is the seat of the most violent physics. The electromagnetic field fluctuates (Chapter 43). Virtual pairs of positive and negative electrons, in effect, are continually being created and annihilated, and likewise pairs of mu mesons, pairs of baryons, and pairs of other particles. All these fluctuations coexist with the quantum fluctuations in the geometry and topology of space. Are they additional to those geometrodynamic zero-point disturbances, or are they, in some sense not now well-understood, mere manifestations of them?

Put the question in other words. Recall Clifford, inspired by Riemann, speaking to the Cambridge Philosophical Society on February 21, 1870, "On the Space Theory of Matter" [Clifford (1879), pp. 244 and 322; and (1882), p. 21], and saying, "I hold in fact (1) That small portions of space *are* in fact of a nature analogous to little hills on a surface which is on the average flat; namely, that the ordinary laws of geometry are not valid in them. (2) That this property of being curved or distorted is continually being passed on from one portion of space to another after the manner of a wave. (3) That this variation of the curvature of space is what really happens in that phenomenon which we call the *motion of matter,* whether ponderable or etherial. (4) That in the physical world nothing else takes place but this variation, subject (possibly) to the law of continuity." Ask if there is a sense in which one can speak of a particle as constructed out of geometry. Or rephrase the question in updated language: "Is a particle a geometrodynamic exciton?" What else is there out of which to build a particle except geometry itself? And what else is there to give discreteness to such an object except the quantum principle?

The Clifford-Einstein space theory of matter has not been forgotten in recent years. "In conclusion," one of the authors wrote a decade ago [Wheeler (1962)], "the vision of Riemann, Clifford, and Einstein, of a purely geometric basis for physics, today has come to a higher state of development, and offers richer prospects—and presents deeper problems—than ever before. The quantum of action adds to this geometrodynamics new features, of which the most striking is the presence of fluctuations of the wormhole type throughout all space. If there is any correspondence at all between this virtual foam-like structure and the physical vacuum as it has come to be known through quantum electrodynamics, then there seems to be no escape from identifying these wormholes with 'undressed electrons.' Completely different from these 'undressed electrons,' according to all available evidence, are the electrons and other particles of experimental physics. For these particles the geometrodynamic picture suggests the model of collective disturbances in a virtual foam-like vacuum, analogous to different kinds of phonons or excitons in a solid.

"The enormous factor from nuclear densities $\sim 10^{14}$ g/cm^3 to the density of field fluctuation energy in the vacuum, $\sim 10^{94}$ g/cm^3, argues that elementary particles represent a percentage-wise almost completely negligible change in the locally violent conditions that characterize the vacuum. ['A particle (10^{14} g/cm^3) means as little

Is matter built out of geometry?

The richness of the physics of the vacuum

to the physics of the vacuum (10^{94} g/cm^3) as a cloud (10^{-6} g/cm^3) means to the physics of the sky (10^{-3} g/cm^3).'] In other words, elementary particles do not form a really basic starting point for the description of nature. Instead, they represent a first-order correction to vacuum physics. That vacuum, that zero-order state of affairs, with its enormous densities of virtual photons and virtual positive-negative pairs and virtual wormholes, has to be described properly before one has a fundamental starting point for a proper perturbation-theoretic analysis."

"These conclusions about the energy density of the vacuum, its complicated topological character, and the richness of the physics which goes on in the vacuum, stand in no evident contradiction with what quantum electrodynamics has to say about the vacuum. Instead the conclusions from the 'small distance' analysis (10^{-33} cm)—sketchy as it is—and from 'larger distance' analysis (10^{-11} cm) would seem to [be able] to reinforce each other in a most natural way.

"The most evident shortcoming of the geometrodynamic model as it stands is this, that *it fails to supply any completely natural place* for spin $\frac{1}{2}$ in general and *for the neutrino* in particular."

Attempts to find a natural place for spin $\frac{1}{2}$ in Einstein's standard geometrodynamics (Box 44.3) founder because there is no natural way for a *change* in connectivity to take place within the context of classical differential geometry.

A uranium nucleus undergoing fission starts with one topology and nevertheless ends up with another topology. It makes this transition in a perfectly continuous way, classical differential geometry notwithstanding.

There are reputed to be two kinds of lawyers. One tells the client what not to do. The other listens to what the client has to do and tells him how to do it. From the first lawyer, classical differential geometry, the client goes away disappointed, still searching for a natural way to describe quantum fluctuations in the connectivity of space. Only in this way can he hope to describe electric charge as lines of electric force trapped in the topology of space. Only in this way does he expect to be able to understand and analyze the final stages of gravitational collapse. Pondering his problems, he comes to the office of a second lawyer, with the name "Pregeometry" on the door. Full of hope, he knocks and enters. What is pregeometry to be and say? Born of a combination of hope and need, of philosophy and physics and mathematics and logic, pregeometry will tell a story unfinished at this writing, but full of incidents of evolution so far as it goes.

No place in geometrodynamics for change of topology; therefore turn to "pregeometry"

§44.4. NOT GEOMETRY, BUT PREGEOMETRY, AS THE MAGIC BUILDING MATERIAL

An early survey (Box 44.4) asked whether geometry can be constructed with the help of the quantum principle out of more basic elements, that do not themselves have any specific dimensionality.

The focus of attention in this 1964 discussion was "dimensionality without dimensionality." However, the prime pressures to ponder pregeometry were and remain

(*continued on page 1206*)

Box 44.3 THE DIFFICULTIES WITH ATTEMPTS TO FIND A NATURAL PLACE FOR SPIN $\frac{1}{2}$ IN EINSTEIN'S STANDARD GEOMETRODYNAMICS

"It is impossible" [Wheeler (1962)] "to accept any description of elementary particles that does not have a place for spin $\frac{1}{2}$. What, then, has any purely geometric description to offer in explanation of spin $\frac{1}{2}$ in general? More particularly and more importantly, what place is there in quantum geometrodynamics for the neutrino—the only entity of half-integral spin that is a pure field in its own right, in the sense that it has zero rest mass and moves with the speed of light? No clear or satisfactory answer is known to this question today. Unless and until an answer is forthcoming, *pure geometrodynamics must be judged deficient as a basis for elementary particle physics*."

A later publication [Wheeler (1968a)] takes up this issue again, noting that, "A new world opens out for analysis in quantum geometrodynamics. The central new concept is space resonating between one foamlike structure and another. For this multiple-connectedness of space at submicroscopic distances no single feature of nature speaks more powerfully than electric charge. Yet at least as impressive as charge is the prevalence of spin $\frac{1}{2}$ throughout the world of elementary particles."

Repeating the statement that "It is impossible to accept any description of elementary particles that does not have a place for spin $\frac{1}{2}$," the article adds to the discussion a new note: "Happily, the concept of spin manifold has come to light, not least through the work of John Milnor [see Lichnerowicz (1961a,b,c) and (1964); Milnor (1962), (1963), and (1965a,b); Hsiang and Anderson (1965); Anderson, Brown, and Peterson (1966a,b); and Penrose (1968a)]. This concept suggests a new and *interesting interpretation of a spinor field* within the context of the resonating microtopology of quantum geometrodynamics, *as the nonclassical two-valuedness* [Pauli's standard term for spin; see, for example, Pauli (1947)] *that attaches to the probability amplitude for otherwise identical 3-geometries endowed with alternative 'spin structures.'*" More specifically: "One does not classify the closed orientable 3-manifold of physics completely

when one gives its topology, its differential structures, and its metric. One must tell which spin structure it has." [On a 3-geometry with the topology of a 3-sphere, one can lay down a continuous field of triads (a triad consisting of three orthonormal vectors). Any other continuous field of triads can be deformed into the first field by a continuous sequence of small readjustments. One says that the 3-sphere admits only one "spin structure," a potentially misleading standard word for what could just as well have been called a "triad structure." In contrast, a 3-sphere with n handles or wormholes admits 2^n "spin structures" (continuous fields of triads) inequivalent to one another under any continuous sequence of small readjustments whatsoever, and distinguished from one another in any convenient way by n "descriptors" $w_1, w_2, \ldots, w_k, \ldots w_n$.] It is natural in quantum geometrodynamics to expect "separate probability amplitudes for a 3-geometry with descriptor $w_k = +1$ and for an otherwise identical 3-geometry with descriptor $w_k = -1$. Does this circumstance imply that quantum geometrodynamics supplies all the machinery one needs to describe fields of spin $\frac{1}{2}$ in general and the neutrino field in particular? . . . That is the only way that has ever turned up within the framework of Einstein's general relativity and Planck's quantum principle. Is this the right path? It is difficult to name any question more decisive than this in one's assessment of 'everything as geometry.'"

Why not spell out these concepts, reduce them to practice, and compare them with what one knows about the behavior of fields of spin $\frac{1}{2}$? There is a central difficulty in this enterprise. It assumes and demands on physical grounds that the topology of the 3-geometry shall be free to change from one connectivity to another. In contrast, classical differential geometry says, in effect, "Once one topology, always that topology." Try a question like this, "When a new handle develops and the number of descriptors rises by one, what boundary condition in superspace connects the probability

amplitude ψ for 3-geometries of the original topology with the probability amplitudes ψ_+ and ψ_- for the two spin structures of the new topology?" Classical differential geometry not only gives one no help in answering this question; it even forbids one to ask it. In other words, one cannot even get the enterprise "on the road" for want of a natural mathematical way to describe the required change in topology. The idea is therefore abandoned here and now that 3-geometry is "the magic building material of the universe." In contrast, pregeometry (see text), far from being endowed with any frozen topology, is to be viewed as not even possessing any dimensionality.

Box 44.4 "A BUCKET OF DUST"—AN EARLY ATTEMPT TO FORMULATE THE CONCEPT OF PREGEOMETRY [Wheeler (1964a)]

"What line of thought could ever be imagined as *leading* to four dimensions—or any dimensionality at all—out of more primitive considerations? In the case of atoms one derives the yellow color of the sodium D-lines by analyzing the quantum dynamics of a system, no part of which is ever endowed with anything remotely resembling the attribute of color. Likewise any derivation of the four-dimensionality of spacetime can hardly *start* with the idea of dimensionality."

"Recall the notion of a Borel set. Loosely speaking, a Borel set is a collection of points ("bucket of dust") which have not yet been assembled into a manifold of any particular dimensionality. . . . Recalling the universal sway of the quantum principle, one can imagine probability amplitudes for the points in a Borel set to be assembled into points with this, that, and the other dimensionality. . . . More conditions have to be imposed on a given number of points—as to which has which for a nearest neighbor—when the points are put together in a five-dimensional array than when these same points are arranged in a two-dimensional pattern. Thus one can think of each dimensionality as having a much higher statistical weight than the next *higher* dimensionality. On the other hand, for manifolds with one, two, and three dimensions, the geometry is too rudimentary—one can suppose—to give anything interesting. Thus Einstein's field equations, applied to a manifold of dimensionality so low, demand flat space; only when the dimensionality is as high as four do really interesting possibilities arise. Can four, therefore, be considered to be the unique dimensionality which is at the same time high enough to give any real physics and yet low enough to have great statistical weight?"

"It is too much to imagine that one has yet made enough mistakes in this domain of thought to explore such ideas with any degree of good judgment."

Consider a handle on the geometry. Let it thin halfway along its length to a point. In other words, let the handle dissolve into two bent prongs that touch at a point. Let these prongs separate and shorten. In this process two points part company that were once immediate neighbors. "However sudden the change is in classical theory, in quantum theory there is a probability amplitude function which falls off in the classically forbidden domain. In other words, there is some residual connection between points which are ostensibly very far away (travel from one 'tip' down one prong, then through the larger space to which these prongs are attached, and then up the other prong to the other tip). But there is nothing distinctive in principle about the two points that have happened to come into discussion. Thus it would seem that there must be a connection . . . between *every* point and every other point. Under these conditions the concept of nearest neighbor would appear no longer to make sense. Thus the tool disappears with the help of which one might otherwise try to speak [un]ambiguously about dimensionality."

two features of nature, spin $\frac{1}{2}$ and charge, that speak out powerfully from every part of elementary particle physics.

A fresh perspective on pregeometry comes from a fresh assessment of general relativity. "Geometrodynamics is neither as important or as simple as it looks. Do not make it the point of departure in searching for underlying simplicity. Look deeper, at elementary particle physics." This is the tenor of interesting new considerations put forward by Sakharov [*the* Sakharov] (1967) and summarized under the heading, "Gravitation as the metric elasticity of space," in Box 17.2. In brief, as elasticity is to atomic physics, so—in Sakharov's view—gravitation is to elementary particle physics. The energy of an elastic deformation is nothing but energy put into the bonds between atom and atom by the deformation. The energy that it takes to curve space is nothing but perturbation in the vacuum energy of fields plus particles brought about by that curvature, according to Sakharov. The energy required for the deformation is governed in the one case by two elastic constants and in the other case by one elastic constant (the Newtonian constant of gravity) but in both cases, he reasons, the constants arise by combination of a multitude of complicated individual effects, not by a brave clean stroke on an empty slate.

One gives all the more favorable reception to Sakharov's view of gravity because one knows today, as one did not in 1915, how opulent in physics the vacuum is. In Einstein's day one had come in a single decade from the ideal God-given Lorentz perfection of flat spacetime to curved spacetime. It took courage to assign even one physical constant to that world of geometry that had always stood so far above physics. The vacuum looked for long as innocent of structure as a sheet of glass emerging from a rolling mill. With the discovery of the positive electron [Anderson (1933)], one came to recognize a little of the life that heat can unfreeze in "empty" space. Each new particle and radiation that was discovered brought a new accretion to the recognized richness of the vacuum. Macadam looks smooth, but a bulldozer has only to cut a single furrow through the roadway to disclose all the complications beneath the surface.

Think of a particle as built out of the geometry of space; think of a particle as a "geometrodynamic exciton"? No model—it would seem to follow from Sakharov's assessment—could be less in harmony with nature, except to think of an atom as built out of elasticity! Elasticity did not explain atoms. Atoms explained elasticity. If, likewise, particles fix the constant in Einstein's geometrodynamic law (Sakharov), must it not be unreasonable to think of the geometrodynamic law as explaining particles?

Carry the comparison between geometry and elasticity one stage deeper (Fig. 44.2). In a mixed solid there are hundreds of distinct bonds, all of which contribute to the elastic constants; some of them arise from Van der Waal's forces, some from ionic coupling, some from homopolar linkage; they have the greatest variety of strengths; but all have their origin in something so fantastically simple as a system of positively and negatively charged masses moving in accordance with the laws of quantum mechanics. In no way was it required or right to meet each complication of the chemistry and physics of a myriad of bonds with a corresponding complication of principle. By going to a level of analysis deeper than bond strengths, one had

Sakharov: gravitation is the "metric elasticity of space"

The stratification of space

Comparison with everyday elasticity

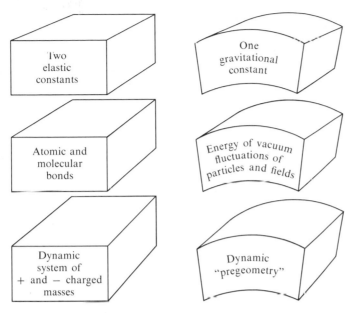

Figure 44.2.
Elasticity and geometrodynamics, as viewed at three levels of analysis. A hundred years of the study of elasticity did not reveal the existence of molecules, and a hundred years of the study of molecular chemistry did not reveal Schrödinger's equation. Revelation moved upward in the diagram, not downward.

emerged into a world of light, where nothing but simplicity and unity was to be seen.

Compare with geometry. The vacuum is animated with the zero-point activity of distinct fields and scores of distinct particles, all of which, according to Sakharov, contribute to the Newtonian G, the "elastic constant of the metric." Some interact via weak forces, some by way of electromagnetic forces, and some through strong forces. These interactions have the greatest variety of strengths. But must not all these particles and interactions have their origin in something fantastically simple? And must not this something, this "pregeometry," be as far removed from geometry as the quantum mechanics of electrons is far removed from elasticity?

If one once thought of general relativity as a guide to the discovery of pregeometry, nothing might seem more dismaying than this comparison with an older realm of physics. No one would dream of studying the laws of elasticity to uncover the principles of quantum mechanics. Neither would anyone investigate the work-hardening of a metal to learn about atomic physics. The order of understanding ran not

$$\text{Work-hardening (1 cm)} \longrightarrow \text{dislocations (10}^{-4}\text{ cm)} \longrightarrow \text{atoms (10}^{-8}\text{ cm),}$$

but the direct opposite,

$$\text{Atoms (10}^{-8}\text{ cm)} \longrightarrow \text{dislocations (10}^{-4}\text{ cm)} \longrightarrow \text{work-hardening (1 cm)}$$

One had to know about atoms to conceive of dislocations, and had to know about dislocations to understand work-hardening. Is it not likewise hopeless to go from the "elasticity of geometry" to an understanding of particle physics, and from particle physics to the uncovering of pregeometry? Must not the order of progress again be the direct opposite? And is not the source of any dismay the apparent loss of guidance that one experiences in giving up geometrodynamics—and not only geometrodynamics, but geometry itself—as a crutch to lean on as one hobbles forward? Yet there is so much chance that this view of nature is right that one must take it seriously and explore its consequences. Never more than today does one have the incentive to explore pregeometry.

§44.5. PREGEOMETRY AS THE CALCULUS OF PROPOSITIONS

Paper in white the floor of the room, and rule it off in one-foot squares. Down on one's hands and knees, write in the first square a set of equations conceived as able to govern the physics of the universe. Think more overnight. Next day put a better set of equations into square two. Invite one's most respected colleagues to contribute to other squares. At the end of these labors, one has worked oneself out into the door way. Stand up, look back on all those equations, some perhaps more hopeful than others, raise one's finger commandingly, and give the order "Fly!" Not one of those equations will put on wings, take off, or fly. Yet the universe "flies."

Search for the central principle of pregeometry

Some principle uniquely right and uniquely simple must, when one knows it, be also so compelling that it is clear the universe is built, and must be built, in such and such a way, and that it could not possibly be otherwise. But how can one discover that principle? If it was hopeless to learn atomic physics by studying work-hardening and dislocations, it may be equally hopeless to learn the basic operating principle of the universe, call it pregeometry or call it what one will, by any amount of work in general relativity and particle physics.

Thomas Mann (1937), in his essay on Freud, utters what Niels Bohr would surely have called a great truth ("A great truth is a truth whose opposite is also a great truth") when he says, "Science never makes an advance until philosophy authorizes and encourages it to do so." If the equivalence principle (Chapter 16) and Mach's principle (§21.9) were the philosophical godfathers of general relativity, it is also true that what those principles do mean, and ought to mean, only becomes clear by study and restudy of Einstein's theory itself. Therefore it would seem reasonable to expect the primary guidance in the search for pregeometry to come from a principle both philosophical and powerful, but one also perhaps not destined to be wholly clear in its contents or its implications until some later day.

Among all the principles that one can name out of the world of science, it is difficult to think of one more compelling than *simplicity;* and among all the simplicities of dynamics and life and movement, none is starker [Werner (1969)] than the *binary choice* yes–no or true–false. It in no way proves that this choice for a starting principle is correct, but it at least gives one some comfort in the choice, that Pauli's "nonclassical two-valuedness" or "spin" so dominates the world of particle physics.

It is one thing to have a start, a tentative construction of pregeometry; but how does one go on? How not to go on is illustrated by Figure 44.3. The "sewing machine" builds objects of one or another definite dimensionality, or of mixed dimensionalities, according to the instructions that it receives on the input tape in yes–no binary code. Some of the difficulties of building up structure on the binary element according to this model, or any one of a dozen other models, stand out at once. (1) Why $N = 10,000$ building units? Why not a different N? And if one feeds in one such arbitrary number at the start, why not fix more features "by hand?" No natural stopping point is evident, nor any principle that would fix such a stopping point. Such arbitrariness contradicts the principle of simplicity and rules out the model. (2) Quantum mechanics is added from outside, not generated from inside (from the model itself). On this point too the principle of simplicity speaks against the model. (3) The passage from pregeometry to geometry is made in a too-literal-minded way, with no appreciation of the need for particles and fields to appear along the way. The model, in the words used by Bohr on another occasion, is "crazy, but not crazy enough to be right."

Noting these difficulties, and fruitlessly trying model after model of pregeometry to see if it might be free of them, one suddenly realizes that a machinery for the combination of yes–no or true–false elements does not have to be invented. It already exists. What else can pregeometry be, one asks oneself, than the calculus of propositions? (Box 44.5.)

A first try at a pregeometry built on the principle of binary choice

A more reasonable picture: pregeometry is the calculus of propositions

§44.6. THE BLACK BOX: THE REPROCESSING OF THE UNIVERSE

No amount of searching has ever disclosed a "cheap way" out of gravitational collapse, any more than earlier it revealed a cheap way out of the collapse of the atom. Physicists in that earlier crisis found themselves in the end confronted with a revolutionary pistol, "Understand nothing—or accept the quantum principle." Today's crisis can hardly force a lesser revolution. One sees no alternative except to say that geometry fails and pregeometry has to take its place to ferry physics through the final stages of gravitational collapse and on into what happens next. No guide is evident on this uncharted way except the principle of simplicity, applied to drastic lengths.

Whether the whole universe is squeezed down to the Planck dimension, or more or less, before reexpansion can begin and dynamics can return to normal, may be irrelevant for some of the questions one wants to consider. Physics has long used the "black box" to symbolize situations where one wishes to concentrate on what goes in and what goes out, disregarding what takes place in between.

At the beginning of the crisis of electric collapse one conceived of the electron as headed on a deterministic path toward a point-center of attraction, and unhappily destined to arrive at a condition of infinite kinetic energy in a finite time. After the advent of quantum mechanics, one learned to summarize the interaction between

The role of the black box in physics

(continued on page 1213)

1210

Figure 44.3.

"Ten thousand rings"; or an example of a way to think of the connection between pregeometry and geometry, wrong because it is too literal-minded, and for other reasons spelled out in the text. The vizier [story by Wheeler, as alluded to by Kilmister (1971)*] speaks: "Take $N = 10,000$ brass rings. Take an automatic fastening device that will cut open a ring, loop it through another ring, and resolder the joint. Pour the brass rings into the hopper that feeds this machine. Take a strip of instruction paper that is long enough to contain $N(N-1)/2$ binary digits. Look at the instruction in the (jk)-th location on this instruction tape $(j,k = 1, 2, \ldots, N; j < k)$. When the binary digit at that location is 0, it is a signal to leave the j-th ring disconnected from the k-th ring. When it is 1, it is an instruction to connect that particular pair of rings. Thread the tape into the machine and press the start button. The clatter begins. Out comes a chain of rings 10,000 links long. It falls on the table and the machine stops. Pour in another 10,000 rings, feed in a new instruction tape, and push the button again. This time it is not a one-dimensional structure that emerges, but a two-dimensional one: a Crusader's coat of mail, complete with neck opening and sleeves. Take still another tape from the library of tapes and repeat. Onto the table thuds a smaller version of the suit of mail, this time filled out internally with a solid network of rings, a three-dimensional structure. Now forego the library and make one's own instruction tape, a random string of 0's and 1's. Guided by it, the fastener produces a "Christmas tree ornament," a collection of segments of one-dimensional chain, two-dimensional surfaces, and three-, four-, five-, and higher-dimensional entities, some joined together, some free-floating. Now turn from a structure deterministically fixed by a tape to a probability amplitude, a complex number,

$$\psi(\text{tape}) = \psi(n_{12}, n_{13}, n_{14}, \ldots \ldots, n_{N-1,N}) \qquad (n_{ij} = 0, 1), \tag{1}$$

defined over the entire range of possibilities for structures built of 10,000 rings. Let these probability amplitudes *not* be assigned randomly. Instead, couple together amplitudes, for structures that differ from each other by the breaking of a single ring, by linear formulas that treat all rings on the same footing. The separate ψ's, no longer entirely independent, will still give non-zero probability amplitudes for "Christmas tree ornaments." Of greater immediate interest than these "unruly" parts of the structures are the following questions about the smoother parts: (1) In what kinds of structures is the bulk of the probability concentrated? (2) What is the dominant dimensionality of these structures in an appropriate correspondence principle limit? (3) In this semiclassical limit, what is the form taken by the dynamic law of evolution of the geometry?" No principle more clearly rules out this model for pregeometry than the principle of simplicity (see text).

*Wheeler's story about the vizier and what the vizier had to say about superspace was told at the May 18, 1970, Gwatt Seminar on the Bearings of Topology upon General Relativity. Kilmister's (1971) published article alludes to the unpublished story, but does not actually contain it.

Box 44.5 "PREGEOMETRY AS THE CALCULUS OF PROPOSITIONS"

A sample proposition taken out of a standard text on logic selected almost at random reads [Kneebone (1963), p. 40]

$$[X \longrightarrow ((X \longrightarrow X) \longrightarrow Y)] \,\&\, (\bar{X} \longrightarrow Z) \,\mathrm{eq}\, (\bar{X} \vee Y \vee Z) \,\&\,$$
$$(\bar{X} \vee Y \vee \bar{Z}) \,\&\, (X \vee Y \vee Z) \,\&\, (X \vee \bar{Y} \vee Z).$$

The symbols have the following meaning:

$$\bar{A}, \qquad \text{Not } A;$$

$$A \vee B, \qquad A \text{ or } B \text{ or both ("}A \text{ vel } B\text{");}$$

$$A \,\&\, B, \qquad A \text{ and } B;$$

$$A \longrightarrow B, \qquad A \text{ implies } B \text{ ("if } A, \text{ then } B\text{");}$$

$$A \longleftrightarrow B, \qquad B \text{ is equivalent to } A \text{ ("}B \text{ if and only if } A\text{").}$$

Propositional formula \mathfrak{A} is said to be equivalent ("eq") to propositional formula \mathfrak{B} if and only if $\mathfrak{A} \longleftrightarrow \mathfrak{B}$ is a tautology. The letters A, B, etc., serve as connectors to "wire together" one proposition with another. Proceeding in this way, one can construct propositions of indefinitely great length.

A switching circuit [see, for example, Shannon (1938) or Hohn (1966)] is isomorphic to a proposition.

Compare a short proposition or an elementary switching circuit to a molecular collision. No idea seemed more preposterous than that of Daniel Bernoulli (1733), that heat is a manifestation of molecular collisions. Moreover, a three-body encounter is difficult to treat, a four-body collision is more difficult, and a five- or more molecule system is essentially intractable. Nevertheless, mechanics acquires new elements of simplicity in the limit in which the number of molecules is very great and in which one can use the concept of density in phase space. Out of statistical mechanics in this limit come such concepts as temperature and entropy. When the temperature is well-defined, the energy of the system is not a well-defined idea; and when the energy is well-defined, the temperature is not. This complementarity is built inescapably into the principles of the subject. Thrust the finger into the flame of a match and experience a sensation like nothing else on heaven or earth; yet what happens is all a consequence of molecular collisions, early critics notwithstanding.

Any individual proposition is difficult for the mind to apprehend when it is long; and still more difficult to grasp is the content of a cluster of propositions. Nevertheless, make a statistical analysis of the calculus of propositions in the limit where the number of propositions is great and most of them are long. Ask if parameters force themselves on one's attention in this analysis (1) analogous in some small measure to the temperature and entropy of statistical mechanics but (2) so much

Box 44.5 (*continued*)

more numerous, and everyday dynamic in character, that they reproduce the continuum of everyday physics.

Nothing could seem so preposterous at first glance as the thought that nature is built on a foundation as ethereal as the calculus of propositions. Yet, beyond the push to look in this direction provided by the principle of simplicity, there are two pulls. First, bare-bones quantum mechanics lends itself in a marvelously natural way to formulation in the language of the calculus of propositions, as witnesses not least the book of Jauch (1968). If the quantum principle were not in this way already automatically contained in one's proposed model for pregeometry, and if in contrast it had to be introduced from outside, by that very token one would conclude that the model violated the principle of simplicity, and would have to reject it. Second, the pursuit of reality seems always to take one away from reality. Who would have imagined describing something so much a part of the here and now as gravitation in terms of curvature of the geometry of spacetime? And when later this geometry came to be recognized as dynamic, who would have dreamed that geometrodynamics unfolds in an arena so ethereal as superspace? Little astonishment there should be, therefore, if the description of nature carries one in the end to logic, the ethereal eyrie at the center of mathematics. If, as one believes, all mathematics reduces to the mathematics of logic, and all physics reduces to mathematics, what alternative is there but for all physics to reduce to the mathematics of logic? Logic is the only branch of mathematics that can "think about itself."

"An issue of logic having nothing to do with physics" was the assessment by many of a controversy of old about the axiom, "parallel lines never meet." Does it follow from the other axioms of Euclidean geometry or is it independent? "Independent," Bolyai and Lobachevsky proved. With this and the work of Gauss as a start, Riemann went on to create Riemannian geometry. Study nature, not Euclid, to find out about geometry, he advised; and Einstein went on to take that advice and to make geometry a part of physics.

"An issue of logic having nothing to do with physics" is one's natural first assessment of the startling limitations on logic discovered by Gödel (1931), Cohen (1966), and others [for a review, see, for example, Kac and Ulam (1968)]. The exact opposite must be one's assessment if the real pregeometry of the real physical world indeed turns out to be identical with the calculus of propositions.

"Physics as manifestation of logic" or "pregeometry as the calculus of propositions" is as yet [Wheeler (1971a)] not an idea, but an idea for an idea. It is put forward here only to make it a little clearer what it means to suggest that the order of progress may not be

$$\text{physics} \longrightarrow \text{pregeometry}$$

but

$$\text{pregeometry} \longrightarrow \text{physics.}$$

Figure 44.4.

The "black-box model" applied (1) to the scattering of an electron by a center of attraction and (2) to the collapse of the universe itself. The deterministic electron world line of classical theory is replaced in quantum theory by a probability amplitude, the wave crests of which are illustrated schematically in the diagram. The catastrophe of classical theory is replaced in quantum theory by a probability distribution of outputs. The same diagram illustrates the "black-box account" of gravitational collapse mentioned in the text. The arena of the diagram is no longer spacetime, but superspace. The incident arrow marks no longer a classical world line of an electron through spacetime, but a classical "leaf of history of geometry" slicing through superspace (Chapter 43). The wave crests symbolize no longer the electron wave function propagating through spacetime, but the geometrodynamic wave function propagating through superspace. The cross-hatched region is no longer the region where the one-body potential goes to infinity, but the region of gravitational collapse where the curvature of space goes to infinity. The outgoing waves describe no longer alternative directions for the new course of the scattered electron, but the beginnings of alternative new histories for the universe itself after collapse and "reprocessing" end the present cycle.

center of attraction and electron in a "black box:" fire in a wave-train of electrons traveling in one direction, and get electrons coming out in this, that, and the other direction with this, that, and the other well-determined probability amplitude (Figure 44.4). Moreover, to predict these probability amplitudes quantitatively and correctly, it was enough to translate the Hamiltonian of classical theory into the language of wave mechanics and solve the resulting wave equation, the key to the "black box."

A similar "black box" view of gravitational collapse leads one to expect a "probability distribution of outcomes." Here, however, one outcome is distinguished from another, one must anticipate, not by a single parameter, such as the angle of scattering of the electron, but by many. They govern, one foresees, such quantities as the size of the system at its maximum of expansion, the time from the start of this new cycle to the moment it ends in collapse, the number of particles present, and a thousand other features. The "probabilities" of these outcomes will be governed by a dynamic law, analogous to (1) the Schrödinger wave equation for the electron, or, to cite another black box problem, (2) the Maxwell equations that couple together, at a wave-guide junction, electromagnetic waves running in otherwise separate wave guides. However, it is hardly reasonable to expect the necessary dynamic law to spring forth as soon as one translates the Hamilton-Jacobi equation of general relativity (Chapter 43) into a Schrödinger equation, simply because geometrodynamics, in both its classical and its quantum version, is built on standard differential geometry. That standard geometry leaves no room for any of those quantum fluctuations in connectivity that seem inescapable at small distances and therefore also inescapable in the final stages of gravitational collapse. Not geometry, but pregeometry, must fill the black box of gravitational collapse.

Probability distribution of the outcomes of collapse

"Reprocessing" the universe

Little as one knows the internal machinery of the black box, one sees no escape from this picture of what goes on: the universe transforms, or transmutes, or transits, or is *reprocessed* probabilistically from one cycle of history to another in the era of collapse.

However straightforwardly and inescapably this picture of the reprocessing of the universe would seem to follow from the leading features of general relativity and the quantum principle, the two overarching principles of twentieth-century physics, it is nevertheless fantastic to contemplate. How can the dynamics of a system so incredibly gigantic be switched, and switched at the whim of probability, from one cycle that has lasted 10^{11} years to one that will last only 10^6 years? At first, only the circumstance that the system gets squeezed down in the course of this dynamics to incredibly small distances reconciles one to a transformation otherwise so unbelievable. Then one looks at the upended strata of a mountain slope, or a bird not seen before, and marvels that the whole universe is incredible:

> mutation of a species,
> metamorphosis of a rock,
> chemical transformation,
> spontaneous transformation of a nucleus,
> radioactive decay of a particle,
> reprocessing of the universe itself.

If it cast a new light on geology to know that rocks can be raised and lowered thousands of meters and hundreds of degrees, what does it mean for physics to think of the universe as being from time to time "squeezed through a knothole," drastically "reprocessed," and started out on a fresh dynamic cycle? Three considerations above all press themselves on one's attention, prefigured in these compressed phrases:

> destruction of all constants of motion in collapse;
> particles, and the physical "constants" themselves, as the
> "frozen-in part of the meteorology of collapse;"
> "the biological selection of physical constants."

All conservation laws transcended in the collapse of the universe

The gravitational collapse of a star, or a collection of stars, to a black hole extinguishes all details of the system (see Chapters 32 and 33) except mass and charge and angular momentum. Whether made of matter or antimatter or radiation, whether endowed with much entropy or little entropy, whether in smooth motion or chaotic turbulence, the collapsing system ends up as seen from outside, according to all indications, in the same standard state. The laws of conservation of baryon number and lepton number are transcended [Chapter 33; also Wheeler (1971b)]. No known means whatsoever will distinguish between black holes of the most different provenance if only they have the same mass, charge, and angular momentum. But for a closed universe, even these constants vanish from the scene. Total charge is automatically zero because lines of force have nowhere to end except upon charge. Total mass and total angular momentum have absolutely no definable meaning whatsoever for a closed universe. This conclusion follows not least because there

is no asymptotically flat space outside where one can put a test particle into Keplerian orbit to determine period and precession.

Of all principles of physics, the laws of conservation of charge, lepton number, baryon number, mass, and angular momentum are among the most firmly established. Yet with gravitational collapse the content of these conservation laws also collapses. The established is disestablished. No determinant of motion does one see left that could continue unchanged in value from cycle to cycle of the universe. Moreover, if particles are dynamic in construction, and if the spectrum of particle masses is therefore dynamic in origin, no option would seem left except to conclude that the mass spectrum is itself reprocessed at the time when "the universe is squeezed through a knot hole." A molecule in this piece of paper is a "fossil" from photochemical synthesis in a tree a few years ago. A nucleus of the oxygen in this air is a fossil from thermonuclear combustion at a much higher temperature in a star a few 10^9 years ago. What else can a particle be but a fossil from the most violent event of all, gravitational collapse?

Three hierarchies of fossils: molecules, nuclei, particles

That one geological stratum has one many-miles long slope, with marvelous linearity of structure, and another stratum has another slope, is either an everyday triteness, taken as for granted by every passerby, or a miracle, until one understands the mechanism. That an electron here has the same mass as an electron there is also a triviality or a miracle. It is a triviality in quantum electrodynamics because it is assumed rather than derived. However, it is a miracle on any view that regards the universe as being from time to time "reprocessed." How can electrons at different times and places in the present cycle of the universe have the same mass if the spectrum of particle masses differs between one cycle of the universe and another?

Inspect the interior of a particle of one type, and magnify it up enormously, and in that interior see one view of the whole universe [compare the concept of monad of Leibniz (1714), "The monads have no window through which anything can enter or depart"]; and do likewise for another particle of the same type. Are particles of the same pattern identical in any one cycle of the universe because they give identically patterned views of the same universe? No acceptable explanation for the miraculous identity of particles of the same type has ever been put forward. That identity must be regarded, not as a triviality, but as a central mystery of physics.

Reason for identity in mass of particles of the same species?

Not the spectrum of particle masses alone, but the physical "constants" themselves, would seem most reasonably regarded as reprocessed from one cycle to another. Reprocessed relative to what? Relative, for example, to the Planck system of units,

Reprocessing of physical constants

$$L^* = (\hbar G/c^3)^{1/2} = 1.6 \times 10^{-33} \text{ cm},$$

$$T^* = (\hbar G/c^5)^{1/2} = 5.4 \times 10^{-44} \text{ sec},$$

$$M^* = (\hbar c/G)^{1/2} = 2.2 \times 10^{-5} \text{ g},$$

the only system of units, Planck (1899) pointed out, free, like black-body radiation itself, of all complications of solid-state physics, molecular binding, atomic constitution, and elementary particle structure, and drawing for its background only on the simplest and most universal principles of physics, the laws of gravitation and black-body radiation. Relative to the Planck units, every constant in every other part of physics is expressed as a pure number.

No pure numbers in physics are more impressive than $\hbar c/e^2 = 137.0360$ and the so-called "big numbers" [Eddington (1931, 1936, 1946); Dirac (1937, 1938); Jordan (1955, 1959); Dicke (1959b, 1961, 1964b); Hayakawa (1965a,b); Carter (1968b)]:

$$\sim 10^{80} \text{ particles in the universe,}^*$$

$$\sim 10^{40} \sim \frac{10^{28} \text{ cm}}{10^{-12} \text{ cm}} \sim \frac{\left(\begin{array}{c}\text{radius of universe at}\\ \text{maximum expansion}\end{array}\right)^*}{\left(\begin{array}{c}\text{"size" of an elementary}\\ \text{particle}\end{array}\right)},$$

$$\sim 10^{40} \sim \frac{e^2}{GmM} \sim \frac{\text{(electric forces)}}{\text{(gravitational forces)}},$$

$$\sim 10^{20} \sim \frac{e^2/mc^2}{(\hbar G/c^3)^{1/2}} \sim \frac{\left(\begin{array}{c}\text{"size" of an elementary}\\ \text{particle}\end{array}\right)}{\text{(Planck length)}},$$

$$\sim 10^{10} \sim \frac{\left(\begin{array}{c}\text{number of photons}\\ \text{in universe}\end{array}\right)}{\left(\begin{array}{c}\text{number of baryons}\\ \text{in universe}\end{array}\right)}.$$

Some understanding of the relationships between these numbers has been won [Carter (1968b)]. Never has any explanation appeared for their enormous magnitude, nor will there ever, if the view is correct that reprocessing the universe reprocesses also the physical constants. These constants on that view are not part of the laws of physics. They are part of the initial-value data. Such numbers are freshly given for each fresh cycle of expansion of the universe. To look for a physical explanation for the "big numbers" would thus seem to be looking for the right answer to the wrong question.

In the week between one storm and the next, most features of the weather are ever-changing, but some special patterns of the wind last the week. If the term "frozen features of the meteorology" is appropriate for them, much more so would it seem appropriate for the big numbers, the physical constants and the spectrum of particle masses in the cycle between one reprocessing of the universe and another.

Values of physical constants as related to the possibilities for life

A per cent or so change one way in one of the "constants," $\hbar c/e^2$, will cause all stars to be red stars; and a comparable change the other way will make all stars be blue stars, according to Carter (1968b). In neither case will any star like the sun be possible. He raises the question whether life could have developed if the determinants of the physical constants had differed substantially from those that characterize this cycle of the universe.

Dicke (1961) has pointed out that the right order of ideas may not be, here is the universe, so what must man be; but here is man, so what must the universe

*Values based on the "typical cosmological model" of Box 27.4; subject to much uncertainty, in the present state of astrophysical distance determinations, not least because the latitude in these numbers is even enough to be compatible with an open universe.

be? In other words: (1) What good is a universe without awareness of that universe? But: (2) Awareness demands life. (3) Life demands the presence of elements heavier than hydrogen. (4) The production of heavy elements demands thermonuclear combustion. (5) Thermonuclear combustion normally requires several 10^9 years of cooking time in a star. (6) Several 10^9 years of time will not and cannot be available in a closed universe, according to general relativity, unless the radius-at-maximum-expansion of that universe is several 10^9 light years or more. So why on this view is the universe as big as it is? Because only so can man be here!

In brief, the considerations of Carter and Dicke would seem to raise the idea of the "biological selection of physical constants." However, to "select" is impossible unless there are options to select between. Exactly such options would seem for the first time to be held out by the only over-all picture of the gravitational collapse of the universe that one sees how to put forward today, the *pregeometry black-box model of the reprocessing of the universe.*

Proceeding with all caution into uncharted territory, one must nevertheless be aware that the conclusions one is reaching and the questions one is asking at a given stage of the analysis may be only stepping stones on the way to still more penetrating questions and an even more remarkable picture. To speak of "reprocessing and selection" may only be a halfway point on the road toward thinking of the universe as Leibniz did, as a world of relationships, not a world of machinery. Far from being brought into its present condition by "reprocessing" from earlier cycles, may the universe in some strange sense be "brought into being" by the participation of those who participate? On this view the concept of "cycles" would even seem to be altogether wrong. Instead the vital act is the act of participation. "Participator" is the incontrovertible new concept given by quantum mechanics; it strikes down the term "observer" of classical theory, the man who stands safely behind the thick glass wall and watches what goes on without taking part. It can't be done, quantum mechanics says. Even with the lowly electron one must participate before one can give any meaning whatsoever to its position or its momentum. Is this firmly established result the tiny tip of a giant iceberg? Does the universe also derive its meaning from "participation"? Are we destined to return to the great concept of Leibniz, of "preestablished harmony" ("Leibniz logic loop"), before we can make the next great advance?

Rich prospects stand open for investigation in gravitation physics, from neutron stars to cosmology and from post-Newtonian celestial mechanics to gravitational waves. Einstein's geometrodynamics exposes itself to destruction on a dozen fronts and by a thousand predictions. No predictions subject to early test are more entrancing than those on the formation and properties of a black hole, "laboratory model" for some of what is predicted for the universe itself. No field is more pregnant with the future than gravitational collapse. No more revolutionary views of man and the universe has one ever been driven to consider seriously than those that come out of pondering the paradox of collapse, the greatest crisis of physics of all time.

Black hole as "laboratory" model for collapse of universe

All of these endeavors are based on the belief that existence should have a completely harmonious structure. Today we have less ground than ever before for allowing ourselves to be forced away from this wonderful belief.

EINSTEIN (1934)

What is now proved was once only imagin'd.

WILLIAM BLAKE

We will first understand
How simple the universe is
When we realize
How strange it is.

ANON.

To some one who could grasp the
universe from a unified standpoint,
the entire creation would appear
as a unique truth and necessity.

J. D'ALEMBERT

Yo ho, it's hot . . . the sun is not
A place where we could live
But here on earth there'd be no life
Without the light it gives

H. ZARET

Probable-Possible, my black hen,
She lays eggs in the Relative When.
She doesn't lay eggs in the Positive Now
Because she's unable to postulate How.

F. WINSOR

(Sur l'air de "J'ai rendez-vous avec vous,"
chanson de Georges Brasseus)

*Le Rayonnement dipolaire
On sait qu'il n'est pas pour nous
C'est pour Maxwell, oui mais Maxwell on s'en fout
Tout est r'latif après tout*

*Un argument qu'on révère
Celui de Synge pour dire le tout
Nous promet le quadrupolaire
Tout est r'latif après tout*

*Les sources quasi stellaires
Disparaissent comme dans un trou
Dans le Schwarzschild, oui mais Schwarzschild on s'en fout
Tout est r'latif après tout*

*Aux solutions singulières
On préfère et de beaucoup
Une métrique partout régulière
Tout est r'latif après tout*

*Les physiciens nucléaires
Comme ils nous aiment pas beaucoup
Y gardent tout l'fric, oui mais le fric on s'en fout
Tout est r'latif après tout*

*Les expériences de Weber
Le gyroscope, ça coûtent des sous
Celles de pensées sont moins chères
Tout est r'latif après tout*

M. A. TONNELAT

''Omnibus ex nihil ducendis sufficit unum!''
(One suffices to create Everything of nothing!)

GOTTFRIED WILHELM VON LEIBNIZ

(Sur l'air de Auprès de ma blonde)

Dans les jardins d'Asnières
La science a refleuri
Tous les savants du monde
Apportent leurs écrits

 Refrain:

 Auprès de nos ondes
 Qu'il fait bon, fait bon, fait bon
 Auprès de nos ondes
 Qu'il fait bon rêver

Tous les savants du monde
Apportent leurs écrits
Loi gravitationnelle
Sans tenseur d'énergie

Loi gravitationnelle
Sans tenseur d'énergie
De ravissants modèles
Pour la cosmologie

De ravissants modèles
Pour la cosmologie
Pour moi ne m'en faut guère
Car j'en ai un joli

Pour moi ne m'en faut guère
Car j'en ai un joli
Il est dans ma cervelle
Voici mon manuscrit

Le champ laisse des plumes
Aux bosses de l'espace-temps
En prendrons quelques unes
Pour décrire le mouvement

C. CATTANEO, J. GÉHÉNIAU
M. MAVRIDES, and M. A. TONNELAT

Reprinted with the kind permission of the authors.

When Arthur Evans began this excavation
neither he nor anyone knew that he
would uncover an unknown world.

PAT (Mrs. Hypatia Vourloumis
at Knossos (1971))

And as imagination bodies forth
The form of things unknown, the poet's pen
Turns them to shapes, and gives to airy nothing
A local habitation and a name.

SHAKESPEARE

Appreciation and farewell to our patient reader.

BIBLIOGRAPHY AND
INDEX OF NAMES

This bibliography, like that of Synge (1960b), merely provides points of entry into the literature, which is far too extensive for a comprehensive listing. For more extensive bibliographies of certain segments of the literature, see, for example, Lecat (1924), Boni, Russ, and Laurence (1960), Chick (1960, 1964), and Combridge (1965). Citations are sometimes not to the earliest publication, but to a later and more accessible source. For ease of reference, conference and summer-school proceedings are cited both under editors' names and under the more familiar place names (Brandeis, Les Houches, Varenna, etc.). Persons mentioned in the text without explicit bibliographical reference are also indexed here. Most doctoral dissertations in the United States are available from University Microfilms, Inc., Ann Arbor, Mich. 48106. Appreciation is expressed to Gregory Cherlin for preparing the initial version of this bibliography, to Nigel Coote for numerous subsequent amendments, and to colleagues without whose help some of the most elusive, and most important, of these references would have escaped capture.

Adler, R., M. Bazin, and M. Schiffer, 1965, *Introduction to General Relativity*, McGraw-Hill, New York. [endpapers]

Aharonov, Y., and L. Susskind, 1967, "Observability of the sign change of spinors under 2π rotations," *Phys. Rev. 158*, 1237–1238. [§41.5]

Ahrens, L. H., F. Press, and S. K. Runcorn, eds., 1964, *Physics and Chemistry of the Earth*, Pergamon Press, London. *Cited under* Bolt 1964.

al Bīrūnī, Abū Raihān, *cited by name in* Box 27.7.

Aldrin, E. E., Jr., *cited by name in* Fig. 40.4.

Alfvén, H., 1971, "Plasma physics applied to cosmology," *Phys. Today 24* (Feb. 1971), 28–33. [§§27.10, 27.11, 28.4]

Alfvén, H., and O. Klein, 1962, "Matter antimatter annihilation and cosmology," *Ark. Fys. 23*, 187–194. [§§27.10, 27.11, 28.4, Box 27.7]

al Khazini, 1115, *Book of the Balance of Wisdom*, partially edited and translated by N. Khanikoff in *J. Am. Oriental Soc. 6* (1860), 25–38. [§1.7]

Allen, C. W., 1963, *Astrophysical Quantities*, University of London, Athlone Press, London. [§36.6, Box 25.1, endpapers]

Allendoerfer, C., and A. Weil, 1943, "The Gauss-Bonnet theorem for Riemannian polyhedra," *Trans. Am. Math. Soc. 53*, 101–129. [Box 15.3]

Alley, C. O., *cited by name in* Fig. 40.4. *See also* Bender *et al.* 1971.

Alpher, R. A., *cited by name in* Box 27.7.

Alpher, R. A., and R. C. Herman 1948a, "Evolution of the universe," *Nature 162*, 774–775. [§28.1]

Alpher, R. A., and R. C. Herman, 1948b, "On the relative abundance of the elements," *Phys. Rev. 74*, 1737–1742. [§28.1]

Alpher, R. A., and R. C. Herman, 1950, "Theory of the origin and relative-abundance distribution of the elements," *Rev. Mod. Phys. 22*, 153–213. [§28.1]

American Ephemeris and Nautical Almanac for 1964, 1962, U.S. Government Printing Office, Washington, D.C. [Fig. 25.1]

Ames, W. L., and K. S. Thorne, 1968, "The optical appearance of a star that is collapsing through its gravitational radius," *Astrophys. J. 151*, 659–670. [§32.3]

Anderson, C. D., 1933, "The positive electron," *Phys. Rev. 43*, 491–494. [§44.4]

Anderson, D. W., E. H. Brown, Jr., and F. P. Peterson, 1966a, "Spin cobordism," *Bull. Am. Math. Soc. 72,* 256–260. [Box 44.3]

Anderson, D. W., E. H. Brown, Jr., and F. P. Peterson, 1966b, "SU-cobordism, KO-characteristic numbers, and the Kervaire invariant," *Ann. Math. 83,* 54. [Box 44.3]

Anderson, J. D., 1973, "Lectures on physical and technical problems posed by precision radio tracking," in Bertotti 1973. [§40.1]

Anderson, J. D., P. B. Esposito, W. L. Martin, and D. O. Muhleman, 1971, "Measurements of the general relativistic time delay with data from Mariners 6 and 7," in Davies 1971, pp. 111–135. [Box 40.2]

Anderson, J. L., 1967, *Principles of Relativity Physics,* Academic Press, New York. [§17.6, endpapers]

Anderson, J. L., 1970, "Relativistic Boltzmann theory and the Grad method of moments," in Carmeli, Fickler, and Witten 1970, pp. 109–124. [Ex. 22.7]

Anonymous, source of quotation at end of Chapter 44: found on beach at Rincon, Puerto Rico, March 21, 1973.

Aristotle, *ca.* 330 B.C., *Physics,* ed. by W. D. Ross, Clarendon Press, Oxford (1950). [Box 1.2] *See also* Guthrie 1939, Warrington 1956.

Armstrong, N. A., *cited by name in* Fig. 40.4.

Arnett, W. D., 1966, "Gravitational collapse and weak interactions," *Can. J. Phys. 44,* 2553–2594. [§32.7]

Arnett, W. D., 1967, "Mass dependence in gravitational collapse of stellar cores," *Can. J. Phys. 45,* 1621–1641. [§32.7]

Arnold, V. I., and A. Avez. 1968, *Ergodic Problems of Classical Mechanics,* W. A. Benjamin, New York. [§8.2]

Arnowitt, R., *cited by name in* §§21.1, 35.8, Box 17.2.

Arnowitt, R., S. Deser, and C. W. Misner, 1962, "The dynamics of general relativity," in Witten 1962, pp. 227–265. [§§21.1, 21.4, 21.5, 21.6, 21.7, 21.8, 35.4]

Arp, H., 1971, "Observational paradoxes in extragalactic astronomy," *Science 174,* 1189–1200. [Box 28.1]

Ash, M. E. *See* Shapiro, Ash, *et. al.* 1971, Shapiro, Smith, *et. al.* 1971, Shapiro *et. al.* 1972.

Auslander, L., and R. E. MacKenzie, 1963, *Introduction to Differentiable Manifolds,* McGraw-Hill, New York. [Box 8.1]

Auslander, L., and L. Markus, 1959, *Flat Lorentz 3-Manifolds,* American Mathematical Society, Providence, R.I. [Box 27.2]

Avez, A., 1960, "Propriétés globales des espace-temps périodiques clos," *Acad. des Sci., Paris, Comptes Rend. 250,* 3583–3587. [§44.1]

Avez, A., 1964, "Le ds^2 de Schwarzschild parmi les ds^2 stationnaires," *Ann. Inst. Henri Poincaré A1,* 291–300. [§23.1]

Avez, A. *See also* Arnold and Avez 1968.

Avicenna, *cited by name in* Box 27.7.

Baade, W., 1942, "The Crab Nebula," *Astrophys. J. 96,* 188–198. [Box 24.1]

Baade, W., 1952, Report to Commission 28 of I.A.U., Rome, September 1952. [Box 27.1]

Baade, W., 1956. "The period-luminosity relation of the cepheids," *Pub. Astron. Soc. Pacific, 28,* 5–16. [Box 27.1]

Baade, W., *cited by name in* Box 27.7.

Baade, W., and F. Zwicky, 1934a, "Cosmic rays from supernovae," *Proc. Nat. Acad. Sci. U.S. 20,* 259–263. [Box 24.1]

Baade, W., and F. Zwicky, 1934b, "On supernovae," *Proc. Nat. Acad. Sci. U.S. 20,* 254–259. [Box 24.1]

Baade, W., and F. Zwicky, 1934c. "Supernovae and cosmic rays," *Phys. Rev. 45,* 138. Abstract of paper presented at Stanford meeting of Amer. Phys. Soc., Dec. 15–16, 1933. [Box 24.1]

Bacon, R., 1268, *Opus Majus,* J. H. Bridges, ed., Clarendon Press, Oxford, Eng. Vols. 1 and 2, 1897; vol. 3, 1900. [Box 27.7]

Bahcall, J. N., W. L. W. Sargent, and M. Schmidt, 1967, "An analysis of the absorption spectrum of 3C191," *Astrophys. J. 149,* 11–15. [§38.6]

Bahcall, J. N., and M. Schmidt, 1967, "Does the fine-structure constant vary with cosmic time?" *Phys. Rev. Lett. 19,* 1294–1295. [§38.6]

Baierlein, R. F., 1967, "Testing general relativity with laser ranging to the moon," *Phys. Rev. 162,* 1275–1288. [§40.6]

Baierlein, R. F., *cited by name in* §21.1. *See also* Krogh and Baierlein 1968.

Baierlein, R. F., D. H. Sharp, and J. A. Wheeler, 1962, "Three-dimensional geometry as carrier of information about time," *Phys. Rev. 126,* 1864–1865. [§§21.9, 43.2]

Bailey, J. *See* Farley *et al.* 1966.

Bailey, J., W. Bartl, G. von Bochmann, R. C. A. Brown, F. J. M. Farley, H. Jostlein, E. Picasso, and R. W. Williams, 1968, "Precision measurement of the anomalous magnetic moment of the muon," *Phys. Lett. 28B,* 287–289. [§44.2]

Bardeen, J. M., 1968, "General relativistic collapse of charged dust (abstract)," *Bull. Am. Phys. Soc. 13,* 41. [§34.6, Fig. 34.4]

Bardeen, J. M., 1970a, "Kerr metric black holes," *Nature 226,* 64–65. [§33.8, Boxes 33.3, 33.5]

Bardeen, J. M., 1970b, "A variational principle for rotating stars in general relativity," *Astrophys. J. 162,* 71–95. [Ex. 33.3]

Bardeen, J. M. *See also* Press and Bardeen 1971.

Bardeen, J. M., B. Carter, and S. W. Hawking, 1973, "The four laws of black-hole mechanics," *Commun. Math. Phys.,* in press. [Box 33.4]

Bardeen, J. M., and W. H. Press, 1972, "Radiation fields in the Schwarzschild background," *J. Math. Phys. 14,* 7–19. [§35.14, Ex. 32.11]

Bardeen, J. M., W. H. Press, and S. A. Teukolsky, 1972, "Rotating black holes: locally nonrotating frames, energy extraction, and scalar synchrotron radiation," *Astrophys. J. 178,* 347–369. [§33.7, Ex. 33.3, 33.14]

Bardeen, J. M., K. S. Thorne, and D. W. Meltzer, 1966, "A catalogue of methods for studying the normal modes of radial pulsation of general-relativistic stellar models," *Astrophys. J. 145,* 505–513. [Box 26.1]

Bardeen, J. M., and R. V. Wagoner, 1971, "Relativistic disks, I: Uniform rotation," *Astrophys. J. 167,* 359–423. [Box 24.1]

Barger, R. L. *See* Evenson *et. al.* 1972.

Bargmann, V., and E. P. Wigner, 1948, "Group theoretical discussion of relativistic wave equations," *Proc. Nat. Acad. Sci. U.S. 34,* 211–222. [Box 32.2]

Barnes, J. A., 1971, "A non-mathematical discussion of some basic concepts of precise time measurement," in *Tracor on Frequency 2,* No. 2 (Tracor Industrial Instruments, Austin, Texas). [Box 1.4]

Barschall, H. H. and W. Haeberli, 1971. "The Madison convention," pp. xxv–xxix in *Polarization Phenomena in Nuclear Reactions,* Univ. of Wisconsin Press, Madison. [§37.7]

Bartl, W. *See* Bailey *et al.* 1968.

Barton, D., S. R. Bourne, and J. P. Fitch, 1970, "An algebra system," *Comput. J. 13,* 32–39. [Box 14.3]

Barton, D., and J. P. Fitch, 1971, "General relativity and the application of algebraic manipulative systems," *Comm. Assoc. Comput. Machinery 14,* 542–547. [Box 14.3]

Battelle Rencontres, Seattle 1967. *See* DeWitt and Wheeler 1968.

Baym, G., 1969, *Lectures on Quantum Mechanics,* W. A. Benjamin, New York. [Box 22.4]

Baym, G., H. A. Bethe, and C. J. Pethick, 1971, "Neutron star matter," *Nuclear Phys A. 175,* 225–271. [§24.2]

Bazin, M. *See* Adler, Bazin, and Schiffer 1965.

Beall, E. F., 1970, "Measuring the gravitational interaction of elementary particles," *Phys. Rev. D 1,* 961–978. [§1.3, Box 1.2]

Beams, J. W., 1971, "Finding a better value for G," *Phys. Today 24* (May 1971), 35–40. [§40.8]

Beams, J. W. *See also* Rose *et al.* 1969.

Beckedorff, D. L., and C. W. Misner, 1962, unpublished; see D. L. Beckedorff's A. B. Senior Thesis, Princeton University. [§32.4]

Bekenstein, J., 1972a, "Nonexistence of baryon number for static black holes," *Phys. Rev. D 5,* 1239–1246. [Boxes 25.2, 32.2, 33.1]

Bekenstein, J., 1972b, "Nonexistence of baryon number for static black holes, II," *Phys. Rev. D 5,* 2403–2412. [Boxes 25.2, 32.2, 33.1]

Bel, L., 1958, "Sur la radiation gravitationelle," *Acad. Sci., Paris, Comptes Rend. 247,* 1094–1096. [Ex. 15.2]

Bel, L., 1959, "Introduction d'un tenseur du quatrième ordre," *Acad. Sci., Paris, Comptes Rend. 248,* 1297–1300. [Ex. 15.2]

Bel, L., 1962, "La radiation gravitationelle," in CNRS 1962, pp. 119–126. [Ex. 15.2]

Belasco, E. P., and H. C. Ohanian, 1969, "Initial conditions in general relativity: Lapse and shift formulation," *J. Math. Phys. 10,* 1503–1507. [§21.9]

Belinfante, F., 1940, "On the current and the density of the electric charge, the energy, the linear momentum, and the angular momentum of arbitrary fields," *Physica 7,* 449–474. [§21.3]

Belinsky, V. A. *See also* Ryan 1972b.

Belinsky, V. A., and I. M. Khalatnikov, 1969a, "On the nature of the singularities in the general solution of the gravitational equations," *Zh. Eksp. & Teor. Fiz. 56,* 1700–1712. English translation in *Sov. Phys.—JETP 29,* 911–917. [§34.6]

Belinsky, V. A., and I. M. Khalatnikov, 1969b, "General solution of the gravitational equations with a physical singularity," *Zh. Eksp. & Teor. Fiz. 57,* 2163–2175, English translation in *Sov. Phys.—JETP 30,* 1174–1180 (1970). [§34.6]

Belinsky, V. A., and I. M. Khalatnikov, 1970, "General solution of the gravitational equations with a physical oscillatory singularity," *Zh. Eksp. & Teor. Fiz. 59,* 314–321. English translation in *Sov. Phys.—JETP 32,* 169–172. [§§30.6, 34.6]

Belinsky, V. A., I. M. Khalatnikov, and E. M. Lifshitz, 1970, "Oscillatory approach to a singular point in the relativistic cosmology," *Usp. Fiz. Nauk 102,* 463–500. English translation in *Advances in Physics 19,* 525–573. [§30.6, Box 30.1]

Belinsky, V. A., E. M. Lifshitz, and I. M. Khalatnikov, 1971, "Oscillatory mode of approach to a singularity in homogeneous cosmological models with rotating axes," *Zh. Eksp. & Teor. Fiz. 60,* 1969–1979. English translation in *Sov. Phys.—JETP 33,* 1061–1066 (1971). [§§30.7, 34.6]

Bell, C. C., R. L. Forward, L. R. Miller, T. D. Beard, and T. M. Baran, 1969, "Static gravitational gradient field detection with a rotating torsional gradiometer," in NASL 1969. [Box 16.5]

Bell, C. C., R. L. Forward, and H. P. Williams, 1970, "Simulated terrain mapping with the rotating gravity gradiometer, in W. T. Kattner, ed., *Advances in Dynamic Gravimetry,* Instrument Society of America, Pittsburgh, 115–129. [Box 16.5]

Bell, S. J., *cited by name in* Box 27.7. *See also* Hewish *et al.* 1968.

Beltran-Lopez, V. B. *See* Hughes, Robinson, and Beltran-Lopez 1960.

Bender, P. L., R. H. Dicke, D. T. Wilkinson, C. O. Alley, D. G. Currie, J. E. Faller, J. D. Mulholland, E. C. Silverberg, H. E. Plotkin, W. M. Kaulla, and G. J. F. McDonald, 1971, "The lunar laser-ranging experiment," in Davies 1971. [§40.9]

Bennett, A. S., *cited by name in* Box 27.7.

Bentley, R., *cited by name in* Box 27.7.

Berger, B. K., D. M. Chitre, V. E. Moncrief, and Y. Nutku, 1972, "Hamiltonian formulation of spherically symmetric gravitational fields," *Phys. Rev. D 5,* 2467–2470. [§21.10]

Berger, M., 1966, "Sur les spectres d'une variété Riemannienne," *Acad. Sci Paris, Comptes Rendus 263,* 13–16. [Box 17.2 part 6]

Berger, M., P. Gauduchon, and E. Mazet, 1971, *Le Spectre d'une variété riemannienne,* Springer, Berlin and New York. [Box 17.2 part 6]

Bergmann, P. G., 1942, *Introduction to the Theory of Relativity,* Prentice-Hall, New York. [endpapers]

Bergmann, P. G., 1968, "Comments on the scalar-tensor theory," *Int. J. Theor. Phys. 1,* 25–36. [§38.2]

Bern, Switzerland, 1955, Jubilee of Relativity Theory, conference held July 11–16, 1955; *see* Mercier and Kerváire 1956.

Bernoulli, D., 1733, *Hydrodynamica,* chapter 10, Archives of Academy of Science, Leningrad; summarized in C. C. Gillispie, ed., *Dictionary of Scientific Biography,* Scribners, New York, vol. 2, 1970. [Box 44.5]

Bers, L., 1952, "Singularities of minimal surfaces," in *Proceedings of the International Congress of Mathematicians, Cambridge, Mass., 1950,* American Mathematical Society, Providence, R.I. [§33.2]

Bers, L., *cited by name in* Box 8.5.

Bertotti, B., 1959, "Uniform electromagnetic field in the theory of general relativity," *Phys. Rev. 116,* 1331–1333. [Ex. 32.1]

Bertotti, B., ed., 1973, proceedings of Course 56 of the International School of Physics Enrico Fermi, dealing with experimental tests of relativistic gravitation theories, Academic Press, New York, in press.

Bertotti, B., D. R. Brill, and R. Krotkov, 1962, "Experiments on gravitation," in Witten 1962. [§38.5, Box 40.1]

Bessel, F. W., *cited by name in* Box 27.7.

Bethe, H. A., *cited by name in* Box 27.7. *See also* Baym, Bethe, and Pethick 1971.

Birkhoff, G. D., 1923, *Relativity and Modern Physics,* Harvard University Press, Cambridge, Mass. [§32.2]

Birkhoff, G. D., 1943, "Matter, electricity and gravitation in flat spacetime," *Proc. Nat. Acad. Sci. U.S. 29,* 231–239. [§39.1]

Bishop, R. L., and S. I. Goldberg, 1968, *Tensor Analysis on Manifolds,* Macmillan, New York. [Box 8.1, Ex. 21.8]

Bisnovatyi-Kogan, G. S., and K. S. Thorne, 1970, "Relativistic gas spheres and clusters of point masses with arbitrarily large central redshifts: Can they be stable?" *Astrophys. J. 160,* 875–885. [Ex. 25.28]

Bisnovatyi-Kogan, G. S., and Ya. B. Zel'dovich, 1969, "Models of clusters of point masses with great central redshift," *Astrofizika 5,* 223–234. English translation in *Astrophys. 5,* 105–109. [Ex. 25.28]

Blackman, R. B., and J. W. Tukey, 1959, *The Measurement of Power Spectra, from the Point of View of Communications Engineering,* Dover, New York. [§37.8]

Blake, W., *quoted at end of* Chapter 44

Blaton, J., 1950, "On a geometrical interpretation of energy and momentum conservation in atomic collisions and disintegration processes," *Kgl. Danske Videnskab. Sels., Mat.-fys. Medd. 24,* no. 20. [§2.10]

Blatt, J. M., and V. F. Weisskopf, 1952, *Theoretical Nuclear Physics,* Wiley, New York. [§37.7]

Blumenthal, L. M., 1953, *Theory and Applications of Distance Geometry,* Clarendon Press, Oxford, England. [Box 13.1]

Bochmann, G. von. *See* Bailey *et al.* 1968.

Bogoliubov, N. N., and D. V. Shirkov, 1959, *Introduction to the Theory of Quantized Fields,* Interscience (now a division of Wiley), New York. [§43.3]

Bogorodskii, A. F., 1962, *Uravneniya Polya Einsteina i ikh Primeneniye v Astronomii (Einstein's Field Equations and Their Application in Astronomy),* Kiev Univ. Press, USSR. [Ex. 25.22]

Bohr, A., and B. R. Mottleson, 1969, *Nuclear Structure, Vol. I: Single-Particle Motion,* W. A. Benjamin, New York. [§23.4]

Bohr, N., 1918, "On the quantum theory of line spectra," *Mém. Acad. R. Sci. & Lett. Danemark 4,* no. 1, parts 1–2; part 1 reprinted in Van der Waerden 1967, pp. 95–137. [Box 17.1]

Bohr, N., *cited by name in* Preface, §§44.1, 44.5, Box 17.1

Bohr, N., and L. Rosenfeld, 1933, "Zur Frage der Messbarkeit der elektromagnetischen Feldgrössen," *Kgl. Danske Videnshab. Sels. Mat.-fys. Medd. 12,* no. 8. [§43.2]

Boissonade, J. F. *See* Westerman and Boissonade 1878.

Bolt, B. A., 1964, "Recent information of the earth's interior from studies of mantle waves and eigenvibrations," in Ahrens, Press, and Runcorn 1964, pp. 55–119. [Ex. 37.14]

Bolton, J. G., *cited by name in* Box 27.7.

Bondi, H., 1947, "Spherically Symmetrical Models in General Relativity," *Mon. Not. R. Astron. Soc. 107,* 410–425. [§30.6]

Bondi, H., 1957, "Plane gravitational waves in general relativity," *Nature 179,* 1072–1073. [Ex. 18.5, Fig. 18.1]

Bondi, H., 1961, *Cosmology,* 2d ed., Cambridge Univ. Press, Cambridge, Eng. [Box 27.1]

Bondi, H., 1965, "Some special solutions of the Einstein equations," in Trautman, Pirani, and Bondi 1965, pp. 375–459, esp. p. 431. [Ex. 18.5]

Bondi, H., *cited by name in* §21.10, Boxes 27.7, 34.2. *See also* Trautman, Pirani, and Bondi 1965.

Bondi, H., and T. Gold, 1948, "The steady-state theory of the expanding universe," *Mon. Not. R. Astron. Soc. 108,* 252–270. [§28.4]

Bondi, H., and W. H. McCrea, 1960, "Energy transfer by gravitation in Newtonian theory," *Proc. Cambridge Phil. Soc. 56,* 410–413. [Ex. 18.5]

Bondi, H., F. A. E. Pirani, and I. Robinson, 1959, "Gravitational waves in general relativity, III: Exact plane waves," *Proc. R. Soc. London A 251,* 519–533. [§35.9]

Boni, N., M. Russ, and D. H. Laurence, 1960, *A Bibliographical Check List and Index to the Published Writings of Albert Einstein,* Pageant Books, Paterson, N.J., cited in preface of present bibliography; based on earlier bibliography of M. C. Shields in Schilpp 1949.

Born, M., 1971, *The Born-Einstein letters: Correspondence Between Albert Einstein and Max and Hedwig Born from 1916 to 1955 with Commentaries by Max Born,* translated by Irene Born, Walker, New York. [Box 1.11]

Born, M., *cited by name in* Box 17.2

Boughn, S. P., D. M. Fram, and R. B. Partridge, 1971, "Isotropy of the microwave background at 8-millimeter wavelength," *Astrophys. J. 165,* 439–444. [§§27.1, 27.2, 30.8]

Bourbaki, N., *cited by name in* Box 8.2; pseudonym for an organization of French mathematicians who, since 1939, have nearly completed the first ten *Books* of a projected larger work: *Éléments de Mathématique,* Hermann, Paris.

Bourne, S. R. *See* Barton, Bourne, and Fitch 1970.

Boyer, R. H., and R. W. Lindquist, 1967, "Maximal analytic extension of the Kerr metric," *J. Math. Phys. 8,* 265–281. [§33.2, Box 33.2, Ex. 33.8]

Braginsky, V. B., 1965, "Gravitational radiation and the prospect of its experimental discovery," *Usp. Fiz. Nauk 86,* 433–446. English translation in *Sov. Phys.—Uspekhi 8,* 513–521. [Box 36.3]

Braginsky, V. B., 1970, *Fizicheskie Eksperimenti s Probnimi Telami (Physics Experiments With Test Bodies),* Nauka, Moscow, Russia. English translation published as NASA-TT F-762, available from National Technical Information Service, Springfield, Va. [§37.8]

Braginsky, V. B., A. B. Manukin, and M. Yu. Tikhonov, 1970, "Investigation of dissipative ponderomotive effects of electromagnetic radiation," *Zh. Eksp. & Teor. Fiz. 58,* 1549–1552, English translation in *Sov. Phys.—JETP 31,* 829–830. [§37.8]

Braginsky, V. B., and M. B. Menskii, 1971, "High-frequency detection of gravitational waves," *Zh. Eksp. & Teor. Fiz. Pis'ma 13,* 585. English translation in *Sov. Phys.—JETP Lett. 13,* 417–419. [§37.9, Box 37.6]

Braginsky, V. B., and V. S. Nazarenko, 1971, "On the detection of gravitational radiation from some astronomical sources with the heterodyne detector," in Davies 1971, pp. 45–46. [§37.3]

Braginsky, V. B., and V. I. Panov, 1971, "Verification of the equivalence of inertial and gravitational mass," *Zh. Eksp. & Teor. Fiz. 61*, 873–879. English translation in *Sov. Phys.—JETP 34*, 464–466. [§§38.3, 40.9, Box 1.2]

Braginsky, V. B., Ya. B. Zel'dovich, and V. N. Rudenko, 1969, "Reception of gravitational radiation of extraterrestrial origin," *Zh. Eksp. & Teor. Fiz. Pis'ma 10*, 437–441. English translation in *Sov. Phys.—JETP Lett. 10*, 280–283. [§37.3, Fig. 37.2]

Brandeis Summer Institute in Theoretical Physics: 1964, *see* Trautman, Pirani, and Bondi 1965; 1968, *see* Chrétian, Deser, and Goldstein 1969.

Brans, C., *cited by name in* §§40.1, 40.6, 40.9.

Brans, C., and R. H. Dicke, 1961, "Mach's principle and a relativistic theory of gravitation," *Phys. Rev. 124*, 925–935. Reprinted in Dicke 1964b. [§§17.6, 28.4, 38.2, 39.2, 40.8, Boxes 38.3, 39.1]

Brasseur, Georges, *quoted in* Fig. 44.5.

Brault, J. W., 1962, *The Gravitational Redshift in the Solar Spectrum,* doctoral dissertation, Princeton Univ. Abstract, "Gravitational Redshift of Solar Lines," in *Bull. Amer. Phys. Soc. 8*, 28 (1963). [§38.5, Fig. 38.2]

Brehme, R. W. *See* DeWitt and Brehme 1960.

Bressanone, Italy: 1970 summer school sponsored by Centro Interzionale Matematico Estivo. *See* Cattaneo 1971.

Brill, D. R., 1959, "On the positive definite mass of the Bondi-Weber-Wheeler time-symmetric gravitational waves," *Ann. Phys. (USA) 7*, 466–483. [§21.10]

Brill, D. R., 1961, *General Relativity: Selected Topics of Current Interest,* report SUI61-4, Physics Department, State University of Iowa, Iowa City, Iowa (multilithed); summarized in DeWitt and DeWitt 1964b, pp. 431–432; published as Brill 1964. [§21.10]

Brill, D. R., 1964, "General relativity: selected topics of current interest," *Suppl. Nuovo Cimento 2*, 1–56. See Brill 1961.

Brill, D. R. *See also* Bertotti, Brill, and Krotkov 1962, Cohen and Brill 1967 *and* 1968, *and* Graves and Brill 1960.

Brill, D. R., P. L. Chrzanowski, C. M. Pereira, E. D. Fackerell, and J. R. Ipser, 1972, "Solution of the scalar wave equation in a Kerr background by separation of variables," *Phys. Rev. D 5*, 1913–1915. [Ex. 33.14]

Brill, D. R., and J. M. Cohen, 1966, "Rotating masses and their effect on inertial frames," *Phys. Rev. 143*, 1011–1015. [§§21.12, 40.7, Ex. 26.1]

Brill, D. R., and J. B. Hartle, 1964, "Method of self-consistent field in general relativity and its application to the gravitational geon," *Phys. Rev. B 135*, 271–278. [§§21.12, 35.13]

Brill, D. R., and J. A. Wheeler, 1957, "Interaction of neutrinos and gravitational fields," *Rev. Mod. Phys. 29*, 465–479. [§§27.10, 41.11]

Brillouin, L., 1970, *Relativity Reexamined,* Academic Press, New York. [Ex. 25.21]

Brodetsky, S. *See* Dobson and Brodetsky 1947.

Brodsky, S. J., and S. D. Drell, 1970, "The present status of quantum electrodynamics," in *Ann. Rev. Nuclear Sci. 20*, 147–194. [§44.2]

Bromberg, V. A., 1958, "Relativistic corrections in the theory of the motion of the moon," *Bull. Inst. Theor. Astron., Akad. Nauk SSSR 6*, 733–756. [§40.6]

Bronshtein, M. P., *cited by name in* §29.2.

Brown, E. H., Jr. *See* Anderson, Brown, and Peterson 1966a *and* 1966b.

Brown, R. C. A. *See* Bailey *et al.* 1968 *and* Farley *et al.* 1966.

Bruhat, Y., 1962, "The Cauchy problem," in Witten 1962. [§§21.11, 21.13]

Bruhat, Y. *See also* Choquet-Bruhat, Y., and Fourès-Bruhat, Y.

Bruno, G. *See* Singer 1950.

Bunsen, R. W., *cited by name in* Box 38.2.

Burbidge, E. M., *cited by name in* Box 27.7. *See also* Burbidge and Burbidge 1969a *and* 1969b *and* Hoyle *et al.* 1964.

Burbidge, G. R., *cited by name in* Box 27.7. *See also* Hoyle *et al.* 1964.

Burbidge, G. R., and E. M. Burbidge, 1969a, "Red-shifts of quasi-stellar objects and related extragalactic systems," *Nature 222*, 735–741. [Ex. 27.14]

Burbidge, G. R., and E. M. Burbidge, 1969b, "Quasi-stellar objects—a progress report," *Nature 224*, 21–24. [Ex. 27.14]

Burke, W. L., 1970, "Runaway solutions: remarks on the asymptotic theory of radiation damping," *Phys. Rev. A 2*, 1501–1505. [§20.6]

Burke, W. L., 1971, "Gravitational radiation damping of slowly moving systems calculated using matched asymptotic expansions," *J. Math. Phys. 12*, 402–418. [§36.8]

Burke, W. L., *cited by name in* §22.5.

Burnett, D. S. *See* Wasserburg and Burnett 1968.

Cairns, S. S., ed., 1965, *Differential and Combinatorial Topology,* Princeton Univ. Press, Princeton, N.J.

Cairns, S. S. *See also* Morse and Cairns 1969.

Callaway, J., 1953, "The equations of motion in Einstein's new unified field theory," *Phys. Rev. 92*, 1567–1570. [§20.6]

Callippus, *cited by name in* Box 27.7.

Caltech-JPL conference on experimental tests of gravitational theories, November 1970. *See* Davies 1971.

Cameron, A. G. W., 1970, "Neutron stars," *Ann. Rev. Astron. & Astrophys. 8*, 179–208. [§24.2]

Cameron, A. G. W., *See also* Langer and Cameron 1969, *and* Truran and Cameron 1971.

Cameron, A. G. W., and J. Truran, 1971, *work cited by* Peebles 1971. [Box 33.3]

Campbell, D. B. *See* Shapiro, Ash, *et al.* 1971 *and* Shapiro *et al.* 1972.

Campbell, L., and Garnett, W., 1882, *Life of James Clerk Maxwell,* Macmillan, London. [Box 4.6]

Campbell, W. W., and R. J. Trumpler, 1928, "Observations made with a pair of five-foot cameras on the light-deflections in the sun's gravitational field at the total solar eclipse of September 21, 1922." *Lick Obs. Bull. 13*, 130–160. [Box 40.1]

Carmeli, M., S. I. Fickler, and L. Witten, 1970, *Relativity,* Plenum, New York. [§40.7]

Carroll, L. *See* Dodgson, C. L.

Cartan, É., 1901, "Sur quelques quadratures dont l'élément différentiel contient des fonctions arbitraires," *Bull. Soc. Math. France 29*, 118–130. [Box 8.2]

Cartan, É., 1922a, "Sur les équations de la gravitation de Einstein," *J. Math. Pures Appl. 1,* 141–203. [§21.1]

Cartan, É., 1922b, "Sur une généralisation de la notion de courbure de Riemann et les espaces à torsion," *Acad. Sci., Paris, Comptes Rend. 174,* 593–595. [§§38.2, 39.2]

Cartan, É., 1923, "Sur les variétés à connexion affine et la théorie de la relativité generalisée (premiére partie)," *Ann. École Norm. Sup. 40,* 325–412. [§§8.2, 12.1, 12.2, 38.2, 39.2]

Cartan, É., 1924, "Sur les variétés à connexion affine et la théorie de la relativité generalisée (suite)," *Ann. École Norm. Sup. 41,* 1–25. [§§8.2, 8.3, 12.1]

Cartan, É., 1925, *La Géométrie des Espaces de Riemann,* Memorial des Sciences Mathématiques, Fascicule IX, Gauthier-Villars, Paris, France. [§14.5]

Cartan, É., 1928 and 1946, *Leçons sur la Géométrie des Espaces de Riemann,* Gauthier-Villars, Paris, France. [§§14.5, 15.3, Box 8.2, Ex. 21.4, endpapers]

Cartan, É., 1935, *La Méthode du Repère Mobile, la Théorie des Groupes Continus et les Espaces Generalisées,* Hermann, Paris, France. [§6.4]

Cartan, É., 1945, "Les systèmes différentiels extérieurs et leurs applications géométriques," *Exposés de Géométrie 14,* Hermann, Paris, France. [§4.1]

Cartan, É., 1946. *See* Cartan 1928.

Cartan, É., 1966, *Theory of Spinors,* M.I.T. Press, Cambridge, Mass. [§41.11]

Cartan, É., *cited by name in* §§9.1, 13.1, 13.4, 15.1, Boxes 10.1, 12.3, Ex. 12.7, 12.8.

Cartan, H., 1970, *Formes Différentiels,* Hermann, Paris, France. [§4.1]

Carter, B., 1966a, "Complete analytic extension of the symmetry axis of Kerr's solution of Einstein's equations," *Phys. Rev. 141,* 1242–1247. [Box 33.2]

Carter, B., 1966b, "The complete analytic extension of the Reissner-Nordstrom metric in the special case $e^2 = m^2$," *Phys. Lett. 21,* 423–424. [Fig. 34.4, Ex. 31.8, 34.3]

Carter, B., 1968a, "Global structure of the Kerr family of gravitational fields," *Phys. Rev. 174,* 1559–1571. [§33.5, Box 33.2, Ex. 33.7]

Carter, B., 1968b, "Large numbers in astrophysics and cosmology," unpublished preprint, Institute of Theoretical Astronomy, Cambridge, Eng. [§44.6]

Carter, B., 1968c, "Hamilton-Jacobi and Schrodinger separable solutions of Einstein's equations," *Commun. Math. Phys. 10,* 280–310. [§33.5]

Carter, B., 1970, "An axisymmetric black hole has only two degrees of freedom," *Phys. Rev. Lett. 26,* 331–333. [Box 33.1]

Carter, B., 1973, "Properties of the Kerr metric," *in* DeWitt and DeWitt 1973. [Box 33.1]

Carter, B., *cited by name in* Box 34.3. *See also* Bardeen, Carter, and Hawking 1973.

Cattaneo, C., ed., 1971, *Relativistic Fluid Dynamics,* proceedings of a summer school sponsored by Centro Internazionale Matematico Estivo, Bressanone, Italy, June 1970, Edizione Cremonese, Rome. [§22.3]

Cattaneo, C., *quoted at end of* Chap. 44.

Cauchy, A. L., *cited by name in* Box 17.2.

Cavendish, H., 1798, "Experiments to determine the density of the Earth" (read June 21, 1798), *Phil. Trans. R. Soc. London, Part II* (1798), pp. 469–526. [§40.8]

Caves, C. M., *cited by name in* §26.1.

Chadwick, J., 1932, "Possible existence of a neutron," *Nature 129,* 312. [Box 24.1]

Chandrasekhar, S., 1931a, "The density of white dwarf stars," *Phil. Mag. 11,* 592–596. [Box 24.1]

Chandrasekhar, S., 1931b, "The maximum mass of ideal white dwarfs," *Astrophys. J. 74,* 81–82. [Box 24.1]

Chandrasekhar, S., 1964a. "Dynamical instability of gaseous masses approaching the Schwarzschild limit in general relativity," *Phys. Rev. Lett. 12,* 114–116 and 437–438. [§26.1, Boxes 24.1, 26.2]

Chandrasekhar, S., 1964b, "The dynamical instability of gaseous masses approaching the Schwarzschild limit in general relativity," *Astrophys. J. 140,* 417–433. [§26.1, Boxes 24.1, 26.2]

Chandrasekhar, S., 1965a, "The post-Newtonian equations of hydrodynamics in general relativity," *Astrophys. J. 142,* 1488–1512. [§39.3, Box 39.3, Ex. 39.14]

Chandrasekhar, S., 1965b, "The post-Newtonian effects of general relativity on the equilibrium of uniformly rotating bodies, I: The Maclaurin spheroids and the virial theorem," *Astrophys. J. 142,* 1513–1518. [§39.3, Box 39.3]

Chandrasekhar, S., 1965c, "The stability of gaseous masses for radial and non-radial oscillations in the post-Newtonian approximation of general relativity," *Astrophys. J. 142,* 1519–1540. [§39.3, Box 26.2]

Chandrasekhar, S., 1968, *Hydrodynamic and Hydromagnetic Stability,* Clarendon Press, Oxford. See especially §108. [Box 32.3]

Chandrasekhar, S., and G. Contopolous, 1967, "On a post-Galilean transformation appropriate to the post-Newtonian theory of Einstein, Infeld, and Hoffmann," *Proc. R. Soc. London A 298,* 123–141. [§39.3, 39.9]

Chandrasekhar, S., and F. P. Esposito, 1970, "The 2½-post-Newtonian equations of hydrodynamics and radiation reaction in general relativity," *Astrophys. J. 160,* 153–180. [§36.8, Box 39.3]

Chandrasekhar, S., and N. R. Lebovitz, 1968, "The pulsations and the dynamical stability of gaseous masses in uniform rotation," *Astrophys. J., 152,* 267–291. [Ex. 39.7]

Chandrasekhar, S., and Y. Nutku, 1969, "The second post-Newtonian equations of hydrodynamics in general relativity," *Astrophys. J. 158,* 55–80. [Box 39.3]

Chapel Hill 1957, Conference on the Role of Gravitation in Physics, Jan. 18–23, 1957. See DeWitt 1957.

Chase, D. M., 1954, "The equations of motion of charged test particles in general relativity," *Phys. Rev. 95,* 243–246. [§20.6]

Cherlin, G., *cited by name in introduction to this bibliography.*

Chern, S.-S., 1955, "On curvature and characteristic classes of a Riemann manifold," *Hamburg Abh. 20,* 117–126. [Box 15.3]

Chern, S.-S., 1960, *Geometric Structures on Manifolds,* colloquium lectures, Amer. Math. Soc., Providence, R.I. [§2.2]

Chern, S.-S., 1962, "Geometry of a quadratic differential

form," *J. Soc. Indust. & Appl. Math. 10,* 751–755, also in *H. P. Robertson January 27, 1903–August 26, 1961, In Memoriam,* Soc. Indust. and Applied Math., Philadelphia 1963, pp. 13–17. [Box 15.3]

Chern, S.-S., 1966, "The geometry of G-structures," *Bull. Am. Math. Soc. 72,* 167–219. [§2.2]

Chern, S.-S., 1971, "Differential geometry: its past and its future," in *Proceedings of the International Congress of Mathematicians, Nice, September 1970,* Gauthier-Villars, Paris, pp. 41–53. [§2.2]

Chern, S.-S., and C. Chevalley, 1952, "Élie Cartan and his mathematical work," *Bull. Am. Math. Soc. 58,* 217–250. [Box 8.2]

Chevalley, C. *See* Chern and Chevalley 1952.

Chick, H. J., 1960, *Theory of Gravitation 1920–Present (Bibliography),* Report LMS-2328 of Los Alamos Scientific Laboratory, Los Alamos, New Mexico, cited in preface of present bibliography.

Chick, H. J., 1964, *Theory of Gravitation 1959–1963 (Bibliography),* Report LMS-3099 of Los Alamos Scientific Laboratory, Los Alamos, New Mexico, cited in preface of present bibliography.

Chinnapared, R. *See* Newman *et al.* 1965.

Chitre, D. M., 1972a, *Investigations of the Vanishing of a Horizon for Bianchi Type IX (Mixmaster) Universe,* doctoral dissertation, Univ. of Maryland. [§30.8]

Chitre, D. M., 1972b, "High-frequency sound waves to eliminate a horizon in the mixmaster universe," *Phys. Rev. D 6,* 3390–3396. [§30.8]

Chitre, D. M. *See also* Berger *et al.* 1972.

Chiu, H.-Y., and W. F. Hoffman, eds., 1964, *Gravitation and Relativity,* W. A. Benjamin, New York.

Choquard, P., 1955, "Traitement semi-classique des forces générales dans la représentation de Feynman," *Helv. Phys. Acta 28,* 89–157. [Box 21.1, Fig. 13.2]

Choquet-Bruhat, Y., 1968a, *Géométrie différentielle et systèmes extérieurs,* Monographies Universitaires de Mathématiques no. 28, Dunod, Paris, France. [§4.1]

Choquet-Bruhat, Y., 1968b, "Espace-temps Einsteiniens généraux, chocs gravitationels," *Ann. Inst. Henri Poincaré 8,* 327–338. [§21.13]

Choquet-Bruhat, Y., 1969, "Construction de solutions radiatives approchées des équations d'Einstein," *Commun. Math. Phys. 12,* 16–35. [§35.13]

Choquet-Bruhat, Y., 1972, "Problème des contraintes sur une variété compacte," *Acad. Sci. Paris, Comptes Rend. A 274,* 682–684; fuller account in "Global solutions of the equations of constraint in general relativity, on open and closed manifolds," *Gen. Rel. & Grav.* 1973, in press. [§21.1]

Choquet-Bruhat, Y. *See also* Bruhat, Y., *and* Fourès-Bruhat, Y.

Chrétien, M., S. Deser, and J. Goldstein, eds., 1969, *Brandeis Summer Institute 1968, Astrophysics and General Relativity,* Gordon and Breach, New York.

Christodolou, D., 1970, "Reversible and irreversible transformations in black-hole physics," *Phys. Rev. Lett. 25,* 1596–1597. [§33.8, Box 33.4]

Christodolou, D., 1971, *Investigations in Gravitational Collapse and the Physics of Black Holes,* doctoral dissertation, Princeton Univ. [§35.8, Boxes 33.3, 33.4, Fig. 33.3]

Christodolou, D., and R. Ruffini, 1971, "Reversible transformations of a charged black hole," *Phys. Rev. D 4,* 3552 3555. [§33.8, Boxes 33.4, 33.5]

Chrzanowski, P. L., and C. W. Misner, 1973, "Geodesic synchrotron radiation in the Kerr geometry by the method of asymptotically factorized Green's functions," technical report of the Dept. of Physics and Astronomy, University of Maryland. [§33.8]

Chrzanowski, P. L. *See also* Brill *et al.* 1972.

Cincinnati 1969, Relativity Conference in The Midwest, June 2–6, 1969, Cincinnati, Ohio. *See* Carmeli, Fickler, and Witten 1970.

Clark, G. *See* Eddington and Clark 1938.

Clarke, C. J. S., 1970, "On the global isometric embedding of pseudo-Riemannian manifolds," *Proc. Roy. Soc. London A 314,* 417–428. [§23.8]

Clayton, D., 1968, *Principles of Stellar Evolution and Nucleosynthesis,* McGraw-Hill, New York. [§28.1]

Clemence, G. M., 1943, "The motion of Mercury, 1765–1937," *Astron. Papers Am. Ephemeris 11,* 1–221. [Box 40.3]

Clemence, G. M., 1947, "The relativity effect in planetary motions," *Rev. Mod. Phys. 19,* 361–364. [Box 40.3]

Clemens, R. *See also* Fletcher *et al.* 1967.

Clemens, R., and R. Matzner, 1967, Technical Report no. 635 of the Univ. of Maryland Department of Physics and Astronomy. [Box 14.3]

Clifford, W. K., 1879, *Lectures and Essays,* L. Stephen and F. Pollock, eds., Macmillan, London, I, 244 and 322. [§44.3]

Clifford, W. K., 1882, *Mathematical Papers,* R. Tucker, ed., Macmillan, London. [§44.3]

CNRS, 1962, *Les Théories relativistes de la gravitation,* Éditions du Centre National de la Recherche Scientifique, Paris, France. *Same as* Royaumont 1959.

Cocke, W. J., 1965, "A maximum-entropy principle in general relativity and the stability of fluid spheres," *Ann. Inst. Henri Poincaré A 2,* 283–306. [§26.1]

Cocke, W. J., H. J. Disney and D. J. Taylor, 1969, "Discovery of optical signals from pulsar NP0532," *Nature 221,* 525–527. [Box 24.1]

Codazzi, D., *cited by name in* §21.5.

Cohen, J. M., 1968, "Angular momentum and the Kerr metric," *J. Math. Phys. 9,* 905–906. [Ex. 26.1]

Cohen, J. M. *See also* Brill and Cohen 1966.

Cohen, J. M., and D. Brill, 1966, "Rotating masses and their effect on inertial frames," *Phys. Rev. 143,* 1011–1015. [§21.12]

Cohen, J. M., and D. R. Brill, 1968, "Further examples of 'Machian effects' of rotating bodies in general relativity," *Nuovo Cim. 56B,* 209–219. [Ex. 26.1]

Cohen, J. M., and R. M. Wald, 1971, "Point charge in the vicinity of a Schwarzschild black hole," *J. Math. Phys. 12,* 1845–1849. [Box 33.1]

Cohen, P., 1966, *Set Theory and the Continuum Hypothesis,* W. A. Benjamin, New York. [Box 44.5]

Cole, J., 1968, *Perturbation Methods in Applied Mathematics,* Ginn-Blaisdell, Waltham, Mass. [§22.5]

Coleman, C. J., 1971, "A simple relativistic theory of gravitation," *J. Phys. A 4,* 611–616. [§40.5]

Coleman, J. A., 1967, *Early Theories of the Universe,* New American Library, New York. [§1.7]

Colgate, S. A., 1967, "Stellar coalescence and the multiple-supernova interpretation of quasi-stellar sources," *Astrophys. J. 150,* 163–192. [§24.5, Box 33.3]

Colgate, S. A., and R. H. White, 1966, "The hydrodynamic behavior of supernovae explosions," *Astrophys. J. 143,* 626–681. [§32.7]

Collins, C. B., and S. W. Hawking, 1973, "Why is the universe isotropic?" *Astrophys. J. 180,* 317–334. [Box 34.3]

Collins, R. A., *cited by name in* Box 27.7. *See also* Hewish *et al.* 1968.

Combridge, J. T., 1965, *Bibliography of Relativity and Gravitation Theory 1921–1937,* Kings College, London, cited in preface of present bibliography.

Comte, A., 1835, *Cours de philosophie positive,* Paris. [Box 27.7]

Conklin, E. K., 1969, "Velocity of the Earth with respect to the cosmic background radiation," *Nature 222,* 971–972. [§30.8]

Contopolous, G. *See* Chandrasekhar and Contopolous 1967.

Coote, N., *cited by name in introduction to this bibliography.*

Copenhagen 1971, Sixth International Conference on General Relativity and Gravitation, held in Copenhagen, July 1971; portions of the proceedings appear as articles in *Gen. Rel. Grav.,* especially vol. 3, 1972.

Copernicus, N. *See* Dobson and Brodetsky 1947.

Corben, H. C., and P. Stehle, 1960, *Classical Mechanics,* 2d ed., Wiley, New York. [Ex. 33.7]

Cornell University, 1965, Summer Seminar on Applied Mathematics, arranged by the American Mathematical Society. Proceedings (vols. 8–10 in Lectures in Applied Mathematics series) published as Ehlers 1967.

Cornford, F. M., 1937, *Plato's Cosmology: The Timaeus of Plato,* Humanities Press, New York. [Box 27.7]

Corson, E. M., 1953, *Introduction to Tensors, Spinors, and Relativistic Wave-Equations,* Hafner, New York. [§21.3]

Couch, W. E. *See also* Newman *et al.* 1965.

Couch, W. E., and E. T. Newman, 1972, "Generalized Lienard-Wiechart fields," *J. Math. Phys. 13,* 929–931. [§36.1]

Couch, W. E., R. J. Torrence, A. I. Janis, and E. T. Newman, 1968, "Tail of a gravitational wave," *J. Math. Phys. 9,* 484–496. [§§35.8, 35.14]

Courant, R., 1937, "Plateau's problem and Dirichlet's principle," *Ann. Math. U.S. 38,* 679–724. [§33.2]

Coxeter, H. S. M., 1963, *Regular Polytopes,* 2d ed., Macmillan, New York. [§42.5, Box 27.2]

Curott, D. R., 1966, "Earth deceleration from ancient solar eclipses," *Astron. J. 71,* 264–269. [Fig. 1.8]

Currie, D. G., *cited by name in* Fig. 40.4. *See also* Bender *et al.* 1971.

Curtis, H. D., *cited by name in* Box 27.7.

Cusanus, N. *See* Heron 1954.

Dabbs, J. W. T., J. A. Harvey, D. Paya, and H. Horstmann, 1965, "Gravitational acceleration of free neutrons," *Phys. Rev. B 139,* 756–760. [§1.3]

D'Alembert, J., *quoted at end of* Chap. 44.

Danielson, B. L. *See* Evenson *et al.* 1972.

Dante Alighieri, *Paradiso,* translated by Philip H. Wicksteed, 1921, Dent, London, England. [Epigraph to Chap. 44]

Danziger, I. J., 1970, "The cosmic abundance of helium," *Ann. Rev. Astron. Astrophys., 8,* 161–178. [§28.1]

Darboux, G., 1941, *Leçons sur la Théorie Générale des Surfaces, Part I,* Gauthier-Villars, Paris, France. [§33.2]

Darmois, G., 1927, *Les équations de la gravitation einsteinienne,* Mémorial des sciences mathematiques XXV, Gauthier-Villars, Paris. [§21.13]

Darwin, C., 1959, "The gravity field of a particle, I," *Proc. R. Soc. London A 249,* 180–194. [§25.5, Fig. 25.7]

Darwin, C., 1961, "The gravity field of a particle, II," *Proc. R. Soc. London A 263,* 39–50. [§25.5, Fig. 25.7]

Datt, B., 1938, "Über eine Klasse von Lösungen der Gravitationsgleichungen der Relativität," *Z. Physik 108,* 314–321. [§30.6, Ex. 32.8]

Davenport, W. B., and W. L. Root, 1958, *An Introduction to the Theory of Random Signals and Noise.* McGraw Hill, New York, [§37.8]

Davies, R. W., ed., 1971, *Proceedings of the Conference on Experimental Tests of Gravitational Theories, November 11–13, 1970, California Institute of Technology,* J.P.L. Technical Memorandum, 33–499.

Davis, M., R. Ruffini, W. H. Press, and R. H. Price, 1971, "Gravitational radiation from a particle falling radially into a Schwarzschild black hole," *Phys. Rev. Lett. 27,* 1466–1469. [§§33.7, 36.5, Box 33.3, Fig. 36.2]

Davis, W. R., 1970, *Classical Theory of Particles and Fields and the Theory of Relativity,* Gordon and Breach, New York. [§21.3, endpapers]

Day, G. W. *See* Evenson *et al.* 1972.

de Broglie, M. *See* Einstein 1912.

Debye, P. J. W., *cited by name in* §22.5.

De Chéseaux, P. L., 1744, *Traité de la comète qui a paru en décembre 1743,* Paris; the relevant material is in the Appendix. [Box 27.7]

De Felice, F., 1968, "Equatorial geodesic motion in the gravitational field of a rotating source," *Nuovo Cimento B 57,* 351–388. [Box 33.5]

Defoe, D., 1722, *The Fortunes and Misfortunes of the Famous Moll Flanders,* reprinted by Fine Editions Press, Cleveland, 1953. [epigraph to Part X]

De Groot, S. R. *See* Kluitenberg and de Groot 1954.

De Hoffmann, F., and E. Teller, 1950, "Magneto-hydrodynamic shocks," *Phys. Rev. 80,* 692–703. [§22.2, Box 22.2]

De la Cruz, V., and W. Israel, 1967, "Gravitational bounce," *Nuovo Cimento A 51,* 744–760. [§34.6, Fig. 34.4]

De Lange, F. G. *See* Refsdal, Stabell, and de Lange 1967.

Dempster, J. R. H., 1963, *Feynman Quantization in Field Theory: Topology of Commutators and Functional Representation of States,* doctoral dissertation, Princeton University. [Box 21.1]

De Rham, G., 1955, *Variétés Différentiables,* Herman, Paris. [§§4.1, 22.4]

Deser, S., 1970, "Self-interaction and gauge invariance," *Gen. Rel. & Grav. 1,* 9–18. [§§7.1, 18.1, Box 17.2 part 5]

Deser, S., *cited by name in* §§21.1, 35.8, Box 17.2, part 2'. *See also* Arnowitt, Deser, and Misner 1962 *and* Chrétien, Deser, and Goldstein 1969.

Deser, S., and K. W. Ford, eds., 1965, *Lectures on General Relativity; 1964 Brandeis Summer Institute in Theoretical Physics I*, Prentice-Hall, Englewood Cliffs, N.J. *Same as* Trautman, Pirani, and Bondi 1965.

De Sitter, W., 1916, "On Einstein's theory of gravitation and its astronomical consequences," *Mon. Not. R. Astron. Soc. 77*, 155–184 and 481. [§§40.6, 40.7]

De Sitter, W., 1917a, "On the relativity of inertia: Remarks concerning Einstein's latest hypothesis," *Proc. Kon. Ned. Akad. Wet. 19*, 1217–1225. [§27.11]

De Sitter, W., 1917b, "On the curvature of space," *Proc. Kon. Ned. Akad. Wet. 20*, 229–243. [§27.11]

De Sitter, W., *cited by name in* §30.3, Box 27.7.

Deutsch, A. J., and W. B. Klemperer, eds., 1962, *Space Age Astronomy*, Academic Press, New York.

De Valera, E., *cited by name in* §41.1.

De Vaucouleurs, G., 1971, "The large-scale distribution of galaxies and clusters of galaxies," *Astron. Soc. Pacific Publ. 83*, 113–143. [§27.11]

DeWitt, B. S., 1953, "Pair production by a curved metric," *Phys. Rev. 90*, 357. [§30.5]

DeWitt, B. S., 1962, "The quantization of geometry," in Witten 1962. [Ex. 40.8]

DeWitt, B. S., 1967a, "Quantum theory of gravity, I," *Phys. Rev. 160*, 1113–1148. [§§21.1, 43.1, 43.3]

DeWitt, B. S., 1967b. "Quantum theory of gravity, II; The manifestly covariant theory." *Phys. Rev. 162*, 1195–1239. [§§43.1, 43.3]

DeWitt, B. S., 1967c, "Quantum theory of gravity, III; Application of the covariant theory," *Phys. Rev. 162*, 1239–1256. [Box 17.2 part 6h]

DeWitt, B. S., 1971, *Lectures on Relativity—Stanford, Fall 1971* (unpublished manuscript). [§36.7]

DeWitt, B. S., *cited by name in* Ex. 35.20. *See also* DeWitt and DeWitt 1964a, 1964b, *and* 1973.

DeWitt, B. S., and R. W. Brehme, 1960, "Radiation damping in a gravitational field," *Ann. Phys. (U.S.A.) 9*, 220–259. [§35.8, Ex. 35.14]

DeWitt, C., ed. 1957, *Conference on the Role of Gravitation in Physics, Proceedings*, W.A.D.C. Technical Report 57-216; ASTIA Document No. AD118180, Wright Air Development Center, Wright-Patterson Air Force Base, Ohio.

DeWitt, C. *See also* Morette, C.

DeWitt, C., and B. S. DeWitt, 1964a, "Falling charges," *Physics 1*, 3–20. [§35.8]

DeWitt, C., and B. S. DeWitt, eds., 1964b, *Relativity, Groups, and Topology*, Gordon and Breach, New York. [§38.2]

DeWitt, C., and B. S. DeWitt, eds., 1973, *Black Holes*, Proceedings of 1972 session of Ecole d'été de physique theorique, Gordon and Breach, New York. [§33.1]

DeWitt, C., E. Schatzman, and P. Veron, eds., 1967, *High-Energy Astrophysics, Vol. III*, Gordon and Breach, New York. [§24.4]

DeWitt, C., and J. A. Wheeler, eds., 1968, *Battelle Rencontres: 1967 Lectures in Mathematics and Physics*, W. A. Benjamin, New York.

Dicke, R. H., 1954, "Coherence in spontaneous radiation processes," *Phys. Rev. 93*, 99–110. [Box 38.3]

Dicke, R. H., 1957, "Coherence and the quantum," *J. Opt. Soc. Am. 47*, 527. [Box 38.3]

Dicke, R. H., 1959a, "Dirac's cosmology and the dating of meteorites," *Nature 183*, 170–171. [§38.6]

Dicke, R. H., 1959b, "New research on old gravitation," *Science 129*, 621–624. [§§38.6, 40.9, 44.6]

Dicke, R. H., 1961, "Dirac's cosmology and Mach's principle," *Nature 192*, 440–441. [§§38.6, 44.6, Box 38.3]

Dicke, R. H., 1962, "Mach's principle and invariance under transformation of units," *Phys. Rev. 125*, 2163–2167. Reprinted in Dicke 1964b. [Box 39.1]

Dicke, R. H., 1964a, "The sun's rotation and relativity," *Nature 202*, 432–435. [Box 38.3]

Dicke, R. H., 1964b, *The Theoretical Significance of Experimental Relativity*, Gordon and Breach, New York; also in DeWitt and DeWitt 1964a. [§§38.2, 38.7, 44.6, Box 40.1]

Dicke, R. H., 1964c, "Remarks on the observational basis of general relativity," in Chiu and Hoffman 1964, pp. 1–16. [§40.9]

Dicke, R. H., 1968, "Scalar-tensor gravitation and the cosmic fireball," *Astrophys. J., 152*, 1–24. [§28.4]

Dicke, R. H., 1969, *Gravitation and the Universe*, American Philosophical Society, Philadelphia. [§§38.6, 40.9]

Dicke, R. H., *cited by name in* §§10.1, 38.3, 38.5, 40.1, 40.6, 40.9, Box 27.7. *See also* Bender *et al.* 1971, Brans and Dicke 1961, *and* Roll, Krotkov, and Dicke 1964.

Dicke, R. H., and H. M. Goldenberg, 1967, "Solar oblateness and general relativity," *Phys. Rev. Lett. 18*, 313–316. [Boxes 38.3, 40.3]

Dicke, R. H., and P. J. E. Peebles, 1962a, "Significance of spatial isotropy," *Phys. Rev. 127*, 629–631. Reprinted in Dicke 1964b. [§38.7]

Dicke, R. H., and P. J. E. Peebles, 1962b, "Cosmology and the radioactive decay ages of terrestrial rocks and meteorites," *Phys. Rev. 128*, 2006–2011. [§38.6, Box 38.3]

Dicke, R. H., and P. J. E. Peebles, 1965, "Gravitation and space science," *Space Sci. Rev. 4*, 419–460. [§40.8]

Dicke, R. H., P. J. E. Peebles, P. G. Roll, and D. T. Wilkinson, 1965, "Cosmic-black-body radiation," *Astrophys. J. 142*, 414–419. [§§27.2, 28.1]

Dieter, N. H., E. E. Epstein, A. E. Lilley, and M. S. Roberts, 1962, "A radio and optical investigation of extragalactic redshifts," *Astron. J. 67*, 270–271. [§29.2]

Dieudonné, J., 1960, *Foundations of Modern Analysis*, Academic Press, New York. [§2.6]

Digges, T., *cited by name in* Box 27.7.

d'Inverno, R. A., 1969, "ALAM—Atlas Lisp algebraic manipulator," *Comput. J. 12*, 124–127. [Box 14.3]

Dirac, P. A. M., 1926, "On the theory of quantum mechanics," *Proc. R. Soc. London A 112*, 661–677. [Box 24.1]

Dirac, P. A. M., 1936, "Relativistic wave equations," *Proc. R. Soc. London A 155*, 447–459. [Box 32.2]

Dirac, P. A. M., 1937, "The cosmological constants," *Nature 139*, 323. [§§38.6, 44.6]

Dirac, P. A. M., 1938, "New basis for cosmology," *Proc. R. Soc. London A 165*, 199–208. [§§38.6, 44.6]

Dirac, P. A. M., 1959, "Fixation of coordinates in the Hamiltonian theory of gravitation," *Phys. Rev. 114*, 924–930. [§21.7]

Dirac, P. A. M. 1964, *Lectures on Quantum Mechanics,* Belfer Graduate School of Science Monograph Series Number Two, Yeshiva University, New York. [§21.7]

Dirac, Paul A. M., *Festschrift* on the occasion of his 70th birthday, Aug. 8, 1972. *See* Salam and Wigner 1972.

Dirac, P. A. M., *cited by name in* §§20.6, 21.1, Box 4.4.

Dirac, P. A. M., V. A. Fock, and B. Podolsky, 1932, "On quantum electrodynamics," *Phys. Z. Sowjetunion 2*, 468–479. [§21.8]

Dobson, J. F., and S. Brodetsky, 1947, *Nicolaus Copernicus's De Revolutionibus* (translation of Preface and Book I), Royal Astronomical Society, London. [Box 27.7]

Dodgson, C. L., 1871, *Through the Looking Glass,* Macmillan, London. [Epigraph to Chap. 32]

Doroshkevich, A. G., 1966, "Change of the angular momentum of a star during accretion," *Astron. Zhur. 43*, 105–107. Terrible English translation in *Sov. Astron.—A. J. 10*, 83–84. [Box 33.3]

Doroshkevich, A. G., V. N. Lukash, and I. D. Novikov, 1971, "Impossibility of mixing in the Bianchi type IX cosmological model," *Zh. Eksp. & Teor. Fiz. 60*, 1201–1205. English translation in *Sov. Phys.—JETP 33*, 649–651. [§30.8]

Doroshkevich, A. G., Ya. B. Zel'dovich, and I. D. Novikov, 1965, "Gravitational collapse of nonsymmetric and rotating masses," *Zh. Eksp. & Teor. Fiz. 49*, 170–181. English translation in *Sov. Phys.—JETP 22*, 122–130. [Box 33.3, Ex. 26.1]

Douglas, K. N., I. Robinson, A. Schild, E. L. Schücking, J. A. Wheeler, and N. J. Woolf, eds., 1969, *Quasars and High-Energy Astronomy,* Proceedings of the Second Texas Symposium on Relativistic Astrophysics, Gordon and Breach, New York.

Douglass, D. H., Jr., 1971, "Properties of 'hollow square' gravitational wave detectors," in Davies 1971. [§37.3, Fig. 37.2]

Douglass, D. H., Jr., *cited by name in* Ex. 37.13.

Douglass, D. H., Jr., and J. A. Tyson, 1971, "New class of gravitational-wave detectors," *Nature 229*, 34–36. [§37.3]

Drabkin, I. E., and S. Drake, 1960, *Galileo Galilei on Motion and on Mechanics,* translations with introductions and notes, University of Wisconsin Press, Madison, Wisc. [Box 1.2]

Drake, S. *See* Drabkin and Drake 1960.

Drell, S. D. *See* Brodsky and Drell 1970.

Drever, R. W., 1961, "A search for anisotropy of inertial mass using a free precession technique," *Phil. Mag. 6*, 683–687. [§38.7]

Dunford, N. and J. T. Schwartz, 1958, *Linear operators,* pp. 467–471, Interscience (now a division of Wiley), New York. *Cited under* Lefschetz 1949.

Durbin, R. P., H. H. Loar, and W. W. Havens, Jr., 1952, "The lifetimes of the pi-plus and pi-minus mesons," *Phys. Rev. 88*, 179–183. [§38.4]

Duyvendak, J. J. L., 1942, "Further data bearing on the identification of the Crab Nebula with the supernova of A.D. 1054, Part I: The ancient oriental chronicles," *Proc. Astron. Soc. Pacific 54*, 91–94. [Box 24.1]

Dyce, R. B. *See* Shapiro, Ash, *et al.* 1971 and Shapiro *et al.* 1972.

Dyson, F. J., 1954, *Advanced Quantum Mechanics,* multilithographed lecture notes, Physics Department, Cornell University. [Box 17.2 part 6]

Dyson, F. J., 1967, "Time variation of the charge of the proton," *Phys. Rev. Lett. 19*, 1291–1293. [§37.3, Fig. 37.2]

Dyson, F. J., 1969, "Seismic response of the Earth to a gravitational wave in the 1-Hz band," *Astrophys. J. 156*, 529–540. [§37.3, Fig. 37.2]

Dyson, F. J., 1972, "The fundamental constants and their time variation," in Salam and Wigner 1972. [§38.6]

Eardley, D., E. Liang, and R. Sachs, 1972, "Velocity-dominated singularities in irrotational dust cosmologies," *J. Math. Phys. 13*, 99–107. [§30.7]

Eckart, C., 1940, "The thermodynamics of irreversible processes, III: Relativistic theory of the simple fluid," *Phys. Rev., 58*, 919–924. [Ex. 22.7]

Ecphantus, *cited by name in* Box 27.7.

Eddington, A. S., 1922, *The Mathematical Theory of Relativity,* Cambridge Univ. Press, Cambridge, Eng. [§§38.2, 39.3]

Eddington, A. S., 1924, "A comparison of Whitehead's and Einstein's formulas," *Nature 113*, 192. [Boxes 31.1, 31.2]

Eddington, A. S., 1931, "Preliminary note on the masses of the electron, the proton, and the universe," *Proc. Cambridge Phil. Soc. 27*, 15–19. [§44.6]

Eddington, A. S., 1936, *Relativity Theory of Protons and Electrons,* Cambridge Univ. Press, Cambridge, Eng. [§44.6]

Eddington, A. S., 1946, *Fundamental Theory,* Cambridge Univ. Press, Cambridge, Eng. [§44.6]

Eddington, A. S., *cited by name in* Box 27.7.

Eddington, A., and G. Clark, 1938, "The problem of *n* bodies in general relativity theory," *Proc. Roy. Soc. A 166*, 465–475. [Ex. 39.15]

Ehlers, J., 1961, "Contributions to the relativistic mechanics of continuous media," *Akad. Wiss. Lit. Mainz Abh. Math.-Nat. Kl.*, 793–837. [§22.3]

Ehlers, J., ed., 1967, *Relativity Theory and Astrophysics: I, Relativity and Cosmology; II, Galactic Structure; III, Stellar Structure,* American Mathematical Society, Providence, R.I.

Ehlers, J., 1971, "General relativity and kinetic theory," in Sachs 1971. [§22.6, Ex. 22.7, endpapers]

Ehlers, J. *See also* Jordan, Ehlers, and Sachs 1961.

Ehlers, J., P. Geren, and R. K. Sachs, 1968, "Isotropic solutions of the Einstein-Liouville equations," *J. Math. Phys. 9*, 1344–1349. [§27.2]

Ehlers, J., and W. Kundt, 1962, "Exact solutions of the gravitational field equations," in Witten 1962. [§§33.6, 35.9, Ex. 35.8, 35.9]

Ehlers, J., F. A. E. Pirani, and A. Schild, 1972, "The geometry of free-fall and light propagation," *in* O'Raifeartaigh 1972, pp. 63–84. [Boxes 10.2, 16.4]

Ehrenfest, P., 1916, "On the adiabatic changes of a system in connection with the quantum theory," *Amsterdam Acad., Proc. 19*, 576–597. [Box 22.4]

Einstein, A., 1905, "Zur elektrodynamik bewegter Körper," *Ann. Phys. (Germany) 17*, 891–921. [Box 1.5]

Einstein, A., 1908, "Über das Relativitätsprinzip und die aus demselben gezogenen Folgerungen," *Jarb. Radioakt. 4*, 411–462. [§§7.4, 17.7]

Einstein, A., 1911, "Über den Einfluss der Schwerkraft auf die Ausbreitung des Lichtes," *Ann. Phys. (Germany) 35*, 898–908, English translation in Lorentz *et al.* 1923. [§§1.3, 7.2, 7.4, 10.1, 16.2]

Einstein, A. 1912, "Si l'on modifie la longueur du pendule . . . ," p. 450 in P. Langevin and M. de Broglie, eds., *La théorie du rayonnement et les quanta,* Gauthier-Villars, Paris, 1912. [Box 22.4]

Einstein, A., 1913a, "Zum gegenwärtigen Stand des Gravitationsproblems," *Phys. Z. 14*, 1249–1262, discussion 1262–1266. [Epigraph to Chap. 35]

Einstein, A., 1913b, letter to Ernst Mach dated June 26. [§21.12, Fig. 21.5]

Einstein, A., 1915a, "Zür allgemeinen Relativitätstheorie," *Preuss. Akad. Wiss. Berlin, Sitzber.,* 778–786 (published November 11). [§17.7]

Einstein, A. 1915b, "Zür allgemeinen Relativitätstheorie (Nachtrag)," *Preuss Akad. Wiss. Berlin, Sitzber.,* 799–801 (published November 18). [§17.7, Box 27.8]

Einstein, A., 1915c. "Erklärung der Perihelbewegung des Merkur aus der allgemeinen Relativitätstheorie," *Preuss. Akad. Wiss. Berlin, Sitzber. 47*, 831–839 (published November 25). [§17.7]

Einstein, A., 1915d, "Die Feldgleichungen der Gravitation," *Preuss. Akad. Wiss. Berlin, Sitzber.,* 844–847 (published December 2). [§17.7, Box 27.7]

Einstein, A., 1916a, "Ernst Mach," *Phys. Z. Sowjetunion 17*, 101–104. [Fig. 21.5]

Einstein, A., 1916b, "Näherungsweise Integration der Feldgleichungen der Gravitation," *Preuss, Akad. Wiss. Berlin, Sitzber.* 688–696. [§17.7]

Einstein, A., 1916c, "Hamiltonsches Prinzip und allgemeine Relativitätstheorie," *Preuss. Akad. Wiss. Berlin, Sitzber.,* 1111–1116. [§17.7]

Einstein, A., 1917, "Kosmologische Betrachtungen zur allgemeinen Relativitätstheorie," *Preuss. Akad. Wiss. Berlin, Sitzber.* 142–152, reprinted in English in Lorentz *et al.* 1923. [§17.3, Boxes 27.1, 27.7]

Einstein, A., 1921, "Geometry and Experience," address to Prussian Academy of Sciences, Berlin, Jan. 27, 1921; published in *Sidelights on Relativity,* Mathuen, London, 1922. [Box 1.11]

Einstein, A., 1931, "Zum kosmologischen Problem der allgemeinen Relativitätstheorie," *Preuss, Akad. Wiss. Berlin, Sitzber,* 235–237. [§27.11]

Einstein, A., 1934, *Essays in Science,* Philosophical Library, New York. Translated from *Mein Weltbild,* Querido Verlag, Amsterdam, 1933. [Epigraph to Chap. 17, §§21.12, 27.11, 27.12, 43.1, 44.6, Box 8.5]

Einstein, A., 1939, "On a stationary system with spherical symmetry consisting of many gravitating masses," *Ann. Math. (U.S.A.) 40*, 922–936. [Ex. 25.29]

Einstein, Albert, *Festschrift,* in commemoration of his seventieth birthday, March 14, 1949; *Rev. Mod. Phys. 21* (July 1949), 343–540.

Einstein, A., 1949a, introduction to Schilpp 1949. [Preface, Epigraph to Chap. 44, §1.2]

Einstein, A., 1949b, "Reply to criticisms," in Schilpp 1949. [Epigraph to Chap. 27]

Einstein, A., 1950, *The Meaning of Relativity,* 3d ed., Princeton Univ. Press, Princeton, N.J. [§21.12, 29.5, 43.1, Box 27.1, endpapers]

Einstein, A., 1954, preface to *Concepts of Space,* M. Jammer, ed., Harper, New York, p. xv. [§1.3]

Einstein, A., 1970, quoted by G. Gamow in *My World Line,* Viking Press, New York (1970), p. 44. [§§17.3, 27.8, Box 27.1]

Einstein, A., *cited by name in* §§1.1, 1.6, 7.1, 7.2, 7.4, 10.1, 12.1, 12.2, 17.3, 20.6, Boxes 12.2, 12.3, 27.7. *See also* Lorentz *et al.* 1923, Born 1971, Holton 1971, Klein 1971.

Einstein, A., and A. D. Fokker, 1914, *Ann. Phys. (Germany), 44*, 321. [§17.6]

Einstein, A., and J. Grommer, 1927, "Allgemeine Relativitätstheorie und Bewegungsgesetze," *Sitzber. Deut. Akad. Wiss. Berlin, Kl. Math. Phys. Tech.* 2–13, 235–245. [§20.6]

Einstein, A., and M. Grossmann, 1913, "Entwurf einer verallgemeinerten Relativitätstheorie und einer Theorie der Gravitation," *Zeit. Math. Phys. 62*, 225–261. [§17.7]

Einstein, A., and M. Grossmann, 1914, "Kovarianzeigenschaften der Feldgleichungen der auf die verallgemeinerte Relativitätstheorie gegründeten Gravitationstheorie," *Zeit. Math. Phys. 63*, 215–225. [§17.7]

Einstein, A., L. Infeld, and B. Hoffmann, 1938, "The gravitational equations and the problem of motion," *Ann. Math. 39*, 65–100. [Ex. 39.15]

Eisenhart, L. P., 1926, *Riemannian Geometry.* Princeton Univ. Press, Princeton, N.J. [§§21.5, 21.12, Ex. 27.4, endpapers]

Ekers, R. D. *See* Muhleman, Ekers, and Fomalont 1970.

Eliot, T. S., 1920, "Tradition and individual talent," in *The Sacred Wood; Essays on Poetry and Criticism,* Methuen, London. [Box 27.7]

Eliot, T. S., 1942, *Little Gidding,* Faber and Faber, London. [§1.7]

Ellis, G. F. R., 1971, "Relativistic cosmology," in Sachs 1971. [§22.3, Ex. 22.6, 29.5]

Ellis., G. F. R., *cited by name in* Boxes 27.7, 34.3, Ex. 32.1. *See also* Hawking and Ellis 1968 *and* 1973, *and* Stewart and Ellis 1968.

Eötvös, R. V., 1889, "Über die Anziehung der Erde auf verschiedene Substanzen," *Math. Naturw. Ber. aus Ungarn 8,* 65–68. [Box 1.2]

Eötvös, R. V., *cited by name in* §§10.1, 38.3, 38.5, 40.9.

Eötvös, R. V., D. Pekár, and E. Fekete, 1922, "Beiträge zum Gesetze der Proportionalität von Trägheit und Gravität," *Ann. Phys. (Germany) 68,* 11–66. [§§10.1, 38.3, 40.9, Boxes 1.2, 38.2]

Epstein, E. E. *See* Dieter *et al.* 1962.

Ernst, F. J., 1968, "7090 FORMAC subroutines: Exterior calculus," report number PORS IIT-9, Illinois Institute of Technology. [Box 14.3]

Esposito, F. P. *See* Chandrasekhar and Esposito 1970.

Esposito, P. B. *See* Anderson *et al.* 1971.

ESRO Conference on Space Physics, Interlaken, Switzerland, September 4, 1969. *See* Ruffini and Wheeler 1971b.

Estabrook, F. B., 1969, "Post-Newtonian *n*-body equations of the Brans-Dicke theory," *Astrophys. J. 158,* 81–83. [Ex. 39.15]

Estermann, I., O. C. Simpson, and O. Stern, 1938, "The free fall of molecules," *Phys. Rev. 53,* 947–948. [§1.3]

Eudoxus, *cited by name in* Box 27.7.

Euler, H. *See* Heisenberg and Euler 1936.

Euler, L., 1760, "Demonstratio theorematis Fermatiani omnem numerum esse summam quatuor quadratorum," *Novi Comment, Petrop. 5, 93, Opera 12,* 369ff. [§41.1]

Evenson, K. M., J. S. Wells, F. R. Petersen, B. L. Danielson, G. W. Day, R. L. Barger, and J. L. Hall, 1972, "Speed of light from direct frequency and wavelength measurements of the methane-stabilized laser", *Phys. Rev. Letters 29,* 1346–1349. [§1.5]

Everitt, C. W. F., 1970, "Maxwell as a student," preprint, Stanford University, Stanford, Calif. [Box 14.6]

Everitt, C. W. F., W. M. Fairbank, and W. O. Hamilton, 1970, "General relativity experiments using low-temperature techniques," in Carmeli, Fickler, and Witten 1970, pp. 145–159. [§40.7]

Exton, A. *See* Newman *et al.* 1965.

Fackerell, E. D., 1966, *Relativistic Stellar Dynamics,* unpublished Ph.D. thesis, University of Sydney. [Box 25.9]

Fackerell, E. D., 1968, "Relativistic stellar dynamics," *Astrophys. J. 153,* 643–660. [Box 25.8]

Fackerell, E. D. *See also* Brill *et al.* 1972.

Fackerell, E. D., and J. R. Ipser, 1972, "Weak electromagnetic fields around a rotating black hole," *Phys. Rev. D5,* 2455–2458. [§41.11]

Fackerell, E. D., J. R. Ipser, and K. S. Thorne, 1969, "Relativistic star clusters," *Comments Astrophys. & Space Phys. 1,* 134–139. [§24.6, Box 25.9]

Fadeev, L. D., 1971, "Symplectic structure and quantization of the Einstein gravitation theory," in *Actes du Congrés International des Mathématiciens, Nice 1–10 Septembre 1970,* Gauthier-Villars, Paris, III, 35–40. [Box 21.1]

Fairbank, W. M. *cited by name in* Ex. 37.13. *See also* Everitt, Fairbank, and Hamilton 1970, Witteborn and Fairbank 1967, *and* Witteborn, Knight, and Fairbank 1967.

Faller, J. E. *See* Bender *et al.* 1971.

Faraday, M., *cited by name in* §5.6.

Farley, F. J. M., 1969, "The status of quantum electrodynamics," *Nuovo Cimento 1,* Numero Speciale, 59–86. [§44.2]

Farley, F. J. M. *See also* Bailey *et al.* 1968.

Farley, F. J. M., J. Bailey, R. C. A. Brown, M. Giesch, H. Jöstlein, S. van der Meer, E. Picasso, and M. Tannenbaum, 1966, "The anomolous magnetic moment of the negative muon," *Nuovo Cimento 45,* 281–286. [§38.4]

Farnsworth, D., J. Fink, J. Porter, and A. Thomson, eds., 1972, *Methods of Local and Global Differential Geometry in General Relativity,* Springer, Berlin.

Faulkner, J., and J. R. Gribbin, 1968, "Stability and radial vibration periods of the Hamada-Salpeter white-dwarf models," *Nature 218,* 734–735. [§24.2, Boxes 24.1, 26.1]

Fejér, L., and S. Mikola, eds., 1918, "Báró Eötvös Loránd Füzet," *Matem. Fiz. Lapok 27,* Nos. 6–7, Oct.–Nov., kindly translated into English for use here by Rev. Stanley Jaki. [Box 38.2]

Fekete, E. *See* Eötvös, Pekar, and Fekete 1922.

Fermi, E., 1922, "Sopra i fenomeni che avvengono in vicinanza di una linea oraria," *Atti Accad. Naz. Lincei Cl. Sci. Fis. Mat. & Nat. 31,* 184–187, 306–309. [§6.3]

Fermi, E., 1932, "Quantum theory of radiation," *Rev. Mod. Phys. 4,* 87–132. [§21.12]

Fermi, E., 1949, "On the origin of the cosmic radiation," *Phys. Rev. 75,* 1169–1174. [Box 27.7].

Fermi, E., and A. Turkevich, 1950, unpublished work described in Alpher and Herman 1950, pp. 194ff. [§28.1]

Ferrari, A., and R. Ruffini, 1969, "Theoretical implications of the second time derivative of the period of the pulsar NP 0532," *Astrophys. J. Letters 158,* L71–L75. [Ex. 36.4]

Feynman, R. P., 1942. *The Principle of Least Action in Quantum Mechanics,* doctoral dissertation, Princeton University. [Boxes 17.2, 21.1]

Feynman, R. P., 1948, "Spacetime approach to nonrelativistic quantum mechanics," *Rev. Mod. Phys. 20,* 367–387. [Box 21.1]

Feynman, R. P., 1949, "Spacetime approach to quantum electrodynamics," *Phys. Rev. 76,* 769–789. [Box 21.1]

Feynman, R. P., 1951, "The concept of probability in quantum mechanics," in Neyman 1951, p. 533. [Box 21.1]

Feynman, R. P., 1955, "Slow electrons in polar crystals," *Phys. Rev. 97,* 660–665. [Box 21.1]

Feynman, R. P., 1963, *Lectures on Gravitation,* unpublished lecture notes prepared by F. B. Moringo and W. G. Wagner, Calif. Institute of Technology. [§7.1, Box 7.1, §18.1]

Feynman, R. P., 1964, unpublished work cited in Fowler 1964. [Box 24.1]

Feynman, R. P. *See also* Wheeler and Feynman 1945 *and* Wheeler and Feynman 1949.

Feynman, R. P., and A. R. Hibbs, 1965, *Quantum Mechanics and Path Integrals,* McGraw-Hill, New York. [§21.1, Fig. 13.3, Boxes 17.2, 21.1]

Feynman, R. P., R. B. Leighton and M. Sands, 1964, *The Feynman Lectures on Physics,* Addison-Wesley, Reading, Mass. [§21.3, Box 16.3]

Fickler, S. I. *See* Carmeli, Fickler, and Witten 1970.

Field, G., 1973, "The origin of galaxies," in A. and M. Sandage, eds., *Stars and Stellar Systems, IX: Galaxies and the Universe,* Univ. of Chicago Press, Chicago (in press). [§30.1]

Fierz, M., and W. Pauli, 1939, "Relativistic wave equations for particles of arbitrary spin in an electromagnetic field," *Proc. R. Soc. London A 173,* 211–232. [§§18.1, 18.3, 39.1, Box 7.1]

Fierz, M., and V. F. Weisskopf, 1960, *Theoretical Physics in the 20th Century: A Memorial Volume to Wolfgang Pauli,* Interscience, New York.

Fink, J. *See* Farnsworth *et al.* 1972.

Finkelstein, D., 1958, "Past-future asymmetry of the gravitational field of a point particle," *Phys. Rev. 110,* 965–967. [Box 31.2]

Fischer, A. E., and J. B. Marsden, 1972, "The Einstein equations of evolution—A geometric approach," *J. Math. Phys. 13*, 546–568. [§§21.5, 21.6]

Fischer, A. E., 1970, "The theory of superspace," in Carmeli, Fickler, and Witten 1970. [§43.1]

Fitch, J. P. *See* Barton and Fitch 1971 *and* Barton, Bourne, and Fitch 1970.

Flamm, L., 1916, "Beitrage zur Einsteinschen Gravitationstheorie," *Physik. Z. 17*, 448–454. [§23.8]

Flanders, H., 1953, "Development of an extended exterior differential calculus," *Trans. Am. Math. Soc. 75*, 311–326. [§14.5, Ex. 14.9, 14.10]

Flanders, H., 1963, *Differential Forms With Applications to the Physical Sciences,* Academic Press, New York. [§4.1, Boxes 4.4, 8.1]

Fletcher, J. G., 1966, "GRAD Assistant—a program for symbolic algebraic manipulation and differentiation," *Commun. Assoc. Comput. Machinery 9*, 552–553. [Box 14.3]

Fletcher, J. G., R. Clemens, R. Matzner, K. S. Thorne, and B. A. Zimmerman, 1967, "Computer programs for calculating general-relativistic curvature tensors," *Astrophys. J. Lett. 148*, L91–L92. [Box 14.3]

Fock, V. A., 1959, *The Theory of Space, Time, and Gravitation,* Pergamon, New York, translated by N. Kemmer. [Box 39.3, Ex. 39.15, Epigraph to Chap. 21, endpapers]

Fock, V. A., chief editor, 1968, *Abstracts, Fifth International Conference on Gravitation and the Theory of Relativity,* Publishing House of Tbilisi University, Tbilisi, USSR.

Fock, V. A. *See also* Dirac, Fock, and Podolsky 1932.

Fokker, A. D., 1965, *Time and Space, Weight and Inertia,* Pergamon Press, London. [endpapers]

Fokker, A. D. *See* Einstein and Fokker 1914.

Fomalont, E. B. *See* Muhleman, Ekers, and Fomalont 1970.

Ford, K. W. *See* Deser and Ford 1967.

Ford, K. W., and J. A. Wheeler, 1959a, "Application of semiclassical scattering analysis," *Ann. Phys. (U.S.A.) 7*, 287–322. [§25.6]

Ford, K. W., and J. A. Wheeler, 1959b, "Semiclassical description of scattering," *Ann. Phys. (U.S.A.) 7*, 259–286. [§25.6]

Forman, F. *See* Gittelman *et al.* 1968.

Forward, R. L., 1972, "Geodesy with orbiting gravity gradiometers," in *The Use of Artificial Satellites for Geodesy,* Geophysical Monograph Series, vol. 15, American Geophysical Union, Washington, 239–243. [Box 16.5]

Fourés-Bruhat, Y. *See* Bruhat, Y., *and* Choquet-Bruhat, Y.

Fowler, R. H., 1926, "Dense matter," *Mon. Not. R. Astron. Soc. 87*, 114–122. [Box 24.1]

Fowler, W. A., 1964, "Massive stars, relativistic polytropes, and gravitational radiation," *Rev. Mod. Phys. 36*, 545–555 and 1104. [Boxes 24.1, 26.2]

Fowler, W. A., 1966, "The stability of supermassive stars," *Astrophys. J. 144*, 180–200. [§24.4, Box 26.2]

Fowler, W. A., 1967, *Nuclear Astrophysics,* American Philosophical Society, Philadelphia, Penn. [§28.1]

Fowler, W. A., 1972, "New observations and old nucleocosmochronologies," in Reines 1972. [§29.7]

Fowler, W. A. *Cited by name in* Box 27.7. *See also* Hoyle and Fowler 1963a *and* 1963b, Hoyle *et al.* 1964, *and* Wagoner, Fowler, and Hoyle 1967.

Fowler, W. A., and F. Hoyle, 1964, "Neutrino processes and pair formation in massive stars and supernovae," *Astrophys. J. Supp. 9*, 201–319. [§24.4]

Fram, D. M. *See* Boughn, Fram, and Partridge 1971.

Friedmann, A., 1922, "Über die Krümmung des Raumes," *Z. Phys. 10*, 377–386. [§§14.1, 27.9, 29.2, Boxes 14.5, 27.6]

Friedmann, A., *cited by name in* §§30.3, 42.1, Boxes 27.6, 27.7.

Frisch, D. H., and J. H. Smith, 1963, "Measurement of the relativistic time dilation using mu-mesons," *Am. J. Phys. 31*, 342–355. [§38.4]

Fröman, N., and P. O. Fröman, *JWKB approximation: contributions to the theory,* North-Holland, Amsterdam, 1965. [§22.5]

Fröman, P. O. *See* Fröman and Fröman 1965.

Fronsdal, C., 1959, "Completion and embedding of the Schwarzschild solution," *Phys. Rev. 116*, 778–781. [§23.8, Box 31.1]

Fujiwara, I., 1962, "The correspondence principle, I," in *Proceedings of the Physics Seminar in Trondheim,* no. 3, Nordita, Copenhagen, Denmark. [Box 21.1]

Fuller, R. W., and J. A. Wheeler, 1962, "Causality and multiply-connected space-time," *Phys. Rev. 128*, 919–929. [§31.6, Fig. 31.5]

Galilei, G., 1592, *De Motu,* unpublished work written at the University of Pisa, 1589–1592; translation in Drabkin and Drake 1961, p. 48. [Box 1.2]

Galilei, G., 1610, *Sidereus Nuncius,* Translated by E. S. Carlos, *The Sidereal messenger,* Rivingtons, London, 1880. [Boxes 1.9, 27.7]

Galilei, G., 1623, *Il Saggiatore,* Accademia dei Lincei, Rome. [epigraph, Chap. 13]

Galilei, G., 1632, *Dialogo dei due massimi sistemi del mundo,* Landini, Florence. Translation by S. Drake, *Galileo Galilei—Dialogue Concerning the Two Chief World Systems—Ptolemaic and Copernican,* Univ. of California Press, Berkeley and Los Angeles, 1953. [Box 1.9]

Galilei, G., 1638, *Discorsi e dimostrazioni matematiche intorno a due nuove scienze,* Elzevir, Leiden. Translation by H. Crew and A. de Salvio, *Dialogues Concerning Two New Sciences,* Macmillan, New York, 1914; reprinted by Dover, New York, 1954. [Boxes 1.2, 1.9, 27.7]

Galilei, G., *cited by name in* §12.2, *quoted in* epigraph of Chap. 38.

Gamow, G., 1948, "The evolution of the universe," *Nature 162*, 680–682. [§28.1]

Gamow, G., 1967, "Electricity, gravity, and cosmology," *Phys. Rev. Lett. 19*, 759–761. [§38.6]

Gamow, G., 1970, *My World Line,* Viking Press, New York. [Box 27.7]

Gamow, George, March 4, 1904–August 20, 1968, Memorial volume. *See* Reines 1972.

Gaposchkin, S., 1958, "The eclipsing binaries," *Handbuch der Physik 50*, 225–242. [Box 36.3]

Gårding, L., 1945, "Relativistic wave equations for zero rest mass," *Proc. Cambridge Phil. Soc. 41*, 49–56. [Box 32.2]

Garnett, W. *See* Campbell and Garnett 1882.

Gauduchon, P. *See* Berger, Gauduchon, and Mazet 1971.

Gasiorowicz, S., 1966, *Elementary Particle Physics,* Wiley, New York. [Box 7.1]

Gauss, K. F., 1817, letter to Olbers in his *Werke,* VIII, 177. [Epigraph to Chap. 8]

Gauss, K. F., 1830, letter to Bessel in his *Werke,* VIII, 201. [Epigraph to Chap. 8]

Gauss, K. F., 1827, "Disquisitiones generales circa superficies curvas," in *Karl Friedrich Gauss Werke,* IV, 217–258; English translation by J. C. Morehead and A. M. Hiltebeitel, *General Investigations of Curved Surfaces of 1827 and 1825,* reprinted by Raven Press, New York, 1965. [§§5.9, 22.2, Boxes 5.3, 5.4, 14.1]

Gauss, K. F., *cited by name in* §41.1.

Géhéniau, J., *quoted at end of* Chap. 44.

Geren, P. *See* Ehlers, Geren, and Sachs 1968.

Gerlach, U., 1968, "The equation of state at supranuclear densities and the existence of a third family of superdense stars," *Phys. Rev. 172,* 1325–1330. [§24.2]

Gerlach, U., 1969, "Derivation of the ten Einstein field equations from the semiclassical approximation to quantum geometrodynamics," *Phys. Rev. 177,* 1929–1941. [§43.3, Box 17.2 part 4]

Geroch, R. P., 1967, *Singularities in the Spacetime of General Relativity: Their Definition, Existence, and Local Characterization,* doctoral dissertation, Princeton University. [§§32.7, 44.1, Epigraph to Chap. 34]

Geroch, R. P., 1968, "What is a singularity in general relativity?" *Ann. Phys. (U.S.A.) 48,* 526–540. [§34.6]

Geroch, R. P., 1971, "Spacetime structure from a global viewpoint," in Sachs 1971. [§34.1]

Geroch, R. P., *cited by name in* §§32.7, 42.1, Box 34.1.

Giacconi, R., *cited by name in* Box 27.7.

Gibbons, J. J., *cited by name in* Box 34.3.

Giesch, M. *See* Farley *et al.* 1966.

Gilbert, R. P., and R. Newton, 1970, *Analytic Methods in Mathematical Physics,* Gordon and Breach, New York. [Box 8.5]

Gillispie, C. C. *See* Bernoulli 1733.

Ginzel, F. K., 1882, "Astronomische Untersuchungen über Finsternisse, I: Über die zwischen 26 und 103 nach Chr. stattgefundenen Sonnenfinsternisse im Allgemeinen und die Finsterniss des Plutarch ins besonders," *Sitzber. Kais. Akad. Wiss. Wien Math.-Nat. Kl. 85II,* 663–747. [Fig. 1.8]

Ginzel, F. K., 1883, "Astronomische Untersuchungen über Finsternisse, II: Grundlagen aus historischen Sonnenfinsternissen zur Ableitung empirischer Correctionen der Mondbahn," *Sitzber. Kais. Akad. Wiss. Wien Math.-Nat. Kl. 88II,* 629–755. [Fig. 1.8]

Ginzel, F. K., 1884, "Astronomische Untersuchungen über Finsternisse, III: Ermittlung empirischer Correction der Bahn des Mondes," *Sitzber. Kais. Akad. Wiss. Wien Math.-Nat. Kl. 89II.* [Fig. 1.8]

Ginzel, F. K., 1899, *Spezieller Kanon der Sonnen- und Mondfinsternisse für das Landergebiet der Klassischen Altertumwissenschaften und den Zeitraum von 900 vor Chr. bis 600 nach Chr.,* Mayer, Berlin. [Fig. 1.8]

Godfrey, B. B., 1970a, "Mach's principle, the Kerr metric, and black-hole physics," *Phys. Rev. D 1,* 2721–2725. [Box 33.3]

Godfrey, B. B., 1970b, *A Survey of the Mathematical Properties*

of Static, Axisymmetric Vacuum Spacetimes, doctoral dissertation, Princeton University. [§34.1]

Goebel, C. J., 1972, "Comments on the 'vibrations' of a black hole," *Astrophys. J. 172,* L95–L96. [Box 33.3]

Gödel, K., 1931, "Über formal unentscheidbare Sätze der Principia Mathematica und verwandte Systeme I," *Monatsh. Math. Phys. 38,* 173–198, English translation by B. Meltzer, *On Formally Undecidable Propositions,* with an introduction by R. O. Braithwaite, Basic Books, New York, 1962; outlined and discussed in Nagel and Newman 1958. [Box 44.5]

Gold, T., 1968, "Rotating neutron stars as the origin of the pulsating radio sources," *Nature 218,* 731–732. [Box 24.1]

Gold, T., 1972, oral remark to J. A. Wheeler. [Epigraph to Chap. 21]

Gold, T., *cited by name in* Box 27.7. *See also* Bondi and Gold 1948.

Goldberg, S. I. *See* Bishop and Goldberg 1968.

Goldenberg, H. M. *See* Dicke and Goldenberg 1967.

Goldreich, P. and W. H. Julian, 1968, "Pulsar electrodynamics," *Astrophys. J. 157,* 869–880. [§24.3, Box 33.3]

Goldstein, H., 1959, *Classical Mechanics,* Addison-Wesley, Reading, Mass. [§§2.1, 9.7, 33.5, Ex. 33.7]

Goldstein, J. *See* Chrétien, Deser, and Goldstein 1969.

Gott, J. R., III, and J. E. Gunn, 1971, "The Coma Cluster as an X-ray source: Some cosmological implications," *Astrophys. J. Lett. 169,* 13–15. [§29.6]

Gowdy, R. H., 1971, "Gravitational waves in closed universes," *Phys. Rev. Lett. 27,* 826–829, and erratum p. 1102. [§30.6]

Gowdy, R. H., 1973, "Vacuum spacetimes with two-parameter spacelike isometry groups and compact invariant hypersurfaces: Topologies and Boundary conditions," University of Maryland technical report 73-089 (unpublished). [§§21.10, 30.6]

Gratton, L., ed., 1966, *High-Energy Astrophysics,* Academic Press, New York.

Graves, J. C., and D. R. Brill, 1960, "Oscillatory character of Reissner-Nordstrom metric for an ideal charged wormhole," *Phys. Rev. 120,* 1507–1513. [Fig. 34.4, Ex. 31.8]

Greenstein, G. S., 1968a, "Brans-Dicke Cosmology," *Astrophys. Letters, 1,* 139–141. [§28.4]

Greenstein, G. S., 1968b, "Brans-Dicke Cosmology, II," *Ap. Space Sci., 2,* 157–165. [§28.4]

Gribbin, J. R. *See* Faulkner and Gribbin 1968.

GR*n, International Conferences on General Relativity and Gravitation. See* Berne 1955 (GR0), Chapel Hill 1957 (GR1), Royaumont 1959 (GR2), Jablonna 1962 (GR3), London 1965 (GR4), Tbilisi 1968 (GR5), Copenhagen 1971 (GR6).

Grommer, J. *See* Einstein and Grommer 1927.

Grossmann, M. *See* Einstein and Grossman 1913 *and* 1914.

Gunn, J. E., 1967, "On the propagation of light in inhomogeneous cosmologies: I, Mean effects," *Astrophys. J. 150,* 737–753. [§29.5]

Gunn, J. E., 1971, "On the distances of the quasistellar objects," *Astrophys. J. Letters, 164,* L133–L118. [Box 28.1]

Gunn, J. E. *Cited by name in* Box 27.7. *See also* Gott and Gunn 1971, Ostriker and Gunn 1969.

Gupta, S. N., 1954, "Gravitation and electromagnetism," *Phys. Rev. 96*, 1683–1685. [§§7.1, 18.1]

Gupta, S. N., 1957, "Einstein's and other theories of gravitation," *Rev. Mod. Phys. 29*, 337–350. [§§7.1, 18.1]

Gupta, S. N., 1962, "Quantum theory of gravitation," in *Recent Developments in General Relativity*, Pergamon, New York, pp. 251–258. [§§7.1, 18.1]

Gurovich, V. Ts., 1965, "The gravitational field of rotating superdense configurations," *Astron. Zhur. 42*, 974–976. English translation in *Sov. Astron.—A.J. 9*, 750–751. [Ex. 26.1]

Gursky, H., *cited by name in* Box 27.7.

Guseynov, O. H. *See* Zel'dovich and Guseynov 1965.

Guthrie, W. K. C., trans., 1939, *Aristotle's On the Heavens*, Harvard Univ. Press, Cambridge, Mass. [Box 27.7]

Haeberli, W. *See* Barschall and Haeberli 1971.

Hagedorn, R., 1964, *Relativistic Kinematics*, W. A. Benjamin, New York. [§2.10]

Hagihara, Y., 1931, "Theory of the relativistic trajectories in a gravitational field of Schwarzschild," *Japan. J. Astron. Geophys., 8*, 67–175. [§25.5]

Hall, D. B. *See* Rossi and Hall 1941.

Hall, J. L., 1972. *See* Evenson *et al.* 1972.

Halley, E., 1693, "Emendationes ac notæ in vetustas Ablatenii observationes astronomicas cum restitutione tabularum lunisolarum," *Roy. Soc., London, Phil. Trans. 17*, 913. [§1.5]

Halley, E. 1720, "Of the infinity of the sphere of fix'd stars" and "Of the number, order, and light of the fix'd stars," *Roy. Soc. London, Phil. Trans. 31*, 22–24 and 24–26. [Box 27.7]

Hamilton, W. O., *cited by name in* Ex. 37.13. *See also* Everitt, Fairbank, and Hamilton 1970.

Hamilton, W. R., *cited by name in* §41.1 Boxes 17.2, 25.3.

Hamoui, A. *See* Papapetrou and Hamoui 1967 *and* 1968.

Hanni, R. S., and R. Ruffini, 1973, paper in preparation. [Box 33.2]

Hardy, E. *See* Sandage, Tamman, and Hardy 1972.

Harrison, B. K., 1959, "Exact three-variable solutions of the field equations of general relativity," *Phys. Rev. 116*, 1285–1296. [§14.1]

Harrison, B. K., K. S. Thorne, M. Wakano, and J. A. Wheeler, 1965, *Gravitation Theory and Gravitational Collapse*, Univ. of Chicago Press, Chicago. [§26.1, Fig. 24.2]

Harrison, B. K., and J. A. Wheeler, 1958, "The Harrison-Wheeler equation of state," first presented in *Onzième Conseil de Physique Solvay: La Structure et l'évolution de l'univers*, Editions Stoops, Brussels, 1959; presented in greater detail in Harrison *et al.* 1965. [§24.2, Fig. 24.2]

Harrison, G. H., 1970, "A Compact Method for Symbolic Computation of the Riemann Tensor," M. S. thesis, Tech. Report, Dept. of Physics and Astronomy, Univ. of Maryland, College Park. [Box 14.3]

Hartle, J. B., 1967, "Slowly rotating relativistic stars, I: Equations of structure," *Astrophys. J. 150*, 1005–1029. [Ex. 26.1]

Hartle, J. B., 1970, "Slowly rotating relativistic stars, IV: Rotational energy and moment of inertia for stars in differential rotation," *Astrophys. J. 161*, 111–118. [§20.5, Ex. 33.13]

Hartle, J. B., 1971, "Long-range neutrino forces exerted by Kerr black holes," *Phys. Rev. D 3*, 2938–2940. [§41.11, Boxes 25.2, 32.2, 33.1]

Hartle, J. B., 1972, "Can a Schwarzschild black hole exert long-range neutrino forces?" in Klauder 1972. [Boxes 25.2, 32.2, 33.1]

Hartle, J. B. *See also* Brill and Hartle 1964, Hawking and Hartle 1972.

Hartle, J. B., and D. H. Sharp 1965, "Variational principle for the hydrostatic equilibrium of a relativistic, rotating fluid," *Phys. Rev. Lett. 15*, 909–911. [Ex. 26.1]

Hartle, J. B., and K. S. Thorne, 1968, "Slowly rotating relativistic stars, II: Models for neutron stars and supermassive stars," *Astrophys. J. 153*, 807–834. [Box 36.1, Fig. 24.2]

Hartle, J. B., and K. S. Thorne, 1974, "Slowly rotating relativistic stars, VII: Emission of gravitational waves due to quasiradial pulsation," *Astrophys. J.*, in preparation. [Box 14.3]

Harvey, J. A. *See* Dabbs *et al.* 1965.

Hauser, W., 1970, "On the fundamental equations of electromagnetism," *Am. J. Phys. 38*, 80–85. [§4.1]

Havens, W. W., Jr. *See* Durbin, Loar, and Havens 1952.

Hawking, S. W., 1966a, "Singularities and the geometry of space-time" Adams Prize Essay, Cambridge University, Cambridge, England. [Box 34.3]

Hawking, S. W., 1966b, "The occurrence of singularities in cosmology," *Proc. R. Soc. London A 294*, 511–521. [Box 34.3]

Hawking, S. W., 1969, "On the rotation of the Universe," *Observatory 89*, 38–39. [Box 34.3]

Hawking, S. W., 1971a, "Gravitationally collapsed objects of very low mass," *Mon. Not. R. Astron. Soc. 152*, 75–78. [Box 33.3]

Hawking, S. W., 1971b, "Gravitational radiation from colliding black holes," *Phys. Rev. Lett. 26*, 1344–1346. [§34.5, Boxes 33.1, 33.4, Ex. 34.4]

Hawking, S. W., 1972a, "Black holes in general relativity," *Commun. Math. Phys. 25*, 152–166. [§34.5, Boxes 33.1, 33.3, 33.4, 34.3, Ex. 34.4]

Hawking, S. W., 1972b, "Gravitational Radiation: The Theoretical Aspect," *Contemp. Phys. 13*, 273–282. [Box 34.3]

Hawking, S. W., 1973, "The Event Horizon," in DeWitt and DeWitt 1973. [§§34.1, 34.5]

Hawking, S. W., *cited by name in* §§32.7, 42.1, Box 27.7. *See also* Bardeen, Carter, and Hawking 1973 *and* Collins and Hawking 1973.

Hawking, S. W., and G. F. R. Ellis, 1968, "The cosmic black-body radiation and the existence of singularities in our universe," *Astrophys. J. 152*, 25–36. [§28.3, Box 34.3]

Hawking, S. W., and G. F. R. Ellis, 1973, *The Large Scale Structure of Space-time*, Cambridge Univ. Press, Cambridge, Eng. [§§34.1, 34.6, Box 33.1, endpapers]

Hawking, S. W., and J. B. Hartle, 1972, "Energy and angular momentum flow into a black hole," *Commun. Math. Phys. 27*, 283–290. [Box 33.3]

Hawking, S. W., and R. Penrose, 1969, "The singularities of gravitational collapse and cosmology.," *Proc. R. Soc. London A 314*, 529–548. [§§34.6, 44.1]

Hayakawa, S., 1965a, "Cosmological interpretation of the weak forces," *Prog. Theor. Phys. 33,* 538–539. [§44.6]

Hayakawa, S., 1965b, "Atomism and cosmology," *Prog. Theor. Phys. Supp.,* special number, Yukawa thirtieth anniversary issue, 532–541. [§44.6]

Hazard, C., *cited by name in* Box 27.7.

Hearn, A. C., 1970, "A program for symbolic algebraic computation," in *Proceedings of SHARE XXXIV,* 4.470–4.481. [Box 14.3]

Heckmann, O. *See* Schücking and Heckmann 1958.

Heisenberg, W., and H. Euler, 1936, "Folgerungen aus der Diracschen Theorie des Positrons," *Z. Phys. 98,* 714–732. [Ex. 40.8]

Helmholtz, H. L. F., *cited by name in* Box 38.2.

Henderson, T., *cited by name in* Box 27.7.

Hepp, K., 1969, *Théorie de la renormalisation,* Springer, Berlin. [Box 17.2 part 6]

Herman, R. C. *See* Alpher and Herman 1948a,b *and* 1950.

Hermann, R., 1966, *Lie Groups for Physicists,* W. A. Benjamin, New York. [§8.2]

Hernandez, W. C., Jr., 1967, "Material sources for the Kerr metric," *Phys. Rev. 159,* 1070–1072. [§33.3]

Heron (Hero of Alexandria), 1899, *Heronis Alexandrini Opera Omnia,* Teubner, Leipzig; *also in* G. Sarton, *A History of Science,* Harvard Univ. Press, Cambridge, Mass., 1952, I, 73. [Box 13.1]

Heron, G., trans., 1954, *Nicolas Cusanus's Of Learned Ignorance,* Yale Univ. Press, New Haven, Conn. [Box 27.7]

Hertz, G. L., *cited by name in* Box 17.3.

Hewish, A., *cited by name in* Box 27.7.

Hewish, A., S. J. Bell, J. D. H. Pilkington, P. F. Scott, and R. A. Collins, 1968, "Observation of a rapidly pulsating radio source," *Nature 217,* 709–713. [Box 24.1]

Hey, J. S., *cited by name in* Box 27.7.

Hibbs, A. R., 1951, "Quantum mechanics," in M. Kac, *Probability and Related Topics in Physical Science,* Interscience, New York, pp. 205–221. [Box 21.1]

Hibbs, A. R. *See also* Feynman and Hibbs 1965.

Hicks, N. J., 1965, *Notes on Differential Geometry,* Van Nostrand, Princeton, N.J. [Box 8.1, endpapers]

Hilbert, D., 1915, "Die Grundlagen der Physik," *Konigl. Gesell. d. Wiss. Göttingen, Nachr., Math.-Phys. Kl.,* 395–407; see also *Math. Annalen 92* (1924), 1–32. [§§17.7, 21.2, Box 17.2]

Hilbert, D., 1917, "Die Grundlagen der Physik (Zweite Mitteilung)," *Konigl. Gesell. d. Wiss. Göttingen, Nachr., Math.-Phys. Kl.,* 53–76. [§17.7]

Hilbert, D., *cited by name in* §21.1.

Hill, H. A. *See* Turner and Hill 1964.

Hill, J. M., 1971, "A Measurement of the gravitational deflection of radio waves by the sun," *Mon. Not. R. Astron. Soc., 154,* 7P-11P. [Box 40.1]

Hill, T., 1967, unpublished work referred to in DeWitt 1967c. [Box 17.2 part 6]

Hiltebeitel, A. M. *See* Gauss 1827

Hipparchus, *cited by name in* Box 16.1.

Hlavatý, Václav, *Festschrift* on the occasion of his 70th birthday, Jan. 27, 1964. *See* Hoffmann 1966.

Hönl, H., 1962, "Allgemeine Relativitätstheorie und Machsches Prinzip," in *Physikertagung Wien,* Physik Verlag, Mosbach/Baden, Germany. [§§21.12, 43.1]

Hoffmann, B., ed., 1966, *Perspectives in Geometry and Relativity: Essays in Honor of Václav Hlavatý,* Indiana Univ. Press, Bloomington.

Hoffman, B. *See also* Einstein, Infeld, and Hoffman 1938.

Hoffman, W. F. *See* Chiu and Hoffman 1964.

Hohn F. E., 1966, *Applied Boolean Algebra: An Elementary Introduction,* 2d ed., Macmillan, New York. [§44.5]

Holden, D. J., 1966, "An investigation of the clustering of radio sources," *Mon. Not. R. Astron. Soc., 133,* 225–228. [§27.1]

Holdridge, D. B. *See* Ohandley *et al.,* 1969.

Holmes, A., *cited by name in* Box 27.7.

Holton, G., 1965, personal communication, London. [Preface]

Holton, G., 1971, *Einstein. The Life and Times,* by R. W. Clark, book review, *New York Times Book Review,* September 5, 1–20. [Box 1.11]

Holton, G. *See also* Kollros 1956.

Horstmann, H. *See* Dabbs *et al.* 1965

Hoyle, F., 1948, "A new model for the expanding universe," *Mon. Not. R. Astron. Soc. 108,* 372–382. [§28.4]

Hoyle, F., *cited by name in* Box 27.7. *See also* Fowler and Hoyle 1964 *and* Wagoner, Fowler, and Hoyle 1967.

Hoyle, F., and W. A. Fowler, 1963a, "On the nature of strong radio sources," *Mon. Not. R. Astron. Soc. 125,* 169–176. [§24.5, Box 24.1]

Hoyle, F., and W. A. Fowler, 1963b, "Nature of strong radio sources," *Nature 197,* 533–535. [§24.5, Box 24.1]

Hoyle, F., W. A. Fowler, G. R. Burbidge, and E. M. Burbidge, 1964, "On relativistic astrophysics," *Astrophys. J. 139,* 909–928. [Box 33.3]

Hsiang, W.-C., and B. J. Sanderson, 1965, "Twist-spinning spheres in spheres," *Illinois J. Math. 9,* 651–659. [Box 44.3]

Hubble, E. P., 1920, "Photographic investigations of faint nebulae" (Ph.D. thesis), *Pub. Yerkes Obs., 4,* 69–85. [Box 29.5]

Hubble, E. P., 1925, "NGC 6822, a remote stellar system," *Astrophys. J. 62,* 409–433. [Box 29.5]

Hubble, E. P., 1926a, "A spiral nebula as a stellar system: Messier 33," *Astrophys. J. 63,* 236–274. [Box 29.5]

Hubble, E. P., 1926b, "Extra-galactic nebulae," *Astrophys. J. 64,* 321–369. [Box 29.5]

Hubble, E. P., 1929, "A relation between distance and radial velocity among extragalactic nebulae," *Proc. Nat. Acad. Sci. U.S. 15,* 169–173. [§§27.9, 29.2, Boxes 27.1, 29.5]

Hubble, E. P., 1934a, *Red-shifts in the spectra of nebulae* (Halley Lecture), Clarendon Press, Oxford, Eng. [Box 29.5]

Hubble, E. P., 1934b, "The distribution of extra-galactic nebulae," *Astrophys. J. 79,* 8–76. [§27.1, Box 29.5]

Hubble, E. P., 1936, "Effects of redshifts on the distribution of nebulae," *Astrophys. J. 84,* 517–554. [§27.1]

Hubble, E. P., *cited by name in* §17.3, Boxes 27.7, 29.5. *See also* Mayall 1970.

Hubble, E. P., and M. L. Humason, 1931, "The velocity-distance relation among extra-galactic nebulae," *Astrophys. J. 74,* 43–80. [Box 29.5]

Hughes, R. G., and M. S. Longair, 1967, "Evidence of the isotropy of faint radio sources," *Mon. Not. R. Astron. Soc. 135*, 131–137. [§27.1]

Hughes, V. W., H. G. Robinson, and V. B. Beltran-Lopez, 1960, "Upper limit for the anisotropy of inertial mass from nuclear resonance experiments," *Phys. Rev. Lett. 4*, 342–344. [§38.7]

Hughston, L., 1969, "Multifluid cosmologies," *Astrophys. J. 158*, 987–989. [§27.9]

Humason, M. L. *See* Hubble and Humason 1931.

Huneke, J. C. *See* Wasserburg, Schramm, and Huneke 1969.

Hunter, C., 1967, "Fragmentation," in Ehlers 1967, pp. 169–194. [Boxes 32.2, 32.3]

Huygens, C., 1698, *Cosmotheoros*, The Hague. English translation, *The Celestial Worlds Discovered or, Conjectures Concerning the Inhabitants, Plants and Productions of the Worlds in the Planets,* London, also 1698. [Box 27.7]

Imshennik, V. S. *See* Ivanova, Imshennik, and Nadezhin 1969.

Infeld, Leopold, 1962, volume dedicated to him in connection with his 60th birthday, Feb. 8, 1958: *Recent Developments in General Relativity,* Pergamon Press, Oxford, and PWN—Polish Scientific Publishers, Warsaw,

Infeld, L., ed., 1964, *Conférence Internationale sur les théories relativistes de la gravitation,* Gauthier-Villars, Paris, France.

Infeld, L. *See also* Einstein, Infeld, and Hoffman 1938.

Infeld, L., and J. Plebanski, 1960, *Motion and Relativity,* Pergamon, New York. [§20.6, endpapers]

Infeld, L., and A. Schild, 1949, "On the motion of test particles in general relativity," *Rev. Mod. Phys. 21*, 408–413. [§20.6]

Ingalls, R. P. *See* Shapiro, Ash *et al.* 1971, Shapiro, Smith *et al.* 1971, Shapiro *et al.* 1972.

Ipser, J. R., 1969, "Relativistic, spherically symmetric star clusters, III: Stability of compact isotropic models," *Astrophys. J. 158*, 17–43. [Boxes 24.1, 25.9]

Ipser, J. R., 1970, "Gravitational radiation from slowly rotating, fully relativistic stars," *Astrophys. J. 166*, 175–196. [Ex. 36.4]

Ipser, J. R., 1971, "Electromagnetic test fields around a Kerr-metric black hole," *Phys. Rev. Lett. 27*, 529–531. [§41.11, Ex. 33.14, Box 33.3]

Ipser, J. R. *See also* Brill *et al.* 1972, Fackerell and Ipser 1972, *and* Fackerell, Ipser, and Thorne 1969.

Ipser, J. R., and K. S. Thorne, 1968, "Relativistic, spherically symmetric star clusters, I: Stability theory for radial perturbations," *Astrophys. J. 154*, 251–270. [Box 24.1]

Isaacson, R. A., 1968a, "Gravitational radiation in the limit of high frequency, I: The linear approximation and geometrical optics," *Phys. Rev. 166*, 1263–1271. [§35.13]

Isaacson, R. A., 1968b, "Gravitational radiation in the limit of high frequency, II: Nonlinear terms and the effective stress tensor," *Phys. Rev. 166*, 1272–1280. [§35.13, Ex. 35.14]

Isenberg, J. A., *cited by name in* Figure 25.7.

Israel, W., 1960, "Relativistic theory of shock waves," *Proc. Roy. Soc. A 259*, 129–143. [§22.2, Box 22.2]

Israel, W., 1966, "Singular hypersurfaces and thin shells in general relativity," *Nuovo Cimento 44 B*, 1–14, and correction in *48 B*, 463. [§§21.5, 21.13]

Israel, W., 1967a, "Event horizons in static vacuum space-times," *Phys. Rev. 164*, 1776–1779. [Box 33.1]

Israel, W., 1967b, "Gravitational collapse and causality," *Phys. Rev. 153*, 1388–1393. [Ex. 21.27]

Israel, W., 1968, "Event horizons in static electrovac space-times," *Commun. Math. Phys. 8*, 245–260. [Box 33.1]

Israel, W., 1970, *Differential Forms in General Relativity,* Communications of the Dublin Institute for Advanced Studies, series A, no. 17. [§4.1]

Israel, W., 1971, personal communication. [Box 33.4]

Israel, W., 1972, "The relativistic Boltzman Equation," in O'Raifeartaigh 1972, pp. 201–241. [§22.6]

Israel, W., *cited by name in* Box 34.3. *See also* de la Cruz and Israel 1967.

Ithaca 1965. *See* Cornell University 1965 *or* Ehlers 1967.

Ivanova, L. N., V. S. Imshennik, and D. K. Nadezhin, 1969, "Investigations into the dynamics of supernova explosions," *Sci. Inf. Astron. Council USSR Acad. Sci. 13*, 3–93. [§32.7]

Jablonna 1962, *Relativistic Theories of Gravitation,* proceedings of a conference held in Warsaw and Jablonna, July 1962, edited by L. Infeld, Pergamon Press, Oxford, 1964.

Jackson, J. D., 1962, *Classical Electrodynamics,* Wiley, New York. [§2.1]

Jacobi, K. G. J., *cited by name in* Box 17.2.

Jaffe, J., 1969, "Collapsing objects and the backward emission of light," *Ann. Phys. U.S.A. 55*, 374–390. [§32.3]

Janis, A. I. *See* Couch *et al.* 1968.

Jauch, J., 1968, *Foundations of Quantum Mechanics,* Addison-Wesley, Reading, Mass. [Box 44.5]

Jeans, J., *cited by name in* Box 27.7.

Jones, B. T. J., and P. J. E. Peebles, 1972, "Chaos in Cosmology," *Comm. Astrophys. Space Sci. 4*, 121–128. [§30.1]

Jöstlein, H. *See* Bailey *et al.* 1968 *and* Farley *et al.* 1966.

Johnson, F. R., 1937, *Astronomical Thought in Renaissance England,* John Hopkins Press, Baltimore, [Box 27.7]

Jordan, P., 1955, *Schwerkraft und Weltall,* Vieweg und Sohn, Braunschweig. [§§38.6, 44.6]

Jordan, P., 1959, "Zum gegenwärtigen Stand der Diracschen kosmologischen Hypothesen," *Z. Phys. 157*, 112–121. [§§38.2, 38.6, 44.6, Box 39.1]

Jordan, P., *cited by name in* §§40.1, 40.6, 40.9.

Jordan, P., J. Ehlers, and R. Sachs, 1961, "Exact solutions of the field equations of general relativity, II: Contributions to the theory of pure gravitational radiation," (in German) *Akad. Wiss. Lit. Mainz Abh. Math.-Nat. Kl. 1*, 3–60. [§22.5]

Julian, W. H. *See* Goldreich and Julian 1968.

Jurgens, R. F. *See* Shapiro, Ash *et al.* 1971.

Kac, M., and S. Ulam, 1968, *Mathematics and Logic: Retrospect and Prospects,* Praeger, New York; abridged paperback edition by Pelican Books, Harmondsworth, Middlesex, 1971. [Box 44.5]

Kant, I., 1754, "Whether the earth has undergone an alteration of its axial rotation," *Wöchentliche Frag- und Anzeigungs-Nachrichten,* Kanter, Königsberg, mid-June issue [§1.5]

Kant, I., 1755, *Allgemeine Naturgeschichte und Theories des Himmels, oder Versuch von der Verfassung und dem Mechanischen Ursprunge des ganzen Welt gebäudes nach Newtonischen Grundsätzen Abgehandelt,* Petersen, Konigsberg and Leipzig. Translation by W. Hastie, with introduction

by M. K. Munitz, *Universal Natural History and Theory of the Heavens,* Univ. of Michigan Press, Ann Arbor, 1969. [Box 27.7]

Kaplan, S. A., 1949a, "Sverkhplotnye' zvezdy ('Superdense' stars)," *Naukovy Zapiski* (*Sci. Notes Univ. Lwow*) *15,* 109–115. [Box 24.1]

Kaplan, S. A., 1949b, "O krugovykh orbitakh v teorii tyagoteniya Einshteina (On circular orbits in Einstein's theory of gravitation)," *Zh. Eksp. & Teor. Fiz. 19,* 951–952. [Box 24.1]

Kasner, E., 1921a, "Geometrical theorems on Einstein's cosmological equations," *Am. J. Math. 43,* 217–221. [§30.2]

Kasner, E., 1921b, "Finite representation of the solar gravitational field in a space of six dimensions," *Am. J. Math 43,* 130–133. [§23.8]

Kasner, E., 1921c, "The impossibility of Einstein fields immersed in flat space of five dimensions," *Am. J. Math 43,* 126–129. [§23.8]

Kasner, E., *cited by name in* Boxes 27.7, 30.1.

Kaula, W. M. *See* Bender *et al.* 1971.

Keller, J. B., R. M. Lewis, and B. D. Seckler, 1956, "Asymptotic solution of some diffraction problems," *Commun. Pure & Appl. Math. 9,* 207–265. [§22.5]

Keller, L. V., *cited by name in* Box 27.6.

Kelley, J. L., 1955, *General Topology,* Van Nostrand, New York. [Box 34.2]

Kelvin. *See* Thomson, W., Lord Kelvin.

Kemble, E. C., 1937, *The Fundamental Principles of Quantum Mechanics,* McGraw-Hill, New York. Later editions available in paperback. [Box 25.3]

Keres, P. *See* Unt and Keres 1972.

Kepler, J., 1609, *Astronomia nova αιτιολογητος, seu Physica coelestis tradita comentariis de motibus stellae Martis,* Prague. [Box 27.7]

Kepler, J., 1619, *De Harmonice Mundi,* Augsburg. [Box 27.7]

Kerr, R. P., 1963, "Gravitational field of a spinning mass as an example of algebraically special metrics," *Phys. Rev. Lett. 11,* 237–238. [§§33.2, 33.6, Box 33.2]

Kerr, R. P., and A. Schild, 1965, "A new class of vacuum solutions of the Einstein field equations," in *Proceedings of the Galileo Galilei Centenary Meeting on General Relativity, Problems of Energy and Gravitational Waves,* G. Barbera, ed., Comitato Nazionale per le Manifestazione Celebrative, Florence, pp. 222–233. [§33.6]

Khalatnikov, I. M., *cited by name in* Box 27.7. *See also* Belinsky and Khalatnikov 1969a, 1969b, *and* 1970, Belinsky, Khalatnikov, and Lifshitz 1970, Belinsky, Lifshitz, and Khalatnikov 1971, Lifshitz and Khalatnikov 1963a *and* 1963b, *and* Ryan 1972b.

Khalatnikov, I. M., and E. M. Lifshitz, 1970, "General cosmological solutions of the gravitational equations with a singularity in time," *Phys. Rev. Lett. 24,* 76–79. [§34.6, Box 30.1]

Khuri, N., 1957, unpublished calculation quoted in Lindquist and Wheeler 1957. [§25.5]

Killing, W., 1892, "Über die Grundlagen der Geometrie," *J. Reine Angew. Math. 109,* 121–186. [§25.2]

Kilmister, C. W., 1971, "More about the King and the Vizier," *Gen. Relativity & Grav. 2,* 35–36. [Fig. 44.3]

Kirchhoff, G.-R., *cited by name in* Box 38.2.

Kittel, C., 1958, *Elementary Statistical Physics,* Wiley, New York. [§23.4]

Klauder, J., ed., 1972, *Magic Without Magic: John Archibald Wheeler,* W. H. Freeman, San Francisco.

Klauder, L. T., Jr., M. Wakano, J. A. Wheeler, and R. Willey, 1959, "Kinematics of a spherical isotropic uniform universe," in *Onzième Conseil de Physique Solvay: La Structure et l'évolution de l'univers,* Éditions Stoops, Brussels, Belgium. [§29.5]

Klein, F., 1872, "Erlangerprogram: Vergleichende Betrachtungen über neuere geometrischen Forschungen," translation in *Bull. Amer. Math. Soc. 2,* 215–249, 1893. [§2.2]

Klein, M. J., 1971, *Einstein: The Life and Times, by R. W. Clark,* book review, *Science, 174,* 1315–1316. [Box 1.11]

Klein, O., 1966, "Boundary conditions and general relativity" in A. de-Shalit, H. Feshbach, and L. Van Hove, eds., *Preludes in Theoretical Physics in Honor of V. F. Weisskopf,* North Holland, Amsterdam, pp. 23–26. [Box 27.7]

Klein, O., 1971, "Arguments concerning relativity and cosmology," *Science 171,* 339–345. [§§27.10, 27.11, Box 27.7]

Klein, O. *See also* Alfvén and Klein 1962.

Klemperer, W. B. *See* Deutsch and Klemperer 1962.

Kline, M., 1954, "Asymptotic solution of linear hyperbolic partial-differential equations," *J. Rat. Mech. & Anal. 3,* 315–342. [§22.5]

Klüber, H. V., 1960, "The determination of Einstein's light-deflection in the gravitational field of the sun," *Vistas in Astronomy 3,* 47–77, London, Pergamon Press. [Box 40.1]

Kluitenberg, G. A., and S. R. deGroot, 1954, "Relativistic thermodynamics of irreversible processes, III," *Physica 20,* 199–209. [§22.2]

Kneebone, G. T., 1963, *Mathematical Logic and the Foundations of Mathematics: An Introductory Survey,* Van Nostrand, Princeton, N. J. [Box 44.5]

Knight, L. V. *See* Witteborn, Knight, and Fairbank 1967.

Kollros, L., 1956, in Carl Seelig, ed. *Helle Zeit- Dunkle Zeit—In Memoriam Albert Einstein,* Europa, Zurich. We thank Professor G. Holton for this quotation. [preface]

Kovács, S., *cited by name in* ex. 35.20.

Kraichnan, R. H., 1955, "Special-relativistic derivation of generally covariant gravitation theory," *Phys. Rev. 55,* 1118–1122. [§§7.1, 18.1]

Kramers, H. A., 1957, *Quantum Mechanics,* North Holland Pub. Co., Amsterdam, Holland. Translated from the German by D. ter Haar. [§21.1]

Krefetz, E., 1967, "The equilibrium of slowly rotating configurations in the post-Newtonian approximation: Corrections to Clairaut's equations," *Astrophys. J. 148,* 589–612. [Ex. 26.1]

Kretschmann, E., 1917, "Über den physikalischen Sinn der Relativitätspostulate, A. Einsteins neue und seine ursprungliche Relativitätstheorie," *Ann. Phys. (Germany) 53,* 575–614. [§§12.5, 17.6]

Kreuzer, L. B., 1968, "Experimental measurement of the

equivalence of active and passive gravitational mass," *Phys. Rev. 169,* 1007–1012. [§40.8]

Krogh, C., and R. Baierlein, 1968, "Lunar laser ranging and the Brans-Dicke theory," *Phys. Rev. 175,* 1576–1579. [§40.6]

Krotkov, R. *See* Bertotti, Brill, and Krotkov 1962 *and* Roll, Krotkov, and Dicke 1964.

Kruskal, M. D., 1960, "Maximal extension of Schwarzschild metric," *Phys. Rev. 119,* 1743–1745. [§31.4, Box 31.1]

Kuchař, K., 1968, "Charged shells in general relativity and their gravitational collapse," *Czech J. Phys. B 18,* 435–463. [§21.13]

Kuchař, K., 1970, "Ground state functional of the linearized gravitational field," *J. Math. Phys. 11,* 3322–3334. [§43.4]

Kuchař, K., 1971a, "Canonical quantization of cylindrical gravitational waves," *Phys. Rev. D 4,* 955–986. [§§21.1, 21.10]

Kuchař, K., 1971b, "Gravitational action functional reexpressed by means of intrinsic and extrinsic curvature," unpublished lecture given in the course *Relativity* at Princeton University. [§§21.6, 21.11, Ex. 21.10]

Kuchař, K., 1972, "A bubble-time canonical formalism for geometrodynamics," *J. Math. Phys. 13,* 768–781. [§§21.1, 21.10, 21.11, Ex. 21.24]

Kuchař, K., *cited by name in* §21.1, Ex. 13.7, 21.10.

Kuhlthau, A. R. *See* Rose *et al.* 1969.

Kundt, W., 1971, "Cosmological density fluctuations during Hadron stage," in Sachs 1971, pp. 365–372. [§28.1]

Kundt, W. *See also* Ehlers and Kundt 1962.

Kundt, W., and E. T. Newman, 1968, "Hyperbolic differential equations in two dimensions," *J. Math. Phys. 9,* 2193–2210. [§§35.8, 38.8]

Künzle, H. P., 1971, "On the Spherical Symmetry of a Static Perfect Fluid," *Commun. Math. Phys. 20,* 85–100. [§23.1]

Kuper, C. G., and A. Peres, eds., 1971, *Relativity and Gravitation,* Gordon and Breach, New York.

Kursunoglu, B., 1962, *Modern Quantum Theory,* W. H. Freeman, San Francisco. [Box 21.1]

Kustaanheimo, P., 1966, "Route dependence of the gravitational red shift," *Phys. Lett. 23,* 75–77. [§39.1]

Lagrange, J. L., *cited by name in* Box 17.2.

Lamb, W. E., Jr., and R. C. Retherford, 1947, "Fine structure of the hydrogen atom by a microwave method," *Phys. Rev. 72,* 241–243. [Fig. 43.3, Box 17.2 part 6]

Lambert, J. H., 1761, *Kosmologischen Briefe,* Augsburg; see his *The System of the World,* translated, with modifications, from the French by J. Jacque, London, 1800; reprinted in part in Munitz 1957. [Box 27.7]

Lanczos, K., 1922, "Bemerkung zur de Sitterschen Welt," *Phys. Z. 23,* 539–543. [§21.13]

Lanczos, K., 1924, "Flächenhafte Verteilung der Materie in der Einsteinschen Gravitationstheorie," *Ann. Phys. (Germany) 74,* 518–540. [§21.13]

Lanczos, K., 1970, *The Variational Principles of Mechanics,* Univ. of Toronto Press, Toronto, 4th ed. [Box 21.1]

Landau, L. D., 1932, "On the theory of stars," *Phys. Z. Sowjetunion 1,* 285–288. [Box 24.1]

Landau, L. D., and E. M. Lifshitz, 1959, *Fluid Mechanics,* Pergamon Press, London. [Ex. 22.7]

Landau, L. D., and E. M. Lifshitz, 1960, *Mechanics,* Addison-Wesley, Reading, Mass. [Box 22.4]

Landau, L. D., and E. M. Lifshitz, 1962, *The Classical Theory of Fields,* translated by M. Hamermesh, Addison-Wesley, Reading, Mass. [§§20.3, 20.5, 36.1, endpapers]

Landau, L. D. and E. M. Lifshitz, 1971, *The Classical Theory of Fields,* 3d rev. English ed., Addison-Wesley, Reading, Massachusetts, and Pergamon, London. [endpapers]

Langer, W. D., and A. G. W. Cameron, 1969, "Effects of hyperons on the vibrations of neutron stars," *Astrophys. & Space Sci. 5,* 213–253. [§24.3]

Lao-tzu, 3d century B.C. *Tao Tê Ching,* Poem 1. Translated as *The Way of Life,* R. B. Blakney, New American Library, 1955; translation revised by Janet Ten Broeck. [Epigraph to Chap. 18]

Langevin, P. *See* Einstein 1912.

Laplace, P. S., 1795, *Le Système du monde,* vol. II, Paris. Published in English as *The System of the World,* W. Flint, London, 1809. [Fig. 24.1, Box 24.1, Epigraph to chap. 33]

Laplace, P. S., *cited by name in* Box 17.2.

Lascoux, J., 1968, "Perturbation theory in quantum field theory and homology," in DeWitt and Wheeler 1968, 354–419. [Ex. 4.5]

Laurence, D. H. *See* Boni, Russ, and Laurence 1960.

Lawson, R. W., *cited by name in* Box 27.7.

Leavitt, H., *cited by name in* Box 27.7.

Lebovitz, N. R. *See* Chandrasekhar and Lebovitz 1968.

Lecat, M., 1924, *"Bibliographie de la relativité,"* Lamertine, Bruxelles, cited in preface of present bibliography.

Ledoux, P., and Th. Walraven, 1958, "Variable stars," *Handbuch der Physik 51,* S, Flügge ed., Springer-Verlag, Berlin, pp. 353–604. [Box 24.2]

Lefschetz, S. 1949, *Introduction to Topology,* Princeton Univ. Press, Princeton, N.J. For a fuller account of fixed-point theorems, see Dunford and Schwartz 1958 [§36.1]

Leibniz, G. W., 1714, *La Monadologie.* English translation available in several books; parts included in P. P. Wiener, ed., *Leibniz Selections,* Scribners, New York, 1951. [§44.6]

Leibniz, G. W., 1961, untitled essay in his *Philosophischen Schriften,* vol. 7, ed. by C. J. Gerhardt, Olms Verlagsbuchhandlung, Hildesheim, p. 200. [Epigraph to Chap. 21]

Leibniz, G. W., *quoted at end of* Chap. 44.

Leighton, R. B., 1959, *Principles of Modern Physics,* McGraw-Hill, New York. [§2.1]

Leighton, R. B. *See also* Feynman, Leighton, and Sands 1964.

Lemaître, G., 1933a, "Formation of nebulae in the expanding universe," *Acad. Sci., Paris, Comptes Rend. 196,* 1085–1087. [§30.6]

Lemaître, G., 1933b, "Spherical condensations in the expanding universe," *Acad. Sci., Paris, Comptes Rend. 196,* 903–904. [§30.6]

Lemaître, G., 1933c, "L'univers en expansion," *Ann. Soc. Sci. Bruxelles I A53,* 51–85. [Box 31.1]

Lemaître, G., 1949, cited in Schilpp 1949, p. 440. [Epigraph to Chap. 36]

Lemaître, G., *cited by name in* Box 27.7.

Lenin, V. I., 1927, *Materialism and Empiriocriticism: Critical Comments on a Reactionary Philosophy,* International Publications, New York, 2d English ed. of *Materializm' i Empiriokrititsizm': Kriticheskia Zametki ob' Odnoi Reaktsionnoi Filosofii,* Zveno Press, Moscow, 1909. [§21.12]

Lense, J. *See* Thirring and Lense 1918.

Les Houches, France, Ecole d'ete de physique theorique: 1963, *see* DeWitt and DeWitt 1964b; 1966, *see* DeWitt, Schatzman, and Veron 1967 (third volume of three); 1972, *see* DeWitt and DeWitt 1973.

Leutwyler, H., 1968, "Boundary conditions for the state functional in the quantum theory of gravity," in DeWitt and Wheeler 1968, pp. 309–317. [§43.3]

Lewis, R. M., 1958, "Asymptotic expansion of steady-state solutions of symmetric hyperbolic linear differential equations," *J. Rat. Mech. & Math. 7,* 593–641. [§22.5]

Lewis, R. M. *See also* Keller, Lewis, and Seckler 1956.

Liang, E. *See* Eardley, Liang, and Sachs 1972.

Lichnerowicz, A., 1944, "L'intégration des équations de la gravitation relativiste et le problème des *n* corps," *J. Math. Pures & Appl. 23,* 37–63. [§21.11]

Lichnerowicz, A., 1955, *Théories relativistes de la gravitation et de l'électromagnétisme,* Masson, Paris. [endpapers]

Lichnerowicz, A., 1961a, "Anticommutateur du champ spinoriel en relativité générale," *Acad. Sci., Paris, Comptes Rend. 252,* 3742–3744. [Box 44.3]

Lichnerowicz, A., 1961b, "Anticommutateur pour le champ correspondant au spin 3/2 sur un espace d'Einstein," *Acad. Sci., Paris, Comptes Rend. 253,* 940–942. [Box 44.3]

Lichnerowicz, A., 1961c, "Théorie de Petiau-Duffin-Kemmeren," *Acad. Sci., Paris, Comptes Rend. 253,* 983–985. [Box 44.3]

Lichnerowicz, A., 1962, "Radiations en relativité générale," especially "Le tenseur de superénergie," in CNRS 1962, pp. 93–106. [Ex. 15.2]

Lichnerowicz, A., 1964, "Propagateurs, commutateurs, et anticommutateurs en relativité générale," in DeWitt and DeWitt 1964. [Box 44.3, Ex. 15.2]

Lichnerowicz, A., 1967, *Relativistic Hydrodynamics and Magnetohydrodynamics; Lectures on the Existence of Solutions,* W. A. Benjamin, New York. [§§22.2, 22.3, Box 22.2]

Lichnerowicz, A., 1971, "Ondes des choc, ondes infinitesimales et rayons en hydrodynamique et magnetohydrodynamique relativistes," in Cattaneo 1971, pp. 88–203. [§22.2, Box 22.2]

Lichtenberg, D. B., 1965, *Meson and Baryon Spectroscopy,* Springer-Verlag, New York. [§38.4]

Lifshitz, E. M., *cited by name in* Box 27.7. *See also* Belinsky, Khalatnikov, and Lifshitz 1970, Belinsky, Lifshitz, and Khalatnikov 1971, Khalatnikov and Lifshitz 1970, *and* Landau and Lifshitz, 1959, 1960, 1962, and 1971.

Lifshitz, E. M., and I. M. Khalatnikov, 1963a, "Investigations in relativistic cosmology," in *Advances in Physics 12,* 185–249; translated from the Russian by J. L. Beeby. [§§30.6, 34.6]

Lifshitz, E. M., and I. M. Khalatnikov, 1963b, "Problems of relativistic cosmology," *Usp. Fiz. Nauk 80,* 391–438. English translation in *Sov. Phys.—Uspekhi 6,* 495–522, 1964. [§§30.6, 34.6]

Lifshitz, E. M., and I. M. Khalatnikov, 1970, "Oscillatory approach to singular point in the open cosmological model," *Zh. Eksp. & Teor. Fiz. Pis'ma., 11,* 200–203. English translation in *Sov. Phys.—JETP Lett. 11,* 123–125 (1971).

Lighthill, M. J., 1958, *Introduction to Fourier Analysis and Generalized Functions,* Cambridge University Press, Cambridge, England. [§4.7]

Lilley, A. E. *See* Dieter *et al.* 1962.

Lin, C. C., L. Mestel, and F. H. Shu, 1965, "The gravitational collapse of a uniform spheroid," *Astrophys. J. 142,* 1431–1446. [Boxes 32.2, 32.3]

Lindquist, R. W., 1960, *The Two-Body Problem in Geometrodynamics,* doctoral dissertation, Princeton University, Princeton, N.J. [Ex. 32.1]

Lindquist, R. W., 1966, "Relativistic transport theory," *Ann. Phys. (USA) 37,* 487–518. [§22.6]

Lindquist, R. W. *See also* Boyer and Lindquist 1967.

Lindquist, R. W., and J. A. Wheeler, 1957, "Dynamics of a lattice universe by the Schwarzschild cell method," *Rev. Mod. Phys. 29,* 432–433. [§27.10, Box 25.2, Fig. 25.3, 27.3]

Liouville, J., 1837, "Second mémoire sur le développement des fonctions ou parties des fonctions en séries ," Liouville's *J. de Math. Pures et Appliquées 2,* 16–35. [§22.5]

Loar, H. H. *See* Durbin, Loar, and Havens 1952.

London 1965, International Conference on Relativistic Theories of Gravitation, King's College, University of London, July 1965. Proceedings multilithed in 2 volumes by King's College.

Longair, M. S. *See* Hughes and Longair 1967.

Lorentz, H. A., *cited by name in* §20.6 *and* Box 16.4.

Lorentz, H. A., A. Einstein, H. Minkowski, and H. Weyl, 1923, *The Principle of Relativity: A Collection of Original Memoirs,* Methuen, London. Paperback reprint, Dover, New York.

Lowry, R. A. *See* Rose *et al.* 1969.

Lukash, V. N. *See* Doroshkevich, Lukash, and Novikov 1971.

Lurie, D., and N. Mukunda, eds., 1963, *Proceedings of Seminar on Unified Theories of Elementary Particles,* University of Rochester report URPA-11, Rochester, N.Y.

Lynden-Bell, D., 1967, "Statistical mechanics of violent relaxation in stellar systems," *Mon. Not. R. Astron. Soc. 136,* 101–121. [Box 33.3]

Lynden-Bell, D., 1969, "Galactic nuclei as collapsed old quasars," *Nature 223,* 690–694. [Box 33.3]

Lynden-Bell, D., and M. Rees, 1971, "On quasars, dust, and the galactic center," *Mon. Not. R. Astron. Soc. 152,* 461–475. [Box 33.3]

Lynds, C. R., *cited by name in* Box 27.7.

MacCallum, M. A. H. *See* Penrose and MacCallum 1973.

MacDonald, G. J. F. *See* Bender *et al.* 1971 *and* Munk and MacDonald 1960.

Mach, E., 1912, *Die Mechanik in Ihrer Entwicklung Historisch-*

Kritisch Dargestellt, Brockhaus, Leipzig. English translation by T. J. McCormack, published in 1960 as *The Science of Mechanics* with an introduction by Karl Menger, Open Court, La Salle, Ill. [§§21.1, 21.12]

Mach, E., *cited by name in* Box 27.1.

Machado, A., 1940, "Proverbios y cantares XXIX," in his *Obras,* Editorial Seneca, Mexico, pp. 235–236. [Epigraph to Chap. 43]

MacKenzie, R. E. *See* Auslander and MacKenzie 1963.

Mackey, M. B., *cited by name in* Box 27.7.

Manasse, F. K., and C. W. Misner, 1963, "Fermi Normal Coordinates and Some Basic Concepts in Differential Geometry," *J. Math. Phys. 4,* 735–745. [§13.6]

Mandelstam, S. *See* Yourgrau and Mandelstam 1968.

Mann, T., 1937, *Freud, Goethe, Wagner,* Knopf, New York. Essay on Freud published originally as *Freud und die Zukunft (Vortrag Gehalten in Wien am 8 Mai 1936 zur Feier von Sigmund Freuds 80 Geburtstag),* Bormann-Fischer, Wien (1936). [§44.5]

Manukin, A. B. *See* Braginsky, Manukin, and Tikhonov 1970.

Marconi, G. W., *cited by name in* Preface.

Marinus of Naples, *cited by name in* Fig. 1.8.

Markus, L. *See* Auslander and Markus 1959.

Marle, C., 1969, "Sur l'établissement des équations de l'hydrodynamique des fluides relativistes dissipatifs, I et II," *Ann. Inst. Henri Poincare, 10,* 67–126 and 127–194. [§22.6, Ex. 22.7]

Marsden, J. B. *See* Fischer and Marsden 1972.

Martin, W. L. *See* Anderson *et al.* 1971.

Marzke, R. F., and J. A. Wheeler, 1964, in Chiu and Hoffman 1964. [Box 16.4]

Mathews, J. *See also* Peters and Mathews 1963.

Mathews, J., and R. Walker, 1965, *Mathematical Methods of Physics,* W. A. Benjamin, New York. [Box 26.1]

Matzner, R. A., 1968, "The present temperature of primordial black-body gravitational radiation," *Astrophys. J. 154,* 1123–1125. [§27.10]

Matzner, R. A. *See also* Clemens and Matzner 1967 *and* Fletcher *et al.* 1967.

Matzner, R. A., and C. W. Misner, 1972, "Dissipative effects in the expansion of the universe," *Astrophys. J. 171,* 415–432. [§§28.2, 30.4]

Maupertuis, P. L. M. de, 1746, "Recherche des loix du mouvement," *Acad. R. Sci. Berlin.* [Epigraph to Chap. 21]

Mavrides, M., *quoted at end of* Chap. 44.

Maxwell, J. C., *cited by name in* Preface, Box 4.6.

May, M., and R. H. White, 1966, "Hydrodynamic calculations of general-relativistic collapse," *Phys. Rev. 141,* 1232–1241. [§§32.7, 42.7, Box 33.3, Ex. 32.7]

May, M., and R. H. White, 1967, "Stellar dynamics and gravitational collapse," *Meth. Computat. Phys. 73,* 219–258. [§22.2, Boxes 22.2, 33.3]

Mayall, N. U., 1970, "Edwin Powell Hubble," *Biograph. Mem. Nat. Acad. Sci., 41,* 176–214. [Box 29.5]

Mayall, N. U., and Oort, J. H., 1942, "Further data bearing on the identification of the Crab Nebula with the supernova of 1054 A.D., part II: The astronomical aspects," *Proc. Astron. Soc. Pacific 54,* 95–104. [Box 24.1]

Mazet, E. *See* Berger, Gauduchon, and Mazet 1971.

McCrea, W. H., *cited by name in* Box 27.7. *See also* Bondi and McCrea 1960.

McDonald, E., *cited by name in* Fig. 41.6.

McVittie, G. C., 1956, *General Relativity and Cosmology,* Chapman and Hall, London. [endpapers]

McWhirter, N., and R. McWhirter, 1971, *The Guinness Book of Records,* Guinness Superlatives, Ltd., Enfield, Middlesex [Chap. 12 epigraph]

Meer, S. van der. *See* Farley *et al.* 1966.

Meixner, J., and H. G. Reik, 1959, "Thermodynamik der irreversiblen Prozesse," *Handbuch der Physik, 3, Teil 2,* 413–505. [§22.2]

Meixner, J., and F. W. Schäfke, 1954, *Mathieusche Funktionen und Sphäroidfunktionen,* Springer-Verlag, Berlin, Germany. [Ex. 33.14]

Melbourne, W. G. *See* Ohandley *et al.* 1969.

Meltzer, D. W. *See* Bardeen, Thorne, and Meltzer 1966.

Meltzer, D. W., and K. S. Thorne, 1966, "Normal modes of radial pulsation of stars at the end point of thermonuclear evolution," *Astrophys. J. 145,* 514–543. [Box 26.1]

Menger, Karl, 1949, see Schilpp 1949, pp. 466–467. [Epigraph to Chap. 2, Epigraph to Chap. 9]

Menskii, M. B. *See* Braginsky and Menskii 1971.

Mercier, A., 1953, *Analytical and Canonical Formalism in Physics,* North-Holland, Amsterdam, Holland. Paperback reprint, Dover, New York, 1963. [Box 21.1]

Mercier, A., and M. Kervaire, 1956, *Jubilee of Relativity Theory,* proceedings of a conference in Bern, Switzerland, July 11–16, 1955, Birkhäuser, Bern.

Merzbacher, E., 1961, *Quantum Mechanics,* Wiley, New York. [§16.3]

Messiah, A., 1961, *Quantum Mechanics,* translated from the French by G. M. Temmer, North-Holland, Amsterdam, Holland. [Ex. 3.12]

Mestel, L. *See* Lin, Mestel, and Shu 1965.

Mielnik, B., and J. Plebanski, 1962, "A study of geodesic motion in the field of Schwarzschild's solution," *Acta Phys. Polon. 21,* 239–268. [§25.5]

Mikola, S. *See* Fejér and Mikola 1918.

Milne, E. A., *cited by name in* Box 27.7.

Milnor, J., 1962, "A survey of cobordism theory," *Enseignement Math. 8,* 16–23. [Box 44.3]

Milnor, J., 1963, "Spin structures on manifolds," *Enseignement Math. 9,* 198–203. [Box 44.3]

Milnor, J., 1965a, "On the Stiefel-Whitney numbers of complex manifolds and of spin manifolds," *Topology 3,* 223–230. [Box 44.3]

Milnor, J., 1965b, "Remarks concerning spin manifolds," in Cairns 1965, p. 55. [Box 44.3]

Milton, J., 1665, *Paradise Lost,* Book V, Lines 623–625, in *The Complete Poems of John Milton,* Union Library Association of New York, 1936. [Epigraph to Chap. 25]

Minkowski, H., 1908, address delivered at the 80th Assembly of German Natural Scientists and Physicians, Cologne. English translation, "Space and time," in *The Principle of Relativity,* Dover, New York, 1923. [§1.5]

Minkowski, H. *See also* Lorentz *et al.* 1923.

Minkowski, R., 1942, "The Crab Nebula," *Astrophys. J. 96*, 199–213. [Box 24.1]

Minkowski, R., *cited by name in* Box 27.7.

Minkowski, R., and O. C. Wilson, 1956, "Proportionality of nebular redshifts to wave length," *Astrophys. J. 123*, 373–376. [§38.6]

Misner, C. W., 1957, "Feynman quantization of general relativity," *Rev. Mod. Phys. 29*, 497–509. [Box 21.1]

Misner, C. W., 1964a, "Differential geometry and differential topology," in DeWitt and DeWitt 1964b, pp. 881–929. [Box 8.1]

Misner, C. W., 1964b, "Waves, Newtonian fields, and coordinate functions," in Infeld 1964. [§35.8]

Misner, C. W., 1965, "Relativistic equations for spherical gravitational collapse with escaping neutrinos," *Phys. Rev. B137*, 1360–1364. [Box 23.1]

Misner, C. W., 1968, "The isotropy of the universe," *Astrophys. J. 151*, 431–457. [§§28.2, 30.3]

Misner, C. W., 1969a, "Gravitational collapse," in Chrétien, Deser, and Goldstein 1969, vol I. [§§12.1, 32.5, Boxes 8.1, 31.2, Ex. 31.8, endpapers]

Misner, C. W., 1969b, "Mixmaster universe," *Phys. Rev. Lett. 22*, 1071–1074. [§§21.10, 28.2, 30.6, 34.6, Box 30.1]

Misner, C. W., 1969c, "Absolute zero of time," *Phys. Rev. 186*, 1328–1333. [§§28.3, 34.6]

Misner, C. W., 1970, "Classical and quantum dynamics of a closed universe," in Carmeli, Fickler, and Witten 1970, pp. 55–79. [Box 30.1]

Misner, C. W., 1972a, "Minisuperspace," in Klauder 1972. [§43.3, Box 30.1]

Misner, C. W., 1972b, "Stability of Kerr black holes against scalar perturbations," *Bull. Am. Phys. Soc. (II) 17*, 472. [Ex. 33.14]

Misner, C. W., 1973, "A minsuperspace example: The Gowdy T^3 cosmology," Caltech Orange Aid Preprint. Submitted to *Phys. Rev.* [§43.3]

Misner, C. W., *cited by name in* §§21.1, 35.8, Boxes 17.2 part 2', 17.3, 27.7. *See also* Arnowitt, Deser, and Misner 1962, Beckedorff and Misner 1962, Chrzanowski and Misner 1973, Manasse and Misner 1963, *and* Matzner and Misner 1972.

Misner, C. W., and D. H. Sharp, 1964, "Relativistic equations for adiabatic, spherically symmetric gravitational collapse," *Phys. Rev. B 136*, 571–576. [§§21.13, 22.2, Ex. 32.7, Box 23.1]

Misner, C. W., and A. H. Taub, 1968, "A singularity-free empty universe," *Zh. Eksp. & Teor. Fiz. 55*, 233. English original in *Sov. Phys.—JETP 28*, 122, 1969. [§§32.5, 34.6]

Misner, C. W., and J. A. Wheeler, 1957, "Classical physics as geometry: Gravitation, electromagnetism, unquantized charge, and mass as properties of curved empty space," *Ann. Phys. (U.S.A.) 2*, 525–603, reprinted in Wheeler 1962. [§§20.6, 21.8, 21.9, Fig. 44.1, Boxes 4.1, 4.4, 15.1, 21.2, Ex. 20.7]

Misner, C. W., and J. A. Wheeler, 1972, "Conservation laws and the boundary of a boundary," in Shelest 1972, pp. 338–351. [§15.3]

Moffet, A., *cited by name in* Box 40.1.

Møller, C., 1952, *The Theory of Relativity*, Oxford Univ. Press, London; 2d ed., 1972. [endpapers]

Møller, C., ed., 1962, *Evidence for Gravitational Theories*, Academic Press, New York.

Moncrief, V. E. *See* Berger *et al.* 1972.

Morduch, G. E. *See* Whitrow and Morduch 1965.

Morehead, J. C. *See* Gauss 1827.

Morette, C., 1951, "On the definition and approximation of Feynman's path integrals," *Phys. Rev. 81*, 848–852. [Box 21.1]

Morette, C. *See also* DeWitt, C.

Morgan, T., and K. S. Thorne, 1973, in preparation. [Box 32.3]

Morganstern, R. E., 1973, "Observational constraints imposed by Brans-Dicke cosmologies," *Phys. Rev. D 7*, 1570–1579. [§28.4]

Moritz, B. K., 1969, *A Study of the General Relativistic Dynamics of the Klein-Alfvén Cosmology*, doctoral dissertation, University of Maryland [§27.11, §28.4]

Morse, M., 1934, *The calculus of variations in the large*, American Mathematical Society, Providence, R.I. [Fig. 13.2]

Morse, M., and S. S. Cairns, 1969, *Critical Point Theory in Global Analysis and Differential Topology: an Introduction*, Academic Press, New York. [Fig. 13.2]

Moser, J. K. *See* Siegel and Moser 1971.

Mottleson, B. R. *See* Bohr and Mottleson 1969.

Muhleman, D. O. *See* Anderson *et al.* 1971.

Muhleman, D. O., R. D. Ekers, and E. B. Fomalont, 1970, "Radio interferometric test of the general relativistic light bending near the sun," *Phys. Rev. Lett. 24*, 1377–1380. [Box 40.1]

Muhlemann, D. O., and P. Reichley, 1964, "Effects of general relativity on planetary radar distance measurements," *JPL Space Programs Summary 4*, no. 37-29, p. 239. [§40.4]

Muhlemann, D. O., and P. Reichley, 1965, "General relativistic effects on radar propagation," *JPL Space Programs Summary 4*, no. 37-31, p. 342. [§40.4]

Mukunda, N. *See* Lurie and Mukunda 1963.

Mulholland, J. D. *See* Bender *et al.* 1971 *and* Ohandley *et al.* 1969.

Munitz, M. K., 1957, *Theories of the Universe*, Free Press, New York. [Box 27.7]

Munk, W. H., and G. J. F. MacDonald, 1960, *The Rotation of the Earth; a Geophysical Discussion*, Cambridge University Press, Cambridge, England. [Fig. 1.8]

Nadezhin, D. K. *See* Ivanova, Imshennik, and Nadezhin 1969.

Nagel, E., and J. R. Newman, 1958, *Gödel's Proof*, New York Univ. Press, New York.

NASL, 1969, *Proceedings of the 1969 Symposium on Unconventional Inertial Sensors (U) Vol. I*, Naval Applied Science Laboratory, Brooklyn, New York.

Nasr, H., 1964, *Three Muslim sages: Avicenna, Suhrawardi, Ibn Arabi*, Harvard University Press, Cambridge, Mass. [Box 27.7]

Navier, L., *cited by name in* Box 17.2 part 6.

Nazarenko, V. S. *See* Braginsky and Nazarenko 1971.

Ne'eman, Y., 1965, "Expansion as an energy source in quasistellar radio sources," *Astrophys. J. 141*, 1303–1305. [§34.3]

Nester, J. M., *cited by name in* §35.1.

Neugebauer, O., 1959, *Greek Horoscopes,* with H. B. van Hösen, American Philosophical Society, Philadelphia, Pa. [Fig. 1.8]

Neugebauer, P. V., 1927, "Zür astronomischen Chronologie," *Astron. Nachr. 230,* 57–64. [Fig. 1.8]

Neugebauer, P. V., 1929, *Astronomische Chronologie,* 2 vols, De Gruyter, Berlin. [Fig. 1.8]

Neugebauer, P. V., 1930, "Schoch Todesanzeige," *Astron. Nachr. 237,* 221–224. [Fig. 1.8]

Neugebauer, P. V., and C. Schoch, 1930, "Zür astronomischen Chronologie, II," *Astronomische Abhandlungen. Ergan-zungshefte zu den Astron. Nachrichten 8,* no. 2, 24–32. [Fig. 1.8]

Newman, E. T. *See also* Couch and Newman 1972, Couch *et al.* 1968, *and* Kundt and Newman 1968.

Newman, E. T., E. Couch, K. Chinnapared, A. Exton, A. Prakash, and R. Torrence, 1965, "Metric of a rotating, charged mass," *J. Math. Phys. 6,* 918–919. [§33.2]

Newman, E. T., and R. Penrose, 1965, "10 exact gravita-tionally-conserved quantities," *Phys. Rev. Lett. 15,* 231–233. [Ex. 32.11]

Newman, J. R. *See* Nagel and Newman 1958.

Newton, I., 1687, *Philosophiae naturalis principia mathematica,* 1st ed., Streater, London; final ed., in English translation by A. Motte, 1729, revised by A. Cajori, *Sir Isaac Newton's Mathematical Principles of Natural Philosophy and His System of the World,* Univ. of California Press, Berkeley and Los Angeles, 1934; paperback, 1962. [Boxes 1.10, 21.1, 27.7]

Newton, I., 1730, *Opticks, or a Treatise of the Reflection, Re-fraction, Inflection, & Colours of Light,* 4th ed.; reprint, ed. by D. H. D. Roller, Dover, New York, 1952. [Box 1.10]

Newton, I., *cited by name in* §§1.1, 1.3, 12.1, 12.2, Boxes 1.2, 12.2, 12.3, 12.6, Ex. 12.7, 12.8, 13.4, 17.2 part 6.

Newton, R. *See* Gilbert and Newton 1970.

Neyman, J., 1951, *Proceedings of the Second Berkeley Sympo-sium on Mathematical Statistics and Probability,* Univ. of California Press, Berkeley, Calif.

Ni, W.-T., 1970, work reported by K. S. Thorne at the NASA-ESRO conference on experimental tests of gravitation theories, California Institute of Technology, November 1970; *see* Thorne, Ni, and Will 1971. [§39.2, Box 39.1]

Ni, W.-T., 1972, "Theoretical frameworks for testing relativistic gravity, IV: A compendium of metric theories of gravity and their post-Newtonian limits," *Astrophys. J. 176,* 769–796. [§§17.6, 38.2, 39.2, 40.1, 40.5, 40.8, Boxes 39.1, 39.2, 39.5]

Ni, W.-T. *See also* Thorne, Ni, and Will 1971.

Ni, W.-T., 1973, "Relativistic stellar stability: an empirical approach," *Astrophys. J. 181,* 939–956. [Box 39.5, Ex. 39.7]

Nicetas, *cited by name in* Box 27.7.

Nickerson, H. K., D. C. Spencer, and N. E. Steenrod, 1959, *Advanced Calculus,* Van Nostrand, Princeton, New Jersey. [§4.1]

Nijenhuis, A., 1952, "Theory of the geometric object," disser-tation, Amsterdam. [§2.2]

Nizer, L., 1972, quoted in the *New York Times* February 7, 1972. [Epigraph to Chap. 40]

Noerdlinger, P. D., and V. Petrosian, 1971, "The effect of cosmological expansion on self-gravitating ensembles of particles," *Astrophys. J. 168,* 1–9. [§27.5]

Noonan, T. W. *See* Robertson and Noonan 1968.

Nordstrøm, G., 1913, "Zur Theorie der Gravitation vom Standpunkt des Relativitätsprinzips," *Ann. Phys. (Ger-many) 42,* 533–554. [§17.6]

Nordstrøm, G., 1918, "On the energy of the gravitational field in Einstein's theory," *Proc. Kon. Ned. Akad. Wet. 20,* 1238–1245. [Ex. 31.8]

Nordstrøm, G., *cited by name in* §§38.2, 42.7.

Nordtvedt, K., 1968a, "Equivalence principle for massive bod-ies, I: Phenomenology," *Phys. Rev. 169,* 1014–1016. [§§38.2, 40.9]

Nordtvedt, K., 1968b, "Equivalence principle for massive bod-ies, II: Theory," *Phys. Rev. 169,* 1017–1025. [§§38.2, 39.3, 40.9, Box 39.5]

Nordtvedt, K., Jr., 1969, "Equivalence principle for massive bodies including rotational energy and radiation pres-sure," *Phys. Rev., 180,* 1293–1298. [§§38.2, 39.3]

Nordtvedt, K., 1970, "Solar system Eötvos experiments," *Icarus 12,* 91–100. [§40.8, Box 39.5]

Nordtvedt, K., 1971a, "Equivalence principle for massive bod-ies, IV: Planetary bodies and modified Eötvos-type exper-iments," *Phys. Rev. D 3,* 1683–1689. [§§40.8, 40.9, Box 39.5]

Nordtvedt, K., 1971b, in Davies 1971, pp. 32–37. [§§39.5, 40.9, Box 39.5]

Nordtvedt, K., Jr. *See also* Will and Nordtvedt 1972.

Nordtvedt, K., Jr., and C. M. Will, 1972, "Conservation laws and preferred frames in relativistic gravity, II: Experi-mental evidence to rule out preferred-frame theories of gravity," *Astrophys. J. 177,* 775–792. [§§39.2, 40.5, 40.8, Box 39.5]

North, J. D., 1965, *The Measure of the Universe,* Clarendon Press, Oxford. [Box 27.7]

Novikov, I. D., 1963, doctoral dissertation, Shternberg Astro-nomical Institute, Moscow. [§31.4]

Novikov, I. D., 1964, "Delayed explosion of a part of the Friedmann universe, and quasars," *Astron. Zh. 41,* 1075–1083. English translation in *Sov. Astron.-Astron. J. 8,* 857–863. [§34.3]

Novikov, I. D., 1966a, "Change of relativistic collapse into anticollapse and kinematics of a charged sphere," *Zh. Eksp. & Teor. Fiz. Pis'ma 3,* 223–227. English trans-lation in *Sov. Phys.—JETP Lett. 3,* 142–144. [§34.6, Fig. 34.4]

Novikov, I. D., 1966b, "The replacement of relativistic gravita-tional contraction by expansion, and the physical singu-larities during contraction," *Astron. Zh. 43,* 911–918. Eng-lish translation in *Sov. Astron.—A. J. 10,* 731–737. [§34.6, Fig. 34.4]

Novikov, I. D. *See also* Doroshkevich, Lukash, and Novikov 1971, Doroshkevich, Zel'dovich, and Novikov 1965, *and* Zel'dovich and Novikov 1966, 1971, *and* 1974.

Novikov, I. D., and K. S. Thorne, 1973, "Astrophysics of black holes," in DeWitt and DeWitt 1973, pp. 343–450. [Box 33.3]

Nutku, Y. *See* Berger *et al.* 1972 *and* Chandrasekhar and Nutku 1969.

O'Connell, R. F., 1972, "Present status of the theory of the relativistic gyroscope experiment," *Gen. Rel. & Grav. 3,* 123–133. [§40.7, Ex. 40.8]

Ogorodnikov, K. F., 1965, *Dynamics of Stellar Systems,* ed. by A. Beer, Macmillan, New York. [§25.7]

Ohandley, D. A., D. B. Holdridge, W. G. Melbourne, and J. D. Mulholland, 1969, *J.P.L. Development Ephemeris No. 69,* J.P.L. Technical report 32-1465, Caltech Jet Propulsion Laboratory, Pasadena. [§40.1]

Ohanian, H. C. *See* Belasco and Ohanian 1969.

Olbers, H. W. M., 1826, "Über die Durchsichtigkcit des Weltraumes" *Bode Jahrbuch, 110;* substantially the same article in *Bibliothèque Universelle* for February 1826, and in *Edinburgh New Philosophical Journal* for April–October 1826, 141. [Box 27.7]

Omnes, R. L., 1969, "Possibility of matter-antimatter separation at high temperature," *Phys. Rev. Lett. 23,* 38–40. [§27.10]

O'Murchadha, N., and J. W. York, Jr., 1973, personal communication; *see also* O'Murchadha, N., "Existence and uniqueness of solutions of the Hamiltonian constraint of general relativity," doctoral dissertation, Princeton Univ., February 5, 1973. [§21.11]

O'Neill, B., 1966, *Elementary Differential Geometry,* Academic Press, New York. [Box 8.1]

Oort, J. H., 1958, "Distribution of galaxies and the density of the universe," in *Onzième Conseil de Physique Solvay: La Structure et l'évolution de l'univers,* Éditions Stoops, Brussels, Belgium. [§29.6]

Oort, J. H., 1970, "Matter far from the galactic plane associated with spiral arms," in *Spiral Structure of our Galaxy, Basel, Switzerland, 29 Aug.–4 Sep. 1969,* D. Reidel, Dordrecht, Netherlands. [§28.2]

Oort, J. H., *cited by name in* Box 27.7. *See also* Mayall and Oort 1942.

Oppenheimer, J. R., and H. Snyder, 1939, "On continued gravitational contraction," *Phys. Rev. 56,* 455–459. [§§32.3, 32.4, 34.6, Fig. 24.1, Box 24.1]

Oppenheimer, J. R., and G. Volkoff, 1939, "On massive neutron cores," *Phys. Rev. 55,* 374–381. [§23.2, Fig. 24.1, Box 24.1, Ex. 23.10]

Oppolzer, Th. V., 1887, "Canon der Finsternisse," *Math.-Naturwiss. Cl. d. Akad. der Wiss. Denkschriften 52,* Wien. [Fig. 1.8]

O'Raifeartaigh, L., ed., 1972, *General Relativity, Papers in Honor of J. L. Synge,* Oxford Univ. Press, London.

Ostriker, J. P., and J. E. Gunn, 1969, "On the nature of pulsars, I: Theory," *Astrophys. J. 157,* 1395–1417. [§24.3, Ex. 33.13]

Pacini, F., 1968, "Rotating neutron stars, pulsars, and supernova remnants," *Nature 219,* 145–146. [§24.3]

Page, L., 1941, lecture at Dec. 17 meeting of the American Institute of Electrical Engineers, New York. [Epigraph to Chap. 3]

Palatini, A., 1919, "Deduzione invariantiva delle equazioni gravitazionali dal principio di Hamilton," *Rend. Circ. Mat. Palermo 43,* 203–212. [§21.2]

Palatini, A., *cited by name in* §§21.1, 21.2.

Panov, V. I. *See* Braginsky and Panov 1971.

Paolini, F., R., *cited by name in* Box 27.7.

Papanastassiou, D. A., and G. J. Wasserburg, 1969, "Initial strontium isotopic abundances and the resolution of small time differences in the formation of planetary objects," *Earth & Planetary Sci. Lett. 5,* 361–376. [Box 29.3]

Papapetrou, A., 1951, "Spinning test-particles in general relativity, I," *Proc. R. Soc. London A 209,* 248–258. [§20.6, Ex. 40.8]

Papapetrou, A., 1954a, "Eine neue Theorie des Gravitationsfeldes, I," *Math. Nachr. 12,* 129–141. [§40.8]

Papapetrou, A., 1954b, "Eine neue Theorie des Gravitationsfeldes, II," *Math. Nachr. 12,* 143–154. [§40.8]

Papapetrou, A., 1954c, "Eine Theorie des Gravitationsfeldes mit einer Feldfunktion," *Z. Phys. 139,* 518–532. [§40.8]

Papapetrou, A., 1958, "Uber zeit abhängige Lösungen der Feldgleichungen der allgemeinen Relativitätstheorie," *Ann. Phys. (Germany) 2,* 87–96. [§35.8]

Papapetrou, A., and A. Hamoui, 1967, "Surfaces caustiques dégénérées dans la solution de Tolman: La singularité physique en relativité générale," *Ann. Inst. Henri Poincaré A 6,* 343–364. [Ex. 32.7]

Papapetrou, A., and A. Hamoui, 1968, "Couches simples de matières en relativité générale," *Ann. Inst. Henri Poincaré A 9,* 179–211. [§21.13]

Papapetrou, A., and H. Treder, 1959, "Das Sprungproblem erster Ordnung in der allgemeinen Relativitätstheorie," *Math. Nachr. 20,* 53–66. [§21.13]

Papapetrou, A., and H. Treder, 1962, "Shock waves in general relativity," in Infeld 1962, pp. 351–360. [§21.13]

Paris 1967, *Fluides et Champs Gravitationnel en relativité générale,* Colloques Internationaux du CNRS, no. 170, organizé au Collège de France, du 19 au 23 juin 1967, Éditions du Centre National de la Recherche Scientifique, 1969.

Parker, H. M. *See* Rose *et al.* 1969.

Parker, L., 1966, *Particle Creation in the Expanding Universe,* Ph.D. thesis, Harvard University, unpublished. [§30.5]

Parker, L., 1969, "Quantized fields and particle creation in expanding universes, I," *Phys. Rev. 183,* 1057–1068. [§30.5]

Parker, L., 1972, "Particle Creation in Isotropic Cosmologies," *Phys. Rev. Lett. 28,* 705–708. [§30.5]

Parsons, S. J., *cited by name in* Box 27.7.

Partridge, R. B., 1969, "The primeval fireball today," *Am. Sci. 57,* 37–74. [§27.2]

Partridge, R. B. *See also* Boughn, Fram, and Partridge 1971.

Pauli, W., 1921, "Relativitätstheorie," in *Enzyklopädie der Mathematischen Wissenschaften,* II, 539–775, Teubner, Leipzig, Germany. [Preface]

Pauli, W., 1934, "Die allgemeinen Prinzipien der Wellenmechanik," *Handbuch der Physik 24,* Part 1, 2d ed. [§16.3]

Pauli, W., 1947, *Exclusion Principle and Quantum Mechanics,* 1945 Nobel prize lecture, Editions Grisson, Neuchatel, Switzerland. Related discussion in Fierz and Weisskopf 1960. [Box 44.3]

Pauli, W., ed., 1955, *Niels Bohr and the Development of Physics,* McGraw-Hill, New York.

Pauli, W., 1958, *Theory of Relativity,* Pergamon Press, London. English translation by Gerard Field of Pauli 1921 with supplementary notes (1956) by the author. [endpapers]

Pauli, Wolfgang, Memorial volume. *See* Fierz and Weisskopf 1960.

Pauli, W., *cited by name in* Preface. *See also* Fierz and Pauli 1939.

Paya, D. *See* Dabbs *et al.* 1965.

Peebles, P. J. E., 1966, "Primordial He abundance and fireball, II," *Astrophys. J. 146,* 542–552. [§28.1]

Peebles, P. J. E., 1969, "Cosmology," *Proc. Roy. Astron. Soc. Canada, 63,* 4–31. [§30.1]

Peebles, P. J. E., 1971, *Physical Cosmology,* Princeton Univ. Press, Princeton, N.J. [§§27.1, 27.2, 28.1, 29.6, Box 27.7]

Peebles, P. J. E., 1972, "Gravitational collapse and related phenomena from an empirical point of view, or, black holes are where you find them," *Gen. Rel. & Grav. 3,* 63–82. [Box 33.3]

Peebles, P. J. E. *Cited by name in* Box 27.7. *See also* Dicke and Peebles 1962a, 1962b, *and* 1965, Dicke *et al.* 1965, *and* Jones and Peebles 1972.

Peebles, P. J. E., and J. T. Yu, 1970, "Primeval adiababic perturbation in an expanding universe," *Astrophys. J. 162,* 815–836. [§30.1]

Pekar, D. *See* Eötvös, Pekar, and Fekete 1922.

Penrose, R., 1959, "The apparent shape of a relativistically moving sphere," *Proc. Cambridge Phil. Soc. 55,* 137–139. [§41.10]

Penrose, R., 1964, "Conformal treatment of infinity," in DeWitt and DeWitt 1964b. [§34.2]

Penrose, R., 1965a, "Zero rest-mass fields including gravitation: Asymptotic behavior," *Proc. R. Soc. London A 284,* 159–203. [§34.2, Box 32.2]

Penrose, R., 1965b, "Gravitational collapse and space-time singularities," *Phys. Rev. Lett. 14,* 57–59. [§§32.7, 34.6, Box 34.2]

Penrose, R., 1966, "An analysis of the structure of spacetime," Adams Prize Essay. [Box 34.2]

Penrose, R., 1968a, "Structure of spacetime," in DeWitt and Wheeler 1968. [§§34.1, 34.4, 41.7, 41.11, Boxes 34.1, 34.2, 44.3]

Penrose, R., 1968b, "Twistor quantization and the curvature of spacetime," *Int. J. Theor. Phys. 1,* 61–99. [Box 34.2]

Penrose, R., 1969, "Gravitational collapse: The role of general relativity," *Nuovo Cimento 1,* special number, 252–276. [§33.7, Boxes 33.2, 33.4, 34.2]

Penrose, R., 1972, "Techniques of differential topology in relativity," in Farnsworth *et al.* (1972). [§34.1]

Penrose, R., *cited by name in* §§32.7, 41.9, 42.1, Box 27.7, Fig. 41.7. *See also* Hawking and Penrose 1969, Infeld 1964, Klauder 1964, Newman and Penrose 1965, *and* Walker and Penrose 1970.

Penrose, R., and M. A. H. MacCallum, 1973, "Twistor Theory: An approach to the quantization of fields and spacetime," *Physics Reports 6C,* no. 4, 242–315. [Box 34.2]

Penzias, A. A., *cited by name in* Box 27.7.

Penzias, A. A., and R. W. Wilson, 1965, "A measurement of excess antenna temperature at 4080 Mc/s," *Astrophys. J. 142,* 419–421. [§27.2]

Pereira, C. M. *See* Brill *et al.* 1972.

Peres, A., 1962, "On the Cauchy problem in general relativity, II," *Nuovo Cimento 26,* 53–62. [§43.3]

Peres, A., and N. Rosen, 1964, "Gravitational radiation damp-

ing of nongravitational motion," *Ann. Phys. (U.S.A.) 10,* 94–99. [§§36.1, 36.8]

Peters, P. C., 1964, "Gravitational radiation and the motion of two point masses," *Phys. Rev. B 136,* 1224–1232. [§§36.1, 36.7, Ex. 36.9]

Peters, P. C., and J. Mathews, 1963, "Gravitational radiation from point masses in a Keplerian orbit," *Phys. Rev. 131,* 435–440. [§36.6, Ex. 36.6]

Peterson, B. A., *cited by name in* Box 27.7.

Peterson, F. P. *See* Anderson, Brown, and Peterson 1966a *and* 1966b.

Peterson, F. R. *See* Evenson *et al.* 1972.

Pethick, C. J. *See* Baym, Bethe, and Pethick 1971.

Petrosian, V. *See* Noerdlinger and Petrosian 1971.

Petrov, A. Z., 1954, *Reports of the State University of Kazan 114,* book 8, 55. [§33.6]

Petrov, A. Z., 1969, *Einstein Spaces,* Pergamon, New York. [§33.6]

Petrov, Aleksei Zinoveivich, Memorial volume. *See* Shelest 1972.

Pettengill, G. H. *See* Shapiro, Ash *et al.* 1971, Shapiro, Smith *et al.* 1971, and Shapiro *et al.* 1972.

Philolaus, *cited by name in* Box 27.7.

Phillips, J. W., *cited by name in* Box 27.7.

Picasso, E. *See* Bailey *et al.* 1968 *and* Farley *et al.* 1966.

Piir, I. I., 1971, "Maxwell equations in the Schwarzschild gravitational field" (in Russian), *Izv. Akad. Nauk Estonian S. S. R., Fiz. Mat.,* 20, 413–419. [Ex. 32.11]

Pilkington, J. D., *cited by name in* Box 27.7. *See also* Hewish *et al.* 1968.

Pirani, F. A. E., 1956, "On the physical significance of the Riemann tensor," *Acta Phys. Pol. 15,* 389–405. [§20.6, Ex. 40.8]

Pirani, F. A. E., 1957, "Invariant formulation of gravitational radiation theory," *Phys. Rev. 105,* 1089–1099. [§§21.13, 33.6, Ex. 15.2]

Pirani, F. A. E., 1965, "Introduction to gravitational radiation theory," in Trautman, Pirani, and Bondi 1965, pp. 249–373. [§§41.7, 41.11, endpapers]

Pirani, F. A. E. *See also* Bondi, Pirani, and Robinson 1959, Ehlers, Pirani, and Schild 1972, *and* Trautman, Pirani, and Bondi 1965.

Pittsburgh 1970, Regional Conference on Relativity, July 13–17, 1970. *See* Farnsworth *et al.* 1972.

Planck, M., 1899, "Über irreversible Strahlungsvorgänge," *Sitzungsber. Deut. Akad. Wiss. Berlin, Kl. Math.—Phys. Tech.,* 440–480. [§§43.4, 44.6]

Plato. *See* Cornford 1937.

Plebanski, J. *See* Infeld and Plebanski 1960 *and* Mielnik and Plebanski 1962.

Plotkin, H. E. *See* Bender *et al.* 1971.

Plutarch, *cited by name in* Box 27.7.

Podolsky, B. *See* Dirac, Fock, and Podolsky 1932.

Podurets, M. A., 1964, "Asymptotic behavior of the optical luminosity of a star in gravitational collapse in terms of general relativity theory," *Astron. Zhur. 41,* 1090–1096, English translation in *Sov. Astron.—A. J. 8,* 868–873 (1965). [§32.3]

Podurets, M. A., 1966, "The character of the singularity in the gravitational collapse of a star," *Dokl. Akad. Nauk SSSR*

167, 1025–1027. English translation in *Sov. Ph.—Doklady 11*, 275–276. [§34.6, Ex. 32.3]

Podurets, M. A. *See also* Zel'dovich and Podurets 1965.

Poincaré, H. *cited by name in* §§3.1, 12.3.

Poisson, S., *cited by name in* Box 17.2 part 6.

Polievktov-Nikoladze, N. M., 1969, "On the theory of the gravitational field," *Zh. Eksp. & Teor. Fiz. 57*, 2010–2020. English translation in *Sov. Phys.—JETP 30* (1970), 1089–1095. [Box 17.2 part 6]

Polkinghorne, J. C., 1955, "On the Feynman principle," *Proc. R. Soc. London A 230*, 272–276. [Box 21.1]

Polubarinova-Kochina, P. Ya., 1963, "Alexander Alexandrovitch Friedmann (In memory of A. A. Friedmann)," *Usp. Fiz. Nauk, 80*, 345–352. English translation in *Sov. Phys.—Uspekhi, XX*, 467–472 (1964). [Box 27.6]

Pope, A., 1733, *Essay on Man*, London [Epigraph to Chap. 24]

Porter, J. *See* Farnsworth *et al.* 1972.

Pound, R. V., and G. A. Rebka, 1960, "Apparent weight of Photons," *Phys. Rev. Lett. 4*, 337–341. [§§7.2, 7.4, 38.5, Fig. 38.1]

Pound, R. V., and J. L. Snider, 1964, "Effect of gravity on nuclear resonance," *Phys. Rev. Lett. 13*, 539–540. [§§7.2, 7.4]

Pound, R. V., and J. L. Snider, 1965, "Effect of gravity on gamma radiation," *Phys. Rev. B 140*, 788–803. [§§7.2, 38.5, Fig. 38.1]

Power, E. A., and J. A. Wheeler, 1957, "Thermal geons," *Rev. Mod. Phys. 29*, 480–495. [§25.6]

Prakash, A. *See* Newman *et al.* 1965.

Press, F., 1965, "Resonant vibrations of the Earth," *Sci. American 213*, no. 5, 28–37. [Ex. 37.14]

Press, F. *See also* Ahrens, Press, and Runcorn 1964.

Press, W. H., 1970, unpublished work described in part in Press and Thorne 1972. [§37.3, Fig. 37.2, Box 37.2]

Press, W. H., 1971, "Long wave trains of gravitational waves from a vibrating black hole," *Astrophys. J. Lett. 170*, 105–108. [Box 33.3]

Press, W. H., 1972, "Time evolution of a rotating black hole immersed in a static scalar field," *Astrophys. J. 175*, 243–252. [Ex. 33.14, Box 33.3]

Press, W. H. *See also* Bardeen and Press 1972, Bardeen, Press, and Teukolsky 1972, *and* Davis *et al.* 1971.

Press, W. H., and J. M. Bardeen, 1971, "Nonconservation of the Newman-Penrose conserved quantities," *Phys. Rev. Lett. 27*, 1303–1306. [Ex. 32.11]

Press, W. H., and S. A. Teukolsky, 1972, "Floating orbits, superradiant scattering and the black-hole bomb," *Nature 238*, 211–212. [Box 33.3]

Press, W. H., and S. A. Teukolsky, 1973, "Perturbations of a rotating black hole, II: Dynamical stability of the Kerr metric," *Astrophys. J.,* in press. [Box 33.3, Ex. 33.14]

Press, W. H., and K. S. Thorne, 1972, "Gravitational-wave astronomy," *Ann. Rev. Astron. Astrophys. 10*, 335–374. [§§36.5, 37.4, 37.8, Box 37.5]

Price, R. H., 1971, *Nonspherical Perturbations of Relativistic Gravitational Collapse*, doctoral dissertation, California Institute of Technology. [Ex. 32.10, Box 32.2]

Price, R. H., 1972a, "Nonspherical perturbations of relativistic gravitational collapse, I: Scalar and gravitational perturbations," *Phys. Rev., D5*, 2419–2438. [§§35.14, 41.11, Ex. 32.10, Box 32.2]

Price, R. H., 1972b, "Nonspherical perturbations of relativistic gravitational collapse, II: Integer-spin, zero-rest-mass fields," *Phys. Rev., D5*, 2439–2454. [§41.11, Box 32.2]

Price, R. H. *See also* Davis *et al.* 1971.

Pringle, J. E., and M. J. Rees, 1972, "Accretion disc models for compact X-ray sources," *Astron. Astrophys. 21*, 1–9. [Box 33.3]

Proclus of Athens, *cited by name in* Fig. 1.8.

Ptolemy, C. *See* Taliaferro 1952.

Ramsey, N. F., 1972, "History of atomic and molecular standards of frequency and time," *I.E.E.E. Trans. IM-21*, 90–98. [Box 16.3]

Rayleigh, Lord, 1912, "On the propagation of waves through a stratified medium, with special reference to the question of reflection," *Proc. R. Soc. London A 86*, 207–226. [§22.5]

Rayzer, Yu. P. *See* Zel'dovich and Rayzer 1967.

Reber, G., *cited by name in* Box 27.7.

Rebka, G. A. *See* Pound and Rebka 1960.

Rees, M. *See* Lynden-Bell and Rees 1971 *and* Pringle and Rees 1972.

Refsdal, S., 1964, "The gravitational lens effect," *Mon. Not. R. Astron. Soc. 128*, 295–306. [§25.6, Fig. 22.2, Box 33.3]

Refsdal, S., R. Stabell, and F. G. de Lange, 1967, "Numerical calculations on relativistic cosmological models," *Mem. Roy. Astron. Soc., 71*, 143–248. [§29.4, Box 29.4]

Regge, T., 1961, "General relativity without coordinates," *Nuovo Cimento 19*, 558–571. [§§42.1, 42.2, 42.4, 42.6]

Regge, T., and J. A. Wheeler, 1957, "Stability of a Schwarzschild singularity," *Phys. Rev. 108*, 1063–1069. [§25.5, Box 33.3]

Reichley, P. *See* Muhlemann and Reichley 1964 *and* 1965.

Reik, H. G. *See* Meixner and Reik 1959.

Reines, F., ed., 1972, *Cosmology, Fusion, and Other Matters,* Univ. of Colorado Press, Boulder.

Reissner, H., 1916, "Über die Eigengravitation des elektrischen Feldes nach der Einsteinschen Theorie," *Ann. Phys. (Germany) 50*, 106–120. [Ex. 31.8]

Reissner, H., *cited by name in* §42.7.

Retherford, R. C. *See* Lamb and Retherford 1947.

Rhoades, C. E. J., 1971, *Investigations in the Physics of Neutron Stars,* doctoral dissertation, Princeton University. [§24.2]

Richelot, *cited by name in* Box 38.2.

Riemann, G. F. B., *achievements and picture in* Box 8.5; for collected works, see Weber 1953.

Riesz, M., 1949, "L'intégrale de Riemann-Liouville et le problème de Cauchy," *Acta Math. 81*, 1–223. [§35.8]

Rindler, W., 1969, *Essential Relativity: Special, General, and Cosmological,* Van Nostrand, New York. [Box 27.7]

Robb, A. A., 1914, *A Theory of Time and Space,* Cambridge Univ. Press, Cambridge, Eng. [Box 13.1]

Robb, A. A., 1936, *Geometry of Time and Space,* Cambridge Univ. Press, Cambridge, Eng. Rev. ed. of Robb 1914. [Boxes 1.3, 13.1]

Robb, A. A., *quoted,* epigraph of Chap. 4.

Roberts, M. S., 1965, "21-cm. hydrogen measurements of NGC5668, a relatively distant galaxy," *Astrophys. J. 142,* 148–153. [§29.2]

Roberts, M. S. *See also* Dieter *et al.* 1962.

Robertson, H. P., 1935, "Kinematics and world structure," *Astrophys. J. 82,* 248–301. [Ex. 27.5]

Robertson, H. P., 1936, "Kinematics and world structure," *Astrophys. J. 83,* 187–201, 257–271. [Ex. 27.5]

Robertson, H. P., 1962, "Relativity and cosmology," in Deutsch and Klemperer 1962, pp. 228–235. [§§38.2, 39.3]

Robertson, H. P., *cited by name in* §30.3, Box 27.7.

Robertson, H. P., and T. W. Noonan, 1968, *Relativity and Cosmology,* W. B. Saunders, Philadelphia, [§?7.11, Ex. 27.4, 29.6, endpapers]

Robinson, H. G. *See* Hughes, Robinson, and Beltran-Lopez 1960.

Robinson, I., 1959a, "A solution of the Maxwell-Einstein equations," *Bull. Acad. Polon. Sci. 7,* 351–352. [Ex. 32.1]

Robinson, I., 1959b, unpublished report at *Colloque Internationale sur les Théories Relativistes de la Gravitation,* Abbaye de Royaumont. [Ex. 15.2]

Robinson, I., 1961, "Null Electromagnetic Fields," *J. Math. Phys., 2,* 290–291. [§22.5]

Robinson, I., A. Schild, and E. Schücking, eds., 1965, *Quasistellar Sources and Gravitational Collapse,* The University of Chicago Press, Chicago.

Robinson, I. *See also* Bondi, Pirani, and Robinson 1959, Douglas *et al.* 1969.

Rohrlich, F., 1965, *Classical Charged Particles; Foundations of Their Theory,* Addison-Wesley, Reading, Mass. [§4.7]

Rohrlich, F., *cited by name in* §20.6.

Roll, P. G., *cited by name in* Box 27.7. *See also* Dicke *et al.* 1965.

Roll, P. G., R. Krotkov, and R. H. Dicke, 1964, "The equivalence of inertial and passive gravitational mass," *Ann. Phys. (U.S.A.) 26,* 442–517. [§§1.3, 10.1, 38.3, Fig. 1.6, Box 1.2]

Root, W. I. *See* Davenport and Root 1958.

Rose, R. D., H. M. Parker, R. A. Lowry, A. R. Kuhlthau, and J. W. Beams, 1969, "Determination of the gravitational constant G, *Phys. Rev. Lett. 23,* 655–658. [§40.8]

Rosen, Nathan, *Festschrift* on The occasion of his 60th birthday, March 22, 1969. *See* Kuper and Peres 1971.

Rosen, N. *See also* Peres and Rosen 1964.

Rosenfeld, L., 1940, "Sur le tenseur d'impulsion-énergie," *Mém. Acad. R. Belg. Sci. 18,* no. 6. [§21.3]

Rosenfeld, L. *See also* Bohr and Rosenfeld 1933.

Rossi, B., *cited by name in* Box 27.7.

Rossi, B., and D. B. Hall, 1941, "Variation of the rate of decay of mesotrons with momentum," *Phys. Rev. 59,* 223–228. [§38.4]

Royaumont 1959, *Les Théories relativistes de la gravitation,* Actes du Colloque International, organizé à Royaumont, du 21 au 27 juin 1959, Centre National de la Recherche Scientifique, Paris, 1962.

Ruark, A. E., and H. C. Urey, 1930, *Atoms, Molecules, and Quanta,* McGraw-Hill, New York. [§§40.7, 41.4]

Rudenko, V. N. *See* Braginsky, Zel'dovich, and Rudenko 1969.

Ruderman, M., 1972, "Pulsars: Structure and dynamics," *Ann. Rev. Astron. Astrophys. 10,* 427–476. [§24.3]

Ruffini, R. *See* Christodolou and Ruffini 1971; Davis *et al.* 1971; Ferrari and Ruffini 1969; *and* Hanni and Ruffini 1973.

Ruffini, R., and J. A. Wheeler, 1970, "Collapse of wave to black hole," *Bull. Am. Phys. Soc. 15,* 76. [§35.8]

Ruffini, R., and J. A. Wheeler, 1971a, "Introducing the black hole," *Phys. Today 24,* 30–36. [§33.7]

Ruffini, R., and J. A. Wheeler, 1971b, "Relativistic cosmology and space platforms," in *Proceedings of the Conference on Space Physics,* European Space Research Organization, Paris, France, pp. 45–174. [Boxes 33.2, 33.3, 33.5, 36.3, Fig. 24.3, Ex. 37.13, 37.14]

Runcorn, S. K. *See* Ahrens, Press, and Runcorn 1964.

Runge, J. *See* Sommerfeld and Runge 1911.

Russ, N. *See* Boni, Russ, and Laurence 1960.

Rutherford, E., *cited by name in* Preface, §44.1.

Ruzmaikin, A. A. *See* Ruzmaikina and Ruzmaikin 1969.

Ruzmaikina, T. V., and A. A. Ruzmaikin, 1969, "Quadratic corrections to the Lagrangian density of the gravitational field and the singularity," *Zh. Eksp. & Teor. Fiz. 57,* 680–685. English translation in *Sov. Phys.—JETP 30,* 372–374. [Box 17.2 part 6]

Ryan, M. P., Jr., 1971, "Qualitative cosmology: Diagrammatic solutions for Bianchi type IX universes with expansion, rotation, and shear, II: The general case," *Ann. Phys. 68,* 541–555. [§30.7]

Ryan, M. P., Jr., 1972a, *Hamiltonian Cosmology,* Springer, Berlin, Germany. [§30.7, Box 30.1]

Ryan, M. P., Jr., 1972b, "The oscillatory regime near the singularity in Bianchi type-IX universes," *Ann. Phys. (U.S.A.) 70,* 301–322. (Includes a report of work by V. A. Belinsky and I. M. Khalatnikov. [§30.7, Box 30.1]

Sachs, R. K., 1961, "Gravitational waves in general relativity, VI: The outgoing radiation condition," *Proc. R. Soc. London A 264,* 309–338. [§22.5]

Sachs, R. K., 1964, "Gravitational radiation," in DeWitt and DeWitt 1964. [§§21.5, 21.13, 33.6, 41.11, Ex. 22.14, endpapers]

Sachs, R. K., ed., 1971, *General Relativity and Cosmology,* proceedings of course 47 of the International School of Physics "Enrico Fermi", Academic Press. [§§27.2, 34.1, 40.7]

Sachs, R. K. *See also* Eardley, Liang, and Sachs 1972, Ehlers, Geren, and Sachs 1968, *and* Jordan, Ehlers, and Sachs 1961.

Sakharov, A. D., 1967, "Vacuum quantum fluctuations in curved space and the theory of gravitation," *Doklady Akad. Nauk S.S.S.R. 177,* 70–71. English translation in *Sov. Phys. Doklady 12,* 1040–1041 (1968). [§44.4, Box 17.2 part 6]

Sakharov, A. D., 1969, unpublished work cited by Braginsky and Nazarenko 1971. [§37.3, Fig. 37.2]

Salam, A., and E. P. Wigner, eds., 1972, *Aspects of Quantum Theory,* Cambridge Univ. Press, Cambridge, Eng.

Salpeter, E. E. and R. V. Wagoner, 1971, "Structure of relativistic disks," *Astrophys. J. 164,* 557–568. [Box 24.1]

Sandage, A., 1958, "Current problems in the extragalactic distance scale," *Astrophys. J. 127,* 513–526. [Box 27.1]

Sandage, A., 1968, "The time scale for creation," in L. Woltjer, ed., *Galaxies and the Universe,* Columbia Univ. Press, New York, pp. 75–112. [§29.7]

Sandage, A., 1970, "Main-sequence photometry, color-magnitude diagrams, and ages for the globular clusters M3, M13, M15, and M92," *Astrophys. J. 162,* 841–870. [§29.7]

Sandage, A., 1972a, "Distances to galaxies: the Hubble constant, the Friedmann time, and the edge of the world," in *Proceedings of the Symposium on the Galaxy and the Distance Scale,* Essex, England, in press. [§§27.1, 29.3, 29.4, Box 29.4]

Sandage, A., 1972b, "The redshift-distance relation, I. Angular diameter of first-ranked cluster galaxies as a function of redshift: the aperture correction to magnitudes," *Astrophys. J. 173,* 485–500. [§29.5, Box 29.4]

Sandage, A., 1972c, "The redshift-distance relation, II. The Hubble diagram and its scatter for first-ranked cluster galaxies: a formal value for q_o," *Astrophys. J. 178,* 1–24. [§29.4, Box 29.4]

Sandage, A., 1972d, "The redshift-distance relation, III. Photometry and the Hubble diagram for radio sources and the possible turn-on time for QSO's," *Astrophys. J. 178,* 25–44. [§29.4, Boxes 28.1, 29.4]

Sandage, A., *cited by name in* Boxes 27.7, 29.5.

Sandage, A., G. A. Tamman, and E. Hardy, 1972, "Limits on the local deviation of the universe from a homogeneous model," *Astrophys. J. 172,* 253–263. [§27.1]

Sanders, R., 1970, "The effects of stellar collisions in dense stellar systems," *Astrophys. J. 162,* 791–809. [§§24.5, 24.6, Box 33.3]

Sanderson, B. J. *See* Hsiang and Sanderson 1965.

Sands, M. *See* Feynman, Leighton, and Sands 1964.

Sard, R. D., 1970, *Relativistic Mechanics,* W. A. Benjamin, New York. [§2.10]

Sargent, W. L. W. *See* Bahcall, Sargent, and Schmidt 1967.

Schäfke, F. W. *See* Meixner and Schäfke 1954.

Schatzman, E. *See* DeWitt, Veron, and Schatzman 1967.

Schiff, L. I., 1962, "General relativity: Theory and experiment," *J. Indust. & Appl. Math 10,* 795–801. [§§38.2, 39.3]

Schiff, L. I., 1967, "Comparison of theory and observation in general relativity," in Ehlers 1967. [§§38.2, 39.3]

Schiffer, M. *See* Adler, Bazin, and Schiffer 1965.

Schild, A., 1960, "Time," *Texas Quarterly, 3,* no. 3, 42–62. [§7.3]

Schild, A., 1962, "Gravitational theories of the Whitehead type and the principle of equivalence," in Møller 1962. [§§7.3, 38.5]

Schild, A., 1967, "Lectures on General Relativity Theory," in Ehlers 1967, pp. 1–105. [§7.3, endpapers]

Schild, A., 1970, "Tearing geometry to pieces: More on conformal geometry," unpublished lecture at Jan. 19 Princeton University relativity seminar. [Box 10.2]

Schild, A., *cited by name in* §10.3, Box 17.3, Ex. 10.6. *See also* Douglas *et al.* 1969, Ehlers, Pirani, and Schild 1972, Infeld and Schild 1949, Kerr and Schild 1965, *and* Robinson, Schild, and Schücking 1965.

Schilpp, P. A., ed., 1949, *Albert Einstein: Philosopher-Scientist,* Library of Living Philosophers, Evanston, Ill.; source of quotations from A. Einstein, K. Menger, and M. von Laue. [Box 1.11, epigraphs to §1.2, Chaps. 2, 9]

Schmidt, B. G., 1971, "A new definition of singular points in general relativity," *Gen. Rel. Grav. 1,* 269–280. [§34.6]

Schmidt, M., 1972, "Statistical studies of the evolution of extragalactic radio sources, I. Quasars," *Astrophys. J. 176,* 273–287. [Box 28.1]

Schmidt, M. *Cited by name in* Box 27.7, Ex. 29.7. *See also* Bahcall, Sargent, and Schmidt 1967 *and* Bahcall and Schmidt 1967.

Schoch, C. *See* Neugebauer and Schoch 1930.

Schouten, J. A., 1954, *Ricci-calculus,* Springer, Berlin, Germany. [§21.5, Ex. 17.8, 21.8, endpapers]

Schramm, D. N. *See* Wasserburg, Schramm, and Huneke 1969.

Schrödinger, E., 1950, *Space-Time Structure,* Cambridge Univ. Press, England. [endpapers]

Schücking, E. *See also* Douglas *et al.* 1969 *and* Robinson, Schild, and Schücking 1965.

Schücking, E., and O. Heckmann, 1958, "World models," in *Onziême Conseil de Physique Solvay,* Éditions Stoops, Brussels, pp. 149–158. [§§30.2, 30.3]

Schutz, B., *cited by name in* Ex. 35.20.

Schutz, S., 1966, statement in January 1966 final examination in course in relativity, Princeton University. [Epigraph to Chap. 16]

Schwartz, J. T. *See* Dunford and Schwartz 1958.

Schwartz, L., 1950–1951, *Théorie des distributions,* vols. 1 and 2, Hermann, Paris. [§4.7]

Schwarzschild, K., 1916a, "Über das Gravitationsfeld eines Massenpunktes nach der Einsteinschen Theorie," *Sitzber. Deut. Akad. Wiss. Berlin, Kl. Math.-Phys. Tech.,* 189–196. [Preface, §§14.1, 23.6]

Schwarzschild, K., 1916b, "Über das Gravitationsfeld einer Kugel aus inkompressibler Flussigkeit nach der Einsteinschen Theorie," *Sitzber. Deut. Akad. Wiss. Berlin, Kl. Math.-Phys. Tech.,* 424–434. [§23.2, Box 23.2]

Schwarzschild, K., *cited by name in* §§42.1, 42.7.

Schwinger, J., 1948, "Quantum electrodynamics, I: A covariant formulation," *Phys. Rev. 74,* 1439–1461. [Box 15.1]

Schwinger, J., ed., 1958, *Selected Papers on Quantum Electrodynamics,* Dover, New York.

Schwinger, J., 1963a, "Energy and momentum density in field theory," *Phys. Rev. 130,* 800–805. [Ex. 40.8]

Schwinger, J., 1963b, "Quantized gravitational field," *Phys. Rev. 130,* 1253–1258. [Ex. 40.8]

Sciama, D. W., 1957, "Inertia," *Sci. American 196,* no. 2, 99–109. [§21.12]

Sciama, D. W., 1961, *The Unity of the Universe,* Anchor, New York. [§21.12]

Sciama, D. W., 1971, *Modern Cosmology,* Cambridge Univ. Press, Cambridge, Eng. [§28.1, Box 27.7]

Sciama, D. W., *cited by name in* Boxes 34.2, 34.3.

Scott, P. F., *cited by name in* Box 27.7. *See also* Hewish *et al.* 1968.

Sears, F. W., 1953, *An Introduction to Thermodynamics, the Kinetic Theory of Gases, and Statistical Mechanics,* Addison-Wesley, Reading, Mass. [§23.4]

Seattle 1967, Battelles Rencontres. *See* DeWitt and Wheeler 1968.

Seckler, B. D. *See* Keller, Lewis, and Seckler 1956.

Seelig, C., ed., 1956. *See* Kollros 1956.

Seielstad, G. A., R. A. Sramek, and K. W. Weiler, 1970, "Measurement of the deflection of 9.602-GHz radiation from 3C279 in the solar gravitational field," *Phys. Rev. Lett. 24,* 1373–1376. [Box 40.1]

Sejnowski, T. J., 1973, personal communication. [Ex. 15.2]

Shakespeare, W., *quoted at end of* Chap. 44.

Shakura, N. I., and R. A. Sunyaev, 1973, "Black holes in binary systems: Observational appearance," *Astron. Astrophys. 24,* 337–355. [Box 33.3]

Shannon, C., 1938, "A symbolic analysis of relay and switching circuits," *Trans. Am. Inst. Elect. Eng. 57,* 713–723. [§44.5]

Shapiro, I. I., 1964, "Fourth test of general relativity," *Phys. Rev. Lett. 13,* 789–791. [§40.4]

Shapiro, I. I., 1967, "New method for the detection of light deflection by solar gravity," *Science 157,* 806–808. [§40.3]

Shapiro, I. I., 1968, "Fourth test of general relativity: preliminary results," *Phys. Rev. Lett. 20,* 1265–1269. [§40.4, Box 40.2]

Shapiro, I. I., 1970, results reported at Third Cambridge Conference on Experimental Relativity, New York City, June 1970. [Box 40.3]

Shapiro, I. I., 1971a. "Testing general relativity: progress, problems, and prospects," in Davies, 1971, pp. 136–147. [Box 40.3]

Shapiro, I. I., 1971b. "Testing general relativity: progress, problems, and prospects," lecture at Sixth International Conference on Gravitation and Relativity, published by NORDITA (Copenhagen) in multilith form for distribution at conference, also published in *Gen. Rel. Grav. 3,* 135–148 (1972). [§39.3, Box 40.3]

Shapiro, I. I., M. E. Ash, D. B. Campbell, R. B. Dyce, R. P. Ingalls, R. F. Jurgens, and G. H. Pettengill, 1971, "Fourth test of general relativity: New radar result," *Phys. Rev. Lett. 26,* 1132–1135. [Box 40.2]

Shapiro, I. I., W. B. Smith, M. E. Ash, R. P. Ingalls, and G. H. Pettengill, 1971, "Gravitational constant: Experimental bound on its time variation," *Phys. Rev. Lett. 26,* 27–30. [§40.8]

Shapiro, I. I., G. H. Pettengill, M. E. Ash, R. P. Ingalls, D. B. Campbell, and R. B. Dyce, 1972, "Mercury's perihelion advance: determination by radar," *Phys. Rev. Lett. 28,* 1594–1597. [Box 40.3]

Shapley, H., *cited by name in* Box 27.7.

Sharp, D. H., 1960, "One- and two-surface formulations of the boundary-value problem for Einstein-Maxwell theory and for 'already unified' field theory," Physics A. B. senior thesis, Princeton University, unpublished; deposited in Fine Library, Princeton University. [§21.9]

Sharp, D. H., *cited by name in* §21.1. *See also* Baierlein, Sharp, and Wheeler 1962, Hartle and Sharp 1965, *and* Misner and Sharp 1964.

Sheldon, S. E., 1971, "A complete redshift-magnitude formula," *Astron. J. 76,* 751–755. [§29.4]

Shelest, V. P., ed., 1972, *Gravitatsiya: Problemi i Perspektivi: pamyati Alekseya Zinovievicha Petrova posvashaetsya,* Naukova Dumka, Kiev.

Shields, M. C. *See* Boni, Russ, and Laurence 1960.

Shimmins, A. J., *cited by name in* Box 27.7.

Shirkov, D. V. *See* Bogoliubov, N. N.

Shu, F. H. *See* Lin, Mestel, and Shu 1965.

Siegel, C. L., 1956, *Vorlesungen über Himmelsmechanik,* Springer, Berlin. [Box 25.4]

Siegel, C. L., and J. K. Moser, 1971, *Lectures on Classical Mechanics,* translated by C. I. Kalme, Springer, Berlin and New York, 1971. [Box 25.4]

Silverberg, E. C. *See* Bender *et al.* 1971.

Simpson, O. C. *See* Estermann, Simpson, and Stern 1938.

Singer, D. W., 1950, *Giordano Bruno, His Life and Thought with Annotated Translation of His Work, On the Infinite Universe and Worlds,* Abelard-Schuman, New York. [Box 27.7]

Slee, O. B., *cited by name in* Box 27.7.

Slipher, V. M., *cited by name in* Box 27.7.

Smith, J. H. *See* Frisch and Smith 1963.

Smith, W. B. *See* Shapiro, Smith *et al.* 1971.

Snider, J. L. *See* Pound and Snider 1964 *and* 1965.

Snyder, H. *See* Oppenheimer and Snyder 1939.

Söding, P., J. Bartels, A. Barbaro-Galtieri, J. E. Enstrom, T. A. Lasinski, A. Rittenberg, A. H. Rosenfeld, T. G. Trippe, N. Barash-Schmidt, C. Bricman, V. Chaloupka, and M. Roos, 1972, "Review of particle properties," *Physics Letters, 39B,* 13. [endpapers]

Solvay 1958, Onzième Conseil de Physique Solvay, held at the Instituts Solvay, Brussels; proceedings: *La Structure et l'evolution de l'univers,* Éditions Stoops, Brussels, Belgium.

Sommerfeld, A., and J. Runge, 1911, "Anwendung der Vektorrechnung auf die Grundlagen der geometrischen Optik," *Ann. Phys. (Germany) 35,* 277–298. [§22.5]

Spencer, D. C. *See* Nickerson, Spencer, and Steenrod 1959.

Spitzer, L., Jr., 1971, "Dynamical evolution of dense spherical star systems," in *Semaine d'étude sur les noyeaux des galaxies 1970 (Pontificae Academiae Scientarum Scripta Varia no. 35).* [§§24.5, 24.6, Box 33.3]

Spivak, M., 1965, *Calculus on Manifolds,* W. A. Benjamin, New York. [§4.1, Box 4.1]

Sramek, R. A., 1971, "A measurement of the gravitational deflection of microwave radiation near the sun, 1970 October," *Astrophys. J. Lett. 167,* L55–L60. [Box 40.1]

Sramek, R. A. *See also* Seielstad, Sramek, and Weiler 1970.

St. Patrick. *See* Stokes 1887.

Stabell, R. *See* Refsdal, Stabell, and de Lange 1967.

Stanley, G. J., *cited by name in* Box 27.7.

Starobinsky, A. A. *See* Zel'dovich and Starobinsky 1971.

Steenrod, N. E. *See* Nickerson, Spencer, and Steenrod 1959.

Stehle, P. *See* Corben and Stehle 1960.

Steigman, G., 1971, "Antimatter and cosmology," in Sachs 1971, pp. 373–382. [§§27.11, 28.4]

Stephenson, F. R., *contributed* Fig. 1.8.

Stern, O. *See* Estermann, Simpson, and Stern 1938.

Sternberg, S., 1969, *Celestial Mechanics,* W. A. Benjamin, New York. [§8.2]

Steuckelberg, E. C. G., and G. Wanders, 1953, "Thermodynamique en relativité général," *Helv. Phys. Act., 26,* 307–316. [§22.2]

Stewart, J. M., 1971, *Non-Equilibrium Relativistic Kinetic Theory,* lecture notes in Physics 10, Springer-Verlag, Berlin. [§22.6, Ex. 22.7]

Stewart, J. M., *cited by name in* §22.3.

Stewart, J. M., and G. F. R. Ellis, 1968, "Solutions of Einstein's equations for a fluid which exhibits local rotational symmetry," *J. Math. Phys. 9,* 1072–1082. [Ex. 32.1]

Stockton, A. N., *cited by name in* Box 27.7.

Stokes, G. C., *cited by name in* Box 4.6.

Stokes, W., 1887, *The Tripartite Life of St. Patrick With Other Documents Relating to That Saint,* Her Majesty's Stationery Office, London. [§19.4]

Stratimyer, H., *cited by name in* Box 16.3.

Struve, F. G. W., *cited by name in* Box 27.7.

Sunyaev, R. A. *See* Shakura and Sunyaev 1973.

Susskind, L. *See* Aharonov and Susskind 1967.

Symanzik, K., 1963, "Aspects of Euclidean field theory and Feynman path integrals," in Lurie and Mukunda 1963. [Box 21.1]

Synge, J. L., 1950, "The gravitational field of a particle," *Proc. R. Irish Acad. A 53,* 83–114. [§31.5, Box 31.1]

Synge, J. L., 1960a, "On certain identities connected with the Einstein tensor," *Proc. R. Irish Acad. A 61,* 29–36. [Ex. 35.14]

Synge, J. L., 1960b, *Relativity: The General Theory,* North-Holland, Amsterdam. [endpapers]

Synge, J. L., 1962, "Tensorial integral conservation laws in general relativity," in CNRS 1962, pp. 75–83. [Box 15.3]

Synge, John Lighton, *Festschrift* on the occasion of his 75th birthday, March 23, 1972. *See* O'Raifeartaigh 1972.

Szekeres, G., 1960, "On the singularities of a Riemannian manifold," *Publ. Mat. Debrecen 7,* 285–301. [§31.4, Box 31.1]

Taliaferro, R. C., trans. 1952, *Ptolemy, The Almagest,* in *Great Books of the Western World,* Encyclopedia Brittanica, Inc., Chicago, Ill., Volume 16. [Box 27.7]

Tamman, G. A., *cited by name in* Box 29.4. *See also* Sandage, Tamman, and Hardy 1972.

Tannenbaum, M. *See* Farley *et al.* 1966.

Tartaglia, N. F., 1560, *General Trattato dei Numeri et Misure,* Cited in *Enzyklopädie der Mathematischen Wissenschaften,* vol. III, part i, p. 1056, Leipzig 1914. [Box 13.1]

Taub, A. H., 1948, "Relativistic Rankine-Hugonoit equations," *Phys. Rev., 74,* 328–334. [§22.2, Box 22.2]

Taub, A. H., 1965, "Equations of motion of test particles," in London 1965. [§40.9]

Taub, A. H., 1971, "Relativistic Hydrodynamics," *Studies in Applied Mathematics—Math. Assoc. Amer. Studies in Math. 7,* 150–180. [§22.3]

Taub, A. H. *See also* Misner and Taub 1968.

Tauber, G. E., and J. W. Weinberg, 1961, "Internal state of a gravitating gas," *Phys. Rev. 122,* 1342–1365. [§22.6]

Taylor, D. J., *cited by name in* Box 34.3. See also Cocke and Taylor 1969.

Taylor, E. F., and J. A. Wheeler, 1966, *Spacetime Physics,* W. H. Freeman, San Francisco, Calif. [§2.1, Fig. 2.7, Ex. 6.9]

Tbilisi, USSR, 1968. Fifth International Conference on Gravitation and the Theory of Relativity, Tbilisi State University, August 1968. *See* Fock 1968.

Technion 1969, *International Seminar on Relativity and Gravi-*

tation held at Technion City, Israel, in July 1969, in honor of the 60th birthday of Professor Nathan Rosen. *See* Kuper and Peres 1971.

Teitelboim, C., 1970, "Splitting of the Maxwell tensor: radiation reaction without advanced fields," *Phys. Rev. D 1,* 1572–1582. [§20.6, Ex. 20.8]

Teitelboim, C., 1971, "Radiation reaction as a retarded self-interaction," *Phys. Rev. D 4,* 345–347. [§20.6]

Teitelboim, C., 1972a, "Nonmeasurability of the baryon number of a black hole," *Nuovo Cimento Lett., II, no. 3,* 326–328. [Box 25.2, Boxes 32.2, 33.1]

Teitelboim, C., 1972b, "Nonmeasurability of the lepton number of a black hole," *Nuovo Cimento, II 3,* 397–400. [Boxes 25.2, 33.1]

Teitelboim, C., 1972c, "Nonmeasurability of the Quantum Numbers of a Black Hole," *Phys. Rev. D 5,* 2941–2954 [Boxes 25.2, 32.2, 33.1]

Teitelboim, C., *cited by name in* §21.1, Fig. 4.6.

Teller, E., 1948, "On the change of physical constants," *Phys. Rev. 73,* 801–802. [§38.6]

Teller, E., *cited by name in* Box 38.2. *See also* De Hoffman and Teller 1950.

Tempier, É., *cited by name in* Box 27.7.

Teukolsky, S. A., 1972. "Rotating black holes: Separable wave equations for gravitational and electromagnetic perturbations," *Phys. Rev. Lett. 29,* 1114–1118. [§41.11, Box 33.3, Ex. 33.14]

Teukolsky, S. A., 1973, "Perturbations of a rotating black hole, I: Fundamental equations for gravitational, electromagnetic, and neutrino-field perturbations," *Astrophys. J.,* in press. [Ex. 33.14]

Teukolsky, S. A. *See also* Bardeen, Press, and Teukolsky 1972 *and* Press and Teukolsky 1972 *and* 1973.

Texas Symposia on Relativistic Astrophysics: 1963, *see* Robinson, Schild, and Schücking 1965 *and* Harrison *et al.* 1965; 1964, *see* Douglas *et al.* 1969; 1972, *see* Trans. *N.Y. Acad. Sci.*

Thirring, H., 1918, "Über die Wirkung rotierender ferner Massen in der Einsteinschen Gravitationstheorie," *Phys. Z. 19,* 33–39. [§21.12]

Thirring, H., 1921, "Berichtigung zu meiner Arbeit: "Über die Wirkung rotierender ferner Massen in der Einsteinschen Gravitationstheorie," *Phys. A. 22,* 29–30. [§21.12]

Thirring, H., and J. Lense, 1918, "Über den Einfluss der Eigenrotation der Zentralkörper auf die Bewegung der Planeten und Monde nach der Einsteinschen Gravitationstheorie," *Phys. Z. 19,* 156–163. [§§21.12, 40.7]

Thirring, W. E., 1961, "An alternative approach to the theory of gravitation," *Ann. Phys. (U.S.A.) 16,* 96–117. [§§7.1, 18.1]

Thomas, L. H., 1927, "The kinematics of an electron with an axis," *Phil. Mag. 3,* 1–22. [§§40.7, 41.4, Ex. 6.9]

Thomson, A. *See* Farnsworth *et al.* 1972.

Thomson, J. J., 1907, *Electricity and Matter,* Archibald Constable, London, Chapter III. [Fig. 4.6]

Thomson, J. J., *cited by name in* Fig. 4.6.

Thomson, W., Lord Kelvin, 1889, *Popular Lectures and Addresses, Vol. I,* Macmillan, London. [Epigraph to Chap. 37]

Thomson, W., Lord Kelvin, 1904, *Baltimore Lectures on Mo-*

lecular Dynamics and the Wave Theory of Light, C. J. Clay, London. [Preface]

Thomson, W., Lord Kelvin, *cited by name in* Box 17.2, *quoted in* epigraph of Chap. 20.

Thorne, K. S., 1967, "The general relativistic theory of stellar structure and dynamics," in DeWitt, Schatzman, and Veron 1967, pp. 259–441. [§24.4, endpapers]

Thorne, K. S., 1969a, "Nonradial pulsation of general-relativistic stellar models, III: Analytic and numerical results for neutron stars," *Astrophys. J. 158,* 1–16. [§24.3, Box 36.1]

Thorne, K. S., 1969b, "Nonradial pulsation of general-relativistic stellar models, IV: The weak-field limit," *Astrophys. J. 158,* 997–1019. [§36.8, Box 36.1]

Thorne, K. S., 1971, "Relativistic stars, black holes, and gravitational waves," in Sachs 1971, pp. 238–283. [§40.7]

Thorne, K. S., 1972, "Nonspherical gravitational collapse, a short review," in Klauder 1972, pp. 231–258. [Box 32.3, Ex. 32.10]

Thorne, K. S., 1973a. "Relativistic shocks: The Taub adiabat," *Astrophys. J. 179,* 897–907. [§22.2, Box 22.2]

Thorne, K. S., 1973b, "Black-hole models for compact x-ray sources," *Astrophys. J.,* in press. [Box 33.3]

Thorne, K. S. *See also* Ames and Thorne 1968, Bardeen, Thorne, and Meltzer 1966, Bisnovatyi-Kogan and Thorne 1970, Fackerell, Ipser, and Thorne 1969, Fletcher *et al.* 1967, Harrison *et al.* 1965, Hartle and Thorne 1968 *and* 1974, Ipser and Thorne 1968, Meltzer and Thorne 1966, Morgan and Thorne 1973, Novikov and Thorne 1973, Press and Thorne 1972, *and* Trimble and Thorne 1969.

Thorne, K. S., W. T. Ni, and C. M. Will, 1971, "Theoretical frameworks for testing relativistic gravity: A review," in Davies 1971, pp. 10–31. [§17.6, Box 39.5]

Thorne, K. S., and C. M. Will, 1971, "Theoretical frameworks for texting relativistic gravity, I: Foundations," *Astrophys. J. 163,* 595–610. [§§38.2, 39.2]

Tikhonov, M. Yu. *See* Braginsky, Manukin, and Tikhonov 1970.

Tobey, R. G., *et al.* 1967, *PL/1 Formac Interpreter, Users Reference Manual* IBM Controlled Program Library 360D 03.3004, Hawthorne, N.Y. [Box 14.3]

Tolman, R. C., 1934a, *Relativity, Thermodynamics, and Cosmology,* Clarendon Press, Oxford. [§44.1, Ex. 22.7, endpapers]

Tolman, R. C., 1934b, "Effect of inhomogeneity on cosmological models," *Proc. Nat. Acad. Sci. U.S. 20,* 169–176. [§30.6, Ex. 32.8]

Tolman, R. C., 1939, "Static solutions of Einstein's field equations for spheres of fluid," *Phys. Rev. 55,* 364–373. [§23.2, Ex. 23.9]

Tomonaga, S., 1946, "On a relativistically invariant formulation of the quantum theory of wave fields," *Prog. Theor. Phys. 1,* 27–42, reprinted in Schwinger 1958. [Boxes 15.1, 21.1, Ex. 43.5]

Tonnelat, M. A., *quoted at end of* Chap. 44.

Torrence, R. J. *See* Couch *et al.* 1968 *and* Newman *et al.* 1965.

Trautman, A., 1965, "Foundations and current problems of general relativity theory," in Trautman, Pirani, and Bondi 1965. [§12.1, Boxes 8.1, 12.4, endpapers]

Trautman, A., 1972, "On the Einstein-Cartan equations, I," *Bulletin de l'Academie Polonaise des Sciences (math., astr., phys.) 20,* 185–190. [§§38.2, 39.2]

Trautman, A., F. A. E. Pirani, and H. Bondi, 1965, *Lectures on General Relativity,* Brandeis 1964 Summer Institute on Theoretical Physics, vol. I, Prentice-Hall, Englewood Cliffs, N.J.

Treder, H. J., 1962, *Gravitative Stosswellen, Nichtanalytischen Wellenlösungen der Einsteinschen Gravitationsgleichungen,* Akademie-Verlag, Berlin. [§21.13]

Treder, H. J. *See also* Papapetrou and Treder 1959 *and* 1962.

Trimble, V. L., and K. S. Thorne, 1969, "Spectroscopic binaries and collapsed stars," *Astrophys. J. 156,* 1013–1019. [Box 33.3]

Trumpler, R. J. *See* Campbell and Trumpler 1928.

Truran, J. *See* Cameron and Truran 1971.

Truran, J., and A. G. W. Cameron, 1971, "Evolutionary models of nucleosynthesis in the galaxy," *Astrophys. & Space Sci. 14,* 179–222. [Box 33.3]

Tukey, J. W. *See* Blackman and Tukey 1959.

Turkevich, A. *See* Fermi and Turkevich 1950.

Turner, K. C., and H. A. Hill, 1964, "New experimental limit on velocity-dependent interactions of clocks and distant matter," *Phys. Rev. B 134,* 252–256. [§38.7, Fig. 38.3]

Tyson, J. A., *cited by name in* Ex. 37.13. *See also* Douglass and Tyson 1971.

Ulam, S. *See* Kac and Ulam 1968.

Unruh, W. G., 1971, "Dirac particles and geometrodynamical charge in curved geometries," doctoral dissertation, physics, Princeton University. [§21.9]

Unt, V., and P. Keres, 1972. "Tails of waves in the Schwarzschild field" (in Russian), *Izv. Akad Nauk Estonian S.S.R., Fiz. Mat., 21,* 17–30. [§35.14]

Urey, H. C. *See* Ruark and Urey 1930.

Van de Kamp, P., 1958, "Visual binaries," *Ann. Phys. (U.S.A.) 50,* 187–224. [Box 36.3]

Van der Waerden, B. L., 1967, *Sources of Quantum Mechanics,* North-Holland, Amsterdam, Holland. Reprinted in 1968 by Dover, New York.

Varenna, International School of Physics "Enrico Fermi," organized at Varenna on Lake Como by the Italian Physical Society; 1961, *see* Møller 1962; 1965, *see* Gratton 1966; 1969, *see* Sachs 1971; 1972, *see* Bertotti 1973.

Vatican 1970, Study Week on the Nuclei of Galaxies, April 13–18, 1970, at the Vatican; proceedings published in *Semain d'etude sur les noyeaux des galaxies, Pontificae Academiae Scientarum Script Varia no. 35,* 1971.

Veblen, O., and J. H. C. Whitehead, 1932, *The Foundations of Differential Geometry,* Cambridge Univ. Press, Cambridge, Eng. Reprinted 1953. [§2.2]

Veron, P. *See* DeWitt, Schatzman, and Veron 1967.

Vishveshwara, C. V., 1970, "Stability of the Schwarzschild Metric," *Phys. Rev. D 1,* 2870–2879. [Box 33.3]

Volkoff, G. *See* Oppenheimer and Volkoff 1939.

Voltaire, François Marie Arouet de, 1738, *Éléments de Philosophie de Newton,* Pt. 3, chap. III; reprinted in vol. 5 of *Oevres de Voltaire,* Bacquenois, Paris, 1836. [epigraph to §1.1]

von Karman, T., *cited by name in* Box 38.2.

von Klüber, H., 1960, "The determination of Einstein's light deflection in the gravitational field of the sun," in Beer 1960, pp. 47–77. [Box 40.1]

von Oppolzer, T. R., 1887, "Canon der Finsternisse," *Sitzber. Kais. Akad. Wiss. Wien Math.—Nat. Kl. 52,* i–xviii, 1–321. [Fig. 1.8]

von Neumann, J., *cited by name in* Box 38.2.

Vourloumis, Hypatia, 1971, statement made at Knossos, May 1971, while serving as guide on Swan's Hellenic tour. *Quoted at end of* Chap. 44.

Vyasa, about 400 A.D., *The Mahabharata,* translated from the Sanskrit by P. Lal, Writers Workshop, Calcutta. [Epigraph to §1.5]

Waage, H., *cited by name in* Fig. 1.7.

Wagoner, R. V., 1967, "Some effects of an intervening galaxy on the radiation from very distant objects," *Astrophys. J. 149,* 465–480. [Ex. 29.8]

Wagoner, R. V., *cited by name in* Box 27.7. *See also* Bardeen and Wagoner 1971 *and* Salpeter and Wagoner 1971.

Wagoner, R. V., W. A. Fowler, and F. Hoyle, 1967, "On the synthesis of elements at very high temperatures," *Astrophys. J. 148,* 3–49. [§28.1]

Wakano, M. *See* Harrison *et al.* 1965 *and* Klauder *et al.* 1959.

Wald, R. M., 1971. *See* Cohen and Wald 1971.

Walker, A. G., 1932, "Relative coordinates," *Proc. R. Soc. Edinburgh 52,* 345–353. [§6.3]

Walker, A. G., 1936, "On Milne's theory of world-structure," *Proc. London Math. Soc. 42,* 90–127. [Ex. 27.5]

Walker, A. G. *cited by name in* §30.3, Box 27.7.

Walker, M., and R. Penrose, 1970, "On quadratic first integrals of the geodesic equations for type [22] spacetimes," *Commun. Math. Phys. 18,* 265–274. [§§33.4, 33.5]

Walker, R. *See* Mathews and Walker 1965.

Wallace, A. H., 1957, *An Introduction to Algebraic Topology,* Pergamon, London. [Box 34.1]

Walraven, Th. *See* Ledoux and Walraven 1958.

Wampler, E. J., *cited by name in* Box 27.7.

Wanders, G. *See* Steuckelberg and Wanders 1953.

Ward, W., 1970, "General-relativistic light deflection for the complete celestial sphere," *Astrophys. J. 162,* 345–348. [§40.3]

Warrington, J., trans. 1956, *Aristotle's Metaphysics,* E. P. Dutton, New York. [Box 27.7]

Wasserburg, G. J. *See* Papanastassiou and Wasserburg 1969.

Wasserburg, G. J., and D. S. Burnett, 1968, "The status of isotopic age determinations on iron and stone meteorites," in P. M. Millman, ed., *Meteorite Research,* Reidel, Dordrecht, Holland, pp. 467–479. [§29.7, Box 29.3]

Wasserburg, G. J., D. N. Schramm, and J. C. Huneke, 1969, "Nuclear chronologies for the galaxy," *Astrophys. J. 157,* L91–L96. [§29.7]

Weber, H., ed., 1953, *B. Riemann: Gesammelte Mathematische Werke,* 2d ed., paperback reprint, Dover, New York. The *Habilitationsvorlesung* of 1854 was translated into English by W. K. Clifford, *Nature, 8,* 14 (1873). [Box 8.5]

Weber, J., 1960, "Detection and generation of gravitational waves," *Phys. Rev. 117,* 307–313. [§37.3]

Weber, J., 1961, *General Relativity and Gravitational Waves,* Wiley-Interscience, New York. [§37.3, endpapers]

Weber, J., 1967, "Gravitational radiation," *Phys. Rev. Lett. 18,* 498–501. [§37.3, Fig. 37.2]

Weber, J., 1969, "Evidence for discovery of gravitational radiation," *Phys. Rev. Lett. 22,* 1320–1324. [§§37.3, 37.10, Fig. 37.2]

Weber, J., 1970a, "Gravitational radiation experiments," *Phys. Rev. Lett. 24,* 276–279. [§§37.3, 37.10]

Weber, J., 1970b, "Anisotropy and polarization in the gravitational-radiation experiments," *Phys. Rev. Lett. 25,* 180–184. [§§37.3, 37.10]

Weber, J., *cited by name in* Preface, Box 17.3.

Weil, A. *See* Allendoerfer and Weil 1943.

Weiler, K. W. *See* Seielstad, Sramek, and Weiler 1970.

Weinberg, J. W. *See* Tauber and Weinberg 1961.

Weinberg, S., 1965, "Photons and gravitons in perturbation theory: Derivation of Maxwell's and Einstein's equations," *Phys. Rev. B 138,* 988–1002. [§§7.1, 18.1]

Weinberg, S., 1972, *Gravitation and Cosmology: Principles and Applications of the General Theory of Relativity,* Wiley, New York. [endpapers]

Weingarten, J., *cited by name in* §21.5.

Weisskopf, V. F. *See* Blatt and Weisskopf 1952 *and* Fierz and Weisskopf 1960.

Wells, J. S. *See* Evenson *et al.* 1972.

Welton, T. A., 1948, "Some observable effects of the quantum-mechanical fluctuations of the electromagnetic field," *Phys. Rev. 74,* 1157–1167. [Box 17.2 part 6]

Wentzel, G., 1949, *Quantum Theory of Fields,* Interscience, New York. [§21.3, Boxes 7.1, 21.2]

Werner, F. G., 1969, private remark to J. A. Wheeler on June 3 at the Cincinnati, Ohio, Relativity Conference in the Midwest. [§44.5]

Westermann, A. and J. F. Boissonade, eds. 1878, "Marini vita Procli," xxxvii, *in Dioginis Laertii de Clarorum Philosophorum,* II, Paris, p. 169. [Fig. 1.8]

Weyl, H., 1922, *Space-Time-Matter,* translated by H. L. Brose, Methuen, London. [endpapers]

Weyl, H., 1924a, "Massenträgheit und Kosmos: Ein Dialog," *Naturwissenschaften 12,* 197–204. [§21.12]

Weyl, H., 1924b, *Was Ist Materie,* Springer, Berlin, Germany. [Box 15.1]

Weyl, H., 1925, "Theorie der Darstellung kontinuierlichen halb-einfacher Gruppen durch lineare Transformationen, I," *Math. Z. 23,* 271–309. [§2.2, Epigraph to Chap. 2]

Weyl, H., 1937, lectures at Princeton University. [§4.5]

Weyl, H., 1946, *Classical Groups: Their Invariants and Representation,* Princeton Univ. Press, Princeton, N.J., 2d ed. [Box 27.2]

Weyl, H., 1949, *Philosophy of Mathematics and Natural Science,* Princeton Univ. Press, Princeton, N.J., p. 116. [Epigraph to Chap. 6, Box 8.5]

Weyl, H., *cited by name in* §4.5, Boxes 16.4, 27.7, *quoted in* epigraph of Chap. 10. *See also* Lorentz *et al.* 1923.

Wheeler, J. A., 1955, "Geons," *Phys. Rev. 97,* 511–536. [§§21.12, 25.5, Box 15.1, Ex. 32.11]

Wheeler, J. A., 1959, "View that the distribution of mass and energy determines the metric," in *Onzième Conseil de Physique Solvay: La Structure et l'évolution de l'univers,* Éditions Stoops, Brussels, Belgium. [§§21.12, 43.1]

Wheeler, J. A., 1961, "Geometrodynamics and the problem of motion," *Rev. Mod. Phys. 33,* 63–78. [§20.6]

Wheeler, J. A., 1962, *Geometrodynamics,* Academic Press, New York. [§§43.4, 44.3, Fig. 44.1, Boxes 33.3, 36.2, 44.3]

Wheeler, J. A., 1964a, "Geometrodynamics and the issue of the final state," in DeWitt and DeWitt 1964b. [§§27.10, 21.11, 34.6, 35.13, 42.6, 43.1, Boxes 27.7, 44.4, Ex. 21.18]

Wheeler, J. A., 1964b, "Gravitation as geometry, II," in Chiu and Hoffman 1964. [§15.3]

Wheeler, J. A., 1964c, "Mach's principle as boundary condition for Einstein's equations," in Chiu and Hoffman 1964, pp. 303–349. [§§21.10, 21.12, 43.1]

Wheeler, J. A., 1966, "Superdense stars," in *Annual Review Astron. and Astrophys. Vol. 4,* L. Goldberg, ed., Palo Alto, Calif. pp. 393–432. [§24.3, Ex. 36.3]

Wheeler, J. A., 1968a, "Superspace and the nature of quantum geometrodynamics," in DeWitt and Wheeler 1968. [§§21.1, 21.8, Box 44.3]

Wheeler, J. A., 1968b, *Einstein's Vision,* Springer, Berlin. Contained in part in Wheeler 1968a. [§§21.8, 21.12, Ex. 21.23]

Wheeler, J. A., 1970, "Superspace," in Gilbert and Newton 1970, pp. 335–378. [§43.1]

Wheeler, J. A., 1971a, notebook entry, "Pregeometry and the calculus of propositions," 9:10 A.M., April 10; seminar, Department of Mathematics, Kings College, London, May 10; letter to L. Thomas, "Pregeometry and propositions," June 11, unpublished. [Box 44.5]

Wheeler, J. A., 1971b, "Transcending the law of conservation of leptons," in *Atti del Convegno Internazionale sul Tema: The Astrophysical Aspects of the Weak Interactions; Quaderno N. 157,* Accademia Nazionale dei Lincei, Roma, pp. 133–164. [§§41.11, 44.6, Box 25.2]

Wheeler, J. A., 1971c, "From Mendeléev's atom to the collapsing star," 189–233 in *Atti del Convegno Mendeléeviano,* Accademia delle Scienze, Torino; reprinted in *Trans. New York Acad. Sci., 33,* 745–779 (1971). *See also* Wyler 1971. [§28.3, Box 25.2]

Wheeler, J. A., 1971d, "Mechanisms for jets," in *Study week on nuclei of galaxies, Pontificiae Academiae Scientarum Scripta Varia no. 35,* 539–567. [Box 33.3]

Wheeler, John Archibald, *Festschrift* on the occasion of his sixtieth birthday, July 9, 1971. *See* Klauder 1972.

Wheeler, J. A., *cited by name in* §21.1, Box 8.5. *See also* Baierlein, Sharp, and Wheeler 1962, Brill and Wheeler 1957, DeWitt and Wheeler 1968, Douglas *et al.* 1969, Ford and Wheeler 1959a *and* 1959b, Fuller and Wheeler 1962, Harrison *et al.* 1965, Harrison and Wheeler 1958, Klauder *et al.* 1959, Lindquist and Wheeler 1957, Marzke and Wheeler 1964, Misner and Wheeler 1957 *and* 1972, Power and Wheeler 1957, Regge and Wheeler 1957, Ruffini and Wheeler 1970, 1971a, *and* 1971b, *and* Taylor and Wheeler 1966.

Wheeler, J. A., and R. P. Feynman, 1945, "Interaction with the absorber as the mechanism of radiation," *Rev. Mod. Phys. 17,* 157–181. [§4.7, Box 2.1]

Wheeler, J. A., and R. P. Feynman, 1949, "Classical electrodynamics in terms of direct inter-particle action," *Rev. Mod. Phys. 21,* 425–434. [§4.7, Box 2.1]

White, R. H. *See* Colgate and White 1966 *and* May and White 1966 *and* 1967.

Whitehead, A. N., 1922, *The Principle of Relativity,* Cambridge Univ. Press, Cambridge, Eng. [§§38.2, 39.1, 39.3, 39.8, Box 17.3]

Whitehead, A. N., 1929, *The Function of Reason,* Princeton Univ. Press, Princeton, N.J. [Epigraph to Chap. 28]

Whitehead, J. H. C. *See* Veblen and Whitehead 1932.

Whitrow, G. J., and G. E. Morduch, 1965, "Relativistic theories of gravitation, a comparative analysis with particular reference to astronomical tests, in *Vistas in Astronomy 6,* 1–67. [§38.2]

Whittaker, E. R., 1904, "On an expression of the electromagnetic field due to electrons by means of two scalar potential functions," *Proc. R. Soc. London 1,* 367–372. [Ex. 4.10]

Wigner, E. P., 1960, "The unreasonable effectiveness of mathematics in the natural sciences," *Commun. Pure & Appl. Math. 13,* 1–14, reprinted in *Symmetries and Reflections: Scientific Essays of Eugene P. Wigner,* W. J. Moore and M. Scriven, eds., Indiana Univ. Press, Bloomington, Ind., 1967. [§44.2]

Wigner, E. P., *cited by name in* Box 38.2. *See also* Bargmann and Wigner 1948 *and* Salam and Wigner 1972.

Wilkins, D., 1972, "Bound geodesics in the Kerr metric," *Phys. Rev. D 5,* 814–822. [Box 33.5]

Wilkinson, D. T., *cited by name in* Box 27.7. *See also* Bender *et al.* 1971 *and* Dicke *et al.* 1965.

Will, C. M., 1971a, "Relativistic gravity in the solar system, I: Effect of an anisotropic gravitational mass on the earth-moon distance," *Astrophys. J. 165,* 409–412. [Box 39.5]

Will, C. M., 1971b, "Relativistic gravity in the solar system, II: Anisotropy in the Newtonian gravitational constant," *Astrophys. J. 169,* 141–156. [§§38.2, 39.1, 40.8, Boxes 17.3, 39.5]

Will, C. M., 1971c, "Theoretical frameworks for testing relativistic gravity, II: Parametrized post-Newtonian hydrodynamics, and the Nordtvedt effect," *Astrophys. J. 163,* 611–628. [§§38.2, 39.3, 40.9, Box 39.5]

Will, C. M., 1971d, "Theoretical frameworks for testing relativistic gravity, III: Conservation laws, Lorentz invariance, and values of the PPN parameters," *Astrophys. J. 169,* 125–140. [§39.9, Box 39.5, Ex. 39.15]

Will, C. M., 1972. "The theoretical tools of experimental gravitation," Caltech Orange-Aid Preprint 289, to be published in Bertotti 1973. [§§38.2, 39.3, Box 39.5]

Will, C. M., 1973, "Relativistic gravity in the solar system, III: Experimental disproof of a class of linear theories of gravitation," *Astrophys. J.,* in press. [§§39.3, 39.8]

Will, C. M. *See also* Nordtvedt and Will 1972, Thorne, Ni, and Will 1971, *and* Thorne and Will 1971.

Will, C. M., and K. Nordtvedt, Jr., 1972, "Conservation laws and preferred frames in relativistic gravity, I: Preferred-frame theories and an extended PPN formalism," *Astrophys. J. 177,* 757–774. [§§38.2, 39.2, 39.3, 39.9, 40.8, Boxes 39.1, 39.5]

Willey, R. *See* Klauder *et al.* 1959.

Williams, R. W. *See* Bailey *et al.* 1968.

Willmore, T. J., 1959, *An Introduction to Differential Geometry,* Clarendon Press, Oxford, [Box 8.1]

Wilson, O. C. *See* Minkowski and Wilson 1956.

Wilson, R. W., *cited by name in* Box 27.7. *See also* Penzias and Wilson 1965.

Winsor, F., *quoted at end of* Chap. 44.

Wintner, A., 1941, *Analytical Foundations of Celestial Mechanics,* Princeton Univ. Press, Princeton, N.J. [Box 25.4]

Witteborn, F. C., and W. M. Fairbank, 1967, "Experimental comparison of the gravitational force on freely falling electrons and metallic electrons," *Phys. Rev. Lett. 19,* 1049–1052. [§1.3]

Witteborn, F. C., L. V. Knight, and W. M. Fairbank, 1965, "Use of low-temperature techniques to measure gravitational forces on charged particles," in *Proceedings of the 9th International Conference on Low-Temperature Physics, Columbus 1964,* Plenum Press, New York, 1965, part B, pp. 1248–1252. [§1.3]

Witten, L., ed., 1962, *Gravitation: An Introduction to Current Research,* Wiley, New York.

Witten, L. *See also* Carmeli, Fickler, and Witten 1970.

Wong, C.-Y., 1971, "Application of Regge calculus to the Schwarzschild and Reissner-Nordstrom geometries at the moment of time-symmetry," *J. Math. Phys. 12,* 70–78. [§§42.6, 42.7]

Woolf, N. *See* Douglas *et al.* 1969.

Wright, J. P., 1964, "General relativistic instability," *Phys. Rev. B 136,* 288–289. [Box 26.2]

Wright, T., 1750, *An Original Theory or New Hypothesis of the Universe,* H. Chapelle, London. Reprinted in Munitz 1957. [Box 27.7]

Wyler, J. A. (pseudonym), 1971, "Rasputin, science, and the transmogrification of destiny," preprint bearing the Princeton University seal.

York, J. W., Jr., 1971, "Gravitational degrees of freedom and the initial-value problem," *Phys. Rev. Lett. 26,* 1656–1658. [§§21.11, 21.12, Ex. 21.22]

York, J. W., Jr., 1972a, personal communication. [Ex. 21.8, 21.9, 21.24]

York, J. W., Jr., 1972b, "The role of conformal 3-geometry in the dynamics of gravitation," *Phys. Rev. Letters 28,* 1082–1085. [§§21.6, 21.11, Ex. 21.24]

York, J. W., Jr., 1973a, "Conformally invariant orthogonal decomposition of symmetric tensors on Riemannian manifolds and the initial-value problem of general relativity," *J. Math. Phys. 14,* in press. [§21.11]

York, J. W., Jr., 1973b, personal communication. [§§21.10, 21.11]

York, J. W., Jr., *cited by name in* §21.1, Ex. 21.8, 21.9. *See also* O'Murchadha and York 1973.

Yourgrau, W., and S. Mandelstam, 1968, *Variational Principles in Dynamics and Quantum Theory,* W. B. Saunders, Philadelphia, Pa. [Box 21.1]

Yu, J. T. *See* Peebles and Yu 1970.

Zapolsky, H. S., 1968, "Can the redshift of quasistellar objects be gravitational?" *Astrophys. J. Lett. 153,* 163–169. [Ex. 25.29]

Zaret, H., *quoted at end of* Chap. 44.

Zeeman, E. C., 1964, "Causality implies the Lorentz group," *J. Math. Phys. 5,* 190–493. [Box 2.1]

Zel'dovich, Ya. B., 1963. "The theory of the expanding universe as originated by A. A. Friedmann," *Usp. Fiz. Nauk, 80,* 357–390. English translation in *Sov. Phys.—Uspekhi. 6,* 475–494 (1964). [§29.2]

Zel'dovich, Ya. B., 1967, "Cosmological constant and elementary particles," *Zh. Eksp. & Teor. Fiz. Pis'ma 6,* 883–884.

English translation in *Sov. Phys.—JETP Lett. 6,* 316–317. [§17.3, Ex. 17.5, Box 17.2 part 6]

Zel'dovich, Ya. B., 1968, "The cosmological constant and the theory of elementary particles," *Usp. Fiz. Nauk 95,* 209–230. English translation in *Sov. Phys.—Uspekhi 11,* 381–393. [Ex. 17.5]

Zel'dovich, Ya. B., 1970, "Particle production in cosmology," *Zh. Eksp. Teor. Fiz.—Pis'ma 12,* 443–447. English translation in *Sov. Phys.—JETP Lett. 12,* 307–311. [§30.5, Ex. 27.12]

Zel'dovich, Ya. B., 1971, "Particle creation by gravitational fields in collapse and singularities," *Comm. Astrophys. & Space Phys. 3,* 179–184. [Ex. 27.12]

Zel'dovich, Ya. B., 1972, "Creation of particles and antiparticles in an electric and gravitational field," in Klauder 1972, pp. 277–288. [§30.8, Ex. 27.12]

Zel'dovich, Ya. B. *See also* Bisnovatyi-Kogan and Zel'dovich 1969, Braginsky, Zel'dovich, and Rudenko 1969, *and* Doroshkevich, Zel'dovich, and Novikov 1965.

Zel'dovich, Ya. B., and O. H. Guseynov, 1965, "Collapsed stars in binaries," *Astrophys. J. 144,* 840–841. [Box 33.3]

Zel'dovich, Ya. B., and I. D. Novikov, 1966, "The hypothesis of cores retarded during expansion and the hot cosmological model," *Astron. Zh. 43,* 758–760. English translation in *Sov. Astron.—A. J. 10,* 602–603. [Box 33.3]

Zel'dovich, Ya. B., and I. D. Novikov, 1971, *Relativistic Astrophysics, Vol. I: Stars and Relativity,* Univ. of Chicago Press, Chicago, Ill. [§§22.2, 23.1, 24.1, 24.3, 24.4, 24.6, 25.6, 32.1, Boxes 24.1, 24.2]

Zel'dovich, Ya. B., and I. D. Novikov, 1974, *Relativistic Astrophysics, Vol. II: The Universe and Relativity,* Univ. of Chicago Press, Chicago, Ill., in press. [§§23.1, 28.1, 28.2, 30.1]

Zel'dovich, Ya. B., and M. A. Podurets, 1965, "The evolution of a system of gravitationally interacting point masses," *Astron. Zh. 42,* 963–973. English translation in *Sov. Astron.—A. J. 9,* 742–749 (1966). [§§24.6, 25.7, Boxes 24.1, 25.9]

Zel'dovich, Ya. B., and Yu. P. Rayzer, 1967, *Physics of Shock Waves and High-temperature Hydrodynamic Phenomena,* Academic Press, New York. [§22.2, Box 22.2]

Zel'dovich, Ya. B., and A. A. Starobinsky, 1971, "Particle production and vacuum polarization in an anisotropic gravitational field," *Zh. Eksp. Teor. Fiz. 61,* 2161–2175. English translation in *Sov. Phys.—JETP 34,* 1159–1166 (1972). [§30.5]

Zeno, *cited by name in* Box 30.1; *see* A. Grünbaum, 1967, *Modern Science and Zeno's Paradoxes,* Wesleyan Univ. Press, Middletown, Conn.

Zerilli, F. J., 1970a, "Effective potential for even-parity Regge-Wheeler gravitational perturbation equations," *Phys. Rev. Lett. 24,* 737–738. [Box 33.3]

Zerilli, F. J., 1970b, "Gravitational field of a particle falling in Schwarzschild geometry analyzed in tensor harmonics," *Phys. Rev. D 2,* 2141–2160. [§36.5, Fig. 36.2, Box 33.3]

Zimmer, H., 1946, *Myths and Symbols in Indian Art and Civilization,* Bollingen Foundation, Washington, D.C.; paperback reprint, Harper and Row, New York, 1962. [Box 27.7]

Zimmerman, B. A. *See* Fletcher *et al.* 1967.

Zwicky, F. *See* Baade and Zwicky 1934a *and* 1934b.

SUBJECT INDEX

photon interpretation of, 580
derived from wave equation and
 Lorentz gauge condition for vector
 potential, 573, 576–577
post-geometric optics corrections, 572f,
 803f
in spinor language, 1165
examples of applications of, 570
geometry of a bundle of rays, 581–582
focusing equation, 582f
breakdown of, related to pair creation,
 803–804
Geometrodynamics (dynamics of geometry)
ideas of, in brief, 4f
built-in plan: initial data plus time
 evolution, 408f, 484f
some history of, 486–488
analogies with electrodynamics, 364,
 367–370
causal propagation of effects in, 554
Arnowitt, Deser, Misner formulation of
 in brief, 486–490
 action principle in, 521
 geometrodynamic field momenta for,
 521
 3-geometry fixed at surfaces in, 522
 split of variables made by, 525–526
 electrodynamic analog, 522–524
Dirac formulation of, 520
subject to standard quantum
 indeterminism, 1182
illustrated in action, for Schwarzschild
 geometry, 528
See also Einstein field equation, General
 relativity, Initial value, Integrating
 forward in time
Geometry. *See specific types:* Spacetime,
 Euclidean, Lorentz, Differential, Affine,
 Riemannian, Prior. *See also* Curvature
Geon, 886
Global techniques of analyzing spacetime
 structure
 in extenso, Chap. 34
 basic references on, 916–917
 examples of, 926–931
 attempt to combine with local methods,
 806
 See also Infinity, regions of, in
 asymptotically flat spacetime; Causal
 structure of spacetime; Singularities
 in spacetime
Globular clusters, 757
 black holes in, 887
"Glory," in particle scattering, 670
Gowdy metrics, 804
Gradient
 of a scalar, in flat spacetime, 59f
 of a tensor, in flat spacetime, 81f
 in a curved manifold, 208–212, 259–261
 See also Exterior derivative
Gradiometer, gravity, 400–403
Gravitation, 13, 163–164
 local description in terms of
 tide-producing acceleration, 29–37

Gravitation-matter "coupling loop," in brief,
 5, 37
Gravitation, theories of
 catalogs of, 429
 criteria for viability of, 1066–1067
 Bergmann's scalar-tensor theories, 1049
 Birkhoff's, 1067
 Cartan's (general relativity plus torsion),
 1049, 1068
 Cartan-Newton. *See* Newton-Cartan
 theory of gravity
 Coleman's, 1114
 completeness of, 1067, 1068
 conservative, 1093
 Dicke-Brans-Jordan, 1048f, 1068f, *esp.*
 1070, 1093, 1098, 1122, 1127, 1129
 cosmological models in, 770
 general relativity, foundations of, Chaps.
 16, 17
 Kustaanheimo's, 1067
 linearized. *See* Linearized theory of
 gravity
 metric. *See* Metric theories of gravity
 metric, not encompassed by the
 10-parameter PPN formalism, 1069
 Newtonian. *See* Newton-Cartan theory of
 gravity
 Ni's, 1068f, *esp.* 1070f, 1083, 1098, 1123,
 1129
 Nordstrøm's, 429ff, 1049
 Papapetrou's, 1124
 post-Newtonian. *See* Post-Newtonian
 approximation; Post-Newtonian
 formalism, parametrized
 preferred-frame, 1083, 1093, 1098,
 1123–1125
 prior-geometric, 429–431, 1068, 1070–1071
 self-consistency of, 1066–1067
 spin-0 field, in flat spacetime, 178f
 spin-1 field, in flat spacetime, 179
 spin-2 field, in flat spacetime. *See*
 Linearized theory of gravity
 Whitehead's, 430, 1049, 1067, 1069, 1124
Gravitational collapse. *See* Collapse,
 gravitational
Gravitational constant
 value of, 29, endpapers
 measurement of, 1121, 1123
 as measure of "metric elasticity of
 space," 426–428
 See also Cavendish gravitational constant
"Gravitational field" in general relativity
 theory
 as term with many meanings and none,
 399f
 spacetime geometry as, 399–400
 metric as, 399f
 covariant derivative and connection
 coefficients as, 387, 399 400
 Riemann curvature as, 399–403
 contribution of, to standard stress-energy
 tensor, specifically excluded, 131
Gravitational lens effect, 589, 887
Gravitational mass, 431

Gravitational potential. *See under*
 Newton-Cartan theory, Post-Newtonian
 formalism
Gravitational radiation reaction. *See*
 Gravitational waves; radiation reaction
Gravitational radius, 820–826. *See also*
 Horizon, Black hole, Schwarzschild
 geometry, Kerr-Newman geometry
Gravitational-wave detectors
 conceivable types of
 Earth-moon separation, 1013, 1014,
 1018
 normal-mode vibrations of Earth and
 moon, 1013, 1015
 oscillations of Earth's crust, 1013, 1015
 normal-mode vibrations of an elastic
 bar, 1013, 1016, 1025, 1035, 1038
 normal-mode vibrations of general
 elastic bodies, 1013, 1016, 1025,
 1028–1035, 1041–1042
 angular accelerations of rotating bars
 ("heterodyne detector"), 1013,
 1016–1017
 angular accelerations of driven
 oscillators, 1013, 1017
 pumping of fluid in a rotating pipe,
 1013, 1018
 idealized vibrator (2 masses on a
 spring), 1022–1028
 beads on stick, 444f
 nonmechanical detectors, 1040
 electromagnetic waves in a toroidal
 wave guide, 1043–1044
 methods of analyzing (for mechanical
 detectors small compared to
 wavelength)
 proper reference frame of detector,
 1005–1006, 1010, 1012
 dynamic analysis: Newtonian equation
 of motion plus wave driving
 forces, 1006–1009
 driving forces of waves, 1006, 1009,
 1010
 line-of-force diagram, 1011–1012
 method of detailed balance, 1028,
 1029–1030, 1033
 for noisy detector, 1019, 1036–1040
 detailed analysis of
 two freely falling bodies, 1018
 idealized vibrator (two masses on a
 spring), 1022–1028
 any resonant vibrator, analyzed by
 detailed balance, 1030, 1033
 any resonant vibrator, analyzed by
 dynamic method, 1031–1034
 noisy resonant vibrator (extraction of
 signal from noise), 1036–1040
 Earth vibrating in quadrupole mode,
 1035–1036
 electromagnetic waves in a toroidal
 wave guide, 1043–1044
 cross sections
 limits on usefulness of concept of cross
 section, 1019, 1022

SOME USEFUL NUMBERS IN CONVENTIONAL AND GEOMETRIZED UNITS*

Fundamental Constants

Gravitation constant
$$G = (6.673 \pm 0.003) \times 10^{-8} \text{ cm}^3/\text{g sec}^2 = 1$$

Speed of light
$$c = (2.997\ 924\ 562 \pm 0.000\ 000\ 011) \times 10^{10} \text{ cm/sec} = 1$$

useful combinations
$$G/c^2 = 0.7425 \times 10^{-28} \text{ cm/g} = 1.476\ 64 \text{ km}/M_\odot = 1$$
$$c^5/G = 3.629 \times 10^{59} \text{ erg/sec} = 2.030 \times 10^5 M_\odot c^2/\text{sec} = 1 \text{ (emission factor)}$$
$$G/c = 2.226 \times 10^{-18} \text{ cm}^2 \text{ Hz/g} = 1 \text{ (receptor factor)}$$
$$c^2/G^{1/2} = 3.479 \times 10^{24} \text{ gauss cm} = 3.479 \times 10^{24} \text{ statvolt} = 1$$

Planck's reduced constant
$$\hbar = (1.054\ 592 \pm 0.000\ 008) \times 10^{-27} \text{ g cm}^2/\text{sec} = 2.612 \times 10^{-66} \text{ cm}^2$$

Planck distance, L^*
$$(\hbar G/c^3)^{1/2} = 1.616 \times 10^{-33} \text{ cm}$$

Planck time, T^*
$$(\hbar G/c^5)^{1/2} = 5.391 \times 10^{-44} \text{ sec}$$

Planck mass, M^*
$$(\hbar c/G)^{1/2} = 2.177 \times 10^{-5} \text{ g}$$

Planck density, M^*/L^{*3}
$$c^5/\hbar G^2 = 5.157 \times 10^{93} \text{ g/cm}^3$$

Quantum of charge
$$e = (4.803\ 25 \pm 0.000\ 02) \times 10^{-10} \text{ (g cm}^3/\text{sec}^2)^{1/2} = 1.381 \times 10^{-34} \text{ cm}$$

Reciprocal fine-structure constant, $1/\alpha$
$$\hbar c/e^2 = 137.0360 \pm 0.0002$$

Electron rest mass
$$m_e = (9.109\ 56 \pm 0.000\ 05) \times 10^{-28} \text{ g} \quad = 8.1873 \times 10^{-7} \text{ erg} = 0.511\ 004 \text{ MeV}$$
$$= 6.764 \times 10^{-56} \text{ cm}$$

Proton rest mass
$$M_p = (1.672\ 614 \pm 0.000\ 012) \times 10^{-24} \text{ g} = 1.503\ 27 \times 10^{-3} \text{ erg} = 0.938\ 259 \text{ GeV}$$
$$= 1.2419 \times 10^{-52} \text{ cm}$$
$$= 5.9301 \times 10^9 \text{ K}$$
$$= 1.088\ 83 \times 10^{13} \text{ K}$$

Bohr radius
$$a_0 = \hbar^2/m_e e^2 = 0.529\ 177 \times 10^{-8} \text{ cm}$$

Reduced Compton wavelength
$$\lambda_e = \hbar/m_e c = 3.861\ 59 \times 10^{-11} \text{ cm}$$

Classical electron radius
$$r_0 = e^2/m_e c^2 = 2.817\ 94 \times 10^{-13} \text{ cm}$$

Atomic energy unit
$$e^2/a_0 = m_e e^4/\hbar^2 = 4.359\ 83 \times 10^{-11} \text{ erg} = 27.2116 \text{ eV} = 3.157\ 86 \times 10^5 \text{ K} = 3.602 \times 10^{-60} \text{ cm}$$

Conversion Factors (see also "fundamental constants," above)

Distance
1 pc $= 3.0856 \times 10^{18}$ cm; 1 lt-yr $= 0.94605 \times 10^{18}$ cm; 1 A.U. $= 1.495\ 985 \times 10^{13}$ cm

Time
1 yr $= 3.155\ 692\ 6 \times 10^7$ sec; 1 day $= 86\ 400.$ sec; 1 sidereal day $= 86\ 164.091$ sec

Mass, energy, temperature
1 eV $= 1.160\ 48 \times 10^4$ K $= 1.602\ 192 \times 10^{-12}$ ergs $= 1.782\ 68 \times 10^{-33}$ g $= 1.324 \times 10^{-61}$ cm

Electromagnetic Radiation

Blackbody radiation

energy density $= aT^4$,
$$a = 7.5647 \times 10^{-15} \text{ erg cm}^{-3} \text{ K}^{-4}$$

emittance $= \sigma T^4$,
$$\sigma = \tfrac{1}{4}ac = 5.6696 \times 10^{-5} \text{ erg cm}^{-2} \text{ sec}^{-1} \text{ K}^{-4}$$

peak of $dE/d\lambda$ spectrum
$$\lambda_{max} T = 0.289\ 79 \text{ cm K} = (0.289\ 79 \text{ Å})(10^8 \text{ K})$$

Free-free emittance of ionized hydrogen,
$$\epsilon = (6 \times 10^{20} \text{ erg/g sec})(\rho_{g\,cm^{-3}}) T_K^{1/2}$$

Free-free opacity of ionized hydrogen,
$$\bar{\kappa}_{ff} = (0.8 \times 10^{23} \text{ cm}^2/\text{g})(\rho_{g\,cm^{-3}}) T_K^{-7/2}$$

Thomson cross section,
$$\sigma_T = (8\pi/3) r_0^2 = 0.665\ 24 \times 10^{-24} \text{ cm}^2$$

Electron scattering opacity of ionized hydrogen,
$$\kappa_{es} = 0.398 \text{ cm}^2/\text{g}$$

The Universe

Solar mass $M_\odot = (1.476\ 64 \pm 0.000\ 02) \times 10^5 \text{ cm} = 1.989 \times 10^{33} \text{ g} = 4.9255 \text{ } \mu\text{sec}$

Solar radius $R_\odot = (6.9598 \pm 0.0007) \times 10^{10} \text{ cm}$

Solar luminosity $L_\odot = (3.90 \pm 0.04) \times 10^{33} \text{ erg/sec} = (1.07 \times 10^{-26})c^5/G$

Earth–Sun distance I A.U. $= (1.495\ 985 \pm 0.000\ 005) \times 10^{13} \text{ cm} = 499.007 \text{ sec}$

Earth mass $M_\oplus = (5.977 \pm 0.004) \times 10^{27} \text{ g} = 0.4438 \text{ cm}$

Earth mean radius $R_\oplus = 6,371\ 03 \times 10^8 \text{ cm}$

Milky Way (Galaxy) mass $M_G = 1.8 \times 10^{11} M_\odot = 2.7 \times 10^{16} \text{ cm} = 0.028 \text{ lt-yr}$

Distance o Galactic center $r_\odot = 10 \text{ kpc} = 30\ 000 \text{ lt-yr}$

Galaxy diameter $D_G \sim /30 \text{ to } 50) \text{ kpc} \sim (100,000 \text{ to } 150,000) \text{ lt-yr}$

Galaxy thickness at Sun's location $h_G \approx 500 \text{ pc} \approx 1500 \text{ lt-yr}$

Nearby galaxies: distance and mass Large Magellanic Cloud, 52 kpc, $10^{10.1} M_\odot$; Small Magellanic Cloud, 54 kpc, $10^{9.2} M_\odot$; Andromeda Nebula (M31), 570 kpc, $10^{11.5} M_\odot$.

Clusters of galaxies: members, Local group, 16 members; $d \sim D \sim 0.4$ Mpc; Virgo cluster, 2500 members, distance, Diameter $d \sim 11$ Mpc, $D \sim 12$ Mpc; "typical" cluster, 130 members, $D \sim 3$ Mpc.

Hubble constant $H = 55 \text{ km sec}^{-1} \text{ Mpc}^{-1} = (1.7 \times 10^{29} \text{ cm})^{-1} = (18 \times 10^9 \text{ yr})^{-1}$

Universal density of luminous matter $\langle \rho \rangle = 2 \times 10^{-31} \text{ g/cm}^3$

* Numerical data based on, or calculated from Söding *et al.* (1971) and Allen (1963), with supplements from the literature.